Math in a Real World Context

Using Signed Numbers in Elevators
When a person enters an elevator and sees that the floor is numbered −1, what does this mean? How many floors are there from Floor −1 to Floor 3? In order to answer these questions, we need to understand integers. See Example 1 on page 88.

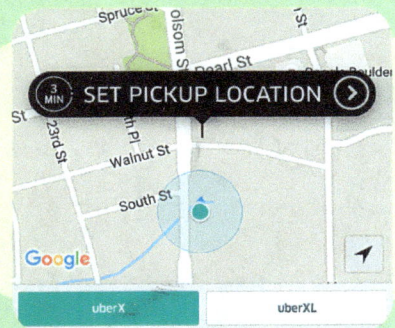

Analyzing Unicorn Companies
Companies that have private venture backing and are worth $1 billion or more (such as Uber) are known as "unicorn" companies. Recently, the total number of unicorn companies in existence during specified months could be modeled by a linear equation. Solving this equation helps us learn more about such companies. See Example 4 on page 545.

Monitoring Heart Rate
When an athlete stops a strenuous exercise routine, his or her heart rate decreases rapidly at first and then more gradually after a few minutes. This situation can be modeled by a polynomial function. Graphs of polynomial functions can be used when straight lines do not apply. See Example 6 on page 986.

Waiting in Traffic
Mathematics shows that in certain conditions, if the number of cars on the road increases even slightly, then the movement of traffic can slow dramatically. To understand this phenomenon more fully, we use a rational equation to model the situation. See Example 6 on page 854.

Taking the Ice Bucket Challenge
During a one-month period when the Ice Bucket Challenge for ALS became a viral fundraising phenomenon, donations increased by more than 20% per *day*. An exponential function can be used to model total donations during the month. See Example 10 on page 1247.

Eating Pizza
Eating your first piece of pizza usually provides a lot of satisfaction. A second piece may be almost as good. After a few more pieces, there is often a point where satisfaction starts to level off, and can even go down. A parabola models this situation. See Example 4 on page 1149.

Posting on Social Networks
When a link is posted on a social network such as Facebook, the majority of those who click on the link do so in the first few hours. After that, the number of people clicking the link decreases by a factor of one-half every few hours. This is an example of exponential decay. See the Chapter 19 Opener on page 1221.

Using the Internet
Of American adults who *don't* use the Internet, about 33% are either "just not interested" or they "don't have a computer." With a little more information, we can use a system of linear equations to learn more about the reasons given in this survey. See Example 7 on page 679.

Owning a Pet
There are both one-time and recurring costs associated with owning a pet. From a table listing such costs for cats and dogs, we can write linear equations that can be used to determine the lifetime cost of ownership for each animal. See the Group Activity on page 641.

Pricing Coffee Drinks
If the cost of a coffee drink is too high, demand will decrease because consumers want to save money. If the price of the drink is too low, supply will decrease because suppliers want to make money. To find a good price for the coffee drink, we can use a system of two linear equations. See Example 6 on page 678.

EDITION 2

Developmental Mathematics
with Applications and Visualization

Prealgebra, Beginning Algebra,
and Intermediate Algebra

Gary K. Rockswold
Minnesota State University, Mankato

Terry A. Krieger
Rochester Community and Technical College

with
Jessica C. Rockswold

Editorial Director *Chris Hoag*
Editor in Chief *Michael Hirsch*
Executive Editor *Cathy Cantin*
Editorial Assistant *Shannon Bushee*
Content Producer *Lauren Morse*
Managing Producer *Scott Disanno*
Product Marketing Manager *Alicia Frankel*
Field Marketing Managers *Lauren Schur and Jenny Crum*
Product Marketing Assistant *Hanna Lafferty*
Media Producer *Erin Carreiro*
Executive Content Manager, MathXL *Rebecca Williams*
MathXL QA Supervisor, TestGen *Mary Durnwald*
Senior Author Support/Technology Specialist *Joe Vetere*
Rights and Permissions Project Manager *Gina Cheselka*
Manufacturing Buyer *Carol Melville, LSC Communications*
Associate Director of Design *Blair Brown*
Text and Cover Design *Tamara Newnam*
Production Coordination and Composition *Cenveo® Publisher Services*

Copyright © 2018, 2013, 2009 by Pearson Education, Inc. All Rights Reserved. Printed in the United States of America. This publication is protected by copyright, and permission should be obtained from the publisher prior to any prohibited reproduction, storage in a retrieval system, or transmission in any form or by any means, electronic, mechanical, photocopying, recording, or otherwise. For information regarding permissions, request forms and the appropriate contacts within the Pearson Education Global Rights & Permissions department, please visit www.pearsoned.com/permissions/.

Acknowledgments of third party content appear on page **C-1**, which constitutes an extension of this copyright page.

PEARSON, ALWAYS LEARNING, and MYMATHLAB are exclusive trademarks owned by Pearson Education, Inc. or its affiliates in the U.S. and/or other countries.

Unless otherwise indicated herein, any third-party trademarks that may appear in this work are the property of their respective owners and any references to third-party trademarks, logos or other trade dress are for demonstrative or descriptive purposes only. Such references are not intended to imply any sponsorship, endorsement, authorization, or promotion of Pearson's products by the owners of such marks, or any relationship between the owner and Pearson Education, Inc. or its affiliates, authors, licensees or distributors.

Library of Congress Cataloging-in-Publication Data
Names: Rockswold, Gary K. | Krieger, Terry A. | Rockswold, Jessica C.
Title: Developmental mathematics with applications and visualization: prealgebra, beginning algebra, and intermediate algebra / Gary K. Rockswold, Minnesota State University, Mankato, Terry A. Krieger, Rochester Community and Technical College; with Jessica C. Rockswold.
Other titles: Prealgebra, beginning algebra, and intermediate algebra
Description: Second edition. | New York : Pearson, 2017. | Includes group activities for students. | Includes index.
Identifiers: LCCN 2017002080 | ISBN 9780134768403 (casebound)
Subjects: LCSH: Algebra–Textbooks. | Algebra–Study and teaching–Activity programs.
Classification: LCC QA152.3 .R634 2017 | DDC 512.9–dc23 LC record available at https://lccn.loc.gov/2017002080

1   17

ISBN 13: 978-0-13-476840-3
ISBN 10: 0-13-476840-X

About the Authors

Gary Rockswold has been a professor and teacher of mathematics, computer science, astronomy, and physical science for over 35 years. Not only has he taught at the undergraduate and graduate college levels, but he has also taught middle school, high school, vocational school, and adult education. He received his BA degree with majors in mathematics and physics from St. Olaf College and his PhD in applied mathematics from Iowa State University. He has been a principal investigator at the Minnesota Supercomputer Institute, publishing research articles in numerical analysis and parallel processing. He is currently an emeritus professor of mathematics at Minnesota State University–Mankato. He is an author for Pearson Education and has numerous textbooks at the developmental and precalculus levels. Making mathematics accessible to students and professing the power of mathematics are special passions for Gary. He frequently gives keynote and invited addresses at regional, national, and international math conferences. In his spare time he enjoys sailing, doing yoga, hiking, and spending time with his family.

Terry Krieger has taught mathematics for over 20 years at the middle school, high school, vocational, community college, and university levels. His undergraduate degree in secondary education is from Bemidji State University in Minnesota, where he graduated summa cum laude. He received his MA in mathematics from Minnesota State University–Mankato. In addition to his teaching experience in the United States, Terry has taught mathematics in Tasmania, Australia, and in a rural school in Swaziland, Africa, where he served as a Peace Corps volunteer. Terry has been involved with various aspects of mathematics textbook publications throughout his career. In his free time, Terry enjoys spending time with his wife and two boys, physical fitness, wilderness camping, and trout fishing.

This book is dedicated to the students.

GR, TK, JR

Contents

Preface xx
Hallmark Features xxiii
Resources for Success xxv
Acknowledgments xxviii
Index of Applications xxx

1 Whole Numbers 1

1.1 Introduction to Whole Numbers 2
Reviewing Natural Numbers and Whole Numbers ▪ Understanding Place Value ▪ Writing Whole Numbers in Word Form ▪ Writing Whole Numbers in Expanded Form ▪ Graphing Whole Numbers on the Number Line ▪ Reading Graphs and Tables

1.2 Adding and Subtracting Whole Numbers; Perimeter 15
Adding Whole Numbers ▪ Using Properties of Addition ▪ Recognizing Words Associated with Addition ▪ Subtracting Whole Numbers ▪ Using Properties of Subtraction ▪ Recognizing Words Associated with Subtraction ▪ Solving Equations Involving Addition and Subtraction ▪ Solving Perimeter and Other Applications Involving Addition and Subtraction
Checking Basic Concepts: Mixed Review of Sections 1.1 and 1.2 29

1.3 Multiplying and Dividing Whole Numbers; Area 30
Multiplying Whole Numbers ▪ Using Properties of Multiplication ▪ Multiplying Larger Whole Numbers ▪ Recognizing Words Associated with Multiplication ▪ Dividing Whole Numbers ▪ Using Properties of Division ▪ Performing Long Division ▪ Recognizing Words Associated with Division ▪ Solving Equations Involving Multiplication and Division ▪ Solving Area and Other Applications Involving Multiplication and Division
GROUP ACTIVITY 45

1.4 Exponents, Variables, and Algebraic Expressions 46
Understanding Exponential Notation ▪ Using Variables ▪ Recognizing Algebraic Expressions ▪ Evaluating Formulas ▪ Translating Words to Expressions and Formulas ▪ Solving Equations
Checking Basic Concepts: Mixed Review of Sections 1.3 and 1.4 56

1.5 Rounding and Estimating; Square Roots 57
Rounding Whole Numbers ▪ Estimating and Approximating ▪ Solving Problems Using Estimation ▪ Estimating Graphically ▪ Finding Square Roots

1.6 Order of Operations 65
Applying the Order of Operations Agreement ▪ Evaluating Algebraic Expressions ▪ Translating Words to Symbols
Checking Basic Concepts: Mixed Review of Sections 1.5 and 1.6 71

1.7 More with Equations and Problem Solving 71
Simplifying Algebraic Expressions ▪ Checking a Solution to an Equation ▪ Applying a Problem-Solving Strategy
Checking Basic Concepts: Review of Section 1.7 79

Summary ▪ Review Exercises ▪ Test 79

2 Integers 87

2.1 Integers and the Number Line 88
Introducing Signed Numbers ▪ Working with Integers and Their Graphs ▪ Comparing Integers ▪ Finding Absolute Value ▪ Solving Applications Involving Integers

2.2 Adding Integers 97
Adding Integers ▪ Recognizing Addition Properties ▪ Adding Integers Visually ▪ Solving Applications Involving Addition of Integers
Checking Basic Concepts: Mixed Review of Sections 2.1 and 2.2 105

2.3 Subtracting Integers 106
Subtracting Integers ▪ Adding and Subtracting Integers ▪ Subtracting Integers Visually ▪ Solving Applications Involving Subtraction of Integers

2.4 Multiplying and Dividing Integers 112
Multiplying Integers ▪ Recognizing Multiplication Properties ▪ Multiplying More Than Two Integer Factors ▪ Dividing Integers ▪ Finding Square Roots of Integers ▪ Solving Applications Involving Multiplication and Division of Integers
Checking Basic Concepts: Mixed Review of Sections 2.3 and 2.4 120

2.5 Order of Operations; Averages 120
Evaluating Exponential Expressions ▪ Using the Order of Operations ▪ Evaluating Algebraic Expressions with Integers ▪ Finding Averages

GROUP ACTIVITY 127

2.6 Solving Equations That Have Integer Solutions 127
Checking a Solution ▪ Solving Equations Using Guess-and-Check ▪ Solving Equations Using Tables of Values ▪ Solving Equations That Have Integer Solutions Graphically
Checking Basic Concepts: Mixed Review of Sections 2.5 and 2.6 136

Summary ▪ Review Exercises ▪ Test 136

Chapters 1–2 Cumulative Review Exercises 144

3 Algebraic Expressions and Linear Equations 145

3.1 Simplifying Algebraic Expressions 146
Reviewing Algebraic Expressions ▪ Combining Expressions That Have Like Terms ▪ Adding Expressions ▪ Subtracting Expressions ▪ Multiplying Expressions ▪ Simplifying Expressions

3.2 Translating Words to Expressions and Equations 154
Translating Words to Expressions ▪ Translating Words to Equations
Checking Basic Concepts: Mixed Review of Sections 3.1 and 3.2 159

3.3 Properties of Equality 160
Finding Solutions and Equivalent Equations ▪ Using the Addition Property of Equality ▪ Using the Multiplication Property of Equality

GROUP ACTIVITY 167

3.4 Solving Linear Equations in One Variable 168
Identifying Linear Equations in One Variable ▪ Solving Linear Equations Symbolically ▪ Solving Linear Equations Numerically ▪ Solving Linear Equations Graphically
Checking Basic Concepts: Mixed Review of Sections 3.3 and 3.4 180

3.5 Applications and Problem Solving 181
Reviewing Steps for Problem Solving ▪ Solving Applications Involving Linear Equations in One Variable
Checking Basic Concepts: Review of Section 3.5 189

Summary ▪ Review Exercises ▪ Test 189

Chapters 1–3 Cumulative Review Exercises 195

4 Fractions 196

4.1 Introduction to Fractions and Mixed Numbers 197
Reviewing Fractions and Fractional Parts ▪ Introducing Rational Numbers ▪ Reviewing Improper Fractions and Mixed Numbers ▪ Graphing Fractions and Mixed Numbers

4.2 Prime Factorization and Lowest Terms 209
Using Divisibility Tests ▪ Finding Prime Factorization of Numbers ▪ Identifying and Writing Equivalent Fractions ▪ Simplifying Fractions to Lowest Terms ▪ Comparing Fractions ▪ Simplifying Rational Expressions
Checking Basic Concepts: Mixed Review of Sections 4.1 and 4.2 223

4.3 Multiplying and Dividing Fractions 223
Multiplying Fractions ▪ Multiplying Rational Expressions ▪ Evaluating Powers and Square Roots of Fractions ▪ Dividing Fractions and Rational Expressions ▪ Solving Applications Involving Multiplication of Fractions

4.4 Adding and Subtracting Fractions—Like Denominators 236
Adding Fractions—Like Denominators ▪ Subtracting Fractions—Like Denominators ▪ Adding and Subtracting Rational Expressions ▪ Solving Applications Involving Addition and Subtraction of Fractions—Like Denominators
Checking Basic Concepts: Mixed Review of Sections 4.3 and 4.4 243

4.5 Adding and Subtracting Fractions—Unlike Denominators 243
Finding the Least Common Multiple (LCM) ▪ Finding the Least Common Denominator (LCD) ▪ Adding and Subtracting Fractions—Unlike Denominators ▪ Solving Applications Involving Addition and Subtraction of Fractions—Unlike Denominators

GROUP ACTIVITY 254

4.6 Operations on Mixed Numbers 254

Writing Mixed Numbers as Improper Fractions ■ Rounding Mixed Numbers to the Nearest Integer ■ Performing Operations on Mixed Numbers ■ Adding and Subtracting Mixed Numbers—Method II
Checking Basic Concepts: Mixed Review of Sections 4.5 and 4.6 262

4.7 Complex Fractions and Order of Operations 263

Simplifying Basic Complex Fractions ■ Simplifying Complex Fractions—Method I ■ Simplifying Complex Fractions—Method II ■ Applying Order of Operations to Complex Fractions

4.8 Solving Equations Involving Fractions 270

Solving Equations Symbolically ■ Solving Equations Numerically ■ Solving Equations Graphically ■ Solving Applications Involving Equations with Fractions
Checking Basic Concepts: Mixed Review of Sections 4.7 and 4.8 284

Summary ■ Review Exercises ■ Test 284

Chapters 1–4 Cumulative Review Exercises 294

5 Decimals 296

5.1 Introduction to Decimals 297

Reviewing Decimal Notation ■ Writing Decimals in Words ■ Writing Decimals as Fractions ■ Graphing Decimals on a Number Line ■ Comparing Decimals ■ Rounding Decimals

5.2 Adding and Subtracting Decimals 307

Estimating Decimal Sums and Differences ■ Adding and Subtracting Positive Decimals ■ Adding and Subtracting Signed Decimals ■ Evaluating and Simplifying Expressions ■ Solving Applications Involving Decimals
Checking Basic Concepts: Mixed Review of Sections 5.1 and 5.2 316

5.3 Multiplying and Dividing Decimals 316

Estimating Decimal Products and Quotients ■ Multiplying Decimals ■ Dividing Decimals ■ Writing Fractions as Decimals ■ Evaluating and Simplifying Expressions ■ Solving Applications Involving Decimals

5.4 Real Numbers, Square Roots, and Order of Operations 330

Introducing Real Numbers ■ Approximating Square Roots ■ Applying the Order of Operations ■ Solving Applications Involving Real Numbers
Checking Basic Concepts: Mixed Review of Sections 5.3 and 5.4 339

5.5 Solving Equations Involving Decimals 340

Solving Equations Symbolically ■ Solving Equations Numerically ■ Solving Equations Graphically ■ Solving Applications Involving Equations with Decimals
GROUP ACTIVITY 350

5.6 Applications from Geometry and Statistics 350

Reviewing the Parts of a Circle ■ Finding the Circumference of a Circle ■ Finding the Area of a Circle ■ Finding the Area of Composite Regions ■ Applying the Pythagorean Theorem ■ Finding the Mean, Median, and Mode ■ Finding a Weighted Mean and GPA
Checking Basic Concepts: Mixed Review of Sections 5.5 and 5.6 363

Summary ■ Review Exercises ■ Test 364

Chapters 1–5 Cumulative Review Exercises 372

6 Ratios, Proportions, and Measurement 374

6.1 Ratios and Rates 375
Writing Ratios as Fractions ▪ Finding Unit Ratios ▪ Writing Rates as Fractions ▪ Finding Unit Rates ▪ Computing Unit Pricing

6.2 Proportions and Similar Figures 383
Identifying Proportions ▪ Solving a Proportion for an Unknown Value ▪ Working with Similar Figures ▪ Solving Applications Involving Proportions and Similar Figures
Checking Basic Concepts: Mixed Review of Sections 6.1 and 6.2 395

6.3 The U.S. System of Measurement 395
Converting U.S. Units of Length ▪ Converting U.S. Units of Area ▪ Converting U.S. Units of Capacity and Volume ▪ Converting U.S. Units of Weight

6.4 The Metric System of Measurement 403
Understanding Metric Prefixes ▪ Converting Metric Units of Length ▪ Converting Metric Units of Area ▪ Converting Metric Units of Capacity and Volume ▪ Converting Metric Units of Mass
Checking Basic Concepts: Mixed Review of Sections 6.3 and 6.4 411

6.5 U.S.–Metric Conversions; Temperature 412
Completing Conversions of Length ▪ Completing Conversions of Capacity and Volume ▪ Completing Conversions of Mass (Weight) ▪ Completing Conversions of Temperature

GROUP ACTIVITY 421

6.6 Time and Speed 421
Converting Units of Time ▪ Converting Units of Speed ▪ Solving an Application Involving Speed
Checking Basic Concepts: Mixed Review of Sections 6.5 and 6.6 425

Summary ▪ Review Exercises ▪ Test 425

Chapters 1–6 Cumulative Review Exercises 432

7 Percent and Probability 434

7.1 Introduction to Percent; Circle Graphs 435
Understanding Percent Notation ▪ Writing a Percent as a Fraction or Decimal ▪ Writing a Fraction or Decimal as a Percent ▪ Reading Circle Graphs ▪ Solving Percent Applications

7.2 Using Equations to Solve Percent Problems 443
Understanding Basic Percent Statements ▪ Translating Percent Statements to Equations ▪ Solving Percent Problems
Checking Basic Concepts: Mixed Review of Sections 7.1 and 7.2 449

7.3 Using Proportions to Solve Percent Problems 449
Reviewing Proportions ▪ Translating Percent Statements to Proportions ▪ Solving Percent Problems

7.4 Applications: Sales Tax, Discounts, and Net Pay 456
Solving General Percent Applications ▪ Finding Sales Tax and Total Amount Paid ▪ Finding Discounts and Sale Price ▪ Calculating Commission and Net Pay ▪ Finding Percent Change
Checking Basic Concepts: Mixed Review of Sections 7.3 and 7.4 465

7.5 Applications: Simple and Compound Interest 466
Introducing Principal, Interest, and Interest Rates ▪ Calculating Simple Interest ▪ Calculating Compound Interest

GROUP ACTIVITY 473

7.6 Probability and Percent Chance 473
Understanding Experiments, Outcomes, and Events ▪ Finding Probability of an Event and Percent Chance ▪ Finding the Complement of an Event
Checking Basic Concepts: Mixed Review of Sections 7.5 and 7.6 479

Summary ▪ Review Exercises ▪ Test 480

Chapters 1–7 Cumulative Review Exercises 487

8 Geometry 489

8.1 Plane Geometry: Points, Lines, and Angles 490
Reviewing Geometric Terms and Concepts ▪ Classifying Angles ▪ Recognizing Parallel, Intersecting, and Perpendicular Lines ▪ Using Properties of Parallel Lines Cut by a Transversal

8.2 Triangles 500
Classifying Triangles ▪ Applying the Sum of the Angle Measures of a Triangle ▪ Using Congruent Triangle Properties
Checking Basic Concepts: Mixed Review of Sections 8.1 and 8.2 507

8.3 Polygons and Circles 508
Understanding Properties of Polygons ▪ Finding Angles in Regular Polygons ▪ Understanding Properties of Quadrilaterals ▪ Finding the Radius and Diameter of a Circle

GROUP ACTIVITY 515

8.4 Perimeter and Circumference 516
Finding the Perimeter of a Polygon ▪ Finding the Circumference of a Circle ▪ Finding the Perimeter of Composite Figures
Checking Basic Concepts: Mixed Review of Sections 8.3 and 8.4 522

8.5 Area, Volume, and Surface Area 523
Finding the Area of Plane Figures ▪ Finding the Volume and Surface Area of Geometric Solids
Checking Basic Concepts: Review of Section 8.5 530

Summary ▪ Review Exercises ▪ Test 531

Chapters 1–8 Cumulative Review Exercises 538

9 Linear Equations and Inequalities in One Variable 540

9.1 Review of Linear Equations in One Variable 541
Identifying Linear Equations in One Variable ▪ Solving Linear Equations ▪ Solving Linear Equations by Applying the Distributive Property ▪ Solving Linear Equations by Clearing Fractions and Decimals ▪ Identifying Equations with No Solutions, One Solution, or Infinitely Many Solutions ▪ Solving a Formula for a Variable

GROUP ACTIVITY 554

9.2 Further Problem Solving 555
Identifying Steps for Solving a Problem ▪ Translating Sentences into Equations ▪ Solving Number Problems and Applications ▪ Solving Distance Problems ▪ Solving Mixture Problems
Checking Basic Concepts: Mixed Review of Sections 9.1 and 9.2 564

9.3 Linear Inequalities in One Variable 565
Graphing Solution Sets on Number Lines ▪ Solving Linear Inequalities with Tables ▪ Writing Solution Sets Using Interval Notation ▪ Applying the Addition Property of Inequalities ▪ Applying the Multiplication Property of Inequalities ▪ Writing Solution Sets Using Set-Builder Notation ▪ Translating Words to Inequalities ▪ Solving Applications Involving Inequalities

Checking Basic Concepts: Review of Section 9.3 579

Summary ▪ Review Exercises ▪ Test 579

Chapters 1–9 Cumulative Review Exercises 583

10 Graphing Equations 585

10.1 Introduction to Graphing 586
Introducing Tables and Graphs ▪ Understanding the Rectangular Coordinate System ▪ Making and Reading Scatterplots ▪ Making and Reading Line Graphs

10.2 Equations in Two Variables 594
Introducing Equations in Two Variables ▪ Recognizing Ordered Pairs as Solutions ▪ Making a Table of Solutions ▪ Graphing Linear Equations in Two Variables ▪ Graphing Nonlinear Equations in Two Variables
Checking Basic Concepts: Mixed Review of Sections 10.1 and 10.2 604

10.3 Intercepts; Horizontal and Vertical Lines 604
Finding Intercepts ▪ Using Intercepts to Graph Lines ▪ Recognizing Horizontal Lines and Their Equations ▪ Recognizing Vertical Lines and Their Equations

GROUP ACTIVITY 615

10.4 Slope and Rates of Change 615
Introducing Slope ▪ Recognizing Positive, Negative, Zero, and Undefined Slope ▪ Finding Slopes of Lines ▪ Understanding Slope as a Rate of Change
Checking Basic Concepts: Mixed Review of Sections 10.3 and 10.4 629

10.5 Slope–Intercept Form 630
  Determining a Line ■ Finding Slope–Intercept Form ■ Working with Parallel and Perpendicular Lines
  GROUP ACTIVITY 641

10.6 Point–Slope Form 642
  Deriving the Point–Slope Form ■ Finding Point–Slope Form ■ Solving Applications Involving Equations of Lines
  Checking Basic Concepts: Mixed Review of Sections 10.5 and 10.6 653

10.7 Introduction to Modeling 653
  Introducing Linear Models ■ Modeling Linear Data
  Checking Basic Concepts: Review of Section 10.7 661

Summary ■ Review Exercises ■ Test 661

Chapters 1–10 Cumulative Review Exercises 670

11 Systems of Linear Equations in Two Variables 672

11.1 Solving Systems of Linear Equations Graphically and Numerically 673
  Solving Equations Graphically and Numerically ■ Introducing Systems of Linear Equations ■ Solving Applications of Systems of Linear Equations

11.2 Solving Systems of Linear Equations by Substitution 686
  Using the Substitution Method ■ Recognizing Systems with No Solutions or Infinitely Many Solutions ■ Solving Applications Using the Substitution Method
  Checking Basic Concepts: Mixed Review of Sections 11.1 and 11.2 695

11.3 Solving Systems of Linear Equations by Elimination 696
  Using the Elimination Method ■ Recognizing Systems with No Solutions or Infinitely Many Solutions ■ Solving Applications Using the Elimination Method
  GROUP ACTIVITY 706

11.4 Systems of Linear Inequalities 706
  Finding Solutions to a Linear Inequality in Two Variables ■ Writing and Graphing a Linear Inequality in Two Variables ■ Finding Solutions to Systems of Linear Inequalities in Two Variables ■ Solving Applications Involving Systems of Linear Inequalities
  Checking Basic Concepts: Mixed Review of Sections 11.3 and 11.4 717

Summary ■ Review Exercises ■ Test 718

Chapters 1–11 Cumulative Review Exercises 724

12 Polynomials and Exponents 726

12.1 Rules for Exponents 727
  Reviewing Bases and Exponents ■ Computing Zero Exponents ■ Applying the Product Rule ■ Applying Power Rules

12.2 Addition and Subtraction of Polynomials 736
Introducing Nonlinear Data and Polynomials ▪ Recognizing Monomials and Polynomials ▪ Adding Polynomials ▪ Subtracting Polynomials ▪ Evaluating Polynomial Expressions
Checking Basic Concepts: Mixed Review of Sections 12.1 and 12.2 746

12.3 Multiplication of Polynomials 746
Multiplying Monomials ▪ Reviewing the Distributive Properties ▪ Multiplying Monomials and Polynomials ▪ Multiplying Polynomials

GROUP ACTIVITY 754

12.4 Special Products 755
Finding the Product of a Sum and Difference ▪ Squaring Binomials ▪ Cubing Binomials
Checking Basic Concepts: Mixed Review of Sections 12.3 and 12.4 762

12.5 Integer Exponents and the Quotient Rule 762
Using Negative Integers as Exponents ▪ Applying the Quotient Rule ▪ Applying Other Rules for Exponents ▪ Reading and Writing Scientific Notation

GROUP ACTIVITY 773

12.6 Division of Polynomials 773
Dividing by a Monomial ▪ Dividing by a Polynomial
Checking Basic Concepts: Mixed Review of Sections 12.5 and 12.6 779

Summary ▪ Review Exercises ▪ Test 780

Chapters 1–12 Cumulative Review Exercises 787

13 Factoring Polynomials and Solving Equations 788

13.1 Introduction to Factoring 789
Introducing Factoring ▪ Finding Common Factors ▪ Finding the Greatest Common Factor ▪ Factoring by Grouping

13.2 Factoring Trinomials I ($x^2 + bx + c$) 798
Reviewing Binomial Multiplication ▪ Factoring Trinomials with Leading Coefficient 1 ▪ Relating Geometry and Visual Factoring
Checking Basic Concepts: Mixed Review of Sections 13.1 and 13.2 806

13.3 Factoring Trinomials II ($ax^2 + bx + c$) 806
Factoring Trinomials by Grouping ▪ Factoring with FOIL in Reverse

13.4 Special Types of Factoring 814
Factoring the Difference of Two Squares ▪ Factoring Perfect Square Trinomials ▪ Factoring the Sum and Difference of Two Cubes ▪ Using Special Methods in General Factoring
Checking БasicConcepts: Mixed Review of Sections 13.3 and 13.4 821

13.5 Summary of Factoring 822
Summarizing General Guidelines for Factoring Polynomials ▪ Factoring Polynomials

GROUP ACTIVITY 827

13.6 Solving Equations by Factoring I (Quadratics) 827

Applying the Zero-Product Property ▪ Solving Quadratic Equations ▪ Solving Applications Involving Quadratic Equations
Checking Basic Concepts: Mixed Review of Sections 13.5 and 13.6 835

13.7 Solving Equations by Factoring II (Higher Degree) 835

Recognizing Nonlinear Data and Polynomial Models ▪ Factoring Higher-Degree Polynomials ▪ Solving Polynomial Equations ▪ Solving Applications Involving Polynomial Equations
Checking Basic Concepts: Review of Section 13.7 841

Summary ▪ Review Exercises ▪ Test 841

Chapters 1–13 Cumulative Review Exercises 847

14 Rational Expressions 849

14.1 Introduction to Rational Expressions 850

Evaluating Rational Expressions ▪ Simplifying Rational Expressions ▪ Solving Applications Involving Rational Expressions
GROUP ACTIVITY 861

14.2 Multiplication and Division of Rational Expressions 861

Reviewing Multiplication and Division of Fractions ▪ Multiplying Rational Expressions ▪ Dividing Rational Expressions
Checking Basic Concepts: Mixed Review of Sections 14.1 and 14.2 867

14.3 Addition and Subtraction with Like Denominators 868

Reviewing Addition and Subtraction of Fractions ▪ Adding and Subtracting Rational Expressions with Like Denominators

14.4 Addition and Subtraction with Unlike Denominators 875

Finding Least Common Multiples ▪ Reviewing Fractions with Unlike Denominators ▪ Adding and Subtracting Rational Expressions with Unlike Denominators ▪ Solving an Application Involving Rational Expressions
Checking Basic Concepts: Mixed Review of Sections 14.3 and 14.4 884

14.5 Complex Fractions 884

Introducing Complex Fractions ▪ Simplifying Complex Fractions

14.6 Rational Equations and Formulas 893

Solving Rational Equations ▪ Recognizing Rational Expressions and Equations ▪ Finding Graphical and Numerical Solutions ▪ Solving an Equation for a Variable ▪ Solving Applications Involving Rational Equations
Checking Basic Concepts: Mixed Review of Sections 14.5 and 14.6 907

14.7 Proportions and Variation 907

Solving Proportions ▪ Solving Direct Variation Problems ▪ Solving Inverse Variation Problems ▪ Analyzing Data ▪ Solving Joint Variation Problems
Checking Basic Concepts: Review of Section 14.7 921

Summary ▪ Review Exercises ▪ Test 922

Chapters 1–14 Cumulative Review Exercises 930

15 Introduction to Functions 932

15.1 Functions and Their Representations 933
Introducing Functions ▪ Recognizing Representations of Functions ▪ Defining a Function ▪ Identifying a Function ▪ Using Graphing Calculators

GROUP ACTIVITY 952

15.2 Linear Functions 953
Recognizing Representations of Linear Functions ▪ Recognizing Graphs of Linear Functions ▪ Modeling Data with Linear Functions ▪ Applying the Midpoint Formula
Checking Basic Concepts: Mixed Review of Sections 15.1 and 15.2 969

15.3 Compound Inequalities and Piecewise-Defined Functions 969
Introducing Compound Inequalities ▪ Finding Symbolic Solutions and Graphing Them on a Number Line ▪ Solving Inequalities and Writing Solutions in Interval Notation ▪ Finding Numerical and Graphical Solutions ▪ Introducing Piecewise-Defined Functions

15.4 Other Functions and Their Properties 983
Introducing Nonlinear Functions ▪ Expressing Domain and Range in Interval Notation ▪ Recognizing Absolute Value Functions and Their Properties ▪ Recognizing Polynomial Functions and Their Properties ▪ Recognizing Rational Functions and Their Properties ▪ Performing Operations on Functions
Checking Basic Concepts: Mixed Review of Sections 15.3 and 15.4 1000

15.5 Absolute Value Equations and Inequalities 1001
Solving Absolute Value Equations ▪ Solving Absolute Value Inequalities
Checking Basic Concepts: Review of Section 15.5 1014

Summary ▪ Review Exercises ▪ Test 1015

Chapters 1–15 Cumulative Review Exercises 1024

16 Systems of Linear Equations 1026

16.1 Systems of Linear Equations in Three Variables 1027
Writing Systems of Linear Equations and Recognizing Solutions ▪ Solving Linear Systems with Substitution and Elimination ▪ Modeling with Linear Systems ▪ Recognizing Linear Systems That Have No Solutions ▪ Recognizing Linear Systems That Have Infinitely Many Solutions

16.2 Matrix Solutions of Linear Systems 1039
Representing Systems of Linear Equations with Matrices ▪ Relating Matrices and Social Networks ▪ Applying Gauss–Jordan Elimination ▪ Using Technology to Solve Systems of Linear Equations
Checking Basic Concepts: Mixed Review of Sections 16.1 and 16.2 1051

GROUP ACTIVITY 1051

16.3 Determinants 1052
Calculating Determinants ▪ Finding the Area of a Region ▪ Applying Cramer's Rule
Checking Basic Concepts: Review of Section 16.3 1058

Summary ▪ Review Exercises ▪ Test 1058

Chapters 1–16 Cumulative Review Exercises 1062

17 Radical Expressions and Functions 1065

17.1 Radical Expressions and Functions 1066
Introducing Radical Expressions ▪ Recognizing Square Root Functions and Their Properties ▪ Recognizing Cube Root Functions and Their Properties

17.2 Rational Exponents 1077
Using Rational Numbers as Exponents ▪ Applying Properties of Rational Exponents
Checking Basic Concepts: Mixed Review of Sections 17.1 and 17.2 1085

17.3 Simplifying Radical Expressions 1086
Applying the Product Rule for Radical Expressions ▪ Applying the Quotient Rule for Radical Expressions
GROUP ACTIVITY 1094

17.4 Operations on Radical Expressions 1095
Adding and Subtracting Radical Expressions ▪ Multiplying Radical Expressions ▪ Rationalizing the Denominator
Checking Basic Concepts: Mixed Review of Sections 17.3 and 17.4 1106

17.5 More Radical Functions 1107
Recognizing Root Functions and Their Properties ▪ Recognizing Power Functions and Their Properties ▪ Modeling with Power Functions
GROUP ACTIVITY 1114

17.6 Equations Involving Radical Expressions 1114
Solving Radical Equations ▪ Deriving and Applying the Distance Formula ▪ Solving Equations of the Form $x^n = k$
Checking Basic Concepts: Mixed Review of Sections 17.5 and 17.6 1127

17.7 Complex Numbers 1128
Introducing the Imaginary Unit and Standard Form ▪ Adding, Subtracting, and Multiplying Complex Numbers ▪ Calculating Powers of i ▪ Finding Complex Conjugates ▪ Dividing Complex Numbers
Checking Basic Concepts: Review of Section 17.7 1135

Summary ▪ Review Exercises ▪ Test 1135

Chapters 1–17 Cumulative Review Exercises 1141

18 Quadratic Functions and Equations 1143

18.1 Quadratic Functions and Their Graphs 1144
Working with Graphs of Quadratic Functions ▪ Solving Min–Max Applications

18.2 Transformations and Translations of Parabolas 1157
Graphing Basic Transformations of $y = ax^2$ ▪ Performing Vertical and Horizontal Translations ▪ Applying Vertex Form ▪ Modeling with Quadratic Functions
Checking Basic Concepts: Mixed Review of Sections 18.1 and 18.2 1170

18.3 Quadratic Equations 1170
Learning the Basics of Quadratic Equations ■ Applying the Square Root Property ■ Completing the Square ■ Solving an Equation for a Variable ■ Solving Applications of Quadratic Equations

GROUP ACTIVITY 1183

18.4 The Quadratic Formula 1184
Solving Quadratic Equations ■ Finding and Applying the Discriminant ■ Using Intercepts to Graph Quadratic Functions ■ Solving Quadratic Equations That Have Complex Solutions
Checking Basic Concepts: Mixed Review of Sections 18.3 and 18.4 1197

18.5 Quadratic Inequalities 1197
Recognizing Quadratic Inequalities ■ Finding Graphical and Numerical Solutions ■ Finding Symbolic Solutions

18.6 Equations in Quadratic Form 1207
Solving Higher Degree Polynomial Equations ■ Solving Equations That Have Rational Exponents ■ Solving Equations That Have Complex Solutions
Checking Basic Concepts: Mixed Review of Sections 18.5 and 18.6 1211

Summary ■ Review Exercises ■ Test 1211

Chapters 1–18 Cumulative Review Exercises 1218

19 Exponential and Logarithmic Functions 1221

19.1 Composite and Inverse Functions 1222
Introducing Composition of Functions ■ Recognizing One-to-One Functions ■ Finding Inverse Functions ■ Using Tables and Graphs to Express Inverse Functions

19.2 Exponential Functions 1238
Introducing Exponential Functions ■ Recognizing Graphs of Exponential Functions ■ Relating Percent Change and Exponential Functions ■ Computing Compound Interest ■ Modeling with Exponential Functions ■ Recognizing the Natural Exponential Function
Checking Basic Concepts: Mixed Review of Sections 19.1 and 19.2 1255

19.3 Logarithmic Functions 1255
Introducing the Common Logarithmic Function ■ Finding the Inverse of the Common Logarithmic Function ■ Working with Logarithms that Have Other Bases

GROUP ACTIVITY 1268

19.4 Properties of Logarithms 1268
Introducing Basic Properties of Logarithms ■ Using the Change of Base Formula
Checking Basic Concepts: Mixed Review of Sections 19.3 and 19.4 1274

19.5 Exponential and Logarithmic Equations 1275
Recognizing Exponential Equations and Models ■ Solving Applications of Exponential Equations ■ Recognizing Logarithmic Equations and Models ■ Solving Applications of Logarithmic Equations
Checking Basic Concepts: Review of Section 19.5 1286

Summary ■ Review Exercises ■ Test 1287

Chapters 1–19 Cumulative Review Exercises 1293

20 Conic Sections 1297

20.1 Parabolas and Circles 1298
Recognizing Types of Conic Sections ▪ Graphing Parabolas with Horizontal Axes of Symmetry and Writing Their Equations ▪ Graphing Circles and Writing Their Equations

GROUP ACTIVITY 1307

20.2 Ellipses and Hyperbolas 1307
Graphing Ellipses and Writing Their Equations ▪ Graphing Hyperbolas and Writing Their Equations
Checking Basic Concepts: Mixed Review of Sections 20.1 and 20.2 1317

20.3 Nonlinear Systems of Equations and Inequalities 1318
Solving Nonlinear Systems of Equations ▪ Solving Nonlinear Systems of Inequalities
Checking Basic Concepts: Review of Section 20.3 1326

Summary ▪ Review Exercises ▪ Test 1327

Chapters 1–20 Cumulative Review Exercises 1331

21 Sequences and Series 1333

21.1 Sequences 1334
Finding Terms of a Sequence ▪ Recognizing Representations of Sequences ▪ Modeling Population Growth with Sequences

21.2 Arithmetic and Geometric Sequences 1342
Recognizing Representations of Arithmetic Sequences ▪ Recognizing Representations of Geometric Sequences ▪ Solving Applications Involving Sequences
Checking Basic Concepts: Mixed Review of Sections 21.1 and 21.2 1351

21.3 Series 1352
Introducing Series ▪ Working with Arithmetic Series ▪ Working with Geometric Series ▪ Using Summation Notation

GROUP ACTIVITY 1361

21.4 The Binomial Theorem 1362
Introducing Pascal's Triangle ▪ Writing Factorial Notation ▪ Finding Binomial Coefficients ▪ Using the Binomial Theorem
Checking Basic Concepts: Mixed Review of Sections 21.3 and 21.4 1368

Summary ▪ Review Exercises ▪ Test 1368

Chapters 1–21 Cumulative Review Exercises 1372

APPENDIX A: Using the Graphing Calculator AP-1
APPENDIX B: Sets AP-12
APPENDIX C: Linear Programming AP-19
APPENDIX D: Synthetic Division AP-26
APPENDIX E: Using a Calculator AP-28
Answers to Selected Exercises A-1
Glossary G-1
Photo Credits C-1
Index I-1

Preface

Developmental Mathematics with Applications and Visualization: Prealgebra, Beginning Algebra, and Intermediate Algebra, Second Edition, gives meaning to the numbers that students encounter by *developing concepts in context* through the use of applications, multiple representations, and visualization. By seeing the concept in context *before* being given the mathematical abstraction, students make math part of their own experiences instead of just memorizing techniques. Research shows that this method is essential to empowering students. Seamlessly integrated real-life connections, graphs, tables, charts, and meaningful data help students deepen understanding and prepare for future math courses—and life—by teaching them critical thinking and problem-solving skills.

By adding new design features and reworking existing content in this text, we increased the amount of content that is delivered in a real-world context. We made the text more accessible by replacing paragraphs of text with visual information. Recognizing that many users want a customizable text, we put new emphasis on clearly identifying objectives and content in both the exposition and the exercise sets.

It is essential that all students achieve competency in mathematics so that they can realize their vocational dreams and contribute to society. A purely traditional/abstract approach to teaching math does not work for most of today's students, and it has led to exclusivity. Our approach promotes inclusivity and diversity within the discipline and beyond. Our goal in *Developmental Mathematics with Applications and Visualization: Prealgebra, Beginning Algebra, and Intermediate Algebra* is to help all students succeed.

This textbook is one part of our comprehensive program:

- *Prealgebra*
- *Beginning Algebra with Applications and Visualization,* Third Edition
- *Beginning & Intermediate Algebra with Applications and Visualization,* Fourth Edition
- *Intermediate Algebra with Applications and Visualization,* Fifth Edition
- *Developmental Mathematics with Applications and Visualization: Prealgebra, Beginning Algebra, and Intermediate Algebra,* Second Edition
- *MyMathLab for Interactive Developmental Mathematics*

New To This Edition

We enriched our program—the text, the supplements, and MyMathLab—in the following areas to enhance our support for conceptual understanding. See also Resources for Success: MyMathLab online course (pp. xxv–xxvii) for additional information.

- **Section Introduction Videos** Every section in the text is now coupled with a short video (in MyMathLab) that introduces the section's concepts in a contextual setting.
- **See the Concept** Throughout the text, we replaced paragraphs of text with clearly labeled infographics that visually—and concisely—walk students through concepts. Every See the Concept in the text has a companion video in MyMathLab to help bring the concept to life.
- **Assignable Content** To assess student understanding when they encounter new material, many text features are assignable in MyMathLab, including every Section Introduction Video and See the Concept Video and the majority of the Reading Checks, Making Connections, and Critical Thinking features.

- **Math in Context** Students are more engaged when mathematics is tied to current and relevant topics. This new feature helps students recognize when and how math is used in the real world.
- **Connecting Concepts with Your Life** Students are often already equipped with the knowledge needed to understand a new concept. Our new Connecting Concepts with Your Life feature gives meaning to mathematics by relating common life experiences that students already understand.
- **Objective List** At the beginning of each section, we list objectives to give a clear outline of the section contents and to make it easier to customize the course.
- **Comment Balloons** When appropriate, comments were inserted next to steps and procedures to make them more (immediately) understandable. Now students don't always have to read a paragraph of text to understand the concept.
- **Modeling Data** The data in hundreds of exercises and examples were updated to keep the applications relevant and fresh. Each chapter opens with a real-data application, and many new examples were added, including discussions about the Internet, social networks, tablet computers, and other contemporary topics.
- *Guided Workbook* Keyed to the text by section and objective, the *Workbook* leads students through the course, giving them the opportunity to record key information, work practice problems, and show and keep their work for reference—as well as taking conceptual understanding one step further by asking students to explain "Why?" after select questions.

Refinement of Content

CHAPTER 1: Interpretation of data that is presented visually has been emphasized with the inclusion of a new subtopic discussing spider charts. Also, new examples and exercises have been added to increase the coverage of properties of addition and multiplication.

CHAPTER 2: A contextual description of signed numbers has been added that relates positive and negative integers to income and debt. Additionally, the process for adding and subtracting integers using a number line has been rewritten and streamlined.

CHAPTER 3: New graphics have been added to help students understand what it means to solve equations numerically and graphically.

CHAPTER 4: Several new infographics have been added throughout the chapter for a more visual coverage of fractions and the arithmetic of fractions. Also, there is now increased coverage of estimation involving fractions.

CHAPTER 5: New infographics have been added that concisely demonstrate graphing decimals, comparing decimals, and the Pythagorean theorem. Also, there is now increased coverage of estimation involving decimals.

CHAPTER 6: New contextual examples involving ratios have been added. Also, coverage of similar figures has been increased.

CHAPTER 7: Dozens of exercises involving applications of percent have been either added or updated to increase the real world relevance of this important topic.

CHAPTER 8: New infographics have been added to streamline the coverage of the most commonly used concepts and formulas from geometry.

CHAPTER 9: Many new exercises have been added throughout the chapter, with an emphasis on linear equations and their solutions.

CHAPTER 10: New examples and exercises were added to cover graphing nonlinear equations by hand. Also, at the request of reviewers, intercepts are now identified as ordered pairs rather than real numbers.

CHAPTER 11: Supply and demand examples and exercises involving systems of linear equations were added to this chapter.

CHAPTER 12: The exercises pertaining to rules for exponents were expanded and more clearly categorized in the exercise sets.

CHAPTER 13: New examples and exercises that require students to rearrange terms before factoring by grouping were added to this chapter.

CHAPTER 14: The description of least common multiple now includes a more visual, contextual explanation involving the listing method.

CHAPTER 15: We put more emphasis on solving absolute value inequalities symbolically. A new objective that more fully discusses the concept of a relation was inserted before the definition of a function, and exercises were added that give students practice differentiating between relations and functions. Also, a new objective on piecewise-defined functions was added in Section 15.3, with corresponding exercises in both Sections 15.3 and 15.4.

CHAPTER 16: New exercises that ask students to represent simple social networks with matrices were added to this chapter.

CHAPTER 17: The coverage of using zeros to write the formula for a function was expanded in Sections 17.5 and 17.7.

CHAPTER 18: A new graphical description for deriving the formula to find the vertex of a parabola is now included. Also, a real-world example was added to highlight a common situation in which a quadratic equation might have zero, one, or two solutions. This chapter additionally includes increased coverage of complex solutions to quadratic equations.

CHAPTER 19: A discussion connecting social network posts with half-life and exponential decay is now included to give students a relevant and accessible example of this important concept. At the request of reviewers, we expanded the coverage of changing between exponential and logarithmic forms of equations.

CHAPTER 20: New exercises involving the equation of a circle were added.

CHAPTER 21: In this final chapter, the coverage of writing a formula for a sequence from given terms of the sequence was increased.

Hallmark Features

Experiencing Math in the Real World

In our program, we emphasize teaching algebra in context. Students typically understand the math best when the concepts and skills are tied to the real world and presented visually. We believe that meaningful applications and visualization of important concepts are pathways to success in mathematics.

- **Chapter Openers** Each chapter opens with a contemporary application that motivates students by offering insights into the relevance of that chapter's mathematical concepts.
- **NEW! Math in Context** Where appropriate, we expand on specific math topics and their connections to everyday life.
- **Applications** We integrate applications (many with actual data) into both the discussions and exercises to help students become more effective problem solvers.
- **Modeling Data** We provide opportunities for students to model real and relevant data with their own functions. **Online Exploration** exercises invite students to find their own data on the Internet and use mathematics to analyze them.
- **NEW! Section Introduction Videos** Every section in the text is now coupled with a short video (in MyMathLab) that introduces the section's concepts in a contextual setting.
- **NEW! Connecting Concepts with Your Life** This new feature helps students understand mathematics by relating it to common life experiences.

Understanding the Concepts

Conceptual understanding is critical to success in mathematics.

- **Learning the Math from Multiple Perspectives** Throughout the text, we present concepts by means of **verbal, graphical, numerical,** and **symbolic representations** to support multiple learning styles and problem-solving methods.
- **New Vocabulary** At the beginning of each section, we direct students' attention to important terms before they are discussed in context.
- **Reading Checks** These questions appear in the text along with important concepts, ensuring that students understand the material they have just read before moving on. Selected Reading Checks (RC) are also available to be assigned in MyMathLab.
- **Making Connections** Throughout the text, we help students understand the relationships between previously learned concepts and new ones. Selected Making Connections (MC) are also available to be assigned in MyMathLab.
- **Critical Thinking** One or more of these exercises appear in most sections, posing questions that can be used for class discussion, group work, or homework assignments. Selected Critical Thinking (CT) features are also available to be assigned in MyMathLab.
- **Putting It All Together** At the end of each section, we summarize the techniques just covered and reinforce the mathematical concepts presented in the section.
- **NEW! See the Concept** Throughout the text, we replaced paragraphs of text with clearly labeled infographics that visually—and concisely—walk students through concepts. Every See the Concept in the text has a companion video in MyMathLab to help bring the concept to life.
- **NEW! Comment Balloons** Comments are included next to steps and procedures to make them more (immediately) understandable.

Practicing the Concepts

A variety of exercise types support the application-based and conceptual nature of this text. Designed to reinforce the skills students need to be able to move on to the next concept, the comprehensive section and chapter exercise sets cover basic concepts, skill-building, writing, applications, and conceptual mastery. These exercise sets are further enhanced by several special types of exercises integrated throughout the text:

- **Now Try** Suggested exercises follow every example for immediate reinforcement of skills and concepts.
- **Concepts and Vocabulary exercises** appear in every section.
- **Checking Basic Concepts** These mixed-review exercise sets appear after every other section and can be used for individual or group review.
- **Thinking Generally** Appearing in most section exercise sets, these open-ended conceptual questions encourage students to synthesize what they have just learned.
- **Writing about Mathematics** At the end of most section exercise sets, students are asked to explain the concepts behind the mathematics procedures they have just learned.
- **NEW! Assignable Content** To assess student understanding when they encounter new material, many text features are assignable in MyMathLab, including every Section Introduction and See the Concept Video and the majority of the Reading Checks, Making Connections, and Critical Thinking features.
- **Group Activities** This feature occurs once or twice per chapter, and provides an opportunity for students to work collaboratively on a problem. Most activities can be completed with limited use of class time.
- **Using a Graphing Calculator**
  - **Graphing Calculator Exercises** The icon denotes an optional exercise that requires students to have access to a graphing calculator.
  - **Technology Notes** Throughout the text, these optional notes offer students guidance, suggestions, and cautions on the use of the graphing calculator.

Mastering the Concepts

By reviewing the material and putting their abilities to the test, students will be able to assess their level of mastery as they complete each chapter.

- **Chapter Summary** In a quick but thorough review, we combine key terms, topics, and procedures with illuminating examples to assist students as they prepare for a test.
- **Chapter Review Exercises** Students can work these exercises to gain confidence that they have mastered the material.
- **Chapter Test** Students can reduce math anxiety by using these tests as a rehearsal for the real thing.
- **Chapter Test Prep Videos** These videos offer further help by showing an instructor's step-by-step solutions to every exercise in each Chapter Test.
- **Cumulative Review Exercises** Starting with Chapter 2 and appearing in all subsequent chapters, Cumulative Review Exercises help students see the big picture of math by reviewing topics and skills they have already learned.

Resources for Success
MyMathLab® Online Course (access code required)

The course for *Developmental Mathematics with Applications and Visualization: Pre-algebra, Beginning Algebra, and Intermediate Algebra,* 2e, includes all of MyMathLab's robust features and functionality, plus these additional new features:

Expanded Video Program
Each section in the text is now supported by a short *new* **Section Introduction Video** that presents the section concepts in context. For key concepts throughout the text, infographics visually—and concisely—walk students through the concepts. Companion **See the Concept Videos** help bring these concepts to life.

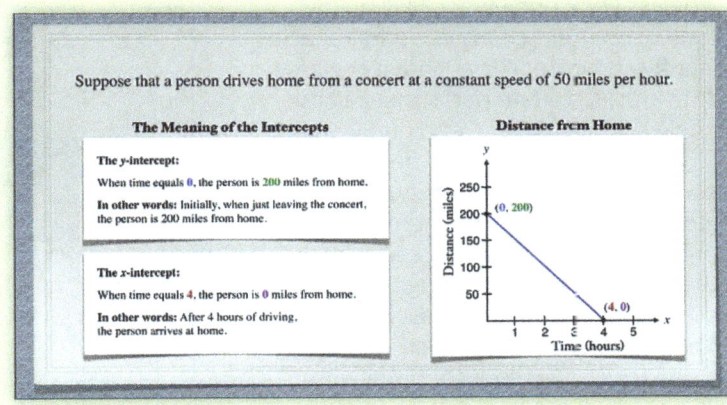

New Workspace Assignments
Workspace Assignments allow students to work through an exercise step-by-step, showing their mathematical reasoning. Students receive immediate feedback after they complete each step, and helpful hints and videos are available for guidance as needed. When students access workspace using a mobile device, handwriting recognition software allows them to write out answers naturally using their fingertip or a stylus.

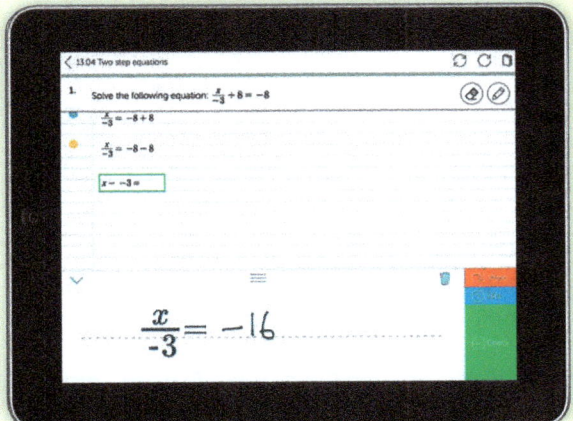

New Learning Catalytics
Learning Catalytics uses students' mobile devices for an engagement, assessment, and classroom intelligence system that gives instructors real-time feedback on student learning.

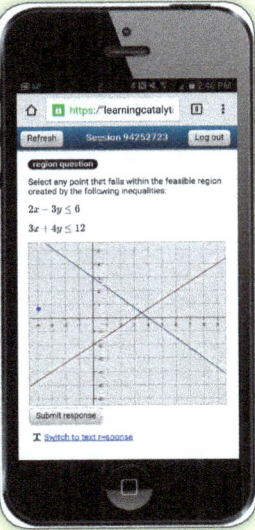

www.mymathlab.com

Resources for Success
MyMathLab® Online Course (access code required)

New Skill Builder Adaptive Practice

When a student is struggling with the assigned homework, Skill Builder exercises offer just-in-time additional adaptive practice. The adaptive engine tracks student performance and delivers questions to each individual that adapt to his or her level of understanding. When the system has determined that the student has a high probability of successfully completing the assigned exercise, it suggests that the student return to the assignment. This new feature allows instructors to assign fewer questions for homework, allowing students to complete as many or as few questions as needed.

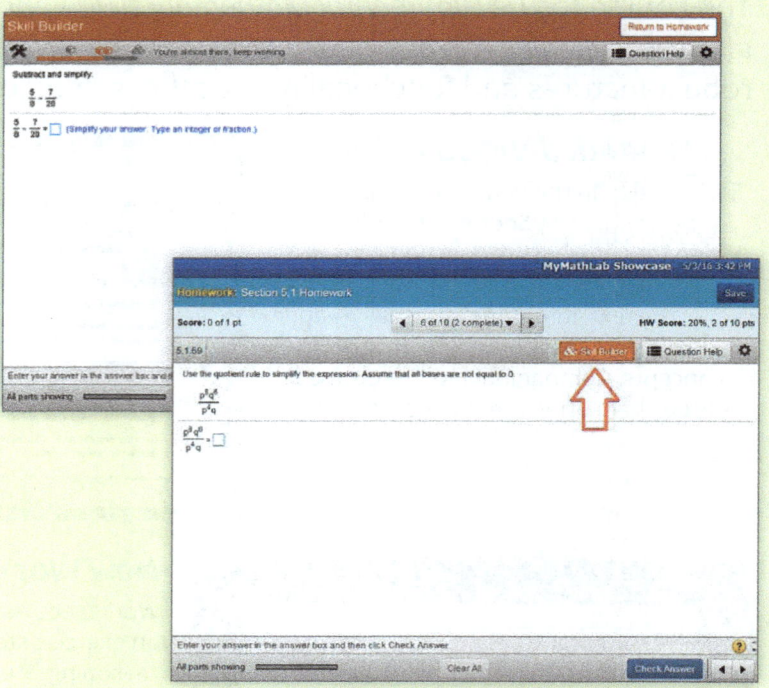

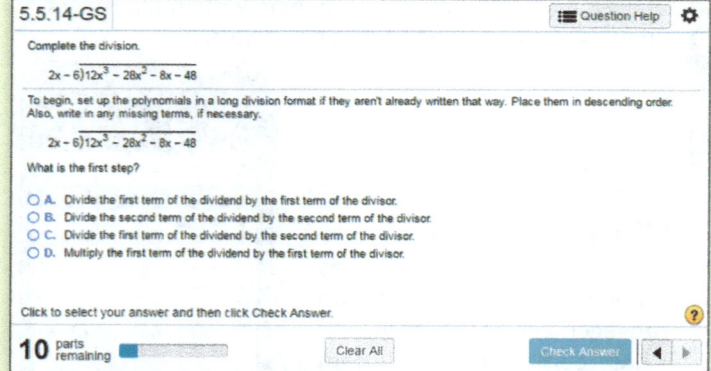

Interactive Exercises

MyMathLab's hallmark interactive exercises help build problem-solving skills and foster conceptual understanding. For this Second Edition, Guided Solutions (GS) exercises were added to reinforce the step-by-step problem-solving process, while the *new* Drag & Drop (DD) functionality was applied to matching exercises throughout the course to better assess a student's understanding of the concepts.

www.mymathlab.com

Resources for Success

Instructor Resources

Additional resources can be downloaded from **www.pearsonhighered.com** or hardcopy resources can be ordered from your sales representative.

Annotated Instructor's Edition
- Contains Teaching Tips, Teaching Examples, and answers to every exercise in the textbook (excluding Writing about Mathematics).
- Answers that do not fit on the same page as the exercises are supplied in the Instructor's Answer Appendix in the back of the textbook.

Instructor's Resource Manual with Tests and Mini-Lectures (Download only)

Includes resources designed to help both new and adjunct faculty with course preparation and classroom management.
- *New!* **Activities** Multi-step Activities for group work, classroom instruction, and homework are available for key concepts.
- Mini-lectures for each text section.
- Notes for presenting graphing calculator topics as well as supplemental activities.
- Teaching Tips and additional exercises for selected content.
- Three free-response alternative test forms and one multiple-choice test form per chapter; one free-response and one multiple-choice final exam.
- Five sets of Cumulative Review Exercises (Chapters 1–3, 1–6, 1–9, 1–12, 1–15, 1–18, and 1–21).

Instructor's Solutions Manual (Download only)

Solutions to section-level exercises (excluding Writing about Mathematics), Checking Basic Concepts, Chapter Review, Chapter Test, and Cumulative Review.

PowerPoint® Lecture Slides

Fully editable slides that correlate to the textbook.

TestGen®

TestGen® (www.pearsoned.com/testgen) enables instructors to build, edit, print, and administer tests using a computerized bank of questions developed to cover all the objectives of the text.

Student Resources

Additional resources to help student success.

Enhanced Video Resources
- Comprehensive video coverage of the course available at the section and objective levels, with new Section Introduction and See the Concept Videos.
- Ideal for variety of course formats, including lecture-based, lab-based, online, self-paced, hybrid, and flipped classroom.
- Chapter Test Prep Videos let students watch an instructor work through step-by-step solutions to all Chapter Test exercises from the text. Available in MyMathLab and on YouTube.

Student's Solutions Manual

Contains solutions to the following text exercises: odd-numbered Section and Chapter Review (excluding Writing about Mathematics), Concepts and Vocabulary, Checking Basic Concepts, Chapter Test, and Cumulative Review.

New Guided Workbook
- Keyed to the text by section and objective.
- Helps the students stay on track and reinforces conceptual understanding throughout the course.
- Promotes active learning with space for students to answer questions about the See the Concept videos, write about key terms, solve exercises and show and keep their work, and reflect about their understanding of the math.
- Binder-ready (3-hole-punched and unbound), it can form the basis of a course notebook.
- **Activities** Multi-step Activities for group work, classroom instruction, and homework are available in MyMathLab for key concepts.

www.mymathlab.com

Acknowledgments

We thank the many individuals who contributed to the development of our program over its two editions by participating in focus groups, completing surveys, completing reviews, and contacting us with suggestions. We thank the following instructors whose comments and suggestions informed our preparation of *Developmental Mathematics with Applications and Visualization: Prealgebra, Beginning Algebra, and Intermediate Algebra*, Second Edition:

Gurdial Arora, *Xavier University of Louisiana*
Susan Barbitta, *Guilford Technical Community College*
Patricia Blus, *National Louis University*
Susan Cooper-Nguyen, *Harrisburg Area Community College*
Laurie DeHerrera, *Pikes Peak Community College*
Sarah Endel, *Rochester Community and Technical College*
Thomas Fitzkee, *Francis Marion University*
Kim Ghiselin, *State College of Florida, Manatee-Sarasota*
Ryan Girard, *Kauai Community College*
Melissa Hedlund, *Christopher Newport University*
Marsha Hodge Cardena, *University of Texas at El Paso*
Roland "Bud" Jenkins, *Hawaii Pacific University*
Nena Kabranski, *Tarrant County College*
Chris Knight, *Walters State Community College*
Doug Mace, *Kirtland Community College*
Marvin F. Mascarenas, *New Mexico Highlands University*
Cindy McCallum, *Tarrant County College*
Alejandro Mena, *University of Texas at El Paso*
Linda Myers, *Harrisburg Area Community College*
David Price, *Tarrant County College, SE*
Elise Price, *Tarrant County College, SE*
Jorge Sarmiento, *County College of Morris*
Salvatore Sciandra, *Niagara County Community College*
Janet E. Teeguarden, *Ivy Tech Community College*
Marissa Wolfe, *Yavapai College*

We would like to welcome David Atwood to our team. He has been instrumental in this revision, providing assistance with answers, proofing, and video creation.

Jennifer Blue, Paul Lorzcak, Hal Whipple, and Mark Rockswold deserve special credit for their help with proofing and accuracy checking. Without the excellent cooperation from the professional staff at Pearson Education, this project would not have been possible. Thanks goes to Michael Hirsch for his support. Particular recognition is due Cathy Cantin, who gave essential advice and assistance. The outstanding contributions of Lauren Morse, Joe Vetere, Alicia Frankel, Erin Carreiro, Rebecca Williams, Eric Gregg, Shannon Bushee, and Alison Oehmen are greatly appreciated. Special thanks go to Chere Bemelmans at Cenveo who was instrumental in the success of this project.

Thanks go to Wendy Rockswold, Carrie Krieger, and Burgess Johns, whose unwavering encouragement and support made this project possible. We also thank the many students and instructors who used the previous editions of this textbook. Their suggestions were insightful and helpful.

Please feel free to send us your comments and questions. You can find our contact information at www.garyrockswold.net. Your opinion is important to us.

Gary Rockswold
Terry Krieger
Jessica Rockswold

Index of Applications

Astronomy

Comets
  orbit of Halley's Comet, 1316
  speed of, 1306–1307
  trajectories of, 1306
Earth, weight near, 920
Jupiter's orbit, 1094
Mars' orbit, 1330
Mars temperatures, 421
Milky Way, 772
Moon
  diameter of, 421
  distance to, 772
  throwing a baseball on, 1155
  weight on, 920
Planets
  alignment of, 884
  orbits of, 1085, 1094, 1114, 1316, 1330
Pluto's orbit, 1316
Satellite orbits, 1316
Stars, magnitude of, 1267
Sun
  distance of, in light-years, 772
  planet orbits and distance from, 1085, 1094, 1114, 1316
  speed of, 772

Business

Advertising
  revenue, 629
  social media, 723
Apple's revenue, 1253
Blackberry shares, 562
Commission, 464, 465, 484
Compensation, female CEO, 1359–1360
Corporate losses, 142
Cost
  average, 874–875
  minimizing, 1127, AP-25
  total production, 867, 875, 968
  of TV ads, 465
Employees
  productivity of, 1076
  working on two projects, 716
  working shifts, 884
Estimated taxes, 329
Interior design, 464
Manufacturing
  container production, 1038
  error analysis, 1014
  mp3 players and digital video players, 716
  tires, 1014
  wheels and trailers, 723
Microsoft income, 105
Microsoft vs. Google, 329
Music shipments, 85
Office supplies, 484
Patents issued, 314
Paypal, 442
Profit, 628, 1000, AP-25
Profit and loss, 126–127
Quality control, 339, 874
Revenues
  advertising, 629
  Apple music, 44
  calculating, 846, 921
  and cost, 578
  digital ad, 463
  DVD and CD, 706
  flash drive, 628
  gaming, 283
  global digital, 950
  Google and Facebook, 866
  Google and Yahoo, 867
  maximizing, 1155, 1216, AP-25, AP-24
  music download, 668, 684–685
  social media, 669, 722
  streaming music, 745
  from ticket sales, 1155
  Tidal music, 44
Sales
  candy and coffee, 716–717
  of DVDs and CDs, 706
  Fitbit, 1285
  personal computer, 999–1000
  street vendor, 153
  ticket, 564, 1038, 1060
Smartwatch market, 1284
Starbucks stores, 55, 64
Street vendor sales, 153
Supply and demand, 685
Ticket prices, 694, 1093
Ticket sales, 564, 1038, 1060
Walgreens locations, 86
Walmart retail sales, 968
Walmart stores, 64
Web browser market, 603
Workforce reduction, 465

Design/Construction

Arch bridge, 1317
Basketball court dimensions, 694
Building a play set, 339
Carpentry, 242
Construction plans, 393
Football field, 694
Garage dimensions, 723
Guy wire length, 1126
Highway curves, 905, 1126
Highway elevations on a sag curve, 1206
Model building, 262
Natural gas line, cost of, 1127
Olympic pool, 394, 411
Open flood channel design, 1076, 1106
Painting, 262
Pen enclosures, 753, 1156, 1205
Raceways, 522
Rain gutters, 1216
Rectangular building, 785
Room ventilation, 1350
Running track, 362, 522
Shed dimensions, 835, 846
Sidewalk dimensions, 761, 840, 1196
Strength of a beam, 918, 919, 920, 928
Width of plot of land, 1182
Wire length, 371

Education

Business majors, 394
Cheating, 463, 484
Classroom population, 242
College, saving for, 464, 484
College degrees
  and pay, 78
  percentage of people with, 1236
  women with pharmacy degrees, 1077
College loans, 563, 1037
College students
  taking classes, 235
  women, 306
College tuition
  cost of, 660–661, 920, 949, 950
  cost of fees and, 968
  historical cost of, 981
  increases in, 660
  modeling, 661
  predicting, 661
Comparing colleges, 382, 430
Comparing schools, 382
Conference fees, 484
Enrollment, 463
Extra credit, 222
Grade average, 578, 583
Graduate school, 464

INDEX OF APPLICATIONS xxxi

High school enrollment, 96
High school grades, 1237
Homeschooling, 262
Homework, completing, 269
Male teachers, 235
Private school tuition, 464
Reading rate, 393
Remembering what you learn, 1000
Scholarships, 242
School children, 382
School districts, 382
School supplies, 464
Study time, 222

Energy

Efficient lighting, 242, 253
Global oil demand, 554
Going green, 242, 393, 431, 484
Nuclear reactors, 193
U.S. energy consumption, 1216–1217
U.S. natural gas consumption, 840
U.S. oil production, 187
U.S. renewable energy, 158, 187
Wind power, 44, 70, 253
Wind turbines, 362, 371

Entertainment

Apple music revenue, 44
CDs
    cost of, 1037
    recording music on, 918
    sales of, 706
Digital entertainment
    global revenue, 950
    music downloads, 668, 684–685
    music subscriptions, 684–685
    streaming music, 745
Digital music, 464
Drive-in theaters, 349
DVD sales, 706
Historical music sales, 1196
Horror movies, 235
Media consumption, 785
Movie tickets, 349
Music sales, 142
Musical revenues, 562
Online videos, 1084, 1125
Recording music, 918
Television
    cable, 293
    classic, 193, 362–363
    diagonal of, 1125
    screen size and distance from, 1182
    width and height of, 1125
Tidal music revenue, 44
Video games, 442

Environment

Air quality, 208, 1022
Antarctic ozone layer, 640
Bamboo growth, 179
Carbon dioxide emissions, 242
Carbon dioxide emitters, 1341
Carbon emissions, 1156
Carbon monoxide emissions, 283
Farm land, 188
Flood control, 1126
Garbage and household size, 1049
Greenhouse gases, 1286
Hazardous waste, 159, 187
Kalahari Desert, 420
Kilimanjaro Glacier melt, 660
National forests, 44
Ocean depth, 96
Old Faithful Geyser, 1049–1050
Ozone and UV radiation, 919
Plant growth, 717
Rainforests, 167, 967, 1370
Rise in sea level, 652
River current speed, 705, 723, 906, 928
Skin cancer and ozone, 1236–1237
Solid waste production, 603, 1340–1341
Underground depth, 104
Underwater depth, 105, 111

General Interest

Ages
    finding, 158, 187, 188, 193, 222, 561
    of siblings, 167
American flag, 382
Antique value, 78
Auditorium seating, 1341, 1350–1351
Baking, 235, 262, 382, 393
Baling twine, 431
Beverage can, 515
Board games, 394
Candy mixture, 1050, 1060
Canoeing, 1195–1196
Cat owners, 463
Checked luggage, 402, 410, 420, 431
Checkerboard, 44
Cleaning solutions, 235, 262, 293
Cleaning windows, 193
Coffee mixture, 564, 694
Coin mixture, 564
Conversion
    feet and yards, 1237
    quarts and gallons, 1237
Counting coins, 78
Currency exchange rates, 293, 329
Cutting hair, 242
Cutting rope, 402
Distance from home, 614
Driving distance, 306, 315, 371
Feeding teenagers, 393
Fencing a property, 45
Fibonacci Sequence, 1351
Filling an aquarium, 402
Fireworks, 78
Fraternal twins, 411
Friendship bracelets, 402
Gambling losses, 981
Game tokens, 104, 111, 119
Gaming revenue, 283
Garden size, 85, 530, 693, 722, 753
Gender, 442, 486
Genetically engineered soybeans, 1077
Going organic, 242, 293
Home prices, 465
Home size, 188
Hotel room prices, 464, 693, 722
Hydraulic lift, 530
Infinite series, 1361
International travel, 242
Joint variation, 920
Keyboarding, 382
Ladder length, 362
Lawn fertilizer mixture, 1038
Longest hair, 420
Making coffee, 919
Making fudge, 918
Model car, 394
Museums, 562
National defense, 515, 521
National Endowment for the Arts, 338
National monuments, 85
NBA basketball court dimensions, 694
New Year's resolutions, 463
Number problems, 45, 56, 85, 561, 694, 705, 840, 1216, 1330
Number puzzles, 78, 159
Paint can, 530
Painting a house, 905
Painting a wall, 153
Pen enclosures, 753, 1156, 1205
Philanthropy, 338
Phone numbers, 293
Photography, 394, 883
Pieces of thread, 411
Pile of sand, 529
Polar plunge, 928
Postage prices, 85, 746, 982
Printer paper, 420
Probability, 761, 860, 866
Pumping water, 905
Raffle tickets, 382
Raising animals, AP-25
Satisfaction from eating pizza, 1154
Serving size, 269
Shoveling snow, 905
Spelunking, 119
Stacking logs, 1360
Standing in line, 859
Stop sign, 537
Super soaker, 420
Swimming pools
    chlorine levels, 1254, 1350
    dimensions, 761, 840
    emptying, 905, 928
    Olympic, 411
    water in, 614
Tall building, 306
Theater seating, 1341, 1350–1351

xxxii   INDEX OF APPLICATIONS

Thermostat setting, 967
Tic-Tac-Toe board, 44
Tightening/loosening a lug nut, 919, 928
Time spent in line, 906, 998
Tree height, 431
Troops in Iraq, 188
Twin prime pairs, 222
Waiting in line at post office, 905, 906, 999
Walking for charities, 628
Wall clock, 515
Wardrobe, 393
Water cup, 530
Water usage in U.S., 668
Weight of milk, 421
Wilderness map, 393
Working together, 283, 293

Geography

Appalachian Trail, 463
Bermuda Triangle, 521
Canadian lakes, 56
Cities, 56, 242
Climbing Kilimanjaro, 235
Elevations, highest/lowest, 64, 96, 105, 111, 1013–1014
European Union, 208, 253
Farming, 382
Great Lakes, 695
Great Pyramid, 530
Languages spoken at home, 262
Longest rivers, 695
Mount Everest and K2, 562
Names of U.S. states, 253
Niagara Falls, flow of water over, 660
State names, 464

Geometry

Angles
    complementary, 693
    supplementary, 693, 705, 722
    in triangles, 693, 722, 1037, 1050, 1060
Area
    of baseball diamonds, 44
    of basketball hoop, 530
    of circles, 745, 746
    of a figure, 745
    maximizing, 1156
    of a pen, 1156
    of rectangles, 44, 45, 85, 153, 159, 193, 262, 338, 348, 745, 753, 760, 779, 785, 797, 845, 847, 883, 1217, 1325–1326, 1330
    of squares, 262, 338, 745, 746, 760, 761, 785, 797, 1341, 1360
    of tennis courts, 44
    of trapezoids, 282, 338, 349
    of triangles, 235, 338, 348, 349, 779, 797
Area and perimeter
    of a desktop, 1330
    of a room, 1325–1326
Base of triangles, 577, 583, 1182

Dimensions
    box, 840, 846
    can, 1325
    cone, 1217, 1326
    cube, 846
    cylinder, 1206, 1330
    of football field, 694
    garage, 723
    garden, 693, 722
    pen, 1205
    picture, 705
    pool, 840
    rectangle, 187, 193, 562, 685, 845, 982
    shed, 835, 846
    triangle, 187–188, 193, 562, 685
Ellipses, 1316, 1330
Height
    antenna, 394
    tree, 394, 919, 928
    triangle, 1182
Length
    of a diagonal, 1106
    of hypotenuse, 1127
    of a side, 45, 55, 167, 193, 845
Perimeter
    of basketball court, 694
    of figures, 27–28, 85
    of football field, 521
    of a pen, 753
    of rectangles, 78, 159, 167, 187, 253, 315, 562, 577, 753, 1105, 1325–1326, 1330
    of squares, 1105, 1341, 1360
    of trapezoids, 253
    of triangles, 78, 187, 253, 315, 562, 1037, 1105
Radius of circle, 846
Surface area
    of a balloon, 1292
    of basketball, 530
    of a box, 753
    of cubes, 754, 761
    of a person, 1112
    of spheres, 754
Volume
    of a balloon, 1236
    of a box, 753, 779, 798
    of cubes, 761
    of spheres, 785

Health and Fitness

Baby's head size, 1085
Blood alcohol level, 1253
Body fat, 442
Body temperature, 420
Bone marrow, 442
Calories
    burning, 55, 329, 371, 694, 705, 723, 968
    counting, 44, 45, 55
    and exercise, 694
Cosmetic surgery, 159, 187

Down Syndrome risk, 1284
Elliptical weight machines, 1316–1317
Exercise
    and calories, 694
    and heart rate, 745
Fat grams, 1022
Heart rate
    animal, 85
    of athletes, 785, 998
    and exercise, 70, 745
    maximum, 716
    resting, 78
    target, 339, 371, 448, 716
Height
    at birth, 262, 293
    and weight, 716, 1126
Human body, 420, 463
Jogging/running
    distance and time, 578, 1182
    distance traveled when, 262, 553, 668, 1182
    heart rate while, 785, 998
    speeds, 563, 564, 906, 928, 1050
    time spent, 705
Life expectancy in U.S., 967
Liquid nitrogen and skin temperature, 111
Liver transplants, 1284
Nutrition, 463, 484
Obesity, 420, 431, 486
Polar plunge, 928
Skin cancer and ozone, 1236–1237
Steroid use, 463
Vegetarian diets, 306
Vegetarian meals, 382
Walking
    distance, 315
    rate, 269
Weight
    gain, 158, 187, 562
    and height, 716, 1126
    lifting, 968
    loss, 158, 187, 562
    on the moon, 920
    obese, 420, 431
    in stones, 402
Working and MS, 442
Workouts, estimating time, 85

Life Sciences

Animal heart rate, 70, 85
Animal pulse rates, 1085, 1113
Animal shelters, 382
Animal speeds, 424
Animal stepping frequency, 1084
Ants, distance and movement of, 629
Bacteria
    growth of, 1253, 1284, 1292
    number of, 15
    size of, 306, 370
Bears, weight of a bear, 1049, 1050

Birds
life span of a robin, 1285
populations of, 921
wing size and weight of, 1085, 1113, 1125
Endangered species, 44, 158, 187, 193, 562
Fawns, predicting number of, 1038
Fish
decline of bluefin tuna population, 1285
populations of, 967, 1285
size and weight of, 578, 967
Frogs
distance and movement of, 629
population of, 859
Genetically engineered soybeans, 1077
Insect populations
of black flies, 629
decline of, 668
density of, 1351
estimating, 1285
growth of, 70, 859, 1341
Large mammals, 55
Mosquitoes and temperature, 1237
Seedling growth, 1156, 1182–1183
Whales
census of humpback whales, 950
and diving, 921
weight of blue whales, 420

Measurement

Fruit drink ounces, 431
Jumbo bottle units, 402
Metric tons, 411
Portage, length of, 402
Relative errors, 1014, 1022
Serving size
in bags of chips, 402
in bottles, 410, 411
in cans, 411
in jugs, 402, 431
Soda bottle units, 402, 410, 420
Water weight, 411

Meteorology

Air temperature, 135
Altitude
and air temperature, 660, 920
and dew point, 578, 981
and temperature, 135, 179, 193, 578, 982
Dew point and altitude, 578, 981
Hurricanes, 1266, 1286
Precipitation averages, 951, 1022
Rainfall
calculating, 640
probability of, 761
Snow, water content in, 394, 918, 928
Sunny days, 64
Temperature
air, 135, 660, 920
and altitude, 135, 179, 193, 982
average, 127, 142, 1013
change in, 104, 111, 142, 651
cold, 96, 119
converting, 85, 126, 142, 269, 283, 349, 371, 982, 1237
and mosquitoes, 1237
Temperature scales, 1022
Tornadoes, 28
Warm day, 431
Wind speeds, 951, 998, 1237, 1292

Personal Finance

Account balances, 96, 105, 111, 142, 315
Airline tickets, 464
Annual income, 15
Annuities, 892
Appreciation of lake property, 1351
ATM fees, 119
Bank loans, 564
Book purchases, 85
Candy mix purchase, 722
Car loans, 472
Car payments, 330
Clearance sale, 464
Commission, 464, 465, 484
Cost
of carpet, 402, 920, 928
of CDs, 1037
of college tuition, 660–661, 920, 950
of different food combinations, 1037, 1060
of dinner, 330, 371, 464
of driving, 640, 949
of fuel, 651
of groceries, 329
of lunch, 314
of pet food, AP-25
of vitamins, AP-25
Credit card
cash back, 463
debt, 564
interest, 472
Depreciation, 1341
Discount, 442
DVD purchases, 45
Electrical rates, 640
Equity loans, 382
Family budget, 235
Federal income taxes, 949
Flash drive purchases, 45
Flea market purchases, 56
Gasoline prices, 306
Golf clubs, 464
Gross pay, 465, 485
Home equity loans, 472
Home loans, 330
Home prices, 78
Interest
compound, 735, 761
earning, 472
on investments, 1050
on savings account, 1284
Investment
growth of, 735, 1284, 1292
interest on, 1050, 1286
mixture, 705, 1038
money, 582
in mutual funds, 1284, 1292
Land line phone plan, 640
Late fees, 119, 142
Lending money, 56
Monthly payments, 371
Movie download account, 55
Net pay, 465, 485
Net worth, 111
Online shopping, 464
Paid vacation days, 393
Parking rates, 578, 583
Pay, 78
Pension savings, 235
Prepaid minutes discount rate, 464, 484
Rental cars, 578, 640, 668, 684, 722
Salary growth, 158, 187, 463, 465, 1253, 1266, 1341, 1350, 1360, 1370
Sale prices, 78
Sales tax calculation, 464, 1292
Sales tax rate, 484
School supplies, 464
Shopping, 486
Snack price increase, 485
Soda price increase, 465
Stock market, 28
Stock prices, 314
Stock value, 735
Swing set, 464
Truck rentals, 684
Unit pricing, 383, 431
Wages, 918
Withholdings, 465, 485
Work hours, 64
Yard sale purchases, 85

Physics

Aging more slowly, 1113
Bouncing ball, 1350, 1360
Circular wave, 1236
Falling objects, 950, 1182, 1341, 1350
Flight
of a ball, 846, 847
of a baseball, 338, 834
of a golf ball, 797, 834
of a projectile, 338, 371
Hang time
football, 1094
golf ball, 1076
jumping person, 1076
Height
reached by a baseball, 785, 1155
reached by a golf ball, 1155
reached by a tennis ball, 1182, 1350, 1360

Hooke's Law, 919
Loudness of sounds, 1286
Musical tone frequencies, 1085
Ohm's Law, 920
Sound levels, 1266, 1274, 1286
Sound speeds, 14
Thrown object height, 1155, 1182, 1216
Toy rockets
 height reached by, 754
 velocity of, 614, 668
Weight when traveling near speed of light, 1113

Science

Acid solution, 563, 694
Air filtration, 1360
Antifreeze mixture, 564, 694
Artifact dating, 1253
Chlorine levels in swimming pool, 1254, 1350
Distance to the horizon, 1076, 1125
Earthquakes, 1267, 1286, 1287, 1292
Electrical current, 920
Electrical resistance, 628, 883, 892, 920
Fuel mixture, 393
Hydrocortisone cream, 564
Light, speed of, 14
Light bulb intensity, 874, 883
Liquid air, 421
Liquid oxygen, 431
Radio telescopes, 1306
Room ventilation, 1350
Saline solution, 563, 582
Temperature and volume, 968
Temperature conversion, 70
Water flow
 from holes, 1196
 over Niagara Falls, 660
 in tanks, 651
 through hoses, 919
Wind power, 920, 928, 1126

Social Networking

Advertising revenue, 629, 669
Facebook
 and Google revenues, 866
 link half-life, 1253
 monthly visitors, 1155, 1183, 1196
 vs. Twitter, 329, 562
 users, 187, 188, 349
Foursquare users, 1195
Global users of, 1360
Google
 revenues, 866, 867
 web searches, 949
Groupon's growth, 1195
MySpace users, 188
StumbleUpon link half-life, 1253

Twitter
 vs. Facebook, 329, 562
 followers, 628
 tweets per month, 1253
Viral emotions online, 448
Yahoo and Google revenue, 867

Social Science

Accidental deaths, 1206
Age at first marriage, 1021
Average household income, 772
Birth rate, 1022
Calorie consumption and land ownership, 1285
Cigarette consumption, 652, 981
Detroit population, 142
Divorce rates, 668
Drug law violations, 1106
Earnings for women, 1094
Federal debt, 1183
Female pharmacists, 222
Female physicians, 222
Fires in United States, 28
Gross domestic product, 772
Heart disease death rates, 1206
HIV/AIDS
 cases in United States, 866
 deaths, 999, 1195
 infections, 554, 603
Home price prediction, 1038
Homeless population, 119
Hospitals, number of, 188, 652, 1196–1197
Immigrant religion, 562
International adoption, 179–180, 652
Learning curves, 1113
Life expectancy, 967
Living in poverty, 394
Marriages, 1022
Median family income, 629
Median household income, 179, 652, 967
Medicare costs, 135, 193, 982
Millionaire earners, 188, 208, 705
Nursing home residents, 242
Pedestrian fatalities, 1060
Population change, 96
Population growth, 64, 78, 1254
Population modeling, 1292
Poverty levels, 208
Prison escapes, 1371
Prison population, 660
Prison time, 394
Residents in a community, 394
Seeking asylum, 142
Smoking doctors, 463
Smoking verdict, 463
State prisoners, 119
Steel imports, 314
Steel production, 314
Union activity, 382
Urban regions population, 1267, 1285
U.S. population, 603, 967, 1183

Violent crimes in U.S., 1022
Voter turnout, 253, 293
Voting, 393
Water consumption, 314
Welfare, 29
Working-age population, 554
World billionaires, 562
World millionaires, 562
World population, 253, 660, 746

Sports

Baseball
 height of MLB players, 28
 stats, 306, 371
Basketball, 393
 hoop, 530
 surface area of, 530
Bench press, 28
Bowling experience, 64
Canoe speed, 1195–1196
Football
 Canadian, 402
 scores, 28
 stats, 104
Golf
 clubs, 464
 pros and percentage of putts made, 1254
Hockey rink, 537
Kentucky Derby, 315
NHL Hockey, 28
Running/jogging
 distance and time, 578, 1182
 distance traveled when, 262, 553, 668, 1182
 heart rate while, 785, 998
 speeds, 563, 564, 906, 928, 1050
 time spent, 705
Sea diving, 112
Sky diving, 1126
Sumo wrestling, 537
Tennis serves, 28
Track and field
 high jump, 306, 410
 long jump, 420
 Olympic shotput, 521–522
 relay race, 329
 sprinters, 306
 triple jump, 420
 Usain Bolt's speed, 424

Technology/Internet

Camera phones, 28
Cell phone
 bending, 28
 complexity, 1156
 international plan for, 348
 path loss for, 1267
 screen size on, 28, 44
 signal strength, 1274
 use, 208, 442, 1206

Computer quality, 442
Computer screen dimensions, 1196, 1217
Computer viruses, 952
Digital pictures
    dimensions of, 834
    number of pixels in, 45, 785
    size of, 761
Downloading, 383
DVD and picture dimensions, 1126
Fitbit sales, 1285
HD video recorder, 464
Indispensable devices, 222
Internet
    access, 208
    ad spending, 554
    online shopping, 464
    real estate sites on, 694
    screen time, 562
    traffic, 142
    unique site visitors, 705
    use, 64, 135, 315
    users, 394, 553, 651
    web browsers, 442, 603
    web searches, 949
iPhone
    colors, 685
    cost, 463
    downloads, 56, 684–685
    models, 329
    prices, 563
    size of, 421
    waiting for, 465
iPod
    application, 464
    capacity, 394
    memory, 14, 15
    music, 45
    photos, 45
Laptop screen diagonal, 362
Online newspaper sites, 381
Phablet growth, 135
School laptop, 464

Skype users, 968
Smartphones
    price of Android, 949
    reselling, 44
    sales of, 652
    sales tax on, 484
    users of, 968
Smartwatch market, 1284
Text messaging, 15, 348, 371, 952, 967
Wearable devices, 235
Wireless networks, 685
Wireless router, 464
Wireless-only households, 968
Worldwide computer sales, 999–1000

Transportation

Airplane(s)
    corrosion in, 1134–1135
    distance and time traveled by, 269, 563
    pilot visibility, 1125
    runway length, 1266, 1285
    speed, 563, 694, 705, 906, 1196
    taxiway speed limits, 1182
Bicycling
    distance between when, 1022
    distance traveled when, 553, 582, 651
    speed, 563, 860, 906
    wheel rotations when, 382
Boat
    occupancy, 411
    speed, 694, 705, 906, 928
Buses, distance and time traveled by, 563
Car(s)
    braking distance, 834, 847, 999, 1182, 1216
    cost of driving, 640, 949
    cruise control and speed, 967
    depreciation of, 485, 1341
    distance and speed of, 651, 668
    distance between, 1014
    distance traveled by, 315

    fuel consumption, 243, 253
    fuel economy, 464
    fuel efficiency, 949
    gas mileage, 135, 188, 329, 339, 660, 669
    gas tank capacity, 421
    historical sales, 968
    hybrid, 442, 485
    lug nut tightening, 919
    NASCAR speeds, 424
    rental, 578, 640, 668, 684, 722
    rolling resistance of, 919
    shipments of connected, 349
    skid marks, 70, 1093, 1126
    speed limits, 424, 425, 431
    speed of, 563, 582, 906, 967
    stopping distance, 70, 846, 1195
    stopping on hills, 866, 906, 999
    stopping on slippery roads, 866, 906
    SUV sales, 339
    time and distance traveled by, 859, 928, 967, 980, 982, 1196
Safety
    of airport taxiway curves, 1182
    of highway curves, 905, 1126, 1206
    speed limit, 982, 1182
    train track, 1306
Toll cost and number of drivers, 919, 928
Traffic
    cones, 537
    fatalities, 722
    flow of, 859, 860, 928, 1254
    roadway congestion, 78
    signs, 515, 537
    speed limits, 906, 982
    waiting in, 905
Trains, 402, 980
Trucks
    renting costs of, 684
    sales of, 208–209
    speed of, 564
    total distance traveled by, 64

1 Whole Numbers

- 1.1 Introduction to Whole Numbers
- 1.2 Adding and Subtracting Whole Numbers; Perimeter
- 1.3 Multiplying and Dividing Whole Numbers; Area
- 1.4 Exponents, Variables, and Algebraic Expressions
- 1.5 Rounding and Estimating; Square Roots
- 1.6 Order of Operations
- 1.7 More with Equations and Problem Solving

The amount of information on the Internet has grown enormously in recent years. Websites such as Facebook, Twitter, and YouTube offer ordinary people the opportunity to add content to the World Wide Web. The extraordinary growth of the Internet and the development of related technologies such as cell phones and tablets are evidence that society is moving quickly into the information age.

Math plays an important role in generating, retrieving, and storing all forms of information. Using math, we can analyze numbers from everyday life and make informed decisions about the future. As we move into the information age, the ability to reason mathematically is becoming increasingly important. In fact, people with a strong math background will be better prepared for careers and life in the 21st century.

In this chapter, we discuss whole numbers. Understanding whole numbers is essential to understanding all types of numbers used in modern society.

1.1 Introduction to Whole Numbers

Objectives

1. Reviewing Natural Numbers and Whole Numbers
2. Understanding Place Value
3. Writing Whole Numbers in Word Form
4. Writing Whole Numbers in Expanded Form
5. Graphing Whole Numbers on the Number Line
   - Graphing Whole Numbers
   - Comparing Whole Numbers
6. Reading Graphs and Tables
   - Bar Graphs
   - Line Graphs
   - Spider Charts
   - Tables

NEW VOCABULARY

☐ Natural numbers
☐ Whole numbers
☐ Period
☐ Standard form
☐ Place value
☐ Word form
☐ Expanded form
☐ Number line
☐ Graph of a whole number
☐ Bar graph
☐ Line graph
☐ Spider chart
☐ Table

READING CHECK 1

- What numbers are the same as the counting numbers?

STUDY TIP

The word "NOTE" is used to draw attention to important concepts that may otherwise be overlooked.

1 Reviewing Natural Numbers and Whole Numbers

When children learn to count, they begin with 1 and follow the *counting numbers* in a way that seems *natural* to us. This thought may help you remember that **natural numbers** are the same as counting numbers and can be expressed as follows.

$$1, 2, 3, 4, 5, 6, \ldots$$

Because there are infinitely many natural numbers, three dots called an *ellipsis* are used to show that the list continues in the same pattern without end. A second set of numbers is called the **whole numbers** and can be expressed as follows.

$$0, 1, 2, 3, 4, 5, \ldots$$

Whole numbers include the natural numbers and the number 0.

STUDY TIP

Bring your book, notebook, and a pen or pencil to every class. Write down major concepts presented by your instructor. Your notes should also include the meaning of words written in bold type in the text. Be sure that you understand the meaning of these important words.

Connecting Concepts with Your Life Natural numbers and whole numbers can be used when data are not broken into fractional parts. For example, **TABLE 1.1** lists the number of Academy Awards won by selected movies. *Avatar*, for example, won 9 Academy Awards, and *The Hunger Games* won 0 Academy Awards. Both natural numbers and whole numbers are appropriate to describe these data because a fraction of an award is not possible.

Academy Awards Won by Selected Movies

Movie	Avatar	Frozen	Gravity	The Hunger Games	Titanic
Awards	9	2	7	0	11

TABLE 1.1 *Source: IMDB*

2 Understanding Place Value

Numbers are written using the *digits* 0, 1, 2, 3, 4, 5, 6, 7, 8, and 9. When a number with more than four digits is written, commas are used to separate the digits of the number into groups called **periods**. For example, the number 18,376,403 has **eight** digits and **three** periods. Numbers written this way are said to be in **standard form**.

18,376,403 — Eight digits — Three periods

NOTE: The period to the left of the first comma in a whole number may contain **one**, **two**, or **three** digits. However, all other periods must contain **three** digits. ∎

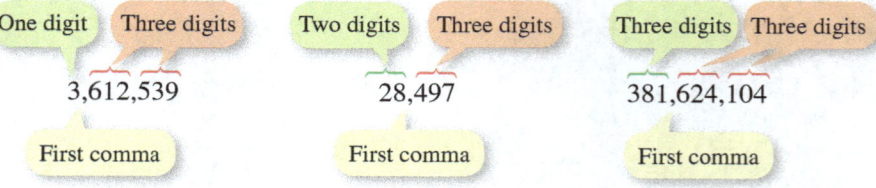

The **place value** of a digit in a number written in standard form is determined by the position that the digit takes in the number. In **FIGURE 1.1**, the number **18,376,403** is shown in a place value chart that gives the first 15 place values.

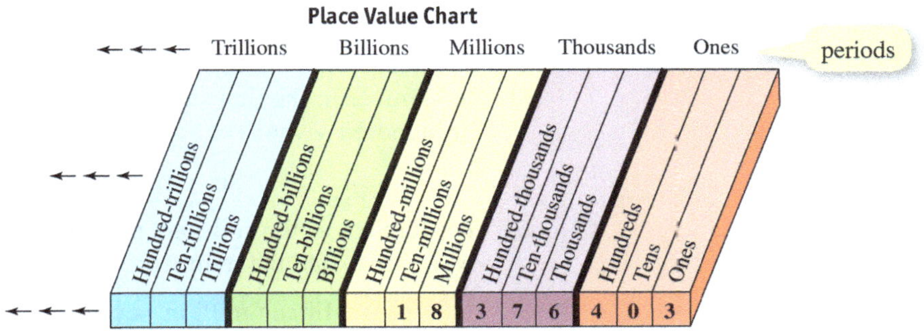

Place Value Chart

[Place value chart showing periods: Trillions, Billions, Millions, Thousands, Ones. Sub-columns: Hundred-trillions, Ten-trillions, Trillions, Hundred-billions, Ten-billions, Billions, Hundred-millions, Ten-millions, Millions, Hundred-thousands, Ten-thousands, Thousands, Hundreds, Tens, Ones. The digits 1, 8, 3, 7, 6, 4, 0, 3 appear in the rightmost eight places.]

FIGURE 1.1

READING CHECK 2

- How many digits are there in any period that is located to the right of the first (left-most) comma?

Arrows to the left of the place value chart in **FIGURE 1.1** indicate that the periods continue indefinitely. As we move to the left from trillions, the next three periods are quadrillions, quintillions, and sextillions.

NOTE: When a four-digit number is written in standard form, the comma that separates the ones period and the thousands period is optional. For example, the number 3,971 can also be written as 3971. In this text, four-digit numbers are written without a comma. ∎

EXAMPLE 1 **Finding the place value of a digit**

For the whole number 3,928,107,465,
(a) Determine the place value of the digit 2.
(b) Name the digit that is in the hundred-thousands place.

Solution
(a) The digit 2 is the eighth digit from the right. It is in the ten-millions place.

Ten-millions place
3,9**2**8,107,465

(b) The hundred-thousands place is the sixth place from the right. That digit is 1.

Hundred-thousands place
3,928,**1**07,465

Now Try Exercises 11, 21

3 Writing Whole Numbers in Word Form

A whole number written in standard form can also be written in words. For example, the whole number 17,024,863 can be written in **word form** as

seventeen million, twenty-four thousand, eight hundred sixty-three.

This number has three periods—millions, thousands, and ones. The names of the millions and thousands periods are used in writing the number in words, but the name of the ones period is not. Regardless of whether a whole number is written in standard form or in word form, commas are used to separate the periods. To write a whole number in word form, use the following procedure.

STUDY TIP

Ideas and procedures written in boxes like the one to the right are major concepts. Be sure that you understand them.

WRITING A WHOLE NUMBER IN WORDS

Starting with the left-most period, write the word form of the number in each period, followed by the period name and a comma. The name of the ones period is commonly not written, and the word *and* is not used when writing a whole number in words.

EXAMPLE 2 Writing whole numbers in word form

Write each whole number in word form.
(a) 62,407   (b) 15,075,410   (c) 2018

Solution
(a) The number in the thousands period is sixty-two, and the number in the ones period is four hundred seven. The word form is

sixty-two thousand, four hundred seven.   ← 62,407

(b) The number in the millions period is fifteen, the number in the thousands period is seventy-five, and the number in the ones period is four hundred ten. The word form is

fifteen million, seventy-five thousand, four hundred ten.   ← 15,075,410

(c) Although the number 2018 does not contain a comma, there are two periods. The number in the thousands period is two, and the number in the ones period is eighteen. The word form is

two thousand, eighteen.   ← 2018

Now Try Exercises 31, 33, 35

 Connecting Concepts with Your Life When writing a check, the dollar amount (without the cents) is a whole number written in word form. This is illustrated in the next example.

EXAMPLE 3 Writing a whole number in standard form

Write the standard form of the dollar amount written on the check in **FIGURE 1.2**.

FIGURE 1.2

Solution
Five hundred thirteen is written in standard form as 513.

Now Try Exercise 37

4 Writing Whole Numbers in Expanded Form

 Connecting Concepts with Your Life A student who cashes a paycheck for $714 receives seven hundred-dollar bills, one ten-dollar bill, and four one-dollar bills. The total received is

700   and   10   and   4

dollars. By replacing each "and" with a *plus sign* (+), we can write the number 714 in **expanded form** as

$$700 + 10 + 4.$$

Note that 700 is the standard form of the number given by the digit 7 in the **hundreds** place, 10 is the standard form of the number given by the digit 1 in the **tens** place, and 4 is the digit in the **ones** place. To write a whole number in expanded form, use the following procedure.

READING CHECK 3

- How many zeros are needed to write the standard form for a digit in the ten-thousands place?

WRITING A WHOLE NUMBER IN EXPANDED FORM

Starting with the left-most digit, write the standard form of the number given by each digit and its corresponding place value. Place a plus sign between each of these results.

EXAMPLE 4 Writing a whole number in expanded form

Write each whole number in expanded form.
(a) 45,923   (b) 709,416

Solution
(a) The digit 4 in the **ten-thousands** place represents 40,000. The digit 5 in the **thousands** place represents 5000. Likewise, the digit 9 represents 900, and the digit 2 represents 20. Finally, the digit 3 is in the ones place. The number can be written in expanded form as

$$40,000 + 5000 + 900 + 20 + 3. \quad \boxed{45,923}$$

(b) The expanded form is

$$700,000 + 9000 + 400 + 10 + 6. \quad \boxed{709,416}$$

NOTE: In part (b), the digit 0 in the ten-thousands place is not used in the expanded form because we do not write 0 ten-thousands as 00,000. Whenever 0 appears in the standard form of a whole number, it will not be included as part of the expanded form. ∎

Now Try Exercises 49, 53

5 Graphing Whole Numbers on the Number Line

GRAPHING WHOLE NUMBERS Sometimes it is helpful to visualize whole numbers using a **number line**. Starting with 0, whole numbers are written below equally spaced *tick marks* as shown in **FIGURE 1.3**.

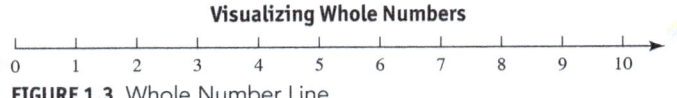

FIGURE 1.3 Whole Number Line

The **graph of a whole number** shows a dot placed on the number line at the whole number's position.

6 CHAPTER 1 WHOLE NUMBERS

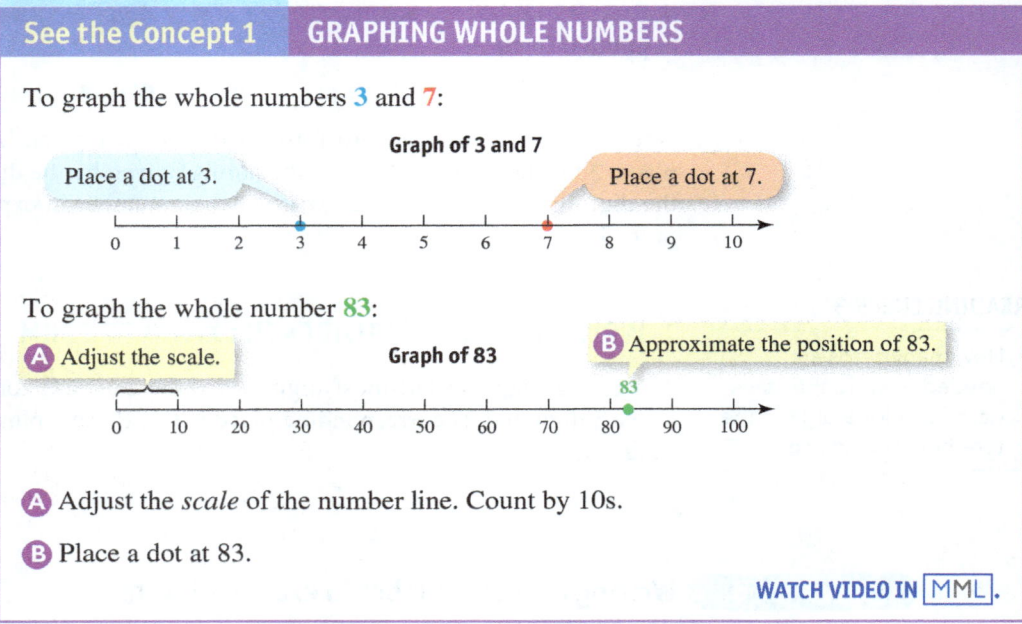

See the Concept 1   **GRAPHING WHOLE NUMBERS**

To graph the whole numbers 3 and 7:

Graph of 3 and 7

Place a dot at 3.
Place a dot at 7.

To graph the whole number 83:

A Adjust the scale.   *Graph of 83*   **B** Approximate the position of 83.

A Adjust the *scale* of the number line. Count by 10s.

B Place a dot at 83.

WATCH VIDEO IN MML.

COMPARING WHOLE NUMBERS When two numbers are graphed on the same number line (see **FIGURE 1.4**), the number to the left is always *less than* the number to the right. The symbol < denotes **less than**. Similarly, the number to the right is always *greater than* the number to the left. The symbol > denotes **greater than**.

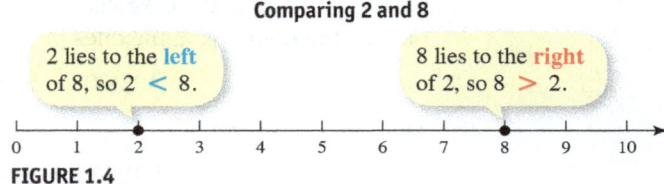

Comparing 2 and 8

2 lies to the **left** of 8, so 2 < 8.

8 lies to the **right** of 2, so 8 > 2.

FIGURE 1.4

EXAMPLE 5   **Graphing and comparing whole numbers**

Graph the whole numbers 12, 6, 19, 2, and 15 on the same number line. Use your graph to compare the following numbers and write the correct symbol, < or >, in the blank.
(a) 6 _____ 15   (b) 19 _____ 12   (c) 6 _____ 2   (d) 12 _____ 15

Solution
The graph is shown in **FIGURE 1.5**. Note that the scale has been adjusted so that the numbers may be more easily represented on the number line.

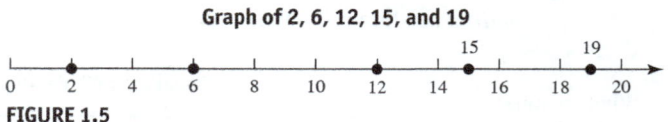

Graph of 2, 6, 12, 15, and 19

FIGURE 1.5

(a) On the number line, the number 6 is located to the **left** of the number 15, so we know that 6 is **less than** 15. We write 6 < 15 and estimate the graph of 15 to be halfway between the tick marks for 14 and 16.
(b) Because 19 is located to the **right** of 12 on the number line, we know that 19 is **greater than** 12, and we write 19 > 12. The graph of 19 is plotted halfway between the tick marks for 18 and 20.
(c) The number 6 is **greater than** the number 2 because it is located to the **right** of 2 on the number line. We write 6 > 2.
(d) We write 12 < 15 because 12 is located to the **left** of 15, or 12 is **less than** 15.

Now Try Exercises 55, 57

6 Reading Graphs and Tables

🌐 Math in Context (Weather) To make it easier to read, compare, and analyze data, numbers are often displayed in graphs and tables.

BAR GRAPHS **FIGURE 1.6** shows a **bar graph** displaying a seven-day temperature forecast. A glance at this bar graph can provide information very quickly.

READING CHECK 4

- Why is it sometimes helpful to display data in graphs and tables?

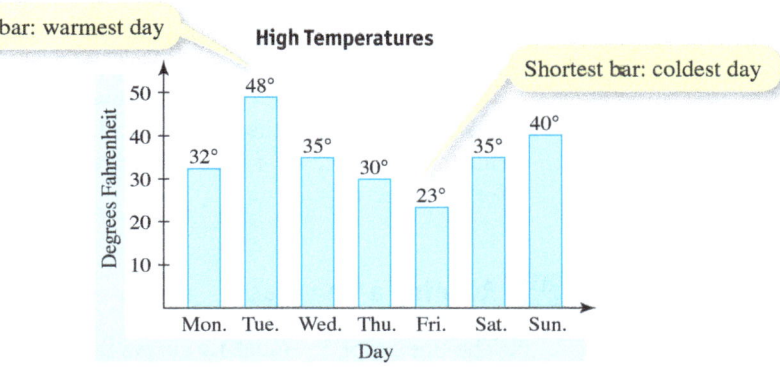

FIGURE 1.6 Seven-Day Temperature Forecast

EXAMPLE 6 Reading a bar graph

The bar graph in **FIGURE 1.7** shows the typical energy produced each month by solar energy panels on an 80 m² home. The units are in kilowatt hours. (*Source*: Woodbrooke Good Lives Project.)

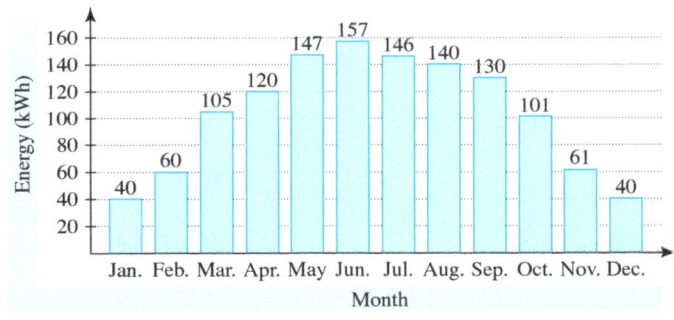

FIGURE 1.7

(a) How much energy is produced in September?
(b) In which month is the largest amount of energy produced?
(c) In which month(s) are 40 kWh of energy produced?

Solution
(a) The bar representing September has a height of 130. Since the units are kilowatt hours (kWh), the amount of energy produced in September is 130 kWh.
(b) The tallest bar corresponds to the largest amount of energy. This occurs in June.
(c) The energy produced is 40 kWh in both January and December, as shown by the two bars with height 40.

Now Try Exercises 75–78

LINE GRAPHS Another type of graph for displaying data is the **line graph**. With this type of graph, we can quickly identify any trends in the data being displayed. The line graph in

FIGURE 1.8 suggests an upward trend in the percentage of global electricity production that is renewable. In this graph, double hash marks // are used on each axis to indicate a break in each scale. The horizontal scale starts at 0 and jumps to 2010 before it shows every 2 years up through 2018. The vertical scale starts at 0 and jumps to 18 (percent) before it shows every 2 percent of electricity production.

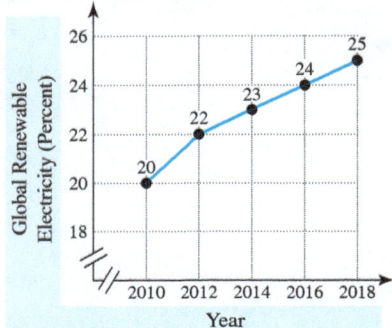

FIGURE 1.8

NOTE: Since the label on the vertical axis states that the data are shown in "percent," each number on the vertical scale is a percentage. For example, the number 22 on the vertical scale represents 22%, meaning that 22% of global electricity production is renewable. ∎

EXAMPLE 7 Reading a line graph

The line graph in FIGURE 1.9 shows budgeted federal income tax receipts in billions of dollars for selected years. (*Source:* Office of Management and Budget.)

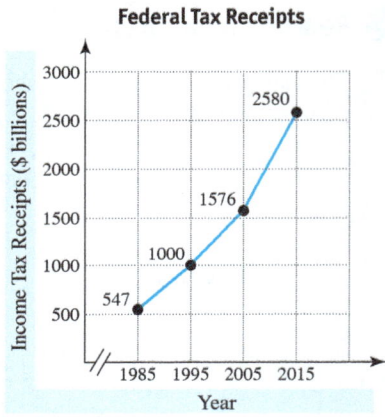

FIGURE 1.9

(a) How much income tax did the federal government collect in 1995?
(b) In which of these years was $547 billion collected?
(c) Comment on the general trend of income tax receipts.

READING CHECK 5

- What type of graph is useful in identifying trends in data?

Solution
(a) In 1995 the line graph reaches $1000 billion, or $1 trillion.
(b) The line graph indicates that $547 billion was collected in 1985.
(c) Income tax receipts increased over this time period.

Now Try Exercises 79–82

SPIDER CHARTS Charts like the ones shown in FIGURES 1.10(a) and 1.10(b) are called **spider charts** and are sometimes used to display data that are divided into several categories. A quick look at these two charts reveals that Student B is a more well-rounded student who *generally* scored higher than Student A on the exam.

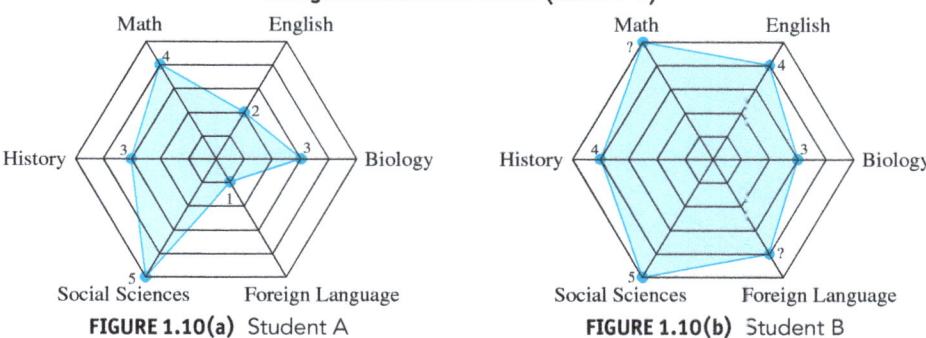

FIGURE 1.10(a) Student A FIGURE 1.10(b) Student B

EXAMPLE 8 **Reading spider charts**

Refer to the spider charts shows in **FIGURES 1.10(a)** and **1.10(b)** to answer the questions.
(a) In which subjects did Student A receive the same scores as Student B?
(b) Which student scored lower on the English portion of the exam?
(c) What is the missing Math score for Student B?

Solution
(a) The students had identical scores of 5 on the Social Sciences portion of the exam and they had identical scores of 3 on the Biology portion of the exam.
(b) Student A scored 2 on the English portion of the exam, while Student B scored 4 on that portion. Student A scored lower.
(c) Starting at the center of the chart, count each "ring" as you move outward toward the Math score. Since the score lies on the fifth ring, the missing Math score is 5.

Now Try Exercise 87–90

TABLES Information can also be displayed visually in a **table**. The renewable energy data shown in **FIGURE 1.8** are also given in **TABLE 1.2**. Even though the data in the line graph are identical to the data in the table, the line graph shows the upward trend in renewable energy more easily than does the table. However, the percentage of electricity that comes from renewable sources can be seen at a glance within the table.

Global Renewable Electricity Production

Year	2010	2012	2014	2016	2018
Renewable Electricity	20%	22%	23%	24%	25%

TABLE 1.2 *Source: Scientific American.*

EXAMPLE 9 **Reading a table**

TABLE 1.3 on the next page shows the number of music albums sold by genre (category) in 2011 and 2012. Use the table to answer the following questions.
(a) How many R&B albums were sold in 2012? Were there fewer or more albums sold in 2012 than in 2011?
(b) Which of these genres sold the fewest albums in 2011?
(c) Which genres increased album sales from 2011 to 2012?

STUDY TIP

Putting It All Together gives a summary of important concepts in each section. Be sure that you have a good understanding of these concepts.

Albums Sold by Genre

	Alternative	Country	R&B	Rap
2011	54,600,000	42,800,000	55,300,000	27,300,000
2012	52,200,000	44,600,000	49,700,000	24,200,000

TABLE 1.3 *Source:* Nielsen Soundscan

Solution

(a) Move downward in the "R&B" column until you find the number listed in the row labeled "2012." There were 49,700,000 R&B albums sold in 2012. This number is fewer than the number sold in 2011, which is 55,300,000.
(b) The smallest number in the "2011" row is 27,300,000. Moving upward, we find that the corresponding type of music is Rap.
(c) The only sales number that increases is in the "Country" column. The number of Country albums sold increased from 42,800,000 in 2011 to 44,600,000 in 2012.

Now Try Exercises 95–98

1.1 Putting It All Together

CONCEPT	COMMENTS	EXAMPLES
Natural Numbers	Sometimes referred to as the *counting numbers*	1, 2, 3, 4, 5, ...
Whole Numbers	Includes the natural numbers and 0	0, 1, 2, 3, 4, ...
Standard Form	A whole number written in digits with commas separating the periods is in standard form.	345,690,274
Place Value	The place value of a digit in a number written in standard form is determined by the position that the digit takes in the number.	In the number 83,451,276, the digit 7 is in the tens place, and the digit 3 is in the millions place.
Word Form	Starting with the left-most period, write the word form of the number in each period followed by the period name and a comma.	In word form, 34,506 is written as thirty-four thousand, five hundred six.
Expanded Form	Starting with the left-most digit, write the standard form of the number given by each digit and its corresponding place value. Place a plus sign between each of these results.	In expanded form, 500,349 is written as $$500{,}000 + 300 + 40 + 9.$$
Number Line	Whole numbers can be visualized on a number line. The graph of a whole number is a dot placed on a number line at the whole number's position.	The numbers 3 and 5 are graphed.
Comparing Whole Numbers	To compare two whole numbers, determine their positions on a number line. Then use the symbols > or < to write an appropriate comparison.	$6 > 3$ and $2 < 5$

CONCEPT	COMMENTS	EXAMPLES
Bar Graph	A bar graph can be used to represent data visually and is helpful when analyzing data.	
Line Graph	A line graph can be used to represent data visually and is helpful when looking for trends in data.	
Table	A table can be used to display data in an at-a-glance way.	Year 2015 2016 2017 / Price $53 $67 $78

1.1 Exercises

MyMathLab®

CONCEPTS AND VOCABULARY

1. The natural numbers are also referred to as the _____ numbers.

2. The whole numbers include the natural numbers and the number _____.

3. Commas are used to separate the digits of a whole number into groups called _____.

4. A whole number expressed in digits with commas separating the periods is in _____ form.

5. The position of a digit in a whole number determines the digit's _____.

6. In a place value chart, the period immediately to the left of billions is _____.

7. The number "six thousand, four hundred seventeen" is written in (word/expanded) form.

8. The number 30,000 + 2000 + 40 + 3 is written in (word/expanded) form.

9. The _____ of a whole number is a dot placed on a number line at the whole number's position.

10. Name two kinds of graphs that can be used to display data visually.

DETERMINING PLACE VALUE

Exercises 11–20: For the given whole number, determine the place value of the digit 8.

11. 18,450
12. 456,981
13. 89,104,765
14. 6,830,142
15. 310,842
16. 4,982,017
17. 3008
18. 48,362,710
19. 890,247,135
20. 38,907,142,516

Exercises 21–30: Name the digit with the given place value in the whole number 3,409,816,725.

21. hundreds
22. ten-thousands
23. hundred-thousands
24. ones

25. billions
26. millions
27. tens
28. thousands
29. hundred-millions
30. ten-millions

WRITING WHOLE NUMBERS IN WORD FORM

Exercises 31–36: Write the whole number in word form.

31. 472,500
32. 79
33. 93,206
34. 10,000,015
35. 1651
36. 632

Exercises 37 and 38: Write the standard form of the dollar amount written on the given check.

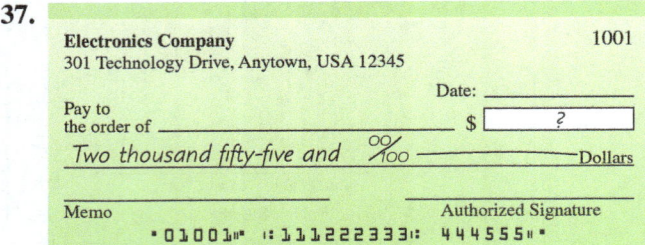

37.
38.

WRITING WHOLE NUMBERS IN STANDARD FORM

Exercises 39 and 40: Write the standard form of the whole number that is expressed in word form in the sentence.

39. A typical student entering college immediately after high school will be five hundred ninety-nine million, six hundred sixteen thousand, four hundred twenty-three seconds old at some point during his or her freshman year.

40. The Nile River is the longest river in the world, at four thousand, one hundred thirty-five miles.

Exercises 41–48: Write the whole number in standard form.

41. Thirty-nine million, four hundred ten thousand
42. Fifty-two thousand, three hundred sixty-seven
43. Eighty-three billion, six hundred thousand, twelve
44. One million, four hundred two thousand, eighty-one

45. $300{,}000 + 40{,}000 + 2000 + 500 + 60 + 3$
46. $5000 + 500 + 50 + 1$
47. $7{,}000{,}000 + 900{,}000 + 5000 + 300 + 70 + 7$
48. $4{,}000{,}000 + 500{,}000 + 7000 + 200 + 9$

WRITING WHOLE NUMBERS IN EXPANDED FORM

Exercises 49–54: Write the whole number in expanded form.

49. 2,510,036
50. 8004
51. 629
52. 63,907
53. 603,138
54. 17

GRAPHING AND COMPARING WHOLE NUMBERS ON THE NUMBER LINE

Exercises 55–60: Graph the given whole numbers on the same number line. Place the correct symbol, $<$ or $>$, in the blank between the given whole numbers.

55. 3, 5, 4, 1
    (a) 3 ___ 5   (b) 4 ___ 3   (c) 5 ___ 1
56. 2, 5, 8, 3
    (a) 3 ___ 2   (b) 5 ___ 8   (c) 2 ___ 5
57. 11, 22, 4, 8
    (a) 22 ___ 4   (b) 8 ___ 11   (c) 11 ___ 22
58. 23, 31, 12, 40
    (a) 12 ___ 31   (b) 23 ___ 40   (c) 31 ___ 40
59. 86, 24, 64, 10
    (a) 86 ___ 64   (b) 64 ___ 24   (c) 24 ___ 10
60. 98, 27, 73, 15
    (a) 27 ___ 73   (b) 98 ___ 15   (c) 15 ___ 73

Exercises 61–70: Place the correct symbol, $<$ or $>$, in the blank between the whole numbers.

61. 34 ___ 0
62. 0 ___ 56
63. 45 ___ 54
64. 72 ___ 27
65. 300 ___ 299
66. 175 ___ 155
67. 30,000 ___ 300,000
68. 2100 ___ 2001
69. 50,101 ___ 51,010
70. 630,020 ___ 632,202

READING BAR GRAPHS

Exercises 71–74: The bar graph shows the 10 countries that had the largest number of Internet users in 2012. (*Source: Internet World Stats.*)

1.1 INTRODUCTION TO WHOLE NUMBERS

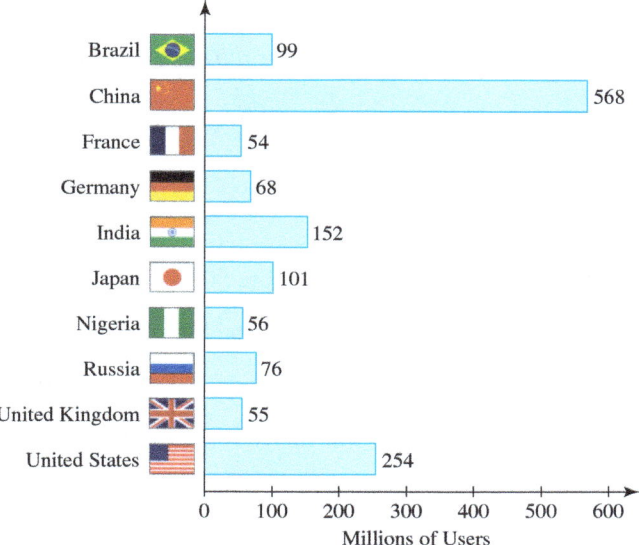

71. Which of the countries shown had 152,000,000 Internet users in 2012?

72. How many Internet users were there in Japan in 2012?

73. Which of these countries had the fewest number of Internet users in 2012?

74. Which country had more Internet users in 2012, Brazil or Russia?

Exercises 75–78: The following bar graph shows the four longest rivers in Canada. (Source: Statistics Canada.)

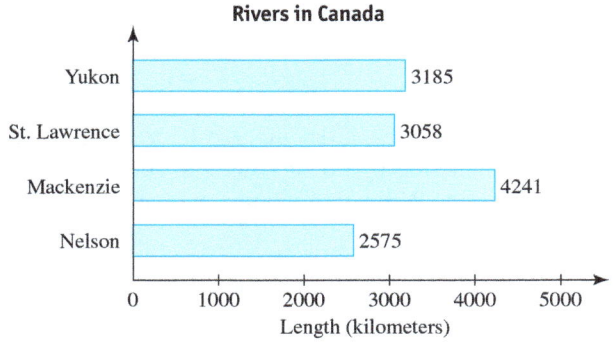

75. What is the longest river in Canada?

76. Which river is 3058 kilometers long?

77. How long is the Nelson River?

78. Which river is longer, the Nelson or the Yukon?

READING LINE GRAPHS

Exercises 79–82: The following line graph shows the box office receipts for top-grossing movies of selected years. These values have not been adjusted for inflation. (Source: MovieWeb.)

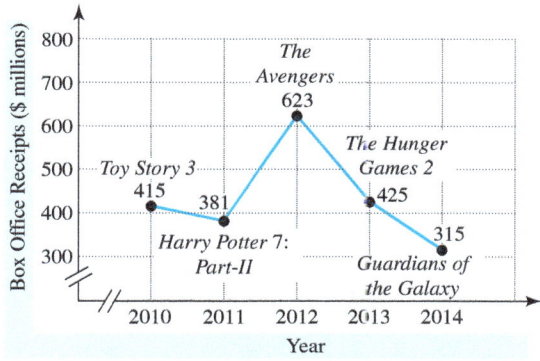

79. In what year did the top-grossing movie have the smallest amount in box office receipts?

80. Which of these movies had box office receipts of $623,000,000?

81. What were the box office receipts of the top-grossing movie of 2014?

82. Which movie grossed more, *The Hunger Games 2* or *Toy Story 3*?

Exercises 83–86: The following line graph shows the federal minimum wage in cents for selected years. (Source: Bureau of Labor Statistics.)

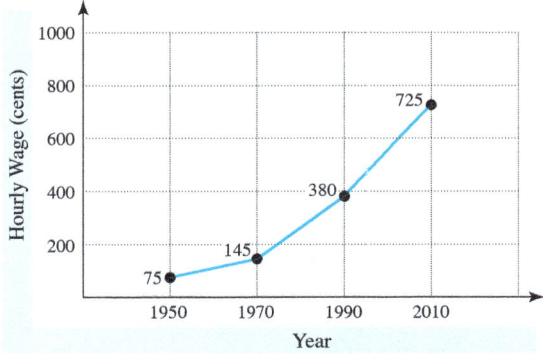

83. Comment on the general trend of the minimum wage over this time period.

84. What was the minimum wage in 1990?

85. What 20-year period had the largest increase in the minimum wage?

86. In what year was the minimum wage 75¢?

READING SPIDER CHARTS

Exercises 87–90: The following spider chart shows the results of a customer satisfaction survey conducted by an Internet shopping site. Higher scores indicate more satisfaction.

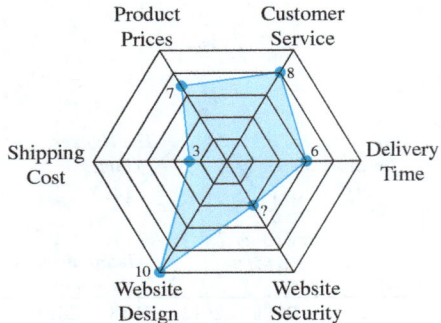

87. In which category does this site have the highest satisfaction score?

88. What is the satisfaction score for product pricing?

89. What is the missing score for website security?

90. Are people generally satisfied with the shipping costs associated with this shopping site?

Exercises 91–94: Answer the following questions by referring to both the spider chart for the previous four exercises and the spider chart shown below.

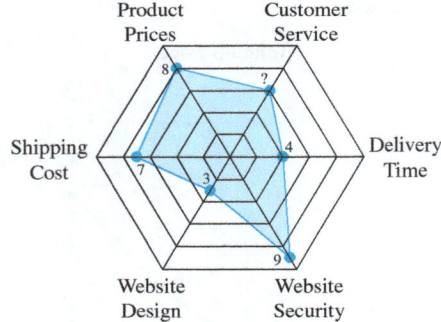

91. Which website has a higher customer service score?

92. In which category does the electronics website have a satisfaction score of 3?

93. By adding the six scores for each website, determine which site has a higher overall score.

94. Are people generally satisfied with the website security associated with the electronics site?

READING TABLES

Exercises 95–98: The following table shows average SAT math scores for males and females during selected years.

	2004	2006	2008	2010	2012
Female	501	502	500	499	499
Male	537	536	533	533	532

Source: The College Board.

95. In what year did females score 502?

96. In what year did males score the highest?

97. Which gender had a higher average score in 2008?

98. In what two years did males record identical average scores?

Exercises 99–102: The following table shows the cost in dollars of tuition and fees at public and private non profit four-year colleges for selected years, in 2013 dollars.

	2009	2010	2011	2012	2013
Public	7672	8174	8557	8821	8893
Private	27,920	28,679	28,830	29,593	30,094

Source: The College Board.

99. In what year did public colleges cost $8174?

100. In what year did private colleges cost $29,593?

101. Which type of college had a cost of $8557 in 2011?

102. Did tuition and fees at public or private colleges ever decrease from one year to the next?

APPLICATIONS INVOLVING WHOLE NUMBERS

103. **Speed of Sound** In dry air at a temperature of 65° Fahrenheit, the speed of sound is about 1124 feet per second. Write this whole number in expanded form.

104. **Speed of Light** The speed of light in a vacuum is about 186,282 miles per second. Write this whole number in expanded form.

105. **iPod Memory** Digital information is stored in units called bytes. A 32-gigabyte iPod Touch holds thirty-four billion, three hundred fifty-nine million, seven hundred thirty-eight thousand, three hundred seventy-eight bytes. Write this whole number in standard form.

106. **iPod Memory** A 1-gigabyte iPod Shuffle holds one billion, seventy-three million, seven hundred forty-one thousand, eight hundred twenty-four bytes. Write this whole number in standard form.

107. **Text Messaging** In June 2012, cell phone users in the U.S. sent over 423,000,000,000 text messages. Write this whole number in word form. (*Source*: CTIA—The Wireless Association.)

108. **Text Messaging** In 2013 there were about 98,700,000 text message–enabled cell phone users. Write this whole number in word form. (*Source*: CTIA—The Wireless Association.)

109. **Annual Income** Who has a greater yearly income, an electrician earning $41,627 per year or a truck driver earning $41,804 per year?

110. **Bacteria** There are 12,678,453 bacteria in a white dish and 12,687,435 bacteria in a black dish. Which dish has fewer bacteria?

WRITING ABOUT MATHEMATICS

111. Explain how to write a whole number in word form. Give an example.

112. Explain how to write a whole number in expanded form. Give an example.

1.2 Adding and Subtracting Whole Numbers; Perimeter

Objectives

1. Adding Whole Numbers
   - Without Regrouping
   - With Regrouping
2. Using Properties of Addition
   - Commutative Property
   - Associative Property
   - Identity Property
3. Recognizing Words Associated with Addition
4. Subtracting Whole Numbers
   - Without Regrouping
   - With Regrouping
5. Using Properties of Subtraction
6. Recognizing Words Associated with Subtraction
7. Solving Equations Involving Addition and Subtraction
8. Solving Perimeter and Other Applications Involving Addition and Subtraction

1 Adding Whole Numbers

Connecting Concepts with Your Life A person planning a Super Bowl party could send out electronic invitations using an online service such as Evite (www.evite.com). The RSVP feature of this service can be used to determine the number of people who will be coming to the party. The total number of guests is found by *adding* the numbers on the electronic RSVP cards.

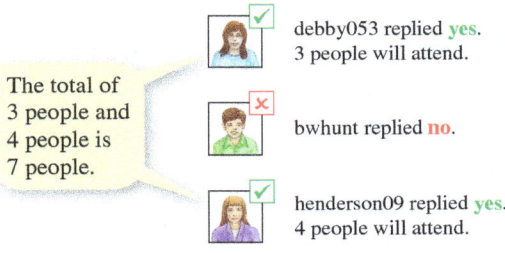

Electronic RSVP Results

The total of 3 people and 4 people is 7 people.

debby053 replied **yes**. 3 people will attend.

bwhunt replied **no**.

henderson09 replied **yes**. 4 people will attend.

FIGURE 1.11

When whole numbers are added, the result is called the **sum**, and the numbers being added are called the **addends**. For **FIGURE 1.11**, the **sum** is **7** and the **addends** are **3** and **4**.

$$3 + 4 = 7$$

Addends   Sum

ADDING WHOLE NUMBERS WITHOUT REGROUPING To add whole numbers with more than one digit, it is often convenient to stack the numbers vertically with corresponding place values aligned. Then we add the digits in each place value, starting with the ones place. For example, the numbers 521 and 6374 are added as follows.

Align digits and add vertically.

```
  5 2 1   ← Addend
+ 6 3 7 4 ← Addend
---------
  6 8 9 5 ← Sum
```

NOTE: The digits in corresponding place values must be aligned vertically when adding numbers. For example, the digit **5** in the top number in the previous example represents 500 and the digit **3** in the second number represents 300. When added, the total is **8**, which represents 800 in the sum. ∎

EXAMPLE 1 Adding whole numbers without regrouping

NEW VOCABULARY

- ☐ Sum
- ☐ Addends
- ☐ Commutative property for addition
- ☐ Associative property for addition
- ☐ Identity property for addition
- ☐ Difference
- ☐ Minuend
- ☐ Subtrahend
- ☐ Equation
- ☐ Solution
- ☐ Solving an equation
- ☐ Perimeter

Add.
(a) $2416 + 332$ (b) $11 + 314 + 5473$

Solution
Start by stacking the addends vertically so that the place values are aligned.

(a) 2416
   $+$ 332
   ──────
    2748

(b)   11
      314
   $+$ 5473
   ──────
     5798

Now Try Exercises 21, 31

ADDING WHOLE NUMBERS WITH REGROUPING Sometimes adding digits within a particular place value results in a sum larger than 9. For example, adding digits in the ones place of the numbers 378 and 607 results in 15. Because **15** is written in expanded form as **10** $+$ **5**, there are **5** ones and **1** ten. We write the **5** in the ones column, and the ten is *regrouped* to the tens column as the digit **1** (because it is 1 ten).

```
    1
  3 7 8         Regroup as 1 ten.
+ 6 0 7
  ─────
  9 8 5
```

EXAMPLE 2 Adding whole numbers with regrouping

Add.
(a) $328 + 4169$ (b) $38 + 367 + 2276$

Solution

(a) The digits in the ones column total 17, so 7 is written as the result in the ones column and **1** ten is regrouped to the tens column. The remaining digits of the sum are found by adding the digits in each column.

```
    1
   328          Regroup as 1 ten.
 + 4169
  ─────
   4497
```

CALCULATOR HELP

To add whole numbers with a calculator, see Appendix E (page AP-28).

READING CHECK 1

- In adding whole numbers, when is it necessary to regroup?

(b) The sum of the digits in the ones column is 21. In this case, a 1 is written as the result in the ones column, and **2** tens are regrouped to the tens column. Adding the digits in the tens column results in 18, meaning that there are 18 tens. An 8 is written as the result in the tens column, and the 10 tens or **1** hundred is regrouped to the hundreds column.

```
    12
    38           Regroup 2 tens
   367           and 1 hundred.
 + 2276
  ─────
   2681
```

Now Try Exercises 23, 33

MAKING CONNECTIONS 1

Regrouping and Expanded Form

To see how regrouping works, we write the addends in expanded form before performing addition. For example, the sum $357 + 876$ is found as follows:

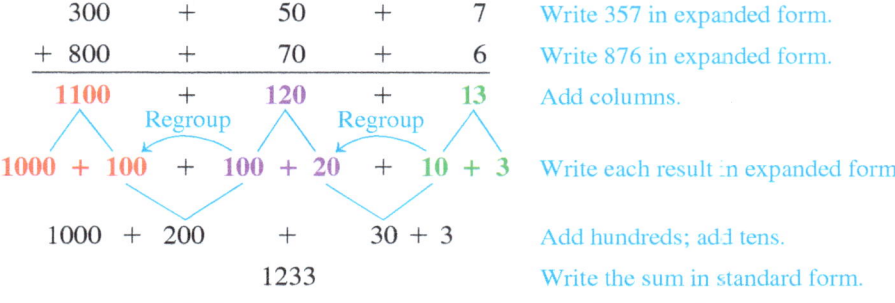

300	+	50	+	7
+ 800	+	70	+	6
1100	+	120	+	13

Write 357 in expanded form.
Write 876 in expanded form.
Add columns.

$1000 + 100 \quad + \quad 100 + 20 \quad + \quad 10 + 3$ — Write each result in expanded form.

$1000 \ + \ 200 \quad + \quad 30 + 3$ — Add hundreds; add tens.

1233 — Write the sum in standard form.

STUDY TIP

The information in Making Connections ties the current concepts to those studied earlier. By reviewing your notes often, you can gain a better understanding of mathematics.

When adding whole numbers, the following procedure can be used.

ADDING WHOLE NUMBERS

To add whole numbers,

1. Stack the numbers vertically with corresponding place values aligned.
2. Add the digits in each place value. Regroup when necessary.

2 Using Properties of Addition

COMMUTATIVE PROPERTY Adding a long list of whole numbers by hand can be time-consuming. However, there are three properties of addition that often make the process easier. The first property, the **commutative property for addition**, states that changing the *order* of the addends does not change the resulting sum.

Example of the Commutative Property

$3 + 2 = 5 \quad \text{and} \quad 2 + 3 = 5$

COMMUTATIVE PROPERTY FOR ADDITION

The *order* in which two addends are written does not affect the sum.

The commutative property for addition can be visualized as shown in **FIGURE 1.12**.

A Visual Representation of the Commutative Property: 3 + 2 = 2 + 3

FIGURE 1.12

ASSOCIATIVE PROPERTY The second property is called the **associative property for addition**. This property allows for the *regrouping* of addends when more than two numbers are being added. Parentheses are used to show which two addends should be grouped (added) first.

Example of the Associative Property

$(3 + 2) + 1 = 5 + 1 = 6 \quad \text{and} \quad 3 + (2 + 1) = 3 + 3 = 6$

ASSOCIATIVE PROPERTY FOR ADDITION

The way in which three or more addends are *grouped* does not affect the sum.

The associative property for addition can be visualized as shown in **FIGURE 1.13**.

A Visual Representation of the Associative Property: $(3 + 2) + 1 = 3 + (2 + 1)$

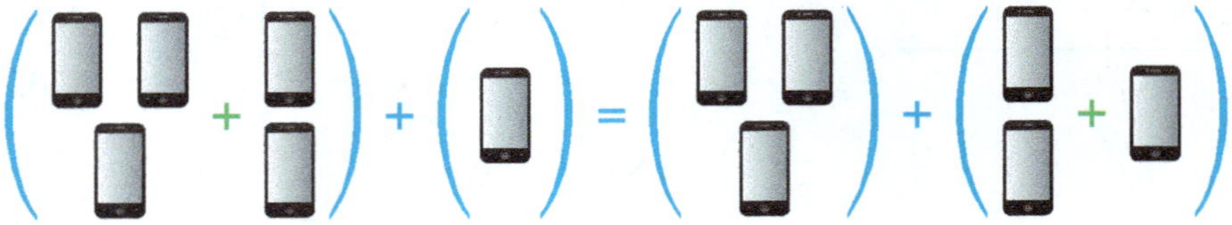

FIGURE 1.13

IDENTITY PROPERTY The **identity property for addition** is the third property. It states that adding 0 to a number does not change the number. If you have $12 and a friend gives you $0, you still have $12.

Example of the Identity Property

$12 + 0 = 12$ and $0 + 12 = 12$

IDENTITY PROPERTY FOR ADDITION

When 0 is added to any number, the result is that number.

EXAMPLE 3 **Identifying properties for addition**

Identify the property of addition that is illustrated in each equation.
(a) $39 + 0 = 39$ (b) $14 + 83 = 83 + 14$ (c) $(3 + 5) + 8 = 3 + (5 + 8)$

Solution
(a) The number 39 remains unchanged when 0 is added to it. This is an example of the identity property for addition.
(b) The sum of 14 and 83 is the same as a sum with the same addends written in a different order. This is an example of the commutative property for addition.
(c) Even though the order of the addends is the same on both sides of the equals sign, the way the addends are grouped differs. This is an example of the associative property for addition.

Now Try Exercises 35, 37, 39

Sometimes the three addition properties can be used to add mentally. Consider the sum $14 + 9 + 0 + 6 + 8 + 1 + 12 + 7$. Note that the addend 0 has no effect on the sum, so it can be ignored. The remaining numbers can be arranged and grouped so that the sum can be computed mentally.

$$14 + 9 + 6 + 8 + 1 + 12 + 7$$
$$(14 + 6) + (9 + 1) + (8 + 12) + 7$$
$$20 + 10 + 20 + 7$$
$$57$$

The sum of 57 is found by adding the numbers 20, 10, 20, and 7.

READING CHECK 2

- Which addition property allows us to add in any order?
- Which addition property allows us to regroup the addends?

EXAMPLE 4 Adding whole numbers mentally

Add mentally.
$16 + 23 + 12 + 8 + 5 + 9 + 7 + 0 + 15 + 4$

Solution
Ignore 0 and regroup to get a sum of 99.

$$16 + 23 + 12 + 8 + 5 + 9 + 7 + 15 + 4 =$$
$$20 + 30 + 20 + 20 + 9 = 99$$

Now Try Exercise 41

3 Recognizing Words Associated with Addition

Many times information is given in words rather than in symbols or numbers. To find a required result, it may be necessary to translate words into a mathematical expression. **TABLE 1.4** shows some words commonly associated with addition, along with sample phrases using the words.

Words Associated with Addition

Words	Sample Phrase
add	add the two temperatures
plus	her age plus his age
more than	10 miles more than the distance
sum	the sum of the length and the width
total	the total of the four prices
increased by	his height increased by 3 inches

TABLE 1.4

EXAMPLE 5 Translating words into a mathematical expression

Translate each phrase into a mathematical expression. Find the result.
(a) The total of 5 inches and 63 inches   (b) 17 medals more than 12 medals already won

Solution
(a) The word *total* suggests that we add 5 and 63. The corresponding mathematical expression is $5 + 63$, which results in 68 inches.
(b) The words *more than* suggest that we add 17 to 12. The corresponding mathematical expression is $12 + 17$, which results in 29 medals.

Now Try Exercises 67, 73

4 Subtracting Whole Numbers

 Connecting Concepts with Your Life On February 1, a student's blog had a cumulative total of 1454 hits, and on March 1, the blog had a cumulative total of 1878 hits. The number of hits that the blog had during the month of February was 424, which can be found by *subtracting* the number 1454 from the number 1878. The result of subtracting one whole number from another is called the **difference**. The number we are subtracting from is called the **minuend**, and the number being subtracted is called the **subtrahend**.

$$1878 \; - \; 1454 \; = \; 424$$

Minuend   Subtrahend   Difference

SUBTRACTING WHOLE NUMBERS WITHOUT REGROUPING To subtract one whole number from another, stack the minuend vertically above the subtrahend with corresponding place values aligned. Then subtract the digits in each place value, starting with the ones place. For example, 431 is subtracted from 7573 as follows.

$$\begin{array}{r} 7\ 5\ 7\ 3 \\ -\ \ \ 4\ 3\ 1 \\ \hline 7\ 1\ 4\ 2 \end{array} \begin{array}{l} \leftarrow \text{Minuend} \\ \leftarrow \text{Subtrahend} \\ \leftarrow \text{Difference} \end{array}$$

NOTE: As with addition, it is important to remember that digits in corresponding place values must be aligned vertically when subtracting one number from another. ∎

EXAMPLE 6 Subtracting whole numbers without regrouping

Subtract.
(a) $1688 - 437$ **(b)** $12{,}877 - 10{,}641$

Solution
Stack the minuend vertically above the subtrahend so that the place values are aligned.

(a) $\begin{array}{r} 1688 \\ -\ 437 \\ \hline 1251 \end{array}$ (b) $\begin{array}{r} 12{,}877 \\ -\ 10{,}641 \\ \hline 2236 \end{array}$

NOTE: In part (b), subtracting the digits in the ten-thousands place results in 0, which is not written as the leading digit in the resulting difference. ∎

Now Try Exercises 49, 51

SUBTRACTING WHOLE NUMBERS WITH REGROUPING Sometimes the digit in a particular place value of the minuend (top number) is smaller than the corresponding digit in the subtrahend (bottom number). When this happens, *regrouping* is necessary. For example, the ones digit in 753 is smaller than the ones digit in 318 (because $3 < 8$). To subtract $753 - 318$, we must regroup. The number **753** has **5** tens and **3** ones. We regroup **1** ten from the tens place to the ones place. After doing this, the number **753** has **4** tens and **13** ones.

$$\begin{array}{r} \overset{4\ 13}{7\ \cancel{5}\ \cancel{3}} \\ -\ 3\ 1\ 8 \\ \hline 4\ 3\ 5 \end{array} \quad \text{Regroup as 10 ones.}$$

EXAMPLE 7 Subtracting whole numbers with regrouping

Subtract.
(a) $3653 - 1481$ **(b)** $4039 - 372$

Solution
(a) Regroup **1** (hundred) from the hundreds place to the tens place, resulting in **15** tens. Then perform the subtraction.

$$\begin{array}{r} \overset{5\ 15}{3\ \cancel{6}\ \cancel{5}\ 3} \\ -\ 1\ 4\ 8\ 1 \\ \hline 2\ 1\ 7\ 2 \end{array} \quad \text{Regroup as 10 tens.}$$

READING CHECK 3

- In subtracting whole numbers, when is it necessary to regroup?

(b) Before we can regroup 1 (hundred) from the hundreds place, which contains a 0, we must first regroup 1 (thousand) from the thousands place to the hundreds place, resulting in 10 hundreds. We may then regroup 1 (hundred) from the hundreds place to the tens place, resulting in 13 tens.

$$\begin{array}{r} \overset{9}{3\,\cancel{10}\,13} \\ \cancel{4}\,\cancel{0}\,\cancel{3}\,9 \\ -\,3\,7\,2 \\ \hline 3\,6\,6\,7 \end{array}$$

Regroup as 10 hundreds and then regroup as 10 tens.

Now Try Exercises 53, 59

CALCULATOR HELP

To subtract whole numbers with a calculator, see Appendix E (page AP-28).

MAKING CONNECTIONS 2

Regrouping and Expanded Form

To see how regrouping works, we can write the minuend in *modified* expanded form before performing subtraction. For example, the difference 847 − 372 is found as follows.

800	+	40	+	7	Write 847 in expanded form.
700 + 100 +		40	+	7	Write 847 in modified expanded form.
	Regroup 100				
700	+ 100 +	40 +		7	Regroup 100 to the tens column.
700	+	140	+	7	Write 100 + 40 as 140.
− (300	+	70	+	2)	Write 372 in expanded form.
400	+	70	+	5	Subtract columns.
		475			Write the difference in standard form.

When subtracting one whole number from another, we can use the following procedure.

SUBTRACTING WHOLE NUMBERS

To subtract one whole number from another,

1. Stack the numbers vertically with corresponding place values aligned.
2. Subtract the digits in each place value. Regroup when necessary.

5 Using Properties of Subtraction

The commutative property for addition does **not** hold true for subtraction. For example, 7 − 3 results in 4, while 3 − 7 does not give a whole number result. Similarly, the associative property for addition does **not** hold true for subtraction, as shown.

$$(10 - 8) - 1 = 2 - 1 = 1, \text{ but } 10 - (8 - 1) = 10 - 7 = 3.$$

 Connecting Concepts with Your Life Subtraction does, however, have two properties related to the identity property for addition. The first property states that subtracting 0 from a number does not change the number. If you have $15 and you give away $0, you still have $15. The second property says that subtracting a number from itself results in 0. If you have $7 and you give away $7, you have no money left. These results can be illustrated as follows.

Examples of the Identity Properties

$$15 - 0 = 15 \quad \text{and} \quad 7 - 7 = 0$$

> **IDENTITY PROPERTIES FOR SUBTRACTION**
>
> 1. When 0 is subtracted from any number, the result is that number.
> 2. When a number is subtracted from itself, the result is 0.

6 Recognizing Words Associated with Subtraction

Just as some words suggest that addition should be used to write a mathematical expression, other words suggest that subtraction is appropriate. **TABLE 1.5** shows some words associated with subtraction, together with sample phrases using the words.

Words Associated with Subtraction

Words	Sample Phrase
subtract	subtract the cost from the revenue
minus	his income minus his taxes
fewer than	18 fewer flowers than shrubs
difference	the difference between their heights
less than	his age is 4 years less than hers
decreased by	the number of boxes decreased by 7
take away	the subtotal take away the cash back

TABLE 1.5

EXAMPLE 8 Translating words into a mathematical expression

Translate each phrase into a mathematical expression. Find the result.
(a) 13 days fewer than 23 days   (b) the difference between 18 cards and 12 cards

Solution
(a) The word *fewer* suggests that we should subtract 13 from 23. The corresponding mathematical expression is $23 - 13$, which results in 10 days.
(b) The word *difference* suggests that we should subtract 12 from 18. The corresponding mathematical expression is $18 - 12$, which results in 6 cards.

Now Try Exercises 69, 75

7 Solving Equations Involving Addition and Subtraction

In mathematics, an **equation** can be written when one quantity is equal to another. Every equation contains an equals sign, $=$. The following words or phrases suggest that an equals sign is needed and an equation can be written.

equals,   is,   gives,   results in,   is the same as

An equation can be true or false. For example, the equation $4 + 8 = 12$ is **true**, and the equation $19 - 4 = 10$ is **false**. However, an equation such as

$$\Box + 4 = 9$$ What number plus 4 equals 9?

READING CHECK 4

- How do we know if a number is a solution to an equation?

contains an *unknown value* and may be either true or false depending on the number that is written in the box. Any such number that makes an equation true is called a **solution** to the equation. In general, **solving an equation** means finding all of its solutions. The whole number 5 is the only solution to the equation above because $5 + 4 = 9$ is a **true** equation, and writing any other number in the box would result in a **false** equation.

EXAMPLE 9 Solving equations

Solve each equation by finding the unknown value.
(a) $14 - \square = 3$ (b) $\square - 10 = 5$ (c) $38 + \square = 68$

Solution
(a) The solution is 11 because $14 - 11 = 3$ is a true equation.
(b) Because $15 - 10 = 5$, the solution is 15.
(c) Add 30 to 38 to obtain a sum of 68. The solution is 30.

Now Try Exercises 79, 81, 87

8 Solving Perimeter and Other Applications Involving Addition and Subtraction

READING CHECK 5
- How is the perimeter of an enclosed region found?

🌐 *Math in Context* (Tennis) The in-bounds playing surface of an official doubles tennis court is a rectangular shape that is 36 feet wide and 78 feet long. The *perimeter* of the court is marked by a line that is the boundary between the in-bounds surface and the out-of-bounds surface. In geometry, the **perimeter** of an enclosed region is the distance around the region. **FIGURE 1.14** shows that the perimeter of a doubles tennis court is the sum of the lengths of its four sides, or $36 + 78 + 36 + 78 = 228$ feet.

Perimeter of a Doubles Tennis Court

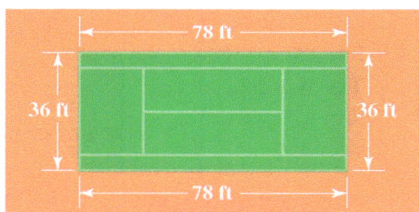

FIGURE 1.14

For some enclosed regions, the length of each side is known, and the perimeter is found by adding the given lengths. Other regions may be missing a length. In this case, we must find the missing length before we can find the perimeter.

EXAMPLE 10 Finding perimeters of shaded regions

Find the perimeter of each shaded region.
(a)
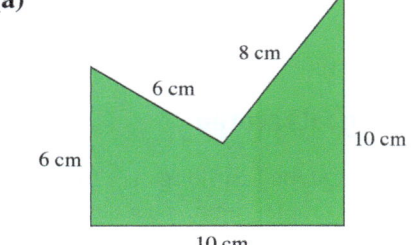

(b)
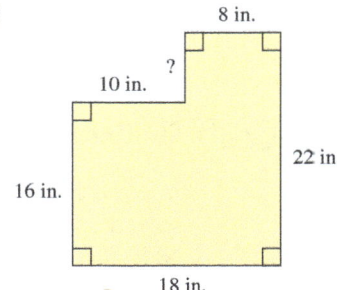

Solution
(a) The length of every side is given. The perimeter is $6 + 6 + 8 + 10 + 10 = 40$ cm.
(b) The missing length can be found by subtraction, as shown in **FIGURE 1.15** on the next page.

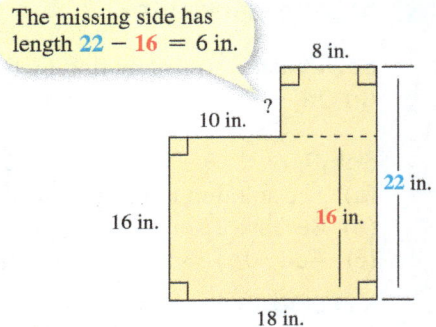

FIGURE 1.15

The missing side has length 22 − 16 = 6 in.

The perimeter is 16 + 10 + 6 + 8 + 22 + 18 = 80 in.

Now Try Exercises 95, 99

In the next two examples, addition and subtraction are used to solve applications involving real-world data.

EXAMPLE 11 **Finding an average temperature in Alaska**

The average summertime high temperature in Nome, Alaska is 11°F more than the average summertime high temperature in Barrow, Alaska. If this temperature is 58 °F in Nome, find the corresponding temperature in Barrow. (*Source: USA Today.*)

Solution
The words

average summertime high temperature in Nome

is 11°*F more than*

the average summertime high temperature in Barrow

translates to an equation of the form

(Nome temperature) = (Barrow temperature) + 11.

Since the average summertime high temperature in Nome is 58 °F, and the corresponding temperature in Barrow is unknown, then the equation to be solved is

58 = □ + 11.   What number plus 11 equals 58?

The solution to this equation is 47 because 58 = 47 + 11. The average summertime high temperature in Barrow is 47 °F.

Now Try Exercise 103

EXAMPLE 12 **Finding the number of goals scored**

In 1920 Joe Malone of the Quebec Bulldogs set the NHL record for the number of goals scored by an *individual* in a single game. That same year, the Montreal Canadiens set the NHL record for the number of goals scored by a *team* in a single game with a 16-3 victory over the Quebec Bulldogs. If the difference between the records is 9, how many goals did Malone score in a single game to set the individual scoring record? (*Source:* NHL.com)

Solution
When the Montreal Canadiens defeated the Quebec Bulldogs, the score was 16 to 3, which means that the team scoring record is 16 goals in a single game. The individual scoring record is unknown.

The phrase

difference between the records *is* 9

suggests subtraction and translates to an equation of the form

(Team record) − (Individual record) = 9.

Because the team record is **16** and the individual record is unknown, the equation is

16 − □ = 9.

The solution to this equation is 7 because 16 − 7 = 9. Therefore, Joe Malone scored 7 goals to set the individual scoring record.

Now Try Exercise 109

1.2 Putting It All Together

CONCEPT	COMMENTS	EXAMPLES
Addition	The numbers being added are the *addends*, and the result is the *sum*.	8 + 7 = 15 **Addend Addend Sum**
Adding Whole Numbers	1. Stack the numbers vertically with corresponding place values aligned. 2. Add the digits in each place value. Regroup when necessary.	$\overset{1}{2}452$ $+6374$ $\overline{8826}$
Properties for Addition	1. Commutative property 2. Associative property 3. Identity property	1. 3 + 6 = 6 + 3 2. (2 + 5) + 4 = 2 + (5 + 4) 3. 4 + 0 = 4 and 0 + 4 = 4
Translating Words to Addition	Words associated with addition include *add, plus, more than, sum, total*, and *increased by*.	7 points more than her score The total of the coins
Subtraction	The number that we subtract from is the *minuend*. The number being subtracted is the *subtrahend*. The result is the *difference*.	24 − 13 = 11 **Minuend Subtrahend Difference**
Subtracting Whole Numbers	1. Stack the numbers vertically with corresponding place values aligned. 2. Subtract the digits in each place value. Regroup when necessary.	$\overset{6\,14}{7\cancel{4}85}$ -2831 $\overline{4654}$
Properties for Subtraction	There are two identity properties, each involving the number 0.	1. 19 − 0 = 19 2. 43 − 43 = 0
Translating Words to Subtraction	Words associated with subtraction include *subtract, minus, fewer than, difference, less than, decreased by,* and *take away*.	The number of bugs decreased by 4 19 days fewer than 10 weeks

continued on next page

26 CHAPTER 1 WHOLE NUMBERS

continued from previous page

CONCEPT	COMMENTS	EXAMPLES
Solving Equations	An *equation* can be written when one quantity is equal to another. A *solution* is any number that makes an equation true when it replaces the unknown value. *Solving an equation* means finding all of its solutions.	The solution to $\square - 13 = 5$ is 18 because $18 - 13 = 5$ is a true equation.
Perimeter	The distance around an enclosed region is called its perimeter.	The perimeter is $3 + 7 + 4 + 6 = 20$ feet.

1.2 Exercises — MyMathLab®

CONCEPTS AND VOCABULARY

1. When adding whole numbers, the numbers being added are called the _____.

2. When whole numbers are being added, the result is called the _____.

3. Is *regrouping* necessary when adding $468 + 215$?

4. The equation $4 + 3 = 3 + 4$ illustrates the _____ property for addition.

5. The equation $(1 + 2) + 6 = 1 + (2 + 6)$ illustrates the _____ property for addition.

6. The _____ property for addition is illustrated by the equation $7 + 0 = 7$.

7. The word *increase* suggests that (addition/subtraction) should be used.

8. When subtracting whole numbers, the number we are subtracting from is the _____ and the number being subtracted is the _____.

9. When whole numbers are being subtracted, the result is called the _____.

10. Is *regrouping* needed to subtract $864 - 521$?

11. The equation $8 - 8 = 0$ illustrates one of the _____ properties for subtraction.

12. The operation (addition/subtraction) should be used for the word *fewer*.

13. A(n) _____ is any number that makes an equation true when it replaces the unknown value.

14. Solving an equation means finding all of its _____.

ADDING WHOLE NUMBERS

Exercises 15–34: Add.

15. $11 + 17$
16. $34 + 21$
17. $65 + 534$
18. $742 + 56$
19. $624 + 261$
20. $322 + 516$
21. 357
    $+7511$
22. 671
    $+2128$
23. 3748
    $+4124$
24. 3352
    $+1539$
25. $16{,}491$
    $+10{,}573$
26. $12{,}458$
    $+23{,}975$
27. $28{,}529 + 53{,}298$
28. $340{,}982 + 72{,}099$
29. $409{,}377 + 654{,}782$
30. $500{,}809 + 499{,}765$

31. 230
5602
+3135

32. 528
6377
+8327

33. 10,669
45,127
+32,255

34. 73,417
56,830
+22,804

USING PROPERTIES OF ADDITION

Exercises 35–40: Identify the property of addition that is illustrated by the given equation.

35. $21 + 8 = 8 + 21$

36. $13 + 54 = 54 + 13$

37. $0 + 87 = 87$

38. $490 + 0 = 490$

39. $2 + (17 + 5) = (2 + 17) + 5$

40. $(22 + 1) + 9 = 22 + (1 + 9)$

Exercises 41–44: Add mentally.

41. $11 + 8 + 13 + 6 + 0 + 7 + 12 + 9$

42. $6 + 3 + 25 + 8 + 0 + 14 + 22 + 5$

43. $0 + 33 + 11 + 6 + 0 + 7 + 9 + 4$

44. $20 + 0 + 44 + 1 + 0 + 6 + 19 + 2$

SUBTRACTING WHOLE NUMBERS

Exercises 45–66: Subtract.

45. $24 - 11$

46. $55 - 31$

47. $468 - 37$

48. $282 - 61$

49. $1769 - 347$

50. $3857 - 554$

51. 3672
− 3521

52. 8175
− 8042

53. 5534
− 3218

54. 6452
− 3327

55. 56,431
− 23,526

56. 81,647
− 58,329

57. $45,832 - 14,399$

58. $184,297 - 98,428$

59. $517,056 - 416,029$

60. $873,870 - 649,335$

61. 3007
− 389

62. 6004
−576

63. 40,063
− 22,378

64. 70,036
− 67,873

65. 100,703
− 89,827

66. 400,102
− 398,516

TRANSLATING WORDS INTO MATH

Exercises 67–78: Translate the phrase into a mathematical expression and then find the result.

67. The sum of $22 and $57

68. 107 songs decreased by 39 songs

69. The difference between 793 photos and 54 photos

70. 873 toothpicks more than 1011 toothpicks

71. Subtract 19 eggs from 62 eggs

72. The total of 13, 89, and 104 cell phone minutes

73. 1200 patients increased by 300 patients

74. 89 degrees fewer than 107 degrees

75. 645 DVDs take away 3 DVDs

76. 58 plates less than 185 plates

77. Add 39 Web pages and 71 Web pages

78. 539 downloads plus 267 downloads

SOLVING EQUATIONS

Exercises 79–92: Solve the given equation by finding the unknown value.

79. $\square - 3 = 6$

80. $5 + \square = 9$

81. $8 - \square = 5$

82. $\square + 7 = 10$

83. $\square + 14 = 34$

84. $24 - \square = 14$

85. $87 - 60 = \square$

86. $31 + 58 = \square$

87. $\square - 10 = 141$

88. $53 + \square = 100$

89. $80 + \square = 107$

90. $84 - 12 = \square$

91. $150 + 379 = \square$

92. $\square - 102 = 14$

APPLICATIONS INVOLVING ADDITION AND SUBTRACTION

Exercises 93–100: Find the perimeter of the shaded region.

93.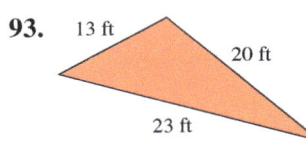

94.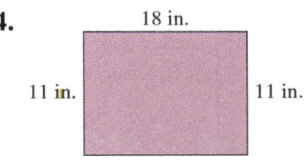

95.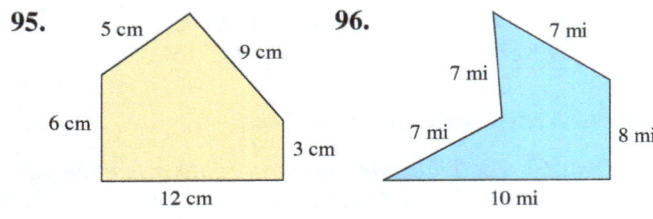

96.

97.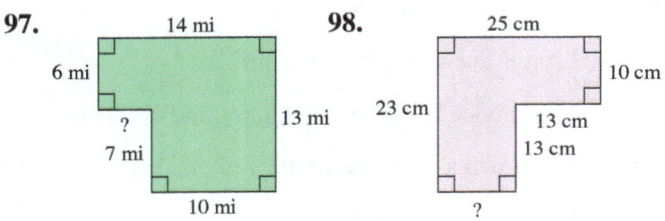

98.

99.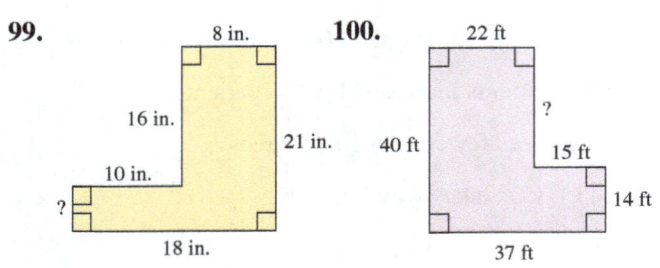

100.

101. **Camera Phones** A Samsung Galaxy S6 has a 16 megapixel camera. The Apple iPhone 6S has a camera that has 4 megapixels less than the Galaxy. How many megapixels does the iPhone camera have?

102. **Mobile Phone Size** A Nokia Asha 501 mobile phone has a screen size of 3 inches. A Samsung Galaxy A9 Pro mobile phone has a screen that is 3 inches larger than the Nokia. What is the screen size of the Samsung phone?

103. **MLB Players** At 83 inches in height, Jon Rauch is the tallest person ever to play major league baseball. The height of the shortest player, Eddie Gaedel, was 40 inches less than Rauchs height. Find Gaedel's height. (*Source:* Major League Baseball.)

104. **Tennis Serves** Samuel Groth has the fastest serve ever recorded in men's tennis. His record serve is 32 miles per hour faster than the fastest serve ever recorded in women's tennis, a 131-mph serve by Sabine Lisicki. How fast is Groth's record serve? (*Source:* U.S. Tennis Association, 2014.)

105. **Deadly Tornados** The following table lists the total number of tornado-related fatalities in the U.S. during selected years. Find the sum of the fatalities from the two deadliest years.

Year	2011	2012	2013	2014
Deaths	553	69	55	44

Source: NOAA, Storm Prediction Center.

106. **U.S. Fires** The following table lists the total number of fires reported in the U.S. during selected years. Find the difference between the largest number and the smallest number of fires.

Year	2010	2011	2012	2013
Fires	1,331,500	1,389,500	1,375,000	2,400,000

Source: National Fire Protection Association.

107. **Bending Phones** The pressure that an iPhone 5 can withstand before bending is 60 pounds more than that of the iPhone 6. If the iPhone 6 can withstand 70 pounds of pressure before it bends, how much pressure can the iPhone 5 withstand? (*Source:* Business Insider.)

108. **Bending Phones** The pressure that an iPhone 6 Plus can withstand before bending is 60 pounds fewer than that of the Samsung Galaxy Note 3. If the Galaxy Note 3 can withstand 150 pounds of pressure before it bends, how much pressure can the iPhone 6 Plus withstand?

109. **NHL Players** There is a 35-year difference between Wayne Gretzky's age at the start of his professional hockey career and Gordie Howe's age at the end of his professional career. If Howe retired at age 52, how old was Gretzky when he began his professional career? (*Source:* National Hockey League.)

110. **NFL Scores** In 1929 Ernie Nevers set the NFL individual scoring record for a single game when he scored *every* point for the Chicago Cardinals in a 40-6 victory over the Chicago Bears. The single-game regular season scoring record for an NFL team, set in 1966 by Washington, is 32 points more than the record set by Nevers. Find the team scoring record. (*Source:* National Football League.)

111. **Bench Press** In 2014 the world record for an unassisted bench press, known as a *raw* bench press, was 722 pounds. A weight lifter wearing a bench shirt was able to bench 380 pounds more to set an equipment-assisted world record. What was the world record for the weight lifter wearing a bench shirt? (*Source:* Powerlifting Watch.)

112. **Stock Market** One of the largest one-day leaps in the history of the Dow Jones Industrial Average occurred on October 13, 2008. That day, the Dow increased by 936 points to close at 9390. What was the value of the Dow when trading began that day? (*Source:* Wall Street Journal.)

Exercises 113–116: **Welfare** *The following bar graph shows the number of people, in millions, who received temporary assistance during selected years.* (*Source:* Administration for Children and Families.)

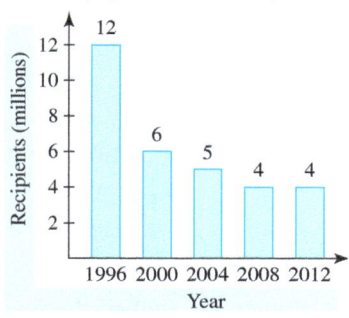

113. Which year had 6,000,000 fewer recipients than there were in 1996?

114. Which year had 1,000,000 more recipients than there were in 2012?

115. To find the number of recipients in 2000, we must decrease the number of recipients in 1996 by what amount?

116. By increasing the number of recipients in 2008 by 2,000,000, we find the number of recipients for which year?

WRITING ABOUT MATHEMATICS

117. Explain what it means to *regroup* when adding whole numbers. Give an example.

118. Explain what it means to *regroup* when subtracting whole numbers. Give an example.

119. A student tells you that the solution to
$$\Box + 14 = 23$$
is 11. How can you convince the student that this answer is incorrect?

120. A student tells you that the solution to
$$29 - \Box = 17$$
is 14. How can you convince the student that this answer is incorrect?

121. Give examples of words associated with addition.

122. Give examples of words associated with subtraction.

SECTIONS 1.1 and 1.2 Checking Basic Concepts

1. Give the place value of the digit 3 in each of the given whole numbers.
   (a) 132,458   (b) 45,267,309

2. Write the number 74,293 in expanded form.

3. Write forty-eight million, two hundred thirty-nine thousand, six hundred ten in standard form.

4. Graph the whole numbers 2, 5, and 1 on the same number line.

5. Place the correct symbol, $<$ or $>$, in the blank between the whole numbers.
   (a) 67 _____ 25   (b) 15 _____ 51

6. Add.
   (a) $581 + 3736$ (b) $204{,}633 + 5897$

7. Subtract.
   (a) $8783 - 124$ (b) $713{,}448 - 112{,}564$

8. Translate each phrase into a mathematical expression and then find the result.
   (a) 97 minus 45   (b) 73 more than 106

9. Solve each equation.
   (a) $3 + \Box = 8$ (b) $\Box - 22 = 7$

10. Find the perimeter of the shaded region.

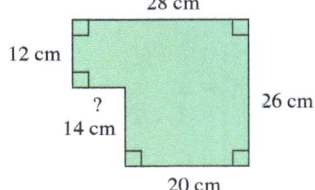

1.3 Multiplying and Dividing Whole Numbers; Area

Objectives

1. Multiplying Whole Numbers
2. Using Properties of Multiplication
   - Commutative Property
   - Associative Property
   - Identity and Zero Properties
   - Distributive Properties
3. Multiplying Larger Whole Numbers
   - Multiplying Whole Numbers Ending in Zeros
4. Recognizing Words Associated with Multiplication
5. Dividing Whole Numbers
   - Remainders
6. Using Properties of Division
7. Performing Long Division
8. Recognizing Words Associated with Division
9. Solving Equations Involving Multiplication and Division
10. Solving Area and Other Applications Involving Multiplication and Division

NEW VOCABULARY

☐ Multiplication
☐ Factors
☐ Product
☐ Multiplication sign
☐ Commutative property for multiplication
☐ Associative property for multiplication
☐ Identity property for multiplication
☐ Zero property for multiplication
☐ Distributive property

continued on page 31

1 Multiplying Whole Numbers

When addends in a sum are the same, we can use **multiplication** as a fast way to perform *repeated addition*. For example, there are seven addends in the sum

$$4 + 4 + 4 + 4 + 4 + 4 + 4 = 28,$$

and each addend is 4. Rather than adding the 4s, the result can be found much faster by using the *multiplication fact*, seven 4s are 28. In multiplication notation, we write

$$7 \cdot 4 = 28.$$

Equivalent to adding seven 4s

Factors   Product

The numbers being multiplied are called **factors**, the result is called the **product**, and the symbol $\cdot$ is called the **multiplication sign**.

NOTE: Multiplication can be written in several ways. Each of the following expressions indicates that we are multiplying 7 and 4. ■

$$7 \cdot 4, \quad 7 \times 4, \quad 7(4), \quad (7)4, \quad \text{and} \quad (7)(4)$$

Before we can multiply larger whole numbers effectively, we must first *memorize* the products that result when multiplying two single-digit numbers. These basic multiplication facts are shown in **TABLE 1.6**, where the product $7 \cdot 4 = 28$ has been highlighted to demonstrate how a product can be found in the table.

Basic Multiplication Facts

$\cdot$	0	1	2	3	4	5	6	7	8	9
0	0	0	0	0	0	0	0	0	0	0
1	0	1	2	3	4	5	6	7	8	9
2	0	2	4	6	8	10	12	14	16	18
3	0	3	6	9	12	15	18	21	24	27
4	0	4	8	12	16	20	24	28	32	36
5	0	5	10	15	20	25	30	35	40	45
6	0	6	12	18	24	30	36	42	48	54
7	0	7	14	21	28	35	42	49	56	63
8	0	8	16	24	32	40	48	56	64	72
9	0	9	18	27	36	45	54	63	72	81

TABLE 1.6

2 Using Properties of Multiplication

Several properties are helpful when performing computations involving multiplication. Some of these properties are similar to those used for addition.

COMMUTATIVE PROPERTY The **commutative property for multiplication** states that changing the *order* of the factors does not change the resulting product.

Example of the Commutative Property

$$5 \cdot 8 = 40 \quad \text{and} \quad 8 \cdot 5 = 40$$

STUDY TIP

Remember that a positive attitude is important. The first step to success is believing in yourself.

CRITICAL THINKING

Rewrite the repeated addition in the following equation as multiplication. What property of multiplication is illustrated by this equation?

$4 + 4 + 4 = 3 + 3 + 3 + 3$

NEW VOCABULARY

continued from page 30
- ☐ Division
- ☐ Quotient
- ☐ Dividend
- ☐ Divisor
- ☐ Division sign
- ☐ Identity properties for division
- ☐ Zero properties for division
- ☐ Undefined
- ☐ Remainder
- ☐ Partial dividend
- ☐ 1 square unit
- ☐ Area

COMMUTATIVE PROPERTY FOR MULTIPLICATION

The *order* in which two factors are written does not affect the product.

ASSOCIATIVE PROPERTY The **associative property for multiplication** allows for the *regrouping* of factors when more than two numbers are multiplied. Parentheses are used to indicate which two factors should be grouped (multiplied) first.

Example of the Associative Property

$(2 \cdot 3) \cdot 8 = 6 \cdot 8 = 48$ and $2 \cdot (3 \cdot 8) = 2 \cdot 24 = 48$

ASSOCIATIVE PROPERTY FOR MULTIPLICATION

The way in which three or more factors are *grouped* does not affect the product.

IDENTITY AND ZERO PROPERTIES Two other properties for multiplication can be illustrated as follows.

One four-can box of energy drinks contains four cans.

Zero four-can boxes of energy drinks contain zero cans.

The first example demonstrates the **identity property for multiplication**, which states that multiplying a number by 1 does not change the number. The second example illustrates the **zero property for multiplication**, which states that multiplying a number by 0 results in 0.

Examples of the Identity Property and Zero Property

$4 \cdot 1 = 4$ and $1 \cdot 4 = 4$ and $0 \cdot 4 = 0$ and $4 \cdot 0 = 0$

IDENTITY PROPERTY FOR MULTIPLICATION

When any number is multiplied by 1, the result is that number.

ZERO PROPERTY FOR MULTIPLICATION

When any number is multiplied by 0, the result is 0.

EXAMPLE 1 Identifying properties of multiplication

Identify the property of multiplication that is illustrated in each equation.
 (a) $9 \cdot 0 = 0$ (b) $4 \cdot 8 = 8 \cdot 4$ (c) $(2 \cdot 5) \cdot 9 = 2 \cdot (5 \cdot 9)$ (d) $7 \cdot 1 = 7$

Solution
(a) When multiplying a number by zero, the result is always zero. This is an example of the zero property for multiplication.
(b) The product of 4 and 8 is the same as a product with the same factors written in a different order. This is an example of the commutative property for multiplication.

READING CHECK 1

- Which multiplication property allows us to multiply in any order?
- Which multiplication property allows us to regroup the factors?

(c) Even though the order of the factors in the equation $(2 \cdot 5) \cdot 9 = 2 \cdot (5 \cdot 9)$ is the same on both sides of the equals sign, the way the factors are grouped differs. This is an example of the **associative property for multiplication**.

(d) When multiplying a number by 1, the result is always the given number. So the equation $7 \cdot 1 = 7$ is an example of the **identity property for multiplication**.

Now Try Exercises 19, 21, 23, 25

DISTRIBUTIVE PROPERTIES One final property of multiplication is illustrated below. This property, called the **distributive property**, allows us to multiply a sum (or difference) by a number.

> **See the Concept 1** — **THE DISTRIBUTIVE PROPERTY**
>
> There are 20 circles on the left side of the equals sign (12 blue and 8 red) and there are also 20 circles on the right side of the equals sign (again, 12 blue and 8 red).
>
> When multiplying $4 \cdot (3 + 2)$, we can *distribute* the 4 *over* the sum in parentheses.
>
>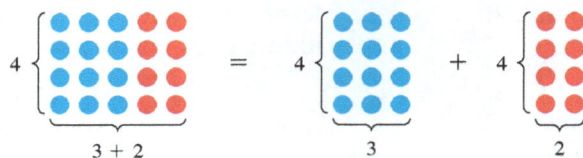
>
> **First Add 3 + 2, and then Multiply**
> $4(3 + 2) = 4(5) = 20$
>
> **First Distribute the 4, and then Add**
> $4(3 + 2) = 4 \cdot 3 + 4 \cdot 2 = 12 + 8 = 20$
>
> The results are the same.
>
> WATCH VIDEO IN MML.

The figure in the See the Concept shows that *multiplication distributes over addition*. The property also holds if the sum is replaced by a difference. That is, *multiplication distributes over subtraction*. Curved arrows are often used to illustrate the distributive properties.

Examples of the Distributive Properties

$3(5 + 7) = 3 \cdot 5 + 3 \cdot 7$ and $4(8 - 5) = 4 \cdot 8 - 4 \cdot 5$

Multiplication over addition   Multiplication over subtraction

DISTRIBUTIVE PROPERTIES

When a sum or difference in parentheses is multiplied by a number outside the parentheses, the outside number can be multiplied by each of the inside numbers *before* the sum or difference is computed.

EXAMPLE 2   **Applying the distributive properties**

Use a distributive property to rewrite each expression. Do not find the product.
(a) $6(3 + 8)$ (b) $5(7 - 2)$

Solution

(a) $6(3 + 8) = 6 \cdot 3 + 6 \cdot 8$

(b) $5(7 - 2) = 5 \cdot 7 - 5 \cdot 2$

Now Try Exercises 29, 31

3 Multiplying Larger Whole Numbers

When multiplying larger whole numbers, we can use the expanded form and a distributive property to find a product. For example, the product 6×47 can be written as $6(40 + 7)$, and a distributive property may be applied as follows:

$$6(40 + 7) = 6 \cdot 40 + 6 \cdot 7$$
$$= 240 + 42$$
$$= 282$$

However, by stacking the numbers vertically and using regrouping when appropriate, we can perform multiplication more efficiently without the need to write any of the factors in expanded form. The product 6×47 can be multiplied using the following two steps:

STEP 1: Multiply 6×7 (ones)   **STEP 2:** Multiply 6×4 (tens)

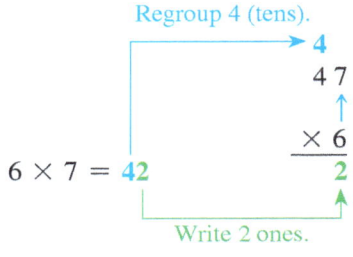
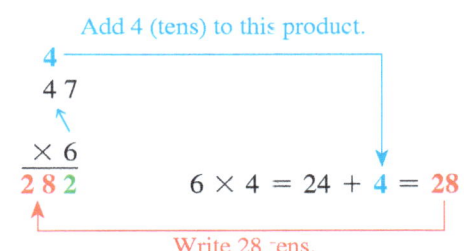

The next example shows how these steps can be combined into a single process in order to multiply larger whole numbers.

EXAMPLE 3   **Multiplying larger whole numbers**

Multiply.
(a) $83(4)$ (b) $92 \cdot 35$ (c) 386×73 (d) $208(867)$

Solution

(a)
```
  1
  83
× 4
 332
```

(b)
```
   1
   92
 × 35
  460    ← 5 × 92
 2760    ← 3 (tens) × 92
 3220    ← add
```

(c)
```
  6 4
  2 1
   386
 × 73
  1 158   ← 3 × 386
 27 020   ← 7 (tens) × 386
 28,178   ← add
```

(d)
```
    6
    4
    5
    208
  × 867
   1 456    ← 7 × 208
  12 480    ← 6 (tens) × 208
 166 400    ← 8 (hundreds) × 208
 180,336    ← add
```

Now Try Exercises 41, 45, 47, 51

MAKING CONNECTIONS 1

The Distributive Property and Expanded Form

The multiplication process used in Example 3 provides a short way to apply the distributive property to the expanded form of each factor. The *partial products* **2760** and **460** found in Example 3(b) are also shown in the following multiplication process.

$92 \times 35 = 92(30 + 5)$	Write 35 in expanded form.
$= 92(30) + 92(5)$	Distribute the 92.
$= (90 + 2)(30) + (90 + 2)(5)$	Write 92 in expanded form.
$= (90)(30) + (2)(30) + (90)(5) + (2)(5)$	Distribute the 30 and the 5.
$= 2700 + 60 + 450 + 10$	Find each product.
$= 3220$	Add.

CALCULATOR HELP

To multiply whole numbers with a calculator, see Appendix E (page AP-28).

MULTIPLYING WHOLE NUMBERS ENDING IN ZEROS

Math in Context *Printing* A *skid* of copier paper contains 40 boxes of paper. Each box contains 10 individually wrapped *reams*, and each ream has 500 sheets of paper. To find the total number of sheets of copier paper in a skid, we multiply

$$40 \times 10 \times 500.$$

Because each of these numbers ends in one or more zeros, we can perform the multiplication mentally. Consider the following products.

$10 \times 10 = 100$ $\qquad$ $20 \times 300 = 6000$

$140 \times 200 = 28{,}000$ $\qquad$ $500 \times 600 = 300{,}000$

In each case, the product is found by counting the total number of zeros in the factors and then writing that number of zeros after the product of the nonzero digits. Using this process, we see that a skid of copier paper has $40 \times 10 \times 500 = 200{,}000$ sheets.

EXAMPLE 4 Multiplying whole numbers that end in zeros

Multiply.
(a) 130×40 $\qquad$ (b) 700×2500

Solution
(a) $130 \times 40 = 5200$ $\qquad$ (b) $700 \times 2500 = 1{,}750{,}000$

Now Try Exercises 55, 57

4 Recognizing Words Associated with Multiplication

Just as there are words associated with addition and subtraction, there are also words associated with multiplication. **TABLE 1.7** shows some of these words, together with sample phrases using the words.

Words Associated with Multiplication

Word	Sample Phrase
multiply	multiply the length and the width
times	the number purchased times the price
product	the product of the measurements
double	double the recipe
triple	triple the score

TABLE 1.7

EXAMPLE 5 Translating words into a mathematical expression

Translate each phrase into a mathematical expression. Find the result.
(a) 15 times 20 cars   (b) the product of 7 and 38 coffee drinks

Solution
(a) The word *times* suggests that we multiply 15 and 20. The corresponding mathematical expression is 15×20, which results in 300 cars.
(b) The word *product* suggests that we multiply 7 and 38. The corresponding mathematical expression is 7×38, which results in 266 coffee drinks.

Now Try Exercises 89, 91

5 Dividing Whole Numbers

Connecting Concepts with Your Life Some gas stations and convenience stores sell music CDs at discounted prices. A customer can use *repeated subtraction* to compute the number of CDs costing $6 each that can be purchased for $24. Each time 6 is subtracted, another CD can be purchased.

$$24 - 6 = 18 \qquad 18 - 6 = 12 \qquad 12 - 6 = 6 \qquad 6 - 6 = 0$$

1st Subtraction   2nd Subtraction   3rd Subtraction   4th Subtraction

Because 6 can be subtracted a total of 4 times, the customer can buy 4 CDs for $24. Just as multiplication is a fast way to perform repeated addition, **division** is a fast way to perform repeated subtraction. We say that 24 divided by 6 is 4 and write $24 \div 6 = 4$. The result of dividing one whole number by another is called the **quotient**. The number we are dividing *into* is called the **dividend**, the number we are dividing *by* is called the **divisor**, and the symbol $\div$ is called the **division sign**. In this example, the dividend is 24, the divisor is 6, and the quotient is 4.

$$24 \div 6 = 4$$

Dividend   Divisor   Quotient

NOTE: Like multiplication, division can be written in several ways. Each of the following expressions represents dividing 24 by 6.

$$24 \div 6, \quad \frac{24}{6}, \quad 24/6, \quad \text{and} \quad 6\overline{)24}$$

Division can be checked by multiplying as follows.

$$\text{Quotient} \times \text{Divisor} = \text{Dividend}$$

From above:
$4 \times 6 = 24$

EXAMPLE 6 Dividing whole numbers

Find each quotient. Check your answers by multiplying.
(a) $48 \div 6$ (b) $\frac{63}{9}$

Solution
(a) The quotient $48 \div 6 = 8$ checks by multiplying $8 \times 6 = 48$. ✓
(b) The quotient $\frac{63}{9} = 7$ checks by multiplying $7 \times 9 = 63$. ✓

Now Try Exercises 65, 73

REMAINDERS Sometimes a divisor does not divide evenly into a dividend. For example, **FIGURE 1.16** shows that 17 books can be divided into 3 stacks of 5 books each, with 2 books remaining. We say that 2 is the **remainder** and write the quotient as 3 r2.

FIGURE 1.16

NOTE: When a remainder exists, division can be checked as follows.

Quotient × Divisor + Remainder = Dividend ∎

READING CHECK 2

• What is a remainder?

6 Using Properties of Division

The commutative properties for addition and multiplication do *not* hold true for division. For example, $10 \div 5$ results in 2, whereas $5 \div 10$ does not give a whole number result. Similarly, the associative properties for addition and multiplication do *not* hold true for division. For example,

$$(24 \div 6) \div 2 = 4 \div 2 = 2, \text{ but } 24 \div (6 \div 2) = 24 \div 3 = 8.$$

Different results

READING CHECK 3

• Which two multiplication properties do not hold true for division?

 Connecting Concepts with Your Life If we think of division as a way to perform repeated subtraction, then we can find several properties for division. Consider the following four questions where "a number" can be any whole number *except* 0.

1. How many times can a number be subtracted from itself?   (Dividing a number by itself)
2. How many times can 1 be subtracted from a number?   (Dividing a number by 1)
3. How many times can a number be subtracted from 0?   (Dividing 0 by a number)
4. How many times can 0 be subtracted from a number?   (Dividing a number by 0)

CALCULATOR HELP

To divide whole numbers with a calculator, see Appendix E (page AP-28).

Questions 1 and 2 illustrate the **identity properties for division**.

1. A person with 5 dimes can give away all 5 dimes exactly 1 time, or $5 \div 5 = 1$.
2. A person with 7 nickels can give away 1 nickel 7 times, or $7 \div 1 = 7$.

Questions 3 and 4 illustrate the **zero properties for division**.

3. A person with 0 pennies can give away 12 pennies 0 times because there are no pennies to give away, or $0 \div 12 = 0$.
4. A person with 6 quarters can give away 0 quarters *any* number of times. We say that $6 \div 0$ is **undefined**.

IDENTITY PROPERTIES FOR DIVISION

1. When any number (except 0) is divided by itself, the result is 1.
2. When any number is divided by 1, the result is the number (dividend).

ZERO PROPERTIES FOR DIVISION

1. When 0 is divided by any number (except 0), the result is 0.
2. When any number is divided by 0, the result is undefined.

EXAMPLE 7 **Applying the division properties**

Use division properties to find each quotient, when possible.

(a) $23 \div 1$ (b) $\dfrac{0}{14}$ (c) $83 \div 83$ (d) $\dfrac{62}{0}$

Solution

(a) Dividing by 1 results in the dividend, 23.
(b) Dividing 0 by a number that is not 0 results in 0.
(c) Dividing a nonzero number by itself results in 1.
(d) Division by 0 is undefined.

Now Try Exercises 61, 63, 67, 69

7 Performing Long Division

When we need to find a quotient involving a larger dividend, we can use a process called *long division*, which allows us to break a large division problem into several smaller division problems. For example, to divide **2461** by **5** using long division, do the following.

$$\text{Divisor} \rightarrow 5\overline{)2461} \leftarrow \text{Dividend}$$

1. Starting at the left end of the dividend, select the fewest digits that give a number that is greater than the divisor. This (highlighted) number is called the **partial dividend**.

$$
\begin{array}{c}
\textbf{1.}\ 5\overline{)2461} \qquad\qquad \textbf{2.}\ \begin{array}{r} 4 \\ 5\overline{)2461} \end{array} \qquad\qquad \textbf{3.}\ \begin{array}{r} 4 \\ 5\overline{)2461} \\ -20 \\ \hline 4 \end{array}
\end{array}
$$

2. The divisor 5 will "go into" the partial dividend 24 at most 4 times. Write **4** above the rightmost digit of the partial dividend.
3. Next, multiply 4 and 5, write the result **20** below the partial dividend, and subtract.
4. Now, "bring down" the first digit in the original dividend that is aligned to the right of the partial dividend. The number 46 is the new partial dividend. We are now ready to begin the process again.

$$
\textbf{4.}\ \begin{array}{r} 4 \\ 5\overline{)2461} \\ -20\downarrow \\ \hline 46 \end{array} \qquad\qquad \textbf{5.}\ \begin{array}{r} 49 \\ 5\overline{)2461} \\ -20 \\ \hline 46 \\ -45\downarrow \\ \hline 11 \end{array} \qquad\qquad \textbf{6.}\ \begin{array}{r} 492 \\ 5\overline{)2461} \\ -20 \\ \hline 46 \\ -45 \\ \hline 11 \\ -10 \\ \hline 1 \end{array}
$$

5. The divisor 5 goes into 46 at most 9 times. Write **9** next to 4 in the quotient. Multiply 9 and 5, write the result **45** below the partial dividend, and subtract. Bring down the 1. The number 11 is the new partial dividend. Go through the process one more time.
6. The divisor 5 goes into 11 at most 2 times. Write **2** next to 9 in the quotient. Multiply 2 and 5, write the result **10** below the partial dividend, and subtract.
7. When there are no more digits to bring down from the original dividend, the process is done. The final difference **1** is the remainder. We write the quotient as **492** r1.

PERFORMING LONG DIVISION

STEP 1: Determine the number of times that the divisor will "go into" the partial dividend. Write this digit above the right-most digit of the partial dividend.

STEP 2: Multiply the digit found in Step 1 by the divisor, and write the product below the partial dividend.

STEP 3: Subtract the product found in Step 2 from the partial dividend.

STEP 4: From the original dividend, "bring down" the first digit aligned to the right of the partial dividend. The number formed becomes the new partial dividend. If there is no digit to bring down, you are done.

READING CHECK 4

- When doing long division, where is the quotient written?

EXAMPLE 8 Performing long division

Divide: $2875 \div 3$. Check your answer.

Solution
The partial dividend is highlighted each time through the steps.

First time through steps

```
          9    ← Step 1
      3)2875
       -27↓   ← Step 2
Step 3 → 17    ← Step 4
```

Second time through steps

```
          95   ← Step 1
      3)2875
       -27
          17
         -15↓  ← Step 2
Step 3 →  25   ← Step 4
```

Third time through steps

```
          958  ← Step 1
      3)2875
       -27
          17
         -15
           25
          -24  ← Step 2
Step 3 →    1  ← Remainder
```

The quotient is 958 with remainder 1, or 958 r1. This result can be checked as follows.

$$958 \times 3 + 1 = 2875$$

Now Try Exercise 75

We do not need to rewrite a division problem every time we go through the steps. Typically, all of the steps used to perform long division are stacked vertically. The next example illustrates this process for divisors with more than 1 digit.

EXAMPLE 9 Performing long division

Divide. Check your answer.
(a) $2511 \div 23$ **(b)** $89{,}285 \div 258$

Solution

In part (a), note that 23 (the divisor) does not divide into the partial dividend 21. As a result, 0 is written in the quotient, and the steps are continued as usual.

(a)
```
              Step 1 (3 times)
         109 r4
    23)2511
       − 23 ↓           ← Step 2
Step 3 →  21            ← Step 4
        −  0 ↓          ← Step 2
Step 3 →  211           ← Step 4
        − 207           ← Step 2
Step 3 →    4           ← Remainder
```

Check:
$109 \times 23 + 4 = 2511$ ✓

(b)
```
                  Step 1 (3 times)
           346 r17
    258)89,285
        − 77 4 ↓         ← Step 2
Step 3 →  11 88          ← Step 4
        − 10 32 ↓        ← Step 2
Step 3 →   1 565         ← Step 4
         − 1 548         ← Step 2
Step 3 →      17         ← Remainder
```

Check:
$346 \times 258 + 17 = 89,285$ ✓

Now Try Exercises 77, 87

8 Recognizing Words Associated with Division

TABLE 1.8 shows some words associated with division, together with sample phrases using the words.

Words Associated with Division

Word	Sample Phrase
divide	divide the area by the length
quotient	the quotient of the pay and the hours worked
per	168 miles per 12 gallons

TABLE 1.8

EXAMPLE 10   Translating words into a mathematical expression

Translate each phrase into a mathematical expression. Find the result.
(a) 126 chairs divided by 14 rows   **(b)** 150 miles per 2 hours

Solution
(a) The words *divided by* suggest that we should divide 126 by 14. The corresponding mathematical expression is $126 \div 14$, which results in 9 chairs per row.
(b) The word *per* suggests that we should divide 150 by 2. The corresponding mathematical expression is $150 \div 2$, which results in 75 miles per hour.

Now Try Exercises 93, 95

9 Solving Equations Involving Multiplication and Division

In the previous section, we solved addition and subtraction equations containing unknown values by finding the value that, when written in the empty box, makes the equation true. Equations involving multiplication or division can be solved in this way also. For example, the solution to the equation

$$\Box \times 4 = 36$$

What number times 4 equals 36?

is **9** because $9 \times 4 = 36$ is a true equation and writing any other number in the box would result in a false equation.

EXAMPLE 11 Solving equations

Solve each equation by finding the unknown value.
(a) $64 \div \square = 16$ (b) $13 \times \square = 65$

Solution
(a) The solution is **4** because $64 \div 4 = 16$ is a true equation.
(b) Because the equation $13 \times 5 = 65$ is true, the solution is **5**.

Now Try Exercises 97, 103

10 Solving Area and Other Applications Involving Multiplication and Division

Connecting Concepts with Your Life Some ceramic floor tiles are available in square pieces that measure 1 foot on each side. A square that measures 1 unit on each side has an *area* of **1 square unit**, as illustrated in **FIGURE 1.17**. To determine the number of square tiles needed to cover a small entryway that is 7 feet long and 5 feet wide, we must find the area of the floor. **Area** is computed by finding the number of square units that are needed to cover a region. **FIGURE 1.18** shows that a 7-feet by 5-feet entryway has an area of 35 square feet.

Square Unit
1 unit
Area is 1 square unit — 1 unit
FIGURE 1.17

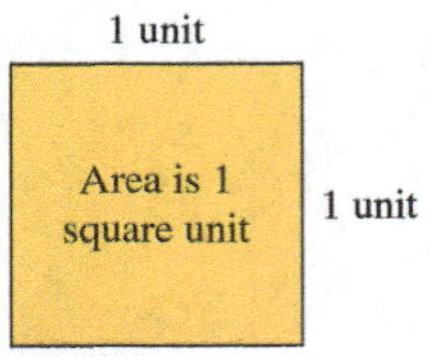

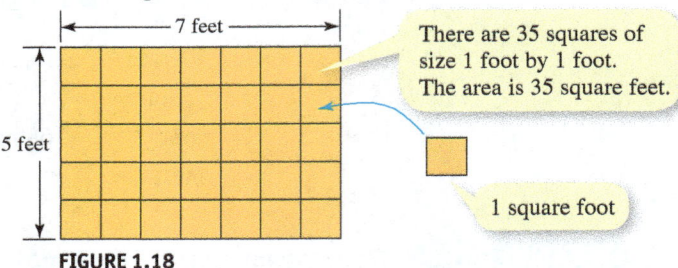

FIGURE 1.18

The area of the rectangle in **FIGURE 1.18** can be found by counting the number of 1-foot by 1-foot tiles that cover the entire region. However, we can also find the area by multiplying the length **7** and the width **5**.

$$7 \text{ feet} \times 5 \text{ feet} = 35 \text{ square feet}$$

To find the area of a rectangle, we use the following formula.

READING CHECK 5
• What kind of units are used to describe area?

AREA OF A RECTANGLE
The area of a rectangle is found by multiplying its length and width.

$$\text{Area} = \text{Length} \times \text{Width}$$

EXAMPLE 12 Finding the area of an Olympic swimming pool's surface

An Olympic swimming pool is 50 meters long and 25 meters wide. (A meter is a unit of length in the metric system.) Find the area of the pool's surface in square meters.

1.3 MULTIPLYING AND DIVIDING WHOLE NUMBERS; AREA

Solution
The area of the pool's surface is found by multiplying the length **50** and the width **25**.

$$50 \times 25 = 1250 \text{ square meters}$$

Now Try Exercise 109

🌐 *Math in Context* (Electricity) The E-126 wind turbine is among the largest in the world. Each rotor blade is approximately 413 feet long. It is estimated that just twenty of these turbines can generate enough electricity to power 35,500 American homes. In the next example, division is used to find the number of homes that can be powered by a single E-126 wind turbine. (*Source: WindBlatt*, Enercon Magazine for Wind Energy.)

EXAMPLE 13 Finding the number of homes powered by a wind turbine

Find the number of American homes that can be powered by a single E-126 wind turbine if 20 such turbines can power 35,500 homes.

Solution
To find the number of homes that can be powered by one E-126 wind turbine, divide 35,500 by 20. As shown to the right, a single E-126 wind turbine can power 1775 American homes.

```
         1 775
    20)35,500
       - 20
         ────
         15 5
       - 14 0
         ────
          1 50
        - 1 40
          ────
           100
         - 100
           ───
             0
```

Now Try Exercise 119

1.3 Putting It All Together

CONCEPT	COMMENTS	EXAMPLES
Multiplication	The numbers being multiplied are the *factors*, and the result is the *product*.	8 × 4 = 32 factor factor product

continued on next page

CONCEPT	COMMENTS	EXAMPLES
Properties for Multiplication	1. Commutative property 2. Associative property 3. Identity property 4. Zero property 5. Distributive properties	1. $4 \times 3 = 3 \times 4$ 2. $(2 \cdot 4) \cdot 5 = 2 \cdot (4 \cdot 5)$ 3. $5 \times 1 = 5$ and $1 \times 5 = 5$ 4. $3 \times 0 = 0$ and $0 \times 3 = 0$ 5. $3(6 + 5) = 3 \cdot 6 + 3 \cdot 5$ $2(7 - 4) = 2 \cdot 7 - 2 \cdot 4$
Words Associated with Multiplication	Words associated with multiplication include *multiply*, *times*, *product*, *double*, and *triple*.	3 times her age Double the number of cups
Division	The number divided into is the *dividend*; the number divided by is the *divisor*, and the result is the *quotient*.	$48 \div 6 = 8$ dividend divisor quotient
Properties for Division	1. Identity properties 2. Zero properties	1. $7 \div 7 = 1$ $9 \div 1 = 9$ 2. $0 \div 3 = 0$ $8 \div 0$ is undefined.
Long Division	When the dividend is large, we use a process called *long division*. When the divisor does not divide perfectly into the dividend, the *remainder* is the amount left over.	$\begin{array}{r} 36 \text{ r}6 \\ 7\overline{)258} \\ -21 \\ \hline 48 \\ -42 \\ \hline 6 \end{array}$ **Check:** $36 \times 7 + 6 = 258$ ✓
Words Associated with Division	Words associated with division include *divide*, *quotient*, and *per*.	Divide the area by the width Student athletes per van
Area of a Rectangle	Area = Length × Width	6 inches 3 inches Area is $3 \times 6 = 18$ square inches.

1.3 Exercises

CONCEPTS AND VOCABULARY

1. Multiplication is a fast way to perform repeated _____.

2. When multiplying, the two numbers being multiplied are called _____.

3. The result when multiplying is called the _____.

4. The equation $13 \cdot 26 = 26 \cdot 13$ is an example of the _____ property for multiplication.

5. The equation $(2 \cdot 5) \cdot 4 = 2 \cdot (5 \cdot 4)$ is an example of the _____ property for multiplication.

6. The _____ property for multiplication states that multiplying any number by 1 results in that number.

7. The _____ property for multiplication states that multiplying any number by 0 results in 0.

8. The equation $3(7 + 5) = 3 \cdot 7 + 3 \cdot 5$ illustrates a(n) _____ property for multiplication.

9. The word *times* indicates that (multiplication/division) should be used.

10. Division is a fast way to perform repeated _____.

11. In a division problem, the number we are dividing into is called the _____ and the number we are dividing by is called the _____.

12. The result when dividing is called the _____.

13. The equation $7 \div 7 = 1$ illustrates one of the _____ properties for division.

14. $0 \div 5 =$ _____, but $5 \div 0$ is _____.

15. A process for finding a quotient involving a larger dividend is called _____.

16. The operation (multiplication/division) should be used when we see the word *per*.

17. A square that measures 1 unit on each side has an area of _____.

18. We compute _____ by finding the number of square units that cover a region.

USING PROPERTIES OF MULTIPLICATION

Exercises 19–28: Identify the property of multiplication that is illustrated by the given equation.

19. $6 \cdot (14 \cdot 2) = (6 \cdot 14) \cdot 2$
20. $(11 \cdot 3) \cdot 7 = 11 \cdot (3 \cdot 7)$
21. $12 \cdot 8 = 8 \cdot 12$
22. $13 \cdot 14 = 14 \cdot 13$
23. $1 \cdot 36 = 36$
24. $71 \cdot 1 = 71$
25. $4(2 + 9) = 4 \cdot 2 + 4 \cdot 9$
26. $3(7 - 1) = 3 \cdot 7 - 3 \cdot 1$
27. $0 \cdot 46 = 0$
28. $73 \cdot 0 = 0$

Exercises 29–34: Use a distributive property to rewrite the expression. Do not find the product.

29. $5(6 + 9)$
30. $7(2 + 5)$
31. $4(8 - 1)$
32. $6(9 - 3)$
33. $(6 - 2)3$
34. $(5 + 7)4$

MULTIPLYING WHOLE NUMBERS

Exercises 35–54: Multiply.

35. 7×1
36. $0 \cdot 9$
37. $0 \cdot 5$
38. 1×12
39. $6(9)$
40. $(4)(8)$
41. $7 \cdot 48$
42. 5×83
43. $(302)6$
44. $(479)(8)$
45. $71(24)$
46. $94 \cdot 18$
47. 172×14
48. $(23)(492)$
49. $35 \cdot 1475$
50. 56×9012
51. $376(754)$
52. $126 \cdot 533$
53. $109 \cdot 1074$
54. $2348(342)$

Exercises 55–60: Multiply mentally.

55. 70×300
56. $30 \cdot 2000$
57. $340 \cdot 2000$
58. 4000×800
59. $30 \cdot 100 \cdot 500$
60. $40 \times 80 \times 20$

DIVIDING WHOLE NUMBERS

Exercises 61–88: Divide, when possible.

61. $9 \div 1$
62. $\dfrac{0}{4}$
63. $\dfrac{17}{17}$
64. $12 \div 0$
65. $88 \div 8$
66. $81 \div 9$
67. $\dfrac{25}{0}$
68. $\dfrac{72}{24}$
69. $0 \div 13$
70. $34 \div 1$
71. $391 \div 391$
72. $\dfrac{354}{6}$
73. $\dfrac{72}{6}$
74. $\dfrac{1026}{1026}$
75. $\dfrac{6729}{7}$
76. $\dfrac{5812}{9}$
77. $2487 \div 31$
78. $4679 \div 53$
79. $6000 \div 30$
80. $8000 \div 20$
81. $9874 \div 0$
82. $\dfrac{0}{5430}$
83. $\dfrac{10{,}651}{84}$
84. $24{,}682 \div 99$
85. $36{,}855 \div 567$
86. $76{,}383 \div 943$
87. $49{,}777 \div 791$
88. $31{,}896 \div 665$

TRANSLATING WORDS INTO MATH

Exercises 89–96: Translate the phrase into a mathematical expression and then find the result.

89. The product of 14 feet and 3 feet

90. Double 50 pounds

91. Multiply $5 by 15

92. 31 text messages times 65

93. 126 miles per 7 gallons

94. Divide 1200 boxes by 20

95. The quotient of 75 days and 15

96. 516 people per 43 tables

SOLVING EQUATIONS

Exercises 97–104: Solve the given equation by finding the unknown value.

97. $12 \div \square = 4$

98. $\square \times 7 = 21$

99. $\square \times 3 = 15$

100. $16 \div \square = 4$

101. $\square \div 10 = 5$

102. $5 \times \square = 100$

103. $6 \times \square = 36$

104. $\square \div 3 = 9$

APPLICATIONS INVOLVING MULTIPLICATION AND DIVISION

Exercises 105–108: **Area** *Find the area of the rectangle.*

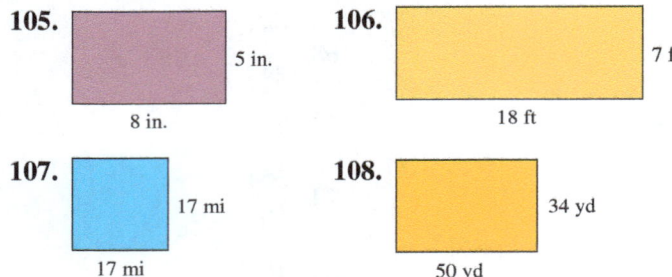

105. 5 in. by 8 in.

106. 7 ft by 18 ft

107. 17 mi by 17 mi

108. 34 yd by 50 yd

109. **Baseball Diamonds** A baseball diamond has the shape of a square measuring 90 feet on each side. Find the area of a baseball diamond.

110. **Tennis Courts** A doubles tennis court is 78 feet long and 36 feet wide. What is the area of a doubles tennis court?

111. **Reselling Phones** The following table lists the resale value of flagship smartphone models by manufacturer. Values are based off a phone traded in 6 months after its launch. What number should be multiplied by the value of a broken Apple phone to find the value of an Apple phone in good condition?

Type	Apple	Apple (broken)	Samsung	Lg
Value	$330	$110	$225	$140

Source: Gazelle.

112. **Endangered Species** The following table shows the number of species listed as threatened or endangered in selected states. Which state has 7 times as many species listed as North Dakota, which lists 10 species?

State	Delaware	Maine	Missouri	Virginia
Species	23	16	40	70

Source: U.S. Fish and Wildlife Service.

113. **Tidal Music** After 12 months, Jay-Z's music streaming app, Tidal, had 3 million subscribers. If each subscriber payed about $10 per month, about how much revenue, in millions, does Tidal make in a month from subscribers? (*Source:* Tidal.)

114. **Apple Music** As of February 2016, Apple Music had 11 million subscribers. If each subscriber payed about $10 per month, about how much revenue, in millions, does Apple Music make in a month from subscribers? (*Source:* Apple.)

115. **Mobile Talk Time** The Nokia Lumia 940 phone has 11 hours of talk time before the battery runs out. If the Samsung Galaxy S7 Edge has triple the talk time of the Nokia Lumia, how many hours of talk time does it have?

116. **Mobile Screen Size** The iPhone SE has a 4-inch screen. If the Asus FonePad 8 has double the screen size of the iPhone, what is the size of its screen?

117. **Checkerboard** A checkerboard is a large square and contains 64 small squares of equal size, half black and half red. If one side of the checkerboard measures 8 small squares, how many squares does the other side measure? How many black squares are there on the board?

118. **Tic-Tac-Toe Board** A tic-tac-toe board is a large square that contains 9 small squares of equal size. If one side of the tic-tac-toe board measures 3 small squares, how many squares does the other side measure? There are 2 players in the game and each fills a small square on their turn, until there is a winner or until all the squares are filled. If the game ends with all squares filled, how many turns does each player get?

119. **Wind Power** Twenty-five E-126 wind turbines can power 125,000 European homes. Find the number of European homes that can be powered by a single E-126 wind turbine. (*Source: WindBlatt*, Enercon Magazine for Wind Energy, 2007.)

120. **National Forests** The number of national forests in Montana is double the number in Utah. If there are 10 national forests in Montana, how many are in Utah? (*Source:* U.S. Forest Service.)

121. **Counting Calories** One bottle of drinking water has 0 calories. How many calories are in 73 bottles of drinking water?

122. Counting Calories One can of grape soda has 100 calories. How many calories are in 3 cans of grape soda?

123. Digital Photos The following digital image was created using a rectangular pattern of small image points called *pixels*. Find the total number of pixels in a photo with a length of 600 pixels and a width of 400 pixels.

124. Digital Photos (Refer to the previous exercise.) Find the total number of pixels in a photo with a length of 400 pixels and a width of 300 pixels.

125. iPod Music A 64-gigabyte iPod holds 16,000 songs. How many songs per gigabyte is this?

126. iPod Photos A 16-gigabyte iPod Touch holds 20,000 iPod-viewable photos. How many photos per gigabyte is this?

127. Fencing Property A homeowner wants to put a fence around a rectangular plot with an area of 150 square feet. If one side of the plot is 10 feet long, find each of the following.

(a) The length of the other side
(b) The total amount of fencing needed

128. Area of a Rectangle A rectangle with an area of 48 square inches has a length of 8 inches. What is the width of the rectangle?

129. A Number Puzzle When a particular number between 2 and 20 is divided by 2, 3, or 4, the remainder is always 1. Find the number.

130. Numbers The product of two numbers is 30. If one of the numbers is 6, find the other number.

131. Flash Drives What is the maximum number of flash drives costing $6 each that a person can buy with $20? How much change will the person receive?

132. DVDs What is the maximum number of DVDs costing $16 each that a person can buy with $80? How much change will the person receive?

WRITING ABOUT MATHEMATICS

133. Explain what a *remainder* is. Give an example.

134. Noting that division can be checked by multiplying

$$\text{quotient} \times \text{divisor} = \text{dividend},$$

explain why dividing by zero is undefined.

135. Give examples of words associated with multiplication.

136. Give examples of words associated with division.

137. Explain how repeated subtraction can be used to show that $79 \div 11 = 7 \, r \, 2$.

138. Explain how repeated addition can be used to show that $15 \cdot 7 = 105$.

Group Activity

Winning the Lottery In the multistate lottery game Powerball, there are 175,223,510 possible number combinations, only one of which is the grand prize winner. The cost of a single ticket (one number combination) is $2. (*Source:* Powerball.com.)

Suppose that a wealthy person decides to buy tickets for every possible number combination, to be assured of winning a $400 million grand prize.

(a) If this strategy is carried out, what would be the total profit?
(b) Suppose this person could purchase one ticket every second. Divide 175,223,510 by 60 to find the number of minutes required to purchase all of the tickets.
(c) Ignoring the remainder in part (b), divide your result by 60 to find the number of hours needed to purchase all of the tickets.
(d) Using the quotient in part (c) without the remainder, divide your result by 24 to find the number of days needed to purchase all of the tickets.
(e) Leaving off the remainder in part (d), divide your result by 365 to find the number of years required to purchase all of the tickets.
(f) If there were a way for this person to buy all possible number combinations quickly, discuss reasons why this strategy would probably lose money.

1.4 Exponents, Variables, and Algebraic Expressions

Objectives

1. Understanding Exponential Notation
   - Squaring and Cubing
   - Powers of 10
2. Using Variables
3. Recognizing Algebraic Expressions
4. Evaluating Formulas
5. Translating Words to Expressions and Formulas
6. Solving Equations

1 Understanding Exponential Notation

Before we begin to study algebra, let's consider one idea from arithmetic. Just as multiplication is a fast way to perform repeated addition and division is a fast way to perform repeated subtraction, there is a mathematical concept called **exponential notation** that is used to perform repeated multiplication. The following equation shows how repeated multiplication can be written in exponential notation.

$$\underbrace{2 \cdot 2 \cdot 2 \cdot 2 \cdot 2}_{\text{5 factors}} = 2^5$$

Repeated multiplication — Exponent — Base

The expression on the right side of the equation is an *exponential expression* with **base 2** and **exponent 5**. It is read as "two to the fifth power." For an exponential expression with a natural number exponent, the base is used as a repeated factor, and the exponent indicates *how many times* the base should be multiplied by itself.

NOTE: An exponent of 1 is usually not written. For example, $9^1 = 9$ and $7^1 = 7$. ∎

STUDY TIP

Do you know your instructor's name and email address? Do you know the location of his or her office and the hours when he or she is available for help? Make sure that you have the answers to these important questions so that you can get help when needed.

EXAMPLE 1   Writing repeated multiplication in exponential notation

NEW VOCABULARY

☐ Exponential notation
☐ Base
☐ Exponent
☐ Base-10 number system
☐ Variable
☐ Algebraic expression
☐ Evaluate
☐ Formula

Write each of the following in exponential notation.
(a) $4 \cdot 4 \cdot 4 \cdot 4 \cdot 4 \cdot 4 \cdot 4$ (b) $2 \cdot 2 \cdot 2 \cdot 5 \cdot 5 \cdot 5 \cdot 5$

Solution
(a) The factor 4 is repeated 7 times. The exponential notation is 4^7.
(b) The factor 2 is repeated 3 times and the factor 5 is repeated 4 times. This product has two bases, each with its own exponent. The exponential notation is $2^3 \cdot 5^4$.

Now Try Exercises 17, 21

The value of an exponential expression with a whole number base and a natural number exponent can be found by first writing the expression as repeated multiplication and then finding the product.

READING CHECK 1

- What is exponential notation used to represent?

EXAMPLE 2   Finding the value of an exponential expression

CALCULATOR HELP

To evaluate exponents with a calculator, see Appendix E (page AP-29).

Find the value of each exponential expression.
(a) 8^2 (b) 2^4 (c) 4^5

Solution
(a) $8^2 = 8 \cdot 8 = 64$
(b) $2^4 = 2 \cdot 2 \cdot 2 \cdot 2 = 16$
(c) $4^5 = 4 \cdot 4 \cdot 4 \cdot 4 \cdot 4 = 1024$

Now Try Exercises 33, 35, 39

SQUARING AND CUBING The word *squared* is commonly used when reading an exponential expression with exponent 2, and the word *cubed* is used when the exponent is 3.

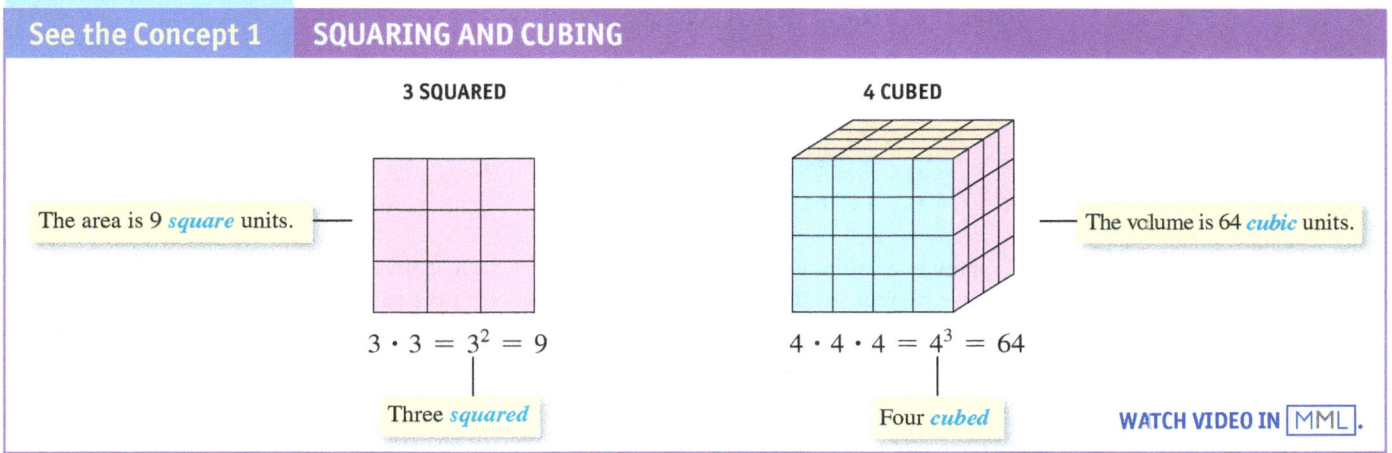

See the Concept 1   SQUARING AND CUBING

3 SQUARED — The area is 9 *square* units. — $3 \cdot 3 = 3^2 = 9$ — Three *squared*

4 CUBED — The volume is 64 *cubic* units. — $4 \cdot 4 \cdot 4 = 4^3 = 64$ — Four *cubed*

WATCH VIDEO IN MML.

POWERS OF 10 A pattern emerges when we look at base 10 raised to natural number powers. **TABLE 1.9** shows that for each power of 10, the **exponent** equals the number of **0**s in the standard form for that power of 10. The table also displays the period name that corresponds to each power of 10. Is it any wonder that our number system is called the **base-10 number system**?

Powers of 10

Power of 10	Repeated Multiplication	Standard Form	Period Name
10^1	10	10	tens
10^2	$10 \cdot 10$	100	hundreds
10^3	$10 \cdot 10 \cdot 10$	1000	thousands
10^4	$10 \cdot 10 \cdot 10 \cdot 10$	10,000	ten-thousands
10^5	$10 \cdot 10 \cdot 10 \cdot 10 \cdot 10$	100,000	hundred-thousands
10^6	$10 \cdot 10 \cdot 10 \cdot 10 \cdot 10 \cdot 10$	1,000,000	millions

TABLE 1.9

READING CHECK 2

- How can you tell how many zeros follow a 1 when finding a power of 10?

NOTE: From **TABLE 1.9**, it may seem reasonable that the standard form of 10^0 should be a 1 followed by **zero** 0s, or simply 1. We will see later that $10^0 = 1$. ■

EXAMPLE 3   Finding the value of a power of 10

Find the value of the expression $4 \cdot 10^5$.

Solution

$4 \cdot 10^5 = 4 \cdot 100,000 = 400,000$

Now Try Exercises 43

2 Using Variables

 Connecting Concepts with Your Life Because there are 3 feet in 1 yard, 4 yards are equal to 4 · 3 = 12 feet and 7 yards are equal to 7 · 3 = 21 feet. **TABLE 1.10** might be useful when converting from yards to feet frequently.

Converting Yards to Feet

Yards	1	2	3	4	5	6	7
Feet	3	6	9	12	15	18	21

Times 3

TABLE 1.10

However, this table is not helpful if we need to convert more than 7 yards. To find the number of feet in 10 yards, for example, we would need to expand **TABLE 1.10** to include 10 · 3 = 30 feet. However, expanding the table to include every possible value for yards would not be possible.

One of the most important ideas in mathematics is the notion of a *variable*. A **variable** is a symbol, typically an italic letter such as x, y, or z, used to represent an unknown quantity. (Uppercase letters such as F, P, and Y can also be variables.) Variables can be used when tables of numbers are inadequate. In the previous example, the number of feet is found by multiplying the corresponding number of **y**ards by 3. This is represented by the product

$$(\text{Yards}) \cdot 3.$$ Changing yards to feet

Any letter can be used as a variable. However, choosing the first letter of the quantity being represented may help give meaning to the variable. For example, if the variable Y is used to represent the number of **y**ards, then the product becomes

$$Y \cdot 3.$$

EXAMPLE 4 Using a variable to express an unknown quantity

Rewrite the given expression using an appropriate variable.
(a) $(\text{Inches}) \div 12$ **(b)** $17 - (\text{Weeks})$

Solution
(a) Using the variable I to represent **i**nches, we can write the expression as $I \div 12$.
(b) Using the variable W to represent **w**eeks, we can write the expression as $17 - W$.

Now Try Exercises 47, 53

3 Recognizing Algebraic Expressions

The product $Y \cdot 3$ is used to find the number of feet in Y yards. This product, $Y \cdot 3$, is an example of a algebraic *expression*. An **algebraic expression** may contain numbers; variables; exponents; operation symbols such as $+$, $-$, $\times$, and $\div$; and grouping symbols, such as parentheses.

If we replace the variable Y in the algebraic expression $Y \cdot 3$ with the number 12, then we can find the number of feet in 12 yards. That is, we **evaluate** the expression for a given value of the variable.

READING CHECK 3

- How is the expression $I \div 12$ evaluated for the value $I = 48$?

EXAMPLE 5 Evaluating algebraic expressions with one variable

Evaluate each algebraic expression for $x = 3$.

(a) $5 + x$ (b) $\dfrac{18}{x}$ (c) $10x$

Solution
(a) Replace x with 3 in the expression $5 + x$ to get $5 + 3 = 8$.
(b) Replace x with 3 in the expression $\dfrac{18}{x}$ to get $\dfrac{18}{3} = 6$.
(c) The expression $10x$ indicates multiplication of 10 and x. So, $10x = 10(3) = 30$.

Now Try Exercises 57, 59, 61

 Connecting Concepts with Your Life Some algebraic expressions contain more than one variable. For example, if a car travels 160 miles on 5 gallons of gasoline, then the car's *mileage* is $\dfrac{160}{5} = 32$ miles per gallon. Generally, if a car travels M miles on G gallons of gasoline, then its mileage is given by the expression $\dfrac{M}{G}$. Note that the expression $\dfrac{M}{G}$ contains two variables, M and G.

EXAMPLE 6 Evaluating algebraic expressions with two variables

Evaluate each algebraic expression for $y = 3$ and $z = 12$.
(a) $5yz$ (b) $z - y$ (c) $z \div y$

Solution
(a) Replace y with 3 and z with 12 to get $5yz = 5 \cdot 3 \cdot 12 = 180$.
(b) $z - y = 12 - 3 = 9$
(c) $z \div y = 12 \div 3 = 4$

Now Try Exercises 65, 67, 69

4 Evaluating Formulas

Recall that an equation is a mathematical statement that two algebraic expressions are equal. Equations *always* contain an equals sign. A **formula** is a special type of equation that expresses a relationship between two or more quantities. The formula $F = Y \cdot 3$ means that to find the number of feet F in Y yards, we multiply Y by 3. A formula usually has a single variable on one side of the equals sign.

The dot ($\cdot$) is often used to indicate multiplication because the symbol ($\times$) can be confused with the variable x. Sometimes the multiplication sign is omitted altogether. The four formulas

$$F = 3 \times Y, \quad F = 3 \cdot Y, \quad F = 3Y, \quad \text{and} \quad F = 3(Y)$$

represent the same relationship between yards and feet. For example, to find the number of feet in 12 yards, we can replace Y in any of these formulas with the number 12. If we let $Y = 12$ in the second formula, we have

$$F = 3 \cdot 12 = 36.$$

This means that there are 36 feet in 12 yards.

Earlier in this chapter, we used the formula

$$\text{Area} = \text{length} \times \text{width}$$

to find the area of a rectangle. By using the variables A, l, and w to represent area, length, and width, respectively, this formula can be written in the form

$$A = lw.$$

FIGURE 1.19 shows three common figures from geometry and their associated formulas.

Formulas Associated with Three Common Geometric Figures

Rectangle

Area: $A = lw$
Perimeter: $P = 2l + 2w$

(a)

Square

Area: $A = s^2$
Perimeter: $P = 4s$

(b)

Triangle

Perimeter: $P = a + b + c$

(c)

FIGURE 1.19

In **FIGURE 1.19**, the variable A represents area, and P represents perimeter. For a rectangle, l and w represent length and width, respectively. The variable s represents the measure of one side of a square, and the variables a, b, and c represent the measures of the three sides of a triangle. (The *area* of a triangle will be discussed later in this text.)

EXAMPLE 7 **Working with geometric formulas**

Use the geometric formulas in **FIGURE 1.19** to complete each of the following.
(a) Find the area of a square with a side length of 6 inches.
(b) Find the perimeter of a triangle with side lengths of 47, 32, and 55 feet.
(c) Find the perimeter of a rectangle with length 7 miles and width 4 miles.

Solution
(a) Substituting 6 for s in the formula $A = s^2$ gives $A = 6^2 = 6 \cdot 6 = 36$ square inches.
(b) Substitute 47, 32, and 55 for a, b, and c in the formula $P = a + b + c$ to find the perimeter of the triangle. $P = 47 + 32 + 55 = 134$ feet
(c) $P = 2l + 2w = 2(7) + 2(4) = 14 + 8 = 22$ miles

Now Try Exercises 71, 73, 75

5 Translating Words to Expressions and Formulas

Many times in mathematics we need to write our own algebraic expressions. To do this, we translate words to symbols. Recall that the symbols $+$, $-$, $\times$, and $\div$ have special mathematical words associated with them. **TABLE 1.11** summarizes many of the words commonly associated with these operations.

READING CHECK 4

- Which arithmetic symbol is associated with the word *double*?

Words Associated with Arithmetic Symbols

Symbol	Associated Words
$+$	add, plus, more than, sum, total, increase by
$-$	subtract, minus, less than, difference, fewer, decrease by
$\times$	multiply, times, twice, double, triple, product
$\div$	divide, divided by, quotient, per

TABLE 1.11

EXAMPLE 8 **Translating words into expressions**

Translate each phrase into a algebraic expression. Explain what each variable represents.
(a) Double the cost of an MP3 player
(b) Eight less than a friend's age
(c) The number of students divided by the number of books
(d) The sum of two different numbers

Solution

(a) If the cost is $79, then twice this cost would be $2 \cdot 79 = \$158$. Generally, if we let C be the cost of an MP3 player, then twice the cost would be $2 \cdot C$, or $2C$.

(b) If a friend's age is 17, then eight less than the age would be $17 - 8 = 9$. If we let A represent the friend's age, then eight less than the age would be $A - 8$.

(c) Let S represent the number of students, and let B represent the number of books. The number of students divided by the number of books is $S \div B$ or $\frac{S}{B}$.

(d) Let x be one of the numbers, and let y be the other number. The sum is $x + y$.

Now Try Exercises 77, 79, 81

🌐 **Math in Fitness Context** When we perform physical activity, the energy used is measured in *calories*. A 175-pound person burns about 8 calories per minute while stacking firewood and about 7 calories per minute while whitewater rafting. We can write a formula that expresses the relationship between the number of minutes spent performing an activity and the number of calories burned. For example, whitewater rafting for 1 minute burns $7 \cdot 1 = 7$ calories, and whitewater rafting for 5 minutes burns $7 \cdot 5 = 35$ calories. If we let C represent the number of calories burned during w minutes of whitewater rafting, then the formula $C = 7w$ expresses the relationship between the number of minutes spent whitewater rafting and the number of calories burned. (*Source:* The Calorie Control Council.)

EXAMPLE 9 **Translating words to a formula**

A 250-pound person burns 2 calories per minute while watching television and 9 calories per minute while gardening.

(a) Write a formula that gives the calories C burned while watching television for t minutes.
(b) Write a formula that gives the calories C burned while gardening for g minutes.
(c) Use your formulas to find the number of calories burned while performing each of these activities for 3 hours.

Solution

(a) Watching television for 1 minute burns $2 \cdot 1 = 2$ calories, and watching television for 10 minutes burns $2 \cdot 10 = 20$ calories. The formula $C = 2t$ gives the number of calories burned while watching television for t minutes.
(b) The formula $C = 9g$ gives the number of calories burned while gardening for g minutes.
(c) Because the formulas require that time is given in minutes, we must convert 3 hours to $3 \cdot 60 = 180$ minutes. For a 250-pound person, watching television for 180 minutes burns $C = 2(180) = 360$ calories, and gardening for that same amount of time burns $C = 9(180) = 1620$ calories.

Now Try Exercise 101

6 Solving Equations

Because a variable can be used to represent an unknown quantity, we can use a variable rather than an empty box when solving equations that contain unknown values. For example, solving an equation of the form

$$\square + 7 = 13 \quad \text{What number plus 7 equals 13?}$$

means that we look for the number that can be written in the box to make the equation true. Similarly, solving an equation of the form

$$n + 7 = 13 \quad \text{What number plus 7 equals 13?}$$

means that we look for the value of the variable n that makes the equation true. For either equation, the solution is 6 because writing 6 in the box or replacing the variable n with 6 results in $6 + 7 = 13$, which is a true equation.

Recall that a number is a *solution* to an equation if it makes the equation true. In the next example, we check a given number to determine if it is a solution to a given equation.

EXAMPLE 10 Checking solutions

Determine if the given number is a solution to the given equation.
(a) Is 3 a solution to $5x = 15$?
(b) Is 11 a solution to $23 - m = 9$?
(c) Is 6 a solution to $y^2 = 12$?

Solution
(a) Replace the variable x in the equation with 3. Because $5(3) = 15$ is a **true** equation, 3 is a solution.
(b) Replacing m in the equation with 11 results in $23 - 11 \stackrel{?}{=} 9$, which is a **false** equation. Subtracting 11 from 23 results in $23 - 11 = 12$, so 11 is not a solution.
(c) Because $6^2 = 36$, replacing the variable y with 6 results in a **false** equation. So, 6 is not a solution.

Now Try Exercises 83, 85, 87

READING CHECK 5

- What math symbol means "not equal to"?

NOTE: In Example 10(b), we could have written $23 - 11 \neq 9$ instead of saying that $23 - 11 \stackrel{?}{=} 9$ is **false**. The symbol $\neq$ means **not equal to**. This notation will be used throughout the remainder of this text. ∎

EXAMPLE 11 Solving equations

Solve each equation.
(a) $114 + y = 125$ (b) $48 \div m = 4$

Solution
(a) Because $114 + 11 = 125$, the solution is 11.
(b) The solution is 12 because $48 \div 12 = 4$ is true.

Now Try Exercises 89, 93

EXAMPLE 12 Reselling Popular Phones

The resale price of a Samsung flagship phone traded in 6 months after its launch is $35 more than the resale price of a similar model Motorola phone traded in 6 months after its launch. If the resale price of a Samsung phone is $225, what is the resale price of a Motorola phone? *(Source: Gazelle.)*

Solution
The question "What is the resale price of a Motorola phone?" indicates that the unknown value is the resale price of a Motorola phone. We will let the variable p represent the resale price of a Motorola phone. Since the resale price of a Samsung phone is **$225**, the phrase

Samsung resale price *is* $35 *more than* the Motorola resale price translates to the equation

$$225 = p + 35.$$

The solution to this equation is 190 because $225 = 190 + 35$ is true. The resale price of a Motorola phone is $190.

Now Try Exercise 103

1.4 EXPONENTS, VARIABLES, AND ALGEBRAIC EXPRESSIONS   53

MAKING CONNECTIONS 1

Equations and Expressions

The words "equation" and "expression" are *not* interchangeable.

Equation	Expression
• Example: $4 + x = 10$	• Example: $4 + x$
• Translates to an English *sentence*	• Translates to an English *phrase*
• Always has an equals sign	• Never has an equals sign
• Can often be *solved*	• Can often be *evaluated*

READING CHECK 6

- Which contains an equals sign, an equation or an expression?

1.4 Putting It All Together

CONCEPT	COMMENTS	EXAMPLES
Exponential Notation	A natural number exponent indicates the number of times that the base is used as a repeated factor.	Exponent ↘ $3^4 = 3 \cdot 3 \cdot 3 \cdot 3$ ↗ Base
Powers of 10	A natural number exponent on 10 equals the number of 0s that follow a 1 in the standard form of the power of 10.	$10^3 = 1000$ $10^5 = 100,000$ $10^{12} = 1,000,000,000,000$
Variable	Used to represent an unknown quantity	P represents the number of pets. n represents an unknown number.
Algebraic Expression	May contain numbers, variables, exponents, operation symbols, and grouping symbols	$3x + 7$, $2(4 - x) + 9$, $2y^2 - 5y + 1$, $2l + 2w$
Formula	A special type of equation that expresses a relationship between two or more quantities	The formula $C = 10D$ gives the number of cents C in D dimes.
Translating Words to Expressions and Formulas	We can translate words into expressions and formulas using math symbols commonly associated with the words.	"More than" means add, while "double" means multiply.
Solving Equations	We can use variables to represent unknown values when solving an equation.	$\square + 5 = 16$ can be written as $x + 5 = 16$. The solution is 11 because $11 + 5 = 16$.

1.4 Exercises

MyMathLab®

CONCEPTS AND VOCABULARY

1. A mathematical concept called _____ can be used to denote repeated multiplication.

2. In the exponential expression 4^7, the base is _____ and the exponent is _____.

3. The word "squared" means that the exponent is _____.

4. The word "cubed" means that the exponent is _____.

5. When writing 10^9 in standard form, how many zeros should be written after a 1?

6. Write 10,000,000 as an exponential expression.

7. A(n) _____ is a symbol or italic letter such as x, y, or z, used to represent an unknown quantity.

8. A(n) _____ may contain numbers; variables; exponents; operation symbols, such as $+$, $-$, $\times$, and $\div$; and grouping symbols, such as parentheses.

9. A(n) _____ is a mathematical statement that two algebraic expressions are equal.

10. A(n) _____ is a special type of equation that expresses a relationship between two or more quantities.

11. Give the formula that is used to find the perimeter of a rectangle.

12. Give the formula that is used to find the area of a square.

13. To _____ an expression, replace the variable with a given value and find the result.

14. To represent an unknown quantity in algebra, we use a _____ rather than an empty box.

15. Is $7x + 21$ an equation or an expression?

16. Is $7x = 21$ an equation or an expression?

USING EXPONENTIAL NOTATION

Exercises 17–24: Use exponential notation to write each repeated multiplication.

17. $8 \cdot 8 \cdot 8$
18. $4 \cdot 4 \cdot 4 \cdot 4 \cdot 4 \cdot 4$
19. $2 \cdot 2 \cdot 2 \cdot 2 \cdot 2$
20. $9 \cdot 9$
21. $2 \cdot 2 \cdot 2 \cdot 5 \cdot 5$
22. $4 \cdot 4 \cdot 6 \cdot 6 \cdot 6 \cdot 6$
23. $5 \cdot 5 \cdot 5 \cdot 7 \cdot 7 \cdot 7$
24. $3 \cdot 9 \cdot 9 \cdot 9$

Exercises 25–32: Write the phrase in exponential notation.

25. Seven squared
26. Five cubed
27. Four to the ninth
28. One to the third
29. Two cubed
30. Ten to the sixth
31. Three to the fifth
32. Eight squared

Exercises 33–46: Evaluate the exponential expression.

33. 9^2
34. 2^3
35. 2^5
36. 3^4
37. 4^4
38. 7^3
39. 6^3
40. 5^3
41. 10^3
42. 10^7
43. $8 \cdot 10^6$
44. $3 \cdot 10^4$
45. $10^2 \cdot 30$
46. $10^4 \cdot 2000$

WORKING WITH VARIABLES AND EXPRESSIONS

Exercises 47–54: Rewrite the given expression using an appropriate variable.

47. $(\text{Age}) - 5$
48. $60 \div (\text{Rate})$
49. $6 \cdot (\text{Goals})$
50. $(\text{Time}) + 6$
51. $(\text{Score}) + 10$
52. $(\text{Laps}) \cdot 4$
53. $(\text{Pieces}) \div 2$
54. $14 - (\text{Days})$

Exercises 55–62: Evaluate the expression for $x = 5$.

55. $25 - x$
56. $x + 13$
57. $\dfrac{0}{x}$
58. $7x$
59. $13x$
60. $x \div 5$
61. $41 + x$
62. $x - 1$

Exercises 63–70: Evaluate the algebraic expression for $x = 8$ and $y = 2$.

63. $y + x$
64. $3xy$
65. $x \div y$
66. $x - 2y$
67. $x - y$
68. $x \div 2y$
69. xy
70. $2x + y$

EVALUATING FORMULAS

Exercises 71–76: Use the appropriate geometric formula from the following list to find the requested measure.

Rectangle: $A = lw$, $P = 2l + 2w$
Square: $A = s^2$, $P = 4s$
Triangle: $P = a + b + c$

71. The perimeter of a square with a 4-inch side

72. The area of a square with a side of length 6 feet

73. The area of a rectangle with a 5-inch length and a 3-inch width

74. The perimeter of a rectangle with width 3 miles and length 8 miles

75. The perimeter of a triangle with sides of length 14, 32, and 23 yards

76. The area of two squares, each with a 5-foot side

TRANSLATING WORDS INTO MATH

Exercises 77–82: Translate the word phrase into an algebraic expression. Explain what each variable represents.

77. Six times an individual's monthly income

78. Seven fewer than a person's age

79. The quotient of the number of pizza slices and 3

80. A child's weight increased by 9

81. The sum of a person's age and heart rate

82. Triple the cost

SOLVING EQUATIONS

Exercises 83–88: Determine if the given number is a solution to the given equation.

83. Is 7 a solution to $21 \div m = 3$?

84. Is 3 a solution to $x^3 = 9$

85. Is 12 a solution to $2y = 6$?

86. Is 15 a solution to $30 - n = 15$?

87. Is 9 a solution to $a^2 = 18$?

88. Is 56 a solution to $x + 8 = 64$?

Exercises 89–96: Solve the equation.

89. $b + 5 = 15$ 90. $3x = 12$

91. $17 - z = 3$ 92. $x + 5 = 27$

93. $72 \div d = 6$ 94. $n \div 9 = 5$

95. $56 = 7x$ 96. $w - 8 = 1$

APPLICATIONS INVOLVING ALGEBRAIC EXPRESSIONS AND EQUATIONS

Exercises 97–100: Use the given information to find the unknown length represented by the variable x.

97. Square with area 100 square feet

98. Rectangle with area 48 square yards

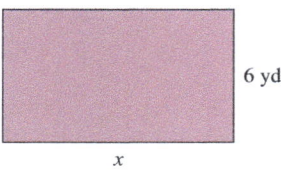

99. Triangle with perimeter 97 inches

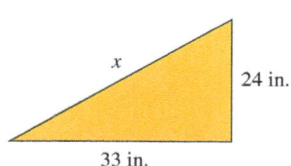

100. Square with perimeter 68 miles

101. **Burning Calories** A 160-pound person burns 6 calories per minute while painting over graffiti and 8 calories per minute while clearing out an illegal dump site. (*Source:* The Calorie Control Council.)
    (a) Write a formula that gives the number of calories C burned while painting over graffiti for p minutes.
    (b) Write a formula that gives the number of calories C burned while clearing out an illegal dump site for d minutes.
    (c) Use your formulas to find the number of calories burned while performing each of these activities for 4 hours.

102. **Counting Calories** Each ounce of a sweetened soft drink contains 15 calories, while each ounce of a "light" version of the same drink contains 2 calories.
    (a) Write a formula that gives the number of calories C in x ounces of sweetened drink.
    (b) Write a formula that gives the number of calories C in y ounces of "light" drink.
    (c) If a person drinks a 12-ounce can of "light" drink rather than a 12-ounce can of sweetened drink, what is the calorie difference?

103. **Starbucks Stores** From 2013 to 2014, the number of Starbucks stores increased by 3538, bringing the worldwide total to 23,305. How many Starbucks stores were there in 2013? (*Source:* Starbucks.)

104. **Large Mammals** The average birth weight of a blue whale is about 30 times the average birth weight of an African elephant. If the average newborn African elephant weighs 232 pounds, find the weight of the average newborn blue whale.

105. **Movie Downloads** A person has $28 in her online movie account. Find the maximum number of movie downloads costing $9 each that this person can purchase. How much is left in the account?

106. **iPhone Downloads** A person has $32 in his online iTunes account. Find the maximum number of iPhone apps costing $3 each that this person can purchase. How much is left in the account?

107. **Flea Market** A person bought 7 items at a flea market and received $6 in change. If each item was priced at $2, how much did the person give the vendor when paying?

108. **Lending Money** A student gave the same amount of money to each of 4 friends and had $9 left over. If the student originally had $49, how much was given to each friend?

109. **Large Cities** Of the 25 largest U.S. cities, 4 are located in California. If California has twice as many large cities as Tennessee, how many large cities does Tennessee have?

110. **Canadian Lakes** The province of Saskatchewan in Canada has 10 times as many lakes as Minnesota. Minnesota is commonly known as "the land of 10,000 lakes." How many lakes are located in Saskatchewan?

111. **Numbers** The quotient of two numbers is 5. If the dividend is 60, what is the divisor?

112. **Numbers** The difference of two numbers is 21. If the subtrahend is 9, what is the minuend?

WRITING ABOUT MATHEMATICS

113. Explain the meaning of a *variable*. Give an example.

114. Explain the difference between an expression and an equation. Give an example of each.

115. Give an example of a *formula*. State what each variable in the formula represents.

116. Explain how to write the standard form of 10 raised to a natural number power.

SECTIONS 1.3 and 1.4 — Checking Basic Concepts

1. Multiply.
   (a) 13×22
   (b) $0 \cdot 11$
   (c) $1(207)$
   (d) $317 \cdot 204$

2. Divide, when possible.
   (a) $35 \div 1$
   (b) $1125 \div 8$
   (c) $\dfrac{34}{0}$
   (d) $\dfrac{12{,}312}{72}$

3. Solve each equation.
   (a) $3 \cdot \square = 36$
   (b) $\square \div 9 = 8$

4. Multiply 4000×30 mentally.

5. Use exponential notation to write each repeated multiplication.
   (a) $7 \cdot 7 \cdot 7 \cdot 7$
   (b) $2 \cdot 2 \cdot 8 \cdot 8$

6. Evaluate each exponential expression.
   (a) $5 \cdot 2^3$
   (b) $3 \cdot 10^5$

7. Evaluate the expression for $x = 2$ and $y = 3$.
   (a) $5y - x$
   (b) $5xy$

8. Is 9 a solution to $17 - n = 8$?

9. Solve each equation.
   (a) $48 \div d = 6$
   (b) $19 - z = 7$

10. Find the perimeter of a rectangle with width 8 inches and length 14 inches.

11. **Numbers** The quotient of two numbers is 7. If the dividend is 49, what is the divisor?

12. **Area** A rectangle is 40 feet long and 20 feet wide. Find the area of the rectangle.

1.5 Rounding and Estimating; Square Roots

Objectives

1. Rounding Whole Numbers
2. Estimating and Approximating
3. Solving Problems Using Estimation
4. Estimating Graphically
5. Finding Square Roots

NEW VOCABULARY

☐ Round
☐ Estimation
☐ Approximation
☐ Perfect square
☐ Square root
☐ Radical
☐ Radicand

READING CHECK 1

- Is a number that lies exactly halfway between two tick marks on a number line rounded up or down?

1 Rounding Whole Numbers

Connecting Concepts with Your Life We **round** a whole number when we approximate it to a given level of accuracy. For example, a person who buys a new car for $28,912 could say that the purchase price was about $29,000 or that the price was about $28,900. The number 29,000 is the result of rounding 28,912 to the nearest *thousand*, and the number 28,900 is the result of rounding 28,912 to the nearest *hundred*. Both results represent an approximation of the price of the car. The difference between these numbers is simply a matter of accuracy.

Number lines can be helpful when rounding whole numbers. To round the number 278 to the nearest **hundred**, plot the number 278 on a number line with tick marks at every **100**, as shown in **FIGURE 1.20**. Because 278 is **closer to 300** than to 200, we round **up to 300**.

Rounding 278 to the Nearest Hundred Results in 300

FIGURE 1.20

Similarly, when rounding 3198 to the nearest thousand, we round **down to 3000** because 3198 is **closer to 3000** than to 4000 on the number line. See **FIGURE 1.21**.

Rounding 3198 to the Nearest Thousand Results in 3000

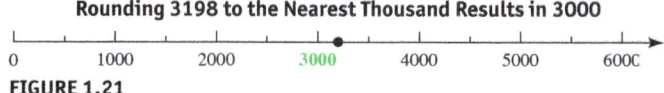

FIGURE 1.21

NOTE: A number that is exactly halfway between two tick marks on the number line will always be rounded up. For example, rounding 55 to the nearest ten results in 60. ∎

Although number lines are convenient for visualizing the rounding process, it is not necessary to draw a number line every time we wish to round a whole number. The following procedure can be used to round whole numbers to a given place value.

> **ROUNDING WHOLE NUMBERS**
>
> **STEP 1:** Identify the first digit to the *right* of the given place value.
>
> **STEP 2:** If this digit is:
>   (a) less than 5, do not change the digit in the given place value.
>   (b) 5 or more, add 1 to the digit in the given place value.
>
> **STEP 3:** Replace each digit to the right of the given place value with 0.

NOTE: Regrouping may sometimes be necessary when rounding whole numbers. For example, when rounding 197 to the nearest tens place, the result is 200 because the 9 rounds up to 10 and ten 10s are regrouped to the hundreds place. ∎

EXAMPLE 1 **Rounding whole numbers**

Round 35,147,289 to the given place value.
(a) ten-thousands (b) millions (c) hundreds

Solution
(a) **STEP 1:** The ten-thousands digit is 4. The first digit to the right of 4 is 7.

$$35,147,289$$

STEP 2: Because 7 is greater than 5, we add 1 to the 4 in the ten-thousands place.

$$35,157,289$$

STEP 3: Replace each digit to the right of the ten-thousands place with 0.

$$35,150,000$$

Rounding 35,147,289 to the ten-thousands place results in 35,150,000.

(b) The millions digit is 5, and the first digit to the right of the millions place is 1 (Step 1). Because 1 is less than 5, we do nothing to the millions place (Step 2), and we replace each digit to the right of the millions place with 0 (Step 3). The result of rounding 35,147,289 to the millions place is 35,000,000.

(c) The first digit to the right of the hundreds place is 8, which is greater than 5. We add 1 to the hundreds place and replace each digit to the right of the hundreds place with 0. The result is 35,147,300.

Now Try Exercises 17, 19, 25

EXAMPLE 2 **Rounding whole numbers**

Round each whole number to the given place value.
(a) 10,517, thousands
(b) 89, hundreds

Solution
(a) The thousands digit is 0. The first digit to the right of the 0 is 5. We round *up* to 11,000.
(b) There is no digit in the hundreds place, which means there are 0 hundreds. The number 89 can be thought of as 089. The first digit to the right of the hundreds place is 8, so we round *up* to 100.

Now Try Exercises 21, 23

2 Estimating and Approximating

 Connecting Concepts with Your Life At a local thrift store, a student buys a pair of jeans for $8, a leather jacket for $29, a pair of dress shoes for $11, and a lava lamp for $9. At checkout, the clerk claims that the total bill is $112. Do you agree with this number? Even without calculating the *exact* total, it should be clear that the clerk made a mistake. Some people can "sense" when a computation is incorrect without performing complicated mental calculations. They *estimate* or *approximate*. Can you see that the total should be about $60?

An **estimation** is a rough calculation used to find a reasonably accurate answer. An estimated answer is usually not exactly accurate and is called an **approximation** of the actual answer. When two numbers are *nearly equal*, we use the **approximately equal symbol**, denoted ≈. For example, 4.01 ≈ 4. The next example demonstrates how rounding is used in the estimating process.

EXAMPLE 3   Estimating a sum and a difference

Round each number to the nearest thousand to estimate the sum or difference. Then give the actual sum or difference.
(a) $3084 + 10{,}987 + 6905$ (b) $13{,}893 - 4019$

Solution
(a) Round each addend to the nearest thousand as follows: **3084** rounds to **3000**, **10,987** rounds to **11,000**, and **6905** rounds to **7000**. Estimate by adding the rounded values, $3000 + 11{,}000 + 7000 = 21{,}000$. The actual value is 20,976.
(b) Round each number as follows: **13,893** rounds to **14,000** and **4019** rounds to **4000**. Estimate by finding the difference of the rounded values, $14{,}000 - 4{,}000 = 10{,}000$. The actual value is 9874.

Now Try Exercises 35, 37

🌐 **Math in Context** Real Estate   When estimating real-world data, a place value for rounding is rarely given. For example, a real estate developer may have an empty plot of land that measures 617 feet by 18,875 feet. If these numbers are rounded to the nearest hundred, we can estimate the area by finding the product $600 \cdot 18{,}900$. Although this product may be easier to find than $617 \cdot 18{,}875$, it is still difficult to do mentally. However, if we round each number to its *highest place value*, then the estimation can be done more easily. Round **617** to the nearest **hundred** to get **600** and **18,875** to the nearest **ten thousand** to get **20,000**. A reasonable estimate of the area is $600 \cdot 20{,}000 = 12{,}000{,}000$ square feet. The actual value is 11,645,875.

EXAMPLE 4   Estimating a product and a quotient

Round each number to its highest place value to estimate the product or quotient.
(a) $83 \cdot 47{,}978$ (b) $798 \div 38$

Solution
(a) Round **83** to the nearest **ten** to get **80** and **47,978** to the nearest **ten thousand** to get **50,000**. An estimate of the product is $80 \cdot 50{,}000 = 4{,}000{,}000$. The actual value is 3,982,174.
(b) Round 798 to **800** and 38 to **40**. An estimate of the quotient is $800 \div 40 = 20$. The actual value is 21.

Now Try Exercises 43, 45

3 Solving Problems Using Estimation
Many real-world problems do not require exact solutions. Sometimes a reasonable estimate will be sufficient. The next two examples illustrate how we can find reasonable estimates of solutions to application problems involving real data.

EXAMPLE 5   Estimating Amazon Acquisitions

TABLE 1.12 shows the price paid in millions of dollars by Amazon to acquire selected companies. Estimate the total price paid for these companies by rounding each price to the nearest hundred million. How does your estimate compare to the actual price?

Amazon Acquisitions

Company	Zappos	Lovefilm	Twitch	Quidsi	Audible
Price ($millions)	1200	312	970	545	300

TABLE 1.12 (*Source:* Business Insider.)

Solution

When rounded to the nearest hundreds million, the values become 1200, 300, 1000, 500 and 300. The sum of these numbers gives 1200 + 300 + 1000 + 500 + 300 = 3300, or $3300 million. This estimate is very close to the actual value of $3327 million.

Now Try Exercise 67

EXAMPLE 6 Estimating blog page hits

During the month of January (31 days), a blog page received an average of 772 daily hits. Estimate the total number of hits that the page received during January. Then, find the actual value.

Solution

Round **31** to the nearest **ten** to get **30** and round **772** to the nearest **hundred** to get **800**. The page received approximately **30** · **800** = 24,000 hits. The actual value is 23,932.

Now Try Exercise 59

4 Estimating Graphically

Math in Context (Digital Music) Newspapers and magazines often illustrate numerical information in the form of a graph. When reading these graphs, we may need to estimate values visually. For example, the graph shown in **FIGURE 1.22** displays the annual sales of digital music in billions of dollars during the early years of downloadable music, from 2000 to 2008. In the next example, we use **FIGURE 1.22** to estimate information regarding digital music sales.
(*Source:* Recording Industry of America.)

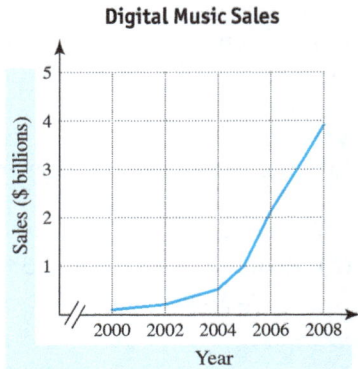

FIGURE 1.22

EXAMPLE 7 Using a graph to estimate digital music sales information

Use the graph in **FIGURE 1.22** to answer each question.
(a) Estimate digital music sales for 2007 to the nearest billion.
(b) Estimate the year when digital music sales hit $1 billion.

Solution
(a) Locate the year 2007 halfway between 2006 and 2008 at the bottom of the graph. Move **vertically** upward to the graphed line. From this position, move **horizontally** to the left to the sales values. See **FIGURE 1.23**. In 2007 digital music sales were about $3 billion.
(b) Find $1 billion among the sales values at the left edge of the graph. Move **horizontally** to the right to the graphed line. From this position, move **vertically** downward to the years. Note that 2005 is located halfway between 2004 and 2006. See **FIGURE 1.24**. Digital music sales reached $1 billion in about 2005.

STUDY TIP

Reading graphs is important in this text. If you have trouble reading a graph, ask your instructor for help.

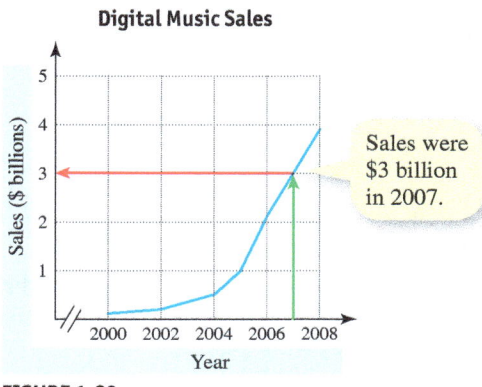

FIGURE 1.23

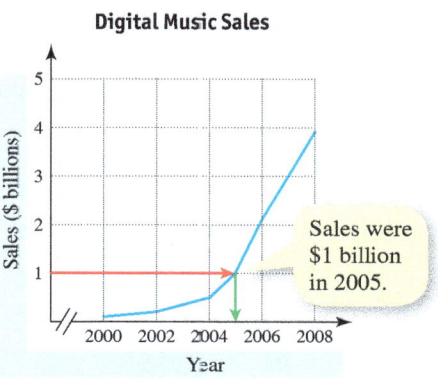

FIGURE 1.24

Now Try Exercises 69, 71

5 Finding Square Roots

In the previous section, we discussed the meaning of the word *squared*. A number is squared when its exponent is 2. For example, 5 squared is written as 5^2 and represents the product $5 \cdot 5 = 25$. Because 25 is the result of a whole number being squared, we say that 25 is a **perfect square**. Other perfect squares include 16, 81, and 289 because $4^2 = 16$, $9^2 = 81$, and $17^2 = 289$, respectively. Note that many numbers, such as 14 and 312, are *not* perfect squares because there is no whole number that can be squared to give 14 or 312. The perfect squares associated with the first 22 whole numbers are shown in **TABLE 1.13**.

READING CHECK 2

- How do you know if a whole number is a perfect square?

Perfect Squares Associated with the First 22 Whole Numbers

Whole Number	0	1	2	3	4	5	6	7	8	9	10
Perfect Square	0	1	4	9	16	25	36	49	64	81	100
Whole Number	11	12	13	14	15	16	17	18	19	20	21
Perfect Square	121	144	169	196	225	256	289	324	361	400	441

TABLE 1.13

In **TABLE 1.13**, each perfect square is the *square* of a whole number, and each whole number is a *square root* of a perfect square. For example, 7 is a square root of 49, and 13 is a square root of 169. A **square root** of a given whole number is a number whose square is the given whole number.

We can use mathematical notation to find or compute a square root. The square root of 121, for example, can be written as the *radical expression*

Radical sign $\sqrt{121}$. Radicand

The symbol $\sqrt{}$ is called the **radical sign** (or simply the **radical**), and the number under the radical (in this case, 121) is called the **radicand**. Because $11^2 = 121$, the square root of 121 is 11, and we write

$$\sqrt{121} = 11. \quad 11^2 = 121$$

EXAMPLE 8 **Computing square roots**

Compute each square root.
(a) $\sqrt{36}$ (b) $\sqrt{256}$ (c) $\sqrt{1156}$

Solution
(a) We look for a number whose square is 36. Because $6^2 = 36$, $\sqrt{36} = 6$.
(b) $\sqrt{256} = 16$ because $16^2 = 256$.
(c) To find a number whose square is 1156, first note that $30^2 = 900$ and $40^2 = 1600$. Because 1156 is between 900 and 1600, $\sqrt{1156}$ is between 30 and 40. By trial and error, $34^2 = 1156$, so $\sqrt{1156} = 34$.

Now Try Exercises 51, 53, 57

1.5 Putting It All Together

CONCEPT	COMMENTS	EXAMPLES
Rounding	Approximating a number to a given level of accuracy	To the nearest thousand, 52,789 rounds to 53,000.
Estimation	A rough calculation used to find a reasonably accurate answer	When rounded to the nearest ten, $52 + 78 + 13$ can be estimated by $50 + 80 + 10$.
Approximation	• The result when estimating or rounding • Not exactly accurate in most cases • Represented by the symbol $\approx$	To the nearest ten, $147 \approx 150$. $52 + 78 + 13 \approx 140$
Perfect Square	A perfect square results when a whole number is squared.	121 is a perfect square because $11^2 = 121$.
Square Root	• The square root of a whole number is a number whose square is that whole number. • The symbol $\sqrt{}$ is called the radical, and the number under it is called the radicand.	$\sqrt{169} = 13$ because $13^2 = 169$. $\sqrt{81} = 9$ because $9^2 = 81$.
Estimating Graphically	Values can be estimated from a graph.	In 2014 sales were about $25,000.

1.5 Exercises

CONCEPTS AND VOCABULARY

1. Approximating a whole number to a given level of accuracy is called _____.

2. When rounding a whole number, we first identify the digit to the (right/left) of the given place value.

3. A(n) _____ is a rough calculation used to find a reasonably accurate answer.

4. A(n) _____ is the result when estimating or rounding. It is usually not exactly accurate.

5. One possible way to estimate a product when no place value for rounding is specified is to round each factor to its _____ place value.

6. The number that results when a whole number is squared is called a(n) _____ square.

7. A(n) _____ of a given whole number is a number whose square is the given whole number.

8. In the expression $\sqrt{64}$, the symbol $\sqrt{}$ is the _____, and the number 64 is the _____.

9. The whole number 49 is a perfect square because it is the result when _____ is squared.

10. The symbol $\approx$ means _____.

ROUNDING WHOLE NUMBERS

Exercises 11–14: Use the given number line to round the whole number to the given place value.

11. 732, hundreds

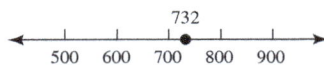

12. 187,654, ten-thousands

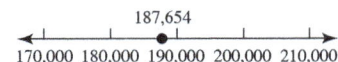

13. 58,923, thousands

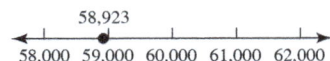

14. 78, tens

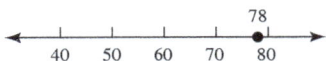

Exercises 15–26: Round the whole number to the given place value.

15. 82, tens
16. 43,903, thousands
17. 850, hundreds
18. 4397, thousands
19. 375,803, ten-thousands
20. 6,702,732, millions
21. 54,208, hundreds
22. 509,982, hundred-thousands
23. 783, tens
24. 74,803, ten-thousands
25. 30,092,441, millions
26. 513,783, thousands

Exercises 27–34: Round the whole number to its highest place value.

27. 3409
28. 347
29. 87,430,933
30. 11,908
31. 68
32. 730,982
33. 15,000
34. 350,000

ESTIMATING AND APPROXIMATING

Exercises 35–42: Round each number to the nearest hundred to estimate the sum or difference.

35. $759 + 311 + 406$
36. $3209 + 287 + 1521$
37. $1739 - 1341$
38. $5866 - 209$
39. $2421 + 576$
40. $399 + 8602$
41. $8289 - 97$
42. $769 - 111$

Exercises 43–50: Round each number to its highest place value to estimate the product or quotient.

43. $72,091 \cdot 68$
44. $311 \cdot 5924$
45. $1007 \div 53$
46. $27,470 \div 510$
47. $391 \cdot 687$
48. $19 \cdot 217$
49. $10,007 \div 38$
50. $593 \div 17$

FINDING SQUARE ROOTS

Exercises 51–58: Compute the square root.

51. $\sqrt{25}$ 52. $\sqrt{9}$ 53. $\sqrt{121}$

54. $\sqrt{100}$ 55. $\sqrt{361}$ 56. $\sqrt{196}$

57. $\sqrt{625}$ 58. $\sqrt{900}$

APPLICATIONS INVOLVING ROUNDING AND ESTIMATION

59. **Yearly Work Hours** There are 52 full weeks in one year. If a person takes no time off and works 38 hours each week, estimate the total number of hours worked in one year.

60. **Career Work Hours** Use your answer from the previous exercise to estimate the total number of hours this person works in 29 years.

61. **Bowling Experience** A woman has 37 years of bowling experience. The other four members of the team have 29, 32, 41, and 43 years of experience, respectively. Estimate the combined number of years of bowling experience for the team by rounding each value to the nearest ten.

62. **Estimating Distance** A truck driver makes three deliveries with distances of 189 miles, 57 miles, and 112 miles. Estimate the total distance traveled by rounding each value to the nearest ten.

63. **Population Growth** In a recent year, there are 15,002 births every hour worldwide but only 6316 deaths. By rounding each value to the nearest thousand, estimate the increase in world population each hour. (*Source:* U.S. Census Bureau, International Data Base.)

64. **Sunny Days** On average, there are 201 cloudless days each year in Bishop, California, but only 59 such days in Kodiak, Alaska. Estimate the difference between these values by rounding to the nearest ten. (*Source:* NOAA, National Climate Data Center.)

65. **Highest Elevations** The highest point in Montana is Granite Peak with an elevation of 12,799 feet. The highest point in Mississippi is Woodall Mountain with an elevation of 806 feet. By rounding to the nearest hundred, estimate the difference between these two points. (*Source:* U.S. Geological Survey.)

66. **Lowest Elevations** The lowest point in Wyoming is Belle Fourche River with an elevation of 3099 feet. The lowest point in New Mexico is Red Bluff Reservoir with an elevation of 2842 feet. By rounding to the nearest hundred, estimate the difference between these two points. (*Source:* U.S. Geological Survey.)

67. **Starbucks Stores** The following table shows the number of Starbucks stores in selected states. Estimate the total number of Starbucks stores in these states by rounding each value to the nearest ten.

State	Texas	Ohio	Hawaii	Idaho
Stores	804	280	80	51

Source: Starbucks.

68. **Walmart Stores** The following table shows the number of Walmart Supercenters in selected states. Estimate the total number of Walmart Supercenters in these states by rounding each value to the nearest ten.

State	Illinois	Iowa	Maine	Utah
Stores	129	56	19	40

Source: Wal-Mart.

Exercises 69–72: **U.S. Internet Use** *The following graph shows the number of U.S. Internet users in millions for selected years.* (*Source:* Department of Commerce.)

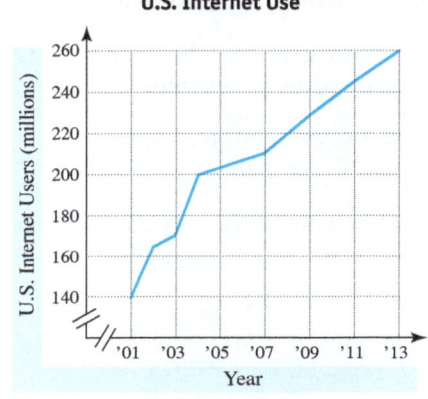

69. Estimate the number of U.S. Internet users in 2008 to the nearest ten-million.

70. Estimate the year when there were 200 million U.S. Internet users.

71. Estimate the year when there were 170 million U.S. Internet users.

72. Estimate the number of U.S. Internet users in 2013 to the nearest ten-million.

WRITING ABOUT MATHEMATICS

73. Explain how a number line can be used as an aid in rounding whole numbers.

74. Create a plan for finding all the whole numbers between 1000 and 1500 that are perfect squares.

1.6 Order of Operations

Objectives

1. Applying the Order of Operations Agreement
2. Evaluating Algebraic Expressions
3. Translating Words to Symbols

NEW VOCABULARY

☐ Order of operations agreement

1 Applying the Order of Operations Agreement

Connecting Concepts with Your Life Retail store employees must occasionally take inventory of items sold in the store. Many times, a particular item is found on the store shelves as well as in boxes in a storage room. Suppose there are 17 light bulbs on the shelf and 4 boxes of 24 light bulbs each in the storage room. The total number of light bulbs can be represented by the expression $17 + 4 \cdot 24$. How would you find the total number of light bulbs? Is the total

$17 + 4 \cdot 24 = 21 \cdot 24 \stackrel{?}{=} 504$ or $17 + 4 \cdot 24 = 17 + 96 \stackrel{?}{=} 113$?
(Add first, then multiply.)   (Multiply first, then add.)

There are $4 \cdot 24 = 96$ light bulbs in the storage room and 17 light bulbs on the shelf. The total number of light bulbs is $17 + 96 = 113$. This result implies that in the expression $17 + 4 \cdot 24$, multiplication should be done before addition. This is just one of the rules stated in the **order of operations agreement**. Because arithmetic expressions may contain parentheses, exponents, radicals, and several arithmetic operations, it is important to evaluate these expressions consistently. To ensure that we all find the same result when evaluating an arithmetic expression, the following rules are used.

READING CHECK 1

- Why is it important to have an order of operations agreement when evaluating $4 - 2 \cdot 2$?

> **ORDER OF OPERATIONS**
>
> 1. Do all calculations within grouping symbols, such as parentheses and radicals, or above and below a fraction bar.
> 2. Evaluate all exponential expressions.
> 3. Do all multiplication and division from *left to right*.
> 4. Do all addition and subtraction from *left to right*.

NOTE: If there is more than one arithmetic operation within grouping symbols, the order of Steps 2, 3, and 4 must be followed when performing Step 1. ∎

EXAMPLE 1 Using the order of operations to evaluate expressions

Evaluate each expression.
(a) $24 - 3 \cdot 6$ (b) $12 \div 3 + 4 \cdot 5$ (c) $3^2 + 8 \div (3 - 1)$

Solution
(a) Perform multiplication before subtraction.

$24 - 3 \cdot 6 = 24 - 18$ Multiply.
(Multiply before subtracting.) $= 6$ Subtract.

CALCULATOR HELP
To use a calculator to evaluate expressions with parentheses, see Appendix E (page AP-29).

(b) Perform multiplication and division from left to right.

$12 \div 3 + 4 \cdot 5 = 4 + 4 \cdot 5$ Divide.
(Divide and multiply before adding.) $= 4 + 20$ Multiply.
$= 24$ Add.

(c) The expression within parentheses is evaluated first.

$$3^2 + 8 \div (3 - 1) = 3^2 + 8 \div 2 \quad \text{Subtract within parentheses.}$$
$$= 9 + 8 \div 2 \quad \text{Evaluate } 3^2.$$
$$= 9 + 4 \quad \text{Divide.}$$
$$= 13 \quad \text{Add.}$$

Calculate within parentheses first.

Now Try Exercises 7, 17, 19

EXAMPLE 2 **Using the order of operations to evaluate expressions**

Evaluate each expression.
(a) $2\sqrt{11 + 14} + (8 - 5)^3$ **(b)** $\dfrac{10 \cdot (2 + 2)}{\sqrt{64} - 3}$

Solution
(a) The expression under the radical and the expression in parentheses are evaluated first.

$$2\sqrt{11 + 14} + (8 - 5)^3 = 2\sqrt{25} + 3^3 \quad \text{Evaluate radicand and parentheses.}$$
$$= 2 \cdot 5 + 27 \quad \text{Evaluate } \sqrt{25} \text{ and } 3^3.$$
$$= 10 + 27 \quad \text{Multiply.}$$
$$= 37 \quad \text{Add.}$$

(b) Evaluate the radical expression and the expression in parentheses first.

$$\frac{10 \cdot (2 + 2)}{\sqrt{64} - 3} = \frac{10 \cdot 4}{8 - 3} \quad \text{Evaluate } \sqrt{64} \text{ and parentheses.}$$
$$= \frac{40}{5} \quad \text{Multiply top; subtract bottom.}$$
$$= 8 \quad \text{Divide.}$$

Now Try Exercises 21, 35

> **STUDY TIP**
>
> Studying with other students can greatly improve your chance of success. Consider exchanging phone numbers or email addresses with some of your classmates so that you can find a time and place to study together. If possible, set up a regular meeting time and invite other classmates to join you.

Sometimes expressions contain grouping symbols *within* grouping symbols, or *nested* grouping symbols, that must be evaluated starting with the innermost grouping and working outward. For example, in the expression

$$(3 + (6 - 1)) \cdot 8$$

the grouping $(6 - 1)$ is evaluated first.

EXAMPLE 3 **Evaluating expressions with nested grouping symbols**

Evaluate each expression.
(a) $(7 - (4 + 2)^2 \div 9) + 3$ **(b)** $(11 - \sqrt{50 - 1} - 3)^3$

Solution
(a) Start by evaluating the innermost grouping, $(4 + 2)$.

$$(7 - (4 + 2)^2 \div 9) + 3 = (7 - 6^2 \div 9) + 3 \quad \text{Evaluate innermost grouping.}$$
$$= (7 - 36 \div 9) + 3 \quad \text{Evaluate } 6^2.$$
$$= (7 - 4) + 3 \quad \text{Divide.}$$
$$= 3 + 3 \quad \text{Evaluate parentheses.}$$
$$= 6 \quad \text{Add.}$$

(b) Start by evaluating the innermost grouping, the radicand $50 - 1$.

$$\left(11 - \sqrt{50 - 1} - 3\right)^3 = \left(11 - \sqrt{49} - 3\right)^3 \quad \text{Evaluate innermost grouping.}$$
$$= (11 - 7 - 3)^3 \quad \text{Evaluate } \sqrt{49}.$$
$$= (4 - 3)^3 \quad \text{Subtract.}$$
$$= 1^3 \quad \text{Evaluate parentheses.}$$
$$= 1 \quad \text{Evaluate } 1^3.$$

Now Try Exercises 29, 39

2 Evaluating Algebraic Expressions

So far, the order of operations agreement has been used to evaluate expressions without variables. Now we will see that the same rules apply when evaluating algebraic expressions for given values of the variables, as shown in the next example.

EXAMPLE 4 Evaluating algebraic expressions

Evaluate each algebraic expression for $x = 3$, $y = 8$, and $z = 4$.

(a) $25 - xy$ (b) $(15 - x) \div z + xy$ (c) $\dfrac{5xz}{y - 4}$

Solution

(a) Start by replacing x with 3 and y with 8 in the given expression. Recall that writing two variables next to each other implies multiplication. Note that z is not used.

$$25 - xy = 25 - 3 \cdot 8 \quad x = 3 \text{ and } y = 8.$$
$$= 25 - 24 \quad \text{Multiply.}$$
$$= 1 \quad \text{Subtract.}$$

(b) This expression contains three variables. Replace x with 3, y with 8, and z with 4.

$$(15 - x) \div z + xy = (15 - 3) \div 4 + 3 \cdot 8 \quad x = 3, y = 8, \text{ and } z = 4.$$
$$= 12 \div 4 + 3 \cdot 8 \quad \text{Evaluate parentheses.}$$
$$= 3 + 24 \quad \text{Divide; multiply.}$$
$$= 27 \quad \text{Add.}$$

(c) Replace x with 3, y with 8, and z with 4.

$$\frac{5xz}{y - 4} = \frac{5 \cdot 3 \cdot 4}{8 - 4} \quad x = 3, y = 8, \text{ and } z = 4.$$
$$= \frac{60}{4} \quad \text{Multiply top; subtract bottom.}$$
$$= 15 \quad \text{Divide.}$$

Now Try Exercises 51, 55, 59

3 Translating Words to Symbols

Often mathematical expressions are not given in application problems. Instead, words are used to describe the mathematics necessary to find a result. When this occurs, we must translate the words into mathematical symbols. However, some phrases may be difficult to translate. Consider the phrase

six plus nine divided by three.

Does the phrase *six plus nine divided by three* translate to

$$6 + 9 \div 3 \quad \text{or} \quad 6 + \frac{9}{3} \quad \text{or} \quad \frac{6 + 9}{3}?$$

Because of the order of operations agreement, the first two expressions are equivalent, and each evaluates to 9. However, the third expression evaluates to 5. It is correct to translate the phrase to either of the first two expressions. A phrase that translates to the third expression must contain a comma, words such as "the quantity," or both. For example, either of the following phrases translates to the third expression above.

six plus nine, divided by three or **the quantity six plus nine, divided by three**

In the next example, we translate words to symbols. Look for the words "the quantity" in each phrase and pay close attention to the placement of any commas.

READING CHECK 2

- If the expression $(3 + 9) \div 4$ were translated into words, would a comma be used?

EXAMPLE 5 Translating words to symbols

Use symbols to write each expression and then evaluate it.
(a) Five squared plus eight
(b) Two less than the quantity three times six
(c) Four plus five, times seven

Solution
(a) The phrase translates to $5^2 + 8$, which evaluates to $25 + 8 = 33$.
(b) Translate "the quantity three times six" by using parentheses and then subtract two. The phrase translates to $(3 \cdot 6) - 2$ and evaluates to $18 - 2 = 16$.
(c) The phrase translates to $(4 + 5) \cdot 7$, which evaluates to $9 \cdot 7 = 63$.

Now Try Exercises 67, 69, 71

1.6 Putting It All Together

CONCEPT	COMMENTS	EXAMPLES
Order of Operations Agreement	1. Do all calculations within grouping symbols, such as parentheses and radicals, or above and below a fraction bar. 2. Evaluate all exponential expressions. 3. Do all multiplication and division from *left to right*. 4. Do all addition and subtraction from *left to right*.	$14 \div 7 + 3 \cdot 6 = 2 + 18$ $= 20$ $20 \div (3 - 1)^2 = 20 \div 2^2$ $= 20 \div 4$ $= 5$ $\sqrt{6 + 10} = \sqrt{16}$ $= 4$
Evaluating Algebraic Expressions	Replace each variable in the algebraic expression with its given value, and then use the order of operations agreement to evaluate.	Evaluate $5 + 2x$ for $x = 4$. $5 + 2x = 5 + 2 \cdot 4$ $= 5 + 8$ $= 13$
Translating Words to Symbols	The words "the quantity" and the placement of any commas are important when translating words to symbols.	The phrase "nine minus two, times six" translates to $(9 - 2) \cdot 6$.

1.6 Exercises

CONCEPTS AND VOCABULARY

1. The _____ agreement must be followed to ensure that we all find the same result when evaluating an arithmetic expression.

2. A fraction bar is an example of a grouping symbol. Name two other examples of grouping symbols from this section.

3. In the expression $3 + 4 \cdot 7$, multiplication should be done (before/after) addition.

4. In the expression $(3 - 1)^2 \cdot 6$, which is done first, subtraction, squaring, or multiplication?

5. When translating words to symbols, parentheses can be used to express the words "_____".

6. In the phrase "six plus two, times ten," which is done first, addition or multiplication?

APPLYING ORDER OF OPERATIONS

Exercises 7–44: Evaluate the expression.

7. $6 + 2 \cdot 9$
8. $23 - 3 \cdot 5$
9. $36 \div 4 + 7$
10. $55 \div 11 - 3$
11. $28 \div 4 \cdot 3$
12. $6 \cdot 8 \div 3$
13. $8 + 10 \cdot 3 \div 5$
14. $15 - 12 \div 4 \cdot 3$
15. $80 - 45 \div 5 + 2$
16. $77 - 12 \cdot 5 + 13$
17. $63 \div 7 + 2 \cdot 8$
18. $2 \cdot 100 - 72 \div 8$
19. $32 - 4^2 \div (5 + 3)$
20. $96 \div (5^2 - 13) + 17$
21. $\sqrt{58 - 9} + (14 - 2)$
22. $110 - \sqrt{99 + 1}$
23. $13^2 + \sqrt{130 - 9}$
24. $\sqrt{3^2 + 16} - 2^2$
25. $\sqrt{256} \cdot \sqrt{300 - 44}$
26. $\sqrt{64} - \sqrt{40 + 24}$
27. $99 - (5 + 4^2 \div 2)$
28. $(72 \div 6^2 + 3) \cdot 12$
29. $((3 + 7)^2 \div 5) - 20$
30. $(3 \cdot (4 + 6) \div 10)^3$
31. $\dfrac{18 - 2}{7 - 3}$
32. $\dfrac{69 + 3}{3 \cdot 8}$
33. $\dfrac{(4 + 10) \cdot 5}{37 - 5 \cdot 6}$
34. $\dfrac{2 \cdot (5 + 1)^2}{5 \cdot 4 - 8}$
35. $\dfrac{(6 - 2) \cdot 9}{\sqrt{49} - 3}$
36. $\dfrac{(\sqrt{81} + 2) \cdot 6}{6^2 - 3}$
37. $(2 + 5)^2 + 3\sqrt{4 \cdot 9}$
38. $(8 - 7)^2 \cdot \sqrt{25 - 16}$
39. $4(15 + 5 \cdot 3) - ((50 \div 5)^2 - 12)$
40. $42 + 5^2 \cdot 2 - 64 \div 2^3 + 8$
41. $35 - (8 \cdot 5 - (20 \div 1^4) + 3)$
42. $35 - \sqrt{60 \div 5 - (2 \cdot 3^2 - 10)}$
43. $\sqrt{49 - 3 + (2 \cdot 6^2 \div 4)} + 14$
44. $17 - 3 \cdot 5 + 70 + 2^2 \cdot 3 - 27 \div 3^3 - 9$

Exercises 45–50: Insert parentheses as needed in the expression to make it equal to 0. More than one set of parentheses may be needed.

45. $14 - 12 \cdot 5 - 10$
46. $18 - 12 - 5 + 1$
47. $36 - 6^2 \div 5 - 1$
48. $39 - 7 + 8^2 \div 2$
49. $32 \div 4^2 - 2 \cdot 9$
50. $5 - 5 \cdot 3^2 \div 3$

EVALUATING ALGEBRAIC EXPRESSIONS

Exercises 51–62: Evaluate the algebraic expression for the given values of the variables.

51. $2 \cdot y + x$, for $x = 4, y = 2$
52. $a - 12 + 5b$, for $a = 13, b = 3$
53. $c(d - 8) + 3$, for $c = 2, d = 10$
54. $12 - g(2 + h)$, for $g = 4, h = 1$
55. $m + (9^2 - n) \div 8$, for $m = 31, n = 1$
56. $w + ((8 - v)^2 \div w)$, for $v = 2, w = 4$
57. $\sqrt{p^2 + 9} - q^2$, for $p = 4, q = 2$
58. $\sqrt{r - 7} + (19 + s)$, for $r = 71, s = 13$
59. $\dfrac{3 \cdot (4 + c)}{d - 2 \cdot 7}$, for $c = 10, d = 16$
60. $\dfrac{g(6 - 2)}{(4 + \sqrt{h} - 3)^2}$, for $g = 8, h = 9$
61. $x + (y^2 - x) \div x$, for $x = 4, y = 8$
62. $m + n \cdot m - n \div m$, for $m = 3, n = 15$

TRANSLATING WORDS TO SYMBOLS

Exercises 63–78: Use symbols to write the expression and then evaluate it.

63. Twelve more than five
64. Three less than forty-five

65. Nine fewer than twenty-one

66. Thirty-eight increased by two

67. Four more than seven squared

68. Three squared minus eight

69. Six plus five, times nine

70. Seven plus four times two

71. The quantity six plus two, times five

72. Ten minus the quantity four plus six

73. Two cubed times three squared

74. The square root of the quantity two plus two

75. The square root of sixteen, plus nine

76. Eight divided by the quantity six minus two

77. Seven times three decreased by two

78. Ten divided by two increased by nine

APPLICATIONS INVOLVING ORDER OF OPERATIONS

79. **Heart Rate** The average heart rate R in beats per minute (bpm) of an animal weighing W pounds can be approximated by

$$R = \frac{885\sqrt{W}}{W}.$$

Find the heart rate for a 25-pound dog.

80. **Heart Rate** If x is the number of minutes that have passed since exercise has stopped, a person's heart rate R in beats per minute can be approximated by

$$R = \frac{4(10 - x)^2}{5} + 80,$$

where $x < 10$. Find the person's heart rate 5 minutes after exercise has stopped.

81. **Wind Power** Electrical power generated by a particular wind turbine is given by

$$W = \frac{5l^2s^3}{32},$$

where W is power in watts, l is the length of a turbine blade in feet, and s is the speed of the wind in miles per hour. How many watts are generated by a wind turbine with a blade length of 10 feet if the wind has a speed of 8 miles per hour?

82. **Skid Marks** Vehicles in accidents often leave skid marks. To determine how fast a vehicle was traveling, officials often use a test vehicle to compare skid marks on the same section of road. If a vehicle involved in a crash left skid marks that are D feet long and a test vehicle traveling at v miles per hour leaves skid marks that are d feet long, then the speed of the vehicle in the crash is given by

$$V = \sqrt{\frac{v^2 D}{d}}.$$

Determine V if $v = 40$ miles per hour, $D = 225$ feet, and $d = 100$ feet.

83. **Insect Population** Suppose that an insect population P, in thousands per acre, is given by

$$P = \frac{10x - 6}{x + 1},$$

where x represents time in months. Find the insect population after 7 months.

84. **Stopping Distance** On dry, level pavement, the stopping distance D in feet for a car traveling at x miles per hour can be estimated by

$$D = \frac{x^2}{11} + \frac{11x}{3}.$$

Find the stopping distance for a car traveling on dry, level pavement at 33 miles per hour.

85. **Converting Temperature** To convert a temperature F given in degrees Fahrenheit to an equivalent temperature C in degrees Celsius, use the formula

$$C = \frac{5(F - 32)}{9}.$$

Find the Celsius temperature that is equivalent to a temperature of 104 °F.

86. **Converting Temperature** To convert a temperature C given in degrees Celsius to an equivalent temperature F in degrees Fahrenheit, use the formula

$$F = \frac{9C}{5} + 32.$$

Find the Fahrenheit temperature that is equivalent to a temperature of 35 °C.

WRITING ABOUT MATHEMATICS

87. Write a paragraph describing how to apply the order of operations to an expression. Give examples.

88. Give an example of a phrase containing the words "the quantity." Translate your phrase into symbols.

SECTIONS 1.5 and 1.6 — Checking Basic Concepts

1. Round 45,277 to the nearest thousand.

2. Round each number to the nearest hundred to estimate the sum or difference.
   (a) $789 + 403$ (b) $5311 - 694$

3. Estimate the product $18 \cdot 314$ by rounding each number to its hightest place value.

4. Compute each square root.
   (a) $\sqrt{81}$ (b) $\sqrt{169}$

5. Evaluate each expression.
   (a) $24 \div 6 + 9$ (b) $\sqrt{5^2 - 9} + 2^3$

6. Evaluate $12 - \left(4 \cdot 2 - (27 \div 3^2) + 5\right)$.

7. Evaluate $x + (3^2 - y) \div 2$ for $x = 7$, $y = 5$.

8. Use symbols to write the expression

   the quantity four plus two, times three

   and then evaluate the expression.

9. **World Population** There are 14,938 births every hour worldwide but only 6459 deaths. Estimate the increase in world population each hour by rounding to the nearest hundred. (*Source:* U.S. Census.)

10. **Building Size** A square garage has a floor area of 400 square feet. Find the length of one side of the garage.

1.7 More with Equations and Problem Solving

Objectives

1. Simplifying Algebraic Expressions
   - Identifying Equations and Expressions
   - Recognizing Like Terms
   - Combining Like Terms
   - Applying Arithmetic Properties
2. Checking a Solution to an Equation
3. Applying a Problem-Solving Strategy

NEW VOCABULARY

☐ Term
☐ Coefficient
☐ Like terms

1 Simplifying Algebraic Expressions

IDENTIFYING EQUATIONS AND EXPRESSIONS There is a difference between an equation and an algebraic expression. An equation *always* contains an equals sign ($=$), but an expression *never* contains an equals sign. We can solve an equation, but we cannot solve an expression. An expression can be evaluated, and as we will see in this section, an expression can sometimes be *simplified*.

Expressions	Equations
$3x + 2$ and $9 - x$	$5x = 10$ and $x - 1 = 7$
Don't have equals signs	Have equals signs

EXAMPLE 1 — Identifying equations and expressions

Identify each of the following as an equation or an expression.
(a) $3x + 5 = 17$ (b) $24y - 34 + 2y$

Solution
(a) Because $3x + 5 = 17$ has an equals sign, it is an equation.
(b) Because there is no equals sign in $24y - 34 + 2y$, it is an expression.

Now Try Exercises 7, 9

RECOGNIZING LIKE TERMS A **term** is a number, a variable, or a product of numbers and variables raised to powers. Examples of terms include

$$7, \quad x, \quad 3y, \quad 12xy, \quad \text{and} \quad 9y^2z^3.$$

STUDY TIP

If you are studying with classmates, make sure that they do not "do the work for you." A classmate with the best intentions may give too many verbal hints while helping you work through a problem. Remember that members of your study group will not be giving hints during an exam.

READING CHECK 1

- How can you tell if two terms are like terms?

Terms do not contain addition or subtraction, but the plus and minus signs in an expression separate terms within an expression.

$3x$ is a term. y is a term. $4z^2$ is a term.

$$3x + y + 4z^2$$

Expression

The number in a term is called the **coefficient**. If no number appears, the coefficient is 1.

The coefficient in this term is 3.   The coefficient in this term is 1.   The coefficient in this term is 4.

$$3x + y + 4z^2$$

If two terms have the same variables raised to the same powers, then we say that the terms are **like terms**. Examples of pairs of like terms include

$3x$ and $5x$, $2y^2$ and $8y^2$,   and $5a^2b$ and a^2b.

EXAMPLE 2 Identifying like terms

Determine whether the terms are like or unlike.
(a) $5w, 9w$ (b) $7x^2, 7y^2$ (c) $4a^2b, 3b^2a$ (d) x, x^2

Solution
(a) The variable in both terms is w (with power 1), so they are like terms.
(b) The variables are different, so they are unlike terms.
(c) Although the variables are the same, the powers on the variables do not match, so they are unlike terms.
(d) The variables are both x, but the powers do not match. They are unlike terms.

Now Try Exercises 15, 17, 19, 21

COMBINING LIKE TERMS We may *combine* like terms by adding or subtracting their coefficients. The following See the Concept illustrates that like terms can be combined but unlike terms cannot.

See the Concept 1   LIKE AND UNLIKE TERMS

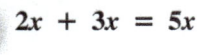 $2x + 3x = 5x$

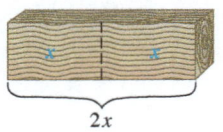

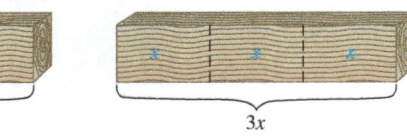

$2x + 3v$ **B** Cannot be combined.

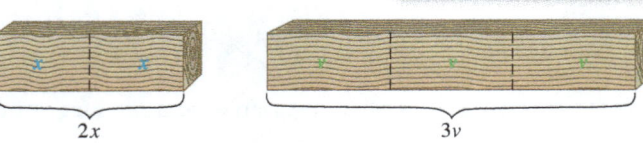

A The terms $2x$ and $3x$ are *like* terms because the length of each segment x is the same. We can find the total length of the two boards by applying the distributive property and *adding like terms*:

$$2x + 3x = (2 + 3)x = 5x.$$

B The lengths x and v might not be equal, so $2x$ and $3v$ are *unlike* terms. We cannot combine unlike terms, so we cannot determine the total length other than to say it is

$$2x + 3v.$$

WATCH VIDEO IN MML.

1.7 MORE WITH EQUATIONS AND PROBLEM SOLVING

EXAMPLE 3 **Combining like terms**

Combine like terms in each expression. Write "not possible" if terms cannot be combined.
(a) $4y + 7y$ (b) $8xy^2 - 3xy^2$ (c) $7m + 3n$

Solution
(a) The variable in both terms is y, so they are like terms and can be combined. Using the distributive property, we have

$$4y + 7y = (4 + 7)y = 11y.$$

We can think of "4 y's plus 7 y's is 11 y's" in the same way as "4 apples plus 7 apples is 11 apples."

(b) The variables and powers match, so the terms are like terms and can be combined. Using the distributive property, we have $8xy^2 - 3xy^2 = (8 - 3)xy^2 = 5xy^2$.

(c) It is not possible to combine $7m$ and $3n$ because the variables are not the same. That is, the terms are unlike.

Now Try Exercises 23, 25, 27

APPLYING ARITHMETIC PROPERTIES Another way to simplify expressions is to apply the commutative, associative, and distributive properties. For example, we can use the associative property to group the terms in the expression

$$(2x + 3) + 5 \quad \text{as} \quad 2x + (3 + 5).$$

READING CHECK 2

- What property allows us to write $2x + 5 + 3x$ as $2x + 3x + 5$?

Now the terms in the parentheses are like terms and can be combined to form the *simplified expression* $2x + 8$.

Similarly, the commutative property can be used to rewrite the expression

$$4x + 1 + 3x \quad \text{as} \quad 4x + 3x + 1.$$

The first two terms can now be combined to form the simplified expression $7x + 1$.

The next example illustrates how algebraic expressions can be simplified by applying the commutative, associative, and distributive properties.

EXAMPLE 4 **Simplifying expressions**

Simplify each expression.
(a) $3 + 5w + 1$ (b) $15x + (2x + y)$ (c) $4y + 5(y - 3)$

Solution
(a) First apply the commutative property to the first two terms.

$$3 + 5w + 1 = 5w + 3 + 1 \qquad \text{Commutative property}$$
$$= 5w + 4 \qquad \text{Add: } 3 + 1 = 4.$$

(b) Apply the associative property to regroup the terms.

$$15x + (2x + y) = (15x + 2x) + y \qquad \text{Associative property}$$
$$= (15 + 2)x + y \qquad \text{Distributive property}$$
$$= 17x + y \qquad \text{Add: } 15 + 2 = 17.$$

(c) First, use the distributive property to multiply 5 and $(y - 3)$.

$$4y + 5(y - 3) = 4y + 5 \cdot y - 5 \cdot 3 \qquad \text{Distributive property}$$
$$= 4y + 5y - 15 \qquad \text{Multiply.}$$
$$= (4 + 5)y - 15 \qquad \text{Distributive property}$$
$$= 9y - 15 \qquad \text{Add: } 4 + 5 = 9.$$

Now Try Exercises 31, 33, 37

2 Checking a Solution to an Equation

To see if a number is a solution to an equation, recall that we replace each occurrence of the variable with the given number. For example, to see if 5 is a solution to the equation $4x + 10 = 6x$, replace each x with 5 and determine if the resulting equation is true or false. A question mark above an equals sign means that we are checking a possible solution.

$$4x + 10 = 6x \qquad \text{Given equation}$$
$$4(5) + 10 \stackrel{?}{=} 6(5) \qquad \text{Replace } x \text{ with 5.}$$
$$20 + 10 \stackrel{?}{=} 30 \qquad \text{Multiply}$$
$$30 = 30 \checkmark \qquad \text{Add; the solution checks.}$$

READING CHECK 3

- How is a solution checked to see if it is correct?

Every equation has an expression on each side of the equals sign. In some equations, it may be possible to simplify one or both of these expressions. In the next example, we simplify the expressions in an equation and then check to see that a given value is a solution to both the given equation and the equation formed by simplifying the expressions.

EXAMPLE 5 Simplifying expressions in an equation and checking a solution

For the equation $5x + 3x = 16 + 2(x + 1)$, do the following.
(a) Simplify the expression on each side of the equals sign.
(b) Check to see if 3 is a solution to both the given equation and the one formed in part (a).

Solution
(a) First apply the distributive property on each side of the equation.

$$5x + 3x = 16 + 2(x + 1) \qquad \text{Given equation}$$
$$(5 + 3)x = 16 + 2x + 2 \qquad \text{Distributive property}$$
$$8x = 2x + 16 + 2 \qquad \text{Add; commutative property}$$
$$8x = 2x + 18 \qquad \text{Add.}$$

(b) To see if 3 is a solution to the given equation, replace each occurrence of x with 3.

$$5x + 3x = 16 + 2(x + 1) \qquad \text{Given equation}$$
$$5(3) + 3(3) \stackrel{?}{=} 16 + 2(3 + 1) \qquad \text{Replace } x \text{ with 3.}$$
$$15 + 9 \stackrel{?}{=} 16 + 2(4) \qquad \text{Multiply; add.}$$
$$15 + 9 \stackrel{?}{=} 16 + 8 \qquad \text{Multiply.}$$
$$24 = 24 \checkmark \qquad \text{Add; the solution checks.}$$

To see if 3 is a solution to the equation formed in part (a), replace each occurrence of x with 3.

$$8x = 2x + 18 \qquad \text{Given equation}$$
$$8(3) \stackrel{?}{=} 2(3) + 18 \qquad \text{Replace } x \text{ with 3.}$$
$$24 \stackrel{?}{=} 6 + 18 \qquad \text{Multiply.}$$
$$24 = 24 \checkmark \qquad \text{Add; the solution checks.}$$

Now Try Exercise 45

3 Applying a Problem-Solving Strategy

Some application problems can be challenging because formulas and equations are not given. To solve such problems, it is often helpful to follow a strategy. The following strategy is based on George Polya's (1888–1985) four-step process for solving problems.

> **STEPS FOR SOLVING A PROBLEM**
>
> **STEP 1:** Read the problem carefully and be sure you understand it. (You may need to read the problem more than once.) Assign a variable to what you are being asked to find.
>
> **STEP 2:** Write an equation that relates the quantities described in the problem. You may need to sketch a diagram or refer to known formulas.
>
> **STEP 3:** Solve the equation. Use the solution to determine the solution(s) to the original problem. Include any necessary units.
>
> **STEP 4:** Look back and check your solution in the original problem. Does your solution seem reasonable?

READING CHECK 4

- What is the first step for solving a problem?

Even when we understand a problem, we may not be able to find a solution if we cannot write an appropriate equation. In the next example, we practice the second step in the four-step process by translating sentences into equations.

EXAMPLE 6 Translating sentences into equations

Translate the sentence into an equation using the variable x. Do not solve the equation.
(a) Four times the number of feet plus 3 times the same number of feet is 28.
(b) A student's age decreased by 7 is 12.
(c) Sixteen thousand, five hundred is 8000 more than the population.

Solution
(a) If x represents the number of feet, then the phrase "**four times** the number of feet" is written $4x$, and the phrase "**3 times** the same number of feet" is written $3x$. The word "**plus**" indicates that these two quantities should be added to get $4x + 3x$. The word "**is**" suggests an equals sign. The entire sentence translates to $4x + 3x = 28$.
(b) If x represents the student's age, then "**decreased by** 7" indicates that 7 should be subtracted from x to get $x - 7$. Again, the word "**is**" suggests an equals sign. The entire sentence translates to $x - 7 = 12$.
(c) If x represents the population, then "8000 **more than** the population" can be written as $x + 8000$. The sentence translates to $16{,}500 = x + 8000$.

Now Try Exercises 49, 51, 53

STUDY TIP

One of the best ways to prepare for class is to read a section *before* it is covered by your instructor. For example, reading ahead about the four steps for solving a problem would give you the chance to consider the process and formulate any questions that you might have about it.

Vintage Television

🌐 Math in Context (New Technology) When a new technology is introduced, it often takes time for the technology to "catch on." For example, thirty-eight years passed between the time that radios were first available to the public and the time when a significant number of people used radios on a regular basis. In the next example, we apply the four-step problem-solving process to a word problem that compares newly introduced technologies.

EXAMPLE 7 Comparing newly introduced technologies

After its introduction to consumers, television took 13 years to catch on in U.S. households. This is eight more years than it took the Internet to catch on. How many years passed between the first availability of the Internet and its widespread use? (*Source:* Internet World Stats.)

Solution

STEP 1: We must find the number of years that it took for the Internet to catch on in the U.S. We assign the variable x to this unknown amount of time.

STEP 2: Reading the paragraph carefully reveals that

13 years is 8 more years than it took for the Internet to catch on.

Because x represents the time it took for the Internet to catch on, the equation is

$$13 = x + 8.$$

STEP 3: To solve the equation in Step 2, we must find the value of x that makes the equation true. Because $13 = 5 + 8$, the solution is 5 years.

STEP 4: Because 13 is 8 more than 5, the solution checks in the original problem. Based on how quickly new technologies become popular in today's society, it seems reasonable that the Internet caught on faster than television.

Now Try Exercise 57

EXAMPLE 8 Analyzing doctorate degrees

In a recent year, there were about 48,000 doctorate degrees awarded in the United States. Twice as many of these doctorates were awarded to U.S. citizens than were awarded to foreign citizens. Find the number of doctorate degrees awarded in the United States to foreign citizens that year. (*Source:* U.S. National Science Foundation.)

Solution

STEP 1: Let x represent the number of doctorates awarded to foreign citizens. Because there were twice as many doctorates awarded to U.S. citizens, $2x$ represents the number of doctorates awarded to U.S. citizens.

STEP 2: The total number of doctorates is found by adding.

U.S. citizen doctorates + *foreign citizen doctorates* = *total number of doctorates*

Because the total is 48,000, the equation can be written as

$$2x + x = 48,000.$$

STEP 3: To solve the equation in Step 2, first we simplify the expression on the left side of the equation. By combining like terms, the equation becomes

$$3x = 48,000.$$

The solution to this equation is 16,000 because $3(16,000) = 48,000$ is a true equation. There were 16,000 doctorates awarded to foreign citizens.

STEP 4: If the number of doctorates awarded to foreign citizens was 16,000, and twice this number of doctorates, or 32,000, were awarded to U.S. citizens, then the total number of doctorates awarded was $16,000 + 32,000 = 48,000$. The solution checks in the original problem.

Now Try Exercise 63

1.7 Putting It All Together

CONCEPT	COMMENTS	EXAMPLES
Equations and Expressions	Equations always contain an equals sign ($=$), but expressions never contain an equals sign.	$3x + 7 = 10$ is an equation. $4y - 19$ is an expression.
Like Terms	• Terms that contain the same variables raised to the same powers • Like terms can be combined.	$4x$ and $7x$ are like terms. $9m$ and $9n$ are unlike terms. $5x - 3x = (5 - 3)x = 2x$
Simplifying an Expression	Use arithmetic properties and combine like terms to write an expression more simply.	$3y + (2y + 4) = (3y + 2y) + 4$ $= (3 + 2)y + 4$ $= 5y + 4$

1.7 Exercises

MyMathLab®

CONCEPTS AND VOCABULARY

1. Is $3x - 4$ an equation or an expression?
2. Is $3x = 4$ an equation or an expression?
3. A(n) _____ is a number, a variable, or the product of numbers and variables raised to powers.
4. The number in a term is called the _____ of the term.
5. The terms $3xy$ and $7xy$ are (like/unlike).
6. The terms $4y$ and $4z$ are (like/unlike).

IDENTIFYING EQUATIONS AND EXPRESSIONS

Exercises 7–14: Identify each of the following as an equation or an expression.

7. $3x + 12$
8. $9 = 3x$
9. $17y + 15 = 49$
10. $38z - 20$
11. $5x = 3x + 10$
12. $2x + (5 - 3x)$
13. $4003 - x$
14. $3m + 2 = 4m - 6$

RECOGNIZING AND COMBINING LIKE TERMS

Exercises 15–22: Determine whether the given terms are like or unlike.

15. $7w, 4w$
16. $2a, 9a$
17. $4bc, 3bc^2$
18. $8x^2y, 17x^2y$
19. $3xy^3, 2xy^3$
20. $9a^2b, 7ab^2$
21. y^2, y
22. $pq, 7p^2q$

Exercises 23–30: Combine like terms in the expression. Write "not possible" if terms cannot be combined.

23. $8x + 3x$
24. $4b + b$
25. $13yz - 6yz$
26. $x - y$
27. $6p - 5q^2$
28. $7z^2 - z^2$
29. $ab + 15ab$
30. $8m^2n - 3m^2n$

SIMPLIFYING EXPRESSIONS

Exercises 31–44: Simplify the expression.

31. $4 + 7x + 5$
32. $2 + 4n + 9$
33. $9y + (2y + 5)$
34. $15m + (3m + 7)$
35. $3a + 4 + 2a$
36. $8x + 7 + 2x$
37. $6z + 2(z - 7)$
38. $10z + 7(z - 3)$
39. $3(x + 2) - 4$
40. $9(q + 1) + 6$
41. $2x + 5 + 3x + 4$
42. $7y + 2 + y + 5$
43. $ab + y + 2ab + 3y$
44. $2x^2 + 3x + 5x^2 + x$

CHECKING SOLUTIONS

Exercises 45–48: For the given equation, do the following:

(a) *Simplify the expression on each side of the equals sign.*
(b) *See if 5 is a solution to both the given equation and the equation formed in part (a).*

45. $3x + 4x = 4(x + 2) + 7$

46. $3x + 2(x + 4) = 8 + 4x + 5$

47. $2(3 + x) + 5 = x + (2x + 6)$

48. $x^2 + 2x^2 = 3x + 12x$

TRANSLATING WORDS TO EQUATIONS

Exercises 49–56: Translate the sentence into an equation using the variable x. Do not solve the equation. State what the variable represents.

49. Six times the number of inches minus two times the same number of inches is 36.

50. Seven pounds less than his weight is 156.

51. Fourteen is 9 less than her score.

52. The total of his age and twice his age is 30.

53. The product of 4 and her shoe size is 28.

54. The total miles divided by 14 is 31.

55. The score is 8 fewer than triple the score.

56. Eight more than 3 times the height is 107.

APPLICATIONS INVOLVING PROBLEM SOLVING

57. **Education and Pay** A recent survey of employed college graduates found that those whose highest degree is a master's earned, on average, $11,000 per year more than graduates whose highest degree is a bachelor's. If those with a master's degree earned $64,000 per year, on average, how much did those with a bachelor's degree earn? (*Source:* National Science Foundation.)

58. **Heart Rate** During exercise, a physically fit male may experience a heart rate that is 3 times his resting heart rate. If this person's heart rate is 186 beats per minute during exercise, what is his resting heart rate?

59. **Roadway Congestion** A recent study found that the average commuter in Atlanta, Georgia, burns 24 gallons of gas each year while sitting in traffic. This is 6 times the amount burned by the average commuter in Cleveland, Ohio. How many gallons of gas does the average Cleveland driver burn each year while stuck in traffic? (*Source:* Federal Highway Administration.)

60. **Roadway Congestion** A recent study found that the average commuter in Houston, Texas, spends 16 more hours each year stuck in traffic than the average commuter in Portland, Oregon. If Houston drivers spend 36 hours each year stuck in traffic, how many hours do Portland drivers spend each year stuck in traffic? (*Source:* Federal Highway Administration.)

61. **Population Growth** The worldwide birth rate is about 14,940 births per hour. If the worldwide death rate is about 6460 deaths per hour, what is the worldwide population increase per hour? (*Source:* U.S. Census Bureau, International Data Base.)

62. **Housing Market** A homeowner recently reduced the price of his home to $218,000. If this represents a $17,000 decrease in price, what was the price before the decrease?

63. **Fireworks** Three friends bought several identical packages of bottle rockets. The first friend bought 2 packages, the second friend bought 4 packages, and the third friend bought 5 packages, giving the three friends a total of 132 bottle rockets. How many bottle rockets are in a single package?

64. **Sale Price** If the sale price of an item is multiplied by 4, then the result is the original price. If the original price is $64, what is the sale price?

65. **A Number Puzzle** If a number is tripled and then added to itself, the result is 12 more than the number. Find the number.

66. **A Number Puzzle** If doubling a natural number has the same result as squaring it, what is the number?

67. **Antique Value** A glass vase purchased at a garage sale has an appraised value of $200. If the appraised value is $179 more than the purchase price, how much did the vase cost at the garage sale?

68. **Counting Coins** A person has 6 coins in his pocket that total 65¢. If 3 of the coins are dimes, what are the other 3 coins?

69. **Perimeter** The rectangle in the following figure has a perimeter of 42 inches. If the length measures $5x$ inches and the width measures $2x$ inches, find x.

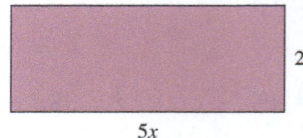

70. **Perimeter** A triangle with sides that measure $3x$, $4x$, and $5x$ feet has a perimeter of 60 feet. Find x.

WRITING ABOUT MATHEMATICS

71. Explain what it means to simplify an expression. Give several examples.

72. Describe in your own words how to use the four-step process to solve word problems.

SECTION 1.7 Checking Basic Concepts

1. Identify each of the following as an equation or an expression.
   (a) $4x = 40$ (b) $y + 16$

2. Determine whether the terms are like or unlike.
   (a) $m, 5m$ (b) $3y, 10x$

3. Combine like terms in each expression. Write "not possible" if terms cannot be combined.
   (a) $8pq - 3pq$ (b) $6m - n^2$

4. Simplify each expression.
   (a) $2(x + 3) - 5$ (b) $6x + 3 + 3x$

5. Translate the displayed sentence into an equation using the variable x. Do not solve the equation. State what the variable represents.

   His age decreased by 7 is 23.

6. **Perimeter** The rectangle in the following figure has a perimeter of 32 feet. If the length measures $3x$ feet and the width measures x feet, find x.

CHAPTER 1 Summary

SECTION 1.1 ■ INTRODUCTION TO WHOLE NUMBERS

Natural Numbers $1, 2, 3, 4, \ldots$

Whole Numbers $0, 1, 2, 3, \ldots$

Place Value   The place value of a digit in a number written in standard form is determined by the position that the digit takes in the number.

Example: The 4 in the number 34,879 is in the thousands place.

Word Form   Starting with the left-most period, write the word form of the number in each period followed by the period name and a comma.

Example: 18,207 is written as eighteen thousand, two hundred seven.

Expanded Form   Starting with the left-most digit, write the standard form of the number given by each digit and its corresponding place value. Place a plus sign ($+$) between each of these results.

Example: 184,079 is written as $100,000 + 80,000 + 4000 + 70 + 9$.

Number Line   A number line can be used to visualize whole numbers. The graph of a whole number is a dot placed at the whole number's position on a number line.

Examples: 2 and 6 are graphed.

Comparing Whole Numbers   To compare two whole numbers, determine their positions on a number line. Then use the symbols $>$ or $<$ to write an appropriate comparison.

Example: From the number line above, $2 < 6$.

80 CHAPTER 1 WHOLE NUMBERS

Bar Graph and Line Graph

A bar graph can be helpful when analyzing data. A line graph can be helpful when looking for trends in data.

Examples:

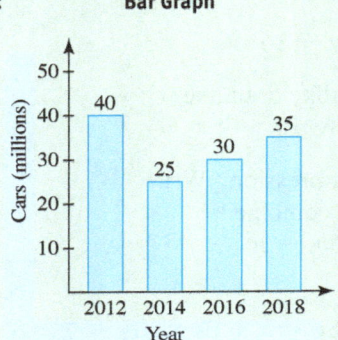

Bar Graph

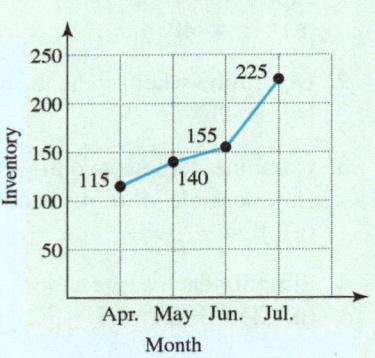

Line Graph

Table

A table can be used to display data in an at-a-glance way.

Example:

Year	2013	2015	2017
Births	207	215	198

SECTION 1.2 ■ ADDING AND SUBTRACTING WHOLE NUMBERS; PERIMETER

Adding Whole Numbers

The numbers being added are the *addends*, and the result is the *sum*.

1. Stack the numbers vertically with corresponding place values aligned.
2. Add the digits in each place value. Regroup when necessary.

Examples:
$$\begin{array}{r} 6532 \\ +\ 2413 \\ \hline 8945 \end{array} \qquad \begin{array}{r} \overset{1\ 1}{2758} \\ +\ 3617 \\ \hline 6375 \end{array}$$

Addition Properties

Property	Examples
1. Commutative property	$4 + 9 = 9 + 4$
2. Associative property	$(3 + 6) + 7 = 3 + (6 + 7)$
3. Identity property	$8 + 0 = 8$ and $0 + 8 = 8$

Translating Words to Addition

Words associated with addition include *add*, *plus*, *more than*, *sum*, *total*, and *increased by*.

Examples: 7 dimes plus 3 dimes;   the number of fish increased by 5.

Subtracting Whole Numbers

The number subtracted from is the *minuend*, the number being subtracted is the *subtrahend*, and the result is the *difference*.

1. Stack the numbers vertically with corresponding place values aligned.
2. Subtract the digits in each place value. Regroup when necessary.

Examples:
$$\begin{array}{r} 8437 \\ -\ 2216 \\ \hline 6221 \end{array} \qquad \begin{array}{r} 5\overset{6\ 13}{7\rlap{/}3}2 \\ -\ 2480 \\ \hline 3252 \end{array}$$

Subtraction Properties

Property	Examples
1. First identity property	$38 - 0 = 38$
2. Second identity property	$22 - 22 = 0$

Translating Words to Subtraction

Words associated with subtraction include *subtract*, *minus*, *fewer than*, *difference*, *less than*, *decreased by*, and *take away*.

Examples: 9 fewer than the number of pies; the price decreased by 5.

Solutions	A *solution* is any number that makes an equation true when it replaces the unknown value. *Solving an equation* means finding all of its solutions. **Example:** The solution to $\square + 3 = 15$ is 12 because $12 + 3 = 15$ is a true equation.
Perimeter	The distance around an enclosed region **Example:** The perimeter of the region shown is $5 + 9 + 6 + 8 = 28$ feet. 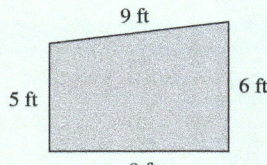

SECTION 1.3 ■ MULTIPLYING AND DIVIDING WHOLE NUMBERS; AREA

Multiplying Whole Numbers	The numbers being multiplied are the *factors*, and the result is the *product*. **Examples:** $$\begin{array}{r} \overset{1}{8}2 \\ \times\ 25 \\ \hline 410 \\ 1640 \\ \hline 2050 \end{array} \begin{array}{l} \\ \leftarrow 5 \times 82 \\ \leftarrow 2\ (\text{tens}) \times 82 \\ \leftarrow \text{add} \end{array} \qquad \begin{array}{r} \overset{4\,2}{\underset{5\,2}{274}} \\ \times\ 67 \\ \hline 1\,918 \\ 16\,440 \\ \hline 18{,}358 \end{array} \begin{array}{l} \\ \leftarrow 7 \times 274 \\ \leftarrow 6\ (\text{tens}) \times 274 \\ \leftarrow \text{add} \end{array}$$
Multiplication Properties	**Property** **Examples** 1. Commutative property $5 \cdot 7 = 7 \cdot 5$ 2. Associative property $(2 \cdot 6) \cdot 5 = 2 \cdot (6 \cdot 5)$ 3. Identity property $3 \times 1 = 3$ and $1 \times 3 = 3$ 4. Zero property $8 \times 0 = 0$ and $0 \times 8 = 0$ 5. Distributive property $3(4 + 2) = 3 \cdot 4 + 3 \cdot 2$ $2(5 - 1) = 2 \cdot 5 - 2 \cdot 1$
Translating Words to Multiplication	Words associated with multiplication include *multiply*, *times*, *product*, *triple*, and *double*. **Examples:** Double the number of cups; the length times the width.
Dividing Whole Numbers	The number divided into is the *dividend*, the number divided by is the *divisor*, and the result is the *quotient*.
Long Division	A process that can be used when the dividend is a large whole number **Example:** $$\begin{array}{r} 301\text{ r}52 \\ 263{\overline{\smash{\big)}\,79{,}215}} \\ \underline{-78\ 9} \\ 31 \\ \underline{-\ \ 0} \\ 315 \\ \underline{-263} \\ 52 \end{array}$$
Division Properties	**Property** **Examples** 1. Identity properties $6 \div 6 = 1$ and $8 \div 1 = 8$ 2. Zero properties $0 \div 7 = 0$ and $5 \div 0$ is undefined.
Translating Words to Division	Words associated with division include *divide*, *quotient*, and *per*. **Examples:** Days per project; the area divided by the width

Area of a Rectangle	Area = Length × Width	
	Example: The area of the region shown is $7 \times 4 = 28$ square inches.	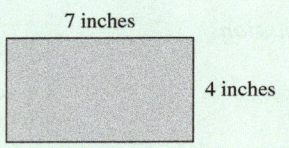

SECTION 1.4 ■ EXPONENTS, VARIABLES, AND ALGEBRAIC EXPRESSIONS

Exponential Notation — Used to represent repeated multiplication

Example: $3^5 = 3 \cdot 3 \cdot 3 \cdot 3 \cdot 3$

Powers of Ten — A natural number power on ten equals the number of 0s that follow a 1 in the standard form of the power of 10.

Example: $10^7 = 10,000,000$

Variable — A symbol or letter used to represent an unknown quantity

Example: W represents the weight of an animal.

Algebraic Expression — May contain numbers, variables, operation symbols, and grouping symbols

Example: $3x + 2$ and $4y$

Formula — An equation that expresses a relationship between two or more quantities

Example: The formula $Q = 4G$ gives the number of quarts Q in G gallons.

SECTION 1.5 ■ ROUNDING AND ESTIMATING; SQUARE ROOTS

Rounding — Approximating a number to a given level of accuracy

Example: To the nearest hundred, 4588 rounds to 4600.

Estimating — A rough calculation used to find a reasonably accurate answer

Example: When rounded to the nearest thousand, $3967 + 2019$ can be estimated by $4000 + 2000$.

Approximation — An approximation is the result when rounding or estimating. It is usually not exactly accurate. The symbol $\approx$ means "is approximately equal to."

Example: To the nearest ten, $87 + 72 \approx 90 + 70 = 160$.

Perfect Square — A perfect square results when a whole number is squared.

Example: 144 is a perfect square because $12^2 = 144$.

Square Root — The square root of a whole number is a number whose square is the given whole number. The symbol $\sqrt{}$ is called the *radical*, and the number under it is called the *radicand*.

Examples: $\sqrt{100} = 10$ because $10^2 = 100$, $\sqrt{49} = 7$ because $7^2 = 49$

SECTION 1.6 ■ ORDER OF OPERATIONS

Order of Operations
1. Do all calculations within grouping symbols, such as parentheses and radicals, or above and below a fraction bar.
2. Evaluate all exponential expressions.
3. Do all multiplication and division from *left to right*.
4. Do all addition and subtraction from *left to right*.

NOTE: If there is more than one arithmetic operation within grouping symbols, the order of Steps 2, 3, and 4 must be followed when performing Step 1. ■

Examples:
$$5 + 36 \div (4 - 2)^2 = 5 + 36 \div 2^2$$
$$= 5 + 36 \div 4$$
$$= 5 + 9$$
$$= 14$$

Evaluating Expressions	Replace each variable in the expression with its given value and then use the order of operations agreement to evaluate.
	Example: Evaluating $13 - 4x$ for $x = 2$ gives $13 - 4(2) = 13 - 8 = 5$.
Translating Words to Symbols	When translating words to symbols, watch for the words "the quantity" and pay special attention to the placement of any commas.
	Example: "four plus seven, times three" translates to $(4 + 7) \cdot 3$.

SECTION 1.7 ■ MORE WITH EQUATIONS AND PROBLEM SOLVING

Equations and Expressions	Equations always contain an equals sign ($=$), but expressions never contain an equals sign. Equations are often solved, and expressions are often simplified.
	Example: $2x - 5 = 9$ is an equation; $5x + 8$ is an expression.
Like Terms	Terms with the same variables raised to the same powers can be combined.
	Example: $7x^2$ and $3x^2$ are like terms; thus $7x^2 + 3x^2 = 10x^2$.
Simplifying Expressions	Use arithmetic properties and combine like terms.
	Example: $4x + 3 + 6x = 4x + 6x + 3 = 10x + 3$

CHAPTER 1   Review Exercises

SECTION 1.1

Exercises 1 and 2: For the given whole number, determine the place value of the digit 3.

1. 25,304
2. 365,719

Exercises 3 and 4: Name the digit with the given place value in the number 2,819,065,347.

3. ten-millions
4. hundred-thousands

Exercises 5 and 6: Write the number in words.

5. 48,309
6. 37

Exercises 7 and 8: Write the number in expanded form.

7. 673
8. 61,004

9. Write *fifty-eight thousand, three hundred forty-five* in standard form.

10. Graph the whole numbers 1 and 4 on a number line.

Exercises 11 and 12: Place the correct symbol, $<$ or $>$, in the blank between the whole numbers.

11. 28 _____ 0
12. 14 _____ 23

SECTION 1.2

Exercises 13–16: Add.

13. $21 + 14$
14. $176 + 949$
15. $378 + 5627$
16. $6952 + 4934$

Exercises 17–20: Subtract.

17. $863 - 97$
18. $2492 - 358$
19. $59{,}415 - 26{,}588$
20. $41{,}637 - 8{,}929$

Exercises 21 and 22: Translate the phrase into a mathematical expression and then find the result.

21. The difference between 83 and 21
22. 48 more than 103

Exercises 23 and 24: Solve the given equation by finding the unknown value.

23. $\square + 17 = 39$
24. $99 - \square = 88$

SECTION 1.3

Exercises 25 and 26: Use the distributive property to rewrite the expression. Do not compute the product.

25. $5(4 + 2)$
26. $7(8 - 5)$

Exercises 27–30: Multiply.

27. $0 \cdot 58$
28. 1×99
29. $43 \cdot 1852$
30. 516×712

Exercises 31–34: Divide, when possible.

31. $84 \div 4$ **32.** $37{,}721 \div 563$

33. $4239 \div 51$ **34.** $132 \div 0$

Exercises 35 and 36: Translate the phrase into a mathematical expression and then find the result.

35. The quotient of 66 and 11

36. 26 times 7

Exercises 37 and 38: Solve the given equation by finding the unknown value.

37. $15 \div \square = 3$ **38.** $\square \times 7 = 21$

SECTION 1.4

Exercises 39 and 40: Use exponential notation to write the repeated multiplication.

39. $8 \cdot 8 \cdot 8 \cdot 8 \cdot 8$ **40.** $9 \cdot 9 \cdot 9$

Exercises 41–44: Evaluate the exponential expression.

41. 7^2 **42.** 5^3

43. $4 \cdot 10^2$ **44.** $9 \cdot 10^5$

Exercises 45 and 46: Evaluate the algebraic expression for $x = 6$ and $y = 3$.

45. $3xy$ **46.** x^y

Exercises 47 and 48: Use the appropriate geometric formulas to find the measure.

47. The perimeter of a triangle with sides of length 7, 12, and 16 feet

48. The area of a rectangle with a 6-inch length and a 20-inch width

Exercises 49–54: Solve the equation.

49. $b + 9 = 19$ **50.** $12x = 36$

51. $33 = 3x$ **52.** $29 - z = 2$

53. $48 \div d = 8$ **54.** $n \div 4 = 7$

SECTION 1.5

55. Round 162 to the nearest ten.

56. Round 978,423 to the nearest ten-thousand.

Exercises 57 and 58: Round the whole number to its highest place value.

57. 52,809 **58.** 393,001

Exercises 59 and 60: Round each number to the nearest hundred to estimate the sum or difference.

59. $689 + 325 + 286$ **60.** $4739 - 3341$

Exercises 61 and 62: Compute the square root.

61. $\sqrt{256}$ **62.** $\sqrt{121}$

SECTION 1.6

Exercises 63–70: Evaluate the expression.

63. $34 - 24 \div 6 + 5$ **64.** $17 - 3 \cdot 5 + 29$

65. $55 \div 11 + 3 \cdot 6$ **66.** $2 \cdot 30 - 72 \div 9$

67. $\dfrac{(6 + 10) \cdot 4}{50 - 7 \cdot 6}$ **68.** $\dfrac{(8 - 2) \cdot 7}{\sqrt{9} - 5 - 1}$

69. $23 - (3 \cdot 5 - (20 \div 2^2) + 3)$

70. $\sqrt{25} - 6 + 4 \cdot 3^2 \div 6 + 8$

Exercises 71 and 72: Evaluate the algebraic expression for the given values of the variables.

71. $7y + 2x$, for $x = 5, y = 3$

72. $a - 8 + 3b$, for $a = 17, b = 4$

Exercises 73 and 74: Use symbols to write an expression and then evaluate it.

73. Nine minus the quantity two plus six

74. Four times three decreased by one

SECTION 1.7

Exercises 75 and 76: Identify each of the following as an equation or an expression.

75. $5x - 17$ **76.** $8 = 3x - 1$

Exercises 77 and 78: Determine whether the given terms are like or unlike.

77. $5xy, 2xy^2$ **78.** $a^2b, 7a^2b$

Exercises 79–82: Combine like terms in the expression. Write "not possible" if terms cannot be combined.

79. $7x + 11x$ **80.** $9b + b$

81. $22mn - 9mn$ **82.** $3x - 2y$

Exercises 83–86: Simplify the expression.

83. $3y + (y + 7)$ **84.** $8z + 2(z - 3)$

85. $7a + 4 + 8a$ **86.** $4y + 3 + y + 6$

Exercises 87 and 88: Translate the sentence into an equation using the variable x. Do not solve the equation. State what the variable represents.

87. Seven inches less than his height is 64.

88. Fourteen is 12 more than her score.

APPLICATIONS

Exercises 89 and 90: **Postage** *The following line graph shows the price of a first class postage stamp in cents for selected years.* (Source: United States Postal Service.)

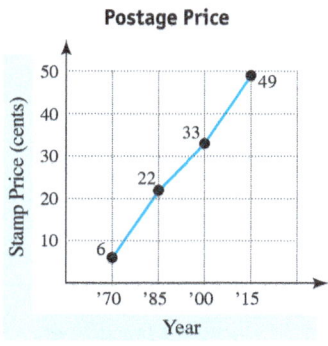

Postage Price

89. Comment on the general trend of the price of a stamp over this time period.

90. What was the price of a stamp in 1985?

91. **National Monuments** There are 11 fewer national monuments in Nebraska than in New Mexico. If New Mexico has 14 national monuments, how many are in Nebraska? (*Source:* National Parks Service.)

92. **Geometry** Find the area of the rectangle.

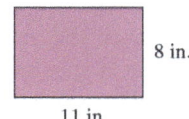

8 in.
11 in.

93. **Heart Rate** The average heart rate R in beats per minute (bpm) of an animal weighing W pounds can be approximated by
$$R = \frac{885\sqrt{W}}{W}.$$
Find the heart rate for a 225-pound bear.

94. **Heart Rate** During exercise, a physically fit female experiences a heart rate that is twice her resting heart rate. If this person's heart rate is 136 beats per minute during exercise, what is her resting heart rate?

95. **Geometry** Find the perimeter of the figure.

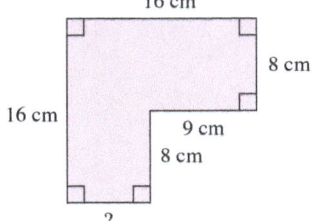

96. **Buying Books** What is the maximum number of books costing $4 that a person can buy with $25? How much change will the person receive?

97. **Yard Sale** A person bought 13 items at a yard sale and received $1 in change. If each item was priced at $3, how much did the person give to the cashier?

98. **Gardening** A garden is being built so that it has the shape of a square with an area of 100 square feet. Determine the length of one side of the garden.

99. **Converting Temperature** To convert a temperature F given in degrees Fahrenheit to an equivalent temperature C in degrees Celsius, use the formula
$$C = \frac{5(F - 32)}{9}.$$
Find the Celsius temperature that is equivalent to a temperature of 77 °F.

100. **A Number Puzzle** If a number is doubled and then added to itself, the result is 10 more than the number. Find the number.

Exercises 101–104: **Music Shipments** *The following line graph shows total shipments in billions of dollars for music in physical form (not downloaded or streamed) during selected years.* (Source: RCIA.)

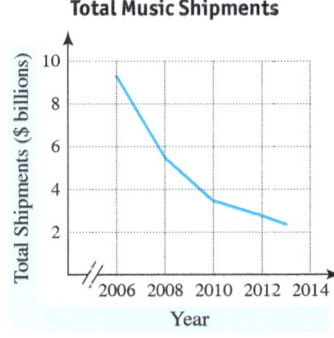

Total Music Shipments

101. Estimate the 2012 value of total shipments in physical form.

102. Estimate the year when the value of total shipments in physical form first fell to $5.5 billion.

103. Estimate the year when the value of total shipments in physical form first fell to $2.4 billion.

104. Estimate the 2010 value of total shipments in physical form.

105. Walgreens Locations There were 110 more Walgreens pharmacy locations in 2014 than there were a year earlier. If there were 8206 Walgreens pharmacies in May 2014, how many were there a year earlier? (*Source:* Walgreens.)

106. Estimating Time An athlete stretches for 13 minutes, jogs for 48 minutes, and then walks for 19 minutes. Estimate the total time for this workout by rounding to the nearest ten.

CHAPTER 1 Test

1. Determine the place value of the digit 5 in the whole number 158,902.

2. Write 7341 in expanded form.

3. Round 78,423 to the nearest thousand.

4. Place the symbol, $<$ or $>$, in the blank: 71____17.

Exercises 5–8: Perform the arithmetic.

5. $3472 + 869$

6. $15{,}902 - 9876$

7. $27 \cdot 4817$

8. $34{,}476 \div 67$

9. Write $3 \cdot 3 \cdot 3 \cdot 3$ using exponential notation.

10. Evaluate $5^2 \cdot 2^3$.

Exercises 11 and 12: Solve the equation.

11. $47 - z = 34$

12. $8x = 48$

Exercises 13–16: Evaluate the expression.

13. $18 - 20 \div 5 + 5$

14. $2 \cdot 10 - 60 \div 4$

15. $\dfrac{(5 + 3) \cdot 3}{20 - 3 \cdot 6}$

16. $29 - (3 \cdot 7 - (4^2 \div 2) + 5)$

17. Evaluate $b - 5 + 6a$ for $a = 2$, $b = 9$.

18. Compute $\sqrt{81}$.

Exercises 19 and 20: Combine like terms in the expression. Write "not possible" if terms cannot be combined.

19. $12p^2 - 7p^2$

20. $4w + 4y$

21. Name the property of multiplication illustrated by the equation $43 \cdot 0 = 0$.

22. Graph the whole numbers 2 and 5 on a number line.

Exercises 23 and 24: Simplify the algebraic expression.

23. $3x + (5x + 2)$

24. $3y + 9 + y + 4$

25. **Geometry** Find the perimeter of the figure.

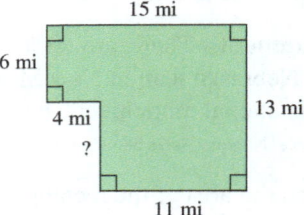

26. **Buying Shirts** What is the maximum number of shirts costing $12 that a person can buy with $80? How much change will the person receive?

27. **College Tuition** A student paid $407 more for tuition this semester than he did last semester. If his tuition bill is $1675 this semester, what was the bill last semester?

28. **Burning Calories** A 180-pound person burns 14 calories each minute while playing handball. Write a formula that gives the total number of calories C burned while playing handball for h minutes. (*Source:* The Calorie Control Council.)

2 Integers

2.1 Integers and the Number Line
2.2 Adding Integers
2.3 Subtracting Integers
2.4 Multiplying and Dividing Integers
2.5 Order of Operations; Averages
2.6 Solving Equations That Have Integer Solutions

The Republic of Maldives, an island nation in the Indian Ocean about 435 miles southwest of Sri Lanka, is considered the "flattest nation on Earth," with a *maximum* natural elevation of less than 8 feet above sea level. Scientists believe that a changing global climate could raise world sea levels. Even a modest sea level rise would be devastating to many coastal locations, including the Maldives.

The following table lists elevations for two locations that are below sea level and two locations that are above sea level.

Elevations for Selected World Locations

Location	Elevation
Amsterdam	13 feet below sea level
Mount Kilimanjaro	19,340 feet above sea level
Death Valley	282 feet below sea level
The Maldives	7 feet above sea level

Source: The World Atlas.

In this chapter, we discuss numbers called *integers*, which can be used to describe elevations above sea level (positive elevation) or below sea level (negative elevation).

2.1 Integers and the Number Line

Objectives

1. Introducing Signed Numbers
   • Finding Opposites
2. Working with Integers and Their Graphs
3. Comparing Integers
4. Finding Absolute Value
5. Solving Applications Involving Integers

NEW VOCABULARY

☐ Positive number
☐ Negative number
☐ Signed numbers
☐ Opposite
☐ Integers
☐ Origin
☐ Absolute value

READING CHECK 1

• What is a positive number?
• What is a negative number?

1 Introducing Signed Numbers

Connecting Concepts with Your Life The following See the Concept illustrates how the need for numbers that are less than zero is a part of everyday life.

See the Concept 1   VALUES LESS THAN ZERO

There are times when numbers that are less than zero are needed.

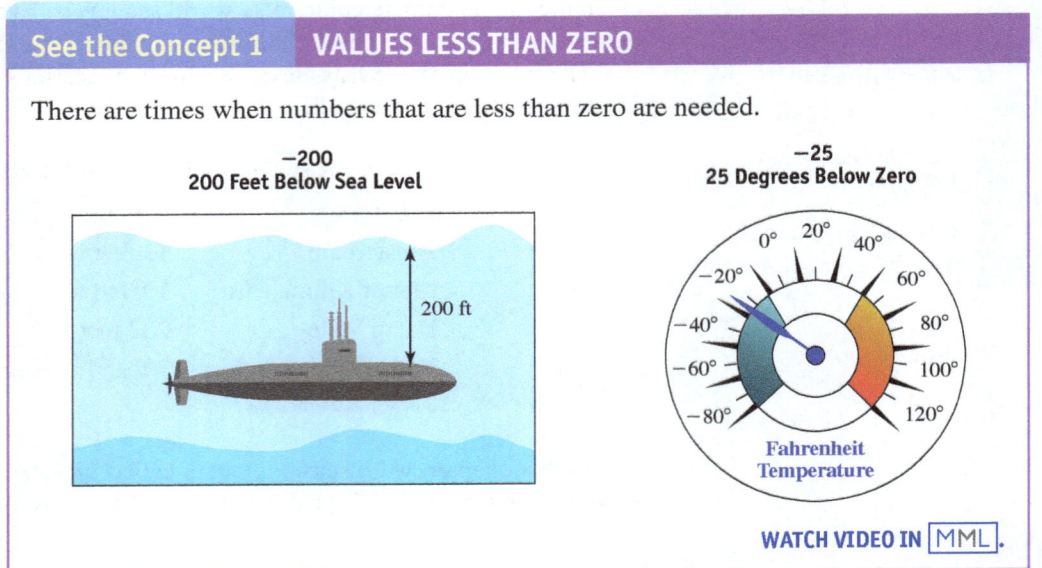

WATCH VIDEO IN MML.

All whole numbers other than zero (the natural numbers) are positive numbers. A **positive number** is a number that is greater than zero. Rather than writing the positive whole numbers as

$$+1, +2, +3, +4, \ldots,$$

we usually omit the positive sign $(+)$ and simply write them as follows.

$$1, 2, 3, 4, \ldots \quad \text{Positive numbers}$$

For every *positive* number, there is a corresponding *negative* number called its *opposite*. A **negative number** is a number that is less than zero. We indicate that a number is negative by placing a negative sign $(-)$ immediately in front of the number.

$$\ldots -4, -3, -2, -1 \quad \text{Negative numbers}$$

Together, positive numbers, negative numbers, and zero are the **signed numbers**.

$$\ldots -4, -3, -2, -1, 0, 1, 2, 3, 4 \ldots \quad \text{Signed numbers}$$

EXAMPLE 1   Using signed numbers in elevators

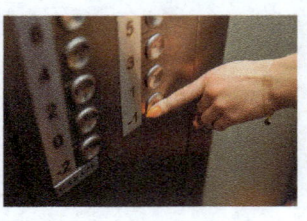

There are buildings that have elevators with buttons labeled with both positive and negative numbers. Refer to the photo of elevator buttons to the left and answer the following.
(a) What might the button labeled -1 represent?
(b) In this building, how many floors do you travel to get from Floor 3 to Floor -1?

Solution
(a) It most likely represents a floor that is one level below ground, such as a parking garage.
(b) Each arrow represents a floor: $3 \rightarrow 2 \rightarrow 1 \rightarrow 0 \rightarrow -1$. The elevator travels 4 floors.

Now Try Exercises 71–74

CALCULATOR HELP

To enter a negative number in a calculator, see Appendix E (page AP-29).

FINDING OPPOSITES Every number has an opposite. For example, the opposite of 3 is -3, and the opposite of 12 is -12. If we let a represent any number, then its **opposite** (also known as the *additive inverse*) is represented by $-a$. See **TABLE 2.1**.

NOTE: Zero is neither positive nor negative, so the opposite of 0 is 0. ∎

Opposites of Signed Numbers

Number	Opposite	Opposite of Opposite
4	-4	$-(-4) = 4$
-7	7	$-(7) = -7$
5	-5	$-(-5) = 5$

The opposite of the opposite is the original number (1st column).

TABLE 2.1

The third column of **TABLE 2.1** suggests the following double negative rule.

DOUBLE NEGATIVE RULE

Let a be any number. Then $-(-a) = a$.

EXAMPLE 2 **Finding opposites (or additive inverses)**

Simplify each of the following.
(a) $-(5)$ (b) $-(-(-14))$ (c) $-(-9)$

Solution
(a) The opposite of 5 is -5.
(b) By the double negative rule, $-(-14) = 14$. So, $-(-(-14)) = -(14) = -14$.
(c) By the double negative rule, $-(-9) = 9$.

Now Try Exercises 19, 21, 27

NOTE: To find the opposite of an exponential expression, evaluate the exponent first. For example, the opposite of 5^2 is ∎

$$-(5^2) = -(5 \cdot 5) = -(25) = -25.$$

MAKING CONNECTIONS 1

Plus, Minus, Positive, and Negative

STUDY TIP

If you have tried to solve a problem but need help, be sure to ask a question in class.

Be sure that you know when the symbols $+$ and $-$ indicate addition and subtraction, and when they indicate that a number is positive or negative.

$4 + 9$	Four plus nine	$+7$	Positive seven
$13 - 8$	Thirteen minus eight	-12	Negative twelve
$-(-6)$	The opposite of negative six		

2 Working with Integers and Their Graphs

The **integers** are a set of numbers that includes the natural numbers, zero, and the opposites of the natural numbers.

Integers

$$\ldots -3, -2, -1, 0, 1, 2, 3, \ldots$$

To graph a negative integer, we must extend (to the left) the number line used for graphing whole numbers so that it can be used for numbers less than 0. The number line in **FIGURE 2.1** can be used to graph integers. The number 0 is called the **origin** and is neither positive nor negative.

READING CHECK 2

- Where are negative numbers found on the number line?

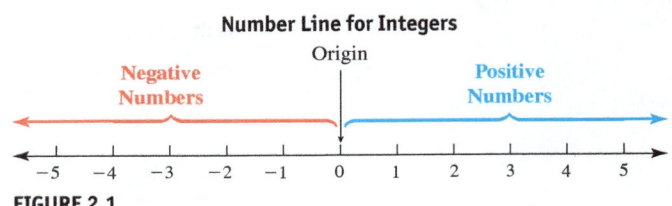

FIGURE 2.1

Just as with whole numbers, the graph of an integer includes a dot placed on the number line at the number's position.

EXAMPLE 3 **Graphing integers**

Graph the integers $-3, 0,$ and 4 on the same number line.

Solution
The integers $-3, 0,$ and 4 are graphed as shown in **FIGURE 2.2**.

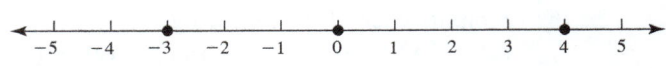

FIGURE 2.2

Now Try Exercise 31

3 Comparing Integers

READING CHECK 3

- On a number line, how do we know if a number is greater than another number?

Recall that when two whole numbers are graphed on the same number line, the number to the left is *less than* the number to the right and the number to the right is *greater than* the number to the left. This method of comparison also holds for integers. As a result, a negative integer is *always less than* a positive integer and a positive integer is *always greater than* a negative integer. See **FIGURE 2.3**.

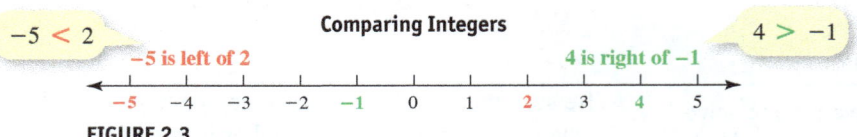

FIGURE 2.3

Connecting Concepts with Your Life When comparing two negative integers, it is helpful to think about temperature. On a cold day in Idaho, the temperature might be -8 degrees Fahrenheit. If the temperature later dips to -13 degrees Fahrenheit, we would say that it got colder. In other words, -13 °F is colder than (less than) -8 °F, and -8 °F is warmer than (greater than) -13 °F. In math symbols, we write

$$-13 < -8 \quad \text{or} \quad -8 > -13.$$

Note that -13 is located to the **left** of -8 on the number line and -8 is located to the **right** of -13 on the number line.

EXAMPLE 4   Comparing two integers

Place the correct symbol, $<$ or $>$, in the blank between each pair of integers.
(a) 5 ____ -9 (b) -3 ____ -12 (c) -7 ____ -6

Solution
(a) Because 5 is located to the **right** of -9 on the number line, $5 > -9$.
(b) Because -3 is located to the **right** of -12 on the number line, $-3 > -12$.
(c) Because -7 is located to the **left** of -6 on the number line, $-7 < -6$.

Now Try Exercises 37, 39, 43

CALCULATOR HELP
To find an absolute value with a calculator, see Appendix E (page AP-29).

4 Finding Absolute Value

The **absolute value** of an integer equals its distance on the number line from 0 (the origin). Because distance is never negative, the absolute value of an integer is *never negative*. If the variable a represents an integer, the absolute value of a is written as $|a|$ and reads as "the absolute value of a." **FIGURE 2.4** shows that $|-3| = 3$ and $|3| = 3$ because both -3 and 3 are located a distance of 3 units from the origin on the number line.

READING CHECK 4

• What is the absolute value of an integer?

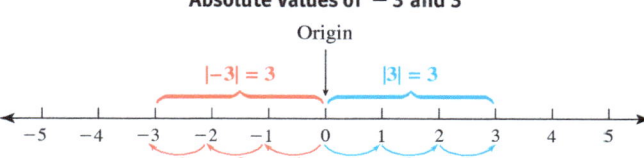

FIGURE 2.4

EXAMPLE 5   Finding absolute value

Evaluate each absolute value.
(a) $|-7|$ (b) $|5|$ (c) $|0|$

Solution
(a) Because -7 is **7** units from the origin, $|-7| = $ **7**.
(b) The integer 5 is **5** units from the origin, so $|5| = $ **5**.
(c) Because 0 is **0** units from the origin, $|0| = $ **0**.

Now Try Exercises 49, 51, 53

MAKING CONNECTIONS 2

Absolute Value and Opposites

Finding the absolute value of an integer is *not* the same as finding its opposite. The following table shows how the absolute values of some integers compare to their opposites.

READING CHECK 5

• Explain why the absolute value of a number may not be the opposite of the number.

Integer	Absolute Value	Opposite
4	4	-4
-2	2	2
0	0	0

The opposite of a positive number is negative.

Absolute value is *never* negative.

NOTE: The vertical lines used to show absolute value are grouping symbols. When evaluating expressions such as $-|21|$ or $-|-16|$, the absolute value should be evaluated first before finding the opposite.

$$-|21| = -(21) = -21 \quad \text{and} \quad -|-16| = -(16) = -16 \quad \blacksquare$$

- Absolute value of 21 is 21.
- Absolute value of −16 is 16.
- Opposite of 21 is −21.
- Opposite of 16 is −16.

EXAMPLE 6 Comparing expressions involving absolute value

Place the correct symbol, $<$, $>$, or $=$, in each blank between the expressions.
(a) $|-7|$ _____ -7 (b) $-|-5|$ _____ $-|5|$ (c) $-|3|$ _____ $|-3|$

Solution
(a) Since $|-7| = 7$ and 7 is **right** of -7 on the number line, $|-7| > -7$.
(b) For the expression on the left, $-|-5| = -(5) = -5$, and for the expression on the right, $-|5| = -(5) = -5$. The expressions are **equal**, so $-|-5| = -|5|$.
(c) Evaluating each expression gives $-|3| = -3$ and $|-3| = 3$. Because -3 is **left** of 3 on the number line, $-|3| < |-3|$.

Now Try Exercises 63, 65, 67

5 Solving Applications Involving Integers

🌐 **Math in Population Context** By analyzing projected (predicted) population growth or decline data, government officials can plan for possible changes in demands on social programs, roadways, and vital resources such as water and electricity. The next example illustrates how both positive and negative integers are used in projecting population changes.

EXAMPLE 7 Analyzing projected population change

TABLE 2.2 lists the projected population change from 2000 to 2030 for selected states.

Projected Population Change: 2000–2030

State	Arkansas	North Dakota	Utah	West Virginia
Change	567,000	−36,000	1,252,000	−88,000

TABLE 2.2 *Source:* U.S. Census Bureau.

(a) Which states have a projected decline in population?
(b) Which state has the largest projected growth in population?

Solution
(a) A decline in population is represented by a negative number. The states with negative population change are North Dakota and West Virginia.
(b) Population growth is represented by a positive number. The largest positive number in the table is 1,252,000, which is the projected population change for Utah.

Now Try Exercise 77

The next example shows a bar graph of positive and negative temperatures.

EXAMPLE 8   Reading a bar graph involving integers

International Falls is a small community in northern Minnesota located on the Canadian border. It is often called "the nation's ice box." **FIGURE 2.5** shows a bar graph of record low temperatures in International Falls by month. (*Source:* NOAA.)

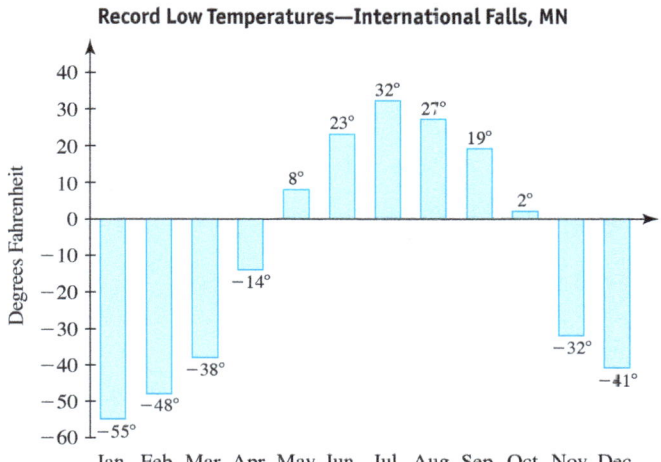

FIGURE 2.5

(a) What is the record low temperature for April?
(b) Which month has the warmest record low temperature?
(c) Does March have a colder or warmer record low when compared to November?

Solution
(a) The bar for April extends below the horizontal line representing 0 °F, so its record low temperature is negative. From the bar graph, the record low for April is −14 °F.
(b) The tallest bar reaches above the horizontal line representing 0 °F and shows a record low temperature of 32 °F in July.
(c) The record low temperature for March is −38 °F, as compared to −32 °F for November. The March temperature is colder.

Now Try Exercise 83

The next example demonstrates how integers can be estimated from a line graph.

EXAMPLE 9   Estimating integer values from a line graph

It is often colder at the top of a mountain than it is at sea level because air temperature decreases as altitude increases. The line graph in **FIGURE 2.6** shows the air temperature in degrees Fahrenheit at various altitudes.
(a) Estimate the air temperature at an altitude of 25,000 feet.
(b) Estimate the altitude where the air temperature is 0 °F.

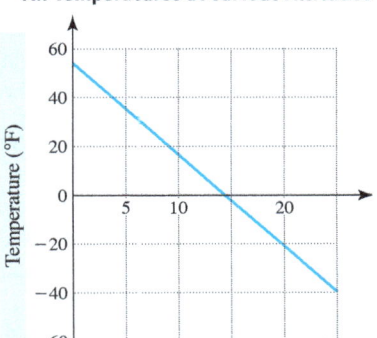

FIGURE 2.6

Solution

(a) To estimate the air temperature at **25** thousand feet, locate **25** on the horizontal scale. Move **vertically** downward to the graphed line. From this position, move **horizontally** to the left to the temperature values. See **FIGURE 2.7**. At an altitude of 25,000 feet, the air temperature is about −40 °F.

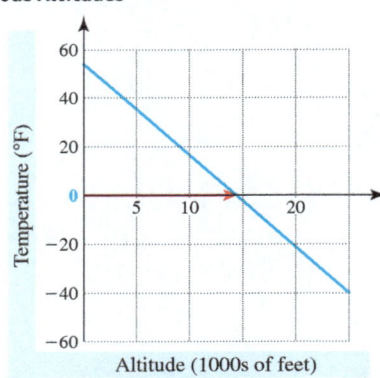

FIGURE 2.7   **FIGURE 2.8**

(b) To estimate the altitude where the air temperature is **0** °F, find **0** °F at the left edge of the graph. Move **horizontally** to the right to the graphed line. There is no need to move vertically upward or downward. See **FIGURE 2.8**. Because altitude is displayed in thousands of feet, the air temperature is 0 °F at an altitude of about 14,000 feet.

Now Try Exercises 85, 87

2.1 Putting It All Together

CONCEPT	COMMENTS	EXAMPLES
Signed Numbers	A positive number is greater than 0. A negative number is less than 0. Zero is neither positive nor negative.	3, +14, 137, and +900 −1, −28, −271, and −1170 0
Opposite (Additive Inverse)	The opposite (additive inverse) of a number a is written as $-a$.	The opposite of 7 is −7. The opposite of −5 is $-(-5) = 5$.
Integers	The integers include the natural numbers, zero, and the opposites of the natural numbers.	… −3, −2, −1, 0, 1, 2, 3 …
Absolute Value	The absolute value of an integer equals its distance on the number line from 0 (the origin). The absolute value of a is written as $\lvert a \rvert$.	$\lvert 9 \rvert = 9$ $\lvert -12 \rvert = 12$ $\lvert 0 \rvert = 0$

2.1 Exercises

CONCEPTS AND VOCABULARY

1. A(n) _____ number is greater than zero.

2. A(n) _____ number is less than zero.

3. If a represents any number, then $-a$ represents the _____ (or additive inverse) of a.

4. The opposite of 0 is _____.

5. If a represents any number, then $-(-a) =$ _____.

6. The _____ include the natural numbers, zero, and the opposites of the natural numbers.

7. On the number line, 0 is called the _____.

8. The _____ of a number equals its distance on the number line from 0.

9. Express the temperature "3 below zero" as an integer.

10. Express an elevation of "17 feet above sea level" as an integer.

FINDING OPPOSITES AND USING THE DOUBLE NEGATIVE RULE

Exercises 11–18: Find the opposite of the given integer.

11. 7
12. 13
13. −43
14. −21
15. −237
16. 452
17. 93,000
18. −3967

Exercises 19–30: Simplify the expression.

19. $-(8)$
20. $-(11)$
21. $-(-26)$
22. $-(-13)$
23. $-(0)$
24. $-(-0)$
25. $-(-(23))$
26. $-(-(39))$
27. $-(-(-5))$
28. $-(-(-9))$
29. $-(-(-(-1)))$
30. $-(-(-(-6)))$

GRAPHING INTEGERS

Exercises 31–36: Graph the integers on a number line.

31. −4, −2, 3
32. −2, 0, 4
33. −16, −8, 12
34. −20, 10, 25
35. −87, 5, 76
36. −92, −63, −12

COMPARING INTEGERS

Exercises 37–48: Place the correct symbol, $<$ or $>$, in the blank between the integers.

37. 4 _____ −7
38. −2 _____ 9
39. −8 _____ −12
40. −17 _____ −1
41. 43 _____ 206
42. 99 _____ 34
43. −34 _____ −29
44. −63 _____ −36
45. 0 _____ −293
46. −349 _____ 0
47. 0 _____ 167
48. 682 _____ 0

FINDING ABSOLUTE VALUE

Exercises 49–56: Evaluate the absolute value.

49. $|10|$
50. $|-8|$
51. $|0|$
52. $|-0|$
53. $|-18|$
54. $|45|$
55. $|-87|$
56. $|-53|$

Exercises 57–62: Simplify the absolute value expression.

57. $-|2|$
58. $-|-3|$
59. $-|-19|$
60. $-|12|$
61. $-|0|$
62. $-|-0|$

Exercises 63–70: Place the correct symbol, $<$, $>$ or $=$, in the blank between the expressions.

63. 2 _____ $-|2|$
64. $-|3|$ _____ -8
65. $|-12|$ _____ $-|12|$
66. $|-3|$ _____ $-|-8|$
67. $-|-29|$ _____ $-|29|$
68. $-|10|$ _____ $|-10|$
69. $-|25|$ _____ 25
70. $-|-46|$ _____ $|-46|$

APPLICATIONS INVOLVING INTEGERS

Exercises 71–74: **Elevation** *Refer to the following table. Express the elevations of the given locations as positive or negative integers.*

Elevations for Selected World Locations

Location	Elevation
Amsterdam	13 feet below sea level
Mount Kilimanjaro	19,340 feet above sea level
Death Valley	282 feet below sea level
The Maldives	7 feet above sea level

Source: The World Atlas.

71. Death Valley
72. The Maldives
73. Mount Kilimanjaro
74. Amsterdam

Exercises 75 and 76: Refer to the photo next to Example 1 in this section.

75. What might the positive numbers in this elevator represent?

76. How many floors do you travel to get from Floor -1 to Floor 1?

77. **Population** The following table lists the population change from 1990 to 2010 for selected countries.

Country	Latvia	Malta	Romania	Tonga
Change	−446,000	48,000	−685,000	31,000

Source: U.S. Census Bureau.

(a) Which of these countries had the largest decline in population from 1990 to 2010?
(b) List the countries that had a growth in population from 1990 to 2010.

78. **High School Enrollment** The following table lists the change in public high school enrollment from 2005 to 2015 for selected states.

State	Alaska	Florida	Nevada	Ohio
Change	−3800	16,000	17,000	−64,000

Source: U.S. National Center for Educational Statistics.

(a) Which of these states had the largest increase in enrollment from 2005 to 2015?
(b) List the states that had a decrease in enrollment from 2005 to 2015.

Finances *Exercises 79–82: Even though* $-\$84$ *is a negative value (a debt), it is a larger financial quantity than a positive balance of* $\$52$ *in a checking account (an asset). The debt is larger than the asset because* $|-84| > |52|$. *Use absolute value to determine which of each pair of financial quantities is larger.*

79. An employee paycheck: $1050
    A credit card debt: −$1745

80. Owing the baby sitter: −$32
    Cash in your wallet: $44

81. A friend owes you: $160
    Club membership dues: −$200

82. Savings account balance: $12,900
    Tuition and fees due: −$7800

83. **Ocean Depth** The following bar graph shows the maximum depth of each ocean. (*Source:* The World Atlas.)

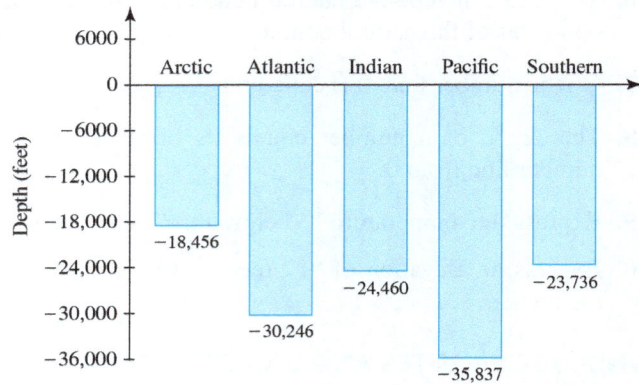

(a) Which ocean is the deepest?
(b) Which ocean is the least deep?
(c) Which ocean is 23,736 feet deep?
(d) Which is deeper, the Indian Ocean or the Southern Ocean?

84. **Cold Temperatures** Refer to the following figure.

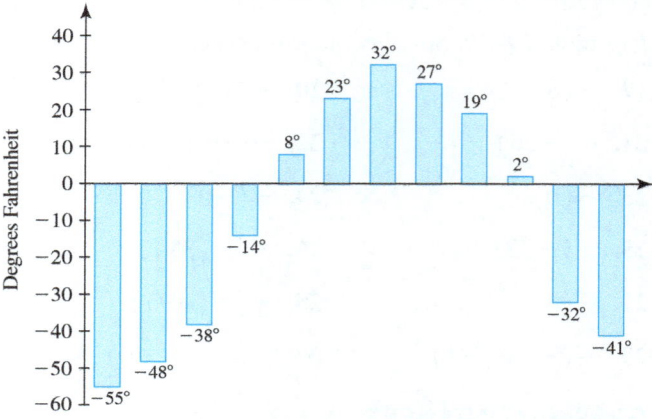

(a) Which month's record low temperature has the largest absolute value?
(b) Which month's record low temperature has the smallest absolute value?

Exercises 85–88: **Music Videos** *The following line graph shows the profit made from selling music videos. A negative profit represents a loss.*

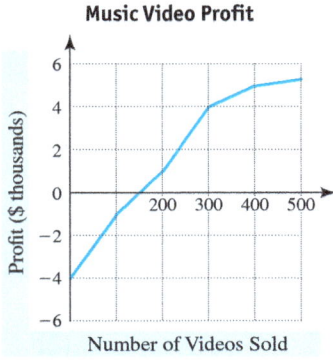

85. Estimate the profit if no videos are sold.

86. Estimate the number of videos that must be sold to make a profit of $1000.

87. Estimate the number of videos that must be sold to make a profit of $5000.

88. Estimate the profit when 150 videos are sold.

WRITING ABOUT MATHEMATICS

89. Explain how absolute value is computed. Give two examples.

90. Sketch a number line and label the locations of the negative numbers, origin, and positive numbers.

91. Explain why a positive number is always greater than a negative number.

92. Explain how the absolute value of a number compares to the absolute value of its opposite.

2.2 Adding Integers

Objectives

1. Adding Integers
   - Adding Integers That Have Like Signs
   - Adding Integers That Have Unlike Signs
2. Recognizing Addition Properties
3. Adding Integers Visually
   - Adding Integers Using a Number Line
   - Adding Integers Using Symbols
4. Solving Applications Involving Addition of Integers

NEW VOCABULARY

☐ Additive inverse
☐ Inverse property for addition

STUDY TIP

Regular and timely practice is one of the keys to having a successful experience in any math class. Learn math by doing math.

1 Adding Integers

Connecting Concepts with Your Life On some football plays, a team may advance the ball toward the opposing team's goal, for a *gain* in yardage. On other plays, the ball may end up farther from the opposing team's goal for a *loss* in yardage. A yardage gain is recorded as a positive number, and a yardage loss is recorded as a negative number. A team's total yardage is found by adding the gains and losses, or more simply, adding positive and negative numbers. In this section, we learn how to add integers.

ADDING INTEGERS THAT HAVE LIKE SIGNS We already know how to add two positive integers from our study of whole numbers.

Math in Context (Football) To understand what it means to add two negative numbers, consider the total yardage of a football team that loses 4 yards on one play and then loses 2 yards on the next play. For the two plays, the team has lost a total of 6 yards. This can be written mathematically as

$$-4 + (-2) = -6.$$

To find the sum of two **negative** numbers, we can *add the absolute values of the numbers* and then keep the **negative** sign in the sum. For example, since $|-4| + |-2| = 4 + 2 = 6$, we can compute the related sum as

$$-4 + (-2) = -6. \quad \text{negative + negative = negative}$$

Because the absolute values of positive numbers are positive, a similar rule can be written for adding two positive numbers. To find the sum of two **positive** numbers, we *add the absolute values of the numbers* and then keep the **positive** sign in the sum.

$$4 + 2 = 6 \quad \text{positive + positive = positive}$$

READING CHECK 1

- Is the sum of two negative numbers a positive or negative number?

ADDING INTEGERS WITH LIKE SIGNS

To add two integers with like signs,

STEP 1: Find the sum of the absolute values of the integers.

STEP 2: Keep the common sign of the two integers as the sign of the sum.

EXAMPLE 1 Finding the sum of two integers with like signs

Find each sum.
(a) $-7 + (-11)$ **(b)** $38 + 9$ **(c)** $-15 + (-26)$

Solution
(a) STEP 1: Start by finding the sum of the absolute values.
$$|-7| + |-11| = 7 + 11 = 18$$
STEP 2: Because the common sign is **negative**, the sum is $-7 + (-11) = -18$.

(b) From our work with whole numbers, we know that $38 + 9 = 47$. However, the rules for adding integers with like signs can be applied to get the same result.
STEP 1: $|38| + |9| = 38 + 9 = 47$
STEP 2: Because the common sign is **positive**, the sum is $38 + 9 = 47$.

(c) Because $|-15| + |-26| = 15 + 26 = 41$ (**STEP 1**), and the common sign is **negative** (**STEP 2**), the sum is $-15 + (-26) = -41$.

Now Try Exercises 11, 13, 15

ADDING INTEGERS THAT HAVE UNLIKE SIGNS To gain an understanding of how to find the sum of two integers with unlike signs, consider the following examples.

🌐 Math in Money Context We can think of both money that we are given (income) and money that we already have (asset) as positive amounts. We can think of money that we owe (a debt) as a negative amount.

A **debt** that is larger than **income** results in a remaining debt.

We owe $6.   We are given $4.
$$-6 + 4 = -2$$ We still owe $2.

A **debt** that is smaller than **income** results in a remaining asset.

We are given $8.   We owe $3.
$$8 + (-3) = 5$$ We still have $5.

READING CHECK 2

- How do we add integers with unlike signs?

ADDING INTEGERS WITH UNLIKE SIGNS

To add two integers with unlike signs,

STEP 1: Find the absolute values of the integers.

STEP 2: Subtract the smaller absolute value from the larger absolute value.

STEP 3: Keep the sign of the integer with the larger absolute value as the sign of the sum.

NOTE: If the absolute values (**STEP 1**) are equal, the sum is 0. For example, $-8 + 8 = 0$. ∎

EXAMPLE 2 Finding the sum of two integers with unlike signs

Find each sum.
(a) $-12 + 4$ (b) $14 + (-9)$

Solution
(a) **STEP 1:** Find the absolute values of -12 and 4.
$$|-12| = 12 \quad \text{and} \quad |4| = 4$$
STEP 2: Subtract the smaller absolute value, 4, from the larger absolute value, 12.
$$12 - 4 = 8$$
STEP 3: Because $|-12| > |4|$, keep the **negative** sign as the sign for the sum.
$$-12 + 4 = -8$$
For example, if you owe \$12 and are given \$4, you still owe \$8.

(b) **STEP 1:** The absolute values are $|14| = 14$ and $|-9| = 9$.
STEP 2: Subtract $14 - 9 = 5$.
STEP 3: Because $|14| > |-9|$, keep the **positive** sign in the sum, $14 + (-9) = 5$.
For example, if you have \$14 and owe \$9, you still have \$5.

Now Try Exercises 17, 19

CALCULATOR HELP
To add integers with a calculator, see Appendix E (page AP-29).

2 Recognizing Addition Properties

In the previous section, we learned that the opposite of an integer is also called the *additive inverse*. More formally, we say that two numbers are **additive inverses** of each other if their sum is 0. For example, -7 is the additive inverse of 7 because $-7 + 7 = 0$, and 7 is the additive inverse of -7 because $7 + (-7) = 0$. The **inverse property for addition** states that the sum of a number and its additive inverse (opposite) is always 0.

Because the addition properties for whole numbers studied earlier also apply to integers, we can create a complete list of addition properties for integers, summarized as follows.

ADDITION PROPERTIES FOR INTEGERS

Let the variables a, b, and c represent integers.

Commutative Property:	$a + b = b + a$
Associative Property:	$(a + b) + c = a + (b + c)$
Identity Property:	$a + 0 = a$ and $0 + a = a$
Inverse Property:	$a + (-a) = 0$ and $-a + a = 0$

EXAMPLE 3 Identifying addition properties

State the addition property illustrated by each equation.
(a) $-99 + 0 = -99$ (b) $4 + (-3) = -3 + 4$
(c) $-65 + 65 = 0$ (d) $(-2 + 4) + 1 = -2 + (4 + 1)$

Solution
(a) The sum of an integer and 0 is that integer. This illustrates the identity property.
(b) The commutative property is illustrated by this equation. Changing the order of the addends does not affect the sum.

(c) The equation $-65 + 65 = 0$ illustrates the inverse property. Adding an integer and its opposite results in 0.

(d) The way in which three or more addends are grouped does not affect the sum. The equation $(-2 + 4) + 1 = -2 + (4 + 1)$ illustrates the associative property.

Now Try Exercises 37, 39, 41, 43

3 Adding Integers Visually

In this subsection, we explore two visual methods that can be used to add integers.

ADDING INTEGERS USING A NUMBER LINE On a number line, we can represent a positive integer with any arrow pointing to the right and a negative integer with any arrow pointing to the left as long as the length of the arrow is equal to the absolute value of the integer. For example, the positive integer 3 can be represented with any arrow that is 3 units long and points to the right. Similarly, the negative integer -4 can be represented with any arrow that is 4 units long and points to the left. See **FIGURE 2.9**.

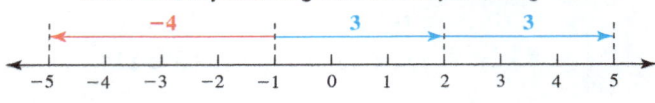

Two Arrows Representing 3 and One Representing -4

FIGURE 2.9

NOTE: The position of an arrow above the number line is not important. Only the length and direction of an arrow are needed to represent an integer. ∎

READING CHECK 3

- Which integers are represented on a number line by arrows pointing to the left?

The following See the Concept illustrates how directed arrows can be used with a number line to add integers.

See the Concept 1 **ADDING ON A NUMBER LINE**

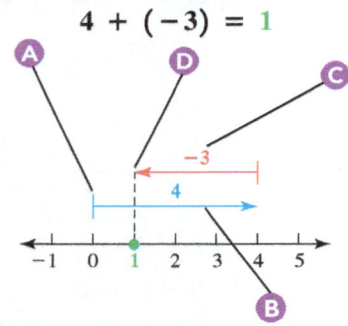

Ⓐ Starting at the origin, complete step B for the first addend.

Ⓑ Draw an arrow with a length equal to the absolute value of this addend. If the addend is **positive**, draw the arrow to the **right**; if it is **negative**, draw the arrow to the **left**.

Ⓒ Starting at the end of the first arrow, repeat step B for the second addend.

Ⓓ The **sum** is located at the end of the second arrow.

WATCH VIDEO IN MML.

EXAMPLE 4 **Adding integers using a number line**

Use a number line to find each sum.
(a) $-2 + 5$ (b) $-1 + (-3)$ (c) $4 + (-4)$

Solution

(a) To add $-2 + 5$, start at 0 and draw an arrow representing -2. From the tip of this arrow, draw a second arrow representing 5. The sum is **3**, as shown in **FIGURE 2.10**.

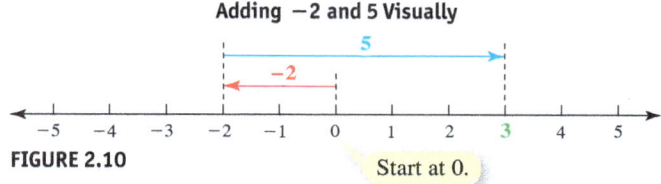

FIGURE 2.10

(b) To add $-1 + (-3)$, start at 0 and draw an arrow representing -1. From the tip of this arrow, draw a second arrow representing -3. The sum is -4, as shown in **FIGURE 2.11**.

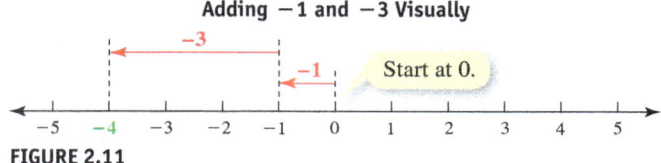

FIGURE 2.11

(c) The sum $4 + (-4)$ is equal to **0**, as shown in **FIGURE 2.12**.

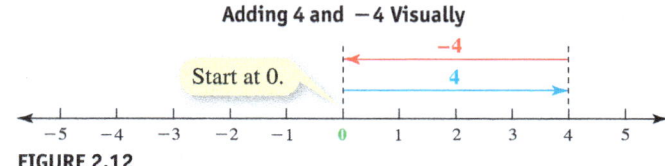

FIGURE 2.12

Now Try Exercises 55, 57, 61

READING CHECK 4

- What number is represented when a positive unit and a negative unit are written together?

ADDING INTEGERS USING SYMBOLS If we use the symbol ⌒ to represent a positive unit and the symbol ⌣ to represent a negative unit, then adding opposites results visually in "zero" as shown.

This is a visual representation of the equation $1 + (-1) = 0$. By combining positive units with negative units, it becomes possible to *see* the sum when two integers are added. The next example illustrates this process.

EXAMPLE 5 Adding integers using symbols

Perform each addition visually, using the symbols ⌒ and ⌣.
(a) $2 + 4$ (b) $-5 + 2$ (c) $7 + (-3)$ (d) $-2 + (-3)$

Solution
(a) Draw two positive units and then draw four more positive units.

Because no "zeros" could be formed, the sum is six positive units, or **6**.

(b) Start by drawing five negative units and then draw two positive units. Remember to form "zeros" when possible.

The "zeros" add no value and can be ignored. The sum is three negative units, or -3.

(c) To add 7 + (−3), draw seven positive units and then draw three negative units. Form "zeros" when possible.

Ignoring the three "zeros" that were formed, the sum is four positive units, or **4**.

(d) To add −2 + (−3), draw two negative units and then draw three negative units.

Because no "zeros" could be formed, the sum is five negative units, or **−5**.

Now Try Exercises 63, 65, 67, 69

4 Solving Applications Involving Addition of Integers

Math in Real World Context At the beginning of this section, we saw how football teams compute total yardage by adding positive and negative integers. The next two examples illustrate other real-world situations that involve adding integers.

EXAMPLE 6 Computing temperature change

One of the quickest and most dramatic temperature changes on record in the United States occurred in Great Falls, Montana on January 11, 1980. In just 7 minutes, the temperature increased by 47 °F. If the temperature was −32 °F before the increase, what was the temperature after the change? (*Source: Montana Almanac.*)

Solution
To find the temperature after a 47 °F increase, we must add 47 to −32.

STEP 1: The absolute values are $|-32| = 32$ and $|47| = 47$.
STEP 2: Subtract $47 - 32 = 15$.
STEP 3: Because $|47| > |-32|$, keep the **positive** sign in the sum $-32 + 47 = $ **15**.

The temperature increased to 15 °F.

Now Try Exercise 71

EXAMPLE 7 Finding underground depth

The activity of exploring caves is known as *spelunking*. If two spelunkers are 135 feet below ground level and they descend an additional 42 feet, what is their new position relative to ground level?

Solution
First, we use the integer 0 to represent ground level. A position of 135 feet **below** ground level can be represented by the integer **−135**. To find the final position after a **descent** of an additional 42 feet, we must add **−42**.

STEP 1: Start by finding the sum of the absolute values.
$$|-135| + |-42| = 135 + 42 = 177$$

STEP 2: Because the common sign is **negative**, the sum is $-135 + (-42) = $ **−177**. The spelunkers are located 177 feet **below** ground level.

Now Try Exercise 75

2.2 Putting It All Together

CONCEPT	COMMENTS	EXAMPLES
Adding Integers with Like Signs	1. Find the sum of the absolute values of the integers. 2. Keep the common sign of the two integers as the sign of the sum.	Because $\|-5\| + \|-6\| = 11$, $-5 + (-6) = -11$. Because $\|3\| + \|12\| = 15$, $3 + 12 = 15$.
Adding Integers with Unlike Signs	1. Find the absolute values of the integers. 2. Subtract the smaller absolute value from the larger absolute value. 3. Keep the sign of the integer with the larger absolute value as the sign of the sum.	To add $-7 + 11$, find the absolute values: $\|-7\| = 7$ and $\|11\| = 11$. Subtract $11 - 7 = 4$. Because $\|11\| > \|-7\|$, the sum is positive. So, $-7 + 11 = 4$.
Addition Properties for Integers	1. Commutative Property 2. Associative Property 3. Identity Property 4. Inverse Property	1. $3 + (-2) = -2 + 3$ 2. $(-2 + 1) + 3 = -2 + (1 + 3)$ 3. $-1 + 0 = -1$ and $0 + 2 = 2$ 4. $2 + (-2) = 0$ and $-5 + 5 = 0$
Adding Integers Using a Number Line	Starting at 0, draw an arrow representing the first addend. From the tip of this arrow, draw an arrow representing the second addend. The sum is located at the tip of the second arrow.	To add $-1 + 3$ using a number line, draw arrows as shown. The sum is 2.
Adding Integers Using Symbols	Using the symbol ⌒ to represent a positive unit and the symbol ⌣ to represent a negative unit, draw the appropriate number of units for each addend. Form "zeros" when possible.	Add $3 + (-5)$ as shown. Ignoring the three "zeros," the sum is -2.

2.2 Exercises

CONCEPTS AND VOCABULARY

1. The first step when adding two integers with like signs is to find the sum of the _____ of the integers

2. When adding two negative integers, what is the sign of the sum?

3. When adding the integers $-1324 + 5678$, what is the sign of the sum?

4. When adding the integers $-32,264 + 11,902$, what is the sign of the sum?

5. Two numbers are _____ of each other if their sum is 0.

6. The sum of a number and its opposite is _____.

7. When adding integers using a number line, an arrow pointing to the _____ represents a positive number, and an arrow pointing to the _____ represents a negative number.

8. When adding two integers visually using symbols, the symbol _____ represents a positive unit, and the symbol _____ represents a negative unit.

ADDING INTEGERS

Exercises 9–30: Find the sum.

9. $3 + 9$
10. $7 + 12$
11. $-5 + (-7)$
12. $-8 + (-2)$
13. $13 + 28$
14. $33 + 21$
15. $-25 + (-17)$
16. $-30 + (-24)$
17. $-28 + 13$
18. $-31 + 17$
19. $35 + (-12)$
20. $50 + (-30)$
21. $39 + (-39)$
22. $47 + (-47)$
23. $-100 + 139$
24. $-75 + 150$
25. $61 + (-62)$
26. $77 + (-79)$
27. $-33 + (-33)$
28. $-41 + (-41)$
29. $-143 + 0$
30. $0 + (-78)$

Exercises 31–36: Evaluate the expression $x + y$ for the given values of the variables.

31. $x = -12, y = -4$
32. $x = -2, y = 19$
33. $x = 27, y = -14$
34. $x = 32, y = 22$
35. $x = 0, y = -93$
36. $x = -65, y = 1$

APPLYING ADDITION PROPERTIES

Exercises 37–44: State the addition property illustrated by the given equation.

37. $13 + (-56) = -56 + 13$
38. $0 + (-1289) = -1289$
39. $347 + (-347) = 0$
40. $13 + (-18 + 47) = (13 + (-18)) + 47$
41. $(-19 + 7) + 43 = -19 + (7 + 43)$
42. $-457 + 457 = 0$
43. $-671 + 0 = -671$
44. $-72 + 561 = 561 + (-72)$

Exercises 45–54: The associative and commutative properties for addition allow for three or more integers to be added in any order. Find the given sum.

45. $-5 + 3 + (-2)$
46. $4 + 8 + (-4)$
47. $-1 + (-9) + (-7)$
48. $-4 + (-8) + 12$
49. $-7 + (-17) + 24$
50. $-11 + 9 + (-7)$
51. $-18 + 53 + 29$
52. $34 + (-51) + 38$
53. $-31 + (-29) + (-47) + 62$
54. $111 + (-15) + (-152) + 68$

ADDING VISUALLY

Exercises 55–62: Use a number line to find the sum.

55. $3 + (-7)$
56. $-6 + 8$
57. $-5 + 9$
58. $-2 + (-3)$
59. $-4 + (-1)$
60. $4 + (-7)$
61. $-5 + 5$
62. $9 + (-9)$

Exercises 63–70: Perform the addition visually, using the symbols ⌒ and ⌣.

63. $3 + 1$
64. $5 + 4$
65. $4 + (-6)$
66. $-7 + (-3)$
67. $-4 + (-5)$
68. $-4 + 2$
69. $-2 + 7$
70. $8 + (-2)$

APPLICATIONS INVOLVING ADDITION OF INTEGERS

71. **Temperature Change** In 1972 Loma, Montana, experienced one of the largest 24-hour temperature swings ever recorded in the United States. On January 14, the temperature was -54 °F and increased 103 °F by the next day. What was the temperature after this change? (*Source: Montana Almanac.*)

72. **Temperature Change** In 1916 a very large 24-hour temperature swing took place in Browning, Montana. On January 23, the temperature was 44 °F and dropped 100 °F by the next day. What was the temperature after this change? (*Source: Montana Almanac.*)

73. **Football Stats** A running back carries the ball four times. Find his total yardage for the four plays if the yardages were $-1, -3, 13,$ and 4 yards.

74. **Game Tokens** A game is played with red and blue tokens. If each red token represents -1 point and each blue token represents $+1$ point, what is the total point value of 7 red tokens and 4 blue tokens?

75. **Underground Depth** If a coal miner is 203 feet below ground level and then descends 816 feet, what is the miner's new position relative to ground level?

76. **Underwater Depth** If a diver who is 97 feet below sea level ascends 56 feet, what is the diver's new position relative to sea level?

77. **Microsoft Income** In September 2010 Microsoft reported online operating income of about −$550 million. By September 2013 the company reported an income $230 million higher than the 2010 value. What was Microsoft's reported online operating income in September 2013?

78. **Geography** The elevation of the highest point in Florida, Britton Hill, is 627 feet higher than the elevation of Death Valley, which has an elevation of −282 feet. What is the elevation of Britton Hill?
(*Source: The World Atlas.*)

79. **Finances** A student's savings account starts the month with a balance of $3534. The following positive entries (deposits) and negative entries (withdrawals) are made in the savings register.

$$-282, 445, 390, \text{ and } -1598$$

What is the ending balance?

80. **Finances** A student's checking account starts the month with a balance of $617. The following positive entries (deposits) and negative entries (withdrawals) are made in the checking register.

$$-17, -120, 200, \text{ and } -40$$

What is the ending balance?

WRITING ABOUT MATHEMATICS

81. Give two examples of real-world situations involving the addition of positive and negative integers.

82. Explain how absolute value is used when adding two negative integers.

83. Explain how to find the sum of two integers with like signs. Give two examples.

84. Explain how to find the sum of two integers with unlike signs. Give two examples.

SECTIONS 2.1 and 2.2 — Checking Basic Concepts

1. Find the opposite of each number.
   (a) 23
   (b) −16

2. Simplify each expression.
   (a) −(−52)
   (b) −(−(−9))

3. Graph the integers −3, 0, and 4 on the same number line.

4. Place the correct symbol, < or >, in each blank between the integers.
   (a) 67 ____ −68
   (b) 0 ____ −10,003

5. Evaluate each absolute value.
   (a) |17|
   (b) |−31|

6. Find each sum.
   (a) −14 + 22
   (b) −27 + (−8)
   (c) 4 + (−25)
   (d) 52 + 31

7. Use a number line to find each sum.
   (a) −4 + 8
   (b) 3 + (−9)

8. Perform the addition visually, using the symbols ∩ and ∪.
   (a) −6 + 8
   (b) 3 + (−8)

9. Use absolute value to determine which of the two given financial quantities is larger.

   A credit card debt: −$420
   Checking account balance: $380

10. **Geography** The elevation of the highest point in Alabama, Cheaha Mountain, is 3783 feet higher than the elevation of the Dead Sea Basin, which is −1378 feet. What is the elevation of Cheaha Mountain? (*Source: The World Atlas.*)

2.3 Subtracting Integers

Objectives

1. Subtracting Integers
2. Adding and Subtracting Integers
3. Subtracting Integers Visually
   - Subtracting Integers Using a Number Line
   - Subtracting Integers Using Symbols
4. Solving Applications Involving Subtraction of Integers

STUDY TIP

If you miss something in class, Video Lectures available in MyMathLab provide a short lecture for each section in this text. These lectures, taught by actual math instructors, offer you the opportunity to review topics that you may not have fully understood before.

READING CHECK 1

- How do we subtract integers?

CALCULATOR HELP

To subtract integers with a calculator, see Appendix E (page AP-30).

1 Subtracting Integers

One of the identity properties for subtraction states that subtracting a number from itself results in 0. For example, $8 - 8 = 0$ and $32 - 32 = 0$. Similarly, the inverse property for addition from the previous section states that the sum of a number and its additive inverse (opposite) is 0. That is, $8 + (-8) = 0$ and $32 + (-32) = 0$. Together, these properties suggest that we can subtract one number from another by adding the first number to the opposite of the second number.

Subtracting a Number Is Equivalent to Adding Its Opposite

$$12 - 4 = 8 \qquad 100 - 45 = 55 \qquad 87 - 85 = 2$$
$$12 + (-4) = 8 \qquad 100 + (-45) = 55 \qquad 87 + (-85) = 2$$

This procedure can be used to find a difference regardless of the signs of the two numbers involved. It is common to say that we subtract integers by *adding the opposite*.

SUBTRACTING INTEGERS

If the variables a and b represent two numbers, then

$$a - b = a + (-b).$$

EXAMPLE 1 Subtracting integers by adding the opposite

Find each difference.
(a) $7 - (-6)$ (b) $-13 - 8$ (c) $-4 - (-11)$

Solution

(a) Rather than subtracting -6, add its opposite, 6.   *Add the opposite of -6.*

$$7 - (-6) = 7 + 6 = 13$$

(b) Instead of subtracting 8, add its opposite, -8.   *Add the opposite of 8.*

$$-13 - 8 = -13 + (-8) = -21$$

(c) Rather than subtracting -11, add its opposite, 11.   *Add the opposite of -11.*

$$-4 - (-11) = -4 + 11 = 7$$

Now Try Exercises 13, 17, 23

Although the procedure used in Example 1 can always be used to find a difference, it is important to remember that some subtraction problems can be done easily without adding the opposite. The next example illustrates this.

EXAMPLE 2 Subtracting integers without adding the opposite

Subtract $22 - 17$.

Solution
Because both numbers are whole numbers with the second number smaller than the first, this difference can be found using simple subtraction of whole numbers. The difference is $22 - 17 = 5$. Adding the opposite is not necessary in this case.

Now Try Exercise 9

2 Adding and Subtracting Integers

Some arithmetic expressions involve a mix of addition and subtraction. When we change each subtraction to the addition of the opposite, the result is an expression that includes only addition. Since both the commutative and associative properties apply to addition, we can rearrange addends. The next example illustrates this process.

EXAMPLE 3 Adding and subtracting integers

Simplify each expression.
(a) $10 - (-3) + 5 - 1$ (b) $4 - 18 + 6 - 2$

Solution
(a) Begin by changing each subtraction to the addition of the opposite.

$$10 - (-3) + 5 - 1 = 10 + 3 + 5 + (-1) \quad \text{Add the opposite.}$$
$$= 13 + 5 + (-1) \quad \text{Add: } 10 + 3 = 13.$$
$$= 18 + (-1) \quad \text{Add: } 13 + 5 = 18.$$
$$= 17 \quad \text{Add.}$$

READING CHECK 2

- After changing every subtraction to addition of the opposite, which properties of addition can be used?

(b) After changing each subtraction to the addition of the opposite, rearrange the addends and find the sum by applying the commutative and associative properties.

$$4 - 18 + 6 - 2 = 4 + (-18) + 6 + (-2) \quad \text{Add the opposite.}$$
$$= 4 + 6 + (-18) + (-2) \quad \text{Rearrange addends.}$$
$$= 10 + (-20) \quad \text{Add.}$$
$$= -10 \quad \text{Add.}$$

Now Try Exercises 39, 43

3 Subtracting Integers Visually

Integers can be subtracted visually in two ways that are similar to those used for addition.

SUBTRACTING INTEGERS USING A NUMBER LINE To subtract integers using a number line, we first change the subtraction to addition of the opposite and then add using the number line.

EXAMPLE 4 Subtracting integers using a number line

Use a number line to find each difference
(a) $-2 - (-5)$ (b) $-1 - 3$

Solution

(a) First change the difference $-2 - (-5)$ to the sum $-2 + 5$. To add $-2 + 5$, start at 0 and draw an arrow representing -2. From the tip of this arrow, draw a second arrow representing 5. The sum is 3, as shown in **FIGURE 2.13**. Thus the difference $-2 - (-5)$ is also 3.

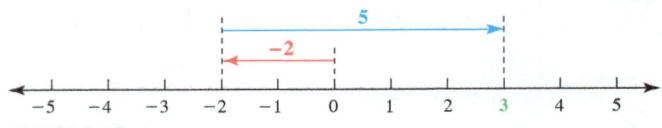

FIGURE 2.13

(b) First change the difference $-1 - 3$ to the sum $-1 + (-3)$. To add $-1 + (-3)$, start at 0 and draw an arrow representing -1. From the tip of this arrow, draw a second arrow representing -3. The sum is -4, as shown in **FIGURE 2.14**. Thus the difference $-1 - 3$ is also -4.

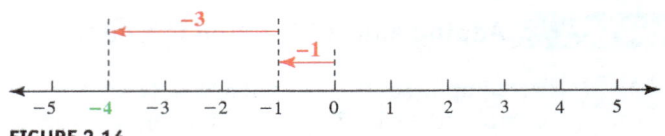

FIGURE 2.14

Now Try Exercises 45, 47

SUBTRACTING INTEGERS USING SYMBOLS To subtract integers visually using symbols, we again use ⌒ to represent a positive unit and ⌣ to represent a negative unit. Subtraction occurs by "taking away" the appropriate number of positive or negative units from the visual representation. We cross out units to indicate that they have been taken away. For example, to show that $7 - 3 = 4$, we start by drawing seven positive units. Crossing out three positive units results in 4 positive units.

Differences that involve integers with unlike signs such as $3 - (-5)$ pose a special problem. If we start by drawing three positive units, how can we cross out five negative units? The difference $3 - (-5)$ is found in part (b) of the next example.

EXAMPLE 5 Subtracting integers using symbols

Perform each subtraction visually, using the symbols ⌒ and ⌣.
(a) $-6 - (-2)$ (b) $3 - (-5)$ (c) $-3 - (-7)$

Solution
(a) To subtract $-6 - (-2)$, draw six negative units and then cross out two of them.

The difference is four negative units, or -4.

(b) To subtract $3 - (-5)$, start by drawing three positive units.

⌒ ⌒ ⌒

Because there are no negative units available to cross out and five are needed, insert five "zeros" in the visual representation of 3. This new representation is also equal to 3.

Now cross out five negative units to result in eight positive units, or 8.

(c) To subtract $-3 - (-7)$, start by drawing three negative units.

Because there are only three negative units available to cross out and seven are needed, insert four "zeros" in the visual representation of -3. This new representation is also equal to -3.

Now cross out seven negative units to result in four positive units, or 4.

Now Try Exercises 55, 57, 59

4 Solving Applications Involving Subtraction of Integers

🌐 *Math in Ocean Context* If a submarine is 235 feet below sea level at a location where the ocean is 640 feet deep, then we can find the distance between the submarine and the bottom of the ocean by subtracting $-235 - (-640)$. We solve this problem in the next example.

EXAMPLE 6   Analyzing the depth of a submarine

If a submarine is 235 feet below sea level at a location where the ocean is 640 feet deep, find the distance between the submarine and the bottom of the ocean.

Solution
The distance is found by subtracting: $-235 - (-640)$. Rather than subtracting -640, add its opposite, or 640. The related sum is $-235 + 640$. Find this sum as follows.

STEP 1: The absolute values are $|-235| = 235$ and $|640| = 640$.
STEP 2: Subtract $640 - 235 = 405$.
STEP 3: Because $|640| > |-235|$, keep the **positive** sign in the sum $-235 + 640 = $ **405**.

The submarine is 405 feet from the bottom of the ocean.

Now Try Exercise 65

EXAMPLE 7   Analyzing an account balance

An overdrawn checking account has a balance of $-\$413$. An amount of money is deposited to bring the balance to $\$776$. How much money has been deposited?

Solution
To determine the amount deposited when the balance changes from −$413 to $776, we must find the difference $776 - (-413)$. Rather than subtracting −413, we add its opposite, or 413. The related sum is $776 + 413$, which is equal to 1189. The amount deposited is $1189.

Now Try Exercise 69

2.3 Putting It All Together

CONCEPT	COMMENTS	EXAMPLES
Subtracting Integers	If the variables a and b represent two numbers, then $a - b = a + (-b)$.	$3 - (-1) = 3 + 1 = 4$ $-6 - 8 = -6 + (-8) = -14$
Subtracting Integers Using a Number Line	To subtract integers using a number line, first change the subtraction to addition of the opposite, and then add using a number line.	To subtract $-1 - (-3)$, first change to the sum $-1 + 3$, and then add. The difference is 2.
Subtracting Integers Using Symbols	Use ◠ to represent a positive unit and ◡ to represent a negative unit.	Subtract: $-2 - (-5)$. The difference is 3.

2.3 Exercises   MyMathLab®

CONCEPTS AND VOCABULARY

1. To subtract $a - b$, add a and the _____ of b.

2. $4 - 7 = 4 +$ _____

3. $-2 - (-9) = -2 +$ _____

4. After changing every subtraction to addition of the opposite in a mix of addition and subtraction, what two properties of addition can be used?

5. To perform subtraction on a number line, first change the subtraction to _____ of the opposite.

6. Is it necessary to add the opposite to find $12 - 5$?

7. In printed text, which has space both before and after it, a subtraction symbol or a negative sign?

8. (True or False?) $8 = 8 + 0 + 0 + 0 + 0$.

SUBTRACTING INTEGERS

Exercises 9–28: Find the difference.

9. $8 - 2$
10. $12 - 5$
11. $13 - 18$
12. $22 - 25$
13. $-10 - 5$
14. $-20 - 7$
15. $-25 - 17$
16. $-24 - 24$
17. $21 - (-6)$
18. $33 - (-10)$
19. $5 - (-24)$
20. $11 - (-29)$
21. $-14 - (-9)$
22. $-40 - (-12)$
23. $-21 - (-29)$
24. $-17 - (-33)$
25. $34 - 0$
26. $-28 - 0$
27. $0 - (-52)$
28. $0 - 75$

Exercises 29–34: Evaluate the expression $x - y$ for the given values of the variables.

29. $x = -8, y = -17$ 30. $x = -3, y = 20$

31. $x = 30, y = -15$ 32. $x = 19, y = 43$

33. $x = -70, y = -3$

34. $x = -48, y = 1$

ADDING AND SUBTRACTING INTEGERS

Exercises 35–44: Simplify the expression.

35. $-3 - 3 - (-4)$

36. $5 - 9 - (-3)$

37. $-1 - (-9) - (-5)$

38. $-4 - (-8) - 12$

39. $-2 + (-3) - (-9)$

40. $4 - (-15) + 10$

41. $-17 - 7 + (-7)$

42. $-14 + 2 - (-8)$

43. $-31 + (-16) - (-40) + 21$

44. $-111 - (-99) + (-55) + 68$

SUBTRACTING INTEGERS VISUALLY

Exercises 45–52: Use a number line to find the difference.

45. $5 - 9$ 46. $4 - 11$

47. $-4 - (-7)$ 48. $-2 - (-8)$

49. $-3 - 3$ 50. $5 - 5$

51. $-4 - (-4)$ 52. $2 - (-2)$

Exercises 53–60: Perform the subtraction visually, using the symbols ⌢ and ⌣.

53. $2 - 7$ 54. $4 - 8$

55. $-9 - (-7)$ 56. $-9 - (-2)$

57. $4 - (-6)$ 58. $1 - (-7)$

59. $-2 - (-9)$ 60. $-5 - (-6)$

APPLICATIONS INVOLVING SUBTRACTION OF INTEGERS

61. **Temperature Change** A traveler leaves South Padre Island, where the temperature is 78 °F, and flies to Chicago, where the temperature is −9 °F. What is the temperature difference for this trip?

62. **Liquid Nitrogen** In some medical procedures, a small amount of liquid nitrogen is applied to the skin. If normal skin temperature is 91 °F and the temperature of liquid nitrogen is −321 °F, then find the temperature difference

63. **Elevations in California** The elevation of Mount Whitney is 14,494 feet, and the elevation of Death Valley is −282 feet. Find the elevation difference between these California landmarks. (*Source:* U.S. Geological Survey.)

64. **Elevations in Louisiana** The elevation of Driskill Mountain is 535 feet, and an elevation found in New Orleans is −8 feet. Find the elevation difference between these Louisiana locations. (*Source:* U.S. Geological Survey.)

65. **Underwater Depth** If a diver is 37 feet below sea level at a location where the ocean is 52 feet deep, find the distance between the diver and the bottom of the ocean.

66. **Underwater Depth** If the highest point on an underwater structure is 26 feet below sea level and the ocean is 73 feet deep at the base of the structure, how tall is the structure?

67. **Net Worth** A student with a net worth of −$35,600 inherits an amount of money so that his net worth becomes $153,800. How much money did he inherit?

68. **Game Tokens** In a family board game, each red token represents −1 point and each blue token represents +1 point. How many more points does a player with 4 blue tokens have compared to a player with 5 red tokens?

69. **Account Balance** An overdrawn checking account has a balance of −$47. An amount of money is deposited to bring the balance to $129. How much money has been deposited?

70. **Account Balance** An overdrawn checking account has a balance of −$55. An amount of money is deposited to bring the balance to $217. How much money has been deposited?

71. **Temperature Change** Over a four-day period in January, a city in Michigan had the following weather changes. The low temperature on Monday was −6 °F. On Tuesday, the low temperature increased by 8 °F. The low temperature then dropped 5 °F on Wednesday. Finally, the low temperature decreased by another 7 °F on Thursday. What was the low temperature recorded on Thursday?

72. Sea Diving A diver located 54 feet below sea level swims up 17 feet and then dives down 26 feet. Write the final position of the diver relative to sea level as a negative integer.

WRITING ABOUT MATHEMATICS

73. Give two examples of real-world situations that involve the subtraction of positive and negative integers.

74. Give an example of a temperature or elevation difference that requires a negative number to be subtracted from a positive number.

75. Explain why subtraction can be accomplished by "adding the opposite."

76. Devise a method for subtracting integers on a number line without first changing the subtraction to addition of the opposite.

2.4 Multiplying and Dividing Integers

Objectives

1. Multiplying Integers
2. Recognizing Multiplication Properties
3. Multiplying More Than Two Integer Factors
4. Dividing Integers
5. Finding Square Roots of Integers
6. Solving Applications Involving Multiplication and Division of Integers

NEW VOCABULARY

☐ Principal square root

STUDY TIP

Is there a math lab or tutor room on your campus? Be sure to take advantage of the resources available at your school.

READING CHECK 1

- When a positive integer and a negative integer are multiplied, is the product positive or negative?

READING CHECK 2

- When two negative integers are multiplied, is the product positive or negative?

1 Multiplying Integers

Finding the product of two integers is similar to finding the product of two whole numbers. Since integers may be positive or negative, the product of two integers may be positive or negative also. To find a general rule for multiplying integers, we can consider the following pattern. What values should replace the blue question marks to continue the pattern?

$$
\begin{aligned}
5 \times 3 &= 15 \\
5 \times 2 &= 10 \\
5 \times 1 &= 5 \\
5 \times 0 &= 0 \\
5 \times (-1) &= ? \\
5 \times (-2) &= ?
\end{aligned}
$$

(Decrease by 1 on the left; Decrease by 5 on the right.)

Continuing the pattern, we find that $5 \times (-1) = -5$ and $5 \times (-2) = -10$. This pattern suggests that *if we multiply a positive integer and a negative integer, the product is negative*.

What can be said about the product of two negative integers? To answer this question, consider the following pattern. This time, what values should replace the blue question marks to continue the pattern?

$$
\begin{aligned}
-4 \times 3 &= -12 \\
-4 \times 2 &= -8 \\
-4 \times 1 &= -4 \\
-4 \times 0 &= 0 \\
-4 \times (-1) &= ? \\
-4 \times (-2) &= ?
\end{aligned}
$$

(Decrease by 1 on the left; Increase by 4 on the right.)

Continuing the pattern results in $-4 \times (-1) = 4$ and $-4 \times (-2) = 8$. This pattern suggests that *if we multiply two negative integers, the product is positive*.

We now have general rules for multiplying integers. To find the product of two integers, multiply their absolute values and use the following rules to find the sign of the product.

SIGNS OF PRODUCTS

The product of two integers with *like* signs is positive.
$(+)(+) = (+)$
$(-)(-) = (+)$

The product of two integers with *unlike* signs is negative.
$(+)(-) = (-)$
$(-)(+) = (-)$

CALCULATOR HELP
To multiply integers with a calculator, see Appendix E (page AP-30).

EXAMPLE 1 Multiplying integers

Multiply.
(a) $-3 \cdot 8$ (b) $-7(-4)$ (c) $2 \times (-9)$

Solution
(a) The factors have *unlike* signs, so the product is negative: $-3 \cdot 8 = -24$.
(b) The factors have *like* signs, so the product is positive: $-7(-4) = 28$.
(c) The factors have *unlike* signs, so the product is negative: $2 \times (-9) = -18$.

Now Try Exercises 13, 21, 23

2 Recognizing Multiplication Properties

Multiplication properties for whole numbers can also be applied to integers.

MULTIPLICATION PROPERTIES FOR INTEGERS

Let the variables a, b, and c represent integers.

Commutative Property: $a \cdot b = b \cdot a$
Associative Property: $(a \cdot b) \cdot c = a \cdot (b \cdot c)$
Identity Property: $a \cdot 1 = a$ and $1 \cdot a = a$
Zero Property: $a \cdot 0 = 0$ and $0 \cdot a = 0$
Distributive Properties: $a(b + c) = a \cdot b + a \cdot c$
$a(b - c) = a \cdot b - a \cdot c$

EXAMPLE 2 Identifying multiplication properties

State the multiplication property illustrated by each of the following.
(a) $-47 \cdot 1 = -47$
(b) $-3(2 - 8) = -3 \cdot 2 - (-3) \cdot 8$
(c) $13 \times (-9) = -9 \times 13$
(d) $(-2 \cdot 7) \cdot 5 = -2 \cdot (7 \cdot 5)$

Solution
(a) The product of an integer and 1 is that integer. This illustrates the identity property.
(b) The distributive property is illustrated by this equation. The integer outside the parentheses is multiplied by each integer inside the parentheses.
(c) This equation illustrates the commutative property. Changing the order of the factors does not affect the product.
(d) The way in which three or more factors are grouped does not affect the product. This equation illustrates the associative property.

Now Try Exercises 25, 27, 29, 31, 33

3 Multiplying More Than Two Integer Factors

Because integer multiplication is both commutative and associative, a product involving three or more factors can be rearranged so that all the positive factors are listed first and all negative factors are listed second. Doing so will help us discover a rule for finding the sign of a product involving several integers.

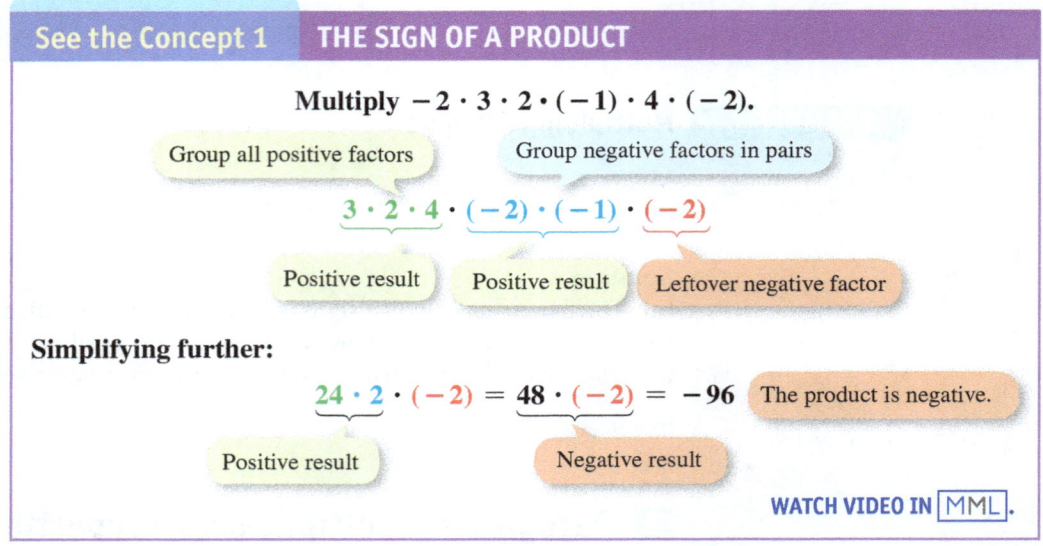

See the Concept 1 — THE SIGN OF A PRODUCT

Multiply $-2 \cdot 3 \cdot 2 \cdot (-1) \cdot 4 \cdot (-2)$.

Group all positive factors: $3 \cdot 2 \cdot 4$ (Positive result)
Group negative factors in pairs: $(-2) \cdot (-1)$ (Positive result), (-2) (Leftover negative factor)

Simplifying further:

$\underbrace{24 \cdot 2}_{\text{Positive result}} \cdot (-2) = \underbrace{48 \cdot (-2)}_{\text{Negative result}} = -96$ The product is negative.

WATCH VIDEO IN MML.

To find the sign of a product involving more than two factors, we need only to determine if there are any *unpaired* negative factors. This means that we determine if there is an odd or even number of negative factors.

READING CHECK 3

- If there is an odd number of negative factors, will the product be positive or negative?

PRODUCTS WITH MORE THAN TWO FACTORS

Begin by multiplying the absolute values of the factors.

- If there is an *even* number of negative factors, the product is positive.
- If there is an *odd* number of negative factors, the product is negative.

EXAMPLE 3 Multiplying more than two factors

Multiply.
(a) $-2 \cdot (-1) \cdot 4 \cdot (-5)$ (b) $-2 \cdot (-4) \cdot (-1) \cdot (-3)$

Solution

(a) Start by multiplying the absolute values: $2 \cdot 1 \cdot 4 \cdot 5 = 40$. Because there are **3** negative factors in the given expression, there is an **odd** number of negative factors and the product is negative.

$$-2 \cdot (-1) \cdot 4 \cdot (-5) = -40$$

(b) Because there are **4** negative factors in the given expression, there is an **even** number of negative factors and the product is positive.

$$-2 \cdot (-4) \cdot (-1) \cdot (-3) = 24$$

Now Try Exercises 41, 47

4 Dividing Integers

Division and multiplication are closely related. For example, the equation $3 \cdot 4 = 12$ can be used to check the equation $12 \div 4 = 3$. Extending this idea to integers reveals the sign rules for quotients. Consider the products and quotients shown in **TABLE 2.3**.

Known Product	Corresponding Quotient	Rule for Quotients
$3 \cdot 4 = 12$	$12 \div 4 = 3$	$(+) \div (+) = (+)$
$3 \cdot (-4) = -12$	$-12 \div (-4) = 3$	$(-) \div (-) = (+)$
$-3 \cdot (-4) = 12$	$12 \div (-4) = -3$	$(+) \div (-) = (-)$
$-3 \cdot 4 = -12$	$-12 \div 4 = -3$	$(-) \div (+) = (-)$

TABLE 2.3

READING CHECK 4

- When a positive integer is divided by a negative integer, is the quotient positive or negative?

When the dividend and divisor have like signs, the quotient is positive, and when the dividend and divisor have unlike signs, the quotient is negative. This result is similar to the rules for the signs of products and is summarized as follows.

SIGNS OF QUOTIENTS

The quotient of two integers with *like* signs is positive. $(+) \div (+) = (+)$
$(-) \div (-) = (+)$

The quotient of two integers with *unlike* signs is negative. $(+) \div (-) = (-)$
$(-) \div (+) = (-)$

EXAMPLE 4 **Dividing integers**

Divide.

(a) $36 \div (-12)$ (b) $\dfrac{-60}{-5}$ (c) $-48 \div 6$

Solution

(a) The numbers have *unlike* signs, so the quotient is negative: $36 \div (-12) = -3$.
(b) The numbers have *like* signs, so the quotient is positive: $\dfrac{-60}{-5} = 12$.
(c) The numbers have *unlike* signs, so the quotient is negative: $-48 \div 6 = -8$.

CALCULATOR HELP

To divide integers with a calculator, see Appendix E (page AP-30).

Now Try Exercises 49, 51, 55

Recall that quotients involving 0 are sometimes undefined. The next example illustrates this and other properties of division.

EXAMPLE 5 **Dividing integers**

Divide, if possible. If a quotient is undefined, state so.

(a) $\dfrac{0}{-7}$ (b) $\dfrac{-12}{-12}$ (c) $-15 \div 0$ (d) $-24 \div 1$

Solution

(a) Dividing 0 by any number (except 0) results in 0: $\dfrac{0}{-7} = 0$.
(b) Dividing a number (except 0) by itself results in 1: $\dfrac{-12}{-12} = 1$.
(c) Any number divided by 0 is undefined: $-15 \div 0$ is undefined.
(d) Dividing a number by 1 results in the number itself: $-24 \div 1 = -24$.

Now Try Exercises 53, 57, 59, 61

READING CHECK 5

- How many square roots does every positive integer have?

5 Finding Square Roots of Integers

In Section 5 of the previous chapter, we discovered that 7 is a square root of 49 because $7^2 = 49$. However, because the product of two negative integers is positive, we also know that $(-7)^2 = 49$. Does this mean that -7 is also a square root of 49? The answer is yes. The number 49 has two integer square roots, -7 and 7. In fact, *every positive integer has two square roots*.

NOTE: Negative integers do not have integer square roots. See part (b) of the next example. ∎

EXAMPLE 6 Finding square roots of integers

Find all integer square roots of the given integer, if possible.
(a) 36   (b) -9

Solution
(a) Because $6^2 = 36$ and $(-6)^2 = 36$, the integer square roots of 36 are -6 and 6.
(b) The product of two integers with like signs is always positive. As a result, the square of any integer (other than 0) is always positive. Therefore, there is no integer whose square is -9. We say that -9 has *no integer square roots*.

Now Try Exercises 65, 69

Recall that the radical sign $\sqrt{}$ means to find the square root of a number. For example, $\sqrt{100} = 10$. However, this notation does not account for the negative square root of 100, or -10. To avoid confusion, we use the notation $\sqrt{100}$ when finding the positive square root or **principal square root** of 100. To find the negative square root of 100, we write $-\sqrt{100}$. For example, the two square roots of 100 are $\sqrt{100} = 10$ and $-\sqrt{100} = -10$.

READING CHECK 6

- What special name is given to the positive square root of an integer?

> **SQUARE ROOTS OF INTEGERS**
>
> A positive integer a has one positive and one negative square root, as shown below.
>
> | Positive Square Root | Negative Square Root |
> |:---:|:---:|
> | $\sqrt{a}$ | $-\sqrt{a}$ |
>
> A negative integer has no integer square roots.

EXAMPLE 7 Finding square roots of integers

Simplify each expression, if possible.
(a) $\sqrt{81}$ (b) $-\sqrt{25}$ (c) $\sqrt{-49}$

Solution
(a) Because $9^2 = 81$ and 9 is positive, $\sqrt{81} = 9$.
(b) The notation $-\sqrt{25}$ means to find the negative square root of 25, or $-\sqrt{25} = -5$.
(c) $\sqrt{-49}$ is not an integer because a negative number has no integer square roots.

Now Try Exercises 73, 75, 79

CALCULATOR HELP

To find the square roots of an integer with a calculator, see Appendix E (page AP-30).

6 Solving Applications Involving Multiplication and Division of Integers

The next two examples illustrate applications involving products and quotients of integers.

EXAMPLE 8　Finding the average surface temperature of a planet

The average surface temperature of Mars is about $-65\,°C$. Because Uranus is much farther from the sun, its average surface temperature is 3 times as cold as that of Mars. What is the average surface temperature of Uranus? (*Source:* NASA.)

Solution
The average surface temperature of Uranus is $-65 \cdot 3 = -195\,°C$.

Now Try Exercise 95

EXAMPLE 9　Calculating investment losses

An investor lost money on six different investments. If the amount lost was $1500 on each investment, write the total loss as a product of integers and find the total loss.

Solution
We can write a loss as a negative number. The total loss can be written as $-1500 \cdot 6$, which equals -9000. The total loss is $9000.

Now Try Exercise 101

2.4　Putting It All Together

CONCEPT	COMMENTS	EXAMPLES
Multiplying Integers	The product of two numbers with *like* signs is positive.	$3 \cdot 12 = 36$ $-4 \cdot (-6) = 24$
	The product of two numbers with *unlike* signs is negative.	$-5 \cdot 7 = -35$ $8 \cdot (-9) = -72$
Multiplication Properties	1. Commutative Property 2. Associative Property 3. Identity Property 4. Zero Property 5. Distributive Properties	1. $4 \cdot (-2) = -2 \cdot 4$ 2. $(-5 \cdot 1) \cdot 7 = -5 \cdot (1 \cdot 7)$ 3. $-2 \cdot 1 = -2$ and $1 \cdot (-2) = -2$ 4. $-7 \cdot 0 = 0$ and $0 \cdot (-7) = 0$ 5. $-5(2 + 4) = -5 \cdot 2 + (-5) \cdot 4$ 　$-2(1 - 5) = -2 \cdot 1 - (-2) \cdot 5$
Products with More Than Two Factors	If there is an *even* number of negative factors, the product is positive. If there is an *odd* number of negative factors, the product is negative.	$3 \cdot (-2) \cdot 1 \cdot (-4)$ is positive. (even number of negative factors) $-3 \cdot (-5) \cdot 1 \cdot (-7)$ is negative. (odd number of negative factors)
Dividing Integers	The quotient of two numbers with *like* signs is positive. The quotient of two numbers with *unlike* signs is negative.	$36 \div 9 = 4$ $-45 \div (-5) = 9$ $-63 \div 9 = -7$ $28 \div (-7) = -4$
Square Roots	Every positive integer a has one positive and one negative square root. The positive square root, or *principal square root*, is written as $\sqrt{a}$. The negative square root is written as $-\sqrt{a}$.	$\sqrt{25} = 5$ $-\sqrt{64} = -8$ $\sqrt{-121}$ is not an integer.

2.4 Exercises

CONCEPTS AND VOCABULARY

1. The product of two numbers with like signs is _____.
2. When two numbers have unlike signs, their product is _____.
3. The equation $1 \cdot (-17) = -17$ is an example of the _____ property of multiplication.
4. The equation $-3(4 - 5) = -3 \cdot 4 - (-3) \cdot 5$ is an example of the _____ property of multiplication.
5. If there is an even number of negative factors, the product is _____.
6. If there is an odd number of negative factors, the product is _____.
7. The quotient of two numbers with like signs is _____.
8. When two numbers have unlike signs, their quotient is _____.
9. The principal square root of 4 is written as _____.
10. The negative square root of 4 is written as _____.

MULTIPLYING INTEGERS

Exercises 11–24: Multiply.

11. $2(-6)$
12. $-7(4)$
13. $-5(-8)$
14. $-8(-7)$
15. $-1 \cdot 18$
16. $-14 \cdot 0$
17. $-10 \cdot (-17)$
18. $-50 \cdot (-2)$
19. $0 \cdot (-21)$
20. $1 \cdot (-34)$
21. $14 \cdot (-3)$
22. $15 \cdot (-4)$
23. $-25 \cdot 6$
24. $-30 \cdot 4$

APPLYING MULTIPLICATION PROPERTIES

Exercises 25–34: State the multiplication property illustrated by the given equation.

25. $12 \cdot (-22 \cdot 41) = (12 \cdot (-22)) \cdot 41$
26. $(-11 \cdot 5) \cdot 23 = -11 \cdot (5 \cdot 23)$
27. $-341 \cdot 0 = 0$
28. $1 \times (-30{,}412) = -30{,}412$
29. $-25 \times (-37) = -37 \times (-25)$
30. $0(-64{,}901) = 0$
31. $-2(-3 + 9) = -2(-3) + (-2)(9)$
32. $-10 \cdot (-19) = -19 \cdot (-10)$
33. $-15{,}400 \cdot 1 = -15{,}400$
34. $3(-1 - 8) = 3(-1) - 3(8)$

MULTIPLYING MORE THAN TWO FACTORS

Exercises 35–48: Multiply.

35. $-2 \cdot 6 \cdot 3$
36. $4 \cdot (-3) \cdot 2$
37. $-3 \cdot 5 \cdot (-2)$
38. $6 \cdot (-4) \cdot (-2)$
39. $-7(-1)(-3)$
40. $-8(-8)(-1)$
41. $5(-2)(-3)(-3)$
42. $5(5)(-1)(-4)$
43. $12(-1)(0)(2)$
44. $9(0)(-2)(-5)$
45. $2(-1)(5)(-2)(-4)$
46. $-2(5)(3)(-2)(-1)$
47. $2(-1)(5)(-2)(-4)(5)(-1)$
48. $-1(-3)(5)(-2)(-3)(5)(-2)$

DIVIDING INTEGERS

Exercises 49–64: Divide, if possible. If a quotient is undefined, state so.

49. $18 \div (-6)$
50. $-48 \div 8$
51. $\dfrac{-40}{-8}$
52. $\dfrac{24}{-3}$
53. $-12 \div (-1)$
54. $-20 \div 1$
55. $\dfrac{-50}{25}$
56. $\dfrac{-72}{-12}$
57. $-35 \div 0$
58. $0 \div (-3)$
59. $-24 \div (-24)$
60. $-10 \div 10$
61. $0 \div (-9)$
62. $-63 \div 0$
63. $\dfrac{72}{-12}$
64. $\dfrac{-64}{16}$

FINDING SQUARE ROOTS OF INTEGERS

Exercises 65–72: Find all integer square roots of the given number, if possible.

65. 25
66. 9
67. 81
68. 100

69. -36
70. -4
71. 0
72. 1

Exercises 73–82: Simplify the expression, if possible.

73. $\sqrt{16}$
74. $\sqrt{49}$
75. $-\sqrt{36}$
76. $-\sqrt{144}$
77. $\sqrt{100}$
78. $-\sqrt{81}$
79. $\sqrt{-121}$
80. $\sqrt{-25}$
81. $-\sqrt{1}$
82. $\sqrt{0}$

EVALUATING EXPRESSIONS

Exercises 83–94: Evaluate the expression for the given value(s) of the variable(s), if possible.

83. $3x$ $x = -7$
84. $-8y$ $y = -2$
85. $\dfrac{x}{6}$ $x = -60$
86. $4xy$ $x = -3, y = 5$
87. $\dfrac{a}{b}$ $a = -30, b = 6$
88. $-ab$ $a = -6, b = 11$
89. $m \cdot (-n)$ $m = -5, n = 5$
90. $2 \cdot (-m) \cdot (-n)$ $m = 4, n = -6$
91. $\sqrt{-x}$ $x = -100$
92. $-\sqrt{y}$ $y = 4$
93. $-\sqrt{a}$ $a = -64$
94. $-\sqrt{-a}$ $a = 81$

APPLICATIONS INVOLVING MULTIPLICATION AND DIVISION OF INTEGERS

95. **Cold Temperatures** Every state in the U.S. has a record low temperature below 0 °C. The warmest of these lows was recorded in Hawaii at Mauna Kea, where the temperature dipped to −11°C. In Iowa, the town of Elkader recorded the state's lowest temperature, which is 4 times as cold as the record in Hawaii. What is Iowa's record low? (*Source:* NOAA.)

96. **Cold Temperatures** The record low temperature for Florida is −19 °C. If the record low temperature for Montana is 3 times as cold as that of Florida, what is the record low for Montana? (*Source:* NOAA.)

97. **Spelunking** A cave explorer descends to the floor of a cave in 5 stages, dropping 107 feet each time. Write the total depth of the cave as a product of integers and find the depth of the cave.

98. **Game Tokens** Each red token in a board game represents −5 points. Write the total number of points in a stack of 4 red tokens as a product of integers and find the total number of points.

99. **State Prisoners** According to a recent study, the number of prisoners in Arkansas decreased by 300 inmates over a one-year period. If the decrease was the same for each of the 12 months, write the monthly decrease as a quotient of integers and find the monthly decrease. (*Source:* Arkansas News Bureau.)

100. **Homeless Population** A recent New York City government study suggested that the number of people living on the streets or in the city's subways dropped by 1089 over a three-year period. Assuming that the decrease was the same for each of the 3 years of the study, write the yearly decrease as a quotient of integers and find the yearly decrease. (*Source:* New York Times.)

101. **Late Fees** Each time that a credit card holder fails to make a credit card payment by the due date, a late fee appears on the monthly statement as −$29. If the card holder missed the payment due date five times over the past year, write the total charges for late fees as a product of integers and find the total charges.

102. **ATM Fees** Each time that a credit card holder uses an ATM for a cash advance, a transaction fee appears on the credit card statement. If a year-end summary shows −$129 for 43 ATM transactions, write the transaction fee as a negative integer.

WRITING ABOUT MATHEMATICS

103. Give an example of a product of three or more integers with a negative result. Give an example of a product of three or more integers with a positive result. Explain how you chose your examples.

104. Give a real-world example of when a negative number is divided by a positive number.

105. Explain why a negative integer cannot have an integer square root.

106. Choose a positive integer that is a perfect square and write both of its square roots using radical notation. Which of your expressions represents the principal square root?

SECTIONS 2.3 and 2.4 — Checking Basic Concepts

1. Find each difference.
   (a) $-11 - 23$
   (b) $-21 - (-7)$
   (c) $3 - (-30)$
   (d) $54 - 39$

2. Use a number line to find each difference.
   (a) $-4 - 4$
   (b) $3 - (-6)$

3. Perform the subtraction visually, using the symbols $\frown$ and $\smile$.
   (a) $-5 - 3$
   (b) $-8 - (-3)$

4. Simplify the expression.
   $-2 - (-7) + 5 + (-6)$

5. Find each product.
   (a) $-11 \cdot 4$
   (b) $-3 \cdot (-13)$
   (c) $6 \cdot (-8)$
   (d) $5 \cdot 10$

6. Simplify the expression.
   $2(-1)(4)(-2)(-3)$

7. Find each quotient.
   (a) $24 \div (-6)$
   (b) $-60 \div 12$
   (c) $\dfrac{-36}{-4}$
   (d) $\dfrac{25}{-1}$

8. Simplify each expression.
   (a) $\sqrt{64}$
   (b) $-\sqrt{16}$

9. **Bank Account** An overdrawn checking account has a balance of $-\$57$. Find the amount of a deposit that would bring the balance to $\$108$.

10. **Ocean Depth** A diver descends in 3 stages, dropping 23 feet each time. Write the total depth of the diver as a product of integers and find the diver's final depth.

2.5 Order of Operations; Averages

Objectives
1. Evaluating Exponential Expressions
2. Using the Order of Operations
3. Evaluating Algebraic Expressions with Integers
4. Finding Averages

NEW VOCABULARY
☐ Average

1 Evaluating Exponential Expressions

Is there a difference in how the expressions -5^2 and $(-5)^2$ are evaluated? To answer this question, consider the following.

The opposite of the square of positive 5: $\quad -5^2 = -(5 \cdot 5) = -25$

Square the integer -5: $\quad (-5)^2 = (-5)(-5) = 25$

The expressions do not give the same result. The base of the exponential expression -5^2 is **positive**, while the base of $(-5)^2$ is **negative**.

EXAMPLE 1 — Evaluating exponential expressions

Evaluate each exponential expression.
(a) $(-2)^4$ (b) -3^3 (c) $(-4)^3$

Solution
(a) The base of the expression is negative. Because the exponent is 4, there is an even number of negative factors: $(-2)^4 = (-2) \cdot (-2) \cdot (-2) \cdot (-2) = 16$.
(b) The base of the expression is positive. First we find the cube of 3 and then find the opposite of the result: $-3^3 = -(3 \cdot 3 \cdot 3) = -27$.
(c) The base of the expression is negative. Because the exponent is 3, there is an odd number of negative factors: $(-4)^3 = (-4) \cdot (-4) \cdot (-4) = -64$.

Now Try Exercises 49, 51, 53, 57

CALCULATOR HELP
To use a calculator to evaluate exponential expressions involving integers, see Appendix E (page AP-30).

2 Using the Order of Operations

READING CHECK 1

- If we want to raise a negative integer to a power, what grouping symbol should be used?

READING CHECK 2

- Is the order of operations agreement for integers essentially the same as or different from the order of operations agreement for whole numbers?

A person getting dressed puts on socks *before* shoes. An airplane can land only *after* it lowers its landing gear. In mathematics, we may need to add, subtract, multiply, divide, take square roots, raise numbers to powers, or find absolute value—all within one problem. When evaluating an expression involving several of these operations, the order in which the operations are performed is important. We now apply the order of operations agreement to expressions involving integers.

> **ORDER OF OPERATIONS**
>
> 1. Do all calculations within grouping symbols, such as parentheses and radicals, or above and below a fraction bar.
> 2. Evaluate all exponential expressions.
> 3. Do all multiplication and division from *left to right*.
> 4. Do all addition and subtraction from *left to right*.

EXAMPLE 2 Using the order of operations to evaluate expressions

Evaluate each expression.
(a) $-16 + 5 \cdot 4$ (b) $24 \div 8 - 2 \cdot 6$ (c) $17 - 4^2 \div 2$

Solution
(a) Perform multiplication before addition.

$$-16 + 5 \cdot 4 = -16 + 20 \qquad \text{Multiply.}$$
$$= 4 \qquad \text{Add.}$$

(b) Perform multiplication and division from left to right.

$$24 \div 8 - 2 \cdot 6 = 3 - 2 \cdot 6 \qquad \text{Divide.}$$
$$= 3 - 12 \qquad \text{Multiply.}$$
$$= -9 \qquad \text{Subtract.}$$

(c) The exponential expression is evaluated first.

$$17 - 4^2 \div 2 = 17 - 16 \div 2 \qquad \text{Evaluate } 4^2.$$
$$= 17 - 8 \qquad \text{Divide.}$$
$$= 9 \qquad \text{Subtract.}$$

Now Try Exercises 5, 7, 11

Parentheses, radical symbols, fraction bars, and absolute value symbols are all grouping symbols. Remember that expressions within grouping symbols are evaluated first.

EXAMPLE 3 Evaluating expressions involving grouping symbols

Evaluate each expression.
(a) $3^2 - (5 + 4)$ (b) $20 \div |6 - 11| - 8$

Solution
(a) The expression within parentheses is evaluated first.

$$3^2 - (5 + 4) = 3^2 - 9 \qquad \text{Add within parentheses.}$$
$$= 9 - 9 \qquad \text{Evaluate } 3^2.$$
$$= 0 \qquad \text{Subtract.}$$

(b) Evaluate the absolute value expression first.

$$20 \div |6 - 11| - 8 = 20 \div |-5| - 8 \quad \text{Subtract within absolute value.}$$
$$= 20 \div 5 - 8 \quad \text{Evaluate } |-5|.$$
$$= 4 - 8 \quad \text{Divide.}$$
$$= -4 \quad \text{Subtract.}$$

Now Try Exercises 17, 23

EXAMPLE 4 Evaluating expressions involving grouping symbols

Evaluate each expression.

(a) $-32 \div \sqrt{17 - 1}$ **(b)** $\dfrac{-8 + 2 \cdot 3}{10 \div 5 - 1}$

Solution
(a) The expression under the radical is evaluated first.

$$-32 \div \sqrt{17 - 1} = -32 \div \sqrt{16} \quad \text{Subtract under radical.}$$
$$= -32 \div 4 \quad \text{Evaluate } \sqrt{16}.$$
$$= -8 \quad \text{Divide.}$$

> **STUDY TIP**
>
> Before visiting your instructor or going to a tutor center for help, be sure that you have tried a problem several times in different ways. Organize your questions so that you can be specific about the part of the problem that is giving you difficulty.

(b) Use the order of operations both above and below the fraction bar.

$$\frac{-8 + 2 \cdot 3}{10 \div 5 - 1} = \frac{-8 + 6}{2 - 1} \quad \text{Multiply above; divide below.}$$
$$= \frac{-2}{1} \quad \text{Add above; subtract below.}$$
$$= -2 \quad \text{Divide.}$$

Now Try Exercises 15, 21

Some expressions contain grouping symbols that are *nested* within other grouping symbols. Expressions of this type are evaluated by starting with the innermost grouping and working outward, as shown in the next example.

EXAMPLE 5 Evaluating expressions with nested grouping symbols

Evaluate each expression.
(a) $7 - \left(4^2 - (2 + 3) \cdot 5\right)$ **(b)** $21 \div \left(5 + |3 - 9| - 8\right)$

Solution
(a) Start by evaluating the innermost grouping, $(2 + 3)$.

$$7 - \left(4^2 - (2 + 3) \cdot 5\right) = 7 - (4^2 - 5 \cdot 5) \quad \text{Evaluate innermost grouping.}$$
$$= 7 - (16 - 5 \cdot 5) \quad \text{Evaluate } 4^2.$$
$$= 7 - (16 - 25) \quad \text{Multiply.}$$
$$= 7 - (-9) \quad \text{Evaluate parentheses.}$$
$$= 16 \quad \text{Subtract.}$$

(b) The absolute value, $|3 - 9|$, is the innermost grouping.

$$21 \div \left(5 + |3 - 9| - 8\right) = 21 \div (5 + |-6| - 8) \quad \text{Subtract within absolute value.}$$
$$= 21 \div (5 + 6 - 8) \quad \text{Evaluate } |-6|.$$
$$= 21 \div (11 - 8) \quad \text{Add.}$$
$$= 21 \div 3 \quad \text{Evaluate parentheses.}$$
$$= 7 \quad \text{Divide.}$$

Now Try Exercises 27, 35

READING CHECK 3

- What grouping symbols are used when substituting a negative value for a variable?

3 Evaluating Algebraic Expressions with Integers

An algebraic expression can be evaluated when particular values are assigned to the variables in the expression. If a negative value is assigned to a variable, then *it is best to place the negative value within parentheses* when replacing the variable. For example, if we let $x = -3$ and $y = -2$ in the expression $5x + y$, then replacing each variable with its given value results in the expression $5(-3) + (-2)$.

EXAMPLE 6 Evaluating algebraic expressions

Evaluate each expression for $x = -4$, $y = 5$, and $z = -2$.

(a) $17 - 2xz$ (b) $6y \div (z + |x| - 7)$ (c) $\dfrac{3yz}{x - 2}$

Solution
(a) Start by replacing x with -4 and z with -2 in the given expression.

$$\begin{aligned} 17 - 2xz &= 17 - 2(-4)(-2) & & x = -4 \text{ and } z = -2. \\ &= 17 - (-8)(-2) & & \text{Multiply } 2(-4). \\ &= 17 - 16 & & \text{Multiply } (-8)(-2). \\ &= 1 & & \text{Subtract.} \end{aligned}$$

(b) This expression contains all three variables. Replace x with -4, y with 5, and z with -2. Note that the positions of x and z in the expression allow for each variable to be replaced by a negative value *without* the need for additional parentheses.

$$\begin{aligned} 6y \div (z + |x| - 7) &= 6(5) \div (-2 + |-4| - 7) & & x = -4, y = 5, \text{ and } z = -2. \\ &= 6(5) \div (-2 + 4 - 7) & & \text{Evaluate } |-4|. \\ &= 6(5) \div (2 - 7) & & \text{Add } -2 + 4. \\ &= 6(5) \div (-5) & & \text{Subtract within parentheses.} \\ &= 30 \div (-5) & & \text{Multiply.} \\ &= -6 & & \text{Divide.} \end{aligned}$$

(c) Replace x with -4, y with 5, and z with -2.

$$\begin{aligned} \dfrac{3yz}{x - 2} &= \dfrac{3(5)(-2)}{-4 - 2} & & x = -4, y = 5, \text{ and } z = -2. \\ &= \dfrac{-30}{-4 - 2} & & \text{Multiply } 3(5)(-2). \\ &= \dfrac{-30}{-6} & & \text{Subtract } -4 - 2. \\ &= 5 & & \text{Divide.} \end{aligned}$$

Now Try Exercises 63, 69, 73

4 Finding Averages

🌐 **Math in Context** (Grading) When graded exams are returned to students, instructors often report the average score for the class. The **average** of a list of numbers is found as follows.

$$\text{Average} = \dfrac{\text{the sum of the numbers in the list}}{\text{the } number \text{ of numbers in the list}}$$

For our example of the average score for the class,

$$\text{Class Average} = \frac{\text{the sum of all student scores}}{\text{the number of students}}.$$

EXAMPLE 7 Finding average monthly profit for an iPhone application

The bar graph in **FIGURE 2.15** shows the monthly profit for a small gaming company after the release of a new iPhone application. A negative profit indicates a loss. Find the average monthly profit for the months shown.

FIGURE 2.15

Solution
To find the average of $-16, -8, -3, 10, 16,$ and 19, add the numbers together and then divide the result by 6 because there are **6** numbers in the list.

$$\text{Average} = \frac{(-16) + (-8) + (-3) + 10 + 16 + 19}{6} = \frac{18}{6} = 3$$

Since the profit is given in thousands of dollars, the average monthly profit for January through June is $3000.

Now Try Exercise 87

2.5 Putting It All Together

CONCEPT	COMMENTS	EXAMPLES
Order of Operations Agreement	1. Do all calculations within grouping symbols, such as parentheses and radicals, or above and below a fraction bar. 2. Evaluate all exponential expressions. 3. Do all multiplication and division from *left to right*. 4. Do all addition and subtraction from *left to right*.	$6 \div (-2) + 3 \cdot 5 = -3 + 3 \cdot 5$ $= -3 + 15$ $= 12$ $18 \div (3 - 6)^2 = 18 \div (-3)^2$ $= 18 \div 9$ $= 2$ $\sqrt{10 - (-6)} = \sqrt{16}$ $= 4$

CONCEPT	COMMENTS	EXAMPLES
Evaluating Algebraic Expressions	Replace each variable in the expression with its given value and then evaluate using the order of operations agreement.	Evaluate $7 + 3x$ for $x = -5$. $$7 + 3x = 7 - 3(-5)$$ $$= 7 - (-15)$$ $$= -8$$
Average	Average is found by using the formula: $$\text{Average} = \frac{\text{the sum of a list of numbers}}{\text{number of numbers in the list}}.$$	The average of $-5, 0, 4, 7,$ and 9 is $$\frac{(-5) + 0 + 4 + 7 + 9}{5} = 3.$$

2.5 Exercises

MyMathLab®

CONCEPTS AND VOCABULARY

1. The _____ agreement ensures that algebraic expressions are evaluated in the same way by everyone.

2. When grouping symbols are nested, we evaluate the _____ grouping first.

3. When replacing a variable with a negative number, we often use _____.

4. When we divide the sum of a list of numbers by the number of numbers in the list, we are finding the _____ of the list.

USING THE ORDER OF OPERATIONS

Exercises 5–40: Evaluate the expression.

5. $2 + (-3) \cdot 4$
6. $-8 - 4(-5)$
7. $-36 \div 3^2 + 7$
8. $48 \div 12 - 11$
9. $60 + (-35) \div 7 - 28$
10. $-55 + 9 \cdot 7 - 18$
11. $49 \div 7 + (-2) \cdot 4$
12. $-2 \cdot (-50) - 8 \div 2$
13. $\dfrac{3 - 25}{7 + 4}$
14. $\dfrac{19 + 16}{5(-1)}$
15. $\dfrac{45 + 9(-10)}{-5}$
16. $\left|\dfrac{19 - 59}{2 \cdot 5}\right|$
17. $36 - 3^2 \div (6 - 9)$
18. $-28 \div (2 - 3^2) + 3$
19. $35 - |3 + 4^2 \div (-2)|$
20. $(36 \div 6^2 - 4) \cdot |-3|$
21. $\sqrt{41 + 8} + (-7)$
22. $-78 - \sqrt{80 + 1}$
23. $-8^2 + |9 \cdot (-8)|$
24. $\sqrt{-5^2 + 61} - 3^2$
25. $\sqrt{100} \cdot \sqrt{|0 - 25|}$
26. $|-43| - |20 - 6^2|$
27. $((2 - 7)^2 \div 5) - 18$
28. $(20 \cdot (6 - 7) \div 10)^3$
29. $\dfrac{(13 - 9) \cdot 6}{|37 - 41| \cdot 2}$
30. $\dfrac{2 \cdot (4 + 1)^2 + 10}{5 \cdot |4 - 7|}$
31. $\dfrac{(6 - 3) \cdot 5}{2^3 - \sqrt{81}}$
32. $\dfrac{(\sqrt{64} + 2) \cdot 5}{|6^2 - 41|}$
33. $(7 - 9)^2 - 3\sqrt{2 \cdot 8}$
34. $(4 - 5)^2 \cdot \sqrt{50 - 25}$
35. $-4|2 - 4 \cdot 3| + ((72 \div 9)^2 + 6)$
36. $50 - 5^2 \cdot 3 - 32 \div 2^3 + 8$
37. $29 - (3 \cdot 9 - (32 \div 2^4) + 3)$
38. $0 - \sqrt{-40 \div 5 - (3 - 3 \cdot 2^2)}$
39. $\sqrt{25 - 3 - (-1 \cdot 6^2 \div 12)} + 11$
40. $-34 - 3 \cdot 7 + 2 + 2^2(-3) - 16 \div 2^3 - 9$

Exercises 41–48: Insert parentheses as needed in the expression in order to make it equal to 0. More than one set of parentheses may be needed.

41. $-20 + 10 \cdot 14 - 12$
42. $-4 - 3 - 8 - 1$
43. $-5^2 \div 3 + 2 + 5$
44. $7 - 10 \cdot 3^2 + 27$
45. $32 \div 4^2 - 2 \cdot 9$
46. $5 - 5 \cdot 3^2 \div 3$
47. $16 - 4^2 \div 4 - 9$
48. $8 - 5 + 6^2 \div 12$

EVALUATING EXPONENTIAL EXPRESSIONS

Exercises 49–62: Evaluate the exponential expression.

49. -2^3
50. $(-3)^2$
51. $(-4)^2$
52. -5^2
53. -9^2
54. $(-8)^2$
55. -1^4
56. -3^4
57. $(-3)^3$
58. -1^5
59. -10^6
60. $(-10)^3$
61. $(-10)^4$
62. -10^5

EVALUATING ALGEBRAIC EXPRESSIONS

Exercises 63–76: Evaluate the expression for the given values of the variables.

63. $4 \cdot y + x$, for $x = -5, y = 1$
64. $3a + 9 - b$, for $a = 5, b = -6$
65. $24v - 6w$, for $v = -2, w = -8$
66. $3c - 5d$, for $c = -5, d = -3$
67. $2m + (4^2 + n) \div 8$, for $m = 7, n = -32$
68. $w + ((3 - v)^2 \div w)$, for $v = 0, w = -3$
69. $2m + |2^3 + n| \div 8$, for $m = 5, n = -16$
70. $w + ((5 - 2v)^2 \div 3w)$, for $v = 4, w = -1$
71. $\sqrt{p^2 - 7} - 2q^2$, for $p = -4, q = 2$
72. $\sqrt{r - 3} - (14 + s)$, for $r = 39, s = -6$
73. $\dfrac{2(7 + c)}{|d - 5| \cdot (-4)}$, for $c = 5, d = 4$
74. $\dfrac{(6 - g) + 7}{(\sqrt{2h} - 3)^2}$, for $g = -5, h = 2$
75. $xy + (y^2 - x) \div y$, for $x = -12, y = -3$
76. $a \cdot b - a \div b + b$, for $a = 0, b = -1$

FINDING AVERAGES

Exercises 77–82: Find the average of the integers.

77. $-13, 8, -1, 12, -5, -7$
78. $-8, 7, 7, -4, 8$
79. $-17, -13, -15, -15, -19, -17$
80. $-3, -5, -2, -6, -4$
81. $-11, -8, -2, 2, 8, 11$
82. $-15, -13, -5, -1, 1, 5, 13, 15$

APPLICATIONS INVOLVING ORDER OF OPERATIONS

83. **Converting Temperature** To convert a temperature F given in degrees Fahrenheit to an equivalent temperature C in degrees Celsius, use the formula
$$C = \frac{5(F - 32)}{9}.$$
Find the Celsius temperature that is equivalent to $-4\,°F$.

84. **Converting Temperature** Use the formula in the previous exercise to find the Celsius temperature that is equivalent to $-40\,°F$.

85. **Converting Temperature** To convert a temperature C given in degrees Celsius to an equivalent temperature F in degrees Fahrenheit, use the formula
$$F = \frac{9C}{5} + 32.$$
Find the Fahrenheit temperature that is equivalent to $-15\,°C$.

86. **Converting Temperature** Use the formula in the previous exercise to find the Fahrenheit temperature that is equivalent to $-40\,°C$.

Exercises 87 and 88: **Profit and Loss** *The following bar graph shows the monthly profit for a jewelry-making company. A negative profit indicates a loss.*

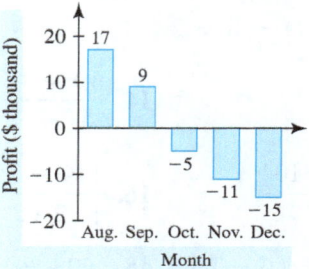

87. Find the average monthly profit for all of the months shown.

88. Find the average monthly profit for the months of August, September, and October only.

89. **Average Temperature** Find the average of the list of Fahrenheit temperatures.

    $-16°, -22°, 1°, 11°, -2°, -21°,$ and $7\ °F$

90. **Average Temperature** Find the average of the list of Celsius temperatures.

    $-6°, -2°, 15°, 18°, -5°,$ and $4°C$

WRITING ABOUT MATHEMATICS

91. Why is it sometimes necessary to insert parentheses when evaluating an expression for a negative value of a variable? Give an example of this situation.

92. Give an example of an expression that contains nested grouping symbols. Explain how to evaluate this expression.

Group Activity

Climate Change If the entire Greenland ice sheet melted completely, it would produce approximately 650,000,000,000,000,000 gallons of water. This means that 650 *quadrillion* gallons of water would enter the ocean, raising sea level significantly. The formula

$$F = \left(\frac{6g}{13}\right) \div 12$$

can be used to estimate the increase in sea level F in feet that would result from g quadrillion gallons of water entering the oceans. (*Source:* U.S. Geological Survey.)

Compute each value of F in the table by replacing g in the formula with each value of g given in the table.

g (quadrillion gallons)	F (feet)
26	
130	
312	
468	
650	

2.6 Solving Equations That Have Integer Solutions

Objectives

1. Checking a Solution
2. Solving Equations Using Guess-and-Check
3. Solving Equations Using Tables of Values
4. Solving Equations That Have Integer Solutions Graphically

NEW VOCABULARY

☐ Equation
☐ Solution

1 Checking a Solution

An **equation** is a mathematical statement that two algebraic expressions are equal. If an equation contains one variable, then any value of the variable that makes the equation true is called a **solution** to the equation. Examples of equations with one variable include

$$4m = 20, \quad 3x - 15 = -12, \quad -80 \div b = 10, \quad \text{and} \quad 1 + y = 37.$$

The solutions to these equations are given by

$$m = 5, \quad x = 1, \quad b = -8, \quad \text{and} \quad y = 36.$$

For example, the solution **1** can be checked in the equation $3x - 15 = -12$ as follows.

$$\begin{aligned}
3x - 15 &= -12 &&\text{Given equation} \\
3(1) - 15 &\stackrel{?}{=} -12 &&\text{Replace } x \text{ with 1.} \\
3 - 15 &\stackrel{?}{=} -12 &&\text{Simplify.} \\
-12 &= -12 \checkmark &&\text{The solution checks.}
\end{aligned}$$

Because the resulting equation is true, the solution checks.

READING CHECK 1

- How do we know if a number is a solution to an equation?

CHECKING A SOLUTION

To check a solution, replace *every* occurrence of the variable in the equation with the proposed solution and check to see if the resulting equation is true.

EXAMPLE 1 Checking solutions

Check each solution.
(a) Is -5 a solution to $3x + 7 = -8$?
(b) Is -1 a solution to $2y^2 = \sqrt{3 - y}$?
(c) Is 2 a solution to $|5w - 11| = -1$?

Solution
(a) Replace the variable x with -5.

$$3x + 7 = -8 \qquad \text{Given equation}$$
$$3(-5) + 7 \stackrel{?}{=} -8 \qquad \text{Replace } x \text{ with } -5.$$
$$-15 + 7 \stackrel{?}{=} -8 \qquad \text{Multiply.}$$
$$-8 = -8 \checkmark \qquad \text{The solution checks.}$$

The solution checks, which means that -5 is a solution to the equation $3x + 7 = -8$.

(b) Replace *every* occurrence of the variable y with -1.

$$2y^2 = \sqrt{3 - y} \qquad \text{Given equation}$$
$$2(-1)^2 \stackrel{?}{=} \sqrt{3 - (-1)} \qquad \text{Replace } y \text{ with } -1.$$
$$2(-1)^2 \stackrel{?}{=} \sqrt{3 + 1} \qquad \text{Add the opposite.}$$
$$2(1) \stackrel{?}{=} \sqrt{4} \qquad \text{Simplify.}$$
$$2 = 2 \checkmark \qquad \text{The solution checks.}$$

The solution checks, which means that -1 is a solution to the equation $2y^2 = \sqrt{3 - y}$.

(c) Replace the variable w with 2.

$$|5w - 11| = -1 \qquad \text{Given equation}$$
$$|5(2) - 11| \stackrel{?}{=} -1 \qquad \text{Replace } w \text{ with 2.}$$
$$|10 - 11| \stackrel{?}{=} -1 \qquad \text{Multiply.}$$
$$|-1| \neq -1 \; \text{✗} \qquad \text{The solution does not check.}$$

Because an absolute value result cannot be negative, the solution does not check and 2 is **not** a solution to the equation $|5w - 11| = -1$.

Now Try Exercises 7, 11, 15

2 Solving Equations Using Guess-and-Check

Some equations can be solved by thinking of the variable as a missing number. For example, the equation $4 + x = 13$ has a missing value for x. To solve this equation, we must find the number that is added to 4 to get 13. Using a *guess-and-check* strategy, we find that the solution is 9 because $4 + 9 = 13$.

EXAMPLE 2 Solving equations using guess-and-check

Solve each equation.
(a) $7x = -63$ (b) $4 - y = -9$ (c) $w \div (-3) = 11$

Solution
(a) This equation has a missing factor. The solution is -9 because $7(-9) = -63$.
(b) To get a result of -9 in this equation, we must subtract 13 from 4. The solution is 13 because $4 - 13 = -9$.
(c) The equation requires us to find a number that can be divided by -3 to get 11. The solution is -33 because it checks in the equation $-33 \div (-3) = 11$.

Now Try Exercises 23, 25, 27

3 Solving Equations Using Tables of Values

Math in Landscaping Context Some equations can be solved by making a table of values and selecting the solution from the table. Suppose that a landscaping crew is able to place 500 cement blocks before noon and 50 blocks each hour afterwards until 5:00 P.M. The total number of blocks B placed by the crew after various elapsed times t is shown in **TABLE 2.4**.

Blocks B Placed t Hours Past Noon

Elapsed Time: t (hours)	0	1	2	3	4	5
Total Blocks Placed: B	500	550	600	650	700	750

TABLE 2.4

(Noon: 500 blocks; 3:00 P.M.: 650 blocks)

A mathematical formula that *describes* or *models* these data is given by

$$B = 500 + 50t.$$

For example, at 2 hours past noon, or 2:00 P.M., a total of

$$B = 500 + 50(2) = 500 + 100 = 600$$

blocks have been placed. By replacing the variable t in the formula with the other values of t in **TABLE 2.4**, we can find the corresponding values of B in the table.

Now suppose that the crew supervisor wants to know the time when 700 blocks had been placed. Replacing B in the formula with 700 results in the equation

$$700 = 500 + 50t.$$

Although symbolic techniques for solving this type of equation have not yet been discussed, we are able to use **TABLE 2.4** to solve it. From the table, the number of blocks B is 700 when t is 4, or at 4:00 P.M. The solution to the equation $700 = 500 + 50t$ is 4.

EXAMPLE 3 Using a table to solve an equation

Complete **TABLE 2.5** for the given values of x and then solve the equation $5x - 13 = -3$.

x	-1	0	1	2	3
$5x - 13$					

TABLE 2.5

Solution

Begin by replacing x in $5x - 13$ with -1. Because $5(-1) - 13$ evaluates to -18, the left side of the equation equals -18 when $x = -1$. We write the result -18 below -1, as shown in **TABLE 2.6**. Similarly, we write -13 below 0 because the expression $5(0) - 13$ evaluates to -13. The remaining values are found by evatuating the expression $5x - 13$ for the last three x-values and are shown in **TABLE 2.6**.

Solving $5x - 13 = -3$ Using a Table

x	-1	0	1	2	3
$5x - 13$	-18	-13	-8	-3	2

TABLE 2.6

An equation is true when its left side equals its right side. **TABLE 2.6** reveals that the left side, $5x - 13$, equals the right side, -3, when $x = 2$. Therefore, the solution to the given equation $5x - 13 = -3$ is 2.

Now Try Exercise 35

Example 3 shows that making a table of values is simply an organized way of selecting and checking possible values of the variable to see if any value makes the equation true. If a value makes an equation true, then that value is a solution.

Math in Context (Twitter) In the next example, a table of values is used to find the year when the number of Twitter users first reached 170 million.

EXAMPLE 4 Analyzing Twitter users

For 14 quarters after Twitter first reached 30 million users, the number of users U in millions could be approximated using the formula

$$U = 14x + 30,$$

where x represents the number of quarters after Twitter reached 30 million users. By replacing U with 170 in the formula, we form the equation $170 = 14x + 30$. The solution represents the quarter when the number of Twitter users reached 170 million. Solve this equation by completing **TABLE 2.7**.

Twitter Users

x	0	2	4	6	8	10	12	14
$14x + 30$								

TABLE 2.7

Solution

In this example, $x = 0$ represents the time when Twitter reached 30 million users, so we can write 30 below the 0 in the table. (See **TABLE 2.8**.) Next replace x in $14x + 30$ with 2 and then evaluate the resulting expression.

$$14(2) + 30 = 28 + 30 = 58$$

Write 58 below 2, as shown in **TABLE 2.8**. The remaining values are found in a similar fashion and are shown in **TABLE 2.8**.

Twitter Users

x	0	2	4	6	8	10	12	14
$14x + 30$	30	58	86	114	142	170	198	226

TABLE 2.8

TABLE 2.8 shows that the right side of the equation, $14x + 30$, equals the left side, 170, when $x = 10$. Therefore, the number of Twitter users reached 170 million 10 quarters after Twitter first reached 30 million users.

Now Try Exercise 56

4 Solving Equations That Have Integer Solutions Graphically

Math in Context (Temperature) If the air temperature at ground level is 80 °F, then the air temperature T at an altitude of x miles can be found using the formula

$$T = 80 - 19x.$$

A table of values corresponding to the expression $80 - 19x$ is shown in **TABLE 2.9**. (Source: Miller, A., and R. Anthes, *Meteorology*, 5th ed.)

Air Temperature at an Altitude of x Miles

x	0	1	2	3	4	5	6
$80 - 19x$	80	61	42	23	4	-15	-34

TABLE 2.9

Note that **TABLE 2.9** does not show that air temperature decreases *gradually* as altitude increases. For example, the temperature is *not* 80 °F from the ground up to an altitude of 1 mile, where it suddenly drops to 61 °F. Rather, between the ground and 1 mile of altitude, air takes on all temperature values from 61 °F to 80 °F. The line graph in **FIGURE 2.16** more accurately displays this situation. The line in **FIGURE 2.16** displays the *graph* of the expression $80 - 19x$. It shows *all* values of the expression $80 - 19x$ for altitudes between 0 miles and 6 miles.

Air Temperatures at Various Altitudes

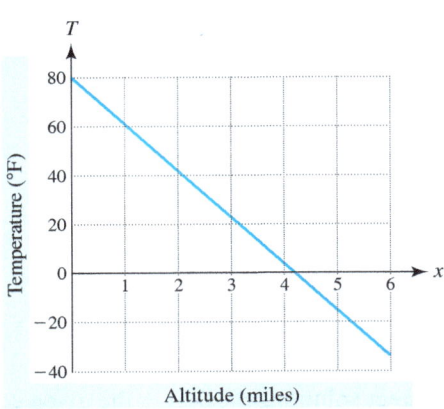

FIGURE 2.16

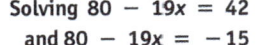

Solving $80 - 19x = 42$ and $80 - 19x = -15$

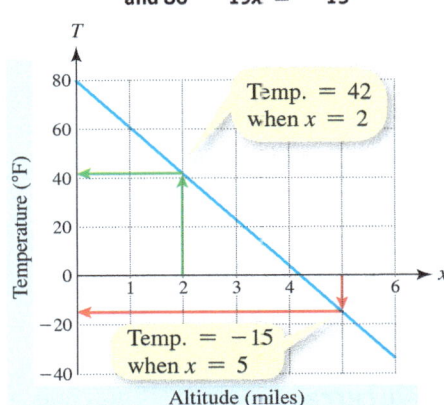

FIGURE 2.17

TABLE 2.9 shows that at an altitude of 2 miles, the temperature is 42 °F, and at an altitude of 5 miles, the temperature is -15 °F. These values are displayed visually in **FIGURE 2.17**. The other values from **TABLE 2.9** can be found visually in a similar way.

When a graph is provided, we can solve some equations graphically.

> **SOLVING AN EQUATION GRAPHICALLY**
>
> To solve an equation graphically,
>
> 1. Use the graph to estimate a solution.
> 2. Check the estimated solution to be sure it is correct. If it is not, try a new value.

NOTE: A graphical solution to an equation is an *estimate* of the actual solution. Be sure to check a graphical solution in the given equation to be sure that it is correct. ■

EXAMPLE 5 **Solving an equation graphically**

The graph of the expression $2x - 21$ is shown in **FIGURE 2.18**. Use **FIGURE 2.18** to solve the equation $2x - 21 = -7$ graphically.

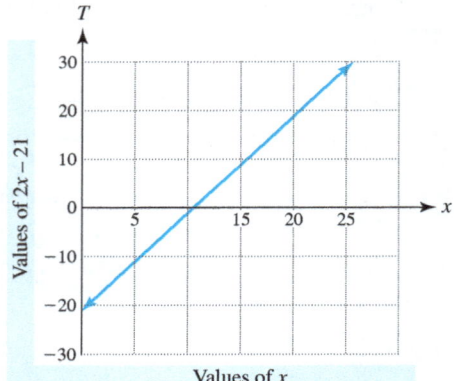

FIGURE 2.18

Solution
Locate -7 at the left edge of the graph. Move **horizontally** to the right to the graphed line. From this position, move **vertically** upward to the value of x. The solution appears to be about 7, as shown in **FIGURE 2.19**.

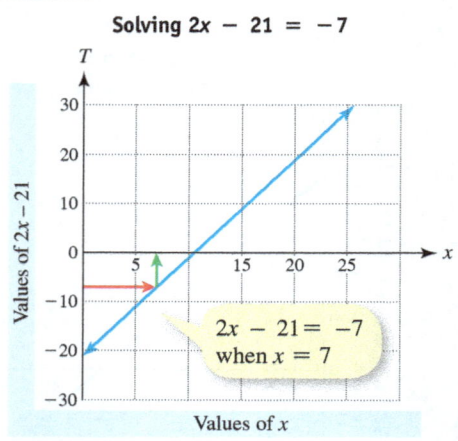

FIGURE 2.19

READING CHECK 2

- Why is it a good idea to check a solution that is found graphically?

To see that 7 is the correct solution, check it in the given equation.

$$2x - 21 = -7 \qquad \text{Given equation}$$
$$2(7) - 21 \stackrel{?}{=} -7 \qquad \text{Replace } x \text{ with 7.}$$
$$14 - 21 \stackrel{?}{=} -7 \qquad \text{Multiply.}$$
$$-7 = -7 \checkmark \qquad \text{The solution checks.}$$

The solution to the equation $2x - 21 = -7$ is 7.

Now Try Exercise 49

2.6 Putting It All Together

CONCEPT	COMMENTS	EXAMPLES
Checking a Solution	To check a solution, replace *every* occurrence of the variable with the proposed solution and see if the resulting equation is true.	The solution to $-4x = 24$ is -6. $-4(-6) \stackrel{?}{=} 24$ Replace x with -6. $24 = 24$ ✓ The solution checks.
Solving Equations Using Guess-and-Check	Determine what portion of the equation is missing: factor, dividend, etc. Test sample values in the equation.	The solution to $x - 4 = 10$ is **14** because $14 - 4 = 10$.
Solving Equations Using Tables of Values	A table of values can be used to solve some equations by completing the table for various values of the variable and then selecting the solution from the table.	The solution to $5x + 3 = 8$ is 1. | x | -1 | 0 | 1 | |---|---|---|---| | $5x + 3$ | -2 | 3 | **8** |
Solving Equations Graphically	1. Use the graph to estimate a solution. 2. Check the estimated solution to be sure it is correct. If it is not, try a new value.	The solution to $3x - 7 = 20$ is 9.

2.6 Exercises

CONCEPTS AND VOCABULARY

1. A(n) _____ is a mathematical statement that two algebraic expressions are equal.

2. Any value of the variable that makes an equation true is a(n) _____ to the equation.

3. To check a solution, replace every occurrence of the _____ in the equation with the proposed solution and see if the resulting equation is true.

4. When solving an equation graphically, be sure to _____ the solution in the given equation.

CHECKING SOLUTIONS

Exercises 5–20: Check the solution as indicated.

5. Is -7 a solution to $2 - x = 9$?

6. Is -12 a solution to $3 + n = -9$?

7. Is -3 a solution to $4x - 2 = -8$?

8. Is 3 a solution to $5a - 12 = -2$?

9. Is -6 a solution to $\frac{3x - 2}{5} = 6$?

10. Is 9 a solution to $\frac{52}{y + 4} = 4$?

11. Is -3 a solution to $\sqrt{1-y} = 3y + 10$?

12. Is 1 a solution to $|2w - 8| = 6$?

13. Is 4 a solution to $6 + b \div 2 = 5$?

14. Is 4 a solution to $\sqrt{6m + 1} = m + 1$?

15. Is -9 a solution to $|w - 8| = 8 - w$?

16. Is -2 a solution to $(3 - x)^2 = 5x + 15$?

17. Is -7 a solution to $x^2 + 2x = 35$?

18. Is -6 a solution to $10 - y \div 2 = y + 19$?

19. Is -2 a solution to $3x^2 - x + 5 = 15$?

20. Is -1 a solution to $x^3 + 5x - 2 = 4$?

SOLVING EQUATIONS USING GUESS-AND-CHECK

Exercises 21–32: Solve the equation.

21. $b + 3 = -12$
22. $3x = 36$
23. $11 - z = 16$
24. $m - (-4) = 3$
25. $-48 \div d = -6$
26. $n \div (-5) = -9$
27. $35 = -7x$
28. $7 + y = -8$
29. $2x + 1 = 11$
30. $3x - 2 = 10$
31. $18 - 2x = 4$
32. $25 - 3m = 7$

SOLVING EQUATIONS USING TABLES

Exercises 33–40: Complete the table. Then solve the given equation.

33. $x + 2 = 1$

x	-2	-1	0	1	2
$x + 2$					

34. $3 - x = 3$

x	-2	-1	0	1	2
$3 - x$					

35. $3x + 5 = 8$

x	-2	-1	0	1	2
$3x + 5$					

36. $7x - 4 = -18$

x	-2	-1	0	1	2
$7x - 4$					

37. $5 + 6x = 17$

x	-2	-1	0	1	2
$5 + 6x$					

38. $3 - 4x = 7$

x	-2	-1	0	1	2
$3 - 4x$					

39. $\sqrt{3 - x} = 3$

x	-13	-6	-1	2	3
$\sqrt{3 - x}$					

40. $\sqrt{x + 7} = 2$

x	-7	-6	-3	2	9
$\sqrt{x + 7}$					

Exercises 41–48: Solve the equation by making a table of values. Use $-3, -2, -1, 0, 1, 2,$ and 3 for the values of x in your table.

41. $x - 2 = -3$
42. $6 - x = 5$
43. $-3x + 8 = 14$
44. $-2x + 5 = -1$
45. $4 - 9x = -14$
46. $4 - 3x = 7$
47. $1 + 3x = 1$
48. $8 + 2x = 2$

SOLVING EQUATIONS GRAPHICALLY

Exercises 49–54: Solve the given equation graphically.

49. $x - 4 = 3$
50. $12 - 2x = -6$

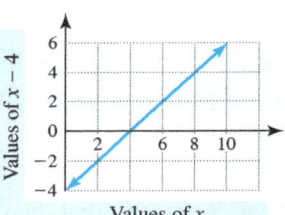
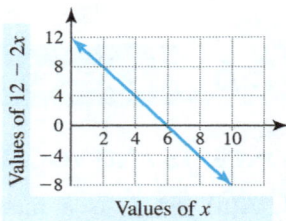

51. $-4x + 27 = 15$
52. $5x - 40 = -10$

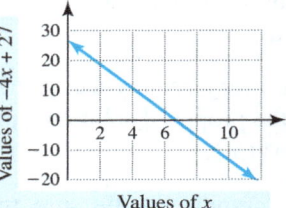
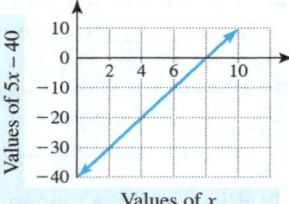

53. $6x - 38 = -20$

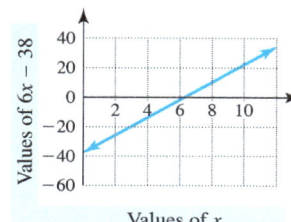

Values of x

54. $-9x + 36 = 0$

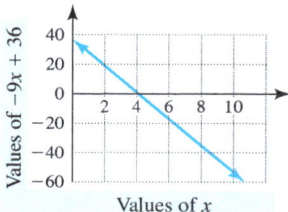

Values of x

APPLICATIONS INVOLVING EQUATIONS

55. Worldwide Internet Users The number of Internet users I in millions during year x, where x is from 2004 to 2014, can be approximated by the formula

$$I = 201x - 401{,}889.$$

Replace I in the formula with 2523 and use a table of values to solve the resulting equation. In what year were there 2523 million Internet users? (*Source:* Internet World Stats.)

56. Phablet Growth The total number of global phablet (tablet phone) shipments P, in millions during year x, where x is from 2014 to 2018, can be approximated by the formula

$$P = 232x - 466{,}848.$$

Replace P in the formula with 1096 and use a table of values to solve the resulting equation. In what year did the number of phablets shipped reach 1096? (*Source:* Business Insider.)

57. Gas Mileage The gas mileage M of a car using G gallons of gasoline to travel a distance of 120 miles is given by

$$M = \frac{120}{G}.$$

Complete the table and find the number of gallons used by a car traveling 120 miles if the car gets $M = 30$ miles per gallon.

G	1	2	3	4	5
$\frac{120}{G}$					

58. Altitude and Temperature If the air temperature on the ground is 75 °F, then the air temperature T at an altitude of x miles is given by the formula

$$T = 75 - 19x.$$

Complete the table and find the altitude where the air temperature is -1 °F.

x	1	2	3	4	5
$75 - 19x$					

59. Medicare Costs Total Medicare costs C in billions of dollars can be found using the formula

$$C = 18x - 35{,}750,$$

where x is a year from 1996 to 2016. Find the year when Medicare costs were \$466 billion by solving the equation $466 = 18x - 35{,}750$ graphically, using the following graph. (*Source:* Office of Management and Budget.)

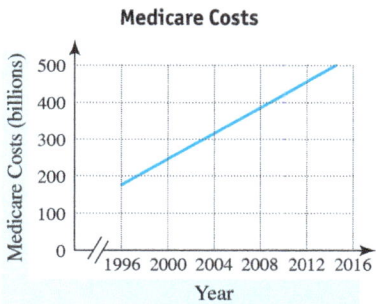

Medicare Costs

60. Air Temperature If the air temperature on the ground is 61 °F, then the air temperature T at an altitude of x miles is given by the formula

$$T = 61 - 19x.$$

Find the altitude where the air temperature is -15 °F by solving the equation $-15 = 61 - 19x$ graphically, using the following graph.

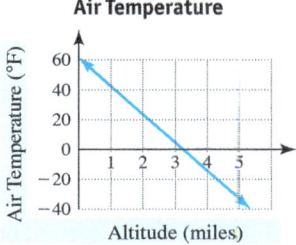

Air Temperature

WRITING ABOUT MATHEMATICS

61. For the equation $3x = 21$, give a value of x that is a solution and a value of x that is not a solution. Explain how to check these values.

62. Explain why the solutions to some equations may be difficult to find using a table of values.

SECTIONS 2.5 and 2.6 — Checking Basic Concepts

1. Evaluate each expression.
   (a) $40 + (-28) \div 4 - 10$
   (b) $4 - |5 + 3^2 \div (-9)|$
   (c) $(5 - 9)^2 - 5\sqrt{2 \cdot 18}$

2. Evaluate the expression for the given values of the variables.
   (a) $6c - 5d$, for $c = -2, d = -4$
   (b) $((7 - v)^2 \div w) + v$, for $v = 1, w = -3$
   (c) $\dfrac{(5 - g) + 14}{(\sqrt{h} - 5)^2}$, for $g = -5, h = 9$

3. Is -2 a solution to $6a - 12 = 0$?

4. Is -3 a solution to $(1 - x)^2 = 1 - 5x$?

5. Solve each equation.
   (a) $b + 2 = -3$ (b) $-24 \div d = -4$

6. Solve $5 - 2x = 7$ by completing the table.

x	-2	-1	0	1	2
$5 - 2x$					

7. **Converting Temperature** To convert a temperature C in degrees Celsius to an equivalent temperature F in degrees Fahrenheit, use the formula
   $$F = \frac{9C}{5} + 32.$$
   Find the Fahrenheit temperature that is equivalent to $-50\,°C$.

CHAPTER 2 Summary

SECTION 2.1 ■ INTEGERS AND THE NUMBER LINE

Positive Numbers A positive number is greater than 0. **Examples:** $7, +17, 152,$ and $+10{,}079$.

Negative Numbers A negative number is less than 0. **Examples:** $-1, -77, -509,$ and -6592.

Signed Numbers Signed numbers are the negative numbers, positive numbers, and zero.

Opposite (Additive Inverse) The opposite (or additive inverse) of a number a is written as $-a$.

Examples: The opposite of 13 is -13, and $13 + (-13) = 0$.
The opposite of -32 is 32, and $-32 + 32 = 0$.

Double Negative Rule Let a be any number. Then $-(-a) = a$.

Examples: $-(-7) = 7$ and $-(-25) = 25$

Integers $\ldots, -3, -2, -1, 0, 1, 2, 3, \ldots$

The integers include the natural numbers, zero, and the opposites of the natural numbers.

Absolute Value The absolute value of an integer equals its distance from 0 on a number line. The absolute value of a is written as $|a|$.

Examples: $|4| = 4, \quad |-17| = 17, \quad$ and $\quad |0| = 0$

SECTION 2.2 ■ ADDING INTEGERS

Adding Integers with Like Signs
1. Find the sum of the absolute value of the integers.
2. Keep the common sign of the two integers as the sign of the sum.

Example: Since $|-2| + |-11| = 13$, we know that $-2 + (-11) = -13$.

Adding Integers with Unlike Signs	1. Find the absolute value of the integers. 2. Subtract the smaller absolute value from the larger absolute value. 3. Keep the sign of the integer with the larger absolute value as the sign of the sum. **Example:** To add $4 + (-9)$, find the absolute values: $\lvert 4 \rvert = 4$ and $\lvert -9 \rvert = 9$. Subtract $9 - 4 = 5$. Because $\lvert -9 \rvert > \lvert 4 \rvert$, the sum is negative. $4 + (-9) = -5$
Addition Properties	**Property** **Examples** 1. Commutative Property $-24 + 9 = 9 + (-24)$ 2. Associative Property $(-3 + 5) + 1 = -3 + (5 + 1)$ 3. Identity Property $-2 + 0 = -2$ and $0 + (-2) = -2$ 4. Inverse Property $5 + (-5) = 0$ and $-5 + 5 = 0$
Adding Integers Using a Number Line	Starting at 0, draw an arrow representing the first addend. From the tip of this arrow, draw an arrow representing the second addend. The sum is located at the tip of the second arrow. **Example:** To add $-2 + 3$ on a number line, draw arrows as shown. The sum is **1**. 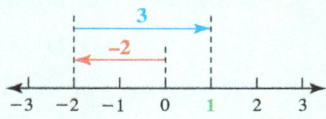
Adding Integers Using Symbols	Using the symbol ⌒ to represent a positive unit and the symbol ⌣ to represent a negative unit, draw the appropriate number of units for each addend. Form "zeros" when possible. **Example:** Add $2 + (-5)$ as shown. Ignoring the two "zeros," the sum is -3. 

SECTION 2.3 ■ SUBTRACTING INTEGERS

Subtracting Integers by Adding the Opposite	If the variables a and b represent two numbers, then $a - b = a - (-b)$. **Examples:** $4 - (-1) = 4 + 1 = 5$ and $-3 - 7 = -3 + (-7) = -10$
Subtracting Integers Using a Number Line	To subtract integers using a number line, first change the subtraction to addition of the opposite, and then add using a number line. **Example:** To subtract $-2 - (-3)$, first change to the sum $-2 + 3$, and then add. The difference is **1**. 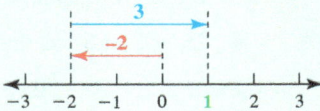
Subtracting Integers Using Symbols	The symbol ⌒ can be used to represent a positive unit and the symbol ⌣ can be used to represent a negative unit. Think of subtraction as "take away." **Example:** Subtract $-1 - (-4)$ as shown. The difference is **3**.

SECTION 2.4 ■ MULTIPLYING AND DIVIDING INTEGERS

Multiplying Integers with Like Signs

The product of two integers with *like* signs is positive.

Examples: $-3(-4) = 12$ and $5 \times 2 = 10$

Multiplying Integers with Unlike Signs

The product of two integers with *unlike* signs is negative.

Examples: $-4 \cdot 7 = -28$ and $3 \times (-6) = -18$

Multiplication Properties

Property	Examples
1. Commutative Property	$5 \cdot (-6) = -6 \cdot 5$
2. Associative Property	$(-3 \cdot 2) \cdot 7 = -3 \cdot (2 \cdot 7)$
3. Identity Property	$-8 \cdot 1 = -8$ and $1 \cdot (-8) = -8$
4. Zero Property	$0 \cdot (-3) = 0$ and $-3 \cdot 0 = 0$
5. Distributive Properties	$-2(1 + 5) = -2(1) + (-2)(5)$ $-3(4 - 7) = -3(4) - (-3)(7)$

Products with More Than Two Factors

If there is an *even* number of negative factors, the product is positive.
If there is an *odd* number of negative factors, the product is negative.

Examples: The product $-4 \cdot 7 \cdot 1 \cdot (-6)$ is positive.

The product $-5 \cdot (-2) \cdot 3 \cdot (-8)$ is negative.

Dividing Integers with Like Signs

The quotient of two integers with *like* signs is positive.

Examples: $-12 \div (-4) = 3$ and $10 \div 5 = 2$

Dividing Integers with Unlike Signs

The quotient of two integers with *unlike* signs is negative.

Examples: $-15 \div 5 = -3$ and $20 \div (-4) = -5$

Number of Square Roots

A positive integer a has one positive and one negative square root. The positive square root, or *principal square root*, is written as $\sqrt{a}$, and the negative square root is written as $-\sqrt{a}$.

Examples: $\sqrt{81} = 9$ and $-\sqrt{49} = -7$

SECTION 2.5 ■ ORDER OF OPERATIONS; AVERAGES

Exponential Expressions with Integers as the Base

Exponential expressions can have integers as the base.

Examples: $(-6)^2 = 36$ and $-6^2 = -36$

Order of Operations

1. Do all calculations within grouping symbols, such as parentheses and radicals, or above and below a fraction bar.
2. Evaluate all exponential expressions.
3. Do all multiplication and division from *left to right*.
4. Do all addition and subtraction from *left to right*.

Example:
$$\begin{aligned} 5 - 36 \div (2 - 5)^2 &= 5 - 36 \div (-3)^2 \\ &= 5 - 36 \div 9 \\ &= 5 - 4 \\ &= 1 \end{aligned}$$

Evaluating Algebraic Expressions

Replace each variable in the expression with its given value and then evaluate using the order of operations agreement.

Example: Evaluating $8 - 3x$ for $x = -7$ gives
$$8 - 3(-7) = 8 - (-21) = 8 + 21 = 29.$$

Average

The average of a list of numbers can be found as follows.

$$\text{Average} = \frac{\text{the sum of a list of numbers}}{\text{the number of numbers in the list}}$$

Example: The average of 3, 6, and 15 is

$$\frac{3 + 6 + 15}{3} = \frac{24}{3} = 8.$$

SECTION 2.6 ■ SOLVING EQUATIONS THAT HAVE INTEGER SOLUTIONS

Checking a Solution

To check a solution, replace every occurrence of the variable with the proposed solution and check to see if the resulting equation is true.

Example: Check that the solution to $3 + 2x = 17$ is 7.

$$3 + 2x = 17 \quad \text{Given equation}$$
$$3 + 2(7) \stackrel{?}{=} 17 \quad \text{Replace } x \text{ with 7.}$$
$$17 = 17 \checkmark \quad \text{Simplify; the solution checks.}$$

Solving Equations Using Guess-and-Check

Guess-and-check is a strategy that can be used to solve some basic equations.

Example: The solution to $5x = -15$ is -3 because $5(-3) = -15$.

Solving Equations Using Tables of Values

We can solve some equations by completing a table of values for the variable and selecting the solution from the table.

Example: The solution to $1 - 3x = -2$ is 1.

x	-1	0	1
$1 - 3x$	4	1	-2

Solving Equations Graphically

To solve an equation graphically,

1. Use the graph to estimate a solution.
2. Check the estimated solution. If it is not correct, try a new value.

Example: The solution to $-4x + 29 = 5$ is 6. Check this solution.

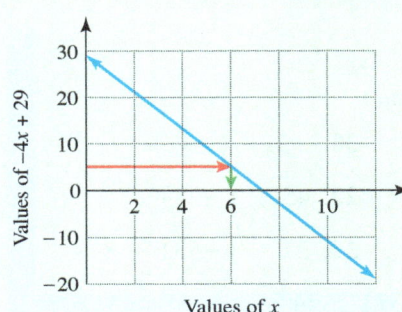

CHAPTER 2 Review Exercises

SECTION 2.1

Exercises 1 and 2: Find the opposite of the given integer.

1. 19
2. −52

Exercises 3 and 4: Simplify the expression.

3. −(−31)
4. −(−(−2))

Exercises 5 and 6: Graph the integers on a number line.

5. −3, −1, 2
6. −4, 0, 3

Exercises 7–10: Place the correct symbol, < or >, in the blank between the integers.

7. −7 ___ 19
8. 2 ___ −5
9. −11 ___ −15
10. −32 ___ −3

Exercises 11–14: Simplify the absolute value expression.

11. $-|6|$
12. $-|-1|$
13. $|-0|$
14. $|-12|$

Exercises 15 and 16: Place the correct symbol, <, >, or =, in the blank between the expressions.

15. 2 ___ $-|2|$
16. $-|8|$ ___ -8

Exercises 17 and 18: The following table lists the change in temperature from 6:00 A.M. to 6:00 P.M. for selected days of the week.

Day	Monday	Tuesday	Friday	Sunday
Change	−4°	15°	19°	−8°

17. Which day had the largest decrease in temperature from 6:00 A.M. to 6:00 P.M.?

18. Was there a temperature increase or decrease between 6:00 A.M and 6:00 P.M. on Friday?

SECTION 2.2

Exercises 19–26: Find the sum.

19. $-14 + 13$
20. $-3 + (-12)$
21. $-21 + (-30)$
22. $45 + (-23)$
23. $-14 + 45 + 22$
24. $27 + (-53) + 8$
25. $-42 + (-21) + (-37) + 54$
26. $105 + (-35) + (-64) + 13$

Exercises 27 and 28: Evaluate the expression $x + y$ for the given values of the variables.

27. $x = 12, y = -7$
28. $x = -2, y = -3$

Exercises 29–32: State the addition property illustrated by the given equation.

29. $48 + (-48) = 0$
30. $1 + (-8 + 7) = (1 + (-8)) + 7$
31. $54 + (-59) = -59 + 54$
32. $-189 + 0 = -189$

Exercises 33 and 34: Use a number line to find the sum.

33. $2 + (-5)$
34. $-5 + 9$

Exercises 35 and 36: Perform the addition visually, using the symbols ⌢ and ⌣.

35. $3 + (-7)$
36. $-6 + 8$

SECTION 2.3

Exercises 37–42: Find the difference.

37. $15 - (-4)$
38. $-16 - 16$
39. $-23 - 7$
40. $11 - 29$
41. $-17 - 0$
42. $0 - 22$

Exercises 43 and 44: Evaluate the expression $x - y$ for the given values of the variables.

43. $x = -5, y = -12$
44. $x = -4, y = 18$

Exercises 45–48: Simplify the expression.

45. $-7 - 13 + (-1)$
46. $8 + (-18) - 2$
47. $-33 - (-15) + (-40) + 9$
48. $101 - (-99) + (-50) + 10$

Exercises 49–52: Use a number line to evaluate.

49. $-2 - (-5)$
50. $5 - 7$
51. $-1 - 3$
52. $2 - (-3)$

Exercises 53 and 54: Perform the subtraction visually, using the symbols ⌢ and ⌣.

53. $-7 - (-4)$ **54.** $1 - (-8)$

SECTION 2.4

Exercises 55–62: Find the product or quotient.

55. $-42 \div 7$ **56.** $-10 \times (-9)$

57. $-3 \cdot 8$ **58.** $-14 \div (-14)$

59. $\dfrac{-75}{5}$ **60.** $\dfrac{-36}{-12}$

61. $3(-1)(3)(-2)(-5)$

62. $-2(4)(3)(2)(-1)$

Exercises 63–66: State the multiplication property illustrated by the given equation.

63. $2 \cdot (-12 \cdot 31) = (2 \cdot (-12)) \cdot 31$

64. $1 \times (-417) = -417$

65. $-6522 \cdot 0 = 0$

66. $-5(-2 + 9) = -5(-2) + (-5)(9)$

Exercises 67 and 68: Simplify the expression if possible.

67. $-\sqrt{16}$ **68.** $\sqrt{-49}$

Exercises 69–72: Evaluate the expression for the given values of the variables, if possible.

69. $\dfrac{2a}{b}$ $a = -20, b = 5$

70. $3 \cdot (-x) \cdot (-y)$ $x = 3, y = -5$

71. $-\sqrt{y}$ $y = 25$

72. $\sqrt{a}$ $a = -36$

SECTION 2.5

Exercises 73 and 74: Evaluate the exponential expression.

73. -7^2 **74.** $(-7)^2$

Exercises 75–80: Evaluate the expression.

75. $-2 \cdot (10) - 12 \div 6$

76. $5 - 3^2 \div (4 - 7)$

77. $\sqrt{52 + 12} + (-7)$

78. $\sqrt{-5^2 + 50} - 3^2$

79. $\dfrac{-39 + 6(-4)}{-3}$ **80.** $-\left|\dfrac{39 - 49}{2 \cdot (-1)}\right|$

Exercises 81–84: Insert parentheses as needed in the expression in order to make it equal to 0.

81. $-10 + 5 \cdot 8 - 6$ **82.** $14 - 16 - 3 - 1$

83. $7 - 11 \cdot 4^2 + 64$ **84.** $-3^2 \div 5 - 2 + 3$

Exercises 85–88: Evaluate the expression for the given values of the variables.

85. $12v - 3w$, for $v = -1, w = -4$

86. $3m + |2^3 + n| \div 4$, for $m = 2, n = -8$

87. $w + ((3 - v)^2 \div w)$, for $v = 9, w = -2$

88. $\sqrt{13 - p^2} - 2q^2$, for $p = -2, q = 2$

SECTION 2.6

Exercises 89–92: Check the solution as indicated.

89. Is 2 a solution to $6a - 14 = -2$?

90. Is -3 a solution to $|3w - 2| = 11$?

91. Is -2 a solution to $\dfrac{5x + 1}{3} = 3$?

92. Is -3 a solution to $\sqrt{1 - y} = 4y + 5$?

Exercises 93–96: Solve the equation.

93. $b + 9 = -2$ **94.** $4 + y = 1$

95. $-5x = -30$ **96.** $n \div (-5) = -4$

Exercises 97 and 98: Complete the table shown. Then solve the given equation.

97. $3x - 5 = 1$

x	-2	-1	0	1	2
$3x - 5$					

98. $7 - 2x = 9$

x	-2	-1	0	1	2
$7 - 2x$					

Exercises 99 and 100: Solve the equation graphically.

99. $-3x + 28 = 10$ **100.** $4x - 30 = -10$

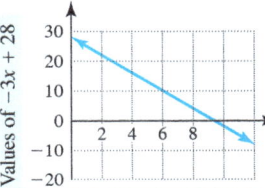
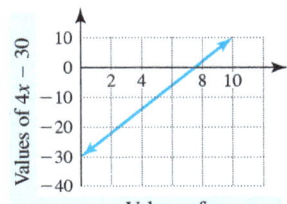

APPLICATIONS

101. Seeking Asylum The following table lists the change from 2010 to 2012 in the number of people from selected countries that sought asylum in other countries. (*Source: The Economist.*)

Country	Serbia	Afghanistan	China	Syria
Change	−6000	10,000	2000	20,000

(a) Which of these countries had the largest increase in the number of people seeking asylum?

(b) Which country saw a decline in the number of people seeking asylum?

102. Finances A checking account starts the month with a balance of $534. The following positive entries (deposits) and negative entries (withdrawals) are made in the checking register.

$$-72, -125, 300, \text{ and } -45$$

What is the ending balance?

103. Music Sales The total music sales S in thousands of dollars for a small band during year x can be computed using the formula

$$S = 7(x - 2012) + 8.$$

Find the year when the music sales were $15,000 by solving the equation $15 = 7(x - 2012) + 8$ graphically, using the following graph.

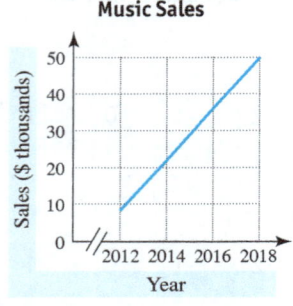

104. Detroit Population From 2000 to 2013, the population of Detroit decreased about 260,000 people. If the decrease was the same for each of these 13 years, write the yearly decrease as a quotient of integers and find the yearly decrease. (*Source: Census Bureau.*)

105. Converting Temperature To convert a temperature F given in degrees Fahrenheit to an equivalent temperature C in degrees Celsius, use the formula

$$C = \frac{5(F - 32)}{9}.$$

What Celsius temperature is equivalent to −13 °F?

106. Internet Traffic The estimated monthly consumer Internet traffic in petabytes during year x can be found using

$$T = 11{,}829(x - 2013) + 21{,}649.$$

Replace T in the formula with 68,965 and use a table of values to solve the resulting equation. For the values of x in your table, use 2014 to 2018. What year has 68,965 petabytes of monthly consumer Internet traffic? (*Source: Statista.*)

107. Temperature Change Over a four-day period in January, a city in Iowa had the following weather changes. The low temperature on Monday was −2 °F. On Tuesday, the low temperature increased by 10 °F. The low temperature then fell 3 °F on Wednesday. Finally, on Thursday, the low temperature decreased by 8 °F. What was the low temperature on Thursday?

108. Late Fees Each time that a credit card holder fails to make a credit card payment by the due date, a late fee appears on the monthly statement as −$19. If the card holder missed the payment due date eight times over the past year, write the total charges for late fees as a product of integers and find the total charges for late fees.

109. Corporate Losses An Internet company reported fourth quarter earnings of −$16 million. During the same quarter, an investment company reported earnings that were $9 million higher than those of the Internet company. What were the investment company's earnings for the fourth quarter?

110. Average Temperature Find the average of the list of Celsius temperatures.

$$-8°, -4°, 13°, 16°, -3°, \text{ and } 4 \text{ °C}$$

Exercises 111 and 112: **Finances** *Absolute value can be used to find the "size" of a financial quantity. Even though* −$63 *is a negative value (a debt), it is a larger financial quantity than a positive balance of* $41 *in a checking account (an asset). The debt is larger than the asset because* $|-63| > |41|$. *Use absolute value to determine which of each pair of financial quantities is larger.*

111. Owing money to a friend: −$64
Cash in savings jar: $32

112. An employee paycheck: $850
A credit card debt: −$753

CHAPTER 2 Test

1. Graph the integers $-3, 0,$ and 4 on a number line.

2. Evaluate $x + y$ for $x = -3$ and $y = 8$.

Exercises 3 and 4: Place the correct symbol, $<, >,$ or $=,$ in the blank between the expressions.

3. -19 _____ -25 4. $-|8|$ _____ -8

Exercises 5–8: Perform the arithmetic.

5. $16 + (-21)$ 6. $-38 - 12$

7. $-12 \cdot (-7)$ 8. $-63 \div 9$

9. Evaluate $3 - 5 + (-1) - (-4)$.

10. Evaluate $2(-1)(-3)(-2)(7)$.

11. Evaluate -11^2.

12. Simplify the expression $-\sqrt{100}$.

Exercises 13–18: Evaluate the expression.

13. $-21 + 4 \cdot 7 - 15$

14. $9 - 4^2 \div 8 - 14$

15. $\sqrt{-13 + 38} - (-3)$

16. $5 - |6^2 \div (-4)|$

17. $\dfrac{(13 - 7) \cdot 5}{|33 - 48| \cdot 2}$ 18. $\dfrac{(-5 - 3) \cdot 3}{2^3 - \sqrt{4}}$

Exercises 19 and 20: Check each solution as specified.

19. Is 10 a solution to $\dfrac{-42}{y + 4} = -3$?

20. Is 7 a solution to $\sqrt{5m + 1} = -m + 1$?

Exercises 21 and 22: Solve the equation.

21. $72 = -12x$

22. $n + (-3) = 6$

23. Complete the table and solve $5 - 4x = -3$.

x	-2	-1	0	1	2
$5 - 4x$					

24. Solve the equation $-3x + 11 = -10$ graphically.

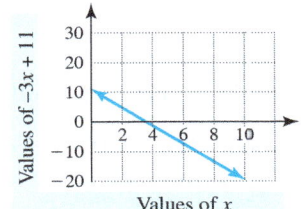

25. **Net Worth** A student with a net worth of $-\$25,700$ inherits an amount of money so that his net worth becomes $\$107,200$. How much money did he inherit?

26. **Altitude and Temperature** If the air temperature on the ground is $70\ °F$, then the air temperature T at an altitude of x miles is given by the formula

$$T = 70 - 19x.$$

Complete the table and determine the altitude where the air temperature is $-6\ °F$.

x	0	1	2	3	4
$70 - 19x$					

CHAPTERS 1–2 Cumulative Review Exercises

1. Identify the digit in the ten-thousands place in the number 9,145,283,705.

2. Write 32,010 in expanded form.

Exercises 3–6: Perform the arithmetic.

3. $\quad 289$
   $\underline{+ 5775}$

4. $\quad 19{,}043$
   $\underline{-7\,938}$

5. $23 \cdot 279$

6. $3672 \div 45$

7. Write $7 \cdot 7 \cdot 7$ using exponential notation.

8. Solve the equation $x - 13 = 6$.

9. Round 32,673,905 to the nearest million.

10. Estimate the sum $789 + 502 + 197$ by rounding each value to the nearest hundred.

11. Evaluate $\sqrt{81}$.

12. Evaluate $3 \cdot 15 - 60 \div 10$.

13. Simplify the expression $4x + (x - 5)$.

14. Place the correct symbol, $<$, $>$, or $=$, in the blank between the whole numbers: $-|5|$ _____ 5.

Exercises 15–18: Perform the arithmetic.

15. $-14 + (-3)$

16. $-3 - (-8)$

17. $-50 \div 5$

18. $-5 \cdot (-20)$

Exercises 19 and 20: Use a number line to evaluate.

19. $3 + (-4)$

20. $-1 - (-5)$

21. Multiply $4(-2)(5)(-1)(-2)$.

22. State the multiplication property illustrated by the equation $-41 \times 0 = 0$.

23. Evaluate $6 - 4^2 \div (3 - 11)$.

24. Evaluate $(w + (2 - v)^2) \div 2$ for $v = 3$, $w = 1$.

25. Is 8 a solution to $4 + b \div 2 = -7$?

26. Complete the table and solve $7x - 12 = -5$.

x	-2	-1	0	1	2
$7x - 12$					

27. **Finances** A checking account starts the month with a balance of $1296. The following positive entries (deposits) and negative entries (withdrawals) are made in the checking register.

$$-504,\ -81,\ 700,\ \text{and}\ -432$$

What is the ending balance?

28. **Converting Temperature** To convert a temperature C given in degrees Celsius to an equivalent temperature F in degrees Fahrenheit, use the formula

$$F = \frac{9C}{5} + 32.$$

Find the Fahrenheit temperature that is equivalent to $-20\,°C$.

29. **Heart Rate** The average heart rate R in beats per minute (bpm) of an animal weighing W pounds can be approximated by

$$R = \frac{885\sqrt{W}}{W}.$$

Find the heart rate for a 25-pound dog.

30. **Geometry** Find the perimeter of the figure.

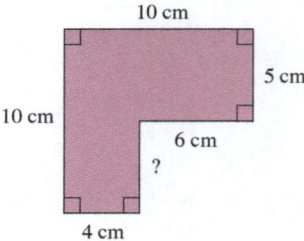

3 Algebraic Expressions and Linear Equations

- 3.1 Simplifying Algebraic Expressions
- 3.2 Translating Words to Expressions and Equations
- 3.3 Properties of Equality
- 3.4 Solving Linear Equations in One Variable
- 3.5 Applications and Problem Solving

Bamboo is the largest plant in the grass family and grows in a wide variety of climates. It is strong and durable with many uses, such as flooring, clothing, paper, medicine, helmets, and musical instruments. Bamboo is an extremely fast-growing plant. Under ideal conditions, bamboo can grow more than 1 inch per hour! To better understand and predict the growth of bamboo plants, we use linear equations. In this chapter, we learn how to write and solve these types of equations.

3.1 Simplifying Algebraic Expressions

Objectives

1. Reviewing Algebraic Expressions
2. Combining Expressions That Have Like Terms
3. Adding Expressions
4. Subtracting Expressions
   - Finding the Opposite of an Expression
   - Using the Opposite to Subtract Expressions
5. Multiplying Expressions
6. Simplifying Expressions

NEW VOCABULARY

☐ Term
☐ Coefficient
☐ Like terms
☐ Unlike terms

STUDY TIP

Try to find a consistent time and place to study your notes and do your homework. When the time comes to study for an exam, do so at your usual study time in your usual place rather than "pulling an all-nighter" in unfamiliar surroundings.

READING CHECK 1

- What is a term?
- How can you tell if two terms are like terms?

1 Reviewing Algebraic Expressions

Earlier, we discussed how to simplify algebraic expressions involving whole numbers and integers. Here, we repeat some of the most important concepts and definitions.

A **term** is a number, a variable, or a product of numbers and variables raised to powers.

Examples of Terms

$$6, \quad -3x, \quad 5y^2, \quad 17ab, \quad -7u^2v$$

The number within a term is called the **coefficient**. If no number appears in a term, then the coefficient is 1 or -1. For example, the coefficients of the terms $2x$, $-a^2$, and $-14y$ are 2, -1, and -14, respectively.

Two terms with the same variables raised to the same powers are called **like terms**.

Examples of Like Terms

$$-5 \text{ and } 8, \quad -2x \text{ and } 93x, \quad y^2 \text{ and } -5y^2, \quad -ab^2 \text{ and } -2ab^2$$

Two terms that are not like terms are **unlike terms**.

Examples of Unlike Terms

$$-7 \text{ and } 2x, \quad -3a \text{ and } 10b, \quad 4y^2 \text{ and } 9y, \quad -4x^2y \text{ and } -3xy^2$$

We can combine (add or subtract) like terms but cannot combine unlike terms.

Several properties of arithmetic are useful when simplifying algebraic expressions. They are listed here with examples corresponding to each property.

Arithmetic Property	Examples
Commutative Property	$3x + (-7x) = (-7x) + 3x$
	$(5x)(-2) = (-2)(5x)$
Associative Property	$(-2y + 4) + 3 = -2y + (4 + 3)$
	$(-b \cdot 4) \cdot 2 = -b \cdot (4 \cdot 2)$
Distributive Property	$5(-7y + 1) = 5(-7y) + 5(1)$
	$6(-2a - 3) = 6(-2a) - 6(3)$

NOTE: The commutative and associative properties allow us to add terms in *any* order or multiply terms in *any* order. ∎

2 Combining Expressions That Have Like Terms

When an expression contains like terms, it can often be simplified using the distributive property. To do this, it is helpful to write the distributive property in a more practical form. For example, when we apply the commutative property for multiplication to the left side and to each term on the right side of the distributive property,

$$c(a + b) = ca + cb, \quad \text{Distributive property}$$

the result is an equation that is equivalent to the distributive property,

$$(a + b)c = ac + bc. \quad \text{Distributive property}$$

Switching the expression on the left side of the equal sign with the expression on the right side results in the following form of the distributive property,

$$ac + bc = (a + b)c. \quad \text{"Reversing" the distributive property}$$

This same process works when using the distributive property with subtraction.

> **DISTRIBUTIVE PROPERTIES FOR COMBINING LIKE TERMS**
>
> If the variables a, b, and c represent numbers, then
> $$ac + bc = (a+b)c \quad \text{and} \quad ac - bc = (a-b)c$$

The next example demonstrates how these distributive properties can be used to simplify algebraic expressions.

EXAMPLE 1   **Combining like terms**

Simplify each expression by combining like terms.
(a) $3x + 11x$ (b) $7y^2 - 13y^2$ (c) $8a - a + 3$

Solution
(a) $3x + 11x = (3 + 11)x = 14x$
(b) $7y^2 - 13y^2 = (7 - 13)y^2 = -6y^2$
(c) The coefficient of the middle term a is understood to be 1. The first two terms are like terms, so $8a - 1a + 3 = (8 - 1)a + 3 = 7a + 3$.

Now Try Exercises 9, 11, 13

Because subtraction is neither commutative nor associative, it is helpful to change each subtraction in an algebraic expression to the addition of its opposite before combining like terms.

EXAMPLE 2   **Combining like terms**

Simplify each expression by combining like terms.
(a) $5x + 9 - 2x - 3$ (b) $3y - 18 + 6y - 2$

Solution
(a) Begin by changing each subtraction to the addition of its opposite.

$\quad 5x + 9 - 2x - 3 = 5x + 9 + (-2x) + (-3)$ Add the opposite.
$\quad\quad\quad\quad\quad\quad\quad\;\; = 5x + (-2x) + 9 + (-3)$ Commutative property
$\quad\quad\quad\quad\quad\quad\quad\;\; = (5 + (-2))x + 9 + (-3)$ Distributive property
$\quad\quad\quad\quad\quad\quad\quad\;\; = 3x + 6$ Simplify.

(b) $3y - 18 + 6y - 2 = 3y + (-18) + 6y + (-2)$ Add the opposite.
$\quad\quad\quad\quad\quad\quad\quad\;\; = 3y + 6y + (-18) + (-2)$ Commutative property
$\quad\quad\quad\quad\quad\quad\quad\;\; = (3 + 6)y + (-18) + (-2)$ Distributive property
$\quad\quad\quad\quad\quad\quad\quad\;\; = 9y + (-20)$ Simplify.
$\quad\quad\quad\quad\quad\quad\quad\;\; = 9y - 20$ Change to subtraction.

Now Try Exercises 17, 19

READING CHECK 2

Which of the following are examples of like terms being combined correctly? Select all that apply.
- $3x + 5x = 8x$
- $-x + 2y = -2xy$
- $3a + 5a - 4a = 4a$
- $-5x + -5y = 0$
- $9y - 13y = -4$

3 Adding Expressions

When two or more expressions are added, we often use parentheses to identify the expressions in the sum. For example, in the sum $(3x - 5) + (2x + 7)$, we are adding the expressions $3x - 5$ and $2x + 7$. However, the parentheses are simply grouping symbols and are not needed to complete the addition.

$$(3x - 5) + (2x + 7) = 3x - 5 + 2x + 7$$

The next example illustrates how to add expressions.

EXAMPLE 3   **Adding expressions**

Simplify each sum.
(a) $(3x - 1) + (4x + 5)$ (b) $(7y^2 + 3) + (y^2 + 5y)$

Solution
(a) Begin by removing the parentheses from the sum.

$$\begin{aligned}
(3x - 1) + (4x + 5) &= 3x - 1 + 4x + 5 && \text{Remove grouping symbols.} \\
&= 3x + (-1) + 4x + 5 && \text{Add the opposite.} \\
&= 3x + 4x + (-1) + 5 && \text{Commutative property} \\
&= (3 + 4)x + (-1) + 5 && \text{Distributive property} \\
&= 7x + 4 && \text{Simplify.}
\end{aligned}$$

(b) The coefficient of the second y^2-term is understood to be **1**.

$$\begin{aligned}
(7y^2 + 3) + (y^2 + 5y) &= 7y^2 + 3 + y^2 + 5y && \text{Remove grouping symbols.} \\
&= 7y^2 + 1y^2 + 5y + 3 && \text{Commutative property} \\
&= (7 + 1)y^2 + 5y + 3 && \text{Distributive property} \\
&= 8y^2 + 5y + 3 && \text{Simplify.}
\end{aligned}$$

Now Try Exercises 21, 25

NOTE: In Example 3(b), since the exponent on y in the term $5y$ is 1, we cannot combine $5y$ with the y^2-terms (which each have an exponent of 2). ■

4 Subtracting Expressions

FINDING THE OPPOSITE OF AN EXPRESSION In the previous chapter, we found that subtracting one integer from another is the same as adding the first integer to the opposite of the second integer. For example, $3 - 9 = 3 + (-9)$. We can extend this idea to include the subtraction of one expression from another. To do so, we must first understand the meaning of *the opposite of an expression*.

> **OPPOSITE OF AN EXPRESSION**
>
> If the variable a represents a number or an expression, then
>
> $$-a = -1 \cdot a.$$

The next example shows us how to find the opposite of an expression by using this general rule.

EXAMPLE 4 Finding the opposite of an expression

Find the opposite of each expression.
(a) $3x - 11$ (b) $-5w + 12$

Solution
(a) Start by writing the opposite of the expression $3x - 11$ as $-(3x - 11)$.

$$
\begin{aligned}
-(3x - 11) &= -1(3x - 11) & -a &= -1 \cdot a \\
&= (-1) \cdot 3x - (-1) \cdot 11 & & \text{Distributive property} \\
&= -3x - (-11) & & \text{Multiply.} \\
&= -3x + 11 & & \text{Add the opposite.}
\end{aligned}
$$

(b) Start by writing the opposite of the expression $-5w + 12$ as $-(-5w + 12)$.

$$
\begin{aligned}
-(-5w + 12) &= -1(-5w + 12) & -a &= -1 \cdot a \\
&= (-1) \cdot (-5w) + (-1) \cdot 12 & & \text{Distributive property} \\
&= 5w + (-12) & & \text{Multiply.} \\
&= 5w - 12 & & \text{Change to subtraction.}
\end{aligned}
$$

Now Try Exercises 33, 37

READING CHECK 3

- What is the opposite of $-2x + 8$?

Example 4 shows that the opposite of the expression $3x - 11$ is $-3x + 11$ and the opposite of the expression $-5w + 12$ is $5w - 12$. These results suggest the following process for finding the opposite of an expression.

> **FINDING THE OPPOSITE OF AN EXPRESSION**
>
> To find the opposite of an expression,
> 1. Change the sign of the first term in the expression from positive to negative or from negative to positive.
> 2. Change every addition to subtraction and change every subtraction to addition.

NOTE: It is common for these rules to be combined into the following single statement.

To find the opposite of an expression, change the sign of each term in the expression. ∎

USING THE OPPOSITE TO SUBTRACT EXPRESSIONS In the next example, we use the opposite of an expression to subtract one expression from another

EXAMPLE 5 Subtracting expressions

Simplify each difference.
(a) $(17x + 6) - (9x - 4)$ (b) $-13a - (3a + 7)$

Solution
(a) Begin by adding the opposite of the second expression. The signs in the first expression do not change.

$$
\begin{aligned}
(17x + 6) - (9x - 4) &= (17x + 6) + (-9x + 4) & & \text{Add the opposite.} \\
&= 17x + 6 + (-9x) + 4 & & \text{Remove grouping symbols.} \\
&= 17x + (-9x) + 6 + 4 & & \text{Commutative property} \\
&= (17 + (-9))x + 6 + 4 & & \text{Distributive property} \\
&= 8x + 10 & & \text{Simplify.}
\end{aligned}
$$

The opposite of $(9x - 4)$ is $(-9x + 4)$.

(b) Begin by adding the opposite of the second expression.

$$-13a - (3a + 7) = -13a + (-3a - 7) \quad \text{Add the opposite.}$$
$$= -13a + (-3a) - 7 \quad \text{Remove grouping symbols.}$$
$$= (-13 + (-3))a - 7 \quad \text{Distributive property}$$
$$= -16a - 7 \quad \text{Simplify.}$$

The opposite of $(3a + 7)$ is $(-3a - 7)$.

Now Try Exercises 41, 47

5 Multiplying Expressions

Algebraic expressions can be multiplied using the commutative, associative, and distributive properties. For now, we focus on the product of a number and an expression.

EXAMPLE 6 **Multiplying a one-term expression by a number**

Simplify each product.
(a) $3(9x)$ **(b)** $(-5w) \cdot 6$

Solution
(a) Apply the associative property of multiplication to regroup the factors.

$$3(9x) = (3 \cdot 9)x \quad \text{Associative property}$$
$$= 27x \quad \text{Multiply.}$$

(b) Apply the commutative and associative properties of multiplication.

$$(-5w) \cdot 6 = 6 \cdot (-5w) \quad \text{Commutative property}$$
$$= (6 \cdot (-5))w \quad \text{Associative property}$$
$$= -30w \quad \text{Multiply.}$$

Now Try Exercises 51, 55

The distributive property can be used to multiply a number and a two-term expression.

EXAMPLE 7 **Multiplying a two-term expression by a number**

Simplify each product.
(a) $7(-3x + 2)$ **(b)** $-5(4y - 9)$

Solution
(a)
$$7(-3x + 2) = 7 \cdot (-3x) + 7 \cdot 2 \quad \text{Distributive property}$$
$$= (7 \cdot (-3))x + 7 \cdot 2 \quad \text{Associative property}$$
$$= -21x + 14 \quad \text{Multiply.}$$

Multiply 7 by both $-3x$ and 2.

(b)
$$-5(4y - 9) = -5 \cdot 4y - (-5) \cdot 9 \quad \text{Distributive property}$$
$$= (-5 \cdot 4)y - (-5) \cdot 9 \quad \text{Associative property}$$
$$= -20y - (-45) \quad \text{Multiply.}$$
$$= -20y + 45 \quad \text{Add the opposite.}$$

Multiply -5 by both $4y$ and 9.

Now Try Exercises 59, 61

6 Simplifying Expressions

Some expressions can be simplified by combining like terms after addition, subtraction, or multiplication has been performed.

EXAMPLE 8 **Simplifying expressions**

Simplify each expression.
(a) $3(2n + 5) - 3n$ (b) $3(5x + 2) + 2(3x - 1)$

Solution
(a) Start with the distributive property.

$$
\begin{aligned}
3(2n + 5) - 3n &= 3(2n) + 3(5) - 3n && \text{Distributive property} \\
&= 6n + 15 - 3n && \text{Multiply.} \\
&= 3n + 15 && \text{Combine like terms.}
\end{aligned}
$$

(b) Start by applying the distributive property twice.

$$
\begin{aligned}
3(5x + 2) + 2(3x - 1) &= 3(5x) + 3(2) + 2(3x) - 2(1) && \text{Distributive property} \\
&= 15x + 6 + 6x - 2 && \text{Multiply.} \\
&= 21x + 4 && \text{Combine like terms.}
\end{aligned}
$$

Now Try Exercises 67, 75

EXAMPLE 9 **Simplifying an expression**

Simplify the expression $-(3x + 8) - 5(x + 2)$.

Solution
Since this expression includes a negative sign and subtraction, extra care should be taken to avoid sign errors.

$$
\begin{aligned}
-(3x + 8) - 5(x + 2) &= -1(3x + 8) + (-5)(x + 2) && \text{Insert 1; add the opposite.} \\
&= -3x + (-8) + (-5x) + (-10) && \text{Distributive property} \\
&= -8x - 18 && \text{Combine like terms.}
\end{aligned}
$$

Now Try Exercise 81

MAKING CONNECTIONS 1

Evaluating and Simplifying Expressions

Although the words "evaluate" and "simplify" occur frequently in mathematics, they are *not* interchangeable. When we *evaluate* an expression, we replace any variables with given values and then find the result (usually a number). When we *simplify* an expression, the result is an equivalent expression often found by applying the distributive property and combining like terms.

READING CHECK 4

- Simplify the expression $3(x - 2) - 2(2x + 1)$, and then evaluate your simplified expression for $x = 2$.

3.1 Putting It All Together

CONCEPT	COMMENTS	EXAMPLES
Like Terms	Terms that contain the same variables raised to the same powers are like terms.	$3x^2$ and $6x^2$ are like terms. $-3p$ and $-3q$ are unlike terms.
Combining Like Terms	To combine like terms, use the distributive properties $$ac + bc = (a+b)c \text{ and}$$ $$ac - bc = (a-b)c.$$	$3w + 5w = (3+5)w = 8w$ $7x - 9x = (7-9)x = -2x$
Opposite of an Expression	To find the opposite of an expression, change the sign of each term in the expression.	$-(3x - 5) = -3x + 5$ $-(-11y + 2) = 11y - 2$
Simplifying an Expression	Use arithmetic properties to remove parentheses and combine like terms.	$2(n + 4) - n = 2n + 8 - n$ $= n + 8$

3.1 Exercises

CONCEPTS AND VOCABULARY

1. A(n) _____ is a number, a variable, or a product of numbers and variables raised to powers.
2. The number that appears in a term is called the _____.
3. The terms $5x^3$ and $-7x^2$ are (like/unlike).
4. The terms $8y$ and $4y$ are (like/unlike).
5. The _____ and _____ properties for addition allow us to add two or more terms in any order.
6. To (evaluate/simplify) an expression, replace any variables with given values and then find the result.

COMBINING LIKE TERMS

Exercises 7–20: Simplify the expression.

7. $2y + 5y$
8. $8x - 3x$
9. $11m - 17m$
10. $-3n - 18n$
11. $3x^3 + 7x^3$
12. $-2y^2 + 7y^2$
13. $3b + b - 8$
14. $t - 4t + 5$
15. $12a + 7 - 12a$
16. $4 - 9r + 9r$
17. $11x + 7 - 3x + 9$
18. $-2y + 3 - y - 8$
19. $-7m + 4 + 9m - 8$
20. $16n - 3 - 7n - 8$

ADDING EXPRESSIONS

Exercises 21–32: Simplify the sum.

21. $(x + 7) + (3x + 1)$
22. $(-x - 5) + (2x + 8)$
23. $(4y - 9) + (-2y - 9)$
24. $(m - 5) + (m + 3)$
25. $(n^2 + 3) + (2n^2 - 7n)$
26. $(t^2 - 2t) + (t^2 + 1)$
27. $(5x - 3) + (-5x + 3)$
28. $(-8y + 1) + (y - 1)$
29. $(4a + 3) + (3a - 3)$
30. $(3b - 2) + (3b - 2)$
31. $(3x + 2y) + (x - y)$
32. $(3p - q) + (-p - q)$

SUBTRACTING EXPRESSIONS

Exercises 33–40: Find the opposite of the expression.

33. $2y - 9$
34. $-5x + 10$
35. $15m + 3$
36. $8n - 3$
37. $-8a + 7$
38. $-2q - 17$

39. $2x^2 - 3x + 1$ **40.** $-5y^2 + y - 9$

Exercises 41–50: Simplify the difference.

41. $(14m + 7) - (4m - 6)$ **42.** $(-x - 3) - (2x + 6)$

43. $(-y + 2) - (-2y + 7)$ **44.** $(4n - 2) - (5n + 2)$

45. $(3a - 5) - (3a - 5)$ **46.** $(2b + 4) - (2b - 4)$

47. $7x - (3x + 12)$ **48.** $-6y - (y - 3)$

49. $8 - (2t + 8)$ **50.** $-13 - (-2w - 13)$

MULTIPLYING EXPRESSIONS

Exercises 51–64: Simplify the product.

51. $5(2x)$ **52.** $-3(8m)$

53. $-10(4n)$ **54.** $-7y(-6)$

55. $(-3p) \cdot 9$ **56.** $(-4w) \cdot 7$

57. $(15x) \cdot 0$ **58.** $(7y)(-1)$

59. $4(-2y - 6)$ **60.** $2(-4w + 3)$

61. $-7(9x - 1)$ **62.** $-2(3m + 4)$

63. $0(-5p + 2)$ **64.** $1(3t - 6)$

SIMPLIFYING EXPRESSIONS

Exercises 65–82: Simplify the expression.

65. $4(3a + 2) - 7$ **66.** $2(3x - 5) - 7$

67. $5(b + 2) - 3b$ **68.** $7(-4x + 5) + 12x$

69. $7 + 2(5x - 4)$ **70.** $11 + 5(x + 2)$

71. $2y - (y + 6) + 9$ **72.** $6x - 5(x + 2) - 8$

73. $17 - 3(2m + 6) + 9m$ **74.** $3 + 7(n - 1) - 5n$

75. $3(x + 2) + 2(4x - 3)$ **76.** $2(3y - 1) + (y - 2)$

77. $-2(3t + 5) + 6(2t - 3)$

78. $-3(4w - 5) + 2(-7w + 2)$

79. $3(5x + 2) - 2(3x - 1)$

80. $-(3y - 1) - 4(y + 6)$

81. $-(2a + 7) - 5(a - 9)$

82. $-4(3b - 5) - 3(-2b - 1)$

APPLICATIONS INVOLVING EXPRESSIONS

83. Painting a Wall A person who can paint 16 square feet of a wall in one minute will paint $16x$ square feet in x minutes. If a second person paints $14x$ square feet in x minutes, complete the following.

(a) Write an expression (as a sum) that gives the total number of square feet painted by both painters in x minutes.

(b) Simplify your expression from part (a).

(c) How many square feet can the two painters cover in 5 minutes?

84. Street Vendors A bagel vendor who can sell 18 bagels every hour sells $18x$ bagels in x hours. If a second vendor sells $22x$ bagels in x hours, complete the following.

(a) Write an expression (as a sum) that gives the total number of bagels sold by both vendors in x hours.

(b) Simplify your expression from part (a).

(c) How many bagels can the two vendors sell in 8 hours?

85. Geometry Refer to the following figure.

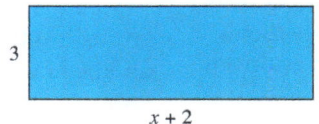

(a) Write an expression (as a product) that gives the area of the rectangle.

(b) Simplify your expression from part (a).

(c) Evaluate the area of the rectangle in square units if $x = 5$.

86. Geometry Refer to the following figure.

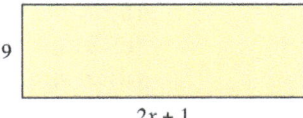

(a) Write an expression (as a product) that gives the area of the rectangle.

(b) Simplify your expression from part (a).

(c) Evaluate the area of the rectangle in square units if $x = 9$.

WRITING ABOUT MATHEMATICS

87. Explain how to determine if two terms are like or unlike.

88. Explain how to combine like terms. What property is used when combining like terms?

89. Explain how to find the opposite of an expression. Give an example.

90. Compare the meanings of the words "evaluate" and "simplify." Give an example of each.

3.2 Translating Words to Expressions and Equations

Objectives

1. Translating Words to Expressions
2. Translating Words to Equations

NEW VOCABULARY

☐ Define the variable

STUDY TIP

Spend some extra time learning words in the language of mathematics. A strong mathematical vocabulary is one of the keys to success in any math course.

1 Translating Words to Expressions

In an earlier chapter, words associated with the arithmetic symbols $+$, $-$, $\cdot$, and $\div$ were discussed. **TABLE 3.1** summarizes these words.

Words Associated with Arithmetic Symbols

Symbol	Associated Words
$+$	add, plus, more, sum, total, increase
$-$	subtract, minus, less, difference, fewer, decrease
$\cdot$	multiply, times, twice, double, triple, product
$\div$	divide, divided by, quotient, per

TABLE 3.1

Paragraphs that describe a mathematical situation often contain one or more of the words in **TABLE 3.1**, but they rarely specify a variable to use when translating the words to an algebraic expression. For example, the phrase "five times your age" contains the word "times" which means that multiplication should be used. However, the variable representing "your age" has not been specified. We need to **define the variable**. Here, we might let the variable a represent "your age."

EXAMPLE 1   Translating words to expressions

Translate each phrase to an algebraic expression. Define each variable.
(a) Seven fewer than the number of peaches
(b) Miles per gallon
(c) Triple the number of bacteria
(d) Four more than a number

Solution
(a) The word "fewer" suggests subtraction. If we let p represent the number of peaches, then the expression $p - 7$ represents seven fewer than the number of peaches.
(b) The word "per" indicates division. Let m represent the number of miles, and let g represent the number of gallons. The expression $m \div g$ represents miles per gallon. This can also be written as $\frac{m}{g}$.
(c) Let b represent the number of bacteria. The word "triple" suggests multiplication by 3. The expression $3b$ represents triple the number of bacteria.
(d) Let n represent a number. The word "more" indicates addition. The expression $n + 4$ represents four more than a number.

Now Try Exercises 7, 9, 11, 13

EXAMPLE 2   Translating words to expressions

Translate each phrase to an algebraic expression. Use x as the variable.
(a) Nine times the sum of a number and 3
(b) Four less than the product of 5 and her score

Solution
(a) The sum of a number and 3 can be written $x + 3$. To find 9 times this sum, we write the sum in parentheses and multiply by 9. The expression is $9(x + 3)$.
(b) The product of 5 and a score can be written $5x$. To find 4 less than this product, we subtract 4 from it. The expression is $5x - 4$.

Now Try Exercises 15, 19

2 Translating Words to Equations

Some written mathematical situations involve equations. To determine whether equations are needed, we look for words associated with an equals sign $(=)$. **TABLE 3.2** summarizes many of these words.

READING CHECK 1

Which of the following words or phrases indicate that a mathematical situation may involve an equation? Select all that apply.

- is
- equals
- results in
- is the same as
- greater than
- less than

Words Associated with an Equals Sign

Symbol	Associated Words
$=$	equals, is, gives, results in, is the same as

TABLE 3.2

Once the word or words associated with an equals sign have been identified, we translate the word or phrase before it to a mathematical expression and also translate the word or phrase after it to a mathematical expression. The equals sign is then placed between the two expressions, forming an equation, as illustrated in the next example.

EXAMPLE 3 **Translating words to equations**

Translate each sentence to an equation using x as the variable. Do not solve the equation.
(a) Three more than a number is nine.
(b) Multiplying a number by five gives -45.
(c) The quotient of a number and 4 is the same as that number minus 9.

Solution
(a) The word associated with the equals sign is "**is**." The phrase

$$\text{Three more than a number}$$

comes before the word "**is**" and translates to $x + 3$. The phrase (word)

$$\text{nine}$$

comes after the word "**is**" and translates to 9. Thus the sentence

$$\text{Three more than a number is nine}$$

translates to the equation $x + 3 = 9$.
(b) The word associated with the equals sign is "**gives**." The sentence

$$\text{Multiplying the number by five gives } -45$$

translates to the equation $x \cdot 5 = -45$ or $5x = -45$.
(c) The words associated with the equals sign are "**is the same as**." The sentence

$$\text{The quotient of a number and 4 is the same as that number minus 9}$$

translates to the equation $x \div 4 = x - 9$.

Now Try Exercises 21, 25, 27

The following box offers a strategy that can be used when translating words to equations.

> **STEPS FOR TRANSLATING WORDS TO EQUATIONS**
>
> **STEP 1:** Read the paragraph carefully for understanding. Identify the unknown quantity and define a variable to represent it.
>
> **STEP 2:** If possible, identify the word or words associated with an equals sign. In some cases, equality may be suggested instead of stated directly.
>
> **STEP 3:** Translate the words and information given in the paragraph to expressions. Two expressions are needed to form an equation.
>
> **STEP 4:** The equation is formed by writing the two expressions found in Step 3 and placing an equals sign between them.

READING CHECK 2

- How many expressions are needed to form an equation?

Math in Traffic Context The U.S. National Highway Safety Administration keeps historical records of the yearly number of fatal traffic accidents for each state. In the next example, we set up an equation relating the numbers of traffic fatalities for two states.

EXAMPLE 4   **Translating traffic information to an equation**

In 2013, there were 3000 traffic fatalities in California. This was 60 less than 60 times the number of traffic fatalities in Alaska that same year. Write an equation for which the solution gives the number of traffic fatalities in Alaska in 2013. Do not solve the equation.

Solution
STEP 1: We want to find the number traffic fatalities in Alaska in 2013. We will use the variable x to represent this unknown number.
STEP 2: The word "was" can be associated with an equals sign. Note that the word "was" is the past tense of the word "is."
STEP 3: The word "This" in the phrase

$$\text{This was } 60 \text{ less than } 60 \text{ times the number}$$

refers to the number of traffic fatalities in California, or 3000. By replacing the word "this" with the number 3000, the phrase becomes

$$3000 \text{ was } 60 \text{ less than } 60 \text{ times the } number.$$

Expression 1: The number before "was" translates to 3000.
Expression 2: The phrase after "was" translates to $60x - 60$.
STEP 4: The equation is $3000 = 60x - 60$.

Now Try Exercise 33

EXAMPLE 5   **Translating population information to an equation**

If the average number of births each minute in 2015 is increased by 73, and then divided by 3, the result is the average number of deaths each minute in 2015. If there were 106 deaths each minute, on average, write an equation for which the solution gives the average number of births each minute. Do not solve the equation. (*Source:* U.S. Census Bureau.)

Solution

STEP 1: We want to find the average number of births each minute in 2015. We use the variable x to represent this unknown number.

STEP 2: The word "**is**" suggests an equals sign.

STEP 3: The paragraph can be paraphrased as

number of births *increased* by 73, then *divided* by 3, *is* 106.

Expression 1: The phrase before "**is**" translates to $(x + 73) \div 3$.
Expression 2: The word after "**is**" translates to 106.

STEP 4: The equation is $(x + 73) \div 3 = 106$.

Now Try Exercise 37

EXAMPLE 6 **Translating a number puzzle to an equation**

When a natural number is tripled, the result equals the natural number squared. Write an equation for which the solution gives the unknown natural number. Do not solve the equation.

Solution

STEP 1: We want to find an unknown natural number. We use the variable x to represent this unknown number.

STEP 2: The word "**equals**" is associated with an equals sign.

STEP 3: The paragraph can be paraphrased as

a *number* *tripled* **equals** the same *number* *squared*.

Expression 1: The phrase before "**equals**" translates to $3x$.
Expression 2: The phrase after "**equals**" translates to x^2.

STEP 4: The equation is $3x = x^2$.

Now Try Exercise 41

3.2 Putting It All Together

CONCEPT	COMMENTS	EXAMPLES
Translating Words to Symbols	Some words or phrases are associated with the math symbols $+, -, \cdot, \div,$ and $=$.	**Symbol** **Associated Words** $+$ add, sum, increase $-$ minus, less, difference $\cdot$ multiply, times, product $\div$ divide, quotient, per $=$ is, equals, results in
Defining a Variable	To define a variable, we assign a variable to represent an unknown quantity.	To "find the total cost," we let the variable c represent the total cost.
Translating Words to Equations	**STEP 1:** Identify the unknown quantity and define a variable. **STEP 2:** Identify the word(s) associated with an equals sign. **STEP 3:** Translate the words to two expressions. **STEP 4:** Write the equation.	The phrase "Increasing a number by 3 results in 9" translates to $x + 3 = 9$.

3.2 Exercises

CONCEPTS AND VOCABULARY

1. The words "decrease" and "fewer" are associated with the mathematical symbol _____.

2. The mathematical symbol _____ is associated with the words "product" and "times."

3. The words "results in" and "is" are associated with the symbol _____.

4. The mathematical symbol _____ is associated with the words "quotient" and "divided."

5. The words "more" and "sum" are associated with the symbol _____.

6. When a variable is not specified, we _____ a variable to represent the unknown quantity.

TRANSLATING WORDS TO EXPRESSIONS

Exercises 7–20: Translate the phrase to an algebraic expression. Define each variable.

7. Double the ticket price

8. The number of pizza slices divided by 4

9. Eight fewer than his height

10. Decrease the count by 9

11. The total of 5 and the number of gallons

12. The product of 3 and the year

13. The quotient of the number of toys and 6

14. Her score increased by 7

15. Seven times the sum of her age and 6

16. Ten fewer than the product of 5 and a number

17. The product of 9 and the quantity 6 less than his age

18. The sum of the price and 4, divided by 7

19. The product of 3 and a number, minus 8

20. Twenty less than the heart rate, times 2

TRANSLATING WORDS TO EQUATIONS

Exercises 21–30: Translate the sentence to an equation using x as the variable. Do not solve the equation.

21. Dividing a number by -10 gives 9.

22. Multiplying a number by 4 results in 28.

23. Sixty-four times a number equals -256.

24. The total of 6 and a number is 19.

25. Three less than a number is 12.

26. Seven is the same as a number decreased by 4.

27. Decreasing a number by 12 is the same as dividing the same number by 2.

28. The sum of a number and 4 is the same as triple the same number.

29. Twice the sum of a number and 6 gives -20.

30. The sum of a number and 10, divided by 5, equals 6.

APPLICATIONS INVOLVING WRITING EQUATIONS

Exercises 31–42: Write an equation as directed. Use the variable x to represent any unknown quantity. Do not solve the equation.

31. **Weight Loss** After losing 30 pounds a person now weighs 135 pounds. Write an equation for which the solution gives the weight of the person before losing weight.

32. **Salary** Before receiving a raise in salary, a worker earned $1100 per week. Write an equation for which the solution gives the amount of the worker's raise if the new salary is $1300 per week.

33. **U.S. Renewable Energy** In 2007, U.S. renewable energy production was 6 quadrillion BTU. This was 12 quadrillion BTU less than twice the 2015 renewable energy production. Write an equation for which the solution gives the renewable energy production level in 2015. (*Source:* U.S. Department of Energy.)

34. **Endangered Species** There were 96 birds on the endangered species list in 2015. This is 24 more than 2 times the number of reptiles on the list. Write an equation for which the solution gives the number of reptiles on the list in 2015. (*Source:* U.S. Fish and Wildlife Service.)

35. **Age** In 10 years, a child's age will be 6 years less than triple her current age. Write an equation for which the solution gives the child's current age.

36. **Weight Gain** After gaining 25 pounds, a person is 115 pounds lighter than double his previous weight. Write an equation for which the solution gives the person's weight before gaining 25 pounds.

37. **Hazardous Waste** By increasing the number of federal hazardous waste sites in California by 2 and then dividing by 2, the result is the number of federal hazardous waste sites in Washington. Write an equation for which the solution gives the number of federal hazardous waste sites in California if there are 13 such sites in Washington. (*Source:* Environmental Protection Agency.)

38. **Cosmetic Surgery** In 2000, there were 1,900,000 cosmetic plastic surgery procedures performed. This was 1,500,000 fewer procedures than twice the number that were performed in 2014. Write an equation for which the solution gives the number of cosmetic plastic surgery procedures performed in 2014. (*Source:* American Society of Plastic Surgeons.)

39. **Geometry** If the perimeter of the rectangle shown is 70 inches, write an equation for which the solution gives the value of x.

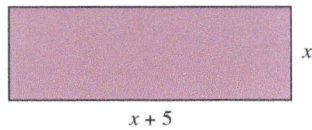

40. **Geometry** If the area of the rectangle shown in Exercise 39 is 300 square feet, write an equation for which the solution gives the value of x.

41. **Number Puzzle** When a positive number is multiplied by 4, the result is the same as the number cubed. Write an equation for which the solution gives the unknown positive number.

42. **Number Puzzle** Seven times a number, increased by 9, results in the same number decreased by 3. Write an equation for which the solution gives the unknown number.

WRITING ABOUT MATHEMATICS

43. Write a paragraph containing information that could be translated to the equation
$$3x + 2 = 11.$$
Explain what the variable x represents.

44. Summarize the process used to translate a paragraph of information to an equation. What are some key words that indicate an equation is appropriate for a given situation?

SECTIONS 3.1 and 3.2 — Checking Basic Concepts

1. Find the opposite of the expression $6y - 5$.

2. Simplify each expression.
   (a) $5a + 9 - 4a$
   (b) $(-6y + 4) - (y - 3)$
   (c) $13 - 2(4m + 8) + 3m$
   (d) $-(2x + 6) - 3(x - 4)$

3. Translate "her score increased by 5" to an expression. Define what your variable represents.

4. Translate the phrase

   Nine less than a number is 15

   to an equation using the variable x.

5. **Bonus Pay** After receiving a $10,000 end-of-year bonus, a manager's annual pay was $76,300. Write an equation for which the solution gives the manager's annual salary before the bonus. Do not solve the equation.

6. **Energy Production** In 2014, the U.S. produced 798 billion kilowatt-hours of electricity at nuclear power plants. This is 45 billion kilowatt-hours more than three times the 1980 nuclear power production level. Write an equation for which the solution gives the production level in 1980. Do not solve the equation. (*Source:* U.S. Department of Energy.)

3.3 Properties of Equality

Objectives

1. Finding Solutions and Equivalent Equations
   - Checking Solutions
   - Recognizing Equivalent Equations
2. Using the Addition Property of Equality
3. Using the Multiplication Property of Equality

NEW VOCABULARY

☐ Solution
☐ Equivalent equations
☐ Addition property of equality
☐ Multiplication property of equality

1 Finding Solutions and Equivalent Equations

CHECKING SOLUTIONS A **solution** to an equation is any value of the variable that makes the equation true. To *check a solution*, replace every occurrence of the variable in the equation with the proposed solution and see if the resulting equation is true.

READING CHECK 1

- Is 5 a solution to $3x - 4 = 11$?

EXAMPLE 1 **Checking solutions**

Check each given solution as specified.
(a) Is -2 a solution to $5x - 1 = -9$?   (b) Is 7 a solution to $23 - 2x = x + 2$?

Solution
(a) Replace the variable x with -2.

$$5x - 1 = -9 \quad \text{Given equation}$$
$$5(-2) - 1 \stackrel{?}{=} -9 \quad \text{Replace } x \text{ with } -2.$$
$$-10 - 1 \stackrel{?}{=} -9 \quad \text{Simplify.}$$
$$-11 \neq -9 \; \text{✗} \quad \text{The solution does not check.}$$

So, -2 is **not** a solution to the equation $5x - 1 = -9$.

(b) Replace the variable x with 7.

$$23 - 2x = x + 2 \quad \text{Given equation}$$
$$23 - 2(7) \stackrel{?}{=} 7 + 2 \quad \text{Replace } x \text{ with } 7.$$
$$23 - 14 \stackrel{?}{=} 9 \quad \text{Simplify.}$$
$$9 = 9 \; \text{✓} \quad \text{The solution checks.}$$

So, 7 **is** a solution to the equation $23 - 2x = x + 2$.

Now Try Exercises 11, 15

RECOGNIZING EQUIVALENT EQUATIONS Until this point in the text, we have solved equations by mentally figuring out what value(s) of the variable make(s) the equation true. This method is often referred to as the *guess-and-check* method. However, for more complicated equations, a step-by-step procedure can be used to transform the equation into a different but *equivalent* equation. Two equations with exactly the same solution(s) are called **equivalent equations**. In the next example, we determine if two equations are equivalent.

EXAMPLE 2 **Determining if two equations are equivalent**

Determine if the equations in each pair are equivalent. Each equation has only one solution.
(a) $2x + 7 = 13$ and $x = 3$ (b) $x + 3 = 4x$ and $x = -2$

Solution

(a) The solution to the second equation is 3. The equations are equivalent if 3 is also the solution to the first equation. Replace the variable x in the first equation with **3**.

$$2x + 7 = 13 \quad \text{Given equation}$$
$$2(3) + 7 \stackrel{?}{=} 13 \quad \text{Replace } x \text{ with 3.}$$
$$6 + 7 \stackrel{?}{=} 13 \quad \text{Simplify.}$$
$$13 = 13 \checkmark \quad \text{The solution checks.}$$

The equations are equivalent because they have the same solution, 3.

(b) The solution to the second equation is -2. However, this is not the solution to the first equation because $-2 + 3 \neq 4(-2)$ or $1 \neq -8$. The equations are not equivalent.

Now Try Exercises 23, 25

READING CHECK 2

- How do we know if two equations are equivalent?

2 Using the Addition Property of Equality

When solving equations, it is often helpful to transform a complicated equation into a simpler, equivalent equation with a more obvious solution. For example, the equation $4x + 9 = 17$ can be transformed into the equivalent equation $x = 2$. Because we can see that the solution to the second equation is 2, we know that the solution to the equation $4x + 9 = 17$ must also be 2. The **addition property of equality** can be used to transform an equation into a simpler, equivalent equation.

> ### ADDITION PROPERTY OF EQUALITY
>
> If the variables a, b, and c represent numbers, then
>
> $$a = b \quad \text{is equivalent to} \quad a + c = b + c.$$
>
> In other words, adding the same number to each side of an equation results in an equivalent equation.

NOTE: Because a subtraction problem can be written as an addition problem, the addition property of equality also works for subtraction. If the same number is subtracted from each side of an equation, then the result is an equivalent equation. ■

STUDY TIP

The addition property of equality is used to solve equations throughout the remainder of the text. Be sure that you have a firm understanding of this important property.

READING CHECK 3

The addition property of equality would be a good first step to solve which of the following equations? Select all that apply.

- $x - 5 = 8$
- $3x = 1$
- $5x + 4 = -8$
- $8x - 5 = 6 - x$

EXAMPLE 3 **Using the addition property of equality**

Use the addition property of equality to solve each equation. Then check the solution.
(a) $x + 9 = 4$ (b) $4 = y - 2$ (c) $73 + a = 9$

Solution
(a) When solving an equation, we isolate the variable on one side of the equation. Because 9 is being **added** to x in the given equation, we can isolate x on the left side of the equation by **subtracting 9** from (or adding its opposite -9 to) each side of the equation.

$$x + 9 = 4 \qquad \text{Given equation}$$
$$x + 9 - 9 = 4 - 9 \qquad \text{Subtract 9 from each side.}$$
$$x + 0 = -5 \qquad \text{Simplify.}$$
$$x = -5 \qquad \text{Identity: } x + 0 = x$$

The solution is -5. To check this solution, replace x with -5 in the *given* equation.

$$x + 9 = 4 \qquad \text{Given equation}$$
$$-5 + 9 \stackrel{?}{=} 4 \qquad \text{Replace } x \text{ with } -5.$$
$$4 = 4 \checkmark \qquad \text{Add; the solution checks.}$$

(b) Because 2 is being **subtracted** from the variable y in the given equation, we can isolate y on the right side of the equation by **adding 2** to each side of the equation.

$$4 = y - 2 \qquad \text{Given equation}$$
$$4 + 2 = y - 2 + 2 \qquad \text{Add 2 to each side.}$$
$$6 = y + 0 \qquad \text{Simplify.}$$
$$6 = y \qquad \text{Identity: } y + 0 = y$$

The solution is 6. To check this solution, replace y with 6 in the *given* equation.

$$4 = y - 2 \qquad \text{Given equation}$$
$$4 \stackrel{?}{=} 6 - 2 \qquad \text{Replace } y \text{ with 6.}$$
$$4 = 4 \checkmark \qquad \text{Add; the solution checks.}$$

(c) Because 73 is being **added** to the variable a in the given equation, we can isolate a on the left side of the equation by **subtracting 73** from each side of the equation.

$$73 + a = 9 \qquad \text{Given equation}$$
$$73 - 73 + a = 9 - 73 \qquad \text{Subtract 73 from each side.}$$
$$0 + a = -64 \qquad \text{Simplify.}$$
$$a = -64 \qquad \text{Identity: } 0 + a = a$$

The solution is -64. To check this solution, replace a with -64 in the *given* equation.

$$73 + a = 9 \qquad \text{Given equation}$$
$$73 + (-64) \stackrel{?}{=} 9 \qquad \text{Replace } a \text{ with } -64.$$
$$9 = 9 \checkmark \qquad \text{Add; the solution checks.}$$

Now Try Exercises 29, 35, 37

See the Concept 1 THE ADDITION PROPERTY OF EQUALITY

We can think of an equation as a scale, with the equals sign at the balancing point. Just as a scale remains *balanced* when the same amount of weight is added to (or removed from) each side, an equation remains *true* (and an equivalent equation results) when the same number is added to (or subtracted from) each side.

True Equation	*Equivalent Equation*	*Not Balanced*
$2 = 2$	$2 + 1 = 2 + 1$	$2 + 1 \neq 2 + 0$
$=$	$=$	$\neq$

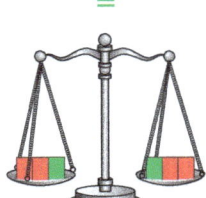

WATCH VIDEO IN MML.

The next example demonstrates how to apply the addition property of equality to "move" a variable term from one side of an equation to the other.

EXAMPLE 4 Using the addition property of equality

Use the addition property of equality to solve each equation.
(a) $5 - x = -12$ (b) $7n = 6n + 13$

Solution
(a) Because this equation can be written as $5 + (-x) = -12$, subtracting 5 from each side of the equation will isolate $-x$ rather than x. To make sure there is not a negative sign in front of the variable, we add x to each side of the equation.

$5 - x = -12$	Given equation
$5 - x + x = -12 + x$	Add x to each side.
$5 - 0 = -12 + x$	Inverses: $-x + x = 0$
$5 = -12 + x$	Identity: $5 - 0 = 5$
$5 + 12 = -12 + 12 + x$	Add 12 to each side.
$17 = x$	Simplify.

The solution is 17.

(b) In this equation, two terms contain the variable—one on each side of the equals sign. To combine these *like terms*, they must be located on the same side of the equals sign. So, we subtract $6n$ from each side of the equation.

$7n = 6n + 13$	Given equation
$7n - 6n = 6n - 6n + 13$	Subtract $6n$ from each side.
$1n = 13$	Combine like terms.
$n = 13$	Simplify.

The solution is 13.

Now Try Exercises 43, 45

3 Using the Multiplication Property of Equality

The **multiplication property of equality** is another property that can be used to transform a complicated equation into a simpler, equivalent equation.

STUDY TIP

The multiplication property of equality is used to solve equations throughout the remainder of the text. Be sure that you have a firm understanding of this important property.

> **MULTIPLICATION PROPERTY OF EQUALITY**
>
> If the variables a, b, and c represent numbers with $c \neq 0$, then
>
> $$a = b \quad \text{is equivalent to} \quad ac = bc.$$
>
> In other words, multiplying each side of an equation by the same *nonzero* number results in an equivalent equation.

NOTE: Because multiplication and division are closely related, the multiplication property of equality also works for division. If each side of an equation is divided by the same nonzero number, then the result is an equivalent equation. ∎

READING CHECK 4

The multiplication property of equality would be a good first step to solve which of the following equations? Select all that apply.

- $x - 5 = 8$
- $3x = 1$
- $5x + 4 = -8$
- $-\frac{2}{9}x = 18$

Before we can use the multiplication property of equality to solve equations, we need to determine how to simplify expressions such as

$$\frac{3x}{3} \quad \text{and} \quad 5 \cdot \frac{x}{5}.$$

In a later chapter, we will learn that the first expression simplifies as

$$\frac{3x}{3} = \frac{3}{3} \cdot x = 1x = x,$$

and the second expression simplifies as

$$5 \cdot \frac{x}{5} = \frac{5}{5} \cdot x = 1x = x.$$

This result leads to the following statement.

Multiplying and dividing a variable by the same (nonzero) number result in the variable.

EXAMPLE 5   Using the multiplication property of equality

Use the multiplication property of equality to solve each equation. Then check the solution.

(a) $12x = 60$ **(b)** $39 = -13x$ **(c)** $\dfrac{x}{-5} = 3$

Solution

(a) Because x is being **multiplied** by 12 in the given equation, we isolate x on the left side of the equation by **dividing** each side of the equation by 12.

$$12x = 60 \qquad \text{Given equation}$$

Divide each side by the coefficient 12.

$$\frac{12x}{12} = \frac{60}{12} \qquad \text{Divide each side by 12.}$$

$$x = 5 \qquad \text{Simplify.}$$

The solution is 5. To check this solution, replace x with 5 in the *given* equation.

$$12x = 60 \qquad \text{Given equation}$$

$$12(5) \stackrel{?}{=} 60 \qquad \text{Replace } x \text{ with 5.}$$

$$60 = 60 \;\checkmark \qquad \text{Multiply; the solution checks.}$$

(b) Because x is being **multiplied** by -13 in the given equation, we isolate x on the right side of the equation by **dividing** each side of the equation by -13.

$$39 = -13x \qquad \text{Given equation}$$

$$\frac{39}{-13} = \frac{-13x}{-13} \qquad \text{Divide each side by } -13.$$

$$-3 = x \qquad \text{Simplify.}$$

The solution is -3. To check this solution, replace x with -3 in the *given* equation.

$$39 = -13x \qquad \text{Given equation}$$

$$39 \stackrel{?}{=} -13(-3) \qquad \text{Replace } x \text{ with } -3.$$

$$39 = 39 \checkmark \qquad \text{Multiply; the solution checks.}$$

(c) Because x is being **divided** by -5 in the given equation, we can isolate x on the left side of the equation by **multiplying** each side of the equation by -5.

$$\frac{x}{-5} = 3 \qquad \text{Given equation}$$

$$-5 \cdot \frac{x}{-5} = -5 \cdot 3 \qquad \text{Multiply each side by } -5.$$

$$x = -15 \qquad \text{Simplify.}$$

The solution is -15. To check this solution, replace x with -15 in the *given* equation.

$$\frac{x}{-5} = 3 \qquad \text{Given equation}$$

$$\frac{-15}{-5} \stackrel{?}{=} 3 \qquad \text{Replace } x \text{ with } -15.$$

$$3 = 3 \checkmark \qquad \text{Divide; the solution checks.}$$

Now Try Exercises 49, 51, 59

3.3 Putting It All Together

CONCEPT	COMMENTS	EXAMPLES
Solution and Equivalent Equations	A solution to an equation is any value of the variable that makes the equation true.	3 is a solution to $4x - 5 = 7$ because $4(3) - 5 = 7$ is true.
	Two equations are equivalent if they have exactly the same solution(s).	$4x - 5 = 7$ and $x = 3$ are equivalent equations (solution: 3).
Addition Property of Equality	If the variables a, b, and c represent numbers, then $a = b$ is equivalent to $a + c = b + c$.	$x - 3 = 1$ Given equation $x - 3 + 3 = 1 + 3$ Add 3 to each side. $x = 4$ Simplify.
Multiplication Property of Equality	If the variables a, b, and c represent numbers with $c \neq 0$, then $a = b$ is equivalent to $ac = bc$.	$\frac{x}{2} = 4$ Given equation $2 \cdot \frac{x}{2} = 2 \cdot 4$ Multiply each side by 2. $x = 8$ Simplify.

3.3 Exercises

CONCEPTS AND VOCABULARY

1. Any value of the variable that makes an equation true is called a(n) _____.

2. To check a solution, replace every occurrence of the _____ with the proposed solution.

3. Two equations with exactly the same solution(s) are called _____ equations.

4. The _____ property of equality states that $a = b$ is equivalent to $a + c = b + c$.

5. Because a subtraction problem can be written as an addition problem, the addition property of equality also works for _____.

6. The _____ property of equality states that if $c \neq 0$, then $a = b$ is equivalent to $ac = bc$.

7. Because multiplication and division are closely related, the multiplication property of equality also works for _____.

8. When we multiply and divide a variable by the same (nonzero) number, the result is the _____ itself.

CHECKING SOLUTIONS

Exercises 9–18: Check each solution as specified.

9. Is 9 a solution to $2 + x = 11$
10. Is -1 a solution to $n - 8 = -9$
11. Is 6 a solution to $3x - 5 = 11$
12. Is -4 a solution to $12 - 5a = 30$
13. Is -9 a solution to $-35 = -8 + 3y$
14. Is 4 a solution to $7 = 5a - 12$
15. Is -2 a solution to $1 - 4m = m + 11$
16. Is -6 a solution to $3w - 7 = -1 + 4w$
17. Is 10 a solution to $3(x + 2) = 4x - 3$
18. Is -5 a solution to $6y + 1 = 5(y - 1)$

RECOGNIZING EQUIVALENT EQUATIONS

Exercises 19–28: Determine if the equations in the given pair are equivalent. Each equation has only one solution.

19. $y + 5 = 11$ and $y = 6$
20. $19 = x - 3$ and $x = -16$
21. $-5w = 30$ and $w = 6$
22. $-24 = 6m$ and $m = -4$
23. $3x + 5 = 17$ and $x = 4$
24. $4 - 7n = -10$ and $n = -2$
25. $x + 7 = 2x$ and $x = -3$
26. $-7b = 4b - 33$ and $b = 3$
27. $4(x + 1) = -3x + 5$ and $x = -2$
28. $-2(a - 3) = a + 9$ and $a = -1$

APPLYING THE ADDITION PROPERTY OF EQUALITY

Exercises 29–48: Use the addition property of equality to solve the equation. Check your solution.

29. $w + 6 = -1$
30. $x + 3 = 12$
31. $-5 + y = 2$
32. $-4 + n = 10$
33. $x - 2 = -5$
34. $-2 = w - 11$
35. $12 = m - 5$
36. $18 + a = -3$
37. $49 + b = 0$
38. $-41 = -10 + p$
39. $-16 = y - 3$
40. $-23 + x = -4$
41. $7 - x = -3$
42. $5 = 2 - y$
43. $12 = 18 - m$
44. $-5 - n = 4$
45. $8w = 7w + 4$
46. $9n - 5 = 10n$
47. $-3p + 1 = -2p$
48. $-5b = -6b + 7$

APPLYING THE MULTIPLICATION PROPERTY OF EQUALITY

Exercises 49–60: Use the multiplication property of equality to solve the equation. Check your solution.

49. $7w = 42$
50. $15x = 45$
51. $50 = -5y$
52. $36 = -9m$
53. $-11p = -77$
54. $-6n = -54$
55. $\dfrac{x}{3} = 9$
56. $\dfrac{a}{5} = 9$
57. $-6 = \dfrac{g}{-8}$
58. $12 = \dfrac{x}{-6}$
59. $\dfrac{p}{-7} = 5$
60. $\dfrac{y}{4} = -7$

APPLYING PROPERTIES OF EQUALITY

Exercises 61–80: Solve the equation.

61. $5 + h = -12$
62. $-6 = -1 + y$
63. $48 = -6y$
64. $4n = -28$
65. $p - 7 = -3$
66. $23 = x - 9$
67. $\frac{y}{4} = -12$
68. $8 = \frac{g}{5}$
69. $6m = 5m + 14$
70. $2 - g = 15$
71. $9n = 81$
72. $27 = -9y$
73. $45 = x - 15$
74. $p - 5 = 11$
75. $3 - g = 11$
76. $7m = 6m - 2$
77. $17 = -2 + y$
78. $12 + h = -3$
79. $22 = \frac{a}{-4}$
80. $\frac{y}{10} = -7$

APPLICATIONS INVOLVING PROPERTIES OF EQUALITY

81. **Geometry** The formula for the area of a rectangle is $A = lw$. Find the width of a rectangle with an area of 90 square inches and a length of 18 inches.

82. **Geometry** The formula for the perimeter of a square is $P = 4s$, where s represents the length of one side of the square. Find the length of one side of a square with a perimeter of 48 feet.

83. **Rainforests** The worldwide change in the number of acres A of rainforest each year is estimated using the formula $A = -49x$, where x represents the number of years and A is in millions. Let $A = -147$ and determine how many years it takes for the world to lose 147 million acres of rainforest. (*Source: New York Times Almanac.*)

84. **Sibling Ages** The ages of two siblings are related using the formula $g = b + 3$, where g represents the girl's age and b represents the boy's age. How old is the boy when the girl is 14?

WRITING ABOUT MATHEMATICS

85. Explain why it is important to check a solution in the *given* equation rather than in one of the equivalent equations found during the solving process.

86. When solving an equation, we can multiply each side of the equation by the same nonzero number. Explain why this number cannot be zero. Give an example.

Group Activity

College Costs Annual average costs of attending a private college in 2008 and 2015 are shown in the following table. These costs include tuition, fees, room, and board.

Year	2008	2015
Cost	$34,200	$46,200

Source: The College Board.

(a) What was the total increase in the cost of attending a private college over this period of time?

(b) Because there are 7 years from 2008 to 2015, divide your answer to part (a) by 7 to find the average yearly increase in the cost of attending private college.

(c) Use your answer from part (b) and the given 2015 cost to predict the cost of attending a private college in 2020.

(d) The formula $C = 1714x + 34,200$ gives the cost C of attending private college x years after 2008. Use both the addition and multiplication properties of equality to find the year when private college costs were $42,770.

3.4 Solving Linear Equations in One Variable

Objectives

1. Identifying Linear Equations in One Variable
2. Solving Linear Equations Symbolically
3. Solving Linear Equations Numerically
4. Solving Linear Equations Graphically

NEW VOCABULARY

☐ Linear equation in one variable
☐ Vertical axis
☐ Horizontal axis

1 Identifying Linear Equations in One Variable

An equation containing a single variable is a *linear equation in one variable* if it can be written in the following form.

> **LINEAR EQUATION IN ONE VARIABLE**
>
> A **linear equation in one variable** is an equation that can be written in the form
> $$ax + b = 0,$$
> where a and b are constants (numbers) and $a \neq 0$.

For example, the equation $3x - 4 = -8$ can be written in the form $ax + b = 0$ by applying the addition property of equality.

$$3x - 4 = -8 \quad \text{Given equation}$$
$$3x - 4 + 8 = -8 + 8 \quad \text{Add 8 to each side.}$$
$$3x + 4 = 0 \quad \text{Simplify.}$$

The equation $3x - 4 = -8$ is a linear equation in one variable because it can be written in the form $ax + b = 0$ with $a = 3$ and $b = 4$.

TABLE 3.3 gives examples of linear equations and corresponding values for a and b.

Linear Equations		
$ax + b = 0$	a	b
$x - 4 = 0$	1	-4
$-2x + 1 = 0$	-2	1
$3x = 0$	3	0

TABLE 3.3

Every linear equation can be written in the form $ax + b = 0$ using only the following properties or processes.

- Using the distributive property to clear any parentheses
- Combining like terms
- Applying the addition property of equality

Also, an equation *cannot* be written in the form $ax + b = 0$ (and is not a linear equation) if after clearing parentheses and combining like terms, any of the following are true.

1. The variable has an exponent other than 1.
2. The variable is in a denominator.
3. The variable is under the symbol $\sqrt{}$ or within absolute value symbols.

READING CHECK 1

Which of the following are linear equations? Select all that apply.

- $x + 3 = -5$
- $4x = 7$
- $\sqrt{2x - 1} = 6$
- $\dfrac{1}{x} - 3 = 2$
- $-4x^2 + 1 = -5$
- $-5x - 7 = 0$

EXAMPLE 1 Determining if an equation is linear

Determine if each equation is linear. If so, give values for a and b.
(a) $8x - 1 = 0$ (b) $4x + 3 = 2$ (c) $\sqrt{x} + 3 = 0$ (d) $-2x^2 + 7 = 0$

Solution

(a) The equation is linear because it is in the form $ax + b = 0$ with $a = 8$ and $b = -1$.

(b) The equation can be written as follows.

$$4x + 3 = 2 \qquad \text{Given equation}$$
$$4x + 3 - 2 = 2 - 2 \qquad \text{Subtract 2 from each side.}$$
$$4x + 1 = 0 \qquad \text{Simplify.}$$

The equation $4x + 3 = 2$ is linear because it can be written in the form $ax + b = 0$ with $a = 4$ and $b = 1$.

(c) The equation $\sqrt{x} + 3 = 0$ is *not* linear because it cannot be written in the form $ax + b = 0$. The variable appears under a square root symbol.

(d) The equation $-2x^2 + 7 = 0$ is *not* linear because it cannot be written in the form $ax + b = 0$. The variable has an exponent of 2.

Now Try Exercises 7, 11, 13, 15

2 Solving Linear Equations Symbolically

When solving an equation using the distributive property and the properties of equality, we are solving the equation *symbolically*. The solution to a linear equation can be found using the following steps.

STUDY TIP

We will be solving equations throughout the remainder of the text. Spend a little extra time practicing these steps so that they are familiar and easy to recall when needed later.

STEPS FOR SOLVING A LINEAR EQUATION

STEP 1: Use the distributive property to clear any parentheses on each side of the equation. Combine any like terms on each side.

STEP 2: Apply the addition property of equality to get all of the terms containing the variable on one side of the equation and all other terms on the other side of the equation. Combine any like terms on each side.

STEP 3: Use the multiplication property of equality to isolate the variable by dividing each side of the equation by the number in front of the variable.

STEP 4: Check the solution by substituting it into the *given* equation.

When a linear equation does not contain parentheses, we can start by combining any like terms on each side of the equation (if needed) and then continue with Step 2.

EXAMPLE 2 Solving linear equations symbolically

Solve each linear equation symbolically. Check your solution.
(a) $5m + 3 = m - 9$ (b) $14 - 3y + 5 = 4y - 1 + 3y$

Solution

(a) The equation does not contain parentheses, and there are no like terms to combine on either side. We begin with Step 2.

$$5m + 3 = m - 9 \qquad \text{Given equation}$$
$$5m + 3 - m = m - 9 - m \qquad \text{Subtract } m \text{ from each side. (Step 2)}$$
$$4m + 3 = -9 \qquad \text{Combine like terms.}$$
$$4m + 3 - 3 = -9 - 3 \qquad \text{Subtract 3 from each side. (Step 2)}$$
$$4m = -12 \qquad \text{Simplify.}$$
$$\frac{4m}{4} = \frac{-12}{4} \qquad \text{Divide each side by 4. (Step 3)}$$
$$m = -3 \qquad \text{Simplify.}$$

READING CHECK 2

- When checking a solution, why should you be sure to check the solution in the *given* equation?

The solution is -3. To check this solution, replace m with -3 in the *given* equation.

$$5m + 3 = m - 9 \qquad \text{Given equation}$$
$$5(-3) + 3 \stackrel{?}{=} -3 - 9 \qquad \text{Replace } m \text{ with } -3. \text{ (Step 4)}$$
$$-15 + 3 \stackrel{?}{=} -12 \qquad \text{Simplify.}$$
$$-12 = -12 \checkmark \qquad \text{The solution checks.}$$

(b) The equation does not contain parentheses, but we can combine like terms on each side before moving on with Step 2.

$$14 - 3y + 5 = 4y - 1 + 3y \qquad \text{Given equation}$$
$$19 - 3y = 7y - 1 \qquad \text{Combine like terms on each side.}$$
$$19 - 3y + 3y = 7y - 1 + 3y \qquad \text{Add } 3y \text{ to each side. (Step 2)}$$
$$19 = 10y - 1 \qquad \text{Combine like terms.}$$
$$19 + 1 = 10y - 1 + 1 \qquad \text{Add 1 to each side. (Step 2)}$$
$$20 = 10y \qquad \text{Simplify.}$$
$$\frac{20}{10} = \frac{10y}{10} \qquad \text{Divide each side by 10. (Step 3)}$$
$$2 = y \qquad \text{Simplify.}$$

The solution is 2. To check this solution, replace y with 2 in the *given* equation.

$$14 - 3y + 5 = 4y - 1 + 3y \qquad \text{Given equation}$$
$$14 - 3(2) + 5 \stackrel{?}{=} 4(2) - 1 + 3(2) \qquad \text{Replace } y \text{ with 2. (Step 4)}$$
$$14 - 6 + 5 \stackrel{?}{=} 8 - 1 + 6 \qquad \text{Simplify.}$$
$$13 = 13 \checkmark \qquad \text{The solution checks.}$$

Now Try Exercises 39, 57

When a linear equation contains parentheses, we begin with Step 1 of the strategy for solving linear equations. The next example shows how to solve linear equations that contain parentheses.

EXAMPLE 3 **Solving linear equations symbolically**

Solve each linear equation symbolically. Check your solution.
(a) $3(x - 2) + 2x = 19$ **(b)** $2(y - 5) + 1 = 3(y + 3)$

Solution
(a) We begin by using the distributive property to clear parentheses.

$$3(x - 2) + 2x = 19 \qquad \text{Given equation}$$
$$3x - 6 + 2x = 19 \qquad \text{Distributive property (Step 1)}$$
$$5x - 6 = 19 \qquad \text{Combine like terms.}$$
$$5x - 6 + 6 = 19 + 6 \qquad \text{Add 6 to each side. (Step 2)}$$
$$5x = 25 \qquad \text{Simplify.}$$
$$\frac{5x}{5} = \frac{25}{5} \qquad \text{Divide each side by 5. (Step 3)}$$
$$x = 5 \qquad \text{Simplify.}$$

The solution is 5. To check this solution, replace x with 5 in the *given* equation.

$$3(x - 2) + 2x = 19 \quad \text{Given equation}$$
$$3(5 - 2) + 2(5) \stackrel{?}{=} 19 \quad \text{Replace } x \text{ with 5. (Step 4)}$$
$$3(3) + 10 \stackrel{?}{=} 19 \quad \text{Simplify.}$$
$$19 = 19 \checkmark \quad \text{The solution checks.}$$

(b) Use the distributive property to clear parentheses on each side of the equation.

$$2(y - 5) + 1 = 3(y + 3) \quad \text{Given equation}$$
$$2y - 10 + 1 = 3y + 9 \quad \text{Distributive property (Step 1)}$$
$$2y - 9 = 3y + 9 \quad \text{Simplify.}$$
$$2y - 9 - 2y = 3y + 9 - 2y \quad \text{Subtract } 2y \text{ from each side. (Step 2)}$$
$$-9 = y + 9 \quad \text{Combine like terms}$$
$$-9 - 9 = y + 9 - 9 \quad \text{Subtract 9 from each side. (Step 2)}$$
$$-18 = y \quad \text{Simplify.}$$

Note that Step 3 is not needed to solve this equation. The solution is -18. To check this solution, replace y with -18 in the *given* equation.

$$2(y - 5) + 1 = 3(y + 3) \quad \text{Given equation}$$
$$2(-18 - 5) + 1 \stackrel{?}{=} 3(-18 + 3) \quad \text{Replace } y \text{ with } -18. \text{ (Step 4)}$$
$$2(-23) + 1 \stackrel{?}{=} 3(-15) \quad \text{Simplify.}$$
$$-45 = -45 \checkmark \quad \text{The solution checks.}$$

Now Try Exercises 61, 65

3 Solving Linear Equations Numerically

Some linear equations can be solved using a table of values. When we solve an equation in this way, we are solving it *numerically*. The next example illustrates this process.

EXAMPLE 4 Solving a linear equation numerically

Complete **TABLE 3.4** for the given values of x. Then solve the equation $4x - 5 = 3$.

x	-1	0	1	2	3
$4x - 5$					

TABLE 3.4

Solution
Begin by replacing x in $4x - 5$ with -1. Because $4(-1) - 5$ evaluates to -9, the left side of the equation is equal to -9 when $x = -1$. We write the result -9 below -1, as shown in **TABLE 3.5**. Similarly, we write -5 below 0 because the expression $4(0) - 5$ evaluates to -5. The remaining values for **TABLE 3.5** are found in a similar manner.

Find an x where $4x - 5$ equals 3.

x	-1	0	1	2	3
$4x - 5$	-9	-5	-1	3	7

TABLE 3.5

TABLE 3.5 reveals that $4x - 5 = 3$ when $x = 2$. Therefore, the solution is **2**.

Now Try Exercise 77

READING CHECK 3

• What do you use to solve an equation when you are asked to solve it numerically?

NOTE: For some equations, making a table that contains the solution could be difficult. For example, the solution to the equation $238x - 7112 = 266$ is 31, but creating a table by hand that shows this solution could take a significant amount of effort. ∎

The following See the Concept summarizes the important points about solving an equation numerically.

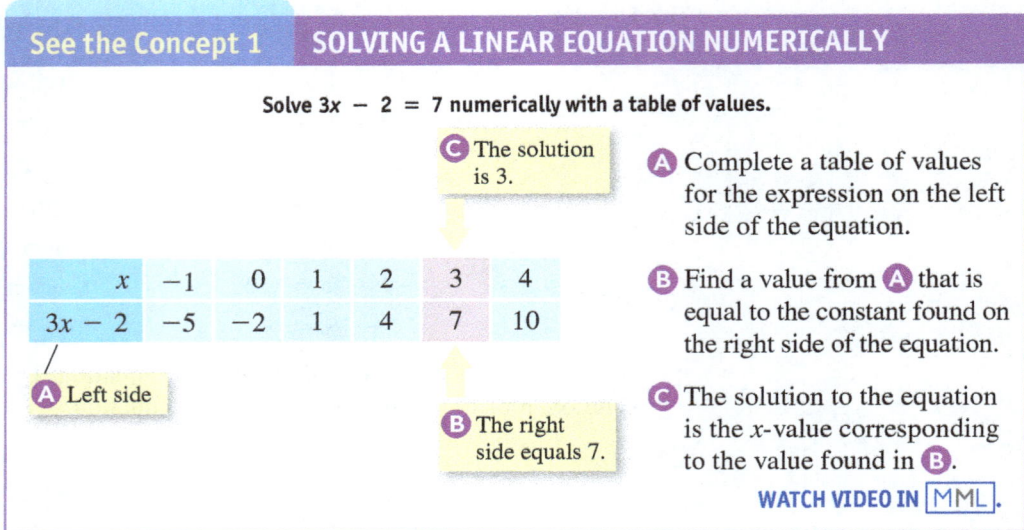

4 Solving Linear Equations Graphically

In an earlier chapter, we solved equations graphically by estimating values from a related graph. As you might expect, the graph associated with a **linear** equation is a straight **line**. When we solve a linear equation by estimating values from its graph, we are solving the equation *graphically*.

Because some linear equations have negative values for solutions, a graph representing a linear equation should show both positive and negative values for the variable. To do this, we can use graphs similar to the one shown in **FIGURE 3.1**.

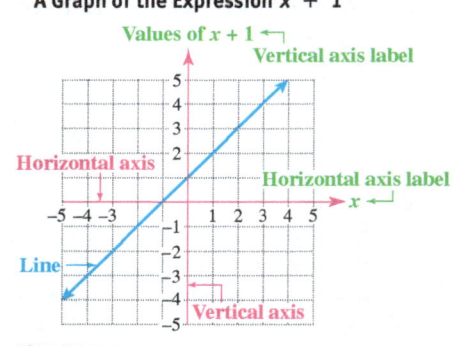

FIGURE 3.1

In this graph, the **vertical axis** shows values of the expression $x + 1$ and the **horizontal axis** shows values of the variable x. Arrows at the ends of the line representing the expression $x + 1$ (shown in blue) are used to show that the line continues in both directions.

NOTE: Although each axis continues in both the positive and negative directions, it is common for an axis to show an arrow pointing in the positive direction only. ∎

In the next example, a linear equation is solved graphically.

EXAMPLE 5 Solving a linear equation graphically

The graph of the expression $-2x - 5$ is shown in **FIGURE 3.2**. Use the graph to solve the equation $-2x - 5 = 3$ graphically.

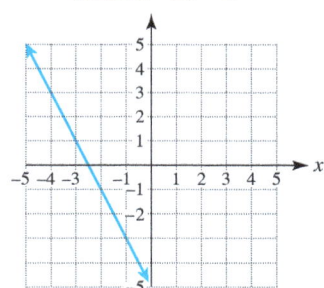

FIGURE 3.2

Solution
Because we want to solve the equation $-2x - 5 = 3$, locate **3** on the vertical axis. Move **horizontally** to the left until the graphed line is reached. From this position, move **vertically** downward to the horizontal axis. The solution appears to be -4. See **FIGURE 3.3**. Remember that a graphical solution is an estimate and must be checked.

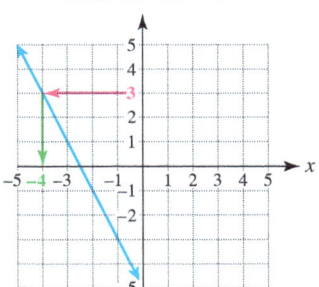

FIGURE 3.3

To see if -4 is the correct solution, check it in the given equation.

$$-2x - 5 = 3 \qquad \text{Given equation}$$
$$-2(-4) - 5 \stackrel{?}{=} 3 \qquad \text{Replace } x \text{ with } -4.$$
$$8 - 5 \stackrel{?}{=} 3 \qquad \text{Simplify.}$$
$$3 = 3 \checkmark \qquad \text{The solution checks.}$$

The solution to the equation $-2x - 5 = 3$ is -4.

Now Try Exercise 81

The following See the Concept summarizes the important points related to solving an equation graphically.

See the Concept 2 — SOLVING A LINEAR EQUATION GRAPHICALLY

Solve $2x - 1 = 3$ graphically.

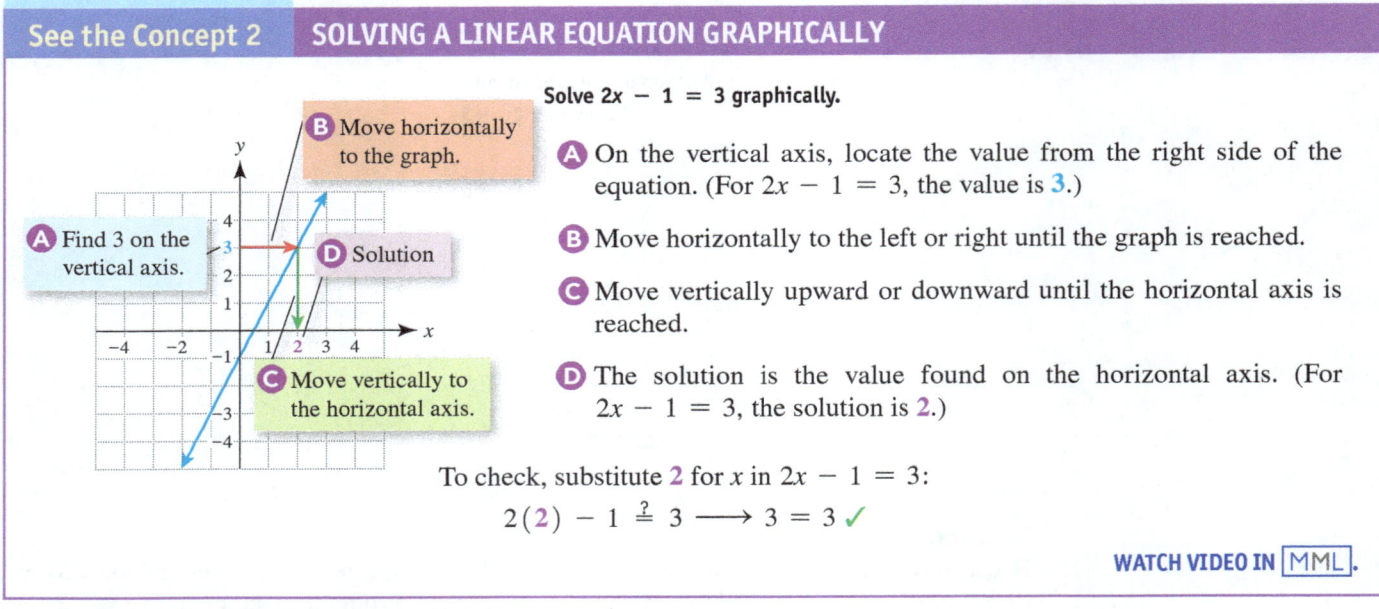

A On the vertical axis, locate the value from the right side of the equation. (For $2x - 1 = 3$, the value is **3**.)

B Move horizontally to the left or right until the graph is reached.

C Move vertically upward or downward until the horizontal axis is reached.

D The solution is the value found on the horizontal axis. (For $2x - 1 = 3$, the solution is **2**.)

To check, substitute **2** for x in $2x - 1 = 3$:
$$2(2) - 1 \stackrel{?}{=} 3 \longrightarrow 3 = 3 \checkmark$$

WATCH VIDEO IN MML.

Math in Energy Context From 2010 to 2035, the total energy production of the Asia-Pacific region is expected to increase substantially. The amount of energy E produced in this region during year x can be estimated using the formula

$$E = 136x - 268{,}360,$$

where E is measured in million metric tons of oil equivalent (Mtoe) and x is any year from 2010 to 2035. By letting $E = 7040$ in this formula, we get the linear equation

$$7040 = 136x - 268{,}360.$$

The solution to this equation gives the year when energy production in the Asia-Pacific region is projected to reach 7040 Mtoe. In the next example, this linear equation is solved symbolically, numerically, and graphically. (*Source:* U.S. Department of Energy.)

EXAMPLE 6 Analyzing energy production

Find the year when total energy production in the Asia-Pacific region is projected to reach 7040 Mtoe by solving the linear equation $7040 = 136x - 268{,}360$.
(a) symbolically,
(b) numerically using **TABLE 3.6**, and
(c) graphically using **FIGURE 3.4**.

Energy Production in the Asia-Pacific Region (Mtoe)

x (year)	2010	2015	2020	2025	2030	2035
$136x - 268{,}360$						

TABLE 3.6

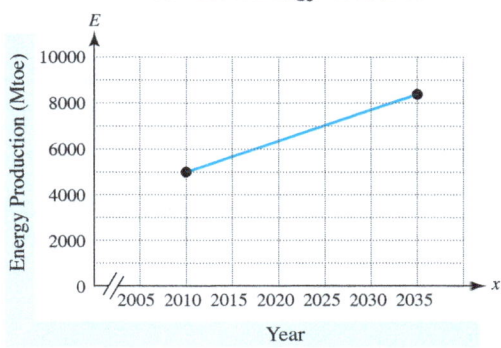

FIGURE 3.4

Solution
(a) **Symbolic Solution**

$$7040 = 136x - 268{,}360 \qquad \text{Given equation}$$
$$7040 + \mathbf{268{,}360} = 136x - 268{,}360 + \mathbf{268{,}360} \qquad \text{Add 268,360 to each side. (Step 2)}$$
$$275{,}400 = 136x \qquad \text{Simplify.}$$
$$\frac{275{,}400}{136} = \frac{136x}{136} \qquad \text{Divide each side by 136. (Step 3)}$$
$$2025 = x \qquad \text{Simplify.}$$

Energy production in the Asia-Pacific region is expected to reach 7040 Mtoe in 2025.

(b) **Numerical Solution** Begin by replacing x in $136x - 268{,}360$ with **2010** to obtain

$$136(\mathbf{2010}) - 268{,}360 = 273{,}360 - 268{,}360 = 5000.$$

Write **5000** below **2010** as shown in **TABLE 3.7** and then complete the rest of the table in a similar fashion.

Energy Production in the Asia-Pacific Region (Mtoe)

x (year)	**2010**	2015	2020	**2025**	2030	2035
$136x - 268{,}360$	**5000**	5680	6360	**7040**	7720	8400

TABLE 3.7

TABLE 3.7 reveals that total energy production in the Asia-Pacific region is expected to reach **7040** Mtoe in **2025**.

(c) **Graphical Solution** Locate 7040 on the vertical axis of **FIGURE 3.4**. Move **horizontally** to the right until the graphed line is reached. From this position, move **vertically** downward to the horizontal axis. The solution appears to be 2025. See **FIGURE 3.5**.

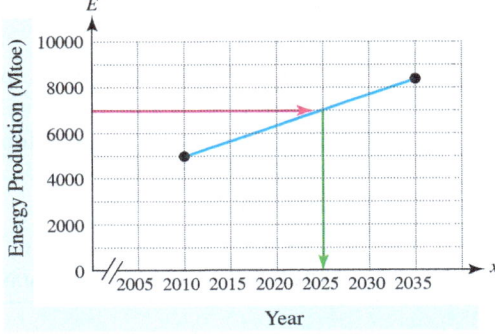

FIGURE 3.5

To see if 2025 is the correct solution, check it in the given equation.

$7040 = 136x - 268{,}360$	Given equation
$7040 \stackrel{?}{=} 136(2025) - 268{,}360$	Replace x with 2025.
$7040 \stackrel{?}{=} 275{,}400 - 268{,}360$	Simplify.
$7040 = 7040$ ✓	The solution checks.

The solution to the equation $7040 = 136x - 268{,}360$ is 2025. Note that symbolic, numerical, and graphical solutions are the same.

Now Try Exercise 89

MAKING CONNECTIONS 1

Why We Solve an Equation Three Different Ways

Because real-world data may be presented in formulas, tables, or graphs, finding answers to real-world questions depends on our ability to work with data in any of these forms. When an equation is too complex to solve symbolically, a graph or table may give a solution. When a table contains millions of numbers, an equation may be more practical. Data that are presented graphically can give us a great deal of information at a glance. No matter which of the three methods we use to solve an equation, the solution is the same.

READING CHECK 4

- Should the solutions that are found symbolically, numerically, and graphically be the same or different?

3.4 Putting It All Together

CONCEPT	COMMENTS	EXAMPLES
Linear Equations in One Variable	A linear equation in one variable is an equation that can be written in the form $$ax + b = 0,$$ where a and b are constants (numbers) and $a \neq 0$.	$x - 8 = 0$ is linear. $(a = 1, b = -8)$ $2x + 3 = 5$ is linear because subtracting 5 from each side gives $2x - 2 = 0$. $(a = 2, b = -2)$ $7x^2 - 2x = 0$ is *not* linear. $\sqrt{2x} - 5 = 10$ is *not* linear.
Solving Linear Equations Symbolically	**STEP 1:** Use the distributive property to clear any parentheses. Combine like terms. **STEP 2:** Apply the addition property of equality to get all of the terms containing the variable on one side of the equation and all other terms on the other side of the equation. Combine like terms. **STEP 3:** Use the multiplication property of equality to isolate the variable by dividing each side of the equation by the number in front of the variable. **STEP 4:** Check the solution by substituting it into the *given* equation.	$2(x+1) = 8$ Given equation $2x + 2 = 8$ Distributive property $2x = 6$ Subtract 2 from each side. $x = 3$ Divide each side by 2. Check this solution. $2(x+1) = 8$ Given equation $2(3+1) \stackrel{?}{=} 8$ Replace x with 3. $2(4) \stackrel{?}{=} 8$ Add. $8 = 8$ ✓ The solution checks.

CONCEPT	COMMENTS	EXAMPLES
Solving Linear Equations Numerically	To solve a linear equation numerically, complete a table for various values of the variable and, if possible, select the solution from the table.	The solution to $2x - 4 = -2$ is 1. | x | -1 | 0 | 1 | |---|---|---|---| | $2x - 4$ | -6 | -4 | -2 |
Solving Linear Equations Graphically	To solve a linear equation graphically, use the graph of the linear equation to estimate a solution. Then check the solution in the given equation.	The solution to $2x - 1 = 3$ is 2.

3.4 Exercises   MyMathLab

CONCEPTS AND VOCABULARY

1. An equation that can be written as $ax + b = 0$, where a and b are constants and $a \neq 0$, is called a(n) _____ equation in one variable.

2. The linear equation $-2x = 0$ can be written in the form $ax + b = 0$ with $a =$ _____ and $b =$ _____.

3. After clearing parentheses and combining like terms, the variable in a linear equation should not have an exponent other than _____.

4. When a table of values is used to solve a linear equation, we are solving the equation _____.

5. When the four-step strategy provided in this section is used to solve a linear equation, we are solving the equation _____.

6. Step 4 of the strategy for solving linear equations states that the solution can be checked by substituting it into the _____ equation.

IDENTIFYING LINEAR EQUATIONS IN ONE VARIABLE

Exercises 7–20: Determine if the equation is linear. If so, give values for a and b in the form $ax + b = 0$.

7. $3x + 7 = 0$
8. $-2x + 5 = 0$
9. $5 - 5 = 0$
10. $-28 + 28 = 0$
11. $4x - 7 = 2$
12. $8x - 3 = -8$
13. $|3x| - 2 = 0$
14. $2\sqrt{x} + 7 = 0$
15. $9x^2 + 3 = 0$
16. $2x^3 - 5 = 0$
17. $3x - 5 = 2x - 4$
18. $x - 3 = -2x + 5$
19. $6(x + 2) = -3$
20. $-3(2x - 1) = -2$

SOLVING EQUATIONS SYMBOLICALLY

Exercises 21–74: Solve the linear equation symbolically.

21. $2a + 8 = 0$
22. $3m - 15 = 0$
23. $-5y - 35 = 0$
24. $-3x + 18 = 0$
25. $6t + (-6) = 12$
26. $7n + (-15) = -1$
27. $6b - 3b = 33$
28. $7k - 9k = -14$
29. $17 - 7 = 2m$
30. $34 + (-9) = 5w$
31. $32p + 9 = 73$
32. $-5 + 15t = 70$
33. $3w = 4 + 11$
34. $16 = 2m - 14$
35. $3(x - 8) = 4x$
36. $-5 + 2y = 7y$
37. $y - 5y = 8 + 4$
38. $4d + d = 1 - 46$
39. $4b - 2 = b - 11$
40. $3b + 4 = 5b - 2$
41. $12 - 2x = 3x - 8$
42. $5 - g = 7 - 2g$

43. $\dfrac{k}{5} = -15 + 18$

44. $14 - 10 = \dfrac{q}{-3}$

45. $3m + 2m = 9m + 4$

46. $5 - b = b - 3b$

47. $a - a - a = 12$

48. $x + x + x = 0$

49. $3d - 7 = d - 19$

50. $4 - n = 2 - 3n$

51. $3t + (-15) = 33$

52. $-r - 23 = -1$

53. $\dfrac{a}{4} = a$

54. $2b = \dfrac{b}{-3}$

55. $3h = -39$

56. $12w - 13 = 11$

57. $11 + 3x - 4 = 5x - 1 + 2x$

58. $2g - 6 - 8g = 9 - g + 5$

59. $2n + 6 - 7n = 33 - 11n + 9$

60. $10 + 3y + 8 = y - 15 - 9y$

61. $2(x - 1) + 7x = 61$

62. $-3(2q + 4) + 7q = 19$

63. $-5(3w + 1) - w = 27$

64. $50 + 4a = 2(5a - 8)$

65. $3(m - 4) + 1 = 2(m + 3)$

66. $4(c - 1) = -2(c + 3) - 28$

67. $-2(k - 1) = 2(k + 1) - 24$

68. $-(d + 3) + 1 = 3(d + 6)$

69. $3x - 4 + 12 = 4x - 1 + 8x$

70. $-2(3w - 1) - w = 23$

71. $2(m - 14) + 1 = 5(m + 7) + 4$

72. $-4(c - 1) = -4c - 11 + 3c$

73. $3(k - 5) - 2(k + 1) - 13 = -22$

74. $3(2y - 5) = 7(3y + 1) - 13 + (-14y)$

SOLVING EQUATIONS NUMERICALLY

Exercises 75–80: Complete the table of values and then solve the given linear equation numerically.

75. $x - 2 = 0$

x	-2	-1	0	1	2
$x - 2$					

76. $3x + 5 = 2$

x	-2	-1	0	1	2
$3x + 5$					

77. $-2x + 7 = 7$

x	-2	-1	0	1	2
$-2x + 7$					

78. $-x - 4 = -5$

x	-2	-1	0	1	2
$-x - 4$					

79. $3(x - 2) = 0$

x	-2	-1	0	1	2
$3(x - 2)$					

80. $-2(x + 1) = 2$

x	-2	-1	0	1	2
$-2(x + 1)$					

SOLVING EQUATIONS GRAPHICALLY

Exercises 81–88: Solve the linear equation graphically.

81. $3x + 2 = -4$

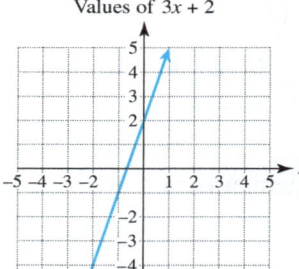

Values of $3x + 2$

82. $-x - 2 = 2$

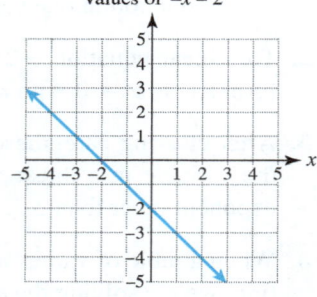

Values of $-x - 2$

83. $-3x - 4 = -4$

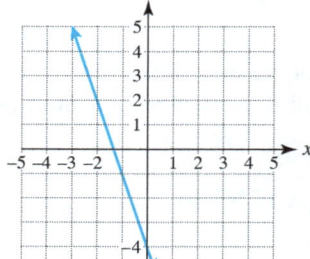

Values of $-3x - 4$

84. $2x - 3 = 3$

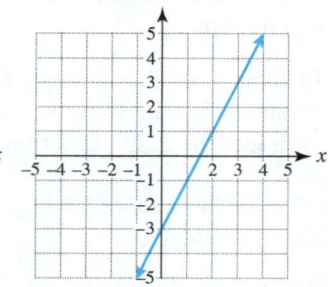

Values of $2x - 3$

85. $10x + 5 = 15$

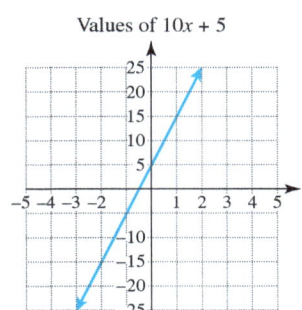

86. $-5x - 10 = 0$

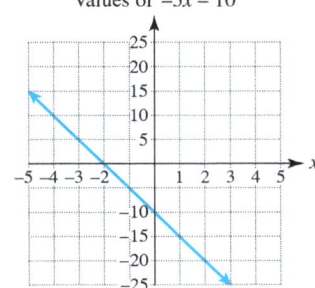

87. $-3x - 6 = -30$

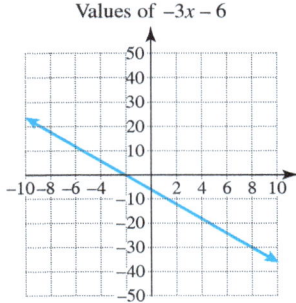

88. $6x + 18 = -30$

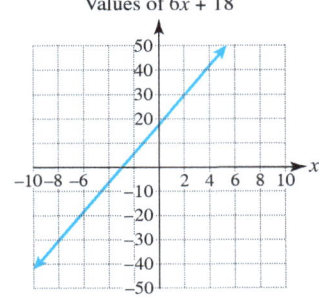

APPLICATIONS INVOLVING EQUATIONS

89. Altitude and Temperature If the temperature on the ground is 70 °F, then the air temperature T at an altitude of x miles is given by $T = 70 - 19x$. Find the altitude where the air temperature is -44 °F by solving the linear equation $-44 = 70 - 19x$
(a) symbolically,
(b) numerically using the given table, and
(c) graphically using the graph shown.
(*Source:* Miller, A., and R. Anthes, *Meteorology*. 5th ed.)

x	2	3	4	5	6
$70 - 19x$					

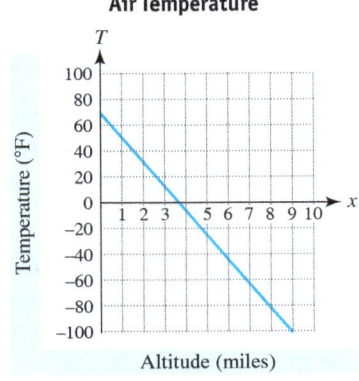

90. Median Household Income The median family income I during year x is approximated by the formula $I = 600x - 1{,}158{,}000$, where I is in dollars and x is any year from 2000 to 2015. Find the year when the median family income was $48,000 by solving $48{,}000 = 600x - 1{,}158{,}000$
(a) symbolically,
(b) numerically using the given table, and
(c) graphically using the given graph.
(*Source:* Department of the Treasury.)

x	2000	2005	2010	2015
$600x - 1{,}158{,}000$				

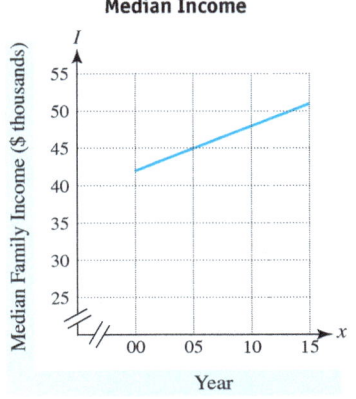

91. Bamboo Growth (Refer to the opener for this chapter.) Under ideal conditions, bamboo can grow at a rate of 3 centimeters per hour. Suppose that a bamboo plant's height H in centimeters after x hours is given by the formula $H = 3x + 210$. Determine how many hours it takes this plant to reach a height of 222 centimeters by solving the linear equation $222 = 3x + 210$
(a) symbolically,
(b) numerically using the given table, and
(c) graphically using the given graph.

x	0	1	2	3	4	5	6
$3x + 210$							

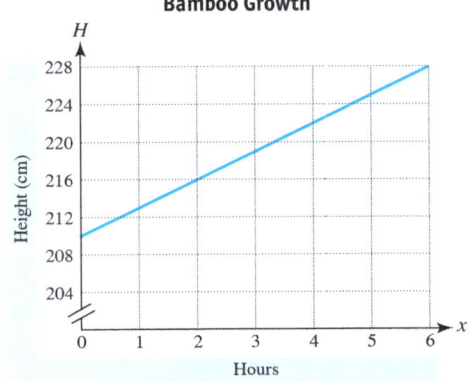

92. International Adoption The number of children C adopted into the U.S. from Ethiopia during year x is approximated by $C = 515x - 1{,}032{,}359$, where x

continued on next page

is any year from 2006 to 2009. Find the year when the number of children adopted into the U.S. from Ethiopia was 1246 by solving the linear equation $1246 = 515x - 1{,}032{,}359$
(a) symbolically,
(b) numerically using the given table, and
(c) graphically using the graph below.

(*Source:* U.S. Department of State.)

x	2006	2007	2008	2009
$515x - 1{,}032{,}359$				

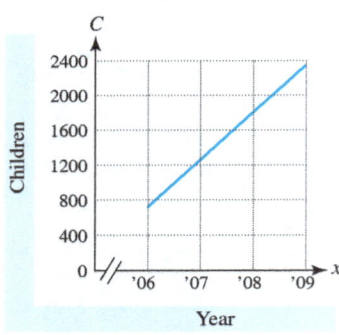

International Adoptions

WRITING ABOUT MATHEMATICS

93. Even though the equation
$$2x^2 - 3x + 1 = 2x^2 + x$$
contains a variable with a power other than 1, it can be written in the form $ax + b = 0$. Explain why this equation is linear.

94. Write the steps used to solve a linear equation in one variable symbolically. Give an example of each step.

SECTIONS 3.3 and 3.4 **Checking Basic Concepts**

1. Is -3 a solution to $n + 5 = 2$?

2. Are $-4w = 8$ and $w = 2$ equivalent equations?

3. Solve each equation.
   (a) $10 = 17 - y$ (b) $45 = -9n$
   (c) $32 + b = 0$ (d) $-7 = \dfrac{g}{8}$

4. Determine if the equation is linear. If so, give values for a and b.
   (a) $4x - 3 = 0$ (b) $5\sqrt{x} - 1 = 0$
   (c) $7x^2 - 4 = 0$ (d) $-2x - 5 = 3$

5. Solve each linear equation symbolically.
   (a) $k - 9k = -24$
   (b) $2n + (-5) = -1$
   (c) $-14 - 3x = 6x + 13$
   (d) $-2(3w + 5) - 2w = -18$

6. Solve $-3(x - 1) = 6$ numerically.

x	-2	-1	0	1	2
$-3(x-1)$					

7. Solve $2x + 3 = -1$ graphically. Be sure to check your answer.

   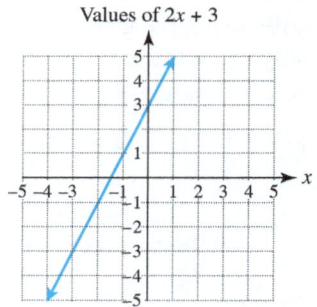

   Values of $2x + 3$

8. **Geometry** The formula for the area of a rectangle is $A = lw$. Find the width of a rectangle with an area of 75 square inches and a length of 15 inches.

9. **Altitude and Temperature** If the temperature on the ground is 80 °F, then the air temperature T at an altitude of x miles is given by $T = 80 - 19x$. Determine the altitude where the air temperature is $T = 4$ °F.

3.5 Applications and Problem Solving

Objectives

1. Reviewing Steps for Problem Solving
2. Solving Applications Involving Linear Equations in One Variable

STUDY TIP

Step 4 in this problem-solving strategy provides good advice for working *any* math problem. Checking your answer, especially when taking an exam, can lead to fewer errors and better scores.

1 Reviewing Steps for Problem Solving

When solving applications involving linear equations, we rely on two skills—translating words to an equation and solving an equation. Together, these skills are used in Steps 2 and 3 of the problem-solving strategy. This strategy is repeated here for convenience.

> **STEPS FOR SOLVING A PROBLEM**
>
> **STEP 1:** Read the problem carefully for understanding. (You may need to read the problem more than once.) Identify key words associated with arithmetic operations. Assign a variable to any unknown quantity.
>
> **STEP 2:** Write an equation that relates the quantities described in the problem. You may need to sketch a diagram or refer to known formulas.
>
> **STEP 3:** Solve the equation. Use the solution to determine the answer to the original problem. Include any necessary units.
>
> **STEP 4:** Check your answer. Does it make sense and is it a reasonable answer to the original problem?

2 Solving Applications Involving Linear Equations in One Variable

The remainder of this section presents examples that demonstrate how we can use the four-step strategy to solve applications involving linear equations.

EXAMPLE 1   Solving a number problem

When a number is added to 18, the result is the same as the product of the number and 4. Find the number.

Solution
STEP 1: We must find an unknown number. Let x represent this number.
STEP 2: The paragraph states that

> *the number **added** to **18** **is the same as** the **product** of the number and **4**.*

Because x represents the number, this means that

$$x + 18 = x \cdot 4.$$

STEP 3: Rewrite and solve the equation from Step 2.

$$\begin{aligned}
x + 18 &= 4x &&\text{Given equation} \\
x - x + 18 &= 4x - x &&\text{Subtract } x \text{ from each side.} \\
18 &= 3x &&\text{Simplify.} \\
\frac{18}{3} &= \frac{3x}{3} &&\text{Divide each side by 3.} \\
6 &= x &&\text{Simplify.}
\end{aligned}$$

STEP 4: Because $6 + 18 = 24$ and $6(4) = 24$, the solution checks in the original problem. The unknown number is 6.

Now Try Exercise 15

Math in Energy Context
From 2000 to 2015, the number of barrels of crude oil imported each year into the United States generally decreased. The next example analyzes this decrease using a linear equation.

EXAMPLE 2 **Analyzing U.S. crude oil imports**

There were 700 million more barrels of crude oil imported into the United States in 2000 than in 2015. Find the number of barrels of crude oil (in millions) that were imported in 2015 if there were 3400 million barrels imported in 2000. (*Source:* U.S. Census Bureau.)

Solution
STEP 1: We must find the number of barrels of crude oil (in millions) that were imported into the U.S. in 2015. Let x represent the number of barrels imported in 2015, where x is expressed in millions of barrels.
STEP 2: Reading the paragraph carefully reveals that

3400 million barrels *is* 700 million *more* barrels than imported in 2015.

Because x represents the number (in millions) of barrels imported in 2015, this means that

$$3400 = x + 700.$$

STEP 3: To solve the equation from Step 2, we use the addition property of equality.

$$x + 700 = 3400 \quad \text{Given equation}$$
$$x + 700 - 700 = 3400 - 700 \quad \text{Subtract 700 from each side.}$$
$$x = 2700 \quad \text{Simplify.}$$

STEP 4: Because 3400 is 700 more than 2700, the solution checks in the original problem. There were 2700 million barrels of crude oil imported into the U.S. in 2015.

Now Try Exercise 31

In an earlier chapter, we used the formula for the perimeter of a rectangle. If we let l represent the length of a rectangle and w represent the width, then the following formula can be used to find the perimeter P of the rectangle.

$$P = 2l + 2w$$

In the next example, this formula is used to find the unknown dimensions of a rectangle.

EXAMPLE 3 **Finding the dimensions of a rectangle**

The length of a rectangle is 4 inches longer than the width. If the perimeter of the rectangle is 28 inches, find the measures of the length and width.

Solution
STEP 1: We must find the length and width of the given rectangle. Because the first sentence in the paragraph gives the length *in terms of* the width, we start by letting w represent the width of the rectangle. A length that is 4 inches longer than the width is represented by the expression $w + 4$.
STEP 2: Sketch a rectangle that represents the given information, as shown in **FIGURE 3.6**. The width is w, and the length is $w + 4$. The paragraph states that the perimeter is 28, so we use the known formula for perimeter, $P = 2l + 2w$. In this formula, replace P with **28** and replace l with the expression $w + 4$. Be sure to use parentheses when replacing l with $w + 4$.

$$28 = 2(w + 4) + 2w$$

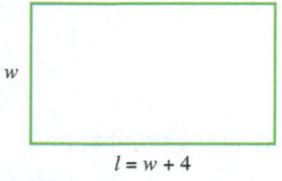

FIGURE 3.6

STEP 3: Rewrite and solve the equation from Step 2.

$$2(w + 4) + 2w = 28 \qquad \text{Given equation}$$
$$2w + 8 + 2w = 28 \qquad \text{Distributive property}$$
$$4w + 8 = 28 \qquad \text{Combine like terms.}$$
$$4w + 8 - 8 = 28 - 8 \qquad \text{Subtract 8 from each side.}$$
$$4w = 20 \qquad \text{Simplify.}$$
$$\frac{4w}{4} = \frac{20}{4} \qquad \text{Divide each side by 4.}$$
$$w = 5 \qquad \text{Simplify.}$$

The width of the rectangle is 5 inches. Because the length is 4 inches longer than the width, the length is $l = w + 4 = 5 + 4 = 9$ inches.

STEP 4: A rectangle with a length of 9 inches and a width of 5 inches has a perimeter of $P = 2(9) + 2(5) = 18 + 10 = 28$. Furthermore, the length is 4 inches longer than the width because $9 - 5 = 4$. The results check in the original problem. The length is 9 inches and the width is 5 inches.

Now Try Exercise 33

🌐 *Math in Context* [Rent] The average rent for an apartment in various cities varies greatly, depending on the city. The next example uses a linear equation to determine rental costs in two cities.

EXAMPLE 4 **Finding the monthly rent for a 1-bedroom apartment**

The monthly rent for an apartment in Oklahoma City is $517 less than the rent for a similar apartment in Fort Lauderdale. If the combined rental cost for both apartments is $1607, what is the monthly rent for an apartment in each city? (*Source:* U.S. Department of Housing and Urban Development.)

Solution

STEP 1: We must find the monthly rent for an apartment in each city. Because the rent in Oklahoma City is given *in terms of* the rent in Fort Lauderdale, we start by letting the variable F represent the monthly rent in Fort Lauderdale. The monthly rent in Oklahoma City can then be represented by the expression $F - 517$.

STEP 2: Because the combined rent is $1607, we add the monthly rent in Oklahoma City to the monthly rent in Fort Lauderdale to get 1607, or $F + (F - 517) = 1607$.

STEP 3: Solve the equation from Step 2.

$$F + (F - 517) = 1607 \qquad \text{Given equation}$$
$$2F - 517 = 1607 \qquad \text{Combine like terms.}$$
$$2F - 517 + 517 = 1607 + 517 \qquad \text{Add 517 to each side.}$$
$$2F = 2124 \qquad \text{Simplify.}$$
$$\frac{2F}{2} = \frac{2124}{2} \qquad \text{Divide each side by 2.}$$
$$F = 1062 \qquad \text{Simplify.}$$

The monthly rent in Fort Lauderdale is $1062, and the monthly rent in Oklahoma City is $F - 517 = 1062 - 517 = \$545$.

STEP 4: The combined rental cost in the two cities is $1062 + 545 = \$1607$, and the monthly rent in Oklahoma City is $1062 - 545 = \$517$ less than the monthly rent in Fort Lauderdale. So, the results check in the original problem.

Now Try Exercise 37

EXAMPLE 5 Finding the ages of two people

The age of a boy's grandfather is 11 times the age of the boy. If the sum of their ages is 72, find the age of each person.

Solution

STEP 1: We must find the age of each person. Because the grandfather's age is given *in terms of* the boy's age, we let b represent the boy's age. The grandfather's age is 11 times the boy's age and is represented by the expression $11b$.

STEP 2: Because the sum of their ages is 72, we add the expressions representing the two ages, which gives the equation $b + 11b = 72$.

STEP 3: Solve the equation from Step 2.

$$b + 11b = 72 \quad \text{Given equation}$$
$$12b = 72 \quad \text{Combine like terms.}$$
$$\frac{12b}{12} = \frac{72}{12} \quad \text{Divide each side by 12.}$$
$$b = 6 \quad \text{Simplify.}$$

The boy's age is 6, and the grandfather's age is $11(6) = 66$.

STEP 4: The sum of the two ages is $6 + 66 = 72$, and the grandfather's age, 66, is 11 times the boy's age, 6. So, the results check in the original problem.

Now Try Exercise 39

Math in Radio Context The total number of radio stations R in the United States from 1950 to 2010 is approximated by the formula $R = 200(x - 1950) + 2770$, where x is any year from 1950 to 2010. This formula is represented graphically in **FIGURE 3.7**. (*Source: National Association of Broadcasters.*)

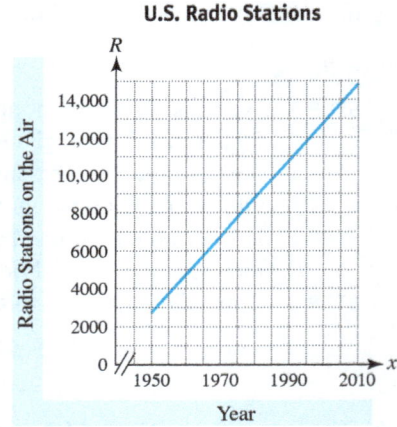

FIGURE 3.7

EXAMPLE 6 Analyzing radio station data

Using the formula $R = 200(x - 1950) + 2770$, find the year when the number of U.S. radio stations was 7170. Find your answer symbolically, numerically, and graphically.

Solution

STEP 1: We must find the year when the number of U.S. radio stations reached 7170. Let x represent this unknown year.

STEP 2: We let $R = 7170$ in the given formula: $7170 = 200(x - 1950) + 2770$.

STEP 3: Solve the equation from Step 2 symbolically, numerically, and graphically.

Symbolic Solution

$$200(x - 1950) + 2770 = 7170 \quad \text{Given equation (rewritten)}$$
$$200x - 390{,}000 + 2770 = 7170 \quad \text{Distributive property}$$
$$200x - 387{,}230 = 7170 \quad \text{Simplify.}$$
$$200x - 387{,}230 + 387{,}230 = 7170 + 387{,}230 \quad \text{Add 387,230 to each side.}$$
$$200x = 394{,}400 \quad \text{Simplify.}$$
$$\frac{200x}{200} = \frac{394{,}400}{200} \quad \text{Divide each side by 200.}$$
$$x = 1972 \quad \text{Simplify.}$$

Numerical Solution First make a table with 10-year intervals, as shown in **TABLE 3.8**.

U.S. Radio Stations

x (year)	1950	1960	1970	1980	1990	2000	2010
$200(x - 1950) + 2770$	2770	4770	6770	8770	10,770	12,770	14,770

TABLE 3.8

Because 7170 is between **6770** and **8770**, we make a new table with years between **1970** and **1980**. **TABLE 3.9** reveals that the solution is **1972**.

U.S. Radio Stations

x (year)	1971	1972	1973	1974	1975	1976
$200(x - 1950) + 2770$	6970	7170	7370	7570	7770	7970

TABLE 3.9

Graphical Solution The solution is shown graphically in **FIGURE 3.8**.

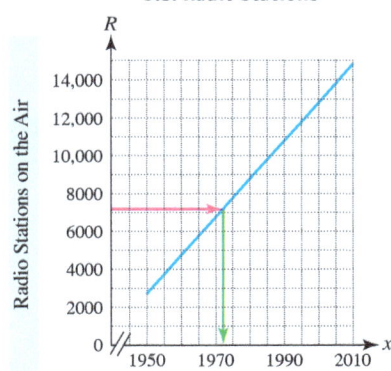

FIGURE 3.8

The solution appears to be 1972. We can check this solution as follows.

$$7170 = 200(x - 1950) + 2770 \quad \text{Given equation}$$
$$7170 \stackrel{?}{=} 200(1972 - 1950) + 2770 \quad \text{Replace } x \text{ with 1972.}$$
$$7170 \stackrel{?}{=} 200(22) + 2770 \quad \text{Subtract.}$$
$$7170 \stackrel{?}{=} 4400 + 2770 \quad \text{Multiply.}$$
$$7170 = 7170 \checkmark \quad \text{The solution checks.}$$

STEP 4: We checked this solution when the graphical solution was found in Step 3. All three methods show that the number of U.S. radio stations reached 7170 in 1972.

Now Try Exercise 43

> **MAKING CONNECTIONS 1**
>
> **The Best Method for Solving a Problem**
>
> Real-world problems come in many different forms. When a formula is given, the best way to solve the problem may be a symbolic method. When data are collected and presented in a table, a numerical solution to the problem may be best. At other times, information may be provided in a graph or bar chart. In these situations, a graphical solution to the problem might be best. Real-world problems can often be solved in more than one correct way. Part of the problem-solving process is choosing which method works best for you.

3.5 Putting It All Together

> **STEPS FOR SOLVING A PROBLEM**
>
> **STEP 1:** Read the problem carefully for understanding. (You may need to read the problem more than once.) Identify key words associated with arithmetic operations. Assign a variable to any unknown quantity.
>
> **STEP 2:** Write an equation that relates the quantities described in the problem. You may need to sketch a diagram or refer to known formulas.
>
> **STEP 3:** Solve the equation. Use the solution to determine the answer to the original problem. Include any necessary units.
>
> **STEP 4:** Check your answer. Does it make sense and is it a reasonable answer to the original problem?

3.5 Exercises

CONCEPTS AND VOCABULARY

1. When solving a problem, the first step involves reading the problem carefully for _____ and assigning a _____ to any unknown quantity.

2. The second step in solving a problem is to write a(n) _____ that relates the quantities described in the problem.

3. The third step in solving a problem is to _____ the equation and determine the answer to the _____ problem.

4. When solving a problem, the fourth (final) step is to _____ your answer.

SOLVING NUMBER PROBLEMS

Exercises 5–20: Solve the number problem by finding the value of the unknown number.

5. Seven more than a number equals 31.

6. Eight fewer than a number is 13.

7. The product of 7 and a number is -21.

8. When -48 is divided by a number, the result is 6.

9. Adding 7 to twice a number gives 13.

10. Fourteen equals 10 less than 6 times a number.

11. A number decreased by 5 gives twice the number.

12. Three times a number equals the number minus 8.

13. When the sum of a number and -10 is multiplied by -5, the result is 60.

14. When the sum of 7 and a number is divided by 3, the result is 6.

15. When a number is subtracted from 32, the result is the same as the product of the number and 3.

16. A number divided by 8 is the same as the number multiplied by 7.

17. Subtracting 3 from the product of 5 and a number results in the number added to 37.

18. If the product of -2 and a number is increased by 11, the result is 4 less than the number.

19. Four times the quantity 5 plus a number yields 16 more than twice the number.

20. The product of -2 and the sum of a number and 7 is the same as the number increased by 1.

APPLICATIONS INVOLVING PROBLEM SOLVING

For Exercises 21–28 refer to Exercises 31–38 in Section 3.2.

21. **Weight Loss** After losing 30 pounds, a person now weighs 135 pounds. Find the weight of the person before losing weight.

22. **Salary** Before receiving a raise in salary, a worker earned $1100 per week. Find the amount of the raise if the new salary is $1300 per week.

23. **U.S. Renewable Energy** In 2007, U.S. renewable energy production was 6 quadrillion BTU. This was 12 quadrillion BTU less than twice the 2015 renewable energy production. Find the renewable energy production level in 2015. (*Source:* U.S. Department of Energy.)

24. **Endangered Species** There were 96 birds on the endangered species list in 2015. This is 24 more than 2 times the number of reptiles on the list. Find the number of reptiles on the list in 2015. (*Source:* U.S. Fish and Wildlife Service.)

25. **Age** In 10 years, a child's age will be 6 years less than triple her current age. Find the child's current age.

26. **Weight Gain** After gaining 25 pounds, a person is 115 pounds lighter than double his previous weight. How much did he weigh before gaining 25 pounds?

27. **Hazardous Waste** By increasing the number of federal hazardous waste sites in California by 2 and then dividing by 2, the result is the number of federal hazardous waste sites in Washington. Find the number of federal hazardous waste sites in California if there are 13 such sites in Washington. (*Source:* Environmental Protection Agency—2010.)

28. **Cosmetic Surgery** In 2000, there were 1,900,000 cosmetic plastic surgery procedures performed. This was 1,500,000 fewer procedures than twice the number that were performed in 2014. Find the number of cosmetic plastic surgery procedures performed in 2014. (*Source:* American Society of Plastic Surgeons.)

29. **Geometry** If the perimeter of the following rectangle is 106 inches, find the value of x.

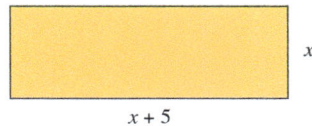

30. **Geometry** If the perimeter of the following triangle is 24 inches, find the value of x.

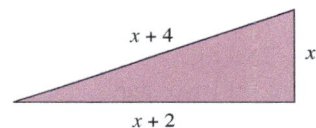

31. **U.S. Oil Production** There were 3000 million fewer barrels of crude oil produced in Alaska in 2014 than there were in the lower 48 states. Find the number of barrels of crude oil (in millions) produced in the lower 48 states in 2014 if there were 181 million barrels produced in Alaska that year. (*Source:* U.S. Energy Information Administration.)

32. **Facebook Users** In February 2015, Facebook had 1 billion more active users than it had in the same month in 2010. If there were 400 million active users in February 2010, how many users were there in February 2015? (*Source:* Facebook.com.)

33. **Rectangle Dimensions** The length of a rectangle is 7 inches longer than the width. If the perimeter of the rectangle is 62 inches, find the measures of the length and width.

34. **Rectangle Dimensions** The width of a rectangle is 9 inches shorter than the length. If the perimeter of the rectangle is 54 inches, find the measures of the length and width.

35. **Triangle Dimensions** The shortest side of a triangle measures 15 feet less than the longest side. If the third side is 6 feet shorter than the longest side and the perimeter is 102 feet, find the measures of the three sides of the triangle.

36. **Triangle Dimensions** A triangle has two sides of the same length and one side that is 5 inches longer than either of the shorter sides. If the perimeter of the triangle is 26 inches, find the measure of the three sides of the triangle

37. **Troops in Iraq** In 2010, there were 52,000 fewer U.S. troops in Iraq than there were in 2003. If the total number of U.S. troops in Iraq for the two years was 248,000, find the number of troops for each year. (*Source: USA Today.*)

38. **Farm Land** In 2015, there were 12 million fewer acres of farm land in Arizona than there were in 1980. If the sum of acreage for these two years is 64 million acres, find the number of farm land acres in Arizona for each of these years. (*Source: U.S. Department of Agriculture, 2007.*)

39. **Finding Ages** A mother is 15 years older than twice her daughter's age. If the difference of their ages is 32, how old is each person?

40. **Millionaire Earners** In 2014, there were 946 more millionaire earners in Kentucky than there were in Rhode Island. Together, a total of 2468 millionaire earners lived in the two states. Find the number of millionaire earners in each of these states. (*Source: Internal Revenue Service.*)

41. **High Mileage Cars** In 2015, the Toyota Prius was one of the highest mileage cars available in the U.S. Its estimated highway mileage was 14 miles per gallon more than the Honda Fit. If the two cars could travel a total of 86 miles on one gallon of gasoline each, find the mileage of each car. (*Source: Environmental Protection Agency 2010 Fuel Economy Guide.*)

42. **Social Networking** In January 2015, the number of active Facebook users was 28 times more than the number of active MySpace users. If the two social networking sites had a total of 1450 million users, how many people used each of the websites? (*Source: comScore.*)

43. **Hospitals** The number of community hospitals H during year x is estimated by $H = -10x + 23{,}040$, where x is any year from 2002 to 2014. This formula is represented graphically in the figure at the top of the next column. In which year were there 3000 hospitals of this type? Find your answer symbolically, numerically, and graphically. (*Source: AHA Hospital Statistics.*)

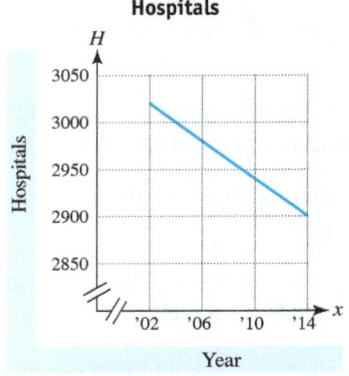

44. **Home Size** The average size F in square feet of new U.S. homes built during year x is estimated by the formula $F = 20x - 37{,}680$, where x is any year from 2004 to 2014. This formula is represented graphically in the following figure. In which year was the average home size 2500 square feet? Find your answer symbolically, numerically, and graphically. (*Source: U.S. Census Bureau.*)

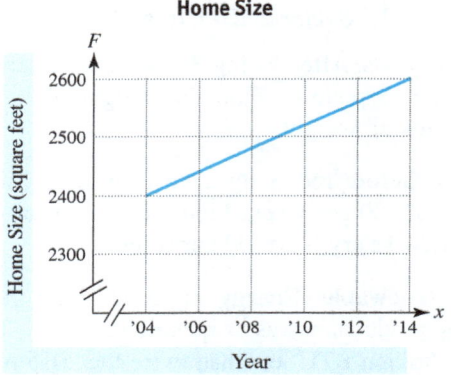

WRITING ABOUT MATHEMATICS

45. In your own words, write the steps in the four-step strategy for solving application problems.

46. In the third step of the problem-solving strategy, we solve the equation formed in the second step. Explain why the solution to this equation may not be the solution to the original problems.

SECTION 3.5 Checking Basic Concepts

Exercises 1–3: Solve the number problem by finding the value of the unknown number.

1. Triple a number plus 9 equals 27.

2. When the sum of a number and -8 is multiplied by -2, the result is 22.

3. Subtracting 4 from the product of -3 and a number results in the number subtracted from -4.

4. **Wind Energy** In 2015, Texas had the capacity to produce 14,000 megawatts of electricity using wind energy. This is 3700 megawatts less than triple the amount of electricity that was being produced using wind energy in California. How many megawatts of electricity could California produce using wind energy in 2015? (*Source:* American Wind Energy Association.)

CHAPTER 3 Summary

SECTION 3.1 ■ SIMPLIFYING ALGEBRAIC EXPRESSIONS

Terms
A number, a variable, or a product of numbers and variables raised to powers
Examples: $4x^2$, $-x^3$, xy, and $4ab$

Combining Like Terms
Terms with the same variables raised to the same powers can be combined.
Example: $8x$ and $-3x$ are like terms, so $8x + (-3x) = 5x$.

Opposite of an Expression
To find the opposite of an expression, change the sign of each term in the expression.
Example: The opposite of $3x^3 - 5x^2 + x - 1$ is $-3x^3 + 5x^2 - x + 1$.

Simplifying an Expression
Use arithmetic properties to clear parentheses and combine like terms.
Example: $3(x + 2) - 2x = 3x + 6 - 2x$ Distributive property
$ = x + 6$ Combine like terms.

SECTION 3.2 ■ TRANSLATING WORDS TO EXPRESSIONS AND EQUATIONS

Translating Words to Symbols
Some words or phrases are associated with the math symbols $+$, $-$, $\cdot$, $\div$, and $=$.
Examples: The word "minus" relates to the symbol " $-$," while the word "twice" relates to " $\cdot$."

Defining a Variable
Assigning a variable to an unknown quantity
Example: If asked to "find the height," let the variable h represent height.

Translating Words to Equations
Use the following steps to translate words to equations.

STEP 1: Read the paragraph carefully for understanding. Identify the unknown quantity and define a variable to represent it.

STEP 2: If possible, identify the word or words associated with an equals sign. In some cases, equality may be suggested instead of stated directly.

STEP 3: Translate the words and information given in the paragraph to expressions. Two expressions are needed to form an equation.

STEP 4: The equation is formed by writing the two expressions found in Step 3 and placing an equals sign between them.

SECTION 3.3 ■ PROPERTIES OF EQUALITY

Solution
A solution to an equation is any value of the variable that makes the equation true.

Example: 2 is a solution to $3x + 5 = 11$ because $3(2) + 5 = 11$ is true.

Equivalent Equations
Two equations are equivalent if they have exactly the same solution(s).

Example: $3x + 5 = 11$ and $x = 2$ are equivalent equations (solution: 2).

Addition Property of Equality
If a, b, and c represent numbers, then $a = b$ is equivalent to $a + c = b + c$.

Example:

$x - 7 = 12$	Given equation
$x - 7 + 7 = 12 + 7$	Add 7 to each side.
$x = 19$	Simplify.

Multiplication Property of Equality
If a, b, and c represent numbers with $c \neq 0$, then $a = b$ is equivalent to $ac = bc$.

Example:

$-3x = 24$	Given equation
$\dfrac{-3x}{-3} = \dfrac{24}{-3}$	Divide each side by -3.
$x = -8$	Simplify.

SECTION 3.4 ■ SOLVING LINEAR EQUATIONS IN ONE VARIABLE

Linear Equations in One Variable
A linear equation in one variable can be written in the form $ax + b = 0$, where a and b are constants (numbers) and $a \neq 0$.

Examples: $5x - 4 = 0$ $(a = 5, b = -4)$; $2x + 7 = 4$ $(a = 2, b = 3)$

Solving Linear Equations Symbolically
The following steps can be used to solve a linear equation symbolically.

STEP 1: Use the distributive property to clear any parentheses on each side of the equation. Combine any like terms on each side.

STEP 2: Apply the addition property of equality to get all of the terms containing the variable on one side of the equation and all other terms on the other side of the equation. Combine any like terms on each side.

STEP 3: Use the multiplication property of equality to isolate the variable by dividing each side of the equation by the number in front of the variable.

STEP 4: Check the solution by substituting it into the given equation.

Example:

$3(x + 2) = 21$	Given equation
$3x + 6 = 21$	Distributive property (Step 1)
$3x = 15$	Subtract 6 from each side. (Step 2)
$x = 5$	Divide each side by 3. (Step 3)

Solving Linear Equations Numerically
To solve a linear equation numerically, complete a table for various values of the variable and then select the solution from the table.

Example: The solution to the linear equation $4x - 1 = 11$ is 3.

x	-1	0	1	2	**3**	4
$4x - 1$	-5	-1	3	7	**11**	15

Solving Linear Equations Graphically

To solve a linear equation graphically, use the graph of the linear equation to estimate a solution. Then check the solution in the given equation.

Example: The solution to the linear equation $2x - 1 = -3$ is -1.

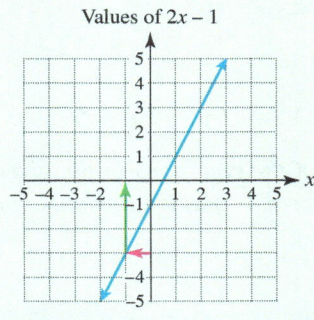

SECTION 3.5 ■ APPLICATIONS AND PROBLEM SOLVING

Solving Applications

STEP 1: Read the problem carefully for understanding. (You may need to read the problem more than once.) Identify key words associated with arithmetic operations. Assign a variable to any unknown quantity.

STEP 2: Write an equation that relates the quantities described in the problem. You may need to sketch a diagram or refer to known formulas.

STEP 3: Solve the equation. Use the solution to determine the answer to the original problem. Include any necessary units.

STEP 4: Check your answer. Does it make sense and is it a reasonable answer to the original problem?

CHAPTER 3 Review Exercises

SECTION 3.1

Exercises 1–12: Simplify the expression.

1. $3y + 6y$
2. $-15n - 2n$
3. $(4x - 2) + (-4x + 2)$
4. $(5b - 2) + (3b - 6)$
5. $(7a - 5) - (-2a + 3)$
6. $-3y - (y - 5)$
7. $7 \cdot (-5p)$
8. $-5(-3w + 1)$
9. $7(b + 3) - 5b$
10. $3x - 6(x + 1) - 9$
11. $-2(5t + 3) + 5(t - 4)$
12. $-(3m + 7) - 2(m - 4)$

SECTION 3.2

Exercises 13–18: Translate the given phrase to an algebraic expression. Define each variable.

13. Double the number of pancakes
14. The product of 3 and her age
15. Five times the sum of her score and 3
16. Eight fewer than the product of 2 and a number
17. The sum of the price and 3, divided by 5
18. The product of 3 and a number, increased by 4

Exercises 19–24: Translate the sentence to an equation using x as the variable. Do not solve the equation.

19. Multiplying a number by 5 results in 45.
20. The total of a number and 7 is 14.
21. Nine equals a number decreased by 11.
22. A number minus 10 gives the same result as triple the number.
23. Twice the sum of a number and 4 gives -12.
24. The sum of a number and 14, divided by 3, equals 7.

SECTION 3.3

Exercises 25–28: Check each solution as specified.

25. Is -1 a solution to $2n - 5 = -7$
26. Is 3 a solution to $-12 - 4a = 0$
27. Is 6 a solution to $2w - 7 = -1 - w$
28. Is -5 a solution to $6y + 1 = 4(y - 1) - 5$

Exercises 29–32: Determine if the equations in the given pair are equivalent. Note that each equation has only one solution.

29. $y + 7 = 0$ and $y = -7$
30. $-26 = 2m$ and $m = -13$
31. $4 - 7t = -8$ and $t = 1$
32. $2(x - 3) = -2x + 5$ and $x = 2$

Exercises 33–44: Solve the equation.

33. $11 = m - 4$
34. $p - 5 = -3$
35. $7w = 6w + 3$
36. $-9n + 1 = -10n$
37. $60 = -10y$
38. $-n = -36$
39. $-5 = \dfrac{g}{-7}$
40. $8 = \dfrac{x}{-9}$
41. $12 - x = 11$
42. $13 = 6 - m$
43. $15 = \dfrac{a}{-3}$
44. $\dfrac{y}{2} = -19$

SECTION 3.4

Exercises 45–48: Determine if the equation is linear. If so, give values for a and b in the form $ax + b = 0$.

45. $2x - 5 = 1$
46. $6(x - 3) = -8$
47. $2x^3 - 5 = 0$
48. $2\sqrt{x} + 7 = 0$

Exercises 49–62: Solve the linear equation symbolically.

49. $-7y - 49 = 0$
50. $-8x + 24 = 0$
51. $27 - 7 = 5m$
52. $15 + (-9) = 3w$
53. $y - 3y = 8 + 6$
54. $4d + 2d = 1 - 7$
55. $10 + 3x = 2x - 3$
56. $4 - g = 8 - 3g$
57. $9 + 6x - 4 = 3x - 5 - 2x$
58. $n + 7 - 4n = 22 - 7n + 5$
59. $1 + 3t + 9 = t - 12 - 9t$
60. $-(3q + 6) + 8q = 19$
61. $46 + 6a = 2(6a - 13)$
62. $-(x + 2) - 6 = 3(x + 4)$

Exercises 63 and 64: Complete the table of values and then solve the given linear equation numerically.

63. $2x - 5 = -7$

x	-2	-1	0	1	2
$2x - 5$					

64. $-3x + 6 = 0$

x	-2	-1	0	1	2
$-3x + 6$					

Exercises 65 and 66: Solve the linear equation graphically.

65. $-x + 2 = 4$

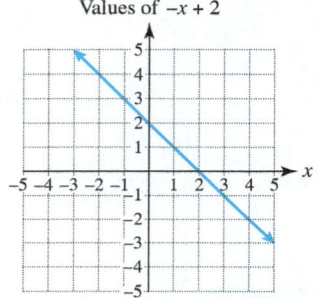

66. $6x + 8 = -40$

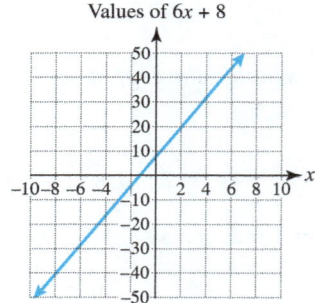

SECTION 3.5

Exercises 67–72: Solve the number problem by finding the value of the unknown number.

67. Nine fewer than a number is 9.
68. Twelve equals 16 more than 2 times a number.
69. Five times a number equals the number minus 12.

70. Subtracting 7 from the product of 4 and a number results in the number added to 8.

71. When the sum of 8 and a number is divided by 3, the result is 6.

72. The product of 4 and the sum of a number and 2 is the same as the number decreased by 1.

APPLICATIONS

73. **Cleaning Windows** A person can clean $8x$ windows in x hours. If a second person can clean $6x$ windows in x hours, complete the following.
    (a) Write an expression (as a sum) that gives the total number of windows that can be cleaned by both people in x hours.
    (b) Simplify your expression from part (a).
    (c) How many windows can the two people clean together in 8 hours?

74. **Finding Age** In 23 years, a person's age will be 7 years less than triple his current age. Write an equation for which the solution gives the person's current age. Use x for your variable. Do not solve the equation.

75. **Geometry** The formula for the area of a rectangle is $A = lw$. Find the width of a rectangle with an area of 80 square inches and a length of 5 inches.

76. **Altitude and Temperature** If the temperature on the ground is 75 °F, then the air temperature T at an altitude of x miles is given by $T = 75 - 19x$. Find the altitude where the air temperature is $T = -1$ °F. (*Source:* Miller, A., and R. Anthes, *Meteorology.* 5th ed.)

77. **Endangered Species** There were 70 insects on the endangered species list in 2015. This is 16 less than two times the number of fish on the list. Find the number of fish on the list in 2015. (*Source:* U.S. Fish and Wildlife Service.)

78. **Rectangle Dimensions** The width of a rectangle is 7 inches shorter than the length. If the perimeter of the rectangle is 74 inches, find the measures of the length and width.

79. **Classic Television** The TV series *Gunsmoke* was on the air for 3 more years than *Lassie*. Together, the two shows aired for 37 years. Find the number of years that each show was on television. (*Source:* Nielsen Media Research.)

80. **Geometry** The formula for the perimeter of a square is $P = 4s$, where s represents the length of one side. Find the length of one side of a square with a perimeter of 60 inches.

81. **Nuclear Reactors** In 2015, the number of operable nuclear reactors in Sweden was 2 less than 6 times the number of operable reactors in Mexico. If there were a total of 12 reactors in these two countries in 2015, write an equation for which the solution gives the number of operable reactors in Mexico in 2015. Use x for your variable. Do not solve the equation. (*Source:* Energy Information Administration.)

82. **Nuclear Reactors** Solve the equation found in the previous exercise. How many operable nuclear reactors were in Mexico in 2015? How many were in Sweden?

83. **Medicare Costs** If x represents a year from 1996 to 2014, then Medicare costs C in billions of dollars is approximated by $C = 18x - 35{,}750$. This formula is represented graphically in the following figure. In which year were Medicare costs \$430 billion? Find your answer symbolically, numerically, and graphically. (*Source:* Office of Management and Budget.)

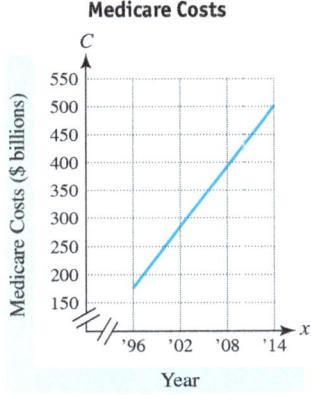

84. **Triangle Dimensions** The longest side of a triangle measures 1 inch less than triple the measure of the shortest side. The measure of the third side is 1 inch more than double the measure of the shortest side. If the perimeter of the triangle is 24 inches, find the measure of the shortest side of the triangle.

CHAPTER 3 Test

Exercises 1–6: Simplify the expression.

1. $(-2x - 3) + (5x + 7)$
2. $(-n - 2) - (2n + 5)$
3. $(-4y)(-6)$
4. $-2(m - 5) + 3$
5. $4(t + 3) - 2(2t - 1)$
6. $3w - (w + 2) - 5$

Exercises 7 and 8: Translate each sentence to an equation using x as the variable. Do not solve the equation.

7. The sum of a number and 8 is the same as triple the the same number.
8. Five times the sum of a number and 3 gives -60.

Exercises 9–14: Solve the equation.

9. $37 = x - 19$
10. $-5 = -3 + y$
11. $9m = 8m - 7$
12. $3n = -42$
13. $\dfrac{t}{6} = -30$
14. $12 = \dfrac{a}{-2}$

Exercises 15 and 16: Determine if the equation is linear. If so, give values for a and b in the form $ax + b = 0$.

15. $6x - 9 = 4$
16. $-2x^2 - x + 5 = 0$

Exercises 17–20: Solve the linear equation symbolically.

17. $3(2x - 1) = 5x$
18. $7 - 3n = 9 - n$
19. $-(3w - 2) - w = 30$
20. $-2(y - 3) = -6y - 12 + 2y$

21. Complete the table and then solve the linear equation $-3x + 1 = -5$ numerically.

x	-2	-1	0	1	2
$-3x + 1$					

22. Solve the equation $-3x - 8 = -20$ graphically.

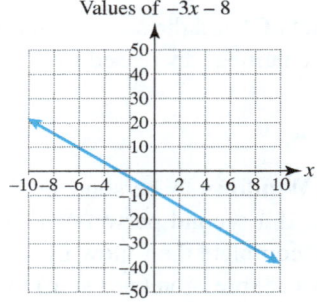

Exercises 23 and 24: Solve the number problem by finding the value of the unknown number.

23. When the sum of 5 and a number is divided by -3, the result is 9.
24. If the product of -7 and a number is increased by 11, the result is 13 less than the number.

25. **Weight Loss** After gaining 15 pounds, a person now weighs 139 pounds. Write an equation for which the solution gives the weight of the person before gaining weight. Use x for your variable. Do not solve the equation.

26. **Rectangle Dimensions** The length of a rectangle is 8 inches longer than the width. If the perimeter of the rectangle is 60 inches, find the measures of the length and width.

27. **Patents for Inventions** In a recent year, residents of South Dakota had 6 less than triple the number of patents granted for their inventions than Alaska residents did. Together, there were a total of 74 patents granted in the two states. Find the number of patents for inventions in each of these states. (*Source:* U.S. Patent and Trademark Office.)

CHAPTERS 1–3 Cumulative Review Exercises

1. Identify the digit in the hundred-thousands place in the number 4,591,083,276.

2. Write $5 \cdot 5 \cdot 5 \cdot 5$ using exponential notation.

3. Round 79,401 to the nearest ten-thousand.

4. Estimate the sum $2989 + 4002 + 997$ by rounding each value to the nearest thousand.

5. Find $\sqrt{121}$.

6. Evaluate $65 \div 5 + 3 \cdot 2$.

7. Simplify the expression $3x - (x - 2)$.

8. Place the correct symbol, $<$, $>$, or $=$, in the blank between the two numbers: $-|7|$ _____ -7.

Exercises 9 and 10: Perform the arithmetic.

9. $-45 \div (-9)$ 10. $-14 - (-6)$

11. Evaluate $4 - 4^2 \div (7 - 3)$.

12. Evaluate $x + (3 - y)^2 \div 2y$ for $x = 7$, $y = -1$.

13. Is -14 a solution to $4 + b \div 2 = -3$?

14. Complete the table and solve $4x - 7 = -3$.

x	-2	-1	0	1	2
$4x - 7$					

Exercises 15 and 16: Simplify the expression.

15. $-4w - (w + 3)$ 16. $(2y - 1) + (5y - 8)$

Exercises 17 and 18: Translate each sentence to an equation using x as the variable. Do not solve the equation.

17. Ten equals a number increased by 7.

18. Triple the sum of a number and 6 equals -18.

Exercises 19–22: Solve the equation.

19. $72 = -8m$ 20. $-5n + 9 = -6n$

21. $9 + 7x - 4 = 2x - 5 - 5x$

22. $3 + (6x - 9) = 7x - (6 + 3x)$

Exercises 23 and 24: Solve the number problem by finding the value of the unknown number.

23. Subtracting 3 from the product of 2 and a number results in 6 more than the number.

24. The product of -2 and the sum of a number and 8 is the same as the number decreased by 1.

25. **Buying DVDs** What is the maximum number of DVDs costing $12 each that a person can buy with $100? How much change will the person receive?

26. **Geometry** Find the perimeter of the figure.

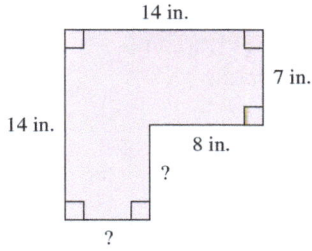

27. **ATM Fees** Each time that a credit card holder uses an ATM machine for a cash advance, a transaction fee appears on the credit card statement. If a year-end summary shows $-\$84$ for 42 ATM transactions, write the fee for a single ATM transaction as a negative integer.

28. **Net Worth** A person with a net worth of $-\$65,200$ wins money in a lottery game so that his net worth becomes $\$134,800$. How much money did the person win?

29. **Salary** Before receiving a raise in salary, a worker earned $1400 per week. Write an equation for which the solution is the amount of the worker's raise if the new salary is $1550 per week.

30. **Farm Land** In a recent year, there were 11 million fewer acres of farm land in Arkansas than there were in California. If the sum of acreage for these two states was 39 million acres, find the number of farm land acres in each state. (*Source:* U.S. Department of Agriculture, 2007.)

4 Fractions

- 4.1 Introduction to Fractions and Mixed Numbers
- 4.2 Prime Factorization and Lowest Terms
- 4.3 Multiplying and Dividing Fractions
- 4.4 Adding and Subtracting Fractions—Like Denominators
- 4.5 Adding and Subtracting Fractions—Unlike Denominators
- 4.6 Operations on Mixed Numbers
- 4.7 Complex Fractions and Order of Operations
- 4.8 Solving Equations Involving Fractions

Throughout history, many calendars have been created to record time accurately. One difficulty with creating an accurate calendar lies in the fact that neither the astronomical month nor year can be measured in a whole number of days. For example, the true length of one year is 365 whole days and part of an additional day. This extra *fraction* of a day is the reason why we have leap years.

Fractions are commonly used in measuring various lengths of time. When a clock reads 5:30 P.M. we say that it is *half* past five. A company's fourth *quarter* earnings refer to the earnings for the last three months of the year. A semester in school is *one-half* of a school year. Before 2001 the New York Stock Exchange used fractions to express stock prices to the nearest *sixteenth* of a dollar. Fractions have played an important role in society.

In this chapter we introduce basic concepts about fractions and then discuss how to perform arithmetic operations with fractions.

4.1 Introduction to Fractions and Mixed Numbers

Objectives

1. Reviewing Fractions and Fractional Parts
2. Introducing Rational Numbers
   - Identifying Rational Numbers
   - Identity and Zero Properties for Fractions
3. Reviewing Improper Fractions and Mixed Numbers
   - Identifying Improper Fractions
   - Writing Improper Fractions as Mixed Numbers
   - Writing Mixed Numbers as Improper Fractions
4. Graphing Fractions and Mixed Numbers

NEW VOCABULARY

☐ Fraction
☐ Numerator
☐ Denominator
☐ Rational numbers
☐ Proper fraction
☐ Improper fraction
☐ Mixed number

READING CHECK 1

- Identify the numerator and the denominator for the fraction $\frac{1}{y}$.

1 Reviewing Fractions and Fractional Parts

In earlier chapters, we used whole numbers and integers to describe data that were not broken into parts. When a situation involves a portion of a whole, we need fractions. A **fraction** is a number that describes a portion of a whole. For example, when a carpenter cuts a board into 4 equal parts, each part is one-fourth of the original board. One-fourth is represented by the fraction $\frac{1}{4}$. See **FIGURE 4.1**.

A Board Cut into Fourths

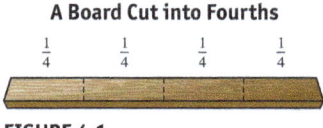

FIGURE 4.1

STUDY TIP

A new concept is often easier to learn when we find a relationship between the concept and our personal experience. List five instances from your life when you have encountered fractions.

If 3 out of the 4 pieces of wood in **FIGURE 4.1** are needed to make shelving, then the carpenter will use $\frac{3}{4}$ of the board for the shelving. The top number in a fraction is called the **numerator**, and the bottom number is called the **denominator**. In the fraction $\frac{3}{4}$, the numerator is 3, and the denominator is 4.

Parts of a Fraction

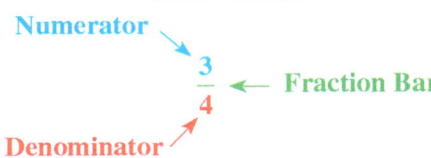

The denominator 4 gives the total number of parts in the whole board. The numerator 3 gives the number of parts needed for the shelving.

EXAMPLE 1 Identifying numerators and denominators

Write the numerator and denominator of each fraction.

(a) $\frac{5}{7}$ (b) $\frac{4x}{19}$

Solution

(a) In the fraction $\frac{5}{7}$, the numerator is 5, and the denominator is 7.
(b) The numerator is $4x$, and the denominator is 19.

Now Try Exercises 7, 13

One way to visualize a fraction is to identify shaded parts of a whole. The diagram shown in **FIGURE 4.2** has a total of 8 equal parts, 5 of which are shaded. Because there are 5 shaded parts out of 8 total parts, the numerator of a fraction representing this shading is 5, and the denominator is 8. The fraction for this shading is $\frac{5}{8}$.

A Shading Representing $\frac{5}{8}$

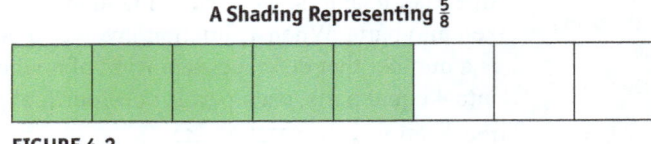

FIGURE 4.2

EXAMPLE 2 Identifying a fraction represented by shading

Write the fraction represented by the shading.

(a)   (b)

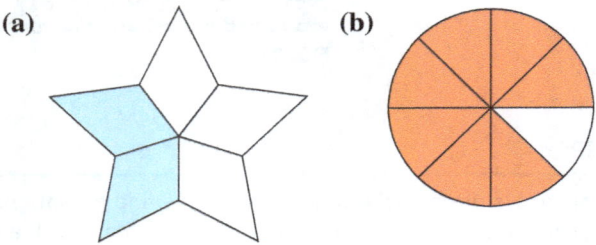

Solution
(a) Because the whole star is divided into 5 equal parts, the denominator of the fraction that represents the shading is 5. Of these parts, 2 are shaded, so the numerator is 2. The fraction for this shading is $\frac{2}{5}$.
(b) There are 7 parts shaded out of a total of 8 equal parts. The fraction for this shading is $\frac{7}{8}$.

Now Try Exercises 15, 19

The following See the Concept summarizes how fractions can be represented by shading.

See the Concept 1   FRACTIONS REPRESENTED BY SHADING

A fraction can be visualized as shaded parts of a whole.

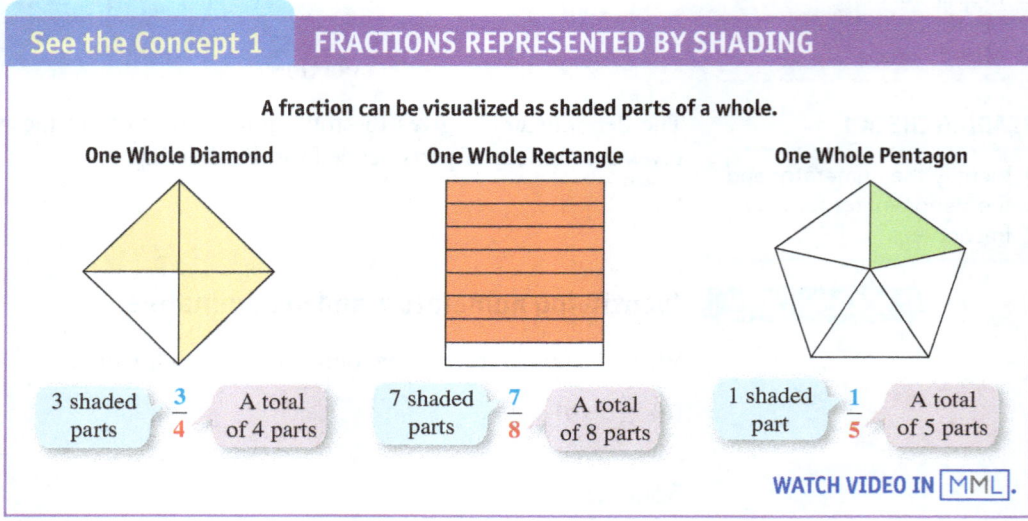

WATCH VIDEO IN MML.

🌐 *Math in Context* (Smoking) Fractions are used to describe many kinds of real data. For example, it is estimated that 21 out of every 100 U.S. men who have 8 or fewer years of education are cigarette smokers. This compares to only 7 smokers out of every 100 U.S. men who have graduate degrees. We can express the fraction of men who smoke and have 8 or fewer years of education as $\frac{21}{100}$, and the fraction of men who smoke and have a graduate degree as $\frac{7}{100}$.
(*Source:* Centers for Disease Control and Prevention.)

The next example illustrates how fractions can be used when part of a whole is needed to represent given information.

EXAMPLE 3 Writing a fraction for given information

The seven continents of the world are Africa, Antarctica, Asia, Australia, Europe, North America, and South America. What fraction of continent names begin with a vowel?

Solution
Because 5 of the 7 continents have names that begin with a vowel, the fraction is $\frac{5}{7}$.

Now Try Exercise 71

2 Introducing Rational Numbers

IDENTIFYING RATIONAL NUMBERS When a number is or *can be* expressed as a fraction where both the numerator and the denominator are integers (with the denominator not equal to 0), the number belongs to the set of numbers known as the **rational numbers**.

> **RATIONAL NUMBER**
>
> A rational number is a number that can be expressed as
> $$\frac{p}{q},$$
> where p and q are integers with $q \neq 0$.

Because every integer can be expressed as a fraction by writing the integer in the numerator and 1 in the denominator, we know that *every integer is a rational number*. For example, the integer -92 can be expressed as $\frac{-92}{1}$, and 0 can be expressed as $\frac{0}{1}$.

Examples of Rational Numbers

$$\frac{3}{11}, \quad -\frac{1}{8}, \quad 5 = \frac{5}{1}, \quad \text{and} \quad -2 = \frac{-2}{1}$$

READING CHECK 2

- Which of the following rational numbers: $\frac{4}{5}, \frac{5}{1}, -7, 0, \frac{13}{7}$, are neither whole numbers nor integers?

> **MAKING CONNECTIONS 1**
>
> **Negative Rational Numbers**
>
> If the variables p and q represent integers with $q \neq 0$, then $\frac{p}{q}$ represents both division and a rational number. Using the rules for signs of quotients, we see that every negative rational number can be written in three ways, each representing the same fraction.
>
> $$\frac{-5}{12} = \frac{5}{-12} = -\frac{5}{12} \qquad \text{Three ways to write negative five-twelfths}$$

IDENTITY AND ZERO PROPERTIES FOR FRACTIONS Because the fraction bar in a rational number represents division, the identity and zero properties for division discussed earlier apply to fractions. These properties can be summarized as follows.

> **IDENTITY AND ZERO PROPERTIES FOR FRACTIONS**
>
> Let the variable n represent any integer.
>
> 1. When $n \neq 0$, the expression $\frac{n}{n}$ simplifies to 1. $\qquad \frac{3}{3} = 1$
>
> 2. The expression $\frac{n}{1}$ simplifies to n. $\qquad \frac{3}{1} = 3$
>
> 3. When $n \neq 0$, the expression $\frac{0}{n}$ simplifies to 0. $\qquad \frac{0}{3} = 0$
>
> 4. The expression $\frac{n}{0}$ is undefined. $\qquad \frac{3}{0}$ is undefined.

The next example demonstrates how the identity and zero properties can be used to evaluate some kinds of fractions. Be sure to pay special attention to any fractions containing zeros. A fraction with a zero in the numerator evaluates differently than a fraction with a zero in the denominator.

EXAMPLE 4 Simplifying fractions

Simplify each fraction, if possible.

(a) $\frac{0}{9}$ (b) $\frac{-13}{-13}$ (c) $\frac{5}{0}$ (d) $\frac{-8}{1}$

Solution
(a) By letting $n = 9$ in Property 3, we see that $\frac{0}{9}$ simplifies to 0.
(b) If $n = -13$ in Property 1, then $\frac{-13}{-13}$ simplifies to 1.
(c) By letting $n = 5$ in Property 4, the fraction $\frac{5}{0}$ is undefined.
(d) If $n = -8$ in Property 2, then $\frac{-8}{1}$ simplifies to -8.

Now Try Exercises 23, 25, 29, 33

3 Reviewing Improper Fractions and Mixed Numbers

IDENTIFYING IMPROPER FRACTIONS When the absolute value of the numerator of a fraction is less than the absolute value of the denominator, the fraction is a **proper fraction**.

Examples of Proper Fractions

$$\frac{3}{7}, \quad \frac{-1}{3}, \quad \frac{5}{9}, \quad \text{and} \quad -\frac{21}{50}$$

A positive proper fraction represents an amount that is less than 1. For example, $\frac{5}{6}$ is less than 1 and is represented by the shading in **FIGURE 4.3**.

When the absolute value of the numerator is greater than or equal to the absolute value of the denominator, the fraction is an **improper fraction**.

Examples of Improper Fractions

$$\frac{9}{4}, \quad \frac{-7}{2}, \quad \frac{13}{13}, \quad \text{and} \quad -\frac{37}{20}$$

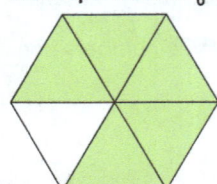

The Proper Fraction $\frac{5}{6}$

FIGURE 4.3

READING CHECK 3

- When is a fraction a proper fraction?
- When is a fraction an improper fraction?

PROPER AND IMPROPER FRACTIONS

The fraction $\frac{a}{b}$ with $b \neq 0$ is a proper fraction whenever $|a| < |b|$, and it is an improper fraction whenever $|a| \geq |b|$.

The following See the Concept summarizes how improper fractions can be represented by shading.

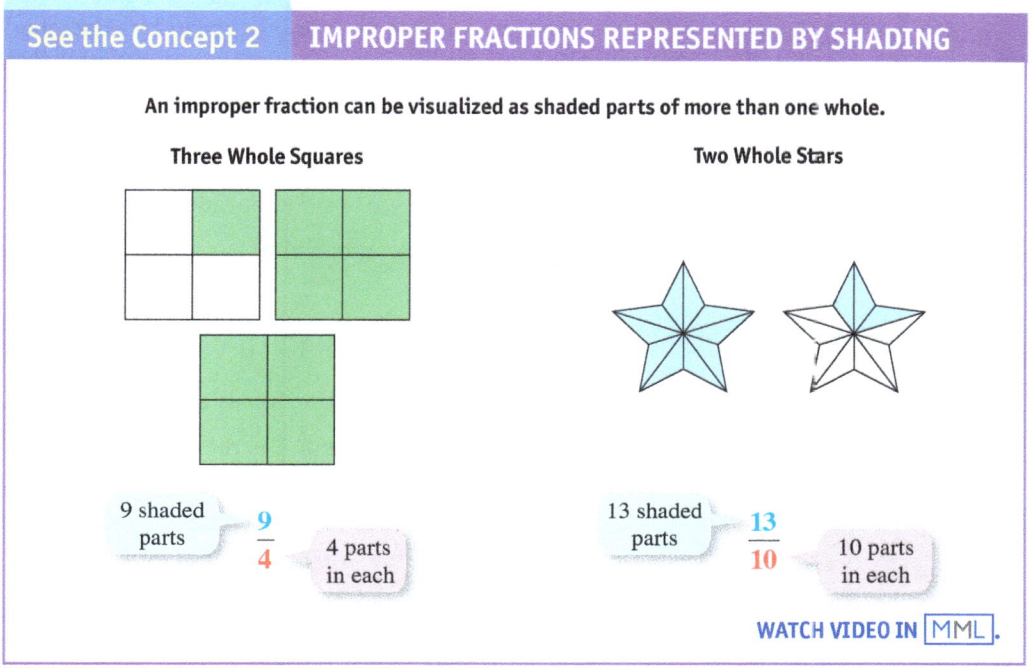

See the Concept 2 — IMPROPER FRACTIONS REPRESENTED BY SHADING

An improper fraction can be visualized as shaded parts of more than one whole.

Three Whole Squares — 9 shaded parts — $\frac{9}{4}$ — 4 parts in each

Two Whole Stars — 13 shaded parts — $\frac{13}{10}$ — 10 parts in each

WATCH VIDEO IN MML.

WRITING IMPROPER FRACTIONS AS MIXED NUMBERS The figure on the left in the above See the Concept suggests that the improper fraction $\frac{9}{4}$ is equal to 2 wholes and $\frac{1}{4}$ more, which can be written as the *mixed number* $2\frac{1}{4}$. A **mixed number** is an integer written with a proper fraction.

Examples of Mixed Numbers

$$7\frac{1}{3}, \quad 11\frac{3}{8}, \quad -5\frac{21}{22}, \quad \text{and} \quad -10\frac{3}{14}$$

EXAMPLE 5   Identifying an improper fraction and mixed number represented by shading

Write the improper fraction and mixed number represented by the shading.

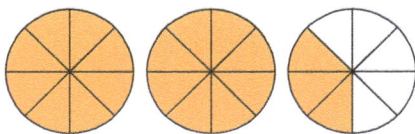

Solution
Each circle is divided into 8 equal parts, giving a denominator of **8**. Because a total of 19 parts are shaded, the numerator is **19**. The improper fraction represented is $\frac{19}{8}$. There are 2 *whole* circles shaded, so the integer part of the mixed number is **2**. Because **3** out of **8** parts of the last circle are shaded, the fraction part is $\frac{3}{8}$. The mixed number is $2\frac{3}{8}$.

Now Try Exercise 35

> **MAKING CONNECTIONS 2**
>
> **Positive and Negative Mixed Numbers**
>
> We read the mixed number $2\frac{1}{3}$ as "two and one-third," which means that there are 2 wholes plus one-third of another whole. Therefore $2\frac{1}{3}$ can be written as $2 + \frac{1}{3}$. Similarly, the negative mixed number $-3\frac{2}{5}$ is "the opposite of the mixed number three and two-fifths." This means that $-3\frac{2}{5}$ can be written as $-\left(3 + \frac{2}{5}\right)$ or $-3 - \frac{2}{5}$.

Example 5 illustrates that an improper fraction can be written as a mixed number. We can write any improper fraction as a mixed number by remembering that a fraction bar indicates division. For example, the improper fraction $\frac{13}{5}$ indicates that we divide 13 by 5 as follows.

$$\text{Divisor} \rightarrow 5\overline{)13} \begin{array}{l} 2 \leftarrow \text{Quotient} \\ \leftarrow \text{Dividend} \\ \underline{10} \\ 3 \leftarrow \text{Remainder} \end{array} \qquad \frac{13}{5} = 2\frac{3}{5}$$

An improper fraction can be written as a mixed number using the following steps.

> **WRITING AN IMPROPER FRACTION AS A MIXED NUMBER**
>
> To write a positive improper fraction as a mixed number, do the following.
>
> **STEP 1:** Divide the denominator into the numerator.
>
> **STEP 2:** The quotient is the integer part of the mixed number. The fraction part of the mixed number has the remainder in its numerator and the divisor in its denominator. Write the mixed number in the form
>
> $$\text{Mixed Number} = \text{Quotient}\,\frac{\text{Remainder}}{\text{Divisor}}.$$
>
> If the remainder is 0, then the improper fraction is equal to the quotient. For example, the quotient is 2 and the remainder is 0 when 6 is divided by 3, so $\frac{6}{3} = 2$.

NOTE: To write a negative improper fraction as a mixed number, perform the same two steps on the absolute value of the fraction and then place a negative sign in front of the resulting mixed number. ∎

EXAMPLE 6 Writing improper fractions as mixed numbers

Write each improper fraction as a mixed number or an integer.

(a) $\frac{15}{8}$ (b) $\frac{24}{4}$ (c) $-\frac{29}{3}$

Solution

(a) Divide 8 into 15.

$$\text{Divisor} \rightarrow 8\overline{)15} \begin{array}{l} 1 \leftarrow \text{Quotient} \\ \leftarrow \text{Dividend} \\ \underline{8} \\ 7 \leftarrow \text{Remainder} \end{array}$$

As a result, $\frac{15}{8} = 1\frac{7}{8}$.

(b) Divide 4 into 24.

$$\text{Divisor} \rightarrow 4\overline{)24} \begin{array}{l} 6 \leftarrow \text{Quotient} \\ \leftarrow \text{Dividend} \\ \underline{24} \\ 0 \leftarrow \text{Remainder} \end{array}$$

The remainder is 0, so $\frac{24}{4} = 6$.

(c) Perform the steps on the absolute value of $-\frac{29}{3}$. Because $\left|-\frac{29}{3}\right| = \frac{29}{3}$, first write $\frac{29}{3}$ as a mixed number by dividing 3 into 29 and then place a negative sign in front of the result.

$$\text{Divisor} \rightarrow 3\overline{)29} \begin{matrix} 9 & \leftarrow \text{Quotient} \\ & \leftarrow \text{Dividend} \end{matrix}$$
$$\underline{27}$$
$$2 \leftarrow \text{Remainder}$$

As a result, $\frac{29}{3} = 9\frac{2}{3}$ and so $-\frac{29}{3} = -9\frac{2}{3}$.

Now Try Exercises 41, 43, 49

WRITING MIXED NUMBERS AS IMPROPER FRACTIONS In Example 6(a), we used *division* to write the improper fraction $\frac{15}{8}$ as the mixed number $1\frac{7}{8}$. If we wish to reverse the process, we can use *multiplication* to write $1\frac{7}{8}$ as $\frac{15}{8}$. First, we note that the numerator of the improper fraction is the dividend and the denominator is the divisor, as follows.

$$\frac{\textbf{Dividend}}{\textbf{Divisor}}$$

Next, recall that we can check the result of a division problem as follows.

$$\textbf{Divisor} \cdot \textbf{Quotient} + \textbf{Remainder} = \textbf{Dividend}$$

As a result, we can replace the numerator **Dividend** in an improper fraction with the expression **Divisor** · **Quotient** + **Remainder** to obtain the expression

$$\frac{\textbf{Divisor} \cdot \textbf{Quotient} + \textbf{Remainder}}{\textbf{Divisor}}.$$

Finally, because the mixed number from Example 6(a) has the form

$$\text{Quotient} \rightarrow 1\frac{7}{8}, \begin{matrix} \leftarrow \text{Remainder} \\ \leftarrow \text{Divisor} \end{matrix}$$

the expression for the improper fraction becomes

$$\frac{\textbf{Divisor} \cdot \textbf{Quotient} + \textbf{Remainder}}{\textbf{Divisor}} = \frac{8 \cdot 1 + 7}{8} = \frac{8 + 7}{8} = \frac{15}{8}.$$

To write a mixed number as an improper fraction, we use the following procedure.

WRITING A MIXED NUMBER AS AN IMPROPER FRACTION

To write a positive mixed number as an improper fraction, do the following.

STEP 1: Multiply the denominator of the fraction by the integer part.

STEP 2: Add the numerator of the fraction to the product from Step 1.

STEP 3: The sum found in Step 2 is the numerator of the improper fraction. The denominator of the improper fraction is the denominator from the original mixed number. So a mixed number of the form

$$\text{Quotient} \frac{\text{Remainder}}{\text{Divisor}}$$

can be written as the improper fraction

$$\frac{\text{Divisor} \cdot \text{Quotient} + \text{Remainder}}{\text{Divisor}}.$$

NOTE: To write a negative mixed number as an improper fraction, perform the same three steps on the absolute value of the mixed number and then place a negative sign in front of the resulting improper fraction. ∎

EXAMPLE 7 — Writing mixed numbers as improper fractions

Write each mixed number as an improper fraction.

(a) $7\frac{2}{3}$ (b) $-5\frac{4}{9}$

Solution

(a) $7\frac{2}{3} = \frac{3 \cdot 7 + 2}{3} = \frac{21 + 2}{3} = \frac{23}{3}$

(b) Because $5\frac{4}{9} = \frac{9 \cdot 5 + 4}{9} = \frac{45 + 4}{9} = \frac{49}{9}$, we know that $-5\frac{4}{9} = -\frac{49}{9}$.

Now Try Exercises 51, 53

4 Graphing Fractions and Mixed Numbers

READING CHECK 4

- When graphing $\frac{5}{6}$, how are the whole distances on a number line divided?

When graphing a fraction, we divide each whole distance on the number line into an equal number of parts determined by the denominator of the fraction to be graphed. For example, to graph the proper fraction $\frac{3}{5}$, divide each whole distance on the number line into **5** equal parts by marking equally spaced tick marks between the whole units on the number line. Then, starting from 0, count **3** tick marks to the right. See **FIGURE 4.4**.

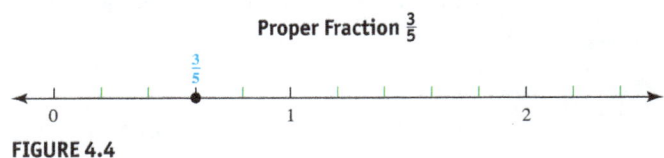

FIGURE 4.4

NOTE: To divide a whole unit on a number line into **5** equal parts, mark **4** equally spaced tick marks between the whole units, as shown in green in **FIGURE 4.4**. ∎

EXAMPLE 8 — Graphing proper and improper fractions

Graph each fraction on a number line.

(a) $\frac{5}{8}$ (b) $-\frac{3}{4}$ (c) $\frac{7}{3}$

Solution

(a) To graph $\frac{5}{8}$, divide each whole unit on the number line into **8** equal parts. Then count **5** tick marks to the right, starting from 0, and place a dot on the number line as shown.

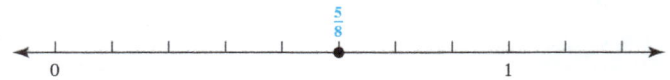

(b) Divide each whole unit on the number line into 4 equal parts. Locate $-\frac{3}{4}$ by counting 3 tick marks to the left, starting from 0. Place a dot on the number line as shown.

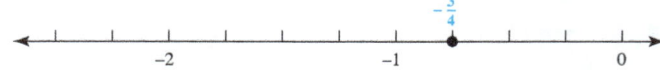

(c) To graph $\frac{7}{3}$, divide each whole unit on the number line into 3 equal parts. Then count 7 tick marks to the right, starting from 0, and place a dot on the number line as shown.

Now Try Exercises 61, 63, 65

The process for graphing a mixed number on a number line is similar to that for graphing a fraction. Mixed numbers are graphed in the next example.

EXAMPLE 9 **Graphing mixed numbers**

Graph each mixed number on a number line.

(a) $2\frac{1}{3}$ (b) $-1\frac{3}{4}$

Solution

(a) To graph $2\frac{1}{3}$, divide each whole unit on the number line into 3 equal parts. Then count 2 whole units and 1 additional tick mark to the right, starting from 0, and place a dot on the number line as shown.

(b) Divide each whole unit on the number line into 4 equal parts. Locate $-1\frac{3}{4}$ by counting 1 whole unit and 3 additional tick marks to the left, starting from 0. Place a dot on the number line as shown.

Now Try Exercises 67, 69

The following See the Concept summarizes how to graph fractions and mixed numbers.

See the Concept 3   GRAPHING FRACTIONS AND MIXED NUMBERS

Graph the fraction $\frac{4}{5}$.

Ⓐ Equal number of parts (denominator in fraction)

Ⓑ Count to the position. (numerator of the fraction)

Graph the mixed number $2\frac{2}{3}$.

Ⓐ Equal number of parts (denominator in mixed number)

Ⓑ Count to the position. (whole number first, and then the numerator of the fraction)

Ⓐ Divide each whole unit on the number line into an equal number of parts determined by the denominator of the fraction.

Ⓑ Starting at 0, count a number of tick marks equal to the numerator of the fraction. Place a dot.

Ⓐ Divide each whole unit on the number line into an equal number of parts determined by the denominator of the fraction in the mixed number.

Ⓑ Starting at 0, first count the whole number and then count a number of tick marks equal to the numerator of the fraction in the mixed number. Place a dot.

WATCH VIDEO IN MML.

4.1 Putting It All Together

CONCEPT	COMMENTS	EXAMPLES
Fraction	A number that can be used to describe a portion of a whole	Numerator → $\frac{2}{5}$ ← Fraction Bar; Denominator ↗
Rational Number	A rational number can be expressed as $\frac{p}{q}$, where p and q are integers with $q \neq 0$.	$\frac{3}{8}, \quad -\frac{7}{5}, \quad 12 = \frac{12}{1}$
Identity and Zero Properties for Fractions	1. When $n \neq 0$, $\frac{n}{n}$ simplifies to 1. 2. $\frac{n}{1}$ simplifies to n. 3. When $n \neq 0$, $\frac{0}{n}$ simplifies to 0. 4. $\frac{n}{0}$ is undefined.	1. $\frac{3}{3} = 1$ 2. $\frac{5}{1} = 5$ 3. $\frac{0}{7} = 0$ 4. $\frac{8}{0}$ is undefined.
Proper and Improper Fractions	The fraction $\frac{a}{b}$ with $b \neq 0$ is a proper fraction whenever $|a| < |b|$. It is an improper fraction whenever $|a| \geq |b|$.	Proper Fractions: $\frac{1}{3}, \quad -\frac{2}{5}, \quad \frac{5}{9}$ Improper Fractions: $\frac{4}{3}, \quad -\frac{8}{5}, \quad \frac{11}{11}$
Mixed Number	An integer written with a proper fraction	$8\frac{1}{4}, \quad -2\frac{3}{5}, \quad 6\frac{5}{9}$
Writing Improper Fractions as Mixed Numbers	1. Divide the denominator into the numerator. 2. The quotient is the integer part of the mixed number. The fraction part of the mixed number has the remainder in its numerator and the divisor in its denominator.	$\frac{8}{3} = 2\frac{2}{3}$ Divisor → $3\overline{)8}$ ← Dividend, Quotient = 2 $\underline{6}$ 2 ← Remainder
Writing Mixed Numbers as Improper Fractions	1. Multiply the denominator of the fraction by the integer part. 2. Add the numerator of the fraction to the product from Step 1. 3. The sum found in Step 2 is the numerator of the improper fraction. The denominator of the improper fraction is the denominator from the original mixed number.	$1\frac{3}{5} = \frac{5(1)+3}{5} = \frac{8}{5}$ $-6\frac{1}{3} = -\left(\frac{3(6)+1}{3}\right) = -\frac{19}{3}$
Graphing Fractions and Mixed Numbers	When graphing a fraction or mixed number, divide each whole distance on the number line into an equal number of parts determined by the denominator of the fraction to be graphed.	The numbers $\frac{1}{3}, \frac{4}{3},$ and $2\frac{2}{3}$ are graphed on the number line.

4.1 Exercises

CONCEPTS AND VOCABULARY

1. A(n) _____ is a number that can be used to describe a portion of a whole.

2. The top number in a fraction is called the _____, and the bottom number is called the _____.

3. A(n) _____ number can be expressed as a fraction where both the numerator and denominator are integers and the denominator does not equal 0.

4. If $|a| < |b|$ in the fraction $\frac{a}{b}$ with $b \neq 0$, then the fraction is a(n) _____ fraction.

5. If $|a| \geq |b|$ in the fraction $\frac{a}{b}$ with $b \neq 0$, then the fraction is a(n) _____ fraction.

6. When an integer is written with a proper fraction, the resulting number is called a(n) _____.

WORKING WITH FRACTIONS AND RATIONAL NUMBERS

Exercises 7–14: Write the numerator and denominator of the given fraction.

7. $\frac{6}{13}$
8. $\frac{3}{7}$
9. $\frac{12}{5}$
10. $\frac{15}{4}$
11. $\frac{x}{y}$
12. $\frac{m}{n}$
13. $\frac{3p}{14}$
14. $\frac{5w}{7}$

Exercises 15–22: Write the fraction that is represented by the given shading.

15.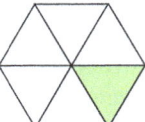
16.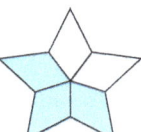
17.
18.
19.
20.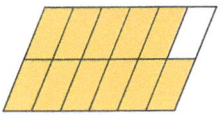
21.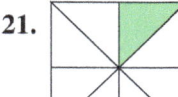
22.

Exercises 23–34: Simplify each fraction, if possible.

23. $\frac{4}{4}$
24. $\frac{-2}{0}$
25. $\frac{0}{-7}$
26. $\frac{10}{10}$
27. $\frac{-13}{1}$
28. $\frac{18}{1}$
29. $\frac{11}{0}$
30. $\frac{0}{-1}$
31. $\frac{-9}{-9}$
32. $\frac{0}{45}$
33. $\frac{53}{1}$
34. $\frac{29}{0}$

WRITING IMPROPER FRACTIONS AND MIXED NUMBERS

Exercises 35–40: Write the improper fraction and mixed number represented by the shading.

35.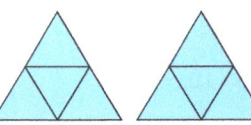

36.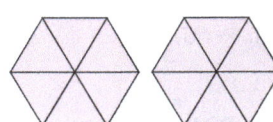

37.

38.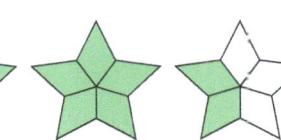

39.

40.

WRITING IMPROPER FRACTIONS AS MIXED NUMBERS

Exercises 41–50: Write the given improper fraction as a mixed number or an integer.

41. $\dfrac{11}{2}$ 42. $\dfrac{-8}{3}$

43. $\dfrac{17}{6}$ 44. $\dfrac{50}{5}$

45. $\dfrac{-16}{4}$ 46. $\dfrac{24}{7}$

47. $\dfrac{91}{8}$ 48. $\dfrac{73}{9}$

49. $-\dfrac{37}{5}$ 50. $\dfrac{115}{-11}$

WRITING MIXED NUMBERS AS IMPROPER FRACTIONS

Exercises 51–60: Write the given mixed number as an improper fraction.

51. $5\dfrac{3}{4}$ 52. $2\dfrac{1}{7}$

53. $-8\dfrac{2}{3}$ 54. $-1\dfrac{7}{8}$

55. $9\dfrac{5}{8}$ 56. $10\dfrac{1}{2}$

57. $-35\dfrac{2}{3}$ 58. $-60\dfrac{4}{5}$

59. $112\dfrac{1}{5}$ 60. $204\dfrac{5}{6}$

GRAPHING FRACTIONS AND MIXED NUMBERS

Exercises 61–70: Graph the given fraction or mixed number on a number line.

61. $\dfrac{2}{3}$ 62. $-\dfrac{1}{4}$

63. $\dfrac{7}{4}$ 64. $\dfrac{3}{5}$

65. $-\dfrac{4}{5}$ 66. $-\dfrac{9}{2}$

67. $1\dfrac{3}{5}$ 68. $-1\dfrac{1}{3}$

69. $-3\dfrac{1}{2}$ 70. $2\dfrac{1}{4}$

APPLICATIONS INVOLVING FRACTIONS AND MIXED NUMBERS

71. **Poverty Level** In 2014, California, New York, Florida, and Texas were the only states with more than 2,000,000 families living below the poverty level. What fraction of all states had more than 2,000,000 families living below the poverty level? (*Source:* U.S. Census Bureau.)

72. **Millionaires** California, Florida, New York, and Texas were the only states with more than 20,000 millionaire earners. What fraction of all states had more than 20,000 millionaire earners? (*Source:* Phoenix Marketing International.)

73. **Air Quality** There were 122 days of unhealthy air quality in Los Angeles in 2014. What fraction of the days in 2014 were *not* unhealthy air quality days in Los Angeles? (*Source:* Environmental Protection Agency.)

74. **Cell Phone Use** In 2015, it was estimated that 28 of every 38 Canadians owned a cell phone. Write the fraction of Canadians that did *not* own a cell phone in 2015. (*Source:* International Telecommunications Union.)

75. **Worldwide Internet Use** It was estimated that 3 of every 7 people had Internet access in 2014. What fraction of the world population did *not* have Internet access in 2014? (*Source:* Internet World Stats.)

76. **European Union** Of the 28 countries that were members of the European Union, only Austria, Estonia, Ireland, Italy, and the United Kingdom have names that begin with a vowel. What fraction of countries in the European Union had names that started with a vowel?

Exercises 77–80: **Truck Sales** *The following bar graph shows the total number of cement trucks made by a truck manufacturer during selected years.*

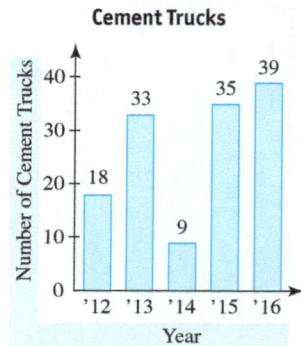

77. For this five-year period, what fraction of the total production occurred in 2016?

78. What fraction of the total production for this five-year period occurred in either 2012 or 2013?

79. Considering only the odd-numbered years, what fraction of the production occurred in 2015?

80. What fraction of production for the first four years occurred in 2012?

WRITING ABOUT MATHEMATICS

81. Give an example of an improper fraction that is not greater than or equal to 1. What must be true about any fraction of this type?

82. Explain how the result of a division problem can be expressed as a mixed number.

83. Describe the steps used to write an improper fraction as a mixed number.

84. Describe the steps used to write a mixed number as an improper fraction.

4.2 Prime Factorization and Lowest Terms

Objectives

1. Using Divisibility Tests
2. Finding Prime Factorization of Numbers
   - Determining If a Number Is Prime
   - Writing Prime Factorization Using a Factor Tree
3. Identifying and Writing Equivalent Fractions
   - Representing Equivalent Fractions
   - Writing Equivalent Fractions
4. Simplifying Fractions to Lowest Terms
   - Finding the Greatest Common Factor (GCF)
   - Simplifying Fractions to Lowest Terms
5. Comparing Fractions
6. Simplifying Rational Expressions
   - Finding the GCF for Rational Expressions
   - Simplifying Rational Expressions

1 Using Divisibility Tests

When two whole numbers are multiplied, we call them *factors*. For example,

$$5 \cdot 4 = 20.$$

factor   factor   product

We multiply two factors to obtain the product.

Because division is closely related to multiplication, we know that dividing 20 by 5 results in a quotient of 4 with no remainder. That is, 20 is *divisible* by 5, and 5 is a *factor* of 20. Similarly, dividing 20 by 4 results in the quotient 5 with no remainder. In this case, we say that 20 is *divisible* by 4, and 4 is a *factor* of 20. A **factor** is any whole number that divides into a second whole number exactly with no remainder. When this occurs, we say that the second number is **divisible** by the first.

NOTE: We often say that a factor divides "evenly" into another number, which means that the remainder is 0. ∎

READING CHECK 1

- Which of the following numbers: 15, 22, 30, 56, or 64, is divisible by 7?

EXAMPLE 1   Checking for divisibility and factors

Using the definitions of "divisible" and "factor," answer each question.
(a) Is 52 divisible by 16?   (b) Is 13 a factor of 39?

Solution
(a) Dividing 52 by 16 results in 3 with remainder 4, not 0, so 52 is not divisible by 16.
(b) Because 13 divides into 39 exactly 3 times with no remainder, 13 is a factor of 39.

Now Try Exercises 15, 17

NEW VOCABULARY

- ☐ Factor
- ☐ Divisible
- ☐ Prime number
- ☐ Composite number
- ☐ Prime factorization
- ☐ Factor tree
- ☐ Equivalent fractions
- ☐ Basic principle of fractions
- ☐ Greatest common factor
- ☐ Lowest terms
- ☐ Cross products
- ☐ Rational expression

It is not always necessary to perform long division to determine if one number is divisible by another. The following rules can be used to test divisibility by 2, 3, or 5.

DIVISIBILITY TESTS

A whole number is divisible by

- 2 if its ones digit is 0, 2, 4, 6, or 8.
- 3 if the sum of its digits is divisible by 3.
- 5 if its ones digit is 0 or 5.

STUDY TIP

If you take a little extra time to memorize basic mathematical concepts such as multiplication facts or these divisibility tests, you can save a great deal of time on homework or on an exam.

EXAMPLE 2   Performing divisibility tests

Use a divisibility test to answer each question.
(a) Is 97 divisible by 2?   (b) Does 3 divide evenly into 144?   (c) Is 5 a factor of 168?

Solution

(a) The number 97 is not divisible by 2 because its ones digit is not 0, 2, 4, 6, or 8.
(b) Because the sum of its digits $1 + 4 + 4 = 9$ is divisible by 3, the number 144 is divisible by 3. So, 3 is a factor of 144.
(c) The number 168 is not divisible by 5 because its ones digit is neither 0 nor 5.

Now Try Exercises 21, 23, 27

In addition to the tests for 2, 3, and 5, the following divisibility tests may also be helpful when determining divisibility by 4, 6, or 9.

MORE DIVISIBILITY TESTS

A whole number is divisible by

- 4 if the number formed by its last two digits is divisible by 4.
- 6 if it is divisible by 2 and 3.
- 9 if the sum of its digits is divisible by 9.

NOTE: There are divisibility tests for 7 and 8; however, they are often more difficult to apply than simply performing long division. ■

EXAMPLE 3   Performing divisibility tests

Use a divisibility test to answer each question.
(a) Can 124 be divided evenly by 4?   (b) Is 6 a factor of 458?
(c) Does 9 divide evenly into 846?

Solution

(a) The number 1**24** divides evenly by 4 because **24** is divisible by 4.

(b) The number 458 is divisible by 2 because its ones digit is 8. However, it is not divisible by 3 because the sum of its digits $4 + 5 + 8 = 17$ is not divisible by 3. So, 6 is not a factor of 458.

(c) Because the sum of its digits $8 + 4 + 6 = 18$ is divisible by 9, we know that 9 divides evenly into 846.

Now Try Exercises 25, 29, 31

2 Finding Prime Factorization of Numbers

DETERMINING IF A NUMBER IS PRIME When we *factor* a number we write it as a product. Some numbers can be factored in many different ways.

Four Factorizations of 12   The *prime* factorization of 12

$12 = 1 \cdot 12, \quad 12 = 2 \cdot 6, \quad 12 = 3 \cdot 4, \quad$ or $\quad 12 = 2 \cdot 2 \cdot 3.$

The last factorization of 12 is important because the factors are all *prime numbers*. A prime number is defined as follows.

> ### PRIME NUMBERS AND COMPOSITE NUMBERS
>
> A **prime number** is a natural number greater than 1 whose *only* factors are 1 and itself.
>
> The first ten prime numbers are 2, 3, 5, 7, 11, 13, 17, 19, 23, and 29.
>
> A **composite number** is a natural number greater than 1 that is not prime.
>
> The first ten composite numbers are 4, 6, 8, 9, 10, 12, 14, 15, 16, and 18.
>
> The natural number 1 is *neither* prime nor composite.

READING CHECK 2

- Which of the following numbers are prime: 1, 2, 3, 21, 31, 109? Select all that apply.

We can determine if a number is prime by using the following property.

Every composite number is divisible by at least one prime number.

For example, the composite number 68 is divisible by the prime number 2. To see if a number is prime, we divide it by each prime number from 2 up to (but not including) the first prime number whose square is greater than the number. This process is illustrated in the next example.

EXAMPLE 4 Determining if a number is prime

Determine if each number is prime, composite, or neither.
(a) 65   (b) 1   (c) 47

Solution

(a) Since $7^2 = 49$ and $11^2 = 121$, the first prime number whose square is greater than 65 is 11. We will check to see if 65 is divisible by the prime numbers 2, 3, 5, or 7.

  2: The ones digit is 5, so we know that 65 is not divisible by 2.

  3: The sum $6 + 5 = 11$ is not divisible by 3, so 65 is not divisible by 3.

  5: Since the ones digit is 5, we know that 65 is divisible by 5.

Since 65 is divisible by the prime number 5, it is a composite number.

(b) As noted in the definition box, the number 1 is neither prime nor composite.

(c) Since $5^2 = 25$ and $7^2 = 49$, the first prime number whose square is greater than 47 is 7. We will check to see if 47 is divisible by the prime numbers 2, 3, or 5.

2: The ones digit is 7, so 47 is not divisible by 2.

3: The sum $4 + 7 = 11$ is not divisible by 3, so 47 is not divisible by 3.

5: Since the ones digit is neither 0 nor 5, we know that 47 is not divisible by 5.

Since 47 is not divisible by 2, 3, or 5, it is a prime number.

Now Try Exercises 33, 37, 39

WRITING PRIME FACTORIZATION USING A FACTOR TREE Every composite number can be factored into a product of prime numbers. This product, known as the **prime factorization**, is unique. For example, the prime factorization of 12 is $12 = 2 \cdot 2 \cdot 3$. No other product of *prime factors* of 12 can be written that is different from this one. (Note that the factorization $12 = 2 \cdot 3 \cdot 2$ does not represent a different product of primes. Only the *order* of the factors is different.)

One way to write a number as the product of primes is to use a **factor tree**. To demonstrate this method, we will find the prime factorization of 120.

See the Concept 1   PRIME FACTORIZATION OF 120

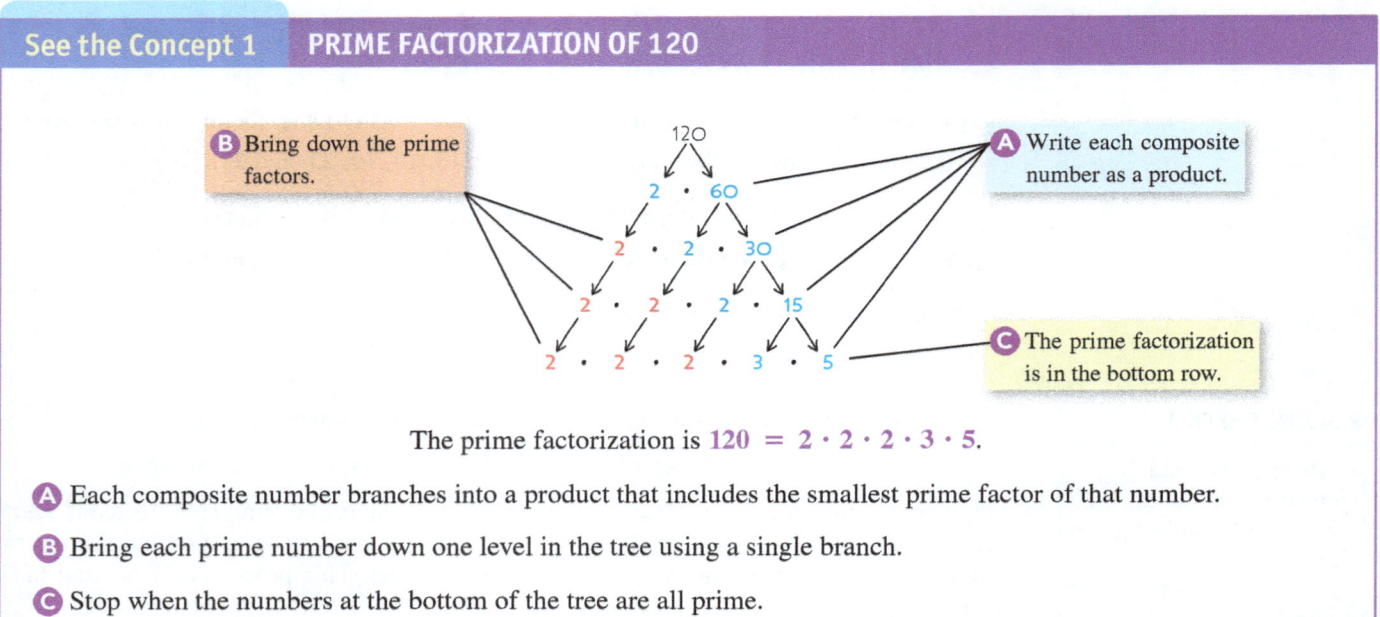

The prime factorization is $120 = 2 \cdot 2 \cdot 2 \cdot 3 \cdot 5$.

A Each composite number branches into a product that includes the smallest prime factor of that number.

B Bring each prime number down one level in the tree using a single branch.

C Stop when the numbers at the bottom of the tree are all prime.

WATCH VIDEO IN MML.

READING CHECK 3

• Why do we use a factor tree?

MAKING CONNECTIONS 1

Factor Trees

Because multiplication is both commutative and associative, there may be several ways to draw a factor tree for a given number. However, all factor trees for a specified number give the same prime factorization. For example, a different factor tree for 120 is shown here.

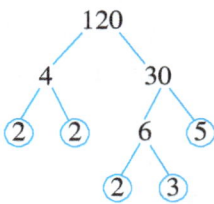

In either case, the prime factorization is the same: $2 \cdot 2 \cdot 2 \cdot 3 \cdot 5$.

EXAMPLE 5 **Writing prime factorizations of composite numbers**

Find the prime factorization of each composite number.
(a) 75   (b) 126

Solution
A factor tree for each number is shown below.

(a)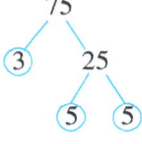

$75 = 3 \cdot 5 \cdot 5$ or $75 = 3 \cdot 5^2$

(b)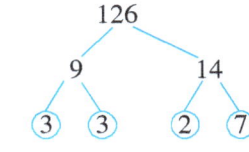

$126 = 2 \cdot 3 \cdot 3 \cdot 7$ or $126 = 2 \cdot 3^2 \cdot 7$

Now Try Exercises 43, 45

3 Identifying and Writing Equivalent Fractions

REPRESENTING EQUIVALENT FRACTIONS

🌐 **Math in Context** (*Inventory*) When the inventory of a particular product sold in a store falls below a certain level, the store manager will order more of the product from the wholesaler. Suppose that it is company policy to reorder a product each time that the inventory reaches one-half of its original amount. In this case, the store manager must understand that there are different ways to express the fraction $\frac{1}{2}$. For Example, when 50 tubes of toothpaste are sold from an inventory of 100 tubes, it is time to reorder because $\frac{50}{100} = \frac{1}{2}$. Similarly, when 15 electric blankets are sold from an inventory of 30 blankets, it is time to reorder because $\frac{15}{30} = \frac{1}{2}$.

This discussion suggests that the fractions $\frac{50}{100}, \frac{15}{30}$, and $\frac{1}{2}$ each represent one-half of a whole amount. To illustrate the equivalence of fractions, the following See the Concept uses both shaded regions and number lines.

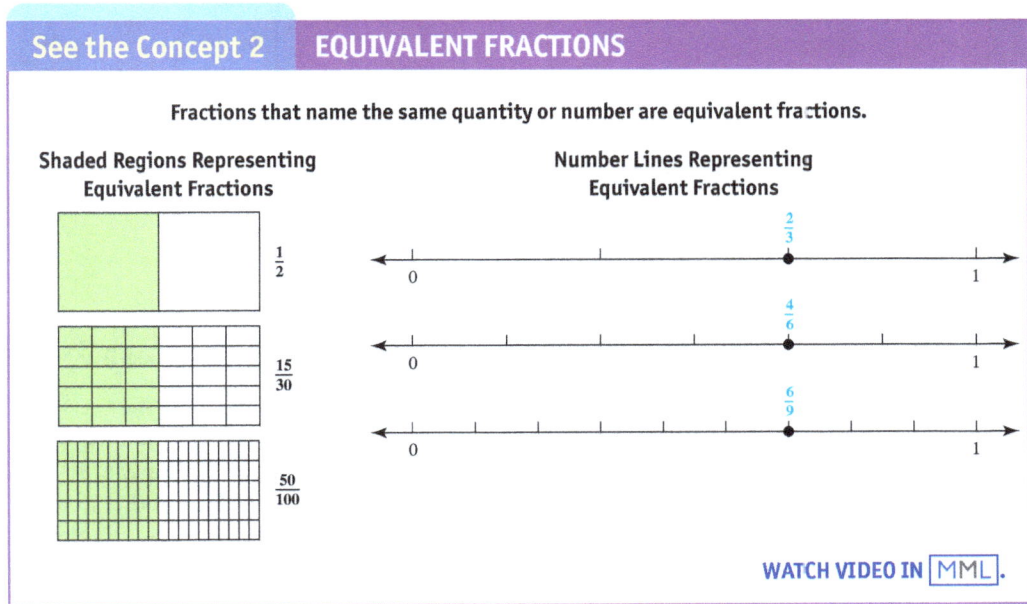

See the Concept 2   EQUIVALENT FRACTIONS

Fractions that name the same quantity or number are equivalent fractions.

WATCH VIDEO IN MML.

WRITING EQUIVALENT FRACTIONS Fractions that name the same number are called **equivalent fractions**. When we write a given fraction as an equivalent fraction, we can use the **basic principle of fractions**. This principle states that if we multiply the numerator and denominator of any fraction by the same nonzero number, the result is an equivalent fraction.

READING CHECK 4

- What principle allows us to write equivalent fractions?

BASIC PRINCIPLE OF FRACTIONS

If the variables a, b, and c represent integers with $b \neq 0$ and $c \neq 0$, then

$$\frac{a \cdot c}{b \cdot c} = \frac{a}{b}.$$

EXAMPLE 6 Writing equivalent fractions

Find a number to replace the question mark so that the fractions are equivalent.

(a) $\dfrac{3}{5} = \dfrac{?}{35}$ (b) $\dfrac{24}{36} = \dfrac{2}{?}$

Solution
(a) For some nonzero number c, the basic principle of fractions gives

$$\frac{3}{5} = \frac{3 \cdot c}{5 \cdot c} = \frac{?}{35}.$$

Because $5 \cdot 7 = 35$ is the value of the denominator in the equivalent fraction, we can conclude that c must be **7**. To find the value of the unknown numerator, we also multiply 3 by c to obtain $3 \cdot c = 3 \cdot 7 = 21$. This method can be visualized as follows, where 3 times **7** gives the unknown value of 21.

$$\frac{3}{5} = \frac{?}{35} \quad \text{×7}$$

The fraction $\frac{3}{5}$ is equivalent to $\frac{21}{35}$.

(b) The unknown value can be found by reversing the arrows used in the visual method shown in part (a). Because $2 \cdot 12 = 24$, the value of c is **12**.

$$\frac{24}{36} = \frac{2}{?} \quad \text{×12}$$

The fraction $\frac{24}{36}$ is equivalent to $\frac{2}{3}$.

Multiplying the unknown value by **12** must result in 36, so the unknown value is 3 because $3 \cdot 12 = 36$.

Now Try Exercises 51, 61

4 Simplifying Fractions to Lowest Terms

FINDING THE GREATEST COMMON FACTOR (GCF) The largest number that divides evenly into two or more given numbers is known as the **greatest common factor** (GCF). For example, 6 is the GCF of 12 and 18 because 6 divides both 12 and 18 evenly and no *larger* number can be found that also divides both 12 and 18. (Note that 2 is a *common factor* of 12 and 18, but it is not the *greatest* common factor.)

One way to find the GCF of two or more numbers is to list all of the factors for each number and then visually search for the largest factor that is common to each list. For example, the GCF of 36, 48, and 72 is 12, as shown in the following lists of factors.

Factors of 36: 1, 2, 3, 4, 6, 9, **12**, 18, 36
Factors of 48: 1, 2, 3, 4, 6, 8, **12**, 16, 24, 48
Factors of 72: 1, 2, 3, 4, 6, 8, 9, **12**, 18, 24, 36, 72

Greatest common factor of 36, 48, and 72 is 12.

EXAMPLE 7 Using the listing method to find the GCF

Use the listing method to find the GCF of 16 and 40.

Solution
First, list the factors of 16 and 40.

Factors of 16: 1, 2, 4, **8**, 16
Factors of 40: 1, 2, 4, 5, **8**, 10, 20, 40

> Greatest common factor of 16 and 40 is **8**.

The common factors are 1, 2, 4, and 8 and the GCF is **8**.

Now Try Exercise 65

Often we find that listing all of the factors for a number is quite difficult. If any factors are missed, we may not be able to find the correct GCF. To avoid this problem, it is common to find the GCF of two or more numbers by using the prime factorization of each number. In the next example, we use this method to verify that 12 is the GCF of 36, 48, and 72.

EXAMPLE 8 Using prime factorization to find the GCF

Find the GCF of 36, 48, and 72.

Solution
First, find the prime factorization of each number. Use a factor tree, if needed.

$36 = \mathbf{2} \cdot \mathbf{2} \cdot \mathbf{3} \cdot 3$
$48 = \mathbf{2} \cdot \mathbf{2} \cdot 2 \cdot 2 \cdot \mathbf{3}$
$72 = \mathbf{2} \cdot \mathbf{2} \cdot 2 \cdot \mathbf{3} \cdot 3$

> 2, 2, and 3 are common prime factors of all three numbers.

The prime factors **2, 2**, and **3** are common to all three factorizations. The GCF is the product of the common prime factors $\mathbf{2} \cdot \mathbf{2} \cdot \mathbf{3} = \mathbf{12}$.

Now Try Exercise 69

READING CHECK 5
- What does it mean to simplify a fraction to lowest terms?

SIMPLIFYING FRACTIONS TO LOWEST TERMS Because a fraction can be written in many equivalent forms, we often want to *simplify* a fraction to *lowest terms*. A fraction is simplified to **lowest terms** if the GCF of its numerator and denominator is 1. In other words, a fraction is in lowest terms if the only common factor of the numerator and the denominator is 1. We can use the GCF of the numerator and the denominator to simplify a fraction to lowest terms. For example, because the GCF of 30 and 45 is **15**, the basic principle of fractions can be used to simplify the fraction $\frac{30}{45}$.

$$\frac{30}{45} = \frac{2 \cdot \mathbf{15}}{3 \cdot \mathbf{15}} = \frac{2}{3}$$

> $\frac{30}{45}$ and $\frac{2}{3}$ are equivalent fractions, but $\frac{2}{3}$ is simplified to lowest terms.

The procedure for simplifying a fraction to lowest terms can be summarized as follows.

CALCULATOR HELP
To simplify fractions with a calculator, see Appendix E (page AP-31).

SIMPLIFYING FRACTIONS TO LOWEST TERMS

To simplify a fraction to lowest terms, do the following.

STEP 1: Determine a number c that is the GCF of the numerator and denominator.

STEP 2: Write the numerator as a product of two factors, a and c.
Then write the denominator as a product of two factors, b and c.

STEP 3: Apply the basic principle of fractions.

$$\frac{a \cdot c}{b \cdot c} = \frac{a}{b}$$

EXAMPLE 9 Simplifying fractions to lowest terms

Simplify each fraction to lowest terms.

(a) $\dfrac{12}{20}$ (b) $-\dfrac{36}{105}$ (c) $\dfrac{84}{120}$

Solution

(a) **STEP 1:** Find the GCF of the numerator and denominator.

$$12 = 2 \cdot 2 \cdot 3$$
$$20 = 2 \cdot 2 \cdot 5$$

The GCF is the product of the common prime factors $2 \cdot 2 = 4$.

STEP 2: Write the numerator as $12 = 3 \cdot 4$ and the denominator as $20 = 5 \cdot 4$.

STEP 3: Apply the basic principle of fractions.

$$\dfrac{12}{20} = \dfrac{3 \cdot 4}{5 \cdot 4} = \dfrac{3}{5}$$

(b) **STEP 1:** The GCF of the numerator and denominator is **3**, as shown below.

$$36 = 2 \cdot 2 \cdot 3 \cdot 3$$
$$105 = 3 \cdot 5 \cdot 7$$

STEP 2: Write the numerator as $36 = 12 \cdot 3$ and the denominator as $105 = 35 \cdot 3$.

STEP 3: Apply the basic principle of fractions.

$$-\dfrac{36}{105} = -\dfrac{12 \cdot 3}{35 \cdot 3} = -\dfrac{12}{35}$$

(c) **STEP 1:** As shown here, the GCF of the numerator and denominator is $2 \cdot 2 \cdot 3 = 12$.

$$84 = 2 \cdot 2 \cdot 3 \cdot 7$$
$$120 = 2 \cdot 2 \cdot 2 \cdot 3 \cdot 5$$

STEP 2: Write the numerator as $84 = 7 \cdot 12$ and the denominator as $120 = 10 \cdot 12$.

STEP 3: Apply the basic principle of fractions.

$$\dfrac{84}{120} = \dfrac{7 \cdot 12}{10 \cdot 12} = \dfrac{7}{10}$$

Now Try Exercises 73, 75, 77

MAKING CONNECTIONS 2

Lowest Terms and Canceling

Have you ever heard the word *canceling* used when simplifying a fraction to lowest terms? Canceling is another way to apply the basic principle of fractions. For example, the fraction in Example 9(c) can be simplified by writing the prime factorizations for the numerator and the denominator and then "canceling" the common prime factors from each factorization.

$$\dfrac{84}{120} = \dfrac{\cancel{2} \cdot \cancel{2} \cdot \cancel{3} \cdot 7}{\cancel{2} \cdot \cancel{2} \cdot 2 \cdot \cancel{3} \cdot 5} = \dfrac{7}{10}$$

NOTE: It is important to remember that a **1** remains after a factor has been canceled. ■

5 Comparing Fractions

When two fractions have denominators that are equal (a common denominator), it is easy to compare the fractions by simply comparing the numerators. For example, $\frac{5}{7} > \frac{2}{7}$ because $5 > 2$. However, when two fractions have different denominators, it can be difficult to compare the sizes of the numbers named by the fractions. To compare fractions with unlike denominators, we write equivalent fractions that have a common denominator and then compare the numerators. This process is demonstrated in the next example.

EXAMPLE 10 **Comparing fractions**

Place the correct symbol, $<$, $>$, or $=$, in the blank for each pair of fractions.

(a) $\frac{7}{9}$ —— $\frac{8}{10}$ (b) $\frac{13}{16}$ —— $\frac{25}{31}$

Solution

(a) To write a fraction that is equivalent to $\frac{7}{9}$, we use the basic principle of fractions with the value of c equal to the denominator of the second fraction, **10**.

$$\frac{7}{9} = \frac{7 \cdot 10}{9 \cdot 10} = \frac{70}{90}$$

Likewise, to write a fraction that is equivalent to $\frac{8}{10}$, use the basic principle of fractions with the value of c equal to the denominator of the first fraction, **9**.

$$\frac{8}{10} = \frac{8 \cdot 9}{10 \cdot 9} = \frac{72}{90}$$

Now we compare the two fractions $\frac{70}{90}$ and $\frac{72}{90}$. Because $\frac{70}{90} < \frac{72}{90}$, the symbol $<$ can be placed in the blank and we write $\frac{7}{9} < \frac{8}{10}$.

(b) To compare the fractions $\frac{13}{16}$ and $\frac{25}{31}$, we use the basic principle of fractions, as demonstrated in part (a). A more concise process is shown here.

$$\frac{13}{16} = \frac{13 \cdot 31}{16 \cdot 31} = \frac{403}{496} \quad \text{and} \quad \frac{25}{31} = \frac{25 \cdot 16}{31 \cdot 16} = \frac{400}{496}$$

Because $\frac{403}{496} > \frac{400}{496}$, the symbol $>$ can be placed in the blank and we write $\frac{13}{16} > \frac{25}{31}$.

Now Try Exercises 85, 87

READING CHECK 6

- How is the cross product used to see if two fractions are equivalent?

In the previous example, we multiplied the numerator of each fraction by the denominator of the other. These products are called the **cross products** and can be used to see if two fractions are equivalent. For example, the cross product method for comparing $\frac{9}{14}$ and $\frac{12}{19}$ is written as follows.

$$19 \cdot 9 \quad \frac{9}{14} \stackrel{?}{=} \frac{12}{19} \quad 14 \cdot 12$$

Since $19 \cdot 9 = 171$ and $14 \cdot 12 = 168$, the cross products are not equal and the fractions $\frac{9}{14}$ and $\frac{12}{19}$ are not equivalent. This result is based on the following rule for cross products.

Two fractions are equivalent if their cross products are equal, and they are not equivalent if their cross products are not equal.

EXAMPLE 11 **Determining if two fractions are equivalent**

Use the cross product method to determine if the two fractions are equivalent.

$$\frac{42}{56} \stackrel{?}{=} \frac{27}{36}$$

Solution
Find the cross product as follows.

> The cross products both equal 1512.

$$36 \cdot 42 \quad \frac{42}{56} \stackrel{?}{=} \frac{27}{36} \quad 27 \cdot 56$$

Since $36 \cdot 42 = 1512$ and $27 \cdot 56 = 1512$, the cross products are equal and the fractions $\frac{42}{56}$ and $\frac{27}{36}$ are equivalent.

Now Try Exercise 91

6 Simplifying Rational Expressions

FINDING THE GCF FOR RATIONAL EXPRESSIONS A *term* is defined as a number, a variable, or a product of numbers and variables raised to powers. Any fraction whose numerator and denominator each contain a term or a sum (or difference) of two or more terms is called a **rational expression**.

Examples of Rational Expressions

$$\frac{5}{9}, \quad \frac{3x}{7}, \quad \frac{2x^2 + 5}{3x - 1}, \quad \frac{10ab}{5c^3}, \quad \text{and} \quad \frac{3x^2 - 3x + 1}{10x^2 + 2x - 6}$$

Here, we will focus on simplifying rational expressions whose numerator and denominator each contain a *single* term. The same basic process that was used earlier to simplify fractions can be used to simplify rational expressions of this type. In the next example, we find the greatest common factor (GCF) for two terms containing variables.

EXAMPLE 12 Finding the GCF of two terms

Find the GCF of $12x^3$ and $18x$.

Solution
First, factor each term completely, as follows.

$$12x^3 = 2 \cdot 2 \cdot 3 \cdot x \cdot x \cdot x$$
$$18x = 2 \cdot 3 \cdot 3 \cdot x$$

> 2, 3, and x are common factors of both terms.

The GCF is the product of the common factors $2 \cdot 3 \cdot x = 6x$.

Now Try Exercise 97

READING CHECK 7
- What is the GCF of $15x^3$ and $25x^2$?

SIMPLIFYING RATIONAL EXPRESSIONS When simplifying rational expressions whose numerator and denominator each contain a single term, we use the same process as that used in Example 9 to simplify fractions. The next example demonstrates this process.

EXAMPLE 13 Simplifying rational expressions

Simplify each rational expression.

(a) $\dfrac{30x^3}{42xy}$ (b) $-\dfrac{55a^3b^2}{33a^2b^2}$

Solution
(a) **STEP 1:** First find the GCF of the numerator and denominator.

$$30x^3 = 2 \cdot 3 \cdot 5 \cdot x \cdot x \cdot x$$
$$42xy = 2 \cdot 3 \cdot 7 \cdot x \cdot y$$

The GCF is the product of the common factors $2 \cdot 3 \cdot x = 6x$

STEP 2: Use the GCF to write the numerator as $30x^3 = 5x^2 \cdot 6x$ and the denominator as $42xy = 7y \cdot 6x$.

STEP 3: Apply the basic principle of fractions.

$$\frac{30x^3}{42xy} = \frac{5x^2 \cdot 6x}{7y \cdot 6x} = \frac{5x^2}{7y}$$

(b) STEP 1: The GCF of the numerator and denominator is $11a^2b^2$, as shown below.

$$55a^3b^2 = 5 \cdot 11 \cdot a \cdot a \cdot a \cdot b \cdot b$$
$$33a^2b^2 = 3 \cdot 11 \cdot a \cdot a \cdot b \cdot b$$

STEP 2: Use the GCF to write the numerator as $55a^3b^2 = 5a \cdot 11a^2b^2$ and the denominator as $33a^2b^2 = 3 \cdot 11a^2b^2$.

STEP 3: Apply the basic principle of fractions.

$$-\frac{55a^3b^2}{33a^2b^2} = -\frac{5a \cdot 11a^2b^2}{3 \cdot 11a^2b^2} = -\frac{5a}{3}$$

Now Try Exercises 111, 113

4.2 Putting It All Together

CONCEPT	COMMENTS	EXAMPLES
Factor	A *factor* is any whole number that divides into a second whole number evenly and results in a quotient with no remainder.	8 is a factor of 32 because $$32 \div 8 = 4 \text{ r } 0.$$ That is, 32 is *divisible* by 8.
Divisibility Tests	A whole number is divisible by • 2 if its ones digit is 0, 2, 4, 6, or 8. • 3 if the sum of its digits is divisible by 3 • 5 if its ones digit is 0 or 5. • 4 if the number formed by its last two digits is divisible by 4 • 6 if it is divisible by 2 and 3. • 9 if the sum of its digits is divisible by 9.	2784 is divisible by 2 because its ones digit is 4. 432 divides evenly by 3 because the digit sum, 9, is divisible by 3. 5 is a factor of 420 because the ones digit in 420 is 0. 873 divides evenly by 9 because the digit sum, 18, is divisible by 9.
Prime and Composite Numbers	A *prime* number is any natural number greater than 1 whose only factors are 1 and itself. A *composite* number is any natural number greater than 1 that is not prime.	Prime numbers include 2, 5, 19, 83, and 773. Composite numbers include 6, 18, 54, 104, and 735.
Prime Factorization	Every composite number can be factored into a unique product of prime numbers.	$54 = 2 \cdot 3 \cdot 3 \cdot 3$ $735 = 3 \cdot 5 \cdot 7 \cdot 7$
Equivalent Fractions	Fractions that name the same number are called *equivalent fractions*. We can write equivalent fractions using the *basic principle of fractions*. $$\frac{a \cdot c}{b \cdot c} = \frac{a}{b}$$	The fractions $$\frac{3}{4} \text{ and } \frac{21}{28}$$ are equivalent because $$\frac{3}{4} = \frac{3 \cdot 7}{4 \cdot 7} = \frac{21}{28}.$$

continued on next page

continued from previous page

CONCEPT	COMMENTS	EXAMPLES
Greatest Common Factor (GCF)	The largest number that divides evenly into two or more numbers is the GCF.	$18 = 2 \cdot 3 \cdot 3$ and $30 = 2 \cdot 3 \cdot 5$ The GCF of 18 and 30 is $2 \cdot 3 = 6$.
Lowest Terms	A fraction is simplified to *lowest terms* if the GCF of its numerator and denominator is 1.	The GCF of 25 and 30 is **5**, so $$\frac{25}{30} = \frac{5 \cdot 5}{6 \cdot 5} = \frac{5}{6}.$$
Cross Product Method	We can determine if two fractions are equivalent by comparing their cross products.	$42 \cdot 5 \quad \frac{5}{14} \stackrel{?}{=} \frac{15}{42} \quad 14 \cdot 15$ The cross products $42 \cdot 5 = 210$ and $14 \cdot 15 = 210$ are equal, so $\frac{5}{14}$ and $\frac{15}{42}$ are equivalent.
Rational Expression	Any fraction whose numerator and denominator each contain a term or a sum (or difference) of two or more terms is a rational expression.	$\frac{14x^2y}{21xy}$ is a rational expression and $$\frac{14x^2y}{21xy} = \frac{2x \cdot 7xy}{3 \cdot 7xy} = \frac{2x}{3}.$$

4.2 Exercises   MyMathLab®

CONCEPTS AND VOCABULARY

1. A(n) _____ is any whole number that divides into a second whole number evenly and results in a quotient with no remainder.

2. We can say that the whole number 24 is _____ by 8 because 8 is a factor of 24.

3. A whole number that has 0 or 5 as its ones digit is always divisible by _____.

4. If the sum of the digits of a whole number is divisible by 3, then the whole number is divisible by _____.

5. A(n) _____ number is a natural number greater than 1 whose only factors are 1 and itself.

6. A(n) _____ number is a natural number greater than 1 that is not prime.

7. Every composite number can be factored into a product of primes called its _____ factorization.

8. One way to find the prime factorization of a composite number is to make a(n) _____ tree.

9. Two fractions that name the same number are called _____ fractions.

10. The letters GCF are used to represent the _____.

11. The GCF of two numbers is the largest number that divides (one/both) of the numbers.

12. A fraction is written in _____ if the GCF of its numerator and denominator is 1.

13. A method that can be used to see if two fractions are equivalent is the _____ product method.

14. A(n) _____ expression is a fraction whose numerator and denominator each contain a term or a sum (or difference) of two or more terms.

DETERMINING DIVISIBILITY

Exercises 15–20: Using the definitions of "divisible" and "factor," answer the given question.

15. Is 64 divisible by 12?

16. Is 46 divisible by 14?

17. Is 13 a factor of 65?

18. Is 8 a factor of 54?

19. Is 19 a factor of 57?

20. Is 72 divisible by 18?

Exercises 21–32: Use a divisibility test to answer the given question.

21. Is 913 divisible by 5?
22. Can 4 be divided evenly into 622?
23. Is 186 divisible by 2?
24. Can 3 be divided evenly into 762?
25. Does 387 divide evenly by 9?
26. Is 5 a factor of 275?
27. Does 691 divide evenly by 3?
28. Can 6 be divided evenly into 533?
29. Is 6 a factor of 834?
30. Can 673 be divided evenly by 9?
31. Does 4 divide evenly into 748?
32. Is 191 divisible by 2?

FINDING PRIME FACTORIZATION

Exercises 33–40: Determine if the given number is prime, composite, or neither.

33. 43
34. 63
35. 57
36. 0
37. 1
38. 65
39. 91
40. 101

Exercises 41–50: Find the prime factorization of the given composite number.

41. 16
42. 54
43. 45
44. 90
45. 140
46. 315
47. 231
48. 390
49. 442
50. 845

FINDING EQUIVALENT FRACTIONS

Exercises 51–62: Find a number to replace the question mark so that the given fractions are equivalent.

51. $\dfrac{7}{9} = \dfrac{?}{27}$
52. $\dfrac{5}{11} = \dfrac{20}{?}$
53. $\dfrac{5}{6} = \dfrac{35}{?}$
54. $\dfrac{?}{15} = \dfrac{12}{45}$
55. $\dfrac{2}{?} = \dfrac{10}{25}$
56. $\dfrac{2}{9} = \dfrac{?}{63}$
57. $\dfrac{?}{4} = \dfrac{36}{48}$
58. $\dfrac{1}{?} = \dfrac{7}{42}$
59. $\dfrac{4}{13} = \dfrac{?}{65}$
60. $\dfrac{8}{17} = \dfrac{24}{?}$
61. $\dfrac{45}{60} = \dfrac{3}{?}$
62. $\dfrac{?}{6} = \dfrac{50}{60}$

SIMPLIFYING FRACTIONS USING THE GCF

Exercises 63–72: Find the GCF of the given numbers.

63. 18 and 40
64. 24 and 60
65. 20 and 70
66. 36 and 63
67. 24 and 72
68. 15 and 75
69. 8, 12, and 20
70. 15, 30, and 55
71. 16, 32, and 48
72. 12, 36, and 72

Exercises 73–82: Simplify the fraction to lowest terms.

73. $\dfrac{16}{24}$
74. $-\dfrac{15}{35}$
75. $\dfrac{26}{65}$
76. $\dfrac{64}{72}$
77. $-\dfrac{42}{77}$
78. $-\dfrac{48}{56}$
79. $\dfrac{105}{350}$
80. $\dfrac{120}{156}$
81. $-\dfrac{200}{450}$
82. $\dfrac{320}{600}$

COMPARING FRACTIONS

Exercises 83–90: Place the correct symbol, <, >, or =, in the blank in the given pair of fractions.

83. $\dfrac{9}{13}$ —— $\dfrac{10}{13}$
84. $\dfrac{8}{11}$ —— $\dfrac{7}{11}$
85. $\dfrac{12}{17}$ —— $\dfrac{22}{31}$
86. $\dfrac{4}{19}$ —— $\dfrac{6}{25}$
87. $\dfrac{27}{34}$ —— $\dfrac{5}{7}$
88. $\dfrac{5}{6}$ —— $\dfrac{75}{90}$
89. $\dfrac{72}{84}$ —— $\dfrac{6}{7}$
90. $\dfrac{33}{47}$ —— $\dfrac{17}{24}$

Exercises 91–96: Use the cross product method to determine if the two fractions are equivalent.

91. $\dfrac{68}{85} \stackrel{?}{=} \dfrac{32}{40}$ **92.** $\dfrac{20}{25} \stackrel{?}{=} \dfrac{28}{35}$

93. $\dfrac{28}{42} \stackrel{?}{=} \dfrac{30}{36}$ **94.** $\dfrac{21}{70} \stackrel{?}{=} \dfrac{9}{30}$

95. $\dfrac{33}{77} \stackrel{?}{=} \dfrac{27}{63}$ **96.** $\dfrac{14}{15} \stackrel{?}{=} \dfrac{17}{18}$

SIMPLIFYING RATIONAL EXPRESSIONS USING THE GCF

Exercises 97–104: Find the GCF of the given terms.

97. $10x^2$ and $16x$ **98.** $15y^2$ and $20y$

99. $14a^2$ and $35a^3$ **100.** $4b$ and $12b^3$

101. $100m^2$ and $50n^2$ **102.** $105x^3$ and $35y^2$

103. $8a^3c^2$ and $24a^2c$ **104.** $10x^2y$ and $25x^4y$

Exercises 105–114: Simplify the rational expression.

105. $\dfrac{6x}{18}$ **106.** $-\dfrac{35a}{55a}$

107. $-\dfrac{12x}{40xy}$ **108.** $\dfrac{18xy}{24y}$

109. $\dfrac{13mn}{26mn}$ **110.** $-\dfrac{5a}{5ab}$

111. $-\dfrac{6x^2y}{3x}$ **112.** $\dfrac{12m^2n^3}{4mn}$

113. $\dfrac{25x^2y}{45xy^3}$ **114.** $-\dfrac{28a^3b^2}{42ab}$

APPLICATIONS INVOLVING FRACTIONS AND PRIMES

115. Study Time A student spends 8 hours studying for a physics exam. What fraction of a 24-hour day is this? Write your answer in lowest terms.

116. Extra Credit In a math class with 31 students, 27 students earned extra credit. In a biology class with 21 students, 17 students earned extra credit. Which class had a larger fraction of its students earning extra credit?

117. Indispensible Devices Of millennial adults (aged 18–35), $\frac{2}{9}$ say their computer would be the hardest device to give up. In Gen X adults (aged 36–50), this number is $\frac{1}{3}$. Do millennial or Gen X adults have a larger fraction who say their computer would be the hardest device to give up. (*Source: Business Insider.*)

118. Indispensible Devices Of baby boomer adults (aged 51–69), $\frac{14}{51}$ say their mobile phone would be the hardest device to give up. In mature adults (aged 70+), this number is $\frac{3}{17}$. Do baby boomers or mature adults have a larger fraction who say their mobile phone would be the hardest device to give up. (*Source: Business Insider.*)

119. Age Puzzle A man who is between the ages of 35 and 60 says that his age is a prime number and that two years ago it was also a prime number. How old is the man?

120. Twin Primes Prime numbers that differ by only 2 are called *twin primes*. For example, 5 and 7 are twin primes. Find all of the twin prime pairs between 50 and 100.

121. Female Physicians In 1970, about 8 out of every 100 physicians were female. By 2016, about 32 out of every 100 physicians were female. Write these values as fractions in lowest terms. (*Source: American Medical Association.*)

122. Female Pharmacists In 1980, about 18 out of every 100 pharmacists were female. By 2016, there were about 68 female pharmacists for every 100 pharmacists. Write these values as fractions in lowest terms. (*Source: American Pharmaceutical Association.*)

WRITING ABOUT MATHEMATICS

123. Explain how to determine if a natural number is prime or composite.

124. Do some research to find a divisibility test for 11. Show how the test works for the number 719,653.

125. Explain how to simplify a fraction to lowest terms.

126. Explain how the cross product method can be used to determine if two fractions are equivalent.

SECTIONS 4.1 and 4.2 Checking Basic Concepts

1. Write the numerator and denominator of $\frac{5}{18}$.

2. Write the improper fraction and mixed number represented by the shading.

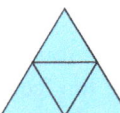 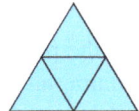 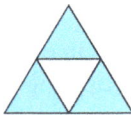

3. Write $\frac{19}{5}$ as a mixed number.

4. Write $6\frac{3}{7}$ as an improper fraction.

5. Is 9762 divisible by 3?

6. Determine if each given number is prime, composite, or neither.
   (a) 91 (b) 103

7. Write the prime factorization of each number.
   (a) 50 (b) 105

8. Find a number to replace the question mark so that the given fractions are equivalent.
$$\frac{5}{12} = \frac{?}{72}$$

9. Simplify each rational expression to lowest terms.
   (a) $\frac{30}{48}$ (b) $-\frac{16xy}{28y}$

10. *United States* The following states have names that begin with vowels: Alabama, Alaska, Arizona, Arkansas, Idaho, Illinois, Indiana, Iowa, Ohio, Oklahoma, Oregon, and Utah. What fraction of the 50 states have names that begin with a vowel? Write your answer in lowest terms.

4.3 Multiplying and Dividing Fractions

Objectives

1. **Multiplying Fractions**
   - Multiplying without Simplification
   - Multiplying with Simplification
2. **Multiplying Rational Expressions**
3. **Evaluating Powers and Square Roots of Fractions**
4. **Dividing Fractions and Rational Expressions**
   - Dividing Fractions
   - Dividing Rational Expressions
5. **Solving Applications Involving Multiplication of Fractions**
   - Finding a Fraction of a Whole and of a Fraction
   - Finding the Area of a Triangle

1 Multiplying Fractions

MULTIPLYING WITHOUT SIMPLIFICATION One way to visualize a product is to draw one vertical and one horizontal number line that each begin at a common location for 0. A shaded rectangle can be used to find a product. For example, a rectangle representing $3 \cdot 4$ is shown in **FIGURE 4.5**. The product, 12, is found by counting the number of squares that appear inside the larger rectangle.

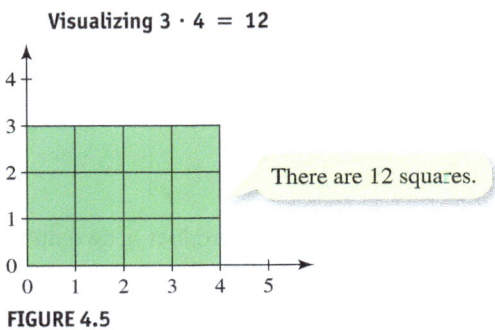

FIGURE 4.5

The product of two fractions can be visualized in a similar manner. The See the Concept on the next page illustrates how to find the product $\frac{2}{3} \cdot \frac{4}{5}$. Because $1 \cdot 1 = 1$ whole, the largest

224 CHAPTER 4 FRACTIONS

NEW VOCABULARY

☐ Reciprocal
☐ Multiplicative inverse

rectangle represents 1 whole. Since this largest rectangle contains 15 smaller rectangles of equal size and 8 of these rectangles are shaded, the product is $\frac{2}{3} \cdot \frac{4}{5} = \frac{8}{15}$.

> **See the Concept 1** — MULTIPLYING FRACTIONS
>
> 8 shaded parts out of 15 total
>
> **Multiplying Fractions**
> - The numerator of the product is found by multiplying the numerators of the factors.
> - The denominator of the product is found by multiplying the denominators of the factors.
>
> Factor → $\frac{2}{3} \cdot \frac{4}{5} = \frac{8}{15}$ ← Product
> Factor
>
> WATCH VIDEO IN [MML].

READING CHECK 1

- Multiply $\frac{3}{4} \cdot \frac{3}{5}$.

STUDY TIP

Much of mathematics builds on previous knowledge. Since we already know how to multiply integers, the only new information needed to multiply fractions is to know that we multiply numerators and denominators separately.

> **MULTIPLYING FRACTIONS**
>
> If the variables a, b, c, and d represent numbers with $b \neq 0$ and $d \neq 0$, then the product of the fractions $\frac{a}{b}$ and $\frac{c}{d}$ is given by
>
> $$\frac{a}{b} \cdot \frac{c}{d} = \frac{ac}{bd}.$$

EXAMPLE 1 **Multiplying fractions**

Multiply.

(a) $\dfrac{3}{7} \cdot \dfrac{2}{5}$ (b) $-\dfrac{1}{6} \cdot \dfrac{5}{8}$ (c) $-\dfrac{2}{3} \cdot \left(-\dfrac{2}{5}\right)$

Solution

(a) Multiply numerators and multiply denominators.

$$\frac{3}{7} \cdot \frac{2}{5} = \frac{3 \cdot 2}{7 \cdot 5} = \frac{6}{35}$$

CALCULATOR HELP

To multiply fractions with a calculator, see Appendix E (page AP-31).

(b) The product of two numbers with unlike signs is negative.

$$-\frac{1}{6} \cdot \frac{5}{8} = -\frac{1 \cdot 5}{6 \cdot 8} = -\frac{5}{48}$$

(c) The product of two numbers with like signs is positive.

$$-\frac{2}{3} \cdot \left(-\frac{2}{5}\right) = \frac{2 \cdot 2}{3 \cdot 5} = \frac{4}{15}$$

Now Try Exercises 7, 11, 15

MULTIPLYING WITH SIMPLIFICATION The three products found in Example 1 are all fractions written in lowest terms. However, this is not necessarily the case with every product of two fractions. Sometimes the resulting fraction needs to be simplified, as illustrated in the next example.

EXAMPLE 2   Multiplying fractions and simplifying the result

Multiply and simplify the result.

(a) $\dfrac{5}{6} \cdot \dfrac{4}{5}$ (b) $-\dfrac{2}{3} \cdot \dfrac{15}{22}$

Solution
In each case, simplify the product by applying the basic principle of fractions.
(a) The GCF of the numerator product 20 and the denominator product 30 is 10.

$$\frac{5}{6} \cdot \frac{4}{5} = \frac{5 \cdot 4}{6 \cdot 5} = \frac{20}{30} = \frac{2 \cdot 10}{3 \cdot 10} = \frac{2}{3}$$

(b) The GCF of 30 and 66 is 6.

$$-\frac{2}{3} \cdot \frac{15}{22} = -\frac{2 \cdot 15}{3 \cdot 22} = -\frac{30}{66} = -\frac{5 \cdot 6}{11 \cdot 6} = -\frac{5}{11}$$

Now Try Exercises 21, 23

> **MAKING CONNECTIONS 1**
>
> **The Basic Principle of Fractions and Multiplying by 1**
>
> The basic principle of fractions is a useful application of the identity property for multiplication. In other words, it is a handy way to multiply by 1. For example, the fraction $\frac{20}{30}$ can be simplified as
>
> $$\frac{20}{30} = \frac{2 \cdot 10}{3 \cdot 10} = \frac{2}{3} \cdot \frac{10}{10} = \frac{2}{3} \cdot 1 = \frac{2}{3}.$$

When a product of fractions contains relatively large numbers, it may be easier to simplify the product *before* multiplying. This method is summarized in the following box and is demonstrated in the next example.

> **SIMPLIFYING BEFORE MULTIPLYING FRACTIONS**
>
> Let the variables a, b, c, and d represent numbers with $b \neq 0$ and $d \neq 0$. To simplify the product $\frac{a}{b} \cdot \frac{c}{d}$ before multiplying, do the following.
>
> **STEP 1:** Write $\frac{a}{b} \cdot \frac{c}{d}$ as $\frac{a \cdot c}{b \cdot d}$. Do not multiply.
>
> **STEP 2:** Replace each composite factor with its prime factorization.
>
> **STEP 3:** Eliminate all factors that are common to the numerator and denominator by applying the basic principle of fractions.
>
> **STEP 4:** Multiply the remaining factors in the numerator and multiply the remaining factors in the denominator.

READING CHECK 2

- Before multiplying, what could the *factors* in the product $\frac{10}{16} \cdot \frac{7}{21}$ be simplified to?

EXAMPLE 3 Simplifying before multiplying fractions

Find each product by simplifying before multiplying.

(a) $\dfrac{12}{25} \cdot \dfrac{35}{36}$ (b) $\dfrac{15}{44} \cdot \dfrac{22}{39}$ (c) $-\dfrac{9}{40} \cdot \left(-\dfrac{8}{27}\right)$

Solution

(a) **STEP 1:** Write the product $\dfrac{12}{25} \cdot \dfrac{35}{36}$ as follows.

$$\dfrac{12 \cdot 35}{25 \cdot 36}$$

STEP 2: Replace every composite factor with its prime factorization.

$$\dfrac{(2 \cdot 2 \cdot 3) \cdot (5 \cdot 7)}{(5 \cdot 5) \cdot (2 \cdot 2 \cdot 3 \cdot 3)}$$

STEP 3: Regroup the factors and apply the basic principle of fractions.

$$\dfrac{7 \cdot (2 \cdot 2 \cdot 3 \cdot 5)}{3 \cdot 5 \cdot (2 \cdot 2 \cdot 3 \cdot 5)} = \dfrac{7}{3 \cdot 5}$$

STEP 4: Multiply the remaining factors in the numerator and in the denominator.

$$\dfrac{7}{3 \cdot 5} = \dfrac{7}{15}$$

(b) The entire four-step process can be written in a single line of equivalent expressions. Note that it is not necessary to regroup the common factors (as in Step 3 above) in order to apply the basic principle of fractions. Such grouping can be done mentally.

$$\dfrac{15}{44} \cdot \dfrac{22}{39} \overset{\text{Step 1}}{=} \dfrac{15 \cdot 22}{44 \cdot 39} \overset{\text{Step 2}}{=} \dfrac{(3 \cdot 5) \cdot (2 \cdot 11)}{(2 \cdot 2 \cdot 11) \cdot (3 \cdot 13)} \overset{\text{Step 3}}{=} \dfrac{5}{2 \cdot 13} \overset{\text{Step 4}}{=} \dfrac{5}{26}$$

(c) The product of two negative fractions is positive. Also, note that the basic principle of fractions eliminates *all* of the factors from the numerator. When this happens, it is important to remember that the remaining factor is **1**.

$$-\dfrac{9}{40} \cdot \left(-\dfrac{8}{27}\right) \overset{\text{Step 1}}{=} \dfrac{9 \cdot 8}{40 \cdot 27} \overset{\text{Step 2}}{=} \dfrac{(3 \cdot 3) \cdot (2 \cdot 2 \cdot 2)}{(2 \cdot 2 \cdot 2 \cdot 5) \cdot (3 \cdot 3 \cdot 3)} \overset{\text{Step 3}}{=} \dfrac{1}{3 \cdot 5} \overset{\text{Step 4}}{=} \dfrac{1}{15}$$

Now Try Exercises 25, 31, 35

2 Multiplying Rational Expressions

In the second section of this chapter, we simplified rational expressions whose numerator and denominator each contained a single term. A process similar to that used in Example 3 can also be used to multiply such rational expressions. The next example illustrates this process.

EXAMPLE 4 Multiplying rational expressions

Multiply the rational expressions. Simplify your result.

(a) $\dfrac{x^2}{3y} \cdot \dfrac{6y^4}{5x}$ (b) $\dfrac{4a^2}{9b^2} \cdot \dfrac{15b}{16a}$ (c) $8m \cdot \left(-\dfrac{3n}{4m^2}\right)$

Solution

(a) **STEP 1:** Write the product $\dfrac{x^2}{3y} \cdot \dfrac{6y^4}{5x}$ as follows.

$$\dfrac{x^2 \cdot 6y^4}{3y \cdot 5x}$$

STEP 2: Replace every factor in $\frac{x^2 \cdot 6y^4}{3y \cdot 5x}$ with its complete factorization.

$$\frac{(x \cdot x) \cdot (2 \cdot 3 \cdot y \cdot y \cdot y \cdot y)}{(3 \cdot y) \cdot (5 \cdot x)}$$

STEP 3: Regroup the factors and apply the basic principle of fractions.

$$\frac{2 \cdot x \cdot y \cdot y \cdot y \cdot (3 \cdot x \cdot y)}{5 \cdot (3 \cdot x \cdot y)} = \frac{2 \cdot x \cdot y \cdot y \cdot y}{5}$$

STEP 4: Multiply the remaining factors in the numerator and in the denominator.

$$\frac{2 \cdot x \cdot y \cdot y \cdot y}{5} = \frac{2xy^3}{5}$$

(b) Once again, the entire four-step process can be written as a single line of equivalent expressions, as shown here.

$$\frac{4a^2}{9b^2} \cdot \frac{15b}{16a} \overset{\text{Step 1}}{=} \frac{4a^2 \cdot 15b}{9b^2 \cdot 16a} \overset{\text{Step 2}}{=} \frac{(2 \cdot 2 \cdot a \cdot a) \cdot (3 \cdot 5 \cdot b)}{(3 \cdot 3 \cdot b \cdot b) \cdot (2 \cdot 2 \cdot 2 \cdot 2 \cdot a)} \overset{\text{Step 3}}{=} \frac{5 \cdot a}{2 \cdot 2 \cdot 3 \cdot b} \overset{\text{Step 4}}{=} \frac{5a}{12b}$$

(c) Begin by writing $8m$ as the rational expression $\frac{8m}{1}$. Note that the product is negative because we are multiplying rational expressions with unlike signs.

$$\frac{8m}{1} \cdot \left(-\frac{3n}{4m^2}\right) \overset{\text{Step 1}}{=} -\frac{8m \cdot 3n}{1 \cdot 4m^2} \overset{\text{Step 2}}{=} -\frac{(2 \cdot 2 \cdot 2 \cdot m) \cdot (3 \cdot n)}{1 \cdot (2 \cdot 2 \cdot m \cdot m)} \overset{\text{Step 3}}{=} -\frac{2 \cdot 3 \cdot n}{1 \cdot m} \overset{\text{Step 4}}{=} -\frac{6n}{m}$$

Now Try Exercises 43, 45, 49

3 Evaluating Powers and Square Roots of Fractions

Recall that a natural number exponent tells us *how many times* to multiply the base of an exponential expression by itself. For example, 5^4 means that we multiply 5 by itself a total of 4 times: $5 \cdot 5 \cdot 5 \cdot 5$. This rule also applies when the base is a fraction; a natural number exponent tells us how many times we multiply the fraction by itself.

EXAMPLE 5 Raising a fraction to a power

Evaluate each exponential expression.

(a) $\left(\frac{1}{2}\right)^3$ **(b)** $\left(-\frac{3}{4}\right)^2$ **(c)** $\left(-\frac{2}{3}\right)^3$

Solution

(a) $\left(\frac{1}{2}\right)^3 = \frac{1}{2} \cdot \frac{1}{2} \cdot \frac{1}{2} = \frac{1 \cdot 1 \cdot 1}{2 \cdot 2 \cdot 2} = \frac{1}{8}$

(b) Because there is an even number of negative factors, the product is positive.

$$\left(-\frac{3}{4}\right)^2 = \left(-\frac{3}{4}\right) \cdot \left(-\frac{3}{4}\right) = \frac{3 \cdot 3}{4 \cdot 4} = \frac{9}{16}$$

(c) With an odd number of negative factors, we have a negative product.

$$\left(-\frac{2}{3}\right)^3 = \left(-\frac{2}{3}\right) \cdot \left(-\frac{2}{3}\right) \cdot \left(-\frac{2}{3}\right) = -\frac{2 \cdot 2 \cdot 2}{3 \cdot 3 \cdot 3} = -\frac{8}{27}$$

Now Try Exercises 57, 59, 61

CALCULATOR HELP

To raise a fraction to a power with a calculator, see Appendix E (page AP-31).

When a fraction is squared, the result is a fraction whose numerator and denominator are both perfect squares. For example, squaring the fraction $\frac{5}{9}$ results in

$$\left(\frac{5}{9}\right)^2 = \frac{5}{9} \cdot \frac{5}{9} = \frac{25}{81},$$

where both 25 and 81 are perfect squares. By definition, the positive square root of $\frac{25}{81}$ is the number (fraction) whose square is $\frac{25}{81}$, or $\frac{5}{9}$. That is,

$$\sqrt{\frac{25}{81}} = \frac{5}{9}.$$

The fact that $\sqrt{25} = 5$ and $\sqrt{81} = 9$ suggests the following procedure for finding the positive square root of a fraction.

SQUARE ROOTS OF FRACTIONS

If the variables a and b represent whole numbers with $b \neq 0$ and the fraction $\frac{a}{b}$ is in lowest terms, then

$$\sqrt{\frac{a}{b}} = \frac{\sqrt{a}}{\sqrt{b}}.$$

READING CHECK 3

- What is the square root of $\frac{36}{64}$ in lowest terms?

NOTE: If $\frac{a}{b}$ is not given in lowest terms, simplify it before applying this procedure. ∎

EXAMPLE 6 Finding the square root of a fraction

Find each square root.

(a) $\sqrt{\frac{9}{49}}$ (b) $\sqrt{\frac{45}{80}}$ (c) $\sqrt{\frac{75}{3}}$

Solution

(a) $\sqrt{\frac{9}{49}} = \frac{\sqrt{9}}{\sqrt{49}} = \frac{3}{7}$

(b) Because the fraction is not given in lowest terms, it should be simplified first.

$$\frac{45}{80} = \frac{9 \cdot 5}{16 \cdot 5} = \frac{9}{16}$$

CALCULATOR HELP

To find the square root of a fraction with a calculator, see Appendix E (page AP-31).

As a result,

$$\sqrt{\frac{45}{80}} = \sqrt{\frac{9}{16}} = \frac{\sqrt{9}}{\sqrt{16}} = \frac{3}{4}.$$

(c) Because $\frac{75}{3} = 25$, the given square root is evaluated as $\sqrt{\frac{75}{3}} = \sqrt{25} = 5$.

Now Try Exercises 65, 69, 71

4 Dividing Fractions and Rational Expressions

DIVIDING FRACTIONS In order to develop a strategy for dividing fractions, we must first understand the concept of a *reciprocal*. Two numbers are **reciprocals** or **multiplicative inverses** of each other if their product is 1. Several numbers and their reciprocals are listed in **TABLE 4.1**.

Numbers and Their Reciprocals

Number	$\frac{2}{3}$	4	$-\frac{5}{2}$	$\frac{1}{7}$	1
Reciprocal	$\frac{3}{2}$	$\frac{1}{4}$	$-\frac{2}{5}$	7	1

TABLE 4.1

NOTE: If neither a nor b is 0, the reciprocal of the fraction $\frac{a}{b}$ is $\frac{b}{a}$. Zero has no reciprocal. ■

🌐 *Math in Education Context* Many college degree programs are designed to be completed in 4 years. Since a semester is $\frac{1}{2}$ of a school year, we can use division to find the number of semesters needed to complete such a program.

$$4 \div \frac{1}{2}$$

Because every school year has 2 semesters, we can also find the number of semesters needed to complete the program, using multiplication.

$$4 \cdot 2$$

Since $4 \cdot 2 = 8$, it is reasonable to conclude that $4 \div \frac{1}{2} = 8$. This discussion suggests a strategy for division by a fraction—*multiply by its reciprocal*.

READING CHECK 4

Which of the following equations show a correct division problem? Select all that apply.

- $\frac{3}{4} \div \frac{3}{4} = 1$
- $\frac{1}{2} \div \frac{1}{4} = \frac{1}{8}$
- $\frac{3}{5} \div \frac{3}{7} = \frac{7}{5}$
- $\frac{10}{3} \div 2 = \frac{5}{3}$
- $5 \div \frac{3}{5} = 3$
- $\frac{14}{3} \div \frac{7}{2} = \frac{4}{3}$

> **DIVIDING FRACTIONS**
>
> If the variables a, b, c, and d represent numbers with $b \neq 0$, $c \neq 0$, and $d \neq 0$, then the quotient found when dividing $\frac{a}{b}$ by $\frac{c}{d}$ is given by
>
> $$\frac{a}{b} \div \frac{c}{d} = \frac{a}{b} \cdot \frac{d}{c}.$$
>
> That is, we multiply the first fraction by the reciprocal of the second fraction.

EXAMPLE 7 **Dividing fractions**

Divide.

(a) $\dfrac{4}{3} \div \dfrac{2}{9}$ (b) $-\dfrac{5}{6} \div \dfrac{5}{8}$ (c) $8 \div \dfrac{4}{3}$ (d) $\dfrac{4}{9} \div 2$

Solution

(a) Multiply the first fraction by the reciprocal of the second fraction.

$$\frac{4}{3} \div \frac{2}{9} = \frac{4}{3} \cdot \frac{9}{2} = \frac{4 \cdot 9}{3 \cdot 2} = \frac{(2 \cdot 2) \cdot (3 \cdot 3)}{3 \cdot 2} = \frac{2 \cdot 3}{1} = \frac{6}{1} = 6$$

(b) The quotient of two numbers with unlike signs is negative.

$$-\frac{5}{6} \div \frac{5}{8} = -\frac{5}{6} \cdot \frac{8}{5} = -\frac{5 \cdot 8}{6 \cdot 5} = -\frac{5 \cdot (2 \cdot 2 \cdot 2)}{(3 \cdot 2) \cdot 5} = -\frac{2 \cdot 2}{3} = -\frac{4}{3}$$

CALCULATOR HELP

To divide fractions with a calculator, see Appendix E (page AP-31).

(c) Start by writing 8 in fraction form as $\frac{8}{1}$.

$$\frac{8}{1} \div \frac{4}{3} = \frac{8}{1} \cdot \frac{3}{4} = \frac{8 \cdot 3}{1 \cdot 4} = \frac{(2 \cdot 2 \cdot 2) \cdot 3}{2 \cdot 2} = \frac{2 \cdot 3}{1} = \frac{6}{1} = 6$$

(d) Because $2 = \frac{2}{1}$, the reciprocal of 2 is $\frac{1}{2}$.

$$\frac{4}{9} \div 2 = \frac{4}{9} \cdot \frac{1}{2} = \frac{4 \cdot 1}{9 \cdot 2} = \frac{2 \cdot 2}{(3 \cdot 3) \cdot 2} = \frac{2}{3 \cdot 3} = \frac{2}{9}$$

Now Try Exercises 81, 83, 87, 91

MAKING CONNECTIONS 2

Simplifying Before Dividing Fractions

Do not simplify a division problem until *after* the problem has been changed to multiplication. For example, we **cannot** eliminate the **2**s from

$$\frac{2}{3} \div \frac{1}{2}.$$

We must first change the division problem to a multiplication problem.

$$\frac{2}{3} \div \frac{1}{2} = \frac{2}{3} \cdot \frac{2}{1} = \frac{2 \cdot 2}{3 \cdot 1} = \frac{4}{3}.$$

Eliminating the 2s before changing to multiplication would give an incorrect answer.

DIVIDING RATIONAL EXPRESSIONS Earlier in this section, we found the product of two rational expressions whose numerator and denominator each contained a single term. We can also find the quotient of two rational expressions of this type. In the next example, we divide rational expressions.

EXAMPLE 8 Dividing rational expressions

Divide.

(a) $\dfrac{x^2}{2y} \div \dfrac{8x}{5y}$ **(b)** $-\dfrac{4n}{3m} \div 2n$

Solution
(a) Multiply the first rational expression by the reciprocal of the second expression.

$$\frac{x^2}{2y} \div \frac{8x}{5y} = \frac{x^2}{2y} \cdot \frac{5y}{8x} = \frac{x^2 \cdot 5y}{2y \cdot 8x} = \frac{(x \cdot x) \cdot (5 \cdot y)}{(2 \cdot y) \cdot (2 \cdot 2 \cdot 2 \cdot x)} = \frac{x \cdot 5}{2 \cdot 2 \cdot 2 \cdot 2} = \frac{5x}{16}$$

(b) Because $2n = \frac{2n}{1}$, the reciprocal of $2n$ is $\frac{1}{2n}$.

$$-\frac{4n}{3m} \div 2n = -\frac{4n}{3m} \cdot \frac{1}{2n} = -\frac{4n \cdot 1}{3m \cdot 2n} = -\frac{2 \cdot 2 \cdot n}{(3 \cdot m) \cdot (2 \cdot n)} = -\frac{2}{3 \cdot m} = -\frac{2}{3m}$$

Now Try Exercises 97, 101

5 Solving Applications Involving Multiplication of Fractions

FINDING A FRACTION OF A WHOLE AND OF A FRACTION In applications involving fractions, we often encounter phrases such as "one-half **of** the total" or "three-fourths **of** the students." In this context, the word "of" indicates multiplication. For example, we can find two-thirds of three-fifths by multiplying $\frac{2}{3} \cdot \frac{3}{5}$. In the next two examples, we solve application problems involving fractions.

EXAMPLE 9 Finding a fraction of a class

In a class of 24 students, two-thirds of the students said that they had studied more than 2 hours for an upcoming exam. Determine the number of students from this class who studied more than 2 hours for the exam.

Solution
Two-thirds **of** 24 can be found by multiplying as follows.

$$\frac{2}{3} \cdot 24 = \frac{2}{3} \cdot \frac{24}{1} = \frac{2 \cdot 24}{3 \cdot 1} = \frac{2 \cdot (2 \cdot 2 \cdot 3)}{3} = \frac{2 \cdot 2 \cdot 2 \cdot 2}{1} = \frac{16}{1} = 16$$

Therefore, a total of 16 students studied more than 2 hours for the exam.

Now Try Exercise 115

EXAMPLE 10 Finding a fraction of a fraction

Suppose that one-half of the fish in a small pond are sunfish. If four-fifths of the sunfish in the pond are under 3 inches in length, what fraction of the fish in the pond are sunfish under 3 inches in length?

Solution
Four-fifths **of** one-half of the fish in the pond are sunfish under 3 inches in length.

$$\frac{4}{5} \cdot \frac{1}{2} = \frac{4 \cdot 1}{5 \cdot 2} = \frac{2 \cdot 2}{5 \cdot 2} = \frac{2}{5}$$

So, $\frac{2}{5}$ of the fish in the pond are sunfish under 3 inches in length.

Now Try Exercise 121

FINDING THE AREA OF A TRIANGLE Sometimes formulas contain fractions. Consider the triangle shown in **FIGURE 4.6**. A formula for its area can be discovered by creating a triangle of identical shape and size that can be turned and placed directly above the first triangle, as shown in **FIGURE 4.7** below.

FIGURE 4.7 is a rectangle with length b and width h. Its area is

$$A = bh.$$

Because the triangle from **FIGURE 4.6** makes up $\frac{1}{2}$ of the area of the rectangle in **FIGURE 4.7**, its area is

$$A = \frac{1}{2}bh.$$

In general, the height of a triangle whose base is positioned horizontally is the vertical distance from the base to the highest point in the triangle, as shown in **FIGURE 4.8**.

Base *b* and Height *h*

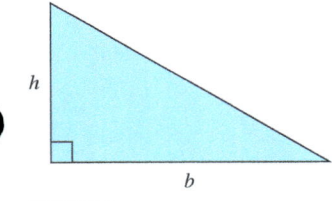

FIGURE 4.6

Length *b* and Width *h*

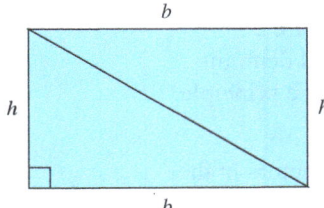

FIGURE 4.7

Heights of Various Triangles

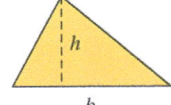 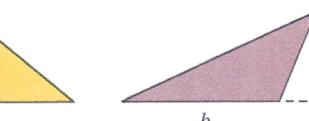

FIGURE 4.8

AREA OF A TRIANGLE

The area A of a triangle with base b and height h is given by the formula

$$A = \frac{1}{2}bh.$$

EXAMPLE 11 Finding the area of a triangle

Find the area of each triangle. Assume that the units are inches.

(a)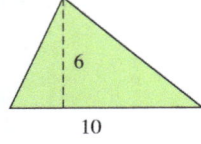

(b)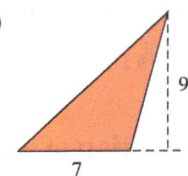

Solution
(a) Replace b with 10 and replace h with 6 in the formula $A = \frac{1}{2}bh$.

$$A = \frac{1}{2}(10)(6) = \frac{1}{2} \cdot 60 = 30$$

The area of the triangle is 30 square inches.

(b) Replace b with 7 and replace h with 9 in the formula $A = \frac{1}{2}bh$.

$$A = \frac{1}{2}(7)(9) = \frac{1}{2} \cdot 63 = \frac{63}{2}$$

The area of the triangle is $\frac{63}{2}$ square inches.

Now Try Exercises 123, 125

4.3 Putting It All Together

CONCEPT	COMMENTS	EXAMPLES
Multiplying Fractions and Rational Expressions	If a, b, c, and d represent numbers (or single terms) with $b \neq 0$ and $d \neq 0$, then $$\frac{a}{b} \cdot \frac{c}{d} = \frac{ac}{bd}.$$	$\frac{1}{4} \cdot \frac{3}{5} = \frac{1 \cdot 3}{4 \cdot 5} = \frac{3}{20}$ $\frac{2x}{y} \cdot \frac{3x}{5y} = \frac{2x \cdot 3x}{y \cdot 5y} = \frac{6x^2}{5y^2}$
Simplifying Before Multiplying Fractions	1. Write $\frac{a}{b} \cdot \frac{c}{d}$ as $\frac{a \cdot c}{b \cdot d}$. Do not multiply. 2. Replace each factor with its prime factorization. 3. Eliminate all factors that are common to the numerator and denominator by applying the basic principle of fractions. 4. Multiply the remaining factors in the numerator and the remaining factors in the denominator.	$\frac{4}{9} \cdot \frac{3}{10} = \frac{4 \cdot 3}{9 \cdot 10}$ Step 1 $= \frac{(2 \cdot 2) \cdot 3}{(3 \cdot 3) \cdot (2 \cdot 5)}$ Step 2 $= \frac{2}{3 \cdot 5}$ Step 3 $= \frac{2}{15}$ Step 4

CONCEPT	COMMENTS	EXAMPLES
Powers of Fractions	A natural number exponent on a fraction indicates how many times to multiply the fraction by itself.	$\left(\dfrac{2}{5}\right)^3 = \dfrac{2}{5} \cdot \dfrac{2}{5} \cdot \dfrac{2}{5} = \dfrac{8}{125}$
Square Roots of Fractions	If a and b represent whole numbers with $b \neq 0$ and the fraction $\dfrac{a}{b}$ is in lowest terms, then $\sqrt{\dfrac{a}{b}} = \dfrac{\sqrt{a}}{\sqrt{b}}.$	$\sqrt{\dfrac{16}{25}} = \dfrac{\sqrt{16}}{\sqrt{25}} = \dfrac{4}{5}$
Dividing Fractions and Rational Expressions	If a, b, c, and d represent numbers (or single terms) with $b \neq 0$, $c \neq 0$ and $d \neq 0$, then $\dfrac{a}{b} \div \dfrac{c}{d} = \dfrac{a}{b} \cdot \dfrac{d}{c}.$	$\dfrac{2}{3} \div \dfrac{3}{4} = \dfrac{2}{3} \cdot \dfrac{4}{3} = \dfrac{2 \cdot 4}{3 \cdot 3} = \dfrac{8}{9}$ $\dfrac{x}{5} \div \dfrac{3}{2y} = \dfrac{x}{5} \cdot \dfrac{2y}{3} = \dfrac{x \cdot 2y}{5 \cdot 3} = \dfrac{2xy}{15}$
Area of a Triangle	A triangle with base b and height h has area $A = \dfrac{1}{2}bh.$	$A = \dfrac{1}{2}(13)(6) = 39$ square units

4.3 Exercises   MyMathLab®

CONCEPTS AND VOCABULARY

1. When multiplying fractions, the numerator of the product is found by multiplying the _____ of the factors, and the denominator of the product is found by multiplying the _____ of the factors.

2. When finding a product of fractions, we can simplify before multiplying by applying the basic principle of _____ to eliminate factors that are common to the numerator and denominator.

3. A natural number exponent on a fraction tells us how many times to _____ the fraction by itself.

4. To find the square root of a fraction, find the square root of the _____ and _____.

5. Fractions $\frac{3}{4}$ and $\frac{4}{3}$ are _____ of each other.

6. In application problems involving fractions, the word "of" often indicates _____.

MULTIPLYING FRACTIONS

Exercises 7–38: Multiply. Simplify, if necessary.

7. $\dfrac{1}{4} \cdot \dfrac{3}{5}$

8. $\dfrac{2}{7} \cdot \dfrac{1}{3}$

9. $\dfrac{11}{9} \cdot \dfrac{2}{3}$

10. $\dfrac{7}{4} \cdot \dfrac{1}{8}$

11. $-\dfrac{2}{3} \cdot \dfrac{5}{3}$

12. $-\dfrac{11}{12} \cdot \dfrac{7}{4}$

13. $\dfrac{7}{5} \cdot \left(-\dfrac{9}{4}\right)$

14. $\dfrac{6}{11} \cdot \left(-\dfrac{8}{5}\right)$

15. $-\dfrac{1}{3} \cdot \left(-\dfrac{1}{8}\right)$

16. $-\dfrac{5}{2} \cdot \left(-\dfrac{7}{9}\right)$

17. $9 \cdot \dfrac{2}{3}$

18. $\dfrac{5}{6} \cdot 24$

19. $-6 \cdot \dfrac{3}{4}$

20. $\dfrac{7}{12} \cdot (-16)$

21. $\dfrac{3}{10} \cdot \dfrac{2}{3}$

22. $\dfrac{4}{5} \cdot \dfrac{5}{12}$

23. $-\dfrac{3}{4} \cdot \dfrac{16}{9}$

24. $-\dfrac{8}{9} \cdot \dfrac{3}{10}$

25. $\dfrac{14}{27} \cdot \dfrac{18}{49}$

26. $\dfrac{10}{9} \cdot \dfrac{3}{25}$

27. $\dfrac{5}{24} \cdot \dfrac{12}{25}$

28. $\dfrac{5}{28} \cdot \dfrac{14}{15}$

29. $-\dfrac{12}{5} \cdot \dfrac{5}{4}$

30. $\dfrac{24}{7} \cdot \dfrac{35}{12}$

31. $\dfrac{14}{39} \cdot \dfrac{26}{35}$

32. $-\dfrac{56}{65} \cdot \dfrac{15}{28}$

33. $\dfrac{27}{55} \cdot \left(-\dfrac{25}{18}\right)$

34. $\dfrac{25}{33} \cdot \left(-\dfrac{44}{75}\right)$

35. $-\dfrac{24}{45} \cdot \left(-\dfrac{55}{48}\right)$

36. $-\dfrac{34}{35} \cdot \left(-\dfrac{15}{17}\right)$

37. $\dfrac{5}{7} \cdot \dfrac{14}{15} \cdot \dfrac{3}{8}$

38. $\dfrac{11}{6} \cdot \dfrac{14}{27} \cdot \dfrac{54}{55}$

MULTIPLYING RATIONAL EXPRESSIONS

Exercises 39–56: Multiply the rational expressions. Simplify, if necessary.

39. $\dfrac{x}{2} \cdot \dfrac{5}{y}$

40. $\dfrac{3m}{5n} \cdot \dfrac{2}{7}$

41. $\dfrac{4p}{3q} \cdot \dfrac{2p}{5q}$

42. $\dfrac{x}{3y} \cdot \dfrac{7x}{y}$

43. $\dfrac{3a^2}{4b} \cdot \dfrac{2b^3}{a}$

44. $\dfrac{4z^3}{9y^2} \cdot \dfrac{3y}{2z^2}$

45. $\dfrac{12x^3}{5y^3} \cdot \dfrac{15y^2}{8x}$

46. $\dfrac{14w}{5x^2} \cdot \dfrac{2x}{21w^3}$

47. $-\dfrac{3u^2}{5v} \cdot \dfrac{10u}{9v}$

48. $\dfrac{4x}{5y} \cdot \left(-\dfrac{5y}{4x}\right)$

49. $-5x \cdot \dfrac{7y}{15x^2}$

50. $\dfrac{2a^2}{3b^2} \cdot (-6b)$

51. $\dfrac{3x}{14y} \cdot \dfrac{7y}{9x^2}$

52. $\dfrac{2a^2}{9b^2} \cdot \dfrac{3b}{10a^3}$

53. $-\dfrac{10m^2}{3n} \cdot \left(-\dfrac{6n}{5m^2}\right)$

54. $-\dfrac{2y^2}{5z} \cdot \left(-\dfrac{5y}{8z^2}\right)$

55. $\dfrac{3x}{4y} \cdot \dfrac{2x}{9y} \cdot \dfrac{6y}{5x}$

56. $\dfrac{7y}{5x} \cdot \dfrac{15}{8x} \cdot \dfrac{4x}{21y}$

EVALUATING POWERS AND SQUARE ROOTS OF FRACTIONS

Exercises 57–72: Evaluate the expression.

57. $\left(\dfrac{1}{4}\right)^2$

58. $\left(\dfrac{3}{2}\right)^3$

59. $\left(-\dfrac{3}{5}\right)^2$

60. $\left(-\dfrac{1}{3}\right)^3$

61. $\left(-\dfrac{3}{4}\right)^3$

62. $\left(\dfrac{2}{5}\right)^3$

63. $\left(\dfrac{8}{11}\right)^2$

64. $\left(-\dfrac{7}{9}\right)^2$

65. $\sqrt{\dfrac{9}{25}}$

66. $\sqrt{\dfrac{16}{81}}$

67. $\sqrt{\dfrac{1}{64}}$

68. $\sqrt{\dfrac{36}{121}}$

69. $\sqrt{\dfrac{80}{5}}$

70. $\sqrt{\dfrac{72}{2}}$

71. $\sqrt{\dfrac{20}{45}}$

72. $\sqrt{\dfrac{27}{48}}$

DIVIDING FRACTIONS AND RATIONAL EXPRESSIONS

Exercises 73–80: Write the reciprocal of the number.

73. $\dfrac{3}{5}$

74. $-\dfrac{7}{13}$

75. $-\dfrac{1}{12}$

76. $\dfrac{1}{9}$

77. 15

78. -8

79. 1

80. -1

Exercises 81–96: Divide. Simplify, if necessary.

81. $\dfrac{3}{4} \div \dfrac{9}{20}$

82. $-\dfrac{4}{7} \div \dfrac{16}{21}$

83. $-\dfrac{9}{2} \div \dfrac{27}{40}$

84. $\dfrac{6}{25} \div \dfrac{42}{15}$

85. $\dfrac{33}{8} \div \left(-\dfrac{55}{64}\right)$

86. $\dfrac{20}{27} \div \left(-\dfrac{40}{81}\right)$

87. $9 \div \dfrac{3}{2}$

88. $-10 \div \dfrac{5}{9}$

89. $-7 \div \dfrac{21}{5}$

90. $5 \div \dfrac{45}{4}$

91. $\dfrac{40}{9} \div 8$

92. $-\dfrac{24}{7} \div 6$

93. $-\dfrac{18}{11} \div 4$

94. $\dfrac{30}{7} \div 8$

95. $\dfrac{35}{72} \div \dfrac{65}{36}$

96. $\dfrac{80}{63} \div \dfrac{100}{81}$

Exercises 97–106: Divide. Simplify, if necessary.

97. $\dfrac{3x^2}{14y} \div \dfrac{3x}{7y^2}$

98. $\dfrac{3u}{5v} \div \dfrac{5u^3}{3v^2}$

99. $-\dfrac{6b^2}{5a^2} \div \dfrac{2b}{15a^2}$

100. $\dfrac{3m^2}{16n^2} \div \dfrac{9m^3}{2n^2}$

101. $-\dfrac{15p^2}{6q} \div 3p$

102. $\dfrac{3y}{10x} \div (-6y)$

103. $12x^3 \div \dfrac{4x^2}{3y^2}$

104. $18w^3 \div \dfrac{6w}{5}$

105. $-\dfrac{3m^2}{7n^2} \div \left(-\dfrac{3m^2}{7n^2}\right)$

106. $\dfrac{a^5}{b^3} \div \left(-\dfrac{a^4}{2b^3}\right)$

Exercises 107–112: Find the indicated fractional part.

107. $\dfrac{2}{3}$ of $\dfrac{5}{4}$

108. $\dfrac{1}{8}$ of $\dfrac{6}{7}$

109. $\dfrac{1}{5}$ of -15

110. $\dfrac{2}{9}$ of -36

111. $\dfrac{7}{6}$ of 20

112. $\dfrac{2}{35}$ of 50

APPLICATIONS INVOLVING MULTIPLICATION OF FRACTIONS

113. Wearable Devices In 2017, there were $\dfrac{22}{15}$ as many wristband devices (like the Fitbit) sold as there were in 2015. If there were 30 million wristbands sold in 2015, how many were sold in 2017? (*Source: Business Insider.*)

114. Wearable Devices In 2017, there were $\dfrac{33}{25}$ as many smartwatches sold as there were in 2016. If there were 50 million smartwatches sold in 2016, how many were sold in 2017? (*Source: Business Insider.*)

115. Climbing Kilimanjaro It is estimated that only $\dfrac{9}{20}$ of the people who attempt to climb Mount Kilimanjaro actually reach the summit. If 180 people attempt the climb, how many are expected to reach the summit? (*Source: Kilitrekker.com*)

116. Male Teachers About $\dfrac{7}{20}$ of teachers in American public secondary schools are male. If a city has 100 secondary teachers, how many are expected to be male? (*Source: National Education Association.*)

117. Family Budget A family spends $\dfrac{3}{20}$ of its monthly budget on utilities. If the family's monthly budget is $3200, how much is spent on the utility bill?

118. Pension Savings An employee has $\dfrac{2}{50}$ of his gross weekly pay withheld for his pension. How much is withheld for his pension if the employee's gross weekly pay is $1250?

119. Baking If a recipe for a batch of cookies calls for $\dfrac{2}{3}$ cup powdered sugar, how much powdered sugar is needed to make only $\dfrac{1}{2}$ of a batch?

120. Cleaning Solution If $\dfrac{3}{4}$ cup of concentrated detergent should be mixed with 1 gallon of water to make a cleaning solution, how much concentrated detergent should be mixed with $\dfrac{4}{5}$ gallon of water?

121. College Students In a particular school, two-thirds of all students are taking a biology class. Of the students taking biology, one-eighth are also taking an art class. What fraction of the students are taking both classes?

122. Horror Movies At a particular theater, one-twelfth of last year's films were horror movies. If two-thirds of the horror films had an R rating, what fraction of last year's films were R-rated horror movies?

Exercises 123–126: Use the formula $A = \dfrac{1}{2}bh$ to find the area of the triangle.

123.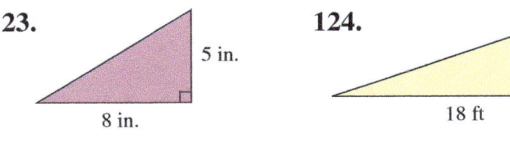

124.

125.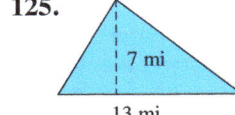

126.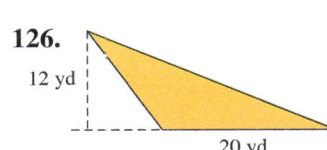

WRITING ABOUT MATHEMATICS

127. Explain the steps necessary to simplify a product of fractions before multiplying.

128. Explain why the number 0 has no reciprocal.

129. A student incorrectly changes a division problem to a multiplication problem by multiplying the reciprocal of the *first* fraction by the second fraction. How does this student's answer compare to the correct answer?

130. Does the commutative property hold when multiplying fractions? Does it hold when dividing fractions? Explain your reasoning and give examples.

4.4 Adding and Subtracting Fractions—Like Denominators

Objectives

1. Adding Fractions—Like Denominators
2. Subtracting Fractions—Like Denominators
3. Adding and Subtracting Rational Expressions
4. Solving Applications Involving Addition and Subtraction of Fractions—Like Denominators

NEW VOCABULARY

☐ Like fractions
☐ Common denominator

READING CHECK 1

- Add $\frac{2}{7} + \frac{1}{7}$.

1 Adding Fractions—Like Denominators

Connecting Concepts with Your Life In the United States, the coin with a value of 25 cents is called a *quarter* because 25 cents is one-quarter (or one-fourth) of a dollar. A person with 3 quarters has $\frac{3}{4}$ of a dollar. Just as a quarter names a fourth of a dollar, a dime names a tenth of a dollar and a nickel names a twentieth of a dollar. In mathematics, we can name the kind of fraction we have by looking at its denominator. For example, $\frac{2}{7}$ is a fraction given in *sevenths*, and $\frac{4}{9}$ is a fraction given in *ninths*. If two fractions have the same denominator, we say that they are **like fractions** and the fractions have a **common denominator**.

If a person with 5 quarters is given 6 more quarters, then the person has a total of 11 quarters. This statement translates to mathematics as

$$\frac{5}{4} + \frac{6}{4} = \frac{11}{4}.$$

Add the numerators and keep the common denominator.

Notice that the numerator of the resulting sum is obtained by adding the numerators of the addends, but the denominator of the resulting sum remains quarters (fourths). When two fractions with a *common denominator* are added, the resulting sum has the same denominator as the fractions being added.

ADDING FRACTIONS WITH LIKE DENOMINATORS

If the variables a, b, and d represent numbers with $d \neq 0$, then

$$\frac{a}{d} + \frac{b}{d} = \frac{a+b}{d}.$$

STUDY TIP

Do you want to know what material will be covered on your next exam? Often, the best place to look is on previously completed assignments and quizzes. If a topic is not discussed in class, is not found on the syllabus, and is not part of your assignments, then your time may be better spent studying other topics.

EXAMPLE 1 Adding fractions with like denominators

Add the fractions.

(a) $\frac{3}{7} + \frac{2}{7}$ (b) $\frac{4}{15} + \frac{4}{15}$ (c) $-\frac{8}{21} + \frac{19}{21}$

Solution

(a) $\frac{3}{7} + \frac{2}{7} = \frac{3+2}{7} = \frac{5}{7}$

(b) $\frac{4}{15} + \frac{4}{15} = \frac{4+4}{15} = \frac{8}{15}$

(c) Begin by writing $-\frac{8}{21}$ as $\frac{-8}{21}$.

$$-\frac{8}{21} + \frac{19}{21} = \frac{-8}{21} + \frac{19}{21} = \frac{-8+19}{21} = \frac{11}{21}$$

Now Try Exercises 7, 11, 13

When finding a sum of fractions, it is often necessary to simplify the resulting fraction to lowest terms. In the next example, we simplify sums of fractions.

EXAMPLE 2 **Adding fractions and simplifying**

Add the fractions. Simplify the result.

(a) $-\dfrac{5}{8} + \dfrac{7}{8}$ (b) $\dfrac{5}{12} + \dfrac{7}{12}$

Solution

(a) $-\dfrac{5}{8} + \dfrac{7}{8} = \dfrac{-5 + 7}{8} = \dfrac{2}{8} = \dfrac{2}{2 \cdot 2 \cdot 2} = \dfrac{1}{2 \cdot 2} = \dfrac{1}{4}$

(b) $\dfrac{5}{12} + \dfrac{7}{12} = \dfrac{5 + 7}{12} = \dfrac{12}{12} = 1$

Now Try Exercises 17, 23

EXAMPLE 3 **Estimating sums of fractions**

(a) Use estimation to determine which of the following sums is greater: $\dfrac{1}{4} + \dfrac{1}{5} + \dfrac{1}{6}$ or $\dfrac{1}{3} + \dfrac{1}{4} + \dfrac{1}{5}$.

(b) Use estimation to determine if the sum $\dfrac{2}{5} + \dfrac{1}{6} + \dfrac{2}{5}$ is greater than or less than 1.

Solution

(a) First, compare the two sums and notice that each sum shares two of its fractional addends, $\dfrac{1}{4}$ and $\dfrac{1}{5}$. Therefore, compare only the differing addends, $\dfrac{1}{6}$ and $\dfrac{1}{3}$, to determine which sum is greater. Because $\dfrac{1}{3} > \dfrac{1}{6}$, then $\dfrac{1}{3} + \dfrac{1}{4} + \dfrac{1}{5} > \dfrac{1}{4} + \dfrac{1}{5} + \dfrac{1}{6}$.

(b) Simplify the sum by adding like fractions. Thus $\dfrac{2}{5} + \dfrac{1}{6} + \dfrac{2}{5} = \dfrac{4}{5} + \dfrac{1}{6}$. Because $\dfrac{4}{5} + \dfrac{1}{5} = 1$ and $\dfrac{1}{6} < \dfrac{1}{5}$, this sum must be less than 1.

Now Try Exercises 25, 29

2 Subtracting Fractions—Like Denominators

 Connecting Concepts with Your Life If a person who has 7 dimes gives away 4 dimes, then the person is left with 3 dimes. Because a dime is one-tenth of a dollar, this statement translates to mathematics as

$$\dfrac{7}{10} - \dfrac{4}{10} = \dfrac{3}{10}.$$

The numerator of the resulting difference is found by subtracting the numerators of the two fractions. Notice that the difference is given in dimes (tenths). When fractions with a common denominator are subtracted, the resulting difference has the same denominator.

SUBTRACTING FRACTIONS WITH LIKE DENOMINATORS

If the variables a, b, and d, represent numbers with $d \neq 0$, then

$$\frac{a}{d} - \frac{b}{d} = \frac{a-b}{d}.$$

EXAMPLE 4 **Subtracting fractions with like denominators**

Subtract.

(a) $\dfrac{8}{11} - \dfrac{3}{11}$ (b) $\dfrac{4}{15} - \dfrac{8}{15}$ (c) $-\dfrac{5}{13} - \dfrac{4}{13}$

Solution

(a) $\dfrac{8}{11} - \dfrac{3}{11} = \dfrac{8-3}{11} = \dfrac{5}{11}$

(b) $\dfrac{4}{15} - \dfrac{8}{15} = \dfrac{4-8}{15} = \dfrac{-4}{15} = -\dfrac{4}{15}$

(c) Begin by writing $-\dfrac{5}{13}$ as $\dfrac{-5}{13}$.

$$-\dfrac{5}{13} - \dfrac{4}{13} = \dfrac{-5}{13} - \dfrac{4}{13} = \dfrac{-5-4}{13} = \dfrac{-9}{13} = -\dfrac{9}{13}$$

Now Try Exercises 33, 37, 39

When finding a difference of two fractions, it is often necessary to simplify the resulting fraction to lowest terms. In the next example, we simplify the resulting difference when one fraction is subtracted from another.

EXAMPLE 5 **Subtracting fractions and simplifying**

Subtract. Simplify the result.

(a) $\dfrac{9}{10} - \dfrac{3}{10}$ (b) $\dfrac{5}{18} - \dfrac{13}{18}$ (c) $\dfrac{5}{7} - \left(-\dfrac{2}{7}\right)$

Solution

(a) $\dfrac{9}{10} - \dfrac{3}{10} = \dfrac{9-3}{10} = \dfrac{6}{10} = \dfrac{2 \cdot 3}{2 \cdot 5} = \dfrac{3}{5}$

(b) $\dfrac{5}{18} - \dfrac{13}{18} = \dfrac{5-13}{18} = \dfrac{-8}{18} = -\dfrac{2 \cdot 2 \cdot 2}{2 \cdot 3 \cdot 3} = -\dfrac{2 \cdot 2}{3 \cdot 3} = -\dfrac{4}{9}$

(c) $\dfrac{5}{7} - \left(-\dfrac{2}{7}\right) = \dfrac{5}{7} - \dfrac{-2}{7} = \dfrac{5-(-2)}{7} = \dfrac{5+2}{7} = \dfrac{7}{7} = 1$

Now Try Exercises 41, 47, 49

READING CHECK 2

Which of the following equations show either a correct addition or subtraction problem? Select all that apply.

- $\frac{1}{5} + \frac{2}{5} = \frac{3}{5}$
- $\frac{6}{7} + \frac{1}{7} = 1$
- $\frac{3}{5} - \frac{3}{5} = 0$
- $\frac{10}{21} - \frac{13}{21} = -\frac{1}{7}$
- $\frac{4}{5} - \frac{2}{5} = \frac{2}{5}$

3 Adding and Subtracting Rational Expressions

The process used to add or subtract rational expressions with like denominators is similar to that used for adding and subtracting fractions with like denominators. We add or subtract the numerators and keep the common denominator. The next two examples illustrate how rational expressions are added and subtracted.

EXAMPLE 6 Adding and subtracting rational expressions

Add or subtract as indicated.

(a) $\dfrac{3x}{5y} + \dfrac{x}{5y}$ (b) $\dfrac{4}{7x^2} - \dfrac{6}{7x^2}$ (c) $\dfrac{2x}{3m} + \left(-\dfrac{4y}{3m}\right)$

Solution

(a) $\dfrac{3x}{5y} + \dfrac{x}{5y} = \dfrac{3x + x}{5y} = \dfrac{4x}{5y}$

(b) $\dfrac{4}{7x^2} - \dfrac{6}{7x^2} = \dfrac{4 - 6}{7x^2} = \dfrac{-2}{7x^2} = -\dfrac{2}{7x^2}$

(c) Begin by writing $-\dfrac{4y}{3m}$ as $\dfrac{-4y}{3m}$.

$$\dfrac{2x}{3m} + \left(-\dfrac{4y}{3m}\right) = \dfrac{2x}{3m} + \dfrac{-4y}{3m} = \dfrac{2x + (-4y)}{3m} = \dfrac{2x - 4y}{3m}$$

Note that the terms $2x$ and $4y$ cannot be combined.

Now Try Exercises 63, 65, 67

EXAMPLE 7 Adding, subtracting, and simplifying rational expressions

Add or subtract as indicated. Simplify the result.

(a) $\dfrac{2y}{6x^3} + \dfrac{y}{6x^3}$ (b) $\dfrac{5a^2}{24b} - \dfrac{7a^2}{24b}$

Solution

(a) $\dfrac{2y}{6x^3} + \dfrac{y}{6x^3} = \dfrac{2y + y}{6x^3} = \dfrac{3y}{6x^3} = \dfrac{3 \cdot y}{2 \cdot 3 \cdot x \cdot x \cdot x} = \dfrac{y}{2 \cdot x \cdot x \cdot x} = \dfrac{y}{2x^3}$

(b) $\dfrac{5a^2}{24b} - \dfrac{7a^2}{24b} = \dfrac{5a^2 - 7a^2}{24b} = \dfrac{-2a^2}{24b} = -\dfrac{2 \cdot a \cdot a}{2 \cdot 2 \cdot 2 \cdot 3 \cdot b} = -\dfrac{a \cdot a}{2 \cdot 2 \cdot 3 \cdot b} = -\dfrac{a^2}{12b}$

Now Try Exercises 71, 73

4 Solving Applications Involving Addition and Subtraction of Fractions—Like Denominators

The next two examples illustrate how we may need to add or subtract fractions in applications involving real-world data.

EXAMPLE 8 Analyzing population data

In 1960, about $\dfrac{17}{20}$ of the U.S. population identified itself as non-Hispanic white. By 2060, it is projected that $\dfrac{9}{20}$ of the population will be non-Hispanic white. Subtract the 2060 value from the 1960 value to find the difference in the fractional portion of the population that identifies itself as non-Hispanic white. (*Source:* Pew Research Center.)

Solution

$$\dfrac{17}{20} - \dfrac{9}{20} = \dfrac{17 - 9}{20} = \dfrac{8}{20} = \dfrac{2 \cdot 2 \cdot 2}{2 \cdot 2 \cdot 5} = \dfrac{2}{5}$$

Over this 100-year period, the fractional portion of the U.S. population that identifies itself as non-Hispanic white will decrease by $\dfrac{2}{5}$.

Now Try Exercise 79

EXAMPLE 9 Constructing trim for cabinetry

If a cabinet maker glues a piece of trim that is $\frac{5}{16}$ inch thick to another piece that is $\frac{7}{16}$ inch thick to make a single, thicker piece of trim, how thick is the newly formed trim piece?

Solution
To find the total thickness, we add $\frac{5}{16}$ and $\frac{7}{16}$.

$$\frac{5}{16} + \frac{7}{16} = \frac{5+7}{16} = \frac{12}{16} = \frac{2 \cdot 2 \cdot 3}{2 \cdot 2 \cdot 2 \cdot 2} = \frac{3}{2 \cdot 2} = \frac{3}{4}$$

The new trim piece is $\frac{3}{4}$ inch thick.

Now Try Exercise 83

4.4 Putting It All Together

CONCEPT	COMMENTS	EXAMPLES
Adding Fractions with Like Denominators	If the variables a, b, and d represent numbers with $d \neq 0$, then $$\frac{a}{d} + \frac{b}{d} = \frac{a+b}{d}.$$	$\frac{3}{11} + \frac{4}{11} = \frac{3+4}{11} = \frac{7}{11}$ $\frac{5}{12} + \frac{3}{12} = \frac{5+3}{12} = \frac{8}{12} = \frac{2}{3}$
Subtracting Fractions with Like Denominators	If the variables a, b, and d represent numbers with $d \neq 0$, then $$\frac{a}{d} - \frac{b}{d} = \frac{a-b}{d}.$$	$\frac{7}{9} - \frac{5}{9} = \frac{7-5}{9} = \frac{2}{9}$ $\frac{7}{10} - \frac{3}{10} = \frac{7-3}{10} = \frac{4}{10} = \frac{2}{5}$
Adding and Subtracting Rational Expressions	The process used to add or subtract rational expressions with like denominators is similar to that used for adding and subtracting fractions.	$\frac{2x}{3y} + \frac{5x}{3y} = \frac{2x+5x}{3y} = \frac{7x}{3y}$ $\frac{7a}{12b^2} - \frac{11a}{12b^2} = \frac{7a-11a}{12b^2} = \frac{-4a}{12b^2}$ $= -\frac{a}{3b^2}$

4.4 Exercises   MyMathLab

CONCEPTS AND VOCABULARY

1. Two fractions with the same denominator are called _____ fractions and we say that the two fractions have a(n) _____ denominator.

2. When two fractions with a common denominator are added, the numerator of the sum is found by adding the _____ of the two fractions.

3. When adding two fractions with a common denominator, the sum has the same _____ as the two fractions being added.

4. When finding the difference of two fractions with a common denominator, the numerator of the difference is found by subtracting the _____ of the two fractions.

5. When finding the difference of two fractions with a common denominator, the difference has the same _____ as the two fractions.

6. The process used to add or subtract rational expressions with like denominators is similar to that used for adding and subtracting _____.

ADDING LIKE FRACTIONS

Exercises 7–24: Add. Simplify, if necessary.

7. $\dfrac{1}{5} + \dfrac{2}{5}$

8. $\dfrac{2}{7} + \dfrac{4}{7}$

9. $\dfrac{2}{9} + \left(-\dfrac{7}{9}\right)$

10. $\dfrac{3}{11} + \left(-\dfrac{10}{11}\right)$

11. $\dfrac{12}{25} + \dfrac{12}{25}$

12. $\dfrac{9}{17} + \dfrac{10}{17}$

13. $-\dfrac{5}{13} + \dfrac{15}{13}$

14. $-\dfrac{14}{19} + \dfrac{7}{19}$

15. $\dfrac{1}{10} + \dfrac{7}{10}$

16. $\dfrac{1}{12} + \dfrac{11}{12}$

17. $\dfrac{8}{9} + \left(-\dfrac{2}{9}\right)$

18. $-\dfrac{8}{15} + \dfrac{13}{15}$

19. $-\dfrac{7}{5} + \left(-\dfrac{3}{5}\right)$

20. $\dfrac{11}{6} + \left(-\dfrac{5}{6}\right)$

21. $-\dfrac{17}{20} + \dfrac{17}{20}$

22. $-\dfrac{3}{8} + \dfrac{3}{8}$

23. $\dfrac{5}{14} + \dfrac{9}{14}$

24. $-\dfrac{5}{12} + \left(-\dfrac{7}{12}\right)$

ESTIMATING SUMS WITH FRACTIONS

Exercises 25–28: Use estimation to determine which sum is greater.

25. $\dfrac{1}{2} + \dfrac{1}{3} + \dfrac{1}{4}$ or $\dfrac{1}{3} + \dfrac{1}{4} + \dfrac{1}{5}$

26. $\dfrac{1}{5} + \dfrac{1}{6} + \dfrac{1}{7}$ or $\dfrac{1}{4} + \dfrac{1}{5} + \dfrac{1}{6}$

27. $\dfrac{1}{3} + \dfrac{1}{3} + \dfrac{1}{10}$ or $\dfrac{1}{10} + \dfrac{1}{2} + \dfrac{1}{2}$

28. $\dfrac{1}{9} + \dfrac{1}{4} + \dfrac{1}{5}$ or $\dfrac{1}{5} + \dfrac{1}{9} + \dfrac{1}{5}$

Exercises 29–32: Use estimation to determine whether the given sum is greater than or less than 1.

29. $\dfrac{1}{3} + \dfrac{1}{3} + \dfrac{1}{4}$

30. $\dfrac{1}{4} + \dfrac{1}{4} + \dfrac{1}{3}$

31. $\dfrac{1}{2} + \dfrac{1}{3} + \dfrac{1}{3}$

32. $\dfrac{1}{4} + \dfrac{1}{4} + \dfrac{1}{4} + \dfrac{1}{3}$

SUBTRACTING LIKE FRACTION

Exercises 33–50: Subtract. Simplify, if necessary.

33. $\dfrac{6}{7} - \dfrac{4}{7}$

34. $\dfrac{4}{5} - \dfrac{1}{5}$

35. $\dfrac{7}{9} - \left(-\dfrac{1}{9}\right)$

36. $-\dfrac{3}{15} - \dfrac{8}{15}$

37. $\dfrac{3}{17} - \dfrac{15}{17}$

38. $\dfrac{19}{21} - \dfrac{11}{21}$

39. $-\dfrac{2}{11} - \dfrac{3}{11}$

40. $-\dfrac{10}{19} - \dfrac{3}{19}$

41. $\dfrac{5}{8} - \dfrac{3}{8}$

42. $\dfrac{19}{24} - \dfrac{7}{24}$

43. $-\dfrac{4}{9} - \left(-\dfrac{4}{9}\right)$

44. $-\dfrac{1}{12} - \dfrac{11}{12}$

45. $\dfrac{1}{10} - \dfrac{3}{10}$

46. $\dfrac{11}{3} - \left(-\dfrac{4}{3}\right)$

47. $-\dfrac{1}{18} - \dfrac{5}{18}$

48. $\dfrac{5}{27} - \dfrac{5}{27}$

49. $\dfrac{11}{16} - \left(-\dfrac{5}{16}\right)$

50. $-\dfrac{5}{14} - \left(-\dfrac{7}{14}\right)$

ADDING AND SUBTRACTING LIKE FRACTION

Exercises 51–62: Add or subtract as indicated. Simplify, if necessary.

51. $\dfrac{22}{27} - \dfrac{13}{27}$

52. $-\dfrac{3}{50} + \dfrac{39}{50}$

53. $\dfrac{7}{36} + \left(-\dfrac{25}{36}\right)$

54. $-\dfrac{13}{45} - \dfrac{8}{45}$

55. $\dfrac{7}{30} - \left(-\dfrac{13}{30}\right)$

56. $\dfrac{28}{75} + \dfrac{8}{75}$

57. $-\dfrac{7}{60} + \dfrac{11}{60}$

58. $-\dfrac{25}{48} - \left(-\dfrac{7}{48}\right)$

59. $\frac{13}{120} - \frac{77}{120}$

60. $-\frac{21}{80} + \left(-\frac{63}{80}\right)$

61. $-\frac{38}{63} + \left(-\frac{25}{63}\right)$

62. $-\frac{77}{90} - \left(-\frac{77}{90}\right)$

ADDING AND SUBTRACTING RATIONAL EXPRESSIONS

Exercises 63–76: Add or subtract as indicated. Simplify, if necessary.

63. $\frac{5x}{9y} + \frac{2x}{9y}$

64. $-\frac{3a}{b} + \frac{5a}{b}$

65. $\frac{8}{5m^2} + \left(-\frac{12}{5m^2}\right)$

66. $-\frac{3p^2}{5q} - \frac{p^2}{5q}$

67. $\frac{7x}{y} - \frac{3w}{y}$

68. $-\frac{8a}{7} + \left(-\frac{4a}{7}\right)$

69. $\frac{2}{3d} - \left(-\frac{1}{3d}\right)$

70. $\frac{3m^2}{10n^3} + \frac{2m^2}{10n^3}$

71. $\frac{4y}{3x^2} + \frac{8y}{3x^2}$

72. $-\frac{3y}{4w^2} - \frac{5y}{4w^2}$

73. $\frac{2k^2}{15c} - \frac{7k^2}{15c}$

74. $-\frac{3x}{2} - \left(-\frac{7x}{2}\right)$

75. $\frac{13x^2}{6} + \frac{5x^2}{6}$

76. $-\frac{4}{x} + \left(-\frac{4}{x}\right)$

APPLICATIONS INVOLVING ADDITION AND SUBTRACTION OF FRACTIONS

77. Carpentry A board that is $\frac{15}{16}$ inch thick is run through a surfacing machine to reduce its thickness by $\frac{3}{16}$ inch. What is the thickness of the board after this process?

78. Cutting Hair A barber cuts hair that is $\frac{7}{2}$ inches long to a length of $\frac{1}{2}$ inch. How many inches of hair are removed?

79. International Travel About $\frac{3}{250}$ of international travelers entering the United States are from Asia and about $\frac{1}{250}$ are from Africa. Subtract the African value from the Asian value to find the difference. (*Source:* U.S. Department of Commerce.)

80. Carbon Emissions The United States accounted for about $\frac{23}{100}$ of worldwide carbon dioxide emissions from the consumption of fossil fuels in 1990. By 2015, this fraction had dropped to $\frac{16}{100}$. Find the difference. (*Source:* U.S. Energy Information Administration.)

81. Nursing Home Residents About $\frac{19}{5000}$ of all Americans are female nursing home residents and about $\frac{7}{5000}$ of all Americans are male nursing home residents. Find the total fraction of the American population residing in nursing homes. (*Source:* Centers for Disease Control and Prevention.)

82. Classroom Population In a large college class, $\frac{57}{300}$ of the students are female and under the age of 24, while $\frac{111}{300}$ of the students are male and under the age of 24. Find the total fraction of students in this class that are under the age of 24.

83. Largest U.S. Cities Texas is home to $\frac{6}{25}$ of the 25 largest U.S. cities, and California is home to $\frac{4}{25}$ of the 25 largest U.S. cities. What fraction of the 25 largest U.S. cities could be found in these two states together? (*Source:* U.S. Census Bureau.)

84. Largest World Cities India and China each have $\frac{3}{20}$ of the world's 20 largest cities. What fraction of the world's 20 largest cities can be found in these two countries together? (*Source:* The World Atlas.)

85. Going Green A company has a goal of replacing $\frac{47}{50}$ of its incandescent light bulbs with energy-efficient LED bulbs. If the company has already replaced $\frac{37}{50}$ of its light bulbs, what fraction of the incandescent bulbs still need to be replaced?

86. Going Organic A farmer plans to convert $\frac{7}{24}$ of his crop land to organic crops. If the farmer has already converted $\frac{5}{24}$ of his crop land, what fraction of his land has yet to be converted?

87. Scholarships A university foundation has set aside $\frac{11}{30}$ of its budget to award scholarships for students with high academic achievement and $\frac{7}{30}$ of its budget to award scholarships for low-income students. What fraction of the foundation's total budget is used for these two groups?

88. Fuel Consumption At the beginning of a trip, a car's gas gauge shows that the car has $\frac{3}{4}$ of a tank of gas. At the end of the trip, the gauge indicates that the car has $\frac{1}{4}$ of a tank of gas. What fraction of the tank was used during the trip?

90. A student says that two fractions can be added by adding the numerators *and* adding the denominators. Give a real-world example that would disprove this student's method.

WRITING ABOUT MATHEMATICS

89. Explain how to determine if a sum or difference of fractions needs to be simplified.

SECTIONS 4.3 and 4.4 — Checking Basic Concepts

1. Multiply or divide as indicated. If necessary, simplify your results.
   (a) $\frac{5}{3} \cdot \frac{12}{15}$
   (b) $-\frac{7}{2} \cdot \frac{10}{7}$
   (c) $\frac{14}{25} \div \frac{7}{10}$
   (d) $\frac{2}{7} \div \left(-\frac{8}{21}\right)$
   (e) $\frac{2x^2}{9y^2} \cdot \frac{3y}{8x^2}$
   (f) $-\frac{15m}{4n} \div \frac{5m}{4n}$
   (g) $-12a \div \frac{6}{5a}$
   (h) $\frac{3x}{8y^2} \cdot 4y$

2. Evaluate the expression.
   (a) $\left(\frac{2}{5}\right)^2$
   (b) $\sqrt{\frac{50}{2}}$

3. Add or subtract as indicated. If necessary, simplify your results.
   (a) $-\frac{5}{19} + \frac{17}{19}$
   (b) $\frac{4}{9} + \left(-\frac{1}{9}\right)$
   (c) $\frac{13}{21} - \frac{7}{21}$
   (d) $-\frac{1}{14} - \frac{13}{14}$
   (e) $\frac{15x^2}{4} + \frac{x^2}{4}$
   (f) $\frac{5m^2}{8n^3} - \frac{3m^2}{8n^3}$

4. **Female Teachers** About $\frac{13}{20}$ of teachers in public secondary schools are female. If a city has 60 secondary teachers, how many are expected to be female? (*Source:* National Education Association.)

4.5 Adding and Subtracting Fractions—Unlike Denominators

Objectives

1. Finding the Least Common Multiple (LCM)
   • Using the Listing Method
   • Using the Prime Factorization Method
2. Finding the Least Common Denominator (LCD)
3. Adding and Subtracting Fractions—Unlike Denominators
4. Solving Applications Involving Addition and Subtraction of Fractions—Unlike Denominators

1 Finding the Least Common Multiple (LCM)

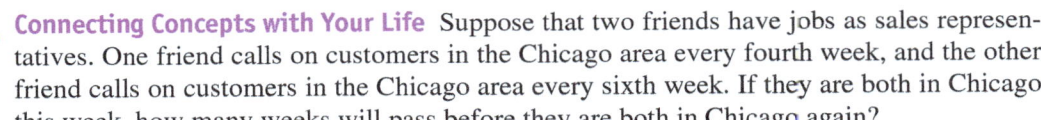 **Connecting Concepts with Your Life** In the previous section, we discussed how different coins represent different fractions of a dollar. Our everyday experience tells us that a person with three dimes and four nickels has $\frac{1}{2}$ of a dollar. Because a dime represents $\frac{1}{10}$ of a dollar and a nickel represents $\frac{1}{20}$ of a dollar, the following sum illustrates that three dimes plus four nickels equals $\frac{1}{2}$ of a dollar.

$$\frac{3}{10} + \frac{4}{20} = \frac{1}{2}$$

The addends in this equation do *not* have a common denominator. Before we can add fractions with unlike denominators, we need to understand least common multiples and least common denominators.

Connecting Concepts with Your Life Suppose that two friends have jobs as sales representatives. One friend calls on customers in the Chicago area every fourth week, and the other friend calls on customers in the Chicago area every sixth week. If they are both in Chicago this week, how many weeks will pass before they are both in Chicago again?

NEW VOCABULARY

☐ Common multiples
☐ Least common multiple
☐ Listing method
☐ Prime factorization method
☐ Least common denominator

We can answer this question by listing the weeks that each person is in Chicago.

First Person: 4, 8, 12, 16, 20, 24, 28, 32,…
Second Person: 6, 12, 18, 24, 30, 36, 42,…

After 12 weeks, the two friends will be in Chicago at the same time. The next time is after 24 weeks. The numbers 12 and 24 are two **common multiples** of 4 and 6. However, 12 is the *least common multiple* (LCM) of 4 and 6. The **least common multiple** of two or more numbers is the smallest number that is divisible by each of the given numbers.

STUDY TIP

Even if we know exactly how to do a math problem correctly, a simple computational error will often cause us to get an incorrect answer. In the above example, if we had listed the multiples of 4 and 6 incorrectly, it is unlikely that we would find the correct LCM. Be sure to take your time on simple calculations.

READING CHECK 1

• Use the listing method to find the LCM of 6 and 9.

USING THE LISTING METHOD The method used above to find the LCM of 4 and 6 is called the **listing method** because multiples of each number are listed and the LCM is chosen from the lists. In the next example, we use the listing method to find the LCM of two numbers.

EXAMPLE 1 Finding the LCM using the listing method

Find the least common multiple of 8 and 10.

Solution

List the multiples of each number by multiplying each number by 1, 2, 3, etc.

Multiples of 8:   8, 16, 24, 32, 40, 48, 56, 64,…
Multiples of 10: 10, 20, 30, 40, 50, 60, 70,…

The LCM is the smallest number that is common to both lists. The LCM of 8 and 10 is 40.

Now Try Exercise 9

USING THE PRIME FACTORIZATION METHOD The listing method is often convenient for small numbers and could be done mentally, but it is not practical for finding the LCM in every case. For example, to find the LCM of 16 and 27 using the listing method, the two lists would contain at least 43 values. A second method for finding the LCM of two or more numbers is called the **prime factorization method**. The steps for this method are summarized as follows.

FINDING THE LCM USING PRIME FACTORIZATION

To find the least common multiple (LCM) of two or more numbers, do the following.

STEP 1: Find the prime factorization of each number.

STEP 2: List each factor that appears in one or more of the factorizations. If a factor is repeated in any of the factorizations, list this factor the maximum number of times that it is repeated in any one of the factorizations.

STEP 3: Find the product of this list of factors. The result is the LCM.

READING CHECK 2

• Use the prime factorization method to find the LCM of 6 and 9.

EXAMPLE 2 Finding the LCM of numbers using the prime factorization method

Find the LCM of the given numbers.
(a) 12 and 18   (b) 4, 6, and 10

Solution
(a) STEP 1: Find the prime factorizations of 12 and 18.

$$12 = 2 \cdot 2 \cdot 3$$
$$18 = 2 \cdot 3 \cdot 3$$

STEP 2: List the factors: 2, 2, 3, 3. Because 2 appears twice in one factorization and one time in the other, we list it twice, which is the maximum number of times that it is repeated in any *one* of the factorizations. For the same reason, 3 is listed twice.

STEP 3: The product of this list is $2 \cdot 2 \cdot 3 \cdot 3 = 36$, so the LCM of 12 and 18 is 36.

(b) STEP 1: Find the prime factorizations of 4, 6, and 10.

$$4 = 2 \cdot 2$$
$$6 = 2 \cdot 3$$
$$10 = 2 \cdot 5$$

STEP 2: List the factors: 2, 2, 3, 5.
STEP 3: The LCM of 4, 6, and 10 is $2 \cdot 2 \cdot 3 \cdot 5 = 60$.

Now Try Exercises 13, 17

NOTE: Because the LCM of two or more numbers is simply the smallest number that is divisible by each of the given numbers, the LCM of 4, 6, and 10 in Example 2(b) is 60 because 60 is the smallest number that is divisible by 4, 6, and 10. ∎

The steps used to find the LCM of two or more numbers can also be used to find the LCM of two or more variable expressions of a single term. In the next example, we find the LCM of two such expressions.

EXAMPLE 3 Finding the LCM of expressions using the factorization method

Find the LCM for the given variable expressions.
(a) $3x^2$ and $4x^3$ (b) $2ab^2$ and $5a^2b$

Solution
(a) STEP 1: Find the complete factorizations of $3x^2$ and $4x^3$.

$$3x^2 = 3 \cdot x \cdot x$$
$$4x^3 = 2 \cdot 2 \cdot x \cdot x \cdot x$$

STEP 2: List the factors: 2, 2, 3, x, x, x.
STEP 3: The LCM of $3x^2$ and $4x^3$ is $2 \cdot 2 \cdot 3 \cdot x \cdot x \cdot x = 12x^3$.

(b) STEP 1: Find the complete factorizations of $2ab^2$ and $5a^2b$.

$$2ab^2 = 2 \cdot a \cdot b \cdot b$$
$$5a^2b = 5 \cdot a \cdot a \cdot b$$

STEP 2: List the factors: 2, 5, a, a, b, b.
STEP 3: The LCM of $2ab^2$ and $5a^2b$ is $2 \cdot 5 \cdot a \cdot a \cdot b \cdot b = 10a^2b^2$.

Now Try Exercises 27, 29

READING CHECK 3

- What is the LCD for $\frac{4}{9}$ and $\frac{5}{12}$?

2 Finding the Least Common Denominator (LCD)

In this section, we will add and subtract fractions with unlike denominators. To do this, we need to write the fractions as equivalent fractions that have a common denominator. Even though *any* common denominator can be used for this purpose, it is often best to find the *least* common denominator (LCD). The **least common denominator** for a list of fractions is the *least common multiple* of the denominators.

EXAMPLE 4 **Finding the LCD for a list of fractions**

Find the LCD for the given fractions.

(a) $\frac{7}{16}$ and $\frac{5}{6}$ (b) $\frac{2}{5x^2}$ and $\frac{3}{4xy}$

Solution

(a) The LCD for the fractions $\frac{7}{16}$ and $\frac{5}{6}$ is the LCM of the denominators 16 and 6. So, we use the three-step process for finding the LCM.

STEP 1: Find the prime factorizations of 16 and 6.

$$16 = 2 \cdot 2 \cdot 2 \cdot 2$$
$$6 = 2 \cdot 3$$

STEP 2: List the factors: **2, 2, 2, 2, 3**.
STEP 3: The LCD is $2 \cdot 2 \cdot 2 \cdot 2 \cdot 3 = 48$.

(b) To find the LCD for $\frac{2}{5x^2}$ and $\frac{3}{4xy}$, we look for the LCM of the denominators $5x^2$ and $4xy$.

STEP 1: Find the complete factorizations of $5x^2$ and $4xy$.

$$5x^2 = 5 \cdot x \cdot x$$
$$4xy = 2 \cdot 2 \cdot x \cdot y$$

STEP 2: List the factors: **2, 2, 5,** x, x, y.
STEP 3: The LCD is $2 \cdot 2 \cdot 5 \cdot x \cdot x \cdot y = 20x^2y$.

Now Try Exercises 35, 41

3 Adding and Subtracting Fractions—Unlike Denominators

When we are faced with a new mathematical problem with no understanding of how to find the solution, it is often helpful to look at similar problems that we *can* solve and then try to relate the new problem to these previously solved problems. For example, we could add and subtract fractions with *unlike* denominators if we could rewrite the fractions as equivalent fractions with *like* denominators. In fact, this is exactly what we do when adding and subtracting fractions with unlike denominators. The entire process is summarized as follows.

READING CHECK 4

Which of the following equations show two fractions with unlike denominators being added correctly? Select all that apply.

- $\frac{1}{2} + \frac{1}{4} = \frac{3}{4}$
- $\frac{2}{3} + \frac{1}{5} = \frac{3}{8}$
- $\frac{1}{4} + \frac{2}{3} = \frac{11}{12}$
- $\frac{1}{3} + \frac{3}{8} = \frac{17}{24}$

ADDING OR SUBTRACTING FRACTIONS WITH UNLIKE DENOMINATORS

To add or subtract fractions with unlike denominators, do the following.

STEP 1: Determine the LCD for all fractions involved.

STEP 2: Write each fraction as an equivalent fraction with the LCD as its denominator.

STEP 3: Add or subtract the newly written *like* fractions and simplify the result.

In the next example, we will practice the important concept in Step 2.

EXAMPLE 5 Writing equivalent fractions using the LCD

Rewrite the fractions as equivalent fractions with the given LCD.
(a) $\frac{1}{6}$ and $\frac{4}{15}$; LCD: 30   (b) $\frac{5}{6}, \frac{1}{4}$, and $\frac{3}{8}$; LCD: 24

Solution
(a) Because the denominator of $\frac{1}{6}$ is 6, we can obtain a denominator of 30 by multiplying $6 \times 5 = 30$. So, we multiply both the numerator and denominator of $\frac{1}{6}$ by **5**.

$$\frac{1}{6} \xrightarrow{\times 5} \frac{?}{30}$$

Since $1 \times 5 = 5$, the unknown numerator is 5. That is, $\frac{1}{6}$ is equivalent to $\frac{5}{30}$.

Because the denominator of $\frac{4}{15}$ is 15, we obtain a denominator of 30 by multiplying $15 \times 2 = 30$. So, we multiply both the numerator and denominator of $\frac{4}{15}$ by **2**.

$$\frac{4}{15} \xrightarrow{\times 2} \frac{?}{30}$$

Since $4 \times 2 = 8$, the unknown numerator is 8. That is, $\frac{4}{15}$ is equivalent to $\frac{8}{30}$.

(b) A convenient and more concise way to write equivalent fractions is shown below. Note that we obtain a denominator of 24 by multiplying the denominator of $\frac{5}{6}$ by **4**, multiplying the denominator of $\frac{1}{4}$ by **6**, and multiplying the denominator of $\frac{3}{8}$ by **3**.

$$\frac{5}{6} \cdot \frac{4}{4} = \frac{20}{24} \qquad \frac{1}{4} \cdot \frac{6}{6} = \frac{6}{24} \qquad \frac{3}{8} \cdot \frac{3}{3} = \frac{9}{24}$$

Now Try Exercises 49, 51

NOTE: The fractions in Example 5(b) were each multiplied by 1, which was written in a form that helped us obtain the desired common denominator. ■

In Example 6, we use the entire three-step process shown above to add or subtract fractions with unlike denominators. Then, Example 7 illustrates how the same three-step process can be written more concisely.

EXAMPLE 6 Adding or subtracting fractions with unlike denominators

Add or subtract as indicated.
(a) $\frac{5}{8} - \frac{2}{3}$ (b) $\frac{2}{3} + \frac{7}{12} + \frac{3}{4}$

Solution
(a) **STEP 1:** Find the LCD for $\frac{5}{8}$ and $\frac{2}{3}$.

$$8 = 2 \cdot 2 \cdot 2$$
$$3 = 3$$

The LCD is $2 \cdot 2 \cdot 2 \cdot 3 = 24$.

STEP 2: Multiply each fraction by 1, written in a form that is appropriate for rewriting each fraction with the LCD.

$$\frac{5}{8} \cdot \frac{3}{3} = \frac{15}{24} \qquad \frac{2}{3} \cdot \frac{8}{8} = \frac{16}{24}$$

STEP 3: Subtract the newly written *like* fractions.

$$\frac{15}{24} - \frac{16}{24} = \frac{15 - 16}{24} = \frac{-1}{24} = -\frac{1}{24}$$

The resulting difference is $-\frac{1}{24}$, which is simplified to lowest terms.

(b) STEP 1: Find the LCD for $\frac{2}{3}$, $\frac{7}{12}$, and $\frac{3}{4}$.

$$3 = 3$$
$$12 = 3 \cdot 2 \cdot 2$$
$$4 = 2 \cdot 2$$

The LCD is $3 \cdot 2 \cdot 2 = 12$.

STEP 2: Multiply each fraction by 1, written in a form that is appropriate for rewriting each fraction with the LCD. Note that the denominator of $\frac{7}{12}$ is already 12.

$$\frac{2}{3} \cdot \frac{4}{4} = \frac{8}{12} \qquad \frac{7}{12} = \frac{7}{12} \qquad \frac{3}{4} \cdot \frac{3}{3} = \frac{9}{12}$$

STEP 3: Add the newly written *like* fractions and simplify the result.

$$\frac{8}{12} + \frac{7}{12} + \frac{9}{12} = \frac{8 + 7 + 9}{12} = \frac{24}{12} = 2$$

The sum is 2.

Now Try Exercises 55, 65

CALCULATOR HELP
To add or subtract fractions with a calculator, See Appendix E (page AP-32).

EXAMPLE 7 **Adding or subtracting fractions with unlike denominators**

Add or subtract as indicated.

(a) $\frac{13}{20} + \frac{1}{8}$ (b) $-\frac{1}{7} - \frac{5}{14}$

Solution

(a) The LCD for $\frac{13}{20}$ and $\frac{1}{8}$ is 40. Verify this. (Step 1)

$$\frac{13}{20} + \frac{1}{8} = \frac{13}{20} \cdot \frac{2}{2} + \frac{1}{8} \cdot \frac{5}{5} \qquad \text{Multiply each fraction by 1.}$$

$$= \frac{26}{40} + \frac{5}{40} \qquad \text{Rewrite fractions with the LCD. (Step 2)}$$

$$= \frac{26 + 5}{40} \qquad \text{Addition of fractions}$$

$$= \frac{31}{40} \qquad \text{Add. (Step 3)}$$

(b) The LCD for $-\frac{1}{7}$ and $\frac{5}{14}$ is 14. Verify this. (Step 1)

$$-\frac{1}{7} - \frac{5}{14} = -\frac{1}{7} \cdot \frac{2}{2} - \frac{5}{14} \qquad \text{Multiply by 1.}$$

$$= -\frac{2}{14} - \frac{5}{14} \qquad \text{Rewrite fractions with the LCD. (Step 2)}$$

$$= \frac{-2-5}{14} \qquad \text{Subtraction of fractions}$$

$$= -\frac{7}{14} \qquad \text{Subtract. (Step 3)}$$

$$= -\frac{1}{2} \qquad \text{Simplify. (Step 3)}$$

Now Try Exercises 59, 63

In the next example, we add and subtract rational expressions with unlike denominators.

EXAMPLE 8 **Adding and subtracting rational expressions with unlike denominators**

Add or subtract as indicated.

(a) $\dfrac{5a}{6} + \dfrac{a}{8}$ **(b)** $\dfrac{3}{4x} - \dfrac{1}{3x^2}$

Solution
(a) The LCD for $\frac{5a}{6}$ and $\frac{a}{8}$ is 24. Verify this. (Step 1)

$$\frac{5a}{6} + \frac{a}{8} = \frac{5a}{6} \cdot \frac{4}{4} + \frac{a}{8} \cdot \frac{3}{3} \qquad \text{Multiply each fraction by 1.}$$

$$= \frac{20a}{24} + \frac{3a}{24} \qquad \text{Rewrite fractions with the LCD. (Step 2)}$$

$$= \frac{20a + 3a}{24} \qquad \text{Addition of fractions}$$

$$= \frac{23a}{24} \qquad \text{Add. (Step 3)}$$

(b) The LCD for $\frac{3}{4x}$ and $\frac{1}{3x^2}$ is $12x^2$. Verify this. (Step 1)

$$\frac{3}{4x} - \frac{1}{3x^2} = \frac{3}{4x} \cdot \frac{3x}{3x} - \frac{1}{3x^2} \cdot \frac{4}{4} \qquad \text{Multiply each fraction by 1.}$$

$$= \frac{9x}{12x^2} - \frac{4}{12x^2} \qquad \text{Rewrite fractions with the LCD. (Step 2)}$$

$$= \frac{9x - 4}{12x^2} \qquad \text{Subtraction of fractions (Step 3)}$$

Note that $\dfrac{9x-4}{12x^2}$ cannot be simplified further.

Now Try Exercises 71, 77

4 Solving Applications Involving Addition and Subtraction of Fractions—Unlike Denominators

Math in Context (Geography) In the next example, we work with unlike fractions to analyze ice on Earth.

EXAMPLE 9 Analyzing ice sheet data

The Antarctic ice sheet contains about $\frac{9}{10}$ of all the ice in the world, while the Greenland ice sheet contains about $\frac{9}{100}$ of the ice. Find the total fraction of Earth's ice that is contained in these two ice sheets. (*Source:* National Science Foundation, Office of Polar Programs.)

Solution
The LCD required for adding $\frac{9}{10}$ and $\frac{9}{100}$ is 100. (Step 1)

$$\frac{9}{10} + \frac{9}{100} = \frac{9}{10} \cdot \frac{10}{10} + \frac{9}{100} \quad \text{Multiply by 1.}$$

$$= \frac{90}{100} + \frac{9}{100} \quad \text{Rewrite fractions with the LCD. (Step 2)}$$

$$= \frac{90 + 9}{100} \quad \text{Addition of fractions}$$

$$= \frac{99}{100} \quad \text{Add. (Step 3)}$$

So, $\frac{99}{100}$ of Earth's ice is contained in the Antarctic and Greenland ice sheets.

Now Try Exercise 81

EXAMPLE 10 Finding the perimeter of a 1-acre plot of land

A 1-acre rectangular plot of land has a length of $\frac{1}{16}$ mile and a width of $\frac{1}{40}$ mile, as shown in the following figure. Find the perimeter of this 1-acre plot.

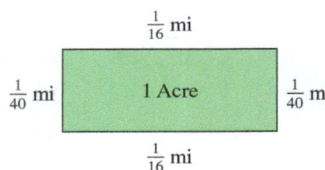

Solution
The perimeter of a rectangle is found by adding the measures of all four sides. The LCD for the fractions $\frac{1}{16}$ and $\frac{1}{40}$ is 80. (Step 1)

$$\frac{1}{16} + \frac{1}{40} + \frac{1}{16} + \frac{1}{40} = \frac{1}{16} \cdot \frac{5}{5} + \frac{1}{40} \cdot \frac{2}{2} + \frac{1}{16} \cdot \frac{5}{5} + \frac{1}{40} \cdot \frac{2}{2} \quad \text{Multiply by 1.}$$

$$= \frac{5}{80} + \frac{2}{80} + \frac{5}{80} + \frac{2}{80} \quad \text{Use the LCD. (Step 2)}$$

$$= \frac{5 + 2 + 5 + 2}{80} \quad \text{Addition of fractions}$$

$$= \frac{14}{80} \quad \text{Add. (Step 3)}$$

$$= \frac{7}{40} \quad \text{Simplify. (Step 3)}$$

The perimeter of the 1-acre plot of land is $\frac{7}{40}$ mile.

Now Try Exercise 91

4.5 Putting It All Together

CONCEPT	COMMENTS	EXAMPLES
Least Common Multiple (LCM)	The least common multiple (LCM) of two or more numbers is the smallest number that is divisible by each of the given numbers.	Common multiples of 3 and 4 include 12, 24, 36, 48, ... The LCM of 3 and 4 is 12.
Prime Factorization Method for LCM	1. Find the prime factorization of each number. 2. List each factor that appears in one or more of the factorizations. If a factor is repeated in any of the factorizations, list this factor the maximum number of times that it is repeated in any one of the factorizations. 3. The product of this list of factors is the LCM.	Find the LCM of 12 and 18. $12 = 2 \cdot 2 \cdot 3$ (Step 1) $18 = 2 \cdot 3 \cdot 3$ (Step 1) 2, 2, 3, 3 (Step 2) $2 \cdot 2 \cdot 3 \cdot 3 = 36$ (Step 3) The LCM of 12 and 18 is 36.
Least Common Denominator (LCD)	The least common denominator (LCD) of two or more fractions is the least common multiple of the denominators.	Because the LCM of 12 and 18 is 36, the LCD of $\frac{7}{12}$ and $\frac{5}{18}$ is 36.
Adding or Subtracting Fractions with Unlike Denominators	1. Determine the LCD for all fractions involved. 2. Write each fraction as an equivalent fraction with the LCD as its denominator. 3. Add or subtract the newly written like fractions and simplify the result.	Add $\frac{7}{12} + \frac{5}{18}$. The LCD of the fractions $\frac{7}{12}$ and $\frac{5}{18}$ is 36. (Step 1) $\frac{7}{12} + \frac{5}{18} = \frac{7}{12} \cdot \frac{3}{3} + \frac{5}{18} \cdot \frac{2}{2}$ $= \frac{21}{36} + \frac{10}{36}$ (Step 2) $= \frac{31}{36}$ (Step 3)

4.5 Exercises

CONCEPTS AND VOCABULARY

1. Although 12, 24, 36, and 48 are all common multiples of 4 and 6, the _____ common multiple is 12.

2. The LCM of a list of numbers is the smallest number that is _____ by each of the given numbers.

3. When the LCM of two numbers is chosen from lists of the multiples of each number, we are using the _____ method for finding the LCM.

4. When the LCM of two numbers is found by first finding the prime factorization of each number, we are using the _____ method for finding the LCM.

5. If a factor is repeated in any of the factorizations used in the prime factorization method for finding the LCM, we list it the _____ number of times that it is repeated in any factorization.

6. The least common denominator (LCD) for a list of fractions is the _____ of the denominators.

252 CHAPTER 4 FRACTIONS

FINDING THE LEAST COMMON MULTIPLE (LCM)

Exercises 7–12: Use the listing method to find the LCM of the given numbers.

7. 4 and 10
8. 6 and 14
9. 12 and 15
10. 9 and 15
11. 6 and 12
12. 8 and 24

Exercises 13–24: Use the prime factorization method to find the LCM of the given numbers.

13. 20 and 16
14. 18 and 24
15. 15 and 90
16. 16 and 80
17. 3, 6, and 15
18. 3, 15, and 21
19. 9, 12, and 45
20. 6, 10, and 27
21. 27 and 45
22. 50 and 75
23. 48 and 81
24. 25 and 42

Exercises 25–32: Use the factorization method to find the LCM of the given variable expressions.

25. $3x$ and $9x$
26. ab^2 and $5ab$
27. $3y^3$ and $8y$
28. $4pq$ and $10p^2q$
29. $6a^3b^2$ and $8ab^2$
30. $4m$ and $12n^2$
31. $4x$, $6y$, and $3z$
32. $8a$, $3ab$, and $4b$

FINDING THE LEAST COMMON DENOMINATOR (LCD)

Exercises 33–46: Find the LCD for the given fractions.

33. $\frac{2}{9}$ and $\frac{5}{12}$
34. $\frac{7}{15}$ and $\frac{1}{6}$
35. $\frac{3}{4}$ and $\frac{9}{14}$
36. $\frac{19}{20}$ and $\frac{5}{8}$
37. $\frac{11}{24}$ and $\frac{17}{30}$
38. $\frac{19}{26}$ and $\frac{13}{24}$
39. $\frac{1}{3}, \frac{4}{5}$, and $\frac{3}{4}$
40. $\frac{1}{2}, \frac{4}{7}$, and $\frac{5}{6}$
41. $\frac{2}{3xy}$ and $\frac{5}{6y^2}$
42. $\frac{7}{10m^2}$ and $\frac{3a}{4mn}$
43. $\frac{5b}{8m}$ and $\frac{3}{16mn^2}$
44. $\frac{x}{13y^2}$ and $\frac{y}{39x}$
45. $\frac{1}{a}, \frac{2}{b}$, and $\frac{3}{c}$
46. $\frac{1}{2x}, \frac{2}{3y}$, and $\frac{3}{4z}$

ADDING AND SUBTRACTING UNLIKE FRACTIONS

Exercises 47–52: Rewrite the fractions as equivalent fractions with the given LCD.

47. $\frac{3}{4}$ and $\frac{1}{6}$ LCD: 12
48. $\frac{5}{8}$ and $\frac{7}{12}$ LCD: 24
49. $\frac{11}{18}$ and $\frac{3}{4}$ LCD: 36
50. $\frac{4}{15}$ and $\frac{7}{9}$ LCD: 45
51. $\frac{1}{2}, \frac{5}{6}$, and $\frac{9}{10}$ LCD: 30
52. $\frac{2}{5}, \frac{11}{12}$, and $\frac{1}{6}$ LCD: 60

Exercises 53–70: Add or subtract as indicated.

53. $\frac{1}{4} + \frac{3}{10}$
54. $\frac{5}{6} + \frac{3}{8}$
55. $\frac{7}{10} - \frac{7}{8}$
56. $\frac{11}{12} - \frac{3}{10}$
57. $-\frac{7}{3} + \frac{8}{9}$
58. $\frac{1}{2} + \left(-\frac{5}{8}\right)$
59. $\frac{1}{10} + \frac{1}{15}$
60. $-\frac{7}{18} + \frac{5}{6}$
61. $\frac{11}{12} + \left(-\frac{1}{4}\right)$
62. $\frac{19}{24} - \frac{5}{8}$
63. $-\frac{5}{12} - \frac{3}{4}$
64. $\frac{17}{30} - \frac{5}{18}$
65. $\frac{3}{8} + \frac{5}{12} + \frac{5}{24}$
66. $\frac{5}{12} + \frac{5}{6} + \frac{3}{4}$
67. $\frac{1}{5} + \frac{3}{10} + \frac{1}{4}$
68. $\frac{1}{6} + \frac{1}{12} + \frac{3}{8}$
69. $\frac{2}{3} + \frac{7}{12} + \frac{5}{4}$
70. $\frac{4}{5} + \frac{7}{10} + \frac{3}{2}$

Exercises 71–80: Add or subtract as indicated.

71. $\frac{3x}{4} + \frac{5x}{6}$
72. $\frac{y}{6} - \frac{7y}{9}$
73. $\frac{7m^2}{14} - \frac{m^2}{6}$
74. $-\frac{7w}{12} + \frac{5w}{8}$

75. $\dfrac{x}{16} + \dfrac{5y}{12}$

76. $\dfrac{a}{9} - \dfrac{b}{12}$

77. $\dfrac{5}{6y} - \dfrac{1}{8y^2}$

78. $\dfrac{3}{a} + \dfrac{4}{b}$

79. $\dfrac{3x}{5y} + \dfrac{5y}{3x}$

80. $\dfrac{m}{2n} - \dfrac{2n}{3m}$

APPLICATIONS INVOLVING ADDITION AND SUBTRACTION OF FRACTIONS

81. **World Population** The two most populous countries in the world are China and India. About $\frac{1}{5}$ of the world's population lives in China, while about $\frac{17}{100}$ of the population lives in India. What is the total fraction of the world's population that lives in these two countries? (*Source:* United Nations.)

82. **World Population** The third and fourth most populous countries in the world are the United States and Indonesia. About $\frac{9}{200}$ of the world's population lives in the United States, while about $\frac{17}{500}$ of the population lives in Indonesia. What is the total fraction of the world's population that lives in these two countries? (*Source:* United Nations.)

83. **United States** The fraction of state names that start with a vowel is $\frac{6}{25}$. If $\frac{9}{50}$ of the state names begin with A, E, I, or U, what fraction of the states have names that begin with O?

84. **European Union** While about $\frac{1}{2}$ of European Union citizens are able to speak English, only about $\frac{3}{20}$ are native English speakers. Determine the fraction of European Union citizens who are non-native English speakers. (*Source:* European Commission.)

85. **Wind Energy** About $\frac{7}{24}$ of U.S. wind power is produced in Texas. California produces about $\frac{1}{8}$ of the country's wind power. What fraction of all U.S wind power is produced in these two states? (*Source:* U.S. Department of Energy.)

86. **Voter Turnout** In a recent presidential election, the nation's highest voter turnout occurred in Minnesota, where $\frac{39}{50}$ of the state's eligible voters went to the polls. In the same presidential election, the national average was $\frac{4}{25}$ lower than the Minnesota fraction. What fraction of the eligible U.S. population voted in this election? (*Source:* Minnesota Secretary of State.)

87. **Fuel Consumption** At the beginning of a trip, a car's gas gauge shows that the car has $\frac{3}{4}$ of a tank of gas. At the end of the trip, the gauge indicates that the car has $\frac{1}{8}$ of a tank of gas. What fraction of the tank was used during the trip?

88. **Efficient Lighting** In an effort to replace $\frac{23}{30}$ of a home's incandescent light bulbs with energy efficient LED bulbs, a contractor has already replaced $\frac{3}{5}$ of the light bulbs. What fraction of the incandescent bulbs still need to be replaced?

Exercises 89–94: **Geometry** *Find the perimeter of the given figure.*

89.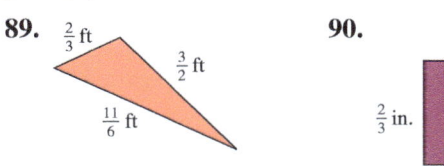

90.

91.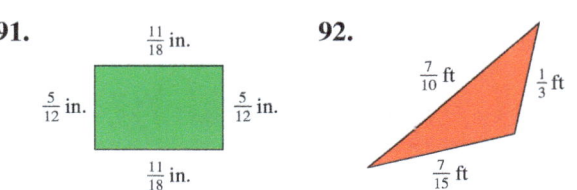

92.

93.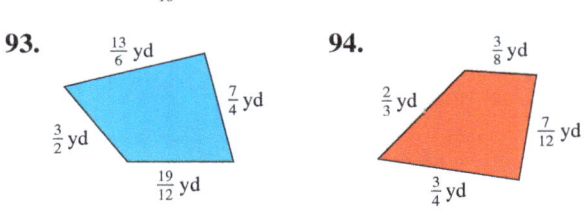

94.

WRITING ABOUT MATHEMATICS

95. Explain how the LCM of two or more numbers could be used to find other multiples of the numbers.

96. Write a detailed paragraph that explains how to add or subtract two fractions with unlike denominators.

97. Use both the listing method and the prime factorization method to find the LCM of 150 and 20. Which method do you prefer?

98. If two fractions are being added and one denominator is a multiple of the other, what can be said about the LCD? Explain your reasoning.

Group Activity

Major U.S. Rivers The mainland United States is made up of all states except Alaska and Hawaii. The two largest rivers within the mainland United States are the Mississippi and the Missouri. The states that each of these rivers passes through (or borders) are listed below.

Mississippi:
Arkansas, Illinois, Iowa, Kentucky, Louisiana, Minnesota, Mississippi, Missouri, Tennessee, and Wisconsin.

Missouri:
Iowa, Kansas, Missouri, Montana, Nebraska, North Dakota, and South Dakota.

(a) Write a simplified fraction that represents the fraction of the mainland states through which the Mississippi river passes.
(b) Write a simplified fraction that represents the fraction of the mainland states through which the Missouri river passes.
(c) Write a simplified fraction that represents the fraction of the mainland states through which both rivers pass.
(d) By adding your answers from parts (a) and (b) and then subtracting your answer from part (c), find the fraction of the mainland states through which either river (or both) passes.

4.6 Operations on Mixed Numbers

Objectives

1. Writing Mixed Numbers as Improper Fractions
2. Rounding Mixed Numbers to the Nearest Integer
3. Performing Operations on Mixed Numbers
4. Adding and Subtracting Mixed Numbers—Method II

1 Writing Mixed Numbers as Improper Fractions

In the first section of this chapter, we discussed how a mixed number can be written as an improper fraction. A summary of the process follows.

> **WRITING A MIXED NUMBER AS AN IMPROPER FRACTION**
>
> A mixed number of the form
>
> $$\text{Quotient} \frac{\text{Remainder}}{\text{Divisor}} \quad \text{can be written as} \quad \frac{\text{Divisor} \cdot \text{Quotient} + \text{Remainder}}{\text{Divisor}}.$$

In the next example, we review how to write mixed numbers as improper fractions.

EXAMPLE 1 Writing mixed numbers as improper fractions

Write each mixed number as an improper fraction.

(a) $5\frac{3}{4}$ (b) $-2\frac{3}{5}$

Solution

(a) $5\frac{3}{4} = \frac{4 \cdot 5 + 3}{4} = \frac{20 + 3}{4} = \frac{23}{4}$

(b) Write the improper fraction for the absolute value of $-2\frac{3}{5}$ and then negate the result. Because $2\frac{3}{5} = \frac{5 \cdot 2 + 3}{5} = \frac{10 + 3}{5} = \frac{13}{5}$, we know that $-2\frac{3}{5} = -\frac{13}{5}$.

Now Try Exercises 7, 13

STUDY TIP

Questions on exams do not always come in the order that they are presented in the text. When studying for an exam, choose review exercises randomly so that the topics are studied in the same random way that they may appear on an exam.

▶ 2 Rounding Mixed Numbers to the Nearest Integer

When performing operations on mixed numbers, it is often helpful to have a rough estimate of the result before completing more complicated computations. To do this, we round the mixed numbers to the nearest integer. For example, the mixed number $3\frac{5}{6}$ can be rounded to 4, and the mixed number $-7\frac{1}{8}$ can be rounded to -7. The following procedure can be used to round a mixed number to the nearest integer.

ROUNDING A MIXED NUMBER TO THE NEAREST INTEGER

To round a mixed number to the nearest integer, do the following.

If the fraction part of the mixed number is

(a) less than $\frac{1}{2}$, drop the fraction part of the mixed number.
(b) $\frac{1}{2}$ or more, drop the fraction part of the mixed number and
  (i) add 1 to the integer part if the mixed number is positive.
  (ii) subtract 1 from the integer part if the mixed number is negative.

NOTE: If twice the numerator of a fraction is greater than or equal to the denominator of the fraction, then the fraction is greater than or equal to $\frac{1}{2}$. For example, the fraction $\frac{7}{13}$ is greater than $\frac{1}{2}$ because $2 \cdot 7 = 14$ and $14 > 13$. Similarly, the fraction $\frac{17}{40}$ is less than $\frac{1}{2}$ because $2 \cdot 17 = 34$ and $34 < 40$. ∎

READING CHECK 1

- Round the mixed numbers $4\frac{3}{4}$ and $-5\frac{1}{3}$ to the nearest integer.

EXAMPLE 2 Rounding mixed numbers to the nearest integer

Round each mixed number to the nearest integer.

(a) $9\frac{3}{7}$ (b) $-6\frac{8}{15}$

Solution
(a) The fraction part of the given mixed number is $\frac{3}{7}$, which is less than $\frac{1}{2}$. To round the mixed number $9\frac{3}{7}$ to the nearest integer, drop the fraction part to obtain 9.
(b) The fraction part is $\frac{8}{15}$, which is greater than $\frac{1}{2}$. Because the mixed number $-6\frac{8}{15}$ is negative, we round it to the nearest integer by dropping the fraction part and subtracting 1 from the integer part to obtain -7.

Now Try Exercises 15, 17

▶ 3 Performing Operations on Mixed Numbers

Because we have already studied operations on fractions, a new process for operations on mixed numbers is not needed. We just need to rewrite any mixed numbers as improper fractions before performing the operations. The following is a process for performing operations on mixed numbers.

OPERATIONS ON MIXED NUMBERS

To perform operations on mixed numbers, do the following.

STEP 1: Rewrite any mixed numbers or integers as improper fractions.

STEP 2: Perform the required arithmetic as usual.

STEP 3: Rewrite the result as a mixed number, when appropriate.

In the next four examples, we use these steps (in a single line of equivalent expressions) to perform operations on mixed numbers and integers. In each example, we first find an estimation of the result in order to reveal any potential computational errors.

EXAMPLE 3   **Adding mixed numbers**

Add.

(a) $2\frac{3}{4} + 6\frac{1}{4}$ (b) $-3\frac{1}{6} + 5\frac{2}{3}$ (c) $\frac{4}{9} + 1\frac{3}{5}$

Solution

(a) Since $2\frac{3}{4}$ rounds to 3 and $6\frac{1}{4}$ rounds to 6, an estimate of the sum is $3 + 6 = 9$.

$$2\frac{3}{4} + 6\frac{1}{4} = \frac{11}{4} + \frac{25}{4} = \frac{11 + 25}{4} = \frac{36}{4} = 9$$

The estimation and the computed sum agree.

(b) Here, $-3\frac{1}{6}$ rounds to -3 and $5\frac{2}{3}$ rounds to 6, so an estimate of the sum is $-3 + 6 = 3$.

$$-3\frac{1}{6} + 5\frac{2}{3} = -\frac{19}{6} + \frac{17}{3} = \frac{-19}{6} + \frac{17}{3} \cdot \frac{2}{2} = \frac{-19}{6} + \frac{34}{6} = \frac{-19 + 34}{6} = \frac{15}{6}$$

Because the problem involves mixed numbers, we express the result as a mixed number.

$$\frac{15}{6} = 2\frac{3}{6} = 2\frac{1}{2}$$

The computed sum is reasonably close to the estimate.

(c) Because $\frac{4}{9}$ rounds to 0 and $1\frac{3}{5}$ rounds to 2, an estimate of the sum is $0 + 2 = 2$.

$$\frac{4}{9} + 1\frac{3}{5} = \frac{4}{9} + \frac{8}{5} = \frac{4}{9} \cdot \frac{5}{5} + \frac{8}{5} \cdot \frac{9}{9} = \frac{20}{45} + \frac{72}{45} = \frac{20 + 72}{45} = \frac{92}{45}$$

Expressing the result as a mixed number, we get

$$\frac{92}{45} = 2\frac{2}{45}.$$

The computed sum is reasonably close to the estimate.

Now Try Exercises 23, 33, 39

EXAMPLE 4   **Subtracting mixed numbers**

Subtract.

(a) $7\frac{2}{15} - 3\frac{1}{6}$ (b) $-14\frac{1}{4} - \frac{7}{8}$

Solution

(a) Because $7\frac{2}{15}$ rounds to 7 and $3\frac{1}{6}$ rounds to 3, the estimated difference is $7 - 3 = \mathbf{4}$.

$$7\frac{2}{15} - 3\frac{1}{6} = \frac{107}{15} - \frac{19}{6} = \frac{107}{15} \cdot \frac{2}{2} - \frac{19}{6} \cdot \frac{5}{5} = \frac{214}{30} - \frac{95}{30} = \frac{214 - 95}{30} = \frac{119}{30}$$

Expressing the result as a mixed number, we get

$$\frac{119}{30} = 3\frac{29}{30}.$$

The computed sum is reasonably close to the estimate.

(b) Here, $-14\frac{1}{4}$ rounds to -14 and $\frac{7}{8}$ rounds to 1, so the estimate is $-14 - 1 = \mathbf{-15}$.

$$-14\frac{1}{4} - \frac{7}{8} = -\frac{57}{4} - \frac{7}{8} = \frac{-57}{4} \cdot \frac{2}{2} - \frac{7}{8} = \frac{-114}{8} - \frac{7}{8} = \frac{-114 - 7}{8} = -\frac{121}{8}$$

As a mixed number, this result is

$$-\frac{121}{8} = -15\frac{1}{8}.$$

The computed sum is reasonably close to the estimate.

Now Try Exercises 27, 35

EXAMPLE 5 Multiplying mixed numbers

Multiply.

(a) $9\frac{4}{7} \cdot 3$ **(b)** $-5\frac{1}{3} \cdot 3\frac{1}{8}$ **(c)** $\frac{3}{4} \cdot 8\frac{2}{3}$

Solution

(a) Because $9\frac{4}{7}$ rounds to 10, the estimated product is $10 \cdot 3 = \mathbf{30}$.

$$9\frac{4}{7} \cdot 3 = \frac{67}{7} \cdot \frac{3}{1} = \frac{201}{7} = 28\frac{5}{7}$$

The computed product is reasonably close to the estimate.

(b) Since $-5\frac{1}{3}$ rounds to -5 and $3\frac{1}{8}$ rounds to 3, the estimated product is $-5 \cdot 3 = \mathbf{-15}$.

$$-5\frac{1}{3} \cdot 3\frac{1}{8} = \frac{-16}{3} \cdot \frac{25}{8} = \frac{-(2 \cdot 2 \cdot 2 \cdot 2) \cdot (5 \cdot 5)}{3 \cdot (2 \cdot 2 \cdot 2)} = \frac{-50}{3} = \mathbf{-16\frac{2}{3}}$$

The computed product is reasonably close to the estimate.

(c) Since $\frac{3}{4}$ rounds to 1 and $8\frac{2}{3}$ rounds to 9, the estimated product is $1 \cdot 9 = \mathbf{9}$.

$$\frac{3}{4} \cdot 8\frac{2}{3} = \frac{3}{4} \cdot \frac{26}{3} = \frac{3 \cdot (2 \cdot 13)}{(2 \cdot 2) \cdot 3} = \frac{13}{2} = \mathbf{6\frac{1}{2}}$$

Although the estimated product may appear to be significantly off from the computed product, the computed result is correct. See the Note following this example.

Now Try Exercises 25, 31, 43

NOTE: The use of rounded values in estimating some products or quotients may give us results that do not seem accurate. This is clear when a fraction rounds to 0. In this case, an estimated product or quotient will be either 0 or undefined. ∎

EXAMPLE 6   Dividing mixed numbers

Divide.

(a) $9\frac{11}{12} \div 2\frac{1}{8}$ (b) $20 \div 3\frac{2}{5}$

Solution
(a) Here, $9\frac{11}{12}$ rounds to 10 and $2\frac{1}{8}$ rounds to 2, so the estimated quotient is $10 \div 2 = $ **5**.

$$9\frac{11}{12} \div 2\frac{1}{8} = \frac{119}{12} \div \frac{17}{8} = \frac{119}{12} \cdot \frac{8}{17} = \frac{(7 \cdot 17) \cdot (2 \cdot 2 \cdot 2)}{(2 \cdot 2 \cdot 3) \cdot 17} = \frac{14}{3} = 4\frac{2}{3}$$

The computed quotient is reasonably close to the estimate.

(b) Because $3\frac{2}{5}$ rounds to 3, the estimated quotient is $20 \div 3 = \frac{20}{3} = 6\frac{2}{3}$.

$$20 \div 3\frac{2}{5} = \frac{20}{1} \div \frac{17}{5} = \frac{20}{1} \cdot \frac{5}{17} = \frac{(2 \cdot 2 \cdot 5) \cdot 5}{17} = \frac{100}{17} = 5\frac{15}{17}$$

The computed quotient is reasonably close to the estimate.

Now Try Exercises 29, 37

4   Adding and Subtracting Mixed Numbers—Method II

Although the method just described for adding and subtracting mixed numbers can always be used, there is a second method that may be more efficient. In this alternate method, we add or subtract the integer parts and the fraction parts separately. The next two examples demonstrate this method.

EXAMPLE 7   Adding mixed numbers—Method II

Add.

(a) $5\frac{3}{4} + 8\frac{2}{5}$ (b) $4 + 7\frac{2}{3}$

Solution
(a) Begin by aligning the integer and fraction parts vertically. The LCD for 4 and 5 is 20.

$$\begin{array}{r} 5\frac{3}{4} \\ + 8\frac{2}{5} \\ \hline \end{array} \quad \text{Rewrite using the LCD} \Rightarrow \quad \begin{array}{r} 5\frac{15}{20} \\ + 8\frac{8}{20} \\ \hline 13\frac{23}{20} \end{array} \quad \text{Add vertically}$$

Since $\frac{23}{20} = 1\frac{3}{20}$, we simplify the sum as follows.

$$13\frac{23}{20} = 13 + 1\frac{3}{20} = 14\frac{3}{20}$$

(b) Method II is particularly efficient for this sum.

$$\begin{array}{r} 4 \\ + 7\frac{2}{3} \\ \hline 11\frac{2}{3} \end{array} \quad \text{Add vertically}$$

Now Try Exercises 47, 51

NOTE: To add mixed numbers with different signs, change the addition problem to an equivalent subtraction problem. Subtraction is illustrated in the next example. ∎

EXAMPLE 8 Subtracting mixed numbers—Method II

Subtract.
(a) $9\frac{7}{8} - 2\frac{1}{6}$ (b) $13\frac{2}{3} - 5\frac{9}{10}$

Solution

(a) Begin by aligning the integer and fraction parts vertically. The LCD for 8 and 6 is 24.

$$9\frac{7}{8} \quad \text{Rewrite using the LCD} \Rightarrow \quad 9\frac{21}{24}$$
$$-2\frac{1}{6} \quad \text{Rewrite using the LCD} \Rightarrow \quad -2\frac{4}{24} \quad \text{Subtract vertically}$$
$$\hline 7\frac{17}{24}$$

(b) Align the integer and fraction parts vertically. The LCD for 3 and 10 is 30.

$$13\frac{2}{3} \quad \text{Rewrite using the LCD} \Rightarrow \quad 13\frac{20}{30}$$
$$-5\frac{9}{10} \quad \text{Rewrite using the LCD} \Rightarrow \quad -5\frac{27}{30}$$

Because subtracting $\frac{27}{30}$ from $\frac{20}{30}$ would result in a negative difference, we regroup 1 from the integer part of $13\frac{20}{30}$, to rewrite the mixed number as follows.

$$\underbrace{13\frac{20}{30} = 12 + 1}_{\text{Regroup 1}} + \frac{20}{30} = 12 + \underbrace{\frac{\mathbf{30}}{\mathbf{30}}}_{\text{Write 1 as } \frac{30}{30}} + \frac{20}{30} = 12 + \frac{30 + 20}{30} = 12\frac{50}{30}$$

Rewriting $13\frac{20}{30}$ as $12\frac{50}{30}$, we complete the given subtraction as follows.

$$13\frac{20}{30} \quad \text{Regroup 1 from 13} \Rightarrow \quad 12\frac{50}{30}$$
$$-5\frac{27}{30} \qquad\qquad\qquad\qquad\qquad -5\frac{27}{30} \quad \text{Subtract vertically}$$
$$\hline 7\frac{23}{30}$$

Now Try Exercises 41, 45

READING CHECK 2

- Change the mixed numbers in the product $2\frac{3}{4} \cdot 5\frac{2}{3}$ to improper fractions and multiply.

> **MAKING CONNECTIONS 1**
>
> **Multiplying Mixed Numbers—Method II**
>
> While we can use Method II for some addition and subtraction, it is not practical for multiplication or division. For example, to multiply $2\frac{3}{4}$ and $5\frac{2}{3}$ using Method II, we must utilize the distributive property as follows.
>
> $$2\frac{3}{4} \cdot 5\frac{2}{3} = \left(2 + \frac{3}{4}\right) \cdot \left(5 + \frac{2}{3}\right) \quad \text{Rewrite the mixed numbers.}$$
>
> $$= \left(2 + \frac{3}{4}\right) \cdot 5 + \left(2 + \frac{3}{4}\right) \cdot \frac{2}{3} \quad \text{Distribute } 2 + \tfrac{3}{4}.$$
>
> $$= 2 \cdot 5 + \frac{3}{4} \cdot 5 + 2 \cdot \frac{2}{3} + \frac{3}{4} \cdot \frac{2}{3} \quad \text{Distribute 5 and } \tfrac{2}{3}.$$
>
> $$= 10 + \frac{15}{4} + \frac{4}{3} + \frac{1}{2} \quad \text{Multiply.}$$
>
> $$= 15\frac{7}{12} \quad \text{Add.}$$

4.6 Putting It All Together

CONCEPT	COMMENTS	EXAMPLES
Writing Mixed Numbers as Improper Fractions	A mixed number of the form $$\text{Quotient}\, \frac{\text{Remainder}}{\text{Divisor}}$$ can be written as $$\frac{\text{Divisor} \cdot \text{Quotient} + \text{Remainder}}{\text{Divisor}}.$$	$9\frac{4}{7} = \frac{7 \cdot 9 + 4}{7} = \frac{67}{7}$ $-4\frac{2}{5} = -\frac{5 \cdot 4 + 2}{5} = -\frac{22}{5}$

Rounding Mixed Numbers to the Nearest Integer	If the fraction part of the mixed number is
	 (a) less than $\frac{1}{2}$, drop the fraction part.
	 (b) $\frac{1}{2}$ or more, drop the fraction part and
	   (i) add 1 to the integer part if the mixed number is positive.
	   (ii) subtract 1 from the integer part if the mixed number is negative.
		$3\frac{2}{9}$ rounds to 3.
		
		 $-4\frac{7}{8}$ rounds to -5.
		
		 $-11\frac{3}{7}$ rounds to -11.

Operations on Mixed Numbers	1. Rewrite any mixed numbers or integers as improper fractions. 2. Perform the required arithmetic as usual. 3. Rewrite the result as a mixed number.	$1\frac{2}{3} \cdot 3\frac{1}{2} = \frac{5}{3} \cdot \frac{7}{2} = \frac{35}{6} = 5\frac{5}{6}$ (Step 1) (Step 2) (Step 3) $5\frac{2}{3} + 2\frac{2}{3} = \frac{17}{3} + \frac{8}{3} = \frac{25}{3} = 8\frac{1}{3}$ (Step 1) (Step 2) (Step 3)
Method II for Adding and Subtracting Mixed Numbers	Add or subtract vertically by adding or subtracting the integer parts and the fraction parts separately. You may have to write the fraction parts with a LCD first.	$\begin{aligned} &4\tfrac{1}{5} \\ +\,&6\tfrac{2}{5} \\ \hline &10\tfrac{3}{5}\end{aligned}$ $\begin{aligned} &11\tfrac{7}{9} \\ -\,&3\tfrac{2}{9} \\ \hline &8\tfrac{5}{9}\end{aligned}$

4.6 Exercises

CONCEPTS AND VOCABULARY

1. To round a mixed number to the nearest integer, we must determine if the fraction part of the mixed number is less than, equal to, or greater than _____.

2. A fraction is less than $\frac{1}{2}$ if twice its numerator is _____ than its denominator.

3. The first step in performing operations on mixed numbers is to rewrite any mixed numbers or integers as _____.

4. When doing arithmetic on mixed numbers, it is often helpful to have a rough _____ of the result before performing more complicated computations.

5. When using Method II for addition or subtraction, we align the mixed numbers _____ before completing the indicated arithmetic.

6. If the fraction in the upper mixed number is less than the fraction in the lower mixed number when subtracting using Method II, we need to _____ 1 from the integer part of the upper mixed number.

WRITING IMPROPER FRACTIONS

Exercises 7–14: Write the given mixed number as an improper fraction.

7. $4\frac{1}{5}$
8. $11\frac{2}{7}$
9. $-3\frac{7}{15}$
10. $-12\frac{3}{8}$
11. $8\frac{9}{10}$
12. $13\frac{2}{3}$
13. $-15\frac{1}{4}$
14. $-20\frac{4}{5}$

ROUNDING MIXED NUMBERS

Exercises 15–22: Round the given mixed number to the nearest integer.

15. $7\frac{8}{11}$
16. $12\frac{1}{8}$
17. $-14\frac{6}{13}$
18. $-3\frac{5}{6}$
19. $11\frac{3}{7}$
20. $6\frac{7}{12}$
21. $-9\frac{5}{9}$
22. $-1\frac{9}{25}$

ADDING, SUBTRACTING, MULTIPLYING, AND DIVIDING MIXED NUMBERS

Exercises 23–60: Perform the indicated arithmetic.

23. $5\frac{1}{6} + 4\frac{5}{6}$
24. $7\frac{1}{2} - \left(-\frac{5}{8}\right)$
25. $9\frac{7}{8} \cdot 4$
26. $16\frac{3}{28} \div 5\frac{6}{7}$
27. $11\frac{5}{12} - 4\frac{3}{8}$
28. $8\frac{2}{9} + \left(-5\frac{1}{12}\right)$
29. $14\frac{5}{8} \div 2\frac{1}{6}$
30. $11\frac{2}{5} \cdot 3$
31. $-7\frac{1}{6} \cdot 4\frac{2}{3}$
32. $\frac{8}{9} \cdot 5\frac{1}{4}$
33. $-7\frac{3}{4} + 6\frac{5}{8}$
34. $6\frac{3}{7} + 5\frac{4}{7}$
35. $-8\frac{2}{5} - \frac{7}{10}$
36. $5\frac{1}{10} + 9\frac{5}{6}$
37. $32 \div 2\frac{3}{8}$
38. $15\frac{7}{18} - 10\frac{1}{2}$
39. $3\frac{4}{5} + \frac{6}{7}$
40. $17\frac{5}{12} - 9\frac{1}{4}$
41. $14\frac{5}{6} - 3\frac{1}{4}$
42. $3\frac{1}{4} \cdot \left(-5\frac{1}{2}\right)$
43. $10\frac{1}{3} \cdot \frac{5}{7}$
44. $18\frac{1}{5} \div 3\frac{1}{5}$
45. $8\frac{1}{8} - 2\frac{5}{6}$
46. $7 + 6\frac{8}{15}$
47. $7\frac{2}{3} + 6\frac{7}{8}$
48. $\frac{4}{7} + 5\frac{9}{14}$
49. $15\frac{1}{3} \div 3\frac{5}{6}$
50. $19\frac{3}{8} - 6\frac{7}{10}$
51. $2\frac{6}{11} + 9$
52. $7\frac{1}{5} \div 9$
53. $-16\frac{2}{9} \div 1\frac{1}{6}$
54. $-5 \cdot \left(-4\frac{2}{9}\right)$
55. $4\frac{1}{8} - \left(-10\frac{6}{8}\right)$
56. $8\frac{4}{5} + \left(-7\frac{13}{15}\right)$

57. $5\frac{11}{12} \cdot (-4)$ **58.** $4\frac{2}{3} \div \left(-6\frac{1}{8}\right)$

59. $-5\frac{7}{15} + 8\frac{1}{3}$ **60.** $-7\frac{1}{12} - 6\frac{2}{3}$

APPLICATIONS INVOLVING MIXED NUMBERS

61. Model Building An industrial model builder needs several pieces of wire that are $6\frac{3}{4}$ inches long. How many whole pieces of wire of this length can be cut from a spool containing 72 inches of wire? How much wire is left over?

62. Jogging An athlete jogs $4\frac{1}{2}$ miles on Tuesday and $7\frac{3}{4}$ miles on Wednesday. What is the total distance for these two days?

63. Baking A recipe calls for $2\frac{2}{3}$ cups of flour. How much flour is needed if the recipe is doubled?

64. Baking A recipe calls for $1\frac{1}{4}$ cups of sugar. How much sugar is needed if the recipe is halved?

65. Languages About 260 thousand Americans speak Hmong as the primary language used in their homes. About $2\frac{3}{5}$ times as many Americans speak Polish in their homes. Find the number of Americans who speak Polish in their homes. (*Source*: U.S. Census Bureau.)

66. Homeschooling About 340 thousand children of parents with annual incomes below the poverty level were homeschooled in 2014. If $2\frac{1}{4}$ times as many children of parents with annual incomes at or above 200% of the poverty level were homeschooled that year, find the number of children at this income level that were homeschooled in 2014. (*Source*: U.S. Department of Education.)

67. Height If the birth announcement for a baby boy indicates that his length at birth was $17\frac{1}{4}$ inches and another birth announcement for a baby girl states that her length at birth was $19\frac{5}{8}$ inches, how much longer was the girl at birth compared to the boy?

68. Height At birth, a baby boy was $18\frac{1}{2}$ inches long. If he grew $5\frac{3}{4}$ inches during his first year, how tall was he on his first birthday?

69. Painting A painter needs $1\frac{1}{4}$ gallons of paint for a project. If there is $\frac{2}{3}$ gallon available in a partially used container, how much paint from a new container must be used?

70. Cleaning Solution If $1\frac{1}{3}$ cups of liquid concentrate are mixed with 1 gallon of water to make a cleaning solution, how many such mixtures can be made from a 16-cup (1-gallon) jug of concentrate?

Exercises 71–74: **Geometry** *Find the area of the given figure.*

71. 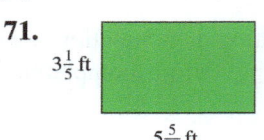 **72.**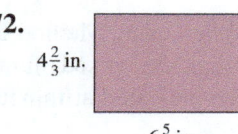

73. 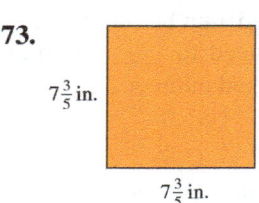 **74.**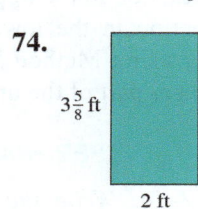

WRITING ABOUT MATHEMATICS

75. A student claims that an easy way to multiply two mixed numbers is to simply multiply the integer parts and then multiply the fraction parts. Give a real-world example that would disprove this student's method.

76. Try to devise a "Method II" for dividing one mixed number by another. Give detailed steps. (Refer to the Making Connections in this section.)

SECTIONS 4.5 and 4.6 **Checking Basic Concepts**

1. Find the least common multiple.
   (a) 12 and 15   (b) $4x$ and $5x^2$

2. Find the least common denominator.
   (a) $\frac{5}{8}$ and $\frac{7}{9}$ (b) $\frac{3}{10xy}$ and $\frac{5}{6x^2}$

3. Add or subtract as indicated. If necessary, simplify your results.
   (a) $\frac{1}{8} + \frac{3}{18}$ (b) $\frac{8}{15} - \frac{7}{12}$
   (c) $\frac{1}{5} + \frac{1}{4} + \frac{3}{8}$ (d) $-\frac{5x}{8} + \frac{7x}{12}$

4. Perform the indicated arithmetic.
   (a) $-5\frac{1}{5} + 4\frac{7}{10}$ (b) $\frac{4}{15} - \left(-6\frac{7}{12}\right)$
   (c) $3\frac{5}{12} \cdot 9$ (d) $16\frac{5}{6} \div 3\frac{5}{32}$

5. **Fuel Consumption** At the beginning of a trip, a car's gas gauge shows that the car has $\frac{7}{8}$ of a tank of gas. At the end of the trip, the gauge indicates that the car has $\frac{1}{3}$ of a tank of gas. What fraction of a tank was used during the trip?

4.7 Complex Fractions and Order of Operations

Objectives
1. Simplifying Basic Complex Fractions
2. Simplifying Complex Fractions—Method I
3. Simplifying Complex Fractions—Method II
4. Applying Order of Operations to Complex Fractions

NEW VOCABULARY

☐ Complex fraction

READING CHECK 1

Which of the following are complex fractions? Select all that apply.

- $\dfrac{\frac{x}{2}}{2}$
- $\dfrac{x+1}{\frac{5}{y}}$
- $\dfrac{1 - \frac{3}{4}}{1 + \frac{3}{4}}$
- $1 + \frac{1}{2}$

1 Simplifying Basic Complex Fractions

Suppose that $3\frac{1}{2}$ watermelons are shared among 14 people so that each person ends up with $\frac{1}{4}$ watermelon. This situation can be represented mathematically by

$$3\frac{1}{2} \div 14 = \frac{1}{4} \quad \text{or} \quad \frac{3\frac{1}{2}}{14} = \frac{1}{4} \quad \text{or} \quad \frac{3 + \frac{1}{2}}{14} = \frac{1}{4}.$$

In the last two equations, the expression on the left side of the equals sign is an example of a *complex fraction*.

A **complex fraction** is a rational expression with fractions in its numerator, denominator, or both. Examples of complex fractions include

$$\frac{1 + \frac{2}{3}}{3 - \frac{5}{6}}, \quad \frac{7}{\frac{x}{3} + \frac{1}{6}}, \quad \text{and} \quad \frac{\frac{3}{x}}{\frac{5x}{7}}. \quad \text{Complex fractions}$$

Typically, we want to rewrite a complex fraction as a standard fraction in the form $\frac{a}{b}$, where the variables a and b represent expressions that are not fractions. For some basic complex fractions, this can be accomplished by noting that fractions are divided using the rule $\frac{a}{b} \div \frac{c}{d} = \frac{a}{b} \cdot \frac{d}{c}$, where b, c, and d are not 0. Because the expression $\frac{a}{b} \div \frac{c}{d}$ can be written as the complex fraction

$$\frac{\frac{a}{b}}{\frac{c}{d}},$$

we can summarize this strategy as follows.

SIMPLIFYING BASIC COMPLEX FRACTIONS

If the variables a, b, c, and d represent numbers with $b \neq 0$, $c \neq 0$, and $d \neq 0$, then

$$\frac{\frac{a}{b}}{\frac{c}{d}} = \frac{a}{b} \cdot \frac{d}{c}.$$

EXAMPLE 1 Simplifying basic complex fractions

Simplify each complex fraction.

(a) $\dfrac{\frac{4}{5}}{\frac{8}{15}}$ (b) $\dfrac{\frac{3}{2x}}{\frac{9}{y}}$

Solution

(a) $\dfrac{\frac{4}{5}}{\frac{8}{15}} = \dfrac{4}{5} \cdot \dfrac{15}{8} = \dfrac{(2 \cdot 2) \cdot (3 \cdot 5)}{5 \cdot (2 \cdot 2 \cdot 2)} = \dfrac{3}{2}$

(b) $\dfrac{\frac{3}{2x}}{\frac{9}{y}} = \dfrac{3}{2x} \cdot \dfrac{y}{9} = \dfrac{3 \cdot y}{(2 \cdot x) \cdot (3 \cdot 3)} = \dfrac{y}{2 \cdot x \cdot 3} = \dfrac{y}{6x}$

Now Try Exercises 5, 13

There are two common methods for simplifying complex fractions. In the first method, we begin by simplifying both the numerator and denominator of the complex fraction and then divide the resulting two fractions. In the second method, we multiply the numerator and denominator of the complex fraction by the LCD of all fractions within the complex fraction.

> **STUDY TIP**
>
> In mathematics, there are often *several correct* ways to perform a particular process. If your instructor does not require you to use a specified method, choose the one that works best for you.

2 Simplifying Complex Fractions—Method I

The following steps outline Method I for simplifying complex fractions.

SIMPLIFYING COMPLEX FRACTIONS—METHOD I

To simplify a complex fraction, do the following.

STEP 1: Write the numerator as a single fraction; do the same to the denominator.

STEP 2: Multiply the fraction in the numerator by the reciprocal of the fraction in the denominator. Simplify the result.

EXAMPLE 2 Simplifying complex fractions using Method I

Simplify each complex fraction.

(a) $\dfrac{\frac{4}{5} + \frac{2}{3}}{\frac{7}{15} - \frac{2}{5}}$ (b) $\dfrac{\frac{3x}{4}}{1 - \frac{7}{16}}$

Solution

(a) **STEP 1:** Add the fractions $\frac{4}{5} + \frac{2}{3}$ in the numerator. The LCD is 15.

$$\dfrac{4}{5} + \dfrac{2}{3} = \dfrac{4}{5} \cdot \dfrac{3}{3} + \dfrac{2}{3} \cdot \dfrac{5}{5} = \dfrac{12}{15} + \dfrac{10}{15} = \dfrac{12 + 10}{15} = \dfrac{22}{15}$$

Subtract the fractions $\frac{7}{15} - \frac{2}{5}$ in the denominator. The LCD here is also 15.

$$\dfrac{7}{15} - \dfrac{2}{5} = \dfrac{7}{15} - \dfrac{2}{5} \cdot \dfrac{3}{3} = \dfrac{7}{15} - \dfrac{6}{15} = \dfrac{7 - 6}{15} = \dfrac{1}{15}$$

STEP 2: Multiply $\frac{22}{15}$ by $\frac{15}{1}$, the reciprocal of $\frac{1}{15}$.

$$\dfrac{\frac{4}{5} + \frac{2}{3}}{\frac{7}{15} - \frac{2}{5}} = \dfrac{\frac{22}{15}}{\frac{1}{15}} = \dfrac{22}{15} \cdot \dfrac{15}{1} = \dfrac{(2 \cdot 11) \cdot (3 \cdot 5)}{(3 \cdot 5)} = \dfrac{2 \cdot 11}{1} = \dfrac{22}{1} = 22$$

(b) **STEP 1:** The numerator of the complex fraction is already written as a single fraction, so we begin by subtracting $1 - \frac{7}{16}$ in the denominator. The LCD is 16.

$$1 - \dfrac{7}{16} = \dfrac{1}{1} \cdot \dfrac{16}{16} - \dfrac{7}{16} = \dfrac{16}{16} - \dfrac{7}{16} = \dfrac{16 - 7}{16} = \dfrac{9}{16}$$

STEP 2: Multiply $\frac{3x}{4}$ by $\frac{16}{9}$, the reciprocal of $\frac{9}{16}$.

$$\frac{\frac{3x}{4}}{1-\frac{7}{16}} = \frac{\frac{3x}{4}}{\frac{9}{16}} = \frac{3x}{4} \cdot \frac{16}{9} = \frac{(3 \cdot x) \cdot (2 \cdot 2 \cdot 2 \cdot 2)}{(2 \cdot 2) \cdot (3 \cdot 3)} = \frac{2 \cdot 2 \cdot x}{3} = \frac{4x}{3}$$

Now Try Exercises 23, 25

3 Simplifying Complex Fractions—Method II

The steps used in Method II for simplifying complex fractions are summarized as follows.

> **SIMPLIFYING COMPLEX FRACTIONS—METHOD II**
>
> To simplify a complex fraction, do the following.
>
> **STEP 1:** Find the LCD of all fractions within the complex fraction.
>
> **STEP 2:** Multiply both the numerator and the denominator of the complex fraction by the LCD found in Step 1. Simplify the result.

READING CHECK 2

- Why isn't the value of a complex fraction changed when we multiply its numerator and denominator by the LCD?

NOTE: Multiplying both the numerator and denominator of a complex fraction by the same nonzero number (the LCD) is equivalent to multiplying the complex fraction by 1. Doing so will *not change the value* of the complex fraction. ∎

EXAMPLE 3 Simplifying complex fractions using Method II

Simplify each complex fraction.

(a) $\dfrac{\frac{5}{8} - \frac{1}{2}}{\frac{1}{4} + \frac{2}{3}}$ (b) $\dfrac{\frac{5x}{4}}{\frac{1}{2} + 7}$

Solution

(a) **STEP 1:** The LCD for the fractions $\frac{5}{8}, \frac{1}{2}, \frac{1}{4}$, and $\frac{2}{3}$ is 24.

STEP 2: Use the distributive property to multiply the numerator and the denominator of the complex fraction by 24. Note that 24 is written as $\frac{24}{1}$.

$$\frac{\frac{5}{8} - \frac{1}{2}}{\frac{1}{4} + \frac{2}{3}} = \frac{24 \cdot \left(\frac{5}{8} - \frac{1}{2}\right)}{24 \cdot \left(\frac{1}{4} + \frac{2}{3}\right)} = \frac{\frac{24}{1} \cdot \frac{5}{8} - \frac{24}{1} \cdot \frac{1}{2}}{\frac{24}{1} \cdot \frac{1}{4} + \frac{24}{1} \cdot \frac{2}{3}}$$

Because 24 is the LCD for all fractions involved, it is divisible by each of the denominators. Dividing each denominator into 24 eliminates the fractions from within the complex fraction. The simplification continues as follows.

$$\frac{\frac{24}{1} \cdot \frac{5}{8} - \frac{24}{1} \cdot \frac{1}{2}}{\frac{24}{1} \cdot \frac{1}{4} + \frac{24}{1} \cdot \frac{2}{3}} = \frac{3 \cdot 5 - 12 \cdot 1}{6 \cdot 1 + 8 \cdot 2} = \frac{15 - 12}{6 + 16} = \frac{3}{22}$$

(b) STEP 1: The LCD for the fractions $\frac{5x}{4}$ and $\frac{1}{2}$ is 4.

STEP 2: Multiply the numerator and the denominator of the complex fraction by 4.

$$\frac{\frac{5x}{4}}{\frac{1}{2} + 7} = \frac{4 \cdot \frac{5x}{4}}{4 \cdot \left(\frac{1}{2} + 7\right)} = \frac{\frac{4}{1} \cdot \frac{5x}{4}}{\frac{4}{1} \cdot \frac{1}{2} + 4 \cdot 7} = \frac{1 \cdot 5x}{2 + 28} = \frac{5x}{30} = \frac{x}{6}$$

Note that the result $\frac{5x}{30}$ was simplified to $\frac{x}{6}$.

Now Try Exercises 29, 37

4 Applying Order of Operations to Complex Fractions

To simplify expressions containing fractions, we use the order of operations agreement.

> **ORDER OF OPERATIONS**
>
> 1. Do all calculations within grouping symbols such as parentheses and radicals, or above and below a fraction bar.
> 2. Evaluate all exponential expressions.
> 3. Do all multiplication and division from *left to right*.
> 4. Do all addition and subtraction from *left to right*.

EXAMPLE 4 **Using the order of operations to simplify expressions with fractions**

Simplify each expression.

(a) $\frac{1}{6} + \frac{3}{4} \cdot \frac{1}{2}$ **(b)** $\left(\frac{5}{6} + \frac{1}{4}\right)\left(\frac{2}{5} - \frac{1}{2}\right)$ **(c)** $\left(\frac{6}{5}\right)^2 \div \left(\frac{16}{25} + \frac{4}{5}\right)$

Solution

(a) Multiply $\frac{3}{4}$ and $\frac{1}{2}$ before adding.

$$\frac{1}{6} + \frac{3}{4} \cdot \frac{1}{2} = \frac{1}{6} + \frac{3}{8} \qquad \text{Multiply.}$$

$$= \frac{1}{6} \cdot \frac{4}{4} + \frac{3}{8} \cdot \frac{3}{3} \qquad \text{The LCD is 24.}$$

$$= \frac{4}{24} + \frac{9}{24} \qquad \text{Multiply.}$$

$$= \frac{13}{24} \qquad \text{Add.}$$

(b) Complete the arithmetic within parentheses before multiplying.

$$\left(\frac{5}{6} + \frac{1}{4}\right)\left(\frac{2}{5} - \frac{1}{2}\right) = \left(\frac{5}{6} \cdot \frac{2}{2} + \frac{1}{4} \cdot \frac{3}{3}\right)\left(\frac{2}{5} \cdot \frac{2}{2} - \frac{1}{2} \cdot \frac{5}{5}\right) \qquad \text{The LCDs are 12 and 10.}$$

$$= \left(\frac{10}{12} + \frac{3}{12}\right)\left(\frac{4}{10} - \frac{5}{10}\right) \qquad \text{Compute four products.}$$

$$= \left(\frac{13}{12}\right)\left(-\frac{1}{10}\right) \qquad \text{Add; subtract.}$$

$$= -\frac{13}{120} \qquad \text{Multiply.}$$

(c) Square $\frac{6}{5}$ and then add within parentheses before performing the division.

$$\left(\frac{6}{5}\right)^2 \div \left(\frac{16}{25} + \frac{4}{5}\right) = \frac{36}{25} \div \left(\frac{16}{25} + \frac{4}{5}\right) \qquad \text{Square } \tfrac{6}{5}.$$

$$= \frac{36}{25} \div \left(\frac{16}{25} + \frac{4}{5} \cdot \frac{5}{5}\right) \qquad \text{The LCD is 25.}$$

$$= \frac{36}{25} \div \left(\frac{16}{25} + \frac{20}{25}\right) \qquad \text{Multiply.}$$

$$= \frac{36}{25} \div \frac{36}{25} \qquad \text{Add.}$$

$$= 1 \qquad a \div a = 1 \text{ when } a \neq 0.$$

Now Try Exercises 41, 45, 51

4.7 Putting It All Together

CONCEPT	COMMENTS	EXAMPLES
Complex Fraction	A complex fraction is a rational expression with fractions in its numerator, denominator, or both.	$\dfrac{3 + \frac{5}{6}}{\frac{1}{2} - \frac{4}{9}}$ and $\dfrac{\frac{3x}{5}}{\frac{2}{3} - 5}$
Simplifying Basic Complex Fractions	If the variables a, b, c, and d represent numbers with $b \neq 0$, $c \neq 0$, and $d \neq 0$, then $$\dfrac{\frac{a}{b}}{\frac{c}{d}} = \dfrac{a}{b} \cdot \dfrac{d}{c}.$$	$\dfrac{\frac{4}{5}}{\frac{5}{6}} = \dfrac{4}{5} \cdot \dfrac{6}{5} = \dfrac{24}{25}$
Simplifying Complex Fractions—Method I	**STEP 1:** Write the numerator as a single fraction; do the same to the denominator. **STEP 2:** Multiply the fraction in the numerator by the reciprocal of the fraction in the denominator. Simplify the result.	$\dfrac{\frac{4}{9} + \frac{1}{9}}{\frac{4}{x} - \frac{2}{x}} = \dfrac{\frac{5}{9}}{\frac{2}{x}} = \dfrac{5}{9} \cdot \dfrac{x}{2} = \dfrac{5x}{18}$
Simplifying Complex Fractions—Method II	**STEP 1:** Find the LCD of all fractions within the complex fraction. **STEP 2:** Multiply both the numerator and the denominator of the complex fraction by the LCD from Step 1. Simplify the result.	$\dfrac{\frac{1}{6} + \frac{2}{3}}{\frac{1}{2} + \frac{5}{6}} = \dfrac{6 \cdot \left(\frac{1}{6} + \frac{2}{3}\right)}{6 \cdot \left(\frac{1}{2} + \frac{5}{6}\right)} = \dfrac{1 + 4}{3 + 5} = \dfrac{5}{8}$

4.7 Exercises

CONCEPTS AND VOCABULARY

1. A(n) _____ fraction is a rational expression with fractions in its numerator, denominator, or both.

2. One strategy for simplifying a complex fraction with a single fraction in the numerator and a single fraction in the denominator is to multiply the fraction in the numerator by the _____ of the fraction in the denominator.

3. If a complex fraction is being simplified by using Method _____, the first step is to write the numerator as a single fraction and write the denominator as a single fraction.

4. To simplify a complex fraction using Method _____, first find the LCD of all fractions within the complex fraction.

SIMPLIFYING COMPLEX FRACTIONS

Exercises 5–16: Simplify the given complex fraction.

5. $\dfrac{\frac{2}{7}}{\frac{4}{5}}$

6. $\dfrac{\frac{9}{10}}{\frac{3}{5}}$

7. $\dfrac{\frac{6}{2}}{3}$

8. $\dfrac{\frac{8}{4}}{5}$

9. $\dfrac{\frac{12}{5}}{\frac{3}{4}}$

10. $\dfrac{\frac{18}{25}}{\frac{4}{5}}$

11. $\dfrac{\frac{3}{4}}{9}$

12. $\dfrac{\frac{4}{9}}{8}$

13. $\dfrac{\frac{5}{4x}}{\frac{y}{10}}$

14. $\dfrac{\frac{x}{7}}{\frac{x}{14}}$

15. $\dfrac{\frac{16}{5x}}{\frac{4}{15x}}$

16. $\dfrac{\frac{w}{18}}{\frac{2}{3w}}$

Exercises 17–28: Simplify the given complex fraction using Method I.

17. $\dfrac{\frac{2}{3}+\frac{5}{6}}{\frac{2}{5}+\frac{1}{2}}$

18. $\dfrac{\frac{3}{7}-\frac{1}{3}}{\frac{7}{8}-\frac{3}{4}}$

19. $\dfrac{\frac{2}{5}+3}{5-\frac{3}{4}}$

20. $\dfrac{6-\frac{3}{8}}{\frac{3}{4}}$

21. $\dfrac{23}{4+\frac{3}{5}}$

22. $\dfrac{7-\frac{3}{5}}{16}$

23. $\dfrac{\frac{5}{3}-\frac{1}{4}}{\frac{9}{10}+\frac{14}{15}}$

24. $\dfrac{\frac{9}{5}+\frac{1}{10}}{\frac{18}{5}-\frac{3}{4}}$

25. $\dfrac{\frac{10x}{7}}{3-\frac{9}{8}}$

26. $\dfrac{\frac{x}{5}+\frac{2x}{3}}{\frac{13}{5}}$

27. $\dfrac{\frac{11x}{3}+x}{\frac{2y}{9}}$

28. $\dfrac{\frac{7b}{12}+\frac{b}{6}}{b}$

Exercises 29–40: Simplify the given complex fraction using Method II.

29. $\dfrac{\frac{5}{12}-\frac{1}{6}}{\frac{7}{4}-\frac{1}{2}}$

30. $\dfrac{\frac{2}{3}+\frac{1}{8}}{\frac{3}{4}+\frac{5}{6}}$

31. $\dfrac{7-\frac{3}{2}}{4}$

32. $\dfrac{26}{3+\frac{7}{2}}$

33. $\dfrac{\frac{7}{10} - \frac{1}{6}}{\frac{9}{5} + \frac{1}{15}}$

34. $\dfrac{\frac{2}{9} + \frac{5}{6}}{\frac{13}{12} - \frac{3}{4}}$

35. $\dfrac{2 - \frac{5}{6}}{\frac{21}{8}}$

36. $\dfrac{\frac{1}{2} + 6}{3 - \frac{2}{5}}$

37. $\dfrac{\frac{13x}{4}}{\frac{4}{3} + 3}$

38. $\dfrac{\frac{11x}{4} - \frac{2x}{3}}{\frac{25}{6}}$

39. $\dfrac{\frac{7a}{5} + a}{\frac{3a}{10}}$

40. $\dfrac{\frac{5w}{9} + \frac{5w}{6}}{\frac{10}{3}}$

APPLYING THE ORDER OF OPERATIONS

Exercises 41–56: Use the order of operations agreement to simplify the given expression.

41. $\dfrac{13}{6} + \dfrac{2}{3} \cdot \dfrac{5}{4}$

42. $\dfrac{11}{8} - \dfrac{1}{3} \cdot \dfrac{3}{2}$

43. $\dfrac{7}{10} \div \dfrac{3}{5} \cdot \dfrac{9}{14}$

44. $\dfrac{3}{8} \cdot \dfrac{4}{9} \div \dfrac{5}{12}$

45. $\left(\dfrac{3}{4} + \dfrac{2}{3}\right)\left(\dfrac{1}{2} - \dfrac{7}{8}\right)$

46. $\left(\dfrac{1}{2} + \dfrac{1}{3}\right)\left(\dfrac{2}{5} - \dfrac{1}{4}\right)$

47. $\dfrac{1}{4^2} \div \left(\dfrac{3}{8} + \dfrac{1}{4}\right)$

48. $3^3 \cdot \left(\dfrac{5}{6} - \dfrac{7}{9}\right)$

49. $5 - 10 \div \sqrt{\dfrac{25}{64}}$

50. $\left(\dfrac{7}{8} - \dfrac{3}{4}\right)^2 + \sqrt{\dfrac{1}{4}}$

51. $\left(\dfrac{3}{2}\right)^2 \cdot \left(\dfrac{7}{12} - \dfrac{5}{6}\right)$

52. $\left(\dfrac{3}{8} - \dfrac{5}{9}\right) \cdot 12$

53. $\dfrac{17}{18} - \dfrac{2}{3} \cdot \left(\dfrac{5}{6} + \dfrac{1}{3}\right)$

54. $\dfrac{3}{8} - 3 \cdot \left(\dfrac{1}{6} - \dfrac{2}{9}\right)$

55. $\dfrac{8}{15} \div \dfrac{4}{5} + \dfrac{5}{6} \cdot \dfrac{9}{25}$

56. $\dfrac{7}{9} - \dfrac{1}{4} \div \dfrac{3}{8} + \dfrac{1}{3}$

APPLICATIONS INVOLVING COMPLEX FRACTIONS

57. **Walking Rate** Use the formula
$$\text{Rate} = \dfrac{\text{Distance}}{\text{Time}}$$
to find the rate for a person who walks $2\frac{1}{2}$ miles in $\frac{4}{5}$ hour. Express your answer as a mixed number.

58. **Flying Distance** A fighter jet is flying at a speed of $\frac{9}{20}$ miles per second. Use the formula
$$\text{Time} = \dfrac{\text{Distance}}{\text{Rate}}$$
to find the time in seconds that it takes for the jet to fly 54 miles.

59. **Average Time** A student completes his math homework in $\frac{5}{6}$ hour and his science homework in $\frac{2}{3}$ hour. Find the average time for completing these two homework assignments.

60. **Serving Size** If a sugar bowl contains $8\frac{1}{2}$ cups of sugar, how many $\frac{1}{4}$-cup scoops of sugar can be taken from the bowl?

61. **Converting Temperature** To convert a temperature F given in degrees Fahrenheit to an equivalent temperature C in degrees Celsius, use the formula
$$C = \dfrac{5}{9}(F - 32).$$
Find the Celsius temperature that is equivalent to a temperature of $82\frac{2}{5}$ °F.

62. **Converting Temperature** To convert a temperature C given in degrees Celsius to an equivalent temperature F in degrees Fahrenheit, use the formula
$$F = \dfrac{9}{5}C + 32.$$
Find the Fahrenheit temperature that is equivalent to a temperature of $18\frac{1}{3}$ °C.

WRITING ABOUT MATHEMATICS

63. Write a paragraph explaining the steps involved in simplifying a complex fraction using Method I.

64. Write a paragraph explaining the steps involved in simplifying a complex fraction using Method II.

4.8 Solving Equations Involving Fractions

Objectives

1. Solving Equations Symbolically
   - Using the Addition Property of Equality
   - Using the Multiplication Property of Equality
   - Clearing Fractions
2. Solving Equations Numerically
3. Solving Equations Graphically
4. Solving Applications Involving Equations with Fractions

NEW VOCABULARY

☐ Trapezoid
☐ Base (of a trapezoid)

1 Solving Equations Symbolically

In a previous chapter, we studied the properties of equality that are used to solve linear equations. These properties are also valid for equations involving fractions. In this subsection, we will review the *addition property of equality* and the *multiplication property of equality* and then apply each property to equations involving fractions.

STUDY TIP

The equations presented in this section are solved symbolically, numerically, and graphically. These are the same methods used to solve many equations throughout the course. Remember these methods when you are trying to solve equations.

USING THE ADDITION PROPERTY OF EQUALITY Because subtraction can be rewritten as addition of the opposite, there are two ways to state the addition property of equality.

1. $a = b$ is equivalent to $a + c = b + c$.
2. $a = b$ is equivalent to $a - c = b - c$.

The requirements for the variables a, b, and c are simply that they represent numbers. Since fractions are numbers, the addition property of equality can be used to solve equations involving fractions, as demonstrated in the next example.

EXAMPLE 1 Using the addition property of equality

Solve each equation and check the solution.

(a) $x + \dfrac{3}{4} = \dfrac{5}{6}$ (b) $2 = y - \dfrac{8}{3}$

Solution

(a) Because $\dfrac{3}{4}$ is being **added** to x in the given equation, we can isolate x on the left side of the equation by **subtracting** $\dfrac{3}{4}$ from each side of the equation.

$$x + \frac{3}{4} = \frac{5}{6} \quad \text{Given equation}$$

$$x + \frac{3}{4} - \frac{3}{4} = \frac{5}{6} - \frac{3}{4} \quad \text{Subtract } \tfrac{3}{4} \text{ from each side.}$$

$$x = \frac{5}{6} \cdot \frac{2}{2} - \frac{3}{4} \cdot \frac{3}{3} \quad \text{Simplify; the LCD is 12.}$$

$$x = \frac{10}{12} - \frac{9}{12} \quad \text{Multiply.}$$

$$x = \frac{1}{12} \quad \text{Subtract.}$$

To check this solution, replace x with $\dfrac{1}{12}$ in the *given* equation.

READING CHECK 1

- Use the addition property of equality to solve the equation $x - \dfrac{3}{2} = \dfrac{1}{2}$.

$$x + \frac{3}{4} = \frac{5}{6}$$ Given equation

$$\frac{1}{12} + \frac{3}{4} \stackrel{?}{=} \frac{5}{6}$$ Replace x with $\frac{1}{12}$.

$$\frac{1}{12} + \frac{3}{4} \cdot \frac{3}{3} \stackrel{?}{=} \frac{5}{6}$$ The LCD is 12.

$$\frac{1}{12} + \frac{9}{12} \stackrel{?}{=} \frac{5}{6}$$ Multiply.

$$\frac{10}{12} \stackrel{?}{=} \frac{5}{6}$$ Add.

$$\frac{5}{6} = \frac{5}{6} \checkmark$$ Simplify; it checks.

(b) Because $\frac{8}{3}$ is being **subtracted** from y in the given equation, we can isolate y on the right side of the equation by **adding** $\frac{8}{3}$ to each side of the equation.

$$2 = y - \frac{8}{3}$$ Given equation

$$2 + \frac{8}{3} = y - \frac{8}{3} + \frac{8}{3}$$ Add $\frac{8}{3}$ to each side.

$$\frac{2}{1} \cdot \frac{3}{3} + \frac{8}{3} = y$$ The LCD is 3; simplify.

$$\frac{6}{3} + \frac{8}{3} = y$$ Multiply.

$$\frac{14}{3} = y$$ Add.

To check this solution, replace y with $\frac{14}{3}$ in the *given* equation.

$$2 = y - \frac{8}{3}$$ Given equation

$$2 \stackrel{?}{=} \frac{14}{3} - \frac{8}{3}$$ Replace y with $\frac{14}{3}$.

$$2 \stackrel{?}{=} \frac{6}{3}$$ Subtract.

$$2 = 2 \checkmark$$ Simplify; it checks.

Now Try Exercises 5, 9

USING THE MULTIPLICATION PROPERTY OF EQUALITY There are two ways to state the multiplication property of equality because division (if defined) can be rewritten as multiplication by the reciprocal.

1. $a = b$ is equivalent to $ac = bc$, where $c \neq 0$.

2. $a = b$ is equivalent to $\frac{a}{c} = \frac{b}{c}$, where $c \neq 0$.

READING CHECK 2

- Use the multiplication property of equality to solve the equation $\frac{1}{2}x = 4$.

The variables a, b, and c must represent numbers with $c \neq 0$. Since fractions are numbers, the multiplication property of equality can be used to solve equations involving fractions, as illustrated in the next example.

EXAMPLE 2 **Using the multiplication property of equality**

Solve each equation and check the solution.

(a) $\dfrac{2}{3}x = 8$ (b) $\dfrac{w}{4} = \dfrac{3}{10}$ (c) $-2y = \dfrac{8}{9}$

Solution

(a) Because x is **multiplied** by $\frac{2}{3}$ in the given equation, we can isolate x on the left side of the equation by **dividing** each side of the equation by $\frac{2}{3}$. Recall that dividing by a fraction is the same as multiplying by its reciprocal. That is, to isolate x, we multiply each side of the equation by the reciprocal of $\frac{2}{3}$, which is $\frac{3}{2}$.

$$\dfrac{2}{3}x = 8 \qquad \text{Given equation}$$

$$\dfrac{3}{2} \cdot \dfrac{2}{3}x = \dfrac{3}{2} \cdot \dfrac{8}{1} \qquad \text{Multiply each side by } \tfrac{3}{2}.$$

$$1x = \dfrac{3 \cdot 8}{2} \qquad \text{Multiply.}$$

$$x = 12 \qquad \text{Simplify.}$$

To check this solution, replace x with 12 in the *given* equation.

$$\dfrac{2}{3}x = 8 \qquad \text{Given equation}$$

$$\dfrac{2}{3} \cdot 12 \stackrel{?}{=} 8 \qquad \text{Replace } x \text{ with 12.}$$

$$\dfrac{2}{3} \cdot \dfrac{12}{1} \stackrel{?}{=} 8 \qquad \text{Write 12 as } \tfrac{12}{1}.$$

$$8 = 8 \checkmark \qquad \text{Simplify; it checks.}$$

(b) Because w is **divided** by 4 in the given equation, we can isolate w on the left side of the equation by **multiplying** each side of the equation by 4, written as $\frac{4}{1}$.

$$\dfrac{w}{4} = \dfrac{3}{10} \qquad \text{Given equation}$$

$$\dfrac{4}{1} \cdot \dfrac{w}{4} = \dfrac{4}{1} \cdot \dfrac{3}{10} \qquad \text{Multiply each side by } \tfrac{4}{1}.$$

$$1w = \dfrac{4 \cdot 3}{1 \cdot 10} \qquad \text{Multiply.}$$

$$w = \dfrac{6}{5} \qquad \text{Simplify.}$$

To check this solution, replace w with $\frac{6}{5}$ in the *given* equation.

$$\dfrac{w}{4} = \dfrac{3}{10} \qquad \text{Given equation}$$

$$\dfrac{\frac{6}{5}}{4} \stackrel{?}{=} \dfrac{3}{10} \qquad \text{Replace } w \text{ with } \tfrac{6}{5}.$$

$$\dfrac{6}{5} \cdot \dfrac{1}{4} \stackrel{?}{=} \dfrac{3}{10} \qquad \text{Invert and multiply.}$$

$$\dfrac{6 \cdot 1}{5 \cdot 4} \stackrel{?}{=} \dfrac{3}{10} \qquad \text{Multiply.}$$

$$\dfrac{3}{10} = \dfrac{3}{10} \checkmark \qquad \text{Simplify; it checks.}$$

(c) Because y is **multiplied** by -2 in the given equation, we can isolate y on the left side of the equation by **dividing** each side of the equation by -2. However, rather than dividing the fraction $\frac{8}{9}$ by -2, we will multiply each side by the reciprocal of -2, or $-\frac{1}{2}$.

$$-2y = \frac{8}{9} \quad \text{Given equation}$$

$$-\frac{1}{2} \cdot (-2y) = -\frac{1}{2} \cdot \frac{8}{9} \quad \text{Multiply each side by } -\tfrac{1}{2}.$$

$$1y = -\frac{1 \cdot 8}{2 \cdot 9} \quad \text{Multiply.}$$

$$y = -\frac{4}{9} \quad \text{Simplify.}$$

To check this solution, replace y with $-\frac{4}{9}$ in the *given* equation.

$$-2y = \frac{8}{9} \quad \text{Given equation}$$

$$-2 \cdot \left(-\frac{4}{9}\right) \stackrel{?}{=} \frac{8}{9} \quad \text{Replace } y \text{ with } -\tfrac{4}{9}.$$

$$\left(-\frac{2}{1}\right) \cdot \left(-\frac{4}{9}\right) \stackrel{?}{=} \frac{8}{9} \quad \text{Write } -2 \text{ as } -\tfrac{2}{1}.$$

$$\frac{2 \cdot 4}{1 \cdot 9} \stackrel{?}{=} \frac{8}{9} \quad \text{Multiply.}$$

$$\frac{8}{9} = \frac{8}{9} \checkmark \quad \text{Simplify; it checks.}$$

Now Try Exercises 13, 17, 25

CLEARING FRACTIONS One of the most efficient ways to solve equations involving fractions is to multiply each side of the equation by the LCD of all fractions within the equation. Doing so "clears fractions," or eliminates fractions, from the equation and produces an equivalent equation.

EXAMPLE 3 Solving equations by clearing fractions

Solve each equation.

(a) $\dfrac{5x}{8} = \dfrac{15}{4}$ (b) $y - \dfrac{1}{6} = \dfrac{4}{9}$

Solution
(a) Multiply each side of the equation by 8, the LCD of all fractions involved. We write 8 in the form $\frac{8}{1}$ in order to simplify computation.

$$\frac{5x}{8} = \frac{15}{4} \quad \text{Given equation}$$

$$\frac{8}{1} \cdot \frac{5x}{8} = \frac{15}{4} \cdot \frac{8}{1} \quad \text{Multiply each side by 8}$$

$$1 \cdot 5x = 15 \cdot 2 \quad \text{Simplify each side.}$$

$$5x = 30 \quad \text{Multiply.}$$

$$\frac{5x}{5} = \frac{30}{5} \quad \text{Divide each side by 5.}$$

$$x = 6 \quad \text{Simplify.}$$

READING CHECK 3

- To clear fractions from $\frac{3x}{4} - 8 = \frac{5}{6}$, multiply each side of the equation by what value?

(b) Multiply each side of the equation by 18, which is the LCD. Multiplying the left side of the equation by 18 requires the use of the distributive property.

$$y - \frac{1}{6} = \frac{4}{9} \qquad \text{Given equation}$$

$$18 \cdot \left(y - \frac{1}{6}\right) = 18 \cdot \frac{4}{9} \qquad \text{Multiply each side by 18.}$$

$$18y - \frac{18}{1} \cdot \frac{1}{6} = \frac{18}{1} \cdot \frac{4}{9} \qquad \text{Distributive property}$$

$$18y - 3 = 2 \cdot 4 \qquad \text{Simplify each side.}$$

$$18y - 3 = 8 \qquad \text{Multiply.}$$

$$18y - 3 + 3 = 8 + 3 \qquad \text{Add 3 to each side.}$$

$$18y = 11 \qquad \text{Simplify.}$$

$$\frac{18y}{18} = \frac{11}{18} \qquad \text{Divide each side by 18.}$$

$$y = \frac{11}{18} \qquad \text{Simplify.}$$

Now Try Exercises 27, 29

The process of multiplying each side of an equation by a number (the LCD) is equivalent to multiplying *each term* in the equation by that number. With this in mind, the following steps can be used to solve equations involving fractions.

SOLVING EQUATIONS INVOLVING FRACTIONS SYMBOLICALLY

To solve an equation involving fractions symbolically, do the following.

STEP 1: Use the distributive property to clear parentheses from the equation.

STEP 2: Multiply each term in the equation by the LCD of all fractions involved.

STEP 3: Simplify each term in the equation and combine any like terms.

STEP 4: Solve the resulting equation.

EXAMPLE 4   **Solving equations involving fractions symbolically**

Solve each equation.

(a) $\dfrac{5}{9}x + \dfrac{1}{3} = \dfrac{5}{6}$ **(b)** $\dfrac{x}{4} - \dfrac{x}{6} = \dfrac{2}{3}$ **(c)** $\dfrac{3}{5}\left(2x - \dfrac{1}{2}\right) = \dfrac{7}{10}$

Solution

(a) Because there are no parentheses in the equation, we start with Step 2. Multiply each term in the equation by 18, the LCD of all fractions involved.

$$\frac{5}{9}x + \frac{1}{3} = \frac{5}{6} \qquad \text{Given equation}$$

$$\frac{18}{1} \cdot \frac{5}{9}x + \frac{18}{1} \cdot \frac{1}{3} = \frac{18}{1} \cdot \frac{5}{6} \qquad \text{Multiply each term by 18. (Step 2)}$$

$$10x + 6 = 15 \qquad \text{Simplify each term. (Step 3)}$$

$$10x + 6 - 6 = 15 - 6 \qquad \text{Subtract 6 from each side. (Step 4)}$$

$$10x = 9 \qquad \text{Simplify.}$$

$$\frac{10x}{10} = \frac{9}{10} \qquad \text{Divide each side by 10. (Step 4)}$$

$$x = \frac{9}{10} \qquad \text{Simplify.}$$

(b) Begin with Step 2. Multiply each term in the equation by 12, the LCD.

$$\frac{x}{4} - \frac{x}{6} = \frac{2}{3} \qquad \text{Given equation}$$

$$\frac{12}{1} \cdot \frac{x}{4} - \frac{12}{1} \cdot \frac{x}{6} = \frac{12}{1} \cdot \frac{2}{3} \qquad \text{Multiply each term by 12. (Step 2)}$$

$$3x - 2x = 8 \qquad \text{Simplify each term.}$$

$$x = 8 \qquad \text{Combine like terms. (Steps 3 and 4)}$$

(c) First use the distributive property to clear the parentheses from the equation.

$$\frac{3}{5}\left(2x - \frac{1}{2}\right) = \frac{7}{10} \qquad \text{Given equation}$$

$$\frac{3}{5} \cdot \frac{2}{1} x - \frac{3}{5} \cdot \frac{1}{2} = \frac{7}{10} \qquad \text{Distributive property (Step 1)}$$

$$\frac{6}{5} x - \frac{3}{10} = \frac{7}{10} \qquad \text{Multiply.}$$

$$\frac{10}{1} \cdot \frac{6}{5} x - \frac{10}{1} \cdot \frac{3}{10} = \frac{10}{1} \cdot \frac{7}{10} \qquad \text{Multiply each term by 10. (Step 2)}$$

$$12x - 3 = 7 \qquad \text{Simplify each term. (Step 3)}$$

$$12x - 3 + 3 = 7 + 3 \qquad \text{Add 3 to each side. (Step 4)}$$

$$12x = 10 \qquad \text{Simplify.}$$

$$\frac{12x}{12} = \frac{10}{12} \qquad \text{Divide each side by 12. (Step 4)}$$

$$x = \frac{5}{6} \qquad \text{Simplify.}$$

Now Try Exercises 37, 41, 47

MAKING CONNECTIONS 1

Clearing Fractions and the Properties of Equality

There are *many correct* ways to solve an equation. For example, the solution to the equation in Example 4(a) can be found without first clearing fractions.

$$\frac{5}{9}x + \frac{1}{3} = \frac{5}{6} \qquad \text{Given equation}$$

$$\frac{5}{9}x + \frac{1}{3} - \frac{1}{3} = \frac{5}{6} - \frac{1}{3} \qquad \text{Subtract } \tfrac{1}{3} \text{ from each side.}$$

$$\frac{5}{9}x = \frac{5}{6} - \frac{1}{3} \cdot \frac{2}{2} \qquad \text{The LCD is 6.}$$

$$\frac{5}{9}x = \frac{1}{2} \qquad \text{Subtract; simplify.}$$

$$\frac{9}{5} \cdot \frac{5}{9}x = \frac{9}{5} \cdot \frac{1}{2} \qquad \text{Multiply each side by } \tfrac{9}{5}.$$

$$x = \frac{9}{10} \qquad \text{Simplify.}$$

2 Solving Equations Numerically

Equations involving fractions can also be solved numerically. In the next example, a table of values is used to solve an equation involving fractions.

EXAMPLE 5 Solving an equation numerically

Complete **TABLE 4.2** for the given values of x and then solve the equation $\frac{3}{4}x - 2 = -\frac{1}{2}$.

x	-1	0	1	2	3	4
$\frac{3}{4}x - 2$						

TABLE 4.2

Solution
Begin by replacing x in $\frac{3}{4}x - 2$ with -1.

$$\frac{3}{4}(-1) - 2 = -\frac{3}{4} - \frac{8}{4} = \frac{-3 - 8}{4} = -\frac{11}{4}$$

We write the result $-\frac{11}{4}$ below -1, as shown in **TABLE 4.3**. The remaining values in **TABLE 4.3** are found as follows.

$\frac{3}{4}(0) - 2 = 0 - 2 = -2$ $\qquad$ $\frac{3}{4}(1) - 2 = \frac{3}{4} - \frac{8}{4} = \frac{3 - 8}{4} = -\frac{5}{4}$

$\frac{3}{4}(2) - 2 = \frac{3}{2} - \frac{4}{2} = \frac{3 - 4}{2} = -\frac{1}{2}$ $\qquad$ $\frac{3}{4}(3) - 2 = \frac{9}{4} - \frac{8}{4} = \frac{9 - 8}{4} = \frac{1}{4}$

$\frac{3}{4}(4) - 2 = 3 - 2 = 1$

> The solution is **2** and is found numerically.

x	-1	0	1	2	3	4
$\frac{3}{4}x - 2$	$-\frac{11}{4}$	-2	$-\frac{5}{4}$	$-\frac{1}{2}$	$\frac{1}{4}$	1

TABLE 4.3

TABLE 4.3 reveals that the left side of the given equation equals $-\frac{1}{2}$ when $x = 2$. Therefore, the solution is **2**.

Now Try Exercise 51

READING CHECK 4

- How does a solution that is found graphically compare to a solution found symbolically or numerically?

3 Solving Equations Graphically

When a graph is provided for an equation involving fractions, an estimate of the solution can be found graphically. As we have seen earlier, solutions found graphically must be checked. In the next example, we estimate and check a solution to an equation involving fractions.

EXAMPLE 6 Solving an equation graphically

A graph of the expression $\frac{3}{2}x + 1$ is shown in **FIGURE 4.9**. Use the graph to solve the equation $\frac{3}{2}x + 1 = 4$ graphically.

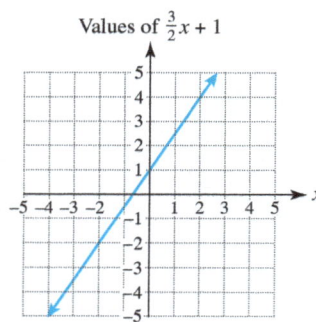

FIGURE 4.9

Solution

Locate **4** on the vertical axis and move **horizontally** to the right to the graphed line. From this position, move **vertically** downward to the horizontal axis, as shown in **FIGURE 4.10**. The solution appears to be **2**.

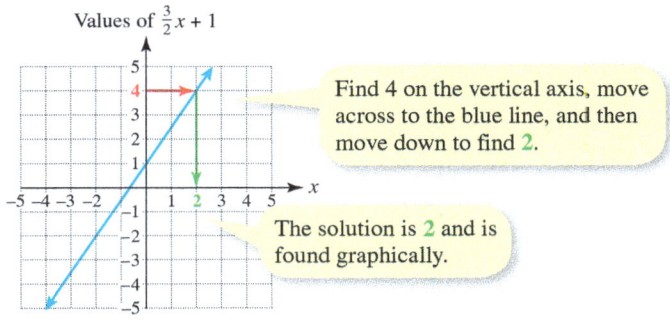

FIGURE 4.10

To see if **2** is the correct solution, check it in the given equation as $\frac{2}{1}$.

$$\frac{3}{2}x + 1 = 4 \qquad \text{Given equation}$$

$$\frac{3}{2}\left(\frac{2}{1}\right) + 1 \stackrel{?}{=} 4 \qquad \text{Replace } x \text{ with } \tfrac{2}{1}.$$

$$3 + 1 \stackrel{?}{=} 4 \qquad \text{Simplify.}$$

$$4 = 4 \;\checkmark \qquad \text{The solution checks.}$$

The solution to the equation $\frac{3}{2}x + 1 = 4$ is 2.

Now Try Exercise 55

4 Solving Applications Involving Equations with Fractions

A **trapezoid** is a four-sided geometric shape with one pair of parallel sides. Each parallel side is called a **base** of the trapezoid. An example of a trapezoid is shown in **FIGURE 4.11**.

Trapezoid with Height *h* and Bases *a* and *b*

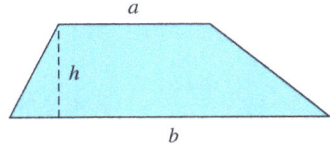

FIGURE 4.11

To compute the area of a trapezoid, we use the following formula.

AREA OF A TRAPEZOID

The area A of a trapezoid with height h and bases a and b is given by

$$A = \frac{1}{2}(a + b)h.$$

In the next example, we use the trapezoid area formula to set up an equation whose solution is an unknown measure for a given trapezoid.

EXAMPLE 7 **Finding an unknown measure for a trapezoid**

Find the length of base *a* for the trapezoid in the following figure if its area is 4 square feet.

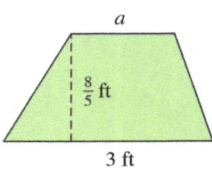

Solution
Let $A = 4$, $b = 3$, and $h = \frac{8}{5}$ in the trapezoid area formula: $A = \frac{1}{2}(a + b)h$.

$$4 = \frac{1}{2}(a + 3)\frac{8}{5} \qquad \text{Equation to solve}$$

$$4 = \frac{1}{2} \cdot \frac{8}{5}(a + 3) \qquad \text{Commutative property}$$

$$4 = \frac{4}{5}(a + 3) \qquad \text{Multiply; simplify.}$$

$$4 = \frac{4}{5}a + \frac{4}{5} \cdot \frac{3}{1} \qquad \text{Distributive property}$$

$$4 = \frac{4}{5}a + \frac{12}{5} \qquad \text{Multiply.}$$

$$5 \cdot 4 = \frac{5}{1} \cdot \frac{4}{5}a + \frac{5}{1} \cdot \frac{12}{5} \qquad \text{Multiply each term by 5.}$$

$$20 = 4a + 12 \qquad \text{Simplify each term.}$$

$$20 - 12 = 4a + 12 - 12 \qquad \text{Subtract 12 from each side.}$$

$$8 = 4a \qquad \text{Simplify.}$$

$$\frac{8}{4} = \frac{4a}{4} \qquad \text{Divide each side by 4.}$$

$$2 = a \qquad \text{Simplify.}$$

The length of base *a* is 2 feet.

Now Try Exercise 73

🌐 **Math in Context** *Temperature* While U.S. temperatures are usually reported using the Fahrenheit temperature scale, many other countries throughout the world use the Celsius temperature scale. In the next example, an equation involving fractions is solved in order to convert a Fahrenheit temperature to its equivalent Celsius temperature.

EXAMPLE 8 **Finding a Celsius temperature**

The formula

$$F = \frac{9}{5}C + 32$$

Use this formula to change degrees Celsius to degrees Fahrenheit.

gives the relationship between *F* degrees Fahrenheit and *C* degrees Celsius. Use the formula to find the Celsius temperature that is equivalent to $57\frac{1}{5}$ °F.

Solution

Write $57\frac{1}{5}$ as the improper fraction $\frac{286}{5}$ and substitute it in the formula for F.

$$\frac{286}{5} = \frac{9}{5}C + 32 \qquad \text{Equation to solve}$$

$$\frac{5}{1} \cdot \frac{286}{5} = \frac{5}{1} \cdot \frac{9}{5}C + 5 \cdot 32 \qquad \text{Multiply each term by 5.}$$

$$286 = 9C + 160 \qquad \text{Simplify each term.}$$

$$286 - 160 = 9C + 160 - 160 \qquad \text{Subtract 160 from each side.}$$

$$126 = 9C \qquad \text{Simplify.}$$

$$\frac{126}{9} = \frac{9C}{9} \qquad \text{Divide each side by 9.}$$

$$14 = C \qquad \text{Simplify.}$$

Therefore, $57\frac{1}{5}$ °F is equivalent to 14 °C.

Now Try Exercise 77

An adult can often perform physical tasks faster than a child. For example, it may take an adult 3 hours to rake leaves in a yard, while the same task may take a child 6 hours to complete. Working together, the adult and child should be able to rake the lawn in less time than it takes the faster person working alone. So, it should take less than 3 hours. To find the time required for two people to rake the lawn while working together, we use an equation involving fractions.

EXAMPLE 9 Finding time to complete a task when working together

If an adult can rake a lawn in 3 hours and the same task takes a child 6 hours to complete, how long will it take for these two people working together to rake the lawn?

Solution

The adult can rake the entire lawn in 3 hours, which means that the adult rakes $\frac{1}{3}$ of the lawn per hour. By the same reasoning, the child rakes $\frac{1}{6}$ of the lawn per hour. If we let x represent the number of hours they work together, then the adult will rake $\frac{1}{3}x$ of the lawn in x hours, and the child will rake $\frac{1}{6}x$ of the lawn in x hours. Together, they will rake

$$\frac{1}{3}x + \frac{1}{6}x$$

of the lawn in x hours. The job is complete when the fraction of the lawn that is raked reaches 1 (one entire lawn is raked). To find out how long this will take the two people working together, solve the equation $\frac{1}{3}x + \frac{1}{6}x = 1$.

$$\frac{1}{3}x + \frac{1}{6}x = 1 \qquad \text{Equation to solve}$$

$$\frac{6}{1} \cdot \frac{1}{3}x + \frac{6}{1} \cdot \frac{1}{6}x = 6 \cdot 1 \qquad \text{Multiply each term by 6.}$$

$$2x + x = 6 \qquad \text{Simplify each term.}$$

$$3x = 6 \qquad \text{Combine like terms.}$$

$$\frac{3x}{3} = \frac{6}{3} \qquad \text{Divide each side by 3.}$$

$$x = 2 \qquad \text{Simplify.}$$

Together, they can rake the lawn in 2 hours.

Now Try Exercise 79

4.8 Putting It All Together

CONCEPT	COMMENTS	EXAMPLES
Properties of Equality	$a = b$ is equivalent to $a + c = b + c$. $a = b$ is equivalent to $a - c = b - c$. $a = b$ is equivalent to $ac = bc$ for $c \neq 0$. $a = b$ is equivalent to $\dfrac{a}{c} = \dfrac{b}{c}$ for $c \neq 0$.	$x - \dfrac{1}{5} = \dfrac{3}{5}$ $x - \dfrac{1}{5} + \dfrac{1}{5} = \dfrac{3}{5} + \dfrac{1}{5}$ $x = \dfrac{4}{5}$
Solving Equations Symbolically	To solve equations symbolically, 1. Use the distributive property to clear parentheses from the equation. 2. Multiply each term in the equation by the LCD of all fractions involved. 3. Simplify each term in the equation and combine any like terms. 4. Solve the resulting equation.	$4\left(x - \dfrac{5}{8}\right) = \dfrac{2}{3}$ $4x - \dfrac{5}{2} = \dfrac{2}{3}$ (Step 1) $6 \cdot 4x - \dfrac{6}{1} \cdot \dfrac{5}{2} = \dfrac{6}{1} \cdot \dfrac{2}{3}$ (Step 2) $24x - 15 = 4$ (Step 3) $24x = 19$ $x = \dfrac{19}{24}$ (Step 4)
Solving Equations Numerically	To solve equations numerically, complete a table of values for various values of the variable and then select the solution from the table, if possible.	The solution to $\frac{1}{2}x - \frac{3}{2} = -1$ is 1. | x | -1 | 0 | 1 | |---|---|---|---| | $\frac{1}{2}x - \frac{3}{2}$ | -2 | $-\frac{3}{2}$ | -1 |
Solving Equations Graphically	To solve equations graphically, use a graph of the equation to estimate a solution and check the estimated solution in the given equation to be sure that it is correct.	The solution to $-\frac{2}{5}x - \frac{4}{5} = -2$ is 3.

4.8 Exercises

CONCEPTS AND VOCABULARY

1. The _____ property of equality states that adding (subtracting) the same number to (from) each side of an equation gives an equivalent equation.

2. The _____ property of equality allows us to multiply or divide each side of an equation by the same nonzero number to obtain an equivalent equation.

3. The _____ property can be used to clear parentheses from an equation.

4. To clear fractions from an equation, multiply each term by the _____ of all fractions involved.

SOLVING EQUATIONS SYMBOLICALLY USING PROPERTIES OF EQUALITY

Exercises 5–26: Use the properties of equality to solve the given equation.

5. $x + \frac{1}{4} = \frac{9}{10}$
6. $m + \frac{5}{3} = \frac{7}{8}$
7. $w - \frac{5}{3} = -\frac{9}{5}$
8. $y - \frac{1}{6} = \frac{3}{4}$
9. $-1 = p - \frac{5}{12}$
10. $x + \frac{9}{7} = 2$
11. $x + 3 = \frac{17}{5}$
12. $-\frac{14}{3} = m - 4$
13. $\frac{3}{7}y = 6$
14. $\frac{9}{2}n = 8$
15. $5 = -\frac{2}{3}x$
16. $10 = -\frac{1}{7}x$
17. $\frac{d}{7} = \frac{9}{28}$
18. $\frac{r}{6} = -\frac{5}{16}$
19. $-\frac{k}{6} = \frac{1}{20}$
20. $-\frac{x}{8} = -\frac{1}{10}$
21. $-\frac{11}{6} = -\frac{c}{6}$
22. $\frac{3}{8} = -\frac{p}{4}$
23. $5x = \frac{10}{9}$
24. $6w = -\frac{12}{5}$
25. $-\frac{10}{3} = 8m$
26. $\frac{16}{5} = -10x$

SOLVING EQUATIONS SYMBOLICALLY BY CLEARING FRACTIONS

Exercises 27–48: Use the LCD to clear fractions and solve the given equation.

27. $\frac{5n}{6} = \frac{3}{8}$
28. $\frac{4q}{15} = -\frac{12}{5}$
29. $y + \frac{2}{9} = \frac{1}{3}$
30. $x + 3 = \frac{2}{5}$
31. $-\frac{7}{2}q = 3$
32. $-\frac{2b}{3} = \frac{8}{15}$
33. $-\frac{14a}{3} = -\frac{7}{9}$
34. $y - \frac{1}{18} = \frac{7}{9}$
35. $\frac{5}{8}x - \frac{1}{2} = \frac{3}{4}$
36. $\frac{7}{12}k + \frac{2}{3} = \frac{1}{4}$
37. $\frac{5}{6} + \frac{2}{3}w = \frac{3}{8}$
38. $\frac{3}{10}d + \frac{5}{6} = \frac{14}{15}$
39. $\frac{13}{18} = \frac{1}{12}q + \frac{2}{9}$
40. $\frac{7}{2} = \frac{1}{6} - \frac{10}{9}p$
41. $\frac{x}{4} + \frac{x}{10} = \frac{3}{5}$
42. $\frac{y}{12} - \frac{y}{4} = \frac{7}{9}$
43. $\frac{13b}{8} - \frac{7b}{12} = \frac{5}{6}$
44. $\frac{9a}{4} - \frac{11a}{16} = \frac{5}{8}$
45. $4\left(\frac{5}{2}n - \frac{2}{3}\right) = \frac{5}{6}$
46. $5\left(\frac{1}{2}r + \frac{3}{5}\right) = \frac{3}{4}$
47. $\frac{4}{5}\left(3x - \frac{1}{3}\right) = \frac{8}{15}$
48. $\frac{1}{3}(6m + 7) = \frac{5}{6}$

SOLVING EQUATIONS NUMERICALLY

Exercises 49–54: Solve the equation numerically by completing the given table of values.

49. $\frac{1}{3}x - 3 = 0$

x	1	3	5	7	9
$\frac{1}{3}x - 3$					

50. $-\frac{2}{3}x + 4 = 2$

x	0	1	2	3	4
$-\frac{2}{3}x + 4$					

51. $-\frac{4}{5}x + 3 = \frac{19}{5}$

x	-2	-1	0	1	2
$-\frac{4}{5}x + 3$					

52. $\frac{3}{4}x - 2 = -\frac{7}{2}$

x	-4	-3	-2	-1	0
$\frac{3}{4}x - 2$					

53. $\frac{1}{2}(x - \frac{2}{3}) = \frac{5}{3}$

x	-4	-2	0	2	4
$\frac{1}{2}(x - \frac{2}{3})$					

54. $-\frac{1}{4}(x + \frac{1}{2}) = \frac{7}{8}$

x	-10	-8	-6	-4	-2
$-\frac{1}{4}(x + \frac{1}{2})$					

SOLVING EQUATIONS GRAPHICALLY

Exercises 55–60: Solve the equation graphically.

55. $-\frac{5}{2}x - 2 = 3$ **56.** $\frac{3}{2}x - 1 = -1$

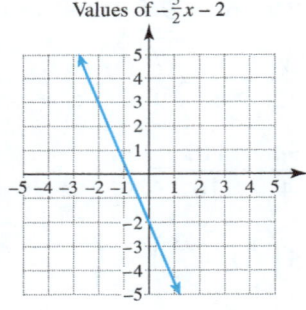 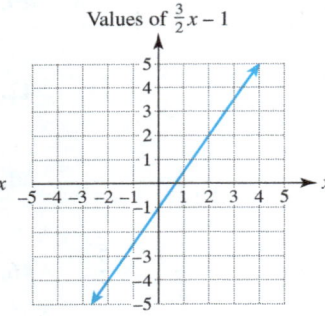

57. $\frac{2}{3}x - \frac{5}{3} = 1$ **58.** $-\frac{3}{4}x - \frac{1}{4} = 2$

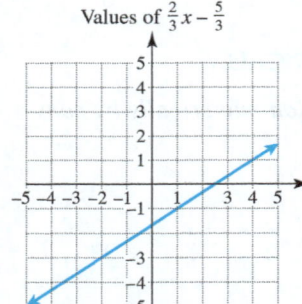 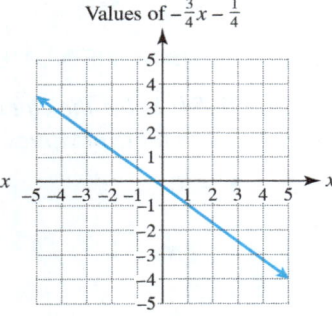

59. $-2x + \frac{5}{2} = \frac{1}{2}$ **60.** $3x + \frac{5}{2} = -\frac{1}{2}$

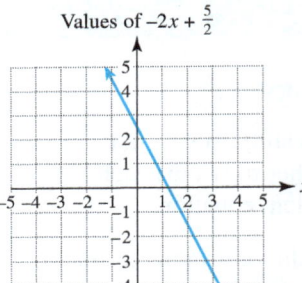 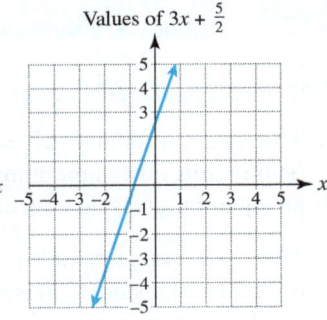

CHOOSING YOUR OWN METHOD

Exercises 61–72: Solve the given equation.

61. $w + \frac{15}{4} = 1$ **62.** $\frac{x}{8} + \frac{x}{12} = \frac{25}{6}$

63. $-\frac{p}{9} = -\frac{7}{6}$ **64.** $8y = \frac{12}{5}$

65. $\frac{1}{4}d + \frac{1}{6} = -\frac{8}{3}$ **66.** $2\left(\frac{5}{4}n - \frac{1}{2}\right) = \frac{5}{4}$

67. $\frac{9a}{20} - \frac{5a}{4} = \frac{7}{10}$ **68.** $\frac{5r}{8} = -\frac{7}{12}$

69. $\frac{3}{2}\left(3x + \frac{2}{3}\right) = \frac{9}{10}$ **70.** $-\frac{9}{7}n = 18$

71. $4\left(3 - \frac{5}{6}m\right) = \frac{32}{3}$ **72.** $7 - \frac{7a}{2} = \frac{7}{6}$

APPLICATIONS INVOLVING EQUATIONS WITH FRACTIONS

Exercises 73–76: **Geometry** *Use the trapezoid area formula, $A = \frac{1}{2}(a + b)h$, to find the unknown measure.*

73. $A = \frac{29}{16}$ square feet **74.** $A = \frac{13}{15}$ square inch

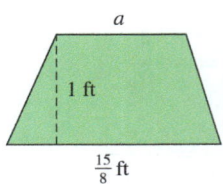 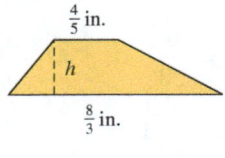

75. $A = \frac{63}{4}$ square inches **76.** $A = \frac{77}{5}$ square feet

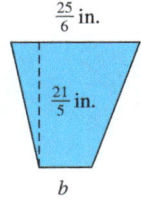 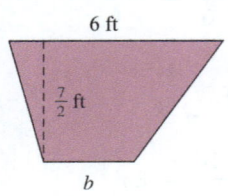

77. **Converting Temperature** The formula
$$F = \frac{9}{5}C + 32$$
gives the relationship between F degrees Fahrenheit and C degrees Celsius. Find the Celsius temperature that is equivalent to $46\frac{2}{5}$ °F.

78. **Converting Temperature** The formula
$$C = \frac{5}{9}(F - 32)$$
gives the relationship between C degrees Celsius and F degrees Fahrenheit. Find the Fahrenheit temperature that is equivalent to $33\frac{1}{3}$ °C.

79. **Working Together** If one pump can inflate an air mattress in 4 minutes and a second pump can inflate the same mattress in 12 minutes, find the amount of time that would be needed for the two pumps working together to inflate the mattress.

80. **Working Together** An experienced painter can paint a house in 6 days, while a less experienced painter can paint the same house in 10 days. How many days would it take the two painters working together to paint the house?

81. **Gaming Revenue** The revenue R, in billions of dollars, for all gaming on the Las Vegas Strip can be estimated by the formula $R = \frac{1}{6}x - 329$, where x is the year from 1984 to 2014. Find the year when gaming revenues were $5 billion by solving the equation $5 = \frac{1}{6}x - 329$. (*Source:* American Gaming Association.)
   (a) Solve symbolically.
   (b) Solve numerically using the table shown.
   (c) Solve graphically using the graph shown.

x	2000	2002	2004	2006	2008
$\frac{1}{6}x - 329$					

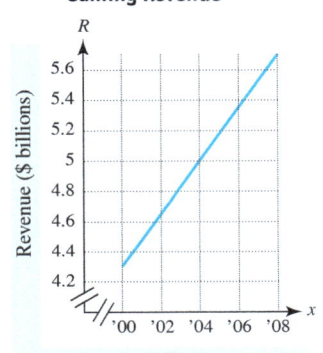

Gaming Revenue

   (d) Compare your symbolic, numerical, and graphical solutions.

82. **Carbon Monoxide Emissions** From 2002 to 2012, the amount of carbon monoxide C, in millions of tons, released into the atmosphere during year x can be estimated using the formula $C = -\frac{5}{2}x + 5120$. Find the year when carbon monoxide emissions reached 95 million tons by solving the equation $95 = -\frac{5}{2}x + 5120$. (*Source:* Environmental Protection Agency.)
   (a) Solve symbolically.
   (b) Solve numerically using the table shown.
   (c) Solve graphically using the graph shown.

x	2004	2006	2008	2010	2012
$-\frac{5}{2}x + 5120$					

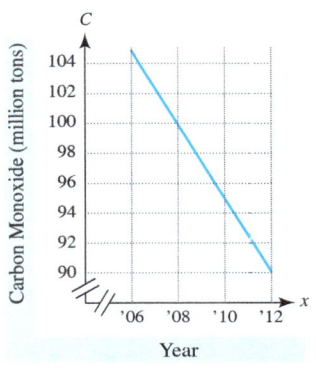

Carbon Monoxide

   (d) Compare your symbolic, numerical, and graphical solutions.

WRITING ABOUT MATHEMATICS

83. Would you rather solve the equation
$$\frac{7}{15}x + \frac{1}{6} = \frac{3}{10}$$
by clearing fractions first, or would using properties of equality work better for you? Explain.

84. Write and explain the steps needed to solve
$$\frac{4}{9}\left(5x - \frac{1}{3}\right) = \frac{7}{6}$$
by clearing fractions first *before* applying the distributive property to clear the parentheses.

SECTIONS 4.7 and 4.8 — Checking Basic Concepts

1. Simplify each complex fraction.

   (a) $\dfrac{\frac{4}{5}}{\frac{8}{15}}$ (b) $\dfrac{\frac{2w}{5} - \frac{w}{3}}{\frac{9}{5}}$

   (c) $\dfrac{\frac{4}{9} + \frac{1}{6}}{\frac{7}{12} - \frac{1}{4}}$ (d) $\dfrac{\frac{11x}{6}}{\frac{3}{4} + 2}$

2. Simplify each expression.

   (a) $\dfrac{5}{8} - \dfrac{2}{3} \cdot \dfrac{1}{2}$ (b) $\dfrac{3}{2} \cdot \left(1 - \dfrac{5}{6}\right)$

3. Solve each equation symbolically.

   (a) $y - \dfrac{3}{8} = \dfrac{1}{4}$ (b) $6 = -\dfrac{1}{5}x$

   (c) $\dfrac{4}{9}d + \dfrac{5}{6} = \dfrac{7}{2}$ (d) $\dfrac{9a}{4} - \dfrac{7a}{12} = \dfrac{5}{3}$

4. Solve $-\dfrac{1}{2}(x-1) = 4$ numerically.

x	-11	-9	-7	-5	-3
$-\frac{1}{2}(x-1)$					

5. Solve $-\dfrac{3}{4}x + \dfrac{1}{2} = 2$ graphically.

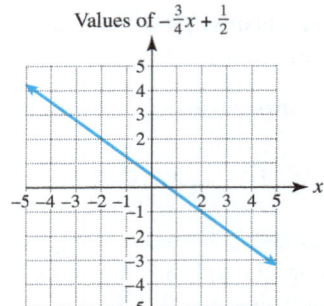

6. **Converting Temperature** The formula
   $$C = \dfrac{5}{9}(F - 32)$$
   gives the relationship between C degrees Celsius and F degrees Fahrenheit. Find the Fahrenheit temperature that is equivalent to $23\frac{1}{3}$ °C.

CHAPTER 4 Summary

SECTION 4.1 ■ INTRODUCTION TO FRACTIONS AND MIXED NUMBERS

Fraction   A fraction is a number that describes a portion of a whole.

Example: $\dfrac{5}{8}$

Rational Number   A rational number can be expressed as $\dfrac{p}{q}$, where p and q are integers with $q \neq 0$.

Examples: $\dfrac{2}{3},\ -\dfrac{4}{9},\ 15 = \dfrac{15}{1}$

Properties of Fractions   The identity and zero properties for fractions are as follows.

1. When $n \neq 0$, $\dfrac{n}{n}$ simplifies to 1.
2. $\dfrac{n}{1}$ simplifies to n.
3. When $n \neq 0$, $\dfrac{0}{n}$ simplifies to 0.
4. $\dfrac{n}{0}$ is undefined.

Examples: 1. $\dfrac{3}{3} = 1$,   2. $\dfrac{7}{1} = 7$,   3. $\dfrac{0}{5} = 0$,   4. $\dfrac{8}{0}$ is undefined.

Proper and Improper Fractions

The fraction $\frac{a}{b}$ with $b \neq 0$ is a proper fraction whenever $|a| < |b|$, and it is an improper fraction whenever $|a| \geq |b|$.

Examples: Proper   Improper

$$\frac{6}{11}, -\frac{3}{8}, \frac{20}{29} \qquad -\frac{7}{2}, \frac{16}{5}, \frac{45}{45}$$

Mixed Number

A mixed number is an integer written with a proper fraction.

Examples: $7\frac{3}{8}, -12\frac{1}{2}, 8\frac{23}{24}$

Writing an Improper Fraction as a Mixed Number

STEP 1: Divide the denominator into the numerator.

STEP 2: The quotient is the integer part of the mixed number. The fraction part of the mixed number has the remainder in its numerator and the divisor in its denominator.

$$\text{Mixed Number} = \text{Quotient} \frac{\text{Remainder}}{\text{Divisor}}$$

Examples: $\frac{5}{2} = 2\frac{1}{2}, \frac{29}{3} = 9\frac{2}{3}$

Writing a Mixed Number as an Improper Fraction

STEP 1: Multiply the denominator of the fraction by the integer.

STEP 2: Add the numerator of the fraction to the product from Step 1.

STEP 3: The sum found in Step 2 is the numerator of the improper fraction. The denominator of the improper fraction is the denominator from the original mixed number. So, a mixed number of the form

$$\text{Quotient} \frac{\text{Remainder}}{\text{Divisor}} \text{ is written as } \frac{\text{Divisor} \cdot \text{Quotient} + \text{Remainder}}{\text{Divisor}}.$$

Examples: $3\frac{7}{9} = \frac{9 \cdot 3 + 7}{9} = \frac{34}{9}, -6\frac{1}{5} = -\left(\frac{5 \cdot 6 + 1}{5}\right) = -\frac{31}{5}$

Graphing Fractions and Mixed Numbers

When graphing a fraction, divide each whole distance on the number line into an equal number of parts determined by the denominator of the fraction.

Examples: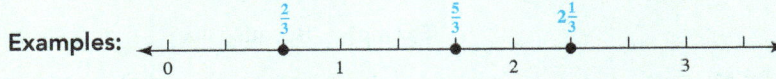

SECTION 4.2 ■ PRIME FACTORIZATION AND LOWEST TERMS

Factor

A factor is any whole number that divides into a second whole number and results in a quotient with no remainder.

Example: 7 is a factor of 21 because $21 \div 7 = 3$ with no remainder. We can also say that 21 is divisible by 7.

Divisibility Tests

A whole number is divisible by

- 2 if its ones digit is 0, 2, 4, 6, or 8.
- 3 if the sum of its digits is divisible by 3.
- 4 if the number formed by its last two digits is divisible by 4.
- 5 if its ones digit is 0 or 5.
- 6 if it is divisible by 2 and 3.
- 9 if the sum of its digits is divisible by 9.

Examples: 5476 is divisible by 2 because its ones digit is 6.
9 is a factor of 621 because the digit sum 9 is divisible by 9.
34,865 divides evenly by 5 because its ones digit is 5.

Prime and Composite Numbers

A prime number is any natural number greater than 1 whose only factors are 1 and itself. A composite number is a number greater than 1 that is not prime.

Examples: **Prime** **Composite**
2, 7, 89, and 101   8, 14, 39, and 117

Prime Factorization

Every composite number can be factored into a unique product of primes.

Examples: $108 = 2 \cdot 2 \cdot 3 \cdot 3 \cdot 3$, $910 = 2 \cdot 5 \cdot 7 \cdot 13$

Equivalent Fractions

Fractions that name the same number are called equivalent fractions. To write equivalent fractions, use the basic principle of fractions.

Example: The fractions $\frac{2}{3}$ and $\frac{10}{15}$ are equivalent because $\frac{2}{3} = \frac{2 \cdot 5}{3 \cdot 5} = \frac{10}{15}$.

Greatest Common Factor (GCF)

The largest number that divides evenly into two or more numbers is the GCF.

Example: The GCF of 25 and 40 is 5.

Lowest Terms

A fraction is in lowest terms if the GCF of its numerator and denominator is 1.

Example: The fraction $\frac{15}{60}$ written in lowest terms is $\frac{1}{4}$.

Simplifying to Lowest Terms

To simplify a fraction to lowest terms,

1. Determine a number c that is the GCF of the numerator and denominator.
2. Write the numerator as a product of two factors, a and c. Then write the denominator as a product of two factors, b and c.
3. Apply the basic principle of fractions.

Example: For $\frac{14}{35}$, the GCF of 14 and 35 is 7, so $c = 7$. In simplest form,

$$\frac{14}{35} = \frac{2 \cdot 7}{5 \cdot 7} = \frac{2}{5}.$$

Cross Product Method

We can determine if two fractions are equivalent by comparing their cross products.

Example: Because $12 \cdot 21 = 28 \cdot 9$, we know that $\frac{21}{28} = \frac{9}{12}$.

Rational Expressions

Any fraction whose numerator and denominator each consist of an expression that can be written as the sum or difference of two or more terms

Examples: $\frac{12x^2}{5}, \frac{3x+7}{2x-1},$ and $\frac{5ab}{3c^2}$

SECTION 4.3 ■ MULTIPLYING AND DIVIDING FRACTIONS

Multiplying Fractions and Rational Expressions

If a, b, c, and d represent numbers (or single terms) with $b \neq 0$ and $d \neq 0$, then

$$\frac{a}{b} \cdot \frac{c}{d} = \frac{ac}{bd}.$$

Examples: $\frac{1}{5} \cdot \frac{3}{4} = \frac{1 \cdot 3}{5 \cdot 4} = \frac{3}{20}$, $\frac{3x}{2y} \cdot \frac{x}{4} = \frac{3x \cdot x}{2y \cdot 4} = \frac{3x^2}{8y}$

Simplifying before Multiplying	**STEP 1:** Write $\frac{a}{b} \cdot \frac{c}{d}$ as $\frac{a \cdot c}{b \cdot d}$. Do not multiply. **STEP 2:** Replace each composite factor with its prime factorization. **STEP 3:** Eliminate all factors that are common to the numerator and denominator by applying the basic principle of fractions. **STEP 4:** Multiply the remaining factors in the numerator and multiply the remaining factors in the denominator. **Example:** $\frac{1}{3} \cdot \frac{6}{7} = \frac{1 \cdot 6}{3 \cdot 7} = \frac{1 \cdot (2 \cdot 3)}{3 \cdot 7} = \frac{1 \cdot 2}{7} = \frac{2}{7}$
Powers of Fractions	For natural number exponent n, multiply the fraction by itself n times. **Example:** $\left(\frac{2}{3}\right)^3 = \frac{2}{3} \cdot \frac{2}{3} \cdot \frac{2}{3} = \frac{8}{27}$
Square Roots of Fractions	If a and b represent whole numbers with $b \neq 0$ and the fraction $\frac{a}{b}$ is in lowest terms, then $$\sqrt{\frac{a}{b}} = \frac{\sqrt{a}}{\sqrt{b}}.$$ **Example:** $\sqrt{\frac{81}{100}} = \frac{\sqrt{81}}{\sqrt{100}} = \frac{9}{10}$
Reciprocal	Two numbers are reciprocals (or multiplicative inverses) of each other if their product is 1. **Example:** $\frac{5}{7}$ and $\frac{7}{5}$ are reciprocals because $\frac{5}{7} \cdot \frac{7}{5} = 1$.
Dividing Fractions and Rational Expressions	If a, b, c, and d represent numbers (or single terms) with $b \neq 0$, $c \neq 0$, and $d \neq 0$, then the quotient found when dividing $\frac{a}{b}$ by $\frac{c}{d}$ is given by $$\frac{a}{b} \div \frac{c}{d} = \frac{a}{b} \cdot \frac{d}{c}.$$ **Examples:** $\frac{2}{7} \div \frac{3}{5} = \frac{2}{7} \cdot \frac{5}{3} = \frac{10}{21}$, $\frac{3x}{2} \div \frac{5}{x} = \frac{3x}{2} \cdot \frac{x}{5} = \frac{3x^2}{10}$
Area of a Triangle	The area A of a triangle with base b and height h is given by $A = \frac{1}{2}bh$. **Example:** For the triangle shown, $$A = \frac{1}{2}(12)(5) = 30 \text{ square units.}$$ 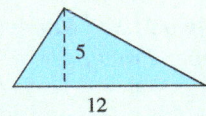

SECTION 4.4 ■ ADDING AND SUBTRACTING FRACTIONS—LIKE DENOMINATORS

Adding or Subtracting Fractions with Like Denominators	If a, b, and d represent numbers with $d \neq 0$, then $$\frac{a}{d} + \frac{b}{d} = \frac{a+b}{d} \quad \text{and} \quad \frac{a}{d} - \frac{b}{d} = \frac{a-b}{d}.$$ **Examples:** $\frac{3}{7} + \frac{2}{7} = \frac{3+2}{7} = \frac{5}{7}$, $\frac{7}{12} - \frac{5}{12} = \frac{7-5}{12} = \frac{2}{12} = \frac{1}{6}$
Adding and Subtracting Rational Expressions	The process used to add or subtract rational expressions with like denominators is similar to that used for adding and subtracting fractions with like denominators. **Examples:** $\frac{2w}{5x} + \frac{w}{5x} = \frac{2w+w}{5x} = \frac{3w}{5x}$, $\frac{3x}{4} - \frac{x}{4} = \frac{3x-x}{4} = \frac{2x}{4} = \frac{x}{2}$

SECTION 4.5 ■ ADDING AND SUBTRACTING FRACTIONS—UNLIKE DENOMINATORS

Least Common Multiple (LCM)

The least common multiple (LCM) of two or more numbers is the smallest number that is divisible by each of the given numbers.

Example: Common multiples of 5 and 6 include 30, 60, 90, 120, ...
The least common multiple is 30.

The Listing Method for Finding the LCM

The LCM of two or more numbers can be found by listing multiples of each number and then finding the smallest common multiple found in all lists.

Example: The LCM of 10 and 15 is 30, which is found as follows.
Multiples of 10: 10, 20, **30**, 40, 50, 60, . . .
Multiples of 15: 15, **30**, 45, 60, 75, 90, . . .

Prime Factorization Method for Finding the LCM

STEP 1: Find the prime factorization of each number.

STEP 2: List each factor that appears in one or more factorizations. If a factor is repeated in any of the factorizations, list this factor the maximum number of times that it is repeated in any one of the factorizations.

STEP 3: Find the product of this list of factors. The result is the LCM.

Example: The LCM of 18 and 24 is found as follows.
$18 = 2 \cdot 3 \cdot 3$
$24 = 2 \cdot 2 \cdot 2 \cdot 3$
The LCM is $2 \cdot 2 \cdot 2 \cdot 3 \cdot 3 = 72$.

Least Common Denominator (LCD)

The least common denominator (LCD) of two or more fractions is the LCM of the denominators.

Example: The LCM of 9 and 12 is 36, so the LCD of $\frac{5}{9}$ and $\frac{11}{12}$ is 36.

Adding or Subtracting Fractions with Unlike Denominators

STEP 1: Determine the LCD for all fractions involved.

STEP 2: Write each fraction as an equivalent fraction with the LCD as its denominator.

STEP 3: Add or subtract the newly written like fractions and simplify.

Example: $\frac{3}{8} + \frac{1}{6} = \frac{3}{8} \cdot \frac{3}{3} + \frac{1}{6} \cdot \frac{4}{4} = \frac{9}{24} + \frac{4}{24} = \frac{9+4}{24} = \frac{13}{24}$

SECTION 4.6 ■ OPERATIONS ON MIXED NUMBERS

Writing a Mixed Number as an Improper Fraction

A mixed number of the form

Quotient $\frac{\text{Remainder}}{\text{Divisor}}$ can be written as $\frac{\text{Divisor} \cdot \text{Quotient} + \text{Remainder}}{\text{Divisor}}$.

Examples: $4\frac{5}{7} = \frac{7 \cdot 4 + 5}{7} = \frac{33}{7}$, $-2\frac{3}{8} = -\frac{8 \cdot 2 + 3}{8} = -\frac{19}{8}$

Rounding a Mixed Number to the Nearest Integer

If the fraction part of the mixed number is

(a) less than $\frac{1}{2}$, drop the fraction part of the mixed number.
(b) $\frac{1}{2}$ or more, drop the fraction part of the mixed number and
  (i) add 1 to the integer part if the mixed number is positive.
  (ii) subtract 1 from the integer part if the mixed number is negative.

Examples: $4\frac{1}{6}$ rounds to 4; $-8\frac{4}{7}$ rounds to -9.

Operations on Mixed Numbers

STEP 1: Rewrite any mixed numbers or integers as improper fractions.
STEP 2: Perform the required arithmetic as usual.
STEP 3: Rewrite the result as a mixed number, when appropriate.

Examples: $1\dfrac{1}{3} \cdot 2\dfrac{1}{2} = \dfrac{4}{3} \cdot \dfrac{5}{2} = \dfrac{20}{6} = \dfrac{10}{3} = 3\dfrac{1}{3}$

$4\dfrac{2}{5} - 2\dfrac{3}{5} = \dfrac{22}{5} - \dfrac{13}{5} = \dfrac{22 - 13}{5} = \dfrac{9}{5} = 1\dfrac{4}{5}$

Method II for Adding or Subtracting Mixed Numbers

Add or subtract the integer and fraction parts vertically.

Examples:
$$\begin{array}{r} 3\dfrac{1}{7} \\ +\,4\dfrac{3}{7} \\ \hline 7\dfrac{4}{7} \end{array} \qquad \begin{array}{r} 13\dfrac{8}{9} \\ -\,5\dfrac{7}{9} \\ \hline 8\dfrac{1}{9} \end{array}$$

SECTION 4.7 ■ COMPLEX FRACTIONS AND ORDER OF OPERATIONS

Complex Fraction

A complex fraction is a rational expression with fractions in its numerator, denominator, or both.

Examples: $\dfrac{4 + \dfrac{5}{6}}{\dfrac{8}{9} - 7}, \quad \dfrac{\dfrac{3x}{7} - \dfrac{1}{2}}{\dfrac{2}{5} + \dfrac{3x}{5}}$

Simplifying Basic Complex Fractions

If a, b, c, and d represent numbers with $b \neq 0$, $c \neq 0$, and $d \neq 0$, then

$$\dfrac{\dfrac{a}{b}}{\dfrac{c}{d}} = \dfrac{a}{b} \cdot \dfrac{d}{c}.$$

Example: $\dfrac{\dfrac{3}{7}}{\dfrac{4}{9}} = \dfrac{3}{7} \cdot \dfrac{9}{4} = \dfrac{27}{28}$

Method I: Simplifying Complex Fractions

To simplify a complex fraction,

1. Write the numerator as a single fraction; do the same to the denominator.
2. Multiply the fraction in the numerator by the reciprocal of the fraction in the denominator. This is called *invert and multiply*. Simplify the result.

Example: $\dfrac{\dfrac{3}{7} + \dfrac{1}{7}}{\dfrac{5}{y} - \dfrac{2}{y}} = \dfrac{\dfrac{4}{7}}{\dfrac{3}{y}} = \dfrac{4}{7} \cdot \dfrac{y}{3} = \dfrac{4y}{21}$

Method II: Simplifying Complex Fractions

To simplify a complex fraction,

1. Find the LCD of all fractions within the complex fraction.
2. Multiply both the numerator and the denominator of the complex fraction by the LCD found in Step 1. Simplify the result.

continued on next page

Example: $\dfrac{\frac{5}{6}-\frac{1}{2}}{\frac{2}{3}+\frac{1}{6}} = \dfrac{6 \cdot \left(\frac{5}{6}-\frac{1}{2}\right)}{6 \cdot \left(\frac{2}{3}+\frac{1}{6}\right)} = \dfrac{5-3}{4+1} = \dfrac{2}{5}$

SECTION 4.8 ■ SOLVING EQUATIONS INVOLVING FRACTIONS

Properties of Equality

$a = b$ is equivalent to $a + c = b + c$.
$a = b$ is equivalent to $a - c = b - c$.
$a = b$ is equivalent to $ac = bc$ for $c \neq 0$.
$a = b$ is equivalent to $\frac{a}{c} = \frac{b}{c}$ for $c \neq 0$.

Example: $x + \dfrac{2}{5} = \dfrac{4}{5}$ is equivalent to $x + \dfrac{2}{5} - \dfrac{2}{5} = \dfrac{4}{5} - \dfrac{2}{5}$, so $x = \dfrac{2}{5}$.

Solving Equations Symbolically

STEP 1: Use the distributive property to clear parentheses from the equation.
STEP 2: Multiply each term by the LCD of all fractions involved.
STEP 3: Simplify each term and combine any like terms.
STEP 4: Solve the resulting equation.

Example:
$3\left(x + \dfrac{5}{6}\right) = \dfrac{7}{8}$

$3x + \dfrac{5}{2} = \dfrac{7}{8}$ (Step 1)

$8 \cdot 3x + \dfrac{8}{1} \cdot \dfrac{5}{2} = \dfrac{8}{1} \cdot \dfrac{7}{8}$ (Step 2)

$24x + 20 = 7$ (Step 3)

$24x = -13$

$x = -\dfrac{13}{24}$ (Step 4)

Solving Equations Numerically

To solve some equations involving fractions, complete a table of values for various values of the variable and then select the solution from the table.

Example: The solution to the equation $\frac{2}{3}x + 4 = 8$ is **6**.

x	-2	0	2	4	6	8
$\frac{2}{3}x + 4$	$\frac{8}{3}$	4	$\frac{16}{3}$	$\frac{20}{3}$	8	$\frac{28}{3}$

Solving Equations Graphically

To solve equations graphically, use a graph of the equation to estimate a solution and then check the estimated solution to be sure that it is correct.

Example: The solution to the equation $-\frac{2}{5}x + \frac{4}{5} = 2$ is -3.

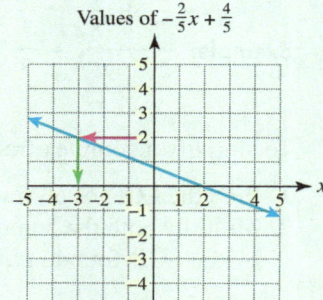

Check:

$-\dfrac{2}{5}(-3) + \dfrac{4}{5} \stackrel{?}{=} 2$

$\dfrac{6}{5} + \dfrac{4}{5} \stackrel{?}{=} 2$

$2 = 2$ ✓

CHAPTER 4   Review Exercises

SECTION 4.1

1. Write the numerator and denominator of each fraction.
   (a) $\dfrac{7}{18}$ (b) $\dfrac{x}{5}$

2. Write the fraction represented by each shading.
   (a) 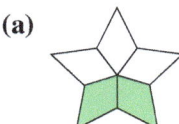   (b)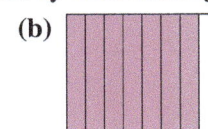

3. Simplify each fraction, if possible.
   (a) $\dfrac{0}{6}$ (b) $\dfrac{-2}{1}$
   (c) $\dfrac{-5}{-5}$ (b) $\dfrac{8}{0}$

4. Write the improper fraction and mixed number represented by each shading.
   (a)
   (b)

5. Write each improper fraction as a mixed number.
   (a) $\dfrac{19}{3}$ (b) $\dfrac{-14}{5}$

6. Write each mixed number as an improper fraction.
   (a) $-6\dfrac{2}{5}$ (b) $7\dfrac{7}{9}$

7. Graph each fraction or mixed number on a number line.
   (a) $-2\dfrac{1}{2}$ (b) $\dfrac{4}{5}$

SECTION 4.2

8. Use a divisibility test to answer each question.
   (a) Is 742 divisible by 3?
   (b) Is 4565 divisible by 5?

9. Determine if each whole number is prime, composite, or neither.
   (a) 39   (b) 59
   (c) 1   (d) 81

10. Find the prime factorization of each number.
    (a) 40   (b) 110

11. Find a number to replace the question mark so that the given fractions are equivalent.
    (a) $\dfrac{5}{?} = \dfrac{15}{33}$ (b) $\dfrac{3}{8} = \dfrac{?}{48}$

12. Find the GCF of the given numbers.
    (a) 12 and 42   (b) 16, 24, and 48

13. Simplify each fraction to lowest terms.
    (a) $-\dfrac{20}{65}$ (b) $\dfrac{64}{88}$

14. Place the correct symbol, $<$, $>$, or $=$, in the blank for each given pair of fractions.
    (a) $\dfrac{5}{8}$ ___ $\dfrac{9}{13}$ (b) $\dfrac{11}{16}$ ___ $\dfrac{21}{32}$

15. Find the GCF of the given terms.
    (a) $3x^2$ and $12x$ (b) $4x^2y^3$ and $6xy^2$

16. Simplify each rational expression.
    (a) $\dfrac{16xy}{20y}$ (b) $-\dfrac{9x^2y}{3y}$

SECTION 4.3

Exercises 17–22: Multiply. Simplify, if necessary.

17. $\dfrac{5}{6} \cdot \dfrac{5}{8}$ 18. $-\dfrac{2}{3} \cdot \left(-\dfrac{5}{8}\right)$

19. $-\dfrac{18}{7} \cdot \dfrac{7}{6}$ 20. $\dfrac{20}{33} \cdot \dfrac{22}{35}$

21. $\dfrac{15w^2}{8x} \cdot \dfrac{4x^2}{25w^3}$ 22. $-4x \cdot \dfrac{9y}{16x^2}$

23. Evaluate each expression.
    (a) $\left(-\dfrac{1}{4}\right)^3$ (b) $\sqrt{\dfrac{49}{100}}$

24. Write the reciprocal of each number.
    (a) $-\dfrac{4}{9}$ (b) $\dfrac{1}{10}$

Exercises 25–28: Divide. Simplify, if necessary.

25. $\dfrac{7}{25} \div \dfrac{21}{10}$ 26. $-\dfrac{7}{2} \div \dfrac{35}{36}$

27. $\dfrac{3x^2}{28y} \div \dfrac{6x}{7y^2}$ 28. $\dfrac{21p^3}{8q} \div 7p$

Exercises 29 and 30: Find the fractional part.

29. $\frac{4}{9}$ of -45

30. $\frac{3}{4}$ of $\frac{5}{18}$

SECTION 4.4

Exercises 31–36: Add or subtract as indicated. Simplify your result if necessary.

31. $\frac{3}{8} + \frac{5}{8}$

32. $-\frac{1}{10} + \frac{9}{10}$

33. $-\frac{2}{3} - \left(-\frac{8}{3}\right)$

34. $-\frac{7}{16} - \frac{3}{16}$

35. $\frac{7y^2}{8x} - \frac{5y^2}{8x}$

36. $\frac{5}{6m} + \left(-\frac{5}{6m}\right)$

SECTION 4.5

37. Find the LCM of the given numbers or expressions.
    (a) 12 and 20
    (b) 6, 8, and 16
    (c) $4x^2$ and $3x$
    (d) $2m$, $4mn$, and $6n$

38. Find the least common denominator for the fractions.
    (a) $\frac{3}{16}$ and $\frac{7}{12}$
    (b) $\frac{x}{3y^2}$ and $\frac{9}{8y}$

Exercises 39 and 40: Rewrite the fractions as equivalent fractions with the given LCD.

39. $\frac{1}{6}$ and $\frac{8}{9}$ LCD: 18

40. $\frac{5}{6}, \frac{2}{7},$ and $\frac{3}{14}$ LCD: 42

Exercises 41–46: Add or subtract as indicated. Simplify your result if necessary.

41. $\frac{3}{4} + \frac{1}{6}$

42. $-\frac{3}{10} - \frac{1}{8}$

43. $\frac{2}{5} - \left(-\frac{4}{15}\right)$

44. $-\frac{7}{18} + \frac{5}{6}$

45. $\frac{5y^2}{2} - \frac{y^2}{6}$

46. $\frac{3}{4m} + \frac{5}{8m}$

SECTION 4.6

Exercises 47 and 48: Round the given mixed number to the nearest integer.

47. $-3\frac{7}{9}$

48. $5\frac{3}{10}$

Exercises 49–56: Perform the indicated arithmetic.

49. $5\frac{1}{4} + 7\frac{5}{6}$

50. $-4\frac{3}{10} \cdot 5$

51. $8\frac{2}{5} - 5\frac{7}{15}$

52. $-3\frac{5}{18} + 8\frac{5}{6}$

53. $-3\frac{1}{8} \cdot 4\frac{2}{5}$

54. $3\frac{3}{8} \div \frac{3}{4}$

55. $6\frac{2}{9} \div 2\frac{1}{3}$

56. $-4\frac{1}{3} - 3\frac{8}{9}$

SECTION 4.7

Exercises 57–60: Simplify the given complex fraction.

57. $\dfrac{\frac{5}{4}}{\frac{15}{8}}$

58. $\dfrac{\frac{4}{3} + \frac{1}{8}}{\frac{5}{6} - \frac{1}{4}}$

59. $\dfrac{\frac{9x}{2} - \frac{3x}{4}}{\frac{15}{8}}$

60. $\dfrac{\frac{12y}{5}}{\frac{8}{3} + 4}$

Exercises 61 and 62: Simplify the given expression.

61. $\frac{7}{9} - \frac{2}{3} \cdot \frac{1}{6}$

62. $1 + \frac{3}{8} \cdot \left(\frac{1}{6} + \frac{1}{2}\right)$

SECTION 4.8

Exercises 63–70: Solve the equation symbolically.

63. $y - \frac{1}{8} = \frac{5}{6}$

64. $8 = -\frac{1}{9}w$

65. $\frac{7}{12}k + \frac{5}{8} = \frac{3}{4}$

66. $\frac{x}{3} + \frac{x}{15} = \frac{4}{5}$

67. $5\left(n - \frac{1}{4}\right) = \frac{5}{3}$

68. $-\frac{16}{3} = m - 2$

69. $\frac{7a}{18} - \frac{5a}{6} = \frac{8}{9}$

70. $\frac{7}{4} = \frac{1}{12} - \frac{5}{9}p$

Exercises 71 and 72: Solve the equation numerically by completing the given table of values.

71. $\frac{1}{5}x - 4 = -3$

x	-3	-1	1	3	5
$\frac{1}{5}x - 4$					

72. $\frac{3}{2}x - 1 = \frac{7}{2}$

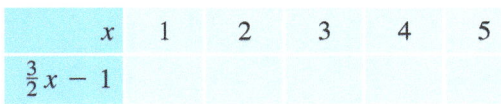

Exercises 73 and 74: Solve the equation graphically.

73. $-\frac{5}{3}x - \frac{1}{3} = 3$ 74. $-\frac{3}{5}x + \frac{2}{5} = -2$

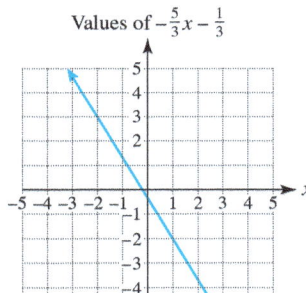 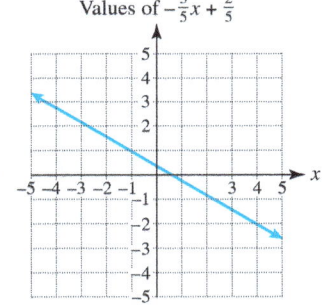

APPLICATIONS

75. **Cable TV** It is estimated that 7 of every 100 households in India have cable TV. Find the fraction of India's households that do *not* have cable TV. (Source: International Telecommunications Union.)

76. **Phone Numbers** Out of 25 students in the red classroom at a preschool, 14 students could recite their parents' phone number. In the blue room, 9 students out of 17 could recite their parents' phone number. Which class had a larger fraction of students who knew their parents' phone number?

77. **Cleaning Solution** If $\frac{2}{3}$ cup of concentrated detergent should be mixed with 1 gallon of water to make a cleaning solution, how much concentrated detergent should be mixed with $\frac{3}{4}$ gallon of water?

78. **Going Organic** A farmer plans to convert $\frac{11}{15}$ of his crop land to organic crops. If the farmer has already converted $\frac{7}{15}$ of his crop land, what fraction of his land has yet to be converted to organic crops?

79. **Voter Turnout** In a recent presidential election, the fraction of Iowa's eligible voters that went to the polls was $\frac{2}{25}$ higher than the national average. If $\frac{31}{50}$ of the nation's voters went to the polls, what fraction of Iowa's eligible voters turned out to vote? (Source: Iowa Secretary of State.)

80. **Height** At birth, a baby boy was $17\frac{3}{4}$ inches long. If he grew $7\frac{1}{2}$ inches during his first year, how tall was he on his first birthday?

81. **Converting Currency** At one time, the formula for converting E euros (€) to D dollars was

$$D = \frac{33}{25}E.$$

Use the formula to find the number of dollars that could be purchased for €100.

82. **Working Together** Experienced roofers can put shingles on a house in 8 hours, while less experienced roofers can shingle the same house in 12 hours. How many hours would it take the two roofing crews working together to shingle the house?

CHAPTER 4 Test

1. Write the improper fraction and mixed number represented by the shading.

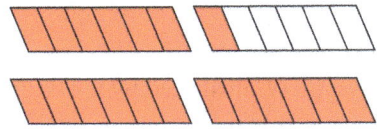

2. Write $4\frac{5}{7}$ as an improper fraction.

3. Write $\frac{22}{3}$ as a mixed number.

4. Determine if the given whole number is prime, composite, or neither.
   (a) 16   (b) 0
   (c) 29   (d) 63

5. Find the prime factorization of 240.

6. Find a number to replace the question mark so that the given fractions are equivalent.
   (a) $\frac{3}{?} = \frac{15}{25}$ (b) $\frac{3}{7} = \frac{?}{63}$

7. Find the GCF of each pair.
   (a) 16 and 72   (b) $14x^2$ and $21x$

8. Simplify to lowest terms.
   (a) $-\frac{45}{60}$ (b) $\frac{28xy^2}{7y^2}$

9. Multiply or divide as indicated. Simplify your result.
   (a) $\dfrac{4}{3} \cdot \left(-\dfrac{9}{8}\right)$
   (b) $\dfrac{6}{25} \div \dfrac{24}{35}$
   (c) $\dfrac{30x^2}{7y} \div 6x$
   (d) $\dfrac{8y}{15x} \cdot (-5x^2)$

10. Evaluate each expression.
    (a) $\sqrt{\dfrac{81}{121}}$
    (b) $\left(-\dfrac{5}{6}\right)^2$

11. Add or subtract as indicated. Simplify your result.
    (a) $\dfrac{5}{24} + \dfrac{3}{8}$
    (b) $-\dfrac{2}{15} - \dfrac{1}{10}$
    (c) $-\dfrac{7}{10} + \left(-\dfrac{5}{6}\right)$
    (d) $\dfrac{9w}{4x} + \dfrac{7w}{4x}$

12. Perform the indicated arithmetic.
    (a) $7\dfrac{1}{4} + 4\dfrac{1}{6}$
    (b) $7 \cdot \left(-3\dfrac{1}{5}\right)$

13. Simplify the expression.
    (a) $\dfrac{\dfrac{1}{6} + \dfrac{7}{8}}{\dfrac{7}{4} - \dfrac{2}{3}}$
    (b) $\dfrac{6}{5} \cdot \left(\dfrac{5}{6} - \dfrac{2}{3}\right) + \dfrac{4}{5}$

14. Solve the equation symbolically.
    (a) $-\dfrac{3}{4}y = 12$
    (b) $\dfrac{3}{4} = 3\left(m - \dfrac{1}{2}\right)$
    (c) $\dfrac{x}{6} + \dfrac{x}{12} = \dfrac{5}{4}$
    (d) $\dfrac{7w}{2} - \dfrac{5w}{6} = \dfrac{4}{3}$

15. Solve the equation $\dfrac{4}{3}x + \dfrac{1}{2} = \dfrac{19}{6}$ numerically by completing the given table of values.

x	-2	-1	0	1	2
$\dfrac{4}{3}x + \dfrac{1}{2}$					

16. Solve the equation $\dfrac{3}{4}x - \dfrac{5}{4} = 1$ graphically.

    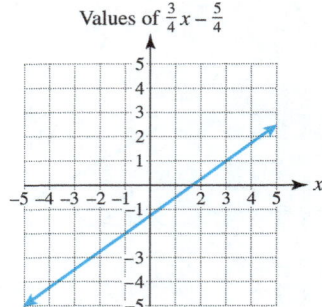
    Values of $\dfrac{3}{4}x - \dfrac{5}{4}$

17. **Family Budget** A family spends $\dfrac{4}{25}$ of its monthly budget on food. If the family has a total of $4200 to spend each month, how much is spent on food?

18. **Gaming Revenue** The revenue R, in billions of dollars, for all gaming on the Las Vegas Strip can be estimated by the formula $R = \dfrac{1}{6}x - 329$, where x is the year from 1984 to 2014. Find the year when gaming revenues were $6.5 billion by solving the equation $6.5 = \dfrac{1}{6}x - 329$. (*Source:* American Gaming Association.)

CHAPTERS 1–4 Cumulative Review Exercises

1. Write *thirty-six thousand, two hundred eighty-five* in standard form.

2. Translate the given word phrase into a mathematical expression and then compute the result.

   The product of 15 and 4

3. Divide $3567 \div 16$.

4. Evaluate the algebraic expression $y \div x$ when $x = 7$ and $y = 35$.

5. Estimate the sum $993 + 2002 + 696$ by rounding each value to the nearest hundred.

6. Evaluate $11 - (7 + 3) \div 2$.

7. Simplify the expression $5x - (2x + 1)$.

8. Simplify $-(-(-(-4)))$.

9. State the addition property illustrated by the equation $3 + (4 + 6) = (3 + 4) + 6$.

10. Simplify the expression $-7 - (-5) + 8$.

11. Multiply $2(-3)(-1)(5)(-2)$.

12. Evaluate $\sqrt{-3^2 + 45} \div (2 + 4)$.

13. Complete the table and solve $3x - 5 = -5$.

x	-2	-1	0	1	2
$3x - 5$					

14. Simplify the expression $-5(2y - 1) + 12y$.

Exercises 15 and 16: Translate each sentence into an equation using x as the variable. Do not solve the equation.

15. Seven equals a number decreased by 13.

16. Double the sum of a number and 3 equals -14.

17. Is -2 a solution to the equation $-x + 3 = -1$?

18. Is the equation $-x^3 + 5 = 2$ linear?

19. Solve the equation $-6y + (-16) = 2y$.

20. Solve the linear equation $-2x - 3 = 3$ graphically.

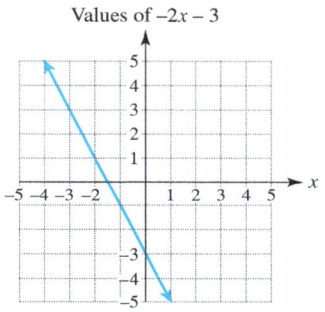

21. Write the improper fraction and mixed number represented by the shading.

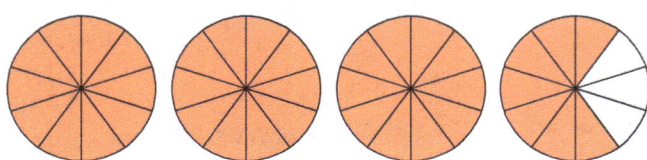

22. Find the prime factorization of 198.

Exercises 23–28: Perform the indicated arithmetic.

23. $-\dfrac{8}{7} \div \dfrac{32}{21}$

24. $\dfrac{3}{14} \cdot \dfrac{7}{9}$

25. $\dfrac{4}{7} + \left(-\dfrac{4}{7}\right)$

26. $\dfrac{13}{10} - \dfrac{5}{6}$

27. $2\dfrac{1}{3} + 6\dfrac{5}{8}$

28. $-\dfrac{1}{6} + \dfrac{2}{3} \cdot \dfrac{9}{10}$

Exercises 29 and 30: Solve the equation symbolically.

29. $-\dfrac{14}{3} = x - 5$

30. $\dfrac{5}{10}y + \dfrac{3}{8} = \dfrac{7}{4}$

31. **Street Vendor** A person bought 9 items from a street vendor and received $13 in change. If each item was priced at $3, how much did the person give to the vendor when paying?

32. **Medicare Costs** Total Medicare costs C, in billions of dollars, can be computed using the formula

$$C = 18x - 35{,}750,$$

where x is a year from 1996 to 2015. Replace C in the formula with 502 and solve the resulting equation to find the year when Medicare costs were $502 billion. (*Source:* Office of Management and Budget.)

33. **Rectangle Dimensions** The length of a rectangle is 5 inches longer than its width. If the perimeter of the rectangle is 26 inches, find its length and width.

34. **Efficient Lighting** In an effort to replace $\frac{17}{30}$ of a home's incandescent light bulbs with energy-efficient LED bulbs, a contractor has already replaced $\frac{1}{6}$ of the light bulbs. What fraction of the incandescent bulbs still need to be replaced?

5 Decimals

- 5.1 Introduction to Decimals
- 5.2 Adding and Subtracting Decimals
- 5.3 Multiplying and Dividing Decimals
- 5.4 Real Numbers, Square Roots, and Order of Operations
- 5.5 Solving Equations Involving Decimals
- 5.6 Applications from Geometry and Statistics

The table below shows Olympic winning times for the men's and women's 100-meter dash. Notice that these times are recorded as decimals, to the hundredth place. In London, Usaine Bolt beat his Beijing time by only 0.06 seconds, or six-hundreths of a second. Six women in the 2012 Olympics finished with times between 10 and 11 seconds, but decimal timing made it possible to identify Shelly-Ann Fraser-Pryce as the clear winner at 10.75 seconds. To analyze Olympic times, we need to understand decimals and be able to add, subtract, multiply, and divide using decimal numbers.

Men's 100m Dash

Year	Sydney 2000	Athens 2004	Beijing 2008	London 2012
Record (sec)	9.87	9.85	9.69	9.63
Runner	Green	Gatlin	Bolt	Bolt

Women's 100m Dash

Year	Sydney 2000	Athens 2004	Beijing 2008	London 2012
Record (sec)	11.12	10.93	10.78	10.75
Runner	Thanou	Siarenka	Fraser-Pryce	Fraser-Pryce

Source: Olympic.org.

5.1 Introduction to Decimals

Objectives

1. Reviewing Decimal Notation
2. Writing Decimals in Words
3. Writing Decimals as Fractions
4. Graphing Decimals on a Number Line
5. Comparing Decimals
6. Rounding Decimals

NEW VOCABULARY

☐ Decimal notation
☐ Decimal point
☐ Decimal number (decimal)
☐ Decimal places

READING CHECK 1

- What two parts of a decimal number are separated by the decimal point?

1 Reviewing Decimal Notation

Connecting Concepts with Your Life It is difficult to imagine just how complicated our modern money system would be if stores and banks used mixed numbers rather than decimals. A bag of chips could cost 3\frac{9}{50}$, a large coffee drink might cost 4\frac{3}{20}$, and a checking account could have a balance of 246\frac{17}{25}$. While computations on mixed numbers are often quite tedious, the same computations on decimal numbers are less time-consuming. In this section, we learn how to write fractions and mixed numbers in decimal notation.

Numbers expressed in **decimal notation** have an integer part and a fractional part separated by a **decimal point**. Numbers written in this way are called **decimal numbers**, or simply **decimals**. Examples of decimals include

$$12.36, \quad -0.198, \quad -545.90132, \quad \text{and} \quad 0.003.$$

In a decimal number, the integer part is written to the left of the decimal point, and the fractional part is written to the right of the decimal point.

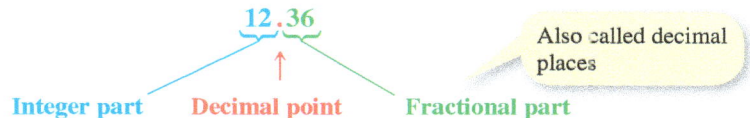

Integer part   Decimal point   Fractional part

Also called decimal places

STUDY TIP

By this time in the semester, it is likely that you know some of your classmates. Have you started or joined a study group? Be sure not to miss the opportunity to study math with your classmates.

Each digit in a decimal has a place value. The place value names for the integer part are identical to those for whole numbers, and the place values for the fractional part have names based on fractions whose denominators are powers of 10. For example, the following See the Concept shows the decimal 617,283.954 written in a place value chart.

See the Concept 1 — PLACE VALUE

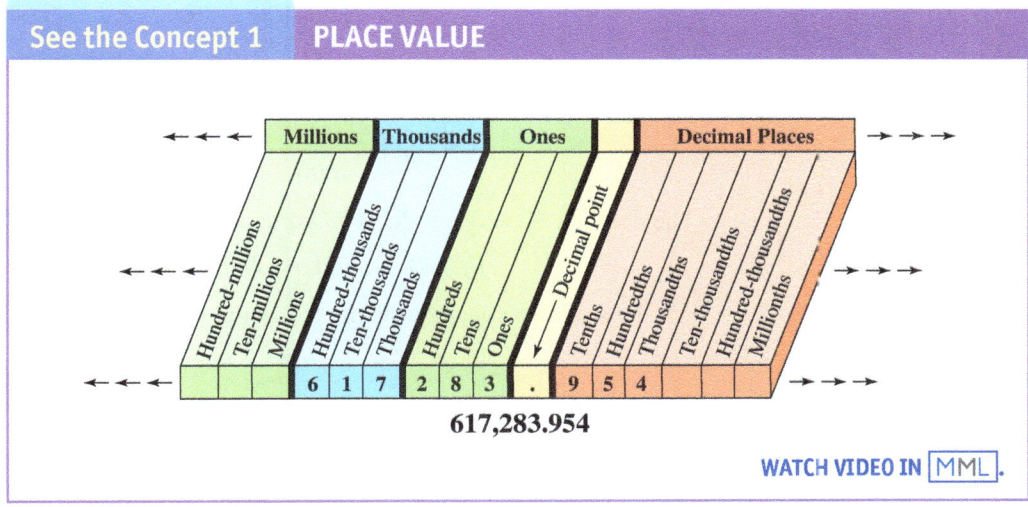

617,283.954

WATCH VIDEO IN MML.

NOTE: There are many ways to write a whole number in decimal form. For example, the following numbers all represent the whole number 72.

$$72 \quad 72.0 \quad 72.00 \quad 72.00000$$

MAKING CONNECTIONS 1

Decimal Place Values and Periods

Except for the ones place value, the names for the place values to the right of the decimal point are similar to the names to the left of the decimal point. This pattern continues indefinitely in both directions. However, we do *not* use commas on the right side of a decimal point.

2 Writing Decimals in Words

 Math in ^Pricing Context Try reading the following sentence aloud.

I bought a new iPhone game for $2.99.

Did you read the price as "two point nine nine dollars?" Probably not. It is more natural to read the price as "two dollars and ninety-nine cents." We separate the whole number 2 from the fractional part 99 (cents) by reading the decimal point as "and." Because 99 cents is the same as 99 hundredths of a dollar, we write the decimal 2.99 in word form as

two and ninety-nine hundredths.

By first writing a decimal in expanded form, we can determine a general procedure for writing a decimal in words. Consider the expanded form of the decimal 57.125.

Tens		Ones		Tenths		Hundredths		Thousandths
$5 \cdot 10$	$+$	$7 \cdot 1$	$+$	$1 \cdot \dfrac{1}{10}$	$+$	$2 \cdot \dfrac{1}{100}$	$+$	$5 \cdot \dfrac{1}{1000}$

We can substitute the word "and" for the plus sign in the position of the decimal point and then simplify the remaining expressions by multiplying and adding.

$5 \cdot 10 + 7 \cdot 1$ and $1 \cdot \dfrac{1}{10} + 2 \cdot \dfrac{1}{100} + 5 \cdot \dfrac{1}{1000}$ Insert the word "and."

$50 + 7$ and $\dfrac{1}{10} + \dfrac{2}{100} + \dfrac{5}{1000}$ Multiply.

$50 + 7$ and $\dfrac{1}{10} \cdot \dfrac{100}{100} + \dfrac{2}{100} \cdot \dfrac{10}{10} + \dfrac{5}{1000}$ The LCD is 1000.

$50 + 7$ and $\dfrac{100}{1000} + \dfrac{20}{1000} + \dfrac{5}{1000}$ Multiply.

57 and $\dfrac{125}{1000}$ Add.

The decimal **57.125** is written in words as

fifty-seven and *one hundred twenty-five thousandths.*

This discussion suggests the following procedure for writing a decimal in words.

READING CHECK 2

- What word is used for the decimal point when writing a decimal in words?

WRITING A DECIMAL IN WORDS

To write a decimal in words, do the following.

1. Write the integer part in words.
2. Include the word "and" for the decimal point.
3. Write the word form of the whole number formed by the fractional part, followed by the place value of the rightmost digit.

NOTE: If the integer part of a decimal is 0, we do not write the words "zero and" before writing the fractional part in words. For example, the decimal 0.922 is written in words as *nine hundred twenty-two thousandths.* ∎

EXAMPLE 1 **Writing decimals in words**

Write each decimal in words.
(a) 3451.92   (b) −3.687   (c) 0.05

Solution
(a) The decimal 3451.92 is written in words as

three thousand four hundred fifty-one and ninety-two hundredths.

(b) We write −3.687 as

negative three and six hundred eighty-seven thousandths.

(c) The integer part and the word "and" are not written. We write 0.05 in words as

five hundredths.

Now Try Exercises 7, 9, 13

🌐 **Math in Checking Context** When writing a check, the whole number part of the amount is written in word form, followed by the word "and" for the decimal point. However, the fractional part of the amount is written as a fraction with a denominator of 100. This is illustrated in the next example.

EXAMPLE 2 **Writing a check**

Fill in the blank line on the check in the following figure.

```
Electronics Company                              1001
301 Technology Drive, Anytown, USA 12345
                                    Date: 2/2/2012
Pay to
the order of  The Power Company    $ 103.47
                                              Dollars
Electric Bill              Gavin Groehler
Memo                       Authorized Signature
  ⑈01001⑈  ⑆111222333⑆  444555⑈
```

Solution
The decimal 103.47 is written on the check in words as shown in the following figure.

```
Electronics Company                              1001
301 Technology Drive, Anytown, USA 12345
                                    Date: 2/2/2012
Pay to
the order of  The Power Company    $ 103.47
                         47
One hundred three and   ⁄100                  Dollars
Electric Bill              Gavin Groehler
Memo                       Authorized Signature
  ⑈01001⑈  ⑆111222333⑆  444555⑈
```

Now Try Exercise 17

3 Writing Decimals as Fractions

In the previous subsection, we saw that the decimal 57.125 could be written as 57 and $\frac{125}{1000}$. By removing the word "and" from this expression and writing the fractional part in lowest terms, we obtain the mixed number $57\frac{1}{8}$. To write a decimal as a mixed number or fraction, use the following procedure.

WRITING A DECIMAL AS A MIXED NUMBER OR FRACTION

To write a decimal as a mixed number or fraction, do the following.

1. Write the integer part of the decimal as the integer part of the mixed number.
2. Take the whole number formed by the fractional part of the decimal and place it in the numerator of a fraction whose denominator is a power of 10 that corresponds to the place value of the rightmost digit.
3. Simplify the fraction to lowest terms.

NOTE: Since we do not write 0 as the integer part of a mixed number, any decimal with 0 as its integer part is written as a proper fraction rather than a mixed number. ∎

EXAMPLE 3 Writing decimals as mixed numbers or fractions

Write each decimal as a mixed number or fraction in lowest terms.
(a) 3.025   (b) −19.48   (c) 0.78

Solution

(a) $3.0\underset{\uparrow}{25} = 3\frac{25}{1000} = 3\frac{1 \cdot 25}{40 \cdot 25} = 3\frac{1}{40}$
Thousandths

(b) $-19.\underset{\uparrow}{48} = -19\frac{48}{100} = -19\frac{12 \cdot 4}{25 \cdot 4} = -19\frac{12}{25}$
Hundredths

(c) $0.\underset{\uparrow}{78} = \frac{78}{100} = \frac{39 \cdot 2}{50 \cdot 2} = \frac{39}{50}$
Hundredths

Now Try Exercises 23, 25, 27

4 Graphing Decimals on a Number Line

Graphing decimals on a number line is similar to graphing fractions and mixed numbers. To get the desired accuracy, we use equally spaced tick marks to break the distance between consecutive values into 10 equal parts. The following See the Concept illustrates how to graph the number 3.68 to various degrees of accuracy on three different number lines.

See the Concept 2   GRAPHING DECIMALS

Graph 3.68 on a number line.

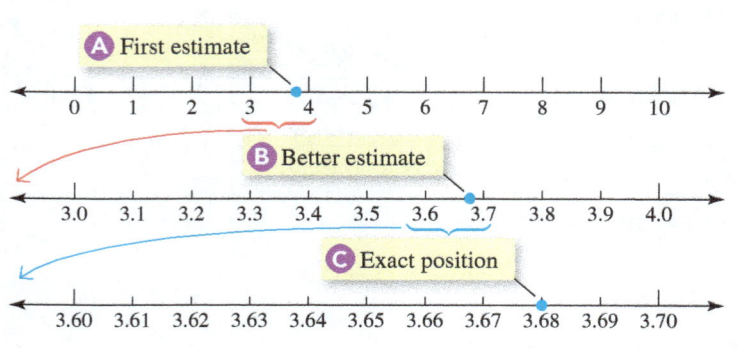

Ⓐ Estimate the position of the decimal between consecutive integers and place a dot on the number line.

Ⓑ Make a new number line with the space between the consecutive integers from Ⓐ divided into 10 equal parts, making tenths.

Ⓒ Make a new number line with the space between the consecutive tenths from Ⓑ divided into 10 equal parts, making hundredths.

Continue this process until the desired accuracy is obtained.

WATCH VIDEO IN MML.

When graphing decimals on a number line, it is *not* necessary to progress through a series of enlarged number lines as demonstrated above. In the next example, decimals are graphed exactly on the first try.

EXAMPLE 4   Graphing decimals on a number line

Graph each decimal on a number line.
(a) 6.43   (b) −2.357

Solution

(a) We look at the next-to-last digit in the given decimal to determine that 6.4**3** is between 6.**4** and 6.**5**. Use equally spaced tick marks to break the distance between 6.**4** and 6.**5** into 10 equal parts and graph the exact position of 6.43, as shown on the number line.

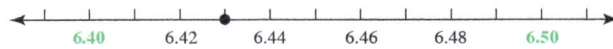

(b) The next-to-last digit in −2.3**5**7 indicates that it is between −2.3**5** and −2.3**6**. Break the distance between −2.**35** and −2.**36** into 10 equal parts and graph the exact position of −2.357. Remember that −2.36 is to the left of −2.35 on the number line.

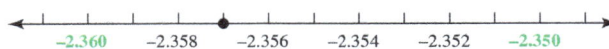

Now Try Exercises 39, 43

READING CHECK 3

- In Example 4(a), between which two *labeled* points on the number line would 6.407 be located?

5   Comparing Decimals

When two decimals are graphed on the same number line, the decimal to the left is *less than* the decimal to the right and the decimal to the right is *greater than* the decimal to the left. However, using a number line to compare decimals such as 23.67843 and 23.68 is not very practical because these numbers are difficult to graph on the same number line. A method for comparing decimals by comparing their digits is summarized as follows.

> **COMPARING TWO POSITIVE DECIMALS**
>
> To compare two positive decimals with the same number of digits to the left of the decimal point, compare digits in corresponding places from left to right until unequal digits are found. The decimal with the greater of these digits is the larger number.

NOTE: This process can also be used to compare two negative decimals as long as you remember that the decimal with the larger of the unequal digits is *less than* the decimal with the smaller of the unequal digits. ■

EXAMPLE 5   Comparing two decimals

Place the correct symbol, < or >, in the blank between the given decimals.
(a) 31.673 _____ 31.689   (b) 7.4569 _____ 7.45   (c) −52.7899 _____ −52.7893

Solution

(a) As we move from left to right, the corresponding digits 3, 1, and 6 are equal. The first unequal digits are in the hundredths place.

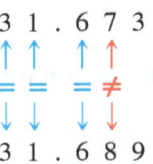

Because $7 < 8$, we conclude that $31.673 < 31.689$.

(b) Begin by writing 7.45 as 7.4500 so that it has the same number of digits to the right of the decimal point as 7.4569. The first unequal digits for 7.4569 and 7.4500 are in the thousandths place. Since $6 > 0$, we know that $7.4569 > 7.45$.

(c) The first unequal digits for -52.7899 and -52.7893 are in the ten-thousandths place. Because $9 > 3$ and the decimals are negative, we have $-52.7899 < -52.7893$.

Now Try Exercises 47, 49, 59

The following See the Concept summarizes how to compare two positive and two negative decimals.

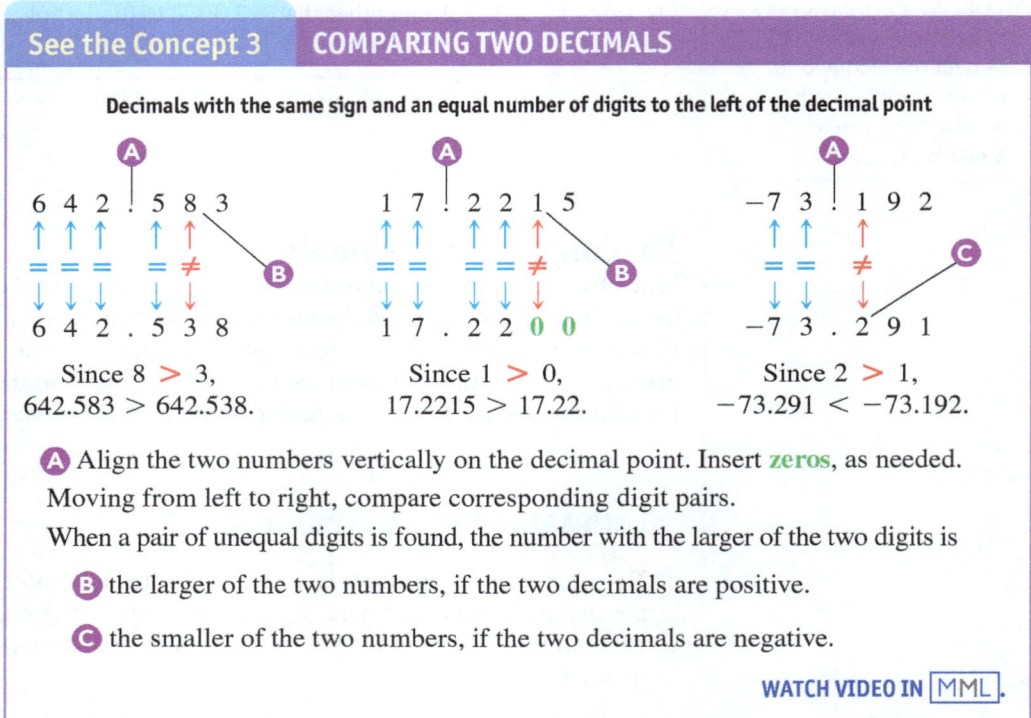

See the Concept 3   COMPARING TWO DECIMALS

Decimals with the same sign and an equal number of digits to the left of the decimal point

Since $8 > 3$, $642.583 > 642.538$.

Since $1 > 0$, $17.2215 > 17.22$.

Since $2 > 1$, $-73.291 < -73.192$.

A Align the two numbers vertically on the decimal point. Insert **zeros**, as needed. Moving from left to right, compare corresponding digit pairs.

When a pair of unequal digits is found, the number with the larger of the two digits is

B the larger of the two numbers, if the two decimals are positive.

C the smaller of the two numbers, if the two decimals are negative.

WATCH VIDEO IN MML

6 Rounding Decimals

Number lines are helpful when rounding decimals to a given place value. For example, we round the decimal 3.568 to the nearest **hundredth** by graphing it on a number line with steps of **0.01** (one-hundredths), as shown in **FIGURE 5.1**. Because 3.568 is **closer to 3.57** than to 3.56, we round **up to 3.57**.

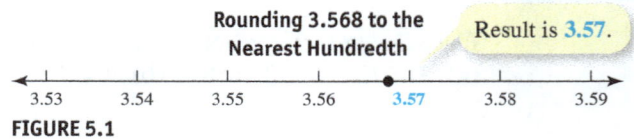

FIGURE 5.1

Since graphing decimals can be tedious, we usually round decimals in a way that is similar to rounding whole numbers.

> **ROUNDING DECIMALS**
>
> To round a decimal to the nearest whole number or to a place value to the right of the decimal point, do the following.
>
> **STEP 1:** Identify the first digit to the *right* of the given place value.
>
> **STEP 2:** If this digit is
>
>   **(a)** less than 5, do not change the digit in the given place value.
>
>   **(b)** 5 or more, add 1 to the digit in the given place value.
>
> **STEP 3:** Drop all digits to the right of the given place value.

READING CHECK 4

- Round 56.107 to the nearest tenth.

EXAMPLE 6   **Rounding decimals**

Round 345.61489 to the given place value.
(a) tenth   **(b)** thousandth

Solution
(a) STEP 1: The digit in the tenths place is 6. The first digit to the right of the 6 is 1.

$$345.61489$$

      STEP 2: Because 1 is less than 5, we do not change the digit in the tenths place.

$$345.61489$$

      STEP 3: Drop all digits to the right of the tenths place.

$$345.6$$

Rounding 345.61489 to the nearest tenth results in 345.6.

(b) The thousandths digit in the decimal 345.61489 is 4, and the first digit to the right of the thousandths place is 8 (Step 1). Because 8 is more than 5, we add 1 to the digit in the thousandths place (Step 2), and we drop all digits to its right (Step 3). The result of rounding 345.61489 to the nearest thousandth is 345.615.

Now Try Exercises 61, 67

In the next example, part (a) shows how to round a decimal to the nearest whole number, and part (b) demonstrates how to round up when a 9 is located in the given place value.

EXAMPLE 7   **Rounding decimals in money**

Round each amount to the given value.
(a) $45.77, dollar   **(b)** $0.498, cent

Solution
(a) Rounding $45.77 to the nearest dollar means that we are rounding to the nearest whole number. The specified place value is the ones place. Because 7 is more than 5, we add 1 to the digit in the ones place. After dropping all digits to the right of the ones place and the decimal point, the result is $46.

(b) Rounding $0.498 to the nearest cent means rounding to the nearest hundredth (of a dollar). Because 8 is more than 5, add 1 to the digit in the hundredths place. However, adding 1 to 9 gives 10, so we must carry 1 to the next digit to the left. The result is $0.50.

Now Try Exercises 73, 75

5.1 Putting It All Together

CONCEPT	COMMENTS	EXAMPLES
Decimal Notation	Numbers represented in decimal notation have an integer part and a fractional part separated by a decimal point. Numbers written in this way are called decimals.	41.78 — Integer part · Decimal point · Fractional part (Also called decimal places)
Writing a Decimal in Words	1. Write the integer part in words. 2. Include the word "and" for the decimal point. 3. Write the word form of the whole number formed by the fractional part, followed by the place value of the rightmost digit.	The decimal 3.79 is written as three and seventy-nine hundredths. The decimal 0.063 is written as sixty-three thousandths.
Writing a Decimal as a Mixed Number or Fraction	1. Write the integer part of the decimal as the integer part of the mixed number. 2. Take the whole number formed by the fractional part of the decimal and place it in the numerator of a fraction whose denominator is a power of 10 that corresponds to the place value of the rightmost digit. 3. Simplify the fraction to lowest terms.	$0.58 = \dfrac{58}{100} = \dfrac{29}{50}$ $76.375 = 76\dfrac{375}{1000} = 76\dfrac{3}{8}$
Graphing Decimals on a Number Line	A decimal can be graphed on a number line by using equally spaced tick marks to break the distance between values into 10 equal parts.	The graph of 3.41 is shown below. 3.40 3.42 3.44 3.46 3.48 3.50
Comparing Decimals	To compare two positive decimals with the same number of digits to the left of the decimal point, compare digits in corresponding places from left to right until unequal digits are found. The decimal with the greater of these digits is the larger number.	$382.792 < 382.795$ $0.29016 > 0.29009$

Rounding Decimals	To round a decimal to the nearest whole number or to a place value to the right of the decimal point:
	1. Identify the first digit to the *right* of the given place value.
	2. If this digit is
	  (a) less than 5, do not change the digit in the given place value.
	  (b) 5 or more, add 1 to the digit in the given place value.
	3. Drop all digits to the right of the given place value.
		Rounding 8.719 to the nearest hundredth results in 8.72.
		
		Rounding 0.5189 to the nearest tenth results in 0.5.

5.1 Exercises

CONCEPTS AND VOCABULARY

1. Numbers in _____ notation have an integer part and a fractional part separated by a decimal point.

2. Numbers written in decimal notation are called _____ numbers, or simply _____.

3. When writing decimals in words, the decimal point is written as the word "_____."

4. When writing decimals as fractions, the denominator of the fractional part is a power of _____ that corresponds to the place value of the rightmost digit.

5. To graph a decimal on a number line, use equally spaced tick marks to break the distance between consecutive values into _____ equal parts.

6. When comparing decimals, compare digits in corresponding places from left to right until _____ digits are found.

WRITING DECIMALS IN WORDS

Exercises 7–16: Write the given decimal in words.

7. 0.56
8. −0.1
9. 7.116
10. 6.0009
11. −58.7
12. 0.000135
13. −2.001003
14. −6.39
15. 501.0012
16. 6002.009

Exercises 17 and 18: Fill in the blank line on the check.

17. [Check image: Electronics Company, 1001, Date 2/2/2012, Pay to the order of The Phone Company $129.68, Memo: Phone Bill, Aaron Groehler]

18. [Check image: Electronics Company, 1001, Date 2/2/2012, Pay to the order of Electronics Superstore $2387.19, Memo: Laptop, Sydney Groehler]

WRITING DECIMALS AS FRACTIONS

Exercises 19–34: Write the given decimal as a proper fraction or mixed number in lowest terms.

19. 0.3
20. 0.25
21. −0.04
22. −0.625
23. 0.85
24. 0.54
25. −8.2
26. 7.075
27. 12.75
28. −14.19
29. 23.205
30. −22.0875
31. −1.028
32. 3.448
33. 6.5125
34. 18.4375

GRAPHING DECIMALS

Exercises 35–44: Graph the decimal on a number line.

35. 8.3
36. 4.7
37. 26.76
38. 54.22
39. 0.315
40. 2.593
41. −2.1
42. −9.9
43. −5.74
44. −7.08

COMPARING DECIMALS

Exercises 45–60: Place the correct symbol, < or >, in the blank between the given decimals.

45. 0.7 ____ 0.9
46. 4.42 ____ 4.402
47. 3.4998 ____ 3.5
48. 0.89 ____ 0.8903
49. 23.654 ____ 23.645
50. 6.0003 ____ 6.003
51. 0.19546 ____ 0.19548
52. 14.0101 ____ 14.0111
53. −3.9 ____ −4.0
54. −5.69 ____ −5.70
55. −1.593 ____ 1.593
56. 23.87 ____ −23.87
57. −7.999 ____ 8.000
58. 5.99 ____ −6.00
59. −14.5903 ____ −14.5913
60. −560.9 ____ −560.1

ROUNDING DECIMALS

Exercises 61–72: Round the given decimal to the indicated place value.

61. 0.3821   tenth
62. 3.7241   thousandth
63. 52.00764   hundredth
64. 265.802   tenth
65. −7.009367   ten-thousandth
66. −0.1111115   millionth
67. 9.00304   thousandth
68. 2.020406   hundred-thousandth
69. −1.1060213   millionth
70. 693.003   hundredth
71. 5.738291   hundred-thousandth
72. −3.016489   ten-thousandth

Exercises 73–78: Round to the given value.

73. $3.78   dollar
74. $23.9087   cent
75. $143.298   cent
76. $8.30   dollar
77. $19.89   dollar
78. $0.9987   cent

APPLICATIONS INVOLVING DECIMALS

79. **Tall Building** The Empire State Building in New York City is among the tallest buildings in the world, with an approximate height of 1453.71 feet. Write this decimal in words. (*Source:* World Almanac, 2008.)

80. **Student Age** A college student calculates that he is about 19.3087 years old. Write this decimal in words.

81. **High Jump** As of 2015, the world record holder in the men's high jump was Javier Sotomayor of Cuba, with a jump of just over 8 feet. In international competition, heights are measured in meters. Write his record jump of 2.45 meters as a mixed number. (*Source:* World Almanac.)

82. **Baseball Stats** Baseball great Harmon Killebrew had a lifetime batting average of 0.256. Write this decimal as a fraction. (*Source:* Major League Baseball.)

83. **Sprinters** In the 1996 Olympic Games, American Michael Johnson ran the 200-meter dash in 19.32 seconds. Twelve years later, Jamaican Usain Bolt ran the 200-meter dash in 19.30 seconds. Which sprinter was faster? (*Source:* Olympics.)

84. **Size of Bacteria** A bacterium in dish A measures 0.0000476 inch across, and a bacterium in dish B measures 0.0000467 inch across. Which of these bacteria is smaller?

85. **Comparing Gasoline Prices** As of April 2016, the average price for regular unleaded gasoline in Illinois was $2.065 per gallon and the average price in Florida was $2.072 per gallon. Which state had a lower price for gasoline at this time? Round the price of gasoline in each state to the nearest cent.

86. **Comparing Gasoline Prices** As of April 2016, the average price for regular unleaded gasoline in North Dakota was $1.978 per gallon and the average price in North Carolina was $1.974 per gallon. Which state had a lower price for gasoline at this time? Round the price of gasoline in each state to the nearest cent.

87. **Gasoline Prices** In April 2016, the national average price of a gallon of regular unleaded gasoline was $2.061. Round this number to the nearest cent. (*Source:* Energy Information Administration.)

88. **Driving Distance** According to MapQuest, the driving distance from a restaurant in East Sandwich, Massachusetts to a hotel in Boston is about 40.78 miles. Round this value to the nearest mile.

89. **Vegetarian Diets** The fraction of U.S. adults that follow a vegetarian diet can be written in decimal form as 0.032. Write this decimal as a fraction in lowest terms. (*Source:* Vegetarian Times)

90. **Female College Students** In 2015, the fraction of college students who were women could be expressed in decimal form as 0.574. Write this decimal in word form. (*Source:* National Center for Education Statistics.)

WRITING ABOUT MATHEMATICS

91. Explain how to round the decimal 12.9995 to the nearest thousandth.

92. Name two numbers that are located between 3.4729 and 3.4730 on a number line.

93. Explain how to compare two decimals that have the same number of digits to the left of the decimal point.

94. Is the process for rounding negative decimals to a given place value different than the process used for rounding positive decimals? Explain.

5.2 Adding and Subtracting Decimals

Objectives

1. Estimating Decimal Sums and Differences
2. Adding and Subtracting Positive Decimals
3. Adding and Subtracting Signed Decimals
4. Evaluating and Simplifying Expressions
5. Solving Applications Involving Decimals

1 Estimating Decimal Sums and Differences

Connecting Concepts with Your Life Have you ever been shopping and suddenly realized that you have only a few dollars with you? When this happens, we often try to keep a running estimate of the total cost of the items we want. If you have only $6, can you purchase a soda for $1.59, two bags of chips for $0.99 each, and a newspaper for $1.50? By estimating $1.59 + 0.99 + 0.99 + 1.50$, we should see that $6 is enough for this purchase. In this subsection, we find estimates of decimal sums and differences before finding the exact results.

Before we find exact values of decimal sums and differences, it is often helpful to first find an estimate. We can do this in several ways. For example, an estimation of the sum

$$148.19 + 21.03$$

can be done by first rounding 148.19 to 148 and 21.03 to 21. With these rounded values, our estimate is found by adding as follows.

$$148 + 21$$ *Round to nearest whole number.*

However, this sum is not as convenient to find mentally as an estimate using the rounded values 150 and 20.

$$\begin{array}{r} 150 \\ +\ 20 \\ \hline 170 \end{array}$$ *Round to nearest ten.*

The actual value of $148.19 + 21.03$ is 169.22. (A procedure for finding an exact sum will be discussed in the next subsection.) In this case, an estimate of 170 is reasonably close to the actual value of 169.22.

In the next example, estimates of a decimal sum and difference are found. In each case, estimates that can be computed mentally are provided.

STUDY TIP

Do you have enough time to study your notes and complete your assignments? One way to manage your time is to make a list of your time commitments and determine the amount of time that each activity requires. Remember to include time for eating, sleeping, and relaxing.

READING CHECK 1

• Why is it useful to estimate a sum or difference?

EXAMPLE 1 Estimating a decimal sum and difference

Estimate the sum or difference.
(a) $38.79 + 408.25$ (b) $307.8 - 89.73$

Solution
(a) If 38.79 is rounded to 40 and 408.25 to 410, the estimated sum to the nearest ten is $40 + 410 = 450$. A different estimate can be found by rounding 38.79 to 40 and 408.25 to 400. In this case, the estimated sum is $40 + 400 = 440$. (The actual sum is 447.04.)

(b) If 307.8 is rounded to 310 and 89.73 to 90, the estimated difference to the nearest ten is $310 - 90 = 220$. A different estimate can be found by rounding 307.8 to 300 and 89.73 to 100. In this case, the estimated sum to the nearest hundred is $300 - 100 = 200$. (The actual difference is 218.07.)

Now Try Exercises 5, 9

NOTE: In Example 1(b), the estimates $310 - 100 = 210$ and $300 - 90 = 210$ are also reasonable. All four of these estimates provide a reasonable approximation. ∎

2 Adding and Subtracting Positive Decimals

In the previous section, decimals were written as mixed numbers, and in an earlier chapter, mixed numbers were added and subtracted. One way to add or subtract decimals is to first change them to mixed numbers. For example, the sum $11.17 + 53.42$ can be found as follows.

$$11.17 + 53.42 = 11\frac{17}{100} + 53\frac{42}{100} = 64\frac{59}{100} = 64.59 \quad \text{Adding as mixed numbers}$$

Although this method works, there is an easier way to find the sum. Consider the same sum when it is stacked vertically.

$$\begin{array}{r} 11.17 \\ + 53.42 \\ \hline 64.59 \end{array} \quad \text{Adding as decimals}$$

The correct sum can be found by adding the digits in corresponding place values. Adding or subtracting decimals is similar to adding or subtracting whole numbers and integers.

CALCULATOR HELP

To add or subtract decimals with a calculator, see Appendix E (page AP-32).

> **ADDING AND SUBTRACTING DECIMALS**
>
> To add or subtract decimals, do the following.
>
> 1. Stack the decimals vertically with the decimal points aligned. Extra zeros may be written to the right of the last digit on any decimal number so that corresponding digits align neatly.
> 2. Add or subtract digits in corresponding place values. Carry or borrow as needed.
> 3. Place a decimal point in the result so that it is aligned vertically with the decimal points in the numbers stacked above it.

EXAMPLE 2 Adding positive decimals

Estimate the sum and then add the decimals.
(a) $123.45 + 56.394$ (b) $17.2 + 134.971 + 1.84$ (c) $68 + 12.75$

Solution
(a) An estimate of the sum is $120 + 60 = 180$. If the actual sum differs significantly from 180, a computational error may have occurred.

$$\begin{array}{r} \overset{1}{}\\ 123.45\textcolor{red}{0} \\ + 56.394 \\ \hline 179.844 \end{array}$$

123.45**0** ← Insert a 0.

179.844 ← The result is reasonably close to the estimate.

(b) An estimate of the sum 17.2 + 134.971 + 1.84 is 20 + 130 + 0 = 150.

$$\begin{array}{r} \overset{1\,2\,1}{}\\ 17.2\mathbf{00} \\ 134.971 \\ +1.84\mathbf{0} \\ \hline 154.011 \end{array}$$ ← Insert two 0s.

← Insert a 0.

← The result is reasonably close to the estimate.

(c) One estimate of the sum 68 + 12.75 is 70 + 10 = 80. Note that in the addition below, a decimal point can be inserted to the right of the ones digit in the whole number 68.

$$\begin{array}{r} \overset{1}{} \\ 68.\mathbf{00} \\ +\,12.75 \\ \hline 80.75 \end{array}$$ ← Insert a decimal point and two 0s.

← The result is reasonably close to the estimate.

Now Try Exercises 19, 21, 27

EXAMPLE 3 Subtracting positive decimals

Estimate the difference and then subtract the decimals.
(a) 687.248 − 35.09   (b) 9274.63 − 510

Solution
(a) One possible estimate of the difference is 690 − 40 = 650.

$$\begin{array}{r} 1\,14 \\ 6\,8\,7.\,\cancel{2}\,\cancel{4}\,8 \\ -3\,5.\,0\,9\,\mathbf{0} \\ \hline 6\,5\,2.\,1\,5\,8 \end{array}$$ ← Insert a 0.

← The result is reasonably close to the estimate.

(b) An estimate of the difference is 9300 − 500 = 8800.

$$\begin{array}{r} 8\,12 \\ \cancel{9}\,\cancel{2}\,7\,4.\,6\,3 \\ -5\,1\,0.\,\mathbf{0\,0} \\ \hline 8\,7\,6\,4.\,6\,3 \end{array}$$

← Insert a decimal point and two 0s.

← The result is reasonably close to the estimate.

Now Try Exercises 23, 25

3 ▸ Adding and Subtracting Signed Decimals

The rules used when adding and subtracting integers are also used when adding and subtracting signed decimals. In the two next examples, we add and subtract signed decimals.

EXAMPLE 4 Adding signed decimals

Estimate the sum and then add the decimals.
(a) −453.62 + 27.119   (b) −99.3 + (−402.597)

Solution
(a) To find the sum of two numbers with unlike signs, subtract the smaller absolute value from the larger absolute value and give the sum the sign of the number with the larger absolute value. One possible estimate of the sum is −450 + 30 = −420.

$$\begin{array}{r} 4\,131\,10 \\ 4\,\cancel{5}\,\cancel{3}.\,6\,\cancel{2}\,\cancel{0} \\ -2\,7.\,1\,1\,9 \\ \hline 4\,2\,6.\,5\,0\,1 \end{array}$$

← Insert a 0 in the absolute value of −453.62.

← Subtract the absolute value of 27.119.

← Difference of the absolute values

Since |−453.62| > |27.119|, the result is negative: −453.62 + 27.119 = −426.501.

(b) To find the sum of two numbers with like signs, add absolute values and give the sum the common sign. One possible estimate of the sum is $-100 + (-400) = -500$.

$$
\begin{array}{r}
11 \\
99.3\mathbf{00} \\
+\,402.597 \\
\hline
501.897
\end{array}
$$

← Insert two 0s in the absolute value of -99.3.
← Add the absolute value of -402.597.
← Sum of the absolute values

The addends are negative, so the sum is negative: $-99.3 + (-402.597) = -501.897$.

Now Try Exercises 31, 33

EXAMPLE 5 Subtracting signed decimals

Estimate the difference and then subtract the decimals.
(a) $-2.7 - 163.902$ (b) $-6044.2 - (-39)$

Solution

(a) Begin by changing the difference to the sum $-2.7 + (-163.902)$. One possible estimate of this sum (and the given difference) is $-3 + (-164) = -167$.

$$
\begin{array}{r}
1 \\
2.7\mathbf{00} \\
+\,163.902 \\
\hline
166.602
\end{array}
$$

← Insert two 0s in the absolute value of -2.7.
← Add the absolute value of -163.902.
← Sum of the absolute values

Since the addends are both negative, this sum is -166.602. The given difference is $-2.7 - 163.902 = -166.602$.

(b) First change the difference to the sum $-6044.2 + 39$. An estimate of this sum (and the given difference) is $-6040 + 40 = -6000$.

$$
\begin{array}{r}
3\,14 \\
6\,0\,\cancel{4}\,\cancel{4}.2 \\
-3\,9.\mathbf{0} \\
\hline
6\,0\,0\,5.2
\end{array}
$$

← Write the absolute value of -6044.2.
← Insert a decimal point and a 0 in the absolute value of 39.
← Difference of the absolute values

Since $|-6044.2| > |39|$, this difference is negative, and the given difference is $-6044.2 - (-39) = -6005.2$.

Now Try Exercises 35, 39

EXAMPLE 6 Estimating sums by comparing decimals

(a) Mentally estimate if the sum $0.44 + 0.5$ is greater than or less than 1.
(b) Mentally estimate if the difference $1.65 - 0.6$ is greater than or less than 1.
(c) Mentally estimate if the difference $-34.56 - (-34.6)$ is greater than or less than 0.

Solution

(a) Because $0.5 + 0.5 = 1$ and $0.44 < 0.5$, the sum $0.44 + 0.5$ is less than 1.
(b) Because $0.65 > 0.6$, the difference $1.65 - 0.6$ is greater than 1.
(c) Rewrite $-34.56 - (-34.6)$ as $-34.56 + 34.6$. Because $0.6 > 0.56$, the difference is greater than 0.

Now Try Exercises 53, 55, 57

READING CHECK 2

- Find the sum and difference of 5.1 and 4.9. Then find the sum and difference of 51 and 49.

4 Evaluating and Simplifying Expressions

 Connecting Concepts with Your Life If the price of a ham sandwich is h dollars and the price of a soft drink is d dollars, then we can find the total cost of the two items by using the algebraic expression given by

$$h + d.$$

To evaluate an algebraic expression for decimal values of the variables, we use the same process as that used to evaluate algebraic expressions for integer or whole number values of the variables. That is, we replace each variable in the expression with its given decimal value. In the next example, we evaluate the expression $h + d$ for given decimal values of h and d.

EXAMPLE 7 **Evaluating an expression**

If a ham sandwich costs \$3.39 and a soft drink costs \$1.79, evaluate the expression $h + d$ for $h = 3.39$ and $d = 1.79$ to find the total cost (excluding tax) of the two items.

Solution
When $h = 3.39$ and $d = 1.79$, the expression $h + d$ becomes $3.39 + 1.79$. The total cost for the ham sandwich and soft drink is \$5.18, as shown here.

$$\begin{array}{r} \overset{1\ 1}{3.39} \\ +\ 1.79 \\ \hline \$5.18 \end{array}$$

Now Try Exercise 67

Recall that like terms in algebraic expressions can be combined. Two or more terms are like terms if they contain the same variables raised to the same powers. In the next example, like terms with decimal coefficients are combined.

EXAMPLE 8 **Simplifying algebraic expressions**

Simplify the expression by combining like terms.
(a) $3.7x + 4 - 2.1x$ (b) $3.68y^2 - 7.2x - 1.4y^2$

Solution
(a) Begin by changing the subtraction to addition of the opposite.

$$\begin{aligned} 3.7x + 4 - 2.1x &= 3.7x + 4 + (-2.1x) &&\text{Add the opposite.} \\ &= 3.7x + (-2.1x) + 4 &&\text{Commutative property} \\ &= (3.7 + (-2.1))x + 4 &&\text{Distributive property} \\ &= 1.6x + 4 &&\text{Simplify.} \end{aligned}$$

(b) First change each subtraction to addition of the opposite.

$$\begin{aligned} 3.68y^2 - 7.2x - 1.4y^2 &= 3.68y^2 + (-7.2x) + (-1.4y^2) &&\text{Add the opposite.} \\ &= 3.68y^2 + (-1.4y^2) + (-7.2x) &&\text{Commutative property} \\ &= (3.68 + (-1.4))y^2 + (-7.2x) &&\text{Distributive property} \\ &= 2.28y^2 - 7.2x &&\text{Simplify.} \end{aligned}$$

Now Try Exercises 71, 75

5 Solving Applications Involving Decimals

Math in Real World Context The next three examples demonstrate everyday situations that require addition and subtraction of decimals. The first of these examples uses decimal addition to analyze changes in the stock market. The second example uses decimal subtraction to calculate the distance traveled by a car. In the last example, both addition and subtraction of decimals are used to find the balance in a checkbook register.

EXAMPLE 9 Analyzing stock market changes

The Dow Jones Industrial Average began a particular trade day at 18,050.17 and finished the day 70.08 points higher. Determine its value at the end of the day.

Solution
To determine the ending value, we need to add 18,050.17 + 70.08.

$$\begin{array}{r} \overset{1}{1}\overset{1}{8},050.17 \\ +70.08 \\ \hline 18,120.25 \end{array}$$

The Dow Jones Industrial Average ended the day at 18,120.25.

Now Try Exercise 81

EXAMPLE 10 Computing miles traveled

At the beginning of a trip, a car's odometer displayed 78,904.7 and at the end of the trip, it displayed 79,351.9. What was the total number of miles traveled on this trip?

Solution
To determine the total number of miles traveled, subtract 79,351.9 − 78,904.7.

$$\begin{array}{r} 7\,\overset{8}{\cancel{9}},\overset{13}{\cancel{3}}\overset{4}{\cancel{5}}\overset{11}{\cancel{1}}.9 \\ -7\,8,9\,0\,4\,.7 \\ \hline 4\,4\,7\,.2 \end{array}$$

A total of 447.2 miles was traveled on the trip.

Now Try Exercise 89

EXAMPLE 11 Finding a checking account balance

Find the final balance in the following checkbook register.

Date	Number	Description	Debits		Credits		Balance	
10/7		Deposit			897	61	2347.83	
10/9	5671	Electric Company	73	92				
10/10	5672	Credit Card Bill	267	14				
10/11		Deposit			207	03		
10/15	5673	Pizza Restaurant	23	76				

Solution

To find the final balance, we must subtract from the initial balance of $2347.83 all debits (payments) and add all credits (deposits). This is done from left to right, as follows.

$$\begin{aligned}\text{Final Balance} &= 2347.83 - 73.92 - 267.14 + 207.03 - 23.76 \\ &= 2273.91 - 267.14 + 207.03 - 23.76 \\ &= 2006.77 + 207.03 - 23.76 \\ &= 2213.80 - 23.76 \\ &= 2190.04 \end{aligned}$$

The final balance in the checking account is $2190.04.

Now Try Exercise 97

5.2 Putting It All Together

CONCEPT	COMMENTS	EXAMPLES
Estimating Decimal Sums and Differences	By rounding decimal numbers to convenient values, a sum or difference can be estimated before the actual computation is performed.	The sum $31.2 + 194.6$ can be estimated as $30 + 190 = 220$ or $30 + 200 = 230$.
Adding or Subtracting Decimals	1. Stack the decimals vertically with the decimal points aligned. Extra zeros may be written to the right of the last digit of any decimal number so that corresponding digits align neatly. 2. Add or subtract digits in corresponding place values. Carry or borrow as needed. 3. Place a decimal point in the result so that it is aligned vertically with the decimal points in the numbers stacked above it.	The difference $458.657 - 25.47$ is found as follows. 458.657 -25.470 433.187

5.2 Exercises   MyMathLab

CONCEPTS AND VOCABULARY

1. Before we find an exact value of a decimal sum or difference, it is often helpful to first _____ the result.

2. When finding an estimate of a decimal sum or difference, the decimals should be _____ so that the estimate can be found mentally when possible.

3. When adding or subtracting, stack decimals vertically with the _____ aligned.

4. A decimal point is placed in the result of a sum or difference so that it is aligned _____ with the decimal points in the numbers stacked above it.

ESTIMATING SUMS AND DIFFERENCES

Exercises 5–16: Estimate the given sum or difference by rounding the decimals. Answers may vary.

5. $22.13 + 397.79$
6. $168.3 + 42.19$
7. $1479.67 - 293.74$
8. $262.78 - 57.9$
9. $689.236 - 90.793$
10. $14.3 - 2008.91$
11. $1652.917 + 349.6$
12. $8003.6 - 261.77$
13. $-302.56 + 73.9$
14. $2012.73 + (-478.22)$
15. $-41.83 - 129.6$
16. $-61.332 - (-39.74)$

ADDING AND SUBTRACTING DECIMALS

Exercises 17–40: Estimate the sum or difference and then add or subtract the decimals. Estimates may vary.

17. $0.458 + 9.93$
18. $16.8 - 0.12$
19. $614.28 + 58.619$
20. $77.94 + 222.71$
21. $52 + 781.62$
22. $1198.2 + 406.39$
23. $786.429 - 28.17$
24. $153.4 - 78.95$

25. 6949 − 401.82 **26.** 570.65 − 310.55

27. 14.38 + 158.9 + 0.2 **28.** 207.8 + 1.3 + 41.35

29. 2.457 + 670.1 + 28.3 **30.** 8.8 + 21.37 + 99.6

31. −48.97 + 211.67 **32.** 309.723 + (−2.67)

33. −178.97 + (−19.4) **34.** −8.83 + (−397.2)

35. −683.01 − 12.6 **36.** −679 − 296.7

37. 19.04 − (−1279.9) **38.** 709.3 − (−7.97)

39. −51.61 − (−107.93) **40.** −293.4 − (−11)

Exercises 41–52: Find the exact sum or difference.

41. −9.623 + 143.83 **42.** 35.97 + (−192.78)

43. 19.2 − (−781.76) **44.** −13 − (−273.42)

45. −360.89 − 2793.4 **46.** 998.99 + 676.99

47. −3.02 + (−58.672) **48.** −16.9 + (−16.9)

49. 27.901 + 1207.998 **50.** 0.987 − (−3.457)

51. 6004.003 − 997.448 **52.** 100.003 − 2997.33

Exercises 53–56: Mentally estimate if the sum or difference is less than or greater than 1.

53. 0.54 + 0.52 **54.** 0.44 + 0.46

55. 1.85 − 0.9 **56.** 1.33 − 0.29

Exercises 57–60: Mentally estimate if the sum or difference is less than or greater than 0.

57. 57.72 + (−57.78) **58.** −6.74 − (−6.78)

59. 20.95 − (−20.98) **60.** −10.15 + (10.14)

EVALUATING AND SIMPLIFYING EXPRESSIONS

Exercises 61–64: Evaluate the expression x − y for the given values of the variables.

61. $x = 4.702, y = 2.104$ **62.** $x = 506.4, y = 27.33$

63. $x = 0.693, y = −0.78$ **64.** $x = −12.8, y = 56$

Exercises 65–68: If a hotdog costs h dollars and a bag of chips costs c dollars, evaluate the expression h + c to find the total cost of the two items.

65. Hotdog: $1.79, chips: $0.99

66. Hotdog: $3.49, chips: $1.39

67. Hotdog: $2.89, chips: $1.25

68. Hotdog: $4.79, chips: $2.55

Exercises 69–78: Simplify the algebraic expression.

69. $4.3y + 6.73y$ **70.** $12.99x − 5.2x$

71. $6.9w − 5 + 2.7w$ **72.** $16.5m + 2b − 8.32m$

73. $8.9n^2 − 7.1n^2$ **74.** $45.3x^3 + 10.7x^3$

75. $0.5p^2 + 13.6 − 3.1p^2$ **76.** $−0.45y^3 − 0.37y^3$

77. $4.3x + 8.1y − 2.7x − 5.4y$

78. $8.6a + 0.3 − 5.7b − 7.9a + 9.2b$

APPLICATIONS INVOLVING DECIMALS

79. Steel Imports The U.S. imported 3.2 million metric tons of steel in April 2015 and 3.1 million metric tons in May 2015. Find the total amount of steel imported in these two years. (*Source:* American Iron and Steel Institute.)

80. Steel Production In 2012, U.S. steel production was 88.6 million metric tons, and in 2014, it was 88.174. Find the total steel production for these two years.

81. Stock Prices If the price of a particular stock begins the day at $18.37 and rises in value by $1.15 for the day, what is the ending price of the stock?

82. Stock Prices If the price of a stock begins the day at $189.32 and decreases by $12.67 for the day, what is the ending price of the stock?

83. Patents There were 127.7 thousand patents issued to U.S. corporations in 2013 and 11.7 thousand fewer similar patents issued a year later. How many patents were issued to U.S. corporations in 2014? (*Source:* U.S. Patent and Trademark Office.)

84. Lunch Cost Find the total cost of a lunch consisting of a sandwich for $2.59, a drink for $0.99, and an apple for $0.59.

Exercises 85–88: **Water Consumption** *The following bar graph shows the daily U.S. per capita (per person) water consumption, in thousands of gallons, during selected years.* (*Source:* U.S. Geological Survey.)

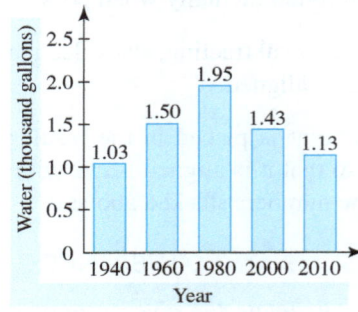

U.S. Water Use per Person

85. Which of the years shown had the highest per capita water consumption?

86. How much lower was the 2010 per capita water consumption compared to 1980?

87. How much higher was the 1980 per capita water consumption compared to 1940?

88. In which 20-year period was the increase in per capita water consumption the largest, from 1940 to 1960 or from 1960 to 1980?

89. **Driving Distance** At the beginning of a trip, a car's odometer displayed 38,410.2 and at the end of the trip, it displayed 38,707.1. What was the total number of miles traveled on this trip?

90. **Walking Distance** A pedometer is a device that measures walking distance. If a person's pedometer reads 2.54 miles at a small bridge and later reads 4.07 miles at a park bench, how far did the person walk between these two locations?

Exercises 91–94: **Perimeter** *Find the perimeter of the given figure.*

91.
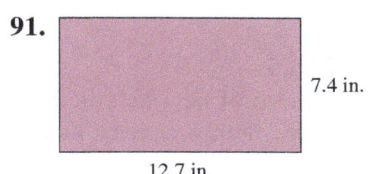
7.4 in.
12.7 in.

92.
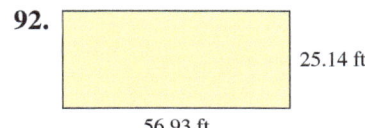
25.14 ft
56.93 ft

93.
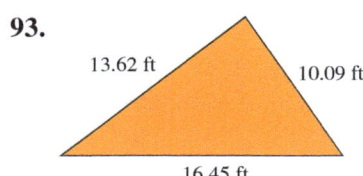
13.62 ft   10.09 ft
16.45 ft

94.
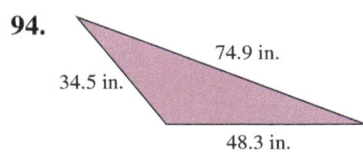
74.9 in.
34.5 in.
48.3 in.

Exercises 95 and 96: **Perimeter** *If the perimeter of the given figure is 52.5 inches, find the length represented by the variable x.*

95.
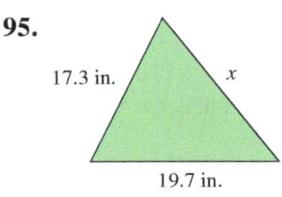
17.3 in. x
19.7 in.

96.
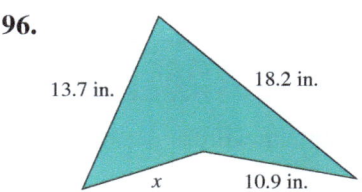
13.7 in.   18.2 in.
x 10.9 in.

Exercises 97 and 98: **Checkbook Register** *Find the final balance in the given checkbook register.*

97.

Date	Num.	Description	Debits		Credits		Balance
1/2		Deposit			153	16	704.93
1/3	8322	Convenience Store	35	61			
1/3	8323	Retail Clothing Store	159	94			
1/7		Deposit			399	83	
1/9	8324	Seafood Restaurant	41	36			

98.

Date	Num.	Description	Debits		Credits		Balance
1/2		Deposit			488	51	2783.06
1/3	5310	Bait & Tackle Shop	79	23			
1/3		Deposit			953	86	
1/7	5311	Health Club	97	22			
1/9	5312	Grocery Store	178	45			

99. **Internet Use** In 2015, the average Internet user spent 1.795 hours each month on Google and 7.764 hours each month on Facebook. How much more time was spent on Facebook? (*Source:* Nielsen/NetRatings.)

100. **Kentucky Derby** The winner of the 2008 Kentucky Derby was Big Brown with a time of 121.82 seconds. In 2015, American Pharoah won the Kentucky Derby with a time of 123.02 seconds. How much faster was Big Brown? (*Source:* Daily Racing Form.)

WRITING ABOUT MATHEMATICS

101. Discuss the difficulties of adding decimals as shown.

$$\begin{array}{r} 1503.98 \\ 1.3964 \\ 0.998 \\ +2.7 \\ \hline \end{array}$$

102. Explain the purpose of estimating a sum or difference before finding its actual value.

103. Give three examples from everyday life when you may need to add or subtract decimals.

104. Is it reasonable to apply the commutative, associative, and distributive properties to decimals? If so, give an example of each property.

SECTIONS 5.1 and 5.2 — Checking Basic Concepts

1. Write the decimal 23.097 in words.

2. Write each decimal as a fraction or mixed number in simplest form.
   (a) -5.6 (b) 0.52

3. Graph the decimal 34.27 on a number line.

4. Place the correct symbol, $<$ or $>$, in the blank between the given decimals.
   (a) $15.47 \_\_ 15.56$ (b) $0.8901 \_\_ 0.08901$

5. Round 0.27839 to the nearest thousandth.

6. Round $8.83 to the nearest dollar.

7. Estimate each sum or difference by rounding the decimals. Answers may vary.
   (a) $149.87 + 21.32$ (b) $6993.7 - 201.6$

8. Find each sum or difference.
   (a) $28.64 + 7.38$ (b) $837.52 - 402.93$
   (c) $-159.2 + 87.54$ (d) $-29 - (-311.62)$

9. Simplify each algebraic expression.
   (a) $0.34x + 4.7x$ (b) $3.9y^2 + 2 - 0.4y^2$

10. **Lunch Cost** Find the total cost of a lunch consisting of a salad for $4.79, a drink for $1.89, and a slice of pie for $1.39.

5.3 Multiplying and Dividing Decimals

Objectives
1. Estimating
2. Multiplying Decimals
   - By a Decimal
   - By a Power of 10
3. Dividing Decimals
   - By a Whole Number
   - By a Decimal
   - By a Power of 10
4. Writing Fractions as Decimals
   - Method I
   - Method II
   - Writing Mixed Numbers as Decimals
5. Evaluating and Simplifying Expressions
6. Solving Applications

NEW VOCABULARY
☐ Repeating decimal
☐ Repeat bar

1 Estimating Decimal Products and Quotients

It is often helpful to first find an estimate of a product or quotient of decimals before computing the actual result. Remember that there may be more than one reasonable way to round decimals when finding an estimate. However, when multiplying and dividing decimals, make sure not to round to zero, as shown in the following Making Connections.

MAKING CONNECTIONS 1
Estimating Products and Quotients

Recall that a decimal such as 0.3 represents a fraction of a whole. If we round 0.3 to the nearest whole number in the product $(2079.6)(0.3)$, then an estimate such as $2100 \cdot 0 = 0$ could result. However, if we write 0.3 as the fraction $\frac{3}{10}$, then a better estimate of $2100 \cdot \frac{3}{10} = 630$ results. The actual product is 623.88.

EXAMPLE 1 Estimating a decimal product and quotient

Estimate the product or quotient.
(a) $(6.93)(11.12)$ (b) $47.4 \div 4.8$

Solution
(a) One possible estimate is obtained if 6.93 is rounded to 7 and 11.12 is rounded to 11. In this case, the estimated product is $7 \cdot 11 = 77$. (The actual product is 77.0616.)
(b) If 47.4 is rounded to 50 and 4.8 to 5, the estimated quotient is $50 \div 5 = 10$. A second estimate can be found by rounding 47.4 to 45 and 4.8 to 5. In this case, the estimated quotient is $45 \div 5 = 9$. (The actual quotient is 9.875.)

Now Try Exercises 9, 11

NOTE: To avoid confusing the multiplication dot with a decimal point, a product such as $38.67 \cdot 13.2$ is written as $38.67(13.2)$ or $(38.67)(13.2)$. ∎

2 Multiplying Decimals

Decimals are multiplied in ways that are similar to how whole numbers are multiplied. One difference is that we must decide where to write the decimal point in the product.

MULTIPLYING A DECIMAL BY A DECIMAL Decimals can be written as fractions, and we already know how to multiply fractions. So, to find a product such as $(0.4)(0.7)$, we can start by writing each factor as a fraction.

1 decimal place
1 decimal place

$$(0.4)(0.7) = \frac{4}{10} \cdot \frac{7}{10} \quad \text{Write the decimals as fractions.}$$
$$= \frac{28}{100} \quad \text{Multiply.}$$
$$= 0.28 \quad \text{Write the fraction as a decimal.}$$

2 decimal places

Note that the factor 0.4 has **1** decimal place, the factor 0.7 has **1** decimal place, and the product 0.28 has $1 + 1 = 2$ decimal places. Next, we multiply $(0.04)(0.007)$.

2 decimal places
3 decimal places

$$(0.04)(0.007) = \frac{4}{100} \cdot \frac{7}{1000} \quad \text{Write the decimals as fractions.}$$
$$= \frac{28}{100{,}000} \quad \text{Multiply.}$$
$$= 0.00028 \quad \text{Write the fraction as a decimal.}$$

5 decimal places

In this case, the factor 0.04 has **2** decimal places, the factor 0.007 has **3** decimal places, and the product 0.00028 has $2 + 3 = 5$ decimal places.

READING CHECK 1

- How does multiplication of (1.1)(2.3) differ from multiplication of (11)(23)?

This discussion suggests that the number of decimal places in the product is equal to the *sum* of the number of decimal places in the factors. The process for multiplying decimals can be summarized as follows.

STUDY TIP

This procedure is another example of how mathematics builds on concepts that have already been studied. Try to get in the regular habit of reviewing topics from earlier parts of the text.

MULTIPLYING DECIMALS

To multiply decimals, do the following.

1. Multiply the decimals as though they were whole numbers. All decimal points may be ignored during computation.
2. Place a decimal point in the product so that the number of decimal places in the product is equal to the sum of the number of decimal places in the given factors.

EXAMPLE 2 Multiplying decimals

Multiply.
(a) $(9.3)(4)$ (b) $0.34(7.3)$ (c) $28.59(1.65)$

Solution

(a) $9.3 \leftarrow$ 1 decimal place
$\underline{\times4} \leftarrow$ 0 decimal places
$37.2 \leftarrow$ 1 decimal place

(b) $0.34 \leftarrow$ 2 decimal places
$\underline{\times 7.3} \leftarrow$ 1 decimal place
102
$2\,380$
$2.482 \leftarrow$ 3 decimal places

(c) $28.59 \leftarrow$ 2 decimal places
$\underline{\times1.65} \leftarrow$ 2 decimal places
$1\,4295$
$17\,1540$
$\underline{28\,5900}$
$47.1735 \leftarrow$ 4 decimal places

Now Try Exercises 17, 21, 25

EXAMPLE 3 Multiplying signed decimals

Multiply.
(a) $(-3.8)(6.7)$ (b) $(-8.2)(-0.51)$

Solution
The product of two decimals with unlike signs is negative, and the product of two decimals with like signs is positive. To avoid confusion, negative signs are omitted during computation and inserted in the final product as necessary.

(a) $3.8 \leftarrow$ 1 decimal place
$\underline{\times 6.7} \leftarrow$ 1 decimal place
$2\,66$
$\underline{22\,80}$
$25.46 \leftarrow$ 2 decimal places
The product is -25.46.

(b) $8.2 \leftarrow$ 1 decimal place
$\underline{\times 0.51} \leftarrow$ 2 decimal places
82
$\underline{4\,100}$
$4.182 \leftarrow$ 3 decimal places
The product is 4.182.

CALCULATOR HELP
To multiply decimals with a calculator, see Appendix E (page AP-32).

Now Try Exercises 29, 31

MULTIPLYING A DECIMAL BY A POWER OF 10 A simple pattern arises when a decimal is multiplied by a power of 10. Consider the resulting products when 3.92 is multiplied by 10, 100, and 1000.

3.92
$\underline{\times10}$
$0\,00$
$\underline{39\,20}$
39.20

3.92
$\underline{\times100}$
$0\,00$
$00\,00$
$\underline{392\,00}$
392.00

3.92
$\underline{\times 1000}$
$0\,00$
$00\,00$
$000\,00$
$\underline{3920\,00}$
3920.00

Note that the number 1**0** has **1** zero, and the product 39.2 is found by moving the decimal point in 3.92 to the right **1** place. Similarly, the number 1**00** has **2** zeros, and the product 392 is found by moving the decimal point in 3.92 to the right **2** places. Finally, the number 1**000** has **3** zeros, and the product 3920 is found by first inserting a zero as a placeholder and then moving the decimal point in 3.920 to the right **3** places. This discussion suggests the following procedure for multiplying a decimal by a power of 10.

READING CHECK 2

- How many places do you move the decimal point when multiplying by a power of 10?

MULTIPLYING A DECIMAL BY A POWER OF 10

The product of a decimal and a (natural number) power of 10 is found by moving the decimal point to the right the same number of places as the number of zeros in the power of 10. If needed, insert zeros at the end of the decimal as placeholders.

NOTE: Powers of 10 that are less than 1 will be discussed later in the text. The procedure presented here works only for powers of 10 such as 10, 100, 1000, etc. ∎

EXAMPLE 4 Multiplying decimals by powers of 10

Multiply.
(a) $10(65.98)$ (b) $0.83(1000)$ (c) $10,000(-2.678452)$

Solution
(a) Because **10** has **1** zero, move the decimal point in 65.98 to the right **1** place.

$$10(65.98) = 659.8$$

10 has 1 zero, so move the decimal point 1 place to the right.

(b) Insert a zero at the end of 0.83 and move the decimal point to the right 3 places.

$$0.830(1000) = 830$$

1000 has 3 zeros, so move the decimal point 3 places to the right.

(c) $10,000(-2.678452) = -26,784.52$

Now Try Exercises 33, 39, 41

3 Dividing Decimals

First we will consider a process for dividing a decimal by a whole number.

DIVIDING A DECIMAL BY A WHOLE NUMBER As with multiplying decimals, we can divide decimals by first writing them as fractions and then performing the division. For example, to find the quotient $54.3 \div 3$, we begin by writing the divisor and dividend as improper fractions.

$$
\begin{aligned}
54.3 \div 3 &= 54\frac{3}{10} \div 3 && \text{Write 54.3 as a mixed number.} \\
&= \frac{543}{10} \div \frac{3}{1} && \text{Write as improper fractions.} \\
&= \frac{543}{10} \cdot \frac{1}{3} && \text{Multiply by the reciprocal.} \\
&= \frac{(181 \cdot 3) \cdot 1}{10 \cdot 3} && \text{Factor before multiplying.} \\
&= \frac{181}{10} && \text{Simplify.} \\
&= 18\frac{1}{10} && \text{Write as a mixed number.} \\
&= 18.1 && \text{Write as a decimal.}
\end{aligned}
$$

After some effort, we have shown that 54.3 ÷ 3 = 18.1. Now consider the results using long division to find this quotient. When the divisor is a whole number, the decimal point in the quotient is written directly above the decimal point in the dividend.

$$\begin{array}{r} 18.1 \\ 3\overline{)54.3} \\ \underline{-3} \\ 24 \\ \underline{-24} \\ 0\,3 \\ \underline{-3} \\ 0 \end{array}$$

When using long division, the correct result is found directly without the need to convert the decimal to a fraction.

DIVIDING A DECIMAL BY A WHOLE NUMBER

To divide a decimal by a whole number, do the following.

1. Using long division, divide as though the dividend is a whole number. The decimal point in the dividend may be ignored during computation.
2. Place a decimal point in the quotient so that it is directly above the decimal point in the original dividend.
3. When needed, insert extra 0s to the right of the last digit in the dividend.

EXAMPLE 5 Dividing a decimal by a whole number

Divide. Check your result using multiplication.
(a) 57 ÷ 6 (b) 3.72 ÷ 5

Solution
The dividend in part (a) is a whole number, so a decimal point is written after its rightmost digit. In part (b), a leading 0 is inserted to the left of the decimal point in the quotient.

(a)
$$\begin{array}{r} 9.5 \\ 6\overline{)57.0} \leftarrow \text{Insert a 0.} \\ \underline{-54} \\ 3\,0 \\ \underline{-3\,0} \\ 0 \end{array}$$

(b) Leading 0
$$\begin{array}{r} 0.744 \\ 5\overline{)3.720} \leftarrow \text{Insert a 0.} \\ \underline{-3\,5} \\ 22 \\ \underline{-20} \\ 20 \\ \underline{-20} \\ 0 \end{array}$$

Check:
$$\begin{array}{r} 9.5 \\ \times\;6 \\ \hline 57.0 \end{array}$$

Check:
$$\begin{array}{r} 0.744 \\ \times\;5 \\ \hline 3.720 \end{array}$$

CALCULATOR HELP
To divide decimals with a calculator, see Appendix E (page AP-32).

Now Try Exercises 43, 45

The digits to the right of the decimal point in some decimals continue without end in a repeating pattern. Such decimals are called **repeating decimals**. Examples include

0.33333…,   235.659659659…,   and   14.0323232….

A **repeat bar** is used to write repeating decimals more concisely. Using repeat bars, the repeating decimals are written as

$$0.\overline{3}, \quad 235.\overline{659}, \quad \text{and} \quad 14.0\overline{32}.$$

The quotient in the next example is a repeating decimal.

EXAMPLE 6 Finding a repeating decimal when dividing

Divide. $64.5 \div 9$

Solution

```
       7.166...
    9)64.500
     −63
      ———
       1 5
      − 9
       ———
         60
       − 54
         ———
          60
        − 54
         ———
           6
```

This pattern continues.

The quotient is $7.1\overline{6}$.

The 6 repeats "forever." That is, the quotient is $7.1666666\ldots$.

Now Try Exercise 51

DIVIDING A DECIMAL BY A DECIMAL When dividing a decimal by a decimal, a small adjustment can be made so that the process for dividing a decimal by a whole number can be used to find the quotient. Consider the following series of equivalent expressions.

$$4.65 \div 2.5 = \frac{4.65}{2.5} = \frac{4.65(10)}{2.5(10)} = \frac{46.5}{25} = 46.5 \div 25$$

READING CHECK 3

- How does division of $2.1 \div 7$ differ from division of $21 \div 7$?

By multiplying both the divisor and the dividend by a power of 10, the quotient $4.65 \div 2.5$ becomes the quotient $46.5 \div 25$. Since multiplying a decimal by a power of 10 simply moves the decimal point to the right, the following procedure can be used to divide a decimal by a decimal.

> **DIVIDING A DECIMAL BY A DECIMAL**
>
> To divide a decimal by a decimal, do the following.
>
> 1. Make the divisor a whole number by moving its decimal point to the right.
> 2. Move the decimal point in the dividend to the right the same number of places.
> 3. Find the quotient using the procedure for dividing a decimal by a whole number.

EXAMPLE 7 Dividing a decimal by a decimal

Divide.
(a) $40.39 \div 3.5$ **(b)** $2.9 \div 0.04$

Solution

(a) $40.39 \div 3.5 = 403.9 \div 35$

Move decimal points one place to the right.

$$\begin{array}{r} 11.54 \\ 35\overline{)403.90} \\ -35 \\ \hline 53 \\ -35 \\ \hline 18\,9 \\ -17\,5 \\ \hline 1\,40 \\ -1\,40 \\ \hline 0 \end{array}$$

(b) $2.90 \div 0.04 = 290 \div 4$

Move decimal points two places to the right.

$$\begin{array}{r} 72.5 \\ 4\overline{)290.0} \\ -28 \\ \hline 10 \\ -8 \\ \hline 2\,0 \\ -2\,0 \\ \hline 0 \end{array}$$

Now Try Exercises 55, 57

DIVIDING A DECIMAL BY A POWER OF 10 When multiplying a decimal by a power of 10, the decimal point moves to the *right*. When dividing a decimal by a power of 10, the decimal point moves to the *left*. To see this, consider the resulting quotients when 54.8 is divided by 10, 100, and 1000.

$$\begin{array}{r} 5.48 \\ 10\overline{)54.80} \\ -50 \\ \hline 4\,8 \\ -4\,0 \\ \hline 80 \\ -80 \\ \hline 0 \end{array} \quad \begin{array}{r} 0.548 \\ 100\overline{)54.800} \\ -50\,0 \\ \hline 4\,80 \\ -4\,00 \\ \hline 800 \\ -800 \\ \hline 0 \end{array} \quad \begin{array}{r} 0.0548 \\ 1000\overline{)54.8000} \\ -50\,00 \\ \hline 4\,800 \\ -4\,000 \\ \hline 8000 \\ -8000 \\ \hline 0 \end{array}$$

The number of places that the decimal point moves to the left is equal to the number of zeros in the power of 10. This process is summarized as follows.

READING CHECK 4

- How do we know how many places to move the decimal point when dividing by a power of 10?

> **DIVIDING A DECIMAL BY A POWER OF 10**
>
> The quotient of a decimal and a (natural number) power of 10 is found by moving the decimal point to the left the same number of places as the number of zeros in the power of 10. If needed, insert zeros at the beginning of the decimal as placeholders.

EXAMPLE 8 Dividing decimals by powers of 10

Divide.
(a) $245.92 \div 10$ (b) $3.9 \div 1000$ (c) $-617 \div 100$

Solution
(a) Because 10 has **1** zero, move the decimal point in 245.92 to the left **1** place.

$$245.92 \div 10 = 24.592$$

10 has 1 zero, so move the decimal point 1 place to the left.

(b) Insert two zeros at the beginning of 3.9 and move the decimal point to the left 3 places.

$$003.9 \div 1000 = 0.0039$$

1000 has 3 zeros, so move the decimal point 3 places to the left.

(c) $-617 \div 100 = -6.17$

Now Try Exercises 63, 67, 69

4　Writing Fractions as Decimals

There are two methods for writing fractions as decimals. The first method involves finding an equivalent fraction whose denominator is a power of 10. The second method uses long division to find a decimal representation.

METHOD I: WRITING THE DENOMINATOR AS A POWER OF 10 Some fractions can be written as decimals by multiplying the numerator and denominator by the same number to result in an equivalent fraction whose denominator is a power of 10. This method is illustrated in the next example.

EXAMPLE 9 Writing a fraction as a decimal—Method I

Write each fraction as a decimal.

(a) $\dfrac{3}{4}$ (b) $-\dfrac{7}{125}$

Solution

(a) Because $4 \cdot 25 = 100$, we multiply the numerator and denominator of $\frac{3}{4}$ by 25 to get an equivalent fraction whose denominator is a power of 10.

$$\frac{3}{4} = \frac{3 \cdot 25}{4 \cdot 25} = \frac{75}{100} = 0.75 \quad \text{Seventy-five hundredths}$$

(b) Because $125 \cdot 8 = 1000$, we multiply the numerator and denominator of $-\frac{7}{125}$ by 8 to get an equivalent fraction whose denominator is a power of 10.

$$-\frac{7}{125} = -\frac{7 \cdot 8}{125 \cdot 8} = -\frac{56}{1000} = -0.056$$

Negative fifty-six thousandths

Now Try Exercises 73, 77

METHOD II: USING LONG DIVISION Method I works well if the denominator of the given fraction is a convenient factor of a power of 10, but it is not practical for writing fractions such as $\frac{14}{37}$ in decimal form. In this case, long division is used to write the decimal form.

EXAMPLE 10 Writing a fraction as a decimal—Method II

Write each fraction as a decimal.

(a) $\dfrac{2}{3}$ (b) $\dfrac{7}{12}$ (c) $-\dfrac{5}{8}$

Solution

(a) Use long division to divide 3 into 2.

```
    0.666...
  3)2.000
   -1 8
     20
    -18
     20
    -18
      2
```
Repeating 6

The decimal form is $0.\overline{6}$.

(b) Use long division to divide 12 into 7.

```
     0.5833...
  12)7.0000
    -6 0
     1 00
     - 96
       40
      -36
       40
      -36
        4
```
Repeating 3

The decimal form is $0.58\overline{3}$.

(c) Ignoring the negative sign, we use long division to divide 8 into 5.

$$\begin{array}{r} 0.625 \\ 8\overline{)5.000} \\ -4\ 8 \\ \hline 20 \\ -16 \\ \hline 40 \\ -40 \\ \hline 0 \end{array}$$

The remainder is 0, so the decimal terminates and is *not* repeating.

Inserting the negative sign results in -0.625.

Now Try Exercises 79, 81, 83

NOTE: We could have used Method I to write the fraction in Example 10(c) as a decimal. ∎

While Method I works for *some* fractions, Method II works for *all* fractions. The two methods can be summarized as follows.

READING CHECK 5

- Which method for writing fractions as decimals works for all fractions?

WRITING FRACTIONS AS DECIMALS

Method I: Multiply the numerator and denominator of the fraction by the same (nonzero) number to get an equivalent fraction whose denominator is a power of 10. Then write the new fraction as a decimal.

Method II: Use long division to divide the fraction's denominator into its numerator.

WRITING MIXED NUMBERS AS DECIMALS A mixed number can be written as a decimal by first writing the integer part of the mixed number to the left of a decimal point. Then Method I or Method II can be used to write the fractional part of the mixed number as the decimal digits to the right of the decimal point. The next example illustrates this process.

EXAMPLE 11 Writing a mixed number as a decimal

Write the mixed number $-11\frac{7}{12}$ as a decimal.

Solution
Because -11 is the integer part of the mixed number, we write -11 to the left of the decimal point and then use Method II to write the fraction $\frac{7}{12}$ as the decimal digits to the right of the decimal point. In Example 10(b), we used Method II to find that $\frac{7}{12} = 0.58\overline{3}$. The decimal form of the mixed number $-11\frac{7}{12}$ is $-11.58\overline{3}$.

Now Try Exercise 87

5 ▶ Evaluating and Simplifying Expressions

Variable expressions that involve multiplication or division can be evaluated by replacing the variable(s) with decimal numbers. The next example illustrates this process.

EXAMPLE 12 Evaluating an expression

Evaluate each expression for $x = 2.5$ and $y = 50.3$.

(a) xy (b) $\dfrac{y}{x}$

Solution

(a) Replacing x with **2.5** and y with **50.3** in the expression xy results in $(2.5)(50.3)$.

$$\begin{array}{r} 50.3 \\ \times\ 2.5 \\ \hline 25\ 15 \\ 100\ 60\ \\ \hline 125.75 \end{array}$$

2.5 and 50.3 each have 1 place after the decimal, so the product 125.75 has 2 places after the decimal.

Evaluating the expression xy for $x = 2.5$ and $y = 50.3$ results in 125.75.

(b) Replacing x with **2.5** and y with **50.3** in the expression $\frac{y}{x}$ results in $\frac{50.3}{2.5}$.

$$50.3 \div 2.5 = 503 \div 25$$

Move decimal points one place right.

$$\begin{array}{r} 20.12 \\ 25\overline{)503.00} \\ -\ 50\ \ \ \ \ \\ \hline 3\ \ \ \ \\ -\ 0\ \ \ \ \\ \hline 3\ 0\ \ \\ -\ 2\ 5\ \ \\ \hline 50 \\ -\ 50 \\ \hline 0 \end{array}$$

The remainder is 0, so the decimal terminates and is *not* repeating.

Evaluating the expression $\frac{y}{x}$ for $x = 2.5$ and $y = 50.3$ results in 20.12.

Now Try Exercises 91, 95

We can also multiply or divide an algebraic expression by a decimal, as illustrated in the next example.

EXAMPLE 13   **Simplifying algebraic expressions**

Simplify each expression.

(a) $3.4(1.8x + 4)$ (b) $\dfrac{4.8y}{3}$

Solution

(a) Use the distributive property first.

$$3.4(1.8x + 4) = 3.4(1.8x) + 3.4(4) = 6.12x + 13.6$$

(b) Divide 3 into 4.8.

$$\frac{4.8y}{3} = \frac{4.8}{3} \cdot \frac{y}{1} = 1.6y$$

Now Try Exercises 99, 103

6 Solving Applications Involving Decimals

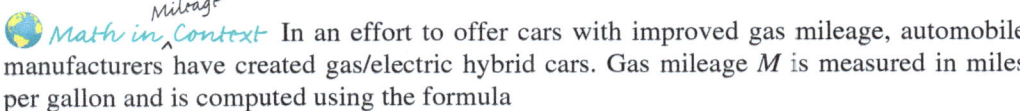

Math in Context — Mileage   In an effort to offer cars with improved gas mileage, automobile manufacturers have created gas/electric hybrid cars. Gas mileage M is measured in miles per gallon and is computed using the formula

$$M = \frac{m}{g},$$

where m is the number of miles driven and g is the number of gallons of gasoline used.

EXAMPLE 14 Computing gas mileage

Find the mileage for a hybrid vehicle that travels 308.2 miles on 6.7 gallons of gas.

Solution
Divide the number of gallons into the number of miles. Move each decimal point one place to the right before using long division.

$$\begin{array}{r} 46 \\ 67\overline{)3082} \\ -268 \\ \hline 402 \\ -402 \\ \hline 0 \end{array}$$

The hybrid vehicle's mileage is 46 miles per gallon.

Now Try Exercise 109

EXAMPLE 15 Finding the cost of catering

A catering company charges $8.89 for each person attending a company dinner. What is the total cost if 76 people attend the dinner?

Solution
Multiply the number of people by the per-person cost.

$$\begin{array}{r} 8.89 \\ \times\ 76 \\ \hline 53\ 34 \\ 622\ 30 \\ \hline 675.64 \end{array}$$

The total cost for the dinner is $675.64.

Now Try Exercise 121

5.3 Putting It All Together

CONCEPT	COMMENTS	EXAMPLES
Estimating Decimal Products and Quotients	By rounding decimal numbers to convenient values, a product or quotient can be estimated before the actual computation is performed.	The product $(7.04)(10.9)$ can be estimated as $7 \cdot 11 = 77$.
Multiplying Decimals	1. Multiply the decimals as though they were whole numbers. All decimal points may be ignored during computation. 2. Place a decimal point in the product so that the number of decimal places in the product is equal to the sum of the number of decimal places in the given factors.	$\begin{array}{r} 13.7 \\ \times\ 8.1 \\ \hline 1\ 37 \\ 109\ 60 \\ \hline 110.97 \end{array}$

CONCEPT	COMMENTS	EXAMPLES
Multiplying a Decimal by a Power of 10	Move the decimal point to the right the same number of places as the number of zeros in the power of 10. If needed, insert zeros at the end of the decimal as placeholders.	$100(5.673) = 567.3$ $10(0.42) = 4.2$
Dividing a Decimal by a Whole Number	1. Divide as though the dividend is a whole number. The decimal point in the dividend may be ignored during computation. 2. Place a decimal point in the quotient so that it is directly above the decimal point in the original dividend. 3. When needed, place extra 0s to the right of the last digit in the dividend.	The quotient $151.9 \div 7$ is found as follows. $$\begin{array}{r} 21.7 \\ 7\overline{)151.9} \\ -14 \\ \hline 11 \\ -7 \\ \hline 4\,9 \\ -4\,9 \\ \hline 0 \end{array}$$
Dividing a Decimal by a Decimal	1. Make the divisor a whole number by moving its decimal point to the right. 2. Move the decimal point in the dividend to the right the same number of places. 3. Find the quotient using the procedure for dividing a decimal by a whole number.	$0.68 \div 0.2 = 6.8 \div 2$ $$\begin{array}{r} 3.4 \\ 2\overline{)6.8} \\ -6 \\ \hline 8 \\ -8 \\ \hline 0 \end{array}$$
Dividing a Decimal by a Power of 10	Move the decimal point to the left the same number of places as the number of zeros in the power of 10. If needed, insert zeros at the beginning of the decimal as placeholders.	$114.2 \div 10 = 11.42$ $27.3 \div 100 = 0.273$
Writing Fractions as Decimals	**Method I:** Multiply the numerator and denominator of the fraction by the same (nonzero) number to get an equivalent fraction whose denominator is a power of 10. Then write the new fraction as a decimal. **Method II:** Use long division to divide the fraction's denominator into its numerator.	**Method I:** Write $\frac{3}{5}$ as a decimal. $$\frac{3}{5} = \frac{3 \cdot 2}{5 \cdot 2} = \frac{6}{10} = 0.6$$ **Method II:** Write $\frac{1}{2}$ as a decimal. $$\begin{array}{r} 0.5 \\ 2\overline{)1.0} \\ -1\,0 \\ \hline 0 \end{array}$$

5.3 Exercises

CONCEPTS AND VOCABULARY

1. Before we find an exact value of a decimal product or quotient, it is often helpful to _____ the result.

2. The number of decimal places in a product for decimal numbers is equal to the _____ of the number of decimal places in the given factors.

3. When multiplying a decimal by a power of 10, move the decimal point to the _____ as many places as there are zeros in the power of 10.

4. When dividing a decimal by a whole number, place the decimal point in the quotient so that it is directly _____ the decimal point in the dividend.

5. When dividing a decimal by a decimal, start by making the divisor a whole number by moving its decimal point to the _____.

6. If the decimal point in the divisor is moved 2 places to make the divisor a whole number, then how many places should the decimal point in the dividend be moved?

7. When dividing a decimal by a power of 10, move the decimal point to the _____ as many places as there are zeros in the power of 10.

8. The method that can *always* be used to write a fraction as a decimal involves using long division to divide the fraction's _____ into its _____.

ESTIMATING PRODUCTS AND QUOTIENTS

Exercises 9–16: Estimate the given product or quotient by rounding the decimals. Answers may vary.

9. $(2.1)(26.97)$
10. $0.92(543.11)$
11. $11.984 \div 4.031$
12. $101.5 \div 19.93$
13. $489.67(5.23)$
14. $(6.99)(10.01)$
15. $87.035 \div 0.996$
16. $52.099 \div 12.9001$

MULTIPLYING DECIMALS

Exercises 17–32: Multiply.

17. $9(13.7)$
18. $6(41.3)$
19. $(0.7)(4.2)$
20. $(0.3)(86.1)$
21. $(5.9)(0.67)$
22. $(6.32)(9.5)$
23. $3.99(4)$
24. $(2.59)(18)$
25. $10.01(3.44)$
26. $(15.02)(3.04)$
27. $-3(14.6)$
28. $11(-7.9)$
29. $(-3.9)(1.8)$
30. $(8.3)(-4.3)$
31. $(-12.3)(-0.17)$
32. $(-7.91)(-9.4)$

Exercises 33–42: Multiply.

33. $10(12.489)$
34. $10(-0.399)$
35. $(4.679)(-100)$
36. $(100)(0.035)$
37. $(100)(4.1)$
38. $(-100)(0.007)$
39. $1000(-0.0098)$
40. $(1000)(1.4)$
41. $10,000(3.4498)$
42. $10,000(-0.006)$

DIVIDING DECIMALS

Exercises 43–62: Divide.

43. $47 \div 5$
44. $98 \div 4$
45. $8.91 \div 3$
46. $56.42 \div 7$
47. $-103.2 \div 8$
48. $-4.96 \div 5$
49. $14 \div 3$
50. $16 \div 9$
51. $22.4 \div 6$
52. $102.7 \div 9$
53. $35.88 \div 26$
54. $377.06 \div 34$
55. $38.35 \div 2.6$
56. $77.184 \div 3.6$
57. $6.9 \div 0.08$
58. $5.1 \div 0.04$
59. $-0.94 \div 2.5$
60. $-6.72 \div 1.44$
61. $720 \div (-1.2)$
62. $350 \div (-2.5)$

Exercises 63–72: Divide.

63. $17.79 \div 10$
64. $-452.6 \div 10$
65. $63.4 \div (-100)$
66. $0.3 \div 100$
67. $7894 \div 100$
68. $-5904 \div 100$
69. $7.6 \div (-1000)$
70. $8.6 \div 1000$
71. $1 \div 10,000$
72. $-7 \div 10,000$

WRITING FRACTIONS AS DECIMALS

Exercises 73–84: Write the fraction as a decimal.

73. $\dfrac{1}{4}$

74. $\dfrac{4}{5}$

75. $\dfrac{3}{8}$

76. $\dfrac{11}{20}$

77. $-\dfrac{6}{25}$

78. $-\dfrac{9}{125}$

79. $\dfrac{1}{3}$

80. $\dfrac{4}{9}$

81. $\dfrac{4}{15}$

82. $\dfrac{19}{30}$

83. $-\dfrac{73}{200}$

84. $-\dfrac{7}{250}$

Exercises 85–90: Write the mixed number as a decimal.

85. $3\dfrac{1}{5}$

86. $-10\dfrac{7}{8}$

87. $-9\dfrac{8}{15}$

88. $92\dfrac{3}{20}$

89. $17\dfrac{2}{3}$

90. $-2\dfrac{2}{9}$

EVALUATING AND SIMPLIFYING EXPRESSIONS

Exercises 91–94: Evaluate the expression xy for the given values of the variables.

91. $x = 4.6, y = 10.8$

92. $x = 103.6, y = 0.7$

93. $x = 0.44, y = -5.31$

94. $x = -19.9, y = 8$

Exercises 95–98: Evaluate the expression $\dfrac{y}{x}$ for the given values of the variables.

95. $x = 0.3, y = 12.66$

96. $x = 13.2, y = 117.48$

97. $x = 1.6, y = -0.08$

98. $x = -1.3, y = 39$

Exercises 99–106: Simplify the algebraic expression.

99. $2.7(1.3x + 8)$

100. $-0.5(6.4y - 7)$

101. $-5.1(4y + 3.6)$

102. $0.8(0.2y + 0.1)$

103. $\dfrac{6.4x}{8}$

104. $\dfrac{33.46y}{1.4}$

105. $\dfrac{-88w}{1.1}$

106. $\dfrac{-36m}{0.9}$

APPLICATIONS INVOLVING DECIMALS

107. **Burning Calories** The average 168-pound person burns 8.9 calories per minute while backpacking. How many calories would a person of this weight burn while backpacking for 65 minutes? (Source: The Calorie Control Council.)

108. **Exchange Rates** In a recent year, a person could exchange 1 U.S. dollar for 1.3 Canadian dollars. How many Canadian dollars could a person get for 180 U.S. dollars?

109. **Gas Mileage** Find the mileage for a Smart car that travels 198.9 miles on 3.9 gallons of gas.

110. **Gas Mileage** Find the mileage for a gas-powered scooter that travels 58.4 miles on 0.8 gallon.

111. **Twitter vs. Facebook** On its 10-year anniversary, Twitter had a quarterly revenue of $710 million. On Facebook's 10-year anniversary, it had a quarterly revenue that was 3.64 times Twitter's revenue. What was Facebook's quarterly revenue on its 10-year anniversary?

112. **Twitter vs. Facebook** On its 10-year anniversary, Twitter had a quarterly net profit of −90 million dollars. On Facebook's 10-year anniversary, it had a quarterly net profit that was −5.81 times Twitter's net profit. What was Facebook's quarterly net profit on its 10-year anniversary?

113. **iPhone Models** As of March 2016, 34.3% of active iPhone users owned the iPhone 6. The percentage of active iPhone users that owned the iPhone 5c was one-seventh the percentage of iPhone 6 users. What percentage of iPhone users owned the 5c model?

114. **Microsoft vs. Google** In 2015, Microsoft held 9% of the market share of cloud infrastructure services, which was $\tfrac{9}{4}$ times the percent of the market share held by Google. Convert $\tfrac{9}{4}$ to a decimal and then determine the percent of the market share held by Google.

115. **Buying Groceries** If the unit price for a box of cereal is $0.24 per ounce, how much does a 19.5-ounce box of the cereal cost?

116. **Running a Relay** Each runner in a relay took 14.2 seconds for her leg of the race. If there were four runners, what was the total time?

117. **Estimated Taxes** A small business owner expects to owe $1056.48 next year in estimated taxes. If estimated taxes are paid in four equal payments, how much is each payment?

118. **Car Payments** An automobile manufacturer offers a zero-interest loan on new car purchases. If a buyer finances $27,891 for 5 years (60 equal payments), how much is the monthly payment?

119. **Home Loans** Most people take out a mortgage when they purchase a home. A mortgage is a loan to be paid back in equal payments that include interest. If the monthly payments for a $235,000 mortgage are $1275.93, how much is paid to the bank over 360 months (30 years)?

120. **Home Loans** If a bank offered a zero-interest loan for the $235,000 mortgage in Exercise 119, what would the equal monthly payments be for a 360-month (30-year) loan?

121. **Feeding the Team** A head coach takes his football team to a buffet restaurant for lunch. What is the total bill for 54 people if the restaurant charges $9.49 per person?

122. **Feeding the Team** A coach pays $515.78 to take 41 people to dinner at a buffet restaurant. How much does the restaurant charge per person for a buffet dinner?

WRITING ABOUT MATHEMATICS

123. Find two methods for estimating the product

$$89.7(0.41).$$

Which of your methods is most accurate?

124. Find a fast way to multiply the decimal 186.34 by each of the following fractions.

$$\frac{1}{10}, \frac{1}{100}, \text{ and } \frac{1}{1000}$$

How does your method compare to that used to multiply by powers of 10 such as 10, 100, and 1000?

5.4 Real Numbers, Square Roots, and Order of Operations

Objectives

1. Introducing Real Numbers
2. Approximating Square Roots
   - The Guess-and-Check Method
   - The Babylonian Algorithm
3. Applying the Order of Operations
   - Evaluating Numerical Expressions
   - Evaluating Algebraic Expressions
4. Solving Applications Involving Real Numbers

NEW VOCABULARY

☐ Irrational number
☐ Real number

1 Introducing Real Numbers

There are infinitely many numbers between any two numbers on a number line. So far in this text, we have used number lines to graph natural numbers, whole numbers, integers, and rational numbers. All of these numbers have decimal representations whose digits either terminate (end) or repeat indefinitely. In this section, we will discuss the *real* numbers, which include all of the sets of numbers listed above and also include a new set called the *irrational* numbers.

As defined in an earlier chapter, a rational number is any number that can be written as a fraction whose numerator and denominator are both integers (and the denominator is not 0). Note that *all* natural numbers, whole numbers, and integers can be written as rational numbers by using 1 as the denominator. Are there numbers that are not rational numbers? The answer is yes. Such numbers are called irrational numbers. Examples include

$$\sqrt{7}, \quad \pi, \quad \text{and} \quad 3.1010010001\ldots.$$

The symbol π is the Greek letter pi. It is used to denote a specific irrational number that is frequently encountered when working with circles. The digits in the decimal representation of π neither terminate nor repeat. In fact, every irrational number has a *nonterminating, nonrepeating* decimal representation. Decimal approximations for the irrational numbers $\sqrt{7}$ and π are

$$\sqrt{7} \approx 2.64575131 \quad \text{and} \quad \pi \approx 3.14159265.$$

An **irrational number** cannot be expressed as a fraction whose numerator and denominator are both integers. In other words, it is *not* rational. Furthermore, its decimal representation is nonterminating and nonrepeating.

5.4 REAL NUMBERS, SQUARE ROOTS, AND ORDER OF OPERATIONS

STUDY TIP

Some computations in this section are quite involved. Students with a firm grasp of multiplication facts are likely to complete these computations with more ease and accuracy than students who lack such knowledge.

READING CHECK 1

- What are the two sets of numbers that include all real numbers?

The **real numbers** are numbers that can be written as decimals. They include all natural numbers, whole numbers, integers, rational numbers, and irrational numbers. The set of real numbers is represented visually in **FIGURE 5.2**.

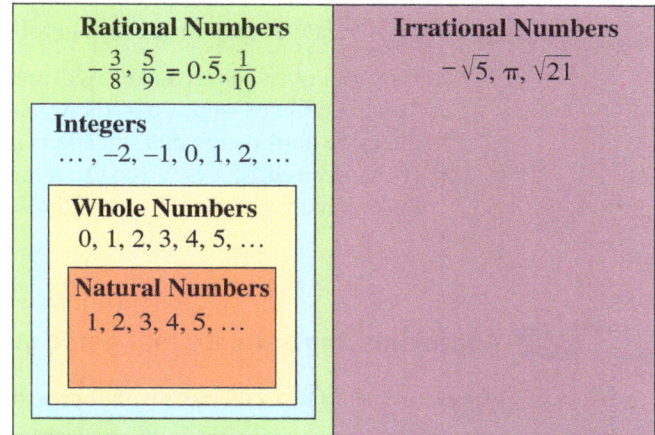

FIGURE 5.2

Note that the numbers $-\sqrt{5}$ and $\sqrt{21}$ are listed as irrational numbers. In fact, if the square root of a whole number is not a whole number, then it is an irrational number. In the next example, real numbers are identified as rational numbers, irrational numbers, integers, whole numbers, and natural numbers.

EXAMPLE 1 Identifying real numbers

Identify the numbers in the given list that belong to each of the following sets of numbers.
(a) natural numbers (b) whole numbers (c) integers
(d) rational numbers (e) irrational numbers

$$\sqrt{11}, \quad -\frac{2}{3}, \quad 0, \quad 4.2, \quad \sqrt{9}, \quad \text{and} \quad -13$$

Solution
(a) Because 9 is a perfect square, $\sqrt{9} = 3$. The only natural number in the list is $\sqrt{9}$.
(b) Whole numbers include 0 and any natural numbers. The whole numbers are 0 and $\sqrt{9}$.
(c) Integers include whole numbers and any numbers that are opposites of whole numbers. The integers are 0, $\sqrt{9}$, and -13.
(d) Rational numbers include the integers and any fractions, terminating decimals, or repeating decimals. The rational numbers are $-\frac{2}{3}$, 0, 4.2, $\sqrt{9}$, and -13.
(e) Any number that is not rational is irrational. The only irrational number is $\sqrt{11}$.

Now Try Exercise 9

2 Approximating Square Roots

Up to this point in the text, square roots such as $\sqrt{10}$ and $\sqrt{23}$ have been approximated to the nearest whole number. There are, however, several methods for computing more accurate approximations for square roots of whole numbers that are not perfect squares. Two such methods are discussed in this subsection.

THE GUESS-AND-CHECK METHOD The process for finding a square root using the guess-and-check method can be summarized as follows.

CALCULATOR HELP
To approximate a square root with a calculator, see Appendix E (page AP-32).

> **APPROXIMATING SQUARE ROOTS BY GUESS-AND-CHECK**
>
> To approximate a square root, do the following.
>
> 1. Use a number line to make a reasonable guess for the square root.
> 2. Make a table of values based on your guess. Compute the square of each value in the table until the given radicand is between consecutive values in the table.
> 3. If the desired accuracy for the approximation has been found, stop. Otherwise, make a more accurate guess of the square root and go back to Step 2.

EXAMPLE 2 Approximating a square root using guess-and-check

Approximate $\sqrt{11}$ to the nearest hundredth.

Solution
The radicand 11 is located on the number line between the perfect squares 9 and 16, as shown in **FIGURE 5.3**.

FIGURE 5.3

READING CHECK 2
- To the nearest tenth, between what two numbers is $\sqrt{11}$?

Because 11 is closer to 9 than to 16, it is reasonable to think that $\sqrt{11}$ is closer to $\sqrt{9} = 3$ than to $\sqrt{16} = 4$. A reasonable initial guess for $\sqrt{11}$ might be 3.3. To check this guess, complete a table such as **TABLE 5.1**. For example, the square of 3.2 is found by multiplying $(3.2)(3.2) = 10.24$. The other squares are found in a similar manner.

Guess	3.2	3.3	3.4	3.5
Guess Squared	10.24	10.89	11.56	12.25

TABLE 5.1

The radicand 11 is between 10.89 and 11.56 but is closer to 10.89. Our next guess for $\sqrt{11}$ might be 3.32. To check this guess, complete a table such as **TABLE 5.2**.

Guess	3.31	3.32	3.33	3.34
Guess Squared	10.9561	11.0224	11.0889	11.1556

TABLE 5.2

TABLE 5.2 reveals that the square of 3.32 is closest to 11. Based on this observation, we can conclude that $\sqrt{11} \approx 3.32$ when rounded to the nearest hundredth.

Now Try Exercise 17

READING CHECK 3
- What is the primary use of the Babylonian algorithm?

THE BABYLONIAN ALGORITHM Nearly 4000 years ago, the Babylonians discovered a technique for finding accurate approximations for square roots. The process involves only basic arithmetic and is outlined as follows.

THE BABYLONIAN ALGORITHM FOR SQUARE ROOTS

To approximate $\sqrt{Q}$ where Q represents a whole number, do the following.

1. Make a reasonable guess for the square root. Assign this value to the variable A.
2. Compute the quotient $Q \div A$.*
3. Add A to the result from Step 2.
4. Divide the result of Step 3 by 2. The result is an approximation for $\sqrt{Q}$.*

For a more accurate approximation, assign the result from Step 4 to the variable A and repeat the process beginning at Step 2.

* To avoid complicated arithmetic, these results may be rounded to four decimal places.

NOTE: When the initial guess is reasonably accurate, the number of correct digits in the approximation for $\sqrt{Q}$ will double each time through the algorithm. ∎

EXAMPLE 3 **Approximating a square root using the Babylonian algorithm**

Go through the Babylonian algorithm two times to approximate the value of $\sqrt{19}$. Give four decimal places in your answer.

Solution
STEP 1: Since the radicand 19 is closer to 16 than to 25, a reasonable initial guess for $\sqrt{19}$ is a value that is slightly more than $\sqrt{16} = 4$. Let A be 4.4.
STEP 2: The quotient $Q \div A$ results in $19 \div 4.4 = 4.3\overline{18}$. For ease of computation, round this value to 4.3182.
STEP 3: Adding A to the result from Step 2 gives $4.3182 + 4.4 = 8.7182$.
STEP 4: Dividing the result from Step 3 by 2 gives $8.7182 \div 2 = 4.3591$.

Let $A = 4.3591$ and repeat the process beginning at Step 2.
STEP 2: The quotient $Q \div A$ results in $19 \div 4.3591 \approx 4.3587$.
STEP 3: Adding A to the result from Step 2 gives $4.3587 + 4.3591 = 8.7178$.
STEP 4: Dividing the result from Step 3 by 2 gives $8.7178 \div 2 = 4.3589$.
By the Babylonian algorithm, $\sqrt{19} \approx 4.3589$ when rounded to four decimal places.

Now Try Exercise 25

NOTE: To nine decimal places, $\sqrt{19} \approx 4.358898944$. Two or three passes through the Babylonian algorithm can give amazingly accurate results. However, to obtain better approximations, results from Steps 2 and 4 should *not* be rounded along the way. ∎

3 Applying the Order of Operations

The order of operations agreement that applies to decimals is the same as the agreement used for whole numbers, integers, and fractions.

ORDER OF OPERATIONS

1. Do all calculations within grouping symbols such as parentheses and radicals, or above and below a fraction bar.
2. Evaluate all exponential expressions. (Do negation after exponents.)
3. Do all multiplication and division from *left to right*.
4. Do all addition and subtraction from *left to right*.

EVALUATING NUMERICAL EXPRESSIONS The next two examples demonstrate how the order of operations agreement applies to numerical expressions involving decimals.

EXAMPLE 4   Evaluating numerical expressions

Evaluate each expression.
(a) $6.8 + 4.1(3) - 2$ (b) $9.83 + 17.8 \div 10$

Solution
(a) Perform multiplication before addition or subtraction.

$$6.8 + \mathbf{4.1(3)} - 2 = 6.8 + \mathbf{12.3} - 2 \quad \text{Multiply.}$$
$$= 19.1 - 2 \quad \text{Add.}$$
$$= 17.1 \quad \text{Subtract.}$$

(b) Perform division before addition.

$$9.83 + \mathbf{17.8 \div 10} = 9.83 + 1.78 \quad \text{Divide.}$$
$$= 11.61 \quad \text{Add.}$$

Now Try Exercises 35, 37

EXAMPLE 5   Evaluating numerical expressions involving grouping symbols

Evaluate each expression.
(a) $-3.2 + 5(4.1 - 0.6)$ (b) $-14.8 \div \sqrt{21.2 - 5.2}$ (c) $\dfrac{0.6 + 2.7 \div 3}{0.03(1000)}$

Solution
(a) The expression within parentheses is evaluated first.

$$-3.2 + 5(\mathbf{4.1 - 0.6}) = -3.2 + 5(\mathbf{3.5}) \quad \text{Subtract within parentheses.}$$
$$= -3.2 + 17.5 \quad \text{Multiply.}$$
$$= 14.3 \quad \text{Add.}$$

(b) The expression under the radical is evaluated first.

$$-14.8 \div \sqrt{\mathbf{21.2 - 5.2}} = -14.8 \div \sqrt{\mathbf{16}} \quad \text{Subtract under radical.}$$
$$= -14.8 \div 4 \quad \text{Evaluate } \sqrt{16}.$$
$$= -3.7 \quad \text{Divide.}$$

(c) Use the order of operations both above and below the fraction bar.

$$\frac{0.6 + \mathbf{2.7 \div 3}}{\mathbf{0.03(1000)}} = \frac{0.6 + \mathbf{0.9}}{\mathbf{30}} \quad \text{Divide above; multiply below.}$$
$$= \frac{1.5}{30} \quad \text{Add.}$$
$$= 0.05 \quad \text{Divide.}$$

Now Try Exercises 39, 41, 45

EVALUATING ALGEBRAIC EXPRESSIONS The order of operations agreement often is needed when algebraic expressions are evaluated for decimal values of the variable(s). The next example demonstrates this situation.

EXAMPLE 6 Evaluating algebraic expressions

Evaluate each expression for $x = 4.2$ and $y = 2.1$.
(a) $-4x + y$ **(b)** $5(y - 0.3) - x$

Solution
(a) Start by replacing x with 4.2 and y with 2.1 in the given expression.

$$\begin{aligned} -4x + y &= -4(4.2) + 2.1 & & x = 4.2 \text{ and } y = 2.1. \\ &= -16.8 + 2.1 & & \text{Multiply } -4(4.2). \\ &= -14.7 & & \text{Add.} \end{aligned}$$

(b) Replace x with 4.2 and y with 2.1 in the given expression.

$$\begin{aligned} 5(y - 0.3) - x &= 5(2.1 - 0.3) - 4.2 & & x = 4.2 \text{ and } y = 2.1. \\ &= 5(1.8) - 4.2 & & \text{Evaluate } (2.1 - 0.3). \\ &= 9 - 4.2 & & \text{Multiply.} \\ &= 4.8 & & \text{Subtract.} \end{aligned}$$

Now Try Exercises 51, 53

4 Solving Applications Involving Real Numbers

Math in Context (Airline) Each year, U.S. airlines transport billions of bags. **TABLE 5.3** lists the number of bags per thousand that were mishandled by selected U.S. airlines during one 10-month period.

Mishandled Bags per Thousand

Airline	American	Continental	Delta	US Airways
Bags	4.35	2.72	4.90	2.97

TABLE 5.3 *Source:* U.S. Department of Transportation.

In the next example, we compute the average number of mishandled bags per thousand for these four airlines.

EXAMPLE 7 Finding the average number of mishandled bags

Find the average number of bags per thousand that were mishandled by the four airlines in **TABLE 5.3**.

Solution
To find the average of 4.35, 2.72, 4.90, and 2.97, add the numbers together and then divide the result by 4 because there are 4 numbers in the list.

$$\text{Average} = \frac{4.35 + 2.72 + 4.90 + 2.97}{4} = \frac{14.94}{4} = 3.735$$

During this period, these four airlines mishandled an average of 3.735 bags per thousand.

Now Try Exercise 59

EXAMPLE 8 **Finding the area of a trapezoid**

Find the area of the trapezoid shown in the following figure.

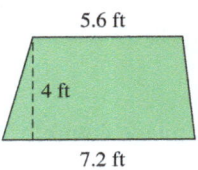

Solution
The formula for the area of a trapezoid is

$$A = \frac{1}{2}(a+b)h,$$

where a and b are the bases and h is the height.

$$A = \frac{1}{2}(5.6 + 7.2)(4) \qquad a = 5.6, b = 7.2, \text{ and } h = 4$$

$$= \frac{1}{2}(12.8)(4) \qquad \text{Add } (5.6 + 7.2).$$

$$= 0.5(12.8)(4) \qquad \text{Write } \tfrac{1}{2} \text{ as } 0.5.$$

$$= 25.6 \qquad \text{Multiply.}$$

The area of the trapezoid is 25.6 square feet.

Now Try Exercise 63

5.4 Putting It All Together

CONCEPT	COMMENTS	EXAMPLES
Irrational Numbers	Irrational numbers cannot be expressed as fractions whose numerator and denominator are both integers. If the square root of a whole number is not a whole number, then it is an irrational number.	$\sqrt{13} \approx 3.605551275$ $\pi \approx 3.141592654$
Real Numbers	Real numbers can be written as decimals. They include natural numbers, whole numbers, integers, rational numbers, and irrational numbers.	$-\sqrt{5}, -2, 0, \frac{5}{7},$ and $4.\overline{9}$

CONCEPT	COMMENTS	EXAMPLES
Approximating Square Roots by Guess-and-Check	1. Use a number line to make a reasonable guess for the square root. 2. Make a table of values based on your guess. Compute the square of each value in the table until the given radicand is between consecutive values in the table. 3. If the desired accuracy for the approximation has been found, stop. Otherwise, make a more accurate guess of the square root and go back to Step 2.	Approximate $\sqrt{8}$ as follows. $\sqrt{4}=2 \quad \sqrt{9}=3$ 0 1 2 3 4 5 6 7 8 9 10 A reasonable guess is 2.8. | Guess | 2.7 | 2.8 | 2.9 | | Guess Squared | 7.29 | 7.84 | 8.41 | To the nearest tenth, $\sqrt{8} \approx 2.8$.
Approximating Square Roots with the Babylonian Algorithm	To approximate $\sqrt{Q}$, do the following. 1. Make a reasonable guess for the square root. Assign this value to the variable A. 2. Compute the quotient $Q \div A$. 3. Add A to the result from Step 2. 4. Divide the result of Step 3 by 2. The result is an approximation for $\sqrt{Q}$. For a more accurate approximation, assign the result from Step 4 to the variable A and repeat the process beginning at Step 2.	Approximate $\sqrt{10}$ as follows. 1. A reasonable guess is 3.2. 2. $10 \div 3.2 = 3.125$ 3. $3.125 + 3.2 = 6.325$ 4. $6.325 \div 2 = 3.1625$ Rounded to two decimal places, $\sqrt{10} \approx 3.16$.

5.4 Exercises   MyMathLab®

CONCEPTS AND VOCABULARY

1. A(n) _____ number cannot be expressed as a fraction whose numerator and denominator are both integers.

2. Numbers that can be written as decimals are called _____ numbers.

3. The _____ numbers include natural numbers, whole numbers, integers, rational numbers, and irrational numbers.

4. Which of the numbers, $\sqrt{35}$ or $\sqrt{36}$, is *not* an irrational number?

5. Which of the methods for approximating square roots involves making tables of values?

6. The _____ algorithm can be used to approximate the square root of a number.

IDENTIFYING REAL NUMBERS

Exercises 7–10: Identify the numbers from the given list that belong to each of the following sets of numbers.

    (a) natural numbers   (b) whole numbers
    (c) integers   (d) rational numbers
    (e) irrational numbers

7. $-\frac{5}{8}, 3, 0, \sqrt{4}, \sqrt{5}$

8. $8, -\frac{2}{3}, 3.\overline{8}, -1.1, \sqrt{6}$

9. $\sqrt{10}, 0, \frac{9}{3}, -1.\overline{2}, 6.4$

10. $-\sqrt{4}, 3, \frac{8}{2}, 7.57, \sqrt{20}$

APPROXIMATING SQUARE ROOTS

Exercises 11–16: Approximate the given square root to the nearest tenth using the guess-and-check method.

11. $\sqrt{5}$
12. $\sqrt{3}$
13. $\sqrt{15}$
14. $\sqrt{22}$
15. $\sqrt{83}$
16. $\sqrt{62}$

Exercises 17–22: Approximate the given square root to the nearest hundredth using the guess-and-check method.

17. $\sqrt{14}$
18. $\sqrt{17}$
19. $\sqrt{42}$
20. $\sqrt{57}$
21. $\sqrt{99}$
22. $\sqrt{88}$

Exercises 23–28: Go through the Babylonian algorithm one time to approximate the given square root. Give two decimal places in your answer.

23. $\sqrt{6}$
24. $\sqrt{8}$
25. $\sqrt{18}$
26. $\sqrt{28}$
27. $\sqrt{78}$
28. $\sqrt{92}$

Exercises 29–34: Go through the Babylonian algorithm two times to approximate the given square root. Give four decimal places in your answer.

29. $\sqrt{2}$
30. $\sqrt{12}$
31. $\sqrt{30}$
32. $\sqrt{44}$
33. $\sqrt{55}$
34. $\sqrt{85}$

APPLYING THE ORDER OF OPERATIONS

Exercises 35–50: Evaluate the numerical expression.

35. $3.9 - 6(0.3) + 7.7$
36. $-5.9 + 9.4(0.8)$
37. $7 - 10(0.37) \div 2$
38. $8.7 + 99 \div 10$
39. $4.6 - 3(2 - 5.1)$
40. $\sqrt{4.1 + 4.9} \div 5$
41. $5 \div \sqrt{18.6 - 2.6}$
42. $|4.2 - 8.7| + 1$
43. $\dfrac{3.2 + 6.9}{5.4 - 4.8}$
44. $\dfrac{100(0.034)}{6}$
45. $\dfrac{1.3 - 2.8 \div 7}{300 \div 1000}$
46. $\dfrac{12.4 - 5(1.6)}{-2.8 - 5.2}$
47. $2 - 0.4^2 + 5.34$
48. $4(5.3 - 4.8)^2$
49. $13.9 - |4.2 - 8.7| + (0.7 + 0.4)^2$
50. $(\sqrt{11.5 + 24.5} - |2.1 - 9.3|) \div 0.4$

Exercises 51–56: Evaluate the algebraic expression for $x = -1.5$ and $y = 4.6$.

51. $3x - y$
52. $2y + 5x$
53. $3.4(5.5 + x) - y$
54. $y - (3x + 1.8)$
55. $\dfrac{10y - 30}{x}$
56. $\dfrac{10x + 32.48}{y}$

APPLICATIONS INVOLVING REAL NUMBERS

57. **Flight of a Baseball** If a baseball is hit upward with a velocity of 66 feet per second, its height h in feet above the ground after t seconds can be approximated using the formula
$$h = -16t^2 + 66t + 3.$$
Find the height of the ball after 0.5 second.

58. **Flight of a Projectile** If a projectile is fired upward with a velocity of 88 feet per second, its height h in feet above the ground after t seconds can be approximated using the formula
$$h = -16t^2 + 88t.$$
Find the projectile's height after 2.5 seconds.

59. **Endowment for the Arts** The following table shows the funding, in millions of dollars, from the National Endowment for the Arts during selected years. Find the average funding during this time.

Year	2012	2013	2014	2015
Funding	146.0	138.4	146.0	146.0

Source: National Endowment for the Arts.

60. **Philanthropy** The following table shows grant funding in billions of dollars from the Bill and Melinda Gates Foundation during selected years. Find the average funding during this time. Round to one decimal place.

Year	2011	2012	2013	2014
Funding	3.2	3.1	3.6	3.9

Source: Bill and Melinda Gates Foundation.

Exercises 61–64: **Geometry** *Use a formula from the list provided to find the area of the figure.*

Rectangle: $A = lw$, Trapezoid: $A = 0.5(a + b)h$,
Triangle: $A = 0.5bh$, Square: $A = s^2$

61.

8.9 in. × 5.2 in.

62.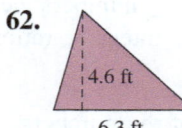

4.6 ft, 6.3 ft

63.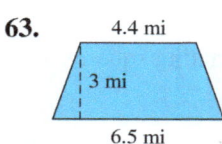

4.4 mi, 3 mi, 6.5 mi

64.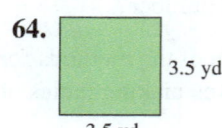

3.5 yd, 3.5 yd

65. Gas Mileage The formula for computing a car's gas mileage M is

$$M = \frac{E - B}{G},$$

where E is the ending odometer reading, B is the beginning odometer reading, and G is the number of gallons of gasoline used for the trip. Find the mileage for a trip where $E = 38{,}989.5$, $B = 38{,}423.1$, and $G = 19.2$.

66. Gas Mileage Refer to Exercise 65. Find the mileage for a trip where $E = 87{,}261.8$, $B = 86{,}658.7$, and $G = 18.5$.

67. SUV Sales The number N in millions of SUVs sold during year x is approximated by

$$N = 0.333x - 667.9.$$

Determine the number of SUVs sold in 2015. (Source: Autodata Corporation.)

68. Target Heart Rate For general health and weight loss, a person who is x years old should maintain a minimum target heart rate of T beats per minute during extended exercise, where

$$T = -0.5x + 120.$$

Determine the minimum target heart rate for a person who is 48 years old.

69. Building a Play Set An engineer who is designing a rope swing for a backyard play set determines that the rope must be $\sqrt{48}$ feet in length. Approximate this length to the nearest hundredth of a foot.

70. Quality Control For safety reasons, the company that makes the play set in Exercise 69 requires that all calculations are checked by a second engineer. The second engineer determines that the rope length should be $4 \cdot \sqrt{3}$ feet. Approximate this length to the nearest hundredth of a foot. Do the engineers agree?

WRITING ABOUT MATHEMATICS

71. A student claims that the number

$$\frac{1.3}{4}$$

is irrational because it is not written as a fraction whose numerator and denominator are both integers. Do you agree with this observation? Explain.

72. Would you rather use the guess-and-check method or the Babylonian algorithm to approximate $\sqrt{12}$ to six decimal places? Explain.

SECTIONS 5.3 and 5.4 — Checking Basic Concepts

1. Multiply.
(a) $(3.6)(2.1)$
(b) $-7(34.9)$
(c) $10(2.63)$
(d) $(3.41)(-5.6)$

2. Divide.
(a) $28.48 \div 8$
(b) $82.4 \div 9$
(c) $63.5 \div 100$
(d) $-0.64 \div 1.6$

3. Write each fraction or mixed number as a decimal.
(a) $\frac{9}{20}$
(b) $\frac{7}{15}$
(c) $-3\frac{1}{8}$
(d) $9\frac{5}{6}$

4. Simplify each expression.
(a) $2.5(7.6x + 3)$
(b) $\frac{-28y}{0.4}$

5. Use the guess-and-check method to approximate the value of $\sqrt{24}$ to the nearest tenth.

6. Go through the Babylonian algorithm one time to approximate $\sqrt{20}$. Give two decimal places in your answer.

7. Simplify each expression.
(a) $6.3 + 4 \div 10$
(b) $\frac{5.2 - 1.8}{0.2 + 1.6}$

8. Evaluate the expression for $x = 4.5$ and $y = 0.3$.

$$100y - (3x + 22.8)$$

9. Estimated Taxes A small business owner must pay \$4678.48 next year in estimated taxes. If the estimated taxes are paid in four equal payments, how much is each payment?

10. Flight of a Golf Ball If a golf ball is hit upward with a velocity of 66 feet per second, its height h in feet above the ground after t seconds can be approximated using the formula

$$h = -16t^2 + 66t.$$

Determine the height of the golf ball 3.5 seconds after it is hit.

5.5 Solving Equations Involving Decimals

Objectives

1. Solving Equations Symbolically
   - Working Directly with Decimals
   - Clearing Decimals
2. Solving Equations Numerically
3. Solving Equations Graphically
4. Solving Applications Involving Equations with Decimals

1 Solving Equations Symbolically

There are two common symbolic methods used to solve equations involving decimals. In the first method, the properties of equality are used directly with all decimals involved. In the second method, decimals are "cleared" before using the properties of equality.

STUDY TIP

Once again, the equations presented in this section are solved symbolically, numerically, and graphically. These are the same methods used in previous chapters to solve equations.

WORKING DIRECTLY WITH DECIMALS In the next example, we use properties of equality to solve equations involving decimals.

EXAMPLE 1   Solving equations by working directly with decimals

Solve each equation and check the solution.
(a) $x + 7.9 = 18.6$ (b) $-4.8y = 31.2$

Solution
(a) Because 7.9 is being **added** to x in the given equation, we can isolate x on the left side of the equation by **subtracting 7.9** from each side of the equation.

$$x + 7.9 = 18.6 \qquad \text{Given equation}$$
$$x + 7.9 - 7.9 = 18.6 - 7.9 \qquad \text{Subtract 7.9 from each side.}$$
$$x = 10.7 \qquad \text{Simplify.}$$

To check this solution, replace x with 10.7 in the given equation.

$$x + 7.9 = 18.6 \qquad \text{Given equation}$$
$$10.7 + 7.9 \stackrel{?}{=} 18.6 \qquad \text{Replace } x \text{ with 10.7.}$$
$$18.6 = 18.6 \checkmark \qquad \text{Add; the solution checks.}$$

(b) Since -4.8 is being **multiplied** with y in the given equation, we isolate y on the left side of the equation by **dividing** each side of the equation by -4.8.

$$-4.8y = 31.2 \qquad \text{Given equation}$$
$$\frac{-4.8y}{-4.8} = \frac{31.2}{-4.8} \qquad \text{Divide each side by } -4.8.$$
$$y = -6.5 \qquad \text{Simplify.}$$

To check this solution, replace y with -6.5 in the given equation.

$$-4.8y = 31.2 \qquad \text{Given equation}$$
$$-4.8(-6.5) \stackrel{?}{=} 31.2 \qquad \text{Replace } y \text{ with } -6.5.$$
$$31.2 = 31.2 \checkmark \qquad \text{Multiply; the solution checks.}$$

Now Try Exercises 7, 11

In the next example, equations are solved by applying *both* the addition and multiplication properties of equality.

EXAMPLE 2 Solving equations by working directly with decimals

Solve each equation and check the solution.
(a) $2.6w + 22.9 = 34.6$ (b) $4.2(m - 7) = 1.7m + 2.1$

Solution
(a) Start by **subtracting 22.9** from each side of the equation.

$2.6w + 22.9 = 34.6$	Given equation
$2.6w + 22.9 - \mathbf{22.9} = 34.6 - \mathbf{22.9}$	Subtract 22.9 from each side.
$2.6w = 11.7$	Simplify.
$\dfrac{2.6w}{\mathbf{2.6}} = \dfrac{11.7}{\mathbf{2.6}}$	Divide each side by 2.6.
$w = 4.5$	Simplify.

To check this solution, replace w with 4.5 in the given equation.

$2.6w + 22.9 = 34.6$	Given equation
$2.6(\mathbf{4.5}) + 22.9 \stackrel{?}{=} 34.6$	Replace w with 4.5.
$11.7 + 22.9 \stackrel{?}{=} 34.6$	Multiply.
$34.6 = 34.6$ ✓	Add; the solution checks.

(b) Start by applying the distributive property on the left side of the equation.

$4.2(m - 7) = 1.7m + 2.1$	Given equation
$4.2m - 29.4 = 1.7m + 2.1$	Distributive property
$4.2m - 29.4 + \mathbf{29.4} = 1.7m + 2.1 + \mathbf{29.4}$	Add 29.4 to each side.
$4.2m = 1.7m + 31.5$	Simplify.
$4.2m - \mathbf{1.7m} = 1.7m - \mathbf{1.7m} + 31.5$	Subtract $1.7m$ from each side.
$2.5m = 31.5$	Simplify.
$\dfrac{2.5m}{\mathbf{2.5}} = \dfrac{31.5}{\mathbf{2.5}}$	Divide each side by 2.5.
$m = 12.6$	Simplify

Check this solution by replacing m with 12.6 in the given equation.

$4.2(m - 7) = 1.7m + 2.1$	Given equation
$4.2(\mathbf{12.6} - 7) \stackrel{?}{=} 1.7(\mathbf{12.6}) + 2.1$	Replace m with 12.6.
$4.2(5.6) \stackrel{?}{=} 1.7(12.6) + 2.1$	Subtract.
$23.52 \stackrel{?}{=} 21.42 + 2.1$	Multiply.
$23.52 = 23.52$ ✓	Add; the solution checks.

Now Try Exercises 13, 25

CLEARING DECIMALS In a previous chapter, the LCD was used to clear fractions from equations. In a similar way, we can multiply each side of an equation by a power of 10 to clear decimals from an equation. Recall that multiplying each *side* of an equation by a (nonzero) number is equivalent to multiplying each *term* in the equation by that number.

> **SOLVING EQUATIONS BY CLEARING DECIMALS**
>
> To solve an equation by clearing decimals, do the following.
>
> **STEP 1:** Use the distributive property to clear parentheses from the equation.
>
> **STEP 2:** Note the maximum number of decimal places in any number in the equation. Multiply every term in the equation by a power of 10 with that many zeros.
>
> **STEP 3:** Simplify each term in the equation and combine any like terms.
>
> **STEP 4:** Solve the resulting equation.

EXAMPLE 3 Solving equations by clearing decimals

Solve each equation.
(a) $5.2(x - 3) = 14.3$ (b) $4.2x - 2.84 = 6.2x + 2.5$

Solution

(a) The maximum number of decimal places in any number is **one**. For Step 2, multiply each term by the power of 10 with **one** zero, or multiply by **10**.

$$5.2(x - 3) = 14.3 \qquad \text{Given equation}$$
$$5.2x - 15.6 = 14.3 \qquad \text{Distributive property (Step 1)}$$
$$10(5.2x) - 10(15.6) = 10(14.3) \qquad \text{Multiply by 10. (Step 2)}$$
$$52x - 156 = 143 \qquad \text{Simplify. (Step 3)}$$
$$52x - 156 + 156 = 143 + 156 \qquad \text{Add 156 to each side. (Step 4)}$$
$$52x = 299 \qquad \text{Simplify.}$$
$$\frac{52x}{52} = \frac{299}{52} \qquad \text{Divide each side by 52. (Step 4)}$$
$$x = 5.75 \qquad \text{Simplify.}$$

(b) The maximum number of decimal places in any number is **two**. For Step 2, multiply each term by **100**. Since there are no parentheses, Step 1 can be skipped.

$$4.2x - 2.84 = 6.2x + 2.5 \qquad \text{Given equation}$$
$$100(4.2x) - 100(2.84) = 100(6.2x) + 100(2.5) \qquad \text{Multiply by 100. (Step 2)}$$
$$420x - 284 = 620x + 250 \qquad \text{Simplify. (Step 3)}$$
$$420x - 284 + 284 = 620x + 250 + 284 \qquad \text{Add 284 to each side. (Step 4)}$$
$$420x = 620x + 534 \qquad \text{Simplify.}$$
$$420x - 620x = 620x - 620x + 534 \qquad \text{Subtract } 620x \text{ from each side. (Step 4)}$$
$$-200x = 534 \qquad \text{Simplify.}$$
$$\frac{-200x}{-200} = \frac{534}{-200} \qquad \text{Divide each side by } -200. \text{ (Step 4)}$$
$$x = -2.67 \qquad \text{Simplify.}$$

Now Try Exercises 27, 29

MAKING CONNECTIONS 1

Clearing Decimals with a Power of 10

To understand why we use powers of 10 to clear decimals, consider that rewriting

$$3.74x + 19.6 = 8.1$$

as an equivalent equation involving fractions results in

$$\frac{374}{100}x + \frac{196}{10} = \frac{81}{10}.$$

We clear fractions from this equation by multiplying each side of the equation by the LCD of all fractions involved. Note that the LCD is 100 and has the same number of zeros as there are decimal places in the number 3.74 in the given equation.

READING CHECK 1

- By what number should we multiply each side of the equation

$$3x + 4.01 = 0.3$$

in order to eliminate decimals?

2 Solving Equations Numerically

Equations involving decimals can also be solved numerically. In the next example, a table of values is used to solve an equation involving decimals.

EXAMPLE 4   **Solving an equation numerically**

Complete **TABLE 5.4** for the given values of x and then solve the equation $2.6x - 3.1 = 4.7$.

x	−1	0	1	2	3	4
$2.6x - 3.1$						

TABLE 5.4

Solution

Begin by replacing x in $2.6x - 3.1$ with -1. Since $2.6(-1) - 3.1$ evaluates to -5.7, the left side of the equation is equal to -5.7 when $x = -1$. We write the result -5.7 below -1, as shown in **TABLE 5.5**. Likewise, we write -3.1 below 0 because the expression $2.6(0) - 3.1$ evaluates to -3.1. The remaining values are found similarly and are shown in **TABLE 5.5**.

x	−1	0	1	2	3	4
$2.6x - 3.1$	−5.7	−3.1	−0.5	2.1	4.7	7.3

TABLE 5.5

TABLE 5.5 reveals that the left side of the given equation equals 4.7 when $x = 3$. Therefore, the solution is 3.

Now Try Exercise 31

When an equation has a decimal solution, the top row of the table of values will need to include decimal values. This is demonstrated in the next example.

EXAMPLE 5 Solving an equation numerically

Complete **TABLE 5.6** for the given values of x and then solve the equation $4x + 5 = 12.2$.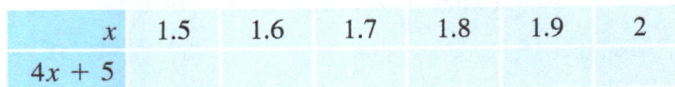

x	1.5	1.6	1.7	1.8	1.9	2
$4x + 5$						

TABLE 5.6

Solution
The completed table is shown in **TABLE 5.7**.

x	1.5	1.6	1.7	1.8	1.9	2
$4x + 5$	11	11.4	11.8	12.2	12.6	13

TABLE 5.7

The left side of the given equation equals 12.2 when $x = 1.8$. The solution is 1.8.

Now Try Exercise 35

READING CHECK 2
- Why do we need to check a solution found graphically?

3 Solving Equations Graphically

When a graph is provided for an equation involving decimals, an estimate of the solution can be found visually. In the next example, we estimate and then check a solution to an equation involving decimals.

EXAMPLE 6 Solving an equation graphically

A graph of the expression $1.7x - 2.3$ is shown in the following figure. Use the graph to solve the equation $1.7x - 2.3 = -5.7$.

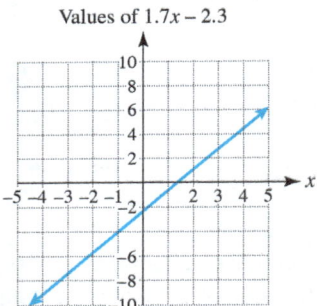

Solution
Approximate the position of -5.7 on the vertical axis and move **horizontally** to the left to the graphed line. From this position, move **vertically** upward to the horizontal axis. The solution appears to be -2, as shown in the figure.

To see if -2 is the correct solution, check it in the given equation.

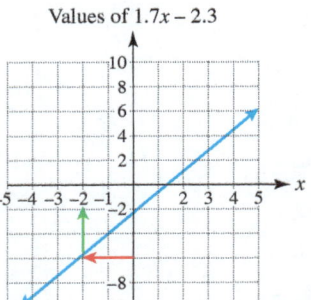

$$1.7x - 2.3 = -5.7 \quad \text{Given equation}$$
$$1.7(-2) - 2.3 \stackrel{?}{=} -5.7 \quad \text{Replace } x \text{ with } -2.$$
$$-3.4 - 2.3 \stackrel{?}{=} -5.7 \quad \text{Multiply.}$$
$$-5.7 = -5.7 \checkmark \quad \text{Subtract; the solution checks.}$$

The solution to the equation $1.7x - 2.3 = -5.7$ is -2.

Now Try Exercise 39

4 Solving Applications Involving Equations with Decimals

🌐 **Math in Context** *Texting* Some cellular phone companies offer text messaging plans that charge a flat fee for a set number of text messages. However, subscribers who go over the designated number are usually charged a fee for *each* additional text message. For example, a plan may charge $2 for up to 1000 text messages, and 0.0075 dollar for each additional text message. A subscriber who sends (or receives) more than 1000 messages can compute the total cost C of x messages using the formula

$$C = 0.0075(x - 1000) + 2.$$

In the next example, we use this formula to analyze the cost of text messaging.

EXAMPLE 7 **Analyzing the cost of text messaging**

Find the number of text messages that correspond to total charges of $13.25 by replacing C in the formula $C = 0.0075(x - 1000) + 2$ with 13.25 and then solving the resulting equation.

Solution

$0.0075(x - 1000) + 2 = 13.25$	Given equation (rewritten)
$0.0075x - 7.5 + 2 = 13.25$	Distributive property (Step 1)
$0.0075x - 5.5 = 13.25$	Combine like terms.
$\mathbf{10{,}000}(0.0075x) - \mathbf{10{,}000}(5.5) = \mathbf{10{,}000}(13.25)$	Multiply by 10,000. (Step 2)
$75x - 55{,}000 = 132{,}500$	Simplify. (Step 3)
$75x - 55{,}000 + \mathbf{55{,}000} = 132{,}500 + \mathbf{55{,}000}$	Add 55,000 to each side. (Step 4)
$75x = 187{,}500$	Simplify.
$\dfrac{75x}{75} = \dfrac{187{,}500}{75}$	Divide each side by 75. (Step 4)
$x = 2500$	Simplify.

The charges are $13.25 for 2500 text messages.

Now Try Exercise 51

🌐 **Math in Context** *Email* Unsolicited bulk email messages known as spam arrive in email inboxes every day. The average *daily* number of spam messages N in billions x years after 2004 can be computed using the formula

$$N = 9.9x + 11.$$

In the next example, we use this formula to analyze email spam. (*Source:* SpamUnit.)

EXAMPLE 8 **Analyzing email spam**

Use the formula $N = 9.9x + 11$ to find the year when the average daily number of email spam messages reached 110 billion.

Solution

Replace N in the formula $N = 9.9x + 11$ with 110 and solve the resulting equation.

$$9.9x + 11 = 110 \qquad \text{Given equation (rewritten)}$$
$$10(9.9x) + 10(11) = 10(110) \qquad \text{Multiply by 10. (Step 2)}$$
$$99x + 110 = 1100 \qquad \text{Simplify. (Step 3)}$$
$$99x + 110 - 110 = 1100 - 110 \qquad \text{Subtract 110 from each side. (Step 4)}$$
$$99x = 990 \qquad \text{Simplify.}$$
$$\frac{99x}{99} = \frac{990}{99} \qquad \text{Divide each side by 99. (Step 4)}$$
$$x = 10 \qquad \text{Simplify.}$$

The average daily number of email spam messages reached 110 billion in 2004 + 10, or 2014.

Now Try Exercise 57

EXAMPLE 9 Finding a Celsius temperature

The formula

$$F = 1.8C + 32$$

gives the relationship between F degrees Fahrenheit and C degrees Celsius. Use the formula to find the Celsius temperature that is equivalent to 98.6 °F.

Solution

Substitute 98.6 in the formula for F and solve the resulting equation.

$$1.8C + 32 = 98.6 \qquad \text{Given equation (rewritten)}$$
$$10(1.8C) + 10(32) = 10(98.6) \qquad \text{Multiply by 10. (Step 2)}$$
$$18C + 320 = 986 \qquad \text{Simplify. (Step 3)}$$
$$18C + 320 - 320 = 986 - 320 \qquad \text{Subtract 320 from each side. (Step 4)}$$
$$18C = 666 \qquad \text{Simplify.}$$
$$\frac{18C}{18} = \frac{666}{18} \qquad \text{Divide each side by 18. (Step 4)}$$
$$C = 37 \qquad \text{Simplify.}$$

Therefore, 98.6 °F is equivalent to 37 °C.

Now Try Exercise 59

5.5 Putting It All Together

CONCEPT	COMMENTS	EXAMPLES
Solving Equations Symbolically by Clearing Decimals	1. Use the distributive property to clear parentheses from the equation. 2. Note the maximum number of decimal places for any number in the equation. Multiply every term in the equation by a power of 10 with that many zeros. 3. Simplify each term in the equation and combine any like terms. 4. Solve the resulting equation.	$2.4(x + 5) = 18.6$ $2.4x + 12 = 18.6$ $24x + 120 = 186$ $24x + 120 - 120 = 186 - 120$ $24x = 66$ $\frac{24x}{24} = \frac{66}{24}$ $x = 2.75$

5.5 SOLVING EQUATIONS INVOLVING DECIMALS

CONCEPT	COMMENTS	EXAMPLES
Solving Equations Numerically	A table of values can be used to solve some equations involving decimals. Complete the table for various values of the variable and then select the solution from the table.	The solution to the equation $5.2x + 3 = -2.2$ is -1. | x | -3 | -2 | -1 | |---|---|---|---| | $5.2x + 3$ | -12.6 | -7.4 | -2.2 |
Solving Equations Graphically	Use a graph of the equation to estimate a solution. Then check the estimated solution in the given equation to be sure that it is correct.	The solution to $0.5x - 1 = 1$ is 4. Values of $0.5x - 1$ **Check:** $0.5(4) - 1 = 1$ ✓

5.5 Exercises   MyMathLab®

CONCEPTS AND VOCABULARY

1. Name the three methods discussed in this section for solving equations.

2. When we multiply each side of an equation by a power of 10, we are clearing _____.

3. The power of 10 used to clear decimals has the same number of zeros as the maximum number of _____ in any number in the equation.

4. When a table of values is used to solve an equation, the equation is being solved _____.

SOLVING EQUATIONS SYMBOLICALLY

Exercises 5–30: Solve the equation symbolically.

5. $x - 4.8 = 11.7$
6. $m + 12.6 = 3.2$
7. $y + 18.1 = 5.7$
8. $w - 0.9 = 13.7$
9. $2.8n = 9.8$
10. $4.25q = -51$
11. $-8.6x = -42.14$
12. $-3.3b = -23.43$
13. $4.5m + 1.1 = 15.5$
14. $8y + 0.1 = 21.7$
15. $16.12 = 3.4w - 7$
16. $3.1k + 7 = 0.8$
17. $-5.7x + 2.1 = 4$
18. $9x - 4.6 = 8$
19. $3.6 - 4b = 15.8$
20. $-6.7 = 5 - 3w$
21. $3(x + 2.1) = 17.4$
22. $4.1(y - 2) = 8.2$
23. $0.6(n - 3.7) = 9$
24. $6(q + 1.6) = 29.4$
25. $2.5(p - 4) = 2.9p + 6.1$
26. $4.8(m + 7) = 7.9m + 2.6$
27. $6.4(y + 12) = 73.6$
28. $10.7(6w - 25) = 2.14$
29. $3.2k - 5.64 = 2.9k + 7.8$
30. $7.8p - 3 = 4.2p + 3.84$

SOLVING EQUATIONS NUMERICALLY

Exercises 31–36: Solve the equation numerically by completing the given table of values.

31. $4.8x - 2.5 = 7.1$

x	0	1	2	3	4
$4.8x - 2.5$					

32. $2.4x + 3.2 = 0.8$

x	−4	−3	−2	−1	0
$2.4x + 3.2$					

33. $1.7 - 3.9x = 9.5$

x	−2	−1	0	1	2
$1.7 - 3.9x$					

34. $3 - 6.8x = -24.2$

x	1	2	3	4	5
$3 - 6.8x$					

35. $7x - 3 = 11.7$

x	2	2.1	2.2	2.3	2.4
$7x - 3$					

36. $5x - 12 = -7.5$

x	0.6	0.7	0.8	0.9	1
$5x - 12$					

SOLVING EQUATIONS GRAPHICALLY

Exercises 37–42: Solve the equation graphically.

37. $-0.8x - 2 = 2$ **38.** $0.6x + 2 = 5$

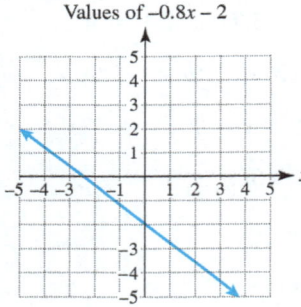
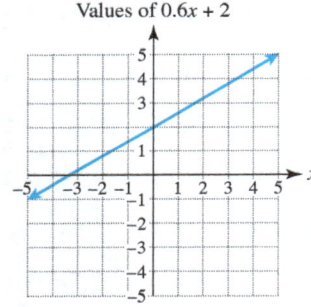

39. $0.7x + 0.4 = 2.5$ **40.** $-0.4x + 1.1 = 1.5$

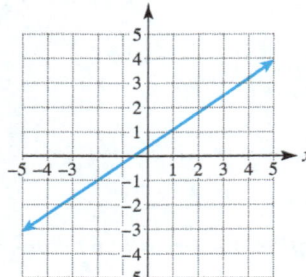
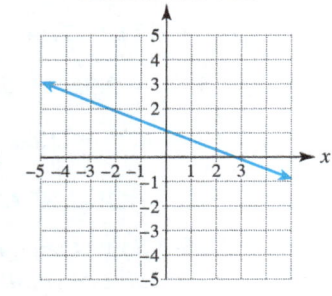

41. $-0.9x - 1.3 = -4$ **42.** $0.6x - 2.2 = -4$

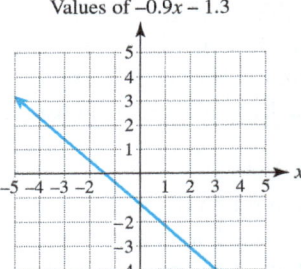
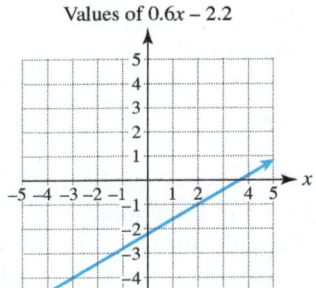

CHOOSING YOUR OWN METHOD

Exercises 43–50: Solve the equation.

43. $x + 13.7 = 22.85$ **44.** $5.7b = -19.38$

45. $8.1y - 3.8 = 20.5$ **46.** $4.9 - 4n = -1.9$

47. $2.6(k - 5) = 9.1$ **48.** $6(w - 1.3) = 24$

49. $8.7(4w + 7) = 19.14$

50. $6.2m + 11.94 = 1.8m - 9.4$

APPLICATIONS INVOLVING DECIMAL EQUATIONS

51. Text Messaging Suppose that the cost C of sending or receiving x text messages is given by the formula

$C = 0.005(x - 800) + 1$, where $x \geq 800$.

Find the number of text messages that correspond to total charges of $2.75 by replacing C in the formula with 2.75 and solving the resulting equation.

52. International Cell Phone Plan Suppose that the cost C of talking internationally for x minutes on a cell phone is given by the formula

$C = 0.25(x - 500) + 30$, where $x \geq 500$.

Find the number of minutes that correspond to total charges of $55.75 by replacing C in the formula with 55.75 and solving the resulting equation.

Exercises 53–56: **Geometry** *Use a formula from the list provided and the given area to find the unknown measure.*

Rectangle: $A = lw$, Trapezoid: $A = 0.5(a + b)h$,

Triangle: $A = 0.5bh$

53. $A = 40.95$ square feet **54.** $A = 11$ square inches

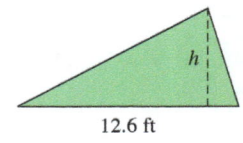
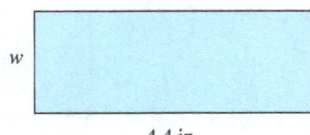

55. $A = 41.7$ square inches **56.** $A = 51.06$ square feet

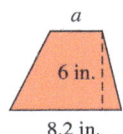
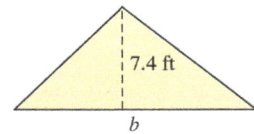

57. Movie Tickets The national average price P (in current dollars) for admission to a movie during year x can be estimated using the formula

$$P = 0.2x - 394.6.$$

Find the year when the average price of a movie ticket was $8.20. (*Source:* Motion Picture Association of America.)

58. Drive-in Theaters The number N of drive-in movie theaters in the United States x years after 2000 can be estimated using the formula

$$N = -6.8x + 443.$$

Find the year when there were 341 drive-in theaters. (*Source:* National Association of Theatre Owners.)

59. Converting Temperature The formula

$$F = 1.8C + 32$$

gives the relationship between F degrees Fahrenheit and C degrees Celsius. Find the Celsius temperature that is equivalent to 57.2 °F.

60. Converting Temperature The formula

$$C = \frac{(F - 32)}{1.8}$$

gives the relationship between C degrees Celsius and F degrees Fahrenheit. Find the Fahrenheit temperature that is equivalent to 46.5 °C.

61. Facebook Users The number of daily Facebook users F, in millions, during year x can be estimated by $F(x) = 156.25x - 313,844$. Find the year when the number of daily Facebook users first reached 900 million by solving $900 = 156.25x - 313,844$. (*Source:* Facebook.)
(a) Solve symbolically.
(b) Solve numerically using the table shown.
(c) Solve graphically using the graph shown at the top of the next column.

x	2011	2012	2013	2014	2015
$156.25x - 313,844$					

(d) Compare the symbolic, numerical, and graphical solutions.

62. Connected Cars Shipments of cars equipped with Internet access are expected to increase in the coming years. The number of connected car shipments C, in millions, during year x can be estimated by $C(x) = 3.226x - 6425.7$. Find the year when there were 84.4 million connected car shipments by solving the equation $84.4 = 3.226x - 6425.7$. (*Source:* Business Insider.)
(a) Solve symbolically.
(b) Solve numerically using the table shown.
(c) Solve graphically using the graph shown.

x	2016	2017	2018	2019	2020
$3.226x - 6425.7$					

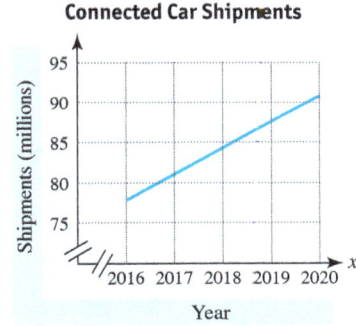

(d) Compare the symbolic, numerical, and graphical solutions.

WRITING ABOUT MATHEMATICS

63. Compare the powers of 10 used to clear decimals and the LCD used to clear fractions. What is the relationship between these two concepts?

64. Would you rather solve the equation

$$4.17x + 6.3 = 22.146$$

by clearing decimals first, or would using the properties of equality directly work better for you? Explain.

65. The solution to the equation

$$9x + 7 = 11$$

is $0.\overline{4}$. Discuss the difficulties that occur when solving this equation numerically with a table of values.

66. The solution to the equation

$$3.2x + 6 = 9.5$$

is 1.09375. Discuss the difficulties that occur when solving this equation graphically.

Group Activity

iPod Capacity The amount of storage space available on an iPod 6 touch is measured in units called gigabytes (GB). Most songs, photos, and applications stored on an iPod have file sizes measured in megabytes (MB). There are 1000 megabytes in 1 gigabyte.

(a) A typical 3-minute music file (one song) requires about 3.46 MB of storage space. How many megabytes are required to store 1794 songs that average 3 minutes in length?

(b) A typical digital photo requires about 0.6 MB of storage space on an iPod. How many megabytes are required to store 673 such photos?

(c) A typical 2-hour movie requires about 1.35 GB of storage. How many gigabytes are required to store 7 such movies?

(d) A student wants to buy an iPod with enough memory to store all of the files listed in parts (a), (b), and (c), as well as 3.74 GB in applications. How many *giga*bytes of storage space are needed to hold all of the files?

(e) If the iPod is available in either a 16-GB, 32-GB, or 64-GB model, which model should the student buy?

5.6 Applications from Geometry and Statistics

Objectives

1. Reviewing the Parts of a Circle
2. Finding the Circumference of a Circle
3. Finding the Area of a Circle
4. Finding the Area of Composite Regions
5. Applying the Pythagorean Theorem
6. Finding the Mean, Median, and Mode
   - Finding the Mean
   - Finding the Median
   - Finding the Mode
7. Finding a Weighted Mean and GPA

1 Reviewing the Parts of a Circle

The following See the Concept summarizes some basic concepts about circles.

See the Concept 1 | **PARTS OF A CIRCLE**

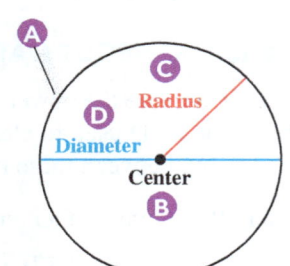

Ⓐ A **circle** is a collection of points that are all the same distance from a central point.

Ⓑ The central point is called the **center** of the circle.

Ⓒ The distance from the center of the circle to any point on the circle is called a **radius**.

Ⓓ The distance across the circle on a straight line through the center of the circle is called a **diameter**.

NOTE: If r represents the radius and d represents the diameter, then the following relationships hold.

$$d = 2r \quad \text{and} \quad r = \frac{1}{2}d$$

WATCH VIDEO IN MML.

NEW VOCABULARY

- ☐ Circle
- ☐ Center
- ☐ Radius
- ☐ Diameter
- ☐ Circumference
- ☐ Composite region
- ☐ Semicircle
- ☐ Right triangle
- ☐ Right angle
- ☐ Legs (of a right triangle)
- ☐ Hypotenuse
- ☐ Measures of central tendency
- ☐ Mean (arithmetic mean)
- ☐ Median
- ☐ Mode
- ☐ Weighted mean
- ☐ Grade point average (GPA)

2 Finding the Circumference of a Circle

The distance around an enclosed region is usually called the perimeter. However, the distance around a circle is called the **circumference**. For *any* circle, dividing the circumference C by the diameter d always results in the irrational number π, where $\pi \approx 3.14159265$. Because the value of π is a nonterminating decimal, we often use either 3.14, or $\frac{22}{7}$ for an approximation to π. We can write this relationship as the equation

$$\frac{C}{d} = \pi.$$

The multiplication property of equality can be used to rewrite this equation so that it gives the circumference of a circle. Because the variable d represents the diameter of the circle, which is a nonzero number, we can multiply each side of the equation by d, as follows.

$$\frac{C}{d} = \pi \qquad \text{Given equation}$$

$$\frac{C}{d} \cdot \frac{d}{1} = \pi \cdot d \qquad \text{Multiply each side by } d.$$

$$C = \pi d \qquad \text{Simplify.}$$

Because $d = 2r$, we have the following formulas for the circumference of a circle.

CIRCUMFERENCE OF A CIRCLE

The circumference C of a circle with radius r and diameter d is

$$C = \pi d \quad \text{or} \quad C = 2\pi r.$$

READING CHECK 1

- How are the radius and diameter related?

EXAMPLE 1 Finding the circumference of a circle

Find the exact circumference of the given circle and then approximate the circumference using 3.14 as an approximation for π.

(a)   (b)

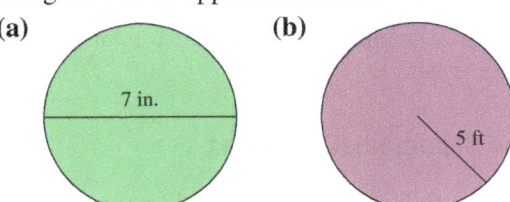

Solution
(a) The exact circumference is $C = \pi d = \pi(7) = 7\pi$ inches. Using 3.14 for π results in an approximate circumference of $C \approx 7(3.14) = 21.98$ inches.
(b) The exact circumference is $C = 2\pi r = 2\pi(5) = 10\pi$ feet. The approximate circumference is $C \approx 10(3.14) = 31.4$ feet.

Now Try Exercises 15, 21

STUDY TIP

This section contains many computations involving decimals. If calculator use is not allowed during exams, do not use one while completing your homework.

3 Finding the Area of a Circle

The formula for finding the area of a circle also involves the irrational number π.

AREA OF A CIRCLE

The area A of a circle with radius r is

$$A = \pi r^2.$$

EXAMPLE 2 Finding the area of a circle

Find the exact area of the given circle and then approximate the area using 3.14 as an approximation for π.

(a)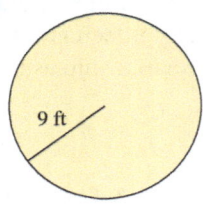

(b)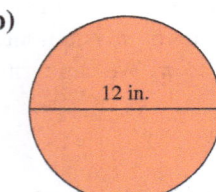

Solution
(a) The exact area is $A = \pi r^2 = \pi(9^2) = 81\pi$ square feet. An approximation for the area of the circle is $A \approx 81(3.14) = 254.34$ square feet.
(b) The radius is half of the diameter, or 6 inches. If $r = 6$, the exact area of the circle is $A = \pi r^2 = \pi(6^2) = 36\pi$ square inches. Using 3.14 for π, we calculate the approximate area as $A \approx 36(3.14) = 113.04$ square inches.

Now Try Exercises 23, 29

4 Finding the Area of Composite Regions

READING CHECK 2

- What is a composite region?

A region that consists of more than one geometric shape is called a **composite region**. For example, the composite region shown in **FIGURE 5.4** includes a rectangle and a half circle. (The geometric name for a half circle is a **semicircle**.)

A Composite Region

FIGURE 5.4

To find the area of a composite region, more than one area formula is usually needed, as shown in the next example.

EXAMPLE 3 Finding areas of composite regions

Find the area of the shaded region. Use 3.14 to approximate π, if needed.

(a)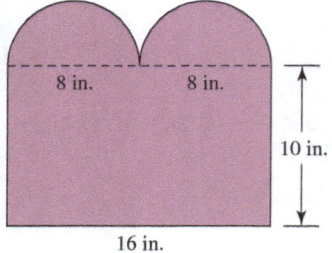

(b)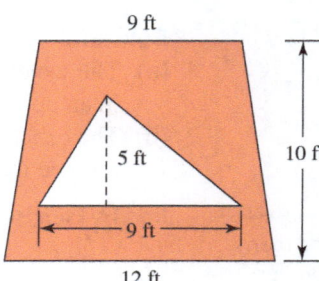

Solution
(a) The region comprises a rectangle and two semicircles. Together, the semicircles make one complete circle with a radius of 4 inches. The total shaded area is the sum of the areas of the rectangle and the two semicircles.

Rectangle: $A = lw = 16(10) = 160$ square inches

Semicircles: $A = \pi r^2 = \pi(4^2) = 16\pi \approx 16(3.14) = 50.24$ square inches

The total area is approximately $160 + 50.24 = 210.24$ square inches.

(b) The area of the shaded region can be found by subtracting the area of the white triangle from the area of the trapezoid.

$$\text{Trapezoid:} \quad A = \frac{1}{2}(a+b)h = \frac{1}{2}(9+12) \cdot 10 = 105 \text{ square feet}$$

$$\text{Triangle:} \quad A = \frac{1}{2}bh = \frac{1}{2}(9)(5) = 22.5 \text{ square feet}$$

The shaded area is exactly $105 - 22.5 = 82.5$ square feet.

Now Try Exercises 33, 35

5 Applying the Pythagorean Theorem

Over 2500 years ago, the Greek mathematician Pythagoras showed that a special relationship exists among the lengths of the sides of a *right triangle*. In a **right triangle**, one of the angles is a **right angle** with a measure of 90° (degrees). To indicate that an angle is a right angle, it is marked with a small square. (Note that each corner of a square forms a right angle.)

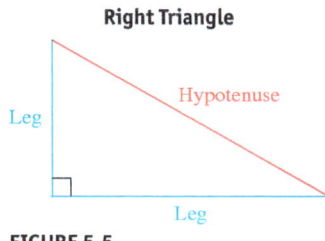

FIGURE 5.5

The two shorter sides that form the right angle are called the **legs** of the right triangle, and the longest side, which is opposite the right angle, is called the **hypotenuse**. See **FIGURE 5.5**. The Pythagorean theorem states that for any right triangle, the sum of the squares of the lengths of the two legs is always equal to the square of the length of the hypotenuse.

The following See the Concept gives a visual illustration of the Pythagorean theorem.

See the Concept 2   THE PYTHAGOREAN THEOREM

For any right triangle with hypotenuse length c and leg lengths a and b:
$$a^2 + b^2 = c^2.$$

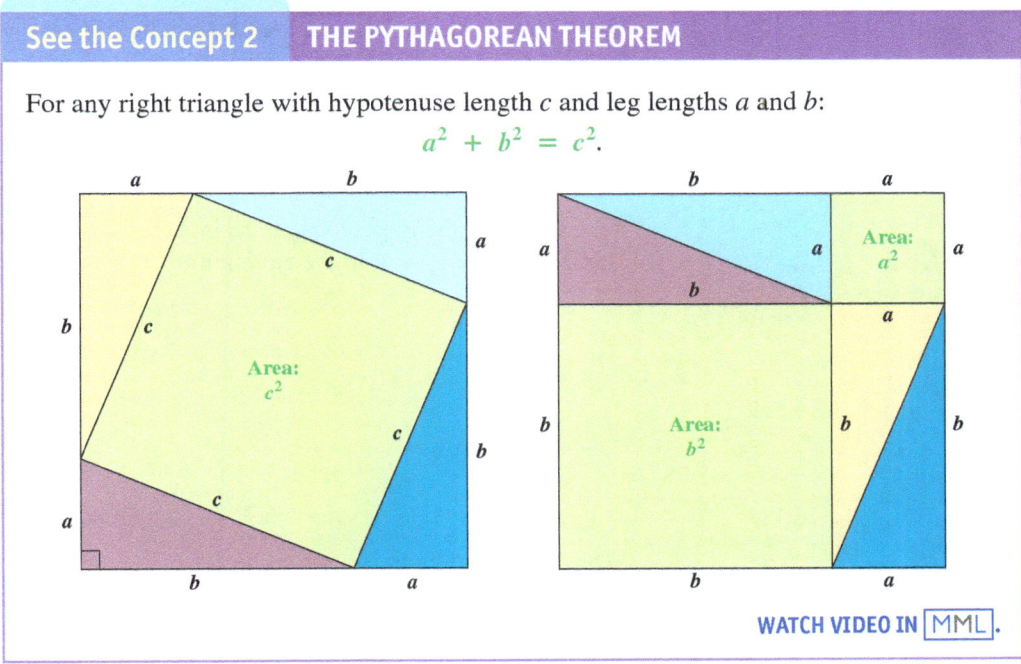

WATCH VIDEO IN MML.

READING CHECK 3

- What kind of triangle must we have in order to apply the Pythagorean theorem?

THE PYTHAGOREAN THEOREM

If a and b represent the lengths of the legs of a right triangle and c represents the length of the hypotenuse, then

$$a^2 + b^2 = c^2.$$

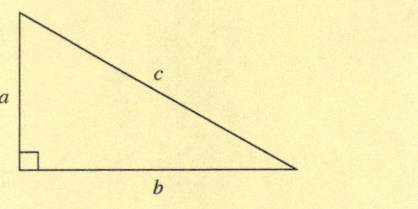

The next example shows how the Pythagorean theorem can be used to find an unknown length of one side of a right triangle when the lengths of the other two sides are known.

EXAMPLE 4 Using the Pythagorean theorem

Find the length of the unknown side of the right triangle. If the answer is not a whole number, approximate it to one decimal place.

(a)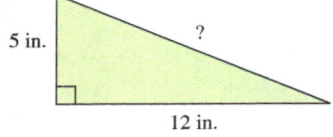

(b)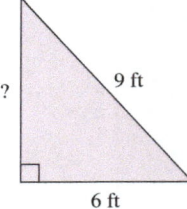

Solution
(a) Let $a = 5$ and $b = 12$ in the equation $a^2 + b^2 = c^2$ and then solve for c.

$a^2 + b^2 = c^2$ Pythagorean theorem
$5^2 + 12^2 = c^2$ Replace a with 5 and b with 12.
$25 + 144 = c^2$ Square 5 and 12.
$169 = c^2$ Add.

Even though the whole number 169 has both a positive and a negative square root, the value of c must be positive because it represents a length. For this reason, we are interested only in the positive (or principal) square root.

$c = \sqrt{169}$ Positive square root
$c = 13$ Simplify.

The length of the hypotenuse is 13 inches.

(b) Let $b = 6$ and $c = 9$ in the equation $a^2 + b^2 = c^2$ and then solve for a.

$a^2 + b^2 = c^2$ Pythagorean theorem
$a^2 + 6^2 = 9^2$ Replace b with 6 and c with 9.
$a^2 + 36 = 81$ Square 6 and 9.
$a^2 = 45$ Subtract 36 from each side.
$a = \sqrt{45}$ Positive square root
$a \approx 6.7$ Approximate.

The unknown length is approximately 6.7 feet.

Now Try Exercises 39, 43

Math in Quilting Context   Quilters often use pinwheel patterns such as the one shown in the margin to create decorative quilts. To make this pattern, right triangles must be cut from selected fabrics and stitched onto the quilt. The two legs of the right triangles used in a pinwheel pattern often have equal measure. The next example illustrates how the length of the legs can be found using the Pythagorean theorem.

EXAMPLE 5   Making a pinwheel pattern

The fabric used in a pinwheel pattern is cut into right triangles with a hypotenuse length of 4 inches. If the two legs must be equal in measure, find the length of one leg of the right triangle to the nearest tenth of an inch.

Solution
Since the legs must have the same length, we will call the unknown length x, as shown in the following triangle.

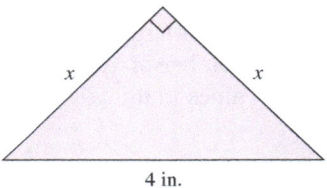

Use the Pythagorean theorem to find x.

$$x^2 + x^2 = c^2 \quad \text{Pythagorean theorem}$$
$$x^2 + x^2 = 4^2 \quad \text{Replace } c \text{ with 4.}$$
$$2x^2 = 16 \quad \text{Combine like terms; simplify.}$$
$$x^2 = 8 \quad \text{Divide each side by 2.}$$
$$x = \sqrt{8} \quad \text{Positive square root}$$
$$x \approx 2.8 \quad \text{Approximate.}$$

Each leg of the triangle is about 2.8 inches long.

Now Try Exercise 73

6 Finding the Mean, Median, and Mode

In statistics, the *mean*, *median*, and *mode* are used to measure the "middle" of a list of numbers. For this reason, they are known as **measures of central tendency**. Each of these measures can be used in a specific way to describe a data set (list of numbers).

FINDING THE MEAN   The **mean** or **arithmetic mean** of a list of numbers is commonly called the *average* of the numbers.

THE MEAN

The mean of a list of numbers is defined as follows.

$$\text{Mean} = \frac{\text{the sum of a list of values}}{\text{the number of values in the list}}$$

EXAMPLE 6 Finding the mean of a data set

TABLE 5.8 lists the hourly wages paid to a school bus driver for selected years. Find the mean hourly wage for the driver over this 6-year period.

Hourly Wages for a Bus Driver

Year	2010	2011	2012	2013	2014	2015
Wages	$13.49	$13.85	$14.39	$14.81	$15.06	$15.58

TABLE 5.8

Solution
To find the mean hourly wage, add the numbers together and then divide the result by 6 because there are **6** numbers in the list.

$$\text{Mean} = \frac{13.49 + 13.85 + 14.39 + 14.81 + 15.06 + 15.58}{6} = \frac{87.18}{6} = 14.53$$

The mean hourly wage over this period of time is $14.53.

Now Try Exercise 75

READING CHECK 4

- Why is it important to order a list of data before finding its median?

FINDING THE MEDIAN If an *ordered* list contains an odd number of values written from smallest to largest, then the **median** is the middle value in the list. If the list contains an even number of values written from smallest to largest, then the median is the mean of the two values in the middle of the list.

> **THE MEDIAN**
>
> The median of an *ordered* list of values is defined as follows.
>
> - For an odd number of values, the median is the middle value in the list.
> - For an even number of values, the median is the mean of the two middle values in the list.

EXAMPLE 7 Finding the median of a data set

Find the median of each data set.
(a) 4.7, 3.9, 2.4, 6.8, 5.2, 7.1, 4.4   (b) 57, 62, 48, 55, 71, 63, 39, 65, 70, 48

Solution
(a) Write the values in order from smallest to largest. Since the list contains an odd number of values, the median is the middle value in the *ordered* list.

$$2.4, 3.9, 4.4, \underbrace{}_{\text{Three values}} \mathbf{4.7}, \underbrace{5.2, 6.8, 7.1}_{\text{Three values}}$$

The median is **4.7**.

(b) Write the values in order from smallest to largest. The list contains an even number of values, so the median is the mean of the middle two values in the *ordered* list.

$$\underbrace{39, 48, 48, 55}_{\text{Four values}}, \mathbf{57, 62}, \underbrace{63, 65, 70, 71}_{\text{Four values}}$$

The median is the mean of **57** and **62**, or

$$\frac{57 + 62}{2} = \frac{119}{2} = 59.5.$$

Now Try Exercises 51, 53

FINDING THE MODE The **mode** in a list of numbers is the value that occurs most often. A data set may have more than one mode or no mode at all. For larger data sets, the mode is often easier to find if the numbers are first written in order from smallest to largest.

> **THE MODE**
>
> The mode of a list of values is the value that occurs most often in the list. It is possible for a data set to have more than one mode. When none of the values in a data set occurs more than once, the data set has no mode.

EXAMPLE 8 **Finding the mode of a data set**

If possible, find the mode(s) of each data set.
(a) 3.8, 2.1, 4.0, 3.8, 2.7, 2.8   (b) 7, 4, 6, 6, 3, 4, 9, 4, 5, 8, 6

Solution
(a) Written in order, the list is 2.1, 2.7, 2.8, 3.8, 3.8, 4.0. The mode is 3.8.
(b) In order, the list is 3, 4, 4, 4, 5, 6, 6, 6, 7, 8, 9. There are two modes, 4 and 6.

Now Try Exercises 59, 63

7 Finding a Weighted Mean and GPA

Math in Grading Context Suppose that a teacher wants to find the mean of the following list of test scores.

74, 74, 74, 74, 76, 76, 76, 80, 80, 85, 85, 85, 85, 88, 88, 88, 90, 90

Finding the sum of 18 scores could take a long time. However, the total can be found more quickly if we note that the score 74 is repeated 4 times, the score 76 is repeated 3 times, and so on. In this way, the mean can be found as follows.

$$\text{Mean} = \frac{4(74) + 3(76) + 2(80) + 4(85) + 3(88) + 2(90)}{4 + 3 + 2 + 4 + 3 + 2}$$

This computation is called a **weighted mean** because each score is *weighted* by the number of times that it occurs. For example, even though 74 is a lower score than 80, it has more *weight* in computing the mean because it occurs more often. This computation gives the correct mean because the numerator is the sum of the 18 test scores and the denominator is the number of test scores.

For college students, one of the most useful applications of the weighted mean is in the computation of **grade point average (GPA)**. Many colleges and universities use the following grade point values when computing GPA.

A: 4 grade points,   B: 3 grade points,   C: 2 grade points,
D: 1 grade point,   and   F: 0 grade points

Based on these grade point values, a student's GPA can be computed as follows.

358 CHAPTER 5 DECIMALS

> **COMPUTING GRADE POINT AVERAGE (GPA)**
>
> If the variables A, B, C, D, and F represent the *number of credits* earned with a grade of A, B, C, D, and F, respectively, then the grade point average (GPA) is given by
>
> $$\text{GPA} = \frac{4A + 3B + 2C + 1D + 0F}{A + B + C + D + F}.$$

READING CHECK 5

- Why is GPA an example of weighted mean?

EXAMPLE 9   **Finding a student's grade point average (GPA)**

A student's grade report is shown in **TABLE 5.9**. Find the student's GPA.

Grade Report

Course	Credits	Grade
Math	4	B
English	3	A
Astronomy	4	D
Psychology	3	A
Soccer	2	B

TABLE 5.9

Solution

The student has **6** credits with a grade of A, **6** credits with a grade of B, and **4** credits with a grade of D. There are **0** credits for both grades C and F.

$$\text{GPA} = \frac{4(6) + 3(6) + 2(0) + 1(4) + 0(0)}{6 + 6 + 0 + 4 + 0} = \frac{24 + 18 + 4}{16} = \frac{46}{16} = 2.875$$

Now Try Exercise 65

5.6   Putting It All Together

CONCEPT	COMMENTS	EXAMPLES
Circle	A *circle* is a collection of points that are all the same distance from a central point called the *center*. The distance between the center and any point on the circle is called the *radius*, and the distance across the circle on a straight line through the center is called the *diameter*.	*(circle with Radius, Diameter, and Center labeled)*
Circumference of a Circle	The distance around a circle is called its *circumference*. For a circle with radius r and diameter d, the circumference is given by $$C = \pi d \quad \text{or} \quad C = 2\pi r.$$	*(circle with radius 4 in.)* $C = 2\pi(4) = 8\pi \approx 25.12$ inches

CONCEPT	COMMENTS	EXAMPLES
Area of a Circle	For a circle with radius r, the area is given by $$A = \pi r^2.$$	$A = \pi(3^2) = 9\pi \approx 28.26$ square feet
Pythagorean Theorem	If a and b are the lengths of the legs of a right triangle and c is the length of the hypotenuse, $$a^2 + b^2 = c^2.$$	$6^2 + 8^2 = 10^2$
The Mean	The mean of a list of numbers is commonly called the average of the numbers. $$\text{Mean} = \frac{\text{the sum of a list of values}}{\text{number of values in the list}}$$	Find the arithmetic mean of the list 3, 6, 2, 7, 6. $$\text{Mean} = \frac{3 + 6 + 2 + 7 + 6}{5} = 4.8$$
The Median	The median of an *ordered* list of values is defined as follows. • For an odd number of values, the median is the middle value in the list. • For an even number of values, the median is the mean of the two middle values in the list.	The median of the list 2, 5, 6, 8, 9 is the middle value, 6. The median of the list 2, 5, 6, 8, 9, 11 is the mean of 6 and 8, $\frac{6+8}{2} = 7$.
The Mode	The mode of a list of values is the value that occurs most often in the list. It is possible for a data set to have more than one mode or no mode.	The mode of the list 3, 6, 4, 7, 3, 5 is 3 because it occurs most often.
Grade Point Average (GPA)	If the variables A, B, C, D, and F represent the *number of credits* earned with a grade of A, B, C, D, and F, respectively, then the grade point average is given by $$\text{GPA} = \frac{4A + 3B + 2C + 1D + 0F}{A + B + C + D + F}.$$	A student who earns 9 credits of As, 5 credits of Bs, and 2 credits of Cs has a GPA of $$\frac{4(9) + 3(5) + 2(2) + 1(0) + 0(0)}{9 + 5 + 2 + 0 + 0}$$ $$= \frac{55}{16} = 3.4375.$$

5.6 Exercises

CONCEPTS AND VOCABULARY

1. A(n) _____ is a collection of points that are all the same distance from a central point.

2. The distance between the center of a circle and any point on the circle is called the _____.

3. The _____ is the distance across a circle on a straight line through its center.

4. The diameter is always twice the _____.

5. The perimeter of a circle is usually called the _____ of the circle.

6. An approximation of the number _____ is 3.14.

7. A half circle is called a(n) _____.

8. The Pythagorean theorem describes the relationship among the legs and hypotenuse of a(n) _____ triangle.

9. The sides of a right triangle that form the right angle are the _____ of the triangle.

10. The _____ is the side that is opposite the right angle in a right triangle.

11. The _____ of a list of numbers is commonly called the average of the numbers.

12. The _____ of an ordered list of numbers is either the middle value in the list or the mean of the two middle values in the list.

13. The _____ of a list of numbers is the value that occurs most often in the list.

14. Grade point average is an example of a(n) _____ mean.

FINDING THE CIRCUMFERENCE OF A CIRCLE

Exercises 15–22: Find the exact circumference of the circle and then approximate the circumference using 3.14 as an approximation for π.

15.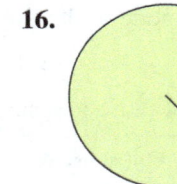

16. 7 ft

17. 25 yd

18. 19 in.

19. 0.6 in.

20. 0.2 yd

21. 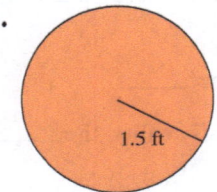 1.5 ft

22. 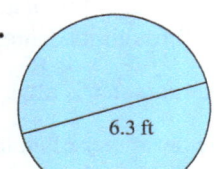 6.3 ft

FINDING THE AREA OF A CIRCLE

Exercises 23–30: Find the exact area of the given circle and then approximate the area using 3.14 as an approximation for π.

23. 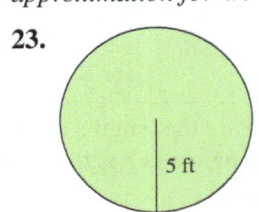 5 ft

24. 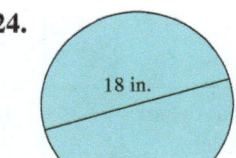 18 in.

25. 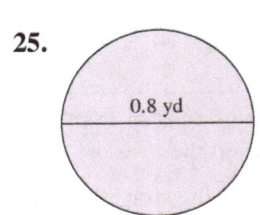 0.8 yd

26. 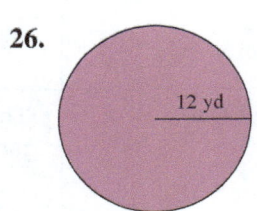 12 yd

27. 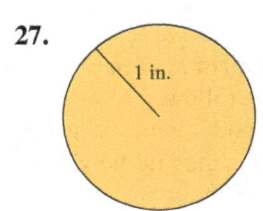 1 in.

28. 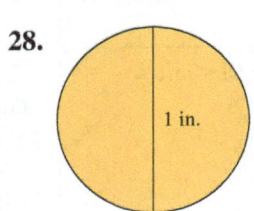 1 in.

29. 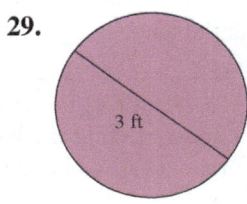 3 ft

30. 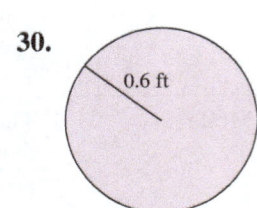 0.6 ft

FINDING THE AREA OF A COMPOSITE REGION

Exercises 31–38: Find the area of the shaded region. Use 3.14 to approximate π, if needed.

31.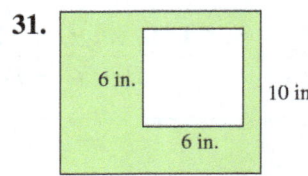

32.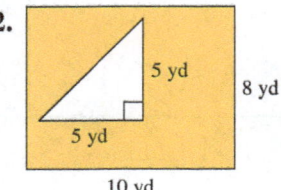

33.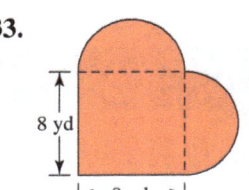

34.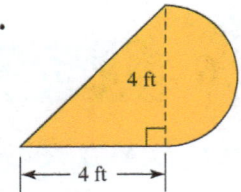

35.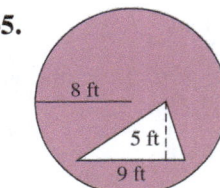

36.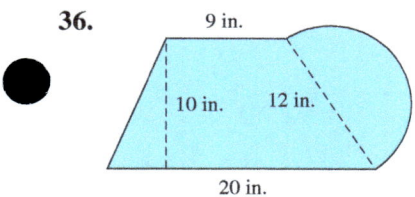

37.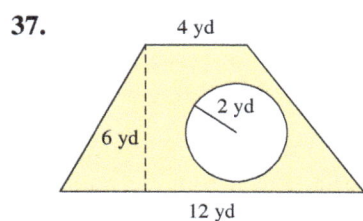

38.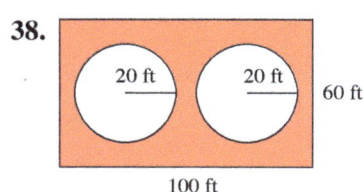

APPLYING THE PYTHAGOREAN THEOREM

Exercises 39–46: Find the length of the unknown side of the right triangle. If the answer is not a whole number, approximate it to one decimal place.

39. **40.**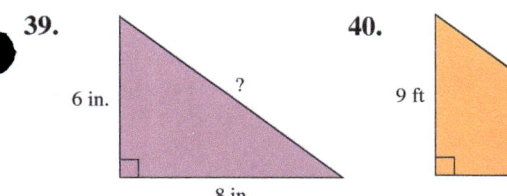

41.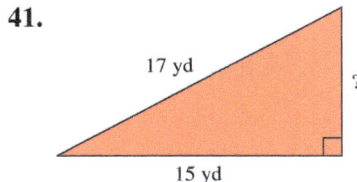

42.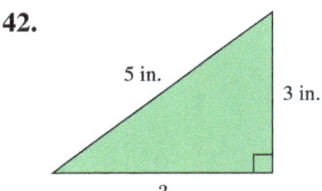

43.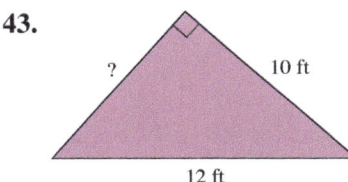

44.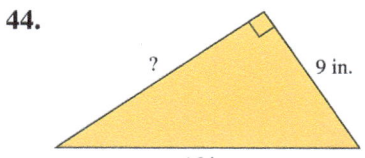

45.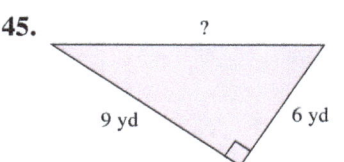

46.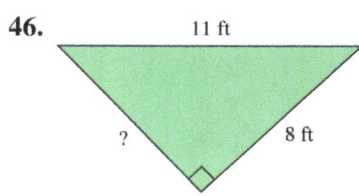

FINDING MEAN, MEDIAN, AND MODE

Exercises 47–50: Find the mean of the data presented in the given table of values.

47.
| Age | 72 | 68 | 70 | 70 | 69 |

48.
| Mileage | 19.7 | 22.6 | 20.8 | 21.5 |

49.
| Score | 27.4 | 32.2 | 30.9 | 26.3 |

50.
| Price | 0.89 | 0.97 | 0.58 | 0.85 | 0.66 |

Exercises 51–58: Find the median of the data set.

51. 2.0, 3.4, 5.4, 1.8, 3.2, 4.4, 6.1

52. 6.98, 7.13, 5.87, 6.53, 7.02, 7.23, 6.74

53. 18, 42, 31, 56, 11, 45, 17, 55, 43, 39, 22, 89

54. 100, 300, 200, 400, 500, 300

55. 25.2, 29.8, 27.3, 24.1, 28.0, 46.9

56. 9, 8, 7, 6, 5, 4, 3, 2, 1

57. 0, 1, 0, 1, 0, 1, 0, 1, 3429

58. 1, 8945, 7356, 6559, 8882

Exercises 59–64: If possible, find the mode(s) of the data set.

59. 3.2, 4.7, 3.3, 3.9, 4.4, 4.7, 3.5

60. 6.99, 5.99, 4.78, 5.99, 6.23, 5.87

61. 17, 14, 23, 11, 15, 19, 22, 28

62. 150, 170, 120, 160, 180, 140, 100, 130

63. 5, 7, 5, 4, 9, 9, 3, 1, 6, 8

64. 11.6, 12.4, 13.2, 13.2, 11.6

Exercises 65–68: Find the GPA for the grade report.

65.

Course	Credits	Grade
Nursing I	3	A
Sociology	3	A
French	3	B
Chemistry	4	A
History	3	D

66.

Course	Credits	Grade
Speech	3	B
Math	4	B
English	4	C
Psychology	3	A
PE	2	C

67.

Course	Credits	Grade
Chemistry	4	A
Calculus II	4	A
Physics	4	A
Biology	3	B

68.

Course	Credits	Grade
Psychology	3	F
English	3	C
Study Skills	3	C
Biology	3	D

APPLICATIONS INVOLVING GEOMETRY AND STATISTICS

69. Wind Turbines The blades on the E-126 wind turbine turn through a large circular region. Approximate the circumference of the circular region if one blade is 206 feet long. Use 3.14 for π. Note that blade length equals the radius. (*Source:* WindBlatt, Enercon Magazine for Wind Energy, 2007.)

70. Wind Turbines Refer to the previous exercise. When turning, the blades of a wind turbine sweep through a circular region. Using 3.14 for π, find the approximate area of the circular region for the blades of the E-126 wind turbine.

71. Running Track A running track is constructed with semicircle curves, as shown in the figure at the top of the next column. Approximate the area of the region enclosed by the track. Use 3.14 for π.

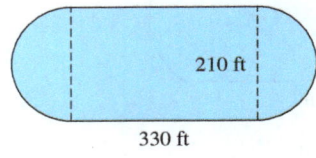

72. Running Track Using 3.14 for π, approximate the distance around the track in the previous exercise.

73. Ladder Length A ladder is leaning against a wall so that the foot of the ladder is 8 feet from the wall and the top of the ladder is 15 feet above the ground, as shown in the figure. How long is the ladder?

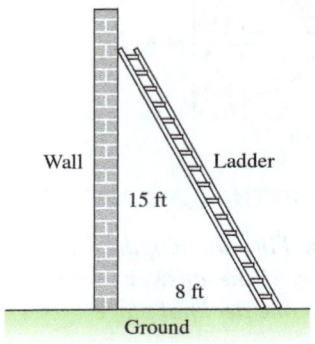

74. Laptop Screen A rectangular laptop screen measures 11 inches by 8 inches. Find the measure of the diagonal. Use 1 decimal place in your answer.

Exercises 75–80: **Television** *The table lists some of the longest-running TV series. Use the information in the table to find the following.*

TV Series	Seasons
60 Minutes	46
Walt Disney Shows	33
Ed Sullivan Show	24
Gunsmoke	20
Red Skelton Show	20
Meet the Press	18
What's My Line?	18
I've Got a Secret	17
Lassie	17
Lawrence Welk Show	17

Source: Nielsen Media Research.

75. The mean number of seasons for the 10 TV series

76. The mean number of seasons for the top 5 TV series

77. The median number of seasons for the 10 TV series

78. The median number of seasons for the top 5 TV series

79. The mode for all 10 TV series

80. The mode for the top 5 TV series

WRITING ABOUT MATHEMATICS

81. Would the mean, median, or mode work best to describe the annual income in a community? Explain your reasoning.

82. Explain why the circumference and diameter of a circle cannot *both* be integers.

83. A student uses the Pythagorean theorem to determine that the longest side of the triangle shown in the right-hand column has a length of 5 feet. Is the student correct? Explain.

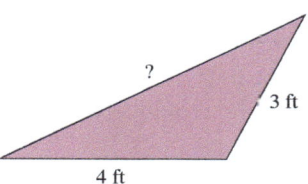

84. A student claims that the median of the following data set is 9.

$$3, 12, 9, 4, 7$$

Explain why this student is not correct.

SECTIONS 5.5 and 5.6 **Checking Basic Concepts**

1. Solve the equation symbolically.
   (a) $6.5w + 7.1 = 27.9$
   (b) $2(x - 4.8) = 26.6$
   (c) $4.2(m + 4) = 6.4m - 1.6$

2. Solve the equation $3.4x + 7.5 = 0.7$ numerically.

x	-4	-3	-2	-1	0
$3.4x + 7.5$					

3. Solve the equation $-0.75x + 1 = -2$ graphically.

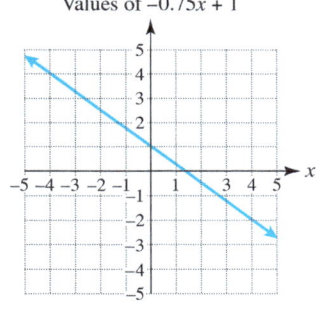

4. Approximate the circumference and the area of the circle shown. Use 3.14 for π.

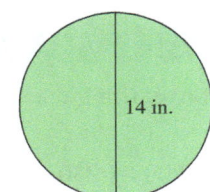

5. Find the length of the unknown side of the right triangle. If the answer is not a whole number, then approximate it to 1 decimal place.

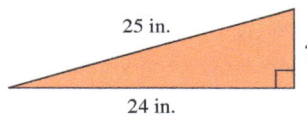

6. Find the mean, median, and mode for the data set.

$$14, 6, 3, 11, 6, 4, 10, 9$$

7. Find the GPA for the given grade report.

Course	Credits	Grade
Physics	4	C
Calculus I	5	C
Ecology	3	A
Biology	4	B

8. **Drive-in Theaters** The number N of drive-in movie theaters in the United States during year x can be estimated using the formula

$$N = -6.8x + 443,$$

where x is years after 2000. Find the year when there might be 307 drive-in movie theaters. (*Source: National Association of Theatre Owners.*)

CHAPTER 5   Summary

SECTION 5.1 ■ INTRODUCTION TO DECIMALS

Decimal Notation Numbers written in decimal notation have an integer part and a fractional part separated by a decimal point. They are called decimals.

Examples: 4.983, 0.64, −3.2

Writing Decimals in Words

1. Write the integer part in words.
2. Include the word "and" for the decimal point.
3. Write the word form of the whole number formed by the fractional part, followed by the place value of the rightmost digit.

Example: The decimal 6.29 is written as "six and twenty-nine hundredths."

Writing Decimals as Mixed Numbers or Fractions

1. Write the integer part of the decimal as the integer part of the mixed number.
2. Take the whole number formed by the fractional part of the decimal and place it in the numerator of a fraction whose denominator is a power of 10 that corresponds to the place value of the rightmost digit.
3. Simplify the fraction to lowest terms.

Examples: $0.65 = \dfrac{65}{100} = \dfrac{13}{20}$, $14.125 = 14\dfrac{125}{1000} = 14\dfrac{1}{8}$

Graphing Decimals A decimal can be graphed on a number line by using equally spaced tick marks to break the distance between values into 10 equal parts.

Example: Graph the decimal 5.28.

Comparing Decimals To compare two positive decimals with the same number of digits to the left of the decimal point, compare digits in corresponding places from left to right until unequal digits are found. The decimal with the greater of these digits is the larger number.

Examples: $652.734 < 652.743$, $0.57401 > 0.57104$

Rounding Decimals To round a decimal to the nearest whole number or to a place value to the right of the decimal point,

STEP 1: Identify the first digit to the *right* of the given place value.

STEP 2: If this digit is:
   (a) less than 5, do not change the digit in the given place value.
   (b) 5 or more, add 1 to the digit in the given place value.

STEP 3: Drop all digits to the right of the given place value.

Examples: Rounding 16.5682 to the nearest hundredth results in 16.57.
Rounding 0.7142 to the nearest tenth results in 0.7.

SECTION 5.2 ■ ADDING AND SUBTRACTING DECIMALS

Estimating Decimal Sums and Differences By rounding decimal numbers to convenient values, a sum or difference can be estimated before the actual computation is performed.

Example: The sum 31.2 + 194.6 can be estimated as 30 + 190 = 220 or as 30 + 200 = 230.

Adding and Subtracting Decimals

1. Stack the decimals vertically with the decimal points aligned. Extra zeros may be written to the right of the last digit of any decimal number so that corresponding digits align neatly.

2. Add or subtract digits in corresponding place values. Carry or borrow as needed.
3. Place a decimal point in the result so that it is aligned vertically with the decimal points in the numbers stacked above it.

Example: The sum $251.873 + 55.48$ is shown.

$$\begin{array}{r} \overset{1\ 1\ 1}{} \\ 251.873 \\ +\ 55.480 \\ \hline 307.353 \end{array}$$

SECTION 5.3 ■ MULTIPLYING AND DIVIDING DECIMALS

Estimating Decimal Products and Quotients By rounding decimal numbers to convenient values, a product or quotient can be estimated before the actual computation is performed.

Example: The quotient $72.15 \div 7.94$ can be estimated as $72 \div 8 = 9$.

Multiplying Decimals

1. Multiply the decimals as though they were whole numbers. All decimal points may be ignored during computation.
2. Place a decimal point in the product so that the number of decimal places in the product equals the sum of the number of decimal places in the factors.

Example: The product $14.2(7.9)$ is shown.

$$\begin{array}{r} 14.2 \\ \times\ 7.9 \\ \hline 12\ 78 \\ 99\ 40 \\ \hline 112.18 \end{array}$$

Multiplying Decimals by Powers of 10 Move the decimal point to the right the same number of places as the number of zeros in the power of 10. If needed, insert zeros at the end of the decimal as placeholders.

Example: $1000(2.6935) = 2693.5$

Dividing Decimals by Whole Numbers

1. Using long division, divide as though the dividend is a whole number. The decimal point in the dividend may be ignored during computation.
2. Place a decimal point in the quotient so that it is directly above the decimal point in the original dividend.
3. When needed, place extra 0s to the right of the last digit in the dividend.

Example: The quotient $259.2 \div 8$ is shown.

$$\begin{array}{r} 32.4 \\ 8\overline{)259.2} \\ -\ 24 \\ \hline 19 \\ -\ 16 \\ \hline 3\ 2 \\ -\ 3\ 2 \\ \hline 0 \end{array}$$

Dividing Decimals by Decimals

1. Make the divisor a whole number by moving its decimal point to the right.
2. Move the decimal point in the dividend to the right the same number of places.
3. Find the quotient using the procedure for dividing a decimal by a whole number.

Example: To find the quotient $0.48 \div 0.6$, move the decimal points one place to the right to result in the quotient $4.8 \div 6$.

$$\begin{array}{r} 0.8 \\ 6\overline{)4.8} \\ -\ 0 \\ \hline 4\ 8 \\ -\ 4\ 8 \\ \hline 0 \end{array}$$

Dividing Decimals by Powers of 10 Move the decimal point to the left the same number of places as the number of zeros in the power of 10. If needed, insert zeros at the beginning of the decimal as placeholders.

Example: $4257.9 \div 100 = 42.579$

Writing Fractions as Decimals

Method I: Multiply the numerator and denominator of the fraction by the same (nonzero) number to get an equivalent fraction whose denominator is a power of 10. Then write the new fraction as a decimal.

Method II: Use long division to divide the denominator into the numerator.

Examples: Method I: Write $\frac{11}{20}$ as a decimal.

$$\frac{11}{20} = \frac{11 \cdot 5}{20 \cdot 5} = \frac{55}{100} = 0.55$$

Method II: Write $\frac{1}{5}$ as a decimal.

$$\begin{array}{r} 0.2 \\ 5\overline{)1.0} \\ -1\,0 \\ \hline 0 \end{array}$$

SECTION 5.4 ■ REAL NUMBERS, SQUARE ROOTS, AND ORDER OF OPERATIONS

Irrational Numbers Irrational numbers cannot be expressed as fractions whose numerator and denominator are both integers.

Examples: $\pi \approx 3.141592654$, $\sqrt{19} \approx 4.358898944$

Real Numbers Real numbers can be written as decimals. They include all natural numbers, whole numbers, integers, rational numbers, and irrational numbers.

Examples: $-\sqrt{7}$, -5, 0, $\frac{3}{4}$, and $6.\overline{3}$

Approximating Square Roots by Guess-and-Check

1. Use a number line to make a reasonable guess for the square root.
2. Make a table of values based on your guess. Compute the square of each value in the table until the given radicand is between consecutive values in the table.
3. If the desired accuracy for the approximation has been found, stop. Otherwise, make a more accurate guess of the square root and go back to Step 2.

Example: The radicand 6 is shown between the perfect squares 4 and 9.

$$\sqrt{4} = 2 \qquad \sqrt{9} = 3$$
$$0\ 1\ 2\ 3\ 4\ 5\ 6\ 7\ 8\ 9\ 10$$

A reasonable guess for $\sqrt{6}$ is 2.5.

Guess	2.4	2.5	2.6
Guess Squared	5.76	6.25	6.76

To the nearest tenth, $\sqrt{6} \approx 2.4$.

Approximating Square Roots with the Babylonian Algorithm To approximate $\sqrt{Q}$, do the following.

1. Make a reasonable guess for the square root. Assign this value to the variable A.
2. Compute the quotient $Q \div A$.
3. Add A to the result from Step 2.
4. Divide the result of Step 3 by 2. The result is an approximation for $\sqrt{Q}$.

For a more accurate approximation, assign the result from Step 4 to the variable A and repeat the process beginning at Step 2.

Example:
1. A reasonable guess for $\sqrt{40}$ is 6.4.
2. $40 \div 6.4 = 6.25$
3. $6.25 + 6.4 = 12.65$
4. $12.65 \div 2 = 6.325$

Rounded to three decimal places, $\sqrt{40} \approx 6.325$.

SECTION 5.5 ■ SOLVING EQUATIONS INVOLVING DECIMALS

Solving Equations Symbolically By Clearing Decimals

1. Use the distributive property to clear parentheses from the equation.
2. Note the maximum number of decimal places in any number in the equation. Multiply every term by a power of 10 with that many zeros.
3. Simplify each term in the equation and combine any like terms.
4. Solve the resulting equation.

Example:
$$3.5x + 9 = 18.1$$
$$35x + 90 = 181$$
$$35x + 90 - 90 = 181 - 90$$
$$35x = 91$$
$$\frac{35x}{35} = \frac{91}{35}$$
$$x = 2.6$$

Solving Equations Numerically A table of values can be used to solve some equations involving decimals. Complete the table for various values of the variable and then select the solution from the table.

Example: The solution to $6.1x + 4 = -8.2$ is -2.

x	-3	-2	-1
$6.1x + 4$	-14.3	-8.2	-2.1

Solving Equations Graphically Use a graph of the equation to estimate a solution. Check the estimated solution in the given equation to be sure that it is correct.

Example: The solution to $0.5x - 2 = -4$ is -4.

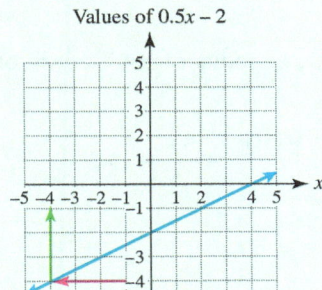

Check: $0.5(-4) - 2 = -4$ ✓

SECTION 5.6 ■ APPLICATIONS FROM GEOMETRY AND STATISTICS

Circumference of a Circle For a circle with radius r and diameter d, the circumference is given by $C = \pi d$ or $C = 2\pi r$.

Example:
$$C = \pi(9)$$
$$= 9\pi$$
$$\approx 28.26 \text{ inches}$$

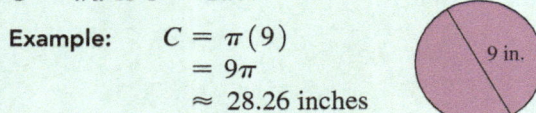

Area of a Circle For a circle with radius r, the area is given by $A = \pi r^2$.

Example:
$$A = \pi(6^2)$$
$$= 36\pi$$
$$\approx 113.04 \text{ square feet}$$

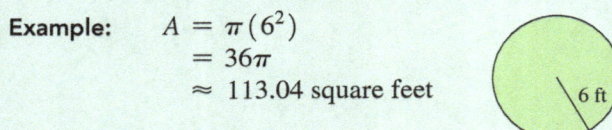

Pythagorean Theorem If a and b are the lengths of the legs of a right triangle and c is the length of the hypotenuse, then $a^2 + b^2 = c^2$.

Example:
$$3^2 + 4^2 = 5^2$$
$$9 + 16 = 25$$
$$25 = 25$$

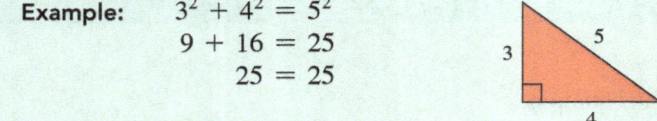

The Mean The mean of a list of numbers is commonly called the average.

$$\text{Mean} = \frac{\text{the sum of a list of values}}{\text{the number of values in the list}}$$

Example: The mean of the numbers 7, 3, 6, 2, 5, and 4 is

$$\text{Mean} = \frac{7 + 3 + 6 + 2 + 5 + 4}{6} = 4.5.$$

The Median The median of an *ordered* list of values is defined as follows.

- For an odd number of values, the median is the middle value in the list.
- For an even number of values, the median is the mean of the two middle values in the list.

Examples: For the list 2, 4, 5, 7, 9, the median is 5.

For the list 3, 5, 8, 9, the median is $\frac{5 + 8}{2} = 6.5$.

The Mode The mode of a list of values is the value that occurs most often in the list. It is possible for a data set to have more than one mode or no mode.

Example: For the list 3, 3, 5, 5, 5, 6, 6, 7, the mode is 5.

Grade Point Average (GPA) If the variables A, B, C, D, and F represent the *number of credits* earned with a grade of A, B, C, D, and F, respectively, then the grade point average is given by

$$\text{GPA} = \frac{4A + 3B + 2C + 1D + 0F}{A + B + C + D + F}.$$

Example: A student who earns 8 credits of As, 4 credits of Bs, 1 credit of Cs, and 2 credits of Ds has a GPA of

$$\frac{4(8) + 3(4) + 2(1) + 1(2) + 0(0)}{8 + 4 + 1 + 2 + 0} = \frac{48}{15} = 3.2.$$

CHAPTER 5 Review Exercises

SECTION 5.1

Exercises 1 and 2: Write the given decimal in words.

1. 0.76
2. −5.206

Exercises 3 and 4: Write the given decimal as a proper fraction or mixed number in lowest terms.

3. −0.08
4. 37.25

Exercises 5 and 6: Graph the decimal on a number line.

5. 7.8
6. −5.13

Exercises 7 and 8: Place the correct symbol, < or >, in the blank between the given decimals.

7. 41.684 ____ 41.648
8. −9.90 ____ −9.89

Exercises 9 and 10: Round to the given place value.

9. −4.008287 to the nearest ten-thousandth
10. 3591.014 to the nearest hundredth

Exercises 11 and 12: Round to the given value.

11. $12.08   dollar
12. $41.807   cent

SECTION 5.2

Exercises 13–18: Estimate the sum or difference and then add or subtract the decimals. Estimates may vary.

13. 290.115 − 39.97
14. 870.38 − 210.4
15. 14.97 + 655.009
16. 57.94 + 782.13
17. 302.41 + (−12.07)
18. −21.3 − (−199.1)

Exercises 19–24: Simplify the algebraic expression.

19. $6.9y + 3.87y$
20. $105.6x^3 + 40.1x^3$
21. $3.8q - 1.1 + 3.2q$
22. $1.3n + 4b + 6.11n$
23. $7.3x + 6.1y - 3.7x - 1.4y$
24. $8.4a + 0.9 - 6.7b - 8.1a + 3.2b$

SECTION 5.3

Exercises 25 and 26: Estimate the product or quotient by rounding the decimals. Answers may vary.

25. (4.1)(29.97)
26. 201.3 ÷ 10.03

Exercises 27–38: Find the given product or quotient.

27. (−3.9)(0.4)
28. −7.56 ÷ 4
29. 583.8 ÷ 3
30. (100)(0.163)
31. 84.072 ÷ 6.2
32. (0.92)(6.8)
33. 6.99(2.5)
34. 9245 ÷ 100
35. 8.4 ÷ 0.06
36. 10.9 ÷ 1.6
37. 1000(−0.0138)
38. (5.05)(2.08)

Exercises 39–42: Write the given fraction or mixed number as a decimal.

39. $-\dfrac{11}{25}$
40. $\dfrac{13}{30}$
41. $8\dfrac{2}{15}$
42. $64\dfrac{17}{20}$

Exercises 43–46: Simplify the algebraic expression.

43. $2.2(3.5x + 8)$
44. $-0.2(7.5y - 3)$
45. $\dfrac{7.2x}{12}$
46. $\dfrac{-48n}{0.8}$

SECTION 5.4

Exercises 47 and 48: Identify the numbers from the given list that belong to each of the following sets of numbers.
  (a) natural numbers   (b) whole numbers
  (c) integers   (d) rational numbers
  (e) irrational numbers

47. $\dfrac{5}{8}, 2, 0, \sqrt{9}, -\sqrt{7}$

48. $7, -\dfrac{2}{5}, 3.\overline{6}, -7.4, \sqrt{5}$

Exercises 49 and 50: Approximate the square root to the nearest hundredth using the guess-and-check method.

49. $\sqrt{13}$
50. $\sqrt{21}$

Exercises 51 and 52: Go through the Babylonian algorithm one time to approximate the given square root. Give two decimal places in your answer.

51. $\sqrt{7}$
52. $\sqrt{28}$

Exercises 53–58: Evaluate the numerical expression.

53. $8 - 10(0.49) \div 4$
54. $\sqrt{8.1 + 16.9} \div 5$
55. $\dfrac{4.6 + 8.6}{9.4 - 7.8}$
56. $\dfrac{12.3 - 1.8 \div 6}{0.004(1000)}$
57. $10.4 - |1.2 - 7.7| + (0.6 + 0.4)^2$
58. $(\sqrt{24.5 + 24.5} - |6.1 - 9.3|) \div 0.4$

Exercises 59 and 60: Evaluate the given expression for $x = -2.5$ *and* $y = 4.9$.

59. $9x - 2y$ **60.** $y - (2x + 0.7)$

SECTION 5.5

Exercises 61–70: Solve the equation symbolically.

61. $x + 6.9 = 23.8$ **62.** $-8.1b = -45.36$

63. $10.12 = 5.2w - 6$ **64.** $6.4w = 3 + 4w$

65. $7.6x = -75.24$ **66.** $2.05(y - 1) = 8.2$

67. $5(x + 2.7) = 37.5$ **68.** $8y - 0.1 = 0.9$

69. $2.2(p - 5) = 2.7p + 4.1$

70. $19.6p - 4 = 14.3p + 7.66$

Exercises 71 and 72: Solve the equation numerically by completing the given table of values.

71. $5.9x - 7.5 = 10.2$

x	0	1	2	3	4
$5.9x - 7.5$					

72. $4x + 8.2 = 10.6$

x	0.2	0.4	0.6	0.8	1.0
$4x + 8.2$					

Exercises 73 and 74: Solve the equation graphically.

73. $-0.8x - 1 = 3$ **74.** $0.7x + 1.4 = 3.5$

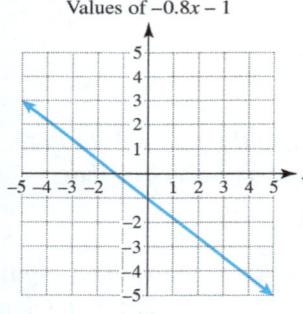
Values of $-0.8x - 1$

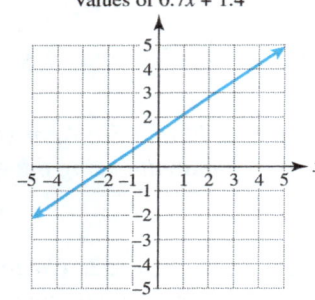
Values of $0.7x + 1.4$

SECTION 5.6

Exercises 75 and 76: Find the exact circumference of the circle and then approximate the circumference using 3.14 as an approximation for π.

75. 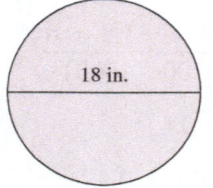 18 in.

76. 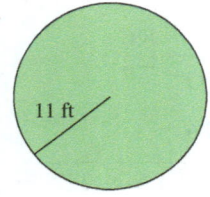 11 ft

Exercises 77 and 78: Find the exact area of the circle and then approximate the area using 3.14 as an approximation for π.

77. 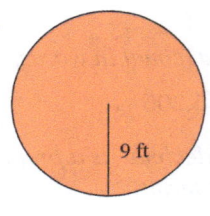 9 ft

78. 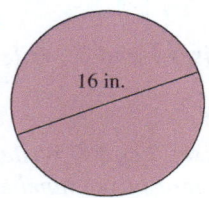 16 in.

Exercises 79 and 80: Find the area of the shaded region. Use 3.14 to approximate π, *if needed.*

79. 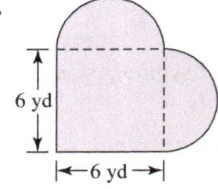 6 yd, 6 yd

80. 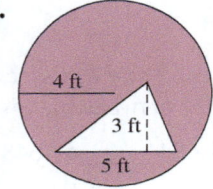 4 ft, 3 ft, 5 ft

Exercises 81 and 82: Find the length of the unknown side of the right triangle. If the answer is not a whole number, approximate it to one decimal place.

81. 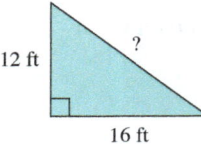 12 ft, ?, 16 ft

82. 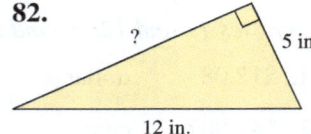 ?, 5 in., 12 in.

Exercises 83 and 84: Find the mean, median, and mode of the given data set.

83. 4.2, 4.7, 4.3, 4.7, 5.4, 5.7, 4.6

84. 9, 13, 12, 13, 15, 10, 15, 12, 10, 15

Exercises 85 and 86: Find the GPA for the grade report.

85.

Course	Credits	Grade
Literature	4	B
Math	3	B
Biology	3	B
Physics	3	C
PE	2	B

86.

Course	Credits	Grade
Chemistry	3	F
English	3	B
Study Skills	3	D
Biology	3	D

APPLICATIONS

87. Size of Bacteria A bacterium in a dish labeled A measures 0.0000635 inch across, and a bacterium in a dish labeled B measures 0.0000653 inch across. Which of these bacteria is smaller?

88. Driving Distance At the beginning of a trip, a car's odometer displayed 96,533.2, and at the end of the trip, it displayed 97,003.6. What was the total number of miles traveled on this trip?

89. Burning Calories The average 175-pound person burns 4.2 calories per minute while playing frisbee. How many calories would a person of this weight burn while playing frisbee for 45 minutes? (*Source:* The Calorie Control Council.)

90. Flight of a Projectile If a projectile is fired upward with a velocity of 72 feet per second, its height h in feet above the ground after t seconds can be approximated using the formula

$$h = -16t^2 + 72t.$$

Find the projectile's height after 1.5 seconds.

91. Converting Temperature The formula

$$C = \frac{(F - 32)}{1.8}$$

gives the relationship between C degrees Celsius and F degrees Fahrenheit. Find the Fahrenheit temperature that is equivalent to 34.5 °C.

92. Wire Length A wire is stretched from the top of a vertical pole to a point on level ground that is 7 feet from the base of the pole. If the wire is 25 feet long, how tall is the pole?

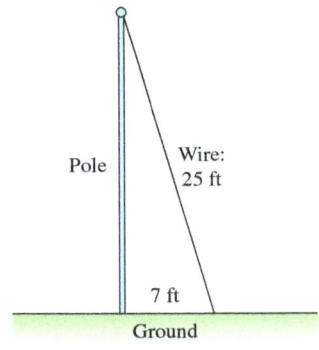

93. Dinner Cost A dinner consists of an entree for $16.59, a drink for $4.99, and dessert for $4.79. Find the total cost of the dinner.

94. Baseball Stats Baseball great Tony Oliva had a lifetime batting average of 0.304. Write this decimal as a fraction. (*Source:* Major League Baseball.)

95. Target Heart Rate For general health and weight loss, a person who is x years old should maintain a minimum target heart rate of T beats per minute during extended exercise, where

$$T = -0.5x + 120.$$

Determine the minimum target heart rate for a person who is 32 years old.

96. Monthly Payments A mail order company offers a zero-interest loan on a new computer purchase. If a buyer finances $967.20 for 4 years (48 equal payments), how much is the monthly payment?

97. Wind Turbines The blades on a wind turbine turn through a large circular region. Approximate the circumference of the circular region if one blade is 86 feet long. Use 3.14 for π. Note that blade length equals the radius.

98. Text Messaging Suppose that the cost C of sending or receiving x international text messages is given by the formula

$$C = 0.1(x - 300) + 7.5, \text{ where } x \geq 300.$$

Find the number of text messages that corresponds to total charges of $26.80 by replacing C in the formula with 26.8 and solving the resulting equation.

CHAPTER 5 Test

Exercises 1 and 2: Write the given decimal as a fraction or mixed number in lowest terms.

1. -0.85 **2.** 13.625

3. Place the correct symbol, $<$ or $>$, in the blank.

153.674 _____ 153.746

4. Round 91.58068 to the nearest thousandth.

Exercises 5 and 6: Add or subtract the decimals.

5. $194.127 - 63.78$ **6.** $156.14 + 3552.09$

Exercises 7–10: Multiply or divide the decimals.

7. $3.49(6.5)$ **8.** $(100)(4.138)$

9. $692.4 \div 10$ **10.** $2.4 \div 0.05$

Exercises 11 and 12: Simplify the expression.

11. $2.5(6x + 0.4)$ **12.** $-\dfrac{75x}{1.5}$

Exercises 13 and 14: Write the given fraction or mixed number as a decimal.

13. $17\dfrac{1}{6}$ **14.** $-\dfrac{11}{40}$

15. Identify the numbers from the given list that belong to each of the following sets of numbers.
(a) natural numbers (b) whole numbers
(c) integers (d) rational numbers
(e) irrational numbers

$$9, -\dfrac{2}{3}, \sqrt{4}, 0, \pi$$

16. Approximate $\sqrt{21}$ to two decimal places.

Exercises 17 and 18: Evaluate the expression.

17. $6.3 - 10(0.9) \div 5$ **18.** $\dfrac{9.7 - 2.4 \div 6}{0.03(100)}$

Exercises 19 and 20: Solve the equation symbolically.

19. $3.25(y - 2.1) = 6.5$ **20.** $1.1w = 6 + 0.3w$

21. Solve the equation $7x - 1.8 = -5.3$ numerically.

x	-0.8	-0.7	-0.6	-0.5	-0.4
$7x - 1.8$					

22. Solve the equation $0.9x + 1 = 2.8$ graphically.

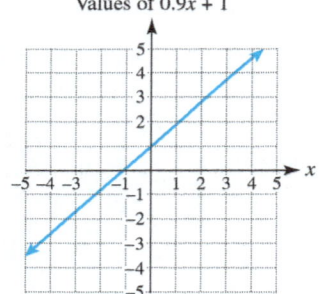

23. Using 3.14 for π, approximate the circumference and the area of the circle shown.

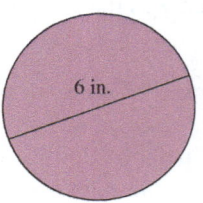

24. Approximate the length of the unknown side of the right triangle to 1 decimal place.

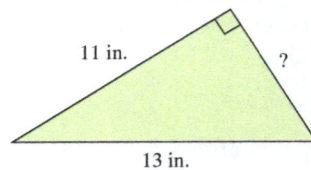

25. Find the mean, median, and mode of the data set.

$$12.6, 10.4, 13.1, 11.8, 12.6, 10.9$$

26. Find the GPA for the given grade report.

Course	Credits	Grade
Earth Science	4	C
Calculus	5	B
English	4	B
PE	3	B

27. Flight of a Baseball If a baseball is hit upward with a velocity of 66 feet per second, its height h in feet above the ground after t seconds can be approximated using the formula

$$h = -16t^2 + 66t + 2.5.$$

Find the height of the ball after 1.5 seconds.

28. Target Heart Rate For general health and weight loss, a person who is x years old should maintain a minimum target heart rate of T beats per minute during extended exercise, where

$$T = -0.5x + 120.$$

Determine the age of a person whose minimum target heart rate is $T = 89$ beats per minute.

CHAPTERS 1–5 Cumulative Review Exercises

1. Write 61,005 in expanded form.

2. Round 48,113 to the nearest thousand.

3. Approximate $\sqrt{65}$ to the nearest whole number.

4. Estimate the sum $904 + 92 + 497$ by rounding each number to the nearest hundred.

5. Evaluate $13 - (8 + 6) \div 7$.

6. Place the correct symbol, <, >, or =, in the blank between the numbers: $-|-9|$ _____ -9.

7. Simplify the expression $-(-(-7))$.

8. State the addition property illustrated by the equation $3 + 6 = 6 + 3$.

9. Simplify the expression $7x - (3x - 8)$.

10. Evaluate $\sqrt{-7^2 + 53} \div (-3 + 2)$.

11. Complete the table and solve $5x - 4 = -9$.

x	-2	-1	0	1	2
$5x - 4$					

12. Is 5 a solution to $4 + w \div 3 = 3$?

Exercises 13 and 14: Translate each sentence into an equation using x as the variable. Do not solve the equation.

13. A number decreased by 12 equals -5.

14. Double the sum of a number and 4 equals -10.

Exercises 15 and 16: Solve the equation.

15. $2 + 3x - 21 = 2x - 3 - 7x$

16. $2 + (5x - 19) = 4x - (1 + 3x)$

17. The product of -3 and the sum of a number and 4 is the same as the number decreased by 4. Find the unknown number.

18. Is the equation $-3x + 11 = 2$ linear?

19. Solve the equation $-x - 2 = -4$ graphically.

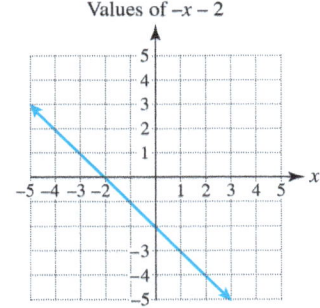

20. Write the improper fraction and mixed number represented by the shading.

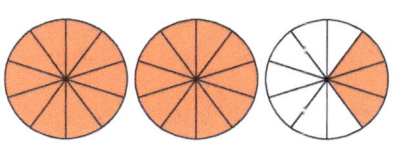

Exercises 21 and 22: Perform the indicated arithmetic.

21. $-\dfrac{4}{7} \div \dfrac{8}{21}$

22. $\dfrac{9}{10} - \dfrac{5}{8}$

23. Write $\frac{4}{15}$ as a decimal.

24. Evaluate the expression $5.3 - 100(0.8) \div 5$.

25. Approximate $\sqrt{45}$ to two decimal places.

26. Place the correct symbol, $<$ or $>$, in the blank between the numbers.

$$87.0254 \underline{\qquad} 87.0524$$

27. Multiply $2.4(6.15)$.

28. Solve the equation $5.1(y - 1) = 11.22$.

29. **Geometry** Find the perimeter of the figure.

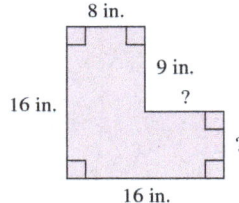

30. **Finances** A checking account starts the month with a balance of $3987. Each of the following positive entries (deposits) and negative entries (withdrawals) are made in the checking register.

$$-467, -26, 900, \text{ and } -1532$$

What is the ending balance?

6 Ratios, Proportions, and Measurement

- 6.1 Ratios and Rates
- 6.2 Proportions and Similar Figures
- 6.3 The U.S. System of Measurement
- 6.4 The Metric System of Measurement
- 6.5 U.S.–Metric Conversions; Temperature
- 6.6 Time and Speed

In the multistate lottery game, Powerball, there are 100 winning tickets (any prize) for every 3511 losing tickets. We say that the *ratio* of winning tickets to losing tickets is 100 to 3511. This ratio is commonly called the *odds* of winning a prize. For the $10,000 prize, there are 25 winning tickets for every 18,078,616 tickets that are not the $10,000 winner. This ratio is expressed as 25 to 18,078,616. In mathematics, ratios are often expressed as fractions in simplest form. In this chapter, we write ratios and *rates* as fractions and discuss their meaning in various real-world situations. *Source:* Powerball.

6.1 Ratios and Rates

Objectives

1. Writing Ratios as Fractions
2. Finding Unit Ratios
3. Writing Rates as Fractions
4. Finding Unit Rates
5. Computing Unit Pricing

NEW VOCABULARY

☐ Ratio
☐ Unit ratio
☐ Rate
☐ Unit rate
☐ Unit pricing

READING CHECK 1

- Write the ratio 6 to 7 as a fraction.

1 Writing Ratios as Fractions

A **ratio** is a comparison of two quantities. The following See the Concept outlines three common ways to write a ratio.

See the Concept 1 — WRITING RATIOS AS FRACTIONS

Comparing Apples to Bananas

A **ratio** is a comparison of two quantities.

Three Common Ways to Write a Ratio:

1. In English using the word "to"
   The ratio of apples to bananas is **3 to 6**.
2. Using a colon
   The ratio of apples to bananas is **3 : 6**.
3. As a fraction
   The ratio of apples to bananas is $\frac{3}{6}$.

NOTE: When a ratio is written as a fraction, we often express the fraction in simplest form. In this case, it is $\frac{1}{2}$. ∎

WATCH VIDEO IN MML.

WRITING A RATIO AS A FRACTION

Write the quantity found before the word "to" (or before the colon) in the numerator of the fraction, and write the quantity found after the word "to" (or after the colon) in the denominator of the fraction. Then simplify the fraction.

EXAMPLE 1 Writing ratios as fractions in simplest form

Write each ratio as a fraction in simplest form.
(a) 10 to 24   (b) $6\frac{1}{2} : 1\frac{2}{3}$ (c) 2.5 to 6.25

Solution
(a) A ratio of 10 to 24 is written as
$$\frac{10}{24} = \frac{5 \cdot 2}{12 \cdot 2} = \frac{5}{12}.$$

(b) For the ratio $6\frac{1}{2} : 1\frac{2}{3}$, first write $6\frac{1}{2}$ as $\frac{13}{2}$ and write $1\frac{2}{3}$ as $\frac{5}{3}$.
$$\frac{\frac{13}{2}}{\frac{5}{3}} = \frac{13}{2} \div \frac{5}{3} = \frac{13}{2} \cdot \frac{3}{5} = \frac{39}{10}.$$

(c) For the ratio 2.5 to 6.25, we start by clearing decimals from the fraction.

$$\frac{2.5}{6.25} = \frac{2.5 \cdot 100}{6.25 \cdot 100} = \frac{250}{625} = \frac{2 \cdot 125}{5 \cdot 125} = \frac{2}{5}$$

Now Try Exercises 9, 15, 21

> Multiplying by $\frac{100}{100}$ clears the decimals.

EXAMPLE 2 **Flying a kite**

When a person flying a kite reels in $8\frac{1}{2}$ feet of string, the kite gets $6\frac{4}{5}$ feet closer to the ground, as illustrated in the figure. Write the ratio of the amount of string reeled in to the decrease in the kite's height as a fraction in simplest form.

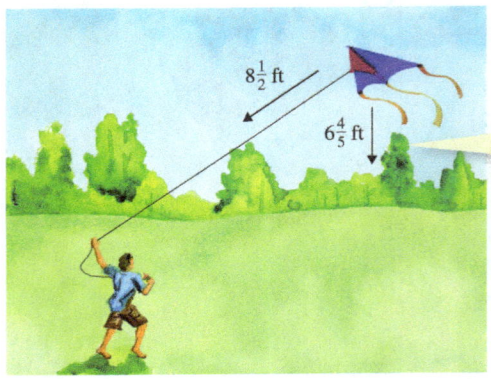

Pulling in $8\frac{1}{2}$ feet of string lowers the kite $6\frac{4}{5}$ feet.

Solution
First write $8\frac{1}{2}$ as $\frac{17}{2}$ and write $6\frac{4}{5}$ as $\frac{34}{5}$. The ratio is written as

"Invert and multiply."

$$\frac{\frac{17}{2}}{\frac{34}{5}} = \frac{17}{2} \cdot \frac{5}{34} = \frac{5}{4}.$$

This means that when 5 feet of string is reeled in, the kite gets 4 feet closer to the ground.

Now Try Exercise 61

EXAMPLE 3 **Analyzing the names of the states**

Of the 50 U.S. states, 12 have names starting with a vowel. Write the ratio of the number of state names starting with a vowel to the number of state names starting with a consonant as a fraction in simplest form.

Solution
There are 12 states having names that begin with a vowel. The rest, $50 - 12 = 38$, have names that begin with a consonant. The ratio 12 to 38 is written as

$$\frac{12}{38} = \frac{6 \cdot 2}{19 \cdot 2} = \frac{6}{19}.$$

So, there are 6 states having names starting with a vowel for every 19 states having names starting with a consonant.

Now Try Exercise 67

READING CHECK 2

- If the student-to-instructor ratio is 30 to 2, what is the unit ratio?

2 Finding Unit Ratios

 Math in Context (college) Students who would like a more personal educational experience may be interested in knowing the student-to-instructor ratio at a college. In order to make comparisons, it is often helpful to compute such ratios as *unit ratios*. A **unit ratio** is a ratio expressed as a fraction with a denominator of 1. For example, at national liberal arts colleges the student-to-instructor unit ratio is about 12 to 1, or 12. This means that there are 12 students for every instructor.

EXAMPLE 4 Using unit ratios to compare two colleges

While researching local schools, a student found that the student-to-instructor ratio at the state college is 945 to 30 and the student-to-instructor ratio at the community college is 2145 to 150. Find the unit ratio at each school and interpret the results.

Solution
One way to find a unit ratio is to write the ratio as a simplified fraction and then divide the denominator into the numerator. We do this for each college, as shown.
For the state college, the unit ratio is

$$\frac{945}{30} = \frac{63 \cdot 15}{2 \cdot 15} = \frac{63}{2} = 31.5.$$

For the community college, the unit ratio is

$$\frac{2145}{150} = \frac{143 \cdot 15}{10 \cdot 15} = \frac{143}{10} = 14.3.$$

This means that there are 31.5 students for every instructor at the state college and there are 14.3 students for every instructor at the community college.

NOTE: Although part of a whole student is not possible, the unit ratios are very helpful in comparing the two colleges. ∎

Now Try Exercise 71

FINDING A UNIT RATIO

To find a unit ratio, divide the denominator into the numerator.

---

3 Writing Rates as Fractions

Connecting Concepts with Your Life A **rate** is a ratio that is used to compare different kinds of quantities. Because the quantities in a rate have different units, the units are expressed as part of the ratio. For example, a car that travels 15 miles in 12 minutes has a rate of

$$\frac{15 \text{ miles}}{12 \text{ minutes}} = \frac{5 \text{ miles}}{4 \text{ minutes}}.$$ Simplify $\frac{15}{12}$ to $\frac{5}{4}$.

EXAMPLE 5　Writing rates as fractions in simplest form

Write each rate as a fraction in simplest form.
(a) A person eats 8 hot dogs in 60 seconds.
(b) A person earns $72 in 9 hours.

Solution
(a) A rate of **8** hot dogs in **60** seconds written as a fraction is

$$\frac{8 \text{ hot dogs}}{60 \text{ seconds}} = \frac{2 \text{ hot dogs}}{15 \text{ seconds}}.$$

Simplify $\frac{8}{60}$ to $\frac{2}{15}$.

The person eats at a rate of 2 hot dogs every 15 seconds.

(b) A rate of **72** dollars in **9** hours written as a fraction is

$$\frac{72 \text{ dollars}}{9 \text{ hours}} = \frac{8 \text{ dollars}}{1 \text{ hour}}.$$

Simplify $\frac{72}{9}$ to $\frac{8}{1}$.

The person's pay rate is $8 per hour.

Now Try Exercises 31, 39

4　Finding Unit Rates

The rate in Example 5(b) is called a *unit rate*. It is common to use the word "per" to indicate division in a unit rate.

 Connecting Concepts with Your Life　As with unit ratios, a **unit rate** is a rate expressed as a fraction with a denominator of 1. Note the use of the word "per" in the following everyday unit rates.

- The speed (rate) of a car is measured in **miles per hour.**
- Heart rate is recorded in **beats per minute.**
- Wages (pay rate) are given in **dollars per hour.**
- Mileage (gas consumption rate) is given in **miles per gallon.**

Unit rates and unit ratios are computed in the same way.

> **FINDING A UNIT RATE**
>
> To find a unit rate, divide the denominator into the numerator.

NOTE: To express a unit rate more concisely, the symbol "/" can be used in place of the word "per" and the units can be abbreviated. For example, we can write the unit rate 17 **miles per gallon** as 17 **mi/gal**. ■

EXAMPLE 6　Writing unit rates

Write each rate as a unit rate.
(a) A person earns $580 in 40 hours.
(b) A person runs a 110-meter hurdle race in 16 seconds.

Solution

(a) $\dfrac{580 \text{ dollars}}{40 \text{ hours}} = \dfrac{29 \text{ dollars}}{2 \text{ hours}} = \$14.50/\text{hr}$

The person earns $14.50 per hour.

(b) $\dfrac{110 \text{ meters}}{16 \text{ seconds}} = \dfrac{55 \text{ meters}}{8 \text{ seconds}} = 6.875 \text{ m/sec}$

The person runs at an *average* rate of 6.875 meters per second.

Now Try Exercises 43, 47

5 Computing Unit Pricing

Math in Context — Pricing One practical use of unit rates occurs in consumer mathematics. Comparison shopping is much easier when unit pricing is involved. For example, it is difficult to tell whether a 4-ounce bag of snack mix for $1.79 is a better buy than a 10.5-ounce bag of the same snack mix for $3.49. To see which is a better buy, we use the following formula for **unit pricing**.

STUDY TIP

Concepts are often easier to remember when they are associated with something that you already do. Unit rates are used in comparison shopping to find unit pricing.

READING CHECK 3

- Is a 4-ounce bag of snack mix for $2.15 a better buy than an 8-ounce bag for $4.25?

FINDING UNIT PRICING

If the price of q units of a product is p, the unit price U is given by

$$U = \dfrac{p}{q}.$$

In the next example, we compare two pricing options using unit pricing.

EXAMPLE 7 Determining the better buy

Use unit pricing to determine the better buy.

16.9 fluid ounces of Red Bull for $3.58   or   12 fluid ounces of Red Bull for $2.68

Solution
Compare the unit prices.

16.9 Ounce Can

$U = \dfrac{p}{q} = \dfrac{\$3.58}{16.9 \text{ fl oz}} \approx \$0.21/\text{fl oz}$

12 Ounce Can

$U = \dfrac{p}{q} = \dfrac{\$2.68}{12 \text{ fl oz}} = \$0.22/\text{fl oz}$

Because the unit price for 16.9 fluid ounces is lower, it is the better buy.

Now Try Exercise 75

In the next example, we analyze a situation in which unit pricing cannot be used to determine a better buy.

EXAMPLE 8 Analyzing unit pricing

For a 16-ounce jar of organic peanut butter that sells for $5.28 and a 7.5-ounce can of pinto beans that sells for $0.90, do the following.
(a) Find the unit pricing for each product.
(b) Explain why the better buy cannot be determined in this case.

Solution

(a) Peanut butter

$$U = \frac{p}{q} = \frac{\$5.28}{16 \text{ ounces}} = \$0.33/\text{oz}$$

Pinto beans

$$U = \frac{p}{q} = \frac{\$0.90}{7.5 \text{ ounces}} = \$0.12/\text{oz}$$

(b) A better buy cannot be determined because we are not comparing size options for the *same* product.

Now Try Exercise 79

CRITICAL THINKING 1

Suppose that the size of a soda can doubles and price is cut in half. What happens to the unit price?

6.1 Putting It All Together

CONCEPT	COMMENTS	EXAMPLES
Ratio	A ratio is a comparison of two quantities. When writing a ratio as a fraction, the first number in the ratio is written as the numerator, and the second number is written as the denominator.	If a bowl contains two pears and four limes, the ratio of pears to limes can be written as $$2 \text{ to } 4, \quad 2:4, \quad \text{or} \quad \frac{2}{4} = \frac{1}{2}.$$
Unit Ratio	A unit ratio is a ratio expressed as a fraction with a denominator of 1. Unit ratios are helpful in making comparisons. To find a unit ratio, divide the denominator into the numerator.	If the student-to-instructor ratio at a college is 496 to 20, the unit ratio is $$\frac{496}{20} = 24.8,$$ or 24.8 students for every instructor.
Rate	A rate is a ratio that is used to compare different kinds of quantities. The units for each quantity are expressed as part of the rate.	A person who earns $19 in 3 hours has an earning rate of $$\frac{19 \text{ dollars}}{3 \text{ hours}}.$$
Unit Rate	A unit rate is a rate expressed as a fraction with a denominator of 1. Unit rates are helpful in making comparisons. To find a unit rate, divide the denominator into the numerator.	If a person earns $67.50 in 9 hours, his hourly pay rate is $$\frac{67.50 \text{ dollars}}{9 \text{ hours}} = \$7.50/\text{hr}.$$
Unit Pricing	If the price of q units of a product is p, then the unit price U is given by $$U = \frac{p}{q}.$$	A 3-ounce bottle of perfume for $17.79 has a unit price of $$U = \frac{p}{q} = \frac{\$17.79}{3 \text{ ounces}} = \$5.93/\text{oz}.$$

6.1 Exercises

CONCEPTS AND VOCABULARY

1. A(n) _____ is a comparison of two quantities.

2. A unit ratio is a ratio expressed as a fraction with a denominator of _____.

3. To find a unit ratio, divide the _____ into the _____.

4. A(n) _____ is a ratio that is used to compare different kinds of quantities.

5. A(n) _____ rate is a rate expressed as a fraction with a denominator of 1.

6. Unit _____ is a special type of unit rate that is helpful in comparison shopping.

WRITING RATIOS AS FRACTIONS

Exercises 7–22: Write the given ratio as a fraction in simplest form.

7. 6 to 18
8. 7 to 42
9. 12 : 32
10. 25 : 60
11. $\frac{1}{2}$ to $\frac{5}{8}$
12. $\frac{4}{3}$ to $\frac{6}{7}$
13. $\frac{12}{25} : \frac{8}{15}$
14. $\frac{16}{9} : \frac{10}{27}$
15. $1\frac{2}{3}$ to $4\frac{1}{6}$
16. $3\frac{3}{4}$ to $1\frac{1}{5}$
17. $6\frac{1}{4} : 1\frac{7}{8}$
18. $2\frac{3}{4} : 1\frac{5}{6}$
19. 8.5 to 6.8
20. 3.2 to 10.4
21. 5.25 : 4.5
22. 8.4 : 2.45

FINDING UNIT RATIOS

Exercises 23–30: Find the unit ratio.

23. 21 to 7
24. 20 to 80
25. 15 : 24
26. 72 : 16
27. 165 to 30
28. 48 to 160
29. 115 : 115
30. 250 : 150

WRITING RATES AS FRACTIONS

Exercises 31–40: Write the given rate as a fraction in simplest form.

31. It rains 6 inches in 9 hours.

32. It snows 16 inches in 12 hours.

33. A person earns $39 in 6 hours.

34. A bicycle travels 18 miles in 48 minutes.

35. A theater has 60 seats in 5 rows.

36. A lab has 10 microscopes for 20 students.

37. An office has 3 copiers for 51 employees.

38. There are 25 calculators for 100 students.

39. There are 8 slices of pizza for 4 people.

40. A car travels 350 miles in 6 hours.

WRITING UNIT RATES

Exercises 41–50: Write the given rate as a unit rate.

41. A person earns $105 in 12 hours.

42. A jet travels 210 miles in 30 minutes.

43. A car goes 128 miles on 5 gallons of gas.

44. It snows 32 inches in 20 hours.

45. It rains 4 inches in 5 hours.

46. There are 12 corn plants for every 12 feet.

47. A vendor makes $40.50 selling 18 drinks.

48. There are 63 players on 7 teams.

49. A heart beats 186 times in 3 minutes.

50. A person pays $100 for 8 gallons of paint.

FINDING UNIT PRICING

Exercises 51–56: Find the unit price.

51. A 3.5-ounce bag of crackers for $1.47

52. A 6-ounce bottle of perfume for $259.50

53. A 2-pound bag of pistachios for $7.90

54. A 3-liter bottle of store-brand soda for $1.89

55. A 2000-pound load of dirt for $8.50

56. A 16-ounce jar of jam for $3.52

APPLICATIONS INVOLVING RATIOS AND RATES

57. **Online Newspapers** In 2014, online newspaper sites had 160 million unique visitors. By 2015, there were 175 million such visitors. Write the ratio of online newspaper visitors in 2014 to the number in 2015 as a fraction in simplest form. (*Source:* Newspaper Association of America.)

58. School Districts There were 155 school districts in Alabama in 2015. By comparison, Arkansas had 285 school districts that year. Write the ratio of the number of school districts in Alabama to the number of school districts in Arkansas as a fraction in simplest form. (*Source:* U.S. Census Bureau.)

59. School Children An elementary school class has 12 boys and 18 girls. Write the ratio of boys to girls as a fraction in simplest form.

60. Animal Shelters An animal shelter has 36 dogs and 16 cats. Write the ratio of dogs to cats as a fraction in simplest form.

61. Bicycle Wheels The wheels on a bicycle rotate $4\frac{1}{2}$ times with $1\frac{1}{4}$ rotations of the pedals. Write the ratio of wheel rotations to pedal rotations as a fraction in simplest form.

62. Baking A recipe calls for $3\frac{2}{3}$ cups of flour and $2\frac{3}{4}$ cups of sugar. Write the ratio of flour to sugar as a fraction in simplest form.

63. American Flag An official American flag has a length of $4\frac{3}{4}$ feet when the width is $2\frac{1}{2}$ feet. Write the ratio of the length to the width as a fraction in simplest form.

64. American Flag The blue rectangle containing the stars on an American flag is called the *union*. The union on an official American flag will have a width of $17\frac{1}{2}$ inches when the flag has a width of $32\frac{1}{2}$ inches. Write the ratio of the width of the union to the width of the flag as a fraction in simplest form.

65. Farming The following table lists the recent number of farms in thousands for selected states. Use the data in the table to write each requested ratio.

State	Minnesota	South Carolina	Texas	Wyoming
Farms	81	27	250	11

Source: U.S. Department of Agriculture.

(a) The number of farms in South Carolina to the number of farms in Minnesota

(b) The number of farms in Wyoming to the number of farms in Texas

(c) The number of farms in South Carolina to the total number of farms in these four states

66. Union Activity The following table lists the number of union-related work stoppages for selected years in companies with 1000 or more workers. Use the data in the table to write each requested ratio.

Year	2011	2012	2013	2014
Stoppages	19	19	15	10

Source: U.S. Bureau of Labor Statistics.

(a) The number of stoppages in 2012 to the number in 2011

(b) The number of stoppages in 2014 to the number in 2013

(c) The number of stoppages in 2013 to the total number of stoppages in these four years

67. Raffle Tickets A math club sold 1000 raffle tickets. If 15 of the tickets are winning tickets, write the ratio of the number of winning tickets to the number of losing tickets as a fraction in simplest form.

68. Vegetarian Meals A banquet is prepared for 500 guests. If 28 of the guests have vegetarian meals, write the ratio of vegetarian meals to non-vegetarian meals as a fraction in simplest form.

69. Equity Loans Some banks will provide an equity loan if the loan-to-value ratio is less than 0.8. Compute the loan-to-value ratio as a unit ratio for a person who owns a $250,000 home and wants an equity loan for $180,000. Does this person qualify for the loan?

70. Equity Loans Some banks will provide an equity loan if the loan-to-value ratio is less than 0.75. Compute the loan-to-value ratio as a unit ratio for a person who owns a $160,000 home and wants an equity loan for $125,000. Does this person qualify for the loan?

71. Comparing Colleges A public university has a student-to-instructor ratio of 692 to 16, while the local community college has a student-to-instructor ratio of 455 to 14. Find the unit ratio at each school. Interpret the results for the community college.

72. Comparing Schools A public high school has a student-to-teacher ratio of 396 to 15, while a private high school has a student-to-teacher ratio of 465 to 25. Find the unit ratio at each school. Interpret the results for the public high school.

73. Keyboarding A receptionist can key a 165-word email in 2.5 minutes, while an office manager can key a 254-word email in 4 minutes. Find the unit rate for each person. Who is faster?

74. **Downloading** A laptop computer can download a 1.8-MB song in 9 seconds, while a desktop computer can download a 3.4-MB song in 16 seconds. Find the unit rate for each computer. Which is faster?

75. **Shopping** Find the unit price for each size option.

    Large jar of jam: 14.5 ounces for $3.19
    Small jar of jam: 8 ounces for $1.64

    Which is the better buy?

76. **Coffee** Find the unit price for each size option.

    Large coffee drink: 26 ounces for $4.81
    Small coffee drink: 14 ounces for $2.66

    Which is the better buy?

77. **Pharmacy** Which option is the better buy?

    Generic allergy pills: 30 pills for $7.68
    Brand name allergy pills: 16 pills for $5.84

78. **Doughnuts** Which option is the better buy?

    Iced doughnuts: 36 doughnuts for $17.64
    Filled doughnuts: 60 doughnuts for $28.20

79. **Unit Pricing** A grocery coupon can be used to buy a 12.5-ounce jar of beets for $1.75 or a 7.8-ounce box of mints for $2.73.
    (a) Find the unit price for each product.
    (b) Explain why we cannot determine the better buy in this situation.

80. **Unit Pricing** A grocery coupon can be used to buy a 6.5-ounce can of peas for $0.52 or a 2.4-ounce bag of chips for $1.08.
    (a) Find the unit price for each product.
    (b) Explain why we cannot determine the better buy in this situation.

WRITING ABOUT MATHEMATICS

81. Give an example other than unit pricing where unit rates can be used for comparison purposes.

82. Give three examples of using rates in everyday life.

6.2 Proportions and Similar Figures

Objectives

1. Identifying Proportions
2. Solving a Proportion for an Unknown Value
3. Working with Similar Figures
4. Solving Applications Involving Proportions and Similar Figures

NEW VOCABULARY

☐ Proportion
☐ Similar figures

STUDY TIP

The mathematical meanings of words such as *proportional* and *similar* may be related to the meanings we know from everyday English. Whenever possible, connect the mathematical meaning to concepts that you already know.

1 Identifying Proportions

A **proportion** is a statement indicating that two ratios are equal. An equation for a proportion can be written and read as follows.

> **PROPORTION**
>
> If $\frac{a}{b}$ and $\frac{c}{d}$ are ratios that are equal in value, then the equation
>
> $$\frac{a}{b} = \frac{c}{d}$$
>
> is a proportion, where $b \neq 0$ and $d \neq 0$. This proportion can be read as
>
> "a is to b as c is to d."

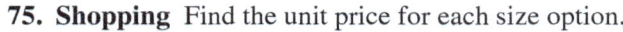

Connecting Concepts with Your Life If 15 inches of snow has the same water content as 2 inches of rain, then we can use a proportion to determine the water content x of 27 inches of the same type of snow. The ratios $\frac{15}{2}$ and $\frac{27}{x}$ must be equal and can be written as an equation in the form of the following proportion.

$$\frac{15}{2} = \frac{27}{x}$$

Solving this proportion for x gives the amount of water in 27 inches of snow.

We sometimes say that "15 is to 2 as 27 is to x" when forming a proportion such as this. Before solving this equation (see Exercise 71), we discuss proportions and cross products in the following See the Concept.

See the Concept 1 — USING CROSS PRODUCTS TO VERIFY A PROPORTION

When two ratios are set equal to each other, the equation may be true or false. If the equation is true, then it is a proportion. Since ratios are written as fractions, we can use *cross products* to determine if two ratios are equal.

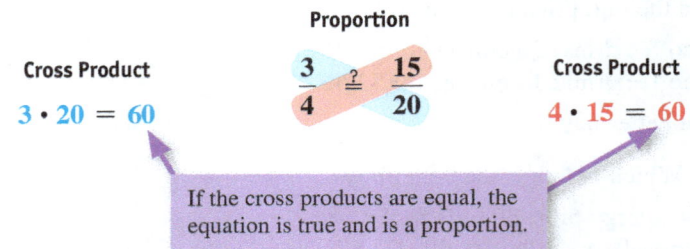

Proportion

Cross Product: $3 \cdot 20 = 60$

$$\frac{3}{4} \stackrel{?}{=} \frac{15}{20}$$

Cross Product: $4 \cdot 15 = 60$

If the cross products are equal, the equation is true and is a proportion.

WATCH VIDEO IN MML.

READING CHECK 1

- If the cross products are not equal, is the equation a proportion?

THE CROSS PRODUCT RULE

For $b \neq 0$ and $d \neq 0$, if the cross products ad and bc are equal, then

$$\frac{a}{b} = \frac{c}{d}$$

is a proportion.

EXAMPLE 1 Determining whether an equation represents a proportion

Determine if the equation is a proportion.

(a) $\dfrac{8}{18} \stackrel{?}{=} \dfrac{4}{9}$ (b) $\dfrac{6}{5.1} \stackrel{?}{=} \dfrac{4}{3.4}$ (c) $\dfrac{3\frac{1}{2}}{4\frac{3}{4}} \stackrel{?}{=} \dfrac{6\frac{2}{3}}{9\frac{1}{4}}$

Solution

(a) Start by finding the cross products.

Equals 72 — $8 \cdot 9$ $\dfrac{8}{18} \stackrel{?}{=} \dfrac{4}{9}$ $18 \cdot 4$ — Equals 72

Because $8 \cdot 9 = 72$ and $18 \cdot 4 = 72$, the cross products are equal, and the equation is a proportion.

(b) Find the cross products.

Equals 20.4 — $6 \cdot 3.4$ $\dfrac{6}{5.1} \stackrel{?}{=} \dfrac{4}{3.4}$ $5.1 \cdot 4$ — Equals 20.4

Because $6 \cdot 3.4 = 20.4$ and $5.1 \cdot 4 = 20.4$, the cross products are equal, and the equation is a proportion.

(c) Write the mixed numbers as improper fractions and then find the cross products.

$$\frac{7}{2} \cdot \frac{37}{4} \qquad \frac{7}{2} \stackrel{?}{=} \frac{20}{3} \qquad \frac{19}{4} \cdot \frac{20}{3}$$
$$\frac{19}{4} \qquad \frac{37}{4}$$

Cross products are not equal.
$$\frac{7}{2} \cdot \frac{37}{4} \neq \frac{19}{4} \cdot \frac{20}{3}$$

Because $\frac{7}{2} \cdot \frac{37}{4} = \frac{259}{8} = 32.375$ and $\frac{19}{4} \cdot \frac{20}{3} = \frac{95}{3} = 31.\overline{6}$, the cross products are *not* equal, and the equation is not a proportion.

Now Try Exercises 7, 11, 19

READING CHECK 2

- What can be used to find an unknown value in a proportion?

2 Solving a Proportion for an Unknown Value

A proportion is an equation with four values: two numerators and two denominators. When one of these values is unknown, we can find its value using cross products. In the next two examples, we solve proportions for unknown values.

EXAMPLE 2 Solving a proportion for an unknown value

Solve each proportion for the unknown value. Check your solutions.

(a) $\frac{3}{5} = \frac{x}{20}$ (b) $\frac{2.4}{3} = \frac{3.6}{w}$

Solution
(a) Use cross products to write an equation.

$$\frac{3}{5} = \frac{x}{20} \qquad \text{Given proportion}$$

$$3 \cdot 20 = 5 \cdot x \qquad \text{Cross products are equal.}$$

$$60 = 5x \qquad \text{Simplify.}$$

$$\frac{60}{5} = \frac{5x}{5} \qquad \text{Divide each side by 5.}$$

$$12 = x \qquad \text{Simplify.}$$

To check this solution, replace x with 12 in the given proportion and use cross products to verify that the resulting equation is a proportion.

Equals 60 — $3 \cdot 20 \qquad \frac{3}{5} \stackrel{?}{=} \frac{12}{20} \qquad 5 \cdot 12$ — Equals 60

Because $3 \cdot 20 = 60$ and $5 \cdot 12 = 60$, the cross products are equal, and the equation is a proportion. The solution checks.

(b) Use cross products to write an equation.

$$\frac{2.4}{3} = \frac{3.6}{w} \qquad \text{Given proportion}$$

$$2.4 \cdot w = 3 \cdot 3.6 \qquad \text{Cross products are equal.}$$

$$2.4w = 10.8 \qquad \text{Simplify.}$$

$$\frac{2.4w}{2.4} = \frac{10.8}{2.4} \qquad \text{Divide each side by 2.4.}$$

$$w = 4.5 \qquad \text{Simplify.}$$

To check this solution, replace *w* with **4.5** in the given proportion and use cross products to verify that the resulting equation is a proportion.

Equals **10.8**   2.4 · 4.5 $\frac{2.4}{3} \stackrel{?}{=} \frac{3.6}{4.5}$ 3 · 3.6   Equals **10.8**

Because 2.4 · 4.5 = 10.8 and 3 · 3.6 = 10.8, the cross products are equal, and the equation is a proportion. The solution checks.

Now Try Exercises 23, 25

EXAMPLE 3 Solving a proportion for an unknown value

Solve each proportion.

(a) $\dfrac{-2}{9} = \dfrac{m}{36}$ (b) $\dfrac{\frac{1}{2}}{n} = \dfrac{12}{\frac{3}{4}}$

Solution
(a)
$$\frac{-2}{9} = \frac{m}{36}$$ Given proportion

Set cross products equal and solve for *m*.

$-2 \cdot 36 = 9 \cdot m$ Cross products are equal.

$-72 = 9m$ Simplify.

$\dfrac{-72}{9} = \dfrac{9m}{9}$ Divide each side by 9.

$-8 = m$ Simplify.

The value of *m* is −8.

(b)
$$\frac{\frac{1}{2}}{n} = \frac{12}{\frac{3}{4}}$$ Given proportion

Set cross products equal and solve for *n*.

$\dfrac{1}{2} \cdot \dfrac{3}{4} = 12 \cdot n$ Cross products are equal.

$\dfrac{3}{8} = 12n$ Simplify.

$\dfrac{1}{12} \cdot \dfrac{3}{8} = \dfrac{1}{12} \cdot 12n$ Multiply each side by $\frac{1}{12}$.

$\dfrac{1}{32} = n$ Simplify.

The value of *n* is $\frac{1}{32}$.

Now Try Exercises 31, 39

3 Working with Similar Figures

When comparing everyday objects, the word "similar" means that the objects resemble each other but are not exactly the same. For example, we might say that two houses are similar. However, in geometry, the word *similar* means that two figures have exactly the same shape, but not necessarily the same size.

🌐 **Math in Context** *Computer Image* When you enlarge or reduce an image on a computer, you usually want to keep the new image *proportional* to the original. For example, if a 6″ by 10″ photo is reduced to a 3″ by 5″ photo, then the images in the two photos will be proportional and similar. However, if the 6″ by 10″ photo is reduced to a 6″ by 6″ photo, then the images in the two photos will neither be proportional nor similar. **FIGURE 6.1** shows two images that are proportional and two that are not.

Proportional and Similar

The dog's image is *not* distorted.

Not Proportional and Not Similar

The dog's image *is* distorted.

FIGURE 6.1

READING CHECK 3

- Suppose two rectangles are similar but not the same size. If the longest side of the larger rectangle is twice as long as the longest side of the smaller rectangle, what can be said about the shorter sides of these rectangles?

Two geometric figures are **similar figures** if the measures of the corresponding angles are equal and the measures of the corresponding sides are proportional. The two images on the left in **FIGURE 6.1** are similar, because their sides are proportional. The two images on the right in **FIGURE 6.1** are not similar, because the rightmost image is a square, while the other is not.

SIMILAR FIGURES

When two geometric figures are similar,

the measures of the corresponding sides are proportional.

The following See the Concept explains similar figures.

See the Concept 2 — SIMILAR FIGURES

When two geometric figures are **similar**, the ratios of the lengths of corresponding sides are equal and form a proportion.

Two Similar Triangles

6 in.   12 in.   4 in.   8 in.
15 in.   10 in.

Corresponding Sides are Proportional

$$\frac{\text{large triangle measure}}{\text{small triangle measure}} = \frac{6}{4} = \frac{12}{8} = \frac{15}{10} = 1.5$$

The lengths of the sides of the larger triangle are **1.5** times the lengths of the corresponding sides of the smaller triangle.

WATCH VIDEO IN MML.

In the next two examples, we use the proportional measures of similar figures to find the length of an unknown side.

EXAMPLE 4 Finding the length of an unknown side

The following triangles are similar. Find the measure of x.

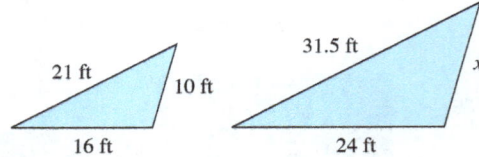

Solution

If the measures from the smaller triangle are written in the numerators of three ratios and the corresponding measures from the larger triangle are written in the corresponding denominators, the following proportions result.

Larger triangle's sides $\rightarrow \dfrac{16}{24} = \dfrac{21}{31.5} = \dfrac{10}{x}$ $\leftarrow$ Smaller triangle's sides

To solve for x, we set $\dfrac{16}{24}$ equal to $\dfrac{10}{x}$.

$\dfrac{16}{24} = \dfrac{10}{x}$ Set up a proportion.

$\dfrac{2}{3} = \dfrac{10}{x}$ Simplify: $\dfrac{16}{24} = \dfrac{2}{3}$.

$2 \cdot x = 3 \cdot 10$ Cross products are equal.

$2x = 30$ Simplify.

$\dfrac{2x}{2} = \dfrac{30}{2}$ Divide each side by 2.

$x = 15$ Simplify.

The measure of the unknown side in the larger triangle is 15 feet.

Now Try Exercise 45

NOTE: It makes no difference which *known* ratio we use when solving for an unknown length. In Example 4, the same result is found by setting up the proportion as $\dfrac{21}{31.5} = \dfrac{10}{x}$. ∎

EXAMPLE 5 Finding the length of an unknown side

The following trapezoids are similar. Find the measure of m.

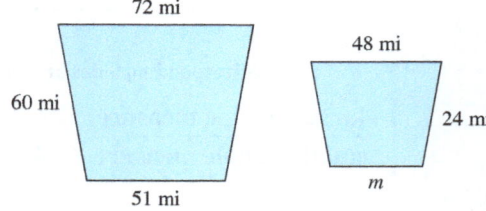

Solution

The only corresponding measures that provide a *known* ratio are the measures of the top lengths of each trapezoid. If the measures from the larger trapezoid are written in the numerators of two ratios and the corresponding measures from the smaller trapezoid are written in the corresponding denominators, then the following proportion results.

$$\frac{72}{48} = \frac{51}{m} \quad \text{Set up a proportion.}$$

$$\frac{3}{2} = \frac{51}{m} \quad \text{In simplest terms, } \tfrac{72}{48} = \tfrac{3}{2}.$$

$$3 \cdot m = 2 \cdot 51 \quad \text{Cross products are equal.}$$

$$3m = 102 \quad \text{Simplify.}$$

$$\frac{3m}{3} = \frac{102}{3} \quad \text{Divide each side by 3.}$$

$$m = 34 \quad \text{Simplify.}$$

The measure of side *m* is 34 miles.

Now Try Exercise 51

4 Solving Applications Involving Proportions and Similar Figures

Math in Mapping Context Some wilderness maps have a *scale* of $1\frac{1}{2}$ inches to 1 mile. This means that objects $1\frac{1}{2}$ inches apart on the map are 1 mile apart on the ground. A canoeist can use the map's scale as a ratio for finding actual distance by setting up a *proportion*. For example, a *portage* (a trail over which a canoe and gear are carried) may have a length of $\frac{3}{4}$ inch on the map. Because this measurement is half the first number in the map's scale, the actual distance must be half the second number in the map's scale. So, the portage is $\frac{1}{2}$ of 1 mile or $\frac{1}{2}$ mile in length. In this subsection, we discuss proportions and how they are used to solve many types of problems.

EXAMPLE 6 Reading a map

Every 2 inches on a map represents a distance of 75 miles on the ground. Find the actual distance between two towns if the corresponding distance on the map is 5 inches.

Solution

If we let *x* represent the unknown ground distance, then the given information can be written as a proportion in word form:

(Map) (Ground) (Map) (Ground)
2 inches is to **75 miles**   as   **5 inches** is to ***x* miles**.

This proportion can be written as an equation by setting the two ratios equal to each other.

$$\frac{2}{75} = \frac{5}{x} \quad \frac{\text{Inches}}{\text{Miles}} = \frac{\text{Inches}}{\text{Miles}}$$

To find the actual distance, solve this proportion for x.

$$\frac{2}{75} = \frac{5}{x} \qquad \text{Proportion to solve}$$

$$2 \cdot x = 5 \cdot 75 \qquad \text{Cross products are equal.}$$

$$2x = 375 \qquad \text{Simplify.}$$

$$\frac{2x}{2} = \frac{375}{2} \qquad \text{Divide each side by 2.}$$

$$x = 187.5 \qquad \text{Simplify.}$$

The actual distance between the two towns is 187.5 miles.

Now Try Exercise 55

EXAMPLE 7 **Providing drinking water for a recreational event**

The organizers of a disc golf tournament found that 40 participants will drink 65 bottles of water. If they are expecting 56 participants at the next tournament, how many bottles of water should be provided?

Solution
If we let x represent the unknown number of water bottles, the given information can be written as a proportion in words:

(Participants)   (Water)   (Participants)   (Water)
40 people is to **65 bottles**   as   **56 people** is to **x bottles**.

This proportion can be written as an equation by setting the two ratios equal to each other.

$$\frac{40}{65} = \frac{56}{x} \qquad \frac{\text{People}}{\text{Bottles}} = \frac{\text{People}}{\text{Bottles}}$$

To find the required number of bottles of water, solve this proportion for x.

$$\frac{40}{65} = \frac{56}{x} \qquad \text{Proportion to solve}$$

$$\frac{8}{13} = \frac{56}{x} \qquad \text{In simplest terms, } \frac{40}{65} = \frac{8}{13}.$$

$$8 \cdot x = 13 \cdot 56 \qquad \text{Cross products are equal.}$$

$$8x = 728 \qquad \text{Simplify.}$$

$$\frac{8x}{8} = \frac{728}{8} \qquad \text{Divide each side by 8.}$$

$$x = 91 \qquad \text{Simplify.}$$

The organizers should provide 91 bottles of water.

Now Try Exercise 61

EXAMPLE 8 **Finding the height of a clock tower**

Similar triangles can be used to find the height of a tall object. A clock tower has a shadow measuring 70 feet, while a nearby sign post has a shadow measuring 3.5 feet. If the sign post is 8 feet tall, how tall is the clock tower?

Solution

To see how similar triangles are used to solve this problem, consider the following picture.

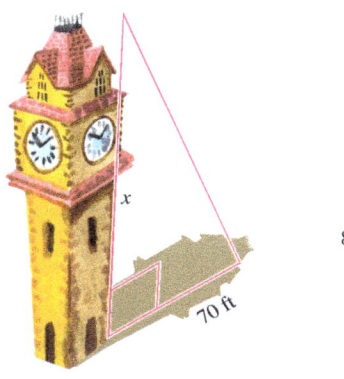

(Not to Scale)

This proportion can be written as

    (Post shadow)   (Tower shadow)   (Post height)   (Tower height)

    3.5 feet is to **70 feet** as **8 feet** is to **x feet**.

Write this proportion as an equation and solve for x.

$$\frac{\text{Post shadow}}{\text{Tower shadow}} = \frac{\text{Post height}}{\text{Tower height}} \qquad \frac{3.5}{70} = \frac{8}{x} \qquad \text{Proportion to solve}$$

$$3.5 \cdot x = 70 \cdot 8 \qquad \text{Cross products are equal.}$$

$$3.5x = 560 \qquad \text{Simplify.}$$

$$\frac{3.5x}{3.5} = \frac{560}{3.5} \qquad \text{Divide each side by 3.5.}$$

$$x = 160 \qquad \text{Simplify.}$$

The clock tower is 160 feet tall.

Now Try Exercise 75

6.2 Putting It All Together

CONCEPT	COMMENTS	EXAMPLES
Proportions	If $\frac{a}{b}$ and $\frac{c}{d}$ are ratios that are equal in value, then $$\frac{a}{b} = \frac{c}{d}$$ is a proportion, where $b \neq 0$ and $d \neq 0$. This proportion is read as "a is to b as c is to d."	The proportion $$\frac{2}{3} = \frac{30}{45}$$ is read as "2 is to 3 as 30 is to 45."
Cross Product Rule	For $b \neq 0$ and $d \neq 0$, if the cross products ad and bc are equal, then $$\frac{a}{b} = \frac{c}{d}$$ represents a proportion.	The equation $$\frac{4}{3} = \frac{20}{15}$$ represents a proportion because $4 \cdot 15 = 3 \cdot 20$.

continued on next page

CONCEPT	COMMENTS	EXAMPLES
Similar Figures	When two figures are similar, the measures of the corresponding sides are proportional.	$\frac{4}{12} = \frac{6}{18} = \frac{3}{9}$

6.2 Exercises

CONCEPTS AND VOCABULARY

1. A(n) _____ is a statement indicating that two ratios are equal.

2. The proportion "a is to b as c is to d" is written as the equation _____ involving two ratios.

3. In the proportion $\frac{a}{b} = \frac{c}{d}$, the products ad and bc are called the _____.

4. The equation $\frac{a}{b} = \frac{c}{d}$ represents a proportion if the cross products are _____.

5. Geometric figures are _____ if they have exactly the same shape but not necessarily the same size.

6. When two geometric figures are similar, the measures of the corresponding sides are _____.

IDENTIFYING PROPORTIONS

Exercises 7–22: Determine if the given equation is a proportion.

7. $\frac{5}{11} \stackrel{?}{=} \frac{15}{33}$

8. $\frac{28}{9} \stackrel{?}{=} \frac{6}{2}$

9. $\frac{18}{10} \stackrel{?}{=} \frac{24}{15}$

10. $\frac{9}{21} \stackrel{?}{=} \frac{6}{14}$

11. $\frac{5.2}{2} \stackrel{?}{=} \frac{15.6}{6}$

12. $\frac{6}{2.4} \stackrel{?}{=} \frac{5}{1.8}$

13. $\frac{7.5}{2.5} \stackrel{?}{=} \frac{12.9}{4.3}$

14. $\frac{1}{6.4} \stackrel{?}{=} \frac{11}{68.2}$

15. $\frac{\frac{1}{3}}{15} \stackrel{?}{=} \frac{\frac{2}{5}}{21}$

16. $\frac{10}{\frac{5}{6}} \stackrel{?}{=} \frac{42}{\frac{7}{2}}$

17. $\frac{\frac{5}{9}}{\frac{3}{5}} \stackrel{?}{=} \frac{\frac{5}{6}}{\frac{9}{10}}$

18. $\frac{\frac{1}{2}}{\frac{3}{4}} \stackrel{?}{=} \frac{\frac{4}{5}}{\frac{5}{6}}$

19. $\frac{4\frac{1}{2}}{2\frac{1}{4}} \stackrel{?}{=} \frac{3\frac{5}{6}}{1\frac{7}{10}}$

20. $\frac{4\frac{2}{3}}{1\frac{1}{6}} \stackrel{?}{=} \frac{5\frac{1}{5}}{1\frac{3}{10}}$

21. $\frac{2\frac{2}{3}}{4\frac{1}{5}} \stackrel{?}{=} \frac{3\frac{1}{3}}{5\frac{2}{5}}$

22. $\frac{3\frac{5}{6}}{1\frac{11}{12}} \stackrel{?}{=} \frac{6\frac{1}{5}}{3\frac{1}{10}}$

SOLVING PROPORTIONS

Exercises 23–28: Solve the proportion for the unknown value. Check your solution.

23. $\frac{2}{7} = \frac{10}{x}$

24. $\frac{y}{6} = \frac{7}{21}$

25. $\frac{4}{2.8} = \frac{9}{m}$

26. $\frac{4.5}{w} = \frac{15}{2.5}$

27. $\frac{\frac{1}{2}}{2} = \frac{k}{\frac{4}{3}}$

28. $\frac{6}{\frac{2}{3}} = \frac{9}{g}$

Exercises 29–44: Solve the proportion.

29. $\frac{9}{2} = \frac{4.5}{m}$

30. $\frac{3.5}{g} = \frac{14}{8}$

31. $\frac{10}{y} = \frac{-5}{6}$

32. $\frac{-16}{8} = \frac{a}{-4}$

33. $\frac{-12}{y} = \frac{-5}{8}$

34. $\frac{-15}{8} = \frac{d}{4}$

35. $\frac{5}{-2.5} = \frac{n}{-1.8}$

36. $\frac{4.8}{b} = \frac{-1.6}{3}$

37. $\frac{-3.6}{-2.4} = \frac{x}{4.6}$

38. $\frac{3.5}{k} = \frac{2.1}{-1.2}$

39. $\frac{x}{\frac{4}{3}} = \frac{\frac{9}{2}}{9}$

40. $\frac{\frac{1}{10}}{-\frac{2}{5}} = \frac{y}{-4}$

41. $\dfrac{\frac{5}{9}}{-\frac{5}{6}} = \dfrac{n}{\frac{9}{10}}$ 42. $\dfrac{\frac{1}{6}}{d} = \dfrac{\frac{4}{5}}{-\frac{6}{7}}$

43. $\dfrac{1\frac{1}{5}}{-6\frac{2}{3}} = \dfrac{a}{4\frac{1}{6}}$ 44. $\dfrac{3\frac{4}{5}}{1\frac{3}{16}} = \dfrac{\frac{4}{5}}{w}$

WORKING WITH SIMILAR FIGURES

Exercises 45–52: Find the measure of x for the similar figures shown.

45.
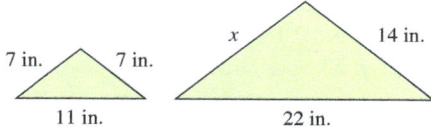

46.
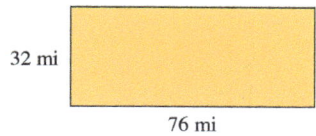

47.
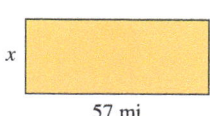

48.
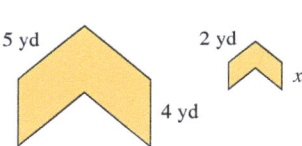

49.
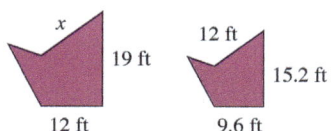

50.
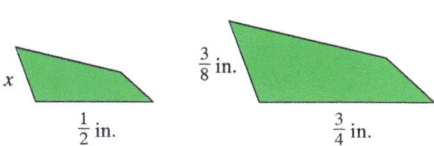

51.
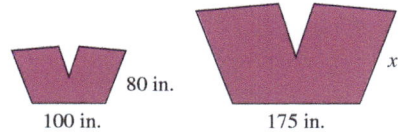

52.
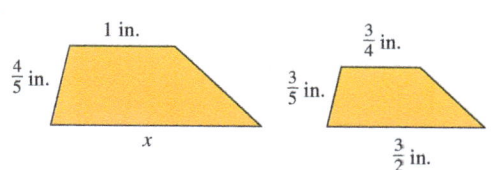

APPLICATIONS INVOLVING PROPORTIONS AND SIMILAR FIGURES

53. **Paid Vacation** An employee earns 6 days of paid vacation after working for 21 weeks. How many days of paid vacation does this employee earn after working for 49 weeks?

54. **Reading Rate** If a student reads 81 pages of a book in 90 minutes, how long will it take this student to read 117 pages?

55. **Wilderness Maps** Every $1\frac{1}{2}$ inches on a wilderness map represents 1 mile. If the actual distance between two lakes is 8 miles, how far apart are the lakes on the map?

56. **Construction Plans** Every $\frac{1}{4}$ inch on a floor plan represents 1 foot. If a wall on the floor plan measures $3\frac{1}{2}$ inches, how long is the actual wall?

57. **Grocery Shopping** If 4 limes sell for $0.76, how much will 6 limes cost?

58. **Wardrobe** A person has 8 shirts for every 5 pairs of pants. If this person has a total of 40 shirts in the closet, how many pairs of pants are there?

59. **Going Green** A survey determined that 80 of 480 students walk to school each day. If this ratio holds in a math class with 24 students, how many of the math students walk to school?

60. **Voting** A survey determined that 6 out of 10 voters intend to vote "no" on a controversial issue. If there are 2400 voters in the district, how many would we expect to vote "no"?

61. **Feeding Teenagers** If 30 teenagers can eat 72 slices of pizza, how many slices are needed to feed 120 teenagers?

62. **Mixing Fuel** The mixing ratio of gas to oil for a lawn mower is 50 to 1. How many gallons of gas should be mixed with 0.05 gallon of oil?

63. **Baking** A recipe that serves 6 people requires $1\frac{1}{2}$ cups of sugar. Using this recipe, how many cups of sugar are needed to serve 10 people?

64. **Basketball** A basketball player makes 16 out of 20 free throws. How many free throws is this player expected to make in 65 attempts?

65. **iPod Capacity** A 32-gigabyte iPod Touch can hold about 7000 songs. How many songs can be stored on an 16-gigabyte iPod Touch?

66. **Model Car** A model car is scaled so that 2 inches on the model are equal to $3\frac{3}{4}$ feet on the actual car. How tall is the model if the actual car is 5 feet tall?

67. **Living in Poverty** Recently, about 21 of every 200 Delaware citizens were living in poverty. How many people would you expect to find living in poverty in a Delaware community with a population of 100,000? (*Source:* U.S. Census Bureau.)

68. **Prison Time** About 9 out of every 200 U.S. citizens who have not been to prison before age 20 will end up serving time during their lifetime. A community has 1000 residents who are 20 years old and have never been to prison. How many of these residents are expected to serve time during their lifetime? (*Source:* Bureau of Justice Statistics.)

69. **Business Majors** About 8 of every 50 college freshmen chose to major in business. How many business majors would you expect to find in a group of 1500 college freshmen? (*Source:* Higher Education Research Institute.)

70. **Education** About 6 out of every 20 people in the U.S. who are over the age of 25 have completed four or more years of college. If a community has 1860 residents over the age of 25 with four or more years of college, how many residents over the age of 25 live there? (*Source:* U.S. Census Bureau.)

71. **Water Content** If 15 inches of snow has the same water content as 2 inches of rain, then find the water content in 27 inches of the same type of snow.

72. **Worldwide Internet Use** In 2005, approximately 1 in 7 people used the Internet, whereas in 2015 this ratio increased to about 3 in 7 people. For every 1000 people, about how many people used the Internet in 2005 and in 2015.

73. **Olympic Pool** A swimming pool that is 16 meters long and 8 meters wide is geometrically similar to an Olympic swimming pool. If an Olympic pool is 50 meters long, how wide is it?

74. **Board Games** The travel version of a popular board game is 8 inches long and 6 inches wide. If the larger version of the game is geometrically similar and has a length of 28 inches, what is its width?

75. **Tree Height** A tree casts a 65-foot shadow, while a nearby tree that is 11 feet tall casts a 5-foot shadow. How tall is the larger tree?

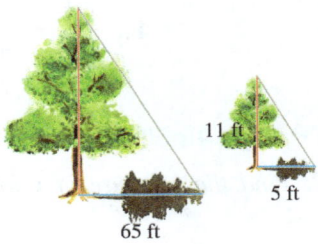

76. **Antenna Height** Refer to the previous exercise. If a 6-foot person casts a 10-foot shadow, how tall is an antenna that casts a 75-foot shadow?

77. **Digital Images** The focal length of a camera is the distance from the lens to the image sensor. A camera with a focal length of 2 inches is used to take a picture of a large letter F that is 24 inches tall. If the letter F is 80 inches from the lens of the camera, as shown in the figure below, how tall is the F on the image sensor?

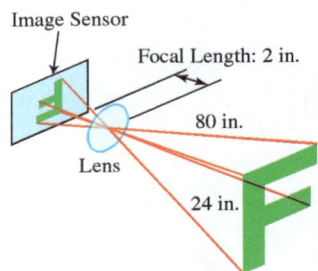

78. **Photography** Refer to the previous exercise. A camera with a 3-inch focal length is used to take a picture of a child standing 16 feet in front of the camera. If the child's image is 0.75 inch tall on the image sensor, how tall is the child?

WRITING ABOUT MATHEMATICS

79. In order to determine how many pages he can read in 1 hour, a student who reads 50 pages in 75 minutes sets up the following proportion.

$$\frac{50}{75} = \frac{x}{1}$$

Is this proportion correct? Explain.

80. Are all squares geometrically similar? Explain.

SECTIONS 6.1 and 6.2 — Checking Basic Concepts

1. Write each ratio as a fraction in simplest form.
   (a) $16:36$
   (b) $\frac{8}{9} : \frac{3}{18}$

2. Write each rate as a fraction in simplest form.
   (a) It rains 6 inches in 15 hours.
   (b) A car travels 582 miles in 8 hours.

3. Write each rate as a unit rate.
   (a) A person earns $148 in 16 hours.
   (b) It snows 30 inches in 20 hours.

4. Compute each unit price.
   (a) A 2-pound bag of chips for $5.90
   (b) A 16-ounce jar of pickles for $2.88

5. Determine if each ratio is a proportion.
   (a) $\frac{4}{26} \stackrel{?}{=} \frac{6}{39}$
   (b) $\frac{5}{2.5} \stackrel{?}{=} \frac{3}{1.6}$

6. Solve each proportion.
   (a) $\frac{18}{y} = \frac{-6}{10}$
   (b) $\frac{\frac{3}{10}}{-\frac{6}{5}} = \frac{m}{-4}$

7. The following figures are similar. Find x.

   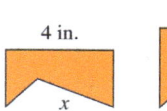 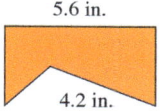

8. **Shopping** Find the unit price for each option.

   Large drink: 28 ounces for $1.96
   Small drink: 12 ounces for $1.14

   Which is the better buy?

9. **Baking** A recipe that serves 8 people requires $1\frac{1}{3}$ cups of milk. How many cups of milk would be needed to serve 10 people?

10. **Tree Height** A tree casts a 35-foot shadow, while a nearby lamp post that is 24 feet tall casts a 10-foot shadow. How tall is the tree?

6.3 The U.S. System of Measurement

Objectives
1. Converting U.S. Units of Length
2. Converting U.S. Units of Area
3. Converting U.S. Units of Capacity and Volume
4. Converting U.S. Units of Weight

NEW VOCABULARY
☐ Inch, foot, yard, mile
☐ Unit fraction
☐ Capacity
☐ Volume
☐ Ounce, cup, pint, quart, gallon
☐ Ounce, pound, ton

1 Converting U.S. Units of Length

The **inch**, **foot**, **yard**, and **mile** are the four most commonly used units of *length* in the U.S. system of measurement. Inches (in.), feet (ft), yards (yd), and miles (mi) are related as follows.

U.S. UNITS OF LENGTH

$$1 \text{ ft} = 12 \text{ in.}$$
$$1 \text{ yd} = 36 \text{ in.} \quad \text{and} \quad 1 \text{ yd} = 3 \text{ ft}$$
$$1 \text{ mi} = 1760 \text{ yd} \quad \text{and} \quad 1 \text{ mi} = 5280 \text{ ft}$$

To convert from one unit of length to another, we use a method called *unit analysis*. This method is based on the fact that multiplying a measurement by 1 does not change the value of the measurement. Since 1 can be written in many forms, it is important to write it in the form of a *unit fraction*. A **unit fraction** is a fraction that is equivalent to 1. Examples include

$$\frac{12 \text{ in.}}{1 \text{ ft}}, \quad \frac{1 \text{ yd}}{36 \text{ in.}}, \quad \frac{3 \text{ ft}}{1 \text{ yd}}, \quad \frac{5280 \text{ ft}}{1 \text{ mi}}, \quad \text{and} \quad \frac{1 \text{ mi}}{1760 \text{ yd}}.$$

STUDY TIP

Unit analysis is used throughout the remainder of this chapter and can be valuable in other classes, especially the sciences. Take some extra time to fully learn this technique. Get extra help and practice if needed.

READING CHECK 1

- Which of the following is a unit fraction?

  $\frac{1 \text{ mi}}{5280 \text{ ft}}$, $\frac{36 \text{ in}}{1 \text{ ft}}$, $\frac{1 \text{ yd}}{5 \text{ mi}}$, and $\frac{1 \text{ yd}}{1760 \text{ mi}}$

For example, because 12 inches equals 1 foot, the fraction $\frac{12 \text{ in.}}{1 \text{ ft}}$ is called a *unit fraction*. If we want to convert 5 feet to inches, we can do so by multiplying 5 feet times this unit fraction.

$$5 \text{ ft} = \frac{5 \text{ ft}}{1} \cdot \frac{12 \text{ in.}}{1 \text{ ft}} = 60 \text{ in.}$$

Notice that the units of feet may be "canceled," much like we have done with numbers. The unit analysis process is especially valuable as the units become more complicated.

> **CONVERTING FROM ONE UNIT OF MEASURE TO ANOTHER**
>
> To convert from one unit of measure to another, do the following.
>
> **STEP 1:** Write the given measure (including its units) over 1.
>
> **STEP 2:** Multiply by a unit fraction whose denominator contains the given (unwanted) units and whose numerator contains the new (desired) units.
>
> It may be necessary to multiply by more than one unit fraction to get the desired units in the final result.

EXAMPLE 1 Converting from one unit of length to another

Convert as indicated.
(a) 78 inches to feet   (b) $\frac{3}{4}$ mile to yards   (c) 6.5 yards to inches

Solution
(a) We would like to determine how many feet there are in 78 inches. To make this conversion, write 78 inches over 1 and then multiply by a unit fraction with inches in the denominator and feet in the numerator. In this way, the inch units will "cancel" and we will be left with feet in the numerator.

$$78 \text{ in.} = \frac{78 \text{ in.}}{1} \cdot \frac{1 \text{ ft}}{12 \text{ in.}} = \frac{78}{12} \text{ ft} = 6\frac{1}{2} \text{ ft, or } 6.5 \text{ ft}$$

(b) We are given miles and want to convert to yards. To do this, write $\frac{3}{4}$ mile over 1 and then multiply by a unit fraction with miles in the denominator and yards in the numerator.

$$\frac{3}{4} \text{ mi} = \frac{\frac{3}{4} \text{ mi}}{1} \cdot \frac{1760 \text{ yd}}{1 \text{ mi}} = \frac{3}{4} \cdot \frac{1760}{1} \text{ yd} = \frac{5280}{4} \text{ yd} = 1320 \text{ yd}$$

Simplify.

(c) The given units are yards, and we want to convert to inches.

$$6.5 \text{ yd} = \frac{6.5 \text{ yd}}{1} \cdot \frac{36 \text{ in.}}{1 \text{ yd}} = 6.5 \cdot 36 \text{ in.} = 234 \text{ in.}$$

Now Try Exercises 11, 15, 19

If you do not know the direct relationship between two units of measure, then it may be necessary to use more than one unit fraction during a conversion. The next example illustrates this process.

EXAMPLE 2 **Converting by using more than one unit fraction**

Convert 1.4 miles to inches by using more than one unit fraction.

Solution

Since most people do not know how many inches are in a mile, we will first convert miles to feet and then convert feet to inches. To convert from miles to feet, we use a unit fraction with feet in the numerator and miles in the denominator. Then, to convert from feet to inches, we use a unit fraction with inches in the numerator and feet in the denominator.

$$1.4 \text{ mi} = \frac{1.4 \text{ mi}}{1} \cdot \frac{5280 \text{ ft}}{1 \text{ mi}} \cdot \frac{12 \text{ in.}}{1 \text{ ft}} = \frac{1.4 \cdot 5280 \cdot 12}{1} \text{ in.} = 88{,}704 \text{ in.}$$

Convert miles to feet.   Convert feet to inches.

Now Try Exercise 21

2 Converting U.S. Units of Area

While there are 3 feet in 1 yard, this does *not* mean that there are 3 square feet in 1 square yard. **FIGURE 6.2** can be used to visualize the correct relationship between square feet and square yards.

Square Feet in a Square Yard

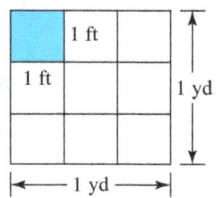

There are $3 \cdot 3 = 9$ square feet in 1 square yard.

FIGURE 6.2

When converting from one square unit to another, we can use the relationships for length that we already know. For example, we can write square yards as either

$$\text{yd}^2 \quad \text{or} \quad \text{yd} \cdot \text{yd}.$$

READING CHECK 2

- When converting from one square unit of area to another, how many times is a unit fraction of length used?

To see symbolically that there are 9 square feet in 1 square yard, we twice use a unit fraction that involves feet and yards.

$$1 \text{ yd}^2 = \frac{1 \text{ yd} \cdot \text{yd}}{1} \cdot \frac{3 \text{ ft}}{1 \text{ yd}} \cdot \frac{3 \text{ ft}}{1 \text{ yd}} = 3 \text{ ft} \cdot 3 \text{ ft} = 9 \text{ ft}^2$$

Unit fraction used twice

EXAMPLE 3 **Converting from one unit of area to another**

Convert as indicated.
(a) 6 square feet to square inches   (b) 108 square feet to square yards

Solution
(a) Use a unit fraction relating feet and inches, twice.

$$6 \text{ ft}^2 = \frac{6 \text{ ft} \cdot \text{ft}}{1} \cdot \frac{12 \text{ in.}}{1 \text{ ft}} \cdot \frac{12 \text{ in.}}{1 \text{ ft}} = 6 \cdot 12 \text{ in.} \cdot 12 \text{ in.} = 864 \text{ in}^2$$

(b) Use a unit fraction relating feet and yards, twice.

$$108 \text{ ft}^2 = \frac{108 \text{ ft} \cdot \text{ft}}{1} \cdot \frac{1 \text{ yd}}{3 \text{ ft}} \cdot \frac{1 \text{ yd}}{3 \text{ ft}} = \frac{108}{3 \cdot 3} \text{ yd} \cdot \text{yd} = 12 \text{ yd}^2$$

Now Try Exercises 25, 29

READING CHECK 3

- How many square inches are there in 5 square feet?

> **MAKING CONNECTIONS 1**
>
> **Converting Units of Area Directly**
>
> We can also derive the appropriate unit fraction for converting units of area directly. For example, a unit fraction for converting square feet to square inches is found by squaring the unit fraction used to convert feet to inches.
>
> $$\left(\frac{12 \text{ in.}}{1 \text{ ft}}\right)^2 = \frac{12 \text{ in.}}{1 \text{ ft}} \cdot \frac{12 \text{ in.}}{1 \text{ ft}} = \frac{12 \text{ in.} \cdot 12 \text{ in.}}{1 \text{ ft} \cdot 1 \text{ ft}} = \frac{144 \text{ in}^2}{1 \text{ ft}^2}$$
>
> So, to convert square feet to square inches directly, use the unit fraction
>
> $$\frac{144 \text{ in}^2}{1 \text{ ft}^2}$$
>
> because there are $12 \cdot 12 = 144$ square inches in 1 square foot.

 **Math in Context** *Home Design* While it is common to measure the floor of a room in square feet, some brands of carpet are sold by the square yard. To find the cost of carpet for a room, we can convert square feet to square yards, as demonstrated in the next example.

EXAMPLE 4 **Computing the cost of carpet for a room**

For a rectangular room that measures 12 feet by 16.5 feet, do the following.
(a) Find the area of the room in square feet.
(b) Convert your answer to square yards.
(c) Determine the cost of carpet for the room if the carpet sells for $38 per square yard.

Solution
(a) The area of the room is

$$A = lw = 16.5 \text{ ft} \cdot 12 \text{ ft} = 198 \text{ ft}^2.$$

(b) Use a unit fraction relating feet and yards, twice.

$$198 \text{ ft}^2 = \frac{198 \cancel{\text{ft}} \cdot \cancel{\text{ft}}}{1} \cdot \frac{1 \text{ yd}}{3 \cancel{\text{ft}}} \cdot \frac{1 \text{ yd}}{3 \cancel{\text{ft}}} = \frac{198}{3 \cdot 3} \text{ yd} \cdot \text{yd} = 22 \text{ yd}^2$$

(c) The total cost of the carpet is

$$\text{Cost} = (\text{yardage})(\text{price}) = \frac{22 \cancel{\text{yd}^2}}{1} \cdot \frac{\$38}{1 \cancel{\text{yd}^2}} = 22 \cdot \$38 = \$836.$$

Now Try Exercise 71

3 Converting U.S. Units of Capacity and Volume

Even though the words *capacity* and *volume* are often used interchangeably, they do not have exactly the same meaning. **Capacity** is a measure of the amount of substance that a container can hold, while **volume** is a measure of the actual amount of the substance present. For example, if a 16-gallon gas tank is half full, the capacity of the tank is 16 gallons, and the volume of gas in the tank is 8 gallons.

In the U.S. system of measurement, the units used to measure *both* capacity and volume of liquids are (fluid) **ounce**, **cup**, **pint**, **quart**, and **gallon**. Ounces (oz), cups (c), pints (pt), quarts (qt), and gallons (gal) are related as follows.

U.S. UNITS OF CAPACITY AND VOLUME

$1 \text{ c} = 8 \text{ oz}$ $1 \text{ pt} = 2 \text{ c}$
$1 \text{ qt} = 2 \text{ pt}$ $1 \text{ gal} = 4 \text{ qt}$

FIGURE 6.3 shows some common containers that hold a cup, pint, quart, and gallon.

Common Containers

1 c   1 pt   1 qt   1 gal

FIGURE 6.3

EXAMPLE 5   **Converting from one unit of capacity to another**

Convert as indicated.
(a) 152 ounces to pints   (b) 2.5 gallons to cups

Solution
(a) First, convert from ounces to cups and then from cups to pints.

$$152 \text{ oz} = \frac{152 \text{ oz}}{1} \cdot \frac{1 \text{ c}}{8 \text{ oz}} \cdot \frac{1 \text{ pt}}{2 \text{ c}} = \frac{152}{8 \cdot 2} \text{ pt} = \frac{19}{2} \text{ pt} = 9\frac{1}{2} \text{ pt or 9.5 pt}$$

Convert ounces to cups.   Convert cups to pints.

(b) Convert from gallons to quarts, then quarts to pints, and finally from pints to cups.

$$2.5 \text{ gal} = \frac{2.5 \text{ gal}}{1} \cdot \frac{4 \text{ qt}}{1 \text{ gal}} \cdot \frac{2 \text{ pt}}{1 \text{ qt}} \cdot \frac{2 \text{ c}}{1 \text{ pt}} = 2.5 \cdot 4 \cdot 2 \cdot 2 \text{ c} = 40 \text{ c}$$

Gallons to quarts.   Quarts to pints.   Pints to cups.

Now Try Exercises 41, 49

4 Converting U.S. Units of Weight

The units used to measure weight in the U.S. system of measurement are the **ounce**, **pound**, and **ton**. The weight of a box of breakfast cereal is measured in ounces, while the weight of a person is measured in pounds. Tons are used for the weight of very heavy objects such as trucks. Ounces (oz), pounds (lb), and tons (T) are related as follows.

U.S. UNITS OF WEIGHT

1 lb = 16 oz   1 T = 2000 lb

MAKING CONNECTIONS 2

Ounces of Capacity and Ounces of Weight

The term *ounce* is used to name two different units of measure: a unit of capacity and a unit of weight. For this reason, we must clearly understand whether a problem involves capacity or weight. To avoid confusion, some people use the term *fluid ounces* (fl oz) for capacity and the term *ounces* (oz) for weight.

EXAMPLE 6 **Converting from one unit of weight to another**

Convert as indicated.
(a) 200 ounces to pounds   (b) 0.04 ton to ounces

Solution
(a) Use a unit fraction to convert ounces to pounds.

$$200 \text{ oz} = \frac{200 \text{ oz}}{1} \cdot \frac{1 \text{ lb}}{16 \text{ oz}} = \frac{200}{16} \text{ lb} = 12\frac{1}{2} \text{ lb or } 12.5 \text{ lb}$$

16 ounces in 1 pound

(b) Convert from tons to pounds and then from pounds to ounces.

$$0.04 \text{ T} = \frac{0.04 \text{ T}}{1} \cdot \frac{2000 \text{ lb}}{1 \text{ T}} \cdot \frac{16 \text{ oz}}{1 \text{ lb}} = 0.04 \cdot 2000 \cdot 16 \text{ oz} = 1280 \text{ oz}$$

2000 pounds in 1 ton   *16 ounces in 1 pound*

Now Try Exercises 53, 57

6.3 Putting It All Together

CONCEPT	COMMENTS	EXAMPLES
Unit Fraction	A unit fraction is a fraction that is equivalent to 1. Unit fractions can be used to convert from one unit of measure to another.	$\frac{3 \text{ ft}}{1 \text{ yd}}, \frac{1 \text{ gal}}{4 \text{ qt}},$ and $\frac{2000 \text{ lb}}{1 \text{ T}}$
Units of Length	Inches, feet, yards, and miles **Relationships:** 1 ft = 12 in. 1 yd = 36 in. and 1 yd = 3 ft 1 mi = 1760 yd and 1 mi = 5280 ft	Convert 24 inches to feet. $\frac{24 \text{ in.}}{1} \cdot \frac{1 \text{ ft}}{12 \text{ in.}} = \frac{24}{12} \text{ ft} = 2 \text{ ft}$
Units of Area	Units of area are squares of the units of length. To convert from one unit of area to another, use the corresponding unit fraction relating units of length, twice.	Convert 2 square yards to square feet. $\frac{2 \text{ yd} \cdot \text{yd}}{1} \cdot \frac{3 \text{ ft}}{1 \text{ yd}} \cdot \frac{3 \text{ ft}}{1 \text{ yd}} = 18 \text{ ft}^2$
Units of Capacity and Volume	Ounce, cup, pint, quart, and gallon **Relationships:** 1 c = 8 oz 1 pt = 2 c 1 qt = 2 pt 1 gal = 4 qt	Convert 4 gallons to quarts. $\frac{4 \text{ gal}}{1} \cdot \frac{4 \text{ qt}}{1 \text{ gal}} = 16 \text{ qt}$
Units of Weight	Ounce, pound, and ton **Relationships:** 1 lb = 16 oz 1 T = 2000 lb	Convert 64 ounces to pounds. $\frac{64 \text{ oz}}{1} \cdot \frac{1 \text{ lb}}{16 \text{ oz}} = \frac{64}{16} \text{ lb} = 4 \text{ lb}$

6.3 Exercises

CONCEPTS AND VOCABULARY

1. The foot, yard, and mile are units of _____.

2. A(n) _____ fraction is a fraction that is equivalent to 1.

3. When using a unit fraction to convert from one unit of measure to another, write the given (unwanted) unit in the _____ of the unit fraction.

4. When using a unit fraction to convert from one unit of measure to another, write the new (desired) unit in the _____ of the unit fraction.

5. When converting from one unit of _____ to another, use the corresponding unit fraction relating units of length, twice.

6. The amount of a substance that a container can hold is called the container's _____.

7. The actual amount of a substance that is present in a container is called the _____ of the substance.

8. The units of _____ and _____ are the ounce, cup, pint, quart, and gallon.

9. The ounce, pound, and ton are units of _____.

10. The term _____ can be used to name two different units of measure: a unit of capacity and a unit of weight.

CONVERTING UNITS OF LENGTH

Exercises 11–24: Convert the length as indicated.

11. 48 inches to feet
12. 6.5 feet to inches
13. 880 yards to miles
14. 4.5 miles to yards
15. $\frac{2}{3}$ yard to inches
16. 156 inches to yards
17. 87 feet to yards
18. $17\frac{1}{3}$ yards to feet
19. 1.2 miles to feet
20. 2112 feet to miles
21. 0.025 mile to inches
22. 47,520 inches to miles
23. 1 inch to feet
24. 1 inch to yards

CONVERTING UNITS OF AREA

Exercises 25–32: Convert the area as indicated.

25. 10 square feet to square inches
26. 216 square inches to square feet
27. 0.01 square mile to square yards
28. 92,928 square yards to square miles
29. 324 square feet to square yards
30. $7\frac{1}{3}$ square yards to square feet
31. 278,784 square feet to square miles
32. 0.02 square mile to square feet

CONVERTING UNITS OF CAPACITY AND VOLUME

Exercises 33–52: Convert the capacity as indicated.

33. 48 ounces to cups
34. $12\frac{1}{2}$ cups to ounces
35. $\frac{1}{2}$ quart to pints
36. 17 pints to quarts
37. 10.25 gallons to quarts
38. 18 quarts to gallons
39. 26 cups to pints
40. $11\frac{1}{2}$ pints to cups
41. 100 ounces to pints
42. 3.25 quarts to ounces
43. $\frac{3}{4}$ gallon to pints
44. 0.5 gallon to ounces
45. 60 cups to gallons
46. 96 ounces to quarts
47. $\frac{1}{4}$ quart to cups
48. 24 cups to quarts
49. 5 gallons to cups
50. 1.25 pints to ounces

51. 800 ounces to quarts

52. 1 pint to gallons

CONVERTING UNITS OF WEIGHT

Exercises 53–64: Convert the weight as indicated.

53. 440 ounces to pounds

54. 0.008 ton to pounds

55. $7\frac{5}{8}$ pounds to ounces

56. 200,000 ounces to tons

57. 0.00025 ton to ounces

58. 15,000 pounds to tons

59. $\frac{1}{100}$ ton to pounds

60. 496 ounces to pounds

61. 1,760,000 ounces to tons

62. 9 pounds to ounces

63. 22,500 pounds to tons

64. 0.4 ton to ounces

APPLICATIONS INVOLVING U.S. UNITS

65. **Soda Bottle** A person buys a 20-ounce bottle of soda from a vending machine. How many pints of soda did the person buy?

66. **Jumbo Bottle** A jumbo bottle of soda contains about 101 ounces of soda. Does this bottle hold more or less than a gallon of soda?

67. **Canadian Football** A standard Canadian football field is 330 feet long. How many yards is this?

68. **Checked Luggage** Many airlines restrict the size of a checked luggage item so that the sum of its length, width, and height can be no more than $5\frac{1}{6}$ feet. What is this restriction in inches?

69. **Weight in Stones** The *stone* is a unit of weight in the Imperial system of measurement. One stone is equivalent to 14 pounds. If a person weighs 154 pounds, find his weight in stones.

70. **Length in Rods** The *rod* is a unit of length often used on wilderness maps to describe the length of a portage (trail between bodies of water). A rod is equivalent to 16.5 feet. Find the length in rods of a portage that is 1 mile long.

71. **Cost of Carpet** For a square room that measures 15 feet on a side, do the following.

    (a) Find the area of the room in square feet.
    (b) Convert your answer to square yards.
    (c) Determine the cost of carpet for the room if the carpet sells for $32 per square yard.

72. **Cost of Carpet** For a rectangular room that measures 15 feet by 18 feet, do the following.

    (a) Find the area of the room in square feet.
    (b) Convert your answer to square yards.
    (c) Determine the cost of carpet for the room if the carpet sells for $46 per square yard.

73. **Serving Size** The label on a 1-gallon jug of fruit juice states that the serving size is 8 oz. How many servings are in the jug?

74. **Serving Size** The label on a 2-pound bag of chips states that the serving size is 4 oz. How many servings are in the bag?

75. **Cutting Rope** A spool of rope contains 75 yards of rope. How many 4-foot pieces of rope can be cut from the spool? How much rope is left over?

76. **Friendship Bracelets** A child uses four 18-inch pieces of string to make a friendship bracelet. If the string is available on a spool with 105 feet of string, how many bracelets can the child make? How much string is left over?

77. **Freight** Freight cars on some trains can each haul 100 tons of cargo. How many 400-pound containers can be hauled by this type of freight car, assuming that all containers fit on the car?

78. **Filling an Aquarium** A person is filling a 30-gallon aquarium by pouring 1 cup of water into the aquarium at a time. How many times will this person need to fill the cup and pour it into the aquarium?

WRITING ABOUT MATHEMATICS

79. A student converts 3 pounds to cups as follows.

$$3 \text{ lb} = \frac{3 \text{ lb}}{1} \cdot \frac{16 \text{ oz}}{1 \text{ lb}} \cdot \frac{1 \text{ c}}{8 \text{ oz}} = 6 \text{ c}$$

Explain the error in this computation.

80. A classmate tells you that there are 36 square inches in 1 square yard because there are 36 inches in 1 yard. How do you convince your classmate that this is incorrect?

81. Explain how to derive a unit fraction used to convert inches to miles directly.

82. Explain how to derive a unit fraction used to convert ounces to gallons directly.

6.4 The Metric System of Measurement

Objectives

1. Understanding Metric Prefixes
2. Converting Metric Units of Length
3. Converting Metric Units of Area
4. Converting Metric Units of Capacity and Volume
5. Converting Metric Units of Mass

NEW VOCABULARY

☐ Base unit
☐ Meter, liter, gram
☐ Mass
☐ Weight

READING CHECK 1

- What are the three base units in the metric system?

1 Understanding Metric Prefixes

Within the metric system of measurement, each type of measure has a **base unit**. For example, the base metric unit for length is the *meter*. All metric measures of length have names that are based on the meter. **TABLE 6.1** summarizes the three base units for the metric system.

Base Units of the Metric System

Measurement	Base Unit
Length	Meter
Capacity and Volume	Liter
Mass	Gram

TABLE 6.1

These base units can be modified in size by applying the following prefixes shown in **TABLE 6.2**.

Metric Prefixes

Prefix	kilo	hecto	deka	deci	centi	milli
Size Relative to Base Unit	$\times 1000$	$\times 100$	$\times 10$	$\times \frac{1}{10}$	$\times \frac{1}{100}$	$\times \frac{1}{1000}$

TABLE 6.2

Different prefixes in **TABLE 6.2** are used depending on a measurement's size *relative* to the base unit. For example, a measurement that is 100 times the length of a meter, or 100 meters, is called a hectometer; a measurement that is $\frac{1}{1000}$ of a gram is called a milligram; and a measurement that is 1000 times the volume of a liter, or 1000 liters, is called a kiloliter.

In the next example, the prefixes from **TABLE 6.2** are used to name metric measurements.

EXAMPLE 1 Naming metric measurements

Write the name of the metric unit of measurement described.
(a) 10 times a meter, or 10 meters   (b) $\frac{1}{10}$ of a liter   (c) $\frac{1}{100}$ of a gram

Solution
(a) The prefix that means to multiply by 10 is "deka." Since the base unit is the meter, the correct name of this unit of length is the dekameter.
(b) The prefix "deci" means multiply by $\frac{1}{10}$. The base unit is the liter, so the correct name of this unit of capacity is the deciliter.
(c) "Centi" is the prefix that means multiply by $\frac{1}{100}$. Since the base unit is the gram, the correct name of this unit of mass is the centigram.

Now Try Exercises 9, 11, 15

2 Converting Metric Units of Length

 Connecting Concepts with Your Life The base unit of length in the metric system is the **meter**. Rather than comparing the meter to a unit of measure in the U.S. system of measurement, it is more helpful to practice holding out your hands a distance of about one meter, as illustrated in **FIGURE 6.4**.

A Meter

FIGURE 6.4

STUDY TIP

Rather than comparing metric units to U.S. units, it is more useful to develop a working knowledge of metric units. Here are some things to try:
- Find the distance from your home to your college campus in kilometers.
- Learn what your weight is in kilograms.
- Take note of the size of a 1-liter bottle of water.

Although any of the metric prefixes can be used with the meter, the three most common measures associated with the meter are *kilometers*, *centimeters*, and *millimeters*. We use kilometers to measure long distances, such as the distance between two towns. Centimeters are used for smaller measurements, such as the height of a dog and the width of a piece of notebook paper. Finally, millimeters are used to measure very small items, such as the length of an insect or a small amount of rainfall. **FIGURE 6.5** shows the relative sizes of the kilometer, centimeter, and millimeter.

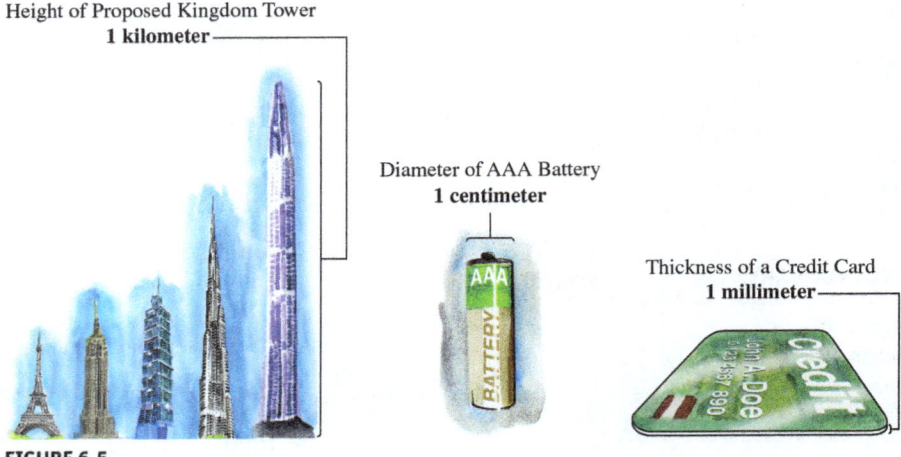

FIGURE 6.5

To convert from one metric unit of length to another, we can first see how each unit relates to the base unit. Kilometers (km), hectometers (hm), dekameters (dam), decimeters (dm), centimeters (cm), and millimeters (mm) relate to the meter (m) as follows.

READING CHECK 2

- How many meters are there in 2 kilometers?

METRIC UNITS OF LENGTH

1 km = 1000 m	1 hm = 100 m	1 dam = 10 m
1 m = 10 dm	1 m = 100 cm	1 m = 1000 mm

The next example illustrates how unit fractions and unit analysis can be used to convert one metric unit of length to another by first converting to the base unit, meters.

EXAMPLE 2 Converting from one unit of length to another

Convert 0.61 hectometer to decimeters.

Solution
First, convert from hectometers to meters and then from meters to decimeters.

$$0.61 \text{ hm} = \frac{0.61 \text{ hm}}{1} \cdot \frac{100 \text{ m}}{1 \text{ hm}} \cdot \frac{10 \text{ dm}}{1 \text{ m}} = 0.61(100)(10) \text{ dm} = 610 \text{ dm}$$

100 meters per hectometer   10 decimeters per meter

Now Try Exercise 17

Notice that the answer in Example 2 can be found by simply moving the decimal point in 0.61 to the right 3 places to get 610. Every unit fraction used in converting metric units has the effect of multiplying or dividing by a power of 10. To determine which direction and the number of places to move the decimal point, consider the following *ordered* list of metric units of length.

Metric Units of Length from Largest to Smallest

km   hm   dam   m   dm   cm   mm

When converting from hectometers to decimeters, we move the decimal point in the given measurement **3** places to the **right**.

$$0.61 \text{ hm} = 0.610 \text{ hm} = 610 \text{ dm}$$

EXAMPLE 3 **Converting from one unit of length to another**

Convert as indicated.
(a) 102,300 millimeters to dekameters   **(b)** 0.045 kilometer to meters

Solution
(a)   km   hm   dam   m   dm   cm   mm   *Convert millimeters to dekameters.*

Move the decimal point in 102,300 mm **4** places to the **left**.

$$102{,}300. \text{ mm} = 10.23 \text{ dam}$$

(b)   km   hm   dam   m   dm   cm   mm   *Convert kilometers to meters.*

Move the decimal point in 0.045 km **3** places to the **right**.

$$0.045 \text{ km} = 45 \text{ m}$$

Now Try Exercises 21, 23

3 Converting Metric Units of Area

When we converted units of area within the U.S. system of measurement, we used unit fractions relating corresponding units of length, *twice*. This concept can be extended to the metric system—when converting metric units of area, the decimal point is moved *twice* the number of places needed to convert the corresponding units of length.

To illustrate this procedure, we will convert 1 square centimeter to square millimeters by first noting the direction and number of places that the decimal point is moved when converting the corresponding units of length, centimeters to millimeters.

km   hm   dam   m   dm   cm   mm

Since the decimal point is moved **1** place to the **right** when centimeters are converted to millimeters, it must be moved twice this distance, or **2** places to the **right**, when square centimeters are converted to *square* millimeters.

$$1 \text{ cm}^2 = 1.00 \text{ cm}^2 = 100 \text{ mm}^2$$

FIGURE 6.6 shows that 1 square centimeter is equivalent to 100 square millimeters.

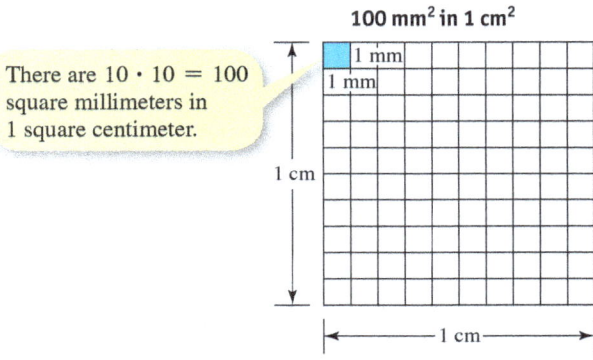

There are $10 \cdot 10 = 100$ square millimeters in 1 square centimeter.

FIGURE 6.6 (not to scale)

EXAMPLE 4 Converting from one unit of area to another

Convert as indicated.
(a) 1,200,000 square centimeters to square meters
(b) 0.00078 square dekameters to square millimeters

Solution

(a)   km   hm   dam   m   dm   cm   mm

First, convert centimeters to meters.

When converting from centimeters to meters, we move the decimal point 2 places to the left. So, when converting from square centimeters to square meters, we move the decimal point twice this amount, or **4** places to the **left**.

$$1,200,000. \text{ cm}^2 = 120 \text{ m}^2$$

Next, convert square centimeters to square meters.

(b)   km   hm   dam   m   dm   cm   mm

First, convert dekameters to millimeters.

When we convert from dekameters to millimeters, we move the decimal point 4 places to the right. So, when converting from square dekameters to square millimeters, we move the decimal point twice this amount, or **8** places to the **right**.

$$0.00078 \text{ dam}^2 = 0.00078000 \text{ dam}^2 = 78,000 \text{ mm}^2$$

Next, convert square dekameters to square millimeters.

Now Try Exercises 33, 35

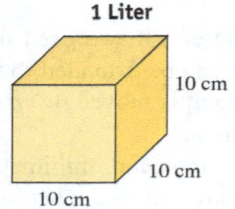

1 Liter
10 cm
10 cm
10 cm

4 Converting Metric Units of Capacity and Volume

The base unit of capacity and volume in the metric system is the **liter**, which is defined as the amount of liquid that would fill a cube measuring 10 centimeters along each edge. As with the meter, any of the metric prefixes can be used with the liter. However, it is most common to measure liquid capacity and volume in either *liters* or *milliliters*. Liters are often used to measure amounts of liquid such as gasoline in a tank, soda in a large bottle, or water in an aquarium. Milliliters can be used to measure very small volumes such as doses of liquid medicine. The relative sizes of the liter and milliliter are shown in **FIGURE 6.7**.

Some Metric Units of Capacity and Volume

Volume of Juice
1 liter

Amount of Milk Shown
1 milliliter

FIGURE 6.7

Kiloliters (kl), hectoliters (hl), dekaliters (dal), deciliters (dl), centiliters (cl), and milliliters (mL) relate to the liter (L) as follows.

METRIC UNITS OF CAPACITY AND VOLUME

| 1 kl = 1000 L | 1 hl = 100 L | 1 dal = 10 L |
| 1 L = 10 dl | 1 L = 100 cl | 1 L = 1000 mL |

READING CHECK 3

- How many cubic centimeters are there in 1 liter?

Because the liter is defined as the volume of a cube measuring 10 centimeters along each edge, there are $10 \cdot 10 \cdot 10 = 1000$ cubic centimeters (cc) in one liter. Based on this observation, we can conclude that 1 cubic centimeter is equivalent to 1 milliliter, because there are also 1000 milliliters in one liter. Since

$$1 \text{ L} = 1000 \text{ mL} = 1000 \text{ cc},$$

the units mL and cc may be used interchangeably. The units of capacity and volume are listed as follows.

Metric Units of Capacity/Volume from Largest to Smallest

kl　　hl　　dal　　L　　dl　　cl　　mL (cc)

EXAMPLE 5　Converting from one unit of capacity to another

Convert as indicated.
(a) 12 deciliters to milliliters　　**(b)** 2500 centiliters to hectoliters

Solution
(a)　kl　　hl　　dal　　L　　dl　　cl　　mL (cc)　　*Convert deciliters to milliliters.*

Move the decimal point in 12 dl **2** places to the **right**.

$$12 \text{ dl} = 12.00 \text{ dl} = 1200 \text{ mL}$$

(b)　kl　　hl　　dal　　L　　dl　　cl　　mL (cc)　　*Convert centiliters to hectoliters.*

Move the decimal point in 2500 cl **4** places to the **left**.

$$2500. \text{ cl} = 0.25 \text{ hl}$$

Now Try Exercises 39, 47

5　Converting Metric Units of Mass

Although the terms *weight* and *mass* are sometimes used interchangeably, they do not have exactly the same meaning. While **mass** is a measure of the amount of matter in an object, **weight** is a measure of the force on an object due to gravity. The mass of an object does not change, but its weight may change depending on its distance from the center of Earth.

The base unit of mass in the metric system is the **gram**, which is defined as the mass of one milliliter of water. While any of the metric prefixes can be used with the gram, it is most common to measure mass in *kilograms*, *grams*, or *milligrams*.

Kilograms are used to measure the mass of heavy objects, such as building materials, people, and pets. Grams are used to measure the mass of relatively light objects, such as snack foods and dry ingredients in recipes. Finally, milligrams are often used to measure the mass of medicine found in small pills or tablets. The relative sizes of the kilogram, gram, and milligram are shown in **FIGURE 6.8** on the next page.

Some Metric Units of Mass

Mass of 1 Liter of Water
1 kilogram

Mass of a Large Paper Clip
1 gram

Mass of 3 to 5 Grains of Table Salt
1 milligram

FIGURE 6.8

Kilogram (kg), hectogram (hg), dekagram (dag), decigram (dg), centigram (cg), and milligram (mg) relate to the gram (g) as follows.

READING CHECK 4

- How many grams are there in 1 kilogram?

METRIC UNITS OF MASS

| 1 kg = 1000 g | 1 hg = 100 g | 1 dag = 10 g |
| 1 g = 10 dg | 1 g = 100 cg | 1 g = 1000 mg |

Units of mass are listed as follows.

Metric Units of Mass from Heaviest to Lightest

kg   hg   dag   g   dg   cg   mg

EXAMPLE 6 Converting from one unit of mass to another

Convert as indicated.
(a) 0.00034 kilogram to decigrams   (b) 10,000 milligrams to grams

Solution

(a)   kg   hg   dag   g   dg   cg   mg

Convert kilograms to decigrams.

Move the decimal point in 0.00034 kg **4** places to the **right**.

$$0.00034 \text{ kg} = 3.4 \text{ dg}$$

(b)   kg   hg   dag   g   dg   cg   mg

Convert milligrams to grams.

Move the decimal point in 10,000 mg **3** places to the **left**.

$$10,000. \text{ mg} = 10 \text{ g}$$

Now Try Exercises 51, 57

6.4 Putting It All Together

CONCEPT	COMMENTS	EXAMPLES
Metric Prefixes	kilo, hecto, deka, deci, centi, and milli. Each prefix indicates the size of a measurement relative to the base unit.	*hecto* means "100 times the base." *milli* means "$\frac{1}{1000}$ of the base." *deka* means "10 times the base."
Units of Length	The base metric unit of length is the *meter*. **Relationships:** 1 km = 1000 m, 1 hm = 100 m 1 dam = 10 m, 1 m = 10 dm 1 m = 100 cm, and 1 m = 1000 mm	Convert 1.8 meters to centimeters. km hm dam m dm cm mm 1.8 m = 1.80 m = 180 cm
Units of Area	Units of area are squares of the units of length. To convert from one unit of area to another, move the decimal point *twice* the number of places needed to convert the corresponding units of length.	Convert 6000 square centimeters to square meters. km hm dam m dm cm mm 6000. cm^2 = 0.6 m^2
Units of Capacity and Volume	The base metric unit of capacity and volume is the *liter*. **Relationships:** 1 kl = 1000 L, 1 hl = 100 L 1 dal = 10 L, 1 L = 10 dl 1 L = 100 cl, and 1 L = 1000 mL	Convert 0.45 hectoliters to liters. kl hl dal L dl cl mL 0.45 hl = 45 L
Units of Mass	The base metric unit of mass is the *gram*. **Relationships:** 1 kg = 1000 g, 1 hg = 100 g 1 dag = 10 g, 1 g = 10 dg 1 g = 100 cg, and 1 g = 1000 mg	Convert 25 grams to hectograms. kg hg dag g dg cg mg 25. g = 0.25 hg

6.4 Exercises

CONCEPTS AND VOCABULARY

1. The prefix *kilo* means _____ times the base.
2. The prefix that means $\frac{1}{10}$ of the base is _____.
3. The base metric unit of length is the _____.
4. Which is the smaller unit of length, the decimeter or the dekameter?
5. The base unit of capacity and volume in the metric system is the _____.
6. One cc is equivalent to one _____.
7. The base metric unit of mass is the _____.
8. Which is the larger unit of mass, the hectogram or the centigram?

WRITING METRIC PREFIXES

Exercises 9–16: Write the name of the metric unit of measurement described.

9. 1000 times a liter
10. $\frac{1}{10}$ of a gram
11. $\frac{1}{100}$ of a meter
12. 100 times a liter

13. 10 times a meter
14. $\frac{1}{100}$ of a liter
15. $\frac{1}{1000}$ of a gram
16. 10 times a gram

CONVERTING UNITS OF LENGTH

Exercises 17–28: Convert the length as indicated.

17. 3400 millimeters to decimeters
18. 8.73 dekameters to meters
19. 121 kilometers to hectometers
20. 53.6 centimeters to millimeters
21. 0.0045 hectometer to decimeters
22. 2007 millimeters to meters
23. 14.59 meters to kilometers
24. 4.9 decimeters to dekameters
25. 0.6 meter to centimeters
26. 13,900 centimeters to dekameters
27. 0.25 meter to decimeters
28. 102,000 millimeters to hectometers

CONVERTING UNITS OF AREA

Exercises 29–36: Convert the area as indicated.

29. 0.004 square meter to square centimeters
30. 655 square decimeters to square meters
31. 7 square hectometers to square kilometers
32. 30 square millimeters to square centimeters
33. 120 square meters to square hectometers
34. 50 square dekameters to square meters
35. 0.0025 square kilometer to square meters
36. 0.6 square decimeter to square millimeters

CONVERTING UNITS OF CAPACITY AND VOLUME

Exercises 37–48: Convert the volume as indicated.

37. 51 hectoliters to deciliters
38. 4730 dekaliters to kiloliters
39. 900 dekaliters to liters
40. 9.63 centiliters to liters
41. 13,000 liters to hectoliters
42. 1.08 liters to milliliters
43. 0.09 deciliter to milliliters
44. 0.67 deciliter to dekaliters
45. 12 kiloliters to deciliters
46. 24,001 centiliters to dekaliters
47. 0.05 milliliter to deciliters
48. 54,600 liters to centiliters

CONVERTING UNITS OF MASS

Exercises 49–60: Convert the mass as indicated.

49. 3.8 decigrams to hectogram
50. 6500 grams to kilograms
51. 0.00095 dekagram to decigrams
52. 13 centigrams to milligrams
53. 0.045 hectogram to grams
54. 2450 grams to centigrams
55. 0.0057 kilogram to centigrams
56. 4.1 dekagrams to kilograms
57. 87.93 hectograms to kilograms
58. 45,900 centigrams to dekagrams
59. 5.5 milligrams to centigrams
60. 5000 grams to hectograms

APPLICATIONS INVOLVING METRIC UNITS

61. **Soda Bottle** A person buys a 500-milliliter bottle of soda from a vending machine. How many liters of soda did the person buy?

62. **Servings in a Bottle** A large bottle contains 3 liters of fruit juice. How many 200-milliliter servings are in this bottle?

63. **Checked Luggage** Some airlines charge a fee for each checked luggage item that weighs more than 23,000 grams. How many kilograms is this?

64. **High Jump** The world record in the high jump was set in 1993 by Javier Sotomayor with a jump of 2.45 meters. How many centimeters high was his jump? (*Source:* International Association of Athletics Federations.)

65. Olympic Pool An Olympic swimming pool measures 25 meters by 50 meters.
(a) Find the area of the surface of the pool.
(b) Convert your answer from square meters to square dekameters.

66. Olympic Pool An Olympic swimming pool typically holds 2500 kiloliters of water. How many 1-liter bottles could be filled with the water in an Olympic swimming pool?

67. Metric Tons The mass of very heavy objects can be measured in metric tons. A metric ton (t) is equal to 1000 kilograms. If an African elephant weighs 5500 kilograms, what is its weight in metric tons?

68. Metric Tons Refer to the previous exercise. Find the number of grams in one metric ton.

69. Serving Size If the label on a 1-kilogram can of mixed nuts states that the serving size is 40 grams, how many servings are in the can?

70. Serving Size If the label on a 2-liter bottle of fruit drink states that the serving size is 250 milliliters, how many servings are in the bottle?

71. Pieces of Thread A spool of thread contains 100 meters of thread. How many 80-centimeter pieces of thread can be cut from the spool?

72. Fraternal Twins A girl and boy are fraternal twins; the birth weight of the girl is 2950 grams and the birth weight of the boy is 2.75 kilograms. Which baby has a higher birth weight?

73. Boat Occupancy A whale-watching tour boat cannot operate if the total passenger weight is more than 6,000,000 grams. What is the maximum number of tourists, weighing an average of 80 kilograms each, that can be allowed to take the tour?

74. Water Weight A milliliter of water weighs 1 gram. How many liters of water weigh the same as a 100-kilogram person?

WRITING ABOUT MATHEMATICS

75. Give two reasons why converting between units in the metric system is easier than converting between units in the U.S. system.

76. The volume of water that flows in a river is sometimes measured in cubic meters. Explain the steps to convert cubic meters to kiloliters.

77. The prefix *mega* means one million times the base unit. Explain how to convert kiloliters to megaliters.

78. Explain why it makes no sense to state that a small swimming pool has a capacity of 500 *cubic liters*.

SECTIONS 6.3 and 6.4 — Checking Basic Concepts

1. Convert each length as indicated.
   (a) 18 feet to yards
   (b) $\frac{1}{3}$ yard to inches

2. Convert 8 square feet to square inches.

3. Convert each capacity as indicated.
   (a) $4\frac{1}{2}$ cups to ounces
   (b) 22 quarts to gallons

4. Convert each weight as indicated.
   (a) 0.002 ton to pounds
   (b) 368 ounces to pounds

5. Convert each length as indicated.
   (a) 4700 millimeters to meters
   (b) 30 meters to hectometers

6. Convert 3 square meters to square centimeters.

7. Convert each capacity as indicated.
   (a) 5.8 liters to milliliters
   (b) 3250 centiliters to dekaliters

8. Convert each mass as indicated.
   (a) 83,000 grams to kilograms
   (b) 6.2 milligrams to centigrams

9. **Serving Size** If the label on a 3-pound bag of chips states that the serving size is 2 oz, how many servings are in the bag?

10. **Lengths of Chain** If there are 30 meters of gold chain on a spool, how many 50-centimeter lengths of chain can be cut from the spool?

6.5 U.S.–Metric Conversions; Temperature

Objectives

1. Completing Conversions of Length
2. Completing Conversions of Capacity and Volume
3. Completing Conversions of Mass (Weight)
4. Completing Conversions of Temperature
   - Symbolically
   - Numerically
   - Graphically

NEW VOCABULARY

☐ Temperature
☐ Fahrenheit
☐ Celsius

1 Completing Conversions of Length

Until the United States moves completely to the metric system, people will need to convert between the U.S. and metric systems. To convert a length in one system to an equivalent length in the other, we use unit fractions and unit analysis. Some common relationships between U.S. and metric units of length are shown below. Recall that the symbol $\approx$ means "approximately equal."

> **CONVERTING UNITS OF LENGTH**
>
> 1 in. = 2.54 cm,   1 m $\approx$ 1.09 yd,   and   1 km $\approx$ 0.62 mi

In the next example, we use unit fractions and unit analysis to convert directly between the two systems of measurement. In each part of the example, only one unit fraction is needed to complete the conversion.

EXAMPLE 1 Completing U.S.–metric conversions of length

Convert as indicated. Round answers to 2 decimal places.
(a) 38 centimeters to inches   (b) 4 kilometers to miles

Solution

(a) $38 \text{ cm} = \dfrac{38 \text{ cm}}{1} \cdot \dfrac{1 \text{ in.}}{2.54 \text{ cm}} = \dfrac{38}{2.54} \text{ in.} \approx 14.96 \text{ in.}$ *There are 2.54 centimeters in 1 inch.*

(b) $4 \text{ km} \approx \dfrac{4 \text{ km}}{1} \cdot \dfrac{0.62 \text{ mi}}{1 \text{ km}} = 4(0.62) \text{ mi} = 2.48 \text{ mi}$ *There is about 0.62 mile in 1 kilometer.*

Now Try Exercises 9, 13

When we do not have a unit fraction to convert directly from one unit to another, it may be necessary to complete some conversions *within* each system first. For example, a measurement in miles could be converted to meters by first converting *miles to kilometers* and then converting *kilometers to meters*.

$$\text{miles} \rightarrow \text{kilometers} \rightarrow \text{meters}$$

U.S. to metric

Another option would be to convert *miles to yards* and then convert *yards to meters*.

$$\text{miles} \rightarrow \text{yards} \rightarrow \text{meters}$$

U.S. to metric

In either case, the green arrow represents a unit fraction that converts U.S. system units to metric system units. In the next example, we convert miles to meters in this manner.

EXAMPLE 2 Completing U.S.–metric conversions of length

Convert 2.5 miles to meters using the given sequence. Round answers to 2 decimal places.
(a) Miles → kilometers → meters   (b) Miles → yards → meters

Solution
(a) First, convert 2.5 miles to kilometers.

$$2.5 \text{ mi} \approx \frac{2.5 \text{ mi}}{1} \cdot \frac{1 \text{ km}}{0.62 \text{ mi}} \approx 4.03226 \text{ km}$$

Next, convert 4.03226 kilometers to meters. Because there are 1000 meters in 1 kilometer, move the decimal point 3 places to the right.

$$4.03226 \text{ km} = 4032.26 \text{ m}$$

(b) Both steps can be combined into one step using the multiplication of 2 unit fractions.

$$2.5 \text{ mi} \approx \frac{2.5 \text{ mi}}{1} \cdot \frac{1760 \text{ yd}}{1 \text{ mi}} \cdot \frac{1 \text{ m}}{1.09 \text{ yd}} = \frac{2.5(1760)}{1.09} \text{ m} \approx 4036.70 \text{ m}$$

Now Try Exercise 15

The answers from parts (a) and (b) are approximately equal.

READING CHECK 1

Why do the results from parts (a) and (b) of Example 2 differ slightly?

NOTE: The two results for Example 2 are not exactly the same because some of the unit fractions used contain approximations that do not give exact results. ■

▶2 Completing Conversions of Capacity and Volume

The process for converting a capacity in one system to an equivalent capacity in the other system is the same process as for converting length from one system to the other. Some common relationships between U.S. and metric units of capacity are shown below.

CONVERTING UNITS OF CAPACITY AND VOLUME

$$1 \text{ L} \approx 1.06 \text{ qt} \quad \text{and} \quad 1 \text{ oz} \approx 29.57 \text{ mL}$$

EXAMPLE 3 Completing U.S.–metric conversions of capacity

Convert as indicated. Round answers to 2 decimal places.
(a) 6.3 liters to quarts   (b) 6 gallons to liters   (c) 250 milliliters to pints

Solution

(a) $\quad 6.3 \text{ L} \approx \dfrac{6.3 \text{ L}}{1} \cdot \dfrac{1.06 \text{ qt}}{1 \text{ L}} = 6.3(1.06) \text{ qt} = 6.678 \text{ qt} \approx 6.68 \text{ qt}$ *(Metric to U.S.)*

(b) We will use the following conversions: gallon → quarts → liters.

$$6 \text{ gal} \approx \frac{6 \text{ gal}}{1} \cdot \frac{4 \text{ qt}}{1 \text{ gal}} \cdot \frac{1 \text{ L}}{1.06 \text{ qt}} = \frac{6(4)}{1.06} \text{ L} \approx 22.64 \text{ L}$$

U.S. to metric

(c) We will use the following conversions: milliliters → ounces → cups → pints.

$$250 \text{ mL} \approx \frac{250 \text{ mL}}{1} \cdot \frac{1 \text{ oz}}{29.57 \text{ mL}} \cdot \frac{1 \text{ c}}{8 \text{ oz}} \cdot \frac{1 \text{ pt}}{2 \text{ c}} = \frac{250}{29.57(8)(2)} \text{ pt} \approx 0.53 \text{ pt}$$

Metric to U.S.

Now Try Exercises 31, 35, 39

3 Completing Conversions of Mass (Weight)

Once again, unit fractions are needed to convert from one system of measurement to the other. Such unit fractions can be derived from the common relationships between U.S. and metric units of mass (weight) shown below.

CONVERTING UNITS OF MASS (WEIGHT)

$$1 \text{ kg} \approx 2.2 \text{ lb} \quad \text{and} \quad 1 \text{ oz} \approx 28.35 \text{ g}$$

EXAMPLE 4 Completing U.S.–metric conversions of mass (weight)

Convert as indicated. Round answers to 2 decimal places.
(a) 135 pounds to kilograms (b) 750 grams to pounds (c) 9 ounces to kilograms

Solution

(a) $135 \text{ lb} \approx \dfrac{135 \text{ lb}}{1} \cdot \dfrac{1 \text{ kg}}{2.2 \text{ lb}} = \dfrac{135}{2.2} \text{ kg} \approx 61.36 \text{ kg}$ *U.S. to metric*

(b) Because there are 1000 grams in 1 kilogram, convert 750 grams to kilograms by moving the decimal point 3 places to the left.

$$750 \text{ g} = 0.75 \text{ kg}$$

Next, convert 0.75 kilograms to pounds.   *Metric to U.S.*

$$0.75 \text{ kg} \approx \frac{0.75 \text{ kg}}{1} \cdot \frac{2.2 \text{ lb}}{1 \text{ kg}} = 0.75(2.2) \text{ lb} = 1.65 \text{ lb}$$

(c) First, convert 9 ounces to grams.   *U.S. to metric*

$$9 \text{ oz} \approx \frac{9 \text{ oz}}{1} \cdot \frac{28.35 \text{ g}}{1 \text{ oz}} = 9(28.35) \text{ g} = 255.15 \text{ g}$$

Next, convert 255.15 grams to kilograms. Because there are 1000 grams in 1 kilogram, move the decimal point 3 places to the left.

$$255.15 \text{ g} = 0.25515 \text{ kg} \approx 0.26 \text{ kg}$$

Now Try Exercises 41, 47, 49

4 Completing Conversions of Temperature

When we are interested in how hot or cold something is, we are considering **temperature**. In the U.S. system, temperature is measured in degrees **Fahrenheit** (°F). In the metric system, temperature is measured in degrees **Celsius** (°C). **FIGURE 6.9** shows how some temperatures compare between the Celsius and Fahrenheit scales.

STUDY TIP

Take time to learn the Celsius temperatures in **FIGURE 6.9** so that you can make sense of a Celsius temperature without having to convert to Fahrenheit.

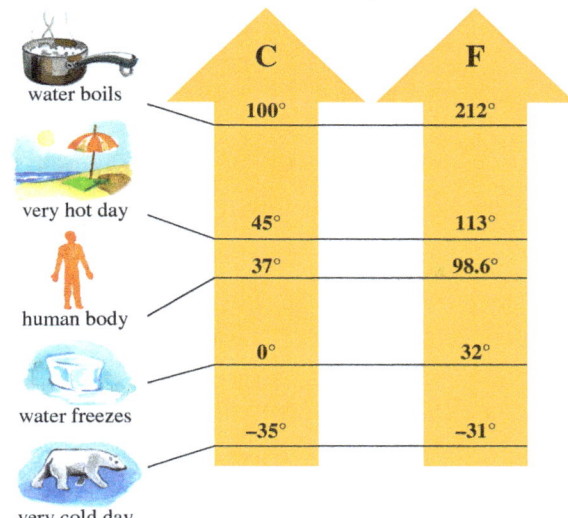

FIGURE 6.9

READING CHECK 2

What are three ways that temperatures can be converted from one scale to the other?

A given temperature can be converted from one scale to the other symbolically using formulas, numerically using a table of values, or graphically using a graph. The method we choose depends on the specific circumstances and the desired accuracy.

CONVERTING TEMPERATURES SYMBOLICALLY When we want the most accurate results, it is often best to use an appropriate formula to convert from one temperature scale to the other. Formulas relating the two scales are stated as follows.

> **FORMULAS FOR CONVERTING TEMPERATURES**
>
> If F and C represent Fahrenheit and Celsius temperatures, respectively, then
>
> $$F = \frac{9}{5}C + 32 \quad \text{and} \quad C = \frac{5}{9}(F - 32).$$
>
> **Convert °C to °F**   **Convert °F to °C**

NOTE: The variables F and C used to represent Fahrenheit and Celsius temperatures are written in italics, while the abbreviations °F and °C are not.

EXAMPLE 5   Converting temperatures symbolically

Convert as indicated. Give exact answers in decimal form.
(a) 127 °F to Celsius   (b) 22 °C to Fahrenheit

Solution
(a) Replace F with 127 in the formula $C = \frac{5}{9}(F - 32)$.

$$C = \frac{5}{9}(\mathbf{127} - 32) \quad \text{Replace } F \text{ with 127.}$$

$$= \frac{5}{9}(95) \quad \text{Subtract.}$$

$$= 52.\overline{7} \quad \text{Multiply and simplify.}$$

The temperature 127 °F is equivalent to $52.\overline{7}$ °C.

(b) Replace C with 22 in the formula $F = \frac{9}{5}C + 32$.

$$F = \frac{9}{5}(22) + 32 \quad \text{Replace } C \text{ with 22.}$$
$$= 39.6 + 32 \quad \text{Multiply and simplify.}$$
$$= 71.6 \quad \text{Add.}$$

So, 22 °C is equivalent to 71.6 °F.

Now Try Exercises 55, 57

CONVERTING TEMPERATURES NUMERICALLY In some cases, temperatures are all relatively close to a specific value. When this happens, a table of values can be used to complete temperature conversions.

🌐 Math in Medical Context Medical professionals who record the body temperatures of patients may do so several times daily. Rather than using a formula, it is sometimes more practical to use a table of values to complete a conversion because human body temperature does not vary greatly from 98.6 °F. In the next example, we use a table of values to convert body temperature from one scale to the other. Note that many of the values in the table have been rounded.

EXAMPLE 6 **Converting temperatures numerically**

Use **TABLE 6.3** to convert as indicated. If there are two choices, use the higher value.
(a) 101.7 °F to Celsius   (b) 37.3 °C to Fahrenheit

Fahrenheit and Celsius Body Temperatures

°F	98.0	98.1	98.2	98.3	98.4	98.5	98.6	98.7	98.8	98.9
°C	36.7	36.7	36.8	36.8	36.9	36.9	37.0	37.1	37.1	37.2
°F	99.0	99.1	99.2	99.3	99.4	99.5	99.6	99.7	99.8	99.9
°C	37.2	37.3	37.3	37.4	37.4	37.5	37.6	37.6	37.7	37.7
°F	100.0	100.1	100.2	100.3	100.4	100.5	100.6	100.7	100.8	100.9
°C	37.8	37.8	37.9	37.9	38.0	38.1	38.1	38.2	38.2	38.3
°F	101.0	101.1	101.2	101.3	101.4	101.5	101.6	101.7	101.8	101.9
°C	38.3	38.4	38.4	38.5	38.6	38.6	38.7	38.7	38.8	38.8

TABLE 6.3

Solution
(a) **TABLE 6.4** shows the two rows from **TABLE 6.3** that can be used to show that **101.7 °F** is about the same as **38.7 °C**.

Fahrenheit and Celsius Body Temperatures

°F	101.0	101.1	101.2	101.3	101.4	101.5	101.6	**101.7**	101.8	101.9
°C	38.3	38.4	38.4	38.5	38.6	38.6	38.7	**38.7**	38.8	38.8

TABLE 6.4

(b) **TABLE 6.5** shows the two rows from **TABLE 6.3** that can be used to show that 37.3 °C is about the same as 99.1 °F or 99.2 °F. Because the instructions tell us to use the higher value if there are two choices, **37.3 °C** is about the same as **99.2 °F**.

Fahrenheit and Celsius Body Temperatures

°F	99.0	99.1	**99.2**	99.3	99.4	99.5	99.6	99.7	99.8	99.9
°C	37.2	37.3	**37.3**	37.4	37.4	37.5	37.6	37.6	37.7	37.7

TABLE 6.5

Now Try Exercises 61, 63

CONVERTING TEMPERATURES GRAPHICALLY When an exact temperature value is unnecessary, it may be convenient to use a graph to convert graphically. In the next example, we use a graph to convert from one scale to the other.

EXAMPLE 7 Converting temperatures graphically

Use the following graph to estimate the requested value. Answers may vary slightly.
(a) 60 °F to Celsius (b) −40 °C to Fahrenheit

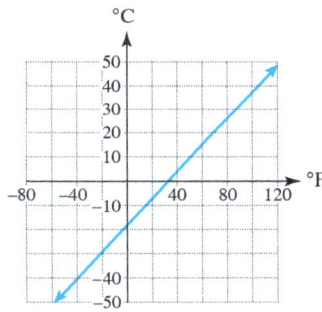

Solution

(a) Locate 60 on the horizontal (°F) axis and move **vertically** upward to the graphed line. Then move **horizontally** to the left until reaching the vertical (°C) axis. It appears that 60 °F is equivalent to about 16 °C. See **FIGURE 6.10**. (The actual value is $15.\overline{5}$ °C.)

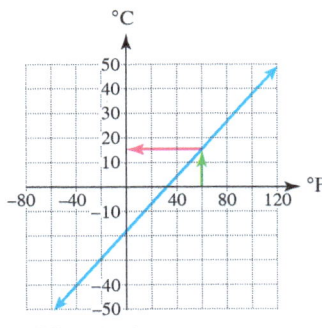

FIGURE 6.10

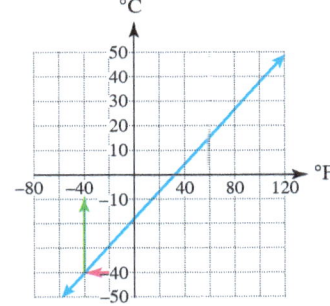

FIGURE 6.11

(b) Locate −40 on the vertical (°C) axis and move **horizontally** to the left to the graphed line. Then move **vertically** upward until reaching the horizontal (°F) axis. It appears that −40 °C is equivalent to about −40 °F. See **FIGURE 6.11**. (The actual value is −40 °F.)

Now Try Exercises 67, 69

6.5 Putting It All Together

CONCEPT	COMMENTS	EXAMPLES
Converting Units of Length	To convert units of length, use the relationships 1 in. = 2.54 cm, 1 m ≈ 1.09 yd, and 1 km ≈ 0.62 mi.	Converting 25 centimeters to inches $\dfrac{25 \text{ cm}}{1} \cdot \dfrac{1 \text{ in.}}{2.54 \text{ cm}} \approx 9.84 \text{ in.}$
Converting Units of Capacity and Volume	To convert units of capacity and volume, use the relationships 1 L ≈ 1.06 qt and 1 oz ≈ 29.57 mL.	Converting 4.2 liters to quarts $\dfrac{4.2 \text{ L}}{1} \cdot \dfrac{1.06 \text{ qt}}{1 \text{ L}} \approx 4.45 \text{ qt}$
Converting Units of Mass (Weight)	To convert units of mass, use the relationships 1 kg ≈ 2.2 lb and 1 oz ≈ 28.35 g.	Converting 0.85 kilogram to pounds $\dfrac{0.85 \text{ kg}}{1} \cdot \dfrac{2.2 \text{ lb}}{1 \text{ kg}} \approx 1.87 \text{ lb}$
Units of Temperature	In the U.S. system, temperature is measured in degrees Fahrenheit, and in the metric system, it is measured in degrees Celsius. To convert from one system to the other, use the formulas $F = \dfrac{9}{5}C + 32$ and $C = \dfrac{5}{9}(F - 32)$. Temperature conversions can also be completed numerically, using a table of values, and graphically, using a graph.	Converting 50 °F to Celsius $C = \dfrac{5}{9}(50 - 32) = \dfrac{5}{9}(18) = 10\,°C$ | °F | 49.9 | **50.0** | 50.1 | 50.2 | | °C | 9.9 | **10.0** | 10.1 | 10.1 |

6.5 Exercises

MyMathLab

CONCEPTS AND VOCABULARY

1. There are _____ centimeters in 1 inch.

2. There are about _____ yards in 1 meter.

3. There is about _____ mile in 1 kilometer.

4. In 1 liter, there are about _____ quarts.

5. In 1 ounce, there are about _____ milliliters.

6. There are about _____ pounds in 1 kilogram.

7. There are about _____ grams in 1 ounce.

8. Temperature is measured in degrees _____ in the U.S. system and in degrees _____ in the metric system.

CONVERTING UNITS OF LENGTH

Exercises 9–14: Convert the length as indicated. Round answers to 2 decimal places.

9. 17 inches to centimeters

10. 98 centimeters to inches

11. 240 yards to meters

12. 35 meters to yards

13. 84 miles to kilometers

14. 316 kilometers to miles

Exercises 15–18: Convert the given length as indicated. Round answers to 2 decimal places.

15. 5800 centimeters to yards
    (a) Centimeters → inches → yards
    (b) Centimeters → meters → yards

16. 8000 yards to kilometers
    (a) Yards → meters → kilometers
    (b) Yards → miles → kilometers

17. 4500 meters to miles
    (a) Meters → kilometers → miles
    (b) Meters → yards → miles

18. 0.5 yard to centimeters
    (a) Yards → inches → centimeters
    (b) Yards → meters → centimeters

Exercises 19–28: Convert the length as indicated. Round answers to 2 decimal places. Answers may vary slightly.

19. 27 feet to meters

20. 180 inches to meters

21. 3400 centimeters to feet

22. 0.335 kilometer to yards

23. 0.25 meter to inches

24. 0.0005 kilometer to inches

25. 15,840 feet to kilometers

26. 5.8 meters to feet

27. 0.003 kilometer to feet

28. 11 feet to centimeters

CONVERTING UNITS OF CAPACITY AND VOLUME

Exercises 29–32: Convert the capacity or volume as indicated. Round answers to 2 decimal places.

29. 16 ounces to milliliters

30. 7.5 liters to quarts

31. 56 quarts to liters

32. 1500 milliliters to ounces

Exercises 33–40: Convert the capacity or volume as indicated. Round answers to 2 decimal places. Answers may vary slightly.

33. 75 liters to gallons

34. 1 cup to milliliters

35. 16 pints to liters

36. 0.4 pint to milliliters

37. 956 milliliters to cups

38. 3.6 gallons to liters

39. 55,000.2 milliliters to gallons

40. 4.5 liters to pints

CONVERTING UNITS OF MASS (WEIGHT)

Exercises 41–44: Convert the mass or weight as indicted. Round answers to 2 decimal places.

41. 420 grams to ounces

42. 98 pounds to kilograms

43. 61.2 kilograms to pounds

44. 3 ounces to grams

Exercises 45–52: Convert the mass or weight as indicated. Round answers to 2 decimal places. Answers may vary slightly.

45. 0.135 ounce to milligrams

46. 3515.4 grams to pounds

47. 2.6 kilograms to ounces

48. 100,000 milligrams to pounds

49. 1.45 pounds to grams

50. 7560 milligrams to ounces

51. 0.0064 pound to milligrams

52. 480 ounces to kilograms

CONVERTING UNITS OF TEMPERATURE

Exercises 53–60: Convert the temperature symbolically. Give exact answers in decimal form.

53. 77 °F to Celsius

54. 55 °C to Fahrenheit

55. 3 °C to Fahrenheit

56. 98 °F to Celsius

57. −15 °F to Celsius

58. −9 °C to Fahrenheit

59. 100 °C to Fahrenheit

60. 98.6 °F to Celsius

Exercises 61–66: Use the following table to convert the temperature numerically. If there are two choices, use the higher value.

°F	98.0	98.1	98.2	98.3	98.4	98.5	98.6	98.7	98.8	98.9
°C	36.7	36.7	36.8	36.8	36.9	36.9	37.0	37.1	37.1	37.2
°F	99.0	99.1	99.2	99.3	99.4	99.5	99.6	99.7	99.8	99.9
°C	37.2	37.3	37.3	37.4	37.4	37.5	37.6	37.6	37.7	37.7
°F	100.0	100.1	100.2	100.3	100.4	100.5	100.6	100.7	100.8	100.9
°C	37.8	37.8	37.9	37.9	38.0	38.1	38.1	38.2	38.2	38.3
°F	101.0	101.1	101.2	101.3	101.4	101.5	101.6	101.7	101.8	101.9
°C	38.3	38.4	38.4	38.5	38.6	38.6	38.7	38.7	38.8	38.8

61. 98.5 °F to Celsius

62. 38.0 °C to Fahrenheit

63. 38.8 °C to Fahrenheit

64. 101.1 °F to Celsius

65. 98.6 °F to Celsius

66. 36.8 °C to Fahrenheit

Exercises 67–72: Use the following graph to convert the temperatures graphically. Answers may vary slightly.

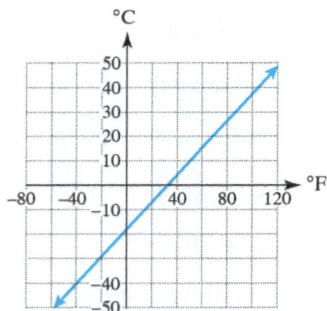

67. 20 °C to Fahrenheit

68. 83 °F to Celsius

69. −10 °F to Celsius

70. −20 °C to Fahrenheit

71. 50 °F to Celsius

72. 46 °C to Fahrenheit

APPLICATIONS INVOLVING CONVERSIONS

73. **Soda Bottle** To the nearest milliliter, how many milliliters are in a 20-ounce soda bottle?

74. **Printer Paper** A standard sheet of printer paper is 8.5 inches by 11 inches. Give these dimensions to the nearest tenth of a centimeter.

75. **Checked Luggage** Some airlines charge a fee for each checked luggage item that weighs more than 22.7 kilograms. To the nearest whole number, how many pounds is this?

76. **Long Jump** The world record in the long jump was set in 1991 by Mike Powell, with a jump of 8.95 meters. Is this jump over 30 feet? (*Source:* International Association of Athletics Federations.)

77. **Body Temperature** The highest body temperature ever recorded for a person surviving occurred in 1980 when Willie Jones had a body temperature of 46.5 °C. Convert this temperature to Fahrenheit. (*Source:* Guinness Book of World Records.)

78. **Human Body** The average adult human has about 6 quarts of blood circulating throughout the body. Find the amount of blood in liters, to the nearest hundredth. (*Source:* NOVA, Cut to the Heart.)

79. **Triple Jump** The world record in the triple jump was set in 1995 by Jonathan Edwards, with a jump of 18.29 meters. Using the sequence

    Meters → centimeters → inches → feet,

    is this jump over 60 feet? (*Source:* International Association of Athletics Federations.)

80. **Blue Whale** An adult blue whale can weigh more than 172,700 kilograms. Find this weight to the nearest ton. (*Source:* Smithsonian National Zoological Park.)

81. **Super Soaker** An Aquashock Hydroblitz Super Soaker has a 101-ounce water reservoir. To the nearest whole number, how many liters of water does this water gun hold? (*Source:* Hasbro.)

82. **Kalahari Desert** Summertime temperatures in the Kalahari Desert can reach 45 °C. Find the equivalent Fahrenheit temperature.

83. **Long Hair** The person with the world's longest hair is Xie Qiuping of China. Her hair is 5.627 meters long. Is her hair over 20 feet long? (*Source:* Guinness Book of World Records.)

84. **Obesity** An adult male who is 5 feet 10 inches tall is considered obese if his weight is above 95.5 kilograms. What is this weight to the nearest pound? (*Source:* Centers for Disease Control.)

85. **iPhone 6** The iPhone 6 is 138.1 millimeters long and 67 millimeters wide. What are these two measurements to the nearest hundredth of an inch? (*Source: Apple.*)

86. **Gas Tank** A car's gas tank holds 18.6 gallons of gasoline. How many liters of gasoline does the tank hold? Round to the nearest tenth.

87. **Mars Temperatures** The temperature at the poles on Mars can dip to −220 °F. What is the equivalent Celsius temperature? (*Source: NASA.*)

88. **The Moon** The average diameter of the moon is about 3474 kilometers. What is this diameter to the nearest mile? (*Source: NASA.*)

89. **Weight of Milk** A gallon of milk weighs about 8.6 pounds. How much does a gallon of milk weigh to the nearest tenth of a kilogram?

90. **Liquid Air** Depending on its chemical makeup, air will liquefy when cooled to about −195 °C. What is the equivalent Fahrenheit temperature?

WRITING ABOUT MATHEMATICS

91. When a conversion is completed in two different ways, the two results may be slightly different. What can be done to be sure results agree more closely?

92. Give two examples of when it might be practical to convert temperatures graphically.

Group Activity

Track and Field Before international standards for track and field were adopted in the United States, races were measured in yards rather than meters. After standards were put in place, many race distances changed.

(a) When international standards were adopted, the 100-yard sprint was replaced by the 100-meter sprint. Which is the shorter race?

(b) The 440-yard "quarter mile" was replaced by a race that is 400 meters long. Did this race get longer or shorter?

(c) The race that corresponds to the mile is now 1500 meters long. Which is longer, the mile race or the 1500-meter race?

(d) The high hurdles race was 120 yards long before standards were adopted. Now the corresponding race is 110 meters long. Using either the sequence

$$\text{meters} \rightarrow \text{centimeters} \rightarrow \text{inches} \rightarrow \text{yards}$$

or

$$\text{yards} \rightarrow \text{inches} \rightarrow \text{centimeters} \rightarrow \text{meters},$$

determine which distance is longer. Note that the relationship 1 m ≈ 1.09 yd is too inaccurate to determine which race is longer.

6.6 Time and Speed

Objectives

1. Converting Units of Time
2. Converting Units of Speed
3. Solving an Application Involving Speed

NEW VOCABULARY

☐ Second, minute, hour, day, week, month, year
☐ Speed

1 Converting Units of Time

The **second** (sec), **minute** (min), **hour** (hr), **day** (d), **week** (wk), **month** (mo), and **year** (yr) are the units most commonly used for measuring time, and these units are related to each other as follows.

UNITS OF TIME

1 min = 60 sec   1 hr = 60 min   1 d = 24 hr

1 wk = 7 d   1 yr = $365\frac{1}{4}$ d (or 365.25 d)

NOTE: While we often hear that there are 52 weeks in 1 year, the relationship is not exactly accurate because a 52-week period is actually $7 \cdot 52 = 364$ days long, which is not equal to 1 year, or 365.25 days. Also, months should not be used in relationships because the number of days in a month can vary.

EXAMPLE 1 Converting from one unit of time to another

Convert as indicated. Express answers in decimal form when needed.
(a) 3.5 minutes to seconds (b) 102 hours to days (c) 5 days to minutes

Solution

(a) $3.5 \text{ min} = \dfrac{3.5 \text{ min}}{1} \cdot \dfrac{60 \text{ sec}}{1 \text{ min}} = 3.5(60) \text{ sec} = 210 \text{ sec}$

(b) $102 \text{ hr} = \dfrac{102 \text{ hr}}{1} \cdot \dfrac{1 \text{ d}}{24 \text{ hr}} = \dfrac{102}{24} \text{ d} = 4.25 \text{ d}$

(c) $5 \text{ d} = \dfrac{5 \text{ d}}{1} \cdot \dfrac{24 \text{ hr}}{1 \text{ d}} \cdot \dfrac{60 \text{ min}}{1 \text{ hr}} = 5 \cdot 24 \cdot 60 \text{ min} = 7200 \text{ min}$

Now Try Exercises 5, 7, 13

STUDY TIP

Look back at your progress so far this semester. Are there parts of your study process that need some adjustment? Are your notes and assignments organized? Are you spending enough time on homework and practice problems?

READING CHECK 1

Speed is a rate that involves what two quantities?

2 Converting Units of Speed

In the first section of this chapter, a *rate* was defined as a ratio used to compare two different kinds of quantities. If one quantity is distance and the other is time, then the rate represents *speed*. **Speed** is a rate that gives a distance traveled in an amount of time. A unit of speed is expressed as a ratio with a unit of distance in the numerator and a unit of time in the denominator.

🌐 **Math in Context** ~Speed Limit~ One recognizable example of speed in everyday life can be found posted along roads and highways on speed limit signs. In the United States, we measure the speed of a vehicle in miles per hour, or **mi/hr** (also abbreviated as mph). In countries using the metric system, the speed of a vehicle is measured in kilometers per hour, or **km/hr**. Other examples of speed include

$$\text{ft/sec}, \quad \text{m/min}, \quad \text{mi/sec}, \quad \text{and} \quad \text{cm/sec}.$$

It is possible to convert from one unit of speed to another by converting the distance unit, the time unit, or both units. The next example illustrates how to convert from one unit of speed to another using unit analysis.

EXAMPLE 2 Converting units of speed

Convert as indicated.
(a) 120 miles per hour to miles per minute
(b) 66 feet per second to miles per hour

READING CHECK 2

Which of the following speeds are equal to 60 miles per hour?

- $88 \frac{\text{ft}}{\text{sec}}$
- $180 \frac{\text{yd}}{\text{hr}}$
- $1 \frac{\text{mi}}{\text{min}}$
- $800 \frac{\text{in}}{\text{sec}}$

Solution

(a) The distance unit does not change, so we only need to convert the time unit from hours to minutes. Since the hours unit is in the denominator of the given speed, we need a unit fraction that relates hours and minutes and has hours in the numerator.

$$120 \text{ mi/hr} = \frac{120 \text{ mi}}{1 \text{ hr}} \cdot \frac{1 \text{ hr}}{60 \text{ min}} = \frac{120 \text{ mi}}{60 \text{ min}} = \frac{2 \text{ mi}}{1 \text{ min}} = 2 \text{ mi/min}$$

(b) The distance unit must be converted from **feet to miles**, and the time unit must be converted from **seconds to hours**.

$$66 \text{ ft/sec} = \frac{66 \text{ ft}}{1 \text{ sec}} \cdot \frac{1 \text{ mi}}{5280 \text{ ft}} \cdot \frac{60 \text{ sec}}{1 \text{ min}} \cdot \frac{60 \text{ min}}{1 \text{ hr}} = \frac{66 \cdot 60 \cdot 60}{5280} \text{ mi/hr} = 45 \text{ mi/hr}$$

Now Try Exercises 17, 25

3 ▸ Solving an Application Involving Speed

Sometimes, speed is more easily interpreted when the units change. The next example illustrates how to convert from one unit of speed to another.

EXAMPLE 3 Analyzing speed limit

After crossing the border into Canada, an American driver uses cruise control set at 60 miles per hour. If the posted speed limit on the Canadian road is 90 kilometers per hour, is the American driving over the speed limit?

Solution
To determine if the driver is speeding, convert 60 miles per hour to kilometers per hour.

$$60 \text{ mi/hr} \approx \frac{60 \text{ mi}}{1 \text{ hr}} \cdot \frac{1 \text{ km}}{0.62 \text{ mi}} = \frac{60}{0.62} \text{ km/hr} \approx 96.8 \text{ km/hr}$$

The American is driving nearly 97 kilometers per hour, which is over the speed limit.

Now Try Exercise 41

6.6 Putting It All Together

CONCEPT	COMMENTS	EXAMPLES
Units of Time	Second (sec), minute (min), hour (hr), day (d), week (wk), month (mo), and year (yr) **Relationships:** 1 min = 60 sec, 1 hr = 60 min, 1 d = 24 hr, 1 wk = 7 d, 1 yr = $365\frac{1}{4}$ d (or 365.25 d)	Convert 3.5 days to hours. $$\frac{3.5 \text{ d}}{1} \cdot \frac{24 \text{ hr}}{1 \text{ d}} = 84 \text{ hr}$$
Units of Speed	A unit of speed is expressed as a ratio with a unit of distance in the numerator and a unit of time in the denominator.	Convert 10,560 feet per hour to miles per hour. $$\frac{10{,}560 \text{ ft}}{1 \text{ hr}} \cdot \frac{1 \text{ mi}}{5280 \text{ ft}} = 2 \text{ mi/hr}$$

6.6 Exercises

CONCEPTS AND VOCABULARY

1. Units of time include seconds, minutes, hours, _____, weeks, months, and years.

2. There are _____ seconds in one minute, and _____ minutes in one hour.

3. A rate that gives distance traveled in an amount of time is called _____.

4. A unit of speed has a unit of _____ in its numerator and a unit of _____ in its denominator.

CONVERTING UNITS OF TIME

Exercises 5–16: Convert as indicated. Express answers in decimal form when needed.

5. 5.25 hours to minutes
6. 3 days to hours
7. 90 minutes to hours
8. 63 days to weeks
9. 4.1 hours to seconds
10. 2 years to days
11. 2.5 weeks to hours
12. 0.25 day to minutes
13. 99,360 seconds to days
14. 19,800 seconds to hours
15. 4383 hours to years
16. 252 hours to weeks

CONVERTING UNITS OF SPEED

Exercises 17–32: Convert as indicated. Express answers in decimal form when needed.

17. 34 yards per second to feet per second
18. 4 kilometers per hour to meters per hour
19. 1200 centimeters per day to meters per day
20. 18 inches per week to feet per week
21. 12 meters per min to meters per second
22. 144 miles per hour to miles per min
23. 3.2 inches per day to inches per week
24. 4 millimeters per hour to millimeters per day
25. 30 miles per hour to feet per second
26. 50 feet per day to inches per hour
27. 7 kilometers per minute to meters per hour
28. 42 meters per hour to centimeters per minute
29. 200 feet per minute to yards per hour
30. 110 feet per second to miles per hour
31. 3 meters per second to kilometers per hour
32. 95 millimeters per minute to meters per hour

Exercises 33–36: Convert as indicated. Round answers to the nearest whole number.

33. 11 meters per second to feet per second
34. 70 miles per hour to kilometers per hour
35. 92 kilometers per hour to miles per hour
36. 190 feet per second to meters per second

APPLICATIONS INVOLVING SPEED

37. **NASCAR Speeds** Some NASCAR drivers reach average speeds of 3 miles per minute. Convert this speed to miles per hour.

38. **NASCAR Speeds** On slower speedways, NASCAR drivers reach average speeds of 120 miles per hour. Convert this speed to miles per minute.

39. **Animal Speeds** With top speeds of about 103 feet per second, the cheetah is the fastest land animal. To the nearest whole number, what is the top speed of a cheetah in miles per hour?

40. **Human Speeds** Usain Bolt set a world record in the 100-meter sprint in 2012 by running an average speed of 34.2 feet per second. To the nearest whole number, what was Bolt's average speed in miles per hour?

41. **Speed Limits** An American drives 30 miles per hour after crossing the border into a Canadian town. If the posted speed limit is 50 kilometers per hour, is the driver exceeding the speed limit?

42. Speed Limits A Canadian drives 75 kilometers per hour after crossing the border into an American town. If the posted speed limit is 45 miles per hour, is the driver exceeding the speed limit?

WRITING ABOUT MATHEMATICS

43. A student writes the following relationship:

$$1 \text{ mo} \stackrel{?}{=} 4 \text{ wk.}$$

Explain why conversions made using this relationship will probably be inaccurate.

44. Without doing a conversion, explain how you would know that a vehicle traveling at 30 miles per hour is moving faster than a vehicle traveling at 30 kilometers per hour.

SECTIONS 6.5 and 6.6 — Checking Basic Concepts

1. Convert as indicated. Answers may vary.
   - (a) 250 inches to meters
   - (b) 0.5 kilometer to feet
   - (c) 5 cups to milliliters
   - (d) 1.5 liters to pints
   - (e) 1250 grams to pounds
   - (f) 704 ounces to kilograms

2. Convert each temperature as indicated.
   - (a) $-5\,°\text{C}$ to Fahrenheit
   - (b) $158\,°\text{F}$ to Celsius

3. Convert 54 meters per minute to centimeters per second.

4. Convert 3.5 yards per week to inches per day.

5. **Human Body** The average adult human brain weighs about 1300 grams. Give the weight of the average adult human brain to the nearest tenth of a pound. (*Source: Brain Facts and Figures.*)

6. **Speed Limits** A Canadian drives 80 kilometers per hour in an American town. If the posted speed limit is 50 miles per hour, is the driver going over the speed limit?

CHAPTER 6 Summary

SECTION 6.1 ■ RATIOS AND RATES

Ratio

Writing Ratios as Fractions

To write a ratio as a fraction, write the quantity found before the word "to" (or before the colon) in the numerator of the fraction, and write the quantity found after the word "to" (or after the colon) in the denominator of the fraction. Then simplify the fraction.

Examples: 3 to 9, 3 : 9, or $\frac{3}{9} = \frac{1}{3}$

Unit Ratio

A unit ratio is a ratio expressed as a fraction with a denominator of 1. To find a unit ratio, divide the denominator into the numerator.

Example: If the student-to-instructor ratio at a college is 420 to 15, the unit ratio is $\frac{420}{15} = 28$, or 28 students for each instructor.

Rate

A rate is a ratio used to compare different kinds of quantities. The units for each quantity are expressed as part of a ratio.

Example: A person spending $23 in 2 hours spends at a rate of $\frac{23 \text{ dollars}}{2 \text{ hours}}$.

Unit Rate

A unit rate is a rate expressed as a fraction with denominator 1. To find a unit rate, divide the denominator into the numerator.

Example: A person earning $90.75 in 11 hours has an hourly pay rate of

$$\frac{90.75 \text{ dollars}}{11 \text{ hours}} = \$8.25/\text{hr.}$$

Unit Pricing	If the price of q units of a product is p, the unit price U is given by $U = \frac{p}{q}$. **Example:** A 20-ounce bottle of soda that sells for $1.60 has a unit price of $$U = \frac{p}{q} = \frac{\$1.60}{20 \text{ oz}} = \$0.08/\text{oz}.$$

SECTION 6.2 ■ PROPORTIONS AND SIMILAR FIGURES

Proportions	If $\frac{a}{b}$ and $\frac{c}{d}$ are ratios that are equal in value, then $\frac{a}{b} = \frac{c}{d}$ is a proportion, where $b \neq 0$ and $d \neq 0$. This proportion is read "a is to b as c is to d." **Example:** The proportion $\frac{4}{5} = \frac{16}{20}$ is read "4 is to 5 as 16 is to 20."
Cross Product Rule	For $b \neq 0$ and $d \neq 0$, if the cross products ad and bc are equal, then $\frac{a}{b} = \frac{c}{d}$ is a proportion. **Example:** $\frac{7}{8} = \frac{14}{16}$ is a proportion because $7 \cdot 16 = 8 \cdot 14$.
Similar Figures	When two geometric figures are similar, the measures of the corresponding sides are proportional. **Example:** 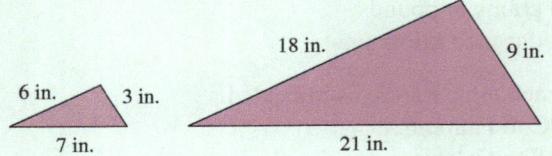

SECTION 6.3 ■ THE U.S. SYSTEM OF MEASUREMENT

U.S. System

U.S. Units of Length	Inches, feet, yards, and miles **Relationships:** 1 ft = 12 in., 1 yd = 36 in., 1 yd = 3 ft, 1 mi = 1760 yd, and 1 mi = 5280 ft **Example:** Convert 48 inches to feet. $48 \text{ in.} = \frac{48 \text{ in.}}{1} \cdot \frac{1 \text{ ft}}{12 \text{ in.}} = \frac{48}{12} \text{ ft} = 4 \text{ ft}$
Unit Fraction	A unit fraction is a fraction that is equivalent to 1. **Examples:** $\frac{12 \text{ in.}}{1 \text{ ft}}, \frac{1 \text{ lb}}{16 \text{ oz}},$ and $\frac{2 \text{ c}}{1 \text{ pt}}$
U.S. Units of Area	Units of area are the squares of units of length. To convert one unit of area to another, use the corresponding unit fraction relating units of length, twice. **Example:** Convert 3 square feet to square inches. $$3 \text{ ft}^2 = \frac{3 \text{ ft} \cdot \text{ft}}{1} \cdot \frac{12 \text{ in.}}{1 \text{ ft}} \cdot \frac{12 \text{ in.}}{1 \text{ ft}} = 432 \text{ in}^2$$
U.S. Units of Capacity and Volume	Ounce, cup, quart, and gallon **Relationships:** 1 c = 8 oz, 1 pt = 2 c, 1 qt = 2 pt, and 1 gal = 4 qt **Example:** Convert 6 cups to pints. $6 \text{ c} = \frac{6 \text{ c}}{1} \cdot \frac{1 \text{ pt}}{2 \text{ c}} = \frac{6}{2} \text{ pt} = 3 \text{ pt}$
U.S. Units of Weight	Ounce, pound, and ton **Relationships:** 1 lb = 16 oz, and 1 T = 2000 lb **Example:** Convert 3.5 tons to pounds. $3.5 \text{ T} = \frac{3.5 \text{ T}}{1} \cdot \frac{2000 \text{ lb}}{1 \text{ T}} = 7000 \text{ lb}$

SECTION 6.4 ■ THE METRIC SYSTEM OF MEASUREMENT

Metric System

Metric Prefixes

Kilo, hecto, deka, deci, centi, and milli
Each prefix designates the size of a measurement relative to the base unit.

Examples: *Deka* means "10 times the base" and *centi* means "$\frac{1}{100}$ of the base."

Metric Units of Length

The base metric unit of length is the *meter*.

Relationships: 1 km = 1000 m,   1 hm = 100 m,   1 dam = 10 m,
1 m = 10 dm,   1 m = 100 cm,   and   1 m = 1000 mm

Example: Convert 2.2 meters to millimeters.

km   hm   dam   m   dm   cm   mm

2.2 m = 2.200 m = 2200 mm

Metric Units of Area

Units of area are the squares of units of length. To convert one unit of area to another, move the decimal point *twice* the number of places needed to convert the corresponding units of length.

Example: Convert 1400 square meters to square hectometers.

km   hm   dam   m   dm   cm   mm

1400. m² = 0.14 hm²

Metric Units of Capacity and Volume

The base metric unit of capacity and volume is the *liter*.

Relationships: 1 kl = 1000 L,   1 hl = 100 L,   1 dal = 10 L,
1 L = 10 dl,   1 L = 100 cl,   and   1 L = 1000 mL

Example: Convert 4.65 deciliters to milliliters.

kl   hl   dal   L   dl   cl   mL

4.65 dl = 465 mL

Metric Units of Mass

The base metric unit of mass is the *gram*.

Relationships: 1 kg = 1000 g,   1 hg = 100 g,   1 dag = 10 g,
1 g = 10 dg,   1 g = 100 cg,   and   1 g = 1000 mg

Example: Convert 0.5 centigram to grams.

kg   hg   dag   g   dg   cg   mg

0.5 cg = 000.5 cg = 0.005 g

SECTION 6.5 ■ U.S.–METRIC CONVERSIONS; TEMPERATURE

Conversions Between Systems

Units of Length

For converting units of length, use the relationships

1 in. = 2.54 cm,   1 m ≈ 1.09 yd,   and   1 km ≈ 0.62 mi.

Example: Convert 7 meters to yards. $7 \text{ m} \approx \frac{7 \text{ m}}{1} \cdot \frac{1.09 \text{ yd}}{1 \text{ m}} = 7.63 \text{ yd}$

Units of Capacity and Volume

For converting units of capacity and volume, use the relationships

1 L ≈ 1.06 qt   and   1 oz ≈ 29.57 mL.

Example: Convert 5 liters to quarts. $5 \text{ L} \approx \frac{5 \text{ L}}{1} \cdot \frac{1.06 \text{ qt}}{1 \text{ L}} = 5.3 \text{ qt}$

428 CHAPTER 6 RATIOS, PROPORTIONS, AND MEASUREMENT

Units of Mass (Weight)	For converting units of mass (weight), use the relationships $1 \text{ kg} \approx 2.2 \text{ lb}$ and $1 \text{ oz} \approx 28.35 \text{ g}$. **Example:** Convert 8.8 pounds to kilograms. $8.8 \text{ lb} \approx \frac{8.8 \text{ lb}}{1} \cdot \frac{1 \text{ kg}}{2.2 \text{ lb}} = 4 \text{ kg}$
Temperature	In the U.S. system, temperature is measured in degrees Fahrenheit, and in the metric system, it is measured in degrees Celsius. To convert from one system to the other, use the formulas $F = \frac{9}{5}C + 32$ and $C = \frac{5}{9}(F - 32)$. Note that temperature conversions can also be completed numerically, using a table of values, and graphically, using a graph. **Example:** Convert 86 °F to Celsius.

Symbolic:

$C = \frac{5}{9}(86 - 32)$

$= \frac{5}{9}(54)$

$= 30 \text{ °C}$

Graphical:

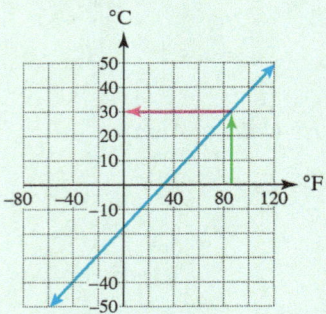

Numerical:

°F	85.9	**86.0**	86.1	86.2
°C	29.9	**30.0**	30.1	30.1

SECTION 6.6 ■ TIME AND SPEED

Units of Time	Second (sec), minute (min), hour (hr), day (d), week (wk), month (mo), and year (yr) **Relationships:** $1 \text{ min} = 60 \text{ sec}$, $1 \text{ hr} = 60 \text{ min}$, $1 \text{ d} = 24 \text{ hr}$, $1 \text{ wk} = 7 \text{ d}$, $1 \text{ yr} = 365\frac{1}{4} \text{ d}$ (or 365.25 d) **Example:** Convert 105 minutes to hours. $105 \text{ min} = \frac{105 \text{ min}}{1} \cdot \frac{1 \text{ hr}}{60 \text{ min}} = 1.75 \text{ hr}$
Units of Speed	A unit of speed is expressed as a ratio with a distance in the numerator and time in the denominator. **Examples:** Convert 450 feet per minute to yards per minute. $450 \text{ ft/min} = \frac{450 \text{ ft}}{1 \text{ min}} \cdot \frac{1 \text{ yd}}{3 \text{ ft}} = 150 \text{ yd/min}$

CHAPTER 6   Review Exercises

SECTION 6.1

Exercises 1–4: Write the given ratio as a fraction in simplest form.

1. 6 to 24
2. 1.2 to 3.3
3. $\frac{15}{8} : \frac{45}{4}$
4. $1\frac{1}{5} : 3\frac{3}{4}$

Exercises 5 and 6: Find the unit ratio.

5. 12 to 15
6. 140 : 30

Exercises 7 and 8: Write the given rate as a fraction in simplest form.

7. An athlete runs 12 miles in 64 minutes.
8. There are 30 laptops for 120 students.

Exercises 9 and 10: Write the given rate as a unit rate.

9. A person earns $58.80 in 8 hours.
10. A heart beats 296 times in 4 minutes.

Exercises 11 and 12: Find the unit price.

11. A 20-pound block of ice for $2.86

12. A 2-liter bottle of store-brand soda for $0.98

SECTION 6.2

Exercises 13–16: Determine if the given equation is a proportion.

13. $\dfrac{10}{80} \stackrel{?}{=} \dfrac{2}{16}$

14. $\dfrac{\frac{1}{4}}{18} \stackrel{?}{=} \dfrac{\frac{1}{6}}{24}$

15. $\dfrac{2}{8.5} \stackrel{?}{=} \dfrac{11}{48.5}$

16. $\dfrac{8\frac{1}{2}}{2\frac{5}{6}} \stackrel{?}{=} \dfrac{5\frac{1}{10}}{1\frac{7}{10}}$

Exercises 17–20: Solve the proportion.

17. $\dfrac{24}{x} = \dfrac{-3}{2}$

18. $\dfrac{1.7}{b} = \dfrac{-5.1}{3}$

19. $\dfrac{\frac{3}{10}}{-\frac{2}{5}} = \dfrac{y}{-4}$

20. $\dfrac{1\frac{4}{5}}{-9} = \dfrac{a}{5\frac{5}{6}}$

Exercises 21 and 22: For the similar figures, find the measure of x.

21.
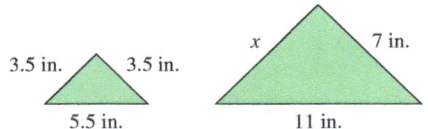

22.
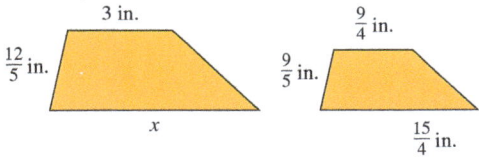

SECTION 6.3

Exercises 23–26: Convert the length as indicated.

23. 12.5 feet to inches

24. 1.8 miles to feet

25. 38,016 inches to miles

26. 144 inches to yards

Exercises 27 and 28: Convert the area as indicated.

27. 270 square feet to square yards

28. 0.04 square mile to square feet

Exercises 29–32: Convert the volume as indicated.

29. 14 gallons to quarts

30. 112 ounces to quarts

31. $8\frac{1}{2}$ pints to cups

32. 2.75 gallons to cups

Exercises 33–36: Convert the weight as indicated.

33. 7.5 pounds to ounces

34. 392 ounces to pounds

35. $\dfrac{1}{250}$ ton to pounds

36. 1,312,000 ounces to tons

SECTION 6.4

Exercises 37 and 38: Write the name of the metric unit of measurement described.

37. $\dfrac{1}{1000}$ of a gram

38. 10 times a liter

Exercises 39–42: Convert the length as indicated.

39. 0.75 meter to centimeters

40. 11.4 decimeters to dekameters

41. 12.6 kilometers to hectometers

42. 0.0019 hectometer to decimeters

Exercises 43 and 44: Convert the area as indicated.

43. 780 square decimeters to square meters

44. 0.00085 square kilometer to square meters

Exercises 45–48: Convert the volume as indicated.

45. 52,530 dekaliters to kiloliters

46. 0.075 deciliter to milliliters

47. 7690 centiliters to dekaliters

48. 0.1 milliliter to deciliters

Exercises 49–52: Convert the mass as indicated.

49. 0.087 dekagram to decigrams

50. 28,000 grams to kilograms

51. 0.00077 kilogram to centigrams

52. 0.021 hectogram to grams

SECTION 6.5

Exercises 53 and 54: Convert the specified length using the given sequences. Round answers to 2 decimal places.

53. 7850 yards to kilometers
    (a) Yards → meters → kilometers
    (b) Yards → miles → kilometers

430 CHAPTER 6 RATIOS, PROPORTIONS, AND MEASUREMENT

54. 2500 meters to miles
    (a) Meters → kilometers → miles
    (b) Meters → yards → miles

Exercises 55–58: Convert the length as indicated. Round answers to 2 decimal places. Answers may vary slightly.

55. 220 inches to meters

56. 0.25 kilometer to yards

57. 36,960 feet to kilometers

58. 9 feet to centimeters

Exercises 59–62: Convert the capacity or volume as indicated. Round answers to 2 decimal places. Answers may vary slightly.

59. 100 liters to gallons

60. 0.2 pint to milliliters

61. 10.25 gallons to liters

62. 39,250 milliliters to gallons

Exercises 63–66: Convert the given mass or weight as indicated. Round answers to 2 decimal places. Answers may vary slightly.

63. 6514 grams to pounds

64. 0.44 pound to milligrams

65. 3780 milligrams to ounces

66. 1.05 pounds to grams

Exercises 67 and 68: Convert the temperature symbolically. Give exact answers in decimal form.

67. 9 °C to Fahrenheit

68. −13 °F to Celsius

Exercises 69 and 70: Use the following table to convert the given temperature numerically. If there are two choices, use the higher value.

°F	98.2	98.3	98.4	98.5	98.6	98.7
°C	36.8	36.8	36.9	36.9	37.0	37.1

69. 98.3 °F to Celsius

70. 36.9 °C to Fahrenheit

Exercises 71 and 72: Use the following graph to convert the given temperature graphically. Answers may vary slightly.

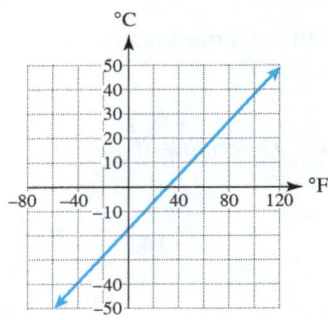

71. −30 °F to Celsius

72. 25 °C to Fahrenheit

SECTION 6.6

Exercises 73–76: Convert as indicated. Express answers in decimal form when needed.

73. 3 days to hours

74. 0.4 day to minutes

75. 16,920 seconds to hours

76. 420 hours to weeks

Exercises 77–80: Convert as indicated. Express answers in decimal form when appropriate.

77. 7.5 kilometers per hour to meters per hour

78. 30 inches per week to feet per week

79. 80 feet per day to inches per hour

80. 8 kilometers per minute to meters per hour

Exercises 81 and 82: Convert as indicated. Round your answers to the nearest whole number.

81. 75 miles per hour to kilometers per hour

82. 120 meters per second to feet per second

APPLICATIONS

83. **Comparing Colleges** A big university has a student-to-instructor ratio of 1548 to 30, while a small college has a student-to-instructor ratio of 764 to 20. Find the unit ratio at each school. Interpret the results for the small college.

84. **Going Green** A survey determined that 60 of 360 students take public transportation to school each day. How many of every 48 students take public transportation to school each day?

85. **Baling Twine** A farmer uses two 80-inch pieces of twine to tie up a bale of hay. If the twine is available on a spool containing 4500 feet of twine, how many bales can be tied up with one spool? How many inches of twine are left over?

86. **Serving Size** If the label on a 4-liter jug of sports drink states that the serving size is 200 milliliters, how many servings are in the jug?

87. **Obesity** An adult male who is 6 feet 2 inches tall is considered obese if his weight is above 106 kilograms. What is this weight to the nearest pound? (*Source:* Centers for Disease Control.)

88. **Speed Limits** An American drives 35 miles per hour after crossing the border into a Mexican town. If the posted speed limit is 55 kilometers per hour, is the driver going over the speed limit?

89. **Tree Height** A tree casts a 54-foot shadow, while a nearby tree that is 13 feet tall casts a 6-foot shadow. See the figure at the top of the next column. How tall is the larger tree?

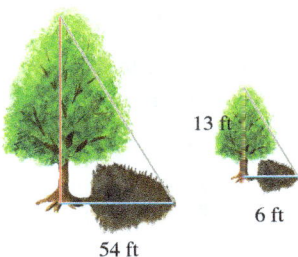

90. **Unit Price** Find the unit price for each size option.

    Large coffee drink: 24 ounces for $4.68

    Small coffee drink: 16 ounces for $3.28

    Which is the better buy?

91. **Checked Luggage** Some airlines charge a fee for each checked luggage item that weighs more than 22.7 kilograms. How many grams is this?

92. **Fruit Drink** A large jug contains 145 ounces of fruit drink. Is this more or less than a gallon?

93. **Warm Day** Temperature in Death Valley can sometimes reach 122 °F. Find an equivalent temperature on the Celsius scale.

94. **Liquid Oxygen** Oxygen will liquefy when cooled to about -183 °C. Find an equivalent temperature on the Fahrenheit scale.

CHAPTER 6 Test

1. Write the ratio 6 : 18 as a fraction in simplest form.

2. An athlete runs 8 miles in 64 minutes. Write this rate as a unit rate.

3. A 16-ounce box of crackers sells for $3.44. Find the unit price of the crackers.

4. Is $\frac{12}{15} \stackrel{?}{=} \frac{20}{24}$ a proportion?

Exercises 5 and 6: Solve the proportion.

5. $\dfrac{-18}{x} = \dfrac{-2}{3}$

6. $\dfrac{\frac{3}{10}}{7} = \dfrac{w}{-\frac{5}{9}}$

7. The triangles are similar. Find the value of x.

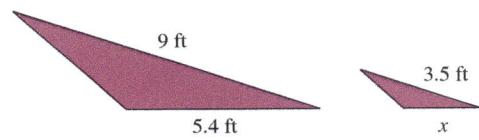

Exercises 8–15: Convert the measurement as indicated.

8. 6.5 yards to inches

9. 540 square inches to square feet

10. $8\frac{1}{2}$ gallons to pints

11. 35,000 pounds to tons

12. 1.54 centimeters to meters

13. 540,000 square meters to square kilometers

14. 0.075 liter to milliliters

15. 2400 decigrams to hectograms

Exercises 16–18: Convert as indicated. Round answers to 2 decimal places. Answers may vary slightly.

16. 350 yards to kilometers

17. 430 milliliters to cups

18. 2.5 pounds to grams

19. Use a formula to convert 15 °C to Fahrenheit.

20. Use the following graph to convert 70 °F to Celsius. Answers may vary slightly.

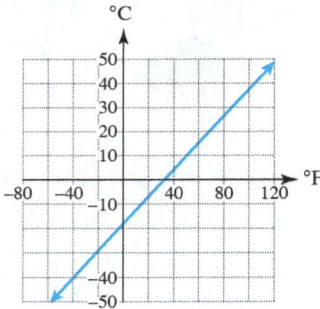

Exercises 21 and 22: Convert as indicated. Express answers in decimal form when appropriate.

21. 41,040 minutes to days

22. 3 days to seconds

Exercises 23 and 24: Convert as indicated. Round your answers to the nearest whole number.

23. 1 inch per second to feet per hour

24. 2 miles per day to meters per day

25. **Unit Price** Find the unit price for each size option.

    Large fruit smoothie: 24 ounces for $3.12

    Small fruit smoothie: 10 ounces for $1.48

    Which is the better buy?

26. **Serving Size** If the label on a 2.2-quart bottle of milk states that the serving size is 250 milliliters, how many full servings are in the bottle?

27. **Temperature** If it was 78 °F on Monday and 27 °C on Tuesday, which day was warmer?

CHAPTERS 1–6   Cumulative Review Exercises

1. Write the whole number 34,206 in word form.

2. Are the terms $5ab$, $3xy$ like or unlike?

3. Graph the whole numbers 2 and 5 on a number line.

4. Round 730,187 to the nearest ten-thousand.

5. Place the correct symbol, $<$ or $>$, in the blank between the integers: -17 _____ -9.

6. Evaluate the expression $x - y$ for the given values of the variables: $x = -5, y = 11$.

7. Find the product: $2(-1)(3)(-2)(-5)$.

8. Solve the equation $-3x + 28 = 25$ graphically.

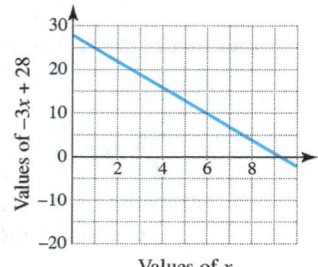

9. Translate the phrase to an algebraic expression using the variable x: *the product of 4 and her age*.

10. Solve symbolically: $-(2q + 3) + 7q = 17$.

11. Solve the linear equation numerically by completing the given table of values: $2x - 5 = -1$.

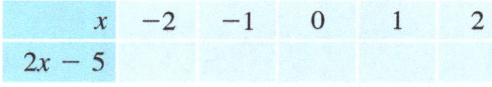

12. When the sum of 6 and a number is divided by 3 the result is 5. Find the number.

13. Write the improper fraction and mixed number represented by the given shading.

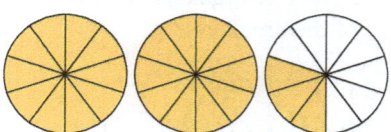

14. Use a divisibility test to answer each question.
    (a) Is 942 divisible by 3?
    (b) Is 4766 divisible by 4?

15. Add and simplify your result, if possible.

    $$-\frac{11}{18} + \frac{5}{6}$$

16. Solve the equation. Simplify your result.

    $$5\left(n - \frac{1}{6}\right) = \frac{8}{3}$$

17. Round 594.032 to the nearest hundredth

18. Solve symbolically: $3(p - 5) = 2.7p + 4.2$.

19. Find the exact area of the circle and then approximate the area using 3.14 as an approximation for π.

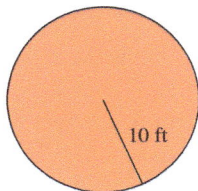

20. Calculate the GPA for the grade report.

Course	Credits	Grade
Chemistry	4	C
English	3	B
Study Skills	1	B
Biology	4	A

21. If a car travels 174 miles in 3 hours, write its speed as a unit rate.

22. Solve the proportion.
$$\frac{15}{x} = \frac{-3}{4}$$

23. **Heart Rate** The average heart rate R, in beats per minute (bpm), of a person weighing W kilograms can be approximated by
$$R = \frac{400\sqrt{W}}{W}.$$
Find the heart rate for a 25-kilogram child.

24. **Music Sales** The total music sales S, in thousands of dollars, for a small band during year x can be found using the formula
$$S = 7(x - 2005) + 8.$$
Find the year when music sales reached $43,000 by solving the equation $43 = 7(x - 2005) + 8$ graphically using the graph provided.

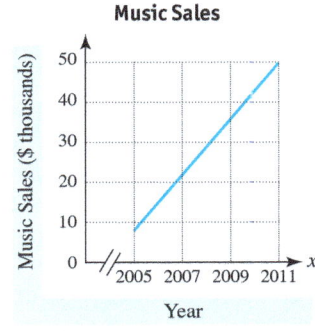

25. **Triangle Dimensions** The longest side of a triangle measures 1 inch less than triple the measure of the shortest side. The measure of the third side is 2 inches more than double the measure of the shortest side. If the perimeter of the triangle is 25 inches, find the measure of the shortest side of the triangle.

26. **Height** At birth, a baby boy was $21\frac{1}{4}$ inches long. If he grew $8\frac{1}{2}$ inches during his first year, how tall was he on his first birthday?

27. **International Calling** Suppose that the cost C of x minutes of international phone calls is given by
$$C = 0.05(x - 500) + 10.5, \text{ where } x \geq 500.$$
Find the number of minutes that correspond to total charges of $22.45 by replacing C in the formula with 22.45 and solving the resulting equation.

28. **Wilderness Maps** Every $1\frac{1}{2}$ inches on a wilderness map represents 1 mile. If the actual distance between two lakes is 18 miles, how far apart are the lakes on the map?

7 Percent and Probability

- 7.1 Introduction to Percent; Circle Graphs
- 7.2 Using Equations to Solve Percent Problems
- 7.3 Using Proportions to Solve Percent Problems
- 7.4 Applications: Sales Tax, Discounts, and Net Pay
- 7.5 Applications: Simple and Compound Interest
- 7.6 Probability and Percent Chance

Health professionals know the importance of proper pacing during exercise. Since fitness levels vary from person to person, no single exercise pace is right for everyone. Regardless of fitness level, maintaining a *target heart rate* during exercise will help maximize the benefits of physical activity.

Target heart rate is a range between 50 and 85 *percent* of a person's maximum heart rate. The proper pace for exercise occurs within this range. Because maximum heart rate depends on age, people can determine their target heart rates before beginning an exercise program. You can find your own target heart rate if you understand percents.

In this chapter, we learn how to solve problems involving percents, which occur in real-world situations such as exercise science, sales tax, interest rates, and exam scores.

Source: The American Heart Association.

7.1 Introduction to Percent; Circle Graphs

Objectives

1. Understanding Percent Notation
2. Writing a Percent as a Fraction or Decimal
3. Writing a Fraction or Decimal as a Percent
4. Reading Circle Graphs
5. Solving Percent Applications

NEW VOCABULARY

☐ Percent
☐ Circle graph

READING CHECK 1

What does percent mean?

1 Understanding Percent Notation

Connecting Concepts with Your Life When you buy an item in a store, there is usually an additional cost due to sales tax. For example, if the purchase price for a jacket is $100, then there may be a 7 *percent*, or 7%, sales tax. This means that there is a $7 sales tax for every $100 spent. The total amount to buy the jacket is $107.

We can remember the meaning of the word *percent* by breaking it into the words *per* and *cent*. Recall that *per* is associated with division and means "divide by." The word *cent* comes from the Latin word *centum*, or 100. (There are 100 *cents* in $1.) So, the word **percent** means "divide by 100" or "out of 100." The symbol for percent is %.

The following See the Concept shows a visualization of the meaning of 67%.

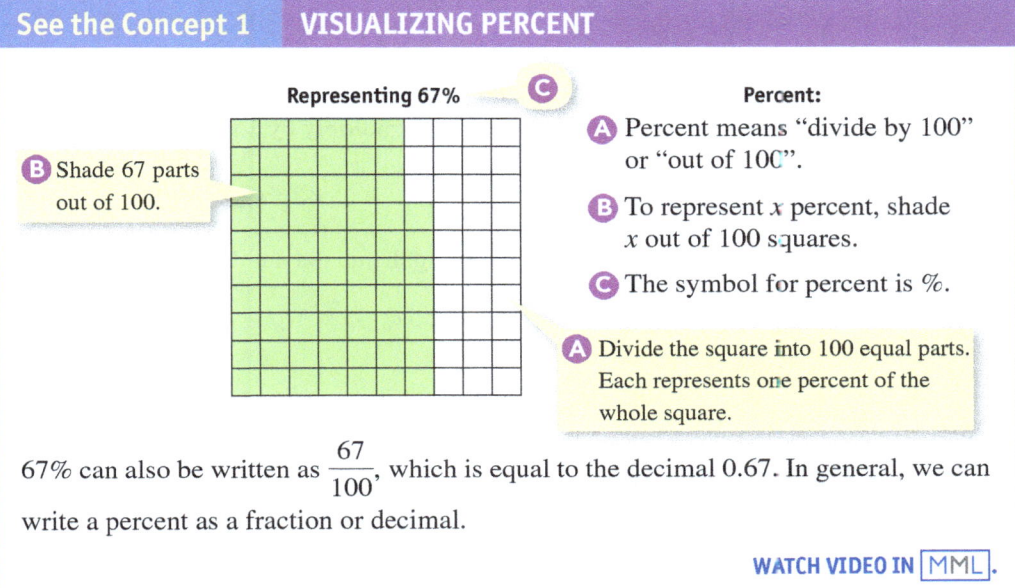

See the Concept 1   **VISUALIZING PERCENT**

Representing 67%

B Shade 67 parts out of 100.

Percent:

A Percent means "divide by 100" or "out of 100".

B To represent x percent, shade x out of 100 squares.

C The symbol for percent is %.

A Divide the square into 100 equal parts. Each represents one percent of the whole square.

67% can also be written as $\frac{67}{100}$, which is equal to the decimal 0.67. In general, we can write a percent as a fraction or decimal.

WATCH VIDEO IN MML.

STUDY TIP

If you have not been studying with other students, consider getting together with classmates so that you can work on your math assignments together.

2 Writing a Percent as a Fraction or Decimal

Recall that dividing by a number is the same as multiplying by its reciprocal. So, dividing by 100 is the same as multiplying by $\frac{1}{100}$ or 0.01. Mathematically, we can write:

the expression $x\%$ represents the fraction $\frac{x}{100}$ or the decimal $0.01x$.

We use the following procedure to write a percent as a fraction or decimal.

> **WRITING A PERCENT AS A FRACTION OR DECIMAL**
>
> To write $x\%$ as a fraction, write $\frac{x}{100}$. Simplify the fraction, if needed.
> To write $x\%$ as a decimal, remove the % symbol and then multiply $0.01x$.

READING CHECK 2

Name two ways that we can write percents.

NOTE: We can write $x\%$ as a decimal by moving the decimal point in the number x two places to the *left* and removing the % symbol. ∎

EXAMPLE 1 Writing percents as fractions

Write each percent as a fraction or mixed number in simplest form.
(a) 24%   (b) 130%

Solution

(a) $24\% = \dfrac{24}{100} = \dfrac{6 \cdot 4}{25 \cdot 4} = \dfrac{6}{25}$ ⟵ Basic principle of fractions

(b) $130\% = \dfrac{130}{100} = \dfrac{13 \cdot 10}{10 \cdot 10} = \dfrac{13}{10}$ or $1\dfrac{3}{10}$

Now Try Exercises 13, 19

EXAMPLE 2 Writing percents as fractions

Write each percent as a fraction in simplest form.
(a) 8.5%   (b) $66\frac{2}{3}\%$

Solution

(a) The fraction $\frac{8.5}{100}$ is not in simplest form because its numerator is a decimal number. To clear the decimal, multiply by 1 in the form $\frac{2}{2}$.

$$8.5\% = \dfrac{8.5}{100} = \dfrac{8.5}{100} \cdot \dfrac{2}{2} = \dfrac{17}{200}$$

(b) Begin by writing the mixed number $66\frac{2}{3}$ as the improper fraction $\frac{200}{3}$.

$$66\tfrac{2}{3}\% = \dfrac{200}{3}\% = \dfrac{\frac{200}{3}}{100} = \dfrac{200}{3} \cdot \dfrac{1}{100} = \dfrac{2 \cdot 100}{3 \cdot 100} = \dfrac{2}{3}$$

Now Try Exercises 17, 23

EXAMPLE 3 Writing percents as decimals

Write each percent as a decimal.
(a) 38%   (b) 119%

Solution   ⟵ The decimal point in 38 moves two places to the left.

(a) $38\% = 0.01(38) = 0.38$
(b) $119\% = 0.01(119) = 1.19$

Now Try Exercises 25, 29

EXAMPLE 4 Writing percents as decimals

Write each percent as a decimal.
(a) 2.9%   (b) $45\frac{1}{4}\%$

Solution   ⟵ The decimal point in 2.9 moves two places to the left.

(a) $2.9\% = 0.01(2.9) = 0.029$
(b) Begin by writing the mixed number $45\frac{1}{4}$ as the decimal 45.25.

$$45\tfrac{1}{4}\% = 45.25\% = 0.01(45.25) = 0.4525$$

Now Try Exercises 27, 39

3 Writing a Fraction or Decimal as a Percent

When we write a percent as a fraction or decimal, we *divide* by 100 and remove the % symbol. To reverse the process and write a fraction or decimal as a percent, we *multiply* by 100 and attach a % symbol. In other words, we multiply by 100%. Since

$$100\% = \frac{100}{100} = 1,$$

Percent means divide by 100.

multiplying by 100% is the same as multiplying by 1, and multiplying a number by 1 does not change its value. For example, $0.75 \cdot 100\% = 75\%$, so 0.75 and 75% are equal.

> **WRITING A FRACTION OR DECIMAL AS A PERCENT**
>
> To write a fraction or decimal as a percent, multiply by 100%.

NOTE: We can write the decimal number x as a percent by moving the decimal point in x two places to the *right* and attaching a % symbol. For example,

$$0.631 = 63.1\%. \quad\blacksquare$$

EXAMPLE 5 Writing fractions as percents

Write each fraction or mixed number as a percent.

(a) $\dfrac{2}{5}$ (b) $2\dfrac{3}{4}$ (c) $\dfrac{5}{6}$

Solution

(a) $\dfrac{2}{5} = \dfrac{2}{5} \cdot 100\% = \dfrac{2}{5} \cdot \dfrac{100}{1}\% = \dfrac{200}{5}\% = 40\%$

(b) Begin by writing the mixed number $2\dfrac{3}{4}$ as the improper fraction $\dfrac{11}{4}$.   *4 · 2 + 3 = 11*

$$2\dfrac{3}{4} = \dfrac{11}{4} = \dfrac{11}{4} \cdot 100\% = \dfrac{11}{4} \cdot \dfrac{100}{1}\% = \dfrac{1100}{4}\% = 275\%$$

(c) $\dfrac{5}{6} = \dfrac{5}{6} \cdot 100\% = \dfrac{5}{6} \cdot \dfrac{100}{1}\% = \dfrac{500}{6}\% = 83\dfrac{1}{3}\% = 83.\overline{3}\%$

Now Try Exercises 41, 45, 49

EXAMPLE 6 Writing decimals as percents

Write each decimal as a percent.
(a) 0.37   (b) 0.9   (c) 0.085   (d) 1.77

Solution *The decimal point in 0.37 moves two places to the right.*

(a) $0.37 = 0.37(100\%) = 37\%$
(b) $0.9 = 0.9(100\%) = 90\%$
(c) $0.085 = 0.085(100\%) = 8.5\%$
(d) $1.77 = 1.77(100\%) = 177\%$

Now Try Exercises 55, 57, 61, 63

CRITICAL THINKING 1

If a decimal number is greater than 1 and written as a percent, what can be said about the percent?

4 Reading Circle Graphs

Another way to visualize percents is to shade parts of a circle that has been divided into 100 equal parts. For example, 17 out of 100 parts are shaded in **FIGURE 7.1**. So, the shaded portion represents 17% of the entire circle. This type of graph is called a **circle graph**. Circle graphs used to compare percent data show only the shaded parts of the graph. They do not show all 100 pieces. For example, the circle graph in **FIGURE 7.2** shows the percentage of students using various modes of transportation to get to a particular college.

READING CHECK 3

How many percent are there in a circle graph?

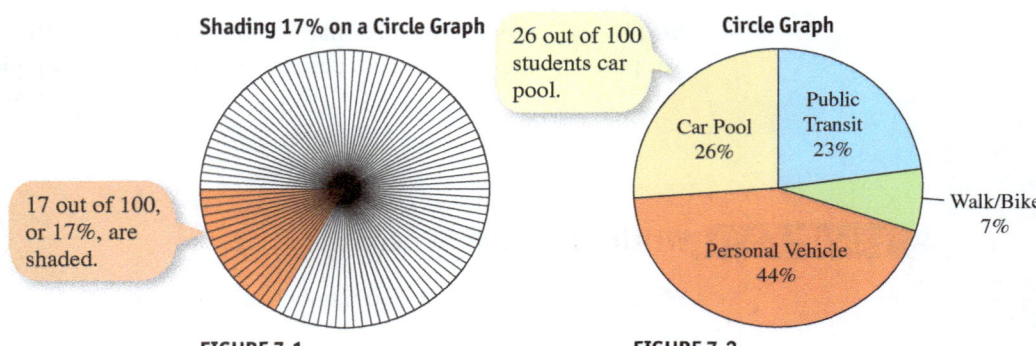

FIGURE 7.1 FIGURE 7.2

EXAMPLE 7 **Reading a circle graph**

FIGURE 7.3 shows the market share for leading Internet search engines during a recent year. (*Source:* Compete.com.)
(a) Which Internet search engine had the largest market share?
(b) Write the market share held by Yahoo! as a fraction in simplest form.
(c) Write the market share held by Other as a decimal.

Solution
(a) The largest market share was held by Google because its shaded region is the largest.
(b) Since $0.8 \cdot 5$ equals a whole number, we can write the market share held by Yahoo! as a fraction by multiplying $\frac{17.8}{100}$ by $\frac{5}{5}$ to clear the decimal.

$$17.8\% = \frac{17.8}{100} = \frac{17.8}{100} \cdot \frac{5}{5} = \frac{89}{500}$$

(c) Written as a decimal, the market share held by Other is $3.5\% = 0.01\,(3.5) = 0.035$.

Now Try Exercise 71

5 Solving Percent Applications

The next two examples illustrate how percents are used regularly in everyday life.

EXAMPLE 8   **Analyzing Internet use**

Of Americans who do not use the Internet, 8% say it is because they are "too old to learn" while 20% are "just not interested". Write each percent as a fraction in simplest form. (*Source: Pew Center for the Internet.*)

Solution

Too old: $\quad 8\% = \dfrac{8}{100} = \dfrac{2 \cdot 4}{25 \cdot 4} = \dfrac{2}{25}$

Not interested: $\quad 20\% = \dfrac{20}{100} = \dfrac{1 \cdot 20}{5 \cdot 20} = \dfrac{1}{5}$

Now Try Exercise 77

EXAMPLE 9   **Analyzing store closings**

In response to the recent economic recession, Starbucks closed $\frac{1}{20}$ of its U.S. stores in 2008. What percent of U.S. Starbucks stores were closed that year? (*Source: Starbucks.*)

Solution

$$\frac{1}{20} = \frac{1}{20} \cdot 100\% = \frac{1}{20} \cdot \frac{100}{1}\% = \frac{100}{20}\% = 5\%$$

Starbucks closed 5% of its U.S. stores in 2008.

Now Try Exercise 75

7.1   Putting It All Together

CONCEPT	COMMENTS	EXAMPLES
Percent	The word *percent* means "divide by 100" or "out of 100."	39% means "39 divided by 100" and is written as 39%, $\frac{39}{100}$, or 0.39.
Writing a Percent as a Fraction or a Decimal	To write $x\%$ as a fraction, write $\frac{x}{100}$ and then simplify the fraction, if needed. To write $x\%$ as a decimal, remove the % symbol and then multiply $0.01x$.	$14\% = \dfrac{14}{100} = \dfrac{7}{50}$ $14\% = 0.01(14) = 0.14$
Writing a Fraction or a Decimal as a Percent	To write a fraction or decimal as a percent, multiply by 100%.	$\dfrac{4}{5} = \dfrac{4}{5} \cdot 100\% = \dfrac{400}{5}\% = 80\%$ $0.64 = 0.64(100\%) = 64\%$

continued on next page

CONCEPT	COMMENTS	EXAMPLES
Circle Graph	A circle graph can be used to display percent data visually.	Employees Who Work Late: Friday 7%, Tuesday 19%, Thursday 11%, Wednesday 22%, Monday 41%

7.1 Exercises   MyMathLab

CONCEPTS AND VOCABULARY

1. The word _____ means "divide by 100."

2. The symbol for percent is _____.

3. To write $x\%$ as a fraction, write _____ and simplify the result, if needed.

4. To write $x\%$ as a decimal, remove the % symbol and then multiply _____.

5. We can write $x\%$ as a decimal by moving the decimal point in the number x two places to the _____ and removing the % symbol.

6. To write a fraction or decimal as a percent, multiply by _____.

7. We can write the decimal number x as a percent by moving the decimal point in the number x two places to the _____ and attaching a % symbol.

8. By shading parts of a circle, percent data can be displayed visually in a(n) _____.

WRITING A PERCENT AS A FRACTION OR DECIMAL

Exercises 9–24: Write the percent as a fraction or mixed number in simplest form.

9. 28%   10. 8%
11. $3\frac{3}{4}\%$ 12. $8\frac{1}{3}\%$
13. 55%   14. 75%
15. 4%   16. 2%
17. 7.5%   18. 16.5%
19. 116%   20. 250%
21. 8.25%   22. 10.4%
23. $33\frac{1}{3}\%$ 24. $9\frac{3}{5}\%$

Exercises 25–40: Write the percent as a decimal.

25. 58%   26. 14%
27. $9\frac{1}{2}\%$ 28. $12\frac{1}{4}\%$
29. 173%   30. 206%
31. 6%   32. 8%
33. $\frac{1}{4}\%$ 34. $\frac{4}{5}\%$
35. 0.3%   36. 0.2%
37. 116%   38. 250%
39. 8.4%   40. 1.3%

WRITING FRACTIONS AND DECIMALS AS A PERCENT

Exercises 41–54: Write the fraction or mixed number as a percent.

41. $\frac{7}{10}$ 42. $\frac{5}{8}$
43. $\frac{9}{40}$ 44. $\frac{6}{25}$

45. $5\frac{1}{2}$ 46. $4\frac{2}{5}$

47. $\frac{9}{50}$ 48. $\frac{3}{16}$

49. $\frac{2}{3}$ 50. $\frac{4}{15}$

51. $\frac{7}{12}$ 52. $\frac{8}{9}$

53. $1\frac{1}{4}$ 54. $2\frac{1}{8}$

Exercises 55–68: Write the decimal as a percent.

55. 0.81   56. 0.57

57. 0.01   58. 0.03

59. 1.6   60. 2.1

61. 0.072   62. 0.049

63. 2.99   64. 3.43

65. 0.005   66. 0.004

67. 0.0401   68. 0.0802

Exercises 69 and 70: Complete the given table by finding the two missing values in each row.

69.
Percent	Decimal	Fraction
80%		
		$\frac{3}{10}$
	0.65	
22.5%		
	2.6	
		$\frac{2}{5}$

70.
Percent	Decimal	Fraction
		$\frac{23}{20}$
108%		
	0.625	
		$\frac{5}{16}$
	1.8	
2.5%		

7.1 INTRODUCTION TO PERCENT; CIRCLE GRAPHS

WORKING WITH CIRCLE GRAPHS

71. **Municipal Solid Waste** The following circle graph shows how Maine recently disposed of its municipal solid waste. (*Source:* Maine State Planning Office.)

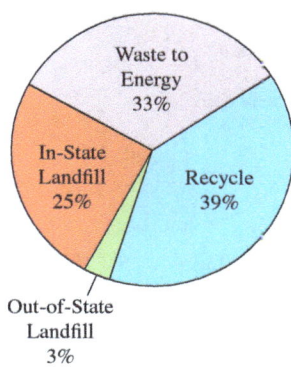

(a) Which disposal method received the largest share of the solid waste?
(b) What fraction of the solid waste went to in-state landfills? Write this fraction in simplest form.
(c) Write the share of solid waste that was converted to energy as a decimal.

72. **Fish and Wildlife Revenue** The following circle graph shows recent revenue sources for the Oregon Department of Fish and Wildlife. (*Source:* Oregon Department of Fish and Wildlife.)

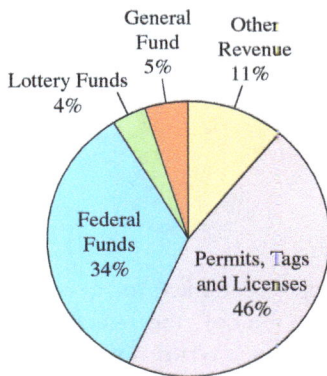

(a) Which source provided the least revenue?
(b) What fraction of the revenue came from the general fund? Write this fraction in simplest form.
(c) Write the share of revenue that came from federal funds as a decimal.

73. **Electricity Use** The following circle graph shows electricity use for a typical Florida home. (*Source:* U.S. Department of Energy.)

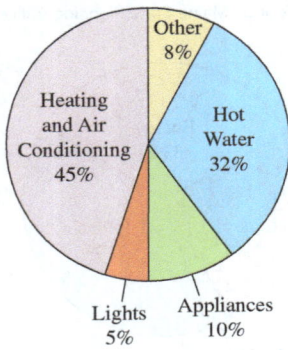

(a) What percent of the electricity was used to power appliances?
(b) What fraction of the electricity was used to heat water? Write this fraction in simplest form.
(c) Write the share of electricity that was used for lighting as a decimal.

74. **Electricity Production** The following circle graph shows recent methods for producing New England's electricity. (*Source:* New England Wind Fund.)

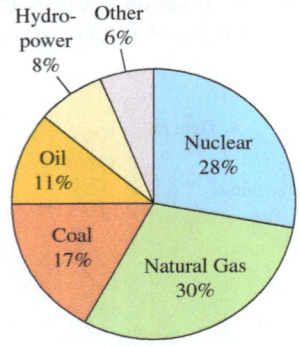

(a) What source accounted for $\frac{3}{10}$ of the electricity?
(b) What fraction of the electricity came from nuclear power? Write this fraction in simplest form.
(c) Write the share of electricity that was produced from hydropower as a decimal.

APPLICATIONS INVOLVING PERCENT

75. **Bone Marrow** About $\frac{7}{300}$ of the American population are registered with the National Marrow Donor Program. Write this fraction as a percent. (*Source:* National Marrow Donor Program.)

76. **PayPal** In 2014, $\frac{13}{50}$ of PayPal's transaction volume was on eBay. Write this fraction as a percent. (*Source:* eBay.)

77. **Shopping** A bargain table at an electronics store advertises "60% off the regular price." Write this percent as a fraction in simplest form.

78. **Online Books** An online book seller claims that it charges "45% less" for its books. Write this percent as a fraction in simplest form.

79. **Cell Phones** The portion of the Swedish population that have cell phone subscriptions can be written in decimal form as 1.1. Write this decimal as a percent. Interpret your answer. (*Source:* Ny Teknik.)

80. **Web Browsers** Recently, the portion of the online population that used Firefox to browse the Internet could be written in decimal form as 0.2251. Write this decimal as a percent. (*Source:* Nielsen Online.)

81. **Hybrid Car** A student claims that her hybrid car gets 23.6% better mileage than her old car. Write this percent as a decimal.

82. **Computer Quality** A computer maker claims that only 0.35% of its hard drives fail within the first year. Write this percent as a decimal.

83. **Working and MS** The percentage of people still working 10–14 years after an MS diagnosis is 16%. What fraction does this percentage represent? Write this fraction in simplest form.

84. **Body Fat** Some athletes have body fat levels as low as 6%. Write this percent as a fraction in simplest form. (*Source:* American Council on Exercise.)

85. **Gender** A class with 30 students has 18 women. What percent of the class are women?

86. **Video Games** A child has completed 36 levels of a video game that has 72 levels. What percent of the levels have been completed?

WRITING ABOUT MATHEMATICS

87. If an improper fraction is written as a percent, what can we say about the result? Give an example.

88. In converting between decimals and percents, when do we move the decimal point to the left and when do we move it to the right?

7.2 Using Equations to Solve Percent Problems

Objectives

1. Understanding Basic Percent Statements
2. Translating Percent Statements to Equations
3. Solving Percent Problems
   - Translating Percent Problems to Equations
   - Finding an Unknown Part
   - Finding an Unknown Whole
   - Finding an Unknown Percent

NEW VOCABULARY

☐ Basic percent statement form
☐ Basic percent equation
☐ Percent problem

1 Understanding Basic Percent Statements

A percent needs to be placed in context in order to have meaning in real-world situations. For example, 25% does not have meaning without the word "of." Phrases such as

25% of the students   or   25% of the price

make more sense. To solve percent problems, we first write the given information in *basic percent statement form*.

BASIC PERCENT STATEMENT FORM

A percent statement is in **basic percent statement form** when it is written as follows.

A percent of the whole is a part.

For example,   20% of 400 is 80.
   Percent Whole Part

The following shows how percent statements can be written in basic percent statement form.

Percent Statement	Basic Percent Statement Form
5 is 10% of 50	10% of 50 is 5
30 out of 200 is 15%	15% of 200 is 30

In each statement, the *whole* always follows the word "of," the *percent* is always followed by the % symbol or the word "percent," and the other number in the statement is always the *part*. In the next example, we write percent statements in basic percent statement form.

EXAMPLE 1   Writing percent statements in basic percent statement form

Write each percent statement in basic percent statement form.
(a) 8 is 25% of 32.   (b) 9 out of 18 is 50%.

Solution

(a) First, write the statement in the following form: A percent of a whole is a part. Note that the number 32 is the *whole* because it follows the word "of," the *percent* is 25, and the *part* is the remaining number, 8. The basic percent statement form is

25% of 32 is 8.

(b) The number 18 is the *whole* because it follows the word "of," the *percent* is 50, and the *part* is 9. The basic percent statement form is

50% of 18 is 9.

Now Try Exercises 9, 13

2 Translating Percent Statements to Equations

Recall that the word "of" means *multiply* when it follows a fraction. The phrase

$$\frac{1}{4} \text{ of the price} \quad \text{means} \quad \frac{1}{4} \cdot \text{the price}.$$

Since the fraction $\frac{1}{4}$ is equal to 25%, we can also write

$$25\% \text{ of the price} \quad \text{means} \quad 25\% \cdot \text{the price}.$$

We also know that the word "is" means *equals*. Therefore, any statement in basic percent statement form can be translated to a **basic percent equation**.

> **READING CHECK 1**
>
> What do the words "of" and "is" mean when translating a statement to a basic percent equation?

BASIC PERCENT EQUATION

The statement *a percent of the whole is a part* translates to the equation

$$\text{Percent} \cdot \text{Whole} = \text{Part}.$$

Be sure to write the percent as either a decimal or a fraction.

> **READING CHECK 2**
>
> How should the percent be written in a basic percent equation?

NOTE: Before a percent equation can be solved for an unknown value, the percent must be written as either a decimal or a fraction. ■

When translating a percent statement to an equation, we should first write the statement in basic percent statement form, as illustrated in the next example.

EXAMPLE 2 Translating percent statements to equations

Translate each percent statement to a basic percent equation.
(a) 24 is 40% of 60.   (b) 11 out of 44 is 25%.

Solution
(a) First, write the statement in basic percent statement form.

$$40\% \text{ of } 60 \text{ is } 24.$$

Using 0.4 for 40%, this statement translates to the equation

$$0.4 \cdot 60 = 24.$$

(b) In basic percent statement form, we have 25% of 44 is 11. Using 0.25 for 25%, this statement translates to the equation $0.25 \cdot 44 = 11$.

Now Try Exercises 19, 23

3 Solving Percent Problems

When the percent, whole, or part of a percent statement is replaced by the words "what percent" or "what number," the result is a *percent problem*. A **percent problem** is a question asking us to find the percent, whole, or part in a percent statement. Each of the following questions is an example of a percent problem.

> **READING CHECK 3**
>
> What is a percent problem?

- 25% of what number is 50?
- What number is 32% of 90?   *Examples of percent problems*
- 18 out of 72 is what percent?

TRANSLATING PERCENT PROBLEMS TO EQUATIONS Percent problems can be translated to equations by using a variable for the unknown value, as shown in the next example.

EXAMPLE 3 — Translating percent problems to equations

Translate each percent problem to a basic percent equation. Do not solve the equation.
(a) What percent of 80 is 6?   (b) What number is 20% of 300?

Solution
(a) The question is already written in basic percent statement (question) form.

$$\text{What percent of } 80 \text{ is } 6?$$

Using x to represent the unknown percent as a decimal, this question translates to the equation

$$x \cdot 80 = 6,$$

which can also be written as $80x = 6$.

(b) First, we write the question in basic percent statement (question) form.

$$20\% \text{ of } 300 \text{ is what number?}$$

Using x for the unknown value and 0.2 for 20%, this question translates to the equation

$$0.2 \cdot 300 = x.$$

Now Try Exercises 29, 31

STUDY TIP

If you are having difficulty with your studies, you may be able to find help at the student support services office on your campus.

FINDING AN UNKNOWN PART The unknown value in a percent problem can be the *percent*, the *whole*, or the *part*. We find the unknown value by solving the corresponding equation. In the next example, we find the value of an unknown *part*.

EXAMPLE 4 — Solving percent problems for an unknown part

Find each unknown value.
(a) 5% of 80 is what number?   (b) What number is 16% of 87.5?

Solution
(a) The question is in the following form: **5%** of **80** is **what number**?

$$5\% \cdot 80 = x \quad \text{Translate to an equation.}$$
$$0.05 \cdot 80 = x \quad \text{Write 5\% as 0.05.}$$
$$4 = x \quad \text{Multiply.}$$

(b) Write the question in the following form: **16%** of **87.5** is **what number**?

$$16\% \cdot 87.5 = x \quad \text{Translate to an equation.}$$
$$0.16 \cdot 87.5 = x \quad \text{Write 16\% as 0.16.}$$
$$14 = x \quad \text{Multiply.}$$

Now Try Exercises 37, 41

FINDING AN UNKNOWN WHOLE In the next example, we find the value of an unknown *whole*.

EXAMPLE 5 Solving percent problems for an unknown whole

Find each unknown value.

(a) 72% of what number is 90? (b) $9\frac{1}{5}$ is 8% of what number?

Solution

(a) The question is in the following form: **72%** of **what number** is **90**?

$$72\% \cdot x = 90 \qquad \text{Translate to an equation.}$$
$$0.72x = 90 \qquad \text{Write 72\% as 0.72.}$$
$$\frac{0.72x}{0.72} = \frac{90}{0.72} \qquad \text{Divide each side by 0.72.}$$
$$x = 125 \qquad \text{Simplify.}$$

(b) Write the question in the following form: **8%** of **what number** is $9\frac{1}{5}$?

$$8\% \cdot x = 9\frac{1}{5} \qquad \text{Translate to an equation.}$$
$$0.08x = 9.2 \qquad \text{Write 8\% as 0.08 and } 9\frac{1}{5} \text{ as 9.2.}$$
$$\frac{0.08x}{0.08} = \frac{9.2}{0.08} \qquad \text{Divide each side by 0.08.}$$
$$x = 115 \qquad \text{Simplify.}$$

Now Try Exercises 45, 47

FINDING AN UNKNOWN PERCENT When finding an unknown *percent*, we must convert the solution to the corresponding equation from a decimal to a percent, as shown in the next example.

EXAMPLE 6 Solving percent problems for an unknown percent

Find each unknown value.
(a) What percent of 84 is 28? (b) 7.8 is what percent of 65?
(c) What percent is 52 of 13?

Solution

(a) The question is in the following form: **What percent** of **84** is **28**?

x is in decimal form.

$$x \cdot 84 = 28 \qquad \text{Translate to an equation.}$$
$$\frac{x \cdot 84}{84} = \frac{28}{84} \qquad \text{Divide each side by 84.}$$
$$x = 0.\overline{3} \qquad \text{Simplify.}$$
$$x = 33.\overline{3}\% \qquad \text{Convert } 0.\overline{3} \text{ to } 33.\overline{3}\%.$$

Note that the last step is completed by multiplying $0.\overline{3}$ by 100%.

(b) Write the question in the following form: **What percent** of **65** is **7.8**?

x is in decimal form.

$$x \cdot 65 = 7.8 \qquad \text{Translate to an equation.}$$
$$\frac{x \cdot 65}{65} = \frac{7.8}{65} \qquad \text{Divide each side by 65.}$$
$$x = 0.12 \qquad \text{Simplify.}$$
$$x = 12\% \qquad \text{Convert 0.12 to 12\%.}$$

(c) Write the question in the following form: **What percent** of **13** is **52**?

$$x \cdot 13 = 52 \quad \text{Translate to an equation.}$$

x is in decimal form.

$$\frac{x \cdot 13}{13} = \frac{52}{13} \quad \text{Divide each side by 13.}$$

$$x = 4 \quad \text{Simplify.}$$

$$x = 400\% \quad \text{Convert 4 to 400\%.}$$

Now Try Exercises 49, 53, 57

7.2 Putting It All Together

CONCEPT	COMMENTS	EXAMPLES
Basic Percent Statement Form	A percent statement is in *basic percent statement form* when it is written as follows. A percent of the whole is a part.	The statement 7 is 10% of 70 can be written as 10% of 70 is 7. Percent Whole Part
Basic Percent Equation	The statement *a percent of a whole is a part* translates to the equation Percent · Whole = Part.	The statement 25% of 60 is 15 translates to the equation $25\% \cdot 60 = 15.$
Percent Problem	A percent problem is a question asking us to find the percent, whole, or part in a percent statement.	What number is 40% of 90? 30% of what number is 12?
Solving a Percent Problem	A percent problem can be solved by solving the corresponding basic percent equation.	20% of 75 is what number? $20\% \cdot 75 = x$ Write an equation. $0.2 \cdot 75 = x$ Write 20% as 0.2. $15 = x$ Multiply.

7.2 Exercises

MyMathLab®

CONCEPTS AND VOCABULARY

1. A percent statement is in basic percent statement form when it is written as: A percent of a(n) _____ is a(n) _____.

2. When writing a statement in basic percent statement form, the *whole* always follows the word _____.

3. In basic percent statement form, the word "of" means _____, and the word "is" means _____.

4. The statement *a percent of a whole is a part* translates to the equation _____.

5. A percent _____ is a question asking us to find the percent, whole, or part in a percent statement.

6. When finding an unknown *percent*, we must convert the solution to the corresponding equation from a(n) _____ to a percent.

WRITING BASIC PERCENT STATEMENTS

Exercises 7–16: Write the given percent statement in basic percent statement form.

7. 15% of 120 is 18.

8. 40% of 50 is 20.

9. 63 is 150% of 42.

10. 55 is 250% of 22.

11. 2.5% is 19 of 760.
12. 12% is 6 of 50.
13. 4 out of 5 is 80%.
14. 7 out of 10 is 70%.
15. 17 of 20 is 85%.
16. 26 of 104 is 25%.

TRANSLATING PERCENT STATEMENTS TO EQUATIONS

Exercises 17–26: Translate the given percent statement to a basic percent equation.

17. 18% of 40 is 7.2.
18. 62% of 130 is 80.6.
19. 49 is 140% of 35.
20. 126 is 210% of 60.
21. 64% is 48 of 75.
22. 67.5% is 54 of 80.
23. 3 out of 24 is 12.5%.
24. 18 out of 20 is 90%.
25. 15 of 24 is 62.5%.
26. 32 of 80 is 40%.

TRANSLATING PERCENT PROBLEMS TO EQUATIONS

Exercises 27–36: Translate the given percent problem to a basic percent equation. Do not solve the equation.

27. 68% of what number is 17?
28. $3\frac{1}{2}$% of 8 is what number?
29. What percent of 95 is 39.9?
30. 44 is what percent of 40?
31. What number is 104% of 70?
32. 198 is 99% of what number?
33. 48 out of 60 is what percent?
34. 76 out of what number is 47.5%?
35. What percent is 3 of 9?
36. 27.5 of 75 is what percent?

SOLVING PERCENT PROBLEMS

Exercises 37–60: Find the unknown value.

37. 6% of 50 is what number?
38. $3\frac{2}{5}$ out of what number is 40%?
39. 3.9 is what percent of 11.7?
40. What percent of 65.4 is 327?
41. What number is 48% of 12.5?
42. 7 of 9 is what percent?
43. 7.6% of 9850 is what number?
44. 3 out of 4 is what percent?
45. 105% of what number is 63?
46. What number is 115% of 60?
47. $4\frac{1}{2}$ is 3% of what number?
48. 96% of what number is 72?
49. What percent of 42 is 28?
50. 66 is 30% of what number?
51. 71 is 2% of what number?
52. 600 out of 960 is what percent?
53. 10.5 is what percent of 60?
54. 45 of 50 is what percent?
55. 66 out of what number is 240%?
56. 18 is what percent of 48?
57. What percent is 13 of 65?
58. 18% of 250 is what number?
59. 19 out of what number is 25%?
60. What number is $8\frac{1}{3}$% of 84?

APPLICATIONS

61. **Target Heart Rate** A 20-year-old student has a maximum heart rate of 200 beats per minute. If the student's target heart rate is 85% of this maximum, find this target rate.

62. **Target Heart Rate** A 50-year-old teacher has a maximum heart rate of 170 beats per minute. If the teacher's target heart rate is 70% of this maximum, find this target rate.

63. **Viral Emotions Online** Of the top 10,000 most shared articles online, 1700 evoke the emotion of laughter. What percent of articles evoke laughter? (*Source:* Business Insider.)

64. **Viral Emotions Online** Of the top 10,000 most shared articles online, 6% evoke the emotion of anger. How many articles evoke the emotion of anger? (*Source:* Business Insider.)

WRITING ABOUT MATHEMATICS

65. If we find 140% of a number, will the result be larger or smaller than the given number? Explain.

66. An advertisement for computer memory claims to cut prices up to 119%. Is this possible? Explain.

SECTIONS 7.1 and 7.2 — Checking Basic Concepts

1. Write each percent as a fraction in simplest form.
   (a) 76%   (b) $6\frac{2}{3}\%$ (c) 9.6%

2. Write each percent as a decimal.
   (a) $6\frac{1}{2}\%$ (b) 214%   (c) 2.8%

3. Write each fraction or mixed number as a percent.
   (a) $\frac{3}{25}$ (b) $3\frac{1}{2}$ (c) $\frac{7}{8}$

4. Write each decimal as a percent.
   (a) 0.29   (b) 0.073   (c) 2.97

5. Write each percent statement in basic percent statement form.
   (a) 49 is 350% of 14.   (b) 11 of 25 is 44%.

6. Translate each percent statement to a basic percent equation.
   (a) 54 is 180% of 30.   (b) 9 of 36 is 25%.

7. Translate each percent problem to a basic percent equation. Do not solve the equation.
   (a) 64 is 75% of what number?
   (b) What percent is 5 of 25?

8. Find each unknown value.
   (a) 12 out of 60 is what percent?
   (b) $2\frac{1}{4}$ is 5% of what number?
   (c) What number is 109% of 120?

9. **Gender in the U.S.** About 51% of Americans are female. What fraction of the U.S. population is female?

10. **Economic Status** The following circle graph shows how Americans were classified in a recent study. (*Source:* Society in Focus.)

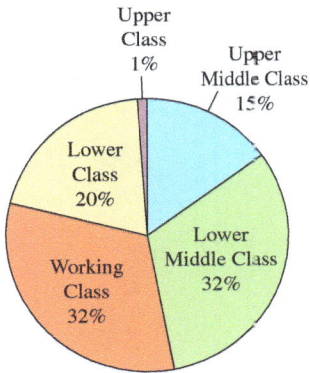

(a) Which class is smallest?
(b) What fraction of Americans were classified as lower middle class? Write this fraction in simplest form.
(c) Write the percent of lower class Americans as a decimal.

7.3 Using Proportions to Solve Percent Problems

Objectives

1. Reviewing Proportions
2. Translating Percent Statements to Proportions
3. Solving Percent Problems
   - Translating Percent Problems to Proportions
   - Finding an Unknown Part
   - Finding an Unknown Whole
   - Finding an Unknown Percent

1 Reviewing Proportions

In the previous chapter, we learned that a proportion is a statement that two ratios are equal. For example, since the ratios $\frac{3}{5}$ and $\frac{18}{30}$ are equal, we can write the proportion

$$\frac{3}{5} = \frac{18}{30}.$$ Proportion

In a proportion, the *cross products* are always equal.

Equals 90. $3 \cdot 30$ $\frac{3}{5} = \frac{18}{30}$ $5 \cdot 18$ Equals 90.

In this case, the cross products $3 \cdot 30$ and $5 \cdot 18$ are both equal to 90.

When a proportion contains a variable for an unknown value, we use the cross products to solve the proportion. For example,

$$\frac{x}{5} = \frac{18}{30} \qquad \text{Given proportion}$$
$$x \cdot 30 = 5 \cdot 18 \qquad \text{Cross products are equal.}$$
$$30x = 90 \qquad \text{Simplify.}$$
$$\frac{30x}{30} = \frac{90}{30} \qquad \text{Divide each side by 30.}$$
$$x = 3 \qquad \text{Simplify.}$$

READING CHECK 1

In a proportion, what can be said about its cross products?

Percent problems can be solved using proportions. To do so, we must first learn how to translate percent statements to proportions.

2 Translating Percent Statements to Proportions

An example of a *basic percent statement* is

$$50\% \text{ of } 60 \text{ is } 30.$$

That is, a *percent* of the *whole* is a *part*. This basic percent statement can be translated into the *basic percent equation*.

$$0.50 \cdot 60 = 30.$$

Consider the general basic percent equation

$$\text{percent} \cdot \text{whole} = \text{part}, \qquad \text{(Decimal or fraction)} \qquad \text{For example, } 0.50 \cdot 60 = 30.$$

where the words "percent," "whole," and "part" represent numbers. Dividing each side of this equation by the (nonzero) number "whole" gives

$$\frac{\text{percent} \cdot \text{whole}}{\text{whole}} = \frac{\text{part}}{\text{whole}}, \qquad \text{Dividing by 60: } \frac{0.50 \cdot 60}{60} = \frac{30}{60}.$$

which simplifies to

$$\text{percent} = \frac{\text{part}}{\text{whole}}. \qquad \text{Simplify: } 0.50 = \frac{30}{60}.$$

Since a percent can be written as a ratio with denominator 100, we get the proportion

$$\frac{\text{percent}}{100} = \frac{\text{part}}{\text{whole}}. \qquad 0.50 = \frac{30}{60} \text{ is equivalent to } \frac{50}{100} = \frac{30}{60}.$$

NOTE: In the two preceding equations, the **percent** (in blue) is a *decimal or fraction*, and the **percent** (in purple) forms a ratio with a denominator of 100 that is equal to the **percent** (in blue). ∎

TRANSLATING PERCENT STATEMENTS TO PROPORTIONS

The percent statement *a percent of the whole is a part* translates to the proportion

$$\frac{\text{percent}}{100} = \frac{\text{part}}{\text{whole}}.$$

Be sure that the percent is written without the % symbol. The % symbol is not needed because we are dividing the percent by 100.

Recall that for any percent statement, the *whole* always follows the word "of," the *percent* is always followed by the % symbol or the word "percent," and the other number in the statement is always the *part*. The next example shows how percent statements are translated to proportions.

EXAMPLE 1 **Translating percent statements to proportions**

Translate each percent statement to a proportion.
(a) 27 is 30% of 90.   (b) 14 out of 56 is 25%.

Solution
(a) We translate the percent statement to the proportion

$$\frac{\text{percent}}{100} = \frac{\text{part}}{\text{whole}}.$$

The number **90** is the *whole* because it follows the word "of," the *percent* is **30**, and the *part* is **27**. The proportion is

$$\frac{30}{100} = \frac{27}{90}.$$

(b) The number **56** is the *whole*, the *percent* is **25**, and the *part* is **14**. The proportion is

$$\frac{25}{100} = \frac{14}{56}.$$

Now Try Exercises 7, 11

3 Solving Percent Problems

TRANSLATING PERCENT PROBLEMS TO PROPORTIONS One way to solve a percent problem is to first translate the problem into a proportion. In the next example, we demonstrate how this is done.

EXAMPLE 2 **Translating percent problems to proportions**

Translate each percent problem to a proportion. Do not solve the proportion.
(a) 12 out of 30 is what percent?   (b) 60 is 35% of what number?

Solution
(a) The number **30** is the *whole* because it follows the word "of," the unknown *percent* can be represented by x, and the *part* is the remaining number, **12**. The proportion is

$$\frac{x}{100} = \frac{12}{30}.$$

(b) The *whole* is unknown. We represent it with x. The *percent* is **35**, and the *part* is **60**. The proportion is

$$\frac{35}{100} = \frac{60}{x}.$$

Now Try Exercises 19, 23

The unknown value in a percent problem can be the *percent*, the *whole*, or the *part*. Its value can be found by solving the corresponding proportion.

The following See the Concept shows how to use cross products to solve a proportion.

See the Concept 1 **USING CROSS PRODUCTS TO SOLVE A PROPORTION**

When a proportion contains a variable for an unknown value, we use cross products to solve the proportion.

Solving a Proportion

Cross Product
$x \cdot 30$

$$\frac{x}{5} = \frac{18}{30}$$

Cross Product
$5 \cdot 18$

$\frac{3}{5} = \frac{18}{30}$

$x \cdot 30 = 5 \cdot 18$ Cross products are equal

$30x = 90$ Simplify.

$\frac{30x}{30} = \frac{90}{30}$ Divide each side by 30.

$x = 3$ Simplify.

WATCH VIDEO IN MML.

FINDING AN UNKNOWN PART In the next example, we find the value of an unknown *part*.

EXAMPLE 3 **Solving percent problems for an unknown part**

Find each unknown value.
(a) What number is 32% of 62.5?   (b) 140% of 75 is what number?

Solution
(a) The *whole* is **62.5**, the *percent* is **32**, and the *part* is **unknown**.

$\frac{32}{100} = \frac{x}{62.5}$ Translate to a proportion.

$32 \cdot 62.5 = 100 \cdot x$ Cross products are equal.

$2000 = 100x$ Simplify.

$\frac{2000}{100} = \frac{100x}{100}$ Divide each side by 100.

$20 = x$ Simplify.

(b) The *whole* is **75**, the *percent* is **140**, and the *part* is **unknown**.

$\frac{140}{100} = \frac{x}{75}$ Translate to a proportion.

$140 \cdot 75 = 100 \cdot x$ Cross products are equal.

$10{,}500 = 100x$ Simplify.

$\frac{10{,}500}{100} = \frac{100x}{100}$ Divide each side by 100.

$105 = x$ Simplify.

Now Try Exercises 25, 31

FINDING AN UNKNOWN WHOLE In the next example, we find the value of an unknown *whole*.

EXAMPLE 4 Solving percent problems for an unknown whole

Find each unknown value.
(a) $7\frac{1}{2}$ is 6% of what number?   (b) 85% of what number is 98.6?

Solution
(a) The *whole* is **unknown**, the *percent* is **6**, and the *part* is $7\frac{1}{2}$, which we write as **7.5** because multiplying 7.5 by 100 can be done mentally.

$$\frac{6}{100} = \frac{7.5}{x} \qquad \text{Translate to a proportion.}$$

$$6 \cdot x = 100 \cdot 7.5 \qquad \text{Cross products are equal.}$$

$$6x = 750 \qquad \text{Simplify.}$$

$$\frac{6x}{6} = \frac{750}{6} \qquad \text{Divide each side by 6.}$$

$$x = 125 \qquad \text{Simplify.}$$

(b) The *whole* is **unknown**, the *percent* is **85**, and the *part* is **98.6**.

$$\frac{85}{100} = \frac{98.6}{x} \qquad \text{Translate to a proportion.}$$

$$85 \cdot x = 100 \cdot 98.6 \qquad \text{Cross products are equal.}$$

$$85x = 9860 \qquad \text{Simplify.}$$

$$\frac{85x}{85} = \frac{9860}{85} \qquad \text{Divide each side by 85.}$$

$$x = 116 \qquad \text{Simplify.}$$

Now Try Exercises 35, 39

FINDING AN UNKNOWN PERCENT The next example shows how to find an unknown *percent*. When finding an unknown *percent*, the solution to the corresponding proportion is already in percent form. Remember to include the % symbol.

EXAMPLE 5 Solving percent problems for an unknown percent

Find each unknown value.
(a) What percent of 93 is 62?   (b) 12.6 is what percent of 90?

Solution
(a) The *whole* is **93**, the *percent* is **unknown**, and the *part* is **62**.

$$\frac{x}{100} = \frac{62}{93} \qquad \text{Translate to a proportion.}$$

$$x \cdot 93 = 100 \cdot 62 \qquad \text{Cross products are equal.}$$

$$93x = 6200 \qquad \text{Simplify.}$$

$$\frac{93x}{93} = \frac{6200}{93} \qquad \text{Divide each side by 93.}$$

$$x = 66.\overline{6}\% \qquad \text{Simplify and include the \% symbol.}$$

(b) The *whole* is **90**, the *percent* is **unknown**, and the *part* is **12.6**.

$$\frac{x}{100} = \frac{12.6}{90} \quad \text{Translate to a proportion.}$$

$$x \cdot 90 = 100 \cdot 12.6 \quad \text{Cross products are equal.}$$

$$90x = 1260 \quad \text{Simplify.}$$

$$\frac{90x}{90} = \frac{1260}{90} \quad \text{Divide each side by 90.}$$

$$x = 14\% \quad \text{Simplify and include the \% symbol.}$$

Now Try Exercises 41, 47

7.3 Putting It All Together

CONCEPT	COMMENTS	EXAMPLES
Proportion	A proportion is a statement that two ratios are equal.	$\frac{45}{100} = \frac{9}{20}$
Cross Products	In a proportion, the cross products are equal.	$\frac{1}{5} = \frac{3}{15}$ $1 \cdot 15$ $5 \cdot 3$ $1 \cdot 15$ and $3 \cdot 5$ both equal 15.
Translating Percent Statements to Proportions	The percent statement *a percent of the whole is a part* translates to the proportion $$\frac{\text{percent}}{100} = \frac{\text{part}}{\text{whole}}.$$	The percent statement 50% of 90 is 45 translates to the proportion $\frac{50}{100} = \frac{45}{90}.$
Solving a Percent Problem	A percent problem can be solved by solving the corresponding proportion.	10% of 40 is what number? $\frac{10}{100} = \frac{x}{40}$ Write a proportion. $10 \cdot 40 = 100 \cdot x$ Equal cross products $400 = 100x$ Simplify. $\frac{400}{100} = \frac{100x}{100}$ Divide by 100. $4 = x$ Simplify.

7.3 Exercises

CONCEPTS AND VOCABULARY

1. A(n) _____ is a statement that two ratios are equal.

2. In a proportion, the cross products are _____.

3. In a percent statement, the _____ is always followed by the % symbol or the word "percent."

4. The percent statement *a percent of a whole is a part* translates to the proportion _____.

TRANSLATING PERCENT STATEMENTS TO PROPORTIONS

Exercises 5–14: Translate the given percent statement to a proportion.

5. 24% of 70 is 16.8.
6. 4% of 755 is 30.2.
7. 45 is 180% of 25.
8. 138 is 230% of 60.
9. 40% is 34 of 85.
10. 13.5% is 27 of 200.
11. 8 out of 64 is 12.5%.
12. 19 out of 20 is 95%.
13. 25 of 40 is 62.5%.
14. 21 of 30 is 70%.

TRANSLATING PERCENT PROBLEMS TO PROPORTIONS

Exercises 15–24: Translate the given percent problem to a proportion. Do not solve the proportion.

15. 48% of what number is 19?
16. $5\frac{1}{2}$% of 22 is what number?
17. What percent of 114 is 38.4?
18. 55 is what percent of 50?
19. 24 out of 80 is what percent?
20. What number is 165% of 20?
21. 39 out of what number is 42.5%?
22. What percent is 2 of 8?
23. 297 is 99% of what number?
24. 13.5 of 92 is what percent?

SOLVING PERCENT PROBLEMS

Exercises 25–48: Find the unknown value.

25. What number is 48% of 37.5?
26. $6\frac{4}{5}$ out of what number is 40%?
27. 14.6 is what percent of 21.9?
28. 16% of 25 is what number?
29. What percent of 90.5 is 362?
30. 5 of 6 is what percent?
31. 108% of 25 is what number?
32. 6 out of 10 is what percent?
33. 130% of what number is 52?
34. What number is 192% of 50?
35. $2\frac{1}{4}$ is 9% of what number?
36. 36 out of what number is 45%?
37. 52 is 40% of what number?
38. 63 is 5% of what number?
39. 88% of what number is 30.8?
40. 696 out of 960 is what percent?
41. What percent of 48 is 16?
42. 32 of 50 is what percent?
43. 52 out of what number is 160%?
44. 49 is what percent of 56?
45. What percent is 19 of 95?
46. 26% of 250 is what number?
47. 14.1 is what percent of 60?
48. What number is $5\frac{1}{3}$% of 225?

APPLICATIONS INVOLVING PERCENT

Exercises 49–52: A survey was given to a group of respondents. They were asked, "How often do you unplug from ALL personal technology?" (including mobile phones, tablets, computers, e-readers, TV, audio players, etc.) (**Source:** CivicScience.)

49. If 3,783 people responded "Never" and they made up 43.4% of the total group, how many people total were given the survey? Round to the nearest whole number.
50. Given your answer for Exercise 49, if 3.7% of the group responded "Once a month", how many people had this response? Round to the nearest whole number.
51. Given your answer for Exercise 49, if 1487 people responded "A few times a year", what percentage of the total group had this response? Round to the nearest percent.
52. Given your answer for Exercise 49, if 1729 people responded "Daily", what percentage of the total group had this response? Round to the nearest percent.

WRITING ABOUT MATHEMATICS

53. Write your own percent problem and solve it using an equation. Then solve it using a proportion. Which method do you prefer? Explain.
54. In your own words, describe how to find the *percent*, the *whole*, and the *part* in a percent statement.

7.4 Applications: Sales Tax, Discounts, and Net Pay

Objectives

1. Solving General Percent Applications
   - Finding a Missing Part
   - Finding a Missing Whole
   - Finding a Missing Percent
2. Finding Sales Tax and Total Amount Paid
3. Finding Discounts and Sale Price
4. Calculating Commission and Net Pay
   - Calculating a Commission
   - Calculating Net Pay
5. Finding Percent Change

NEW VOCABULARY

☐ Sales tax
☐ Total amount paid
☐ Discount
☐ Sale price
☐ Commission
☐ Net pay
☐ Withholdings
☐ Gross pay
☐ Percent change
☐ Percent increase
☐ Percent decrease

1 Solving General Percent Applications

When solving general percent applications, we may need to find the *part*, the *whole*, or the *percent*. Each of these types of percent applications is discussed in this section, beginning with finding the *part*.

FINDING A MISSING PART As we learned in Section 2 of this chapter, the *whole* in a percent problem follows the word "of," and the *percent* is followed by the % symbol or the word "percent." In the next example, we identify the *whole* and the *percent* so that an equation can be used to find the missing *part*.

EXAMPLE 1 Finding the number of downloaded country music songs

If country music accounts for 32% of all downloads from an online music store, how many country songs are sold on a day with a total of 1425 songs downloaded?

Solution
To use an equation to solve this percent problem, we write the question in basic percent statement form.

$$32\% \text{ of } 1425 \text{ is } \text{what number}?$$

$$\text{Percent} \quad \text{Whole} \quad \text{Part}$$

$$32\% \cdot 1425 = x \quad \text{Translate to an equation.}$$
$$0.32 \cdot 1425 = x \quad \text{Write 32\% as 0.32.}$$
$$456 = x \quad \text{Multiply.}$$

On a day when 1425 songs are downloaded, we can predict that 456 are country songs.

Now Try Exercise 13

FINDING A MISSING WHOLE In the previous example, we used an equation to find a missing *part*. In the next example, we set up and solve a proportion to find the missing *whole* in a percent application.

EXAMPLE 2 Finding a selected portion of the Hawaiian population

A recent survey found that 25% of Hawaii's population over the age of 4 spoke a language other than English at home. If 297 thousand Hawaiians fit this description, how many Hawaiians were over the age of 4? (*Source*: U.S. Census Bureau.)

Solution
Begin by writing the given information in the following form:

$$25\% \text{ of } \text{what number} \text{ is } 297 \text{ thousand}?$$

The *whole* is **unknown**, the *percent* is **25**, and the *part* is **297** thousand.

$$\frac{25}{100} = \frac{297}{x} \qquad \text{Translate to a proportion.}$$

$$25 \cdot x = 100 \cdot 297 \qquad \text{Cross products are equal.}$$

$$25x = 29{,}700 \qquad \text{Simplify.}$$

$$\frac{25x}{25} = \frac{29{,}700}{25} \qquad \text{Divide each side by 25.}$$

$$x = 1188 \qquad \text{Simplify.}$$

Since the information is given in thousands, the number of Hawaiians over age 4 was $1188 \cdot 1000 = 1{,}188{,}000$.

Now Try Exercise 19

FINDING A MISSING PERCENT In the next example, we will solve a proportion to find the missing *percent* in a percent application.

EXAMPLE 3 **Finding the percent of teens who recycle**

In a survey of 420 teenagers, 35 said that they *never* recycle. What percent of teenagers surveyed did some kind of recycling?

Solution
Since 35 teens stated that they never recycle, the remaining $420 - 35 = 385$ do some kind of recycling. Write this information in the following form:

What percent of **420** is **385**?

The *whole* is **420**, the *percent* is **unknown**, and the *part* is **385**.

$$\frac{x}{100} = \frac{385}{420} \qquad \text{Translate to a proportion.}$$

$$x \cdot 420 = 100 \cdot 385 \qquad \text{Cross products are equal.}$$

$$420x = 38{,}500 \qquad \text{Simplify.}$$

$$\frac{420x}{420} = \frac{38{,}500}{420} \qquad \text{Divide each side by 420.}$$

$$x = 91.\overline{6}\% \qquad \text{Simplify.}$$

Of the teens surveyed, $91.\overline{6}\%$ did some kind of recycling.

Now Try Exercise 25

READING CHECK 1

How does sales tax affect the total amount paid for an item?

2 Finding Sales Tax and Total Amount Paid

Most states require retailers and service providers to collect sales tax when certain kinds of items or services are sold. **Sales tax** is a percent of the purchase price, which is added to the purchase price. The **total amount paid** is the sum of the purchase price and the sales tax.

🌐 *Math in Context* (Tax) Many states have a sales tax rate of **6%**. If the purchase price of an item is **$24.50**, then the sales tax is **6%** of **$24.50**, or

$$0.06(24.50) = \$1.47.$$

The total amount paid for the item is the sum of the price and tax, or

$$24.50 + 1.47 = \$25.97.$$

Sales tax and total amount paid can be found using the following equations.

STUDY TIP

The box to the right is the first of four formula boxes in this section. Try to avoid simply memorizing the formulas. Instead, study and practice the formulas until you have a firm grasp of the situations that require their use.

> **SALES TAX AND TOTAL AMOUNT PAID**
>
> If P represents the purchase price of an item and r (written as a decimal) represents the sales tax rate, then the amount of sales tax S is given by
>
> $$S = rP.$$
>
> The total amount paid T for the item is
>
> $$T = P + S.$$

NOTE: Taxes and dollar amounts are rounded to the nearest cent. ∎

EXAMPLE 4 Finding sales tax and total amount paid

If a cell phone costs $149 and the sales tax rate is 4%, find the sales tax and the total amount paid.

Solution
Written as a decimal, the sales tax rate is $r = 0.04$, and the purchase price is $P = 149$. The sales tax is $S = 0.04(149) = \$5.96$ and the total amount paid is $149 + 5.96 = \$154.96$.

Now Try Exercise 29

> **MAKING CONNECTIONS 1**
>
> **Calculating Total Amount Paid Directly**
>
> In Example 4, the buyer pays the purchase price, which is 100% of $149, and then pays an additional 4% of $149 as sales tax. In other words, the buyer pays 100% + 4%, or 104% of the purchase price. To calculate the total amount paid directly, add the sales tax rate to 100% to get 1.04 and then multiply by the purchase price, 149, to get the total amount paid of $154.96.
>
> $$1.04(149) = \$154.96$$

EXAMPLE 5 Finding the sales tax rate

When a student bought a new Nintendo Wii U console with a purchase price of $300, the sales tax was $22.50. Find the sales tax rate.

Solution
The purchase price is $P = 300$, and the sales tax is $S = 22.50$. Let r represent the unknown sales tax rate.

$$S = rP \qquad \text{Sales tax equation}$$
$$22.50 = r \cdot 300 \qquad S = 22.5 \text{ and } P = 300.$$
$$\frac{22.50}{300} = \frac{r \cdot 300}{300} \qquad \text{Divide each side by 300.}$$
$$0.075 = r \qquad \text{Simplify.}$$

The sales tax rate is 0.075 or 7.5%.

Now Try Exercise 33

READING CHECK 2

How does a discount affect the price paid for an item?

3 Finding Discounts and Sale Price

To attract customers, retailers offer discounted prices. A **discount** is a percent of the original (or regular) price that is subtracted from the original price to give the **sale price**.

A discount amount and the sale price can be found using the following equations.

> **DISCOUNTS AND SALE PRICE**
>
> If O represents the *original* price of an item and r (written as a decimal) represents the discount rate, then the discount D is given by
>
> $$D = rO.$$
>
> The sale price is the new purchase price P, where
>
> $$P = O - D.$$

EXAMPLE 6 Finding the discount and sale price

If new shoes priced at $120 are on sale for 30% off, find the discount and the sale price.

Solution
Written as a decimal, the discount rate is $r = 0.3$, and the original price is $O = 120$. The discount is $D = 0.3(120) = \$36$, and the sale price is $120 - 36 = \$84$.

Now Try Exercise 37

CRITICAL THINKING 1

Can a discount be greater than 100%?

> **MAKING CONNECTIONS 2**
>
> **Calculating Sale Price Directly**
>
> If the price of an item is discounted 30%, then a buyer pays 70% of the original (or regular) price. In Example 6, the buyer pays 70% of $120, or
>
> $$0.70(120) = \$84.$$

4 Calculating Commission and Net Pay

CALCULATING A COMMISSION Workers can be paid in several ways. Some earn hourly wages, while others receive pay based on an annual salary. Some people who work in sales earn a **commission**, which is a percent of total sales. A commission amount can be found as follows.

READING CHECK 3

How is a salesperson's commission calculated?

> **COMMISSION**
>
> If T represents the total sales and r (written as a decimal) represents the commission rate, then the commission C is given by
>
> $$C = rT.$$

EXAMPLE 7 Finding commission

A real estate broker earns a commission of 6% on total sales. If the broker sells a home for $212,000, find her commission.

Solution
Written as a decimal, the commission rate is $r = 0.06$, and total sales are $T = 212,000$. The commission is $C = 0.06(212,000) = \$12,720$.

Now Try Exercise 45

EXAMPLE 8 Finding total sales

A salesperson receives a commission of $876. If his commission rate is 3%, find the total sales.

Solution
Written as a decimal, the commission rate is $r = 0.03$, and the commission is $C = 876$. Let T represent the unknown sales tax rate.

$$\begin{align} C &= rT &&\text{Commission equation} \\ 876 &= 0.03T &&C = 876 \text{ and } r = 0.03. \\ \frac{876}{0.03} &= \frac{0.03T}{0.03} &&\text{Divide each side by 0.03.} \\ 29{,}200 &= T &&\text{Simplify.} \end{align}$$

Total sales are $29,200.

Now Try Exercise 49

READING CHECK 4

Which amount is larger, net pay or gross pay?

CALCULATING NET PAY Most paycheck amounts are less than hourly wages, salary, or commission would suggest. The final dollar amount on a paycheck is called the *net pay* or *take-home pay*. **Net pay** is the amount remaining after **withholdings** such as taxes, insurance, and other deductions have been subtracted from the **gross pay**. Withholdings are calculated as a percent of gross pay.

NET PAY

If G represents gross pay and r (written as a decimal) represents the total withholding rate, then the total withholdings W are given by

$$W = rG.$$

The net pay N is

$$N = G - W.$$

EXAMPLE 9 Finding withholdings and net pay

A factory worker has 26% of her weekly pay withheld for taxes, insurance, and a pension plan. If her weekly gross pay is $860, find her total withholdings and net pay.

Solution
Written as a decimal, the withholding rate is $r = 0.26$, and the gross pay is $G = 860$. The withholdings are $W = 0.26(860) = \$223.60$, and the net pay is $860 - 223.60 = \$636.40$.

Now Try Exercise 53

5 Finding Percent Change

 Connecting Concepts with Your Life When prices increase (or decrease), the actual amount of the increase (or decrease) is often not as significant as the *percent change* in the price. For example, consider the $1000 increases shown in **FIGURE 7.4**.

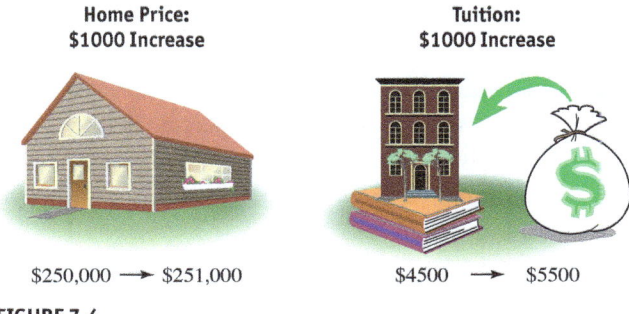

FIGURE 7.4

Even though both prices increased by $1000, the percent increase in the price of the home is four-tenths of 1% and the percent increase in cost of tuition is more than 22%. The percent increase in tuition is much more dramatic than the percent increase in the price of the home.

If a quantity changes from an **old amount** to a **new amount**, the **percent change** is given by

$$\text{Percent Change} = \frac{\text{new amount} - \text{old amount}}{\text{old amount}} \cdot 100.$$

Multiply by 100 to make it a percent.

READING CHECK 5

What does it mean when a percent change is negative?

NOTE: A positive percent change corresponds to an increase and a negative percent change corresponds to a decrease. ∎

EXAMPLE 10 Calculating percent increase in Skype international calls

From 2005 to 2013, the number of international Skype-to-Skype calling minutes increased from 6 billion to 54 billion. Calculate the percent increase in Skype calling minutes from 2005 to 2013. (*Source:* Skype SEC filling.)

Solution
The **old amount** is **6** and the **new amount** is **54**. The percent **increase** is

$$\frac{\text{new amount} - \text{old amount}}{\text{old amount}} \cdot 100 = \frac{54 - 6}{6} \cdot 100 = 800\%.$$

Note that the number of minutes increased 9 times, which corresponds to a percent increase of 800%.

Now Try Exercise 61

EXAMPLE 11 Finding percent decrease

A CEO reduces her company's workforce from 250 to 220. Find the percent decrease.

Solution
The **old amount** is **250**, and the **new amount** is **220**. The percent change is

$$\frac{220 - 250}{250} \cdot 100 = \frac{-30}{250} \cdot 100 = -0.12 \cdot 100 = -12\%.$$

Since the percent change is negative, we have a 12% decrease.

Now Try Exercise 63

NOTE: In Example 11, we write that either the percent *change* is -12%, or that the percent *decrease* is 12%. It is *not* correct to say that the percent decrease is -12%. ∎

7.4 Putting It All Together

CONCEPT	COMMENTS	EXAMPLES
General Percent Applications	When solving a basic percent application, we may need to find a missing *part*, *whole*, or *percent*.	If 8% of 150 students surveyed say that they like tofu, then $$0.08(150) = 12$$ students like tofu.
Sales Tax and Total Amount Paid	If S is sales tax, P is purchase price, and r (written as a decimal) is the sales tax rate, then $$S = rP.$$ If T is total amount paid, then $$T = P + S.$$	If there is a 5% sales tax rate, then an $800 canoe has $$S = 0.05(800) = \$40$$ in sales tax, and the total amount paid is $$T = 800 + 40 = \$840.$$
Discounts and Sale Price	If D is a discount, O is the original price, and r (written as a decimal) is the discount rate, then $$D = rO.$$ If P is purchase price, then $$P = O - D.$$	If a $200 camera is marked 25% off, then the discount is $$D = 0.25(200) = \$50,$$ and the purchase price is $$P = 200 - 50 = \$150.$$
Commission	If C is commission, T is total sales, and r (written as a decimal) is the commission rate, then $$C = rT.$$	If a person has a 4% commission rate on $7000 in total sales, then the commission is $$C = 0.04(7000) = \$280.$$
Net Pay	If W is withholdings, G is gross pay, and r (written as a decimal) is the withholding rate, then $$W = rG.$$ If N is net pay, then $$N = G - W.$$	If 18% of a worker's $1200 gross pay is withheld, then withholdings are $$W = 0.18(1200) = \$216,$$ and the net pay is $$N = 1200 - 216 = \$984.$$
Percent Change	If an amount changes from an **old amount** to a **new amount**, then the percent change is $$\frac{\text{new amount} - \text{old amount}}{\text{old amount}} \cdot 100.$$	When a price increases from $40 to $50, the percent change is $$\frac{50 - 40}{40} \cdot 100 = 25\%.$$ When a price decreases from $50 to $40, the percent change is $$\frac{40 - 50}{50} \cdot 100 = -20\%.$$

7.4 Exercises

CONCEPTS AND VOCABULARY

1. When solving basic percent applications, we may be asked to find the *part*, the *whole*, or the _____.

2. A percent of the purchase price that is added to the purchase price is called _____ tax.

3. The _____ is the sum of the purchase price and the sales tax.

4. A percent of the original price that is subtracted from the original price is called a(n) _____.

5. The _____ is the result of subtracting the discount from the original price.

6. When a person's pay is a percent of the total sales, then the person is earning a(n) _____.

7. The amount of pay before taxes, insurance, and other deductions are subtracted is called _____.

8. Taxes, insurance, and other deductions that are subtracted from gross pay are called _____.

9. A worker's _____ is the amount remaining after withholdings are subtracted from gross pay.

10. A positive percent change means that we have a percent _____, whereas a negative percent change means that we have a percent _____.

APPLICATIONS INVOLVING PERCENT

11. **Cash Back** A credit card company offers a 1.5% annual cash back rebate on all purchases. If a person uses this credit card for $18,600 in purchases, how much is the rebate?

12. **Nutrition** If a daily diet should include 35 grams of dietary fiber, how many grams of dietary fiber are in a bowl of whole grain cereal with 6% of the recommended amount?

13. **Digital Ad Revenue** Digital video accounted for 7% of the total digital ad revenue for the first half of 2013. If the total was $20 billion, what was the revenue from digital video?

14. **iPhone Cost** The touchscreen display made up 21% of the manufacturing cost of the iPhone 6 Plus. The total manufacturing cost of this phone was $242.50. What was the manufacturing cost of the touchscreen display? (*Source:* Teardown.com.)

15. **Smoking Verdict** A jury awarded $5.3 million to a man whose wife died from lung cancer after smoking two packs of cigarettes every day for more than 50 years. However, the tobacco company Phillip Morris USA was found to be only 36.5% responsible. How much of the total award was the tobacco company required to pay? (*Source:* CNN.)

16. **Cheating** According to a recent survey of 12,000 high school students, 74% admitted to cheating on an exam at some point during the past year. How many of the students surveyed admitted to cheating? (*Source:* Josephson Institute of Ethics.)

17. **The Human Body** An adult male's body is about 60% water. If a man's body has 111 pounds of water, what is his total body weight?

18. **Cat Owners** A recent survey found that 34% of U.S. households (or 38.42 million) own at least one cat. Find the total number of U.S. households at the time of the survey. (*Source:* American Pet Products Manufacturers Association.)

19. **Steroid Use** In a recent survey, 1.5% of high school seniors admitted to using steroids within the last year. If 300 high school seniors admitted to steroid use, how many seniors were surveyed? (*Source:* National Institute on Drug Abuse.)

20. **Smoking Doctors** About 23% of Chinese doctors recently surveyed were smokers. If 817 doctors were smokers, how many doctors were surveyed? Round to the nearest whole number. (*Source:* University of California–Los Angeles.)

21. **Salary** A worker's salary increases by 3.5%. If the raise amounts to $2450, how much was the worker being paid before the raise?

22. **Enrollment** The spring semester's enrollment at a small college was 138 students more than the fall enrollment. If this represents an increase of 12%, how many students were there in the fall?

23. **New Year's Resolutions** On New Year's Eve, 200 people made New Year's resolutions. By the end of January, 14 people had kept their resolutions. What percent of the people kept their resolutions?

24. **Appalachian Trail** In a survey of 8000 hikers, 224 had hiked on the Appalachian Trail. What percent of the hikers surveyed had hiked on the Appalachian Trail?

25. **Saving for College** The parents of 800 college freshmen were asked how much they had saved for their child's college education. The parents of 520 freshmen had saved less than $5000. What percent of parents saved less than $5000? (*Source:* College Savings Foundation.)

26. **Fuel Economy** Of 1540 cars and trucks in a parking lot, only 77 had mileage ratings of more than 30 miles per gallon. What percent of the vehicles had mileage ratings of more than 30 miles per gallon?

27. **Graduate School** At a small private college, 124 of 465 seniors intend to go to graduate school. What percent is this?

28. **State Names** The states with names that begin with a vowel are Alabama, Alaska, Arizona, Arkansas, Idaho, Illinois, Indiana, Iowa, Ohio, Oklahoma, Oregon, and Utah. What percent of the 50 U.S. states have names that begin with a vowel?

29. **Digital Music** If the purchase price of a portable MP3 player is $64 and the sales tax rate is 6.5%, find the sales tax and the total amount paid.

30. **Online Shopping** A hammock company offers free shipping on orders over $150. If a hanging hammock chair sells for $199 and the sales tax rate is 7%, find the sales tax and the total amount paid.

31. **Dining Out** The bill for two diners at an upscale restaurant comes to $148.80. If the restaurant is in a state with a 7.5% sales tax rate, find the sales tax and the total amount paid (without tip).

32. **iPod Application** A business application for an iPod Touch has a purchase price of $9.75. If the sales tax rate is 4%, find the sales tax and the total amount paid for the application.

33. **HD Video Recorder** A shopper paid $12.75 sales tax on an HD video recorder with a purchase price of $318.75. Find the sales tax rate.

34. **School Laptop** A student bought a new laptop computer with a purchase price of $980.50. If the sales tax was $58.83, find the sales tax rate.

35. **Wireless Router** A homeowner buys a wireless router for a home computer network and pays $10.40 in sales tax. If the sales tax rate is 6.5%, find the purchase price of the router.

36. **Hotel Rooms** In some states, higher tax rates are charged for hotel rooms. If a traveler pays $35.70 in taxes for a one-night stay and the tax rate is 17%, find the purchase price for one night at the hotel.

37. **Golf Clubs** A set of golf clubs is marked at 25% off. If the regular price is $560, find the discount and the sale price.

38. **Online Shopping** A toy company offers free shipping on orders over $100. If an electric train that sells for $189 is marked at 15% off, find the discount and the sale price.

39. **Clearance Sale** A lamp on a clearance table is marked at 70% off. If the regular (or original) price is $76.50, find the discount and the sale price.

40. **Swing Set** A backyard swing set is marked at 75% off. If the regular price is $858.80, find the discount and the sale price.

41. **Prepaid Minutes** A cellular customer receives a $33.75 discount when she buys 1500 prepaid minutes. If the regular (or original) price for 1500 minutes is $187.50, what is the discount rate?

42. **Airline Tickets** An airline offers a $50 discount on tickets regularly priced at over $375. If a customer buys a ticket with a $400 regular price, what is the discount rate?

43. **School Supplies** A student got a $10.50 discount on a new backpack. If the discount rate is 30%, find the original price of the backpack.

44. **Private School** A parent gets a $225 discount for prepaying his child's entire annual tuition at a private elementary school. If this represents a 5% discount, find the original cost of annual tuition.

45. **Real Estate** A realtor earns a commission of 3% on total sales. If the realtor sells a home for $380,000, find the commission.

46. **Interior Design** An interior designer receives 8% commission on all sales related to a design project. If a corporate design project has total sales of $198,500, find the commission.

47. **Electronics** If a home electronics salesperson earns 5% commission on total sales, find the commission for a month with $87,400 in total sales.

48. **Home Delivery** A frozen-food company pays its sales representatives 11% commission on all home delivery sales. Find the commission for $3200 in home delivery sales.

49. **Telemarketing** A telemarketing company pays its employees 9% commission. Find total sales for an employee who gets $783 in commission.

50. **Home Furnishings** A person selling home furnishings makes $2370 in commission. If the commission rate is 7.5%, find the total sales.

51. **Livestock Feed** If a livestock feed salesperson earns a $990 commission for $16,500 in total sales, find the commission rate.

52. **Building Supplies** If a building supplies salesperson earns a $1340 commission for $26,800 in total sales, find the commission rate.

53. **Net Pay** A worker has 24% of her weekly pay withheld for taxes, insurance, and a pension plan. If her weekly gross pay is $740, find her total withholdings and net pay.

54. **Net Pay** Find the total withholdings and net pay for a worker who has 18% of his $520 weekly pay withheld for taxes and insurance.

55. **Withholdings** What percent of gross pay is being withheld for a worker who has $320 withheld from $2560 in gross pay?

56. **Withholdings** A worker has $472 withheld from $2950 in gross pay. What percent of gross pay is being withheld?

57. **Gross Pay** A worker has $858 withheld from her pay every two weeks. If this represents 22% of her gross pay, what is her gross pay?

58. **Gross Pay** A worker has 28% of his salary withheld each month. Find his gross pay if he has $1820 withheld each month.

59. **Soda Prices** The cost of a 16-ounce soda from a vending machine increases from $1.25 to $1.50. Find the percent increase.

60. **Home Prices** If the average price of a 3-bedroom home decreases from $220,000 to $187,000, find the percent decrease.

61. **Waiting for iPhones** When the iPhone 4 was launched, there were 1300 customers waiting in line at 8 A.M. at the flagship store in New York City. When the iPhone 6 and 6 Plus launched, there were 1880 customers waiting. To the nearest percent, find the percent increase in customers waiting. (*Source:* Piper Jaffray.)

62. **Salary** If a worker's annual salary increases from $64,200 to $67,410, what is the percent increase in the worker's salary?

63. **Workforce** A company reduces its global workforce from 18,500 employees to 16,280 employees. Find the percent decrease.

64. **Cost of TV Ads** In 2009, the cost per 30-second primetime TV ad spot was $8800. By 2013, this cost decreased to $7700. Find this percent decrease in cost. (*Source:* Nielsen.)

WRITING ABOUT MATHEMATICS

65. What happens to a price that increases by 100%? Is it possible for a price to decrease by 100%? If so, explain what happens to the price.

66. If the price of an item is discounted by x%, explain how to find the sale price without first finding the discount amount.

SECTIONS 7.3 and 7.4 **Checking Basic Concepts**

1. Translate each percent problem to a proportion. Do not solve the proportion.
   (a) 64% of what number is 144?
   (b) 80 is what percent of 60?
   (c) What number is 99% of 70?

2. Find each unknown value.
   (a) What percent of 18.5 is 37?
   (b) 12 out of what number is 80%?
   (c) $2\frac{3}{4}$ is 11% of what number?
   (d) 1 of 6 is what percent?

3. **Nutrition** If a woman's daily diet should include 75 milligrams of vitamin C, how many milligrams of vitamin C are in a cup of yogurt with 20% of the recommended amount for women?

4. **Enrollment** The spring semester's enrollment at a college was 722 students more than the fall enrollment. If this represents an increase of 19%, how many students were there in the fall?

5. **Online Shopping** If accounting software sells for $130 and the sales tax rate is 5.5%, find the sales tax and the total amount paid.

6. **Clearance Sale** A book on a clearance table is marked at 80% off. If the regular price is $15.20, find the discount and the sale price.

7. **Home Theater Sales** If a home theater salesperson earns 8% commission on total sales, what is her commission for a month with $63,900 in total sales?

8. **Hybrid Car Price** If the price of a hybrid car decreases from $24,500 to $24,010, find the percent decrease.

7.5 Applications: Simple and Compound Interest

Objectives

1. Introducing Principal, Interest, and Interest Rates
2. Calculating Simple Interest
   - Calculating Simple Interest
   - Total Value of an Investment or Loan
3. Calculating Compound Interest

NEW VOCABULARY

☐ Principal
☐ Interest
☐ Interest rate
☐ Simple interest
☐ Annual interest rate
☐ Total value
☐ Compound interest
☐ Annual percentage rate (APR)

1 Introducing Principal, Interest, and Interest Rates

When people borrow money, they *pay interest* for the use of the money over the loan period. An amount of money that is borrowed or invested is called the **principal**. Lenders or investors are paid a fee for use of the principal. The fee, which is a percent of the principal, is called **interest**. The percent used to calculate the fee is called the **interest rate**. These three terms are summarized as follows.

PRINCIPAL, INTEREST, AND INTEREST RATE

Principal: The initial amount of an investment or loan

Interest: A fee paid to the lender or investor for use of the principal

Interest Rate: The percent used to calculate interest

STUDY TIP

If you are unfamiliar with the new vocabulary in this section, spend extra time learning the meaning of new words. Try writing the definitions in your own words or discussing their meanings with a classmate.

2 Calculating Simple Interest

CALCULATING SIMPLE INTEREST Interest that is based only on the original principal is called **simple interest**. Such interest is paid at the end of a loan period, which is written in *years*. Because the interest rate is *per year*, we say that it is an **annual interest rate**.

🌐 **Math in Context** (Interest) If a person borrows $100 from a friend and agrees to pay 4% simple interest, then the borrower pays $100 \cdot 0.04 = \$4$ in interest after one year. If the loan period is changed to two years, the borrower pays twice the $4 interest, or $100 \cdot 0.04 \cdot 2 = \$8$. *Interest is found by multiplying the principal, interest rate, and time.*

READING CHECK 1

Describe in words how simple interest is found.

SIMPLE INTEREST

The total amount of simple interest I is given by

$$I = Prt,$$

where

P = principal,
r = annual interest rate (written as a decimal), and
t = time (in years).

EXAMPLE 1 Finding simple interest

Find the simple interest for the given values of P, r, and t.
(a) $P = \$120$, $r = 6\%$, and $t = 1$ year
(b) $P = \$1600$, $r = 4.5\%$, and $t = 9$ months
(c) $P = \$9520$, $r = 3\%$, and $t = 5$ years

Solution
(a) Substitute the values $P = 120$, $r = 0.06$, and $t = 1$ into the simple interest formula.
$$I = Prt = 120 \cdot 0.06 \cdot 1 = \$7.20$$
(b) Since 9 months is $\frac{9}{12} = \frac{3}{4}$ year, we have $P = 1600$, $r = 0.045$, and $t = \frac{3}{4}$.
$$I = Prt = 1600 \cdot 0.045 \cdot \tfrac{3}{4} = \$54$$
(c) $P = 9520$, $r = 0.03$, and $t = 5$, so
$$I = Prt = 9520 \cdot 0.03 \cdot 5 = \$1428.$$

Now Try Exercises 11, 13, 15

CALCULATOR HELP

To use a calculator to find simple interest, see Appendix E (page AP-33).

EXAMPLE 2 Finding simple interest

If a student borrows $3400 at 5% simple interest for 6 months to pay for tuition, how much will the student pay in interest?

Solution
The amount of the loan is $P = 3400$, the interest rate is $r = 0.05$, and the time is 6 months, or $t = \frac{6}{12} = \frac{1}{2}$ year.
$$I = Prt = 3400 \cdot 0.05 \cdot \tfrac{1}{2} = \$85$$

Now Try Exercise 19

TOTAL VALUE OF AN INVESTMENT OR LOAN At the end of a loan period, a borrower repays the original principal and also pays the interest. Similarly, an investor receives both the original principal and the interest earned by the investment. In either case, the **total value** of the loan or investment is the sum of the principal and the interest.

$$\text{Total Value} = P + I \qquad \text{Principal plus interest}$$

EXAMPLE 3 Finding the total value of an investment

If a person invests $1250 in an account that pays 5% simple interest, find the total value of the investment after 18 months.

Solution
First, find the amount of interest received after 18 months. The amount of the investment is $P = 1250$, the interest rate is $r = 0.05$, and the time is 18 months, or $t = \frac{18}{12} = \frac{3}{2}$ year.
$$I = Prt = 1250 \cdot 0.05 \cdot \tfrac{3}{2} = \$93.75$$
The total value of the investment is the sum of the principal and the interest, or
$$P + I = 1250 + 93.75 = \$1343.75.$$

Now Try Exercise 23

If the amount of simple interest is known, then we can find an unknown principal, interest rate, or time, as shown in the next example.

EXAMPLE 4 **Finding an unknown amount of time**

A person pays $216 in interest when borrowing $2400 at 4.5% simple interest. Find the length of time for the loan (the loan period).

Solution
The interest is $I = 216$, the principal is $P = 2400$, and the interest rate is $r = 0.045$. Replace these variables in the equation $I = Prt$ and solve for t.

$$I = Prt \qquad \text{Simple interest formula}$$
$$216 = 2400 \cdot 0.045 \cdot t \qquad I = 216, P = 2400, r = 0.045$$
$$216 = 108t \qquad \text{Simplify.}$$
$$\frac{216}{108} = \frac{108t}{108} \qquad \text{Divide each side by 108.}$$
$$2 = t \qquad \text{Simplify.}$$

The loan period or time of the loan is 2 years.

Now Try Exercise 29

3 Calculating Compound Interest

Simple interest is paid once at the end of the loan period. **Compound interest** is paid at the end of each compounding period and is usually paid several times over the loan period. As a result, compound interest is computed on the original principal *and* the interest already earned.

For example, suppose $1000 is invested with an annual interest rate of 6% that is compounded semiannually, or every 6 months. After 6 months, the interest earned is

$$I = Prt = 1000 \cdot 0.06 \cdot \tfrac{1}{2} = \$30,$$

so there is $1000 + 30 = \$1030$ in the account after 6 months. During the second 6 months, the interest earned is

$$I = Prt = 1030 \cdot 0.06 \cdot \tfrac{1}{2} = \$30.90,$$

so there is now $1030.00 + 30.90 = \$1060.90$ in the account after 1 year. Simple interest would have earned $60 after 1 year. The "extra $0.90" is due to compounding. Compounding pays more interest because a person is paid "interest on his or her interest."

TABLE 7.1 lists the number of times that interest is compounded (added to the principal) each year for common compounding periods.

Compounding Periods

Compounding	Number of Times Compounded per Year
Annually	1
Semiannually	2
Quarterly	4
Monthly	12
Daily	365

TABLE 7.1

> Compounded monthly means that interest is calculated and paid after each month.

The interest rate for compound interest is usually given as an **annual percentage rate**, or **APR**. **TABLE 7.2** shows how compound interest is calculated for $1000 invested at 5% APR compounded *annually*.

$1000 Invested at 5% Compounded Annually for 4 Years

Year	Beginning Principal	Interest $I = Prt$	Ending Amount Becomes New Principal for Next Year
1	$1000.00	$1000.00 \cdot 0.05 \cdot 1 = 50.00$	$1000.00 + 50.00 = $**$1050.00**
2	$1050.00	$1050.00 \cdot 0.05 \cdot 1 = 52.50$	$1050.00 + 52.50 = $**$1102.50**
3	$1102.50	$1102.50 \cdot 0.05 \cdot 1 \approx 55.13$	$1102.50 + 55.13 = $**$1157.63**
4	$1157.63	$1157.63 \cdot 0.05 \cdot 1 \approx 57.88$	$1157.63 + 57.88 = $**$1215.51**

TABLE 7.2

Interest earned each year increases.

READING CHECK 2

If $1000 is invested at 5% compounded annually, how much interest is earned the second year?

NOTE: In **TABLE 7.2**, the total interest after 4 years is $1215.51 - 1000 = \$215.51$. However, simple interest for the same situation, $1000 invested at 5% for 4 years, is only $I = 1000 \cdot 0.05 \cdot 4 = \$200$. In general, compound interest earns more than simple interest because it *earns interest on previously earned interest*. ∎

The process used to find the final amount in **TABLE 7.2** is not practical for most compound interest problems. For example, compounding interest *monthly* for 6 years would require a table with $12 \cdot 6 = 72$ rows of computation! For such problems, the following compound interest formula is used to find the final amount directly.

COMPOUND INTEREST

The final amount A in an account paying compound interest is given by

$$A = P\left(1 + \frac{r}{n}\right)^{nt},$$

where

P = principal,
r = annual interest rate (written as a decimal),
t = time (in years), and
n = number of compounding periods per year.

NOTE: For compound interest, the final amount A is the *total value* of the account. There is no need to add the principal to this amount as we did for simple interest. ∎

MAKING CONNECTIONS 1

Simple Interest and Compound Interest

Whether we are working with simple interest or compound interest, we may want to find either the *total value* (final amount) of the loan or investment, or just the *interest* that is earned or paid. To do this, we use the following formulas.

	Simple Interest	Compound Interest	
Interest only	$I = Prt$	$A = P\left(1 + \frac{r}{n}\right)^{nt}$	Total value
Principal plus Interest	Total Value $= P + I$	Interest $= A - P$	Interest only

In the next example, we use the compound interest formula to find the final amount (total value) that was computed earlier in **TABLE 7.2**.

EXAMPLE 5 Finding a total amount after compounding interest

If $1000 is invested in an account that pays 5% APR compounded annually, find the amount in the account after 4 years.

Solution
The principal is $P = 1000$, the interest rate is $r = 0.05$, and the time is $t = 4$. Because the interest is compounded annually (one time per year), $n = 1$.

CALCULATOR HELP
To use a calculator to find compound interest, see Appendix E (page AP-33).

$$A = P\left(1 + \frac{r}{n}\right)^{nt} \qquad \text{Compound interest formula}$$

$$A = 1000\left(1 + \frac{0.05}{1}\right)^{1 \cdot 4} \qquad P = 1000, r = 0.05, t = 4, n = 1$$

$$A = 1000(1 + 0.05)^4 \qquad \text{Divide 0.05 by 1, and multiply 1 and 4.}$$

$$A = 1000(1.05)^4 \qquad \text{Add 1 and 0.05.}$$

$$A \approx 1215.51 \qquad \text{Evaluate the exponent and then multiply.}$$

The total amount in the account after 4 years is $1215.51, which agrees with **TABLE 7.2**.

Now Try Exercise 35

EXAMPLE 6 Finding a total amount after compounding interest

If $2500 is invested in an account that pays 6% APR compounded quarterly, find the amount in the account after 2 years.

Solution
The principal is $P = 2500$, the interest rate is $r = 0.06$, and the time is $t = 2$. Because the interest is compounded quarterly (four times per year), $n = 4$.

Evaluate the exponent first.

$$A = P\left(1 + \frac{r}{n}\right)^{nt} \qquad \text{Compound interest formula}$$

$$A = 2500\left(1 + \frac{0.06}{4}\right)^{4 \cdot 2} \qquad P = 2500, r = 0.06, t = 2, n = 4$$

$$A = 2500(1 + 0.015)^8 \qquad \text{Divide 0.06 by 4 and multiply 4 and 2.}$$

$$A = 2500(1.015)^8 \qquad \text{Add 1 and 0.015.}$$

$$A \approx 2816.23 \qquad \text{Evaluate the exponent and then multiply.}$$

The total amount in the account after 2 years is $2816.23.

Now Try Exercise 37

NOTE: Before using a calculator to evaluate the compound interest formula, it is often helpful to first find the value of the exponent nt by multiplying n and t. ∎

7.5 Putting It All Together

CONCEPT	COMMENTS	EXAMPLES
Principal, Interest, and Interest Rate	The *principal* is the initial amount of an investment or loan, *interest* is a fee paid to the lender or investor for use of the principal, and the *interest rate* is the percent used to calculate interest.	If $800 is invested in an account that pays 5% simple interest for 1 year, then the principal is $800, the interest is $800 \cdot 0.05 \cdot 1 = \$40$, and the interest rate is 5%.
Simple Interest	The amount of simple interest I is given by $$I = Prt,$$ where P = principal, r = annual interest rate (written as a decimal), and t = time (in years).	If $200 is borrowed at 4% simple interest for 3 years, then the interest is $$I = 200 \cdot 0.04 \cdot 3 = \$24.$$
Compound Interest	The amount A in an account paying compound interest is given by $$A = P\left(1 + \frac{r}{n}\right)^{nt},$$ where P = principal, r = annual interest rate (written as a decimal), t = time (in years), and n = the number of compounding periods per year.	If $300 is invested in an account that pays 2% interest compounded monthly, then the amount in the account after 5 years is $$A = 300\left(1 + \frac{0.02}{12}\right)^{12 \cdot 5}$$ $$\approx \$331.52.$$

7.5 Exercises

CONCEPTS AND VOCABULARY

1. An amount of money that is borrowed or invested is called _____.

2. The fee that lenders or investors are paid for use of the principal is called _____.

3. The percent used to calculate the interest on a loan or investment is called the _____.

4. Interest that is based only on the original principal is called _____ interest.

5. An interest rate that is *per year* is called a(n) _____ interest rate.

6. When using the formula $I = Prt$, the interest rate must be written as a(n) _____, and the time must be given in _____.

7. The total value of a loan or investment is the sum of the _____ and _____.

8. Interest that is paid or earned at the end of each compounding period is called _____ interest.

9. The interest rate for compound interest is usually given as a(n) _____, or APR.

10. When using the formula $A = P\left(1 + \frac{r}{n}\right)^{nt}$, the interest rate must be written as a(n) _____, and the time must be given in _____.

CALCULATING SIMPLE INTEREST

Exercises 11–16: Find the simple interest for the given values of P, r, and t.

11. $P = \$400$, $r = 8\%$, and $t = 1$ year

12. $P = \$600$, $r = 5\%$, and $t = 1$ year

13. $P = \$1200$, $r = 3.5\%$, and $t = 6$ months

14. $P = \$840$, $r = 6.5\%$, and $t = 3$ months

15. $P = \$3250$, $r = 2\%$, and $t = 8$ years

16. $P = \$4750$, $r = 7\%$, and $t = 9$ years

Exercises 17–22: Find the simple interest.

17. A mechanic borrows $2500 at 7% simple interest for 9 months to pay for tools.

18. A homeowner borrows $15,000 at 6.5% simple interest for 2 years to pay for a home theater.

19. A student lends $300 to a friend for 1 month at 4% simple interest.

20. A truck driver borrows $18,000 at 4.5% simple interest for 15 months to pay for new equipment.

21. A student borrows $40,000 from a wealthy aunt for 10 years at 3.5% simple interest.

22. A businessman lends $100,000 to a start-up company for 4 years at 7.5% simple interest.

Exercises 23–28: For the given initial investment, simple interest rate, and time of investment, find the total value of the investment.

23. $640 at 8% for 21 months

24. $12,000 at 3% for 4 months

25. $1600 at 5% for 9 years

26. $80,000 at 2.5% for 8 years

27. $900 at 1.5% for 20 months

28. $4000 at 6.5% for 11 years

Exercises 29–34: Find the unknown value.

29. $504 interest is paid when $3600 is borrowed at 3.5% simple interest. Find the time for the loan.

30. When money is invested at 8% simple interest for 3 years, the interest is $528. Find the principal.

31. $150 simple interest is paid when $5000 is borrowed for 6 months. Find the simple interest rate.

32. $90 interest is paid when $4800 is borrowed at 2.5% simple interest. Find the time for the loan.

33. When money is invested at 7.5% simple interest for 4 months, the interest is $175. Find the principal.

34. $162 simple interest is paid when $1800 is borrowed for 6 years. Find the simple interest rate.

CALCULATING COMPOUND INTEREST

Exercises 35–44: Find the final amount.

35. $1650 invested at 4% APR compounded annually for 3 years

36. $22,000 invested at 5% APR compounded annually for 4 years

37. $1000 invested at 3% APR compounded quarterly for 10 years

38. $8000 invested at 2.5% APR compounded quarterly for 2 years

39. $200 invested at 6.5% APR compounded semiannually for 5 years

40. $12,000 invested at 6% APR compounded semiannually for 15 years

41. $4000 invested at 7% APR compounded daily for 6 years

42. $2500 invested at 2% APR compounded daily for 13 years

43. $1500 invested at 8% APR compounded monthly for 54 years

44. $10,000 invested at 4.5% APR compounded monthly for 9 years

APPLICATIONS INVOLVING INTEREST

45. **Car Loans** In the fall of 2014, interest rates for a 36-month used car loan were 4.73%, where interest is compounded quarterly. If you received a loan for $5000 and paid nothing on it for 2 years, what would be the total amount owed on the loan at that time? Assume no penalties.

46. **Credit Card Interest** In the fall of 2014, fixed credit card interest rates were 13.02%, compounded daily. If you had a balance of $250 on this card and paid nothing for 3 years, what would be the balance at the end of this time period? Assume no penalties.

47. **Earning Interest** In the fall of 2014, interest rates for a 5-year CD were 0.84%. If you put $500 into CDs at this time and interest is compounded monthly, what would be the total amount earned in interest after 5 years?

48. **Home Equity Loans** In the fall of 2014, the interest rate for a home equity loan was 6.15%, compounded monthly. If you received a loan for $50,000 and paid nothing on the balance for a year, what would be the total balance at the end of this time period? Assume no penalties.

WRITING ABOUT MATHEMATICS

49. Would you rather invest money in an account that pays 5% APR compounded daily or one that pays 5% APR compounded quarterly? Explain.

50. When $100 is invested in an account that pays 6% APR compounded quarterly for 5 years, a student determined that the final amount in the account is

$$A = 100\left(1 + \frac{6}{4}\right)^{4 \cdot 5} \approx \$9{,}094{,}947{,}018.$$

How do you know that this is not correct? Find the error in the calculation.

Group Activity

Federal Debt In 2014, the U.S. public held $12.5 trillion in federal debt. At the same time, 30-year treasury bonds were paying 3.2% interest.

(a) The U.S. population was about 317 million in 2014. How much debt was held by each U.S. citizen in 2014?

(b) If interest was compounded daily and the government intended to pay back the entire amount, including interest, at the end of 30 years, how much would be paid back?

(c) If the U.S. population were to grow to 400 million by 2044, when the debt was to be paid, how much would the government owe each U.S. citizen in 2044?

(d) Suppose that the interest rate was 4.2%, rather than 3.2%, in 2014. How much would the federal government owe after 30 years?

(e) Rework part (c) if the interest rate was 4.2% as computed in part (d).

7.6 Probability and Percent Chance

Objectives

1. Understanding Experiments, Outcomes, and Events
2. Finding Probability of an Event and Percent Chance
3. Finding the Complement of an Event

NEW VOCABULARY

☐ Experiment
☐ Outcome
☐ Event
☐ Probability
☐ Percent chance
☐ Impossible event
☐ Certain event
☐ Complement

READING CHECK 1

What is an outcome?

1 Understanding Experiments, Outcomes, and Events

We often think of the word *experiment* as it relates to science. In mathematics, however, an **experiment** is an activity with an observable result.

Examples of Experiments

1. Tossing a coin and observing whether heads or tails results
2. Observing the weather and noting any precipitation
3. Rolling a six-sided die and observing the resulting number (see **FIGURE 7.5**)

Six-Sided Die

FIGURE 7.5

A result of an experiment is called an **outcome** of the experiment. Any outcome or group of outcomes for an experiment is called an **event**. Examples of events that are related to the experiments listed above include the following.

Examples of Events from the above Experiments

1. A result of tails when the coin is tossed
2. No precipitation on a given day
3. A result of 1, 3, or 5 when the six-sided die is rolled

EXAMPLE 1 Listing all possible outcomes for an experiment

When a four-sided die such as the one shown in **FIGURE 7.6** is rolled, the result is not determined by the side facing upward; rather, by the side facing downward. The result in **FIGURE 7.6** is 4. List all of the possible outcomes for an experiment in which a four-sided die is rolled.

Four-Sided Die

This four-sided die is read by looking at the face of the die that is resting on the table. This die is showing a 4.

FIGURE 7.6

Solution
A four-sided die has four possible outcomes: 1, 2, 3, and 4.

Now Try Exercise 11

STUDY TIP

If the vocabulary for this section is new to you, try to come up with your own examples for each new word. By listing several examples of your own, the words will have more meaning for you.

2 Finding Probability of an Event and Percent Chance

The **probability** of an event is a measure of its likelihood of occurring. The probability of an event must be a number from 0 to 1, which can be written as a decimal, fraction, or percent. A probability written as a percent is called a **percent chance**. For example, the percent chance of tossing a coin and observing a tail is 50%, or $\frac{1}{2}$. Numbers that could represent a probability include

$$0, \quad \frac{1}{2}, \quad 0.65, \quad 30\%, \quad \text{and} \quad 1. \qquad \text{Probabilities}$$

Because a probability must be a number from 0 to 1, numbers that could *not* represent a probability include

$$10, \quad 6\frac{1}{3}, \quad -0.33, \quad 160\%, \quad \text{and} \quad -1. \qquad \text{Not probabilities}$$

An event with probability 0 is called an **impossible event** because it cannot occur. For example, the probability of rolling a 13 on a six-sided die is 0 (or 0%) because there is no way to get a 13 on a six-sided die. An event with probability 1 is called a **certain event** because it is certain to occur. For example, the probability of getting heads *or* tails when tossing a coin is 1 (or 100%) because a coin is certain to land as either heads or tails.

READING CHECK 2

- What is the probability of an impossible event?
- What is the probability of a certain event?

EXAMPLE 2 Determining whether an event is impossible or certain

State whether each event is impossible or certain and then give its probability.
(a) Choosing a white marble from a bag that contains only black marbles
(b) Getting a result that is less than 9 when rolling a six-sided die

Solution
(a) Since it is impossible to choose a white marble from a bag that contains only black marbles, the event is impossible, and its probability is 0.
(b) Since every number on a six-sided die is less than 9, the event is certain to occur, and its probability is 1.

Now Try Exercises 17, 19

Many events have probabilities that are neither 0 nor 1. To compute the probability of such events, use the following formula.

> **THE PROBABILITY OF AN EVENT**
>
> $$\text{probability of an event} = \frac{\text{number of ways the event can occur}}{\text{number of possible outcomes for the experiment}}$$

NOTE: Probabilities computed using this formula are called *theoretical* probabilities, based on the idea that all outcomes for the experiment are *equally likely* to occur. ∎

EXAMPLE 3 Computing probability when tossing a coin

When a coin is tossed, the result is either heads or tails, as shown in **FIGURE 7.7**. Find the probability of getting heads when a coin is tossed. Express your answer as a percent chance.

FIGURE 7.7

Solution

When a coin is tossed, only **1** of the possible outcomes is heads. Since a coin can land as either heads or tails, the total number of possible outcomes is **2**.

$$\text{probability of heads} = \frac{\text{number of ways the event can occur}}{\text{number of possible outcomes}} = \frac{1}{2}$$

Since $\frac{1}{2} = 50\%$, there is a 50% chance of getting heads when a coin is tossed.

Now Try Exercise 21

EXAMPLE 4 Computing probability when rolling a die

Find the probability that a number greater than 4 results when a six-sided die is rolled. Express your answer as a fraction in simplest form.

Solution

Only **2** of the possible outcomes on a six-sided die are greater than 4, namely, 5 and 6. The total number of possible outcomes for a six-sided die is **6**.

$$\text{probability the result is greater than 4} = \frac{\text{number of ways the event can occur}}{\text{number of possible outcomes}} = \frac{2}{6} = \frac{1}{3}$$

The probability that the result is greater than 4 when rolling a six-sided die is $\frac{1}{3}$.

Now Try Exercise 27

A standard deck of 52 playing cards contains 4 suits with 13 cards each. The two black suits are Clubs and Spades, and the two red suits are Hearts and Diamonds. Each suit includes an Ace, 2, 3, 4, 5, 6, 7, 8, 9, 10, Jack, Queen, and King. The Jack, Queen, and King of each suit are known as *face cards*. See **FIGURE 7.8** on the next page.

A Standard Deck of 52 Playing Cards

FIGURE 7.8

EXAMPLE 5 Computing probabilities involving playing cards

A single card is randomly drawn from a standard deck of playing cards. Expressing your answer as a fraction in simplest form, what is the probability of getting a
(a) King of Clubs?   (b) red 5?   (c) face card?

Solution
(a) Only 1 card is the King of Clubs. The total number of possible outcomes is 52 because there are 52 cards in a standard deck.

$$\text{probability of King of Clubs} = \frac{\text{number of ways the event can occur}}{\text{number of possible outcomes}} = \frac{1}{52}$$

The probability of drawing the King of Clubs is $\frac{1}{52}$.

(b) There are 2 red 5s (5 of Hearts and 5 of Diamonds) out of a total of 52 cards.

$$\text{probability of red 5} = \frac{\text{number of ways the event can occur}}{\text{number of possible outcomes}} = \frac{2}{52} = \frac{1}{26}$$

The probability of drawing a red 5 is $\frac{1}{26}$.

(c) There are 12 face cards (Jack, Queen, and King in 4 suits) out of 52 cards.

$$\text{probability of face card} = \frac{\text{number of ways the event can occur}}{\text{number of possible outcomes}} = \frac{12}{52} = \frac{3}{13}$$

The probability of drawing a face card is $\frac{3}{13}$.

Now Try Exercises 33, 35, 37

3 Finding the Complement of an Event

 Connecting Concepts with Your Life If there are 36 chairs in a classroom and we know that 3 of the chairs are not occupied, then without counting students in the classroom, we know that there are 33 students in class. In other words, the number of students in class is found by counting the ones that are *not* in class. This example illustrates the idea of the *complement*.

In mathematics, the **complement** of an event includes all outcomes of an experiment that are *not* part of the given event. Together, an event and its complement make up all possible outcomes for an experiment. **TABLE 7.3** lists some events and their complements.

Events and Their Complements

Event	Complement
Rolling an odd number on a die	Rolling an even number on a die
Getting tails on a coin toss	Getting heads on a coin toss
Drawing a Club, Spade, or Heart	Drawing a Diamond

Rolling 1, 3, or 5 *Rolling 2, 4, or 6*

TABLE 7.3

Because an event and its complement make up all possible outcomes for an experiment, the sum of the probability of an event and the probability of its complement is 100% or 1. This idea is used in the following formulas for the probability of the complement.

> **THE PROBABILITY OF THE COMPLEMENT**
>
> probability of an event's complement = 1 − probability of the event
>
> or
>
> probability of an event's complement = 100% − percent chance of the event

READING CHECK 3

If you know the probability of an event, how is the complement of the event computed?

EXAMPLE 6   Finding the probability of the complement

Use the given information to find the requested probability or percent chance.
(a) If there is a 30% chance of rain for a given day, what is the chance of no rain?
(b) The probability that a student will pass a test is $\frac{43}{50}$. What is the probability that the student will not pass the test?

Solution
(a) The event "no rain" is the complement of the event "rain."

$$\text{probability of no rain} = 100\% - \text{probability of rain} = 100\% - 30\% = 70\%$$

There is a 70% chance that it will *not* rain.

(b) The event "not pass" is the complement of the event "pass."

$$\text{probability of not pass} = 1 - \text{probability of pass} = 1 - \frac{43}{50} = \frac{7}{50}$$

The probability that the student will not pass the test is $\frac{7}{50}$.

Now Try Exercises 43, 45

7.6   Putting It All Together

CONCEPT	COMMENTS	EXAMPLES
Experiment, Outcome, and Event	An *experiment* is an activity with an observable result. A result of an experiment is called an *outcome* of the experiment. Any outcome or group of outcomes for an experiment is called an *event*.	Rolling a six-sided die is an example of an experiment with possible outcomes of 1, 2, 3, 4, 5, and 6. Getting an odd number when rolling a die is an example of an event.
Probability and Percent Chance	The *probability* of an event is a measure of its likelihood of occurring. A probability written as a percent is called a *percent chance*. $$\text{probability} = \frac{\text{number of ways event can occur}}{\text{number of possible outcomes}}$$	Because 3 numbers are even on a six-sided die, the probability of rolling an even number is $$\text{probability of even} = \frac{3}{6} = \frac{1}{2}.$$

continued on next page

478   CHAPTER 7   PERCENT AND PROBABILITY

continued from previous page

CONCEPT	COMMENTS	EXAMPLES
Impossible and Certain Events	An *impossible* event has probability 0, and a *certain* event has probability 1.	Getting a 9 on a six-sided die is an impossible event, while getting a result less than 8 is a certain event.
Complement of an Event	The complement of an event includes all outcomes of an experiment that are *not* part of the given event. The probability of an event's complement is equal to 1 − probability of the event or 100% − percent chance of the event.	If there is a 40% chance of rain on a particular day, then there is a 100% − 40% = 60% chance that it will not rain.

7.6 Exercises   MyMathLab®

CONCEPTS AND VOCABULARY

1. A(n) _____ is an activity with an observable result.

2. The result of an experiment is called a(n) _____ of the experiment.

3. A(n) _____ is any outcome or group of outcomes for an experiment.

4. The _____ of an event is a measure of its likelihood of occurring.

5. When a probability is written as a percent, it is called a(n) _____.

6. An event with probability 0 is called a(n) _____.

7. An event with probability 1 is called a(n) _____.

8. The _____ of an event includes all outcomes that are *not* part of the given event.

LISTING OUTCOMES FOR AN EXPERIMENT

Exercises 9–14: List all possible outcomes for the given experiment.

9. An answer is picked randomly on a true/false test.

10. A woman guesses the gender of her unborn child.

11. An eight-sided die is rolled.

12. A marble is chosen from a bag containing only red, yellow, and green marbles.

13. A teacher randomly chooses a weekday for the next pop quiz.

14. A prime number less than 10 is randomly chosen.

FINDING PROBABILITY AND PERCENT CHANCE

Exercises 15–20: State whether the event is impossible or certain and then give its probability.

15. Randomly choosing a month with more than 20 days

16. Naming a day of the week that starts with the letter *b*

17. Rolling a 14 on a six-sided die

18. Getting either heads or tails when tossing a coin

19. Choosing a red marble from a bag that contains only red marbles

20. Drawing the 23 of Clubs from a standard deck

Exercises 21–26: Find the probability of the given event. Express your answer as a percent chance.

21. Guessing correctly on a true/false test question

22. Getting a number that is 3 or less when rolling a four-sided die

23. Choosing a heart from a standard deck of cards

24. Guessing correctly on a multiple choice test question with five possible answers

25. Choosing a white sock from a drawer that contains 8 white socks and 12 black socks

26. Choosing a blue marble from a bag that contains 3 blue marbles and 7 red marbles

Exercises 27–32: Find the probability of the given event. Express your answer as a fraction in simplest form.

27. Rolling a 1 or 2 on a six-sided die

28. Getting a number that is 7 or more when rolling an eight-sided die

29. Guessing *incorrectly* on a true/false test question

30. Guessing *incorrectly* on a multiple choice test question with four possible answers

31. Choosing a green marble from a bag that contains 4 blue marbles and 6 green marbles

32. Rolling a prime number on a twelve-sided die

Exercises 33–38: A card is drawn from a standard deck of cards. Expressing your answer as a fraction in simplest form, what is the probability of getting the specified card?

33. The 9 of Spades   34. A 9

35. A black face card   36. An even-numbered card

37. A black card   38. A red Ace

Exercises 39–42: The following spinner is used to play a board game. Expressing your answer as a fraction in simplest form, what is the probability that the spinner will land on the specified space?

39. Lose Turn

40. A number

41. A number greater than 200

42. A number less than 200

WORKING WITH COMPLEMENTS

Exercises 43–48: Use the given information to find the requested probability or percent chance.

43. There is an 80% chance of snow for a given day. What is the chance that it will not snow?

44. There is a 70% chance that a flight will be on time. What is the chance that it will not be on time?

45. The probability that a ski resort will have enough snow to open on or before November 1st is $\frac{7}{10}$. What is the probability that the ski resort will open after November 1st?

46. The probability of drawing a black King from a standard deck is $\frac{1}{26}$. What is the probability of drawing a card that is not a black King?

47. If there is a 60% chance that a baseball pitcher will strike out a batter, what is the chance that the pitcher will not strike out the batter?

48. The probability that a bus is early or on time is $\frac{7}{12}$. What is the probability that the bus is late?

WRITING ABOUT MATHEMATICS

49. What is the complement of the following event?

    At least one window is open

    Explain your reasoning.

50. Give a real-world example of one impossible event and one certain event.

SECTIONS 7.5 and 7.6 Checking Basic Concepts

1. Find the simple interest when lending $200 to a friend for 3 months at 5% simple interest.

2. If $1200 is invested in an account that pays 4.5% simple interest, what is the total amount in the account after 6 months?

3. If $90 interest is paid when $8000 is borrowed at 1.5% simple interest, find the time for the loan.

4. If $18,000 is invested in an account that pays 6% APR compounded monthly, find the amount in the account after 6 years.

5. Find the probability of getting an 8 when rolling a six-sided die.

6. Find the probability of choosing a face card that is a Spade when selecting a card from a standard deck. Write your answer as a fraction in simplest form.

7. Find the probability of guessing *incorrectly* on a multiple choice test question with five possible answers. Give your answer as a percent.

8. The probability that a student will pass a biology exam is $\frac{9}{11}$. What is the probability that the student will fail the exam?

CHAPTER 7 Summary

SECTION 7.1 ■ INTRODUCTION TO PERCENT; CIRCLE GRAPHS

Percent Terminology

Percent Notation The word *percent* means "divide by 100" or "out of 100."

Example: 7% means "7 divided by 100," and is written as 7%, $\frac{7}{100}$, or 0.07.

Percents as Fractions To write x% as a fraction, write $\frac{x}{100}$ and then simplify the fraction, if needed.

Example: $26\% = \frac{26}{100} = \frac{13}{50}$

Percents as Decimals To write x% as a decimal, remove the % symbol and then multiply $0.01x$.

Example: $89\% = 0.01(89) = 0.89$

Fractions or Decimals as Percents To write a fraction or decimal as a percent, multiply by 100%.

Examples: $\frac{2}{5} = \frac{2}{5} \cdot 100\% = \frac{200}{5}\% = 40\%$

$0.99 = 0.99(100\%) = 99\%$

Circle Graph A circle graph can be used to visually display percent data.

Examples: The following circle graph shows the percentage of sales for each pizza variety sold at a small pizza vendor.

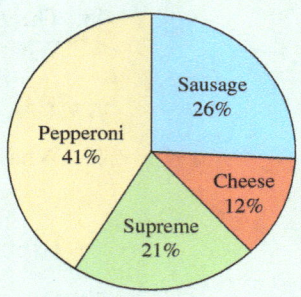

SECTION 7.2 ■ USING EQUATIONS TO SOLVE PERCENT PROBLEMS

Percent Statements, Equations, and Problems

Basic Percent Statement Form The following percent statement is written in basic percent statement form.

A percent of the whole is a part.

Example: The percent statement 12 is 15% of 80 is written in basic percent statement form as 15% of 80 is 12.

Basic Percent Equation *A percent of a whole is a part* translates to the equation

percent · whole = part.

Example: 25% of 60 is 15 translates to the equation $25\% \cdot 60 = 15$.

Percent Problem A percent problem is a question asking us to find the percent, whole, or part in a percent statement.

Example: What number is 40% of 90?

Solving a Percent Problem Solve a percent problem by solving the corresponding basic percent equation.

Example: 30% of 90 is what number?

$$30\% \cdot 90 = x \quad \text{Write an equation.}$$
$$0.30 \cdot 90 = x \quad \text{Write 30\% as 0.30.}$$
$$27 = x \quad \text{Multiply.}$$

SECTION 7.3 ■ USING PROPORTIONS TO SOLVE PERCENT PROBLEMS

Important Terms

Proportion A proportion is a statement that two ratios are equal.

Example: $\dfrac{35}{100} = \dfrac{7}{20}$

Cross Products In a proportion, the cross products are equal.

Example: $1 \cdot 18 \quad \dfrac{1}{2} = \dfrac{9}{18} \quad 2 \cdot 9$

$1 \cdot 18$ and $2 \cdot 9$ both equal 18.

Percent Statements and Problems

Translating Percent Statements to Proportions A *percent of the whole is a part* translates to the proportion

$$\dfrac{\text{percent}}{100} = \dfrac{\text{part}}{\text{whole}}.$$

Example: 40% of 50 is 20 translates to the proportion

$$\dfrac{40}{100} = \dfrac{20}{50}.$$

Solving a Percent Problem A percent problem can be solved by solving the corresponding proportion.

Example: 20% of 60 is what number?

$$\dfrac{20}{100} = \dfrac{x}{60} \quad \text{Write a proportion.}$$
$$20 \cdot 60 = 100 \cdot x \quad \text{Cross products are equal.}$$
$$1200 = 100x \quad \text{Simplify.}$$
$$\dfrac{1200}{100} = \dfrac{100x}{100} \quad \text{Divide by 100.}$$
$$12 = x \quad \text{Simplify.}$$

SECTION 7.4 ■ APPLICATIONS: SALES TAX, DISCOUNTS, AND NET PAY

Important Concepts

Basic Percent Applications When solving a basic percent application, we may need to find a missing *part*, *whole*, or *percent*.

Example: If 12% of 150 students surveyed say that they like carrots, then $0.12(150) = 18$ students like carrots.

Sales Tax and Total Amount Paid If S is sales tax, P is purchase price, and r (written as a decimal) is sales tax rate, then $S = rP$. If T is total amount paid, then $T = P + S$.

Example: If there is a 6% sales tax rate, then a $400 digital camera requires $S = 0.06(400) = \$24$ in sales tax, and the total amount paid for the digital camera is $T = 400 + 24 = \$424$.

Discounts and Sale Price   If D is a discount, O is original price, and r (written as a decimal) is discount rate, then $D = rO$. If P is purchase price, then $P = O - D$.

Example: If a $900 HDTV is marked at 25% off, then the discount is $D = 0.25(900) = \$225$, and the purchase price is $P = 900 - 225 = \$675$.

Commission   If C is commission, T is total sales, and r (written as a decimal) is commission rate, then $C = rT$.

Example: If a person has a 3% commission rate on $18,000 in total sales, then the commission is $C = 0.03(18,000) = \$540$.

Net Pay   If W is withholdings, G is gross pay, and r (written as a decimal) is the withholding rate, then $W = rG$. If N is net pay, then $N = G - W$.

Example: If 22% of a factory worker's $1400 gross pay is withheld, then the withholdings are $W = 0.22(1400) = \$308$, and the net pay is $N = 1400 - 308 = \$1092$.

Percent Change   If an amount changes from an **old amount** to a **new amount**, then the percent change is

$$\frac{\text{new amount} - \text{old amount}}{\text{old amount}} \cdot 100.$$

Examples: When a price increases from $60 to $72, the percent change is

$$\frac{72 - 60}{60} \cdot 100 = 20\%.$$

When a price decreases from $60 to $45, the percent change is

$$\frac{45 - 60}{60} \cdot 100 = -25\%.$$

SECTION 7.5 ■ APPLICATIONS: SIMPLE AND COMPOUND INTEREST

Interest Terminology

Principal, Interest, and Interest Rate   The *principal* is the initial amount of an investment or loan, *interest* is a fee paid to the lender or investor for use of the principal, and the *interest rate* is the percent used to calculate interest.

Example: If $100 is invested in an account that pays 4% simple interest for 1 year, then the principal is $100, the interest is $100 \cdot 0.04 \cdot 1 = \$4$, and the interest rate is 4%.

Simple Interest   The amount of simple interest I is given by $I = Prt$, where $P =$ principal, $r =$ annual interest rate (written as a decimal), and $t =$ time (in years).

Example: If $300 is borrowed at 5% simple interest for 6 years, then the interest is $I = 300 \cdot 0.05 \cdot 6 = \$90$.

Compound Interest   The amount A in an account paying compound interest is given by

$$A = P\left(1 + \frac{r}{n}\right)^{nt},$$

where $P =$ principal, $r =$ annual interest rate (written as a decimal), $t =$ time (in years), and $n =$ the number of compounding periods per year.

Example: If $1600 is invested in an account that pays 8% interest compounded monthly, then the amount in the account after 4 years is

$$A = 1600\left(1 + \frac{0.08}{12}\right)^{12 \cdot 4} \approx \$2201.07.$$

SECTION 7.6 ■ PROBABILITY AND PERCENT CHANCE

Important Probability Terms

Experiment, Outcome, and Event

An *experiment* is an activity with an observable result. A result of an experiment is called an *outcome* of the experiment. Any outcome or group of outcomes for an experiment is called an *event*.

Example: Rolling a 12-sided die is an example of an experiment that has possible outcomes of 1, 2, 3, 4, 5, 6, 7, 8, 9, 10, 11, and 12. Getting a number less than 5 when rolling this die is an example of an event.

Probability and Percent Chance

The *probability* of an event is a measure of its likelihood of occurring. A probability written as a percent is called a *percent chance*.

$$\text{probability} = \frac{\text{number of ways event can occur}}{\text{number of possible outcomes}}$$

Example: Because 3 numbers are odd on a six-sided die, the probability of rolling an odd number is

$$\text{probability of odd} = \frac{3}{6} = \frac{1}{2}. \text{ (The percent chance is 50\%.)}$$

Impossible and Certain Events

An *impossible* event has probability 0, and a *certain* event has probability 1.

Example: Getting a 5 on a four-sided die is an impossible event, while getting a result less than 5 is a certain event.

Complement of an Event

The complement of an event includes all outcomes of an experiment that are *not* part of the given event. The probability of an event's complement equals

1 − probability of the event   or   100% − percent chance of the event.

Example: If there is a 15% chance of fog on a particular day, then there is a 100% − 15% = 85% chance that there will be no fog on that day.

CHAPTER 7 Review Exercises

SECTION 7.1

Exercises 1–4: Write the percent as a fraction or mixed number in simplest form.

1. 68%
2. 180%
3. $6\frac{1}{4}\%$
4. 7.2%

Exercises 5–8: Write the percent as a decimal.

5. 29%
6. 0.4%
7. 625%
8. $35\frac{3}{4}\%$

Exercises 9–12: Write the fraction or mixed number as a percent.

9. $\frac{17}{25}$
10. $3\frac{1}{5}$
11. $\frac{7}{9}$
12. $\frac{5}{16}$

Exercises 13–16: Write the decimal as a percent.

13. 0.16
14. 0.049
15. 7.02
16. 0.0305

Exercises 17 and 18: The following circle graph shows expenses for a student at a private college.

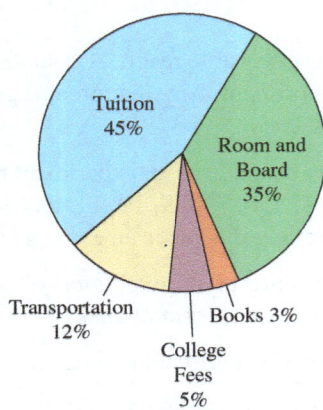

17. Which expense was greatest?

18. What fraction of the student's expenses went toward transportation?

SECTION 7.2

Exercises 19 and 20: Write the given percent statement in basic percent statement form.

19. 52 is 130% of 40.
20. 9 out of 12 is 75%.

Exercises 21 and 22: Translate the given percent statement to a basic percent equation.

21. 77.5% is 62 of 80.
22. 50% of 184 is 92.

Exercises 23 and 24: Translate the given percent problem to a basic percent equation. Do not solve the equation.

23. 75% of what number is 46.5?
24. 48 out of 72 is what percent?

Exercises 25–28: Use a basic percent equation to find the unknown value.

25. What number is 16% of 85?
26. $10\frac{1}{2}$ out of what number is 30%?
27. 4.1 is what percent of 12.3?
28. 3 out of 5 is what percent?

SECTION 7.3

Exercises 29 and 30: Translate the given percent statement to a proportion.

29. 96 is 240% of 40.
30. 28 of 35 is 80%.

Exercises 31 and 32: Translate the given percent problem to a proportion. Do not solve the proportion.

31. What number is 165% of 20?
32. What percent is 3 of 9?

Exercises 33–36: Use a proportion to find the unknown value.

33. What number is 13% of 60?
34. 560 out of 840 is what percent?
35. 34.6 is what percent of 69.2?
36. 4.5% of what number is 198?

SECTION 7.4

37. **Nutrition** If a daily diet should include 35 grams of dietary fiber, how many grams of dietary fiber are in a serving of whole grain bread with 8% of the recommended amount?

38. **Cheating** Of 12,000 high school students surveyed, 26% did *not* cheat on an exam during the past year. How many of the students surveyed did not cheat? (*Source:* Josephson Institute of Ethics.)

39. **Going Green** Only 82 of the 2050 students on a college campus used a bicycle to get to campus. What percent of the students rode a bike to school?

40. **Saving for College** The parents of 800 college freshmen were asked how much they had saved for their child's college education. The parents of 280 freshmen had saved more than $5000. What percent of parents saved more than $5000? (*Source:* College Savings Foundation.)

41. **Smartphone** If the price of a smartphone is $198 and the sales tax rate is 5.5%, find the sales tax and the total amount paid.

42. **Tax Rate** A student bought a car stereo with a purchase price of $646.50. If the sales tax was $25.86, find the sales tax rate.

43. **Microwave Oven** A microwave oven is marked at 20% off. If the regular price is $380, find the discount and the sale price.

44. **Prepaid Minutes** An international cellular customer receives a $42.50 discount when she buys 2000 prepaid minutes. If the regular price for 2000 minutes is $250, what is the discount rate?

45. **Conference Fees** A professor gets a $15 discount for prepaying conference fees at least 60 days in advance. If this represents an 8% discount, find the original cost of the conference.

46. **Clothing** A salesperson at a women's clothing store earns 7.5% commission on total sales. Find the commission for a month with $7,400 in total sales.

47. **Office Supplies** If an office supplies salesperson earns a $126 commission for $1050 in total sales, find the commission rate.

48. Net Pay Find the total withholdings and net pay for a worker who has 24% of his $820 weekly pay withheld for taxes and insurance.

49. Withholdings What percent of gross pay is being withheld for a worker who has $420 withheld from $1890 in gross pay?

50. Gross Pay A worker has $456 withheld from her pay every two weeks. If this represents a 20% withholding rate, find her gross pay.

51. Snack Prices The cost of a candy bar from a vending machine increases from $0.75 to $1.00. Find the percent increase.

52. Depreciation If driving a new car off the lot decreases its value from $22,600 to $18,080, find the percent decrease.

SECTION 7.5

Exercises 53 and 54: Find the simple interest for the given values of P, r, and t.

53. $P = \$480$, $r = 4.5\%$, and $t = 4$ months

54. $P = \$38,650$, $r = 4\%$, and $t = 6$ years

Exercises 55 and 56: Find the simple interest.

55. A homeowner borrows $28,000 at 4.5% simple interest for 2 years to pay for a swimming pool.

56. A student borrows $64,000 for 12 years at 3.5% simple interest.

Exercises 57 and 58: For the given initial investment, simple interest rate, and time of investment, find the total value of the investment.

57. $4500 at 6% for 8 years.

58. $1200 at 2.5% for 18 months.

Exercises 59 and 60: Find the unknown value.

59. When money is invested at 6% simple interest for 5 years, the interest is $1170. Find the principal.

60. $45 interest is paid when $2400 is borrowed at 2.5% simple interest. Find the time for the loan.

Exercises 61–64: Find the final amount.

61. $2500 invested at 8% APR compounded monthly for 48 years

62. $7000 invested at 3.5% APR compounded quarterly for 7 years

63. $34,000 invested at 5% APR compounded semiannually for 11 years

64. $400 invested at 7% APR compounded daily for 19 years

SECTION 7.6

Exercises 65 and 66: List all possible outcomes for the given experiment.

65. A marble is chosen from a bag containing only red, black, and blue marbles.

66. A prime number less than 20 is randomly chosen.

Exercises 67 and 68: State whether the event is impossible or certain and then give its probability.

67. Rolling a number less than 14 on a six-sided die.

68. Naming a U.S. state that starts with the letter z.

Exercises 69 and 70: Find the probability of the given event. Express your answer as a percent chance.

69. Guessing correctly on a multiple choice test question with four possible answers

70. Choosing a white marble from a jar that contains 9 white marbles and 15 black marbles

Exercises 71 and 72: Find the probability of the given event. Express your answer as a fraction in simplest form.

71. Rolling a 5 or 6 on a six-sided die

72. Guessing *incorrectly* on a multiple choice test question with five possible answers

Exercises 73 and 74: A card is drawn from a standard deck of cards. Expressing your answer as a fraction in simplest form, what is the probability of getting the specified card?

73. a Jack

74. a red, even-numbered card

Exercises 75 and 76: Use the given information to find the requested probability or percent chance.

75. There is a 75% chance that farmers will be able to harvest crops in the next week. What is the chance that they will not be able to harvest crops in the next week?

76. The probability that it will be warm enough for a water park to open on or before June 1st is $\frac{8}{11}$. What is the probability that the park will open after June 1st?

APPLICATIONS

77. Obesity A recent study found that $\frac{33}{50}$ of the American population was *not* obese. Write this fraction as a percent. (*Source:* National Center for Health Statistics.)

78. Shopping A bargain table at a sporting goods store advertises "40% off the regular price." Write this percent as a fraction.

79. Hybrid Car A car dealer claims that a hybrid car gets 42.8% better mileage than a large SUV. Write this percent as a decimal.

80. Gender A class with 32 students has 20 women. What percent of the class do women represent?

CHAPTER 7 Test

Exercises 1 and 2: Write the given percent as a fraction or mixed number in simplest form.

1. 55%
2. 260%

Exercises 3 and 4: Write the given fraction or decimal as a percent.

3. $\frac{19}{20}$
4. 0.078

5. Translate the following percent problem to a basic percent equation. Do not solve the equation.

   What percent of 160 is 32?

6. Translate the following percent problem to a proportion. Do not solve the proportion.

   What number is 12 percent of 90?

Exercises 7 and 8: Use a basic percent equation to find the unknown value.

7. 30% of what number is 42?
8. What percent of 225 is 75?

Exercises 9 and 10: Use a proportion to find the unknown value.

9. 38 out of 40 is what percent?
10. 10.5% of what number is 84?

11. Find the simple interest when $1400 is borrowed for 6 months at 4.5% interest.

12. If $20,000 is invested in an account that pays 6% simple interest, find the total value of the investment after 5 years.

13. If an investment of $1500 earns $90 simple interest after 2 years, find the interest rate.

14. If $2000 is invested in an account that pays 6% APR compounded quarterly, find the total amount in the account after 3 years.

15. List all possible outcomes when a positive even number less than 11 is randomly selected.

16. State whether the following event is an impossible or certain event and give its probability.

    Naming a month that starts with the letter q.

17. Find the probability of getting a 3, 4, or 5 when rolling a six-sided die. Express your answer as a fraction in simplest form.

18. There is a 38% chance that a student has blonde hair. What is the chance that a student is not blonde?

19. **The Human Body** An adult female's body is about 55% water. If a woman's body has 77 pounds of water, what is her total body weight?

20. **Salary** A worker's salary increases by 4.5%. If the amount of the raise amounts to $2880, how much was the worker being paid before the raise?

21. **Car Stereo** If speakers for a car stereo sell for $349 and the sales tax rate is 5%, find the sales tax and the total amount paid.

22. **Clearance Sale** A video game on a clearance table is marked at 60% off. If the regular price is $19.25, find the discount and the sale price.

23. **Home Design** An interior designer receives 12% commission on all sales related to a design project. If a home design project has total sales of $38,500, find the commission.

24. **Stock Price** If a stock price increases from $58.40 per share to $67.16 per share, find the percent increase.

CHAPTERS 1–7 Cumulative Review Exercises

1. Write *forty-six thousand, three hundred ninety-one* in standard form.

2. Round 658,255 to the nearest thousand.

3. Evaluate $18 - (3 \cdot 4 - (24 \div 2^2) + 1)$.

4. Simplify $7w + 2(w - 5)$.

5. Simplify $-7 - 13 + (-1)$.

6. Find the product $3(-1)(3)(-2)(-4)$.

7. Evaluate the expression $w + ((3 - v)^2 \div 2w)$, when $v = 3$ and $w = -2$.

8. Use the following graph to solve $-3x + 28 = 25$ graphically.

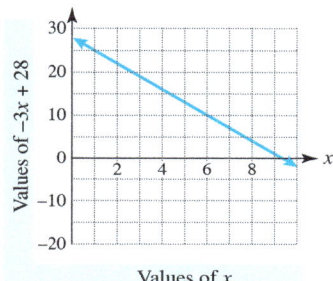

9. Simplify $-2(5t + 1) + 3(t - 2)$.

10. Translate the following sentence into an equation using x as the variable. Do not solve the equation.

    Twice the sum of a number and 5 gives -14.

11. Solve $10 + 5x = 4x - 6$ symbolically.

12. Complete the following table of values to solve the equation $-3x + 6 = 3$ numerically.

x	-2	-1	0	1	2
$-3x + 6$					

13. Find the prime factorization of 105.

14. Add $-\frac{5}{7} + \frac{9}{14}$.

15. Subtract $7\frac{3}{5} - 1\frac{4}{15}$.

16. Solve $4\left(x - \frac{2}{3}\right) = \frac{5}{4}$ symbolically.

17. Round 468.049 to the nearest hundredth.

18. For the given list of numbers, identify any of the following types of numbers.
    (a) natural numbers
    (b) whole numbers
    (c) integers
    (d) rational numbers
    (e) irrational numbers

    $$9, -\frac{1}{5}, 4.\overline{6}, -2.1, \sqrt{7}$$

19. Solve $5(x + 4.7) = 53.5$ symbolically.

20. Find the exact area of the circle and then approximate the area using 3.14 as an approximation for π.

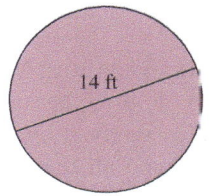

21. Convert 2.6 miles to feet.

22. Convert 0.089 deciliter to milliliters.

23. Convert $-31°$F to Celsius.

24. Convert 5 kilometers per minute to meters per hour.

25. Write 137% as a decimal.

26. What percent of 144 is 48?

27. What number is 15% of 840?

28. A card is randomly selected from a standard deck. Find the probability that the card is a red card with the number 2, 7, or 9. Express your answer as a fraction in simplest form.

29. **Heart Rate** The average heart rate R in beats per minute (bpm) of an animal weighing W pounds can be approximated by

    $$R = \frac{885\sqrt{W}}{W}.$$

    Find the heart rate for a 25-pound coyote.

30. **Music Sales** The total music sales S in thousands of dollars for a small band during year x can be computed using the formula
$$S = 7(x - 2009) + 8.$$
Find the year when the music sales were $29,000 by solving the equation $29 = 7(x - 2009) + 8$ graphically, using the following graph.

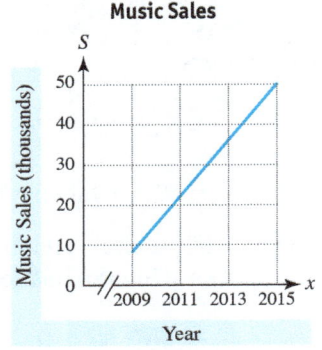

31. **Altitude and Temperature** If the temperature on the ground is 75 °F, then the air temperature T at an altitude of x miles is given by $T = 75 - 19x$. Find the altitude where the air temperature is $T = 18$ °F.
(*Source:* Miller, A., and R. Anthes, *Meteorology.* 5th ed.)

32. **Height** At birth, a baby boy was $19\frac{3}{4}$ inches long. If he grew $6\frac{1}{2}$ inches during his first year, how tall was he on his first birthday?

33. **Text Messaging** Suppose that the cost C of sending or receiving x international text messages is given by the formula
$$C = 0.1(x - 400) + 7.5, \text{ where } x \geq 400.$$
Find the number of text messages that correspond to total charges of $19.60 by replacing C in the formula with 19.6 and solving the resulting equation.

34. **Unit Price** Find the unit price for each size option.

   Large fruit drink: 24 ounces for $4.20
   Small fruit drink: 16 ounces for $2.96

   Which is the better buy?

35. **Coffee Maker** A coffee maker is marked at 25% off. If the regular price is $38, find the discount and the sale price.

36. **Going Green** A company reduces solid waste from 4600 tons per year to 4048 tons per year. Find the percent decrease.

8 Geometry

- 8.1 **Plane Geometry: Points, Lines, and Angles**
- 8.2 **Triangles**
- 8.3 **Polygons and Circles**
- 8.4 **Perimeter and Circumference**
- 8.5 **Area, Volume, and Surface Area**

One of the most noted landmark buildings ever constructed is the Beijing National Stadium, the venue for the Olympic Games. Because of its seemingly random web of twisting steel, the building is commonly called the *Bird's Nest*. Its 110,000 tons of steel make the Bird's Nest the largest steel structure in the world.

From a distance, the outer surface of the stadium appears smooth and simple, but the *geometry* used in its design is very complex.

Without geometry, building structures as remarkable as the Bird's Nest or as simple as a backyard shed would be impossible. In this chapter, we discuss many of the basic concepts in the branch of mathematics known as geometry. (*Source:* Design Build Network, *Beijing National Stadium, China.*)

8.1 Plane Geometry: Points, Lines, and Angles

Objectives

1. Reviewing Geometric Terms and Concepts
2. Classifying Angles
   - Right, Straight, Acute, and Obtuse Angles
   - Congruent, Supplementary, and Complementary Angles
3. Recognizing Parallel, Intersecting, and Perpendicular Lines
4. Using Properties of Parallel Lines Cut by a Transversal

NEW VOCABULARY

- ☐ Plane
- ☐ Plane geometry
- ☐ Vertex
- ☐ Side (of an angle)
- ☐ Degree
- ☐ Right angle
- ☐ Straight angle
- ☐ Acute angle
- ☐ Obtuse angle
- ☐ Congruent angles
- ☐ Complementary angles
- ☐ Supplementary angles
- ☐ Parallel lines
- ☐ Intersecting lines
- ☐ Perpendicular
- ☐ Vertical angles
- ☐ Adjacent angles
- ☐ Transversal
- ☐ Corresponding angles
- ☐ Alternate interior angles
- ☐ Alternate exterior angles

1 Reviewing Geometric Terms and Concepts

Connecting Concepts with Your Life A **plane** is a flat surface that continues without end. When a child draws on a sidewalk with chalk, the sidewalk represents (a portion of) a plane. Other plane surfaces include a white board in a classroom and a flat sheet of paper. A plane is *two-dimensional*—it has length and width but no height.

Geometric figures with two or fewer dimensions are the focus of **plane geometry**. **TABLE 8.1** shows descriptions and examples of terms used in plane geometry.

Terms Used in Plane Geometry

Term	Description	Example(s)	Notation
Point	A location in space having no length, width, or height	• P	Point P
Line	A straight figure representing a set of points extending indefinitely in two directions	$\overleftrightarrow{AB}$; line m	Line AB or $\overleftrightarrow{AB}$ Line m
Line Segment	A straight figure representing a set of points extending between two endpoints	segment from A to B	Segment AB or $\overline{AB}$
Ray	A straight figure representing a set of points extending indefinitely in one direction	ray from A through B	Ray AB or $\overrightarrow{AB}$
Angle	A figure formed by two rays with a common endpoint	angle with vertex B, sides to A and C, measure x	$\angle ABC$ or $\angle CBA$ or $\angle B$ or $\angle x$

TABLE 8.1

STUDY TIP

The terms listed in **TABLE 8.1** are the "building blocks" of plane geometry. Spend extra time learning the meanings of these terms.

EXAMPLE 1 Identifying geometric figures

Identify the figure and name it using the labels shown.

(a) • A (b) ray PQ (c) segment CD (d) angle x (e) line MN

Solution

(a) Point A is a dot representing a location.
(b) This figure is straight, has one endpoint, and extends in one direction. It is a ray. To name this ray, we write ray PQ or $\overrightarrow{PQ}$.

(c) The figure contains all points extending between two endpoints. It is a line segment. To name this line segment, we write segment CD or $\overline{CD}$.
(d) An angle is formed by two rays with a common endpoint. It is $\angle x$.
(e) A line is straight and extends in two directions. This line is line MN or $\overleftrightarrow{MN}$.

Now Try Exercises 23, 25, 27, 29, 31

When two rays form an angle, the common endpoint is the angle's **vertex** and the rays are the angle's **sides**. Some angles can be named using only the vertex letter. However, when three letters are used to name an angle, the vertex letter should always be in the middle. For example, the angle in **FIGURE 8.1** can be named as follows.

Naming an Angle

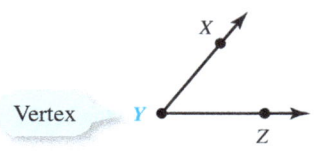

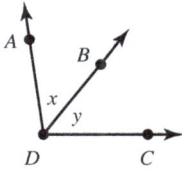

FIGURE 8.1

In the next example, we analyze the angles in a figure and discuss why it is not always appropriate to name an angle using only its vertex letter.

EXAMPLE 2 Analyzing angles in a figure

Refer to **FIGURE 8.2**.
(a) Name the vertex of $\angle x$.
(b) Name the two sides of $\angle y$.
(c) Use three-letter naming to name three *different* angles that have vertex D.
(d) Explain why it is not appropriate to name an angle as $\angle D$.

FIGURE 8.2

Solution
(a) The common endpoint of the rays that form $\angle x$ is the vertex. It is D.
(b) The two sides of $\angle y$ are the rays that form the angle. They are $\overrightarrow{DB}$ and $\overrightarrow{DC}$.
(c) The angles are $\angle ADB$, $\angle BDC$, and $\angle ADC$.
(d) Because there are three different angles with vertex D, naming an angle as $\angle D$ would not clearly define a single angle. We cannot tell which of the angles is being named.

Now Try Exercise 33

2 Classifying Angles

Angles are often measured in *degrees*. There are 360 degrees in one revolution, as shown in **FIGURE 8.3**, so a **degree** is a measure representing $\frac{1}{360}$ of a revolution.

360° in One Direction

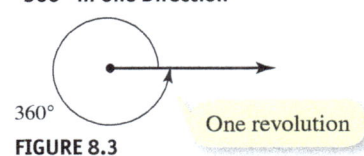

FIGURE 8.3

RIGHT, STRAIGHT, ACUTE, AND OBTUSE ANGLES An angle that measures 90° is called a **right angle**. A right angle is $\frac{1}{4}$ of a revolution. A small square near its vertex, as shown in **FIGURE 8.4**, tells us that the angle is a right angle and measures 90°.

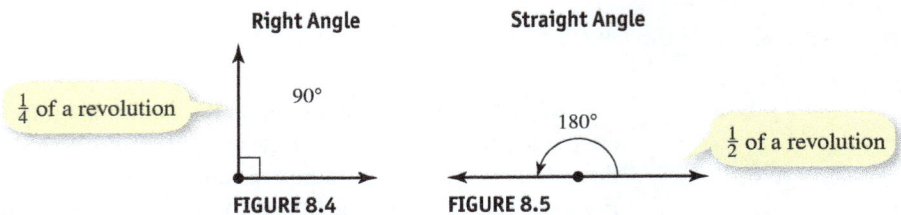

READING CHECK 1

What is the difference between an acute angle and an obtuse angle?

An angle that measures 180° is called a **straight angle**. A straight angle is $\frac{1}{2}$ of a revolution, as shown in **FIGURE 8.5**.

An angle that measures between 0° and 90° is called an **acute angle**, while an angle that measures between 90° and 180° is called an **obtuse angle**. Examples of acute and obtuse angles are shown in **FIGURES 8.6** and **8.7**, respectively.

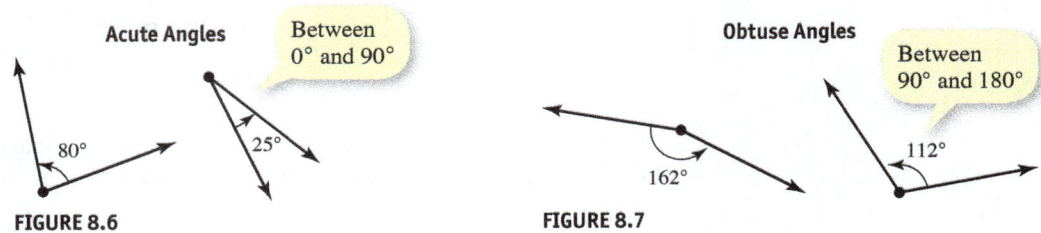

EXAMPLE 3   Classifying angles as acute, right, obtuse, or straight

Classify each angle as acute, right, obtuse, or straight.

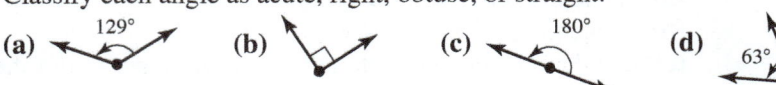

Solution
(a) The measure of the angle is between 90° and 180°. It is an obtuse angle.
(b) The square near the vertex indicates that the angle is a right angle.
(c) The angle forms a straight line and has measure 180°. It is a straight angle.
(d) The measure of the angle is between 0° and 90°. It is an acute angle.

Now Try Exercises 35, 37, 39, 41

CONGRUENT, SUPPLEMENTARY, AND COMPLEMENTARY ANGLES There are three special ways to classify two angles considered together.

1. **Congruent angles**: Two angles whose measures are equal.
2. **Complementary angles**: Two angles whose measures sum to 90°.
3. **Supplementary angles**: Two angles whose measures sum to 180°.

NOTE: If two angles are complementary, then each angle is the **complement** of the other. If two angles are supplementary, then each angle is the **supplement** of the other. ∎

FIGURES 8.8, **8.9**, and **8.10** show examples of congruent, complementary, and supplementary angles, respectively.

Congruent Angles

$31° = 31°$

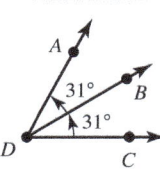

FIGURE 8.8

Complementary Angles

$35° + 55° = 90°$

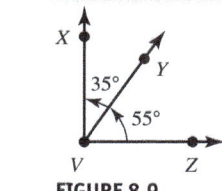

FIGURE 8.9

Supplementary Angles

$58° + 122° = 180°$

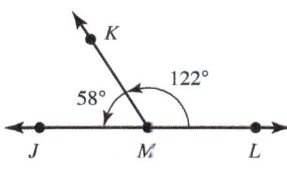

FIGURE 8.10

NOTE: The symbol ($\cong$) is used to show that two angles are congruent. For the angles shown in **FIGURE 8.8**, we write $\angle ADB \cong \angle BDC$. ∎

EXAMPLE 4 Finding congruent, complementary, and supplementary angles

Find the measure of an angle that is
(a) the complement of an angle with measure 71°.
(b) the supplement of an angle with measure 32°.
(c) congruent to an angle with measure 49°.

Solution
(a) Two angles are complements of each other if the sum of their angle measures is 90°. If one angle measures 71°, then the other must measure **19°** because

$$71° + 19° = 90°.\quad \text{Complementary angles sum to 90°.}$$

(b) The supplement of an angle with measure 32° is an angle with measure **148°** because

$$32° + 148° = 180°.\quad \text{Supplementary angles sum to 180°.}$$

(c) Congruent angles have equal measure, so an angle that is congruent to an angle with measure 49° also has measure 49°.

Now Try Exercises 43, 45, 47

3 Recognizing Parallel, Intersecting, and Perpendicular Lines

If two lines in a plane never touch, they are **parallel lines**. We can indicate that the lines in **FIGURE 8.11** are parallel by using the symbol ($\parallel$), as in $m \parallel n$. When two lines meet, they are called **intersecting lines**. The lines in **FIGURE 8.12** intersect at the point I.

Perpendicular Lines

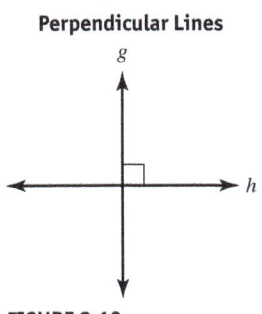

FIGURE 8.13

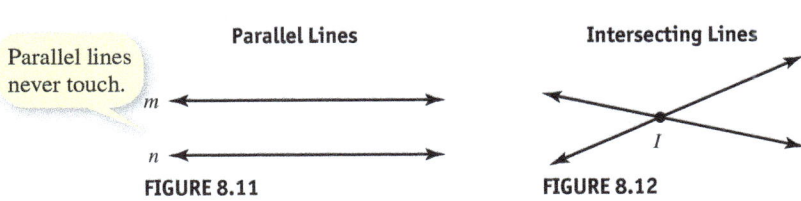

FIGURE 8.11   FIGURE 8.12

When two lines intersect to form right angles, we say that the lines are **perpendicular**. We can indicate that the intersecting lines in **FIGURE 8.13** are perpendicular by using the symbol ($\perp$), as in $g \perp h$.

READING CHECK 2

- What symbol indicates that two lines are parallel?
- What symbol indicates that two lines are perpendicular?

See the Concept 1 — ANGLE PAIRS FOR INTERSECTING LINES

Two intersecting lines that are not perpendicular form two *pairs* of congruent angles and four *pairs* of supplementary angles. Two angles in a congruent pair are called **vertical angles** and two angles in a supplementary pair are called **adjacent angles**.

Vertical Angles	Intersecting Lines	Adjacent Angles
∠w and ∠y		∠w and ∠x
∠x and ∠z		∠x and ∠y
Congruent Pairs		∠y and ∠z
		∠z and ∠w
		Supplementary Pairs

WATCH VIDEO IN MML.

We can use mathematical notation to say that an angle has a specified measure. For example, if ∠b has measure 79°, we can write $m\angle b = 79°$. The m in front of the angle symbol tells us that we are talking about the *measure of the angle*.

EXAMPLE 5 Finding angle measures

If $m\angle w = 36°$ in **FIGURE 8.14**, find the measures of ∠x, ∠y, and ∠z.

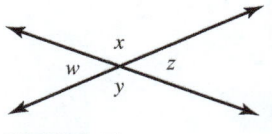

FIGURE 8.14

Solution
Since ∠x and ∠w are adjacent angles, they are supplementary. So, $m\angle x = 144°$ because $36° + 144° = 180°$. Similarly, since ∠y and ∠w are also adjacent angles, $m\angle y = 144°$. Since ∠w and ∠z are vertical angles, they have the same measure. So, $m\angle z = 36°$.

Now Try Exercise 49

NOTE: After determining that $m\angle x = 144°$ in Example 5, we could have determined that $m\angle y = 144°$ because ∠x and ∠y are vertical angles, and are therefore congruent. ■

4 Using Properties of Parallel Lines Cut by a Transversal

The following See the Concept shows parallel lines m and n "cut" by the transversal t.

See the Concept 2 — PARALLEL LINES CUT BY A TRANSVERSAL

A line that intersects two other lines in the same plane (at different points) is called a **transversal**. If a transversal intersects two *parallel* lines, then the eight angles formed have special relationships.

Corresponding Angles	Alternate Interior Angles	Alternate Exterior Angles
∠a and ∠e	∠c and ∠f	∠a and ∠h
∠b and ∠f	∠d and ∠e	∠b and ∠g
∠c and ∠g		
∠d and ∠h		

WATCH VIDEO IN MML.

In **TABLE 8.2** we define the special angle pairs shown in the See the Concept.

Key Definitions When Parallel Lines Are Cut by a Transversal

Definition	Angle Pairs and Illustration	
Corresponding angles are angles in the same relative position with respect to the parallel lines.	∠a and ∠e ∠b and ∠f ∠c and ∠g ∠d and ∠h	
Alternate interior angles are angles that are between the parallel lines, on opposite sides of the transversal, and are *not* adjacent angles.	∠c and ∠f ∠d and ∠e	
Alternate exterior angles are angles that are outside of the parallel lines, on opposite sides of the transversal, and are *not* adjacent angles.	∠a and ∠h ∠b and ∠g	

TABLE 8.2

These definitions help us to define the following properties of angles when two parallel lines are cut by a transversal.

> ### TWO PARALLEL LINES CUT BY A TRANSVERSAL
>
> When two parallel lines are cut by a transversal, the following properties apply.
>
> 1. Vertical angles are congruent.
> 2. Corresponding angles are congruent.
> 3. Alternate interior angles are congruent.
> 4. Alternate exterior angles are congruent.
>
> Two angles in any other angle pair are supplementary.

EXAMPLE 6 Finding angle measures

If $p \parallel q$ and $m\angle h = 78°$ in **FIGURE 8.15**, find the measures of $\angle a$, $\angle c$, and $\angle f$.

Parallel Lines Cut by a Transversal

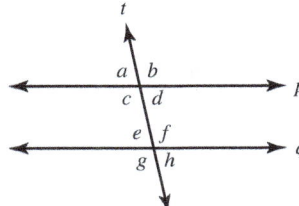

FIGURE 8.15

Solution
Since $\angle h$ and $\angle a$ are alternate exterior angles, they are congruent. So, $m\angle a = 78°$. Since $\angle c$ and $\angle a$ are adjacent angles, the sum of their measures is 180°. That is, $m\angle c = \mathbf{102°}$

because 78° + 102° = 180°. Finally, ∠c and ∠f are congruent because they are alternate interior angles. So, m∠f = 102°.

Now Try Exercise 65

READING CHECK 3

- When parallel lines are cut by a transversal, how many angle measures result?

NOTE: In Example 6, each of the angles has a measure of either 78° or 102°. In general, when a transversal cuts two parallel lines, all angles have one of two measures and the two measures sum to 180°. ∎

8.1 Putting It All Together

CONCEPT	EXAMPLES
A **point** is a location in space having no length, width, or height.	• P
A **line** is a straight figure representing a set of points extending indefinitely in two directions.	 A B
A **line segment** is a straight figure representing a set of points extending between two endpoints.	 A B
A **ray** is a straight figure representing a set of points extending indefinitely in one direction.	 A B
An **angle** is a figure formed by two rays with a common endpoint.	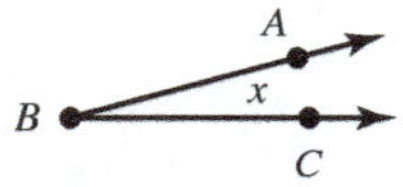 A x B C
The common endpoint of the rays that form an angle is called the **vertex** of the angle. The two rays are called the **sides** of the angle.	The vertex is B, and the sides are $\overrightarrow{BA}$ and $\overrightarrow{BC}$. 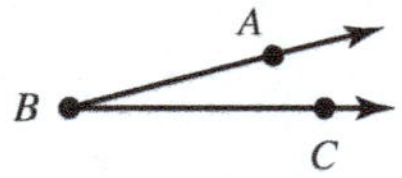 A B C
A **degree** is a unit of measure that represents $\frac{1}{360}$ of a revolution.	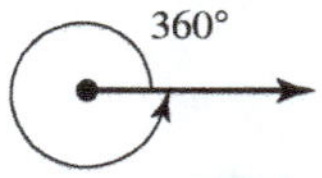 360°
A **right angle** has measure 90°. A small square near the vertex indicates a right angle.	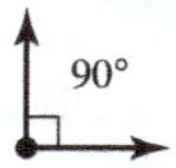 90°
A **straight angle** has measure 180°.	 180°
An **acute angle** measures between 0° and 90°.	 82°
An **obtuse angle** measures between 90° and 180°.	 159°
Angles with equal measures are **congruent angles**.	 29° 29°

CONCEPT	EXAMPLES
The sum of the measures of two **complementary** angles is 90°. Each angle is the complement of the other.	37° 53°
The sum of the measures of two **supplementary** angles is 180°. Each angle is the supplement of the other.	56° 124°
Parallel lines never touch. We indicate that two lines are parallel by using the symbol ($\parallel$). When two lines meet, they are **intersecting lines**.	**Parallel Lines** **Intersecting Lines** m n l
Two lines are perpendicular if they intersect to form right angles. We use the symbol ($\perp$) to represent **perpendicular lines**.	

Two **intersecting lines** that are not perpendicular form two pairs of congruent angles called **vertical angles** and four pairs of supplementary angles called **adjacent angles**.	w x y z
	
	Vertical **Adjacent**
	$\angle w$ and $\angle y$ $\angle w$ and $\angle x$
	$\angle x$ and $\angle z$ $\angle x$ and $\angle y$
	                       $\angle y$ and $\angle z$
	                       $\angle z$ and $\angle w$

For **parallel lines cut by a transversal**, the following properties apply. 1. Vertical angles are congruent. 2. Corresponding angles are congruent. 3. Alternate interior angles are congruent. 4. Alternate exterior angles are congruent. Any other angle pairs are supplementary.	a b c d e f g h **Corresponding** **Alternate Interior** $\angle a$ and $\angle e$ $\angle c$ and $\angle f$ $\angle b$ and $\angle f$ $\angle d$ and $\angle e$ $\angle c$ and $\angle g$ **Alternate Exterior** $\angle d$ and $\angle h$ $\angle a$ and $\angle h$ $\angle b$ and $\angle g$

8.1 Exercises

MyMathLab®

CONCEPTS AND VOCABULARY

1. A position in space having no length, width, or height is called a(n) _____.

2. A(n) _____ is a straight figure representing a set of points extending indefinitely in two directions.

3. A(n) _____ is a straight figure representing a set of points extending between two endpoints.

4. A(n) _____ is a straight figure representing a set of points extending indefinitely in one direction.

5. A figure formed by two rays that share a common endpoint is called a(n) _____.

6. The shared endpoint of the rays that form an angle is called the _____ of the angle.

7. An angle that measures 90° is a(n) _____ angle.

8. A(n) _____ angle measures 180°.

9. An angle that measures between 0° and 90° is called a(n) _____ angle.

10. An angle that measures between 90° and 180° is a(n) _____ angle.

11. Two angles with the same measure are _____ angles.

12. Two angles whose measures sum to 90° are called _____ angles.

13. Two angles whose measures sum to 180° are called _____ angles.

14. If two lines in a plane never touch, they are called _____ lines.

15. If two lines in a plane meet at a point, they are called _____ lines.

16. A pair of congruent angles formed when two lines intersect are called _____ angles.

17. A pair of supplementary angles formed when two lines intersect are called _____ angles.

18. Two lines are _____ if their intersection forms right angles.

19. When parallel lines are cut by a transversal, _____ angles are in the same position with respect to the parallel lines.

20. When parallel lines are cut by a transversal, _____ angles are between the parallel lines, on opposite sides of the transversal, and are not adjacent angles.

21. When parallel lines are cut by a transversal, _____ angles are outside of the parallel lines, on opposite sides of the transversal, and are not adjacent angles.

22. When parallel lines are cut by a transversal, any two angles that are *not* vertical, corresponding, alternate interior, or alternate exterior angles are _____.

IDENTIFYING GEOMETRIC TERMS

Exercises 23–32: Identify the figure and name it using the labels shown.

23. A B

24. • D

25. m

26. R S

27. C D

28. X Y

29. • K

30. G H

31. P Q

32. y

33. Refer to the following figure.

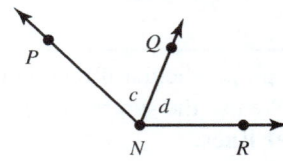

(a) Name the vertex of ∠c.
(b) Name the two sides of ∠d.
(c) Use three-letter naming to name three *different* angles that have vertex N.
(d) Explain why it is not appropriate to name an angle as ∠N.

34. Refer to the following figure.

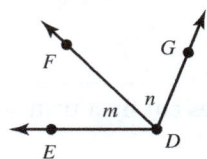

(a) Name the vertex of ∠n.
(b) Name the two sides of ∠m.
(c) Use three-letter naming to name three *different* angles that have vertex D.
(d) Explain why it is not appropriate to name an angle as ∠D.

CLASSIFYING ANGLES

Exercises 35–42: Classify the angle as acute, right, obtuse, or straight.

35.

36.

37.

38.

39.

40.

41.

42.

Exercises 43–48: Find the measure of an angle that has the specified description.

43. The supplement of an angle with measure 14°

44. The complement of an angle with measure 43°

45. Congruent to an angle with measure 109°
46. The supplement of an angle with measure 154°
47. The complement of an angle with measure 9°
48. Congruent to an angle with measure 13°

WORKING WITH INTERSECTING LINES

49. If $m\angle c = 49°$ in the following figure, find the measures of $\angle a$, $\angle b$, and $\angle d$.

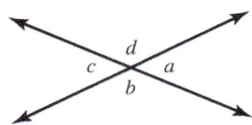

50. If $m\angle y = 158°$ in the following figure, find the measures of $\angle x$, $\angle w$, and $\angle z$.

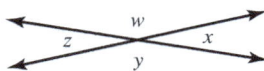

51. If $m\angle k = 84°$ in the following figure, find the measures of $\angle j$, $\angle l$, and $\angle m$.

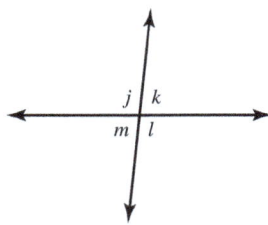

52. If $m\angle f = 102°$ in the following figure, find the measures of $\angle c$, $\angle d$, and $\angle e$.

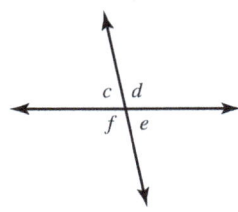

WORKING WITH PARALLEL LINES CUT BY A TRANSVERSAL

Exercises 53–64: In the following figure, $m \parallel n$. Determine whether the given angles are congruent or supplementary.

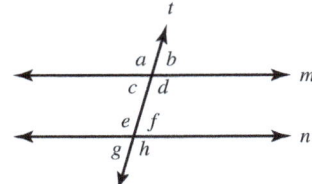

53. $\angle d$ and $\angle a$
54. $\angle e$ and $\angle g$
55. $\angle c$ and $\angle f$
56. $\angle a$ and $\angle e$
57. $\angle g$ and $\angle h$
58. $\angle b$ and $\angle g$
59. $\angle f$ and $\angle a$
60. $\angle e$ and $\angle d$
61. $\angle c$ and $\angle g$
62. $\angle b$ and $\angle h$
63. $\angle h$ and $\angle a$
64. $\angle e$ and $\angle h$

65. If $m \parallel n$ and $m\angle c = 119°$ in the following figure, find the measures of $\angle b$, $\angle g$, and $\angle h$.

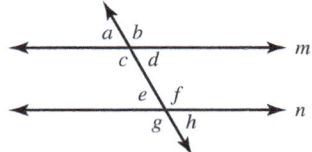

66. If $m \parallel n$ and $m\angle g = 63°$ in the following figure, find the measures of $\angle a$, $\angle b$, and $\angle c$.

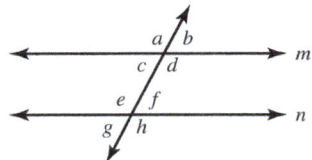

67. If $m \parallel n$ and $m\angle e = 92°$ in the following figure, find the measures of $\angle a$, $\angle c$, and $\angle h$.

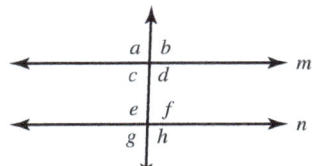

68. If $m \parallel n$ and $m\angle b = 142°$ in the following figure, find the measures of $\angle d$, $\angle e$, and $\angle g$.

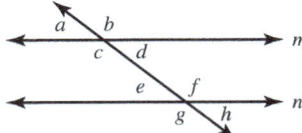

WRITING ABOUT MATHEMATICS

69. Suppose that two angles are supplementary. Can both of the angles be acute? Explain.

70. If two parallel lines are cut by a transversal that is perpendicular to one of the lines, explain how you know that the transversal is also perpendicular to the other line.

8.2 Triangles

Objectives
1. Classifying Triangles
2. Applying the Sum of the Angle Measures of a Triangle
3. Using Congruent Triangle Properties

NEW VOCABULARY
☐ Acute triangle
☐ Obtuse triangle
☐ Right triangle
☐ Scalene triangle
☐ Isosceles triangle
☐ Equilateral triangle
☐ Congruent triangles
☐ Angle-side-angle (ASA)
☐ Side-angle-side (SAS)
☐ Side-side-side (SSS)

1 Classifying Triangles

We can classify a triangle as *acute*, *obtuse*, or *right* by looking at its angles. We can classify a triangle as *scalene*, *isosceles*, or *equilateral* by looking at its sides. **TABLE 8.3** shows three types of triangles that are classified by looking at angles.

Classifying Triangles by Looking at the Angles

Type of Triangle	Description	Example
Acute	A triangle in which every angle measures less than 90°	
Obtuse	A triangle in which one angle measures between 90° and 180°	
Right	A triangle in which one angle measures exactly 90°	90° angle

TABLE 8.3

TABLE 8.4 shows three types of triangles that are classified by looking at sides.

Classifying Triangles by Looking at the Sides

Type of Triangle	Description	Example
Scalene	A triangle with no sides of the same length	
Isosceles	A triangle with at least two sides of the same length	
Equilateral	A triangle with three sides of the same length	

TABLE 8.4

NOTE: *Every* triangle must be one of the types in **TABLE 8.3** and, at the same time, must be one of the types in **TABLE 8.4**. For example, one triangle might be *both* acute and scalene, while a different triangle might be *both* right and isosceles. ∎

EXAMPLE 1 Classifying triangles as acute, obtuse, or right

Classify each triangle as acute, obtuse, or right.

(a) (b) (c)

Solution
(a) Since one angle of the triangle measures between 90° and 180°, it is an obtuse triangle.
(b) The triangle has one angle that measures exactly 90°, so it is a right triangle.
(c) Every angle in the triangle measures less than 90°, so it is an acute triangle.

Now Try Exercises 13, 17, 19

EXAMPLE 2 Classifying triangles as scalene, isosceles, or equilateral

Classify each triangle as scalene, isosceles, or equilateral.

(a) 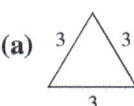   (b) 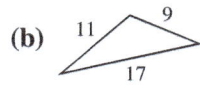   (c)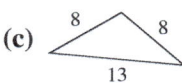

Solution
(a) Since the triangle has three sides of the same length, it is an equilateral triangle.
(b) The triangle has no sides of the same length, so it is a scalene triangle.
(c) Since the triangle has two sides of the same length, it is an isosceles triangle.

Now Try Exercises 25, 29, 31

2 Applying the Sum of the Angle Measures of a Triangle

No matter what type of triangle is being considered, one property is always true—the sum of the measures of the angles is 180°. To show that this property is true, we position the triangle between two parallel lines as shown in **FIGURE 8.16**, where $m \parallel n$ and the three angles of the triangle are $\angle 1$, $\angle 2$, and $\angle 3$.

A Triangle Between Parallel Lines

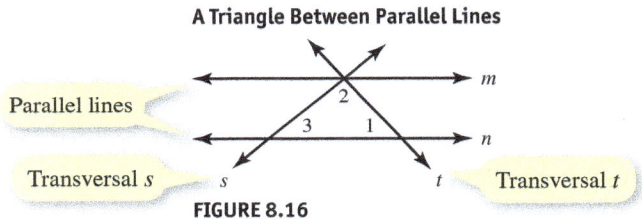

FIGURE 8.16

Look at the See the Concept below. Since transversal t forms one side of the triangle, the measure of $\angle 1$ is equal to that of its corresponding angle, $\angle 1$. Since transversal s forms another side of the triangle, the measure of $\angle 3$ is equal to that of its corresponding angle, $\angle 3$. Finally, transversals s and t intersect to form vertical angles with the measure of $\angle 2$ equal to that of $\angle 2$.

See the Concept 1   SUM OF THE ANGLE MEASURES OF A TRIANGLE

The sum of the measures of the three angles in any triangle is 180°.

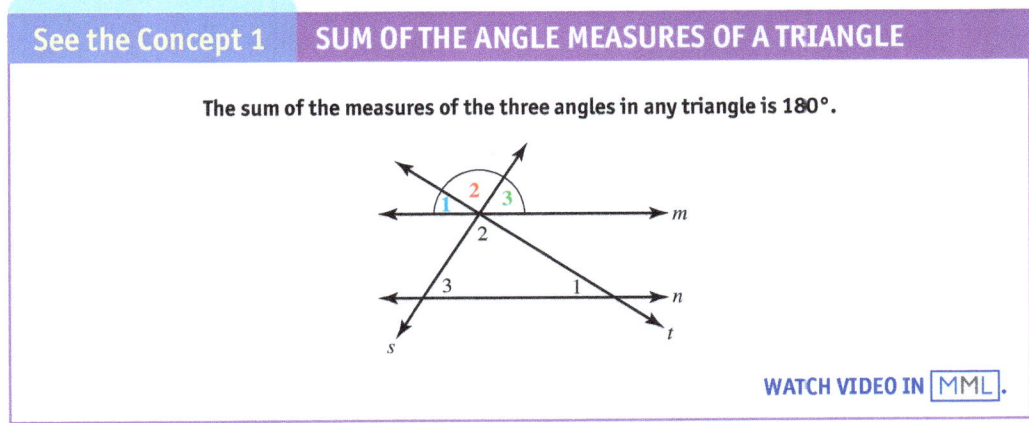

WATCH VIDEO IN MML.

READING CHECK 1

What is the sum of the angle measures for any triangle?

Together, $\angle 1$, $\angle 2$, and $\angle 3$ form a straight angle with measure 180°. Since the measures of $\angle 1$, $\angle 2$, and $\angle 3$ are equal to the measures of $\angle 1$, $\angle 2$, and $\angle 3$, respectively, the three angles of the triangle also sum to 180°.

THE SUM OF THE ANGLE MEASURES

For any triangle, the sum of the measures of the three angles is 180°.

EXAMPLE 3 Finding a missing angle measure in a triangle

Find the measure of $\angle x$ in **FIGURE 8.17**.

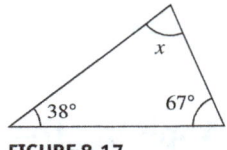

FIGURE 8.17

Solution
For simplicity, the degree symbol can be left out during the computation.

$$38 + 67 + x = 180 \qquad \text{Angle measures sum to 180.}$$
$$105 + x = 180 \qquad \text{Add.}$$
$$105 + x - 105 = 180 - 105 \qquad \text{Subtract 105 from each side.}$$
$$x = 75 \qquad \text{Simplify.}$$

The measure of $\angle x$ is 75°. $38° + 67° + 75° = 180°$

Now Try Exercise 37

EXAMPLE 4 Finding the value of a variable in a triangle

Find the value of x in **FIGURE 8.18**. Assume that all angle measures are given in degrees.

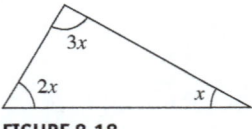

FIGURE 8.18

Solution

$$x + 2x + 3x = 180 \qquad \text{Angle measures sum to 180.}$$
$$6x = 180 \qquad \text{Combine like terms.}$$
$$\frac{6x}{6} = \frac{180}{6} \qquad \text{Divide each side by 6.}$$
$$x = 30 \qquad \text{Simplify.}$$

The value of x is 30°. We can check this by finding the sum of the measures of the three angles. Here, $x = 30°$, $2x = 60°$, and $3x = 90°$, so the sum is $30° + 60° + 90° = 180°$.

Now Try Exercise 45

3 Using Congruent Triangle Properties

In the previous section, we learned that two angles are *congruent angles* if they have the same measure. Two triangles are **congruent triangles** if they have exactly the same shape and size. In other words, two triangles are congruent if the corresponding angles are congruent and the lengths of the corresponding sides are equal. **FIGURE 8.19** shows congruent triangles.

Congruent Triangles

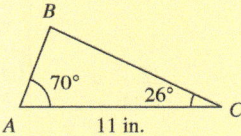

FIGURE 8.19

To show that the corresponding sides or angles of two triangles have the same measure, we use single, double, or triple hash marks, as shown in **FIGURE 8.19**.

When comparing two triangles, we do not need to know the measures of every side and angle to know if the triangles are congruent. Any of the following properties can be used.

CONGRUENT TRIANGLE PROPERTIES

The Angle-Side-Angle Property (ASA)

If two angles and the included side of one triangle are congruent to two angles and the included side of another triangle, then the triangles are congruent.

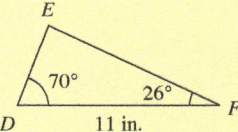

The Side-Angle-Side Property (SAS)

If two sides and the included angle of one triangle are congruent to two sides and the included angle of another triangle, then the triangles are congruent.

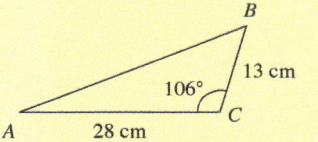

The Side-Side-Side Property (SSS)

If three sides of one triangle are congruent to three sides of another triangle, then the triangles are congruent.

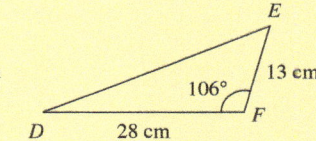

READING CHECK 2

Name the three properties that are used to determine when two triangles are congruent.

NOTE: Even though SSS can be used to show that two triangles are congruent, there is **not** an AAA property. Two triangles of different sizes can have three pairs of congruent angles, as shown in **FIGURE 8.20**.

Triangles That Are Not Congruent

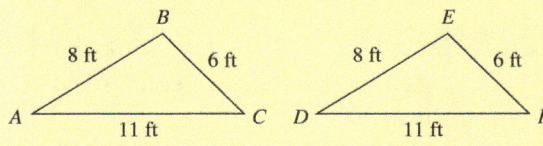

FIGURE 8.20 ■

EXAMPLE 5 Stating the property that shows triangles are congruent

For each pair of triangles, state the property that shows the triangles are congruent.

(a) 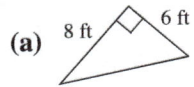 (b)

Solution
(a) Two sides and the included angle of the first triangle are congruent to two sides and the included angle of the second triangle. The triangles are congruent by SAS.
(b) Since two angles and the included side of the first triangle are congruent to two angles and the included side of the second triangle, the triangles are congruent by ASA.

Now Try Exercises 53, 55

EXAMPLE 6 Determining whether triangles are congruent

Determine whether the triangles in each pair are congruent. If so, state the property that shows that the triangles are congruent.

(a) 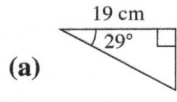 (b)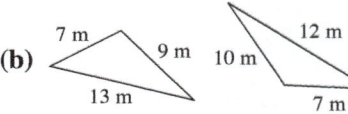

Solution
(a) Each triangle has a 19-cm side included between a 90° angle and a 29° angle. The two triangles are congruent by ASA.
(b) The two triangles are not congruent because the corresponding sides do not have equal measures.

Now Try Exercises 59, 61

8.2 Putting It All Together

CONCEPT	COMMENTS	EXAMPLES
Types of Triangles	**Acute:** Every angle measures less than 90°. **Obtuse:** One angle measures between 90° and 180°. **Right:** One angle measures exactly 90°. **Scalene:** No sides have the same length. **Isosceles:** At least two sides have the same length. **Equilateral:** Three sides have the same length.	Acute, Obtuse, Right, Scalene, Isosceles, Equilateral
The Sum of the Angle Measures	For any triangle, the sum of the measures of the three angles is 180°.	37° + 68° + 75° = 180°

CONCEPT	COMMENTS	EXAMPLES
Congruent Triangles	Any of the following properties can be used to determine whether two triangles are congruent.	
	Angle-Side-Angle (ASA) If two angles and the included side of one triangle are congruent to two angles and the included side of another triangle, then the triangles are congruent.	Angle-Side-Angle (ASA)
	Side-Angle-Side (SAS) If two sides and the included angle of one triangle are congruent to two sides and the included angle of another triangle, then the triangles are congruent.	Side-Angle-Side (SAS)
	Side-Side-Side (SSS) If three sides of one triangle are congruent to three sides of another triangle, then the triangles are congruent.	Side-Side-Side (SSS)

8.2 Exercises

CONCEPTS AND VOCABULARY

1. If every angle of a triangle measures less than 90°, then the triangle is a(n) _____ triangle.

2. If one angle of a triangle measures between 90° and 180°, then the triangle is a(n) _____ triangle.

3. If one angle of a triangle measures exactly 90°, then the triangle is a(n) _____ triangle.

4. A(n) _____ triangle has no sides of the same length.

5. A(n) _____ triangle has two sides of the same length.

6. A(n) _____ triangle has three sides of the same length.

7. For any triangle, the sum of the measures of the three angles is _____.

8. Two triangles are _____ triangles if they have exactly the same shape and size.

9. According to the _____ property, if two angles and the included side of one triangle are congruent to two angles and the included side of another triangle, then the triangles are congruent.

10. According to the _____ property, if two sides and the included angle of one triangle are congruent to two sides and the included angle of another triangle, then the triangles are congruent.

11. According to the _____ property, if three sides of one triangle are congruent to three sides of another triangle, then the two triangles are congruent.

12. There is not a(n) _____ property to determine if two triangles are congruent because triangles of different sizes can have three pairs of congruent angles.

CLASSIFYING TRIANGLES

Exercises 13–24: Classify the triangle as acute, obtuse, or right.

13. 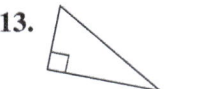   14.

15.    16.

17.    18.

19. 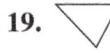   20.

506 CHAPTER 8 GEOMETRY

21. 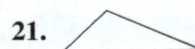   22.

23. 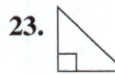   24.

Exercises 25–36: Classify the triangle as scalene, isosceles, or equilateral.

25. 6, 6, 10   26. 57, 29, 33

27. 12, 12, 12   28. 19, 7, 19

29. 14, 8, 19   30. 99, 99, 99

31. 7, 7, 7   32. 10, 10, 18

33. 14, 22, 15   34. 13, 13, 13

35. 85, 30, 85   36. 31, 15, 40

FINDING THE SUM OF THE ANGLE MEASURES

Exercises 37–44: Find the measure of ∠x in the triangle.

37. 82°, 47°, x   38. x, 68°, 38°

39. x, 73°, 73°   40. x, 44°, 48°

41. 27°, x, 35°   42. 123°, 32°, x

43. 60°, x, 60°   44. x, 45°, 45°

Exercises 45–52: Find the value of x. Assume that all angle measures are given in degrees.

45. 6x, 3x, 3x   46. 3x, 5x, x

47. 2x, x+36, x   48. 3x, x+20, x

49. 2x, 4x, 4x   50. 3x, x, x

51. 30, 2x, x   52. x, x, x

IDENTIFYING CONGRUENT TRIANGLES

Exercises 53–58: For the triangle pair, state the property that shows the triangles are congruent.

53.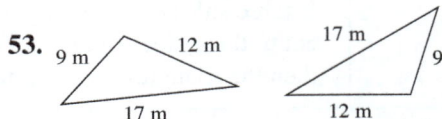

54.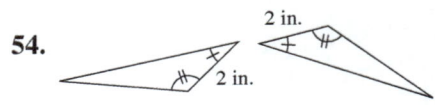

55.

56.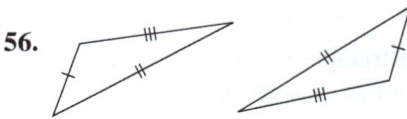

57.

58.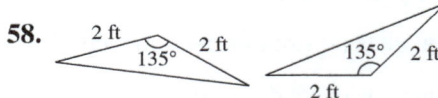

Exercises 59–64: Determine whether the triangles in the given pair are congruent. If so, state the property that shows that the triangles are congruent.

59.

60.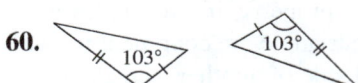

61.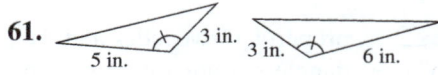

62.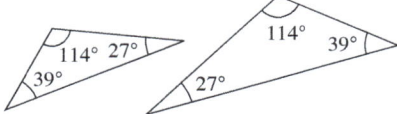

63.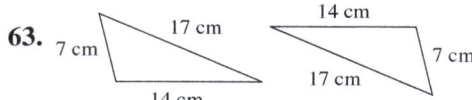

64.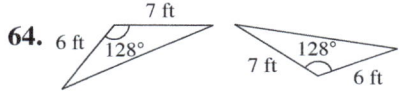

WRITING ABOUT MATHEMATICS

65. Can a right triangle also be obtuse? Explain.

66. For every triangle, the sum of the lengths of any two sides must be greater than the length of the third side. Use this fact to explain why a triangle with side lengths of 2, 7, and 4 inches cannot exist.

67. Can an equilateral triangle also be right? Explain.

68. If two angles and a non-included side of one triangle are congruent to two angles and a non-included side of a second triangle, then the two triangles are congruent. This property is known as AAS. Explain why this property is equivalent to the ASA property.

SECTIONS 8.1 and 8.2 — Checking Basic Concepts

1. Identify each of the given figures and name it using the labels shown.
   (a)    (b)

2. Classify each of the given angles as acute, right, obtuse, or straight.
   (a)    (b)

3. Find the measure of the complement of an angle with measure 19°.

4. Find the measure of the supplement of an angle with measure 114°.

5. If $m\angle y = 161°$ in the following figure, find the measures of $\angle x$, $\angle w$, and $\angle z$.

   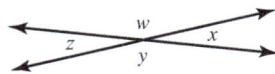

6. If $m \parallel n$ and $m\angle b = 139°$ in the following figure, find the measures of $\angle d$, $\angle e$, and $\angle g$.

   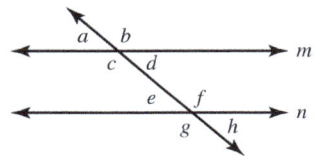

7. Classify each triangle as acute, obtuse, or right.
   (a)    (b)

8. Classify each of the given triangles as scalene, isosceles, or equilateral.
   (a)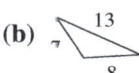

9. Find the measure of $\angle x$ in the triangle.

   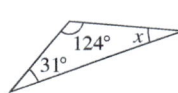

10. Determine whether the triangles in the given pair are congruent. If so, state the property that shows that the triangles are congruent.

8.3 Polygons and Circles

Objectives

1. Understanding Properties of Polygons
2. Finding Angles in Regular Polygons
3. Understanding Properties of Quadrilaterals
4. Finding the Radius and Diameter of a Circle

1 Understanding Properties of Polygons

A **polygon** is a closed plane figure determined by three or more line segments. Each line segment is called a **side** of the polygon, where two sides never touch except at a common endpoint called a **vertex**. FIGURE 8.21 shows several examples of polygons.

Examples of Polygons

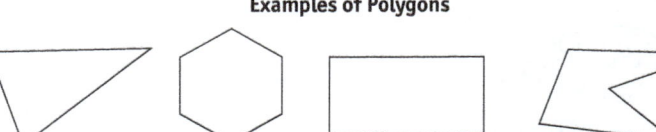

FIGURE 8.21

PROPERTIES OF POLYGONS

Every polygon has the following three properties.

1. All sides must be *straight* line segments.
2. Sides cannot intersect, except at endpoints.
3. The figure must be closed.

NEW VOCABULARY

☐ Polygon
☐ Side (of a polygon)
☐ Vertex
☐ Triangle, quadrilateral, pentagon, hexagon, heptagon, octagon
☐ Regular polygon
☐ Square, rectangle, parallelogram, trapezoid, rhombus, kite
☐ Circle
☐ Center (of a circle)
☐ Radius
☐ Diameter

EXAMPLE 1   Explaining why figures are not polygons

Explain why each figure is not a polygon.

(a)   (b)   (c)

Solution

(a) The figure is not closed.
(b) One side is curved.
(c) The sides intersect.

Now Try Exercises 11, 13, 15

Every polygon is named by the number of its sides. TABLE 8.5 lists the names of polygons with 3 to 8 sides.

Names of Polygons

Polygon Name	Triangle	Quadrilateral	Pentagon	Hexagon	Heptagon	Octagon
Number of Sides	3	4	5	6	7	8
Example	△	▱	⬠	⬡	⌂	⌂

TABLE 8.5

EXAMPLE 2   Determining the name of a polygon

Determine the name of each polygon.

(a)    (b)    (c)

Solution

(a) The polygon has four sides, so it is a quadrilateral.
(b) Since the polygon has six sides, it is a hexagon.
(c) A polygon with seven sides is a heptagon.

Now Try Exercises 17, 23, 27

2 Finding Angles in Regular Polygons

When all sides of a polygon have equal length and all angles have equal measure, the polygon is a **regular polygon**. **FIGURE 8.22** shows a regular quadrilateral (square), a regular hexagon, and a regular octagon.

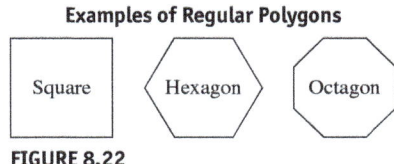

FIGURE 8.22

The sum of the angles of any regular polygon can be found using the fact that the angle measures of any triangle add up to 180°. For example, the regular pentagon in **FIGURE 8.23** is divided into 3 triangles. The sum of the angle measures of the pentagon is equal to the sum of the angle measures of 3 triangles. Since we know that the sum of the angle measures of a triangle is equal to 180°, we know that the sum of the angle measures of a pentagon is $3 \cdot 180° = 540°$.

We can find the sum of the angle measures of *any* regular polygon by simply counting the number of triangles inside and multiplying by 180°. The regular polygons in **FIGURE 8.24** show that the number of triangles is always 2 less than the number of sides.

3 Triangles in a Pentagon

Angle measure of pentagon:
$3 \cdot 180° = 540°$

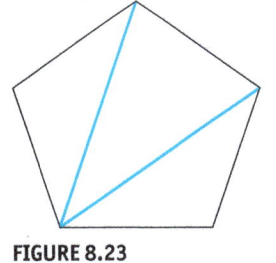

FIGURE 8.23

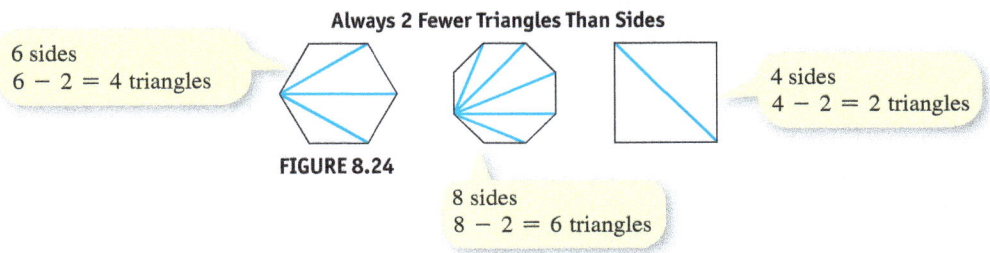

FIGURE 8.24

READING CHECK 1

If there are 4 triangles in a regular polygon, what product gives the sum of the angle measures in the polygon?

This discussion leads to the following rule for finding the sum of the angle measures of any regular polygon.

THE SUM OF THE ANGLE MEASURES OF A REGULAR POLYGON

For a regular polygon with *n* sides, the sum of the angle measures is

$$(n - 2) \cdot 180°.$$

EXAMPLE 3 Finding the sum of the angle measures of a regular polygon

Find the sum of the angle measures for the regular polygon shown.

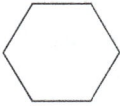

Solution
Since $n = 6$, the sum of the angle measures is $(6 - 2) \cdot 180° = 4 \cdot 180° = 720°$.

Now Try Exercise 33

Because all angles of a regular polygon have equal measure, we can find the measure of a *single* angle by dividing the angle measure sum by the number of angles. For a regular polygon, the number of angles equals the number of sides and the following rule can be used to find the measure of one angle of a regular polygon.

THE MEASURE OF ONE ANGLE OF A REGULAR POLYGON

For a regular polygon with n sides, the measure of one angle is
$$\frac{(n - 2) \cdot 180°}{n}.$$

EXAMPLE 4 Finding the measure of one angle of a regular polygon

For the following regular polygon, find the measure of one angle.

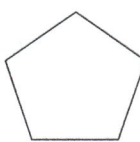

Solution
Since $n = 5$, the measure of one angle is
$$\frac{(5 - 2) \cdot 180°}{5} = \frac{3 \cdot 180°}{5} = \frac{540°}{5} = 108°.$$

Now Try Exercise 37

3 Understanding Properties of Quadrilaterals

Recall that a quadrilateral is a 4-sided polygon. In the previous section, we learned that there are several special classifications for triangles. The same is true for quadrilaterals. **TABLE 8.6** lists different types of quadrilaterals.

Types of Quadrilaterals

Quadrilateral	Defining Feature	Example
Square	A regular quadrilateral	
Rectangle	All angles measure 90°	
Parallelogram	Two pairs of parallel sides	
Trapezoid	One pair of parallel sides	
Rhombus	Four sides of equal length	
Kite	Two pairs of adjacent sides equal in length	

TABLE 8.6

NOTE: Small arrowheads are used along the sides of the parallelogram and trapezoid shown in **TABLE 8.6** to indicate which sides are parallel. However, such arrowheads are not commonly shown on a square, rectangle, or rhombus, which can each be classified as a parallelogram. ∎

EXAMPLE 5 Naming quadrilaterals

Name each quadrilateral using every classification that applies.

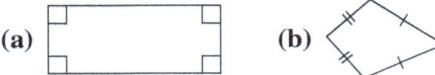

(a)   (b)

Solution
(a) Since all angles measure 90°, the quadrilateral is a rectangle. It is also a parallelogram because it has two pairs of parallel sides.
(b) The quadrilateral has two pairs of adjacent sides that are equal in length, so it is a kite.

Now Try Exercises 41, 43

4 Finding the Radius and Diameter of a Circle

A **circle** is a closed plane figure consisting of all points that are equally distant from a single point called the **center** of the circle. A line segment with one endpoint on the circle and the other endpoint at the center is called a **radius**. A line segment that contains the center and has both endpoints on the circle is called a **diameter**. **FIGURE 8.25** shows these terms labeled on a circle.

A Circle

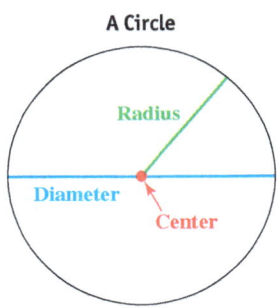

FIGURE 8.25

NOTE: Not only are the terms radius and diameter used to name the segments that they represent, but they are also used to refer to the length of those segments. For example, we may say that the radius is 5 feet, or the diameter is 10 feet. ■

The radius and diameter of a circle are related in the following ways.

READING CHECK 2

How are the radius and diameter of a circle related?

THE RADIUS AND DIAMETER OF A CIRCLE

If r represents the radius and d represents the diameter of a circle, then

$$d = 2r \quad \text{and} \quad r = \frac{d}{2}.$$

EXAMPLE 6 Finding the radius when given the diameter

Find the radius of the following circle.

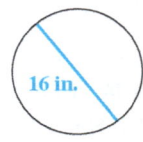

Solution
The radius is half the diameter, or

$$r = \frac{d}{2} = \frac{16}{2} = 8 \text{ in.}$$

Now Try Exercise 51

EXAMPLE 7 Finding the diameter when given the radius

Find the diameter of the following circle.

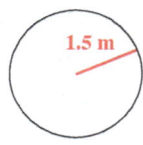

Solution
The diameter is twice the radius, or

$$d = 2r = 2(1.5) = 3 \text{ m.}$$

Now Try Exercise 59

8.3 Putting It All Together

CONCEPT	COMMENTS	EXAMPLES
Polygon	A polygon is a closed plane figure determined by three or more line segments. To be a polygon, the following must be true. 1. All sides must be *straight* line segments. 2. Sides cannot intersect, except at endpoints. 3. The figure must be closed.	

CONCEPT	COMMENTS	EXAMPLES
Names of Polygons	A polygon is named by the number of its sides. 3 sides: **Triangle** 4 sides: **Quadrilateral** 5 sides: **Pentagon** 6 sides: **Hexagon** 7 sides: **Heptagon** 8 sides: **Octagon**	Triangle, Quadrilateral, Pentagon, Hexagon, Heptagon, Octagon
Regular Polygon	In a regular polygon, all sides have equal length and all angles have equal measure.	
Angle Measures for a Regular Polygon	For a regular polygon with n sides, $$(n-2) \cdot 180°$$ gives the sum of the angle measures, and $$\frac{(n-2) \cdot 180°}{n}$$ gives the measure of one angle.	For a regular polygon with 6 sides, the sum of the angle measures is $$(6-2) \cdot 180° = 720°.$$ The measure of one angle is $$\frac{(6-2) \cdot 180°}{6} = \frac{720°}{6} = 120°.$$
Quadrilaterals	There are six kinds of quadrilaterals. **Square:** A regular quadrilateral **Rectangle:** All angles measure 90° **Parallelogram:** Two pairs of parallel sides **Trapezoid:** One pair of parallel sides **Rhombus:** Four sides of equal length **Kite:** Two pairs of adjacent sides equal in length	Square, Rectangle, Parallelogram, Trapezoid, Rhombus, Kite
Circle	A circle is a closed plane figure that consists of all points that are equally distant from a central point (center). A radius is a line segment with one endpoint on the circle and the other endpoint at the center. A diameter is a line segment that contains the center and has both endpoints on the circle.	Radius r, Diameter d, Center $d = 2r$ and $r = \dfrac{d}{2}$

8.3 Exercises

CONCEPTS AND VOCABULARY

1. A(n) _____ is a closed plane figure determined by three or more line segments.

2. A polygon with 6 sides is called a(n) _____.

3. A polygon with 7 sides is called a(n) _____.

4. In a(n) _____ polygon, all sides have the same length and all angles have the same measure.

5. For a regular polygon with n sides, the sum of the angle measures is given by _____.

6. For a regular polygon with n sides, the measure of one angle is given by _____.

7. A(n) _____ is a closed plane figure consisting of all points that are equally distant from a single point called the _____.

8. A line segment with one endpoint on a circle and the other endpoint at the center is a(n) _____.

9. A line segment that contains the center and has both endpoints on the circle is a(n) _____.

10. If r represents the radius and d represents the diameter of a circle, then $d =$ _____ and $r =$ _____.

IDENTIFYING POLYGONS

Exercises 11–16: Explain why the figure is not a polygon.

11.  12.

13.  14.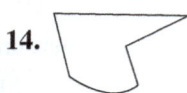

15.  16.

Exercises 17–28: Determine the name of the polygon.

17. 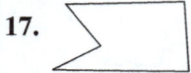 18.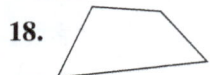

19.  20.

21.  22.

23.  24.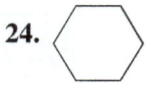

25. 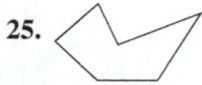 26.

27.  28.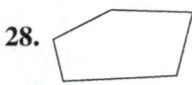

WORKING WITH REGULAR POLYGONS

Exercises 29–34: Find the sum of the angle measures for the given regular polygon.

29. 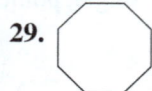 30.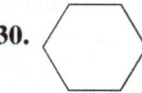

31. 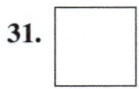 32.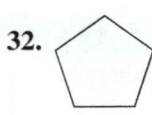

33. 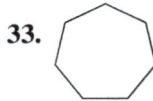 34.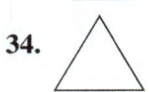

Exercises 35–40: For the given regular polygon, find the measure of one angle.

35. 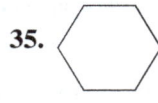 36.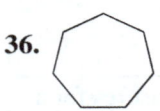

37. 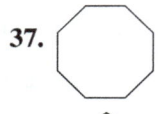 38.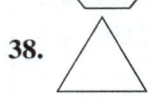

39. 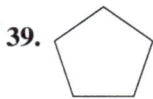 40.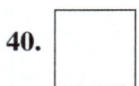

IDENTIFYING QUADRILATERALS

Exercises 41–50: Name the quadrilateral using every classification that applies.

41. 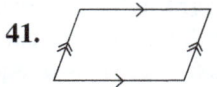 42.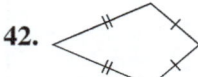

43. 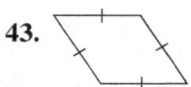 44.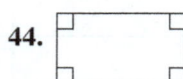

45. 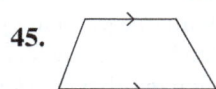 46.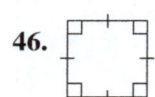

47. 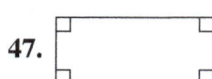 48.

49.  50.

FINDING RADIUS AND DIAMETER

Exercises 51–56: Find the radius of the given circle.

51. 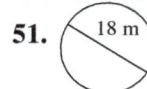 52.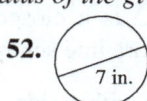

53. 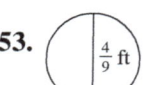 54.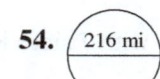

55.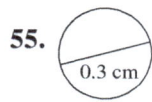
56.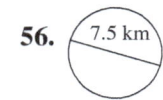

Exercises 57–62: Find the diameter of the given circle.

57.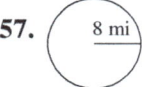
58.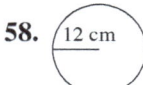
59.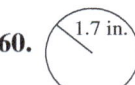
60.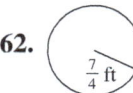
61. $\frac{5}{2}$ m
62. $\frac{7}{4}$ ft

APPLICATIONS INVOLVING POLYGONS AND CIRCLES

63. **Traffic Sign** A stop sign has the shape of a regular octagon. Find the measure of one of its angles.

64. **Wall Clock** A circular clock has a 5.5-inch radius. Find the diameter of the clock.

65. **National Defense** The U.S. Department of Defense is housed in a building called the Pentagon. The building has the shape of a regular five-sided polygon. Find the sum of the angles of the Pentagon.

66. **Beverage Can** The top of a soda can is a circle with a diameter of 5.4 cm. Find the radius of the top of the can.

WRITING ABOUT MATHEMATICS

67. How are the radius and diameter of a circle related?

68. Give three examples of real-world things with polygon shapes.

69. A student draws a rhombus with four right angles. What are some other quadrilateral names that describe what the student has drawn? Explain.

70. Explain why it is necessary to have a *regular* polygon when finding the measure of one angle.

Group Activity

The Honeycomb Tessellation Honey bees can construct a honeycomb that fits regular polygons together perfectly to make the pattern shown.

A tiling pattern made of non-overlapping polygons with no gaps is called a *tessellation*.

(a) Name the regular polygon that honey bees use to make a honeycomb.
(b) What is the measure of one angle of this regular polygon?
(c) How many polygons share any given vertex?
(d) For the angles that share a vertex, what is the sum of the angle measures?
(e) Repeat parts (b), (c), and (d) if squares made up a honeycomb.
(f) Repeat parts (b), (c), and (d) if equilateral triangles made up a honeycomb.

8.4 Perimeter and Circumference

Objectives

1. Finding the Perimeter of a Polygon
2. Finding the Circumference of a Circle
3. Finding the Perimeter of Composite Figures

NEW VOCABULARY

☐ Perimeter
☐ Circumference
☐ Composite figure

1 Finding the Perimeter of a Polygon

The **perimeter** of a polygon is the distance around the polygon. We can find the perimeter of a polygon by adding the lengths of its sides.

> **THE PERIMETER OF A POLYGON**
>
> To find the perimeter of a polygon, add the lengths of all sides of the polygon.

Math in Context (Materials) City planners often include *green space* in new city development projects. To determine the amount of material needed to create a walkway around a city park, the perimeter of the park is measured. The perimeter of the rectangular park in **FIGURE 8.26** is 780 feet because

$$200 + 190 + 200 + 190 = 780.$$

Add the lengths of the four sides.

Perimeter of a Rectangular Park

FIGURE 8.26

EXAMPLE 1 **Finding the perimeter of polygons**

Find the perimeter of each polygon.

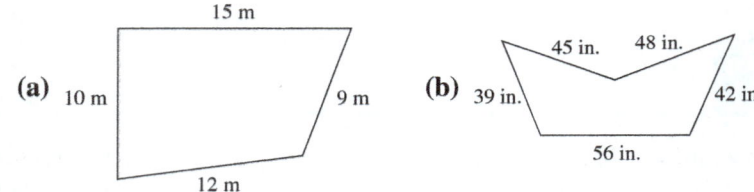

Solution

(a) The perimeter is $10 + 15 + 9 + 12 = 46$ meters.
(b) The perimeter is $39 + 45 + 48 + 42 + 56 = 230$ inches.

Now Try Exercises 9, 11

We use multiplication rather than addition to find the perimeter of a *regular* polygon because all sides of a regular polygon have the same measure.

> **THE PERIMETER OF A REGULAR POLYGON**
>
> To find the perimeter of a regular polygon, multiply the number of sides by the length of one side.

EXAMPLE 2 Finding the perimeter of a regular polygon

Find the perimeter of the regular polygon in **FIGURE 8.27**.

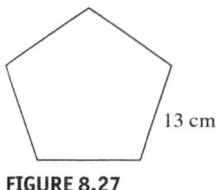

FIGURE 8.27

Solution
A regular pentagon has 5 sides with the same length. The perimeter is $5 \cdot 13 = 65$ cm.

Now Try Exercise 15

Earlier, we discussed the perimeter of a region in which the measure of one or more of the sides was unknown. When this happens, we must find any missing lengths before we can find the perimeter, as demonstrated in the next example.

EXAMPLE 3 Finding the perimeter of a polygon with missing measures

Find the perimeter of the polygon in **FIGURE 8.28**.

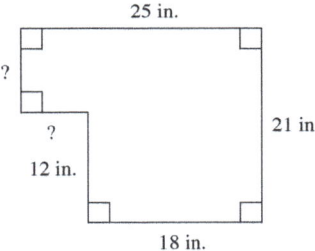

FIGURE 8.28

Solution
The lengths of the missing sides can be found by subtraction, as shown in **FIGURE 8.29**.

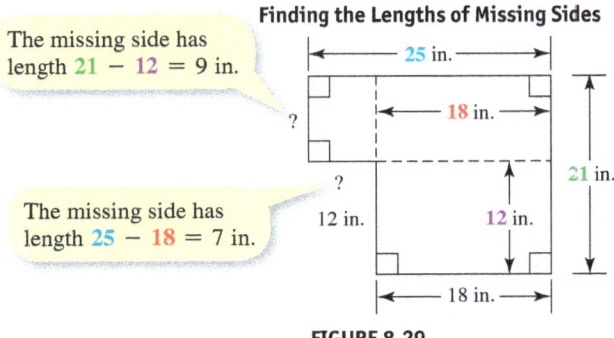

FIGURE 8.29

The perimeter is $25 + 21 + 18 + 12 + 7 + 9 = 92$ in.

Now Try Exercise 21

2 Finding the Circumference of a Circle

The perimeter of a circle is called the **circumference** of the circle. A formula involving the number π (pi) is needed to find the circumference of a circle. Recall that π is an irrational number whose decimal representation neither terminates nor repeats.

$$\pi \approx 3.1415926536$$

The ratio of the circumference to the diameter of *any* circle is always equal to the number π. In other words,

$$\frac{\text{Circumference}}{\text{diameter}} = \pi.$$

Using C for circumference and d for diameter, we have

$$\frac{C}{d} = \pi.$$

Multiplying each side of this equation by d gives the formula

$$C = \pi \cdot d.$$

Since diameter is twice the radius, the circumference formula can be written in two ways.

READING CHECK 1

What two values are used as approximations for π?

THE CIRCUMFERENCE OF A CIRCLE

The circumference C of a circle with radius r and diameter d is found using

$$C = \pi d \quad \text{or} \quad C = 2\pi r,$$

where π can be approximated as $\pi \approx 3.14$ or $\pi \approx \frac{22}{7}$.

EXAMPLE 4 Finding the circumference of a circle using the diameter

Refer to the circle in **FIGURE 8.30**.

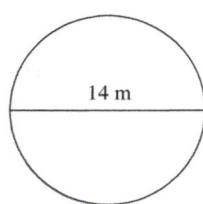

FIGURE 8.30

(a) Find the exact circumference of the circle.
(b) Approximate the circumference using 3.14 as an approximation for π.

Solution
(a) Using the formula $C = \pi d$ gives $C = \pi \cdot 14 = 14\pi$ m.
(b) Using $\pi \approx 3.14$ gives $C \approx 14 \cdot 3.14 = 43.96$ m.

Now Try Exercise 27

EXAMPLE 5   Finding the circumference of a circle using the radius

Refer to the circle in **FIGURE 8.31**.

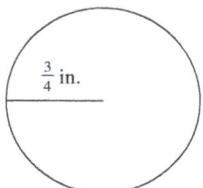

FIGURE 8.31

(a) Find the exact circumference of the circle.
(b) Approximate the circumference, using $\frac{22}{7}$ as an approximation for π.

Solution
(a) Using the formula $C = 2\pi r$ gives $C = 2 \cdot \pi \cdot \frac{3}{4} = \frac{3}{2}\pi$ in.
(b) Using $\pi \approx \frac{22}{7}$ gives $C \approx \frac{3}{2} \cdot \frac{22}{7} = \frac{33}{7}$ in.

Now Try Exercise 37

3 Finding the Perimeter of Composite Figures

An enclosed geometric region made up of polygons and semicircles (half circles) is called a **composite figure**. Examples of composite figures can be found in everyday situations.

🌐 *Math in Sports Context* On a basketball court, the *key* is an enclosed region on the floor at each end of the court. The key is a composite figure made up of a rectangle and a semicircle. In the next example, we find the perimeter of the key in **FIGURE 8.32**.

Basketball Court Key

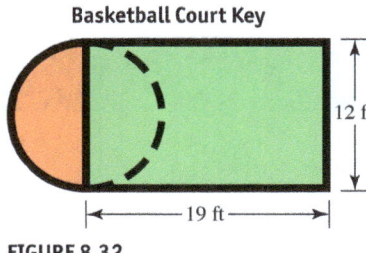

FIGURE 8.32

EXAMPLE 6   Finding the perimeter of a basketball court key

Find the perimeter of the key in **FIGURE 8.32**. Use 3.14 as an approximation for π.

Solution
The key is enclosed by three line segments and a semicircle. The segments have lengths 19, 12, and 19 feet. The length of the semicircle is one-half the circumference of a circle with a 12-foot diameter. Since the circumference of the circle is $C = \pi \cdot 12 \approx 3.14 \cdot 12 = 37.68$ feet, the length of the semicircle is approximately $0.5 \cdot 37.68 = 18.84$ feet. The total perimeter is approximately $19 + 12 + 19 + 18.84 = 68.84$ feet.

Now Try Exercise 53

8.4 Putting It All Together

CONCEPT	COMMENTS	EXAMPLES
Perimeter of a Polygon	To find the perimeter of a polygon, add the lengths of all sides of the polygon.	Triangle with sides 15 cm, 8 cm, 11 cm. $15 + 11 + 8 = 34$ cm
Perimeter of a Regular Polygon	To find the perimeter of a regular polygon, multiply the number of sides by the length of one side.	Equilateral triangle with side 7 ft. $3 \cdot 7 = 21$ ft
Circumference of a Circle	The circumference C of a circle with radius r and diameter d is found using $$C = \pi d \quad \text{or} \quad C = 2\pi r,$$ where $\pi \approx 3.14$ or $\pi \approx \frac{22}{7}$.	Circle with diameter 4 mi: $C = 4\pi \approx 4 \cdot 3.14 = 12.56$ mi. Circle with radius $\frac{7}{11}$ m: $C = 2 \cdot \frac{7}{11}\pi \approx 2 \cdot \frac{7}{11} \cdot \frac{22}{7} = 4$ m

8.4 Exercises

CONCEPTS AND VOCABULARY

1. The distance around a polygon is called the _____ of the polygon.

2. To find the perimeter of a polygon, add the lengths of all _____ of the polygon.

3. The perimeter of a circle is its _____.

4. To find the circumference of a circle with a given diameter, use the formula _____.

5. To find the circumference of a circle with a given radius, use the formula _____.

6. An enclosed geometric region made up of polygons and semicircles is called a(n) _____ figure.

FINDING PERIMETER OF A POLYGON

Exercises 7–14: Find the perimeter of the polygon.

7. 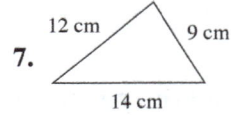 (Triangle with sides 12 cm, 9 cm, 14 cm)

8. 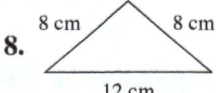 (Triangle with sides 8 cm, 8 cm, 12 cm)

9. 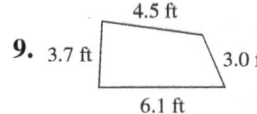 (Quadrilateral with sides 3.7 ft, 4.5 ft, 3.0 ft, 6.1 ft)

10. 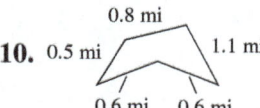 (Polygon with sides 0.5 mi, 0.8 mi, 1.1 mi, 0.6 mi, 0.6 mi)

11. 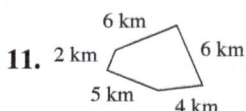 (Polygon with sides 2 km, 6 km, 6 km, 4 km, 5 km)

12. 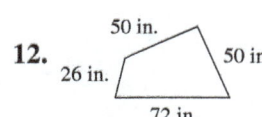 (Polygon with sides 26 in., 50 in., 50 in., 72 in.)

13. Triangle with sides $\frac{1}{2}$ m, $\frac{7}{8}$ m, 1 m

14. Triangle with sides $\frac{9}{4}$ in., $\frac{9}{4}$ in., $\frac{5}{2}$ in.

Exercises 15–20: Find the perimeter of the regular polygon.

15. 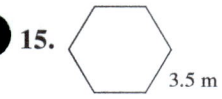 3.5 m

16. 13 in.

17. 52 ft

18. 16.4 mi

19. $\frac{7}{10}$ cm

20. $\frac{9}{2}$ ft

Exercises 21–26: Find the perimeter of the polygon.

21. 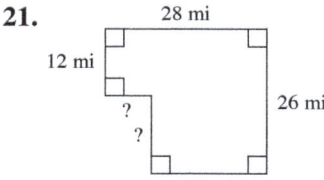 28 mi, 12 mi, 26 mi, 20 mi, ?, ?

22. 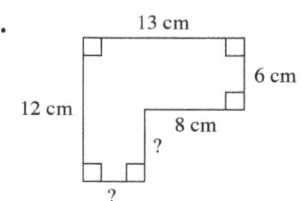 13 cm, 6 cm, 12 cm, 8 cm, ?, ?

23. 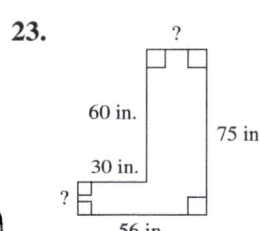 ?, 60 in., 75 in., 30 in., 56 in., ?

24. 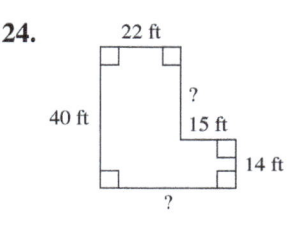 22 ft, ?, 40 ft, 15 ft, 14 ft, ?

25. 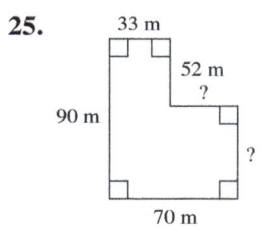 33 m, 52 m, ?, 90 m, ?, 70 m

26. 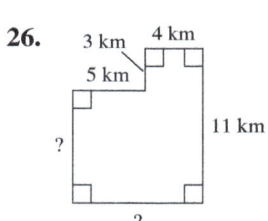 3 km, 4 km, 5 km, 11 km, ?, ?

FINDING CIRCUMFERENCE OF A CIRCLE

Exercises 27–34: Do the following.

(a) Find the exact circumference of the circle.
(b) Approximate the circumference, using 3.14 as an approximation for π.

27. 12 m

28. 4 mi

29. 5 in.

30. 9 ft

31. 2.5 ft

32. 0.7 m

33. 1.4 cm

34. 6.5 mi

Exercises 35–42: Do the following.

(a) Find the exact circumference of the circle.
(b) Approximate the circumference, using $\frac{22}{7}$ as an approximation for π.

35. 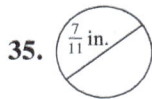 $\frac{7}{11}$ in.

36. 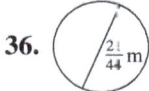 $\frac{21}{44}$ m

37. 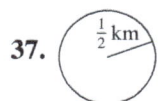 $\frac{1}{2}$ km

38. 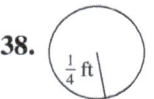 $\frac{1}{4}$ ft

39. 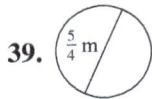 $\frac{5}{4}$ m

40. 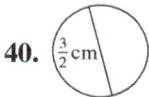 $\frac{3}{2}$ cm

41. 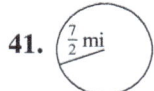 $\frac{7}{2}$ mi

42. 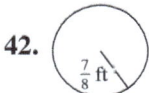 $\frac{7}{8}$ ft

FINDING THE PERIMETER OF COMPOSITE FIGURES

Exercises 43–48: Find the perimeter of the composite figure. Use 3.14 as an approximation for π.

43. 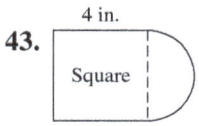 4 in., Square

44. 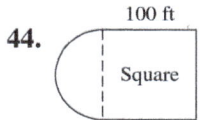 100 ft, Square

45. 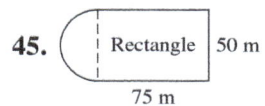 Rectangle, 50 m, 75 m

46. 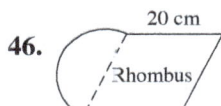 20 cm, Rhombus

47. 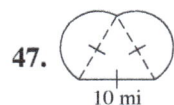 10 mi

48. 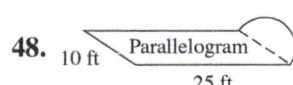 10 ft, Parallelogram, 25 ft

APPLICATIONS INVOLVING PERIMETER AND CIRCUMFERENCE

49. **Football Field** An official NFL football field is a rectangle that measures 160 feet by 360 feet. Find the perimeter of an official NFL football field.

50. **National Defense** The U.S. Department of Defense building has the shape of a regular pentagon with each wall measuring 921 feet. Find the perimeter of the Pentagon building.

51. **Bermuda Triangle** A triangular region with vertices at Bermuda, Puerto Rico, and southern Florida is known as the Bermuda Triangle. It is roughly an equilateral triangle measuring 1580 km on each side. Find the perimeter of the Bermuda Triangle.

522 CHAPTER 8 GEOMETRY

52. **Olympic Shotput** An Olympic shotput ring is a circle with a diameter of 7 feet. Find the circumference of an Olympic shotput ring, using 3.14 as an approximation for π.

53. **Track and Field** A running track is constructed with the dimensions shown in the following figure. Find the perimeter of the track, using 3.14 as an approximation for π. Round to the nearest meter.

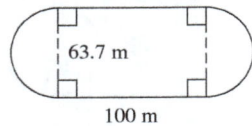

54. **Raceways** A racing track is constructed with the dimensions shown in the following figure. Find the perimeter of the track, using 3.14 as an approximation for π. Round to the nearest tenth of a mile.

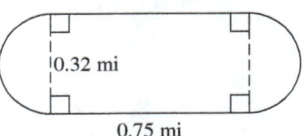

WRITING ABOUT MATHEMATICS

55. Which number, 3.14 or $\frac{22}{7}$, is a more accurate approximation for π? Explain your reasoning.

56. If the perimeter of a regular polygon is known, explain how you can find the length of one side.

SECTIONS 8.3 and 8.4 — Checking Basic Concepts

1. Name each polygon.

   (a) 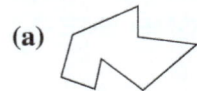   (b)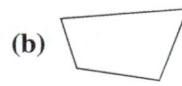

2. Find the sum of the angle measures for each regular polygon.

   (a) 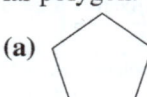   (b)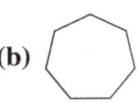

3. For each regular polygon, find the measure of one angle.

   (a)    (b)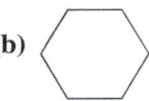

4. Name each quadrilateral, using every classification that applies.

   (a) 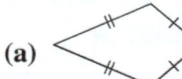   (b)

5. Find the radius of the circle shown.

   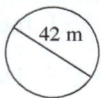

6. Find the diameter of the circle shown.

   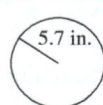

7. Find the perimeter of the following polygon.

   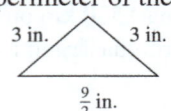

8. Find the perimeter of the following regular polygon.

   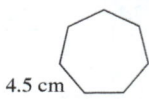

9. Find the perimeter of the polygon.

   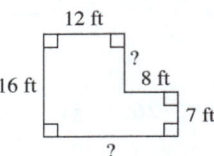

10. Approximate the circumference, using 3.14 as an approximation for π.

    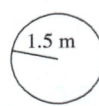

11. Approximate the circumference, using $\frac{22}{7}$ as an approximation for π.

    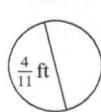

12. Approximate the perimeter of the given composite figure, using 3.14 as an approximation for π.

    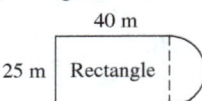

13. **Bicycle Tire** A bicycle tire has a 9.5-inch radius. Find the diameter of the tire.

14. **Flower Garden** A flower garden is constructed in the shape of a rhombus, measuring 16 feet on each side. Find the perimeter of the garden.

8.5 Area, Volume, and Surface Area

Objectives

1. Finding the Area of Plane Figures
   - Area of Common Polygons
   - Area of a Circle
2. Finding the Volume and Surface Area of Geometric Solids
   - Volume
   - Surface Area

NEW VOCABULARY

☐ Volume
☐ Surface area
☐ Cube
☐ Rectangular prism
☐ Circular cylinder
☐ Cone
☐ Square-based pyramid
☐ Sphere

1 Finding the Area of Plane Figures

The *area* of a region is computed by finding the number of square units that are needed to cover the region. In this section, we are interested in regions that are common polygons or circles.

AREA OF COMMON POLYGONS Earlier in this text, area formulas were presented for squares, rectangles, and triangles. For convenience, these formulas are provided again in **FIGURE 8.33**.

Area of Squares, Rectangles, and Triangles

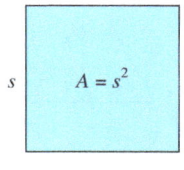 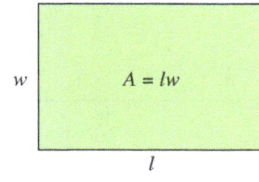 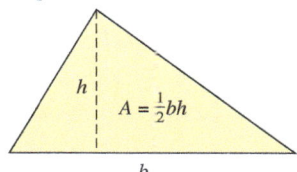

FIGURE 8.33

STUDY TIP

Organize your completed notes for this course and keep them ready as an aid for your studies in your next math course.

EXAMPLE 1 Finding the area of a square, rectangle, or triangle

Find the area of each figure.

(a) 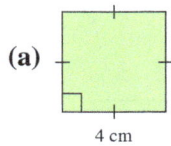 (b) 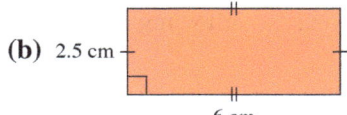 (c)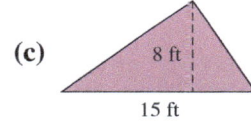

Solution
(a) The area of the square is $A = s^2 = 4^2 = 16 \text{ cm}^2$.
(b) The area of the rectangle is $A = lw = 6(2.5) = 15 \text{ cm}^2$.
(c) The area of the triangle is $A = \frac{1}{2}bh = \frac{1}{2}(15)(8) = 60 \text{ ft}^2$.

Now Try Exercises 7, 9, 11

The area formula for a parallelogram can be found by "cutting" a right triangle off one end of the parallelogram and connecting it to the other end. The area formula for a trapezoid can be found by connecting the trapezoid to a second, identical but rotated, trapezoid to form a parallelogram. These cases are shown in the following See the Concept.

See the Concept 1 — AREA OF A PARALLELOGRAM AND TRAPEZOID

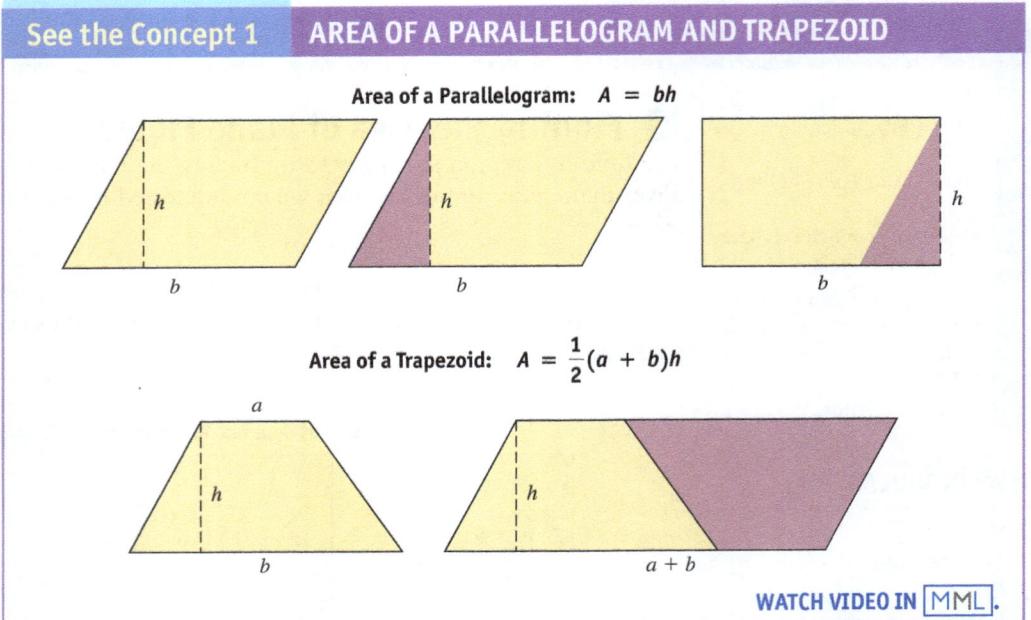

AREA OF A PARALLELOGRAM

The area A of a parallelogram with base b and height h is

$$A = bh.$$

EXAMPLE 2 **Finding the area of a parallelogram**

Find the area of the following parallelogram.

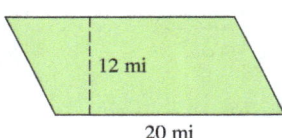

Solution
The area of the parallelogram is $A = bh = 20(12) = 240$ mi^2.

Now Try Exercise 13

AREA OF A TRAPEZOID

The area A of a trapezoid with parallel sides of lengths a and b and height h is

$$A = \frac{1}{2}(a + b)h.$$

EXAMPLE 3 Finding the area of a trapezoid

Find the area of the following trapezoid.

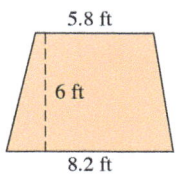

Solution
The area of the trapezoid is $A = \frac{1}{2}(a+b)h = \frac{1}{2}(5.8 + 8.2) \cdot 6 = 42 \text{ ft}^2$.

Now Try Exercise 15

AREA OF A CIRCLE To find the area formula for a circle, we will take the circle apart in a way that allows us to reassemble it as a figure whose shape closely resembles a parallelogram, as shown in the following See the Concept.

See the Concept 2   AREA OF A CIRCLE

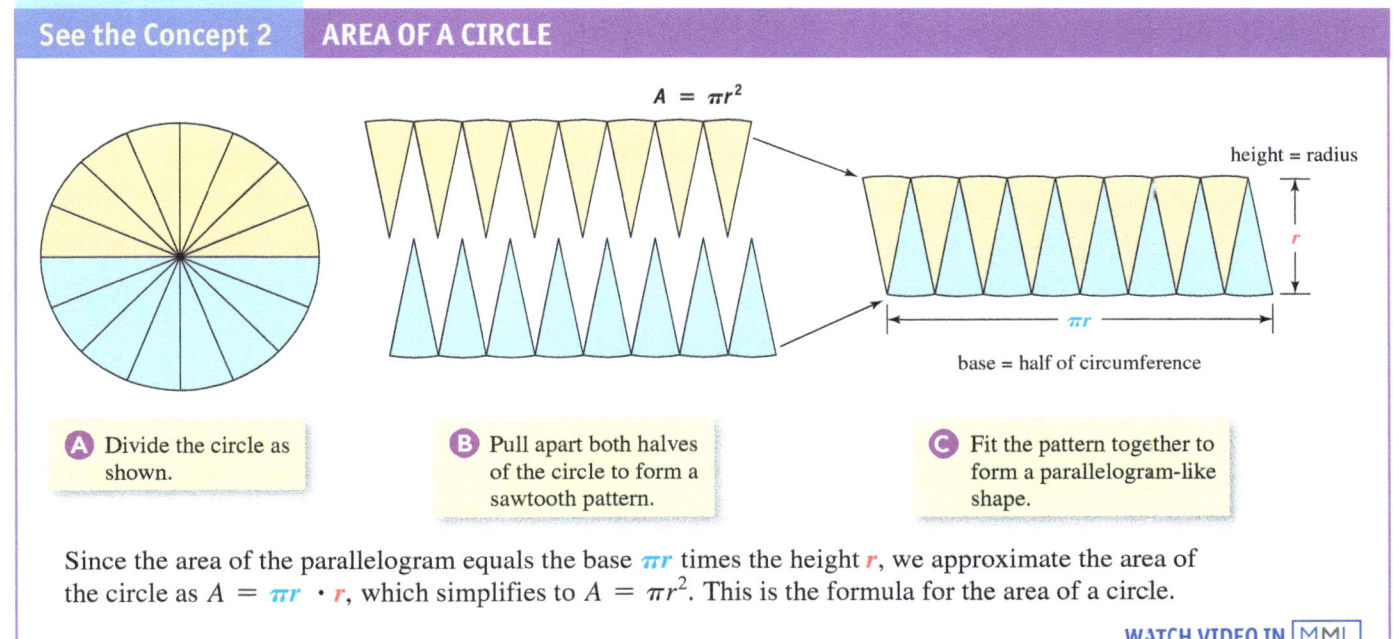

A Divide the circle as shown.

B Pull apart both halves of the circle to form a sawtooth pattern.

C Fit the pattern together to form a parallelogram-like shape.

Since the area of the parallelogram equals the base πr times the height r, we approximate the area of the circle as $A = \pi r \cdot r$, which simplifies to $A = \pi r^2$. This is the formula for the area of a circle.

WATCH VIDEO IN MML.

READING CHECK 1

After transforming a circle into a parallelogram shape, why is the base equal to πr?

AREA OF A CIRCLE

The area A of a circle with radius r is found using

$$A = \pi r^2,$$

where $\pi \approx 3.14$ or $\pi \approx \frac{22}{7}$.

EXAMPLE 4 Finding the area of circles

Find the approximate area of each circle, using 3.14 for π.

(a) 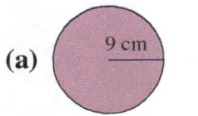 (b)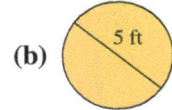

Solution
(a) The area of the circle is $A = \pi r^2 \approx 3.14(9^2) = 254.34 \text{ cm}^2$.
(b) The radius of the circle is half the diameter, or $0.5(5) = 2.5$ feet. So, the area of the circle is $A = \pi r^2 \approx 3.14(2.5)^2 = 19.625 \text{ ft}^2$.

Now Try Exercises 17, 19

2 Finding the Volume and Surface Area of Geometric Solids

VOLUME For geometric solids (three-dimensional figures), we are interested in finding *volume* and *surface area*. As noted earlier, **volume** is a measure of the amount of a substance needed to fill a defined space. Volume is measured in *cubic units*. **FIGURE 8.34** shows a cubic centimeter and a cubic inch.

A Cubic Centimeter and a Cubic Inch

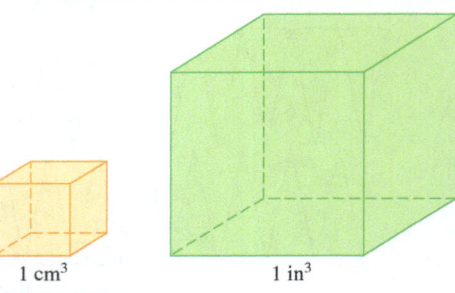

FIGURE 8.34

SURFACE AREA **Surface area** measures the amount of exposed area for a given solid. For example, a simple die from a board game has the shape of a cube with six square surfaces called *faces*, as shown in **FIGURE 8.35**.

A Die with Six Faces

FIGURE 8.35

The surface area of a die is the sum of the areas of the six exposed faces. This discussion suggests that surface area is found by adding the areas of the exposed faces or surfaces.

Several common geometric solids are shown in **TABLE 8.7**, along with formulas for the volume and surface area. In the table, we use V to represent volume, S to represent surface area, and B to represent the *area of the base* of a solid.

Volume and Surface Area Formulas for Common Solids

Solid		Formulas
Cube	side, side, side	$V = s^3$ (or $V = Bh$), $S = 6s^2$ (or $S = 6B$), where s = side length, $B = s^2$, and $h = s$ = height
Rectangular Prism	height, length, width	$V = lwh$ (or $V = Bh$), $S = 2lw + 2lh + 2wh$, where l = length, w = width, h = height, and $B = lw$
Circular Cylinder	height, radius	$V = \pi r^2 h$ (or $V = Bh$), $S = 2\pi rh + 2\pi r^2$, where r = radius, h = height, and $B = \pi r^2$
Cone	height, radius	$V = \frac{1}{3}\pi r^2 h$ (or $V = \frac{1}{3}Bh$), $S = \pi r\sqrt{r^2 + h^2} + \pi r^2$, where r = radius, h = height, and $B = \pi r^2$
Square-Based Pyramid	slant height, height, side, side	$V = \frac{1}{3}s^2 h$ (or $V = \frac{1}{3}Bh$), $S = B + 2sl$, where s = side, h = height, $B = s^2$, and l = slant height
Sphere	radius	$V = \frac{4}{3}\pi r^3$, $S = 4\pi r^2$, where r = radius

TABLE 8.7

READING CHECK 2

What does the variable B represent when finding the volume of geometric solids?

EXAMPLE 5 Finding the volume of geometric solids

Find the volume of each geometric solid. If needed, use 3.14 as an approximation for π and round answers to the nearest tenth.
(a) Circular cylinder: $r = 3$ ft and $h = 12$ ft
(b) Rectangular prism: $l = 14$ m, $w = 10$ m, and $h = 3$ m

Solution
(a) $V = \pi r^2 h = (3.14)(3^2)(12) \approx 339.1$ ft^3
(b) $V = lwh = (14)(10)(3) = 420$ m^3

Now Try Exercises 27, 29

528 CHAPTER 8 GEOMETRY

EXAMPLE 6 **Finding the surface area of geometric solids**

Find the surface area of each geometric solid. If needed, use 3.14 as an approximation for π and round answers to the nearest tenth.
(a) Square-based pyramid: $s = 6$ ft, $h = 4$ ft and $l = 5$ ft
(b) Sphere: $r = 6$ cm

Solution
(a) $S = B + 2sl = 6^2 + 2(6)(5) = 96$ ft^2
(b) $S = 4\pi r^2 \approx 4(3.14)(6^2) \approx 452.2$ cm^2

Now Try Exercises 45, 47

8.5 Putting It All Together

CONCEPT	COMMENTS	EXAMPLES
Area of Plane Figures	**Square:** $A = s^2$, ($s =$ side) **Rectangle:** $A = lw$, ($l =$ length; $w =$ width) **Triangle:** $A = \frac{1}{2}bh$, ($b =$ base; $h =$ height) **Parallelogram:** $A = bh$, ($b =$ base; $h =$ height) **Trapezoid:** $A = \frac{1}{2}(a+b)h$, ($h =$ height; a and b are lengths of parallel sides) **Circle:** $A = \pi r^2$, ($r =$ radius)	The area of a rectangle with $l = 9$ ft and $w = 5$ ft is $A = 9(5) = 45$ ft^2. The area of a circle with $r = 7$ cm is $A \approx 3.14(7^2) \approx 153.9$ cm^2.
Volume and Surface Area of Geometric Solids	A full summary of the volume and surface area formulas for the cube, rectangular prism, circular cylinder, cone, square-based pyramid, and sphere can be found in **TABLE 8.7**.	The volume of a sphere with $r = 2$ m is $V \approx \left(\frac{4}{3}\right)(3.14)(2^3) \approx 33.5$ m^3. The surface area of a cube with $s = 9$ mi is $S = 6(9^2) = 486$ mi^2.

8.5 Exercises

MyMathLab®

CONCEPTS AND VOCABULARY

1. When finding the area of a plane figure, the units for the result are (square/cubic) units.

2. When finding the volume of a geometric solid, the units for the result are (square/cubic) units.

3. When finding the surface area of a geometric solid, the units for the result are (square/cubic) units.

4. (True or False?) When finding the area of a circle, the units for the result are *circular* units.

5. When we find the amount of a substance needed to fill a defined space, we are finding _____.

6. When we find the amount of exposed area on a solid, we are finding _____.

FINDING THE AREA OF PLANE FIGURES

Exercises 7–16: Find the area of the polygon.

7.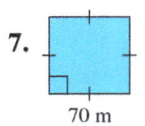
8.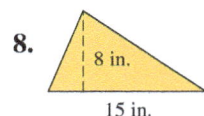
9.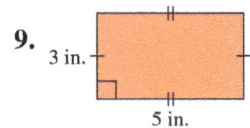
10.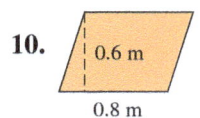
11.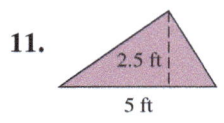
12.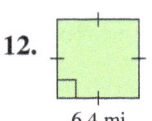
13.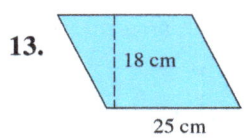
14.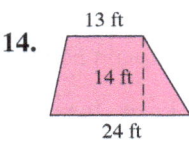
15.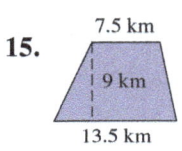
16.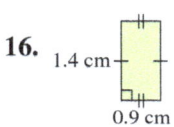

Exercises 17–24: Determine the approximate area of the given circle, using 3.14 for π. Round your answers to the nearest tenth.

17.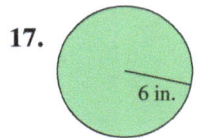
18.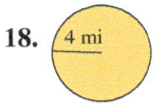
19.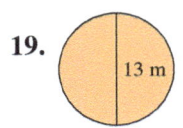
20.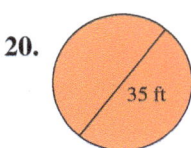
21.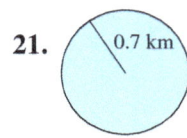
22.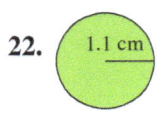
23.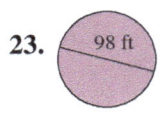
24.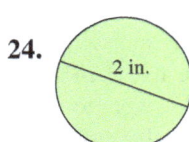

FINDING THE VOLUME OF GEOMETRIC SOLIDS

Exercises 25–36: Find the volume of the geometric solid. If needed, use 3.14 as an approximation for π and round answers to the nearest tenth.

25. Cube: $s = 7$ in.
26. Cube: $s = 30$ m
27. Rectangular prism: $l = 7$ ft, $w = 3$ ft, $h = 2$ ft
28. Rectangular prism: $l = 16$ mi, $w = 5$ mi, $h = 4$ mi
29. Circular cylinder: $r = 8$ cm, $h = 15$ cm
30. Circular cylinder: $r = 12$ mm, $h = 56$ mm
31. Cone: $r = 3$ in., $h = 4$ in.
32. Cone: $r = 14$ m, $h = 50$ m
33. Square-based pyramid: $s = 9$ ft, $h = 10$ ft
34. Square-based pyramid: $s = 4$ in., $h = 6$ in.
35. Sphere: $r = 2$ cm
36. Sphere: $r = 10$ m

FINDING THE SURFACE AREA OF GEOMETRIC SOLIDS

Exercises 37–48: Find the surface area of the geometric solid. If needed, use 3.14 as an approximation for π and round answers to the nearest tenth.

37. Cube: $s = 6$ in.
38. Cube: $s = 56$ m
39. Rectangular prism: $l = 8$ ft, $w = 4$ ft, $h = 3$ ft
40. Rectangular prism: $l = 10$ mi, $w = 3$ mi, $h = 1$ mi
41. Circular cylinder: $r = 4$ cm, $h = 9$ cm
42. Circular cylinder: $r = 20$ mm, $h = 80$ mm
43. Cone: $r = 5$ in., $h = 10$ in.
44. Cone: $r = 7$ m, $h = 4$ m
45. Square-based pyramid: $s = 3$ ft, $l = 8$ ft
46. Square-based pyramid: $s = 12$ in., $l = 9$ in.
47. Sphere: $r = 7$ cm
48. Sphere: $r = 25$ m

APPLICATIONS INVOLVING VOLUME AND SURFACE AREA

49. **Pile of Sand** A pile of sand has the shape of a cone with radius 6 feet and height 5 feet. Find the volume of sand in the pile. Use 3.14 as an approximation for π and round to the nearest tenth.

50. **Great Pyramid** The Great Pyramid of Giza has a square base measuring about 756 feet. Approximate the volume of the Great Pyramid if the height is about 480 feet. (*Source: National Geographic.*)

51. **Basketball** A basketball has a 4.7-inch radius. Find the surface area of the ball. Use 3.14 as an approximation for π and round to the nearest tenth.

52. **Basketball Hoop** A basketball hoop has an 18-inch diameter. Find the area enclosed by the hoop. Use 3.14 as an approximation for π and round to the nearest tenth.

53. **Heavy Lifting** A hydraulic lift has the shape of a parallelogram. Find the area of the parallelogram if the base measures 3.5 feet and the lift is extended to a height of 2 feet.

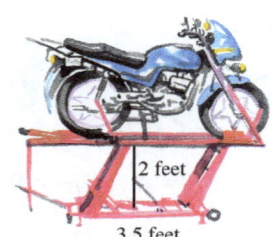

54. **Geometric Garden** A garden is constructed in the shape of a trapezoid. If the parallel sides are 20 feet apart and measure 18 feet and 32 feet, find the area of the garden.

55. **Paint Can** A paint can has the shape of a cylinder with a 6-inch diameter and a 7-inch height. Find the surface area of the can. Use 3.14 as an approximation for π and round to the nearest tenth.

56. **Water Cup** How much water fits in a cone-shaped cup with a 4-centimeter radius and an 8-centimeter height? Use 3.14 as an approximation for π and round to the nearest cubic centimeter.

WRITING ABOUT MATHEMATICS

57. Half of a sphere is called a semi-sphere. Write the formulas for the volume and surface area of a semi-sphere. Explain how you arrived at your answers.

58. If you knew the distance around Earth's equator, explain how you could compute Earth's radius.

59. Suppose a box (rectangular prism) has no top. Write a formula for the surface area for the box. Explain how you arrived at your answer.

60. Explain how you could find the volume of wood that remains after a circular hole is drilled through a wooden cube.

SECTION 8.5 Checking Basic Concepts

1. Find the area of each polygon.

   (a) 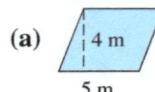   (b)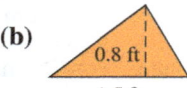

2. Find the area of each circle, using 3.14 as an approximation for π. Round your answers to the nearest tenth.

   (a) 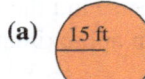   (b)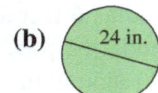

3. Find the volume of each geometric solid. If needed, use 3.14 as an approximation for π and round answers to the nearest tenth.
   (a) Cube: $s = 14$ km
   (b) Cone: $r = 6$ ft, $h = 15$ ft

4. Find the surface area of each geometric solid. If needed, use 3.14 as an approximation for π and round answers to the nearest tenth.
   (a) Circular cylinder: $r = 1$ m, $h = 3$ m
   (b) Sphere: $r = 60$ cm

CHAPTER 8 Summary

SECTION 8.1 ■ PLANE GEOMETRY: POINTS, LINES, AND ANGLES

Geometric Terms

Point
A point is a location in space having no length, width, or height.

Example: • P

Line
A line is a straight figure representing a set of points extending indefinitely in two directions.

Example: ←•————•→
         A   B

Line Segment
A line segment is a straight figure representing a set of points extending between two endpoints.

Example: •————————•
         A   B

Ray
A ray is a straight figure representing a set of points extending indefinitely in one direction.

Example: •————•——→
         A   B

Angle
An angle is a figure formed by two rays with a common endpoint. The common endpoint is called the *vertex*, and the two rays are called the *sides* of the angle.

Example: The vertex is B and the sides are $\vec{BA}$ and $\vec{BC}$.

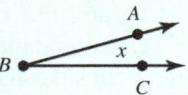

Classifying Angles

Types of Angles
A *right* angle measures 90°. A *straight* angle measures 180°. An *acute* angle measures between 0° and 90°. An *obtuse* angle measures between 90° and 180°.

Examples:
90° Right Angle   180° Straight Angle   77° Acute Angle   154° Obtuse Angle

Pairs of Angles
Angles with equal measures are *congruent*. The sum of the measures of two *complementary* angles is 90°. The sum of the measures of two *supplementary* angles is 180°.

Examples:
31°/31° Congruent Angles   39°/51° Complementary Angles   59°/121° Supplementary Angles

Perpendicular Lines
Two intersecting lines are *perpendicular* if they form right angles.

Example:

Vertical and Adjacent Angles
Two intersecting lines that are not perpendicular form two pairs of congruent angles called vertical angles and four pairs of supplementary angles called adjacent angles. (See example on the next page.)

Example: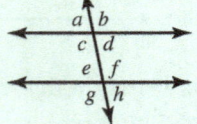

Vertical: ∠w and ∠y, ∠x and ∠z
Adjacent: ∠w and ∠x, ∠x and ∠y, ∠y and ∠z, ∠z and ∠w

Parallel Lines Cut by a Transversal

When two parallel lines are cut by a transversal, the following properties apply.

1. Vertical angles are congruent.
2. Corresponding angles are congruent.
3. Alternate interior angles are congruent.
4. Alternate exterior angles are congruent.

Any other angle pairs are supplementary.

Example: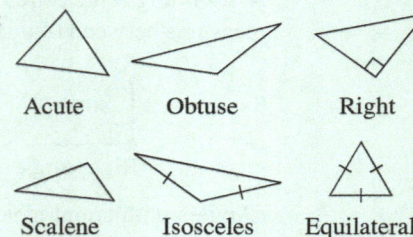

Corresponding: ∠a and ∠e, ∠b and ∠f, ∠c and ∠g, ∠d and ∠h
Alternate Interior: ∠c and ∠f, ∠d and ∠e
Alternate Exterior: ∠a and ∠h, ∠b and ∠g

SECTION 8.2 ■ TRIANGLES

Types of Triangles

Acute: Every angle measures less than 90°.
Obtuse: One angle measures between 90° and 180°.
Right: One angle measures exactly 90°.
Scalene: No sides have the same length.
Isosceles: At least two sides have the same length.
Equilateral: Three sides have the same length.

Examples:

Acute Obtuse Right

Scalene Isosceles Equilateral

The Sum of the Angle Measures

For any triangle, the sum of the measures of the three angles is 180°.

Example: 38° + 66° + 76° = 180°

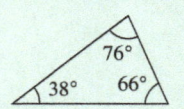

Congruent Triangles

Any of the following properties can be used to determine whether two triangles are congruent. (See examples on the next page.)

Angle-Side-Angle (ASA): If two angles and the included side of one triangle are congruent to two angles and the included side of another triangle, then the triangles are congruent.

Side-Angle-Side (SAS): If two sides and the included angle of one triangle are congruent to two sides and the included angle of another triangle, then the triangles are congruent.

Side-Side-Side (SSS): If three sides of one triangle are congruent to three sides of another triangle, then the triangles are congruent.

Examples: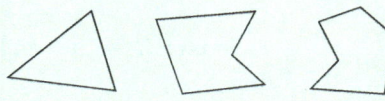

Angle-Side-Angle   Side-Angle-Side

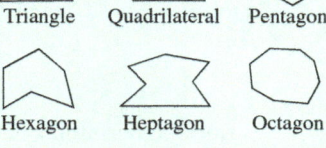

Side-Side-Side

SECTION 8.3 ■ POLYGONS AND CIRCLES

Polygons

Polygons A polygon is a closed plane figure determined by three or more line segments.

Examples:

Names of Polygons A polygon is named by the number of its sides.

3 sides: Triangle   4 sides: Quadrilateral   5 sides: Pentagon
6 sides: Hexagon   7 sides: Heptagon   8 sides: Octagon

Example:

Triangle   Quadrilateral   Pentagon

Hexagon   Heptagon   Octagon

Regular Polygons In a regular polygon, all sides have the same length and all angles are congruent.

Examples:

Angle Measures for a Regular Polygon For a regular polygon with n sides, the sum of the angle measures is given by

$$(n - 2) \cdot 180°,$$

and the measure of one angle is given by

$$\frac{(n - 2) \cdot 180°}{n}.$$

Examples: For a regular 4-sided polygon, the sum of the angle measures is $(4 - 2) \cdot 180° = 360°$ and the measure of one angle is

$$\frac{(4 - 2) \cdot 180°}{4} = \frac{360°}{4} = 90°.$$

Quadrilaterals Quadrilaterals are 4-sided polygons that include the following. (See examples on the next page.)

Square: A regular quadrilateral
Rectangle: All angles measure 90°
Parallelogram: Two pairs of parallel sides
Trapezoid: One pair of parallel sides
Rhombus: Four sides of equal length
Kite: Two pairs of adjacent sides equal in length

534 CHAPTER 8 GEOMETRY

Examples: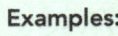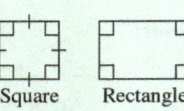

Square Rectangle Parallelogram

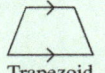 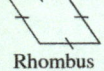 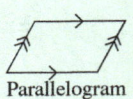

Trapezoid Rhombus Kite

Circles

A circle is a closed plane figure that consists of all points that are equally distant from a central point (called the center of the circle). A radius is a line segment with one endpoint on the circle and the other endpoint at the center. A diameter is a line segment that contains the center and has both endpoints on the circle.

Example: $d = 2r$ and $r = \dfrac{d}{2}$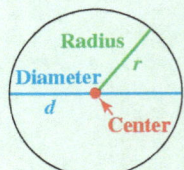

SECTION 8.4 ■ PERIMETER AND CIRCUMFERENCE

Perimeter

Perimeter of a Polygon

To find the perimeter of a polygon, add the lengths of all sides of the polygon.

Example: $9 + 14 + 17 = 40$ cm

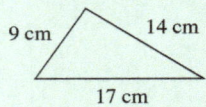

Perimeter of a Regular Polygon

To find the perimeter of a regular polygon, multiply the number of sides by the length of one side.

Example: $4 \cdot 7 = 28$ ft

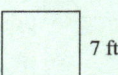

Circumference

Circumference of a Circle

The circumference C of a circle with radius r and diameter d is found using $C = \pi d$ or $C = 2\pi r$, where $\pi \approx 3.14$ or $\pi \approx \frac{22}{7}$.

Example: $C = 16\pi \approx 16 \cdot 3.14 = 50.24$ cm

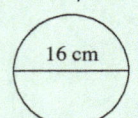

SECTION 8.5 ■ AREA, VOLUME, AND SURFACE AREA

Area

Area of Plane Figures

The following formulas can be used to find the area of common plane figures.

Square: $A = s^2$, ($s =$ side)
Rectangle: $A = lw$, ($l =$ length; $w =$ width)
Triangle: $A = \frac{1}{2}bh$, ($b =$ base; $h =$ height)
Parallelogram: $A = bh$, ($b =$ base; $h =$ height)
Trapezoid: $A = \frac{1}{2}(a + b)h$, ($h =$ height; a and b are lengths of parallel sides)
Circle: $A = \pi r^2$, ($r =$ radius)

Examples: The area of a parallelogram with $b = 10$ ft and $h = 13$ ft is $A = 10(13) = 130$ ft^2.

The area of a circle with $r = 2$ cm is $A \approx 3.14(2^2) = 12.56$ cm^2.

Volume and Surface Area

Volume and Surface Area of Plane Figures

A full summary of the volume and surface area formulas for the cube, rectangular prism, circular cylinder, cone, square-based pyramid, and sphere can be found in **TABLE 8.7**.

Example: The volume of a circular cylinder with $r = 2$ m and $h = 4$ m is
$$V \approx (3.14)(2^2)(4) = 50.24 \text{ m}^3.$$

The surface area of a rectangular prism with $l = 8$ ft, $w = 5$ ft, and $h = 3$ ft is $S = 2(8)(5) + 2(8)(3) + 2(5)(3) = 158$ ft^2.

CHAPTER 8 Review Exercises

SECTION 8.1

Exercises 1–4: Identify the figure and name it using the labels shown.

1. $\xleftrightarrow{X \quad Y}$
2. $\overset{\bullet}{P} \xrightarrow{\quad Q}$
3. $\overset{\bullet}{A} \quad \overset{\bullet}{B}$
4. • T

Exercises 5–8: Classify the angle as acute, right, obtuse, or straight.

5. 6. 7. 8.

Exercises 9 and 10: Find the measure of an angle that has the specified description.

9. The complement of an angle with measure 24°

10. The supplement of an angle with measure 33°

11. If $m\angle d = 104°$ in the following figure, find the measures of $\angle c$, $\angle f$, and $\angle e$.

12. If $m \| n$ and $m\angle f = 65°$ in the following figure, find the measures of $\angle a$, $\angle b$, and $\angle c$.

Exercises 13–16: In the following figure, $m \| n$. Determine whether the given angles are congruent or supplementary.

13. $\angle b$ and $\angle e$
14. $\angle a$ and $\angle h$
15. $\angle b$ and $\angle f$
16. $\angle c$ and $\angle h$

SECTION 8.2

Exercises 17–20: Classify the given triangle as acute, obtuse, or right.

17. 18. 19. 20.

Exercises 21–24: Classify the given triangle as scalene, isosceles, or equilateral.

21. (15, 8, 21)
22. (29, 29, 29)
23. (15, 41, 41)
24. (11, 7, 8)

Exercises 25 and 26: Find the measure of $\angle x$ in the triangle.

25.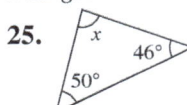
26.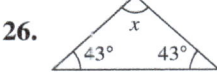

Exercises 27 and 28: Find the value of x. Assume that all angle measures are given in degrees.

27.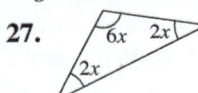
28.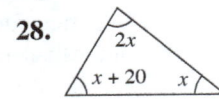

Exercises 29 and 30: For the triangle pair, state the property that shows the triangles are congruent.

29.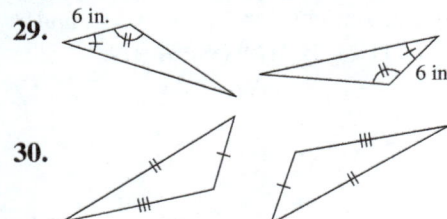

30.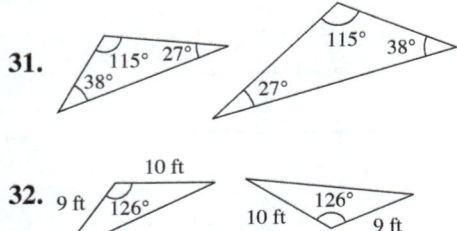

Exercises 31 and 32: Determine whether the triangles in the given pair are congruent. If so, state the property that shows the triangles are congruent.

31.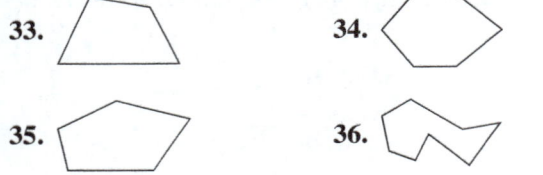

32.

SECTION 8.3

Exercises 33–36: Determine the name of the polygon.

33. 34.

35. 36.

Exercises 37 and 38: Find the sum of the angle measures for the given regular polygon.

37. 38.

Exercises 39 and 40: For the given regular polygon, find the measure of one angle.

39. 40.

Exercises 41–44: Name the quadrilateral using every classification that applies.

41.  42.

43. 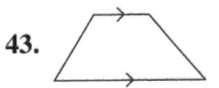 44.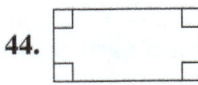

Exercises 45 and 46: Find the radius of the given circle.

45.  46.

Exercises 47 and 48: Find the diameter of the given circle.

47. 48.

SECTION 8.4

Exercises 49 and 50: Find the perimeter of the polygon.

49.  50.

Exercises 51 and 52: Find the perimeter of the given regular polygon.

51. 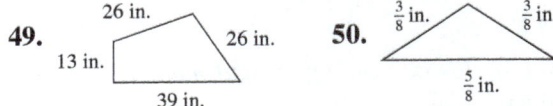 52.

Exercises 53 and 54: Do the following.
  (a) Find the exact circumference of the circle.
  (b) Approximate the circumference, using 3.14 as an approximation for π.

53.  54.

Exercises 55 and 56: Do the following.
  (a) Find the exact circumference of the circle.
  (b) Approximate the circumference, using $\frac{22}{7}$ as an approximation for π.

55.  56.

57. Find the perimeter of the polygon.

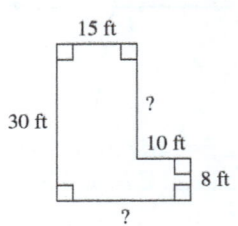

58. Find the perimeter of the composite figure. Use 3.14 as an approximation for π.

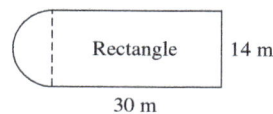

SECTION 8.5

Exercises 59–62: Find the area of the given polygon.

59. 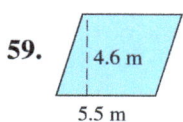 **60.**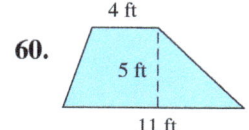

61. 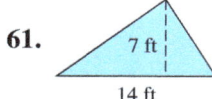 **62.**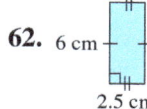

Exercises 63 and 64: Find the approximate area of the circle using 3.14 as an approximation for π. Round your answers to the nearest tenth.

63. 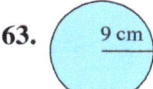 **64.**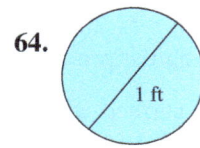

Exercises 65–68: Find the volume of the geometric solid. If needed, use 3.14 as an approximation for π and round answers to the nearest tenth.

65. Cube: $s = 5$ ft

66. Sphere: $r = 9$ cm

67. Circular cylinder: $r = 2$ in., $h = 4$ in.

68. Cone: $r = 7$ m, $h = 5$ m

Exercises 69–72: Find the surface area of the geometric solid. If needed, use 3.14 as an approximation for π and round answers to the nearest tenth.

69. Square-based pyramid: $s = 9$ in., $l = 7$ in.

70. Cone: $r = 10$ m, $h = 3$ m

71. Circular cylinder: $r = 2$ cm, $h = 6$ cm

72. Rectangular prism: $l = 9$ ft, $w = 5$ ft, $h = 2$ ft

APPLICATIONS

73. Hockey Rink The center circle on a hockey rink has a diameter that measures 30 feet. Find the radius of the center circle on a hockey rink.

74. Stop Sign A standard stop sign has the shape of a regular octagon with one side measuring about 16.24 inches. Find the perimeter of a stop sign.

75. Sumo Wrestling Sumo wrestlers compete within a circle with a diameter of 4.55 meters. Approximate the area of the circle to the nearest tenth, using 3.14 for π.

76. Traffic Cone An orange traffic cone has a radius of 5 inches and a height of 24 inches. Approximate the volume of the cone. Use 3.14 for π and round to the nearest whole number.

CHAPTER 8 Test

1. Name the figure using the labels shown.

2. Classify the angle as acute, right, obtuse, or straight.

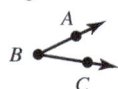

3. Find the measure of the complement of a 71° angle.

4. Find the measure of the supplement of a 118° angle.

5. If $m \| n$ and $m\angle c = 72°$ in the following figure, find the measures of $\angle a$, $\angle f$, and $\angle h$.

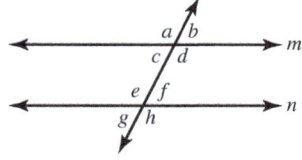

6. Classify the given triangle as acute, obtuse, or right.

7. Classify the given triangle as scalene, isosceles, or equilateral.

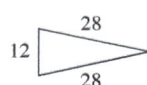

8. Find the measure of $\angle x$ in the triangle.

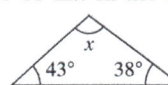

9. Determine whether the triangles in the given pair are congruent. If so, state the property that shows the triangles are congruent.

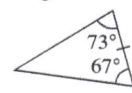 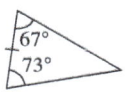

10. Determine the name of the polygon.

11. For the regular polygon shown, do the following.

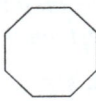

(a) Find the sum of the angle measures.
(b) Find the measure of one angle.

12. Name the given quadrilateral using every classification that applies.

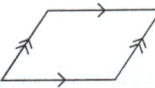

13. Find the diameter of the given circle.

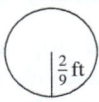

14. Find the perimeter of the polygon.

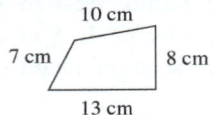

15. For the circle shown, do the following.
(a) Find the exact circumference of the circle.
(b) Approximate the circumference, using 3.14 for π.

16. Find the perimeter of the polygon.

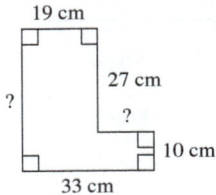

17. Find the area of each polygon.
(a) 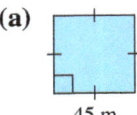 (b)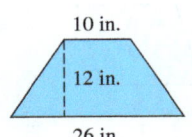

18. Find the approximate area of the circle, using 3.14 for π. Round your answer to the nearest tenth.

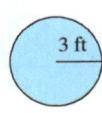

19. Find the volume of the geometric solid. If needed, use 3.14 as an approximation for π and round answers to the nearest tenth.
(a) Sphere: $r = 2$ in.
(b) Rectangular prism: $l = 14$ ft, $w = 6$ ft, $h = 4$ ft

20. Find the surface area of the geometric solid. If needed, use 3.14 as an approximation for π and round answers to the nearest tenth.
(a) Cube: $s = 16$ cm
(b) Circular cylinder: $r = 3$ ft, $h = 8$ ft

CHAPTERS 1–8   Cumulative Review Exercises

1. Write 18,020 in expanded form.

2. Write $9 \cdot 9 \cdot 9 \cdot 9$ using exponential notation.

3. Round 73,987 to the nearest thousand.

4. Evaluate $4 \cdot 8 - 50 \div 10$.

5. Graph the integers $-4, 0,$ and 3 on a number line.

6. Evaluate $6 - 8 + (-3) - (-9)$.

7. Evaluate $25 - 3^2 \div (4 - 7)$.

8. Complete the table and solve $5x - 2 = 3$.

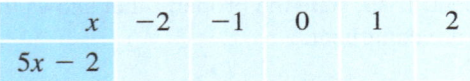

9. Is -2 a solution to the equation $-4x + 3 = 11$?

10. Solve the equation $-3y + (-8) = y$.

11. Solve the linear equation $-2x - 3 = -5$ graphically.

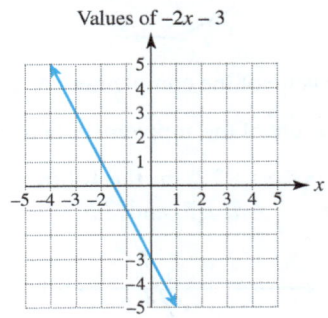

12. Translate the given sentence to an equation using *x* as the variable. Do not solve the equation.

Four equals a number increased by 9.

13. Give both the improper fraction and mixed number represented by the shading.

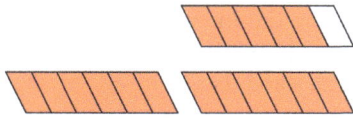

14. Determine whether the given whole number is prime, composite or neither.
(a) 39   (b) 1
(c) 17   (d) 75

Exercises 15 and 16: Perform the indicated arithmetic.

15. $-\dfrac{8}{7} \div \dfrac{4}{35}$ **16.** $-\dfrac{5}{6} + \dfrac{1}{3} \cdot \dfrac{2}{5}$

17. Approximate the length of the unknown side of the right triangle to 1 decimal place.

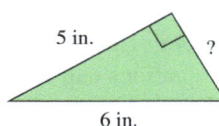

18. Round 841.5706 to the nearest hundredth.

19. Solve the equation $2.1y = 6 - 0.4y$.

20. For the numbers in the given list, identify any of the following types of numbers:
(a) natural numbers   (b) whole numbers
(c) integers   (d) rational numbers
(e) irrational numbers

$$8, -\frac{1}{5}, \sqrt{9}, 0, \pi$$

21. The triangles are similar. Find the value of *x*.

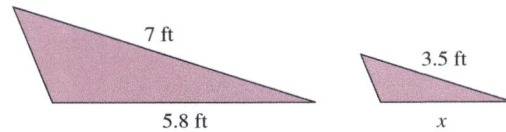

22. A 16-ounce box of crackers sells for $3.68. Find the unit price of the crackers.

23. Convert $7\frac{1}{2}$ gallons to pints.

24. Convert 7 inches per second to feet per hour.

25. Translate the following percent problem to a basic percent equation. Do not solve the equation.

What percent of 80 is 16?

26. If an investment of $1200 earns $144 simple interest after 3 years, find the interest rate.

27. There is a 78% chance that a student likes pie. What is the chance a student does not like pie?

28. 14.5% of what number is 116?

29. If $m \| n$ and $m\angle g = 71°$ in the following figure, find the measures of $\angle a$, $\angle b$, and $\angle h$.

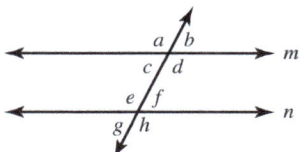

30. Find the measure of $\angle x$ in the triangle.

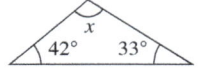

31. For the regular polygon shown, do the following.

(a) Find the sum of the angle measures.
(b) Find the measure of one angle.

32. Find the surface area of the sphere with radius 6 cm. Use 3.14 as an approximation for π and round your answer to the nearest tenth.

33. Gardening A garden is built in the shape of a square with an area of 330 square feet. Estimate the length of one side of the garden to the nearest foot.

34. Finding Age In 17 years, a person's age will be 6 years less than double his current age. Write an equation whose solution gives the person's current age. Use *x* for your variable. Do not solve the equation.

35. Converting Temperature The formula

$$C = \frac{(F - 32)}{1.8}$$

gives the relationship between *C* degrees Celsius and *F* degrees Fahrenheit. Find the Fahrenheit temperature that is equivalent to 26.5 °C.

36. Smart Phone If the price of a smart phone is $178 and the sales tax rate is 6.5%, find the sales tax and the total amount paid.

9 Linear Equations and Inequalities in One Variable

9.1 Review of Linear Equations in One Variable

9.2 Further Problem Solving

9.3 Linear Inequalities in One Variable

Mathematics lets us use numbers to describe the intensity of ultraviolet light from the sun. The table shows the maximum ultraviolet intensity measured in milliwatts per square meter for various latitudes and dates.

The Sun's Ultraviolet Intensity (mW/m²)

Latitude	Mar. 21	June 21	Sept. 21	Dec. 21
0°	325	254	325	272
10°	311	275	280	220
20°	249	292	256	143
30°	179	248	182	80
40°	99	199	127	34
50°	57	143	75	13

If a student from Chicago, located at a latitude of 42°, spends spring break in Hawaii with a latitude of 20°, the sun's ultraviolet rays in Hawaii will be approximately $\frac{249}{99} \approx 2.5$ times as intense as they are in Chicago. Equations can be used to describe, or model, the intensity of the sun at various latitudes. In this chapter, we will focus on *linear equations* and the related concept of *linear inequalities*. Source: J. Williams, *The USA Today Weather Almanac*.

9.1 Review of Linear Equations in One Variable

Objectives

1. Identifying Linear Equations in One Variable
2. Solving Linear Equations
   - Solving with Tables
   - Solving Symbolically
3. Solving Linear Equations by Applying the Distributive Property
4. Solving Linear Equations by Clearing Fractions and Decimals
5. Identifying Equations with No Solutions, One Solution, or Infinitely Many Solutions
6. Solving a Formula for a Variable

NEW VOCABULARY

☐ Linear equation
☐ Identity
☐ Contradiction

1 Identifying Linear Equations in One Variable

Connecting Concepts with Your Life Suppose that a bicyclist is 5 miles from home, riding *away* from home at 10 miles per hour, as shown in **FIGURE 9.1**. The distance between the bicyclist and home for various elapsed times is shown in **TABLE 9.1**.

Riding Away from Home

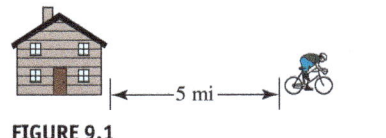

FIGURE 9.1

Distance from Home

Elapsed Time (hours)	0	1	2	3
Distance (miles)	5	15	25	35

TABLE 9.1

10 mi   10 mi   10 mi

The bicyclist is moving at a constant speed, so the distance increases by 10 miles every hour. The distance D from home after x hours can be calculated by the equation

$$D = 10x + 5.$$

For example, after 2 hours the distance is

$$D = 10(2) + 5 = 25 \text{ miles.}$$

TABLE 9.1 verifies that the bicyclist is 25 miles from home after 2 hours. However, the table is less helpful if we want to find the elapsed time when the bicyclist is 18 miles from home. To answer this question, we could begin by substituting 18 for D in the distance equation to obtain

$$18 = 10x + 5.$$

The equation $18 = 10x + 5$ can be written in a different form by applying the addition property of equality. Subtracting 18 from each side gives an equivalent equation.

Equivalent equations
$$\begin{cases} 18 - 18 = 10x + 5 - 18 & \text{Subtract 18 from each side.} \\ 0 = 10x - 13 & \text{Simplify.} \\ 10x - 13 = 0 & \text{Rewrite the equation.} \end{cases}$$

Even though these steps did not result in a solution to the equation $18 = 10x + 5$, applying the addition property of equality allowed us to rewrite the equation as $10x - 13 = 0$, which is an example of a *linear equation*. (See Example 3(a) for a solution to the equation $10x - 13 = 0$.)

> **LINEAR EQUATION IN ONE VARIABLE**
>
> A **linear equation** in one variable is an equation that can be written in the form
>
> $$ax + b = 0,$$
>
> where a and b are constants with $a \neq 0$.

NOTE: Linear equations can model applications in which things move or change at a *constant rate*. ∎

If an equation is linear, writing it in the form $ax + b = 0$ should not require any properties or processes other than the following.

- Using the distributive property to clear any parentheses
- Combining like terms
- Applying the addition property of equality

TABLE 9.2 gives examples of linear equations and values for a and b.

Examples of Linear Equations

Equation	$ax + b = 0$ Form	a	b
$x = 1$	$x - 1 = 0$	1	-1
$-5x + 4 = 3$	$-5x + 1 = 0$	-5	1
$2.5x = 0$	$2.5x + 0 = 0$	2.5	0

TABLE 9.2

> **MAKING CONNECTIONS 1**
>
> **Equations That Are Not Linear**
>
> An equation *cannot* be written in the form $ax + b = 0$ (and is not a *linear* equation) if after clearing parentheses and combining like terms, any of the following statements are true.
>
> 1. The variable has an exponent other than 1.
> 2. The variable appears in a denominator of a fraction.
> 3. The variable appears under the symbol $\sqrt{}$ or within an absolute value.

READING CHECK 1

Name three things that tell you that an equation is not a linear equation.

EXAMPLE 1 **Determining whether an equation is linear**

Determine whether the equation is linear. If the equation is linear, give values for a and b that result when the equation is written in the form $ax + b = 0$.
(a) $4x + 5 = 0$ **(b)** $5 = -\frac{3}{4}x$ **(c)** $4x^2 + 6 = 0$ **(d)** $\frac{3}{x} + 5 = 0$

Solution
(a) The equation is linear because it is in the form $ax + b = 0$ with $a = 4$ and $b = 5$.
(b) The equation can be rewritten as follows.

$$5 = -\frac{3}{4}x \qquad \text{Given equation}$$

$$\frac{3}{4}x + 5 = \frac{3}{4}x + \left(-\frac{3}{4}x\right) \qquad \text{Add } \tfrac{3}{4}x \text{ to each side.}$$

$$\frac{3}{4}x + 5 = 0 \qquad \text{Additive inverse}$$

The given equation is linear because it can be written in the form $ax + b = 0$ with $a = \frac{3}{4}$ and $b = 5$.

NOTE: If 5 had been subtracted from each side, the result would be $0 = -\frac{3}{4}x - 5$, which is an equivalent linear equation with $a = -\frac{3}{4}$ and $b = -5$. ∎

(c) The equation is *not* linear because it cannot be written in the form $ax + b = 0$. The variable has exponent 2.
(d) The equation is *not* linear because it cannot be written in the form $ax + b = 0$. The variable appears in the denominator of a fraction.

Now Try Exercises 9, 11, 13, 15

2 Solving Linear Equations

Every linear equation $(ax + b = 0, a \neq 0)$ has *exactly one* solution. Showing that this is true is left as an exercise (see Exercise 59). Solving a linear equation means finding the (one) value of the variable that makes the equation true.

SOLVING WITH TABLES One way to solve a linear equation is to make a table of values. A table provides an organized way of checking possible values of the variable to see if there is a value that makes the equation true. For example, if we want to solve the equation

$$2x - 5 = -7,$$

we substitute various values for x in the left side of the equation. If one of these values results in -7, then the value makes the equation true and is the solution. In the next example, a table of values is used to solve this equation.

EXAMPLE 2 Using a table to solve an equation

Complete **TABLE 9.3** for the given values of x. Then solve the equation $2x - 5 = -7$.

Listing Possible Values for $2x - 5$

x	-3	-2	-1	0	1	2	3
$2x - 5$	-11						

TABLE 9.3

Solution
To complete the table, substitute $x = -2, -1, 0, 1, 2$, and 3 into the expression $2x - 5$. For example, if $x = -2$, then $2x - 5 = 2(-2) - 5 = -9$. The other values shown in **TABLE 9.4** can be found similarly.

Solving $2x - 5 = -7$ Numerically

x	-3	-2	-1	0	1	2	3
$2x - 5$	-11	-9	-7	-5	-3	-1	1

TABLE 9.4

We see in **TABLE 9.4** that $2x - 5$ equals -7 when $x = -1$. The solution to $2x - 5 = -7$ is -1.

Now Try Exercise 25

SOLVING SYMBOLICALLY Although tables can be used to solve some linear equations, the process of creating a table that contains the solution can take a significant amount of time. For example, the solution to the equation $9x - 4 = 0$ is $\frac{4}{9}$. However, creating a table that reveals this solution would be quite challenging.

The following strategy, which involves the addition and multiplication properties of equality, is a method for solving linear equations symbolically.

SOLVING A LINEAR EQUATION SYMBOLICALLY

STEP 1: Use the distributive property to clear any parentheses on each side of the equation. Combine any like terms on each side.

STEP 2: Use the addition property of equality to get all of the terms containing the variable on one side of the equation and all other terms on the other side of the equation. Combine any like terms on each side.

STEP 3: Use the multiplication property of equality to isolate the variable by multiplying each side of the equation by the reciprocal of the number in front of the variable (or divide each side by that number).

STEP 4: Check the solution by substituting it in the given equation.

TECHNOLOGY NOTE

Graphing Calculators and Tables
Many graphing calculators have the capability to make tables. Table 9.4 is shown in the accompanying figure.

CALCULATOR HELP
To make a table on a calculator, see Appendix A (pages AP-2 and AP-3).

STUDY TIP
We will be solving equations throughout the remainder of the text. Spend a little extra time practicing these steps so that they are easy to recall when needed later.

EXAMPLE 3 Solving linear equations

Solve each linear equation. Check the answer for part (b).
(a) $10x - 13 = 0$ (b) $\frac{1}{2}x + 3 = 6$ (c) $5x + 7 = 2x + 3$

Solution
(a) First, isolate the x-term on the left side of the equation by adding 13 to each side.

$10x - 13 = 0$	Given equation
$10x - 13 + 13 = 0 + 13$	Add 13 to each side. (STEP 2)
$10x = 13$	Add the real numbers.

To obtain a coefficient of 1 on the x-term, divide each side by 10.

$\dfrac{10x}{10} = \dfrac{13}{10}$	Divide each side by 10. (STEP 3)
$x = \dfrac{13}{10}$	Simplify.

The solution is $\frac{13}{10}$.

(b) Start by subtracting 3 from each side.

$\dfrac{1}{2}x + 3 = 6$	Given equation
$\dfrac{1}{2}x + 3 - 3 = 6 - 3$	Subtract 3 from each side. (STEP 2)
$\dfrac{1}{2}x = 3$	Subtract the real numbers.
$2 \cdot \dfrac{1}{2}x = 2 \cdot 3$	Multiply each side by 2. (STEP 3)
$x = 6$	Multiply the real numbers.

The solution is 6. To check it, substitute 6 for x in the equation, $\frac{1}{2}x + 3 = 6$.

$\dfrac{1}{2} \cdot 6 + 3 \stackrel{?}{=} 6$	Replace x with 6. (STEP 4)
$3 + 3 \stackrel{?}{=} 6$	Multiply.
$6 = 6$ ✓	Add; the answer checks.

(c) Since this equation has two x-terms, we need to get all x-terms on one side of the equation and all real numbers on the other side. To do this, begin by subtracting $2x$ from each side.

$5x + 7 = 2x + 3$	Given equation
$5x - 2x + 7 = 2x - 2x + 3$	Subtract $2x$ from each side. (STEP 2)
$3x + 7 = 3$	Combine like terms.
$3x + 7 - 7 = 3 - 7$	Subtract 7 from each side. (STEP 2)
$3x = -4$	Simplify.
$\dfrac{3x}{3} = \dfrac{-4}{3}$	Divide each side by 3. (STEP 3)
$x = -\dfrac{4}{3}$	Simplify the fractions.

The solution is $-\frac{4}{3}$.

Now Try Exercises 31, 33, 37

9.1 REVIEW OF LINEAR EQUATIONS IN ONE VARIABLE   545

🌐 Math in Financial Context The number of private venture-backed companies worth $1 billion or more is increasing. The rise of these "unicorns" has been dramatic. One such example is Uber, a popular ride-sharing service worth billions. (*Source:* Statista.)

EXAMPLE 4 Solving an application involving a linear equation

The number N of "unicorn" companies worth $1 billion or more can be modeled by the equation $N = 3.625x + 42$, where $0 \leq x \leq 20$ and x is the number of months after January 2014. Use this equation to determine the x-value when N was equal to 100. Interpret the meaning of this solution.

Solution
Begin by solving for x when $N = 100$. That is, solve $3.625x + 42 = 100$.

$$3.625x + 42 = 100 \qquad \text{Equation to solve}$$
$$3.625x = 58 \qquad \text{Subtract 42 from each side.}$$
$$x = \frac{58}{3.625} \qquad \text{Divide each side by 3.625.}$$
$$x = 16 \qquad \text{Simplify}$$

The value $x = 16$ represents 16 months after January 2014, or May 2015. Thus there were 100 venture-backed private companies worth $1 billion or more in May 2015.

Now Try Exercise 93

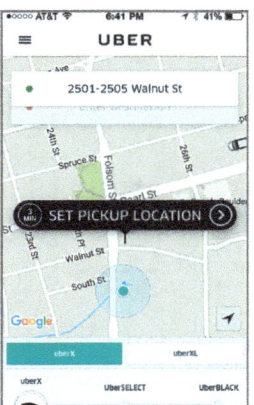

3 Solving Linear Equations by Applying the Distributive Property

Sometimes the distributive property is helpful in solving linear equations. The next example demonstrates how to apply the distributive property in such situations. Use of the distributive property appeared in STEP 1 of the strategy for solving linear equations discussed earlier.

EXAMPLE 5 Applying the distributive property

Solve each linear equation. Check the answer for part (a).
(a) $4(x - 3) + x = 0$ **(b)** $2(3z - 4) + 1 = 3(z + 1)$

Solution
(a) Begin by applying the distributive property.

$$4(x - 3) + x = 0 \qquad \text{Given equation}$$

$4 \cdot x = 4x$
$4 \cdot 3 = 12$

$$4x - 12 + x = 0 \qquad \text{Distributive property (STEP 1)}$$
$$5x - 12 = 0 \qquad \text{Combine like terms.}$$
$$5x - 12 + 12 = 0 + 12 \qquad \text{Add 12 to each side. (STEP 2)}$$
$$5x = 12 \qquad \text{Add the real numbers.}$$
$$\frac{5x}{5} = \frac{12}{5} \qquad \text{Divide each side by 5. (STEP 3)}$$
$$x = \frac{12}{5} \qquad \text{Simplify.}$$

READING CHECK 2

In the strategy for solving linear equations symbolically, which step involves the use of the distributive property?

To see if $\frac{12}{5}$ is the solution, substitute $\frac{12}{5}$ for x in the equation $4(x - 3) + x = 0$.

$$4\left(\frac{12}{5} - 3\right) + \frac{12}{5} \stackrel{?}{=} 0 \qquad \text{Replace } x \text{ with } \tfrac{12}{5}. \text{ (STEP 4)}$$

$$4\left(\frac{12}{5} - \frac{15}{5}\right) + \frac{12}{5} \stackrel{?}{=} 0 \qquad \text{Common denominator}$$

$$4\left(-\frac{3}{5}\right) + \frac{12}{5} \stackrel{?}{=} 0 \qquad \text{Subtract within parentheses.}$$

$$-\frac{12}{5} + \frac{12}{5} \stackrel{?}{=} 0 \qquad \text{Multiply.}$$

$$0 = 0 \checkmark \qquad \text{Add; the answer checks.}$$

(b) Begin by applying the distributive property to each side of the equation. Then get all z-terms on the left side and terms containing only real numbers on the right side.

$$2(3z - 4) + 1 = 3(z + 1) \qquad \text{Given equation}$$
$$6z - 8 + 1 = 3z + 3 \qquad \text{Distributive property (STEP 1)}$$
$$6z - 7 = 3z + 3 \qquad \text{Add the real numbers.}$$
$$3z - 7 = 3 \qquad \text{Subtract } 3z \text{ from each side. (STEP 2)}$$
$$3z = 10 \qquad \text{Add 7 to each side. (STEP 2)}$$
$$\frac{3z}{3} = \frac{10}{3} \qquad \text{Divide each side by 3. (STEP 3)}$$
$$z = \frac{10}{3} \qquad \text{Simplify.}$$

The solution is $\frac{10}{3}$.

Now Try Exercises 43, 45

4 Solving Linear Equations by Clearing Fractions and Decimals

Some people prefer to do calculations without fractions or decimals. For this reason, clearing an equation of fractions or decimals before solving it can be helpful. To clear fractions or decimals, multiply each side of the equation by the least common denominator (LCD).

EXAMPLE 6 Solving a linear equation by clearing fractions

Solve each linear equation.
(a) $\frac{1}{7}x - \frac{5}{7} = \frac{3}{7}$ **(b)** $\frac{2}{3}x - \frac{1}{6} = x$

Solution
(a) Multiply each side of the equation by the LCD 7 to clear (remove) fractions.

$$\frac{1}{7}x - \frac{5}{7} = \frac{3}{7} \qquad \text{Given equation}$$

The LCD is 7.
$$7\left(\frac{1}{7}x - \frac{5}{7}\right) = 7 \cdot \frac{3}{7} \qquad \text{Multiply each side by 7.}$$

$$\frac{7}{1} \cdot \frac{1}{7}x - \frac{7}{1} \cdot \frac{5}{7} = \frac{7}{1} \cdot \frac{3}{7} \qquad \text{Distributive property}$$

$$x - 5 = 3 \qquad \text{Simplify.}$$

$$x = 8 \qquad \text{Add 5 to each side.}$$

The solution is 8.

(b) The LCD for 3 and 6 is 6. Multiply each side of the equation by **6**.

$$\frac{2}{3}x - \frac{1}{6} = x \qquad \text{Given equation}$$

The LCD is 6.
$$6\left(\frac{2}{3}x - \frac{1}{6}\right) = 6 \cdot x \qquad \text{Multiply each side by 6.}$$

$$\frac{6}{1} \cdot \frac{2}{3}x - \frac{6}{1} \cdot \frac{1}{6} = 6 \cdot x \qquad \text{Distributive property}$$

$$4x - 1 = 6x \qquad \text{Simplify.}$$

$$-1 = 2x \qquad \text{Subtract } 4x \text{ from each side.}$$

$$-\frac{1}{2} = x \qquad \text{Divide each side by 2.}$$

The solution is $-\frac{1}{2}$.

Now Try Exercises 55, 57

EXAMPLE 7 Solving a linear equation by clearing decimals

Solve each linear equation.
(a) $0.2x - 0.7 = 0.4$
(b) $0.01x - 0.42 = -0.2x$

Solution
(a) The least common denominator for 0.2, 0.7, and 0.4 $\left(\text{or } \frac{2}{10}, \frac{7}{10}, \text{ and } \frac{4}{10}\right)$ is **10**. Multiply each side by 10. When multiplying by 10, move the decimal point 1 place to the right.

$$0.2x - 0.7 = 0.4 \qquad \text{Given equation}$$

The LCD is 10.
$$10(0.2x - 0.7) = 10(0.4) \qquad \text{Multiply each side by 10.}$$

$$10(0.2x) - 10(0.7) = 10(0.4) \qquad \text{Distributive property}$$

$$2x - 7 = 4 \qquad \text{Simplify.}$$

$$2x = 11 \qquad \text{Add 7 to each side.}$$

$$x = \frac{11}{2} \qquad \text{Divide each side by 2.}$$

The solution is $\frac{11}{2}$, or 5.5.

(b) The least common denominator for 0.01, 0.42, and 0.2 $\left(\text{or } \frac{1}{100}, \frac{42}{100}, \text{ and } \frac{2}{10}\right)$ is **100**. Multiply each side by 100. To do this, move the decimal point 2 places to the right.

$$0.01x - 0.42 = -0.2x \qquad \text{Given equation}$$

The LCD is 100.
$$100(0.01x - 0.42) = 100(-0.2x) \qquad \text{Multiply each side by 100.}$$

$$100(0.01x) - 100(0.42) = 100(-0.2x) \qquad \text{Distributive property}$$

$$x - 42 = -20x \qquad \text{Simplify.}$$

$$x - 42 + 20x + 42 = -20x + 20x + 42 \qquad \text{Add } 20x \text{ and } 42.$$

$$21x = 42 \qquad \text{Combine like terms.}$$

$$x = 2 \qquad \text{Divide each side by 21.}$$

The solution is 2.

Now Try Exercises 51, 53

> **MAKING CONNECTIONS 2**
>
> **Clearing Decimals and the LCD**
>
> When clearing decimals from a linear equation, the LCD is always a power of 10 determined by the maximum number of decimal places in the coefficients of the terms in the equation. For example, for the equation in Example 7(b), the maximum number of such decimal places is 2.
>
> 2 decimal places — $0.01x - 0.42 = -0.2x$ — 1 decimal place
>
> The LCD is 10^2, or 100. That is, decimals are cleared from the equation by multiplying each term by 100. As a result, the decimal point in each term is moved 2 places to the right.

5 Identifying Equations with No Solutions, One Solution, or Infinitely Many Solutions

Some equations that appear to be linear are not because when they are written in the form $ax + b = 0$ the value of a is 0 and no x-term appears. This type of equation can have *no solutions* or *infinitely many solutions*. The following See the Concept explains how we can tell if an equation has no solutions, one solution, or infinitely many solutions.

> **See the Concept 1   THE NUMBER OF SOLUTIONS**
>
> | | *No Solutions* | *One Solution* | *Infinitely Many Solutions* |
> |---|---|---|---|
> | Equation | $x + 1 = x$ | $4x - 1 = 7$ | $5x = 2x + 3x$ |
> | $ax + b = 0$ form | $0x + 1 = 0$ Ⓐ | $4x - 8 = 0$ Ⓐ | $0x + 0 = 0$ Ⓐ |
> | Simplified form | $1 = 0$ Ⓑ | $x = 2$ Ⓑ | $0 = 0$ Ⓑ |
>
> **Equations with *No Solutions*:**
>
> Ⓐ When the equation is written in $ax + b = 0$ form, the value of a is 0.
> Ⓑ The simplified equation is *always false* for any value of the variable. An equation that is always false is called a **contradiction**.
>
> **Equations with *One Solution*:**
>
> Ⓐ When the equation is written in $ax + b = 0$ form, the value of a is *not* 0.
> Ⓑ The simplified equation is true for *only one* value of the variable. This is called a **conditional equation**.
>
> **Equations with *Infinitely Many Solutions*:**
>
> Ⓐ When the equation is written in $ax + b = 0$ form, the value of a is 0.
> Ⓑ The simplified equation is *always true* for any value of the variable. An equation that is always true is called an **identity**.
>
> WATCH VIDEO IN MML.

READING CHECK 3

How can you tell when an equation will have infinitely many solutions?

EXAMPLE 8   Determining numbers of solutions

Determine whether the equation has no solutions, one solution, or infinitely many solutions.
(a) $3x = 2(x + 1) + x$ (b) $2x - (x + 1) = x - 1$ (c) $5x = 2(x - 4)$

Solution

(a) Start by applying the distributive property.

$$3x = 2(x + 1) + x \qquad \text{Given equation}$$
$$3x = 2x + 2 + x \qquad \text{Distributive property}$$
$$3x = 3x + 2 \qquad \text{Combine like terms.}$$
$$0 = 2 \qquad \text{Subtract } 3x \text{ from each side.}$$

The equation $0 = 2$ is always false, so the given equation is a contradiction and it has no solutions.

(b) Start by applying the distributive property.

$$2x - (x + 1) = x - 1 \qquad \text{Given equation}$$
$$2x - x - 1 = x - 1 \qquad \text{Distributive property} \quad [-1(x+1)]$$
$$x - 1 = x - 1 \qquad \text{Combine like terms.}$$
$$x = x \qquad \text{Add 1 to each side.}$$
$$0 = 0 \qquad \text{Subtract } x \text{ from each side.}$$

The equation $0 = 0$ is always true, so the given equation is an identity that has infinitely many solutions. Note that the solution set contains all real numbers.

(c) Start by applying the distributive property.

$$5x = 2(x - 4) \qquad \text{Given equation}$$
$$5x = 2x - 8 \qquad \text{Distributive property}$$
$$3x = -8 \qquad \text{Subtract } 2x \text{ from each side.}$$
$$x = -\frac{8}{3} \qquad \text{Divide each side by 3.}$$

Thus there is one solution, $-\frac{8}{3}$.

Now Try Exercises 61, 63, 65

CRITICAL THINKING 1

What must be true about b and d for the equation

$$bx - 2 = dx + 7$$

to have no solutions? What must be true about b and d for this equation to have exactly one solution?

---

6 Solving a Formula for a Variable

Sometimes a formula must be rewritten to solve for the needed variable. For example, if the area A and the width w of a rectangular region are given, then its length l can be found by solving the formula $A = lw$ for l.

$$A = lw \qquad \text{Area formula}$$
$$\frac{A}{w} = \frac{lw}{w} \qquad \text{Divide each side by } w.$$
$$\frac{A}{w} = l \qquad \text{Simplify the fraction.}$$
$$l = \frac{A}{w} \qquad \text{Rewrite the equation.}$$

If the area A of a rectangle is 400 square inches and its width w is 16 inches, then the rectangle's length l is

$$l = \frac{A}{w} = \frac{400}{16} = 25 \text{ inches.}$$

Once the area formula has been solved for l, the resulting formula can be used to find the length of *any* rectangle whose area and width are known.

EXAMPLE 9 **Finding the base of a trapezoid**

The area of a trapezoid is given by

$$A = \frac{1}{2}(a+b)h,$$

where a and b are the bases of the trapezoid and h is the height.
(a) Solve the formula for b.
(b) A trapezoid has area $A = 36$ square inches, height $h = 4$ inches, and base $a = 8$ inches. Find b.

STUDY TIP

If you are studying with classmates, make sure that they do not "do the work for you." A classmate with the best intentions may give too many verbal hints while helping you work through a problem. Remember that members of your study group will not be giving hints during an exam.

Solution
(a) To clear the equation of the fraction, multiply each side by 2.

$$A = \frac{1}{2}(a+b)h \quad \text{Trapezoid area formula}$$

$$2A = (a+b)h \quad \text{Multiply each side by 2.}$$

$$\frac{2A}{h} = a + b \quad \text{Divide each side by } h.$$

$$\frac{2A}{h} - a = b \quad \text{Subtract } a \text{ from each side.}$$

$$b = \frac{2A}{h} - a \quad \text{Rewrite the formula.}$$

(b) Let $A = 36$, $h = 4$, and $a = 8$ in $b = \frac{2A}{h} - a$. Then

$$b = \frac{2(36)}{4} - 8 = 18 - 8 = 10 \text{ inches.}$$

Now Try Exercise 79

EXAMPLE 10 **Solving for a variable**

Solve each equation for the indicated variable.
(a) $c = \frac{a+b}{2}$ for b (b) $ab - bc = ac$ for c

Solution
(a) To clear the equation of the fraction, multiply each side by 2.

$$c = \frac{a+b}{2} \quad \text{Given equation}$$

$$2c = a + b \quad \text{Multiply each side by 2.}$$

$$2c - a = b \quad \text{Subtract } a.$$

The equation solved for b is $b = 2c - a$.

(b) In $ab - bc = ac$, the variable c appears in two terms. We will combine the terms containing c by using the distributive property. Begin by moving the term on the left side containing c to the right side of the equation.

$$ab - bc = ac \quad \text{Given equation}$$
$$ab - bc + bc = ac + bc \quad \text{Add } bc \text{ to each side.}$$
$$ab = (a + b)c \quad \text{Combine terms; distributive property.}$$
$$\frac{ab}{a + b} = \frac{(a + b)c}{(a + b)} \quad \text{Divide each side by } (a + b).$$
$$\frac{ab}{a + b} = c \quad \text{Simplify the fraction.}$$

The equation solved for c is $c = \frac{ab}{a + b}$.

Now Try Exercises 83, 87

CRITICAL THINKING 2

Are the equations $c = \frac{1}{a - b}$ and $c = \frac{-1}{b - a}$ equivalent? Why?

9.1 Putting It All Together

CONCEPT	COMMENTS	EXAMPLES
Linear Equation	Can be written as $$ax + b = 0,$$ where $a \neq 0$; has one solution	The equation $5x - 8 = 0$ is linear, with $a = 5$ and $b = -8$. The equation $2x^2 + 4 = 0$ is not linear.
Solving Linear Equations Numerically	To solve a linear equation numerically, complete a table for various values of the variable and then select the solution from the table, if possible.	The solution to $2x - 4 = -2$ is 1. | x | -1 | 0 | 1 | |---|---|---|---| | $2x - 4$ | -6 | -4 | -2 |
Solving Linear Equations Symbolically	Use the addition and multiplication properties of equality to isolate the variable. See the four-step approach to solving a linear equation in this section.	$5x - 8 = 0$ Given equation $5x = 8$ Add 8 to each side. $x = \frac{8}{5}$ Divide each side by 5.
Equations with No Solutions	Some equations that appear to be linear have no solutions. Solving will result in an equivalent equation that is always false.	The equation $$x = x + 5$$ has no solutions because it is equivalent to the equation $0 = 5$, which is always false.
Equations with Infinitely Many Solutions	Some equations that appear to be linear have infinitely many solutions. Solving will result in an equivalent equation that is always true.	The equation $$2x = x + x$$ has infinitely many solutions because it is equivalent to the equation $0 = 0$, which is always true.

9.1 Exercises

CONCEPTS AND VOCABULARY

1. Linear equations can model applications in which things move or change at a(n) _____ rate.

2. A linear equation can be written in the form _____ with $a \neq 0$.

3. How many solutions does a linear equation in one variable have?

4. When a table of values is used to solve a linear equation, the equation is being solved _____.

5. What two properties of equality are frequently used to solve linear equations?

6. To clear fractions or decimals from an equation, multiply each side by the _____.

7. If solving an equation results in $0 = 4$, how many solutions does it have?

8. If solving an equation results in $0 = 0$, how many solutions does it have?

IDENTIFYING LINEAR EQUATIONS IN ONE VARIABLE

Exercises 9–22: (Refer to Example 1.) Is the equation linear? If it is linear, give values for a and b that result when the equation is written in the form $ax + b = 0$.

9. $3x - 7 = 0$
10. $-2x + 1 = 4$
11. $\frac{1}{2}x = 0$
12. $-\frac{3}{4}x = 0$
13. $4x^2 - 6 = 11$
14. $-2x^2 + x = 4$
15. $\frac{6}{x} - 4 = 2$
16. $2\sqrt{x} - 1 = 0$
17. $1.1x + 0.9 = 1.8$
18. $-5.7x - 3.4 = -6.8$
19. $2(x - 3) = 0$
20. $\frac{1}{2}(x + 4) = 0$
21. $|3x| + 2 = 1$
22. $3x = 4x^3$

SOLVING LINEAR EQUATIONS NUMERICALLY

Exercises 23–28: Complete the table for the given values of x. Then solve the given equation numerically.

23. $x - 3 = -1$

x	-1	0	1	2	3
$x - 3$	-4				

24. $-2x = 0$

x	-2	-1	0	1	2
$-2x$	4				

25. $-3x + 7 = 1$

x	0	1	2	3	4
$-3x + 7$	7				

26. $5x - 2 = 3$

x	-1	0	1	2	3
$5x - 2$	-7				

27. $4 - 2x = 6$

x	-2	-1	0	1	2
$4 - 2x$	8				

28. $9 - (x + 3) = 4$

x	-2	-1	0	1	2
$9 - (x + 3)$	8				

SOLVING LINEAR EQUATIONS SYMBOLICALLY

Exercises 29–40: Solve and check the solution.

29. $11x = 3$
30. $-5x = 15$
31. $x - 18 = 5$
32. $8 = 5 + x$
33. $\frac{1}{2}x - 1 = 13$
34. $\frac{1}{4}x + 3 = 9$
35. $-6 = 5x + 5$
36. $31 = -7x - 4$
37. $3z + 2 = z - 5$
38. $z - 5 = 5z - 3$
39. $12y - 6 = 33 - y$
40. $-13y + 2 = 22 - 3y$

SOLVING LINEAR EQUATIONS BY APPLYING THE DISTRIBUTIVE PROPERTY

Exercises 41–50: Solve and check the solution.

41. $4(x - 1) = 5$
42. $-2(2x + 7) = 1$
43. $1 - (3x + 1) = 5 - x$
44. $6 + 2(x - 7) = 10 - 3(x - 3)$
45. $5t - 6 = 2(t + 1) + 2$

46. $-2(t - 7) - (t + 5) = 5$

47. $3(4z - 1) - 2(z + 2) = 2(z + 1)$

48. $-(z + 4) + (3z + 1) = -2(z + 1)$

49. $4y - 2(y + 1) = 0$

50. $(15y + 20) - 5y = 5 - 10y$

SOLVING LINEAR EQUATIONS BY CLEARING DECIMALS

Exercises 51–54: Solve and check the solution.

51. $7.3x - 1.7 = 5.6$
52. $5.5x + 3x = 51$

53. $-9.5x - 0.05 = 10.5x + 1.05$

54. $0.04x + 0.03 = 0.02x - 0.1$

SOLVING LINEAR EQUATIONS BY CLEARING FRACTIONS

Exercises 55–58: Solve and check the solution.

55. $\frac{1}{2}x - \frac{3}{2} = \frac{5}{2}$

56. $-\frac{1}{4}x + \frac{5}{4} = \frac{3}{4}$

57. $-\frac{3}{8}x + \frac{1}{4} = \frac{1}{8}$

58. $\frac{1}{3}x + \frac{1}{4} = \frac{1}{6} - x$

59. **Thinking Generally** A linear equation has exactly one solution. Find the solution to the equation $ax + b = 0$, where $a \neq 0$, by solving for x.

60. **Thinking Generally** Solve the linear equation given by $\frac{1}{a}x - b = 0$ for x.

IDENTIFYING EQUATIONS WITH NO SOLUTIONS, ONE SOLUTION, OR INFINITELY MANY SOLUTIONS

Exercises 61–70: Determine whether the equation has no solutions, one solution, or infinitely many solutions.

61. $5x = 5x + 1$
62. $4x = 5(x + 3) - x$

63. $8x = 0$
64. $9x = x + 1$

65. $2(x - 3) = 2x - 6$

66. $5x = 15 - 2(x + 7)$

67. $x - (3x + 2) = 15 - 2x$

68. $4(x + 2) - 2(2x + 3) = 10$

69. $5(2x + 7) - (10x + 5) = 30$

70. $2x - (x + 5) = x - 5$

SOLVING A FORMULA FOR A VARIABLE

Exercises 71–88: Solve the formula for the given variable.

71. $9x + 3y = 6$ for y
72. $-2x - 2y = 10$ for y

73. $4x + 3y = 12$ for y
74. $5x - 2y = 22$ for y

75. $A = lw$ for w
76. $A = \frac{1}{2}bh$ for b

77. $V = \pi r^2 h$ for h
78. $V = \frac{1}{3}\pi r^2 h$ for h

79. $A = \frac{1}{2}(a + b)h$ for a
80. $C = 2\pi r$ for r

81. $V = lwh$ for w
82. $P = 2l + 2w$ for w

83. $s = \dfrac{a + b + c}{2}$ for b
84. $t = \dfrac{x - y}{3}$ for x

85. $\dfrac{a}{b} - \dfrac{c}{b} = 1$ for b
86. $\dfrac{x}{y} + \dfrac{z}{y} = 5$ for z

87. $ab = cd + ad$ for a

88. $S = 2lw + 2lh + 2wh$ for w

89. **Perimeter of a Rectangle** If the width of a rectangle is 5 inches and its perimeter is 40 inches, find the length of the rectangle.

90. **Perimeter of a Triangle** Two sides of a triangle have lengths of 5 feet and 7 feet. If the triangle's perimeter is 21 feet, what is the length of the third side?

APPLICATIONS INVOLVING LINEAR EQUATIONS

91. **Distance Traveled** A bicyclist is 4 miles from home, riding away from home at 8 miles per hour.
    (a) Make a table that shows the bicyclist's distance D from home after 0, 1, 2, 3, and 4 hours.
    (b) Write an equation that calculates the distance D after x hours.
    (c) Determine D when $x = 3$ hours. Does your answer agree with the value found in your table?
    (d) Find x when $D = 22$ miles. Interpret the result.

92. **Distance Traveled** An athlete is 16 miles from home, running *toward* home at 6 miles per hour.
    (a) Write an equation that calculates the distance D the athlete is from home after x hours.
    (b) Determine D when $x = 1.5$ hours.
    (c) Find x when $D = 5.5$ miles. Interpret the result.

93. **Internet Users** The number of Internet users I in millions during year x, where $x \geq 2007$, can be approximated by the equation
$$I = 241x - 482{,}440.$$
Approximate the year in which there were 1730 million (1.73 billion) Internet users.

94. HIV Infections The cumulative number of HIV infections N in thousands for the United States in year x can be approximated by the equation

$$N = 42x - 83{,}197,$$

where $x \geq 2000$. Approximate the year when this number reached 970 thousand. (*Source:* CDC)

95. Working-Age Population The U.S. working age population P in millions during year x can be approximated by $P = 1.2x - 2205$, where x is any year between 1970 and 2050. Find the year when the U.S. working age population was 207 million. (*Source:* UN)

96. Working-Age Population The sub-Saharan African working age population P in millions during year x can be approximated by $P = 19.44x - 38{,}661$, where x is any year between 2014 and 2050. Estimate the year when the sub-Saharan African working age population will be 705 million. (*Source:* UN)

97. Internet Ad Spending The amount A spent worldwide on Internet advertising in billions of dollars during year x can be approximated by $A = 1.885x - 3774$, where x is any year between 2006 and 2015. Determine the year when Internet advertising spending was $20.5 billion.

98. Global Oil Demand The global demand D for oil in millions of barrels per day during year x can be approximated by $D = 0.75x - 1420$, where x is any year from 2012 to 2020. Estimate the year when global demand for oil was 92 million barrels per day.

WRITING ABOUT MATHEMATICS

99. A student says that the equation $4x - 1 = 1 - x$ is not a linear equation because it is not in the form $ax + b = 0$. Is the student correct? Explain.

100. A student solves a linear equation as follows.

$$4(x + 3) = 5 - (x + 3)$$
$$4x + 3 \stackrel{?}{=} 5 - x + 3$$
$$4x + 3 \stackrel{?}{=} 8 - x$$
$$5x \stackrel{?}{=} 5$$
$$x \stackrel{?}{=} 1$$

Identify and explain the errors that the student made. What is the correct answer?

Group Activity

Exercises 1–5: **Climate Change** If the global climate were to warm significantly, the Arctic ice cap in Greenland would start to melt. This ice cap contains the equivalent of about 680,000 cubic miles of water. More than 200 million people live on land that is less than 3 feet above sea level. In the United States, several large cities have low average elevations. Three examples are Boston (14 feet), New Orleans (4 feet), and San Diego (13 feet). In the following exercises, you are to estimate the rise in the sea level if the Arctic ice cap were to melt and to determine whether this event would have a significant impact on people living in coastal areas.

1. The surface area of a sphere is given by the formula $4\pi r^2$, where r is its radius. Although the shape of Earth is not exactly spherical, it has an average radius of 3960 miles. Use the formula to estimate the surface area of Earth.

2. Oceans cover approximately 71% of the total surface area of Earth. How many square miles of Earth's surface area are covered by oceans?

3. Approximate the potential rise in sea level by dividing the total volume of the water from the ice cap by the surface area of the oceans. Convert your answer from miles to feet.

4. Discuss the implications of your calculation. How would cities such as Boston, New Orleans, and San Diego be affected?

5. The Antarctic ice cap contains about 6,300,000 cubic miles of water. Estimate how much the sea level would rise if this ice cap melted. (*Source:* Department of the Interior Geological Survey.)

9.2 Further Problem Solving

Objectives

1. Identifying Steps for Solving a Problem
2. Translating Sentences into Equations
3. Solving Number Problems and Applications
4. Solving Distance Problems
5. Solving Mixture Problems

1 Identifying Steps for Solving a Problem

Word problems can be challenging because formulas and equations are not usually given. To solve such problems, we need a strategy. The following steps are based on George Polya's (1888–1985) four-step process for problem solving.

STEPS FOR SOLVING A PROBLEM

STEP 1: Read the problem carefully and be sure that you understand it. (You may need to read the problem more than once.) Assign a variable to what you are being asked to find. If necessary, write other quantities in terms of this variable.

STEP 2: Write an equation that relates the quantities described in the problem. You may need to sketch a diagram or refer to known formulas.

STEP 3: Solve the equation. Use the solution to determine the solution(s) to the original problem. Include any necessary units.

STEP 4: Check your solution in the original problem. Does it seem reasonable?

READING CHECK 1

- Why doesn't the problem-solving strategy stop after the equation is solved in STEP 3?

STUDY TIP

STEP 4 in the problem-solving strategy provides good advice for working any math problem. Checking your answer, especially when taking an exam, can lead to fewer errors and better scores.

2 Translating Sentences into Equations

Even if we understand the problem that we are trying to solve (STEP 1), we may not be able to find a solution if we cannot write an appropriate equation (STEP 2). **TABLE 9.5** lists common words that are associated with the math symbols needed to write equations.

Math Symbols and Associated Words

Symbol	Associated Words
+	add, plus, more, sum, total, increase
−	subtract, minus, less, difference, fewer, decrease
·	multiply, times, twice, double, triple, product
÷	divide, divided by, quotient, per
=	equals, is, gives, results in, is the same as

TABLE 9.5

EXAMPLE 1 Translating sentences into equations

Translate the sentence into an equation using the variable x. Then solve the resulting equation.
(a) Three times a number minus 6 is equal to 18.
(b) The sum of half a number and 5 is zero.
(c) Sixteen is 4 less than twice a number.

Solution

(a) The phrase "Three times a number" indicates that we multiply x by 3 to get $3x$. The word "minus" indicates that we then subtract 6 from $3x$ to get $3x - 6$. This expression "equals" 18, so the equation is $3x - 6 = 18$. The solution is 8 as shown.

$$3x - 6 = 18 \qquad \text{Equation to be solved}$$
$$3x = 24 \qquad \text{Add 6 to each side.}$$
$$\frac{3x}{3} = \frac{24}{3} \qquad \text{Divide each side by 3.}$$
$$x = 8 \qquad \text{Simplify the fractions.}$$

(Three times a number minus 6 is equal to 18.)

(b) The word "sum" indicates that we add "half a number" and 5 to get $\frac{1}{2}x + 5$. The word "is" implies equality, so the equation is $\frac{1}{2}x + 5 = 0$. The solution is -10 as shown.

> The sum of half a number and 5 is zero.

$$\frac{1}{2}x + 5 = 0 \quad \text{Equation to be solved}$$
$$\frac{1}{2}x = -5 \quad \text{Subtract 5 from each side.}$$
$$x = -10 \quad \text{Multiply each side by 2.}$$

(c) To translate "4 less than twice a number" into a mathematical expression, we write $2x - 4$. If this seems backwards, consider how you would calculate "4 less than your age." The equation is $16 = 2x - 4$ and the solution is 10 as shown.

> Sixteen is 4 less than twice a number.

$$16 = 2x - 4 \quad \text{Equation to be solved}$$
$$20 = 2x \quad \text{Add 4 to each side.}$$
$$\frac{20}{2} = \frac{2x}{2} \quad \text{Divide each side by 2.}$$
$$x = 10 \quad \text{Simplify and rewrite.}$$

Now Try Exercises 7, 9, 13

3 Solving Number Problems and Applications

In the next example, we apply the four-step process to a word problem that involves three consecutive numbers. Two integers (or natural numbers) are *consecutive* if they differ by 1. For example, 21 and 22 are consecutive, as are -137 and -136.

EXAMPLE 2 Solving a number problem

The sum of three consecutive natural numbers is 81. Find the three numbers.

Solution
STEP 1: Start by assigning a variable n to an unknown quantity.

n: smallest of the three natural numbers

Next, write the other two natural numbers in terms of n.

$n + 1$: next consecutive natural number

$n + 2$: largest of the three consecutive natural numbers

STEP 2: Write an equation that relates these unknown quantities. The sum of the three consecutive natural numbers is **81**, so the needed equation is

$$n + (n + 1) + (n + 2) = 81.$$

STEP 3: Solve the equation in STEP 2.

$$n + (n + 1) + (n + 2) = 81 \quad \text{Equation to be solved}$$
$$(n + n + n) + (1 + 2) = 81 \quad \text{Commutative and associative properties}$$
$$3n + 3 = 81 \quad \text{Combine like terms.}$$
$$3n = 78 \quad \text{Subtract 3 from each side.}$$
$$n = 26 \quad \text{Divide each side by 3.}$$

The smallest of the three numbers is 26, so the three numbers are 26, 27, and 28.
STEP 4: To check this solution, we can add the three numbers to see if their sum is 81.

$$26 + 27 + 28 = 81 \checkmark$$

Now Try Exercise 15

Math in Context (World Health) The infant mortality rate for a country measures the number of deaths of infants under one year of age per 1000 live births in a given year. A high infant mortality rate may indicate a lack of good quality health care for infants. In the next example, a number problem is solved to find the infant mortality rate in Iceland, which has one of the lowest rates in the world.

EXAMPLE 3 Finding an infant mortality rate

Sierra Leone, in West Africa, has one of the highest infant mortality rates in the world at 75. This rate is 15 more than 20 times the rate in Iceland. Find the infant mortality rate in Iceland. (*Source:* World Population Prospectus.)

Solution
STEP 1: Let r represent the unknown infant mortality rate in Iceland.
STEP 2: Since the rate 75 in Sierra Leone is 15 more than 20 times the rate r in Iceland, the equation to solve is $75 = 15 + 20r$.
STEP 3: Rewrite the equation in STEP 2 and then solve it.

$$20r + 15 = 75 \quad \text{Equation to be solved (rewritten)}$$
$$20r = 60 \quad \text{Subtract 15 from each side.}$$
$$r = 3 \quad \text{Divide each side by 20.}$$

The infant mortality rate in Iceland is 3 deaths per 1000 live births.

STEP 4: To check this solution, we verify that 15 more than 20 times the Iceland infant mortality rate of 3 gives the Sierra Leone infant mortality rate of 75.

$$20(3) + 15 = 60 + 15 = 75 \checkmark$$

Now Try Exercise 27

4 Solving Distance Problems

Connecting Concepts with Your Life If a person drives on an interstate highway at 70 miles per hour for 3 hours, then the total distance traveled is $70 \cdot 3 = 210$ miles. In general,

$$d = rt,$$

where d is the distance traveled, r is the rate (or speed), and t is time. In this example, the distance is in miles, the time is in hours, and the rate is expressed in miles per hour. In general, the rate in a distance problem is expressed in units of distance per unit of time.

EXAMPLE 4 Solving a distance problem

A person drives for 2 hours and 30 minutes at a constant speed and travels 180 miles. See **FIGURE 9.2**. Find the speed of the car in miles per hour.

Traveling 180 Miles in 2 Hours and 30 Minutes

FIGURE 9.2

Solution
STEP 1: Let r represent the car's rate, or speed, in miles per hour.
STEP 2: The rate is to be given in miles per hour, so change 2 hours and 30 minutes to 2.5 or $\frac{5}{2}$ hours. Because $d = 180$ and $t = \frac{5}{2}$, the equation $d = rt$ becomes

$$180 = r \cdot \frac{5}{2}.$$

STEP 3: Solve the equation in STEP 2 for r by multiplying each side of the equation by $\frac{2}{5}$, which is the reciprocal of $\frac{5}{2}$.

$$180 = \frac{5}{2} \cdot r \qquad \text{Equation to be solved}$$

Reciprocal of $\frac{5}{2}$

$$\frac{2}{5} \cdot 180 = r \cdot \frac{5}{2} \cdot \frac{2}{5} \qquad \text{Multiply each side by } \frac{2}{5}.$$

$$72 = r \qquad \text{Simplify.}$$

The speed of the car is 72 miles per hour.

STEP 4: Because 2 hours and 30 minutes is equivalent to $\frac{5}{2}$ hours, traveling for 2 hours and 30 minutes at a constant rate of **72** miles per hour results in a distance of

$$d = rt = 72 \cdot \frac{5}{2} = 180 \text{ miles.} \checkmark$$

Now Try Exercise 49

EXAMPLE 5 **Solving a distance problem**

An athlete jogs at two speeds, covering a distance of 7 miles in $\frac{3}{4}$ hour. If the athlete runs $\frac{1}{4}$ hour at 8 miles per hour, find the second speed.

Solution
STEP 1: Let r represent the second speed of the jogger in miles per hour.
STEP 2: The total time spent jogging is $\frac{3}{4}$ hour, so the time spent jogging at the second speed must be $\frac{3}{4} - \frac{1}{4} = \frac{1}{2}$ hour. See **FIGURE 9.3**.

Jogging 7 Miles in $\frac{3}{4}$ Hour

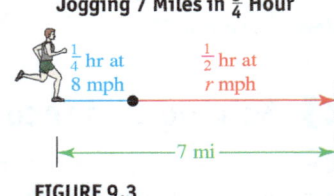

FIGURE 9.3

The **distance** traveled at 8 miles per hour for $\frac{1}{4}$ hour is given by $8 \cdot \frac{1}{4}$ and the **distance** traveled at r miles per hour for $\frac{1}{2}$ hour is given by $r \cdot \frac{1}{2}$. The sum of these distances must equal **7** miles. Thus

$$8 \cdot \frac{1}{4} + r \cdot \frac{1}{2} = 7.$$

STEP 3: Solve the equation in STEP 2 for r.

$$8 \cdot \frac{1}{4} + r \cdot \frac{1}{2} = 7 \qquad \text{Equation to be solved}$$

$$2 + \frac{r}{2} = 7 \qquad \text{Simplify.}$$

$$\frac{r}{2} = 5 \qquad \text{Subtract 2 from each side.}$$

$$r = 10 \qquad \text{Multiply each side by 2.}$$

The athlete's second speed is 10 miles per hour.

STEP 4: Jogging at 8 miles per hour for $\frac{1}{4}$ hour results in a distance of $8 \cdot \frac{1}{4} = 2$ miles. Jogging at 10 miles per hour for $\frac{1}{2}$ hour results in a distance of $10 \cdot \frac{1}{2} = 5$ miles. The total distance is $2 + 5 = 7$ miles and the total time is $\frac{1}{4} + \frac{1}{2} = \frac{3}{4}$ hour. $\checkmark$

Now Try Exercise 53

5 Solving Mixture Problems

🌐 Math in Context (Health Science) Many applied problems involve linear equations. For example, right after people have their wisdom teeth pulled, they may need to rinse their mouth with salt water. In the next example, we use linear equations to determine how much water must be added to dilute a concentrated saline solution.

EXAMPLE 6 **Diluting a saline solution**

A solution contains 4% salt. How much pure water should be added to 30 ounces of the solution to dilute it to a 1.5% solution?

Solution
STEP 1: FIGURE 9.4 illustrates this situation. Assign a variable x as follows.

x: ounces of pure water (0% salt solution)   30: ounces of 4% salt solution

$x + 30$: ounces of 1.5% salt solution

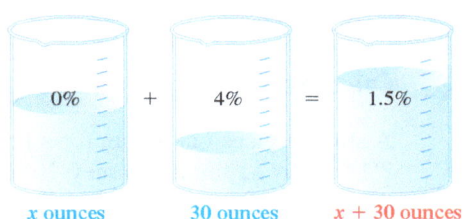

Mixing a Saline Solution

x ounces   30 ounces $x + 30$ ounces

FIGURE 9.4

STEP 2: Note that the amount of salt in the first two beakers must equal the amount of salt in the third beaker. TABLE 9.6 can be used to organize our calculations. The amount of salt in a solution equals the concentration times the solution amount, as shown in the last column of the table.

Mixing a Saline Solution

Solution Type	Concentration (as a decimal)	Solution Amount (ounces)	Salt (ounces)
Pure Water	0% = 0.00	x	$0.00x$
Initial Solution	4% = 0.04	30	$0.04(30)$
Diluted Solution	1.5% = 0.015	$x + 30$	$0.015(x + 30)$

TABLE 9.6

The amount of salt in the first two beakers is
$$0.00x + 0.04(30) = 0 + 1.2 = 1.2 \text{ ounces.}$$
The amount of salt in the final beaker is
$$0.015(x + 30) \text{ ounces.}$$
Because the amounts of salt in the solutions before and after mixing must be equal, the following equation must hold.
$$0.015(x + 30) = 1.2$$

STEP 3: Solve the equation in STEP 2.

$$0.015(x + 30) = 1.2 \qquad \text{Equation to be solved}$$
$$0.015x + 0.45 = 1.2 \qquad \text{Distributive property}$$
$$0.015x = 0.75 \qquad \text{Subtract 0.45 from each side.}$$
$$\frac{0.015x}{0.015} = \frac{0.75}{0.015} \qquad \text{Divide each side by 0.015.}$$
$$x = 50 \qquad \text{Simplify fractions.}$$

The amount of water that should be added is 50 ounces.

STEP 4: Adding 50 ounces of water will yield $50 + 30 = 80$ ounces of water containing $0.04(30) = 1.2$ ounces of salt. The concentration is $\frac{1.2}{80} = 0.015$ or 1.5%. ✓

Now Try Exercise 61

🌐 *Math in Context* — Financial Many times interest rates for student loans vary. In the next example, we present a situation in which a student has to borrow money at two different interest rates. This problem can be thought of as a mixture problem where the mixture is made up of two loans with different rates.

EXAMPLE 7 **Calculating interest on college loans**

A student takes out a loan for a limited amount of money at 5% interest and then must pay 7% for any additional money. If the student borrows $2000 more at 7% than at 5%, then the total interest for one year is $440. How much does the student borrow at each rate?

Solution

STEP 1: Assign a variable x as follows.

$$x: \text{loan amount at 5\% interest}$$
$$x + 2000: \text{loan amount at 7\% interest}$$

STEP 2: The amount of interest paid for the 5% loan is 5% of x, or $0.05x$. The amount of interest paid for the 7% loan is 7% of $x + 2000$, or $0.07(x + 2000)$. The total interest equals $440, so we solve the equation

$$0.05x + 0.07(x + 2000) = 440.$$

STEP 3: Solve the equation in STEP 2 for x.

$$0.05x + 0.07(x + 2000) = 440 \qquad \text{Equation to be solved}$$
$$0.05x + 0.07x + 140 = 440 \qquad \text{Distributive property}$$
$$0.12x + 140 = 440 \qquad \text{Combine like terms.}$$
$$0.12x = 300 \qquad \text{Subtract 140 from each side.}$$
$$x = 2500 \qquad \text{Divide each side by 0.12.}$$

The student borrows $2500 at 5% and $2500 + 2000 = \$4500$ at 7%.

STEP 4: The amount of interest on $2500 borrowed at 5% is $0.05(2500) = \$125$ and the amount of interest on $4500 borrowed at 7% is $0.07(4500) = \$315$. Thus the total interest is $125 + 315 = \$440$. Furthermore, the amount borrowed at 7% is $2000 more than the amount borrowed at 5%. ✓

Now Try Exercise 63

9.2 Putting It All Together

In this section, we presented a four-step approach to problem solving. However, because no approach works in every situation, solving mathematical problems takes time, effort, and creativity.

STEPS FOR SOLVING A PROBLEM

STEP 1: Read the problem carefully and be sure that you understand it. (You may need to read the problem more than once.) Assign a variable to what you are being asked to find. If necessary, write other quantities in terms of this variable.

STEP 2: Write an equation that relates the quantities described in the problem. You may need to sketch a diagram or refer to known formulas.

STEP 3: Solve the equation. Use the solution to determine the solution(s) to the original problem. Include any necessary units.

STEP 4: Check your solution in the original problem. Does it seem reasonable?

9.2 Exercises   MyMathLab®

CONCEPTS AND VOCABULARY

1. When you are solving a word problem, what is the last step?

2. The words *more*, *sum*, and *increase* are associated with the symbol _____.

3. The words *is*, *gives*, and *results in* are associated with the symbol _____.

4. Given an integer n, what are the next two consecutive integers?

5. If a car travels at speed r for time t, then distance d is given by $d =$ _____.

6. In general, the rate in a distance problem is expressed in units of _____ per unit of _____.

TRANSLATING SENTENCES INTO EQUATIONS

Exercises 7–14: Using the variable x, translate the sentence into an equation. Solve the resulting equation.

7. The sum of 2 and a number is 12.

8. A number divided by 5 equals the number decreased by 24.

9. Twice a number plus 7 equals 9.

10. 25 times a number is 125.

11. A number subtracted from 7 is -2.

12. A number subtracted from 8 is 5.

13. The quotient of a number and 2 is 17.

14. The product of 5 and a number equals 95.

SOLVING NUMBER PROBLEMS AND APPLICATIONS

Exercises 15–22: Find the number or numbers.

15. The sum of three consecutive natural numbers is 96.

16. The sum of three consecutive integers is -123.

17. Three times a number equals 102.

18. A number plus 18 equals twice the number.

19. Five times a number is 24 more than twice the number.

20. Three times a number is 18 less than the number.

21. Six times a number divided by 7 equals 18.

22. Two less than twice a number, divided by 5, equals 4.

23. **Finding Age** In 10 years, a child will be 3 years older than twice her current age. What is the current age of the child?

24. **Finding Age** A mother is 15 years older than twice her daughter's age. If the mother is 49 years old, how old is the daughter?

25. **Weight Loss** After losing 30 pounds, an individual weighs 110 pounds more than one-third his previous weight. Find the previous weight of the individual.

26. **Weight Gain** After gaining 25 pounds, a person is 115 pounds lighter than double his previous weight. How much did the person weigh before gaining 25 pounds?

27. **World Billionaires** In 2012, New York City was home to the most billionaires. This is 10 more than 5 times the number of billionaires in Tokyo. How many billionaires were in Tokyo if there were 70 billionaires in New York City? (*Source:* The Economist.)

28. **World Millionaires** In 2012, Tokyo was home to the most millionaires. This is 317 thousand less than twice the number of millionaires in New York City. How many millionaires were in New York City if there were 461 thousand in Tokyo? (*Source:* The Economist.)

29. **Endangered Species** There were 80 birds on the endangered species list in 2014. This is 10 more than 5 times the number of reptiles on the list. How many reptiles were on the endangered species list in 2014? (*Source:* U.S. Fish & Wildlife Service.)

30. **Immigrant Religion** Of those who immigrated to the United States between 1992 and 2012, 12.7% self-identified as Christian. This is 0.8% more than 7 times the percentage that self-identified as Muslim. What percentage of immigrants self-identified as Muslim? (*Source:* Pew Research Center.)

31. **Musical Revenues** From their opening dates, *Phantom of the Opera* and *Les Miserables* have together made $8.2 billion worldwide. *Phantom of the Opera* has made $0.4 billion more than twice what *Les Miserables* has made. Find the revenue of each musical (*Source:* The Economist.)

32. **Museums** The number of museums in Canada and the United States combined is 20 thousand. The number of museums in the United States is 2.5 thousand less than 8 times the number in Canada. Find the number of museums in each country. (*Source:* The Economist.)

33. **Highest Mountains** By 2013, there were 5656 successful ascents of Mount Everest. This is 148 more than 18 times the number of successful ascents of K2. How many successful ascents of K2 were there? (*Source:* The Economist.)

34. **Highest Mountains** By 2013, there were 29 deaths on K2 for every 100 safe returns. This is 1 more than 7 times the number of deaths for every 100 safe returns on Mount Everest. How many deaths for every 100 safe returns were there on Mount Everest? (*Source:* The Economist.)

35. **Geometry** If the perimeter of the rectangle shown is 106 inches, find the value of x.

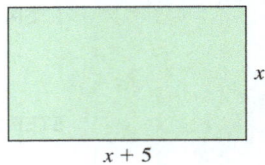

36. **Geometry** If the perimeter of the following triangle is 24 inches, find the value of x.

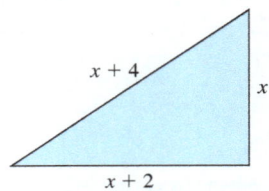

37. **Rectangle Dimensions** The length of a rectangle is 7 inches longer than the width. If the perimeter of the rectangle is 62 inches, find the measures of the length and width.

38. **Triangle Dimensions** The shortest side of a triangle measures 15 feet less than the longest side. If the third side is 6 feet shorter than the longest side and the perimeter is 102 feet, find the measures of the three sides of the triangle.

39. **Facebook vs. Twitter** In June 2013, the number of monthly active Facebook users in the U.S. was 2 million more than 4 times the number of monthly active Twitter users. If the two sites had a total of 247 million active monthly users, how many people used each site? (*Source:* Company Reports.)

40. **Blackberry Shares** In June 2008, Blackberry share prices (in Canadian dollars) peaked at $149.90. This is $4.46 less than 17 times what the share price was in September 2013. What was the share price in September 2013? (*Source:* Reuters.)

41. **Screen Time** In 2013, the monthly time spent watching traditional TV was 10 hours less than 6 times the monthly time spent using the Internet on a computer. If the time spent watching TV was 146 hours, what was the time spent using the Internet on a computer? (*Source:* Nielsen.)

42. **iPhone Prices** At launch, the price in Brazil of the 16GB iPhone 5S was $1200. This is $214 less than twice the price of the same phone in the United States. Find the price in the United States? (*Source:* Mobile Unlocked.)

SOLVING DISTANCE PROBLEMS

Exercises 43–48: Use the formula $d = rt$ to find the value of the missing variable.

43. $r = 4$ mph, $t = 2$ hours

44. $r = 70$ mph, $t = 2.5$ hours

45. $d = 1000$ feet, $t = 50$ seconds

46. $d = 1250$ miles, $t = 5$ days

47. $d = 200$ miles, $r = 40$ mph

48. $d = 1700$ feet, $r = 10$ feet per second

49. **Driving a Car** A person drives a car at a constant speed for 4 hours and 15 minutes, traveling 255 miles. Find the speed of the car in miles per hour.

50. **Flying an Airplane** A pilot flies a plane at a constant speed for 5 hours and 30 minutes, traveling 715 miles. Find the speed of the plane in miles per hour.

51. **Jogging Speeds** One runner passes another runner traveling in the same direction on a hiking trail. The faster runner is jogging 2 miles per hour faster than the slower runner. Determine how long it will be before the faster runner is $\frac{3}{4}$ mile ahead of the slower runner.

52. **Distance Running** An athlete runs 8 miles, first at a slower speed and then at a faster speed. The total time spent running is 1 hour. If the athlete runs $\frac{1}{3}$ hour at 6 miles per hour, find the second speed.

53. **Jogging Speeds** At first an athlete jogs at 5 miles per hour and then jogs at 8 miles per hour, traveling 7 miles in 1.1 hours. How long does the athlete jog at each speed? (*Hint:* Let t represent the amount of time the athlete jogs at 5 mph. Then $1.1 - t$ represents the amount of time the athlete jogs at 8 mph.)

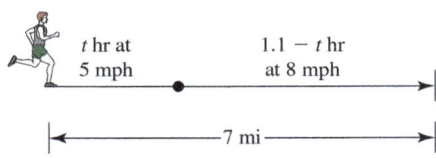

54. **Distance and Time** A bus is 160 miles east of the North Dakota–Montana border and is traveling west at 70 miles per hour. How long will it take for the bus to be 295 miles west of the border?

55. **Finding Speeds** Two cars pass on a straight highway while traveling in opposite directions. One car is traveling 6 miles per hour faster than the other car. After 1.5 hours, the two cars are 171 miles apart. Find the speed of each car.

56. **Distance and Time** A plane is 300 miles west of Chicago, Illinois, and is flying west at 500 miles per hour. How long will it take for the plane to be 2175 miles west of Chicago?

57. **Average Speed** A driver travels at 50 mph for the first hour and then travels at 70 mph for the second hour. What is the average speed of the car?

58. **Average Speed** A bicyclist rides 1 mile uphill at 5 mph and then rides 1 mile downhill at 10 mph. Find the average speed of the bicyclist. Does your answer agree with what you expected?

59. **Average Speed** At a 3-mile cross-country race, an athlete runs 2 miles at 8 mph and 1 mile at 10 mph. What is the athlete's average speed?

60. **Average Speed** A pilot flies an airplane between two cities and travels half the distance at 200 mph and the other half at 100 mph. Find the average speed of the airplane.

SOLVING MIXTURE PROBLEMS

61. **Saline Solution** (Refer to Example 6.) A solution contains 3% salt. How much water should be added to 20 ounces of this solution to make a 1.2% solution?

62. **Acid Solution** A solution contains 15% hydrochloric acid. How much water should be added to 50 milliliters of this solution to dilute it to a 2% solution?

63. **College Loans** (Refer to Example 7.) A student takes out two loans, one at 5% interest and the other at 6% interest. The 5% loan is $1000 more than the 6% loan, and the total interest for 1 year is $215. How much is each loan?

64. **Bank Loans** Two bank loans, one for $5000 and the other for $3000, cost a total of $550 in interest for one year. The $5000 loan has an interest rate 3% lower than the interest rate for the $3000 loan. Find the interest rate for each loan.

65. **Coffee Mixture** A more expensive brand of organic coffee costs $12 per pound and a cheaper brand is $7 per pound. How many pounds of each coffee would it take to make an 8-pound bag of coffee that costs $83.50?

66. **Coin Mixture** A sample of 23 dimes and nickels has a value of $1.45. How many of each type of coin are there?

67. **Running Speeds** A jogger runs a total distance of 9.9 miles at 8 miles per hour and at 9 miles per hour. If the jogger runs for 1 hour and 12 minutes, how many minutes did the jogger run at each speed?

68. **Traveling Speeds** A truck driver on an interstate highway travels a total distance of 285 miles at 65 miles per hour and at 75 miles per hour. If it takes 4 hours, how many hours did the truck driver travel at each speed?

69. **Ticket Sales** At a show, student tickets cost $7 and adult tickets cost $12. If 151 tickets were sold for a total of $1532, how many tickets of each type were sold?

70. **Ticket Sales** At a ball game, student tickets cost $15 and adult tickets cost $20. If 260 tickets were sold for a total of $4575, how many tickets of each type were sold?

71. **Mixing Antifreeze** How many gallons of 70% antifreeze should be mixed with 10 gallons of 30% antifreeze to obtain a 45% antifreeze mixture?

72. **Mixing Antifreeze** How many gallons of 65% antifreeze and how many gallons of 20% antifreeze should be mixed to obtain 50 gallons of a 56% mixture of antifreeze? (*Hint:* Let x represent the number of gallons of 65% antifreeze. Then $50 - x$ represents the amount of 20% antifreeze.)

73. **Hydrocortisone Cream** A pharmacist needs to make a 1% hydrocortisone cream. How many grams of 2.5% hydrocortisone cream should be added to 15 grams of cream base (0% hydrocortisone) to make the 1% cream?

74. **Credit Card Debt** A person carries a balance on two credit cards, one with a monthly interest rate of 1.5% and the other with a monthly rate of 1.75%. The balance on the 1.5% card is $600 less than the balance on the 1.75% card. If the total interest for the month is $49.50, what is the balance on each card?

WRITING ABOUT MATHEMATICS

75. State the four steps for solving a word problem.

76. The cost of living has increased about 600% during the past 50 years. Does this percent change correspond to a cost of living increase of 6 times? Explain.

SECTIONS 9.1 and 9.2 Checking Basic Concepts

1. Determine whether the equation is linear.
   (a) $4x^3 - 2 = 0$ (b) $2(x + 1) = 4$

2. Complete the table for each value of x. Then use the table to solve $4x - 3 = 13$.

x	3	3.5	4	4.5	5
$4x - 3$	9				17

3. Solve each equation and check your answer.
   (a) $x - 12 = 6$
   (b) $\frac{3}{4}z = \frac{1}{8}$
   (c) $0.6t + 0.4 = 2$
   (d) $5 - 2(x - 2) = 3(4 - x)$

4. Determine whether each equation has no solutions, one solution, or infinitely many solutions.
   (a) $x - 5 = 6x$
   (b) $-2(x - 5) = 10 - 2x$
   (c) $-(x - 1) = -x - 1$

5. **Distance Traveled** A driver is 300 miles from home and is traveling toward home on a freeway at a constant speed of 75 miles per hour.
   (a) Write an formula to calculate the distance D that the driver is from home after x hours.
   (b) Write an equation whose solution gives the hours needed for the driver to reach home.
   (c) Solve the equation from part (b).

6. **Number Problem** When three consecutive integers are added, the sum is -93. Find the three integers.

7. **Driving a Car** How many hours does it take the driver of a car to travel 390 miles at 60 miles per hour?

8. **College Loans** A student takes out two loans, one at 6% and the other at 7%. The 6% loan is $2000 more than the 7% loan, and the total interest for one year is $510. Find the amount of each loan.

9.3 Linear Inequalities in One Variable

Objectives

1. Graphing Solution Sets on Number Lines
   • Checking Solutions
2. Solving Linear Inequalities with Tables
3. Writing Solution Sets Using Interval Notation
4. Applying the Addition Property of Inequalities
5. Applying the Multiplication Property of Inequalities
6. Writing Solution Sets Using Set-Builder Notation
7. Translating Words to Inequalities
8. Solving Applications Involving Inequalities

NEW VOCABULARY

☐ Linear inequality
☐ Solution
☐ Solution set
☐ Interval notation
☐ Set-builder notation

1 Graphing Solution Sets on Number Lines

At an amusement park, a particular ride might be restricted to people at least 48 inches tall. A child who is x inches tall may go on the ride if $x \geq 48$ but may not go on the ride if $x < 48$. A height of 48 inches represents the *boundary* between being allowed on the ride and being denied access to the ride.

Solving linear inequalities is closely related to solving linear equations because equality is the boundary between *greater than* and *less than*. In this section, we discuss techniques used to solve linear inequalities.

A **linear inequality** results whenever the equals sign in a linear equation is replaced with any one of the symbols $<, \leq, >,$ or $\geq$. Examples of linear equations include

$$x = 5, \quad 2x + 1 = 0, \quad 1 - x = 6, \quad \text{and} \quad 5x + 1 = 3 - 2x.$$

Therefore examples of linear inequalities include

$$x > 5, \quad 2x + 1 < 0, \quad 1 - x \geq 6, \quad \text{and} \quad 5x + 1 \leq 3 - 2x.$$

TABLE 9.7 shows how each of the inequality symbols is read.

Inequality Symbols

Symbol	How the Symbol Is Read
$>$	greater than
$<$	less than
$\geq$	greater than or equal to
$\leq$	less than or equal to

TABLE 9.7

Inequalities often have infinitely many solutions. A **solution** to an inequality is a value of the variable that makes the inequality statement true. The set of all solutions is called the **solution set**. Two inequalities are *equivalent* if they have the same solution sets.

The following See the Concept shows how to graph inequalities on number lines.

See the Concept 1 — GRAPHING THE SOLUTION SET TO AN INEQUALITY

$x < 2$ (A A Solution, B Solution Set, C Not Included)

$x \leq 2$ (A A Solution, B Solution Set, D Included)

A Since $-1 < 2$ is a true statement, -1 is *one* of the infinitely many solutions to $x < 2$.
Since $0 \leq 2$ is a true statement, 0 is *one* of the infinitely many solutions to $x \leq 2$.

B The solution set (*all solutions*) can be graphed on a number line.

C A *parenthesis* is used to show that 2 is *not included* in the solution set.

D A *bracket* is used to show that 2 is *included* in the solution set.

WATCH VIDEO IN MML.

EXAMPLE 1 Graphing inequalities on a number line

Use a number line to graph the solution set to each inequality.
(a) $x > 0$ (b) $x \geq 0$ (c) $x \leq -1$ (d) $x < 3$

Solution
(a) First locate $x = 0$ (or the origin) on a number line. Numbers greater than 0 are located to the right of the origin, so shade the number line to the right of the origin. Because the inequality is $x > 0$, the number 0 is not included, so place a parenthesis "(" at 0, as shown in **FIGURE 9.5(a)**.
(b) **FIGURE 9.5(b)** is similar to the graph in part (a) except that a bracket "[" is placed at the origin because the inequality symbol is $\geq$ and 0 is included in the solution set.

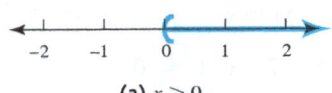 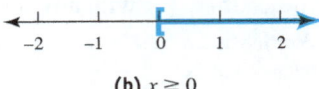

(a) $x > 0$ (b) $x \geq 0$

FIGURE 9.5

(c) First locate $x = -1$ on the number line. Numbers less than -1 are located to the left of -1. Because -1 is included, a bracket "]" is placed at -1, as shown in **FIGURE 9.5(c)**.
(d) Real numbers less than 3 are graphed in **FIGURE 9.5(d)**.

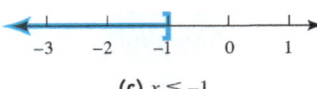 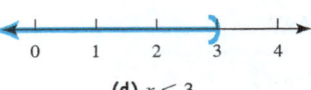

(c) $x \leq -1$ (d) $x < 3$

FIGURE 9.5

Now Try Exercises 19, 21, 23, 25

CHECKING SOLUTIONS We can check possible solutions to an inequality in the same way that we checked possible solutions to an equation. For example, to check whether 5 is a solution to the equation $2x + 3 = 13$, we substitute **5** for x in the equation.

$$2(5) + 3 \stackrel{?}{=} 13 \quad \text{Replace } x \text{ with 5.}$$
$$13 = 13 \checkmark \quad \text{A true statement}$$

Thus 5 is a solution to this equation. Similarly, to check whether 7 is a solution to the inequality $2x + 3 > 13$, we substitute **7** for x in the inequality.

$$2(7) + 3 \stackrel{?}{>} 13 \quad \text{Replace } x \text{ with 7.}$$
$$17 > 13 \checkmark \quad \text{A true statement}$$

Thus 7 is a solution to the inequality.

READING CHECK 1

How many solutions do inequalities often have?

EXAMPLE 2 Checking possible solutions

Determine whether the given value of x is a solution to the inequality.
(a) $3x - 4 < 10$, $x = 6$ (b) $4 - 2x \leq 8$, $x = -2$

Solution
(a) Substitute **6** for x and simplify.

$$3(6) - 4 \stackrel{?}{<} 10 \quad \text{Replace } x \text{ with 6.}$$
$$14 < 10 \; \text{✗} \quad \text{A false statement}$$

Thus 6 is *not* a solution to the inequality.

(b) Substitute -2 for x and simplify.

$$4 - 2(-2) \stackrel{?}{\leq} 8 \quad \text{Replace } x \text{ with } -2.$$
$$8 \leq 8 \checkmark \quad \text{A true statement}$$

Thus -2 is a solution to the inequality.

Now Try Exercises 13, 15

2 Solving Linear Inequalities with Tables

Just as solving a linear equation means finding the value of the variable that makes the equation true, solving a linear inequality means finding the *values* of the variable that make the inequality true.

Making a table is an organized way of checking possible values of the variable to see if there are values that make an inequality true. In the next example, we use a table to find solutions to an equation and related inequalities.

EXAMPLE 3 — Finding solutions to equations and inequalities

In **TABLE 9.8**, the expression $2x - 6$ has been evaluated for several values of x. Use the table to determine any solutions to each equation or inequality.

(a) $2x - 6 = 0$ **(b)** $2x - 6 > 0$ **(c)** $2x - 6 \geq 0$ **(d)** $2x - 6 < 0$

Solving an Inequality with a Table

x	0	1	2	3	4	5	6
$2x - 6$	-6	-4	-2	0	2	4	6

TABLE 9.8

Solution
(a) From **TABLE 9.8**, $2x - 6$ equals **0** when $x = 3$.
(b) The values of x in the table that make the expression $2x - 6$ greater than 0 are 4, 5, and 6. These values are all greater than 3, which is the solution found in part (a). It follows that $2x - 6 > 0$ when $x > 3$.
(c) The values of x in the table that make the expression $2x - 6$ greater than or equal to 0 are 3, 4, 5, and 6. It follows that $2x - 6 \geq 0$ when $x \geq 3$.
(d) The expression $2x - 6$ is less than 0 when $x < 3$.

Now Try Exercise 33

3 Writing Solution Sets Using Interval Notation

Rather than always graphing a solution set on a number line, a convenient notation called **interval notation** can be used to represent the solutions to an inequality. The following See the Concept shows how interval notation is written.

See the Concept 2: INTERVAL NOTATION

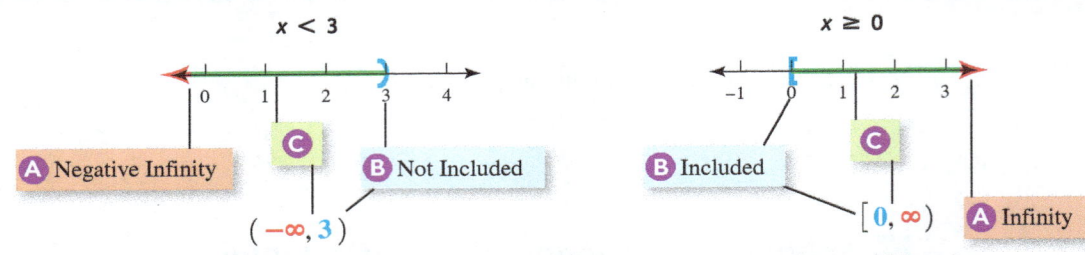

Interval Notation

A The left arrow corresponds to negative infinity, denoted $-\infty$ in interval notation. The right arrow corresponds to infinity, denoted ∞ in interval notation.

B The *parenthesis* used in both the graph and interval notation shows that 3 is *not included*. The *bracket* used in both the graph and interval notation shows that 0 is *included*.

C A comma is used in interval notation to separate the two ends of the interval.

WATCH VIDEO IN MML

NOTE: In interval notation, a *parenthesis* is always used with ∞ or $-\infty$. ∎

EXAMPLE 4 Writing solution sets in interval notation

Write the solution set to each inequality in interval notation.
(a) $x > 4$ (b) $y \leq -3$ (c) $z \geq -1$

Solution
(a) Real numbers greater than 4 are represented by the interval $(4, \infty)$.
(b) Real numbers less than or equal to -3 are represented by the interval $(-\infty, -3]$.
(c) The solution set is represented by the interval $[-1, \infty)$.

Now Try Exercises 41, 43, 45

4 Applying the Addition Property of Inequalities

Connecting Concepts with Your Life Suppose that the speed limit on a country road is 55 miles per hour, and this is 25 miles per hour faster than the speed limit in town. If x represents lawful speeds in town, then x satisfies the inequality

$$x + 25 \leq 55.$$

To solve this inequality, we **add** -25 to (or subtract 25 from) each side of the inequality.

$$x + 25 + (-25) \leq 55 + (-25) \quad \text{Add } -25 \text{ to each side.}$$
$$x \leq 30 \quad \text{Add the real numbers.}$$

Thus drivers are obeying the speed limit in town when they travel at 30 miles per hour or less. To solve this inequality, the addition property of inequalities was used.

ADDITION PROPERTY OF INEQUALITIES

Let a, b, and c be expressions that represent real numbers. The inequalities

$$a < b \quad \text{and} \quad a + c < b + c$$

are equivalent. That is, the same number may be added to (or subtracted from) each side of an inequality. Similar properties exist for $>$, $\leq$, and $\geq$.

To solve some inequalities, we apply the addition property of inequalities to obtain a simpler, equivalent inequality.

EXAMPLE 5   Applying the addition property of inequalities

Solve each inequality. Then graph the solution set.
(a) $x - 1 > 4$ (b) $3 + 2x \leq 5 + x$

Solution
(a) Begin by adding 1 to each side of the inequality.

$x - 1 > 4$ Given inequality
$x - 1 + 1 > 4 + 1$ Add 1 to each side.
$x > 5$ Add the real numbers.

The solution set is given by $x > 5$ and is graphed as follows.

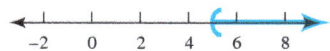

(b) Begin by subtracting x from (or adding $-x$ to) each side of the inequality.

$3 + 2x \leq 5 + x$ Given inequality
$3 + 2x - x \leq 5 + x - x$ Subtract x from each side.
$3 + x \leq 5$ Combine like terms.
$3 + x - 3 \leq 5 - 3$ Subtract 3 from each side.
$x \leq 2$ Subtract the real numbers.

The solution set is given by $x \leq 2$ and is graphed as follows.

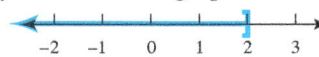

Now Try Exercises 47, 51

The following See the Concept illustrates the addition property of inequalities by using a scale.

See the Concept 3   THE ADDITION PROPERTY OF INEQUALITIES

To understand the addition property of inequalities, we can think of an inequality as a scale that is not balanced. Just as a scale remains *in the same position* when the same amount of weight is added to (or subtracted from) each side, an inequality remains *true* (and an equivalent inequality results) when the same number is added to (or subtracted from) each side.

True Inequality	*Equivalent Inequality*	*Different Position*
$1 < 2$	$1 + 1 < 2 + 1$	$1 + 2 > 2 - 0$
<	<	>

WATCH VIDEO IN

EXAMPLE 6   Applying the addition property of inequalities

Solve $5 + \frac{1}{2}x \leq 3 - \frac{1}{2}x$. Then graph the solution set.

Solution
Begin by subtracting 5 from each side of the inequality.

$$5 + \frac{1}{2}x \leq 3 - \frac{1}{2}x \qquad \text{Given inequality}$$

$$5 + \frac{1}{2}x - 5 \leq 3 - \frac{1}{2}x - 5 \qquad \text{Subtract 5 from each side.}$$

$$\frac{1}{2}x \leq -\frac{1}{2}x - 2 \qquad \text{Subtract real numbers.}$$

$$\frac{1}{2}x + \frac{1}{2}x \leq -\frac{1}{2}x + \frac{1}{2}x - 2 \qquad \text{Add } \tfrac{1}{2}x \text{ to each side.}$$

$$x \leq -2 \qquad \text{Combine like terms.}$$

The solution set is given by $x \leq -2$ and is graphed as follows.

▸ **Now Try Exercise 53**

STUDY TIP
If you miss something in class, section video lectures provide a short lecture for each section in this text. These videos are available as streaming videos and video podcasts within MyMathLab.

5 ▸ Applying the Multiplication Property of Inequalities

The multiplication property of inequalities is not exactly the same as the multiplication property of *equality*. When we multiply each side of an inequality by the same nonzero number, we may need to reverse the inequality symbol to make sure that the resulting inequality remains true. **TABLE 9.9** shows various results that occur when each side of a true inequality is multiplied by the same nonzero number.

Determining When the Inequality Symbol Should Be Reversed

True Statement	Multiply Each Side By	Resulting Inequality	Is the Result True or False?	Reverse the Inequality Symbol
$-3 < 5$	4	$-12 \stackrel{?}{<} 20$	True	Not needed
$7 > -1$	-2	$-14 \stackrel{?}{>} 2$	False	$-14 < 2$
$-2 > -5$	3	$-6 \stackrel{?}{>} -15$	True	Not needed
$4 < 9$	-11	$-44 \stackrel{?}{<} -99$	False	$-44 > -99$

TABLE 9.9

TABLE 9.9 indicates that the inequality symbol must be reversed when each side of the given inequality is *multiplied by a negative number*. This result is summarized below.

> ### MULTIPLICATION PROPERTY OF INEQUALITIES
>
> Let a, b, and c be expressions that represent real numbers with $c \neq 0$.
>
> 1. If $c > 0$, then the inequalities $a < b$ and $ac < bc$ are equivalent. That is, each side of an inequality may be multiplied (or divided) by the same positive number.
> 2. If $c < 0$, then the inequalities $a < b$ and $ac > bc$ are equivalent. That is, each side of an inequality may be multiplied (or divided) by the same negative number, provided the inequality symbol is reversed.
>
> Note that similar properties exist for $\leq$ and $\geq$.

READING CHECK 2
When solving an inequality, when does it become necessary to reverse the inequality symbol?

NOTE: Remember to reverse the inequality symbol when either multiplying or *dividing* by a negative number. ∎

EXAMPLE 7 Applying the multiplication property of inequalities

Solve each inequality. Then graph the solution set.

(a) $3x < 18$ **(b)** $-7 \leq -\frac{1}{2}x$

Solution
(a) To solve for x, divide each side by 3.

$$3x < 18 \qquad \text{Given inequality}$$
$$\frac{3x}{3} < \frac{18}{3} \qquad \text{Divide each side by 3.}$$
$$x < 6 \qquad \text{Simplify fractions.}$$

The solution set is given by $x < 6$ and is graphed as follows.

(b) To isolate x in $-7 \leq -\frac{1}{2}x$, multiply each side by -2 and reverse the inequality symbol.

$$-7 \leq -\frac{1}{2}x \qquad \text{Given inequality}$$
$$-2(-7) \geq -2\left(-\frac{1}{2}\right)x \qquad \text{Multiply by } -2; \text{ reverse the inequality.}$$
$$14 \geq 1 \cdot x \qquad \text{Multiply the real numbers.}$$
$$x \leq 14 \qquad \text{Rewrite the inequality.}$$

The solution set is given by $x \leq 14$ and is graphed as follows.

Now Try Exercises 55, 57

6 Writing Solution Sets Using Set-Builder Notation

Because $x \leq 14$ is an inequality with infinitely many solutions and is not itself a set of solutions, a notation called **set-builder notation** has been devised for writing the solutions to an inequality as a set. For example, the solution set consisting of "all real numbers x such that x is less than or equal to 14" can be written as $\{x | x \leq 14\}$. The vertical line segment in this notation "|" is read "such that."

In the next example, the solution sets are expressed in set-builder notation. However, this notation is not widely used throughout this text.

EXAMPLE 8 Applying both properties of inequalities

Solve each inequality. Write the solution set in set-builder notation.
(a) $4x - 7 \geq -6$ **(b)** $-8 + 4x \leq 5x + 3$ **(c)** $0.4(2x - 5) < 1.1x + 2$

Solution

(a) Start by adding 7 to each side.

$4x - 7 \geq -6$	Given inequality
$4x - 7 + 7 \geq -6 + 7$	Add 7 to each side.
$4x \geq 1$	Add real numbers.
$\dfrac{4x}{4} \geq \dfrac{1}{4}$	Divide each side by 4.
$x \geq \dfrac{1}{4}$	Simplify.

In set-builder notation, the solution set is $\left\{x \mid x \geq \frac{1}{4}\right\}$.

(b) Begin by adding 8 to each side.

$-8 + 4x \leq 5x + 3$	Given inequality
$-8 + 4x + 8 \leq 5x + 3 + 8$	Add 8 to each side.
$4x \leq 5x + 11$	Add real numbers.
$4x - 5x \leq 5x + 11 - 5x$	Subtract $5x$ from each side.
$-x \leq 11$	Combine like terms.
$-1 \cdot (-x) \geq -1 \cdot 11$	Multiply by -1; reverse the inequality.
$x \geq -11$	Simplify.

The solution set is $\{x \mid x \geq -11\}$.

(c) Begin by applying the distributive property.

$0.4(2x - 5) < 1.1x + 2$	Given inequality
$0.8x - 2 < 1.1x + 2$	Distributive property
$0.8x < 1.1x + 4$	Add 2 to each side.
$-0.3x < 4$	Subtract $1.1x$ from each side.
$x > -\dfrac{4}{0.3}$	Divide by -0.3; reverse the inequality.

Since $-\frac{4}{0.3} = -\frac{40}{3}$, the solution set is $\left\{x \mid x > -\frac{40}{3}\right\}$.

Now Try Exercises 89, 91, 93

CRITICAL THINKING 1

Solve

$$-5 - 3x > -2x + 7$$

without having to reverse the inequality symbol.

7 Translating Words to Inequalities

To solve applications involving inequalities, we often have to translate words or phrases to mathematical statements. **TABLE 9.10** lists words and phrases that are associated with each inequality symbol.

Words and Phrases Associated with Inequality Symbols

Symbol	Associated Words and Phrases
>	greater than, more than, exceeds, above, over
<	less than, fewer than, below, under
≥	greater than or equal to, at least, is not less than
≤	less than or equal to, at most, does not exceed

TABLE 9.10

EXAMPLE 9 Translating words to inequalities

Translate each phrase to an inequality. Let the variable be x.
(a) A number that is more than 30
(b) An age that is at least 18
(c) A grade point average that is at most 3.25

Solution
(a) The inequality $x > 30$ represents a number x that is more than 30.
(b) The inequality $x \geq 18$ represents an age x that is at least 18.
(c) The inequality $x \leq 3.25$ represents a grade point average x that is at most 3.25.

Now Try Exercises 95, 97, 101

8 Solving Applications Involving Inequalities

  Weather

Math in Context In the atmosphere, the air temperature generally becomes colder as the altitude increases. One mile above Earth's surface the temperature is about 19 °F colder than the ground-level temperature. As the air cools, there is an increased chance of clouds forming. In the next example, we estimate the altitudes where clouds may form. (*Source:* A. Miller and R. Anthes, *Meteorology*.)

EXAMPLE 10 Finding the altitude of clouds

If the ground temperature is 79 °F, then the temperature T above Earth's surface is given by the formula $T = 79 - 19x$, where x is the altitude in miles. Suppose that clouds form only where the temperature is 3 °F or colder. Solve an inequality to determine the heights at which clouds may form.

Solution
Clouds may form when $T \leq 3$ degrees. Since $T = 79 - 19x$, we can substitute the expression $79 - 19x$ for T to obtain the inequality $79 - 19x \leq 3$.

$$79 - 19x \leq 3 \quad \text{Inequality to be solved}$$
$$-19x \leq -76 \quad \text{Subtract 79 from each side.}$$
$$\frac{-19x}{-19} \geq \frac{-76}{-19} \quad \text{Divide by } -19; \text{ reverse the inequality.}$$
$$x \geq 4 \quad \text{Simplify the fractions.}$$

Clouds may form at 4 miles or higher.

Now Try Exercise 115

EXAMPLE 11 Calculating revenue, cost, and profit

For a computer company, the cost to produce one high-end laptop computer (variable cost) is $1320 plus a one-time cost (fixed cost) of $200,000 for research and development. The revenue received from selling one laptop computer is $1850.

(a) Write a formula that gives the cost C of producing x laptop computers.
(b) Write a formula that gives the revenue R from selling x laptop computers.
(c) Profit equals revenue minus cost. Write a formula that calculates the profit P from selling x laptop computers.
(d) How many computers need to be sold to yield a positive profit?

Solution
(a) The cost of producing the first laptop is

$$1320 \times 1 + 200{,}000 = \$201{,}320.$$

The cost of producing two laptops is

$$1320 \times 2 + 200{,}000 = \$202{,}640.$$

And, in general, the cost of producing x laptops is

$$1320 \times x + 200{,}000 = 1320x + 200{,}000.$$

Thus $C = 1320x + 200{,}000$.

(b) Because the company receives $1850 for each laptop, the revenue for x laptops is given by $R = 1850x$.

(c) Profit equals revenue minus cost, so

$$\begin{aligned} P &= R - C \\ &= 1850x - (1320x + 200{,}000) \\ &= 530x - 200{,}000. \end{aligned}$$

Thus $P = 530x - 200{,}000$.

(d) To determine how many laptops need to be sold to yield a positive profit, we must solve the inequality $P > 0$.

$$530x - 200{,}000 > 0 \quad \text{Inequality to be solved}$$
$$530x > 200{,}000 \quad \text{Add 200,000 to each side.}$$
$$x > \frac{200{,}000}{530} \quad \text{Divide each side by 530.}$$

Because $\frac{200{,}000}{530} \approx 377.4$, the company must sell at least 378 laptops. Note that the company cannot sell a fraction of a laptop.

Now Try Exercise 111

9.3 Putting It All Together

CONCEPT	COMMENTS	EXAMPLES
Linear Inequality	If the equals sign in a linear equation is replaced with $<$, $>$, $\leq$, or $\geq$, a linear inequality results.	*Linear Equation* *Linear Inequality* $4x - 1 = 0$ $4x - 1 > 0$ $2 - x = 3x$ $2 - x \leq 3x$ $4(x + 3) = 1 - x$ $4(x + 3) < 1 - x$

CONCEPT	COMMENTS	EXAMPLES
Solution to an Inequality	A value for the variable that makes the inequality a true statement	6 is a solution to $2x > 5$ because $2(6) > 5$ is a true statement.
Solution Set to an Inequality	The set of all solutions to an inequality	The solution set to $x + 1 > 5$ is given by $x > 4$ and can be written in set-builder notation as $\{x \mid x > 4\}$.
Number Line Graphs	The solutions to an inequality can be graphed on a number line.	$x < 2$ is graphed as follows. $x \geq -1$ is graphed as follows.
Addition Property of Inequalities	$a < b$ is equivalent to $a + c < b + c$, where a, b, and c represent real number expressions.	$x - 5 \geq 6$ Given inequality $x \geq 11$ Add 5. $3x > 5 + 2x$ Given inequality $x > 5$ Subtract $2x$.
Multiplication Property of Inequalities	$a < b$ is equivalent to $ac < bc$ when $c > 0$, and is equivalent to $ac > bc$ when $c < 0$.	$\frac{1}{2}x \geq 6$ Given inequality $x \geq 12$ Multiply by 2. $-3x > 5$ Given inequality $x < -\frac{5}{3}$ Divide by -3; reverse the inequality symbol.
Interval Notation	A notation that can be used to identify the solution set to an inequality as an interval of real numbers	The inequality $x < 5$ is written in interval notation as $(-\infty, 5)$. The inequality $x \geq -1$ is written in interval notation as $[-1, \infty)$.
Set-Builder Notation	A notation that can be used to identify the solution set to an inequality as a set of real numbers	The solution set for $x - 2 < 5$ can be written as $\{x \mid x < 7\}$ and is read "the set of real numbers x such that x is less than 7."

9.3 Exercises

CONCEPTS AND VOCABULARY

1. A linear inequality results whenever the equals sign in a linear equation is replaced by any one of the symbols _____, _____, _____ or _____.

2. Equality is the boundary between _____ and _____.

3. A(n) _____ is a value of the variable that makes an inequality statement true.

4. Two linear inequalities are _____ if they have the same solution set.

5. (True or False?) When a linear equation is solved, the solution set contains one solution.

6. (True or False?) When a linear inequality is solved, the solution set contains infinitely many solutions.

7. The solution set to a linear inequality can be graphed by using a(n) _____.

8. The addition property of inequalities states that if $a > b$, then $a + c$ _____ $b + c$.

9. The multiplication property of inequalities states that if $a < b$ and $c > 0$, then ac _____ bc.

10. The multiplication property of inequalities states that if $a < b$ and $c < 0$, then ac _____ bc.

GRAPHING SOLUTION SETS ON NUMBER LINES

Exercises 11–18: Determine whether the given value of the variable is a solution to the inequality.

11. $x + 5 > 5$, $x = 4$

12. $-3x \leq -8$, $x = -2$

13. $5x \geq 25$, $x = 3$

14. $4y - 3 \leq 5$, $y = -3$

15. $3y + 5 \geq -8$, $y = -3$

16. $-(z + 7) > 3(6 - z)$, $z = 2$

17. $5(z + 1) < 3z - 7$, $z = -4$

18. $\frac{3}{2}t - \frac{1}{2} \geq 1 - t$, $t = \frac{3}{5}$

Exercises 19–26: Use a number line to graph the solution set to the inequality.

19. $x < 0$

20. $x > -2$

21. $x > 1$

22. $x < -\frac{5}{2}$

23. $x \leq 1.5$

24. $x \geq -3$

25. $z \geq -2$

26. $z \leq -\pi$

Exercises 27–32: Express the set of real numbers graphed on the number line as an inequality.

27.

28.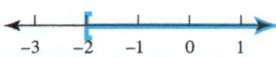

29.

30.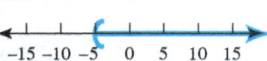

31.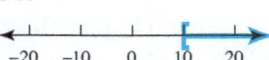

32.

SOLVING LINEAR INEQUALITIES WITH TABLES

Exercises 33–36: Use the table to solve the inequality.

33. $3x + 6 > 0$

x	−4	−3	−2	−1	0
$3x + 6$	−6	−3	0	3	6

34. $6 - 3x \leq 0$

x	1	2	3	4	5
$6 - 3x$	3	0	−3	−6	−9

35. $-2x + 7 > 5$

x	−1	0	1	2	3
$-2x + 7$	9	7	5	3	1

36. $5(x - 3) \leq 4$

x	3.2	3.4	3.6	3.8	4
$5(x - 3)$	1	2	3	4	5

Exercises 37–40: Complete the table. Then use the table to solve the inequality.

37. $-2x + 6 \leq 0$

x	1	2	3	4	5
$-2x + 6$	4				

38. $3x - 1 < 8$

x	0	1	2	3	4
$3x - 1$	−1				

39. $5 - x > x + 7$

x	−3	−2	−1	0	1
$5 - x$	8				4
$x + 7$	4				8

40. $2(3 - x) \geq -3(x - 2)$

x	−2	−1	0	1	2
$2(3 - x)$					
$-3(x - 2)$					

WRITING INTERVAL NOTATION

Exercises 41–46: Write the solution set to the inequality in interval notation.

41. $x \geq 6$ 42. $x < 3$

43. $y > -2$ 44. $y \geq 1$

45. $z \leq 7$ 46. $z < -5$

APPLYING THE ADDITION PROPERTY OF INEQUALITIES

Exercises 47–54: Use the addition property of inequalities to solve the inequality. Graph the solution set.

47. $x - 3 > 0$
48. $x + 6 < 3$
49. $3 - y \leq 5$
50. $8 - y \geq 10$
51. $12 + z < 4 + 2z$
52. $2z \leq z + 17$
53. $5 - 2t \geq 10 - t$
54. $-2t > -3t + 1$

APPLYING THE MULTIPLICATION PROPERTY OF INEQUALITIES

Exercises 55–62: Solve the inequality using the multiplication property of inequalities. Graph the solution set.

55. $2x < 10$
56. $3x > 9$
57. $-\frac{1}{2}t \geq 1$
58. $-5t \leq -6$
59. $\frac{3}{4} > -5y$
60. $10 \geq -\frac{1}{7}y$
61. $-\frac{2}{3} \leq \frac{1}{7}z$
62. $-\frac{3}{10}z < 11$

SOLVING LINEAR INEQUALITIES SYMBOLICALLY

Exercises 63–88: Solve the linear inequality. First write all solutions for x as an inequality. Then use interval notation to write to the solution set.

63. $3x + 1 < 22$
64. $4 + 5x \leq 9$
65. $5 - \frac{3}{4}x \geq 6$
66. $10 - \frac{2}{5}x > 0$
67. $45 > 6 - 2x$
68. $69 \geq 3 - 11x$
69. $5x - 2 \leq 3x + 1$
70. $12x + 1 < 25 - 3x$
71. $-x + 24 < x + 23$
72. $6 - 4x \leq x + 1$
73. $-(x + 1) \geq 3(x - 2)$
74. $5(x + 2) > -2(x - 3)$
75. $3(2x + 1) > -(5 - 3x)$
76. $4x \geq -3(7 - 2x) + 1$
77. $1.6x + 0.4 \leq 0.4x$
78. $-5.1x + 1.1 < 0.1 - 0.1x$
79. $0.8x - 0.5 < x + 1 - 0.5x$
80. $0.1(x + 1) - 0.1 \leq 0.2x - 0.5$
81. $-\frac{1}{2}\left(\frac{2}{3}x + 4\right) \geq x$
82. $-5x > \frac{4}{5}\left(\frac{10}{3}x + 10\right)$
83. $\frac{3}{7}x + \frac{2}{7} > -\frac{1}{7}x - \frac{5}{14}$
84. $\frac{5}{6} - \frac{1}{3}x \geq -\frac{1}{3}\left(\frac{5}{6}x - 1\right)$
85. $\frac{x}{3} + \frac{5x}{6} \leq \frac{2}{3}$
86. $\frac{3x}{4} - \frac{x}{2} < 1$
87. $\frac{6x}{7} < \frac{1}{3}x + 1$
88. $\frac{5x}{8} - \frac{3x}{4} \leq 8$

WRITING SOLUTION SETS USING SET-BUILDER NOTATION

Exercises 89–94: Solve the linear inequality and write the solution in set-builder notation.

89. $x + 6 > 7$
90. $x + 4 < 1$
91. $-3x \leq 21$
92. $4x \geq -20$
93. $2x - 3 < 9$
94. $-5x + 4 < 44$

TRANSLATING WORDS TO INEQUALITIES

Exercises 95–102: Translate each phrase to an inequality. Let x be the variable.

95. A speed that is greater than 60 miles per hour
96. A speed that is at most 60 miles per hour
97. An age that is at least 21 years old
98. An age that is less than 21 years old
99. A salary that is more than $40,000
100. A salary that is less than or equal to $40,000
101. A speed that does not exceed 70 miles per hour
102. A speed that is not less than 70 miles per hour

APPLICATIONS INVOLVING INEQUALITIES

103. **Geometry** Find all values for x so that the perimeter of the rectangle is less than 50 feet.

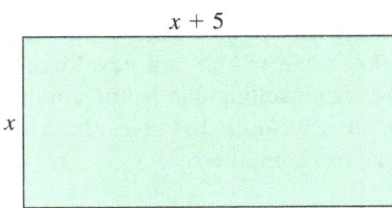

104. **Geometry** A triangle with height 12 inches is to have area less than 120 square inches. What must be true about the base of the triangle?

105. **Grade Average** A student scores 74 out of 100 on a test. If the maximum score on the next test is also 100 points, what score does the student need to maintain at least an average of 80?

106. **Grade Average** A student scores 65 and 82 on two different 100-point tests. If the maximum score on the next test is also 100 points, what score does the student need to maintain at least an average of 70?

107. **Parking Rates** Parking in a student lot costs $2 for the first half hour and $1.25 for each hour thereafter. A partial hour is charged the same as a full hour. What is the longest time that a student can park in this lot for $8?

108. **Parking Rates** Parking in a student lot costs $2.50 for the first hour and $1 for each hour thereafter. A nearby lot costs $1.25 for each hour. In both lots a partial hour is charged as a full hour. In which lot can a student park the longest for $5? For $11?

109. **Car Rental** A rental car costs $25 per day plus $0.20 per mile. If someone has $200 to spend and needs to drive the car 90 miles each day, for how many days can that person rent the car? Assume that the car cannot be rented for part of a day.

110. **Car Rental** One car rental agency charges $20 per day plus $0.25 per mile. A different agency charges $37 per day with unlimited mileage. For what mileages is the second rental agency a better deal?

111. **Revenue and Cost** (Refer to Example 11.) The cost to produce one compact disc is $1.50 plus a one-time fixed cost of $2000. The revenue received from selling one compact disc is $12.
    (a) Write a formula that gives the cost C of producing x compact discs. Be sure to include the fixed cost.
    (b) Write a formula that gives the revenue R from selling x compact discs.
    (c) Profit equals revenue minus cost. Write a formula that calculates the profit P from selling x compact discs.
    (d) How many compact discs need to be sold to yield a positive profit?

112. **Revenue and Cost** The cost to produce one laptop computer is $890 plus a one-time fixed cost of $100,000 for research and development. The revenue received from selling one laptop computer is $1520.
    (a) Write a formula that gives the cost C of producing x laptop computers.
    (b) Write a formula that gives the revenue R from selling x laptop computers.
    (c) Profit equals revenue minus cost. Write a formula that calculates the profit P from selling x laptop computers.
    (d) How many computers need to be sold to yield a positive profit?

113. **Distance and Time** Two athletes are jogging in the same direction along an exercise path. After x minutes, the first athlete's distance in miles from a parking lot is given by $\frac{1}{6}x$ and the second athlete's distance is given by $\frac{1}{8}x + 2$.
    (a) When are the athletes the same distance from the parking lot?
    (b) When is the first athlete farther from the parking lot than the second?

114. **Altitude and Dew Point** If the dew point on the ground is 65 °F, then the dew point x miles high is given by $D = 65 - 5.8x$. Determine the altitudes at which the dew point is greater than 36 °F. (Source: A. Miller.)

115. **Altitude and Temperature** (Refer to Example 10.) If the temperature on the ground is 90 °F, then the air temperature x miles high is given by $T = 90 - 19x$. Determine the altitudes at which the air temperature is less than 4.5 °F. (Source: A. Miller.)

116. **Size and Weight of a Fish** If the length of a bass is between 20 and 25 inches, its weight W in pounds can be estimated by $W = 0.96x - 14.4$, where x is the length of the fish. (Source: Minnesota Department of Natural Resources.)
    (a) What length of bass is likely to weigh 7.2 pounds?
    (b) What lengths of bass are likely to weigh less than 7.2 pounds?

WRITING ABOUT MATHEMATICS

117. Explain each of the terms and give an example.
    (a) Linear equation
    (b) Linear inequality

118. Suppose that a student says that a linear equation and a linear inequality can be solved the same way. How would you respond?

SECTION 9.3 Checking Basic Concepts

1. Use a number line to graph the solution set to the inequality $x + 1 \geq -1$.

2. Express the set of real numbers graphed on the number line by using an inequality.

3. Determine whether -3 is a solution to the inequality $4x - 5 \leq -15$.

4. Complete the table. Then use the table to solve the inequality $5 - 2x \leq 7$.

x	-2	-1	0	1	2
$5 - 2x$				1	

5. Solve each inequality.
   (a) $x + 5 > 8$
   (b) $-\frac{5}{7}x \leq 25$
   (c) $3x \geq -2(1 - 2x) + 3$

6. Translate the phrase "a price that is not more than $12" to an inequality using the variable x.

7. **Geometry** The length of a rectangle is 5 inches longer than twice its width. If the perimeter of the rectangle is more than 88 inches, find all possible widths for the rectangle.

CHAPTER 9 Summary

SECTION 9.1 ■ REVIEW OF LINEAR EQUATIONS IN ONE VARIABLE

Linear Equation Can be written in the form $ax + b = 0$, where $a \neq 0$

Examples: $3x - 5 = 0$ is linear, whereas $5x^2 + 2x = 0$ is *not* linear.

Solving Linear Equations with a Table To solve a linear equation numerically, complete a table for various values of the variable and then select the solution from the table.

Example: The solution to $3x - 4 = -1$ is 1.

x	-1	0	1
$3x - 4$	-7	-4	-1

Solving Linear Equations Symbolically The following steps can be used as a guide for solving linear equations symbolically.

STEP 1: Use the distributive property to clear any parentheses on each side of the equation. Combine any like terms on each side.

STEP 2: Use the addition property of equality to get all of the terms containing the variable on one side of the equation and all other terms on the other side of the equation. Combine any like terms on each side.

STEP 3: Use the multiplication property of equality to isolate the variable by multiplying each side of the equation by the reciprocal of the number in front of the variable (or divide each side by that number).

STEP 4: Check the solution by substituting it in the given equation.

Distributive Properties $a(b + c) = ab + ac$ or $a(b - c) = ab - ac$

Examples: $5(2x + 3) = 10x + 15$ and $5(2x - 3) = 10x - 15$

Clearing Fractions and Decimals When fractions or decimals appear in an equation, multiplying each side by the least common denominator can be helpful.

Examples: Multiply each side of $\frac{1}{3}x - \frac{1}{6} = \frac{2}{3}$ by 6 to obtain $2x - 1 = 4$.

Multiply each side of $0.04x + 0.1 = 0.07$ by 100 to obtain $4x + 10 = 7$.

Number of Solutions Equations that can be written in the form $ax + b = 0$, where a and b are *any* real number, can have no solutions, one solution, or infinitely many solutions.

Examples: $x + 3 = x$ is equivalent to $3 = 0$. (No solutions)

$2y + 1 = 9$ is equivalent to $y = 4$. (One solution)

$z + z = 2z$ is equivalent to $0 = 0$. (Infinitely many solutions)

SECTION 9.2 ■ FURTHER PROBLEM SOLVING

Steps for Solving a Problem The following steps can be used for solving word problems.

STEP 1: Read the problem carefully and be sure that you understand it. (You may need to read the problem more than once.) Assign a variable to what you are being asked to find. If necessary, write other quantities in terms of this variable.

STEP 2: Write an equation that relates the quantities described in the problem. You may need to sketch a diagram or refer to known formulas.

STEP 3: Solve the equation. Use the solution to determine the solution(s) to the original problem. Include any necessary units.

STEP 4: Check your solution in the original problem. Does it seem reasonable?

Translating Sentences into Equations The following table lists common words that are associated with the math symbols needed to write equations.

Symbol	Associated Words
$+$	add, plus, more, sum, total, increase
$-$	subtract, minus, less, difference, fewer, decrease
$\cdot$	multiply, times, twice, double, triple, product
$\div$	divide, divided by, quotient, per
$=$	equals, is, gives, results in, is the same as

Example: Translate the following sentence into an equation using the variable x.

Four times a number plus 7 equals 15.

The phrase "four times a number" indicates that we multiply 4 and x to get $4x$. The word "plus" indicates that we then add $4x$ and 7 to get $4x + 7$. This expression "equals 15," so the equation is $4x + 7 = 15$.

Distance Problems If an object travels at speed (rate) r for time t, then the distance d traveled is calculated by $d = rt$.

Example: A car moving at 65 mph for 2 hours travels

$$d = rt = 65 \cdot 2 = 130 \text{ miles.}$$

SECTION 9.3 ■ LINEAR INEQUALITIES IN ONE VARIABLE

Linear Inequality When the equals sign in a linear equation is replaced with any one of the symbols $<, \leq, >,$ or $\geq$, a linear inequality results.

Examples: $x > 0, \quad 6 - \frac{2}{3}x \leq 7,$ and $4(x - 1) < 3x - 1$

Solution to an Inequality Any value for the variable that makes the inequality a true statement

Example: 3 is a solution to $2x < 9$ because $2(3) < 9$ is a true statement.

Number Line Graphs A number line can be used to graph the solution set to a linear inequality.

Example: The graph of $x \leq 1$ is shown in the figure.

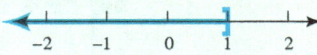

Properties of Inequalities

Addition Property $\qquad a < b$ is equivalent to $a + c < b + c$.

$\qquad$ **Example:** $x - 3 < 0$ and $x - 3 + 3 < 0 + 3$ are equivalent inequalities.

Multiplication Property $\qquad$ When $c > 0$, $a < b$ is equivalent to $ac < bc$.

$\qquad$ When $c < 0$, $a < b$ is equivalent to $ac > bc$.

$\qquad$ **Examples:** $2x < 6$ is equivalent to $2x\left(\frac{1}{2}\right) < 6\left(\frac{1}{2}\right)$ or $x < 3$.

$\qquad\qquad\qquad\quad -2x < 6$ is equivalent to $-2x\left(-\frac{1}{2}\right) > 6\left(-\frac{1}{2}\right)$ or $x > -3$.

CHAPTER 9 Review Exercises

SECTION 9.1

Exercises 1 and 2: Decide whether the equation is linear. If the equation is linear, give values for a and b so that it can be written in the form $ax + b = 0$.

1. $-4x + 3 = 2$
2. $\frac{3}{8}x^2 - x = \frac{1}{4}$

Exercises 3–12: Solve the equation. Check the solution.

3. $4x - 5 = 3$
4. $7 - \frac{1}{2}x = -4$
5. $5(x - 3) = 12$
6. $3 + x = 2x - 4$
7. $2(x - 1) = 4(x + 3)$
8. $1 - (x - 3) = 6 + 2x$
9. $3.4x - 4 = 5 - 0.6x$
10. $2.1x - 0.3(2 - x) = 0.5 + x$
11. $\frac{2}{3}x - \frac{1}{6} = \frac{5}{12}$
12. $-\frac{1}{3}(3 - 6x) = -(x + 2) + 1$

Exercises 13–16: Determine whether the equation has no solutions, one solution, or infinitely many solutions.

13. $4(3x - 2) = 2(6x + 5)$
14. $5(3x - 1) = 15x - 5$
15. $8x = 5x + 3x$
16. $9x - 2 = 8x - 2$

Exercises 17 and 18: Complete the table. Then use the table to solve the given equation.

17. $-2x + 3 = 0$

x	0.5	1.0	1.5	2.0	2.5
$-2x + 3$	2				

18. $-(x + 1) + 3 = 2$

x	-2	-1	0	1	2
$-(x + 1) + 3$					

Exercises 19–24: Solve the given formula for the specified variable.

19. $3x = 5 + y$ for y
20. $16 = 2x + 2y$ for y
21. $z = 2xy$ for y
22. $S = \dfrac{a + b + c}{3}$ for b

23. $T = \dfrac{a}{3} + \dfrac{b}{4}$ for b

24. $cd = ab + bc$ for c

SECTION 9.2

Exercises 25–28: Using the variable x, translate the sentence into an equation. Solve the resulting equation.

25. The product of a number and 6 is 72.

26. The sum of a number and 18 is -23.

27. Twice a number minus 5 equals the number plus 4.

28. The sum of a number and 4 equals the product of the number and 3.

Exercises 29 and 30: Find the number or numbers.

29. Five times a number divided by 3 equals 15.

30. The sum of three consecutive integers is -153.

Exercises 31–34: Use the formula $d = rt$ to find the value of the missing variable.

31. $r = 8$ miles per hour, $t = 3$ hours

32. $r = 70$ feet per second, $t = 55$ seconds

33. $d = 500$ yards, $t = 20$ seconds

34. $d = 125$ miles, $r = 15$ miles per hour

SECTION 9.3

Exercises 35–38: Use a number line to graph the solution set to the inequality.

35. $x < 2$

36. $x > -1$

37. $y \geq -\dfrac{3}{2}$

38. $y \leq 2.5$

Exercises 39 and 40: Express the set of real numbers graphed on the number line with an inequality.

39.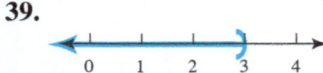

40.

Exercises 41 and 42: Determine whether the given value of x is a solution to the inequality.

41. $1 - (x + 3) \geq x$ $x = -2$

42. $4(x + 1) < -(5 - x)$ $x = -1$

Exercises 43 and 44: Complete the table and then use the table to solve the inequality.

43. $5 - x > 3$

x	0	1	2	3	4
$5 - x$	5				

44. $2x - 5 \leq 0$

x	1	1.5	2	2.5	3
$2x - 5$					

Exercises 45–50: Solve the inequality.

45. $x - 3 > 0$

46. $-2x \leq 10$

47. $5 - 2x \geq 7$

48. $3(x - 1) < 20$

49. $5x \leq 3 - (4x + 2)$

50. $3x - 2(4 - x) \geq x + 1$

Exercises 51–54: Translate the phrase to an inequality. Let x be the variable.

51. A speed that is less than 50 miles per hour

52. A salary that is at most $45,000

53. An age that is at least 16 years old

54. A year before 1995

APPLICATIONS

55. **Distance Traveled** At noon, a bicyclist is 50 miles from home, riding toward home at 10 miles per hour.
    (a) Make a table that shows the bicyclist's distance D from home after 1, 2, 3, 4, and 5 hours.
    (b) Write a formula that calculates the distance D from home after x hours.
    (c) Use your formula to determine D when $x = 3$ hours. Does your answer agree with the value shown in your table?
    (d) For what times was the bicyclist at least 20 miles from home? Assume that $0 \leq x \leq 5$.

56. **Car Speeds** One car passes another car on a freeway. The faster car is traveling 12 miles per hour faster than the slower car. Determine how long it will be before the faster car is 2 miles ahead of the slower car.

57. **Saline Solution** A saline solution contains 3% salt. How much water should be added to 100 milliliters of this solution to dilute it to a 2% solution?

58. **Investment Money** A student invests two sums of money, $500 and $800, at different interest rates, receiving a total of $55 in interest after one year. The $500 investment receives an interest rate 2% lower than the interest rate for the $800 investment. Find the interest rate for each investment.

59. Geometry A triangle with height 8 inches is to have an area that is not more than 100 square inches. What lengths are possible for the base of the triangle?

60. Grade Average A student scores 75 and 91 on two different tests of 100 points. If the maximum score on the next test is also 100 points, what score does the student need to maintain an average of at least 80?

61. Parking Rates Parking in a lot costs $2.25 for the first hour and $1.25 for each hour thereafter. A partial hour is charged the same as a full hour. What is the longest time that someone can park for $9?

62. Profit The cost to produce one DVD player is $85 plus a one-time fixed cost of $150,000. The revenue received from selling one DVD player is $225.
  (a) Write a formula that gives the cost C of producing x DVD players.
  (b) Write a formula that gives the revenue R from selling x DVD players.
  (c) Profit equals revenue minus cost. Write a formula that calculates the profit P from selling x DVD players.
  (d) What numbers of DVD players sold will result in a loss? (*Hint:* A loss corresponds to a negative profit.)

CHAPTER 9 Test

Exercises 1–4: Solve the equation. Check your solution.

1. $9 = 3 - x$
2. $4x - 3 = 7$
3. $4x - (2 - x) = -3(2x + 6)$
4. $\frac{1}{12}x - \frac{2}{3} = \frac{1}{2}\left(\frac{3}{4} - \frac{1}{3}x\right)$

Exercises 5 and 6: Determine the number of solutions to the given equation.

5. $6(2x - 1) = -4(3 - 3x)$
6. $4(2x - 1) = 8x - 4$

Exercises 7 and 8: Solve the formula for x.

7. $z = y - 3xy$
8. $R = \dfrac{x}{4} + \dfrac{y}{5}$

9. Complete the table. Then use the table to solve the equation $6 - 2x = 0$.

x	0	1	2	3	4
$6 - 2x$	6				

Exercises 10 and 11: Translate the sentence into an equation, using the variable x. Then solve the resulting equation.

10. The sum of a number and -7 is 6.

11. Twice a number plus 6 equals the number minus 7.

12. The sum of three consecutive natural numbers is 336. Find the three numbers.

13. Use the formula $d = rt$ to find r when $d = 200$ feet and $t = 4$ seconds.

14. Express the set of real numbers graphed on the number line with an inequality.

Exercises 15 and 16: Solve the inequality.

15. $-3x + 9 \geq x - 15$
16. $3(6 - 5x) < 20 - x$

17. **Snowfall** Suppose 5 inches of snow fall before noon and 2 inches per hour fall thereafter until 10 P.M.
  (a) Write a formula that calculates the snowfall S in inches, x hours past noon.
  (b) Use your formula to calculate the total snowfall at 8 P.M.
  (c) How much snow had fallen at 6:15 P.M.?

18. **Mixing Acid** A solution is 45% hydrochloric acid. How much water should be added to 1000 milliliters of this solution to dilute it to a 15% solution?

CHAPTERS 1–9 Cumulative Review Exercises

1. Identify the digit in the hundred-thousands place in the number 4,612,350,798.

2. Write $9 \cdot 9 \cdot 9 \cdot 9 \cdot 9$ using exponential notation.

3. Estimate the sum $698 + 401 + 197$ by rounding each value to the nearest hundred.

4. Evaluate $(x + (5 - y)^2) \div x$ for $x = 3$, $y = 2$.

5. Is 9 a solution to $4 + b \div 3 = -7$?

6. Complete the table and solve $-3x + 5 = -1$.

x	-2	-1	0	1	2
$-3x + 5$					

7. Translate the following sentence to an equation using x as the variable. Do not solve the equation.

Eleven equals a number increased by 4.

8. Solve $-9 + 2x - 4 = 6x - 5$.

9. Solve the linear equation $-2x - 3 = -3$ graphically.

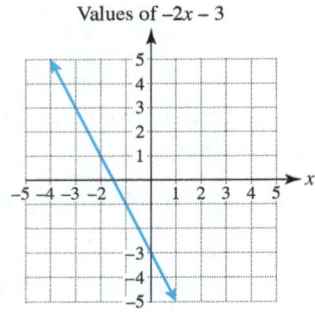

10. Write $\frac{7}{20}$ as a decimal.

Exercises 11 and 12: Perform the indicated arithmetic.

11. $\frac{11}{10} - \frac{1}{4}$

12. $\frac{3}{16} \cdot \frac{8}{9}$

13. Place the correct symbol, $<$ or $>$, in the blank.

13.0263 _____ 13.0623

14. If a car travels 171 miles in 3 hours, write its speed as a unit rate.

15. Solve the proportion.

$$\frac{6}{x} = \frac{-3}{10}$$

16. Convert 180 hours to days.

17. What percent of 120 is 15?

18. What number is 15% of 700?

Exercises 19 and 20: Solve the equation.

19. $5(6y + 2) = 25$

20. $11 - (y + 2) = 3y + 5$

Exercises 21 and 22: Determine whether the equation has no solutions, one solution, or infinitely many solutions.

21. $6x + 2 = 2(3x + 1)$

22. $2(3x - 4) = 6(x - 1)$

23. Find three consecutive integers whose sum is 90.

24. Find the average speed of a car that travels 325 miles in 5 hours.

Exercises 25 and 26: Solve the formula for x.

25. $a = 3xy - 4$

26. $A = \dfrac{x + y + z}{3}$

Exercises 27 and 28: Solve the inequality and graph the solution set.

27. $7 - 3x > 4$

28. $6x \leq 5 - (x - 9)$

29. **Going Organic** A farmer plans to convert $\frac{13}{15}$ of his crop land to organic crops. If the farmer has already converted $\frac{8}{15}$ of his crop land, what fraction of his land has yet to be converted to organic crops?

30. **Music Sales** The total music sales S in thousands of dollars for a small band during year x can be computed using the formula

$$S = 7(x - 2005) + 8.$$

Find the year when the music sales were \$36,000 by solving the equation $36 = 7(x - 2005) + 8$ graphically using the graph below.

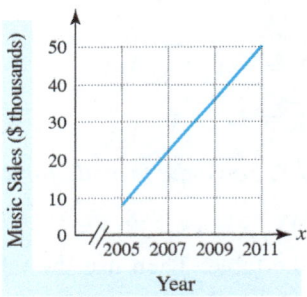

31. **Text Messaging** Suppose that the cost C of sending or receiving x text messages is given by the formula

$$C = 0.05(x - 400) + 10.5, \text{ where } x > 400.$$

Find the number of text messages that correspond to total charges of \$18.50 by replacing C in the formula with 18.5 and solving the resulting equation.

32. **HD Video** If the purchase price of an HD video recorder is \$178 and the sales tax rate is 4.5%, find the sales tax and the total price.

33. **Acid Solution** How much of a 4% acid solution should be added to 150 milliliters of a 10% acid solution to dilute it to a 6% acid solution?

34. **Energy Production** In 1997, U.S. hydroelectric power production hit an all-time high of 356 billion kilowatt-hours. This is 188 billion kilowatt-hours less than twice the 2007 production level. Find the production level in 2007. (*Source:* U.S. Department of Energy.)

10 Graphing Equations

- **10.1** Introduction to Graphing
- **10.2** Equations in Two Variables
- **10.3** Intercepts; Horizontal and Vertical Lines
- **10.4** Slope and Rates of Change
- **10.5** Slope–Intercept Form
- **10.6** Point–Slope Form
- **10.7** Introduction to Modeling

Since entering the 21st century, the U.S. economy has experienced both incredible growth and substantial decline. One way that economists monitor changes in the economy is by tracking median home prices. The *line graph* in the accompanying figure shows median new home prices displayed in two-year intervals. Graphs such as this one provide an excellent way to visualize data trends. In this chapter, we discuss line graphs and other types of graphs that can be used to represent mathematical information visually. *Source:* U.S. Census Bureau.

10.1 Introduction to Graphing

Objectives

1. Introducing Tables and Graphs
2. Understanding the Rectangular Coordinate System
   - Plotting and Reading Ordered Pairs
3. Making and Reading Scatterplots
   - Choosing a Scale for a Graph
4. Making and Reading Line Graphs

NEW VOCABULARY

- ☐ Rectangular coordinate system (xy-plane)
- ☐ x-axis
- ☐ y-axis
- ☐ Origin
- ☐ Quadrants
- ☐ Ordered pair
- ☐ Scatterplot
- ☐ x-coordinate
- ☐ y-coordinate
- ☐ Line graph

STUDY TIP

Have you been completing all of the assigned homework on time? Regular and timely practice is one of the keys to having a successful experience in any math class.

1 Introducing Tables and Graphs

When data are displayed in a table, it is often difficult to recognize any trends. In order to visualize information provided by a set of data, it can be helpful to create a graph. In math, we visualize data by plotting points to make *scatterplots* and *line graphs*.

Math in Financial Context TABLE 10.1 lists the per capita (per person) income for the United States for selected years. In TABLE 10.1, we can easily see that income first increased, and then remained relatively flat before it increased again. However, if a table contained 1000 data values, determining trends would be extremely difficult.

Rather than always using tables to display data, presenting data on a graph is often more useful. For example, the data in TABLE 10.1 are graphed in FIGURE 10.1. This line graph is more visual than the table and shows the trend at a glance. Line graphs will be discussed in more detail later in this section.

U.S. per Capita Income (2014 Dollars)

Year	Amount
2004	$41,929
2006	$46,444
2008	$48,407
2010	$48,358
2012	$51,749

TABLE 10.1 *Source:* World Bank

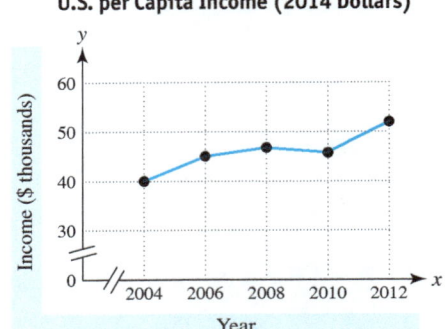

FIGURE 10.1

NOTE: In FIGURE 10.1, the double hash marks // on the graph indicate a break in data values. The years before 2004 and incomes below $30,000 have been skipped. ∎

2 Understanding the Rectangular Coordinate System

One common way to graph data is to use the **rectangular coordinate system**, or **xy-plane**. The following See the Concept shows important information about the xy-plane.

See the Concept 1 — THE xy-PLANE

Two axes divide the xy-plane into four regions called **quadrants**, which are numbered **I**, **II**, **III**, and **IV** counterclockwise, as shown in the figure.

Ⓐ The horizontal axis is called the x-axis.

Ⓑ The vertical axis is called the y-axis.

Ⓒ The axes intersect at the origin, which is the point associated with zero on each axis.

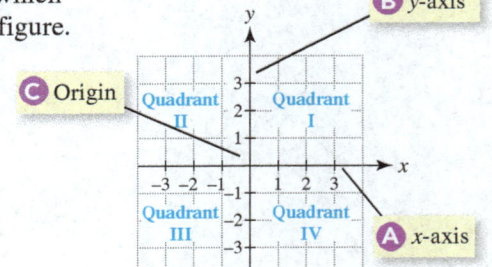

Negative values are located to the left of the origin along the x-axis and below the origin along the y-axis.

Positive values are located to the right of the origin along the x-axis and above the origin along the y-axis.

WATCH VIDEO IN MML.

PLOTTING AND READING ORDERED PAIRS Before we can plot data, we must understand the concept of an **ordered pair** (x, y). In **TABLE 10.1**, we can let x-values correspond to the year and y-values correspond to the per capita income. Then the fact that the per capita income in 2004 was $41,929 can be summarized by the ordered pair (2004, 41929). Similarly, the ordered pair (2008, 48407) indicates that the per capita income was $48,407 in 2008.

Order is important in an ordered pair. The ordered pairs given by (1950, 2025) and (2025, 1950) are different. The first ordered pair indicates that the per capita income in 1950 was $2025, whereas the second ordered pair indicates that the per capita income in 2025 will be $1950 (which is probably not correct).

To plot an ordered pair such as $(-2, 3)$ in the xy-plane, begin at the origin and move left to locate $x = -2$ on the x-axis. Then move upward until a height of $y = 3$ is reached. Thus the point $(-2, 3)$ is located 2 units left of the origin and 3 units above the origin. In **FIGURE 10.2**, the point $(-2, 3)$ is plotted in quadrant II.

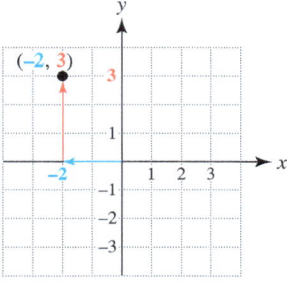

FIGURE 10.2

NOTE: A point that lies on an axis is not located in a quadrant. ■

READING CHECK 1

- Is the order of the numbers in an ordered pair important? Explain.

EXAMPLE 1 Plotting points

Plot the following ordered pairs on the same xy-plane. State the quadrant in which each point is located, if possible.
(a) $(3, 2)$ (b) $(-2, -3)$ (c) $(-3, 0)$

Solution
(a) The point $(3, 2)$ is located in quadrant I, 3 units to the right of the origin and 2 units above the origin. See **FIGURE 10.3**.
(b) The point $(-2, -3)$ is located 2 units to the left of the origin and 3 units below the origin. **FIGURE 10.3** shows the point $(-2, -3)$ in quadrant III.
(c) The point $(-3, 0)$ is not in any quadrant because it is located 3 units left of the origin on the x-axis, as shown in **FIGURE 10.3**.

Now Try Exercise 13

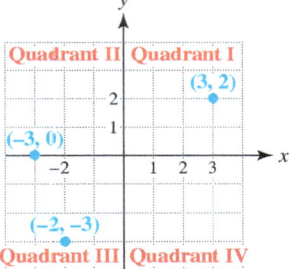

FIGURE 10.3

🌐 Math in Business Context In the next example, a scatterplot is used to analyze historical data concerning the valuation of the social network Twitter.

EXAMPLE 2 Reading a graph containing Twitter data

FIGURE 10.4 shows the value of Twitter in billions of dollars during its first 6 years. Use the graph to estimate the value of Twitter in 2009 and 2013. *(Source: Business Insider.)*

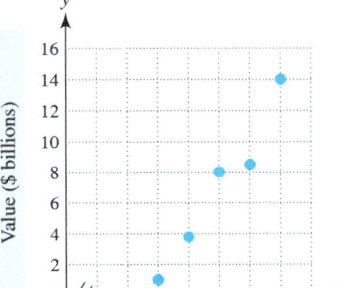

FIGURE 10.4

Solution

To find the value of Twitter in 2009, start by locating 2009 on the *x*-axis and move **vertically** upward to the data point. From the data point, move **horizontally** to the *y*-axis. **FIGURE 10.5(a)** shows that Twitter's value was about $1 billion in 2009. Similarly, the value was about $14 billion by 2013, as shown in **FIGURE 10.5(b)**.

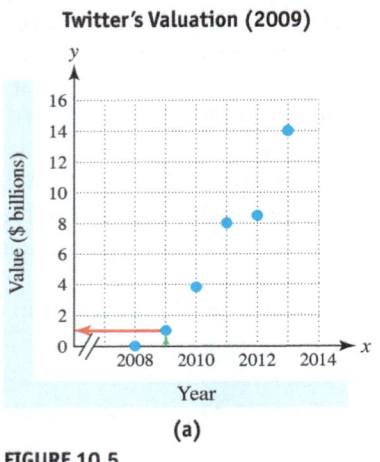
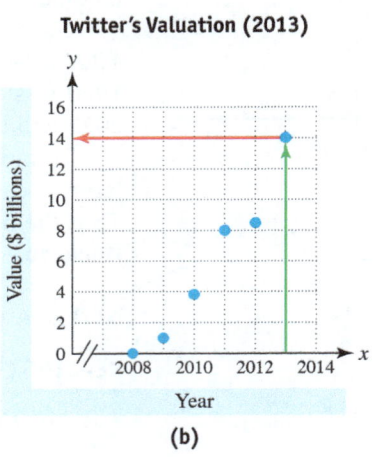

(a) (b)

FIGURE 10.5

Now Try Exercise 35

3 Making and Reading Scatterplots

If distinct points are plotted in the *xy*-plane, then the resulting graph is called a **scatterplot**. **FIGURE 10.5(a)** is an example of a scatterplot that displays information about Twitter. A different scatterplot is shown in **FIGURE 10.6**, in which the points $(1, 3)$, $(2, 2)$, $(3, 1)$, $(4, 4)$, and $(5, 1)$ are plotted.

CHOOSING A SCALE FOR A GRAPH Choosing appropriate scales for the axes is important when plotting points and making graphs. This can be accomplished by looking at the *coordinates* of the ordered pairs to be plotted. When plotting in the *xy*-plane, the first value in an ordered pair is called the **x-coordinate** and the second value is called the **y-coordinate**. In **FIGURE 10.6**, the *x*-coordinates of the points are 1, 2, 3, 4, and 5 and the *y*-coordinates are 1, 2, 3, and 4. Because no coordinate is more than 6 units from the origin, the scale shown in **FIGURE 10.6** is appropriate.

A Scatterplot

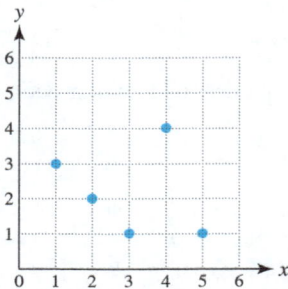

FIGURE 10.6

READING CHECK 2

- Which value in the ordered pair $(1, -2)$ is the *x*-coordinate, and which is the *y*-coordinate?

🌐 **Math in Context** *Economics* While gasoline prices can vary from day to day, historical data show a general upward trend in gas prices. The next example shows how a scatterplot can be used to visualize this trend.

EXAMPLE 3 Making a scatterplot of gasoline prices

TABLE 10.2 lists the average price of a gallon of gasoline for selected years. Make a scatterplot of the data. These prices have *not* been adjusted for inflation.

Average Price of Gasoline

Year	1960	1970	1980	1990	2000	2010
Cost (per gal)	31¢	36¢	119¢	115¢	156¢	273¢

TABLE 10.2 *Source:* Department of Energy.

Solution

The data point $(1960, 31)$ can be used to indicate that the average cost of a gallon of gasoline in 1960 was 31¢. Plot the six data points $(1960, 31)$, $(1970, 36)$, $(1980, 119)$,

(1990, 115), (2000, 156), and (2010, 273) in the *xy*-plane. The *x*-values vary from 1960 to 2010, so label the *x*-axis from 1960 to 2020 every 10 years. The *y*-values vary from 31 to 273, so label the *y*-axis from 0 to 350 every 50¢. Note that the *x*- and *y*-scales must be large enough to accommodate every data point. **FIGURE 10.7** shows the scatterplot.

In **FIGURE 10.7**, the *increment* on the *x*-axis is 10 because each step from one vertical grid line to the next represents a change of 10 years. Similarly, the increment on the *y*-axis is 50 because each step from one horizontal grid line to the next represents a 50-cent change in price. This example demonstrates that the scale and increment on one axis are not always the same as those on the other axis.

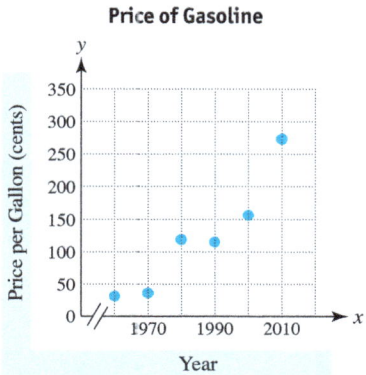

FIGURE 10.7

Now Try Exercise 25

4 Making and Reading Line Graphs

Sometimes it is helpful to connect consecutive data points in a scatterplot with line segments. This type of graph visually emphasizes changes in the data and is called a **line graph**. When making a line graph, be sure to plot all of the given data points *before* connecting the points with line segments. The points should be connected consecutively from left to right on the scatterplot, even if the data are given "out of order" in a table.

EXAMPLE 4 Making a line graph

Use the data in **TABLE 10.3** to make a line graph.

Data Points for a Line Graph

x	-2	-1	0	1	2
y	1	2	-2	-1	1

TABLE 10.3

Solution
The data in **TABLE 10.3** are represented by the five ordered pairs $(-2, 1)$, $(-1, 2)$, $(0, -2)$, $(1, -1)$, and $(2, 1)$. Plot these points and then connect consecutive points with line segments, as shown in **FIGURE 10.8**.

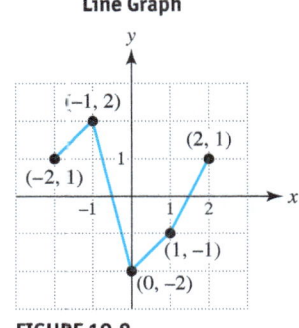

FIGURE 10.8

Now Try Exercise 37

CALCULATOR HELP

To make a scatterplot or a line graph, see Appendix A (pages AP-3 and AP-4).

TECHNOLOGY NOTE

Scatterplots and Line Graphs
Graphing calculators are capable of creating line graphs and scatterplots. The line graph in **FIGURE 10.8** is shown (below) to the left, and the corresponding scatterplot is shown to the right.

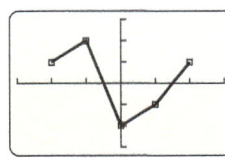 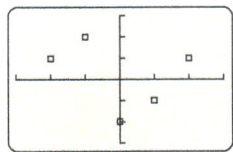

Math in Energy Context

The next example illustrates how a line graph can be used to analyze U.S. energy use over a 50-year period.

EXAMPLE 5 Analyzing a line graph

The line graph in **FIGURE 10.9** shows the per capita energy consumption in the United States from 1960 to 2010. Units are in millions of Btu, where 1 Btu equals the amount of energy necessary to heat 1 pound of water 1 °F. (*Source:* Department of Energy.)

(a) Did energy consumption ever decrease during this time period? If so, when?
(b) Estimate the per capita energy consumption in 1960 and in 2000.
(c) Estimate the percent change in energy consumption from 1960 to 2000.

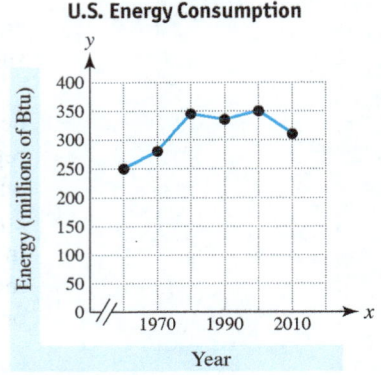

FIGURE 10.9

READING CHECK 3

- What is the main difference between a scatterplot and a line graph?

CRITICAL THINKING 1

When analyzing data, do you prefer a table of values, a scatterplot, or a line graph? Explain your answer.

Solution

(a) Yes, energy consumption decreased slightly between 1980 and 1990, and again between 2000 and 2010.

(b) From the graph, per capita energy consumption in 1960 was about 250 million Btu. In 2000 it was about 350 million Btu.

NOTE: When different people read a graph, the values they obtain may vary slightly. ∎

(c) The percent change from 1960 to 2000 was

$$\frac{350 - 250}{250} \cdot 100 = 40,$$

so the increase was 40%.

Now Try Exercise 49

10.1 Putting It All Together

CONCEPT	EXPLANATION	EXAMPLES
Ordered Pair	Has the form (x, y), where the order of x and y is important	$(1, 2)$, $(-2, 3)$, $(2, 1)$ and $(-4, -2)$ are distinct ordered pairs.
Rectangular Coordinate System, or xy-plane	Consists of a horizontal x-axis and a vertical y-axis that intersect at the origin. Has four quadrants, which are numbered counterclockwise as I, II, III, and IV	Quadrant II, Quadrant I, Origin, Quadrant III, Quadrant IV

CONCEPT	EXPLANATION	EXAMPLES
Scatterplot	Individual points that are plotted in the *xy*-plane	The points $(1, 1)$, $(2, 3)$, $(3, 2)$, $(4, 5)$, and $(5, 4)$ are plotted in the graph.
Line Graph	Similar to a scatterplot except that line segments are drawn between consecutive data points	

10.1 Exercises

MyMathLab®

CONCEPTS AND VOCABULARY

1. Another name for the rectangular coordinate system is the _____.

2. The point where the *x*-axis and *y*-axis intersect is called the _____.

3. In the *xy*-plane, the origin corresponds to the ordered pair _____.

4. How many quadrants are there in the *xy*-plane?

5. (True or False?) Every point in the *xy*-plane is located in one of the quadrants.

6. In the *xy*-plane, the first value in an ordered pair is called the _____ -coordinate and the second value is called the _____ -coordinate.

7. If distinct points are plotted in the *xy*-plane, the resulting graph is called a(n) _____.

8. If the consecutive points in a scatterplot are connected with line segments, the resulting graph is called a(n) _____ graph.

UNDERSTANDING THE RECTANGULAR COORDINATE SYSTEM (*xy*-PLANE)

Exercises 9–12: Identify the coordinates of each point in the graph.

9.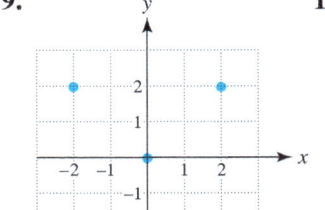

10.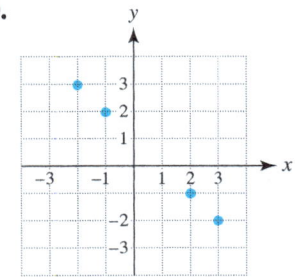

11.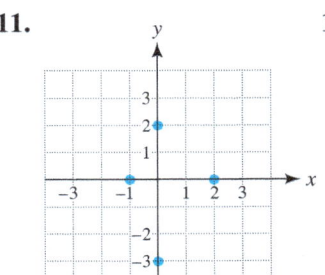

12.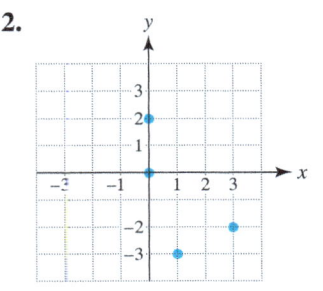

Exercises 13–16: Plot the given ordered pairs in the same xy-plane. If possible, state the quadrant in which each of the points is located.

13. $(1, 3)$, $(0, -3)$, and $(-2, 2)$

14. $(4, 0)$, $(-3, -4)$, and $(2, -3)$

15. $(0, 6)$, $(8, -4)$, and $(-6, -6)$

16. $(-4, 8)$, $(6, 8)$, and $(-8, 0)$

Exercises 17–22: If possible, identify the quadrant in which each point is located.

17. (a) $(1, 4)$ **(b)** $(-1, -4)$

18. (a) $(-2, -3)$ **(b)** $(2, -3)$

19. (a) $(7, 0)$ **(b)** $(0.1, 7)$

20. (a) $(100, -3)$ **(b)** $(-100, 3)$

21. (a) $\left(-\frac{1}{2}, \frac{3}{4}\right)$ **(b)** $\left(\frac{3}{4}, -\frac{1}{2}\right)$

22. (a) $(1.2, 0)$ **(b)** $(0, -1.2)$

23. Thinking Generally Which of the four quadrants contain points whose x- and y-coordinates have the same sign?

24. Thinking Generally Which of the four quadrants contain points whose x- and y-coordinates have different signs?

MAKING AND READING SCATTERPLOTS

Exercises 25–30: Make a scatterplot by plotting the given points. Be sure to label each axis.

25. $(0, 0)$, $(1, 2)$, $(-3, 2)$, $(-1, -2)$

26. $(0, -3)$, $(-2, 1)$, $(2, 2)$, $(-4, -4)$

27. $(-1, 0)$, $(4, -3)$, $(0, -1)$, $(3, 4)$

28. $(1, 1)$, $(-2, 2)$, $(-3, -3)$, $(4, -4)$

29. $(2, 4)$, $(-4, 4)$, $(0, -4)$, $(-6, 2)$

30. $(4, 8)$, $(8, 4)$, $(-8, -4)$, $(-4, 0)$

Exercises 31–34: Use the table to make a scatterplot.

31.

x	5	5	-10	10
y	0	-5	-20	-10

32.

x	10	-20	40	-30
y	30	10	0	-10

33.

x	0	0.2	-0.1
y	0.1	-0.3	0.4

34.

x	1.5	-1	-2.5
y	2.5	-1.5	0

Exercises 35 and 36: **Real Data Graphs** *Identify the coordinates of each point in the graph. Then explain what the coordinates of the first point indicate.*

35. Cigarette consumption in the United States (*Source:* U.S. Department of Health and Human Services.)

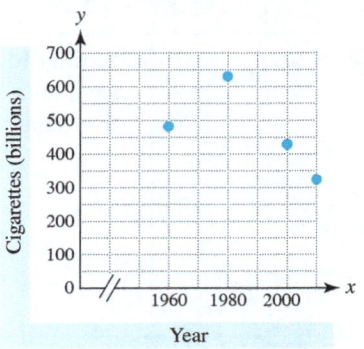

36. Number of vinyl albums sold (*Source:* Recording Industry Association of America.)

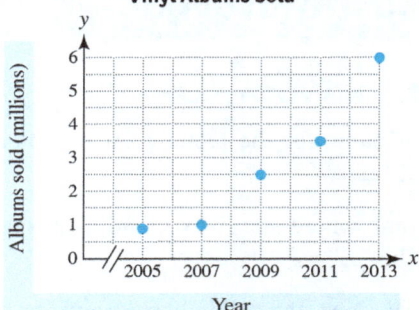

MAKING AND READING LINE GRAPHS

Exercises 37–42: Use the table to make a line graph.

37.

x	-2	-1	0	1	2
y	2	1	0	-1	-2

38.

x	-4	-2	0	2	4
y	4	-2	3	-1	2

39.

x	-10	-5	0	5	10
y	20	-10	10	0	-20

40.

x	1	2	3	4	5
y	2	3	1	5	4

41.

x	−5	5	−10	10	0
y	10	20	30	20	40

42.

x	3	−2	2	1	−3
y	4	3	3	−2	−3

GRAPHING REAL DATA

Exercises 43–48: **Graphing Real Data** *The table contains real data.*

(a) *Make a line graph of the data. Be sure to label the axes.*
(b) *Comment on any trends in the data.*

43. Average annual cost C of a four-year residential college degree in thousands of dollars during year t

t	1980	1990	2000	2010	2015
C	10	12	16	21	24

Source: National Center for Education Statistics.

44. Federal income tax receipts I in billions during year t

t	1970	1980	1990	2000	2010
I	90	244	467	1003	2165

Source: Office of Management and Budget.

45. Welfare beneficiaries B in millions during year t

t	1970	1980	1990	2000	2010
B	7	11	12	6	4

Source: Administration for Children and Families.

46. Detroit city population P in millions during year t

t	1900	1920	1940	1960	1980	2000
P	0.3	1.0	1.6	1.7	1.2	0.9

Source: Census Bureau.

47. U.S. Internet users y in millions during year x

x	2010	2011	2012	2013	2014
y	221	229	237	244	251

Source: eMarketer.

48. Number of farms in Iowa F in thousands during year x

x	2003	2005	2007	2009	2011
F	90	89	92	92	92

Source: U.S. Department of Agriculture.

49. Infant Mortality The line graph shows the U.S. infant mortality rate for selected years.

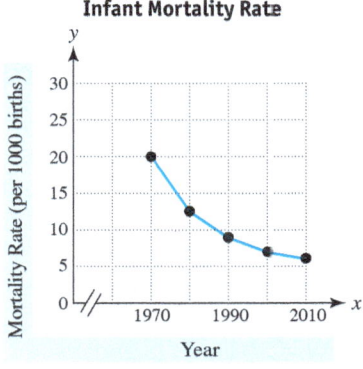

(a) Comment on any trends in the data.
(b) Estimate the infant mortality rate in 1990.
(c) Estimate the percent change in the infant mortality rate from 1970 to 2010.

50. Population The line graph shows the population of the midwestern states, in millions, for selected years.

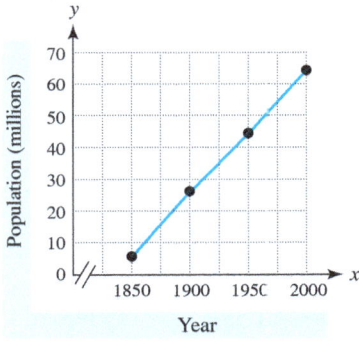

(a) Comment on any trends in the data.
(b) Estimate this population in 1900.
(c) Estimate the percent change in population from 1850 to 2000.

WRITING ABOUT MATHEMATICS

51. Explain how to identify the quadrant that a point lies in if it has coordinates (x, y).

52. Explain the difference between a scatterplot and a line graph. Give an example of each.

10.2 Equations in Two Variables

Objectives

1. Introducing Equations in Two Variables
2. Recognizing Ordered Pairs as Solutions
3. Making a Table of Solutions
4. Graphing Linear Equations in Two Variables
   - Standard Form for a Linear Equation
   - Solving Standard Form for a Variable
5. Graphing Nonlinear Equations in Two Variables

NEW VOCABULARY

☐ Standard form (of a linear equation in two variables)
☐ Linear equation in two variables

STUDY TIP

If you have tried to solve a homework problem but need help, ask a question in class. Other students will likely have the same question.

READING CHECK 1

- How is a solution to an equation in two variables expressed?

1 Introducing Equations in Two Variables

Connecting Concepts with Your Life FIGURE 10.10 shows a scatterplot of average college tuition and fees at public colleges and universities (in 2010 dollars), together with a line that models the data. If we could find an equation for this line, then we could use it to estimate tuition and fees for years without data points, such as 2015. In this section, we discuss linear equations, whose graphs are lines. Linear equations and lines are often used to approximate data. (*Source:* The College Board.)

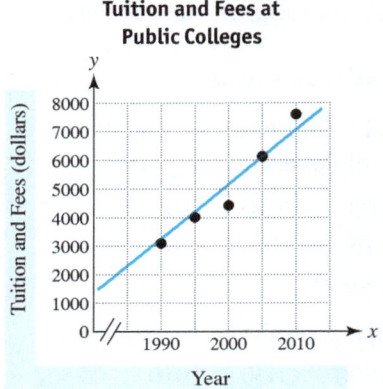

FIGURE 10.10

NOTE: The graph shown in FIGURE 10.10 is *not* a line graph that has consecutive points connected with line segments. Rather, it is a scatterplot together with a single straight line that closely models the data. ∎

Equations can have any number of variables. Previously, we solved equations with *one* variable. The following are examples of equations with *two* variables.

$$y = 2x, \quad 3x + 2y = 4, \quad z = t^2, \quad \text{and} \quad a - b = 1$$
Equations with two variables

2 Recognizing Ordered Pairs as Solutions

A solution to an equation with two variables consists of two numbers, one for each variable, which can be expressed as an ordered pair. For example, one solution to the equation $y = 5x$ is given by $x = 1$ and $y = 5$ because $5 = 5(1)$ is a true statement. This solution can be expressed as the ordered pair $(1, 5)$. TABLE 10.4 lists several ordered pairs and shows whether each ordered pair is a solution to the equation $y = 5x$.

Checking Ordered-Pair Solutions to $y = 5x$

Ordered Pair	Check: $y = 5x$	Is it a solution?
$(1, 5)$	$5 \stackrel{?}{=} 5(1)$	Yes, because $5 = 5$
$(2, 12)$	$12 \stackrel{?}{=} 5(2)$	No, because $12 \neq 10$
$(-2, -10)$	$-10 \stackrel{?}{=} 5(-2)$	Yes, because $-10 = -10$
$(4, 20)$	$20 \stackrel{?}{=} 5(4)$	Yes, because $20 = 20$

TABLE 10.4

EXAMPLE 1   **Testing solutions to equations**

Determine whether the given ordered pair is a solution to the given equation.

(a) $y = x + 3$, $(1, 4)$ **(b)** $2x - y = 5$, $\left(\frac{1}{2}, -4\right)$ **(c)** $-4x - 5y = 20$, $(-5, 1)$

Solution
(a) Let $x = 1$ and $y = 4$ in the given equation.

$$y = x + 3 \quad \text{Given equation}$$
$$4 \stackrel{?}{=} 1 + 3 \quad \text{Substitute.}$$
$$4 = 4 \checkmark \quad \text{The solution checks.}$$

The ordered pair $(1, 4)$ is a solution.

(b) Let $x = \frac{1}{2}$ and $y = -4$ in the given equation.

$$2x - y = 5 \quad \text{Given equation}$$
$$2\left(\frac{1}{2}\right) - (-4) \stackrel{?}{=} 5 \quad \text{Substitute.}$$
$$1 + 4 \stackrel{?}{=} 5 \quad \text{Simplify the left side.}$$
$$5 = 5 \checkmark \quad \text{The solution checks}$$

The ordered pair $\left(\frac{1}{2}, -4\right)$ is a solution.

(c) Let $x = -5$ and $y = 1$ in the given equation.

$$-4x + 5y = 20 \quad \text{Given equation}$$
$$-4(-5) + 5(1) \stackrel{?}{=} 20 \quad \text{Substitute.}$$
$$20 + 5 \stackrel{?}{=} 20 \quad \text{Simplify the left side.}$$
$$25 \neq 20 \text{ ✗} \quad \text{The solution does } not \text{ check.}$$

The ordered pair $(-5, 1)$ is *not* a solution.

Now Try Exercises 11, 15, 17

3 Making a Table of Solutions

A table can be used to list solutions to an equation. For example, **TABLE 10.5** lists solutions to $x + y = 5$, where the sum of each xy-pair equals 5.

$x + y = 5$

x	-2	-1	0	1	2
y	7	6	5	4	3

Sum equals 5

TABLE 10.5

READING CHECK 2

- When is it helpful to have a table that lists a few solutions to an equation?

Most equations in two variables have infinitely many solutions, so it is impossible to list all solutions in a table. However, when you are graphing an equation, having a table that lists a few solutions to the equation is often helpful. The next two examples demonstrate how to complete a table for a given equation.

EXAMPLE 2 Completing a table of solutions

Complete the table for the equation $y = 2x - 3$.

x	-4	-2	0	2
y				

Solution
Start by determining the corresponding y-value for each x-value in the table. For example, when $x = -2$, the equation $y = 2x - 3$ implies that $y = 2(-2) - 3 = -4 - 3 = -7$. Filling in the y-values results in **TABLE 10.6**.

$y = 2x - 3$

x	-4	-2	0	2
y	-11	-7	-3	1

TABLE 10.6

Now Try Exercise 21

EXAMPLE 3 Making a table of solutions

Use $y = 0, 5, 10,$ and 15 to make a table of solutions to $5x + 2y = 10$.

Solution
Begin by listing the required y-values in the table. Next determine the corresponding x-values for each y-value by using the equation $5x + 2y = 10$.

When $y = 0$,	When $y = 5$,	When $y = 10$,	When $y = 15$,
$5x + 2(0) = 10$	$5x + 2(5) = 10$	$5x + 2(10) = 10$	$5x + 2(15) = 10$
$5x + 0 = 10$	$5x + 10 = 10$	$5x + 20 = 10$	$5x + 30 = 10$
$5x = 10$	$5x = 0$	$5x = -10$	$5x = -20$
$x = 2$	$x = 0$	$x = -2$	$x = -4$

Filling in the x-values results in **TABLE 10.7**.

$5x + 2y = 10$

x	2	0	-2	-4
y	0	5	10	15

TABLE 10.7

Now Try Exercise 23

Math in Context (Health Care) Newspapers, magazines, and books often list numbers in a table rather than presenting a formula for the reader to use. The next example illustrates a situation in health care in which a table might be preferable to a formula.

EXAMPLE 4 Calculating appropriate lengths of crutches

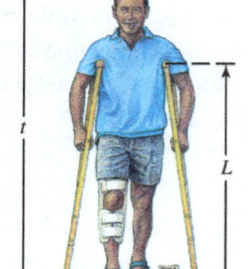

People with leg injuries often need crutches. An appropriate crutch length L in inches for an injured person who is t inches tall is estimated by $L = 0.72t + 2$. (*Source: Journal of the American Physical Therapy Association.*)

(a) Complete the table. Round values to the nearest inch.

t	60	65	70	75	80
L					

(b) Use the table to determine the appropriate crutch length for a person 5 feet 10 inches tall.

Solution

(a) In the formula $L = 0.72t + 2$, if $t = 60$, then $L = 0.72(60) + 2 = 43.2 + 2 = 45.2$, or about 45 inches. If $t = 65$, then $L = 0.72(65) + 2 = 46.8 + 2 = 48.8 \approx 49$. Other values in **TABLE 10.8** are found similarly.

Crutch Lengths

t	60	65	70	75	80
L	45	49	52	56	60

TABLE 10.8

(b) A person who is 5 feet 10 inches tall is $5 \cdot 12 + 10 = 70$ inches tall. **TABLE 10.8** shows that a person 70 inches tall needs crutches that are about 52 inches long.

Now Try Exercise 79

4 Graphing Linear Equations in Two Variables

Many times graphs are used in mathematics to make concepts easier to visualize and understand. Graphs can be either curved or straight; however, graphs of *linear* equations in two variables are always *straight* lines.

The following See the Concept shows how to graph the equation $y = 2x$.

See the Concept 1 — GRAPHING $y = 2x$

Start by making a table of values. A table can be either vertical or horizontal. We will use a vertical table.

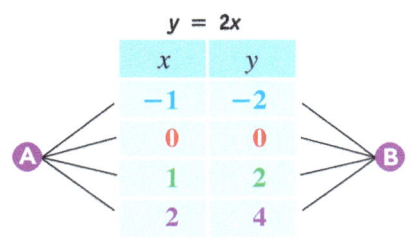
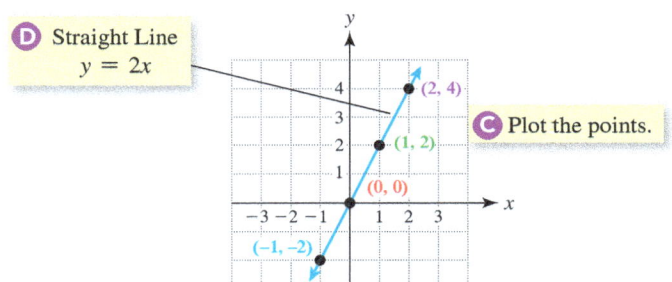

Ⓐ Select a few convenient values for x, such as $x = -1, 0, 1$ and 2. *Any* numbers may be chosen for x-values. In this example, we choose values that are easy to evaluate in the equation $y = 2x$.

Ⓑ Calculate the y-values for each x-value. Since $y = 2x$, each y-value is found by doubling the x-value.

Ⓒ The table indicates that the points $(-1, -2)$, $(0, 0)$, $(1, 2)$, and $(2, 4)$ lie on the line. Plot these points. Since $y = 2x$ is a *linear equation*, the points should lie on a *straight line*.

Ⓓ Draw a straight line that passes through the plotted points. This line is the graph of $y = 2x$.

WATCH VIDEO IN MML.

NOTE: Because the graph of a linear equation in two variables is a straight line, you may want to use a ruler or other straight edge when graphing this kind of equation. ∎

EXAMPLE 5 Graphing linear equations

Graph each linear equation.
(a) $y = \frac{1}{2}x - 1$ (b) $x + y = 4$

Solution

(a) The equation $y = \frac{1}{2}x - 1$ is a linear equation in two variables and its graph is a line. Two points determine a line. However, it is a good idea to plot three points to be sure that the line is graphed correctly. Start by choosing three values for x and then calculate the corresponding y-values, as shown in **TABLE 10.9**. In **FIGURE 10.11**, the points $(-2, -2)$, $(0, -1)$, and $(2, 0)$ are plotted and the line passing through these points is drawn.

$y = \frac{1}{2}x - 1$

x	y
-2	-2
0	-1
2	0

TABLE 10.9

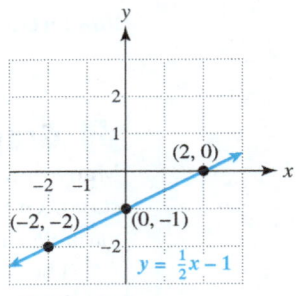

FIGURE 10.11

(b) The equation $x + y = 4$ is a linear equation in two variables, so its graph is a line. If an ordered pair (x, y) is a solution to the given equation, then the sum of x and y is 4. **TABLE 10.10** shows three examples. In **FIGURE 10.12**, the points $(0, 4)$, $(2, 2)$, and $(4, 0)$ are plotted with the line passing through each one.

$x + y = 4$

x	y
0	4
2	2
4	0

TABLE 10.10

The sum is 4.

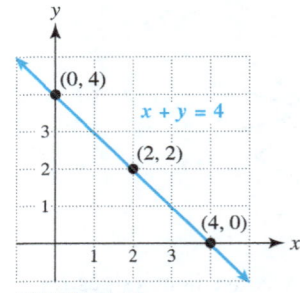

FIGURE 10.12

Now Try Exercises 43, 57

TECHNOLOGY NOTE

CALCULATOR HELP
To graph an equation, see Appendix A (page AP-5).

Graphing Equations
Graphing calculators can be used to graph equations. Before graphing the equation $x + y = 4$, solve the equation for y to obtain $y = 4 - x$. A calculator graph of **FIGURE 10.12** is shown, except that the three points have not been plotted.

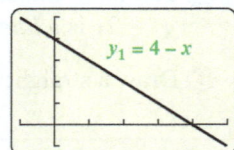

MAKING CONNECTIONS 1

Line Graphs and Graphs of Linear Equations

Do not confuse a line graph with the graph of a linear equation. A *line graph* results when consecutive points from a table of values are connected with line segments. However, the *graph of a linear equation* is a single, straight line that passes through points from a table of values and continues indefinitely in two directions.

READING CHECK 3

- Is a line graph the same as a graph of a linear equation?

STANDARD FORM FOR A LINEAR EQUATION Not every equation with two variables has a graph that is a straight line. Those equations whose graphs are straight lines are called *linear* equations in two variables. Every linear equation in two variables can be written in the following **standard form**.

> **LINEAR EQUATION IN TWO VARIABLES**
>
> A **linear equation in two variables** can be written as
>
> $$Ax + By = C,$$
>
> where A, B, and C are fixed numbers (constants) and A and B are not both equal to 0. The graph of a linear equation in two variables is a line.

The equation $y = 2x$ was presented earlier in this section in See the Concept 1. This equation is a linear equation in two variables because it can be written in standard form by adding $-2x$ to each side.

$$y = 2x \quad \text{Given equation}$$
$$-2x + y = -2x + 2x \quad \text{Add } -2x \text{ to each side.}$$
$$-2x + y = 0 \quad \text{Simplify.}$$

The equation $-2x + y = 0$ is a linear equation in two variables because it is in the form $Ax + By = C$ with $A = -2$, $B = 1$, and $C = 0$.

SOLVING STANDARD FORM FOR A VARIABLE When a linear equation in two variables is given in standard form, it is sometimes difficult to create a table of solutions. Solving an equation for y often makes it easier to select x-values for the table that will make the y-values simpler to calculate. This is demonstrated in the next example.

EXAMPLE 6 Solving for y and then graphing

Graph the linear equation $4x - 3y = 12$ by solving for y first.

Solution
First solve the given equation for y.

$$4x - 3y = 12 \quad \text{Given equation (standard form)}$$
$$-3y = -4x + 12 \quad \text{Subtract } 4x \text{ from each side.}$$
$$\frac{-3y}{-3} = \frac{-4x + 12}{-3} \quad \text{Divide each side by } -3.$$
$$\frac{-3y}{-3} = \frac{-4x}{-3} + \frac{12}{-3} \quad \text{Property of fractions, } \frac{a+b}{c} = \frac{a}{c} + \frac{b}{c}$$
$$y = \frac{4}{3}x - 4 \quad \text{Simplify fractions.}$$

Divide each term by -3.

Select *multiples of 3* (the denominator of $\frac{4}{3}$) as x-values for the table of solutions. For example, if $x = 6$ is chosen, $y = \frac{4}{3}(6) - 4 = \frac{24}{3} - 4 = 8 - 4 = 4$. **TABLE 10.11** lists the solutions $(0, -4)$, $(3, 0)$, and $(6, 4)$, which are plotted in **FIGURE 10.13** on the next page with the line passing through each one.

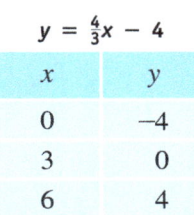

$y = \frac{4}{3}x - 4$

x	y
0	-4
3	0
6	4

TABLE 10.11

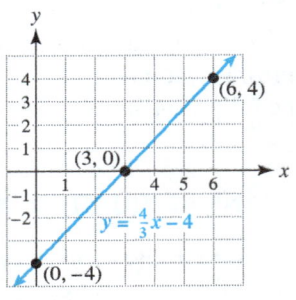

FIGURE 10.13

Now Try Exercise 59

> **MAKING CONNECTIONS 2**
>
> **Graphs and Solution Sets**
>
> A graph visually depicts the set of solutions to an equation. Each point (x, y) on the graph represents one solution to the equation. In a linear equation, each x-value determines a unique y-value. Because there are infinitely many x-values, there are infinitely many points located on the graph of a linear equation. Thus the graph of a linear equation is a *continuous* line with no breaks.

5 Graphing Nonlinear Equations in Two Variables

Some equations in two variables have graphs that are *not* straight lines. Such equations are called **nonlinear equations**. It is possible to graph some nonlinear equations by completing a table of values and plotting points. However, because these graphs can take on many different shapes, it is important to plot enough points so that an accurate graph can be obtained.

EXAMPLE 7 **Graphing a nonlinear equation**

Graph the nonlinear equation $y = x^2$.

Solution
Complete a table of values as shown in **TABLE 10.12**. Be sure to include both positive and negative x-values to get an accurate graph. Plot the points from the table and draw a smooth curve that passes through each point, as shown in **FIGURE 10.14**.

$y = x^2$

x	y
-2	4
-1	1
0	0
1	1
2	4

TABLE 10.12

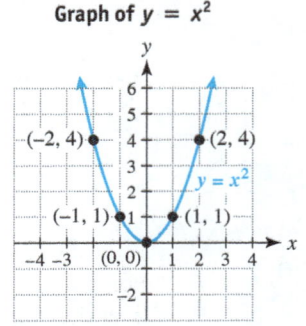

FIGURE 10.14

Now Try Exercise 71

10.2 Putting It All Together

CONCEPT	EXPLANATION	EXAMPLES
Equation in Two Variables	An equation that has two variables	$y = 4x + 5$, $4x - 5y = 20$, and $u - v = 100$
Solution to an Equation in Two Variables	An ordered pair (x, y) whose x- and y-values satisfy the equation. Equations in two variables often have infinitely many solutions.	$(1, 3)$ is a solution to $3x + y = 6$ because $3(1) + 3 = 6$ is a true statement. The equation $y = 2x$ has infinitely many solutions, such as $(1, 2), (2, 4), (3, 6)$, and so on.
Table of Solutions	A table can be used to list solutions to an equation in two variables.	The following table lists solutions to the equation $y = 3x - 1$. | x | -1 | 0 | 1 | 2 | |---|---|---|---|---| | y | -4 | -1 | 2 | 5 |
Linear Equation in Two Variables (Standard Form)	Can be written as $$Ax + By = C,$$ where A, B, and C are fixed numbers and A and B are not both equal to 0	$3x + 4y = 5$, $\quad y = 4 - 3x$, $\quad x = 2y + 1$ $\qquad\qquad\quad$ (or $3x + y = 4$) $\;$ (or $x - 2y = 1$) The graph of each equation is a line.

10.2 Exercises

MyMathLab®

CONCEPTS AND VOCABULARY

1. (True or False?) The equation $4x + 6y = 24$ is an equation in two variables.

2. (True or False?) The equation $4x - 3x = 10$ is a linear equation in two variables.

3. A solution to an equation with two variables consists of two numbers expressed as a(n) _____.

4. A(n) _____ equation in two variables can be written in the form $Ax + By = C$.

5. (True or False?) A table of solutions lists all of the solutions to an equation in two variables.

6. (True or False?) A linear equation in two variables has infinitely many solutions.

7. An equation's _____ visually depicts its solution set.

8. The graph of a linear equation in two variables is a(n) _____

9. (True or False?) Every equation in two variables has a graph that is a straight line.

10. (True or False?) A line graph and the graph of a linear equation in two variables are the same thing.

RECOGNIZING ORDERED PAIRS AS SOLUTIONS

Exercises 11–20: Determine whether the ordered pair is a solution to the given equation.

11. $y = x + 1$, $(5, 6)$

12. $y = 4 - x$, $(6, 2)$

13. $y = 4x + 7$, $(2, 13)$

14. $y = -3x + 2$, $(-2, 8)$

15. $4x - y = -13$, $(-2, 3)$

16. $3y + 2x = 0$, $(-2, 3)$

17. $y - 6x = -1$, $\left(\frac{1}{2}, 2\right)$

18. $\frac{1}{2}x + \frac{3}{2}y = 0$, $\left(-\frac{3}{2}, \frac{1}{2}\right)$

19. $0.31x - 0.42y = -9$, $(100, 100)$

20. $0.5x - 0.6y = 4$, $(20, 10)$

MAKING A TABLE OF SOLUTIONS

Exercises 21–28: Complete the table for the given equation.

21. $y = 4x$

x	-2	-1	0	1	2
y	-8				

22. $y = \frac{1}{2}x - 1$

x	0	1	2	3	4
y	-1				

23. $3y + 2x = 6$

x					
y	-2	0	2	4	8

24. $3x - 5y = 30$

x					
y	-9	-6	-3	0	3

25. $y = x + 4$

x	y
-8	-4
	0
4	
8	
12	

26. $2x - y = 1$

x	y
-1	-3
	-1
	0
	1
	3

27. $2x - 6y = 12$

x	y
-6	
0	
6	
12	
18	

28. $4x + 3y = 12$

x	y
9	
6	
3	
0	
-3	

Exercises 29–32: Use the given values of the variable to make a horizontal table of solutions for the equation.

29. $y = 3x$ $x = -3, 0, 3, 6$

30. $y = 1 - 2x$ $x = 0, 1, 2, 3$

31. $x + y = 6$ $y = -2, 0, 2, 4$

32. $2x - 3y = 9$ $y = -3, 0, 1, 2$

Exercises 33–36: Use the given values of the variable to make a vertical table of solutions for the equation.

33. $y = \frac{x + 4}{2}$ $x = -8, -4, 0, 4$

34. $y = \frac{x}{3} - 1$ $x = 0, 2, 4, 6$

35. $y - 4x = 0$ $y = -2, -1, 0, 1$

36. $-4x = 6y - 4$ $y = -1, 0, 1, 2$

37. Thinking Generally If a student wishes to avoid fractional y-values when making a table of solutions for $y = \frac{3}{5}x - 7$, what must be true about any selected integer x-values?

38. Thinking Generally If a student wishes to avoid fractional y-values when making a table of solutions for $y = \frac{a}{b}x$, where a and b are natural numbers, what must be true about any selected integer x-values?

GRAPHING LINEAR EQUATIONS IN TWO VARIABLES

Exercises 39–44: Make a table of solutions for the equation, and then use the table to graph the equation.

39. $y = -2x$ **40.** $y = 2x - 1$

41. $x = 3 - y$ **42.** $x = y + 1$

43. $x + 2y = 4$ **44.** $2x - y = 1$

Exercises 45–58: Graph the equation.

45. $y = x$ **46.** $y = \frac{1}{2}x$

47. $y = \frac{1}{3}x$ **48.** $y = -2x$

49. $y = x + 3$ **50.** $y = x - 2$

51. $y = 2x + 1$ **52.** $y = \frac{1}{2}x - 1$

53. $y = 4 - 2x$ **54.** $y = 2 - 3x$

55. $y = 7 + x$ **56.** $y = 2 + 2x$

57. $y = -\frac{1}{2}x + \frac{1}{2}$ **58.** $y = -\frac{3}{4}x + 2$

SOLVING STANDARD FORM FOR A VARIABLE

Exercises 59–70: Graph the linear equation by solving for y first.

59. $2x + 3y = 6$ **60.** $3x + 2y = 6$

61. $x + 4y = 4$ **62.** $4x + y = -4$

63. $-x + 2y = 8$ **64.** $-2x + 6y = 12$

65. $y - 2x = 7$
66. $3y - x = 2$
67. $5x - 4y = 20$
68. $4x - 5y = -20$
69. $3x + 5y = -9$
70. $5x - 3y = 10$

GRAPHING NONLINEAR EQUATIONS IN TWO VARIABLES

Exercises 71–78: (See Example 7.) Make a convenient table of values for the given equation. Then plot the points and graph the equation.

71. $y = x^2 - 1$
72. $y = x^2 + 2$
73. $y = 2x^2$
74. $y = \frac{1}{2}x^2$
75. $y = |x| - 1$
76. $y = |x - 1|$
77. $y = \sqrt{x}$
78. $y = 2\sqrt{x}$

APPLICATIONS INVOLVING EQUATIONS IN TWO VARIABLES

79. **U.S. Population** For the years 2010 to 2050, the projected percentage P of the U.S. population that will be over the age of 65 during year t is estimated by $P = 0.178t - 344.6$. (*Source*: U.S. Census Bureau.)
    (a) Complete the table. Round each resulting value to the nearest tenth.

t	2010	2020	2030	2040	2050
P					

    (b) Use the table to find the year when the percentage of the population over the age of 65 is expected to reach 16.7%.

80. **U.S. Population** For the years 2010 to 2050, the projected percentage P of the U.S. population that will be 18 to 24 years old during year t is estimated by $P = -0.025t + 60.35$. (*Source*: U.S. Census Bureau.)
    (a) Complete the table. Round each resulting value to the nearest tenth.

t	2010	2020	2030	2040	2050
P					

    (b) Use the table to find the year when the percentage of the population that is 18 to 24 years old is expected to be 9.4%.

81. **Solid Waste in the Past** In 1960, the amount A of garbage in pounds produced after t days by the average American is given by $A = 2.7t$. (*Source*: Environmental Protection Agency.)
    (a) Graph the equation for $t \geq 0$.
    (b) How many days did it take for the average American in 1960 to produce 100 pounds of garbage?

82. **Solid Waste Today** (See Exercise 81.) Today the amount A of garbage in pounds produced after t days by the average American is given by $A = 4.5t$. (*Source*: Environmental Protection Agency.)
    (a) Graph the equation for $t \geq 0$.
    (b) How many days does it take for the average American to produce 100 pounds of garbage today?

83. **Web Browsers** From 2008 to 2014, the percent P of the global desktop web-browser market held by Internet Explorer can be modeled by $P = -7.333t + 14,793$, where t is the year. (*Source*: Business Insider.)
    (a) Evaluate P for $t = 2008$ and $t = 2014$. Interpret your answers.
    (b) Use your results from part (a) to graph the equation for P from 2008 to 2014.
    (c) In what year was $P = 50\%$?

84. **U.S. HIV Infections** The cumulative number of HIV infections I in thousands during year t is modeled by the equation $I = 42t - 83,197$, where $t \geq 2000$. (*Source*: Department of Health and Human Services.)
    (a) Evaluate I for $t = 2000$ and for $t = 2005$.
    (b) Use your results from part (a) to graph the equation from 2000 to 2005.
    (c) In what year was $I = 971$?

WRITING ABOUT MATHEMATICS

85. The number of welfare beneficiaries B in millions during year t is shown in the table. Discuss whether a linear equation might work to approximate these data during this period of time.

t	1970	1980	1990	2000	2010
B	7	11	12	6	4

    Source: Administration for Children and Families.

86. The Asian-American population P in millions during year t is shown in the table. Discuss whether a linear equation might model these data.

t	2002	2004	2006	2008	2010
P	12.0	12.8	13.6	14.4	15.2

    Source: U.S. Census Bureau.

SECTIONS 10.1 and 10.2 Checking Basic Concepts

1. Identify the coordinates of the four points in the graph. State the quadrant, if any, in which each point lies.

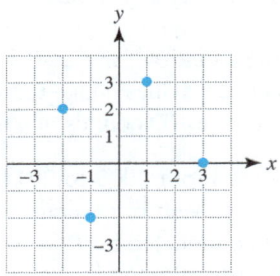

2. Make a scatterplot of the five points $(-2, -2)$, $(-1, -3)$, $(0, 0)$, $(1, 2)$, and $(2, 3)$.

3. **U.S. Population** The table gives the percentage P of the U.S. population that was over the age of 85 during year t. Make a line graph of the data in the table, and then comment on any trends. (*Source:* U.S. Census Bureau.)

t	1970	1980	1990	2000	2010
P	0.7	1.0	1.2	1.5	2.1

4. Determine whether $(-2, -3)$ is a solution to the equation $-2x - y = 7$.

5. Complete the table for the equation $y = -2x + 1$.

x	-2	-1	0	1	2
y	5				

6. Graph the equation.
   (a) $y = \frac{1}{2}x$ (b) $4x + 6y = 12$

7. **Windows 8 Performance** The percent P of the global market share of Windows 8 in the first 12 months after its release can be approximated by $P = 0.73t + 0.27$, where t is months after the release date.
   (a) Find P when $t = 1$ and $t = 12$. Interpret each result.
   (b) Use your results from part (a) to graph the equation for P from month 1 to month 12. Be sure to label the axes.
   (c) How many months after its launch did Windows 8 represent 4% of the global market share?

10.3 Intercepts; Horizontal and Vertical Lines

Objectives
1. Finding Intercepts
2. Using Intercepts to Graph Lines
3. Recognizing Horizontal Lines and Their Equations
4. Recognizing Vertical Lines and Their Equations

NEW VOCABULARY
☐ x-intercept
☐ y-intercept

1 Finding Intercepts

When things move or change at a *constant* rate, linear equations can often be used to describe or model the situation. For example, when a car moves at a constant speed, a graph of a linear equation can be used to visualize its driver's distance from a particular location. In real-world situations such as this, the *x*- and *y-intercepts* of the graph often provide important information.

Connecting Concepts with Your Life Suppose that someone leaves a family gathering at a state park and drives home at a constant speed of 50 miles per hour. The graph in **FIGURE 10.15** shows the distance of the driver from home at various times. The graph intersects the *y*-axis at $(0, 200)$, which is called the *y-intercept*. In this situation, the *y*-intercept tells us that the initial distance (when $x = 0$) between the driver and home is 200 miles. The graph also intersects the *x*-axis at $(4, 0)$, which is called the *x-intercept*. This intercept tells us that the elapsed time is 4 hours when the distance of the driver from home is 0 miles (when $y = 0$.) In other words, the driver arrived at home after 4 hours of driving.

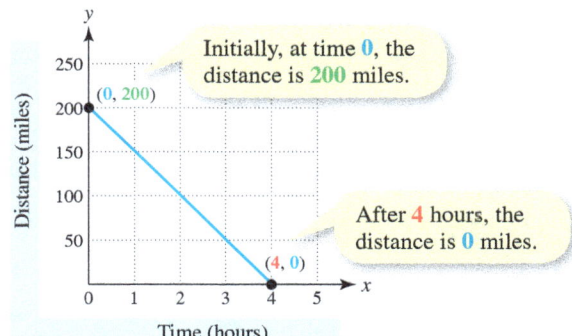

FIGURE 10.15

STUDY TIP

Before getting help, be sure that you have tried a problem. Organize your questions so that you can be specific about the part of the problem that is giving you difficulty.

READING CHECK 1

- Where are the intercepts located on a graph?

FINDING x- AND y-INTERCEPTS

A point where a graph intersects the x-axis is an **x-intercept**. To find an x-intercept, let $y = 0$ in the equation and solve for x.

A point where a graph intersects the y-axis is a **y-intercept**. To find a y-intercept, let $x = 0$ in the equation and solve for y.

EXAMPLE 1 Using a table to find intercepts

Complete the table for the equation $x - y = -1$. Then determine the x-intercept and y-intercept for the graph of the equation $x - y = -1$.

x	-2	-1	0	1	2
y					

Solution
Substitute -2 for x in $x - y = -1$ to find the corresponding y-value.

$$-2 - y = -1 \quad \text{Let } x = -2.$$
$$-y = 1 \quad \text{Add 2 to each side.}$$
$$y = -1 \quad \text{Multiply each side by } -1.$$

The other y-values can be found similarly. See **TABLE 10.13**.

The x-intercept corresponds to a point on the graph whose y-coordinate is 0. **TABLE 10.13** shows that the y-coordinate is 0 when $x = -1$. So the x-intercept is $(-1, 0)$. Similarly, the y-intercept corresponds to a point on the graph whose x-coordinate is 0. **TABLE 10.13** shows that the x-coordinate is 0 when $y = 1$. So the y-intercept is $(0, 1)$.

$$x - y = -1$$

x	-2	-1	0	1	2
y	-1	0	1	2	3

TABLE 10.13

READING CHECK 2

- How can the intercepts of a graph be found using a table of values?

Now Try Exercise 17

CRITICAL THINKING 1

If a line has no x-intercept, what can you say about the line?
If a line has no y-intercept, what can you say about the line?

2 Using Intercepts to Graph Lines

It is often convenient to use intercepts to graph lines. The following See the Concept shows how to graph a linear equation using its intercepts.

See the Concept 1 USING INTERCEPTS TO GRAPH $3x + 2y = 6$

To graph the linear equation $3x + 2y = 6$, start by finding the intercepts.

A Find the x-intercept.

$3x + 2(0) = 6$ Let $y = 0$.
$3x = 6$ Simplify.
$x = 2$ Divide each side by 3.

B Find the y-intercept.

$3(0) + 2y = 6$ Let $x = 0$.
$2y = 6$ Simplify.
$y = 3$ Divide each side by 2.

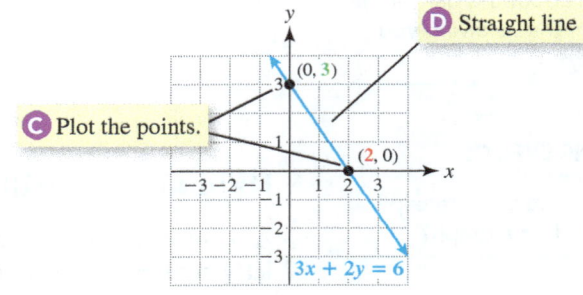

C Plot the points.

D Straight line

A Find the x-intercept: Let $y = 0$ in the given equation and solve for x. The line passes through $(2, 0)$.

B Find the y-intercept: Let $x = 0$ in the given equation and solve for y. The line passes through $(0, 3)$.

C Plot the points: Plot the points $(2, 0)$ and $(0, 3)$ on the xy-plane.

D Draw the line: Draw a straight line passing through the intercepts.

WATCH VIDEO IN MML.

EXAMPLE 2 Using intercepts to graph a line

Use intercepts to graph $2x - 6y = 12$.

Solution
The x-intercept is found by letting $y = 0$.

$2x - 6(0) = 12$ Let $y = 0$.
$x = 6$ Solve for x.

The y-intercept is found by letting $x = 0$.

$2(0) - 6y = 12$ Let $x = 0$.
$y = -2$ Solve for y.

Therefore the graph passes through the points $(6, 0)$ and $(0, -2)$, as shown in **FIGURE 10.16**.

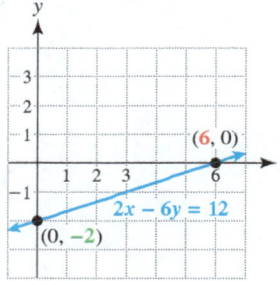

FIGURE 10.16

Now Try Exercise 21

EXAMPLE 3 Modeling the velocity of a toy rocket

A toy rocket is shot vertically into the air. Its velocity v in feet per second after t seconds is given by $v = 160 - 32t$. Assume that $t \geq 0$ and $t \leq 5$.
(a) Graph the equation by finding the intercepts. Let t correspond to the horizontal axis (x-axis) and v correspond to the vertical axis (y-axis).
(b) Interpret each intercept.

Solution

(a) To find the *t*-intercept, let $v = 0$.

$$0 = 160 - 32t \quad \text{Let } v = 0.$$
$$32t = 160 \quad \text{Add } 32t \text{ to each side.}$$
$$t = 5 \quad \text{Divide each side by 32.}$$

To find the *v*-intercept, let $t = 0$.

$$v = 160 - 32(0) \quad \text{Let } t = 0.$$
$$v = 160 \quad \text{Simplify.}$$

Therefore the graph passes through the intercepts $(5, 0)$ and $(0, 160)$, as shown in **FIGURE 10.17**.

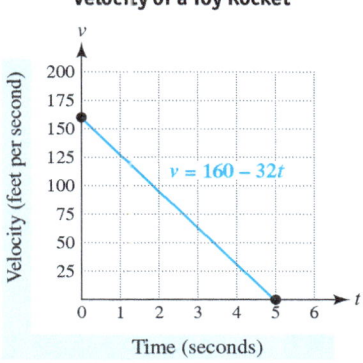

Velocity of a Toy Rocket

FIGURE 10.17

(b) The *t*-intercept indicates that the rocket had a velocity of 0 feet per second after 5 seconds. The *v*-intercept indicates that the rocket's initial velocity was 160 feet per second.

Now Try Exercise 77

3 Recognizing Horizontal Lines and Their Equations

 Connecting Concepts with Your Life Suppose that someone drives a car on a freeway at a constant speed of 70 miles per hour. **TABLE 10.14** shows the speed *y* after *x* hours.

Speed of a Car

x	1	2	3	4	5
y	70	70	70	70	70

TABLE 10.14

We can make a scatterplot of the data by plotting the points $(1, 70)$, $(2, 70)$, $(3, 70)$, $(4, 70)$, and $(5, 70)$, as shown in **FIGURE 10.18(a)**. The speed is always 70 miles per hour and the graph of the car's speed is a horizontal line, as shown in **FIGURE 10.18(b)**. The equation of this line is $y = 70$ with *y*-intercept $(0, 70)$. There are no *x*-intercepts.

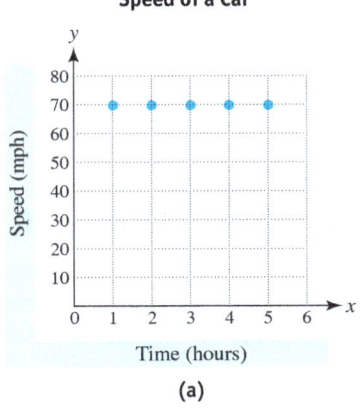
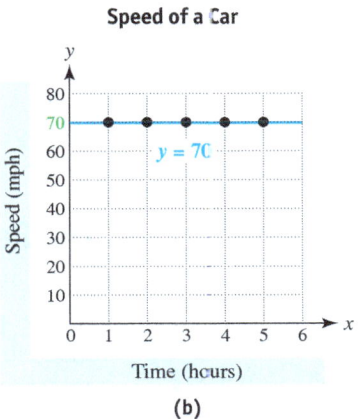

FIGURE 10.18

In general, the equation of a horizontal line is $y = b$, where *b* is a constant that corresponds to the *y*-coordinate of the *y*-intercept. Examples of horizontal lines are shown in **FIGURE 10.19** on the next page. Note that every point on the graph of $y = 3$ in **FIGURE 10.19(a)** has a *y*-coordinate of 3, and that every point on the graph of $y = -2$ in **FIGURE 10.19(b)** has a *y*-coordinate of -2.

Horizontal Lines: $y = b$

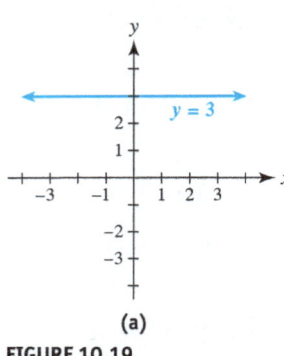
(a)

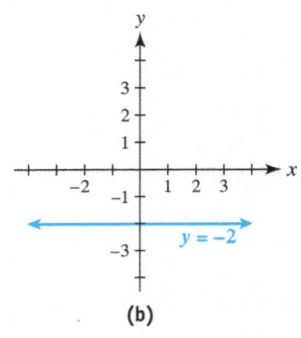
(b)

FIGURE 10.19

READING CHECK 3

- Which variable, *x* or *y*, is *not* present in the equation for a horizontal line?

HORIZONTAL LINE

The equation of a horizontal line with *y*-intercept $(0, b)$ is $y = b$.

EXAMPLE 4   **Graphing a horizontal line**

Graph the equation $y = -1$ and identify its *y*-intercept.

Solution
The graph of $y = -1$ is a horizontal line passing through the point $(0, -1)$, as shown in **FIGURE 10.20**. Its *y*-intercept is $(0, -1)$, and it has no *x*-intercept.

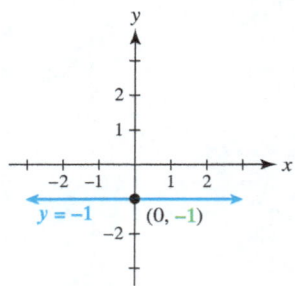

FIGURE 10.20

Now Try Exercise 45(a)

4   Recognizing Vertical Lines and Their Equations

Connecting Concepts with Your Life A parent looks at the number of text messages sent in one month by each of the four people on the family's data plan. **TABLE 10.15** shows the results, where *x* represents the number of months and *y* represents the number of text messages.

We can make a scatterplot of the data by plotting the points $(1, 1631)$, $(1, 12)$, $(1, 2314)$, and $(1, 359)$, as shown in **FIGURE 10.21(a)**. In each case the time is always 1 month and each point lies on the graph of a vertical line, as shown in **FIGURE 10.21(b)**. This vertical line has the equation $x = 1$ because each point on the line has an *x*-coordinate of 1 and there are no restrictions on the *y*-coordinate. This line has *x*-intercept $(1, 0)$ but no *y*-intercept.

Text Messages

x (months)	*y* (texts)
1	1631
1	12
1	2314
1	359

TABLE 10.15

10.3 INTERCEPTS; HORIZONTAL AND VERTICAL LINES

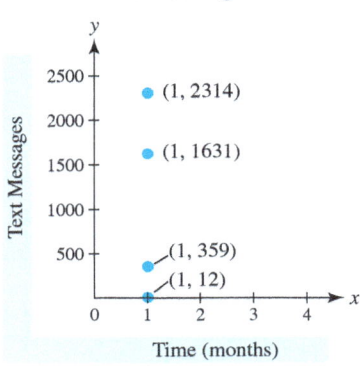
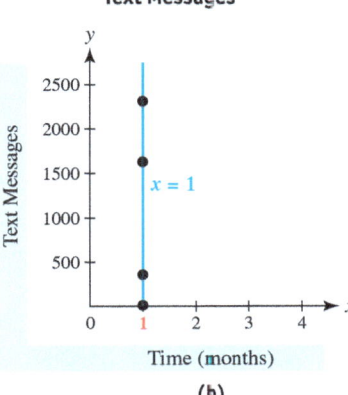

FIGURE 10.21

In general, the graph of a vertical line is $x = k$, where k is a constant that corresponds to the x-coordinate of the x-intercept. Examples of vertical lines are shown in **FIGURE 10.22**. Note that every point on the graph of $x = 3$ in **FIGURE 10.22(a)** has an x-coordinate of 3 and that every point on the graph of $x = -2$ shown in **FIGURE 10.22(b)** has an x-coordinate of -2.

Vertical Lines: $x = k$

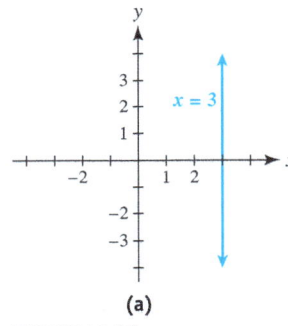
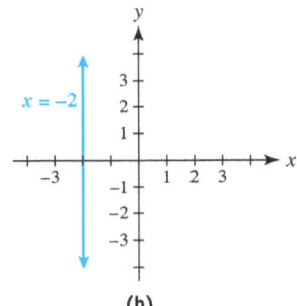

FIGURE 10.22

CALCULATOR HELP

To graph a vertical line, see Appendix A (page AP-5).

READING CHECK 4

- Which variable, x or y, is not present in the equation for a vertical line?

VERTICAL LINE

The equation of a vertical line with x-intercept $(k, 0)$ is $x = k$.

EXAMPLE 5 Graphing a vertical line

Graph the equation $x = -3$, and identify its x-intercept.

Solution
The graph of $x = -3$ is a vertical line passing through the point $(-3, 0)$, as shown in **FIGURE 10.23**. Its x-intercept is $(-3, 0)$ and it has no y-intercept.

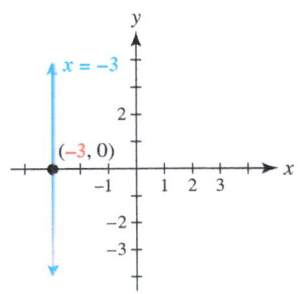

FIGURE 10.23

Now Try Exercise 45(b)

EXAMPLE 6 Writing equations of horizontal and vertical lines

Write the equation of the line shown in each graph.

(a)
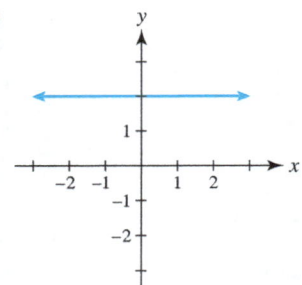

(b)
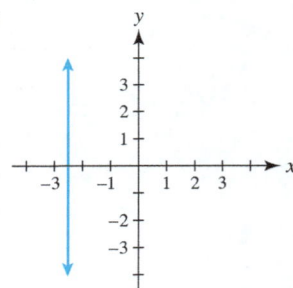

Solution
(a) The graph is a horizontal line with y-intercept $(0, 2)$. Its equation is $y = 2$.
(b) The graph is a vertical line with x-intercept $(-2.5, 0)$. Its equation is $x = -2.5$.

Now Try Exercises 51, 53

EXAMPLE 7 Writing equations of horizontal and vertical lines

Find an equation for a line satisfying the given conditions.
(a) Vertical, passing through $(2, -3)$
(b) Horizontal, passing through $(3, 1)$
(c) Perpendicular to $x = 3$, passing through $(-1, 2)$

Solution
(a) A vertical line passing through $(2, -3)$ has x-intercept $(2, 0)$ as shown in **FIGURE 10.24(a)**. The equation of a vertical line with x-intercept $(2, 0)$ is $x = 2$.
(b) A horizontal line passing through $(3, 1)$ has y-intercept $(0, 1)$, as shown in **FIGURE 10.24(b)**. The equation of a horizontal line with y-intercept $(0, 1)$ is $y = 1$.
(c) Because the line $x = 3$ is vertical, a line that is perpendicular to this line is horizontal, as shown in **FIGURE 10.24(c)**. The equation of a horizontal line passing through the point $(-1, 2)$ is $y = 2$.

Horizontal and Vertical Lines

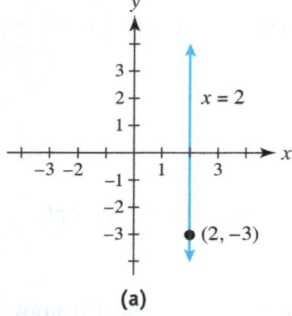
(a)

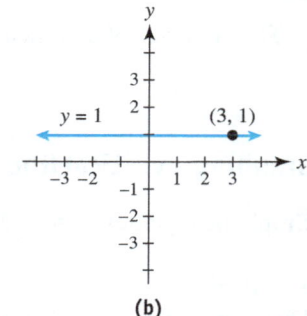
(b)

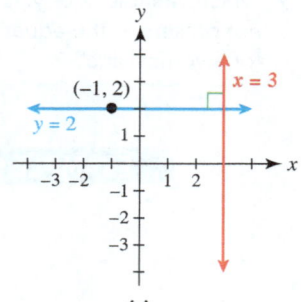
(c)

FIGURE 10.24

Now Try Exercises 65, 67, 69

The following See the Concept shows how any line can be written in the standard form $Ax + By = C$.

See the Concept 2 — STANDARD FORM: $Ax + By = C$

Every straight line in the xy-plane must be exactly one of the following.

Horizontal line ($A = 0, B \neq 0$)	Vertical line ($A \neq 0, B = 0$)	General (Slanted) Line ($A \neq 0, B \neq 0$)
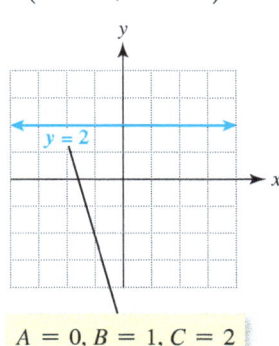 $A = 0, B = 1, C = 2$	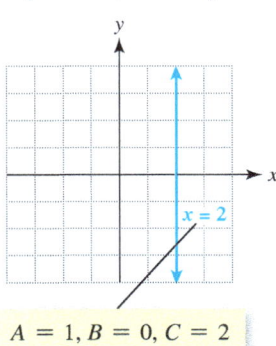 $A = 1, B = 0, C = 2$	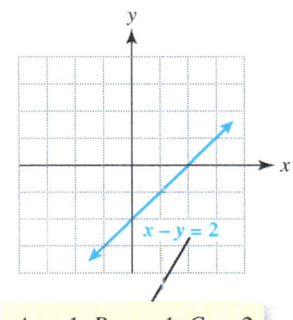 $A = 1, B = -1, C = 2$

The equation of *any* line can be written in the standard form $Ax + By = C$.

WATCH VIDEO IN MML.

10.3 Putting It All Together

CONCEPT	EXPLANATION	EXAMPLES
x- and y-Intercepts	A point at which a graph intersects the x-axis is called an x-intercept. A point at which a graph intersects the y-axis is called a y-intercept.	Graph showing y-intercept at $(0, 2)$ and x-intercept at $(-3, 0)$.
Finding Intercepts	To find x-intercepts, let $y = 0$ in the equation and solve for x. To find y-intercepts, let $x = 0$ in the equation and solve for y.	Let $4x - 5y = 20$. x-intercept: $\quad 4x - 5(0) = 20$ $\qquad\qquad\qquad x = 5$ y-intercept: $\quad 4(0) - 5y = 20$ $\qquad\qquad\qquad y = -4$ The x-intercept is $(5, 0)$ and the y-intercept is $(0, -4)$.
Horizontal Line	A horizontal line has equation $y = b$, where b is a constant. It also has y-intercept $(0, b)$ and no x-intercept when $b \neq 0$.	y-intercept $(0, b)$; line $y = b$.

continued on next page

612 CHAPTER 10 GRAPHING EQUATIONS

continued from previous page

CONCEPT	EXPLANATION	EXAMPLES
Vertical Line	A vertical line has equation $x = k$, where k is a constant. It also has x-intercept $(k, 0)$ and no y-intercept when $k \neq 0$.	*x*-intercept $(k, 0)$; graph of $x = k$

10.3 Exercises — MyMathLab

CONCEPTS AND VOCABULARY

1. A point where a graph intersects the x-axis is called a(n) _____.

2. To find an x-intercept, let $y =$ _____ and solve for x.

3. A point where a graph intersects the y-axis is called a(n) _____.

4. To find a y-intercept, let $x =$ _____ and solve for y.

5. The graph of the linear equation $Ax + By = C$ with $A = 0$ and $B = 1$ is a(n) _____ line.

6. A horizontal line with y-intercept $(0, b)$ has equation _____.

7. The graph of the linear equation $Ax + By = C$ with $A = 1$ and $B = 0$ is a(n) _____ line.

8. A vertical line with x-intercept $(k, 0)$ has equation _____.

FINDING INTERCEPTS

Exercises 9–16: Identify any x-intercepts and y-intercepts in the graph.

9.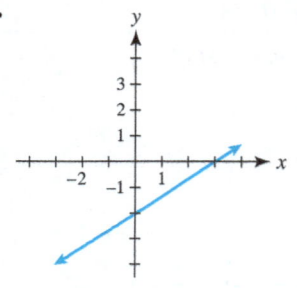

10.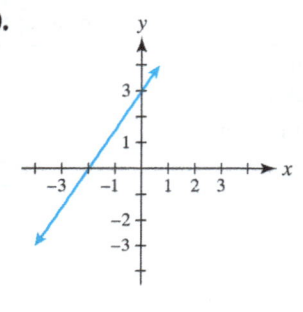

11.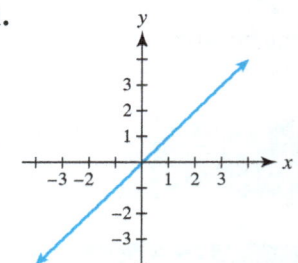

12.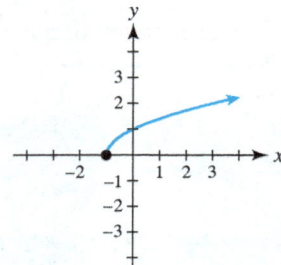

13.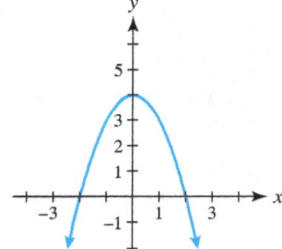

14.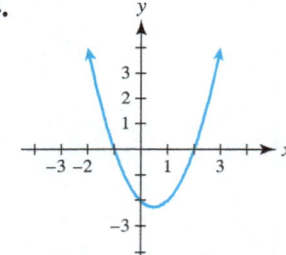

15.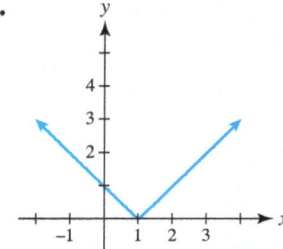

16.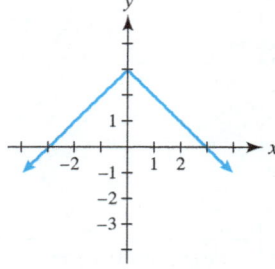

Exercises 17–20: Complete the table. Then determine the x-intercept and the y-intercept for the graph of the equation.

17. $y = x + 2$

x	−2	−1	0	1	2
y					

18. $y = 2x - 4$

x	−2	−1	0	1	2
y					

19. $-x + y = -2$

x	−4	−2	0	2	4
y					

20. $x + y = 1$

x	−2	−1	0	1	2
y					

USING INTERCEPTS TO GRAPH LINES

Exercises 21–38: Find any intercepts for the graph of the linear equation and then graph the equation.

21. $-2x + 3y = -6$ **22.** $4x + 3y = 12$

23. $x - 3y = 6$ **24.** $5x + y = -5$

25. $6x - y = -6$ **26.** $5x + 7y = -35$

27. $3x + 7y = 21$ **28.** $-3x + 8y = 24$

29. $40y - 30x = -120$ **30.** $10y - 20x = 40$

31. $\frac{1}{2}x - y = 2$ **32.** $x - \frac{1}{2}y = 4$

33. $-\frac{x}{4} + \frac{y}{3} = 1$ **34.** $\frac{x}{3} - \frac{y}{4} = 1$

35. $\frac{x}{3} + \frac{y}{2} = 1$ **36.** $\frac{x}{5} - \frac{y}{4} = 1$

37. $0.6y - 1.5x = 3$ **38.** $0.5y - 0.4x = 2$

39. Thinking Generally Find any intercepts for the graph of $Ax + By = C$.

40. Thinking Generally Find any intercepts for the graph of $\frac{x}{A} + \frac{y}{B} = 1$.

RECOGNIZING HORIZONTAL AND VERTICAL LINES

Exercises 41–44: Write an equation for the line that passes through the points shown in the table.

41.

x	−2	−1	0	1	2
y	1	1	1	1	1

42.

x	0	1	2	3	4
y	−10	−10	−10	−10	−10

43.

x	−6	−6	−6	−6	−6
y	5	4	3	2	1

44.

x	20	20	20	20	20
y	−2	−1	0	1	2

Exercises 45–50: Graph each equation.

45. (a) $y = 2$ (b) $x = 2$

46. (a) $y = -4$ (b) $x = -4$

47. (a) $y = 0$ (b) $x = 0$

48. (a) $y = -\frac{1}{2}$ (b) $x = -\frac{1}{2}$

49. (a) $y = \frac{3}{2}$ (b) $x = \frac{3}{2}$

50. (a) $y = -1.5$ (b) $x = -1.5$

Exercises 51–58: Write an equation for the line shown in the graph.

51. 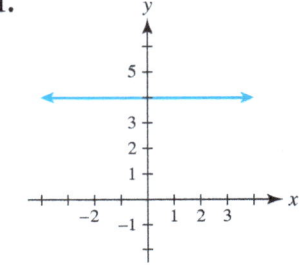 **52.**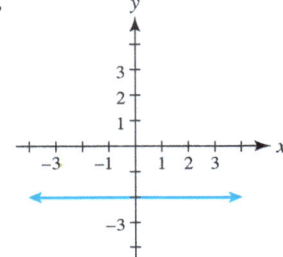

53. 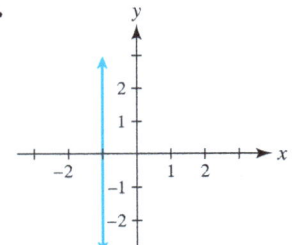 **54.**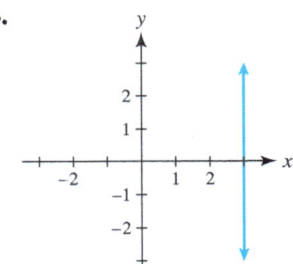

55. 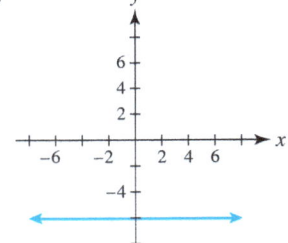 **56.**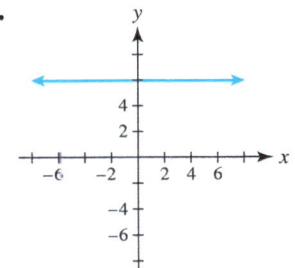

57. 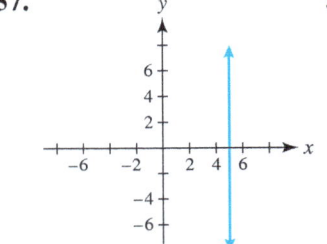 **58.**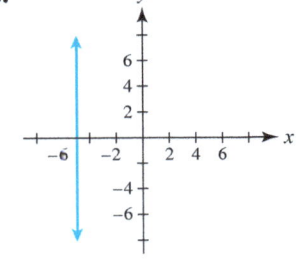

Exercises 59–64: Write the equations of a horizontal line and a vertical line that pass through the given point. (Hint: Make a sketch.)

59. $(1, 2)$ **60.** $(-3, 4)$

61. $(20, -45)$ **62.** $(-5, 12)$

63. $(0, 5)$ **64.** $(-3, 0)$

Exercises 65–72: Find an equation for a line satisfying the following conditions.

65. Vertical, passing through $(-1, 6)$

66. Vertical, passing through $(2, -7)$

67. Horizontal, passing through $\left(\frac{3}{4}, -\frac{5}{6}\right)$

68. Horizontal, passing through $(5.1, 6.2)$

69. Perpendicular to $y = \frac{1}{2}$, passing through $(4, -9)$

70. Perpendicular to $x = 2$, passing through $(3, 4)$

71. Parallel to $x = 4$, passing through $\left(-\frac{2}{3}, \frac{1}{2}\right)$

72. Parallel to $y = -2.1$, passing through $(7.6, 3.5)$

73. Thinking Generally Write the equation of the x-axis. (*Hint:* The x-axis is a horizontal line.)

74. Thinking Generally Write the equation of the y-axis. (*Hint:* The y-axis is a vertical line.)

APPLICATIONS INVOLVING LINES

Exercises 75 and 76: **Distance** *The distance of a driver from home is illustrated in the graph.*

(a) Find the intercepts.
(b) Interpret each intercept.

75. Distance from Home

76. Distance from Home

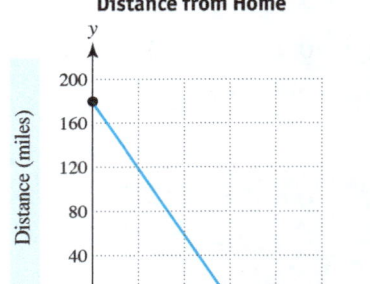

Exercises 77 and 78: **Modeling a Toy Rocket** *(Refer to Example 3.) The velocity v of a toy rocket in feet per second after t seconds of flight is given. Assume that $t \geq 0$ and t does not take on values greater than the t-intercept.*

(a) Find the intercepts and then graph the equation.
(b) Interpret each intercept.

77. $v = 128 - 32t$ **78.** $v = 96 - 32t$

Exercises 79 and 80: **Water in a Pool** *The amount of water in a swimming pool is depicted in the graph.*

(a) Find the intercepts.
(b) Interpret each intercept.

79. Water in a Pool

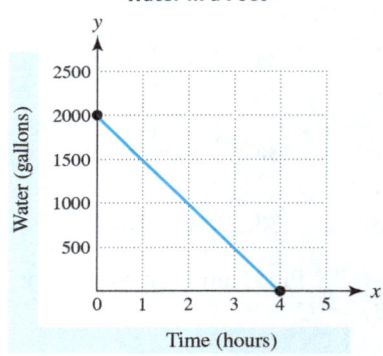

80. Water in a Pool

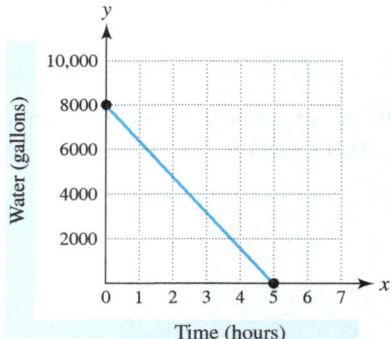

WRITING ABOUT MATHEMATICS

81. Given an equation, explain how to find an x-intercept and a y-intercept.

82. The form $\frac{x}{a} + \frac{y}{b} = 1$ is called the **intercept form** of a linear equation. Explain how you can use this equation to find the intercepts. (*Hint:* Graph $\frac{x}{2} + \frac{y}{3} = 1$ and find its intercepts.)

Group Activity

1. **Radio Stations** The approximate number of radio stations on the air for each decade from 1960 to 2010 is shown in the table.

x (year)	1960	1970	1980
y (stations)	4100	6800	8600

x (year)	1990	2000	2010
y (stations)	10,800	12,600	14,500

   Source: M. Street Corporation.

   Make a line graph of the data. Be sure to label both axes.

2. **Estimation** Discuss ways to estimate the number of radio stations on the air in 1975. Compare your estimates with the actual value of 7700 stations. Repeat this estimate for 1985 and compare it to the actual value of 10,400. Discuss your results.

3. **Modeling Equation** Substitute each x-value from the table into the equation $y = 220x - 427{,}100$ and determine the corresponding y-value. Do these y-values give reasonable approximations to the y-values in the table? Explain your answer.

4. **Making Estimates** Use $y = 220x - 427{,}100$ to estimate the number of radio stations on the air in 1975 and 1985. Compare the results to your answer in Exercise 2.

10.4 Slope and Rates of Change

Objectives

1. Introducing Slope
2. Recognizing Positive, Negative, Zero, and Undefined Slope
3. Finding Slopes of Lines
   - Using Slope to Sketch a Line
   - Using Slope to Complete a Table
4. Understanding Slope as a Rate of Change

NEW VOCABULARY

☐ Rise
☐ Run
☐ Slope
☐ Positive slope
☐ Negative slope
☐ Zero slope
☐ Undefined slope
☐ Rate of change

1 Introducing Slope

Take a moment to look at the graphs in **FIGURE 10.25**, where the horizontal axis represents time.

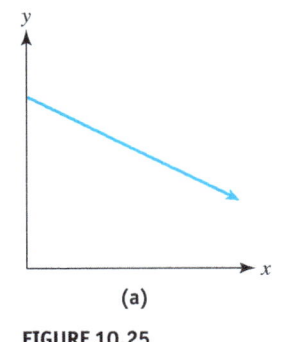
(a)

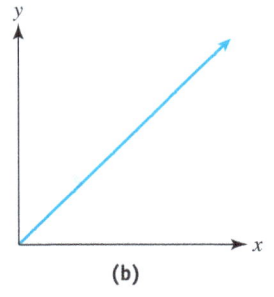
(b)

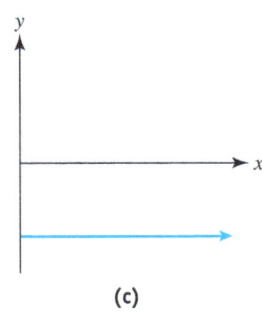
(c)

FIGURE 10.25

Which graph might represent:

- the distance traveled by you if you are walking at a constant rate?
- the temperature in your freezer?
- the value of a laptop computer after it is purchased?

To be able to answer these questions, you probably used the concept of slope. In mathematics, slope is a real number that measures the "tilt" of a line.

Connecting Concepts with Your Life Suppose you are using a stairway where each tread is 12 inches wide and the rise between steps is 6 inches high, as illustrated in **FIGURE 10.26(a)**. We say that this stairway has a *rise* of 6 inches for each 12 inches of *run*. Next, if we draw a blue line along this stairway and include an *x*- and *y*-axis, as shown in **FIGURE 10.26(b)**, then this blue line has a *slope* of $\frac{6}{12}$, or $\frac{1}{2}$. That is, the slope of a line along this stairway equals *rise over run*.

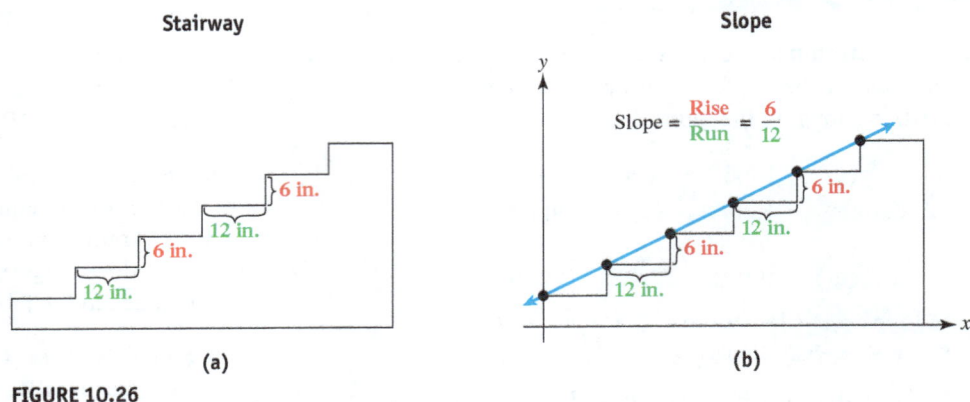

FIGURE 10.26

NOTE: For a line, the ratio $\frac{\text{rise}}{\text{run}}$ equals the slope and is always the same regardless of where one is on the line. ∎

2 Recognizing Positive, Negative, Zero, and Undefined Slope

By visually scanning across the graph of a line from *left to right*, we can determine if the graph has a positive slope, a negative slope, a slope of zero, or an undefined slope. The following See the Concept shows how this is done.

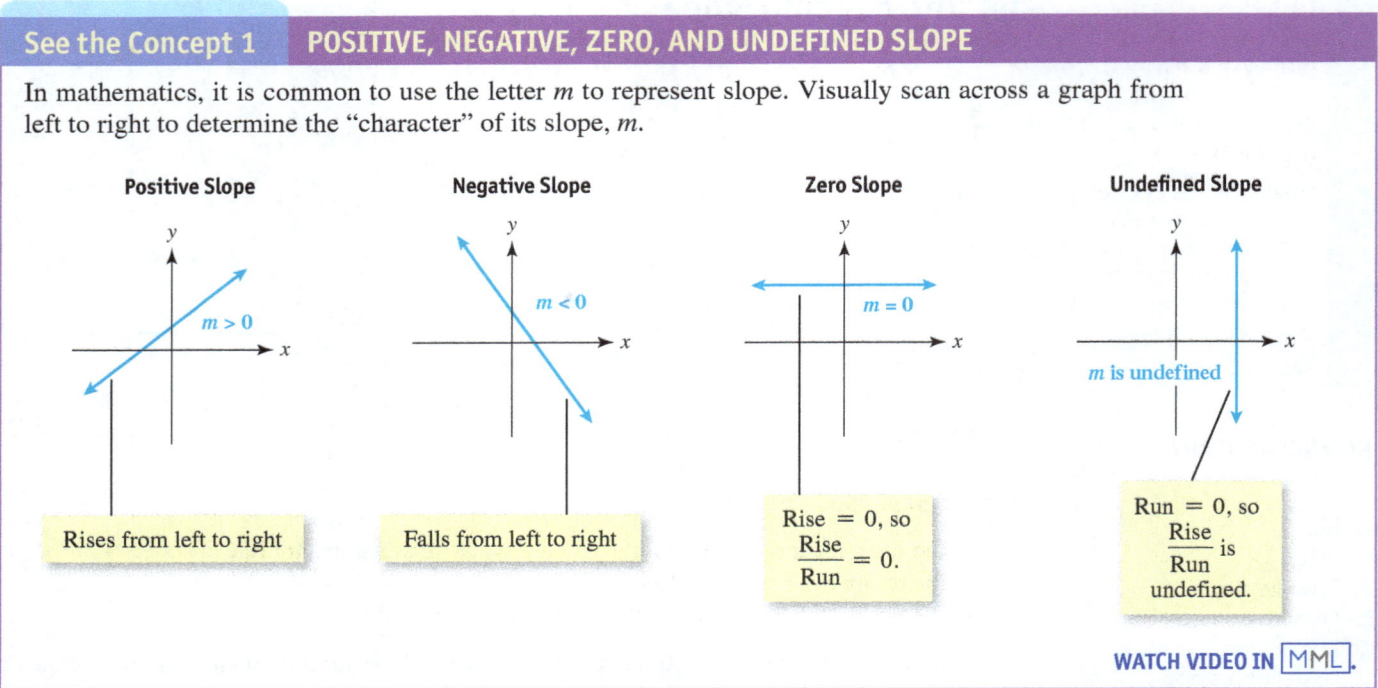

10.4 SLOPE AND RATES OF CHANGE

> **SLOPE OF A LINE**
>
> 1. A line that rises from left to right has **positive slope**.
> 2. A line that falls from left to right has **negative slope**.
> 3. A horizontal line has **zero slope**.
> 4. A vertical line has **undefined slope**.

3 Finding Slopes of Lines

 Connecting Concepts with Your Life The graph shown in **FIGURE 10.27** illustrates the cost of parking for x hours. The graph tilts upward from left to right, which indicates that the cost increases as the number of hours increases.

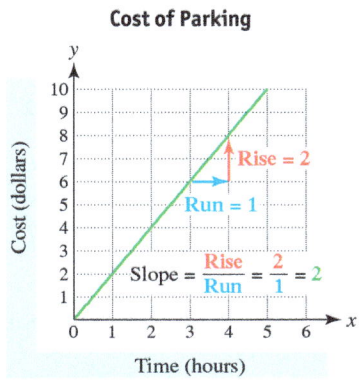

FIGURE 10.27

The graph *rises* 2 units vertically for every horizontal unit of *run*, and the ratio $\frac{\text{rise}}{\text{run}}$ equals the *slope* of the line. The slope m of this line is 2, which indicates that the cost of parking is $2 per hour. In applications, slope indicates a *rate of change*.

If we know the coordinates of any two points on a (non-vertical) line, we can calculate the slope of the line. A general case is shown in the following See the Concept, where a line passes through the points (x_1, y_1) and (x_2, y_2).

See the Concept 2 — **CALCULATING SLOPE USING TWO POINTS**

Any two *distinct* points on a *non-vertical* line can be used to find the line's slope.

Equivalent Slope Formulas

$$m = \frac{\text{Rise}}{\text{Run}}$$

$$m = \frac{\text{Change in } y}{\text{Change in } x}$$

$$m = \frac{y_2 - y_1}{x_2 - x_1} \quad (\text{where } x_1 \neq x_2)$$

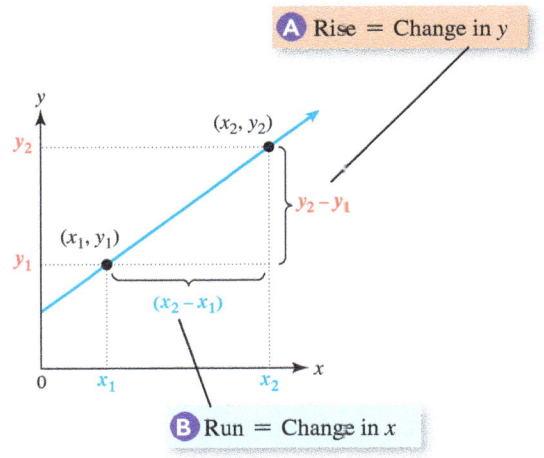

A The **rise** is found by calculating the **change in y**. To do this, subtract $y_2 - y_1$.

B The **run** is found by calculating the **change in x**. To do this, subtract $x_2 - x_1$.

WATCH VIDEO IN MML.

NOTE: The symbol x_1 has a *subscript* of 1 and is read "x sub one" or "x one". Thus x_1 and x_2 are used to denote two different x-values. Similar comments apply to y_1 and y_2. ∎

READING CHECK 1

- By looking at the x-coordinates of two points on a line, how can you tell if the line has undefined slope?

SLOPE

The **slope** m of the line passing through the points (x_1, y_1) and (x_2, y_2) is

$$m = \frac{\text{Rise}}{\text{Run}} = \frac{y_2 - y_1}{x_2 - x_1},$$

where $x_1 \neq x_2$. That is, slope equals *rise over run*.

NOTE: If $x_1 = x_2$, the line is vertical and the slope is undefined. ∎

EXAMPLE 1   Calculating the slope of a line

Use the two points labeled in **FIGURE 10.28** to find the slope of the line. What are the rise and run between these two points? Interpret the slope in terms of rise and run.

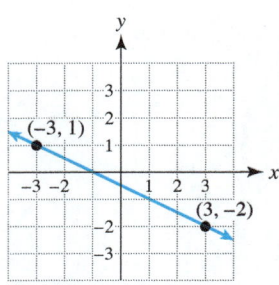

FIGURE 10.28

Solution
The line passes through the points $(-3, 1)$ and $(3, -2)$, so let $(x_1, y_1) = (-3, 1)$ and $(x_2, y_2) = (3, -2)$. The slope is

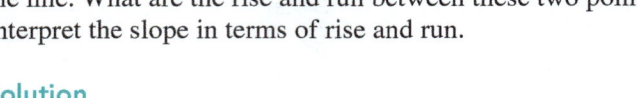
$$m = \frac{y_2 - y_1}{x_2 - x_1} = \frac{-2 - 1}{3 - (-3)} = \frac{-3}{6} = -\frac{1}{2}.$$

Starting at the point $(-3, 1)$, count 3 units downward and then 6 units to the right to return to the graph at the point $(3, -2)$. Thus the "rise" is -3 units and the run is 6 units. See **FIGURE 10.29(a)**. The ratio $\frac{\text{rise}}{\text{run}}$ is $\frac{-3}{6}$, or $-\frac{1}{2}$.

Finding Slope Moving Left to Right

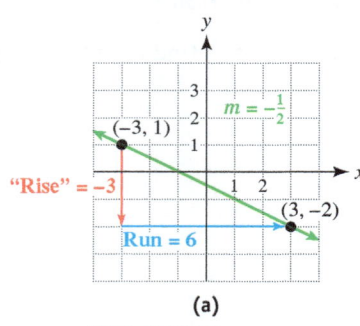

(a)

Finding Slope Moving Right to Left

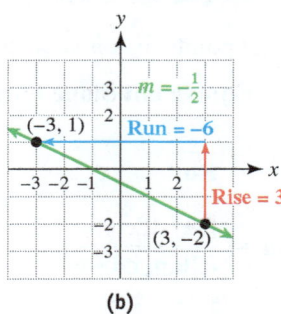

(b)

FIGURE 10.29

FIGURE 10.29 (b) shows an alternate way of finding this slope. Starting at the point $(3, -2)$, count 3 units upward and then 6 units to the left to return to the graph at the point $(-3, 1)$. Here, the rise is 3 units and the run is -6 units so that the ratio $\frac{\text{rise}}{\text{run}}$ is $\frac{3}{-6}$, or $-\frac{1}{2}$. In either case, the slope is $-\frac{1}{2}$.

Now Try Exercise 27

NOTE: In Example 1, the same slope would result if we let $(x_1, y_1) = (3, -2)$ and $(x_2, y_2) = (-3, 1)$. In this case, the calculation would be

$$m = \frac{y_2 - y_1}{x_2 - x_1} = \frac{1 - (-2)}{-3 - 3} = \frac{3}{-6} = -\frac{1}{2}.$$ ∎

A graph is not needed to find slope. Any two points on a line can be used to find the line's slope, as demonstrated in the next example.

EXAMPLE 2 Calculating the slope of a line

Calculate the slope of the line passing through each pair of points. Graph the line.
(a) $(-2, 3), (2, 1)$ (b) $(-1, 3), (2, 3)$ (c) $(-3, 3), (-3, -2)$

Solution

(a) $m = \frac{y_2 - y_1}{x_2 - x_1} = \frac{1 - 3}{2 - (-2)} = \frac{-2}{4} = -\frac{1}{2}$. This slope indicates that the line falls 1 unit for every 2 units of horizontal run, as shown in **FIGURE 10.30 (a)**.

(b) $m = \frac{y_2 - y_1}{x_2 - x_1} = \frac{3 - 3}{2 - (-1)} = \frac{0}{3} = 0$. The line is horizontal, as shown in **FIGURE 10.30 (b)**.

(c) Because $x_1 = x_2 = -3$, the slope formula does not apply. If we try to use it, we obtain $m = \frac{y_2 - y_1}{x_2 - x_1} = \frac{-2 - 3}{-3 - (-3)} = \frac{-5}{0}$, which is an **undefined** expression. The line has undefined slope and is vertical, as shown in **FIGURE 10.30 (c)**.

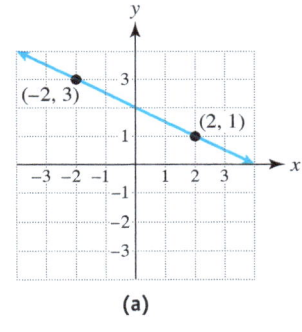
(a)

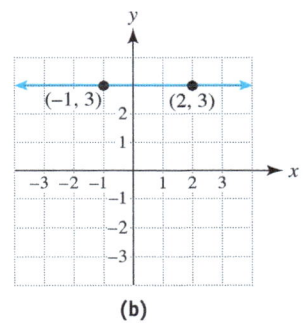
(b)

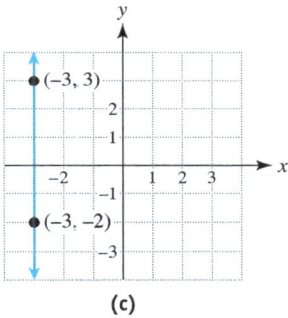
(c)

FIGURE 10.30

Now Try Exercises 41, 43, 45

EXAMPLE 3 Finding slope from a graph

Find the slope of each line.

(a)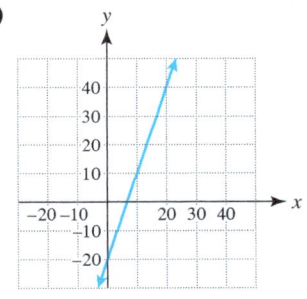

(b)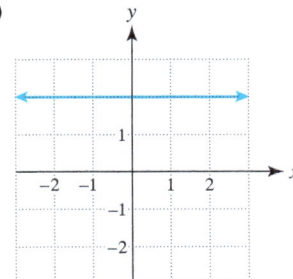

Solution

(a) The graph rises 30 units for every 10 units of run. For example, the graph passes through $(10, 10)$ and $(20, 40)$, so the line rises $40 - 10 = 30$ units while the run is $20 - 10 = 10$ units, as shown in **FIGURE 10.31**.

$$m = \frac{\text{Rise}}{\text{Run}} = \frac{30}{10} = 3$$

(b) The line is horizontal, so the slope is 0.

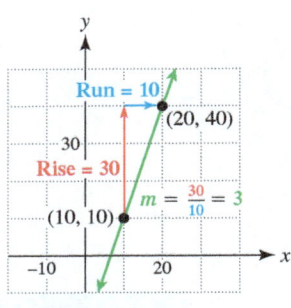

FIGURE 10.31

Now Try Exercises 19, 29

USING SLOPE TO SKETCH A LINE A point and a slope also determine a line, as illustrated in the next example.

EXAMPLE 4 Sketching a line with a given slope

Sketch a line that passes through the point $(1, 4)$ and has slope $-\frac{2}{3}$.

Solution
Start by plotting the point $(1, 4)$. A slope of $-\frac{2}{3}$ can be written as $\frac{-2}{3}$, which indicates that the line *falls* 2 units for every 3-unit increase in the run. Because the line passes through $(1, 4)$, a 2-unit decrease in y and a 3-unit increase in x result in the line passing through the point $(1 + 3, 4 - 2)$ or $(4, 2)$. See **FIGURE 10.32**.

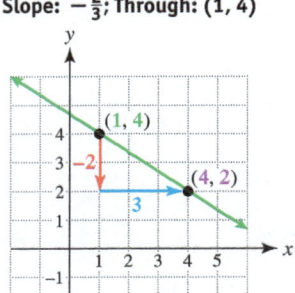

FIGURE 10.32

Now Try Exercise 59

EXAMPLE 5 Sketching a line with a given y-intercept

Sketch a line with slope -2 and y-intercept $(0, 3)$.

Solution
First plot the y-intercept, $(0, 3)$. The slope of -2 can be written as $\frac{-2}{1}$, so the y-values decrease 2 units for each unit increase in x. Increasing the x-value in the point $(0, 3)$ by 1 and decreasing the y-value by 2 result in the point $(0 + 1, 3 - 2)$ or $(1, 1)$. Plot $(1, 1)$ and then sketch the line, as shown in **FIGURE 10.33**.

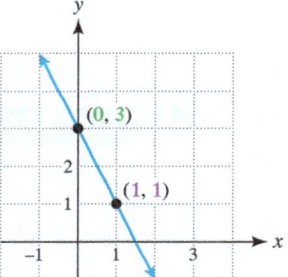

FIGURE 10.33

Now Try Exercise 55

USING SLOPE TO COMPLETE A TABLE If we know the slope of a line and a point on the line, we can complete a table of values that gives the coordinates of other points on the line, as demonstrated in the next example.

EXAMPLE 6 Completing a table of values

A line has slope 2 and passes through the first point listed in the table. Complete the table so that each point lies on the line.

x	−2	−1	0	1
y	1			

Complete the table so that the slope between consecutive points is 2.

Solution
Slope 2 indicates that $\frac{\text{Rise}}{\text{Run}} = 2$. Because consecutive x-values in the table increase by 1 unit, the run from one point in the table to the next is 1 unit. Substituting **1** for the **run** in the slope equation results in $\frac{\text{Rise}}{1} = 2$. Thus the **rise** is **2** and consecutive y-values shown in **TABLE 10.16** increase by 2 units.

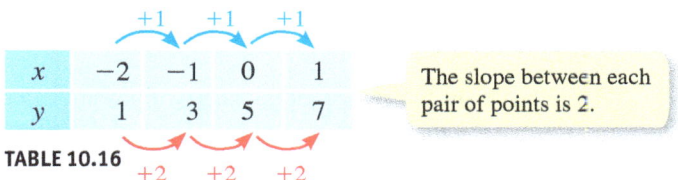

x	−2	−1	0	1
y	1	3	5	7

TABLE 10.16

The slope between each pair of points is 2.

Now Try Exercise 63

4 Understanding Slope as a Rate of Change

When lines are used to model physical quantities in applications, their slopes provide important information. Slope measures the **rate of change** in a quantity.

EXAMPLE 7 Interpreting slope

The distance y in miles that an athlete training for a marathon is from home after x hours is shown in **FIGURE 10.34**.
(a) Find the y-intercept. What does it represent?
(b) The graph passes through the point $(1, 10)$. Discuss the meaning of this point.
(c) Find the slope of this line. Interpret the slope as a rate of change.

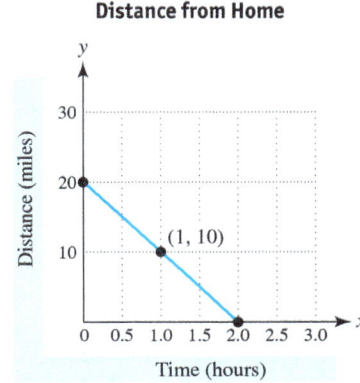

FIGURE 10.34

Solution
(a) The y-intercept is $(0, 20)$, so the athlete is initially 20 miles from home.
(b) The point $(1, 10)$ means that after 1 hour the athlete is 10 miles from home.

(c) The line passes through the points $(0, 20)$ and $(1, 10)$. Its slope is

$$m = \frac{10 - 20}{1 - 0} = -10.$$

Slope -10 means that the athlete is running *toward* home at 10 miles per hour. A *negative* slope indicates that the distance between the runner and home is decreasing.

Now Try Exercise 87

NOTE: The units for a rate of change are determined by putting the *y*-axis units over the *x*-axis units. In Example 7, the *y*-units are **miles** and the *x*-units are **hours**, so the units for the rate of change are $\frac{\text{miles}}{\text{hour}}$, or miles per hour. ∎

READING CHECK 2

- When a line models a real-world situation, how are the units of the slope found?

MAKING CONNECTIONS 1

Average Rates of Change

In real-world applications, lines are often used to model data that may not be perfectly linear. In such cases, the slope of the line that models the situation gives an *average* rate of change for the quantities involved. For example, **FIGURE 10.35** illustrates the actual number of words typed as a student writes a term paper. Because the student does not type at a *constant* rate, a line does not fit the situation perfectly. However, the line shown in **FIGURE 10.36** can be used to *model* the student's typing and its slope gives the *average* number of words typed per minute.

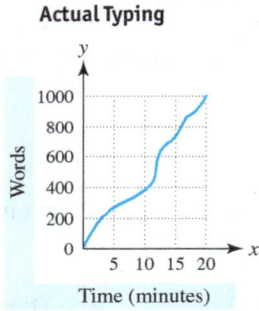

FIGURE 10.35   **FIGURE 10.36**

The student types at an *average* rate of **50** words per minute.

EXAMPLE 8   **Interpreting slope**

When a company manufactures 2000 MP3 players, its profit is $10,000, and when it manufactures 4500 MP3 players, its profit is $35,000.
(a) Find the slope of the line passing through $(2000, 10000)$ and $(4500, 35000)$.
(b) Interpret the slope as a rate of change.

Solution

(a) $m = \frac{35{,}000 - 10{,}000}{4500 - 2000} = \frac{25{,}000}{2500} = 10$

(b) When 2000 to 4500 MP3 players are manufactured, profit is, *on average*, $10 per MP3 player made and sold.

Now Try Exercise 91

🌐 *Math in Medical Context* Slope can be used to estimate *average rates of change* in the occurrences of a disease. In the next example, this concept is applied to tetanus.

EXAMPLE 9 Analyzing tetanus cases

TABLE 10.17 lists numbers of reported cases of tetanus in the United States for selected years.

Cases of Tetanus in the United States

Year	1960	1970	1980	1990	2000	2010
Cases of Tetanus	368	148	95	64	45	18

TABLE 10.17 *Source:* Department of Health and Human Services.

(a) Make a line graph of the data.
(b) Find the slope of each line segment.
(c) Interpret each slope as a rate of change.

Solution

(a) A line graph connecting the points $(1960, 368)$, $(1970, 148)$, $(1980, 95)$, $(1990, 64)$, $(2000, 45)$, and $(2010, 18)$ is shown in **FIGURE 10.37**.

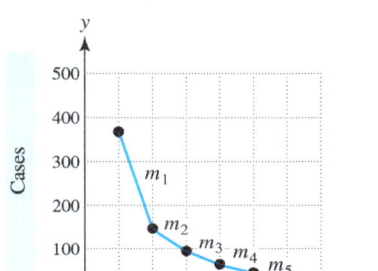

FIGURE 10.37

(b) The slope of each line segment may be calculated as follows.

$$m_1 = \frac{148 - 368}{1970 - 1960} = -22.0, \quad m_2 = \frac{95 - 148}{1980 - 1970} = -5.3,$$

$$m_3 = \frac{64 - 95}{1990 - 1980} = -3.1, \quad m_4 = \frac{45 - 64}{2000 - 1990} = -1.9$$

$$m_5 = \frac{18 - 45}{2010 - 2000} = -2.7$$

(c) Slope $m_1 = -22.0$ indicates that, *on average*, the number of tetanus cases *decreased* by 22.0 cases per year from 1960 to 1970. The other four slopes can be interpreted similarly.

NOTE: The number of tetanus cases did not decrease by *exactly* 22.0 cases each year between 1960 and 1970. However, the yearly *average* decrease was 22.0 cases. ∎

Now Try Exercise 89

EXAMPLE 10 Sketching a model

During a storm, rain falls at the average rates of 2 inches per hour from 1 A.M. to 3 A.M., 1 inch per hour from 3 A.M. to 4 A.M., and $\frac{1}{2}$ inch per hour from 4 A.M. to 6 A.M.
(a) Sketch a graph that shows the total accumulation of rainfall from 1 A.M. to 6 A.M.
(b) What does the slope of each line segment represent?

Solution

(a) At 1 A.M. the accumulated rainfall is 0, so place a point at $(1, 0)$. Rain falls at an average rate of 2 inches per hour for the next 2 hours, so at 3 A.M. the total rainfall is 4 inches. Place a point at $(3, 4)$. Sketch a line segment from $(1, 0)$ to $(3, 4)$, as shown in **FIGURE 10.38**. Similarly, during the next hour 1 inch of rain falls, so draw a line segment from $(3, 4)$ to $(4, 5)$. Finally, 1 inch of rain falls from 4 A.M. to 6 A.M., so draw a line segment from $(4, 5)$ to $(6, 6)$.

(b) The slope of each line segment represents the average rate at which rain is falling. For example, the first segment has slope 2 because rain falls at an average rate of 2 inches per hour during this period of time.

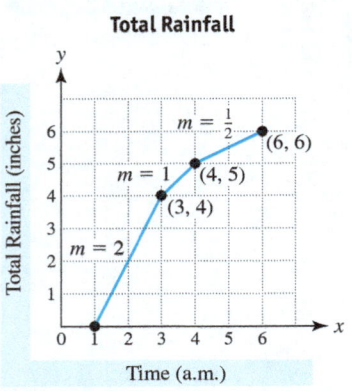

FIGURE 10.38

Now Try Exercise 81

10.4 Putting It All Together

CONCEPT	COMMENTS	EXAMPLE
Rise, Run, and Slope	Rise is a vertical change in a line, and run is a horizontal change in a line. The ratio $\frac{\text{rise}}{\text{run}}$ is the slope m when run is nonzero. 1. A line that rises from left to right has positive slope. 2. A line that falls from left to right has negative slope. 3. A horizontal line has zero slope. 4. A vertical line has undefined slope.	$m = \frac{\text{Rise}}{\text{Run}} = \frac{1}{2}$
Calculating Slope	For any two points (x_1, y_1) and (x_2, y_2), slope m is $$m = \frac{y_2 - y_1}{x_2 - x_1},$$ where $x_1 \neq x_2$.	The slope of the line passing through $(-2, 3)$ and $(1, 5)$ is $$m = \frac{5 - 3}{1 - (-2)} = \frac{2}{3}.$$ The line rises vertically 2 units for every 3 horizontal units of run.
Slope as a Rate of Change	Slope measures the "tilt" of a line. In applications, slope measures the rate of change in a quantity.	Slope $m = -66\frac{2}{3}$ indicates that water is *leaving* the pool at the rate of $66\frac{2}{3}$ gallons per hour. **Water in a Pool**

10.4 Exercises

CONCEPTS AND VOCABULARY

1. (True or False?) The change in the horizontal distance along a line is called run.

2. (True or False?) The change in the vertical distance along a line is called rise.

3. Slope m of a line equals _____ over _____.

4. Slope 0 indicates that a line is _____.

5. Undefined slope indicates that a line is _____.

6. If a line passes through (x_1, y_1) and (x_2, y_2) where $x_1 \neq x_2$, then $m =$ _____.

7. A line that rises from left to right has _____ slope.

8. A line that falls from left to right has _____ slope.

9. When a line models a physical quantity in an application, its slope measures the _____ of change in the quantity.

10. If a line that models the distance between a hiker and camp has a negative slope, is the hiker moving away from or toward camp?

RECOGNIZING SLOPE

Exercises 11–18: State whether the slope of the line is positive, negative, zero, or undefined.

11. 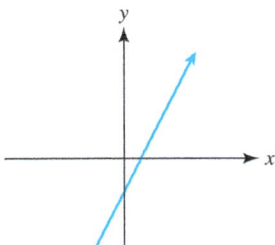 12.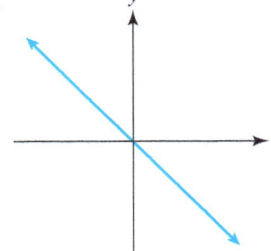

13. 14.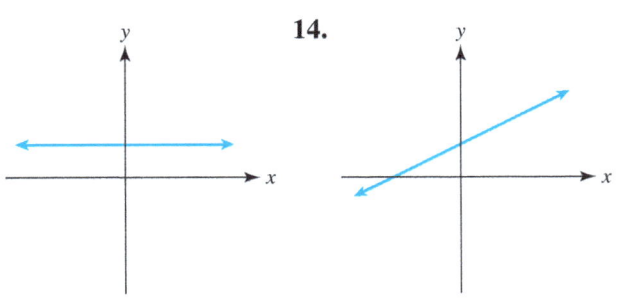

15. 16.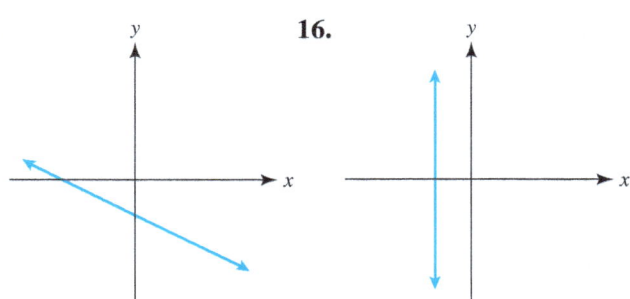

17. 18.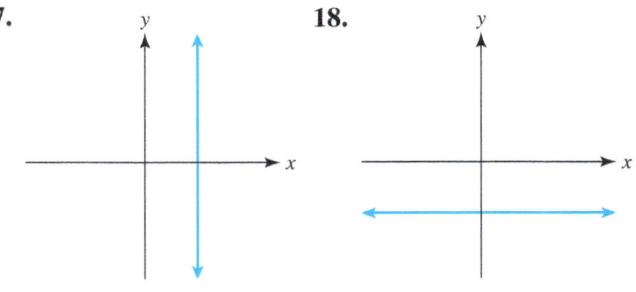

FINDING SLOPES OF LINES

Exercises 19–30: If possible, find the slope of the line. Interpret the slope in terms of rise and run.

19. 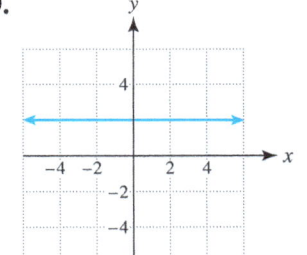 20.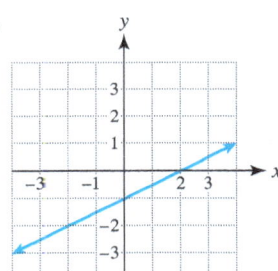

21. 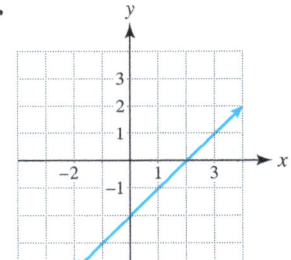 22.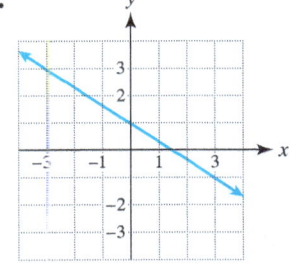

23. 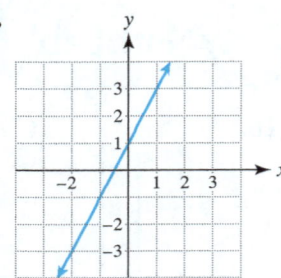 **24.**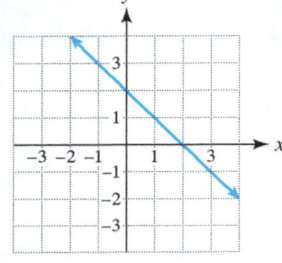

25. 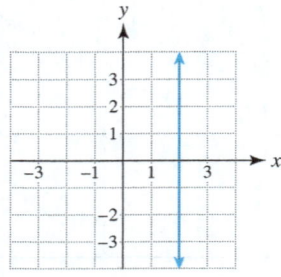 **26.**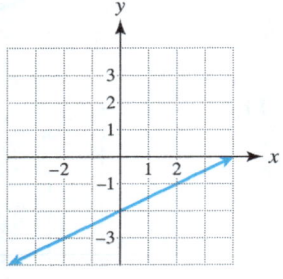

27. 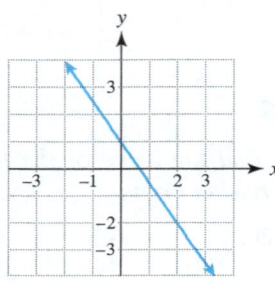 **28.**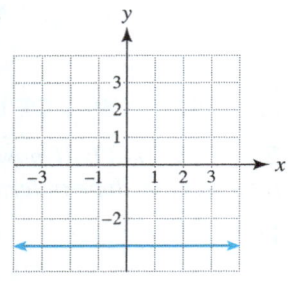

29. 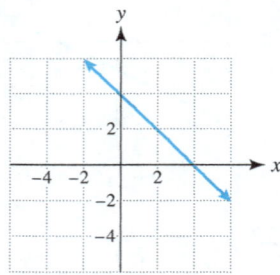 **30.**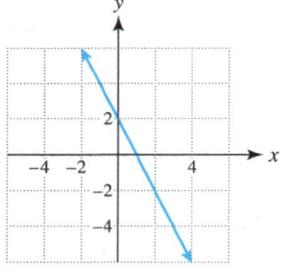

Exercises 31–38: If possible, find the slope of the line passing through the two points. Graph the line.

31. $(1, 2), (2, 4)$ **32.** $(-4, 7), (7, -4)$

33. $(2, 1), (2, 4)$ **34.** $(-1, 3), (-1, -1)$

35. $(1, 3), (-2, 5)$ **36.** $(0, -4), (-4, 6)$

37. $(2, -1), (-2, -1)$ **38.** $(-2, 3), (1, 3)$

Exercises 39–54: If possible, find the slope of the line passing through the two points.

39. $(4, -2), (-3, -9)$ **40.** $(15, -3), (20, 9)$

41. $(-3, 4), (4, -2)$ **42.** $(1, -3), (3, -5)$

43. $(-3, 5), (2, 5)$ **44.** $(-3, 3), (-5, 3)$

45. $(-1, 6), (-1, -4)$ **46.** $\left(\frac{1}{2}, -\frac{2}{7}\right), \left(\frac{1}{2}, \frac{13}{17}\right)$

47. $(1980, 5), (2000, 18)$

48. $(1989, 10), (1999, 16)$

49. $(1950, 6.1), (2000, 10.6)$

50. $(1900, 10), (1950, 35)$

51. $\left(\frac{1}{3}, -\frac{2}{7}\right), \left(-\frac{2}{3}, \frac{3}{7}\right)$

52. $(-1.3, 5.6), (-2.6, -2.5)$

53. $(12, -34), (14, 64)$

54. $(-25, 105), (60, 55)$

SKETCHING LINES

Exercises 55–62: (Refer to Example 4.) Sketch a line that passes through the point and has slope m.

55. $(0, 2), m = -1$ **56.** $(0, -1), m = 2$

57. $(1, 1), m = 3$ **58.** $(1, -1), m = -2$

59. $(-2, 3), m = -\frac{1}{2}$ **60.** $(-1, -2), m = \frac{3}{4}$

61. $(-3, 1), m = \frac{1}{2}$ **62.** $(-2, 2), m = -3$

COMPLETING TABLES

Exercises 63–68: (Refer to Example 6.) A line has the given slope m and passes through the first point listed in the table. Complete the table so that each point in the table lies on the line.

63. $m = 2$

x	0	1	2	3
y	-4			

64. $m = -\frac{1}{2}$

x	0	1	2	3
y	2			

65. $m = -3$

x	1	2	3	4
y	4			

66. $m = -1$

x	-1	0	1	2
y	10			

67. $m = \frac{3}{2}$

x	-4	-2	0	2
y	0			

68. $m = 3$

x	−2	0	2	4
y	−4			

UNDERSTANDING SLOPE AS A RATE OF CHANGE

Exercises 69–72: **Modeling** *Choose the graph (a.–d.) that models the situation best.*

69. Cost of buying x gum balls at a price of 25¢ each

70. Total number of computers with CD/DVD drives purchased during the past 5 years

71. Average cost of a new car over the past 30 years

72. Height of the Empire State Building after x people have entered it

a.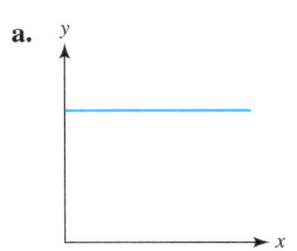

b.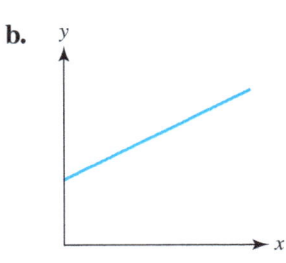

c.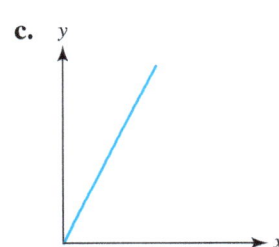

d.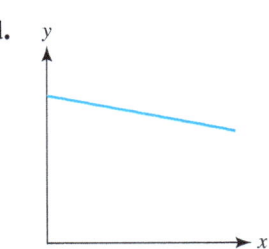

Exercises 73 and 74: **Modeling** *The line graph represents the gallons of water in a small swimming pool after x hours. Assume that there is a pump that can either add water to or remove water from the pool at a constant rate.*

(a) *Estimate the slope of each line segment.*
(b) *Interpret each slope as a rate of change.*
(c) *Describe what happened to the amount of water in the pool.*

73.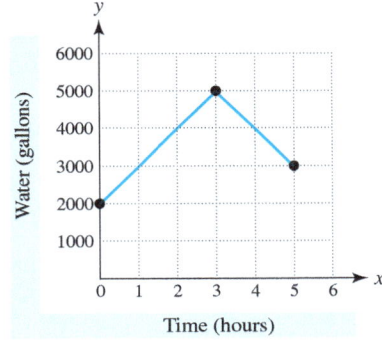

74.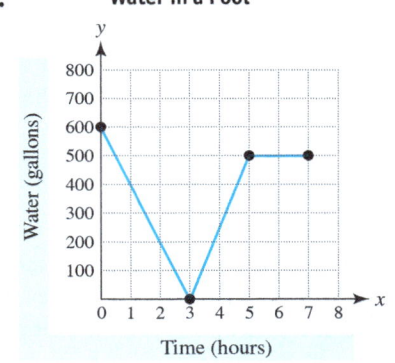

Exercises 75 and 76: **Modeling** *An individual is driving a car at a constant speed along a straight road. The graph shows the distance that the driver is from home after x hours.*

(a) *Find the slope of each line segment in the graph.*
(b) *Interpret each slope as a rate of change.*
(c) *Describe both the motion of the car and its distance from home.*

75.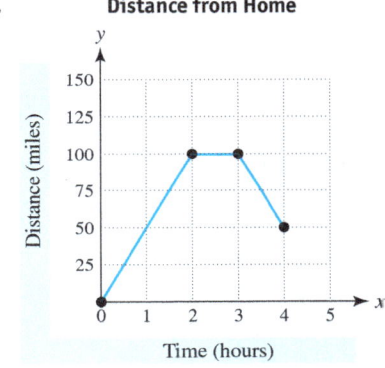

76.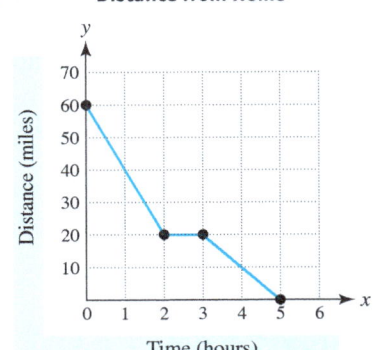

77. Thinking Generally If a line is used to model physical data where the x-axis is labeled "Time (minutes)" and the y-axis is labeled "Volume (cubic meters)," what are the units for the rate of change represented by the slope of the line?

78. Thinking Generally If a line is used to model physical data where the *x*-axis is labeled "Cookies" and the *y*-axis is labeled "Chocolate Chips," what are the units for the rate of change represented by the slope of the line?

Exercises 79–82: **Sketching a Model** *Sketch a graph that models the given situation.*

79. The distance that a boat is from a harbor if the boat is initially 6 miles from the harbor and arrives at the harbor after sailing at a constant speed for 3 hours

80. The distance that a person is from home if the person starts at home, walks away from home at 4 miles per hour for 90 minutes, and then walks back home at 3 miles per hour

81. The distance that an athlete is from home if the athlete jogs for 1 hour to a park that is 7 miles away, stays for 30 minutes, and then jogs home at the same pace

82. The amount of oil in a 55-gallon drum that is initially full, then is drained at a rate of 5 gallons per minute for 4 minutes, is left for 6 minutes, and then is emptied at a rate of 7 gallons per minute

83. Online Exploration Search the "Fast Facts" pages on the U.S. Census Bureau's Web site to complete the following.
(a) Rounding to the nearest million, what was the U.S. population in 1900?
(b) Rounding to the nearest million, what was the U.S. population in 2000?
(c) Using the concept of slope, find the *average* rate of change in the U.S. population over this time.

84. Online Exploration Search the "Fast Facts" pages on the U.S. Census Bureau's Web site to complete the following.
(a) Find the population of New York City in 1900. Round to the nearest hundred-thousand.
(b) Find the population of New York City in 2000. Round to the nearest hundred-thousand.
(c) Using the concept of slope, find the *average* rate of change in the New York City population over this time.

APPLICATIONS INVOLVING SLOPE

85. Twitter Followers One year after a celebrity first began posting comments on Twitter, she had 25,600 followers. Four years after this celebrity first began posting comments, she had 148,000 followers.
(a) Find the slope of the line that passes through the points $(1, 25600)$ and $(4, 148000)$.
(b) Interpret the slope as an average rate of change.

86. Profit from Tablet Computers When a company manufactures 500 tablet computers, its profit is $100,000, and when it manufactures 1500 tablet computers, its profit is $400,000.
(a) Find the slope of the line passing through the points $(500, 100000)$ and $(1500, 400000)$.
(b) Interpret the slope as an average rate of change.

87. Revenue The graph shows revenue received from selling *x* flash drives.

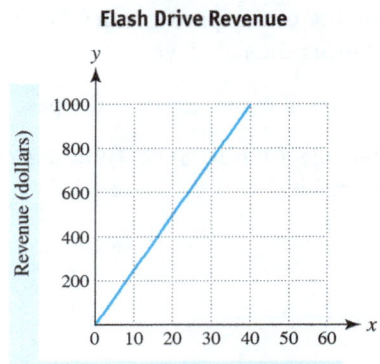

Flash Drive Revenue

(a) Find the slope of the line shown.
(b) Interpret the slope as a rate of change.

88. Electricity The graph shows how voltage is related to amperage in an electrical circuit. The slope corresponds to the resistance in ohms. Find the resistance in this electrical circuit.

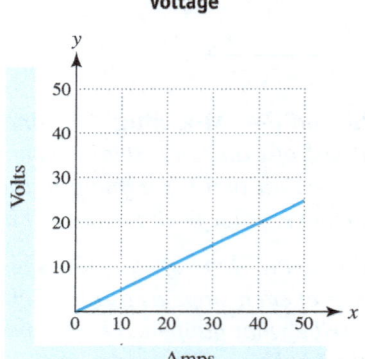

Voltage

89. Walking for Charities The table lists the amount of money *M* in dollars raised for walking various distances *x* in miles for a charity.

x	0	5	10	15
M	0	100	250	450

(a) Make a line graph of the data.
(b) Calculate the slope of each line segment.
(c) Interpret each slope as an average rate of change.

90. Insect Population The table lists the number N of black flies in thousands per acre after x weeks.

x	0	2	4	6
N	3	4	10	18

(a) Make a line graph of the data.
(b) Calculate the slope of each line segment.
(c) Interpret each slope as an average rate of change.

91. Median Family Income In 2000, median family income was about $42,000, and in 2012, it was about $51,000. (*Source: Department of the Treasury.*)
(a) Find the slope of the line passing through the points $(2000, 42000)$ and $(2012, 51000)$.
(b) Interpret the slope as an average rate of change.
(c) If this trend continues, estimate the median family income in 2018.

92. Social Media Ads In 2012, revenue from social media ads was $4.3 billion, and by 2017, it is projected to be $12 billion. (*Source: BIA/Kelsey.*)
(a) Find the slope of the line that passes through the points $(2012, 4.3)$ and $(2017, 12)$.
(b) Interpret the slope as an average rate of change.
(c) If this trend continues, estimate the revenue from ads in 2020.

93. Rate of Change Suppose that $y = -2x + 10$ is graphed in the first quadrant of the xy-plane where the x-axis is labeled "Time (minutes)" and the y-axis is labeled "Distance (feet)." If this graph represents the distance y that an ant is from a stone after x minutes, answer each of the following.
(a) Is the ant moving toward or away from the stone?
(b) Initially, how far from the stone is the ant?
(c) At what rate is the ant moving?
(d) What is the value of x (time) when the ant reaches the stone?

94. Rate of Change Suppose that we graph $y = 15x + 8$ in the first quadrant of the xy-plane where the x-axis is labeled "Time (minutes)" and the y-axis is labeled "Distance (feet)." If this graph represents the distance y that a frog is from a tree after x minutes, answer each of the following.
(a) Is the frog moving toward or away from the tree?
(b) Initially, how far from the tree is the frog?
(c) At what rate is the frog moving?
(d) What is the value of x when the frog is 53 feet from the tree?

WRITING ABOUT MATHEMATICS

95. If you are given two points and the slope formula $m = \frac{y_2 - y_1}{x_2 - x_1}$, does it matter which point is (x_1, y_1) and which point is (x_2, y_2)? Explain.

96. Suppose that a line approximates the distance y in miles that a person drives in x hours. What does the slope of the line represent? Give an example.

97. Describe the information that the slope m gives about a line. Be as complete as possible.

98. Could one line have two slopes? Explain.

SECTIONS 10.3 and 10.4 — Checking Basic Concepts

1. Identify the x- and y-intercepts in the graph.

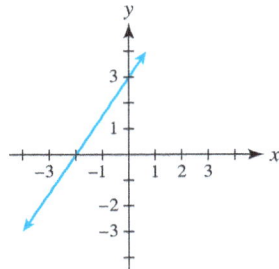

2. Complete the table for the equation $2x - y = 2$. Then determine the x- and y-intercepts.

x	-2	-1	0	1	2
y	-6				

3. Find any intercepts for the graphs of the equations and then graph each linear equation.
(a) $x - 2y = 6$ (b) $y = 2$ (c) $x = -1$

4. Write the equations of a horizontal line and a vertical line that pass through the point $(-2, 4)$.

5. If possible, find the slope of the line passing through each pair of points.
(a) $(-2, 3), (2, 6)$
(b) $(-5, 3), (0, 3)$
(c) $(1, 5), (1, 8)$

6. Find the slope of the line shown.

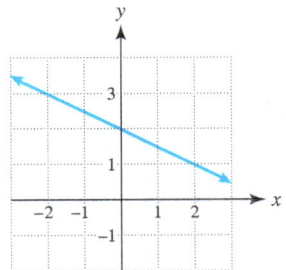

continued on next page

7. Sketch a line that passes through the point $(-3, 1)$ and has slope 2.

8. **Modeling** The line graph to the right shows the depth of water in a small pond before and after a rain storm.
   (a) Estimate the slope of each line segment.
   (b) Interpret each slope as a rate of change.
   (c) Describe what happened to the amount of water in the pond.

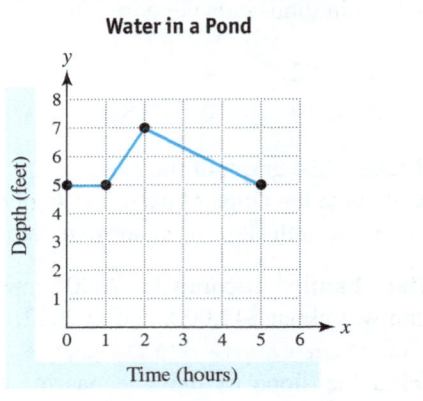

Water in a Pond

10.5 Slope–Intercept Form

Objectives

1. Determining a Line
2. Finding Slope–Intercept Form
3. Working with Parallel and Perpendicular Lines
   - Parallel Lines
   - Perpendicular Lines

1 Determining a Line

Connecting Concepts with Your Life When a line models a real-world situation, slope represents a rate of change and the y-intercept often represents an initial value. For example, if there are initially 30 gallons of water in a small wading pool, and the pool is being filled by a garden hose at a constant rate of 2 gallons per minute, a line representing this situation has a slope of 2 and the y-coordinate of its y-intercept is 30.

For any two points in the xy-plane, we can draw a unique line passing through them, as illustrated in **FIGURE 10.39(a)**. Another way we can determine a unique line is to know the y-intercept and the slope. For example, if a line has y-intercept $(0, 2)$ and slope $m = 1$, then the resulting line is shown in **FIGURE 10.39(b)**.

NEW VOCABULARY

☐ Slope–intercept form
☐ Negative reciprocals

STUDY TIP

A new concept is often easier to learn when we can find a relationship between the concept and our personal experience. Try to list two real-life situations that have an initial value and a constant rate of change.

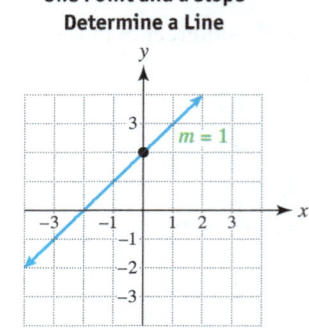

Two Points Determine a Line (a)

One Point and a Slope Determine a Line (b)

FIGURE 10.39

READING CHECK 1

- How many points does it take to determine a line?

2 Finding Slope–Intercept Form

The graph of $y = 2x + 3$ passes through $(0, 3)$ and $(1, 5)$, as shown in **FIGURE 10.40**. The slope of this line is

$$m = \frac{5 - 3}{1 - 0} = 2.$$

Slope 2, y-intercept (0, 3)

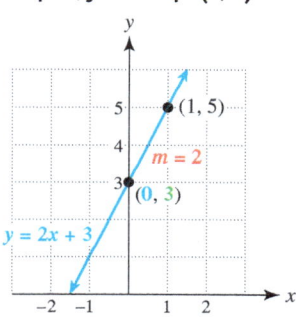

FIGURE 10.40

If $x = 0$ in $y = 2x + 3$, then $y = 2(0) + 3 = 3$. Thus the graph of $y = 2x + 3$ has slope 2 and y-intercept $(0, 3)$. In general, the graph of $y = mx + b$ has slope m and y-intercept $(0, b)$. The form $y = mx + b$ is called the *slope–intercept form*.

> ### SLOPE–INTERCEPT FORM
>
> The line with slope m and y-intercept $(0, b)$ is given by
>
> $$y = mx + b,$$
>
> the **slope–intercept form** of a line.

TABLE 10.18 shows several equations in the form $y = mx + b$ and lists the corresponding slope and y-intercept for the graph associated with each.

READING CHECK 2

- In $y = mx + b$, which value is the slope and what is the y-intercept?

	$y = mx + b$ Form	
Equation	Slope	y-intercept
$y = 4x - 3$	4	$(0, -3)$
$y = 12$	0	$(0, 12)$
$y = -x - \frac{5}{8}$	-1	$(0, -\frac{5}{8})$
$y = -\frac{2}{3}x$	$-\frac{2}{3}$	$(0, 0)$

TABLE 10.18

EXAMPLE 1 Using a graph to write the slope–intercept form

For each graph, write the slope–intercept form of the line.

(a)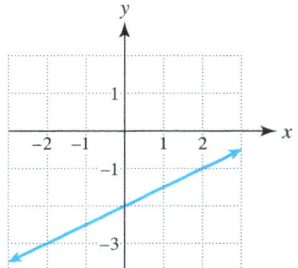

(b)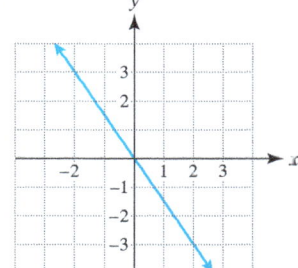

Solution

(a) The graph intersects the y-axis at the y-intercept, $(0, -2)$. Because the graph rises 1 unit for each 2-unit increase in x, the slope is $\frac{1}{2}$. The slope–intercept form of the line is $y = \frac{1}{2}x - 2$.

(b) The graph intersects the y-axis at the y-intercept, $(0, 0)$. Because the graph falls 3 units for each 2-unit increase in x, the slope is $-\frac{3}{2}$. The slope–intercept form of the line is $y = -\frac{3}{2}x + 0$, or $y = -\frac{3}{2}x$.

Now Try Exercises 15, 19

EXAMPLE 2 Sketching a line

Sketch a line with slope $-\frac{1}{2}$ and y-intercept $(0, 1)$. Write its slope–intercept form.

Solution

For the y-intercept, plot the point $(0, 1)$. Slope $-\frac{1}{2}$ indicates that the graph falls **1** unit for each **2**-unit increase in x. Thus the line passes through the point $(0 + 2, 1 - 1)$, or $(2, 0)$, as shown in **FIGURE 10.41**. The slope–intercept form of this line is $y = -\frac{1}{2}x + 1$.

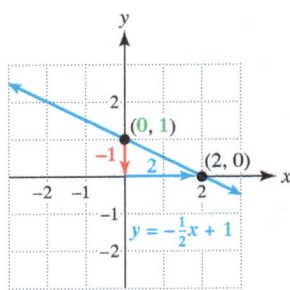

FIGURE 10.41

Now Try Exercise 23

When a linear equation is not given in slope-intercept form, the coefficient of the x-term may not represent the slope and the constant term may not represent the y-intercept. For example, the graph of $2x + 3y = 12$ does *not* have slope 2 and it does *not* have y-intercept $(0, 12)$. To find the correct slope and y-intercept, the equation can first be written in slope–intercept form. This is demonstrated in the next example.

EXAMPLE 3 Writing an equation in slope–intercept form

Write each equation in slope–intercept form. Then give the slope and y-intercept of the line.
(a) $2x + 3y = 12$ (b) $x = 2y + 4$

Solution
(a) To write the equation in slope–intercept form, solve for y.

$$2x + 3y = 12 \qquad \text{Given equation (standard form)}$$
$$3y = -2x + 12 \qquad \text{Subtract } 2x \text{ from each side.}$$
$$y = -\frac{2}{3}x + 4 \qquad \text{Divide each side by 3.}$$

The slope of the line is $-\frac{2}{3}$, and the y-intercept is $(0, 4)$.

(b) This equation is *not* in slope–intercept form because it is solved for x, not y.

$$x = 2y + 4 \qquad \text{Given equation}$$
$$x - 4 = 2y \qquad \text{Subtract 4 from each side.}$$
$$\frac{1}{2}x - 2 = y \qquad \text{Divide each side by 2.}$$
$$y = \frac{1}{2}x - 2 \qquad \text{Rewrite the equation.}$$

The slope of the line is $\frac{1}{2}$, and the y-intercept is $(0, -2)$.

Now Try Exercises 39, 41

EXAMPLE 4 Writing an equation in slope–intercept form and graphing it

Write the equation $2x + y = 3$ in slope–intercept form and then graph it.

Solution
First write the given equation in slope–intercept form.

$2x + y = 3$ Given equation (standard form)

$y = -2x + 3$ Subtract $2x$ from each side.

The slope–intercept form is $y = -2x + 3$, with slope -2 and y-intercept $(0, 3)$. To graph this equation, plot the y-intercept as the point $(0, 3)$. The line falls **2** units for each **1**-unit increase in x, so plot the point $(0 + 1, 3 - 2) = (1, 1)$. Sketch a line passing through $(0, 3)$ and $(1, 1)$, as shown in **FIGURE 10.42**.

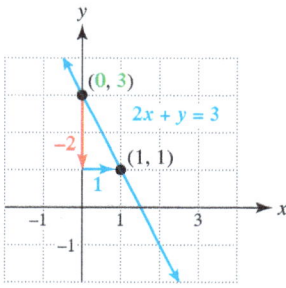

FIGURE 10.42

Now Try Exercise 51

MAKING CONNECTIONS 1

Two Ways to Graph the Same Line

In Example 4, the line $2x + y = 3$ was graphed by finding its slope–intercept form. A second way to graph this line is to find the x- and y-intercepts of the line. If $y = 0$, then $x = \frac{3}{2}$ makes this equation true, and if $x = 0$, then $y = 3$ makes this equation true. Thus the x-intercept is $(\frac{3}{2}, 0)$ and the y-intercept is $(0, 3)$. Note that the line in **FIGURE 10.42** passes through these points, which could be used to graph the line.

EXAMPLE 5 Modeling cell phone costs

Suppose that international calls with a cell phone cost $5 for the initial connection and $0.50 per minute.
(a) If someone talks for 23 minutes, what is the charge?
(b) Write the slope–intercept form that gives the cost of talking for x minutes.
(c) If the charge is $8.50, how long did the person talk?

Solution
(a) The charge for 23 minutes at $**0.50** per minute plus $**5** would be

$$0.50 \cdot 23 + 5 = \$16.50.$$

(b) The rate of increase is $0.50 per minute with an initial cost of $5. Let $y = 0.5x + 5$, where the slope or rate of change is **0.5** and the y-intercept is $(0, 5)$.

(c) To determine how long a person can talk for $8.50, we solve the following equation.

$$0.5x + 5 = 8.5 \quad \text{Equation to solve}$$
$$0.5x = 3.5 \quad \text{Subtract 5 from each side.}$$
$$x = \frac{3.5}{0.5} \quad \text{Divide each side by 0.5.}$$
$$x = 7 \quad \text{Simplify.}$$

The person talked for 7 minutes. Note that this solution is based on the assumption that the phone company did not round up a fraction of a minute.

Now Try Exercise 79

3 Working with Parallel and Perpendicular Lines

Slope is important for determining whether two lines are parallel or perpendicular.

PARALLEL LINES If two lines have the same slope, they are parallel. For example, the lines $y = 2x$ and $y = 2x - 1$ are parallel because they both have slope 2, as shown in **FIGURE 10.43**.

Parallel Lines Have the Same Slope

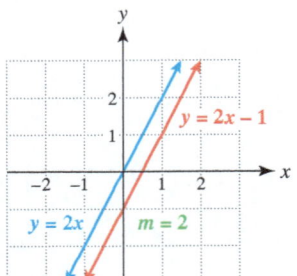

FIGURE 10.43

> **PARALLEL LINES**
>
> Two lines with the same slope are parallel.
> Two nonvertical parallel lines have the same slope.

NOTE: Two vertical lines are parallel and the slope of each is undefined. ∎

EXAMPLE 6 Finding parallel lines

Find the slope–intercept form of a line parallel to $y = -2x + 3$ and passing through the point $(-2, 3)$. Sketch each line in the same xy-plane.

Solution
Because the line $y = -2x + 3$ has slope -2, any parallel line also has slope -2 with slope–intercept form $y = -2x + b$ for some value of b. The value of b can be found by substituting the point $(-2, 3)$ in the slope–intercept form.

$$y = -2x + b \quad \text{Slope–intercept form}$$
$$3 = -2(-2) + b \quad \text{Let } x = -2 \text{ and } y = 3.$$
$$3 = 4 + b \quad \text{Multiply}$$
$$-1 = b \quad \text{Subtract 4 from each side.}$$

The y-intercept is $(0, -1)$, and so the slope–intercept form is $y = -2x - 1$. The graphs of the equations $y = -2x + 3$ and $y = -2x - 1$ are shown in **FIGURE 10.44**. Note that they are parallel lines, both with slope -2 but with different y-intercepts, $(0, 3)$ and $(0, -1)$.

Now Try Exercise 69

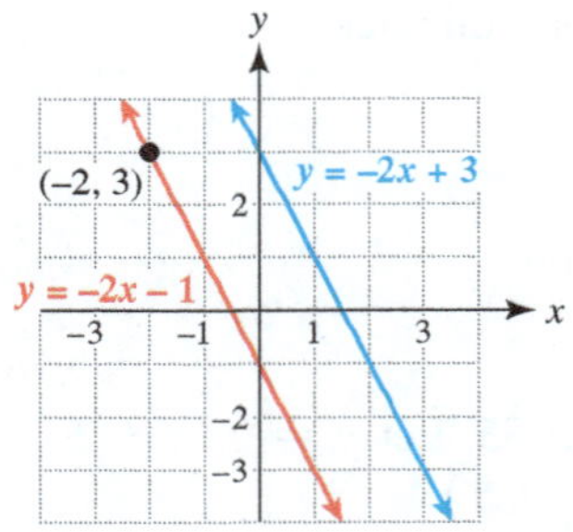

FIGURE 10.44

PERPENDICULAR LINES The lines shown in **FIGURE 10.45** are perpendicular because they intersect at a 90° angle. Rather than measure the angle between two intersecting lines, we can determine whether two lines are perpendicular from their slopes. The slopes of perpendicular lines satisfy the properties given in the box below **FIGURE 10.45**.

Perpendicular Lines

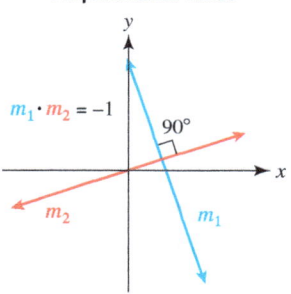

FIGURE 10.45

PERPENDICULAR LINES

If two perpendicular lines have nonzero slopes m_1 and m_2, then $m_1 \cdot m_2 = -1$.

If two lines have slopes m_1 and m_2 such that $m_1 \cdot m_2 = -1$, then they are perpendicular lines.

NOTE: A vertical line and a horizontal line are perpendicular. ∎

The following See the Concept shows how to determine if two lines are perpendicular by looking at their slopes.

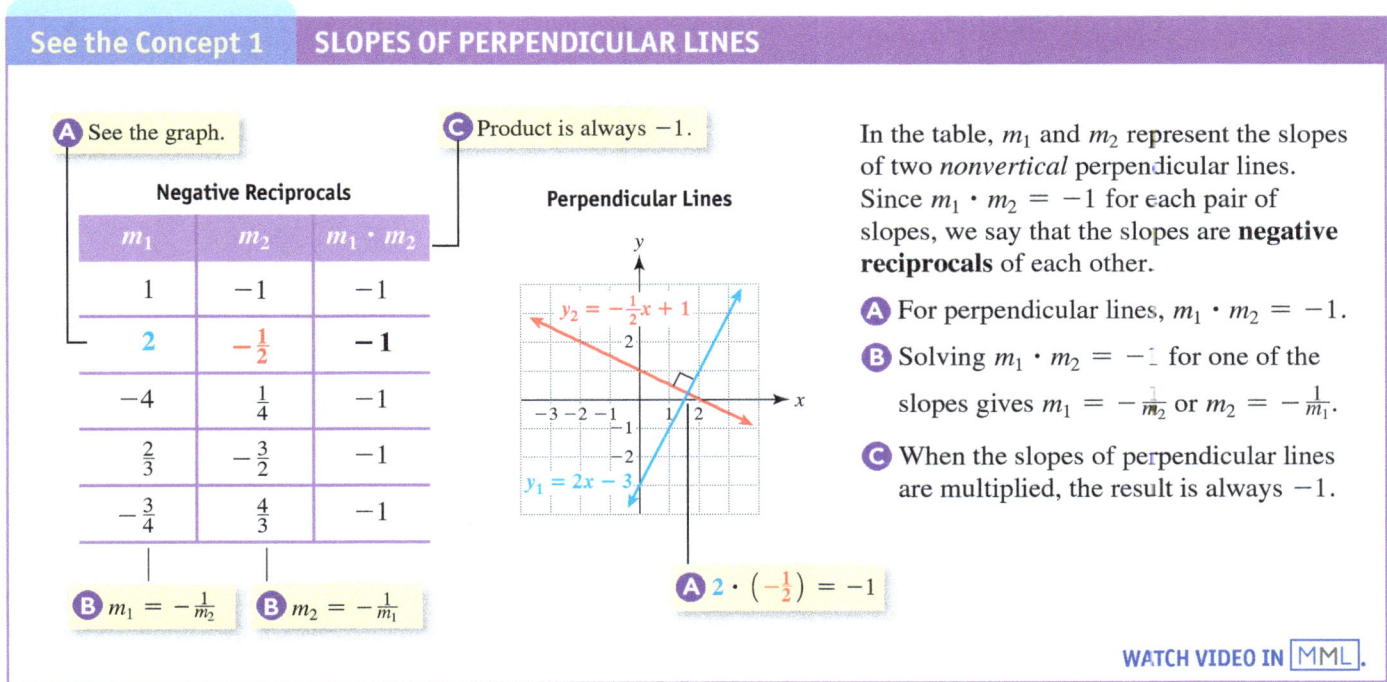

See the Concept 1 — SLOPES OF PERPENDICULAR LINES

Ⓐ See the graph.

Negative Reciprocals

m_1	m_2	$m_1 \cdot m_2$
1	−1	−1
2	$-\frac{1}{2}$	−1
−4	$\frac{1}{4}$	−1
$\frac{2}{3}$	$-\frac{3}{2}$	−1
$-\frac{3}{4}$	$\frac{4}{3}$	−1

Ⓑ $m_1 = -\frac{1}{m_2}$ Ⓑ $m_2 = -\frac{1}{m_1}$

Ⓒ Product is always −1.

Perpendicular Lines

$y_2 = -\frac{1}{2}x + 1$

$y_1 = 2x - 3$

Ⓐ $2 \cdot \left(-\frac{1}{2}\right) = -1$

In the table, m_1 and m_2 represent the slopes of two *nonvertical* perpendicular lines. Since $m_1 \cdot m_2 = -1$ for each pair of slopes, we say that the slopes are **negative reciprocals** of each other.

Ⓐ For perpendicular lines, $m_1 \cdot m_2 = -1$.

Ⓑ Solving $m_1 \cdot m_2 = -1$ for one of the slopes gives $m_1 = -\frac{1}{m_2}$ or $m_2 = -\frac{1}{m_1}$.

Ⓒ When the slopes of perpendicular lines are multiplied, the result is always −1.

WATCH VIDEO IN MML.

READING CHECK 3

- What does it mean for one number to be a negative reciprocal of another?

EXAMPLE 7 Finding negative reciprocals

Find the negative reciprocal of each number.
(a) $\frac{2}{3}$ (b) -6 (c) $-\frac{1}{4}$

Solution
(a) The negative reciprocal is found by inverting the fraction $\frac{2}{3}$ and finding its opposite, which results in $-\frac{3}{2}$. We can verify this result by showing that $\frac{2}{3} \cdot \left(-\frac{3}{2}\right) = -1$.
(b) After first writing -6 in fraction form as $-\frac{6}{1}$, invert the fraction and find its opposite. The reciprocal of $-\frac{6}{1}$ is $\frac{1}{6}$. Note that $-\frac{6}{1} \cdot \frac{1}{6} = -1$.
(c) Since $-\frac{1}{4} \cdot \frac{4}{1} = -1$, the negative reciprocal of $-\frac{1}{4}$ is $\frac{4}{1}$ or **4**.

Now Try Exercises 61, 63

We can use the concept of negative reciprocals to find equations of perpendicular lines, as illustrated in the next two examples.

EXAMPLE 8 Finding equations of perpendicular lines

For each of the given lines, find the slope–intercept form of a line passing through the origin that is perpendicular to the given line.
(a) $y = 3x$ (b) $y = -\frac{2}{5}x + 5$ (c) $-3x + 4y = 24$

Solution
(a) If a line passes through the origin, then its y-intercept is $(0, 0)$ and its slope–intercept form is $y = mx$. The given line $y = 3x$ has slope $m_1 = 3$, so a line perpendicular to it has a slope that is the negative reciprocal of 3.

$$m_2 = -\frac{1}{m_1} = -\frac{1}{3} = -\frac{1}{3}$$

The required slope–intercept form is $y = -\frac{1}{3}x$.

(b) The given line $y = -\frac{2}{5}x + 5$ has slope $m_1 = -\frac{2}{5}$, so a line perpendicular to it has slope $m_2 = \frac{5}{2}$ because $-\frac{2}{5} \cdot \frac{5}{2} = -1$. The required slope–intercept form is $y = \frac{5}{2}x$.

(c) To find the slope of the given line, write $-3x + 4y = 24$ in slope–intercept form.

$$-3x + 4y = 24 \qquad \text{Given equation}$$
$$4y = 3x + 24 \qquad \text{Add } 3x \text{ to each side.}$$
$$y = \frac{3}{4}x + 6 \qquad \text{Divide each side by 4.}$$

The slope of the given line is $m_1 = \frac{3}{4}$, so a line perpendicular to it has slope $m_2 = -\frac{4}{3}$. The required slope–intercept form is $y = -\frac{4}{3}x$.

Now Try Exercise 73

EXAMPLE 9 Finding a perpendicular line equation

Find the slope–intercept form of the line perpendicular to $y = -\frac{1}{2}x + 1$ and passing through the point $(1, -1)$. Sketch each line in the same xy-plane.

Solution

The line $y = -\frac{1}{2}x + 1$ has slope $m_1 = -\frac{1}{2}$. Any line perpendicular to it has slope $m_2 = 2$ (because $-\frac{1}{2} \cdot 2 = -1$) with slope–intercept form $y = 2x + b$ for some value of b. The value of b can be found by substituting the point $(1, -1)$ in the slope–intercept form.

$$y = 2x + b \qquad \text{Slope–intercept form}$$
$$-1 = 2(1) + b \qquad \text{Let } x = 1 \text{ and } y = -1.$$
$$-1 = 2 + b \qquad \text{Multiply.}$$
$$-3 = b \qquad \text{Subtract 2 from each side.}$$

The slope–intercept form is $y = 2x - 3$. The graphs of $y = -\frac{1}{2}x + 1$ and $y = 2x - 3$ are shown in **FIGURE 10.46**. Note that the point $(1, -1)$ lies on the graph of $y = 2x - 3$.

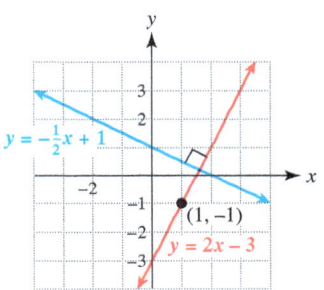

FIGURE 10.46

Now Try Exercise 75

10.5 Putting It All Together

CONCEPT	COMMENTS	EXAMPLE
Slope–Intercept Form $y = mx + b$	A unique equation for a line, determined by the slope m and the y-intercept $(0, b)$	The equation of the line with slope $m = 3$ and y-intercept $(0, -5)$ is $y = 3x - 5$.
Parallel Lines	$y = m_1x + b_1$ and $y = m_2x + b_2$, where $m_1 = m_2$ Nonvertical parallel lines have the same slope. Two vertical lines are parallel.	The lines $y = 2x - 1$ and $y = 2x + 2$ are parallel because they both have slope 2.

continued on next page

continued from previous page

CONCEPT	COMMENTS	EXAMPLE
Perpendicular Lines	$y = m_1 x + b_1$ and $y = m_2 x + b_2$, where $m_1 m_2 = -1$ Perpendicular lines that are neither vertical nor horizontal have slopes whose product equals -1. A vertical line and a horizontal line are perpendicular.	The lines $y = 3x - 1$ and $y = -\frac{1}{3}x + 2$ are perpendicular because $m_1 m_2 = 3\left(-\frac{1}{3}\right) = -1.$

10.5 Exercises

CONCEPTS AND VOCABULARY

1. The slope–intercept form of a line is _____.

2. In the slope–intercept form of a line, m represents the _____ of the line.

3. In the slope–intercept form of a line, b represents the _____ of the line.

4. If $b = 0$ in the slope–intercept form of a line, then its graph passes through the _____.

5. Two lines with the same slope are _____.

6. Two nonvertical parallel lines have the same _____.

7. If m_1 and m_2 are the slopes of two lines where $m_1 \cdot m_2 = -1$, the lines are _____.

8. If two perpendicular lines have nonzero slopes, the slopes are negative _____ of each other.

DETERMINING A LINE

Exercises 9–14: Match the description of the line with its graph (a.–f.).

9. A line with positive slope and negative y-coordinate of the y-intercept

10. A line with positive slope and positive y-coordinate of the y-intercept

11. A line with negative slope and y-coordinate of the y-intercept 0

12. A line with negative slope and nonzero y-coordinate of the y-intercept

13. A line with no x-intercept

14. A line with no y-intercept

a.
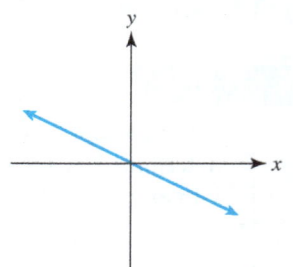

b.
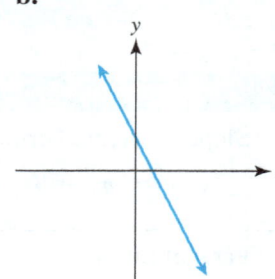

c.
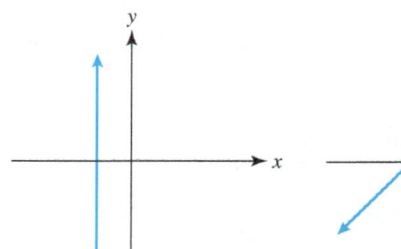

d.
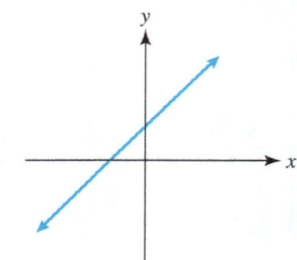

10.5 SLOPE–INTERCEPT FORM

e.

f.

Exercises 23–32: Sketch a line with the given slope m and y-coordinate of the y-intercept b. Write the slope–intercept form of the line.

23. $m = 1, b = 2$ **24.** $m = -1, b = 3$

25. $m = 2, b = -1$ **26.** $m = -3, b = 2$

27. $m = -\frac{1}{2}, b = -2$ **28.** $m = -\frac{2}{3}, b = 0$

29. $m = \frac{1}{3}, b = 0$ **30.** $m = 3, b = -3$

31. $m = 0, b = 3$ **32.** $m = 0, b = -3$

FINDING SLOPE–INTERCEPT FORM

Exercises 15–22: Write the slope–intercept form for the line shown in the graph.

Exercises 33–44: Do the following.

(a) Write the equation in slope–intercept form.
(b) Give the slope and y-intercept of the line.

33. $x + y = 4$ **34.** $x - y = 6$

15. 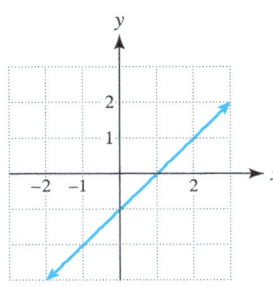 **16.**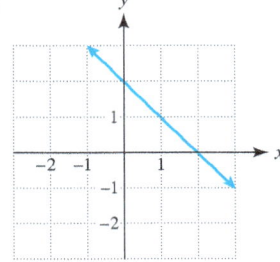

35. $2x + y = 4$ **36.** $-4x + y = 8$

37. $x - 2y = -4$ **38.** $x + 3y = -9$

39. $2x - 3y = 6$ **40.** $4x + 5y = 20$

41. $x = 4y - 6$ **42.** $x = -3y + 2$

43. $\frac{1}{2}x + \frac{3}{2}y = 1$ **44.** $-\frac{3}{4}x + \frac{1}{2}y = \frac{1}{2}$

Exercises 45–56: Graph the equation.

17. 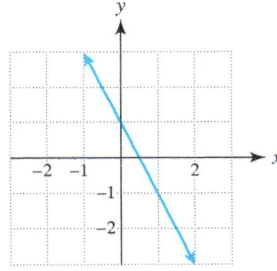 **18.**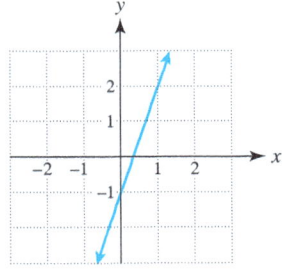

45. $y = -3x + 2$ **46.** $y = \frac{1}{2}x - 1$

47. $y = \frac{1}{3}x$ **48.** $y = -2x$

49. $y = 2$ **50.** $y = -3$

51. $x + y = 3$ **52.** $-\frac{2}{3}x + y = -2$

53. $x + 2y = 2$ **54.** $-2x - y = -2$

19. 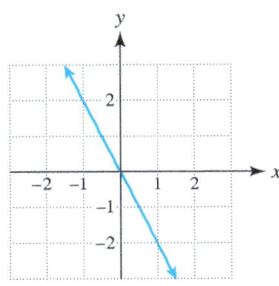 **20.**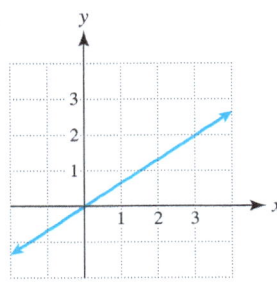

55. $x = 2 - y$ **56.** $x = -\frac{1}{3}y + \frac{2}{3}$

Exercises 57–60: The table shows points that all lie on the same line. Find the slope–intercept form for the line.

57.

x	0	1	2
y	2	4	6

58.

x	-1	0	1
y	4	8	12

59.

x	-2	0	2
y	-4	-2	0

60.

x	0	2	4
y	6	3	0

21. 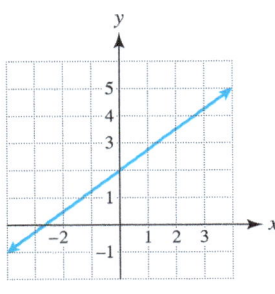 **22.**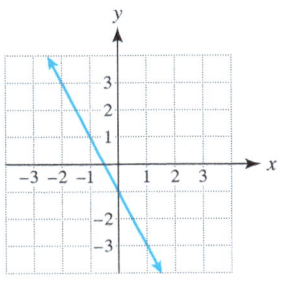

WORKING WITH PARALLEL AND PERPENDICULAR LINES

Exercises 61–66: Find the negative reciprocal of the number.

61. 5 **62.** −4

63. $-\frac{4}{5}$ **64.** $\frac{7}{5}$

65. −0.25 **66.** 0.4

Exercises 67–76: Find the slope–intercept form of the line satisfying the given conditions.

67. Slope $\frac{4}{7}$, y-intercept $(0, 3)$

68. Slope $-\frac{1}{2}$, y-intercept $(0, -7)$

69. Parallel to $y = 3x + 1$, passing through $(0, 0)$

70. Parallel to $y = -2x$, passing through $(0, 1)$

71. Parallel to $2x + 4y = 5$, passing through $(1, 2)$

72. Parallel to $-x - 3y = 9$, passing through $(-3, 1)$

73. Perpendicular to $y = -\frac{1}{2}x - 3$, passing through $(0, 0)$

74. Perpendicular to $y = \frac{3}{4}x - \frac{1}{2}$, passing through $(3, -2)$

75. Perpendicular to $x = -\frac{1}{3}y$, passing through $(-1, 0)$

76. Perpendicular to $6x - 3y = 18$, passing through $(4, -3)$

APPLICATIONS INVOLVING SLOPE–INTERCEPT FORM

77. **Rental Cars** Driving a rental car x miles costs $y = 0.25x + 25$ dollars.
    (a) How much would it cost to rent the car but not drive it?
    (b) How much does the cost increase for each additional mile driven?
    (c) What is the y-intercept of $y = 0.25x + 25$? What does it represent?
    (d) What is the slope of $y = 0.25x + 25$? What does it represent?

78. **Calculating Rainfall** The total rainfall y in inches that fell x hours past noon is given by $y = \frac{1}{2}x + 3$.
    (a) How much rainfall was there at noon?
    (b) At what rate was rain falling in the afternoon?
    (c) What is the y-intercept of $y = \frac{1}{2}x + 3$? What does it represent?
    (d) What is the slope of $y = \frac{1}{2}x + 3$? What does it represent?

79. **Land Line Phone Plan** A phone plan in a rural area costs $3.95 per month plus $0.07 per minute. (Assume that a partial minute is not rounded up.)
    (a) During July, a person talks a total of 50 minutes. What is the charge?
    (b) Write an equation in slope–intercept form that gives the monthly cost C of talking on this plan for x minutes.
    (c) If the charge for one month amounts to $8.64, how much time did the person spend talking on the phone?

80. **Electrical Rates** Electrical service costs $8 per month plus $0.10 per kilowatt-hour of electricity used. (Assume that a partial kilowatt-hour is not rounded up.)
    (a) If the resident of an apartment uses 650 kilowatt-hours in 1 month, what is the charge?
    (b) Write an equation in slope–intercept form that gives the cost C of using x kilowatt-hours in 1 month.
    (c) If the monthly electrical bill for the apartment's resident is $43, how many kilowatt-hours of electricity were used?

81. **Cost of Driving** The cost of driving a car includes both fixed costs and mileage costs. Assume that it costs $164.30 per month for insurance and car payments and $0.35 per mile for gasoline, oil, and routine maintenance.
    (a) Find values for m and b so that the equation $y = mx + b$ models the monthly cost of driving the car x miles.
    (b) What does the value of b represent?

82. **Antarctic Ozone Layer** The ozone layer occurs in Earth's atmosphere between altitudes of 12 and 18 miles and is an important filter of ultraviolet light from the sun. The thickness of the ozone layer is frequently measured in Dobson units. An average value is 300 Dobson units. In 2013, the reported minimum in the antarctic *ozone hole* was about 133 Dobson units. (*Source:* NASA.)
    (a) The equation $T = 0.01D$ describes the thickness T in millimeters of an ozone layer that is D Dobson units. How many millimeters thick was the ozone layer over the antarctic in 2013?
    (b) What is the average thickness of the ozone layer in millimeters?

WRITING ABOUT MATHEMATICS

83. Explain how the values of m and b can be used to graph the equation $y = mx + b$.

84. Explain how to find the value of b in the equation $y = 2x + b$ if the point $(3, 4)$ lies on the line.

Group Activity

Exercises 1–4: **Pet Ownership** *In following table, you will see different costs associated with having a dog or a cat. Use the table to complete each of the following.*

1. Find the annual costs and the fixed costs for each type of pet.

2. Write a linear equation for each type of pet that models the total cost of owning the pet over a period of x years.

3. How much will the first year of owning a dog cost? How much will the first year of owning a cat cost?

4. If the average dog is expected to live 13 years, and the average cat is expected to live 15 years, what is the total expense of owning each type of animal over its entire lifetime?

Expenses	Dog Expenses ($)	Cat Expenses ($)
Food	120	115
Recurring medical	235	160
Spay/neuter	200	145
Other initial medical	70	130
Litter	0	165
Toys/treats	55	25
Yearly health insurance	225	175
Scratching post	0	15
Yearly licence	15	0
Initial training class	110	0
Carrier bag/crate	95	40
Collar/leash	30	10
Litter box	0	25
Other recurring costs	45	30

Source: ASPCA.

10.6 Point–Slope Form

Objectives

1. Deriving the Point–Slope Form
2. Finding Point-Slope Form
3. Solving Applications Involving Equations of Lines

1 Deriving the Point–Slope Form

If we know the slope and y-intercept of a line, we can write its slope–intercept form, $y = mx + b$, which is an example of an **equation of a line**. The point–slope form is a different type of equation of a line.

Suppose that a (nonvertical) line with slope m passes through the point (x_1, y_1). If (x, y) is a different point on this line, then $m = \frac{y - y_1}{x - x_1}$. See **FIGURE 10.47**. Using this slope formula, we can find the point–slope form of the equation of a line.

$$m = \frac{y - y_1}{x - x_1} \qquad \text{Slope formula}$$

$$m \cdot (x - x_1) = \frac{y - y_1}{x - x_1} \cdot (x - x_1) \qquad \text{Multiply each side by } (x - x_1).$$

$$m(x - x_1) = y - y_1 \qquad \text{Simplify.}$$

$$y - y_1 = m(x - x_1) \qquad \text{Rewrite the equation.}$$

Slope of a Line

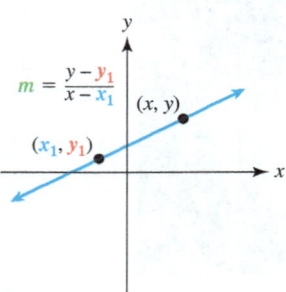

FIGURE 10.47

NEW VOCABULARY

☐ Equation of a line
☐ Point–slope form

POINT–SLOPE FORM

The line with slope m passing through the point (x_1, y_1) is given by

$$y - y_1 = m(x - x_1),$$

the **point–slope form** of a line.

NOTE: The equation $y - y_1 = m(x - x_1)$ is traditionally called the *point–slope form*. By adding y_1 to each side of this equation, we get $y = m(x - x_1) + y_1$, which is an equivalent form that is helpful when graphing. ■

2 Finding Point–Slope Form

In the next example, we find a point–slope form for a line. Note that *any* point that lies on the line can be used in its point–slope form.

EXAMPLE 1 Finding a point–slope form

Use the labeled point in each figure to write a point–slope form for the line and then simplify it to the slope–intercept form.

(a)
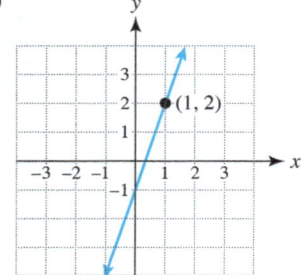

(b)
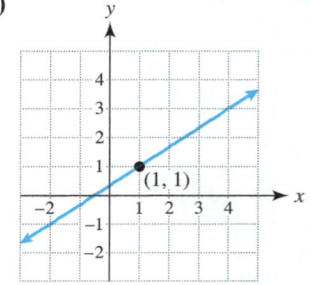

Solution

(a) The graph rises 3 units for each unit of horizontal run, so the slope is $\frac{3}{1}$, or 3. Let $m = 3$ and $(x_1, y_1) = (1, 2)$ in the point–slope form.

$$y - y_1 = m(x - x_1) \quad \text{Point–slope form}$$

Point–slope form $\quad y - 2 = 3(x - 1) \quad$ Let $m = 3, x_1 = 1$ and $y_1 = 2$.

To obtain slope–intercept form, apply the distributive property to $y - 2 = 3(x - 1)$.

$$y - 2 = 3x - 3 \quad \text{Distributive property}$$

Slope–intercept form $\quad y = 3x - 1 \quad$ Add 2 to each side.

(b) The graph rises 2 units for each 3 units of horizontal run, so the slope is $\frac{2}{3}$. Let $m = \frac{2}{3}$ and $(x_1, y_1) = (1, 1)$ in the point–slope form.

$$y - y_1 = m(x - x_1) \quad \text{Point–slope form}$$

Point–slope form $\quad y - 1 = \frac{2}{3}(x - 1) \quad$ Let $m = \frac{2}{3}, x_1 = 1,$ and $y_1 = 1$.

To obtain slope–intercept form, apply the distributive property to $y - 1 = \frac{2}{3}(x - 1)$.

$$y - 1 = \frac{2}{3}x - \frac{2}{3} \quad \text{Distributive property}$$

Slope–intercept form $\quad y = \frac{2}{3}x + \frac{1}{3} \quad$ Add 1, or $\frac{3}{3}$, to each side.

Now Try Exercises 13, 15

EXAMPLE 2 **Finding a point–slope form**

Find a point–slope form for a line passing through the point $(-2, 3)$ with slope $-\frac{1}{2}$. Does the point $(2, 1)$ lie on this line?

Solution
Let $m = -\frac{1}{2}$ and $(x_1, y_1) = (-2, 3)$ in the point–slope form.

$$y - y_1 = m(x - x_1) \quad \text{Point–slope form}$$

$$y - 3 = -\frac{1}{2}(x - (-2)) \quad x_1 = -2, y_1 = 3, \text{ and } m = -\frac{1}{2}$$

Point–slope form $\quad y - 3 = -\frac{1}{2}(x + 2) \quad$ Simplify.

To determine whether $(2, 1)$ lies on the line, substitute $x = 2$ and $y = 1$ in the equation.

$$1 - 3 \stackrel{?}{=} -\frac{1}{2}(2 + 2) \quad \text{Let } x = 2 \text{ and } y = 1.$$

Yes, $(2, 1)$ lies on the line. $\quad -2 \stackrel{?}{=} -\frac{1}{2}(4) \quad$ Simplify.

$$-2 = -2 \checkmark \quad \text{A true statement}$$

Now Try Exercises 9, 19

In the next example, we use the point–slope form to find an equation of a line passing through two points.

644 CHAPTER 10 GRAPHING EQUATIONS

EXAMPLE 3 Finding an equation of a line passing through two points

Use the point–slope form to find an equation of the line passing through the points $(1, -4)$ and $(-2, 5)$.

Solution
Before we can apply the point–slope form, we must find the slope.

$$m = \frac{y_2 - y_1}{x_2 - x_1} \quad \text{Slope formula}$$

$$= \frac{5 - (-4)}{-2 - 1} \quad x_1 = 1, y_1 = -4, x_2 = -2, \text{ and } y_2 = 5$$

$$= -3 \quad \text{Simplify.}$$

We can let either the point $(1, -4)$ or the point $(-2, 5)$ be (x_1, y_1) in the point–slope form. If $(x_1, y_1) = (1, -4)$, then the equation of the line becomes the following.

$$y - y_1 = m(x - x_1) \quad \text{Point–slope form}$$

$$y - (-4) = -3(x - 1) \quad x_1 = 1, y_1 = -4, \text{ and } m = -3$$

$$y + 4 = -3(x - 1) \quad \text{Simplify.}$$

If we let $(x_1, y_1) = (-2, 5)$, the point–slope form becomes

$$y - 5 = -3(x + 2).$$

Now Try Exercise 25

NOTE: Although the two point–slope forms in Example 3 might appear to be different, they actually are equivalent because they simplify to the *same* slope–intercept form.

$y + 4 = -3(x - 1)$	$y - 5 = -3(x + 2)$	Point–slope forms
$y = -3(x - 1) - 4$	$y = -3(x + 2) + 5$	Addition property
$y = -3x + 3 - 4$	$y = -3x - 6 + 5$	Distributive property
$y = -3x - 1$	$y = -3x - 1$	Same slope–intercept form ∎

A line can have only one slope–intercept form.

MAKING CONNECTIONS 1
Slope–Intercept and Point–Slope Forms

READING CHECK 1

- Is the point–slope form of the equation of a line unique?

The slope–intercept form, $y = mx + b$, is *unique* because any nonvertical line has one slope m and one y-intercept $(0, b)$. The point–slope form, $y - y_1 = m(x - x_1)$, is *not unique* because (x_1, y_1) can be any point that lies on the line. However, any point–slope form can be simplified to a unique slope–intercept form.

EXAMPLE 4 Finding equations of lines

Find the slope–intercept form for the line that satisfies the conditions.
(a) Slope $\frac{1}{2}$, passing through $(-2, 4)$
(b) x-intercept $(-3, 0)$, y-intercept $(0, 2)$
(c) Perpendicular to $y = -\frac{2}{3}x$, passing through $(\frac{2}{3}, 3)$

Solution

(a) Substitute $m = \frac{1}{2}$, $x_1 = -2$, and $y_1 = 4$ in the point–slope form.

$$y - y_1 = m(x - x_1) \quad \text{Point–slope form}$$

$$y - 4 = \frac{1}{2}(x + 2) \quad \text{Substitute and simplify.}$$

$$y - 4 = \frac{1}{2}x + 1 \quad \text{Distributive property}$$

$$y = \frac{1}{2}x + 5 \quad \text{Add 4 to each side.}$$

(b) The line passes through the points $(-3, 0)$ and $(0, 2)$. The slope of the line is

$$m = \frac{2 - 0}{0 - (-3)} = \frac{2}{3}.$$

Because the line has slope $\frac{2}{3}$ and y-intercept $(0, 2)$, the slope–intercept form is

$$y = \frac{2}{3}x + 2.$$

(c) The slope of $y = -\frac{2}{3}x$ is $m_1 = -\frac{2}{3}$, so the slope of a line perpendicular to it is the negative reciprocal of $-\frac{2}{3}$, or $\frac{3}{2}$. Let $m = \frac{3}{2}$, $x_1 = \frac{2}{3}$, and $y_1 = 3$ in the point–slope form.

$$y - y_1 = m(x - x_1) \quad \text{Point–slope form}$$

$$y - 3 = \frac{3}{2}\left(x - \frac{2}{3}\right) \quad \text{Substitute.}$$

$$y - 3 = \frac{3}{2}x - 1 \quad \text{Distributive property}$$

$$y = \frac{3}{2}x + 2 \quad \text{Add 3 to each side.}$$

Now Try Exercises 43, 47, 51

In the next example, the point–slope form is used to find the slope–intercept form of a line that passes through several points given in a table.

EXAMPLE 5 Using a table to find slope–intercept form

The points in **TABLE 10.19** lie on a line. Find the slope–intercept form of the line.

x	2	4	6	8
y	2	1	0	-1

TABLE 10.19

Solution

The y-values in the table decrease 1 unit for every 2-unit increase in the x-values, so the line has a "rise" of -1 when the run is 2. The slope is $m = \frac{\text{Rise}}{\text{Run}} = \frac{-1}{2} = -\frac{1}{2}$. Any point from the table can be used to obtain a point–slope form of the line, which can then be simplified to slope–intercept form. Letting $(x_1, y_1) = (2, 2)$ and $m = -\frac{1}{2}$ in the point–slope form yields the following result.

$$y - y_1 = m(x - x_1) \qquad \text{Point–slope form}$$

$$y - 2 = -\frac{1}{2}(x - 2) \qquad \text{Substitute.}$$

$$y - 2 = -\frac{1}{2}x + 1 \qquad \text{Distributive property}$$

$$y = -\frac{1}{2}x + 3 \qquad \text{Add 2 to each side.}$$

This result can be checked by substituting each x-value from the table in the equation. For example, when $x = 4$, the corresponding y-value is $y = -\frac{1}{2}(4) + 3 = -2 + 3 = 1$. This agrees with the table.

Now Try Exercise 57

MAKING CONNECTIONS 2

Finding Slope–Intercept Form Without Using Point–Slope Form

The equation in Example 5 can be obtained without using the point–slope form. The value of b can be found by letting $(x, y) = (2, 2)$ and $m = -\frac{1}{2}$ in the slope–intercept form.

$$y = mx + b \qquad \text{Slope–intercept form}$$

$$2 = -\frac{1}{2}(2) + b \qquad \text{Substitute.}$$

$$2 = -1 + b \qquad \text{Multiply.}$$

$$3 = b \qquad \text{Add 1 to each side.}$$

Because $m = -\frac{1}{2}$ and $b = 3$, the slope–intercept form is $y = -\frac{1}{2}x + 3$.

In the next example, we use the process shown in the above Making Connections to find the slope–intercept form of a line equation without using the point–slope form.

EXAMPLE 6 Finding slope–intercept form directly using $y = mx + b$

If a line has slope 4 and passes through the point $(2, 6)$, find its slope–intercept form *without* using the point–slope equation.

Solution
Find the value of b in the slope–intercept form by letting $(x, y) = (2, 6)$ and $m = 4$ in the equation $y = mx + b$.

$$y = mx + b \qquad \text{Slope–intercept form}$$

$$6 = 4(2) + b \qquad x = 2, y = 6 \text{ and } m = 4.$$

$$6 = 8 + b \qquad \text{Multiply.}$$

$$-2 = b \qquad \text{Subtract 8 from each side.}$$

Since $m = 4$ and $b = -2$, the slope–intercept form is $y = 4x - 2$.

Now Try Exercise 61

3 Solving Applications Involving Equations of Lines

🌐 *Math in Context* (Gender ^) In 1995, there were 690 female officers in the Marine Corps, and by 2010, this number had increased to about 1110. This growth is illustrated in **FIGURE 10.48**, where the line passes through the points $(1995, 690)$ and $(2010, 1110)$. Because two points determine a unique line, we can find the equation of this line and use it to *estimate* the number of female officers in other years. (*Source:* Department of Defense.)

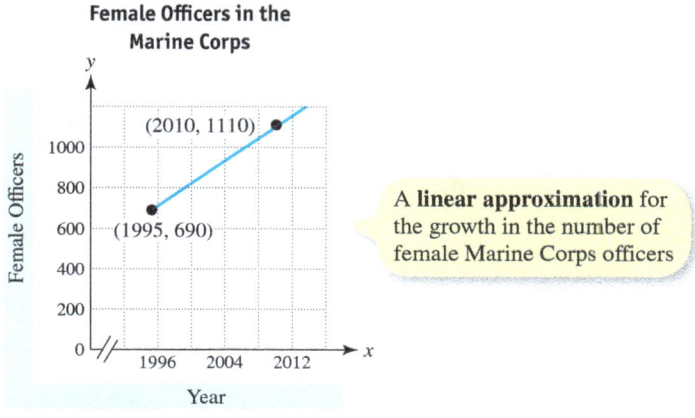

FIGURE 10.48

EXAMPLE 7   **Modeling numbers of female officers**

Refer to **FIGURE 10.48** above.
(a) Use the point $(1995, 690)$ to find a point–slope form of the line.
(b) Interpret the slope as a rate of change.
(c) Use **FIGURE 10.48** to estimate the number of female officers in 2006. Then use your equation from part (a) to approximate this number. How do your answers compare?

Solution
(a) The slope of the line passing through $(1995, 690)$ and $(2010, 1110)$ is
$$m = \frac{1110 - 690}{2010 - 1995} = 28.$$
If we let $x_1 = 1995$ and $y_1 = 690$, then the point–slope form becomes
$$y - 690 = 28(x - 1995).$$

(b) Slope $m = 28$ indicates that the number of female officers increased, *on average*, by about 28 officers per year.

(c) From **FIGURE 10.48**, it appears that the number of female officers in 2006 was about 1000. To estimate this value, let $x = 2006$ in the equation found in part (a).
$$y - 690 = 28(2006 - 1995) \quad \text{or} \quad y = 998$$

Although the graphical estimate and calculated answers are not exactly equal, they are approximately equal. Estimations made from a graph usually are not exact.

Now Try Exercise 73

In the next example, we review several concepts related to lines.

EXAMPLE 8 Modeling water in a pool

A small swimming pool is being emptied by a pump that removes water at a constant rate. After 1 hour, the pool contains 5000 gallons, and after 3 hours, it contains 3000 gallons.
(a) How fast is the pump removing water?
(b) Find the slope–intercept form of a line that models the amount of water in the pool. Interpret the slope.
(c) Find the y-intercept and the x-intercept. Interpret each.
(d) Sketch a graph of the amount of water in the pool during the first 6 hours.
(e) The point $(2, 4000)$ lies on the graph. Explain its meaning.

Solution

(a) The pump removes $5000 - 3000 = 2000$ gallons of water in 2 hours, or 1000 gallons per hour.
(b) The line passes through the points $(1, 5000)$ and $(3, 3000)$, so the slope is

$$m = \frac{3000 - 5000}{3 - 1} = -1000.$$

One way to find the slope–intercept form is to use the point–slope form.

$y - y_1 = m(x - x_1)$ Point–slope form
$y - 5000 = -1000(x - 1)$ $m = -1000$, $x_1 = 1$, and $y_1 = 5000$
$y - 5000 = -1000x + 1000$ Distributive property
$y = -1000x + 6000$ Add 5000 to each side.

Slope -1000 means that the pump is *removing* 1000 gallons of water per hour.

(c) The y-intercept is $(0, 6000)$ and indicates that the pool initially contained 6000 gallons of water. To find the x-intercept, let $y = 0$ in the slope–intercept form.

$0 = -1000x + 6000$ Let $y = 0$.
$1000x = 6000$ Add $1000x$ to each side.
$x = \dfrac{6000}{1000}$ Divide by 1000.
$x = 6$ Simplify.

An x-intercept of $(6, 0)$ indicates that the pool is empty after 6 hours.

(d) Sketch a line from the x-intercept $(6, 0)$ to the y-intercept $(0, 6000)$, as shown in **FIGURE 10.49**.

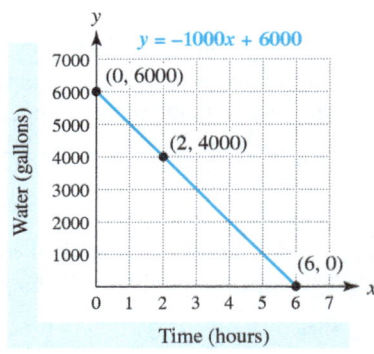

FIGURE 10.49

(e) The point $(2, 4000)$ indicates that after 2 hours the pool contains 4000 gallons of water.

Now Try Exercise 67

CRITICAL THINKING 1

Suppose that a line models the amount of water in a swimming pool. What does a negative slope indicate?

10.6 Putting It All Together

CONCEPT	COMMENTS	EXAMPLE
Point–Slope Form $$y - y_1 = m(x - x_1)$$	Used to find an equation of a line, given two points or one point and the slope Can always be simplified to slope–intercept form Does *not* provide a unique equation for a line because *any* point on the line can be used An equivalent form that is useful for graphing is $$y = m(x - x_1) + y_1.$$	For two points $(1, 2)$ and $(3, 5)$, first compute $m = \frac{5-2}{3-1} = \frac{3}{2}$. An equation of this line is $$y - 2 = \frac{3}{2}(x - 1).$$

10.6 Exercises

MyMathLab®

CONCEPTS AND VOCABULARY

1. (True or False?) One line is determined by two distinct points.

2. (True or False?) One line is determined by a point and a slope.

3. Give the slope–intercept form of a line.

4. Give the point–slope form of a line.

5. If the point–slope form is written for a line passing through $(1, 3)$, then $x_1 = $ _____ and $y_1 = $ _____ .

6. To write a point–slope equation in slope–intercept form, use the _____ property to clear the parentheses.

7. Is the slope–intercept form of a line unique? Explain.

8. Is the point–slope form of a line unique? Explain.

WORKING WITH POINT–SLOPE FORM

Exercises 9–12: Determine whether the given point lies on the line.

9. $(3, -11)$ $y + 1 = -2(x + 3)$

10. $(1, 4)$ $y - 3 = -(x - 1)$

11. $(0, 4)$ $y = \frac{1}{2}(x + 4) + 2$

12. $(2, -5)$ $y = -\frac{1}{3}(x - 5) - 6$

Exercises 13–18: Use the labeled point to write a point–slope form for the line.

13.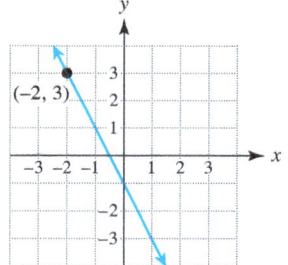

14.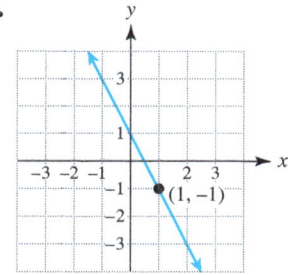

15.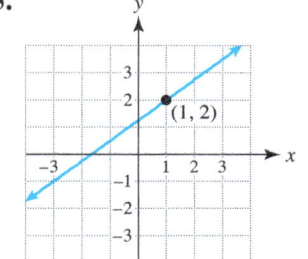

16.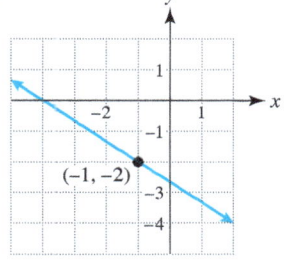

17.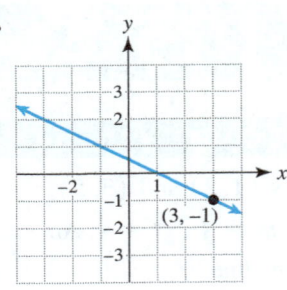
18.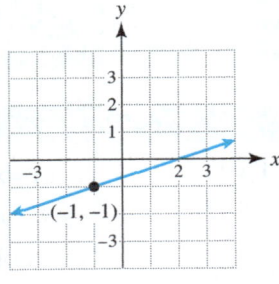

Exercises 19–30: Find a point–slope form for the line that satisfies the stated conditions. When two points are given, use the first point in the point–slope form.

19. Slope 4, passing through $(-3, 1)$

20. Slope -3, passing through $(1, -2)$

21. Slope $\frac{1}{2}$, passing through $(-5, -3)$

22. Slope $-\frac{2}{3}$, passing through $(-3, 6)$

23. Slope 1.5, passing through $(2010, 30)$

24. Slope -10, passing through $(2014, 100)$

25. Passing through $(2, 4)$ and $(-1, -3)$

26. Passing through $(3, -1)$ and $(1, -4)$

27. Passing through $(5, 0)$ and $(0, -3)$

28. Passing through $(-2, 0)$ and $(0, -1)$

29. Passing through $(2003, 15)$ and $(2013, 65)$

30. Passing through $(2009, 5)$ and $(2014, 30)$

Exercises 31–42: Write the point–slope form in slope–intercept form.

31. $y - 4 = 3(x - 2)$
32. $y - 3 = -2(x + 1)$
33. $y + 2 = \frac{1}{3}(x + 6)$
34. $y - 1 = \frac{2}{5}(x + 10)$
35. $y - \frac{3}{4} = \frac{2}{3}(x - 1)$
36. $y + \frac{2}{3} = -\frac{1}{6}(x - 2)$
37. $y = -2(x - 2) + 5$
38. $y = 4(x + 3) - 7$
39. $y = \frac{3}{5}(x - 5) + 1$
40. $y = -\frac{1}{2}(x + 4) - 6$
41. $y = -16(x + 1.5) + 5$
42. $y = -15(x - 1) + 100$

Exercises 43–56: Find the slope–intercept form for the line satisfying the conditions.

43. Slope -2, passing through $(4, -3)$

44. Slope $\frac{1}{5}$, passing through $(-2, 5)$

45. Passing through $(3, -2)$ and $(2, -1)$

46. Passing through $(8, 3)$ and $(-7, 3)$

47. x-intercept $(3, 0)$, y-intercept $(0, \frac{1}{3})$

48. x-intercept $(2, 0)$, y-intercept $(0, -3)$

49. Parallel to $y = 2x - 1$, passing through $(2, -3)$

50. Parallel to $y = -\frac{3}{2}x$, passing through $(0, 20)$

51. Perpendicular to $y = -\frac{1}{2}x + 3$, passing through the point $(6, -3)$

52. Perpendicular to $y = \frac{3}{5}(x + 1) + 3$, passing through the point $(1, -2)$

53. Passing through $(2, 6)$ and parallel to the line passing through $(4, 1)$ and $(-2, -2)$

54. Passing through $(1, 3)$ and parallel to the line passing through $(2, 1)$ and $(4, 5)$

55. Passing through $(-1, -5)$ and perpendicular to the line passing through $(2, -3)$ and $(-2, -1)$

56. Passing through $(-3, 2)$ and perpendicular to the line passing through $(0, -1)$ and $(3, 8)$

Exercises 57–60: The points in the table lie on a line. Find the slope–intercept form of the line.

57.
x	1	2	3	4
y	-3	-5	-7	-9

58.
x	2	3	4	5
y	5	8	11	14

59.
x	-1	1	3	5
y	-3	-2	-1	0

60.
x	-1	5	11	17
y	1	-3	-7	-11

Exercises 61–64: (Refer to Example 6.) Without using point–slope form, find the slope–intercept form of the line that has slope m and passes through the given point.

61. $m = 2$, $(1, -3)$
62. $m = -3$, $(-1, 1)$
63. $m = -\frac{1}{2}$, $(-1, -1)$
64. $m = \frac{2}{3}$, $(4, 3)$

65. **Thinking Generally** Find the y-intercept of the line given by $y - y_1 = m(x - x_1)$.

66. **Thinking Generally** Find the x-intercept of the line given by $y - y_1 = m(x - x_1)$.

INTERPRETING GRAPHS OF LINES

67. Change in Temperature The outside temperature was 40 °F at 1 A.M. and 15 °F at 6 A.M. Assume that the temperature changed at a constant rate.
(a) At what rate did the temperature change?
(b) Find the slope–intercept form of a line that models the temperature T at x A.M. Interpret the slope as a rate of change.
(c) Assuming that your equation is valid for times after 6 A.M., find and interpret the x-intercept.
(d) Sketch a graph that shows the temperature from 1 A.M. to 9 A.M.
(e) The point $(4, 25)$ lies on the graph. Explain its meaning.

68. Cost of Fuel The cost of buying 5 gallons of fuel oil is $12 and the cost of buying 15 gallons of fuel oil is $36.
(a) What is the cost of a gallon of fuel oil?
(b) Find the slope–intercept form of a line that models the cost of buying x gallons of fuel oil. Interpret the slope as a rate of change.
(c) Find and interpret the x-intercept.
(d) Sketch a graph that shows the cost of buying 20 gallons or less of fuel oil.
(e) The point $(11, 26.40)$ lies on the graph. Explain its meaning.

69. Water and Flow The graph shows the amount of water y in a 500-gallon tank after x minutes have elapsed.
(a) Is water entering or leaving the tank? How much water is in the tank after 4 minutes?
(b) Find the y-intercept. Explain its meaning.
(c) Find the slope–intercept form of the line. Interpret the slope as a rate of change.
(d) After how many minutes will the tank be full?

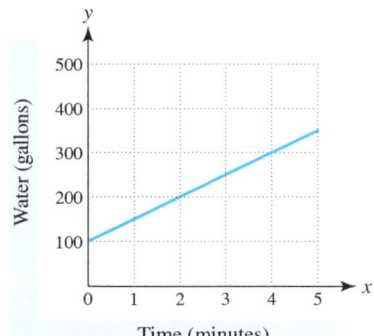

70. Water and Flow A hose is used to fill a 100-gallon barrel. If the hose delivers 5 gallons of water per minute, sketch a graph of the amount A of water in the barrel during the first 20 minutes.

71. Distance and Speed A person is driving a car along a straight road. The graph shows the distance y in miles that the driver is from home after x hours.
(a) Is the person traveling toward or away from home?
(b) The graph passes through $(1, 250)$ and $(4, 100)$. Discuss the meaning of these points.
(c) How fast is the driver traveling?
(d) Find the slope–intercept form of the line. Interpret the slope as a rate of change.

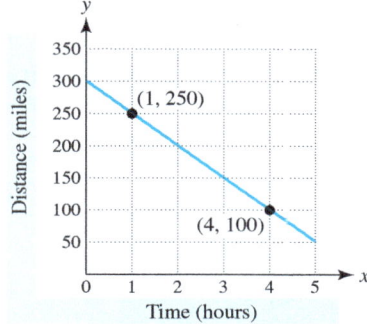

72. Distance and Speed A person rides a bicycle at 10 miles per hour, first away from home for 1 hour and then toward home for 1 hour. Sketch a graph that shows the distance d between the bicyclist and home after x hours.

SOLVING APPLICATIONS INVOLVING EQUATIONS OF LINES

73. Mobile Internet Users The graph models the number of mobile Internet users worldwide in millions from 2009 to 2015. (*Source:* ComScore.)
(a) The line passes through the points $(2009, 700)$ and $(2015, 1900)$. Explain the meaning of the first point.
(b) Use the first point to write a point–slope form for the equation of this line.
(c) Use the graph to estimate the number of mobile Internet users in 2013. Then use your equation from part (b) to estimate the number of mobile Internet users in 2013.
(d) Interpret the slope as a rate of change.

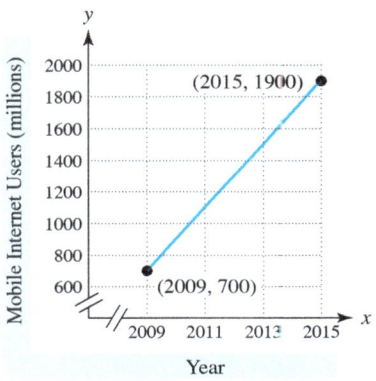

74. **Wyoming Tuition** The graph models average tuition at Wyoming public 2-year colleges after the year 2000, where 1 corresponds to 2001, 2 corresponds to 2002, and so on. (*Source:* SSTI *Weekly Digest.*)
   (a) The line passes through the points $(3, 1557)$ and $(13, 2607)$. What is the meaning of the first of these two points?
   (b) Use the first point to write a point–slope form for the equation of this line.
   (c) Use the graph to estimate average tuition in 2010. Then use your equation from part (b) to estimate tuition in 2010.
   (d) Interpret the slope as a rate of change.

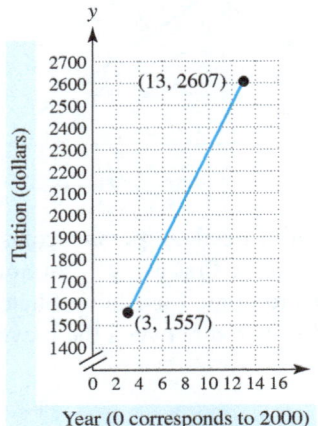

75. **International Adoption** In 2012, there were 1567 children adopted into the United States from Ethiopia, and this number was decreasing by 165 children per year. (*Source:* U.S. Department of State.)
   (a) Determine a point–slope equation of a line that approximates the number of children who were adopted into the United States from Ethiopia during year x, where $x \geq 2012$.
   (b) Estimate the number of children adopted into the United States from Ethiopia in 2014.

76. **Median Household Income** In 2012, median income for U.S. households was $50,099, and this number was increasing at a rate of $941 per year. (*Source:* Department of the Treasury.)
   (a) Determine a point–slope equation of a line that approximates median family income during year x, where $x \geq 2012$.
   (b) Estimate median family income in 2022.

77. **Cigarette Consumption** For any year x from 1975 to 2015, the number of cigarettes y consumed in the United States is modeled by $y = -10.33x + 21{,}087$, where y is in billions. Interpret the slope as a rate of change. (*Source:* The Tobacco Outlook Report.)

78. **Worldwide Smartphones** If x represents a year from 2010 to 2016, then the number y of smartphones sold worldwide in millions is approximated by $y = 216.67x - 435{,}200$. (*Source:* Business Insider.)
   (a) Interpret the slope as a rate of change.
   (b) Estimate the number of smartphones sold worldwide in 2015.

79. **Rise in Sea Level** The highest projection for the rise in sea level by year 2100 is that it will be 1.5 meters higher than the level in 1940. (*Source:* The Economist.)
   (a) Find the slope–intercept equation of a line that approximates the rise in sea level above the 1940 level for year x, where x is between 1940 and 2100.
   (b) Interpret the slope.
   (c) Estimate the rise in sea level by 2020.

80. **Hospitals** In 2002, there were 3039 U.S. hospitals with more than 100 beds, and this number was decreasing at a rate of 33 hospitals per year. (*Source:* AHA Hospital Statistics.)
   (a) Determine a slope–intercept equation of a line that approximates the number of U.S. hospitals with more than 100 beds during year x, where $x \geq 2002$.
   (b) Estimate the number of hospitals with more than 100 beds in 2012.

WRITING ABOUT MATHEMATICS

81. Explain how to find the equation of a line passing through two points with coordinates (x_1, y_1) and (x_2, y_2). Give an example.

82. Explain how slope is related to rate of change. Give an example.

SECTIONS 10.5 and 10.6 Checking Basic Concepts

1. Write the slope–intercept form for the line shown in the graph.

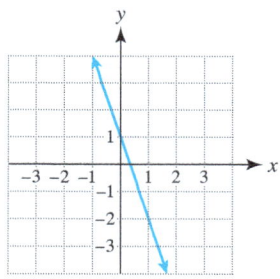

2. Write $4x - 5y = 20$ in slope–intercept form. Give the slope and y-intercept.

3. Graph $y = \frac{1}{2}x - 3$

4. Write the slope–intercept form of a line that satisfies the given conditions.
   (a) Slope 3, passing through $(0, -2)$
   (b) Perpendicular to $y = \frac{2}{3}x$, passing through the point $(-2, 3)$
   (c) Passing through $(1, -4)$ and $(-2, 3)$

5. Write a point–slope form for a line with slope -2, passing through $(-1, 3)$.

6. Write the equation $y + 3 = -2(x - 2)$ in slope–intercept form.

7. Find the slope–intercept form of the line passing through the points in the table.

x	−3	−1	1	3
y	−3	1	5	9

8. **Distance and Speed** A bicyclist is riding at a constant speed and is 36 miles from home at 1 P.M. Two hours later, the bicyclist is 12 miles from home.
   (a) Find the slope–intercept form of a line passing through $(1, 36)$ and $(3, 12)$.
   (b) How fast is the bicyclist traveling?
   (c) When will the bicyclist arrive home?
   (d) How many miles was the bicyclist from home at noon?

9. **Snowfall** The total amount of snowfall S in inches t hours past noon is given by $S = 2t + 5$.
   (a) How many inches of snow fell by noon?
   (b) At what rate did snow fall in the afternoon?
   (c) What is the S-intercept for the graph of this equation? What does it represent?
   (d) What is the slope for the graph of this equation? What does it represent?

10.7 Introduction to Modeling

Objectives

1. Introducing Linear Models
2. Modeling Linear Data

1 Introducing Linear Models

For centuries people have tried to understand the world around them by creating models. A model is an *abstraction* of something that people observed. Not only should a good model describe *known* data, but it should also be able to predict *future* data.

🌐 Math in Real-World Context **FIGURE 10.50(a)** on the next page shows a scatterplot of the number of inmates in the federal prison system from 2007 to 2013. The four points in the graph appear almost to lie on the same line. An example of such a line is shown in **FIGURE 10.50(b)** on the next page. Using mathematical modeling, we can find an equation for such a line. Once we have found this equation, we can use it to make estimates about the federal inmate population. (See Exercise 49 at the end of this section.)

STUDY TIP

There is often more than one way to accurately model data. Don't be afraid to try different methods. Part of the modeling process is trying to justify the reasoning behind a particular modeling choice.

READING CHECK 1

- What are some uses for a mathematical model?

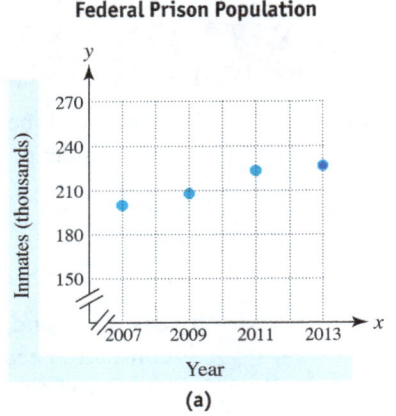

FIGURE 10.50 *Source:* U.S. Department of Justice.

Generally, mathematical models are not *exact* representations of data. Although the line in **FIGURE 10.50(b)** appears to touch every point, it does not pass through all four points exactly. **FIGURE 10.51(a)** shows data modeled *exactly* by a line, whereas **FIGURE 10.51(b)** shows data modeled *approximately* by a line. In applications, a model is more likely to be approximate than exact.

READING CHECK 2

- Does a mathematical model always touch every point in a scatterplot?

FIGURE 10.51

EXAMPLE 1 Determining whether a model is exact

TABLE 10.20 shows the worldwide spending on Internet advertising S in billions dollars for selected years x. Does the equation $S = 2x - 4006$ model the data exactly? Explain.

Spending on Internet Advertising

x	2009	2011	2013	2015
S	13	16	20	24

TABLE 10.20

Solution
To determine whether the equation models the data exactly, let $x = 2009, 2011, 2013,$ and 2015 in the given equation.

$$x = 2009: \quad S = 2(2009) - 4006 = 12$$
$$x = 2011: \quad S = 2(2011) - 4006 = 16$$
$$x = 2013: \quad S = 2(2013) - 4006 = 20$$
$$x = 2015: \quad S = 2(2015) - 4006 = 24$$

The model is *not exact*. It does not predict a spending of $13 billion in 2009.

Now Try Exercise 13

2 Modeling Linear Data

A line can model linear data. In the next example, we use a line to model gas mileage.

EXAMPLE 2 **Determining gas mileage**

TABLE 10.21 shows the number of miles y traveled by an SUV on x gallons of gasoline.

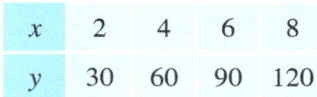

x	2	4	6	8
y	30	60	90	120

TABLE 10.21

(a) Plot the data in the xy-plane. Be sure to label each axis.
(b) Sketch a line that models the data. (You may want to use a ruler.)
(c) Find the equation of the line and interpret the slope of the line.
(d) How far could this SUV travel on 11 gallons of gasoline?

Solution
(a) Plot the points $(2, 30)$, $(4, 60)$, $(6, 90)$, and $(8, 120)$, as shown in FIGURE 10.52(a).

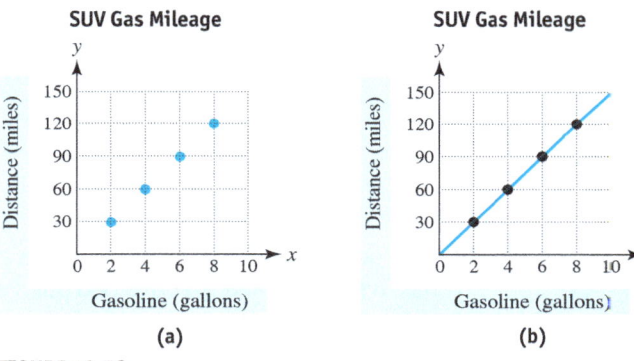

FIGURE 10.52

(b) Sketch a line similar to the one shown in FIGURE 10.52(b). This particular line passes through each data point.

(c) First find the slope m of the line by choosing two points that the line passes through, such as $(2, 30)$ and $(8, 120)$.

$$m = \frac{120 - 30}{8 - 2} = \frac{90}{6} = 15$$

Now find the equation of the line passing through $(2, 30)$ with slope 15.

$y - y_1 = m(x - x_1)$ Point–slope form

$y - 30 = 15(x - 2)$ $x_1 = 2, y_1 = 30$, and $m = 15$

$y - 30 = 15x - 30$ Distributive property

Linear Model → $y = 15x$ Add 30 to each side.

The data are modeled by the equation $y = 15x$. Slope 15 indicates that the mileage of this SUV is 15 miles per gallon.

(d) On 11 gallons of gasoline the SUV could go $y = 15(11) = 165$ miles.

Now Try Exercise 51

EXAMPLE 3 Modeling linear data

TABLE 10.22 contains ordered pairs that can be modeled *approximately* by a line.
(a) Plot the data. Could a line pass through all five points?
(b) Sketch a line that models the data and then determine its equation.

x	1	2	3	4	5
y	3	5	6	10	11

TABLE 10.22

Solution
(a) Plot the ordered pairs $(1, 3)$, $(2, 5)$, $(3, 6)$, $(4, 10)$, and $(5, 11)$, as shown in FIGURE 10.53(a). The points are not collinear, so it is impossible to sketch a single line that passes through *all* five points.

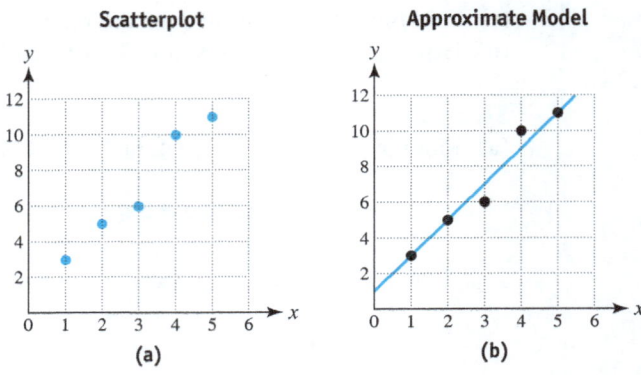

FIGURE 10.53

(b) One possibility for a line is shown in FIGURE 10.53(b). This line passes through three of the five points, and is above one point and below another point. To determine the equation of this line, pick two points that the line passes through. For example, the points $(1, 3)$ and $(5, 11)$ lie on the line. The slope of this line is

$$m = \frac{11 - 3}{5 - 1} = \frac{8}{4} = 2.$$

The equation of the line passing through $(1, 3)$ with slope 2 can be found as follows.

$y - y_1 = m(x - x_1)$ Point–slope form
$y - 3 = 2(x - 1)$ $x_1 = 1, y_1 = 3,$ and $m = 2$
$y - 3 = 2x - 2$ Distributive property
Linear Model $y = 2x + 1$ Add 3 to each side.

The equation of this line is $y = 2x + 1$.

Now Try Exercise 33

NOTE: The equation found in Example 3 represents *one* possible linear model. Other linear models are possible. ∎

World Health
Math in Context When a quantity increases at a constant rate, it can be modeled with the linear equation $y = mx + b$. This concept is illustrated in the next example.

EXAMPLE 4 Modeling worldwide HIV/AIDS cases in children

In 2012, a total of 3.3 million children (under age 15) were living with HIV/AIDS. The rate of new infections was 0.26 million per year.
(a) Write a linear equation $C = mx + b$ that models the total number of children C in millions that were living with HIV/AIDS, x years after 2012.
(b) Estimate C in 2015.

Solution
(a) In the equation $C = mx + b$, the rate of change in HIV infections (**0.26** million per year) corresponds to the slope m, and the initial number of cases (**3.3** million) in 2012 corresponds to b. Therefore the equation $C = \mathbf{0.26}x + \mathbf{3.3}$ models the data.
(b) Since 2015 is 3 years after 2012, let $x = 3$.

$$C = 0.26(3) + 3.3 = 4.08 \text{ million}$$

Now Try Exercise 47

The following box offers a strategy for modeling data that have a constant rate of change.

MODELING WITH A LINEAR EQUATION

To model a quantity y that has a constant rate of change, use the equation

$$y = mx + b,$$

where m = (constant rate of change) and b = (initial amount).

EXAMPLE 5 Modeling with linear equations

Find a linear equation in the form $y = mx + b$ that models the quantity y after x days.
(a) A quantity y is initially 500 and increases at a rate of 6 per day.
(b) A quantity y is initially 1800 and decreases at a rate of 25 per day.
(c) A quantity y is initially 10,000 and remains constant.

Solution
(a) In the equation $y = mx + b$, the y-intercept $(0, b)$ represents the initial amount and the slope m represents the rate of change. Therefore $y = 6x + 500$.
(b) The quantity y is decreasing at the rate of 25 per day with an initial amount of 1800, so $y = -25x + 1800$.
(c) The quantity is constant, so $m = 0$. The equation is $y = 10,000$.

Now Try Exercises 25, 27, 29

10.7 Putting It All Together

CONCEPT	COMMENTS	EXAMPLE
Linear Model	Used to model a quantity that has a constant rate of change	If a total of 2 inches of rain falls before noon, and if rain falls at the rate of $\frac{1}{2}$ inch per hour, then $y = \frac{1}{2}x + 2$ models the total rainfall x hours past noon.

continued on next page

CONCEPT	COMMENTS	EXAMPLE
Modeling Linear Data with a Line	1. Plot the data. 2. Sketch a line that passes either through or nearly through the points. 3. Pick two points on the line and find the equation of the line.	To model $(0, 4)$, $(1, 3)$, and $(2, 2)$, plot the points and sketch a line as shown in the accompanying figure. Many times one line cannot pass through all the points. The equation of the line is $y = -x + 4$.

10.7 Exercises

CONCEPTS AND VOCABULARY

1. Linear data are modeled by a(n) _____ equation.

2. If a line passes through all the data points, it is a(n) _____ model.

3. If a line passes near but not through each data point, it is a(n) _____ model.

4. Linear models are used to describe data that have a(n) _____ rate of change.

5. If a quantity is modeled by the equation $y = mx + b$, then m represents the _____.

6. If a quantity is modeled by the equation $y = mx + b$, then b represents the _____.

UNDERSTANDING LINEAR MODELS

Exercises 7–12: **Modeling** *Match the situation to the graph (a.–f.) that models it best.*

7. College tuition from 2000 to 2015

8. Average temperature in a freezer in degrees Celsius.

9. Profit from selling boxes of candy if it costs $200 to make the candy

10. Height of Mount Hood in Oregon over time

11. Total amount of water delivered by a garden hose if it flows at a constant rate

12. Sales of music on CDs from 2000 to 2010

a.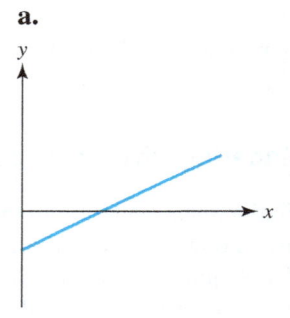

b.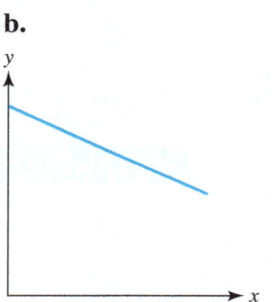

c.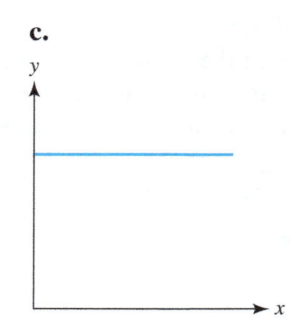

d.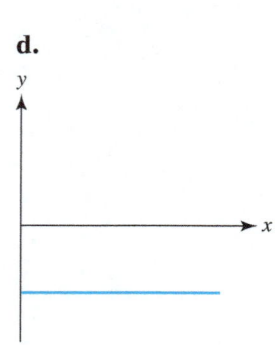

e.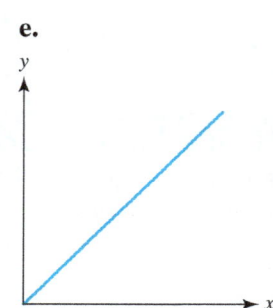

f.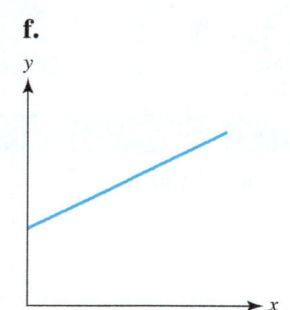

Exercises 13–18: State whether the ordered pairs in the table are modeled exactly by the linear equation.

13. $y = 2x + 2$

x	0	1	2
y	2	4	6

14. $y = -2x + 5$

x	0	1	2
y	5	3	0

15. $y = -4x$

x	-1	0	1
y	4	0	-8

16. $y = 5 - x$

x	-2	1	4
y	7	4	1

17. $y = 1.4x - 4$

x	0	5	10
y	-4	3	9

18. $y = -\frac{4}{3}x - \frac{13}{3}$

x	-7	-4	-1
y	5	1	-3

Exercises 19–24: State whether the linear model in the graph is exact or approximate. Then find the equation of the line.

19.

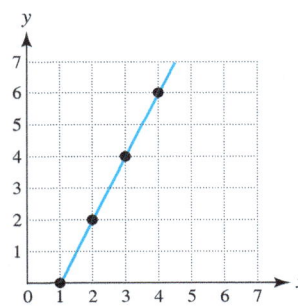

20.

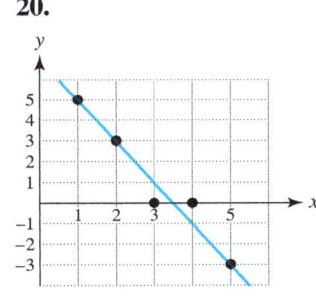

21.

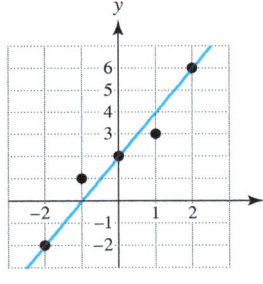

22.

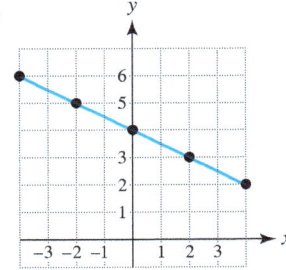

23.

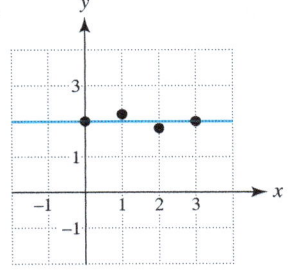

24.

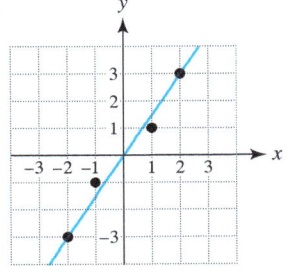

MODELING LINEAR DATA

Exercises 25–30: Find an equation $y = mx + b$ that models the quantity y after x units of time.

25. A quantity y is initially 40 and increases at a rate of 5 per minute.

26. A quantity y is initially -60 and increases at a rate of 1.7 per minute.

27. A quantity y is initially -5 and decreases at a rate of 20 per day.

28. A quantity y is initially 5000 and decreases at a rate of 35 per day.

29. A quantity y is initially 8 and remains constant.

30. A quantity y is initially -45 and remains constant.

Exercises 31–38: **Modeling Data** *For the ordered pairs in the table, do the following.*

(a) Plot the data. Could a line pass through all five points?
(b) Sketch a line that models the data.
(c) Determine an equation of the line. For data that are not exactly linear, answers may vary.

31.

x	0	1	2	3	4
y	4	2	0	-2	-4

32.

x	0	1	2	2	3
y	7	6	5	4	3

33.

x	-2	-1	0	1	2
y	4	1	0	-1	-4

34.

x	-2	-1	0	1	2
y	7	5	1	-3	-5

35.

x	-6	-4	-2	0	2
y	1	0	-1	-2	-3

36.

x	-4	-2	0	2	4
y	1	2	3	4	5

37.

x	-6	-3	0	3	6
y	-3	-2	-0.5	0	0.5

38.

x	-3	-2	-1	0	1
y	1	2	3	4	5

APPLICATIONS INVOLVING LINEAR MODELS

Exercises 39–46: Write an equation that models the described quantity. Specify what each variable represents.

39. A barrel contains 200 gallons of water and is being filled at a rate of 5 gallons per minute.

40. A barrel contains 40 gallons of gasoline and is being drained at a rate of 3 gallons per minute.

41. An athlete has run 5 miles and is jogging at 6 miles per hour.

42. A new car has 141 miles on its odometer and is traveling at 70 miles per hour.

43. A worker has already earned $200 and is being paid $8 per hour.

44. A gambler has lost $500 and is losing money at a rate of $150 per hour.

45. A carpenter has already shingled 5 garage roofs and is shingling roofs at a rate of 1 per day.

46. A hard drive has been spinning for 2 minutes and is spinning at 7200 revolutions per minute.

47. **Kilimanjaro Glacier** Mount Kilimanjaro is located in Tanzania, Africa, and has an elevation of 19,340 feet. In 1912, the glacier on its peak covered 5 acres. By 2002, this glacier had melted to only 1 acre. (*Source:* NBC News.)
    (a) Assume that this glacier melted at a constant rate each year. Find this *yearly* rate.
    (b) Use your answer in part (a) to write a linear equation that gives the acreage A of this glacier t years past 1912.

48. **World Population** In 1987, the world's population reached 5 billion people, and by 2012, the world's population reached 7 billion people. (*Source:* U.S. Census Bureau.)
    (a) Find the average yearly increase in the world's population from 1987 to 2012.
    (b) Write a linear equation that estimates the world's population P in billions x years after 1987.

49. **Prison Population** The points shown in **FIGURE 10.50** include $(2007, 200)$, $(2009, 208)$, $(2011, 218)$, and $(2013, 219)$, where the *y*-coordinates are in thousands.
    (a) Use the first and last data points to determine a line that models the data. Write the equation in slope–intercept form.
    (b) Use the line to estimate the population in 2017.

50. **Niagara Falls** The average flow of water over Niagara Falls is 212,000 cubic feet of water per second.
    (a) Write an equation that gives the number of cubic feet of water F that flow over Niagara Falls in x seconds.
    (b) How many cubic feet of water flow over Niagara Falls in 1 minute?

51. **Gas Mileage** The table shows the number of miles y traveled by a car on x gallons of gasoline.

x (gallons)	3	6	9	12
y (miles)	60	120	180	240

   (a) Plot the data in the *xy*-plane. Label each axis.
   (b) Sketch a line that models these data. (You may want to use a ruler.)
   (c) Calculate the slope of the line. Interpret the slope.
   (d) Find an equation of the line.
   (e) How far could this car travel on 7 gallons of gasoline?

52. **Air Temperature** Generally, the air temperature becomes colder as the altitude above the ground increases. The table lists typical air temperatures x miles high when the ground temperature is 80 °F.

x (miles)	0	1	2	3
y (°F)	80	61	42	43

   (a) Plot the data in the *xy*-plane. Label each axis.
   (b) Sketch a line that models these data. (You may want to use a ruler.)
   (c) Calculate the slope of the line. Interpret the slope.
   (d) Find the slope–intercept form of the line.
   (e) Estimate the air temperature 5 miles high.

Exercises 53–56: In this set of exercises, you are to use your knowledge of equations of lines to model the average annual cost of tuition and fees.

53. **Cost of Tuition** In 2005, the average cost of tuition and fees at *private* four-year colleges was $21,235, and in 2013 it was $30,083. Sketch a line that passes through the points $(2005, 21235)$ and $(2013, 30083)$. (*Source:* The College Board.)

54. **Rate of Change in Tuition** Calculate the slope of the line in your graph. Interpret this slope as a rate of change.

55. **Modeling Tuition** Find the slope–intercept form of the line in your sketch. What is the y-intercept and does it have meaning in this situation?

56. **Predicting Tuition** Use your equation to estimate tuition and fees in 2010 and compare it to the known value of $27,290. Estimate tuition and fees in 2015.

WRITING ABOUT MATHEMATICS

57. In Example 2, the gas mileage of an SUV is modeled with a linear equation. Explain why it is reasonable to use a linear equation to model this situation.

58. Explain the steps for finding the equation of a line that models data points in a table.

SECTION 10.7 Checking Basic Concepts

1. State whether the ordered pairs shown in the table are modeled exactly by $y = -5x + 10$.

x	-2	-1	0	1
y	20	15	10	5

2. State whether the linear model in the graph is exact or approximate. Find the equation of the line.

   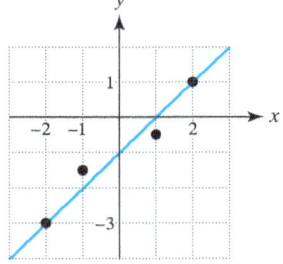

3. Find an equation, $y = mx + b$, that models the quantity y after x units of time.
   (a) A quantity y is initially 50 pounds and increases at a rate of 10 pounds per day.
   (b) A quantity y is initially 200 °F and decreases at a rate of 2 °F per minute.

4. The table contains ordered pairs.
   (a) Plot the data.
   (b) Sketch a line that models the data.
   (c) Determine the equation of the line.

x	-2	0	2	4
y	2	1	0	-1

5. **Climate Change** Since 1945, the average annual recorded temperature on the Antarctic Peninsula has increased by 0.075 °F per year.
   (a) Write an equation that models the average temperature increase T, x years after 1945.
   (b) Use your equation to calculate the temperature increase between 1945 and 2017.

CHAPTER 10 Summary

SECTION 10.1 ■ INTRODUCTION TO GRAPHING

The Rectangular Coordinate System (xy-plane)

Points Plotted as (x, y) ordered pairs (see the xy-plane below).

Four Quadrants The x- and y-axes divide the xy-plane into quadrants I, II, III, and IV.

NOTE: A point on an axis, such as $(1, 0)$, does not lie in a quadrant. ■

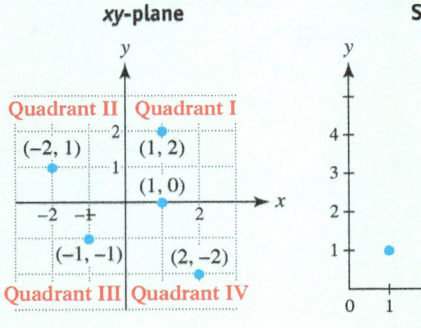

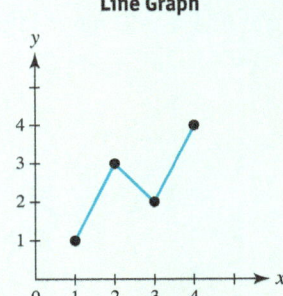

SECTION 10.2 ■ EQUATIONS IN TWO VARIABLES

Equations in Two Variables An equation with two variables and possibly some constants

Examples: $y = 3x + 7$ and $x + y = 100$

Solution to an Equation in Two Variables The solution to an equation in two variables is an ordered pair that makes the equation a true statement.

Example: $(1, 2)$ is a solution to $2x + y = 4$ because $2(1) + 2 = 4$ is true.

Graphing a Linear Equation in Two Variables

Linear Equation $\qquad\qquad y = mx + b$ or $Ax + By = C$ (standard form)

Graphing $\qquad\qquad\qquad$ Plot at least three points and sketch a line passing through each one.

Example: $y = 2x - 1$

x	y
0	−1
1	1
2	3

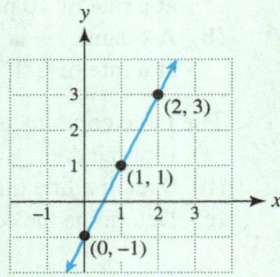

SECTION 10.3 ■ INTERCEPTS; HORIZONTAL AND VERTICAL LINES

Intercepts

x-Intercept $\qquad\qquad$ A point at which a graph intersects the x-axis; to find an x-intercept, let $y = 0$ in the equation and solve for x.

y-Intercept $\qquad\qquad$ A point at which a graph intersects the y-axis; to find a y-intercept, let $x = 0$ in the equation and solve for y.

Example: $x + 3y = 3$

x-intercept: Solve $x + 3(0) = 3$ to find the x-intercept of $(3, 0)$.
y-intercept: Solve $0 + 3y = 3$ to find the y-intercept of $(0, 1)$.

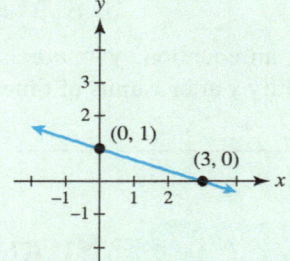

Horizontal and Vertical Lines

The equation of a horizontal line with y-intercept $(0, b)$ is $y = b$.

The equation of a vertical line with x-intercept $(k, 0)$ is $x = k$.

Example: The horizontal line $y = -1$ has y-intercept $(0, -1)$ and no x-intercepts.
$\qquad\qquad$ The vertical line $x = 2$ has x-intercept $(2, 0)$ and no y-intercepts.

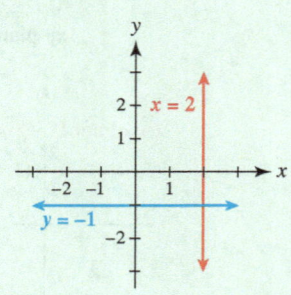

SECTION 10.4 ■ SLOPE AND RATES OF CHANGE

Slope The ratio $\frac{rise}{run}$, or $\frac{change\ in\ y}{change\ in\ x}$, is the slope m of a line when run (change in x) is nonzero. A positive slope indicates that a line rises from left to right, and a negative slope indicates that a line falls from left to right.

Example: Slope $\frac{2}{3}$ indicates that a line rises 2 units for every 3 units of horizontal run.

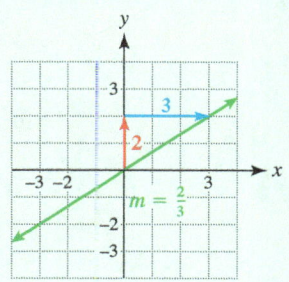

Calculating Slope   A line passing through (x_1, y_1) and (x_2, y_2) has slope

$$m = \frac{y_2 - y_1}{x_2 - x_1}, \text{ where } x_1 \neq x_2.$$

Example: The line through $(-2, 2)$ and $(3, 4)$ has slope

$$m = \frac{4 - 2}{3 - (-2)} = \frac{2}{5}.$$

Horizontal Line   Has zero slope

Vertical Line   Has undefined slope

Slope as a Rate of Change   Slope measures the "tilt" of a line. In applications, slope measures the rate of change in a quantity.

Example: The line shown in the graph has slope -2 and depicts an initial outside temperature of $6\,°F$. Slope -2 indicates that the temperature is *decreasing* at a rate of $2\,°F$ per hour.

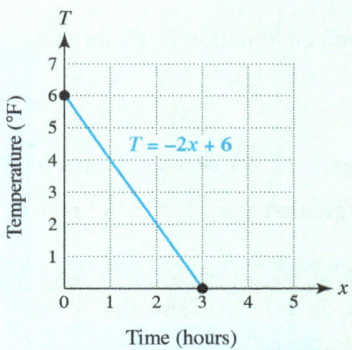

SECTION 10.5 ■ SLOPE–INTERCEPT FORM

Slope–Intercept Form

For the line given by $y = mx + b$, the slope is m and the y-intercept is $(0, b)$.

Example: $y = -\frac{1}{2}x + 2$ has slope $-\frac{1}{2}$ and y-intercept $(0, 2)$, as shown in the graph on the following page.

Parallel Lines
Lines with the same slope are parallel; nonvertical parallel lines have the same slope. Two vertical lines are parallel.

Example: The equations $y = -2x + 1$ and $y = -2x$ determine two parallel lines because $m_1 = m_2 = -2$. See the graph on the following page.

Perpendicular Lines

If two perpendicular lines have nonzero slopes m_1 and m_2, then $m_1 \cdot m_2 = -1$.

If two lines have nonzero slopes satisfying $m_1 \cdot m_2 = -1$, then they are perpendicular.

A vertical line and a horizontal line are perpendicular.

Example: The equations $y = -\frac{1}{2}x$ and $y = 2x - 2$ determine perpendicular lines because $m_1 \cdot m_2 = -\frac{1}{2} \cdot 2 = -1$. See the graph shown on the following page.

Slope–intercept Form	Parallel Lines	Perpendicular Lines
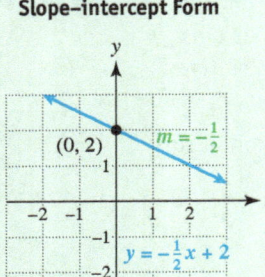	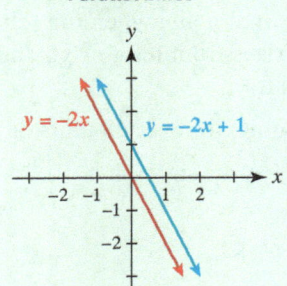	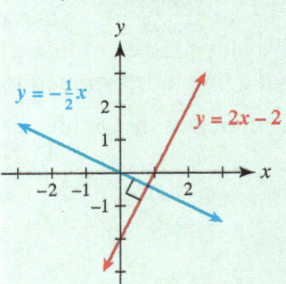

SECTION 10.6 ■ POINT–SLOPE FORM

Point–Slope Form An equation of the line passing through (x_1, y_1) with slope m is $y - y_1 = m(x - x_1)$.

Example: If $m = -2$ and $(x_1, y_1) = (-2, 3)$, then the point–slope form is
$$y - 3 = -2(x + 2).$$

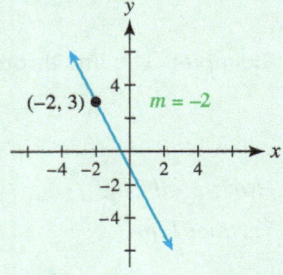

Example: To find an equation of a line passing through $(-2, 5)$ and $(4, 2)$, first find the slope.
$$m = \frac{2-5}{4-(-2)} = \frac{-3}{6} = -\frac{1}{2}$$

Either $(-2, 5)$ or $(4, 2)$ may be used in the point–slope form. The point $(4, 2)$ results in the equation $y - 2 = -\frac{1}{2}(x - 4)$.

SECTION 10.7 ■ INTRODUCTION TO MODELING

Mathematical Modeling Mathematics can be used to describe or approximate the behavior of real-world phenomena.

Exact Model   The equation describes the data precisely without error.

Example: The equation $y = 3x$ models the data in the table exactly.

x	0	1	2	3
y	0	3	6	9

Approximate Model   The modeling equation describes the data approximately. An approximate model occurs most often in applications.

Example: The line in the graph models the data approximately.

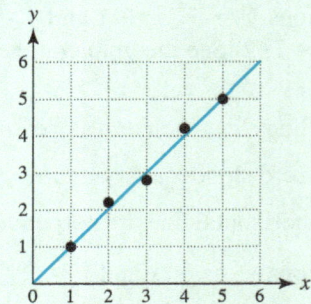

Modeling with a Linear Equation To model a quantity y that has a constant rate of change, use the equation $y = mx + b$, where

$$m = (\text{constant rate of change}) \quad \text{and} \quad b = (\text{initial amount}).$$

Example: If the temperature is initially $100\,°F$ and cools at $5\,°F$ per hour, then

$$T = -5x + 100$$

models the temperature T after x hours.

CHAPTER 10 Review Exercises

SECTION 10.1

1. Identify the coordinates of each point in the graph. Identify the quadrant, if any, in which each point lies.

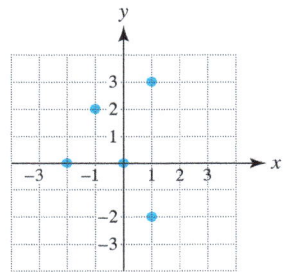

2. Make a scatterplot by plotting the following four points: $(-2, 3)$, $(-1, -1)$, $(0, 3)$, and $(2, -1)$.

Exercises 3 and 4: If possible, identify the quadrant in which each point is located.

3. (a) $(-4, 3)$ (b) $(\tfrac{1}{3}, -\tfrac{1}{2})$

4. (a) $(0, 3.2)$ (b) $(-5, -1.7)$

Exercises 5 and 6: Use the table of xy-values to make a line graph.

5.
x	-2	-1	0	1	2
y	-3	2	-1	-2	3

6.
x	-10	-5	0	5	10
y	5	-10	10	-5	0

SECTION 10.2

Exercises 7 and 8: Determine whether the ordered pair is a solution for the given equation.

7. $y = x - 3$ $(6, 3)$

8. $3x - y = 3$ $(-1, 6)$

Exercises 9 and 10: Complete the table for the given equation.

9. $y = -3x$

x	-2	-1	0	1	2
y					

10. $2x + y = 5$

x					
y	-3	-1	0	1	3

Exercises 11 and 12: Use the given values of the variable to make a horizontal table of solutions for the equation.

11. $y = 3x + 2$ $x = -2, 0, 2, 4$

12. $2y + x = 1$ $y = 1, 2, 3, 4$

Exercises 13–16: Graph the equation.

13. $y = 2x$ 14. $y = -3x + 2$

15. $x + y = 2$ 16. $3x - 2y = 6$

SECTION 10.3

Exercises 17 and 18: Identify the x- and y-intercepts.

17. 18.

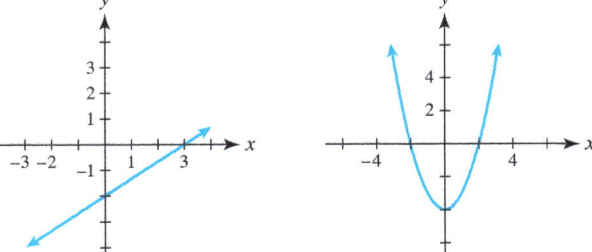

Exercises 19 and 20: Complete the table. Then determine the x- and y-intercepts for the graph of the equation.

19. $y = 2 - x$

x	−2	−1	0	1	2
y					

20. $x - 2y = 4$

x	−4	−2	0	2	4
y					

Exercises 21–24: Find any intercepts for the graph of the equation and then graph the linear equation.

21. $2x - 3y = 6$

22. $5x - y = 5$

23. $0.1x - 0.2y = 0.4$

24. $\dfrac{x}{2} + \dfrac{y}{3} = 1$

Exercises 25 and 26: Write an equation for the line that passes through the points shown in the table.

25.

x	−2	−1	0	1	2
y	1	1	1	1	1

26.

x	3	3	3	3	3
y	−2	−1	0	1	2

27. Graph each equation.
   (a) $y = 1$ (b) $x = -3$

28. Write an equation for each line shown in the graph.

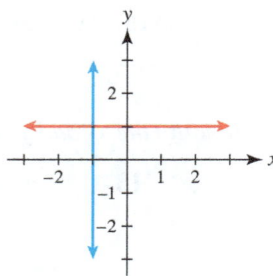

29. Write the equations of a horizontal line and a vertical line that pass through the point $(-2, 3)$.

30. Write the equation of a line that is perpendicular to $y = -\frac{1}{2}$ and passes through $(4, 1)$.

31. Write the equation of a line that is parallel to $y = 3$ and passes through $(-6, -5)$.

32. Distance The distance a driver is from home is illustrated in the graph.
   (a) Find the intercepts.
   (b) Interpret each intercept.

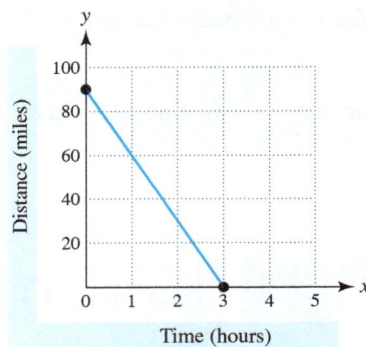

SECTION 10.4

Exercises 33–36: Find the slope, if possible, of the line passing through the two points.

33. $(2, 3), (4, 7)$ **34.** $(-3, 1), (2, -1)$

35. $(2, 1), (5, 1)$ **36.** $(-5, 6), (-5, 10)$

Exercises 37 and 38: Find the slope of the line shown in the graph.

37. 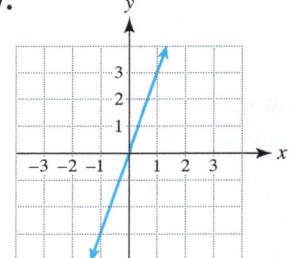   **38.**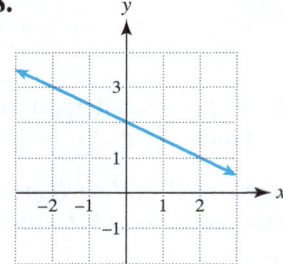

Exercises 39 and 40: Do the following.

   (a) *Graph the linear equation.*
   (b) *What are the slope and y-intercept of the line?*

39. $y = -2x$ **40.** $2x - 3y = -6$

Exercises 41 and 42: Sketch a line that passes through the given point and has slope m.

41. $(0, -3), m = 2$ **42.** $(2, 2), m = 1$

43. A line with slope $\frac{1}{2}$ passes through the first point shown in the table. Complete the table so that each point in the table lies on the line.

x	0	1	2	3
y	1			

44. If a line models the cost of buying x coffee drinks at $3.49 each, does the line have positive or negative slope?

SECTION 10.5

Exercises 45 and 46: Write the slope–intercept form for the line shown in the graph.

45. 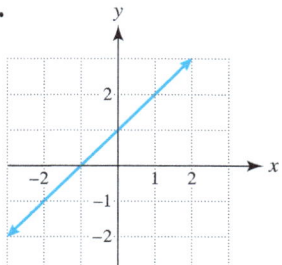   **46.**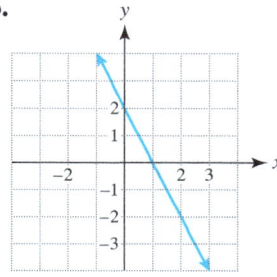

Exercises 47 and 48: Sketch a line with slope m and y-coordinate of the y-intercept b. Write the slope–intercept form of the equation of the line.

47. $m = 2, b = -2$ **48.** $m = -\frac{3}{4}, b = 3$

Exercises 49 and 50: Do the following.

 (a) Write the equation in slope–intercept form.
 (b) Give the slope and y-intercept of the line.

49. $x + y = 3$

50. $5x - 6y = 30$

Exercises 51–54: Graph the equation.

51. $y = \frac{1}{2}x + 1$ **52.** $y = 3x - 2$

53. $y = 4 - x$ **54.** $y = 2 - \frac{2}{3}x$

Exercises 55 and 56: All the points shown in the table lie on the same line. Find the slope–intercept form for the line.

55.

x	0	1	2
y	−5	0	5

56.

x	−1	0	1
y	2	0	−2

Exercises 57–60: Find the slope–intercept form for the line satisfying the given conditions.

57. Slope $-\frac{5}{6}$, y-intercept $(0, 2)$

58. Parallel to $y = -2x + 1$, passing through $(1, -5)$

59. Perpendicular to $y = -\frac{3}{2}x$, passing through $(3, 0)$

60. Perpendicular to $y = 5x - 3$, passing through the point $(0, -2)$

SECTION 10.6

Exercises 61 and 62: Determine whether the given point lies on the line.

61. $(-3, 1)$ $y - 1 = 2(x + 3)$

62. $(3, -8)$ $y = -3(x - 1) + 2$

Exercises 63–68: Find a point–slope form for the line that satisfies the conditions given. When two points are given, use the first point in the point–slope form.

63. Slope 5, passing through $(1, 2)$

64. Passing through $(20, -30)$ and $(40, 30)$

65. x-intercept $(3, 0)$, y-intercept $(0, -4)$

66. x-intercept $(\frac{1}{2}, 0)$, y-intercept $(0, -1)$

67. Parallel to $y = 2x$, passing through $(5, 7)$

68. Perpendicular to $y - 4 = \frac{3}{2}(x + 1)$, and passing through the point $(-1, 0)$

Exercises 69 and 70: Write the given point–slope form in slope–intercept form.

69. $y - 2 = 3(x + 1)$ **70.** $y = -\frac{1}{4}(x - 8) + 1$

SECTION 10.7

71. State whether the ordered pairs shown in the table are modeled exactly by $y = -x + 4$.

x	0	1	2	2
y	4	3	2	1

72. State whether the linear model shown in the graph is exact or approximate. Find the equation of the line.

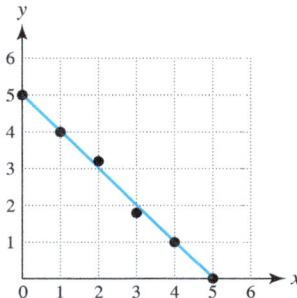

Exercises 73–76: Find an equation $y = mx + b$ that models y after x units of time.

73. y is initially 40 pounds and decreases at a rate of 2 pounds per minute.

74. y is initially 200 gallons and increases at 20 gallons per hour.

75. y is initially 50 and remains constant.

76. y is initially 20 feet *below* sea level and rises at 5 feet per second.

Exercises 77 and 78: For the ordered pairs in the table, do the following.

(a) Plot the data. Could a line pass through all five points?
(b) Sketch a line that models the data.
(c) Determine an equation of the line. For data that are not exactly linear, answers may vary.

77.

x	0	1	2	3	4
y	10	6	2	-2	-6

78.

x	-4	-2	0	2	4
y	1	2.1	3	3.9	5

APPLICATIONS

79. Graphing Real Data The table contains real data on the number of divorces D in millions during year t.
(a) Make a line graph of the data. Label the axes.
(b) Comment on any trends in the data.

t	1970	1980	1990	2000	2010
D	0.7	1.2	1.2	1.2	1.0

Source: National Center for Health Statistics.

80. Water Usage The average American uses 100 gallons of water each day.
(a) Write an equation that gives the gallons G of water that a person uses in t days.
(b) Graph the equation for $t \geq 0$.
(c) How many days does it take for the average American to use 5000 gallons of water?

81. Modeling a Toy Rocket The velocity v of a toy rocket in feet per second after t seconds of flight is given by $v = 160 - 32t$, where $t \geq 0$.
(a) Graph the equation.
(b) Interpret each intercept.

82. Modeling The line graph at the top of the next column represents the insect population on 1 acre of land after x weeks. During this time, a farmer sprayed pesticides on the land.
(a) Estimate the slope of each line segment.
(b) Interpret each slope as a rate of change.
(c) Describe what happened to the insect population.

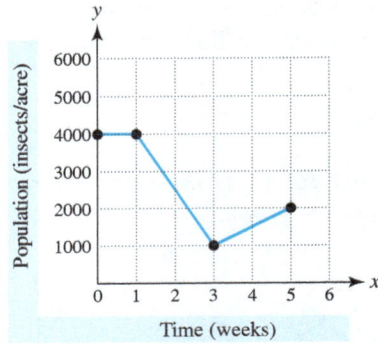

83. Sketching a Graph An athlete jogs 4 miles away from home at a constant rate of 8 miles per hour and then turns around and jogs back home at the same speed. Sketch a graph that shows the distance that the athlete is from home. Be sure to label each axis.

84. Music Downloads In 2008, global recorded-music revenue from downloads was $3 billion, and it was projected to be $9 billion by 2017. (Source: The Economist.)
(a) Calculate the slope of the line passing through $(2008, 3)$ and $(2017, 9)$.
(b) Interpret the slope as a rate of change.

85. Rental Cars The cost C in dollars for driving a rental car x miles is $C = 0.2x + 35$.
(a) How much would it cost to rent the car but not drive it?
(b) How much does the cost increase for each additional mile driven?
(c) Determine the C-intercept for the graph of $C = 0.2x + 35$. What does this C-intercept represent?
(d) What is the slope of the graph of $C = 0.2x + 35$? What does it represent?

86. Distance and Speed A person is driving a car along a straight road. The graph at the top of the next page shows the distance y in miles that the driver is from home after x hours.
(a) Is the person traveling toward or away from home? Why?
(b) The graph passes through $(1, 200)$ and $(3, 100)$. Discuss the meaning of these points.
(c) Find the slope–intercept form of the line. Interpret the slope as a rate of change.
(d) Use the graph to estimate the distance from home after 2 hours. Then check your answer by using your equation from part (c).

87. Social Ad Revenue Social display advertising is advertising seen next to content on social media pages, such as Facebook. The table gives the U.S. ad revenue R in billions of dollars for social display ads during year t.

t	2013	2014	2015	2016
R	4.3	4.9	5.5	6.1

(a) Make a scatterplot of the data.

(b) Find the slope–intercept form of a line that models the revenue R in year t. Interpret the slope of this line.

(c) Is the line you found in part (b) an exact model for the data in the table?

(d) If this trend continues, what will the revenue be in 2018?

88. Gas Mileage The table shows the number of miles y traveled by a car on x gallons of gasoline.

x	2	4	8	10
y	40	79	161	200

(a) Plot the data in the xy-plane. Be sure to label each axis.

(b) Sketch a line that models these data. (You may want to use a ruler.)

(c) Calculate the slope of the line. Interpret the slope of this line.

(d) Find the equation of the line. Is your line an *exact* model?

(e) How far could this car travel on 9 gallons of gasoline?

CHAPTER 10 Test

1. Identify the coordinates of each point in the graph. State the quadrant, if any, in which each point lies.

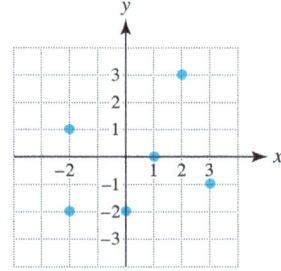

2. Make a scatterplot by plotting the four points $(0, 0)$, $(-2, -2)$, $(3, 0)$, and $(3, -2)$.

3. Complete the table for the equation $y = 2x - 4$. Then determine the x- and y-intercepts for the graph of the equation.

x	-2	-1	0	1	2
y					

4. Determine whether the ordered pair $(1, -3)$ is a solution for the equation $2x - y = 5$.

5. Sketch a line passing through the point $(2, 1)$ and having slope $-\frac{1}{2}$.

6. Find the x- and y-intercepts for the graph of the equation $5x - 3y = 15$.

Exercises 7–10: Graph the equation.

7. $y = 2$ **8.** $x = -3$

9. $y = -3x + 3$ **10.** $4x - 3y = 12$

11. Write the slope–intercept form for the line shown in the graph.

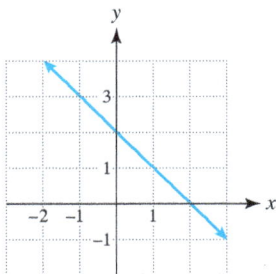

12. Write the equation $-4x + 2y = 1$ in slope–intercept form. Give the slope and the y-intercept.

13. Write the equation of a horizontal line and a vertical line passing through the point $(1, -5)$.

14. Find the slope of a line passing through the points $(-4, 3)$ and $(5, 1)$.

Exercises 15–18: Find the slope–intercept form for the line that satisfies the given conditions.

15. Slope $-\frac{4}{3}$, y-intercept $(0, -5)$

16. Parallel to $y = 3x - 1$, passing through $(2, -5)$

17. Perpendicular to $y = \frac{1}{3}x$, passing through $(1, 2)$

18. Passing through $(-4, 2)$ and $(2, -1)$

19. Write $y - 3 = \frac{1}{2}(x + 4)$ in slope–intercept form.

20. Do all of the points shown in the table lie on the same line. If so, find the slope–intercept form for the line.

x	-2	-1	0	1	2
y	-8	-5	-2	1	4

21. Write a point–slope form for a line that is parallel to $y = -3x$ and passes through $(-2, 7)$.

22. State whether the linear model shown is exact or approximate. Then find the equation of the line.

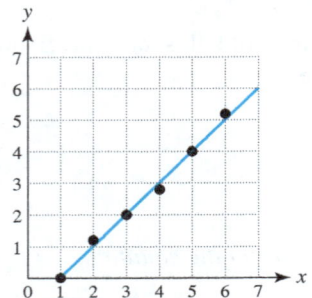

23. **Sketching a Graph** A cyclist rides a bicycle at 10 miles per hour for 2 hours and then at 8 miles per hour for 1 hour. Sketch a graph that shows the total distance d traveled after x hours. Be sure to label each axis.

24. **Modeling** The line graph represents the total fish population P in a small lake after x years. One winter the lake almost froze solid.
    (a) Estimate the slope of each line segment.
    (b) Interpret each slope as a rate of change.
    (c) Describe what happened to the fish population.

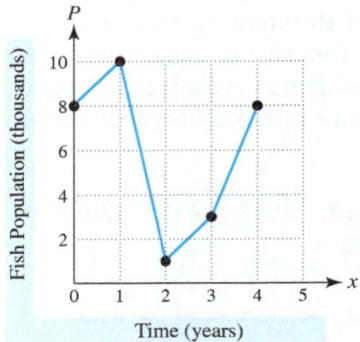

25. **Modeling Insect Population** Write an equation in slope–intercept form that models the number of insects N after x days if there are initially 2000 insects and the population increases at a rate of 100 insects per day.

CHAPTERS 1–10 Cumulative Review Exercises

Exercises 1 and 2: Classify the number as prime or composite. If a number is composite, write it as a product of prime numbers.

1. 40
2. 61

Exercises 3 and 4: Translate the phrase into an algebraic expression using the variable n.

3. Ten more than a number

4. A number squared decreased by 2

Exercises 5 and 6: Evaluate by hand and then simplify to lowest terms.

5. $\frac{3}{4} \div \frac{9}{8}$
6. $\frac{7}{10} - \frac{2}{15}$

Exercises 7 and 8: Evaluate the expression by hand.

7. $20 - 2 \cdot 3$
8. -3^2

Exercises 9 and 10: Classify the number as one or more of the following: natural number, whole number, integer, rational number, or irrational number.

9. $-\frac{4}{5}$
10. $\sqrt{3}$

Exercises 11 and 12: Simplify the expression.

11. $3 + 4x - 2 + 3x$
12. $2(x - 1) - (x + 2)$

Exercises 13 and 14: Solve the equation. Check your solution.

13. $3t - 5 = 1$
14. $2(x - 3) = -6 - x$

Exercises 15 and 16: Find the area of the figure shown.

15.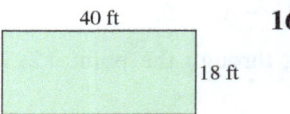
16.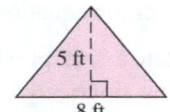

17. Complete the table. Then use the table to solve the equation $6 - 2x = 4$.

x	-2	-1	0	1	2
$6 - 2x$					

18. Translate the sentence "Twice a number increased by 2 equals the number decreased by 5" to an equation, using the variable n. Then solve the equation.

19. If $r = 10$ mph and $d = 80$ miles, use the formula $d = rt$ to find t.

20. Use an inequality to express the set of real numbers graphed.

Exercises 21 and 22: Solve the inequality. Write the solution set in set-builder notation.

21. $3 - 6x < 3$

22. $2x \le 1 - (2x - 1)$

23. Identify the coordinates of each point in the graph. State the quadrant, if any, in which each point lies.

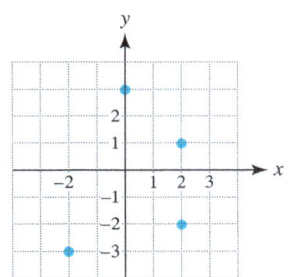

24. Complete the table for the equation $x + 2y = 4$.

x	-2	-1	0	1	2
y					

Exercises 25 and 26: Graph the equation.

25. $2x + 3y = -6$ 26. $y = -\frac{3}{2}$

27. Determine the intercepts for the graph of the equation $-4x + 5y = 40$.

28. Write an equation for each line shown in the graph.

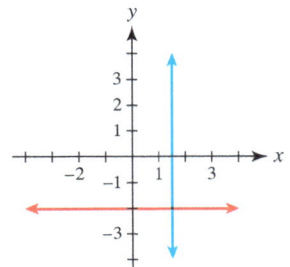

29. The table lists points located on a line. Write the slope–intercept form of the line.

x	-1	0	1	2
y	-6	-3	0	3

30. Write the equation $3x - 5y = 15$ in slope–intercept form. Graph the equation.

Exercises 31 and 32: Find the slope–intercept form for the line satisfying the given conditions.

31. Perpendicular to $3x - 2y = 6$, passing through the point $(0, -3)$

32. Passing through $(-1, 3)$ and $(2, -3)$

33. Let the initial value of y be 100 pounds, increasing at 5 pounds per hour. Find an equation in slope–intercept form that models y after x hours.

34. An insect population is initially 20,000 and is increasing at 5000 per day. Write an equation that gives the number of insects I after x days.

APPLICATIONS

35. **Pizzas** The table lists the cost C of buying x large pepperoni pizzas. Write an equation that relates C and x.

x	2	3	4
C	$16	$24	$32

36. **Photo-Sharing Sites** In 2013, among the top four photo-sharing sites, Snapchat and Instagram made up about $\frac{14}{25}$ of the daily photo uploads. For every Instagram upload, there were 7 Snapchat uploads. Estimate the fraction of the uploads that were on Snapchat.

37. **Investment Money** A student invests two sums of money at 3% and 4% interest, receiving a total of $110 in interest after 1 year. Twice as much money is invested at 4% as at 3%. Find the amount invested at each interest rate.

38. **Rental Cars** The cost C in dollars for driving a rental car x miles is $C = 0.3x + 25$.
    (a) How much does it cost to rent the car and drive it 200 miles?
    (b) How much does it cost to rent the car but not drive it?
    (c) How much does the cost increase for each additional mile driven?

11 Systems of Linear Equations in Two Variables

- **11.1** Solving Systems of Linear Equations Graphically and Numerically
- **11.2** Solving Systems of Linear Equations by Substitution
- **11.3** Solving Systems of Linear Equations by Elimination
- **11.4** Systems of Linear Inequalities

With a land-line telephone, information is typically sent along a single wire, or path. However, if a person is talking on a cell phone in a car, any *one* path could fail at any time because the location of the person's cell phone is always changing. As a result, information must be sent simultaneously along several paths so that this information can be reconstructed even when one or more paths fail during transmission. Mathematics, and *systems of equations* in particular, play an important role in determining how information is sent. (*Source:* Effros, M., and A. Goldsmith, *Scientific American*.)

In this chapter we discuss graphical, numerical, and symbolic methods for solving systems of linear equations.

11.1 Solving Systems of Linear Equations Graphically and Numerically

Objectives

1. **Solving Equations Graphically and Numerically**
   - Solving Graphically
   - Solving Numerically with a Table of Values
2. **Introducing Systems of Linear Equations**
   - Types of Equations and Number of Solutions
   - Checking a Solution to a System of Equations
   - Solving Systems Graphically and Numerically
3. **Solving Applications of Systems of Linear Equations**
   - A Four-step Process for Solving Applications

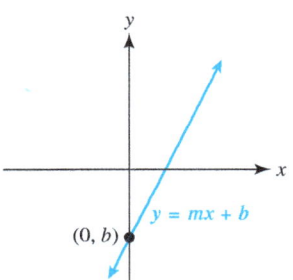

FIGURE 11.1

NEW VOCABULARY

☐ Intersection-of-graphs method
☐ System of linear equations in two variables
☐ Solution to a system
☐ Inconsistent system
☐ Consistent system with independent equations
☐ Consistent system with dependent equations

1 Solving Equations Graphically and Numerically

SOLVING GRAPHICALLY In the previous chapter, we showed that the graph of $y = mx + b$ is a line with slope m and y-intercept $(0, b)$, as illustrated in **FIGURE 11.1**. Each point on this line represents a solution to the equation $y = mx + b$. Because there are infinitely many points on a line, there are infinitely many solutions to this equation. However, many applications require that we find one particular solution to a linear equation. One way to find such a solution is to graph a second line in the same xy-plane and determine the point of intersection (if one exists).

Connecting Concepts with Your Life Consider the following application of a line. If renting a moving truck for one day costs $25 plus $0.50 per mile driven, then the equation $C = 0.5x + 25$ represents the cost C in dollars of driving the rental truck x miles. The graph of this line is shown in **FIGURE 11.2(a)** for $x \geq 0$.

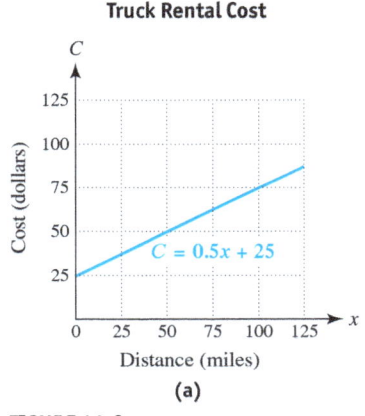
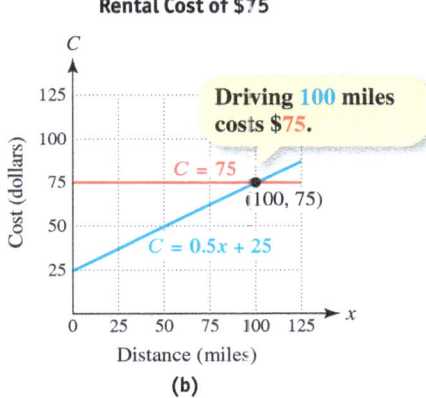

FIGURE 11.2

Suppose that we want to determine the number of miles that the truck is driven when the rental cost is $75. One way to solve this problem *graphically* is to graph both $C = 0.5x + 25$ and $C = 75$ in the same coordinate plane, as shown in **FIGURE 11.2(b)**. The lines intersect at the point $(100, 75)$, which is a solution to $C = 0.5x + 25$ and to $C = 75$. That is, if the rental cost is $75, then the mileage must be 100 miles. This graphical technique for solving two equations is sometimes called the **intersection-of-graphs method**. To find a solution with this method, we locate a point where two graphs intersect.

STUDY TIP

Try to find a consistent time and place to study your notes and do your homework. When the time comes to study for an exam, do so at your usual study time in your usual place rather than "pulling an all-nighter" in unfamiliar surroundings.

EXAMPLE 1 Solving an equation graphically

Use a graph to find the x-value when $y = 3$.
(a) $y = 2x - 1$ (b) $-3x + 2y = 12$

Solution

(a) Begin by graphing the equations $y = 2x - 1$ and $y = 3$. The graph of $y = 2x - 1$ is a line with slope 2 and y-intercept $(0, -1)$. The graph of $y = 3$ is a horizontal line with y-intercept $(0, 3)$. In **FIGURE 11.3(a)**, the graphs intersect at the point $(2, 3)$. Therefore an x-value of 2 corresponds to a y-value of 3.

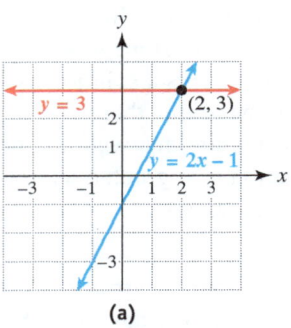

FIGURE 11.3

(b) One way to graph $-3x + 2y = 12$ is to write this equation in slope–intercept form.

$$-3x + 2y = 12 \quad \text{Given equation}$$
$$2y = 3x + 12 \quad \text{Add } 3x \text{ to each side.}$$
$$y = \frac{3}{2}x + 6 \quad \text{Divide each side by 2.}$$

The line has slope $\frac{3}{2}$ and y-intercept $(0, 6)$. Its graph and the graph of $y = 3$ are shown in **FIGURE 11.3(b)**. The graphs intersect at $(-2, 3)$. Therefore an x-value of -2 corresponds to a y-value of 3.

Now Try Exercises 15, 17

EXAMPLE 2 Solving an equation graphically

The equation $P = 10x$ calculates an employee's pay for working x hours at \$10 per hour. Use the intersection-of-graphs method to find the number of hours that the employee worked if the amount paid is \$40.

Solution

Begin by graphing the equations $P = 10x$ and $P = 40$, as illustrated at right.

The graphs intersect at the point $(4, 40)$. Since the x-coordinate represents the number of hours worked and the P-coordinate represents pay, the point $(4, 40)$ indicates that the employee must work 4 hours to earn $\$40$.

Now Try Exercise 73(a), (b)

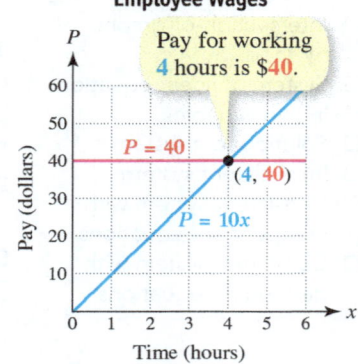

READING CHECK 1

- How is the intersection-of-graphs method used to find a solution to two equations?

SOLVING NUMERICALLY WITH A TABLE OF VALUES Equations can be solved in more than one way. In Example 2, we determined graphically that $P = 10x$ is equal to 40 when $x = 4$. We could also solve this problem by making a table of values, as illustrated by **TABLE 11.1**. Note that, when $x = 4$, $P = \$40$. A table of values provides a *numerical solution*.

Wages Earned at $10 per Hour

x (hours)	0	1	2	3	4	5	6
P (pay)	$0	$10	$20	$30	**$40**	$50	$60

TABLE 11.1

Find the x-value where $P = 40$.

NOTE: Although graphical and numerical methods are different, both methods should give the same solution. However, slight variations may occur because reading a graph precisely may be difficult, or a needed value may not appear in a table. ■

TECHNOLOGY NOTE

Intersection of Graphs and Table of Values
A graphing calculator can be used to find the intersection of the two graphs shown in Example 2. It can also be used to create **TABLE 11.1**. The accompanying figures illustrate how a calculator can be used to determine that $y_1 = 10x$ equals $y_2 = 40$ when $x = 4$.

CALCULATOR HELP

To find a point of intersection, see Appendix A (page AP-6).
To make a table, see Appendix A (pages AP-2 and AP-3).

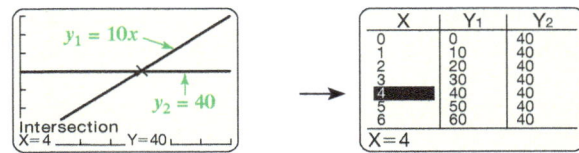

2 Introducing Systems of Linear Equations

In Example 1(b), we determined the x-value when $y = 3$ in the equation $-3x + 2y = 12$. This problem can be thought of as solving the following *system of two equations in two variables*.

$$-3x + 2y = 12$$
$$y = 3$$

READING CHECK 2

• How do we express the solution to a system of equations in two variables?

The solution is the ordered pair $(-2, 3)$, which indicates that when $x = -2$ and $y = 3$, each equation is a true statement.

$-3(-2) + 2(3) = 12$ ✓ A true statement

$3 = 3$ ✓ A true statement

Suppose that the sum of two numbers is 10 and that their difference is 4. If we let x and y represent the two numbers, then the equations

$x + y = 10$ Sum is 10.

$x - y = 4$ Difference is 4.

describe this situation. Each equation is a linear equation in two variables, so we call these equations a **system of linear equations in two variables**. Its graph typically consists of two lines. A **solution to a system** of two equations is an ordered pair (x, y) that makes *both* equations true. If a single solution exists, the ordered pair gives the coordinates of a point where the two lines intersect.

NOTE: For two distinct lines, there can be no more than one intersection point. If such an intersection point exists, the ordered pair corresponding to it represents the only solution to the system of linear equations. In this case, we say the ordered pair is *the* solution to the system of equations. ■

TYPES OF EQUATIONS AND NUMBER OF SOLUTIONS When a system of two linear equations in two variables is graphed, exactly one of the following situations will result.

See the Concept 1 — THREE TYPES OF LINEAR SYSTEMS

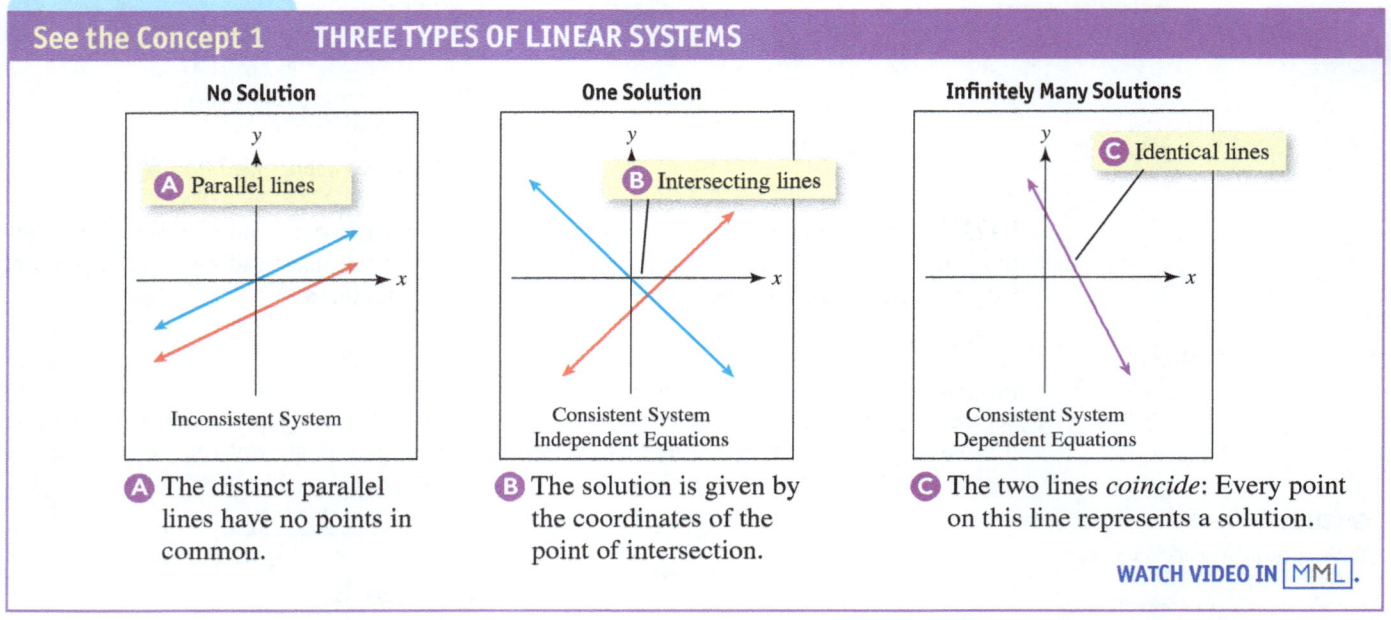

A The distinct parallel lines have no points in common.

B The solution is given by the coordinates of the point of intersection.

C The two lines *coincide*: Every point on this line represents a solution.

WATCH VIDEO IN MML.

In the first situation, there are no solutions, and it is an **inconsistent system**. In the second situation, there is exactly one solution, and it is a **consistent system** with **independent equations**. In the third situation, there are infinitely many solutions, and it is a **consistent system** with **dependent equations**. This information is summarized in **TABLE 11.2**.

Types of Systems of Equations

Type of Graph	Number of Solutions	Type of System	Type of Equations
Parallel Lines	0	Inconsistent	—
Intersecting Lines	1	Consistent	Independent
Identical Lines	Infinitely many	Consistent	Dependent

TABLE 11.2

READING CHECK 3

- How many solutions are possible for a system of linear equations in two variables?

EXAMPLE 3 Identifying types of equations

Graphs of two equations are shown. State the number of solutions to each system of equations. Then state whether the system is consistent or inconsistent. If it is consistent, state whether the equations are dependent or independent.

(a)   (b)   (c)

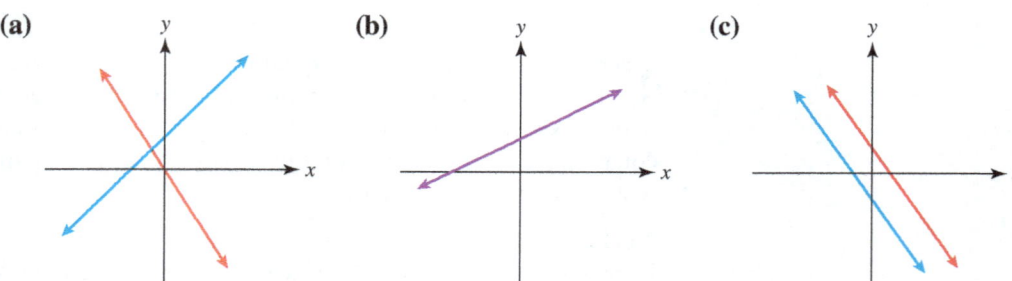

Solution
(a) The lines intersect at one point, so there is one solution. The system is consistent, and the equations are independent.
(b) There is only one line, which indicates that the graphs are identical, or coincide, so there are infinitely many solutions. The system is consistent and the equations are dependent.
(c) The lines are parallel, so there are no solutions. The system is inconsistent.

Now Try Exercises 25, 27, 29

CHECKING A SOLUTION TO A SYSTEM OF EQUATIONS For the remainder of this section, we focus on systems of linear equations in two variables that have exactly one solution. We will solve the systems by graphing and by making a table of values. Methods for solving two equations symbolically will be discussed later in the chapter.

Recall that an ordered pair is a solution to a system of two equations if its coordinates make *both* equations true. In the next example, we test possible solutions.

EXAMPLE 4 Checking possible solutions

Determine whether $(4, 6)$ or $(7, 3)$ is the solution to the system of equations

$$x + y = 10$$
$$x - y = 4.$$

Solution
To determine whether $(4, 6)$ is the solution, substitute $x = 4$ and $y = 6$ in each equation. It must make *both* equations true.

$x + y = 10$	$x - y = 4$	Given equations
$4 + 6 \stackrel{?}{=} 10$	$4 - 6 \stackrel{?}{=} 4$	Let $x = 4, y = 6$.
$10 = 10$ (True) ✓	$-2 = 4$ (False) ✗	Second equation is false.

Because $(4, 6)$ does not satisfy *both* equations, it is not the solution for the system of equations. Next let $x = 7$ and $y = 3$ to determine whether $(7, 3)$ is the solution.

$x + y = 10$	$x - y = 4$	Given equations
$7 + 3 \stackrel{?}{=} 10$	$7 - 3 \stackrel{?}{=} 4$	Let $x = 7, y = 3$.
$10 = 10$ (True) ✓	$4 = 4$ (True) ✓	Both are true.

Because $(7, 3)$ makes *both* equations true, it is the solution to the system of equations.

Now Try Exercise 31

SOLVING SYSTEMS GRAPHICALLY AND NUMERICALLY In the next example, we find the solution to a system of linear equations graphically and numerically.

EXAMPLE 5 Solving a system graphically and numerically

Solve the system of linear equations

$$x + 2y = 4$$
$$2x - y = 3$$

with a graph and with a table of values.

Solution
Graphically Begin by writing each equation in slope–intercept form.

$x + 2y = 4$	First equation		$2x - y = 3$	Second equation
$2y = -x + 4$	Subtract x.		$-y = -2x + 3$	Subtract $2x$.
$y = -\dfrac{1}{2}x + 2$	Divide by 2.		$y = 2x - 3$	Multiply by -1.

The graphs of $y = -\frac{1}{2}x + 2$ and $y = 2x - 3$ are shown in **FIGURE 11.4**. The graphs intersect at the point $(2, 1)$; thus, $(2, 1)$ is the solution.

Numerically **TABLE 11.3** on the next page shows the equations $y = -\frac{1}{2}x + 2$ and $y = 2x - 3$ evaluated for various values of x. Note that when $x = 2$, both equations have a y-value of 1. Thus $(2, 1)$ is the solution.

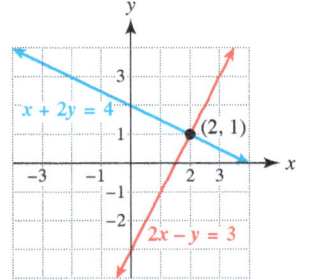

FIGURE 11.4

A Numerical Solution

x	−1	0	1	2	3
$y = -\frac{1}{2}x + 2$	2.5	2	1.5	1	0.5
$y = 2x - 3$	−5	−3	−1	1	3

TABLE 11.3

Find an x-value where the y-values are equal.

Now Try Exercise 51

🌐 Math in Business Context In business, linear equations are sometimes used to model supply and demand for a product. For example, if the price of a coffee drink is too high, the demand for the drink will decrease because consumers are interested in saving money. Similarly, if the price of the coffee drink is too low, supply will decrease because suppliers are interested in making money. To find an appropriate price for the coffee drink, a system of linear equations can be solved.

EXAMPLE 6 Solving a supply and demand problem graphically

The quantity of coffee drinks a company will supply at price P is $Q_s = 3P$ and the quantity of coffee drinks customers will demand at price P is $Q_d = -2P + 10$, where quantity is in hundreds of cups of coffee and price is in dollars. Graph each equation and determine the number of cups of coffee the company should sell and the corresponding price per cup.

Solution
The equation $Q_s = 3P$ is a line passing through the origin with slope 3. The equation $Q_d = -2P + 10$ is a line passing through the point $(0, 10)$ with slope -2. As shown in **FIGURE 11.5**, their intersection point is $(2, 6)$, meaning that the company should sell 600 cups of coffee at a price of $2 per cup.

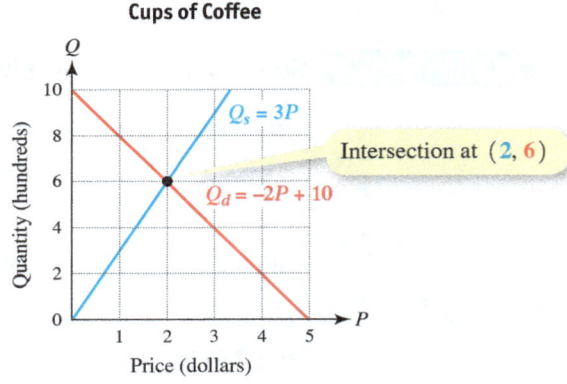

FIGURE 11.5

Now Try Exercise 77

3 Solving Applications of Systems of Linear Equations

A FOUR-STEP PROCESS FOR SOLVING APPLICATIONS In the next example, we use a four-step process to solve an application involving a system of linear equations. These steps are based on the four-step process discussed earlier in this text, with Step 3 split into two parts to emphasize the importance of using the solution to the system of equations to determine the solution to the given problem.

EXAMPLE 7 **Reasons for not using the Internet**

Of American adults who don't use the Internet, about 33% are either "Just not interested" or "Don't have a computer." There are 7% more adults who are "Just not interested" than those who "Don't have a computer." Find the percentage of adults that are "Just not interested" and the percentage that "Don't have a computer." (*Source:* Pew Center for the Internet.)

Solution

STEP 1: *Identify each variable.*

x: percentage of adults who are "Just not interested"

y: percentage of adults who "Don't have a computer"

STEP 2: *Write a system of equations.* Of those who don't use the Internet, the total percentage of adults who are either "Just not interested" or "Don't have a computer" is 33%, so $x + y = 33$. Because 7% more adults were "Just not interested" than those who "Don't have a computer," we also know that $x - y = 7$. Thus a system of equations representing this situation is

$$x + y = 33$$
$$x - y = 7.$$

STEP 3A: *Solve the system of equations.* To solve this system graphically, write each equation in slope–intercept form.

$$y = -x + 33$$
$$y = x - 7$$

Their graphs intersect at the point $(20, 13)$, as shown in **FIGURE 11.6**.

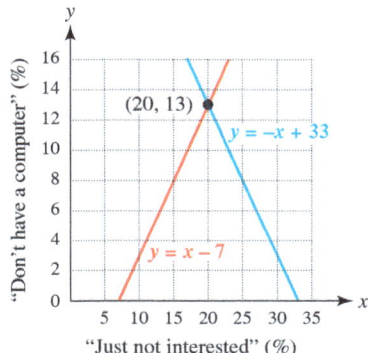

FIGURE 11.6

STEP 3B: *Determine the solution to the problem.* The point $(20, 13)$ corresponds to $x = 20$ and $y = 13$. Thus about 20% of adults who don't use the Internet are "Just not interested" and 13% "Don't have a computer."

STEP 4: *Check your solution.* Note that $20 + 13 = 33\%$ of adults who don't use the Internet are either "Just not interested" or "Don't have a computer," and $20 - 13 = 7\%$ more stated that they are "Just not interested" compared to those who "Don't have a computer."

Now Try Exercise 75

680 CHAPTER 11 SYSTEMS OF LINEAR EQUATIONS IN TWO VARIABLES

CALCULATOR HELP

To find a point of intersection, see Appendix A (page AP-6).

TECHNOLOGY NOTE

Checking Solutions
The solution to a system can be checked with a graphing calculator, as shown in the accompanying figure where the graphs of $y_1 = -x + 50$ and $y_2 = x - 10$ intersect at the point $(30, 20)$. However, a graphing calculator cannot read an application problem and write down the system of equations. A human mind is needed for these tasks.

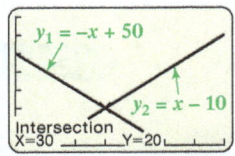

🌐 *Math in Context* (Communications) In the opening discussion for this chapter, we considered how systems of equations are used to transmit information reliably over wireless networks. To better understand how this is done, consider the following simple example.

Suppose we want to send the number 35 from one cell phone to another. Begin by letting $x = 3$ and $y = 5$. Instead of sending both 3 and 5 over one path, which could fail, we will use four paths. Over the first path we send $x = 3$, over the second $y = 5$, over the third $x + y = 8$, and over the fourth $2x + y = 11$. See the figure below.

Systems of Equations in Wireless Networks

Sending the Number 35

$x + y = 8$ and $x = 3$, so $y = 5$.

In the figure, any two paths can fail and the receiving phone can still determine the two numbers, 3 and 5. For example, suppose that the second and fourth paths shown fail. Then the receiving phone gets the information only from the first and third paths, $x = 3$ and $x + y = 8$. However, it can still determine the values x and y by solving the following system of linear equations.

$$x = 3 \quad \text{First path information}$$
$$x + y = 8 \quad \text{Third path information}$$

The solution is given by $x = 3$ and $y = 5$, so the number is 35.

NOTE: Wireless network systems are often designed so that any pair of equations is consistent and independent, and thus the solution (information) is unique. ∎

EXAMPLE 8 Sending cell phone messages

A person's age is being transmitted over a wireless network in a manner similar to that illustrated in the figure above. Let x be the tens digit and y be the ones digit. Suppose that the first and second paths fail, resulting in the following system that must be solved.

$$x + y = 3 \quad \text{Third path information}$$
$$2x + y = 5 \quad \text{Fourth path information}$$

Solve this system graphically and find the person's age.

Decoding Cell Phone Message

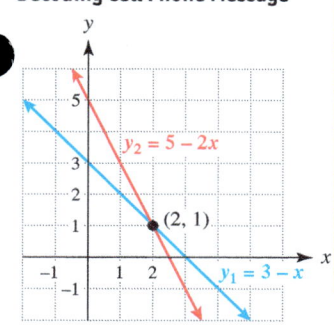

FIGURE 11.7 Person's Age: 21

Solution
Begin by solving each equation for y.

$$y = 3 - x \quad \text{Subtract } x \text{ in third path equation.}$$
$$y = 5 - 2x \quad \text{Subtract } 2x \text{ in fourth path equation.}$$

Graph $y_1 = 3 - x$ and $y_2 = 5 - 2x$, as in **FIGURE 11.7**. Their graphs intersect at $(2, 1)$. It follows that $x = 2$, $y = 1$, and the person's age is 21.

Now Try Exercise 81

11.1 Putting It All Together

CONCEPT	EXPLANATION	EXAMPLE
System of Linear Equations in Two Variables	Can be written as $$Ax + By = C$$ $$Dx + Ey = F$$	$x + y = 8$ $x - y = 2$
Solution to a System of Equations	An ordered pair (x, y) that satisfies *both* equations	The solution to the preceding system is $(5, 3)$ because, when $x = 5$ and $y = 3$ are substituted, both equations are true. $5 + 3 \stackrel{?}{=} 8$ ✓ True $5 - 3 \stackrel{?}{=} 2$ ✓ True
Graphical Solution to a System of Equations	Graph each equation. A point of intersection represents a solution.	The graphs of $y = -x + 8$ and $y = x - 2$ intersect at $(5, 3)$.
Numerical Solution to a System of Equations	Make a table for each equation. A solution occurs when one x-value gives the same y-values in both equations.	Make a table for $y = -x + 8$ and $y = x - 2$. When $x = 5$, $y = 3$ in both equations, so $(5, 3)$ is the solution. | x | 4 | 5 | 6 | | $y = -x + 8$ | 4 | 3 | 2 | | $y = x - 2$ | 2 | 3 | 4 |

continued on next page

continued from previous page

THREE TYPES OF LINEAR SYSTEMS

No Solution	One Solution	Infinitely Many Solutions
Parallel lines	Intersecting lines	Identical lines
Inconsistent System	Consistent System / Independent Equations	Consistent System / Dependent Equations
The distinct parallel lines have no points in common.	The solution is given by the coordinates of the point of intersection.	The two lines *coincide*: Every point on this line represents a solution.

11.1 Exercises

CONCEPTS AND VOCABULARY

1. A solution to a system of two equations in two variables is a(n) _____ pair.

2. A graphical technique for solving a system of two equations in two variables is the _____ method.

3. A system of linear equations can have _____, _____, or _____ solutions.

4. If a system of linear equations has at least one solution, then it is a(n) (consistent/inconsistent) system.

5. If a system of linear equations has no solutions, then it is a(n) (consistent/inconsistent) system.

6. If a system of linear equations has exactly one solution, then the equations are (dependent/independent).

7. If a system of linear equations has infinitely many solutions, then the equations are (dependent/independent).

8. To find a numerical solution to a system, start by creating a(n) _____ of values for the equations.

9. If a graphical method and a numerical method (table of values) are used to solve the same system of equations, then the two solutions should be (the same/different).

10. One way to graph a line is to write its equation in slope–intercept form. A second method is to find the x- and y- _____.

SOLVING AN EQUATION GRAPHICALLY

Exercises 11–18: Determine graphically the x-value when $y = 2$ in the given equation.

11. $y = 2x$
12. $y = \frac{1}{3}x$
13. $y = 4 - x$
14. $y = -2 - x$
15. $y = -\frac{1}{2}x + 1$
16. $y = 3x - 1$
17. $-3x + 4y = 11$
18. $2x + y = 6$

SOLVING AN EQUATION NUMERICALLY WITH A TABLE OF VALUES

Exercises 19–24: Complete the table. Then use it to solve the given equation.

19. $y = 4$

x	0	1	2	3
$y = x + 3$				

20. $y = -1$

x	-1	0	1	2
$y = 2x - 1$				

21. $y = -3$

x	-4	-2	2	4
$y = 3 - 3x$				

22. $y = -42$

x	-10	-5	5	10
$y = 8 - 5x$				

23. $y = 0$

x	-3	0	3	6
$2x + 3y = 6$				

24. $y = -4$

x	-3	0	3	6
$4x - 3y = 12$				

DETERMINING TYPES OF EQUATIONS AND NUMBER OF SOLUTIONS

Exercises 25–30: The graphs of two equations are shown. State the number of solutions to each system of equations. Then state whether the system is consistent or inconsistent. If it is consistent, state whether the equations are dependent or independent.

CHECKING A SOLUTION TO A SYSTEM OF EQUATIONS

Exercises 31–36: Determine which ordered pair is a solution to the system of equations.

31. $(0, 0), (1, 1)$
$x + y = 2$
$x - y = 0$

32. $(-1, 2), (1, -2)$
$2x + y = 0$
$x - 2y = -5$

33. $(-1, -1), (2, -3)$
$2x + 3y = -5$
$4x - 5y = 23$

34. $(2, -1), (-2, -2)$
$-x + 4y = -6$
$6x - 7y = 19$

35. $\left(\frac{1}{5}, \frac{4}{5}\right), \left(\frac{1}{4}, \frac{1}{3}\right)$
$-5x + 5y = 3$
$4x + 9y = 8$

36. $\left(\frac{1}{2}, \frac{3}{2}\right), \left(\frac{3}{4}, \frac{5}{4}\right)$
$x + y = 2$
$3x - y = 0$

SOLVING SYSTEMS GRAPHICALLY AND NUMERICALLY

Exercises 37–42: The graphs of two equations are shown. Use the intersection-of-graphs method to identify the solution to both equations. Then check your answer.

25.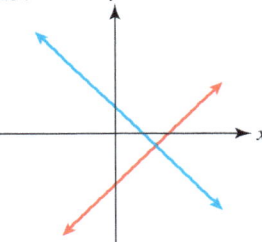

26.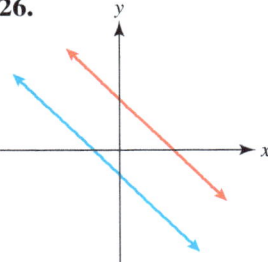

37.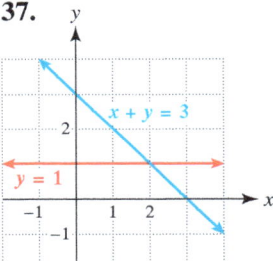

38.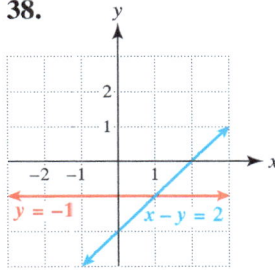

27.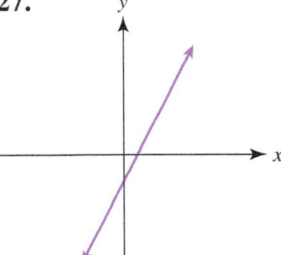

28.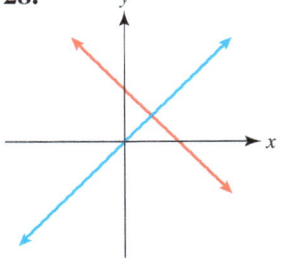

39.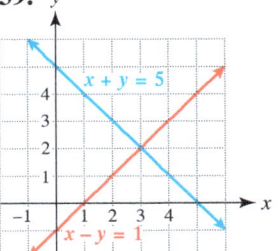

40.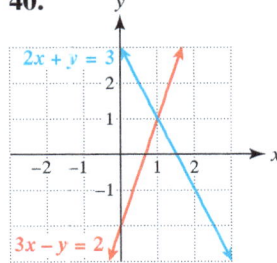

29.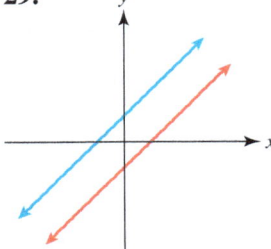

30.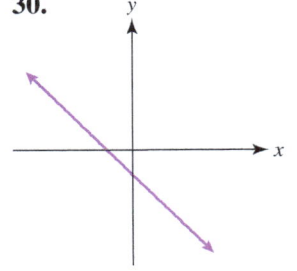

41.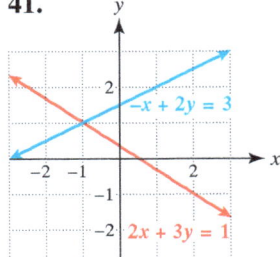

42.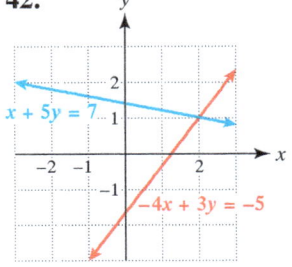

Exercises 43–46: A table for two equations is given. Identify the solution to both equations.

43.
x	1	2	3	4
$y = 2x$	2	4	6	8
$y = 4$	4	4	4	4

44.
x	2	3	4	5
$y = 6 - x$	4	3	2	1
$y = x - 2$	0	1	2	3

45.
x	1	2	3	4
$y = 4 - x$	3	2	1	0
$y = x - 2$	-1	0	1	2

46.
x	1	2	3	4
$y = 6 - 3x$	3	0	-3	-6
$y = 2 - x$	1	0	-1	-2

Exercises 47 and 48: Complete the table for each equation. Then identify the solution to both equations.

47.
x	0	1	2	3
$y = x + 2$				
$y = 4 - x$				

48.
x	-5	-4	-3	-2
$y = 2x + 1$				
$y = x - 1$				

Exercises 49–54: Use the specified method to solve the system of equations.

(a) Graphically
(b) Numerically (table of values)

49. $y = 2x + 3$
    $y = 1$

50. $y = 2 - x$
    $y = 0$

51. $y = 4 - x$
    $y = x - 2$

52. $y = 2x$
    $y = -\frac{1}{2}x$

53. $y = 3x$
    $y = x + 2$

54. $y = 2x - 3$
    $y = -x + 3$

Exercises 55–66: Solve the system of equations graphically.

55. $x + y = -3$
    $x - y = 1$

56. $x - y = 3$
    $2x + y = 3$

57. $2x - y = 3$
    $3x + y = 2$

58. $x + 2y = 6$
    $-x + 3y = 4$

59. $-4x + 2y = 0$
    $x - y = -1$

60. $4x - y = 2$
    $y = 2x$

61. $2x - y = 4$
    $x + 2y = 7$

62. $x - y = 2$
    $\frac{1}{2}x + y = 4$

63. $x = -y + 4$
    $x = 3y$

64. $x = 2y$
    $y = -\frac{1}{2}x$

65. $x + y = 3$
    $x = \frac{1}{2}y$

66. $2x - 4y = 8$
    $\frac{1}{2}x + y = -4$

WRITING AND SOLVING SYSTEMS

Exercises 67–72: **Number Problems** *For each problem, complete each of the following.*

(a) Write a system of equations for the problem.
(b) Find the unknown numbers by solving the system of equations graphically.

67. The sum of two numbers is 4 and their difference is 0.

68. The sum of two numbers is -5 and their difference is 1.

69. The sum of twice a number and another number is 7. Their difference is 2.

70. Three times a number subtracted from another number results in 1. Their sum is 5.

71. One number is triple another number. Their difference is 4.

72. Half of a number added to another number equals 5. Their difference is 1.

APPLICATIONS INVOLVING SYSTEMS OF EQUATIONS

73. **Renting a Truck** A rental truck costs $50 plus $0.50 per mile.
    (a) Write an equation that gives the cost C of driving the truck x miles.
    (b) Use the intersection-of-graphs method to determine the number of miles that the truck is driven if the rental cost is $80.
    (c) Solve part (b) numerically with a table of values.

74. **Renting a Car** A rental car costs $25 plus $0.25 per mile.
    (a) Write an equation that gives the cost C of driving the car x miles.
    (b) Use the intersection-of-graphs method to determine the number of miles that the car is driven if the rental cost is $100.
    (c) Solve part (b) numerically with a table of values.

75. **Recorded Music** In 2013, digital downloads and streaming subscriptions accounted for 60% of all music revenues. Digital downloads revenues were double the

streaming subscriptions revenues. (*Source:* Recording Industry Association of America.)

(a) Let x be the percentage of revenues due to digital downloads and y be the percentage of revenues due to streaming subscriptions. Write a system of two equations that describes the given information.

(b) Solve your system graphically and interpret your solution.

76. **iPhone Colors** About 70% of iPhone 5S users have a phone that is either gold or space gray. About 16% more users have a phone that is space gray than a phone that is gold.

(a) Let x be the percentage of users who have a space gray phone and y be the percentage of users who have a gold phone. Write a system of two equations that describes the given information.

(b) Solve your system graphically and interpret your solution.

77. **Supply and Demand** The quantity of e-readers a company supplies per week at price P is $Q_s = 1.5P$ and the quantity of e-readers their customers demand at price P is $Q_d = 300 - P$, where P is in dollars. Determine graphically the number of e-readers the company should sell and the corresponding price per e-reader.

78. **Supply and Demand** The quantity of donuts a bakery supplies per week at price P is $Q_s = P$ and the quantity of donuts their customers demand at price P is $Q_d = 9 - 2P$, where price is in dollars and quantity is in hundreds of donuts. Determine graphically the number of donuts the bakery should sell and the corresponding price per donut.

79. **Dimensions of a Rectangle** A rectangle is 4 inches longer than it is wide. Its perimeter is 28 inches.

(a) Write a system of two equations in two variables that describes this information. Be sure to specify what each variable means.

(b) Solve your system graphically. Interpret your results.

80. **Dimensions of a Triangle** An isosceles triangle has a perimeter of 17 inches, with its two shorter sides equal in length. The longest side measures 2 inches more than either of the shorter sides.

(a) Write a system of two equations in two variables that describes this information. Be sure to specify what each variable means.

(b) Solve your system graphically. Explain what your results mean.

Exercises 81–84: **Wireless Networks** (*Refer to Example 8.*) *The system of four equations represents a wireless transmission of a person's age.*

$$x = 2 \quad (1)$$
$$y = 3 \quad (2)$$
$$x + y = 5 \quad (3)$$
$$3x + y = 9 \quad (4)$$

Solve the given pair of equations and find the age of the person. Do you get the same answer in each case?

81. Equations (1) and (3)   82. Equations (2) and (3)

83. Equations (2) and (4)   84. Equations (3) and (4)

WRITING ABOUT MATHEMATICS

85. Use the intersection-of-graphs method to help explain why you typically expect a linear system in two variables to have one solution.

86. Could a system of two linear equations in two variables have exactly two solutions? Explain your reasoning.

87. Give one disadvantage of using a table to solve a system of equations. Explain your answer.

88. Do the equations $y = 2x + 1$ and $y = 2x - 1$ have a common solution? Explain your answer.

11.2　Solving Systems of Linear Equations by Substitution

Objectives

1. Using the Substitution Method
   - Solving for a Variable Before Using Substitution
2. Recognizing Systems with No Solutions or Infinitely Many Solutions
3. Solving Applications Using the Substitution Method

NEW VOCABULARY

☐ Method of substitution

READING CHECK 1

- What is an advantage of using the method of substitution rather than using a graph or table to solve a system of linear equations?

STUDY TIP

Questions on exams do not always come in the order that they are presented in the text. When studying for an exam, choose review exercises randomly so that the topics are studied in the same random way they may appear on an exam.

1 Using the Substitution Method

In the previous section, we solved systems of linear equations by using graphs and tables. A disadvantage of a graph is that reading the graph precisely can be difficult. A disadvantage of using a table is that locating the solution can be difficult when it is either a fraction or a large number. In this subsection, we introduce the *method of substitution*, in which we solve systems of equations symbolically. The advantage of this method is that the *exact* solution can always be found (provided it exists).

Connecting Concepts with Your Life Suppose that you and a friend earned $120 together. If x represents how much your friend earned and y represents how much you earned, then the equation $x + y = 120$ describes this situation. Now, if we also know that you earned twice as much as your friend, then we can include a second equation, $y = 2x$. The amount that each of you earned can now be determined by *substituting* $2x$ for y in the first equation.

$$x + y = 120 \quad \text{First equation}$$
$$x + 2x = 120 \quad \text{Substitute } 2x \text{ for } y.$$
$$3x = 120 \quad \text{Combine like terms.}$$
$$x = 40 \quad \text{Divide each side by 3.}$$

So your friend earned $40, and you earned twice as much, or $80.

This technique of substituting an expression for a variable and solving the resulting equation is called the **method of substitution**. The following four-step process can be used when using the method of substitution.

STEPS FOR USING THE METHOD OF SUBSTITUTION

STEP 1: Solve one of the equations for one variable.

STEP 2: Substitute the expression found in Step 1 in the other equation and solve the resulting equation to find the value of one variable.

STEP 3: Use the solution from Step 2 to find the value of the other variable. Write the solution to the system of equations as an ordered pair.

STEP 4: Check the solution.

EXAMPLE 1 Using the method of substitution

Solve each system of equations.

(a) $2x + y = 10$
    $y = 3x$

(b) $-2x + 3y = -8$
    $x = 3y + 1$

Solution

(a) **STEP 1:** The second equation is already solved for y.

STEP 2: From the second equation, substitute $3x$ for y in the first equation.

$$2x + y = 10 \quad \text{First equation}$$
$$2x + 3x = 10 \quad \text{Substitute } 3x \text{ for } y.$$
$$5x = 10 \quad \text{Combine like terms.}$$
$$x = 2 \quad \text{Divide each side by 5.}$$

STEP 3: The solution to this system is an *ordered pair*, so we must also find y. Because $y = 3x$ and $x = 2$, it follows that $y = 3(2) = 6$. The solution is $(2, 6)$.

STEP 4: To check the solution, substitute 2 for x and 6 for y in each equation.

$2x + y = 10$	$y = 3x$	Given equations
$2(2) + 6 \stackrel{?}{=} 10$	$6 \stackrel{?}{=} 3(2)$	Let $x = 2$ and $y = 6$.
$10 = 10$ ✓	$6 = 6$ ✓	Both are true.

(b) STEP 1: The second equation, $x = 3y + 1$, is solved for x.

STEP 2: Substitute $(3y + 1)$ for x in the first equation. Include parentheses around the expression $3y + 1$ since this entire expression is to be multiplied by -2.

$-2x + 3y = -8$	First equation
$-2(3y + 1) + 3y = -8$	Substitute $(3y + 1)$ for x.
$-6y - 2 + 3y = -8$	Distributive property
$-3y - 2 = -8$	Combine like terms.
$-3y = -6$	Add 2 to each side.
$y = 2$	Divide each side by -3.

Be sure to include parentheses.

STEP 3: To find x, substitute 2 for y in $x = 3y + 1$ to obtain $x = 3(2) + 1 = 7$. The solution is $(7, 2)$.

STEP 4: Substitute 7 for x and 2 for y in each equation.

$-2x + 3y = -8$	$x = 3y + 1$	Given equations
$-2(7) + 3(2) \stackrel{?}{=} -8$	$7 \stackrel{?}{=} 3(2) + 1$	Let $x = 7$ and $y = 2$.
$-8 = -8$ ✓	$7 = 7$ ✓	Both are true.

Now Try Exercises 5, 11

READING CHECK 2

- When substituting an expression with two or more terms for a single variable, why is it important to use parentheses?

NOTE: When an expression contains two or more terms, it is usually best to place parentheses around it when substituting it for a single variable in an equation. In Example 1(b), the distributive property would not have been applied correctly without the parentheses. ∎

SOLVING FOR A VARIABLE BEFORE USING SUBSTITUTION Sometimes it is necessary to solve one of the equations for a variable before substitution can be used, as demonstrated in the next example.

EXAMPLE 2 Using the method of substitution

Solve each system of equations.
(a) $x + y = 8$
$2x - 3y = 6$
(b) $3a - 2b = 2$
$a + 4b = 3$

Solution
(a) STEP 1: Neither equation is solved for a variable, but we can easily solve the first equation for y.

$x + y = 8$	First equation
$y = 8 - x$	Subtract x from each side.

STEP 2: Now we can substitute $(8 - x)$ for y in the second equation.

$2x - 3y = 6$	Second equation
$2x - 3(8 - x) = 6$	Substitute $(8 - x)$ for y.
$2x - 24 + 3x = 6$	Distributive property
$5x = 30$	Combine like terms; add 24.
$x = 6$	Divide each side by 5.

(Include parentheses.)

STEP 3: Because $y = 8 - x$ and $x = 6$, we know that $y = 8 - 6 = 2$. The solution is $(6, 2)$.

STEP 4: To check the solution, substitute 6 for x and 2 for y in each equation.

$x + y = 8$	$2x - 3y = 6$	Given equations
$6 + 2 \stackrel{?}{=} 8$	$2(6) - 3(2) \stackrel{?}{=} 6$	Let $x = 6$ and $y = 2$.
$8 = 8$ ✓	$6 = 6$ ✓	Both are true.

(b) STEP 1: Although we could solve either equation for either variable, solving the second equation for a is easiest because the coefficient of a is 1 and we can avoid fractions.

$a + 4b = 3$	Second equation
$a = 3 - 4b$	Subtract $4b$ from each side.

STEP 2: Now substitute $(3 - 4b)$ for a in the first equation.

$3a - 2b = 2$	First equation
$3(3 - 4b) - 2b = 2$	Substitute $(3 - 4b)$ for a.
$9 - 12b - 2b = 2$	Distributive property
$-14b = -7$	Combine like terms; subtract 9.
$b = \dfrac{1}{2}$	Divide each side by -14.

STEP 3: Substitute $b = \frac{1}{2}$ in $a = 3 - 4b$ to obtain $a = 1$. The solution is $\left(1, \frac{1}{2}\right)$.

STEP 4: Substitute 1 for a and $\frac{1}{2}$ for b in each equation.

$3a - 2b = 2$	$a + 4b = 3$	Given equations
$3(1) - 2\left(\dfrac{1}{2}\right) \stackrel{?}{=} 2$	$1 + 4\left(\dfrac{1}{2}\right) \stackrel{?}{=} 3$	Let $a = 1$ and $b = \frac{1}{2}$.
$2 = 2$ ✓	$3 = 3$ ✓	Both are true.

Now Try Exercises 17, 23

NOTE: When a system of equations contains variables other than x and y, we will list them alphabetically in an ordered pair. ■

2 Recognizing Systems with No Solutions or Infinitely Many Solutions

A system of linear equations typically has exactly one solution. However, in the next example we see how the method of substitution can be used on systems that have no solutions or infinitely many solutions.

EXAMPLE 3 Solving other types of systems

If possible, use substitution to solve the system of equations. Then use graphing to help explain the result.

(a) $3x + y = 4$
$6x + 2y = 2$

(b) $x + y = 2$
$2x + 2y = 4$

Solution

(a) Solve the first equation for y to obtain $y = 4 - 3x$. Next substitute $(4 - 3x)$ for y in the second equation.

$$6x + 2y = 2 \qquad \text{Second equation}$$
$$6x + 2(4 - 3x) = 2 \qquad \text{Substitute } (4 - 3x) \text{ for } y.$$
$$6x + 8 - 6x = 2 \qquad \text{Distributive property}$$
$$8 = 2 \text{ (False)} \qquad \text{Combine like terms.}$$

The equation $8 = 2$ is *always false*, which indicates that there are *no solutions*. One way to graph each equation is to write the equations in slope–intercept form.

$3x + y = 4$ First equation | $6x + 2y = 2$ Second equation

$y = -3x + 4$ Subtract $3x$. | $y = -3x + 1$ Subtract $6x$; divide by 2.

The graphs of these equations are parallel lines with slope -3, as shown in **FIGURE 11.8(a)**. Because the lines *do not intersect*, there are *no solutions* to the system of equations.

(b) Solve the first equation for y to obtain $y = 2 - x$. Now substitute $(2 - x)$ for y in the second equation.

$$2x + 2y = 4 \qquad \text{Second equation}$$
$$2x + 2(2 - x) = 4 \qquad \text{Substitute } (2 - x) \text{ for } y.$$
$$2x + 4 - 2x = 4 \qquad \text{Distributive property}$$
$$4 = 4 \text{ (True)} \qquad \text{Combine like terms.}$$

The equation $4 = 4$ is *always true*, which means that there are *infinitely many solutions*. One way to graph these equations is to write them in slope–intercept form.

$x + y = 2$ First equation | $2x + 2y = 4$ Second equation

$y = -x + 2$ Subtract x. | $y = -x + 2$ Subtract $2x$; divide by 2.

Because the equations have the same slope–intercept form, their graphs are identical, resulting in a single line, as shown in **FIGURE 11.8(b)**. *Every point on this line represents a solution* to the system of equations, so there are infinitely many solutions.

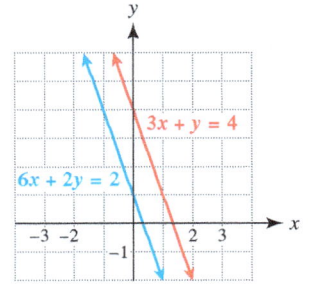

(a) No Solutions

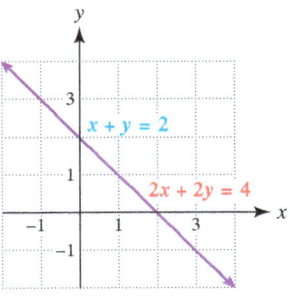

(b) Infinitely Many Solutions
FIGURE 11.8

Now Try Exercises 33, 35

3 Solving Applications Using the Substitution Method

Math in Context *Messaging* Suppose that a boy and a girl sent a total of 760 text messages in one month. With only this information, it is impossible to know how many messages were sent by each person. The boy may have sent 432 messages while the girl sent 328. However, 11 and 749 also total 760, as do many other possibilities. If we are told that the boy sent three times as many messages as the girl, a system of linear equations can be written and solved to find the answer.

EXAMPLE 4 Finding numbers of text messages

A boy and a girl sent a total of 760 text messages in one month. If the boy sent three times as many messages as the girl, how many text messages did each person send?

Solution

If x represents the number of text messages sent by the girl and y represents the number of text messages sent by the boy, then the equation $x + y = 760$ can be written because the total number of messages is 760. Also, the equation $y = 3x$ can be included because the boy sent three times as many messages as the girl. The system is

$$x + y = 760$$
$$y = 3x.$$

Substitute $3x$ for y in the first equation.

$x + y = 760$	First equation
$x + 3x = 760$	Substitute $3x$ for y.
$4x = 760$	Combine like terms.
$x = 190$	Divide each side by 4.

The girl sent 190 text messages and the boy sent 3 times as many, or $3(190) = 570$. This answer checks because $190 + 570 = 760$ and 570 is three times 190.

Now Try Exercise 71

The next two examples illustrate how the method of substitution can be used to solve applications. In these examples, we apply the same four-step process that was used to solve application problems in the previous section.

EXAMPLE 5 Determining time spent online

In December 2013, the total time Americans spent using tablets or smartphones to access the Internet was 566 billion minutes. The amount of online time for smartphones was about 3.565 times the amount of online time for tablets. Find the amount of time spent accessing the Internet on each type of device. (*Source: ComScore.*)

Solution

STEP 1: *Identify each variable.* Clearly identify what each variable represents.

x: tablet time online, in billions of minutes

y: smartphone time online, in billions of minutes

STEP 2: *Write a system of equations.*

$x + y = 566$ Total time is 566 billion minutes.

$y = 3.565x$ Time on smartphones is 3.565 times the amount of time on tablets.

STEP 3A: *Solve the system of linear equations.* Substitute $3.565x$ for y in the first equation.

$x + y = 566$	First equation
$x + 3.565x = 566$	Substitute $3.565x$ for y.
$4.565x = 566$	Combine like terms.
$x = \frac{566}{4.565}$	Divide each side by 4.565.
$x \approx 124$	Simplify.

Because $y = 3.565x$, it follows that $y = 3.565(124) \approx 442$.

STEP 3B: *Determine the solution to the problem.* The solution $x = 124$ and $y = 422$ indicates that online time for tablets was 124 billion minutes and online time for smartphones was 442 billion minutes in December 2013.

STEP 4: *Check the solution.* The sum of these times is $124 + 442 = 566$ billion minutes, and the amount of time spent on smartphones is $\frac{442}{124} \approx 3.565$ times the amount of time spent on tablets. The answer checks.

Now Try Exercise 57

EXAMPLE 6 Determining airplane speed and wind speed

An airplane flies 2400 miles into (or against) the wind in 8 hours. The return trip takes 6 hours. Find the speed of the airplane with no wind and the speed of the wind.

Solution

STEP 1: *Identify each variable.*

x: the speed of the airplane without wind

y: the speed of the wind

STEP 2: *Write a system of equations.* If we solve the distance formula $d = rt$ for the variable r, the result is $r = \frac{d}{t}$. Using this formula, the speed (rate) of the airplane against the wind is $\frac{2400}{8} = 300$ miles per hour, because it traveled 2400 miles in 8 hours. The wind slowed the plane, so $x - y = 300$. Similarly, the airplane flew $\frac{2400}{6} = 400$ miles per hour with the wind because it traveled 2400 miles in 6 hours. The wind made the plane fly faster, so $x + y = 400$.

$$x - y = 300 \quad \text{Speed against the wind}$$
$$x + y = 400 \quad \text{Speed with the wind}$$

STEP 3A: *Solve the system of linear equations.* Solve the first equation for x to obtain $x = y + 300$. Substitute $(y + 300)$ for x in the second equation.

$$x + y = 400 \quad \text{Second equation}$$
$$\mathbf{(y + 300)} + y = 400 \quad \text{Substitute } (y + 300) \text{ for } x.$$
$$2y = 100 \quad \text{Combine like terms; subtract 300.}$$
$$y = 50 \quad \text{Divide each side by 2.}$$

Because $x = y + 300$, it follows that $x = 50 + 300 = 350$.

STEP 3B: *Determine the solution to the problem.* The solution $x = 350$ and $y = 50$ indicates that the airplane can fly 350 miles per hour with no wind, and the wind speed is 50 miles per hour.

STEP 4: *Check the solution.* The plane flies $350 - 50 = 300$ miles per hour into the wind, taking $\frac{2400}{300} = 8$ hours. The plane flies $350 + 50 = 400$ miles per hour with the wind, taking $\frac{2400}{400} = 6$ hours. The answers check.

Now Try Exercise 63

CRITICAL THINKING 1

A boat travels 10 miles per hour upstream and 16 miles per hour downstream. How fast is the current?

11.2 Putting It All Together

CONCEPT	EXPLANATION	EXAMPLE
Method of Substitution	Can be used to solve a system of equations Gives the exact solution, provided one exists STEP 1: Solve one equation for one variable. STEP 2: Substitute the result in the other equation and solve. STEP 3: Use the solution for the first variable to find the other variable. STEP 4: Check the solution.	$x + y = 5$ Sum is 5. $x - y = 1$ Difference is 1. STEP 1: Solve for x in the second equation. $$x = y + 1$$ STEP 2: Substitute $(y + 1)$ in the first equation for x. $$(y + 1) + y = 5$$ $$2y = 4$$ $$y = 2$$ STEP 3: $x = 2 + 1 = 3$ STEP 4: $3 + 2 = 5$ ✓ $3 - 2 = 1$ ✓ $(3, 2)$ checks.

11.2 Exercises

CONCEPTS AND VOCABULARY

1. One advantage of solving a linear system using the method of substitution rather than graphical or numerical methods is that the _____ solution can always be found (provided it exists).

2. When substituting an expression that contains two or more terms for a single variable in an equation, it is usually best to place _____ around it.

3. Suppose that the method of substitution results in the equation $1 = 1$. What does this indicate about the number of solutions to the system of equations?

4. Suppose that the method of substitution results in the equation $0 = 1$. What does this indicate about the number of solutions to the system of equations?

USING THE METHOD OF SUBSTITUTION

Exercises 5–32: Use the method of substitution to solve the system of linear equations.

5. $x + y = 9$
   $y = 2x$

6. $x + y = -12$
   $y = -3x$

7. $x + 2y = 4$
   $x = 2y$

8. $-x + 3y = -12$
   $x = 5y$

9. $2x + y = -2$
   $y = x + 1$

10. $-3x + y = -10$
    $y = x - 2$

11. $x + 3y = 3$
    $x = y + 3$

12. $x - 2y = -5$
    $x = 4 - y$

13. $3x + 2y = \frac{3}{2}$
    $y = 2x - 1$

14. $-3x + 5y = 4$
    $y = 2 - 3x$

15. $2x - 3y = -12$
    $x = 2 - \frac{1}{2}y$

16. $\frac{3}{4}x + \frac{1}{4}y = -\frac{7}{4}$
    $x = 1 - 2y$

17. $2x - 3y = -4$
    $3x - y = 1$

18. $\frac{1}{2}x - y = -1$
    $2x - \frac{1}{2}y = \frac{13}{2}$

19. $x - 5y = 26$
    $2x + 6y = -12$

20. $4x - 3y = -4$
    $x + 7y = -63$

21. $\frac{1}{2}y - z = 5$
    $y - 3z = 13$

22. $3y - 7z = -2$
    $5y - z = 2$

23. $10r - 20t = 20$
    $r + 60t = -29$

24. $-r + 10t = 22$
    $-10r + 5t = 30$

25. $3x + 2y = 9$
    $2x - 3y = -7$

26. $5x - 2y = -5$
    $2x - 5y = 19$

27. $2a - 3b = 6$
    $-5a + 4b = -8$

28. $-5a + 7b = -1$
    $3a + 2b = 13$

29. $-\frac{1}{2}x + 3y = 5$
    $2x - \frac{1}{2}y = 3$

30. $3x - \frac{1}{2}y = 2$
    $-\frac{1}{2}x + 5y = \frac{19}{2}$

31. $3a + 5b = 16$
    $-8a + 2b = 34$

32. $5a - 10b = 20$
    $10a + 5b = 15$

RECOGNIZING OTHER TYPES OF SYSTEMS

Exercises 33–50: Use the method of substitution to solve the system of linear equations. These systems may have no solutions, one solution, or infinitely many solutions.

33. $x + y = 9$
    $x + y = 7$

34. $x - y = 8$
    $x - y = 4$

35. $x - y = 4$
    $2x - 2y = 8$

36. $2x + y = 5$
    $4x + 2y = 10$

37. $x - y = 3$
    $2x - y = 7$

38. $x + y = 4$
    $x - y = 2$

39. $x - y = 7$
    $-x + y = -7$

40. $2u - v = 6$
    $-4u + 2v = -12$

41. $u - 2v = 5$
    $2u - 4v = -2$

42. $3r + 3t = 9$
    $2r + 2t = 4$

43. $5x - y = -1$
    $2x - 7y = 7$

44. $2r + 3t = 1$
    $r - 3t = -5$

45. $y = 5x$
    $y = -3x$

46. $a = b + 1$
    $a = b - 1$

47. $5a = 4 - b$
    $5a = 3 - b$

48. $3y = x$
    $3y = 2x$

49. $2x + 4y = 0$
    $3x + 6y = 5$

50. $-5x + 10y = 3$
    $\frac{1}{2}x - y = 1$

51. **Thinking Generally** If the method of substitution results in an equation that is always true, what can be said about the graphs of the two equations?

52. **Thinking Generally** If the method of substitution results in an equation that is always false, what can be said about the graphs of the two equations?

SOLVING APPLICATIONS USING THE METHOD OF SUBSTITUTION

53. **Rectangle** A rectangular garden is 10 feet longer than it is wide. Its perimeter is 72 feet.
    (a) Let W be the width of the garden and L be the length. Write a system of linear equations whose solution gives the width and length of the garden.
    (b) Use the method of substitution to solve the system. Check your answer.

54. **Angles** The measures of the two smaller angles in a triangle are equal and their sum equals the largest angle.
    (a) Let x be the measure of each of the two smaller angles and y be the measure of the largest angle. Write a system of linear equations whose solution gives the measures of these angles.
    (b) Use the method of substitution to solve the system. Check your answer.

55. **Complementary Angles** The smaller of two angles that sum to 90° is half the measure of the larger angle.
    (a) Let x be the measure of the smaller angle and y be the measure of the larger angle. Write a system of linear equations whose solution gives the measures of these angles.
    (b) Use the method of substitution to solve your system.
    (c) Use graphing to solve the system.

56. **Supplementary Angles** The smaller of two angles that sum to 180° is one-fourth the measure of the larger angle.
    (a) Let x be the measure of the smaller angle and y be the measure of the larger angle. Write a system of linear equations whose solution gives the measures of these angles.
    (b) Use the method of substitution to solve the system.

57. **Average Room Prices** In 2013, the average room price for a hotel in Cancun, Mexico, was $30 more than it was in 2012. The 2012 room price was 86.4% of the 2013 room price. (*Source:* Hotel Price Index.)
    (a) Let x be the average room price in 2013 and y be this price in 2012. Write a system of equations whose solution gives the average room prices for each year.
    (b) Use the method of substitution to solve the system to the nearest cent.

58. **Ticket Prices** Two hundred tickets were sold for a baseball game, which amounted to $840. Student tickets cost $3, and adult tickets cost $5.
    (a) Let x be the number of student tickets and y be the number of adult tickets. Write a system of linear equations whose solution gives the number of each type of ticket.
    (b) Use the method of substitution to solve the system of equations.

59. **NBA Basketball Court** An official NBA basketball court is 44 feet longer than it is wide. If its perimeter is 288 feet, find its dimensions.

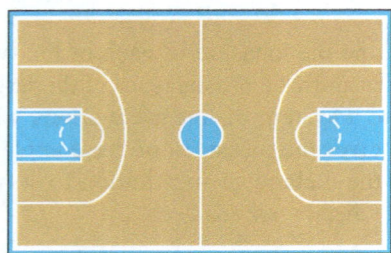

60. **Football Field** A U.S. football field is 139.5 feet longer than it is wide. If its perimeter is 921 feet, find its dimensions.

61. **Number Problem** The sum of two numbers is 70. The larger number is two more than three times the smaller number. Find the two numbers.

62. **Number Problem** The difference of two numbers is 12. The larger number is one less than twice the smaller number. Find the two numbers.

63. **Speed Problem** A tugboat goes 120 miles upstream in 15 hours. The return trip downstream takes 10 hours. Find the speed of the tugboat without a current and the speed of the current.

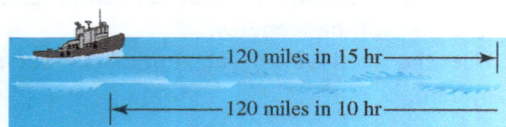

64. **Speed Problem** An airplane flies 1200 miles into the wind in 3 hours. The return trip takes 2 hours. Find the speed of the airplane without a wind and the speed of the wind.

65. **Mixture Problem** A chemist has 20% and 50% solutions of acid available. How many liters of each solution should be mixed to obtain 10 liters of a 40% acid solution?

66. **Mixture Problem** A mechanic needs a radiator to have a 40% antifreeze solution. The radiator currently is filled with 4 gallons of a 25% antifreeze solution. How much of the antifreeze mixture should be drained from the car if the mechanic replaces it with pure antifreeze?

67. **Mixture Problem** A 15-pound mixture of coffee that contains both decaffeinated and regular coffee costs $53. The decaffeinated coffee costs $3 per pound and the regular coffee costs $4 per pound.
    (a) Let x be the weight of the decaffeinated coffee and y be the weight of the regular coffee. Write a system of equations whose solution gives the amount of each type of coffee in the mixture.
    (b) Use the method of substitution to solve the system of equations.

68. **Real Estate Sites** In 2014, the top two real estate websites were Zillow and Trulia. Combined, they made up 25.4% of the U.S. market share of visits. Zillow had 7.6% more market share than Trulia. (*Source:* Experian Marketing Services.)
    (a) Let x be the Zillow market share of visits and y be the Trulia market share of visits. Write a system of equations whose solution gives the market share of visits for each site.
    (b) Use the method of substitution to solve the system of equations.

69. **Exercise and Calories** In a 48-minute workout, a student burns a total of 450 calories on a tread climber and a stationary bicycle. The student burns 12 calories per minute on the tread climber and 5 calories per minute on the stationary bicycle.
    (a) Let x be the time spent on the tread climber and y be the time spent on the stationary bicycle. Write a system of equations whose solution gives the number of minutes spent on each machine.
    (b) Use the method of substitution to solve the system of equations.
    (c) How many calories were burned on each machine?

70. **Exercise and Calories** In 1.2 hours, a student travels 10.2 miles by running at 8 miles per hour and at 10 miles per hour.
    (a) Let x be the time spent running at 8 miles per hour and y be the time spent running at 10 miles per hour. Write a system of equations whose solution gives the time spent running at each speed.
    (b) Use the method of substitution to solve the system of equations.

71. Great Lakes Together, Lake Superior and Lake Michigan cover 54 thousand square miles. Lake Superior is approximately 10 thousand square miles larger than Lake Michigan. Find the size of each lake. (*Source:* National Oceanic and Atmospheric Administration.)

72. Longest Rivers The two longest rivers in the world are the Nile and the Amazon. Together, they are 8145 miles, with the Amazon being 145 miles shorter than the Nile. Find the length of each river. (*Source:* National Oceanic and Atmospheric Administration.)

WRITING ABOUT MATHEMATICS

73. State one advantage that the method of substitution has over the intersection-of-graphs method. Explain your answer.

74. When applying the method of substitution, how do you know that there are no solutions?

75. When applying the method of substitution, how do you know that there are infinitely many solutions?

76. When applying the intersection-of-graphs method, how do you know that there are no solutions?

SECTIONS 11.1 and 11.2 Checking Basic Concepts

1. Determine graphically the x-value in each equation when $y = 2$.
   (a) $y = 1 - \frac{1}{2}x$ (b) $2x - 3y = 6$

2. Determine whether $(-1, 0)$ or $(4, 2)$ is a solution to the system
$$2x - 5y = -2$$
$$3x + 2y = 16.$$

3. Solve the system of equations graphically. Check your answer.
$$x - y = 1$$
$$2x + y = 5$$

4. If possible, use the method of substitution to solve each system of equations.
   (a) $x + y = -1$
       $y = 2 - x$
   (b) $4x - y = 5$
       $-x + y = -2$
   (c) $x + 2y = 3$
       $-x - 2y = -3$

5. **Room Prices** A hotel rents single and double rooms for $150 and $200, respectively. The hotel receives $55,000 for renting 300 rooms.
   (a) Let x be the number of single rooms rented and let y be the number of double rooms rented. Write a system of linear equations whose solution gives the values of x and y.
   (b) Use the method of substitution to solve the system. Check your answer.

11.3 Solving Systems of Linear Equations by Elimination

Objectives

1. Using the Elimination Method
   - Multiplying Before Applying Elimination
   - Solving a System Symbolically, Graphically, and Numerically
2. Recognizing Systems with No Solutions or Infinitely Many Solutions
3. Solving Applications Using the Elimination Method

NEW VOCABULARY

☐ Elimination Method

STUDY TIP

Much of mathematics builds on previous knowledge. We learned about the addition property of equality in Chapter 2. In this section, we see how we can use this property to solve a system of linear equations.

READING CHECK 1

- What is eliminated from both of the equations when the elimination method is used to solve a system of linear equations?

1 Using the Elimination Method

The **elimination method** is based on the addition property of equality. If

$$a = b \quad \text{and} \quad c = d,$$

then

$$a + c = b + d.$$

For example, if the sum of two numbers is 20 and their difference is 4, then the system of equations

$$x + y = 20$$
$$x - y = 4$$

describes these two numbers. By the addition property of equality, the sum of the left sides of these equations equals the sum of their right sides.

$$(x + y) + (x - y) = 20 + 4$$

Add the left sides.   Add the right sides.

$$2x = 24 \quad \text{Combine terms.}$$

Note that the y-variable is eliminated when the left sides are added. The resulting equation, $2x = 24$, simplifies to $x = 12$. Thus the value of x in the solution to the system of equations is 12. The value of y can be found by substituting 12 for x in either of the given equations. Substituting 12 for x in the first equation, $x + y = 20$, results in $12 + y = 20$ or $y = 8$. The solution to the system of equations is $x = 12, y = 8$, which can be written as the ordered pair $(12, 8)$.

To organize the elimination method better, we can carry out the addition vertically.

$$x + y = 20$$
$$x - y = 4$$

Add the right sides.

$$2x + 0y = 24 \quad \text{Add the equations.}$$
$$2x = 24 \quad \text{Simplify.}$$

Add the left sides.

$$x = 12 \quad \text{Divide each side by 2.}$$

Once the value of one variable is known, in this case $x = 12$, don't forget to find the value of the other variable by substituting this known value in either of the *given* equations. By substituting 12 for x in the second equation, we obtain $12 - y = 4$ or $y = 8$. The solution is the ordered pair $(12, 8)$.

EXAMPLE 1 **Applying the elimination method**

Solve each system of equations. Check each solution.

(a) $2x + y = 1$
    $3x - y = 9$

(b) $-2a + b = -3$
    $2a + 3b = 7$

Solution

(a) Adding these two equations eliminates the y-variable.

$$2x + y = 1 \quad \text{First equation}$$
$$3x - y = 9 \quad \text{Second equation}$$
$$5x = 10 \quad \text{or} \quad x = 2 \quad \text{Add and solve for } x.$$

To find y, substitute 2 for x in either of the *given* equations. We will use $2x + y = 1$.

$$2(2) + y = 1 \qquad \text{Let } x = 2 \text{ in first equation.}$$
$$y = -3 \qquad \text{Subtract 4 from each side.}$$

The solution is the *ordered pair* $(2, -3)$, which can be checked by substituting 2 for x and -3 for y in each of the given equations.

$2x + y = 1$	$3x - y = 9$	Given equations
$2(2) + (-3) \stackrel{?}{=} 1$	$3(2) - (-3) \stackrel{?}{=} 9$	Let $x = 2$ and $y = -3$.
$4 - 3 \stackrel{?}{=} 1$	$6 + 3 \stackrel{?}{=} 9$	Simplify.
$1 = 1$ ✓	$9 = 9$ ✓	The solution checks.

(b) Adding these two equations eliminates the a-variable.

$$\begin{aligned} -2a + b &= -3 \qquad &\text{First equation} \\ 2a + 3b &= 7 \qquad &\text{Second equation} \\ \hline 4b &= 4 \quad \text{or} \quad b = 1 \qquad &\text{Add and solve for } b. \end{aligned}$$

To find a, substitute 1 for b in either of the *given* equations. We will use $2a + 3b = 7$.

$$2a + 3(1) = 7 \qquad \text{Let } b = 1 \text{ in second equation.}$$
$$2a = 4 \qquad \text{Subtract 3 from each side.}$$
$$a = 2 \qquad \text{Divide each side by 2.}$$

The solution is the *ordered pair* $(2, 1)$, which can be checked by substituting 2 for a and 1 for b in each of the given equations.

$-2a + b = -3$	$2a + 3b = 7$	Given equations
$-2(2) + 1 \stackrel{?}{=} -3$	$2(2) + 3(1) \stackrel{?}{=} 7$	Let $a = 2$ and $b = 1$.
$-4 + 1 \stackrel{?}{=} -3$	$4 + 3 \stackrel{?}{=} 7$	Simplify.
$-3 = -3$ ✓	$7 = 7$ ✓	The solution checks.

Now Try Exercises 15, 17

MULTIPLYING BEFORE APPLYING ELIMINATION Adding two equations does not always eliminate a variable. For example, adding the following equations eliminates neither variable.

$$\begin{aligned} 3x - 2y &= 11 \qquad &\text{First equation} \\ 4x + y &= 11 \qquad &\text{Second equation} \\ \hline 7x - y &= 22 \qquad &\text{Add the equations.} \end{aligned}$$

READING CHECK 2

- In using the elimination method, when is it necessary to apply the multiplication property of equality?

However, by the multiplication property of equality, we can multiply the second equation by 2. Then adding the equations eliminates the y-variable.

$$\begin{aligned} 3x - 2y &= 11 \qquad &\text{First equation} \\ 8x + 2y &= 22 \qquad &\text{Multiply the second equation by 2.} \\ \hline 11x &= 33 \quad \text{or} \quad x = 3 \qquad &\text{Add and solve for } x. \end{aligned}$$

Opposites

EXAMPLE 2 Multiplying before applying elimination

Solve each system of equations.
(a) $5x - y = -11$
$\phantom{\textbf{(a)}\ }2x + 3y = -1$

(b) $3x + 2y = 1$
$\phantom{\textbf{(b)}\ }2x - 3y = 5$

Solution

(a) We multiply $5x - y = -11$ by 3 and then add to eliminate the y-variable.

Opposites
$$\begin{array}{rl} 15x - 3y = -33 & \text{Multiply the first equation by 3.} \\ 2x + 3y = -1 & \text{Second equation} \\ \hline 17x = -34 \quad \text{or} \quad x = -2 & \text{Add and solve for } x. \end{array}$$

We can find y by substituting -2 for x in the second equation, $2x + 3y = -1$.

$$\begin{aligned} 2(-2) + 3y &= -1 & \text{Let } x = -2 \text{ in second equation.} \\ 3y &= 3 & \text{Add 4 to each side.} \\ y &= 1 & \text{Divide each side by 3.} \end{aligned}$$

The solution is $(-2, 1)$.

(b) We must apply the multiplication property to both equations. If we multiply the first equation, $3x + 2y = 1$, by 3, and the second equation, $2x - 3y = 5$, by 2, then the coefficients of the y-variables will be opposites. Adding eliminates the y-variable.

Opposites
$$\begin{array}{rl} 9x + 6y = 3 & \text{Multiply the first equation by 3.} \\ 4x - 6y = 10 & \text{Multiply the second equation by 2.} \\ \hline 13x = 13 \quad \text{or} \quad x = 1 & \text{Add and solve for } x. \end{array}$$

To find y, substitute 1 for x in the first *given* equation, $3x + 2y = 1$.

$$\begin{aligned} 3(1) + 2y &= 1 & \text{Let } x = 1 \text{ in first equation.} \\ 2y &= -2 & \text{Subtract 3 from each side.} \\ y &= -1 & \text{Divide each side by 2.} \end{aligned}$$

The solution is $(1, -1)$.

Now Try Exercises 29, 33

In practice, it is possible to eliminate *either* variable from a system of linear equations. It is often best to choose the variable that requires the least amount of computation to complete the elimination. In the next example, we solve a system of equations twice—first by using the multiplication property of equality to eliminate the x-variable and then by using it to eliminate the y-variable.

EXAMPLE 3 Multiplying before applying elimination

Solve the system of equations two times, first by eliminating x and then by eliminating y.

$$2y = -6 - 5x$$
$$2x = -5y + 6$$

Solution

It is best to write each equation in the standard form: $Ax + By = C$.

$$\begin{aligned} 5x + 2y &= -6 & \text{First equation in standard form} \\ 2x + 5y &= 6 & \text{Second equation in standard form} \end{aligned}$$

Eliminate x If we multiply the first equation in standard form by -2 and the second equation in standard form by 5, then we can eliminate the x-variable by adding.

$$\begin{array}{rl} -10x - 4y = 12 & \text{Multiply the first equation by } -2. \\ 10x + 25y = 30 & \text{Multiply the second equation by 5.} \\ \hline 21y = 42 \quad \text{or} \quad y = 2 & \text{Add and solve for } y. \end{array}$$

To find x, substitute 2 for y in the first *given* equation, $2y = -6 - 5x$.

$$2(2) = -6 - 5x \qquad \text{Let } y = 2 \text{ in first equation.}$$
$$10 = -5x \qquad \text{Add 6 to each side.}$$
$$-2 = x \qquad \text{Divide each side by } -5.$$

The solution is $(-2, 2)$.

Eliminate y If we multiply the first equation in standard form by -5 and the second equation in standard form by 2, then we can eliminate the y-variable by adding.

$$-25x - 10y = 30 \qquad \text{Multiply the first equation by } -5.$$
$$\underline{4x + 10y = 12} \qquad \text{Multiply the second equation by 2.}$$
$$-21x = 42 \quad \text{or} \quad x = -2 \qquad \text{Add and solve for } x.$$

To find y, substitute -2 for x in the second given equation, $2x = -5y + 6$.

$$2(-2) = -5y + 6 \qquad \text{Let } x = -2 \text{ in second equation.}$$
$$-10 = -5y \qquad \text{Subtract 6 from each side.}$$
$$2 = y \qquad \text{Divide each side by } -5.$$

The solution is $(-2, 2)$.

Now Try Exercise 31

SOLVING A SYSTEM SYMBOLICALLY, GRAPHICALLY, AND NUMERICALLY In the next example, we use three different methods to solve a system of equations.

EXAMPLE 4 Solving a system with different methods

Solve the system of equations symbolically, graphically, and numerically.

$$x + y = 2$$
$$x - 3y = 6$$

Solution

Symbolic Solution Both the method of substitution and the elimination method are symbolic methods. The elimination method is used here. We can solve the system by multiplying the second equation by -1 and adding to eliminate the x-variable.

$$x + y = 2 \qquad \text{First equation}$$
$$\underline{-x + 3y = -6} \qquad \text{Multiply the second equation by } -1.$$
$$4y = -4 \quad \text{or} \quad y = -1 \qquad \text{Add and solve for } y.$$

We can find x by substituting -1 for y in the first equation, $x + y = 2$.

$$x + (-1) = 2 \qquad \text{Let } y = -1 \text{ in first equation.}$$
$$x = 3 \qquad \text{Add 1 to each side.}$$

The solution is $(3, -1)$.

Graphical Solution For a graphical solution, we solve each equation for y to obtain the slope–intercept form.

$x + y = 2$	First equation	$x - 3y = 6$	Second equation
$y = -x + 2$	Subtract x.	$-3y = -x + 6$	Subtract x.
		$y = \dfrac{1}{3}x - 2$	Divide by -3.

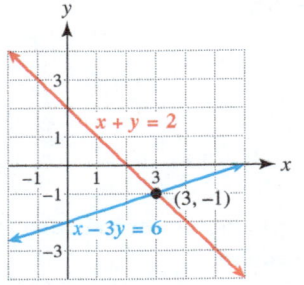

FIGURE 11.9

The graphs of $y = -x + 2$ and $y = \frac{1}{3}x - 2$ are shown in **FIGURE 11.9**. They intersect at $(3, -1)$. (These graphs could also be obtained by finding the x- and y-intercepts for each equation.)

Numerical Solution A numerical solution consists of a table of values, as shown in **TABLE 11.4**. Note that when $x = 3$, both y-values equal -1. Therefore the solution is $(3, -1)$.

x	0	1	2	3	4
$y = -x + 2$	2	1	0	-1	-2
$y = \frac{1}{3}x - 2$	-2	$-\frac{5}{3}$	$-\frac{4}{3}$	-1	$-\frac{2}{3}$

TABLE 11.4

Find the x-value where the two y-values are equal.

Now Try Exercise 41

MAKING CONNECTIONS 1

Solving Systems Using Different Methods

Systems of linear equations can be solved symbolically, graphically, and numerically. Symbolic solutions are *always exact*, whereas graphical and numerical solutions are *sometimes approximate*. The following example shows how to use each method to solve the system

$$-x + 2y = 3$$
$$x + 6y = 5.$$

For the graphical and numerical solutions, solve each equation for y to obtain

$$y = \frac{1}{2}x + \frac{3}{2}$$
$$y = -\frac{1}{6}x + \frac{5}{6}.$$

Symbolic Solution (elimination)

$$-x + 2y = 3$$
$$\underline{x + 6y = 5}$$
$$8y = 8 \quad \text{or} \quad y = 1$$
$$-x + 2(1) = 3$$
$$-x = 1$$
$$x = -1$$

The solution is $(-1, 1)$.

Graphical Solution

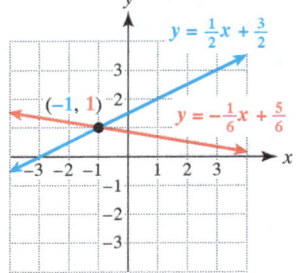

The solution is $(-1, 1)$.

Numerical Solution

x	-2	-1	0	1
$y = \frac{1}{2}x + \frac{3}{2}$	$\frac{1}{2}$	1	$\frac{3}{2}$	2
$y = -\frac{1}{6}x + \frac{5}{6}$	$\frac{7}{6}$	1	$\frac{5}{6}$	$\frac{2}{3}$

The solution is $(-1, 1)$.

---

2 Recognizing Systems with No Solutions or Infinitely Many Solutions

In the first section of this chapter, we discussed how a system of linear equations can have no solutions, one solution, or infinitely many solutions. Elimination can also be used on systems that have no solutions or infinitely many solutions.

EXAMPLE 5 Solving other types of systems

Solve each system of equations by using the elimination method. Then graph the system.

(a) $\quad x - 2y = 4$
$\quad\;\; -2x + 4y = -8$

(b) $\quad 3x + 3y = 6$
$\quad\;\;\; x + y = 1$

Solution

(a) We multiply the first equation by 2 and then add, which eliminates both variables.

$$\begin{aligned} 2x - 4y &= 8 \quad &&\text{Multiply the first equation by 2.} \\ -2x + 4y &= -8 \quad &&\text{Second equation} \\ \hline 0 &= 0 \quad \text{(True)} \quad &&\text{Add.} \end{aligned}$$

The equation $0 = 0$ is *always true*, which indicates that the system has *infinitely many solutions*. A graph of the two equations is shown in **FIGURE 11.10**. The two lines are identical so there actually is only one line, and *every point on this line represents a solution*. For example, $(0, -2)$ and $(4, 0)$ lie on the line and are both solutions.

(b) We multiply the second equation by -3 and add, eliminating both variables.

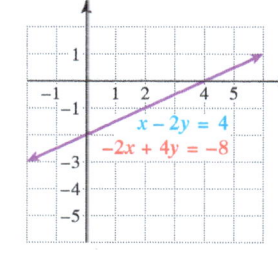

FIGURE 11.10

$$\begin{aligned} 3x + 3y &= 6 \quad &&\text{First equation} \\ -3x - 3y &= -3 \quad &&\text{Multiply the second equation by } -3. \\ \hline 0 &= 3 \quad \text{(False)} \quad &&\text{Add.} \end{aligned}$$

The equation $0 = 3$ is *always false*, which indicates that the system has *no solutions*. A graph of the two equations is shown in **FIGURE 11.11**. Note that the two lines are parallel and thus do not intersect.

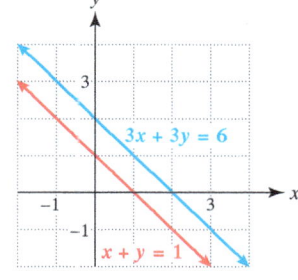

FIGURE 11.11

Now Try Exercises 51, 55

3 Solving Applications Using the Elimination Method

Math in Real World Context In the next two examples, we use elimination to solve applications relating to connected devices and to burning calories during exercise.

EXAMPLE 6 Determining numbers of Android and Windows devices

In 2015, there were 1737 million (1.737 billion) connected devices shipped globally that ran either a Windows or Android operating system. There were 979 million more Android devices shipped than Windows devices. How many connected devices were shipped with each operating system?

Solution
STEP 1: *Identify each variable.*

x: Android devices, in millions

y: Windows devices, in millions

STEP 2: *Write a system of equations.*

$x + y = 1737$ The total number of devices is 1737 million.

$x - y = 979$ Android devices outnumber Windows devices by 979 million.

STEP 3A: *Solve the system of linear equations.* Add the two equations to eliminate the y-variable.

$$\begin{aligned} x + y &= 1737 \\ x - y &= 979 \\ \hline 2x &= 2716 \quad \text{or} \quad x = 1358 \end{aligned}$$

Substituting 1358 for x in the first equation results in $1358 + y = 1737$ or $y = 379$. The solution is $x = 1358$ and $y = 379$.

STEP 3B: *Determine the solution to the problem.* A total of 1358 million connected devices shipped running Android and 379 million shipped running Windows.

STEP 4: *Check the solution.* The total number of devices using these operating systems is $1358 + 379 = 1737$ million. The number of Android devices exceeds the number of Windows devices by $1358 - 379 = 979$ million. The answer checks.

Now Try Exercise 57

EXAMPLE 7   **Burning calories during exercise**

During strenuous exercise, an athlete can burn 10 calories per minute on a rowing machine and 11.5 calories per minute on a stair climber. If an athlete uses both machines and burns 433 calories in a 40-minute workout, how many minutes does the athlete spend on each machine? (*Source: Runner's World.*)

Solution

STEP 1: *Identify each variable.*

x: number of minutes on a rowing machine

y: number of minutes on a stair climber

STEP 2: *Write a system of equations.* The total workout takes 40 minutes, so $x + y = 40$. The athlete burns $10x$ calories on the rowing machine and $11.5y$ calories on the stair climber. Because the total number of calories equals 433, it follows that $10x + 11.5y = 433$.

$x + y = 40$ Workout is 40 minutes.

$10x + 11.5y = 433$ Total calories is 433.

STEP 3A: *Solve the system of linear equations.* Multiply the first equation by -10 and add the two equations.

$$\begin{aligned} -10x - 10y &= -400 \\ 10x + 11.5y &= 433 \\ \hline 1.5y &= 33 \quad \text{or} \quad y = \frac{33}{1.5} = 22 \end{aligned}$$

Multiply first equation by -10.

Second equation

Add and solve for y.

Because $x + y = 40$ and $y = 22$, it follows that $x = 18$.

STEP 3B: *Determine the solution to the problem.* The athlete spends 18 minutes on the rowing machine and 22 minutes on the stair climber.

STEP 4: *Check your answer.* Because $18 + 22 = 40$, the athlete works out for 40 minutes. Also, the athlete burns $10(18) + 11.5(22) = 433$ calories. The answer checks.

Now Try Exercise 59

11.3 Putting It All Together

CONCEPT	EXPLANATION	EXAMPLE
Elimination Method	Is based on the addition property of equality If $a = b$ and $c = d$, then $$a + c = b + d.$$ May be used to solve systems of equations	$x + y = 5$ $x - y = -1$ $2x = 4$ or $x = 2$ Add. Because $x + y = 5$ and $x = 2$, it follows that $y = 3$. The solution is $(2, 3)$.
Systems with No Solutions	Elimination can be used to recognize systems that have no solutions.	$-x - y = -4$ $x + y = 2$ $ 0 = -2$ Add. Because $0 = -2$ is always false, there are no solutions.
Systems with Infinitely Many Solutions	Elimination can be used to recognize systems that have infinitely many solutions.	$x + y = 4$ $2x + 2y = 8$ Multiply the first equation by -2. $-2x - 2y = -8$ $2x + 2y = 8$ $ 0 = 0$ Add. Because $0 = 0$ is always true, there are infinitely many solutions.

11.3 Exercises

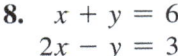

CONCEPTS AND VOCABULARY

1. Name two symbolic methods for solving a system of linear equations.

2. The elimination method is based on the _____ property of equality.

3. The addition property of equality states that if $a = b$ and $c = d$, then $a + c$ _____ $b + d$.

4. The multiplication property of equality states that if $a = b$, then ca _____ cb.

5. Suppose that the elimination method results in the equation $0 = 0$. What does this indicate about the number of solutions to the system of equations?

6. Suppose that the elimination method results in the equation $0 = 1$. What does this indicate about the number of solutions to the system of equations?

USING THE ELIMINATION METHOD

Exercises 7–14: If possible, use the given graph to solve the system of equations. Then use the elimination method to verify your answer.

7. $x - y = 0$
   $x + y = 2$

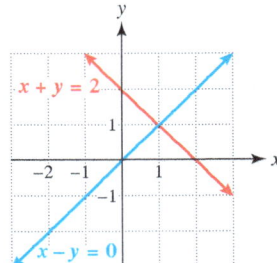

8. $x + y = 6$
   $2x - y = 3$

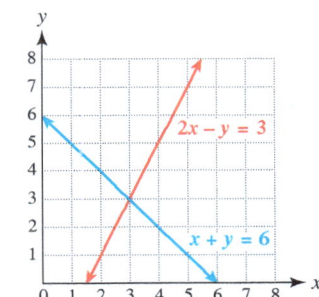

9. $2x + 3y = -1$
   $2x - 3y = -7$

10. $-2x + y = -3$
    $4x - 3y = 7$

27. $5a - 6b = -2$
    $5a + 5b = 9$

28. $-r + 2t = 0$
    $3r + 2t = 8$

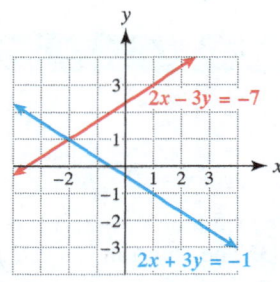
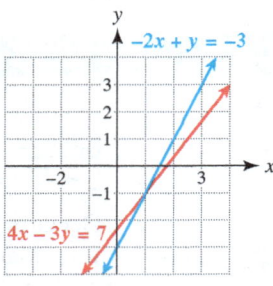

29. $3u + 2v = -16$
    $2u + v = -9$

30. $5u - v = 0$
    $3u + 3v = -18$

31. $5x - 7y = 5$
    $-2x + 2y = -2$

32. $2x + 7y = 6$
    $4x - 3y = -22$

33. $5x - 3y = 4$
    $3x + 2y = 10$

34. $-3x - 8y = 1$
    $2x + 5y = 0$

35. $-5x - 10y = -22$
    $10x + 15y = 35$

36. $-15x + 4y = -20$
    $5x + 7y = 90$

11. $x + y = 3$
    $x + y = -1$

12. $2x - y = 4$
    $-2x + y = -4$

SOLVING A SYSTEM USING A TABLE OF VALUES

Exercises 37–40: A table of values is given for two linear equations. Use the table to solve this system.

37.
x	0	1	2	3	4
$y = -x + 5$	5	4	3	2	1
$y = 2x - 4$	-4	-2	0	2	4

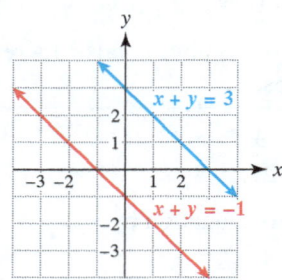
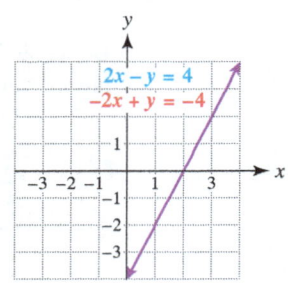

38.
x	-3	-2	-1	0	1
$y = x + 1$	-2	-1	0	1	2
$y = -x - 3$	0	-1	-2	-3	-4

13. $2x + 2y = 6$
    $x + y = 3$

14. $-x + 3y = 4$
    $x - 3y = 3$

39.
x	-2	-1	0	1	2
$y = 3x + 1$	-5	-2	1	4	7
$y = -x + 1$	3	2	1	0	-1

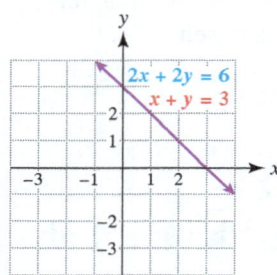
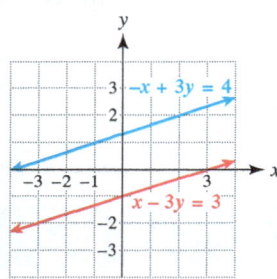

40.
x	-2	-1	0	1	2
$y = 2x$	-4	-2	0	2	4
$y = -x$	2	1	0	-1	-2

Exercises 15–36: Use the elimination method to solve the system of equations.

15. $x + y = 7$
    $x - y = 5$

16. $x - y = 8$
    $x + y = 4$

SOLVING A SYSTEM SYMBOLICALLY, GRAPHICALLY, AND NUMERICALLY

17. $-x + y = 5$
    $x + y = 3$

18. $x - y = 10$
    $-x - y = 20$

Exercises 41–46: Solve the system of equations

(a) *symbolically,*
(b) *graphically, and*
(c) *numerically.*

19. $2x + y = 8$
    $3x - y = 2$

20. $-x + 2y = 3$
    $x + 6y = 5$

21. $-2x + y = -3$
    $2x - 4y = 0$

22. $2x + 6y = -5$
    $7x - 6y = -4$

41. $2x + y = 5$
    $x - y = 1$

42. $-x + y = 2$
    $3x + y = -2$

23. $\frac{1}{2}x - y = 3$
    $\frac{3}{2}x + y = 5$

24. $x - \frac{1}{4}y = 4$
    $-4x + \frac{1}{4}y = -9$

43. $2x + y = 5$
    $x + y = 1$

44. $-x + y = 2$
    $3x - y = -2$

25. $a + 6b = 2$
    $a + 3b = -1$

26. $3r - t = 7$
    $2r - t = 2$

45. $6x + 3y = 6$
    $-2x + 2y = -2$

46. $-x + 2y = 5$
    $2x + 2y = 8$

RECOGNIZING OTHER TYPES OF SYSTEMS

Exercises 47–56: Use elimination to determine whether the system of equations has no solutions, one solution, or infinitely many solutions. Then graph the system.

47. $2x - 2y = 4$
    $-x + y = -2$

48. $-2x + y = 4$
    $4x - 2y = -8$

49. $x - y = 0$
    $x + y = 0$

50. $x - y = 2$
    $x + y = 2$

51. $x - y = 4$
    $x - y = 1$

52. $-2x + 3y = 5$
    $4x - 6y = 10$

53. $x - 3y = 2$
    $-x + 3y = 4$

54. $6x + 9y = 18$
    $4x + 6y = 12$

55. $4x - 8y = 24$
    $6x - 12y = 36$

56. $x - y = 5$
    $2x - y = 4$

SOLVING APPLICATIONS USING THE ELIMINATION METHOD

57. **Unique Site Visitors** In March 2014, the websites healthcare.gov and ncaa.com saw large percentage increases in unique visitors over the previous month. There were a combined 20.4 million unique visitors to ncaa.com and healthcare.gov. There were 3 million more visitors to healthcare.gov than ncaa.com. How many unique visitors were there to each site in March 2014? (*Source:* Compete.com.)

58. **Millionaires** In 2012, there were 210,000 millionaires in Chicago and Houston. If there were 3,600 more millionaires in Chicago than Houston, find the number of millionaires in each city that year. (*Source:* The Economist.)

59. **Burning Calories** During strenuous exercise, an athlete can burn 9 calories per minute on a stationary bicycle and 11.5 calories per minute on a stair climber. In a 30-minute workout, an athlete burns 300 calories. How many minutes does the athlete spend on each type of exercise equipment? (*Source:* Runner's World.)

60. **Distance Running** An athlete runs at 9 mph and then at 12 mph, covering 10 miles in 1 hour. How long does the athlete run at each speed?

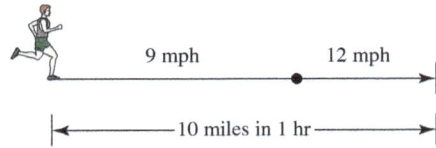

61. **River Current** A riverboat takes 8 hours to travel 64 miles downstream and 16 hours for the return trip. Find the speed of the current and the speed of the riverboat in still water?

62. **Airplane Speed** An airplane travels 3000 miles with the wind in 5 hours and takes 6 hours for the return trip into the wind. Find the speed of the wind and the speed of the airplane without any wind?

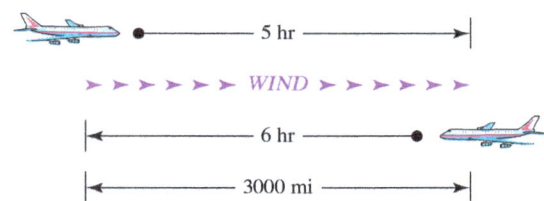

63. **Investments** A total of $5000 is invested at 3% and 5% annual interest. After 1 year, the total interest equals $210. How much money is invested at each interest rate?

64. **Mixing Antifreeze** A car radiator holds 2 gallons of fluid and initially is empty. If a mixture of water and antifreeze contains 70% antifreeze and another mixture contains 15% antifreeze, how much of each should be combined to fill the radiator with a 50% antifreeze mixture?

65. **Number Problem** The sum of two integers is -17, and their difference is -69. Find the two integers.

66. **Supplementary Angles** The measures of two supplementary angles differ by 74°. Find the two angles.

67. **Picture Dimensions** The figure below shows a red graph that gives possible dimensions for a rectangular picture frame with perimeter 120 inches. The blue graph shows possible dimensions for a rectangular frame whose length L is twice its width W.

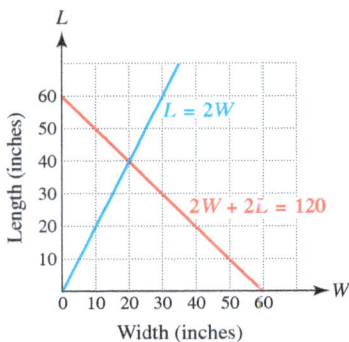

(a) Use the figure to determine the dimensions of a frame with a perimeter of 120 inches and a length that is twice the width.

(b) Solve this problem symbolically.

68. Sales of DVDs and CDs A company sells DVDs d and CDs c. The figure shows a red graph of $d + c = 2000$. The blue graph shows a revenue of $15,000 received from selling d DVDs at $12 each and c CDs at $6 each.

Digital Media

(a) If the total number of DVDs and CDs sold is 2000, determine how many of each were sold to obtain a revenue of $15,000.

(b) Solve this problem symbolically.

WRITING ABOUT MATHEMATICS

69. Suppose that a system of linear equations is solved symbolically, numerically, and graphically. How do the solutions from each method compare? Explain your answer.

70. When you are solving a system of linear equations by elimination, how can you recognize that the system has no solutions?

Group Activity

Exercises 1–4: **Facebook Apps** In early 2014, the two Facebook apps with the highest daily active users were Candy Crush Saga and Farm Heroes Saga. The total number of daily active users for the two apps was 73.8 million. The number of Candy Crush Saga users exceeded the number of Farm Heroes Saga users by 40.2 million. (*Source:* Inside Facebook.)

1. Set up a system of equations whose solution gives the number of daily active users in millions for each app. Identify what each variable represents.

2. Use substitution to solve this system. Interpret the result.

3. Use elimination to solve this system.

4. Solve this system graphically. Do all your answers agree?

11.4 Systems of Linear Inequalities

Objectives

1. Finding Solutions to a Linear Inequality in Two Variables
2. Writing and Graphing a Linear Inequality in Two Variables
3. Finding Solutions to Systems of Linear Inequalities in Two Variables
4. Solving Applications Involving Systems of Linear Inequalities

1 Finding Solutions to a Linear Inequality in Two Variables

Connecting Concepts with Your Life Suppose that a college student works at both the library and a department store. The library pays $10 per hour, and the department store pays $8 per hour. The equation $A = 10L + 8D$ calculates the amount of money earned from working L hours at the library and D hours at the department store. If the cost of one college credit is $80, then solutions to the equation

$$10L + 8D = 80$$

are ordered pairs (L, D) that result in the student earning (exactly) enough to pay for one credit. The equation's graph is the line shown in **FIGURE 11.12(a)**. The point $(4, 5)$ lies on this line, which indicates that, if the student works 4 hours at the library and 5 hours at the department store, then the pay is $80.

NEW VOCABULARY

☐ Test point
☐ Linear inequality in two variables
☐ System of linear inequalities in two variables

STUDY TIP

If you have not been studying with other students, consider getting together with classmates so that you can work on your math homework together.

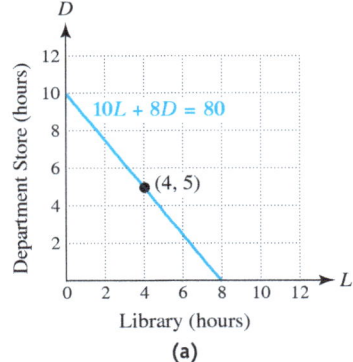
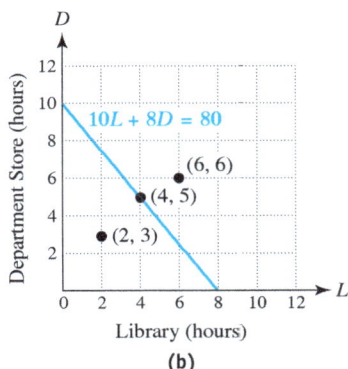

FIGURE 11.12

There are many situations in which the student can make more than $80. For example, the **test point** $(6, 6)$ lies above the line in **FIGURE 11.12(b)**, indicating that, if the student works 6 hours at both the library and the department store, then the pay is more than $80. In fact, any point *above* the line results in pay *greater than* $80. The region above the line is described by the inequality

$$10L + 8D > 80.$$

The test point $(6, 6)$ represents earnings of $108 and *satisfies* this inequality because

$$10(6) + 8(6) > 80$$

is a true statement. Similarly, any point *below* the line gives an ordered pair (L, D) that results in earnings *less than* $80. The point $(2, 3)$ in **FIGURE 11.12(b)** lies below the line and represents earnings of $44. That is,

$$10(2) + 8(3) < 80.$$

The region below the line is described by the inequality

$$10L + 8D < 80.$$

2 Writing and Graphing a Linear Inequality in Two Variables

Any linear equation in two variables can be written in standard form as

$$Ax + By = C,$$

where A, B, and C are constants. When the equals sign is replaced with $<$, $>$, $\leq$, or $\geq$, a **linear inequality in two variables** results. Examples of linear *equations* in two variables include

$$2x + 3y = 10 \quad \text{and} \quad y = \frac{1}{2}x - 5,$$

and so examples of linear *inequalities* in two variables include

$$2x + 3y < 10 \quad \text{and} \quad y \geq \frac{1}{2}x - 5.$$

A *solution* to a linear inequality in two variables is an ordered pair (x, y) that makes the inequality a true statement. The *solution set* is the set of all solutions to the inequality. The solution set to an inequality in two variables is typically a region in the xy-plane, which means that there are infinitely many solutions.

EXAMPLE 1   Writing a linear inequality

Write a linear inequality that describes each shaded region.

(a)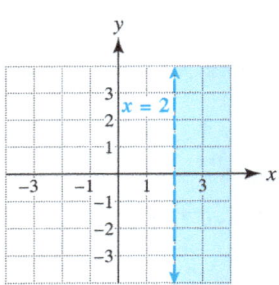

(b)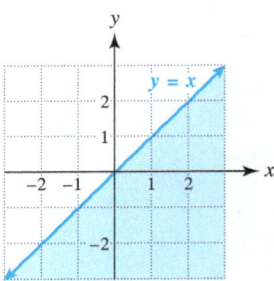

Solution
(a) The shaded region is bounded by the line $x = 2$. The *dashed* line indicates that the line is not included in the solution set. Only points with *x*-coordinates **greater than** 2 are shaded. Thus every point in the shaded region satisfies $x > 2$.

(b) The solution set includes all points that are **on or below** the line $y = x$. An inequality that describes this region is $y \leq x$, which can also be written as $-x + y \leq 0$.

Now Try Exercises 27, 29

TECHNOLOGY NOTE

Shading an Inequality
Graphing calculators can be used to shade a solution set to an inequality. The left-hand screen shows how to enter the equation from Example 1(b), and the right-hand screen shows the resulting graph.

CALCULATOR HELP
To shade an inquality, see Appendix A (pages AP-6 and AP-7).

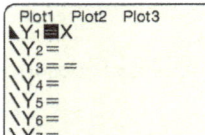 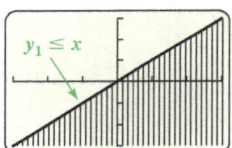

READING CHECK 1
- How do you know whether to graph a solid line or a dashed line for an inequality?

GRAPHING A LINEAR INEQUALITY IN TWO VARIABLES

STEP 1: Replace the inequality symbol with an equals sign and graph the resulting line. If the inequality is $<$ or $>$, use a dashed line, and if it is $\leq$ or $\geq$, use a solid line.

STEP 2: Pick a *test point* that does *not* lie on the line. Substitute this point in the given inequality. Determine whether the resulting statement is true or false.

STEP 3: If the statement is true, shade the region containing the test point. If the statement is false, shade the region that does not contain the test point.

EXAMPLE 2   Graphing a linear inequality

Shade the solution set for each inequality.
(a) $y \leq 1$ (b) $x + y < 3$ (c) $-x + 2y \geq 2$

Solution
(a) **STEP 1:** Graph the horizontal line $y = 1$, where a solid line indicates that the line is included in the solution set. See **FIGURE 11.13(a)**.

STEP 2: Select a test point either above or below the horizontal line. For example, the point $(0, 0)$ satisfies the inequality $y \leq 1$, because $0 \leq 1$ is a true statement.

STEP 3: Shade the region that contains $(0, 0)$, **below** the line. See **FIGURE 11.13(a)**.

(b) STEP 1: Graph the line $x + y = 3$, as shown in **FIGURE 11.13(b)**. Because the inequality is $<$ and not $\leq$, the line is not included and is dashed rather than solid.

STEP 2: Select a test point in either region. For example, the point $(0, 0)$ satisfies the inequality $x + y < 3$, because $0 + 0 < 3$ is a true statement.

STEP 3: Shade the region that contains $(0, 0)$, **below** the line. See **FIGURE 11.13(b)**.

(c) STEP 1: Graph $-x + 2y = 2$ as a solid line, as shown in **FIGURE 11.13(c)**.

STEP 2: To decide whether to shade above or below the solid line, use the test point $(0, 0)$ again. (Note that other test points can be used.) The point $(0, 0)$ does not satisfy the inequality $-x + 2y \geq 2$ because $-0 + 2(0) \geq 2$ is a false statement.

STEP 3: Shade the region that does not contain $(0, 0)$, **above** the line. See **FIGURE 11.13(c)**.

READING CHECK 1

- How is a test point used when shading a solution set for an inequality?

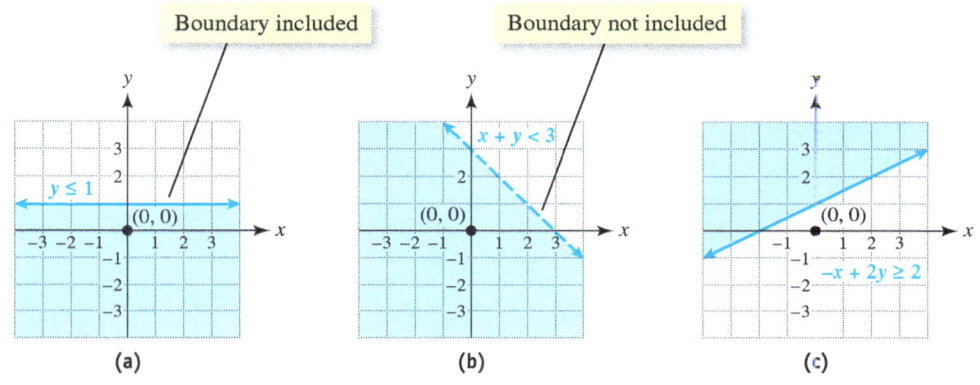

FIGURE 11.13

Now Try Exercises 35, 41, 43

NOTE: In all three parts of Example 2, the test point $(0, 0)$ was used to determine which region should be shaded. However, *any point that is not on* the solid or dashed line can be used as a test point. When the line does not pass through the origin, it is often convenient to use $(0, 0)$ as a test point because substituting 0 for both x and y in an inequality results in a very simple computation. ∎

See the Concept 1 — GRAPHING A LINEAR INEQUALITY

To graph $3x - 2y \leq 6$:

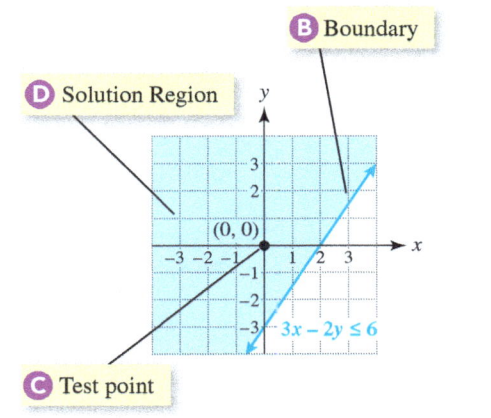

Ⓐ Solve the inequality for y.

$3x - 2y \leq 6$:   Given inequality

$-2y \leq -3x + 6$ Subtract $3x$ from each side

$y \geq \dfrac{3}{2}x - 3$ Divide each side by -2.

Ⓑ Graph the *equation* $y = \dfrac{3}{2}x - 3$.
(Use a **solid** line for $\geq$ or $\leq$. Use a **dashed** line for $>$ or $<$.)

Ⓒ Choose a test point that is not on the line.

Ⓓ Substitute the test point $(0, 0)$ in the given inequality. Since $3(0) - 2(0) \leq 6$ is true, shade the region containing the test point.

WATCH VIDEO IN MML.

3 Finding Solutions to Systems of Linear Inequalities in Two Variables

Sometimes a solution set must satisfy two inequalities in a **system of linear inequalities in two variables**. For example, the point $(2, 1)$ satisfies both of the inequalities in the system of inequalities given by

$$x > 1$$
$$y < 2.$$

FIGURE 11.14(a) shows the solution set to $x > 1$ in blue, and **FIGURE 11.14(b)** shows the solution set to $y < 2$ in red. The two regions are shaded together in **FIGURE 11.14(c)**, where the blue and red regions intersect to form a purple region. The solution set to the system of inequalities includes all points in the purple region, as shown in **FIGURE 11.14(d)**. Points in this region satisfy *both* inequalities.

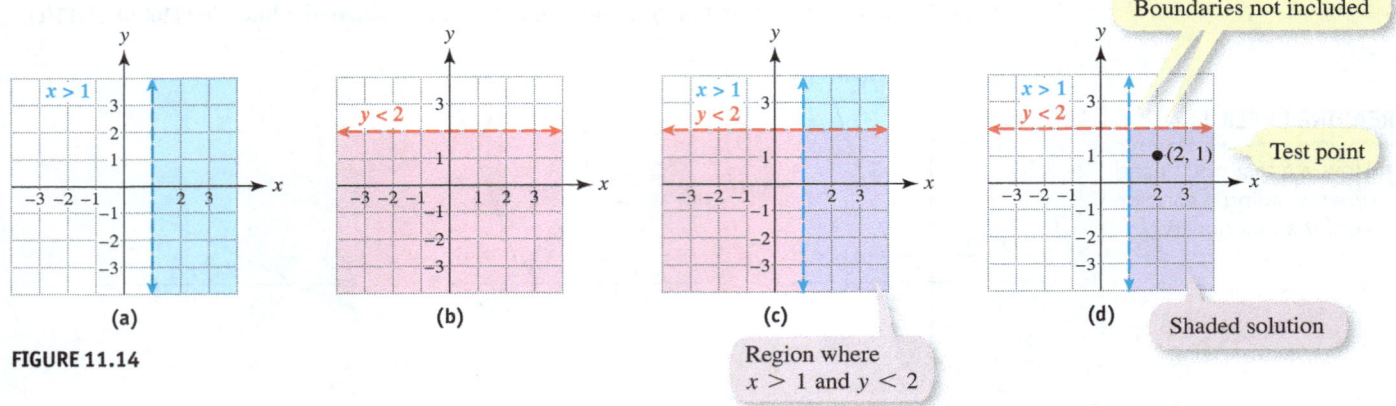

FIGURE 11.14

EXAMPLE 3 Graphing a system of linear inequalities

Shade the solution set to the system of inequalities.

$$x > -1$$
$$x + y \leq 1$$

Solution
Start by graphing the solution set to each inequality. The solution set to $x > -1$ is the blue region to the **right** of the dashed vertical line $x = -1$, as shown in **FIGURE 11.15(a)**. The solution set to $x + y \leq 1$ includes the solid line $x + y = 1$ and the red region that lies **below** it, as shown in **FIGURE 11.15(b)**. For a point to satisfy the *system* of inequalities, it must satisfy *both* inequalities. Therefore the solution set is the *intersection* of the blue and red regions, shown as the purple region in **FIGURE 11.15(c)**. Note that the test point $(0, 0)$, located in the shaded region, satisfies both inequalities.

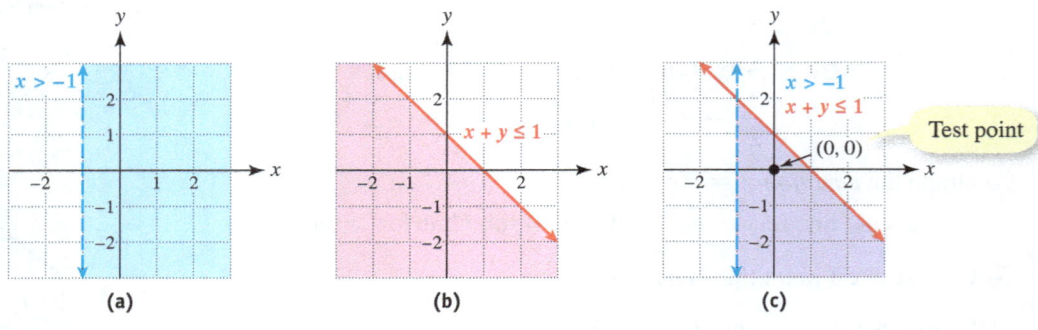

FIGURE 11.15

Now Try Exercise 67

GRAPHING A SYSTEM OF LINEAR INEQUALITIES IN TWO VARIABLES

STEP 1: For each inequality in the system, perform Step 1 from Graphing a Linear Inequality in Two Variables. See the box just before Example 2 in this section.

STEP 2: Pick a test point from one region and substitute it in the given inequalities.

STEP 3: If both of the resulting statements are true, shade the region containing the test point. If not, pick a test point from a different region and substitute it in the given inequalities. Repeat this step until the region to be shaded is found.

EXAMPLE 4 Graphing a system of linear inequalities

Shade the solution set to the system of inequalities.

$$x + 2y < -2$$
$$2x + y \geq 2$$

Solution

STEP 1: Start by graphing the dashed line $x + 2y = -2$ and the solid line $2x + y = 2$, as shown in **FIGURE 11.16(a)**. Note that these two lines divide the xy-plane into 4 regions, numbered **1, 2, 3,** and **4**.

STEP 2: If we let $(0, 0)$ be a test point, it does not satisfy either inequality. Therefore we do not shade region **2**, which contains $(0, 0)$. However, there are still 3 possible regions. If we try the test point $(4, -4)$ in region **4**, it satisfies both the given inequalities.

$$4 + 2(-4) < -2 \checkmark \quad \text{A true statement}$$
$$2(4) + (-4) \geq 2 \checkmark \quad \text{A true statement}$$

STEP 3: Shade region **4**, as shown in **FIGURE 11.16(b)**. Once a test point is found that makes both inequalities true, there is no need to check test points in other regions.

Boundaries and Regions

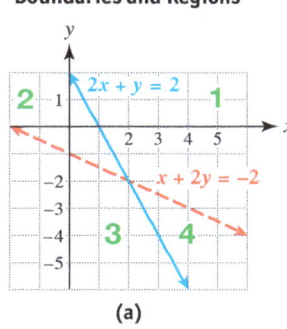

Shaded Solution

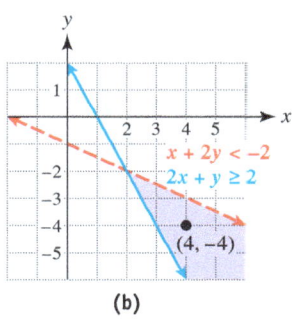

(a)   (b)

FIGURE 11.16

CRITICAL THINKING 1

Does the solution set in **FIGURE 11.16(b)** include the point of intersection, $(2, -2)$? Explain your reasoning.

Now Try Exercise 63

MAKING CONNECTIONS 1

Solving for y and shading inequalities

Another way to solve the system of inequalities in Example 4 that does not involve test points is to solve each inequality for y to obtain

$$y < -\frac{1}{2}x - 1$$
$$y \geq -2x + 2.$$

The solution set is the region in **FIGURE 11.16(b)** that lies **below** the line $y = -\frac{1}{2}x - 1$ and **above and including** the line $y = -2x + 2$.

4 Solving Applications Involving Systems of Linear Inequalities

Math in Context (Manufacturing) The next example illustrates an application from business and manufacturing that involves systems of inequalities.

EXAMPLE 5 Manufacturing MP3 players and digital video players

A business manufactures MP3 players and digital video players. Because every digital video player contains an MP3 player, it must produce at least as many MP3 players as digital video players. In addition, the total number of MP3 players and digital video players produced each day cannot exceed 50 because of limited resources. Shade the region that shows the numbers of MP3 players M and digital video players V that can be produced within these restrictions. Label the horizontal axis M and the vertical axis V.

Solution
Because the company must produce *at least* as many MP3 players M as digital video players V, we have $M \geq V$, which can also be written as $V \leq M$. The total number of MP3 players and digital video players *cannot exceed* 50, so $M + V \leq 50$. To shade the solution set for

$$V \leq M$$
$$M + V \leq 50,$$

we first graph the lines $V = M$ and $M + V = 50$, as shown in **FIGURE 11.17(a)**. Because the number of MP3 players and digital video players cannot be negative, the graph includes only quadrant I. These lines divide this quadrant into four regions, and we can determine the correct region to shade by selecting one test point from each region. The region containing the test point satisfying both inequalities is the one to be shaded. For example, the test point $(20, 10)$ with $M = 20$, $V = 10$ satisfies both inequalities.

$$10 \leq 20 \checkmark \quad \text{A true statement; } V \leq M$$
$$20 + 10 \leq 50 \checkmark \quad \text{A true statement; } M + V \leq 50$$

The solution set is shaded in **FIGURE 11.17(b)**.

NOTE: An alternative solution is to write the inequalities as $V \leq M$ and $V \leq -M + 50$. Then the solution set lies *below both lines*. This region is shaded in **FIGURE 11.17(b)**. ∎

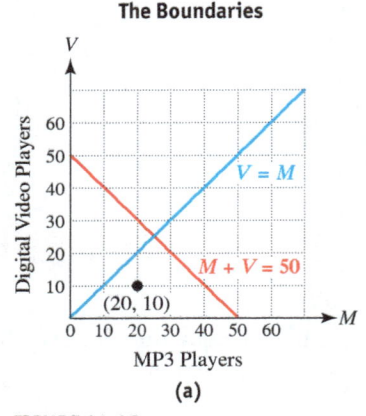
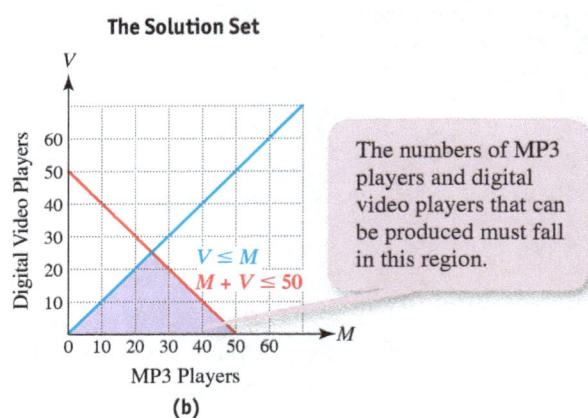

The numbers of MP3 players and digital video players that can be produced must fall in this region.

FIGURE 11.17

Now Try Exercise 73

Math in Context (Health) Although there is no *ideal* weight for a person, government agencies and insurance companies sometimes recommend a *range* of weights for various heights. *Inequalities* are used with these recommendations.

EXAMPLE 6 Finding weight–height combinations

FIGURE 11.18 shows a shaded region containing recommended weights w for heights h. (*Source: Department of Agriculture.*)

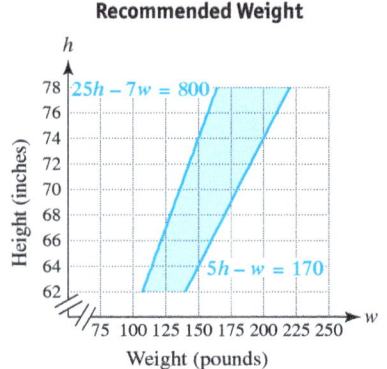

FIGURE 11.18

(a) What does this graph indicate about a 68-inch person who weighs 150 pounds?
(b) The shaded region in **FIGURE 11.18** is determined by the following system of inequalities.

$$25h - 7w \leq 800$$
$$5h - w \geq 170$$

Verify that $h = 68$ and $w = 150$ satisfies the system of inequalities.
(c) What do ordered pairs (w, h) to the left of the shaded region indicate?

Solution
(a) The point $(150, 68)$ lies in the shaded region. Therefore someone who is 68 inches tall and weighs 150 pounds falls within the recommended guidelines.
(b) Both inequalities are satisfied by $h = 68$, $w = 150$.

$$25(68) - 7(150) = 650 \leq 800 \checkmark$$
$$5(68) - 150 = 190 \geq 170 \checkmark$$

(c) To the left of the shaded region are ordered pairs (w, h) that represent smaller weights and larger heights. These ordered pairs correspond to people who weigh less than recommended.

Now Try Exercise 77

11.4 Putting It All Together

CONCEPT	EXPLANATION	EXAMPLES
Linear Inequality in Two Variables	An inequality that can be written as $$Ax + By < C,$$ where $<$ can also be $\leq$, $>$, or $\geq$	$3x + y \geq 10$, $-x + 3y < 5$, $y \leq 5 - x$, and $x > 5$

continued on next page

CHAPTER 11 SYSTEMS OF LINEAR EQUATIONS IN TWO VARIABLES

continued from previous page

CONCEPT	EXPLANATION	EXAMPLES
Solution	A solution (x, y) makes the inequality a true statement.	The ordered pair $(0, 0)$ satisfies $$2x - y < 2,$$ so it is a solution to the inequality.
Solution Set	The set of all solutions Usually a region in the xy-plane	The solution set to $x + y > 2$ is all points above the line $x + y = 2$.
System of Linear Inequalities in Two Variables	Solutions to systems must satisfy both inequalities. The solution set usually includes infinitely many solutions.	The ordered pair $(0, 0)$ is a solution to $$x + y \leq 2$$ $$2x - y > -4,$$ because both inequalities are true when $x = 0$ and $y = 0$.

11.4 Exercises

CONCEPTS AND VOCABULARY

1. When the equals sign in $Ax + By = C$ is replaced with $<, >, \leq,$ or $\geq$, a linear _____ in two variables results.

2. A solution to a linear inequality in two variables is a(n) _____ that makes the inequality a true statement.

3. Describe the graph of the solution set to $y \leq k$ for some number k.

4. Describe the graph of the solution set to $x > k$ for some number k.

5. Describe the graph of the solution set to $y \geq x$.

6. When graphing the solution set to a linear inequality, one way to determine which region to shade is to use a(n) _____ point.

7. When graphing a linear inequality containing either $<$ or $>$, use a (dashed/solid) line.

8. When graphing a linear inequality containing either $\leq$ or $\geq$, use a (dashed/solid) line.

9. When graphing the linear inequality $Ax + By < C$, a first step is to graph the line _____.

10. A solution to a system of two inequalities must make (both inequalities/one inequality) true.

11. If two shaded regions represent the solution sets for two inequalities in a system, then the solution set for the system is where these two shaded regions _____.

12. If a test point is found that satisfies both inequalities in a system, do other test points still need to be checked?

TESTING SOLUTIONS TO ONE LINEAR INEQUALITY

Exercises 13–24: Determine whether the test point is a solution to the linear inequality.

13. $(3, 1), x > 2$
14. $(-3, 4), x \leq -3$
15. $(0, 0), y \geq 2$
16. $(0, 0), y < -3$
17. $(5, 4), y \geq x$
18. $(-1, 2), y < x$
19. $(3, 0), y < x - 1$
20. $(0, 5), y > 2x + 4$
21. $(-2, 6), x + y \leq 4$
22. $(2, -4), x - y \geq 7$
23. $(-1, -1), 2x + y \geq -1$
24. $(0, 1), -x - 5y \geq -1$

WRITING AND GRAPHING A LINEAR INEQUALITY

Exercises 25–32: Write a linear inequality that describes the shaded region.

25.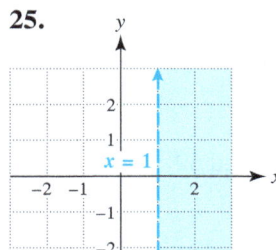
26.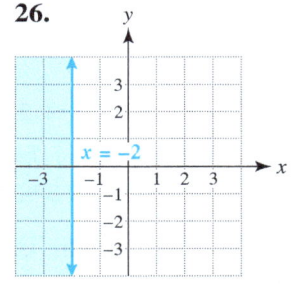
27.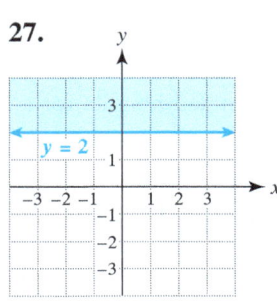
28.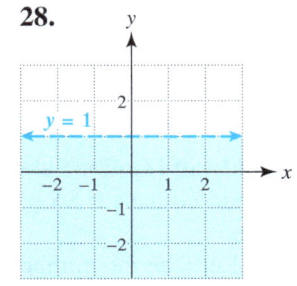
29.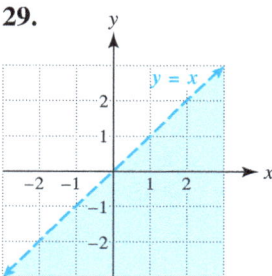
30.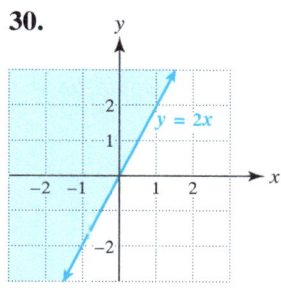
31.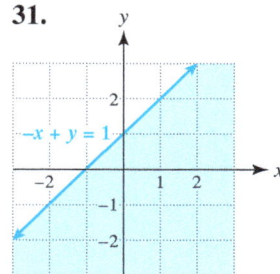
32.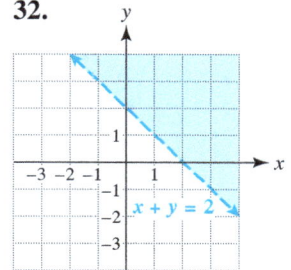

Exercises 33–44: Shade the solution set to the inequality.

33. $x \leq -1$
34. $x > 3$
35. $y < -2$
36. $y \geq 0$
37. $y > x$
38. $y \leq x$
39. $y \geq 3x$
40. $y < -2x$
41. $x + y \leq 1$
42. $x + y \geq -2$
43. $2x - y > 2$
44. $-x - y < 1$

TESTING SOLUTIONS TO SYSTEMS OF LINEAR INEQUALITIES

Exercises 45–50: Determine if the test point is a solution to the system of linear inequalities.

45. $(3, 1)$
$x - y < 3$
$x + y > 3$

46. $(0, 0)$
$x - 2y < 1$
$2x - y > -1$

47. $(-2, 3)$
$3x - 2y \geq 1$
$-x + 3y > 3$

48. $(\_, 2)$
$2x - 2y < 5$
$x - y > -1$

49. $(4, -2)$
$x - 2y \geq 8$
$-2x - 5y > 0$

50. $(-1, -2)$
$x + y < 0$
$-2x - 3y \leq -1$

Exercises 51–54: The graphs of two equations are shown with four test points labeled. Use these points to decide which region should be shaded to solve the given system of inequalities.

51. $x \leq 2$
$x + y \geq 2$

52. $y \geq 1$
$2x - y \geq 3$

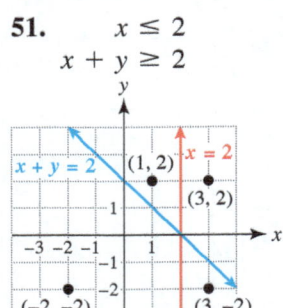
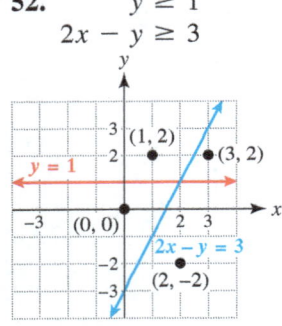

53. $x + y \leq 3$
$y \leq 2x$

54. $y \leq x$
$y \geq -x$

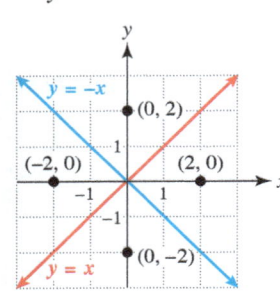

GRAPHING SOLUTIONS TO SYSTEMS OF LINEAR INEQUALITIES

Exercises 55–72: Shade the solution set to the system of inequalities.

55. $x > 2$
$y < 3$

56. $x \leq -1$
$y \geq 3$

57. $x \leq -2$
$y < 2x$

58. $y > 2$
$y \geq -x$

59. $y \leq x$
$y > -x$

60. $y \leq \frac{1}{2}x$
$y \geq -2x$

61. $x + y \leq 3$
$-x + y \leq 1$

62. $x + y > 2$
$x - y < 2$

63. $2x + y > -3$
$x + y \leq -1$

64. $-x + y \geq 3$
$2x - y \geq -2$

65. $2x + y \geq -3$
$x + y > -1$

66. $-x + y \geq 2$
$3x - y \geq -2$

67. $y > -2$
$x + 2y \leq -4$

68. $x \geq 2$
$3y < x - 3$

69. $x + 2y > -4$
$2x + y \leq 3$

70. $x + 3y \geq 3$
$3x - 2y \geq 6$

71. $3x + 4y \leq 12$
$5x + 3y \geq 15$

72. $-2x + y \geq 6$
$x - 2y \geq -4$

APPLICATIONS INVOLVING SYSTEMS OF INEQUALITIES

73. **Manufacturing** (Refer to Example 5.) A business manufactures at least two MP3 players for each digital video player. The total number of MP3 players and digital video players must be less than 90. Shade the region that represents the number of MP3 players M and digital video players V that can be produced within these restrictions. Put V on the horizontal axis.

74. **Working on Two Projects** An employee must spend more time on project X than on project Y. The employee can work at most 40 hours on the two projects. Shade the region in the xy-plane that represents the number of hours the employee can spend on each project.

75. **Maximum Heart Rate** When exercising, people often try to maintain heart rates that are a percentage of their maximum heart rate. Maximum heart rate R is given by $R = 220 - A$, where A is the person's age in years and R is the heart rate in beats per minute.
    (a) Find R for a person 20 years old; 70 years old.
    (b) Sketch a graph of $R \leq 220 - A$. Assume that A is between 20 and 70 and put A on the horizontal axis of your graph.
    (c) Interpret your graph.

76. **Target Heart Rate** (Refer to the preceding exercise.) A target heart rate T that is half a person's maximum heart rate is given by $T = 110 - \frac{1}{2}A$, where A is a person's age in years.
    (a) What is T for a person 30 years old? 50 years old?
    (b) Sketch a graph of the system of inequalities.
    $$T \geq 110 - \frac{1}{2}A$$
    $$T \leq 220 - A$$
    Assume that A is between 20 and 60.
    (c) Interpret your graph.

77. **Height and Weight** Use Figure 11.18 in Example 6 to determine the range of recommended weights for a person who is 74 inches tall.

78. **Height and Weight** Use Figure 11.18 in Example 6 to determine the range of recommended heights for a person who weighs 150 pounds.

79. **Candy and Coffee Sales** A small business sells freshly ground coffee for $6 per pound and candy for $4 per pound.
    (a) Let x be the number of pounds of coffee sold and y be the number of pounds of candy sold. Sketch a line that represents all possible values of x and y that result in total sales of $240. Label your graph and assume that $x \geq 0$ and $y \geq 0$. (*Hint:* Find the x- and y-intercepts.)
    (b) Shade the region in your graph that represents all sales (x, y) that result in total sales of $240 or less.

80. Candy and Coffee Sales (Continuation of Exercise 79) Suppose that the business sells at least as many pounds of candy as coffee and that total sales are $240 or less. Shade the region in your graph that represents all sales (x, y) that satisfy these two conditions.

Exercises 81–86: **Plant Growth** *The following figure illustrates the relationships among forests, grasslands, and deserts, suggested by annual Temperature T in degrees Fahrenheit and precipitation P in inches.*

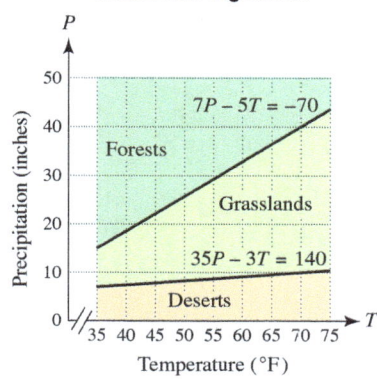

81. Use the graph to describe in terms of precipitation and temperature the regions where forests grow.

82. Use the graph to describe in terms of precipitation and temperature the regions where deserts exist.

83. The equation of the line that separates grasslands and forests is
$$7P - 5T = -70.$$
Write an inequality that describes temperatures and amounts of precipitation that correspond to forested regions. (Include the line.)

84. The equation of the line that separates grasslands and deserts is
$$35P - 3T = 140.$$
Write an inequality that describes temperatures and amounts of precipitation that correspond to desert regions. (Include the line.)

85. Using the information from Exercises 83 and 84, write a system of inequalities that describes temperatures and amounts of precipitation that correspond to grassland regions. (Include the lines.)

86. Cheyenne, Wyoming, has an average annual temperature of about 50°F and an average annual precipitation of about 14 inches. Use the graph to predict the type of plant growth you might expect near Cheyenne. Then check to determine whether $T = 50$ and $P = 14$ satisfy the proper inequalities.

WRITING ABOUT MATHEMATICS

87. What is the solution set to the following system of inequalities? Explain your reasoning.
$$y > x$$
$$y < x - 1$$

88. Write a system of linear inequalities whose solution set is the entire xy-plane. Explain your reasoning.

SECTIONS 11.3 and 11.4 — Checking Basic Concepts

1. Use elimination to solve the system of equations.
$$2x + 3y = 5$$
$$x - 7y = -6$$

2. Use elimination to solve each system of equations. How many solutions are there in each case?
   (a) $\quad x + y = -1$
   $\quad\quad x - 2y = 2$
   (b) $\quad 5x - 6y = 4$
   $\quad\quad -5x + 6y = 1$
   (c) $\quad x - 3y = 0$
   $\quad\quad 2x - 6y = 0$

3. Solve the given system of equations symbolically, graphically, and numerically. How many solutions are there?
$$-2x + y = 0$$
$$y = 2x$$

4. Shade the solution set to each inequality.
   (a) $y < -1$ (b) $x + y < 1$

5. Shade the solution set to the following system of inequalities.
$$x \leq -1$$
$$-2x + y > -3$$

6. **Large U.S. Cities** The combined population of New York and Chicago was 11.1 million people in 2013. The population of New York exceeded the population of Chicago by 5.7 million people.
   (a) Let x be the population of New York and y be the population of Chicago. Write a system of equations whose solution gives the population of each city in 2013.
   (b) Solve the system of equations.

CHAPTER 11   Summary

SECTION 11.1 ■ SOLVING SYSTEMS OF LINEAR EQUATIONS GRAPHICALLY AND NUMERICALLY

System of Linear Equations

Solution — An ordered pair (x, y) that satisfies *both* equations

Solution Set — The set of all solutions

Graphical Solution — Graph each equation. A point of intersection is a solution. (Sometimes determining the exact answer when estimating from a graph may be difficult.)

Numerical Solution — Solve each equation for y and make a table for each equation. A solution occurs when two y-values are equal for a given x-value.

Example: The ordered pair $(3, 1)$ is the solution to the following system.

$$x + y = 4 \qquad 3 + 1 = 4 \text{ is a true statement.}$$
$$x - y = 2 \qquad 3 - 1 = 2 \text{ is a true statement.}$$

A Graphical Solution
The point of intersection $(3, 1)$ is the solution to the system of equations.

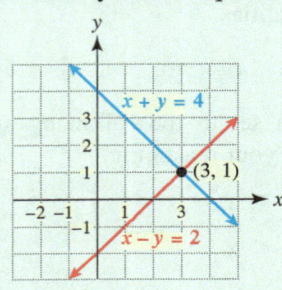

A Numerical Solution
The ordered pair $(3, 1)$ is the solution. When $x = 3$, both y-values equal 1.

x	1	2	3	4
$y = 4 - x$	3	2	1	0
$y = x - 2$	-1	0	1	2

Types of Systems of Linear Equations   A system of linear equations can have no solutions, one solution, or infinitely many solutions.

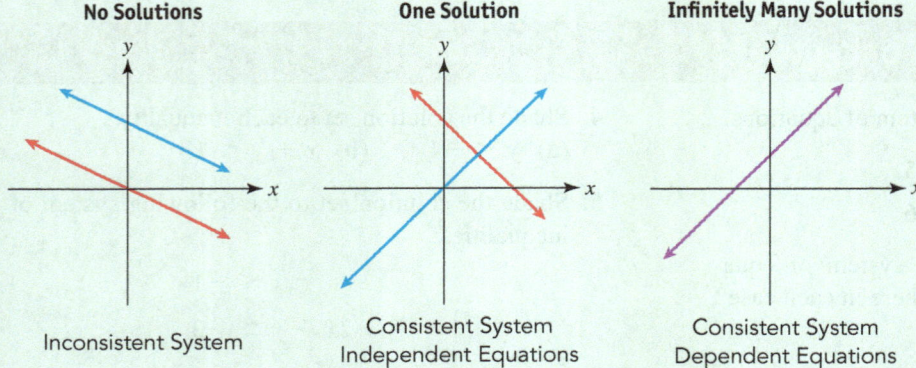

No Solutions — Inconsistent System

One Solution — Consistent System, Independent Equations

Infinitely Many Solutions — Consistent System, Dependent Equations

SECTION 11.2 ■ SOLVING SYSTEMS OF LINEAR EQUATIONS BY SUBSTITUTION

Method of Substitution   This method can be used to solve a system of equations symbolically and always gives the exact solution, provided one exists.

Example: $-2x + y = -3$
$\phantom{\text{Example: }}x + y = 3$

STEP 1: Solve one of the equations for a convenient variable.

$$x + y = 3 \quad \text{becomes} \quad y = 3 - x.$$

STEP 2: Substitute this result in the other equation and then solve.

$-2x + (3 - x) = -3$ Substitute $(3 - x)$ for y.

$-3x = -6$ Combine like terms; subtract 3.

$x = 2$ Divide each side by -3.

STEP 3: Find the value of the other variable. Because $y = 3 - x$ and $x = 2$, it follows that $y = 3 - 2 = 1$.

STEP 4: Check to determine that $(2, 1)$ is the solution.

$-2(2) + (1) \stackrel{?}{=} -3$ ✓   A true statement

$2 + 1 \stackrel{?}{=} 3$ ✓   A true statement

The solution $(2, 1)$ checks.

Recognizing Types of Systems

No solutions	The final equation is always false, such as $0 = 1$.
One solution	The final equation has one solution, such as $x = 1$.
Infinitely many solutions	The final equation is always true, such as $0 = 0$.

SECTION 11.3 ■ SOLVING SYSTEMS OF LINEAR EQUATIONS BY ELIMINATION

Method of Elimination This method can be used to solve a system of linear equations symbolically and always gives the exact solution, provided one exists.

Example: $\begin{aligned} x + 3y &= 1 \\ \underline{-x + y} &= 3 \\ 4y &= 4 \quad \text{or} \quad y = 1 \end{aligned}$ Add and solve for y.

Substitute $y = 1$ in either of the given equations: $x + 3(1) = 1$ implies that $x = -2$, so $(-2, 1)$ is the solution.

NOTE: To eliminate a variable, it may be necessary to multiply one or both equations by a constant before adding. ■

Recognizing Types of Systems

No solutions	The final equation is always false, such as $0 = 1$.
One solution	The final equation has one solution, such as $x = 1$.
Infinitely many solutions	The final equation is always true, such as $0 = 0$.

SECTION 11.4 ■ SYSTEMS OF LINEAR INEQUALITIES

Graphing a Linear Inequality in Two Variables

1. Replace the inequality symbol with an equals sign and graph the resulting line. If the inequality is $<$ or $>$, use a dashed line; if it is $\leq$ or $\geq$, use a solid line.
2. Pick a *test point* that does *not* lie on the line. Substitute this point in the given inequality. Determine whether the resulting statement is true or false.
3. If the statement is true, shade the region containing the test point. If the statement is false, shade the region that does not contain the test point.

Solving a System of Linear Inequalities in Two Variables

1. For each inequality in the system, perform Step 1 from Graphing a Linear Inequality in Two Variables (above).
2. Pick a test point from one region and substitute it in the given inequalities.
3. If both of the resulting statements are true, shade the region containing the test point. If not, pick a test point from a different region and substitute it in the given inequalities. Repeat this step until the region to be shaded is found.

Example: $x \leq 1$
$x + y \leq 2$

Graph the lines $x = 1$ and $x + y = 2$. Then pick a test point, such as $(0, 0)$, and substitute it in each inequality.

$0 \leq 1$ ✓   A true statement
$0 + 0 \leq 2$ ✓   A true statement

Because $(0, 0)$ satisfies *both* inequalities, shade the region containing $(0, 0)$. See the graph.

NOTE: When shading the solution set to a *system* of inequalities, you may need to try more than one test point. ∎

NOTE: An alternative way to determine the solution set is to shade the region that is both to the **left** of the line $x = 1$ (because $x \leq 1$) and **below** the line $y = -x + 2$ (because $y \leq -x + 2$). ∎

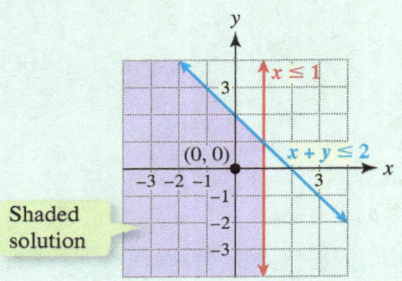
Shaded solution

CHAPTER 11   Review Exercises

SECTION 11.1

Exercises 1 and 2: Determine graphically the x-value for the equation when $y = 3$.

1. $y = 2x - 3$
2. $y = \frac{3}{2}x$

Exercises 3–6: The graphs of two equations are shown.
(a) State the number of solutions to the system of equations.
(b) Is the system consistent or inconsistent? If the system is consistent, state whether the equations are dependent or independent.

3.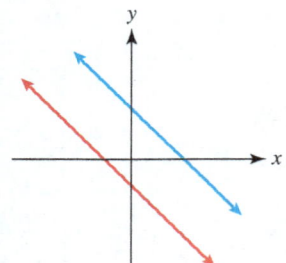
4.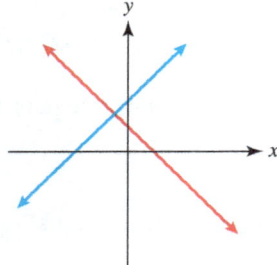
5.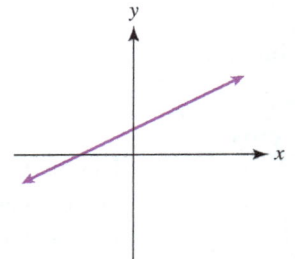
6.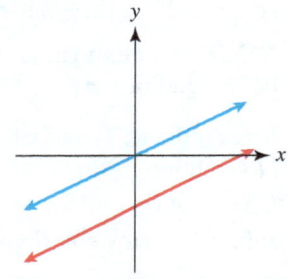

Exercises 7–10: Determine which ordered pair is a solution to the system of equations.

7. $(0, 1), (1, 2)$
$x + 2y = 5$
$x - y = -1$

8. $(5, 2), (4, 0)$
$2x - y = 8$
$x + 3y = 11$

9. $(2, 2), (4, 3)$
$\frac{1}{2}x = y - 1$
$2x = 3y - 1$

10. $(2, -4), (-1, 2)$
$5x - 2y = 18$
$y = -2x$

Exercises 11 and 12: The graphs for two equations are shown. Use the intersection-of-graphs method to identify the solution to both equations. Then check your result.

11.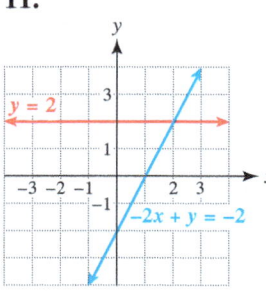

12.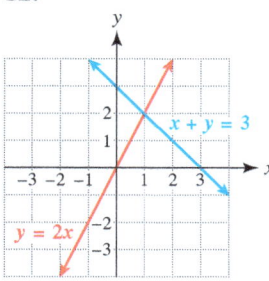

Exercises 13 and 14: A table for two equations is given. Identify the solution to both equations.

13.

x	1	2	3	4
$y = 3x$	3	6	9	12
$y = 6$	6	6	6	6

14.

x	-1	0	1	2
$y = 2x - 1$	-3	-1	1	3
$y = 2 - x$	3	2	1	0

Exercises 15–20: Solve the system of linear equations graphically.

15. $y = -3$
$x + y = 1$

16. $x = 1$
$x - y = -1$

17. $2x + y = 3$
$-x + y = 0$

18. $y = 2x$
$2x + y = 4$

19. $x + 2y = 3$
$2x + y = 3$

20. $-3x - y = 7$
$2x + 3y = -7$

SECTION 11.2

Exercises 21–28: Use the method of substitution to solve the system of linear equations. These systems may have no solutions, one solution, or infinitely many solutions.

21. $x + y = 8$
$y = 3x$

22. $x - 2y = 22$
$y = -5x$

23. $x + 3y = 1$
$-2x + 2y = 6$

24. $3x - 2y = -4$
$2x - y = -4$

25. $x + y = 2$
$y = -x$

26. $x + y = -2$
$x + y = 3$

27. $-x + 2y = 2$
$x - 2y = -2$

28. $-x - y = -2$
$2x - y = 1$

SECTION 11.3

Exercises 29 and 30: Use the graph to solve the system of equations. Then use the elimination method to verify your answer.

29. $x + y = 3$
$x - y = 1$

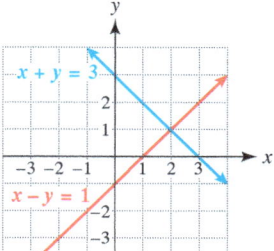

30. $2x + 3y = 4$
$x - 2y = -5$

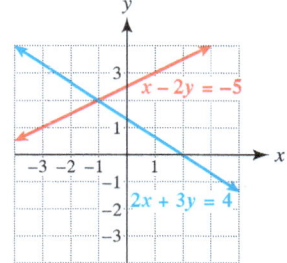

Exercises 31–38: Use the elimination method to solve the system of equations.

31. $x + y = 10$
$x - y = 12$

32. $2x - y = 2$
$3x + y = 3$

33. $-2x + 2y = -1$
$x - 3y = -3$

34. $2x - 5y = 0$
$2x + 4y = 9$

35. $2a + b = 3$
$-3a - 2b = -1$

36. $a - 3b = 2$
$3a + b = 26$

37. $5r + 3t = -1$
$-2r - 5t = -11$

38. $5r + 2t = 5$
$3r - 7t = 3$

Exercises 39 and 40: Solve the system of equations (a) symbolically, (b) graphically, and (c) numerically.

39. $3x + y = 6$
$x - y = -2$

40. $2x + y = 3$
$-x + 2y = -4$

Exercises 41–44: Use elimination to determine whether the system of equations has no solutions, one solution, or infinitely many solutions.

41. $x - y = 5$
$-x + y = -5$

42. $3x - 3y = 0$
$-x + y = 0$

43. $-2x + y = 3$
$2x - y = 3$

44. $-2x + y = 2$
$3x - y = 3$

SECTION 11.4

Exercises 45–48: Determine whether the test point is a solution to the linear inequality.

45. $(5, -3)$ $y \leq 2$

46. $(-1, 3)$ $x > -1$

47. $(1, 2)$ $x + y < -2$

48. $(1, -4)$ $2x - 3y \geq 2$

Exercises 49 and 50: Write a linear inequality that describes the shaded region.

49.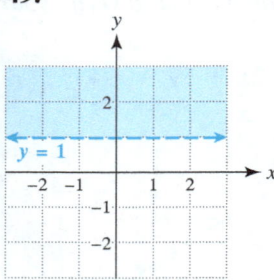

50.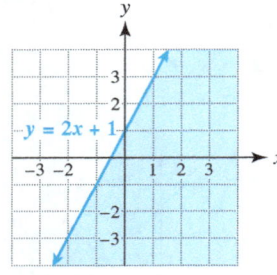

Exercises 51–56: Shade the solution set for the inequality.

51. $x \leq 1$
52. $y > 2$
53. $y > 3x$
54. $x \geq 2y$
55. $y < x + 1$
56. $2x + y \geq -2$

Exercises 57 and 58: Determine whether the test point is a solution to the system of linear inequalities.

57. $(1, -2)$
    $x - 2y > 3$
    $2x + y < 3$

58. $(4, -3)$
    $x - y \geq 1$
    $4x + 3y \leq 4$

Exercises 59 and 60: The graphs of two equations are shown with four test points labeled. Use these points to decide which region should be shaded to solve the system of inequalities.

59. $y \leq 1$
    $2x + y \geq -1$

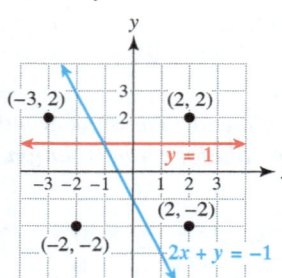

60. $y \geq x$
    $x + y \geq 2$

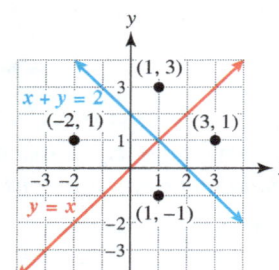

Exercises 61–66: Shade the solution set for the system of inequalities.

61. $x > -1$
    $y < -2$

62. $y \leq x$
    $y \geq -2x$

63. $x + y \leq 3$
    $y \geq -x$

64. $2x + y < 3$
    $y > x$

65. $\frac{1}{2}x + y \geq 2$
    $x - 2y \leq 0$

66. $2x - y < 3$
    $4x + 2y > -6$

APPLICATIONS

67. **Traffic Fatalities** The number of traffic fatalities increased by a factor of 10.9 from 1912 to 2012. There were 30,690 more deaths in 2012 than in 1912. Find the number of traffic fatalities in each of the two years. Note that the number of vehicles on the road increased from 1 million to 253 million between 1912 and 2012. (*Source:* Department of Transportation)

68. **Social Ad Revenue** By 2017, U.S. total social ad revenue reached $11.8 billion. Social display ad revenue was $1.8 billion more than social native ad revenue. What was the revenue for social display ads and for social native ads?

69. **Renting a Car** A rental car costs $40 plus $0.20 per mile that it is driven.
    (a) Write an equation that gives the cost C of driving the car x miles.
    (b) Use the intersection-of-graphs method to determine the number of miles that the car is driven if the rental cost is $90.
    (c) Solve part (b) numerically with a table of values.

70. **Supplementary Angles** The smaller of two angles that sum to 180° is 30° less than the measure of the larger angle. Find each angle.

71. **Triangle** In a triangle, the measures of the two smaller angles are equal and their sum is 40° more than the larger angle.
    (a) Let x be the measure of each of the two smaller angles and y be the measure of the larger angle. Write a system of linear equations whose solution gives the measures of these angles.
    (b) Use the method of substitution to solve the system.
    (c) Use the method of elimination to solve the system.

72. **Garden Dimensions** A rectangular garden has 88 feet of fencing around it. The garden is 4 feet longer than it is wide. Find the dimensions of the garden.

73. **Room Prices** Ten rooms are rented at rates of $80 and $120 per night. The total collected for the 10 rooms is $920.
    (a) Write a system of linear equations whose solution gives the number of each type of room rented. Be sure to state what each variable represents.
    (b) Solve the system of equations.

74. **Mixture Problem** One type of candy sells for $2 per pound, and another type sells for $3 per pound. An order for 18 pounds of candy costs $47. How much of each type of candy was bought?

75. **Burning Calories** An athlete burns 9 calories per minute on a stationary bicycle and 11 calories per minute on a stair climber. In a 60-minute workout, the athlete burns 590 calories. How many minutes does the athlete spend on each type of exercise equipment? (*Source:* Runner's World.)

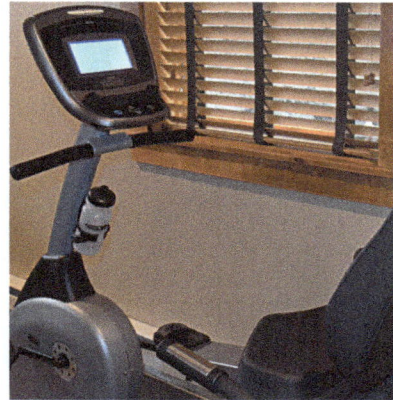

76. **River Current** A riverboat travels 140 miles downstream in 10 hours, and the return trip takes 14 hours. What is the speed of the current?

77. **Garage Dimensions** The blue graph shown in the figure gives possible dimensions for a rectangular garage with perimeter 80 feet. The red graph shows possible dimensions for a garage that has width W two-thirds of its length L.

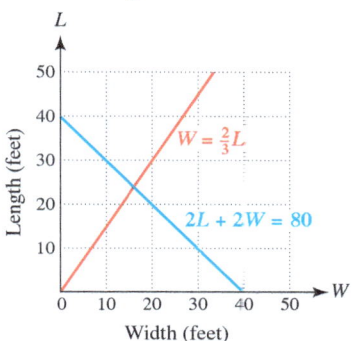

(a) Use the graph to estimate the dimensions of a garage with perimeter 80 feet and width two-thirds its length.
(b) Solve this problem symbolically.

78. **Wheels and Trailers** A business manufactures at least two wheels for each trailer it makes. The total number of trailers and wheels manufactured cannot exceed 30 per week. Shade the region that represents numbers of wheels W and trailers T that can be produced each week within these restrictions. Label the horizontal axis W and the vertical axis T.

CHAPTER 11 Test

1. Determine which ordered pair is a solution to the system of equations.

$$(3, -1), (1, 2)$$
$$3x + 2y = 7$$
$$2x - y = 0$$

2. The graphs for two equations are shown. Use the intersection-of-graphs method to identify the solution to both equations. Then check your solution.

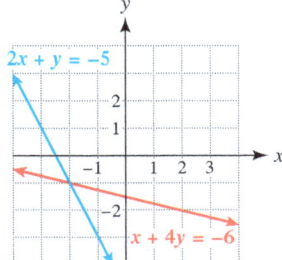

3. A table for two equations is given. Identify the solution to both equations.

x	-2	-1	0	1
$y = 2x$	-4	-2	0	2
$y = 3x + 1$	-5	-2	1	4

4. Solve the system of equations graphically.

$$x + 2y = 4$$
$$x + y = 1$$

5. Use the method of substitution to solve the system of linear equations.

$$3x + 2y = 9$$
$$y = 3x$$

6. Use the method of substitution to solve the system of linear equations. How many solutions are there? Is the system consistent or inconsistent?
(a) $x + 3y = 5$
$\quad\;\,3x - 2y = 4$
(b) $-x + \frac{1}{2}y = 12$
$\quad\;\,2x - y = -4$

Exercises 7 and 8: The graphs of two equations are shown.

(a) State the number of solutions to the system of equations.
(b) Is the system consistent or inconsistent? If the system is consistent, state whether the equations are dependent or independent.

7. 8.

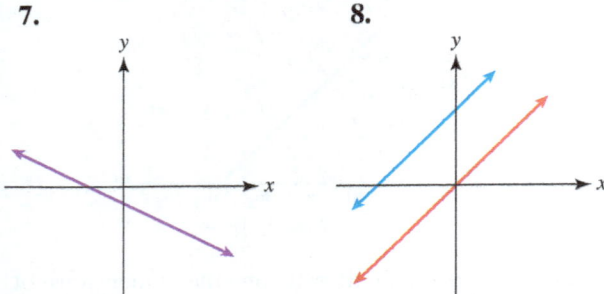

Exercises 9–12: Use the elimination method to solve the system of equations. Note that these systems may have no solutions, one solution, or infinitely many solutions.

9. $x + 2y = 5$
   $3x - 2y = -17$

10. $2x - 2y = 3$
    $-x + y = 5$

11. $x - 2y = 3$
    $-3x + 6y = -9$

12. $4x + 3y = 5$
    $3x - 2y = -9$

13. Write a linear inequality that describes the shaded region.

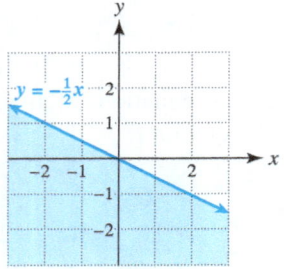

14. Determine whether the test point $(4, -3)$ is a solution to the system of linear inequalities.

$$2x + y > 3$$
$$x - y \geq 7$$

Exercises 15 and 16: Shade the solution set for the given inequality.

15. $x \leq 4$
16. $x + y > 2$

17. Determine whether region **1**, **2**, **3**, or **4** should be shaded to solve the system of inequalities.

$$x + y \leq 5$$
$$x - y \geq 1$$

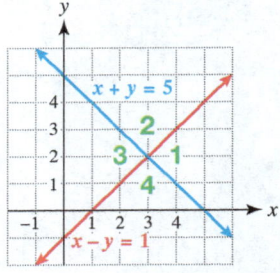

Exercises 18 and 19: Shade the solution set for the given system of inequalities.

18. $x > 2$
    $y < 2x$

19. $2x + y \leq 3$
    $x - y \geq 0$

20. **Music Subscriptions** The combined number of paid music subscriptions in 2012 and 2013 was 9.5 million. There were 2.7 million more subscriptions in 2013 than in 2012. How many paid music subscriptions were there in each year?

21. **Mixture Problem** A chemist has 20% and 60% solutions of acid available. How many liters of each solution should be mixed to obtain 10 liters of a 30% acid solution?

22. **Jogging Speed** An athlete jogs at 6 miles per hour and at 9 miles per hour for a total time of 1 hour, covering a distance of 7 miles. How long does the athlete jog at each speed?

CHAPTERS 1–11 Cumulative Review Exercises

1. Write 120 as a product of prime numbers.

2. Evaluate each expression by hand.
   (a) $2^3 \div \frac{5 + 7}{9 - 3}$
   (b) $-\frac{2}{5} \cdot (5 - 25)$

3. Classify each number as rational or irrational.
   (a) -6.9 (b) $\sqrt{14}$

4. Insert $>$ or $<$ to make each statement true.
   (a) -5 _____ $|-5|$ (b) $|7|$ _____ $|-1|$

5. Simplify each expression.
   (a) $5x^2 - x^2$ (b) $3 - 2x + 7x - 5$

Exercises 6 and 7: Solve the equation.

6. $5(2x + 1) = 7 + x$

7. $1 - (x + 1) = x - 1$

8. Determine whether the linear equation given by $2(5x + 1) = 10x - 3$ has no solutions, one solution, or infinitely many solutions.

9. Find four consecutive integers whose sum is 50.

10. Find the area of a rectangle that has a 36-inch length and a 1-foot width.

11. Solve the formula $W = 3x - 7y$ for x.

12. Solve the inequality $3 - (2x - 7) \leq 8x$.

13. Use the table of xy-values to make a line graph.

x	-2	-1	0	1	2
y	-2	2	0	3	-1

14. Graph each equation.
    (a) $y = -2x + 2$ (b) $3y + 2x = 6$

15. Find the x- and y-intercepts for the graph of the equation $4x - y = 8$.

16. Sketch a line with the given slope that passes through the given point.
    (a) $m = -2$, $(1, 1)$ (b) $m = \frac{4}{3}$, $(-3, 2)$

17. Write the slope–intercept form for each line.
    (a)   (b)

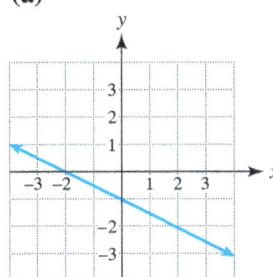

18. Find the slope–intercept form of the line passing through $(4, 2)$ and $(-4, 4)$.

19. Write the slope–intercept form of the line perpendicular to $2x - 6y = 7$, passing through $(1, 1)$.

20. Determine whether $(3, 1)$ or $(4, 4)$ is a solution to the system of equations.

$$3x - y = 8$$
$$2x + y = 12$$

Exercises 21–24: Solve the system of equations. Note that these systems may have no solutions, one solution, or infinitely many solutions.

21. $x - y = 4$
    $-2x - y = 1$

22. $2x + 3y = 4$
    $-4x - 6y = 7$

23. $3x - 4y = 8$
    $-15x + 20y = -40$

24. $7x + 2y = -3$
    $-5x - 3y = -1$

Exercises 25 and 26: Shade the solution set for the system of inequalities.

25. $y < x + 1$
    $x + y \leq 3$

26. $3x + y \geq 6$
    $x - 3y \leq 3$

27. **Rods to Feet** There are 16.5 feet in 1 rod. Write a formula that converts R rods to F feet.

28. **Temperature Change** Find the temperature change for a package of frozen carrots that is removed from a freezer at $-11\,°C$ and placed in water at $83\,°C$.

29. **Cost of a Digital Camera** A 7% sales tax on a digital camera amounts to $17.15. Find the cost of the digital camera.

30. **Tuition Increase** If tuition is currently $145 per credit and it is going to be increased by 9%, what will the new tuition per credit be?

31. **Gasoline Consumption** Write an equation in slope–intercept form that models the number of gallons G of gas in a truck's tank after x hours if the tank initially contains 30 gallons of gas and the truck uses 3 gallons every hour.

32. **Bank Loans** An individual has two loans totaling $2400. One loan charges 5% interest and the other charges 6% interest. If the interest for one year is $132, how much money is borrowed at each rate?

12 Polynomials and Exponents

- 12.1 Rules for Exponents
- 12.2 Addition and Subtraction of Polynomials
- 12.3 Multiplication of Polynomials
- 12.4 Special Products
- 12.5 Integer Exponents and the Quotient Rule
- 12.6 Division of Polynomials

Digital images were first sent between New York and London by cable in the early 1920s. Unfortunately, the transmission time was 3 hours and the quality was poor. Digital photography was developed further by NASA in the 1960s because ordinary pictures were subject to interference when transmitted through space. Today, digital pictures remain crystal clear even if they travel millions of miles.

Whether they are taken with a webcam, with a smartphone, or by a Mars rover, digital images consist of tiny units called pixels, which are represented by numbers. As a result, mathematics plays an important role in digital images. In this chapter, we illustrate some of the ways mathematics is used to describe digital pictures (see Example 4 and Exercise 80 in the Special Products section of this chapter). We also discuss how mathematics is used to model things such as heart rate, computer sales, motion of the planets, and interest on money. *Source:* NASA.

12.1 Rules for Exponents

Objectives

1. Reviewing Bases and Exponents
2. Computing Zero Exponents
3. Applying the Product Rule
4. Applying Power Rules

1 Reviewing Bases and Exponents

The expression 5^3 is an exponential expression with *base* 5 and *exponent* 3. Its value is

$$5 \cdot 5 \cdot 5 = 125.$$

In general, b^n is an exponential expression with base b and exponent n. If n is a natural number, it indicates the number of times the base b is to be multiplied with itself.

$$b^n = \underbrace{b \cdot b \cdot b \cdot \cdots \cdot b}_{n \text{ factors of } b}$$

TABLE 12.1 contains several examples of exponential expressions.

Exponential Expressions

Equal Expressions	Base	Exponent
$2 \cdot 2 \cdot 2 = 2^3$	2	3
$6 \cdot 6 \cdot 6 \cdot 6 = 6^4$	6	4
$7 = 7^1$	7	1
$0.5 \cdot 0.5 = 0.5^2$	0.5	2
$x \cdot x \cdot x = x^3$	x	3

TABLE 12.1

STUDY TIP

Exponents occur throughout mathematics. Because exponents are so important, this section is essential for your success in mathematics. It takes practice, so set aside some extra time.

EVALUATING EXPRESSIONS

When evaluating expressions, use the following order of operations. First perform all calculations within parentheses and absolute values, or above and below a fraction bar.

1. Evaluate exponents.
2. Perform negation.
3. Do multiplication and division from left to right.
4. Do addition and subtraction from left to right.

NOTE: Evaluate exponents *before* performing negation. See parts (c) and (d) in the next example. ∎

EXAMPLE 1 Evaluating exponential expressions

Evaluate each expression.

(a) $1 + \dfrac{2^4}{4}$ (b) $3\left(\dfrac{1}{3}\right)^2$ (c) -2^4 (d) $(-2)^4$

Solution

(a) Evaluate the exponent first.

$$1 + \frac{2^4}{4} = 1 + \frac{\overbrace{2 \cdot 2 \cdot 2 \cdot 2}^{4 \text{ factors}}}{4} = 1 + \frac{16}{4} = 1 + 4 = 5$$

(b) $3\left(\dfrac{1}{3}\right)^2 = 3\left(\overbrace{\dfrac{1}{3} \cdot \dfrac{1}{3}}^{\text{2 factors}}\right) = 3 \cdot \dfrac{1}{9} = \dfrac{3}{9} = \dfrac{1}{3}$

(c) Because exponents are evaluated before negation is performed,

$-\underset{\text{Base is 2.}}{2}^4 = -(\overbrace{2 \cdot 2 \cdot 2 \cdot 2}^{\text{4 factors}}) = -16.$

(d) $(\underset{\text{Base is }-2.}{-2})^4 = \overbrace{(-2)(-2)(-2)(-2)}^{\text{4 factors}} = 16$

Now Try Exercises 9, 11, 13, 15

READING CHECK 1

- Simplify -4^2 and $(-4)^2$.

NOTE: Parts (c) and (d) of Example 1 appear to be very similar. However, the negation sign is inside the parentheses in part (d), which means that the base for the exponential expression is -2. In part (c), no parentheses are used, indicating that the base of the exponential expression is 2. ∎

TECHNOLOGY NOTE

Evaluating Exponents
Exponents can often be evaluated on calculators by using the ∧ key. The four expressions from Example 1 are evaluated with a calculator and the results are shown in the following two figures. When evaluating the last two expressions on your calculator, remember to use the negation key rather than the subtraction key.

CALCULATOR HELP
To evaluate exponents, see Appendix A (page AP-1).

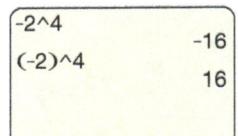

2 Computing Zero Exponents

So far we have discussed natural number exponents. What if an exponent is 0? What does 2^0 equal? To answer these questions, consider **TABLE 12.2**, which shows values for decreasing powers of 2. Note that each time the power of 2 decreases by 1, the resulting value is divided by 2. For this pattern to continue, we need to define 2^0 to be 1 because dividing 2 by 2 results in 1.

Powers of 2

Power of 2	Value
2^3	8
2^2	4
2^1	2
2^0	?

Decrease exponent by 1 → Divide by 2
Decrease exponent by 1 → Divide by 2
Decrease exponent by 1 → Divide by 2

TABLE 12.2

This discussion suggests that $2^0 = 1$, and is generalized as follows.

> **ZERO EXPONENT**
>
> For any nonzero real number b,
> $$b^0 = 1.$$
>
> The expression 0^0 is undefined.

EXAMPLE 2 **Evaluating zero exponents**

Evaluate each expression. Assume that all variables represent nonzero numbers.

(a) 7^0 (b) $3\left(\dfrac{4}{9}\right)^0$ (c) $\left(\dfrac{x^2 y^5}{3z}\right)^0$

Solution
(a) Since 7 is a nonzero base, $7^0 = 1$.
(b) $3\left(\dfrac{4}{9}\right)^0 = 3(1) = 3$. (Note that the exponent 0 does not apply to 3.)
(c) All variables are nonzero, so the expression inside the parentheses is also nonzero. Thus $\left(\dfrac{x^2 y^5}{3z}\right)^0 = 1$.

Now Try Exercises 17, 19, 21

3 Applying the Product Rule

We can use a special rule to calculate products of exponential expressions *provided their bases are the same.* For example,

$$4^3 \cdot 4^2 = \underbrace{(4 \cdot 4 \cdot 4) \cdot (4 \cdot 4)}_{\text{5 factors of 4}} = 4^5.$$

The expression $4^3 \cdot 4^2$ has $3 + 2 = 5$ factors of 4, so the result is $4^{3+2} = 4^5$. To multiply exponential expressions with the *same base*, add exponents and keep the base.

READING CHECK 2

- Use the product rule to simplify the expression $3^2 \cdot 3^3$.

> **THE PRODUCT RULE**
>
> For any real number a and natural numbers m and n,
> $$a^m \cdot a^n = a^{m+n}.$$

NOTE: The product $2^4 \cdot 3^5$ cannot be simplified by using the product rule because the exponential expressions have different bases: **2** and **3**. ∎

EXAMPLE 3 **Using the product rule**

Multiply and simplify.
(a) $2^3 \cdot 2^2$ (b) $x^4 x^5$ (c) $2x^2 \cdot 5x^6$ (d) $x^3(2x + 3x^2)$

Product rule: Add exponents.

Solution
(a) $2^3 \cdot 2^2 = 2^{3+2} = 2^5 = 32$ (b) $x^4 x^5 = x^{4+5} = x^9$

(c) Begin by applying the commutative property of multiplication to write the product in a more convenient order.

$$2x^2 \cdot 5x^6 = 2 \cdot 5 \cdot x^2 \cdot x^6 = 10x^{2+6} = 10x^8$$

(d) To simplify this expression, first apply the distributive property.

$$x^3(2x + 3x^2) = x^3 \cdot 2x + x^3 \cdot 3x^2 = 2x^4 + 3x^5$$

(Exponent is 1.)

Now Try Exercises 25, 27, 31, 75

NOTE: If an exponent does not appear in an expression, it is assumed to be 1. For example, x can be written as x^1 and $(x + y)$ can be written as $(x + y)^1$. ∎

EXAMPLE 4 Applying the product rule

Multiply and simplify.
(a) $x \cdot x^3$ (b) $(a + b)(a + b)^4$

Solution
(a) Begin by writing x as x^1. Then $x^1 \cdot x^3 = x^{1+3} = x^4$.
(b) First write $(a + b)$ as $(a + b)^1$. Then

$$(a + b)^1 \cdot (a + b)^4 = (a + b)^{1+4} = (a + b)^5.$$

Now Try Exercises 23, 67

4 Applying Power Rules

How should $(4^3)^2$ be evaluated? To answer this question, consider how the product rule can be used in evaluating

$$(4^3)^2 = \underbrace{4^3 \cdot 4^3}_{\text{Product rule}} = 4^{3+3} = 4^6.$$

Similarly,

$$(a^5)^3 = \underbrace{a^5 \cdot a^5 \cdot a^5}_{\text{Product rule}} = a^{5+5+5} = a^{15}.$$

This discussion suggests that to raise a power to a power, we multiply the exponents.

READING CHECK 3

- Use the raising a power to a power rule to simplify the expression $(2^2)^3$.

RAISING A POWER TO A POWER

For any real number a and natural numbers m and n,

$$(a^m)^n = a^{mn}.$$

EXAMPLE 5 Raising a power to a power

Simplify each expression.
(a) $(3^2)^4$ **(b)** $(a^3)^2$

Power rule: Multiply exponents

Solution
(a) $(3^2)^4 = 3^{2 \cdot 4} = 3^8$ **(b)** $(a^3)^2 = a^{3 \cdot 2} = a^6$

Now Try Exercises 35, 37

To decide how to simplify the expression $(2x)^3$, consider

$$(2x)^3 = \underbrace{2x \cdot 2x \cdot 2x}_{\text{3 factors}} = \underbrace{(2 \cdot 2 \cdot 2)}_{\text{3 factors}} \cdot \underbrace{(x \cdot x \cdot x)}_{\text{3 factors}} = 2^3 x^3.$$

To raise a product to a power, we raise each factor to the power.

READING CHECK 4

- Use the raising a product to a power rule to rewrite the expression $(4x)^3$.

RAISING A PRODUCT TO A POWER

For any real numbers a and b and natural number n,

$$(ab)^n = a^n b^n.$$

EXAMPLE 6 Raising a product to a power

Simplify each expression.
(a) $(3z)^2$ **(b)** $(-2x^2)^3$ **(c)** $4(x^2y^3)^5$ **(d)** $(-2^2 a^5)^3$

Solution
(a) $(3z)^2 = 3^2 z^2 = 9z^2$
(b) $(-2x^2)^3 = (-2)^3 (x^2)^3 = -8x^6$
(c) $4(x^2 y^3)^5 = 4(x^2)^5 (y^3)^5 = 4x^{10} y^{15}$
(d) $(-2^2 a^5)^3 = (-4a^5)^3 = (-4)^3 (a^5)^3 = -64 a^{15}$

Now Try Exercises 41, 43, 47, 49

The following equation illustrates another power rule.

$$\left(\frac{2}{3}\right)^4 = \underbrace{\frac{2}{3} \cdot \frac{2}{3} \cdot \frac{2}{3} \cdot \frac{2}{3}}_{\text{4 factors}} = \frac{2 \cdot 2 \cdot 2 \cdot 2}{3 \cdot 3 \cdot 3 \cdot 3} = \frac{2^4}{3^4}$$

To raise a quotient to a power, raise both the numerator and the denominator to the power.

READING CHECK 5

- Use the raising a quotient to a power rule to rewrite the expression $\left(\frac{x}{2}\right)^5$.

RAISING A QUOTIENT TO A POWER

For any real numbers a and b and natural number n,

$$\left(\frac{a}{b}\right)^n = \frac{a^n}{b^n}. \quad b \neq 0$$

EXAMPLE 7   Raising a quotient to a power

Simplify each expression.

(a) $\left(\dfrac{2}{3}\right)^3$ (b) $\left(\dfrac{a}{b}\right)^9$ (c) $\left(\dfrac{a+b}{5}\right)^2$

Solution

(a) $\left(\dfrac{2}{3}\right)^3 = \dfrac{2^3}{3^3} = \dfrac{8}{27}$ (b) $\left(\dfrac{a}{b}\right)^9 = \dfrac{a^9}{b^9}$

(c) Because the numerator is an expression with more than one term, we must place parentheses around it before raising it to the power 2.

$$\left(\dfrac{a+b}{5}\right)^2 = \dfrac{(a+b)^2}{5^2} = \dfrac{(a+b)^2}{25}$$

Now Try Exercises 51, 53, 57

MAKING CONNECTIONS 1

Raising a Sum or Difference to a Power

Although there are power rules for products and quotients, similar rules for sums and differences do not exist. In general,

$$(a+b)^n \neq a^n + b^n \qquad \text{and} \qquad (a-b)^n \neq a^n - b^n.$$

$(3+4)^2 = 7^2 = 49$, but
$3^2 + 4^2 = 9 + 16 = 25$.

$(4-1)^3 = 3^3 = 27$, but
$4^3 - 1^3 = 64 - 1 = 63$.

The five rules for exponents discussed in this section are summarized as follows.

RULES FOR EXPONENTS

The following rules hold for real numbers a and b, and natural numbers m and n.

Description	Rule	Example
Zero Exponent	$b^0 = 1$, for $b \neq 0$	$(-13)^0 = 1$
The Product Rule	$a^m \cdot a^n = a^{m+n}$	$5^4 \cdot 5^3 = 5^{4+3} = 5^7$
Power to a Power	$(a^m)^n = a^{m \cdot n}$	$(y^2)^5 = y^{2 \cdot 5} = y^{10}$
Product to a Power	$(ab)^n = a^n b^n$	$(pq)^7 = p^7 q^7$
Quotient to a Power	$\left(\dfrac{a}{b}\right)^n = \dfrac{a^n}{b^n}$, for $b \neq 0$	$\left(\dfrac{x}{y}\right)^3 = \dfrac{x^3}{y^3}$, for $y \neq 0$

Simplification of some expressions may require the application of more than one rule of exponents. This is demonstrated in the next example.

EXAMPLE 8   Combining rules for exponents

Simplify each expression.

(a) $(2a)^2(3a)^3$ (b) $\left(\dfrac{a^2 b^3}{c}\right)^4$ (c) $(2x^3 y)^2(-4x^2 y^3)^3$

Solution

(a) $(2a)^2(3a)^3 = 2^2 a^2 \cdot 3^3 a^3$ Raising a product to a power
$= 4 \cdot 27 \cdot a^2 \cdot a^3$ Evaluate powers; commutative property
$= 108 a^5$ Product rule

(b) $\left(\dfrac{a^2 b^3}{c}\right)^4 = \dfrac{(a^2)^4 (b^3)^4}{c^4}$ Raising a quotient to a power; raising a product to a power
$= \dfrac{a^8 b^{12}}{c^4}$ Raising a power to a power

(c) $(2x^3 y)^2 (-4x^2 y^3)^3 = 2^2 (x^3)^2 y^2 (-4)^3 (x^2)^3 (y^3)^3$ Raising a product to a power
$= 4 x^6 y^2 (-64) x^6 y^9$ Raising a power to a power
$= 4(-64) x^6 x^6 y^2 y^9$ Commutative property
$= -256 x^{12} y^{11}$ Product rule

Now Try Exercises 59, 61, 65

🌐 **Math in Financial Context** Exponents occur frequently in calculations involving yearly percent increases, such as the increase in property value illustrated in the next example.

EXAMPLE 9 **Calculating growth in property value**

If a parcel of property increases in value by about 11% each year for 20 years, then its value will double three times.
(a) Write an exponential expression that represents "doubling three times."
(b) If the property is initially worth $25,000, how much will it be worth after it doubles three times?

Solution
(a) Doubling three times is represented by 2^3.
(b) $2^3 (25,000) = 8(25,000) = \$200,000$

Now Try Exercise 89

12.1 Putting It All Together

CONCEPT	EXPLANATION	EXAMPLES
Bases and Exponents	In the expression b^n, b is the base and n is the exponent. If n is a natural number, then $b^n = \underbrace{b \cdot b \cdot \cdots \cdot b}_{n \text{ factors}}.$	2^3 has base 2 and exponent 3. $9^1 = 9,$ $3^2 = 3 \cdot 3 = 9,$ $4^3 = 4 \cdot 4 \cdot 4 = 64,$ and $-6^2 = -(6 \cdot 6) = -36$
Zero Exponents	For any nonzero number b, $b^0 = 1$.	$5^0 = 1$, $x^0 = 1$, and $(xy^3)^0 = 1$

continued on next page

continued from previous page

CONCEPT	EXPLANATION	EXAMPLES
The Product Rule	For any real number a and natural numbers m and n, $$a^m \cdot a^n = a^{m+n}.$$	$2^4 \cdot 2^3 = 2^{4+3} = 2^7$, $x \cdot x^2 \cdot x^6 = x^{1+2+6} = x^9$, and $(x+1) \cdot (x+1)^2 = (x+1)^3$
Raising a Power to a Power	For any real number a and natural numbers m and n, $$(a^m)^n = a^{mn}.$$	$(2^4)^2 = 2^{4 \cdot 2} = 2^8$, $(x^2)^5 = x^{2 \cdot 5} = x^{10}$, and $(a^4)^3 = a^{4 \cdot 3} = a^{12}$
Raising a Product to a Power	For any real numbers a and b and natural number n, $$(ab)^n = a^n b^n.$$	$(3x)^3 = 3^3 x^3 = 27x^3$, $(x^2 y)^4 = (x^2)^4 y^4 = x^8 y^4$, and $(-xy)^6 = (-x)^6 y^6 = x^6 y^6$
Raising a Quotient to a Power	For any real numbers a and b and natural number n, $$\left(\frac{a}{b}\right)^n = \frac{a^n}{b^n}. \quad b \neq 0$$	$\left(\frac{x}{y}\right)^5 = \frac{x^5}{y^5}$ and $\left(\frac{a^2 b}{d^3}\right)^4 = \frac{(a^2)^4 b^4}{(d^3)^4} = \frac{a^8 b^4}{d^{12}}$

12.1 Exercises

MyMathLab®

CONCEPTS AND VOCABULARY

1. In the expression b^n, b is the _____ and n is the _____.
2. The expression $b^0 =$ _____ for any nonzero number b.
3. $a^m \cdot a^n =$ _____
4. $(a^m)^n =$ _____
5. $(ab)^n =$ _____
6. $\left(\frac{a}{b}\right)^n =$ _____

REVIEWING BASES AND EXPONENTS

Exercises 7–16: Evaluate the expression.

7. 8^2
8. 4^3
9. $(-2)^3$
10. $(-3)^4$
11. -2^3
12. -3^4
13. $3 + \frac{4^2}{2}$
14. $6 - \left(\frac{-4}{2}\right)^2$
15. $4\left(\frac{1}{2}\right)^3$
16. $16\left(\frac{1}{4}\right)^2$

COMPUTING ZERO EXPONENTS

Exercises 17–22: Simplify the expression. Assume that all variables represent nonzero numbers.

17. 6^0
18. $(-0.5)^0$
19. $5(-4)^0$
20. $-9(45)^0$
21. $\left(\frac{xy^3}{z^2}\right)^0$
22. $(x^2 y^3)^0$

APPLYING THE PRODUCT RULE

Exercises 23–34: Simplify the expression. Assume that all variables represent nonzero numbers.

23. $3 \cdot 3^2$
24. $5^3 \cdot 5^3$
25. $4^2 \cdot 4^6$
26. $10^4 \cdot 10^3$
27. $x^3 x^6$
28. $a^5 a^2$
29. $x^2 x^2 x^2$
30. $y^7 y^3 y^0$
31. $4x^2 \cdot 5x^5$
32. $-2y^6 \cdot 5y^2$
33. $3(-xy^3)(x^2 y)$
34. $(a^2 b^3)(-ab^2)$

APPLYING POWER RULES

Exercises 35–58: Simplify the expression. Assume that all variables represent nonzero numbers.

35. $(2^3)^2$
36. $(10^3)^4$
37. $(n^3)^4$
38. $(z^7)^3$
39. $x(x^3)^2$
40. $(z^3)^2 (5z^5)$
41. $(-7b)^2$
42. $(-4z)^3$

43. $(ab)^3$
44. $(xy)^8$
45. $(2x^2)^0$
46. $(3a^2)^4$
47. $(-4b^2)^3$
48. $(-3r^4t^3)^2$
49. $(x^2y^3)^7$
50. $(rt^2)^5$
51. $\left(\dfrac{1}{3}\right)^3$
52. $\left(\dfrac{5}{2}\right)^2$
53. $\left(\dfrac{a}{b}\right)^5$
54. $\left(\dfrac{x}{2}\right)^4$
55. $\left(\dfrac{x-y}{3}\right)^3$
56. $\left(\dfrac{4}{x+y}\right)^2$
57. $\left(\dfrac{5}{a+b}\right)^2$
58. $\left(\dfrac{a-b}{2}\right)^3$

COMBINING RULES

Exercises 59–78: Simplify the expression. Assume that all variables represent nonzero numbers.

59. $(y^3)^2(x^4y)^3$
60. $(ab^3)^2(ab)^3$
61. $(a^2b)^2(a^2b^2)^3$
62. $(x^3y)(x^2y^4)^2$
63. $\left(\dfrac{2x}{5}\right)^3$
64. $\left(\dfrac{3y}{2}\right)^4$
65. $\left(\dfrac{3x^2}{5y^4}\right)^3$
66. $\left(\dfrac{a^2b^3}{3}\right)^5$
67. $(x+y)(x+y)^3$
68. $(a-b)^2(a-b)$
69. $(a+b)^2(a+b)^3$
70. $(x-y)^5(x-y)^4$
71. $6(x^4y^6)^0$
72. $\left(\dfrac{xy}{z^2}\right)^0$
73. $a(a^2+2b^2)$
74. $x^3(3x-5y^4)$
75. $3a^3(4a^2+2b)$
76. $2x^2(5-4y^3)$
77. $(x-y)(x^2y^3)$
78. $(r+t)(rt)$

79. **Thinking Generally** Students sometimes mistakenly apply the "rule" $a^m \cdot b^n \stackrel{?}{=} (ab)^{m+n}$. In general, this equation is *not true*. Find values for a, b, m, and n with $a \ne b$ and $m \ne n$ that will make this equation true.

80. **Thinking Generally** Students sometimes mistakenly apply the "rule" $(a+b)^n \stackrel{?}{=} a^n + b^n$. In general, this equation is *not true*. Find values for a, b, and n with $a \ne b$ that will make this equation true.

APPLICATIONS INVOLVING EXPONENTS

Exercises 81–84: Write a simplified expression for the area of each figure at the top of the next column.

81.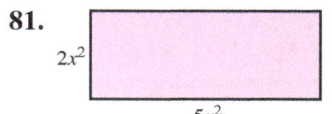
82.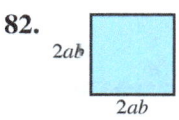
83.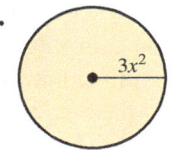
84.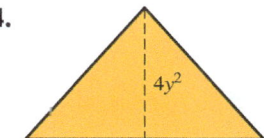

Exercises 85 and 86: Write a simplified expression for the volume of the given figure.

85.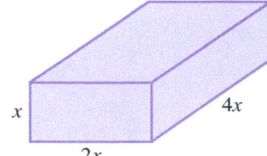
86.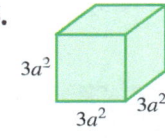

87. **Compound Interest** If P dollars are deposited in an account that pays 5% annual interest, then the amount of money in the account after 3 years is $P(1+0.05)^3$. Find the amount when $P = \$1000$.

88. **Compound Interest** If P dollars are deposited in an account that pays 9% annual interest, then the amount of money in the account after 4 years is $P(1+0.09)^4$. Find the amount when $P = \$500$.

89. **Investment Growth** If an investment increases in value by about 10% each year for 22 years, then its value will triple two times.
    (a) Write an exponential expression that represents "tripling two times."
    (b) If the investment has an initial value of $8000, how much will it be worth if it triples two times?

90. **Stock Value** If a stock decreases in value by about 23% each year for 9 years, then its value will be halved three times.
    (a) Write an exponential expression that represents "halved three times."
    (b) If the stock is initially worth $88 per share, how much will it be worth if it is halved three times?

WRITING ABOUT MATHEMATICS

91. Are the expressions $(4x)^2$ and $4x^2$ equal in value? Explain your answer.

92. Are the expressions $3^3 \cdot 2^3$ and 6^6 equal in value? Explain your answer.

12.2 Addition and Subtraction of Polynomials

Objectives

1. Introducing Nonlinear Data and Polynomials
2. Recognizing Monomials and Polynomials
3. Adding Polynomials
4. Subtracting Polynomials
5. Evaluating Polynomial Expressions

1 Introducing Nonlinear Data and Polynomials

Connecting Concepts with Your Life If you have ever exercised strenuously and then taken your pulse immediately afterward, you may have discovered that your pulse slowed quickly at first and then gradually leveled off. A typical scatterplot of this phenomenon is shown in **FIGURE 12.1(a)**. These data points cannot be modeled accurately with a line because the data are nonlinear. A new expression, called a *polynomial*, is needed to model them. A graph of a polynomial that models these data is shown in **FIGURE 12.1(b)** and discussed in Exercise 75.
(*Source:* V. Thomas, *Science and Sport.*)

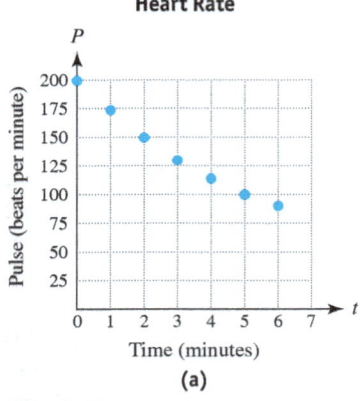
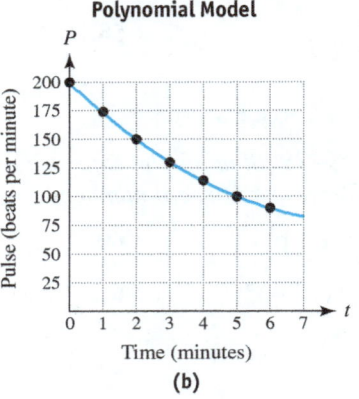

FIGURE 12.1

NEW VOCABULARY

☐ Monomial
☐ Degree of a monomial
☐ Coefficient of a monomial
☐ Polynomial
☐ Polynomial in one variable
☐ Binomial
☐ Trinomial
☐ Degree of a polynomial
☐ Like terms

READING CHECK 1

- What is the degree of the monomial $-5x^2y^4$?

2 Recognizing Monomials and Polynomials

A **monomial** is a number, a variable, or a product of numbers and variables raised to natural number powers. Examples of monomials include

$$-3, \quad xy^2, \quad 5a^2, \quad -z^3, \quad \text{and} \quad -\frac{1}{2}xy^3. \quad \text{Monomials}$$

A monomial may contain more than one variable, but monomials do not contain division by variables. For example, the expression $\frac{3}{z}$ is not a monomial. If an expression contains addition or subtraction signs, it is *not* a monomial.

The **degree of a monomial** is the sum of the exponents of the variables. If the monomial has only one variable, its degree is the exponent of that variable. Remember, when a variable does not have a written exponent, the exponent is implied to be 1. A nonzero number has degree 0, and the number 0 has *undefined* degree. The number in a monomial is called the **coefficient of the monomial**. **TABLE 12.3** contains the degree and coefficient of several monomials.

Properties of Monomials

Monomial	-5	$6a^3b$	$-xy$	$7y^3$
Degree	0	4	2	3
Coefficient	-5	6	-1	7

TABLE 12.3

A **polynomial** is a monomial or the sum of two or more monomials. Each monomial is called a *term* of the polynomial. Addition or subtraction signs separate terms. For example, the expression $2x^2 - 3x + 5$ is a **polynomial in one variable** with three terms. Other examples of polynomials in one variable include

$-2x$, $3x + 1$, $4y^2 - y + 7$,   and $x^5 - 3x^3 + x - 7$.   *Polynomials*

1 term   2 terms   3 terms   4 terms

These polynomials have 1, 2, 3, and 4 terms, respectively. A polynomial with *two terms* is called a **binomial**, and a polynomial with *three terms* is called a **trinomial**.

A polynomial can have more than one variable, as in

x^2y^2, $2xy^2 + 5x^2y - 1$,   and $a^2 + 2ab + b^2$.   *2-variable polynomials*

Note that all variables in a polynomial are raised to natural number powers. The **degree of a polynomial** is the degree of the term (or monomial) with greatest degree. See **TABLE 12.4**.

Properties of Polynomials

Polynomial	$3x^2 + 7$	$x^3 + 3x^2 + 5x + 9$	$6xy - x^2y^2$
Degree	2	3	4
Number of terms	2	4	2
Number of variables	1	1	2

TABLE 12.4

READING CHECK 2

- What is the degree of the polynomial $-2 + x - x^2 - 4x^5$?

EXAMPLE 1   Identifying properties of polynomials

Determine whether the expression is a polynomial. If it is, state how many terms and variables the polynomial contains and give its degree.

(a) $7x^2 - 3x + 1$ (b) $5x^3 - 3x^2y^3 + xy^2 - 2y^3$ (c) $4x^2 + \dfrac{5}{x+1}$

Solution
(a) The expression $7x^2 - 3x + 1$ is a polynomial with three terms and one variable. The first term $7x^2$ has degree 2 because the exponent on the variable is 2. The second term $-3x$ has degree 1 because the exponent on the variable is implied to be 1. The third term 1 has degree 0 because it is a nonzero number. The term with greatest degree is $7x^2$, so the polynomial has degree **2**.
(b) The expression $5x^3 - 3x^2y^3 + xy^2 - 2y^3$ is a polynomial with four terms and two variables, x and y. The first term has degree 3 because the exponent on the variable is 3. The second term has degree 5 because the *sum* of the exponents on the variables x and y is 5. Likewise, the third term has degree 3 and the fourth term has degree 3. The term with greatest degree is $-3x^2y^3$, so the polynomial has degree **2 + 3 = 5**.
(c) The expression $4x^2 + \dfrac{5}{x+1}$ is not a polynomial because it contains division by the polynomial $x + 1$.

Now Try Exercises 21, 23, 27

3   Adding Polynomials

Suppose that we have 2 identical rectangles with length l and width w. Then the area of one rectangle is lw and the total area is

$$lw + lw.$$

This area is equivalent to 2 times *lw*, which can be expressed as 2*lw*. See **FIGURE 12.2**.

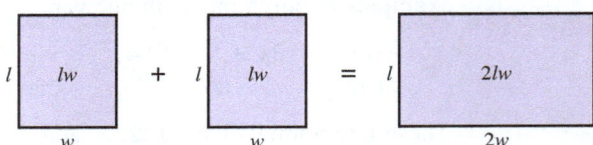

FIGURE 12.2

If two monomials contain the same variables raised to the same powers, we call them **like terms**. We can add or subtract (combine) *like* terms but cannot combine *unlike* terms. The terms *lw* and 2*lw* are like terms and can be combined geometrically, as shown in **FIGURE 12.3**. If we joined one of the small rectangles with area *lw* and a larger rectangle with area 2*lw*, then the total area is 3*lw*.

READING CHECK 3

Are $5xy^2$ and $5x^2y$ like terms? Are $-3xy^2$ and $9xy^2$ like terms?

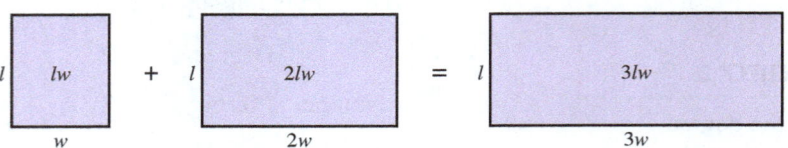

FIGURE 12.3

STUDY TIP

Do you want to know what material will be covered on your next exam? Often, the best place to look is on previously completed assignments and quizzes.

The *distributive property* justifies combining like terms.

$$1lw + 2lw = (1 + 2)lw = 3lw$$

The rectangles shown in **FIGURE 12.4** have areas of *ab* and *xy*. Together, their area is the sum, $ab + xy$. However, because these monomials are unlike terms, they cannot be combined into one term.

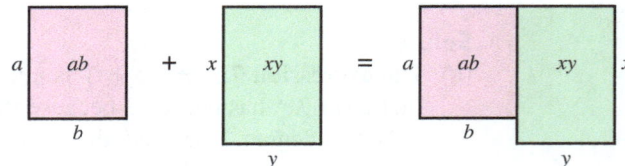

FIGURE 12.4

EXAMPLE 2 Adding like terms

State whether each pair of expressions contains like terms or unlike terms. If they are like terms, add them.

(a) $5x^2, -x^2$ (b) $7a^2b, 10ab^2$ (c) $4rt^2, \frac{1}{2}rt^2$

Solution

(a) The terms $5x^2$ and $-x^2$ have the same variable raised to the same power, so they are like terms. To add like terms, add their coefficients. Note that the coefficient of $-x^2$ is -1.

$$5x^2 + (-x^2) = (5 + (-1))x^2 \quad \text{Distributive property}$$
$$= 4x^2 \quad \text{Add.}$$

(b) The terms $7a^2b$ and $10ab^2$ have the same variables, but these variables are not raised to the same powers. They are unlike terms and cannot be added.

(c) The terms $4rt^2$ and $\frac{1}{2}rt^2$ have the same variables raised to the same powers, so they are like terms. We add them as follows.

$$4rt^2 + \frac{1}{2}rt^2 = \left(4 + \frac{1}{2}\right)rt^2 \quad \text{Distributive property}$$
$$= \frac{9}{2}rt^2 \quad \text{Add.}$$

Now Try Exercises 29, 31, 33

To add two polynomials, combine like terms, as illustrated in the next example.

EXAMPLE 3 Adding polynomials

Add by combining like terms.
(a) $(3x + 4) + (-4x + 2)$
(b) $(y^2 - 2y + 1) + (3y^2 + y + 11)$

Solution
(a) $(3x + 4) + (-4x + 2) = 3x + (-4x) + 4 + 2$
$\qquad = (3 - 4)x + (4 + 2)$
$\qquad = -x + 6$

(b) $(y^2 - 2y + 1) + (3y^2 + y + 11) = y^2 + 3y^2 - 2y + y + 1 + 11$
$\qquad = (1 + 3)y^2 + (-2 + 1)y + (1 + 11)$
$\qquad = 4y^2 - y + 12$

Now Try Exercises 37, 39

Recall that the commutative and associative properties of addition allow us to rearrange a sum in any order. For example, if we write each subtraction in $2x - 5 - 4x + 10$ as addition of the opposite, we have

$$2x - 5 - 4x + 10 = 2x + (-5) + (-4x) + 10,$$

and the terms can be rearranged as

$$2x + (-4x) + (-5) + 10 = 2x - 4x - 5 + 10 = -2x + 5.$$

If we pay attention to the sign in front of each term in a polynomial, the like terms can be combined without rearranging the terms, as demonstrated in the next example.

EXAMPLE 4 Adding polynomials

Add $(x^3 - 3x^2 + 7x - 4) + (4x^3 - 5x + 9)$ by combining like terms.

Solution
Remove parentheses and identify like terms with their signs as shown.

$$x^3 - 3x^2 + 7x - 4 + 4x^3 - 5x + 9$$

When like terms (of the same color) are added, the resulting sum is

$$5x^3 - 3x^2 + 2x + 5.$$

Now Try Exercise 41

Polynomials can also be added vertically, as demonstrated in the next example.

EXAMPLE 5   **Adding polynomials vertically**

Simplify $(3x^2 - 3x + 5) + (-x^2 + x - 6)$.

Solution
Write the polynomials in a vertical format and then add each column of like terms.

$$
\begin{array}{r}
3x^2 - 3x + 5 \\
-x^2 + x - 6 \\
\hline
2x^2 - 2x - 1
\end{array}
$$
Add like terms in each column.

Regardless of the method used, the same answer should be obtained. However, adding vertically requires that *like terms be placed in the same column.*

Now Try Exercise 47

4 Subtracting Polynomials

To subtract one integer from another, add the first integer and the *additive inverse* or *opposite* of the second integer. For example, $3 - 5$ is evaluated as follows.

$$3 - 5 = 3 + (-5) \quad \text{Add the opposite.}$$
$$= -2 \quad \text{Simplify.}$$

READING CHECK 4

- How do you subtract one polynomial from another?

Similarly, to subtract one polynomial from another, add the first polynomial and the *opposite* of the second polynomial. To find the opposite of a polynomial, simply negate each term. **TABLE 12.5** lists some polynomials and their opposites.

CRITICAL THINKING 1

What is the result when a polynomial and its opposite are added?

Opposites of Polynomials

Polynomial	Opposite
$2x - 4$	$-2x + 4$
$-x^2 - 2x + 9$	$x^2 + 2x - 9$
$6x^3 - 12$	$-6x^3 + 12$
$-3x^4 - 2x^2 - 8x + 3$	$3x^4 + 2x^2 + 8x - 3$

TABLE 12.5

EXAMPLE 6   **Subtracting polynomials**

Simplify each expression.
(a) $(3x - 4) - (5x + 1)$
(b) $(5x^2 + 2x - 3) - (6x^2 - 7x + 9)$
(c) $(6x^3 + x^2) - (-3x^3 - 9)$

Solution
(a) To subtract $(5x + 1)$ from $(3x - 4)$, we add the binomial $(3x - 4)$ and the opposite of the binomial $(5x + 1)$, or $(-5x - 1)$.

$$(3x - 4) - (5x + 1) = (3x - 4) + (-5x - 1)$$
$$= (3 - 5)x + (-4 - 1)$$
$$= -2x - 5$$

Add the opposite.

(b) The opposite of $(6x^2 - 7x + 9)$ is $(-6x^2 + 7x - 9)$.
$$(5x^2 + 2x - 3) - (6x^2 - 7x + 9) = (5x^2 + 2x - 3) + (-6x^2 + 7x - 9)$$
$$= (5 - 6)x^2 + (2 + 7)x + (-3 - 9)$$
$$= -x^2 + 9x - 12$$

Add the opposite.

(c) The opposite of $(-3x^3 - 9)$ is $(3x^3 + 9)$.
$$(6x^3 + x^2) - (-3x^3 - 9) = (6x^3 + x^2) + (3x^3 + 9)$$
$$= (6 + 3)x^3 + x^2 + 9$$
$$= 9x^3 + x^2 + 9$$

Now Try Exercises 57, 59, 61

NOTE: Some students prefer to subtract one polynomial from another by noting that a subtraction sign in front of parentheses changes the signs of all of the terms within the parentheses. For example, part (a) of the previous example could be worked as follows.

$$(3x - 4) - (5x + 1) = 3x - 4 - 5x - 1$$
$$= (3 - 5)x + (-4 - 1)$$
$$= -2x - 5 \blacksquare$$

EXAMPLE 7   Subtracting polynomials vertically

Simplify $(5x^2 - 2x + 7) - (-3x^2 + 3)$.

Solution
To subtract one polynomial from another vertically, simply add the first polynomial and the opposite of the second polynomial. No x-term occurs in the second polynomial, so insert $0x$.

$$5x^2 - 2x + 7$$
$$\underline{3x^2 + 0x - 3}\quad \text{The opposite of } -3x^2 + 3 \text{ is } 3x^2 - 3 \text{ or } 3x^2 + 0x - 3.$$
$$8x^2 - 2x + 4 \quad \text{Add like terms in each column.}$$

Now Try Exercise 69

5 Evaluating Polynomial Expressions
Frequently, monomials and polynomials represent formulas that may be evaluated. We illustrate such applications in the next two examples.

EXAMPLE 8   Writing and evaluating a monomial

Write the monomial that represents the volume of the box having a square bottom, as shown in **FIGURE 12.5**. Find the volume of the box if $x = 3$ feet and $y = 2$ feet.

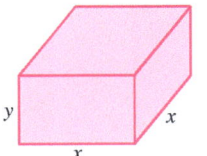

FIGURE 12.5

Solution
The volume of a box is found by multiplying the length, width, and height together. Because the length and width are both x and the height is y, the monomial xxy represents the volume

of the box. This can be written as x^2y. To calculate the volume, let $x = 3$ and $y = 2$ in the monomial x^2y.

$$x^2y = 3^2 \cdot 2 = 9 \cdot 2 = 18 \text{ cubic feet}$$

Now Try Exercise 77

Math in Business Context Polynomials can be used to model trends that occur in business. In the next example, a polynomial is used to model computer sales.

EXAMPLE 9 Modeling sales of personal computers

Worldwide sales of personal computers increased dramatically during the first 10 years of the 21st century, as illustrated in **FIGURE 12.6**. The polynomial

$$0.7868x^2 + 16.72x + 122.58$$

approximates the number of computers sold in millions, where $x = 0$ corresponds to 2000, $x = 1$ to 2001, and so on. Estimate the number of personal computers sold in 2008 by using both the graph and the polynomial. (*Source:* International Data Corporation.)

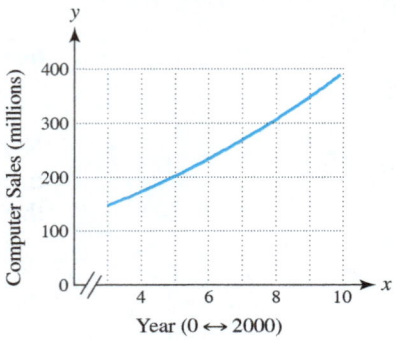

FIGURE 12.6

Solution

From the graph shown in **FIGURE 12.7**, it appears that personal computer sales were slightly more than 300 million, or about 310 million, in 2008.

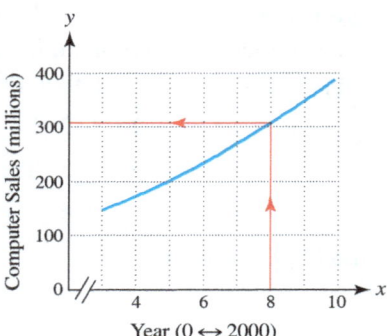

FIGURE 12.7

The year 2008 corresponds to $x = 8$ in the given polynomial, so substitute **8** for x and evaluate the resulting expression.

$$0.7868x^2 + 16.72x + 122.58 = 0.7868(8)^2 + 16.72(8) + 122.58$$
$$\approx 307 \text{ million}$$

The graph and the polynomial give similiar results.

Now Try Exercise 75

12.2 Putting It All Together

CONCEPT	EXPLANATION	EXAMPLES
Monomial	A number, variable, or product of numbers and variables raised to natural number powers The degree is the sum of the exponents. The coefficient is the number in a monomial.	$4x^2y$ Degree: 3; coefficient: 4 $-6x^2$ Degree: 2; coefficient: -6 $-a^4$ Degree: 4; coefficient: -1 x Degree: 1; coefficient: 1 -8 Degree: 0; coefficient: -8
Polynomial	A monomial or the sum of two or more monomials	$4x^2 + 8xy^2 + 3y^2$ Trinomial $-9x^4 + 100$ Binomial $-3x^2y^3$ Monomial
Like Terms	Monomials containing the same variables raised to the same powers	$10x$ and $-2x$, $4x^2$ and $3x^2$ $5ab^2$ and $-ab^2$, $5z$ and $\frac{1}{2}z$
Addition of Polynomials	To add polynomials, combine like terms.	$(x^2 + 3x + 1) + (2x^2 - 2x + 7)$ $= (1 + 2)x^2 + (3 - 2)x + (1 + 7)$ $= 3x^2 + x + 8$ $3xy + 5xy = (3 + 5)xy = 8xy$
Opposite of a Polynomial	To obtain the opposite of a polynomial, negate each term.	*Polynomial* *Opposite* $-2x^2 + x - 6$ $2x^2 - x + 6$ $a^2 - b^2$ $-a^2 + b^2$ $-3x - 18$ $3x + 18$
Subtraction of Polynomials	To subtract one polynomial from another, add the first polynomial and the opposite of the second polynomial.	$(x^2 + 3x) - (2x^2 - 5x)$ $= (x^2 + 3x) + (-2x^2 + 5x)$ $= (1 - 2)x^2 + (3 + 5)x$ $= -x^2 + 8x$
Evaluating a Polynomial	To evaluate a polynomial in x, substitute a value for x in the expression and simplify.	To evaluate the polynomial $$3x^2 - 2x + 1 \quad \text{for} \quad x = 2,$$ substitute 2 for x and simplify. $$3(2)^2 - 2(2) + 1 = 9$$

12.2 Exercises

CONCEPTS AND VOCABULARY

1. A(n) _____ is a number, a variable, or a product of numbers and variables raised to a natural number power.

2. A(n) ____ is a monomial or a sum of monomials.

3. The ____ of a monomial is the sum of the exponents of the variables.

4. The _____ of a polynomial is the degree of the term with the greatest degree.

5. A polynomial with two terms is called a(n) ____.

6. A polynomial with three terms is called a(n) ____.

7. Two monomials with the same variables raised to the same powers are ____ terms.

8. To add two polynomials, combine ____ terms.

9. To subtract two polynomials, add the first polynomial to the ____ of the second polynomial.

10. Polynomials can be added horizontally or ____.

RECOGNIZING MONOMIALS AND POLYNOMIALS

Exercises 11–18: Identify the degree and coefficient of the monomial.

11. $3x^2$
12. y
13. $-ab$
14. $-2xy$
15. $-5rt$
16. $8x^2y^5$
17. 6
18. $-\frac{1}{2}$

Exercises 19–28: Determine whether the expression is a polynomial. If it is, state how many terms and variables the polynomial contains. Then state its degree.

19. $-x$
20. $7z$
21. $4x^2 - 5x + 9$
22. $x^3 - 9$
23. $x + \frac{1}{x}$
24. $\frac{5}{xy + 1}$
25. $3x^{-2}y^{-3}$
26. $5^2 a^3 b^4$
27. $-2^3 a^4 bc + b^2 c$
28. $-7y^{-1} z^{-3}$

Exercises 29–36: State whether the given pair of expressions are like terms. If they are like terms, add them.

29. $5x, -4x$
30. $x^2, 8x^2$
31. $x^3, -6x^3$
32. $4xy, -9xy$
33. $9x, -xy$
34. $5x^2 y, -3xy^2$
35. ab, ba
36. $rt^2, -2t^2 r$

ADDING POLYNOMIALS

Exercises 37–46: Add the polynomials.

37. $(3x + 5) + (-4x + 4)$
38. $(-x + 5) + (2x - 5)$
39. $(3x^2 + 4x + 1) + (x^2 + 4x)$
40. $(-x^2 - x) + (2x^2 + 3x - 1)$
41. $(y^3 + 3y^2 - 5) + (3y^3 + 4y - 4)$
42. $(4z^4 + z^2 - 10) + (-z^4 + 4z - 5)$
43. $(-xy + 5) + (5xy - 4)$
44. $(2a^2 + b^2) + (3a^2 - 5b^2)$
45. $(a^3 b^2 + a^2 b^3) + (a^2 b^3 - a^3 b^2)$
46. $(a^2 + ab + b^2) + (a^2 - ab + b^2)$

Exercises 47–50: Add the polynomials vertically.

47. $4x^2 - 2x + 1$
    $\underline{5x^2 + 3x - 7}$

48. $8x^2 + 3x + 5$
    $\underline{-x^2 - 3x - 9}$

49. $-x^2 + x$
    $\underline{2x^2 - 8x - 1}$

50. $a^3 - 3a^2 b + 3ab^2 - b^3$
    $\underline{a^3 + 3a^2 b + 3ab^2 + b^3}$

SUBTRACTING POLYNOMIALS

Exercises 51–56: Write the opposite of the polynomial.

51. $5x^2$
52. $17x + 12$
53. $3a^2 - a + 4$
54. $-b^3 + 3b$
55. $-2t^2 - 3t + 4$
56. $7t^2 + t - 10$

Exercises 57–66: Subtract the polynomials.

57. $(3x + 1) - (-x + 3)$

58. $(-2x + 5) - (x + 7)$

59. $(-x^2 + 6x) - (2x^2 + x - 2)$

60. $(2y^2 + 3y - 2) - (y^2 - y)$

61. $(z^3 - 2z^2 - z) - (4z^2 + 5z + 1)$

62. $(3z^4 - z) - (-z^4 + 4z^2 - 5)$

63. $(4xy + x^2y^2) - (xy - x^2y^2)$

64. $(a^2 + b^2) - (-a^2 + b^2)$

65. $(ab^2) - (ab^2 + a^3b)$

66. $(x^2 + 3xy + 4y^2) - (x^2 - xy + 4y^2)$

Exercises 67–70: Subtract the polynomials vertically.

67. $(x^2 + 2x - 3) - (2x^2 + 7x + 1)$

68. $(5x^2 - 9x - 1) - (x^2 - x + 3)$

69. $(3x^3 - 2x) - (5x^3 + 4x + 2)$

70. $(a^2 + 3ab + 2b^2) - (a^2 - 3ab + 2b^2)$

EVALUATING POLYNOMIAL EXPRESSIONS

Exercises 71–74: Evaluate the polynomial expression for each value of x.

71. $x^2 - x + 3$ for $x = -2, 3$

72. $2x^2 - x$ for $x = -1, 2$

73. $4x^3 + x^2 - 2x$ for $x = -3, 1$

74. $-2x^4 - x^3$ for $x = -2, 3$

APPLICATIONS INVOLVING POLYNOMIALS

75. Exercise and Heart Rate The polynomial given by $1.6t^2 - 28t + 200$ calculates the heart rate shown in **FIGURE 12.1(b)** in this section, where t represents the elapsed time in minutes since exercise stopped.
(a) What is the heart rate when the athlete first stops exercising?
(b) What is the heart rate after 5 minutes?
(c) Describe what happens to the heart rate after exercise stops.

76. Music Streaming The proportion of total streaming music revenues after 2008 can be modeled by the polynomial given by $0.7x^2 - 0.4x + 4.2$, where $x = 0$ corresponds to 2008, $x = 1$ to 2009, and so on. The graph at the top of the next column illustrates this growth.

Streaming Music Revenues

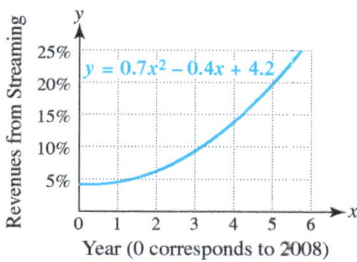

(a) Use the graph to estimate the proportion of streaming music revenues in 2012.
(b) Use the polynomial to estimate the proportion of streaming music revenues in 2012.
(c) Do your answers from parts (a) and (b) agree?

77. Areas of Squares Write a monomial that equals the sum of the areas of the squares. Then calculate this sum for $z = 10$ inches.

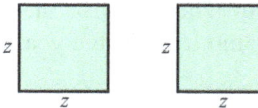

78. Areas of Rectangles Find a monomial that equals the sum of the areas of the three rectangles. Find this sum for $a = 5$ yards and $b = 3$ yards.

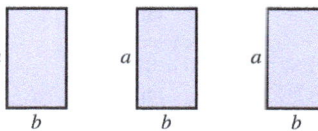

79. Area of a Figure Find a polynomial that equals the area of the figure. Calculate its area for $x = 6$ feet.

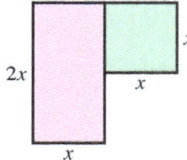

80. Area of a Rectangle Write a polynomial that gives the area of the rectangle. Calculate its area for $x = 3$ feet.

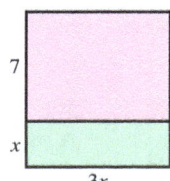

81. Areas of Circles Write a polynomial that gives the sum of the areas of two circles, one with radius x and the other with radius y. Find this sum for $x = 2$ feet and $y = 3$ feet. Leave your answer in terms of π.

82. Squares and Circles Write a polynomial that gives the sum of the areas of a square having sides of length x and a circle having diameter x. Approximate this sum to the nearest hundredth of a square foot for $x = 6$ feet.

83. World Population The table lists world population P in billions for selected years t.

t	1974	1987	1999	2012
P	4	5	6	7

Source: U.S. Census Bureau.

(a) Find the slope of each line segment connecting consecutive data points in the table. Can these data be modeled with a line? Explain.
(b) Does the polynomial $0.077t - 148$ give good estimates for the world population in year t? Explain how you decided.

84. Price of a Stamp The table lists the price P of a first-class postage stamp for selected years t.

t	1963	1975	1987	2002	2007	2011
P	5¢	13¢	25¢	37¢	41¢	44¢

Source: U.S. Postal Service.

(a) Does the polynomial $0.835t - 1635$ model the data in the table exactly?
(b) Does it give approximations that are within 1.5¢ of the actual values?

WRITING ABOUT MATHEMATICS

85. Explain what the terms monomial, binomial, trinomial, and polynomial mean. Give an example of each.

86. Explain how to determine the degree of a polynomial that has one variable. Give an example.

87. Explain how to obtain the opposite of a polynomial. Give an example.

88. Explain how to subtract two polynomials. Give an example.

SECTIONS 12.1 and 12.2 — Checking Basic Concepts

1. Evaluate each expression.
   (a) -5^2 (b) $3^2 - 2^3$

2. Simplify each expression.
   (a) $10^3 \cdot 10^5$ (b) $(3x^2)(-4x^5)$
   (c) $(a^3 b)^2$ (d) $\left(\dfrac{x}{z^3}\right)^4$

3. Simplify each expression.
   (a) $(4y^3)^0$ (b) $(x^3)^2 (3x^4)^2$
   (c) $2a^2(5a^3 - 7)$

4. State the number of terms and variables in the polynomial $5x^3y - 2x^2y + 5$. What is its degree?

5. A box has a rectangular bottom twice as long as it is wide.
   (a) If the bottom has width w and the box has height h, write a monomial that gives the volume of the box.
   (b) Find the volume of the box for $w = 12$ inches and $h = 10$ inches.

6. Simplify each expression.
   (a) $(2a^2 + 3a - 1) + (a^2 - 3a + 7)$
   (b) $(4z^3 + 5z) - (2z^3 - 2z + 8)$
   (c) $(x^2 + 2xy + y^2) - (x^2 - 2xy + y^2)$

12.3 Multiplication of Polynomials

Objectives
1. Multiplying Monomials
2. Reviewing the Distributive Property
3. Multiplying Monomials and Polynomials
4. Multiplying Polynomials

1 Multiplying Monomials

A monomial is a number, a variable, or a product of numbers and variables raised to natural number powers. To multiply monomials, we often use the product rule for exponents.

READING CHECK 1

- Which rule for exponents is commonly used to multiply monomials?

EXAMPLE 1 Multiplying monomials

Multiply.
(a) $-5x^2 \cdot 4x^3$ (b) $(7xy^4)(x^3y^2)$

Solution

(a) $\quad -5x^2 \cdot 4x^3 = (-5)(4)x^2x^3 \qquad$ Commutative property
$\qquad\qquad\qquad = -20x^{2+3} \qquad\qquad$ The product rule
$\qquad\qquad\qquad = -20x^5 \qquad\qquad\quad$ Simplify.

(b) $(7xy^4)(x^3y^2) = 7xx^3y^4y^2 \qquad\quad$ Commutative property
$\qquad\qquad\qquad = 7x^{1+3}y^{4+2} \qquad$ The product rule
$\qquad\qquad\qquad = 7x^4y^6 \qquad\qquad$ Simplify.

Now Try Exercises 9, 13

2 Reviewing the Distributive Properties

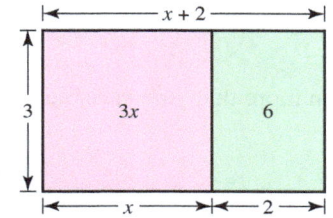

Distributive Property
$3(x + 2) = 3x + 6$

FIGURE 12.8 Area: $3x + 6$

Distributive properties are used frequently for multiplying monomials and polynomials. For all real numbers a, b, and c,

$$a(b + c) = ab + ac \quad \text{and} \quad a(b - c) = ab - ac.$$

The first distributive property above can be visualized geometrically. For example,

$$3(x + 2) = 3x + 6$$

is illustrated in **FIGURE 12.8**. The dimensions of the large rectangle are 3 by $x + 2$, and its area is $3(x + 2)$. The areas of the two small rectangles, $3x$ and 6, equal the area of the large rectangle. Therefore $3(x + 2) = 3x + 6$.

In the next example, we use the distributive properties to multiply expressions.

EXAMPLE 2 Using distributive properties

Multiply.
(a) $2(3x + 4)$ (b) $(3x^2 + 4)5$ (c) $-x(3x - 6)$

Solution

(a) $2(3x + 4) = 2 \cdot 3x + 2 \cdot 4 = 6x + 8$

(b) $(3x^2 + 4)5 = 3x^2 \cdot 5 + 4 \cdot 5 = 15x^2 + 20$

(c) $-x(3x - 6) = -x \cdot 3x + x \cdot 6 = -3x^2 + 6x$

Now Try Exercises 15, 19, 21

READING CHECK 2

- What properties are commonly used to multiply a monomial and a polynomial?

3 Multiplying Monomials and Polynomials

A monomial consists of one term, whereas a polynomial consists of one or more terms separated by $+$ or $-$ signs. To multiply a monomial by a polynomial, we apply the distributive properties and the product rule.

EXAMPLE 3 Multiplying monomials and polynomials

Multiply.
(a) $9x(2x^2 - 3)$
(b) $(5x - 8)x^2$
(c) $-7(2x^2 - 4x + 6)$
(d) $4x^3(x^4 + 9x^2 - 8)$

Solution

(a) $9x(2x^2 - 3) = 9x \cdot 2x^2 - 9x \cdot 3$ Distributive property
$= 18x^3 - 27x$ The product rule

(b) $(5x - 8)x^2 = 5x \cdot x^2 - 8 \cdot x^2$ Distributive property
$= 5x^3 - 8x^2$ The product rule

(c) $-7(2x^2 - 4x + 6) = -7 \cdot 2x^2 + 7 \cdot 4x - 7 \cdot 6$ Distributive property
$= -14x^2 + 28x - 42$ Simplify.

(d) $4x^3(x^4 + 9x^2 - 8) = 4x^3 \cdot x^4 + 4x^3 \cdot 9x^2 - 4x^3 \cdot 8$ Distributive property
$= 4x^7 + 36x^5 - 32x^3$ The product rule

Now Try Exercises 23, 25, 27, 29

We can also multiply monomials and polynomials that contain more than one variable.

EXAMPLE 4 Multiplying monomials and polynomials

Multiply.
(a) $2xy(7x^2y^3 - 1)$
(b) $-ab(a^2 - b^2)$

Solution

(a) $2xy(7x^2y^3 - 1) = 2xy \cdot 7x^2y^3 - 2xy \cdot 1$ Distributive property
$= 14xx^2yy^3 - 2xy$ Commutative property
$= 14x^3y^4 - 2xy$ The product rule

(b) $-ab(a^2 - b^2) = -ab \cdot a^2 + ab \cdot b^2$ Distributive property
$= -aa^2b + abb^2$ Commutative property
$= -a^3b + ab^3$ The product rule

Now Try Exercises 31, 35

4 Multiplying Polynomials

Monomials, binomials, and trinomials are examples of polynomials. Recall that a monomial has one term, a binomial has two terms, and a trinomial has three terms. In the next example, we multiply two binomials, using both geometric and symbolic techniques.

EXAMPLE 5 Multiplying binomials

Multiply $(x + 4)(x + 2)$
(a) geometrically and (b) symbolically.

Solution

(a) To multiply $(x + 4)(x + 2)$ geometrically, draw a rectangle $x + 4$ long and $x + 2$ wide, as shown in **FIGURE 12.9(a)**. The area of this rectangle equals length times width, or $(x + 4)(x + 2)$. The large rectangle can be divided into four smaller rectangles, which have areas of x^2, $4x$, $2x$, and 8, as shown in **FIGURE 12.9(b)**. Thus

$$(x + 4)(x + 2) = x^2 + 4x + 2x + 8$$
$$= x^2 + 6x + 8.$$

(b) To multiply $(x + 4)(x + 2)$ symbolically, apply the distributive property two times.

$$(x + 4)(x + 2) = (x + 4)(x) + (x + 4)(2)$$
$$= x \cdot x + 4 \cdot x + x \cdot 2 + 4 \cdot 2$$
$$= x^2 + 4x + 2x + 8$$
$$= x^2 + 6x + 8$$

Now Try Exercise 39

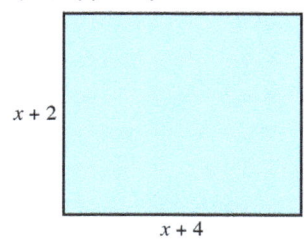

$(x + 2)(x + 4) = x^2 + 6x + 8$

(a) Area $= (x + 4)(x + 2)$

(b) Area $= x^2 + 4x + 2x + 8$

FIGURE 12.9

MAKING CONNECTIONS 1

Multiplying Binomials Using FOIL

The distributive properties used in part (b) of the previous example show that if we want to multiply $(x + 4)$ by $(x + 2)$, we should multiply *every* term in the binomial $x + 4$ by *every* term in the binomial $x + 2$.

$$(x + 4)(x + 2) = x^2 + 2x + 4x + 8$$
$$= x^2 + 6x + 8$$

When multiplying *binomials*, this process is sometimes called *FOIL*. This acronym may be used to remind us to multiply the first terms (*F*), outside terms (*O*), inside terms (*I*), and last terms (*L*). The *FOIL* process is a shortcut for the process used in Example 5(b).

Multiply the *First terms* to obtain x^2. $(x + 4)(x + 2)$

Multiply the *Outside terms* to obtain $2x$. $(x + 4)(x + 2)$

Multiply the *Inside terms* to obtain $4x$. $(x + 4)(x + 2)$

Multiply the *Last terms* to obtain 8. $(x + 4)(x + 2)$

NOTE: The *FOIL* acronym is useful for remembering how to multiply two *binomials*, but it is not helpful, in general, for multiplying any two *polynomials*. ∎

The following statement summarizes how to multiply two polyncmials in general.

MULTIPLYING POLYNOMIALS

The product of two polynomials may be found by multiplying every term in the first polynomial by every term in the second polynomial and then combining like terms.

EXAMPLE 6 Multiplying binomials

Multiply. Draw arrows to show how each term is found.
(a) $(3x + 2)(x + 1)$ (b) $(1 - x)(1 + 2x)$ (c) $(4x - 3)(x^2 - 2x)$

Solution

(a) $(3x + 2)(x + 1) = 3x \cdot x + 3x \cdot 1 + 2 \cdot x + 2 \cdot 1$
$= 3x^2 + 3x + 2x + 2$
$= 3x^2 + 5x + 2$

(b) $(1 - x)(1 + 2x) = 1 \cdot 1 + 1 \cdot 2x - x \cdot 1 - x \cdot 2x$
$= 1 + 2x - x - 2x^2$
$= 1 + x - 2x^2$

(c) $(4x - 3)(x^2 - 2x) = 4x \cdot x^2 - 4x \cdot 2x - 3 \cdot x^2 + 3 \cdot 2x$
$= 4x^3 - 8x^2 - 3x^2 + 6x$
$= 4x^3 - 11x^2 + 6x$

Now Try Exercises 51, 53, 59

The *FOIL* process cannot be used for every product of polynomials. In the next example, the general process for multiplying polynomials is used.

EXAMPLE 7 Multiplying polynomials

Multiply.
(a) $(2x + 3)(x^2 + x - 1)$ (b) $(a - b)(a^2 + ab + b^2)$
(c) $(x^4 + 2x^2 - 5)(x^2 + 1)$

Solution
(a) Multiply every term in $(2x + 3)$ by every term in $(x^2 + x - 1)$.

$(2x + 3)(x^2 + x - 1) = 2x \cdot x^2 + 2x \cdot x - 2x \cdot 1 + 3 \cdot x^2 + 3 \cdot x - 3 \cdot 1$
$= 2x^3 + 2x^2 - 2x + 3x^2 + 3x - 3$
$= 2x^3 + 5x^2 + x - 3$

(b) $(a - b)(a^2 + ab + b^2) = a \cdot a^2 + a \cdot ab + a \cdot b^2 - b \cdot a^2 - b \cdot ab - b \cdot b^2$
$= a^3 + a^2b + ab^2 - a^2b - ab^2 - b^3$
$= a^3 - b^3$

(c) $(x^4 + 2x^2 - 5)(x^2 + 1) = x^4 \cdot x^2 + x^4 \cdot 1 + 2x^2 \cdot x^2 + 2x^2 \cdot 1 - 5 \cdot x^2 - 5 \cdot 1$
$= x^6 + x^4 + 2x^4 + 2x^2 - 5x^2 - 5$
$= x^6 + 3x^4 - 3x^2 - 5$

Now Try Exercises 63, 67, 69

> **STUDY TIP**
>
> Even if you know exactly how to do a math problem correctly, a simple computational error will often cause you to get an incorrect answer. Be sure to take your time on simple calculations.

Polynomials can be multiplied vertically in a manner similar to multiplication of real numbers. For example, multiplication of 123 times 12 is performed as follows.

$$\begin{array}{r} 1\ 2\ 3 \\ \times\ 1\ 2 \\ \hline 2\ 4\ 6 \\ 1\ 2\ 3 \\ \hline 1\ 4\ 7\ 6 \end{array}$$

A similar method can be used to multiply polynomials vertically.

EXAMPLE 8 Multiplying polynomials vertically

Multiply $2x^2 - 4x + 1$ and $x + 3$ vertically.

Solution
Write the polynomials vertically. Then multiply every term in the first polynomial by each term in the second polynomial. Arrange the results so that *like terms are in the same column.*

$$\begin{array}{r} 2x^2 - 4x + 1 \\ x + 3 \\ \hline 6x^2 - 12x + 3 \\ 2x^3 - 4x^2 + x \\ \hline 2x^3 + 2x^2 - 11x + 3 \end{array}$$

Multiply top polynomial by 3.
Multiply top polynomial by x.
Add each column.

Now Try Exercise 71

MAKING CONNECTIONS 2

Vertical and Horizontal Formats

Whether you decide to add, subtract, or multiply polynomials vertically or horizontally, remember that the *same* answer is obtained either way.

EXAMPLE 9 Finding the volume of a box

A box has a width 3 inches less than its height and a length 4 inches more than its height.
(a) If h represents the height of the box, write a polynomial that represents the volume of the box.
(b) Use this polynomial to calculate the volume of the box if $h = 10$ inches.

Solution
(a) If h is the height, then $h - 3$ is the width and $h + 4$ is the length, as illustrated in **FIGURE 12.10**. Its volume equals the product of these three expressions.

$$h(h - 3)(h + 4) = (h^2 - 3h)(h + 4)$$
$$= h^2 \cdot h + h^2 \cdot 4 - 3h \cdot h - 3h \cdot 4$$
$$= h^3 + 4h^2 - 3h^2 - 12h$$
$$= h^3 + h^2 - 12h$$

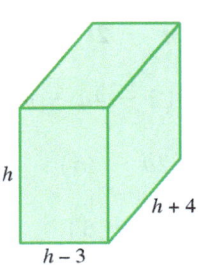

FIGURE 12.10

(b) If $h = 10$, then the volume is

$$10^3 + 10^2 - 12(10) = 1000 + 100 - 120 = 980 \text{ cubic inches.}$$

Now Try Exercises 79

12.3 Putting It All Together

CONCEPT	EXPLANATION	EXAMPLES
Distributive Properties	For all real numbers a, b, and c, $$a(b + c) = ab + ac \text{ and}$$ $$a(b - c) = ab - ac.$$	$5(x + 3) = 5x + 15$, $3(x - 6) = 3x - 18$, and $-2x(3 - 5x^3) = -6x + 10x^4$
Multiplying Polynomials	The product of two polynomials may be found by multiplying every term in the first polynomial by every term in the second polynomial and then combining like terms.	$3x(5x^2 + 2x - 7)$ $= 3x \cdot 5x^2 + 3x \cdot 2x - 3x \cdot 7$ $= 15x^3 + 6x^2 - 21x$ $(x + 2)(7x - 3)$ $= x \cdot 7x - x \cdot 3 + 2 \cdot 7x - 2 \cdot 3$ $= 7x^2 - 3x + 14x - 6$ $= 7x^2 + 11x - 6$

12.3 Exercises

CONCEPTS AND VOCABULARY

1. The equation $x^2 \cdot x^3 = x^5$ illustrates what rule of exponents?

2. The equation $3(x - 2) = 3x - 6$ illustrates what property?

3. The product of two polynomials may be found by multiplying every ____ in the first polynomial by every ____ in the second polynomial and then combining like terms.

4. Polynomials can be multiplied horizontally or ____.

MULTIPLYING MONOMIALS

Exercises 5–14: Multiply.

5. $x^2 \cdot x^5$
6. $-a \cdot a^5$
7. $-3a \cdot 4a$
8. $7x \cdot 5x$
9. $4x^3 \cdot 5x^2$
10. $6b^6 \cdot 3b^5$
11. $xy^2 \cdot 4xy$
12. $3ab \cdot ab^2$
13. $(-3xy^2)(4x^2y)$
14. $(-r^2t^2)(-r^3t)$

MULTIPLYING MONOMIALS AND POLYNOMIALS

Exercises 15–36: Multiply and simplify the expression.

15. $3(x + 4)$
16. $-7(4x - 1)$
17. $-5(9x + 1)$
18. $10(1 - 6x)$
19. $(4 - z)z$
20. $(2y - 8)2y$
21. $-y(5 + 3y)$
22. $3z(1 - 5z)$
23. $3x(5x^2 - 4)$
24. $-6x(2x^3 + 1)$
25. $(6x - 6)x^2$
26. $(1 - 2x^2)3x^2$
27. $-8(4t^2 + t + 1)$
28. $7(3t^2 - 2t - 5)$
29. $n^2(-5n^2 + n - 2)$
30. $6n^3(2 - 4n + n^2)$
31. $xy(x + y)$
32. $ab(2a - 3b)$
33. $x^2(x^2y - xy^2)$
34. $2y^2(xy - 5)$
35. $-ab(a^3 - 2b^3)$
36. $5rt(r^2 + 2rt + t^2)$

MULTIPLYING POLYNOMIALS

Exercises 37–42: (Refer to Example 5.) Multiply the given expression (a) geometrically and (b) symbolically.

37. $x(x + 3)$
38. $2x(x + 5)$
39. $(x + 2)(x + 2)$
40. $(x + 1)(x + 3)$
41. $(x + 3)(x + 6)$
42. $(x + 5)(x + 2)$

Exercises 43–70: Multiply and simplify the expression.

43. $(x + 3)(x + 5)$
44. $(x - 4)(x - 7)$
45. $(x - 8)(x - 9)$
46. $(x + 10)(x + 10)$
47. $(3z - 2)(2z - 5)$
48. $(z + 6)(2z - 1)$
49. $(8b - 1)(8b + 1)$
50. $(3t + 2)(3t - 2)$
51. $(10y + 7)(y - 1)$
52. $(y + 6)(2y + 7)$
53. $(5 - 3a)(1 - 2a)$
54. $(4 - a)(5 + 3a)$
55. $(1 - 3x)(1 + 3x)$
56. $(10 - x)(5 - 2x)$
57. $(x - 1)(x^2 + 1)$
58. $(x + 2)(x^2 - x)$
59. $(x^2 + 4)(4x - 3)$
60. $(3x^2 - 1)(3x^2 + 1)$
61. $(2n + 1)(n^2 + 3)$
62. $(2 - n^2)(1 + n^2)$
63. $(m + 1)(m^2 + 3m + 1)$
64. $(m - 2)(m^2 - m + 5)$
65. $(3x - 2)(2x^2 - x + 4)$
66. $(5x + 4)(x^2 - 3x + 2)$
67. $(x + 1)(x^2 - x + 1)$
68. $(x - 2)(x^2 + 4x + 4)$
69. $(4b^2 + 3b + 7)(b^2 + 3)$
70. $(-3a^2 - 2a + 1)(3a^2 - 3)$

Exercises 71–76: Multiply the polynomials vertically.

71. $(x + 2)(x^2 - 3x + 1)$
72. $(2y - 3)(3y^2 - 2y - 2)$
73. $(a - 2)(a^2 + 2a + 4)$
74. $(b - 3)(b^2 + 3b + 9)$
75. $(3x^2 - x + 1)(2x^2 + 1)$
76. $(2x^2 - 3x - 5)(2x^2 + 3)$

77. **Thinking Generally** If a polynomial with m terms and a polynomial with n terms are multiplied, how many terms are there in the product before like terms are combined?

78. **Thinking Generally** When a polynomial with m terms is multiplied by a second polynomial, the product contains k terms before like terms are combined. How many terms does the second polynomial contain?

APPLICATIONS INVOLVING MULTIPLYING POLYNOMIALS

79. **Volume of a Box** (Refer to Example 9.) A box has a width 4 inches less than its height and a length 2 inches more than its height.
    (a) If h is the height of the box, write a polynomial that represents the volume of the box.
    (b) Use this polynomial to calculate the volume for $h = 25$ inches.

80. **Surface Area of a Box** Use the drawing of the box to write a polynomial that represents each of the following.

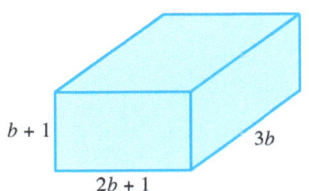

   (a) The area of its bottom
   (b) The area of its front
   (c) The area of its right side
   (d) The total area of its six sides

81. **Perimeter of a Pen** A rectangular pen for a pet has a perimeter of 100 feet. If one side of the pen has length x, then its area is given by $x(50 - x)$.
    (a) Multiply this expression.
    (b) Evaluate the expression obtained in part (a) for $x = 25$.

82. **Rectangular Garden** A rectangular garden has a perimeter of 500 feet.
    (a) If one side of the garden has length x, then write a polynomial expression that gives its area. Multiply this expression completely.
    (b) Evaluate the expression for $x = 50$ and interpret your answer.

83. **Surface Area of a Cube** Write a polynomial that represents the total area of the six sides of the cube having edges with length $x + 1$.

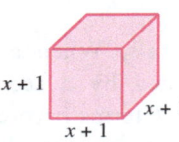

84. **Surface Area of a Sphere** The surface area of a sphere with radius r is $4\pi r^2$. Write a polynomial that gives the surface area of a sphere with radius $x + 2$. Leave your answer in terms of π.

85. **Toy Rocket** A toy rocket is shot straight up into the air. Its height h in feet above the ground after t seconds is represented by the expression $t(64 - 16t)$.
    (a) Multiply this expression.
    (b) Evaluate the expression obtained in part (a) and the given expression for $t = 2$.
    (c) Are your answers in part (b) the same? Should they be the same?

86. **Toy Rocket on the Moon** (Refer to the preceding exercise.) If the same toy rocket were flown on the moon, then its height h in feet after t seconds would be $t(64 - \frac{5}{2}t)$.
    (a) Multiply this expression.
    (b) Evaluate the expression obtained in part (a) and the given expression for $t = 2$. Did the rocket go higher on the moon?

WRITING ABOUT MATHEMATICS

87. Explain how the acronym FOIL relates to multiplying two binomials, such as $x + 3$ and $2x + 1$.

88. Does the FOIL method work for multiplying a binomial and a trinomial? Explain.

89. Explain in words how to multiply any two polynomials. Give an example.

90. Give two properties of real numbers that are used for multiplying $3x(5x^2 - 3x + 2)$. Explain your answer.

Group Activity

Exercises 1 and 2: **Constructing a Box** *A box is constructed from a rectangular piece of metal by cutting squares from the corners and folding up the sides. The square, cutout corners are x inches by x inches.*

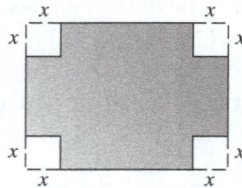 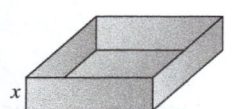

1. Suppose that the dimensions of the metal piece are 20 inches by 30 inches.
    (a) Write a polynomial that gives the volume of the box.
    (b) Find the volume of the box for $x = 4$ inches.

2. Suppose that the metal piece is square with sides of length 25 inches.
    (a) Write a polynomial expression that gives the outside surface area of the box. (Assume that the box does not have a top.)
    (b) Find this area for $x = 3$ inches.

12.4 Special Products

Objectives
1. Finding the Product of a Sum and Difference
2. Squaring Binomials
3. Cubing Binomials

STUDY TIP

In mathematics, there are often several correct ways to perform a particular process. If your instructor does not require you to use a specified method, choose the one that works best for you.

READING CHECK 1

- Multiply $(x - 4)(x + 4)$.

1 Finding the Product of a Sum and Difference

Products of the form $(a + b)(a - b)$ occur frequently in mathematics. Other examples include the products

$$(x + y)(x - y) \quad \text{and} \quad (2r + 3t)(2r - 3t).$$

These products can always be multiplied by using the techniques discussed in the previous section. However, there is a faster way to multiply these special products.

$$(a + b)(a - b) = a \cdot a - a \cdot b + b \cdot a - b \cdot b$$
$$= a^2 - ab + ba - b^2$$
$$= a^2 - b^2$$

In words, the product of a sum of two numbers and their difference equals the difference of their squares. We generalize this method as follows.

> **PRODUCT OF A SUM AND DIFFERENCE**
>
> For any real numbers a and b,
>
> $$(a + b)(a - b) = a^2 - b^2.$$

EXAMPLE 1 Finding products of sums and differences

Multiply.
(a) $(x + y)(x - y)$
(b) $(z - 2)(z + 2)$
(c) $(2r + 3t)(2r - 3t)$
(d) $(5m^2 - 4n^2)(5m^2 + 4n^2)$

Solution
(a) If we let $a = x$ and $b = y$, then we can apply the rule

$$(a + b)(a - b) = a^2 - b^2.$$

Thus

$$(x + y)(x - y) = (x)^2 - (y)^2$$
$$= x^2 - y^2.$$

(b) Because the expressions $(z + 2)(z - 2)$ and $(z - 2)(z + 2)$ are equal by the commutative property, we can apply the formula for the product of a sum and difference.

$$(z - 2)(z + 2) = (z)^2 - (2)^2$$
$$= z^2 - 4$$

(c) Let $a = 2r$ and $b = 3t$. Then the product can be evaluated as follows.

$$(2r + 3t)(2r - 3t) = (2r)^2 - (3t)^2$$
$$= 4r^2 - 9t^2$$

(d) $(5m^2 - 4n^2)(5m^2 + 4n^2) = (5m^2)^2 - (4n^2)^2 = 25m^4 - 16n^4$

Now Try Exercises 7, 9, 13, 17

The next example demonstrates how the product of a sum and difference can be used to find some products of numbers mentally.

EXAMPLE 2 **Finding a product**

Use the product of a sum and difference to find $22 \cdot 18$.

Solution
Because $22 = 20 + 2$ and $18 = 20 - 2$, rewrite and evaluate $22 \cdot 18$ as follows.

$$\begin{aligned} 22 \cdot 18 &= (20 + 2)(20 - 2) &&\text{Rewrite 22 as } 20 + 2 \text{ and 18 as } 20 - 2. \\ &= 20^2 - 2^2 &&\text{Product of a sum and difference} \\ &= 400 - 4 &&\text{Evaluate exponents.} \\ &= 396 &&\text{Subtract.} \end{aligned}$$

Now Try Exercise 21

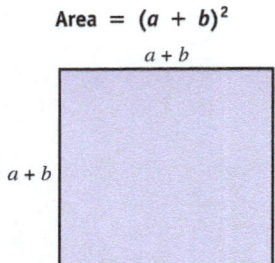

Area = $(a + b)^2$

FIGURE 12.11

2 Squaring Binomials

Because each side of the square shown in **FIGURE 12.11** has length $(a + b)$, its area equals
$$(a + b)(a + b),$$
which can be written as $(a + b)^2$. We can multiply this expression as follows.
$$\begin{aligned} (a + b)^2 &= (a + b)(a + b) \\ &= a^2 + ab + ba + b^2 \\ &= a^2 + 2ab + b^2 \end{aligned}$$

This result is illustrated geometrically in **FIGURE 12.12**, where the area of the large square is $(a + b)^2$. This area can also be found by adding the areas of the four small rectangles.

$$a^2 + ab + ba + b^2 = a^2 + 2ab + b^2$$

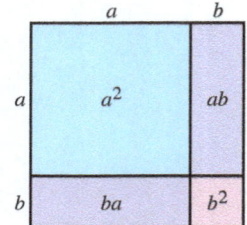

$(a + b)^2 = a^2 + 2ab + b^2$

FIGURE 12.12

The geometric and symbolic results are the same. Note that to obtain the middle term, $2ab$, we can multiply the two terms, a and b, in the binomial and *double* the result.

A similar product that is also the square of a binomial can be calculated as
$$\begin{aligned} (a - b)^2 &= (a - b)(a - b) \\ &= a^2 - ab - ba + b^2 \\ &= a^2 - 2ab + b^2. \end{aligned}$$

These results are summarized as follows.

READING CHECK 2

- Multiply $(a - b)^2$.

SQUARING A BINOMIAL

For any real numbers a and b,
$$(a + b)^2 = a^2 + 2ab + b^2 \quad \text{and}$$
$$(a - b)^2 = a^2 - 2ab + b^2.$$

That is, the square of a binomial equals the square of the first term, plus (or minus) twice the product of the two terms, plus the square of the last term.

NOTE: $(a + b)^2 \neq a^2 + b^2$. Don't forget the middle term when squaring a binomial. ∎

EXAMPLE 3   Squaring a binomial

Multiply.
(a) $(x + 3)^2$ (b) $(2x - 5)^2$ (c) $(1 - 5y)^2$ (d) $(7a^2 + 3b)^2$

Solution
(a) Let $a = x$ and $b = 3$ in the formula $(a + b)^2 = a^2 + 2ab + b^2$.
$$(x + 3)^2 = (x)^2 + 2(x)(3) + (3)^2$$
$$= x^2 + 6x + 9$$

(b) Apply the formula $(a - b)^2 = a^2 - 2ab + b^2$ with $a = 2x$ and $b = 5$.
$$(2x - 5)^2 = (2x)^2 - 2(2x)(5) + (5)^2$$
$$= 4x^2 - 20x + 25$$

(c) $(1 - 5y)^2 = (1)^2 - 2(1)(5y) + (5y)^2 = 1 - 10y + 25y^2$
(d) $(7a^2 + 3b)^2 = (7a^2)^2 + 2(7a^2)(3b) + (3b)^2 = 49a^4 + 42a^2b + 9b^2$

Now Try Exercises 27, 29, 35, 39

MAKING CONNECTIONS 1

Multiplying Binomials and Special Products

If you forget these special products, you can still multiply polynomials by using earlier techniques. For example, the binomial in Example 3(b) can be multiplied as

$$(2x - 5)^2 = (2x - 5)(2x - 5)$$
$$= 2x \cdot 2x - 2x \cdot 5 - 5 \cdot 2x + 5 \cdot 5$$
$$= 4x^2 - 10x - 10x + 25$$
$$= 4x^2 - 20x + 25.$$

Digital Photo

FIGURE 12.13

Math in Digital Context NASA first developed digital photos because they were easy to transmit through space and because they provided clear images. A digital image from outer space is shown in **FIGURE 12.13**.

Today, digital cameras are readily available, and the Internet uses digital images exclusively. The next example shows how polynomials relate to digital pictures.

EXAMPLE 4   Calculating the size of a digital picture

A digital picture is made up of tiny square units called *pixels*. Shading individual pixels creates a picture. A simplified version of a digital picture of the letter T is shown in **FIGURE 12.14**. This picture includes an image of the letter T that measures 3 pixels by 3 pixels and a 1-pixel border.
(a) If a square digital image measures x pixels by x pixels, then a picture that includes the image and a 1-pixel border will measure $x + 2$ pixels by $x + 2$ pixels. Find a polynomial that gives the total number of pixels in the picture, including the border.
(b) Let $x = 3$ and evaluate the polynomial. Does it agree with **FIGURE 12.14**?

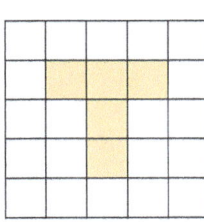

FIGURE 12.14

Solution
(a) The total number of pixels equals $(x + 2)$ times $(x + 2)$, or $(x + 2)^2$.
$$(x + 2)^2 = x^2 + 4x + 4$$

(b) For $x = 3$, the polynomial $x^2 + 4x + 4$ evaluates to $3^2 + 4 \cdot 3 + 4 = 25$, the total number of pixels. This result agrees with **FIGURE 12.14** on the previous page, which has a total of $5 \cdot 5 = 25$ pixels with a 3 pixel by 3 pixel image of the letter T inside.

Now Try Exercise 79

3 Cubing Binomials

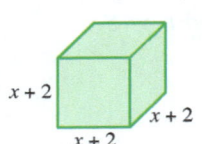

Volume = $(x + 2)^3$

FIGURE 12.15

To calculate the volume of the cube shown in **FIGURE 12.15**, we find the product of its length, width, and height. Because all sides have the same measure, its volume is $(x + 2)^3$. That is, the volume equals the *cube* of $x + 2$.

To multiply the expression $(x + 2)^3$, we proceed as follows.

$(x + 2)^3 = (x + 2)(x + 2)^2$ $a^3 = a \cdot a^2$

$\quad\quad\quad = (x + 2)(x^2 + 4x + 4)$ Square the binomial.

$\quad\quad\quad = x \cdot x^2 + x \cdot 4x + x \cdot 4 + 2 \cdot x^2 + 2 \cdot 4x + 2 \cdot 4$ Multiply the polynomials.

$\quad\quad\quad = x^3 + 4x^2 + 4x + 2x^2 + 8x + 8$ Simplify terms.

$\quad\quad\quad = x^3 + 6x^2 + 12x + 8$ Combine like terms.

EXAMPLE 5   **Cubing a binomial**

Multiply $(2z - 3)^3$.

Solution

$(2z - 3)^3 = (2z - 3)(2z - 3)^2$ $a^3 = a \cdot a^2$

$\quad\quad\quad\quad = (2z - 3)(4z^2 - 12z + 9)$ Square the binomial.

$\quad\quad\quad\quad = 8z^3 - 24z^2 + 18z - 12z^2 + 36z - 27$ Multiply the polynomials.

$\quad\quad\quad\quad = 8z^3 - 36z^2 + 54z - 27$ Combine like terms.

Now Try Exercise 47

CRITICAL THINKING 1

Suppose that a student is convinced that the expressions

$(x + y)^3$ and $x^3 + y^3$

are equal. How could you convince the student otherwise?

NOTE: In Example 5, it is not correct to find the result by simply cubing each term in the binomial. That is,

$$(2z - 3)^3 \neq (2z)^3 - (3)^3 = 8z^3 - 27. \blacksquare$$

🌐 **Math in Business Context** Powers of expressions often occur in business when annual interest is calculated. The next example illustrates how this is done.

EXAMPLE 6   **Calculating interest**

If a savings account pays x percent annual interest, where x is expressed as a decimal, then after 3 years a sum of money will grow by a factor of $(1 + x)^3$.
(a) Multiply this expression.
(b) Evaluate the expression for $x = 0.10$ (or 10%), and interpret the result.

Solution

(a) $(1 + x)^3 = (1 + x)(1 + x)^2$ $a^3 = a \cdot a^2$

$= (1 + x)(1 + 2x + x^2)$ Square the binomial.

$= 1 + 2x + x^2 + x + 2x^2 + x^3$ Multiply the polynomials.

$= 1 + 3x + 3x^2 + x^3$ Combine like terms.

(b) Let $x = 0.1$ in the expression $1 + 3x + 3x^2 + x^3$.

$$1 + 3(0.1) + 3(0.1)^2 + (0.1)^3 = 1.331$$

The sum of money will increase by a factor of 1.331. For example, $1000 deposited in this account will grow to $1331 in 3 years.

Now Try Exercise 75

12.4 Putting It All Together

CONCEPT	EXPLANATION	EXAMPLES
Product of a Sum and Difference	For any real numbers x and y, $(x + y)(x - y) = x^2 - y^2.$	$(x + 6)(x - 6) = x^2 - 36,$ $(2x - 3)(2x + 3) = 4x^2 - 9,$ and $(x^2 + y^2)(x^2 - y^2) = x^4 - y^4$
Squaring a Binomial	For all real numbers x and y, $(x + y)^2 = x^2 + 2xy + y^2$ and $(x - y)^2 = x^2 - 2xy + y^2.$	$(x + 4)^2 = x^2 + 8x + 16,$ $(5x - 2)^2 = 25x^2 - 20x + 4,$ and $(x^2 + y^2)^2 = x^4 + 2x^2y^2 + y^4$
Cubing a Binomial	Multiply the binomial by its square.	$(x + 3)^3$ $= (x + 3)(x + 3)^2$ $= (x + 3)(x^2 + 6x + 9)$ $= x^3 + 6x^2 + 9x + 3x^2 + 18x + 27$ $= x^3 + 9x^2 + 27x + 27$

12.4 Exercises

CONCEPTS AND VOCABULARY

1. $(a + b)(a - b) = $ _____

2. $(a + b)^2 = $ _____

3. $(a - b)^2 = $ _____

4. $(a + b)^3 = (a + b) \cdot$ _____

5. (True or False?) The two expressions $(x + y)^2$ and $x^2 + y^2$ are equal for all real numbers x and y.

6. (True or False?) The two expressions $(r - t)^2$ and $r^2 - t^2$ are equal for all real numbers r and t.

FINDING THE PRODUCT OF A SUM AND DIFFERENCE

Exercises 7–20: Multiply.

7. $(x - 3)(x + 3)$
8. $(x + 6)(x - 6)$
9. $(4x - 1)(4x + 1)$
10. $(10x + 3)(10x - 3)$
11. $(1 + 2a)(1 - 2a)$
12. $(4 - 9b)(4 + 9b)$
13. $(2x + 3y)(2x - 3y)$
14. $(5r - 6t)(5r + 6t)$
15. $(ab - 5)(ab + 5)$
16. $(2xy + 7)(2xy - 7)$
17. $(a^2 - b^2)(a^2 + b^2)$
18. $(3x^2 + y^2)(3x^2 - y^2)$
19. $(x^3 - y^3)(x^3 + y^3)$
20. $(2a^4 + b^4)(2a^4 - b^4)$

Exercises 21–26: (Refer to Example 2.) Use the product of a sum and a difference to evaluate the expression.

21. $101 \cdot 99$
22. $52 \cdot 48$
23. $23 \cdot 17$
24. $29 \cdot 31$
25. $90 \cdot 110$
26. $38 \cdot 42$

SQUARING BINOMIALS

Exercises 27–40: Multiply.

27. $(a - 2)^2$
28. $(x - 7)^2$
29. $(2x + 3)^2$
30. $(7x - 2)^2$
31. $(3b + 5)^2$
32. $(7t + 10)^2$
33. $\left(\frac{3}{4}a - 4\right)^2$
34. $\left(\frac{1}{5}a + 1\right)^2$
35. $(1 - b)^2$
36. $(1 - 4a)^2$
37. $(5 + y^3)^2$
38. $(9 - 5x^2)^2$
39. $(a^2 + b)^2$
40. $(x^3 - y^3)^2$

CUBING BINOMIALS

Exercises 41–48: Multiply.

41. $(a + 1)^3$
42. $(b + 4)^3$
43. $(x - 2)^3$
44. $(y - 7)^3$
45. $(2x + 1)^3$
46. $(4z + 3)^3$
47. $(6u - 1)^3$
48. $(5v + 3)^3$

MULTIPLYING POLYNOMIALS

Exercises 49–66: Multiply, using any appropriate method.

49. $4(5x + 9)$
50. $(2x + 1)(3x - 5)$
51. $(x - 5)(x + 7)$
52. $(x + 10)(x + 10)$
53. $(3x - 5)^2$
54. $(x - 3)(x + 9)$
55. $(5x + 3)(5x + 4)$
56. $-x^3(x^2 - x + 1)$
57. $(4b - 5)(4b + 5)$
58. $(x + 5)^3$
59. $-5x(4x^2 - 7x + 2)$
60. $(4x^2 - 5)(4x^2 + 5)$
61. $(4 - a)^3$
62. $2x(x - 3)^3$
63. $x(x + 3)^2$
64. $(x - 1)^2(x + 1)$
65. $(x + 2)(x - 2)(x + 1)(x - 1)$
66. $(x - y)(x + y)(x^2 + y^2)$
67. **Thinking Generally** Multiply $(a^n + b^n)(a^n - b^n)$.
68. **Thinking Generally** Multiply $(a^n + b^n)^2$.

APPLICATIONS INVOLVING SPECIAL PRODUCTS

Exercises 69–72: Do each part and verify that your answers are the same.

(a) Find the area of the large square by multiplying its length and width.
(b) Find the sum of the areas of the smaller rectangles inside the large square.

69.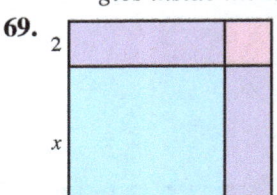

70.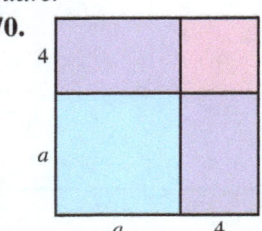

71.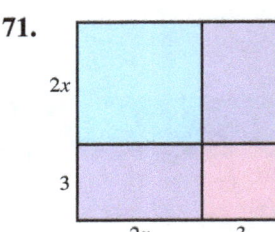

72.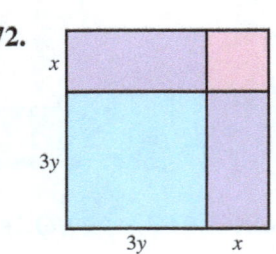

Exercises 73 and 74: Find a polynomial that represents the following.

(a) The outside surface area given by the six sides of the cube
(b) The volume of the cube

73.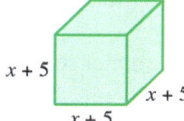

74.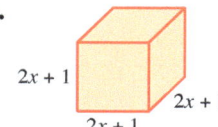

75. **Compound Interest** (Refer to Example 6.) If a sum of money is deposited in a savings account that is paying x percent annual interest (expressed as a decimal), then this sum of money increases by a factor of $(1 + x)^2$ after 2 years.
    (a) Multiply this expression.
    (b) Evaluate the polynomial expression found in part (a) for an annual interest rate of 10%, or $x = 0.10$, and interpret the answer.

76. **Compound Interest** If a sum of money is deposited in a savings account that is paying x percent annual interest, then this sum of money increases by a factor of $\left(1 + \frac{1}{100}x\right)^3$ after 3 years.
    (a) Multiply this expression.
    (b) Evaluate the polynomial expression in part (a) for an annual interest rate of 8%, or $x = 8$, and interpret the answer.

77. **Probability** If there is an x percent chance of rain on each of two consecutive days, then the expression $(1 - x)^2$ gives the percent chance that neither day will have rain. Assume that all percentages are expressed as decimals.
    (a) Multiply this expression.
    (b) Evaluate the polynomial expression in part (a) for a 50% chance of rain, or $x = 0.50$, and interpret the answer.

78. **Probability** If there is an x percent chance of rolling a 6 with one die, then the expression $(1 - x)^3$ gives the percent chance of not rolling a 6 with three dice. Assume that all percentages are expressed as decimals or fractions.
    (a) Multiply this expression.
    (b) Evaluate the polynomial expression found in part (a) for a $16.\overline{6}\%$ chance of rolling a 6, or $x = \frac{1}{6}$, and interpret the answer.

79. **Swimming Pool** A square swimming pool has an 8-foot-wide sidewalk around it.

    (a) If the sides of the pool have length z, as shown in the accompanying figure, find a polynomial that gives the area of the sidewalk.
    (b) Evaluate the polynomial in part (a) for $z = 60$ and interpret the answer.

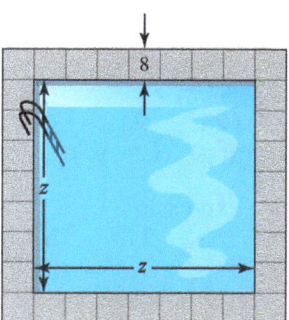

80. **Digital Picture** (Refer to Example 4.) Suppose that a digital picture, including its border, is $x + 2$ pixels by $x + 2$ pixels and that the actual image inside the border is $x - 2$ pixels by $x - 2$ pixels, as shown in the following figure.

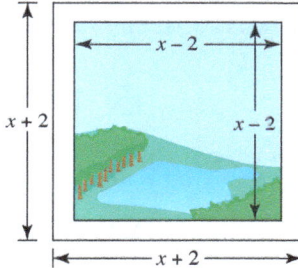

    (a) Find a polynomial that gives the number of pixels in the border.
    (b) Evaluate the polynomial in part (a) for $x = 5$.
    (c) Sketch a digital picture of the letter H with $x = 5$. Does the picture agree with the answer in part (b)?
    (d) Digital pictures typically have large values for x. If a picture has $x = 500$, find the total number of pixels in its border.

WRITING ABOUT MATHEMATICS

81. Explain why $(a + b)^2$ does not equal $a^2 + b^2$ in general for real numbers a and b.

82. Explain how to find the cube of a binomial.

SECTIONS 12.3 and 12.4 — Checking Basic Concepts

1. Multiply each expression.
   (a) $(-3xy^4)(5x^2y)$
   (b) $-x(6-4x)$
   (c) $3ab(a^2 - 2ab + b^2)$

2. Multiply each expression.
   (a) $(x+3)(4x-3)$
   (b) $(x^2-1)(2x^2+2)$
   (c) $(x+y)(x^2-xy+y^2)$

3. Multiply each expression.
   (a) $(5x+2)(5x-2)$
   (b) $(x+3)^2$
   (c) $(2-7x)^2$
   (d) $(t+2)^3$

4. Complete each part and verify that your answers are the same.
   (a) Find the area of the large square by squaring the length of one of its sides.
   (b) Find the sum of the areas of the smaller rectangles inside the large square.

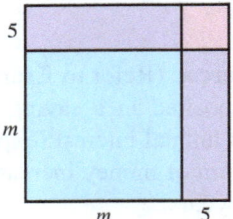

12.5 Integer Exponents and the Quotient Rule

Objectives
1. Using Negative Integers as Exponents
2. Applying the Quotient Rule
3. Applying Other Rules for Exponents
4. Reading and Writing Scientific Notation

NEW VOCABULARY
☐ Scientific notation

STUDY TIP

Mathematics often builds on concepts that have already been studied. Try to get in the regular habit of reviewing topics from earlier parts of the text.

READING CHECK 1
- Simplify 3^{-2}.

1 ▸ Using Negative Integers as Exponents

So far, we have defined exponents that are whole numbers. For example,

$$5^0 = 1 \quad \text{and} \quad 2^3 = 2 \cdot 2 \cdot 2 = 8.$$

What if an exponent is a negative integer? To answer this question, consider **TABLE 12.6**, which shows values for decreasing powers of 2. Note that each time the exponent on 2 decreases by 1, the resulting value is divided by 2.

Powers of 2

Power of 2	Value
2^1	2
2^0	1
2^{-1}	$\frac{1}{2} = \frac{1}{2^1}$
2^{-2}	$\frac{1}{4} = \frac{1}{2^2}$

Decrease exponent by 1 — Divide by 2
Decrease exponent by 1 — Divide by 2
Decrease exponent by 1 — Divide by 2

TABLE 12.6

TABLE 12.6 shows that $2^{-1} = \frac{1}{2^1}$ and $2^{-2} = \frac{1}{2^2}$. In other words, if the exponent on the base 2 is negative, then the expression is equal to the reciprocal of the corresponding expression with a positive exponent on the base 2. This discussion suggests the following definition for negative integer exponents.

NEGATIVE INTEGER EXPONENTS

Let a be a nonzero real number and n be a positive integer. Then
$$a^{-n} = \frac{1}{a^n}.$$
That is, a^{-n} is the reciprocal of a^n.

EXAMPLE 1 **Evaluating negative exponents**

Simplify each expression.
(a) 2^{-3} (b) 7^{-1} (c) x^{-2} (d) $(x + y)^{-8}$

Solution

(a) Because $a^{-n} = \frac{1}{a^n}$, $2^{-3} = \frac{1}{2^3} = \frac{1}{2 \cdot 2 \cdot 2} = \frac{1}{8}$.

(b) $7^{-1} = \frac{1}{7^1} = \frac{1}{7}$ (c) $x^{-2} = \frac{1}{x^2}$ (d) $(x + y)^{-8} = \frac{1}{(x + y)^8}$

Now Try Exercises 7, 19, 25(b)

TECHNOLOGY NOTE

Negative Exponents
Calculators can evaluate negative exponents. The figure shows how a calculator evaluates the expressions in parts (a) and (b) of Example 1.

```
2^(-3)▶Frac
              1/8
7^(-1)▶Frac
              1/7
```

CALCULATOR HELP

To use the fraction feature (Frac), see Appendix A (pages AP-1 and AP-2).

The rules for exponents discussed in this chapter so far also apply to expressions that have negative exponents. For example, we can apply the product rule, $a^m \cdot a^n = a^{m+n}$, as follows.

$$2^{-3} \cdot 2^2 = 2^{-3+2} = 2^{-1} = \frac{1}{2}$$ — Add

We can check this result by evaluating the expression without using the product rule.

$$2^{-3} \cdot 2^2 = \frac{1}{2^3} \cdot 2^2 = \frac{1}{8} \cdot 4 = \frac{4}{8} = \frac{1}{2} \checkmark$$

EXAMPLE 2 **Using the product rule with negative exponents**

Evaluate each expression.
(a) $5^2 \cdot 5^{-4}$ (b) $3^{-2} \cdot 3^{-1}$

Solution

Product rule: Add exponents.

(a) $5^2 \cdot 5^{-4} = 5^{2+(-4)} = 5^{-2} = \frac{1}{5^2} = \frac{1}{25}$

(b) $3^{-2} \cdot 3^{-1} = 3^{-2+(-1)} = 3^{-3} = \frac{1}{3^3} = \frac{1}{27}$

Now Try Exercise 9

EXAMPLE 3 Using the rules of exponents

Simplify the expression. Write the answer using positive exponents.
(a) $x^2 \cdot x^{-5}$ (b) $(y^3)^{-4}$ (c) $(rt)^{-5}$ (d) $(ab)^{-3}(a^{-2}b)^3$

Solution
(a) Using the product rule, $a^m \cdot a^n = a^{m+n}$, gives

$$x^2 \cdot x^{-5} = x^{2+(-5)} = x^{-3} = \frac{1}{x^3}.$$

(b) Using the power rule, $(a^m)^n = a^{mn}$, gives

$$(y^3)^{-4} = y^{3(-4)} = y^{-12} = \frac{1}{y^{12}}.$$

(c) Using the power rule, $(ab)^n = a^n b^n$, gives

$$(rt)^{-5} = r^{-5}t^{-5} = \frac{1}{r^5} \cdot \frac{1}{t^5} = \frac{1}{r^5 t^5}.$$

This expression could also be simplified as follows.

$$(rt)^{-5} = \frac{1}{(rt)^5} = \frac{1}{r^5 t^5}$$

(d) $(ab)^{-3}(a^{-2}b)^3 = a^{-3}b^{-3}a^{-6}b^3$
$= a^{-3+(-6)}b^{-3+3}$
$= a^{-9}b^0$
$= \frac{1}{a^9} \cdot 1$
$= \frac{1}{a^9}$

Now Try Exercises 21, 27, 29(a)

2 Applying the Quotient Rule

Consider the division problem

$$\frac{3^4}{3^2} = \frac{3 \cdot 3 \cdot 3 \cdot 3}{3 \cdot 3} = \frac{3}{3} \cdot \frac{3}{3} \cdot 3 \cdot 3 = 1 \cdot 1 \cdot 3^2 = 3^2.$$

Because there are two more 3s in the numerator than in the denominator, the result is

$$3^{4-2} = 3^2.$$

(Subtract)

That is, to divide exponential expressions that have the *same base*, subtract the exponent of the denominator from the exponent of the numerator and keep the same base. This rule is called the *quotient rule*, which we express in symbols as follows.

READING CHECK 2

Use the quotient rule to simplify $\frac{5^2}{5^4}$.

THE QUOTIENT RULE

For any nonzero number a and integers m and n,

$$\frac{a^m}{a^n} = a^{m-n}.$$

EXAMPLE 4 Using the quotient rule

Simplify each expression. Write the answer using positive exponents.

(a) $\dfrac{4^3}{4^5}$ (b) $\dfrac{6a^7}{3a^4}$ (c) $\dfrac{xy^7}{x^2y^5}$

Solution

> Quotient rule: Subtract exponents.

(a) $\dfrac{4^3}{4^5} = 4^{3-5} = 4^{-2} = \dfrac{1}{4^2} = \dfrac{1}{16}$

(b) $\dfrac{6a^7}{3a^4} = \dfrac{6}{3} \cdot \dfrac{a^7}{a^4} = 2a^{7-4} = 2a^3$

(c) $\dfrac{xy^7}{x^2y^5} = \dfrac{x^1}{x^2} \cdot \dfrac{y^7}{y^5} = x^{1-2}y^{7-5} = x^{-1}y^2 = \dfrac{y^2}{x}$

Now Try Exercises 13(b), 31(b), 33(a)

MAKING CONNECTIONS 1

The Quotient Rule and Simplifying Quotients

Some quotients can be simplified mentally. Because

$$\dfrac{x^5}{x^3} = \dfrac{x \cdot x \cdot x \cdot x \cdot x}{x \cdot x \cdot x},$$

the quotient $\dfrac{x^5}{x^3}$ has five factors of x in the numerator and three factors of x in the denominator. There are two more factors of x in the numerator than in the denominator, so this expression simplifies to $\dfrac{x^2}{1}$, or simply x^2. Similarly,

$$\dfrac{x^3}{x^5} = \dfrac{x \cdot x \cdot x}{x \cdot x \cdot x \cdot x \cdot x}$$

has two more factors of x in the denominator than in the numerator. This quotient $\dfrac{x^3}{x^5}$ simplifies to $\dfrac{1}{x^2}$.

---

3 Applying Other Rules for Exponents

Other rules can be used to simplify expressions with negative exponents.

QUOTIENTS AND NEGATIVE EXPONENTS

The following three rules hold for any nonzero real numbers a and b and positive integers m and n.

1. $\dfrac{1}{a^{-n}} = a^n$ 2. $\dfrac{a^{-n}}{b^{-m}} = \dfrac{b^m}{a^n}$ 3. $\left(\dfrac{a}{b}\right)^{-n} = \left(\dfrac{b}{a}\right)^n$

We demonstrate the validity of the first rule as follows.

1. $\dfrac{1}{a^{-n}} = \dfrac{1}{\frac{1}{a^n}} = 1 \cdot \dfrac{a^n}{1} = a^n$

We demonstrate the validity of rules 2 and 3 from the previous page as follows.

2. $\dfrac{a^{-n}}{b^{-m}} = \dfrac{\frac{1}{a^n}}{\frac{1}{b^m}} = \dfrac{1}{a^n} \cdot \dfrac{b^m}{1} = \dfrac{b^m}{a^n}$

3. $\left(\dfrac{a}{b}\right)^{-n} = \dfrac{a^{-n}}{b^{-n}} = \dfrac{\frac{1}{a^n}}{\frac{1}{b^n}} = \dfrac{1}{a^n} \cdot \dfrac{b^n}{1} = \dfrac{b^n}{a^n} = \left(\dfrac{b}{a}\right)^n$

EXAMPLE 5 Working with quotients and negative exponents

Simplify each expression. Write the answer using positive exponents.

(a) $\dfrac{1}{2^{-5}}$ (b) $\dfrac{3^{-3}}{4^{-2}}$ (c) $\dfrac{5x^{-4}y^2}{10x^2y^{-4}}$ (d) $\left(\dfrac{2}{z^2}\right)^{-4}$

Solution

(a) $\dfrac{1}{2^{-5}} = 2^5 = 2 \cdot 2 \cdot 2 \cdot 2 \cdot 2 = 32$ (b) $\dfrac{3^{-3}}{4^{-2}} = \dfrac{4^2}{3^3} = \dfrac{16}{27}$

(c) $\dfrac{5x^{-4}y^2}{10x^2y^{-4}} = \dfrac{y^2 y^4}{2x^2 x^4} = \dfrac{y^6}{2x^6}$ (d) $\left(\dfrac{2}{z^2}\right)^{-4} = \left(\dfrac{z^2}{2}\right)^4 = \dfrac{z^8}{2^4} = \dfrac{z^8}{16}$

Now Try Exercises 15(b), 17, 37, 47

The rules for natural number exponents that are summarized in the first section of this chapter also hold for integer exponents. Additional rules for integer exponents are summarized as follows.

RULES FOR INTEGER EXPONENTS

The following rules hold for nonzero real numbers a and b, and positive integers m and n.

Description	Rule	Example
Negative Exponents	$a^{-n} = \dfrac{1}{a^n}$	$9^{-2} = \dfrac{1}{9^2} = \dfrac{1}{81}$
The Quotient Rule	$\dfrac{a^m}{a^n} = a^{m-n}$	$\dfrac{2^3}{2^{-2}} = 2^{3-(-2)} = 2^5$
Negative Exponents	$\dfrac{1}{a^{-n}} = a^n$	$\dfrac{1}{7^{-5}} = 7^5$
Negative Exponents	$\dfrac{a^{-n}}{b^{-m}} = \dfrac{b^m}{a^n}$	$\dfrac{4^{-3}}{2^{-5}} = \dfrac{2^5}{4^3}$
Negative Exponents	$\left(\dfrac{a}{b}\right)^{-n} = \left(\dfrac{b}{a}\right)^n$	$\left(\dfrac{1}{5}\right)^{-2} = \left(\dfrac{5}{1}\right)^2 = 25$

READING CHECK 3

- What kinds of numbers are often expressed in scientific notation?

TECHNOLOGY NOTE

Powers of 10
Calculators make use of scientific notation, as illustrated in the accompanying figure. The letter E denotes a power of 10. That is,
$2.5\text{E}13 = 2.5 \times 10^{13}$ and
$5\text{E}{-}6 = 5 \times 10^{-6}$.

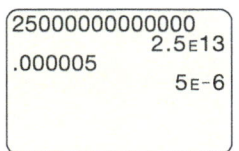

NOTE: The calculator has been set in *scientific mode*. ∎

4 Reading and Writing Scientific Notation

Powers of 10 are important because they are used in science to express numbers that are either very small or very large in absolute value. **TABLE 12.7** lists some powers of 10. Note that if the power of 10 decreases by 1, the result decreases by a factor of $\frac{1}{10}$, or equivalently, the decimal point is moved one place to the left. **TABLE 12.8** shows the names of some important powers of 10.

Powers of 10

Power of 10	Value
10^3	1000
10^2	100
10^1	10
10^0	1
10^{-1}	$\frac{1}{10} = 0.1$
10^{-2}	$\frac{1}{100} = 0.01$
10^{-3}	$\frac{1}{1000} = 0.001$

TABLE 12.7

Important Powers of 10

Power of 10	Name
10^{12}	Trillion
10^9	Billion
10^6	Million
10^3	Thousand
10^{-1}	Tenth
10^{-2}	Hundredth
10^{-3}	Thousandth
10^{-6}	Millionth

TABLE 12.8

Recall that numbers written in decimal notation are sometimes said to be in *standard form*. Decimal numbers that are either very large or very small in absolute value can be expressed in *scientific notation*.

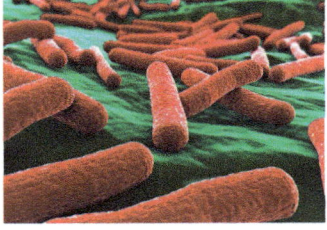

🌐 Math in Context [*Science*] The distance to the planet WASP-17b, discovered in 2009, is about 5,880,000,000,000,000 miles. This distance can be written in scientific notation as 5.88×10^{15} because

$$5{,}880{,}000{,}000{,}000{,}000 = 5.88 \times 10^{15}.$$

15 decimal places

The 10^{15} indicates that the decimal point in 5.88 should be moved **15** places to the **right**.

A typical virus is about 0.00000468 inch in diameter, which can be written in scientific notation as 4.68×10^{-6} because

$$0.00000468 = 4.68 \times 10^{-6}.$$

6 decimal places

The 10^{-6} indicates that the decimal point in 4.68 should be moved **6** places to the **left**.
The following definition provides a more complete explanation of scientific notation.

SCIENTIFIC NOTATION

A real number a is in **scientific notation** when a is written in the form $b \times 10^n$, where $1 \leq |b| < 10$ and n is an integer.

EXAMPLE 6   Converting scientific notation to standard form

Write each number in standard form.
(a) 5.23×10^4 (b) 8.1×10^{-3} (c) 6×10^{-2}

Solution
(a) The positive exponent 4 indicates that the decimal point in 5.23 is to be moved **4** places to the **right**.

$$5.23 \times 10^4 = 5.2\,3\,0\,0. = 52{,}300$$

(b) The negative exponent -3 indicates that the decimal point in 8.1 is to be moved **3** places to the **left**.

$$8.1 \times 10^{-3} = 0.0\,0\,8.1 = 0.0081$$

(c) $6 \times 10^{-2} = 0.0\,6. = 0.06$

Now Try Exercises 61, 63

The following steps can be used for writing a positive number a in scientific notation.

WRITING A POSITIVE NUMBER IN SCIENTIFIC NOTATION

For a positive, rational number a expressed as a decimal, if $1 \leq a < 10$, then $a = a \times 10^0$. Otherwise, use the following process to write a in scientific notation.

1. Move the decimal point in a until it becomes a number b such that $1 \leq b < 10$.
2. Let the positive integer n be the number of places the decimal point was moved.
3. Write a in scientific notation as follows.
   - If $a \geq 10$, then $a = b \times 10^n$.
   - If $a < 1$, then $a = b \times 10^{-n}$.

NOTE: The scientific notation for a negative number a is the opposite of the scientific notation of $|a|$. For example, $450 = 4.5 \times 10^2$ and $-450 = -4.5 \times 10^2$. ■

EXAMPLE 7   Writing a number in scientific notation

Write each number in scientific notation.
(a) 321,000,000   (U.S. population in 2015)
(b) 0.001   (Approximate time in seconds for sound to travel one foot)

Solution
(a) Move the assumed decimal point in 321,000,000 eight places to obtain 3.21.

$$3.2\,1\,0\,0\,0\,0\,0\,0.$$

Since $321{,}000{,}000 \geq 10$, the scientific notation is 3.21×10^8.

(b) Move the decimal point in 0.001 three places to obtain 1.

$$0.\underset{1\ 2\ 3}{001.}$$

Since $0.001 < 1$, the scientific notation is 1×10^{-3}.

Now Try Exercises 75, 79

Numbers in scientific notation can be multiplied by applying properties of real numbers and properties of exponents.

$$(6 \times 10^4) \cdot (3 \times 10^3) = (6 \cdot 3) \times (10^4 \cdot 10^3) \quad \text{Properties of real numbers}$$
$$= 18 \times 10^7 \quad \text{Product rule}$$
$$= 1.8 \times 10^8 \quad \text{Scientific notation}$$

Division can also be performed with scientific notation.

$$\frac{6 \times 10^4}{3 \times 10^3} = \frac{6}{3} \times \frac{10^4}{10^3} \quad \text{Property of fractions}$$
$$= 2 \times 10^1 \quad \text{Quotient rule}$$

These results are supported in **FIGURE 12.16**, where the calculator is in scientific mode.

```
(6*10^4)(3*10^3)
                1.8E8
(6*10^4)/(3*10^3
)
                  2E1
```

FIGURE 12.16

CALCULATOR HELP

To display numbers in scientific notation, see Appendix A (page AP-2).

Math in Context (Advertising) In the next example, we show how to use scientific notation to calculate the per person spending on advertising in the United States.

EXAMPLE 8 Analyzing the cost of advertising

In 2016, a total of $\$1.969 \times 10^{11}$ was spent on advertising in the United States. At that time, the population of the United States was 3.26×10^8. Determine how much was spent per person on advertising. (*Source: eMarketer.*)

Solution
To determine the amount spent per person, divide $\$1.969 \times 10^{11}$ by 3.26×10^8.

$$\frac{1.969 \times 10^{11}}{3.26 \times 10^8} = \frac{1.969}{3.26} \times 10^{11-8} \approx 0.604 \times 10^3 = 604$$

In 2016, about $604 per person was spent on advertising in the United States.

Now Try Exercise 97

CRITICAL THINKING 1

Estimate the number of seconds that you have been alive. Write your answer in scientific notation.

12.5 Putting It All Together

For the rules for integer exponents in this table, assume that a and b are nonzero real numbers and that m and n are integers.

CONCEPT	EXPLANATION	EXAMPLES
Negative Integer Exponents	$a^{-n} = \dfrac{1}{a^n}$	$2^{-4} = \dfrac{1}{2^4} = \dfrac{1}{16}$, $a^{-8} = \dfrac{1}{a^8}$, and $(xy)^{-2} = \dfrac{1}{(xy)^2} = \dfrac{1}{x^2y^2}$

continued on next page

770 CHAPTER 12 POLYNOMIALS AND EXPONENTS

continued from previous page

CONCEPT	EXPLANATION	EXAMPLES
Quotient Rule	$\dfrac{a^m}{a^n} = a^{m-n}$	$\dfrac{7^2}{7^4} = 7^{2-4} = 7^{-2} = \dfrac{1}{7^2} = \dfrac{1}{49}$ and $\dfrac{x^6}{x^3} = x^{6-3} = x^3$
Quotients and Negative Integer Exponents	1. $\dfrac{1}{a^{-n}} = a^n$ 2. $\dfrac{a^{-n}}{b^{-m}} = \dfrac{b^m}{a^n}$ 3. $\left(\dfrac{a}{b}\right)^{-n} = \left(\dfrac{b}{a}\right)^n$	1. $\dfrac{1}{5^{-2}} = 5^2 = 25$ 2. $\dfrac{x^{-4}}{y^{-2}} = \dfrac{y^2}{x^4}$ 3. $\left(\dfrac{2}{3}\right)^{-3} = \left(\dfrac{3}{2}\right)^3 = \dfrac{3^3}{2^3} = \dfrac{27}{8}$
Scientific Notation	Write a as $b \times 10^n$, where $1 \le |b| < 10$ and n is an integer.	$23{,}500 = 2.35 \times 10^4$, $0.0056 = 5.6 \times 10^{-3}$, and $1000 = 1 \times 10^3$

12.5 Exercises   MyMathLab®

CONCEPTS AND VOCABULARY

Exercises 1–5: Complete the given rule for integer exponents m and n, where a and b are nonzero real numbers.

1. $a^{-n} = $ _____
2. $\dfrac{1}{a^{-n}} = $ _____
3. $\dfrac{a^m}{a^n} = $ _____
4. $\dfrac{a^{-n}}{b^{-m}} = $ _____
5. $\left(\dfrac{a}{b}\right)^{-n} = $ _____

6. To write a positive number a in scientific notation as $b \times 10^n$, the number b must satisfy _____.

USING NEGATIVE INTEGERS AS EXPONENTS

Exercises 7–18: Simplify the expression.

7. (a) 4^{-1} (b) $\left(\dfrac{1}{3}\right)^{-2}$
8. (a) 6^{-2} (b) 2.5^{-1}
9. (a) $2^3 \cdot 2^{-2}$ (b) $10^{-1} \cdot 10^{-2}$
10. (a) $3^{-4} \cdot 3^2$ (b) $10^4 \cdot 10^{-2}$
11. (a) $3^{-2} \cdot 3^{-1} \cdot 3^{-1}$ (b) $(2^3)^{-1}$
12. (a) $2^{-3} \cdot 2^5 \cdot 2^{-4}$ (b) $(3^{-2})^{-2}$
13. (a) $(3^2 4^3)^{-1}$ (b) $\dfrac{4^5}{4^2}$
14. (a) $(2^{-2} 3^2)^{-2}$ (b) $\dfrac{5^5}{5^3}$
15. (a) $\dfrac{1^9}{1^7}$ (b) $\dfrac{1}{4^{-3}}$
16. (a) $\dfrac{-6^4}{6}$ (b) $\dfrac{1}{6^{-2}}$
17. (a) $\dfrac{5^{-2}}{5^{-4}}$ (b) $\left(\dfrac{2}{7}\right)^{-2}$
18. (a) $\dfrac{7^{-3}}{7^{-1}}$ (b) $\left(\dfrac{3}{4}\right)^{-3}$

APPLYING RULES FOR INTEGER EXPONENTS

Exercises 19–50: Simplify the expression. Write the answer using positive exponents.

19. (a) x^{-1} (b) a^{-4}
20. (a) y^{-2} (b) z^{-7}
21. (a) $x^{-2} \cdot x^{-1} \cdot x$ (b) $a^{-5} \cdot a^{-2} \cdot a^{-1}$

22. (a) $y^{-3} \cdot y^4 \cdot y^{-5}$ (b) $b^5 \cdot b^{-3} \cdot b^{-6}$

23. (a) $x^2 y^{-3} x^{-5} y^6$ (b) $(xy)^{-3}$

24. (a) $a^{-2} b^{-6} b^3 a^{-1}$ (b) $(ab)^{-1}$

25. (a) $(2t)^{-4}$ (b) $(x+1)^{-7}$

26. (a) $(8c)^{-2}$ (b) $(a+b)^{-9}$

27. (a) $(a^{-2})^{-4}$ (b) $(rt^3)^{-2}$

28. (a) $(4x^3)^{-3}$ (b) $(xy^{-3})^{-2}$

29. (a) $(ab)^2 (a^2)^{-3}$ (b) $\dfrac{x^4}{x^2}$

30. (a) $(x^3)^{-2} (xy)^4 y^{-5}$ (b) $\dfrac{y^9}{y^5}$

31. (a) $\dfrac{a^{10}}{a^{-3}}$ (b) $\dfrac{4z}{2z^4}$

32. (a) $\dfrac{b^5}{b^{-2}}$ (b) $\dfrac{12x^2}{24x^7}$

33. (a) $\dfrac{-4xy^5}{6x^3 y^2}$ (b) $\dfrac{x^{-4}}{x^{-1}}$

34. (a) $\dfrac{12a^6 b^2}{8ab^3}$ (b) $\dfrac{y^{-2}}{y^{-7}}$

35. (a) $\dfrac{10b^{-4}}{5b^{-5}}$ (b) $\left(\dfrac{a}{b}\right)^3$

36. (a) $\dfrac{8a^{-2}}{2a^{-3}}$ (b) $\left(\dfrac{2x}{y}\right)^5$

37. (a) $\dfrac{6x^2 y^{-4}}{18x^{-5} y^4}$ (b) $\dfrac{16a^{-3} b^{-5}}{4a^{-8} b}$

38. (a) $\dfrac{m^2 n^4}{3m^{-5} n^4}$ (b) $\dfrac{7x^{-3} y^{-5}}{x^{-3} y^{-2}}$

39. (a) $\dfrac{1}{y^{-5}}$ (b) $\dfrac{4}{2t^{-3}}$

40. (a) $\dfrac{1}{z^{-6}}$ (b) $\dfrac{5}{10b^{-5}}$

41. (a) $\dfrac{3a^4}{(2a^{-2})^3}$ (b) $\dfrac{(2b^5)^{-3}}{4b^{-6}}$

42. (a) $\dfrac{(2x^4)^{-2}}{5x^{-2}}$ (b) $\dfrac{2y^5}{(3y^{-4})^{-2}}$

43. (a) $\dfrac{1}{(xy)^{-2}}$ (b) $\dfrac{1}{(a^2 b)^{-3}}$

44. (a) $\dfrac{1}{(ab)^{-1}}$ (b) $\dfrac{1}{(rt^4)^{-2}}$

45. (a) $\dfrac{(3m^4 n)^{-2}}{(2mn^{-2})^3}$ (b) $\dfrac{(-4x^4 y)^2}{(xy^{-5})^{-3}}$

46. (a) $\dfrac{(x^4 y^2)^2}{(-2x^2 y^{-2})^3}$ (b) $\dfrac{(m^2 n^{-6})^{-2}}{(4m^2 n^{-4})^{-3}}$

47. (a) $\left(\dfrac{a}{b}\right)^{-2}$ (b) $\left(\dfrac{u}{4v}\right)^{-1}$

48. (a) $\left(\dfrac{2x}{y}\right)^{-3}$ (b) $\left(\dfrac{5u}{3v}\right)^{-2}$

49. (a) $\left(\dfrac{3a^4 b}{2ab^{-2}}\right)^{-2}$ (b) $\left(\dfrac{4m^4 n}{5m^{-3} n^2}\right)^2$

50. (a) $\left(\dfrac{2x^4 y^2}{3x^3 y^{-3}}\right)^3$ (b) $\left(\dfrac{a^{-5} b^2}{2ab^{-2}}\right)^{-2}$

51. **Thinking Generally** For positive integers m and n, show that $\dfrac{a^n}{a^m} = \dfrac{1}{a^{m-n}}$.

52. **Thinking Generally** For positive integers m and n, show that $\dfrac{a^{-n}}{a^{-m}} = a^{m-n}$.

READING AND WRITING SCIENTIFIC NOTATION

Exercises 53–58: (Refer to Table 12.8.) Write the value of the power of 10 in words.

53. 10^3 54. 10^6

55. 10^9 56. 10^{-1}

57. 10^{-2} 58. 10^{-6}

Exercises 59–70: Write the expression in standard form.

59. -2×10^{-3} 60. -5×10^{-2}

61. 4.5×10^4 62. 7.1×10^6

63. 8×10^{-3} 64. 9×10^{-1}

65. 4.56×10^{-4} 66. 9.4×10^{-2}

67. 3.9×10^7 68. 5.27×10^6

69. -5×10^5 70. -9.5×10^3

Exercises 71–82: Write the number in scientific notation.

71. 2000   72. 11,000

73. -567 74. -9300

75. 12,000,000
76. 600,000
77. 0.004
78. 0.0008
79. 0.000895
80. 0.0123
81. −0.05
82. −0.934

Exercises 83–90: Evaluate the expression. Write the answer in standard form.

83. $(5 \times 10^3)(3 \times 10^2)$
84. $(2.1 \times 10^2)(2 \times 10^4)$
85. $(-3 \times 10^{-3})(5 \times 10^2)$
86. $(4 \times 10^2)(1 \times 10^3)(5 \times 10^{-4})$
87. $\dfrac{4 \times 10^5}{2 \times 10^2}$
88. $\dfrac{9 \times 10^2}{3 \times 10^6}$
89. $\dfrac{8 \times 10^{-6}}{4 \times 10^{-3}}$
90. $\dfrac{6.3 \times 10^2}{2 \times 10^{-3}}$

APPLICATIONS INVOLVING INTEGER EXPONENTS

91. **Light-year** The distance that light travels in 1 year is called a *light-year*. Light travels at 1.86×10^5 miles per second, and there are about 3.15×10^7 seconds in 1 year.
    (a) Estimate the number of miles in 1 light-year.
    (b) Except for the sun, Alpha Centauri is the nearest star, and its distance is 4.27 light-years from Earth. Estimate its distance in miles. Write your answer in scientific notation.

92. **Milky Way** It takes 2×10^8 years for the sun to make one orbit around the Milky Way galaxy. Write this number in standard form.

93. **Speed of the Sun** (Refer to the two previous exercises.) Assume that the sun's orbit in the Milky Way galaxy is circular with a diameter of 10^5 light-years. Estimate how many miles the sun travels in 1 year.

94. **Distance to the Moon** The moon is about 240,000 miles from Earth.
    (a) Write this number in scientific notation.
    (b) If a rocket traveled at 4×10^4 miles per hour, how long would it take for it to reach the moon?

95. **Online Exploration** In 1997, the creators of the Internet search engine BackRub renamed it Google. This new name is a play on the word *googol*, which is a very large number. Look up a googol and write it in scientific notation.

96. **Online Exploration** An astronomical unit (AU) is based on the distance from Earth to the sun. Look up the distance in kilometers from Earth to the sun.
    (a) Write an astronomical unit in standard form to the nearest million kilometers.
    (b) Convert your rounded answer from part (a) to scientific notation.

97. **Gross Domestic Product** The gross domestic product (GDP) is the total national output of goods and services valued at market prices *within* the United States. The GDP of the United States in 2013 was $15,684,000,000,000. (*Source:* World Bank.)
    (a) Write this number in scientific notation.
    (b) In 2013, the U.S. population was 3.17×10^8. On average, how many dollars of goods and services were produced by each individual?

98. **Average Household Income** Recently, there were 1.22×10^8 households in the United States and the average household income was $51,371. Calculate the total household income in the United States. (*Source:* U.S. Census Bureau.)

WRITING ABOUT MATHEMATICS

99. Explain what a negative exponent is and how it is different from a positive exponent. Give an example.

100. Explain why scientific notation is helpful for writing some numbers.

Group Activity

Water in a Lake East Battle Lake in Minnesota covers an area of about 1950 acres, or 8.5×10^7 square feet, and its average depth is about 3.2×10^1 feet.
(a) Estimate the cubic feet of water in the lake. (*Hint:* volume = area × average depth.)
(b) One cubic foot of water equals about 7.5 gallons. How many gallons of water are in this lake?
(c) The population of the United States is about 3.1×10^8, and the average American uses about 1.5×10^2 gallons of water per day. Could this lake supply the American population with water for 1 day?

12.6 Division of Polynomials

Objectives

1. Dividing by a Monomial
2. Dividing by a Polynomial

1 Dividing by a Monomial

To add two fractions with like denominators, we use the property

$$\frac{a}{d} + \frac{b}{d} = \frac{a+b}{d}.$$

For example, $\frac{1}{7} + \frac{3}{7} = \frac{1+3}{7} = \frac{4}{7}$.

To divide a polynomial by a monomial, we use the same property, only in reverse. That is,

$$\frac{a+b}{d} = \frac{a}{d} + \frac{b}{d}.$$

Note that each term in the numerator is divided by the monomial in the denominator. The next example shows how to divide a polynomial by a monomial.

EXAMPLE 1 Dividing a polynomial by a monomial

Divide.

(a) $\dfrac{a^5 + a^3}{a^2}$ (b) $\dfrac{5x^4 + 10x}{10x}$ (c) $\dfrac{3y^2 + 2y - 12}{6y}$

Solution

(a) $\dfrac{a^5 + a^3}{a^2} = \dfrac{a^5}{a^2} + \dfrac{a^3}{a^2} = a^{5-2} + a^{3-2} = a^3 + a$

(b) $\dfrac{5x^4 + 10x}{10x} = \dfrac{5x^4}{10x} + \dfrac{10x}{10x} = \dfrac{x^3}{2} + 1$

(c) $\dfrac{3y^2 + 2y - 12}{6y} = \dfrac{3y^2}{6y} + \dfrac{2y}{6y} - \dfrac{12}{6y} = \dfrac{y}{2} + \dfrac{1}{3} - \dfrac{2}{y}$

Now Try Exercises 17, 19, 21

MAKING CONNECTIONS 1

Division and Simplification

A common mistake made when dividing expressions is to "cancel" incorrectly. Note in Example 1(b) that

$$\frac{5x^4 + 10x}{10x} \neq 5x^4 + \frac{10x}{10x}.$$

We cannot "cancel" the $10x$ terms.

The monomial must be divided into *every* term in the numerator.

When dividing two natural numbers, we can check our work by multiplying. For example, $\frac{10}{5} = 2$, and we can check this result by finding the product $5 \cdot 2 = 10$. Similarly, to check

$$\frac{a^5 + a^3}{a^2} = a^3 + a$$

in Example 1(a), we can multiply a^2 and $a^3 + a$.

$$a^2(a^3 + a) = a^2 \cdot a^3 + a^2 \cdot a \quad \text{Distributive property}$$
$$= a^5 + a^3 \checkmark \quad \text{It checks.}$$

EXAMPLE 2 Dividing and checking

Divide the expression $\frac{8x^3 - 4x^2 + 6x}{2x^2}$ and then check the result.

Solution
Be sure to divide $2x^2$ into *every* term in the numerator.

$$\frac{8x^3 - 4x^2 + 6x}{2x^2} = \frac{8x^3}{2x^2} - \frac{4x^2}{2x^2} + \frac{6x}{2x^2} = 4x - 2 + \frac{3}{x}$$

Check:

$$2x^2\left(4x - 2 + \frac{3}{x}\right) = 2x^2 \cdot 4x - 2x^2 \cdot 2 + 2x^2 \cdot \frac{3}{x}$$
$$= 8x^3 - 4x^2 + 6x \checkmark$$

Now Try Exercise 13

EXAMPLE 3 Finding the length of a rectangle

The rectangle in **FIGURE 12.17** has an area $A = x^2 + 2x$ and width x. Write an expression for its length l in terms of x.

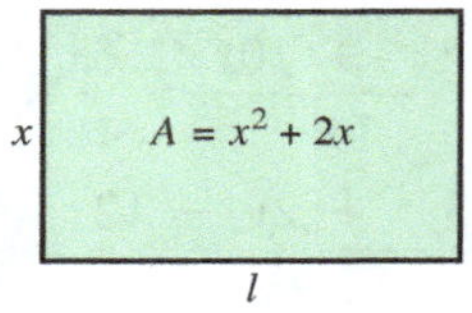

FIGURE 12.17

Solution

The area A of a rectangle equals length l times width w, or $A = lw$. Solving for l gives

$$l = \frac{A}{w}.$$

Thus to find the length of the given rectangle, divide the area by the width.

$$l = \frac{x^2 + 2x}{x} = \frac{x^2}{x} + \frac{2x}{x} = x + 2$$

The length of the rectangle is $x + 2$. The answer checks because $x(x + 2) = x^2 + 2x$.

Now Try Exercise 49

2 Dividing by a Polynomial

To understand division by a polynomial better, we first need to review some terminology related to long division of natural numbers. To compute $271 \div 4$, we complete long division as follows.

$$\begin{array}{r} \text{Quotient} \longrightarrow\ \ 67\text{ R }3 \longleftarrow \text{Remainder} \\ \text{Divisor} \longrightarrow 4\overline{)271} \longleftarrow \text{Dividend} \\ \underline{24}\ \ \ \ \ \\ 31\ \ \\ \underline{28}\ \ \\ 3\ \ \end{array}$$

To check this result, we find the product of the quotient and divisor and then add the remainder. Because $67 \cdot 4 + 3 = 271$, the answer checks. The quotient and remainder can also be expressed as $67\frac{3}{4}$. Division of polynomials is similar to long division of natural numbers.

EXAMPLE 4 Dividing polynomials

Divide $\frac{6x^2 + 13x + 3}{3x + 2}$.

Solution

Begin by dividing the first term of $3x + 2$ into the first term of $6x^2 + 13x + 3$. That is, divide $3x$ into $6x^2$ to obtain $2x$. Then find the product of $2x$ and $3x + 2$, or $6x^2 + 4x$, place it below $6x^2 + 13x$, and subtract. Bring down the 3.

$$\begin{array}{r} 2x \\ 3x + 2\overline{)6x^2 + 13x + 3} \\ \underline{6x^2 + 4x} \\ 9x + 3 \end{array} \quad \begin{array}{l} \frac{6x^2}{3x} = 2x \\ 2x(3x + 2) = 6x^2 + 4x \\ \text{Subtract: } 13x - 4x = 9x. \text{ Bring down the 3.} \end{array}$$

In the next step, divide $3x$ into the first term of $9x + 3$ to obtain 3. Then find the product of 3 and $3x + 2$, or $9x + 6$, place it below $9x + 3$, and subtract.

$$\begin{array}{r} 2x + 3 \\ 3x + 2\overline{)6x^2 + 13x + 3} \\ \underline{6x^2 + 4x} \\ 9x + 3 \\ \underline{9x + 6} \\ -3 \end{array} \quad \begin{array}{l} \frac{9x}{3x} = 3 \\ \\ \\ \\ 3(3x + 2) = 9x + 6 \\ \text{Subtract: } 3 - 6 = -3. \end{array}$$

The quotient is $2x + 3$ with remainder -3. This result can also be written as

$$2x + 3 + \frac{-3}{3x + 2}, \quad \text{Quotient} + \frac{\text{Remainder}}{\text{Divisor}}$$

in the same manner that 67 R 3 was written as $67\frac{3}{4}$.

Now Try Exercise 27

> **MAKING CONNECTIONS 2**
>
> **Checking Polynomial Division**
>
> Check polynomial division by adding the remainder to the product of the divisor and the quotient. That is,
>
> $$(\text{\color{blue}Divisor})(\text{\color{purple}Quotient}) + \text{\color{red}Remainder} = \text{\color{teal}Dividend}.$$
>
> For Example 4, the equation becomes
>
> $$(3x + 2)(2x + 3) + (-3) = 3x \cdot 2x + 3x \cdot 3 + 2 \cdot 2x + 2 \cdot 3 - 3$$
> $$= 6x^2 + 9x + 4x + 6 - 3$$
> $$= 6x^2 + 13x + 3. \checkmark \quad \text{It checks.}$$

EXAMPLE 5 Dividing polynomials having a missing term

Simplify $(3x^3 + 2x - 4) \div (x - 2)$.

Solution
Because the dividend does not have an x^2-term, insert $0x^2$ as a "place holder." Then begin by dividing x into $3x^3$ to obtain $3x^2$.

$$\begin{array}{r} 3x^2 \\ x - 2 \overline{\smash{\big)}\, 3x^3 + 0x^2 + 2x - 4} \\ \underline{3x^3 - 6x^2} \\ 6x^2 + 2x \end{array}$$

$\dfrac{3x^3}{x} = 3x^2$

$3x^2(x - 2) = 3x^3 - 6x^2$

Subtract $0x^2 - (-6x^2) = 6x^2$. Bring down $2x$.

In the next step, divide x into $6x^2$.

$$\begin{array}{r} 3x^2 + 6x \\ x - 2 \overline{\smash{\big)}\, 3x^3 + 0x^2 + 2x - 4} \\ \underline{3x^3 - 6x^2} \\ 6x^2 + 2x \\ \underline{6x^2 - 12x} \\ 14x - 4 \end{array}$$

$\dfrac{6x^2}{x} = 6x$

$6x(x - 2) = 6x^2 - 12x$

Subtract: $2x - (-12x) = 14x$. Bring down -4.

Now divide x into $14x$.

$$\begin{array}{r} 3x^2 + 6x + 14 \\ x - 2 \overline{\smash{\big)}\, 3x^3 + 0x^2 + 2x - 4} \\ \underline{3x^3 - 6x^2} \\ 6x^2 + 2x \\ \underline{6x^2 - 12x} \\ 14x - 4 \\ \underline{14x - 28} \\ 24 \end{array}$$

$\dfrac{14x}{x} = 14$

$14(x - 2) = 14x - 28$

Subtract: $-4 - (-28) = 24$.

The quotient is $3x^2 + 6x + 14$ with remainder 24. This result can also be written as

$$3x^2 + 6x + 14 + \frac{24}{x - 2}.$$

Now Try Exercise 37

EXAMPLE 6 **Dividing when the divisor is not linear**

Divide $x^3 - 3x^2 + 3x + 2$ by $x^2 + 1$.

Solution
Begin by writing $x^2 + 1$ as $x^2 + 0x + 1$.

$$\begin{array}{r} x - 3 \\ x^2 + 0x + 1 \overline{\smash{)}x^3 - 3x^2 + 3x + 2} \\ \underline{x^3 + 0x^2 + x} \\ -3x^2 + 2x + 2 \\ \underline{-3x^2 + 0x - 3} \\ 2x + 5 \end{array}$$

The quotient is $x - 3$ with remainder $2x + 5$. This result can also be written as

$$x - 3 + \frac{2x + 5}{x^2 + 1}.$$

Now Try Exercise 41

12.6 Putting It All Together

CONCEPT	EXPLANATION	EXAMPLES
Division by a Monomial	Use the property $$\frac{a + b}{d} = \frac{a}{d} + \frac{b}{d}.$$ Be sure to divide the denominator into every term in the numerator.	$\frac{2x^3 + 4x}{2x^2} = \frac{2x^3}{2x^2} + \frac{4x}{2x^2} = x + \frac{2}{x}$ and $\frac{a^2 - 2a}{4a} = \frac{a^2}{4a} - \frac{2a}{4a} = \frac{a}{4} - \frac{1}{2}$
Division by a Polynomial	Is done similarly to the way long division of natural numbers is performed. If either the divisor or the dividend is missing a term, be sure to insert as a "place holder" the missing term with coefficient 0.	Divide $x^2 + 3x + 3$ by $x + 1$. $$\begin{array}{r} x + 2 \\ x + 1 \overline{\smash{)}x^2 + 3x + 3} \\ \underline{x^2 + x} \\ 2x + 3 \\ \underline{2x + 2} \\ 1 \end{array}$$ The quotient is $x + 2$ with remainder 1, which can be expressed as $$x + 2 + \frac{1}{x + 1}.$$

continued on next page

12.6 Exercises

CONCEPT	EXPLANATION	EXAMPLES
Checking a Result	Dividend = (Divisor)(Quotient) + Remainder	When $x^2 + 3x + 3$ is divided by $x + 1$, the quotient is $x + 2$ with remainder 1. Thus $(x + 1)(x + 2) + 1 = x^2 + 3x + 3$, and the answer checks.

CONCEPTS AND VOCABULARY

1. $\frac{a + b}{d} =$ _____

2. $\frac{a + b - c}{d} =$ _____

3. When dividing a polynomial by a monomial, the monomial must be divided into every _____ of the polynomial.

4. (True or False?) The expressions $\frac{5x^2 + 2x}{2x}$ and $5x^2 + 1$ are equal.

5. (True or False?) The expressions $\frac{5x^2 + 2x}{2x}$ and $\frac{5x^2}{2x}$ are equal.

6. Because $\frac{37}{9} = 4$ with remainder 1, it follows that $37 =$ _____ · _____ + _____.

7. Because $2x^3 - x + 5$ divided by $x + 1$ equals $2x^2 - 2x + 1$ with remainder 4, it follows that $2x^3 - x + 5 =$ _____ · _____ + _____.

8. When dividing $2x^3 + 3x - 1$ by $x - 1$, insert _____ into the dividend as a "place holder" for the missing x^2-term.

DIVIDING BY A MONOMIAL

Exercises 9–16: Divide and check.

9. $\frac{6x^2}{3x}$

10. $\frac{-5x^2}{10x^4}$

11. $\frac{z^4 + z^3}{z}$

12. $\frac{t^3 - t}{t}$

13. $\frac{a^5 - 6a^3}{2a^3}$

14. $\frac{b^4 - 4b}{4b^2}$

15. $\frac{y + 6y^2}{3y^3}$

16. $\frac{8z^2 - z}{4z^2}$

Exercises 17–26: Divide.

17. $\frac{4x - 7x^4}{x^2}$

18. $\frac{1 + 6x^4}{3x^3}$

19. $\frac{6y^2 + 3y}{3y^3}$

20. $\frac{5z^2 - 10z^3}{5z^4}$

21. $\frac{9x^4 - 3x + 6}{3x}$

22. $\frac{y^3 - 4y + 6}{y}$

23. $\frac{12y^4 - 3y^2 + 6y}{3y^2}$

24. $\frac{2x^2 - 6x + 9}{12x}$

25. $\frac{15m^4 - 10m^3 + 20m^2}{5m^2}$

26. $\frac{n^8 - 8n^6 + 4n^4}{2n^5}$

DIVIDING BY A POLYNOMIAL

Exercises 27–34: Divide and check.

27. $\frac{2x^2 - 3x + 1}{x - 2}$

28. $\frac{4x^2 - x + 3}{x + 2}$

29. $\frac{x^2 + 2x + 1}{x + 1}$

30. $\frac{4x^2 - 4x + 1}{2x - 1}$

31. $\frac{x^3 - x^2 + x - 2}{x - 1}$

32. $\frac{2x^3 + 3x^2 + 3x - 1}{2x + 1}$

33. $\frac{x^3 + x^2 - 7x + 2}{x - 2}$

34. $\frac{x^3 + x^2 - 2x + 12}{x + 3}$

Exercises 35–46: Divide.

35. $\frac{4x^3 - 3x^2 + 7x + 3}{4x + 1}$

36. $\frac{10x^3 - x^2 - 17x - 7}{5x + 2}$

37. $\frac{x^3 - x + 2}{x - 2}$

38. $\frac{6x^3 + 8x^2 + 4}{3x + 4}$

39. $(3x^3 + 2) \div (x - 1)$

40. $(-3x^3 + 8x^2 + x) \div (3x + 4)$

41. $(x^3 + 3x^2 + 1) \div (x^2 + 1)$

42. $(x^4 - x^3 + x^2 - x + 1) \div (x^2 - 1)$

43. $\dfrac{x^3 + 1}{x^2 - x + 1}$

44. $\dfrac{4x^3 + 3x + 2}{2x^2 - x + 1}$

45. $\dfrac{x^3 + 8}{x + 2}$

46. $\dfrac{x^4 - 16}{x - 2}$

47. **Thinking Generally** If the quotient in a polynomial division problem is an integer, what must be true about the degrees of the dividend and divisor?

48. **Thinking Generally** If the quotient in a polynomial division problem is a polynomial of degree 1, what must be true about the degrees of the dividend and divisor?

APPLICATIONS INVOLVING DIVISION OF POLYNOMIALS

Exercises 49 and 50: **Area of a Rectangle** *The area of a rectangle and its width are given. Find an expression for the length l.*

49.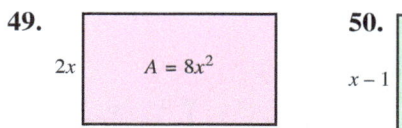

50.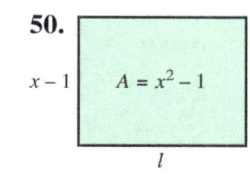

51. **Volume of a Box** The volume V of a box is $2x^3 + 4x^2$, and the area of its bottom is $2x^2$. Find the height of the box in terms of x. Make a possible sketch of the box, and label the length of each side.

52. **Area of a Triangle** A triangle has height h and area $A = 2h^2 - 4h$. Find its base b in terms of h. Make a possible sketch of the triangle, and label the height and base. (*Hint:* $A = \frac{1}{2}bh$.)

WRITING ABOUT MATHEMATICS

53. Suppose that one polynomial is divided into another polynomial and the remainder is 0. What does the product of the divisor and quotient equal? Explain.

54. A student simplifies the expression $\dfrac{4x^3 - 1}{4x^2}$ to $x - 1$. Explain the student's error.

SECTIONS 12.5 and 12.6 — **Checking Basic Concepts**

1. Simplify each expression. Write the result with positive exponents.
   (a) 9^{-2} (b) $\dfrac{3x^{-3}}{6x^4}$ (c) $(4ab^{-4})^{-2}$

2. Simplify each expression. Write the result with positive exponents.
   (a) $\dfrac{1}{z^{-5}}$ (b) $\dfrac{x^{-3}}{y^{-6}}$ (c) $\left(\dfrac{3}{x^2}\right)^{-3}$

3. Write each number in scientific notation.
   (a) 45,000   (b) 0.000234   (c) 0.01

4. Write each expression in standard form.
   (a) 4.71×10^4 (b) 6×10^{-3}

5. Simplify $\dfrac{25a^4 - 15a^3}{5a^3}$.

6. Divide $3x^2 - x - 4$ by $x - 1$. State the quotient and remainder.

7. Divide $x^4 + 2x^3 - 2x^2 - 5x - 2$ by $x^2 - 3$. State the quotient and remainder.

8. **Distance to the Sun** The distance to the sun is approximately 93 million miles.
   (a) Write this distance in scientific notation.
   (b) Light travels at 1.86×10^5 miles per second. How long does it take for the sun's light to reach Earth?

CHAPTER 12 Summary

SECTION 12.1 ▪ RULES FOR EXPONENTS

Bases and Exponents The expression b^n has base b and exponent n and equals the expression $\underbrace{b \cdot b \cdot b \cdot \cdots \cdot b}_{n \text{ times}}$, when n is a natural number.

Example: 2^3 has base 2 and exponent 3 and equals $2 \cdot 2 \cdot 2 = 8$.

Evaluating Expressions When evaluating expressions, use the following order of operations. First perform all calculations within parentheses and absolute values, or above and below a fraction bar.

1. Evaluate exponents.
2. Perform negation.
3. Do multiplication and division from left to right.
4. Do addition and subtraction from left to right.

Example: $-3^2 + 3 \cdot 4 = -9 + 3 \cdot 4 = -9 + 12 = 3$

Zero Exponents For any nonzero number b, $b^0 = 1$. Note that 0^0 is undefined.

Examples: $5^0 = 1$ and $\left(\dfrac{x}{y}\right)^0 = 1$, where x and y are nonzero.

Product Rule For any real number a and natural numbers m and n,
$$a^m \cdot a^n = a^{m+n}.$$

Examples: $3^4 \cdot 3^2 = 3^6$ and $x^3 x^2 x^4 = x^9$

Power Rules For any real numbers a and b and natural numbers m and n,
$$(a^m)^n = a^{mn}, \quad (ab)^n = a^n b^n, \quad \text{and} \quad \left(\dfrac{a}{b}\right)^n = \dfrac{a^n}{b^n}, \quad b \neq 0.$$

Examples: $(x^2)^3 = x^6$, $(3x)^4 = 3^4 x^4 = 81 x^4$, and $\left(\dfrac{2}{y}\right)^3 = \dfrac{2^3}{y^3} = \dfrac{8}{y^3}$

SECTION 12.2 ▪ ADDITION AND SUBTRACTION OF POLYNOMIALS

Terms Related to Polynomials

Monomial	A number, variable, or product of numbers and variables raised to natural number powers
Degree of a Monomial	Sum of the exponents of the variables
Coefficient of a Monomial	The number in a monomial

Example: The monomial $-3x^2 y^3$ has degree 5 and coefficient -3.

Polynomial	A monomial or the sum of two or more monomials
Term of a Polynomial	Each monomial is a term of the polynomial.
Binomial	A polynomial with two terms
Trinomial	A polynomial with three terms
Degree of a Polynomial	The degree of the term with highest degree
Opposite of a Polynomial	The opposite is found by negating each term.

Example: $2x^3 - 4x + 5$ is a trinomial with degree 3. Its opposite is $-2x^3 + 4x - 5$.

Like Terms   Two monomials with the same variables raised to the same powers

Examples: $3xy^2$ and $-xy^2$ are like terms.

$5x^3$ and $3x^3$ are like terms.

$5x^2$ and $5x$ are *unlike* terms.

Addition of Polynomials Combine like terms, using the distributive property.

Example: $(2x^2 - 4x) + (-x^2 - x) = (2 - 1)x^2 + (-4 - 1)x$
$= x^2 - 5x$

Subtraction of Polynomials Add the first polynomial to the opposite of the second polynomial.

Example: $(4x^4 - 5x) - (7x^4 + 6x) = (4x^4 - 5x) + (-7x^4 - 6x)$
$= (4 - 7)x^4 + (-5 - 6)x$
$= -3x^4 - 11x$

SECTION 12.3 ■ MULTIPLICATION OF POLYNOMIALS

Multiplication of Monomials Use the commutative property and the product rule.

Examples: $-2x^3 \cdot 3x^2 = -2 \cdot 3 \cdot x^3 \cdot x^2 = -6x^5$

$(2xy^2)(3x^2y^3) = 2 \cdot 3 \cdot x \cdot x^2 \cdot y^2 \cdot y^3 = 6x^3y^5$

         ↑ Assumed exponent of 1

Distributive Properties

$$a(b + c) = ab + ac \quad \text{and} \quad a(b - c) = ab - ac$$

Examples: $4x(3x + 6) = 4x \cdot 3x + 4x \cdot 6 = 12x^2 + 24x$

$ab(a^2 - b^2) = ab \cdot a^2 - ab \cdot b^2 = a^3b - ab^3$

Multiplication of Monomials and Polynomials Apply the distributive properties. Be sure to multiply every term in the polynomial by the monomial.

Example: $-2x^2(4x^2 - 5x - 3) = -8x^4 + 10x^3 + 6x^2$

Multiplication of Polynomials The product of two polynomials may be found by multiplying every term in the first polynomial by every term in the second polynomial. Be sure to combine like terms.

Examples: $(x + 3)(2x - 5) = 2x^2 - 5x + 6x - 15$
$= 2x^2 + x - 15$

$(2x + 1)(x^2 - 5x + 2) = 2x^3 - 10x^2 + 4x + x^2 - 5x + 2$
$= 2x^3 - 9x^2 - x + 2$

SECTION 12.4 ■ SPECIAL PRODUCTS

Product of a Sum and Difference

$$(a + b)(a - b) = a^2 - b^2$$

Examples: $(x + 4)(x - 4) = x^2 - 16$

$(2r - 3t)(2r + 3t) = (2r)^2 - (3t)^2 = 4r^2 - 9t^2$

Squaring Binomials

$$(a + b)^2 = a^2 + 2ab + b^2 \quad \text{and} \quad (a - b)^2 = a^2 - 2ab + b^2$$

Examples: $(2x + 1)^2 = (2x)^2 + 2(2x)1 + 1^2 = 4x^2 + 4x + 1$

$(z^2 - 2)^2 = (z^2)^2 - 2z^2(2) + 2^2 = z^4 - 4z^2 + 4$

Cubing Binomials
To multiply $(a + b)^3$, write it as $(a + b)(a + b)^2$.

Example:
$$\begin{aligned}
(x + 4)^3 &= (x + 4)(x + 4)^2 \\
&= (x + 4)(x^2 + 8x + 16) && \text{Square the binomial.} \\
&= x^3 + 8x^2 + 16x + 4x^2 + 32x + 64 && \text{Distributive property} \\
&= x^3 + 12x^2 + 48x + 64 && \text{Combine like terms.}
\end{aligned}$$

SECTION 12.5 ■ INTEGER EXPONENTS AND THE QUOTIENT RULE

Negative Integers as Exponents For any nonzero real number a and positive integer n,

$$a^{-n} = \frac{1}{a^n}.$$

Examples: $5^{-2} = \dfrac{1}{5^2}$ and $x^{-4} = \dfrac{1}{x^4}$

The Quotient Rule For any nonzero real number a and integers m and n,

$$\frac{a^m}{a^n} = a^{m-n}.$$

Examples: $\dfrac{6^4}{6^2} = 6^{4-2} = 6^2 = 36$ and $\dfrac{xy^3}{x^4y^2} = x^{1-4}y^{3-2} = x^{-3}y^1 = \dfrac{y}{x^3}$

Other Rules For any nonzero real numbers a and b and positive integers m and n,

$$\frac{1}{a^{-n}} = a^n, \quad \frac{a^{-n}}{b^{-m}} = \frac{b^m}{a^n}, \quad \text{and} \quad \left(\frac{a}{b}\right)^{-n} = \left(\frac{b}{a}\right)^n.$$

Examples: $\dfrac{1}{4^{-3}} = 4^3$, $\dfrac{x^{-3}}{y^{-2}} = \dfrac{y^2}{x^3}$, and $\left(\dfrac{4}{5}\right)^{-2} = \left(\dfrac{5}{4}\right)^2$

Scientific Notation A real number a written as $b \times 10^n$, where $1 \leq |b| < 10$ and n is an integer

Examples: $2.34 \times 10^3 = 2340$ Move the decimal point 3 places to the right.

$2.34 \times 10^{-3} = 0.00234$ Move the decimal point 3 places to the left.

SECTION 12.6 ■ DIVISION OF POLYNOMIALS

Division of a Polynomial by a Monomial Divide the monomial into *every* term of the polynomial.

Example: $\dfrac{5x^3 - 10x^2 + 15x}{5x} = \dfrac{5x^3}{5x} - \dfrac{10x^2}{5x} + \dfrac{15x}{5x} = x^2 - 2x + 3$

Division of a Polynomial by a Polynomial Division of polynomials is performed similarly to long division of natural numbers.

Example: Divide $2x^3 + 4x^2 - 3x + 1$ by $x + 1$.

$$\begin{array}{r}
2x^2 + 2x - 5 \\
x+1\overline{\smash{\big)}2x^3 + 4x^2 - 3x + 1}\\
\underline{2x^3 + 2x^2}\\
2x^2 - 3x\\
\underline{2x^2 + 2x}\\
-5x + 1\\
\underline{-5x - 5}\\
6
\end{array}$$

The quotient is $2x^2 + 2x - 5$ with remainder 6, which can be written as

$$2x^2 + 2x - 5 + \frac{6}{x+1}.$$

CHAPTER 12 Review Exercises

SECTION 12.1

Exercises 1–6: Evaluate the expression.

1. 5^3
2. -3^4
3. $4(-2)^0$
4. $3 + 3^2 - 3^0$
5. $\dfrac{-5^2}{5}$
6. $\left(\dfrac{-5}{5}\right)^2$

Exercises 7–24: Simplify the expression.

7. $6^2 \cdot 6^3$
8. $10^5 \cdot 10^7$
9. $z^4 \cdot z^5$
10. $y^2 \cdot y \cdot y^3$
11. $5x^2 \cdot 6x^7$
12. $(ab^3)(a^3b)$
13. $(2^5)^2$
14. $(m^4)^5$
15. $(ab)^3$
16. $(x^2y^3)^4$
17. $(xy)^3(x^2y^4)^2$
18. $(a^2b^9)^0$
19. $(r-t)^4(r-t)^5$
20. $(a+b)^2(a+b)^4$
21. $\left(\dfrac{3}{x-y}\right)^2$
22. $\left(\dfrac{x+y}{2}\right)^3$
23. $2x^2(3x - 5)$
24. $3x(4x + x^3)$

SECTION 12.2

Exercises 25 and 26: Identify the degree and coefficient of the monomial.

25. $6x^7$
26. $-x^2y^3$

Exercises 27–30: Determine whether the expression is a polynomial. If it is, state how many terms and variables the polynomial contains. Then state its degree.

27. $8y$
28. $8x^3 - 3x^2 + x - 5$
29. $a^2 + 2ab + b^2$
30. $\dfrac{1}{xy}$

31. Add the polynomials vertically.

$$\begin{array}{r} 3x^2 + 4x + 8 \\ \underline{2x^2 - 5x - 5} \end{array}$$

32. Write the opposite of $6x^2 - 3x - 7$.

Exercises 33–40: Simplify.

33. $(4x - 3) + (-x + 7)$
34. $(3x^2 - 1) - (5x^2 + 12)$
35. $(x^2 + 5x + 6) - (3x^2 - 4x + 1)$
36. $(x^2 + 3x - 5) + (2x^2 - 5x - 1)$

37. $(a^3 + 4a^2) + (a^3 - 5a^2 + 7a)$

38. $(4x^3 - 2x + 6) - (4x^3 - 6)$

39. $(xy + y^2) + (4y^2 - 4xy)$

40. $(7x^2 + 2xy + y^2) - (7x^2 - 2xy + y^2)$

SECTION 12.3

Exercises 41–54: Multiply and simplify.

41. $-x^2 \cdot x^3$
42. $-(r^2t^3)(rt)$
43. $-3(2t - 5)$
44. $2y(1 - 6y)$
45. $6x^3(3x^2 + 5x)$
46. $-x(x^2 - 2x + 9)$
47. $-ab(a^2 - 2ab + b^2)$
48. $(a - 2)(a + 5)$
49. $(8x - 3)(x + 2)$
50. $(2x - 1)(1 - x)$
51. $(y^2 + 1)(2y + 1)$
52. $(y^2 - 1)(2y^2 + 1)$
53. $(z + 1)(z^2 - z + 1)$
54. $(4z - 3)(z^2 - 3z + 1)$

Exercises 55 and 56: Multiply the expression
(a) *geometrically and*
(b) *symbolically.*

55. $z(z + 1)$
56. $2x(x + 2)$

SECTION 12.4

Exercises 57–72: Multiply.

57. $(z + 2)(z - 2)$
58. $(5z - 9)(5z + 9)$
59. $(1 - 3y)(1 + 3y)$
60. $(5x + 4y)(5x - 4y)$
61. $(rt + 1)(rt - 1)$
62. $(2m^2 - n^2)(2m^2 + n^2)$
63. $(x + 1)^2$
64. $(4x + 3)^2$
65. $(y - 3)^2$
66. $(2y - 5)^2$
67. $(4 + a)^2$
68. $(4 - a)^2$
69. $(x^2 + y^2)^2$
70. $(xy - 2)^2$
71. $(z + 5)^3$
72. $(2z - 1)^3$

Exercises 73 and 74: Use the product of a sum and a difference to evaluate the expression.

73. $59 \cdot 61$
74. $22 \cdot 18$

SECTION 12.5

Exercises 75–82: Simplify the expression.

75. 9^{-1}
76. 3^{-2}
77. $4^3 \cdot 4^{-2}$
78. $10^{-6} \cdot 10^3$
79. $\dfrac{1}{6^{-2}}$
80. $\dfrac{5^7}{5^9}$
81. $(3^{-1} 2^2)^{-2}$
82. $(2^{-4} 5^3)^0$

Exercises 83–98: Simplify the expression. Write the answer using positive exponents.

83. z^{-2}
84. y^{-4}
85. $a^{-4} \cdot a^2$
86. $x^2 \cdot x^{-5} \cdot x$
87. $(2t)^{-2}$
88. $(ab^2)^{-3}$
89. $(xy)^{-2}(x^{-2}y)^{-1}$
90. $\dfrac{x^6}{x^2}$
91. $\dfrac{4x}{2x^4}$
92. $\dfrac{20x^5y^3}{30xy^6}$
93. $\left(\dfrac{a}{b}\right)^5$
94. $\dfrac{4}{t^{-4}}$
95. $\dfrac{(3m^3n)^{-2}}{(2m^2n^{-3})^3}$
96. $\left(\dfrac{x^{-4}y^2}{3xy^{-3}}\right)^{-2}$
97. $\left(\dfrac{x}{y}\right)^{-2}$
98. $\left(\dfrac{3u}{2v}\right)^{-1}$

Exercises 99–102: Write the expression in standard form.

99. 6×10^2
100. 5.24×10^4
101. -3.7×10^{-3}
102. -6.234×10^{-2}

Exercises 103–106: Write the number in scientific notation.

103. 10,000
104. 56,100,000
105. 0.000054
106. 0.001

Exercises 107 and 108: Evaluate the expression. Write the result in standard form.

107. $(4 \times 10^2)(6 \times 10^4)$
108. $\dfrac{8 \times 10^3}{4 \times 10^4}$

SECTION 12.6

Exercises 109–116: Divide and check.

109. $\dfrac{5x^2 + 3x}{3x}$
110. $\dfrac{6b^4 - 4b^2 + 2}{2b^2}$
111. $\dfrac{3x^2 - x + 2}{x - 1}$
112. $\dfrac{9x^2 - 6x - 2}{3x + 2}$

113. $\dfrac{4x^3 - 11x^2 - 7x - 1}{4x + 1}$

114. $\dfrac{2x^3 - x^2 - 1}{2x - 1}$

115. $\dfrac{x^3 - x^2 - x + 1}{x^2 + 1}$

116. $\dfrac{x^4 + 3x^3 + 8x^2 + 7x + 5}{x^2 + x + 1}$

APPLICATIONS

117. **Heart Rate** An athlete starts running and continues for 10 seconds. The polynomial $\frac{1}{2}t^2 + 60$ calculates the heart rate of the athlete in beats per minute t seconds after beginning the run, where $t \leq 10$.
    (a) What is the athlete's heart rate when the athlete first starts to run?
    (b) What is the athlete's heart rate after 10 seconds?
    (c) What happens to the athlete's heart rate while the athlete is running?

118. **Areas of Rectangles** Find a monomial equal to the sum of the areas of the rectangles. Calculate this sum for $x = 3$ feet and $y = 4$ feet.

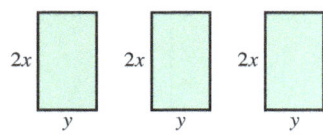

119. **Area of a Rectangle** Write a polynomial that gives the area of the rectangle. Calculate its area for $z = 6$ inches.

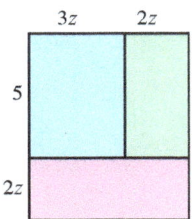

120. **Area of a Square** Find the area of the square whose sides have length x^2y.

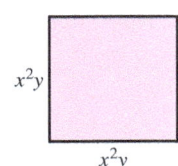

121. **Media Consumption** Recently, there were 240 million people in the United States over the age of 18. They spent an average of 11 hours per day with electronic media. Use scientific notation to estimate the total number of hours of media consumed by this group.

122. **Volume of a Sphere** The expression for the volume of a sphere with radius r is $\frac{4}{3}\pi r^3$. Find a polynomial that gives the volume of a sphere with radius $x + 2$. Leave your answer in terms of π.

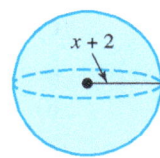

123. **Height Reached by a Baseball** A baseball is hit straight up. Its height h in feet above the ground after t seconds is given by $t(96 - 16t)$.
    (a) Multiply this expression.
    (b) Evaluate both the expression in part (a) and the given expression for $t = 2$. Interpret the result.

124. **Rectangular Building** A rectangular building has a perimeter of 1200 feet.
    (a) If one side of the building has length L, write a polynomial expression that gives its area. (Be sure to multiply your expression.)
    (b) Evaluate the expression in part (a) for $L = 50$ and interpret the answer.

125. **Geometry** Complete each part and verify that your answers are equal.
    (a) Find the area of the large square by multiplying its length and width.
    (b) Find the sum of the areas of the smaller rectangles inside the large square.

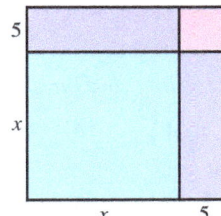

126. **Digital Picture** A digital picture, including its border, is $x + 4$ pixels by $x + 4$ pixels, and the actual picture inside the border is $x - 4$ pixels by $x - 4$ pixels.

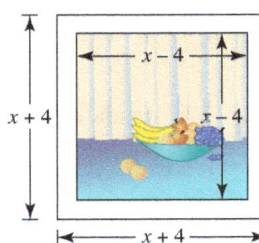

(a) Find a polynomial that gives the number of pixels in the border.
(b) Let $x = 100$ and evaluate the polynomial.

CHAPTER 12 Test

1. Simplify each expression.
   (a) -5^0 (b) -9^2

2. Evaluate each expression by hand.
   (a) $-4^2 + 10$ (b) 8^{-2}
   (c) $\dfrac{1}{2^{-3}}$ (d) $-3x^0$

3. State how many terms and variables the polynomial $5x^2 - 3xy - 7y^3$ contains. Then state its degree.

4. Write the opposite of $-x^3 + 4x - 8$.

Exercises 5–8: Simplify.

5. $(-3x + 4) + (7x + 2)$

6. $(y^3 - 2y + 6) - (4y^3 + 5)$

7. $(5x^2 - x + 3) - (4x^2 - 2x + 10)$

8. $(a^3 + 5ab) + (3a^3 - 3ab)$

Exercises 9–16: Write the given expression with positive exponents.

9. $6y^4 \cdot 4y^7$ 10. $(a^2b^3)^2(ab^2)$

11. $x^7 \cdot x^{-3}$ 12. $(a^{-1}b^2)^{-3}$

13. $ab(a^2 - b^2)$ 14. $\left(\dfrac{3a^2}{2b^{-3}}\right)^{-2}$

15. $\dfrac{12xy^4}{6x^2y}$ 16. $\left(\dfrac{2}{a+b}\right)^4$

Exercises 17–22: Multiply and simplify.

17. $3x^2(4x^3 - 6x + 1)$ 18. $(z - 3)(2z + 4)$

19. $(7y^2 - 3)(7y^2 + 3)$ 20. $(3x - 2)^2$

21. $(m + 3)^3$

22. $(y + 2)(y^2 - 2y + 3)$

23. Evaluate $78 \cdot 82$ using the product of a sum and a difference.

24. Write 6.1×10^{-3} in standard form.

25. Write 5410 in scientific notation.

Exercises 26 and 27: Divide.

26. $\dfrac{9x^3 - 6x^2 + 3x}{3x^2}$ 27. $\dfrac{x^3 + x^2 - x + 1}{x + 2}$

28. **Concert Tickets** Tickets for a concert are sold for $20 each.
    (a) Write a polynomial that gives the revenue from selling t tickets.
    (b) Putting on the concert costs management $2000 to hire the band plus $2 for each ticket sold. What is the total cost of the concert if t tickets are sold?
    (c) Subtract the polynomial that you found in part (b) from the polynomial that you found in part (a). What does the resulting polynomial represent?

29. **Areas of Rectangles** Find a polynomial representing the sum of the areas of two identical rectangles that have width $2x$ and length $3x$. Calculate this sum for $x = 10$ feet.

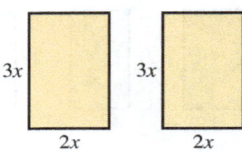

30. **Volume of a Box** Write a polynomial that represents the volume of the box. Be sure to multiply your answer completely.

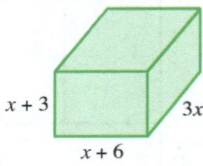

31. **Height Reached by a Golf Ball** When a golf ball is hit into the air, its height in feet above the ground after t seconds is given by $t(88 - 16t)$.
    (a) Multiply this expression.
    (b) Evaluate the expression in part (a) for $t = 3$. Interpret the result.

CHAPTERS 1–12   Cumulative Review Exercises

Exercises 1 and 2: Evaluate each expression by hand.

1. (a) $18 - 2 \cdot 5$ (b) $42 \div 7 + 2$

2. (a) $21 - (-8)$ (b) $-\frac{7}{3} \div \left(-\frac{14}{9}\right)$

Exercises 3 and 4: Solve the equation. Note that these equations may have no solutions, one solution, or infinitely many solutions.

3. (a) $(x - 3) + x = 4 + x$
   (b) $2(5x - 4) = 1 + 10x$

4. (a) $2 + 6x = 2(3x + 1)$
   (b) $11x - 9 = -31$

5. Find the average speed of a car that travels 306 miles in 4 hours 30 minutes.

6. Write each value as a fraction in lowest terms.
   (a) 42%   (b) 0.076

7. Graph the equation $4x - 5y = 20$.

8. Sketch a line with slope $-\frac{2}{3}$ that passes through the point $(1, 1)$.

9. Write the slope–intercept form for the line shown.

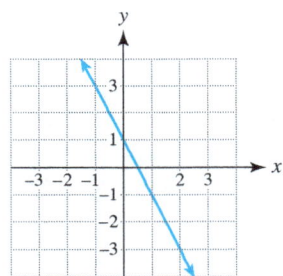

10. Find the x- and y-intercepts for the graph of the equation $2y = 3x - 6$.

Exercises 11 and 12: Write the slope–intercept form of a line that satisfies the given information.

11. Parallel to $3x - 6y = 7$, passing through $(2, -3)$

12. Passing through $(-2, -5)$ and $(1, 4)$

Exercises 13–16: Solve the system of equations. Note that these systems may have no solutions, one solution, or infinitely many solutions.

13. $4x + 3y = -6$
    $8x + 6y = 12$

14. $x - 3y = 5$
    $3x + y = 5$

15. $x + 4y = -8$
    $-3x - 12y = 24$

16. $x - 5y = 30$
    $2x + y = -6$

Exercises 17 and 18: Shade the solution set for the system of inequalities.

17. $x + y < 3$
    $y \geq x + 2$

18. $x - 2y > 4$
    $3x + y < 6$

19. Simplify the expression.
    (a) $3x^2 \cdot 5x^3$ (b) $(x^3 y)^2 (x^4 y^5)$

20. Simplify.
    (a) $(5x^2 - 3x + 4) - (3x^2 - 2x + 1)$
    (b) $(7a^3 - 4a^2 - 5) + (5a^3 + 4a^2 + a)$

21. Multiply and simplify.
    (a) $(2x + 3)(x - 7)$
    (b) $(y + 3)(y^2 - 3y - 1)$
    (c) $(4x + 7)(4x - 7)$
    (d) $(5a + 3)^2$

22. Simplify the expression. Write the answer using positive exponents.
    (a) $x^{-5} \cdot x^3 \cdot x$ (b) $\left(\dfrac{2}{x^3}\right)^{-3}$
    (c) $\dfrac{3x^2 y^{-1}}{6x^{-2} y}$ (d) $(xy^{-2})^3 (x^{-2} y)^{-2}$

23. Write 24,000,000,000 in scientific notation.

24. Write 4.71×10^{-7} in standard form.

25. Divide.
    (a) $\dfrac{8x^3 - 2x}{2x}$ (b) $\dfrac{2x^2 + x - 14}{x + 3}$

26. **Price Decrease** If the price of a computer is reduced from \$1200 to \$900, find the percent change.

27. **Mixing an Acid Solution** How many milliliters of a 3% acid solution should be added to 400 milliliters of a 6% acid solution to dilute it to a 5% acid solution?

28. **Surface Area of a Box** Use the drawing of the box to write a polynomial that represents the area of each of the following.

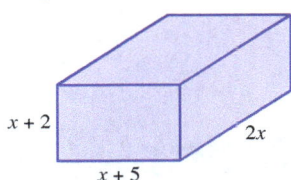

(a) The bottom
(b) The front
(c) The right side
(d) All six sides

13 Factoring Polynomials and Solving Equations

- 13.1 Introduction to Factoring
- 13.2 Factoring Trinomials I ($x^2 + bx + c$)
- 13.3 Factoring Trinomials II ($ax^2 + bx + c$)
- 13.4 Special Types of Factoring
- 13.5 Summary of Factoring
- 13.6 Solving Equations by Factoring I (Quadratics)
- 13.7 Solving Equations by Factoring II (Higher Degree)

Most cars are designed so that their exteriors are curved and smooth. This characteristic is especially true for solar cars. In fact, a side view of a solar car often resembles the cross section of an airplane wing, as illustrated in the accompanying figure. This design reduces drag from air resistance and increases fuel efficiency. As a result, solar cars are capable of traveling at speeds as high as 100 kilometers per hour.

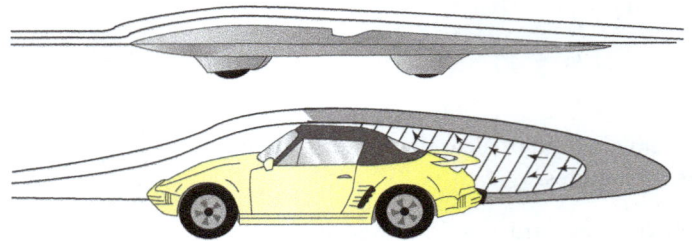

Views of a Solar Car and a Standard Car

Mathematics plays an important role in the design of cars. A special type of third-degree polynomial is used extensively by engineers to model smooth shapes for new cars. In this chapter, we introduce many concepts necessary for understanding polynomials and polynomial equations. *Source:* R. Burden and J. Faires. *Numerical Analysis.*

13.1 Introduction to Factoring

Objectives

1. Introducing Factoring
2. Finding Common Factors
3. Finding the Greatest Common Factor
4. Factoring by Grouping

NEW VOCABULARY

☐ Factoring a polynomial
☐ Greatest common factor (GCF)

1 Introducing Factoring

When two or more numbers are multiplied, each number is called a *factor*. For example, in the equation $4 \cdot 6 = 24$, the numbers 4 and 6 are factors and we say that 24 can be *factored* into the product $4 \cdot 6$. Other ways to factor 24 are shown below.

Factors of 24

$2 \cdot 12$, $3 \cdot 8$, $2 \cdot 2 \cdot 6$,   and $2 \cdot 2 \cdot 2 \cdot 3$

When we factor a positive number, we reverse the multiplication process and write the number as a product of two or more smaller numbers. Similarly, we **factor a polynomial** by writing the polynomial as a product of two or more *lower-degree* polynomials. For example, possible ways to factor the polynomial $12x^5 + 6x^4$ are shown below.

Factors of $12x^5 + 6x^4$

$2x(6x^4 + 3x^3)$, $x^2(12x^3 + 6x^2)$,   and $6x^4(2x + 1)$

The second of these factorizations shows that the **fifth**-degree polynomial $12x^5 + 6x^4$ can be written as the product of the **second**-degree polynomial x^2 and the **third**-degree polynomial $12x^3 + 6x^2$, where each factor has a lower degree than the given polynomial. Similar statements can be made about the first and third factorizations.

2 Finding Common Factors

When factoring a polynomial, we first look for factors that are common to each term. By applying the distributive property, we can often write a polynomial as a product. For example, each term in the polynomial $8x^2 + 6x$ has a factor of $2x$ because

$$8x^2 = 2x \cdot 4x \quad \text{and} \quad 6x = 2x \cdot 3.$$

Therefore by the distributive property,

$$8x^2 + 6x = 2x(4x + 3).$$

READING CHECK 1

• What property allows us to write a polynomial as a product?

Thus the product $2x(4x + 3)$ equals $8x^2 + 6x$. We check this result by multiplying.

$$2x(4x + 3) = 2x \cdot 4x + 2x \cdot 3$$
$$= 8x^2 + 6x \checkmark \quad \text{It checks.}$$

This factorization is shown visually in **FIGURE 13.1**, where possible dimensions for a rectangle with an area of $8x^2 + 6x$ are $2x$ by $4x + 3$.

Geometric Illustration of Factoring:
$8x^2 + 6x = 2x(4x + 3)$

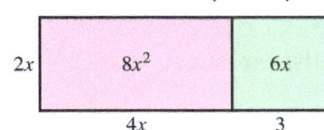

FIGURE 13.1

CALCULATOR HELP

To make a table of values, see Appendix A (pages AP-2 and AP-3).

The expressions $8x^2 + 6x$ and $2x(4x + 3)$ are equal for all values of x. **FIGURE 13.2** illustrates this fact with a partial table of values for each expression.

Numerical Illustration: $8x^2 + 6x = 2x(4x + 3)$

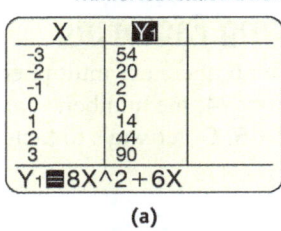
(a)

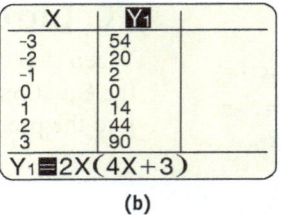
(b)

FIGURE 13.2

EXAMPLE 1   Factoring an expression

Factor the expression and sketch a rectangle that illustrates the factorization.
(a) $10x + 6$ (b) $6x^2 + 15x$

Solution
(a) Each term in the polynomial $10x + 6$ has a factor of 2 because

$$10x = 2 \cdot 5x \quad \text{and} \quad 6 = 2 \cdot 3.$$

Common factor is 2.

By the distributive property, $10x + 6 = 2(5x + 3)$. This factorization is illustrated visually in **FIGURE 13.3(a)**.

(b) Each term in the polynomial $6x^2 + 15x$ has a factor of $3x$ because

$$6x^2 = 3x \cdot 2x \quad \text{and} \quad 15x = 3x \cdot 5.$$

Common factor is $3x$.

By the distributive property, $6x^2 + 15x = 3x(2x + 5)$. This factorization is illustrated visually in **FIGURE 13.3(b)**.

Geometric Illustrations of Factoring:

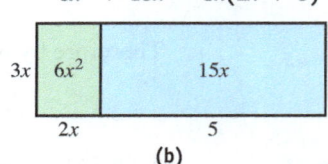

FIGURE 13.3

Now Try Exercises 9, 11

EXAMPLE 2   Finding common factors

Factor.
(a) $15x^2 + 10x$ (b) $6y^3 - 2y^2$ (c) $3z^3 + 9z^2 - 6z$ (d) $2x^2y^2 + 4xy^3$

Solution
(a) In the expression $15x^2 + 10x$, the terms $15x^2$ and $10x$ contain a common factor of $5x$.

$$15x^2 = 5x \cdot 3x \quad \text{and} \quad 10x = 5x \cdot 2$$

Therefore this polynomial can be factored as

$$15x^2 + 10x = 5x(3x + 2).$$

(b) In the expression $6y^3 - 2y^2$, the terms $6y^3$ and $2y^2$ contain a common factor of $2y^2$.

$$6y^3 = 2y^2 \cdot 3y \quad \text{and} \quad 2y^2 = 2y^2 \cdot 1$$

Therefore this polynomial can be factored as

$$6y^3 - 2y^2 = 2y^2(3y - 1).$$

(c) In $3z^3 + 9z^2 - 6z$, the terms $3z^3$, $9z^2$, and $6z$ contain a common factor of $3z$.

$$3z^3 = 3z \cdot z^2, \quad 9z^2 = 3z \cdot 3z, \quad \text{and} \quad 6z = 3z \cdot 2$$

Therefore this polynomial can be factored as

$$3z^3 + 9z^2 - 6z = 3z(z^2 + 3z - 2).$$

(d) In $2x^2y^2 + 4xy^3$, the terms $2x^2y^2$ and $4xy^3$ contain a common factor of $2xy^2$.

$$2x^2y^2 = 2xy^2 \cdot x \quad \text{and} \quad 4xy^3 = 2xy^2 \cdot 2y$$

Thus $2x^2y^2 + 4xy^3 = 2xy^2(x + 2y)$.

Now Try Exercises 15, 17, 19, 21

READING CHECK 2

- What is the greatest common factor of $18b^4 + 12b^3$?

STUDY TIP

In general, the ability to factor polynomials is important in solving many types of equations. Factoring out the GCF is a first step in preparing a polynomial for other factoring techniques discussed in this chapter.

3 Finding the Greatest Common Factor

In most situations, we factor out the *greatest common factor* (GCF). For example, the polynomial $12b^3 + 8b^2$ has a common factor of $2b$. We could factor this polynomial as

$$12b^3 + 8b^2 = 2b(6b^2 + 4b).$$

However, we can factor out $4b^2$ instead.

$$12b^3 + 8b^2 = 4b^2(3b + 2)$$

Because $4b^2$ is the common factor with the greatest (integer) coefficient and highest degree, we say that $4b^2$ is the **greatest common factor** of $12b^3 + 8b^2$. In Examples 1 and 2, we factored out the greatest common factor for each expression.

The following See the Concept illustrates a method for factoring the greatest common factor (GCF) from a polynomial.

See the Concept 1   FACTORING OUT THE GCF

To factor the polynomial $12x^4 + 9xy^3$:

A Complete factorization

$$12x^4 = 2 \cdot 2 \cdot 3 \cdot x \cdot x \cdot x \cdot x$$
$$= 3x \cdot 4x^3$$

Common factors   Noncommon factors

A Complete factorization

$$9xy^3 = 3 \cdot 3 \cdot x \cdot y \cdot y \cdot y$$
$$= 3x \cdot 3y^3$$

Common factors   Noncommon factors

B GCF times the sum of the noncommon factors

$$12x^4 + 9xy^3 = 3x(4x^3 + 3y^3)$$

A Completely factor each term of the polynomial by writing its coefficient as the product of prime numbers and writing any powers of variables as repeated multiplication.

B The given polynomial can be factored into the product of the common factors (GCF) times the sum (or difference) of the noncommon factors.

WATCH VIDEO IN MML.

EXAMPLE 3 Finding the greatest common factor

Find the greatest common factor for each expression. Then factor the expression.
(a) $9x^2 + 6x$ (b) $4z^4 + 8z^2$ (c) $8a^2b^3 - 16a^3b^2$

Solution
(a) Because

$$9x^2 = 3 \cdot 3 \cdot x \cdot x \quad \text{and} \quad 6x = 3 \cdot 2 \cdot x,$$

the terms have common factors of **3** and x. The GCF is the product of these two factors, or $3 \cdot x = 3x$. Thus the expression $9x^2 + 6x$ can be factored as $3x(3x + 2)$.

(b) Because

$$4z^4 = 2 \cdot 2 \cdot z \cdot z \cdot z \cdot z \quad \text{and}$$
$$8z^2 = 2 \cdot 2 \cdot 2 \cdot z \cdot z,$$

the terms have common factors of **2, 2,** z, and z. The GCF is the product of these four factors, or $2 \cdot 2 \cdot z \cdot z = 4z^2$. Thus $4z^4 + 8z^2$ can be factored as $4z^2(z^2 + 2)$.

(c) Because

$$8a^2b^3 = 2 \cdot 2 \cdot 2 \cdot a \cdot a \cdot b \cdot b \cdot b \quad \text{and}$$
$$16a^3b^2 = 2 \cdot 2 \cdot 2 \cdot 2 \cdot a \cdot a \cdot a \cdot b \cdot b,$$

the terms have common factors of **2, 2, 2,** a, a, b, and b. The GCF is the product of these seven factors, or $2 \cdot 2 \cdot 2 \cdot a \cdot a \cdot b \cdot b = 8a^2b^2$. Here, $8a^2b^3 - 16a^3b^2$ can be factored as $8a^2b^2(b - 2a)$.

Now Try Exercises 23, 25, 37

NOTE: With practice, you may find that you can determine the GCF mentally without factoring each term, as was done in Example 3. ■

MAKING CONNECTIONS 1

Checking Common Factors with Multiplication

When factoring, we can check our work by multiplying. For example, if we are uncertain whether the equation

$$6y^3 - 2y^2 = 2y^2(3y - 1)$$

is correct, we can apply the distributive property to the right side of the above equation to obtain

$$2y^2(3y - 1) = 2y^2 \cdot 3y - 2y^2 \cdot 1$$
$$= 6y^3 - 2y^2. \checkmark \quad \text{It checks.}$$

🌐 **Math in Context** (Sports) In the next example, we factor an expression that occurs in an application involving the flight of a golf ball.

EXAMPLE 4 Modeling the flight of a golf ball

If a golf ball is hit upward at 66 feet per second (45 miles per hour), then its height in feet after t seconds is approximated by $66t - 16t^2$. Factor this expression.

Solution
The GCF for $66t$ and $16t^2$ is $2t$ because

$$66t = 2t \cdot 33 \quad \text{and} \quad 16t^2 = 2t \cdot 8t.$$

Therefore this polynomial can be factored as

$$66t - 16t^2 = 2t(33 - 8t).$$

Now Try Exercise 81

4 Factoring by Grouping

Factoring by grouping is a technique that makes use of the associative and distributive properties. The next example illustrates one step in this factoring technique.

EXAMPLE 5 **Factoring out binomials**

Factor.
(a) $5x(x + 3) + 6(x + 3)$ (b) $x^2(2x - 5) - 4x(2x - 5)$

Solution
(a) Each term in the expression $5x(x + 3) + 6(x + 3)$ contains the binomial $(x + 3)$. Therefore the distributive property can be used to factor this expression.

$$5x(x + 3) + 6(x + 3) = (5x + 6)(x + 3)$$

(b) Each term in $x^2(2x - 5) - 4x(2x - 5)$ contains the binomial $(2x - 5)$. Therefore the distributive property can be used to factor this expression.

$$x^2(2x - 5) - 4x(2x - 5) = (x^2 - 4x)(2x - 5)$$
$$= x(x - 4)(2x - 5) \quad \text{Factor out the GCF.}$$

Now Try Exercises 41, 45

Now consider the polynomial

> The middle arithmetic symbol is $-$, and it separates the polynomial into two binomials.

$$x^3 + x^2 + 2x + 2.$$

We can factor this polynomial by first grouping it into two binomials. The first binomial has a common factor of x^2, and the second binomial has a common factor of 2.

$$(x^3 + x^2) + (2x + 2) \quad \text{Associative property}$$
$$x^2(x + 1) + 2(x + 1) \quad \text{Factor out common factors.}$$
$$(x^2 + 2)(x + 1) \quad \text{Factor out } (x + 1).$$

When factoring by grouping, we factor out a common factor more than once.

The first step in factoring a *four-term* polynomial by grouping requires the use of the associative property to write the polynomial as the *sum* of two binomials. However, this property must be applied carefully to avoid sign errors. The next two examples illustrate that the middle arithmetic symbol ($+$ or $-$) in a four-term polynomial determines how the associative property is applied.

READING CHECK 3

- What factoring technique is sometimes used to factor four-term polynomials?

EXAMPLE 6 Factoring by grouping when the middle symbol is (+)

Factor each polynomial.
(a) $2x^3 - 4x^2 + 3x - 6$ (b) $3x + 3y + ax + ay$

Solution
(a) Use the associative property to write the polynomial as the *sum* of two binomials.

$$2x^3 - 4x^2 + 3x - 6 = (2x^3 - 4x^2) + (3x - 6) \quad \text{Associative property}$$
$$= 2x^2(x - 2) + 3(x - 2) \quad \text{Factor out common factors.}$$
$$= (2x^2 + 3)(x - 2) \quad \text{Factor out } (x - 2).$$

The middle arithmetic symbol is +.

(b) Group the polynomial into the *sum* of two binomials.

$$3x + 3y + ax + ay = (3x + 3y) + (ax + ay) \quad \text{Associative property}$$
$$= 3(x + y) + a(x + y) \quad \text{Factor out common factors.}$$
$$= (3 + a)(x + y) \quad \text{Factor out } (x + y).$$

The middle arithmetic symbol is +.

Now Try Exercises 47, 65

EXAMPLE 7 Factoring by grouping when the middle symbol is (−)

Factor each polynomial.
(a) $3y^3 - y^2 - 9y + 3$ (b) $z^3 + 4z^2 - 5z - 20$

Solution
(a) Begin by changing the middle subtraction to addition by adding the opposite of $9y$. Then apply the associative property to write the result as the *sum* of two binomials.

$$3y^3 - y^2 - 9y + 3 = 3y^3 - y^2 + (-9y) + 3 \quad \text{Add the opposite of } 9y.$$
$$= (3y^3 - y^2) + (-9y + 3) \quad \text{Associative property}$$
$$= y^2(3y - 1) - 3(3y - 1) \quad \text{Factor out common factors.}$$
$$= (y^2 - 3)(3y - 1) \quad \text{Factor out } (3y - 1).$$

The middle arithmetic symbol is −.

Note that in the third step, -3 was factored from the second binomial.

(b) Begin by changing the middle subtraction to addition by adding the opposite of $5z$. Then apply the associative property to write the result as the *sum* of two binomials.

$$z^3 + 4z^2 - 5z - 20 = z^3 + 4z^2 + (-5z) - 20 \quad \text{Add the opposite of } 5z.$$
$$= (z^3 + 4z^2) + (-5z - 20) \quad \text{Associative property}$$
$$= z^2(z + 4) - 5(z + 4) \quad \text{Factor out common factors.}$$
$$= (z^2 - 5)(z + 4) \quad \text{Factor out } (z + 4).$$

The middle arithmetic symbol is −.

Factor out −5.

Now Try Exercises 55, 61

When factoring some polynomials, it may be necessary to factor out the greatest common factor before completing other factoring techniques such as factoring by grouping. In the next example, the GCF is factored out before grouping is applied.

EXAMPLE 8 Factoring out the GCF before grouping

Completely factor each polynomial.
(a) $6x^3 - 12x^2 - 3x + 6$
(b) $2x^5 - 8x^4 + 6x^3 - 24x^2$

Solution
(a) The GCF of $6x^3 - 12x^2 - 3x + 6$ is 3, so factor out 3 before factoring the remaining polynomial by grouping.

$$\begin{align*}
6x^3 - 12x^2 - 3x + 6 &= 3(2x^3 - 4x^2 - x + 2) &&\text{Factor out the GCF.}\\
&= 3(2x^3 - 4x^2 + (-x) + 2) &&\text{Add the opposite of } x.\\
&= 3((2x^3 - 4x^2) + (-x + 2)) &&\text{Associative property}\\
&= 3(2x^2(x - 2) - 1(x - 2)) &&\text{Factor out common factors.}\\
&= 3(2x^2 - 1)(x - 2) &&\text{Factor out } (x - 2).
\end{align*}$$

Factor out -1.

(b) The GCF of $2x^5 - 8x^4 + 6x^3 - 24x^2$ is $2x^2$, so factor out $2x^2$ before factoring the remaining polynomial by grouping.

$$\begin{align*}
2x^5 - 8x^4 + 6x^3 - 24x^2 &= 2x^2(x^3 - 4x^2 + 3x - 12) &&\text{Factor out the GCF.}\\
&= 2x^2((x^3 - 4x^2) + (3x - 12)) &&\text{Associative property}\\
&= 2x^2(x^2(x - 4) + 3(x - 4)) &&\text{Factor out common factors.}\\
&= 2x^2(x^2 + 3)(x - 4) &&\text{Factor out } (x - 4).
\end{align*}$$

Now Try Exercises 67, 71

Sometimes we need to rearrange the terms before we can apply factoring by grouping as shown in the next example.

EXAMPLE 9 Rearranging terms before factoring by grouping

Factor the polynomial $2x^3 - 15 - 3x + 10x^2$.

Solution
The terms of the first binomial, $2x^3 - 15$, have no factors in common. However, if we rearrange the terms as $2x^3 + 10x^2 - 3x - 15$, we can factor as follows.

$$\begin{align*}
2x^3 + 10x^2 - 3x - 15 &= 2x^3 + 10x^2 + (-3x) - 15\\
&= (2x^3 + 10x^2) + (-3x - 15)\\
&= 2x^2(x + 5) - 3(x + 5)\\
&= (2x^2 - 3)(x + 5)
\end{align*}$$

Now Try Exercise 57

13.1 Putting It All Together

CONCEPT	EXPLANATION	EXAMPLES
Common Factor	Factor out a monomial common to each term in a polynomial.	$6z^2 - 6z = 6z(z - 1)$ $4y^3 - 6y^2 = 2y^2(2y - 3)$ $5x^3 - 10x^2 + 15x = 5x(x^2 - 2x + 3)$ $2a^3b^3 - 4a^2b^3 = 2a^2b^3(a - 2)$
Greatest Common Factor (GCF)	The common factor with the greatest (integer) coefficient and highest degree	The GCF of $10x^4 + 15x^2$ is $5x^2$. Common factors include $1, 5, x, 5x, x^2$, and $5x^2$. However, $5x^2$ is *the greatest* common factor.
Factoring by Grouping	Factoring by grouping is a method that can be used to factor some *four-term* polynomials into a product of two binomials. It makes use of the associative and distributive properties.	$2x^3 + 3x^2 + 2x + 3$ $= (2x^3 + 3x^2) + (2x + 3)$ $= x^2(2x + 3) + 1(2x + 3)$ $= (x^2 + 1)(2x + 3)$ $4x^3 - 24x^2 - 3x + 18$ $= 4x^3 - 24x^2 + (-3x) + 18$ $= (4x^3 - 24x^2) + (-3x + 18)$ $= 4x^2(x - 6) - 3(x - 6)$ $= (4x^2 - 3)(x - 6)$

13.1 Exercises

CONCEPTS AND VOCABULARY

1. To _____ a polynomial, write it as a product of two or more lower degree polynomials.

2. A common factor in the expression $ab + ac$ is _____.

3. When factoring, we can check our work by _____.

4. When finding the GCF for a polynomial, it is often helpful to completely factor each term by writing its coefficient as the product of _____ numbers and writing any powers of variables as repeated _____.

5. The _____ of a polynomial is the common factor with the greatest (integer) coefficient and highest degree.

6. Factoring by _____ can be used to factor some four-term polynomials into a product of two binomials by using the associative and distributive properties.

7. Identify four common factors of $2x^2 + 4x$.

8. Identify the greatest common factor (GCF) of the expression $2x^2 + 4x$.

FINDING COMMON FACTORS

Exercises 9–14: Factor the expression. Then make a sketch of a rectangle that illustrates this factorization.

9. $2x + 4$
10. $6 + 3x$
11. $z^2 + 4z$
12. $a^2 + 5a$
13. $3y^2 + 12y$
14. $2z^2 + 10z$

Exercises 15–22: Factor the expression.

15. $3x^2 + 9x$
16. $10y^2 + 2y$
17. $4y^3 - 2y^2$
18. $6x^4 + 9x^2$
19. $2z^3 + 8z^2 - 4z$
20. $5x^4 - 15x^3 - 10x^2$

21. $6x^2y - 3xy^2$
22. $7x^3y^3 + 14x^2y^2$

FINDING THE GREATEST COMMON FACTOR

Exercises 23–40: Identify the greatest common factor. Then factor the expression.

23. $6x - 18x^2$
24. $16x^2 - 24x^3$
25. $8y^3 - 12y^2$
26. $12y^3 - 8y^2 + 4y$
27. $6z^3 + 3z^2 + 9z$
28. $16z^3 - 24z^2 - 36z$
29. $x^4 - 5x^3 - 4x^2$
30. $2x^4 + 8x^2$
31. $5y^5 + 10y^4 - 15y^3 + 10y^2$
32. $7y^4 - 14y^3 - 21y^2 + 7y$
33. $xy + xz$
34. $ab - bc$
35. $ab^2 - a^2b$
36. $4x^2y + 6xy^2$
37. $5x^2y^4 + 10x^3y^3$
38. $3r^3t^3 - 6r^4t^2$
39. $a^2b + ab^2 + ab$
40. $6ab^2 - 9ab + 12b^2$

FACTORING BY GROUPING

Exercises 41–46: Factor.

41. $x(x + 1) - 2(x + 1)$
42. $5x(3x - 2) + 2(3x - 2)$
43. $(z + 5)z + (z + 5)4$
44. $3y^2(y - 2) + 5(y - 2)$
45. $4x^3(x - 5) - 2x(x - 5)$
46. $8x^2(x + 3) + (x + 3)$

Exercises 47–66: Factor by grouping.

47. $x^3 + 2x^2 + 3x + 6$
48. $x^3 + 6x^2 + x + 6$
49. $2y^3 + y^2 + 2y + 1$
50. $4y^3 + 10y^2 + 2y + 5$
51. $15z^3 - 5z^2 + 6z - 2$
52. $2z^3 - 6z^2 + 5z - 15$
53. $4t^3 - 20t^2 + 3t - 15$
54. $4t^3 - 12t^2 + 3t - 9$
55. $9r^3 + 6r^2 - 6r - 4$
56. $3r^3 + 12r^2 - 2r - 8$
57. $7x^3 - 6 - 2x + 21x^2$
58. $6x^3 - 5 - 10x + 3x^2$
59. $2y^3 - 7y^2 - 4y + 14$
60. $y^3 - 5y^2 - 3y + 15$
61. $z^3 - 4z^2 - 7z + 28$
62. $12z^3 - 18z^2 - 10z + 15$
63. $2x^4 - 3x^3 + 4x - 6$
64. $x^4 + x^3 + 5x + 5$
65. $ax + bx + ay + by$
66. $ax - bx + ay - by$

Exercises 67–78: Completely factor the polynomial.

67. $3x^3 + 6x^2 + 3x + 6$
68. $5x^3 - 5x^2 + 5x - 5$
69. $6y^4 - 24y^3 - 2y^2 + 8y$
70. $6x^4 - 12x^3 + 3x^2 - 6x$
71. $x^5 + 2x^4 - 3x^3 - 6x^2$
72. $y^6 + 3y^5 - 2y^4 - 6y^3$
73. $4x^5 + 2x^4 - 12x^3 - 6x^2$
74. $18y^5 + 27y^4 + 12y^3 + 18y^2$
75. $x^3y + x^2y^2 - 2x^2y - 2xy^2$
76. $6x^3y - 3x^2y^2 + 18x^2y - 9xy^2$
77. $2x^3y^3 - 2x^4y^2 + 4x^2y^3 - 4x^3y^2$
78. $4x^2y^6 - 4x^2y^5 - 8xy^7 + 8xy^6$

79. **Thinking Generally** Factor a from the polynomial expression $ax^2 + bx + c$.

80. **Thinking Generally** Factor c from the polynomial expression $ax^2 + bx + c$.

APPLICATIONS INVOLVING FACTORING

81. **Flight of a Golf Ball** The height of a golf ball in feet after t seconds is given by $80t - 16t^2$.
    (a) Identify the greatest common factor.
    (b) Factor this expression.

82. **Flight of a Golf Ball** Repeat the previous exercise if the height of a golf ball in feet after t seconds is given by $128t - 16t^2$.

Exercises 83–86: **Geometry** *Use the information in the figure to write a polynomial expression that represents the area of the shaded region. Then factor the expression.*

83.

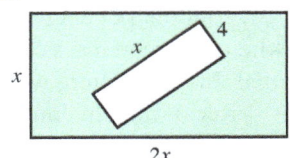

84.

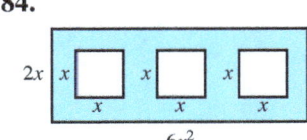

85.

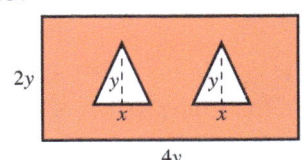

86.

87. Volume of a Box A box is constructed by cutting out square corners of a rectangular piece of cardboard and folding up the sides. If the cutout corners have sides with length x, then the volume of the box is given by the polynomial $4x^3 - 60x^2 + 200x$.

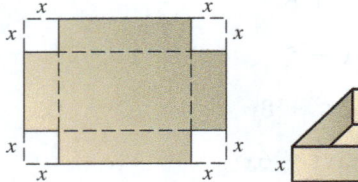

(a) Determine the volume of the box when $x = 3$ inches.
(b) Factor out the greatest common factor for this expression.

88. Volume of a Box (Refer to the preceding exercise.) A box is constructed from a square piece of metal that is 20 inches on a side.
(a) If the square corners of length x are cut out, write a polynomial that gives the volume of the box.
(b) Find the volume when $x = 4$ inches.
(c) Factor out the greatest common factor for this polynomial expression.

WRITING ABOUT MATHEMATICS

89. Use an example to explain the difference between a common factor and the greatest common factor.

90. Use an example to explain how to factor a polynomial by grouping. What two properties of real numbers did you use?

13.2 Factoring Trinomials I ($x^2 + bx + c$)

Objectives
1 Reviewing Binomial Multiplication
2 Factoring Trinomials with Leading Coefficient 1
3 Relating Geometry and Visual Factoring

NEW VOCABULARY
☐ Leading coefficient
☐ Prime polynomial

1 Reviewing Binomial Multiplication

Recall that a *trinomial* is a polynomial that has three terms. We begin by reviewing products of binomials that result in trinomials. For example, we multiply the binomials $(x + 2)$ and $(x + 3)$ as follows.

$$(x + 2)(x + 3) = x \cdot x + x \cdot 3 + 2 \cdot x + 2 \cdot 3$$
$$= x^2 + 5x + 6$$

Note that the first term, x^2, in the trinomial results from multiplying the *first* terms of each binomial. The middle term, $5x$, results from adding the product of the *outside* terms and the product of the *inside* terms. Finally, the last term, 6, results from multiplying the *last* terms of each binomial. We discussed this method of multiplying binomials, called FOIL, in Section 3 of the previous chapter and illustrate it as follows.

$$(x + 2)(x + 3) = x^2 + 5x + 6$$

← The middle term checks.

2 Factoring Trinomials with Leading Coefficient 1

Any trinomial of degree 2 in the variable x can be written in *standard form* as $ax^2 + bx + c$, where a, b, and c are constants. The constant a is called the **leading coefficient** of the trinomial. In this section, we focus on trinomials where $a = 1$ and b and c are integers.
Recall that the binomials $(x + m)$ and $(x + n)$ are multiplied as follows.

$$(x + m)(x + n) = x^2 + nx + mx + mn$$
$$= x^2 + (m + n)x + mn$$

READING CHECK 1
• How are the coefficient of the x-term b and the constant term c related to the numbers m and n?

Note that the coefficient of the x-term is the sum of m and n and that the constant (or third) term is the product of m and n. Thus to factor a trinomial in the form $x^2 + bx + c$, we start by finding two numbers, m and n, such that when they are multiplied, $m \cdot n = c$, and when they are added, $m + n = b$. In the next example, we find integer pairs that have a specified product and sum.

EXAMPLE 1 Finding integer pairs with a given product and sum

For each of the following, find an integer pair that has the given product and sum.
(a) Product: 18; Sum: 11   (b) Product: -20; Sum: 1

Solution
(a) For two integers to have a product of (positive) 18, the integers must have the same sign. Because the specified sum is positive and any sum of two negative integers is negative, the two integers must be positive. **TABLE 13.1** lists positive integer factor pairs for 18 along with the corresponding sum for each pair.

Factor Pairs for 18

Factors	1, 18	2, 9	3, 6
Sum	19	11	9

TABLE 13.1

From the table, we see that the integers 2 and 9 have a product of 18 and a sum of 11.

(b) For two integers to have a negative product, they must have unlike signs. Integer factor pairs for -20 and the corresponding sum for each pair are listed in **TABLE 13.2**.

Factor Pairs for -20

Factors	1, -20	2, -10	4, -5	-1, 20	-2, 10	-4, 5
Sum	-19	-8	-1	19	8	1

TABLE 13.2

Here, we see that the integers -4 and 5 have a product of -20 and a sum of 1.

Now Try Exercises 11, 15

STUDY TIP

Spend extra time practicing the process illustrated in Example 1. Pay special attention to factor pairs in which at least one of the integers is negative.

The ability to find integer pairs as demonstrated in Example 1 is essential for factoring trinomials of the form $x^2 + bx + c$. To illustrate the factoring process, we will find an integer pair that can be used to factor $x^2 + 6x + 8$.

Standard Form	*Example*	**Factor Pairs for 8**		
$x^2 + bx + c$	$x^2 + 6x + 8$	Factors	1, 8	2, 4	$m \cdot n = 8$
$m \cdot n = c$	$m \cdot n = 8$	Sum	9	6
$m + n = b$	$m + n = 6$	**TABLE 13.3**			$m + n = 6$

To determine possible values for m and n, we list factor pairs for 8 and search for a pair whose sum is 6, as in **TABLE 13.3**.

Because $2 \cdot 4 = 8$ and $2 + 4 = 6$, we let $m = 2$ and $n = 4$. We then factor the given trinomial as

$$x^2 + 6x + 8 = (x + 2)(x + 4).$$

Note that, if you can find this factor pair mentally, making a table is not necessary. We check the result by multiplying the two binomials.

$$(x + 2)(x + 4) = x^2 + 6x + 8$$

with middle term $2x + 4x = 6x$ ✓ ← The middle term checks.

> **FACTORING $x^2 + bx + c$**
>
> To factor the trinomial $x^2 + bx + c$, find two numbers m and n that satisfy
>
> $$m \cdot n = c \quad \text{and} \quad m + n = b.$$
>
> Then $x^2 + bx + c = (x + m)(x + n)$.

EXAMPLE 2 **Factoring a trinomial with only positive coefficients**

Factor each trinomial.
(a) $x^2 + 7x + 12$ (b) $x^2 + 13x + 30$ (c) $z^2 + 9z + 20$

Solution
(a) To factor $x^2 + 7x + 12$, we need to find a factor pair for 12 whose sum is 7. To do so, we make **TABLE 13.4**.

Factor Pairs for 12

Factors	1, 12	2, 6	3, 4
Sum	13	8	7

TABLE 13.4

The required factor pair is 3 and 4 because $3 \cdot 4 = 12$ and $3 + 4 = 7$. Therefore the given trinomial can be factored as

$$x^2 + 7x + 12 = (x + 3)(x + 4).$$

(b) To factor $x^2 + 13x + 30$, we need to find a factor pair for 30 whose sum is 13. The required pair is 3 and 10. Thus

$$x^2 + 13x + 30 = (x + 3)(x + 10).$$

(c) To factor $z^2 + 9z + 20$, we need to find a factor pair for 20 whose sum is 9. The required pair is 4 and 5. Thus

$$z^2 + 9z + 20 = (z + 4)(z + 5).$$

Now Try Exercises 25, 27, 29

In the next example, the coefficients of the middle terms are negative.

EXAMPLE 3 Factoring trinomials with a negative middle coefficient

Factor each trinomial.
(a) $x^2 - 7x + 10$ (b) $x^2 - 8x + 15$ (c) $y^2 - 9y + 18$

Solution
(a) To factor $x^2 - 7x + 10$, we need to find a factor pair for 10 whose sum equals -7. To have a positive product *and* a negative sum, *both* numbers must be negative, as shown in **TABLE 13.5**.

Factor Pairs for 10

Factors	$-1, -10$	$-2, -5$
Sum	-11	-7

TABLE 13.5

The required pair is -2 and -5 because $-2 \cdot (-5) = 10$ and $-2 + (-5) = -7$. Therefore the given trinomial can be factored as
$$x^2 - 7x + 10 = (x - 2)(x - 5).$$

(b) To factor $x^2 - 8x + 15$, we need to find a factor pair for 15 whose sum is -8. The required pair is -3 and -5. Thus
$$x^2 - 8x + 15 = (x - 3)(x - 5).$$

(c) To factor $y^2 - 9y + 18$, we need to find a factor pair for 18 whose sum is -9. The required pair is -3 and -6. Thus
$$y^2 - 9y + 18 = (y - 3)(y - 6).$$

Now Try Exercises 33, 35, 37

In Examples 2 and 3, the coefficient of the last term was always positive. In the next example, this coefficient is negative and the coefficient of the middle term is either positive or negative.

EXAMPLE 4 Factoring trinomials with a negative constant term

Factor each trinomial.
(a) $x^2 - 3x - 4$ (b) $x^2 + 7x - 8$ (c) $t^2 - 2t - 24$

Solution
(a) To factor $x^2 - 3x - 4$, we need to find a factor pair for -4 whose sum is -3. To have a negative product, one factor must be positive and the other factor must be negative, as shown in **TABLE 13.6**.

Factor Pairs for -4

Factors	$-1, 4$	$1, -4$	$-2, 2$
Sum	3	-3	0

TABLE 13.6

The required pair is 1 and -4 because $1 \cdot (-4) = -4$ and $1 + (-4) = -3$. Therefore the given trinomial can be factored as
$$x^2 - 3x - 4 = (x + 1)(x - 4),$$
which can be checked by multiplying $(x + 1)(x - 4)$.

(b) To factor $x^2 + 7x - 8$, we need to find a factor pair for -8 whose sum is 7. The required pair is -1 and 8. Thus
$$x^2 + 7x - 8 = (x - 1)(x + 8).$$

(c) To factor $t^2 - 2t - 24$, we need to find a factor pair for -24 whose sum is -2. The required pair is -6 and 4. Thus
$$t^2 - 2t - 24 = (t - 6)(t + 4).$$

Now Try Exercises 43, 55, 59

READING CHECK 2

- What is a prime polynomial?

A polynomial with integer coefficients that cannot be factored by using integer coefficients is called a **prime polynomial**. The next example illustrates that some trinomials of the form $x^2 + bx + c$ cannot be factored into the product of two binomials.

EXAMPLE 5 Discovering that a trinomial is prime

Factor each trinomial, if possible.
(a) $x^2 + 9x + 12$ (b) $x^2 + 5x - 4$

Solution
(a) To factor $x^2 + 9x + 12$, we need to find a factor pair for 12 whose sum is 9. **TABLE 13.7** reveals that no such factor pair exists.

Factor Pairs for 12

Factors	1, 12	2, 6	3, 4
Sum	13	8	7

TABLE 13.7

The trinomial $x^2 + 9x + 12$ is prime.

(b) At first glance, it may appear that the required factor pair is 4 and 1 because $4 \cdot 1 = 4$ and $4 + 1 = 5$. However, it is important to pay close attention to the signs of the coefficients. To factor $x^2 + 5x - 4$, we need to find a factor pair for -4 whose sum is 5. No such factor pair exists. The trinomial $x^2 + 5x - 4$ is prime.

Now Try Exercises 23, 47

MAKING CONNECTIONS 1

The Signs in the Binomial Factors

If a trinomial of the form $x^2 + bx + c$ can be factored, the signs of the coefficients in the trinomial can be used to determine the signs in the binomial factors. If b and c represent positive numbers, this can be summarized as follows.

Form of the Trinomial	Signs in the Binomial Factors	Examples
$x^2 + bx + c$	$(+)(+)$	$x^2 + 3x + 2 = (x + 1)(x + 2)$
$x^2 - bx + c$	$(-)(-)$	$x^2 - 3x + 2 = (x - 1)(x - 2)$
$x^2 + bx - c$	$(-)(+)$	$x^2 + 4x - 5 = (x - 1)(x + 5)$
$x^2 - bx - c$	$(-)(+)$	$x^2 - 4x - 5 = (x - 5)(x + 1)$

When factoring some trinomials, it may be necessary to factor out the greatest common factor before attempting to factor the trinomial into the product of two binomials. The next example illustrates this process.

EXAMPLE 7 Factoring out the GCF before factoring further

Factor each trinomial completely.
(a) $7x^2 + 35x + 42$ (b) $2x^4 - 4x^3 - 6x^2$

Solution
(a) Because the GCF of $7x^2 + 35x + 42$ is 7, factor out **7** first.
$$7x^2 + 35x + 42 = 7(x^2 + 5x + 6)$$
Now, to factor $x^2 + 5x + 6$, we need to find a factor pair for 6 whose sum is 5. The required pair is 2 and 3. Thus
$$7x^2 + 35x + 42 = 7(x+2)(x+3).$$

(b) Because the GCF of $2x^4 - 4x^3 - 6x^2$ is $2x^2$, factor out $2x^2$ first.
$$2x^4 - 4x^3 - 6x^2 = 2x^2(x^2 - 2x - 3)$$
Now, to factor $x^2 - 2x - 3$, we need to find a factor pair for -3 whose sum is -2. The required pair is -3 and 1. Thus
$$2x^4 - 4x^3 - 6x^2 = 2x^2(x-3)(x+1).$$

Now Try Exercises 65, 77

CRITICAL THINKING 1

A cube has a surface area of $6x^2 + 24x + 24$. What is the length of each side?

3 Relating Geometry and Visual Factoring

Trinomials of the form $x^2 + bx + c$ can be factored visually using rectangles, as illustrated in the next example and in the following Making Connections.

EXAMPLE 7 Finding the dimensions of a rectangle

Find one possibility for the dimensions of a rectangle that has an area of $x^2 + 6x + 5$.

Solution
The area of a rectangle equals length times width. If we can factor $x^2 + 6x + 5$, then the factors can represent its length and width. Because
$$x^2 + 6x + 5 = (x+1)(x+5),$$
one possibility for the rectangle's dimensions is width $x + 1$ and length $x + 5$, as illustrated in **FIGURE 13.4**.

Now Try Exercise 93

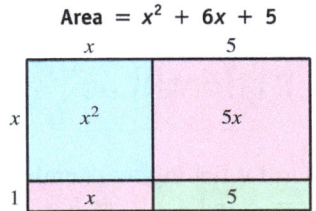

FIGURE 13.4

MAKING CONNECTIONS 2

Factoring Trinomials Visually

A special grid can be used to factor trinomials visually. Like an *xy*-plane, the grid has four quadrants. But this grid is not an *xy*-plane because the axes are not labeled and the grid is not used to plot ordered pairs. The area of any region located in quadrants I or III represents a positive term and the area of any region in quadrants II or IV represents a negative term.

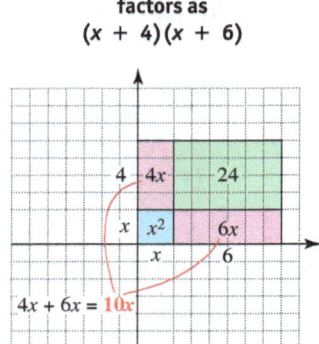

$x^2 + 10x + 24$
factors as
$(x + 4)(x + 6)$

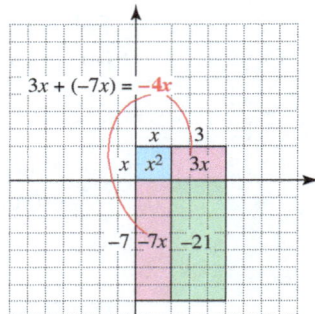

$x^2 - 4x - 21$
factors as
$(x - 7)(x + 3)$

See Exercises 99–104.

13.2 Putting It All Together

CONCEPT	EXPLANATION	EXAMPLES
Factoring Trinomials of the Form $$x^2 + bx + c$$	Find two numbers m and n that satisfy $mn = c$ and $m + n = b$. Then $$x^2 + bx + c = (x + m)(x + n).$$	$x^2 + 9x + 20 = (x + 4)(x + 5)$ because $4 \cdot 5 = 20$ and $4 + 5 = 9$. $x^2 + x - 6 = (x - 2)(x + 3)$ because $-2 \cdot 3 = -6$ and $-2 + 3 = 1$. $x^2 - 8x + 12 = (x - 6)(x - 2)$ because $-6 \cdot (-2) = 12$ and $-6 + (-2) = -8$.

13.2 Exercises   MyMathLab®

CONCEPTS AND VOCABULARY

1. In the trinomial $x^2 + bx + c$, the leading coefficient is _____.

2. Multiply $(x + m)(x + n)$. What is the coefficient of the *x*-term? What is the constant term?

3. To factor $x^2 + bx + c$, start by finding two numbers m and n that satisfy _____ $= c$ and _____ $= b$.

4. A trinomial with integer coefficients that cannot be factored using integer coefficients is _____.

REVIEWING BINOMIAL MULTIPLICATION

Exercises 5–10: Multiply and simplify.

5. $(x + 2)(x + 5)$
6. $(x + 1)(x + 7)$
7. $(x - 3)(x - 6)$
8. $(x - 5)(x - 2)$
9. $(x + 8)(x - 7)$
10. $(x - 4)(x + 9)$

FINDING INTEGER PAIRS

Exercises 11–18: Find the integer pair that has the given product and sum.

11. Product: 28   Sum: 11
12. Product: 35   Sum: 12
13. Product: −30   Sum: −7
14. Product: −15   Sum: −2
15. Product: −50   Sum: 5
16. Product: −100   Sum: 21
17. Product: 28   Sum: −11
18. Product: 80   Sum: −42

FACTORING TRINOMIALS

Exercises 19–64: Factor the trinomial. If the trinomial cannot be factored, write "prime."

19. $x^2 + 3x + 2$
20. $x^2 + 5x + 4$
21. $y^2 + 4y + 4$
22. $y^2 + 8y + 7$
23. $z^2 + 3z + 7$
24. $z^2 + 4z + 5$
25. $x^2 + 8x + 15$
26. $x^2 + 9x + 14$
27. $m^2 + 13m + 36$
28. $m^2 + 15m + 36$
29. $n^2 + 20n + 100$
30. $n^2 + 52n + 100$
31. $x^2 - 6x + 5$
32. $x^2 - 6x + 8$
33. $y^2 - 7y + 12$
34. $y^2 - 12y + 27$
35. $z^2 - 13z + 40$
36. $z^2 - 15z + 54$
37. $a^2 - 16a + 63$
38. $a^2 - 82a + 81$
39. $y^2 - 6y + 10$
40. $y^2 - 2y + 3$
41. $b^2 - 30b + 125$
42. $b^2 - 19b + 90$
43. $x^2 + 13x - 90$
44. $x^2 + 15x - 100$
45. $m^2 + 4m - 45$
46. $m^2 + 4m - 60$
47. $a^2 + 16a - 63$
48. $a^2 + 13a - 42$
49. $n^2 + 10n - 200$
50. $n^2 + 2n - 120$
51. $x^2 + 22x - 23$
52. $x^2 + 18x - 19$
53. $a^2 + 4a - 32$
54. $a^2 + 9a - 36$
55. $b^2 - b - 20$
56. $b^2 - b - 12$
57. $m^2 - 14m - 22$
58. $m^2 - 11m - 24$
59. $x^2 - x - 72$
60. $x^2 - 2x - 80$
61. $y^2 - 15y - 34$
62. $y^2 - 10y - 39$
63. $z^2 - 5z - 66$
64. $z^2 - 6z - 55$

Exercises 65–80: Factor the trinomial completely.

65. $5x^2 - 10x - 40$
66. $2x^2 + 8x - 10$
67. $-3m^2 - 9m + 12$
68. $-4n^2 + 20n - 24$
69. $y^3 - 7y^2 + 10y$
70. $z^3 + 9z^2 + 20z$
71. $-x^3 - 2x^2 + 15x$
72. $-y^3 + 9y^2 - 14y$
73. $3a^3 + 21a^2 + 18a$
74. $5b^3 - 5b^2 - 60b$
75. $-2x^3 + 6x^2 - 8x$
76. $-4y^3 + 20y^2 - 32y$
77. $2m^4 - 10m^3 - 28m^2$
78. $6n^4 - 18n^3 + 12n^2$
79. $-3x^4 + 3x^3 + 6x^2$
80. $-5y^4 + 25y^3 - 30y^2$

Exercises 81–84: Factor the trinomial. (Hint: Write the expression in standard form.)

81. $5 + 6x + x^2$
82. $8 + 6x + x^2$
83. $3 - 4x + x^2$
84. $10 - 7x + x^2$

Exercises 85–88: Factor the trinomial. (Hint: Write $(m - x)(n + x)$ and find m and n.)

85. $12 + 4x - x^2$
86. $28 + 3x - x^2$
87. $32 - 4x - x^2$
88. $40 - 3x - x^2$

89. **Thinking Generally** Factor the trinomial expression $x^2 + (k + 1)x + k$.

90. **Thinking Generally** Factor the trinomial expression $x^2 + (k - 2)x - 2k$.

RELATING GEOMETRY AND VISUAL FACTORING

91. A square has an area of $x^2 + 2x + 1$. Find the length of a side. Make a sketch of the square.

92. A square has an area of $x^2 + 6x + 9$. Find the length of a side. Make a sketch of the square.

93. A rectangle has an area of $x^2 + 3x + 2$. Find one possibility for its width and length. Make a sketch of the rectangle.

94. A rectangle has an area of $x^2 + 9x + 8$. Find one possibility for its width and length. Make a sketch of the rectangle.

95. A cube has a surface area of $6x^2 + 12x + 6$. Find the length of a side. (*Hint:* First factor out the GCF.)

96. A cube has a surface area of $6x^2 + 36x + 54$. Find the length of a side.

97. Write a polynomial in factored form that represents the total area of the figure.

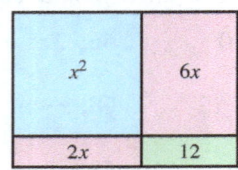

98. Write a polynomial in factored form that represents the total area of the figure.

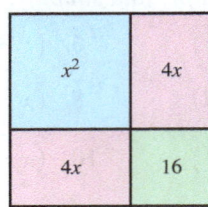

Exercises 99–104: Factor the trinomial visually, as is discussed in Making Connections 2 at the end of this section.

99. $x^2 + 5x + 6$ 100. $x^2 + 9x + 20$

101. $x^2 - 11x + 30$ 102. $x^2 - 3x - 10$

103. $x^2 + 4x - 12$ 104. $x^2 - 8x + 16$

WRITING ABOUT MATHEMATICS

105. Explain how to determine whether a trinomial has been factored correctly. Give an example.

106. Factoring $x^2 + bx + c$ involves finding two integers m and n such that $mn = c$ and $m + n = b$. Is it better first to determine values of m and n so that the product is c or the sum is b? Explain your reasoning.

SECTIONS 13.1 and 13.2 — Checking Basic Concepts

1. What is the greatest common factor for the expression $8x^3 - 12x^2 + 24x$?

2. Factor $12z^3 - 18z^2$.

3. Factor each expression completely.
   (a) $6y(y - 2) + 5(y - 2)$
   (b) $2x^3 + x^2 + 10x + 5$
   (c) $4z^3 - 12z^2 + 4z - 12$

4. Factor each trinomial completely, if possible.
   (a) $x^2 + 6x + 8$ (b) $x^2 - x - 42$

   (c) $a^2 + 3a - 5$ (d) $4a^3 + 20a^2 + 24a$

5. Write a polynomial in factored form that represents the total area of the figure.

13.3 Factoring Trinomials II ($ax^2 + bx + c$)

Objectives

1. Factoring Trinomials by Grouping
2. Factoring with FOIL in Reverse

1 Factoring Trinomials by Grouping

To factor the trinomial $x^2 + 6x + 5$, we find two numbers, m and n, such that $mn = 5$ and $m + n = 6$. For this trinomial, we let $m = 1$ and $n = 5$, which gives

$$x^2 + 6x + 5 = (x + 1)(x + 5).$$

READING CHECK 1

- What are the values of m and n for $6x^2 + 11x + 3$?

To factor the trinomial $2x^2 + 7x + 3$, which has a leading coefficient of 2, we must find two numbers m and n such that $mn = 2 \cdot 3 = 6$ and $m + n = 7$. One solution is $m = 1$ and $n = 6$. Using grouping, we can now factor this trinomial by writing $7x$ as the sum $1x + 6x$.

$$
\begin{aligned}
2x^2 + 7x + 3 &= 2x^2 + \overbrace{x + 6x}^{7x} + 3 && \text{Write } 7x \text{ as } x + 6x. \\
&= (2x^2 + x) + (6x + 3) && \text{Associative property} \\
&= x(2x + 1) + 3(2x + 1) && \text{Factor out common factors.} \\
&= (x + 3)(2x + 1) && \text{Factor out } (2x + 1).
\end{aligned}
$$

This technique of factoring trinomials by grouping is summarized as follows.

FACTORING $ax^2 + bx + c$ BY GROUPING

To factor $ax^2 + bx + c$, perform the following steps. (Assume that a, b, and c are integers and have no common factors.)

1. Find two numbers, m and n, such that $mn = ac$ and $m + n = b$.
2. Write the trinomial as $ax^2 + mx + nx + c$.
3. Use grouping to factor this expression into two binomials.

STUDY TIP

Be sure that you understand that the values of m and n are used in different ways, depending on whether or not the leading coefficient is 1.

MAKING CONNECTIONS 1

The Values of m and n

The values of m and n found when factoring $ax^2 + bx + c$, where $a \neq 1$, are not used in the same way as the values of m and n found when factoring $x^2 + bx + c$.

Factoring $x^2 + bx + c$	*Factoring* $ax^2 + bx + c$, *where* $a \neq 1$
$mn = c$ and $m + n = b$	$mn = ac$ and $m + n = b$
$x^2 + bx + c = (x + m)(x + n)$	$ax^2 + bx + c = ax^2 + mx + nx + c$
	Factor the resulting four-term polynomial by grouping.

EXAMPLE 1 Factoring $ax^2 + bx + c$ by grouping

Factor each trinomial.
(a) $2x^2 + 5x + 2$ (b) $3z^2 + z - 2$ (c) $10t^2 - 11t + 3$

Solution

(a) To factor $2x^2 + 5x + 2$, we need to find m and n that satisfy $mn = 2 \cdot 2 = 4$ and $m + n = 5$. **TABLE 13.8** shows that two such numbers are $m = 1$ and $n = 4$.

$$
\begin{aligned}
2x^2 + 5x + 2 &= 2x^2 + \overbrace{x + 4x}^{5x} + 2 && \text{Write } 5x \text{ as } x + 4x. \\
&= (2x^2 + x) + (4x + 2) && \text{Associative property} \\
&= x(2x + 1) + 2(2x + 1) && \text{Factor out common factors.} \\
&= (x + 2)(2x + 1) && \text{Factor out } (2x + 1).
\end{aligned}
$$

Factor Pairs for 4

Factors	1, 4	2, 2
Sum	5	4

TABLE 13.8

(b) To factor $3z^2 + 1z - 2$, we need to find m and n that satisfy $mn = 3 \cdot (-2) = -6$ and $m + n = 1$. Two such numbers are $m = 3$ and $n = -2$.

$$\begin{aligned} 3z^2 + z - 2 &= 3z^2 + 3z - 2z - 2 & &\text{Write } z \text{ as } 3z - 2z. \\ &= (3z^2 + 3z) + (-2z - 2) & &\text{Associative property} \\ &= 3z(z + 1) - 2(z + 1) & &\text{Factor out common factors.} \\ &= (3z - 2)(z + 1) & &\text{Factor out } (z + 1). \end{aligned}$$

(c) To factor $10t^2 - 11t + 3$, we need to find m and n that satisfy $mn = 10 \cdot 3 = 30$ and $m + n = -11$. Two such numbers are $m = -5$ and $n = -6$.

$$\begin{aligned} 10t^2 - 11t + 3 &= 10t^2 - 5t - 6t + 3 & &\text{Write } -11t \text{ as } -5t - 6t. \\ &= (10t^2 - 5t) + (-6t + 3) & &\text{Associative property} \\ &= 5t(2t - 1) - 3(2t - 1) & &\text{Factor out common factors.} \\ &= (5t - 3)(2t - 1) & &\text{Factor out } (2t - 1). \end{aligned}$$

Now Try Exercises 15, 25, 37

MAKING CONNECTIONS 2

Different Ways to Factor by Grouping

In Example 1(c), we could have written $-11t$ as $-6t - 5t$, rather than $-5t - 6t$. Then the factoring could have been written as

$$\begin{aligned} 10t^2 - 11t + 3 &= 10t^2 - 6t - 5t + 3 \\ &= (10t^2 - 6t) + (-5t + 3) \\ &= 2t(5t - 3) - 1(5t - 3) \\ &= (2t - 1)(5t - 3), \end{aligned}$$

which gives the same result.

In the previous section, we showed that some trinomials of the form $x^2 + bx + c$ are prime and cannot be factored into the product of two binomials with integer coefficients. The next example illustrates that some trinomials of the form $ax^2 + bx + c$, with $a \neq 1$, may also be prime.

EXAMPLE 2 **Discovering that a trinomial is prime**

Factor each trinomial.
(a) $3x^2 + 5x + 4$ (b) $2x^2 - 8x - 3$

Solution

(a) To factor $3x^2 + 5x + 4$, we need to find integers m and n such that $mn = 3 \cdot 4 = 12$ and $m + n = 5$. Because the middle term is positive, we consider only positive factors of 12. **TABLE 13.9** reveals that no such integers exist.

Factor Pairs for 12

Factors	1, 12	2, 6	3, 4
Sum	13	8	7

No sum equals 5.

TABLE 13.9

The trinomial $3x^2 + 5x + 4$ is prime.

(b) To factor $2x^2 - 8x - 3$, we must find integers m and n such that $mn = 2 \cdot (-3) = -6$ and $m + n = -8$. **TABLE 13.10** reveals that no such integers exist.

Factor Pairs for −6

Factors	−1, 6	−2, 3	2, −3	1, −6
Sum	5	1	−1	−5

No sum equals −8.

TABLE 13.10

The trinomial $2x^2 - 8x - 3$ is prime.

Now Try Exercises 17, 41

Although some trinomials may look as though they can be factored using the process discussed in Example 1, it is important to remember to first factor out the greatest common factor whenever possible. In the next example, we factor out the GCF before factoring the trinomial further.

EXAMPLE 3 **Factoring out the GCF before factoring further**

Factor each trinomial completely.
(a) $15x^2 - 50x - 40$ **(b)** $4x^3 - 22x^2 + 30x$

Solution
(a) Because the GCF of $15x^2 - 50x - 40$ is 5, factor out **5** before factoring the remaining trinomial.

$$15x^2 - 50x - 40 = 5(3x^2 - 10x - 8)$$

To factor $3x^2 - 10x - 8$, we need numbers m and n such that $mn = 3 \cdot (-8) = -24$ and $m + n = -10$. Two such numbers are -12 and 2.

$$\begin{aligned} 15x^2 - 50x - 40 &= 5(3x^2 - 12x + 2x - 8) && \text{Write } -10x \text{ as } -12x + 2x. \\ &= 5((3x^2 - 12x) + (2x - 8)) && \text{Associative property} \\ &= 5(3x(x - 4) + 2(x - 4)) && \text{Factor out common factors.} \\ &= 5(3x + 2)(x - 4) && \text{Factor out } (x - 4). \end{aligned}$$

(b) Because the GCF of $4x^3 - 22x^2 + 30x$ is $2x$, factor out $2x$ before factoring the remaining trinomial.

$$4x^3 - 22x^2 + 30x = 2x(2x^2 - 11x + 15)$$

To factor the trinomial $2x^2 - 11x + 15$, we need to find numbers m and n such that $mn = 2 \cdot 15 = 30$ and $m + n = -11$. Two such numbers are -6 and -5.

$$\begin{aligned} 4x^3 - 22x^2 + 30x &= 2x(2x^2 - 6x - 5x + 15) && \text{Write } -11x \text{ as } -6x - 5x. \\ &= 2x((2x^2 - 6x) + (-5x + 15)) && \text{Associative property} \\ &= 2x(2x(x - 3) - 5(x - 3)) && \text{Factor out common factors.} \\ &= 2x(2x - 5)(x - 3) && \text{Factor out } (x - 3). \end{aligned}$$

Now Try Exercises 55, 59

2 Factoring with FOIL in Reverse

Rather than factoring a trinomial by grouping, we can use FOIL in reverse to determine its binomial factors. For example, the factors of $2x^2 + 5x + 2$ are two binomials:

$$2x^2 + 5x + 2 = (\underline{} + \underline{})(\underline{} + \underline{}),$$

where the expressions to be placed in the four blanks are yet to be found. By the FOIL method, we know that the product of the first terms of these two binomials is $2x^2$. Because $2x^2 = 2x \cdot x$, we can write

$$2x^2 + 5x + 2 = (\underline{2x} + \underline{})(\underline{x} + \underline{}).$$

By FOIL, the product of the last terms in these binomials must be 2. Because $2 = 1 \cdot 2$, we could put 1 and 2 in the blanks. However, we must be sure to place them correctly so that the product of the outside terms plus the product of the inside terms is $5x$.

$$(2x + 1)(x + 2) = 2x^2 + 5x + 2$$

$1x$
$+ 4x$
$5x$ ✓ ⟵ Middle term checks.

If we had interchanged the 1 and 2, we would have obtained an incorrect result.

$$(2x + 2)(x + 1) = 2x^2 + 4x + 2$$

$2x$
$+2x$
$4x$ ✗ ⟵ Middle term is *not* $5x$.

> **READING CHECK 2**
>
> - Does $6x^2 + 17x + 5$ factor as $(3x + 1)(2x + 5)$ or as $(3x + 5)(2x + 1)$?

MAKING CONNECTIONS 3

The Signs in the Binomial Factors

Let a, b, and c represent positive integers. If a trinomial of the form $ax^2 + bx + c$ can be factored, the signs in the binomial factors can be summarized as follows.

Form of the Trinomial	Signs in the Binomial Factors	Examples
$ax^2 + bx + c$	$(\,+\,)(\,+\,)$	$2x^2 + 5x + 3 = (2x + 3)(x + 1)$
$ax^2 - bx + c$	$(\,-\,)(\,-\,)$	$2x^2 - 5x + 3 = (2x - 3)(x - 1)$
$ax^2 + bx - c$	$(\,-\,)(\,+\,)$	$3x^2 + 5x - 2 = (3x - 1)(x + 2)$
$ax^2 - bx - c$	$(\,-\,)(\,+\,)$	$3x^2 - 2x - 5 = (3x - 5)(x + 1)$

In the next example, we factor expressions of the form $ax^2 + bx + c$, where $a \neq 1$. We use FOIL in reverse and *trial and error* to find the correct factors.

EXAMPLE 4   Factoring the form $ax^2 + bx + c$

Factor each trinomial.
(a) $3x^2 + 5x + 2$ (b) $6x^2 + 7x - 3$ (c) $6 + 4y^2 - 11y$

Solution
(a) To factor $3x^2 + 5x + 2$, we start by finding factors of $3x^2$, which include $3x$ and x, so we write

$$3x^2 + 5x + 2 = (3x + \underline{})(x + \underline{}).$$

The factors of the last term 2 are **1** and **2**. (Because the middle term is positive, we do not consider -1 and -2.) If we place values of 1 and 2 in the binomial as follows, the middle term becomes $7x$ rather than $5x$.

$$(3x + 1)(x + 2) = 3x^2 + 7x + 2$$

$$\begin{array}{c} 1x \\ +6x \\ \hline 7x \; \text{✗} \end{array} \longleftarrow \text{Middle term is } not \; 5x.$$

By reversing the positions of 1 and 2, we obtain a correct factorization.

$$(3x + 2)(x + 1) = 3x^2 + 5x + 2$$

$$\begin{array}{c} 2x \\ +3x \\ \hline 5x \; \checkmark \end{array} \longleftarrow \text{Middle term checks.}$$

(b) To factor $6x^2 + 7x - 3$, we start by finding factors of $6x^2$, which include $2x$ and $3x$ or $6x$ and x. Factors of the last term, -3, include -1 and 3 or 1 and -3. The following two factorizations give incorrect results.

$$(2x + 1)(3x - 3) = 6x^2 - 3x - 3 \qquad (6x + 1)(x - 3) = 6x^2 - 17x - 3$$

$$\begin{array}{c} 3x \\ -6x \\ \hline -3x \; \text{✗} \end{array} \longleftarrow \text{Middle term is } not \; 7x. \qquad \begin{array}{c} 1x \\ -18x \\ \hline -17x \; \text{✗} \end{array} \longleftarrow \text{Middle term is } not \; 7x.$$

To obtain a middle term of $7x$, we use the following arrangement.

$$(3x - 1)(2x + 3) = 6x^2 + 7x - 3$$

$$\begin{array}{c} -2x \\ +9x \\ \hline 7x \; \checkmark \end{array} \longleftarrow \text{Middle term checks.}$$

To find the correct factorization, we often need to try more than one arrangement.

(c) To factor $6 + 4y^2 - 11y$, we start by writing the trinomial in standard form as $4y^2 - 11y + 6$. Then we find possible factors of the first term, $4y^2$, which include $2y$ and $2y$ or $4y$ and y. Factors of the last term, 6, include either -1 and -6 or -2 and -3. (Because the middle term is negative, we do not use the positive factors of 1 and 6 or 2 and 3.) To obtain a middle term of $-11y$, we use the following arrangement.

$$(4y - 3)(y - 2) = 4y^2 - 11y + 6$$

$$\begin{array}{c} -3y \\ -8y \\ \hline -11y \; \checkmark \end{array} \longleftarrow \text{Middle term checks.}$$

READING CHECK 3

- What is a good first step when factoring $-x^2 - 2x + 3$, which has a negative leading coefficient?

Now Try Exercises 19, 33, 69

In the next example, we demonstrate how to factor trinomials that have a negative leading coefficient. This task is sometimes accomplished by first factoring out -1.

EXAMPLE 5   **Factoring trinomials that have a negative leading coefficient**

Factor each trinomial.
(a) $-6x^2 + 17x - 5$ (b) $1 - x - 2x^2$

Solution
(a) One way to factor $-6x^2 + 17x - 5$ is to start by factoring out -1. Then we can apply FOIL in reverse.

$$-6x^2 + 17x - 5 = \overbrace{-1(6x^2 - 17x + 5)}^{\text{Opposite of } -6x^2 + 17x - 5} \quad \text{Factor out } -1.$$
$$= -1(3x - 1)(2x - 5) \quad \text{Factor the trinomial.}$$
$$= -(3x - 1)(2x - 5) \quad \text{Rewrite.}$$

(b) To factor the trinomial $1 - x - 2x^2$, write it in standard form, factor out -1, and then apply FOIL in reverse.

$$1 - x - 2x^2 = -2x^2 - x + 1 \quad \text{Standard form}$$
$$= -1(2x^2 + x - 1) \quad \text{Factor out } -1.$$
$$= -(2x - 1)(x + 1) \quad \text{Factor the trinomial.}$$

Now Try Exercises 71, 75

13.3 Putting It All Together

CONCEPT	EXPLANATION	EXAMPLES
Factoring Trinomials by Grouping	To factor $ax^2 + bx + c$, find two numbers, m and n, such that $mn = ac$ and $m + n = b$. Then write $$ax^2 + mx + nx + c.$$ Use grouping to factor this expression into two binomials.	For $3x^2 + 10x + 8$, let $m = 6$ and $n = 4$ because $mn = 24$ and $m + n = 10$. $$\begin{aligned}3x^2 + 10x + 8 &= 3x^2 + 6x + 4x + 8 \\ &= (3x^2 + 6x) + (4x + 8) \\ &= 3x(x + 2) + 4(x + 2) \\ &= (3x + 4)(x + 2)\end{aligned}$$
Factoring Trinomials by Using FOIL in Reverse	To factor $ax^2 + bx + c$, find factors of ax^2 and of c. Choose and arrange these factors in two binomials so that the middle term is bx.	For $3x^2 + 10x + 8$, the factors of $3x^2$ are $3x$ and x. The positive factors of 8 are either 2 and 4, or 1 and 8. A middle term of $10x$ can be obtained as follows. $$(3x + 4)(x + 2) = 3x^2 + 10x + 8$$ $4x$ $+6x$ $10x$ ✓ ← Middle term checks.
Choosing Signs When Factoring $ax^2 + bx + c$	For $ax^2 + bx + c$ with $a > 0$, 1. If $c > 0$ and $b > 0$, use $$(\_\_ + \_\_)(\_\_ + \_\_).$$ 2. If $c > 0$ and $b < 0$, use $$(\_\_ - \_\_)(\_\_ - \_\_).$$ 3. If $c < 0$, use $$(\_\_ - \_\_)(\_\_ + \_\_) \text{ or}$$ $$(\_\_ + \_\_)(\_\_ - \_\_).$$	1. $2x^2 + 7x + 3 = (2x + 1)(x + 3)$ $(c = 3, b = 7)$ 2. $2x^2 - 7x + 3 = (2x - 1)(x - 3)$ $(c = 3, b = -7)$ 3. $2x^2 + 5x - 3 = (2x - 1)(x + 3)$ $(c = -3, b = 5)$ $2x^2 - 5x - 3 = (2x + 1)(x - 3)$ $(c = -3, b = -5)$

13.3 Exercises

CONCEPTS AND VOCABULARY

1. To factor the polynomial $ax^2 + bx + c$ by grouping, you first find two numbers, m and n, such that $mn =$ _____ and $m + n =$ _____.

2. To factor the polynomial $ax^2 + bx + c$ with FOIL in reverse, you first find possible factors for _____ and for _____.

Exercises 3–6: Insert the symbol + or − in each binomial factor to make the equation true.

3. $3x^2 + 5x + 2 = (3x \_ 2)(x \_ 1)$.
4. $3x^2 - x - 2 = (3x \_ 2)(x \_ 1)$.
5. $3x^2 - 5x + 2 = (3x \_ 2)(x \_ 1)$.
6. $3x^2 + x - 2 = (3x \_ 2)(x \_ 1)$.

Exercises 7–14: Fill in the blank in each binomial factor to make the equation true.

7. $4x^2 + 11x + 6 = (4x + \_)(\_ + 2)$.
8. $4x^2 - 5x - 6 = (\_ - 2)(\_ + 3)$.
9. $4x^2 + 4x - 3 = (2x - \_)(2x + \_)$.
10. $4x^2 - 8x + 3 = (2x - \_)(\_ - 3)$.
11. $6x^2 + 11x - 7 = (\_ + 7)(2x - \_)$.
12. $6x^2 - 31x + 28 = (\_ - 7)(\_ - 4)$.
13. $6x^2 - x - 15 = (3x - \_)(2x + \_)$.
14. $6x^2 - 53x + 40 = (\_ - 5)(x - \_)$.

FACTORING TRINOMIALS

Exercises 15–54: Factor the trinomial. If the trinomial cannot be factored, write "prime."

15. $2x^2 + 7x + 3$
16. $2x^2 + 3x + 1$
17. $3y^2 + 2y + 4$
18. $2y^2 + 5y + 1$
19. $3x^2 + 4x + 1$
20. $3x^2 + 10x + 3$
21. $6x^2 + 11x + 3$
22. $6x^2 + 17x + 5$
23. $5x^2 - 11x + 2$
24. $7x^2 - 8x + 1$
25. $2y^2 - 7y + 5$
26. $2y^2 - 11y + 12$
27. $3m^2 - 11m - 6$
28. $5m^2 - 7m - 2$
29. $7z^2 - 37z + 10$
30. $3z^2 - 11z + 6$
31. $3t^2 - 7t - 6$
32. $8t^2 - 6t - 9$
33. $15r^2 + r - 6$
34. $12r^2 + r - 6$
35. $24m^2 - 23m - 12$
36. $24m^2 + 29m - 4$
37. $25x^2 + 5x - 2$
38. $30x^2 + 7x - 2$
39. $6x^2 + 11x - 2$
40. $12x^2 + 28x - 5$
41. $15y^2 - 7y + 2$
42. $14y^2 - 5y + 1$
43. $21n^2 + 4n - 1$
44. $21n^2 + 10n + 1$
45. $14y^2 + 23y + 3$
46. $28y^2 + 25y + 3$
47. $28z^2 - 25z + 3$
48. $15z^2 - 19z + 6$
49. $30x^2 - 29x + 6$
50. $50x^2 - 55x + 12$
51. $20a^2 + 18a - 5$
52. $40a^2 + 21a - 2$
53. $18t^2 + 23t - 6$
54. $33t^2 + 7t - 10$

Exercises 55–64: Factor the trinomial completely.

55. $12a^2 + 12a - 9$
56. $21b^2 - 14b - 56$
57. $10z^3 + 19z^2 + 6z$
58. $12y^3 - 11y^2 + 2y$
59. $24x^3 - 30x^2 + 9x$
60. $8y^3 - 16y^2 + 6y$
61. $8x^4 - 6x^3 + 2x^2$
62. $10y^3 + 15y^2 - 5y$
63. $28x^4 + 56x^3 + 21x^2$
64. $20y^4 + 42y^3 - 20y^2$

65. **Thinking Generally** Factor the trinomial expression $3x^2 + (3k + 1)x + k$.

66. **Thinking Generally** Factor the trinomial expression $3x^2 + (3k - 2)x - 2k$.

Exercises 67–76: Factor the trinomial.

67. $2 + 15x + 7x^2$
68. $3 + 16x + 5x^2$
69. $2 - 5x + 2x^2$
70. $5 - 6x + x^2$
71. $3 - 2x - 8x^2$
72. $5 - 3x - 2x^2$
73. $-2x^2 - 7x + 15$
74. $-5x^2 - 19x + 4$
75. $-5x^2 + 14x + 3$
76. $-6x^2 + 17x + 14$

814 CHAPTER 13 FACTORING POLYNOMIALS AND SOLVING EQUATIONS

RELATING GEOMETRY AND VISUAL FACTORING

77. A rectangle has an area of $6x^2 + 7x + 2$. Find possible dimensions for this rectangle. Make a sketch of the rectangle.

78. A rectangle has an area of $2x^2 + 5x + 3$. Find possible dimensions for the rectangle. Make a sketch of the rectangle.

79. Write a polynomial in factored form that represents the total area of the figure.

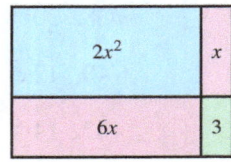

80. Write a polynomial in factored form that represents the total area of the figure.

WRITING ABOUT MATHEMATICS

81. Explain how the sign of the third term in the trinomial $ax^2 + bx + c$ affects how it is factored.

82. Explain the steps to be used to factor $ax^2 + bx + c$ by grouping.

13.4 Special Types of Factoring

Objectives
1. Factoring the Difference of Two Squares
2. Factoring Perfect Square Trinomials
3. Factoring the Sum and Difference of Two Cubes
4. Using Special Methods in General Factoring

NEW VOCABULARY
☐ Perfect square trinomial

STUDY TIP
The special types of factoring discussed in this section are helpful for factoring some (but not all) types of polynomials. Don't forget to review factoring techniques that were discussed earlier in this chapter.

1 Factoring the Difference of Two Squares

In Section 4 of the previous chapter, we showed that $(a - b)(a + b) = a^2 - b^2$. If we rewrite this equation as

$$a^2 - b^2 = (a - b)(a + b),$$

then we can use it to factor a difference of two squares. For example, to factor the expression $x^2 - 25$, we first write it in the form $a^2 - b^2$, where $a = x$ and $b = 5$. Doing so results in expressing $x^2 - 25$ as $x^2 - 5^2$, and the equation

$$a^2 - b^2 = (a - b)(a + b) \quad \text{General rule}$$

becomes

$$x^2 - 5^2 = (x - 5)(x + 5). \quad \text{Substitute: } a = x, b = 5$$

This discussion suggests the following rule for factoring a difference of two squares.

DIFFERENCE OF TWO SQUARES

For any real numbers a and b,

$$a^2 - b^2 = (a - b)(a + b).$$

In the next example, we apply this method to other expressions.

EXAMPLE 1 Factoring the difference of two squares

Factor each difference of two squares.
(a) $x^2 - 36$ (b) $4x^2 - 9$ (c) $100 - 16t^2$ (d) $49y^2 - 64z^2$

Solution
(a) Write $x^2 - 36$ as $x^2 - 6^2$ and then substitute x for a and 6 for b. The equation

$$a^2 - b^2 = (a - b)(a + b) \quad \text{General rule}$$

becomes
$$x^2 - 6^2 = (x - 6)(x + 6).$$ Substitute: $a = x, b = 6$

(b) The expression $4x^2 - 9$ can be written as $(2x)^2 - 3^2$. Thus
$$4x^2 - 9 = (2x - 3)(2x + 3).$$ $a = 2x, b = 3$

(c) The expression $100 - 16t^2$ can be written as $(10)^2 - (4t)^2$. Thus
$$100 - 16t^2 = (10 - 4t)(10 + 4t).$$ $a = 10, b = 4t$

(d) The expression $49y^2 - 64z^2$ can be written as $(7y)^2 - (8z)^2$. Thus
$$49y^2 - 64z^2 = (7y - 8z)(7y + 8z).$$ $a = 7y, b = 8z$

Now Try Exercises 17, 19, 25, 29

MAKING CONNECTIONS 1

Sum of Squares versus Difference of Squares

The sum of two squares, $a^2 + b^2$, cannot be factored by using real numbers. However, the difference of two squares, $a^2 - b^2$, can be factored.

$$x^2 + 4 \qquad\qquad x^2 - 4 = (x - 2)(x + 2)$$

Sum of squares cannot be factored.   Difference of squares can be factored.

READING CHECK 1

- If possible, factor $4x^2 + 9$.

2 Factoring Perfect Square Trinomials

In Section 4 of the previous chapter, we also showed how to compute $(a + b)^2$ and $(a - b)^2$ as
$$(a + b)^2 = a^2 + 2ab + b^2 \quad \text{and}$$
$$(a - b)^2 = a^2 - 2ab + b^2.$$

The expressions $a^2 + 2ab + b^2$ and $a^2 - 2ab + b^2$ are called **perfect square trinomials**. If we can recognize a perfect square trinomial, we can use the following formulas to factor it.

PERFECT SQUARE TRINOMIALS

For any real numbers a and b,
$$a^2 + 2ab + b^2 = (a + b)^2 \quad \text{and}$$
$$a^2 - 2ab + b^2 = (a - b)^2.$$

READING CHECK 2

- Is either $4x^2 + 4x + 1$ or $4x^2 + 9x + 16$ a perfect square trinomial?

When factoring a trinomial as a perfect square trinomial, we must first verify that the middle term is correct. For example, to factor $x^2 + 14x + 49$, we start by rewriting the expression as $x^2 + 14x + 7^2$. In order to factor this trinomial as a perfect square trinomial, the middle term $14x$ must be equal to twice the product of x and 7.

$$x^2 + 14x + 7^2$$
$$\downarrow$$
$$2(x)(7) \quad \checkmark \text{ The middle term checks.}$$

When $a = x$ and $b = 7$, the equation $a^2 + 2ab + b^2 = (a + b)^2$ allows us to factor the given polynomial as $x^2 + 14x + 49 = (x + 7)^2$.

EXAMPLE 2   **Factoring perfect square trinomials**

If possible, factor each trinomial as a perfect square trinomial.
(a) $x^2 + 10x + 25$ (b) $4x^2 - 4x + 1$ (c) $9z^2 + 18z + 4$ (d) $x^2 - 4xy + 4y^2$

Solution
(a) Start by writing $x^2 + 10x + 25$ as $x^2 + 10x + 5^2$ and then check the middle term.

$$x^2 + 10x + 5^2$$
$$\downarrow$$
$$2(x)(5) \quad \checkmark \quad \text{The middle term checks.}$$

When $a = x$ and $b = 5$, the equation $a^2 + 2ab + b^2 = (a + b)^2$ allows us to factor the given polynomial as $x^2 + 10x + 25 = (x + 5)^2$.

(b) The polynomial $4x^2 - 4x + 1$ can be written as $(2x)^2 - 4x + 1^2$.

$$(2x)^2 - 4x + 1^2$$
$$\downarrow$$
$$2(2x)(1) \quad \checkmark \quad \text{The middle term checks.}$$

When $a = 2x$ and $b = 1$, the equation $a^2 - 2ab + b^2 = (a - b)^2$ allows us to factor the given polynomial as $4x^2 - 4x + 1 = (2x - 1)^2$.

(c) Write $9z^2 + 18z + 4$ as $(3z)^2 + 18z + 2^2$ and then check the middle term.

$$(3z)^2 + 18z + 2^2$$
$$\downarrow$$
$$2(3z)(2) \quad \text{✗} \quad \text{The middle term is } 12z \text{ and does not equal } 18z.$$

The expression $9z^2 + 18z + 4$ cannot be factored as a perfect square trinomial.

(d) The polynomial $x^2 - 4xy + 4y^2$ can be written as $x^2 - 4xy + (2y)^2$.

$$x^2 - 4xy + (2y)^2$$
$$\downarrow$$
$$2(x)(2y) \quad \checkmark \quad \text{The middle term checks.}$$

When $a = x$ and $b = 2y$, the equation $a^2 - 2ab + b^2 = (a - b)^2$ allows us to factor the given polynomial as $x^2 - 4xy + 4y^2 = (x - 2y)^2$.

Now Try Exercises 37, 41, 43, 51

MAKING CONNECTIONS 2

Special Factoring and General Techniques

If you do not recognize a polynomial as the difference of two squares or a perfect square trinomial, you can still factor the polynomial by using the methods discussed in earlier sections.

---

3 Factoring the Sum and Difference of Two Cubes

The sum or difference of two cubes may be factored—a result of the two equations

$$(a + b)(a^2 - ab + b^2) = a^3 + b^3 \quad \text{and}$$
$$(a - b)(a^2 + ab + b^2) = a^3 - b^3.$$

These equations can be verified by multiplying the left side to obtain the right side. For example, multiplying the polynomials on the left side of the first equation results in

$$(a + b)(a^2 - ab + b^2) = a \cdot a^2 - a \cdot ab + a \cdot b^2 + b \cdot a^2 - b \cdot ab + b \cdot b^2$$
$$= a^3 - a^2b + ab^2 + a^2b - ab^2 + b^3$$
$$= a^3 + b^3.$$

SUM AND DIFFERENCE OF TWO CUBES

For any real numbers a and b,

$$a^3 + b^3 = (a + b)(a^2 - ab + b^2) \text{ and}$$
$$a^3 - b^3 = (a - b)(a^2 + ab + b^2).$$

(Opposite Signs)

Any binomial whose terms can be expressed as cubes can be factored as a sum or difference of cubes. We demonstrate this method in the next example.

EXAMPLE 3 Factoring the sum and difference of two cubes

Factor each polynomial.
(a) $z^3 + 8$ (b) $x^3 - 27$ (c) $8x^3 - 1$

Solution
(a) To factor $z^3 + 8$, we let $a^3 = z^3$ and $b^3 = 2^3$ so that $a = z$ and $b = 2$. Then

$$a^3 + b^3 = (a + b)(a^2 - ab + b^2) \quad \text{General rule}$$

becomes

$$z^3 + 2^3 = (z + 2)(z^2 - z \cdot 2 + 2^2) \quad \text{Substitute: } a = z, b = 2$$
$$= (z + 2)(z^2 - 2z + 4).$$

(b) To factor $x^3 - 27$, we let $a^3 = x^3$ and $b^3 = 3^3$ so that $a = x$ and $b = 3$. Then

$$a^3 - b^3 = (a - b)(a^2 + ab + b^2) \quad \text{General rule}$$

becomes

$$x^3 - 3^3 = (x - 3)(x^2 + x \cdot 3 + 3^2) \quad \text{Substitute: } a = x, b = 3$$
$$= (x - 3)(x^2 + 3x + 9).$$

(c) To factor $8x^3 - 1$, we let $a^3 = (2x)^3$ and $b^3 = 1^3$ so that $a = 2x$ and $b = 1$. Be sure to write $2x$ in *parentheses* when substituting for a in the term a^2.

Write $2x$ in parentheses.

$$(2x)^3 - 1^3 = (2x - 1)((2x)^2 + 2x \cdot 1 + 1^2) \quad a = 2x, b = 1$$
$$= (2x - 1)(4x^2 + 2x + 1)$$

Now Try Exercises 57, 59, 63

4 Using Special Methods in General Factoring

In this section, we have discussed special methods for factoring polynomials that can be identified as the difference of two squares, perfect square trinomials, the sum of two cubes, or the difference of two cubes. The next example demonstrates how to recognize and factor such polynomials.

EXAMPLE 4 Recognizing polynomials that can be factored with special methods

Factor each polynomial.
(a) $8x^3 + 27$ (b) $4y^2 - 20y + 25$ (c) $9z^2 - 64$

Solution
(a) Because this polynomial has only two terms, it cannot be factored as a perfect square trinomial. Since it is not a difference, it cannot be factored as the difference of two squares. We will try to factor the polynomial as the sum of two cubes. To factor $8x^3 + 27$, we note that $8x^3 = (2x)^3$ and $27 = 3^3$. Then

$$8x^3 + 27 = (2x)^3 + 3^3 \quad \text{Sum of cubes}$$
$$= (2x + 3)((2x)^2 - 2x \cdot 3 + 3^2)$$
$$= (2x + 3)(4x^2 - 6x + 9).$$

(b) Because this polynomial has three terms, we will try to factor it as a perfect square trinomial. Begin by writing $4y^2 - 20y + 25$ as $(2y)^2 - 20y + 5^2$ and then verify that the middle term checks.

$$(2y)^2 - 20y + 5^2 \quad \text{Perfect square trinomial}$$
$$\downarrow$$
$$2(2y)(5) \quad \checkmark \text{ The middle term checks.}$$

When $a = 2y$ and $b = 5$, the equation $a^2 - 2ab + b^2 = (a - b)^2$ allows us to factor the given polynomial as $4y^2 - 20y + 25 = (2y - 5)^2$.

(c) Because this polynomial is a difference of two terms that appear to be square terms, we will try to factor the polynomial as the difference of two squares. The expression $9z^2 - 64$ can be written as $(3z)^2 - 8^2$. Thus

Difference of squares $9z^2 - 64 = (3z - 8)(3z + 8).$

Now Try Exercises 31, 47, 67

When using the special factoring methods discussed in this section, it is important to remember to first factor out the greatest common factor whenever possible. In the next example, we factor out the GCF before factoring further.

EXAMPLE 5 Factoring out the GCF before factoring further

Factor each polynomial completely.
(a) $27x^3 + 72x^2 + 48x$ (b) $18a^3 - 8ab^2$

Solution
(a) Because the GCF of $27x^3 + 72x^2 + 48x$ is $3x$, factor out $3x$ before factoring the remaining trinomial.

$$27x^3 + 72x^2 + 48x = 3x(9x^2 + 24x + 16)$$

The expression $9x^2 + 24x + 16$ is a perfect square trinomial that can be written as $(3x)^2 + 24x + 4^2$.

$$(3x)^2 + 24x + 4^2 \quad \text{Perfect square trinomial}$$
$$\downarrow$$
$$2(3x)(4) \quad \checkmark \text{ The middle term checks.}$$

Thus $9x^2 + 24x + 16 = (3x + 4)^2$. As a result, the given polynomial can be factored as

$$27x^3 + 72x^2 + 48x = 3x(3x + 4)^2.$$

(b) Because the GCF of $18a^3 - 8ab^2$ is $2a$, factor out $2a$ before factoring the remaining polynomial.

$$18a^3 - 8ab^2 = 2a(9a^2 - 4b^2) \quad \text{Difference of squares}$$

The expression $9a^2 - 4b^2$ is the difference of two squares and can be written as $(3a)^2 - (2b)^2$. Thus

$$18a^3 - 8ab^2 = 2a(3a - 2b)(3a + 2b).$$

Now Try Exercises 71, 79

13.4 Putting It All Together

FACTORING	EXPLANATION	EXAMPLES
Difference of Two Squares	$a^2 - b^2 = (a - b)(a + b)$ **NOTE:** The *sum* of two squares, $a^2 + b^2$, cannot be factored by using real numbers. ∎	$y^2 - 49 = (y - 7)(y + 7)$ $81 - z^2 = (9 - z)(9 + z)$ $4r^2 - 25t^2 = (2r - 5t)(2r + 5t)$ $16a^2 + b^2$ cannot be factored.
Perfect Square Trinomial	$a^2 + 2ab + b^2 = (a + b)^2$ $a^2 - 2ab + b^2 = (a - b)^2$ Be sure to verify that the given middle term equals $2ab$ before factoring.	$m^2 + 2m + 1 = (m + 1)^2$ $25y^2 - 30y + 9 = (5y - 3)^2$ $36r^2 + 12rt + t^2 = (6r + t)^2$ $x^2 + 5x + 4$ is *not* a perfect square trinomial because $$2ab = 2 \cdot x \cdot 2 = 4x \neq 5x.$$
Sum and Difference of Two Cubes	$a^3 + b^3 = (a + b)(a^2 - ab + b^2)$ $a^3 - b^3 = (a - b)(a^2 + ab + b^2)$	$y^3 + 27 = (y + 3)(y^2 - y \cdot 3 + 3^2)$ $\qquad = (y + 3)(y^2 - 3y + 9)$ $27r^3 - 64t^3$ $= (3r - 4t)((3r)^2 + 3r \cdot 4t + (4t)^2)$ $= (3r - 4t)(9r^2 + 12rt + 16t^2)$

13.4 Exercises

CONCEPTS AND VOCABULARY

1. $a^2 - b^2 =$ _____
2. If the expression $36x^2 - 49y^2$ is written in the form $a^2 - b^2$, then $a =$ _____ and $b =$ _____.
3. (True or False?) The expression $a^2 + b^2$ can be factored by using real numbers.
4. (True or False?) Using only integer coefficients, the expression $3x^2 - 5y^2$ can be factored as a difference of two squares.
5. $a^2 + 2ab + b^2 =$ _____
6. $a^2 - 2ab + b^2 =$ _____
7. $x^2 +$ _____ $+ 9$ is a perfect square trinomial.
8. $4r^2 -$ _____ $+ 25t^2$ is a perfect square trinomial.
9. $a^3 + b^3 =$ _____
10. $a^3 - b^3 =$ _____
11. If the expression $8x^3 + 27y^3$ is written in the form $a^3 + b^3$, then $a =$ _____ and $b =$ _____.
12. If the expression $x^3 - 1$ is written in the form $a^3 - b^3$, then $a =$ _____ and $b =$ _____.
13. $y^3 - 8 = (y\_\_\_2)(y^2\_\_\_2y + 4)$
14. $64z^3 + 27 = (4z\_\_\_3)(16z^2\_\_\_12z + 9)$

FACTORING THE DIFFERENCE OF TWO SQUARES

Exercises 15–32: Factor.

15. $x^2 - 1$
16. $x^2 - 16$
17. $z^2 - 100$
18. $z^2 - 81$
19. $4y^2 - 1$
20. $9y^2 - 16$
21. $36z^2 - 25$
22. $49z^2 - 64$
23. $9 - x^2$
24. $25 - x^2$
25. $1 - 9y^2$
26. $49 - 16y^2$
27. $4a^2 - 9b^2$
28. $16a^2 - b^2$
29. $36m^2 - 25n^2$
30. $49m^2 - 100n^2$
31. $81r^2 - 49t^2$
32. $625r^2 - 121t^2$

FACTORING PERFECT SQUARE TRINOMIALS

Exercises 33–54: Factor as a perfect square trinomial whenever possible.

33. $x^2 + 8x + 16$
34. $x^2 + 4x + 4$
35. $z^2 + 12z + 25$
36. $z^2 - 18z + 36$
37. $x^2 - 6x + 9$
38. $x^2 - 10x + 25$
39. $9y^2 + 6y + 1$
40. $16y^2 + 8y + 1$
41. $4z^2 - 4z + 1$
42. $25z^2 - 12z + 1$
43. $9t^2 + 16t + 4$
44. $4t^2 + 12t + 9$
45. $9x^2 + 30x + 25$
46. $25x^2 + 60x + 36$
47. $4a^2 - 36a + 81$
48. $9a^2 - 60a + 100$
49. $x^2 + 2xy + y^2$
50. $x^2 - 6xy + 9y^2$
51. $r^2 - 10rt + 25t^2$
52. $15r^2 + 10rt + t^2$
53. $4y^2 - 10yz + 9z^2$
54. $25y^2 - 20yz + 4z^2$

FACTORING THE SUM AND DIFFERENCE OF TWO CUBES

Exercises 55–68: Factor.

55. $z^3 + 1$
56. $z^3 + 8$
57. $x^3 + 64$
58. $x^3 + 125$
59. $y^3 - 8$
60. $y^3 - 27$
61. $n^3 - 1$
62. $n^3 - 64$
63. $8x^3 + 1$
64. $27x^3 - 1$
65. $m^3 - 64n^3$
66. $m^3 + 8n^3$
67. $8x^3 + 125y^3$
68. $27x^3 + 64y^3$

USING SPECIAL METHODS IN GENERAL FACTORING

Exercises 69–86: Factor the expression completely.

69. $4x^2 - 16$
70. $12x^2 - 60x + 75$

71. $2y^2 - 28y + 98$
72. $y^3 - 9y$
73. $5z^3 + 40$
74. $4z^3 + 36z^2 + 100z$
75. $x^3y - xy^3$
76. $8m^3 - 8$
77. $2m^3 - 10m^2 + 18m$
78. $2a^3b - 18ab^3$
79. $700x^4 - 63x^2y^2$
80. $135r^3 - 5t^3$
81. $16a^3 + 2b^3$
82. $192x^2y^2 - 3y^4$
83. $4b^4 + 24b^3 + 36b^2$
84. $2y^4 + 24y^3 + 72y^2$
85. $500r^3 - 32t^3$
86. $8r^3 - 64t^3$

RELATING GEOMETRY AND VISUAL FACTORING

87. A square has an area of $4x^2 + 12x + 9$. Find the length of a side. Make a sketch of the square.

88. A square has an area of $9x^2 + 30x + 25$. Find the length of a side. Make a sketch of the square.

89. **Difference of Two Squares** Write the difference in the areas between the large square with sides of length x and a smaller square with sides of length 5, using factored form.

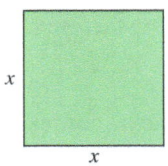
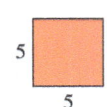

90. **Difference of Two Cubes** Write the difference in the volumes between a large cube with sides of length 2 and a smaller cube with sides of length x, using factored form.

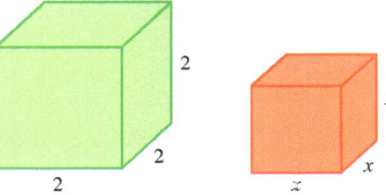

WRITING ABOUT MATHEMATICS

91. Explain how factoring $x^3 + y^3$ is different from factoring $x^3 - y^3$.

92. Using the techniques discussed in this section, can you factor the expression $4x^2 + 9y^2$ into two binomials? Explain your reasoning.

SECTIONS 13.3 and 13.4 — Checking Basic Concepts

1. Factor each trinomial.
   (a) $2x^2 - 5x - 12$ (b) $6x^2 + 17x - 14$

2. Factor completely when possible.
   (a) $3y^2 + 4y - 2$ (b) $6y^3 - 10y^2 - 4y$

3. Write a polynomial in factored form that represents the total area of the figure.

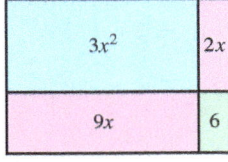

4. Factor each polynomial.
   (a) $z^2 - 64$ (b) $9r^2 - 4t^2$

5. Factor each trinomial.
   (a) $x^2 + 12x + 36$ (b) $9a^2 - 12ab + 4b^2$

6. Factor.
   (a) $m^3 - 27$ (b) $125n^3 + 27$

7. Factor completely.
   (a) $16x^2 - 4$ (b) $3y^4 + 24y$

13.5 Summary of Factoring

Objectives

1. Summarizing General Guidelines for Factoring Polynomials
2. Factoring Polynomials

STUDY TIP

This section provides an opportunity to practice the material from several sections. In general, it is a good idea to regularly review topics that have been covered earlier.

1 Summarizing General Guidelines for Factoring Polynomials

So far in this chapter, we have discussed several useful techniques for factoring polynomials. But in most factoring problems, the specific method that should be used is not stated. Instead, we must look carefully at each factoring problem and decide which approach is best. The following guidelines can be used to factor polynomials in general.

> **FACTORING POLYNOMIALS**
>
> **STEP 1:** Factor out the greatest common factor (GCF), if possible.
>
> **STEP 2:** **A.** If the polynomial has *four terms*, try factoring by grouping.
>
> **B.** If the polynomial is a *binomial*, try one of the following.
>   1. $a^2 - b^2 = (a - b)(a + b)$ Difference of two squares
>   2. $a^3 - b^3 = (a - b)(a^2 + ab + b^2)$ Difference of two cubes
>   3. $a^3 + b^3 = (a + b)(a^2 - ab + b^2)$ Sum of two cubes
>
> **C.** If the polynomial is a *trinomial*, check for a perfect square.
>   1. $a^2 + 2ab + b^2 = (a + b)^2$ Perfect square trinomial
>   2. $a^2 - 2ab + b^2 = (a - b)^2$ Perfect square trinomial
>
> Otherwise, try to factor the trinomial by grouping or apply FOIL in reverse, as described in Sections 2 and 3 of this chapter.
>
> **STEP 3:** Check to make sure that the polynomial is *completely* factored.

READING CHECK 1

- Why is it important to first factor out the greatest common factor?

NOTE: Always perform Step 1 first. Factoring out the greatest common factor usually makes it easier to factor the resulting polynomial. After a polynomial has been factored, remember to perform Step 3 so that you are sure the given polynomial is completely factored. ∎

2 Factoring Polynomials

In the first example, we apply Step 1 to a polynomial with a common factor.

EXAMPLE 1   Factoring out a common factor

Factor $5x^3 - 10x^2 + 15x$.

Solution

STEP 1: The greatest common factor is **$5x$**.   *Prime trinomial*

$$5x^3 - 10x^2 + 15x = 5x(x^2 - 2x + 3)$$

STEP 2C: The trinomial $x^2 - 2x + 3$ is prime and cannot be factored further.

STEP 3: The completely factored polynomial is $5x(x^2 - 2x + 3)$.

Now Try Exercise 11

When factoring polynomials completely, it is often necessary to apply more than one factoring technique. In several of the next examples, we factor polynomials that require more than one method of factoring.

EXAMPLE 2　Factoring a difference of squares

Factor $3x^4 - 48x^2$.

Solution
STEP 1: The greatest common factor is $3x^2$.

$$3x^4 - 48x^2 = 3x^2(x^2 - 16) \quad \text{Difference of squares}$$

STEP 2B: The binomial $x^2 - 16$ can be factored as a difference of two squares.

$$3x^2(x^2 - 16) = 3x^2(x - 4)(x + 4)$$

STEP 3: The completely factored polynomial is $3x^2(x - 4)(x + 4)$.

Now Try Exercise 39

MAKING CONNECTIONS 1

Factoring Polynomials

We can often determine how a polynomial should be factored by considering the number of terms in the polynomial. This is summarized as follows.

Type of Polynomial	*Factoring Technique*
4-term Polynomial	Grouping
Trinomial	Perfect square trinomial
	FOIL in reverse or grouping
Binomial	Difference of two squares
	Sum or difference of two cubes

EXAMPLE 3　Factoring a perfect square trinomial

Factor $36y^3 - 24y^2 + 4y$.

Solution
STEP 1: The greatest common factor is $4y$.

$$36y^3 - 24y^2 + 4y = 4y(9y^2 - 6y + 1) \quad \text{Perfect square trinomial}$$

STEP 2C: We can factor $9y^2 - 6y + 1$ as a perfect square trinomial.

$$4y(9y^2 - 6y + 1) = 4y(3y - 1)(3y - 1)$$

STEP 3: The completely factored polynomial is $4y(3y - 1)^2$.

Now Try Exercise 41

EXAMPLE 4　Factoring a sum of cubes

Factor $27z^3 + 64$.

Solution
STEP 1: There are no common factors.
STEP 2B: The binomial $27z^3 + 64$ can be written as $(3z)^3 + 4^3$ and factored as a sum of two cubes.

$$\text{Sum of cubes}\quad 27z^3 + 64 = (3z + 4)\left((3z)^2 - 3z \cdot 4 + 4^2\right)$$
$$= (3z + 4)(9z^2 - 12z + 16)$$

NOTE: The trinomial $9z^2 - 12z + 16$ is prime and cannot be factored further. ∎

STEP 3: The completely factored polynomial is $(3z + 4)(9z^2 - 12z + 16)$.

Now Try Exercise 15

EXAMPLE 5 Factoring a trinomial

Factor $14x^4 + 7x^3 - 42x^2$.

Solution
STEP 1: The greatest common factor is $7x^2$.   *Factor this trinomial further.*

$$14x^4 + 7x^3 - 42x^2 = 7x^2(2x^2 + x - 6)$$

STEP 2C: We can factor $2x^2 + x - 6$ using FOIL in reverse.

$$7x^2(2x^2 + x - 6) = 7x^2(2x - 3)(x + 2)$$

STEP 3: The completely factored polynomial is $7x^2(2x - 3)(x + 2)$.

Now Try Exercise 35

EXAMPLE 6 Factoring by grouping

Factor $15x^3 + 10x^2 - 60x - 40$.

Solution   *Factor by grouping.*
STEP 1: The greatest common factor is 5.

$$15x^3 + 10x^2 - 60x - 40 = 5(3x^3 + 2x^2 - 12x - 8)$$

STEP 2A: Because the resulting polynomial has four terms, we apply grouping.

$$\begin{aligned} 5(3x^3 + 2x^2 - 12x - 8) &= 5\big((3x^3 + 2x^2) + (-12x - 8)\big) &&\text{Associative property} \\ &= 5\big(x^2(3x + 2) - 4(3x + 2)\big) &&\text{Factor out common factors.} \\ &= 5(x^2 - 4)(3x + 2) &&\text{Factor out } (3x + 2). \end{aligned}$$

STEP 2B: The binomial $x^2 - 4$ can now be factored as a difference of two squares.

$$5(x^2 - 4)(3x + 2) = 5(x - 2)(x + 2)(3x + 2)$$

Difference of squares

STEP 3: The completely factored polynomial is $5(x - 2)(x + 2)(3x + 2)$.

Now Try Exercise 33

EXAMPLE 7 Factoring a polynomial that has two variables

Factor $18x^3y - 8xy^3$.

Solution
STEP 1: The greatest common factor is $2xy$.

$$18x^3y - 8xy^3 = 2xy(9x^2 - 4y^2)$$

STEP 2B: The binomial $9x^2 - 4y^2$ can be written as $(3x)^2 - (2y)^2$ and can be factored as a difference of two squares.

Difference of squares $\quad 9x^2 - 4y^2 = (3x - 2y)(3x + 2y)$

STEP 3: The completely factored polynomial is $2xy(3x - 2y)(3x + 2y)$.

Now Try Exercise 55

EXAMPLE 8 Applying several techniques

Factor $3x^5 - 3x^3 - 24x^2 + 24$.

Solution
STEP 1: The greatest common factor is 3.
$$3x^5 - 3x^3 - 24x^2 + 24 = 3(x^5 - x^3 - 8x^2 + 8)$$

STEP 2A: The resulting four-term polynomial can be factored by grouping.

$$3(x^5 - x^3 - 8x^2 + 8) = 3((x^5 - x^3) + (-8x^2 + 8)) \quad \text{Associative property}$$
$$= 3(x^3(x^2 - 1) - 8(x^2 - 1)) \quad \text{Factor out common factors.}$$
$$= 3(x^3 - 8)(x^2 - 1) \quad \text{Factor out } x^2 - 1.$$

Factor by grouping.

STEP 2B: Both binomials in this expression can be factored further. The binomial $x^3 - 8$ can be factored as a difference of two cubes, and the binomial $x^2 - 1$ can be factored as a difference of two squares.

$$3(x^3 - 8)(x^2 - 1) = 3(x - 2)(x^2 + 2x + 4)(x - 1)(x + 1)$$

Difference of cubes $\qquad$ Difference of squares

NOTE: The trinomial $x^2 + 2x + 4$ is prime and cannot be factored further. ■

STEP 3: The completely factored form is $3(x - 2)(x^2 + 2x + 4)(x - 1)(x + 1)$.

Now Try Exercise 45

13.5 Putting It All Together

CONCEPT	EXPLANATION	EXAMPLES
Greatest Common Factor	Factor out the greatest common factor, or monomial, that occurs in each term.	$2x^2 - 4x + 10 = 2(x^2 - 2x + 5)$ $3x^3 + 6x = 3x(x^2 + 2)$ $7xy - x^2y = xy(7 - x)$
Factoring by Grouping	Use the associative and distributive properties to factor a polynomial with four terms.	$x^3 - 3x^2 + 2x - 6 = (x^3 - 3x^2) + (2x - 6)$ $\qquad = x^2(x - 3) + 2(x - 3)$ $\qquad = (x^2 + 2)(x - 3)$
Factoring Binomials	Use the difference of two squares, the difference of two cubes, or the sum of two cubes.	$9x^2 - 4 = (3x - 2)(3x + 2)$ $x^3 - 27 = (x - 3)(x^2 + 3x + 9)$ $x^3 + 27 = (x + 3)(x^2 - 3x + 9)$

continued on next page

CONCEPT	EXPLANATION	EXAMPLES
Factoring Trinomials	Use FOIL in reverse or grouping.	$x^2 + 5x - 6 = (x + 6)(x - 1)$ Check middle term: $-x + 6x = 5x.$ ✓ $4x^2 + 4x - 3 = (4x^2 - 2x) + (6x - 3)$ $= 2x(2x - 1) + 3(2x - 1)$ $= (2x + 3)(2x - 1)$

13.5 Exercises

CONCEPTS AND VOCABULARY

1. What do the letters GCF mean?

2. A good first step for factoring polynomials is to factor out the _____ .

3. If a polynomial has four terms, what factoring method might be appropriate?

4. If a polynomial is a binomial, we can try to factor it as a difference of two _____, a difference of two _____, or a sum of two _____.

5. Can $x^2 + 1$ be factored? Explain.

6. Can $x^3 + 1$ be factored? Explain.

7. If a polynomial is a trinomial, we can try to factor it as a perfect _____ trinomial. Otherwise, try factoring it by _____ or apply _____ in reverse.

8. The last step for factoring is to be sure that the polynomial is _____ factored.

PERFORMING BASIC FACTORING

Exercises 9–24: Factor completely, if possible.

9. $4x - 2$
10. $x^2 + 3x$
11. $2y^2 - 4y + 4$
12. $5y^2 - 25y + 10$
13. $z^2 - 4$
14. $9z^2 - 25$
15. $a^3 + 8$
16. $8a^3 - 1$
17. $4b^2 - 12b + 9$
18. $b^2 + 4b + 4$
19. $m^2 + 9$
20. $4m^2 + 49$
21. $x^3 - x^2 + 5x - 5$
22. $3x^3 + 6x^2 + x + 2$
23. $y^2 - 5y + 4$
24. $y^2 - 3y - 10$

PERFORMING GENERAL FACTORING

Exercises 25–62: Factor completely.

25. $x^3 + 4x^2 - 9x - 36$
26. $6x^2 - 19x + 15$
27. $8a^3 - 64$
28. $ab^2 - 4a$
29. $12x^4 - 18x^3 + 4x^2 - 6x$
30. $3x^2y + 24xy + 48y$
31. $54t^4 + 16t$
32. $3t^3 + 18t^2 - 48t$
33. $2r^3 + 6r^2 - 2r - 6$
34. $3r^4 + 3r^3 - 24r - 24$
35. $6z^4 - 21z^3 - 45z^2$
36. $3x^4y + 24xy^4$
37. $12b^4 - 10b^3 + 2b^2$
38. $6a^4b + 4a^3b + 18a^2b + 12ab$
39. $6z - 24z^3$
40. $6y^3z - 48z^4$
41. $3x^2y - 30xy + 75y$
42. $8x^3 + y^3$
43. $27m^3 - 8n^3$
44. $45m^3 - 69m^2 + 12m$
45. $3x^5 - 12x^3 - 3x^2 + 12$
46. $8x^3 - 8$
47. $5a^2 - 27a - 18$
48. $2a^2 - 6ab + 3a - 9b$
49. $3rt^2 + 33rt + 90r$
50. $9t^2 + 24t + 16$
51. $9b^3 + 6b^2 + 12b + 8$
52. $5b^3 - 55b^2 - 60b$
53. $6n^3 + 2n^2 - 10n$
54. $7n^4 + 28n^3 - 63n^2$
55. $4x^2 - 36y^2$

56. $64x^2 - 25y^2$

57. $2a^3 - 16a^2 + 32a$

58. $24a^3 + 72a^2 + 54a$

59. $32xy^3 + 4x$

60. $24x^3 - 4x^2 - 160x$

61. $8b^4 + 24b^3 - 2b^2 - 6b$

62. $3z^3 - 6z^2 - 27z + 54$

RELATING GEOMETRY AND VISUAL FACTORING

63. **Dimensions of a Square** If three identical squares have a total area of $27x^2 + 18x + 3$, find the length of one side of one of the squares.

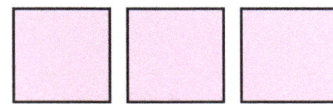

64. **Dimensions of a Cube** If three identical cubes have a total volume of $3x^3 + 18x^2 + 36x + 24$, find the length of one side of one of the cubes.

WRITING ABOUT MATHEMATICS

65. Explain how the number of terms in a polynomial can help determine what method should be used to factor it.

66. Describe a method for determining whether a polynomial has been factored correctly.

Group Activity

Exercises 1–6: **Difference of Two Squares** *The difference of two squares can be factored by using*

$$a^2 - b^2 = (a - b)(a + b).$$

This equation can also be used in some situations where an expression may not appear to be the difference of two squares. For example, because $(\sqrt{3})^2 = 3$, the binomial $x^2 - 3$ can be written and then factored as

$$x^2 - 3 = x^2 - (\sqrt{3})^2$$
$$= (x - \sqrt{3})(x + \sqrt{3}).$$

Use this concept to factor the following expressions as the difference of two squares.

1. $x^2 - 5$
2. $y^2 - 7$
3. $3z^2 - 25$
4. $7t^2 - 2$
5. $x - 4$ for $x \geq 0$ (Hint: $(\sqrt{x})^2 = x$.)
6. $x - 7$ for $x \geq 0$

13.6 Solving Equations by Factoring I (Quadratics)

Objectives

1. Applying the Zero-Product Property
2. Solving Quadratic Equations
3. Solving Applications Involving Quadratic Equations

NEW VOCABULARY

☐ Zero-product property
☐ Zeros (of a polynomial)
☐ Quadratic polynomial
☐ Quadratic equation
☐ Standard form

1 Applying the Zero-Product Property

To solve equations, we often use the **zero-product property**, which states that if the product of two numbers (or expressions) is 0, then at least one of the numbers (or expressions) must equal 0.

ZERO-PRODUCT PROPERTY

For all real numbers a and b, if $ab = 0$, then $a = 0$ or $b = 0$ (or both).

NOTE: The zero-product property works only for 0. If $ab = 1$, then it does *not* follow that $a = 1$ or $b = 1$. For example, $a = \frac{1}{3}$ and $b = 3$ satisfy the equation $ab = 1$. ∎

After factoring an expression, we can use the zero-product property to solve an equation. The left side of the equation $3t^2 - 9t = 0$ may be factored to obtain

$$3t(t - 3) = 0.$$

Note that the product of $3t$ and $t - 3$ is 0. By the zero-product property, either

$$3t = 0 \quad \text{or} \quad t - 3 = 0.$$

Solving each equation for t results in *Zero-product property*

$$t = 0 \quad \text{or} \quad t = 3.$$

These values can be checked by substituting them in the given equation $3t^2 - 9t = 0$.

$$3(0)^2 - 9(0) = 0 \checkmark \quad \text{Let } t = 0. \text{ It checks.}$$
$$3(3)^2 - 9(3) = 0 \checkmark \quad \text{Let } t = 3. \text{ It checks.}$$

The t-values of 0 and 3 are called **zeros** of the polynomial $3t^2 - 9t$, because when either is substituted in this polynomial, the result is 0.

STUDY TIP

The zero-product property is used extensively in mathematics for solving many types of equations. Be sure that you learn how to apply this important property correctly.

READING CHECK 1

- Which of the following are zeros of $4x^2 - 8x$: $-8, -4, -2, 0, 2, 4, 8$?

EXAMPLE 1   Applying the zero-product property

Solve each equation.
(a) $x(x - 1) = 0$
(b) $2z^2 = 0$
(c) $(t + 3)(t + 2) = 0$
(d) $x(x - 2)(2x + 1) = 0$

Solution
(a) By the zero-product property, $x(x - 1) = 0$ when $x = 0$ or $x - 1 = 0$. The solutions are 0 and 1.
(b) $2z^2 = 2 \cdot z \cdot z$ and $2 \neq 0$, so $2z^2 = 0$ when $z = 0$.
(c) $(t + 3)(t + 2) = 0$ implies that $t + 3 = 0$ or $t + 2 = 0$. The solutions to the given equation are -3 and -2.
(d) We can apply the zero-product property to $x(x - 2)(2x + 1) = 0$. Thus $x = 0$, or $x - 2 = 0$, or $2x + 1 = 0$. The solutions are $-\frac{1}{2}$, 0, and 2.

Now Try Exercises 13, 15, 19, 23

2 Solving Quadratic Equations

Any **quadratic polynomial** in the variable x can be written as $ax^2 + bx + c$ with $a \neq 0$. Any **quadratic equation** in the variable x can be written as $ax^2 + bx + c = 0$ with $a \neq 0$. This form of quadratic equation is called the **standard form** of a quadratic equation. **TABLE 13.11** shows examples of quadratic polynomials along with related quadratic equations that can be expressed in standard form.

Quadratic Polynomials and Equations

Quadratic Polynomial	Quadratic Equation	Standard Form
$x^2 - 2x + 1$	$x^2 - 2x + 1 = 9$	$x^2 - 2x - 8 = 0$
$3x^2 + 7x$	$3x^2 + 7x = 4$	$3x^2 + 7x - 4 = 0$
$x^2 - 9$	$x^2 - 9 = 2$	$x^2 - 11 = 0$
$6x - x^2 + 2$	$6x - x^2 + 2 = 1$	$-x^2 + 6x + 1 = 0$

TABLE 13.11

READING CHECK 2

- What does it mean for a polynomial to be written so that it contains descending powers of x?

NOTE: When a quadratic polynomial in the variable x is in standard form and we read it from left to right, the terms contain *descending powers* of x. In other words, the first term contains x^2, the second term contains x, and the third term is a constant (the exponent on x is 0). ∎

To solve a quadratic equation, we often use factoring and the zero-product property. This method is summarized by the following steps. Although it is not necessary to label each step in the solution to a quadratic equation, it is important to keep these steps in mind.

SOLVING QUADRATIC EQUATIONS

To solve a quadratic equation by factoring, follow these steps.

STEP 1: If necessary, write the equation in standard form as $ax^2 + bx + c = 0$.

STEP 2: Factor the left side of the equation using any method.

STEP 3: Apply the zero-product property.

STEP 4: Solve each of the resulting equations. Check any solutions.

EXAMPLE 2 Solving equations by factoring

Solve each quadratic equation. Check your answers.
(a) $x^2 + 2x = 0$ (b) $y^2 = 16$ (c) $z^2 - 3z + 2 = 0$ (d) $2x^2 = 5 - 9x$

Solution
(a) Because $x^2 + 2x = 0$ is in standard form, Step 1 is unnecessary. We begin by factoring out the GCF, x.

$x^2 + 2x = 0$ Given equation
$x(x + 2) = 0$ Factor out x. (Step 2)
$x = 0$ or $x + 2 = 0$ Zero-product property (Step 3)
$x = 0$ or $x = -2$ Solve for x. (Step 4)

To check these values, substitute -2 and 0 for x in the given equation.

$(-2)^2 + 2(-2) \stackrel{?}{=} 0$ $(0)^2 + 2(0) \stackrel{?}{=} 0$ Substitute -2 and 0.
$0 = 0$ ✓ $0 = 0$ ✓   Both answers check.

Therefore the solutions are -2 and 0.

(b) To write $y^2 = 16$ in standard form, we begin by subtracting 16 from each side to obtain 0 on the right side.

$y^2 = 16$ Given equation
$y^2 - 16 = 0$ Subtract 16. (Step 1)
$(y - 4)(y + 4) = 0$ Difference of squares (Step 2)
$y - 4 = 0$ or $y + 4 = 0$ Zero-product property (Step 3)
$y = 4$ or $y = -4$ Solve for y. (Step 4)

To check these values, substitute -4 and 4 for y in the given equation.

$(-4)^2 \stackrel{?}{=} 16$ $(4)^2 \stackrel{?}{=} 16$ Substitute -4 and 4.
$16 = 16$ ✓ $16 = 16$ ✓   Both answers check.

The solutions are -4 and 4.

(c) Step 1 is unnecessary because the equation is already in standard form. We begin by factoring the left side of the equation, $z^2 - 3z + 2$.

$$z^2 - 3z + 2 = 0 \qquad \text{Given equation}$$
$$(z - 1)(z - 2) = 0 \qquad \text{Factor. (Step 2)}$$
$$z - 1 = 0 \quad \text{or} \quad z - 2 = 0 \qquad \text{Zero-product property (Step 3)}$$
$$z = 1 \quad \text{or} \quad z = 2 \qquad \text{Solve for } z. \text{ (Step 4)}$$

To check these values, substitute 1 and 2 for z in the given equation.

$$1^2 - 3(1) + 2 \stackrel{?}{=} 0 \qquad 2^2 - 3(2) + 2 \stackrel{?}{=} 0 \qquad \text{Substitute 1 and 2.}$$
$$0 = 0 \checkmark \qquad 0 = 0 \checkmark \qquad \text{Both answers check.}$$

The solutions are 1 and 2.

(d) We write $2x^2 = 5 - 9x$ in standard form by adding -5 and $9x$ to each side.

$$2x^2 = 5 - 9x \qquad \text{Given equation}$$
$$2x^2 + 9x - 5 = 0 \qquad \text{Add } -5 \text{ and } 9x. \text{ (Step 1)}$$
$$(2x - 1)(x + 5) = 0 \qquad \text{Factor. (Step 2)}$$
$$2x - 1 = 0 \quad \text{or} \quad x + 5 = 0 \qquad \text{Zero-product property (Step 3)}$$
$$x = \frac{1}{2} \quad \text{or} \quad x = -5 \qquad \text{Solve for } x. \text{ (Step 4)}$$

To check these values, substitute -5 and $\frac{1}{2}$ for x in the given equation.

$$2(-5)^2 \stackrel{?}{=} 5 - 9(-5) \qquad 2\left(\frac{1}{2}\right)^2 \stackrel{?}{=} 5 - 9\left(\frac{1}{2}\right) \qquad \text{Substitute } -5 \text{ and } \tfrac{1}{2}.$$
$$50 = 50 \checkmark \qquad \frac{1}{2} = \frac{1}{2} \checkmark \qquad \text{Both answers check.}$$

The solutions are -5 and $\frac{1}{2}$.

Now Try Exercises 27, 35, 43, 49

EXAMPLE 3 Solving an equation by factoring

Solve $6x^2 - x = 12$.

Solution
We *cannot* solve the equation $6x^2 - x = 12$ by factoring out the common factor of x in $6x^2 - x$ and setting each factor equal to 12. Instead, we apply the zero-product property by first writing the given equation in the standard form as $ax^2 + bx + c = 0$.

$$6x^2 - x = 12 \qquad \text{Given equation}$$
$$6x^2 - x - 12 = 0 \qquad \text{Subtract 12. (Step 1)}$$
$$(2x - 3)(3x + 4) = 0 \qquad \text{Factor. (Step 2)}$$
$$2x - 3 = 0 \quad \text{or} \quad 3x + 4 = 0 \qquad \text{Zero-product property (Step 3)}$$
$$x = \frac{3}{2} \quad \text{or} \quad x = -\frac{4}{3} \qquad \text{Solve for } x. \text{ (Step 4)}$$

The solutions are $-\frac{4}{3}$ and $\frac{3}{2}$.

Now Try Exercise 51

READING CHECK 3

Which of the following are expressions: $x^2 - 4x + 8$, $\sqrt{x+1} = x$, $\frac{x+4}{5} - x^2$, $x + 5 = x^2$?

> **MAKING CONNECTIONS 1**
>
> **Equations and Expressions**
>
> The words "equation" and "expression" occur frequently in mathematics. However, they are *not* interchangeable. We often want to *solve* equations to find the values of the variable that make the equation true. We *factor* and *simplify* expressions. An equation is a statement that two expressions are equal. For example, $2x^2 - 3 = 5x + 1$ is an equation where $2x^2 - 3$ and $5x + 1$ are each expressions.

3 ▸ Solving Applications Involving Quadratic Equations

🌐 **Math in Sports Context** If a golf ball is hit upward at 132 feet per second, or 90 miles per hour, then its height h in feet above the ground after t seconds is given by $h = 132t - 16t^2$. The expression $132t - 16t^2$ is an example of a *quadratic polynomial*. To determine the elapsed time between when the ball is hit and when it strikes the ground (or when $h = 0$), we solve the quadratic equation

$$132t - 16t^2 = 0.$$

One method for solving this equation is by factoring.

EXAMPLE 4 **Modeling the flight of a golf ball**

If a golf ball is hit upward at 132 feet per second, or 90 miles per hour, then its height h in feet after t seconds is $h = 132t - 16t^2$. Find the time when the golf ball strikes the ground.

Solution
The golf ball strikes the ground when its height is 0.

$132t - 16t^2 = 0$		Let $h = 0$.
$4t(33 - 4t) = 0$		Factor out $4t$.
$4t = 0$ or $33 - 4t = 0$		Zero-product property
$t = 0$ or $-4t = -33$		Divide by 4; subtract 33.
$t = 0$ or $t = \frac{33}{4}$		Solve for t.

The ball strikes the ground after $\frac{33}{4} = 8.25$ seconds. The solution of 0 is not used in this problem because it corresponds to the time when the ball is hit from ground level.

Now Try Exercise 67 (a)

🌐 **Math in Highway Safety Context** When you try to stop a car, the faster you are driving, the further the stopping distance will be. In fact, if you drive twice as fast, the braking distance will be about four times as much, and if you drive three times faster, the braking distance will be about nine times as much.

EXAMPLE 5 **Modeling braking distance**

The braking distance D in feet required to stop a car traveling at x miles per hour on dry, level pavement can be approximated by $D = \frac{1}{11}x^2$. (*Source:* L. Haefner.)
(a) Calculate the braking distance for a car traveling 70 miles per hour.
(b) If the braking distance is 44 feet, calculate the speed of the car.
(c) If you have a calculator available, use it to solve part (b) numerically with a table of values.

Solution

(a) If $x = 70$, then $D = \frac{1}{11}(70)^2 = \frac{4900}{11} \approx 445$ feet.

(b) **Symbolic Solution** Let $D = 44$ in the given equation and solve.

$$\frac{1}{11}x^2 = 44 \qquad \text{Let } D = 44.$$

$$\frac{1}{11}x^2 - 44 = 0 \qquad \text{Subtract 44.}$$

$$x^2 - 484 = 0 \qquad \text{Multiply by 11.}$$

$$(x - 22)(x + 22) = 0 \qquad \text{Difference of two squares } (22^2 = 484)$$

$$x - 22 = 0 \quad \text{or} \quad x + 22 = 0 \qquad \text{Zero-product property}$$

$$x = 22 \quad \text{or} \quad x = -22 \qquad \text{Solve for } x.$$

The car is traveling at (approximately) 22 miles per hour. (Note that $x = -22$ has no physical meaning in this problem.)

(c) **Numerical Solution** Let $Y_1 = X^2/11$ and make a table of values. Scroll through the table, as shown in **FIGURE 13.5**, to where $x = 22$ when $y_1 = 44$. Thus the solution is 22 miles per hour.

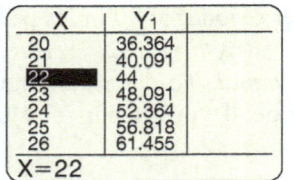

FIGURE 13.5

Now Try Exercise 69

EXAMPLE 6 Finding the dimensions of a digital photograph

A small digital photograph is 20 pixels longer than it is wide, as illustrated in **FIGURE 13.6**. It has a total of 2400 pixels. Find the dimensions of this photograph.

Dimensions of a Photograph

FIGURE 13.6

Solution
From **FIGURE 13.6** the rectangular photograph has an area of 2400 pixels.

$$x(x + 20) = 2400 \qquad \text{Area = width} \times \text{length}$$

$$x^2 + 20x = 2400 \qquad \text{Distributive property}$$

$$x^2 + 20x - 2400 = 0 \qquad \text{Subtract 2400.}$$

$$(x - 40)(x + 60) = 0 \qquad \text{Factor.}$$

$$x - 40 = 0 \quad \text{or} \quad x + 60 = 0 \qquad \text{Zero-product property}$$

$$x = 40 \quad \text{or} \quad x = -60 \qquad \text{Solve for } x.$$

The only valid solution is 40. Thus the dimensions of the photograph are 40 pixels by $40 + 20 = 60$ pixels.

Now Try Exercise 71

CRITICAL THINKING 1

Are the two solutions to $x(2x + 1) = 1$ found by setting both factors equal to 1? Explain your reasoning. What are the solutions to this equation?

13.6 Putting It All Together

CONCEPT	EXPLANATION	EXAMPLES
Zero-Product Property	If the product of two or more expressions is 0, then at least one of the expressions must equal 0.	$ab = 0$ implies that $a = 0$ or $b = 0$. $x(x + 1) = 0$ implies that $x = 0$ or $x + 1 = 0$. $z(z - 1)(z + 2) = 0$ implies that $z = 0$ or $z - 1 = 0$ or $z + 2 = 0$.
Solving Quadratic Equations by Factoring	1. Write the equation as $$ax^2 + bx + c = 0.$$ 2. Factor the left side of this equation. 3. Apply the zero-product property. 4. Solve each resulting equation. Check any solutions.	$2x^2 + 11x = 6$ $2x^2 + 11x - 6 = 0$ Step 1 $(2x - 1)(x + 6) = 0$ Step 2 $2x - 1 = 0$ or $x + 6 = 0$ Step 3 $x = \dfrac{1}{2}$ or $x = -6$ Step 4

13.6 Exercises

CONCEPTS AND VOCABULARY

1. If $ab = 0$, then either $a =$ _____ or $b =$ _____.

2. Can the zero-product property be used to state that if $(x - 1)(x - 2) = 3$, then either $x - 1 = 3$ or $x - 2 = 3$? Explain your answer.

3. If $2x(x + 6) = 0$, then either _____ or _____.

4. What is a good first step when you are solving the equation $4x^2 + 1 = 4x$ by factoring?

5. What is the next step when you are solving the equation $(x + 5)(x - 4) = 0$?

6. Factoring is an important method for ____ equations.

7. Because $2(4) - 8 = 0$, the value 4 is called a(n) _____ of the polynomial $2x - 8$.

8. What is the zero of the polynomial $3x - 6$?

9. Any quadratic equation in the variable x can be written in standard form as _____.

10. Standard form for $x^2 + 1 = 6x$ is _____.

11. When written in standard form and read from left to right, a quadratic polynomial in the variable x contains ____ powers of x.

12. For the constant term in a quadratic polynomial in the variable x, the exponent on x is _____.

APPLYING THE ZERO-PRODUCT PROPERTY

Exercises 13–24: Solve the equation.

13. $x^2 = 0$
14. $5m^2 = 0$
15. $2x(x + 8) = 0$
16. $x(x + 10) = 0$
17. $(y - 1)(y - 2) = 0$
18. $(y + 4)(y - 3) = 0$
19. $(6z + 5)(z - 7) = 0$
20. $(2z - 1)(4z - 3) = 0$
21. $(1 - 3n)(3 - 7n) = 0$
22. $(5 - n)(5 + n) = 0$
23. $x(x - 5)(x - 8) = 0$
24. $x(x + 1)(x - 6) = 0$

SOLVING QUADRATIC EQUATIONS

Exercises 25–60: Solve and check.

25. $x^2 - x = 0$
26. $2x^2 + 4x = 0$
27. $z^2 - 5z = 0$
28. $6z^2 - 3z = 0$

29. $10y^2 + 15y = 0$
30. $2y^2 + 3y = 0$
31. $x^2 - 1 = 0$
32. $x^2 - 9 = 0$
33. $4n^2 - 1 = 0$
34. $9n^2 - 4 = 0$
35. $z^2 + 3z + 2 = 0$
36. $z^2 - 2z - 3 = 0$
37. $x^2 - 12x + 35 = 0$
38. $x^2 - x - 20 = 0$
39. $2b^2 + 3b - 2 = 0$
40. $3b^2 + b - 2 = 0$
41. $6y^2 + 19y + 10 = 0$
42. $4y^2 - 25y - 21 = 0$
43. $x^2 = 25$
44. $x^2 = 81$
45. $t^2 = 5t$
46. $10t^2 = -5t$
47. $3m^2 = -9m$
48. $4m^2 = 9$
49. $x^2 = 5x + 6$
50. $12z^2 = 5 - 4z$
51. $2x^2 + 3x = 14$
52. $12z^2 + 11z = 15$
53. $t(t + 1) = 2$
54. $t(t - 7) = -12$
55. $x(2x + 5) = 3$
56. $x(3x + 2) = 5$
57. $12x^2 + 12x = -3$
58. $18x^2 + 2 = 12x$
59. $30y^2 + 50y + 20 = 0$
60. $30y^2 - 25y + 5 = 0$

RELATING GEOMETRY AND EQUATIONS

61. **Dimensions of a Square** A square has an area of 144 square feet. Find the length of a side.

62. **Dimensions of a Cube** A cube has a surface area of 96 square feet. Find the length of a side.

63. **Radius of a Circle** The numerical difference between the area and the circumference of a circle is 8π. Find the radius of the circle. (*Hint:* First factor out π in your equation.)

64. **Dimensions of a Rectangle** A rectangle is 5 feet longer than it is wide and has an area of 126 square feet. What are its dimensions?

Exercise 65 and 66: **The Pythagorean Theorem** *Suppose that a right triangle has legs a and b and hypotenuse c, as illustrated in the figure. Then these values satisfy $a^2 + b^2 = c^2$.*

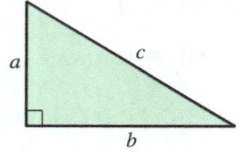

Use the Pythagorean theorem to find the value of x in the figure.

65.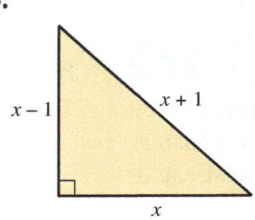

66.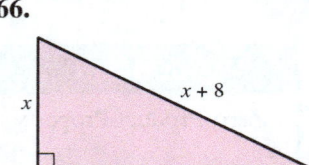

APPLICATIONS INVOLVING QUADRATIC EQUATIONS

67. **Flight of a Golf Ball** (Refer to Example 4.) The height h in feet of a golf ball after t seconds is given by $h = 96t - 16t^2$.
    (a) How long does it take for the golf ball to hit the ground?
    (b) Make a table of h for $t = 0, 1, 2, \ldots, 6$. After how many seconds does the golf ball reach its maximum height?

68. **Flight of a Baseball** The height h in feet of a baseball after t seconds is given by $h = -16t^2 + 88t + 3$. At what values of t is the height of the baseball 75 feet?

69. **Braking Distance** (Refer to Example 5.) The braking distance D in feet required to stop a car traveling x miles per hour on dry, level pavement can be approximated by $D = \frac{1}{11}x^2$.
    (a) Calculate the braking distance for 30 miles per hour and 60 miles per hour. How do your answers compare?
    (b) If the braking distance is 33 feet, estimate the speed of the car.
    (c) If you have a calculator, use it to solve part (b) numerically. Do your answers agree?

70. **Braking Distance** The braking distance D in feet required to stop a car traveling x miles per hour on wet, level pavement is approximated by $D = \frac{1}{9}x^2$.
    (a) Calculate the braking distance for 36 miles per hour and 72 miles per hour. How do your answers compare?
    (b) If the braking distance is 49 feet, estimate the speed of the car.
    (c) If you have a calculator, use it to solve part (b) numerically. Do your answers agree?

71. **Digital Photographs** (Refer to Example 6.) A digital photograph is 10 pixels longer than it is wide and has a total area of 2000 pixels. Find the dimensions of this rectangular photograph.

72. **Dimensions of a Building** The rectangular floor of a shed has a length 4 feet longer than its width, and its area is 140 square feet. Let x be the width of the floor.
    (a) Write a quadratic equation whose solution gives the width of the floor.
    (b) Solve this equation.

WRITING ABOUT MATHEMATICS

73. List four steps for solving a quadratic equation by factoring.

74. Explain why factoring is important.

SECTIONS 13.5 and 13.6 Checking Basic Concepts

1. Factor out the greatest common factor.
   (a) $9a^2 - 18a + 27$ (b) $7xy^2 + 28x$

2. Factor completely.
   (a) $6z^4 - 28z^3 + 16z^2$
   (b) $2r^2t^2 - 18r^2$

3. Factor completely.
   (a) $36x^3 - 48x^2 + 16x$
   (b) $24b^3 - 81$

4. Solve each quadratic equation.
   (a) $4y^2 - 6y = 0$ (b) $5z^2 + 2z = 3$

5. Solve $x^2 + 2x - 3 = 0$ symbolically and numerically with a table of values.

6. If a golf ball is hit upward at 60 miles per hour, then its height h in feet after t seconds is given by $h = 88t - 16t^2$. Use factoring to determine when the golf ball strikes the ground.

13.7 Solving Equations by Factoring II (Higher Degree)

Objectives

1. Recognizing Nonlinear Data and Polynomial Models
2. Factoring Higher-Degree Polynomials
3. Solving Polynomial Equations
4. Solving Applications Involving Polynomial Equations

1 Recognizing Nonlinear Data and Polynomial Models

In this section, we discuss factoring polynomials that have higher degree. Polynomials of degree 2 or higher are often used in applications. For example, the polynomial

$$0.0013x^3 - 0.085x^2 + 1.6x + 12$$

models natural gas consumption in the United States in trillions of cubic feet, where $x = 0$ corresponds to 1960, $x = 10$ corresponds to 1970, and so on until $x = 40$ corresponds to 2000, as shown in **FIGURE 13.7**. (This trend did not continue after 2000.) In Exercises 81–84 we discuss further the consumption of natural gas. (*Source:* Department of Energy.)

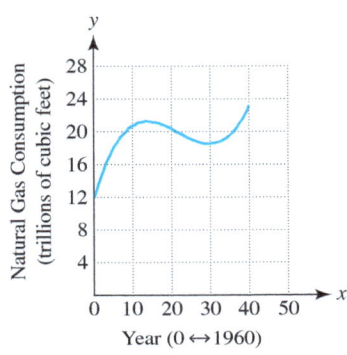

FIGURE 13.7

READING CHECK 1

- What is a good first step in factoring a polynomial?

2 Factoring Higher-Degree Polynomials

The first step in factoring a polynomial is to factor out the greatest common factor (GCF). Once this is accomplished, the resulting expression should be factored further by using any of the factoring methods discussed earlier in this chapter.

EXAMPLE 1   Factoring trinomials with common factors

Factor each trinomial completely.
(a) $-4x^2 + 28x - 40$ (b) $10x^3 + 28x^2 - 6x$

Solution
(a) Each term in $-4x^2 + 28x - 40$ has a factor of -4, so we start by factoring out -4. (In this example, we factor out the *opposite* of the GCF to obtain a positive leading coefficient in the resulting expression.)

$$-4x^2 + 28x - 40 = -4(x^2 - 7x + 10) \quad \text{Factor out } -4.$$
$$= -4(x-5)(x-2) \quad \text{Factor the trinomial.}$$

(b) We start by factoring out the GCF for $10x^3 + 28x^2 - 6x$, which is $2x$.

$$10x^3 + 28x^2 - 6x = 2x(5x^2 + 14x - 3) \quad \text{Factor out } 2x.$$
$$= 2x(5x-1)(x+3) \quad \text{Factor the trinomial.}$$

Now Try Exercises 13, 21

EXAMPLE 2   Factoring higher-degree polynomials

Factor each polynomial completely.
(a) $x^4 - 16$ (b) $y^4 + 5y^2 + 4$
(c) $x^4 + 2x^2y^2 + y^4$ (d) $r^4 - t^4$

Solution
(a) We view this polynomial as the difference of two squares, where $x^4 = (x^2)^2$ and $16 = 4^2$. Then we factor twice.

$$x^4 - 16 = (x^2)^2 - 4^2 \quad \text{Rewrite.}$$
$$= (x^2 - 4)(x^2 + 4) \quad \text{Difference of two squares}$$
$$= (x-2)(x+2)(x^2+4) \quad \text{Difference of two squares}$$

Note that $x^2 + 4$ does not factor.

(b) Because the trinomial $a^2 + 5a + 4$ factors as $(a+1)(a+4)$, we let $a = y^2$ and then factor the given trinomial.

$$y^4 + 5y^2 + 4 = (y^2)^2 + 5(y^2) + 4$$
$$= (y^2 + 1)(y^2 + 4)$$

Note that neither $y^2 + 1$ nor $y^2 + 4$ can be factored further.

(c) Because the perfect square trinomial $a^2 + 2ab + b^2$ factors as $(a+b)^2$, we let $a = x^2$ and $b = y^2$ and then factor the given trinomial.

$$x^4 + 2x^2y^2 + y^4 = (x^2)^2 + 2x^2y^2 + (y^2)^2$$
$$= (x^2 + y^2)^2$$

(d) Because the difference of squares $a^2 - b^2$ factors as $(a-b)(a+b)$, we let $a = r^2$ and $b = t^2$ and then factor the given binomial.

$$r^4 - t^4 = (r^2)^2 - (t^2)^2$$
$$= (r^2 - t^2)(r^2 + t^2)$$
$$= (r-t)(r+t)(r^2 + t^2)$$

Note that $r^2 + t^2$ cannot be factored.

Now Try Exercises 29, 37, 43, 45

3 Solving Polynomial Equations

READING CHECK 2

- When solving a polynomial equation by factoring, to which factors can we apply the zero-product property?

Many equations involving higher-degree polynomials can be solved using the four-step process discussed in the previous section. It is important to remember that the zero-product property applies to *all* factors that contain the variable. In the next example, we apply the zero-product property to three factors, resulting in three solutions.

EXAMPLE 3   Solving higher-degree polynomial equations

Solve each equation.
(a) $x^3 - x^2 - 6x = 0$ (b) $4x^4 + 10x^3 = 6x^2$

Solution
(a) We start by factoring out the GCF, which is x.

$$x^3 - x^2 - 6x = 0 \quad \text{Given equation}$$
$$x(x^2 - x - 6) = 0 \quad \text{Factor out } x.$$
$$x(x - 3)(x + 2) = 0 \quad \text{Factor the trinomial.}$$
$$x = 0 \quad \text{or} \quad x - 3 = 0 \quad \text{or} \quad x + 2 = 0 \quad \text{Zero-product property}$$
$$x = 0 \quad \text{or} \quad x = 3 \quad \text{or} \quad x = -2 \quad \text{Solve for } x.$$

The solutions are -2, 0, and 3.

(b) We start by subtracting $6x^2$ from each side to obtain 0 on one side of the equation.

$$4x^4 + 10x^3 = 6x^2 \quad \text{Given equation}$$
$$4x^4 + 10x^3 - 6x^2 = 0 \quad \text{Subtract } 6x^2.$$
$$2x^2(2x^2 + 5x - 3) = 0 \quad \text{Factor out the GCF, } 2x^2.$$
$$2x^2(2x - 1)(x + 3) = 0 \quad \text{Factor the trinomial.}$$
$$2x \cdot x = 0 \quad \text{or} \quad 2x - 1 = 0 \quad \text{or} \quad x + 3 = 0 \quad \text{Zero-product property}$$
$$x = 0 \quad \text{or} \quad x = \frac{1}{2} \quad \text{or} \quad x = -3 \quad \text{Solve for } x.$$

The solutions are -3, 0, and $\frac{1}{2}$.

Now Try Exercises 59, 65

STUDY TIP

If you are having difficulty with your studies, you may be able to find help at the student support services office on your campus.

EXAMPLE 4   Solving a higher-degree equation

Solve $x^5 - 81x = 0$.

Solution
We start by factoring out the common factor of x.

$$x^5 - 81x = 0 \quad \text{Given equation}$$
$$x(x^4 - 81) = 0 \quad \text{Factor out } x.$$
$$x(x^2 - 9)(x^2 + 9) = 0 \quad \text{Difference of two squares}$$
$$x(x - 3)(x + 3)(x^2 + 9) = 0 \quad \text{Difference of two squares}$$
$$x = 0 \quad \text{or} \quad x - 3 = 0 \quad \text{or} \quad x + 3 = 0 \quad \text{or} \quad x^2 + 9 = 0 \quad \text{Zero-product property}$$
$$x = 0 \quad \text{or} \quad x = 3 \quad \text{or} \quad x = -3 \quad \text{Solve for } x.$$

Note that $x^2 + 9 = 0$ has no real-number solutions because the square of a number plus 9 is never 0. The solutions are -3, 0, and 3.

Now Try Exercise 71

4 Solving Applications Involving Polynomial Equations

CRITICAL THINKING 1

Calculate the outside surface area A of the box shown in **FIGURE 13.8** two different ways.

🌐 **Math in Context** Construction The corners of a square piece of metal are cut out to form a box, as shown in **FIGURE 13.8**. This square piece of metal has sides with length 10 inches, and the cutout corners are squares with length x. The outside surface area A of this box, including the bottom and the sides but *not* the top, is $A = 100 - 4x^2$. (See Critical Thinking 1 in the margin.)

Construction of a Box

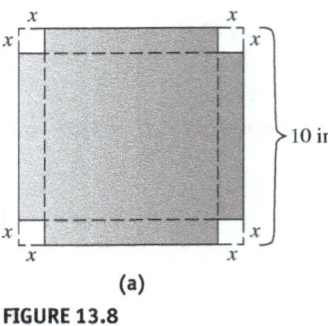
(a)

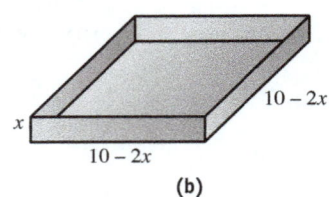
(b)

FIGURE 13.8

EXAMPLE 5 Finding a dimension of a box

Find the value of x in **FIGURE 13.8**, if the outside surface area A is 84 square inches.

Solution
Let $A = 84$ in the equation $A = 100 - 4x^2$ and solve for x.

$$100 - 4x^2 = 84 \quad \text{Let } A = 84.$$
$$16 - 4x^2 = 0 \quad \text{Subtract 84.}$$
$$4(4 - x^2) = 0 \quad \text{Factor out 4.}$$
$$4(2 - x)(2 + x) = 0 \quad \text{Difference of two squares}$$
$$2 - x = 0 \quad \text{or} \quad 2 + x = 0 \quad \text{Zero-product property}$$
$$x = 2 \quad \text{or} \quad x = -2 \quad \text{Solve for } x.$$

If squares measuring 2 inches on a side are cut out of each corner, the surface area of the box is 84 square inches. (Note that $x = -2$ has no physical meaning in this problem.)

Now Try Exercise 77

13.7 Putting It All Together

CONCEPT	EXPLANATION	EXAMPLES
Common Factors	A first step when factoring polynomials is to factor out the GCF. Once this is done, factor the resulting expression, if possible.	$x^3 - 4x^2 - 5x = x(x^2 - 4x - 5)$ $= x(x - 5)(x + 1)$ $4x^4 - 16x^2 = 4x^2(x^2 - 4)$ $= 4x^2(x - 2)(x + 2)$

CONCEPT	EXPLANATION	EXAMPLES
Factoring Higher Degree Polynomials	Some higher degree polynomials can be factored by using the same methods that we use to factor quadratic polynomials.	$x^4 + 6x^2 + 5 = (x^2 + 5)(x^2 + 1)$ $4y^4 - 25 = (2y^2 - 5)(2y^2 + 5)$ $2z^4 - 32 = 2(z^4 - 16)$ $ = 2(z^2 - 4)(z^2 + 4)$ $ = 2(z - 2)(z + 2)(z^2 + 4)$
Solving Equations by Factoring	Use factoring and the zero-product property to solve polynomial equations.	$3x^3 - 12x^2 + 9x = 0$ $3x(x^2 - 4x + 3) = 0$ $3x(x - 3)(x - 1) = 0$ $3x = 0$ or $x - 3 = 0$ or $x - 1 = 0$ $x = 0$ or $x = 3$ or $x = 1$ The solutions are 0, 1, and 3.

13.7 Exercises

CONCEPTS AND VOCABULARY

1. When you are factoring polynomials, a good first step is to factor out the _____.

2. When you are solving an equation by factoring, the _____ property is used.

3. The zero-product property applies to all _____ containing the variable.

4. Because $a^2 - 2ab + b^2 = (a - b)^2$, it follows that $x^4 - 2x^2y^2 + y^4 =$ _____.

5. Because $x^2 + 3x + 2 = (x + 1)(x + 2)$, it follows that $z^4 + 3z^2 + 2 =$ _____.

6. If $x^4 - 1$ is factored as $(x^2 - 1)(x^2 + 1)$, is it factored *completely*? Explain.

7. If $x^3 - x^2 + x - 1$ is factored as $(x - 1)(x^2 + 1)$, is it factored *completely*? Explain.

8. When you are solving the equation $x^3 = x$, what is a good first step?

9. When you are solving $x^4 - x^2 = 0$, what is a good first step?

10. How many real-number solutions does the equation $(x - 6)(x^2 + 4) = 0$ have?

FACTORING HIGHER-DEGREE POLYNOMIALS

Exercises 11–48: Factor the polynomial completely.

11. $5x^2 - 5x - 30$
12. $3x^2 - 15x + 12$
13. $-4y^2 - 32y - 48$
14. $-7y^2 + 14y + 21$
15. $-20z^2 - 110z - 50$
16. $-12z^2 - 54z + 30$
17. $60 - 64t - 28t^2$
18. $18 - 45t - 27t^2$
19. $r^3 - r$
20. $r^3 + 2r^2 - 3r$
21. $3x^3 + 3x^2 - 18x$
22. $6x^3 - 26x^2 - 20x$
23. $72z^3 + 12z^2 - 24z$
24. $6z^3 - 4z^2 - 42z$
25. $x^4 - 4x^2$
26. $4x^4 - 36x^2$
27. $t^4 + t^3 - 2t^2$
28. $t^4 + 5t^3 - 24t^2$
29. $x^4 - 5x^2 + 6$
30. $x^4 - 3x^2 - 10$
31. $2x^4 + 7x^2 + 3$
32. $3x^4 - 8x^2 + 5$
33. $y^4 + 6y^2 + 9$
34. $y^4 - 10y^2 + 25$
35. $x^4 - 9$
36. $x^4 - 25$

37. $x^4 - 81$
38. $4x^4 - 64$
39. $z^5 + 2z^4 + z^3$
40. $6z^5 - 47z^4 + 35z^3$
41. $2x^2 + xy - y^2$
42. $2x^2 + 5xy + 2y^2$
43. $a^4 - 2a^2b^2 + b^4$
44. $a^3 + 2a^2b + ab^2$
45. $x^3 - xy^2$
46. $2x^2y - 2y^3$
47. $4x^3 + 4x^2y + xy^2$
48. $x^2y - 6xy^2 + 9y^3$

SOLVING POLYNOMIAL EQUATIONS

Exercises 49–54: Do the following.

49. (a) Factor $x^3 - 4x$.
    (b) Solve $x^3 - 4x = 0$.
50. (a) Factor $4x^3 - 16x$.
    (b) Solve $4x^3 - 16x = 0$.
51. (a) Factor $2y^3 - 6y^2 - 36y$.
    (b) Solve $2y^3 - 6y^2 - 36y = 0$.
52. (a) Factor $z^4 - 13z^2 + 36$.
    (b) Solve $z^4 - 13z^2 + 36 = 0$.
53. (a) Factor $x^3 - x^2 + 4x - 4$.
    (b) Solve $x^3 - x^2 + 4x - 4 = 0$.
54. (a) Factor $y^4 - 8y^2 + 16$.
    (b) Solve $y^4 - 8y^2 + 16 = 0$.

Exercises 55–76: Solve.

55. $3x^2 + 33x + 72 = 0$
56. $4x^2 - 16x - 20 = 0$
57. $25x^2 = 50x + 75$
58. $10x^2 = 20x + 80$
59. $y^3 - 3y^2 - 4y = 0$
60. $y^3 - 3y^2 + 2y = 0$
61. $3z^3 + 6z^2 = 72z$
62. $4z^3 = 4z^2 + 24z$
63. $x^4 - 36x^2 = 0$
64. $4x^4 = 100x^2$
65. $r^4 + 6r^3 = 7r^2$
66. $r^4 + 30r^2 = 11r^3$
67. $x^4 - 13x^2 = -36$
68. $x^4 - 17x^2 + 16 = 0$
69. $x^4 + 1 = 2x^2$
70. $x^4 - 8x^2 + 16 = 0$
71. $a^5 = 81a$
72. $b^3 = -8$
73. $x^3 - 2x^2 - x + 2 = 0$
74. $x^3 - x^2 + 4x - 4 = 0$
75. $x^3 - 5x^2 + x - 5 = 0$
76. $3x^3 + 2x^2 - 27x - 18 = 0$

APPLICATIONS

77. **Dimensions of a Box** (Refer to Example 5) A box is made from a rectangular piece of metal with length 20 inches and width 15 inches. The box has no top.
    (a) What are the limitations on the size of x? Explain your answer.
    (b) Write an expression that gives the outside surface area of the box. (*Hint:* Consider the size of the metal sheet and how much was cut out.)
    (c) If the outside surface area of the box is 275 square inches, find x.

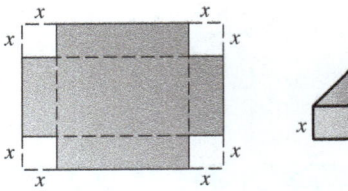 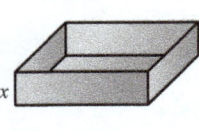

78. **Dimensions of a Box** Refer to the previous exercise.
    (a) Find a polynomial that gives the volume of the box for a given x.
    (b) Factor your polynomial.
    (c) What are the zeros of your polynomial? What do they represent in this problem?

79. **Numbers** If a positive number n is increased by 3, its square equals 121. Find n.

80. **Sidewalk around a Pool** A 5-foot-wide side walk around a rectangle swimming pool has a total area of 900 square feet. Find the dimensions of the swimming pool if the pool is 20 feet longer than it is wide.

Exercises 81–84: U.S. Natural Gas Consumption The polynomial

$$0.0013x^3 - 0.085x^2 + 1.6x + 12$$

models natural gas consumption in trillions of cubic feet, where $x = 0$ corresponds to 1960, $x = 1$ corresponds to 1961, and so on until $x = 40$ corresponds to 2000.

81. How much natural gas was consumed in 1990?

82. In which year between 1970 and 1990 was natural gas consumption about 20.4 trillion cubic feet?

83. Explain any difficulties encountered when you try to solve the equation

$$0.0013x^3 - 0.085x^2 + 1.6x + 12 = 23.2.$$

84. How might you solve the equation in Exercise 83 without factoring? If you were to find the solution to this equation, what would it represent?

86. Suppose that a polynomial can be factored. Explain how its factors can be used to find the zeros of the polynomial. Give an example.

WRITING ABOUT MATHEMATICS

85. Compare factoring the polynomial $x^2 + 6x + 5$ with factoring the polynomial $z^4 + 6z^2 + 5$.

SECTION 13.7 **Checking Basic Concepts**

1. Factor the trinomial completely.
   (a) $3x^2 - 6x - 24$ **(b)** $-10y^2 + 5y + 5$

2. Factor the binomial completely.
   (a) $z^4 - 25$ **(b)** $7t^4 - 7$

3. Factor completely.
   (a) $x^4 - 8x^2 + 16$ **(b)** $2y^3 + 17y^2 - 30y$

4. Solve $t^4 + t^3 = 12t^2$.

5. Solve $x^3 - 3x^2 + 2x - 6 = 0$.

CHAPTER 13 Summary

SECTION 13.1 ■ INTRODUCTION TO FACTORING

Terms Related to Factoring Polynomials

Factoring Writing a polynomial as a product, usually of lower-degree polynomials

Common Factor An expression that is a factor of each term in a polynomial

Example: Some common factors of $4x^4 + 8x^2$ are $2x$, x^2, and $4x^2$.

Greatest Common Factor (GCF) The common factor with the greatest (integer) coefficient and the highest degree

Example: The GCF of $4x^4 + 8x^2$ is $4x^2$.
$$4x^4 + 8x^2 = 4x^2(x^2 + 2)$$

Factoring by Grouping Used to factor a four-term polynomial into two binomials

Example: $x^3 + 5x^2 + 3x + 15 = x^2(x + 5) + 3(x + 5)$
$$= (x^2 + 3)(x + 5)$$

SECTION 13.2 ■ FACTORING TRINOMIALS I $(x^2 + bx + c)$

Review of the FOIL Method A method used for multiplying two binomials

First Terms $(2x + 3)(5x + 4)$: $2x \cdot 5x = 10x^2$

Outside Terms $(2x + 3)(5x + 4)$: $2x \cdot 4 = 8x$

Inside Terms $(2x + 3)(5x + 4)$: $3 \cdot 5x = 15x$

Last Terms $(2x + 3)(5x + 4)$: $3 \cdot 4 = 12$

The product is the sum of these four terms:
$$(2x + 3)(5x + 4) = 10x^2 + 8x + 15x + 12 = 10x^2 + 23x + 12.$$

Factoring Trinomials with a Leading Coefficient of 1 To factor the trinomial $x^2 + bx + c$, find two numbers, m and n, that satisfy

$$m \cdot n = c \quad \text{and} \quad m + n = b.$$

Then $x^2 + bx + c = (x + m)(x + n)$.

Example: Because $-3 \cdot 5 = -15$ and $-3 + 5 = 2$,

$$x^2 + 2x - 15 = (x - 3)(x + 5).$$

SECTION 13.3 ■ FACTORING TRINOMIALS II ($ax^2 + bx + c$)

Factoring Trinomials by Grouping To factor $ax^2 + bx + c$, perform the following steps. (Assume that a, b, and c are integers and have no factor in common.)

1. Find two numbers, m and n, such that $mn = ac$ and $m + n = b$.
2. Write the trinomial as $ax^2 + mx + nx + c$.
3. Use grouping to factor this expression into two binomials.

Example: To factor $3x^2 + 10x - 8$, find two numbers whose product is -24 and whose sum is 10. These two numbers are $m = 12$ and $n = -2$, so write $10x$ as $12x - 2x$.

$$
\begin{aligned}
3x^2 + 10x - 8 &= 3x^2 + 12x - 2x - 8 &&\text{Write } 10x \text{ as } 12x - 2x.\\
&= (3x^2 + 12x) + (-2x - 8) &&\text{Associative property}\\
&= 3x(x + 4) - 2(x + 4) &&\text{Factor out common factors.}\\
&= (3x - 2)(x + 4) &&\text{Factor out } (x + 4).
\end{aligned}
$$

Factoring with FOIL in Reverse Use trial and error and FOIL in reverse to find the factors of a trinomial.

Example: To factor $3x^2 + 10x - 8$, first find factors of $3x^2$.

$$(3x + \underline{})(x + \underline{})$$

Then place factors of -8 so that the resulting middle term is $10x$.

$(3x + -2) \cdot (x + 4) = 3x^2 + 10x - 8$
$\llcorner -2x \lrcorner$
$\llcorner + 12x \lrcorner$
$\overline{10x} \checkmark \longleftarrow$ Middle term checks.

SECTION 13.4 ■ SPECIAL TYPES OF FACTORING

Difference of Two Squares

$$a^2 - b^2 = (a - b)(a + b)$$

Examples: $x^2 - 16 = (x - 4)(x + 4) \quad (a = x, b = 4)$
$\phantom{\text{Examples:}}\ 4r^2 - 9t^2 = (2r - 3t)(2r + 3t) \quad (a = 2r, b = 3t)$

Perfect Square Trinomials

$$a^2 + 2ab + b^2 = (a + b)^2 \quad \text{and} \quad a^2 - 2ab + b^2 = (a - b)^2$$

Examples: $4x^2 + 4x + 1 = (2x)^2 + 2(2x)1 + 1^2 = (2x + 1)^2 \quad (a = 2x, b = 1)$
$\phantom{\text{Examples:}}\ x^2 - 10x + 25 = x^2 - 2 \cdot x \cdot 5 + 5^2 = (x - 5)^2 \quad (a = x, b = 5)$

Sums and Differences of Two Cubes

$$a^3 + b^3 = (a + b)(a^2 - ab + b^2) \quad \text{and} \quad a^3 - b^3 = (a - b)(a^2 + ab + b^2)$$

Examples: $x^3 + 8 = (x + 2)(x^2 - 2x + 4)$ $\quad (a = x, b = 2)$

$27x^3 - 1 = (3x - 1)(9x^2 + 3x + 1)$ $\quad (a = 3x, b = 1)$

SECTION 13.5 ■ SUMMARY OF FACTORING

Guidelines for Factoring Polynomials The following guidelines can be used to factor polynomials in general.

STEP 1: Factor out the greatest common factor, if possible.

STEP 2: **A.** If the polynomial has *four terms*, try factoring by grouping.
 B. If the polynomial is a *binomial,* try one of the following.
  1. $a^2 - b^2 = (a - b)(a + b)$ Difference of two squares
  2. $a^3 - b^3 = (a - b)(a^2 + ab + b^2)$ Difference of two cubes
  3. $a^3 + b^3 = (a + b)(a^2 - ab + b^2)$ Sum of two cubes
 C. If the polynomial is a *trinomial*, check for a perfect square.
  1. $a^2 + 2ab + b^2 = (a + b)^2$ Perfect square trinomial
  2. $a^2 - 2ab + b^2 = (a - b)^2$ Perfect square trinomial

 Otherwise, try to factor the trinomial by grouping or apply FOIL in reverse, as described in Sections 2 and 3 of this chapter.

STEP 3: Check to make sure that the polynomial is *completely* factored.

Examples: $12x^3 - 12x^2 + 3x = 3x(4x^2 - 4x + 1) = 3x(2x - 1)^2$ Steps 1, 2C, and 3

$9x^3 - 6x^2 + 18x - 12 = 3(3x^3 - 2x^2 + 6x - 4) = 3(x^2 + 2)(3x - 2)$ Steps 1, 2A, and 3

$16x^3 - 100x = 4x(4x^2 - 25) = 4x(2x - 5)(2x + 5)$ Steps 1, 2B, and 3

SECTION 13.6 ■ SOLVING EQUATIONS BY FACTORING I (QUADRATICS)

Zero-Product Property
For any real numbers a and b, if $ab = 0$, then $a = 0$ or $b = 0$ (or both). The zero-product property is used to solve equations.

Examples: $xy = 0$ implies that $x = 0$ or $y = 0$.

$(x + 5)(x - 3) = 0$ implies $x + 5 = 0$ or $x - 3 = 0$.

Zero of a Polynomial A number a is a zero of a polynomial if the result is 0 when a is substituted in that polynomial.

Example: The number -2 is a zero of $x^2 - 4$ because $(-2)^2 - 4 = 0$.

Solving Quadratic Equations by Factoring To solve a quadratic equation by factoring, follow these steps.

STEP 1: If necessary, write the equation in standard form as $ax^2 + bx + c = 0$.

STEP 2: Factor the left side of the equation using any method.

STEP 3: Apply the zero-product property.

STEP 4: Solve each of the resulting equations. Check any solutions.

Example:

$$x^2 + 7x = 8 \quad \text{Given equation}$$
$$x^2 + 7x - 8 = 0 \quad \text{Step 1}$$
$$(x + 8)(x - 1) = 0 \quad \text{Step 2}$$
$$x + 8 = 0 \quad \text{or} \quad x - 1 = 0 \quad \text{Step 3}$$
$$x = -8 \quad \text{or} \quad x = 1 \quad \text{Step 4}$$

SECTION 13.7 ■ SOLVING EQUATIONS BY FACTORING II (HIGHER DEGREE)

Factoring Polynomials of Higher Degree The distributive property and the techniques for factoring quadratic polynomials can also be applied to polynomials of higher degree.

Examples: $10r^3 + 15r = 5r(2r^2 + 3)$ (To check, multiply the right side.)

Because $2x^2 + x - 1 = (x + 1)(2x - 1)$,

it follows that $2z^4 + z^2 - 1 = (z^2 + 1)(2z^2 - 1)$.

Solving Equations by Factoring Use algebra to obtain 0 on one side of the equation. Factor the other side and apply the zero-product property.

Example:

$$x^3 = 4x \quad \text{Given equation}$$
$$x^3 - 4x = 0 \quad \text{Subtract } 4x.$$
$$x(x^2 - 4) = 0 \quad \text{Factor out the GCF, } x.$$
$$x(x - 2)(x + 2) = 0 \quad \text{Difference of two squares}$$
$$x = 0 \quad \text{or} \quad x - 2 = 0 \quad \text{or} \quad x + 2 = 0 \quad \text{Zero-product property}$$
$$x = 0 \quad \text{or} \quad x = 2 \quad \text{or} \quad x = -2 \quad \text{Solve for } x.$$

CHAPTER 13 Review Exercises

SECTION 13.1

Exercises 1–6: Identify the greatest common factor for the expression and then factor the expression.

1. $5x - 15$
2. $y^2 + 2y$
3. $8z^3 - 4z^2$
4. $6x^4 + 3x^3 - 12x^2$
5. $9xy + 15yz^2$
6. $a^2b^3 + a^3b^2$

Exercises 7–14: Use grouping to factor the given polynomial completely.

7. $x(x + 2) - 3(x + 2)$
8. $y^2(x - 5) + 3y(x - 5)$
9. $z^3 - 2z^2 + 5z - 10$
10. $t^3 + t^2 + 8t + 8$
11. $x^3 - 3x^2 + 6x - 18$
12. $ax + bx - ay - by$
13. $x^5 + 3x^4 - 2x^3 - 6x^2$
14. $2y^4 + 6y^3 + 2y^2 + 6y$

SECTION 13.2

Exercises 15–18: Find an integer pair that has the given product and sum.

15. Product: 20   Sum: 9
16. Product: -21 Sum: 4
17. Product: 36   Sum: -13
18. Product: -100 Sum: -21

Exercises 19–32: Factor the trinomial completely. If the trinomial cannot be factored, write "prime."

19. $x^2 - x - 12$
20. $x^2 + 10x + 24$
21. $x^2 + 6x - 16$
22. $x^2 - x - 42$
23. $x^2 + 4x - 6$
24. $x^2 - 5x + 8$
25. $x^2 + 2x - 3$
26. $x^2 + 22x + 120$
27. $2x^3 + 6x^2 - 20x$
28. $x^4 - 3x^3 - 28x^2$

29. $10 - 7x + x^2$
30. $24 + 2x - x^2$
31. $-2x^2 - 4x + 30$
32. $-x^3 - 9x^2 + 10x$

SECTION 13.3

Exercises 33–44: Factor the trinomial completely. If the trinomial cannot be factored, write "prime."

33. $9x^2 + 3x - 2$
34. $2x^2 + 3x - 5$
35. $3x^2 + 14x + 15$
36. $35x^2 - 2x - 1$
37. $3x^2 + 4x - 5$
38. $4x^2 - 12x - 5$
39. $24x^2 - 7x - 5$
40. $4x^2 + 33x - 27$
41. $12x^3 + 48x^2 + 21x$
42. $8x^4 + 14x^3 - 30x^2$
43. $12 - 5x - 2x^2$
44. $1 + 3x - 10x^2$

SECTION 13.4

Exercises 45–58: Factor completely.

45. $z^2 - 4$
46. $9z^2 - 64$
47. $36 - y^2$
48. $100a^2 - 81b^2$
49. $x^2 + 14x + 49$
50. $x^2 - 10x + 25$
51. $4x^2 - 12x + 9$
52. $9x^2 + 48x + 64$
53. $8t^3 - 1$
54. $27r^3 + 8t^3$
55. $2x^3 - 50x$
56. $24x^3 + 81$
57. $2x^3 + 28x^2 + 98x$
58. $2x^4 - 128x$

SECTION 13.5

Exercises 59–68: Factor completely.

59. $12x - 8$
60. $6x^3 + 9x^2$
61. $9y^2 - 6y + 6$
62. $yz^2 - 9y$
63. $x^4 + 7x^3 - 4x^2 - 28x$
64. $12x^3 + 36x^2 + 27x$
65. $3ab^3 - 24a$
66. $5x^3 + 20x$
67. $24x^3 - 6xy^2$
68. $x^3y + 27y$

SECTION 13.6

Exercises 69–80: Solve the equation.

69. $mn = 0$
70. $y^2 = 0$
71. $(4x - 3)(x + 9) = 0$
72. $(1 - 4x)(6 + 5x) = 0$
73. $z(z - 1)(z - 2) = 0$
74. $z^2 - 7z = 0$
75. $y^2 - 64 = 0$
76. $y^2 + 9y + 14 = 0$
77. $x^2 = x + 6$
78. $10x^2 + 11x = 6$
79. $t(t - 14) = 72$
80. $t(2t - 1) = 10$

SECTION 13.7

Exercises 81–90: Factor completely.

81. $5x^2 - 15x - 50$
82. $-3x^2 - 6x + 45$
83. $y^3 - 4y$
84. $3y^3 + 6y^2 - 9y$
85. $2z^4 + 14z^3 + 20z^2$
86. $8z^4 - 32z^2$
87. $x^4 - 6x^2 + 9$
88. $2x^4 - 15x^2 - 27$
89. $a^2 + 10ab + 25b^2$
90. $x^3 - xy^2$

Exercises 91–98: Solve.

91. $16x^2 - 72x - 40 = 0$
92. $2x^3 - 11x^2 + 15x = 0$
93. $t^3 = 25t$
94. $t^5 - 7t^3 + 12t^2 = 0$
95. $z^4 + 16 = 8z^2$
96. $z^4 - 256 = 0$
97. $y^3 = -64$
98. $y^3 - y^2 - y + 1 = 0$

APPLICATIONS

99. A square has area $9x^2 + 42x + 49$. Find the length of a side. Make a sketch of the square.

100. A rectangle has area $x^2 + 6x + 5$. Find possible dimensions for the rectangle. Make a sketch of the rectangle.

101. A cube has surface area $6x^2 + 12x + 6$. Find the length of a side.

102. Write a polynomial in factored form that represents the total area of the rectangle.

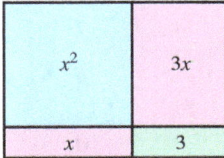

103. Write a polynomial in factored form that represents the total area of the rectangle.

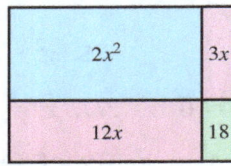

104. **Radius of a Circle** The area and the circumference of a circle are numerically equal. Find the radius of the circle.

105. **Dimensions of a Shed** The floor of a rectangular shed is 7 feet longer than it is wide and has an area of 120 square feet. What are its dimensions?

106. **Flight of a Ball** A ball is hit upward. Its height h in feet after t seconds is $h = -16t^2 + 80t + 4$. At what times is the ball 100 feet in the air?

107. **Stopping Distance** The distance D in feet that it takes to stop a car traveling x miles per hour on wet, level pavement can be approximated by $D = \frac{1}{9}x^2 + \frac{11}{3}x$.
   (a) Estimate the distance required for the car to stop when it is traveling 45 miles per hour.
   (b) If the stopping distance is 80 feet, what is the speed of the car?
   (c) If you have a calculator, use it to solve part (b) numerically with a table of values. Do your answers agree?

108. **Revenue** A company makes tops for the boxes of pickup trucks. The total revenue R in dollars from selling the tops for p dollars each is given by $R = p(200 - p)$, where $p \le 200$.
   (a) Find R when $p = \$100$.
   (b) Find p when $R = \$7500$.
   (c) If you have a calculator, use it to solve part (b) numerically with a table of values. Do your answers agree?

109. **Dimensions of a Box** A box is made from a rectangular piece of metal with length 50 inches and width 40 inches by cutting out square corners of length x and folding up the sides.
   (a) Write an expression that gives the surface area of the inside of the box.
   (b) If the surface area of the box is 1900 square inches, find x.

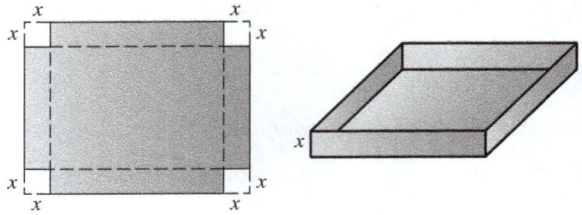

110. **Dimensions of a Cube** If two identical cubes have a total surface area of $12x^2 + 48x + 48$, find the length of one side of one of the cubes.

CHAPTER 13 Test

Exercises 1 and 2: Identify the greatest common factor for the expression. Then factor the expression.

1. $4x^2y - 20xy^2 + 12xy$ 2. $9a^3b^2 + 3a^2b^2$

Exercises 3 and 4: Factor by grouping.

3. $ay + by + az + bz$ 4. $3x^3 + x^2 - 15x - 5$

Exercises 5–10: Factor the trinomial completely. If the trinomial cannot be factored, write "prime."

5. $y^2 + 4y - 12$ 6. $4x^2 + 20x + 25$

7. $4z^2 - 19z + 12$ 8. $21 - 17t + 2t^2$

9. $x^2 + 7x - 10$ 10. $3y^2 + 4y + 2$

Exercises 11–16: Factor completely.

11. $6x^3 + 3x^2 - 3x$ 12. $2z^4 - 12z^2 - 54$

13. $36y^3 - 100y$ 14. $7x^4 + 56x$

15. $16a^4 + 24a^3 + 9a^2$ 16. $2b^4 - 32$

Exercises 17–22: Solve the equation.

17. $x^2 - 16 = 0$ 18. $y^2 = y + 20$

19. $9z^2 + 16 = 24z$ 20. $x(x - 5) = 66$

21. $y^3 = 9y$ 22. $x^4 - 5x^2 + 4 = 0$

23. A square has area $9x^2 + 30x + 25$. Find the length of a side in terms of x.

24. Write a polynomial in factored form that represents the total area of the rectangle.

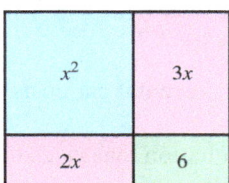

25. **Braking Distance** The braking distance D in feet required for a car traveling at x miles per hour to stop on dry, level pavement can be modeled by $D = \frac{1}{11}x^2$.
    (a) Calculate the distance required for the car to stop when it is traveling 55 miles per hour.
    (b) If the braking distance is 99 feet, estimate the speed of the car.

26. **Flight of a Ball** A ball is hit upward. Its height h in feet after t seconds is given by $h = -16t^2 + 48t + 4$. At what times is the ball 36 feet in the air?

CHAPTERS 1–13 Cumulative Review Exercises

Exercises 1 and 2: Evaluate by hand and then simplify to lowest terms.

1. $\frac{3}{5} \cdot \frac{15}{21}$
2. $\frac{4}{5} - \frac{1}{10}$

Exercises 3 and 4: Evaluate by hand.

3. $26 - 3 \cdot 6 \div 2$
4. $-2^2 + \frac{3+2}{8+2}$

5. Complete the table. Then use the table to solve the equation $2x + 3 = 5$.

x	−2	−1	0	1	2
$2x + 3$					

6. Translate the sentence "Triple a number decreased by 5 equals the number decreased by 7" into an equation using the variable n. Then solve the equation.

7. Convert 5.7% to fraction and decimal notation.

8. Solve $P = 2W + 2L$ for W.

9. Solve $5 - 3z < -1$.

10. Make a scatterplot having the following five points: $(-2, 3)$, $(-1, 2)$, $(0, -1)$, $(1, 1)$, and $(2, 2)$.

Exercises 11 and 12: Graph the given equation. Determine any intercepts.

11. $y = 3x - 2$
12. $y = -2$

Exercises 13 and 14: Find the slope–intercept form for the line satisfying the given conditions.

13. Perpendicular to $2x - 3y = -6$, passing through the point $(1, 2)$

14. Passing through the points $(-2, 1)$ and $(1, 5)$

15. Identify the x-intercept and the y-intercept. Then write the slope–intercept form of the line.

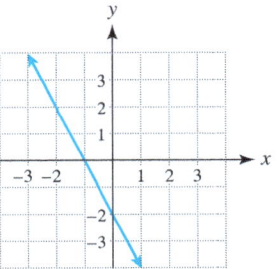

16. The graphs of two equations are shown. Use the graphs to identify the solution to the system of equations. Then check your answer.

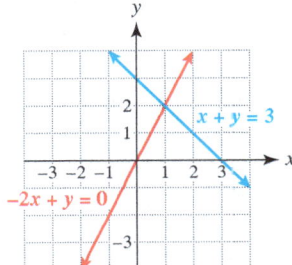

Exercises 17 and 18: Solve the system of equations.

17. $y = -1$
    $2x + y = 1$

18. $5x + y = -5$
    $-x + 2y = 12$

Exercises 19 and 20: Shade the solution set to the system of inequalities.

19. $x > -1$
    $y < x$

20. $2x - y \leq 4$
    $x + 2y \geq 2$

Exercises 21–24: Simplify the expression.

21. -2^4 **22.** $(xy)^0$

23. $(xy)^4(x^3y^{-4})^2$ **24.** $7x^3(-2x^2 + 3x)$

Exercises 25–28: Simplify and write the expression using positive exponents.

25. $a^{-4} \cdot a^2$ **26.** $(2t^3)^{-2}$

27. $(xy)^{-3}(x^{-1}y^2)^{-1}$ **28.** $\left(\dfrac{2x}{y^{-2}}\right)^5$

Exercises 29 and 30: Divide and check.

29. $\dfrac{6x^3 + 12x^2}{3x}$ **30.** $\dfrac{3x^3 - x + 1}{x^2 + 1}$

Exercises 31–36: Factor completely.

31. $x^2 + 3x - 28$ **32.** $6y^2 + y - 12$

33. $25x^2 - 4y^2$ **34.** $64x^2 - 16x + 1$

35. $27t^3 - 8$ **36.** $-4x^2 + 4x + 24$

Exercises 37–40: Solve the equation.

37. $y^4 = 25y^2$ **38.** $8z^2 + 8z - 16 = 0$

39. $4z^3 = 49z$ **40.** $x^4 - 18x^2 + 81 = 0$

APPLICATIONS

41. Shoveling a Driveway Two people are shoveling snow from a driveway. The first person shovels 10 square feet per minute, while the second person shovels 8 square feet per minute.
 (a) Write and simplify an expression that gives the total square feet that the two people shovel in x minutes.
 (b) How many minutes would it take for them to clear a driveway with an area of 900 square feet?

42. Running Distance A person runs the following distances over three days: $1\frac{3}{4}$ miles, $2\frac{1}{2}$ miles, and $2\frac{2}{3}$ miles. How far does the person run altogether?

43. Burning Calories An athlete can burn 12 calories per minute while cross-country skiing and 9 calories per minute while running at 5 miles per hour. If the athlete burns 615 calories in 60 minutes, how long is spent on each activity?

44. Renting a Car A rental car costs $20 plus $0.25 per mile driven.
 (a) Write an equation that gives the cost C of driving the car x miles.
 (b) Determine the number of miles that the car is driven if the rental cost is $100.

45. Triangle In an isosceles triangle, the measures of the two smaller angles are equal and their sum is 20° more than the larger angle.
 (a) Let x be the measure of one of the two smaller angles and y be the measure of the larger angle. Write a system of linear equations whose solution gives the measures of these angles.
 (b) Solve your system.

46. Areas of Rectangles Find a monomial equal to the sum of the areas of the rectangles. Calculate this sum for $x = 2$ yards and $y = 3$ yards.

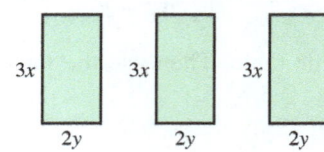

47. Area of a Square A square has area $x^2 + 12x + 36$. Find the length of a side.

48. Flight of a Golf Ball A golf ball is hit upward. Its height h in feet after t seconds is given by the formula $h = 64t - 16t^2$.
 (a) How long does it take for the ball to hit the ground?
 (b) At what times is the ball 48 feet in the air?

14 Rational Expressions

- 14.1 Introduction to Rational Expressions
- 14.2 Multiplication and Division of Rational Expressions
- 14.3 Addition and Subtraction with Like Denominators
- 14.4 Addition and Subtraction with Unlike Denominators
- 14.5 Complex Fractions
- 14.6 Rational Equations and Formulas
- 14.7 Proportions and Variation

One of the most significant problems facing the U.S. transportation system is chronic highway congestion. According to a newly released "Highway Statistics," Americans drove about 4.6 trillion miles in 2014. Our ability to keep traffic moving smoothly and safely is key to keeping our economy strong, and traffic congestion costs motorists more than $121 billion annually in wasted fuel. A recent study found that, collectively, Americans spend as many as 5.5 billion hours stuck in traffic each year.

If the amount of traffic doubles on a highway, the time spent waiting in traffic may more than double. At times, only a slight increase in the traffic rate can result in a dramatic increase in the time spent waiting. (See Example 6 in the first section of this chapter.) Mathematics can describe this effect by using rational expressions. In this chapter, we introduce rational expressions and some of their applications. *Source:* Texas A&M Transportation Institute.

14.1 Introduction to Rational Expressions

Objectives

1. Evaluating Rational Expressions
   - Undefined Expressions
2. Simplifying Rational Expressions
3. Solving Applications Involving Rational Expressions
   - An Application Involving Probability

NEW VOCABULARY

☐ Rational expression
☐ Lowest terms
☐ Vertical asymptote
☐ Probability

1 Evaluating Rational Expressions

Recall that a *rational number* is any number that can be expressed as a ratio of two integers $\frac{p}{q}$, where $q \neq 0$. In this chapter, we discuss *rational expressions*, which can be written as the ratio of two polynomials. Because examples of polynomials include

$$3, \quad 2x, \quad x^2 + 4, \quad \text{and} \quad x^3 - 1, \quad \leftarrow \text{Polynomials}$$

it follows that examples of rational expressions include

$$\frac{3}{2x}, \quad \frac{2x}{x^2 + 4}, \quad \frac{x^2 + 4}{3}, \quad \text{and} \quad \frac{x^3 - 1}{x^2 + 4}. \quad \leftarrow \text{Rational expressions}$$

> **RATIONAL EXPRESSION**
>
> A **rational expression** can be written as $\frac{P}{Q}$, where P and Q are polynomials. A rational expression is defined whenever $Q \neq 0$.

We can evaluate polynomials for different values of a variable. For example, for $x = 2$, the polynomial $x^2 - 3x + 1$ evaluates to

$$(2)^2 - 3(2) + 1 = -1.$$

Rational expressions can be evaluated similarly.

EXAMPLE 1 **Evaluating rational expressions**

If possible, evaluate each expression for the given value of the variable.

(a) $\dfrac{1}{x+1} \quad x = 2$

(b) $\dfrac{y^2}{2y - 1} \quad y = -4$

(c) $\dfrac{5z + 8}{z^2 - 2z + 1} \quad z = 1$

(d) $\dfrac{2 - x}{x - 2} \quad x = -3$

Solution

(a) If $x = 2$, then $\dfrac{1}{x+1} = \dfrac{1}{2+1} = \dfrac{1}{3}$.

(b) If $y = -4$, then $\dfrac{y^2}{2y - 1} = \dfrac{(-4)^2}{2(-4) - 1} = -\dfrac{16}{9}$.

(c) If $z = 1$, then $\dfrac{5z + 8}{z^2 - 2z + 1} = \dfrac{5(1) + 8}{1^2 - 2(1) + 1}$, or $\dfrac{13}{0}$, which is undefined because division by 0 is not possible.

(d) If $x = -3$, then $\dfrac{2 - x}{x - 2} = \dfrac{2 - (-3)}{-3 - 2} = \dfrac{5}{-5} = -1$.

Now Try Exercises 7, 11, 13, 17

READING CHECK 1

- When is a rational expression undefined?

UNDEFINED EXPRESSIONS Division by 0 is undefined. As a result, rational expressions are different from polynomials because they are *undefined* whenever their *denominators are 0*. For example, the expression in Example 1(c) is undefined when $z = 1$.

EXAMPLE 2 **Determining when a rational expression is undefined**

Find all values of the variable for which each expression is undefined.

(a) $\dfrac{1}{x}$ (b) $\dfrac{4t}{t-3}$ (c) $\dfrac{1-6r}{r^2-4}$ (d) $\dfrac{4}{x^2+1}$

Solution

(a) A rational expression is undefined when its denominator is 0. Thus $\dfrac{1}{x}$ is undefined when $x=0$.

(b) The expression $\dfrac{4t}{t-3}$ is undefined when its denominator, $t-3$, is 0, or when $t=3$.

(c) The expression $\dfrac{1-6r}{r^2-4}$ is undefined when its denominator, r^2-4, is 0. Here,

$$r^2-4=(r-2)(r+2)=0$$

implies that the denominator is 0 when $r=-2$ or $r=2$.

(d) In the expression $\dfrac{4}{x^2+1}$, the denominator x^2+1 is never 0 because any real number squared plus 1 is always greater than or equal to 1. Thus this rational expression is defined for all real numbers x.

Now Try Exercises 25, 27, 31, 33

STUDY TIP

You may want to review your notes on fractions. Many of the mathematical concepts that apply to fractions also apply to rational expressions.

2 Simplifying Rational Expressions

Earlier in this text, we used the *basic principle of fractions*,

$$\dfrac{a \cdot c}{b \cdot c} = \dfrac{a}{b}.$$

For example, this basic principle allows us to simplify the fraction $\dfrac{8}{12}$ as

$$\dfrac{8}{12} = \dfrac{2 \cdot 4}{3 \cdot 4} = \dfrac{2}{3}.$$

EXAMPLE 3 **Simplifying fractions**

Simplify each fraction by applying the basic principle of fractions.

(a) $\dfrac{5}{10}$ (b) $-\dfrac{36}{48}$

Solution

(a) $\dfrac{5}{10} = \dfrac{1 \cdot 5}{2 \cdot 5} = \dfrac{1}{2}$ (b) $-\dfrac{36}{48} = -\dfrac{3 \cdot 12}{4 \cdot 12} = -\dfrac{3}{4}$

Now Try Exercises 39, 43

We can also apply this basic principle to rational expressions. For example,

$$\dfrac{x(x-1)}{4(x-1)} = \dfrac{x}{4},$$

provided that $x \neq 1$.

NOTE: The above simplification is not valid when $x=1$ because the expression is undefined for this x-value. When simplifying a rational expression, *we assume that values of the variable that make the rational expression undefined are excluded*, unless stated otherwise. ∎

BASIC PRINCIPLE OF RATIONAL EXPRESSIONS

The following property can be used to simplify rational expressions, where P, Q, and R are polynomials.

$$\frac{P \cdot R}{Q \cdot R} = \frac{P}{Q} \qquad Q \text{ and } R \text{ are nonzero.}$$

NOTE: $\frac{P \cdot R}{Q \cdot R} = \frac{P}{Q} \cdot \frac{R}{R} = \frac{P}{Q} \cdot 1 = \frac{P}{Q}$, provided that $Q \neq 0$ and $R \neq 0$. ∎

Like fractions, rational expressions can be written in *lowest terms*. For example, the rational expression $\frac{x^2 - 1}{x^2 + 2x + 1}$ can be written in lowest terms by factoring the numerator and the denominator and then applying the basic principle of rational expressions.

READING CHECK 2

- How do you know when a rational expression is written in lowest terms?

$$\frac{x^2 - 1}{x^2 + 2x + 1} = \frac{(x - 1)(x + 1)}{(x + 1)(x + 1)} \qquad \text{Factor the numerator and the denominator.}$$

$$= \frac{x - 1}{x + 1} \qquad \text{Apply } \frac{PR}{QR} = \frac{P}{Q} \text{ with } R = x + 1.$$

Because the basic principle of rational expressions cannot be applied further to $\frac{x-1}{x+1}$, we say that this expression is written in **lowest terms**.

EXAMPLE 4 Simplifying rational expressions

Simplify each expression.

(a) $\dfrac{8y}{4y^2}$ **(b)** $\dfrac{2x + 6}{3x + 9}$ **(c)** $\dfrac{(z + 1)(z - 5)}{(z - 5)(z + 3)}$ **(d)** $\dfrac{x^2 - 9}{2x^2 + 7x + 3}$

Solution

(a) Factor out the greatest common factor, $4y$, in the numerator and the denominator.

$$\frac{8y}{4y^2} = \frac{2 \cdot 4y}{y \cdot 4y} = \frac{2}{y} \qquad \text{Apply } \frac{PR}{QR} = \frac{P}{Q} \text{ with } R = 4y.$$

(b) Start by factoring the numerator and denominator.

$$\frac{2x + 6}{3x + 9} = \frac{2(x + 3)}{3(x + 3)} = \frac{2}{3} \qquad \text{Apply } \frac{PR}{QR} = \frac{P}{Q} \text{ with } R = x + 3.$$

(c) The commutative property allows us to write $\frac{PR}{RQ}$ as $\frac{PR}{QR}$.

$$\frac{(z + 1)(z - 5)}{(z - 5)(z + 3)} = \frac{(z + 1)(z - 5)}{(z + 3)(z - 5)} = \frac{z + 1}{z + 3} \qquad \text{Apply } \frac{PR}{RQ} = \frac{P}{Q} \text{ with } R = z - 5.$$

(d) Start by factoring the numerator and the denominator.

$$\frac{x^2 - 9}{2x^2 + 7x + 3} = \frac{(x - 3)(x + 3)}{(2x + 1)(x + 3)} = \frac{x - 3}{2x + 1} \qquad \text{Apply } \frac{PR}{QR} = \frac{P}{Q} \text{ with } R = x + 3.$$

Now Try Exercises 51, 55, 61, 79

READING CHECK 3

- How do rational expressions and rational equations differ?

> **MAKING CONNECTIONS 1**
>
> **Rational Expressions and Equations**
>
> $$\frac{x}{x+4} \text{ and } \frac{2}{x} \qquad \frac{x}{x+4} = \frac{2}{x}$$
>
> Expressions do not contain an equals sign.
>
> An equation must contain an equals sign.
>
> We simplify and evaluate expressions. We solve equations by finding x-values that make the equation true.

A negative sign can be placed in a fraction in a number of ways.

$$-\frac{5}{7} = \frac{-5}{7} = \frac{5}{-7} \qquad \text{Three equal fractions}$$

This property about negative signs can also be applied to rational expressions, as demonstrated in the next example.

EXAMPLE 5 **Distributing a negative sign**

Simplify each expression.

(a) $\dfrac{-x-6}{2x+12}$ (b) $\dfrac{10-z}{z-10}$ (c) $-\dfrac{5-x}{x-5}$

Solution

(a) Factor -1 out of the numerator and 2 out of the denominator.

$$\frac{-x-6}{2x+12} = \frac{-1(x+6)}{2(x+6)} = -\frac{1}{2}$$

(b) Factor -1 out of the numerator.

$$\frac{10-z}{z-10} = \frac{-1(-10+z)}{z-10} = \frac{-1(z-10)}{z-10} = -1$$

(c) Rewrite the expression with the negative sign in the numerator and then apply the distributive property. Be sure to include parentheses around the numerator.

$$-\frac{5-x}{x-5} = \frac{-(5-x)}{x-5} = \frac{-5+x}{x-5} = \frac{x-5}{x-5} = 1$$

The same answer can be obtained by distributing the negative sign in the denominator.

$$-\frac{5-x}{x-5} = \frac{5-x}{-(x-5)} = \frac{5-x}{-x+5} = \frac{5-x}{5-x} = 1$$

Now Try Exercises 63, 67, 71

NOTE: The result for Example 5(b) becomes more obvious if we substitute a number for z. For example, if we let $z = 6$, then

$$\frac{10-z}{z-10} = \frac{10-6}{6-10} = \frac{4}{-4} = -1.$$

> **MAKING CONNECTIONS 2**
>
> **Negative Signs and Rational Expressions**
>
> In general, $(b - a)$ equals $-1(a - b)$. Thus if $a \neq b$, then $\frac{b - a}{a - b} = -1$.

3 Solving Applications Involving Rational Expressions

Math in Context *Highway Traffic* — Have you ever been moving smoothly in traffic, only to come to a sudden halt? Mathematics shows that in certain conditions, if the number of cars on a road increases even slightly, then the movement of traffic can slow dramatically. To understand why this occurs, we will consider how *rational equations* can be used to model traffic flow.

EXAMPLE 6   Modeling traffic flow

Suppose that 10 cars per minute can pass through a construction zone. If traffic arrives *randomly* at an average rate of x cars per minute, the average time T in minutes spent waiting in line and passing through the construction zone is given by

$$T = \frac{1}{10 - x},$$

where $x < 10$. (*Source:* N. Garber and L. Hoel, *Traffic and Highway Engineering*.)

(a) Complete **TABLE 14.1** by finding T for each value of x.

Waiting in Traffic

x (cars/minute)	5	7	9	9.5	9.9	9.99
T (minutes)						

TABLE 14.1

(b) Interpret the results.

Solution
(a) When $x = 5$ cars per minute, $T = \frac{1}{10 - 5} = \frac{1}{5}$ minute. Other values are found similarly and are shown in **TABLE 14.2**.

Waiting in Traffic

x (cars/minute)	5	7	9	9.5	9.9	9.99
T (minutes)	$\frac{1}{5}$	$\frac{1}{3}$	1	2	10	100

TABLE 14.2

(b) As the average traffic rate increases from 9 cars per minute to 9.9 cars per minute, the time needed to pass through the construction zone increases from 1 minute to 10 minutes. As x nears 10 cars per minute, a small increase in x increases the waiting time dramatically.

Now Try Exercise 103

This nonlinear effect for traffic congestion in Example 6 is shown in **FIGURE 14.1**, where points from **TABLE 14.2** have been plotted and a curve passing through them has been sketched. A vertical dashed line has also been sketched at $x = 10$. This dashed line is called a **vertical asymptote** and indicates that the rational expression is undefined at this value of x. Near the left side of the vertical asymptote, the waiting time T increases dramatically for small increases in x. The graph of T does not intersect or cross this vertical asymptote.

Length of Wait in Traffic

$T = \dfrac{1}{10 - x}$

The wait gets very long as x increases between 9 and 10 cars per minute.

The dashed line $x = 10$ is a *vertical asymptote*.

FIGURE 14.1

TECHNOLOGY NOTE

Making Tables
TABLE 14.2 can also be created with a graphing calculator by using the Ask feature, as illustrated in the following displays.

CALCULATOR HELP
To make a table of values, see Appendix A (pages 2 and 3).

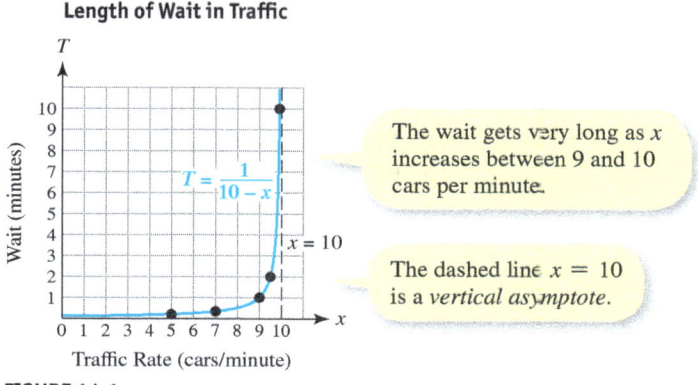

Math in Context (Biology) When a new species of animal is introduced into an area that it did not previously inhabit, its population may grow quickly at first, and then level off over time. Rational equations can be used to model such situations, as demonstrated in the next example.

EXAMPLE 7 Modeling a fish population

Suppose that a small fish species is introduced into a pond that had not previously held this type of fish, and that its population P in thousands is modeled by the equation

$$P = \frac{3x + 1}{x + 4},$$

where $x \geq 0$ represents time in months.
(a) Complete **TABLE 14.3** by finding P for each value of x. Round to 3 decimal places.

Fish Population

x (months)	0	6	12	36	72
P (thousands)					

TABLE 14.3

(b) How many fish were initially introduced into the pond?
(c) Interpret the results shown in your completed table.

Solution

(a) If we evaluate $P = \frac{3x + 1}{x + 4}$ when $x = 0$, the population is $P = \frac{3(0) + 1}{0 + 4} = \frac{1}{4} = 0.25$ thousand fish. The other values are found similarly and are shown in **TABLE 14.4**.

Fish Population

x (months)	0	6	12	36	72
P (thousands)	0.25	1.9	2.313	2.725	2.855

TABLE 14.4

(b) **TABLE 14.4** shows that initially (when $x = 0$) there were 0.25 thousand, or 250 fish.
(c) The fish population increased quickly at first but then leveled off. This population growth is shown graphically in **FIGURE 14.2**.

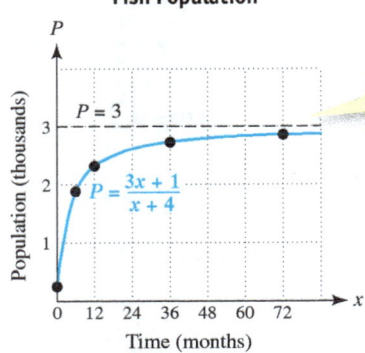

The horizontal dashed line $P = 3$ is called a **horizontal asymptote**. The fish population levels off at 3 thousand fish.

FIGURE 14.2

Now Try Exercise 105

AN APPLICATION INVOLVING PROBABILITY If 10 marbles, one blue and nine red, are placed in a jar, then the *probability*, or *likelihood*, of picking the blue marble at random is 1 *chance* in 10, or $\frac{1}{10}$. The probability of drawing a red marble at random is 9 chances in 10, or $\frac{9}{10}$. **Probability** is a real number from 0 to 1. A probability of 0, or 0%, indicates that an event is impossible, whereas a probability of 1, or 100%, indicates that an event is certain. Rational expressions are often used to describe probability.

EXAMPLE 8 Calculating probability

Suppose that n balls, numbered 1 to n, are placed in a container and only one ball has the winning number.
(a) What is the probability of drawing the winning ball at random?
(b) Calculate this probability for $n = 100, 1000,$ and $10,000$.
(c) What happens to the probability of drawing the winning ball as the number of balls increases?

Solution

(a) There is 1 chance in n of drawing the winning ball, so the probability is $\frac{1}{n}$.
(b) For $n = 100, 1000,$ and $10,000$, the probabilities are $\frac{1}{100}, \frac{1}{1000},$ and $\frac{1}{10,000}$.
(c) As the number of balls increases, the probability of picking the winning ball decreases.

Now Try Exercise 113

14.1 Putting It All Together

CONCEPT	EXPLANATION	EXAMPLES
Rational Expression	An expression of the form $\frac{P}{Q}$, where P and Q are polynomials with $Q \neq 0$	$\frac{1}{x}, \frac{x-3}{2x^2-1}, \frac{2x+9}{5x}$, and $\frac{x^2+3x-5}{1}$
Undefined Rational Expressions	A rational expression is undefined for any value of the variable that makes the denominator equal to 0.	$\frac{1}{x-3}$ is undefined when $x = 3$. $\frac{5y}{y^2-1}$ is undefined when $y = 1$ or when $y = -1$.
Basic Principle of Rational Expressions	Factor the numerator and the denominator completely. Then apply $$\frac{P \cdot R}{Q \cdot R} = \frac{P}{Q}.$$	$\frac{4xy^2}{6xy^3} = \frac{2(2xy^2)}{3y(2xy^2)} = \frac{2}{3y}$ $\frac{4x(x-4)}{(x+1)(x-4)} = \frac{4x}{x+1}$

14.1 Exercises

CONCEPTS AND VOCABULARY

1. A rational expression can be written as _____, where P and Q are _____ with $Q \neq 0$.

2. Is $\frac{x}{2x^2+1}$ a rational expression? Why or why not?

3. A rational expression is undefined whenever the _____ is equal to 0.

4. The rational expression $\frac{1}{x-a}$ is undefined whenever $x =$ _____.

5. The basic principle of fractions states that a fraction can be simplified by using $\frac{a \cdot c}{b \cdot c} =$ _____.

6. The basic principle of rational expressions can be used to simplify _____ expressions.

EVALUATING RATIONAL EXPRESSIONS

Exercises 7–20: If possible, evaluate the expression for the given value of the variable.

7. $\frac{3}{x}$; $x = -7$

8. $\frac{3}{x+3}$; $x = 0$

9. $-\frac{x}{x-5}$; $x = -4$

10. $-\frac{4x}{5x+1}$; $x = 1$

11. $\frac{y+1}{y^2}$; $y = -2$

12. $\frac{3y-1}{y^2+1}$; $y = -1$

13. $\frac{7z}{z^2-4}$; $z = -2$

14. $\frac{5}{z^2-3z+2}$; $z = -1$

15. $\frac{5}{3t+6}$; $t = -2$

16. $\frac{4t}{2t+5}$; $t = -\frac{5}{2}$

17. $\frac{4-x}{x-4}$; $x = -2$

18. $\frac{x-7}{7-x}$; $x = 4$

19. $-\frac{6-x}{x-6}$; $x = 0$

20. $\frac{8-2x}{2x-8}$; $x = -5$

Exercises 21–24: Complete the table for the given expression. If a value is undefined, place a dash in the table.

21.
x	-2	-1	0	1	2
$\frac{x}{x+1}$					

22.
x	-2	-1	0	1	2
$\frac{2x}{3x-1}$					

23.

x	-2	-1	0	1	2
$\dfrac{3x}{2x^2+1}$					

24.

x	-2	-1	0	1	2
$\dfrac{2x-1}{x^2-1}$					

DETERMINING UNDEFINED EXPRESSIONS

Exercises 25–38: Find any values of the variable that make the expression undefined.

25. $-\dfrac{8}{x}$

26. $\dfrac{7}{x+1}$

27. $\dfrac{4}{z-3}$

28. $\dfrac{7-z}{z-7}$

29. $\dfrac{4y}{5y+4}$

30. $\dfrac{3+y}{3y-7}$

31. $\dfrac{5t+2}{t^2+1}$

32. $\dfrac{8t}{t^2+25}$

33. $\dfrac{8x}{x^2-25}$

34. $\dfrac{x+4}{x^2-36}$

35. $\dfrac{x^2+3x+2}{x^2+5x+6}$

36. $\dfrac{2x-1}{x^2-7x+10}$

37. $\dfrac{8z^2+z+1}{2z^2-7z+5}$

38. $\dfrac{4n^2+17n-15}{3n^2-8n+4}$

SIMPLIFYING RATIONAL EXPRESSIONS

Exercises 39–46: Simplify the fraction to lowest terms.

39. $\dfrac{12}{18}$

40. $\dfrac{24}{32}$

41. $\dfrac{24}{48}$

42. $\dfrac{8}{22}$

43. $-\dfrac{6}{15}$

44. $-\dfrac{22}{33}$

45. $-\dfrac{25}{75}$

46. $-\dfrac{36}{42}$

Exercises 47–50: First simplify the fraction in part (a), then simplify the rational expression in part (b).

47. (a) $\dfrac{8}{16}$ (b) $\dfrac{x+2}{2x+4}$

48. (a) $\dfrac{6}{9}$ (b) $\dfrac{4x+12}{6x+18}$

49. (a) $\dfrac{7-3}{3-7}$ (b) $\dfrac{7-x}{x-7}$

50. (a) $-\dfrac{8-5}{5-8}$ (b) $-\dfrac{x-5}{5-x}$

Exercises 51–88: Simplify the expression.

51. $\dfrac{5x^4}{10x^6}$

52. $\dfrac{6y^2}{9y}$

53. $\dfrac{8xy^3}{6x^2y^2}$

54. $\dfrac{36x^2y^5}{6x^5y}$

55. $\dfrac{x+4}{2x+8}$

56. $\dfrac{5x-10}{x-2}$

57. $\dfrac{3z-9}{5z-15}$

58. $\dfrac{4z+8}{10+5z}$

59. $\dfrac{(x+1)(x-1)}{(x+6)(x-1)}$

60. $\dfrac{(2x+1)(x+9)}{(4x-3)(x+9)}$

61. $\dfrac{(5y+3)(2y-1)}{(2y-1)(y+2)}$

62. $\dfrac{(4y-1)(5y+7)}{(5y+7)(1-4y)}$

63. $\dfrac{x-7}{7-x}$

64. $\dfrac{5-x}{x-5}$

65. $\dfrac{a-b}{b-a}$

66. $\dfrac{2t-3r}{3r-2t}$

67. $\dfrac{-6-x}{18+3x}$

68. $\dfrac{-2x-6}{x+3}$

69. $\dfrac{x+1}{-2x-2}$

70. $\dfrac{3x+21}{-7-x}$

71. $-\dfrac{9-x}{x-9}$

72. $-\dfrac{4-2x}{x-2}$

73. $\dfrac{(3x+5)(x-1)}{(3x-5)(1-x)}$

74. $\dfrac{(2-x)(x-2)}{(x-2)(2-x)}$

75. $\dfrac{n^2-n}{n^2-5n}$

76. $\dfrac{3n^2-4n}{n^2+4n}$

77. $\dfrac{x^2-3x}{6x-18}$

78. $\dfrac{4x^2+16x}{5x^2+20x}$

79. $\dfrac{z^2-3z+2}{z^2-4z+3}$

80. $\dfrac{z^2-3z-10}{z^2-2z-8}$

81. $\dfrac{2x^2+7x-4}{6x^2+x-2}$

82. $\dfrac{5x^2+3x-2}{5x^2+13x-6}$

83. $\dfrac{x-3}{3x^2-11x+6}$

84. $\dfrac{2x-1}{4x^2+6x-4}$

85. $-\dfrac{a-9}{9-a}$

86. $-\dfrac{-b-6}{b+6}$

87. $\dfrac{-2x-1}{4x+2}$

88. $\dfrac{4x+3}{-8x-6}$

Exercises 89 and 90: **Thinking Generally** *Complete the statement involving a rational expression.*

89. The expression $\frac{x-a}{a-x}$ simplifies to _____.

90. The expression $-\frac{x-a}{x+a}$ simplifies to $\frac{?}{x+a}$.

Exercises 91–98: Refer to Making Connections 1 in this section.
   (a) *Decide whether you are given an expression or an equation.*
   (b) *If you are given an expression, simplify it. If you are given an equation, solve it.*

91. $x + 1 = 7$

92. $x^2 - 4 = 0$

93. $\dfrac{x}{x(x+1)}$

94. $\dfrac{x-2}{(x-2)(x-8)}$

95. $\dfrac{x^2 - 4}{x+2}$

96. $\dfrac{x-4}{8-2x}$

97. $\dfrac{x}{2(1+3)} = 1$

98. $\dfrac{x}{2} + 2 = \dfrac{8}{2}$

Exercises 99–102: **Graphing Rational Equations** *Complete the following.*

   (a) *Use the given equation to complete the table of values for y.*

x	−4	−3	−2	−1	0	1	2	3	4
y									

   (b) *Determine any x-value that will make the expression undefined.*
   (c) *Sketch a dashed, vertical line (asymptote) in the xy-plane at any undefined values of x.*
   (d) *Plot the points from the table.*
   (e) *Sketch a graph of the equation. Do not let your graph cross the vertical, dashed line.*

99. $y = \dfrac{1}{x-1}$

100. $y = \dfrac{1}{x+1}$

101. $y = \dfrac{x}{x+1}$

102. $y = \dfrac{x}{x-1}$

APPLICATIONS INVOLVING RATIONAL EXPRESSIONS

103. **Traffic Flow** (Refer to Example 6.) Five vehicles per minute can pass through a construction zone. If the traffic arrives randomly at an average rate of x vehicles per minute, the average time T in minutes spent waiting in line and passing through the construction zone is given by $T = \frac{1}{5-x}$ for $x < 5$. (*Source:* N. Garber.)
   (a) Evaluate T for $x = 3$ and interpret the result.
   (b) Complete the table and interpret the results.

x	2	4	4.5	4.9	4.99
T					

104. **Standing in Line** A worker at a poolside store can serve 20 customers per hour. If children arrive randomly at an average rate of x per hour, then the average number of children N waiting in line is given by $N = \dfrac{x^2}{400 - 20x}$ for $x < 20$. (*Source:* N. Garber.)
   (a) Complete the table.

x	5	10	18	19	20
N					

   (b) Compare the number of children waiting in line if the average rate increases from 18 to 19 children per hour.

105. **Frog Population** Suppose that a frog species is introduced into a wetland area and its population in hundreds is modeled by
$$P = \frac{7x + 3}{x + 6},$$
where $x \geq 0$ is time in months.
   (a) Complete the table by finding P for each given value of x. Round to 2 decimal places.

x (months)	0	12	36	72
P (hundreds)				

   (b) What was the initial frog population?
   (c) Interpret the results in your completed table.

106. **Insect Population** Suppose that an insect population in thousands per acre is modeled by
$$P = \frac{5x + 2}{x + 1},$$
where $x \geq 0$ is time in months.
   (a) Complete the table by finding P for each given value of x. Round to 3 decimal places.

x (months)	0	12	36	60
P (thousands)				

   (b) What was the initial insect population?
   (c) Interpret the results in your completed table.

107. **Distance and Time** A car is traveling at 60 miles per hour.
   (a) How long does it take the car to travel 360 miles?
   (b) Write a rational expression that gives the time that it takes the car to travel M miles.

108. Distance and Time A bicyclist rides uphill at 10 miles per hour for 5 miles and then rides downhill at 20 miles per hour for 5 miles. What is the bicyclist's average speed? (*Hint:* Average speed equals distance divided by time.)

Exercises 109 and 110: **Traffic Flow** *(Refer to Example 6.) The figure shows a graph of the waiting time T in minutes at a construction zone when cars are arriving randomly at an average rate of x cars per minute.*
  (a) Give the equation of the vertical asymptote.
  (b) Explain how the graph relates to traffic flow.

109.

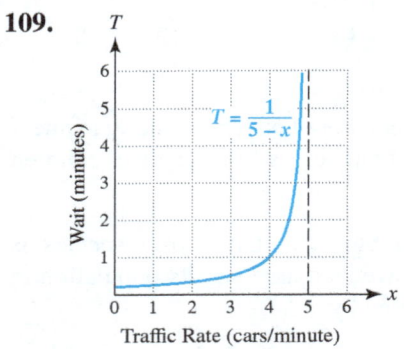

110.

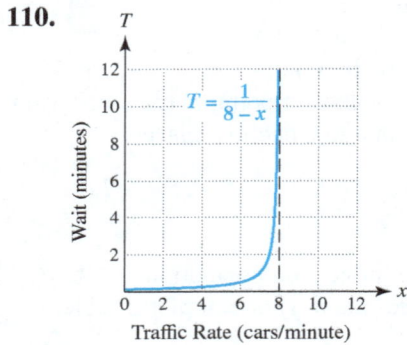

APPLICATIONS INVOLVING PROBABILITY

111. Probability Suppose that a coin is flipped. What is the probability that a head appears?

112. Probability A die shows the numbers 1, 2, 3, 4, 5, and 6. If each number has an equal chance of appearing on any given roll, what is the probability that a 2 or 4 appears?

113. Probability (Refer to Example 8.) Suppose that there are n balls in a container and that three balls have a winning number. If a ball is drawn randomly, do each of the following.
  (a) Write a rational expression that gives the probability of drawing a winning ball.
  (b) Write a rational expression that gives the probability of not drawing a winning ball. Evaluate your expression for $n = 100$ and interpret the result.

114. Probability Suppose that there are n balls numbered 1 to n in a container and that every even-numbered ball is green while every odd-numbered ball is red. If a ball is drawn randomly, answer each of the following.
  (a) If n is an odd number, are there more green balls or more red balls?
  (b) If n is 39, what is the probability of drawing a green ball?

WRITING ABOUT MATHEMATICS

115. What is a rational expression? When is a rational expression undefined?

116. Does the rational expression $\frac{5x + 2}{10x + 4}$ equal $\frac{5x}{10x} + \frac{2}{4}$? Explain your answer.

Group Activity

Students per Computer In the early years of personal computers, school districts could not afford to buy a computer for every student. As the price of computers decreased, more and more school districts were able to move toward this goal. The following table lists numbers of students per computer during these early years.

Year	1983	1985	1987	1989
Students/Computer	125	50	32	22

Year	1991	1993	1995	1997
Students/Computer	18	14	10	6

Source: Quality Education Data, Inc.

(a) Make a scatterplot of the data in the table. Would a straight line model the data accurately? Explain.
(b) Discuss how well the formula

$$S = \frac{125}{1 + 0.7(x - 1983)}, \quad x \geq 1983$$

models these data, where S represents the students per computer and x represents the year.
(c) In what year does the formula suggest that there were about 17 students per computer?

14.2 Multiplication and Division of Rational Expressions

Objectives

1. Reviewing Multiplication and Division of Fractions
2. Multiplying Rational Expressions
3. Dividing Rational Expressions

1 Reviewing Multiplication and Division of Fractions

To multiply two fractions, we use the property

$$\frac{a}{b} \cdot \frac{c}{d} = \frac{ac}{bd}.$$

In the next example, we review multiplication of fractions.

EXAMPLE 1 Multiplying fractions

Multiply and simplify your answers to lowest terms.
(a) $\frac{3}{7} \cdot \frac{4}{5}$ (b) $2 \cdot \frac{3}{4}$ (c) $\frac{4}{21} \cdot \frac{7}{8}$

Solution

(a) $\frac{3}{7} \cdot \frac{4}{5} = \frac{12}{35}$ (b) $2 \cdot \frac{3}{4} = \frac{2}{1} \cdot \frac{3}{4} = \frac{6}{4} = \frac{3}{2}$

(c) $\frac{4}{21} \cdot \frac{7}{8} = \frac{7 \cdot 4}{21 \cdot 8} = \frac{7}{21} \cdot \frac{4}{8} = \frac{1}{3} \cdot \frac{1}{2} = \frac{1}{6}$

Now Try Exercises 5, 7, 9

To divide two fractions, we change the division problem to a multiplication problem by using the property

$$\frac{a}{b} \div \frac{c}{d} = \frac{a}{b} \cdot \frac{d}{c}.$$

Invert $\frac{c}{d}$ to $\frac{d}{c}$ and multiply.

EXAMPLE 2　Dividing fractions

Divide and simplify your answers to lowest terms.

(a) $\frac{1}{3} \div \frac{5}{7}$　(b) $\frac{4}{5} \div 8$　(c) $\frac{8}{9} \div \frac{10}{3}$

Solution

Invert $\frac{5}{7}$ to $\frac{7}{5}$ and multiply.

(a) $\frac{1}{3} \div \frac{5}{7} = \frac{1}{3} \cdot \frac{7}{5} = \frac{7}{15}$

(b) $\frac{4}{5} \div 8 = \frac{4}{5} \cdot \frac{1}{8} = \frac{4}{40} = \frac{1}{10}$

(c) $\frac{8}{9} \div \frac{10}{3} = \frac{8}{9} \cdot \frac{3}{10} = \frac{24}{90} = \frac{4 \cdot 6}{15 \cdot 6} = \frac{4}{15}$

Now Try Exercises 13, 15, 17

2　Multiplying Rational Expressions

Multiplying rational expressions is similar to multiplying fractions.

> **PRODUCTS OF RATIONAL EXPRESSIONS**
>
> To multiply two rational expressions, multiply the numerators and multiply the denominators. That is,
>
> $$\frac{A}{B} \cdot \frac{C}{D} = \frac{AC}{BD},$$
>
> where B and D are nonzero.

READING CHECK 1

- How do we multiply rational expressions?

EXAMPLE 3　Multiplying rational expressions

Multiply and simplify to lowest terms. Leave your answers in factored form.

(a) $\dfrac{3}{x} \cdot \dfrac{2x - 5}{x - 1}$　(b) $\dfrac{x - 1}{4x} \cdot \dfrac{x + 3}{x - 1}$

(c) $\dfrac{x^2 - 4}{x + 3} \cdot \dfrac{x + 3}{x + 2}$　(d) $\dfrac{4}{x^2 + 3x + 2} \cdot \dfrac{x^2 + 2x + 1}{8}$

Solution

(a) $\dfrac{3}{x} \cdot \dfrac{2x - 5}{x - 1} = \dfrac{3(2x - 5)}{x(x - 1)}$　Multiply the numerators and the denominators.

(b) $\dfrac{x - 1}{4x} \cdot \dfrac{x + 3}{x - 1} = \dfrac{(x - 1)(x + 3)}{4x(x - 1)}$　Multiply the numerators and the denominators.

$= \dfrac{x + 3}{4x}$　Simplify.

(c) $\dfrac{x^2 - 4}{x + 3} \cdot \dfrac{x + 3}{x + 2} = \dfrac{(x - 2)(x + 2)}{x + 3} \cdot \dfrac{x + 3}{x + 2}$　Factor.

$= \dfrac{(x - 2)(x + 2)(x + 3)}{(x + 3)(x + 2)}$　Multiply the numerators and the denominators.

$= x - 2$　Simplify.

STUDY TIP

Check with your instructor about your progress in the class to be sure that your work is complete and your grades are up-to-date.

(d) $\dfrac{4}{x^2 + 3x + 2} \cdot \dfrac{x^2 + 2x + 1}{8} = \dfrac{4(x^2 + 2x + 1)}{8(x^2 + 3x + 2)}$ Multiply the numerators and the denominators.

$= \dfrac{4(x + 1)(x + 1)}{8(x + 2)(x + 1)}$ Factor.

$= \dfrac{x + 1}{2(x + 2)}$ Simplify.

Now Try Exercises 29, 31, 41, 45

Math in Context — Highway Safety   Stopping distance for a car can vary depending on the road conditions. If the road is slippery, more distance is needed to stop. Also, cars that are traveling downhill require additional stopping distance. Rational expressions are frequently used by highway engineers to estimate the stopping distance of a car on slippery surfaces or on hills.

EXAMPLE 4 **Estimating stopping distance**

If a car is traveling at 60 miles per hour on a slippery road, then its stopping distance D in feet can be calculated by

$$D = \dfrac{3600}{30} \cdot \dfrac{1}{x},$$

where x is the coefficient of friction between the tires and the road and $0 < x \leq 1$. The more slippery the road is, the smaller the value of x. (Source: L. Haefner, *Introduction to Transportation Systems*.)

(a) Multiply and simplify the formula for D.
(b) Compare the stopping distance on an icy road with $x = 0.1$ to the stopping distance on dry pavement with $x = 0.4$.

Solution
(a) Because

$$\dfrac{3600}{30} \cdot \dfrac{1}{x} = \dfrac{3600}{30x} = \dfrac{120 \cdot 30}{x \cdot 30} = \dfrac{120}{x},$$

it follows that $D = \dfrac{120}{x}$.

(b) When $x = 0.1$, $D = \dfrac{120}{0.1} = 1200$ feet, and when $x = 0.4$, $D = \dfrac{120}{0.4} = 300$ feet. A car traveling at 60 miles per hour on an icy road requires a stopping distance that is 4 times that of a car traveling at the same speed on dry pavement.

Now Try Exercise 73

3 Dividing Rational Expressions

Dividing rational expressions is similar to dividing fractions.

READING CHECK 2

- When we divide the rational expressions $\dfrac{A}{B} \div \dfrac{C}{D}$, what do we multiply $\dfrac{A}{B}$ by?

QUOTIENTS OF RATIONAL EXPRESSIONS

To divide two rational expressions, multiply by the reciprocal of the divisor. That is,

$$\dfrac{A}{B} \div \dfrac{C}{D} = \dfrac{A}{B} \cdot \dfrac{D}{C},$$

where B, C, and D are nonzero.

NOTE: When we "multiply by the reciprocal of the divisor" in a division problem, we will sometimes refer to this step as "invert and multiply." ∎

EXAMPLE 5 Dividing rational expressions

Divide and simplify to lowest terms.

(a) $\dfrac{5}{2x} \div \dfrac{10}{x-4}$ (b) $\dfrac{x^2-9}{x^2+4} \div (x-3)$ (c) $\dfrac{x^2-x}{x^2-x-2} \div \dfrac{x}{x-2}$

Solution

(a)
$$\dfrac{5}{2x} \div \dfrac{10}{x-4} = \dfrac{5}{2x} \cdot \dfrac{x-4}{10}$$ Multiply by the reciprocal of the divisor.

Invert $\dfrac{10}{x-4}$ to $\dfrac{x-4}{10}$ and multiply.

$$= \dfrac{5(x-4)}{20x}$$ Multiply the numerators and the denominators.

$$= \dfrac{x-4}{4x}$$ Simplify. Note that $\dfrac{5}{20} = \dfrac{1}{4}$.

(b)
$$\dfrac{x^2-9}{x^2+4} \div (x-3) = \dfrac{x^2-9}{x^2+4} \cdot \dfrac{1}{x-3}$$ Multiply by the reciprocal of the divisor.

Invert $\dfrac{x-3}{1}$ to $\dfrac{1}{x-3}$ and multiply.

$$= \dfrac{x^2-9}{(x^2+4)(x-3)}$$ Multiply the numerators and the denominators.

$$= \dfrac{(x+3)(x-3)}{(x^2+4)(x-3)}$$ Factor the numerator.

$$= \dfrac{x+3}{x^2+4}$$ Simplify.

(c)
$$\dfrac{x^2-x}{x^2-x-2} \div \dfrac{x}{x-2} = \dfrac{x^2-x}{x^2-x-2} \cdot \dfrac{x-2}{x}$$ Multiply by the reciprocal of the divisor.

Invert $\dfrac{x}{x-2}$ to $\dfrac{x-2}{x}$ and multiply.

$$= \dfrac{(x^2-x)(x-2)}{(x^2-x-2)x}$$ Multiply the numerators and the denominators.

$$= \dfrac{x(x-1)(x-2)}{x(x+1)(x-2)}$$ Factor the numerator and the denominator.

$$= \dfrac{x-1}{x+1}$$ Simplify.

Now Try Exercises 49, 57, 65

14.2 Putting It All Together

CONCEPT	EXPLANATION	EXAMPLES
Multiplication of Rational Expressions	To multiply two rational expressions, multiply the numerators and multiply the denominators. $\dfrac{A}{B} \cdot \dfrac{C}{D} = \dfrac{AC}{BD}$ B and D are nonzero Then simplify the result to lowest terms.	$\dfrac{4x}{x-1} \cdot \dfrac{x-1}{x+1} = \dfrac{4x(x-1)}{(x-1)(x+1)}$ $= \dfrac{4x}{x+1}$

CONCEPT	EXPLANATION	EXAMPLES
Division of Rational Expressions	To divide two rational expressions, multiply the first expression by the reciprocal of the second expression. $$\frac{A}{B} \div \frac{C}{D} = \frac{A}{B} \cdot \frac{D}{C} \quad B, C, \text{ and } D \text{ are nonzero}$$ Then simplify the result to lowest terms.	$$\frac{x+1}{x-3} \div \frac{x+1}{x-5} = \frac{x+1}{x-3} \cdot \frac{x-5}{x+1}$$ $$= \frac{(x+1)(x-5)}{(x-3)(x+1)}$$ $$= \frac{x-5}{x-3}$$

14.2 Exercises — MyMathLab

CONCEPTS AND VOCABULARY

1. To multiply rational expressions, multiply the _____ and multiply the _____.

2. To divide two rational expressions, multiply the first expression by the _____ of the second expression.

3. $\frac{A}{B} \cdot \frac{C}{D} =$ _____   4. $\frac{A}{B} \div \frac{C}{D} =$ _____

REVIEWING MULTIPLICATION AND DIVISION OF FRACTIONS

Exercises 5–20: Multiply or divide as indicated. Simplify to lowest terms.

5. $\frac{1}{2} \cdot \frac{4}{5}$ 6. $\frac{6}{7} \cdot \frac{7}{18}$

7. $\frac{3}{7} \cdot 4$ 8. $5 \cdot \frac{4}{5}$

9. $\frac{5}{4} \cdot \frac{8}{15}$ 10. $\frac{3}{10} \cdot \frac{5}{9}$

11. $\frac{1}{3} \cdot \frac{2}{3} \cdot \frac{9}{11}$ 12. $\frac{2}{5} \cdot \frac{10}{11} \cdot \frac{1}{4}$

13. $\frac{2}{3} \div \frac{1}{6}$ 14. $\frac{5}{7} \div \frac{5}{8}$

15. $\frac{8}{9} \div 3$ 16. $\frac{7}{3} \div 6$

17. $8 \div \frac{4}{5}$ 18. $7 \div \frac{14}{5}$

19. $\frac{4}{5} \div \frac{2}{3} \div \frac{1}{2}$ 20. $\frac{1}{2} \div \frac{5}{4} \div \frac{2}{5}$

SIMPLIFYING RATIONAL EXPRESSIONS

Exercises 21–28: Simplify the expression.

21. $\frac{x+5}{x+5}$ 22. $\frac{2x-3}{2x-3}$

23. $\frac{(z+1)(z+2)}{(z+4)(z+2)}$ 24. $\frac{(2z-7)(z+5)}{(3z+5)(z+5)}$

25. $\frac{8y(y+7)}{12y(y+7)}$ 26. $\frac{6(y+1)}{12(y+1)}$

27. $\frac{x(x+2)(x+3)}{x(x-2)(x+3)}$ 28. $\frac{2(x+1)(x-1)}{4(x+1)(x-1)}$

MULTIPLYING RATIONAL EXPRESSIONS

Exercises 29–48: Multiply and simplify to lowest terms. Leave your answers in factored form.

29. $\frac{8}{x} \cdot \frac{x+1}{x}$ 30. $\frac{7}{2x} \cdot \frac{x}{x-1}$

31. $\frac{8+x}{x} \cdot \frac{x-3}{x+8}$ 32. $\frac{5x^2+x}{2x-1} \cdot \frac{1}{x}$

33. $\frac{z+3}{z+4} \cdot \frac{z+4}{z-7}$ 34. $\frac{2z+1}{3z} \cdot \frac{3z}{z+2}$

35. $\frac{5x+1}{3x+2} \cdot \frac{3x+2}{5x+1}$ 36. $\frac{x+1}{x+3} \cdot \frac{x+3}{x+1}$

37. $\frac{(t+1)^2}{t+2} \cdot \frac{(t+2)^2}{t+1}$ 38. $\frac{(t-1)^2}{(t+5)^2} \cdot \frac{t+5}{t-1}$

39. $\frac{x^2}{x^2+4} \cdot \frac{x+4}{x}$ 40. $\frac{x-1}{x^2} \cdot \frac{x^2}{x^2+1}$

41. $\frac{z^2-1}{z^2-4} \cdot \frac{z-2}{z+1}$ 42. $\frac{z^2-9}{z-5} \cdot \frac{z-5}{z+3}$

43. $\frac{y^2-2y}{y^2-1} \cdot \frac{y+1}{y-2}$ 44. $\frac{y^2-4y}{y+1} \cdot \frac{y+1}{y-4}$

45. $\frac{2x^2-x-3}{3x^2-8x-3} \cdot \frac{3x+1}{2x-3}$

46. $\frac{6x^2+11x-2}{3x^2+11x-4} \cdot \frac{3x-1}{6x-1}$

47. $\dfrac{(x-3)^3}{x^2-2x+1} \cdot \dfrac{x-1}{(x-3)^2}$

48. $\dfrac{x^2+4x+4}{x^2-2x+1} \cdot \dfrac{(x-1)^2}{(x+2)^2}$

DIVIDING RATIONAL EXPRESSIONS

Exercises 49–70: Divide and simplify to lowest terms. Leave your answers in factored form.

49. $\dfrac{2}{x} \div \dfrac{2x+3}{x}$

50. $\dfrac{6}{2x} \div \dfrac{x+2}{2x}$

51. $\dfrac{x-2}{3x} \div \dfrac{2-x}{6x}$

52. $\dfrac{x+1}{2x-1} \div \dfrac{x+1}{x}$

53. $\dfrac{z+2}{z+1} \div \dfrac{z+2}{z-1}$

54. $\dfrac{z+7}{z-4} \div \dfrac{z+7}{z-4}$

55. $\dfrac{3y+4}{2y+1} \div \dfrac{3y+4}{y+2}$

56. $\dfrac{y+5}{y-2} \div \dfrac{y}{y+3}$

57. $\dfrac{t^2-1}{t^2+1} \div \dfrac{t+1}{4}$

58. $\dfrac{4}{2t^3} \div \dfrac{8}{t^2}$

59. $\dfrac{y^2-9}{y^2-25} \div \dfrac{y+3}{y+5}$

60. $\dfrac{y+1}{y-4} \div \dfrac{y^2-1}{y^2-16}$

61. $\dfrac{2x^2-4x}{2x-1} \div \dfrac{x-2}{2x-1}$

62. $\dfrac{x-4}{x^2+x} \div \dfrac{5}{x+1}$

63. $\dfrac{2z^2-5z-3}{z^2+z-20} \div \dfrac{z-3}{z-4}$

64. $\dfrac{z^2+12z+27}{z^2-5z-14} \div \dfrac{z+3}{z+2}$

65. $\dfrac{t^2-1}{t^2+5t-6} \div (t+1)$

66. $\dfrac{t^2-2t-3}{t^2-5t-6} \div (t-3)$

67. $\dfrac{a-b}{a+b} \div \dfrac{a-b}{2a+3b}$

68. $\dfrac{x^3-y^3}{x^2-y^2} \div \dfrac{x^2+xy+y^2}{x-y}$

69. $\dfrac{x-y}{x^2+2xy+y^2} \div \dfrac{1}{(x+y)^2}$

70. $\dfrac{a^2-b^2}{4a^2-9b^2} \div \dfrac{a-b}{2a+3b}$

71. **Thinking Generally** Simplify $\dfrac{a-b}{b-c} \cdot \dfrac{c-b}{b-a}$.

72. **Thinking Generally** Simplify $\dfrac{a-b}{b-c} \div \dfrac{b-a}{a-b}$.

APPLICATIONS INVOLVING RATIONAL EXPRESSIONS

73. **Stopping on Slippery Roads** (Refer to Example 4.) If a car is traveling at 30 miles per hour on a slippery road, then its stopping distance D in feet can be calculated by

$$D = \dfrac{900}{30} \cdot \dfrac{1}{x},$$

where x is the coefficient of friction between the tires and the road and $0 < x \leq 1$. (*Source:* L. Haefner.)
(a) Multiply and simplify the formula for D.
(b) Compare the stopping distance on an icy road with $x = 0.1$ and on dry pavement with $x = 0.4$.

74. **Stopping on Hills** If a car is traveling at 50 miles per hour on a hill with wet pavement, then its stopping distance D is given by

$$D = \dfrac{2500}{30} \cdot \dfrac{1}{x+0.3},$$

where x equals the slope of the hill. (*Source:* L. Haefner.)
(a) Multiply and simplify the formula for D.
(b) Compare the stopping distance for an uphill slope of $x = 0.1$ to a downhill slope of $x = -0.1$.

75. **Probability** Suppose that one jar holds n balls and that a second jar holds $n+1$ balls. Each jar contains one winning ball.
(a) The probability, or chance, of drawing the winning ball from the first jar and *not* drawing it from the second jar is

$$\dfrac{1}{n} \cdot \dfrac{n}{n+1}.$$

Simplify this expression.
(b) Find this probability for $n = 99$.

76. **U.S. AIDS Cases** The cumulative number of AIDS cases C in the United States from 1982 to 1994 can be modeled by $C = 3200x^2 + 1586$, and the cumulative number of AIDS deaths D from 1982 to 1994 can be modeled by $D = 1900x^2 + 619$. In these equations, $x = 0$ corresponds to 1982, $x = 1$ corresponds to 1983, and so on until $x = 12$ corresponds to 1994. (*Source:* U.S. Department of Health.)
(a) Write the rational expression $\dfrac{D}{C}$ in terms of x.
(b) Evaluate your expression for $x = 4, 7$, and 10. Round your answers to the nearest thousandth. Interpret the results.
(c) Explain what the rational expression $\dfrac{D}{C}$ represents.

77. **Google and Facebook Revenues** The equation given by $G = 350(x-3)^2 + 50$ gives Google's revenue x years after its startup, and $F = 206(x-3)^2 + 150$ gives Facebook's revenue x years after its startup. Results are in millions of dollars and $x \geq 3$.
(a) Write a formula for $R = \dfrac{G}{F}$.
(b) Evaluate R when $x = 6$ and interpret your result.

78. Google and Yahoo Revenues The equation given by $G = 350(x - 3)^2 + 50$ gives Google's revenue x years after its startup, and $Y = 117(x - 3)^2 + 50$ gives Yahoo's revenue x years after its startup. Results are in millions of dollars and $x \geq 3$.
(a) Write a formula for $R = \frac{G}{Y}$.
(b) Evaluate R when $x = 5$ and interpret your result.

79. Total Production Cost The average cost per item in dollars for a company to produce x digital cameras is given by
$$C = \frac{50x + 20{,}000}{x}.$$
(a) If the company sells 5000 cameras, evaluate C when $x = 5000$ and interpret the result.
(b) Write an expression T that gives the *total* cost of producing x cameras.
(c) Evaluate T when $x = 5000$ and interpret the result.

80. Total Production Cost The average cost per item in dollars for a company to produce x flash drives is given by
$$C = \frac{18x + 36{,}000}{x}.$$
(a) The company expects to sell 9000 flash drives. Evaluate C when $x = 9000$ and interpret the result.
(b) Write an expression for T that gives the *total* cost of producing x flash drives. Simplify this expression.
(c) Evaluate T when $x = 9000$ and interpret the result.

WRITING ABOUT MATHEMATICS

81. Explain how to multiply two rational expressions.

82. Explain how to divide two rational expressions.

SECTIONS 14.1 and 14.2 Checking Basic Concepts

1. If possible, evaluate the expression $\frac{3}{x^2 - 1}$ for $x = -1$ and $x = 3$.

2. Simplify to lowest terms.
   (a) $\dfrac{6x^3y^2}{15x^2y^3}$ (b) $\dfrac{5x - 15}{x - 3}$
   (c) $\dfrac{x^2 - x - 6}{x^2 + x - 12}$

3. Multiply and simplify to lowest terms.
   (a) $\dfrac{4}{3x} \cdot \dfrac{2x}{6}$ (b) $\dfrac{2x + 4}{x^2 - 1} \cdot \dfrac{x + 1}{x + 2}$

4. Divide and simplify to lowest terms.
   (a) $\dfrac{7}{3z^2} \div \dfrac{14}{5z^3}$ (b) $\dfrac{x^2 + x}{x - 3} \div \dfrac{x}{x - 3}$

5. **Waiting in Line** Customers are waiting in line at a department store. They arrive randomly at an average rate of x per minute. If the clerk can wait on 2 customers per minute, then the average time in minutes spent waiting in line is given by $T = \frac{1}{2 - x}$ for $x < 2$. (*Source:* N. Garber, *Traffic and Highway Engineering*.)
   (a) Complete the table.

x	0.5	1.0	1.5	1.9
T				

   (b) What happens to the waiting time as x increases but remains less than 2?

14.3 Addition and Subtraction with Like Denominators

Objectives

1. Reviewing Addition and Subtraction of Fractions
2. Adding and Subtracting Rational Expressions with Like Denominators

1 ▶ Reviewing Addition and Subtraction of Fractions

Previously, we demonstrated how the property

$$\frac{a}{c} + \frac{b}{c} = \frac{a+b}{c}$$

can be used to add fractions with like denominators. For example,

$$\frac{3}{7} + \frac{2}{7} = \frac{3+2}{7} = \frac{5}{7}. \qquad \text{Add the numerators.}$$

(Like denominators)

To subtract two fractions with like denominators, the property

$$\frac{a}{c} - \frac{b}{c} = \frac{a-b}{c}$$

is used. For example,

$$\frac{2}{5} - \frac{4}{5} = \frac{2-4}{5} = -\frac{2}{5}. \qquad \text{Subtract the numerators.}$$

(Like denominators)

STUDY TIP

By this time in the semester, it is likely that you know some of your classmates. Have you started or joined a study group? Be sure not to miss the opportunity to study math with your classmates.

EXAMPLE 1   Adding and subtracting fractions with like denominators

Simplify each expression to lowest terms.

(a) $\frac{3}{8} + \frac{4}{8}$ (b) $\frac{5}{9} + \frac{1}{9}$ (c) $\frac{12}{5} - \frac{7}{5}$ (d) $\frac{23}{20} - \frac{13}{20}$

Solution

(a) $\frac{3}{8} + \frac{4}{8} = \frac{3+4}{8} = \frac{7}{8}$

(b) $\frac{5}{9} + \frac{1}{9} = \frac{5+1}{9} = \frac{6}{9} = \frac{2}{3}$

(c) $\frac{12}{5} - \frac{7}{5} = \frac{12-7}{5} = \frac{5}{5} = 1$

(d) $\frac{23}{20} - \frac{13}{20} = \frac{23-13}{20} = \frac{10}{20} = \frac{1}{2}$

Now Try Exercises 7, 9, 11, 13

TECHNOLOGY NOTE

Arithmetic of Fractions

Many calculators have the capability to perform addition and subtraction of fractions, as illustrated in the following figures. Compare these results with those from Example 1.

CALCULATOR HELP

To express a result as a fraction on a calculator, see Appendix A (pages AP-1 and AP-2).

```
(3/8)+(4/8)▶Frac
              7/8
(5/9)+(1/9)▶Frac
              2/3
```

```
(12/5)-(7/5)▶Frac
              1
(23/20)-(13/20)▶
Frac
              1/2
```

2 Adding and Subtracting Rational Expressions with Like Denominators

Addition and subtraction of rational expressions with like denominators are similar to addition and subtraction of fractions. The following property can be used to add two rational expressions with like denominators.

READING CHECK 1
- How do we add rational expressions with like denominators?

> **SUMS OF RATIONAL EXPRESSIONS**
>
> To add two rational expressions that have like denominators, add their numerators. Keep the same denominator.
>
> $$\frac{A}{C} + \frac{B}{C} = \frac{A+B}{C} \quad C \text{ is nonzero.}$$

When we add rational expressions with like denominators, we add the numerators. Then we combine like terms and simplify the resulting expression by applying the basic principle of rational expressions. For example, we can add $\frac{2x}{x+1}$ and $\frac{1-x}{x+1}$ as follows.

$$\frac{2x}{x+1} + \frac{1-x}{x+1} = \frac{2x + 1 - x}{x+1} \quad \text{Add the numerators.}$$

$$= \frac{2x - x + 1}{x+1} \quad \text{Commutative property}$$

$$= \frac{x+1}{x+1} \quad \text{Combine like terms.}$$

$$= 1 \quad \text{Simplify.}$$

(Like denominators)

It is important to understand that the expressions $\frac{2x}{x+1} + \frac{1-x}{x+1}$ and 1 are *equivalent expressions*. That is, they are equal for *every* value of x except -1, for which the first expression is undefined.

In the next example, we add rational expressions with like denominators and simplify the result to lowest terms.

EXAMPLE 2   Adding rational expressions with like denominators

Add and simplify to lowest terms.

(a) $\dfrac{3}{b} + \dfrac{2}{b}$ (b) $\dfrac{z}{z+2} + \dfrac{2}{z+2}$

(c) $\dfrac{x-1}{x^2+x} + \dfrac{1}{x^2+x}$ (d) $\dfrac{t^2+t}{t-1} + \dfrac{1-3t}{t-1}$

Solution

(a) $\dfrac{3}{b} + \dfrac{2}{b} = \dfrac{3+2}{b} = \dfrac{5}{b}$ Add the numerators.

(b) $\dfrac{z}{z+2} + \dfrac{2}{z+2} = \dfrac{z+2}{z+2}$ Add the numerators.

$\phantom{(b)\ \dfrac{z}{z+2} + \dfrac{2}{z+2}} = 1$ Simplify.

(c) $\dfrac{x-1}{x^2+x} + \dfrac{1}{x^2+x} = \dfrac{x-1+1}{x^2+x}$ Add the numerators.

(Add numerator; keep the same denominator.)

$\phantom{(c)\ \dfrac{x-1}{x^2+x} + \dfrac{1}{x^2+x}} = \dfrac{x}{x(x+1)}$ Factor the denominator.

$\phantom{(c)\ \dfrac{x-1}{x^2+x} + \dfrac{1}{x^2+x}} = \dfrac{1}{x+1}$ Simplify.

(d) $\dfrac{t^2 + t}{t - 1} + \dfrac{1 - 3t}{t - 1} = \dfrac{t^2 + t + 1 - 3t}{t - 1}$ Add the numerators.

$\phantom{(d) \dfrac{t^2 + t}{t - 1} + \dfrac{1 - 3t}{t - 1}} = \dfrac{t^2 - 2t + 1}{t - 1}$ Combine like terms.

$\phantom{(d) \dfrac{t^2 + t}{t - 1} + \dfrac{1 - 3t}{t - 1}} = \dfrac{(t - 1)(t - 1)}{t - 1}$ Factor the numerator.

$\phantom{(d) \dfrac{t^2 + t}{t - 1} + \dfrac{1 - 3t}{t - 1}} = t - 1$ Simplify.

Now Try Exercises 19, 25, 33, 35

EXAMPLE 3   Adding rational expressions with two variables

Add and simplify to lowest terms.

(a) $\dfrac{4}{xy} + \dfrac{5}{xy}$ (b) $\dfrac{a}{a^2 - b^2} + \dfrac{b}{a^2 - b^2}$ (c) $\dfrac{1}{x - y} + \dfrac{-1}{y - x}$

Solution

(a) $\dfrac{4}{xy} + \dfrac{5}{xy} = \dfrac{4 + 5}{xy} = \dfrac{9}{xy}$ Add the numerators.

(b) $\dfrac{a}{a^2 - b^2} + \dfrac{b}{a^2 - b^2} = \dfrac{a + b}{a^2 - b^2}$ Add the numerators.

$\phantom{(b) \dfrac{a}{a^2 - b^2} + \dfrac{b}{a^2 - b^2}} = \dfrac{a + b}{(a - b)(a + b)}$ Factor the denominator.

$\phantom{(b) \dfrac{a}{a^2 - b^2} + \dfrac{b}{a^2 - b^2}} = \dfrac{1}{a - b}$ Simplify.

(c) First write $\dfrac{1}{x - y} + \dfrac{-1}{y - x}$ with a common denominator. Note that if we multiply the second term by 1, written in the form $\dfrac{-1}{-1}$, it becomes

$$\dfrac{-1}{y - x} \cdot \dfrac{-1}{-1} = \dfrac{(-1)(-1)}{(y - x)(-1)} = \dfrac{1}{-y + x} = \dfrac{1}{x - y}.$$

Thus the given sum can be simplified as follows.

$\dfrac{1}{x - y} + \dfrac{-1}{y - x} = \dfrac{1}{x - y} + \dfrac{1}{x - y}$ Rewrite the second term.

$\phantom{\dfrac{1}{x - y} + \dfrac{-1}{y - x}} = \dfrac{2}{x - y}$ Add the numerators.

Now Try Exercises 51, 53, 55

Next we consider subtraction of rational expressions with like denominators.

READING CHECK 2

- How do we subtract rational expressions with like denominators?

DIFFERENCES OF RATIONAL EXPRESSIONS

To subtract two rational expressions that have like denominators, subtract their numerators. Keep the same denominator.

$$\dfrac{A}{C} - \dfrac{B}{C} = \dfrac{A - B}{C} \quad \text{C is nonzero.}$$

Subtraction of rational expressions with like denominators is similar to addition except that instead of adding numerators, we subtract them. For example, the expressions $\frac{3x}{x-4}$ and $\frac{2x}{x-4}$ have like denominators and can be subtracted as follows.

$$\underbrace{\frac{3x}{x-4} - \frac{2x}{x-4}}_{\text{Like denominators}} = \frac{3x - 2x}{x-4} \quad \text{Subtract the numerators.}$$
$$= \frac{x}{x-4} \quad \text{Combine like terms.}$$

In the next example, we subtract rational expressions with like denominators and simplify the result to lowest terms.

EXAMPLE 4 **Subtracting rational expressions with like denominators**

Subtract and simplify to lowest terms.

(a) $\dfrac{a+1}{a} - \dfrac{1}{a}$ (b) $\dfrac{2y}{3y-1} - \dfrac{3y}{3y-1}$ (c) $\dfrac{1+x}{2x^2+5x-3} - \dfrac{-2}{2x^2+5x-3}$

Solution

(a) $\dfrac{a+1}{a} - \dfrac{1}{a} = \dfrac{a+1-1}{a} = \dfrac{a}{a} = 1$

(b) $\dfrac{2y}{3y-1} - \dfrac{3y}{3y-1} = \dfrac{2y-3y}{3y-1} = \dfrac{-y}{3y-1}$ or $-\dfrac{y}{3y-1}$

(c) $\dfrac{1+x}{2x^2+5x-3} - \dfrac{-2}{2x^2+5x-3} = \dfrac{1+x-(-2)}{2x^2+5x-3}$

(Subtract numerators; keep the same denominator.)

$$= \dfrac{x+3}{(2x-1)(x+3)}$$
$$= \dfrac{1}{2x-1}$$

Now Try Exercises 21, 27, 67

If the numerator of the second fraction in a difference has more than one term, it is important to put parentheses around the second numerator.

$$\dfrac{x+6}{2x+1} - \dfrac{3-x}{2x+1} = \dfrac{x+6-(3-x)}{2x+1} \quad \text{Subtract the numerators; insert parentheses.}$$
$$= \dfrac{x+6-3-(-x)}{2x+1} \quad \text{Distributive property}$$
$$= \dfrac{x+6-3+x}{2x+1} \quad \text{Double negative property}$$
$$= \dfrac{2x+3}{2x+1} \quad \text{Combine like terms.}$$

NOTE: If parentheses were not inserted in the previous calculation, the numerator would be

$$x + 6 - 3 - x = 3,$$

which would give an incorrect result. ∎

EXAMPLE 5 Subtracting rational expressions with like denominators

Subtract and simplify to lowest terms.

(a) $\dfrac{2x}{x+1} - \dfrac{x-1}{x+1}$ (b) $\dfrac{x+y}{3y} - \dfrac{x-y}{3y}$

Solution

(a) $\dfrac{2x}{x+1} - \dfrac{x-1}{x+1} = \dfrac{2x-(x-1)}{x+1}$ Subtract the numerators.

$= \dfrac{2x-x+1}{x+1}$ Distributive property

$= \dfrac{x+1}{x+1}$ Simplify the numerator.

$= 1$ Simplify.

(b) $\dfrac{x+y}{3y} - \dfrac{x-y}{3y} = \dfrac{x+y-(x-y)}{3y}$ Subtract the numerators.

$= \dfrac{x+y-x+y}{3y}$ Distributive property

$= \dfrac{2y}{3y}$ Simplify the numerator.

$= \dfrac{2}{3}$ Simplify.

Now Try Exercises 31, 59

🌐 **Math in Context** Quality Control When companies manufacture a large number of items, *quality control* is important. For example, suppose that a company makes computer flash drives. Because it is not practical to check every flash drive to make sure that it works properly, inspectors often check a random sample. By using mathematics and rational expressions, this technique helps determine the likelihood that all the flash drives are good.

EXAMPLE 6 Analyzing quality control

A container holds a mixture of 8-GB and 16-GB computer flash drives. In this container, there is a total of n flash drives, including 2 defective 8-GB flash drives and 4 defective 16-GB flash drives. If a flash drive is picked at random by a quality control inspector, then the probability, or chance, of one of the defective flash drives being chosen is given by the expression $\dfrac{2}{n} + \dfrac{4}{n}$.

(a) Simplify this expression.   (b) Interpret the result.

Solution

(a) Because the denominators are the same, we simply add the numerators.

$$\dfrac{2}{n} + \dfrac{4}{n} = \dfrac{2+4}{n} = \dfrac{6}{n}$$ Add the numerators.

(b) There are 6 chances in n that a defective flash drive is chosen.

Now Try Exercise 73

14.3 Putting It All Together

CONCEPT	EXPLANATION	EXAMPLES
Addition of Rational Expressions	For polynomials A, B, and C, where C is nonzero, $$\frac{A}{C} + \frac{B}{C} = \frac{A+B}{C}.$$	$\frac{x}{x+1} + \frac{1-x}{x+1} = \frac{x+1-x}{x+1} = \frac{1}{x+1}$ $\frac{2x}{x^2-1} + \frac{x}{x^2-1} = \frac{2x+x}{x^2-1} = \frac{3x}{x^2-1}$
Subtraction of Rational Expressions	For polynomials A, B, and C, where C is nonzero, $$\frac{A}{C} - \frac{B}{C} = \frac{A-B}{C}.$$ If B consists of more than one term, put parentheses around B and apply the distributive property.	$\frac{2x}{x^2-4} - \frac{x+2}{x^2-4} = \frac{2x-(x+2)}{x^2-4}$ $= \frac{2x-x-2}{x^2-4}$ $= \frac{x-2}{(x+2)(x-2)}$ $= \frac{1}{x+2}$

14.3 Exercises   MyMathLab®

CONCEPTS AND VOCABULARY

1. When adding two rational expressions that have like denominators, add their _____ . The _____ do not change.

2. When subtracting two rational expressions that have like denominators, subtract their _____ . The _____ do not change.

3. $\frac{A}{C} + \frac{B}{C} = $ _____

4. $\frac{A}{C} - \frac{B}{C} = $ _____

ADDING AND SUBTRACTING FRACTIONS THAT HAVE LIKE DENOMINATORS

Exercises 5–18: Simplify to lowest terms.

5. $\frac{1}{2} + \frac{1}{2}$
6. $\frac{3}{7} + \frac{2}{7}$
7. $\frac{4}{5} + \frac{2}{5}$
8. $\frac{3}{11} + \frac{5}{11}$
9. $\frac{1}{6} + \frac{5}{6}$
10. $\frac{3}{10} + \frac{5}{10}$
11. $\frac{4}{7} - \frac{1}{7}$
12. $\frac{5}{13} - \frac{7}{13}$
13. $\frac{7}{8} - \frac{3}{8}$
14. $\frac{9}{16} - \frac{5}{16}$
15. $\frac{11}{12} - \frac{5}{12}$
16. $\frac{7}{24} - \frac{3}{24}$
17. $\frac{7}{15} + \frac{4}{15} - \frac{1}{15}$
18. $\frac{11}{36} - \frac{5}{36} + \frac{1}{36}$

ADDING AND SUBTRACTING RATIONAL EXPRESSIONS THAT HAVE LIKE DENOMINATORS

Exercises 19–70: Simplify to lowest terms.

19. $\frac{2}{x} + \frac{1}{x}$
20. $\frac{9}{x} - \frac{7}{x}$
21. $\frac{7+2x}{4x} - \frac{7}{4x}$
22. $\frac{x-1}{5x} + \frac{2x+1}{5x}$
23. $\frac{y+3}{y-3} + \frac{2y-12}{y-3}$
24. $\frac{5-y}{y+2} + \frac{3y-1}{y+2}$
25. $\frac{x}{x-3} + \frac{-3}{x-3}$
26. $\frac{2x}{2x+1} + \frac{1}{2x+1}$
27. $\frac{5z}{4z+3} - \frac{z}{4z+3}$
28. $\frac{z}{2z+1} - \frac{1-z}{2z+1}$
29. $\frac{t+5}{t+6} + \frac{t+7}{t+6}$
30. $\frac{4t-13}{t-4} + \frac{1-t}{t-4}$
31. $\frac{5x}{2x+3} - \frac{3x-3}{2x+3}$
32. $\frac{x}{5-x} - \frac{2x-5}{5-x}$
33. $\frac{x-4}{x^2-x} + \frac{4}{x^2-x}$
34. $\frac{2x-2}{4x^2-1} + \frac{1}{4x^2-1}$
35. $\frac{z^2-1}{z-2} + \frac{3-3z}{z-2}$
36. $\frac{x^2+2}{x+1} + \frac{3x}{x+1}$

37. $\dfrac{x^2 + 4x - 1}{4x + 2} - \dfrac{x^2 - 4x - 5}{4x + 2}$

38. $\dfrac{2x^2 - x + 5}{x^2 - 9} - \dfrac{x^2 - x + 14}{x^2 - 9}$

39. $\dfrac{3y}{5} + \dfrac{2y - 5}{5}$

40. $\dfrac{3y - 22}{11} + \dfrac{8y}{11}$

41. $\dfrac{x + y}{4} + \dfrac{x - y}{4}$

42. $\dfrac{x + y}{4} - \dfrac{x - y}{4}$

43. $\dfrac{z^2 + 4}{z - 2} - \dfrac{4z}{z - 2}$

44. $\dfrac{z^2 + 2z}{z + 1} + \dfrac{1}{z + 1}$

45. $\dfrac{2x^2 - 5x}{2x + 1} - \dfrac{3}{2x + 1}$

46. $\dfrac{2x^2}{x + 2} + \dfrac{9x + 10}{x + 2}$

47. $\dfrac{3n}{2n^2 - n + 5} + \dfrac{4n}{2n^2 - n + 5}$

48. $\dfrac{n}{n^2 + n + 1} - \dfrac{1}{n^2 + n + 1}$

49. $\dfrac{1}{x + 3} + \dfrac{2}{x + 3} + \dfrac{3}{x + 3}$

50. $\dfrac{x}{2x - 5} - \dfrac{1}{2x - 5} + \dfrac{2x + 1}{2x - 5}$

51. $\dfrac{8}{ab} + \dfrac{1}{ab}$

52. $\dfrac{6}{xy} + \dfrac{9}{xy}$

53. $\dfrac{x}{(x + y)^2} + \dfrac{y}{(x + y)^2}$

54. $\dfrac{x - 2y}{x^2 - y^2} + \dfrac{y}{x^2 - y^2}$

55. $\dfrac{5}{x - y} + \dfrac{-5}{y - x}$

56. $\dfrac{-4}{y - x} + \dfrac{4}{x - y}$

57. $\dfrac{8}{a - b} + \dfrac{8}{b - a}$

58. $\dfrac{6}{x - y} + \dfrac{6}{y - x}$

59. $\dfrac{a + b}{4a} - \dfrac{a - b}{4a}$

60. $\dfrac{x - y}{5x} - \dfrac{x + 9y}{5x}$

61. $\dfrac{x}{x + y} + \dfrac{y}{x + y}$

62. $\dfrac{x}{x - y} - \dfrac{y}{x - y}$

63. $\dfrac{a^2}{a + b} - \dfrac{b^2}{a + b}$

64. $\dfrac{a^2}{a + b} + \dfrac{2ab + b^2}{a + b}$

65. $\dfrac{2x - 5}{2x^2 + 5x + 2} + \dfrac{6}{2x^2 + 5x + 2}$

66. $\dfrac{4x}{3x^2 + 5x - 2} + \dfrac{2 - 3x}{3x^2 + 5x - 2}$

67. $\dfrac{3x + 7}{3x^2 - 2x - 5} - \dfrac{2x + 6}{3x^2 - 2x - 5}$

68. $\dfrac{5x - 4}{2x^2 - 7x + 6} - \dfrac{3x - 1}{2x^2 - 7x + 6}$

69. $\dfrac{4x^2}{2x + 3y} - \dfrac{9y^2}{2x + 3y}$

70. $\dfrac{x^3}{x^2 + xy + y^2} - \dfrac{y^3}{x^2 + xy + y^2}$

71. **Thinking Generally** If $\dfrac{2}{3 + x} + \dfrac{3}{3 + x}$ equals $\dfrac{5}{10}$, what must be true about x?

72. **Thinking Generally** If $\dfrac{8}{6 + x} - \dfrac{4}{3 + y}$ equals 0, what must be true about x and y?

APPLICATIONS INVOLVING RATIONAL EXPRESSIONS

73. **Quality Control** (Refer to Example 6.) A container holds a total of $n + 1$ batteries. In this container, there are 6 defective AA batteries, 5 defective C batteries, and 3 defective D batteries. If a battery is chosen at random by a quality control inspector, the probability, or chance, of one of the defective batteries being chosen is

$$\dfrac{6}{n + 1} + \dfrac{5}{n + 1} + \dfrac{3}{n + 1}.$$

(a) Simplify this expression.
(b) Evaluate the simplified expression for $n = 99$ and interpret the result.

74. **Intensity of a Light Bulb** The farther a person is from a light bulb, the less intense its light. The equation $I = \dfrac{19}{4d^2}$ approximates the light intensity from a 60-watt light bulb at a distance of d meters, where I is measured in watts per square meter. (*Source:* R. Weidner.)
(a) Find I for $d = 2$ meters and interpret the result.
(b) The intensity of light from a 100-watt light bulb is about $I = \dfrac{32}{4d^2}$. Find an expression for the sum of the intensities of light from a 100-watt bulb and a 60-watt bulb.

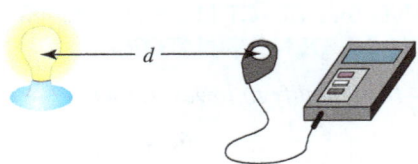

75. **Average Cost** The average cost per item in dollars to make x video games is given by

$$M = \dfrac{5x + 1{,}000{,}000}{x}$$

and the average cost per item in dollars to package the same video game is given by

$$P = \dfrac{2x + 500{,}000}{x}.$$

Write an equation C that gives the total cost per item of making and packaging x video games.

76. **Total Cost** Use the result from the previous exercise to determine the total cost per item when 50,000 video games are made and packaged.

WRITING ABOUT MATHEMATICS

77. Explain how to add two rational expressions with like denominators. Give an example.

78. Explain how to subtract two rational expressions with like denominators. Give an example.

14.4 Addition and Subtraction with Unlike Denominators

Objectives

1. Finding Least Common Multiples
   - Using a Step Diagram to Find the LCM
2. Reviewing Fractions with Unlike Denominators
3. Adding and Subtracting Rational Expressions with Unlike Denominators
4. Solving an Application Involving Rational Expressions

1 Finding Least Common Multiples

Connecting Concepts with Your Life Two friends work part-time at a store. The first person works every sixth day, and the second person works every eighth day. If they both work today, how many days will pass before they work on the same day again?

We can answer this question by listing the days that each person works.

Days the First Person Works: 0, 6, 12, 18, 24, 30, 36, 42, 48, 54
(every sixth day)

Both worked today. *Both will work on day 24.* *Both will work on day 48.*

Days the Second Person Works: 0, 8, 16, 24, 32, 40, 48, 56, 64, 72
(every eighth day)

After 24 days, the two friends work on the same day. The next time is after 48 days. The numbers 24 and 48 are *common multiples* of 6 and 8. (Find another.) However, 24 is the *least* common multiple (LCM) of 6 and 8 because it is the smallest common multiple.

Another way to find the least common multiple of 6 and 8 is to factor each number into prime numbers.

Prime factors of 6   *Prime factors of 8*

$6 = 2 \cdot 3$ and $8 = 2 \cdot 2 \cdot 2$

STUDY TIP

The methods discussed earlier in this text for finding the LCM of two numbers can also be applied to find the LCM of two polynomials.

To find the least common multiple, first list each factor the greatest number of times that it occurs in either factorization. Then find the product of these numbers. For this example, the factor 2 occurs three times in the factorization of 8 and only once in the factorization of 6, so list 2 three times. The factor 3 appears only once in the factorization of 6, so list it once:

$2, 2, 2, 3.$

The least common multiple is their product: $2 \cdot 2 \cdot 2 \cdot 3 = 24.$

This same procedure can also be used to find the least common multiple of two or more polynomials.

READING CHECK 1

- What is the first step in finding the least common multiple of two polynomials?

FINDING THE LEAST COMMON MULTIPLE

The least common multiple (LCM) of two or more polynomials can be found as follows.

STEP 1: Factor each polynomial completely.

STEP 2: List each factor the greatest number of times that it occurs in any factorization.

STEP 3: Find the product of this list of factors. The result is the LCM.

EXAMPLE 1　Finding least common multiples

Find the least common multiple of each pair of expressions.
(a) $2x, 5x^2$
(b) $x^2 - x, x - 1$
(c) $x + 2, x - 3$
(d) $x^2 + 2x + 1, x^2 + 3x + 2$

Solution

(a) **STEP 1:** Factor $2x$ and $5x^2$ completely.

$$2x = 2 \cdot x \quad \text{and} \quad 5x^2 = 5 \cdot x \cdot x$$

STEP 2: In either factorization, the factor 2 occurs at most once, the factor 5 occurs at most once, and the factor x occurs at most twice. The list of factors is $2, 5, x, x$.

STEP 3: The LCM equals the product

$$2 \cdot 5 \cdot x \cdot x = 10x^2.$$

(b) **STEP 1:** Factor $x^2 - x$ and $x - 1$ completely. Note that $x - 1$ cannot be factored.

$$x^2 - x = x(x - 1) \quad \text{and} \quad x - 1 = x - 1$$

STEP 2: Both factors, x and $x - 1$, occur at most once in either factorization. The list of factors is $x, (x - 1)$.

STEP 3: The LCM is the product $x(x - 1)$, or $x^2 - x$.

(c) **STEP 1:** Neither $x + 2$ nor $x - 3$ can be factored.

STEP 2: The list of factors is $(x + 2), (x - 3)$.

STEP 3: The LCM is the product $(x + 2)(x - 3)$, or $x^2 - x - 6$.

(d) **STEP 1:** Factor $x^2 + 2x + 1$ and $x^2 + 3x + 2$ completely.

$$x^2 + 2x + 1 = (x + 1)(x + 1) \quad \text{and} \quad x^2 + 3x + 2 = (x + 1)(x + 2)$$

STEP 2: In either factorization, the factor $(x + 1)$ occurs at most twice and the factor $(x + 2)$ occurs at most once. The list is $(x + 1), (x + 1), (x + 2)$.

STEP 3: The LCM is the product $(x + 1)^2(x + 2)$.

Now Try Exercises 15, 19, 27, 29

USING A STEP DIAGRAM TO FIND THE LCM The least common multiple for two polynomials can be found using a step diagram similar to that used earlier in this text for finding the least common multiple of two numbers. The next example illustrates this process.

EXAMPLE 2　Using a step diagram to find the LCM

Use a step diagram to find the LCM of $2x^3 - 8x^2$ and $2x^3 - 2x^2 - 24x$.

Solution
Start by factoring $2x^3 - 8x^2$ and $2x^3 - 2x^2 - 24x$ completely.

$$2x^3 - 8x^2 = 2x^2(x - 4) \quad \text{and} \quad 2x^3 - 2x^2 - 24x = 2x(x + 3)(x - 4)$$

Write the factored form of each expression on the top step of the diagram. We find the expressions in each of the following steps of the diagram by factoring out *any* common factor from the expressions in the previous step. The process continues until no common factor can be found, as shown in **FIGURE 14.3**.

Factor Step Diagram for Finding the LCM

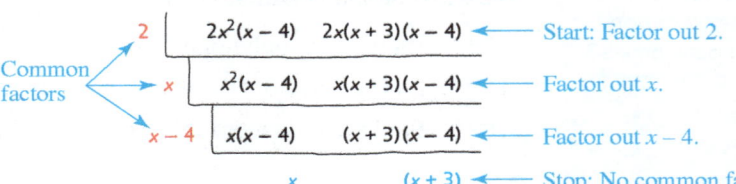

FIGURE 14.3

We find the LCM by multiplying the expressions along the side and at the bottom of the diagram. The LCM is $2 \cdot x \cdot (x - 4) \cdot x \cdot (x + 3) = 2x^2(x - 4)(x + 3)$.

Now Try Exercise 31

2 ▶ Reviewing Fractions with Unlike Denominators

Before we can find the sum $\frac{1}{2} + \frac{1}{3}$ by hand, we need to rewrite these fractions by using their least common denominator. The *least common denominator* (LCD) of $\frac{1}{2}$ and $\frac{1}{3}$ corresponds to the *least common multiple* of the denominators 2 and 3, which is 6. As a result, we can rewrite these fractions as

$$\frac{1}{2} \cdot \frac{3}{3} = \frac{3}{6} \quad \text{and} \quad \frac{1}{3} \cdot \frac{2}{2} = \frac{2}{6}.$$

> Multiplying a fraction by 1 in the form of $\frac{2}{2}$ or $\frac{3}{3}$ does not change its value.

Their sum is $\frac{1}{2} + \frac{1}{3} = \frac{3}{6} + \frac{2}{6} = \frac{5}{6}$.

EXAMPLE 3 **Adding and subtracting fractions with unlike denominators**

Simplify each expression.
(a) $\frac{3}{10} + \frac{4}{15}$ (b) $\frac{7}{8} - \frac{1}{6}$

Solution
(a) The LCD for $\frac{3}{10}$ and $\frac{4}{15}$ equals the LCM of 10 and 15, which is 30. We rewrite these fractions as

$$\frac{3}{10} \cdot \frac{3}{3} = \frac{9}{30} \quad \text{and} \quad \frac{4}{15} \cdot \frac{2}{2} = \frac{8}{30}.$$

Their sum is

$$\frac{3}{10} + \frac{4}{15} = \frac{9}{30} + \frac{8}{30} = \frac{17}{30}.$$

(b) The LCD for $\frac{7}{8}$ and $\frac{1}{6}$ is the LCM of 8 and 6, which is 24. We rewrite these fractions as

$$\frac{7}{8} \cdot \frac{3}{3} = \frac{21}{24} \quad \text{and} \quad \frac{1}{6} \cdot \frac{4}{4} = \frac{4}{24}.$$

Their difference is

$$\frac{7}{8} - \frac{1}{6} = \frac{21}{24} - \frac{4}{24} = \frac{17}{24}.$$

Now Try Exercises 33, 35

3 ▶ Adding and Subtracting Rational Expressions with Unlike Denominators

The first step in adding or subtracting rational expressions with unlike denominators is to rewrite each expression by using the least common denominator. Then the sum or difference can be found by using the techniques discussed in Section 3 of this chapter.

NOTE: The LCD of two or more rational expressions equals the LCM of their denominators. ∎

To add or subtract rational expressions, we often rewrite a rational expression with a different denominator. This technique is demonstrated in the next example and is used in future examples.

EXAMPLE 4 **Rewriting rational expressions**

Rewrite each rational expression so it has the given denominator D.
(a) $\frac{3}{2x}$, $D = 8x^2$ (b) $\frac{1}{x + 1}$, $D = x^2 - 1$

Solution

(a) We need to write $\frac{3}{2x}$ so that it is equivalent to $\frac{?}{8x^2}$.

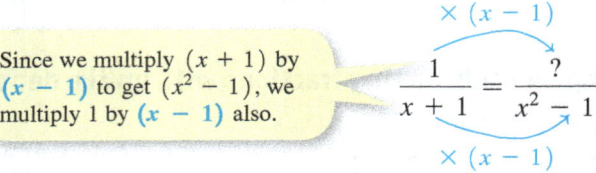

Since we multiply $2x$ by $4x$ to get $8x^2$, we multiply 3 by $4x$ also.

Because $8x^2 = 2x \cdot 4x$, we can multiply $\frac{3}{2x}$ by 1 in the form $\frac{4x}{4x}$ as follows.

$$\frac{3}{2x} \cdot \frac{4x}{4x} = \frac{12x}{8x^2} \qquad \text{Multiply rational expressions.}$$

(b) We must write $\frac{1}{x+1}$ so that it is equivalent to $\frac{?}{x^2-1}$.

Since we multiply $(x+1)$ by $(x-1)$ to get (x^2-1), we multiply 1 by $(x-1)$ also.

$$\frac{1}{x+1} \xrightarrow{\times (x-1)} \frac{?}{x^2-1}$$

Because $x^2 - 1 = (x+1)(x-1)$, we can multiply $\frac{1}{x+1}$ by 1 in the form $\frac{x-1}{x-1}$ as follows.

$$\frac{1}{x+1} \cdot \frac{x-1}{x-1} = \frac{x-1}{x^2-1} \qquad \text{Multiply rational expressions.}$$

Now Try Exercises 45, 47

EXAMPLE 5 Adding rational expressions with unlike denominators

Find each sum and leave your answer in factored form.

(a) $\dfrac{5}{8y} + \dfrac{7}{4y^2}$ **(b)** $\dfrac{1}{x-1} + \dfrac{1}{x+1}$ **(c)** $\dfrac{x}{x^2+2x+1} + \dfrac{1}{x+1}$

Solution

(a) First find the LCM of $8y$ and $4y^2$.

$$8y = 2 \cdot 2 \cdot 2 \cdot y \quad \text{and} \quad 4y^2 = 2 \cdot 2 \cdot y \cdot y$$

Thus the LCM is $2 \cdot 2 \cdot 2 \cdot y \cdot y = 8y^2$. Now, because

$$8y^2 = 8y \cdot y \quad \text{and} \quad 8y^2 = 4y^2 \cdot 2,$$

we multiply the first expression by $\frac{y}{y}$ and the second expression by $\frac{2}{2}$.

$$\frac{5}{8y} + \frac{7}{4y^2} = \frac{5}{8y} \cdot \frac{y}{y} + \frac{7}{4y^2} \cdot \frac{2}{2} \qquad \text{Rewrite by using the LCD.}$$

The LCD is $8y^2$.

$$= \frac{5y}{8y^2} + \frac{14}{8y^2} \qquad \text{Multiply rational expressions.}$$

$$= \frac{5y + 14}{8y^2} \qquad \text{Add the numerators.}$$

(b) The LCM for $x-1$ and $x+1$ is their product, $(x-1)(x+1)$.

$$\frac{1}{x-1} + \frac{1}{x+1} = \frac{1}{x-1} \cdot \frac{x+1}{x+1} + \frac{1}{x+1} \cdot \frac{x-1}{x-1} \qquad \text{Rewrite by using the LCD.}$$

The LCD is $(x-1)(x+1)$.

$$= \frac{x+1}{(x-1)(x+1)} + \frac{x-1}{(x+1)(x-1)} \qquad \text{Multiply rational expressions.}$$

$$= \frac{x+1+x-1}{(x-1)(x+1)} \qquad \text{Add the numerators.}$$

$$= \frac{2x}{(x-1)(x+1)} \qquad \text{Simplify the numerator.}$$

(c) First find the LCM for $x^2 + 2x + 1$ and $x + 1$. Because
$$x^2 + 2x + 1 = (x+1)(x+1),$$
their LCM is $(x+1)(x+1) = (x+1)^2$.

$$\frac{x}{x^2+2x+1} + \frac{1}{x+1} = \frac{x}{(x+1)^2} + \frac{1}{x+1} \cdot \frac{x+1}{x+1} \qquad \text{Rewrite by using the LCD.}$$

The LCD is $(x+1)^2$.
$$= \frac{x}{(x+1)^2} + \frac{x+1}{(x+1)^2} \qquad \text{Multiply rational expressions.}$$

$$= \frac{2x+1}{(x+1)^2} \qquad \text{Add the numerators.}$$

Now Try Exercises 53, 65, 71

Subtraction of rational expressions is performed in a manner similar to addition and is illustrated in the next example.

EXAMPLE 6 **Subtracting rational expressions with unlike denominators**

Simplify each expression. Write your answer in lowest terms and leave it in factored form.

(a) $\dfrac{5}{z} - \dfrac{z}{z-1}$
(b) $\dfrac{5}{x+1} - \dfrac{1}{x^2-1}$
(c) $\dfrac{x}{x^2-2x} - \dfrac{1}{x^2+2x}$
(d) $\dfrac{3}{x-1} - \dfrac{3}{x^2-x} + \dfrac{5}{x}$

Solution

(a) The LCD is $z(z-1)$.

$$\frac{5}{z} - \frac{z}{z-1} = \frac{5}{z} \cdot \frac{z-1}{z-1} - \frac{z}{z-1} \cdot \frac{z}{z} \qquad \text{Rewrite by using the LCD.}$$

$$= \frac{5(z-1)}{z(z-1)} - \frac{z^2}{z(z-1)} \qquad \text{Multiply rational expressions.}$$

$$= \frac{5(z-1) - z^2}{z(z-1)} \qquad \text{Subtract the numerators.}$$

$$= \frac{-z^2 + 5z - 5}{z(z-1)} \qquad \text{Simplify the numerator.}$$

(b) The LCD is $(x-1)(x+1)$. Note that $x^2 - 1 = (x-1)(x+1)$.

$$\frac{5}{x+1} - \frac{1}{x^2-1} = \frac{5}{x+1} \cdot \frac{x-1}{x-1} - \frac{1}{(x-1)(x+1)} \qquad \text{Rewrite by using the LCD.}$$

$$= \frac{5(x-1) - 1}{(x-1)(x+1)} \qquad \text{Subtract the numerators.}$$

$$= \frac{5x - 5 - 1}{(x-1)(x+1)} \qquad \text{Distributive property}$$

$$= \frac{5x - 6}{(x-1)(x+1)} \qquad \text{Simplify the numerator.}$$

(c) Start by factoring each denominator in $\frac{x}{x^2 - 2x} - \frac{1}{x^2 + 2x}$. Because

$$x^2 - 2x = x(x - 2) \quad \text{and} \quad x^2 + 2x = x(x + 2),$$

the LCD is $x(x - 2)(x + 2)$.

$$\begin{aligned}
\frac{x}{x^2 - 2x} - \frac{1}{x^2 + 2x} &= \frac{x}{x(x-2)} \cdot \frac{x+2}{x+2} - \frac{1}{x(x+2)} \cdot \frac{x-2}{x-2} && \text{Rewrite by using the LCD.} \\
&= \frac{x(x+2)}{x(x-2)(x+2)} - \frac{x-2}{x(x-2)(x+2)} && \text{Multiply rational expressions.} \\
&= \frac{x(x+2) - (x-2)}{x(x-2)(x+2)} && \text{Subtract the numerators.} \quad \text{Insert parentheses.} \\
&= \frac{x^2 + 2x - x + 2}{x(x-2)(x+2)} && \text{Distributive property} \\
&= \frac{x^2 + x + 2}{x(x-2)(x+2)} && \text{Simplify the numerator.}
\end{aligned}$$

(d) The given expression $\frac{3}{x-1} - \frac{3}{x^2 - x} + \frac{5}{x}$ contains three rational expressions. Begin by finding the LCM of the three denominators: $x - 1$, $x^2 - x$, and x. Because $x^2 - x = x(x - 1)$, the LCM is $x(x - 1)$.

$$\begin{aligned}
\frac{3}{x-1} - \frac{3}{x^2 - x} + \frac{5}{x} &= \frac{3}{x-1} \cdot \frac{x}{x} - \frac{3}{x(x-1)} + \frac{5}{x} \cdot \frac{x-1}{x-1} && \text{Rewrite by using the LCD.} \\
&= \frac{3x}{x(x-1)} - \frac{3}{x(x-1)} + \frac{5(x-1)}{x(x-1)} && \text{Multiply rational expressions.} \\
&= \frac{3x - 3 + 5x - 5}{x(x-1)} && \text{Combine the expressions.} \\
&= \frac{8x - 8}{x(x-1)} && \text{Simplify the numerator.} \\
&= \frac{8(x-1)}{x(x-1)} && \text{Factor the numerator.} \\
&= \frac{8}{x} && \text{Simplify to lowest terms.}
\end{aligned}$$

Now Try Exercises 63, 77, 79, 97

4 Solving an Application Involving Rational Expressions

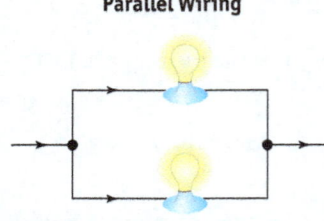

Parallel Wiring

FIGURE 14.4

🌐 **Math in Context** Electrical Sums of rational expressions occur in applications involving electricity. Suppose that two light bulbs are wired in parallel so that electricity can flow through either light bulb, as illustrated in **FIGURE 14.4**. If the individual resistances of the light bulbs are R and S, then the total resistance of the circuit is found by adding the expression

$$\frac{1}{R} + \frac{1}{S},$$

and then taking the *reciprocal*. (*Source:* R. Weidner and R. Sells, *Elementary Classical Physics*, vol. 2.)

EXAMPLE 7 Modeling electrical resistance

Add $\frac{1}{R} + \frac{1}{S}$, and then find the reciprocal of the result.

Solution
The LCD for $\frac{1}{R}$ and $\frac{1}{S}$ is RS.

$$\frac{1}{R} + \frac{1}{S} = \frac{1}{R} \cdot \frac{S}{S} + \frac{1}{S} \cdot \frac{R}{R} \qquad \text{Rewrite by using the LCD.}$$

$$= \frac{S}{RS} + \frac{R}{RS} \qquad \text{Multiply rational expressions.}$$

$$= \frac{S + R}{RS} \qquad \text{Add the numerators.}$$

In general, the reciprocal of $\frac{a}{b}$ is $\frac{b}{a}$, so the reciprocal of $\frac{S+R}{RS}$ is

$$\frac{RS}{S + R}.$$

This final expression can be used to find the total resistance of the circuit.

Now Try Exercise 101

> **CRITICAL THINKING 1**
>
> Find the reciprocal of the sum $x + \frac{1}{x}$.

14.4 Putting It All Together

CONCEPT	EXPLANATION	EXAMPLES
Least Common Multiple (LCM) of Polynomials	1. Factor each polynomial completely. 2. List each factor the greatest number of times that it occurs in any factorization. 3. The LCM is the product of this list.	For $x^2 - 2x$ and $x^2 - 4x + 4$, 1. $x^2 - 2x = x(x - 2)$ $\quad x^2 - 4x + 4 = (x - 2)(x - 2)$ 2. $x, (x - 2), (x - 2)$ 3. LCM $= x(x - 2)(x - 2)$.
Least Common Denominator (LCD)	The LCD of two or more rational expressions equals the least common multiple (LCM) of their denominators.	The LCD of $\frac{2}{x^2 - 2x}$ and $\frac{3}{x^2 - 4x + 4}$ is $x(x - 2)(x - 2)$ because the LCM of $x^2 - 2x$ and $x^2 - 4x + 4$ is $x(x - 2)(x - 2)$, as shown above.
Addition and Subtraction of Rational Expressions with Unlike Denominators	First rewrite each expression by using the LCD. Then add or subtract the expressions. Finally, write your answer in lowest terms. It is often helpful to leave the result in factored form.	The LCM of x and $x + 2$ is $x(x + 2)$. $\frac{2}{x} + \frac{5}{x + 2} = \frac{2}{x} \cdot \frac{x + 2}{x + 2} + \frac{5}{x + 2} \cdot \frac{x}{x}$ $= \frac{2x + 4}{x(x + 2)} + \frac{5x}{x(x + 2)}$ $= \frac{7x + 4}{x(x + 2)}$

14.4 Exercises

CONCEPTS AND VOCABULARY

1. Give a common multiple of 6 and 9 that is not the *least* common multiple.

2. The LCM of x and y is ____.

3. To rewrite $\frac{3}{4}$ with denominator 12, multiply $\frac{3}{4}$ by 1 written as the fraction ____.

4. To rewrite $\frac{4}{x-1}$ with denominator $x^2 - 1$, multiply $\frac{4}{x-1}$ by 1 written as the rational expression ____.

FINDING LEAST COMMON MULTIPLES

Exercises 5–12: Find the least common multiple.

5. 4, 6
6. 6, 9
7. 2, 3
8. 5, 4
9. 10, 15
10. 8, 12
11. 24, 36
12. 32, 40

Exercises 13–32: Find the least common multiple. Leave your answer in factored form.

13. $4x, 6x$
14. $6x, 9x$
15. $5x, 10x^2$
16. $4x^2, 12x$
17. $x, x+1$
18. $4x, x-1$
19. $2x+1, x+3$
20. $5x+3, x+9$
21. $x^2 - x, x^2 + x$
22. $x^2 + 2x, x^2$
23. $(x-8)^2, (x-8)(x+1)$
24. $(2x-1)^3, (2x-1)(x+3)$
25. $4x^2 - 1, 2x+1$
26. $x^2 + 4x + 3, x+3$
27. $x^2 - 1, x+1$
28. $x^2 - 4, x-2$
29. $2x^2 + 7x + 6, x^2 + 5x + 6$
30. $x^2 - 3x + 2, x^2 + 2x - 3$
31. $3y^2 + 6y, 3y^3 + 3y^2 - 6y$
32. $y^2 + 3y + 2, y^4 - 4y^2$

REVIEWING FRACTIONS WITH UNLIKE DENOMINATORS

Exercises 33–40: Evaluate and simplify. Write your answer in lowest terms.

33. $\frac{4}{5} + \frac{1}{2}$
34. $\frac{3}{8} + \frac{1}{4}$
35. $\frac{5}{9} - \frac{1}{3}$
36. $\frac{7}{10} - \frac{3}{15}$
37. $\frac{4}{25} + \frac{2}{5}$
38. $\frac{7}{9} - \frac{1}{3}$
39. $\frac{1}{5} + \frac{3}{4} - \frac{1}{2}$
40. $\frac{6}{7} - \frac{8}{9} + \frac{2}{3}$

ADDING AND SUBTRACTING RATIONAL EXPRESSIONS WITH UNLIKE DENOMINATORS

Exercises 41–52: (Refer to Example 4.) Rewrite the rational expression so it has the given denominator D.

41. $\frac{1}{3}, D = 9$
42. $\frac{3}{4}, D = 24$
43. $\frac{5}{7}, D = 21$
44. $\frac{4}{5}, D = 30$
45. $\frac{1}{4x}, D = 8x^3$
46. $\frac{5}{3x}, D = 9x^2$
47. $\frac{1}{x+2}, D = x^2 - 4$
48. $\frac{3}{x-3}, D = x^2 - 9$
49. $\frac{1}{x+1}, D = x^2 + x$
50. $\frac{3}{x-3}, D = x^2 - 3x$
51. $\frac{2x}{x+1}, D = x^2 + 2x + 1$
52. $\frac{x}{2x-1}, D = 2x^2 + 11x - 6$

Exercises 53–100: Simplify the expression. Write your answer in lowest terms and leave it in factored form.

53. $\frac{1}{3x} + \frac{3}{4x}$
54. $\frac{4}{2x^2} + \frac{7}{3x}$
55. $\frac{5}{z^2} - \frac{7}{z^3}$
56. $\frac{8}{z} - \frac{3}{2z}$
57. $\frac{1}{x} - \frac{1}{y}$
58. $\frac{1}{xy} - \frac{4}{y}$
59. $\frac{a}{b} + \frac{b}{a}$
60. $\frac{3}{x} - \frac{4}{y}$
61. $\frac{1}{2x+4} + \frac{3}{x+2}$
62. $\frac{1}{5x-10} - \frac{x}{x-2}$
63. $\frac{2}{t-2} - \frac{1}{t}$
64. $\frac{7}{2t} + \frac{1}{t+5}$
65. $\frac{5}{n-1} + \frac{n}{n+1}$
66. $\frac{4n}{3n-2} + \frac{n}{n+1}$

67. $\dfrac{3}{x-3} + \dfrac{6}{3-x}$

68. $\dfrac{x}{x-8} + \dfrac{x}{8-x}$

69. $\dfrac{1}{5k-1} + \dfrac{1}{1-5k}$

70. $\dfrac{4}{4-3k} + \dfrac{3k}{3k-4}$

71. $\dfrac{2x}{(x-1)^2} + \dfrac{4}{x-1}$

72. $\dfrac{5}{(x+5)} - \dfrac{x}{(x+5)^2}$

73. $\dfrac{2y}{y(2y-1)} + \dfrac{1}{2y-1}$

74. $\dfrac{5y}{y(y+1)} - \dfrac{5}{y+1}$

75. $\dfrac{1}{x+2} - \dfrac{1}{x^2+2x}$

76. $\dfrac{1}{x-3} - \dfrac{2}{x^2-3x}$

77. $\dfrac{3}{x-2} - \dfrac{1}{x^2-4}$

78. $\dfrac{x}{9-x^2} - \dfrac{1}{3-x}$

79. $\dfrac{2}{x^2-3x} - \dfrac{1}{x^2+3x}$

80. $\dfrac{3}{x^2+4x} - \dfrac{2}{x^2-4x}$

81. $\dfrac{1}{x-2} - \dfrac{1}{x+2} + \dfrac{1}{x}$

82. $\dfrac{1}{x^2} - \dfrac{2}{x} + \dfrac{2}{x-1}$

83. $\dfrac{x}{x^2+4x+4} + \dfrac{1}{x+2}$

84. $\dfrac{1}{x^2-3x-4} - \dfrac{1}{x+1}$

85. $\dfrac{x}{(x+1)(x+2)} - \dfrac{1}{(x+2)(x+3)}$

86. $\dfrac{2x}{(x-1)(x-2)} - \dfrac{5}{x-2}$

87. $\dfrac{1}{a+b} - \dfrac{1}{a-b}$

88. $\dfrac{x}{x^2-y^2} - \dfrac{1}{x+y}$

89. $\dfrac{r}{r-t} + \dfrac{t}{t-r} - 1$

90. $\dfrac{1}{a-b} - \dfrac{a}{a^2-b^2}$

91. $\dfrac{1}{x} + \dfrac{2}{x^2-2x} + \dfrac{5}{x-2}$

92. $\dfrac{1}{b} + \dfrac{1}{b+1} + \dfrac{1}{b+2}$

93. $\dfrac{2}{x-y} + \dfrac{3}{y-x} + \dfrac{1}{x-y}$

94. $\dfrac{a}{a-b} + \dfrac{b}{b-a} + \dfrac{3}{b}$

95. $\dfrac{3}{x-3} - \dfrac{3}{x^2-3x} - \dfrac{6}{x(x-3)}$

96. $\dfrac{3}{2a-4} + \dfrac{5}{2a} - \dfrac{3}{a^2-2a}$

97. $x + \dfrac{1}{x-1} - \dfrac{1}{x+1}$

98. $\dfrac{x}{x^2-4} - \dfrac{1}{x^2+4x+4}$

99. $\dfrac{2x+1}{x-1} - \dfrac{3}{x+1} + \dfrac{x}{x-1}$

100. $\dfrac{1}{x-3} - \dfrac{2}{x+3} + \dfrac{x}{x^2-9}$

APPLICATIONS INVOLVING RATIONAL EXPRESSIONS

101. **Electricity** In Example 7, we demonstrated that the expressions

$$\dfrac{1}{R} + \dfrac{1}{S} \quad \text{and} \quad \dfrac{S+R}{RS}$$

are equivalent. Evaluate both expressions by using $R = 120$ and $S = 200$. Are your answers the same?

102. **Intensity of a Light Bulb** The formula $I = \dfrac{32}{4d^2}$ approximates the intensity of light from a 100-watt light bulb at a distance of d meters, where I is in watts per square meter. For light from a 40-watt bulb, the equation for intensity becomes $I = \dfrac{16}{5d^2}$. (Source: R. Weidner.)
   (a) Find an expression for the sum of the intensities of light from the two light bulbs.
   (b) Find the combined intensity of their light at $d = 5$ meters.

103. **Photography** A lens in a camera has a focal length, which is important for focusing the camera. If an object is at a distance D from a lens that has a focal length F, then to be in focus, the distance S between the lens and the film should satisfy the equation

$$\dfrac{1}{S} = \dfrac{1}{F} - \dfrac{1}{D},$$

as illustrated in the figure. Write the difference $\dfrac{1}{F} - \dfrac{1}{D}$ as one term.

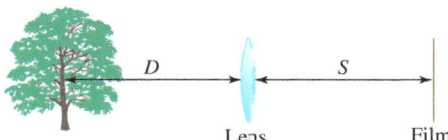

104. **Geometry** Find the sum of the areas of the two rectangles shown in the figure. Write your answer in factored form.

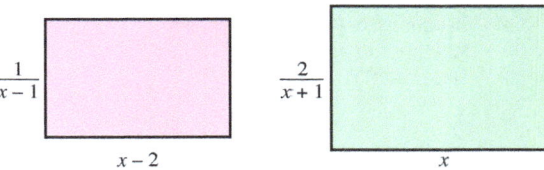

105. Planet Alignment Suppose that Saturn and Jupiter are in alignment with respect to the sun. If Jupiter orbits the sun every 12 years and Saturn orbits the sun every 30 years, how many years will it be before these two planets are back in the same location with the same alignment again?

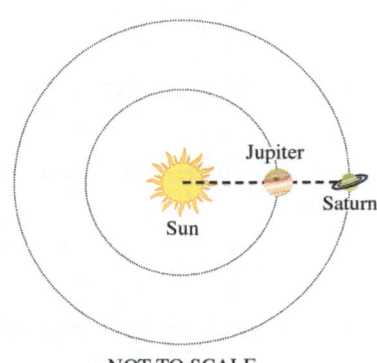

NOT TO SCALE

106. Working Shifts Two friends work part-time at a store. That first person works every sixth day and the second person works every tenth day. If they are both working today, how many days pass before they both work on the same day again?

WRITING ABOUT MATHEMATICS

107. Explain how to find the least common multiple of two polynomials.

108. Explain how to subtract two rational expressions with unlike denominators.

SECTIONS 14.3 and 14.4 — Checking Basic Concepts

1. Simplify each expression.
(a) $\dfrac{x}{x+2} + \dfrac{2}{x+2}$
(b) $\dfrac{2}{3x} - \dfrac{x}{3x}$
(c) $\dfrac{z^2 + z}{z+2} + \dfrac{z}{z+2}$

2. Find the least common multiple of each pair of expressions.
(a) $3x, 5x$
(b) $4x, x^2 + x$
(c) $x + 1, x - 1$

3. Simplify each expression.
(a) $\dfrac{1}{x+1} + \dfrac{5}{x}$
(b) $\dfrac{5}{x-3} + \dfrac{1}{3-x}$
(c) $\dfrac{-4}{4x+2} - \dfrac{x+2}{2x+1}$

4. Simplify the expression $\dfrac{a}{a-b} - \dfrac{b}{a+b}$.

14.5 Complex Fractions

Objectives

1 Introducing Complex Fractions
2 Simplifying Complex Fractions
 • Simplifying the Numerator and Denominator (Method I)
 • Multiplying by the LCD (Method II)

1 Introducing Complex Fractions

 **Connecting Concepts with Your Life** If two and a half pizzas are cut so that each piece equals one-eighth of a pizza, then there are 20 pieces of pizza. The quotient two and a half divided by one-eighth can be written as

$$\dfrac{2\tfrac{1}{2}}{\tfrac{1}{8}} \quad \text{or} \quad \dfrac{2 + \tfrac{1}{2}}{\tfrac{1}{8}}.$$

Complex fractions

NEW VOCABULARY

☐ Complex fraction
☐ Basic complex fraction

These expressions are examples of *complex* fractions. Typically, we want to simplify a complex fraction by rewriting it as a standard fraction in the form $\frac{a}{b}$.

A **complex fraction** is a rational expression that contains fractions in its numerator, denominator, or both. Examples of complex fractions include

$$\frac{1+\frac{1}{2}}{1-\frac{1}{3}}, \quad \frac{5z}{\frac{7}{z}-\frac{4}{2z}}, \quad \text{and} \quad \frac{\frac{2x}{3}+\frac{1}{x}}{x-\frac{1}{x+1}}. \quad \text{Complex fractions}$$

STUDY TIP

Two methods for simplifying complex fractions are discussed in this section. If your instructor does not require you to use a specific method, choose the one that works best for you.

The expression $\frac{a}{b} \div \frac{c}{d}$ can be written as a **basic complex fraction**, where both the numerator and denominator of the complex fraction are single fractions.

$$\frac{\frac{a}{b}}{\frac{c}{d}} \quad \text{Basic complex fraction}$$

Because

$$\frac{a}{b} \div \frac{c}{d} = \frac{a}{b} \cdot \frac{d}{c},$$

Invert $\frac{c}{d}$ to $\frac{d}{c}$ and multiply.

we can simplify basic complex fractions by multiplying the numerator and the reciprocal of the denominator. This strategy is summarized as follows.

READING CHECK 1

- Does the complex fraction $\frac{\frac{a}{b}}{\frac{c}{d}}$ equal $\frac{ad}{bc}$?

SIMPLIFYING BASIC COMPLEX FRACTIONS

For any real numbers a, b, c, and d,

$$\frac{\frac{a}{b}}{\frac{c}{d}} = \frac{a}{b} \cdot \frac{d}{c},$$

where b, c, and d are nonzero.

2 Simplifying Complex Fractions

There are two methods of simplifying a complex fraction. Method I is to simplify both the numerator and the denominator and then divide the resulting two fractions. Method II is to multiply the numerator and denominator by the least common denominator of all fractions within the complex fraction.

SIMPLIFYING THE NUMERATOR AND DENOMINATOR (METHOD I) The following steps outline Method I for simplifying complex fractions.

SIMPLIFYING COMPLEX FRACTIONS (METHOD I)

To simplify a complex fraction, perform the following steps.

STEP 1: Write the numerator as a single fraction; write the denominator as a single fraction.

STEP 2: Divide the denominator into the numerator by multiplying the numerator and the reciprocal of the denominator. (Invert and multiply.)

STEP 3: Simplify the result to lowest terms.

READING CHECK 2

- What is the first step when using Method I to simplify complex fractions?

Sometimes a complex fraction already has both its numerator and denominator written as single fractions. As a result, we can start with Step 2. This situation is shown in Example 1.

EXAMPLE 1 **Simplifying basic complex fractions (Method I)**

Simplify each basic complex fraction.

(a) $\dfrac{\frac{3}{4}}{\frac{9}{8}}$ (b) $\dfrac{3\frac{3}{4}}{1\frac{1}{2}}$ (c) $\dfrac{\frac{x}{4}}{\frac{3}{2y}}$ (d) $\dfrac{\frac{(x-1)^2}{4}}{\frac{x-1}{8}}$

Solution

(a) We can skip Step 1. To simplify $\frac{3}{4} \div \frac{9}{8}$, multiply $\frac{3}{4}$ by the reciprocal of $\frac{9}{8}$, or $\frac{8}{9}$.

$\dfrac{\frac{3}{4}}{\frac{9}{8}} = \dfrac{3}{4} \cdot \dfrac{8}{9}$ Multiply by the reciprocal of $\frac{9}{8}$ (Step 2).

Invert $\frac{9}{8}$ to $\frac{8}{9}$ and multiply.

$= \dfrac{24}{36}$ Multiply the fractions.

$= \dfrac{2}{3}$ Simplify (Step 3).

(b) Start by writing $3\frac{3}{4}$ and $1\frac{1}{2}$ as improper fractions.

$\dfrac{3\frac{3}{4}}{1\frac{1}{2}} = \dfrac{\frac{15}{4}}{\frac{3}{2}}$ Write as improper fractions (Step 1).

Invert $\frac{3}{2}$ to $\frac{2}{3}$ and multiply.

$= \dfrac{15}{4} \cdot \dfrac{2}{3}$ Multiply by the reciprocal of $\frac{3}{2}$ (Step 2).

$= \dfrac{5}{2}$ Multiply and simplify (Step 3).

(c) We can skip Step 1. To simplify $\frac{x}{4} \div \frac{3}{2y}$, multiply $\frac{x}{4}$ by the reciprocal of $\frac{3}{2y}$, or $\frac{2y}{3}$.

$\dfrac{\frac{x}{4}}{\frac{3}{2y}} = \dfrac{x}{4} \cdot \dfrac{2y}{3}$ Multiply by the reciprocal of $\frac{3}{2y}$ (Step 2).

Invert $\frac{3}{2y}$ to $\frac{2y}{3}$ and multiply.

$= \dfrac{2xy}{12}$ Multiply the fractions.

$= \dfrac{xy}{6}$ Simplify (Step 3).

(d) We can skip Step 1. Multiply $\frac{(x-1)^2}{4}$ by the reciprocal of $\frac{x-1}{8}$, or $\frac{8}{x-1}$.

$\dfrac{\frac{(x-1)^2}{4}}{\frac{x-1}{8}} = \dfrac{(x-1)^2}{4} \cdot \dfrac{8}{x-1}$ Multiply by the reciprocal of $\frac{x-1}{8}$ (Step 2).

Invert $\frac{x-1}{8}$ to $\frac{8}{x-1}$ and multiply.

$= \dfrac{8(x-1)(x-1)}{4(x-1)}$ Multiply the fractions.

$= 2(x-1)$ Simplify (Step 3).

Now Try Exercises 13, 15, 19, 29

EXAMPLE 2 — Simplifying complex fractions (Method I)

Simplify. Write your answer in lowest terms.

(a) $\dfrac{\dfrac{1}{a} + \dfrac{1}{b}}{\dfrac{1}{a} - \dfrac{1}{b}}$ (b) $\dfrac{x - \dfrac{4}{x}}{x + \dfrac{4}{x}}$ (c) $\dfrac{\dfrac{1}{x} + \dfrac{2}{x-1}}{\dfrac{2}{x} - \dfrac{1}{x-1}}$ (d) $\dfrac{\dfrac{1}{x} + \dfrac{1}{y}}{\dfrac{1}{x^2} - \dfrac{1}{y^2}}$

Solution

(a) In Step 1, write the numerator as one fraction by using the LCD, ab.

$$\frac{1}{a} + \frac{1}{b} = \frac{1}{a} \cdot \frac{b}{b} + \frac{1}{b} \cdot \frac{a}{a} = \frac{b}{ab} + \frac{a}{ab} = \frac{b+a}{ab}$$

Write the denominator as one fraction by using the LCD, ab.

$$\frac{1}{a} - \frac{1}{b} = \frac{1}{a} \cdot \frac{b}{b} - \frac{1}{b} \cdot \frac{a}{a} = \frac{b}{ab} - \frac{a}{ab} = \frac{b-a}{ab}$$

Finally, use these results to simplify the given complex fraction.

$$\frac{\dfrac{1}{a} + \dfrac{1}{b}}{\dfrac{1}{a} - \dfrac{1}{b}} = \frac{\dfrac{b+a}{ab}}{\dfrac{b-a}{ab}} \qquad \text{Simplify the numerator and the denominator (Step 1).}$$

$$= \frac{b+a}{ab} \cdot \frac{ab}{b-a} \qquad \text{Multiply by the reciprocal of } \tfrac{b-a}{ab} \text{ (Step 2).}$$

$$= \frac{ab(b+a)}{ab(b-a)} \qquad \text{Multiply.}$$

$$= \frac{b+a}{b-a} \qquad \text{Simplify (Step 3).}$$

(b) The LCD for both the numerator and the denominator is x.

$$\frac{x - \dfrac{4}{x}}{x + \dfrac{4}{x}} = \frac{\dfrac{x^2}{x} - \dfrac{4}{x}}{\dfrac{x^2}{x} + \dfrac{4}{x}} \qquad \text{Write by using the LCD (Step 1).}$$

$$= \frac{\dfrac{x^2 - 4}{x}}{\dfrac{x^2 + 4}{x}} \qquad \text{Combine terms.}$$

$$= \frac{x^2 - 4}{x} \cdot \frac{x}{x^2 + 4} \qquad \text{Multiply by the reciprocal of } \tfrac{x^2+4}{x} \text{ (Step 2).}$$

$$= \frac{x(x^2 - 4)}{x(x^2 + 4)} \qquad \text{Multiply.}$$

$$= \frac{x^2 - 4}{x^2 + 4} \qquad \text{Simplify (Step 3).}$$

(c) The LCD for the numerator and the denominator is $x(x-1)$.

$$\dfrac{\dfrac{1}{x}+\dfrac{2}{x-1}}{\dfrac{2}{x}-\dfrac{1}{x-1}} = \dfrac{\dfrac{x-1}{x(x-1)}+\dfrac{2x}{x(x-1)}}{\dfrac{2(x-1)}{x(x-1)}-\dfrac{x}{x(x-1)}}$$ Write by using the LCD (Step 1).

$$= \dfrac{\dfrac{x-1+2x}{x(x-1)}}{\dfrac{2(x-1)-x}{x(x-1)}}$$ Combine terms.

$$= \dfrac{\dfrac{3x-1}{x(x-1)}}{\dfrac{x-2}{x(x-1)}}$$ Simplify.

$$= \dfrac{3x-1}{x(x-1)} \cdot \dfrac{x(x-1)}{x-2}$$ Multiply by the reciprocal of $\dfrac{x-2}{x(x-1)}$ (Step 2).

$$= \dfrac{3x-1}{x-2}$$ Multiply and simplify (Step 3).

(d) The LCD for the numerator is xy, and the LCD for the denominator is x^2y^2.

$$\dfrac{\dfrac{1}{x}+\dfrac{1}{y}}{\dfrac{1}{x^2}-\dfrac{1}{y^2}} = \dfrac{\dfrac{y}{xy}+\dfrac{x}{xy}}{\dfrac{y^2}{x^2y^2}-\dfrac{x^2}{x^2y^2}}$$ Write by using the LCD (Step 1).

$$= \dfrac{\dfrac{y+x}{xy}}{\dfrac{y^2-x^2}{x^2y^2}}$$ Combine terms.

$$= \dfrac{y+x}{xy} \cdot \dfrac{x^2y^2}{y^2-x^2}$$ Multiply by the reciprocal of $\dfrac{y^2-x^2}{x^2y^2}$ (Step 2).

$$= \dfrac{x^2y^2(y+x)}{xy(y^2-x^2)}$$ Multiply.

$$= \dfrac{x^2y^2(y+x)}{xy(y-x)(y+x)}$$ Factor the denominator (Step 3).

$$= \dfrac{xy}{y-x}$$ Simplify.

Now Try Exercises 31, 37, 41, 45

MULTIPLYING BY THE LCD (METHOD II) In Method II for simplifying a complex fraction, we multiply the numerator and denominator of the complex fraction by the least common denominator of all the fractions within the complex fraction.

The following steps outline Method II for simplifying complex fractions.

READING CHECK 3

- When using Method II, by what do we multiply the numerator and denominator of the complex fraction?

SIMPLIFYING COMPLEX FRACTIONS (METHOD II)

To simplify a complex fraction, perform the following steps.

STEP 1: Find the LCD of all fractions within the complex fraction.

STEP 2: Multiply the numerator and the denominator of the complex fraction by the LCD.

STEP 3: Simplify the result to lowest terms.

We can use this method to simplify the complex fraction in Example 2(a). Because the LCD for the numerator and the denominator is ab, we multiply the complex fraction by 1, expressed in the form $\frac{ab}{ab}$. This method is equivalent to multiplying the numerator and the denominator by ab.

See Example 2(a).

$$\frac{\frac{1}{a} + \frac{1}{b}}{\frac{1}{a} - \frac{1}{b}} = \frac{\left(\frac{1}{a} + \frac{1}{b}\right)ab}{\left(\frac{1}{a} - \frac{1}{b}\right)ab} \quad \text{Multiply by } \frac{ab}{ab} = 1.$$

$$= \frac{\frac{ab}{a} + \frac{ab}{b}}{\frac{ab}{a} - \frac{ab}{b}} \quad \text{Distributive property}$$

$$= \frac{b + a}{b - a} \quad \text{Simplify the fractions.}$$

EXAMPLE 3 Simplifying complex fractions (Method II)

Simplify.

(a) $\dfrac{2z}{\dfrac{4}{z} + \dfrac{3}{z}}$ (b) $\dfrac{\dfrac{1}{x - 3}}{\dfrac{1}{x} + \dfrac{3}{x - 3}}$ (c) $\dfrac{\dfrac{1}{a} - \dfrac{1}{b}}{\dfrac{1}{2b^2} - \dfrac{1}{2a^2}}$

Solution

(a) The LCD for all fractions within the complex fraction is z, so multiply the numerator and denominator of the complex fraction by z (Step 1).

$$\frac{2z}{\frac{4}{z} + \frac{3}{z}} = \frac{(2z)z}{\left(\frac{4}{z} + \frac{3}{z}\right)z} \quad \text{Multiply by } \frac{z}{z} = 1 \text{ (Step 2).}$$

$$= \frac{2z^2}{\frac{4z}{z} + \frac{3z}{z}} \quad \text{Distributive property}$$

$$= \frac{2z^2}{4 + 3} \quad \text{Simplify the fractions (Step 3).}$$

$$= \frac{2z^2}{7} \quad \text{Add.}$$

(b) The LCD for all fractions within the complex fraction is the product $x(x-3)$ (Step 1).

$$\frac{\dfrac{1}{x-3}}{\dfrac{1}{x}+\dfrac{3}{x-3}} = \frac{\dfrac{1}{x-3}}{\left(\dfrac{1}{x}+\dfrac{3}{x-3}\right)} \cdot \frac{x(x-3)}{x(x-3)} \qquad \text{Multiply by } \frac{x(x-3)}{x(x-3)}=1 \text{ (Step 2).}$$

$$= \frac{\dfrac{x(x-3)}{x-3}}{\dfrac{x(x-3)}{x}+\dfrac{3x(x-3)}{x-3}} \qquad \text{Distributive property}$$

$$= \frac{x}{(x-3)+3x} \qquad \text{Simplify the fractions (Step 3).}$$

$$= \frac{x}{4x-3} \qquad \text{Simplify the denominator.}$$

(c) The LCD for the numerator *and* the denominator is $2a^2b^2$ (Step 1).

$$\frac{\dfrac{1}{a}-\dfrac{1}{b}}{\dfrac{1}{2b^2}-\dfrac{1}{2a^2}} = \frac{\dfrac{1}{a}-\dfrac{1}{b}}{\dfrac{1}{2b^2}-\dfrac{1}{2a^2}} \cdot \frac{2a^2b^2}{2a^2b^2} \qquad \text{Multiply by } \frac{2a^2b^2}{2a^2b^2}=1 \text{ (Step 2).}$$

$$= \frac{\dfrac{2a^2b^2}{a}-\dfrac{2a^2b^2}{b}}{\dfrac{2a^2b^2}{2b^2}-\dfrac{2a^2b^2}{2a^2}} \qquad \text{Distributive property}$$

$$= \frac{2ab^2-2a^2b}{a^2-b^2} \qquad \text{Simplify the fractions (Step 3).}$$

$$= \frac{2ab(b-a)}{(a-b)(a+b)} \qquad \text{Factor.}$$

$$= -\frac{2ab}{a+b} \qquad \text{Simplify.}$$

Now Try Exercises 33, 35, 49

CRITICAL THINKING 1

Are the expressions $\dfrac{\frac{a}{b}}{\frac{a}{b}+1}$ and $\dfrac{1}{1+1}$ equal? Explain.

Are the expressions $\dfrac{\frac{a}{b}+1}{\frac{a}{b}}$ and $1+\dfrac{b}{a}$ equal? Explain.

14.5 Putting It All Together

CONCEPT	EXPLANATION	EXAMPLES
Complex Fraction	A rational expression that contains fractions in its numerator, denominator, or both	$\dfrac{3+\frac{1}{x+1}}{3-\frac{1}{x+1}}$ and $\dfrac{\frac{x}{y}-\frac{y}{x}}{\frac{x}{y}+\frac{y}{x}}$
Simplifying Basic Complex Fractions	When b, c, and d are nonzero, $\dfrac{\frac{a}{b}}{\frac{c}{d}} = \dfrac{a}{b} \cdot \dfrac{d}{c}.$	$\dfrac{\frac{2}{x}}{\frac{4}{x-1}} = \dfrac{2}{x} \cdot \dfrac{x-1}{4} = \dfrac{x-1}{2x}$

14.5 COMPLEX FRACTIONS   891

CONCEPT	EXPLANATION	EXAMPLES
Method I: Simplifying the Numerator and Denominator	Combine the terms in the numerator, combine the terms in the denominator, and then multiply the numerator by the reciprocal of the denominator.	$\dfrac{\dfrac{1}{x}+\dfrac{3}{x}}{\dfrac{5}{y}-\dfrac{4}{y}} = \dfrac{\dfrac{4}{x}}{\dfrac{1}{y}} = \dfrac{4}{x}\cdot\dfrac{y}{1} = \dfrac{4y}{x}$
Method II: Multiplying the Numerator and Denominator by the LCD	Multiply the numerator *and* the denominator by the LCD of *all* fractions within the expression.	$\dfrac{\dfrac{2}{x}+\dfrac{1}{y}}{\dfrac{4}{y}-\dfrac{1}{x}} = \dfrac{\left(\dfrac{2}{x}+\dfrac{1}{y}\right)xy}{\left(\dfrac{4}{y}-\dfrac{1}{x}\right)xy} = \dfrac{2y+x}{4x-y}$ Note that the LCD is xy.

14.5 Exercises

MyMathLab®

CONCEPTS AND VOCABULARY

1. $\dfrac{\frac{1}{2}}{\frac{3}{4}} = $ _____

2. $\dfrac{\frac{a}{b}}{\frac{c}{d}} = $ _____

3. A complex fraction is a rational expression that contains _____ in its numerator, denominator, or both.

4. Write the expression $\dfrac{x}{2} \div \dfrac{1}{x-1}$ as a complex fraction.

5. Write the expression $\dfrac{a}{b} \div \dfrac{c}{d}$ as a complex fraction.

6. What is the LCD for $\dfrac{1}{x+2}$ and $\dfrac{1}{x}$?

SIMPLIFYING COMPLEX FRACTIONS

Exercises 7–12: For the complex fraction, determine the LCD of all the fractions appearing within the expression.

7. $\dfrac{\dfrac{x}{5}-\dfrac{1}{6}}{\dfrac{2}{15}-3x}$

8. $\dfrac{\dfrac{1}{2}-\dfrac{1}{x}}{\dfrac{1}{2}+\dfrac{1}{x}}$

9. $\dfrac{\dfrac{2}{x+1}-x}{\dfrac{2}{x-1}+x}$

10. $\dfrac{\dfrac{1}{4x}-\dfrac{4}{x}}{\dfrac{1}{2x}+\dfrac{1}{3x}}$

11. $\dfrac{\dfrac{1}{2x-1}-\dfrac{1}{2x+1}}{\dfrac{x+1}{x}}$

12. $\dfrac{\dfrac{1}{4x^2}-\dfrac{1}{2x^3}}{\dfrac{1}{x-1}+\dfrac{1}{x-1}}$

Exercises 13–52: Simplify the complex fraction.

13. $\dfrac{\frac{2}{3}}{\frac{5}{6}}$

14. $\dfrac{\frac{8}{9}}{\frac{5}{4}}$

15. $\dfrac{2\frac{1}{2}}{1\frac{3}{4}}$

16. $\dfrac{1\frac{2}{3}}{3\frac{1}{2}}$

17. $\dfrac{1\frac{1}{2}}{2\frac{1}{3}}$

18. $\dfrac{2\frac{1}{5}}{2\frac{1}{10}}$

19. $\dfrac{\frac{r}{t}}{\frac{2r}{t}}$

20. $\dfrac{\frac{8}{p}}{\frac{4}{p}}$

21. $\dfrac{\frac{6}{x}}{\frac{2}{y}}$

22. $\dfrac{\frac{3}{14x}}{\frac{6}{7x}}$

23. $\dfrac{\dfrac{6}{m-2}}{\dfrac{2}{m-2}}$

24. $\dfrac{\dfrac{3}{n+1}}{\dfrac{6}{n+1}}$

25. $\dfrac{\dfrac{p+1}{p}}{\dfrac{p+2}{p}}$

26. $\dfrac{\dfrac{2p}{2p+5}}{\dfrac{1}{4p+10}}$

27. $\dfrac{\dfrac{5}{z^2-1}}{\dfrac{z}{z^2-1}}$

28. $\dfrac{\dfrac{z}{z-2}}{\dfrac{z}{z-2}}$

29. $\dfrac{\dfrac{y}{y^2-9}}{\dfrac{1}{y+3}}$

30. $\dfrac{\dfrac{2y}{2y-1}}{\dfrac{1}{4y^2-1}}$

31. $\dfrac{4-\dfrac{1}{x}}{4+\dfrac{1}{x}}$

32. $\dfrac{x-\dfrac{1}{x}}{x+\dfrac{1}{x}}$

33. $\dfrac{x}{\dfrac{2}{x}+\dfrac{1}{x}}$

34. $\dfrac{5x}{1+\dfrac{1}{x}}$

35. $\dfrac{\dfrac{3}{x+1}}{\dfrac{4}{x+1}-\dfrac{1}{x+1}}$

36. $\dfrac{\dfrac{5}{2x-3}-\dfrac{4}{2x-3}}{\dfrac{7}{2x-3}+\dfrac{8}{2x-3}}$

37. $\dfrac{\dfrac{1}{m^2n}+\dfrac{1}{mn^2}}{\dfrac{1}{m^2n}-\dfrac{1}{mn^2}}$

38. $\dfrac{\dfrac{1}{x^2y}-\dfrac{1}{xy^2}}{\dfrac{1}{x^2y}+\dfrac{1}{xy^2}}$

39. $\dfrac{\dfrac{1}{2x}+\dfrac{1}{y}}{\dfrac{1}{y}-\dfrac{1}{2x}}$

40. $\dfrac{\dfrac{3}{x}-\dfrac{2}{y}}{\dfrac{3}{x}+\dfrac{2}{y}}$

41. $\dfrac{\dfrac{1}{ab}+\dfrac{1}{a}}{\dfrac{1}{ab}-\dfrac{1}{b}}$

42. $\dfrac{\dfrac{1}{a}+\dfrac{2}{3b}}{\dfrac{1}{a}-\dfrac{5}{2b}}$

43. $\dfrac{\dfrac{2}{q}-\dfrac{1}{q+1}}{\dfrac{1}{q+1}}$

44. $\dfrac{\dfrac{5}{p}+\dfrac{4}{p-5}}{\dfrac{5}{p}-\dfrac{5}{p-5}}$

45. $\dfrac{\dfrac{1}{x+1}+\dfrac{1}{x+2}}{\dfrac{1}{x+1}-\dfrac{1}{x+2}}$

46. $\dfrac{\dfrac{1}{x-3}-\dfrac{1}{x+3}}{1-\dfrac{1}{x^2-9}}$

47. $\dfrac{\dfrac{1}{2x-1}-\dfrac{1}{2x+1}}{\dfrac{x+1}{x}}$

48. $\dfrac{\dfrac{1}{4x^2}-\dfrac{1}{x^3}}{\dfrac{1}{x-1}+\dfrac{1}{x-1}}$

49. $\dfrac{\dfrac{1}{ab^2}-\dfrac{1}{a^2b}}{\dfrac{1}{b}-\dfrac{1}{a}}$

50. $\dfrac{\dfrac{1}{x^2}-\dfrac{1}{y^2}}{\dfrac{1}{x}-\dfrac{1}{y}}$

51. $\dfrac{1}{a^{-1}+b^{-1}}$

52. $\dfrac{a^2-b^2}{a^{-2}-b^{-2}}$

APPLICATIONS INVOLVING COMPLEX FRACTIONS

53. **Annuity** If P dollars are deposited every 2 weeks in an account paying an annual interest rate r expressed as a decimal, then the amount in the account after 2 years can be approximated by

$$\left(P\left(1+\dfrac{r}{26}\right)^{52}-P\right) \div \dfrac{r}{26}.$$

Write this expression as a complex fraction.

54. **Annuity** (Continuation of the preceding exercise) Use a calculator to evaluate the expression when $r = 0.026$ (2.6%) and $P = \$100$. Interpret the result.

55. **Resistance in Electricity** Light bulbs are often wired so that electricity can flow through either bulb, as illustrated in the accompanying figure.

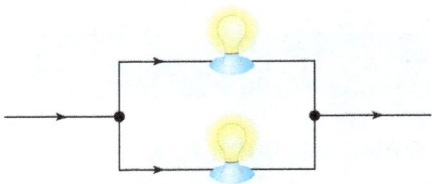

In this way, if one bulb burns out, the other bulb still works. If two light bulbs have resistances T and S, their combined resistance R is

$$R = \dfrac{1}{\dfrac{1}{T}+\dfrac{1}{S}}.$$

Simplify this formula.

56. **Resistance in Electricity** (Refer to the preceding exercise.) Evaluate the formula

$$R = \dfrac{1}{\dfrac{1}{T}+\dfrac{1}{S}}$$

when $T = 100$ and $S = 200$.

WRITING ABOUT MATHEMATICS

57. A student simplifies a complex fraction as shown below. Explain the student's mistake and how you would simplify the complex fraction correctly.

$$\dfrac{\dfrac{1}{x}+1}{\dfrac{1}{x}} \stackrel{?}{=} \dfrac{\dfrac{1}{x}}{\dfrac{1}{x}}+1=2$$

58. Explain one method for simplifying a complex fraction.

14.6 Rational Equations and Formulas

Objectives

1. Solving Rational Equations
2. Recognizing Rational Expressions and Equations
   - Expressions
   - Equations
3. Finding Graphical and Numerical Solutions
4. Solving an Equation for a Variable
5. Solving Applications Involving Rational Equations

NEW VOCABULARY

☐ Rational equation
☐ Basic rational equation
☐ Extraneous solutions

1 Solving Rational Equations

🌐 **Math in Traffic Context** In the first section of this chapter, we demonstrated that if cars enter a construction zone at random at an average rate of x cars per minute and if 10 cars per minute can pass through the zone, then the average time T that a driver spends waiting in line and passing through the construction zone is

$$T = \frac{1}{10 - x},$$

where T is in minutes and $x < 10$. If the highway department wants to limit the average wait for a car to $\frac{1}{2}$ minute, then mathematics can be used to determine the corresponding value of x. (Source: N. Garber and L. Hoel, *Traffic and Highway Engineering*.)

If an equation contains one or more rational expressions, it is called a **rational equation**. Rational equations occur in mathematics whenever a rational expression is set equal to a constant. For example, the rational expression

$$\frac{1}{10 - x} \quad \text{Rational expression}$$

can be used to estimate the average time that drivers wait to get through a construction site. If the wait is $\frac{1}{2}$ minute, we determine x by solving the *rational equation*

$$\frac{1}{10 - x} = \frac{1}{2}. \quad \text{Rational equation}$$

To solve this equation, we multiply each side by the LCD: $2(10 - x)$.

$$\frac{1 \cdot 2(10 - x)}{10 - x} = \frac{1 \cdot 2(10 - x)}{2} \quad \text{Multiply by the LCD.}$$

$$2 = 10 - x \quad \text{Simplify.}$$

$$x = 8 \quad \text{Add } x\text{; subtract 2.}$$

The average wait is $\frac{1}{2}$ minute when cars arrive randomly at an average rate of 8 cars per minute. In general, if a rational equation is in the form

$$\frac{a}{b} = \frac{c}{d}, \quad \text{Basic rational equation}$$

we can multiply each side of this equation by the common denominator bd to obtain

$$\frac{a(bd)}{b} = \frac{c(bd)}{d}, \quad \text{Multiply each side by the common denominator } bd.$$

which simplifies to $ad = cb$. This technique can be used to solve some types of **basic rational equations**, which have a single rational expression on each side of the equals sign.

SOLVING BASIC RATIONAL EQUATIONS

The equations

$$\frac{a}{b} = \frac{c}{d} \quad \text{and} \quad ad = bc$$

are equivalent, provided that b and d are nonzero. Note that converting the first equation to the second equation is sometimes called *cross multiplying*.

READING CHECK 1

- When do we cross multiply?

EXAMPLE 1 Solving rational equations

Solve each equation.

(a) $\dfrac{5}{3} = \dfrac{4}{x}$ (b) $\dfrac{x+1}{5} = \dfrac{3x}{2}$ (c) $x = \dfrac{4}{3x-4}$ (d) $\dfrac{1}{x} + \dfrac{2}{x} = \dfrac{3}{7}$

Solution

(a)
$$\dfrac{5}{3} = \dfrac{4}{x} \qquad \text{Given equation}$$
$$5x = 12 \qquad \text{Cross multiply.}$$
$$x = \dfrac{12}{5} \qquad \text{Divide by 5.}$$

The solution is $\dfrac{12}{5}$.

(b)
$$\dfrac{x+1}{5} = \dfrac{3x}{2} \qquad \text{Given equation}$$
$$2(x+1) = 15x \qquad \text{Cross multiply.}$$
$$2x + 2 = 15x \qquad \text{Distributive property}$$
$$2 = 13x \qquad \text{Subtract } 2x.$$
$$\dfrac{2}{13} = x \qquad \text{Divide by 13.}$$

The solution is $\dfrac{2}{13}$.

(c)
$$\dfrac{x}{1} = \dfrac{4}{3x-4} \qquad \text{Write } x \text{ as } \dfrac{x}{1}.$$
$$x(3x-4) = 4 \cdot 1 \qquad \text{Cross multiply.}$$
$$3x^2 - 4x = 4 \qquad \text{Distributive property}$$
$$3x^2 - 4x - 4 = 0 \qquad \text{Subtract 4.}$$
$$(3x+2)(x-2) = 0 \qquad \text{Factor.}$$
$$3x + 2 = 0 \quad \text{or} \quad x - 2 = 0 \qquad \text{Zero-product property}$$
$$x = -\dfrac{2}{3} \quad \text{or} \quad x = 2 \qquad \text{Solve each equation.}$$

The solutions are $-\dfrac{2}{3}$ and 2.

(d)
> Add before cross multiplying; the equation must be in basic form.

$$\dfrac{1}{x} + \dfrac{2}{x} = \dfrac{3}{7} \qquad \text{Given equation}$$
$$\dfrac{3}{x} = \dfrac{3}{7} \qquad \text{Basic form}$$
$$21 = 3x \qquad \text{Cross multiply.}$$
$$7 = x \qquad \text{Divide by 3.}$$

The solution is 7.

Now Try Exercises 9, 15, 31, 41

Another technique for solving rational equations is to multiply each side by the least common denominator. Unlike cross multiplying, this technique can *always* be used.

EXAMPLE 2 Multiplying each side by the LCD

Solve each equation. Check your answer.

(a) $\dfrac{1}{x-1} - \dfrac{1}{x} = \dfrac{1}{9x}$

(b) $\dfrac{1}{x-1} + \dfrac{1}{x+1} = \dfrac{12}{x^2-1}$

Solution

(a) Start by multiplying each term by the LCD, $9x(x-1)$.

$$\dfrac{1}{x-1} - \dfrac{1}{x} = \dfrac{1}{9x} \qquad \text{Given equation}$$

$$\dfrac{1 \cdot 9x(x-1)}{x-1} - \dfrac{1 \cdot 9x(x-1)}{x} = \dfrac{1 \cdot 9x(x-1)}{9x} \qquad \text{Multiply each term by the LCD.}$$

$$9x - 9(x-1) = x - 1 \qquad \text{Simplify each rational expression.}$$

$$9 = x - 1 \qquad \text{Distributive property; simplify.}$$

$$10 = x \qquad \text{Add 1.}$$

Check:

$$\dfrac{1}{10-1} - \dfrac{1}{10} \stackrel{?}{=} \dfrac{1}{9(10)} \qquad \text{Substitute 10 for } x.$$

$$\dfrac{10}{90} - \dfrac{9}{90} \stackrel{?}{=} \dfrac{1}{90} \qquad \text{The LCD is 90.}$$

$$\dfrac{1}{90} = \dfrac{1}{90} \checkmark \qquad \text{The answer checks.}$$

(b) Start by multiplying each term by the LCD, $x^2 - 1 = (x-1)(x+1)$.

$$\dfrac{1}{x-1} + \dfrac{1}{x+1} = \dfrac{12}{x^2-1} \qquad \text{Given equation}$$

$$\dfrac{1(x-1)(x+1)}{x-1} + \dfrac{1(x-1)(x+1)}{x+1} = \dfrac{12(x-1)(x+1)}{(x-1)(x+1)} \qquad \text{Multiply each term by the LCD.}$$

$$(x+1) + (x-1) = 12 \qquad \text{Simplify.}$$

$$2x = 12 \qquad \text{Combine like terms.}$$

$$x = 6 \qquad \text{Divide by 2.}$$

Check:

$$\dfrac{1}{6-1} + \dfrac{1}{6+1} \stackrel{?}{=} \dfrac{12}{6^2-1} \qquad \text{Substitute 6 for } x.$$

$$\dfrac{1}{5} + \dfrac{1}{7} \stackrel{?}{=} \dfrac{12}{35} \qquad \text{Simplify.}$$

$$\dfrac{7}{35} + \dfrac{5}{35} \stackrel{?}{=} \dfrac{12}{35} \qquad \text{The LCD is 35.}$$

$$\dfrac{12}{35} = \dfrac{12}{35} \checkmark \qquad \text{The answer checks.}$$

Now Try Exercises 49, 55

In Example 2, we checked our answers. Although the answer checked in both cases, it may not always check. When multiplying a rational equation by the LCD, it is possible to obtain **extraneous solutions** that do not satisfy the *given* equation. This situation is demonstrated in Example 3. Before solving Example 3, we present a step-by-step strategy for solving rational equations.

> **STEPS FOR SOLVING A RATIONAL EQUATION**
>
> **STEP 1:** Find the LCD of the terms in the equation.
> **STEP 2:** Multiply each side of the equation by the LCD.
> **STEP 3:** Simplify each term.
> **STEP 4:** Solve the resulting equation.
> **STEP 5:** Check each answer in the *given* equation. Any value that makes a denominator equal 0 should be rejected because it is an extraneous solution.

READING CHECK 2

- What is the first step in solving a rational equation?

NOTE: If the rational equation is in the form $\frac{a}{b} = \frac{c}{d}$, or "fraction equals fraction," cross multiplying to obtain $ad = bc$ may be helpful. However, be sure to check your answers. ∎

EXAMPLE 3   **Solving an equation that has an extraneous solution**

If possible, solve $\dfrac{1}{x-2} + \dfrac{1}{x+2} = \dfrac{4}{x^2-4}$.

Solution
The LCD is $x^2 - 4 = (x-2)(x+2)$ (Step 1).

$$\dfrac{1}{x-2} + \dfrac{1}{x+2} = \dfrac{4}{x^2-4} \qquad \text{Given equation}$$

$$\dfrac{1(x-2)(x+2)}{x-2} + \dfrac{1(x-2)(x+2)}{x+2} = \dfrac{4(x^2-4)}{x^2-4} \qquad \text{Multiply by the LCD (Step 2).}$$

$$(x+2) + (x-2) = 4 \qquad \text{Simplify each term (Step 3).}$$

$$2x = 4 \qquad \text{Combine like terms (Step 4).}$$

$$x = 2 \qquad \text{Divide by 2.}$$

Check:

$$\underbrace{\dfrac{1}{2-2}} + \underbrace{\dfrac{1}{2+2} \stackrel{?}{=} \dfrac{4}{2^2-4}} \quad ✗ \qquad \text{Substitute 2 for } x \text{ (Step 5).}$$

Undefined terms

Note that terms on both sides of the equation are undefined because it is not possible to divide by 0. Therefore 2 is not a solution; rather, it is an *extraneous solution*. That is, there are no solutions to the *given* equation.

Now Try Exercise 53

2 Recognizing Rational Expressions and Equations

EXPRESSIONS Rational expressions and rational equations are not the same concepts.

$$\dfrac{3}{x} \quad \text{and} \quad \dfrac{x}{x^2 - x} \qquad \text{Two rational expressions}$$

Neither expression contains an equals sign. We often simplify or evaluate rational expressions. By factoring the denominator and applying the basic principle of rational expressions, we can simplify the second expression.

$$\frac{x}{x^2 - x} \text{ is equivalent to } \frac{x}{x(x-1)} \text{ and}$$

$$\frac{x}{x(x-1)} \text{ is equivalent to } \frac{1}{x-1} \quad (\text{if } x \neq 0).$$

Thus the expressions $\frac{x}{x^2-x}$ and $\frac{1}{x-1}$ are equal for every value of x (except 0 and 1). Expressions can also be evaluated. For example, if we replace x with 3, the expression $\frac{3}{x}$ equals $\frac{3}{3}$, or 1.

EQUATIONS When two expressions are set equal to each other, an equation is formed that is typically true for only a limited number of x-values. For example, $\frac{3}{x} = \frac{1}{x-1}$ is a rational equation that is true for only one x-value.

$\dfrac{3}{x} = \dfrac{1}{x-1}$	Given rational equation
$3(x - 1) = x \cdot 1$	Cross multiply.
$3x - 3 = x$	Distributive property
$2x = 3$	Add 3; subtract x.
$x = \dfrac{3}{2}$	Divide each side by 2.

> When solving an equation, we find values of x that make the equation true.

The only solution is $\frac{3}{2}$. (Check this solution.) If we replace x with any value other than $\frac{3}{2}$, the equation $\frac{3}{x} = \frac{1}{x-1}$ is a false statement.

EXAMPLE 4 Identifying expressions and equations

Determine whether you are given an expression or an equation. If it is an expression, *simplify* it and then *evaluate* it for $x = 5$. If it is an equation, *solve* it.

(a) $\dfrac{x-1}{x+1} = \dfrac{x}{x+3}$ (b) $\dfrac{x^2 - 2x}{x - 1} + \dfrac{1}{x - 1}$

Solution
(a) There is an equals sign, so it is an equation that can be solved.

$\dfrac{x-1}{x+1} = \dfrac{x}{x+3}$	Given equation
$(x-1)(x+3) = x(x+1)$	Cross multiply.
$x^2 + 2x - 3 = x^2 + x$	Multiply.
$2x - 3 = x$	Subtract x^2.
$x = 3$	Subtract x; add 3.

Check:

$$\frac{3-1}{3+1} \stackrel{?}{=} \frac{3}{3+3} \quad \text{Replace } x \text{ with 3 in given equation.}$$

$$\frac{1}{2} = \frac{1}{2} \checkmark \quad \text{The answer checks.}$$

The only solution to the equation is 3.

(b) There is no equals sign, so $\frac{x^2 - 2x}{x - 1} + \frac{1}{x - 1}$ is an expression that can be simplified. The common denominator is $x - 1$, so add the numerators.

$$\underbrace{\frac{x^2 - 2x}{x - 1} + \frac{1}{x - 1}}_{\text{Given expression}} = \frac{x^2 - 2x + 1}{x - 1}$$
$$= \frac{(x - 1)(x - 1)}{x - 1}$$
$$= x - 1$$

These expressions are all equivalent to the given expression when $x \neq 1$.

The given expression is equivalent to, or simplifies to, $x - 1$ when $x \neq 1$. When x equals 5, the expression evaluates to $5 - 1$, or 4.

Now Try Exercises 63 and 67

MAKING CONNECTIONS 1

Expressions versus Equations

1. If a problem does not contain an equals sign, you are probably adding, subtracting, multiplying, dividing, or otherwise simplifying an expression. Your answer will be an expression, *not* a value for x.
2. If the problem has an equals sign, it is an equation to be solved. Your answer will be a value (or values) for x that makes the equation a true statement. Check all your answers.

3 Finding Graphical and Numerical Solutions

Like other types of equations, rational equations can also be solved graphically and numerically. Graphs of rational expressions are not lines. They are typically curves that can be graphed by plotting several points and then sketching a graph or by using a graphing calculator.

EXAMPLE 5 Solving a rational equation graphically and numerically

Solve $\frac{2}{x} = x + 1$ graphically and numerically.

Solution
Graphical Solution To solve $\frac{2}{x} = x + 1$, graph $y_1 = \frac{2}{x}$ and $y_2 = x + 1$. The graph of y_2 is a line with slope 1 and y-intercept $(0, 1)$. To graph y_1, make a table of values, as shown in **TABLE 14.5**. Then plot the points and connect them with a smooth curve. Note that y_1 is undefined for $x = 0$, so $x = 0$ is a vertical asymptote. Generally, it is a good idea to plot at least three points on each side of an asymptote. These points are plotted and the curves sketched in **FIGURE 14.5**. Note that the graphs of y_1 and y_2 intersect at $(-2, -1)$ and $(1, 2)$. The solutions to the equation are -2 and 1. Check these solutions.

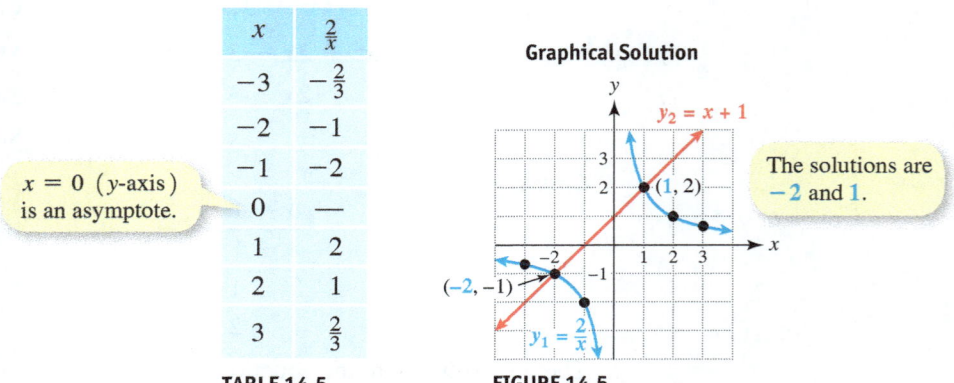

TABLE 14.5   FIGURE 14.5

CRITICAL THINKING 1

If you solve Example 5 symbolically, what will be the result? Verify your answer.

Numerical Solution Make a table of values for $y_1 = \frac{2}{x}$ and $y_2 = x + 1$, as shown in **TABLE 14.6**. Note that $\frac{2}{x} = x + 1$ when $x = -2$ or $x = 1$.

Numerical Solution

x	-3	-2	-1	0	1	2	3
$y_1 = \frac{2}{x}$	$-\frac{2}{3}$	-1	-2	—	2	1	$\frac{2}{3}$
$y_2 = x + 1$	-2	-1	0	1	2	3	4

TABLE 14.6

Solutions indicated at $x=-2$ and $x=1$.

Now Try Exercise 75

A graphing calculator can be used to create **FIGURE 14.5** and **TABLE 14.6**, as illustrated in **FIGURE 14.6**.

CALCULATOR HELP

To make a graph or table, see Appendix A (pages 2, 3, and 4).

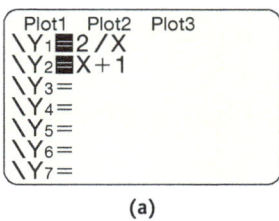
(a)

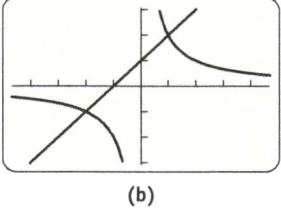
(b)

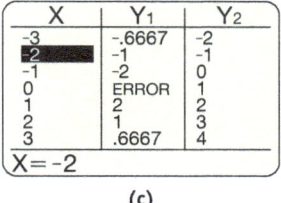
(c)

FIGURE 14.6

In the following Making Connections, we solve the rational equation $\frac{x+3}{2x} = x$ using symbolic, numerical, and graphical methods.

MAKING CONNECTIONS 2

Symbolic, Numerical, and Graphical Solutions to a Rational Equation

Symbolic Solution	Numerical Solution	Graphical Solution
$\frac{x+3}{2x} = 2$		
$2x \cdot \frac{x+3}{2x} = 2 \cdot 2x$		
$x + 3 = 4x$		
$3 = 3x$		
$x = 1$		

Numerical Solution table:

x	$(x+3)/(2x)$
-2	-0.25
-1	-1
0	—
1	2
2	1.25

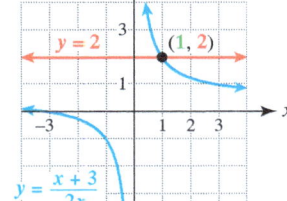

The solution is **1**.   When $x = 1$, $\frac{x+3}{2x} = 2$.   The graphs intersect at $(1, 2)$.

STUDY TIP

It can be helpful to talk and study with other students. However, make sure that others are not "doing the work for you."

4 Solving an Equation for a Variable

🌐 **Math in Context** (Science) Equations and formulas in applications often involve both the use of rational expressions and the need to solve an equation for a variable. This idea is illustrated in the next two examples, which involve formulas from geometry, physics, and chemistry.

EXAMPLE 6 Finding time in a distance problem

If a person travels at a speed (rate) r for time t, then the distance d traveled is $d = rt$.
(a) How far does a person travel in 2 hours when traveling at 60 miles per hour?
(b) Solve the formula $d = rt$ for t.
(c) How long does it take a person to go 250 miles when traveling at 40 miles per hour?

Solution
(a) A person traveling at 60 miles per hour for 2 hours travels
$$d = rt = 60 \cdot 2 = 120 \text{ miles.}$$
(b) To solve $d = rt$ for t, divide each side by r.

$$d = rt \quad \text{Given equation}$$
$$\frac{d}{r} = t \quad \text{Divide each side by } r.$$

(c) At 40 miles per hour, a person can travel 250 miles in
$$t = \frac{d}{r} = \frac{250}{40} = \frac{25}{4} = 6.25 \text{ hours.}$$

Note that 6.25 hours equals 6 hours and 15 minutes.

Now Try Exercise 87

EXAMPLE 7 Solving a formula for a variable

Solve each equation for the specified variable.
(a) $A = \frac{bh}{2}$ for b (b) $P = \frac{nRT}{V}$ for V (c) $S = 2\pi rh + \pi r^2$ for h

Solution
(a) First, multiply each side by 2.

$$A = \frac{bh}{2} \quad \text{Given equation}$$
$$2A = bh \quad \text{Multiply each side by 2.}$$
$$\frac{2A}{h} = b \quad \text{Divide each side by } h.$$

(b) Begin by multiplying each side by V.

$$P = \frac{nRT}{V} \quad \text{Given equation}$$
$$PV = nRT \quad \text{Multiply each side by } V.$$
$$V = \frac{nRT}{P} \quad \text{Divide each side by } P.$$

(c) Begin by subtracting πr^2 from each side.

$$S = 2\pi rh + \pi r^2 \quad \text{Given equation}$$
$$S - \pi r^2 = 2\pi rh \quad \text{Subtract } \pi r^2 \text{ from each side.}$$
$$\frac{S - \pi r^2}{2\pi r} = h \quad \text{Divide each side by } 2\pi r.$$

Now Try Exercises 89, 91, 97

5 Solving Applications Involving Rational Equations

Math in Context (Speed and Time) Rational equations sometimes occur in time and rate problems, as demonstrated in the next two examples.

EXAMPLE 8 Mowing a lawn

Two people are mowing a large lawn. One person has a riding mower, and the other person has a push mower. The person with the riding mower can cut the lawn alone in 4 hours, and the person with the push mower can cut the lawn alone in 9 hours. How long does it take them, working together, to cut the lawn?

Solution
The first person can cut the entire lawn in 4 hours, so this person can cut $\frac{1}{4}$ of the lawn in 1 hour, $\frac{2}{4}$ of the lawn in 2 hours, and in general, $\frac{t}{4}$ of the lawn in t hours. The second person can cut the lawn in 9 hours, so (using similar reasoning) this person can cut $\frac{t}{9}$ of the lawn in t hours. Together, they can cut

$$\frac{t}{4} + \frac{t}{9}$$

of the lawn in t hours. The job is complete when the fraction of the lawn cut reaches **1**. To find out how long this task takes, solve the equation

$$\frac{t}{4} + \frac{t}{9} = 1.$$

Begin by multiplying each side by the LCD, or **36**.

$\frac{t}{4} + \frac{t}{9} = 1$	Equation to be solved
$\frac{36t}{4} + \frac{36t}{9} = 36(1)$	Multiply each term by 36.
$9t + 4t = 36$	Simplify.
$13t = 36$	Combine like terms.
$t = \frac{36}{13}$	Divide each side by 13.

Working together they can cut the lawn in $\frac{36}{13} \approx 2.8$ hours.

Now Try Exercise 103

CRITICAL THINKING 2

If one person can mow a lawn in x hours and another person can mow it in y hours, how long does it take them, working together, to mow the lawn?

EXAMPLE 9 Solving a distance problem

Suppose that the winner of a 600-mile car race finishes 12 minutes ahead of the second place finisher. If the winner averages 5 miles per hour faster than the second racer, find the average speed of each racer.

Solution
Here, we apply the four-step method for solving an application problem.

STEP 1: *Identify any variables.*

x: speed of slower car in miles per hour
$x + 5$: speed of faster car in miles per hour

STEP 2: *Write an equation.* To determine the time required for each car to finish the race, we use the equation $t = \frac{d}{r}$. Because the race is 600 miles, the time for the slower car is $\frac{600}{x}$ and the time for the faster car is $\frac{600}{x+5}$. The difference between these times is 12 minutes, or $\frac{12}{60} = \frac{1}{5}$ hour. Thus to determine x, we solve

$$\frac{600}{x} - \frac{600}{x+5} = \frac{1}{5}.$$

NOTE: Speeds are in miles per *hour*, so time must be in hours, not minutes. ■

STEP 3: *Solve the equation.* Start by multiplying by the LCD, $5x(x+5)$.

$\dfrac{600}{x} - \dfrac{600}{x+5} = \dfrac{1}{5}$	Equation to be solved
$\dfrac{600 \cdot 5x(x+5)}{x} - \dfrac{600 \cdot 5x(x+5)}{x+5} = \dfrac{1 \cdot 5x(x+5)}{5}$	Multiply each term by the LCD.
$3000(x+5) - 3000x = x(x+5)$	Simplify.
$3000x + 15{,}000 - 3000x = x^2 + 5x$	Distributive property
$15{,}000 = x^2 + 5x$	Combine like terms.
$x^2 + 5x - 15{,}000 = 0$	Rewrite the equation.
$(x - 120)(x + 125) = 0$	Factor.
$x - 120 = 0 \quad \text{or} \quad x + 125 = 0$	Zero-product property
$x = 120 \quad \text{or} \quad x = -125$	Solve.

The slower car travels at 120 miles per hour, and the faster car travels 5 miles per hour faster, or 125 miles per hour. (The solution -125 has no meaning in this problem.)

STEP 4: *Check your answer.* The slower car travels 600 miles at 120 miles per hour, which takes $\frac{600}{120} = 5$ hours. The faster car travels 600 miles at 125 miles per hour, which requires $\frac{600}{125} = 4.8$ hours. The difference between their times is $5 - 4.8 = 0.2$ hour, or $0.2 \times 60 = 12$ minutes, so the answer checks.

Now Try Exercise 105

TECHNOLOGY NOTE

Solving Rational Equations Numerically
Tables can be used to find solutions to rational equations. The following displays show the positive solution of 120 from Example 9. Note the use of parentheses for entering the formula for Y_1.

CALCULATOR HELP

To make a table, see Appendix A (pages 2 and 3).

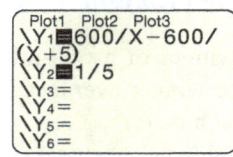
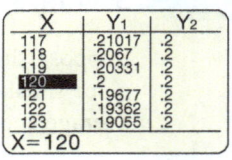

14.6 Putting It All Together

CONCEPT	EXPLANATION	EXAMPLES
Basic Rational Equations	To solve an equation of the form $$\frac{a}{b} = \frac{c}{d}, \quad b \neq 0, d \neq 0$$ cross multiply to obtain $ad = bc$.	$\frac{3}{6} = \frac{5}{2x}$ is equivalent to $6x = 30$. $\frac{x}{2} = \frac{8}{x}$ is equivalent to $x^2 = 16$. (Cross multiplication works only for equations with *one* rational expression on each side: "fraction equals fraction.")
Rational Equations	To solve a rational equation, use the following steps. **STEP 1:** Find the LCD. **STEP 2:** Multiply each side of the equation by the LCD. **STEP 3:** Simplify each term. **STEP 4:** Solve the resulting equation. **STEP 5:** Check each possible solution.	Solve $\frac{1}{x} - \frac{2}{3x} = \frac{5}{6}$. **STEP 1:** The LCD is $6x$. **STEP 2:** $\frac{1(6x)}{x} - \frac{2(6x)}{3x} = \frac{5(6x)}{6}$ **STEP 3:** $6 - 4 = 5x$ **STEP 4:** $\frac{2}{5} = x$ **STEP 5:** $\frac{1}{\frac{2}{5}} - \frac{2}{3(\frac{2}{5})} = \frac{5}{6}$ ✓ It checks.

14.6 Exercises

CONCEPTS AND VOCABULARY

1. If an equation contains one or more rational expressions, it is called a(n) _____ equation.

2. Give an example of a rational expression and an example of a rational equation.

3. The equation $\frac{a}{b} = \frac{c}{d}$ is equivalent to _____, provided that _____ and _____ are nonzero.

4. Are the equations $\frac{2}{x-1} = 5$ and $5(x-1) = 2$ equivalent provided that $x \neq 1$?

5. One way to solve the equation $\frac{5}{3x} + \frac{3}{4x} = 1$ is to multiply each side by the LCD, which is _____.

6. To solve the equation $T = \frac{R}{SV}$ for V, multiply each side by the variable _____ and then divide each side by the variable _____.

SOLVING RATIONAL EQUATIONS

Exercises 7–62: Solve and check your answer.

7. $\frac{x}{2} = \frac{3}{4}$

8. $\frac{2x}{3} = \frac{2}{5}$

9. $\frac{3}{z} = \frac{6}{5}$

10. $\frac{2}{7} = \frac{1}{z}$

11. $\frac{3y}{4} = \frac{7y}{2}$

12. $\frac{y}{6} = \frac{5y}{3}$

13. $\frac{2}{3} = \frac{1}{2x+1}$

14. $\frac{1}{x+4} = \frac{3}{5}$

15. $\frac{5}{2x} = \frac{8}{x+2}$

16. $\frac{1}{x-1} = \frac{5}{3x}$

17. $\frac{1}{z-1} = \frac{2}{z+1}$

18. $\frac{4}{z+3} = \frac{2}{z-2}$

19. $\dfrac{3}{n+5} = \dfrac{2}{n-5}$

20. $\dfrac{4}{3n+2} = \dfrac{1}{n-1}$

21. $\dfrac{m}{m-1} = \dfrac{5}{4}$

22. $\dfrac{5m}{2m-1} = \dfrac{3}{2}$

23. $\dfrac{5x}{5-x} = \dfrac{1}{3}$

24. $\dfrac{x+2}{3x} = \dfrac{4}{3}$

25. $\dfrac{6}{5-2x} = 2$

26. $\dfrac{x+1}{x} = 6$

27. $\dfrac{2x}{2x+1} = \dfrac{-1}{2x+1}$

28. $\dfrac{x}{x-4} = \dfrac{4}{x-4}$

29. $\dfrac{1}{1-x} = \dfrac{3}{1+x}$

30. $\dfrac{2x}{1-2x} = \dfrac{1}{2}$

31. $\dfrac{1}{z+2} = -z$

32. $\dfrac{1}{z-2} = \dfrac{z}{3}$

33. $\dfrac{-1}{2x+5} = \dfrac{x}{3}$

34. $\dfrac{x}{2} = \dfrac{1}{3x+5}$

35. $\dfrac{x}{2} + \dfrac{x}{4} = 3$

36. $\dfrac{x}{4} - \dfrac{x}{3} = 1$

37. $\dfrac{3x}{4} - \dfrac{x}{2} = 1$

38. $\dfrac{2x}{3} + \dfrac{x}{3} = 6$

39. $\dfrac{4}{t+1} + \dfrac{1}{t+1} = -1$

40. $\dfrac{2}{t-5} - \dfrac{5}{t-5} = 3$

41. $\dfrac{1}{x} + \dfrac{2}{x} = \dfrac{1}{2}$

42. $\dfrac{1}{2x} - \dfrac{2}{x} = -3$

43. $\dfrac{2}{x-1} + 1 = \dfrac{4}{x^2-1}$

44. $\dfrac{1}{x} + 2 = \dfrac{1}{x^2+x}$

45. $\dfrac{1}{x+2} = \dfrac{4}{4-x^2} - 1$

46. $\dfrac{1}{x-3} + 1 = \dfrac{6}{x^2-9}$

47. $\dfrac{5}{4z} - \dfrac{2}{3z} = 1$

48. $\dfrac{3}{z+1} - \dfrac{1}{z+1} = 2$

49. $\dfrac{4}{y-1} + \dfrac{1}{y} = \dfrac{6}{5}$

50. $\dfrac{6}{y+1} + \dfrac{6}{y} = 5$

51. $\dfrac{1}{2x} - \dfrac{1}{x+3} = 0$

52. $\dfrac{2}{x} - \dfrac{6}{2x-1} = -1$

53. $\dfrac{1}{x-1} + \dfrac{1}{x+1} = \dfrac{2}{x^2-1}$

54. $\dfrac{1}{2x+1} + \dfrac{1}{2x-1} = \dfrac{2}{4x^2-1}$

55. $\dfrac{1}{x-2} + \dfrac{1}{x+2} = \dfrac{6}{x^2-4}$

56. $\dfrac{2}{x+3} - \dfrac{1}{x-3} = \dfrac{1}{x^2-9}$

57. $\dfrac{1}{p+1} + \dfrac{1}{p+2} = \dfrac{1}{p^2+3p+2}$

58. $\dfrac{1}{p-1} - \dfrac{1}{p+3} = \dfrac{1}{p^2+2p-3}$

59. $\dfrac{1}{x-2} + \dfrac{3}{2x-4} = \dfrac{6}{3x-6}$

60. $\dfrac{4}{x+1} - \dfrac{4}{2x+2} = \dfrac{1}{(x+1)^2}$

61. $\dfrac{1}{r^2-r-2} + \dfrac{2}{r^2-2r} = \dfrac{1}{r^2+r}$

62. $\dfrac{3}{r^2-1} + \dfrac{1}{r^2+r} = \dfrac{3}{r^2-r}$

RECOGNIZING RATIONAL EXPRESSIONS AND EQUATIONS

Exercises 63–70: (Refer to Example 4.) Determine if you are given an expression or an equation. If it is an expression, simplify (if possible) and evaluate it for $x = 2$. If it is an equation, solve it.

63. $\dfrac{1}{x} - \dfrac{1-x}{x}$

64. $\dfrac{1}{x} - x = 0$

65. $\dfrac{1}{2x} - \dfrac{1}{4x} = \dfrac{1}{8}$

66. $\dfrac{1}{2x} - \dfrac{1}{4x}$

67. $\dfrac{x+1}{x-1} = \dfrac{2x-3}{2x-5}$

68. $\dfrac{2x-1}{4x+1} = \dfrac{x+1}{2x-1}$

69. $\dfrac{4x+4}{x+2} + \dfrac{x^2}{x+2}$

70. $\dfrac{x^2-2}{x-2} - \dfrac{x}{x-2}$

FINDING GRAPHICAL AND NUMERICAL SOLUTIONS

Exercises 71–74: Use the graph to solve the given equation. Check your answers.

71. $\dfrac{1}{x} = 4x$

72. $\dfrac{x}{x-1} = x$

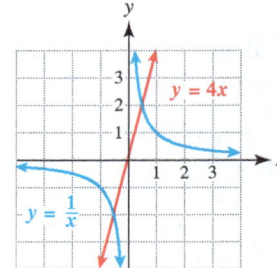

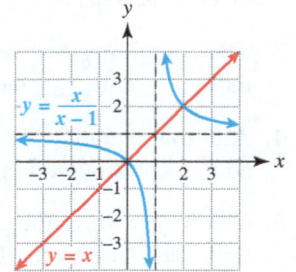

73. $\dfrac{x-1}{x} = -2x$ 74. $\dfrac{3}{x^2-1} = 1$

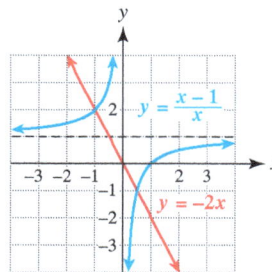
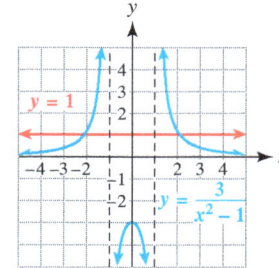

Exercises 75–82: (Refer to Example 5.) Solve the equation
(a) *graphically and*
(b) *numerically.*

75. $\dfrac{3}{x} = x + 2$ 76. $-\dfrac{2}{x} = 1 - x$

77. $\dfrac{3x}{2} = \dfrac{1}{2}x - 1$ 78. $\dfrac{x}{3} = 2 - \dfrac{2}{3}x$

79. $\dfrac{3}{x-1} = 3$ 80. $\dfrac{2}{x+5} = 1$

81. $\dfrac{4}{x^2} = 1$ 82. $\dfrac{-18}{x^2} = -2$

Exercises 83–86: Solve the rational equation graphically to the nearest thousandth.

83. $\dfrac{1}{\pi x - 2} + \dfrac{\sqrt{2}}{2.1x} = 1.3$ 84. $\dfrac{6}{x^2-4} + \dfrac{\pi}{2x} = \dfrac{3}{2}$

85. $\dfrac{1}{x^3} - \dfrac{\pi}{4} = 0$ 86. $\dfrac{1}{2\pi x^2} - \dfrac{x}{2} = 1$

SOLVING AN EQUATION FOR A VARIABLE

Exercises 87–98: (Refer to Examples 6 and 7.) Solve the equation for the specified variable.

87. $m = \dfrac{F}{a}$ for a 88. $m = \dfrac{2K}{v^2}$ for K

89. $I = \dfrac{V}{R+r}$ for r 90. $\dfrac{1}{T} = \dfrac{r}{R-r}$ for R

91. $h = \dfrac{2A}{b}$ for b 92. $h = \dfrac{2A}{b_1+b_2}$ for b_1

93. $\dfrac{3}{k} = \dfrac{z}{z+5}$ for z 94. $\dfrac{5}{r} = \dfrac{t+r}{t}$ for t

95. $T = \dfrac{ab}{a+b}$ for b 96. $A = \dfrac{2b}{a-b}$ for b

97. $\dfrac{3}{k} = \dfrac{1}{x} - \dfrac{2}{y}$ for x 98. $\dfrac{1}{R} = \dfrac{1}{R_1} + \dfrac{1}{R_2}$ for R_1

APPLICATIONS INVOLVING RATIONAL EQUATIONS

99. **Waiting in Traffic** (Refer to the introduction to this section.) Solve the equation
$$\dfrac{1}{10-x} = 1$$
to determine the traffic rate x in cars per minute corresponding to an average waiting time of 1 minute.

100. **Waiting in Line** At a post office, customers arrive at random at an average rate of x people per minute. The clerk can wait on 4 customers per minute. The average time T in minutes spent waiting in line is given by
$$T = \dfrac{1}{4-x},$$
where $x < 4$.
(a) Evaluate T for $x = 3$, 3.9, and 3.99. What happens to waiting time as the arrival rate nears 4 people per minute?
(b) Find x when the waiting time is 5 minutes.

101. **Shoveling a Sidewalk** It takes a less experienced employee 4 hours to shovel the snow from a sidewalk, but a more experienced employee can shovel the same sidewalk in 3 hours. How long will it take them to clear the walk if they work together?

102. **Pumping Water** One pump can empty a pool in 5 days, whereas a second pump can empty the pool in 7 days. How long will it take the two pumps, working together, to empty the pool?

103. **Painting a House** One painter can paint a house in 8 days, yet a more experienced painter can paint the house in 4 days. How long will it take the two painters, working together, to paint the house?

104. **Highway Curves** To make a highway curve safe, highway engineers often bank it, as shown in the figure. If a curve is designed for a speed of 50 miles per hour and is banked with positive slope m, then a minimum radius R in feet for the curve is given by
$$R = \dfrac{2500}{15m+2}.$$

(*Source:* N. Garber and L. Hoel, *Traffic and Highway Engineering.*)

(a) Find R for $m = 0.1$. Interpret the result.
(b) If $R = 500$, find m. Interpret the result.

105. Bicycle Race The winner of a 6-mile bicycle race finishes 2 minutes ahead of a teammate and travels, on average, 2 miles per hour faster than the teammate. Find the average speed of each racer.

106. Freeway Travel Two drivers travel 150 miles on a freeway and then stop at a wayside rest area. The first driver travels 5 miles per hour faster and arrives $\frac{1}{7}$ hour ahead of the second. Find the average speed of each car.

107. Braking Distance If a car is traveling *downhill* at 30 miles per hour on wet pavement, then the braking distance B in feet for this car is given by

$$B = \frac{30}{0.3 + m},$$

where $m < 0$ is the slope of the hill. (*Source:* L. Haefner, *Introduction to Transportation Systems.*)
   (a) Find the braking distance for $m = -0.05$ and interpret the result.
   (b) Find m if the braking distance is 150 feet.

108. Slippery Roads If a car is traveling at 30 miles per hour on a level road, then its braking distance in feet is $\frac{30}{x}$, where x is the coefficient of friction between the road and the tires. The variable x is positive and satisfies $x \leq 1$. The closer the value of x is to 0, the more slippery the road is. (*Source:* L. Haefner.)
   (a) Evaluate the expression for $x = 1, 0.5$, and 0.1. Interpret the results.
   (b) Find x for a braking distance of 150 feet.

109. River Current A boat can travel 36 miles upstream in the same time that it can travel 54 miles downstream. If the speed of the current is 3 miles per hour, find the speed of the boat without a current.

110. Airplane Speed An airplane can travel 380 miles into the wind in the same time that it can travel 420 miles with the wind. If the wind speed is 10 miles per hour, find the speed of the airplane without any wind.

111. Airplane Speed An airplane can travel 450 miles into the wind in the same time that it can travel 750 miles with the wind. If the wind speed is 50 miles per hour, find the speed of the airplane without any wind

112. River Current A boat can travel 114 miles upstream in the same time that it can travel 186 miles downstream. If the speed of the current is 6 miles per hour, find the speed of the boat without a current.

113. Running and Jogging An athlete runs 10 miles and then jogs home. The trip home takes 1 hour longer than it took to run that distance. If the athlete runs 5 miles per hour faster than she jogs, what are her average running and jogging speeds?

114. Speed Limit A person drives 390 miles on a stretch of road. Half the distance is driven traveling 5 miles per hour below the speed limit, and half the distance is driven traveling 5 miles per hour above the speed limit. If the time spent traveling at the slower speed exceeds the time spent traveling at the faster speed by 24 minutes, find the speed limit.

115. Time Spent in Line If a parking lot attendant can wait on 5 vehicles per minute and vehicles are leaving the lot randomly at an average rate of x vehicles per minute, then the average time T in minutes spent waiting in line *and* paying the attendant is given by

$$T = \frac{1}{5 - x},$$

where $x < 5$. (*Source:* N. Garber.)
   (a) Evaluate T when $x = 4$ and interpret the result.
   (b) A graph of T is shown in the figure. Interpret the graph as x increases from 0 to 5. Does this result agree with your intuition?

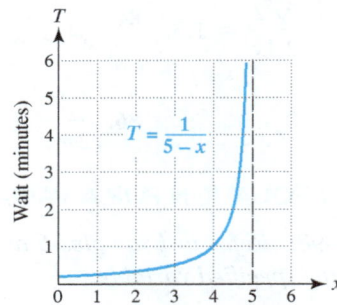

   (c) Find x if the waiting time is 3 minutes.

116. People Waiting in Line At a post office, workers can wait on 50 people per hour. If people arrive randomly at an average rate of x per hour, then the average number of people N waiting in line is given by

$$N = \frac{x^2}{2500 - 50x},$$

where $x < 50$. (*Source:* N. Garber.)
   (a) Evaluate N when $x = 30$ and interpret the result.

(b) A graph of N is shown in the figure. Interpret the graph as x increases from 0 to 50. Does this result agree with your intuition?

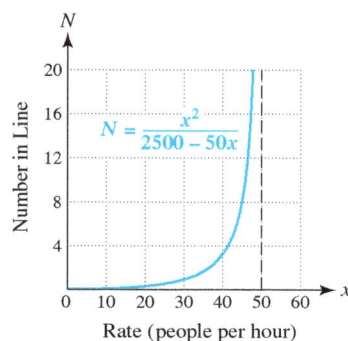

(c) Find x if $N = 8$.

WRITING ABOUT MATHEMATICS

117. Do all rational equations have solutions? Explain.

118. Why is it important to check your answer when solving rational equations? Explain.

SECTIONS 14.5 and 14.6 — Checking Basic Concepts

1. Simplify each complex fraction.

(a) $\dfrac{\frac{x}{3}}{\frac{2x}{5}}$

(b) $\dfrac{\frac{2}{2x} - \frac{1}{3x}}{6x}$

(c) $\dfrac{\frac{1}{a} - \frac{1}{b}}{\frac{1}{a} + \frac{1}{b}}$

(d) $\dfrac{\frac{1}{r^2} - \frac{1}{t^2}}{\frac{2}{r} - \frac{2}{t}}$

2. Solve each equation. Check your answer.

(a) $\dfrac{1}{2x} = \dfrac{3}{x+1}$

(b) $\dfrac{x}{2x+3} = \dfrac{4}{5}$

(c) $\dfrac{1}{2x} + \dfrac{3}{2x} = 1$

(d) $\dfrac{3}{x+1} - \dfrac{2}{x} = -2$

(e) $\dfrac{1}{x-1} = \dfrac{2}{x^2-1} - \dfrac{1}{2}$

3. Solve each equation for the specified variable.

(a) $\dfrac{ax}{2} - 3y = b$ for x

(b) $\dfrac{1}{2m-1} = \dfrac{k}{m}$ for m

4. Braking Distance If a car is traveling *uphill* at 60 miles per hour on wet pavement, then the braking distance D in feet for this car is given by

$$D = \dfrac{120}{0.3 + m},$$

where $m > 0$ is the slope of the hill. (*Source:* L. Haefner, *Introduction to Transportation Systems.*)

(a) Find D for $m = 0.1$ and interpret the result.

(b) Find the slope of the road if D is 200 feet. Interpret the result.

14.7 Proportions and Variation

Objectives

1. Solving Proportions • Similar Figures
2. Solving Direct Variation Problems
3. Solving Inverse Variation Problems
4. Analyzing Data
5. Solving Joint Variation Problems

1 Solving Proportions

Connecting Concepts with Your Life A **ratio** is a comparison of two quantities. For example, a math class might have 7 boys for every 8 girls. Thus the boy–girl ratio in this class is *7 to 8*, or $\frac{7}{8}$. In mathematics, ratios are typically expressed as fractions.

Ratios and proportions are sometimes used to find how much space remains for music on a compact disc (CD). A 700-megabyte CD can store about 80 minutes of music. Suppose that some music has been recorded on the CD and that 256 megabytes are still available. Using ratios and proportions, we can find how many more minutes of music could be recorded. A **proportion** is a statement that two *ratios* are equal.

908 CHAPTER 14 RATIONAL EXPRESSIONS

NEW VOCABULARY

☐ Ratio
☐ Proportion
☐ Directly proportional (varies directly)
☐ Constant of proportionality/variation
☐ Inversely proportional (varies inversely)
☐ Varies jointly

Let x represent the number of minutes available on a CD. Then 80 minutes are to 700 megabytes as x minutes are to 256 megabytes. By setting the *ratios* $\frac{80}{700}$ and $\frac{x}{256}$ equal to each other, we obtain the *proportion*

$$\frac{80}{700} = \frac{x}{256}. \qquad \frac{\text{Minutes}}{\text{Megabytes}} = \frac{\text{Minutes}}{\text{Megabytes}}$$

Solving this equation for x gives

$$700x = 80(256) \qquad \text{Cross multiply.}$$

$$x = \frac{80 \cdot 256}{700} \approx 29.3 \text{ minutes.} \qquad \text{Divide by 700.}$$

About 29 minutes are available to record on the CD.

🌐 Math in Context (Weather) Flooding in Fargo, North Dakota, has been an annual problem in recent years. With snow drifts reaching 10 to 15 feet high in some areas, an accurate estimate of the water content in the snow is essential for predicting spring flood levels. In the next example, we estimate the water content in snow.

EXAMPLE 1 Calculating the water content in snow

Six inches of light, fluffy snow are equivalent to about half an inch of rain in terms of water content. If 15 inches of this type of snow fall, estimate the water content.

Solution
Let x be the equivalent amount of rain. Then 6 inches of snow are to $\frac{1}{2}$ inch of rain as 15 inches of snow are to x inches of rain, which can be written as the proportion

$$\frac{6}{\frac{1}{2}} = \frac{15}{x}. \qquad \frac{\text{Snow}}{\text{Rain}} = \frac{\text{Snow}}{\text{Rain}}$$

Solving this equation gives

$$6x = \frac{15}{2} \quad \text{or} \quad x = \frac{15}{12} = 1.25.$$

Thus 15 inches of light, fluffy snow are equivalent to about 1.25 inches of rain.

Now Try Exercise 65

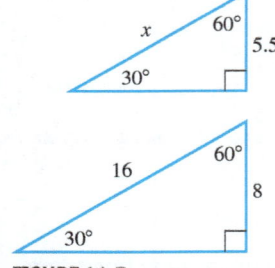

Similar Triangles

FIGURE 14.7

SIMILAR FIGURES Proportions frequently occur in geometry when we work with similar figures. Two triangles are similar if the measures of their corresponding angles are equal. Corresponding sides of similar triangles are proportional. **FIGURE 14.7** shows two right triangles that are similar because each has angles of 30°, 60°, and 90°.

We can find the length of side x by using proportions. Side x is to 16 as 5.5 is to 8, which can be written as the proportion

$$\frac{x}{16} = \frac{5.5}{8}. \qquad \frac{\text{Hypotenuse}}{\text{Hypotenuse}} = \frac{\text{Shorter leg}}{\text{Shorter leg}}$$

Solving yields the following equation.

$$8x = 5.5(16) \qquad \text{Cross multiply.}$$
$$x = 11 \qquad \text{Divide each side by 8.}$$

NOTE: Proportions can be set up in different ways and still produce the correct result. For example, we could say that x is to 5.5 in the smaller triangle as 16 is to 8 in the larger triangle.

$$\frac{x}{5.5} = \frac{16}{8} \qquad \frac{\text{Hypotenuse}}{\text{Shorter leg}} = \frac{\text{Hypotenuse}}{\text{Shorter leg}}$$

Solving, we obtain $8x = 5.5(16)$, or $x = 11$, which is the same answer. ∎

READING CHECK 1

• Solve the proportion $\frac{4}{x} = \frac{6}{7}$.

EXAMPLE 2 Calculating the height of a tree

A 6-foot-tall person casts a 4-foot-long shadow. If a nearby tree casts a 36-foot-long shadow, estimate the height of the tree. See **FIGURE 14.8**.

Forming a Proportion from Heights and Shadows

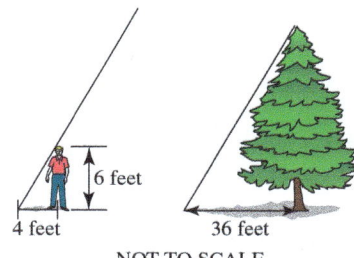

NOT TO SCALE
FIGURE 14.8

Similar Triangles

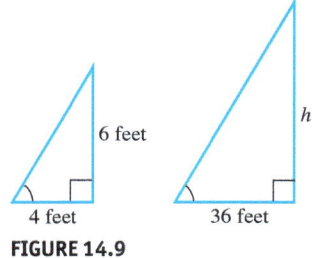

FIGURE 14.9

Solution
The triangles shown in **FIGURE 14.9** represent the situation shown in **FIGURE 14.8**. These triangles are similar because the measures of the corresponding angles are equal. Therefore their sides are proportional. Let h be the height of the tree.

$$\frac{h}{6} = \frac{36}{4} \qquad \frac{\text{Height}}{\text{Height}} = \frac{\text{Shadow length}}{\text{Shadow length}}$$

$$4h = 6(36) \qquad \text{Cross multiply.}$$

$$h = \frac{6(36)}{4} \qquad \text{Divide each side by 4.}$$

$$h = 54 \qquad \text{Simplify.}$$

The tree is 54 feet tall.

Now Try Exercise 69

2 Solving Direct Variation Problems

 Connecting Concepts with Your Life If your wage is $12 per hour, then your total pay P is proportional to the number of hours H you work, which can be represented by the equation

$$\frac{P}{H} = \frac{12}{1}, \qquad \frac{\text{Pay}}{\text{Hours}} = \frac{\text{Pay}}{\text{Hours}}$$

or, equivalently,

$$P = 12H.$$

We say that your pay P is *directly proportional* to the number of hours H that you work. The *constant of proportionality* is 12.

> **DIRECT VARIATION**
>
> Let x and y denote two quantities. Then y is **directly proportional** to x, or y **varies directly** with x, if there is a nonzero number k such that
>
> $$y = kx.$$
>
> The number k is called the **constant of proportionality**, or the **constant of variation**.

The following 4-step process is helpful when solving variation applications. This process is used in Examples 3 and 4 and can also be used for other types of variation problems.

READING CHECK 2

- What is the first step in solving a variation problem?

SOLVING A VARIATION APPLICATION

When solving a variation problem, the following steps can be used.

STEP 1: Write the general equation for the type of variation problem that you are solving.

STEP 2: Substitute given values in this equation so the constant of variation k is the only unknown value in the equation. Solve for k.

STEP 3: Substitute the value of k in the general equation in Step 1.

STEP 4: Use this equation to find the requested quantity.

EXAMPLE 3   Solving a direct variation problem

Let y be directly proportional to x, or vary directly with x. Suppose $y = 7$ when $x = 5$. Find y when $x = 11$.

Solution
STEP 1: The general equation for direct variation is $y = kx$.
STEP 2: Substitute **7** for y and **5** for x in $y = kx$. Solve for k.

$$7 = k(5) \qquad \text{Let } y = 7 \text{ and } x = 5.$$
$$\frac{7}{5} = k \qquad \text{Divide each side by 5.}$$

STEP 3: Replace k with $\frac{7}{5}$ in the equation $y = kx$ to obtain $y = \frac{7}{5}x$.
STEP 4: To find y when $x = 11$, let $x = 11$ in $y = \frac{7}{5}x$.

$$y = \frac{7}{5}(11) = \frac{77}{5} = 15.4$$

Now Try Exercise 33

EXAMPLE 4   Solving a direct variation application

The amount of weight that a beam of wood can support varies directly with its width. A beam that is 2.5 inches wide can support 800 pounds. How much weight can a similar beam support if its width is 3.2 inches?

Solution
STEP 1: The general equation for direct variation is $y = kx$, where y is the weight and x is the width.
STEP 2: Substitute **800** for y and **2.5** for x in $y = kx$. Solve for k.

$$800 = k(2.5) \qquad \text{Let } y = 800 \text{ and } x = 2.5.$$
$$\frac{800}{2.5} = k \qquad \text{Divide each side by 2.5.}$$
$$k = 320 \qquad \text{Simplify; rewrite the equation.}$$

STEP 3: Replace k with **320** in $y = kx$ to obtain $y = 320x$.
STEP 4: To find the weight y when $x = 3.2$, substitute **3.2** for x in $y = 320x$.

$$y = 320(3.2) = 1024 \text{ pounds}$$

Now Try Exercise 67

The following See the Concept shows how the graph of a direct variation equation relates to the constant of variation k.

See the Concept 1 — GRAPHS AND DIRECT VARIATION

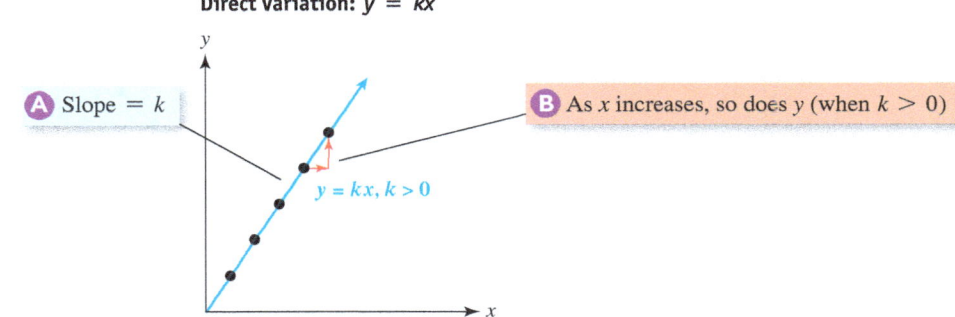

A The slope of the line equals k, which is the constant of variation. (The value of k could be negative.)

B These data lie on the line $y = kx$, so y varies directly with x.

WATCH VIDEO IN MML.

READING CHECK 3

- How does the constant of proportionality for data that represent direct variation compare to the slope of a line that passes through the data?

EXAMPLE 5 Modeling college tuition

TABLE 14.7 lists the tuition for taking various numbers of credits.

(a) A scatterplot of the data is shown in **FIGURE 14.10**. Could the data be modeled using a line?

Credits	Tuition
3	$189
5	$315
8	$504
11	$693
17	$1071

TABLE 14.7

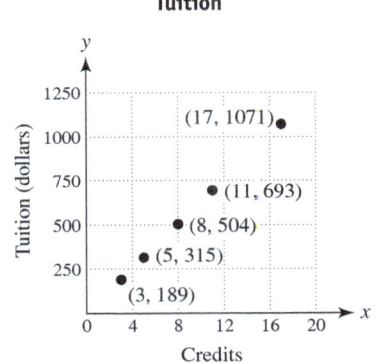

FIGURE 14.10

(b) Explain why tuition is directly proportional to the number of credits taken.
(c) Find the constant of proportionality. Interpret your result.
(d) Predict the cost of taking 16 credits.

Solution
(a) The data are linear and suggest a line passing through the origin.
(b) Because the data can be modeled by a line passing through the origin, tuition is directly proportional to the number of credits taken. Hence doubling the credits will double the tuition and tripling the credits will triple the tuition.
(c) The slope of the line equals the constant of proportionality k. If we use the first and last data points $(3, 189)$ and $(17, 1071)$, the slope is

$$k = \frac{1071 - 189}{17 - 3} = 63.$$

That is, tuition is $63 per credit. If we graph the line $y = 63x$, it models the data, as shown in **FIGURE 14.11** on the next page.

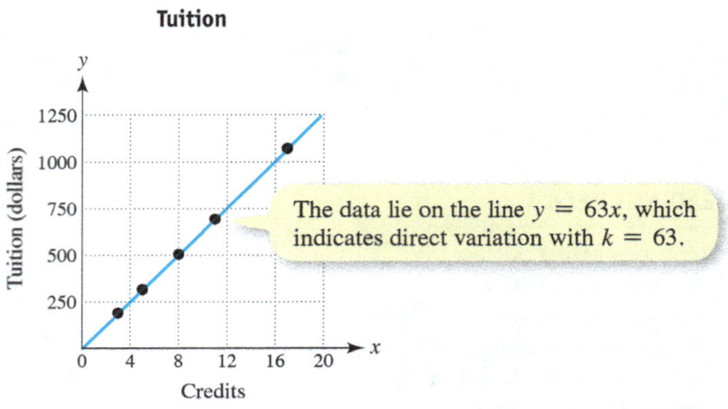

Tuition

The data lie on the line $y = 63x$, which indicates direct variation with $k = 63$.

FIGURE 14.11

(d) If y represents tuition and x represents the credits taken, 16 credits would cost

$$y = 63(16) = \$1008.$$

Now Try Exercise 73

MAKING CONNECTIONS 1

Ratios and the Constant of Proportionality

The constant of proportionality in Example 5 can be found by calculating the ratios $\frac{y}{x}$.

x	3	5	8	11	17
y	189	315	504	693	1071
$\frac{y}{x}$	63	63	63	63	63

Each ratio $\frac{y}{x}$ equals 63, which indicates direct variation and that $k = 63$.

3 Solving Inverse Variation Problems

Connecting Concepts with Your Life When two quantities vary inversely, an increase in one quantity results in a decrease in the second quantity. For example, at 30 miles per hour, a car travels 120 miles in 4 hours, whereas at 60 miles per hour, the car travels 120 miles in 2 hours. Doubling the speed (or rate) decreases the travel time by half. Distance equals rate times time, so $d = rt$. Thus

$$120 = rt, \quad \text{or equivalently,} \quad t = \frac{120}{r}.$$

We say that the time t to travel 120 miles is *inversely proportional* to the speed or rate r. The constant of proportionality or constant of variation is 120.

READING CHECK 4

- When two quantities, x and y, are inversely proportional with positive k, how does an increase in x affect y?

INVERSE VARIATION

Let x and y denote two quantities. Then y is **inversely proportional** to x, or y **varies inversely** with x, if there is a nonzero number k such that

$$y = \frac{k}{x}.$$

The following See the Concept shows how the graph of an inverse variation equation relates to points (x, y) on the graph and to the constant of variation k.

See the Concept 2 — GRAPHS AND INVERSE VARIATION

Inverse Variation: $y = \frac{k}{x}$

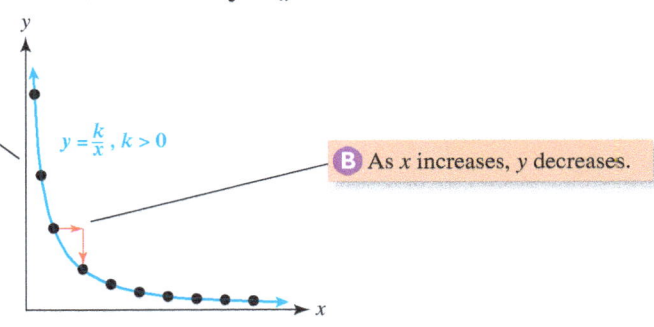

Ⓐ For any data point (x, y), $xy = k$

Ⓑ As x increases, y decreases.

Ⓐ If (x, y) is any point on the curve, then the product xy always equals k, which is the constant of variation.

Ⓑ These data lie on the curve $y = \frac{k}{x}$, so y varies inversely with x.

WATCH VIDEO IN MML.

EXAMPLE 6 Solving an inverse variation problem

Let y be inversely proportional to x, or vary inversely with x. Suppose $y = 5$ when $x = 6$. Find y when $x = 21$.

Solution
STEP 1: The general equation for inverse variation is $y = \frac{k}{x}$.
STEP 2: Because $y = 5$ when $x = 6$, substitute **5** for y and **6** for x in $y = \frac{k}{x}$. Solve for k.

$$5 = \frac{k}{6} \quad \text{Let } y = 5 \text{ and } x = 6.$$

$$30 = k \quad \text{Multiply each side by 6.}$$

STEP 3: Replace k with **30** in the equation $y = \frac{k}{x}$ to obtain $y = \frac{30}{x}$.
STEP 4: To find y, let $x = 21$. Then $y = \frac{30}{21} = \frac{10}{7}$.

Now Try Exercise 39

 Connecting Concepts with Your Life A wrench is commonly used to loosen a nut on a bolt. See **FIGURE 14.12**. If the nut is difficult to loosen, a wrench with a longer handle is often helpful.

Loosening a Nut with a Wrench

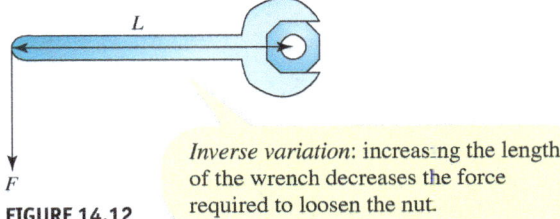

Inverse variation: increasing the length of the wrench decreases the force required to loosen the nut.

FIGURE 14.12

EXAMPLE 7 Illustrating inverse variation with a wrench

TABLE 14.8 on the next page lists the force F necessary to loosen a particular nut with wrenches of different lengths L.

L (inches)	F (pounds)
6	12
8	9
12	6
18	4
24	3

TABLE 14.8

(a) Make a scatterplot of the data and discuss the graph. Are the data linear?
(b) Explain why the force F is inversely proportional to the handle length L. Find k so that $F = \frac{k}{L}$ models the data.
(c) Predict the force needed to loosen the nut with a 15-inch wrench.

Solution
(a) The scatterplot shown in **FIGURE 14.13** reveals that the data are nonlinear. As the length L of the wrench increases, the force F necessary to loosen the nut decreases.
(b) If F is inversely proportional to L, then $F = \frac{k}{L}$, or $FL = k$. That is, the product of F and L equals the constant of proportionality k. In **TABLE 14.8**, the product of F and L always equals 72 for each data point. Thus F is inversely proportional to L with constant of proportionality $k = 72$, so $F = \frac{72}{L}$.
(c) If $L = 15$, then $F = \frac{72}{15} = 4.8$. A wrench with a 15-inch handle requires a force of 4.8 pounds to loosen the nut.

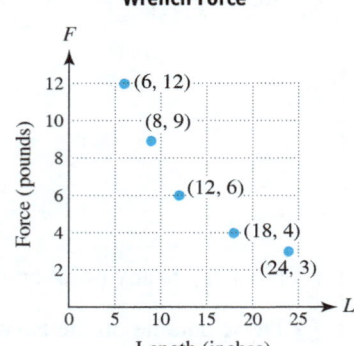

FIGURE 14.13

Now Try Exercises 75

TECHNOLOGY NOTE

Scatterplots and Graphs
A graphing calculator can be used to create scatterplots and graphs. A scatterplot of the data in **TABLE 14.8** is shown in the first figure. In the second figure, the data and the equation $y = \frac{72}{x}$ are graphed. Note that each tick mark represents 5 units.

CALCULATOR HELP
To make a scatterplot, see Appendix A (pages 3 and 4).

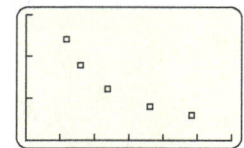 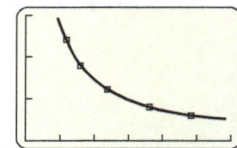

4 Analyzing Data

So far in this section, we have discussed *direct* and *inverse* variation. **TABLE 14.9** gives a summary of these two types of variation.

Direct and Inverse Variation

Type of Variation	Equation	Constant of Variation
y varies *directly* with x	$y = kx$	$k = \frac{y}{x}$
y varies *inversely* with x	$y = \frac{k}{x}$	$k = xy$

TABLE 14.9

READING CHECK 5
- How can the constant of variation be used to determine whether data are directly or inversely proportional?

The last column in **TABLE 14.9** shows that a set of data represents direct variation when the quotients $\frac{y}{x}$ equal a constant, and it represents inverse variation when the products xy equal a constant. In the next example, we determine if tables of data represent direct variation, inverse variation, or neither.

EXAMPLE 8   Analyzing data

Determine whether the data in each table represent direct variation, inverse variation, or neither.

(a)
x	4	5	10	20
y	40	32	16	8

(b)
x	2	5	9	11
y	18	45	81	99

(c)
x	2	4	6	8
y	8	20	30	56

Solution
(a) As x increases, y decreases. Because $xy = 160$ for each data point in the table, the equation $y = \frac{160}{x}$ models the data. The data represent inverse variation.
(b) Because $\frac{y}{x} = 9$ for each data point in the table, the equation $y = 9x$ models the data in the table. These data represent direct variation.
(c) Neither the products xy nor the ratios $\frac{y}{x}$ are constant for the data in the table. Therefore these data represent neither direct variation nor inverse variation.

Now Try Exercises 51(a), 53(a), 55(a)

5 Solving Joint Variation Problems

Math in Context (Construction) In many applications, a quantity depends on more than one variable. In *joint variation*, a quantity varies with the product of more than one variable. For example, the formula for the area A of a rectangle is given by

$$A = WL,$$

where W and L are the width and length, respectively. Thus the area of a rectangle varies jointly with the width and length.

> **JOINT VARIATION**
>
> Let x, y, and z denote three quantities. Then z **varies jointly** with x and y if there is a nonzero number k such that
>
> $$z = kxy.$$

Sometimes joint variation can involve a power of a variable. For example, the volume V of a cylinder is given by $V = \pi r^2 h$, where r is its radius and h is its height, as illustrated in **FIGURE 14.14**. In this case, we say that the volume varies jointly with the height and the *square* of the radius. The constant of variation is $k = \pi$.

Cylinder
$V = \pi r^2 h,$

FIGURE 14.14

EXAMPLE 9   Finding the strength of a rectangular beam

The strength S of a rectangular beam varies jointly with its width w and the square of its thickness t. See **FIGURE 14.15**. If a beam 3 inches wide and 5 inches thick supports 750 pounds, how much can a similar beam 2 inches wide and 6 inches thick support?

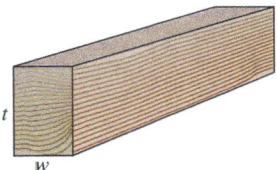

FIGURE 14.15

Solution

STEP 1: The strength of a beam is modeled by $S = kwt^2$, where k is a constant of variation.

STEP 2: We can find k by substituting $S = 750$, $w = 3$, and $t = 5$ in the formula.

$$750 = k \cdot 3 \cdot 5^2 \quad \text{Substitute in } S = kwt^2.$$

$$k = \frac{750}{3 \cdot 5^2} \quad \text{Solve for } k; \text{ rewrite.}$$

$$= 10 \quad \text{Simplify.}$$

STEP 3: The equation $S = 10wt^2$ models the strength of this type of beam.

STEP 4: When $w = 2$ and $t = 6$, the beam can support

$$S = 10 \cdot 2 \cdot 6^2 = 720 \text{ pounds.}$$

Now Try Exercises 83

CRITICAL THINKING 1

Compare the increased strength of a beam if the width doubles and if the thickness doubles. What happens to the strength of a beam if both the width and thickness triple?

14.7 Putting It All Together

CONCEPT	EXPLANATION	EXAMPLES
Proportion	A statement that two ratios are equal	$\dfrac{8}{17} = \dfrac{49}{x}$ and $\dfrac{x}{6} = \dfrac{3}{14}$
Direct Variation	Two quantities x and y vary according to the equation $y = kx$, where k is a nonzero constant. The constant of proportionality (or variation) is k.	$y = 4x$ or $\dfrac{y}{x} = 4$ | x | 1 | 2 | 4 | |---|---|---|---| | y | 4 | 8 | 16 | Note that if x doubles, then y also doubles.
Inverse Variation	Two quantities x and y vary according to the equation $y = \frac{k}{x}$, where k is a nonzero constant. The constant of proportionality (or variation) is k.	$y = \dfrac{3}{x}$ or $xy = 3$ | x | 1 | 3 | 6 | |---|---|---|---| | y | 3 | 1 | $\frac{1}{2}$ | Note that if x doubles from 3 to 6, then y decreases by half.
Joint Variation	Three quantities x, y, and z vary according to the equation $z = kxy$, where k is a nonzero constant.	The area A of a triangle varies jointly with b and h according to the equation $A = \frac{1}{2}bh$, where b is its base and h is its height. The constant of variation is $k = \frac{1}{2}$.

14.7 Exercises

CONCEPTS AND VOCABULARY

1. What is a proportion?

2. If 5 is to 6 as x is to 7, write a proportion that allows you to find x.

3. Suppose that y is directly proportional to x. If x doubles, what happens to y?

4. Suppose that y is inversely proportional to x. If x doubles, what happens to y?

5. If y varies directly with x, then $\frac{y}{x}$ equals a(n) _____.

6. If y varies inversely with x, then xy equals a(n) _____.

7. Would the food bill B generally vary directly or inversely with the number of people N being fed? Explain your reasoning.

8. Would the time T needed to paint a building vary directly or inversely with the number of painters N working on the job? Explain your reasoning.

9. If xy equals a constant for every data point (x, y) in a table, then the data represent _____ variation.

10. If $\frac{y}{x}$ equals a constant for every data point (x, y) in a table, then the data represent _____ variation.

SOLVING PROPORTIONS

Exercises 11–22: Solve the proportion.

11. $\dfrac{x}{24} = \dfrac{5}{8}$

12. $\dfrac{x}{5} = \dfrac{3}{7}$

13. $\dfrac{14}{x} = \dfrac{2}{3}$

14. $\dfrac{4}{9} = \dfrac{9}{x}$

15. $\dfrac{3}{16} = \dfrac{h}{256}$

16. $\dfrac{20}{a} = \dfrac{15}{4}$

17. $\dfrac{3}{4} = \dfrac{2x}{7}$

18. $\dfrac{7}{3z} = \dfrac{5}{4}$

19. $\dfrac{x}{6} = \dfrac{8}{3x}$

20. $\dfrac{4}{x} = \dfrac{4x}{9}$

21. $\dfrac{x}{7} = \dfrac{7}{4x}$

22. $\dfrac{2}{3x} = \dfrac{27x}{8}$

23. **Thinking Generally** Solve $\dfrac{a}{b} = \dfrac{c}{d}$ for b.

24. **Thinking Generally** Solve $\dfrac{a+b}{c^2} = \dfrac{1}{2}$ for b.

Exercises 25–32: Do the following.
(a) Write a proportion that models the situation.
(b) Solve the proportion for x.

25. 5 is to 8 as 9 is to x.

26. x is to 11 as 7 is to 4.

27. A triangle has sides of 4, 7, and 10. In a similar triangle, the shortest side is 8 and the longest side is x.

28. A rectangle has sides of 5 and 12. In a similar rectangle, the longer side is 10 and the shorter side is x.

29. If you earn \$98 in 7 hours, then you can earn x dollars in 11 hours.

30. If 14 gallons of gasoline contain 1.4 gallons of ethanol, then 22 gallons of gasoline contain x gallons of ethanol.

31. If 3 MP3 players can hold 750 songs, then 7 similar MP3 players can hold x songs.

32. If a gas pump fills a 25-gallon tank in 6 minutes, it can fill a 14-gallon tank in x minutes.

SOLVING DIRECT VARIATION PROBLEMS

Exercises 33–38: Suppose that y is directly proportional to x.
(a) Use the given information to find the constant of proportionality k.
(b) Then use $y = kx$ to find y for $x = 6$.

33. $y = 4$ when $x = 2$

34. $y = 5$ when $x = 10$

35. $y = 3$ when $x = 2$

36. $y = 11$ when $x = 55$

37. $y = -60$ when $x = 8$

38. $y = -17$ when $x = 68$

SOLVING INVERSE VARIATION PROBLEMS

Exercises 39–44: Suppose that y is inversely proportional to x.
(a) Use the given information to find the constant of proportionality k.
(b) Then use $y = \dfrac{k}{x}$ to find y for $x = 8$.

39. $y = 6$ when $x = 4$

40. $y = 2$ when $x = 24$

41. $y = 80$ when $x = \frac{1}{2}$ **42.** $y = \frac{1}{4}$ when $x = 32$

43. $y = 20$ when $x = 20$ **44.** $y = \frac{8}{3}$ when $x = 12$

SOLVING JOINT VARIATION PROBLEM

Exercises 45–50: Let z vary jointly with x and y.
  (a) *Find the constant of variation k.*
  (b) *Use* $z = kxy$ *to find z when* $x = 5$ *and* $y = 7$.

45. $z = 6$ when $x = 3$ and $y = 8$

46. $z = 135$ when $x = 2.5$ and $y = 9$

47. $z = 5775$ when $x = 25$ and $y = 21$

48. $z = 1530$ when $x = 22.5$ and $y = 4$

49. $z = 25$ when $x = \frac{1}{2}$ and $y = 5$

50. $z = 12$ when $x = \frac{1}{4}$ and $y = 12$

ANALYZING DATA

Exercises 51–56: (Refer to Example 8.)
  (a) *Determine whether the data represent direct variation, inverse variation, or neither.*
  (b) *If the data represent either direct or inverse variation, find an equation that models the data.*
  (c) *Graph the equation and the data when possible.*

51.

x	2	3	4	5
y	3	4.5	6	7.5

52.

x	1	5	9	15
y	6	30	54	90

53.

x	3	6	9	12
y	12	6	4	3

54.

x	2	6	10	14
y	105	35	21	15

55.

x	4	6	12	20
y	10	20	30	40

56.

x	10	20	30	40
y	12	6	5	4

Exercises 57–62: Use the graph to determine whether the data represent direct variation, inverse variation, or neither. Find the constant of variation whenever possible.

57. 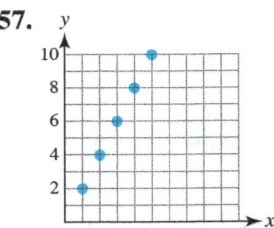 **58.**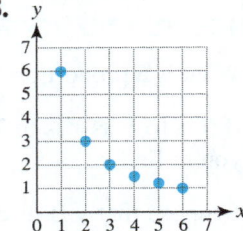

59. 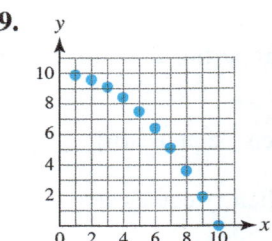 **60.**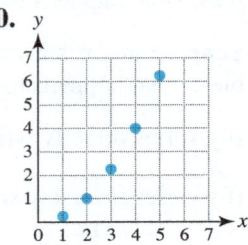

61. 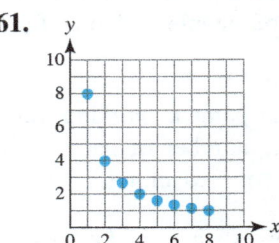 **62.**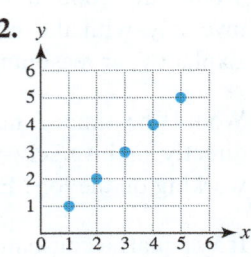

APPLICATIONS INVOLVING PROPORTIONS AND VARIATION

63. Recording Music A 750-megabyte CD can record 85 minutes of music. How many minutes can be recorded on 420 megabytes?

64. Making Fudge If $2\frac{2}{3}$ cups of sugar can make 14 pieces of fudge, how much sugar is needed to make 49 pieces of fudge?

65. Water Content in Snow (Refer to Example 1.) Eight inches of heavy, wet snow are equivalent to an inch of rain. Estimate the water content in 13 inches of heavy, wet snow.

66. Wages If a person working for an hourly wage earns $143 in 13 hours, how much will that person earn in 15 hours?

67. Strength of a Beam (Refer to Example 4.) The strength of a metal beam varies directly with its width. A beam that is 6.2 inches wide can support 2800 pounds. How much weight can a similar beam support if it is 4.7 inches wide?

68. **Strength of a Beam** The strength of a wood beam varies inversely with its length. A beam that is 33 feet long can support 1200 pounds. How much weight can a similar beam support if it is 23 feet long?

69. **Height of a Tree** (Refer to Example 2.) A 5-foot-tall person casts an 8-foot-long shadow, and a nearby tree casts a 30-foot-long shadow. See the figure below. Estimate the height of the tree.

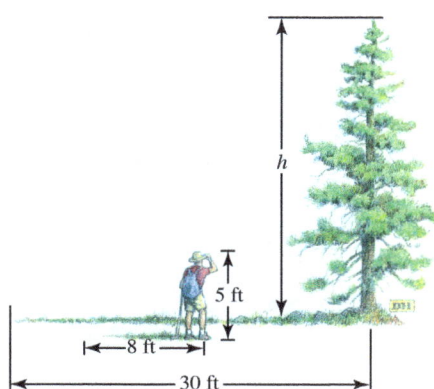

70. **Making Coffee** If 6 tablespoons of coffee grounds make 10 cups of coffee, how many tablespoons of coffee grounds are needed to make 35 cups of coffee?

71. **Rolling Resistance of Cars** If you were to try to push a car, you would experience *rolling resistance*. This resistance equals the force necessary to keep the car moving slowly in neutral gear. The following table shows the rolling resistance R for passenger cars of different gross weights W. (*Source:* N. Garber and L. Hoel, *Traffic and Highway Engineering.*)

W (pounds)	2000	2500	3000	3500
R (pounds)	24	30	36	42

(a) Do the data represent direct or inverse variation? Explain.
(b) Find an equation that models the data. Graph the equation with the data.
(c) Estimate the rolling resistance of a 3200-pound car.

72. **Transportation Costs** The use of a particular toll bridge varies inversely according to the toll. If the toll is $0.50, then 8000 vehicles use the bridge. Estimate the number of users if the toll is $0.80. (*Source:* N. Garber.)

73. **Flow of Water** The gallons of water G flowing in 1 minute through a hose with a cross-sectional area A are shown in the table.

A (square inch)	0.2	0.3	0.4	0.5
G (gallons)	5.4	8.1	10.8	13.5

(a) Do the data represent direct or inverse variation? Explain.
(b) Find an equation that models the data. Graph the equation with the data.
(c) Interpret the constant of variation k.

74. **Hooke's Law** The table shows the distance D that a spring stretches when a weight W is hung on it.

W (pounds)	2	6	9	15
D (inches)	1.6	4.8	7.2	12

(a) Do the data represent direct or inverse variation? Explain.
(b) Find an equation that models the data.
(c) How far will the spring stretch if an 11-pound weight is hung on it?

75. **Tightening Lug Nuts** (Refer to Example 7.) When a tire is mounted on a car, the lug nuts should not be over-tightened. The table below shows the maximum force used with wrenches of different lengths.

L (inches)	8	10	16
F (pounds)	150	120	75

Source: Tires plus.

(a) Model the data, using the equation $F = \frac{k}{L}$.
(b) How much force should be used with a wrench 15 inches long?

76. **Ozone and UV Radiation** Ozone in the atmosphere filters out approximately 90% of the harmful ultraviolet (UV) rays from the sun. Depletion of the ozone layer increases the amount of UV radiation reaching Earth's surface. An increase in UV radiation is associated with skin cancer. The following graph shows the percentage increase y in UV radiation for a decrease in the ozone layer of x percent. (*Source:* R. Turner, D. Pearce, and I. Bateman, *Environmental Economics.*)

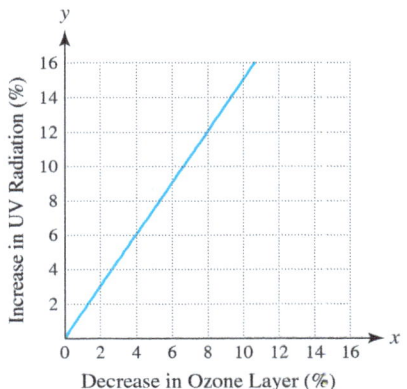

Ozone Depletion

continued on next page

(a) Does this graph represent direct or inverse variation?
(b) Find an equation for the line in the graph.
(c) Estimate the percentage increase in UV radiation if the ozone layer decreases by 5%.

77. **Air Temperature and Altitude** In the first 6 miles of Earth's atmosphere, air cools as the altitude increases. The following graph shows the temperature change y in degrees Fahrenheit at an altitude of x miles. (*Source:* A. Miller and R. Anthes, *Meteorology*.)

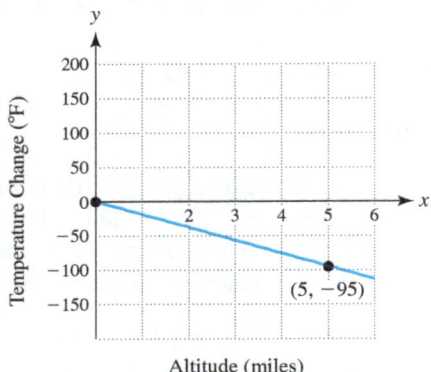

(a) Does this graph represent direct variation or inverse variation?
(b) Find an equation that models the data in the graph.
(c) Is the constant of proportionality k positive or negative? Interpret k.
(d) Find the change in air temperature 2.5 miles high.

78. **Cost of Tuition** (Refer to Example 5.) The cost of tuition is directly proportional to the number of credits taken. If 6 credits cost $483, find the cost of 13 credits. What does the constant of proportionality represent?

79. **Electrical Resistance** The electrical resistance of a wire is directly proportional to its length. If a 30-foot-long wire has a resistance of 3 ohms, find the resistance of an 18-foot-long wire.

80. **Resistance and Current** The current that flows through an electrical circuit is inversely proportional to the resistance. When the resistance R is 180 ohms, the current I is 0.6 amp. Find the current when the resistance is 54 ohms.

81. **Joint Variation** The variable z varies jointly with the second power of x and the third power of y. Write a formula for z if $z = 31.9$ when $x = 2$ and $y = 2.5$.

82. **Wind Power** The electrical power generated by a windmill varies jointly with the square of the diameter of the area swept out by the blades and the cube of the wind velocity. If a windmill with a 10-foot diameter and a 16-mile-per-hour wind generates 15,392 watts, how much power would be generated if the blades swept out an area 12 feet in diameter and the wind speed was 15 miles per hour?

83. **Strength of a Beam** (Refer to Example 9.) If a wood beam 5 inches wide and 3 inches thick supports 300 pounds, how much can a similar beam 5 inches wide and 2 inches thick support?

84. **Carpeting** The cost of carpet for a rectangular room varies jointly with its width and length. If a room 11 feet wide and 14 feet long costs $539 to carpet, find the cost to carpet a room 17 feet by 19 feet. Interpret the constant of variation k.

85. **Weight on the Moon** The weight of a person on the moon is directly proportional to the weight of the person on Earth. If a 175-pound person weighs 28 pounds on the moon, how much will a 220-pound person weigh on the moon?

86. **Weight Near Earth** The weight W of a person near Earth is inversely proportional to the square of the person's distance d from the *center* of Earth. If a person weighs 200 pounds when $d = 4000$ miles, how much does the same person weigh when $d = 7000$ miles? (*Note:* The radius of Earth is about 4000 miles.)

87. **Ohm's Law** The voltage V in an electrical circuit varies jointly with the amperage I and resistance R. If $V = 220$ when $I = 10$ and $R = 22$, find V when $I = 15$ and $R = 50$.

88. **Revenue** The revenue R from selling x items at price p varies jointly with x and p. If $R = \$24{,}000$ when $x = 3000$ and $p = \$8$, find the number of items x sold when $R = \$30{,}000$ and $p = \$6$.

89. **Whales and Diving** At a depth of 1800 feet, a sperm whale experiences water pressure equal to 800 pounds per square inch. Find the pressure at 2700 feet. (*Hint:* Water pressure is proportional to water depth.)

90. **Eagle Population** Twenty-two bald eagles are tagged and released into the wilderness. Later, an observed sample of 56 bald eagles contains 7 eagles that are tagged. Estimate the bald eagle population in this wilderness area. (*Hint:* The number of tagged birds in the sample is proportional to the number of tagged birds in the population)

WRITING ABOUT MATHEMATICS

91. Explain what it means for a quantity y to be directly proportional to a quantity x.

92. Explain what it means for a quantity y to be inversely proportional to a quantity x.

SECTION 14.7 Checking Basic Concepts

1. Solve each proportion.
   (a) $\dfrac{x}{9} = \dfrac{2}{5}$ (b) $\dfrac{4}{3} = \dfrac{5}{b}$

2. Write a proportion that models each situation. Then solve it.
   (a) 4 is to 6 as 8 is to x.
   (b) If 2 compact discs can record 148 minutes of music, then 5 compact discs can record x minutes of music.

3. Suppose that y is inversely proportional to x. If $y = 4$ when $x = 15$, find the constant of proportionality k. Find y when $x = 10$.

4. Decide whether the data in the table represent direct or inverse variation. Explain your reasoning. Find the constant of variation.

   (a)
x	2	4	6	8
y	3	6	9	12

   (b)
x	2	4	6	8
y	12	6	4	3

5. **Wages** If a person working for an hourly wage earns $272 in 17 hours, how much will that person earn in 10 hours?

CHAPTER 14 Summary

SECTION 14.1 ■ INTRODUCTION TO RATIONAL EXPRESSIONS

Rational Expression
A rational expression can be written in the form $\frac{P}{Q}$, where P and Q are polynomials, and is defined whenever $Q \neq 0$.

Example: $\frac{x^2}{x-5}$ is a rational expression that is defined for all real numbers except $x = 5$.

Undefined Rational Expression A rational expression is undefined for any value of the variable that makes the denominator equal to 0.

Examples: $\frac{1}{x-5}$ is undefined when $x = 5$.

$\frac{3y}{y^2 - 4}$ is undefined when $y = -2$ or when $y = 2$.

Simplifying a Rational Expression To simplify a rational expression, factor the numerator and the denominator. Then apply the basic principle of rational expressions,

$$\frac{PR}{QR} = \frac{P}{Q}. \quad Q \text{ and } R \text{ nonzero}$$

Examples: $\frac{x^2 - 4}{x^2 - 3x + 2} = \frac{(x+2)(x-2)}{(x-1)(x-2)} = \frac{x+2}{x-1}, \quad \frac{x-3}{3-x} = -\frac{x-3}{x-3} = -1$

SECTION 14.2 ■ MULTIPLICATION AND DIVISION OF RATIONAL EXPRESSIONS

Multiplying Rational Expressions
To multiply two rational expressions, multiply the numerators and multiply the denominators.

$$\frac{A}{B} \cdot \frac{C}{D} = \frac{AC}{BD} \quad B \text{ and } D \text{ nonzero}$$

Example: $\frac{3}{x-1} \cdot \frac{4}{x+1} = \frac{12}{(x-1)(x+1)}$

Dividing Rational Expressions To divide two rational expressions, multiply by the reciprocal of the divisor.

$$\frac{A}{B} \div \frac{C}{D} = \frac{A}{B} \cdot \frac{D}{C} = \frac{AD}{BC} \quad B, C, \text{ and } D \text{ nonzero}$$

Example: $\frac{x+2}{x^2+3x} \div \frac{x+2}{x} = \frac{x+2}{x(x+3)} \cdot \frac{x}{x+2} = \frac{x(x+2)}{x(x+3)(x+2)} = \frac{1}{x+3}$

SECTION 14.3 ■ ADDITION AND SUBTRACTION WITH LIKE DENOMINATORS

Addition of Rational Expressions That Have Like Denominators To add two rational expressions that have like denominators, add their numerators. Keep the same denominator.

$$\frac{A}{C} + \frac{B}{C} = \frac{A+B}{C} \quad C \text{ nonzero}$$

Example: $\frac{5x}{x+4} + \frac{1}{x+4} = \frac{5x+1}{x+4}$

Subtraction of Rational Expressions That Have Like Denominators To subtract two rational expressions that have like denominators, subtract their numerators. Keep the same denominator.

$$\frac{A}{C} - \frac{B}{C} = \frac{A-B}{C} \quad C \text{ nonzero}$$

Example: $\dfrac{6}{2x+1} - \dfrac{2}{2x+1} = \dfrac{4}{2x+1}$

SECTION 14.4 ■ ADDITION AND SUBTRACTION WITH UNLIKE DENOMINATORS

Finding Least Common Multiples The least common multiple (LCM) of two or more polynomials can be found as follows.

STEP 1: Factor each polynomial completely.

STEP 2: List each factor the greatest number of times that it occurs in any factorization.

STEP 3: Find the product of this list of factors. It is the LCM.

Example: $4x^2(x+1) = 2 \cdot 2 \cdot x \cdot x \cdot (x+1)$
$2x(x^2-1) = 2 \cdot x \cdot (x+1) \cdot (x-1)$

Listing each factor the greatest number of times and multiplying give the LCM.

$$2 \cdot 2 \cdot x \cdot x \cdot (x+1)(x-1) = 4x^2(x^2-1)$$

Finding the Least Common Denominator The least common denominator (LCD) is the least common multiple (LCM) of the denominators.

Example: From the preceding example, the LCD for $\dfrac{1}{4x^2(x+1)}$ and $\dfrac{1}{2x(x^2-1)}$ is $4x^2(x^2-1)$.

Addition and Subtraction of Rational Expressions That Have Unlike Denominators First write each rational expression by using the LCD. Then add or subtract the resulting rational expressions. Finally, write your answer in lowest terms.

Example: $\dfrac{1}{x-1} - \dfrac{1}{x} = \dfrac{x}{x(x-1)} - \dfrac{x-1}{x(x-1)} = \dfrac{x-(x-1)}{x(x-1)} = \dfrac{1}{x(x-1)}$

Note that the LCD is $x(x-1)$.

SECTION 14.5 ■ COMPLEX FRACTIONS

Complex Fractions A complex fraction is a rational expression that contains fractions in its numerator, denominator, or both. The following equation can be used to simplify basic complex fractions.

$$\frac{\frac{a}{b}}{\frac{c}{d}} = \frac{a}{b} \cdot \frac{d}{c} \quad b, c, \text{ and } d \text{ nonzero}$$

Example: $\dfrac{\frac{x}{3}}{\frac{x}{x-1}} = \dfrac{x}{3} \cdot \dfrac{x-1}{x} = \dfrac{x(x-1)}{3x} = \dfrac{x-1}{3}$

Simplifying Complex Fractions

Method I Combine terms in the numerator, combine terms in the denominator, and then multiply the numerator by the reciprocal of the denominator.

Method II Multiply the numerator and denominator by the LCD of all fractions within the expression and simplify the resulting expression.

Example: *Method I*

$$\frac{\dfrac{1}{a} - \dfrac{1}{b}}{\dfrac{1}{a} + \dfrac{1}{b}} = \frac{\dfrac{b-a}{ab}}{\dfrac{b+a}{ab}} = \frac{b-a}{ab} \cdot \frac{ab}{b+a} = \frac{b-a}{b+a}$$

Method II The least common denominator is ab.

$$\frac{\dfrac{1}{a} - \dfrac{1}{b}}{\dfrac{1}{a} + \dfrac{1}{b}} = \frac{\left(\dfrac{1}{a} - \dfrac{1}{b}\right)ab}{\left(\dfrac{1}{a} + \dfrac{1}{b}\right)ab} = \frac{\dfrac{ab}{a} - \dfrac{ab}{b}}{\dfrac{ab}{a} + \dfrac{ab}{b}} = \frac{b-a}{b+a}$$

SECTION 14.6 ■ RATIONAL EQUATIONS AND FORMULAS

Solving Rational Equations One way to solve the equation $\frac{a}{b} = \frac{c}{d}$ is to cross multiply to obtain $ad = bc$. (Check each answer.) A general way to solve rational equations is to use the following steps.

STEP 1: Find the LCD of the terms in the equation.
STEP 2: Multiply each side of the equation by the LCD.
STEP 3: Simplify each term.
STEP 4: Solve the resulting equation.
STEP 5: Check each answer in the *given* equation. Reject any value that makes a denominator equal 0.

Examples: $\frac{1}{2x} = \frac{2}{x+3}$ implies that $x + 3 = 4x$, or $x = 1$. This answer checks.

To solve $\frac{5}{x} - \frac{1}{3x} = \frac{7}{3}$, multiply each term by the LCD, $3x$:

$$\frac{5(3x)}{x} - \frac{1 \cdot (3x)}{3x} = \frac{7(3x)}{3},$$

which simplifies to $15 - 1 = 7x$, or $x = 2$. This answer checks.

Solving for a Variable Many formulas contain more than one variable. To solve for a particular variable, use the rules of algebra to isolate the variable.

Example: To solve $S = \frac{2\pi}{r}$ for r, multiply each side by r to obtain $Sr = 2\pi$ and then divide each side by S to obtain $r = \frac{2\pi}{S}$.

SECTION 14.7 ■ PROPORTIONS AND VARIATION

Proportions A proportion is a statement that two ratios are equal.

Example: $\frac{5}{x} = \frac{4}{7}$

Similar Triangles Two triangles are similar if the measures of their corresponding angles are equal. Corresponding sides of similar triangles are proportional.

Example: The following triangles are similar.

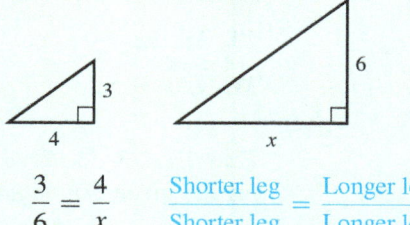

$$\frac{3}{6} = \frac{4}{x} \qquad \frac{\text{Shorter leg}}{\text{Shorter leg}} = \frac{\text{Longer leg}}{\text{Longer leg}}$$

By solving this proportion, we see that $x = 8$.

Direct Variation A quantity y is *directly proportional* to a quantity x, or y *varies directly* with x, if there is a nonzero constant k such that $y = kx$. The number k is called the *constant of proportionality* or the *constant of variation*.

Example: If y varies directly with x, then the ratios $\frac{y}{x}$ always equal k. The following data satisfy $\frac{y}{x} = 4$, so the constant of variation is 4. Thus $y = 4x$.

x	1	2	3	4
y	4	8	12	16

Inverse Variation A quantity y is *inversely proportional* to a quantity x, or y *varies inversely* with x, if there is a nonzero constant k such that $y = \frac{k}{x}$.

Example: If y varies inversely with x, then the products xy always equal k. The following data satisfy $xy = 12$, so the constant of variation is 12. Thus $y = \frac{12}{x}$.

x	1	2	4	6
y	12	6	3	2

Joint Variation The quantity z *varies jointly* with x and y if $z = kxy$, $k \neq 0$.

Example: The area A of a rectangle varies jointly with the width W and length L because $A = LW$. Note that $k = 1$ in this example.

CHAPTER 14 Review Exercises

SECTION 14.1

Exercises 1–4: If possible, evaluate the expression for the given value of x.

1. $\dfrac{3}{x - 3}$, $x = -2$

2. $\dfrac{4x}{5 - x^2}$, $x = 3$

3. $\dfrac{-x}{7 - x}$, $x = 7$

4. $\dfrac{4x}{x^2 - 3x + 2}$, $x = 2$

5. Complete the table for the rational expression. If a value is undefined, place a dash in the table.

x	-2	-1	0	1	2
$\dfrac{3x}{x - 1}$					

6. Find the x-values that make $\dfrac{8}{x^2 - 4}$ undefined.

Exercises 7–10: Simplify to lowest terms.

7. $\dfrac{25x^3y^4}{15x^5y}$

8. $\dfrac{x^2 - 36}{x + 6}$

9. $\dfrac{2x^2 + 5x - 3}{2x^2 + x - 1}$

10. $\dfrac{3x^2 + 10x - 8}{3x^2 + x - 2}$

Exercises 11 and 12: Do the following.
  (a) Decide whether you are given an expression or an equation.
  (b) If you are given an expression, simplify it. If you are given an equation, solve it.

11. $\dfrac{x + 1}{(x - 3)(x + 1)}$

12. $\dfrac{x}{3(4 - 1)} = 2$

SECTION 14.2

Exercises 13–16: Multiply and write in lowest terms.

13. $\dfrac{x - 3}{x + 1} \cdot \dfrac{2x + 2}{x - 3}$

14. $\dfrac{2x + 5}{(x + 5)(x - 1)} \cdot \dfrac{x - 1}{2x + 5}$

15. $\dfrac{z + 3}{z - 4} \cdot \dfrac{z - 4}{(z + 3)^2}$

16. $\dfrac{x^2}{x^2 - 4} \cdot \dfrac{x + 2}{x}$

Exercises 17–20: Divide and write in lowest terms.

17. $\dfrac{x + 1}{2x} \div \dfrac{3x + 3}{5x}$

18. $\dfrac{4}{x^3} \div \dfrac{x + 1}{2x^2}$

19. $\dfrac{x - 5}{x + 2} \div \dfrac{2x - 10}{x + 2}$

20. $\dfrac{x^2 - 6x + 5}{x^2 - 25} \div \dfrac{x - 1}{x + 5}$

SECTION 14.3

Exercises 21–28: Add or subtract and write in lowest terms.

21. $\dfrac{2}{x + 10} + \dfrac{8}{x + 10}$

22. $\dfrac{9}{x - 1} - \dfrac{8}{x - 1}$

23. $\dfrac{x + 2y}{2x} + \dfrac{x - 2y}{2x}$

24. $\dfrac{x}{x - 3} - \dfrac{3}{x - 3}$

25. $\dfrac{x}{x^2 - 1} - \dfrac{1}{x^2 - 1}$

26. $\dfrac{2x}{x^2 - 25} + \dfrac{10}{x^2 - 25}$

27. $\dfrac{3}{xy} - \dfrac{1}{xy}$

28. $\dfrac{x + y}{2y} + \dfrac{x - y}{2y}$

SECTION 14.4

Exercises 29–32: Find the least common multiple for the expressions. Leave your answer in factored form.

29. $3x,\ 5x$

30. $5x^2,\ 10x$

31. $x,\ x - 5$

32. $10x^2,\ x^2 - x$

Exercises 33–36: Rewrite the rational expression by using the given denominator D.

33. $\dfrac{3}{8},\ D = 24$

34. $\dfrac{4}{3x},\ D = 12x$

35. $\dfrac{3x}{x - 2},\ D = x^2 - 4$

36. $\dfrac{2x}{2x - 3},\ D = 2x^2 + x - 6$

Exercises 37–46: Simplify the expression. Write your answer in lowest terms.

37. $\dfrac{5}{8} + \dfrac{1}{6}$

38. $\dfrac{3}{4x} + \dfrac{1}{x}$

39. $\dfrac{5}{9x} - \dfrac{2}{3x}$

40. $\dfrac{7}{x - 1} - \dfrac{3}{x}$

41. $\dfrac{1}{x + 1} + \dfrac{1}{x - 1}$

42. $\dfrac{4}{3x^2} - \dfrac{3}{2x}$

43. $\dfrac{1 + x}{3x} - \dfrac{3}{2x}$

44. $\dfrac{x}{x^2 - 1} - \dfrac{1}{x - 1}$

45. $\dfrac{2}{x - y} - \dfrac{3}{x + y}$

46. $\dfrac{x}{y - x} + \dfrac{y}{x - y}$

SECTION 14.5

Exercises 47–54: Simplify the complex fraction.

47. $\dfrac{\frac{3}{4}}{\frac{7}{11}}$

48. $\dfrac{\frac{x}{5}}{\frac{2x}{7}}$

49. $\dfrac{\frac{m}{n}}{\frac{2m}{n^2}}$

50. $\dfrac{\frac{3}{p - 1}}{\frac{1}{p + 1}}$

51. $\dfrac{\frac{1}{2x} - \frac{1}{3x}}{\frac{2}{3x} - \frac{1}{6x}}$

52. $\dfrac{\frac{2}{xy} - \frac{1}{y}}{\frac{2}{xy} + \frac{1}{y}}$

53. $\dfrac{\frac{1}{x} - \frac{1}{x+1}}{\frac{x}{x+1}}$

54. $\dfrac{\frac{2}{x-1} - \frac{1}{x+1}}{\frac{1}{x^2 - 1}}$

SECTION 14.6

Exercises 55–58: Solve and check your answer.

55. $\dfrac{x}{5} = \dfrac{4}{7}$

56. $\dfrac{4}{x} = \dfrac{3}{2}$

57. $\dfrac{3}{z+1} = \dfrac{1}{2z}$

58. $\dfrac{x+2}{x} = \dfrac{3}{5}$

Exercises 59–70: If possible, solve. Check your answer.

59. $\dfrac{1}{5x} + \dfrac{3}{5x} = \dfrac{1}{5}$

60. $\dfrac{1}{x-1} + \dfrac{2x}{x-1} = 1$

61. $\dfrac{1}{x} + \dfrac{2}{3x} = \dfrac{1}{3}$

62. $\dfrac{1}{x+3} + \dfrac{2x}{x+3} = \dfrac{3}{2}$

63. $\dfrac{5}{x} - \dfrac{3}{x+1} = \dfrac{1}{2}$

64. $\dfrac{1}{x-1} - \dfrac{1}{x+1} = \dfrac{1}{4}$

65. $\dfrac{4}{p} - \dfrac{5}{p+2} = 0$

66. $\dfrac{1}{x-3} - \dfrac{1}{x+3} = \dfrac{1}{x^2 - 9}$

67. $\dfrac{1}{x+1} = \dfrac{-x}{x+1}$

68. $\dfrac{2}{x} = \dfrac{2}{x^2 + x} - 4$

69. $\dfrac{2}{x^2 - 2x} + \dfrac{1}{x^2 - 4} = \dfrac{1}{x^2 + 2x}$

70. $\dfrac{3}{x^2 - 3x} - \dfrac{1}{x^2 - 9} = \dfrac{1}{x^2 + 3x}$

Exercises 71 and 72: Do the following.
(a) Decide whether you are given an expression or an equation.
(b) If you are given an expression, simplify it. If you are given an equation, solve it.

71. $\dfrac{4}{x} - x = 0$

72. $\dfrac{x^2}{x-3} - \dfrac{9}{x-3}$

Exercises 73 and 74: Solve for the specified variable.

73. $\dfrac{1}{a} + \dfrac{2}{b} = \dfrac{3}{c}$ for b

74. $y = \dfrac{x}{x-1}$ for x

SECTION 14.7

Exercises 75 and 76: Solve the proportion.

75. $\dfrac{x}{6} = \dfrac{1}{3}$

76. $\dfrac{5}{x} = \dfrac{7}{3}$

Exercises 77 and 78: **Proportions** *Do the following.*
(a) Write a proportion that models the situation.
(b) Solve the proportion for x.

77. A rectangle has sides of 6 and 13. In a similar rectangle, the longer side is 20 and the shorter side is x.

78. If you earn $341 in 11 hours, then you can earn x dollars in 8 hours.

Exercises 79 and 80: **Direct Variation** *Suppose that y is directly proportional to x.*
(a) Use the given information to find the constant of proportionality k.
(b) Then use $y = kx$ to find y for $x = 5$.

79. $y = 8$ when $x = 2$

80. $y = 21$ when $x = 7$

Exercises 81 and 82: **Inverse Variation** *Suppose that y is inversely proportional to x.*
(a) Use the given information to find the constant of proportionality k.
(b) Then use $y = \frac{k}{x}$ to find y for $x = 5$.

81. $y = 2.5$ when $x = 4$

82. $y = 7$ when $x = 3$

83. **Joint Variation** Suppose that z varies jointly with x and y. If $z = 483$ when $x = 23$ and $y = 7$, find the constant of variation k.

84. **Joint Variation** Suppose that z varies jointly with x and the square of y. If $z = 891$ when $x = 22$ and $y = 3$, find z when $x = 10$ and $y = 4$.

Exercises 85 and 86: Do the following.
(a) Determine whether the data represent direct or inverse variation.
(b) Find an equation that models the data.
(c) Graph the data and your equation.

85.

x	2	3	4	5
y	30	20	15	12

86.

x	2	4	6	8
y	6	12	18	24

Exercises 87 and 88: Use the graph to determine whether the data represent direct or inverse variation. Find the constant of variation.

87.

88.

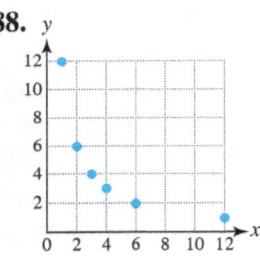

APPLICATIONS

89. **Modeling Traffic Flow** Fifteen vehicles per minute can pass through an intersection. If vehicles arrive randomly at an average rate of x per minute, the average waiting time T in minutes is given by $T = \frac{1}{15 - x}$ for $x < 15$. (*Source:* N. Garber, *Traffic and Highway Engineering*.)
    (a) Find T when $x = 10$ and interpret the result.
    (b) Complete the table.

x	5	10	13	14	14.9
T					

    (c) What happens to the waiting time as the traffic rate x approaches 15 vehicles per minute?

90. **Distance and Time** A car traveled at 50 miles per hour for 150 miles and then traveled at 75 miles per hour for 150 miles. What was the car's average speed?

91. **Emptying a Swimming Pool** A large pump can empty a swimming pool in 100 hours, whereas a small pump can empty the pool in 160 hours. How long will it take to empty the pool if both pumps are used?

92. **Running** Two athletes run 10 miles. One of the athletes runs 2 miles per hour faster and finishes 10 minutes ahead of the other athlete. Find the average speed of each athlete.

93. **River Current** A boat can travel 16 miles upstream in the same time that it can travel 48 miles downstream. If the speed of the current is 4 miles per hour, find the boat's speed.

94. **Height of a Tree** A 5-foot-tall person has a 6-foot-long shadow, and a nearby tree has a 40-foot-long shadow. Estimate the tree's height.

95. **Transportation Costs** Use of a toll road varies inversely with the toll. If the toll is $0.25, then 400 vehicles use the road. Estimate the number of users for a toll of $0.50.

96. **Cost of Carpet** The cost of carpet is directly proportional to the amount of carpet purchased. If 17 square yards cost $612, find the cost of 13 square yards.

97. **Tightening a Bolt** The torque exerted on a nut by a wrench is inversely proportional to the length of the wrench's handle. Suppose that a 12-inch wrench can be used to tighten a nut by using 30 pounds of force. How much force is necessary to tighten the same nut by using a 10-inch wrench?

98. **Water Content in Snow** Twenty inches of extremely dry snow are equivalent to an inch of rain. Estimate the water content in 32 inches of this type of snow.

99. **Polar Plunge** When a person swims in extremely cold water, the water removes body heat 25 times faster than air at the same temperature. To be safe, a person has between 30 and 90 seconds to get out of the water. How long could a person safely remain in air at the same temperature?

100. **Strength of a Beam** The strength of a wood beam varies inversely with its length. A beam that is 18 feet long can support 900 pounds. How much weight can a similar beam support if its length is 21 feet?

101. **Wind Power** The electric power generated by a windmill varies jointly with the square of the diameter of the area swept out by the blades and the cube of the wind velocity. If a windmill with 6-foot-diameter blades and a 20-mile-per-hour wind generates 10,823 watts, how much power would be generated if the blades were 10 feet in diameter and the wind speed were 12 miles per hour?

102. **Strength of a Beam** The strength of a beam varies jointly with its width w and the square of its thickness t. If a beam 8 inches wide and 5 inches thick supports 650 pounds, how much can a similar beam 6 inches wide and 6 inches thick support?

CHAPTER 14 Test

1. Evaluate the expression $\frac{3x}{2x-1}$ for $x = 3$.

2. Find any x-value that makes $\frac{x-1}{x+2}$ undefined.

Exercises 3 and 4: Simplify the expression.

3. $\frac{x^2 - 25}{x - 5}$

4. $\frac{3x^2 - 15x}{3x}$

Exercises 5–10: Simplify the expression. Write your answer in lowest terms.

5. $\frac{x - 2}{x + 4} \cdot \frac{3x + 12}{x - 2}$

6. $\frac{z + 1}{z + 3} \cdot \frac{2z + 6}{z + 1}$

7. $\frac{x + 1}{5x} \div \frac{2x + 2}{x - 1}$

8. $\frac{2}{x^2} \div \frac{x + 3}{3x}$

9. $\frac{x}{x + 4} + \frac{3x + 1}{x + 4}$

10. $\frac{4t + 1}{2t - 3} - \frac{3t - 6}{2t - 3}$

11. Find the least common multiple for
$$6x^2 \text{ and } 3x^2 - 3x.$$

12. Rewrite the rational expression $\frac{4}{7x}$ by using the denominator $7x^2 - 7x$.

Exercises 13 and 14: Simplify the expression. Write your answer in lowest terms.

13. $\frac{1}{y^2 + y} - \frac{y - 1}{y^2 - y}$

14. $\frac{1}{xy} + \frac{x}{y} - \frac{1}{y^2}$

Exercises 15 and 16: Simplify the complex fraction.

15. $\dfrac{\frac{a}{3b}}{\frac{5a}{b^2}}$

16. $\dfrac{1 + \frac{1}{p - 1}}{1 - \frac{1}{p - 1}}$

Exercises 17–24: Solve the equation and check your answer.

17. $\frac{2}{7} = \frac{5}{x}$

18. $\frac{x + 3}{2x} = 1$

19. $\frac{1}{2x} + \frac{2}{5x} = \frac{9}{10}$

20. $\frac{1}{x - 1} + \frac{2}{x + 2} = \frac{3}{2}$

21. $\frac{1}{x^2 - 1} - \frac{4}{x + 1} = \frac{3}{x - 1}$

22. $\frac{1}{x^2 - 4x} + \frac{2}{x^2 - 16} = \frac{2}{x^2 + 4x}$

23. $\frac{x}{2x - 1} = \frac{1 - x}{2x - 1}$

24. $\frac{x}{x - 5} + \frac{x}{x + 5} = \frac{10x}{x^2 - 25}$

Exercises 25 and 26: Solve the equation for the specified variable.

25. $y = \frac{2}{3x - 5}$ for x

26. $\frac{a + b}{ab} = 1$ for b

27. Suppose that y is directly proportional to x.
    (a) If $y = 14$ when $x = 4$, find k so that $y = kx$.
    (b) Use $y = kx$ to find y for $x = 6$.

28. Use the table to determine whether y varies directly or inversely with x. Find the constant of variation.

x	2	4	8	16
y	16	8	4	2

29. **Emptying a Swimming Pool** It takes a large pump 40 hours to empty a swimming pool, whereas a small pump can empty the pool in 60 hours. How long will it take to empty the pool if both pumps are used?

30. **Height of a Building** A 5-foot-tall post has a 4-foot-long shadow, and a nearby building has a 54-foot-long shadow. Estimate the height of the building.

31. **Standing in Line** A department store clerk can wait on 30 customers per hour. If people arrive randomly at an average rate of x per hour, then the average number of customers N waiting in line is given by
$$N = \frac{x^2}{900 - 30x},$$
for $x < 30$. Evaluate the expression for $x = 24$ and interpret your result.

CHAPTERS 1–14 Cumulative Review Exercises

1. Evaluate $\pi r^2 h$ when $r = 2$ and $h = 6$.

2. Translate the phrase "two less than twice a number" into an algebraic expression using the variable x.

Exercises 3 and 4: Evaluate and simplify to lowest terms.

3. $\frac{1}{2} \div \frac{5}{4}$

4. $\frac{5}{8} + \frac{1}{8}$

Exercises 5 and 6: Simplify the expression.

5. $-2 + 7x + 4 - 5x$

6. $-4(4 - y) + (5 - 3y)$

7. Solve $-2x + 11 = 13$ and check the solution.

8. Solve $-3x + 1 \geq x$.

Exercises 9 and 10: Graph the equation. Determine any intercepts.

9. $2x - 3y = 6$

10. $x = 1$

11. Sketch a line with slope $m = 3$ passing through $(-2, -1)$. Write its slope–intercept form.

12. The table lists points located on a line. Write the slope–intercept form of the line.

x	-2	-1	0	1
y	-5	-3	-1	1

Exercises 13 and 14: Find the slope–intercept form for the line satisfying the given conditions.

13. Parallel to $y = -\frac{2}{3}x + 1$, passing through $(2, -1)$

14. Passing through $(-1, 2)$ and $(2, 4)$

15. Determine which ordered pair is a solution to the system of equations: $(2, -6)$ or $(1, -2)$.
    $$4x + y = 2$$
    $$x - 4y = 9$$

16. Solve the system of equations.
    $$-3r - t = 2$$
    $$2r + t = -4$$

Exercises 17 and 18: Determine if the system of linear equations has no solutions, one solution, or infinitely many solutions.

17. $2x - y = 5$
    $-2x + y = -5$

18. $4x - 6y = 12$
    $-6x + 9y = 18$

Exercises 19–22: Simplify the expression.

19. $3z^2 \cdot 5z^6$

20. $(ab)^3$

21. $(2y - 3)(5y + 2)$

22. $(x^2 - y^2)^2$

Exercises 23 and 24: Simplify and write the expression using positive exponents.

23. $(3x^2)^{-3}$

24. $\frac{4x^2}{2x^4}$

25. Write 0.00123 in scientific notation.

26. Divide and check: $\frac{2x^2 - x + 3}{x - 1}$.

Exercises 27–30: Factor completely.

27. $6 + 13x - 5x^2$

28. $9z^2 - 4$

29. $t^2 + 16t + 64$

30. $x^3 - 16x$

Exercises 31 and 32: Solve the equation.

31. $y^2 + 5y - 14 = 0$

32. $x^3 = 4x$

Exercises 33 and 34: Simplify.

33. $\frac{3x}{4y} \cdot \frac{y}{9x^2}$

34. $\frac{x}{x^2 - 4} \div \frac{2x}{x - 2}$

35. Simplify $\dfrac{1 + \frac{2}{x}}{1 - \frac{2}{x}}$.

36. Solve $z = 3x - 2y$ for x.

Exercises 37 and 38: Solve.

37. $\dfrac{4}{3x} - \dfrac{3}{4x} = 1$

38. $\dfrac{1}{x - 1} + \dfrac{2}{x + 2} = \dfrac{3}{2}$

39. Suppose that y is directly proportional to x and that $y = 7$ when $x = 14$. Find y when $x = 11$.

40. Determine whether the data in the table represent direct or inverse variation. Find an equation that models the data.

x	1	2	4	10
y	20	10	5	2

APPLICATIONS

41. Shoveling a Driveway Two people are shoveling snow from a driveway. The first person shovels 12 square feet per minute, while the second person shovels 9 square feet per minute.
  (a) Write a simplified expression that gives the total square feet that the two people shovel together in x minutes.
  (b) How long would it take them to clear a driveway with 1890 square feet?

42. Burning Calories An athlete can burn 10 calories per minute while running and 4 calories per minute while walking. If the athlete burns 450 calories in 60 minutes, how long is spent on each activity?

43. Triangle In an isosceles triangle, the measures of the two smaller angles are equal and their sum is 32° more than the largest angle.
  (a) Let x be the measure of one of the two smaller angles and y be the measure of the largest angle. Write a system of linear equations whose solution gives the measures of these angles.
  (b) Solve your system.

44. Strength of a Beam The strength of a wood beam varies inversely with its length. A beam that is 10 feet long can support 1100 pounds. How much weight can a similar beam support if it is 22 feet long?

15 Introduction to Functions

- 15.1 Functions and Their Representations
- 15.2 Linear Functions
- 15.3 Compound Inequalities and Piecewise-Defined Functions
- 15.4 Other Functions and Their Properties
- 15.5 Absolute Value Equations and Inequalities

The trend today is to use more green building material. As a result, demand for green building material is expected to increase dramatically, particularly bamboo. Bamboo doesn't need pesticides or much water, it pulls large amounts of carbon dioxide out of the air, and it can be harvested every 5–10 years. One essential reason why bamboo is valuable is because it grows at an astonishing rate, as much as 2 inches per hour.

In Examples 6 and 7, in the second section of this chapter, the mathematical concept of a *function* is used to describe these facts about bamboo and green building material. However, functions can also be used to describe other things like human behavior and social networks, which are discussed in Example 10, in the fourth section of this chapter.

15.1 Functions and Their Representations

Objectives

1. Introducing Functions
2. Recognizing Representations of Functions
   - Verbal Representation (Words)
   - Numerical Representation (Table of Values)
   - Symbolic Representation (Formula)
   - Graphical Representation (Graph)
   - Diagrammatic Representation (Diagram)
   - Evaluating Representations of Functions
3. Defining a Function
   - Relations
   - Functions
   - Domain and Range
4. Identifying a Function
   - Ordered Pairs and Tables as Functions
   - Vertical Line Test
5. Using Graphing Calculators
   - Scatterplots
   - Graphs and Tables

NEW VOCABULARY

☐ Function
☐ Function notation
☐ Input/output
☐ Name of the function
☐ Dependent variable
☐ Independent variable
☐ Verbal representation
☐ Numerical representation
☐ Symbolic representation
☐ Graphical representation
☐ Diagrams/diagrammatic representation
☐ Relation
☐ Domain/range
☐ Nonlinear functions
☐ Vertical line test

1 Introducing Functions

Connecting Concepts with Your Life Functions are used to calculate many important quantities. For example, suppose that a person works for $7 per hour. Then we could use a function *named f* to calculate the amount of money the person earns after working x hours simply by multiplying the *input x* by 7. The result y is called the *output*. This concept is shown visually in the following diagram.

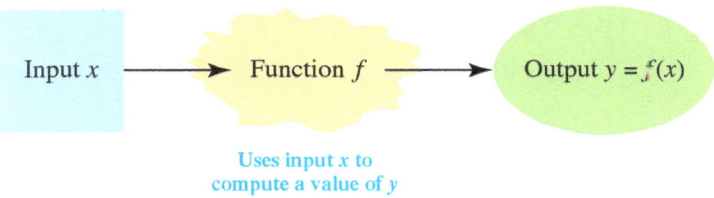

Uses input x to compute a value of y

For each valid input x, a function computes *exactly one* output y. This can be represented by the ordered pair (x, y). If the input is 5 hours, f outputs $7 \cdot 5 = \$35$; if the input is 8 hours, f outputs $7 \cdot 8 = \$56$. These results can be represented by the ordered pairs $(5, 35)$ and $(8, 56)$. Sometimes an input may not be valid. For example, if $x = -3$, there is no reasonable output because a person cannot work -3 hours. The set of all valid inputs is called the **domain** of the function and the set of corresponding outputs is called the **range** of the function.

We say that *y is a function of x* because the output y is determined by and *depends* on the input x. As a result, y is called the *dependent variable* and x is the *independent variable*. To emphasize that y is a function of x, we use the notation $y = f(x)$. The symbol $f(x)$ does not represent multiplication of a variable f and a variable x. The notation $y = f(x)$ is called *function notation*, is read "y equals f of x," and means that function f with input x produces output y. For example, if $x = 3$ hours, $y = f(3) = 7 \cdot 3 = \$21$.

The following See the Concept summarizes these basic concepts about a function.

See the Concept 1 — **FUNCTION NOTATION**

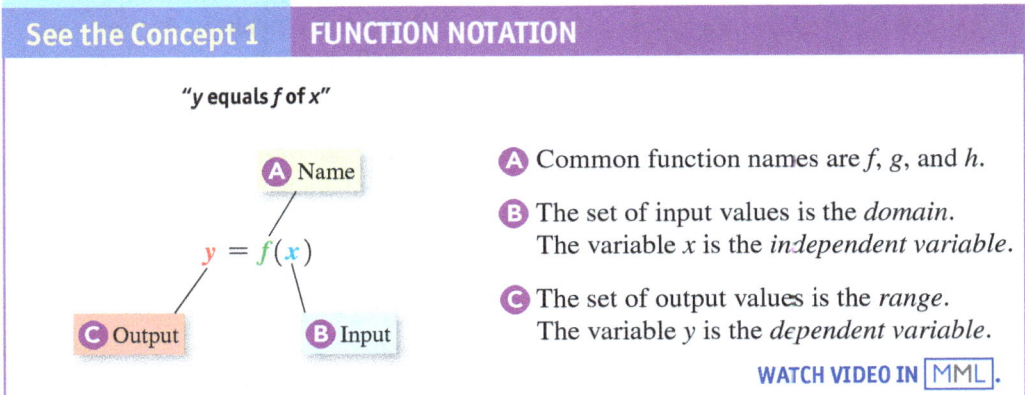

Ⓐ Common function names are f, g, and h.

Ⓑ The set of input values is the *domain*.
The variable x is the *independent variable*.

Ⓒ The set of output values is the *range*.
The variable y is the *dependent variable*.

WATCH VIDEO IN MML.

NOTE: Functions can be given *meaningful* names and variables. For example, function f could have been defined by $P(h) = 7h$, where function P calculates the pay after working h hours for $7 per hour. ■

Connecting Concepts with Your Life Functions can be used to compute a variety of quantities. For example, suppose that a boy has a sister who is exactly 5 years older than he is. If the age of the boy is x, then a function g can calculate the age of his sister by adding 5 to x. Thus $g(4) = 4 + 5 = 9$, $g(10) = 10 + 5 = 15$, and in general $g(x) = x + 5 = y$. That is, function g adds 5 to input x to obtain the output $y = g(x)$.

Functions can be represented by an input–output machine, as illustrated in **FIGURE 15.1**. This machine represents function g and receives input $x = 4$, adds 5 to this value, and then outputs $g(4) = 4 + 5 = 9$.

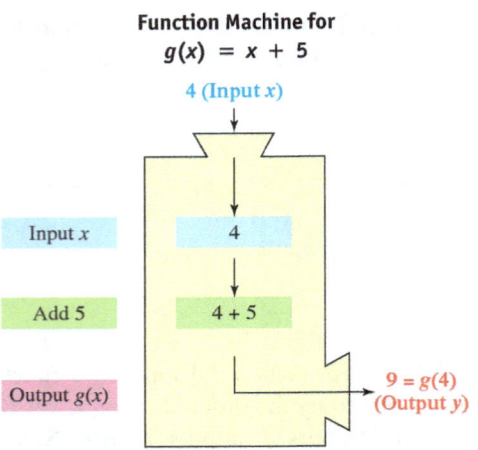

FIGURE 15.1

2 Recognizing Representations of Functions

 Connecting Concepts with Your Life A function f forms a relation between inputs x and outputs y that can be represented verbally, numerically, symbolically, and graphically. Functions can also be represented with diagrams. We begin by considering a function f that converts yards to feet.

VERBAL REPRESENTATION (WORDS) To convert x yards to y feet, we multiply x by 3. Therefore, if function f computes the number of feet in x yards, a **verbal representation** of f is "Multiply the input x in yards by 3 to obtain the output y in feet."

NUMERICAL REPRESENTATION (TABLE OF VALUES) A function f that converts yards to feet is shown in **TABLE 15.1**, where $y = f(x)$.

A *table of values* is called a **numerical representation** of a function. Many times it is impossible to list all valid inputs x in a table. On the one hand, if a table does not contain every x-input, it is a *partial* numerical representation. On the other hand, a *complete* numerical representation includes *all* valid inputs. **TABLE 15.1** is a partial numerical representation of f because many valid inputs, such as $x = 10$ or $x = 5.3$, are not shown in it. Note that for each valid input x, there is exactly one output y. *For a function, inputs are not listed more than once in a table.*

Yards to Feet

x (yards)	y (feet)
1	3
2	6
3	9
4	12
5	15
6	18
7	21

TABLE 15.1

SYMBOLIC REPRESENTATION (FORMULA) A *formula* can be used to provide a **symbolic representation** of a function. The computation performed by f to convert x yards to y feet is expressed by $y = 3x$. A formula for f is $f(x) = 3x$, where $y = f(x)$. We say that function f is *defined by* or *given by* $f(x) = 3x$. Thus $f(2) = 3 \cdot 2 = 6$.

GRAPHICAL REPRESENTATION (GRAPH) A **graphical representation**, or **graph**, visually associates an x-input with a y-output. The ordered pairs

$(1, 3), (2, 6), (3, 9), (4, 12), (5, 15), (6, 18),$ and $(7, 21)$

from **TABLE 15.1** are plotted in **FIGURE 15.2(a)**. This scatterplot suggests a line for the graph of f. For each real number x, there is exactly one real number y determined by $y = 3x$. If we restrict inputs to $x \geq 0$ and plot all ordered pairs $(x, 3x)$, then a line with no breaks will appear, as shown in **FIGURE 15.2(b)**.

STUDY TIP

Be sure that you understand what verbal, numerical, graphical, and symbolic representations are.

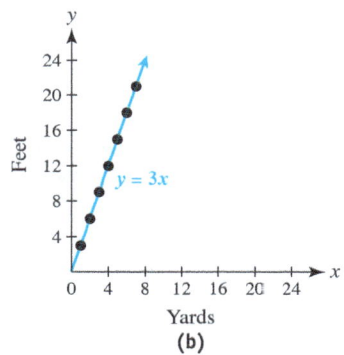

FIGURE 15.2

Because $f(1) = 3$, it follows that the point $(1, 3)$ lies on the graph of f, as shown in **FIGURE 15.3**. Graphs can sometimes be used to define a function f. For example, because the point $(1, 3)$ lies on the graph of f in **FIGURE 15.3**, we can conclude that $f(1) = 3$. That is, each point on the graph of f defines an input–output pair for f. Because there are infinitely many points on the graph of f, there are also infinitely many input-output pairs for function f.

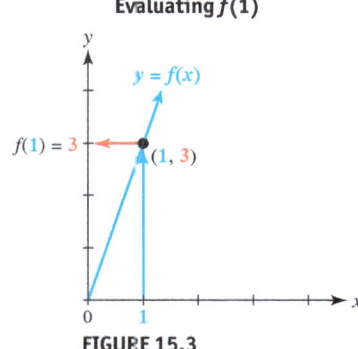

FIGURE 15.3

MAKING CONNECTIONS 1

Functions, Points, and Graphs

If $f(a) = b$, then the point (a, b) lies on the graph of f. Conversely, if the point (a, b) lies on the graph of f, then $f(a) = b$. See **FIGURE 15.4(a)**. Thus each point on the graph of f can be written in the form $(a, f(a))$. See **FIGURE 15.4(b)**.

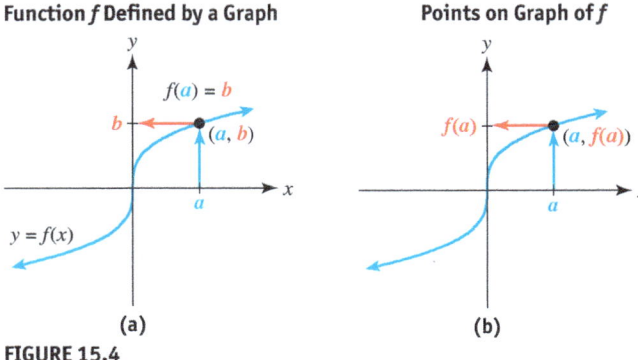

FIGURE 15.4

In the next example, we find a formula and then sketch a graph of a function.

EXAMPLE 1 Finding representations of a function

Let function f square the input x and then subtract 1 to obtain the output y.
(a) Write a formula, or symbolic representation, for f.
(b) Make a table of values, or numerical representation, for f. Use $x = -2, -1, 0, 1, 2$.
(c) Sketch a graph, or graphical representation, of f.

Solution
(a) **Symbolic Representation** If we square x and then subtract 1, we obtain $x^2 - 1$. Thus a formula for f is $f(x) = x^2 - 1$.
(b) **Numerical Representation** Make a table of values for $f(x)$, as shown in **TABLE 15.2**. For example,
$$f(-2) = (-2)^2 - 1 = 4 - 1 = 3.$$
(c) **Graphical Representation** To obtain a graph of $f(x) = x^2 - 1$, plot the points from **TABLE 15.2** and then connect them with a smooth curve, as shown in **FIGURE 15.5**. Note that we need to plot enough points so that we can determine the overall shape of the graph.

$f(x) = x^2 - 1$

x	$f(x)$
-2	3
-1	0
0	-1
1	0
2	3

TABLE 15.2

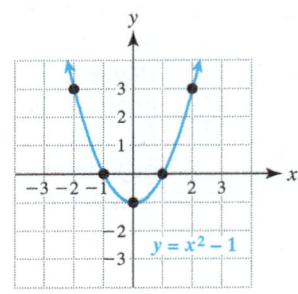

FIGURE 15.5

READING CHECK 1

• Give a verbal and a symbolic representation of a function that calculates the number of days in a given number of weeks. Choose meaningful variables.

Now Try Exercise 63

The following See the Concept illustrates four basic representations of a function.

See the Concept 2 **FOUR REPRESENTATIONS OF A FUNCTION**

Symbolic	Numerical		Graphical	Verbal
$f(x) = x^2 + 1$	x	y		Function f squares the input x and then adds 1 to produce the output y.
	-2	5		
	-1	2		
	0	1		
	1	2		
	2	5		

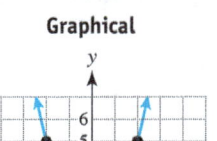

To get the table, evaluate $f(x) = x^2 + 1$ for x equal to $-2, -1, 0, 1,$ and 2.

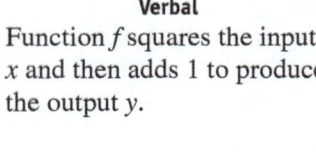

Plot the points from the table, and sketch a smooth curve.

WATCH VIDEO IN MML.

Yards to Feet

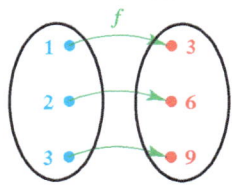

Function
FIGURE 15.6

DIAGRAMMATIC REPRESENTATION (DIAGRAM) Functions may be represented by **diagrams**. **FIGURE 15.6** is a diagram of a function where an arrow is used to identify the output y associated with input x. For example, an arrow is drawn from input **2** to output **6**, which is written in function notation as $f(2) = 6$. That is, **2** yards are equivalent to **6** feet.

FIGURE 15.7(a) shows a (different) function f even though $f(1) = 4$ and $f(2) = 4$. Although two inputs for f have the same output, each valid input has exactly one output. In contrast, **FIGURE 15.7(b)** is *not* a function because input 2 results in two different outputs, 5 and 6.

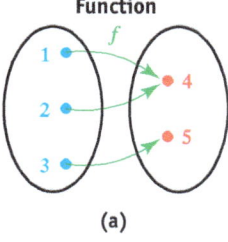

(a)   (b)
FIGURE 15.7

EVALUATING REPRESENTATIONS OF FUNCTIONS In the next three examples, we evaluate several representations of functions.

EXAMPLE 2   **Evaluating symbolic representations (formulas)**

Evaluate each function f at the given value of x.
(a) $f(x) = 3x - 7$ $x = -2$
(b) $f(x) = \dfrac{x}{x + 2}$ $x = 0.5$
(c) $f(x) = \sqrt{x - 1}$ $x = 10$

Solution
(a) $f(-2) = 3(-2) - 7 = -6 - 7 = -13$
(b) $f(0.5) = \dfrac{0.5}{0.5 + 2} = \dfrac{0.5}{2.5} = 0.2$
(c) $f(10) = \sqrt{10 - 1} = \sqrt{9} = 3$

Now Try Exercises 21, 23, 31

Math in Context (Tax) In the next example, we calculate sales tax by evaluating different representations of a function.

EXAMPLE 3   **Calculating sales tax**

Let a function f compute a sales tax of 7% on a purchase of x dollars. Use the given representation to evaluate $f(2)$.
(a) **Verbal Representation** Multiply a purchase of x dollars by 0.07 to obtain a sales tax of y dollars.
(b) **Numerical Representation (partial)** Shown in **TABLE 15.3** on the next page
(c) **Symbolic Representation** $f(x) = 0.07x$
(d) **Graphical Representation** Shown in **FIGURE 15.8** on the next page
(e) **Diagrammatic Representation** Shown in **FIGURE 15.9** on the next page

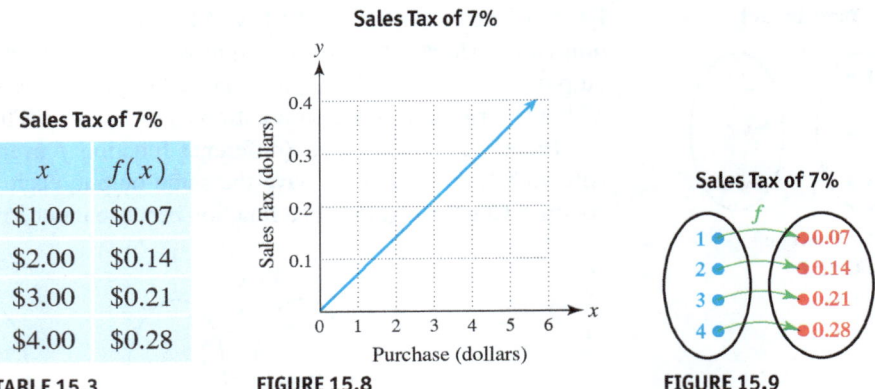

Sales Tax of 7%

x	$f(x)$
$1.00	$0.07
$2.00	$0.14
$3.00	$0.21
$4.00	$0.28

TABLE 15.3

FIGURE 15.8

FIGURE 15.9

Solution

(a) **Verbal** To evaluate $f(2)$, multiply the input 2 by 0.07 to obtain 0.14. The sales tax on a $2.00 purchase is $0.14.

(b) **Numerical** From **TABLE 15.3**, $f(2) = \$0.14$.

(c) **Symbolic** Because $f(x) = 0.07x$, $f(2) = 0.07(2) = 0.14$, or $0.14.

(d) **Graphical** To evaluate $f(2)$ with a graph, first find **2** on the x-axis in **FIGURE 15.10**. Then move vertically upward until you reach the graph of f. The point on the graph may be estimated as $(2, 0.14)$, meaning that $f(2) = 0.14$. Note that it may not be possible to find the exact answer from a graph. For example, one might estimate $f(2)$ to be 0.13 or 0.15 instead of 0.14.

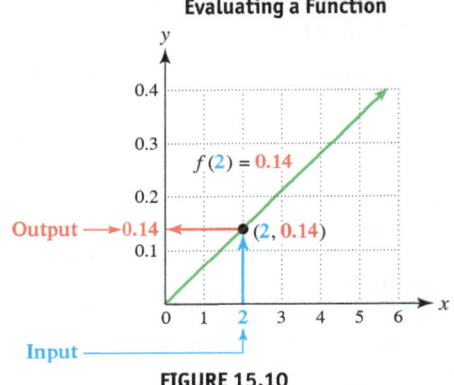

FIGURE 15.10

(e) **Diagrammatic** In **FIGURE 15.9**, follow the arrow from **2** to **0.14**. Thus $f(2) = 0.14$.

Now Try Exercises 25, 33, 53, 59, 61

The following See the Concept summarizes how to evaluate a function graphically.

See the Concept 3   EVALUATING A FUNCTION GRAPHICALLY

If $f(x) = x^2 - 2x$, evaluate $f(-1)$ graphically.

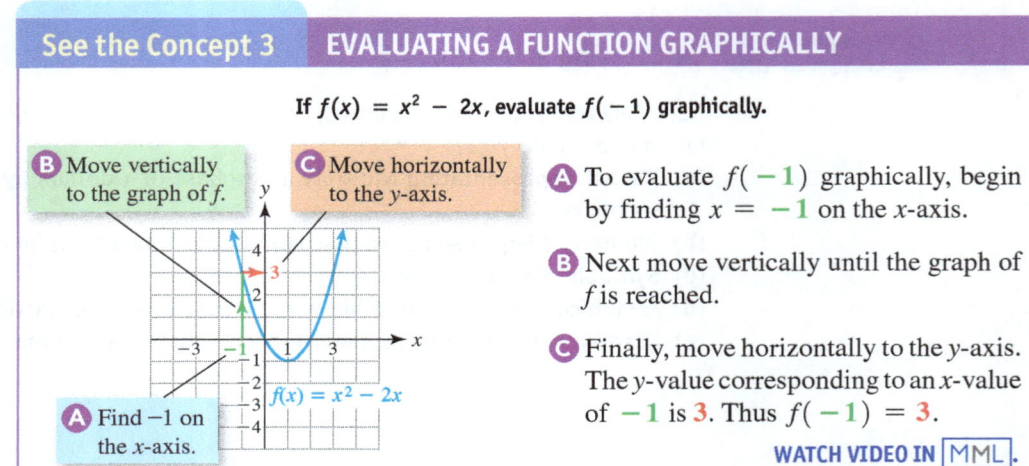

A Find -1 on the x-axis.

B Move vertically to the graph of f.

C Move horizontally to the y-axis.

A To evaluate $f(-1)$ graphically, begin by finding $x = -1$ on the x-axis.

B Next move vertically until the graph of f is reached.

C Finally, move horizontally to the y-axis. The y-value corresponding to an x-value of -1 is **3**. Thus $f(-1) = 3$.

WATCH VIDEO IN MML.

Math in Context (Medical) There are many examples of functions. To give more meaning to a function, sometimes we change both its name and its input variable. For instance, if we know the radius r of a circle, we can calculate its circumference by using $C(r) = 2\pi r$. The next example illustrates how functions are used in physical therapy.

EXAMPLE 4 **Computing crutch length**

People who sustain leg injuries often require crutches. A proper crutch length can be estimated without using trial and error. The function L, given by $L(t) = 0.72t + 2$, outputs an appropriate crutch length L in inches for a person t inches tall. (*Source: Journal of the American Physical Therapy Association.*)

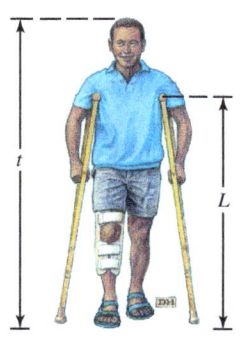

(a) Find $L(60)$ and interpret the result.
(b) If one person is 70 inches tall and another person is 71 inches tall, what should be the difference in their crutch lengths?

Solution
(a) $L(60) = 0.72(60) + 2 = \mathbf{45.2}$. Thus a person 60 inches tall needs crutches that are about 45.2 inches long.
(b) From the formula $L(t) = 0.72t + 2$, we can see that each 1-inch increase in t results in a 0.72-inch increase in $L(t)$. For example,

$$L(71) - L(70) = 53.12 - 52.4 = 0.72.$$

Now Try Exercise 75

3 Defining a Function

RELATIONS Sometimes there is a relationship between two quantities or variables, such as the cost y of parking x hours. We can represent this relationship by using a set of ordered pairs. For example, it might cost $2 to park 1 hour, $4 to park 2 hours, and $8 to park from 3 to 5 hours. Thus, the ordered pairs $(1, 2)$, $(2, 4)$, $(3, 8)$, $(4, 8)$, and $(5, 8)$ are elements or members of this relation.

A **relation** is a set of ordered pairs. The sets C and S are examples of relations.

$$C = \{(1, 2), (2, 4), (3, 8), (4, 8), (5, 8)\} \quad \text{and} \quad S = \{(1, 1), (3, 4), (3, 5)\}$$

The *domain* of a relation is the set of x-values and the *range* of a relation is the set of y-values. For example, the domain of S is $\{1, 3\}$ and the range is $\{1, 4, 5\}$.

FUNCTIONS A function is a fundamental concept in mathematics. Its definition should allow for all representations of a function. *A function receives an input x and produces exactly one output y*, which can be expressed as an ordered pair:

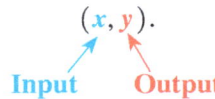

A *relation* is a set of ordered pairs, and a function is a special type of relation.

FUNCTION

A **function** f is a set of ordered pairs (x, y), where each x-value corresponds to exactly one y-value.

The *domain* of f is the set of all x-values, and the *range* of f is the set of all y-values. For example, a function f that converts 1, 2, 3, and 4 yards to feet could be expressed as

$$f = \{(1, 3), (2, 6), (3, 9), (4, 12)\}.$$

The domain of f is $D = \{1, 2, 3, 4\}$, and the range of f is $R = \{3, 6, 9, 12\}$.

> **MAKING CONNECTIONS 2**
>
> **Relations and Functions**
>
> A relation can be thought of as a set of input–output pairs. A function is a special type of relation whereby each input results in *exactly one output*.

Math in Education Context In the next example, we see how education can improve a person's chances for earning a higher income.

EXAMPLE 5 Computing average income

A function f computes the median individual income in dollars in relation to educational attainment. This function is defined by $f(N) = 25{,}376$, $f(H) = 34{,}736$, $f(B) = 57{,}252$, and $f(M) = 68{,}952$, where N denotes no diploma, H a high school diploma, B a bachelor's degree, and M a master's degree. (*Source:* Bureau of Labor.)

(a) Write f as a set of ordered pairs.
(b) Give the domain and range of f.
(c) Discuss the relationship between education and income.

Solution
(a) $f = \{(N, 25376), (H, 34736), (B, 57252), (M, 68952)\}$
(b) The domain of function f is given by $D = \{N, H, B, M\}$, and the range of function f is given by $R = \{25376, 34736, 57252, 68952\}$.
(c) Education pays—the greater the educational attainment, the greater annual earnings are.

Now Try Exercise 101

DOMAIN AND RANGE We can determine the domain and range of a function from either a graph or a formula, as shown in the next two examples.

EXAMPLE 6 Finding the domain and range graphically

Use the graphs of f shown in **FIGURES 15.11** and **15.12** to find each function's domain and range.

(a)

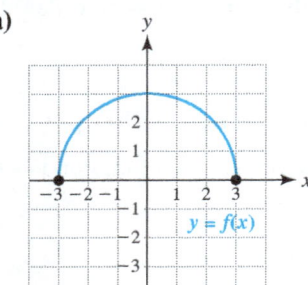

FIGURE 15.11

(b)

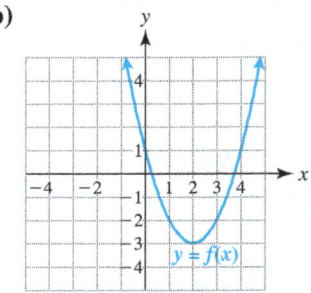

FIGURE 15.12

READING CHECK 2

- Use the graph in FIGURE 15.12 to evaluate $f(3)$.

Solution

(a) The domain is the set of all x-values that correspond to points on the graph of f. **FIGURE 15.13** shows that the domain D includes all x-values satisfying $-3 \leq x \leq 3$. (Recall that the symbol $\leq$ is read "*less than or equal to*.") Because the graph is a semicircle with no breaks, the domain includes all real numbers between and including -3 and 3. The range R is the set of y-values that correspond to points on the graph of f. Thus R includes all y-values satisfying $0 \leq y \leq 3$.

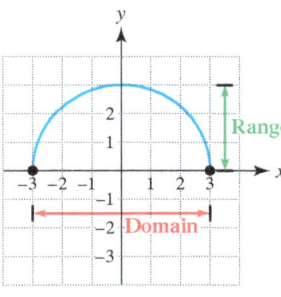

FIGURE 15.13

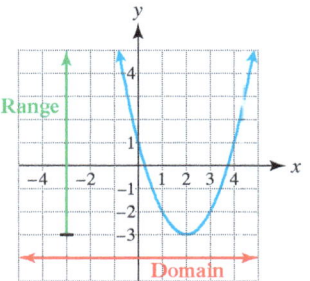

FIGURE 15.14

(b) The arrows on the ends of the graph in **FIGURE 15.12** indicate that the graph extends indefinitely left and right, as well as upward. Thus D includes **all real numbers**. See **FIGURE 15.14**. The smallest y-value on the graph is $y = -3$, which occurs when $x = 2$. Thus the range R is $y \geq -3$. (Recall that the symbol $\geq$ is read "*greater than or equal to*.")

Now Try Exercises 77, 81

CRITICAL THINKING 1

Suppose that a car travels at 50 miles per hour to a city that is 250 miles away. Sketch a graph of a function f that gives the distance y traveled after x hours. Identify the domain and range of f.

The domain of a function is the set of all valid inputs. To determine the domain of a function from a formula, we must find x-values for which the formula is defined. To do this, we must determine if we can substitute any real number in the formula for $f(x)$. If we can, then the domain of f is *all real numbers*. However, there are situations in which we must limit the domain of f. For example, the domain must often be limited when there is either division or a square root in the formula for f. When division occurs, we must be careful to avoid values of the variable that result in division by 0, which is undefined. When a square root occurs, we must be careful to avoid values of the variable that result in the square root of a negative number, which is not a real number. This concept is demonstrated in the next example.

EXAMPLE 7 Finding the domain of a function

Use $f(x)$ to find the domain of f.

(a) $f(x) = 5x$ (b) $f(x) = \dfrac{1}{x - 2}$ (c) $f(x) = \sqrt{x}$

Solution

(a) Because we can always multiply a real number x by 5, $f(x) = 5x$ is defined for all real numbers. Thus the domain of f includes all real numbers.
(b) Because we cannot divide by 0, input $x = 2$ is not valid for $f(x) = \dfrac{1}{x-2}$. The expression for $f(x)$ is defined for all other values of x. Thus the domain of f includes all real numbers except 2, or $x \neq 2$.
(c) Because square roots of negative numbers are not real numbers, inputs for $f(x) = \sqrt{x}$ cannot be negative. Thus the domain of f includes all nonnegative numbers, or $x \geq 0$.

Now Try Exercises 87, 91, 95

Symbolic, numerical, and graphical representations of three common functions are shown in FIGURE 15.15. Note that their graphs are not lines. For this reason, they are called **nonlinear functions**. Use the graphs to find the domain and range of each function.

Absolute Value: $f(x) = |x|$

x	-2	-1	0	1	2
$|x|$	2	1	0	1	2

Square: $f(x) = x^2$

x	-2	-1	0	1	2
x^2	4	1	0	1	4

Square Root: $f(x) = \sqrt{x}$

x	0	1	4	9
$\sqrt{x}$	0	1	2	3

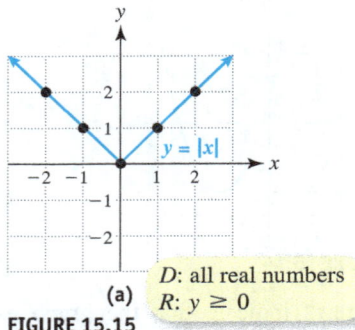

(a) D: all real numbers
$R: y \geq 0$

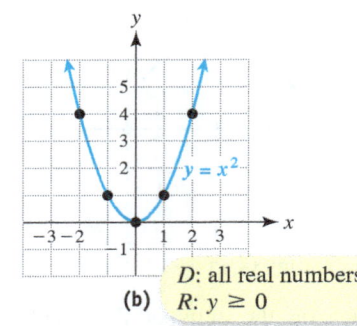

(b) D: all real numbers
$R: y \geq 0$

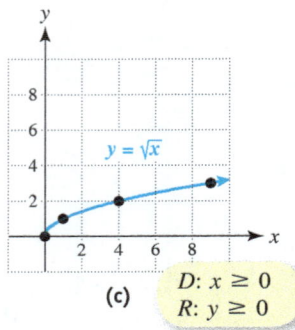

(c) $D: x \geq 0$
$R: y \geq 0$

FIGURE 15.15

4 Identifying a Function

ORDERED PAIRS AND TABLES AS FUNCTIONS Recall that for a function each valid input x produces exactly one output y. In the next three examples, we demonstrate techniques for identifying a function.

EXAMPLE 8 Determining whether a set of ordered pairs is a function

The set S of ordered pairs (x, y) represents the number of mergers and acquisitions y in a recent year for selected technology companies x.

$$S = \{(\text{IBM}, 12), (\text{HP}, 7), (\text{Oracle}, 5), (\text{Apple}, 5), (\text{Microsoft}, 0)\}$$

Determine if S is a function. (*Source:* cbinsights.)

Solution
The input x is the name of the technology company, and the output y is the number of mergers and acquisitions associated with that company. The set S *is* a function because each company x is associated with exactly one number y. Note that even though there were 5 mergers and acquisitions corresponding to both Oracle and Apple, S is nonetheless a function.

Now Try Exercise 131

EXAMPLE 9 Determining whether a table of values represents a function

Determine whether the relation in **TABLE 15.4** represents a function.

x	y
1	-4
2	8
3	2
1	5
4	-6

TABLE 15.4

Solution
The table does not represent a function because input $x = 1$ produces two outputs: -4 and 5. That is, the following two ordered pairs both belong to this relation.

Same input x

$(1, -4) \qquad (1, 5) \qquad \leftarrow$ Not a function

Different outputs y

Now Try Exercise 133

VERTICAL LINE TEST To determine whether a graph represents a function, we must be convinced that it is impossible for an input x to have two or more outputs y. If two distinct points have the *same* x-coordinate on a graph, then the graph cannot represent a function. For example, the ordered pairs $(-1, 1)$ and $(-1, -1)$ could not be on the graph of a function because input -1 results in *two* outputs: 1 and -1. When the points $(-1, 1)$ and $(-1, -1)$ are plotted, they lie on the same vertical line, as shown in **FIGURE 15.16(a)**. A graph passing through these points intersects the vertical line twice, as illustrated in **FIGURE 15.16(b)**.

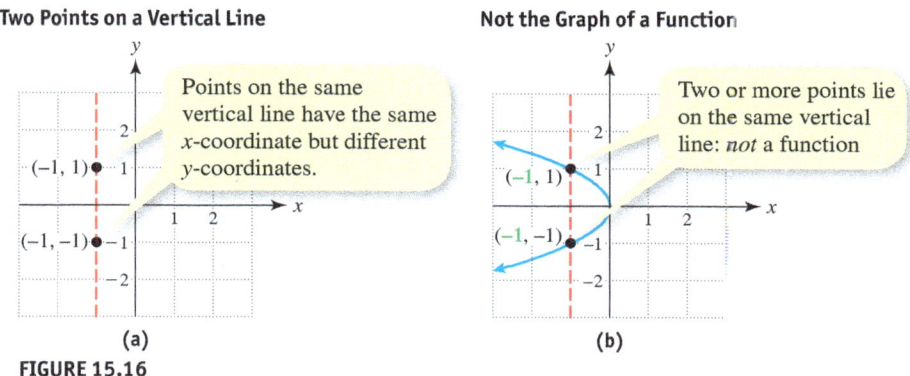

FIGURE 15.16

To determine whether a graph represents a function, visualize vertical lines moving across the xy-plane. If each vertical line intersects the graph *at most once*, then it is a graph of a function. This test is called the **vertical line test**. Note that the graph in **FIGURE 15.16(b)** fails the vertical line test and therefore does not represent a function.

READING CHECK 3

- What is the vertical line test used for?

VERTICAL LINE TEST

If every vertical line intersects a graph at no more than one point, then the graph represents a function.

EXAMPLE 10 Determining whether a graph represents a function

Determine whether the graphs shown in **FIGURE 15.17** represent functions.

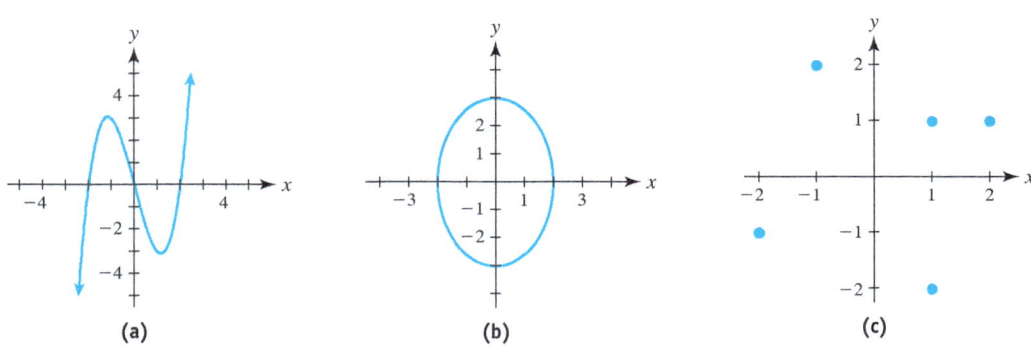

FIGURE 15.17

Solution
(a) Visualize vertical lines moving across the xy-plane from left to right. Any (red) vertical line will intersect the graph at most once, as depicted in **FIGURE 15.18(a)** on the next page. Therefore the graph *does* represent a function.

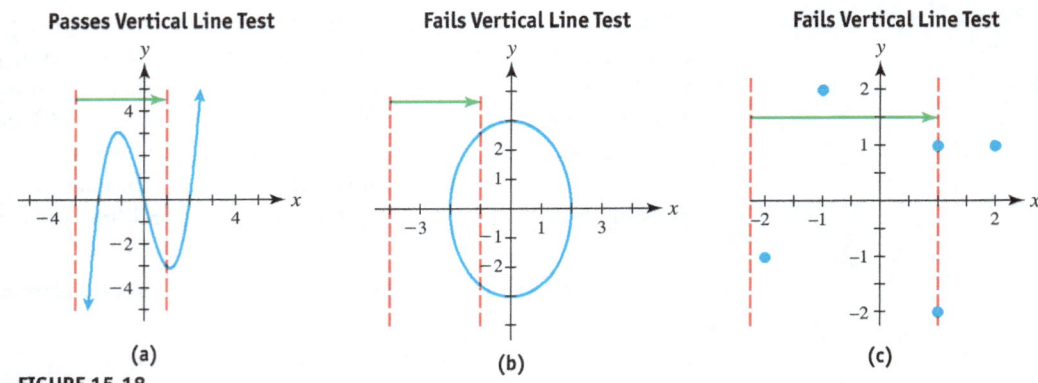

FIGURE 15.18

(b) Visualize vertical lines moving across the *xy*-plane from left to right. The graph *does not* represent a function because there exist (red) vertical lines that can intersect the graph twice. One such line is shown in **FIGURE 15.18(b)**.

(c) Visualize vertical lines moving across the *xy*-plane from left to right. The graph is a scatterplot and *does not* represent a function because there exists one (red) vertical line that intersects two points: $(1, 1)$ and $(1, -2)$. These points have the same *x*-coordinate but different *y*-coordinates. See **FIGURE 15.18(c)**.

Now Try Exercises 119, 121, 127

5 Using Graphing Calculators

Graphing calculators provide several features beyond those found on scientific calculators. For example, graphing calculators have additional keys that can be used to create tables, scatterplots, and graphs.

 Connecting Concepts with Your Life The **viewing rectangle**, or **window**, on a graphing calculator is similar to the viewfinder in a camera. A camera cannot take a picture of an entire scene. The camera must be centered on some object and can photograph only a portion of the available scenery. A camera can capture different views of the same scene by zooming in and out, as can graphing calculators. The *xy*-plane is infinite, but the calculator screen can show only a finite, rectangular region of the *xy*-plane. The viewing rectangle must be specified by setting minimum and maximum values for both the *x*- and *y*-axes before a graph can be drawn.

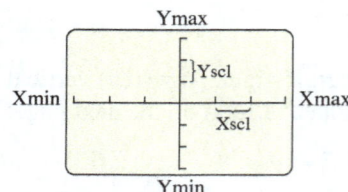

FIGURE 15.19

We use the following terminology regarding the size of a viewing rectangle. **Xmin** is the minimum *x*-value along the *x*-axis, and **Xmax** is the maximum *x*-value. Similarly, **Ymin** is the minimum *y*-value along the *y*-axis, and **Ymax** is the maximum *y*-value. Most graphs show an *x*-scale and a *y*-scale with tick marks on the respective axes. Sometimes the distance between consecutive tick marks is 1 unit, but at other times it might be 5 or 10 units. The distance represented by consecutive tick marks on the *x*-axis is called **Xscl**, and the distance represented by consecutive tick marks on the *y*-axis is called **Yscl** (see **FIGURE 15.19**).

If we want to represent this information about the viewing rectangle, we can write [Xmin, Xmax, Xscl] by [Ymin, Ymax, Yscl]. For example, $[-10, 10, 1]$ by $[-10, 10, 1]$ indicates that Xmin = -10, Xmax = 10, Xscl = 1, Ymin = -10, Ymax = 10, and Yscl = 1. This setting is referred to as the **standard viewing rectangle**. The window in **FIGURE 15.19** is $[-3, 3, 1]$ by $[-3, 3, 1]$.

EXAMPLE 11 Setting the viewing rectangle

Show the viewing rectangle $[-2, 3, 0.5]$ by $[-100, 200, 50]$ on your calculator.

Solution
The window setting and viewing rectangle are displayed in **FIGURE 15.20**. Note that in **FIGURE 15.20(b)**, there are 6 tick marks on the positive x-axis because its length is 3 units and the distance between consecutive tick marks is 0.5 unit.

CALCULATOR HELP
To set a viewing rectangle, see Appendix A (page AP-3).

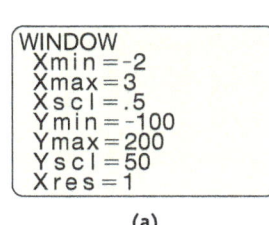
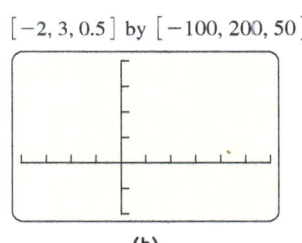

(a)   (b)
FIGURE 15.20

Now Try Exercise 141

SCATTERPLOTS Many graphing calculators have the capability to create scatterplots and line graphs. The next example illustrates how to make a scatterplot with a graphing calculator.

EXAMPLE 12 Making a scatterplot with a graphing calculator

Plot the points $(-2, -2)$, $(-1, 3)$, $(1, 2)$, and $(2, -3)$ in $[-4, 4, 1]$ by $[-4, 4, 1]$.

Solution
We enter the points $(-2, -2)$, $(-1, 3)$, $(1, 2)$, and $(2, -3)$ as shown in **FIGURE 15.21(a)** using the STAT EDIT feature. The variable L1 represents the list of x-values, and the variable L2 represents the list of y-values. In **FIGURE 15.21(b)**, we set the graphing calculator to make a scatterplot with the STATPLOT feature, and in **FIGURE 15.21(c)**, the points have been plotted. If you have a different model of calculator, you may need to consult your owner's manual.

CALCULATOR HELP
To make a scatterplot, see Appendix A (pages AP-3 and AP-4).

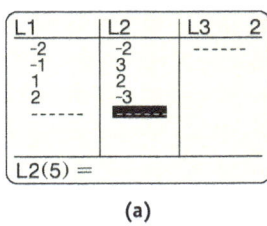
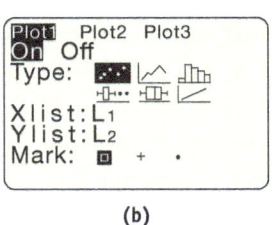
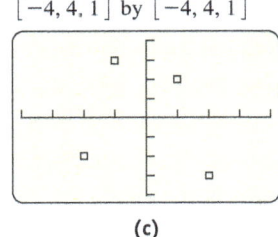

(a)   (b)   (c)
FIGURE 15.21

Now Try Exercise 145

GRAPHS AND TABLES We can use graphing calculators to create graphs and tables, usually more efficiently and reliably than with pencil-and-paper techniques. However, a graphing calculator uses the same techniques that we might use to sketch a graph.

For example, one way to sketch a graph of $y = 2x - 1$ is first to make a table of values, as shown in **TABLE 15.5**. We can plot these points in the xy-plane, as shown in **FIGURE 15.22**. Next we might connect the points, as shown in **FIGURE 15.23**.

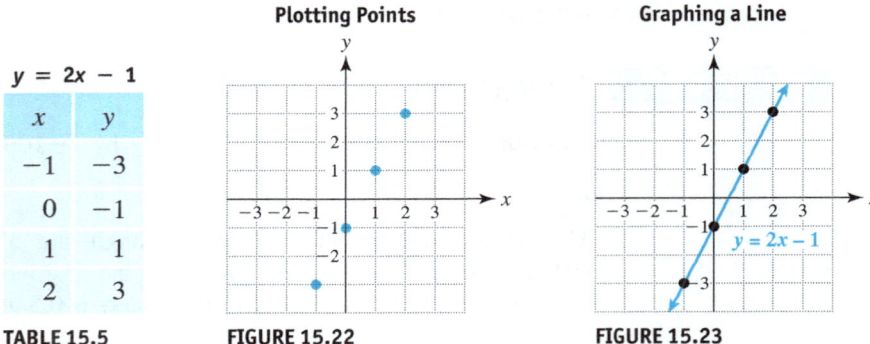

TABLE 15.5 **FIGURE 15.22** **FIGURE 15.23**

In a similar manner, a graphing calculator plots numerous points and connects them to make a graph. To create a similar graph with a graphing calculator, we enter the formula $Y_1 = 2X - 1$, set an appropriate viewing rectangle, and graph as shown in **FIGURES 15.24** and **15.25**. A table of values can also be generated, as illustrated in **FIGURE 15.26**.

CALCULATOR HELP

To make a graph, see Appendix A (page AP-3). To make a table, see Appendix A (pages AP-2 and AP-3).

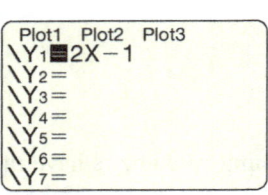
FIGURE 15.24

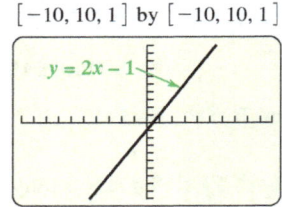
FIGURE 15.25

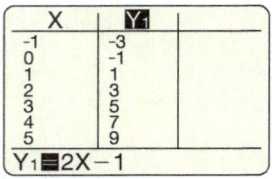
FIGURE 15.26

15.1 Putting It All Together

CONCEPT	EXPLANATION	EXAMPLES
Function	A set of ordered pairs (x, y), where each x-value corresponds to exactly one y-value	$f = \{(1, 3), (2, 3), (3, 1)\}$ $f(x) = 2x$ A graph of $y = x + 2$ A table of values for $y = 4x$
Independent Variable	The *input* variable for a function	*Function* *Independent Variable* $f(x) = 2x$ x $A(r) = \pi r^2$ r $V(s) = s^3$ s
Dependent Variable	The *output* variable of a function. There is exactly one output for each valid input.	*Function* *Dependent Variable* $y = f(x)$ y $T = F(r)$ T $V = g(r)$ V
Domain and Range of a Function	The domain D is the set of all valid inputs. The range R is the set of all outputs.	For $S = \{(-1, 0), (3, 4), (5, 0)\}$, $D = \{-1, 3, 5\}$ and $R = \{0, 4\}$. For $f(x) = \frac{1}{x}$, the domain includes all real numbers except 0, or $x \neq 0$.

CONCEPT	EXPLANATION	EXAMPLES
Vertical Line Test	If every vertical line intersects a graph at no more than one point, the graph represents a function.	This graph does *not* pass this test and thus does not represent a function. Two points lie on the same vertical line: *not* a function

A function can be represented verbally, symbolically, numerically, and graphically.

REPRESENTATION	EXPLANATION	COMMENTS
Verbal	Precise word description of what is computed	May be oral or written Must be stated *precisely*
Symbolic	Mathematical formula	Efficient and concise way of representing a function (e.g., $f(x) = 2x - 3$)
Numerical	List of specific inputs and their outputs	May be in the form of a table or an explicit set of ordered pairs
Graphical, diagrammatic	Shows inputs and outputs visually	No words, formulas, or tables Many types of graphs and diagrams are possible.

15.1 Exercises

MyMathLab®

CONCEPTS AND VOCABULARY

1. The notation $y = f(x)$ is called _____ notation.

2. The notation $y = f(x)$ is read _____.

3. The notation $f(x) = x^2 + 1$ is a(n) _____ representation of a function.

4. A table of values is a(n) _____ representation of a function.

5. The set of valid inputs for a function is the _____.

6. The set of outputs for a function is the _____.

7. A function computes ____ output for each valid input.

8. (True or False?) The vertical line test is used to identify graphs of relations.

9. (True or False?) Four ways to represent functions are verbal, numerical, symbolic, and graphical.

10. If $f(3) = 4$, the point _____ is on the graph of f. If $(3, 6)$ is on the graph of f, then $f(\_\_\_\_) = \_\_\_\_$.

11. **Thinking Generally** If $f(a) = b$, the point ____ is on the graph of f.

12. **Thinking Generally** If (c, d) is on the graph of g, then $g(c) = \_\_\_\_$.

13. **Thinking Generally** If a is in the domain of f, then $f(a)$ represents how many outputs?

14. **Thinking Generally** If $f(x) = x$ for every x in the domain of f, then the domain and range of f are ____.

RECOGNIZING FUNCTIONS

Exercises 15–20: Determine whether the phrase describes a function.

15. Calculating the square of a number

16. Calculating the low temperature for a day

17. Listing the students who passed a given math exam

18. Listing the children of parent x

19. Finding sales tax on a purchase

20. Naming the people in your class

REPRESENTING AND EVALUATING FUNCTIONS

Exercises 21–32: Evaluate $f(x)$ at the given values of x.

21. $f(x) = 4x - 2$ $\quad x = -1, 0$

22. $f(x) = 5 - 3x$ $\quad x = -4, 2$

23. $f(x) = \sqrt{x}$ $\quad x = 0, \frac{9}{4}$

24. $f(x) = \sqrt[3]{x}$ $\quad x = -1, 27$

25. $f(x) = x^2$ $\quad x = -5, \frac{3}{2}$

26. $f(x) = x^3$ $\quad x = -2, 0.1$

27. $f(x) = 3$ $\quad x = -8, \frac{7}{3}$

28. $f(x) = 100$ $\quad x = -\pi, \frac{1}{3}$

29. $f(x) = 5 - x^3$ $\quad x = -2, 3$

30. $f(x) = x^2 + 5$ $\quad x = -\frac{1}{2}, 6$

31. $f(x) = \dfrac{2}{x + 1}$ $\quad x = -5, 4$

32. $f(x) = \dfrac{x}{x - 4}$ $\quad x = -3, 1$

Exercises 33–38: Do the following.

(a) Write a formula for the function described.
(b) Evaluate the function for input 10 and interpret the result.

33. Function I computes the number of inches in x yards.

34. Function A computes the area of a circle with radius r.

35. Function M computes the number of miles in x feet.

36. Function C computes the circumference of a circle with radius r.

37. Function A computes the square feet in x acres. (*Hint:* There are 43,560 square feet in one acre.)

38. Function K computes the number of kilograms in x pounds. (*Hint:* There are about 2.2 pounds in one kilogram.)

Exercises 39–42: Write each function f as a set of ordered pairs. Give the domain and range of f.

39. $f(1) = 3, f(2) = -4, f(3) = 0$

40. $f(-1) = 4, f(0) = 6, f(1) = 4$

41. $f(a) = b, f(c) = d, f(e) = a, f(d) = b$

42. $f(a) = 7, f(b) = 7, f(c) = 7, f(d) = 7$

Exercises 43–52: Sketch a graph of f.

43. $f(x) = -x + 3$ **44.** $f(x) = -2x + 1$

45. $f(x) = 2x$ **46.** $f(x) = \frac{1}{2}x - 2$

47. $f(x) = 4 - x$ **48.** $f(x) = 6 - 3x$

49. $f(x) = x^2$ **50.** $f(x) = \sqrt{x}$

51. $f(x) = \sqrt{x + 1}$ **52.** $f(x) = \frac{1}{2}x^2 - 1$

Exercises 53–58: Use the graph of f to evaluate the given expressions.

53. $f(0)$ and $f(2)$ **54.** $f(-2)$ and $f(2)$

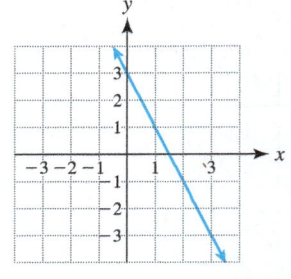 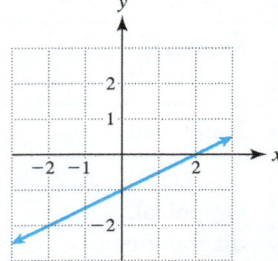

55. $f(-2)$ and $f(1)$ **56.** $f(-1)$ and $f(0)$

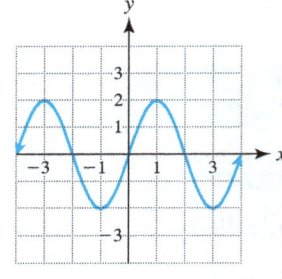 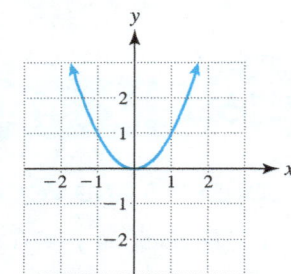

57. $f(1)$ and $f(2)$ **58.** $f(-1)$ and $f(4)$

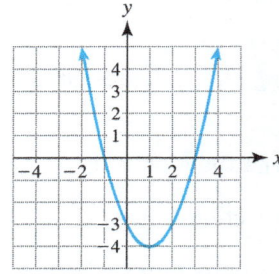 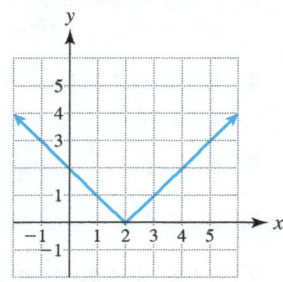

Exercises 59 and 60: Use the table to evaluate the given expressions.

59. $f(0)$ and $f(2)$

x	0	1	2	3	4
$f(x)$	5.5	4.3	3.7	2.5	1.9

60. $f(-10)$ and $f(5)$

x	-10	-5	0	5	10
$f(x)$	23	96	-45	-33	23

Exercises 61 and 62: Use the diagram to evaluate $f(1990)$. Interpret your answer.

61. Fuel Efficiency The function f computes average fuel efficiency of new U.S. passenger cars in miles per gallon during year x. (*Source: Department of Transportation.*)

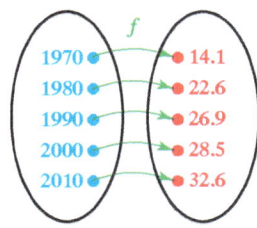

62. Cost of Tuition The function f computes average cost of tuition at public colleges and universities during academic year x. (*Source: The College Board.*)

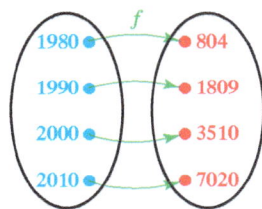

Exercises 63–66: Express the verbal representation for the function f numerically, symbolically, and graphically. Let $x = -3, -2, -1, \ldots, 3$ for the numerical representation (table), and let $-3 \leq x \leq 3$ for the graph.

63. Add 5 to the input x to obtain the output y.

64. Square the input x to obtain the output y.

65. Multiply the input x by 5 and then subtract 2 to obtain the output y.

66. Divide the input x by 2 and then add 3 to obtain the output y.

Exercises 67–72: Give a verbal representation for $f(x)$.

67. $f(x) = x - \frac{1}{2}$ **68.** $f(x) = \frac{3}{4}x$

69. $f(x) = \frac{x}{3}$ **70.** $f(x) = x^2 + 1$

71. $f(x) = \sqrt{x-1}$ **72.** $f(x) = 1 - 3x$

73. Cost of Driving In 2015, the average cost of driving a new car in the United States was about 60 cents per mile. Symbolically, graphically, and numerically represent a function f that computes the cost in dollars of driving x miles. For the numerical representation (table), let $x = 10, 20, 30, \ldots, 70$. (*Source: AAA.*)

74. Federal Income Taxes In 2015, the lowest U.S. income tax rate was 10 percent. Symbolically, graphically, and numerically represent a function f that computes the tax on a taxable income of x dollars. For the numerical representation (table) let $x = 1000, 2000, 3000, \ldots, 7000$. For the graphical representation let $0 \leq x \leq 10{,}000$. (*Source: Internal Revenue Service.*)

75. Google Web Searches The number of Google searches S in billions during year x can be approximated by $S(x) = 120x - 240{,}280$ from 2009 to 2015. Evaluate $S(2014)$ and interpret the result. (*Source: DMR.*)

76. Price of Android Smartphones The average price P in dollars for an Android smartphone during year x is projected to be $P(x) = -9.75x + 19{,}890.5$, where $2014 \leq x \leq 2019$. (*Source: Computerworld.*)
(a) Evaluate $P(2018)$ and interpret your answer.
(b) What does function P indicate about how the price of Android smartphones is changing?

IDENTIFYING DOMAINS AND RANGES

Exercises 77–84: Use the graph of f to identify its domain and range.

77. 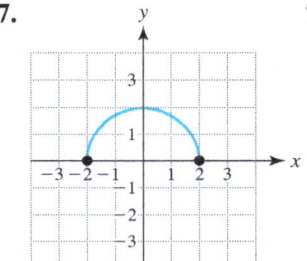 **78.**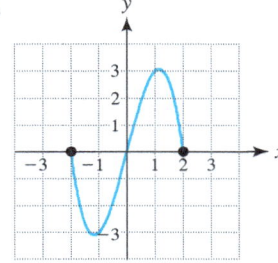

79. 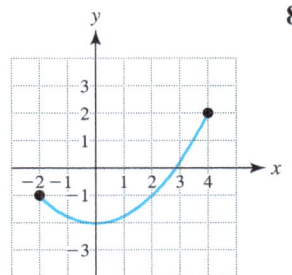 **80.**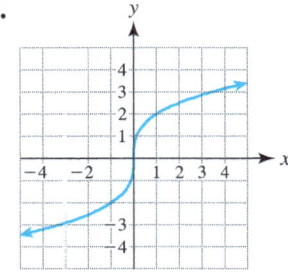

81. 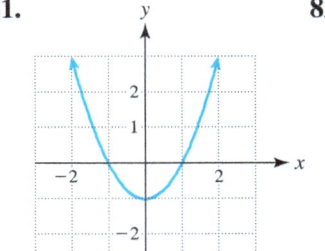 **82.**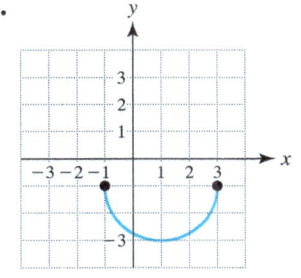

83. 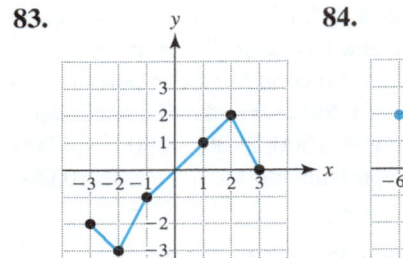 **84.**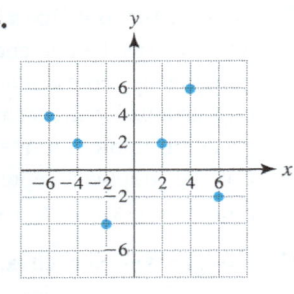

Exercises 85 and 86: Use the diagram to find the domain and range of f.

85. 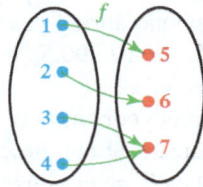 **86.**

Exercises 87–100: Find the domain.

87. $f(x) = 10x$

88. $f(x) = 5 - x$

89. $f(x) = x^2 - 3$

90. $f(x) = \frac{1}{2}x^2$

91. $f(x) = \dfrac{3}{x - 5}$

92. $f(x) = \dfrac{x}{x + 1}$

93. $f(x) = \dfrac{2x}{x^2 + 1}$

94. $f(x) = \dfrac{6}{1 - x}$

95. $f(x) = \sqrt{x - 1}$

96. $f(x) = |x|$

97. $f(x) = |x - 5|$

98. $f(x) = \sqrt{2 - x}$

99. $f(x) = \dfrac{1}{x}$

100. $f(x) = 1 - 3x^2$

101. Humpback Whales The number of humpback whales W sighted in Maui's annual whale census for year x is given by $W(2011) = 1607$, $W(2012) = 1054$, $W(2013) = 1126$, $W(2014) = 1331$, and finally $W(2015) = 1488$. (*Source:* Pacific Whale Foundation.)
(a) Evaluate $W(2015)$ and interpret the result.
(b) Identify the domain and range of W.
(c) Describe the pattern in the data.

102. Global Digital Revenues Revenue R in billions of dollars from global sales of digital music during year x is given by $R(2010) = 4.6$, $R(2011) = 5.1$, $R(2012) = 5.6$, $R(2013) = 6.4$, and $R(2014) = 6.9$. (*Source:* IFPI.)
(a) Evaluate $R(2014)$ and interpret the result.
(b) Identify the domain and range of function R.
(c) Describe the pattern in the data.

103. Cost of Tuition Suppose that a student can take from 1 to 20 credits at a college and that each credit costs \$200. If function C calculates the cost of taking x credits, determine the domain and range of C.

104. Falling Ball Suppose that a ball is dropped from a window that is 64 feet above the ground and that the ball strikes the ground after 2 seconds. If function H calculates the height of the ball after t seconds, determine a domain and range for H, while the ball is falling.

IDENTIFYING RELATIONS AND FUNCTIONS

Exercises 105–112: Identify whether set S is a relation, function, or both. If it is neither, state so. Give its domain and range, if possible.

105. $S = \{(2, 3), (3, 5), (2, -4)\}$

106. $S = \{(-1, 4), (-1, 6), (-1, 7)\}$

107. $S = \{(4, 5), (-7, 5), (3, 5)\}$

108. $S = \{(-5, 0), (0, 0), (5, 0)\}$

109. $S = \{1, 2, 3, 4\}$

110. $S = \{a, b, c, d, e\}$

111. $S = \{(-5, -5), (-2, -2), (0, 0), (3, 3)\}$

112. $S = \{(-1, 3), (1, 2), (1, 7), (-1, -2)\}$

IDENTIFYING A FUNCTION

Exercises 113–116: Determine whether the diagram could represent a function.

113. 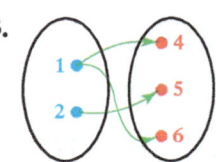 **114.**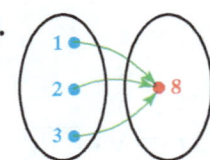

115. 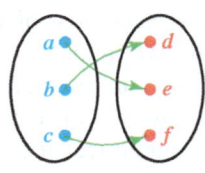 **116.**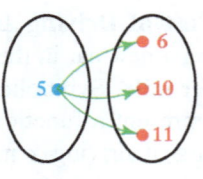

117. **Average Precipitation** The table lists the monthly average precipitation P in Las Vegas, Nevada, where $x = 1$ corresponds to January and $x = 12$ corresponds to December.

x (month)	1	2	3	4	5	6
P (inches)	0.5	0.4	0.4	0.2	0.2	0.1

x (month)	7	8	9	10	11	12
P (inches)	0.4	0.5	0.3	0.2	0.4	0.3

Source: J. Williams.

(a) Determine the value of P during May.
(b) Is P a function of x? Explain.
(c) If $P = 0.4$, find x.

118. **Wind Speeds** The table lists the monthly average wind speed W in Louisville, Kentucky, where $x = 1$ corresponds to January and $x = 12$ corresponds to December.

x (month)	1	2	3	4	5	6
W (mph)	10.4	12.7	10.4	10.4	8.1	8.1

x (month)	7	8	9	10	11	12
W (mph)	6.9	6.9	6.9	8.1	9.2	9.2

Source: J. Williams.

(a) Determine the month with the highest average wind speed.
(b) Is W a function of x? Explain.
(c) If $W = 6.9$, find x.

Exercises 119–130: Determine whether the graph represents a function. If it does, identify the domain and range.

119.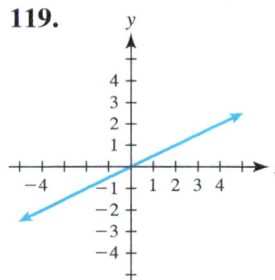

120.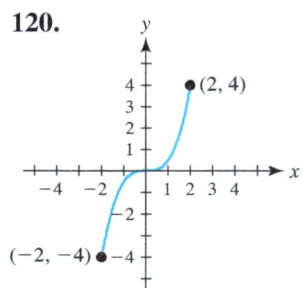

121.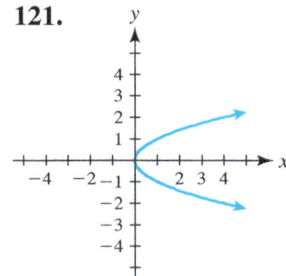

122.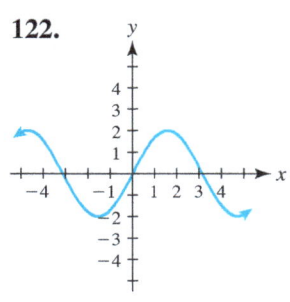

123.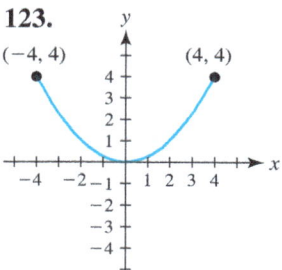

124.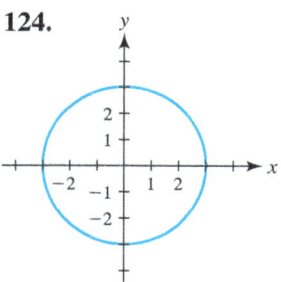

125.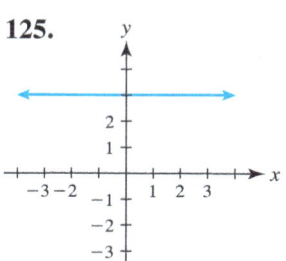

126.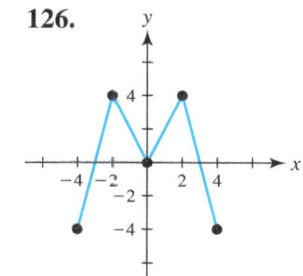

127. 128.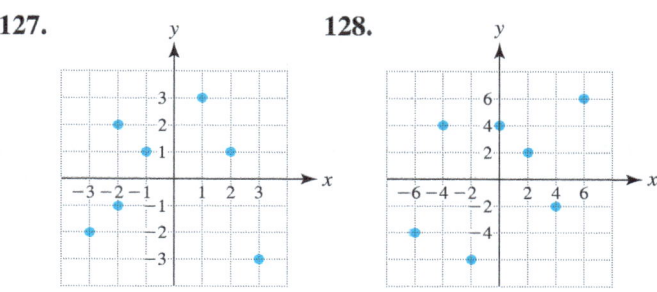

129. 130.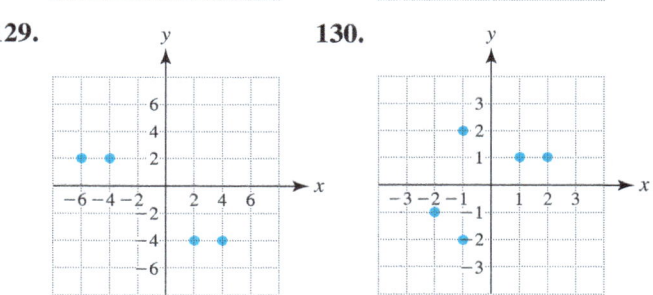

Exercises 131–134: Determine whether S is a function.

131. $S = \{(1, 2), (4, 5), (7, 8), (5, 4), (2, 2)\}$

132. $S = \{(4, 7), (-2, 1), (3, 8), (4, 9)\}$

133. S is given by the table.

x	5	10	5
y	2	1	0

134. S is given by the table.

x	-3	-2	-1
y	10	10	10

INTERPRETING GRAPHS

Exercises 135 and 136: The graph represents the distance that a person is from home while walking on a straight path. The x-axis represents time and the y-axis represents distance. Interpret the graph.

135.   136.

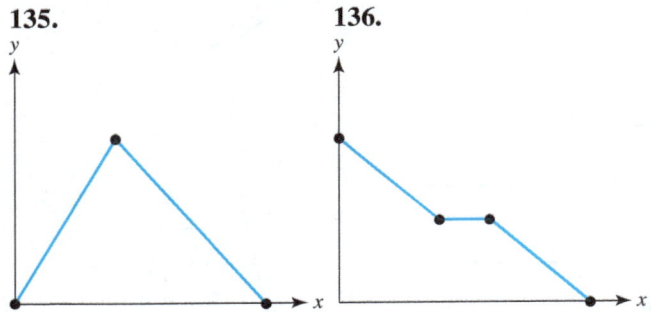

137. **Texting** The average 18- to 24-year-old person texts about 2000 messages per month. Sketch a graph that shows the total number of text messages sent over a period of 4 months. Assume that the same number of texts is sent each day. (*Source: The Nielsen Company.*)

138. **Computer Viruses** In 2000, there were about 50 thousand computer viruses. In 2015, there were about 27 million computer viruses. Sketch a graph of this increase from 2000 to 2015. Answers may vary. (*Source: Symantec.*)

USING GRAPHING CALCULATORS

Exercises 139–144: Show the given viewing rectangle on your graphing calculator. Predict the number of tick marks on the positive x-axis and the positive y-axis.

139. Standard viewing rectangle

140. $[-12, 12, 2]$ by $[-8, 8, 2]$

141. $[0, 100, 10]$ by $[-50, 50, 10]$

142. $[-30, 30, 5]$ by $[-20, 20, 5]$

143. $[1980, 1995, 1]$ by $[12000, 16000, 1000]$

144. $[1900, 1990, 10]$ by $[1700, 2800, 100]$

Exercises 145–150: Use your calculator to make a scatterplot of the relation after determining an appropriate viewing rectangle.

145. $\{(4, 3), (-2, 1), (-3, -3), (5, -2)\}$

146. $\{(5, 5), (2, 0), (-2, 7), (2, -8), (-1, -5)\}$

147. $\{(20, 40), (-25, -15), (-20, 25), (15, -25)\}$

148. $\{(-13, 12), (3, 10), (-15, -4), (12, -9)\}$

149. $\{(100, -100), (50, 200), (-150, -140), (-30, 80)\}$

150. $\{(-125, 75), (45, 65), (-53, -67), (150, -80)\}$

Exercises 151–154: Make a table and graph of $y = f(x)$. Let $x = -3, -2, -1, \ldots, 3$ for your table and use the standard window for your graph.

151. $f(x) = \sqrt{x + 3}$ 152. $f(x) = x^3 - \frac{1}{2}x^2$

153. $f(x) = \dfrac{5 - x}{5 + x}$ 154. $f(x) = |2 - x| + \sqrt[3]{x}$

WRITING ABOUT MATHEMATICS

155. Give an example of a function. Identify the domain and range of your function.

156. Explain in your own words what a function is. How is a function different from other relations?

157. Explain how to evaluate a function by using a graph. Give an example.

158. Give one difficulty that may occur when you use a table of values to evaluate a function.

Group Activity

Digital Music Album Downloads The following table lists the number of digital music album downloads in the United States for selected years.

Year	Downloads (millions)
2004	5
2007	45
2009	75
2012	115

(*Source: Statista 2015.*)

(a) Make a scatterplot of the data. Let x represent the number of years after 2004. Discuss any trend in numbers of album downloads.
(b) Estimate the slope of a line that could be used to model the data.
(c) Find an equation of a line $y = mx + b$ that models the data.
(d) Interpret the slope as a rate of change.
(e) Use your results to estimate the number of album downloads in 2014. Compare your answer with the true value of 118 million. Comment on the prediction made by your equation.

15.2 Linear Functions

Objectives

1. Recognizing Representations of Linear Functions
   - Numerical Representation (Table of Values)
   - Graphical Representation (Graph)
   - Verbal Representation (Words)
   - Symbolic Representation (Formula)
   - Identifying Linear Functions
2. Recognizing Graphs of Linear Functions
3. Modeling Data with Linear Functions
4. Applying the Midpoint Formula
   - Midpoint Formula on the Real Number Line
   - Midpoint Formula in the xy-Plane

STUDY TIP

Be sure you understand how functions can be represented.

1 Recognizing Representations of Linear Functions

Connecting Concepts with Your Life Suppose that the air conditioner is turned on when the temperature inside a house is 80 °F. The resulting temperatures are listed in **TABLE 15.6** for various elapsed times. Note that for each 1-hour increase in elapsed time, the temperature decreases by 2 °F.

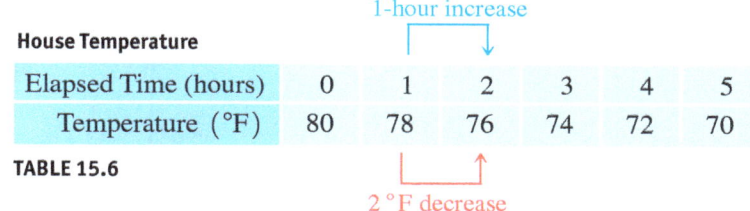

House Temperature

Elapsed Time (hours)	0	1	2	3	4	5
Temperature (°F)	80	78	76	74	72	70

TABLE 15.6

We want to determine a function f that models, or calculates, the temperature inside the house after x hours. To do this, we will find numerical, graphical, verbal, and symbolic representations of f.

NUMERICAL REPRESENTATION (TABLE OF VALUES) We can think of **TABLE 15.6** as a numerical representation (table of values) for the function f. A similar numerical representation that uses x and $f(x)$ is shown in **TABLE 15.7**.

Numerical Representation of $f(x)$

x	0	1	2	3	4	5
$f(x)$	80	78	76	74	72	70

TABLE 15.7

GRAPHICAL REPRESENTATION (GRAPH) To graph $y = f(x)$, we begin by plotting the points in **TABLE 15.7**, as shown in **FIGURE 15.27**. This scatterplot suggests that a line models these data, as shown in **FIGURE 15.28**. We call f a *linear function* because its graph is a *line*.

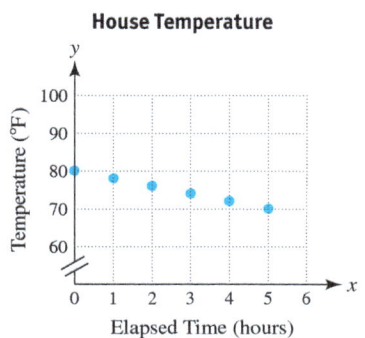

FIGURE 15.27 A Scatterplot

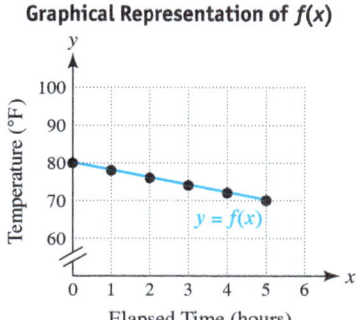

FIGURE 15.28 A Linear Function

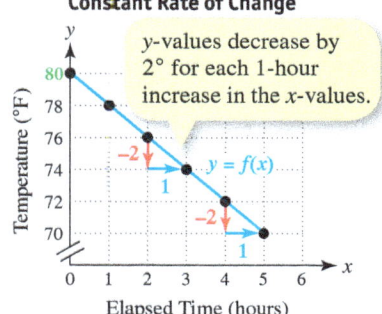

FIGURE 15.29

Another graph of $y = f(x)$ with a different y-scale is shown in **FIGURE 15.29**. Because the y-values always decrease by the same amount for each 1-hour increase on the x-axis, we say that function f has a *constant rate of change*. In this example, the constant rate of change is -2 °F per hour.

VERBAL REPRESENTATION (WORDS) Over a 5-hour period, the air conditioner lowers the initial temperature of 80 °F by 2 °F for each elapsed hour x. Thus a description of how to calculate the temperature is written as follows.

Verbal representation of $f(x)$

"Multiply x by -2 °F and then add 80 °F."

SYMBOLIC REPRESENTATION (FORMULA) Our verbal representation of $f(x)$ makes it straightforward for us to write a formula.

Symbolic representation of $f(x)$

$$f(x) = -2x + 80$$

(Elapsed Hours; Rate of Change; Initial Temperature)

For example,

$$f(2.5) = -2(2.5) + 80 = 75$$

means that the temperature is 75 °F after the air conditioner has run for 2.5 hours. In this instance, it might be appropriate to *limit the domain* of f to x-values between 0 and 5, inclusive.

> **LINEAR FUNCTION**
>
> A function f defined by $f(x) = mx + b$, where m and b are constants, is a **linear function**.

For $f(x) = -2x + 80$, we have $m = -2$ and $b = 80$. The constant m represents the rate at which the air conditioner cools the building, and the constant b represents the initial temperature.

NOTE: The value of m represents the slope of the graph of $f(x) = mx + b$, and $(0, b)$ is the y-intercept. ∎

 Connecting Concepts with Your Life In general, a linear function defined by $f(x) = mx + b$ changes by m units for each unit increase in x. This **rate of change** is an increase if $m > 0$ and a decrease if $m < 0$. For example, if new carpet costs $20 per square yard, then the linear function defined by $C(x) = 20x$ gives the cost of buying x square yards of carpet. The value of $m = 20$ gives the cost (rate of change) for each additional square yard of carpet. For function C, the value of b is 0 because it costs $0 to buy 0 square yards of carpet.

READING CHECK 1

- Explain what a linear function is and what its graph looks like.

NOTE: If f is a linear function, then $f(0) = m(0) + b = b$. Thus b can be found by evaluating $f(x)$ at $x = 0$. ∎

IDENTIFYING LINEAR FUNCTIONS We can easily identify the graph of a linear function because it is always a line. The next two examples illustrate how to identify a linear function from its formula and from a table of values.

EXAMPLE 1 Identifying linear functions from formulas

Determine whether f is a linear function. If f is a linear function, find values for m and b so that $f(x) = mx + b$.

(a) $f(x) = 4 - 3x$ (b) $f(x) = 8$ (c) $f(x) = 2x^2 + 8$

Solution
(a) Let $m = -3$ and $b = 4$. Then $f(x) = -3x + 4$, and f is a linear function.
(b) Let $m = 0$ and $b = 8$. Then $f(x) = 0x + 8$, and f is a linear function.
(c) Function f is not linear because its formula contains x^2. The formula for a linear function cannot contain a variable with an exponent other than 1.

Now Try Exercises 11, 13, 15

EXAMPLE 2 Identifying linear functions from tables of values

Use each table of values to determine whether $f(x)$ could represent a linear function. If f could be linear, write a formula for f in the form $f(x) = mx + b$.

(a)
x	0	1	2	3
$f(x)$	10	15	20	25

(b)
x	−2	0	2	4
$f(x)$	4	2	0	−2

(c)
x	0	1	2	3
$f(x)$	1	2	4	7

(d)
x	−2	0	3	5
$f(x)$	7	7	7	7

Solution
(a) For each unit increase in x, $f(x)$ increases by 5 units, so $f(x)$ could be linear with $m = 5$. Because $f(0) = 10$, $b = 10$. Thus $f(x) = 5x + 10$.
(b) For each 2-unit increase in x, $f(x)$ decreases by 2 units. Equivalently, each unit increase in x results in a 1-unit decrease in $f(x)$, so $f(x)$ could be linear with $m = -1$. Because $f(0) = 2$, $b = 2$. Thus $f(x) = -x + 2$.
(c) Each unit increase in x does not result in a constant change in $f(x)$. Thus $f(x)$ does *not* represent a linear function.
(d) For any change in x, $f(x)$ does *not* change, so $f(x)$ could be linear with $m = 0$. Because $f(0) = 7$, let $b = 7$. Thus $f(x) = 0x + 7$, or $f(x) = 7$. (When $m = 0$, we say that f is a *constant function*. See Example 8.)

Now Try Exercises 23, 25, 27, 31

2 Recognizing Graphs of Linear Functions

The graph of a linear function is a line. To graph a linear function f, we can start by making a table of values and then plotting three or more points. We can then sketch the graph of f by drawing a line through these points, as demonstrated in the next example.

EXAMPLE 3 Graphing a linear function by hand

Sketch a graph of $f(x) = x - 1$. Use the graph to evaluate $f(-2)$.

Solution
Begin by making a table of values containing at least three points. Pick convenient values of x, such as $x = -1, 0, 1$.

$$f(-1) = -1 - 1 = -2$$
$$f(0) = 0 - 1 = -1$$
$$f(1) = 1 - 1 = 0$$

Display the results, as shown in **TABLE 15.8** on the next page.

Plot the points $(-1, -2)$, $(0, -1)$, and $(1, 0)$. Sketch a line through these points to obtain the graph of f. A graph of a line results when *infinitely* many points are plotted, as shown in **FIGURE 15.30**.

To evaluate $f(-2)$, first find $x = -2$ on the x-axis. See **FIGURE 15.31**. Then move downward to the graph of f. By moving across to the y-axis, we see that the corresponding y-value is -3. Thus $f(-2) = -3$.

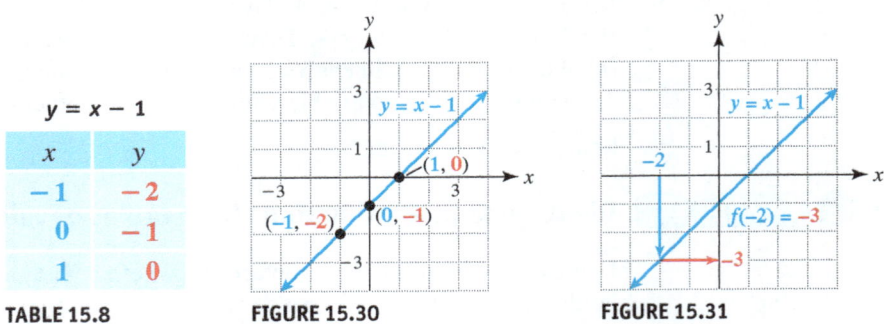

TABLE 15.8   FIGURE 15.30   FIGURE 15.31

Now Try Exercises 39, 57

In the next example, a graphing calculator is used to create a graph and table.

EXAMPLE 4 Using a graphing calculator

Give numerical and graphical representations of $f(x) = \frac{1}{2}x - 2$.

Solution
Numerical Representation To make a numerical representation, construct the table for $Y_1 = 0.5X - 2$, starting at $x = -3$ and incrementing by 1, as shown in **FIGURE 15.32(a)**. (Other tables are possible.)
Graphical Representation Graph Y_1 in the standard viewing rectangle, as shown in **FIGURE 15.32(b)**. (Other viewing rectangles may be used.)

CALCULATOR HELP
To make a table, see Appendix A (pages AP-2 and AP-3). To make a graph, see Appendix A (page AP-3).

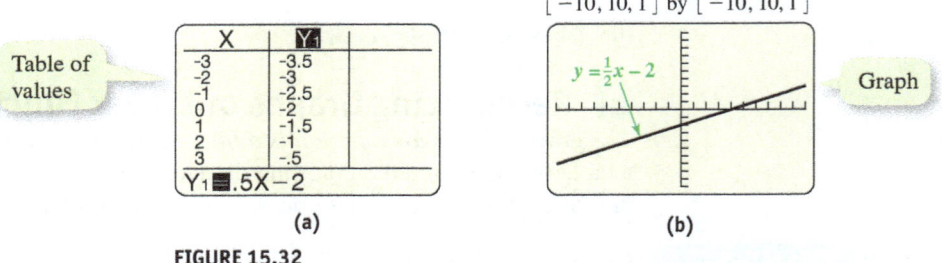

FIGURE 15.32

Now Try Exercise 75

CRITICAL THINKING 1

Two points determine a line. Why is it a good idea to plot at least three points when graphing a linear function by hand?

EXAMPLE 5 Representing a linear function

A linear function is given by $f(x) = -3x + 2$.
(a) Give a verbal representation (description) of f.
(b) Make a numerical representation (table) of f by letting $x = -1, 0, 1$.
(c) Plot the points listed in the table from part (b). Then sketch a graph of $y = f(x)$.

Solution
(a) **Verbal Representation** Multiply the input x by -3 and add 2 to obtain the output.
(b) **Numerical Representation** Evaluate the formula $f(x) = -3x + 2$ at $x = -1, 0, 1$, which results in **TABLE 15.9**. Note that $f(-1) = 5$, $f(0) = 2$, and $f(1) = -1$.
(c) **Graphical Representation** To make a graph of f by hand without a graphing calculator, plot the points $(-1, 5)$, $(0, 2)$, and $(1, -1)$ from **TABLE 15.9**. Then draw a line passing through these points, as shown in **FIGURE 15.33**.

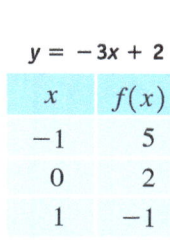
$y = -3x + 2$

x	$f(x)$
-1	5
0	2
1	-1

TABLE 15.9

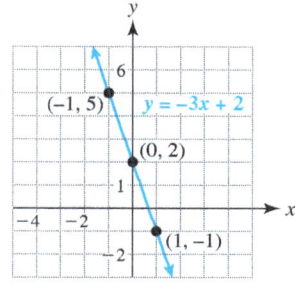

FIGURE 15.33

Now Try Exercise 71

NOTE: To graph $y = -3x + 2$ in Example 5, we could also graph a line with slope -3 and y-intercept $(0, 2)$. ∎

MAKING CONNECTIONS 1

Mathematics in Newspapers

Think of the mathematics that you see in newspapers or in online publications. Often, percentages are described *verbally*, numbers are displayed in *tables*, and data are shown in *graphs*. Seldom are *formulas* given, which is an important reason to study verbal, numerical, and graphical representations.

3 Modeling Data with Linear Functions

🌐 *Math in Context* (Wireless) A distinguishing feature of a linear function is that when the input x increases by 1 unit, the output $f(x) = mx + b$ always changes by an amount equal to m. For example, the percentage of wireless-only households during year x from 2005 to 2014 can be modeled by the linear function

$$f(x) = 4x - 8013,$$

where x is the year. The value of $m = 4$ indicates that the percentage of wireless-only households has increased, on average, by 4% per year. (*Source:* National Center for Health Statistics.)

The following are other examples of quantities that are modeled by linear functions. Try to determine the value of the constant m.

- The wages earned by an individual working x hours at $8 per hour
- The distance traveled by a jet airliner in x hours if the speed of the jet is 500 miles per hour
- The cost of tuition and fees when registering for x credits if each credit costs $200 and the fees are fixed at $300

When we are modeling data with a linear function defined by $f(x) = mx + b$, the following concepts are helpful to determine m and b.

MODELING DATA WITH A LINEAR FUNCTION

The formula $f(x) = mx + b$ may be interpreted as follows.

$$f(x) = mx + b$$

(New amount) = (Change) + (Fixed amount)

When x represents time, *change* equals (rate of change) × (time).

$$f(x) = m \times x + b$$

(Future amount) = (Rate of change) × (Time) + (Initial amount)

Math in Context (*Biology*) These concepts are applied in the next example where a linear function is used to model the height of a fast-growing bamboo plant.

EXAMPLE 6   **Modeling growth of bamboo**

Bamboo is gaining popularity as a *green* building material because of its fast-growing, regenerative characteristics. Under ideal conditions, some species of bamboo grow at an astonishing 2 inches per hour. Suppose a bamboo plant is initially 6 inches tall. (*Source:* Cali Bamboo.)

(a) Find a function H that models the plant's height in inches under ideal conditions after t hours.

(b) Find $H(3)$ and interpret the result.

Solution

(a) The initial height is **6** inches and the rate of change is **2** inches per hour.

$$H(t) = 2 \times t + 6,$$

(Future height) = (Rate of change) × (Time) + (Initial height)

or $H(t) = 2t + 6$.

(b) $H(3) = 2(3) + 6 = 12$. After **3** hours, the bamboo plant is **12** inches tall.

Now Try Exercise 117

Math in Context (*Green*) People are becoming more energy conscious, so there is an increase in the demand for green building material. In the next example, this increase in demand is modeled with a linear function.

EXAMPLE 7 Modeling demand for green building material

TABLE 15.10 lists actual and projected sales of green building material in billions of dollars.

Green Building Material Sales ($ billions)

Year	2015	2016	2017	2018
Sales	72	79	86	93

TABLE 15.10 *Source: Freedonia Group, Green Building Material.*

(a) Make a scatterplot of the data and sketch the graph of a function f that models these data. Let x represent years after 2015. That is, let $x = 0$ correspond to 2015, $x = 1$ to 2016, and so on.
(b) What were sales in 2015? What was the annual increase in sales each year?
(c) Find a formula for $f(x)$.
(d) Use your formula to estimate sales in 2019.

READING CHECK 2

- How can you determine whether data in a table can be modeled by a linear function?

Solution

(a) In **FIGURE 15.34** the scatterplot suggests that a linear function models the data. A line has been sketched with the data.

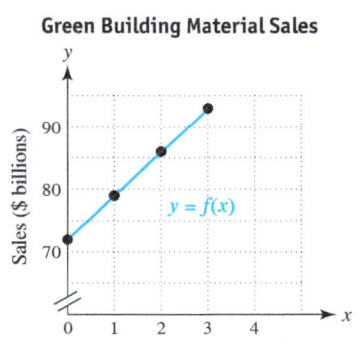

FIGURE 15.34 A Linear Model

(b) From **TABLE 15.10**, sales for green building material totaled $72 billion in 2015, with sales increasing at a *constant rate* of $7 billion per year.

(c) From part (b) initial sales ($x = 0$) were $**72** billion, and sales increased by $**7** billion per year. Thus

$$f(x) = 7 \times x + 72$$
$$(\text{Future sales}) = (\text{Rate of change in sales}) \times (\text{Time}) + (\text{Initial sales})$$

or $f(x) = 7x + 72$.

(d) Because $x = 4$ corresponds to 2019, evaluate $f(4)$.

$$f(4) = 7(4) + 72 = \mathbf{100}$$

This model projects sales of green building material to be $100 billion in 2019.

Now Try Exercise 119

🌎 **Math in Context** *Driving* In the next example, we consider a simple function that models the speed of a car.

EXAMPLE 8 Modeling with a constant function

A car travels on a freeway with its speed recorded at regular intervals, as listed in **TABLE 15.11**.

Speed of a Car

Elapsed Time (hours)	0	1	2	3	4
Speed (miles per hour)	70	70	70	70	70

TABLE 15.11

(a) Discuss the speed of the car during this time interval.
(b) Find a formula for a function f that models these data.
(c) Sketch a graph of f together with the data.

Solution
(a) The speed of the car appears to be constant at 70 miles per hour.
(b) Because the speed is constant, the rate of change is 0. Thus

$$f(x) \quad = \quad 0x \quad + \quad 70$$

(Future speed) = (Change in speed) + (Initial speed)

and $f(x) = 70$. We call f a *constant function*.

(c) Because $y = f(x)$, graph $y = 70$ with the data points

$(0, 70), (1, 70), (2, 70), (3, 70),$ and $(4, 70)$

to obtain **FIGURE 15.35**.

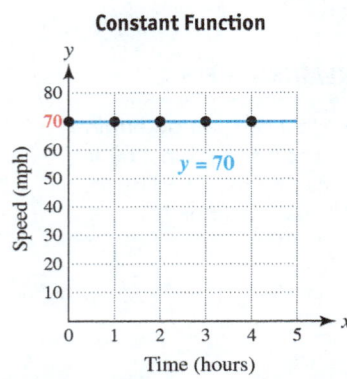

FIGURE 15.35 Speed of a Car

Now Try Exercise 113

The function defined by $f(x) = 70$ is an example of a *constant function*. A **constant function** *is a linear function* with $m = 0$ and can be written as $f(x) = b$. Regardless of the input, a constant function always outputs the same value, b. Its graph is a horizontal line. In general, its domain is all real numbers and its range is $R = \{b\}$.

CRITICAL THINKING 2

Find a formula for a function D that calculates the *distance* traveled by the car in Example 8 after x hours. What is the rate of change for $D(x)$?

 Math in Context *(Real-World)* The following are three applications of constant functions.

- A thermostat calculates a constant function regardless of the weather outside by maintaining a set temperature.
- A cruise control in a car calculates a constant function by maintaining a fixed speed, regardless of the type of road or terrain.
- A constant function calculates the 1250-foot height of the Empire State Building.

4 Applying the Midpoint Formula

Connecting Concepts with Your Life A common way to make estimations is to average data items. For example, in 2000 the average cost of tuition and fees at public two-year colleges was about $1700, and in 2014 it was about $3300. (*Source:* The College Board.) To estimate tuition and fees in 2007, we could average the 2000 and 2014 amounts.

$$\frac{1700 + 3300}{2} = \$2500 \quad \text{Finding the average}$$

This technique predicts that the average cost of tuition and fees was $2500 in 2007 and is referred to as finding the *midpoint*.

15.2 LINEAR FUNCTIONS

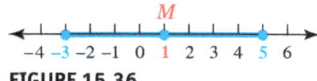

FIGURE 15.36

MIDPOINT FORMULA ON THE REAL NUMBER LINE The **midpoint** of a line segment is the unique point on the line segment that is an equal distance from the endpoints. For example, in **FIGURE 15.36**, the midpoint M of -3 and 5 on the real number line is 1.

We can calculate the value of M as follows.

$$M = \frac{x_1 + x_2}{2} = \frac{-3 + 5}{2} = \frac{2}{2} = 1$$

Average the x-values to find the midpoint.

MIDPOINT FORMULA IN THE xy-PLANE The midpoint of a line segment in the xy-plane can be found in a similar way. **FIGURE 15.37(a)** shows the midpoint on the line segment connecting the points (x_1, y_1) and (x_2, y_2). The x-coordinate of M is equal to the average of x_1 and x_2, and the y-coordinate of M is equal to the average of y_1 and y_2. For example, the line segment shown in **FIGURE 15.37(b)** has endpoints $(-2, 1)$ and $(4, -3)$. The coordinates of the midpoint are

$$M = \left(\frac{-2 + 4}{2}, \frac{1 + (-3)}{2}\right) = (1, -1).$$

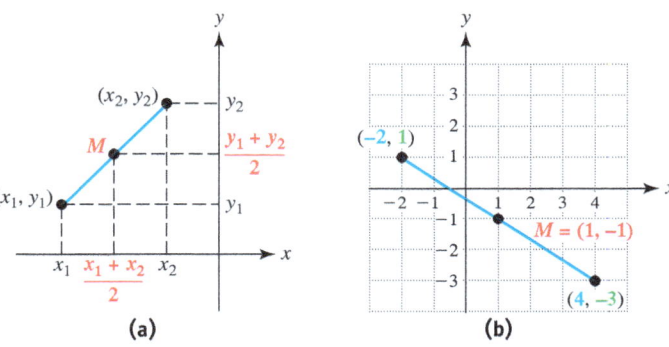

FIGURE 15.37

This discussion is summarized as follows.

MIDPOINT FORMULA IN THE xy-PLANE

The midpoint of the line segment with endpoints (x_1, y_1) and (x_2, y_2) in the xy-plane is

$$\left(\frac{x_1 + x_2}{2}, \frac{y_1 + y_2}{2}\right).$$

EXAMPLE 9 Finding the midpoint

Find the midpoint of the line segment connecting the points $(-3, -2)$ and $(4, 1)$.

Solution
In the midpoint formula, let $(-3, -2)$ be (x_1, y_1) and $(4, 1)$ be (x_2, y_2).

$$M = \left(\frac{x_1 + x_2}{2}, \frac{y_1 + y_2}{2}\right) \quad \text{Midpoint formula}$$

$$= \left(\frac{-3 + 4}{2}, \frac{-2 + 1}{2}\right) \quad \text{Substitute.}$$

$$= \left(\frac{1}{2}, -\frac{1}{2}\right) \quad \text{Simplify.}$$

The midpoint of the line segment is $\left(\frac{1}{2}, -\frac{1}{2}\right)$.

Now Try Exercise 93

EXAMPLE 10 Estimating the U.S. divorce rate

~Social Science~
Math in Context In the next example, we use the midpoint formula to estimate the divorce rate in the United States in 2007.

The divorce rate per 1000 people in 2000 was 4.2, and in 2014 it was 3.6. (*Source:* Statistical Abstract of the United States.)

(a) Use the midpoint formula to estimate the divorce rate in 2007.
(b) Could the midpoint formula be used to estimate the divorce rate in 2003? Explain.

Solution
(a) In the midpoint formula, let $(2000, 4.2)$ be (x_1, y_1) and let $(2014, 3.6)$ be (x_2, y_2).

$$M = \left(\frac{x_1 + x_2}{2}, \frac{y_1 + y_2}{2}\right) \quad \text{Midpoint formula}$$

$$= \left(\frac{2000 + 2014}{2}, \frac{4.2 + 3.6}{2}\right) \quad \text{Substitute.}$$

$$= (2007, 3.9) \quad \text{Simplify.}$$

The midpoint formula estimates that the divorce rate was **3.9** per 1000 people in **2007**.

(b) No, the midpoint formula can only be used to estimate data that are exactly halfway between two given data points. Because the year 2003 is not exactly halfway between 2000 and 2014, the midpoint formula cannot be used.

Now Try Exercise 107

NOTE: An estimate obtained from the midpoint formula is equal to an estimate obtained from a linear function whose graph passes through the endpoints of the line segment. This fact is illustrated in the next example. ■

EXAMPLE 11 Relating midpoints to linear functions

The graph of the linear function f shown in **FIGURE 15.38** is a straight line passing through the points $(-1, 3)$ and $(2, -3)$.

(a) Find a formula for $f(x)$.
(b) Evaluate $f\left(\frac{1}{2}\right)$. Does your answer agree with the graph?
(c) Find the midpoint M of the line segment connecting the points $(-1, 3)$ and $(2, -3)$. Comment on your result.

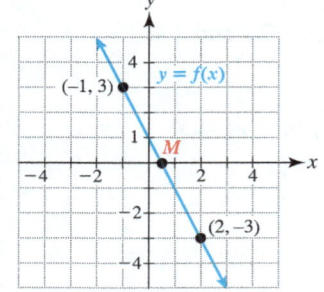

FIGURE 15.38

Solution
(a) The graph of f is a line that passes through $(-1, 3)$ and $(2, -3)$. The slope m of the line is

$$m = \frac{-3 - 3}{2 - (-1)} = -\frac{6}{3} = -2,$$

and from the graph, the y-intercept is $(0, 1)$. Thus $f(x) = -2x + 1$.

(b) $f\left(\frac{1}{2}\right) = -2\left(\frac{1}{2}\right) + 1 = 0$. Yes, they agree because the point $\left(\frac{1}{2}, 0\right)$ lies on the graph of $y = f(x)$ in **FIGURE 15.38**.

(c) The midpoint of the line segment connecting $(-1, 3)$ and $(2, -3)$ is

$$M = \left(\frac{-1+2}{2}, \frac{3+(-3)}{2}\right) = \left(\frac{1}{2}, 0\right).$$

Finding the midpoint $M = \left(\frac{1}{2}, 0\right)$ of the line segment with endpoints $(-1, 3)$ and $(2, -3)$ is equivalent to evaluating the linear function f, whose graph passes through $(-1, 3)$ and $(2, -3)$, at $x = \frac{1}{2}$.

Now Try Exercise 103

15.2 Putting It All Together

CONCEPT	EXPLANATION	EXAMPLES
Linear Function	Can be represented by $f(x) = mx + b$. Its graph is a line with slope m and y-intercept $(0, b)$.	$f(x) = 2x - 6$, $m = 2$ and $b = -6$ $f(x) = 10$, $m = 0$ and $b = 10$
Constant Function	Can be represented by $f(x) = b$. Its graph is a horizontal line.	$f(x) = -7$, $b = -7$ $f(x) = 22$, $b = 22$
Rate of Change for a Linear Function	The output of a linear function changes by a constant amount for each unit increase in the input.	$f(x) = -3x + 8$ decreases 3 units for each unit increase in x. $f(x) = 5$ neither increases nor decreases. The rate of change is 0.
Midpoint Formula	The midpoint of the line segment connecting (x_1, y_1) and (x_2, y_2) is $\left(\dfrac{x_1 + x_2}{2}, \dfrac{y_1 + y_2}{2}\right).$	The midpoint of the line segment connecting $(-2, 3)$ and $(4, 5)$ is $\left(\dfrac{-2 + 4}{2}, \dfrac{3 + 5}{2}\right) = (1, 4).$

REPRESENTATION	COMMENTS	EXAMPLE
Symbolic	Mathematical formula in the form $f(x) = mx + b$	$f(x) = 2x + 1$, where $m = 2$ and $b = 1$
Verbal	Multiply the input x by m and add b.	Multiply the input x by 2 and then add 1 to obtain the output.
Numerical (table of values)	For each unit increase in x in the table, the output of $f(x) = mx + b$ changes by an amount equal to m.	1-unit increase | x | 0 | 1 | 2 | | $f(x)$ | 1 | 3 | 5 | 2-unit increase
Graphical	The graph of a linear function is a line. Plot at least 3 points and then sketch the line. If $f(x) = mx + b$, then the graph of f has slope m and y-intercept $(0, b)$.	Graph of $y = 2x + 1$

15.2 Exercises

CONCEPTS AND VOCABULARY

1. The formula for a linear function is $f(x) =$ ___.

2. The formula for a constant function is $f(x) =$ ___.

3. The graph of a linear function is a(n) ___.

4. The graph of a constant function is a(n) ___ line.

5. If $f(x) = 7x + 5$, each time x increases by 1 unit, $f(x)$ increases by ___ units.

6. If $f(x) = 5$, each time x increases by 1 unit, $f(x)$ increases by ___ units.

7. (True or False?) Every constant function is a linear function.

8. (True or False?) Every linear function is a constant function.

9. If $C(x) = 2x$ calculates the cost in dollars of buying x square feet of carpet, what does 2 represent in the formula? Interpret the fact that the point $(10, 20)$ lies on the graph of C.

10. If $G(x) = 100 - 4x$ calculates the number of gallons of water in a tank after x minutes, what does -4 represent in the formula? Interpret the fact that the point $(5, 80)$ lies on the graph of G.

IDENTIFYING LINEAR FUNCTIONS

Exercises 11–18: Determine whether f is a linear function. If f is linear, give values for m and b so that f may be expressed as $f(x) = mx + b$.

11. $f(x) = \dfrac{1}{2}x - 6$

12. $f(x) = x$

13. $f(x) = \dfrac{5}{2} - x^2$

14. $f(x) = \sqrt{x} + 3$

15. $f(x) = -9$

16. $f(x) = 1.5 - 7.3x$

17. $f(x) = -9x$

18. $f(x) = \dfrac{1}{x}$

Exercises 19–22: Determine whether the graph represents a linear function.

19.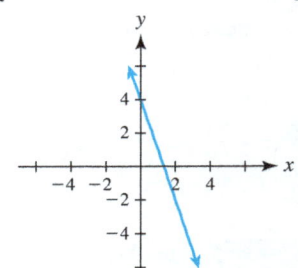

20.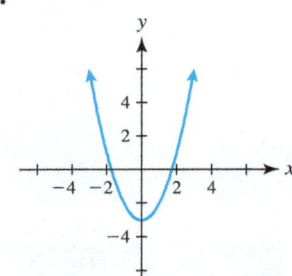

21.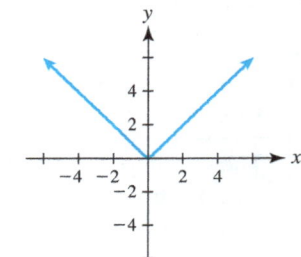

22.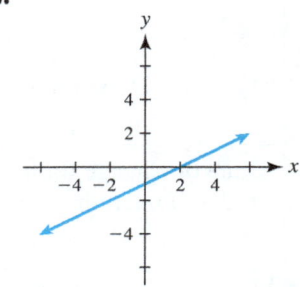

Exercises 23–32: (Refer to Example 2.) Use the table to determine whether $f(x)$ could represent a linear function. If it could, write $f(x)$ in the form $f(x) = mx + b$.

23.
x	0	1	2	3
$f(x)$	-6	-3	0	3

24.
x	0	2	4	6
$f(x)$	-2	2	6	10

25.
x	-2	0	2	4
$f(x)$	6	3	0	-3

26.
x	0	3	6	9
$f(x)$	8	4	2	1

27.
x	-2	-1	0	1
$f(x)$	-5	0	20	40

28.
x	-2	-1	0	1
$f(x)$	6	3	0	-3

29.
x	0	2	3	4
$f(x)$	0	4	6	8

30.

x	1	2	3	4
f(x)	0	1	3	7

31.

x	−1	0	1	2
f(x)	−4	−4	−4	−4

32.

x	2	5	6	8
f(x)	5	5	5	5

EVALUATING LINEAR FUNCTIONS

Exercises 33–38: Evaluate $f(x)$ at the given values of x.

33. $f(x) = 4x$ $x = -4, 5$

34. $f(x) = -2x + 1$ $x = -2, 3$

35. $f(x) = 5 - x$ $x = -\frac{2}{3}, 3$

36. $f(x) = \frac{1}{2}x - \frac{1}{4}$ $x = 0, \frac{1}{2}$

37. $f(x) = -22$ $x = -\frac{3}{4}, 13$

38. $f(x) = 9x - 7$ $x = -1.2, 2.8$

Exercises 39–44: Use the graph of f to evaluate the given expressions.

39. $f(-1)$ and $f(0)$ 40. $f(-2)$ and $f(2)$

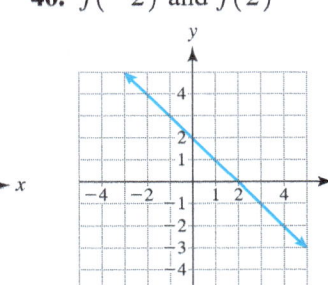

41. $f(-2)$ and $f(4)$ 42. $f(0)$ and $f(3)$

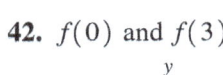
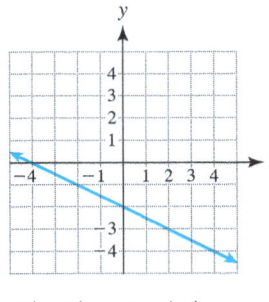
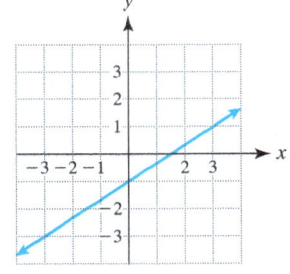

43. $f(-3)$ and $f(1)$ 44. $f(1.5)$ and $f(0.5\pi)$

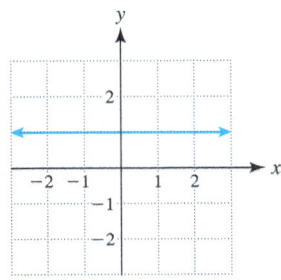
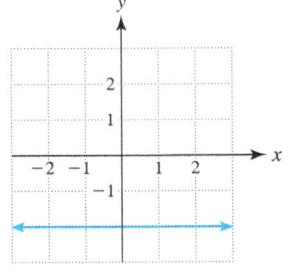

Exercises 45–48: Use the verbal description to write a formula for $f(x)$. Then evaluate $f(3)$.

45. Multiply the input by 6.

46. Multiply the input by −3 and add 7.

47. Divide the input by 6 and subtract $\frac{1}{2}$.

48. Output 8.7 for every input.

REPRESENTING LINEAR FUNCTIONS

Exercises 49–52: Match $f(x)$ with its graph (a–d).

49. $f(x) = 3x$ 50. $f(x) = -2x$

51. $f(x) = x - 2$ 52. $f(x) = 2x + 1$

a. 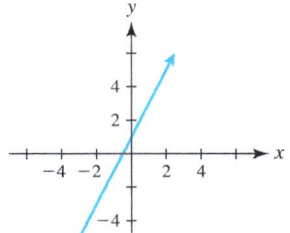 b.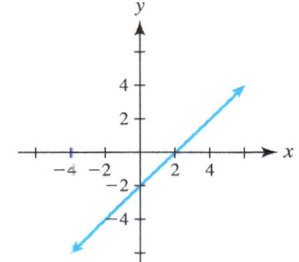

c. 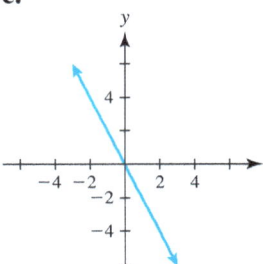 d.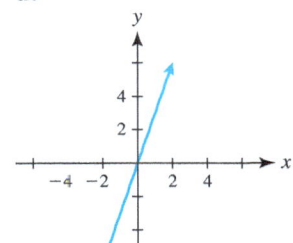

Exercises 53–62: Sketch a graph of $y = f(x)$.

53. $f(x) = 2$ 54. $f(x) = -1$

55. $f(x) = -\frac{1}{2}x$ 56. $f(x) = 2x$

57. $f(x) = x + 1$ 58. $f(x) = x - 2$

59. $f(x) = 3x - 3$ 60. $f(x) = -2x + 1$

61. $f(x) = 3 - x$ 62. $f(x) = \frac{1}{4}x + 2$

Exercises 63–68: Write a symbolic representation (formula) for a linear function f that calculates the following.

63. The number of pounds in x ounces

64. The number of dimes in x dollars

65. The distance traveled by a car moving at 65 miles per hour for t hours

66. The international phone bill *in dollars* for calling t minutes at 10 cents per minute and a fixed fee of $4.95

67. The total number of hours in a day during day x

68. The total cost of downhill skiing x times with a $500 season pass

69. Thinking Generally For each 1-unit increase in x with $y = ax + b$ and $a > 0$, y increases by ___ units.

70. Thinking Generally For each 1-unit decrease in x with $y = cx + d$ and $c < 0$, y increases by ___ units.

Exercises 71–74: Do the following.

(a) Give a verbal representation of f.
(b) Make a numerical representation (table) of f for $x = -2, 0, 2$.
(c) Plot the points listed in the table from part (b), then sketch a graph of f.

71. $f(x) = -2x + 1$ **72.** $f(x) = 1 - x$

73. $f(x) = \frac{1}{2}x - 1$ **74.** $f(x) = \frac{3}{4}x$

Exercises 75–78: Do the following.

(a) Make a numerical representation (table) of f for $x = -3, -2, -1, \ldots, 3$.
(b) Graph f in the window $[-6, 6, 1]$ by $[-4, 4, 1]$.

75. $f(x) = \frac{1}{3}x + \sqrt{2}$ **76.** $f(x) = -\frac{2}{3}x - \sqrt{3}$

77. $f(x) = \dfrac{x + 2}{5}$ **78.** $f(x) = \dfrac{2 - 3x}{7}$

MODELING WITH LINEAR FUNCTIONS

Exercises 79–82: Match the situation with the graph (a–d) that models it best, where x-values represent time from 2000 to 2010.

79. The cost of college tuition

80. The cost of 1 gigabyte of computer memory

81. The distance between Chicago and Denver

82. The total distance traveled by a satellite that is orbiting Earth if the satellite was launched in 2000

a.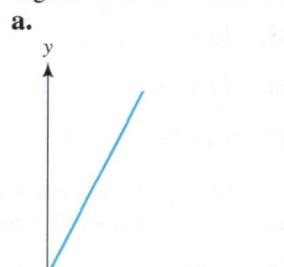
b.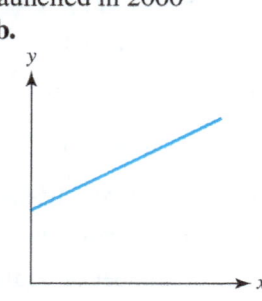
c.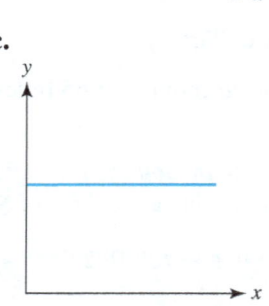
d.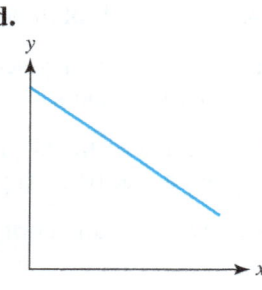

83. Online Exploration Look up the fuel efficiency E in miles per gallon for one of your favorite cars. (Answers will vary.)
(a) Find a function G that calculates the number of gallons required to travel x miles.
(b) If the cost of gasoline is $3 per gallon, find function C that calculates the cost of fuel to travel x miles.

84. Online Exploration Suppose that you would like to drive to Miami for spring break (if it is possible) in the car that you chose in Exercise 83. Calculate the gallons of gasoline needed for the trip.

APPLYING THE MIDPOINT FORMULA

Exercises 85–92: Find the midpoint of the line segment shown.

85.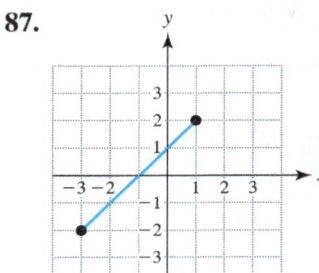

86.

87.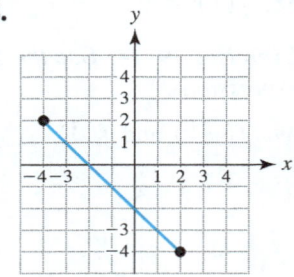

88.

89.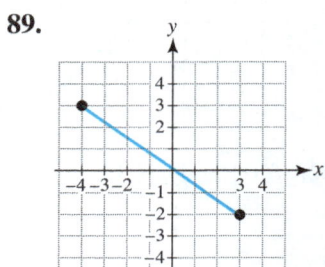

90.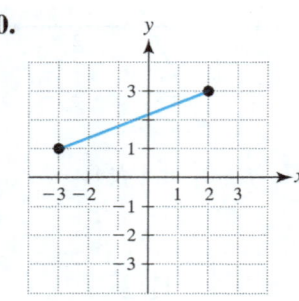

91.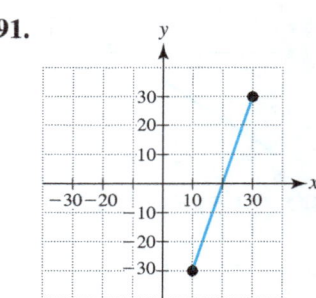

92.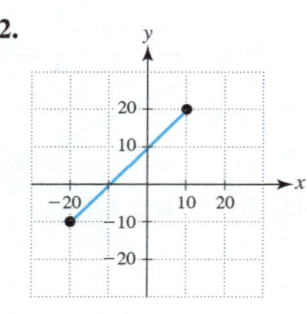

Exercises 93–102: Find the midpoint of the line segment connecting the given points.

93. $(-9, -3), (-7, 1)$ **94.** $(7, -2), (-5, 8)$

95. $\left(\frac{1}{2}, \frac{1}{3}\right), \left(-\frac{5}{2}, -\frac{2}{3}\right)$ **96.** $\left(-\frac{3}{5}, -\frac{1}{4}\right), \left(\frac{1}{10}, \frac{1}{2}\right)$

97. $(-0.3, 0.1)$, $(0.7, 0.4)$

98. $(0.8, -0.4)$, $(0.9, -0.1)$

99. $(2000, 5)$, $(2010, 13)$ 100. $(2005, 9)$, $(2011, 3)$

101. **Thinking Generally** $(a, -b)$, $(3a, 5b)$

102. **Thinking Generally** $(-a, b)$, $(a, -b)$

Exercises 103–106: (Refer to Example 11.) The graph of a linear function f passes through the two given points.

(a) *Find a formula for* $f(x)$. *Determine* $f(2)$.
(b) *Determine* $f(2)$ *by finding the midpoint of the line segment connecting the given points.*
(c) *Compare your answers for parts (a) and (b).*

103. $(0, 5)$, $(4, -3)$

104. $(0, 2)$, $(4, 10)$

105. $(-3, -1)$, $(7, 3)$

106. $(-1, 3)$, $(5, -5)$

Exercises 107–112: Use the midpoint formula to make the requested estimation.

107. **U.S. Life Expectancy** The life expectancy of a female born in 1990 was 78.8 years, and the life expectancy of a female born in 2010 rose to 80.8 years. Estimate the life expectancy of a female born in 2000. (*Source:* Centers for Disease Control and Prevention.)

108. **U.S. Life Expectancy** The life expectancy of a male born in 1990 was 71.8 years, and the life expectancy of a male born in 2010 rose to 75.6 years. Estimate the life expectancy of a male born in 2000. (*Source:* Centers for Disease Control and Prevention.)

109. **U.S. Population** The population of the United States in 1970 was 205 million, and in 2010 it was 308 million. Estimate the population in 1990. (*Source:* U.S. Census Bureau.)

110. **Distance Traveled** A car is moving at a constant speed on an interstate highway. After 1 hour, the car passes the 103-mile marker, and after 5 hours, the car passes the 391-mile marker. What mile marker does the car pass after 3 hours?

111. **U.S. Median Income** In 2009, the median family income was $49,800, and in 2013, it was $51,900. Estimate the median family income in 2011.

112. **Estimating Fish Populations** In 2008, there were approximately 3200 large-mouth bass in a lake. This number increased to 3800 in 2012. Estimate the number of large-mouth bass in the lake in 2010.

APPLICATIONS INVOLVING LINEAR FUNCTIONS

113. **Thermostat** Let $y = f(x)$ describe the temperature y of a room that is kept at $70\,°F$ for x hours.

(a) Represent f symbolically and graphically over a 24-hour period for $0 \leq x \leq 24$.
(b) Construct a table of f for $x = 0, 4, 8, 12, \ldots, 24$.
(c) What type of function is f?

114. **Cruise Control** Let $y = f(x)$ describe the speed y of an automobile after x minutes if the cruise control is set at 60 miles per hour.

(a) Represent f symbolically and graphically over a 15-minute period for $0 \leq x \leq 15$.
(b) Construct a table of f for $x = 0, 1, 2, \ldots, 6$.
(c) What type of function is f?

115. **Distance** A car is initially 50 miles south of the Minnesota—Iowa border, traveling south on Interstate 35. Distances D between the car and the border are recorded in the table for various elapsed times t. Find a linear function D that models these data.

t (hours)	0	2	3	5
D (miles)	50	170	230	350

116. **Estimating the Weight of a Bass** Sometimes the weight of a fish can be estimated by measuring its length. The table lists typical weights of bass that have various lengths.

Length (inches)	12	14	16	18	20	22
Weight (pounds)	1.0	1.7	2.5	3.6	5.0	6.6

Source: Minnesota Department of Natural Resources.

(a) Let x be the length and y be the weight. Make a line graph of the data.
(b) Could the data be modeled accurately with a linear function? Explain your answer.

117. **Texting** In 2014, the average American between the ages of 18 and 24 sent approximately 130 texts per day, whereas the average adult over age 65 sent approximately 5 texts per day.

(a) Find a formula for a function K that calculates the number of texts sent in x days by the average person between the ages of 18 and 24.
(b) Find a formula for a function A that calculates the number of texts sent in x days by the average person over age 65.
(c) Evaluate $K(365)$ and $A(365)$. Interpret your results.

118. **Rain Forests** Rain forests are forests that grow in regions receiving more than 70 inches of rain per year. The world is losing about 49 million acres of rain forest each year. (*Source:* New York Times Almanac.)

(a) Find a linear function f that calculates the change in the acres of rain forest in millions in x years.
(b) Evaluate $f(7)$ and interpret the result.

119. **Historical Car Sales** The table shows the number of U.S. Toyota vehicles sold in millions for past years.

Year	2000	2001	2002	2003	2004
Vehicles	1.6	1.7	1.8	1.9	2.0

Source: Autodata.

(a) What were the sales in 2000?
(b) What was the annual increase in sales?
(c) Find a linear function f that models these data. Let $x = 0$ correspond to 2000, $x = 1$ to 2001, and so on.
(d) Use f to estimate sales in 2006.

120. **Tuition and Fees** Suppose tuition costs $300 per credit and that student fees are fixed at $100.
(a) Find a formula for a linear function T that models the cost of tuition and fees for x credits.
(b) Evaluate $T(16)$ and interpret the result.

121. **Skype Users** The number of active Skype users S in millions x years after 2010 can be modeled by the formula $S(x) = 11x + 145$.
(a) How many active users were there in 2014?
(b) What does the number 145 indicate in the formula?
(c) What does the number 11 indicate in the formula?

122. **Temperature and Volume** If a sample of a gas such as helium is heated, it will expand. The formula $V(T) = 0.147T + 40$ calculates the volume V in cubic inches of a sample of gas at temperature T in degrees Celsius.
(a) Evaluate $V(0)$ and interpret the result.
(b) If the temperature increases by 10 °C, by how much does the volume increase?
(c) What is the volume of the gas when the temperature is 100 °C?

123. **Temperature and Volume** (Refer to the preceding exercise.) A sample of gas at 0 °C has a volume V of 137 cubic centimeters, which increases in volume by 0.5 cubic centimeter for every 1 °C increase in temperature T.
(a) Write a formula $V(T) = aT + b$ that gives the volume of the gas at temperature T.
(b) Find the volume of the gas when $T = 50$ °C.

124. **Cost** To make a music video, it costs $750 to rent a studio plus $5 for each copy produced.
(a) Write a formula $C(x) = ax + b$ that calculates the cost of producing x videos.
(b) Find the cost of producing 2500 videos.

125. **Weight Lifting** Lifting weights can increase a person's muscle mass. Each additional pound of muscle burns an extra 40 calories per day. Write a linear function that models the number of calories burned each day by x pounds of muscle. By burning an extra 3500 calories, a person can lose 1 pound of fat. How many pounds of muscle are needed to burn 1 pound of fat in 30 days? (Source: Runner's World.)

126. **Wireless-Only Households** The percentage P of wireless-only households x years after 2005 can be modeled by the formula $P(x) = 4x + 7$, where $0 \le x \le 7$.
(a) Evaluate $P(0)$ and $P(3)$. Interpret your results.
(b) Explain the meaning of 4 and 7 in the formula.

127. **Smartphone Users** The table lists the percentage P of people with cell phones who have smartphones.

Year	2015	2016	2017
Percentage	43%	46%	49%

(a) What was this percentage in 2015?
(b) By how much did this percentage change each year?
(c) Write a function P that models these data. Let x be years after 2015.
(d) Estimate this percentage in 2019.

128. **Past Walmart Sales** The table shows Walmart's share as a percentage of overall U.S. retail sales for past years. (This percentage excludes restaurants and motor vehicles.)

Year	1998	1999	2000	2001	2002
Share(%)	6	6.5	7	7.5	8

Source: Commerce Department, Walmart.

(a) What was Walmart's share in 1998?
(b) By how much (percent) did Walmart's share increase each year?
(c) Find a linear function f that models these data. Let $x = 0$ correspond to 1998.
(d) Use f to estimate Walmart's share in 2014. Compare your answer to the actual value of 11.3%.

WRITING ABOUT MATHEMATICS

129. Explain how you can determine whether a function is linear by using its
(a) symbolic representation,
(b) graphical representation, and
(c) numerical representation.

130. Describe one way to determine whether data can be modeled by a linear function.

SECTIONS 15.1 and 15.2 Checking Basic Concepts

1. Find a formula and sketch a graph for a function f that squares the input x and then subtracts 1.

2. Use the graph of f to do the following.
   (a) Find the domain and range of f.
   (b) Evaluate $f(0)$ and $f(2)$.
   (c) Is f a linear function? Explain.

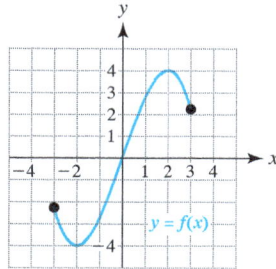

3. Determine whether f is a linear function.
   (a) $f(x) = 4x - 2$
   (b) $f(x) = 2\sqrt{x} - 5$
   (c) $f(x) = -7$
   (d) $f(x) = 9 - 2x + 5x$

4. Graph $f(x) = 4 - 3x$. Evaluate $f(-2)$.

5. Find a formula for a linear function that models the data.

x	0	1	2	3	4
$f(x)$	-1	$-\frac{1}{2}$	0	$\frac{1}{2}$	1

6. The median age in the United States from 1970 to 2015 can be approximated by
$$f(x) = 0.2x + 27.7,$$
where $x = 0$ corresponds to 1970, $x = 1$ to 1971, and so on.
   (a) Evaluate $f(45)$ and interpret the result.
   (b) Interpret the numbers 0.2 and 27.7.

7. Find the midpoint of the line segment connecting the points $(-3, 4)$ and $(5, -6)$.

15.3 Compound Inequalities and Piecewise-Defined Functions

Objectives
1. Introducing Compound Inequalities
2. Finding Symbolic Solutions and Graphing Them on a Number Line
3. Solving Inequalities and Writing Solutions in Interval Notation
4. Finding Numerical and Graphical Solutions
5. Introducing Piecewise-Defined Functions

1 Introducing Compound Inequalities

A person weighing 143 pounds and needing to purchase a life vest for white-water rafting is not likely to find one designed exactly for this weight. Life vests are manufactured to support a range of body weights. A vest approved for weights between 100 and 160 pounds might be appropriate for this person. In other words, if a person's weight is w, this life vest is safe if $w \geq 100$ and $w \leq 160$. This example illustrates the concept of a *compound inequality*.

A **compound inequality** consists of two inequalities joined by the words *and* or *or*. The following are two examples of compound inequalities.

$2x \geq -3$ **and** $2x < 5$ $x + 2 \geq 3$ **or** $x - 1 < -5$
First compound inequality   Second compound inequality

If a compound inequality contains the word *and*, a solution must satisfy *both* inequalities. For example, 1 is a solution to the first compound inequality because

$2(1) \geq -3$ and $2(1) < 5$ First compound inequality with $x = 1$
True   True

are *both* true statements.

Both parts are true so this compound inequality containing "and" is also true.

If a compound inequality contains the word *or*, a solution must satisfy *at least one* of the two inequalities. Thus 5 is a solution to the second compound inequality, because the first statement is true.

> One part is true so this compound inequality containing "or" is also true.

$5 + 2 \geq 3$ or $5 - 1 < -5$ Second compound inequality with $x = 5$
True   False

Note that 5 does not need to satisfy both statements for this compound inequality to be true.

EXAMPLE 1 Determining solutions to compound inequalities

NEW VOCABULARY

☐ Compound inequality
☐ Intersection
☐ Three-part inequality
☐ Union
☐ Interval notation
☐ Infinity
☐ Negative infinity
☐ Piecewise-defined function
☐ Piecewise-defined linear function

Determine whether the given x-values are solutions to the compound inequalities.
(a) $x + 1 < 9$ and $2x - 1 > 8$ $x = 5, -5$
(b) $5 - 2x \leq -4$ or $5 - 2x \geq 4$ $x = 2, -3$

Solution
(a) Substitute $x = 5$ in the compound inequality $x + 1 < 9$ and $2x - 1 > 8$.

$5 + 1 < 9$ and $2(5) - 1 > 8$ Check to see if 5 satisfies *both* inequalities.
True   True

Both inequalities are true, so 5 is a solution. Now substitute $x = -5$.

$-5 + 1 < 9$ and $2(-5) - 1 > 8$ Check to see if -5 satisfies *both* inequalities.
True   False

To be a solution, both inequalities must be true, so -5 is *not* a solution.

(b) Substitute $x = 2$ into the compound inequality $5 - 2x \leq -4$ or $5 - 2x \geq 4$.

$5 - 2(2) \leq -4$ or $5 - 2(2) \geq 4$ Check to see if 2 satisfies *at least one* inequality.
False   False

Neither inequality is true, so 2 is *not* a solution. Now substitute $x = -3$.

$5 - 2(-3) \leq -4$ or $5 - 2(-3) \geq 4$ Check to see if -3 satisfies *at least one* inequality.
False   True

At least one of the two inequalities is true, so -3 is a solution.

Now Try Exercises 7, 9

2 Finding Symbolic Solutions and Graphing Them on a Number Line

We can use a number line to graph solutions to compound inequalities, such as

$$x \leq 6 \text{ and } x > -4.$$

The solution set for $x \leq 6$ is shaded to the left of 6, with a bracket placed at $x = 6$, as shown in **FIGURE 15.39** on the next page. The solution set for $x > -4$ can be shown by shading a different number line to the right of -4 and placing a left parenthesis at -4. Because the inequalities are connected by *and*, the solution set consists of all numbers that are shaded on *both* number lines. The final number line represents the intersection of the two solution sets. That is, the solution set includes real numbers where the graphs "overlap." For any two sets A and B, the **intersection** of A and B, denoted $A \cap B$, is defined by

$$A \cap B = \{x \mid x \text{ is an element of } A \text{ and an element of } B\}.$$

15.3 COMPOUND INEQUALITIES AND PIECEWISE-DEFINED FUNCTIONS

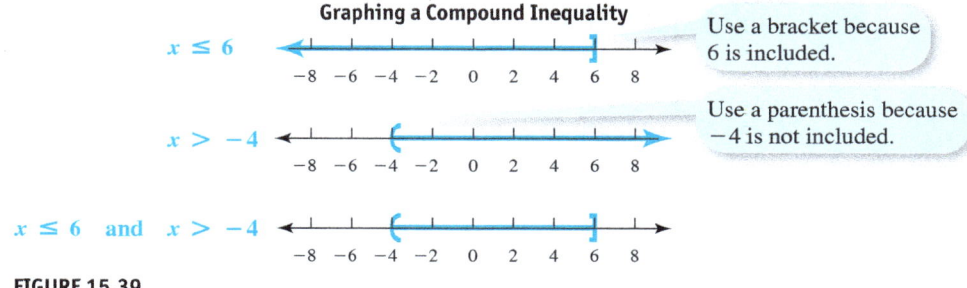

FIGURE 15.39

NOTE: A bracket, either [or], is used when an inequality contains $\leq$ or $\geq$. A parenthesis, either (or), is used when an inequality contains $<$ or $>$. This notation makes clear whether or not an endpoint is included in the inequality. ∎

EXAMPLE 2 Solving a compound inequality containing "and"

Solve $2x + 4 > 8$ and $5 - x < 9$. Graph the solution set.

Solution
First solve each linear inequality separately.

$$2x + 4 > 8 \quad \text{and} \quad 5 - x < 9$$

Subtract 4 and divide by 2. $2x > 4$ and $-x < 4$ Subtract 5 and multiply by -1.

$$x > 2 \quad \text{and} \quad x > -4$$

Graph $x > 2$ and $x > -4$ on two different number lines. On a third number line, shade solutions that appear on both of the first two number lines. As shown in **FIGURE 15.40**, the solution set written in set-builder notation is $\{x \mid x > 2\}$.

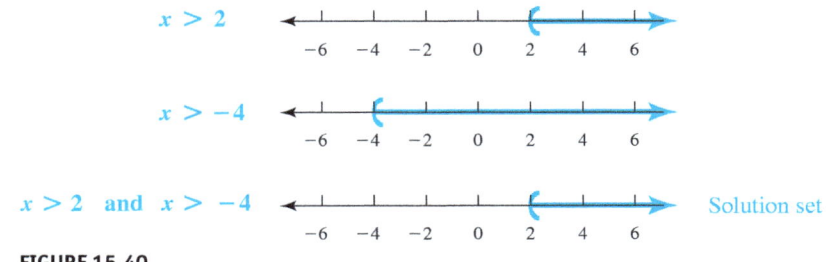

FIGURE 15.40

Now Try Exercise 43

We can also solve compound inequalities containing the word *or*. To write the solution to such an inequality, we sometimes use *union* notation. For any two sets A and B, the **union** of A and B, denoted $A \cup B$, is defined by

$$A \cup B = \{x \mid x \text{ is an element of } A \textit{ or } \text{an element of } B\}.$$

If the solution to an inequality is $\{x \mid x < 1\}$ or $\{x \mid x \geq 3\}$, then it can also be written as

$$\{x \mid x < 1\} \cup \{x \mid x \geq 3\}.$$

That is, we can replace the word *or* with the $\cup$ symbol.

EXAMPLE 3   Solving a compound inequality containing "or"

Solve $x + 2 < -1$ or $x + 2 > 1$. Graph the solution set.

Solution
We first solve each linear inequality.

$$x + 2 < -1 \quad \text{or} \quad x + 2 > 1 \qquad \text{Given compound inequality}$$
$$x < -3 \quad \text{or} \quad x > -1 \qquad \text{Subtract 2.}$$

We can graph the simplified inequalities on different number lines, as shown in **FIGURE 15.41**. A solution must satisfy at least one of the two inequalities. Thus the solution set for the compound inequality results from taking the *union* of the first two number lines. We can write the solution, using set-builder notation, as $\{x \mid x < -3\} \cup \{x \mid x > -1\}$ or as $\{x \mid x < -3 \text{ or } x > -1\}$.

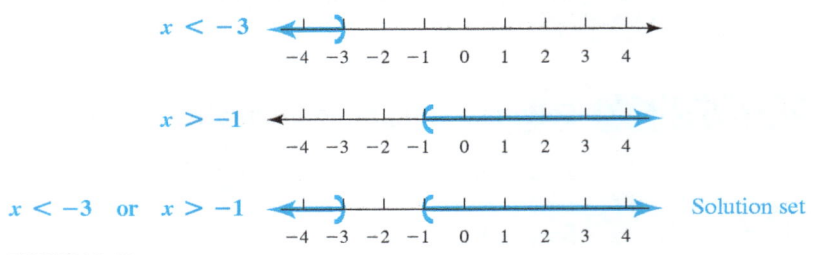

FIGURE 15.41

Now Try Exercise 47

3 Solving Inequalities and Writing Solutions in Interval Notation

READING CHECK 1

- Write the compound inequality $x < 2$ and $x \geq -1$ as a three-part inequality.

Sometimes a compound inequality containing the word *and* can be combined into a three-part inequality. For example, rather than writing

$$x > 5 \quad \text{and} \quad x \leq 10,$$

we could write the **three-part inequality**

$$5 < x \leq 10.$$

This three-part inequality is represented by the number line shown in **FIGURE 15.42**.

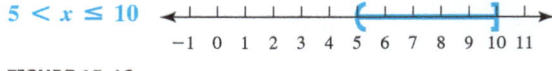

FIGURE 15.42

A convenient notation for number line graphs is called **interval notation**. Instead of drawing the entire number line as in **FIGURE 15.42**, the solution set can be expressed as $(5, 10]$ in interval notation. The solution set does not include 5, but does include 10. To denote this, a parenthesis is used next to 5 and a bracket is used next to 10, as is done on the number line. A solution set that includes all real numbers satisfying $-2 < x < 3$ can be expressed as $(-2, 3)$. Parentheses indicate that the endpoints are *not* included. The interval $0 \leq x \leq 4$ is represented by $[0, 4]$.

MAKING CONNECTIONS 1

Points and Intervals

The expression $(1, 2)$ may represent a point in the xy-plane or the interval $1 < x < 2$. To alleviate confusion, phrases such as "the point $(1, 2)$" or "the interval $(1, 2)$" are used.

TABLE 15.12 provides examples of interval notation. The symbol ∞ refers to **infinity**, and it does not represent a real number. The interval $(5, \infty)$ represents $x > 5$, which has no maximum x-value, so ∞ is used for the right endpoint. The symbol $-\infty$ may be used similarly and denotes **negative infinity**. Real numbers are denoted $(-\infty, \infty)$.

Interval Notation

Inequality	Interval Notation	Number Line Graph
$-1 < x < 3$	$(-1, 3)$	
$-3 < x \leq 2$	$(-3, 2]$	
$-2 \leq x \leq 2$	$[-2, 2]$	
$x < -1$ or $x > 2$	$(-\infty, -1) \cup (2, \infty)$ ($\cup$ is the union symbol.)	
$x > -1$	$(-1, \infty)$	
$x \leq 2$	$(-\infty, 2]$	

TABLE 15.12

EXAMPLE 4 Writing inequalities in interval notation

Write each expression in interval notation.
(a) $-2 \leq x < 5$ (b) $x \geq 3$ (c) $x < -5$ or $x \geq 2$
(d) $\{x \mid x > 0 \text{ and } x \leq 3\}$ (e) $\{x \mid x \leq 1 \text{ or } x \geq 3\}$

Solution
(a) $[-2, 5)$ (b) $[3, \infty)$ (c) $(-\infty, -5) \cup [2, \infty)$
(d) $(0, 3]$ (e) $(-\infty, 1] \cup [3, \infty)$

Now Try Exercises 13, 17, 23, 27, 37

EXAMPLE 5 Solving three-part inequalities

Solve each inequality and graph each solution set. Write the solution set in interval notation.
(a) $4 < t + 2 \leq 8$ (b) $-3 \leq 3z \leq 6$ (c) $-\dfrac{5}{2} < \dfrac{1 - m}{2} < 4$

Solution
(a) To solve a three-part inequality, isolate the variable by applying properties of inequalities to each part of the inequality.

$$4 < t + 2 \leq 8 \quad \text{Given three-part inequality}$$
$$4 - 2 < t + 2 - 2 \leq 8 - 2 \quad \text{Subtract 2 from each part.}$$
$$2 < t \leq 6 \quad \text{Simplify each part.}$$

The solution set is $(2, 6]$. See **FIGURE 15.43**.

$2 < t \leq 6$  Solution set

FIGURE 15.43

CRITICAL THINKING

Graph the following inequalities and discuss your results.
1. $x < 2$ and $x > 5$
2. $x > 2$ or $x < 5$

(b) To simplify, divide each part by 3.

$$-3 \leq 3z \leq 6 \qquad \text{Given three-part inequality}$$
$$\frac{-3}{3} \leq \frac{3z}{3} \leq \frac{6}{3} \qquad \text{Divide each part by 3.}$$
$$-1 \leq z \leq 2 \qquad \text{Simplify each part.}$$

The solution set is $[-1, 2]$. See **FIGURE 15.44**.

$-1 \leq z \leq 2$ Solution set

FIGURE 15.44

(c) Multiply each part by 2 to clear (eliminate) fractions.

$$-\frac{5}{2} < \frac{1-m}{2} < 4 \qquad \text{Given three-part inequality}$$
$$2 \cdot \left(-\frac{5}{2}\right) < 2 \cdot \left(\frac{1-m}{2}\right) < 2 \cdot 4 \qquad \text{Multiply each part by 2.}$$
$$-5 < 1 - m < 8 \qquad \text{Simplify each part.}$$
$$-5 - 1 < 1 - m - 1 < 8 - 1 \qquad \text{Subtract 1 from each part.}$$
$$-6 < -m < 7 \qquad \text{Simplify each part.}$$
$$-1 \cdot (-6) > -1 \cdot (-m) > -1 \cdot 7 \qquad \text{Multiply each part by } -1;$$
$$\qquad \textit{reverse} \text{ inequality symbols.}$$
$$6 > m > -7 \qquad \text{Simplify each part.}$$
$$-7 < m < 6 \qquad \text{Rewrite inequality.}$$

The solution set is $(-7, 6)$. See **FIGURE 15.45**.

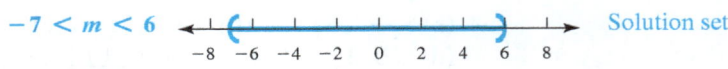

$-7 < m < 6$ Solution set

FIGURE 15.45

STUDY TIP

When simplifying a three-part inequality, be sure to perform the same step on each of the three parts.

NOTE: Either $6 > m > -7$ or $-7 < m < 6$ is a correct way to write a three-part inequality. However, we usually write the smaller number on the left side and the larger number on the right side. ∎

Now Try Exercises 59, 63, 79

Math in Context *Meteorology* Three-part inequalities occur frequently in applications. In the next example, we find altitudes at which the air temperature is within a certain range.

EXAMPLE 6   **Solving a three-part inequality**

If the ground-level temperature is 80 °F, the air temperature x miles above Earth's surface is cooler and can be modeled by $T(x) = 80 - 19x$. Find the altitudes at which the air temperature ranges from 42 °F down to 23 °F. (*Source:* A. Miller and R. Anthes, *Meteorology*.)

Solution

We write and solve the three-part inequality $23 \leq T(x) \leq 42$.

$$23 \leq 80 - 19x \leq 42 \quad \text{Substitute for } T(x).$$
$$-57 \leq -19x \leq -38 \quad \text{Subtract 80 from each part.}$$
$$\frac{-57}{-19} \geq x \geq \frac{-38}{-19} \quad \text{Divide by } -19; \text{ } reverse \text{ inequality symbols.}$$
$$3 \geq x \geq 2 \quad \text{Simplify.}$$
$$2 \leq x \leq 3 \quad \text{Rewrite inequality.}$$

The air temperature ranges from 42 °F to 23 °F for altitudes between 2 and 3 miles. In interval notation, the solution set can be expressed as $[2, 3]$.

Now Try Exercise 123

MAKING CONNECTIONS 2

Writing Three-Part Inequalities

The inequality $-2 < x < 1$ means that $x > -2$ *and* $x < 1$. A three-part inequality should *not* be used when *or* connects a compound inequality. Writing $x < -2$ or $x > 1$ as $1 < x < -2$ is **incorrect** because it states that x must be both greater than 1 *and* less than -2. It is impossible for any value of x to satisfy this statement.

CRITICAL THINKING 1

Carbon dioxide is emitted when human beings breathe. In one study of college students, the amount of carbon dioxide exhaled in grams per hour was measured during both lectures and exams. The average amount exhaled during lectures L satisfied $25.33 \leq L \leq 28.17$, whereas the average amount exhaled during exams E satisfied $36.58 \leq E \leq 40.92$. What do these results indicate? Explain. (*Source:* T. Wang, ASHRAE Trans.)

EXAMPLE 7 Solving an inequality

Solve $2x + 1 \leq -1$ or $2x + 1 \geq 3$. Write the solution set in interval notation.

Solution
First solve each inequality.

$$2x + 1 \leq -1 \quad \text{or} \quad 2x + 1 \geq 3 \quad \text{Given compound inequality}$$
$$2x \leq -2 \quad \text{or} \quad 2x \geq 2 \quad \text{Subtract 1.}$$
$$x \leq -1 \quad \text{or} \quad x \geq 1 \quad \text{Divide by 2.}$$

The solution set may be written in interval notation as $(-\infty, -1] \cup [1, \infty)$.

Now Try Exercise 55

---

4 Finding Numerical and Graphical Solutions

Compound inequalities can also be solved graphically and numerically, as illustrated in the next example.

EXAMPLE 8 Estimating numbers of Internet users

The number of U.S. Internet users in millions during year x can be modeled by the formula $f(x) = 11.6(x - 2000) + 124$. Estimate the years when the number of users was between 240 and 275 million. (*Source:* The Nielsen Company.)

Solution
Numerical Solution Let $y_1 = 11.6(x - 2000) + 124$. Make a table of values, as shown in **FIGURE 15.46(a)**. In 2010, the number of Internet users was 240 million, and in 2013, this number was about 275 million. Thus from 2010 to about 2013, the number of Internet users was between 240 million and 275 million.

Graphical Solution The graph of $y_1 = 11.6(x - 2000) + 124$ is shown between the graphs of $y_2 = 240$ and $y_3 = 275$ in **FIGURES 15.46(b)** and **15.46(c)** from 2010 to about 2013, or when $2010 \leq x \leq 2013$.

CALCULATOR HELP
To find a point of intersection, see Appendix A (page AP-6).

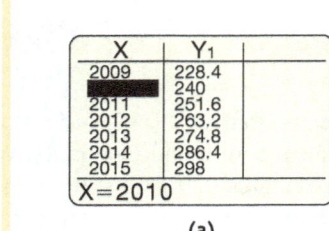
(a)

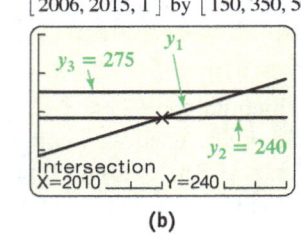
(b)

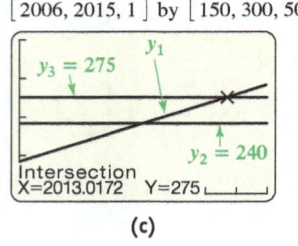
(c)

FIGURE 15.46

Now Try Exercise 117(a) and (b)

5 Introducing Piecewise-Defined Functions

When a function f models data, there may not be one formula for $f(x)$ that works. In this case, the function is sometimes defined on pieces of its domain and is therefore called a **piecewise-defined function**. If each piece is linear, the function is a **piecewise-defined linear function**.

An example of a piecewise-defined function is the *Fujita scale*, which classifies tornadoes by intensity. If a tornado has wind speeds between 40 and 72 miles per hour, it is an F1 tornado. Tornadoes with wind speeds *greater than* 72 miles per hour but not more than 112 miles per hour are F2 tornadoes. The Fujita scale is represented by the following function F, where the input x represents the maximum wind speed of a tornado and the output is the F-scale number from 1 to 5.

Fujita Scale (F1 → F5)

F is defined in five pieces over intervals of its domain.

$$F(x) = \begin{cases} 1 & \text{if } 40 \leq x \leq 72 \\ 2 & \text{if } 72 < x \leq 112 \\ 3 & \text{if } 112 < x \leq 157 \\ 4 & \text{if } 157 < x \leq 206 \\ 5 & \text{if } 206 < x \leq 260 \end{cases}$$

$F(180) = 4$

For example, if the maximum wind speed is 180 miles per hour, then $F(180) = 4$ because 180 is between 157 and 206; that is, $157 < 180 \leq 206$. Thus a tornado with a maximum wind speed of 180 miles per hour is an F4 tornado.

A graph of $y = f(x)$ is shown in **FIGURE 15.47**. It is composed of horizontal line segments. Because each piece is constant, F is sometimes called a **piecewise-constant function** or **step function**. A solid dot occurs at the point $(72, 1)$ and an open circle occurs at the point $(72, 2)$, because technically a tornado with 72-mile-per-hour winds is an F1 tornado, not an F2 tornado.

Tornado Intensities

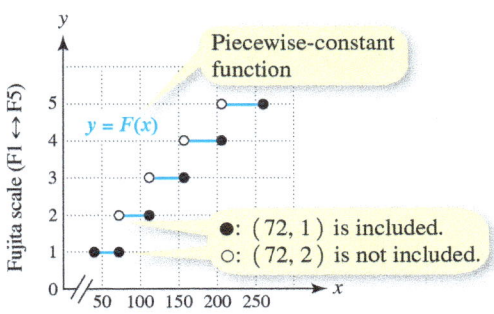

FIGURE 15.47

EXAMPLE 9 Evaluating and graphing a piecewise-defined function

Use $f(x)$ to complete the following.

$$f(x) = \begin{cases} x - 1 & \text{if } -4 \leq x < 2 \\ -2x & \text{if } 2 \leq x \leq 4 \end{cases}$$

(a) What is the domain of f?
(b) Evaluate $f(-3), f(2), f(4),$ and $f(5)$.
(c) Sketch a graph of f.
(d) What is the range of f?

Solution
(a) Function f is defined for x-values satisfying either $-4 \leq x < 2$ or $2 \leq x \leq 4$. Thus the domain of f is $-4 \leq x \leq 4$.
(b) For x-values satisfying $-4 \leq x < 2$, $f(x) = x - 1$ and so $f(-3) = -4$. Similarly, if $2 \leq x \leq 4$, then $f(x) = -2x$. Thus $f(2) = -4$ and $f(4) = -8$. The expression $f(5)$ is undefined because 5 is not in the domain of f.
(c) Because each piece of $f(x)$ is linear, the graph of $y = f(x)$ consists of two line segments. Therefore we can find the endpoints of each line segment and then sketch the two line segments together to form the graph of f.

A Graph the First Piece Evaluate $y_1 = x - 1$ at $x = -4$ and $x = 2$. Place a dot at $(-4, -5)$ and an open circle at $(2, 1)$, because 2 is not included in $-4 \leq x < 2$. Sketch a line segment between these points. See **FIGURE 15.48**.

B Graph the Second Piece Evaluate $y_2 = -2x$ at $x = 2$ and $x = 4$. Place dots at $(2, -4)$ and $(4, -8)$. Sketch a line segment between these points. See **FIGURE 15.49**.

Graph the First Piece

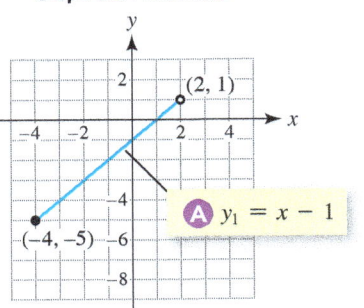

FIGURE 15.48

Graph the Second Piece

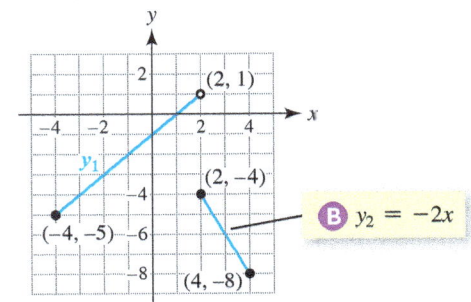

FIGURE 15.49

(d) The y-values in **FIGURE 15.49** vary from $y = -8$ up to but *not* including $y = 1$. The range is $-8 \leq x < 1$.

Now Try Exercise 111

15.3 Putting It All Together

CONCEPT	EXPLANATION	EXAMPLES
Compound Inequality	Two inequalities joined by *and* or *or*	$2x \geq 10$ and $x + 2 < 16$; $x < -1$ or $x > 2$
Three-Part Inequality	Can be used to write some types of compound inequalities involving *and*	$x > -2$ and $x \leq 3$ is equivalent to $-2 < x \leq 3$.
Interval Notation	Notation used to write sets of real numbers rather than using number lines or inequalities	$-2 \leq z \leq 4$ is equivalent to $[-2, 4]$. $x < 4$ is equivalent to $(-\infty, 4)$. $x \leq -2$ or $x > 0$ is equivalent to $(-\infty, -2] \cup (0, \infty)$.
Piecewise-Defined Function	A function that is defined by using different formulas on different intervals of its domain	$f(x) = \begin{cases} x^2 & \text{if } x \leq 2 \\ 2x & \text{if } x > 2 \end{cases}$

TYPE OF INEQUALITY	METHOD USED TO SOLVE THE INEQUALITY
Solving a Compound Inequality Containing *and*	**STEP 1:** First solve each inequality individually. **STEP 2:** The solution set includes values that satisfy *both* inequalities from Step 1.
Solving a Compound Inequality Containing *or*	**STEP 1:** First solve each inequality individually. **STEP 2:** The solution set includes values that satisfy *at least one* of the inequalities from Step 1.
Solving a Three-Part Inequality	Work on all three parts at the same time. Be sure to perform the same step on each part. Continue until the variable is isolated in the middle part.

15.3 Exercises

CONCEPTS AND VOCABULARY

1. Give an example of a compound inequality containing the word *and*.

2. Give an example of a compound inequality containing the word *or*.

3. Is 1 a solution to $x > 3$ and $x \leq 5$?

4. Is 1 a solution to $x < 3$ or $x \geq 5$?

5. Is the compound inequality $x \geq -5$ and $x \leq 5$ equivalent to $-5 \leq x \leq 5$?

6. Name three ways to solve a compound inequality.

TESTING SOLUTIONS TO COMPOUND INEQUALITIES

Exercises 7–12: Determine whether the given values of x are solutions to the compound inequality.

7. $x - 1 < 5$ and $2x > 3$ $x = 2, x = 6$

8. $2x + 1 \geq 4$ and $1 - x \leq 3$ $x = -2, x = 3$

9. $3x < -5$ or $2x \geq 3$ $x = 0, x = 3$

10. $x + 1 \leq -4$ or $x + 1 \geq 4$ $x = -5, x = 2$

11. $2 - x > -5$ and $2 - x \leq 4$ $x = -3, x = 0$

12. $x + 5 \geq 6$ or $3x \leq 3$ $x = -1, x = 1$

WRITING INTERVAL NOTATION

Exercises 13–38: Write the inequality in interval notation.

13. $2 \leq x \leq 10$
14. $-1 < x < 5$
15. $5 < x \leq 8$
16. $-\frac{1}{2} \leq x \leq \frac{5}{6}$
17. $x < 4$
18. $x \leq -3$
19. $x > -2$
20. $x \geq 6$
21. $x \geq -2$ and $x < 5$
22. $x \leq 6$ and $x \geq 2$
23. $x \leq 8$ and $x > -8$
24. $x \geq -4$ and $x < 3$
25. $x \geq 6$ or $x > 3$
26. $x \leq -4$ or $x < -3$
27. $x \leq -2$ or $x \geq 4$
28. $x \leq -1$ or $x > 6$
29. $x < 1$ or $x \geq 5$
30. $x < -3$ or $x > 3$

31. [number line graph: -2 to 5]
32. [number line graph: from 2 rightward]
33. [number line graph: leftward to -2]
34. [number line graph: -4 to 5]

35. $\{x \mid x < 4\}$
36. $\{x \mid -1 \leq x < 4\}$
37. $\{x \mid x < 1 \text{ or } x > 2\}$
38. $\{x \mid -\infty < x < \infty\}$

FINDING SYMBOLIC SOLUTIONS

Exercises 39–48: Solve the compound inequality. Graph the solution set on a number line.

39. $x \leq 3$ and $x \geq -1$
40. $x \geq 5$ and $x > 6$
41. $2x < 5$ and $2x > -4$
42. $2x + 1 < 3$ and $x - 1 \geq -5$
43. $x + 2 > 5$ and $3 - x < 10$
44. $x + 2 > 5$ or $3 - x < 10$
45. $x \leq -1$ or $x \geq 2$
46. $2x \leq -6$ or $x \geq 6$
47. $5 - x > 1$ or $x + 3 \geq -1$
48. $1 - 2x > 3$ or $2x - 4 \geq 4$

Exercises 49–58: Solve the compound inequality. Write your answer in interval notation.

49. $x - 3 \leq 4$ and $x + 5 \geq -1$
50. $2z \geq -10$ and $z < 8$
51. $3t - 1 > -1$ and $2t - \frac{1}{2} > 6$
52. $2(x + 1) < 8$ and $-2(x - 4) > -2$
53. $x - 4 \geq -3$ or $x - 4 \leq 3$
54. $1 - 3n \geq 6$ or $1 - 3n \leq -4$
55. $-x < 1$ or $5x + 1 < -10$
56. $7x - 6 > 0$ or $-\frac{1}{2}x \leq 6$
57. $1 - 7x < -48$ and $3x + 1 \leq -9$
58. $3x - 4 \leq 8$ or $4x - 1 \leq 13$

Exercises 59–80: Solve the three-part inequality. Write your answer in interval notation.

59. $-2 \leq t + 4 < 5$
60. $5 < t - 7 < 10$
61. $-\frac{5}{8} \leq y - \frac{3}{8} < 1$
62. $-\frac{1}{2} < y - \frac{3}{2} < \frac{1}{2}$
63. $-27 \leq 3x \leq 9$
64. $-4 < 2y < 22$
65. $\frac{1}{2} < -2y \leq 8$
66. $-16 \leq -4x \leq 8$
67. $-4 < 5z + 1 \leq 6$
68. $-3 \leq 3z + 6 < 9$
69. $3 \leq 4 - n \leq 6$
70. $-1 < 3 - n \leq 1$
71. $-1 < 2z - 1 < 3$
72. $2 \leq 4z + 5 \leq 6$
73. $-2 \leq 5 - \frac{1}{3}m < 2$
74. $-\frac{3}{2} < 4 - 2m < \frac{7}{2}$
75. $100 \leq 10(5x - 2) \leq 200$
76. $-15 < 5(x - 1990) < 30$
77. $-3 < \dfrac{3z + 1}{4} < 1$
78. $-3 < \dfrac{z - 1}{2} < 5$
79. $-\dfrac{5}{2} \leq \dfrac{2 - m}{4} \leq \dfrac{1}{2}$
80. $\dfrac{4}{5} \leq \dfrac{4 - 2m}{10} \leq 2$

FINDING NUMERICAL AND GRAPHICAL SOLUTIONS

Exercises 81–84: Use the table to solve the three-part inequality. Write your answer in interval notation.

81. $-3 \leq 3x \leq 6$

X	Y₁
-2	-6
-1	-3
0	0
1	3
2	6
3	9
4	12

Y₁■3X

82. $-5 \leq 2x - 1 \leq 1$

X	Y₁
-4	-9
-3	-7
-2	-5
-1	-3
0	-1
1	1
2	3

Y₁■2X−1

83. $-1 < 1 - x < 2$ **84.** $-2 \leq -2x < 4$

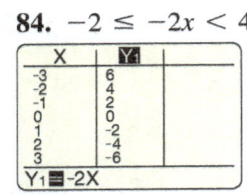

Exercises 85–88: Use the graph to solve the compound inequality. Write your answer in interval notation.

85. $-2 \leq y_1 \leq 2$ **86.** $1 \leq y_1 < 3$

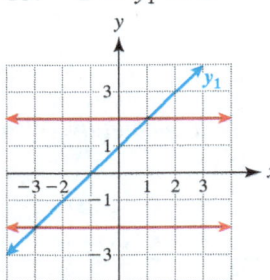 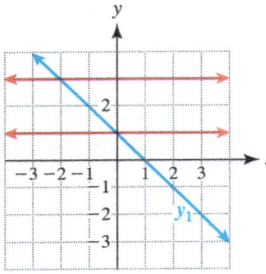

87. $y_1 < -2$ or $y_1 > 2$ **88.** $y_1 \leq -2$ or $y_1 \geq 4$

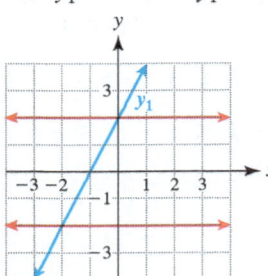 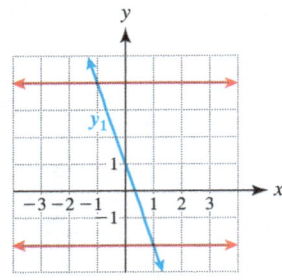

89. Distance The function f, shown in the figure, gives the distance y in miles between a car and Omaha, Nebraska, after x hours, where $0 \leq x \leq 6$.
 (a) Is the car moving toward or away from Omaha? Explain.
 (b) Determine the times when the car is 100 miles or 200 miles from Omaha.
 (c) When is the car from 100 to 200 miles from Omaha?
 (d) When is the car's distance from Omaha greater than or equal to 200 miles?

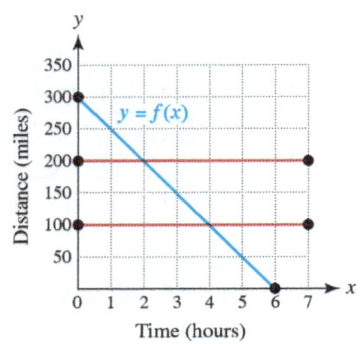

90. Distance The function g, shown in the figure, gives the distance y in miles between a train and Seattle after x hours, where $0 \leq x \leq 5$.

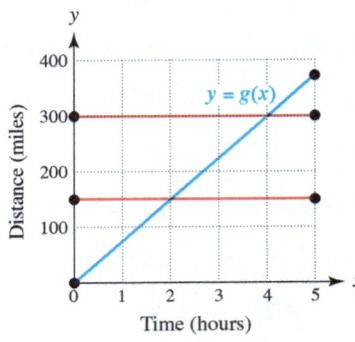

 (a) Is the train moving toward or away from Seattle? Explain.
 (b) Determine the times when the train is 150 miles or 300 miles from Seattle.
 (c) When is the train from 150 to 300 miles from Seattle?
 (d) When is the train's distance from Seattle less than or equal to 150 miles?

91. Use the figure to solve each equation or inequality. Let the domains of y_1, y_2, and y_3 be $0 \leq x \leq 8$.
 (a) $y_1 = y_2$ (b) $y_2 = y_3$
 (c) $y_1 \leq y_2 \leq y_3$ (d) $y_2 < y_1$

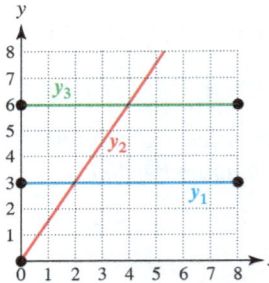

92. Use the figure to solve each equation or inequality. Let the domains of y_1, y_2, and y_3 be $0 \leq x \leq 5$.
 (a) $y_1 = y_2$ (b) $y_2 = y_3$
 (c) $y_1 \leq y_2 \leq y_3$ (d) $y_2 < y_3$

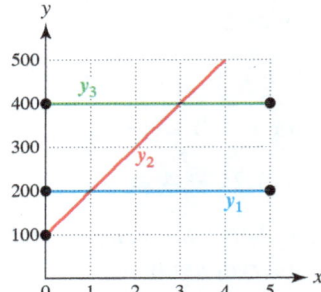

Exercises 93–98: Solve numerically or graphically. Write your answer in interval notation.

93. $-2 \leq 2x - 4 \leq 4$ 94. $-1 \leq 1 - x \leq 3$

95. $x + 1 < -1$ or $x + 1 > 1$

96. $2x - 1 < -3$ or $2x - 1 > 5$

97. $95 \leq 25(x - 2000) + 45 \leq 295$

98. $42 \leq -13(x - 2005) + 120 \leq 94$

USING MORE THAN ONE METHOD

Exercises 99–104: Solve symbolically, graphically, and numerically. Write the solution set in interval notation.

99. $4 \leq 5x - 1 \leq 14$ 100. $-4 < 2x < 4$

101. $4 - x \geq 1$ or $4 - x < 3$

102. $x + 3 \geq -2$ or $x + 3 \leq 1$

103. $2x + 1 < 3$ or $2x + 1 \geq 7$

104. $3 - x \leq 4$ or $3 - x > 8$

105. **Thinking Generally** Solve $c < x + b \leq d$ for x.

106. **Thinking Generally** Solve $c \leq ax + b \leq d$ for x, if $a < 0$.

WORKING WITH PIECEWISE-DEFINED FUNCTIONS

Exercises 107–110: Graph each piecewise-defined function.

107. $f(x) = \begin{cases} x - 1 & \text{if } x \leq 3 \\ 2 & \text{if } x > 3 \end{cases}$

108. $f(x) = \begin{cases} 6 - x & \text{if } x \leq 3 \\ 3x - 6 & \text{if } x > 3 \end{cases}$

109. $f(x) = \begin{cases} 4 - x & \text{if } x < 2 \\ 1 + 2x & \text{if } x \geq 2 \end{cases}$

110. $f(x) = \begin{cases} 2x + 1 & \text{if } x \geq 0 \\ x & \text{if } x < 0 \end{cases}$

Exercises 111–116: Complete the following for $f(x)$.

(a) Determine the domain of f.
(b) Evaluate $f(-2)$, $f(0)$, and $f(3)$.
(c) Graph f.
(d) What is the range of f?

111. $f(x) = \begin{cases} 2 & \text{if } -5 \leq x \leq -1 \\ x + 3 & \text{if } -1 < x \leq 5 \end{cases}$

112. $f(x) = \begin{cases} 2x + 1 & \text{if } -3 \leq x < 0 \\ x - 1 & \text{if } 0 \leq x \leq 3 \end{cases}$

113. $f(x) = \begin{cases} 3x & \text{if } -1 \leq x < 1 \\ x + 1 & \text{if } 1 \leq x \leq 2 \end{cases}$

114. $f(x) = \begin{cases} -2 & \text{if } -6 \leq x < -2 \\ 0 & \text{if } -2 \leq x < 0 \\ 3x & \text{if } 0 \leq x \leq 4 \end{cases}$

115. $f(x) = \begin{cases} x & \text{if } -3 \leq x \leq -1 \\ 1 & \text{if } -1 < x < 1 \\ 2 - x & \text{if } 1 \leq x \leq 3 \end{cases}$

116. $f(x) = \begin{cases} 3 & \text{if } -4 \leq x \leq -1 \\ x - 2 & \text{if } -1 < x \leq 2 \\ 0.5x & \text{if } 2 < x \leq 4 \end{cases}$

APPLICATIONS INVOLVING COMPOUND INEQUALITIES

117. **Worldwide Gambling Losses** Global betting losses in billions of dollars can be estimated and projected to the year 2018 by $L(x) = 20x + 250$, where x represents years after 2003. Use each of the following methods to estimate when losses ranged from $450 billion to $530 billion. (*Source:* The Economist.)
(a) Numerical
(b) Graphical
(c) Symbolic

118. **College Tuition** From 1980 to 2000, college tuition and fees at private colleges could be modeled by the linear function $f(x) = 575(x - 1980) + 3600$. (Since 2000, the increase in tuition and fees has been more than linear growth.) Use each method to estimate when the average tuition and fees ranged from $8200 to $10,500. (*Source:* The College Board.)
(a) Numerical
(b) Graphical
(c) Symbolic

119. **Altitude and Dew Point** If the dew point D on the ground is 60 °F, then the dew point x miles high is given by $D(x) = 60 - 5.8x$. Find the altitudes where the dew point ranges from 57.1 °F to 51.3 °F. (*Source:* A. Miller.)

120. **Cigarette Consumption** Worldwide cigarette consumption in trillions from 1950 to 2015 can be modeled by $C(x) = 0.09x - 173.8$, where x is the year. Estimate the years when cigarette consumption was between 5.3 and 6.2 trillion. (*Source:* Department of Agriculture.)

121. **Geometry** For what values of x is the perimeter of the rectangle from 40 to 60 feet?

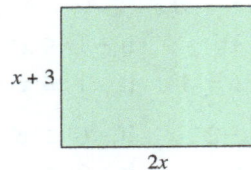

122. **Geometry** A rectangle is three times as long as it is wide. If the perimeter ranges from 100 to 160 inches, what values for the width are possible?

123. **Altitude and Temperature** If the air temperature at ground level is 70 °F, the air temperature x miles high is given by $T(x) = 70 - 19x$. Determine the altitudes at which the air temperature is from 41.5 °F to 22.5 °F. (*Source:* A. Miller and R. Anthes, Meteorology.)

124. **Distance** A car's distance in miles from a rest stop after x hours is given by $f(x) = 70x + 50$.
  (a) Make a table for $f(x)$ for $x = 4, 5, 6, \ldots, 10$ and then use the table to solve the inequality $470 \le f(x) \le 680$. Explain your result.
  (b) Solve the inequality in part (a) symbolically.

125. **Medicare Costs** In 2000, Medicare cost taxpayers $250 billion, and in 2010, it cost $500 billion. (*Source:* Department of Health and Human Services.)
  (a) Find a linear function M that models these data x years after 2000.
  (b) Estimate when Medicare costs might be from $600 billion to $700 billion.

126. **Temperature Conversion** Water freezes at 32 °F, or 0 °C, and boils at 212 °F, or 100 °C.
  (a) Find a linear function $C(F)$ that converts Fahrenheit temperature to Celsius temperature.
  (b) The greatest temperature ranges on Earth are recorded in Siberia, where temperature has varied from about -70 °C to 35 °C. Find this temperature range in Fahrenheit.

127. **Speed Limits** The graph of $y = f(x)$ in the next column gives the speed limit y along a rural highway x miles from its starting point.
  (a) What are the maximum and minimum speed limits along this stretch of highway?
  (b) Estimate the miles of highway with a speed limit of 55 miles per hour.
  (c) Evaluate $f(4)$, $f(12)$, and $f(18)$.

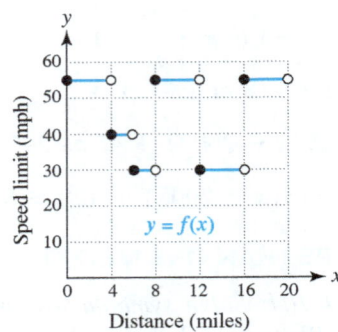

128. **First-Class Mail** Recently, the retail flat rate in dollars for first-class mail weighing up to 5 ounces could be computed by the piecewise-constant function P, where x is the number of ounces.

$$P(x) = \begin{cases} 0.90 & \text{if } 0 < x \le 1 \\ 1.10 & \text{if } 1 < x \le 2 \\ 1.30 & \text{if } 2 < x \le 3 \\ 1.50 & \text{if } 3 < x \le 4 \\ 1.70 & \text{if } 4 < x \le 5 \end{cases}$$

  (a) Evaluate $P(1.5)$ and $P(3)$. Interpret your results.
  (b) Sketch a graph of P. What is the domain of P?

WRITING ABOUT MATHEMATICS

129. Suppose that the solution set for a compound inequality can be written as $x < -3$ or $x > 2$. A student writes it as $2 < x < -3$. Is the student's three-part inequality correct? Explain your answer.

130. How can you determine whether an x-value is a solution to a compound inequality containing the word *and*? Give an example. Repeat the question for a compound inequality containing the word *or*.

15.4 Other Functions and Their Properties

Objectives

1. Introducing Nonlinear Functions
2. Expressing Domain and Range in Interval Notation
3. Recognizing Absolute Value Functions and Their Properties
4. Recognizing Polynomial Functions and Their Properties
5. Recognizing Rational Functions and Their Properties
6. Performing Operations on Functions

NEW VOCABULARY

☐ Absolute value function
☐ Polynomial function of one variable
☐ Linear function
☐ Quadratic function
☐ Cubic function
☐ Rational function

1 Introducing Nonlinear Functions

Many quantities in applications cannot be modeled with linear functions and equations. If data points do not lie on a line, we say that the data are *nonlinear*. For example, a scatterplot of the *cumulative* number of AIDS deaths from 1981 through 2015 is nonlinear, as shown in **FIGURE 15.50**. To model such data, we often use *nonlinear functions*, whose graphs are *not* lines. Because scatterplots of nonlinear data can have a variety of shapes, mathematicians use many different types of nonlinear functions, such as polynomial functions, which we discuss in this section. See Exercise 133. (*Source:* U.S. Department of Health and Human Services.)

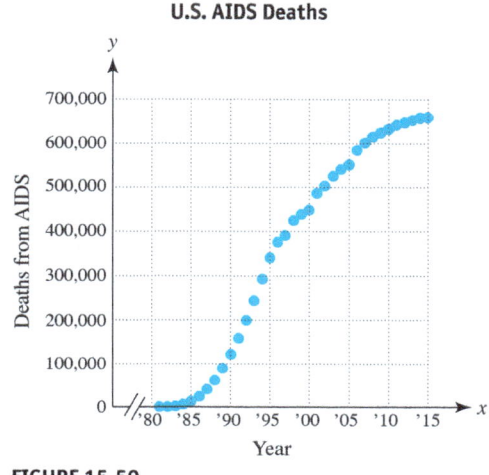

FIGURE 15.50

2 Expressing Domain and Range in Interval Notation

The set of all valid inputs for a function is called the *domain*, and the set of all outputs from a function is called the *range*. For example, all real numbers are valid inputs for $f(x) = x^2$. Rather than writing "the set of all real numbers" for the domain of f, we can use *interval notation* to express the domain as $(-\infty, \infty)$. The symbol ∞ represents infinity and is not a real number. Because $x^2 \geq 0$ for every real number x, the output from $f(x) = x^2$ is never negative. Therefore the range of f is $[0, \infty)$, which denotes all nonnegative real numbers, or $x \geq 0$. Note that 0 is in the range of f because $f(0) = 0$, and a bracket "[" is used to indicate that 0 is included in the range of f.

EXAMPLE 1 Writing domains in interval notation

Write the domain for each function in interval notation.
(a) $f(x) = 4x$ (b) $g(t) = \sqrt{t-1}$ (c) $h(v) = \dfrac{1}{v+3}$

Solution
(a) The expression $4x$ is defined for all real numbers x. Thus the domain of f is $(-\infty, \infty)$.
(b) The square root $\sqrt{t-1}$ is defined only when $t - 1$ is *not* negative. Thus the domain of g includes all real numbers satisfying $t - 1 \geq 0$ or $t \geq 1$. In interval notation this inequality is written as $[1, \infty)$.
(c) The expression $\frac{1}{v+3}$ is defined except when $v + 3 = 0$ or $v = -3$. Thus the domain of h includes all real numbers except -3 and can be written as $(-\infty, -3) \cup (-3, \infty)$. Parentheses are used because -3 is not included in the domain of h.

Now Try Exercises 13, 21, 25

In the next example, we determine the domain and range of a function from its graph. Note that dots placed at each end of a graph indicate that the endpoints are included.

EXAMPLE 2   **Writing the domain and range in interval notation**

Use the graph of f in **FIGURE 15.51** to write its domain and range in interval notation.

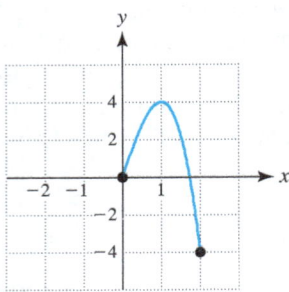

FIGURE 15.51

Solution
Because dots are placed at $(0, 0)$ and $(2, -4)$, the endpoints are included in the graph of f. Thus the graph in **FIGURE 15.51** includes x-values from $x = 0$ to $x = 2$. In interval notation, the domain of f is $[0, 2]$. The range of f includes y-values from -4 to 4 and can be expressed in interval notation as $[-4, 4]$.

Now Try Exercise 39

3   Recognizing Absolute Value Functions and Their Properties

In an earlier chapter, we discussed the absolute value of a number. We can define a function called the **absolute value function** as $f(x) = |x|$. We evaluate f as follows.

$$f(11) = |11| = 11, \quad f(-4) = |-4| = 4, \quad \text{and} \quad f(-\pi) = |-\pi| = \pi$$

To graph $y = |x|$, we begin by making a table of values, as shown in **TABLE 15.13**. Next we plot these points and sketch the graph, as shown in **FIGURE 15.52**. Note that the graph is V-shaped and never lies below the x-axis because the absolute value of a number cannot be negative.

$y = |x|$

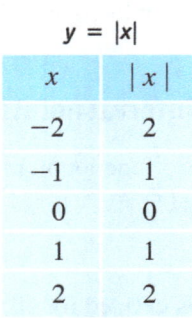

TABLE 15.13

Absolute Value Function

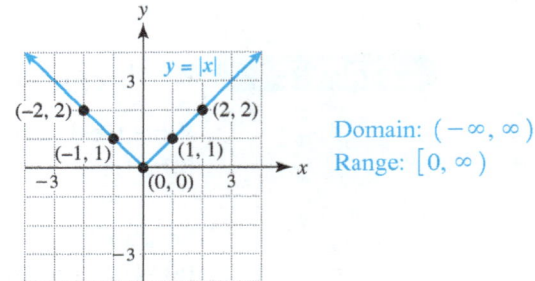

FIGURE 15.52

Domain: $(-\infty, \infty)$
Range: $[0, \infty)$

Because the input for $f(x) = |x|$ is any real number, the domain of f is all real numbers, or $(-\infty, \infty)$. The graph of the absolute value function shows that the output y (range) is any real number greater than or equal to 0. That is, the range is $[0, \infty)$.

4 Recognizing Polynomial Functions and Their Properties

In a previous chapter, we introduced polynomials and defined their degrees.

Polynomials of One Variable

$$1 - 5x, \quad 3t^2 - 5t + 1, \quad \text{and} \quad z^3 + 5$$

(The exponents on variables in polynomials must be nonnegative integers.) Recall that the *degree* of a polynomial of one variable equals the largest exponent on the variable. Thus the degree of $1 - 5x$ is 1, the degree of $3t^2 - 5t + 1$ is 2, and the degree of $z^3 + 5$ is 3.

Polynomial Functions of One Variable

$$f(x) = 1 - 5x, \quad g(t) = 3t^2 - 5t + 1, \quad \text{and} \quad h(z) = z^3 + 5$$

The equations shown above define three **polynomial functions of one variable**. Function f is a **linear function** because it has degree 1, function g is a **quadratic function** because it has degree 2, and function h is a **cubic function** because it has degree 3.

NOTE: The domain of *every* polynomial function is *all* real numbers, $(-\infty, \infty)$. ∎

EXAMPLE 3 Identifying polynomial functions

Determine whether $f(x)$ represents a polynomial function. If possible, identify the type of polynomial function and its degree.
(a) $f(x) = 5x^3 - x + 10$
(b) $f(x) = x^{-2.5} + 1$
(c) $f(x) = 1 - 2x$
(d) $f(x) = \dfrac{3}{x - 1}$

Solution
(a) The expression $5x^3 - x + 10$ is a cubic polynomial, so $f(x)$ represents a cubic polynomial function. It has degree 3.
(b) $f(x) = x^{-2.5} + 1$ does not represent a polynomial function because the variables in a polynomial must have *nonnegative integer* exponents.
(c) $f(x) = 1 - 2x$ represents a polynomial function that is linear. It has degree 1.
(d) $f(x) = \dfrac{3}{x - 1}$ does not represent a polynomial function because $\dfrac{3}{x - 1}$ is not a polynomial.

Now Try Exercises 43, 45, 47, 51

Frequently, polynomials represent functions or formulas that can be evaluated. This situation is illustrated in the next two examples.

EXAMPLE 4 Evaluating a polynomial function graphically and symbolically

A graph of $f(x) = 4x - x^3$ is shown in **FIGURE 15.53**, where $y = f(x)$. Evaluate $f(-1)$ graphically and check your result symbolically.

Solution
Graphical Evaluation To evaluate $f(-1)$ graphically, find -1 on the x-axis and move down until the graph of f is reached. Then move horizontally to the y-axis, as shown in **FIGURE 15.54** on the next page. Thus when $x = -1, y = -3$ and $f(-1) = -3$.

Symbolic Evaluation When $x = -1$, evaluation of $f(x) = 4x - x^3$ is performed as follows.

$$f(-1) = 4(-1) - (-1)^3 = -4 - (-1) = -3$$

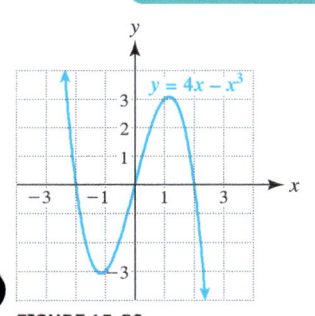

FIGURE 15.53

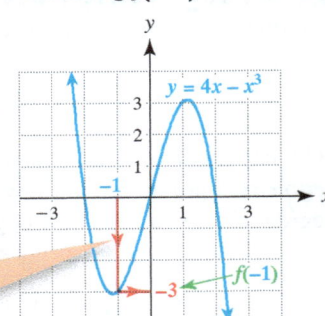

Evaluating $f(-1) = -3$

Find -1 on the x-axis. Move downward to the graph of f and then across to the y-axis.

FIGURE 15.54

Now Try Exercise 73

EXAMPLE 5 Evaluating a polynomial function symbolically

Evaluate $f(x)$ at the given value of x.
(a) $f(x) = -3x^4 - 2, \quad x = 2$
(b) $f(x) = -2x^3 - 4x^2 + 5, \quad x = -3$

Solution
(a) Be sure to evaluate exponents before multiplying.
$$f(2) = -3(2)^4 - 2 = -3 \cdot 16 - 2 = -50$$
(b) $f(-3) = -2(-3)^3 - 4(-3)^2 + 5 = -2(-27) - 4(9) + 5 = 23$

Now Try Exercises 61, 63

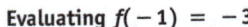

Math in Context Sports A well-conditioned athlete's heart rate can reach 200 beats per minute during strenuous physical activity. Upon quitting, a typical heart rate decreases rapidly at first and then more gradually after a few minutes, as illustrated in the next example.

EXAMPLE 6 Modeling heart rate of an athlete

Let $P(t) = 1.875t^2 - 30t + 200$ model an athlete's heart rate (or pulse P) in beats per minute (bpm) t minutes after strenuous exercise has stopped, where $0 \leq t \leq 8$. (*Source:* V. Thomas, *Science and Sport.*)
(a) What is the initial heart rate when the athlete stops exercising?
(b) What is the heart rate after 8 minutes?
(c) A graph of P is shown in **FIGURE 15.55**. Interpret this graph.

Athlete's Heart Rate

FIGURE 15.55

Solution
(a) To find the initial heart rate, evaluate $P(t)$ at $t = 0$, or
$$P(0) = 1.875(0)^2 - 30(0) + 200 = 200.$$

When the athlete stops exercising, the heart rate is **200** beats per minute. (This result agrees with the graph.)

(b) $P(8) = 1.875(8)^2 - 30(8) + 200 = 80$ beats per minute.
(c) The heart rate does not drop at a constant rate; rather, it drops rapidly at first and then gradually begins to level off.

Now Try Exercise 127

5 Recognizing Rational Functions and Their Properties

A rational expression is formed when a polynomial is divided by a polynomial.

Rational Expressions

$$\frac{2x-1}{x}, \quad \frac{5}{x^2-1}, \quad \text{and} \quad \frac{2x-5}{x^2-9} \quad \text{Polynomial divided by a polynomial}$$

Rational expressions can be used to define *rational functions*.

RATIONAL FUNCTIONS

Let $p(x)$ and $q(x)$ be polynomials. Then a **rational function** is given by

$$f(x) = \frac{p(x)}{q(x)}.$$

The domain of f includes all x-values such that $q(x) \neq 0$.

Using Rational Expressions to Define Rational Functions

$$f(x) = \frac{2x-1}{x}, \quad g(x) = \frac{5}{x^2+1}, \quad \text{and} \quad h(x) = \frac{2x-5}{x^2-9}$$

The domain of f includes all real numbers except 0, the domain of g includes all real numbers because $x^2 + 1 \neq 0$ for any x-value, and the domain of h includes all real numbers except ± 3.

Formulas for linear and polynomial functions are *defined* for all x-values. However, formulas for a rational function are *undefined* for x-values that make the denominator equal to 0. (Division by 0 is undefined.) For example, if $f(x) = \frac{1}{x-2}$, then $f(2) = \frac{1}{2-2} = \frac{1}{0}$ is undefined because the denominator equals 0. A graph of $y = f(x)$ is shown in **FIGURE 15.56**. The graph does not cross the dashed vertical line $x = 2$, because $f(x)$ is *undefined* at $x = 2$. The red vertical dashed line $x = 2$ is called a *vertical asymptote*, and is used as an aid for sketching a graph of f. It is not actually part of the graph of the function.

READING CHECK 1

- How can you determine the domain of a rational function?

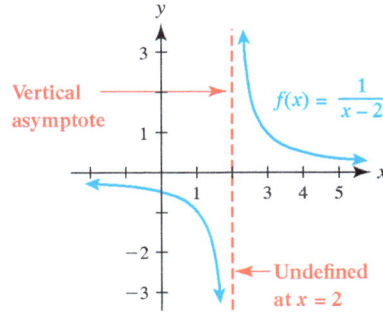

FIGURE 15.56

EXAMPLE 7 Identifying the domains of rational functions

Write the domain of each function in interval notation.

(a) $f(x) = \dfrac{1}{x+2}$ (b) $g(x) = \dfrac{2x}{x^2 - 3x + 2}$ (c) $h(t) = \dfrac{4}{t^3 - t}$

Solution

(a) The domain of f includes all x-values except when the denominator equals 0.

$x + 2 = 0$ Set the denominator equal to 0.

$x = -2$ Subtract 2.

Thus $f(-2)$ is undefined and -2 must be excluded from the domain of f. In interval notation, the domain of f is $(-\infty, -2) \cup (-2, \infty)$.

(b) The domain of g includes all real numbers except when $x^2 - 3x + 2 = 0$.

$x^2 - 3x + 2 = 0$ Set the denominator equal to 0.

$(x - 1)(x - 2) = 0$ Factor.

$x = 1 \quad \text{or} \quad x = 2$ Zero-product property

Because $g(1)$ and $g(2)$ are both undefined, 1 and 2 must be excluded from the domain of g. In interval notation, the domain of g is $(-\infty, 1) \cup (1, 2) \cup (2, \infty)$.

(c) The domain of h includes all real numbers except when $t^3 - t = 0$.

$t^3 - t = 0$ Set the denominator equal to 0.

$t(t^2 - 1) = 0$ Factor out t.

$t(t - 1)(t + 1) = 0$ Difference of squares

$t = 0 \quad \text{or} \quad t = 1 \quad \text{or} \quad t = -1$ Zero-product property

In interval notation, the domain of h is $(-\infty, -1) \cup (-1, 0) \cup (0, 1) \cup (1, \infty)$.

Now Try Exercises 27, 33, 35

To graph a rational function by hand, we usually start by making a table of values, as demonstrated in the next example. Because the graphs of rational functions are typically nonlinear, it is a good idea to plot at least 3 points on each side of an x-value where the formula is undefined—that is, where the denominator equals 0.

EXAMPLE 8 Graphing a rational function

Graph $f(x) = \dfrac{1}{x}$. State the domain of f.

Solution
Make a table of values for $f(x) = \dfrac{1}{x}$, as shown in **TABLE 15.14**. Notice that $x = 0$ is not in the domain of f, and a dash can be used to denote this undefined value. The domain of f is all real numbers such that $x \neq 0$. Start by picking three x-values on each side of 0.

x	-2	-1	$-\frac{1}{2}$	0	$\frac{1}{2}$	1	2
$\frac{1}{x}$	$-\frac{1}{2}$	-1	-2	—	2	1	$\frac{1}{2}$

TABLE 15.14   Undefined

A table of values to help graph $y = \dfrac{1}{x}$

Plot the points shown in **TABLE 15.14** and then connect the points with a smooth curve, as shown in **FIGURE 15.57**. Because $f(0)$ is undefined, the graph of $f(x) = \frac{1}{x}$ does not cross the line $x = 0$ (the y-axis). The line $x = 0$ is a vertical asymptote.

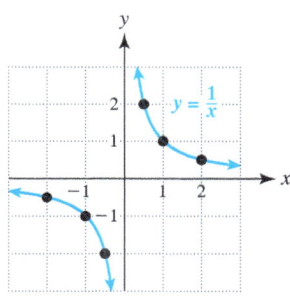

FIGURE 15.57

Now Try Exercise 89

In the next example, we evaluate a rational function in three ways.

EXAMPLE 9 **Evaluating a rational function**

Use **TABLE 15.15**, the formula for $f(x)$, and **FIGURE 15.58** to evaluate $f(-1), f(1),$ and $f(2)$.

(a)

x	$f(x)$
-3	$\frac{3}{2}$
-2	$\frac{4}{3}$
-1	1
0	0
1	—
2	4
3	3

TABLE 15.15

(b) $f(x) = \dfrac{2x}{x - 1}$

(c)

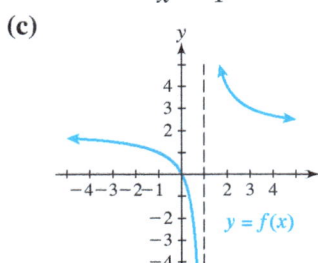

FIGURE 15.58

Solution
(a) **Numerical Evaluation** **TABLE 15.15** shows that

$$f(-1) = 1, \quad f(1) \text{ is undefined}, \quad \text{and} \quad f(2) = 4.$$

(b) **Symbolic Evaluation** Let $f(x) = \dfrac{2x}{x - 1}$.

$$f(-1) = \frac{2(-1)}{-1 - 1} = 1$$

$$f(1) = \frac{2(1)}{1 - 1} = \frac{2}{0}, \text{ which is undefined. Input 1 is } \textit{not} \text{ in the domain of } f.$$

$$f(2) = \frac{2(2)}{2 - 1} = 4$$

(c) **Graphical Evaluation** To evaluate $f(-1)$ graphically, find $x = -1$ on the x-axis and move upward to the graph of f. The y-value is 1 at the point of intersection, so

$f(-1) = 1$, as shown in **FIGURE 15.59(a)**. In **FIGURE 15.59(b)**, the red, dashed vertical line $x = 1$ is a vertical asymptote. Because the graph of f does not intersect this line, $f(1)$ is undefined. **FIGURE 15.59(c)** reveals that $f(2) = 4$.

Evaluating a Rational Function Graphically

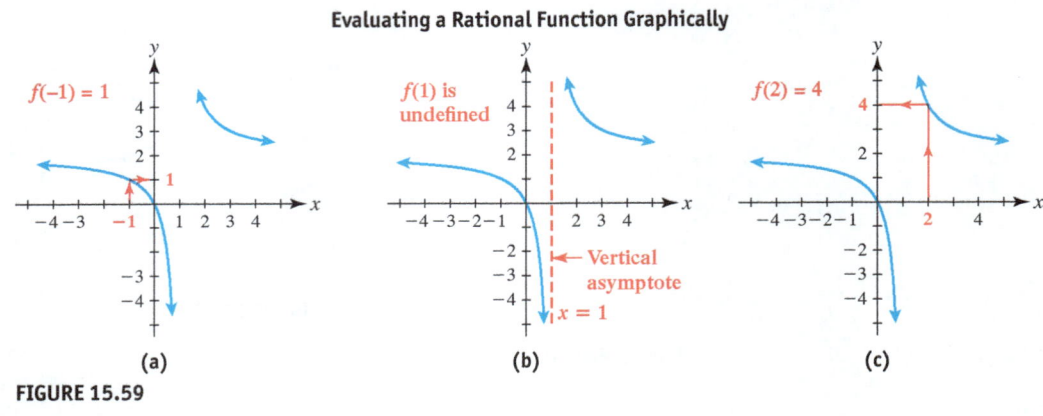

FIGURE 15.59

Now Try Exercises 67, 77, 79

The following See the Concept explains more about vertical asymptotes and graphs of rational functions.

See the Concept 1   RATIONAL FUNCTIONS AND VERTICAL ASYMPTOTES

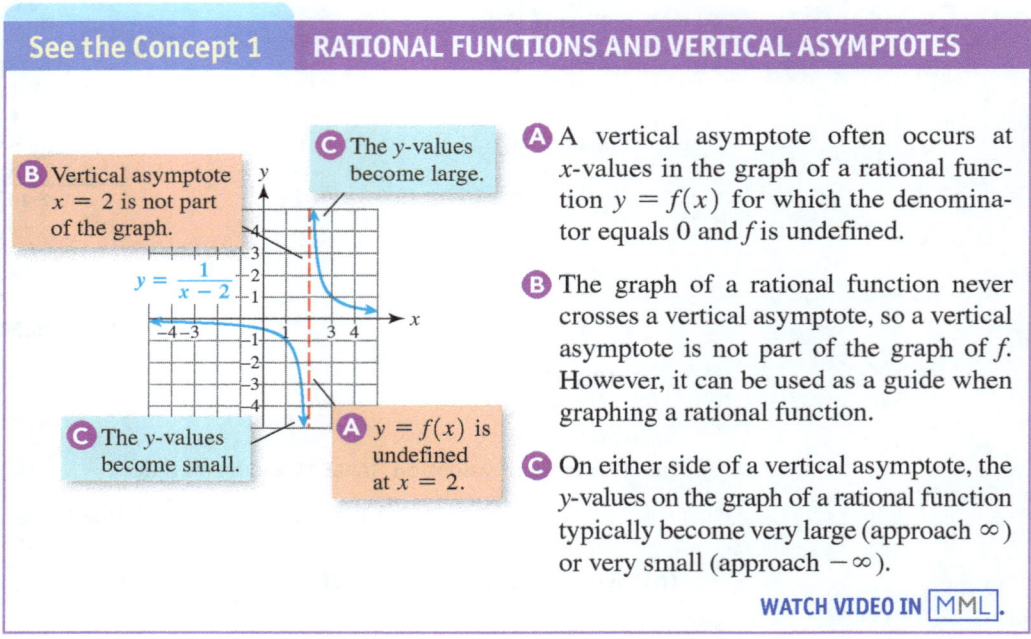

A A vertical asymptote often occurs at x-values in the graph of a rational function $y = f(x)$ for which the denominator equals 0 and f is undefined.

B The graph of a rational function never crosses a vertical asymptote, so a vertical asymptote is not part of the graph of f. However, it can be used as a guide when graphing a rational function.

C On either side of a vertical asymptote, the y-values on the graph of a rational function typically become very large (approach ∞) or very small (approach $-\infty$).

WATCH VIDEO IN MML.

Connecting Concepts with Your Life You may have noticed that a relatively small percentage of people do the vast majority of postings on social networks, such as Facebook and Twitter. This phenomenon is called *participation inequality*. That is, a vast majority of the population falls under the category of "lurkers," who are on the network but are not posting material. This characteristic of a social network can be modeled approximately by a rational function, as illustrated in the next example. (*Source:* Wu, Michael, *The Economics of 90–9–1.*)

EXAMPLE 10 Modeling social network participation

The rational function given by

$$f(x) = \frac{100}{101 - x}, \quad 5 \le x \le 100,$$

models participation inequality in a social network. In this formula, $f(x)$ outputs the percentage of the postings done by the least active (bottom) x percent of the population.
(a) Evaluate $f(95)$. Interpret your answer.
(b) A graph of $y = f(x)$ is shown in **FIGURE 15.60**. Interpret the graph.

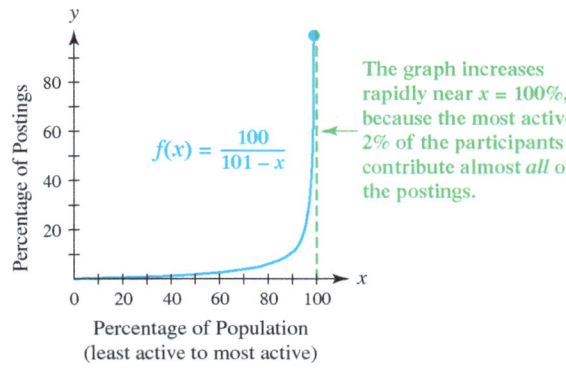

FIGURE 15.60 Participation Inequality

(c) Solve the rational equation $\frac{100}{101 - x} = 9$. Interpret your answer.

Solution

(a) $f(95) = \dfrac{100}{101 - 95} = \dfrac{100}{6} \approx 16.7\%$ Let $x = 95$.

This means that the least active 95% of the population contributes only 16.7% of the postings, so the most active 5% of the population is responsible for the remaining 83.3% of the postings.

(b) The graph shows participation inequality visually. The graph remains at a relatively low percentage until $x = 90\%$. This means that the bottom 90% of the population does very few postings. For $x \ge 90\%$, the graph rises rapidly because the top 10% contributes a vast majority of the postings.

(c) To solve this equation, we begin by multiplying each side by $(101 - x)$.

$$\frac{100}{101 - x} = 9 \qquad \text{Given equation}$$

$$(101 - x) \cdot \frac{100}{101 - x} = 9(101 - x) \qquad \text{Multiply by } (101 - x).$$

$$100 = 9(101 - x) \qquad \text{Simplify left side.}$$

$$100 = 909 - 9x \qquad \text{Distributive property}$$

$$9x + 100 - 100 = 909 - 100 + 9x - 9x \qquad \text{Add } 9x.\text{ Subtract } 100.$$

$$9x = 809 \qquad \text{Simplify}$$

$$x = \frac{809}{9} \approx 90\% \qquad \text{Simplify}$$

This result indicates that the least active 90% of the population contributes only 9% of the postings.

Now Try Exercise 129

CRITICAL THINKING 1

Suppose that a social network had *participation equality*, in which every member contributed an equal number of postings. Sketch a graph similar to **FIGURE 15.60** that describes this social network.

TECHNOLOGY NOTE

Asymptotes, Dot Mode, and Decimal Windows

When a rational function is graphed on a graphing calculator in connected mode, pseudo-asymptotes often occur because the calculator is simply connecting dots to draw a graph. The accompanying figures show the graph of $y = \frac{2}{x-2}$ in connected mode, in dot mode, and with a *decimal*, or *friendly*, window. In dot mode, pixels in the calculator screen are not connected. With dot mode (and sometimes with a decimal window) pseudo-asymptotes do not appear. To learn more about these features, consult your owner's manual.

CALCULATOR HELP

To set a calculator in dot mode or to set a decimal window, see Appendix A (pages AP-10 and AP-11).

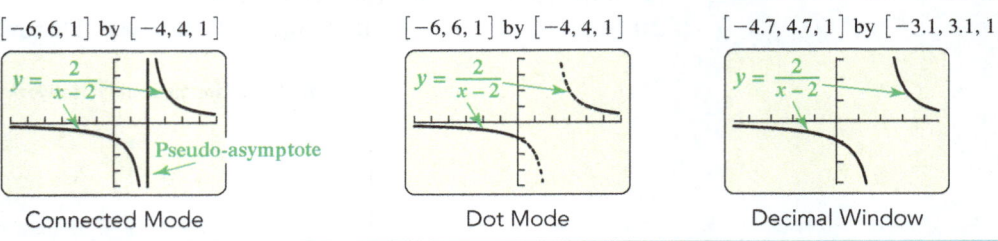

Connected Mode   Dot Mode   Decimal Window

6 Performing Operations on Functions

🌐 **Math in Business Context** A business incurs a *cost* to make its product and then it receives *revenue* from selling this product. For example, suppose a small business reconditions motorcycles. The graphs of its cost and of its revenue for reconditioning and selling x motorcycles are shown in **FIGURE 15.61**.

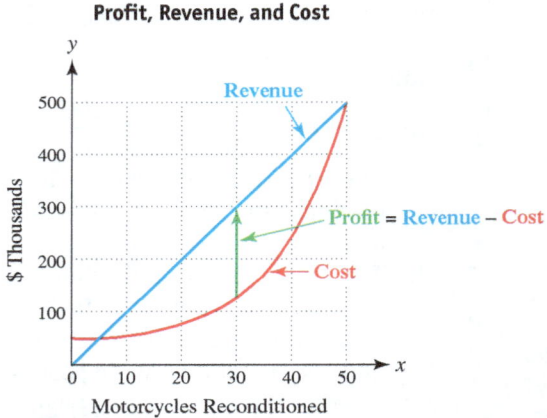

FIGURE 15.61

In general, *profit equals revenue minus cost*. In **FIGURE 15.61**, profit is shown visually as the length of the vertical green arrow between the graphs of revenue and cost. For any x-value, the distance by which revenue is above cost is called the profit for reconditioning and selling x motorcycles. *Maximum profit* for the company occurs at the x-value where the *length of the vertical green arrow is greatest*.

If we let $C(x), R(x)$, and $P(x)$ be functions that calculate the cost, revenue, and profit, respectively, for reconditioning and selling x motorcycles, then

$$P(x) = R(x) - C(x).$$

Profit equals Revenue minus Cost.

This example helps explain why we subtract functions in the real world. See Exercise 137. Functions can be added, multiplied, and divided in a similar manner.

Given two functions f and g, we define the sum $f + g$, difference $f - g$, product fg, and quotient $\frac{f}{g}$, as follows.

> **OPERATIONS ON FUNCTIONS**
>
> If $f(x)$ and $g(x)$ are both defined, then the sum, difference, product, and quotient of two functions f and g are defined by
>
> $(f + g)(x) = f(x) + g(x)$ Sum
>
> $(f - g)(x) = f(x) - g(x)$ Difference
>
> $(fg)(x) = f(x) \cdot g(x)$ Product
>
> $\left(\dfrac{f}{g}\right)(x) = \dfrac{f(x)}{g(x)}$, where $g(x) \neq 0$. Quotient

EXAMPLE 11 Performing arithmetic on functions

Use $f(x) = x^2$ and $g(x) = 2x - 4$ to evaluate each of the following.

(a) $(f + g)(3)$ (b) $(fg)(-1)$ (c) $\left(\dfrac{f}{g}\right)(0)$ (d) $(f/g)(2)$

Solution
(a) $(f + g)(3) = f(3) + g(3) = 3^2 + (2 \cdot 3 - 4) = 9 + 2 = 11$
(b) $(fg)(-1) = f(-1) \cdot g(-1) = (-1)^2 \cdot (2 \cdot (-1) - 4) = 1 \cdot (-6) = -6$
(c) $\left(\dfrac{f}{g}\right)(0) = \dfrac{f(0)}{g(0)} = \dfrac{0^2}{2 \cdot 0 - 4} = \dfrac{0}{-4} = 0$
(d) Note that $(f/g)(2)$ is equivalent to $\left(\dfrac{f}{g}\right)(2)$.

$$(f/g)(2) = \dfrac{f(2)}{g(2)} = \dfrac{2^2}{2 \cdot 2 - 4} = \dfrac{4}{0},$$

which is not possible because division by 0 is undefined. Thus $(f/g)(2)$ is *undefined*.

Now Try Exercise 101

In the next example, we find the sum, difference, product, and quotient of two functions for a general x.

EXAMPLE 12 Performing arithmetic on functions

Use $f(x) = 4x - 5$ and $g(x) = 3x + 1$ to evaluate each of the following.

(a) $(f + g)(x)$ (b) $(f - g)(x)$ (c) $(fg)(x)$ (d) $\left(\dfrac{f}{g}\right)(x)$

Solution
(a) $(f + g)(x) = f(x) + g(x) = (4x - 5) + (3x + 1) = 7x - 4$
(b) $(f - g)(x) = f(x) - g(x) = (4x - 5) - (3x + 1) = x - 6$
(c) $(fg)(x) = f(x) \cdot g(x) = (4x - 5)(3x + 1) = 12x^2 - 11x - 5$
(d) $\left(\dfrac{f}{g}\right)(x) = \dfrac{f(x)}{g(x)} = \dfrac{4x - 5}{3x + 1}$

Now Try Exercise 105

15.4 Putting It All Together

CONCEPT	COMMENTS	EXAMPLES
Writing Domain and Range in Interval Notation	Interval notation can be used to specify the domain and range of a function.	If $f(x) = x^2 + 1$, the domain of f is $(-\infty, \infty)$, and the range of f is $[1, \infty)$.
Absolute Value Function	Defined by $$f(x) = |x|$$ and has a V-shaped graph	$f(-5) = |-5| = 5$ $f(0) = |0| = 0$ $f(4) = |4| = 4$
Polynomial Function of One Variable	Can be defined by a polynomial; its degree equals the largest exponent of the variable.	Because $x^3 - 4x^2 + 6$ is a polynomial with degree 3, $$f(x) = x^3 - 4x^2 + 6$$ defines a polynomial function of degree 3 and is called a cubic function.
Rational Function	A rational function can be written as $$f(x) = \frac{p(x)}{q(x)},$$ where $p(x)$ and $q(x)$ are polynomials. Note that $q(x) \neq 0$.	Because $2x - 3$ and $x + 1$ are polynomials, $$f(x) = \frac{2x - 3}{x + 1}$$ defines a rational function. Because $f(x)$ is undefined at $x = -1$, the domain of f is $(-\infty, -1) \cup (-1, \infty)$
Operations on Functions	$(f + g)(x) = f(x) + g(x)$ Sum $(f - g)(x) = f(x) - g(x)$ Difference $(fg)(x) = f(x)g(x)$ Product $\left(\dfrac{f}{g}\right)(x) = \dfrac{f(x)}{g(x)}, g(x) \neq 0$ Quotient	Let $f(x) = x^2$ and $g(x) = 1 - x^2$. $(f + g)(x) = f(x) + g(x)$ $ = x^2 + (1 - x^2)$ $ = 1$ $(f - g)(x) = f(x) - g(x)$ $ = x^2 - (1 - x^2)$ $ = 2x^2 - 1$ $(fg)(x) = f(x)g(x)$ $ = x^2(1 - x^2)$ $ = x^2 - x^4$ $\left(\dfrac{f}{g}\right)(x) = \dfrac{f(x)}{g(x)}$ $\phantom{\left(\dfrac{f}{g}\right)(x)} = \dfrac{x^2}{1 - x^2}, x \neq -1, x \neq 1$

15.4 Exercises

CONCEPTS AND VOCABULARY

1. The set of all valid inputs for a function is called its _____.

2. The set of all outputs for a function is called its _____.

3. The set of all real numbers can be written in interval notation as _____.

4. If the domain of a function includes all real numbers except 5, then its domain can be written in interval notation as _____.

5. The graph of the _____ function is V-shaped.

6. The degree of a polynomial of one variable equals the largest _____ of the variable.

7. A quadratic function has degree _____.

8. If a function is linear, then its degree is _____.

9. If $f(x) = \frac{x}{2x+1}$, then f is a(n) _____ function.

10. If $f(x) = \frac{x}{2x+1}$, then the domain of f includes all real numbers except _____.

11. Which of the following expressions (a.–d.) is not a rational function?
    a. $f(x) = \frac{1}{x}$
    b. $f(x) = x^2 + 1$
    c. $f(x) = \sqrt{x}$
    d. $f(x) = \frac{2x^2}{x-1}$

12. Which (a.–d.) is the domain of $f(x) = \frac{2x}{2x-1}$?
    a. $\{x \mid x \neq \frac{1}{2}\}$
    b. $\{x \mid x \neq 1\}$
    c. $\{x \mid x \neq 0\}$
    d. $\{x \mid x = 1\}$

FINDING DOMAIN AND RANGE OF FUNCTIONS

Exercises 13–24: Write the domain and the range of the function in interval notation. (Hint: You may want to consider the graph of the function.)

13. $f(x) = -2x$
14. $f(x) = -\frac{1}{4}x + 1$
15. $g(t) = \frac{2}{3}t - 3$
16. $g(t) = 9t$
17. $h(z) = z^2 + 2$
18. $h(z) = z^2 - 1$
19. $f(z) = -z^2$
20. $f(z) = -\frac{1}{4}z^2$
21. $g(x) = \sqrt{x+1}$
22. $g(x) = \sqrt{x-2}$
23. $h(x) = |x-1|$
24. $h(x) = |2x|$

Exercises 25–36: Write the domain of the rational function in interval notation.

25. $f(x) = \frac{1}{x-1}$
26. $f(x) = \frac{6}{x}$
27. $f(x) = \frac{x}{6-3x}$
28. $f(x) = \frac{3x}{2x-4}$
29. $g(t) = \frac{2}{t^2-4}$
30. $g(t) = \frac{5}{1-t^2}$
31. $g(t) = \frac{5t}{t^2-2t}$
32. $g(t) = \frac{-t}{2t^2-3t}$
33. $h(z) = \frac{2-z}{z^3-1}$
34. $h(z) = \frac{z+1}{z^3-z^2}$
35. $f(x) = \frac{4}{x^2-2x-3}$
36. $f(x) = \frac{1}{x^2+4x-5}$

Exercises 37–42: A graph of a function is shown. Write the domain and range of the function in interval notation.

37.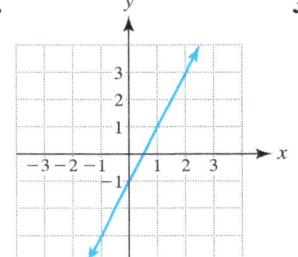

38.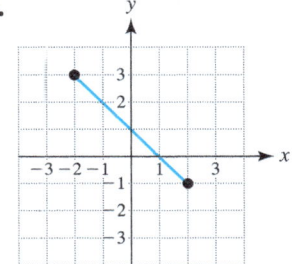

39.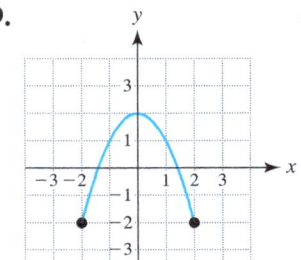

40.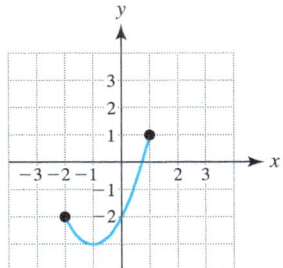

41. 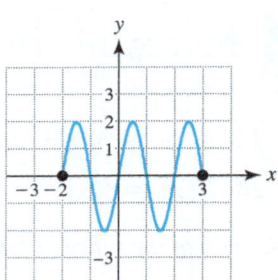 **42.**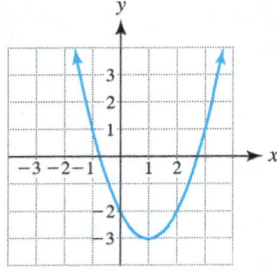

RECOGNIZING POLYNOMIAL FUNCTIONS

Exercises 43–54: Determine whether $f(x)$ represents a polynomial function. If possible, identify the degree and type of polynomial function.

43. $f(x) = 5x - 11$
44. $f(x) = 9 - x$
45. $f(x) = x^3$
46. $f(x) = x^2 + 3$
47. $f(x) = \dfrac{6}{x + 5}$
48. $f(x) = |x|$
49. $f(x) = 1 + 2x - x^2$
50. $f(x) = \frac{1}{4}x^3 - x$
51. $f(x) = 5x^{-2}$
52. $f(x) = x^2 + x^{-1}$
53. $f(x) = x^4 + 2x^2$
54. $f(x) = x^5 - 3x^3$

EVALUATING NONLINEAR FUNCTIONS

Exercises 55–70: If possible, evaluate $g(t)$ for the given values of t.

55. $g(t) = |4t|$ $t = 3, t = 0$
56. $g(t) = |t + 12|$ $t = 18, t = -15$
57. $g(t) = |t - 2|$ $t = 1, t = -\frac{3}{4}$
58. $g(t) = |2t + 1|$ $t = 2, t = -\frac{1}{2}$
59. $g(t) = t^2 - t - 6$ $t = 3, t = -3$
60. $g(t) = 3t^2 - 2t$ $t = -2, t = 4$
61. $g(t) = -2t^3 + t$ $t = 2, t = -2$
62. $g(t) = \frac{1}{3}t^3$ $t = 1, t = -3$
63. $g(t) = t^2 - 2t - 6$ $t = 0, t = -3$
64. $g(t) = 2t^3 - t^2 + 4$ $t = 2, t = -1$
65. $g(t) = \dfrac{1}{t}$ $t = 11, t = -7$
66. $g(t) = \dfrac{2}{3 - t}$ $t = 10, t = 3$
67. $g(t) = -\dfrac{t}{t + 1}$ $t = 5, t = -1$
68. $g(t) = -\dfrac{2 - t}{4t}$ $t = 4, t = -1$
69. $g(t) = \dfrac{t^2}{t^2 - t}$ $t = -5, t = 1$
70. $g(t) = \dfrac{t - 3}{t^2 - 3t + 2}$ $t = -2, t = 1$

Exercises 71–78: If possible, use the graph to evaluate each expression. Then use the formula for $f(x)$ to check your results.

71. $f(0)$ and $f(1)$ 72. $f(-1)$ and $f(2)$

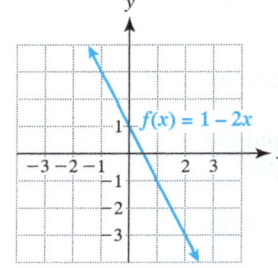 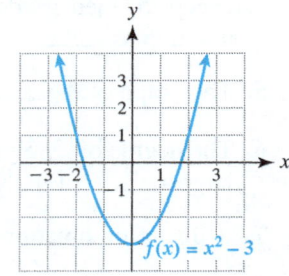

73. $f(-1)$ and $f(2)$ 74. $f(0)$ and $f(-2)$

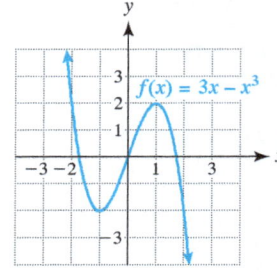 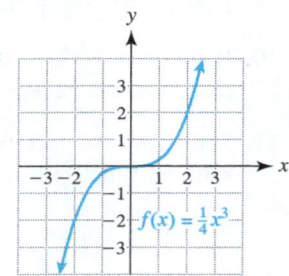

75. $f(-2)$ and $f(2)$ 76. $f(-1)$ and $f(0)$

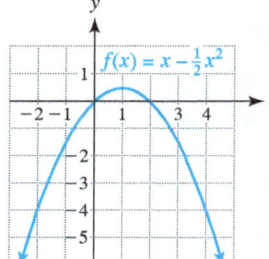 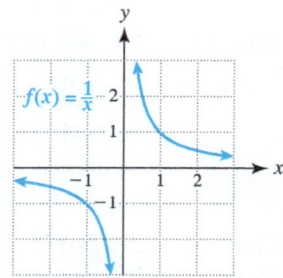

77. $f(-3)$ and $f(-1)$ 78. $f(0)$ and $f(1)$

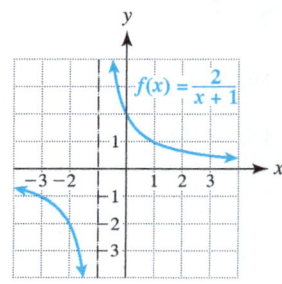 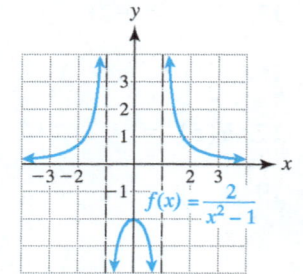

Exercises 79 and 80: Complete the table. Then evaluate $f(1)$.

79.

x	−2	−1	0	1	2
$f(x) = \frac{1}{x-1}$					

80.

x	−2	−1	0	1	2
$f(x) = \frac{2x}{x+2}$					

GRAPHING NONLINEAR FUNCTIONS

Exercises 81–100: Graph $y = f(x)$.

81. $f(x) = |2x|$ **82.** $f(x) = \left|\frac{1}{2}x\right|$

83. $f(x) = |x + 2|$ **84.** $f(x) = |x − 2|$

85. $f(x) = 1 − 2x^2$ **86.** $f(x) = \frac{1}{2}x^2 + 1$

87. $f(x) = \frac{1}{2}x^2$ **88.** $f(x) = x^2 − 2$

89. $f(x) = \frac{1}{x-1}$ **90.** $f(x) = \frac{1}{x+1}$

91. $f(x) = \frac{1}{2x}$ **92.** $f(x) = \frac{2}{x}$

93. $f(x) = \frac{1}{x+2}$ **94.** $f(x) = \frac{1}{x-2}$

95. $f(x) = \frac{4}{x^2+1}$ **96.** $f(x) = \frac{6}{x^2+2}$

97. $f(x) = \frac{3}{2x-3}$ **98.** $f(x) = \frac{1}{3x+2}$

99. $f(x) = \frac{1}{x^2-1}$ **100.** $f(x) = \frac{4}{4-x^2}$

PERFORMING OPERATIONS ON FUNCTIONS

Exercises 101–104: Use $f(x)$ and $g(x)$ to evaluate each of the following.

(a) $(f + g)(3)$ (b) $(f − g)(−2)$
(c) $(fg)(5)$ (d) $(f/g)(0)$

101. $f(x) = 5x, g(x) = x + 1$

102. $f(x) = x^2 + 2, g(x) = −2x$

103. $f(x) = 2x − 1, g(x) = 4x^2$

104. $f(x) = x^2 − 1, g(x) = x + 2$

Exercises 105–108: Use $f(x)$ and $g(x)$ to find each of the following.

(a) $(f + g)(x)$ (b) $(f − g)(x)$
(c) $(fg)(x)$ (d) $(f/g)(x)$

105. $f(x) = x + 1, g(x) = x + 2$

106. $f(x) = −3x, g(x) = x − 1$

107. $f(x) = 1 − x, g(x) = x^2$

108. $f(x) = x^2 + 4, g(x) = 6x$

109. Thinking Generally If $f(x) = x^2 − 2x$, then it follows that $f(a) =$ _____.

110. Thinking Generally If $f(x) = 2x − 1$, then it follows that $f(a + 2) =$ _____.

WORKING WITH PIECEWISE-DEFINED FUNCTIONS

Exercises 111–114: Match each piecewise-defined function with its graph (a–d).

111. $f(x) = \begin{cases} x^2 - 4 & \text{if } x \geq 0 \\ -x + 5 & \text{if } x < 0 \end{cases}$

112. $g(x) = \begin{cases} |x - 2| & \text{if } x \geq -1 \\ -x^2 & \text{if } x < -1 \end{cases}$

113. $h(x) = \begin{cases} 4 & \text{if } x \geq 0 \\ -4 & \text{if } x < 0 \end{cases}$

114. $k(x) = \begin{cases} \sqrt{x} & \text{if } x \geq 0 \\ -x^2 & \text{if } x < 0 \end{cases}$

a.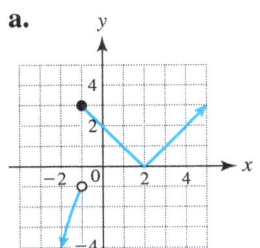

b.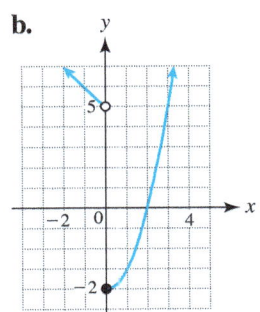

c.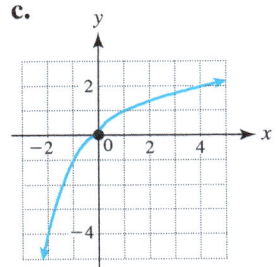

d.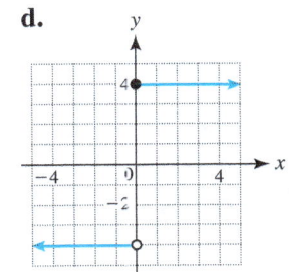

Exercises 115–120: Evaluate $f(x)$ at the given values of x.

115. $x = -2$ and 1

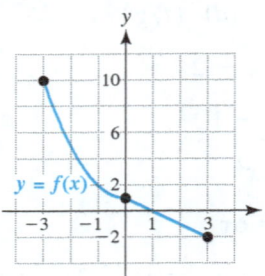

116. $x = -1, 0,$ and 3

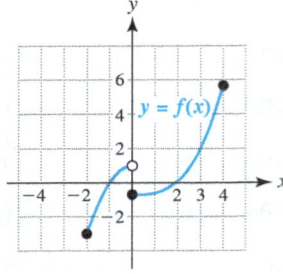

117. $x = -1, 1,$ and 2

118. $x = -2, 0,$ and 2

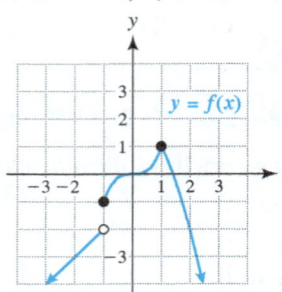
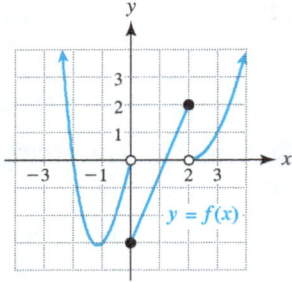

119. $x = -3, 1,$ and 4

$$f(x) = \begin{cases} x^3 - 4x^2 & \text{if } x \leq -3 \\ 3x^2 & \text{if } -3 < x < 4 \\ x^3 - 54 & \text{if } x \geq 4 \end{cases}$$

120. $x = -4, 0,$ and 4

$$f(x) = \begin{cases} -4x & \text{if } x \leq -4 \\ x^3 + 2 & \text{if } -4 < x \leq 2 \\ 4 - x^2 & \text{if } x > 2 \end{cases}$$

Exercises 121–122: Graph the piecewise-defined function and state its domain in interval notation.

121. $f(x) = \begin{cases} |x| & \text{if } -3 \leq x \leq 0 \\ x^2 & \text{if } 0 < x \leq 3 \end{cases}$

122. $f(x) = \begin{cases} x^2 & \text{if } -2 \leq x < 0 \\ x + 1 & \text{if } 0 \leq x \leq 2 \end{cases}$

APPLICATIONS INVOLVING NONLINEAR FUNCTIONS

Exercises 123–126: **Graphical Interpretation** *Match the physical situation with the graph of the rational function (a.–d.) in the next column that models it best.*

123. A population of fish that increases and then levels off

124. An insect population that dies out

125. The length of a ticket line as the rate at which people arrive in line increases

126. The wind speed during a day that is initially calm, becomes windy, and then is calm again

a.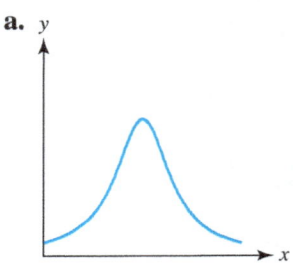

b.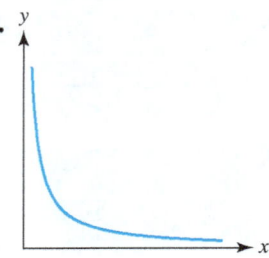

c.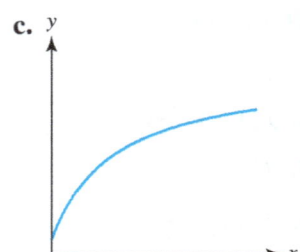

d.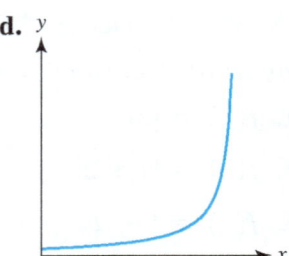

127. Heart Rate of an Athlete The following table lists the heart rate of an athlete running a 100-meter race. The race lasts 10 seconds.

Time (minutes)	0	2	4	6	8	10
Heart Rate (bpm)	90	100	113	127	143	160

(a) Does $P(t) = 0.2t^2 + 5t + 90$ model the data in the table exactly? Explain.
(b) Does P provide a reasonable model for the athlete's heart rate?
(c) Does $P(12)$ have significance in this situation? What should be the domain of P?

128. Heart Rate of an Athlete The following table lists an athlete's heart rate after the athlete finishes exercising strenuously.

Time (minutes)	0	2	4	6
Heart Rate (bpm)	180	137	107	90

(a) Does $P(t) = \frac{5}{3}t^2 - 25t + 180$ model the data in the table exactly? Explain.
(b) Does P provide a reasonable model for the athlete's heart rate?
(c) Does $P(12)$ have significance in this situation? What should be the domain of P?

129. Time Spent in Line If a parking lot attendant can wait on 5 vehicles per minute and vehicles are leaving the lot randomly at an average rate of x vehicles per minute, then the average time T in minutes spent waiting in line *and* paying the attendant is given by

$$T(x) = \frac{1}{5 - x},$$

where $x < 5$. (*Source:* N. Garber.)

(a) Evaluate $T(4)$ and interpret the result.
(b) A graph of T is shown in the figure. Interpret the graph as x increases from 0 to 5. Does this result agree with your intuition?

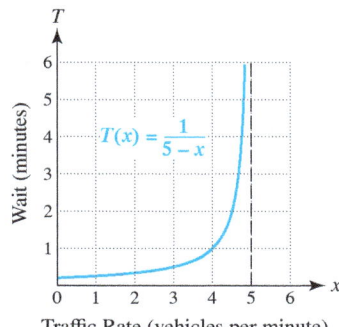
Traffic Rate (vehicles per minute)

(c) Find x if the waiting time is 3 minutes.

130. People Waiting in Line At a post office, workers can wait on 50 people per hour. If people arrive randomly at an average rate of x per hour, then the average number of people N waiting in line is given by

$$N(x) = \frac{x^2}{2500 - 50x},$$

where $x < 50$. (*Source:* N. Garber.)
(a) Evaluate $N(30)$ and interpret the result.
(b) A graph of N is shown in the figure. Interpret the graph as x increases from 0 to 50. Does this result agree with your intuition?

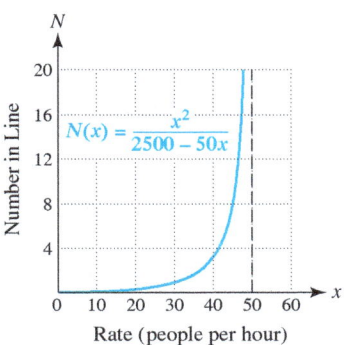
Rate (people per hour)

(c) Find x if $N = 8$.

131. Uphill Highway Grade The *grade* x of a hill is a measure of its steepness and corresponds to the slope of the road. For example, if a road rises 10 feet for every 100 feet of horizontal distance, it has an uphill grade of $x = \frac{10}{100}$, or 10%, as illustrated in the figure.

The braking distance for a car traveling 30 miles per hour on a wet, *uphill* grade x is given by

$$D(x) = \frac{900}{10.5 + 30x}.$$

(*Source:* N. Garber.)
(a) Evaluate $D(0.05)$ and interpret the result.
(b) If the braking distance for this car is 60 feet, find the uphill grade x.

132. Downhill Highway Grade (See Exercise 131.) The braking distance for a car traveling 30 miles per hour on a wet, *downhill* grade x is given by

$$S(x) = \frac{900}{10.5 - 30x}.$$

(a) Evaluate $S(0.05)$ and interpret the result.
(b) Make a table for $D(x)$ from Exercise 131 and $S(x)$, starting at $x = 0$ and incrementing by 0.05.
(c) How do the braking distances for uphill and downhill grades compare? Does this result agree with your driving experience?

133. U.S. AIDS Deaths Historically The following scatterplot shows the cumulative number of reported AIDS deaths. The data may be modeled x years after 1980 by $f(x) = 2.4x^2 - 14x + 23$, where the output is in thousands of deaths. (Note that after 1994, the number of deaths started to level off because of new drug therapy.)

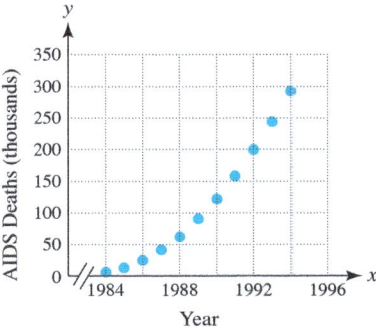
Year

(a) Use $f(x)$ to estimate the cumulative total of AIDS deaths in 1990. Compare it with the actual value of 121.6 thousand.
(b) In 1997, the cumulative number of AIDS deaths was 390 thousand. What estimate does $f(x)$ give? Discuss your result.

134. A PC for All? Worldwide sales of computers have climbed as prices have continued to drop. The function $f(x) = 0.29x^2 + 8x + 19$ models the number of personal computers sold in millions during year x, where $x = 0$ corresponds to 1990, $x = 1$ corresponds to 1991, and so on until $x = 25$ corresponds to 2015. Estimate the number of personal computers sold in 2010, using both the graph and the polynomial. (*Source:* eTForcasts.)

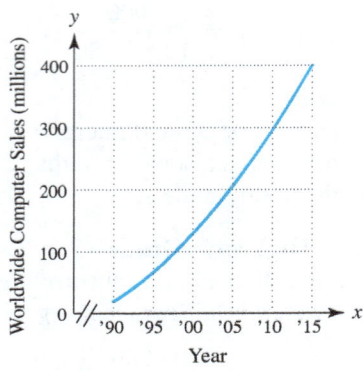

Exercises 135 and 136: **Remembering What You Learn**
After a test, students often forget what they learned. The rational function

$$R(x) = \frac{100}{1.2x + 1}, 0 \le x \le 5,$$

gives an estimate of the percentage of the material a student remembers x days after a test.

135. Evaluate $R(1)$ and $R(3)$. Interpret your results.

136. If a student takes notes in class, these percentages increase by 30% for $1 \le x \le 5$. Write another function $N(x)$ that models this result. Evaluate $N(3)$.

137. **Profit** A company makes and sells notebook computers. The company's cost function in thousands of dollars is $C(x) = 0.3x + 100$, and the revenue function in thousands of dollars is $R(x) = 0.75x$, where x is the number of notebook computers.
  (a) Evaluate and interpret $C(100)$.
  (b) Interpret the y-intercepts on the graphs of C and R.
  (c) Give the profit function $P(x)$.
  (d) How many computers need to be sold to make a profit?

138. **Profit** A company makes and sells sailboats. The company's cost function in thousands of dollars is $C(x) = 2x + 20$, and the revenue function in thousands of dollars is $R(x) = 4x$, where x is the number of sailboats.
  (a) Evaluate and interpret $C(5)$.
  (b) Interpret the y-intercepts on the graphs of C and R.
  (c) Give the profit function $P(x)$.
  (d) How many sailboats need to be sold to break even?

WRITING ABOUT MATHEMATICS

139. Name two functions. Give their formulas, sketch their graphs, and state their domains and ranges.

140. Explain the difference between the domain and the range of a function.

SECTIONS 15.3 and 15.4 Checking Basic Concepts

1. (a) Is 3 a solution to the compound inequality $x + 2 < 4$ or $2x - 1 \ge 3$?
   (b) Is 3 a solution to the compound inequality $x + 2 < 4$ and $2x - 1 \ge 3$?

2. Solve the following compound inequalities. Write your answers in interval notation.
   (a) $-5 \le 2x + 1 \le 3$
   (b) $1 - x \le -2$ or $1 - x \ge 2$
   (c) $-2 < \dfrac{4 - 3x}{2} \le 6$

3. Write the domain of each function in interval notation.
   (a) $f(x) = x^2$
   (b) $g(t) = \dfrac{1}{t - 1}$
   (c) $h(z) = \sqrt{z}$

4. Use the graph of f to do the following.
   (a) Write the domain and range of f in interval notation.
   (b) Evaluate $f(0)$ and $f(-2)$.

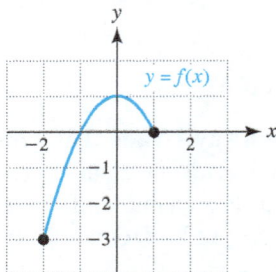

5. Graph $f(x) = |x - 3|$.

15.5 Absolute Value Equations and Inequalities

Objectives

1. **Solving Absolute Value Equations**
   - Solving $|x| = k$
   - Solving $|ax + b| = k$
   - Solving $|ax + b| = |cx + d|$
2. **Solving Absolute Value Inequalities**
   - Finding Numerical and Graphical Solutions
   - Finding Symbolic Solutions Based on Visualization (Method I)
   - Finding Symbolic Solutions (Method II)

NEW VOCABULARY

☐ Absolute value equation
☐ Absolute value inequality

1 Solving Absolute Value Equations

SOLVING $|x| = k$ An equation that contains an absolute value expression is called an **absolute value equation**.

Absolute Value Equations

$$|x| = 2, \quad |2x - 1| = 5, \quad |5 - 3x| - 3 = 1$$

Consider the absolute value equation $|x| = 2$. This equation has *two* solutions, -2 and 2, because $|-2| = 2$ and $|2| = 2$. We can also demonstrate this result with a table of values or a graph. In **TABLE 15.16**, $|x| = 2$ when $x = -2$ or $x = 2$. In **FIGURE 15.62**, the graph of $y_1 = |x|$ intersects the graph of $y_2 = 2$ at the points $(-2, 2)$ and $(2, 2)$ The x-values of these points of intersection correspond to the solutions -2 and 2.

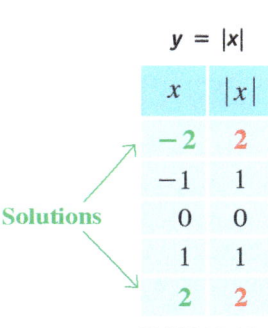

TABLE 15.16

FIGURE 15.62

We generalize this discussion in the following manner.

> **SOLVING $|x| = k$**
>
> 1. If $k > 0$, then $|x| = k$ is equivalent to $x = k$ or $x = -k$.
> 2. If $k = 0$, then $|x| = k$ is equivalent to $x = 0$.
> 3. If $k < 0$, then $|x| = k$ has no solutions.

EXAMPLE 1 Solving absolute value equations

Solve each equation.
(a) $|x| = 20$ (b) $|x| = -5$ (c) $|x| = 0$

Solution
(a) The solutions are -20 and 20.
(b) There are no solutions because $|x|$ is never negative.
(c) The only solution is 0.

Now Try Exercises 21, 23, 24

EXAMPLE 2 Graphing an absolute value equation

Graph the equation $y = |3x + 3|$.

Solution
Begin by making a table of values for $y = |3x + 3|$, as shown in **TABLE 15.17**. Plot these points in the xy-plane, and sketch the resulting V-shaped graph, as shown in **FIGURE 15.63**.

$y = |3x + 3|$

x	$|3x + 3|$
-3	6
-2	3
-1	0
0	3
1	6

TABLE 15.17

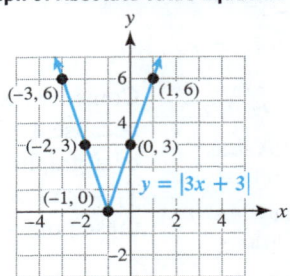

Graph of Absolute Value Equation

FIGURE 15.63

Now Try Exercise 15

MAKING CONNECTIONS 1

Graphs and the Absolute Value

The graph of $y = |ax + b|$ is a reflection of the line $y = ax + b$ across the x-axis whenever $ax + b < 0$, which is shown in red. Otherwise, when $ax + b \geq 0$, their graphs are identical, which is shown in blue. Note that the graph of $y = |ax + b|$ is *always* V-shaped and *never* goes below the x-axis, whereas the graph of $y = ax + b$ is a line with slope $m = a$.

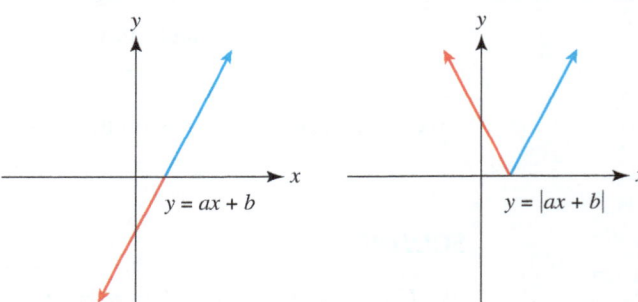

SOLVING $|ax + b| = k$ The next example illustrates how to solve an absolute value equation symbolically, numerically, and graphically.

EXAMPLE 3 Solving an absolute value equation

Solve $|2x - 5| = 3$
(a) symbolically, (b) numerically, and (c) graphically.

Solution
(a) **Symbolic Solution** If $|2x - 5| = 3$, then either $2x - 5 = 3$ or $2x - 5 = -3$. Solve each equation separately.

$2x - 5 = 3$ or $2x - 5 = -3$ Equations to be solved
$2x = 8$ or $2x = 2$ Add 5.
$x = 4$ or $x = 1$ Divide by 2.

The solutions are **1** and **4**.

(b) Numerical Solution A table of values can be used to solve the equation $|2x - 5| = 3$. **TABLE 15.18**, shows that $|2x - 5| = 3$ when $x = 1$ or $x = 4$.

$|2x - 5| = 3$

x	0	1	2	3	4	5	6
$|2x - 5|$	5	3	1	1	3	5	7

TABLE 15.18 Solutions are 1 and 4.

(c) Graphical Solution The equation $|2x - 5| = 3$ can be solved by graphing the equations $y_1 = |2x - 5|$ and $y_2 = 3$. To graph y_1, first plot some of the points from **TABLE 15.18**. Its graph is V-shaped, as shown in **FIGURE 15.64**. Note that the x-coordinate of the "point" or vertex of the V is found by solving the equation $2x - 5 = 0$ to obtain $\frac{5}{2}$. The graph of y_1 intersects the graph of y_2 at the points $(1, 3)$ and $(4, 3)$, giving the solutions 1 and 4. The graphical solutions agree with the numerical and symbolic solutions.

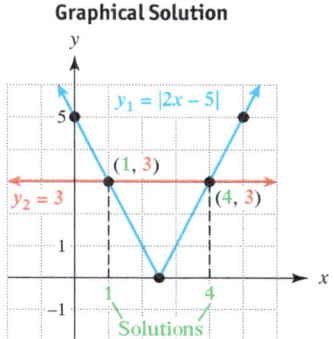

FIGURE 15.64

Now Try Exercises 29, 89(a), 91

This discussion leads to the following result.

ABSOLUTE VALUE EQUATIONS

If $k > 0$, then

$$|ax + b| = k$$

is equivalent to

$$ax + b = k \quad \text{or} \quad ax + b = -k.$$

EXAMPLE 4 Solving absolute value equations

Solve.
(a) $|5 - x| - 2 = 8$ **(b)** $\left|\frac{1}{2}(x - 6)\right| = \frac{3}{4}$

Solution
(a) Start by adding 2 to each side to obtain $|5 - x| = 10$. This new equation is satisfied by the solution from either of the following equations.

$$5 - x = 10 \quad \text{or} \quad 5 - x = -10 \quad \text{Equations to be solved}$$
$$-x = 5 \quad \text{or} \quad -x = -15 \quad \text{Subtract 5.}$$
$$x = -5 \quad \text{or} \quad x = 15 \quad \text{Multiply by } -1.$$

The solutions are -5 and 15.

(b) The solutions to $\left|\frac{1}{2}(x-6)\right| = \frac{3}{4}$ are found by solving the following equations.

$$\frac{1}{2}(x-6) = \frac{3}{4} \quad \text{or} \quad \frac{1}{2}(x-6) = -\frac{3}{4} \quad \text{Equations to be solved}$$

$$4 \cdot \frac{1}{2}(x-6) = 4 \cdot \frac{3}{4} \quad \text{or} \quad 4 \cdot \frac{1}{2}(x-6) = 4\left(-\frac{3}{4}\right) \quad \text{Multiply by 4 to clear fractions.}$$

$$2(x-6) = 3 \quad \text{or} \quad 2(x-6) = -3 \quad \text{Simplify.}$$

$$2x - 12 = 3 \quad \text{or} \quad 2x - 12 = -3 \quad \text{Distributive property}$$

$$2x = 15 \quad \text{or} \quad 2x = 9 \quad \text{Add 12.}$$

$$x = \frac{15}{2} \quad \text{or} \quad x = \frac{9}{2} \quad \text{Divide by 2.}$$

The solutions are $\frac{9}{2}$ and $\frac{15}{2}$.

Now Try Exercises 31, 35

EXAMPLE 5 Solving absolute value equations that have no solutions or one solution

Solve.
(a) $|2x - 1| = -2$ **(b)** $|4 - 2x| = 0$

Solution
(a) Because an absolute value is never negative, there are no solutions. **FIGURE 15.65** shows that the graph of $y_1 = |2x - 1|$ never intersects the graph of $y_2 = -2$.
(b) If $|y| = 0$, then $y = 0$. Thus $|4 - 2x| = 0$ when $4 - 2x = 0$ or when $x = 2$. The only solution is 2.

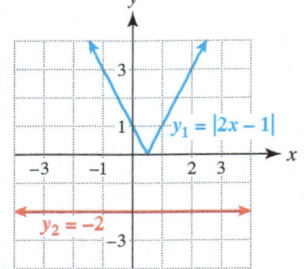

FIGURE 15.65

Now Try Exercises 33, 37

SOLVING $|ax + b| = |cx + d|$ Sometimes an equation can have an absolute value on each side. An example would be $|2x| = |x - 3|$. In this situation, either $2x = x - 3$ (the two expressions are equal) or $2x = -(x - 3)$ (the two expressions are opposites).

These concepts are summarized as follows.

SOLVING $|ax + b| = |cx + d|$

Let a, b, c, and d be constants. Then $|ax + b| = |cx + d|$ is equivalent to
$$ax + b = cx + d \quad \text{or} \quad ax + b = -(cx + d).$$

EXAMPLE 6 Solving an absolute value equation

Solve $|2x| = |x - 3|$.

Solution
Solve the following equations.

$$2x = x - 3 \quad \text{or} \quad 2x = -(x - 3)$$
$$x = -3 \quad \text{or} \quad 2x = -x + 3$$
$$3x = 3$$
$$x = 1$$

The solutions are -3 and 1.

Now Try Exercise 41

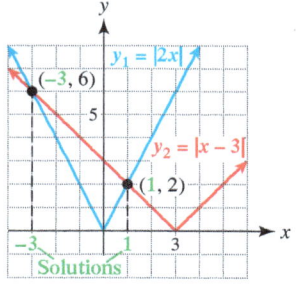

FIGURE 15.66

Example 6 is solved graphically in **FIGURE 15.66**. The graphs of $y_1 = |2x|$ and $y_2 = |x - 3|$ are V-shaped and intersect at $(-3, 6)$ and $(1, 2)$. The solutions are -3 and 1.

2 Solving Absolute Value Inequalities

FINDING NUMERICAL AND GRAPHICAL SOLUTIONS We can solve absolute value inequalities numerically and graphically. For example, to solve $|x| < 3$ graphically, let $y_1 = |x|$ and $y_2 = 3$ (see **FIGURE 15.67**). Their graphs intersect at $(-3, 3)$ and $(3, 3)$. The V-shaped graph of y_1 is *below* the graph of y_2 for x-values *between*, but not including, $x = -3$ and $x = 3$. The solution set for $|x| < 3$ is $(-3, 3)$ in interval notation and is shaded on the x-axis.

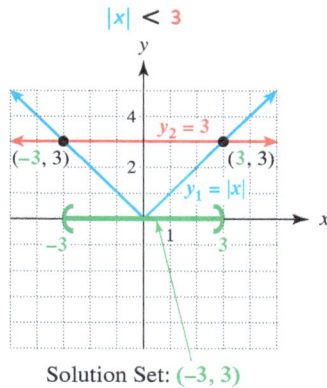

Solution Set: $(-3, 3)$
FIGURE 15.67

In the next example, we solve absolute value inequalities both numerically and graphically.

EXAMPLE 7 Solving inequalities numerically and graphically

Solve each inequality numerically and graphically.
(a) $|2x - 1| < 3$ (b) $|2x - 1| > 3$

Solution
(a) **Numerical Solution** **TABLE 15.19** shows a table of values for $y_1 = |2x - 1|$. Note that when $x = -1$ and $x = 2$, the absolute value $|2x - 1| = 3$. From this table, we see that $|2x - 1| < 3$ when $-1 < x < 2$. The solution set is $(-1, 2)$.

$y = |2x - 1|$

x	$|2x - 1|$	
-3	7	$\left.\begin{matrix}\\\\\end{matrix}\right\} \leftarrow |2x - 1| > 3$
-2	5	
-1	3	$\leftarrow |2x - 1| = 3$
0	1	$\left.\begin{matrix}\\\\\end{matrix}\right\} \leftarrow |2x - 1| < 3$
1	1	
2	3	$\leftarrow |2x - 1| = 3$
3	5	$\left.\begin{matrix}\\\\\end{matrix}\right\} \leftarrow |2x - 1| > 3$
4	7	

TABLE 15.19

Graphical Solution In **FIGURE 15.68**, the V-shaped graph of $y_1 = |2x - 1|$ lies *below* the horizontal line $y_2 = 3$ **between** $x = -1$ and $x = 2$, or when $-1 < x < 2$. Thus the solution set to $|2x - 1| < 3$ is $(-1, 2)$ in interval notation, and the graphical and numerical solutions agree.

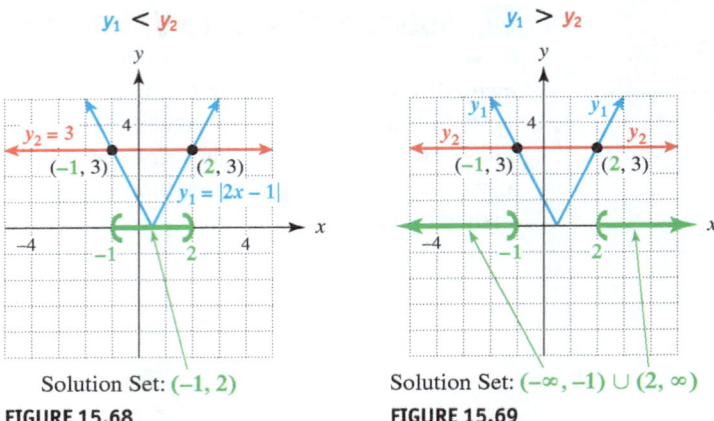

Solution Set: $(-1, 2)$
FIGURE 15.68

Solution Set: $(-\infty, -1) \cup (2, \infty)$
FIGURE 15.69

(b) Numerical Solution From **TABLE 15.19** on the previous page, we see that $|2x - 1| > 3$ when $x < -1$ or when $x > 2$. Thus the solution set is $(-\infty, -1) \cup (2, \infty)$.

Graphical Solution In **FIGURE 15.69**, the V-shaped graph of $y_1 = |2x - 1|$ lies *above* the horizontal line $y_2 = 3$ to the **left of** $x = -1$ or to the **right of** $x = 2$. Thus the solution set to $|2x - 1| > 3$ is $(-\infty, -1) \cup (2, \infty)$ in interval notation, and the graphical and numerical solutions agree.

Now Try Exercises 107(b), 107(c)

SYMBOLIC SOLUTIONS BASED ON VISUALIZATION (METHOD I) By understanding graphical solutions to absolute value inequalities, we can use visualization to solve absolute value inequalities symbolically. If we find the solutions to $|ax + b| = k$, then we can write the solutions to $|ax + b| < k$ and $|ax + b| > k$ directly without further work. This technique is summarized as follows.

> **ABSOLUTE VALUE INEQUALITIES**
>
> Let the solutions to $|ax + b| = k$ be c and d, where $c < d$ and $k > 0$.
>
> 1. $|ax + b| < k$ is equivalent to $c < x < d$.
> 2. $|ax + b| > k$ is equivalent to $x < c$ or $x > d$.
>
> Similar statements can be made for inequalities involving $\leq$ or $\geq$.

EXAMPLE 8 Solve absolute value equations and inequalities

Solve each absolute value equation or inequality.
(a) $|2 - 3x| = 4$ (b) $|2 - 3x| < 4$ (c) $|2 - 3x| > 4$

Solution
(a) The given equation is equivalent to the following equations.

$$2 - 3x = 4 \quad \text{or} \quad 2 - 3x = -4 \quad \text{Equations to be solved}$$
$$-3x = 2 \quad \text{or} \quad -3x = -6 \quad \text{Subtract 2.}$$
$$x = -\frac{2}{3} \quad \text{or} \quad x = 2 \quad \text{Divide by } -3.$$

The solutions are $-\frac{2}{3}$ and 2.

STUDY TIP

Be sure you understand how to write the solution to Example 8(c).

(b) Solutions to $|2 - 3x| < 4$ include x-values **between**, but not including, $-\frac{2}{3}$ and 2. Thus the solution set is $\{x \mid -\frac{2}{3} < x < 2\}$, or in interval notation, $\left(-\frac{2}{3}, 2\right)$.

(c) Solutions to $|2 - 3x| > 4$ include x-values to the **left** of $x = -\frac{2}{3}$ **or** to the **right** of $x = 2$. Thus the solution set is $\{x \mid x < -\frac{2}{3} \text{ or } x > 2\}$, or in interval notation, $\left(-\infty, -\frac{2}{3}\right) \cup (2, \infty)$.

Now Try Exercise 55

FIGURE 15.70(a) can be used to visualize the solution to $|2 - 3x| = 4$ in Example 8(a). Similarly, **FIGURES 15.70(b)** and **15.70(c)** can be used to visualize the solutions to $|2 - 3x| < 4$ and $|2 - 3x| > 4$ in parts (b) and (c) of Example 8.

Visualizing Solutions to Equations and Inequalities

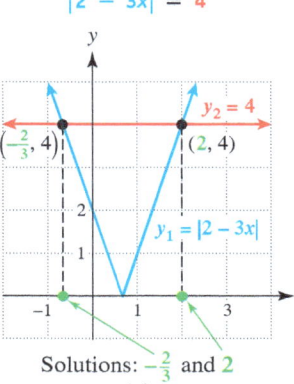
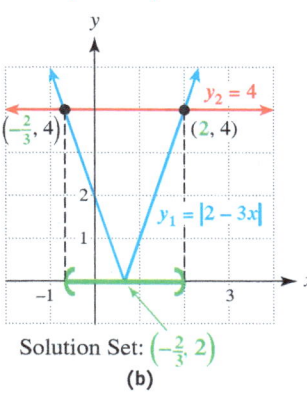
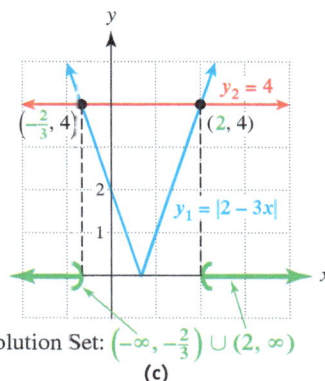

Solutions: $-\frac{2}{3}$ and 2 Solution Set: $\left(-\frac{2}{3}, 2\right)$ Solution Set: $\left(-\infty, -\frac{2}{3}\right) \cup (2, \infty)$
(a)   (b)   (c)

FIGURE 15.70

EXAMPLE 9   **Solving an absolute value inequality**

Solve $\left|\frac{2x - 5}{3}\right| > 3$. Write the solution set in interval notation.

Solution
Start by solving $\left|\frac{2x - 5}{3}\right| = 3$ as follows.

$$\frac{2x - 5}{3} = 3 \quad \text{or} \quad \frac{2x - 5}{3} = -3 \quad \text{\color{blue}Equations to be solved}$$

$$2x - 5 = 9 \quad \text{or} \quad 2x - 5 = -9 \quad \text{\color{blue}Multiply by 3.}$$

$$2x = 14 \quad \text{or} \quad 2x = -4 \quad \text{\color{blue}Add 5.}$$

$$x = 7 \quad \text{or} \quad x = -2 \quad \text{\color{blue}Divide by 2.}$$

Because the inequality symbol is $>$, the solution set is given by $x < -2$ or $x > 7$, or in interval notation, $(-\infty, -2) \cup (7, \infty)$.

Now Try Exercise 83

FINDING SYMBOLIC SOLUTIONS (METHOD II) The absolute value of a number gives its distance from the origin on a number line. For example $|3| = 3$ and $|-3| = 3$ because both 3 and -3 are located 3 units from the origin, as shown in **FIGURE 15.71**.

Absolute Values of -3 and 3

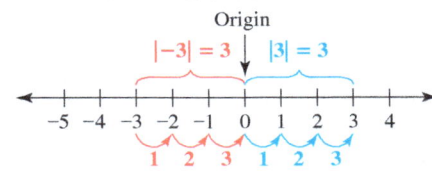

FIGURE 15.71

Now suppose we want to find the solution set to $|x| < 3$. An x-value that satisfies this inequality has a distance that is less than 3 from the origin. For example, both $x = 1$ and $x = -2$ are solutions to $|x| < 3$ because $|1| < 3$ and $|-2| < 3$. In general, solutions to $|x| < 3$ satisfy $-3 < x < 3$, as shown in **FIGURE 15.72**.

Next suppose we want to find the solution set to $|x| > 3$. An x-value that satisfies this inequality has a distance that is greater than 3 from the origin. For example, both $x = 4$ and $x = -4$ are solutions to $|x| > 3$ because $|4| > 3$ and $|-4| > 3$. In general, solutions to $|x| > 3$ satisfy $x < -3$ or $x > 3$, as shown in **FIGURE 15.73**.

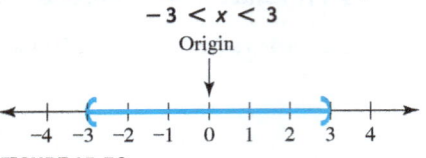

Distance from Origin is Less Than 3: $|x| < 3$
$-3 < x < 3$

FIGURE 15.72

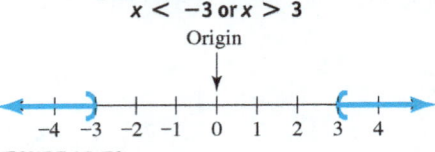

Distance from Origin is Greater Than 3: $|x| > 3$
$x < -3$ or $x > 3$

FIGURE 15.73

From the preceding discussion, it follows that if $|ax + b| < 3$, then $-3 < ax + b < 3$. Similarly, if $|ax + b| > 3$, then either $ax + b < -3$ or $ax + b > 3$. This method for solving absolute value inequalities symbolically is summarized as follows.

CRITICAL THINKING 1

How many solutions are there to $|2x + 1| < 4$? How many solutions are there to $|2x + 1| < -4$?

ABSOLUTE VALUE INEQUALITIES

Let k be a positive number.

1. $|ax + b| < k$ is equivalent to $-k < ax + b < k$.
2. $|ax + b| > k$ is equivalent to $ax + b < -k$ or $ax + b > k$.

Similar statements can be made for inequalities involving $\leq$ or $\geq$.

The next two examples, use this symbolic technique to solve absolute value inequalities.

EXAMPLE 10 Solving absolute value inequalities symbolically

Solve each absolute value inequality. Write your answer in interval notation.
(a) $|4 + x| < 5$ (b) $|4 + x| > 5$

Solution
(a) The solution set to $|4 + x| < 5$ is found by solving the following three-part inequality.

$-5 < 4 + x < 5$ Three-part inequality
$-9 < x < 1$ Subtract 4 from each part.

Thus the solution set in interval notation is $(-9, 1)$.

(b) The solution set to $|4 + x| > 5$ is found by solving the following compound inequality.

$4 + x < -5$ or $4 + x > 5$ Compound inequality
$x < -9$ or $x > 1$ Subtract 4 from each inequality.

Thus the solution set in interval notation is $(-\infty, -9) \cup (1, \infty)$.

Now Try Exercises 53(b), 53(c)

EXAMPLE 11 Solving absolute value inequalities symbolically

Solve each absolute value inequality. Write your answer in interval notation.
(a) $|4 - 5x| \leq 3$ (b) $|-4x - 6| > 2$

Solution
(a) $|4 - 5x| \leq 3$ is equivalent to the following three-part inequality.

$$-3 \leq 4 - 5x \leq 3 \quad \text{Equivalent inequality}$$
$$-7 \leq -5x \leq -1 \quad \text{Subtract 4 from each part.}$$
$$\frac{7}{5} \geq x \geq \frac{1}{5} \quad \text{Divide each part by } -5;\ \text{reverse the inequality.}$$

In interval notation, the solution is $\left[\frac{1}{5}, \frac{7}{5}\right]$.

(b) $|-4x - 6| > 2$ is equivalent to the following compound inequality.

$$-4x - 6 < -2 \quad \text{or} \quad -4x - 6 > 2 \quad \text{Equivalent compound inequality}$$
$$-4x < 4 \quad \text{or} \quad -4x > 8 \quad \text{Add 6 to each side.}$$
$$x > -1 \quad \text{or} \quad x < -2 \quad \text{Divide each by } -4;\ \text{reverse the inequality.}$$

In interval notation, the solution set is $(-\infty, -2) \cup (-1, \infty)$.

Now Try Exercises 65, 67

 *Math in Context* (Manufacturing) Absolute value inequalities often occur in real life when error in manufacturing occurs. The next example illustrates the fact that a circular cover cannot be made with an *exact* diameter. There is always some error.

EXAMPLE 12 Analyzing error

An engineer is designing a circular cover for a container. The diameter d of the cover is to be 4.25 inches and must be accurate to within 0.01 inch. Write an absolute value inequality that gives acceptable values for d.

Solution
The diameter d must satisfy $4.24 \leq d \leq 4.26$. Subtracting 4.25 from each part gives

$$-0.01 \leq d - 4.25 \leq 0.01,$$

which is equivalent to $|d - 4.25| \leq 0.01$. The "distance" or difference between 4.25 and the diameter is less than or equal to 0.01.

CALCULATOR HELP
To graph an absolute value, see Appendix A (page AP-8).

Now Try Exercise 127

 Connecting Concepts with Your Life Monthly average temperatures can vary greatly from one month to another, whereas yearly average temperatures remain fairly constant from one year to the next. In Boston, Massachusetts, the yearly average temperature is 50 °F, but monthly average temperatures can vary from 28 °F to 72 °F. Because 50 °F − 28 °F = 22 °F and 72 °F − 50 °F = 22 °F, the monthly average temperatures are always within 22 °F of the yearly average temperature. If T represents a monthly average temperature, we can model this situation by using the *absolute value inequality*

$$|T - 50| \leq 22.$$

The absolute value is necessary because a monthly average temperature T can be either greater than or less than 50 °F by as much as 22 °F.

EXAMPLE 13 Modeling temperature in Boston

The inequality $|T - 50| \leq 22$ models the range for the monthly average temperatures T in Boston.
(a) Solve this inequality and interpret the result.
(b) Give graphical support for part (a).

Solution
(a) **Symbolic Solution** To solve $|T - 50| \leq 22$, we can solve the three-part inequality $-22 \leq T - 50 \leq 22$.

$$-22 \leq T - 50 \leq 22 \quad \text{Inequality to be solved}$$
$$28 \leq T \leq 72 \quad \text{Add 50 to each part.}$$

Thus the solution set to $|T - 50| \leq 22$ is $\{T \,|\, 28 \leq T \leq 72\}$. Monthly average temperatures in Boston vary from 28 °F to 72 °F.

(b) **Graphical Solution** The graphs of $y_1 = |x - 50|$ and $y_2 = 22$ intersect at $(28, 22)$ and $(72, 22)$, as shown in **FIGURES 15.74(a)** and **(b)**. The V-shaped graph of y_1 intersects the horizontal graph of y_2, or is below it, when $28 \leq x \leq 72$. Thus the solution set is $\{T \,|\, 28 \leq T \leq 72\}$. This result agrees with the symbolic result.

Now Try Exercise 119

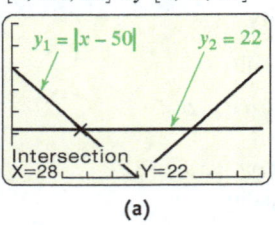
$[0, 100, 10]$ by $[0, 70, 10]$
(a)

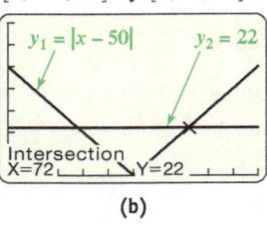
$[0, 100, 10]$ by $[0, 70, 10]$
(b)

FIGURE 15.74

Sometimes the solution set to an absolute value inequality can be either empty or the set of all real numbers. These two situations are illustrated in the next example.

EXAMPLE 14 Solving absolute value inequalities

Solve if possible.
(a) $|2x - 5| > -1$ (b) $|5x - 1| + 3 \leq 2$

Solution
(a) Because the absolute value of an expression cannot be negative, $|2x - 5|$ is greater than -1 for every x-value. The solution set is all real numbers, or $(-\infty, \infty)$.
(b) Subtracting 3 from each side results in $|5x - 1| \leq -1$. Because the absolute value is always greater than or equal to 0, no x-values satisfy this inequality. There are no solutions.

Now Try Exercises 79, 81

15.5 Putting It All Together

PROBLEM	SYMBOLIC SOLUTION	GRAPHICAL SOLUTION
$\|ax + b\| = k, k > 0$	Solve the equations $$ax + b = k$$ and $$ax + b = -k.$$	Graph $y_1 = \|ax + b\|$ and $y_2 = k$. Find the x-values of the two points of intersection.

PROBLEM	SYMBOLIC SOLUTION	GRAPHICAL SOLUTION
$\lvert ax+b \rvert < k, k > 0$	**Method I:** If the solutions to $$\lvert ax+b \rvert = k$$ are c and d, $c < d$, then the solutions to $$\lvert ax+b \rvert < k$$ satisfy $$c < x < d.$$ **Method II:** The solutions to $\lvert ax+b \rvert < k$ can be found by solving $$-k < ax+b < k.$$	Graph $y_1 = \lvert ax+b \rvert$ and $y_2 = k$. Find the x-values of the two points of intersection. The solutions are between these x-values on the number line, where the graph of y_1 lies below the graph of y_2.
$\lvert ax+b \rvert > k, k > 0$	**Method I:** If the solutions to $$\lvert ax+b \rvert = k$$ are c and d, $c < d$, then the solutions to $$\lvert ax+b \rvert > k$$ satisfy $$x < c \quad \text{or} \quad x > d.$$ **Method II:** The solutions to $\lvert ax+b \rvert > k$ can be found by solving $$ax+b < -k \text{ or } ax+b > k.$$	Graph $y_1 = \lvert ax+b \rvert$ and $y_2 = k$. Find the x-values of the two points of intersection. The solutions are "outside" (left or right of) these x-values on the number line, where the graph of y_1 is above the graph of y_2.

15.5 Exercises   MyMathLab®

CONCEPTS AND VOCABULARY

1. Give an example of an absolute value equation.
2. Give an example of an absolute value inequality.
3. Is -3 a solution to $\lvert x \rvert = 3$?
4. Is -4 a solution to $\lvert x \rvert > 3$?
5. Is the solution set to $\lvert x \rvert = 5$ equal to $\{-5, 5\}$?
6. Is $\lvert x \rvert < 3$ equivalent to $x < -3$ or $x > 3$? Explain.
7. How many times does the graph of $y = \lvert 2x - 1 \rvert$ intersect the graph of $y = 5$?
8. How many times does the graph of $y = \lvert 2x - 1 \rvert$ intersect the graph of $y = -5$?

Exercises 9–14: Determine whether the given values of x are solutions to the absolute value equation or inequality.

9. $\lvert 2x - 5 \rvert = 1$ $x = -3, x = 3$
10. $\lvert 5 - 6x \rvert = 1$ $x = 1, x = 0$
11. $\lvert 7 - 4x \rvert \leq 5$ $x = -2, x = 2$
12. $\lvert 2 + x \rvert < 2$ $x = -4, x = -1$
13. $\lvert 7x + 4 \rvert > -1$ $x = -\frac{4}{7}, x = 2$
14. $\lvert 12x + 3 \rvert \geq 3$ $x = -\frac{1}{4}, x = 2$

GRAPHING ABSOLUTE VALUE EQUATIONS

Exercises 15–20: Graph the absolute value equation.

15. $y = \lvert 2x + 2 \rvert$
16. $y = \lvert x + 1 \rvert$
17. $y = \lvert -x - 3 \rvert$
18. $y = \lvert 3x - 3 \rvert$
19. $y = \lvert 3 - x \rvert$
20. $y = \lvert 4 - 2x \rvert$

FINDING SYMBOLIC SOLUTIONS

Exercises 21–46: Solve the absolute value equation.

21. $\lvert x \rvert = 7$
22. $\lvert x \rvert = 4$
23. $\lvert x \rvert = -6$
24. $\lvert x \rvert = 0$

25. $|4x| = 9$ 26. $|-3x| = 7$
27. $|-2x| - 6 = 2$ 28. $|5x| + 1 = 5$
29. $|2x + 1| = 11$ 30. $|1 - 3x| = 4$
31. $|-2x + 3| + 3 = 4$
32. $|6x + 2| - 2 = 6$
33. $|3 - 4x| = 0$ 34. $|5x - 3| = -1$
35. $\left|\frac{1}{2}x - 1\right| = 5$ 36. $\left|6 - \frac{3}{4}x\right| = 3$
37. $|2x - 6| = -7$ 38. $\left|1 - \frac{2}{3}x\right| = 0$
39. $\left|\frac{2}{3}z - 1\right| - 3 = 8$ 40. $|1 - 2z| + 5 = 10$
41. $|z - 1| = |2z|$ 42. $|2z + 3| = |2 - z|$
43. $|3t + 1| = |2t - 4|$
44. $\left|\frac{1}{2}t - 1\right| = \left|3 - \frac{3}{2}t\right|$ 45. $\left|\frac{1}{4}x\right| = \left|3 + \frac{1}{4}x\right|$
46. $|2x - 1| = |2x + 2|$

Exercises 47–56: Solve each equation or inequality. Use interval notation for inequalities.

47. $|x| = 3$ 48. $|x| = 5$
49. $|x| < 3$ 50. $|x| < 5$
51. $|x| > 3$ 52. $|x| > 5$
53. (a) $|2x| = 8$ (b) $|2x| < 8$
    (c) $|2x| > 8$
54. (a) $|3x - 9| = 6$
    (b) $|3x - 9| \le 6$
    (c) $|3x - 9| \ge 6$
55. (a) $|5 - 4x| = 3$
    (b) $|5 - 4x| \le 3$
    (c) $|5 - 4x| \ge 3$
56. (a) $\left|\frac{x - 5}{2}\right| = 2$ (b) $\left|\frac{x - 5}{2}\right| < 2$
    (c) $\left|\frac{x - 5}{2}\right| > 2$

Exercises 57–88: Solve the absolute value inequality. Write your answer in interval notation.

57. $|x| \le 3$ 58. $|x| < 2$
59. $|k| > 4$ 60. $|k| \ge 5$
61. $|t| \le -3$ 62. $|t| < -1$
63. $|z| > 0$ 64. $|2z| \ge 0$
65. $|2x| > 7$ 66. $|-12x| < 30$
67. $|-4x + 4| < 16$ 68. $|-5x - 8| > 2$
69. $2|x + 5| \ge 8$ 70. $-3|x - 1| \ge -9$
71. $|8 - 6x| - 1 \le 2$ 72. $4 - \left|\frac{2x}{3}\right| < -7$
73. $5 + \left|\frac{2 - x}{3}\right| \le 9$ 74. $\left|\frac{x + 3}{5}\right| \le 12$
75. $|2x - 1| \le -3$ 76. $|x + 6| \ge -5$
77. $|x + 1| - 1 > -3$ 78. $-2|1 - 7x| \ge 2$
79. $|2z - 4| + 2 \le 1$ 80. $|4 - z| \le 0$
81. $|3z - 1| > -3$ 82. $|2z| \ge -2$
83. $\left|\frac{2 - t}{3}\right| \ge 5$ 84. $\left|\frac{2t + 3}{5}\right| \ge 7$
85. $|t - 1| \le 0.1$
86. $|t - 2| \le 0.01$
87. $|b - 10| > 0.5$
88. $|b - 25| \ge 1$

FINDING NUMERICAL AND GRAPHICAL SOLUTIONS

Exercises 89 and 90: Use the table of $y = |ax + b|$ to solve each equation or inequality. Write your answers in interval notation for parts (b) and (c).

89. (a) $y = 2$ (b) $y < 2$ (c) $y > 2$

x	-2	-1	0	1	2	3	4
y	3	2	1	0	1	2	3

90. (a) $y = 6$ (b) $y \le 6$ (c) $y \ge 6$

x	-12	-6	0	6	12	18	24
y	9	6	3	0	3	6	9

Exercises 91 and 92: Use the graph to solve the equation.

91. $|x - 2| = 2$ 92. $|2x + 1| = 3$

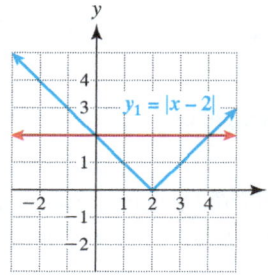
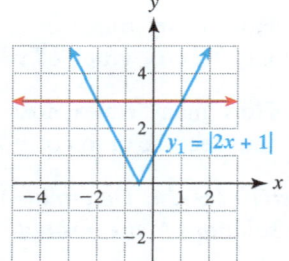

Exercises 93 and 94: Use the graph of y_1 to solve each equation or inequality. Write your answers in interval notation for parts (b) and (c).

93. (a) $|2x + 1| = 1$
    (b) $|2x + 1| \leq 1$
    (c) $|2x + 1| \geq 1$

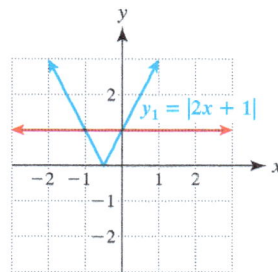

94. (a) $|x - 1| = 3$ (b) $|x - 1| < 3$
    (c) $|x - 1| > 3$

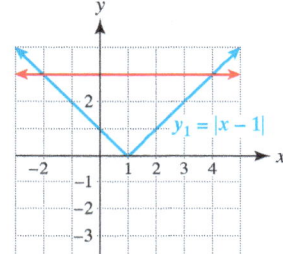

Exercises 95–104: Solve the inequality graphically. Write your answer in interval notation.

95. $|x| \geq 1$
96. $|x| < 2$
97. $|x - 1| \leq 3$
98. $|x + 5| \geq 2$
99. $|4 - 2x| > 2$
100. $|1.5x - 3| \geq 6$
101. $|10 - 3x| < 4$
102. $|7 - 4x| \leq 2.5$
103. $|8.1 - x| > -2$
104. $\left|\dfrac{5x - 9}{2}\right| \leq -1$

USING MORE THAN ONE METHOD

Exercises 105–108: Solve the absolute value inequality
(a) symbolically,
(b) graphically, and
(c) numerically.
Write your answer in set-builder notation.

105. $|3x| \leq 9$
106. $|5 - x| \geq 3$
107. $|2x - 5| > 1$
108. $|-8 - 4x| < 6$

WRITING ABSOLUTE VALUE INEQUALITIES

Exercises 109–116: Write each compound inequality as an absolute value inequality. Do not simplify Exercises 113–116.

109. $-4 \leq x \leq 4$
110. $-0.1 < y < 0.1$
111. $y < -2$ or $y > 2$
112. $-0.1 \leq x \leq 0.1$
113. $-0.3 \leq 2x + 1 \leq 0.3$
114. $4x < -5$ or $4x > 5$
115. $\pi x \leq -7$ or $\pi x \geq 7$
116. $-0.9 \leq x - \sqrt{2} \leq 0.9$

117. **Thinking Generally** If $a \neq 0$ and $k > 0$, then the graph of $y = |ax + b|$ intersects the graph of $y = k$ at _____ points.

118. **Thinking Generally** If a and k are positive, then the solution set to $|ax + b| < k$ is _____.

APPLICATIONS INVOLVING ABSOLUTE VALUES

Exercises 119–122: **Average Temperatures** *(Refer to Example 13.)* The given inequality models the range for the monthly average temperatures T in degrees Fahrenheit at the location specified.
(a) Solve the inequality.
(b) Give a possible interpretation of the inequality.

119. $|T - 43| \leq 24$, Marquette, Michigan
120. $|T - 62| \leq 19$, Memphis, Tennessee
121. $|T - 10| \leq 36$, Chesterfield, Canada
122. $|T - 61.5| \leq 12.5$, Buenos Aires, Argentina

123. **Highest Elevations** The table lists the highest elevation on each continent.

Continent	Elevation (feet)
Asia	29,028
S. America	22,834
N. America	20,320
Africa	19,340
Europe	18,510
Antarctica	16,066
Australia	7,310

Source: National Geographic.

(a) Calculate the average A of these elevations.
(b) Which continents have their highest elevations within 1000 feet of A?

continued on next page

continued from previous page

(c) Which continents have their highest elevations within 5000 feet of A?

(d) Let E be an elevation. Write an absolute value inequality that says E is within 5000 feet of A.

124. **Distance** Suppose that two cars, both traveling at a constant speed of 60 miles per hour, approach each other on a straight highway.
   (a) If the two cars are initially 4 miles apart, sketch a graph of the distance between the two cars after x minutes, where $0 \leq x \leq 4$. (*Hint:* 60 miles per hour = 1 mile per minute.)
   (b) Write an absolute value equation whose solution gives the times when the cars are 2 miles apart.
   (c) Solve your equation from part (b).

125. **Error in Measurements** (Refer to Example 12.) The maximum error in the diameter of a container is restricted to 0.002 inch, so an acceptable diameter d must satisfy the absolute value inequality
$$|d - 2.5| \leq 0.002.$$
Solve this inequality for d and interpret the result.

126. **Error in Measurements** Suppose that a person can operate a stopwatch accurately to within 0.02 second. If Byron Dyce's time in the 800-meter race is recorded as 105.30 seconds, write an absolute value inequality that gives the possible values for the actual time t.

127. **Error in Measurements** A circular lid is being designed for a container. The diameter d of the lid is to be 3.8 inches and must be accurate to within 0.03 inch. Write an absolute value inequality that gives acceptable values for d.

128. **Manufacturing a Tire** An engineer is designing a tire for a truck. The diameter d of the tire is to be 36 inches and the *circumference* must be accurate to within 0.1 inch. Write an absolute value inequality that gives acceptable values for d.

129. **Relative Error** If a quantity is measured to be x and the true value is t, then the relative error in the measurement is $\left|\frac{x-t}{t}\right|$. If the true measurement is $t = 20$ and you want the relative error to be less than 0.05 (5%), what values for x are possible?

130. **Relative Error** (Refer to the preceding exercise.) The volume V of a box is 50 cubic inches. How accurately must you measure the volume of the box for the relative error to be less than 3%?

WRITING ABOUT MATHEMATICS

131. If $a \neq 0$, how many solutions are there to the equation $|ax + b| = k$ when
   (a) $k > 0$, (b) $k = 0$, and (c) $k < 0$?
   Explain each answer.

132. Suppose that you know two solutions to the equation $|ax + b| = k$. How can you use these solutions to solve the inequalities $|ax + b| < k$ and $|ax + b| > k$? Give an example.

SECTION 15.5 Checking Basic Concepts

Write answers in interval notation whenever possible.

1. Solve $\left|\frac{3}{4}x - 1\right| - 3 = 5$.

2. Solve the absolute value equation and inequalities.
   (a) $|3x - 6| = 8$
   (b) $|3x - 6| < 8$
   (c) $|3x - 6| > 8$

3. Solve the inequality $|-2(3 - x)| < 6$, and then solve $|-2(3 - x)| \geq 6$.

4. Use the graph to solve the equation and inequalities.

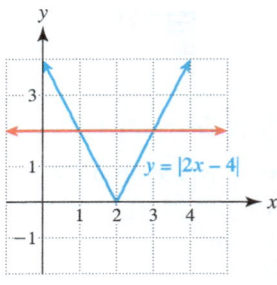

   (a) $|2x - 4| = 2$ (b) $|2x - 4| \leq 2$
   (c) $|2x - 4| \geq 2$

CHAPTER 15 Summary

SECTION 15.1 ■ FUNCTIONS AND THEIR REPRESENTATIONS

Function A function is a set of ordered pairs (x, y), where each x-value corresponds to exactly one y-value. A function takes a valid input x and computes exactly one output y, forming the ordered pair (x, y).

Domain and Range of a Function The domain D is the set of all valid inputs, or x-values, and the range R is the set of all outputs, or y-values.

Examples: $f = \{(1, 2), (2, 3), (3, 3)\}$ has $D = \{1, 2, 3\}$ and $R = \{2, 3\}$.

$f(x) = x^2$ has domain all real numbers and range $y \geq 0$. (See the graph below.)

Function Notation $y = f(x)$ and is read "y equals f of x."

Example: $f(x) = \frac{2x}{x-1}$ implies that $f(3) = \frac{2 \cdot 3}{3-1} = \frac{6}{2} = 3$. Thus the point $(3, 3)$ is on the graph of f.

Function Representations A function can be represented symbolically, numerically, graphically, or verbally.

Symbolic Representation (Formula) $f(x) = x^2$

Numerical Representation (Table)

x	y
−2	4
−1	1
0	0
1	1
2	4

Graphical Representation (Graph)

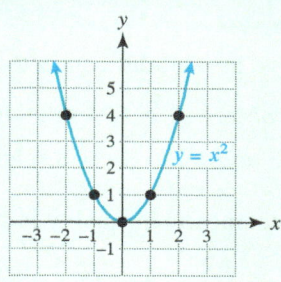

Verbal Representation (Words) f computes the square of the input x.

Vertical Line Test If every vertical line intersects a graph at most once, then the graph represents a function.

SECTION 15.2 ■ LINEAR FUNCTIONS

Linear Function A linear function can be represented by $f(x) = mx + b$. Its graph is a (straight) line, where m is the slope and $(0, b)$ is the y-intercept. For each unit increase in x, $f(x)$ changes by an amount equal to m.

Example: $f(x) = 2x - 1$ represents a linear function with $m = 2$ and $b = -1$.

Numerical Representation

x	f(x)
−1	−3
0	−1
1	1
2	3

1 unit → 2 units
1 unit → 2 units
1 unit → 2 units

Graphical Representation

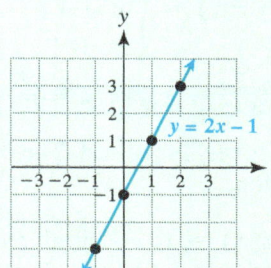

Each 1-unit increase in x results in a 2-unit increase in $f(x)$; thus $m = 2$.

NOTE: A numerical representation is a table of values of $f(x)$. ■

Modeling Data with Linear Functions When data have a constant rate of change, they can be modeled by $f(x) = mx + b$. The constant m represents the *rate of change*, and the constant b represents the *initial amount* or the value when $x = 0$. That is,

$$f(x) = (\text{Rate of change})x + (\text{Initial amount}).$$

Example: In the following table, the y-values decrease by 3 units for each 1-unit increase in x. When $x = 0, y = 4$. Thus the data are modeled by $f(x) = -3x + 4$.

x	-2	-1	0	1	2
y	10	7	4	1	-2

Midpoint Formula The midpoint of the line segment connecting (x_1, y_1) and (x_2, y_2) is

$$\left(\frac{x_1 + x_2}{2}, \frac{y_1 + y_2}{2}\right).$$

Example: The midpoint of the line segment connecting $(-5, 8)$ and $(9, 4)$ is

$$\left(\frac{-5 + 9}{2}, \frac{8 + 4}{2}\right) = (2, 6).$$

SECTION 15.3 ■ COMPOUND INEQUALITIES

Compound Inequality Two inequalities connected by *and* or *or*.

Examples: For $x + 1 < 3$ or $x + 1 > 6$, a solution satisfies *at least* one of the inequalities.

For $2x + 1 < 3$ and $1 - x > 6$, a solution satisfies *both* inequalities.

Three-Part Inequality A compound inequality in the form $x > a$ and $x < b$ can be written as the three-part inequality $a < x < b$.

Example: $1 \leq x < 7$ means $x \geq 1$ *and* $x < 7$.

Interval Notation Can be used to identify intervals on the real number line

Examples: $-2 < x \leq 3$ is equivalent to $(-2, 3]$.

$x < 5$ is equivalent to $(-\infty, 5)$.

All real numbers are denoted $(-\infty, \infty)$.

Piecewise-Defined Function A function that is defined using different formulas over different intervals of its domain

Example: $f(0) = 0 + 6 = 6$, because when $x = 0, f(x) = x + 6$.

$$f(x) = \begin{cases} x^2 + 2x + 6 & \text{if } -5 \leq x < 0 \\ x + 6 & \text{if } 0 \leq x < 2 \\ x^3 + 1 & \text{if } 2 \leq x \leq 5 \end{cases}$$

SECTION 15.4 ■ OTHER FUNCTIONS AND THEIR PROPERTIES

Domain and Range in Interval Notation The domain and range of a function can often be expressed in interval notation.

Example: The domain of $f(x) = x^2 - 2$ is all real numbers, or $(-\infty, \infty)$, and its range is real numbers greater than or equal to -2, or $[-2, \infty)$.

Absolute Value Function The domain of $f(x) = |x|$ is $(-\infty, \infty)$, and the range is $[0, \infty)$.

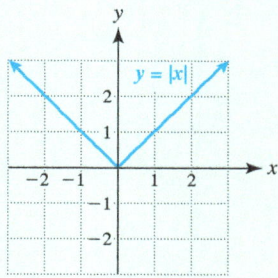

Polynomial Functions The degree of a polynomial function (of one variable) equals the largest exponent of a variable. The graphs of polynomial functions with degree greater than 1 are not lines. The domain of a polynomial function is $(-\infty, \infty)$.

Examples: $f(x) = 4x - 1$ defines a linear function with degree 1.

$g(x) = 4x^2 + x - 4$ defines a quadratic function with degree 2.

$h(x) = x^3 + 0.7x - 1$ defines a cubic function with degree 3.

Rational Functions If $f(x) = \frac{p(x)}{q(x)}$, where $p(x)$ and $q(x)$ are polynomials, f is a rational function. The domain of a rational function includes all real numbers, except x-values for which $q(x) = 0$.

Examples: $f(x) = \frac{1}{x}$ has domain $(-\infty, 0) \cup (0, \infty)$, or $x \neq 0$.

$g(x) = \frac{x}{x^2 - 9}$ has domain $(-\infty, -3) \cup (-3, 3) \cup (3, \infty)$, or $x \neq -3, x \neq 3$.

Operations on Functions If $f(x)$ and $g(x)$ are both defined, then the sum, difference, product, and quotient of two functions f and g are defined by

$$(f + g)(x) = f(x) + g(x) \quad \text{Sum}$$
$$(f - g)(x) = f(x) - g(x) \quad \text{Difference}$$
$$(fg)(x) = f(x) \cdot g(x) \quad \text{Product}$$
$$\left(\frac{f}{g}\right)(x) = \frac{f(x)}{g(x)}, \text{ where } g(x) \neq 0. \quad \text{Quotient}$$

Examples: Let $f(x) = x^2 - 1$ and $g(x) = x^2 + 1$.

$$(f + g)(x) = f(x) + g(x) = (x^2 - 1) + (x^2 + 1) = 2x^2$$
$$(f - g)(x) = f(x) - g(x) = (x^2 - 1) - (x^2 + 1) = -2$$
$$(fg)(x) = f(x) \cdot g(x) = (x^2 - 1)(x^2 + 1) = x^4 - 1$$
$$\left(\frac{f}{g}\right)(x) = \frac{f(x)}{g(x)} = \frac{x^2 - 1}{x^2 + 1}$$

SECTION 15.5 ■ ABSOLUTE VALUE EQUATIONS AND INEQUALITIES

Absolute Value Equations The graph of $y = |ax + b|$, $a \neq 0$, is V-shaped and intersects the horizontal line $y = k$ twice if $k > 0$. In this case, there are two solutions to the equation $|ax + b| = k$ determined by $ax + b = k$ or $ax + b = -k$.

Example: The equation $|2x - 1| = 5$ has two solutions.

Symbolic Solution

$2x - 1 = 5$ or $2x - 1 = -5$

$\quad 2x = 6$ or $\quad 2x = -4$ Add 1.

$\quad\ \ x = 3$ or $\quad\ \ x = -2$ Divide by 2.

Graphical Solution

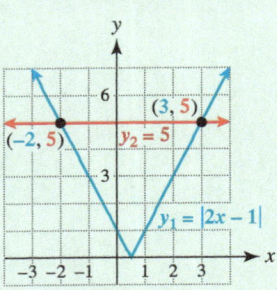

The solutions are -2 and 3.

Numerical Solution

x	-3	-2	-1	0	1	2	3
$\|2x - 1\|$	7	5	3	1	1	3	5

The solutions are -2 and 3.

Absolute Value Inequalities

Method I If the solutions to $|ax + b| = k$ are c and d with $c < d$, then the solution set for $|ax + b| < k$ is $\{x \mid c < x < d\}$, and the solution set for $|ax + b| > k$ is $\{x \mid x < c \text{ or } x > d\}$.

Examples: The solutions to the equation $|2x - 1| = 5$ are -2 and 3, so the solution set for $|2x - 1| < 5$ is $\{x \mid -2 < x < 3\}$, and the solution set for $|2x - 1| > 5$ is $\{x \mid x < -2 \text{ or } x > 3\}$.

Method II The solution set to $|ax + b| < k$ can be found by solving the three-part inequality

$$-k < ax + b < k.$$

The solution set to $|ax + b| > k$ can be found by solving the compound inequality

$$ax + b < -k \text{ or } ax + b > k.$$

Examples: $|3 - x| < 5$ is equivalent to $-5 < 3 - x < 5$ and
$|3 - x| > 5$ is equivalent to $3 - x < -5$ or $3 - x > 5$.

CHAPTER 15 Review Exercises

SECTION 15.1

Exercises 1–4: Evaluate $f(x)$ for the given values of x.

1. $f(x) = 3x - 1$ $x = -2, \frac{1}{3}$

2. $f(x) = 5 - 3x^2$ $x = -3, 1$

3. $f(x) = \sqrt{x} - 2$ $x = 0, 9$

4. $f(x) = 5$ $x = -5, \frac{7}{5}$

Exercises 5 and 6: Do the following.
(a) Write a symbolic representation (formula) for the function described.
(b) Evaluate the function for input 5 and interpret the result.

5. Function P computes the number of pints in q quarts.

6. Function f computes 3 less than 4 times a number x.

7. If $f(3) = -2$, then the point _____ lies on the graph of f.

8. If the point $(4, -6)$ lies on the graph of function f, then $f(\text{\_\_\_\_}) = \text{\_\_\_\_}$.

Exercises 9–12: Sketch a graph of f.

9. $f(x) = -2x$ 10. $f(x) = \frac{1}{2}x - \frac{3}{2}$

11. $f(x) = x^2 - 1$ 12. $f(x) = \sqrt{x + 1}$

Exercises 13 and 14: Use the graph of f to evaluate the given expressions.

13. $f(0)$ and $f(-3)$ 14. $f(-2)$ and $f(1)$

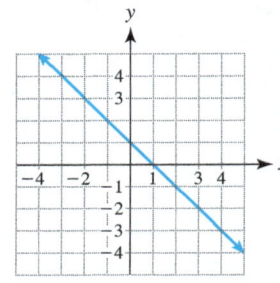 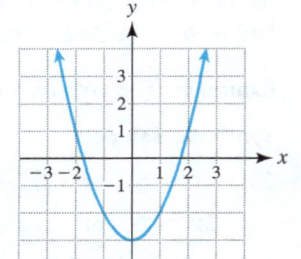

15. Use the table to evaluate $f(-1)$ and $f(3)$.

x	-1	1	3	5
$f(x)$	7	3	-1	-5

16. A function f is represented verbally by "Multiply the input x by 3 and then subtract 2." Give numerical, symbolic, and graphical representations for f. Let $x = -3, -2, -1, \ldots, 3$ in the table of values, and let $-3 \le x \le 3$ for the graph.

Exercises 17 and 18: Use the graph of f to estimate its domain and range.

17. 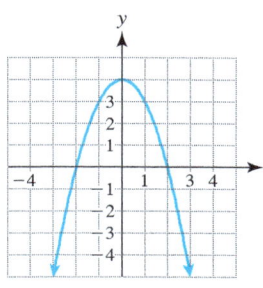 **18.**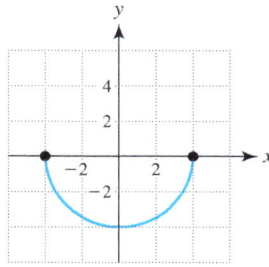

Exercises 19 and 20: Does the graph represent a function?

19. 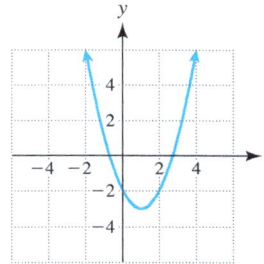 **20.**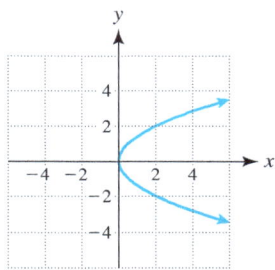

Exercises 21 and 22: Find the domain and range of S. Then state whether S defines a function.

21. $S = \{(-3, 4), (-1, 4), (2, 3), (4, -1)\}$

22. $S = \{(-1, 5), (0, 3), (1, -2), (-1, 2), (2, 4)\}$

Exercises 23–30: Find the domain.

23. $f(x) = -3x + 7$ **24.** $f(x) = \sqrt{x}$

25. $f(x) = \dfrac{3}{x}$ **26.** $f(x) = x^2 + 2$

27. $f(x) = \sqrt{5 - x}$ **28.** $f(x) = \dfrac{x}{x + 2}$

29. $f(x) = |2x + 1|$ **30.** $f(x) = x^3$

SECTION 15.2

Exercises 31 and 32: Does the graph represent a linear function?

31. 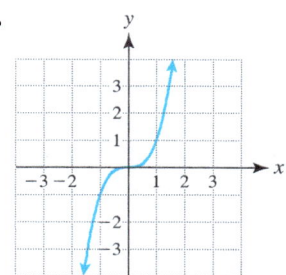 **32.**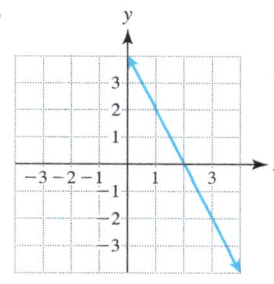

Exercises 33–36: Determine whether f is a linear function. If f is linear, give values for m and b so that f may be expressed as $f(x) = mx + b$.

33. $f(x) = -4x + 5$ **34.** $f(x) = 7 - x$

35. $f(x) = \sqrt{x}$ **36.** $f(x) = 6$

Exercises 37 and 38: Use the table to determine whether $f(x)$ could represent a linear function. If it could, write the formula for f in the form $f(x) = mx + b$.

37.

x	0	2	4	6
$f(x)$	-3	0	3	6

38.

x	-1	0	1	2
$f(x)$	-5	0	10	15

39. Evaluate $f(x) = \tfrac{1}{2}x + 3$ at $x = -4$.

40. Use the graph to evaluate $f(-2)$ and $f(1)$.

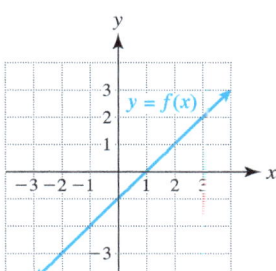

Exercises 41–44: Sketch a graph of $y = f(x)$.

41. $f(x) = x + 1$ **42.** $f(x) = 1 - 2x$

43. $f(x) = -\tfrac{1}{3}x$ **44.** $f(x) = -1$

45. Write a symbolic representation (formula) for a linear function H that calculates the number of hours in x days. Evaluate $H(2)$ and interpret the result.

46. Let $f(x) = \sqrt{x+2} - x^2$.
 (a) Make a numerical representation (table) for the function f with $x = 1, 2, 3, \ldots, 7$.
(b) Graph f in the standard window. What is the domain of f?

Exercises 47–48: Find the midpoint of the line segment connecting the given points.

47. $(-5, 3), (6, -9)$ **48.** $\left(\frac{2}{3}, -\frac{3}{4}\right), \left(\frac{1}{6}, \frac{3}{2}\right)$

SECTION 15.3

Use interval notation whenever possible for the remaining exercises.

Exercises 49–52: Solve the compound inequality. Graph the solution set on a number line.

49. $x + 1 \leq 3$ and $x + 1 \geq -1$

50. $2x + 7 < 5$ and $-2x \geq 6$

51. $5x - 1 \leq 3$ or $1 - x < -1$

52. $3x + 1 > -1$ or $3x + 1 < 10$

53. Use the table to solve $-2 \leq 2x + 2 \leq 4$.

x	-3	-2	-1	0	1	2	3
$2x + 2$	-4	-2	0	2	4	6	8

54. Use the following figure to solve each equation or inequality.
(a) $y_1 = y_2$ **(b)** $y_2 = y_3$
(c) $y_1 \leq y_2 \leq y_3$ **(d)** $y_2 < y_3$

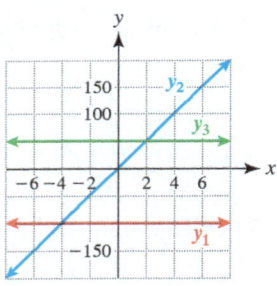

55. The graphs of y_1 and y_2 are shown in the figure. Solve each equation or inequality.
(a) $y_1 = y_2$ **(b)** $y_1 < y_2$
(c) $y_1 > y_2$

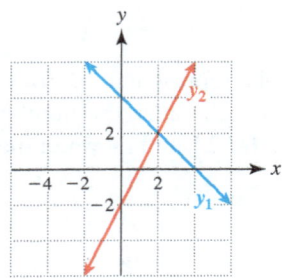

56. The graphs of three linear functions f, g, and h are shown in the following figure. Solve each equation or inequality.
(a) $f(x) = g(x)$ **(b)** $g(x) = h(x)$
(c) $f(x) < g(x) < h(x)$

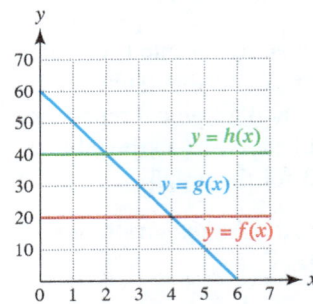

Exercises 57–62: Write the given inequality in interval notation.

57. $-3 \leq x \leq \frac{2}{3}$ **58.** $-6 < x \leq 45$

59. $x < \frac{7}{2}$ **60.** $x \geq 1.8$

61. $x > -3$ and $x < 4$ **62.** $x < 4$ or $x > 10$

Exercises 63–66: Solve the three-part inequality. Write the solution set in interval notation.

63. $-4 < x + 1 < 6$ **64.** $20 \leq 2x + 4 \leq 60$

65. $-3 < 4 - \frac{1}{3}x < 7$

66. $30 \leq \dfrac{2x - 6}{5} - 4 < 50$

Exercises 67 and 68: Let f be defined as follows.

$$f(x) = \begin{cases} 3x - 1 & \text{if } -5 \leq x < 1 \\ 4 & \text{if } 1 \leq x \leq 3 \\ 6 - x & \text{if } 3 < x \leq 5 \end{cases}$$

Use $f(x)$ to evaluate each expression.

67. $f(1)$ and $f(4)$ **68.** $f(-3)$ and $f(6)$

SECTION 15.4

Exercises 69 and 70: Write the domain and the range of the function.

69. $f(t) = \frac{1}{2}t^2$ **70.** $f(x) = |x + 2|$

71. Write the domain of $f(x) = \dfrac{x + 1}{2x - 8}$.

72. Write the domain and range of the function shown in the graph.

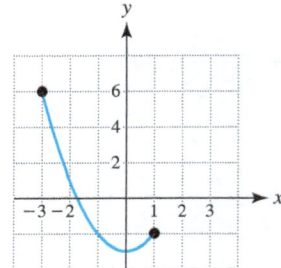

Exercises 73–76: Determine whether $f(x)$ represents a polynomial function. If possible, identify the degree and type of polynomial function.

73. $f(x) = 1 + 2x - 3x^2$

74. $f(x) = 5 + 7x$

75. $f(x) = x^3 + 2x$ **76.** $f(x) = |2x - 1|$

Exercises 77 and 78: If possible, evaluate $g(t)$ for the given values of t.

77. $g(t) = |1 - 4t|$ $t = 3, t = -\frac{1}{4}$

78. $g(t) = \dfrac{4}{4 - t^2}$ $t = 3, t = -2$

Exercises 79–82: Graph $y = f(x)$.

79. $f(x) = |x + 3|$ **80.** $f(x) = x^2 + 1$

81. $f(x) = \dfrac{1}{x}$ **82.** $f(x) = -3x$

83. Use $f(x) = 2x^2 - 3x$ and $g(x) = 2x - 3$ to find each of the following.
(a) $(f + g)(3)$ (b) $(fg)(3)$

84. Use $f(x) = x^2 - 1$ and $g(x) = x - 1$ to find each of the following.
(a) $(f - g)(x)$ (b) $(f/g)(x)$

SECTION 15.5

Exercises 85–88: Determine whether the given values of x are solutions to the absolute value equation or inequality.

85. $|12x - 24| = 24$ $x = -3; x = 2$

86. $|5 - 3x| > 3$ $x = \frac{4}{3}; x = 0$

87. $|3x - 6| \le 6$ $x = -3; x = 4$

88. $|2 + 3x| + 4 < 11$ $x = -3; x = \frac{2}{3}$

89. Use the table to solve each equation or inequality.
(a) $y_1 = 2$ (b) $y_1 < 2$
(c) $y_1 > 2$

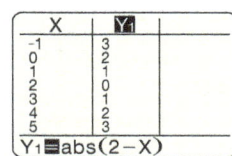

90. Use the graph of $y = |2x + 2|$ at the top of the next column to solve each equation or inequality.
(a) $|2x + 2| = 4$ (b) $|2x + 2| \le 4$
(c) $|2x + 2| \ge 4$

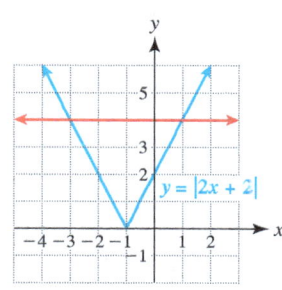

Exercises 91–96: Solve the absolute value equation.

91. $|x| = 22$ **92.** $|2x - 9| = 7$

93. $|4 - \frac{1}{2}x| = 17$ **94.** $\frac{1}{3}|3x - 1| + 1 = 9$

95. $|2x - 5| = |5 - 3x|$

96. $|-3 + 3x| = |-2x + 6|$

Exercises 97 and 98: Solve each absolute value equation or inequality.

97. (a) $|x + 1| = 7$ (b) $|x + 1| \le 7$
(c) $|x + 1| \ge 7$

98. (a) $|1 - 2x| = 6$ (b) $|1 - 2x| \le 6$
(c) $|1 - 2x| \ge 6$

Exercises 99–106: Solve the absolute value inequality.

99. $|x| > 3$ **100.** $|-5x| < 20$

101. $|4x - 2| \le 14$ **102.** $|1 - \frac{4}{5}x| \ge 3$

103. $|t - 4.5| \le 0.1$ **104.** $-2|13t - 5| \ge -4$

105. $|5 - 4x| > -5$ **106.** $|2t - 3| \le 0$

Exercises 107 and 108: Solve the inequality graphically.

107. $|2x| \ge 3$ **108.** $|\frac{1}{2}x - 1| \le 2$

Exercises 109 and 110: Write each compound inequality as an absolute value inequality.

109. $-0.05 \le x \le 0.05$

110. $5x - 1 < -4$ or $5x - 1 > 4$

APPLICATIONS

111. Age at First Marriage The median age at the first marriage for men from 1890 to 1960 can be modeled by $f(x) = -0.0492x + 119.1$, where x is the year. (Source: National Center of Health Statistics.)
(a) Find the median age in 1910.
(b) Graph f in $[1885, 1965, 10]$ by $[22, 26, 1]$. What happened to the median age?
(c) What is the slope of the graph of f? Interpret the slope as a rate of change.

112. **Marriages** From 2009 to 2014, the number of U.S. marriages in millions could be modeled by the formula $f(x) = 2.1$, where x is the year.
    (a) Estimate the number of marriages in 2012.
    (b) What information does f give about the number of marriages from 2009 to 2014?

113. **Fat Grams** A cup of milk contains 8 grams of fat.
    (a) Give a formula for $f(x)$ that calculates the number of fat grams in x cups of milk.
    (b) What is the slope of the graph of f?
    (c) Interpret the slope as a rate of change.

114. **Birth Rate** The U.S. birth rate per 1000 people for selected years is shown in the table.

Year	1950	1970	1990	2010
Birth Rate	24.1	18.4	16.7	13.5

    Source: U.S. Census Bureau.

    (a) Make a scatterplot of the data.
    (b) Model the data with $f(x) = mx + b$, where x is the year. Answers may vary.
    (c) Use f to estimate the birth rate in 2000.

115. **Unhealthy Air Quality** The Environmental Protection Agency (EPA) monitors air quality in U.S. cities. The function f gives the annual number of days with unhealthy air quality in Los Angeles, California, for selected years.

x	2006	2008	2010	2012	2014
$f(x)$	148	145	128	122	120

    Source: Environmental Protection Agency.

    (a) Find $f(2010)$ and interpret your result.
    (b) Identify the domain and range of f.
    (c) Discuss the trend of air pollution in Los Angeles.

116. **Temperature Scales** The table shown below gives equivalent temperatures in degrees Celsius and degrees Fahrenheit.

°C	−40	0	15	35	100
°F	−40	32	59	95	212

    (a) Plot the data. Let the x-axis correspond to the Celsius temperature and the y-axis correspond to the Fahrenheit temperature. What type of relation exists between the data?
    (b) Find $f(x) = mx + b$ so that f receives the Celsius temperature x as input and outputs the corresponding Fahrenheit temperature. Interpret the slope of the graph of f.
    (c) If the temperature is 20 °C, what is the equivalent temperature in degrees Fahrenheit?

117. **Distance between Bicyclists** The following graph shows the distance between two bicyclists traveling toward each other along a straight road after x hours.
    (a) After how long did the bicycle riders meet?
    (b) When were they 20 miles apart?
    (c) Find the times when they were less than 20 miles apart.
    (d) Estimate the sum of the speeds of the bicyclists.

    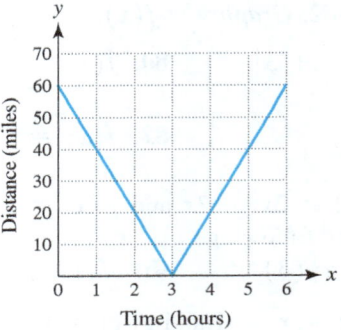

118. **Violent Crimes in the U.S.** The number of violent crimes reported has dropped from 1.3 million in 2009 to 1.1 million in 2013.
    (a) Find a linear function $f(x) = mx + b$ that models the data x years after 2009.
    (b) Use $f(x)$ to estimate the number of violent crimes in 2012.

119. **Average Precipitation** The average rainfall in Houston, Texas, is 3.9 inches per month. Each month's average A is within 1.7 inches of 3.9 inches. (*Source:* J. Williams, The Weather Almanac)
    (a) Write an absolute value inequality that models this situation.
    (b) Solve the inequality.

120. **Relative Error** If a quantity is measured to be T and the actual value is A, then the relative error in this measurement is $\left|\frac{T-A}{A}\right|$. If $A = 35$ and the relative error is to be less than 0.08 (8%), what values for T are possible?

CHAPTER 15 Test

1. Evaluate $f(4)$ if $f(x) = 3x^2 - \sqrt{x}$. Give a point on the graph of f.

2. Write a symbolic representation (formula) for a function C that calculates the cost of buying x pounds of candy at $4 per pound. Evaluate $C(5)$ and interpret your result.

3. Sketch a graph of f.
   (a) $f(x) = -2x + 1$ (b) $f(x) = x^2 + 1$
   (c) $f(x) = \sqrt{x+3}$ (d) $f(x) = |x+1|$

4. Use the graph of f to evaluate $f(-3)$ and $f(0)$. Determine the domain and range of f.

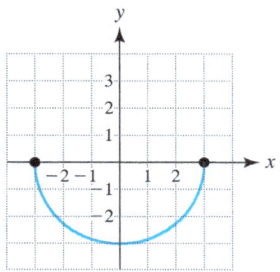

5. A function f is represented verbally by "Square the input x and then subtract 5." Give symbolic, numerical, and graphical representations of f. Let $x = -3, -2, -1, \ldots, 3$ in the numerical representation (table) and let $-3 \le x \le 3$ for the graph.

6. Determine whether the graph represents a function. Explain your reasoning.

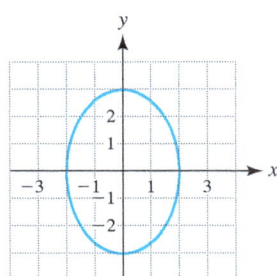

7. Find the domain of function f.
   (a) $f = \{(-2, 3), (-1, 5), (0, 3), (5, 7)\}$
   (b) $f(x) = \frac{3}{4}x - 5$
   (c) $f(x) = \sqrt{x+4}$
   (d) $f(x) = 2x^2 - 1$
   (e) $f(x) = \frac{3x}{5-x}$

8. Determine if $f(x) = 6 - 8x$ is a linear function. If it is, write it in the form $f(x) = mx + b$.

9. Graph the solution set to $2x + 6 < 2$ and $-3x \ge 3$ on a number line.

10. Use the given table to solve the compound inequality $-3x < -3$ or $-3x > 6$. Use interval notation.

x	-3	-2	-1	0	1	2	3
$-3x$	9	6	3	0	-3	-6	-9

11. Use the following figure to solve each equation or inequality. Write your answers for parts (c) and (d) in interval notation.
    (a) $y_1 = y_2$ (b) $y_2 = y_3$
    (c) $y_1 \le y_2 \le y_3$ (d) $y_2 < y_3$

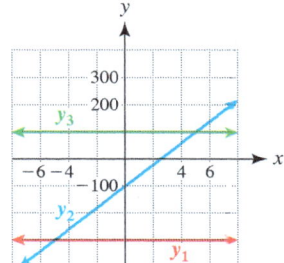

12. Solve $-2 < 2 + \frac{1}{2}x < 2$ and write the solution set in interval notation.

13. Solve the equation $|2 - \frac{1}{3}x| = 6$.

14. Solve each inequality. Write your answer in interval notation.
    (a) $|x| \le 5$ (b) $|x| > 0$

15. Determine whether $f(x) = 1 - 2x + x^3$ represents a polynomial function. If possible, identify the degree and type of polynomial function.

16. Evaluate $h(t) = -\frac{4t}{5-t}$ at $t = -2$. Write the domain of h in interval notation.

17. Let $f(x) = x^2 + 1$ and $g(x) = 2x$. Find each of the following.
    (a) $(f - g)(-2)$ (b) $(fg)(x)$

18. **Drinking Fluids and Exercise** To determine the number of ounces of fluid that a person should drink in a day, divide his or her weight in pounds by 2 and then add 0.4 ounce for every minute of exercise.
    (a) Write a function that gives the fluid requirements for a person weighing 150 pounds and exercising x minutes a day.
    (b) If a 150-pound runner needs 89 ounces of fluid each day, determine the runner's daily minutes of exercise.

19. Heart Rate of an Athlete The table below lists the heart rate or pulse of an athlete running a 400-meter race. The race lasts 50 seconds.

Time (seconds)	0	20	30	50
Heart Rate (bpm)	100	134	150	180

(a) Does $P(t) = -\frac{1}{300}t^2 + \frac{53}{30}t + 100$ model the data in the table exactly? Explain.

(b) Does $P(60)$ have significance in this situation? What should be the domain of P?

20. Time Spent in Line Suppose that parking lot attendants can wait on 25 vehicles per minute and vehicles are arriving randomly at an average rate of x vehicles per minute. Then the average time T in minutes spent waiting in line *and* paying the attendant is given by

$$T(x) = \frac{1}{25 - x},$$

where $x < 25$. (*Source:* N. Garber.)

(a) Graph T in $[0, 25, 5]$ by $[0, 2, 0.5]$. Identify any vertical asymptotes.

(b) If the wait is 1 minute, how many vehicles are arriving on average?

CHAPTERS 1–15 Cumulative Review Exercises

1. Write 120 as a product of prime numbers.

2. Translate the sentence "Triple a number decreased by 4 equals the number" into an equation using the variable n. Solve the equation.

Exercises 3 and 4: Simplify completely.

3. $\frac{1}{4} \div \frac{3}{4} - \frac{1}{2}$

4. $12 - 3^2 \div 3 \cdot 2$

5. Solve $-3(3 - x) - 6 = 2x$.

6. Convert 0.075 to a percentage.

7. Solve $A = \frac{1}{3}(2a - b)$ for a.

8. Solve $5 - 3t < 1 - t$. Write the solution set in set-builder notation.

Exercises 9 and 10: Graph the equation.

9. $3x - 2y = -6$

10. $y = |2x - 4|$

11. Write the slope–intercept form of a line that passes through the points $(-4, 4)$ and $(2, 1)$.

12. Solve the system of equations. Write your answer as an ordered pair.

$$2x + 3y = 5$$
$$3x - 2y = 1$$

Exercises 13 and 14: Shade the solution set for the given inequality.

13. $x \geq -1$

14. $x - y \geq 2$

15. Shade the solution set for the system of inequalities.

$$x - y \geq 1$$
$$2x + y \leq 0$$

16. Simplify $(3x^2 + 2x - 4) - (4x^2 + 5)$.

Exercises 17–20: Simplify and use positive exponents.

17. $\dfrac{x^{-4}}{x^{-3}}$

18. $(3b^{-3})(2b^4)$

19. $3(2t)^3$

20. $\left(\dfrac{2x^3}{x^2 y^{-1}}\right)^{-2}$

Exercises 21–24: Multiply the expression.

21. $2x^2(x^3 - 4x^2 - 5)$

22. $(5x + 1)(2x - 7)$

23. $(y - 3)(y + 3)$

24. $(x - 4y)^2$

25. Write 2.5×10^4 in standard form.

26. Write 0.028 in scientific notation.

Exercises 27 and 28: Divide.

27. $\dfrac{6x^3 - 4x^2 + 8x}{2x}$

28. $(3x^3 + 2x^2 + 1) \div (x - 1)$

Exercises 29–34: Factor completely.

29. $10x^2y^3 - 15x^3y^2$

30. $x^3 + 3x^2 - x - 3$

31. $2z^2 + z - 3$

32. $16x^2 - 25$

33. $a^3 - 8$

34. $z^4 + 7z^2 + 6$

Exercises 35–38: Solve the equation.

35. $x(x + 5) = 0$
36. $4x^2 = 0$
37. $2x^2 + 5x = 3$
38. $x^3 = x$

39. Simplify $\frac{x^2 + 4x + 4}{x + 2}$ to lowest terms.

40. If possible, evaluate $\frac{x^2 + 1}{x - 1}$ for $x = -2$ and $x = 1$.

Exercises 41 and 42: Simplify to lowest terms.

41. $\frac{x^2 - 3x + 2}{x + 2} \div \frac{x - 1}{2x + 4}$

42. $\frac{1}{x + 3} + \frac{2}{x + 1}$

43. Solve $\frac{x + 2}{x - 1} = \frac{2}{3}$.

44. Suppose that y is directly proportional to x and that $y = 5$ when $x = 10$. Find y when $x = 20$.

45. Evaluate $f(x) = x^2 - 4x$ for $x = -2$.

46. Graph $f(x) = x^2 - 2$ and identify the domain and range of f. Write your answer in interval notation.

47. Solve $x - 2 < 3$ or $x - 2 > 6$. Write your answer in interval notation.

48. Solve $|2x - 4| = 6$.

49. Solve $|2x - 4| \leq 6$. Write your answer in interval notation.

50. Solve $|x - 4| > 2$. Write your answer in interval notation.

APPLICATIONS

51. **Modeling Motion** The table lists the distance d in miles traveled by a car for various elapsed times t in hours. Find an equation that models these data.

t (hours)	2	3	4	6
d (miles)	144	216	288	432

52. **Rainfall** At noon, 2 inches of rain had fallen. For the next 6 hours, rain fell at $\frac{1}{4}$ inch per hour.
    (a) Find an equation in the form $y = mx + b$ that calculates the number of inches of rain that fell x hours past noon.
    (b) What is the slope of the graph of f?
    (c) Interpret the slope as a rate of change.
    (d) How much rain had fallen by 4 P.M.?

53. **Interest** A total of $5000 is deposited in two accounts paying 5% and 7% annual interest. If total interest received at the end of the year is $308, determine how much is invested at each interest rate.

54. **Height of a Building** An 8-foot-tall stop sign casts a 5-foot-long shadow, while a nearby building casts a 65-foot-long shadow. Find the height of the building.

55. **Shoveling the Driveway** Two people are shoveling snow from a driveway. The first person shovels 9 square feet per minute, while the second person shovels 11 square feet per minute.
    (a) Write a simplified expression that gives the total number of square feet the two people can shovel in x minutes.
    (b) How many minutes would it take for them to clear a driveway with 1000 square feet?

56. **Cost of a Television** A 5% sales tax on a television amounts to $82.50. Find the cost of the television.

16 Systems of Linear Equations

- 16.1 Systems of Linear Equations in Three Variables
- 16.2 Matrix Solutions of Linear Systems
- 16.3 Determinants

In 1940, a physicist named John Atanasoff at Iowa State University needed to solve 29 equations with 29 variables simultaneously. This task was too difficult to do by hand, so he and a graduate student invented the first fully electronic digital computer. Thus the desire to solve a mathematical problem led to one of the most important inventions of the twentieth century. Today, people can solve thousands of equations with thousands of variables. Solutions to such equations have resulted in better airplanes, cars, electronic devices, weather forecasts, and medical equipment.

Equations are even used in biology. The following table contains the weight W, neck size N, and chest size C for three black bears. Suppose that park rangers find a bear with a neck size of 22 inches and a chest size of 38 inches. Can they use the data in the table to estimate the bear's weight? Using systems of linear equations, they *can* estimate the bear's weight. See Example 8 and Exercise 61 in the second section of this chapter.

Sources: A. Tucker, *Fundamentals of Computing*; M. Triola, *Elementary Statistics*; Minitab, Inc.

Black Bear Measurements

W (pounds)	N (inches)	C (inches)
80	16	26
344	28	45
416	31	54
?	22	38

16.1 Systems of Linear Equations in Three Variables

Objectives

1. Writing Systems of Linear Equations and Recognizing Solutions
   - Checking for Solutions
   - Setting up a System
2. Solving Linear Systems with Substitution and Elimination
3. Modeling with Linear Systems
4. Recognizing Linear Systems That Have No Solutions
5. Recognizing Linear Systems That Have Infinitely Many Solutions

1 Writing Systems of Linear Equations and Recognizing Solutions

When we solve a linear system in two variables, we can express a solution as an ordered pair (x, y). A linear equation in two variables can be represented graphically by a line. A system of two linear equations with a unique solution can be represented graphically by two lines intersecting at a point, as shown in **FIGURE 16.1**.

When solving **linear systems in three variables,** we often use the variables $x, y,$ and z. A solution is expressed as an **ordered triple** (x, y, z), rather than an ordered pair (x, y). For example, if the ordered triple $(1, 2, 3)$ is a solution, $x = 1, y = 2,$ and $z = 3$ satisfy each equation. A linear equation in three variables can be represented by a flat plane in space. If the solution is unique, we can represent a linear system of three equations in three variables graphically by three planes intersecting at a single point. (See **B** below.)

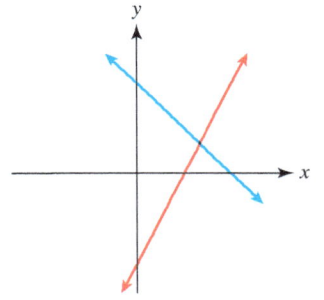

FIGURE 16.1

NEW VOCABULARY

☐ Linear system in three variables
☐ Ordered triple

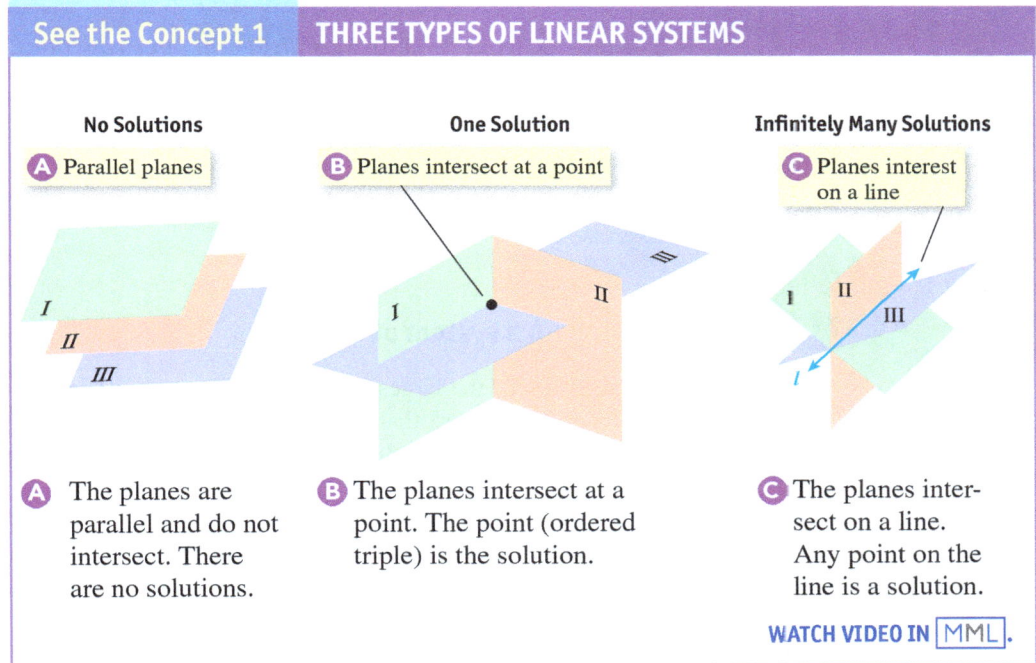

See the Concept 1 — THREE TYPES OF LINEAR SYSTEMS

No Solutions
Ⓐ Parallel planes

One Solution
Ⓑ Planes intersect at a point

Infinitely Many Solutions
Ⓒ Planes interest on a line

Ⓐ The planes are parallel and do not intersect. There are no solutions.

Ⓑ The planes intersect at a point. The point (ordered triple) is the solution.

Ⓒ The planes intersect on a line. Any point on the line is a solution.

WATCH VIDEO IN MML.

NOTE: The three planes in Ⓑ above all intersect at right angles. In general, three planes can intersect at a point even if they are not at right angles to each other. ■

READING CHECK 1

- What is the possible number of solutions to a system of linear equations in three variables?

CHECKING FOR SOLUTIONS The next example shows how to check whether an ordered triple is a solution to a system of linear equations in three variables.

EXAMPLE 1 **Checking for solutions to a system of three equations**

Determine whether $(4, 2, -1)$ or $(-1, 0, 3)$ is a solution to the system.

$$2x - 3y + z = 1$$
$$x - 2y + 2z = 5$$
$$2y + z = 3$$

Solution
To check $(4, 2, -1)$, substitute $x = 4$, $y = 2$, and $z = -1$ in each equation.

$$2(4) - 3(2) + (-1) \stackrel{?}{=} 1 \checkmark \text{ True}$$
$$4 - 2(2) + 2(-1) \stackrel{?}{=} 5 \text{ ✗ False}$$
$$2(2) + (-1) \stackrel{?}{=} 3 \checkmark \text{ True}$$

A solution must satisfy all equations.

The ordered triple $(4, 2, -1)$ *does not satisfy all three equations, so it is not a solution.* Next substitute $x = -1$, $y = 0$, and $z = 3$ in each equation.

$$2(-1) - 3(0) + 3 \stackrel{?}{=} 1 \checkmark \text{ True}$$
$$-1 - 2(0) + 2(3) \stackrel{?}{=} 5 \checkmark \text{ True}$$
$$2(0) + 3 \stackrel{?}{=} 3 \checkmark \text{ True}$$

The ordered triple $(-1, 0, 3)$ *satisfies all three equations, so it is a solution.*

Now Try Exercise 9

SETTING UP A SYSTEM In the next example, we set up a system of three equations in three variables that involves the angles of a triangle. You are asked to solve this system in Exercise 51.

EXAMPLE 2 **Setting up a system of equations**

The measure of the largest angle in a triangle is 40° greater than the sum of the two smaller angles and 90° more than the smallest angle. Set up a system of three linear equations in three variables whose solution gives the measure of each angle.

Solution
Let x, y, and z be the measures of the three angles from largest to smallest. Because the sum of the measures of the angles in a triangle equals 180°, we have

$$x + y + z = 180. \quad (1)$$

The measure of the largest angle x is 40° greater than the sum of the measures of the two smaller angles $y + z$, so

$$x - (y + z) = 40 \quad \text{or} \quad x - y - z = 40. \quad (2)$$

The measure of the largest angle x is 90° more than the measure of the smallest angle z, so

$$x - z = 90. \quad (3)$$

Thus the required system of equations can be written as follows.

$$x + y + z = 180 \quad (1)$$
$$x - y - z = 40 \quad (2)$$
$$x - z = 90 \quad (3)$$

Now Try Exercise 49(a)

Math in Biology Context In the next example, we show how a linear system involving three equations and three variables can be used to model a real-world situation. We solve this system of equations in Example 7.

EXAMPLE 3 Modeling real data with a linear system

The Bureau of Land Management studies antelope populations in Wyoming. It monitors the number of adult antelope, the number of fawns each spring, and the severity of the winter. The first two columns of **TABLE 16.1** contain counts of fawns and adults for three representative winters. The third column shows the severity of each winter. The severity of the winter is measured from 1 to 5, with 1 being mild and 5 being severe.

Antelope Numbers and Winter Severity

Fawns (F)	Adults (A)	Winter (W)
405	870	3
414	848	2
272	684	5
?	750	4

TABLE 16.1

We want to use the data in the first three rows of the table to estimate the number of fawns F in the fourth row when the number of adults is 750 and the severity of the winter is 4. To do so, we use the formula

$$F = a + bA + cW,$$

where a, b, and c are constants. Write a system of linear equations whose solution gives appropriate values for a, b, and c.

Solution
From the first row in the table, when $F = 405$, $A = 870$, and $W = 3$, the formula

$$F = a + bA + cW$$

becomes

$$405 = a + b(870) + c(3). \quad (1)$$

Similarly, $F = 414$, $A = 848$, and $W = 2$ gives

$$414 = a + b(848) + c(2), \quad (2)$$

and $F = 272$, $A = 684$, and $W = 5$ yields

$$272 = a + b(684) + c(5). \quad (3)$$

To find values for a, b, and c, we can solve the following system of linear equations.

System of equations
$$\begin{cases} 405 = a + 870b + 3c & (1) \\ 414 = a + 848b + 2c & (2) \\ 272 = a + 684b + 5c & (3) \end{cases}$$

We can also write these equations as a linear system in the following form.

$$a + 870b + 3c = 405 \quad (1)$$
$$a + 848b + 2c = 414 \quad (2)$$
$$a + 684b + 5c = 272 \quad (3)$$

Finding values for a, b, and c will allow us to use the formula $F = a + bA + cW$ to predict the number of fawns F when the number of adults A is 750 and the severity of the winter W is 4 (see Example 7).

Now Try Exercise 53(a)

NOTE: Linear systems of two equations can have no solutions, one solution, or infinitely many solutions. The same is true for larger linear systems. In the following subsection, we focus on linear systems that have one solution. ∎

2 Solving Linear Systems with Substitution and Elimination

When solving systems of linear equations with more than two variables, we usually use both substitution and elimination. However, in the next example, we use only substitution to solve a particular type of linear system in three variables.

EXAMPLE 4 Using substitution to solve a linear system of equations

Solve the following system.

$$2x - y + z = 7$$
$$3y - z = 1$$
$$z = 2$$

Solution
From the third equation, we know z is 2. Let $z = 2$ in the second equation and determine y.

	$3y - z = 1$	Second equation
Let $z = 2$ and then find y.	$3y - 2 = 1$	Substitute $z = 2$.
	$3y = 3$	Add 2 to each side.
	$y = 1$	Divide each side by 3.

Knowing that $y = 1$ and $z = 2$ allows us to find x by using the first equation.

	$2x - y + z = 7$	First equation
Let $y = 1$ and $z = 2$ and then find x.	$2x - 1 + 2 = 7$	Let $y = 1$ and $z = 2$.
	$2x = 6$	Simplify and subtract 1.
	$x = 3$	Divide each side by 2.

Thus $x = 3$, $y = 1$, and $z = 2$, and the solution is $(3, 1, 2)$.

Now Try Exercise 13

In the next example, we use elimination and substitution in a four-step method to solve a system of linear equations. This method is summarized by the following.

STUDY TIP

Be sure to follow this four-step method. It will help you get the correct answer consistently.

SOLVING A LINEAR SYSTEM IN THREE VARIABLES

STEP 1: Eliminate one variable, such as x, from two of the equations.

STEP 2: Use the two resulting equations in two variables to eliminate one of the variables, such as y. Solve for the remaining variable, z.

STEP 3: Substitute z in one of the two equations from Step 2. Solve for the unknown variable y.

STEP 4: Substitute values for y and z in one of the given equations and find x. The solution is (x, y, z).

EXAMPLE 5 Solving a linear system in three variables

Solve the following system.

$$x - y + 2z = 6$$
$$2x + y - 2z = -3$$
$$-x - 2y + 3z = 7$$

Solution

STEP 1: We begin by eliminating the variable x from the second and third equations. To eliminate x from the second equation, we multiply the first equation by -2 and then add it to the second equation. To eliminate x from the third equation, we add the first and third equations.

$-2x + 2y - 4z = -12$	First equation times -2		$x - y + 2z = 6$	First equation
$2x + y - 2z = -3$	Second equation		$-x - 2y + 3z = 7$	Third equation
$3y - 6z = -15$	Add.		$-3y + 5z = 13$	Add.

STEP 2: Take the two resulting equations from Step 1 and eliminate either variable. Here, we add the two equations to eliminate the variable y.

$$3y - 6z = -15$$
$$-3y + 5z = 13$$

Add to eliminate y.

$$-z = -2 \quad \text{Add the equations.}$$
$$z = 2 \quad \text{Multiply by } -1.$$

STEP 3: Now we can use substitution to find the values of x and y. We let $z = 2$ in either equation used in Step 2 to find y.

Substitute 2 for z.

$$3y - 6z = -15 \quad \text{Equation from Step 2}$$
$$3y - 6(2) = -15 \quad \text{Substitute } z = 2.$$
$$3y - 12 = -15 \quad \text{Multiply.}$$
$$3y = -3 \quad \text{Add 12.}$$
$$y = -1 \quad \text{Divide by 3.}$$

STEP 4: Finally, we substitute $y = -1$ and $z = 2$ in one of the given equations to find x.

Substitute -1 for y and 2 for z.

$$x - y + 2z = 6 \quad \text{First given equation}$$
$$x - (-1) + 2(2) = 6 \quad \text{Let } y = -1 \text{ and } z = 2.$$
$$x + 1 + 4 = 6 \quad \text{Simplify.}$$
$$x = 1 \quad \text{Subtract 5.}$$

The solution is $(1, -1, 2)$. Check this solution.

Now Try Exercise 25

🌐 **Math in Sales Context** In the next example, we solve a system of linear equations to determine the number of tickets sold at a play.

EXAMPLE 6 Finding the number of tickets sold

One thousand tickets were sold for a play, which generated $3800 in revenue. The prices of the tickets were $3 for children, $4 for students, and $5 for adults. There were 100 fewer student tickets sold than adult tickets. Find the number of each type of ticket sold.

Solution
Let x be the number of tickets sold to children, y be the number of tickets sold to students, and z be the number of tickets sold to adults. The total number of tickets sold was 1000, so

$$x + y + z = 1000. \quad (1)$$

Each child's ticket cost $3, so the revenue generated from selling x tickets was $3x$. Similarly, the revenue generated from student tickets was $4y$, and the revenue from adult tickets was $5z$. Total ticket revenue was $3800, so

$$3x + 4y + 5z = 3800. \quad (2)$$

The equation $z - y = 100$, or

$$y - z = -100, \quad (3)$$

must also be satisfied, because 100 fewer tickets were sold to students than to adults.

To find the number of each type of ticket sold, we need to solve the following system of linear equations.

$$\left. \begin{array}{r} x + y + z = 1000 \quad (1) \\ 3x + 4y + 5z = 3800 \quad (2) \\ y - z = -100 \quad (3) \end{array} \right\} \text{System to be solved}$$

STEP 1: We begin by eliminating the variable x from the second equation. To do so, we multiply the first equation by -3 and add the second equation.

Add equations to eliminate x.

$$\begin{array}{rl} -3x - 3y - 3z = -3000 & \text{Equation (1) times } -3 \\ \underline{3x + 4y + 5z = 3800} & \text{Equation (2)} \\ y + 2z = 800 & \text{Add Equations to get equation (4).} \end{array}$$

STEP 2: We then multiply the resulting equation from Step 1 by -1 and add the third given equation to eliminate y.

Add equations to eliminate y.

$$\begin{array}{rl} -y - 2z = -800 & \text{Equation (4) times } -1 \\ \underline{y - z = -100} & \text{Equation (3)} \\ -3z = -900 & \text{Add Equations.} \\ z = 300 & \text{Divide by } -3 \text{ to find the value of } z. \end{array}$$

STEP 3: To find y, we can substitute $z = 300$ in one of the equations from Step 2.

$$\begin{array}{rl} y - z = -100 & \text{Equation (3)} \\ y - 300 = -100 & \text{From Step 2, let } z = 300. \\ y = 200 & \text{Add 300 to find the value of } y. \end{array}$$

STEP 4: Finally, we substitute $y = 200$ and $z = 300$ in the first given equation.

$$\begin{array}{rl} x + y + z = 1000 & \text{Equation (1)} \\ x + 200 + 300 = 1000 & \text{From Steps 2 and 3, let } y = 200 \text{ and } z = 300. \\ x = 500 & \text{Subtract 500 to find the value of } x. \end{array}$$

Thus **500** tickets were sold to children, **200** to students, and **300** to adults.

Now Try Exercise 47

3 Modeling with Linear Systems

In the next example, we solve the system of equations that we discussed in Example 3.

EXAMPLE 7 Predicting fawns in the spring

Solve the following linear system for a, b, and c.

$$a + 870b + 3c = 405$$
$$a + 848b + 2c = 414$$
$$a + 684b + 5c = 272$$

Then use $F = a + bA + cW$ to predict the number of fawns when there are 750 adults and the severity of the winter is 4.

Solution

STEP 1: We begin by eliminating the variable a from the second and third equations. To do so, we add the second and third equations to the first equation times -1.

$$\begin{aligned}-a - 870b - 3c &= -405 \\ a + 848b + 2c &= 414 \\ \hline -22b - c &= 9\end{aligned}$$ $$\begin{aligned}-a - 870b - 3c &= -405 \quad \text{First times } -1\\ a + 684b + 5c &= 272 \quad \text{Second/third equation}\\ \hline -186b + 2c &= -133 \quad \text{Add.}\end{aligned}$$

STEP 2: We use the two resulting equations from Step 1 to eliminate c. To do so, we multiply the equation $-22b - c = 9$ by 2 and add it to the other equation.

Add equations to eliminate c.

$$\begin{aligned}-44b - 2c &= 18 \quad (-22b - c = 9) \text{ times 2}\\ -186b + 2c &= -133 \\ \hline -230b &= -115 \quad \text{Add the equations.}\\ b &= 0.5 \quad \text{Divide by } -230.\end{aligned}$$

STEP 3: To find c, we substitute $b = 0.5$ in either equation used in Step 2.

Let $b = 0.5$ and find c.

$$\begin{aligned}-44b - 2c &= 18 \quad \text{First equation in Step 2}\\ -44(0.5) - 2c &= 18 \quad \text{Let } b = 0.5.\\ -22 - 2c &= 18 \quad \text{Multiply.}\\ -2c &= 40 \quad \text{Add 22.}\\ c &= -20 \quad \text{Divide by } -2.\end{aligned}$$

STEP 4: Finally, we substitute $b = 0.5$ and $c = -20$ in one of the given equations to find a.

Let $b = 0.5$ and $c = -20$ and find a.

$$\begin{aligned}a + 870b + 3c &= 405 \quad \text{First given equation}\\ a + 870(0.5) + 3(-20) &= 405 \quad \text{Let } b = 0.5 \text{ and } c = -20.\\ a + 435 - 60 &= 405 \quad \text{Multiply.}\\ a &= 30 \quad \text{Solve for } a.\end{aligned}$$

The solution is $a = 30$, $b = 0.5$, and $c = -20$. Thus we may write

$$F = a + bA + cW$$
$$= 30 + 0.5A - 20W. \quad \text{Modeling equation}$$

If there are 750 adults and the winter has a severity of 4, this model predicts

$$F = 30 + 0.5(750) - 20(4)$$
$$= 325 \text{ fawns.}$$

Now Try Exercise 53

CRITICAL THINKING 1

Give reasons why the coefficient for A is positive and the coefficient for W is negative in the formula $F = 30 + 0.5A - 20W$.

4 Recognizing Linear Systems That Have No Solutions

It is possible for a system of three linear equations in three variables to be inconsistent and have no solutions. If we apply substitution and elimination to this type of system, we arrive at a contradiction. This case is demonstrated in the next example.

EXAMPLE 8 **Recognizing an inconsistent system**

Solve the system, if possible.

$$x + y + z = 4$$
$$-x + y + z = 2$$
$$y + z = 1$$

Solution
STEP 1: If we add the first two equations, we can eliminate x. The variable x is already eliminated from the third equation.

Add equations to eliminate x.

$$\begin{array}{ll} x + y + z = 4 & \text{First equation} \\ -x + y + z = 2 & \text{Second equation} \\ \hline 2y + 2z = 6 & \text{Add.} \end{array}$$

STEP 2: If we multiply the third *given* equation by -2 and add it to the resulting equation in Step 1, we arrive at a contradiction.

The equation $0 = 4$ is never true.

$$\begin{array}{ll} -2y - 2z = -2 & \text{Third equation times } -2 \\ 2y + 2z = 6 & \text{Equation from Step 1} \\ \hline 0 = 4 \;\text{✗} & \text{Add. (Contradiction)} \end{array}$$

Because $0 = 4$ is a contradiction, there are no solutions to the given system of equations.

Now Try Exercise 31

5 Recognizing Linear Systems That Have Infinitely Many Solutions

It is possible for a system of linear equations in three variables to have infinitely many solutions. If we apply substitution and elimination to this type of system, we arrive at an identity. This case is demonstrated in the next example.

EXAMPLE 9 **Solving a system with infinitely many solutions**

Solve the system.

$$x + y + z = 2$$
$$x - y + z = 4$$
$$3x - y + 3z = 10$$

Solution
STEP 1: To eliminate y from the second equation, add the first equation to the second. To eliminate y from the third equation, add the first equation to the third equation.

$$\begin{array}{ll} x + y + z = 2 & \text{First equation} \\ x - y + z = 4 & \text{Second equation} \\ \hline 2x + 2z = 6 & \text{Add.} \end{array} \qquad \begin{array}{ll} x + y + z = 2 & \text{First equation} \\ 3x - y + 3z = 10 & \text{Third equation} \\ \hline 4x + 4z = 12 & \text{Add.} \end{array}$$

STEP 2: If we multiply the first resulting equation in Step 1 by -2 and add it to the second resulting equation in Step 1, we arrive at an identity.

$$-4x - 4z = -12 \quad (2x + 2z = 6) \text{ times } -2$$
$$\underline{4x + 4z = 12} \quad \text{Second equation from Step 1}$$
$$0 = 0 \quad \text{Add. (Identity)}$$

The equation $0 = 0$ is always true.

The variable x can be written in terms of z by solving $2x + 2z = 6$ for x.

$$2x + 2z = 6 \quad \text{Equation from Step 1}$$
$$2x = 6 - 2z \quad \text{Subtract } 2z.$$
$$x = 3 - z \quad \text{Divide by 2.}$$

STEP 3: To find y in terms of z, substitute $3 - z$ for x in the first *given* equation.

$$x + y + z = 2 \quad \text{First equation}$$
$$(3 - z) + y + z = 2 \quad \text{Let } x = 3 - z.$$
$$y = -1 \quad \text{Solve for } y.$$

All solutions have the form $(3 - z, -1, z)$, where z can be any real number. For example, if $z = 1$, then $(2, -1, 1)$ is one of infinitely many solutions to the system of equations. Note that in this particular system, y must always equal -1.

Now Try Exercise 33

16.1 Putting It All Together

In this section, we discussed how to solve a system of three linear equations in three variables. Systems of linear equations can have no solutions, one solution, or infinitely many solutions.

CONCEPT	EXPLANATION/EXAMPLE
System of Linear Equations in Three Variables	The following is a system of three linear equations in three variables. Each equation is linear. $$x - 2y + z = 0$$ $$-x + y + z = 4$$ $$-y + 4z = 10$$
Solution to a Linear System in Three Variables	The solution to a linear system in three variables is an ordered triple, expressed as (x, y, z). The solution to the preceding system is $(1, 2, 3)$ because substituting $x = 1$, $y = 2$, and $z = 3$ in each equation results in a true statement. We can check solutions this way. $$(1) - 2(2) + (3) = 0 \checkmark \text{ True}$$ $$-(1) + (2) + (3) = 4 \checkmark \text{ True}$$ $$-(2) + 4(3) = 10 \checkmark \text{ True}$$

continued on next page

CONCEPT	EXPLANATION
Solving a Linear System with Substitution and Elimination	**STEP 1:** Eliminate one variable, such as x, from two of the equations.
	STEP 2: Use the two resulting equations in two variables to eliminate one of the variables, such as y. Solve for the remaining variable z.
	STEP 3: Substitute z in one of the two equations from Step 2. Solve for the unknown variable y.
	STEP 4: Substitute values for y and z in one of the given equations and find x. The solution is (x, y, z).

16.1 Exercises

CONCEPTS AND VOCABULARY

1. Can a system of three linear equations in three variables have exactly two solutions? Explain.

2. Give an example of a system of three linear equations in three variables.

3. Does the ordered triple $(1, 2, 3)$ satisfy the equation $x + y + z = 6$?

4. Does $(3, 4)$ represent a solution to the equation $x + y + z = 7$? Explain.

5. To solve uniquely for two variables, how many equations do you usually need?

6. To solve uniquely for three variables, how many equations do you usually need?

7. If a contradiction occurs while solving a system of linear equations, then there is/are _____ solution(s).

8. If an identity occurs while solving a linear system, how many solutions are there?

RECOGNIZING SOLUTIONS TO LINEAR SYSTEMS

Exercises 9–12: Determine which ordered triple is a solution to the linear system.

9. $(1, 2, 3)$, $(0, 2, 4)$
$$x + y + z = 6$$
$$x - y - z = -4$$
$$-x - y + z = 0$$

10. $(-1, 0, 2)$, $(0, 4, 4)$
$$2x + y - 3z = -8$$
$$x - 3y + 2z = -4$$
$$3x - 2y + z = -4$$

11. $(1, 0, 3)$, $(-1, 1, 2)$
$$3x - 2y + z = -3$$
$$-x + 3y - 2z = 0$$
$$x + 4y + 2z = 7$$

12. $(\tfrac{1}{2}, \tfrac{3}{2}, -\tfrac{1}{2})$, $(-1, 0, -2)$
$$x + 3y - 4z = 7$$
$$-x + 5y + 3z = \tfrac{11}{2}$$
$$3x - 2y - 7z = 2$$

SOLVING LINEAR SYSTEMS WITH SUBSTITUTION

Exercises 13–18: (Refer to Example 4.) Use substitution to solve the system of linear equations. Check your solution.

13. $x + y - z = 1$
$2y + z = -1$
$z = 1$

14. $2x + y - 3z = 1$
$y + 4z = 0$
$z = -1$

15. $-x - 3y + z = -2$
$2y + 3z = 3$
$z = 2$

16. $3x + 2y - 3z = -4$
$-y + 2z = 4$
$z = 0$

17. $a - b + 2c = 3$
$-3b + c = 4$
$c = -2$

18. $5a + 2b - 3c = 10$
$5b - 2c = -4$
$c = 3$

SOLVING LINEAR SYSTEMS

Exercises 19–44: Solve the system, if possible.

19. $x + y - z = 11$
$-x + 2y + 3z = -1$
$2z = 4$

20. $x + 2y - 3z = -7$
$-2x + y + z = -1$
$3z = 9$

21. $x + y - z = -2$
$-x + z = 1$
$y + 2z = 3$

22. $x + y - 3z = 11$
$-2x + y + 2z = 1$
$-3y + 3z = -21$

23. $x + y - 2z = -7$
$y + z = -1$
$-y + 3z = 9$

24. $2x + 3y + z = 5$
$y + 2z = 4$
$-2y + z = 2$

25. $x + 2y + 2z = 1$
$x + y + z = 0$
$-x - 2y + 3z = -11$

26. $x + y - z = 0$
    $x - 3y + z = -2$
    $x - y + 3z = 8$

27. $x + y + z = 5$
    $y + z = 6$
    $x + z = 3$

28. $x + y + z = 0$
    $x - y - z = 6$
    $-x + y - z = 4$

29. $x + 2y + 3z = 24$
    $-x + y + 2z = 1$
    $x + y - 2z = 9$

30. $5x - 15y + z = 22$
    $-10x + 12y - 2z = -8$
    $4x - 2y - 3z = 9$

31. $x + y + z = 2$
    $x - y + z = 1$
    $x + z = 3$

32. $4x - y + 3z = 3$
    $2x + y + z = 2$
    $x - y + z = 1$

33. $x + y + z = 6$
    $x - y + z = 2$
    $-x + 5y - z = 6$

34. $x - y + z = 3$
    $2x - y + z = 2$
    $-x - y + z = 5$

35. $2x + y + z = 3$
    $2x - y - z = 9$
    $x + y - z = 0$

36. $x + 3y + z = -8$
    $x - 2y = 11$
    $2y - z = -16$

37. $2x + 6y - 2z = 47$
    $2x + y + 3z = -28$
    $-x + y + z = -\frac{7}{2}$

38. $x + y + 2z = 23$
    $3x - y + 3z = 8$
    $2x + 2y + z = 13$

39. $x + 3y - 4z = \frac{13}{2}$
    $-2x + 3y - z = \frac{1}{2}$
    $3x + z = 4$

40. $x - 2y + z = \frac{9}{2}$
    $4x - y + 3z = 9$
    $x + 2y = -\frac{3}{2}$

41. $2x - 2y + z = 4$
    $x - y + z = 1$
    $x - y + z = 3$

42. $x - 2y + 3z = 1$
    $x + 3y + z = 2$
    $3x + 4y + 5z = 7$

43. $x + y + z = 5$
    $x - y + z = 3$
    $2x + y + 2z = 9$

44. $x + y - 2z = 0$
    $x + 2y - 4z = -1$
    $y - 2z = -1$

Exercises 45 and 46: **Thinking Generally** *Solve the system of linear equations, if $a \neq 0$.*

45. $x + y + z = a$
    $x + y + z = 2a$
    $-x + y + z = 0$

46. $x + y + z = a$
    $-x - y - z = -a$
    $y - z = 0$

APPLICATIONS OF LINEAR SYSTEMS

47. **Finding Costs** The accompanying table shows the costs of purchasing different combinations of hamburgers, fries, and soft drinks.

Hamburgers	Fries	Soft Drinks	Total Cost
1	2	4	$10
1	4	6	$15
0	3	2	$6

(a) Let x be the cost of a hamburger, y the cost of fries, and z the cost of a soft drink. Write a system of three linear equations that represents the data in the table.

(b) Solve the system and interpret your answer.

48. **Cost of CDs** The accompanying table shows the total cost of purchasing combinations of differently priced CDs. The types of CDs are labeled A, B, and C.

A	B	C	Total Cost
1	1	1	$37
3	2	1	$69
1	1	4	$82

(a) Let x be the cost of a CD of type A, y the cost of a CD of type B, and z the cost of a CD of type C. Write a system of three linear equations that represents the data in the table.

(b) Solve the system and interpret your answer.

49. **Geometry** The largest angle in a triangle is 55° more than the smallest angle. The sum of the measures of the two smaller angles is 10° more than the measure of the largest angle.
    (a) Let x, y, and z be the measures of the three angles from largest to smallest. Write a system of three linear equations whose solution gives the measure of each angle.
    (b) Solve the system.
    (c) Check your solution.

50. **Geometry** The perimeter of a triangle is 90 inches. The longest side is 20 inches longer than the shortest side and 10 inches longer than the remaining side.
    (a) Let x, y, and z be the lengths of the three sides from largest to smallest. Write a system of three linear equations whose solution gives the lengths of each side.
    (b) Solve the system.
    (c) Check your solution.

51. **Geometry** (Refer to Example 2.) Solve the system to find the measure of each angle.

$$x + y + z = 180$$
$$x - y - z = 40$$
$$x - z = 90$$

52. **Loan Mixture** A student takes out a total of $5000 in three loans: one subsidized, one unsubsidized, and one from the parents of the student. The subsidized loan is $200 more than the combined total of the unsubsidized and parent loans. The unsubsidized loan is twice the amount of the parent loan. Find the amount of each loan.

53. **Predicting Fawns** (Refer to Examples 3 and 7.) The accompanying table shows counts for fawns and adult deer and the severity of the winter. These data may be modeled by the equation $F = a + bA + cW$.

Fawns (F)	Adults (A)	Winter (W)
525	600	4
365	400	2
805	900	5
?	500	3

(a) Use the first three rows to write a system of three linear equations in three variables whose solution gives values for a, b, and c.
(b) Solve the system.
(c) Predict the number of fawns when there are 500 adults and the winter has severity 3.

54. **Business Production** A business has three machines that manufacture containers. Together, they make 100 containers per day, whereas the two fastest machines together can make 80 containers per day. The fastest machine makes 34 more containers per day than the slowest machine.
(a) Let x, y, and z be the number of containers that the machines make from fastest to slowest. Write a system of three equations whose solution gives the number of containers that each machine can make.
(b) Solve the system.

55. **Mixture Problem** One type of lawn fertilizer consists of a mixture of nitrogen, N, phosphorus, P, and potassium, K. An 80-pound sample contains 8 more pounds of nitrogen and phosphorus than potassium. There is 9 times as much potassium as phosphorus.
(a) Write a system of three equations whose solution gives the amount of nitrogen, phosphorus, and potassium in this sample.
(b) Solve the system.

56. **Predicting Home Prices** Selling prices of homes can depend on several factors such as size and age. The accompanying table shows the selling price for three homes. In this table, price P is given in thousands of dollars, age A in years, and home size S in thousands of square feet. These data may be modeled by the equation $P = a + bA + cS$.

Price (P)	Age (A)	Size (S)
190	20	2
320	5	3
50	40	1
?	10	2.5

(a) Write a system of linear equations whose solution gives a, b, and c.
(b) Solve the system.
(c) Predict the price of a home that is 10 years old and has 2500 square feet.

57. **Investment Mixture** A sum of $30,000 was invested in three mutual funds. In one year, the first fund grew by 4%, the second by 5%, and the third by 7.5%. Total earnings were $1775. The amount invested in the third fund was $2000 less than the combined amount invested in the other two funds. Use a linear system of equations to determine the amount invested in each fund.

58. **Football Tickets** A total of 2500 tickets to a football game were sold. Prices were $2 for children, $3 for students, and $5 for adults. Twice as many tickets were sold to students as to children, and ticket revenues were $7250. Use a system of linear equations to determine how many of each type of ticket were sold.

WRITING ABOUT MATHEMATICS

59. Earlier in this text, we solved problems with two variables; to obtain a unique solution, we needed two linear equations. In this section, we solved problems with three variables; to obtain a unique solution, we needed three equations. Try to generalize these results. In the design of aircraft, problems commonly involve 100,000 variables. How many equations are required to solve such problems? Can such problems be solved by hand? If not, how are such problems solved? Explain your answers.

60. In Exercise 56, the price of a home was estimated by its age and size. What other factors might affect the price of a home? Explain how these factors might affect the number of variables and equations in the linear system.

16.2 Matrix Solutions of Linear Systems

Objectives

1. Representing Systems of Linear Equations with Matrices
2. Relating Matrices and Social Networks
3. Applying Gauss–Jordan Elimination
4. Using Technology to Solve Systems of Linear Equations

NEW VOCABULARY

☐ Matrix
☐ Element
☐ Dimension of a matrix
☐ Square matrix
☐ Augmented matrix
☐ Main diagonal
☐ Reduced row–echelon form
☐ Gauss–Jordan elimination
☐ Matrix row transformations

1 Representing Systems of Linear Equations with Matrices

Arrays of numbers are used frequently in many different real-world situations. Spreadsheets often make use of arrays. A **matrix** is a rectangular array of numbers. Each number in a matrix is called an **element**. The following are examples of *matrices* (plural of matrix), with their dimensions written below them.

Matrices and Their Dimensions

$$\begin{bmatrix} 2 & 0 \\ 3 & 1 \end{bmatrix} \quad \begin{bmatrix} -1.2 & 5 & 0 \\ 1 & 0 & 1 \\ 4 & -5 & 7 \end{bmatrix} \quad \begin{bmatrix} 3 & -6 & 0 & \sqrt{3} \\ 1 & 4 & 0 & 9 \\ -3 & 1 & 1 & 18 \\ -10 & -4 & 5 & -1 \end{bmatrix} \quad \begin{bmatrix} 4 & 2 \\ 0 & 1 \\ 1 & 0 \end{bmatrix} \quad \begin{bmatrix} 1 & 5 & -1 \\ 3 & 4 & 2 \end{bmatrix}$$

$2 \times 2 \qquad 3 \times 3 \qquad\qquad 4 \times 4 \qquad\qquad 3 \times 2 \qquad 2 \times 3$

Rows × Columns

Connecting Concepts with Your Life The dimension of a matrix is stated much like the dimensions of a rectangular room. We might say that a room is m feet long and n feet wide. Similarly, the **dimension of a matrix** is $m \times n$ (m by n) if it has m rows and n columns. For example, the last matrix in the preceding group has a dimension of 2×3 because it has 2 rows and 3 columns. If the number of rows equals the number of columns, the matrix is a **square matrix**. The first three matrices in that group are square matrices.

Matrices can be used to represent a system of linear equations.

See the Concept 1 — REPRESENTING A LINEAR SYSTEM WITH A MATRIX

A System of Three Equations

$$\begin{aligned} 3x - y + 2z &= 7 \\ x - 2y + z &= 0 \\ 2x + 5y - 7z &= -9 \end{aligned}$$

C Main diagonal

B Augmented Matrix

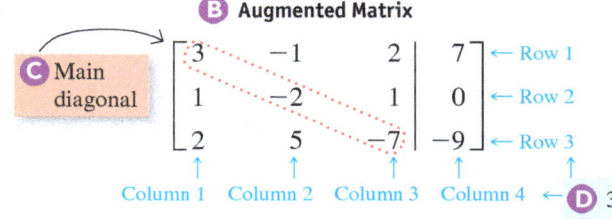

← Row 1
← Row 2
← Row 3

↑ Column 1 ↑ Column 2 ↑ Column 3 ↑ Column 4 ← **D** 3×4

A Write the system of equations as shown, with all variables to the left of the equals sign, constants to the right of the equals sign, and the matching variables aligned vertically in columns.

B The coefficients and constants are represented by the **augmented matrix**. The vertical line in the matrix corresponds to the position of the equals sign in each equation.

C The elements of the **main diagonal** of the augmented matrix are circled.

D The matrix has 3 rows and 4 columns, so its dimension is 3×4.

WATCH VIDEO IN MML.

EXAMPLE 1 Representing a linear system

Represent each linear system with an augmented matrix. State the dimension of the matrix.

(a) $\quad x - 2y = 9$
$\quad\quad 6x + 7y = 16$

(b) $\quad x - 3y + 7z = 4$
$\quad\quad 2x + 5y - z = 15$
$\quad\quad 2x + y = 8$

Solution

(a) This system can be represented by the following 2×3 augmented matrix.

2 rows
3 columns

$$\begin{bmatrix} 1 & -2 & | & 9 \\ 6 & 7 & | & 16 \end{bmatrix}$$

(b) This system can be represented by the following 3×4 augmented matrix.

3 rows
4 columns

$$\begin{bmatrix} 1 & -3 & 7 & | & 4 \\ 2 & 5 & -1 & | & 15 \\ 2 & 1 & 0 & | & 8 \end{bmatrix}$$

Now Try Exercises 11, 13

2 Relating Matrices and Social Networks

Math in Context *Social Networking* — Today, there are several different types of online social networks, such as Facebook and Twitter. Mathematics is essential to the success of these social networks. Consider **FIGURE 16.2**, which represents a simple social network of four people.

STUDY TIP

Matrices are very important because they have a wide variety of applications in the real world.

Simple Social Network

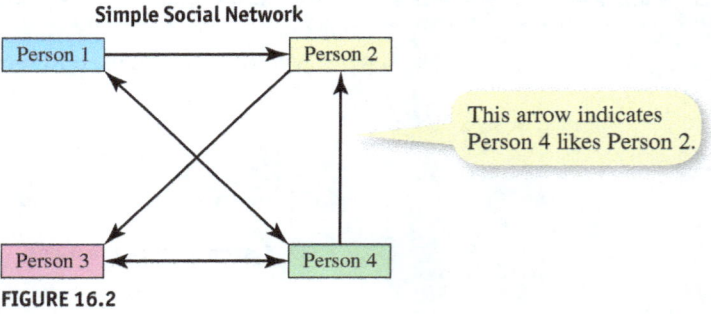

This arrow indicates Person 4 likes Person 2.

FIGURE 16.2

In **FIGURE 16.2**, the arrow from Person 1 to Person 2 indicates that Person 1 likes Person 2, but Person 2 does not like Person 1 because there is no arrow from Person 2 to Person 1. On the other hand, Person 3 and Person 4 like each other because there is a double arrow between them. In the next example, we can see how a matrix can be used to represent this simple social network.

EXAMPLE 2 Representing social networks

Use a matrix to model the social network shown in **FIGURE 16.2**.

Solution

To model a social network with four people, we use a square 4×4 matrix, where rows 1, 2, 3, and 4 represents the "likes" of each person, respectively. In this matrix, a 1 represents a connection (or "like") and a 0 represents no connection. Because Person 1 likes Person 2, we put a **1** in row 1 column 2. Because Person 1 also likes Person 4, we put a **1** in row 1 column 4. Continuing in the same manner, we arrive at the following matrix.

A Social Network

$$\begin{bmatrix} 0 & 1 & 0 & 1 \\ 0 & 0 & 1 & 0 \\ 0 & 0 & 0 & 1 \\ 1 & 1 & 1 & 0 \end{bmatrix} \begin{matrix} \leftarrow \text{Person 1's "likes"} \\ \leftarrow \text{Person 2's "likes"} \\ \leftarrow \text{Person 3's "likes"} \\ \leftarrow \text{Person 4's "likes"} \end{matrix}$$

Person 4 "likes" Person 1

Now Try Exercise 21

3 ▶ Applying Gauss–Jordan Elimination

A convenient matrix form for representing a system of linear equations is **reduced row–echelon form**. The following augmented matrices are examples of reduced row–echelon form. Note that there are 1s on the main diagonal with 0s above and below the 1s.

Reduced Row–Echelon Form

$$\begin{bmatrix} 1 & 0 & | & 3 \\ 0 & 1 & | & -2 \end{bmatrix} \quad \begin{bmatrix} 1 & 0 & 0 & | & 3 \\ 0 & 1 & 0 & | & 1 \\ 0 & 0 & 1 & | & -1 \end{bmatrix} \quad \begin{bmatrix} 1 & 0 & 0 & | & 8 \\ 0 & 1 & 0 & | & 2 \\ 0 & 0 & 1 & | & 3 \end{bmatrix}$$

If an augmented matrix representing a linear system is in reduced row–echelon form, we can usually determine the solution easily.

EXAMPLE 3 **Determining a solution from reduced row–echelon form**

Each matrix represents a system of linear equations. Find the solution.

(a) $\begin{bmatrix} 1 & 0 & 0 & | & 2 \\ 0 & 1 & 0 & | & -3 \\ 0 & 0 & 1 & | & 5 \end{bmatrix}$ (b) $\begin{bmatrix} 1 & 0 & | & 10 \\ 0 & 1 & | & -4 \end{bmatrix}$

Solution
(a) The top row $1\ 0\ 0\ 2$ represents $1x + 0y + 0z = 2$ or $x = 2$. The second and third rows tell us that $y = -3$ and $z = 5$. The solution is $(2, -3, 5)$.
(b) The system involves two equations in two variables. The solution is $(10, -4)$.

Now Try Exercise 19

In the previous section, we solved systems of three linear equations in three variables by using elimination and substitution. In real life, systems of equations often contain thousands of variables. To solve a large system of equations, we need an efficient method. Long before the invention of the computer, Carl Friedrich Gauss (1777–1855) developed a method called *Gaussian elimination* to solve systems of linear equations. Even though it was developed more than 150 years ago, it is still used today in modern computers and calculators.

We will use a numerical method called **Gauss–Jordan elimination** to solve a linear system. It makes use of the following **matrix row transformations**.

READING CHECK 1

- Why do we use matrix row transformations?

MATRIX ROW TRANSFORMATIONS

For any augmented matrix representing a system of linear equations, the following row transformations result in an equivalent system of linear equations.

1. Any two rows may be interchanged.
2. The elements of any row may be multiplied by a nonzero constant.
3. Any row may be changed by adding to (or subtracting from) its elements a nonzero multiple of the corresponding elements of another row.

Gauss–Jordan elimination can be used to transform an augmented matrix into reduced row–echelon form. Its objective is to use the matrix row transformations to obtain a matrix that has the following reduced row–echelon form, where (a, b) represents the solution.

$$\begin{bmatrix} 1 & 0 & | & a \\ 0 & 1 & | & b \end{bmatrix}$$

This method is illustrated in the next example.

EXAMPLE 4 Transforming a matrix into reduced row–echelon form

Use Gauss–Jordan elimination to transform the augmented matrix of the linear system into reduced row–echelon form. Find the solution.

$$x + y = 5$$
$$-x + y = 1$$

Solution
Both the linear system and the augmented matrix are shown.

Linear System
$$x + y = 5$$
$$-x + y = 1$$

Augmented Matrix
$$\begin{bmatrix} 1 & 1 & | & 5 \\ -1 & 1 & | & 1 \end{bmatrix}$$

First, we want to obtain a 0 in the second row, where the -1 is highlighted. To do so, we add row 1 to row 2 and place the result in row 2. This step is denoted $R_2 + R_1$ and eliminates the x-variable from the second equation.

$$x + y = 5$$
$$2y = 6 \qquad R_2 + R_1 \rightarrow \begin{bmatrix} 1 & 1 & | & 5 \\ 0 & 2 & | & 6 \end{bmatrix}$$

1 1 5
−1 1 1
0 2 6 Add.

To obtain a 1 where the 2 in the second row is located, we multiply the second row by $\frac{1}{2}$, denoted $\frac{1}{2}R_2$.

$$x + y = 5$$
$$y = 3 \qquad \frac{1}{2}R_2 \rightarrow \begin{bmatrix} 1 & 1 & | & 5 \\ 0 & 1 & | & 3 \end{bmatrix}$$

Half of
0 2 6 is
0 1 3.

Next, we need to obtain a 0 where the 1 is highlighted. We do so by subtracting row 2 from row 1 and placing the result in row 1, denoted $R_1 - R_2$.

$$x = 2 \qquad R_1 - R_2 \rightarrow \begin{bmatrix} 1 & 0 & | & 2 \\ 0 & 1 & | & 3 \end{bmatrix}$$
$$y = 3$$

1 1 5
0 1 3
1 0 2 Subtract.

This matrix is in reduced row–echelon form. The solution is $(2, 3)$.

Now Try Exercise 25

NOTE: When using Gauss–Jordan elimination to solve a system, it is not necessary to write the system alongside the matrix solution. ∎

In the next example, we use Gauss–Jordan elimination to solve a system with three linear equations and three variables. To do so, we transform the matrix into the following reduced row–echelon form, where (a, b, c) represents the solution.

$$\begin{bmatrix} 1 & 0 & 0 & | & a \\ 0 & 1 & 0 & | & b \\ 0 & 0 & 1 & | & c \end{bmatrix}$$

EXAMPLE 5 Transforming a matrix into reduced row–echelon form

Use Gauss–Jordan elimination to transform the augmented matrix of the linear system into reduced row–echelon form. Find the solution.

$$x + y + 2z = 1$$
$$-x + z = -2$$
$$2x + y + 5z = -1$$

Solution
The linear system and the augmented matrix are both shown.

Linear System
$$x + y + 2z = 1$$
$$-x + z = -2$$
$$2x + y + 5z = -1$$

Augmented Matrix
$$\begin{bmatrix} 1 & 1 & 2 & | & 1 \\ -1 & 0 & 1 & | & -2 \\ 2 & 1 & 5 & | & -1 \end{bmatrix}$$

First, we want to put 0s in the second and third rows, where the -1 and 2 are highlighted. To obtain a 0 in the first position of the second row, we add row 1 to row 2 and place the result in row 2, denoted $R_2 + R_1$. To obtain a 0 in the first position of the third row, we subtract 2 times row 1 from row 3 and place the result in row 3, denoted $R_3 - 2R_1$. Row 1 does not change. These steps eliminate the x-variable from the second and third equations.

$$x + y + 2z = 1$$
$$y + 3z = -1 \quad R_2 + R_1 \rightarrow$$
$$-y + z = -3 \quad R_3 - 2R_1 \rightarrow$$

$$\begin{bmatrix} 1 & 1 & 2 & | & 1 \\ 0 & 1 & 3 & | & -1 \\ 0 & -1 & 1 & | & -3 \end{bmatrix}$$

To eliminate the y-variable in row 1, we subtract row 2 from row 1. To eliminate the y-variable from row 3, we add row 2 to row 3.

$$x - z = 2 \quad R_1 - R_2 \rightarrow$$
$$y + 3z = -1$$
$$4z = -4 \quad R_3 + R_2 \rightarrow$$

$$\begin{bmatrix} 1 & 0 & -1 & | & 2 \\ 0 & 1 & 3 & | & -1 \\ 0 & 0 & 4 & | & -4 \end{bmatrix}$$

To obtain a 1 in row 3, where the highlighted 4 is located, we multiply row 3 by $\frac{1}{4}$.

$$x - z = 2$$
$$y + 3z = -1$$
$$z = -1 \quad \frac{1}{4}R_3 \rightarrow$$

$$\begin{bmatrix} 1 & 0 & -1 & | & 2 \\ 0 & 1 & 3 & | & -1 \\ 0 & 0 & 1 & | & -1 \end{bmatrix}$$

For the matrix to be in reduced row–echelon form, we need 0s in the highlighted locations. We first add row 3 to row 1 and then subtract 3 times row 3 from row 2.

$$x = 1 \quad R_1 + R_3 \rightarrow$$
$$y = 2 \quad R_2 - 3R_3 \rightarrow$$
$$z = -1$$

$$\begin{bmatrix} 1 & 0 & 0 & | & 1 \\ 0 & 1 & 0 & | & 2 \\ 0 & 0 & 1 & | & -1 \end{bmatrix}$$

This matrix is now in reduced row–echelon form. The solution is $(1, 2, -1)$.

Now Try Exercise 35

CRITICAL THINKING 1

An *inconsistent* system of linear equations has no solutions, and a system of *dependent* linear equations has infinitely many solutions. Suppose that an augmented matrix row reduces to either of the following matrices. Explain what each matrix indicates about the given system of linear equations.

$$\begin{bmatrix} 1 & 0 & 0 & | & 2 \\ 0 & 1 & 0 & | & 3 \\ 0 & 0 & 0 & | & 1 \end{bmatrix} \quad \begin{bmatrix} 1 & 0 & 1 & | & 2 \\ 0 & 1 & 2 & | & 3 \\ 0 & 0 & 0 & | & 0 \end{bmatrix}$$

In the next example, we find the amounts invested in three mutual funds.

EXAMPLE 6 Determining investment amounts in mutual funds

A total of $8000 was invested in three funds that grew at a rate of 5%, 10%, and 20% over 1 year. After 1 year, the combined value of the three funds had grown by $1200. Five times as much money was invested at 20% as at 10%. Find the amount invested in each fund.

Solution
Let x be the amount invested at 5%, y be the amount invested at 10%, and z be the amount invested at 20%. The total amount invested was $8000, so

$$x + y + z = 8000.$$

The growth in the first mutual fund, paying 5% of x, is given by $0.05x$. Similarly, the growths in the other mutual funds are given by $0.10y$ and $0.20z$. As the total growth was $1200, we can write

$$0.05x + 0.10y + 0.20z = 1200.$$

Multiplying each side of this equation by 20 to eliminate decimals results in

$$x + 2y + 4z = 24{,}000.$$

Five times as much was invested at 20% as at 10%, so $z = 5y$, or $5y - z = 0$.

These three equations can be written as a system of linear equations and as an augmented matrix.

Linear System
$$\begin{aligned} x + y + z &= 8{,}000 \\ x + 2y + 4z &= 24{,}000 \\ 5y - z &= 0 \end{aligned}$$

Augmented Matrix
$$\begin{bmatrix} 1 & 1 & 1 & | & 8{,}000 \\ 1 & 2 & 4 & | & 24{,}000 \\ 0 & 5 & -1 & | & 0 \end{bmatrix}$$

A 0 can be obtained in the highlighted position by subtracting row 1 from row 2.

$$\begin{aligned} x + y + z &= 8{,}000 \\ y + 3z &= 16{,}000 \quad R_2 - R_1 \to \\ 5y - z &= 0 \end{aligned} \qquad \begin{bmatrix} 1 & 1 & 1 & | & 8{,}000 \\ 0 & 1 & 3 & | & 16{,}000 \\ 0 & 5 & -1 & | & 0 \end{bmatrix}$$

Zeros can be obtained in the highlighted positions by subtracting row 2 from row 1 and by subtracting 5 times row 2 from row 3.

$$\begin{aligned} x - 2z &= -8{,}000 \quad R_1 - R_2 \to \\ y + 3z &= 16{,}000 \\ -16z &= -80{,}000 \quad R_3 - 5R_2 \to \end{aligned} \qquad \begin{bmatrix} 1 & 0 & -2 & | & -8{,}000 \\ 0 & 1 & 3 & | & 16{,}000 \\ 0 & 0 & -16 & | & -80{,}000 \end{bmatrix}$$

To obtain a 1 in the highlighted position, multiply row 3 by $-\frac{1}{16}$.

$$\begin{aligned} x - 2z &= -8{,}000 \\ y + 3z &= 16{,}000 \\ z &= 5{,}000 \quad -\tfrac{1}{16} R_3 \to \end{aligned} \qquad \begin{bmatrix} 1 & 0 & -2 & | & -8{,}000 \\ 0 & 1 & 3 & | & 16{,}000 \\ 0 & 0 & 1 & | & 5{,}000 \end{bmatrix}$$

To obtain a 0 in each of the highlighted positions, add twice row 3 to row 1 and subtract three times row 3 from row 2.

$$\begin{aligned} x &= 2{,}000 \quad R_1 + 2R_3 \to \\ y &= 1{,}000 \quad R_2 - 3R_3 \to \\ z &= 5{,}000 \end{aligned} \qquad \begin{bmatrix} 1 & 0 & 0 & | & 2{,}000 \\ 0 & 1 & 0 & | & 1{,}000 \\ 0 & 0 & 1 & | & 5{,}000 \end{bmatrix}$$

Thus $2000 was invested at 5%, $1000 at 10%, and $5000 at 20%.

Now Try Exercise 67

4 Using Technology to Solve Systems of Linear Equations

Examples 5 and 6 involve a lot of arithmetic. Trying to solve a large system of equations by hand is an enormous—if not impossible—task. In the real world, people use technology to solve large systems. In the next example, we solve the linear systems from Examples 4 and 5 with a graphing calculator.

EXAMPLE 7 Using technology

Use a graphing calculator to solve the following systems of equations.
(a) $\quad x + y = 5 \qquad$ (b) $\quad x + y + 2z = 1$
$\quad\;\; -x + y = 1 \qquad\qquad\;\; -x + \; z = -2$
$\qquad\qquad\qquad\qquad\qquad 2x + y + 5z = -1$

Solution
(a) Enter the 2×3 augmented matrix from Example 4 in a graphing calculator, as shown in **FIGURE 16.3(a)**. Then transform the matrix into reduced row–echelon form (rref), as shown in **FIGURE 16.3(b)**. The solution is $(2, 3)$.

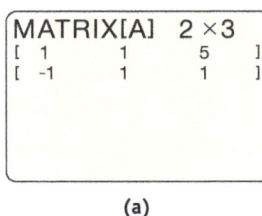
(a)
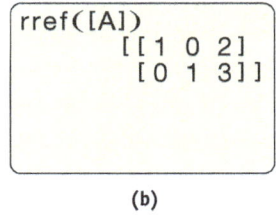
(b)

FIGURE 16.3

(b) Enter the 3×4 augmented matrix from Example 5, as shown in **FIGURE 16.4(a)**. (The fourth column of A can be seen by scrolling right.) Then transform the matrix into reduced row–echelon form (rref), as shown in **FIGURE 16.4(b)**. The solution is $(1, 2, -1)$.

CALCULATOR HELP
To enter a matrix and put it in reduced row–echelon form, see Appendix A (pages AP-8 and AP-9).

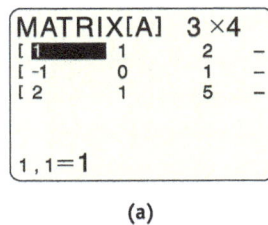
(a)
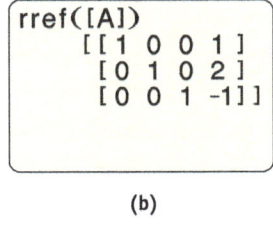
(b)

FIGURE 16.4

Now Try Exercises 43, 45

🌐 *Math in Context* — Biology Suppose that the size of a bear's head and its overall length are known. Can its weight be estimated from these variables? In the next example, we show that a system of linear equations can be used to make this estimate.

EXAMPLE 8 Modeling the weight of male bears

The data shown in **TABLE 16.2** on the next page give the weight W, head length H, and overall length L of three bears. These data can be modeled with the equation $W = a + bH + cL$, where a, b, and c are constants that we need to determine. (*Sources:* M. Triola, *Elementary Statistics*; Minitab, Inc.)

Bear Measurements

W (pounds)	H (inches)	L (inches)
362	16	72
300	14	68
147	11	52
?	13	65

TABLE 16.2

(a) Set up a system of equations whose solution gives values for constants a, b, and c.
(b) Solve the system.
(c) Predict the weight of a bear with $H = 13$ inches and $L = 65$ inches.

Solution
(a) Substitute each row of **TABLE 16.2** in the equation $W = a + bH + cL$.

$$362 = a + b(16) + c(72)$$
$$300 = a + b(14) + c(68)$$
$$147 = a + b(11) + c(52)$$

Rewrite this system as

$$a + 16b + 72c = 362$$
$$a + 14b + 68c = 300$$
$$a + 11b + 52c = 147$$

and represent it as the following augmented matrix.

$$A = \begin{bmatrix} 1 & 16 & 72 & | & 362 \\ 1 & 14 & 68 & | & 300 \\ 1 & 11 & 52 & | & 147 \end{bmatrix}$$

(b) Enter A and put it in reduced row–echelon form, as shown in **FIGURES 16.5(a)** and **(b)**, respectively. The solution is $a = -374$, $b = 19$, and $c = 6$.

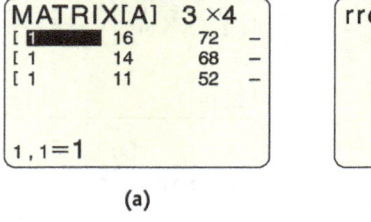

(a) (b)

FIGURE 16.5

(c) For $W = a + bH + cL$, use

$$W = -374 + 19H + 6L$$

to predict the weight of a bear with head length $H = 13$ and overall length $L = 65$.

$$W = -374 + 19(13) + 6(65) = 263 \text{ pounds}$$

Now Try Exercise 63

16.2 Putting It All Together

A matrix is a rectangular array of numbers. An augmented matrix may be used to represent any system of linear equations. One common method for solving a system of linear equations is Gauss–Jordan elimination. Matrix row operations may be used to transform an augmented matrix to reduced row–echelon form.

CONCEPT	EXPLANATION/EXAMPLE
Augmented Matrix	A linear system can be represented by an augmented matrix. The following matrix has dimension 3×4. **Linear System** $x + 2y - z = 6$ $-2x + y - z = 7$ $2x + 3z = -11$ **Augmented Matrix** $\begin{bmatrix} 1 & 2 & -1 & 6 \\ -2 & 1 & -1 & 7 \\ 2 & 0 & 3 & -11 \end{bmatrix}$
Reduced Row–Echelon Form	The following augmented matrix is in reduced row–echelon form, which results from transforming the preceding system to reduced row–echelon form. There are 1s along the main diagonal and 0s elsewhere in the first three columns. The solution to the linear system is $(-1, 2, -3)$. $\begin{bmatrix} 1 & 0 & 0 & -1 \\ 0 & 1 & 0 & 2 \\ 0 & 0 & 1 & -3 \end{bmatrix}$

16.2 Exercises

CONCEPTS AND VOCABULARY

1. What is a matrix?

2. Give an example of a matrix and state its dimension.

3. Give an example of an augmented matrix and state its dimension.

4. If an augmented matrix is used to solve a system of three linear equations in three variables, what will be its dimension?

5. Give an example of a matrix that is in reduced row–echelon form.

6. Identify the elements on the main diagonal in the augmented matrix.

$\begin{bmatrix} 4 & -6 & -1 & 3 \\ 6 & 2 & -2 & 9 \\ 7 & 5 & -3 & 1 \end{bmatrix}$

REPRESENTING SYSTEMS WITH MATRICES

Exercises 7–10: State the dimension of the matrix.

7. $\begin{bmatrix} 3 & -3 & 7 \\ 2 & 6 & -2 \\ 4 & 2 & 5 \end{bmatrix}$

8. $\begin{bmatrix} -2 & 3 & 0 \\ 1 & -8 & 4 \end{bmatrix}$

9. $\begin{bmatrix} 1 & 7 \\ 0 & 2 \\ 2 & -5 \end{bmatrix}$

10. $\begin{bmatrix} 4 & 2 & -3 & -1 \\ 4 & -3 & 2 & -7 \\ 14 & 6 & 4 & 0 \end{bmatrix}$

Exercises 11–14: Represent the linear system as an augmented matrix.

11. $x - 3y = 1$
    $-x + 3y = -1$

12. $4x + 2y = -5$
    $5x + 8y = 2$

13. $2x - y + 2z = -4$
    $x - 2y = 2$
    $-x + y - 2z = -6$

14. $3x - 2y + z = 5$
    $-x + 2z = -4$
    $x - 2y + z = -1$

Exercises 15–20: Write the system of linear equations that the augmented matrix represents. Use the variables x, y, and z.

15. $\begin{bmatrix} 1 & 2 & | & -6 \\ 5 & -1 & | & 4 \end{bmatrix}$

16. $\begin{bmatrix} 1 & -5 & | & 7 \\ 0 & -3 & | & 6 \end{bmatrix}$

17. $\begin{bmatrix} 1 & -1 & 2 & | & 6 \\ 2 & 1 & -2 & | & 1 \\ -1 & 2 & -1 & | & 3 \end{bmatrix}$

18. $\begin{bmatrix} 3 & -1 & 2 & | & -1 \\ 2 & -2 & 2 & | & 4 \\ 1 & 7 & -2 & | & 2 \end{bmatrix}$

19. $\begin{bmatrix} 1 & 0 & 0 & | & 4 \\ 0 & 1 & 0 & | & -2 \\ 0 & 0 & 1 & | & 7 \end{bmatrix}$

20. $\begin{bmatrix} 1 & 0 & 0 & | & 6 \\ 0 & 1 & 0 & | & -2 \\ 0 & 0 & 1 & | & 4 \end{bmatrix}$

RELATING MATRICES AND SOCIAL NETWORKS

Exercises 21 and 24: (Refer to Example 2.) Use a 4 × 4 matrix to represent the social network.

21.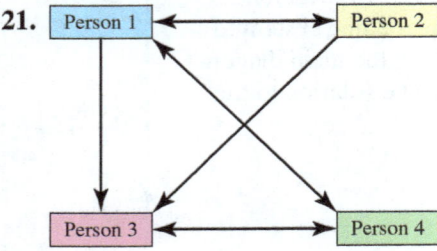

22.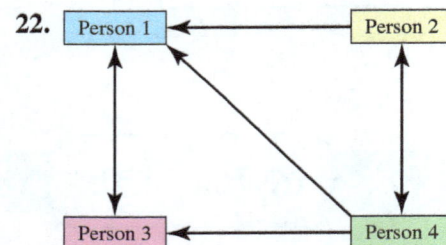

23.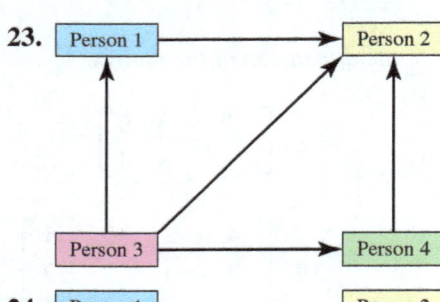

24.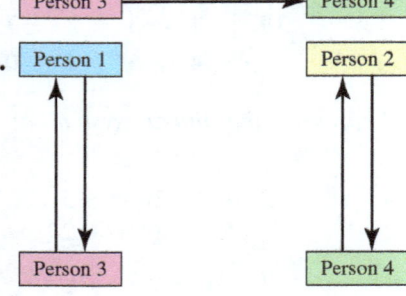

APPLYING GAUSS–JORDAN ELIMINATION

Exercises 25–42: Use Gauss–Jordan elimination to find the solution. Write the solution as an ordered pair or ordered triple and check the solution.

25. $x + y = 4$
    $x + 3y = 10$

26. $x - 3y = -7$
    $2x + y = 0$

27. $2x + 3y = 3$
    $-2x + 2y = 7$

28. $x + 3y = -14$
    $2x + 5y = -24$

29. $x - y = 5$
    $x + 3y = -1$

30. $x + 4y = 1$
    $3x - 2y = 10$

31. $4x - 8y = -10$
    $x + y = 2$

32. $x - 7y = -16$
    $4x + 10y = 50$

33. $x + y + z = 6$
    $2y - z = 1$
    $y + z = 5$

34. $x + y + z = 3$
    $x + y - z = 2$
    $y + z = 2$

35. $x + 2y + 3z = 6$
    $-x + 3y + 4z = 0$
    $x + y - 2z = -6$

36. $2x - 4y + 2z = 10$
    $-x + 3y - 4z = -19$
    $2x - y - 6z = -28$

37. $x + y + z = 0$
    $2x + y + 2z = -1$
    $x + y = 0$

38. $x + y - 2z = 5$
    $x + 2y - 2z = 4$
    $-x - y + z = -4$

39. $x + y + z = 3$
    $-x - z = -2$
    $x + y + 2z = 4$

40. $x + 2y - z = 3$
    $-x - y + z = 0$
    $x + 2y = 5$

41. $x + 2y + z = 3$
    $2x + y - z = -6$
    $-x - y + 2z = 5$

42. $x + y + z = -3$
    $x - y - z = -1$
    $-2x + y + 4z = 4$

USING TECHNOLOGY TO SOLVE SYSTEMS

Exercises 43–52: **Technology** *Use a graphing calculator to solve the system of linear equations.*

43. $x + 4y = 13$
    $5x - 3y = -50$

44. $9x - 11y = 7$
    $5x + 6y = 16$

45. $2x - y + 3z = 9$
    $-4x + 5y + 2z = 12$
    $2x + 7z = 23$

46. $3x - 2y + 4z = 29$
    $2x + 3y - 7z = -14$
    $5x - y + 11z = 59$

47. $6x + 2y + z = 4$
    $-2x + 4y + z = -3$
    $2x - 8y = -2$

48. $-x - 9y + 2z = -28.5$
    $2x - y + 4z = -17$
    $x - y + 8z = -9$

49. $4x + 3y + 12z = -9.25$
    $15y + 8z = -4.75 + x$
    $7z = -5.5 - 6y$

50. $5x + 4y = 13.3 + z$
    $7y + 9z = 16.9 - x$
    $x - 3y + 4z = -4.1$

51. $1.2x - 0.9y + 2.7z = 5.37$
    $3.1x - 5.1y + 7.2z = 14.81$
    $1.8y + 6.38 = 3.6z - 0.2x$

52. $11x + 13y - 17z = 380$
    $5x - 14y - 19z = 24$
    $-21y + 46z = -676 + 7x$

Exercises 53–58: (Refer to Critical Thinking 1 in this section.) Row-reduce the matrix associated with the given system to determine whether the system of linear equations is inconsistent or has dependent equations.

53. $x + 2y = 4$
    $-2x - 4y = -8$

54. $x - 5y = 4$
    $-2x + 10y = 8$

55. $x + y + z = 3$
    $x + y - z = 1$
    $x + y = 3$

56. $x + y + z = 5$
    $x - y - z = 8$
    $2x + 2y + 2z = 6$

57. $x + 2y + 3z = 14$
    $2x - 3y - 2z = -10$
    $3x - y + z = 4$

58. $x + 2y + 3z = 6$
    $-x + 3y + 4z = 6$
    $5y + 7z = 12$

Exercises 59 and 60: **Thinking Generally** *The matrix represents a linear system of equations. Solve the system, if a and b are nonzero constants.*

59. $\begin{bmatrix} a & 0 & 0 & | & 1 \\ 0 & b & 0 & | & 1 \\ 0 & 0 & ab & | & 2 \end{bmatrix}$

60. $\begin{bmatrix} a & 0 & 0 & | & 1 \\ 0 & b & 0 & | & 2 \\ 0 & 0 & 0 & | & a \end{bmatrix}$

APPLICATIONS INVOLVING MATRICES

61. **Weight of a Bear** Use the results of Example 8 to estimate the weight of a bear with a head length of 12 inches and an overall length of 60 inches.

62. **Garbage and Household Size** A larger household produces more garbage, on average, than a smaller household. If we know the amount of metal M and plastic P waste produced each week, we can estimate the household size H from $H = a + bM + cP$. The table contains representative data for three households. (*Source:* M. Triola, *Elementary Statistics.*)

H (people)	M (pounds)	P (pounds)
3	2.00	1.40
2	1.50	0.65
6	4.00	3.40

(a) Set up a system of equations whose solution gives values for the constants a, b, and c.
(b) Solve this system.
(c) Predict the size of a household that produces 3 pounds of metal waste and 2 pounds of plastic waste each week.

63. **Weight of a Bear** (Refer to Example 8.) Head length and overall length are not the only variables that can be used to estimate the weight of a bear. The data in the accompanying table list the weight W, neck size N, and chest size C of three bears. These data can be modeled by $W = a + bN + cC$. (*Sources:* M. Triola, *Elementary Statistics*; Minitab, Inc.)

W (pounds)	N (inches)	C (inches)
80	16	26
344	28	45
416	31	54

(a) Set up a system of equations whose solution gives values for the constants a, b, and c.
(b) Solve this system. Round each value to the nearest tenth.
(c) Predict the weight of a bear with neck size $N = 22$ inches and chest size $C = 38$ inches.

64. **Old Faithful Geyser** In Yellowstone National Park, Old Faithful Geyser has been a favorite attraction for decades. Although this geyser erupts about every 80 minutes, this time interval varies, as do the duration and height of the eruptions. The accompanying table shows the height H, duration D, and time interval T for three representative eruptions. (*Source:* National Park Service.)

H (feet)	D (seconds)	T (minutes)
160	276	94
125	203	84
140	245	79

continued on next page

continued from previous page

(a) Assume that these data can be modeled by $H = a + bD + cT$. Set up a system of equations whose solution gives values for the constants a, b, and c.
(b) Solve this system. Round each value to the nearest thousandth.
(c) Use this equation to estimate H when $D = 220$ and $T = 81$.

65. Jogging Speeds A runner in preparation for a marathon jogs at 5, 6, and 8 miles per hour. The runner travels a total distance of 12.5 miles in 2 hours and jogs the same length of time at 5 miles per hour and at 8 miles per hour. How long does the runner jog at each speed?

66. Mixture Problem Three types of candy that cost $2, $3, and $4 per pound are to be mixed to produce a 5-pound bag of candy that costs $14.50. If there are to be equal amounts of the $3-per-pound candy and the $4-per-pound candy, how much of each type of candy should be included in the mixture?

67. Interest and Investments (Refer to Example 6.) A total of $3000 is invested at 3%, 4%, and 6% annual interest. The interest earned after 1 year equals $145. The amount invested at 6% is triple the amount invested at 3%. Find the amount invested at each rate.

68. Geometry The measure of the largest angle in a triangle is twice the measure of the smallest angle. The remaining angle is 10° less than the largest angle. Find the measure of each angle.

69. Online Exploration Go to your Facebook profile and select three or four of your friends plus yourself. Answers will vary.
(a) Draw a social network that shows the friendships between each pair of people. Use a double arrow to represent this. (See **FIGURE 16.2**.)
(b) Use a matrix to represent your social network.

70. Online Exploration Go online and look up a diagram of the molecular structure of water (H_2O).

(a) Draw double arrows wherever the hydrogen and oxygen are connected.
(b) Use a matrix to represent the structure of water. Let the first row represent oxygen.

SOLVING AN EQUATION IN FOUR VARIABLES

71. Weight of a Bear Earlier in this section, we estimated the weight of a bear by using three variables. We may be able to make more accurate estimates by using four variables. The accompanying table shows the weight W, neck size N, overall length L, and chest size C for four bears. (*Sources:* M. Triola, *Elementary Statistics*, Minitab, Inc.)

W (pounds)	N (inches)	L (inches)	C (inches)
125	19	57.5	32
316	26	65	42
436	30	72	48
514	30.5	75	54
?	24	63	39

(a) We can model the data in the table with the equation $W = a + bN + cL + dC$, where a, b, c, and d are constants. To do so, represent a system of linear equations by a 4×5 augmented matrix whose solution gives values for a, b, c, and d.
(b) Solve the system with a graphing calculator. Round each value to the nearest thousandth.

72. Refer to the previous exercise. Predict the weight of a bear with $N = 24$, $L = 63$, and $C = 39$. Interpret the result.

WRITING ABOUT MATHEMATICS

73. Explain what the dimension of a matrix means. What is the difference between a matrix that has dimension 3×4 and one that has dimension 4×3?

74. Discuss the advantages of using technology to transform an augmented matrix to reduced row–echelon form. Are there any disadvantages? Explain.

SECTIONS 16.1 and 16.2 Checking Basic Concepts

1. Determine which ordered triple is the solution to the system of equations: $(5, -4, 0)$ or $(1, 3, -1)$.

$$x - y + 7z = -9$$
$$2x - 2y + 5z = -9$$
$$-x + 3y - 2z = 10$$

2. Solve the system of equations by using elimination and substitution.

$$x - y + z = 2$$
$$2x - 3y + z = -1$$
$$-x + y + z = 4$$

3. Use an augmented matrix to represent the system of equations. Solve the system by using
   (a) Gauss–Jordan elimination and
   (b) technology.

$$x + 2y + z = 1$$
$$x + y + z = -1$$
$$y + z = 1$$

4. A total of $1500 is invested at 1%, 2%, and 4% annual interest. The interest after 1 year equals $46. The amount invested at 2% is double the amount invested at 1%. Find the amount invested at each rate.

Group Activity

Adjacency Matrix *A matrix A can be used to represent a map showing distances between cities. Let a_{ij} denote the number in row i and column j of a matrix A. Now consider the following map illustrating freeway distances in miles between four cities. Each city has been assigned a number. For example, there is a direct route from Denver, Colorado (city 1), to Colorado Springs, Colorado (city 2), of approximately 60 miles. Therefore $a_{12} = 60$ in the accompanying matrix A. (Note that a_{12} is the number in row 1 and column 2.) The distance from Colorado Springs to Denver is also 60 miles, so $a_{21} = 60$. As there is no direct freeway connection between Las Vegas, Nevada (city 4), and Colorado Springs (city 2), we let $a_{24} = a_{42} = *$. The matrix A is called an* **adjacency matrix**. (Source: S. Baase, Computer Algorithms: Introduction to Design and Analysis.)

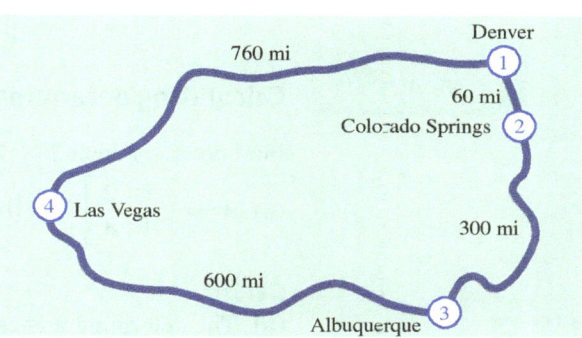

$$A = \begin{bmatrix} 0 & 60 & * & 760 \\ 60 & 0 & 300 & * \\ * & 300 & 0 & 600 \\ 760 & * & 600 & 0 \end{bmatrix}$$

1. Explain how to use A to find the freeway distance from Denver to Las Vegas.

2. Explain how to use A to find the shortest freeway distance from Denver to Albuquerque.

3. If a map shows 20 cities, what would be the dimension of the adjacency matrix? How many elements would there be in this matrix?

4. Why are there only zeros on the main diagonal of A?

5. What does $a_{14} + a_{41}$ equal?

6. What does $a_{11} + a_{44}$ equal?

16.3 Determinants

Objectives
1. Calculating Determinants
2. Finding the Area of a Region
3. Applying Cramer's Rule

NEW VOCABULARY
☐ Determinants
☐ Expansion by minors
☐ Minors
☐ Cramer's rule

1 Calculating Determinants

The concept of determinants originated with the Japanese mathematician Seki Kowa (1642–1708), who used them to solve systems of linear equations. Later, Gottfried Leibniz (1646–1716) formally described determinants and also used them to solve systems of linear equations. (*Source: Historical Topics for the Mathematical Classroom, NCTM.*)

We begin by defining a determinant of a 2×2 matrix.

DETERMINANT OF A 2×2 MATRIX

The **determinant** of

$$A = \begin{bmatrix} a & b \\ c & d \end{bmatrix}$$

is a *real number* defined by

$$\det A = ad - cb.$$

EXAMPLE 1 Calculating determinants

Find det A for each 2×2 matrix.

(a) $A = \begin{bmatrix} 1 & 2 \\ 3 & 4 \end{bmatrix}$ (b) $A = \begin{bmatrix} -1 & -3 \\ 2 & -8 \end{bmatrix}$

Solution

(a) The determinant is calculated as follows.

$$\det A = \det \begin{bmatrix} 1 & 2 \\ 3 & 4 \end{bmatrix} = (1)(4) - (3)(2) = -2$$

(b) Similarly,

$$\det A = \det \begin{bmatrix} -1 & -3 \\ 2 & -8 \end{bmatrix} = (-1)(-8) - (2)(-3) = 14.$$

Now Try Exercises 5, 7

READING CHECK 1

- How do you calculate the determinant of a 2×2 matrix?

We can use determinants of 2×2 matrices to find determinants of 3×3 matrices. This method is called **expansion of a determinant by minors**.

DETERMINANT OF A 3×3 MATRIX

$$\det A = \det \begin{bmatrix} a_1 & b_1 & c_1 \\ a_2 & b_2 & c_2 \\ a_3 & b_3 & c_3 \end{bmatrix}$$

$$= a_1 \cdot \det \begin{bmatrix} b_2 & c_2 \\ b_3 & c_3 \end{bmatrix} - a_2 \cdot \det \begin{bmatrix} b_1 & c_1 \\ b_3 & c_3 \end{bmatrix} + a_3 \cdot \det \begin{bmatrix} b_1 & c_1 \\ b_2 & c_2 \end{bmatrix}$$

The 2×2 matrices in this equation are called **minors**.

EXAMPLE 2 Calculating 3 × 3 determinants

Evaluate det A.

(a) $A = \begin{bmatrix} 2 & 1 & -1 \\ -1 & 3 & 2 \\ 4 & -3 & -5 \end{bmatrix}$
(b) $A = \begin{bmatrix} 5 & -2 & 4 \\ 0 & 2 & 1 \\ -1 & 4 & -4 \end{bmatrix}$

Solution

(a) We evaluate the determinant as follows.

$$\det \begin{bmatrix} 2 & 1 & -1 \\ -1 & 3 & 2 \\ 4 & -3 & -5 \end{bmatrix} = 2 \cdot \det \begin{bmatrix} 3 & 2 \\ -3 & -5 \end{bmatrix} - (-1) \cdot \det \begin{bmatrix} 1 & -1 \\ -3 & -5 \end{bmatrix}$$

$$+ 4 \cdot \det \begin{bmatrix} 1 & -1 \\ 3 & 2 \end{bmatrix}$$

$$= 2(-9) + 1(-8) + 4(5)$$

$$= -6$$

(b) We evaluate the determinant as follows.

$$\det \begin{bmatrix} 5 & -2 & 4 \\ 0 & 2 & 1 \\ -1 & 4 & -4 \end{bmatrix} = 5 \cdot \det \begin{bmatrix} 2 & 1 \\ 4 & -4 \end{bmatrix} - (0) \cdot \det \begin{bmatrix} -2 & 4 \\ 4 & -4 \end{bmatrix}$$

$$+ (-1) \cdot \det \begin{bmatrix} -2 & 4 \\ 2 & 1 \end{bmatrix}$$

$$= 5(-12) - 0(-8) + (-1)(-10)$$

$$= -50$$

Now Try Exercises 13, 15

Many graphing calculators can evaluate the determinant of a matrix, as illustrated in the next example, where we evaluate the determinants from Example 2.

EXAMPLE 3 Using technology to find determinants

Find each determinant of A, using a graphing calculator.

(a) $A = \begin{bmatrix} 2 & 1 & -1 \\ -1 & 3 & 2 \\ 4 & -3 & -5 \end{bmatrix}$
(b) $A = \begin{bmatrix} 5 & -2 & 4 \\ 0 & 2 & 1 \\ -1 & 4 & -4 \end{bmatrix}$

Solution

(a) Begin by entering the matrix and then evaluate the determinant, as shown in **FIGURE 16.6**. The result is det $A = -6$, which agrees with our earlier calculation.

```
MATRIX[A]  3×3
[  2    1   -1  ]
[ -1    3    2  ]
[  4   -3   -5  ]
```

```
[A]
  [[2   1  -1]
   [-1  3   2]
   [4  -3  -5]]
det([A])
              -6
```

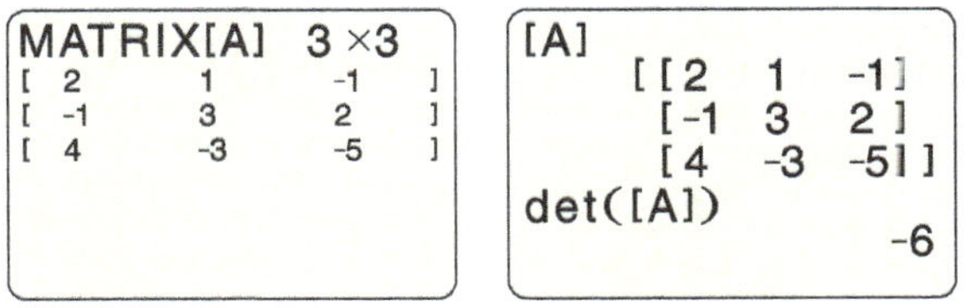

(a)   (b)

FIGURE 16.6

CALCULATOR HELP

To find a determinant, see Appendix A (pages AP-9 and AP-10).

(b) The determinant of A evaluates to -50. See **FIGURE 16.7**.

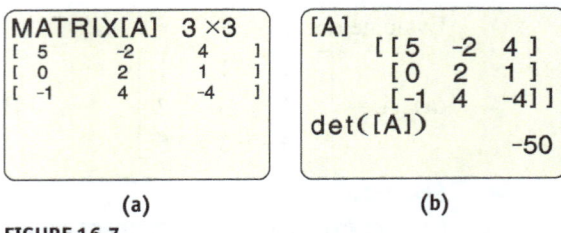

(a)   (b)

FIGURE 16.7

Now Try Exercise 21

2 Finding the Area of a Region

Math in Context Surveying A determinant may be used in surveying to find the area of a triangular region. For example, if a triangle has vertices (a_1, a_2), (b_1, b_2), and (c_1, c_2), its area equals the absolute value of D, where

$$D = \frac{1}{2} \det \begin{bmatrix} a_1 & b_1 & c_1 \\ a_2 & b_2 & c_2 \\ 1 & 1 & 1 \end{bmatrix}.$$

If the vertices are entered in the columns of D counterclockwise as they appear in the xy-plane, D will be positive. (*Source:* W. Taylor, *The Geometry of Computer Graphics.*)

EXAMPLE 4 Computing the area of a triangular parcel of land

A triangular parcel of land is shown in **FIGURE 16.8**. If all units are miles, find the area of the parcel of land by using a determinant.

Solution
The vertices of the triangular parcel of land are $(2, 2)$, $(5, 4)$, and $(3, 8)$. The area of the triangle is

$$D = \frac{1}{2} \det \begin{bmatrix} 2 & 5 & 3 \\ 2 & 4 & 8 \\ 1 & 1 & 1 \end{bmatrix} = \frac{1}{2} \cdot 16 = 8.$$

The area of the triangle is 8 square miles.

Now Try Exercise 27

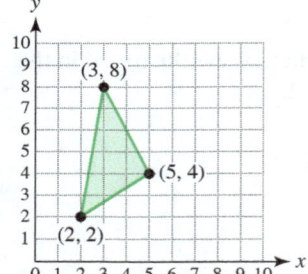

FIGURE 16.8

3 Applying Cramer's Rule

The use of determinants was expanded by Gabriel Cramer (1704–1752). His work, published in 1750, provided a method called **Cramer's rule** for solving systems of linear equations.

CRITICAL THINKING 1

Suppose that you are given three distinct points in the xy-plane and $D = 0$. What must be true about the three points?

CRAMER'S RULE FOR LINEAR SYSTEMS IN TWO VARIABLES

The solution to the system of linear equations

$$a_1 x + b_1 y = c_1$$
$$a_2 x + b_2 y = c_2$$

is given by $x = \frac{E}{D}$ and $y = \frac{F}{D}$, where

$$E = \det \begin{bmatrix} c_1 & b_1 \\ c_2 & b_2 \end{bmatrix}, \quad F = \det \begin{bmatrix} a_1 & c_1 \\ a_2 & c_2 \end{bmatrix}, \quad \text{and} \quad D = \det \begin{bmatrix} a_1 & b_1 \\ a_2 & b_2 \end{bmatrix} \neq 0.$$

NOTE: If $D = 0$, the system has either no solutions or infinitely many solutions. ∎

EXAMPLE 5 Using Cramer's rule

Use Cramer's rule to solve the following linear systems.
(a) $3x - 4y = 18$
$7x + 5y = -1$
(b) $-4x + 9y = -24$
$6x + 17y = -25$

Solution

(a) $E = \det\begin{bmatrix} c_1 & b_1 \\ c_2 & b_2 \end{bmatrix} = \det\begin{bmatrix} 18 & -4 \\ -1 & 5 \end{bmatrix} = (18)(5) - (-1)(-4) = 86$

$F = \det\begin{bmatrix} a_1 & c_1 \\ a_2 & c_2 \end{bmatrix} = \det\begin{bmatrix} 3 & 18 \\ 7 & -1 \end{bmatrix} = (3)(-1) - (7)(18) = -129$

$D = \det\begin{bmatrix} a_1 & b_1 \\ a_2 & b_2 \end{bmatrix} = \det\begin{bmatrix} 3 & -4 \\ 7 & 5 \end{bmatrix} = (3)(5) - (7)(-4) = 43$

Because $x = \frac{E}{D} = \frac{86}{43} = 2$ and $y = \frac{F}{D} = \frac{-129}{43} = -3$, the solution is $(2, -3)$.

(b) $E = \det\begin{bmatrix} c_1 & b_1 \\ c_2 & b_2 \end{bmatrix} = \det\begin{bmatrix} -24 & 9 \\ -25 & 17 \end{bmatrix} = (-24)(17) - (-25)(9) = -183$

$F = \det\begin{bmatrix} a_1 & c_1 \\ a_2 & c_2 \end{bmatrix} = \det\begin{bmatrix} -4 & -24 \\ 6 & -25 \end{bmatrix} = (-4)(-25) - (6)(-24) = 244$

$D = \det\begin{bmatrix} a_1 & b_1 \\ a_2 & b_2 \end{bmatrix} = \det\begin{bmatrix} -4 & 9 \\ 6 & 17 \end{bmatrix} = (-4)(17) - (6)(9) = -122$

Because $x = \frac{E}{D} = \frac{-183}{-122} = 1.5$ and $y = \frac{F}{D} = \frac{244}{-122} = -2$, the solution is $(1.5, -2)$.

Now Try Exercises 33, 35

Cramer's rule can be applied to systems that have any number of linear equations. Cramer's rule for three linear equations is discussed in Exercises 39–44.

16.3 Putting It All Together

CONCEPT	EXPLANATION
Determinant of a 2×2 Matrix	The determinant of a 2×2 matrix A is given by $$\det A = \det\begin{bmatrix} a & b \\ c & d \end{bmatrix} = ad - cb.$$
Determinant of a 3×3 Matrix	The determinant of a 3×3 matrix A is given by $$\det A = \det\begin{bmatrix} a_1 & b_1 & c_1 \\ a_2 & b_2 & c_2 \\ a_3 & b_3 & c_3 \end{bmatrix}$$ $$= a_1 \cdot \det\begin{bmatrix} b_2 & c_2 \\ b_3 & c_3 \end{bmatrix} - a_2 \cdot \det\begin{bmatrix} b_1 & c_1 \\ b_3 & c_3 \end{bmatrix}$$ $$+ a_3 \cdot \det\begin{bmatrix} b_1 & c_1 \\ b_2 & c_2 \end{bmatrix}.$$

continued on next page

CONCEPT	EXPLANATION
Area of a Triangle	If a triangle has vertices (a_1, a_2), (b_1, b_2), and (c_1, c_2), its area equals the absolute value of D, where $$D = \frac{1}{2} \det \begin{bmatrix} a_1 & b_1 & c_1 \\ a_2 & b_2 & c_2 \\ 1 & 1 & 1 \end{bmatrix}.$$
Cramer's Rule for Linear Systems in Two Variables	The solution to the linear system $$a_1 x + b_1 y = c_1$$ $$a_2 x + b_2 y = c_2$$ is given by $x = \frac{E}{D}$ and $y = \frac{F}{D}$ where $$E = \det \begin{bmatrix} c_1 & b_1 \\ c_2 & b_2 \end{bmatrix}, \quad F = \det \begin{bmatrix} a_1 & c_1 \\ a_2 & c_2 \end{bmatrix}, \quad \text{and}$$ $$D = \det \begin{bmatrix} a_1 & b_1 \\ a_2 & b_2 \end{bmatrix} \neq 0.$$ **NOTE:** If $D = 0$, then the system has either no solutions or infinitely many solutions. ■

16.3 Exercises

CONCEPTS AND VOCABULARY

1. We can find the determinant of a(n) _____ matrix.

2. If we find a determinant, the answer is a(n) _____.

3. Cramer's rule can be used to solve a(n) _____.

4. If the first column of a matrix is all 0s, then its determinant equals _____.

CALCULATING DETERMINANTS

Exercises 5–20: Evaluate det A by hand where A is the given matrix.

5. $\begin{bmatrix} 1 & -2 \\ 3 & -8 \end{bmatrix}$

6. $\begin{bmatrix} 5 & -1 \\ 3 & 7 \end{bmatrix}$

7. $\begin{bmatrix} -3 & 7 \\ 8 & -1 \end{bmatrix}$

8. $\begin{bmatrix} 0 & -7 \\ -3 & 1 \end{bmatrix}$

9. $\begin{bmatrix} 23 & 4 \\ 6 & -13 \end{bmatrix}$

10. $\begin{bmatrix} 44 & -51 \\ -9 & 32 \end{bmatrix}$

11. $\begin{bmatrix} 1 & -1 & 2 \\ 0 & 1 & -3 \\ 0 & -4 & 7 \end{bmatrix}$

12. $\begin{bmatrix} 2 & -1 & -5 \\ -1 & 4 & -2 \\ 0 & 1 & 4 \end{bmatrix}$

13. $\begin{bmatrix} 2 & -1 & 0 \\ 1 & -2 & 6 \\ 0 & 1 & 8 \end{bmatrix}$

14. $\begin{bmatrix} 0 & 1 & -4 \\ 3 & -6 & 10 \\ 4 & -2 & 7 \end{bmatrix}$

15. $\begin{bmatrix} -1 & 3 & 5 \\ 3 & -3 & 5 \\ 2 & -3 & 7 \end{bmatrix}$

16. $\begin{bmatrix} 6 & -1 & 9 \\ 7 & 0 & -3 \\ 2 & 5 & -1 \end{bmatrix}$

17. $\begin{bmatrix} 5 & 0 & 0 \\ 0 & -2 & 0 \\ 0 & 0 & 5 \end{bmatrix}$

18. $\begin{bmatrix} 1 & 2 & 3 \\ 2 & 4 & 6 \\ 3 & 6 & 9 \end{bmatrix}$

19. $\begin{bmatrix} 0 & 2 & -3 \\ 0 & 3 & -9 \\ 0 & 5 & 9 \end{bmatrix}$

20. $\begin{bmatrix} 3 & -1 & 2 \\ 0 & 5 & 7 \\ 0 & 0 & -1 \end{bmatrix}$

Exercises 21–24: Use technology to calculate det A, where A is the given matrix.

21. $\begin{bmatrix} 2 & -5 & 13 \\ 10 & 15 & -10 \\ 17 & -19 & 22 \end{bmatrix}$

22. $\begin{bmatrix} 1.6 & 3.1 & 5.7 \\ 2.1 & 6.7 & 8.1 \\ -0.4 & -0.8 & -3.1 \end{bmatrix}$

23. $\begin{bmatrix} 17 & 0 & 4 \\ -9 & 14 & 1.5 \\ 13 & 67 & -11 \end{bmatrix}$

24. $\begin{bmatrix} 121 & 45 & -56 \\ -45 & 87 & 32 \\ -14 & -34 & 67 \end{bmatrix}$

Exercises 25 and 26: **Thinking Generally** *Find det A, if a, b, and c are nonzero constants.*

25. $A = \begin{bmatrix} a & 0 & 0 \\ 0 & b & 0 \\ 0 & 0 & c \end{bmatrix}$ **26.** $A = \begin{bmatrix} 0 & 0 & 0 \\ a & b & 0 \\ 0 & 0 & c \end{bmatrix}$

FINDING THE AREA OF A REGION

Exercises 27–32: Find the area of the figure by using a determinant. Assume that units are feet.

27.
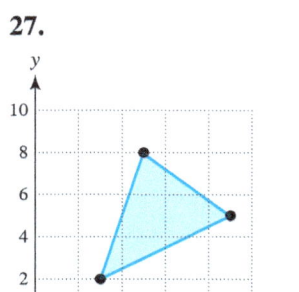

28.
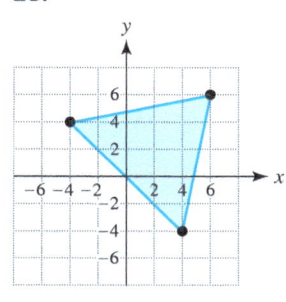

29.
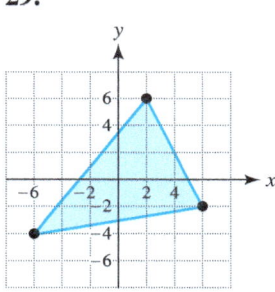

30.
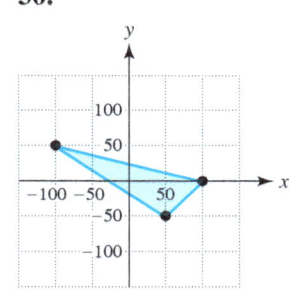

31.
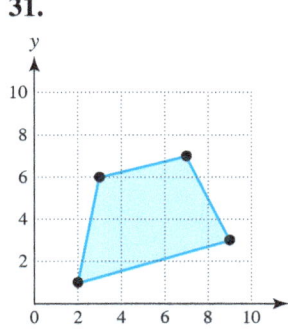

32.
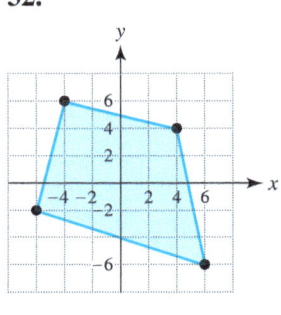

APPLYING CRAMER'S RULE IN TWO VARIABLES

Exercises 33–38: Solve the system of equations by using Cramer's rule.

33. $5x + 3y = 4$
$6x - 4y = 20$

34. $-5x + 4y = -5$
$4x + 4y = -32$

35. $7x - 5y = -3$
$-4x + 6y = -8$

36. $-4x - 9y = -17$
$8x + 4y = 9$

37. $8x = 3y - 61$
$-x = 4y - 23$

38. $15y = -188 - 22x$
$23y = -173 - 16x$

APPLYING CRAMER'S RULE IN THREE VARIABLES

Exercises 39–44: Cramer's rule can be applied to systems of three equations in three variables. For the system of equations

$$a_1 x + b_1 y + c_1 z = d_1$$
$$a_2 x + b_2 y + c_2 z = d_2$$
$$a_3 x + b_3 y + c_3 z = d_3,$$

the solution can be written as follows.

$$D = \det \begin{bmatrix} a_1 & b_1 & c_1 \\ a_2 & b_2 & c_2 \\ a_3 & b_3 & c_3 \end{bmatrix}, \quad E = \det \begin{bmatrix} d_1 & b_1 & c_1 \\ d_2 & b_2 & c_2 \\ d_3 & b_3 & c_3 \end{bmatrix}$$

$$F = \det \begin{bmatrix} a_1 & d_1 & c_1 \\ a_2 & d_2 & c_2 \\ a_3 & d_3 & c_3 \end{bmatrix}, \quad G = \det \begin{bmatrix} a_1 & b_1 & d_1 \\ a_2 & b_2 & d_2 \\ a_3 & b_3 & d_3 \end{bmatrix}$$

If $D \neq 0$, a unique solution exists and is given by

$$x = \frac{E}{D}, \; y = \frac{F}{D}, \; z = \frac{G}{D}.$$

Use Cramer's rule to solve the system of equations.

39. $x + y + z = 6$
$2x + y + 2z = 9$
$y + 3z = 9$

40. $y + z = 1$
$2x - y - z = -1$
$x + y - z = 3$

41. $x + z = 2$
$x + y = 0$
$y + 2z = 1$

42. $x + y + 2z = 1$
$-x - 2y - 3z = -2$
$y - 3z = 5$

43. $x + 2z = 7$
$-x + y + z = 5$
$2x - y + 2z = 6$

44. $x + 2y + 3z = -1$
$2x - 3y - z = 12$
$x + 4y - 2z = -12$

WRITING ABOUT MATHEMATICS

45. Suppose that one row of a 3 × 3 matrix *A* is all 0s. What is the value of det *A*? Give an example and explain your answer.

46. Explain how to evaluate a determinant of a 2 × 2 matrix. Can you find the determinant of a 2 × 3 matrix? Explain your answer.

SECTION 16.3 Checking Basic Concepts

1. Evaluate det A.

    (a) $A = \begin{bmatrix} -3 & 4 \\ -2 & 3 \end{bmatrix}$

    (b) $A = \begin{bmatrix} 1 & -2 & 3 \\ 5 & 1 & 1 \\ 0 & 2 & -1 \end{bmatrix}$

2. Use Cramer's rule to solve the system of linear equations.

    $2x - y = -14$
    $3x - 4y = -36$

3. Find the area of a triangle that has the vertices $(-1, 2)$, $(5, 6)$, and $(2, -3)$.

CHAPTER 16 Summary

SECTION 16.1 ■ SYSTEMS OF LINEAR EQUATIONS IN THREE VARIABLES

Solution to a System of Linear Equations in Three Variables An ordered triple (x, y, z) that satisfies all three equations

Example: $x - y + 2z = 3$
$2x - y + z = 5$
$x + y + z = 6$

The solution is $(3, 2, 1)$ because these values for (x, y, z) satisfy all three equations.

$3 - 2 + 2(1) = 3$ ✓ True
$2(3) - 2 + 1 = 5$ ✓ True
$3 + 2 + 1 = 6$ ✓ True

Elimination and Substitution Systems of linear equations in three variables can be solved by elimination and substitution, using the following steps.

STEP 1: Eliminate one variable, such as x, from two of the given equations.

STEP 2: Use the two resulting equations in two variables to eliminate one of the variables, such as y. Solve for the remaining variable z.

STEP 3: Substitute z in one of the two equations from Step 2. Solve for the unknown variable y.

STEP 4: Substitute values for y and z in one of the given equations. Then find x. The solution is the ordered triple (x, y, z).

SECTION 16.2 ■ MATRIX SOLUTIONS OF LINEAR SYSTEMS

Matrix A rectangular array of numbers is a matrix. If a matrix has m rows and n columns, it has dimension $m \times n$.

Example: Matrix $A = \begin{bmatrix} 3 & -1 & 7 \\ 0 & 6 & -2 \end{bmatrix}$ has dimension 2×3.

Augmented Matrix Any linear system can be represented with an augmented matrix.

Linear System
$4x - 3y = 5$
$x + 2y = 4$

Augmented Matrix
$\begin{bmatrix} 4 & -3 & | & 5 \\ 1 & 2 & | & 4 \end{bmatrix}$

Gauss–Jordan Elimination A numerical method that uses matrix row transformations to transform a matrix into reduced row–echelon form

Example: The matrix $\begin{bmatrix} 4 & -3 & | & 5 \\ 1 & 2 & | & 4 \end{bmatrix}$ reduces to $\begin{bmatrix} 1 & 0 & | & 2 \\ 0 & 1 & | & 1 \end{bmatrix}$.

The solution to the system is $(2, 1)$.

SECTION 16.3 ■ DETERMINANTS

Determinant for a 2 × 2 Matrix A determinant is a *real number*. The determinant of a 2 × 2 matrix is

$$\det A = \det \begin{bmatrix} a & b \\ c & d \end{bmatrix} = ad - cb.$$

Example: $\det \begin{bmatrix} 2 & 3 \\ 4 & 5 \end{bmatrix} = (2)(5) - (4)(3) = -2$

Determinant for a 3 × 3 Matrix

$$\det A = \det \begin{bmatrix} a_1 & b_1 & c_1 \\ a_2 & b_2 & c_2 \\ a_3 & b_3 & c_3 \end{bmatrix}$$

$$= a_1 \cdot \det \begin{bmatrix} b_2 & c_2 \\ b_3 & c_3 \end{bmatrix} - a_2 \cdot \det \begin{bmatrix} b_1 & c_1 \\ b_3 & c_3 \end{bmatrix} + a_3 \cdot \det \begin{bmatrix} b_1 & c_1 \\ b_2 & c_2 \end{bmatrix}$$

Example: $\det \begin{bmatrix} 2 & 3 & 2 \\ 3 & 7 & -3 \\ 0 & 0 & -1 \end{bmatrix} = 2 \det \begin{bmatrix} 7 & -3 \\ 0 & -1 \end{bmatrix} - 3 \det \begin{bmatrix} 3 & 2 \\ 0 & -1 \end{bmatrix} + 0 \det \begin{bmatrix} 3 & 2 \\ 7 & -3 \end{bmatrix}$

$$= 2(-7) - 3(-3) + 0(-23) = -5$$

Cramer's rule uses determinants to solve linear systems of equations. Determinants can also be used to find areas of triangles. See Putting It All Together for this section.

CHAPTER 16 Review Exercises

SECTION 16.1

1. Is $(3, -4, 5)$ a solution for $x + y + z = 4$?

2. Decide whether either ordered triple is a solution: $(1, -1, 2)$ or $(1, 0, 5)$.
$$2x - 3y + z = 7$$
$$-x - y + 3z = 6$$
$$3x - 2y + z = 7$$

Exercises 3–8: Use elimination and substitution to solve the system of linear equations, if possible.

3. $x - y - 2z = -11$
$-x + 2y + 3z = 16$
$3z = 6$

4. $x + y = 4$
$-2x + y + 3z = -2$
$x - 2y + 5z = -26$

5. $2x - y = -5$
$x + 2y + z = 7$
$-2x + y + z = 7$

6. $2x + 3y + z = 6$
$-x + 2y + 2z = 3$
$x + y + 2z = 4$

7. $x - y + 3z = 2$
$2x + y + 4z = 3$
$x + 2y + z = 5$

8. $x - y + 3z = 3$
$x + y - z = 1$
$x + z = 2$

SECTION 16.2

Exercises 9–12: Write the system of linear equations as an augmented matrix. Then use Gauss–Jordan elimination to solve the system, writing the solution as an ordered triple. Check your solution.

9. $x + y + z = -6$
$x + 2y + z = -8$
$y + z = -5$

10. $x + y + z = -3$
$-x + y = 5$
$y + z = -1$

11. $x + 2y - z = 1$
$-x + y - 2z = 5$
$2y + z = 10$

12. $2x + 2y - 2z = -14$
$-2x - 3y + 2z = 12$
$x + y - 4z = -22$

Exercises 13 and 14: Use a graphing calculator to solve the system of linear equations.

13. $3x - 2y + 6z = -17$
$-2x - y + 5z = 20$
$4y + 7z = 30$

14. $19x - 13y - 7z = 7.4$
$22x + 33y - 8z = 110.5$
$10x - 56y + 9z = 23.7$

SECTION 16.3

Exercises 15–18: Evaluate det A, if A is the given matrix.

15. $\begin{bmatrix} 6 & -5 \\ -4 & 2 \end{bmatrix}$

16. $\begin{bmatrix} 0 & -6 \\ 5 & 9 \end{bmatrix}$

17. $\begin{bmatrix} 3 & -5 & -3 \\ 1 & 4 & 7 \\ 0 & -3 & 1 \end{bmatrix}$

18. $\begin{bmatrix} -2 & -1 & -7 \\ 2 & 1 & -3 \\ 3 & -5 & 8 \end{bmatrix}$

Exercises 19 and 20: Use technology to calculate det A, where A is the given matrix.

19. $\begin{bmatrix} 22 & -45 & 3 \\ 15 & -12 & -93 \\ 5 & 81 & -21 \end{bmatrix}$

20. $\begin{bmatrix} 0.5 & -7.3 & 9.6 \\ 0.1 & 3.1 & 9.2 \\ -0.5 & -1.9 & 5.4 \end{bmatrix}$

Exercises 21–22: Use a determinant to find the area of the triangle. Assume that the units are feet.

21.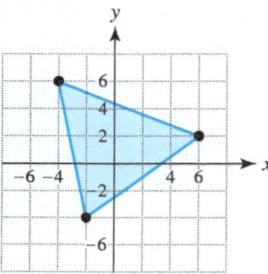

22.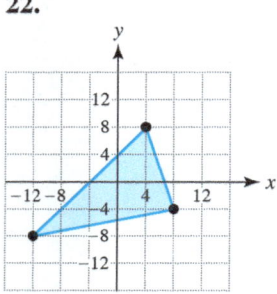

Exercises 23 and 24: Use Cramer's rule to solve the system.

23. $7x + 6y = 8$
$5x - 8y = 18$

24. $-2x + 5y = 25$
$3x + 4y = -3$

APPLICATIONS

25. **Pedestrian Fatalities** Forty-seven percent of pedestrian fatalities occur on Friday and Saturday nights. The combined total of pedestrian fatalities in 1994 and 2004 was 10,130. There were 848 more fatalities in 1994 than in 2004. Find the number of pedestrian fatalities during each year. (*Source:* National Highway Traffic Safety Administration.)

26. **Tickets** Tickets for a football game cost $8 and $12. If 480 tickets were sold for total receipts of $4620, how many of each type of ticket were sold?

27. **Determining Costs** The accompanying table shows the costs for purchasing different combinations of malts, cones, and ice cream bars.

Malts	Cones	Bars	Total Cost
1	3	5	$14
1	2	4	$11
0	1	3	$5

(a) Let m be the cost of a malt, c the cost of a cone, and b the cost of an ice cream bar. Write a system of equations that represents the data in the table.
(b) Solve this system.

28. **Geometry** The largest angle in a triangle is 20° more than the sum of the two smaller angles. The measure of the largest angle is 85° more than the smallest angle. Find the measure of each angle in the triangle.

29. **Mixture Problem** Three types of candy that cost $1.50, $2.00, and $2.50 per pound are to be mixed to produce 12 pounds of candy worth $26.00. If there are to be 2 pounds more of the $2.50 candy than the $2.00 candy, how much of each type of candy should be used in the mixture?

30. **Estimating the Chest Size of a Bear** The accompanying table shows the chest size C, weight W, and overall length L of three bears. These data can be modeled with the formula $C = a + bW + cL$. (*Sources:* M. Triola, Elementary Statistics; Minitab, Inc.)

C (inches)	W (pounds)	L (inches)
40	202	63
50	365	70
55	446	77
?	300	68

(a) Set up a system of linear equations whose solution gives values for the constants a, b, and c.

(b) Solve this system. Round each value to the nearest thousandth.

(c) Predict the chest size of a bear weighing 300 pounds and having a length of 68 inches.

CHAPTER 16 Test MyMathLab®, YouTube

1. Can a system of three equations and three variables have exactly three solutions? Explain.

2. Determine which ordered triple is a solution to the linear system.

$$(-4, 3, 3), (1, -2, -2)$$
$$x - 3y + 4z = -1$$
$$2x + y - 3z = 6$$
$$x - y + z = 1$$

Exercises 3–6: Use substitution and elimination to solve the system, if possible.

3. $x + 3y = 2$
$-2x + y + z = 5$
$ y + z = -3$

4. $x + y - z = 1$
$2x - 3y + z = 0$
$x - 4y + 2z = 2$

5. $x - y + 2z = 5$
$-x + y - 3z = -8$
$x - 2y + 2z = 3$

6. $x - y + z = 1$
$x + y - z = 1$
$3x + y - z = 3$

7. Consider the system of linear equations.

$$2x - 4y = -10$$
$$-3x - 2y = 7$$

(a) Write the system of linear equations as an augmented matrix.
(b) Use Gauss–Jordan elimination to solve the system, writing the solution as an ordered pair.

8. Consider the system of linear equations.

$$x + y + z = 2$$
$$x - y - z = 3$$
$$2x + 2y + z = 6$$

(a) Write the system as an augmented matrix.

(b) Use Gauss–Jordan elimination to solve the system, writing the solution as an ordered triple.

9. Evaluate det A if $A = \begin{bmatrix} -1 & 2 \\ -5 & 4 \end{bmatrix}$.

10. Evaluate det A if $A = \begin{bmatrix} 3 & 2 & -1 \\ 6 & 2 & -6 \\ 0 & 8 & -3 \end{bmatrix}$.

11. Solve the system of equations using Cramer's rule.

$$5x - 3y = 7$$
$$-4x + 2y = 11$$

12. **Jogging Speeds** A runner in preparation for a race jogs at 6, 7, and 9 miles per hour and travels a total of 7.1 miles in 1 hour. The runner jogs 12 minutes longer at 6 miles per hour than at 9 miles per hour. How many minutes did the runner jog at each speed?

13. **Geometry** The largest angle in a triangle is 50° more than the smallest angle. The sum of the measures of the smaller two angles is 10° more than the largest angle. Find the measure of each angle.

14. **Weight of a Bear** The data in the accompanying table list the weight W, neck size N, and chest size C of three bears. These data can be modeled by $W = a + bN + cC$.

W (pounds)	N (inches)	C (inches)
168	20	25
270	24	40
405	30	50

(a) Set up a system of equations whose solution gives values for the constants a, b, and c.
(b) Solve this system.
(c) Predict the weight W of a bear with neck size $N = 26$ and chest size $C = 44$.

CHAPTERS 1–16 Cumulative Review Exercises

1. Write 360 as a product of prime numbers.

2. Translate the sentence "Double a number increased by 7 equals the number decreased by 2" into an equation by using the variable n. Then solve the equation.

Exercises 3 and 4: Simplify to lowest terms.

3. $\frac{2}{3} + \frac{4}{7} \cdot \frac{21}{28}$

4. $\frac{3}{5} \div \frac{6}{5} - \frac{2}{3}$

Exercises 5 and 6: Evaluate by hand.

5. $30 - 4 \div 2 \cdot 6$

6. $\frac{3^2 - 2^3}{20 - 5 \cdot 2}$

7. Solve $2(x + 1) - 6x = x - 4$.

8. Solve $4 - 3x = -2$ graphically. Check your answer.

9. Convert 124% to fraction and decimal notation.

10. If $A = 30$ square miles and $h = 10$ miles, use the formula $A = \frac{1}{2}bh$ to find b.

11. Solve $6t - 1 < 3 - t$.

12. Make a scatterplot with the points $(-1, 2)$, $(1, -2)$, $(0, 3)$, $(-2, 0)$, and $(2, 3)$.

Exercises 13 and 14: Graph the equation and determine any intercepts.

13. $-3x + 4y = 12$

14. $x = -2$

15. Use the graph to identify the x-intercept and the y-intercept. Then write the slope–intercept form of the line.

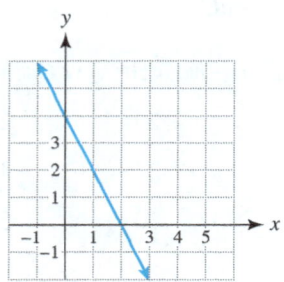

16. Write the slope–intercept form of a line that passes through the point $(-1, 3)$ with slope $m = -2$.

Exercises 17 and 18: Find the slope–intercept form of the line satisfying the given conditions.

17. Passing through the points $(-3, 5)$ and $(2, 8)$

18. Perpendicular to $x + 2y = 5$, passing through the point $(-1, 1)$

19. An insect population P is initially 4000 and increases by 500 insects per day. Write an equation in slope–intercept form that calculates P after x days.

20. Solve the system of equations. Write your answer as an ordered pair.

$$-2a + b = -5$$
$$4a - 3b = 0$$

Exercises 21 and 22: Shade the solution set.

21. $y \geq -1$

22. $x - y < 3$
$2x + y \geq 6$

Exercises 23–28: Simplify and use positive exponents, when appropriate, to write the expression.

23. $5 - 3^4$

24. $(8t^{-3})(3t^2)(t^5)$

25. $\frac{2^{-4}}{4^{-2}}$

26. $(2t^3)^{-2}$

27. $(4a^2b^3)^2(2ab)^{-3}$

28. $\left(\frac{2a^{-1}}{ab^{-2}}\right)^{-3}$

Exercises 29–32: Multiply the expression.

29. $2a^2(a^2 - 2a + 3)$

30. $(5x + 1)(x - 7)$

31. $(2x + 3y)^2$

32. $(a + b)(a - b)$

Exercises 33 and 34: Divide.

33. $\frac{4x^3 - 8x^2 + 6x}{2x}$

34. $(x^4 - 9x^3 + 23x^2 - 17x + 11) \div (x - 5)$

Exercises 35–42: Factor completely.

35. $10ab^2 - 25a^3b^5$

36. $y^3 - 3y^2 + 2y - 6$

37. $6z^2 + 7z - 3$

38. $4z^2 - 9$

39. $4y^2 - 20y + 25$

40. $a^3 - 27$

41. $4z^4 - 17z^2 + 15$

42. $2a^3b + a^2b^2 - ab^3$

Exercises 43 and 44: Solve the equation.

43. $(x - 1)(x + 2) = 0$

44. $6y^2 - 7y = 3$

Exercises 45 and 46: Simplify to lowest terms.

45. $\dfrac{x^2 - 16}{x + 4}$

46. $\dfrac{2x^2 - 11x - 6}{6x^2 - 5x - 4}$

Exercises 47 and 48: Simplify and write in lowest terms.

47. $\dfrac{x^2 - 3x + 2}{x + 7} \div \dfrac{x - 2}{2x + 14}$

48. $\dfrac{x}{2x + 3} + \dfrac{x + 3}{2x + 3}$

Exercises 49 and 50: Solve the equation.

49. $\dfrac{x + 2}{5} = \dfrac{x}{4}$

50. $\dfrac{1}{3x} + \dfrac{5}{2x} = 2$

51. Suppose that y is inversely proportional to x and that $y = 25$ when $x = 4$. Find y for $x = 10$.

52. Evaluate $f(x) = 1 - 4x$ for $x = -3$.

53. Graph $f(x) = x^2 - 2$ and identify the domain and range of f. Use interval notation.

54. Use the graph to evaluate $f(0)$ and $f(-2)$.

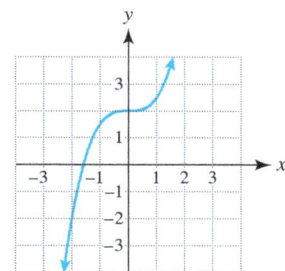

Exercises 55 and 56: Solve the inequality symbolically and write the solution set in interval notation.

55. $\dfrac{4x - 9}{6} > \dfrac{1}{2}$

56. $\dfrac{2}{3}z - 2 \le \dfrac{1}{4}z - (2z + 2)$

Exercises 57 and 58: Solve the compound inequality and graph the solution set on a number line.

57. $x + 2 > 1$ and $2x - 1 \le 9$

58. $4x + 7 < 1$ or $3x + 2 \ge 11$

Exercises 59 and 60: Solve the three-part inequality and write the solution set in interval notation.

59. $-7 \le 2x - 3 \le 5$

60. $-8 \le -\frac{1}{2}x - 3 \le 5$

61. Use the graph to solve the equation and inequalities.
(a) $y_1 = 2$
(b) $y_1 \le 2$
(c) $y_1 \ge 2$

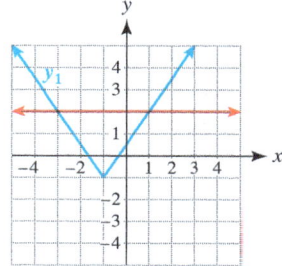

62. Solve the absolute value equation $\left|\frac{2}{3}x - 4\right| = 8$.

Exercises 63 and 64: Solve the absolute value inequality. Write the solution set in interval notation.

63. $|3x + 5| > 13$

64. $-3|2t - 11| \ge -9$

65. Use elimination and substitution to solve the system of linear equations.

$$2x + 3y - z = 3$$
$$3x - y + 4z = 10$$
$$2x + y - 2z = -1$$

66. Write the system of linear equations as an augmented matrix. Then use Gauss–Jordan elimination to solve the system. Write the solution as an ordered triple.

$$x + y - z = 4$$
$$-x - y - z = 0$$
$$x - 2y + z = -9$$

APPLICATIONS

67. Graphical Model An athlete rides a bicycle on a straight road away from home at 20 miles per hour for 1.5 hours. The athlete then turns around and rides home at 15 miles per hour. Sketch a graph that depicts the athlete's distance d from home after t hours.

68. Ticket Prices The price of admission to a baseball game is $15 for children and $25 for adults. If a group of 8 people pays $170 to enter the game, find the number of children and the number of adults in the group.

69. Working Together One person can shovel the snow from a sidewalk in 2 hours and another person can shovel the same sidewalk in 1.5 hours. How long will it take for them to shovel the snow from the sidewalk if they work together?

70. Flight of a Golf Ball If a golf ball is hit upward with a velocity of 88 feet per second (60 miles per hour), then its height h in feet after t seconds can be approximated by
$$h(t) = 88t - 16t^2.$$
(a) What is the height of the golf ball after 2 seconds?
(b) After how long does the golf ball strike the ground?

71. Height of a Building A 7-foot-tall stop sign casts a 4-foot-long shadow, while a nearby building casts a 35-foot-long shadow. Find the height of the building.

72. Determining Costs The accompanying table shows the costs for purchasing different combinations of burgers, fries, and malts.

Burgers	Fries	Malts	Total Cost
4	3	4	$23
1	2	1	$7
3	1	2	$13

(a) Let b be the cost of a burger, f be the cost of an order of fries, and m be the cost of a malt. Write a system of three linear equations that represents the data in the table.
(b) Solve this system and find the price of each item.

17 Radical Expressions and Functions

- 17.1 Radical Expressions and Functions
- 17.2 Rational Exponents
- 17.3 Simplifying Radical Expressions
- 17.4 Operations on Radical Expressions
- 17.5 More Radical Functions
- 17.6 Equations Involving Radical Expressions
- 17.7 Complex Numbers

Wind power is rapidly becoming a larger portion of America's total energy production. Recently, the United States had a wind capacity that was able to produce 61,000 megawatts of energy, which was enough to power about 15.5 million homes. At that time, there were over 500 new wind turbines installed and their placement was critical in maximizing their efficiency. For example, suppose location A has an average wind velocity that is twice the velocity of location B. Does this mean that a wind turbine at location A will be able to produce twice the wind power as a wind turbine at location B? To better understand the mathematics behind wind energy, we need to learn more about radical expressions and functions. (See Section 6, Example 11, of this chapter.) *Source:* American Wind Energy Association.

17.1 Radical Expressions and Functions

Objectives

1. Introducing Radical Expressions
   - Square Roots
   - Square Roots of Negative Numbers
   - Cube Roots
   - *n*th Roots
   - Absolute Value
2. Recognizing Square Root Functions and Their Properties
3. Recognizing Cube Root Functions and Their Properties

NEW VOCABULARY

☐ Radical sign
☐ Radicand
☐ Radical expression
☐ *n*th root
☐ Index
☐ Odd root
☐ Even root
☐ Principal *n*th root
☐ Square root function
☐ Cube root function

1 Introducing Radical Expressions

SQUARE ROOTS Recall the definition of the square root of a number a.

> **SQUARE ROOT**
>
> The number b is a *square root* of a if $b^2 = a$.

For example, since
$$6^2 = 36 \quad \text{and} \quad (-6)^2 = 36,$$
we know that both 6 and -6 are square roots of 36. From this discussion, we can see that every positive number has two square roots: one positive and one negative.

EXAMPLE 1 **Finding square roots**

Find the square roots of 100.

Solution
The square roots of 100 are 10 and -10, because $10^2 = 100$ and $(-10)^2 = 100$.

Now Try Exercise 1

Recall that the *positive square root* of a positive number a is called the *principal square root* and is denoted $\sqrt{a}$. The *negative square root* is denoted $-\sqrt{a}$. To identify both square roots, we write $\pm \sqrt{a}$. The symbol $\pm$ is read "plus or minus." The symbol $\sqrt{}$ is called the **radical sign**. The expression under the radical sign is called the **radicand**, and an expression containing a radical sign is called a **radical expression**.

Radical Expression

Radical sign $\sqrt{6}$ Radicand

EXAMPLE 2 **Finding principal square roots**

Evaluate each square root.

(a) $\sqrt{25}$ (b) $\sqrt{0.49}$ (c) $\sqrt{\frac{4}{9}}$ (d) $\sqrt{c^2}, c > 0$

Solution
(a) Because $5 \cdot 5 = 25$, the principal, or *positive*, square root of 25 is $\sqrt{25} = 5$.
(b) Because $(0.7)(0.7) = 0.49$, the principal square root of 0.49 is $\sqrt{0.49} = 0.7$.
(c) Because $\frac{2}{3} \cdot \frac{2}{3} = \frac{4}{9}$, the principal square root of $\frac{4}{9}$ is $\sqrt{\frac{4}{9}} = \frac{2}{3}$.
(d) The principal square root of c^2 is $\sqrt{c^2} = c$, as c is positive.

Now Try Exercises 15, 17, 19, 21

The square roots of many real numbers, such as $\sqrt{17}$, $\sqrt{1.2}$, and $\sqrt{\frac{5}{7}}$, cannot be conveniently evaluated (or approximated) by hand. In these cases, we sometimes use a calculator to give a decimal *approximation*, as demonstrated in the next example.

EXAMPLE 3 Approximating a square root

Approximate $\sqrt{17}$ to the nearest thousandth.

Solution
FIGURE 17.1 shows that $\sqrt{17} \approx 4.123$, rounded to the nearest thousandth. This result means that

$$4.123 \times 4.123 \approx 17.$$

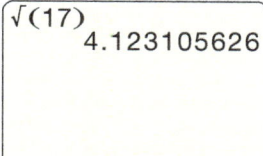

FIGURE 17.1

Now Try Exercise 39

READING CHECK 1

Are square roots of negative numbers real numbers?

SQUARE ROOTS OF NEGATIVE NUMBERS The square root of a *negative* number is *not* a real number. For example, $\sqrt{-4} \neq 2$ because $2 \cdot 2 \neq -4$ and $\sqrt{-4} \neq -2$ because $(-2)(-2) \neq -4$. (Later in this chapter, we will use the complex numbers to identify square roots of negative numbers.)

 Connecting Concepts with Your Life Have you ever noticed that if you climb up a hill or a tower, you can see farther to the horizon? This phenomenon can be described by a formula containing a square root, as demonstrated in the next example.

EXAMPLE 4 Seeing the horizon

A formula for calculating the distance d in miles that one can see to the horizon on a clear day is approximated by $d = 1.22\sqrt{x}$, where x is the elevation, in feet, of a person.
(a) Approximate how far a 6-foot-tall person can see to the horizon.
(b) Approximate how far a person can see from an 8000-foot mountain.
(c) If a person's elevation doubles, can this person see twice as far to the horizon?

Solution
(a) Let $x = 6$ in the formula $d = 1.22\sqrt{x}$.

$$d = 1.22\sqrt{6} \approx 3, \text{ or about 3 miles.}$$

(b) If $x = 8000$, then $d = 1.22\sqrt{8000} \approx 109$, or about 109 miles.
(c) No, this person can see $\sqrt{2} \approx 1.4$ times as far, which is less than twice as far.

Now Try Exercise 107

CUBE ROOTS The cube root of a number a is denoted  $\sqrt[3]{a}$.

CALCULATOR HELP

To calculate a cube root, see Appendix A (page AP-1).

CUBE ROOT

The number b is a *cube root* of a if $b^3 = a$.

Although the square root of a negative number is *not* a real number, the cube root of a negative number is a negative real number. *Every real number has one real cube root.*

EXAMPLE 5 Finding cube roots

Evaluate the cube root. Approximate your answer to the nearest hundredth when appropriate.

(a) $\sqrt[3]{8}$ (b) $\sqrt[3]{-27}$ (c) $\sqrt[3]{\frac{1}{64}}$ (d) $\sqrt[3]{d^6}$ (e) $\sqrt[3]{16}$

Solution
(a) $\sqrt[3]{8} = 2$ because $2^3 = 2 \cdot 2 \cdot 2 = 8$.
(b) $\sqrt[3]{-27} = -3$ because $(-3)^3 = (-3)(-3)(-3) = -27$.
(c) $\sqrt[3]{\frac{1}{64}} = \frac{1}{4}$ because $\left(\frac{1}{4}\right)^3 = \frac{1}{4} \cdot \frac{1}{4} \cdot \frac{1}{4} = \frac{1}{64}$.
(d) $\sqrt[3]{d^6} = d^2$ because $(d^2)^3 = d^2 \cdot d^2 \cdot d^2 = d^{2+2+2} = d^6$.
(e) $\sqrt[3]{16}$ is not an integer. **FIGURE 17.2** shows that $\sqrt[3]{16} \approx 2.52$.

Now Try Exercises 23, 25, 27, 31, 41

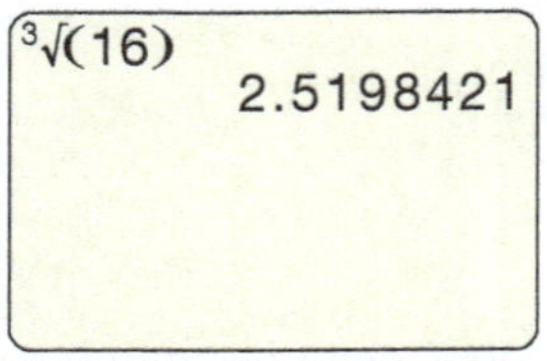

```
³√(16)
         2.5198421
```

FIGURE 17.2

NOTE: $\sqrt[3]{-b} = -\sqrt[3]{b}$ for any real number b. That is, the cube root of a negative number is the negative of the cube root of the number. For example, $\sqrt[3]{-8} = -\sqrt[3]{8} = -2$. ∎

TABLE 17.1 illustrates how to evaluate both square roots and cube roots. If the radical expression is undefined, a dash is used.

Evaluating Radical Expressions

Expression	$\sqrt{64}$	$-\sqrt{64}$	$\sqrt{-64}$	$\sqrt[3]{64}$	$-\sqrt[3]{64}$	$\sqrt[3]{-64}$
Evaluated	8	−8	— (Undefined)	4	−4	−4

TABLE 17.1

nth ROOTS We can generalize square roots and cube roots to include *n*th roots of a number *a*. The number *b* is an **nth root** of *a* if $b^n = a$, where *n* is a positive integer. For example, $2^5 = 32$, so the **5**th root of **32** is **2** and can be written as $\sqrt[5]{32} = 2$.

> ### THE NOTATION $\sqrt[n]{a}$
>
> The equation $\sqrt[n]{a} = b$ means that $b^n = a$, where *n* is a natural number called the **index**. If *n* is odd, we are finding an **odd root** and if *n* is even, we are finding an **even root**.
>
> 1. If $a > 0$, then $\sqrt[n]{a}$ is a positive number. $\sqrt[4]{16} = 2$ and is positive.
> 2. If $a < 0$ and
>   (a) *n* is odd, then $\sqrt[n]{a}$ is a negative number. $\sqrt[3]{-8} = -2$ and is negative.
>   (b) *n* is even, then $\sqrt[n]{a}$ is *not* a real number. $\sqrt[4]{-8}$ is not a real number.

If $a > 0$ and *n* is even, then *a* has two real *n*th roots: one positive and one negative. The positive root is denoted $\sqrt[n]{a}$ and called the **principal nth root** of *a*. For example, $(-3)^4 = 81$ *and* $3^4 = 81$, but $\sqrt[4]{81} = 3$ in the same way *principal square roots* are calculated.

EXAMPLE 6 Finding nth roots

Find each root, if possible.
(a) $\sqrt[4]{16}$ (b) $\sqrt[5]{-32}$ (c) $\sqrt[4]{-81}$ (d) $-\sqrt[4]{81}$

Solution
(a) $\sqrt[4]{16} = 2$ because $2^4 = 2 \cdot 2 \cdot 2 \cdot 2 = 16$.
(b) $\sqrt[5]{-32} = -2$ because $(-2)^5 = (-2)(-2)(-2)(-2)(-2) = -32$.
(c) An *even* root of a *negative* number is *not* a real number.
(d) $-\sqrt[4]{81} = -3$ because $\sqrt[4]{81} = 3$.

Now Try Exercises 33, 35, 37

ABSOLUTE VALUE Consider the calculations

$$\sqrt{3^2} = \sqrt{9} = 3, \quad \sqrt{(-4)^2} = \sqrt{16} = 4, \quad \text{and} \quad \sqrt{(-6)^2} = \sqrt{36} = 6.$$

In general, the expression $\sqrt{x^2}$ equals $|x|$. See **FIGURE 17.3**.

CRITICAL THINKING 1

Evaluate $\sqrt[6]{(-2)^6}$ and $\sqrt[3]{(-2)^3}$.

Identical Graphs

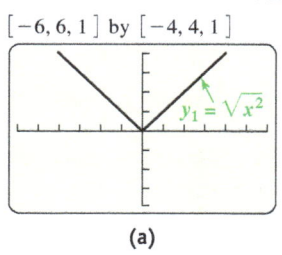
(a)

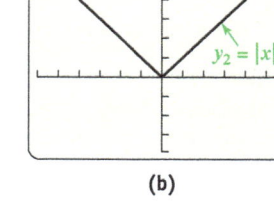
(b)

The graphs of $y_1 = \sqrt{x^2}$ and $y_2 = |x|$ are identical.

FIGURE 17.3

THE EXPRESSION $\sqrt{x^2}$

For every real number x, $\sqrt{x^2} = |x|$.

EXAMPLE 7 Simplifying expressions

Write each expression in terms of an absolute value.
(a) $\sqrt{(-3)^2}$ (b) $\sqrt{(x+1)^2}$ (c) $\sqrt{z^2 - 4z + 4}$

Solution
(a) $\sqrt{x^2} = |x|$, so $\sqrt{(-3)^2} = |-3|$
(b) $\sqrt{(x+1)^2} = |x+1|$
(c) $\sqrt{z^2 - 4z + 4} = \sqrt{(z-2)^2} = |z-2|$

Now Try Exercises 45, 49, 51

2 Recognizing Square Root Functions and Their Properties

The important properties of the **square root function** are shown in the following See the Concept.

> **See the Concept 1**   **THE SQUARE ROOT FUNCTION**
>
> **Formula**
>
> $f(x) = \sqrt{x},\ x \geq 0$
>
> **Table of Values**
>
> | x | $\sqrt{x}$ |
> |---|---|
> | 0 | 0 |
> | 1 | 1 |
> | 4 | 2 |
>
> **Graph**
>
>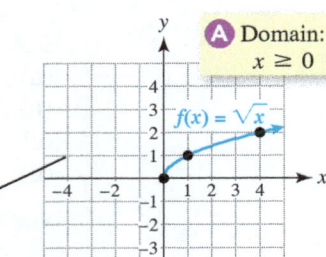
>
> **A** Domain: $x \geq 0$
>
> **B** No graph left of origin
>
> **A** The domain of the square root function is $[0, \infty)$.
>
> **B** $f(x) = \sqrt{x}$ is defined for only nonnegative real numbers, so its graph does not appear to the left of the origin.
>
> WATCH VIDEO IN MML.

TECHNOLOGY NOTE

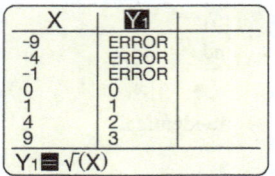

Square Roots of Negative Numbers
If a table of values for $y_1 = \sqrt{x}$ includes both negative and positive values for x, then many calculators give error messages when x is negative, as shown in the figure in the margin.

EXAMPLE 8   **Evaluating functions involving square roots**

If possible, evaluate $f(1)$ and $f(-2)$ for each $f(x)$.
(a) $f(x) = \sqrt{2x - 1}$ (b) $f(x) = \sqrt{4 - x^2}$

Solution
(a) $f(1) = \sqrt{2(1) - 1} = \sqrt{1} = 1$
$f(-2) = \sqrt{2(-2) - 1} = \sqrt{-5}$, which does not equal a real number.
(b) $f(1) = \sqrt{4 - (1)^2} = \sqrt{3}$
$f(-2) = \sqrt{4 - (-2)^2} = \sqrt{0} = 0$

Now Try Exercises 61, 63

🌐 **Math in Context** (Sports)   A good punter can kick a football so that the ball has a long *hang time*. Hang time is the length of time that the ball is in the air, and a long hang time gives the kicking team time to run down the field and stop the punt return. By using a function involving a square root, we can estimate hang time.

EXAMPLE 9   **Calculating hang time**

If a football is kicked x feet high, then the time T in seconds that the ball is in the air is given by the function

$$T(x) = \frac{1}{2}\sqrt{x}.$$

(a) Find the hang time if the ball is kicked 50 feet into the air.
(b) Does the hang time double if the ball is kicked 100 feet in the air?

CRITICAL THINKING 2

How high would a football have to be kicked to have twice the hang time of a football kicked 50 feet into the air?

Solution

(a) The hang time is $T(50) = \frac{1}{2}\sqrt{50} \approx 3.5$ seconds.

(b) The hang time is $T(100) = \frac{1}{2}\sqrt{100} = 5$ seconds. The time does *not* double when the height doubles.

Now Try Exercise 105

EXAMPLE 10 Finding the domain of a square root function

Let $f(x) = \sqrt{x - 1}$.
(a) Find the domain of f. Write your answer in interval notation.
(b) Graph $y = f(x)$ and compare it to the graph of $y = \sqrt{x}$.

Solution

(a) For $f(x)$ to be defined, $x - 1$ cannot be negative. Thus valid inputs for x must satisfy

$$x - 1 \geq 0 \quad \text{or} \quad x \geq 1.$$

To solve, add 1 to each side.

The domain is $[1, \infty)$.

(b) **TABLE 17.2** lists points that lie on the graph of $y = \sqrt{x - 1}$. Note in **FIGURE 17.4** that the graph appears only when $x \geq 1$. Compared to the graph of $y = \sqrt{x}$, the graph of $y = \sqrt{x - 1}$ is shifted one unit to the right.

$y = \sqrt{x - 1}$

x	$\sqrt{x - 1}$
1	0
2	1
5	2

TABLE 17.2

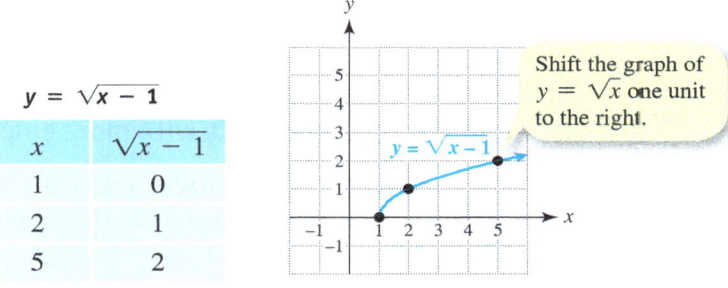

Shift the graph of $y = \sqrt{x}$ one unit to the right.

FIGURE 17.4

Now Try Exercises 75, 89

EXAMPLE 11 Finding the domains of square root functions

Find the domain of each function. Write your answer in interval notation.
(a) $f(x) = \sqrt{4 - 2x}$ (b) $g(x) = \sqrt{x^2 + 1}$

Solution

(a) To determine when $f(x) = \sqrt{4 - 2x}$ is defined, we must solve $4 - 2x \geq 0$.

The radicand must be nonnegative for f to be defined.

$4 - 2x \geq 0$	Inequality to be solved
$4 \geq 2x$	Add $2x$ to each side.
$2 \geq x$	Divide each side by 2.

The domain is $(-\infty, 2]$.

(b) Regardless of the value of x in $g(x) = \sqrt{x^2 + 1}$, the expression $x^2 + 1$ is always positive because $x^2 \geq 0$. Thus $g(x)$ is defined for all real numbers, and its domain is $(-\infty, \infty)$.

Now Try Exercises 83, 85

> **MAKING CONNECTIONS 1**
>
> **Domains of Functions and Their Graphs**
>
> The graphs of f and g from Example 11 are shown below and can be used to find their domains.
>
>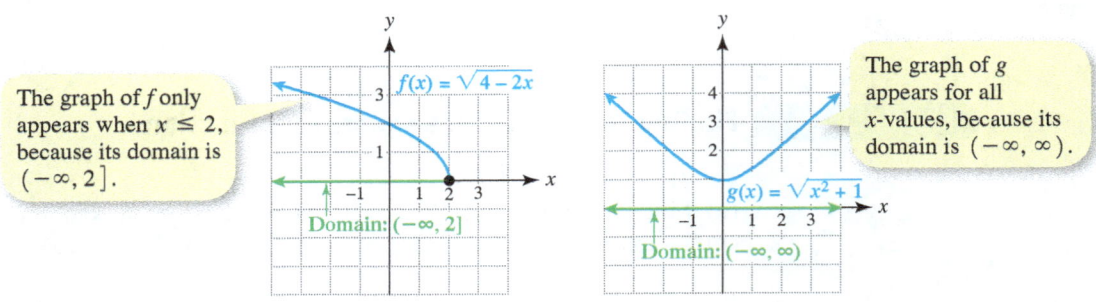
>
> The graph of f only appears when $x \leq 2$, because its domain is $(-\infty, 2]$.
>
> The graph of g appears for all x-values, because its domain is $(-\infty, \infty)$.

🌐 **Math in Context** *Business* Suppose you start a summer painting business, and initially you are the only employee. Because business is good, you hire another employee and productivity goes up, more than enough to pay for the new employee. As time goes on, you hire more employees. Eventually, productivity starts to level off because there are too many employees to keep busy. This situation is common in business and can be modeled by using a function involving a square root. In the next example, we analyze the benefit of adding employees to a small business.

EXAMPLE 12 **Increasing productivity with more employees**

The function $R(x) = 108\sqrt{x}$ gives the total revenue per year in thousands of dollars generated by a small business that has x employees. A graph of $y = R(x)$ is shown in **FIGURE 17.5**.

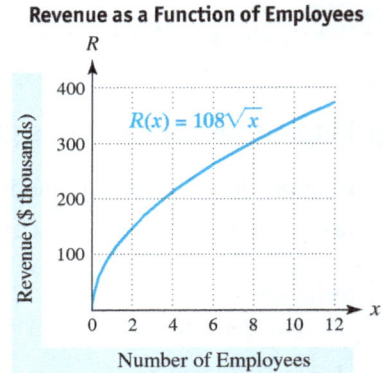

FIGURE 17.5

(a) Approximate values for $R(4)$, $R(8)$, and $R(12)$ by using both the formula and the graph.
(b) Evaluate $R(12) - R(8)$ and $R(8) - R(4)$ using the formula. Interpret your answer.

Solution
(a) $R(4) = 108\sqrt{4} = \$216$ thousand, $R(8) = 108\sqrt{8} \approx \$305$ thousand, and finally $R(12) = 108\sqrt{12} \approx \$374$ thousand.

We can graphically evaluate $R(4) \approx \$215$ thousand, $R(8) \approx \$305$ thousand, and $R(12) \approx \$375$ thousand, as shown in **FIGURE 17.6**.

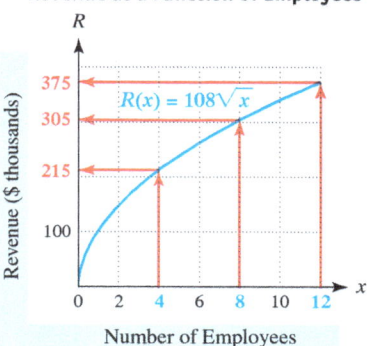

FIGURE 17.6

(b) $R(8) - R(4) \approx 305 - 216 = \$89$ thousand
$R(12) - R(8) \approx 374 - 305 = \$69$ thousand

There is more revenue gained when increasing from 4 to 8 employees than there is when increasing from 8 to 12 employees. Because the graph of $R(x) = 108\sqrt{x}$ starts to level off, there is a limited benefit to adding employees.

Now Try Exercise 109

3 Recognizing Cube Root Functions and Their Properties

The important properties of the cube root function are shown in the following See the Concept. Note that, unlike the square root function, the cube root function is defined for negative numbers.

See the Concept 2 — THE CUBE ROOT FUNCTION

Formula

$f(x) = \sqrt[3]{x}$

Table of Values

x	$\sqrt[3]{x}$
-8	-2
-1	-1
0	0
1	1
8	2

Graph

A Domain: $-\infty < x < \infty$

B Graph is both left and right of the origin

A The domain of the cube root function is $(-\infty, \infty)$.

B $f(x) = \sqrt[3]{x}$ is defined for all real numbers, so its graph appears both to the left and to the right of the origin.

WATCH VIDEO IN MML.

EXAMPLE 13   Evaluating functions involving cube roots

Evaluate $f(1)$ and $f(-3)$ for each $f(x)$.
(a) $f(x) = \sqrt[3]{x^2 - 1}$ (b) $f(x) = \sqrt[3]{2 - x^2}$

Solution
(a) $f(1) = \sqrt[3]{1^2 - 1} = \sqrt[3]{0} = 0$; $f(-3) = \sqrt[3]{(-3)^2 - 1} = \sqrt[3]{8} = 2$
(b) $f(1) = \sqrt[3]{2 - 1^2} = \sqrt[3]{1} = 1$; $f(-3) = \sqrt[3]{2 - (-3)^2} = \sqrt[3]{-7}$ or $-\sqrt[3]{7}$

Now Try Exercises 65, 69

17.1   Putting It All Together

CONCEPT	EXPLANATION	EXAMPLES
nth Root of a Real Number	An nth root of a real number a is b if $b^n = a$, and the (principal) nth root is denoted $\sqrt[n]{a}$. If $a < 0$ and n is even, $\sqrt[n]{a}$ is not a real number.	The square roots of 25 are 5 and -5. The principal square root is $\sqrt{25} = 5$. $\sqrt[3]{-125} = -5$ because $(-5)^3 = -125$. $\sqrt[4]{-9}$ is not a real number.
Square Root and Cube Root Functions	$f(x) = \sqrt{x}$ and $g(x) = \sqrt[3]{x}$ The cube root function g is defined for all inputs, whereas the square root function f is defined only for nonnegative inputs. **Square Root Function** $y = \sqrt{x}$, Domain: $[0, \infty)$ **Cube Root Function** $y = \sqrt[3]{x}$, Domain: $(-\infty, \infty)$	$f(64) = \sqrt{64} = 8$ $f(-64) = \sqrt{-64}$ is *not* a real number. $g(64) = \sqrt[3]{64} = 4$ $g(-64) = \sqrt[3]{-64} = -4$

17.1   Exercises

CONCEPTS AND VOCABULARY

1. What are the square roots of 9?
2. What is the principal square root of 9?
3. What is the cube root of 8?
4. Does every real number have a cube root?
5. If $b^n = a$ and $b > 0$, then $\sqrt[n]{a} = $ _____.
6. What is $\sqrt{x^2}$ equal to?
7. Evaluate $\sqrt{-25}$, if possible.
8. Evaluate $\sqrt[3]{-27}$, if possible.

9. Which of the following (a–d) equals -4?
   a. $\sqrt{16}$
   b. $\sqrt{-16}$
   c. $-\sqrt[3]{16}$
   d. $\sqrt[3]{-64}$

10. Which of the following (a–d) equals $|2x + 1|$?
    a. $\sqrt{(2x+1)^2}$
    b. $\sqrt[3]{(2x+1)^3}$
    c. $(\sqrt{2x+1})^2$
    d. $\sqrt{(2x)^2 + 1}$

11. Sketch a graph of the square root function.

12. Sketch a graph of the cube root function.

13. What is the domain of the square root function?

14. What is the domain of the cube root function?

EVALUATING RADICAL EXPRESSIONS

Exercises 15–38: Evaluate the expression by hand, if possible. Variables represent any real number.

15. $\sqrt{9}$
16. $\sqrt{121}$
17. $\sqrt{0.36}$
18. $\sqrt{0.64}$
19. $\sqrt{\frac{16}{25}}$
20. $\sqrt{\frac{9}{49}}$
21. $\sqrt{x^2}, x > 0$
22. $\sqrt{(x-1)^2}, x > 1$
23. $\sqrt[3]{27}$
24. $\sqrt[3]{64}$
25. $\sqrt[3]{-64}$
26. $-\sqrt[3]{-1}$
27. $\sqrt[3]{\frac{8}{27}}$
28. $\sqrt[3]{-\frac{1}{125}}$
29. $-\sqrt[3]{x^9}$
30. $\sqrt[3]{(x+1)^6}$
31. $\sqrt[3]{27x^3}$
32. $\sqrt[3]{(2x)^6}$
33. $\sqrt[4]{81}$
34. $\sqrt[5]{-1}$
35. $\sqrt[5]{-243}$
36. $\sqrt[4]{625}$
37. $\sqrt[4]{-16}$
38. $\sqrt[6]{-64}$

Exercises 39–44: Approximate to the nearest hundredth.

39. $\sqrt{11}$
40. $-\sqrt{5}$
41. $\sqrt[3]{5}$
42. $\sqrt[3]{-13}$
43. $\sqrt[5]{-7}$
44. $\sqrt[4]{6}$

CONNECTING ABSOLUTE VALUE AND RADICALS

Exercises 45–58: Simplify the expression. Assume that all variables are real numbers.

45. $\sqrt{(-4)^2}$
46. $\sqrt{9^2}$
47. $\sqrt{y^2}$
48. $\sqrt{z^4}$
49. $\sqrt{(x-5)^2}$
50. $\sqrt{(2x-1)^2}$
51. $\sqrt{x^2 - 2x + 1}$
52. $\sqrt{4x^2 + 4x + 1}$
53. $\sqrt[4]{y^4}$
54. $\sqrt[4]{z^4}$
55. $\sqrt[4]{x^{12}}$
56. $\sqrt[6]{x^6}$
57. $\sqrt[5]{x^5}$
58. $\sqrt[5]{32(x+4)^5}$

EVALUATING SQUARE AND CUBE ROOT FUNCTIONS

Exercises 59–74: If possible, evaluate the function at the given value(s) of the variable.

59. $f(x) = \sqrt{x - 1}$ $x = 10, 0$
60. $f(x) = \sqrt{4 - 3x}$ $x = -4, 1$
61. $f(x) = \sqrt{3 - 3x}$ $x = -1, 5$
62. $f(x) = \sqrt{x - 5}$ $x = -1, 5$
63. $f(x) = \sqrt{x^2 - x}$ $x = -4, 3$
64. $f(x) = \sqrt{2x^2 - 3}$ $x = -1, 2$
65. $f(x) = \sqrt[3]{x^2 - 8}$ $x = -3, 4$
66. $f(x) = \sqrt[3]{2x^2}$ $x = -2, 2$
67. $f(x) = \sqrt[3]{5x - 2}$ $x = -5, 2$
68. $f(x) = \sqrt[3]{x - 9}$ $x = 1, 10$
69. $f(x) = \sqrt[3]{3 - x^2}$ $x = -2, 3$
70. $f(x) = \sqrt[3]{-1 - x^2}$ $x = 0, 3$
71. $T(h) = \frac{1}{2}\sqrt{h}$ $h = 64$
72. $L(k) = 2\sqrt{k + 2}$ $k = 23$
73. $f(x) = \sqrt{x + 5} + \sqrt{x}$ $x = 4$
74. $f(x) = \dfrac{\sqrt{x - 5} - \sqrt{x}}{2}$ $x = 9$

FINDING THE DOMAIN OF ROOT FUNCTIONS

Exercises 75–88: Find the domain of f. Write your answer in interval notation.

75. $f(x) = \sqrt{x + 2}$
76. $f(x) = \sqrt{x - 1}$
77. $f(x) = \sqrt{x - 2}$
78. $f(x) = \sqrt{x + 1}$
79. $f(x) = \sqrt{2x - 4}$
80. $f(x) = \sqrt{4x + 2}$
81. $f(x) = \sqrt{1 - x}$
82. $f(x) = \sqrt{6 - 3x}$
83. $f(x) = \sqrt{8 - 5x}$
84. $f(x) = \sqrt{3 - 2x}$
85. $f(x) = \sqrt{3x^2 + 4}$
86. $f(x) = \sqrt{1 + 2x^2}$
87. $f(x) = \dfrac{1}{\sqrt{2x + 1}}$
88. $f(x) = \dfrac{1}{\sqrt{x - 1}}$

APPLYING TRANSLATIONS OF GRAPHS

Exercises 89–94: Graph the function and give its domain in interval notation. Compare the graph of $y = f(x)$ to either $y = \sqrt{x}$ or $y = \sqrt[3]{x}$.

89. $f(x) = \sqrt{x+2}$
90. $f(x) = \sqrt{x-1}$
91. $f(x) = \sqrt{x}+2$
92. $f(x) = \sqrt[3]{x}+2$
93. $f(x) = \sqrt[3]{x+2}$
94. $f(x) = \sqrt[3]{x}-1$

REPRESENTING ROOT FUNCTIONS

Exercises 95–102: Give symbolic, numerical, and graphical representations for the function f.

95. Function f takes the square root of x and then adds 1 to the result.
96. Function f takes the square root of x and then subtracts 2 from the result.
97. Function f takes the square root of the quantity three times x.
98. Function f takes the square root of the quantity x plus 1.
99. Function f takes the cube root of x and then multiplies the result by 2.
100. Function f takes the cube root of the quantity 4 times x.
101. Function f takes the cube root of the quantity x minus 1.
102. Function f takes the cube root of x and then adds 1.

FINDING AREA OF A TRIANGLE

Exercises 103 and 104: **Heron's Formula** Suppose the lengths of the sides of a triangle are a, b, and c as illustrated in the figure.

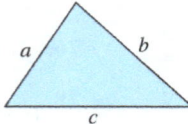

If the **semiperimeter** (half of the perimeter) of the triangle is $s = \frac{1}{2}(a + b + c)$, then the area of the triangle is
$$A = \sqrt{s(s-a)(s-b)(s-c)}.$$

Find the area A of the triangle with the given sides.

103. $a = 3, b = 4, c = 5$
104. $a = 5, b = 9, c = 10$

APPLICATIONS INVOLVING RADICAL EXPRESSIONS

105. **Jumping** (Refer to Example 9.) If a person jumps 4 feet off the ground, estimate how long the person is in the air.

106. **Hang Time** (Refer to Example 9.) Find the hang time for a golf ball hit 80 feet into the air.

107. **Distance to the Horizon** (Refer to Example 4.) Use the formula $d = 1.22\sqrt{x}$ to estimate how many miles a person can see from a jet airliner at 10,000 feet.

108. **Distance to the Horizon** (Refer to Example 4.) Use the formula $d = 1.22\sqrt{x}$ to estimate how many miles a 5-foot-tall person can see standing on the deck of a ship that is 50 feet above the ocean.

109. **Increasing Productivity** (Refer to Example 12.) Use $R(x) = 108\sqrt{x}$ to evaluate $R(16) - R(15)$. If the salary for the sixteenth employee is \$25,000, is it a good decision to hire the sixteenth employee?

110. **Increasing Productivity** (Refer to Example 12.) Use $R(x) = 108\sqrt{x}$ to evaluate $R(2) - R(1)$. If the salary for the second employee is \$25,000, is it a good decision to hire the second employee?

111. **Productivity of Workers** If workers are given more equipment, or *physical capital*, they are often more productive. For example, a carpenter definitely needs a hammer to be productive, but probably does not need 20 hammers. There is a leveling off in a worker's productivity as more is spent on equipment. The function $P(x) = 400\sqrt{x} + 8000$ approximates the worth of the goods produced by a typical U.S. worker in dollars when x dollars are spent on equipment per worker.
    (a) Evaluate $P(25,000)$ and interpret the result.
    (b) Sketch a graph of $y = P(x)$.
    (c) Use the graph from (b) and the formula to evaluate $P(50,000) - P(25,000)$ and also to evaluate $P(75,000) - P(50,000)$. Interpret the result.

112. **Design of Open Channels** To protect cities from flooding during heavy rains, open channels are sometimes constructed to handle runoff. The rate R at which water flows through the channel is modeled by $R = k\sqrt{m}$, where m is the slope of the channel and k is a constant determined by the shape of the channel. (*Source:* N. Garber and L. Hoel, *Traffic and Highway Design.*)
    (a) Suppose that a channel has a slope of $m = 0.01$ (or 1%) and a runoff rate of $R = 340$ cubic feet per second (cfs). Find k.
    (b) If the slope of the channel increases to $m = 0.04$ (or 4%), what happens to R? Be specific.

113. Genetically Engineered Soybeans The following graph shows the percentage y of planted acres of genetically engineered soybeans from 2000 to 2012, where x is the number of years after 2000. This growth is modeled by $y = 13\sqrt{x} + 50$.

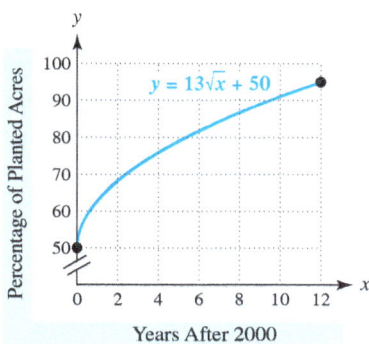

(a) Find y when $x = 0, 6,$ and 12.
(b) Interpret your result when $x = 12$.
(c) What happens if you use this equation to estimate this percentage in 1990?

114. Women in Pharmacy From 1975 to 2005, the percentage y of pharmacy degrees granted to women x years after 1975 could be modeled by $y = 7.5\sqrt{x} + 30$.

(a) Find y when $x = 0$ and 30.
(b) Interpret each answer.
(c) What happens if you use this equation to estimate this percentage in 1970?

WRITING ABOUT MATHEMATICS

115. Try to calculate $\sqrt{-7}$, $\sqrt[4]{-56}$, and $\sqrt[6]{-10}$ with a calculator. Describe what happens when you evaluate an even root of a negative number. Does the same difficulty occur when you evaluate an odd root of a negative number? Try to evaluate $\sqrt[3]{-7}$, $\sqrt[5]{-56}$, and $\sqrt[7]{-10}$. Explain.

116. Explain the difference between a root and a positive integer power of a number. Give examples.

17.2 Rational Exponents

Objectives

1. Using Rational Numbers as Exponents
   - The Expression $a^{1/n}$
   - The Expression $a^{m/n}$
   - The Expression $a^{-m/n}$
2. Applying Properties of Rational Exponents

1 Using Rational Numbers as Exponents

THE EXPRESSION $a^{1/n}$ When m and n are integers, the product rule states that $a^m a^n = a^{m+n}$. This rule can be extended to include exponents that are fractions. For example,

$$4^{1/2} \cdot 4^{1/2} = 4^{1/2 + 1/2} = 4^1 = 4.$$

That is, if we multiply the expression $4^{1/2}$ by itself, the result is 4. Because we also know that $\sqrt{4} \cdot \sqrt{4} = 4$, this discussion suggests that $4^{1/2} = \sqrt{4}$ and leads to the following definition.

THE EXPRESSION $a^{1/n}$

If n is an integer greater than 1 and a is a real number, then

$$a^{1/n} = \sqrt[n]{a}. \qquad 7^{1/5} = \sqrt[5]{7}$$

NOTE: If $a < 0$ and n is an even positive integer, then $a^{1/n}$ is *not* a real number. ∎

In the next two examples, we show how to interpret rational exponents.

EXAMPLE 1 Interpreting rational exponents

Write each expression in radical notation. Then evaluate the expression and round to the nearest hundredth when appropriate.
(a) $36^{1/2}$ (b) $23^{1/5}$ (c) $x^{1/3}$ (d) $(5x)^{1/2}$

Solution
(a) The exponent $\frac{1}{2}$ indicates a square root. Thus $36^{1/2} = \sqrt{36}$, which evaluates to 6.
(b) The exponent $\frac{1}{5}$ indicates a fifth root. Thus $23^{1/5} = \sqrt[5]{23}$, which is not an integer. **FIGURE 17.7** shows this expression approximated in both exponential and radical notation. In either case, $23^{1/5} \approx 1.87$.
(c) The exponent $\frac{1}{3}$ indicates a cube root, so $x^{1/3} = \sqrt[3]{x}$.
(d) The exponent $\frac{1}{2}$ indicates a square root, so $(5x)^{1/2} = \sqrt{5x}$.

Now Try Exercises 17, 45, 59, 63

```
23^(1/5)
      1.872171231
5ˣ√(23)
      1.872171231
```

FIGURE 17.7

CALCULATOR HELP
To calculate other roots, see Appendix A (pages AP-7 and AP-8).

THE EXPRESSION $a^{m/n}$ Suppose that we want to define the expression $8^{2/3}$. On the one hand, using properties of exponents, we have

$$8^{1/3} \cdot 8^{1/3} = 8^{1/3 + 1/3} = 8^{2/3}.$$

On the other hand, we have

$$8^{1/3} \cdot 8^{1/3} = \sqrt[3]{8} \cdot \sqrt[3]{8} = 2 \cdot 2 = 4.$$

Thus $8^{2/3} = 4$, and that value is obtained whether we interpret $8^{2/3}$ as either

$$8^{2/3} = (8^{1/3})^2 = (\sqrt[3]{8})^2 = 2^2 = 4$$

or

$$8^{2/3} = (8^2)^{1/3} = \sqrt[3]{8^2} = \sqrt[3]{64} = 4.$$

This result suggests the following definition.

STUDY TIP

Be sure that you understand how rational exponents relate to radical notation.

> **THE EXPRESSION** $a^{m/n}$
>
> If m and n are positive integers with $\frac{m}{n}$ in lowest terms, then
>
> $$a^{m/n} = \sqrt[n]{a^m} = (\sqrt[n]{a})^m. \qquad 2^{3/4} = \sqrt[4]{2^3} = (\sqrt[4]{2})^3$$
>
> **NOTE:** If $a < 0$ and n is an even integer, then $a^{m/n}$ is *not* a real number. ∎

The exponent $\frac{m}{n}$ indicates that we either take the nth root and then calculate the mth power of the result or calculate the mth power and then take the nth root. For example, $4^{3/2}$ means that we can either take the **square root** of 4 and then **cube** the result or we can **cube** 4 and then take the **square root** of the result. In either case, the result is the same: $4^{3/2} = 8$. This concept is illustrated in the next example.

EXAMPLE 2 Interpreting rational exponents

Write each expression in radical notation. Evaluate the expression by hand when possible.
(a) $(-27)^{2/3}$ (b) $12^{3/5}$

Solution

(a) The exponent $\frac{2}{3}$ indicates either that we take the **cube root** of -27 and then **square** it or that we **square** -27 and then take the **cube root**. Thus

$$(-27)^{2/3} = (\sqrt[3]{-27})^2 = (-3)^2 = 9$$

or

$$(-27)^{2/3} = \sqrt[3]{(-27)^2} = \sqrt[3]{729} = 9.$$

Same result

(b) The exponent $\frac{3}{5}$ indicates either that we take the **fifth root** of 12 and then **cube** it or that we **cube** 12 and then take the **fifth root**. Thus

$$12^{3/5} = (\sqrt[5]{12})^3 \quad \text{or} \quad 12^{3/5} = \sqrt[5]{12^3}.$$

This result cannot be evaluated by hand.

Now Try Exercises 47, 51

TECHNOLOGY NOTE

Rational Exponents
When evaluating expressions with rational (fractional) exponents, be sure to put parentheses around the fraction. For example, most calculators will evaluate 8^(2/3) and 8^2/3 differently. The accompanying figure shows that evaluating $8^{2/3}$ correctly as 8^(2/3) results in 4, but evaluating $8^{2/3}$ incorrectly as 8^2/3 results in $\frac{8^2}{3} = 21.\overline{3}$.

Correct → 8^(2/3)   4
Incorrect → 8^2/3   21.33333333

Math in Context (Online Video) In today's society, people are accustomed to having a lot of choices when it comes to online videos. In fact, if a typical online video has 1,000,000 viewers, then about 200,000 of them, or 20%, have abandoned the video in 10 seconds or less. This average viewer abandonment rate can be modeled by a new function A, given by

$$A(x) = 7.3x^{7/16},$$

where A computes the percentage of viewers who abandon an online video after x seconds. (*Source: Business insider.*)

EXAMPLE 3 Calculating abandonment rates for online videos

The function $A(x) = 7.3x^{7/16}$ computes the percentage of viewers who abandon an online video after x seconds.
(a) Find $A(10)$ and $A(90)$. Interpret your answers.
(b) Use radical notation to write $A(x)$.

Solution

(a) To approximate these expressions, we will use a calculator, as shown in **FIGURE 17.8**.

$$A(10) = 7.3(10)^{7/16} \approx 20\%$$
$$A(90) = 7.3(90)^{7/16} \approx 52\%$$

After **10** seconds, about **20%** of the viewers have abandoned the online video, and after **90** seconds, over half, or **52%**, of the viewers have abandoned the online video.

(b) The exponent $\frac{7}{16}$ means we raise x to the **seventh** power and then take the **sixteenth** root. That is, $A(x) = 7.3\sqrt[16]{x^7}$.

```
7.3*10^(7/16)
       19.99046333
7.3*90^(7/16)
       52.27619364
```

FIGURE 17.8

Now Try Exercise 97

THE EXPRESSION $a^{-m/n}$ From properties of exponents, we know that $a^{-n} = \frac{1}{a^n}$, where n is a positive integer. We now define this property for negative rational exponents.

> **THE EXPRESSION $a^{-m/n}$**
>
> If m and n are positive integers with $\frac{m}{n}$ in lowest terms, then
>
> $$a^{-m/n} = \frac{1}{a^{m/n}}, \qquad a \neq 0. \qquad 2^{-3/4} = \frac{1}{2^{3/4}}$$

READING CHECK 1

- Write the expression $8^{-2/3}$ in radical notation and then evaluate it.

Examples of changing rational exponents to radical expressions are shown in **TABLE 17.3**.

Converting Rational Exponents to Radical Notation

Expression	$a^{3/4}$	$5^{1/7}$	$x^{-5/3}$
Radical Form	$\sqrt[4]{a^3}$	$\sqrt[7]{5}$	$\dfrac{1}{\sqrt[3]{x^5}}$

TABLE 17.3

EXAMPLE 4 **Interpreting negative rational exponents**

Write each expression in radical notation and then evaluate.
(a) $64^{-1/3}$ (b) $81^{-3/4}$

Solution

(a) $64^{-1/3} = \dfrac{1}{64^{1/3}} = \dfrac{1}{\sqrt[3]{64}} = \dfrac{1}{4}.$

(b) $81^{-3/4} = \dfrac{1}{81^{3/4}} = \dfrac{1}{(\sqrt[4]{81})^3} = \dfrac{1}{3^3} = \dfrac{1}{27}.$

Now Try Exercises 53, 55

EXAMPLE 5 **Converting to rational exponents**

Use rational exponents to write each radical expression.
(a) $\sqrt[5]{x^4}$ (b) $\dfrac{1}{\sqrt{b^5}}$ (c) $\sqrt[3]{(x-1)^2}$ (d) $\sqrt[3]{a^2 + b^2}$

Solution

(a) $\sqrt[5]{x^4} = x^{4/5}$

(b) $\dfrac{1}{\sqrt{b^5}} = b^{-5/2}$

(c) $\sqrt[3]{(x-1)^2} = (x-1)^{2/3}$

(d) $\sqrt[3]{a^2 + b^2} = (a^2 + b^2)^{1/3}$ Note that $\sqrt[3]{a^2 + b^2} \neq a^{2/3} + b^{2/3}$.

Now Try Exercises 35, 37, 41, 43

🌐 **Math in Context** (Biology) In the next example, we use a formula from biology that involves a rational exponent.

EXAMPLE 6 Analyzing stepping frequency

When smaller (four-legged) animals walk, they tend to take faster, shorter steps, whereas larger animals tend to take slower, longer steps. If an animal is h feet high at the shoulder, then the number N of steps per second that the animal takes *while walking* can be estimated by $N(h) = 1.6h^{-1/2}$. (*Source:* C. Pennycuick, *Newton Rules Biology*.)
(a) Use radical notation to write $N(h)$.
(b) Estimate the stepping frequency for an adult African elephant that is 10 feet high at the shoulders and a baby elephant that is only 33 inches high.

Solution
(a) We can rewrite the formula as $N(h) = \dfrac{1.6}{\sqrt{h}}$.

(b) We can evaluate N as follows. Note that 33 inches is $\dfrac{33}{12} = 2.75$ feet.

$$N(10) = \frac{1.6}{\sqrt{10}} \approx 0.51 \quad \text{and} \quad N(2.75) = \frac{1.6}{\sqrt{2.75}} \approx 0.96$$

While walking, the adult elephant takes about $\frac{1}{2}$ step every second, or 1 step every 2 seconds, and the baby elephant takes about 1 step every second.

Now Try Exercise 99

2 Applying Properties of Rational Exponents

Any rational number can be written as a ratio of two integers. That is, if p is a rational number, then $p = \frac{m}{n}$, where m and n are integers. Properties for integer exponents also apply to rational exponents, with one exception. If n is even in the expression $a^{m/n}$ and $\frac{m}{n}$ is written in lowest terms, then a must be nonnegative for the result to be a real number.

For example, the expression $(-8)^{3/2}$ is not a real number because -8 is negative and 2 is even. That is, $(-8)^{3/2} = (\sqrt{-8})^3$ is not a real number because the square root of a negative number is not a real number.

PROPERTIES OF EXPONENTS

Let p and q be rational numbers written in lowest terms. For all real numbers a and b for which the expressions are real numbers, the following properties hold.

1. $a^p \cdot a^q = a^{p+q}$ Product rule for exponents $3^{1/7} \cdot 3^{2/7} = 3^{3/7}$

2. $a^{-p} = \dfrac{1}{a^p}, \dfrac{1}{a^{-p}} = a^p$ Negative exponents $7^{-1/4} = \dfrac{1}{7^{1/4}}, \dfrac{1}{7^{-1/4}} = 7^{1/4}$

3. $\left(\dfrac{a}{b}\right)^{-p} = \left(\dfrac{b}{a}\right)^p$ Negative exponents for quotients $\left(\dfrac{3}{7}\right)^{-1/3} = \left(\dfrac{7}{3}\right)^{1/3}$

4. $\dfrac{a^p}{a^q} = a^{p-q}$ Quotient rule for exponents $\dfrac{5^{4/7}}{5^{1/7}} = 5^{3/7}$

5. $(a^p)^q = a^{pq}$ Power rule for exponents $(2^{1/3})^{1/2} = 2^{1/6}$

6. $(ab)^p = a^p b^p$ Power rule for products $(5x)^{2/3} = 5^{2/3} x^{2/3}$

7. $\left(\dfrac{a}{b}\right)^p = \dfrac{a^p}{b^p}$ Power rule for quotients $\left(\dfrac{5}{3}\right)^{3/4} = \dfrac{5^{3/4}}{3^{3/4}}$

In the next two examples, we apply these properties.

> **EXAMPLE 7** **Applying properties of exponents**
>
> Write each expression using rational exponents and simplify. Write the answer with a positive exponent. Assume that all variables are positive numbers.
>
> (a) $\sqrt{x} \cdot \sqrt[3]{x}$ (b) $\sqrt[3]{27x^2}$ (c) $\dfrac{\sqrt[4]{16x}}{\sqrt[3]{x}}$ (d) $\left(\dfrac{x^2}{81}\right)^{-1/2}$
>
> **Solution**
> (a) $\sqrt{x} \cdot \sqrt[3]{x} = x^{1/2} \cdot x^{1/3}$ Use rational exponents.
> $\phantom{\sqrt{x} \cdot \sqrt[3]{x}} = x^{1/2 + 1/3}$ Product rule for exponents
> $\phantom{\sqrt{x} \cdot \sqrt[3]{x}} = x^{5/6}$ Simplify: $\frac{1}{2} + \frac{1}{3} = \frac{3}{6} + \frac{2}{6} = \frac{5}{6}$.
>
> (b) $\sqrt[3]{27x^2} = (27x^2)^{1/3}$ Use rational exponents.
> $\phantom{\sqrt[3]{27x^2}} = 27^{1/3}(x^2)^{1/3}$ Power rule for products
> $\phantom{\sqrt[3]{27x^2}} = 3x^{2/3}$ Simplify; power rule for exponents.
>
> (c) $\dfrac{\sqrt[4]{16x}}{\sqrt[3]{x}} = \dfrac{(16x)^{1/4}}{x^{1/3}}$ Use rational exponents.
> $\phantom{\dfrac{\sqrt[4]{16x}}{\sqrt[3]{x}}} = \dfrac{16^{1/4}x^{1/4}}{x^{1/3}}$ Power rule for products
> $\phantom{\dfrac{\sqrt[4]{16x}}{\sqrt[3]{x}}} = 16^{1/4}x^{1/4 - 1/3}$ Quotient rule for exponents
> $\phantom{\dfrac{\sqrt[4]{16x}}{\sqrt[3]{x}}} = 2x^{-1/12}$ Simplify: $\frac{1}{4} - \frac{1}{3} = \frac{3}{12} - \frac{4}{12} = -\frac{1}{12}$.
> $\phantom{\dfrac{\sqrt[4]{16x}}{\sqrt[3]{x}}} = \dfrac{2}{x^{1/12}}$ Negative exponents
>
> (d) $\left(\dfrac{x^2}{81}\right)^{-1/2} = \left(\dfrac{81}{x^2}\right)^{1/2}$ Negative exponents for quotients
> $\phantom{\left(\dfrac{x^2}{81}\right)^{-1/2}} = \dfrac{(81)^{1/2}}{(x^2)^{1/2}}$ Power rule for quotients
> $\phantom{\left(\dfrac{x^2}{81}\right)^{-1/2}} = \dfrac{9}{x}$ Simplify; power rule for exponents.
>
> **Now Try Exercises** 73, 79, 87, 93

> **EXAMPLE 8** **Applying properties of exponents**
>
> Write each expression with positive rational exponents and simplify, if possible.
>
> (a) $\sqrt[3]{\sqrt{x+1}}$ (b) $\sqrt[5]{c^{15}}$ (c) $\dfrac{y^{-1/2}}{x^{-1/3}}$ (d) $\sqrt{x}(\sqrt{x} - 1)$
>
> **Solution**
> (a) $\sqrt[3]{\sqrt{x+1}} = ((x+1)^{1/2})^{1/3} = (x+1)^{1/6}$
> (b) $\sqrt[5]{c^{15}} = c^{15/5} = c^3$
> (c) $\dfrac{y^{-1/2}}{x^{-1/3}} = \dfrac{x^{1/3}}{y^{1/2}}$
> (d) $\sqrt{x}(\sqrt{x} - 1) = x^{1/2}(x^{1/2} - 1) = x^{1/2}x^{1/2} - x^{1/2} = x - x^{1/2}$
>
> **Now Try Exercises** 81, 85, 91

17.2 Putting It All Together

CONCEPT	EXPLANATION	EXAMPLES
Rational Exponents	If m and n are positive integers with $\frac{m}{n}$ in lowest terms, $$a^{m/n} = \sqrt[n]{a^m} = (\sqrt[n]{a})^m.$$ If $a < 0$ and n is even, $a^{m/n}$ is not a real number.	$8^{4/3} = (\sqrt[3]{8})^4 = 2^4 = 16$ $(-27)^{3/4} = (\sqrt[4]{-27})^3$ is *not* a real number.
Properties of Exponents	Let p and q be rational numbers. 1. $a^p \cdot a^q = a^{p+q}$ 2. $a^{-p} = \frac{1}{a^p}, \frac{1}{a^{-p}} = a^p$ 3. $\left(\frac{a}{b}\right)^{-p} = \left(\frac{b}{a}\right)^p$ 4. $\frac{a^p}{a^q} = a^{p-q}$ 5. $(a^p)^q = a^{pq}$ 6. $(ab)^p = a^p b^p$ 7. $\left(\frac{a}{b}\right)^p = \frac{a^p}{b^p}$	1. $2^{1/3} \cdot 2^{2/3} = 2^{1/3 + 2/3} = 2^1 = 2$ 2. $2^{-1/2} = \frac{1}{2^{1/2}}, \frac{1}{3^{-1/4}} = 3^{1/4}$ 3. $\left(\frac{3}{4}\right)^{-4/5} = \left(\frac{4}{3}\right)^{4/5}$ 4. $\frac{7^{2/3}}{7^{1/3}} = 7^{2/3 - 1/3} = 7^{1/3}$ 5. $(8^{2/3})^{1/2} = 8^{(2/3) \cdot (1/2)} = 8^{1/3} = 2$ 6. $(2x)^{1/3} = 2^{1/3} x^{1/3}$ 7. $\left(\frac{x}{y}\right)^{1/6} = \frac{x^{1/6}}{y^{1/6}}$

17.2 Exercises

CONCEPTS AND VOCABULARY

1. Simplify $4^{1/2}$.
2. Simplify $8^{1/3}$.
3. Simplify $4^{-1/2}$.
4. Simplify $8^{-1/3}$.
5. Use a rational exponent to write $\sqrt{x}$.
6. Use a rational exponent to write $\sqrt[3]{a^4}$.
7. Write $a^{1/n}$ in radical notation.
8. Write $a^{m/n}$ in radical notation.

Exercises 9–16: Match the given expression to the expression (a.–h.) that it equals.

9. $\sqrt{x^3}$
10. $\sqrt[3]{8^2}$
11. $25^{-1/2}$
12. $27^{-2/3}$
13. $x^{1/5}$
14. $(x^{1/3})^2$
15. $\sqrt{x} \cdot \sqrt[3]{x}$
16. $\dfrac{x^{1/2}}{x^{-1/3}}$

a. $\sqrt[6]{x^5}$
b. $\sqrt[6]{x}$
c. $x^{3/2}$
d. $\sqrt[5]{x}$
e. $\sqrt[3]{x^2}$
f. 4
g. $\frac{1}{5}$
h. $\frac{1}{9}$

EVALUATING RATIONAL EXPONENTS

Exercises 17–24: Approximate to the nearest hundredth.

17. $16^{1/5}$
18. $7^{1/4}$
19. $5^{1/3}$
20. $11^{1/2}$
21. $9^{3/5}$
22. $13^{5/4}$
23. $4^{-3/7}$
24. $2^{-3/4}$

CONVERTING BETWEEN RATIONAL EXPONENTS AND RADICAL NOTATION

Exercises 25–34: Use radical notation to write each expression.

25. $7^{1/2}$
26. $10^{1/2}$
27. $a^{1/3}$
28. $y^{1/4}$
29. $x^{5/6}$
30. $x^{3/2}$
31. $(x+y)^{1/2}$
32. $(x+y)^{2/3}$
33. $b^{-2/3}$
34. $b^{-3/4}$

Exercise 35–44: Use rational exponents to write each expression.

35. $\sqrt{t}$
36. $\sqrt[3]{t}$
37. $\sqrt[3]{(x+1)}$
38. $\sqrt{y-3}$
39. $\dfrac{1}{\sqrt{x+1}}$
40. $\dfrac{1}{\sqrt[3]{x-3}}$
41. $\sqrt{a^2-b^2}$
42. $\sqrt[3]{a^3+b^3}$
43. $\dfrac{1}{\sqrt[3]{x^7}}$
44. $\dfrac{1}{\sqrt[3]{y^4}}$

Exercises 45–64: Write each expression in radical notation. Evaluate the expression by hand when possible.

45. $9^{1/2}$
46. $100^{1/2}$
47. $8^{1/3}$
48. $(-8)^{1/3}$
49. $\left(\tfrac{4}{9}\right)^{1/2}$
50. $\left(\tfrac{64}{27}\right)^{1/3}$
51. $(-8)^{2/3}$
52. $(16)^{3/2}$
53. $\left(\tfrac{1}{8}\right)^{-1/3}$
54. $\left(\tfrac{1}{81}\right)^{-1/4}$
55. $16^{-3/4}$
56. $(32)^{-2/5}$
57. $(4^{1/2})^{-3}$
58. $(8^{1/3})^{-2}$
59. $z^{1/4}$
60. $b^{1/3}$
61. $y^{-2/5}$
62. $z^{-3/4}$
63. $(3x)^{1/3}$
64. $(5x)^{1/2}$

USING PROPERTIES OF RATIONAL EXPONENTS

Exercises 65–94: Use positive rational exponents to simplify the expression. Assume that all variables are positive.

65. $\sqrt{x} \cdot \sqrt{x}$
66. $\sqrt[3]{x} \cdot \sqrt[3]{x}$
67. $\sqrt[3]{8x^2}$
68. $\sqrt[3]{27z}$
69. $\dfrac{\sqrt{49x}}{\sqrt[3]{x^2}}$
70. $\dfrac{\sqrt[3]{8x}}{\sqrt[4]{x^3}}$
71. $(x^2)^{3/2}$
72. $(y^4)^{1/2}$
73. $\sqrt[3]{x^3y^6}$
74. $\sqrt{16x^4}$
75. $\sqrt{y^3} \cdot \sqrt[3]{y^2}$
76. $\left(\dfrac{x^6}{81}\right)^{1/4}$
77. $\left(\dfrac{x^6}{27}\right)^{2/3}$
78. $\left(\dfrac{1}{x^8}\right)^{-1/4}$
79. $\left(\dfrac{x^2}{y^6}\right)^{-1/2}$
80. $\dfrac{\sqrt{x}}{\sqrt[3]{27x^6}}$
81. $\sqrt{\sqrt{y}}$
82. $\sqrt{\sqrt[3]{(3x)^2}}$
83. $(a^{-1/2})^{4/3}$
84. $(x^{-3/2})^{2/3}$
85. $\dfrac{(k^{1/2})^{-3}}{(k^2)^{1/4}}$
86. $\dfrac{(b^{3/4})^4}{(b^{4/5})^{-5}}$
87. $\sqrt[3]{b} \cdot \sqrt[4]{b}$
88. $\sqrt[3]{t} \cdot \sqrt[5]{t}$
89. $p^{1/2}(p^{3/2}+p^{1/2})$
90. $d^{3/4}(d^{1/4}-d^{-1/4})$
91. $\sqrt[3]{x}(\sqrt{x}-\sqrt[3]{x^2})$
92. $\tfrac{1}{2}\sqrt{x}(\sqrt{x}+\sqrt[4]{x^2})$
93. $\dfrac{\sqrt[3]{27x}}{\sqrt{x}}$
94. $\dfrac{\sqrt[5]{32x^2}}{\sqrt[3]{8x}}$

Exercises 95 and 96: **Thinking Generally** *Simplify the expression. Assume that a and b are positive integers.*

95. $(b^{1/a}b^{2/a})^a$
96. $b^{(b-1)/b} \cdot \sqrt[b]{b}$

APPLICATIONS INVOLVING RATIONAL EXPONENTS

97. **Online Videos** (Refer to Example 3.) The function given by $A(x) = 7.3x^{7/16}$ computes the percentage of viewers who abandon an online video after x seconds for $0 \le x \le 120$.
    (a) Make a table of values for $x = 0, 20, 40, 60,$ and 80. Round values to the nearest percent.
    (b) Interpret the table in terms of how people watch online videos.

98. **Online Videos** (Refer to Example 3.) Suppose that the function given by $A(x) = 7.3x^{4/3}$ computes the percentage of viewers who abandon an online video after x seconds. Evaluate $A(10)$. Is this function realistic?

99. **Animal Stepping Frequency** (Refer to Example 6.) Use the formula $N(h) = 1.6h^{-1/2}$ to estimate the stepping frequency of a dog that is 2.5 feet high at the shoulders. (*Source:* C. Pennycuick.)

100. Animal Pulse Rate According to one model, an animal's heart rate varies according to its weight. The formula $N(w) = 885w^{-1/2}$ gives an estimate for the average number N of beats per minute for an animal that weighs w pounds. Use the formula to estimate the heart rate for a horse that weighs 800 pounds. (*Source:* C. Pennycuick.)

101. Bird Wings Heavier birds tend to have larger wings than lighter birds do. For some birds, the relationship between the surface area A of the bird's wings in square inches and its weight W in pounds can be modeled by $A = 100\sqrt[3]{W^2}$. (*Source:* C. Pennycuick, *Newton Rules Biology*.)
(a) Find the area of the wings when the weight is 8 pounds.
(b) Write this formula with rational exponents.

102. Planet Orbits and Distance Johannes Kepler (1571–1630) discovered a relationship between a planet's distance D from the sun and the time T it takes to orbit the sun. This formula is $T = \sqrt{D^3}$, where T is in Earth years and $D = 1$ corresponds to the distance between Earth and the sun, or 93,000,000 miles.
(a) The planet Neptune is 30 times farther from the sun than Earth ($D = 30$). Estimate the number of years required for Neptune to orbit the sun.
(b) Write this formula with rational exponents.

103. Baby's Head Size If a female baby's head circumference is 35.2 centimeters at birth, then the function given by $H(x) = 35.2x^{3/40}$ gives the infant's head circumference at x months for $1 \leq x \leq 36$.
(a) Evaluate $H(12)$ and $H(24)$.
(b) Does an infant's head circumference increase the most in the first year or the second year?

104. Musical Tones One octave on a piano contains 12 keys (including both the black and white keys). The frequency of each successive key increases by a factor of $2^{1/12}$. For example, middle C is two keys below the first D above it. Therefore the frequency of this D is

$$2^{1/12} \cdot 2^{1/12} = 2^{1/6} \approx 1.12$$

times the frequency of middle C.
(a) If two tones are one octave apart, how do their frequencies compare?
(b) The A tone below middle C on a piano has a frequency of 220 cycles per second. Middle C is 3 keys above this A note. Estimate the frequency of middle C.

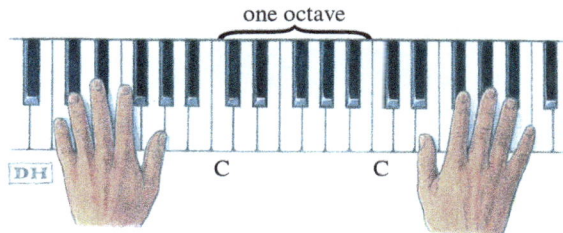

105. Online Exploration Go online and look up the shoulder height of three animals of different sizes. Use the formula in Exercise 99 to calculate their stepping frequencies while walking.

106. Online Exploration Go online and look up the weight of three animals of different sizes. Use the formula in Exercise 100 to calculate their heart rates.

WRITING ABOUT MATHEMATICS

107. Explain the meaning of the rational exponent in $x^{3/5}$.

108. Explain the meaning of each expression: $\sqrt[3]{x^5}$ and $(\sqrt[3]{x})^5$. Are these two expressions equal?

SECTIONS 17.1 and 17.2 — Checking Basic Concepts

1. Find the following.
   (a) The square roots of 49
   (b) The principal square root of 49

2. Evaluate.
   (a) $\sqrt[3]{-8}$
   (b) $-\sqrt[4]{81}$

3. Write the expression in radical notation.
   (a) $x^{3/2}$
   (b) $x^{2/3}$
   (c) $x^{-2/5}$

4. Simplify $\sqrt{(x-1)^2}$ for any real number x.

5. Evaluate each function at the given value of x.
   (a) $f(x) = \sqrt{x}$, $x = 9$
   (b) $g(x) = \sqrt[3]{x}$, $x = 125$
   (c) $h(x) = x^{7/12}$, $x = \frac{12}{7}$

17.3 Simplifying Radical Expressions

Objectives

1. Applying the Product Rule for Radical Expressions
   - Multiplying Radical Expressions
   - Simplifying Radicals
2. Applying the Quotient Rule for Radical Expressions

NEW VOCABULARY

☐ Perfect nth power
☐ Perfect square
☐ Perfect cube

1 Applying the Product Rule for Radical Expressions

MULTIPLYING RADICAL EXPRESSIONS Consider the following examples of multiplying radical expressions. The equations

$$\sqrt{4} \cdot \sqrt{25} = 2 \cdot 5 = 10 \quad \text{and} \quad \sqrt{4 \cdot 25} = \sqrt{100} = 10$$

imply that

$$\sqrt{4} \cdot \sqrt{25} = \sqrt{4 \cdot 25} \quad \text{(see FIGURE 17.9(a))}.$$

Similarly, the equations

$$\sqrt[3]{8} \cdot \sqrt[3]{27} = 2 \cdot 3 = 6 \quad \text{and} \quad \sqrt[3]{8 \cdot 27} = \sqrt[3]{216} = 6$$

imply that

$$\sqrt[3]{8} \cdot \sqrt[3]{27} = \sqrt[3]{8 \cdot 27} \quad \text{(see FIGURE 17.9(b))}.$$

Illustrating $\sqrt[n]{a} \cdot \sqrt[n]{b} = \sqrt[n]{a \cdot b}$

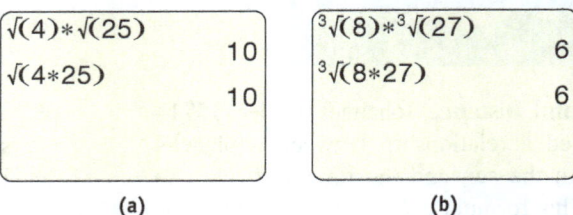

(a) (b)

FIGURE 17.9

These examples suggest that *the product of the roots is equal to the root of the product.*

PRODUCT RULE FOR RADICAL EXPRESSIONS

Let a and b be real numbers, where $\sqrt[n]{a}$ and $\sqrt[n]{b}$ are both defined. Then

$$\sqrt[n]{a} \cdot \sqrt[n]{b} = \sqrt[n]{a \cdot b}. \qquad \sqrt[4]{2} \cdot \sqrt[4]{8} = \sqrt[4]{2 \cdot 8}$$

READING CHECK 1

- Can the product rule be used to simplify a product of radicals with different indexes?

NOTE: The product rule works only when the radicals have the *same index*. For example, the product $\sqrt{2} \cdot \sqrt[3]{4}$ cannot be simplified because the indexes are 2 and 3. (However, by using rational exponents, we can simplify this product. See Example 6(b).) ∎

We apply the product rule in the next two examples.

EXAMPLE 1 Multiplying radical expressions

Multiply each radical expression.

(a) $\sqrt{5} \cdot \sqrt{20}$ (b) $\sqrt[3]{-3} \cdot \sqrt[3]{9}$ (c) $\sqrt[4]{\frac{1}{3}} \cdot \sqrt[4]{\frac{1}{9}} \cdot \sqrt[4]{\frac{1}{3}}$

Solution

(a) $\sqrt{5} \cdot \sqrt{20} = \sqrt{5 \cdot 20} = \sqrt{100} = 10$

(b) $\sqrt[3]{-3} \cdot \sqrt[3]{9} = \sqrt[3]{-3 \cdot 9} = \sqrt[3]{-27} = -3$

(c) The product rule can also be applied to three or more factors. Thus

$$\sqrt[4]{\frac{1}{3}} \cdot \sqrt[4]{\frac{1}{9}} \cdot \sqrt[4]{\frac{1}{3}} = \sqrt[4]{\frac{1}{3} \cdot \frac{1}{9} \cdot \frac{1}{3}} = \sqrt[4]{\frac{1}{81}} = \frac{1}{3},$$

because $\frac{1}{3} \cdot \frac{1}{3} \cdot \frac{1}{3} \cdot \frac{1}{3} = \frac{1}{81}$.

Now Try Exercises 13, 15, 21

EXAMPLE 2 Multiplying radical expressions containing variables

Multiply each radical expression. Assume that all variables are positive.

(a) $\sqrt{x} \cdot \sqrt{x^3}$ **(b)** $\sqrt[3]{2a} \cdot \sqrt[3]{5a}$ **(c)** $\sqrt{11} \cdot \sqrt{xy}$ **(d)** $\sqrt[5]{\frac{2x}{y}} \cdot \sqrt[5]{\frac{16y}{x}}$

Solution

(a) $\sqrt{x} \cdot \sqrt{x^3} = \sqrt{x \cdot x^3} = \sqrt{x^4} = x^2$

(b) $\sqrt[3]{2a} \cdot \sqrt[3]{5a} = \sqrt[3]{2a \cdot 5a} = \sqrt[3]{10a^2}$

(c) $\sqrt{11} \cdot \sqrt{xy} = \sqrt{11xy}$

(d) $\sqrt[5]{\frac{2x}{y}} \cdot \sqrt[5]{\frac{16y}{x}} = \sqrt[5]{\frac{2x}{y} \cdot \frac{16y}{x}}$ Product rule

$= \sqrt[5]{\frac{32xy}{xy}}$ Multiply fractions.

$= \sqrt[5]{32}$ Simplify.

$= 2$ $2^5 = 32$

Now Try Exercises 23, 51, 57, 61

SIMPLIFYING RADICALS An integer a is a **perfect nth power** if there exists an integer b such that $b^n = a$. Thus 36 is a **perfect square** because $6^2 = 36$. Similarly, 8 is a **perfect cube** because $2^3 = 8$, and 81 is a *perfect fourth power* because $3^4 = 81$. **TABLE 17.4** lists examples of perfect squares and perfect cubes.

First Six Perfect Squares and Cubes

Perfect Squares	1	4	9	16	25	36
Perfect Cubes	1	8	27	64	125	216

TABLE 17.4

To simplify a square root, we sometimes need to recognize the *largest* perfect square factor of the radicand. For example, 50 has several factors.

$$1, \ 2, \ 5, \ 10, \ 25, \ 50 \quad \text{Factors of 50}$$

From **TABLE 17.4**, the perfect square factors of 50 are 1 and 25. The *largest* perfect square factor of 50 is 25.

The product rule for radicals can be used to simplify radical expressions. For example, because the largest perfect square factor of 50 is 25, the expression $\sqrt{50}$ can be simplified as

$$\sqrt{50} = \sqrt{25 \cdot 2} = \sqrt{25} \cdot \sqrt{2} = 5\sqrt{2}.$$

This procedure is generalized as follows.

READING CHECK 2

- Simplify the expression $\sqrt{40}$.

SIMPLIFYING RADICALS (nth ROOTS)

STEP 1: Determine the largest perfect nth power factor of the radicand.

STEP 2: Use the product rule to factor out and simplify this perfect nth power.

EXAMPLE 3 Simplifying radical expressions

Simplify each expression.
(a) $\sqrt{300}$ (b) $\sqrt[3]{16}$ (c) $\sqrt{54}$ (d) $\sqrt[4]{512}$

Solution
(a) First note that $300 = 100 \cdot 3$ and that 100 is the largest perfect square factor of 300.
$$\sqrt{300} = \sqrt{100 \cdot 3} = \sqrt{100} \cdot \sqrt{3} = 10\sqrt{3}$$
(b) The largest perfect cube factor of 16 is 8. (See **TABLE 17.4** on the previous page.)
$$\sqrt[3]{16} = \sqrt[3]{8} \cdot \sqrt[3]{2} = 2\sqrt[3]{2}$$
(c) $\sqrt{54} = \sqrt{9} \cdot \sqrt{6} = 3\sqrt{6}$
(d) $\sqrt[4]{512} = \sqrt[4]{256} \cdot \sqrt[4]{2} = 4\sqrt[4]{2}$ because $4^4 = 256$.

Now Try Exercises 75, 77, 79, 81

🌐 Math in Context Driving When a car stops suddenly, it often leaves skid marks. If the car is involved in an accident, authorities often measure the length of the skid marks M and use this information to determine the minimum speed S of the car. One formula that is sometimes used when the road surface is both wet and level is given by $S = \sqrt{9M}$. In the next example, we simplify this radical expression and estimate the speed of a car based on the length of its skid marks.

EXAMPLE 4 Calculating minimum speed

A car is in an accident and leaves skid marks on wet, level pavement that are 121 feet long. The formula $S = \sqrt{9M}$ calculates the minimum speed S in miles per hour for a car that leaves skid marks M feet long.
(a) Simplify this formula. (b) Determine the minimum speed of the car.

Solution
(a) Because 9 is a perfect square, we can factor it out.
$$\sqrt{9M} = \sqrt{9} \cdot \sqrt{M} = 3\sqrt{M}$$
Thus $S = 3\sqrt{M}$.
(b) $S = 3\sqrt{121} = 3 \cdot 11 = 33$ miles per hour.

Now Try Exercise 111

READING CHECK 3

• Simplify $\sqrt[3]{-54}$.

NOTE: To simplify a cube root of a negative number, we factor out the negative of the largest perfect cube factor. For example, $-16 = -8 \cdot 2$, so $\sqrt[3]{-16} = \sqrt[3]{-8} \cdot \sqrt[3]{2} = -2\sqrt[3]{2}$. This procedure can be used with any odd root of a negative number. (See Example 5(c).) ∎

EXAMPLE 5 Simplifying radical expressions

Simplify each expression. Assume that all variables are positive.
(a) $\sqrt{25x^4}$ (b) $\sqrt{32n^3}$ (c) $\sqrt[3]{-16x^3y^5}$ (d) $\sqrt[3]{2a} \cdot \sqrt[3]{4a^2b}$

Solution
(a) $\sqrt{25x^4} = 5x^2$ Perfect square: $(5x^2)^2 = 25x^4$
(b) $\sqrt{32n^3} = \sqrt{(16n^2)2n}$ $16n^2$ is the largest perfect square factor.
$\phantom{\sqrt{32n^3}} = \sqrt{16n^2} \cdot \sqrt{2n}$ Product rule
$\phantom{\sqrt{32n^3}} = 4n\sqrt{2n}$ $(4n)^2 = 16n^2$

(c) $\sqrt[3]{-16x^3y^5} = \sqrt[3]{(-8x^3y^3)2y^2}$ $8x^3y^3$ is the largest perfect cube factor.
$= \sqrt[3]{-8x^3y^3} \cdot \sqrt[3]{2y^2}$ Product rule
$= -2xy\sqrt[3]{2y^2}$ $(-2xy)^3 = -8x^3y^3$

(d) $\sqrt[3]{2a} \cdot \sqrt[3]{4a^2b} = \sqrt[3]{(2a)(4a^2b)}$ Product rule
$= \sqrt[3]{(8a^3)b}$ $8a^3$ is the largest perfect cube factor.
$= \sqrt[3]{8a^3} \cdot \sqrt[3]{b}$ Product rule
$= 2a\sqrt[3]{b}$ $(2a)^3 = 8a^3$

Now Try Exercises 45, 85, 89, 91

The product rule for radical expressions cannot be used if the radicals do not have the same indexes. In this case, we use rational exponents, as illustrated in the next example.

EXAMPLE 6 Multiplying radicals with different indexes

Simplify each expression. Write your answer in radical notation.
(a) $\sqrt{5} \cdot \sqrt[4]{5}$ (b) $\sqrt{2} \cdot \sqrt[3]{4}$ (c) $\sqrt[3]{x} \cdot \sqrt[4]{x}$

Solution
(a) Because $\sqrt{5} = 5^{1/2}$ and $\sqrt[4]{5} = 5^{1/4}$,
$$\sqrt{5} \cdot \sqrt[4]{5} = 5^{1/2} \cdot 5^{1/4} = 5^{1/2+1/4} = 5^{3/4}.$$
In radical notation, $5^{3/4} = \sqrt[4]{5^3} = \sqrt[4]{125}$.

(b) Because $\sqrt[3]{4} = \sqrt[3]{2^2} = 2^{2/3}$,
$$\sqrt{2} \cdot \sqrt[3]{4} = 2^{1/2} \cdot 2^{2/3} = 2^{1/2+2/3} = 2^{7/6}.$$
In radical notation, $2^{7/6} = \sqrt[6]{2^7} = \sqrt[6]{2^6 \cdot 2^1} = \sqrt[6]{2^6} \cdot \sqrt[6]{2} = 2\sqrt[6]{2}$.

(c) $\sqrt[3]{x} \cdot \sqrt[4]{x} = x^{1/3} \cdot x^{1/4} = x^{7/12} = \sqrt[12]{x^7}$

Now Try Exercises 101, 103, 107

2 Applying the Quotient Rule for Radical Expressions

Consider the following examples of dividing radical expressions. The equations
$$\sqrt{\frac{4}{9}} = \sqrt{\frac{2}{3} \cdot \frac{2}{3}} = \frac{2}{3} \quad \text{and} \quad \frac{\sqrt{4}}{\sqrt{9}} = \frac{2}{3}$$
imply that
$$\sqrt{\frac{4}{9}} = \frac{\sqrt{4}}{\sqrt{9}} \quad \text{(see FIGURE 17.10).}$$

```
√(4/9)▶Frac
              2/3
√(4)/√(9)▶Frac
              2/3
```

FIGURE 17.10

CALCULATOR HELP

To use the Frac feature, see Appendix A (pages AP-1 and AP-2).

These examples suggest that *the root of a quotient is equal to the quotient of the roots.*

QUOTIENT RULE FOR RADICAL EXPRESSIONS

Let a and b be real numbers, where $\sqrt[n]{a}$ and $\sqrt[n]{b}$ are both defined and $b \neq 0$. Then
$$\sqrt[n]{\frac{a}{b}} = \frac{\sqrt[n]{a}}{\sqrt[n]{b}}. \qquad \sqrt[3]{\frac{23}{5}} = \frac{\sqrt[3]{23}}{\sqrt[3]{5}}$$

EXAMPLE 7 Simplifying quotients

Simplify each radical expression. Assume that all variables are positive.

(a) $\sqrt[3]{\dfrac{5}{8}}$ (b) $\sqrt[4]{\dfrac{x}{16}}$ (c) $\sqrt{\dfrac{16}{y^2}}$

Solution

(a) $\sqrt[3]{\dfrac{5}{8}} = \dfrac{\sqrt[3]{5}}{\sqrt[3]{8}} = \dfrac{\sqrt[3]{5}}{2}$ (b) $\sqrt[4]{\dfrac{x}{16}} = \dfrac{\sqrt[4]{x}}{\sqrt[4]{16}} = \dfrac{\sqrt[4]{x}}{2}$

(c) $\sqrt{\dfrac{16}{y^2}} = \dfrac{\sqrt{16}}{\sqrt{y^2}} = \dfrac{4}{y}$ because $y > 0$.

Now Try Exercises 25, 27, 29

EXAMPLE 8 Simplifying quotients

Simplify each radical expression. Assume that all variables are positive.

(a) $\dfrac{\sqrt{40}}{\sqrt{10}}$ (b) $\dfrac{\sqrt[3]{2}}{\sqrt[3]{16}}$ (c) $\dfrac{\sqrt{x^2 y}}{\sqrt{y}}$

Solution

(a) $\dfrac{\sqrt{40}}{\sqrt{10}} = \sqrt{\dfrac{40}{10}} = \sqrt{4} = 2$

(b) $\dfrac{\sqrt[3]{2}}{\sqrt[3]{16}} = \sqrt[3]{\dfrac{2}{16}} = \sqrt[3]{\dfrac{1}{8}} = \dfrac{1}{2}$ because $\dfrac{1}{2} \cdot \dfrac{1}{2} \cdot \dfrac{1}{2} = \dfrac{1}{8}$.

(c) $\dfrac{\sqrt{x^2 y}}{\sqrt{y}} = \sqrt{\dfrac{x^2 y}{y}} = \sqrt{x^2} = x$ because $x > 0$.

Now Try Exercises 33, 39, 41

MAKING CONNECTIONS 1

Rules for Radical Expressions and Rational Exponents

The rules for radical expressions are a result of the properties of rational exponents.

$\sqrt[n]{a \cdot b} = \sqrt[n]{a} \cdot \sqrt[n]{b}$ is equivalent to $(a \cdot b)^{1/n} = a^{1/n} \cdot b^{1/n}$.

$\sqrt[n]{\dfrac{a}{b}} = \dfrac{\sqrt[n]{a}}{\sqrt[n]{b}}$ is equivalent to $\left(\dfrac{a}{b}\right)^{1/n} = \dfrac{a^{1/n}}{b^{1/n}}$.

READING CHECK 4

- Use rational exponents to show that $\sqrt[3]{x} \cdot \sqrt[3]{x} \cdot \sqrt[3]{x} = x$.

EXAMPLE 9 Simplifying radical expressions

Simplify each radical expression. Assume that all variables are positive.

(a) $\sqrt[4]{\dfrac{16x^3}{y^4}}$ (b) $\sqrt{\dfrac{5a^2}{8}} \cdot \sqrt{\dfrac{5a^3}{2}}$

Solution

(a) To simplify this expression, we first use the quotient rule for radical expressions and then apply the product rule for radical expressions.

$$\sqrt[4]{\frac{16x^3}{y^4}} = \frac{\sqrt[4]{16x^3}}{\sqrt[4]{y^4}} \qquad \text{Quotient rule}$$

$$= \frac{\sqrt[4]{16}\sqrt[4]{x^3}}{\sqrt[4]{y^4}} \qquad \text{Product rule}$$

$$= \frac{2\sqrt[4]{x^3}}{y} \qquad \text{Evaluate 4th roots.}$$

(b) To simplify this expression, we use both the product and quotient rules.

$$\sqrt{\frac{5a^2}{8}} \cdot \sqrt{\frac{5a^3}{2}} = \sqrt{\frac{5a^2 \cdot 5a^3}{8 \cdot 2}} \qquad \text{Product rule}$$

$$= \sqrt{\frac{25a^5}{16}} \qquad \text{Multiply.}$$

$$= \frac{\sqrt{25a^5}}{\sqrt{16}} \qquad \text{Quotient rule}$$

$$= \frac{\sqrt{25a^4 \cdot a}}{\sqrt{16}} \qquad \text{Factor out largest perfect square.}$$

$$= \frac{\sqrt{25a^4} \cdot \sqrt{a}}{\sqrt{16}} \qquad \text{Product rule}$$

$$= \frac{5a^2\sqrt{a}}{4} \qquad (5a^2)^2 = 25a^4$$

Now Try Exercises 95, 97

EXAMPLE 10   Simplifying products and quotients of roots

Simplify each expression. Assume all radicands are positive.

(a) $\sqrt{x-3} \cdot \sqrt{x+3}$ **(b)** $\dfrac{\sqrt[3]{x^2 + 3x + 2}}{\sqrt[3]{x+1}}$

Solution

(a) Start by applying the product rule for radical expressions.

$$\sqrt{x-3} \cdot \sqrt{x+3} = \sqrt{(x-3)(x+3)} \qquad \text{Product rule}$$
$$= \sqrt{x^2 - 9} \qquad \text{Multiply binomials.}$$

NOTE: The expression $\sqrt{x^2 - 9}$ does *not* simplify further. It is important to realize that
$$\sqrt{x^2 - 9} \neq \sqrt{x^2} - \sqrt{9} = x - 3.$$
For example, $\sqrt{5^2 - 3^2} = \sqrt{16} = 4$, but $\sqrt{5^2} - \sqrt{3^2} = 5 - 3 = 2$. ■

(b) Start by applying the quotient rule for radical expressions.

$$\frac{\sqrt[3]{x^2 + 3x + 2}}{\sqrt[3]{x+1}} = \sqrt[3]{\frac{x^2 + 3x + 2}{x+1}} \qquad \text{Quotient rule}$$

$$= \sqrt[3]{\frac{(x+1)(x+2)}{x+1}} \qquad \text{Factor trinomial.}$$

$$= \sqrt[3]{x+2} \qquad \text{Simplify quotient.}$$

Now Try Exercises 63, 67

17.3 Putting It All Together

CONCEPT	EXPLANATION	EXAMPLES
Product Rule for Radical Expressions	Let a and b be real numbers, where $\sqrt[n]{a}$ and $\sqrt[n]{b}$ are both defined. Then $\sqrt[n]{a} \cdot \sqrt[n]{b} = \sqrt[n]{a \cdot b}.$	$\sqrt{2} \cdot \sqrt{32} = \sqrt{64} = 8$ $\sqrt{500} = \sqrt{100} \cdot \sqrt{5} = 10\sqrt{5}$
Simplifying Radicals (nth roots)	**STEP 1:** Find the largest perfect nth power factor of the radicand. **STEP 2:** Factor out and simplify this perfect nth power.	$\sqrt{12} = \sqrt{4 \cdot 3} = \sqrt{4} \cdot \sqrt{3} = 2\sqrt{3}$ $\sqrt[3]{81} = \sqrt[3]{27 \cdot 3} = \sqrt[3]{27} \cdot \sqrt[3]{3} = 3\sqrt[3]{3}$ $\sqrt[4]{x^5} = \sqrt[4]{x^4 \cdot x} = \sqrt[4]{x^4} \cdot \sqrt[4]{x} = x\sqrt[4]{x},$ provided $x \geq 0.$
Quotient Rule for Radical Expressions	Let a and b be real numbers, where $\sqrt[n]{a}$ and $\sqrt[n]{b}$ are both defined and $b \neq 0.$ Then $\sqrt[n]{\dfrac{a}{b}} = \dfrac{\sqrt[n]{a}}{\sqrt[n]{b}}.$	$\dfrac{\sqrt{60}}{\sqrt{15}} = \sqrt{\dfrac{60}{15}} = \sqrt{4} = 2$ $\sqrt[3]{\dfrac{x^2}{-27}} = \dfrac{\sqrt[3]{x^2}}{\sqrt[3]{-27}} = \dfrac{\sqrt[3]{x^2}}{-3} = -\dfrac{\sqrt[3]{x^2}}{3}$

17.3 Exercises

CONCEPTS AND VOCABULARY

1. Does $\sqrt{2} \cdot \sqrt{3}$ equal $\sqrt{6}$?
2. Does $\sqrt{5} \cdot \sqrt[3]{5}$ equal 5?
3. $\sqrt[3]{a} \cdot \sqrt[3]{b} =$ _____
4. $\dfrac{\sqrt{a}}{\sqrt{b}} = \sqrt{?}$
5. $\dfrac{\sqrt[n]{a}}{\sqrt[n]{b}} = \sqrt[n]{?}$
6. $\dfrac{\sqrt{3}}{\sqrt{27}} =$ _____
7. Does $\sqrt{50} = \sqrt{25} + \sqrt{25}$?
8. Does $\sqrt{50} = 5\sqrt{2}$?
9. Is $\sqrt[3]{3}$ equal to 1? Explain.
10. Is 64 a perfect cube? Explain.

MULTIPLYING AND DIVIDING RADICAL EXPRESSIONS

Exercises 11–62: Simplify the expression. Assume that all variables are positive.

11. $\sqrt{3} \cdot \sqrt{3}$
12. $\sqrt{2} \cdot \sqrt{18}$
13. $\sqrt{2} \cdot \sqrt{50}$
14. $\sqrt[3]{-2} \cdot \sqrt[3]{-4}$
15. $\sqrt[3]{4} \cdot \sqrt[3]{16}$
16. $\sqrt[3]{x} \cdot \sqrt[3]{x^2}$
17. $\sqrt{\dfrac{9}{25}}$
18. $\sqrt[3]{\dfrac{x}{8}}$
19. $\sqrt{\dfrac{1}{2}} \cdot \sqrt{\dfrac{1}{8}}$
20. $\sqrt{\dfrac{5}{3}} \cdot \sqrt{\dfrac{1}{3}}$
21. $\sqrt[3]{\dfrac{2}{3}} \cdot \sqrt[3]{\dfrac{4}{3}} \cdot \sqrt[3]{\dfrac{1}{3}}$
22. $\sqrt[4]{\dfrac{8}{3}} \cdot \sqrt[4]{\dfrac{4}{9}} \cdot \sqrt[4]{\dfrac{8}{3}}$
23. $\sqrt{x^3} \cdot \sqrt{x^3}$
24. $\sqrt{z} \cdot \sqrt{z^7}$
25. $\sqrt[3]{\dfrac{7}{27}}$
26. $\sqrt[3]{\dfrac{9}{64}}$
27. $\sqrt[4]{\dfrac{x}{81}}$
28. $\sqrt[4]{\dfrac{16x}{81}}$
29. $\sqrt{\dfrac{x^2}{81}}$
30. $\sqrt{\dfrac{9}{z^2}}$
31. $\sqrt{\dfrac{x}{2}} \cdot \sqrt{\dfrac{x}{8}}$
32. $\sqrt{\dfrac{4}{y}} \cdot \sqrt{\dfrac{y}{5}}$
33. $\dfrac{\sqrt{45}}{\sqrt{5}}$
34. $\dfrac{\sqrt{7}}{\sqrt{28}}$
35. $\sqrt[3]{-4} \cdot \sqrt[3]{-16}$
36. $\sqrt[3]{9} \cdot \sqrt[3]{3}$
37. $\sqrt[4]{9} \cdot \sqrt[4]{9}$
38. $\sqrt[5]{16} \cdot \sqrt[5]{-2}$
39. $\dfrac{\sqrt[5]{64}}{\sqrt[5]{-2}}$
40. $\dfrac{\sqrt[4]{324}}{\sqrt[4]{4}}$
41. $\dfrac{\sqrt{4xy^2}}{\sqrt{x}}$
42. $\dfrac{\sqrt{a^2b}}{\sqrt{b}}$
43. $\dfrac{\sqrt[3]{54}}{\sqrt[3]{2}}$
44. $\dfrac{\sqrt[3]{x^3y^7}}{\sqrt[3]{y^4}}$

45. $\sqrt{4x^4}$
46. $\sqrt[3]{-8y^3}$
47. $\sqrt[3]{-5a^6}$
48. $\sqrt{9x^2y}$
49. $\sqrt[4]{16x^4y}$
50. $\sqrt[3]{8xy^3}$
51. $\sqrt{6x^5} \cdot \sqrt{6x}$
52. $\sqrt{3x} \cdot \sqrt{12x}$
53. $\sqrt{16x^4y^6}$
54. $\sqrt[3]{8x^6y^3z^9}$
55. $\sqrt[4]{\frac{3}{4}} \cdot \sqrt[4]{\frac{27}{4}}$
56. $\sqrt[5]{-\frac{4}{9}} \cdot \sqrt[5]{-\frac{8}{27}}$
57. $\sqrt[3]{12} \cdot \sqrt[3]{ab}$
58. $\sqrt{5x} \cdot \sqrt{5z}$
59. $\sqrt[4]{25z} \cdot \sqrt[4]{25z}$
60. $\sqrt[5]{3z^2} \cdot \sqrt[5]{7z}$
61. $\sqrt[5]{\frac{7a}{b^2}} \cdot \sqrt[5]{\frac{b^2}{7a^6}}$
62. $\sqrt[3]{\frac{8m}{n}} \cdot \sqrt[3]{\frac{n^4}{m^2}}$

Exercises 63–68: Use properties of polynomials to simplify the expression. Assume all radicands are positive.

63. $\sqrt{x+4} \cdot \sqrt{x-4}$
64. $\sqrt[3]{x-1} \cdot \sqrt[3]{x^2+x+1}$
65. $\sqrt[3]{a+1} \cdot \sqrt[3]{a^2-a+1}$
66. $\sqrt{b-1} \cdot \sqrt{b+1}$
67. $\dfrac{\sqrt{x^2+2x+1}}{\sqrt{x+1}}$
68. $\dfrac{\sqrt{x^2-4x+4}}{\sqrt{x-2}}$

SIMPLIFYING RADICAL EXPRESSIONS

Exercises 69–74: Complete the equation.

69. $\sqrt{500} = \underline{} \sqrt{5}$
70. $\sqrt{28} = \underline{} \sqrt{7}$
71. $\sqrt{8} = \underline{} \sqrt{2}$
72. $\sqrt{99} = \underline{} \sqrt{11}$
73. $\sqrt{45} = \underline{} \sqrt{5}$
74. $\sqrt{243} = \underline{} \sqrt{3}$

Exercises 75–98: Simplify the radical expression by factoring out the largest perfect nth power. Assume that all variables are positive.

75. $\sqrt{200}$
76. $\sqrt{72}$
77. $\sqrt[3]{81}$
78. $\sqrt[3]{256}$
79. $\sqrt[4]{64}$
80. $\sqrt[5]{27 \cdot 81}$
81. $\sqrt[5]{-64}$
82. $\sqrt[3]{-81}$
83. $\sqrt{b^5}$
84. $\sqrt{t^3}$
85. $\sqrt{8n^3}$
86. $\sqrt{32a^2}$
87. $\sqrt{12a^2b^5}$
88. $\sqrt{20a^3b^2}$
89. $\sqrt[3]{-125x^4y^5}$
90. $\sqrt[3]{-81a^5b^2}$
91. $\sqrt[3]{5t} \cdot \sqrt[3]{125t}$
92. $\sqrt[4]{4bc^3} \cdot \sqrt[4]{64ab^3c^2}$
93. $\sqrt[4]{\frac{9t^5}{r^8}} \cdot \sqrt[4]{\frac{9r}{5t}}$
94. $\sqrt[5]{\frac{4t^6}{r}} \cdot \sqrt[5]{\frac{8t}{r^6}}$
95. $\sqrt[3]{\frac{27x^2}{y^3}}$
96. $\sqrt[4]{\frac{32x^8}{z^4}}$
97. $\sqrt{\frac{7a^2}{27}} \cdot \sqrt{\frac{7a}{3}}$
98. $\sqrt{\frac{8a}{125}} \cdot \sqrt{\frac{2a}{5}}$

Exercises 99 and 100: **Thinking Generally** *Let m and n be positive integers and assume each expression exists.*

99. Simplify $\left(\sqrt[mn]{a^m b^m}\right)^n$ so that it equals ab.

100. Simplify $\left(\sqrt[mn]{a^n b^m}\right) \cdot \left(\sqrt[mn]{a^m b^n}\right)$ so that it equals $\sqrt[m]{ab} \cdot \sqrt[n]{ab}$.

Exercises 101–110: Simplify the expression. Let all variables be positive and write your answer in radical notation.

101. $\sqrt{3} \cdot \sqrt[3]{3}$
102. $\sqrt{5} \cdot \sqrt[3]{5}$
103. $\sqrt[4]{8} \cdot \sqrt[3]{4}$
104. $\sqrt[5]{16} \cdot \sqrt{2}$
105. $\sqrt[4]{27} \cdot \sqrt[3]{9} \cdot \sqrt{3}$
106. $\sqrt[5]{16} \cdot \sqrt[3]{16}$
107. $\sqrt[4]{x^3} \cdot \sqrt[3]{x}$
108. $\sqrt[4]{x^3} \cdot \sqrt{x}$
109. $\sqrt[4]{rt} \cdot \sqrt[3]{r^2t}$
110. $\sqrt[3]{a^3b^2} \cdot \sqrt{a^2b}$

APPLICATIONS INVOLVING RADICAL EXPRESSIONS

111. **Minimum Speed** (Refer to Example 4.) If the pavement is dry cement, then $S = \sqrt{25M}$ is an estimate for a car's speed S in miles per hour when it leaves skid marks M feet long.
   (a) Simplify this formula.
   (b) Determine the minimum speed of the car if the skid marks are 100 feet.

112. **Minimum Speed** (Refer to Example 4.) If the road surface is gravel, then $S = \sqrt{15M}$ is an estimate for a car's speed S in miles per hour when it leaves skid marks M feet long in the gravel. Determine the minimum speed of the car if the skid marks are 30 feet.

113. **Baseball Ticket Prices** The formula $y = \sqrt{36x}$ gives the percent price markup of a Major League Baseball ticket x days prior to game day. (*Source:* StubHub.com)
   (a) Find y when $x = 4$ and interpret your result.
   (b) Simplify the equation $y = \sqrt{36x}$
   (c) Find y when $x = 4$ using your equation from part (b). Do you get the same answer as in part (a)?

114. Earnings for Women The equation $I = 710\sqrt{25x}$ gives an estimate for the median annual earnings in thousands of dollars for *all* women x years after 1975. This equation has been adjusted to 2009 dollars. (*Source:* Current Populations Survey.)
(a) Find I when $x = 25$ and interpret your result.
(b) Simplify the equation $I = 710\sqrt{25x}$.
(c) Find I when $x = 25$ using your equation from part (b). Do you get the same answer as in part (a)?

115. Hang Time of a Football A good punter can kick a football so that it stays in the air for a relatively long time. This time is called *hang time*. If a football is kicked h feet high, then its hang time T in seconds is

$$T = \sqrt{\frac{h}{4}}.$$

(a) Calculate the exact hang time for a football kicked 81 feet into the air.
(b) Use the quotient rule to simplify the formula.

116. Orbits and Distance Johannes Kepler (1571–1630) discovered a relationship between a planet's distance D from the sun and the time T it takes the planet to orbit the sun. This formula is

$$T = \sqrt{D} \cdot \sqrt{D} \cdot \sqrt{D},$$

where T is in Earth years and $D = 1$ corresponds to the average distance between Earth and the sun, or 93,000,000 miles.
(a) If $D = 5.2$ for Jupiter, estimate the number of years required for Jupiter to orbit the sun.
(b) Use the product rule to simplify the formula.

WRITING ABOUT MATHEMATICS

117. Explain what it means for a positive integer to be a *perfect square* or a *perfect cube*. List the positive integers that are perfect squares and less than 101. List the positive integers that are perfect cubes and less than 220.

118. Explain how the product and quotient rules for radical expressions are the result of properties of rational exponents.

Group Activity

Designing a Paper Cup A paper drinking cup is being designed in the shape shown in the accompanying figure. The amount of paper needed to manufacture the cup is determined by the surface area S of the cup, which is given by

$$S = \pi r \sqrt{r^2 + h^2},$$

where r is the radius and h is the height.

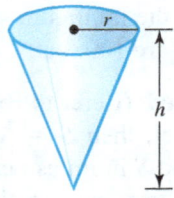

(a) Approximate S to the nearest hundredth when $r = 1.5$ inches and $h = 4$ inches.
(b) Could the formula for S be simplified as follows?
$$\pi r \sqrt{r^2 + h^2} \stackrel{?}{=} \pi r (\sqrt{r^2} + \sqrt{h^2})$$
$$\stackrel{?}{=} \pi r (r + h)$$

Try evaluating this second formula with $r = 1.5$ inches and $h = 4$ inches. Is your answer the same as in part (a)? Explain.
(c) Discuss why evaluating real-world formulas correctly is important.
(d) In general, does $\sqrt{a + b}$ equal $\sqrt{a} + \sqrt{b}$? Justify your answer by completing the following table. Approximate answers to the nearest hundredth when appropriate.

a	b	$\sqrt{a + b}$	$\sqrt{a} + \sqrt{b}$
0	4		
4	0		
5	4		
9	7		
4	16		
25	100		

17.4 Operations on Radical Expressions

Objectives

1. Adding and Subtracting Radical Expressions
   - Adding Like Radicals
   - Finding Like Radicals and Adding
   - Subtracting Like Radicals
2. Multiplying Radical Expressions
3. Rationalizing the Denominator

1 Adding and Subtracting Radical Expressions

ADDING LIKE RADICALS We can add $2x^2$ and $5x^2$ to obtain $7x^2$ because they are *like* terms. That is,

$$2x^2 + 5x^2 = (2 + 5)x^2 = 7x^2.$$

Like terms

We can also add and subtract *like radicals*. **Like radicals** *have the same index and the same radicand.* For example, we can add $3\sqrt{2}$ and $5\sqrt{2}$ because they are like radicals.

$$3\sqrt{2} + 5\sqrt{2} = (3 + 5)\sqrt{2} = 8\sqrt{2}$$

Like radicals

EXAMPLE 1 Adding like radicals

If possible, add the expressions and simplify.
(a) $10\sqrt{11} + 4\sqrt{11}$ (b) $5\sqrt[3]{6} + \sqrt[3]{6}$
(c) $4 + 5\sqrt{3}$ (d) $\sqrt{7} + \sqrt{11}$

Solution
(a) These terms are like radicals because they have the same index, 2, and the same radicand, 11.

$$10\sqrt{11} + 4\sqrt{11} = (10 + 4)\sqrt{11} = 14\sqrt{11}$$

(b) These terms are like radicals because they have the same index, 3, and the same radicand, 6. Note that the coefficient on the second term is understood to be 1.

$$5\sqrt[3]{6} + 1\sqrt[3]{6} = (5 + 1)\sqrt[3]{6} = 6\sqrt[3]{6}$$

(c) The expression $4 + 5\sqrt{3}$ can be written as $4\sqrt{1} + 5\sqrt{3}$. These terms cannot be added because they are not like radicals.

NOTE: $4 + 5\sqrt{3} \neq 9\sqrt{3}$ because multiplication is performed before addition. ∎

(d) The expression $\sqrt{7} + \sqrt{11}$ contains unlike radicals that *cannot* be simplified.

Now Try Exercises 19, 21, 23, 25

READING CHECK 1
- What are like radicals?

NEW VOCABULARY
☐ Like radicals
☐ Rationalizing the denominator
☐ Conjugate

FINDING LIKE RADICALS AND ADDING Sometimes two radical expressions that are not alike can be added by changing them to like radicals. For example, $\sqrt{20}$ and $\sqrt{5}$ are unlike radicals. However,

because $\sqrt{20} = \sqrt{4 \cdot 5} = \sqrt{4} \cdot \sqrt{5} = 2\sqrt{5}$,
we have $\sqrt{20} + \sqrt{5} = 2\sqrt{5} + 1\sqrt{5} = 3\sqrt{5}$.

We cannot combine $x + x^2$ because they are unlike terms. Similarly, we cannot combine $\sqrt{2} + \sqrt{5}$ because they are unlike radicals. When combining radicals, the first step is to see if we can write pairs of terms as like radicals, as demonstrated in the next example.

EXAMPLE 2 Finding like radicals

Write each pair of terms as like radicals, if possible.
(a) $\sqrt{45}, \sqrt{20}$ (b) $\sqrt{27}, \sqrt{5}$ (c) $5\sqrt[3]{16}, 4\sqrt[3]{54}$

Solution

(a) The expressions $\sqrt{45}$ and $\sqrt{20}$ are unlike radicals. However, they can be changed to like radicals, as follows.

Write as like radicals.
$$\sqrt{45} = \sqrt{9 \cdot 5} = \sqrt{9} \cdot \sqrt{5} = 3\sqrt{5} \text{ and}$$
$$\sqrt{20} = \sqrt{4 \cdot 5} = \sqrt{4} \cdot \sqrt{5} = 2\sqrt{5}$$

The expressions $3\sqrt{5}$ and $2\sqrt{5}$ are like radicals.

(b) Because $\sqrt{27} = \sqrt{9 \cdot 3} = \sqrt{9} \cdot \sqrt{3} = 3\sqrt{3}$, the given expressions $\sqrt{27}$ and $\sqrt{5}$ are unlike radicals and cannot be written as like radicals.

(c) $5\sqrt[3]{16} = 5\sqrt[3]{8 \cdot 2} = 5\sqrt[3]{8} \cdot \sqrt[3]{2} = 5 \cdot 2 \cdot \sqrt[3]{2} = 10\sqrt[3]{2}$ and
$$4\sqrt[3]{54} = 4\sqrt[3]{27 \cdot 2} = 4\sqrt[3]{27} \cdot \sqrt[3]{2} = 4 \cdot 3 \cdot \sqrt[3]{2} = 12\sqrt[3]{2}$$

The expressions $10\sqrt[3]{2}$ and $12\sqrt[3]{2}$ are like radicals.

Now Try Exercises 9, 11, 13

We use these techniques to add radical expressions in the next two examples.

EXAMPLE 3 Adding radical expressions

Add the expressions and simplify.
(a) $\sqrt{12} + 7\sqrt{3}$ (b) $\sqrt[3]{16} + \sqrt[3]{2}$ (c) $3\sqrt{2} + \sqrt{8} + \sqrt{18}$

Solution
(a)
$$\sqrt{12} + 7\sqrt{3} = \sqrt{4 \cdot 3} + 7\sqrt{3}$$
$$= \sqrt{4} \cdot \sqrt{3} + 7\sqrt{3}$$
$$= 2\sqrt{3} + 7\sqrt{3} \quad \text{Like radicals}$$
$$= 9\sqrt{3}$$

(b)
$$\sqrt[3]{16} + \sqrt[3]{2} = \sqrt[3]{8 \cdot 2} + \sqrt[3]{2}$$
$$= \sqrt[3]{8} \cdot \sqrt[3]{2} + \sqrt[3]{2}$$
$$= 2\sqrt[3]{2} + 1\sqrt[3]{2} \quad \text{Like radicals}$$
$$= 3\sqrt[3]{2}$$

READING CHECK 2

Which pairs of terms can be written as like radicals.
- $\sqrt{45}, \sqrt{80}$
- $\sqrt{5}, \sqrt{20}$
- $\sqrt{18}, \sqrt{6}$
- $\sqrt{2}, \sqrt{32}$

(c)
$$3\sqrt{2} + \sqrt{8} + \sqrt{18} = 3\sqrt{2} + \sqrt{4 \cdot 2} + \sqrt{9 \cdot 2}$$
$$= 3\sqrt{2} + \sqrt{4} \cdot \sqrt{2} + \sqrt{9} \cdot \sqrt{2}$$
$$= 3\sqrt{2} + 2\sqrt{2} + 3\sqrt{2} \quad \text{Like radicals}$$
$$= 8\sqrt{2}$$

Now Try Exercises 29, 31, 43

EXAMPLE 4 Adding radical expressions

Add the expressions and simplify. Assume that all variables are positive.
(a) $\sqrt[4]{32} + 3\sqrt[4]{2}$ (b) $-2\sqrt{4x} + \sqrt{x}$ (c) $3\sqrt{3k} + 5\sqrt{12k} + 9\sqrt{48k}$

Solution
(a) Because $\sqrt[4]{32} = \sqrt[4]{16 \cdot 2} = \sqrt[4]{16} \cdot \sqrt[4]{2} = 2\sqrt[4]{2}$, we can add and simplify as follows.
$$\sqrt[4]{32} + 3\sqrt[4]{2} = 2\sqrt[4]{2} + 3\sqrt[4]{2} = 5\sqrt[4]{2}$$

(b) Note that $\sqrt{4x} = \sqrt{4} \cdot \sqrt{x} = 2\sqrt{x}$.
$$-2\sqrt{4x} + \sqrt{x} = -2(2\sqrt{x}) + \sqrt{x} = -4\sqrt{x} + 1\sqrt{x} = -3\sqrt{x}$$

(c) Note that $\sqrt{12k} = \sqrt{4} \cdot \sqrt{3k} = 2\sqrt{3k}$ and that $\sqrt{48k} = \sqrt{16} \cdot \sqrt{3k} = 4\sqrt{3k}$.

$$3\sqrt{3k} + 5\sqrt{12k} + 9\sqrt{48k} = 3\sqrt{3k} + 5(2\sqrt{3k}) + 9(4\sqrt{3k})$$
$$= (3 + 10 + 36)\sqrt{3k}$$
$$= 49\sqrt{3k}$$

Now Try Exercises 45, 47, 49

SUBTRACTING LIKE RADICALS Subtraction of radical expressions is similar to addition, as illustrated in the next four examples.

EXAMPLE 5 Subtracting like radicals

Simplify the expressions.
(a) $5\sqrt{7} - 3\sqrt{7}$ **(b)** $8\sqrt[3]{5} - 3\sqrt[3]{5} + \sqrt[3]{11}$ **(c)** $5\sqrt{z} + \sqrt[3]{z} - 2\sqrt{z}$

Solution
(a) $5\sqrt{7} - 3\sqrt{7} = (5 - 3)\sqrt{7} = 2\sqrt{7}$
(b) $8\sqrt[3]{5} - 3\sqrt[3]{5} + \sqrt[3]{11} = (8 - 3)\sqrt[3]{5} + \sqrt[3]{11} = 5\sqrt[3]{5} + \sqrt[3]{11}$
(c) $5\sqrt{z} + \sqrt[3]{z} - 2\sqrt{z} = 5\sqrt{z} - 2\sqrt{z} + \sqrt[3]{z}$ Commutative property
$\qquad\qquad\qquad\qquad\quad = (5 - 2)\sqrt{z} + \sqrt[3]{z}$ Distributive property
$\qquad\qquad\qquad\qquad\quad = 3\sqrt{z} + \sqrt[3]{z}$ Subtract.

NOTE: We cannot combine $3\sqrt{z} + \sqrt[3]{z}$ because their indexes are different. That is, one term is a square root and the other is a cube root. ∎

Now Try Exercises 33, 35, 39

EXAMPLE 6 Subtracting radical expressions

Subtract and simplify. Assume that all variables are positive.
(a) $3\sqrt[3]{xy^2} - 2\sqrt[3]{xy^2}$ **(b)** $\sqrt{16x^3} - \sqrt{x^3}$ **(c)** $\sqrt[3]{\dfrac{5x}{27}} - \dfrac{\sqrt[3]{5x}}{6}$

Solution
(a) $3\sqrt[3]{xy^2} - 2\sqrt[3]{xy^2} = (3 - 2)\sqrt[3]{xy^2} = \sqrt[3]{xy^2}$
(b) $\sqrt{16x^3} - \sqrt{x^3} = \sqrt{16x^2} \cdot \sqrt{x} - \sqrt{x^2} \cdot \sqrt{x}$ Factor out perfect squares.
$\qquad\qquad\qquad = 4x\sqrt{x} - x\sqrt{x}$ Simplify.
$\qquad\qquad\qquad = (4x - x)\sqrt{x}$ Distributive property
$\qquad\qquad\qquad = 3x\sqrt{x}$ Subtract.

(c) $\sqrt[3]{\dfrac{5x}{27}} - \dfrac{\sqrt[3]{5x}}{6} = \dfrac{\sqrt[3]{5x}}{\sqrt[3]{27}} - \dfrac{\sqrt[3]{5x}}{6}$ Quotient rule for radical expressions

$\qquad\qquad\qquad = \dfrac{\sqrt[3]{5x}}{3} - \dfrac{\sqrt[3]{5x}}{6}$ Evaluate $\sqrt[3]{27} = 3$.

$\qquad\qquad\qquad = \dfrac{2\sqrt[3]{5x}}{6} - \dfrac{\sqrt[3]{5x}}{6}$ Find a common denominator.

$\qquad\qquad\qquad = \dfrac{2\sqrt[3]{5x} - \sqrt[3]{5x}}{6}$ Subtract numerators.

$\qquad\qquad\qquad = \dfrac{\sqrt[3]{5x}}{6}$ Simplify.

Now Try Exercises 55, 63, 65

🌐 Math in Context (Business) Our productivity as a nation has increased significantly during the past 70 years because of new technology. For example, if a typical business bought $10,000 in equipment per worker in 2005, it would have generated about $48,000 per worker in revenue. This same investment by a business in 1935 would have generated only $23,000 in revenue (adjusted to 2005 dollars). This situation can be described by the following two functions, which approximate the increase in revenue resulting from investing x dollars in equipment (per worker).

$$N(x) = 400\sqrt{x} + 8000 \quad \text{and} \quad O(x) = 195\sqrt{x} + 3500$$
$$\text{2005 technology (new)} \qquad \text{1935 technology (old)}$$

What does the difference of these two functions represent? In the next example, we learn how to simplify and interpret this difference.

EXAMPLE 7   **Increasing productivity with new technology**

For the functions given by

$$N(x) = 400\sqrt{x} + 8000 \quad \text{and} \quad O(x) = 195\sqrt{x} + 3500,$$
$$\text{2005 technology (new)} \qquad \text{1935 technology (old)}$$

complete each of the following.
(a) Find their difference $D(x) = N(x) - O(x)$. Simplify your answer.
(b) Evaluate $D(40{,}000)$ and interpret the result.

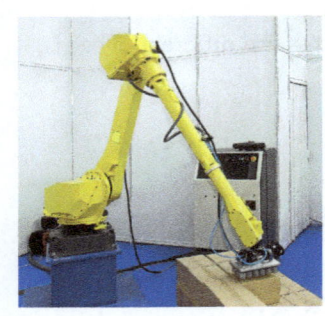

Solution
(a) $D(x) = N(x) - O(x)$

$\qquad = (400\sqrt{x} + 8000) - (195\sqrt{x} + 3500)$ Substitute.
$\qquad = 400\sqrt{x} - 195\sqrt{x} + 8000 - 3500$ Rewrite terms.
$\qquad = (400 - 195)\sqrt{x} + 4500$ Distributive property
$\qquad = 205\sqrt{x} + 4500$ Simplify.

The difference is given by $D(x) = 205\sqrt{x} + 4500$.

(b) $D(40{,}000) = 205\sqrt{40{,}000} + 4500 = \$45{,}500$. An investment of $40,000 per worker resulted in $45,500 *more* revenue per worker in 2005 than in 1935 (adjusted to 2005 dollars).

Now Try Exercises 81, 135

EXAMPLE 8   **Subtracting radical expressions**

Subtract and simplify. Assume that all variables are positive.

(a) $\dfrac{5\sqrt{2}}{3} - \dfrac{2\sqrt{2}}{4}$ (b) $\sqrt[4]{81a^5b^6} - \sqrt[4]{16ab^2}$ (c) $3\sqrt[3]{\dfrac{n^5}{27}} - 2\sqrt[3]{n^2}$

Solution

(a) $\dfrac{5\sqrt{2}}{3} - \dfrac{2\sqrt{2}}{4} = \dfrac{5\sqrt{2}}{3} \cdot \dfrac{4}{4} - \dfrac{2\sqrt{2}}{4} \cdot \dfrac{3}{3}$ LCD is 12.

$\qquad = \dfrac{20\sqrt{2}}{12} - \dfrac{6\sqrt{2}}{12}$ Multiply fractions.

$\qquad = \dfrac{14\sqrt{2}}{12}$ Subtract numerators.

$\qquad = \dfrac{7\sqrt{2}}{6}$ Simplify.

(b) $\sqrt[4]{81a^5b^6} - \sqrt[4]{16ab^2} = \sqrt[4]{81a^4b^4} \cdot \sqrt[4]{ab^2} - \sqrt[4]{16} \cdot \sqrt[4]{ab^2}$ Factor out perfect powers.
$= 3ab\sqrt[4]{ab^2} - 2\sqrt[4]{ab^2}$ Simplify.
$= (3ab - 2)\sqrt[4]{ab^2}$ Distributive property

(c) $3\sqrt[3]{\dfrac{n^5}{27}} - 2\sqrt[3]{n^2} = 3\sqrt[3]{\dfrac{n^3}{27}} \cdot \sqrt[3]{n^2} - 2\sqrt[3]{n^2}$ Factor out perfect cube.
$= \dfrac{3\sqrt[3]{n^3}}{\sqrt[3]{27}} \cdot \sqrt[3]{n^2} - 2\sqrt[3]{n^2}$ Quotient rule
$= n\sqrt[3]{n^2} - 2\sqrt[3]{n^2}$ Simplify: $\sqrt[3]{n^3} = n$ and $\sqrt[3]{27} = 3$.
$= (n - 2)\sqrt[3]{n^2}$ Distributive property

Now Try Exercises 69, 77, 79

Radicals often occur in geometry. In the next example, we find the perimeter of a triangle by adding radical expressions.

EXAMPLE 9 **Finding the perimeter of a triangle**

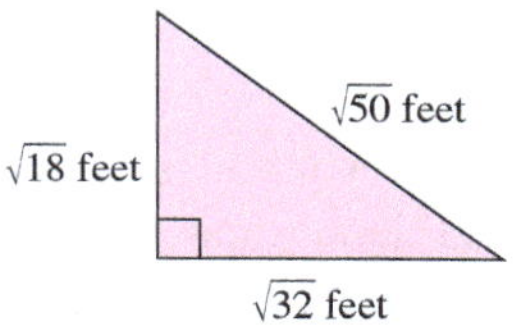

Find the *exact* perimeter of the right triangle shown in **FIGURE 17.11**. Then approximate your answer to the nearest hundredth of a foot.

FIGURE 17.11

Solution
The sum of the lengths of the sides of the triangle is
$$\sqrt{18} + \sqrt{32} + \sqrt{50} = 3\sqrt{2} + 4\sqrt{2} + 5\sqrt{2} = 12\sqrt{2}.$$
The perimeter is $12\sqrt{2} \approx 16.97$ feet.

Now Try Exercise 125

2 Multiplying Radical Expressions

Some types of radical expressions can be multiplied like binomials. For example, because
$$(x + 1)(x + 2) = x^2 + 3x + 2,$$
we have
$$(\sqrt{x} + 1)(\sqrt{x} + 2) = (\sqrt{x})^2 + 2\sqrt{x} + 1\sqrt{x} + 2 = x + 3\sqrt{x} + 2,$$
provided that $x \geq 0$. The next example demonstrates this technique.

EXAMPLE 10 **Multiplying radical expressions**

Multiply and simplify.
(a) $(\sqrt{b} - 4)(\sqrt{b} + 5)$ (b) $(4 + \sqrt{3})(4 - \sqrt{3})$

Solution
(a) This expression can be multiplied and then simplified.

$$(\sqrt{b} - 4)(\sqrt{b} + 5) = \sqrt{b} \cdot \sqrt{b} + 5\sqrt{b} - 4\sqrt{b} - 4 \cdot 5$$
$$= b + \sqrt{b} - 20$$

Multiply in the same way as we multiply $(b - 4)(b + 5)$.

Compare this product with $(b - 4)(b + 5) = b^2 + b - 20$.

(b) $(4 + \sqrt{3})(4 - \sqrt{3}) = (4)^2 - (\sqrt{3})^2$

Multiply as a sum and difference: $(a + b)(a - b) = a^2 - b^2$.

$= 16 - 3$
$= 13$

Now Try Exercises 85, 87

NOTE: Example 10(b) illustrates a special case for multiplying radicals. In general,

$$(\sqrt{a} + \sqrt{b})(\sqrt{a} - \sqrt{b}) = (\sqrt{a})^2 - (\sqrt{b})^2 = a - b,$$

provided a and b are nonnegative. ∎

3 Rationalizing the Denominator

In mathematics, it is common to write expressions without radicals in the denominator. Quotients containing radical expressions in the numerator or denominator can appear to be different but actually be equal. For example, $\frac{1}{\sqrt{3}}$ and $\frac{\sqrt{3}}{3}$ represent the same real number even though they look as though they are unequal. To show that they are equal, we multiply the first quotient by 1 in the form $\frac{\sqrt{3}}{\sqrt{3}}$.

$$\frac{1}{\sqrt{3}} \cdot \frac{\sqrt{3}}{\sqrt{3}} = \frac{1 \cdot \sqrt{3}}{\sqrt{3} \cdot \sqrt{3}} = \frac{\sqrt{3}}{3}$$

READING CHECK 3

Rationalize the denominator of $\frac{1}{\sqrt{5}}$.

If the denominator of a quotient contains only one term with one square root, then we can rationalize the denominator by multiplying the numerator and denominator by this square root. For example, the denominator of $\frac{1}{\sqrt{3}}$ contains one term, which is $\sqrt{3}$. Therefore, we multiplied $\frac{1}{\sqrt{3}}$ by $\frac{\sqrt{3}}{\sqrt{3}}$ to rationalize the denominator.

NOTE: $\sqrt{b} \cdot \sqrt{b} = \sqrt{b^2} = b$ for any *positive* number b. ∎

One way to *standardize* radical expressions is to remove any radical expressions from the denominator. This process is called **rationalizing the denominator**. Exercise 139 suggests one reason why people rationalized denominators before calculators were invented. The next example demonstrates how to rationalize the denominator of several quotients.

EXAMPLE 11 Rationalizing the denominator

Rationalize each denominator. Assume that all variables are positive.

(a) $\frac{1}{\sqrt{2}}$ **(b)** $\frac{3}{5\sqrt{3}}$ **(c)** $\sqrt{\frac{x}{24}}$ **(d)** $\frac{xy}{\sqrt{y^3}}$

Solution

(a) We start by multiplying this expression by 1 in the form $\frac{\sqrt{2}}{\sqrt{2}}$.

$$\frac{1}{\sqrt{2}} \cdot \frac{\sqrt{2}}{\sqrt{2}} = \frac{\sqrt{2}}{\sqrt{4}} = \frac{\sqrt{2}}{2}$$

Note that the expression $\frac{\sqrt{2}}{2}$ does not have a radical in the denominator.

(b) We multiply $\frac{3}{5\sqrt{3}}$ by 1 in the form $\frac{\sqrt{3}}{\sqrt{3}}$.

$$\frac{3}{5\sqrt{3}} \cdot \frac{\sqrt{3}}{\sqrt{3}} = \frac{3\sqrt{3}}{5\sqrt{9}} = \frac{3\sqrt{3}}{5 \cdot 3} = \frac{\sqrt{3}}{5}$$

(c) Because $\sqrt{24} = \sqrt{4} \cdot \sqrt{6} = 2\sqrt{6}$, we start by simplifying $\sqrt{\dfrac{x}{24}}$.

$$\sqrt{\dfrac{x}{24}} = \dfrac{\sqrt{x}}{\sqrt{24}} = \dfrac{\sqrt{x}}{2\sqrt{6}}$$

To rationalize the denominator, we multiply this expression by **1** in the form $\dfrac{\sqrt{6}}{\sqrt{6}}$.

$$\dfrac{\sqrt{x}}{2\sqrt{6}} = \dfrac{\sqrt{x}}{2\sqrt{6}} \cdot \dfrac{\sqrt{6}}{\sqrt{6}} = \dfrac{\sqrt{6x}}{12}$$

(d) Because $\sqrt{y^3} = \sqrt{y^2} \cdot \sqrt{y} = y\sqrt{y}$, we start by simplifying $\dfrac{xy}{\sqrt{y^3}}$.

$$\dfrac{xy}{\sqrt{y^3}} = \dfrac{xy}{y\sqrt{y}} = \dfrac{x}{\sqrt{y}}$$

To rationalize the denominator, we multiply by **1** in the form $\dfrac{\sqrt{y}}{\sqrt{y}}$.

$$\dfrac{x}{\sqrt{y}} \cdot \dfrac{\sqrt{y}}{\sqrt{y}} = \dfrac{x\sqrt{y}}{y}$$

Now Try Exercises 97, 101, 103, 105

EXAMPLE 12 **Rationalizing a denominator in geometry**

A square with a diagonal of length x has sides of length $\dfrac{x}{\sqrt{2}}$. Find the perimeter and rationalize the denominator. See **FIGURE 17.12**.

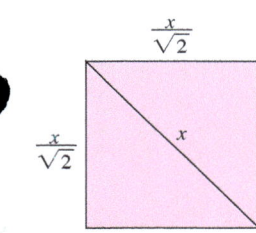

FIGURE 17.12

Solution
The perimeter is 4 times the length of one side.

$$4 \cdot \dfrac{x}{\sqrt{2}} = \dfrac{4x}{\sqrt{2}}$$

To rationalize the denominator, we multiply the ratio by **1** in the form $\dfrac{\sqrt{2}}{\sqrt{2}}$.

$$\dfrac{4x}{\sqrt{2}} \cdot \dfrac{\sqrt{2}}{\sqrt{2}} = \dfrac{4x\sqrt{2}}{2} = 2x\sqrt{2}$$

The perimeter is $2x\sqrt{2}$.

Now Try Exercise 127

When the denominator is either a sum or difference containing a square root, we rationalize the denominator by multiplying the numerator and denominator by the *conjugate* of the denominator. In this case, the **conjugate** of the denominator is found by changing a $+$ sign to a $-$ sign or vice versa. **TABLE 17.5** lists examples of conjugates.

Expressions and Their Conjugates

Expression	$1 + \sqrt{2}$	$\sqrt{3} - 2$	$\sqrt{x} + 7$	$\sqrt{a} - \sqrt{b}$
Conjugate	$1 - \sqrt{2}$	$\sqrt{3} + 2$	$\sqrt{x} - 7$	$\sqrt{a} + \sqrt{b}$

TABLE 17.5

In the next two examples, we use this method to rationalize a denominator.

EXAMPLE 13 Using a conjugate to rationalize the denominator

Rationalize the denominator of $\dfrac{1}{1 + \sqrt{2}}$.

Solution
From **TABLE 17.5**, on the previous page, the conjugate of $1 + \sqrt{2}$ is $1 - \sqrt{2}$.

$$\dfrac{1}{1 + \sqrt{2}} = \dfrac{1}{1 + \sqrt{2}} \cdot \dfrac{(1 - \sqrt{2})}{(1 - \sqrt{2})} \quad \text{Multiply numerator and denominator by the conjugate.}$$

$$= \dfrac{1 - \sqrt{2}}{(1)^2 - (\sqrt{2})^2} \quad (a + b)(a - b) = a^2 - b^2$$

$$= \dfrac{1 - \sqrt{2}}{1 - 2} \quad \text{Simplify.}$$

$$= \dfrac{1 - \sqrt{2}}{-1} \quad \text{Subtract.}$$

$$= \dfrac{1}{-1} - \dfrac{\sqrt{2}}{-1} \quad \dfrac{a - b}{c} = \dfrac{a}{c} - \dfrac{b}{c}$$

$$= -1 + \sqrt{2} \quad \text{Simplify.}$$

READING CHECK 4

- Rationalize the denominator of $\dfrac{1}{1 - \sqrt{3}}$.

Now Try Exercise 107

EXAMPLE 14 Rationalizing the denominator

Rationalize the denominator.

(a) $\dfrac{3 + \sqrt{5}}{2 - \sqrt{5}}$ (b) $\dfrac{\sqrt{x}}{\sqrt{x} - 2}$

Solution
(a) The conjugate of the denominator is $2 + \sqrt{5}$.

$$\dfrac{3 + \sqrt{5}}{2 - \sqrt{5}} = \dfrac{(3 + \sqrt{5})}{(2 - \sqrt{5})} \cdot \dfrac{(2 + \sqrt{5})}{(2 + \sqrt{5})} \quad \text{Multiply by 1.}$$

$$= \dfrac{6 + 3\sqrt{5} + 2\sqrt{5} + (\sqrt{5})^2}{(2)^2 - (\sqrt{5})^2} \quad \text{Multiply.}$$

$$= \dfrac{11 + 5\sqrt{5}}{4 - 5} \quad \text{Combine terms.}$$

$$= -11 - 5\sqrt{5} \quad \text{Simplify.}$$

(b) The conjugate of the denominator is $\sqrt{x} + 2$.

$$\dfrac{\sqrt{x}}{\sqrt{x} - 2} = \dfrac{\sqrt{x}}{(\sqrt{x} - 2)} \cdot \dfrac{(\sqrt{x} + 2)}{(\sqrt{x} + 2)} \quad \text{Multiply by 1.}$$

$$= \dfrac{x + 2\sqrt{x}}{x - 4} \quad \text{Multiply.}$$

Now Try Exercises 111, 115

EXAMPLE 15 **Rationalizing a denominator that has a cube root**

Rationalize the denominator of $\frac{5}{\sqrt[3]{x}}$.

Solution

The expression $\frac{5}{\sqrt[3]{x}}$ is equal to $\frac{5}{x^{1/3}}$. To rationalize the denominator, $x^{1/3}$, we can multiply it by $x^{2/3}$ because $x^{1/3} \cdot x^{2/3} = x^{1/3+2/3} = x^1$.

$$\frac{5}{x^{1/3}} = \frac{5}{x^{1/3}} \cdot \frac{x^{2/3}}{x^{2/3}} \quad \text{Multiply by 1.}$$

$$= \frac{5x^{2/3}}{x^{1/3+2/3}} \quad \text{Product rule}$$

$$= \frac{5\sqrt[3]{x^2}}{x} \quad \text{Add; write in radical notation.}$$

Now Try Exercise 121

17.4 Putting It All Together

CONCEPT	EXPLANATION	EXAMPLES
Like Radicals	Like radicals have the same index and the same radicand.	$7\sqrt{5}$ and $3\sqrt{5}$ are like radicals. $5\sqrt[3]{ab}$ and $\sqrt[3]{ab}$ are like radicals. $\sqrt[3]{5}$ and $\sqrt[3]{4}$ are unlike radicals. $\sqrt[3]{7}$ and $\sqrt{7}$ are unlike radicals.
Adding and Subtracting Radical Expressions	Combine like radicals when adding or subtracting. We cannot combine unlike radicals such as $\sqrt{2}$ and $\sqrt{5}$. But sometimes we can rewrite radicals and then combine them.	$6\sqrt{13} + \sqrt{13} = (6+1)\sqrt{13} = 7\sqrt{13}$ $\sqrt{40} - \sqrt{10} = \sqrt{4} \cdot \sqrt{10} - \sqrt{10}$ $= 2\sqrt{10} - \sqrt{10}$ $= \sqrt{10}$
Multiplying Radical Expressions	Radical expressions can sometimes be multiplied like binomials.	$(\sqrt{a}-5)(\sqrt{a}+5) = a - 25$ and $(\sqrt{x}-3)(\sqrt{x}+1) = x - 2\sqrt{x} - 3$
Conjugate	The conjugate is found by changing a $+$ sign to a $-$ sign or vice versa.	Expression Conjugate $\sqrt{x}+7$ $\sqrt{x}-7$ $\sqrt{a}-2\sqrt{b}$ $\sqrt{a}+2\sqrt{b}$
Rationalizing a Denominator that has One Term	Write the quotient without a radical expression in the denominator.	To rationalize $\frac{5}{\sqrt{7}}$, multiply the expression by 1 in the form $\frac{\sqrt{7}}{\sqrt{7}}$. $\frac{5}{\sqrt{7}} \cdot \frac{\sqrt{7}}{\sqrt{7}} = \frac{5\sqrt{7}}{\sqrt{49}} = \frac{5\sqrt{7}}{7}$
Rationalizing a Denominator that has Two Terms	Multiply the numerator and denominator by the conjugate of the denominator.	$\frac{1}{2-\sqrt{3}} = \frac{1}{2-\sqrt{3}} \cdot \frac{(2+\sqrt{3})}{(2+\sqrt{3})}$ $= \frac{2+\sqrt{3}}{4-3} = 2+\sqrt{3}$

17.4 Exercises

CONCEPTS AND VOCABULARY

1. $\sqrt{a} + \sqrt{a} =$ _____
2. $\sqrt[3]{b} + \sqrt[3]{b} + \sqrt[3]{b} =$ _____
3. You cannot simplify $\sqrt[3]{4} + \sqrt[3]{7}$ because they are not _____ radicals.
4. Can you simplify $4\sqrt{15} - 3\sqrt{15}$? Explain.
5. Does $6 + 3\sqrt{5}$ equal $9\sqrt{5}$?
6. To rationalize the denominator of $\frac{2}{\sqrt{7}}$, multiply this expression by _____.
7. What is the conjugate of $\sqrt{t} - 5$?
8. To rationalize the denominator of $\frac{1}{5 - \sqrt{2}}$, multiply this expression by _____.

WRITING LIKE RADICALS

Exercises 9–18: (Refer to Example 2.) Write the terms as like radicals, if possible. Assume that all variables are positive.

9. $\sqrt{12}, \sqrt{24}$
10. $\sqrt{18}, \sqrt{27}$
11. $\sqrt{7}, \sqrt{28}, \sqrt{63}$
12. $\sqrt{200}, \sqrt{300}, \sqrt{500}$
13. $\sqrt[3]{16}, \sqrt[3]{-54}$
14. $\sqrt[3]{80}, \sqrt[3]{10}$
15. $\sqrt{x^2 y}, \sqrt{4y^2}$
16. $\sqrt{x^5 y^3}, \sqrt{9xy}$
17. $\sqrt[3]{8xy}, \sqrt[3]{x^4 y^4}$
18. $\sqrt[3]{64x^4}, \sqrt[3]{-8x}$

ADDING AND SUBSTRACTING RADICAL EXPRESSIONS

Exercises 19–80: If possible, simplify the expression. Assume that all variables are positive.

19. $2\sqrt{3} + 7\sqrt{3}$
20. $8\sqrt{7} + 2\sqrt{7}$
21. $4\sqrt[3]{5} + 2\sqrt[3]{5}$
22. $\sqrt[3]{13} + 3\sqrt[3]{13}$
23. $7 + 4\sqrt{7}$
24. $8 - 4\sqrt{3}$
25. $2\sqrt{3} + 3\sqrt{2}$
26. $\sqrt{6} + \sqrt{17}$
27. $\sqrt{3} + \sqrt[3]{3}$
28. $\sqrt{6} + \sqrt[4]{6}$
29. $\sqrt[3]{16} + 3\sqrt[3]{2}$
30. $\sqrt[3]{24} + \sqrt[3]{81}$
31. $\sqrt{2} + \sqrt{18} + \sqrt{32}$
32. $2\sqrt{3} + \sqrt{12} + \sqrt{27}$
33. $11\sqrt{11} - 5\sqrt{11}$
34. $9\sqrt{5} + \sqrt{2} - \sqrt{5}$
35. $\sqrt{x} + \sqrt{x} - \sqrt{y}$
36. $\sqrt{xy^2} - \sqrt{x}$
37. $\sqrt[3]{z} + \sqrt[3]{z}$
38. $\sqrt[3]{y} - \sqrt[3]{y}$
39. $2\sqrt[3]{6} - 7\sqrt[3]{6}$
40. $18\sqrt[3]{3} + 3\sqrt[3]{3}$
41. $\sqrt[3]{y^6} - \sqrt[3]{y^3}$
42. $2\sqrt{20} + 7\sqrt{5} + 3\sqrt{2}$
43. $3\sqrt{28} + 3\sqrt{7}$
44. $9\sqrt{18} - 2\sqrt{8}$
45. $\sqrt[4]{48} + 4\sqrt[4]{3}$
46. $\sqrt[4]{32} + \sqrt[4]{16}$
47. $\sqrt{9x} + \sqrt{16x}$
48. $-3\sqrt{x} + 5\sqrt{x}$
49. $3\sqrt{2k} + \sqrt{8k} + \sqrt{18k}$
50. $3\sqrt{k} + 2\sqrt{4k} + \sqrt{9k}$
51. $\sqrt{44} - 4\sqrt{11}$
52. $\sqrt[4]{5} + 2\sqrt[4]{5}$
53. $2\sqrt[3]{16} + \sqrt[3]{2} - \sqrt{2}$
54. $5\sqrt[3]{x} - 3\sqrt[3]{x}$
55. $\sqrt[3]{xy} - 2\sqrt[3]{xy}$
56. $3\sqrt{x^3} - \sqrt{x}$
57. $\sqrt{4x + 8} + \sqrt{x + 2}$
58. $\sqrt{2a + 1} + \sqrt{8a + 4}$
59. $\sqrt{9x + 18} - \sqrt{4x + 8}$
60. $\sqrt{25x - 25} - \sqrt{4x - 4}$
61. $\sqrt{x^3 + x^2} - \sqrt{x + 1}, x \geq 0$
62. $\sqrt{x^3 - x^2} - \sqrt{4x - 4}, x \geq 1$
63. $\sqrt{25x^3} - \sqrt{x^3}$
64. $\sqrt{36x^5} - \sqrt{25x^5}$
65. $\sqrt[3]{\frac{7x}{8}} - \frac{\sqrt[3]{7x}}{3}$
66. $\sqrt[3]{\frac{8x^2}{27}} - \sqrt[3]{\frac{x^2}{8}}$
67. $\frac{4\sqrt{3}}{3} + \frac{\sqrt{3}}{6}$
68. $\frac{8\sqrt{5}}{7} + \frac{4\sqrt{5}}{2}$
69. $\frac{15\sqrt{8}}{4} - \frac{2\sqrt{2}}{5}$
70. $\frac{23\sqrt{11}}{2} - \frac{\sqrt{44}}{8}$
71. $2\sqrt[4]{64} - \sqrt[4]{324} + \sqrt[4]{4}$
72. $2\sqrt[3]{16} - 5\sqrt[3]{54} + 10\sqrt[3]{2}$
73. $5\sqrt[4]{x^5} - \sqrt[4]{x}$
74. $20\sqrt[3]{b^4} - 4\sqrt[3]{b}$
75. $\sqrt{64x^3} - \sqrt{x} + 3\sqrt{x}$
76. $2\sqrt{3z} + 3\sqrt{12z} + 3\sqrt{48z}$
77. $\sqrt[4]{81a^5 b^5} - \sqrt[4]{ab}$
78. $\sqrt[4]{xy^5} - \sqrt[4]{x^5 y}$
79. $5\sqrt[3]{\frac{n^4}{125}} - 2\sqrt[3]{n}$
80. $\sqrt[3]{\frac{8x}{27}} - \frac{2\sqrt[3]{x}}{3}$

PERFORMING OPERATIONS ON RADICAL FUNCTIONS

Exercises 81–84: Find $(f + g)(x)$ and $(f - g)(x)$.

81. $f(x) = 5\sqrt{x} - 2$, $g(x) = -2\sqrt{x} + 3$
82. $f(x) = 3\sqrt{4x - 4}$, $g(x) = 5\sqrt{x - 1}$
83. $f(x) = \sqrt[3]{8x} + 1$, $g(x) = 2\sqrt[3]{x} - 1$
84. $f(x) = \sqrt[3]{27x}$, $g(x) = \sqrt[3]{64x}$

MULTIPLYING BINOMIALS CONTAINING RADICALS

Exercises 85–96: Multiply and simplify.

85. $(\sqrt{x} - 3)(\sqrt{x} + 2)$
86. $(2\sqrt{x} + 1)(\sqrt{x} + 4)$
87. $(3 + \sqrt{7})(3 - \sqrt{7})$
88. $(5 - \sqrt{5})(5 + \sqrt{5})$
89. $(11 - \sqrt{2})(11 + \sqrt{2})$
90. $(6 + \sqrt{3})(6 - \sqrt{3})$
91. $(\sqrt{x} + 8)(\sqrt{x} - 8)$
92. $(\sqrt{ab} - 3)(\sqrt{ab} + 3)$
93. $(\sqrt{ab} - \sqrt{c})(\sqrt{ab} + \sqrt{c})$
94. $(\sqrt{2x} + \sqrt{3y})(\sqrt{2x} - \sqrt{3y})$
95. $(\sqrt{x} - 7)(\sqrt{x} + 8)$
96. $(\sqrt{ab} - 1)(\sqrt{ab} - 2)$

RATIONALIZING THE DENOMINATOR

Exercises 97–124: Rationalize the denominator.

97. $\frac{1}{\sqrt{7}}$
98. $\frac{1}{\sqrt{23}}$
99. $\frac{4}{\sqrt{3}}$
100. $\frac{8}{\sqrt{2}}$
101. $\frac{5}{3\sqrt{5}}$
102. $\frac{6}{11\sqrt{3}}$
103. $\sqrt{\frac{b}{12}}$
104. $\sqrt{\frac{5b}{72}}$
105. $\frac{rt}{2\sqrt{r^3}}$
106. $\frac{m^2 n}{2\sqrt{m^5}}$
107. $\frac{1}{3 - \sqrt{2}}$
108. $\frac{1}{\sqrt{3} - 2}$
109. $\frac{\sqrt{2}}{\sqrt{5} + 2}$
110. $\frac{\sqrt{3}}{\sqrt{3} + 2}$
111. $\frac{\sqrt{7} - 2}{\sqrt{7} + 2}$
112. $\frac{\sqrt{3} - 1}{\sqrt{3} + 1}$
113. $\frac{1}{\sqrt{7} - \sqrt{6}}$
114. $\frac{1}{\sqrt{8} - \sqrt{7}}$
115. $\frac{\sqrt{z}}{\sqrt{z} - 3}$
116. $\frac{2\sqrt{z}}{2 - \sqrt{z}}$
117. $\frac{\sqrt{a} + \sqrt{b}}{\sqrt{a} - \sqrt{b}}$
118. $\frac{\sqrt{x} - 2\sqrt{y}}{\sqrt{x} + 2\sqrt{y}}$
119. $\frac{1}{\sqrt{x + 1} - \sqrt{x}}$
120. $\frac{1}{\sqrt{a + 1} + \sqrt{a}}$
121. $\frac{3}{\sqrt[3]{x}}$
122. $\frac{6}{5\sqrt[3]{x}}$
123. $\frac{1}{\sqrt[3]{x^2}}$
124. $\frac{2}{\sqrt[3]{(x - 2)^2}}$

RELATING GEOMETRY AND RADICALS

125. **Perimeter** (Refer to Example 9.) Find the exact perimeter of the right triangle. Then approximate your answer.

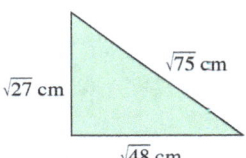

126. **Perimeter** Find the exact perimeter of the rectangle. Then approximate your answer.

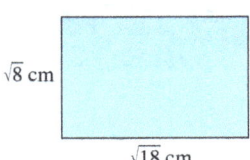

127. **Geometry** (Refer to Example 12.) A square has a diagonal with length $\sqrt{3}$. Find the perimeter and rationalize the denominator.

128. **Geometry** A rectangle has sides of length $\sqrt{8}$ and $\sqrt{32}$. Find the perimeter of the rectangle.

129. **Geometry** (Refer to Example 12.) A square has a diagonal that is 60 feet long. Find the exact perimeter of the rectangle and simplify your answer.

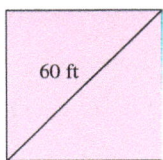

130. **Geometry** A square has a diagonal that is 10 feet long. Find the exact perimeter of the rectangle and simplify your answer.

131. Geometry A square has an area of x square feet. Find the length of the diagonal of the square.

132. Geometry A square has an area of 16 square feet. Find the length of the diagonal of the square.

APPLICATIONS INVOLVING RADICAL EXPRESSIONS

Exercises 133 and 134. **Flood Channels** *The rate R at which water flows through a channel is given by the formula $R = k\sqrt{m}$, where m is the slope of the channel and k is a constant determined by the shape of the channel. Suppose that two open flood channels have flow rates R_1 and R_2 given by*

$$R_1 = 1500\sqrt{m_1} \quad \text{and} \quad R_2 = 2000\sqrt{m_2},$$

where R_1 and R_2 are in cubic feet per second and m_1 and m_2 are the slopes of each channel, respectively.

133. Find $R_1 + R_2$ if $m_1 = 0.09$ and $m_2 = 0.04$.

134. Find $R_1 + R_2$ if both channels have slope m.

Exercises 135–138: **Drug Law Violations** *The number of adults in thousands incarcerated in state or federal prisons for drug law violations x years after 1984 can be estimated by the formula $P = 70\sqrt{x} + 45$. Similarly, the formula $J = 20\sqrt{x} + 25$ gives the number of adults in thousands incarcerated in jails for drug law violations x years after 1984.*

135. Evaluate P and J for $x = 16$. Interpret each.

136. Add your results from Exercise 135. Interpret the result.

137. Simplify $T = (70\sqrt{x} + 45) + (20\sqrt{x} + 25)$.

138. What does T represent?

WRITING ABOUT MATHEMATICS

139. Suppose that a student knows that $\sqrt{3} \approx 1.73205$ and does not have a calculator. Which of the following expressions would be easier to evaluate by hand? Why?

$$\frac{1}{\sqrt{3}} \quad \text{or} \quad \frac{\sqrt{3}}{3}$$

140. A student simplifies an expression *incorrectly*:

$$\sqrt{8} + \sqrt[3]{16} \stackrel{?}{=} \sqrt{4 \cdot 2} + \sqrt[3]{8 \cdot 2}$$
$$\stackrel{?}{=} \sqrt{4} \cdot \sqrt{2} + \sqrt[3]{8} \cdot \sqrt[3]{2}$$
$$\stackrel{?}{=} 2\sqrt{2} + 2\sqrt[3]{2}$$
$$\stackrel{?}{=} 4\sqrt{4}$$
$$\stackrel{?}{=} 8.$$

Explain any errors that the student made. What would you do differently?

SECTIONS 17.3 and 17.4 — Checking Basic Concepts

1. Simplify each expression. Assume that all variables are positive.
  (a) $(64^{-3/2})^{1/3}$
  (b) $\sqrt{5} \cdot \sqrt{20}$
  (c) $\sqrt[3]{-8x^4y}$
  (d) $\sqrt{\dfrac{4b}{5}} \cdot \sqrt{\dfrac{4b^3}{5}}$

2. Simplify $\sqrt[3]{7} \cdot \sqrt{7}$.

3. Simplify each expression.
  (a) $\sqrt{3} \cdot \sqrt{12}$
  (b) $\dfrac{\sqrt[3]{81}}{\sqrt[3]{3}}$
  (c) $\sqrt{36x^6}, x > 0$

4. Simplify each expression.
  (a) $5\sqrt{6} + 2\sqrt{6} + \sqrt{7}$
  (b) $8\sqrt[3]{x} - 3\sqrt[3]{x}$
  (c) $\sqrt{9x} - \sqrt{4x}$

5. Simplify each expression.
  (a) $\sqrt[3]{xy^4} - \sqrt[3]{x^4y}$
  (b) $(4 - \sqrt{2})(4 + \sqrt{2})$

6. Rationalize the denominator of $\dfrac{6}{2\sqrt{6}}$.

7. Rationalize the denominator of $\dfrac{2}{\sqrt{5} - 1}$.

17.5 More Radical Functions

Objectives

1. Recognizing Root Functions and Their Properties
2. Recognizing Power Functions and Their Properties
3. Modeling with Power Functions

1 Recognizing Root Functions and Their Properties

Earlier in this chapter, we discussed the square root and cube root functions. There are also functions for higher roots. The following See the Concept shows graphs of four common root functions.

See the Concept 1   THE ROOT FUNCTIONS

Square Root Function

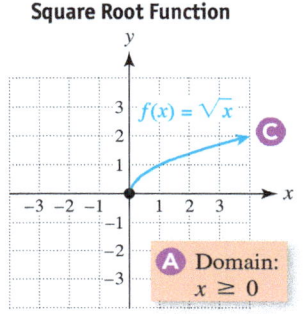

A Domain: $x \geq 0$

Cube Root Function

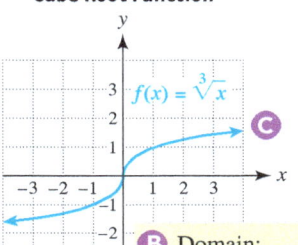

B Domain: $-\infty < x < \infty$

Fourth Root Function

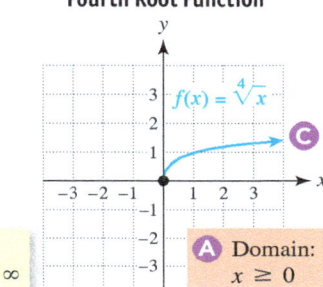

A Domain: $x \geq 0$

Fifth Root Function

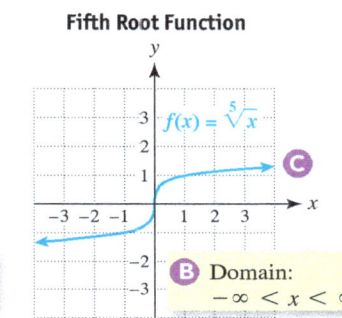

B Domain: $-\infty < x < \infty$

A The domain for an even root function is $[0, \infty)$.

B The domain for an odd root function is $(-\infty, \infty)$.

C The higher the root, the more slowly the graph increases for $x \geq 1$.

WATCH VIDEO IN MML.

NEW VOCABULARY

☐ Root function
☐ Power function

EXAMPLE 1   Evaluating root functions

If possible, evaluate each root function f at the given x-values.
(a) $f(x) = \sqrt{x + 1}, \quad x = -5, x = 3$
(b) $f(x) = \sqrt[3]{1 - x}, \quad x = -7, x = 28$
(c) $f(x) = \sqrt[4]{x - 7}, \quad x = 5, x = 88$
(d) $f(x) = \sqrt[5]{x}, \quad x = -1, x = 32$

Solution
(a) $f(-5) = \sqrt{-5 + 1} = \sqrt{-4}$, which is *not* a real number.
  $f(3) = \sqrt{3 + 1} = \sqrt{4} = 2$ because $2^2 = 4$.
(b) $f(-7) = \sqrt[3]{1 - (-7)} = \sqrt[3]{8} = 2$ because $2^3 = 8$.
  $f(28) = \sqrt[3]{1 - 28} = \sqrt[3]{-27} = -3$ because $(-3)^3 = -27$.

(c) For $f(x) = \sqrt[4]{x - 7}$, $f(5) = \sqrt[4]{5 - 7} = \sqrt[4]{-2}$, which is not a real number.
$f(88) = \sqrt[4]{88 - 7} = \sqrt[4]{81} = 3$ because $3^4 = 81$.

(d) For $f(x) = \sqrt[5]{x}$, $f(-1) = \sqrt[5]{-1} = -1$ because $(-1)^5 = -1$.
$f(32) = \sqrt[5]{32} = 2$ because $2^5 = 32$.

Now Try Exercises 9, 11, 13, 15

2 Recognizing Power Functions and Their Properties

STUDY TIP

Power functions are important in many areas of study. Be sure to become familiar with them.

Power functions are a generalization of root functions.

Power Functions

$$f(x) = x^{1/2}, g(x) = x^{2/3}, \text{ and } h(x) = x^{-3/5}.$$

The exponent for a power function is frequently a rational number p, written in lowest terms. That is, we usually write $f(x) = x^{1/2}$, rather than $f(x) = x^{3/6}$. The power function $f(x) = x^{m/n}$ can also be written as $f(x) = \sqrt[n]{x^m}$. A power function f is defined for all real numbers when n is odd and defined for only nonnegative numbers when n is even.

> **POWER FUNCTION**
>
> If a function f can be represented by
>
> $$f(x) = x^p,$$
>
> where p is a rational number, then f is a **power function**. If $p = \frac{1}{n}$, where $n \geq 2$ is an integer, then f is also a **root function**, which is given by
>
> $$f(x) = \sqrt[n]{x}.$$

EXAMPLE 2 **Evaluating power functions**

If possible, evaluate $f(x)$ at the given value of x.
(a) $f(x) = x^{0.75}$ at $x = 16$ (b) $f(x) = x^{1/4}$ at $x = -81$

Solution
(a) $0.75 = \frac{3}{4}$, so $f(x) = x^{3/4}$. Thus $f(16) = 16^{3/4} = (16^{1/4})^3 = 2^3 = 8$.

NOTE: $16^{1/4} = \sqrt[4]{16} = 2$. ∎

(b) $f(-81) = (-81)^{1/4} = \sqrt[4]{-81}$, which is not a real number. There is no real number a such that $a^4 = -81$ because a^4 is never negative.

Now Try Exercises 23, 27

🌐 *Math in Context* (Biology) The surface area of the skin covering the human body is influenced by both the height and weight of a person. A taller person tends to have a larger surface area, as does a heavier person. In the next example, we use a power function from biology to model this situation. Also, see the Group Activity at the end of this section.

EXAMPLE 3 **Modeling surface area of the human body**

The surface area of a person who is 66 inches tall and weighs w pounds can be estimated by $S(w) = 327w^{0.425}$, where S is in square inches. (*Source:* H. Lancaster, *Quantitative Methods in Biological and Medical Sciences.*)

(a) Find S if this person weighs 130 pounds.
(b) If the person gains 20 pounds, by how much does the person's surface area increase?
(c) A graph of $y = S(w)$ is shown in **FIGURE 17.13**. Suppose that a person's weight is more than 50 pounds and it doubles. Use the graph to determine if the person's surface area also doubles.

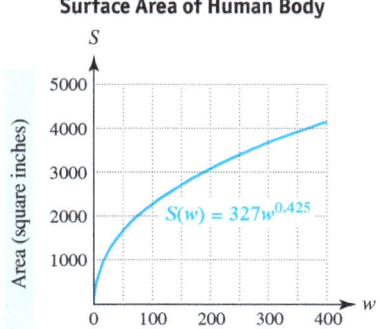

FIGURE 17.13

Solution
(a) $S(130) = 327(130)^{0.425} \approx 2588$ square inches
(b) Because $S(150) = 327(150)^{0.425} \approx 2750$ square inches, the surface area of the person increases by about $2750 - 2588 = 162$ square inches.
(c) For $w \geq 50$, the graph increases more slowly. This means that the person's surface area (S-values) will not double even if his or her weight doubles (w-values). For example, $S(100) \approx 2300$ square inches and $S(200) \approx 3100$ square inches. The surface area did not double.

Now Try Exercise 57

In the next example, we investigate the graph of $y = x^p$ for different values of p.

EXAMPLE 4 Graphing power functions

The graphs of three power functions,

$$f(x) = x^{1/3}, \, g(x) = x^{0.75}, \quad \text{and} \quad h(x) = x^{1.4},$$

are shown in **FIGURE 17.14**, where the two decimal exponents have been written as fractions. Discuss how the value of p affects the graph of $y = x^p$ when $x > 1$ and when $0 < x < 1$.

Power Functions

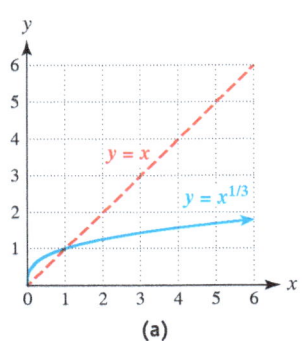
(a)

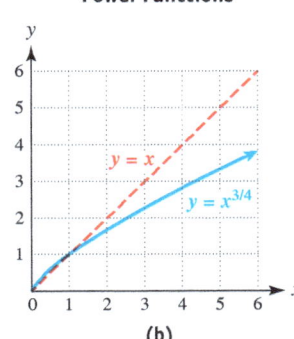
(b)

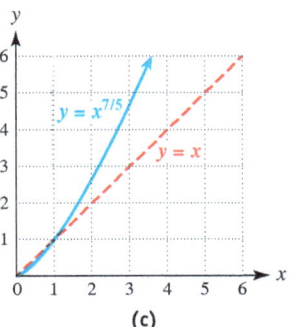
(c)

FIGURE 17.14

Solution
The graphs of these power functions have the following characteristics.
1. If $x > 1$, then $x^{1/3} < x^{3/4} < x^{7/5}$. Larger exponents result in graphs that increase (rise) *faster* for $x > 1$.
2. If $0 < x < 1$, then $x^{7/5} < x^{3/4} < x^{1/3}$. Smaller exponents result in graphs that have larger y-values for $0 < x < 1$.
3. All graphs pass through the points $(0, 0)$ and $(1, 1)$.

Now Try Exercise 35

3 Modeling with Power Functions

Math in Context (Biology) Allometry is the study of the relative sizes of different characteristics of an organism. For example, the weight of a bird is related to the surface area of its wings; heavier birds tend to have larger wings. Allometric relations are often modeled with $f(x) = kx^p$, where k and p are constants. (*Source*: C. Pennycuick, *Newton Rules Biology*.)

EXAMPLE 5 Modeling surface area of wings

Weights w of various birds and the corresponding surface area of their wings A are shown in **TABLE 17.6**.

Weight and Wing Size

w (kilograms)	0.5	2.0	3.5	5.0
A (square meters)	0.069	0.175	0.254	0.325

TABLE 17.6

CALCULATOR HELP
To make a scatterplot, see Appendix A (pages AP-3 and AP-4).

(a) Make a scatterplot of the data. Discuss any trends in the data.
(b) Biologists modeled the data with $A(w) = kw^{2/3}$, where k is a constant. Find k.
(c) Graph A and the data in the same viewing rectangle.
(d) Estimate the area of the wings of a 3-kilogram bird.

Solution
(a) A scatterplot of the data is shown in **FIGURE 17.15(a)**. As the weight of a bird increases so does the surface area of its wings.
(b) To determine k, substitute one of the data points into $A(w)$. We use $(2.0, 0.175)$.

$$A(w) = kw^{2/3} \quad \text{Given formula}$$
$$0.175 = k(2)^{2/3} \quad \text{Let } w = 2 \text{ and } A(w) = 0.175$$
$$\quad \text{(any data point could be used).}$$
$$k = \frac{0.175}{2^{2/3}} \quad \text{Solve for } k; \text{ rewrite equation.}$$
$$k \approx 0.11 \quad \text{Approximate } k.$$

Thus $A(w) = 0.11w^{2/3}$.

(c) The data and graph of $y_1 = 0.11x^{2/3}$ are shown in **FIGURE 17.15(b)**. Note that the graph appears to pass through each data point.
(d) $A(3) = 0.11(3)^{2/3} \approx 0.23$ square meter

Now Try Exercise 65

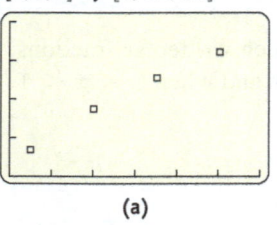

$[0, 6, 1]$ by $[0, 0.4, 0.1]$

(a)

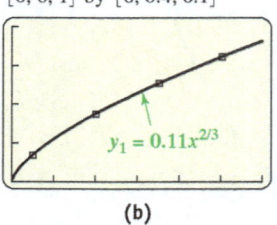

$[0, 6, 1]$ by $[0, 0.4, 0.1]$

$y_1 = 0.11x^{2/3}$

(b)

FIGURE 17.15

17.5 Putting It All Together

FUNCTION	EXPLANATION	EXAMPLES
Root	Odd root functions are defined for all real numbers, and even root functions are defined for nonnegative real numbers.	$f(x) = \sqrt[3]{x}$, $g(x) = \sqrt[4]{x}$, and $h(x) = \sqrt[5]{x}$
Power	$f(x) = x^p$, where p is a rational number, is a power function. If $p = \frac{1}{n}$, where $n \geq 2$ is an integer, f is also a root function, given by $f(x) = \sqrt[n]{x}$.	$f(x) = x^{5/3}$ Power function $g(x) = x^{1/4} = \sqrt[4]{x}$ Root function

17.5 Exercises

CONCEPTS AND VOCABULARY

1. Sketch a graph of $y = x^{1/2}$.
2. Sketch a graph of $y = x^{1/3}$.
3. What is the domain of $f(x) = x^{1/2}$?
4. What is the domain of $f(x) = x^{1/3}$?
5. Give a symbolic representation for a power function.
6. Give a symbolic representation for a root function.
7. What is the domain of $f(x) = \sqrt[4]{x}$?
8. What is the domain of $f(x) = \sqrt[5]{x}$?

EVALUATING ROOT FUNCTIONS

Exercises 9–16: If possible, evaluate each root function f at the given x-values. When the result is not an integer, approximate it to the nearest hundredth.

9. $f(x) = \sqrt{x^2 - 1}$, $x = -2, x = 0$
10. $f(x) = \sqrt{2x - 5}$, $x = 0, x = 3$
11. $f(x) = \sqrt[4]{1 - x}$, $x = -5, x = 2$
12. $f(x) = \sqrt[4]{x + 1}$, $x = -15, x = 15$
13. $f(x) = \sqrt[5]{4 - 3x}$, $x = -3, x = 1$
14. $f(x) = \sqrt[5]{1 - x^2}$, $x = 3, x = 1$
15. $f(x) = \sqrt[3]{1 - x}$, $x = -5, x = 2$
16. $f(x) = \sqrt[3]{(x + 1)^2}$, $x = -2, x = 7$

WORKING WITH POWER FUNCTIONS

Exercises 17–22: Use radical notation to write $f(x)$.

17. $f(x) = x^{1/2}$
18. $f(x) = x^{1/3}$
19. $f(x) = x^{2/3}$
20. $f(x) = x^{3/4}$
21. $f(x) = x^{-1/5}$
22. $f(x) = x^{-2/5}$

Exercises 23–30: If possible, evaluate $f(x)$ at the given values of x. When appropriate, approximate the answer to the nearest hundredth.

23. $f(x) = x^{5/2}$ $x = 4, x = 5$
24. $f(x) = x^{-3/4}$ $x = 1, x = 3$
25. $f(x) = x^{-7/5}$ $x = -32, x = 10$
26. $f(x) = x^{4/3}$ $x = -8, x = 27$
27. $f(x) = x^{1/4}$ $x = 256, x = -10$
28. $f(x) = x^{3/4}$ $x = 16, x = -1$
29. $f(x) = x^{2/5}$ $x = 32, x = -32$
30. $f(x) = x^{5/6}$ $x = -5, x = 64$

Exercises 31–34: Graph $y = f(x)$. Write the domain of f in interval notation.

31. $f(x) = x^{1/4}$
32. $f(x) = x^{1/5}$
33. $f(x) = x^{2/3}$
34. $f(x) = (x + 1)^{1/4}$

Exercises 35–38: Graph f and g in the window $[0, 6, 1]$ by $[0, 6, 1]$. Which function is greater when $x > 1$?

35. $f(x) = x^{1/5}, g(x) = x^{1/3}$

36. $f(x) = x^{4/5}, g(x) = x^{5/4}$

37. $f(x) = x^{1.2}, g(x) = x^{0.45}$

38. $f(x) = x^{-1.4}, g(x) = x^{1.4}$

Exercises 39 and 40: **Thinking Generally** *Let $0 < q < p$, where p and q are rational numbers.*

39. For $x > 1$, is $x^p > x^q$ or $x^p < x^q$?

40. For $0 < x < 1$, is $x^p > x^q$ or $x^p < x^q$?

PERFORMING OPERATIONS ON ROOT FUNCTIONS

Exercises 41 and 42: For the given $f(x)$ and $g(x)$, evaluate each expression.

(a) $(f + g)(2)$ (b) $(f - g)(x)$
(c) $(fg)(x)$ (d) $(f/g)(x)$

41. $f(x) = \sqrt{8x}, g(x) = \sqrt{2x}$

42. $f(x) = \sqrt{9x + 18}, g(x) = \sqrt{4x + 8}$

RECOGNIZING GRAPHS OF POWER FUNCTIONS

Exercises 43–46: **Graphical Interpretation** *Match the situation with the graph of the power function (a–d) that models it best.*

43. Amount of water in a barrel that is initially full and has a hole near the bottom

44. The average weight of a type of bird as its wing span increases (*Hint:* The exponent is greater than 1.)

45. Money made after x hours by a person who works for a fixed hourly wage

46. Rapid growth of an insect population that eventually slows down

a.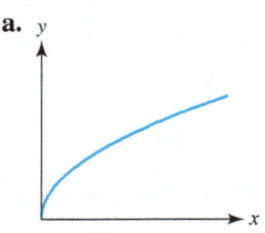

b.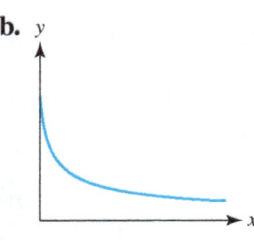

c.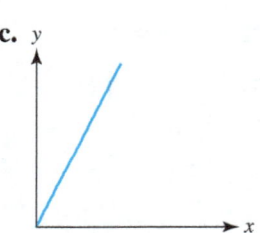

d.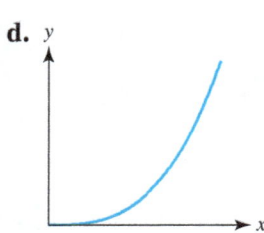

Exercises 47–50: **Graphical Recognition** *Match the equation with its graph (a–d). Do not use a calculator.*

47. $y = x^{7/4}$ 48. $y = x^{4/3}$

49. $y = x^{1/3}$ 50. $y = x^{1/2}$

a.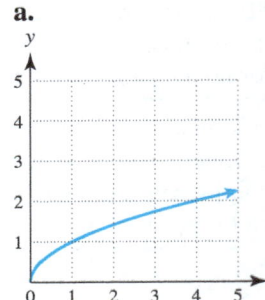

b.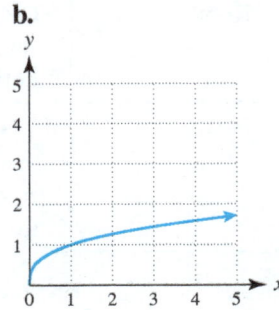

c.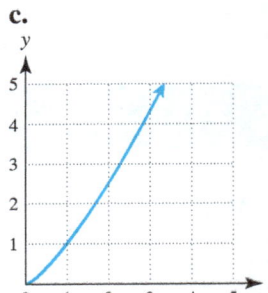

d.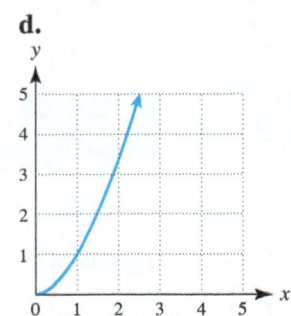

USING ZEROS TO WRITE A FORMULA FOR A FUNCTION

Exercises 51–56: A quadratic function f has the given zeros and leading coefficient 1.

(a) Write the formula for $f(x)$ in factored form.
(b) Write the expanded form for $f(x)$ by multiplying the factor form.

(*Hint: If -5 and 5 are the two zeros of $f(x)$, then $f(x)$ can be written in factored form using the formula given by $f(x) = (x + 5)(x - 5)$.*)

51. $-2, 2$

52. $-3, 3$

53. $-\sqrt{7}, \sqrt{7}$

54. $-\sqrt{3}, \sqrt{3}$

55. $-\sqrt{6}, \sqrt{6}$

56. $-\sqrt{11}, \sqrt{11}$

APPLICATIONS INVOLVING ROOT AND POWER FUNCTIONS

Exercises 57 and 58: **Surface Area** *(Refer to Example 3.) The surface area of the skin of a person who is 70 inches tall and weighs w pounds can be estimated by $S(w) = 342w^{0.425}$, where S is in square inches. Evaluate each of the following.*

57. $S(150)$ 58. $S(200)$

Exercises 59–62: **Learning Curves** *The graph shows the percentage P of times that individuals completed a simple task correctly after practicing it x times.*

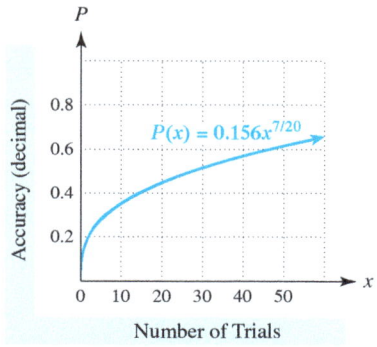

Number of Trials

59. If a group of people does 10 practice trials each, what percentage of the time is the task done correctly?

60. How many trials need to be performed to reach 50% accuracy?

61. If a person doubles the number of practice trials, does the accuracy always double? Explain.

62. What happens to a person's accuracy after a large number of trials?

63. Aging More Slowly In his theory of relativity, Albert Einstein showed that if a person travels at nearly the speed of light, then time slows down significantly. Suppose that there are twins; one remains on Earth and the other leaves in a very fast spaceship that has velocity v. If the twin on Earth ages T_0 years, then according to Einstein, the twin in the spaceship ages T years, where

$$T(v) = T_0\sqrt{1 - (v/c)^2}.$$

In this formula, c represents the speed of light, which is 186,000 miles per second.
(a) Evaluate T when $v = 0.8c$ (eight-tenths the speed of light) and $T_0 = 10$ years. (*Hint:* Simplify $\frac{v}{c}$ without using 186,000 miles per second.)
(b) Interpret your result.

64. Increasing Your Weight (Refer to Exercise 63.) Albert Einstein also showed that the weight (mass) of an object increases when traveling near the speed of light. If a person's weight on Earth is W_0, then the same person's weight W in a spaceship traveling at velocity v is

$$W(v) = \frac{W_0}{\sqrt{1 - (v/c)^2}}.$$

(a) Evaluate function W when $v = 0.6c$ (six-tenths the speed of light) and $W_0 = 220$ pounds (100 kilograms).
(b) Interpret your result.

65. Modeling Wing Size (Refer to Example 5.) The surface area of wings for a species of bird with weight w is shown in the table.

w (kilograms)	2	3	4
A (square meters)	0.254	0.333	0.403

(a) Make a scatterplot of the data.
(b) The data can be modeled with $A(w) = kw^{2/3}$. Find k.
(c) Graph A and the data in the same viewing rectangle. Does the graph of A pass through the data points?
(d) Estimate the area of the wings of a 2.5-kilogram bird.

66. Pulse Rate in Animals The following table lists typical pulse rates R in beats per minute (bpm) for animals with various weights W in pounds. (*Source:* C. Pennycuick.)

W (pounds)	20	150	500	1500
R (beats per minute)	198	72	40	23

(a) Describe what happens to the pulse rate as the weight of the animal increases.
(b) Plot the data in $[0, 1600, 400]$ by $[0, 220, 20]$.
(c) If $R = kW^{-1/2}$, find k.
(d) Find R when $W = 700$, and interpret the result.

67. Modeling Wing Span (Refer to Example 5.) Biologists have found that the weight W of a bird and the length L of its wing span are related by $L = kW^{1/3}$, where k is a constant. The following table lists L and W for one species of bird. (*Source:* C. Pennycuick.)

W (kilograms)	0.1	0.4	0.8	1.1
L (meters)	0.422	0.670	0.844	0.938

(a) Use the data to approximate the value of k.
(b) Graph L and the data in the same viewing rectangle. What happens to L as W increases?
(c) Find the wing span of a 0.7-kilogram bird.
(d) Find L when $W = 0.65$, and interpret the result.

68. Orbits and Distances The time T in years for a planet to orbit the sun is given by $T = D^{3/2}$, where D is the planet's distance from the sun and $D = 1$ corresponds to Earth's distance from the sun. If a planet is twice the distance from the sun as Earth, does the time to orbit the sun also double?

WRITING ABOUT MATHEMATICS

69. Explain why a root function is an example of a power function.

70. Discuss the shape of the graph of $y = x^p$ as p increases. Assume that p is a positive rational number and that x is a positive real number.

Group Activity

Area of Skin The surface area of the skin covering the human body is a function of more than one variable. Both height and weight influence the surface area of a person's body. Hence a taller person tends to have a larger surface area, as does a heavier person. A formula to determine the area of a person's skin in square inches is $S = 15.7 w^{0.425} h^{0.725}$, where w is weight in pounds and h is height in inches. (*Source:* H. Lancaster, *Quantitative Methods in Biological and Medical Sciences.*)

(a) Use S to estimate the area of skin covering a person who is 65 inches tall and weighs 154 pounds.
(b) If a person's weight doubles, what happens to the area of the person's skin? Explain.
(c) If a person's height doubles, what happens to the area of the person's skin? Explain.

17.6 Equations Involving Radical Expressions

Objectives

1 **Solving Radical Equations**
   - Power Rule
   - Extraneous Solutions
   - Squaring Twice
   - Equations Involving Cube Roots

2 **Deriving and Applying the Distance Formula**
   - Pythagorean Theorem
   - Distance Between Two Points

3 **Solving Equations of the Form $x^n = k$**

NEW VOCABULARY

☐ Extraneous solutions
☐ Pythagorean theorem
☐ Distance

1 Solving Radical Equations

POWER RULE Many times, equations contain either radical expressions or rational exponents. Examples include

$$\sqrt{x} = 6, \quad 5x^{1/2} = 1, \quad \text{and} \quad \sqrt[3]{x-1} = 3. \quad \text{— Radical equations}$$

One strategy for solving an equation containing a square root is to isolate the square root (if necessary) and then square each side of the equation. This technique is an example of the *power rule for solving equations*.

POWER RULE FOR SOLVING EQUATIONS

If each side of an equation is raised to the same positive integer power, then any solutions to the given equation are among the solutions to the new equation. That is, the solutions to the equation $a = b$ are among the solutions to $a^n = b^n$.

NOTE: After applying the power rule, the new equation can have solutions that do not satisfy the given equation, which are sometimes called **extraneous solutions**. We must *always* check our answers in the *given* equation after applying the power rule. ∎

READING CHECK 1

- When you apply the power rule to solve an equation, what must you do with your solutions?

We illustrate the power rule in the next example.

EXAMPLE 1   **Solving a radical equation symbolically**

Solve $\sqrt{2x-1} = 3$. Check your solution.

Solution
Begin by squaring each side of the equation. That is, apply the power rule with $n = 2$.

$\sqrt{2x-1} = 3$ Given equation
$(\sqrt{2x-1})^2 = 3^2$ Square each side.
$2x - 1 = 9$ Simplify.
$2x = 10$ Add 1.
$x = 5$ Divide by 2.

To check this answer, we substitute $x = 5$ in the given equation.

$\sqrt{2(5) - 1} \stackrel{?}{=} 3$
$3 = 3$ ✓   It checks.

Now Try Exercise 21

NOTE: To simplify $(\sqrt{2x-1})^2$ in Example 1, we used the fact that
$$(\sqrt{a})^2 = \sqrt{a} \cdot \sqrt{a} = a. \quad \blacksquare$$

The following steps can be used to solve a radical equation.

SOLVING A RADICAL EQUATION

STEP 1: Isolate a radical term on one side of the equation.

STEP 2: Apply the power rule by raising each side of the equation to the power equal to the index of the isolated radical term.

STEP 3: Solve the equation. If it still contains a radical, repeat Steps 1 and 2.

STEP 4: Check your answers by substituting each result in the *given* equation.

MAKING CONNECTIONS 1

Expressions and Equations

An equation is a statement that two expressions are equal. For example,

$\sqrt{x+1}$ and $\sqrt{5-x}$ Expressions do not have an equals sign.

are two different expressions, and

$\sqrt{x+1} = \sqrt{5-x}$ Equations contain an equals sign.

is an equation. We often *solve equations*, whereas we *simplify* and *evaluate* expressions.

In the next example, we apply these steps to solve a radical equation.

EXAMPLE 2   **Isolating the radical term before applying the power rule**

Solve $\sqrt{4-x} + 5 = 8$.

Solution
STEP 1: To isolate the radical term, we subtract 5 from each side of the equation.

$$\sqrt{4-x} + 5 = 8 \qquad \text{Given equation}$$
$$\sqrt{4-x} = 3 \qquad \text{Subtract 5.}$$

STEP 2: The isolated term involves a square root, so we must **square** each side.

$$(\sqrt{4-x})^2 = 3^2 \qquad \text{Square each side.}$$

STEP 3: Next we solve the resulting equation. (It is not necessary to repeat Steps 1 and 2 because the resulting equation does not contain any radical expressions.)

$$4 - x = 9 \qquad \text{Simplify.}$$
$$-x = 5 \qquad \text{Subtract 4.}$$
$$x = -5 \qquad \text{Multiply by } -1.$$

STEP 4: To check this answer, we substitute $x = -5$ in the given equation.

$$\sqrt{4-(-5)} + 5 \stackrel{?}{=} 8$$
$$\sqrt{9} + 5 \stackrel{?}{=} 8$$
$$8 = 8 \checkmark \qquad \text{It checks.}$$

Now Try Exercise 23

EXTRANEOUS SOLUTIONS The next example shows that we must check our answers to identify extraneous solutions when squaring each side of an equation.

EXAMPLE 3 Solving a radical equation

Solve $\sqrt{3x+3} = 2x - 1$ symbolically. Check your results and then solve the equation graphically.

Solution
Symbolic Solution Begin by squaring each side of the equation.

$$\sqrt{3x+3} = 2x - 1 \qquad \text{Given equation}$$
$$(\sqrt{3x+3})^2 = (2x-1)^2 \qquad \text{Square each side.}$$
$$3x + 3 = 4x^2 - 4x + 1 \qquad \text{Multiply.}$$
$$0 = 4x^2 - 7x - 2 \qquad \text{Subtract } 3x + 3.$$
$$0 = (4x + 1)(x - 2) \qquad \text{Factor.}$$
$$x = -\frac{1}{4} \quad \text{or} \quad x = 2 \qquad \text{Solve for } x.$$

To check these values, start by substituting $x = -\frac{1}{4}$ in the given equation.

$$\sqrt{3\left(-\frac{1}{4}\right) + 3} \stackrel{?}{=} 2\left(-\frac{1}{4}\right) - 1$$
$$\sqrt{2.25} \stackrel{?}{=} -1.5$$
$$1.5 \neq -1.5 \; \text{✗} \qquad \text{It does not check.}$$

Thus $-\frac{1}{4}$ is an *extraneous solution*. Next substitute $x = 2$ in the given equation.

$$\sqrt{3(2) + 3} \stackrel{?}{=} 2(2) - 1$$
$$\sqrt{9} \stackrel{?}{=} 3$$
$$3 = 3 \checkmark \qquad \text{It checks.}$$

The only solution is 2. Next we find a graphical solution.

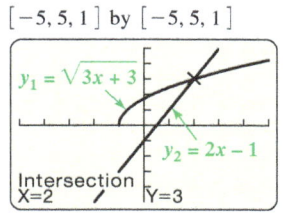

FIGURE 17.16

CALCULATOR HELP
To find a point of intersection, see Appendix A (page AP-6).

Graphical Solution The solution 2 to the equation $\sqrt{3x+3} = 2x - 1$ is supported graphically in **FIGURE 17.16**, where the graphs of $y_1 = \sqrt{3x+3}$ and $y_2 = 2x - 1$ intersect at the point $(2, 3)$. *Note that the graphical solution does not give an extraneous solution.*

Now Try Exercises 27, 79(a), (b)

Example 3 demonstrates that *checking a solution is essential when you are squaring each side of an equation.* Squaring may introduce extraneous solutions, which are solutions to the resulting equation but are not solutions to the given equation.

CRITICAL THINKING 1

Will a numerical solution give extraneous solutions?

SQUARING TWICE When an equation contains two or more terms with square roots, it may be necessary to square each side of the equation more than once. In these situations, isolate one of the square roots and then square each side of the equation. If a radical term remains after simplifying, repeat these steps. We apply this technique in the next example.

EXAMPLE 4 Solving a radical equation by squaring twice

Solve $\sqrt{2x} - 1 = \sqrt{x+1}$.

Solution
Begin by squaring each side of the equation.

$$\sqrt{2x} - 1 = \sqrt{x+1} \quad \text{Given equation (Step 1)}$$
$$(\sqrt{2x} - 1)^2 = (\sqrt{x+1})^2 \quad \text{Square each side. (Step 2)}$$
$$(\sqrt{2x})^2 - 2(\sqrt{2x})(1) + 1^2 = x + 1 \quad (a-b)^2 = a^2 - 2ab + b^2$$
$$2x - 2\sqrt{2x} + 1 = x + 1 \quad \text{Simplify.}$$
$$2x - 2\sqrt{2x} = x \quad \text{Subtract 1.}$$
$$x = 2\sqrt{2x} \quad \text{Subtract } x \text{ and add } 2\sqrt{2x}. \text{ (Step 1)}$$
$$x^2 = 4(2x) \quad \text{Square each side again. (Step 2)}$$
$$x^2 - 8x = 0 \quad \text{Subtract } 8x.$$
$$x(x-8) = 0 \quad \text{Factor.}$$
$$x = 0 \quad \text{or} \quad x = 8 \quad \text{Solve. (Step 3)}$$

To check these answers, substitute $x = 0$ and $x = 8$ in the given equation (Step 4).

$$\sqrt{2(0)} - 1 \stackrel{?}{=} \sqrt{0+1} \qquad \sqrt{2(8)} - 1 \stackrel{?}{=} \sqrt{8+1}$$
$$-1 \neq 1 \quad \text{✗ It does not check.} \qquad 3 = 3 \quad \text{✓ It checks.}$$

Because 0 is an extraneous solution, the only solution is 8.

Now Try Exercise 43

EQUATIONS INVOLVING CUBE ROOTS In the next example, we apply the power rule to an equation that contains a cube root.

EXAMPLE 5 Solving an equation containing a cube root

Solve $\sqrt[3]{4x - 7} = 4$.

Solution

STEP 1: The cube root term is already isolated, so we proceed to Step 2.

STEP 2: Because the index is 3, we **cube** each side of the equation.

$\sqrt[3]{4x - 7} = 4$ Given equation

$(\sqrt[3]{4x - 7})^3 = (4)^3$ Cube each side.

STEP 3: We solve the resulting equation.

$4x - 7 = 64$ Simplify.

$4x = 71$ Add 7 to each side.

$x = \dfrac{71}{4}$ Divide each side by 4.

STEP 4: To check this answer, we substitute $x = \frac{71}{4}$ in the given equation.

$\sqrt[3]{4\left(\frac{71}{4}\right) - 7} \stackrel{?}{=} 4$

$\sqrt[3]{64} \stackrel{?}{=} 4$

$4 = 4$ ✓   It checks.

Now Try Exercise 31

Math in Context *Biology* Larger birds tend to have larger wings. This relationship can sometimes be modeled by $A(W) = 100\sqrt[3]{W^2}$, where W is the weight in pounds of the bird and A is the area of the wings in square inches. In the next example, we use the formula to determine the weight of a bird that has wings with a surface area of 600 square inches.
(*Source:* C. Pennycuick, *Newton Rules Biology*.)

EXAMPLE 6 Finding the weight of a bird

Solve the equation $600 = 100\sqrt[3]{W^2}$ to determine the weight in pounds of a bird that has wings with an area of 600 square inches.

Solution

Begin by dividing each side of the equation $600 = 100\sqrt[3]{W^2}$ by 100 to isolate the radical term on the right side of the equation.

$\dfrac{600}{100} = \sqrt[3]{W^2}$ Divide each side by 100.

$(6)^3 = (\sqrt[3]{W^2})^3$ Cube each side.

$216 = W^2$ Simplify.

$\sqrt{216} = W$ Take principal square root, $W > 0$.

$W \approx 14.7$ Approximate.

The weight of the bird is approximately 14.7 pounds.

Now Try Exercise 117

TECHNOLOGY NOTE

Graphing Radical Expressions

The equation in Example 6 can be solved graphically. Sometimes it is more convenient to use rational exponents than radical notation. Thus $y = 100\sqrt[3]{W^2}$ can be entered as $y_1 = 100x^{2/3}$. The accompanying figure shows y_1 intersecting the line $y_2 = 600$ near the point $(14.7, 600)$, which supports our symbolic result.

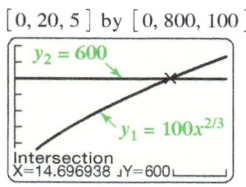

EXAMPLE 7 Solving an equation with rational exponents graphically

Solve $x^{2/3} = 3 - x^2$ graphically.

Solution
Graph $y_1 = x^{2/3}$ and $y_2 = 3 - x^2$. The graphs intersect near $(-1.34, 1.21)$ and $(1.34, 1.21)$, as shown in **FIGURE 17.17**. Thus the solutions are given by $x \approx \pm 1.34$.

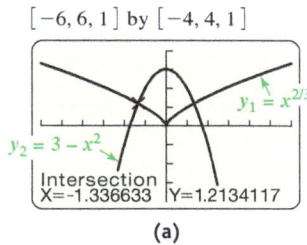
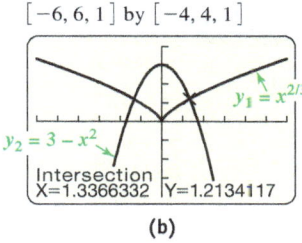

(a) (b)

FIGURE 17.17

Now Try Exercise 73

The following See the Concept illustrates how the equation $\sqrt{x + 2} = x$ can be solved symbolically, numerically, and graphically.

See the Concept 1   WAYS TO SOLVE AN EQUATION

Solve the equation $\sqrt{x + 2} = x$ symbolically, numerically, and graphically.

Symbolic Solution

$$\sqrt{x + 2} = x$$
$$x + 2 = x^2$$
$$x^2 - x - 2 = 0$$
$$(x - 2)(x + 1) = 0$$
$$x = 2 \quad \text{or} \quad x = -1$$

Check: $\sqrt{2 + 2} = 2$ ✓
$\sqrt{-1 + 2} \neq -1$ ✗

The only solution is 2.

Numerical Solution

X	Y₁	Y₂
-2	0	-2
-1	1	-1
0	1.4142	0
1	1.7321	1
2	2	2
3	2.2361	3
4	2.4495	4

X=2

$y_1 = y_2$ when $x = 2$, so 2 is a solution. Note that -1 is *not* a solution.

Graphical Solution

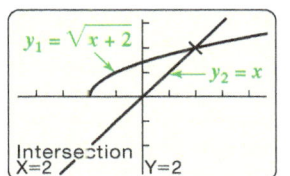

The graphs intersect at $(2, 2)$. Note that there is no point of intersection when $x = -1$.

Note: The solution is 2 for each method. ∎

WATCH VIDEO IN MML.

2 Deriving and Applying the Distance Formula

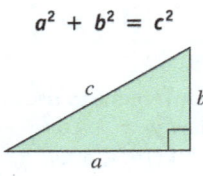

FIGURE 17.18

PYTHAGOREAN THEOREM One of the most famous theorems in mathematics is the **Pythagorean theorem**. It states that if a right triangle has legs a and b with hypotenuse c (see **FIGURE 17.18**), then

$$a^2 + b^2 = c^2. \quad \text{Pythagorean theorem}$$

For example, if the legs of a right triangle are $a = 3$ and $b = 4$, the hypotenuse is $c = 5$ because $3^2 + 4^2 = 5^2$. Also, if the sides of a triangle satisfy $a^2 + b^2 = c^2$, it is a right triangle.

EXAMPLE 8 Applying the Pythagorean theorem

An HDTV has a width of 60 inches and a height of 32 inches. Find the diagonal of the television. Why is it called a 68-inch television?

Solution
In **FIGURE 17.19**, let $a = 60$ and $b = 32$. Then the diagonal of the television corresponds to the hypotenuse of a right triangle with legs of 60 inches and 32 inches.

$$c^2 = a^2 + b^2 \quad \text{Pythagorean theorem}$$
$$c = \sqrt{a^2 + b^2} \quad \text{Take the principal square root, } c > 0.$$
$$c = \sqrt{60^2 + 32^2} \quad \text{Substitute } a = 60 \text{ and } b = 32.$$
$$c = 68 \quad \text{Simplify.}$$

A 68-inch television has a diagonal of 68 inches.

Now Try Exercise 125

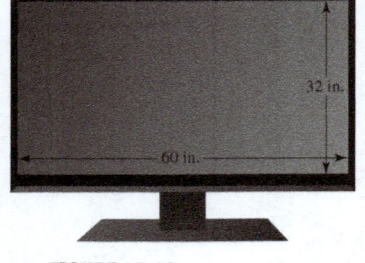

FIGURE 17.19

DISTANCE BETWEEN TWO POINTS The Pythagorean theorem can be used to determine the distance between two points, as illustrated in the following See the Concept.

See the Concept 2   **DISTANCE BETWEEN TWO POINTS**

A The length of the vertical leg is $y_2 - y_1$.

B The length of the horizontal leg is $x_2 - x_1$.

C By the Pythagorean theorem:
$$d^2 = (x_2 - x_1)^2 + (y_2 - y_1)^2$$

Thus, $d = \sqrt{(x_2 - x_1)^2 + (y_2 - y_1)^2}$.

Distance is nonnegative.   **WATCH VIDEO IN** MML.

DISTANCE FORMULA

The **distance** d between the points (x_1, y_1) and (x_2, y_2) in the xy-plane is

$$d = \sqrt{(x_2 - x_1)^2 + (y_2 - y_1)^2}.$$

EXAMPLE 9 Finding distance between points

Find the distance between the points $(-2, 3)$ and $(1, -4)$.

Solution
Start by letting $(x_1, y_1) = (-2, 3)$ and $(x_2, y_2) = (1, -4)$. Then substitute these values into the distance formula.

$$\begin{aligned} d &= \sqrt{(x_2 - x_1)^2 + (y_2 - y_1)^2} & \text{Distance formula} \\ &= \sqrt{(1 - (-2))^2 + (-4 - 3)^2} & \text{Substitute.} \\ &= \sqrt{9 + 49} & \text{Simplify.} \\ &= \sqrt{58} & \text{Add.} \\ &\approx 7.62 & \text{Approximate.} \end{aligned}$$

The distance between the points, as shown in **FIGURE 17.20**, is exactly $\sqrt{58}$ units, or about 7.62 units. Note that we would obtain the same result if we let $(x_1, y_1) = (1, -4)$ and $(x_2, y_2) = (-2, 3)$.

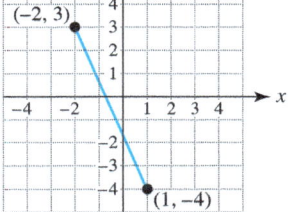

FIGURE 17.20

Now Try Exercise 109

NOTE: In Example 9, $\sqrt{9 + 49} \neq \sqrt{9} + \sqrt{49} = 3 + 7 = 10$. In general, for any a and b, $\sqrt{a^2 + b^2} \neq a + b$. ∎

3 Solving Equations of the Form $x^n = k$

The equation $x^n = k$, where n is a *positive integer*, can be solved by taking the nth root of each side of the equation. The following technique allows us to find all *real* solutions to this equation.

> **SOLVING THE EQUATION $x^n = k$**
>
> Take the nth root of each side of $x^n = k$ to obtain $\sqrt[n]{x^n} = \sqrt[n]{k}$.
>
> 1. If n is odd, then $\sqrt[n]{x^n} = x$ and the equation becomes $x = \sqrt[n]{k}$.
> 2. If n is *even* and $k > 0$, then $\sqrt[n]{x^n} = |x|$ and the equation becomes $|x| = \sqrt[n]{k}$. (If $k < 0$, there are no real solutions.)

To understand this technique better, consider the following examples. First let $x^3 = 8$ so that n is odd. Taking the cube root of each side gives

$$\sqrt[3]{x^3} = \sqrt[3]{8}, \text{ which is equivalent to } x = \sqrt[3]{8} \text{ or } x = 2.$$

Next let $x^2 = 4$ so that n is even. Taking the square root of each side gives

$$\sqrt{x^2} = \sqrt{4}, \text{ which is equivalent to } |x| = \sqrt{4} \text{ or } |x| = 2.$$

The solutions to $|x| = 2$ are -2 or 2, which can be written as ± 2.

EXAMPLE 10 Solving equations of the form $x^n = k$

Solve each equation.
(a) $x^3 = -64$ (b) $x^2 = 12$ (c) $2(x - 1)^4 = 32$

Solution
(a) Taking the cube root of each side of $x^3 = -64$ gives

$$\sqrt[3]{x^3} = \sqrt[3]{-64} \quad \text{or} \quad x = -4.$$

(b) Taking the square root of each side of $x^2 = 12$ gives
$$\sqrt{x^2} = \sqrt{12} \quad \text{or} \quad |x| = \sqrt{12}.$$
The equation $|x| = \sqrt{12}$ is equivalent to $x = \pm\sqrt{12}$.

(c) First divide each side of $2(x-1)^4 = 32$ by 2 to isolate the power of $x - 1$.

$$\begin{aligned}
(x-1)^4 &= 16 &&\text{Divide each side by 2.} \\
\sqrt[4]{(x-1)^4} &= \sqrt[4]{16} &&\text{Take the 4th root of each side.} \\
|x-1| &= 2 &&\text{Simplify: } \sqrt[4]{y^4} = |y|. \\
x - 1 = -2 \quad &\text{or} \quad x - 1 = 2 &&\text{Solve the absolute value equation.} \\
x = -1 \quad &\text{or} \quad x = 3 &&\text{Add 1 to each side.}
\end{aligned}$$

Now Try Exercises 47, 55, 67

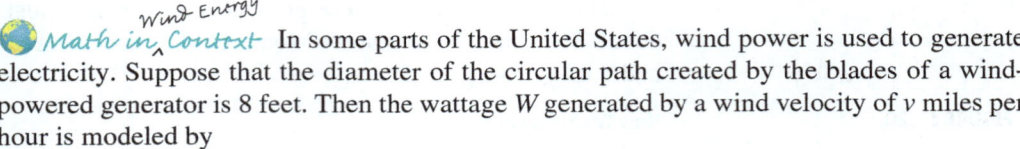

Math in Context   Wind Energy   In some parts of the United States, wind power is used to generate electricity. Suppose that the diameter of the circular path created by the blades of a wind-powered generator is 8 feet. Then the wattage W generated by a wind velocity of v miles per hour is modeled by

$$W(v) = 2.4v^3.$$

If the wind blows at 10 miles per hour, the generator can produce about

$$W(10) = 2.4 \cdot 10^3 = 2400 \text{ watts.}$$

(*Source: Conquering the Sciences, Sharp Electronics.*)

EXAMPLE 11   Modeling a wind generator

The formula $W(v) = 2.4v^3$ is used to calculate the watts generated when there is a wind velocity of v miles per hour.
(a) Find a function f that calculates the wind velocity when W watts are being produced.
(b) If the wattage doubles, has the wind velocity also doubled? Explain.

Solution
(a) Given W we need a formula to find v, so solve $W = 2.4v^3$ for v.

$$\begin{aligned}
W &= 2.4v^3 &&\text{Given formula} \\
\frac{W}{2.4} &= v^3 &&\text{Divide by 2.4.} \\
\sqrt[3]{\frac{W}{2.4}} &= \sqrt[3]{v^3} &&\text{Take the cube root of each side.} \\
v &= \sqrt[3]{\frac{W}{2.4}} &&\text{Simplify and rewrite equation.}
\end{aligned}$$

Thus $f(W) = \sqrt[3]{\frac{W}{2.4}}$.

(b) Suppose that the power generated is 1000 watts. Then the wind speed is

$$f(1000) = \sqrt[3]{\frac{1000}{2.4}} \approx 7.5 \text{ miles per hour.}$$

If the power doubles to 2000 watts, then the wind speed is

$$f(2000) = \sqrt[3]{\frac{2000}{2.4}} \approx 9.4 \text{ miles per hour.}$$

Thus for the wattage to double, the wind speed does not need to double.

Now Try Exercise 133

17.6 Putting It All Together

CONCEPT	EXPLANATION	EXAMPLES
Power Rule for Solving Equations	If each side of an equation is raised to the same positive integer power, any solutions to the given equation are among the solutions to the new equation.	$\sqrt{2x} = x$ $2x = x^2$ Square each side. $x^2 - 2x = 0$ Rewrite equation. $x = 0$ or $x = 2$ Factor and solve. Be sure to check any solutions.
Pythagorean Theorem	If c is the hypotenuse of a right triangle and a and b are its legs, then $a^2 + b^2 = c^2$.	If the sides of the right triangle are $a = 5$, $b = 12$, and $c = 13$, then they satisfy $a^2 + b^2 = c^2$ or $5^2 + 12^2 = 13^2$.
Distance Formula	The distance d between the points (x_1, y_1) and (x_2, y_2) is $d = \sqrt{(x_2 - x_1)^2 + (y_2 - y_1)^2}$.	The distance between the points $(2, 3)$ and $(-3, 4)$ is $d = \sqrt{(-3 - 2)^2 + (4 - 3)^2}$ $= \sqrt{(-5)^2 + (1)^2} = \sqrt{26}$.

Solving the Equation $x^n = k$, Where n is a Positive Integer	Take the nth root of each side to obtain $\sqrt[n]{x^n} = \sqrt[n]{k}$. Then
	1. $x = \sqrt[n]{k}$, if n is odd.
	2. $x = \pm\sqrt[n]{k}$, if n is even and $k \geq 0$.
		1. n odd:
		   If $x^5 = 32$, then $x = \sqrt[5]{32} = 2$.
		2. n even:
		   If $x^4 = 81$, then $x = \pm\sqrt[4]{81} = \pm 3$.

17.6 Exercises

CONCEPTS AND VOCABULARY

1. What is a good first step for solving $\sqrt{4x - 1} = 5$?

2. What is a good first step for solving $\sqrt[3]{x + 1} = 6$?

3. Can an equation involving rational exponents have more than one solution?

4. When you square each side of an equation to solve for an unknown, what must you do with any answers?

5. What is the Pythagorean theorem used for?

6. If the legs of a right triangle are 3 and 4, what is the length of the hypotenuse?

7. What formula can you use to find the distance d between two points?

8. Write the equation $\sqrt{x} + \sqrt[4]{x^3} = 2$ with rational exponents.

SIMPLIFYING RADICALS

Exercises 9–16: Simplify. Assume radicands of square roots are positive.

9. $\sqrt{2} \cdot \sqrt{2}$ 10. $\sqrt{5} \cdot \sqrt{5}$

11. $\sqrt{x} \cdot \sqrt{x}$ 12. $\sqrt{2x} \cdot \sqrt{2x}$

13. $(\sqrt{2x + 1})^2$ 14. $(\sqrt{7x})^2$

15. $(\sqrt[3]{5x^2})^3$ 16. $(\sqrt[3]{2x - 5})^3$

FINDING SYMBOLIC SOLUTIONS

Exercises 17–46: Solve the equation symbolically. Check your results.

17. $\sqrt{x} = 8$ 18. $\sqrt{3z} = 6$

19. $\sqrt[4]{x} = 3$ 20. $\sqrt[5]{x - 4} = 2$

21. $\sqrt{2t + 4} = 4$ 22. $\sqrt{y + 4} = 3$

23. $\sqrt{x+1} - 3 = 4$
24. $\sqrt{2x+5} + 2 = 5$
25. $2\sqrt{x-2} + 1 = 5$
26. $-\sqrt{x+7} - 1 = -7$
27. $\sqrt{x+6} = x$
28. $\sqrt{z} + 6 = z$
29. $\sqrt[3]{x} = 3$
30. $\sqrt[3]{x+10} = 4$
31. $\sqrt[3]{2z-4} = -2$
32. $\sqrt[3]{z-1} = -3$
33. $\sqrt[4]{t+1} = 2$
34. $\sqrt[4]{5t} = 5$
35. $\sqrt{5z-1} = \sqrt{z+1}$
36. $y = \sqrt{y+1} + 1$
37. $\sqrt{1-x} = 1-x$
38. $\sqrt[3]{4x} = x$
39. $\sqrt{b^2-4} = b-2$
40. $\sqrt{b^2-2b+1} = b$
41. $\sqrt{1-2x} = x+7$
42. $\sqrt{4-y} = y-2$
43. $\sqrt{x} = \sqrt{x-5} + 1$
44. $\sqrt{x-1} = \sqrt{x+4} - 1$
45. $\sqrt{2t-2} + \sqrt{t} = 7$
46. $\sqrt{x+1} - \sqrt{x-6} = 1$

SOLVING THE EQUATION $x^n = k$

Exercises 47–68: (Refer to Example 10.) Solve.

47. $x^2 = 49$
48. $x^2 = 9$
49. $2z^2 = 200$
50. $3z^2 = 48$
51. $(t+1)^2 = 16$
52. $(t-5)^2 = 81$
53. $(3x-6)^2 = 25$
54. $(4-2x)^2 = 100$
55. $b^3 = 64$
56. $a^3 = 1000$
57. $2t^3 = -128$
58. $3t^3 = -81$
59. $(x+1)^3 = 8$
60. $(4-x)^3 = -1$
61. $(2-5z)^3 = -125$
62. $(2x+4)^3 = 125$
63. $x^4 = 16$
64. $x^4 = 7$
65. $x^5 = 12$
66. $x^5 = -32$
67. $2(x+2)^4 = 162$
68. $\frac{1}{2}(x-1)^5 = 16$

FINDING GRAPHICAL SOLUTIONS

Exercises 69–78: Solve graphically. Approximate solutions to the nearest hundredth when appropriate.

69. $\sqrt[3]{x+5} = 2$
70. $\sqrt[3]{x} + \sqrt{x} = 3.43$
71. $\sqrt{2x-3} = \sqrt{x} - \frac{1}{2}$
72. $x^{4/3} - 1 = 2$
73. $x^{5/3} = 2 - 3x^2$
74. $x^{3/2} = \sqrt{x+2} - 2$
75. $z^{1/3} - 1 = 2 - z$
76. $z^{3/2} - 2z^{1/2} - 1 = 0$
77. $\sqrt{x+1} - \sqrt{x-1} = 4$
78. $\sqrt{y+2} + \sqrt{3y+2} = 2$

USING MORE THAN ONE METHOD TO SOLVE EQUATIONS

Exercises 79–82: Solve the equation
 (a) symbolically, *(b) graphically, and*
 (c) numerically.

79. $2\sqrt{x} = 8$
80. $\sqrt[3]{5-x} = 2$
81. $\sqrt{6z-2} = 8$
82. $\sqrt{y+4} = \dfrac{y}{3}$

SOLVING AN EQUATION FOR A VARIABLE

Exercises 83–86: Solve for the indicated variable.

83. $T = 2\pi\sqrt{\dfrac{L}{32}}$ for L
84. $Z = \sqrt{L^2 + R^2}$ for R
85. $r = \sqrt{\dfrac{A}{\pi}}$ for A
86. $F = \dfrac{1}{2\pi\sqrt{LC}}$ for C

APPLYING THE PYTHAGOREAN THEOREM

Exercises 87–94: If the sides of a triangle are a, b, and c and they satisfy $a^2 + b^2 = c^2$, the triangle is a right triangle. Determine whether the triangle with the given sides is a right triangle.

87. $a = 6$ $b = 8$ $c = 10$
88. $a = 5$ $b = 12$ $c = 13$
89. $a = \sqrt{5}$ $b = \sqrt{9}$ $c = \sqrt{14}$
90. $a = 4$ $b = 5$ $c = 7$
91. $a = 7$ $b = 24$ $c = 25$
92. $a = 1$ $b = \sqrt{3}$ $c = 2$
93. $a = 8$ $b = 8$ $c = 16$
94. $a = 11$ $b = 60$ $c = 61$

Exercises 95–98: Find the length of the missing side in the right triangle.

95.
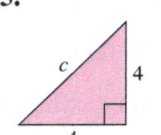

96.
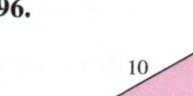

97.
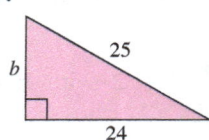

98.
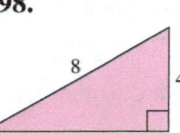

Exercises 99–104: A right triangle has legs a and b with hypotenuse c. Find the length of the missing side.

99. $a = 3, b = 4$
100. $a = 4, b = 7$
101. $a = \sqrt{3}, c = 8$
102. $a = \sqrt{6}, c = \sqrt{10}$
103. $b = 48, c = 50$
104. $b = 10, c = 26$

APPLYING THE DISTANCE FORMULA

Exercises 105–108: Find the length of the line segment.

105.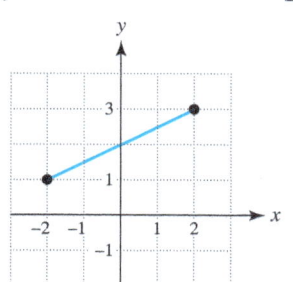
106.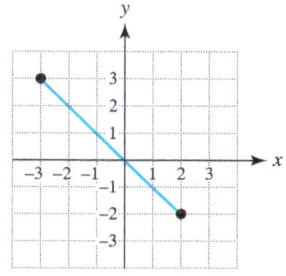
107.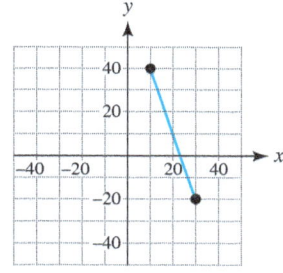
108.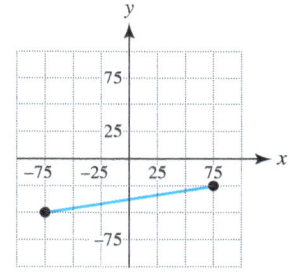

Exercises 109–112: Find the distance between the points.

109. $(-1, 2), (4, 10)$
110. $(5, -40), (-6, 20)$
111. $(0, -3), (4, 0)$
112. $(3, 9), (-4, 2)$

Exercises 113–116: Find x if the distance between the points is d. Assume that $x \geq 0$.

113. $(x, 3), (0, 6) \quad d = 5$
114. $(x, -1), (6, 11) \quad d = 13$
115. $(x, -5), (62, 6) \quad d = 61$
116. $(x, 3), (12, -4) \quad d = 25$

APPLICATIONS INVOLVING RADICAL EQUATIONS

Exercises 117 and 118: **Weight of a Bird** *(Refer to Example 6.) Estimate the weight W of a bird that has wings with area A. Let $A = 100\sqrt[3]{W^2}$.*

117. $A = 400$ square inches
118. $A = 1000$ square inches

Exercises 119 and 120: **Abandonment Rate** *In a fast-paced society, people do not spend time watching videos that they find boring. For example, in Example 3 of the second section of this chapter, we learned how the function*

$$A(x) = 7.3\sqrt[16]{x^7}$$

computes the percentage of viewers who abandon an online video after x seconds. Use $A(x) = 7.3\sqrt[16]{x^7}$ to find the average number of seconds x that it takes for A percent of the viewers to abandon an online movie.

119. $A = 50$
120. $A = 20$

Exercises 121–124: **Distance to the Horizon** *Because of Earth's curvature, a person can see a limited distance to the horizon. The higher the location of the person, the farther that person can see. The distance D in miles to the horizon can be estimated by $D(h) = 1.22\sqrt{h}$, where h is the height of the person above the ground in feet.*

121. Find D for a 6-foot-tall person standing on level ground.

122. Find D for a person on top of Mount Everest with a height of 29,028 feet.

123. What height allows a person to see 20 miles?

124. How high does a plane need to fly for the pilot to be able to see 100 miles?

125. **Diagonal of a Television** (Refer to Example 8.) A rectangular television screen is 11.4 inches by 15.2 inches. Find the diagonal of the television screen.

126. **Dimensions of a Television** The height of a television with a 13-inch diagonal is $\frac{3}{4}$ of its width. Find the width and height of the television set.

127. DVD and Picture Dimensions If the picture shown on a television set is h units high and w units wide, the *aspect ratio* of the picture is $\frac{w}{h}$ (see the accompanying figure). Digital video discs (DVDs) support the newer aspect ratio of $\frac{16}{9}$ rather than the older ratio of $\frac{4}{3}$. If the width of a picture with an aspect ratio of $\frac{16}{9}$ is 29 inches, approximate the height and diagonal of the rectangular picture. (*Source:* J. Taylor, *DVD Demystified.*)

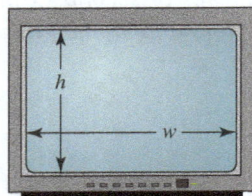

128. Flood Control The spillway capacity of a dam is important in flood control. Spillway capacity Q in cubic feet of water per second flowing over the spillway depends on the width W and the depth D of the spillway, as illustrated in the accompanying figure. If W and D are measured in feet, capacity can be modeled by $Q = 3.32WD^{3/2}$. (*Source:* D. Callas, Project Director, *Snapshots of Applications in Mathematics.*)
(a) Find the capacity of a spillway with $W = 20$ feet and $D = 5$ feet.
(b) A spillway with a width of 30 feet is to have a capacity of $Q = 2690$ cubic feet per second. Estimate to the nearest foot the appropriate depth of the spillway.

129. Sky Diving When sky divers initially fall from an airplane, their velocity v in miles per hour after free falling d feet can be approximated by $v = \frac{60}{11}\sqrt{d}$. (Because of air resistance, they will eventually reach a terminal velocity.) How far do sky divers need to fall to attain the following velocities? (These values for d represent minimum distances.)
(a) 60 miles per hour
(b) 100 miles per hour

130. Guy Wire A guy wire attached to the top of a 30-foot-long pole is anchored 10 feet from the base of the pole, as illustrated in the figure at the top of the next column. Find the length of the guy wire to the nearest tenth of a foot.

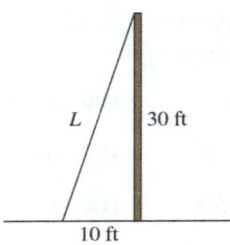

131. Skid Marks Vehicles involved in accidents often leave skid marks, which can be used to determine how fast a vehicle was traveling. To determine this speed, officials often use a test vehicle to compare skid marks on the same section of road. Suppose that a vehicle in a crash left skid marks D feet long and that a test vehicle traveling at v miles per hour leaves skid marks d feet long. Then the speed V of the vehicle involved in the crash is given by

$$V = v\sqrt{\frac{D}{d}}.$$

(*Source:* N. Garber and L. Hoel, *Traffic and Highway Engineering.*)
(a) Find V if $v = 30$ mph, $D = 285$ feet, and $d = 178$ feet. Interpret your result.
(b) A test vehicle traveling at 45 mph leaves skid marks 255 feet long. How long would the skid marks be for a vehicle traveling 60 miles per hour?

132. Highway Curves If a circular highway curve without any banking has a radius of R feet, the speed limit L in miles per hour for the curve is $L = 1.5\sqrt{R}$. (*Source:* N. Garber.)
(a) Find the speed limit for a curve that has a radius of 400 feet.
(b) If the radius of a curve doubles, what happens to the speed limit?
(c) A curve with a 40-mile-per-hour speed limit is being designed. What should be its radius?

133. Wind Power (Refer to Example 11.) If a wind-powered generator has blades that create a circular path with a diameter of 10 feet, then the wattage W generated by a wind velocity of v miles per hour is modeled by $W(v) = 3.8v^3$.
(a) If the wind velocity doubles, what happens to the wattage generated?
(b) Solve $W = 3.8v^3$ for v.
(c) If the wind generator is producing 30,400 watts, find the wind speed.

134. Height and Weight Suppose that the weight of a person is directly proportional to the cube of the person's height. If one person weighs twice as much as a (similarly proportioned) second person, by what factor is the heavier person's height greater than the shorter person's height?

135. 45°–45° Right Triangle Suppose that the legs of a right triangle with angles of 45° and 45° both have length a, as depicted in the accompanying figure. Find the length of the hypotenuse.

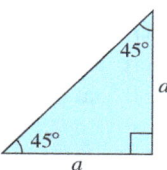

136. 30°–60° Right Triangle In a right triangle with angles of 30° and 60°, the shortest side is half the length of the hypotenuse (see the accompanying figure). If the hypotenuse has length c, find the length of the other two sides, a and b, in terms of c.

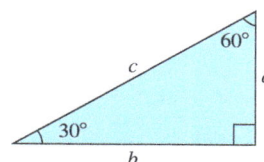

137. Minimizing Cost A natural gas line running along a river is to be connected from point A to a cabin on the other bank located at point D, as illustrated in the figure. The width of the river is 500 feet, and the distance from point A to point C is 1000 feet. The cost of running the pipe along the shoreline is $30 per foot, and the cost of running it underwater is $50 per foot. The cost of connecting the gas line from A to D is to be minimized.

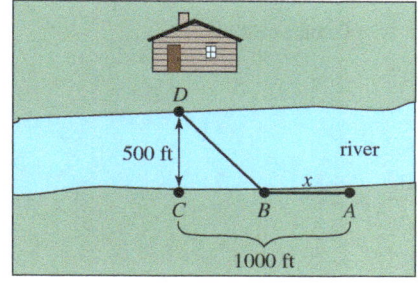

(a) Write an expression that gives the cost of running the line from A to B if the distance between these points is x feet.
(b) Find the distance from B to D in terms of x.
(c) Write an expression that gives the cost of running the line from B to D.
(d) Use your answer from parts (a) and (c) to write an expression that gives the cost of running the line from A to B to D.
(e) Graph your expression from part (d) in the window $[0, 1000, 100]$ by $[40000, 60000, 5000]$ to determine the value of x that minimizes the cost of the line going from A to D. What is the minimum cost?

138. Minimizing Cost Repeat Exercise 137 if the width of the river is 600 feet and the distance from point A to point C is 1500 feet. For the window in part (e), use $[0, 1500, 500]$ by $[50000, 90000, 10000]$.

WRITING ABOUT MATHEMATICS

139. A student solves an equation *incorrectly* as follows.
$$\sqrt{3-x} = \sqrt{x} - 1$$
$$(\sqrt{3-x})^2 \stackrel{?}{=} (\sqrt{x})^2 - (1)^2$$
$$3 - x \stackrel{?}{=} x - 1$$
$$-2x \stackrel{?}{=} -4$$
$$x \stackrel{?}{=} 2$$

(a) How could you convince the student that the answer is wrong?
(b) Discuss where any errors were made.

140. When each side of an equation is squared, you must check your results. Explain why.

SECTIONS 17.5 and 17.6 — Checking Basic Concepts

1. Sketch a graph of each function and then evaluate $f(-1)$, if possible.
   (a) $f(x) = \sqrt{x}$
   (b) $f(x) = \sqrt[3]{x}$
   (c) $f(x) = \sqrt{x^2}$

2. Evaluate $f(x) = 0.2x^{2/3}$ when $x = 64$.

3. Find the domain of $f(x) = \sqrt{x - 4}$. Write your answer in interval notation.

4. Solve each equation. Check your answers.
   (a) $\sqrt{2x - 4} = 2$
   (b) $\sqrt[3]{x - 1} = 3$
   (c) $\sqrt{3x} = 1 + \sqrt{x + 1}$

5. Find the distance between $(-3, 5)$ and $(2, -7)$.

6. A 16-inch diagonal television set has a rectangular picture with a width of 12.8 inches. Find the height of the picture.

7. Solve $(x + 1)^4 = 16$.

17.7 Complex Numbers

Objectives

1. Introducing the Imaginary Unit and Standard Form
2. Adding, Subtracting, and Multiplying Complex Numbers
   - Addition and Subtraction
   - Multiplication
3. Calculating Powers of i
4. Finding Complex Conjugates
5. Dividing Complex Numbers

NEW VOCABULARY

☐ Imaginary unit
☐ Complex number
☐ Standard form
☐ Real part
☐ Imaginary part
☐ Nonreal complex number
☐ Imaginary number
☐ Pure imaginary number
☐ Complex conjugate

1 Introducing the Imaginary Unit and Standard Form

A graph of $y = x^2 + 1$ is shown in **FIGURE 17.21**. There are no x-intercepts, so the quadratic equation $x^2 + 1 = 0$ has no real number solutions.

If we try to solve $x^2 + 1 = 0$ by subtracting 1 from each side, the result is $x^2 = -1$. Because $x^2 \geq 0$ for any real number x, there are no real solutions. However, mathematicians have invented (or discovered) solutions.

$$x^2 = -1$$
$$x = \pm\sqrt{-1} \quad \text{Solve for } x.$$

We now define a number called the **imaginary unit**, denoted i.

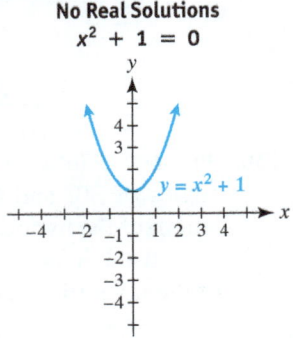

No Real Solutions
$x^2 + 1 = 0$

FIGURE 17.21

PROPERTIES OF THE IMAGINARY UNIT i

$$i = \sqrt{-1} \quad \text{and} \quad i^2 = -1$$

By creating the number i, the solutions to the equation $x^2 + 1 = 0$ are i and $-i$. Using the real numbers and the imaginary unit i, we can define a new set of numbers called the *complex numbers*. A **complex number** can be written in **standard form**, as $a + bi$, where a and b are real numbers. The **real part** is a and the **imaginary part** is b. Every real number a is also a complex number because it can be written $a + 0i$. A complex number $a + bi$ with $b \neq 0$ is a **nonreal complex number**, or **imaginary number**. A complex number $a + bi$ with $a = 0$ and $b \neq 0$ is sometimes called a **pure imaginary number**. Examples of pure imaginary numbers include $4i$ and $-2i$. **TABLE 17.7** lists several complex numbers with their real and imaginary parts.

Real and Imaginary Parts of Complex Numbers

Complex Number: $a + bi$	$-3 + 2i$	5	$-3i$	$-1 + 7i$	$-5 - 2i$	$4 + 6i$
Real Part: a	-3	5	0	-1	-5	4
Imaginary Part: b	2	0	-3	7	-2	6

TABLE 17.7

The following See the Concept illustrates how different sets of numbers are related.

See the Concept 1 — **SETS OF NUMBERS**

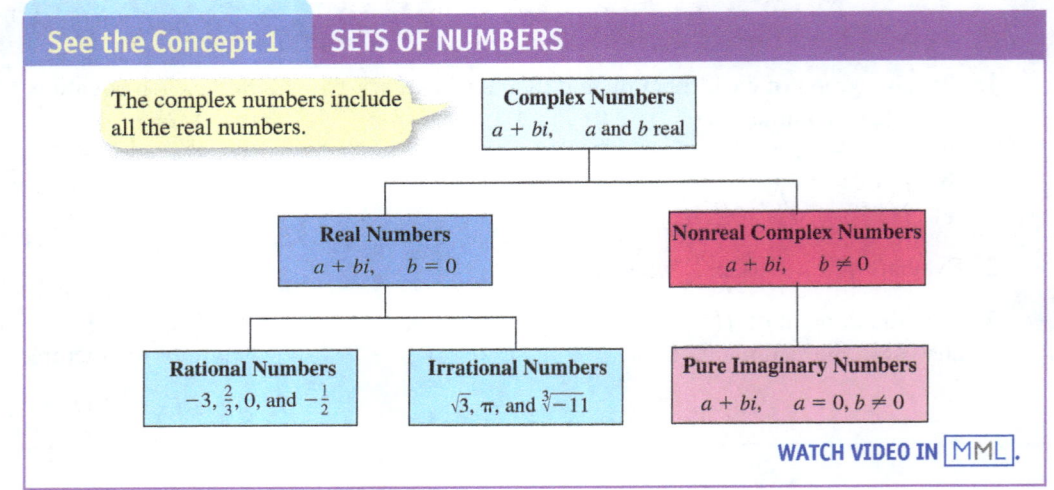

WATCH VIDEO IN MML.

Using the imaginary unit i, we may write the square root of a negative number as a complex number. For example, $\sqrt{-2} = i\sqrt{2}$, and $\sqrt{-4} = i\sqrt{4} = 2i$. This method is summarized as follows.

CALCULATOR HELP

To set your calculator in $a + bi$ mode, see Appendix A (page AP-7).

THE EXPRESSION $\sqrt{-a}$

If $a > 0$, then $\sqrt{-a} = i\sqrt{a}$.

NOTE: Although it is standard for a complex number to be expressed as $a + bi$, we often write $\sqrt{2}i$ as $i\sqrt{2}$ so that it is clear i is not under the square root. Similarly, $\frac{1}{2}\sqrt{2}i$ is sometimes written as $\frac{1}{2}i\sqrt{2}$ or $\frac{i\sqrt{2}}{2}$. ∎

EXAMPLE 1 Writing the square root of a negative number

Write each square root using the imaginary unit i.
(a) $\sqrt{-25}$ (b) $\sqrt{-7}$ (c) $\sqrt{-20}$

Solution
(a) $\sqrt{-25} = i\sqrt{25} = 5i$
(b) $\sqrt{-7} = i\sqrt{7}$
(c) $\sqrt{-20} = i\sqrt{20} = i\sqrt{4}\sqrt{5} = 2i\sqrt{5}$

Now Try Exercises 13, 15, 19

2 Adding, Subtracting, and Multiplying Complex Numbers

Addition can be defined for complex numbers in a manner similar to how we add binomials. For example,

$$(-3 + 2x) + (2 - x) = (-3 + 2) + (2x - x) = -1 + x.$$

ADDITION AND SUBTRACTION To add the complex numbers $(-3 + 2i)$ and $(2 - i)$, add the real parts and then add the imaginary parts.

$$(-3 + 2i) + (2 - i) = (-3 + 2) + (2i - i)$$
$$= (-3 + 2) + (2 - 1)i$$
$$= -1 + i$$

This same process works for subtraction.

$$(6 - 3i) - (2 + 5i) = (6 - 2) + (-3i - 5i)$$
$$= (6 - 2) + (-3 - 5)i$$
$$= 4 - 8i$$

This method is summarized as follows.

READING CHECK 1

Add $(1 - 2i) + (3 + i)$.

SUM OR DIFFERENCE OF COMPLEX NUMBERS

Let $a + bi$ and $c + di$ be two complex numbers. Then

$(a + bi) + (c + di) = (a + c) + (b + d)i$ Sum

and

$(a + bi) - (c + di) = (a - c) + (b - d)i$.   Difference

EXAMPLE 2   Adding and subtracting complex numbers

Write each sum or difference in standard form.
(a) $(-7 + 2i) + (3 - 4i)$ (b) $3i - (5 - i)$

Solution
(a) $(-7 + 2i) + (3 - 4i) = (-7 + 3) + (2 - 4)i = -4 - 2i$
(b) $3i - (5 - i) = 3i - 5 + i = -5 + (3 + 1)i = -5 + 4i$

Now Try Exercises 23, 29

CALCULATOR HELP
To access the imaginary unit i, see Appendix A (page AP-7).

MULTIPLICATION We multiply two complex numbers in the same way that we multiply binomials and then we apply the property $i^2 = -1$. For example,

$$(2 - 3x)(1 + 4x) = 2 + 8x - 3x - 12x^2 = 2 + 5x - 12x^2.$$

In the next example, we find the product of $2 - 3i$ and $1 + 4i$ in a similar manner.

EXAMPLE 3   Multiplying complex numbers

Write each product in standard form.
(a) $(2 - 3i)(1 + 4i)$ (b) $(5 - 2i)(5 + 2i)$

Solution
(a) Multiply the complex numbers like binomials.

$$(2 - 3i)(1 + 4i) = (2)(1) + (2)(4i) - (3i)(1) - (3i)(4i)$$
$$= 2 + 8i - 3i - 12i^2$$
$$= 2 + 5i - 12(-1) \quad i^2 = -1$$
$$= 14 + 5i$$

Complex Multiplication
```
(2-3i)(1+4i)
                14+5i
(5-2i)(5+2i)
                   29
```

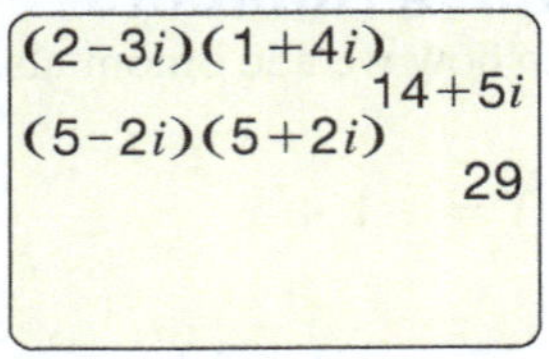

FIGURE 17.22

(b) $(5 - 2i)(5 + 2i) = (5)(5) + (5)(2i) - (2i)(5) - (2i)(2i)$
$$= 25 + 10i - 10i - 4i^2$$
$$= 25 - 4(-1) \quad i^2 = -1$$
$$= 29$$

TECHNOLOGY NOTE

Complex Numbers
Many calculators can perform arithmetic with complex numbers. The figure shows a calculator display for the results in Example 2.

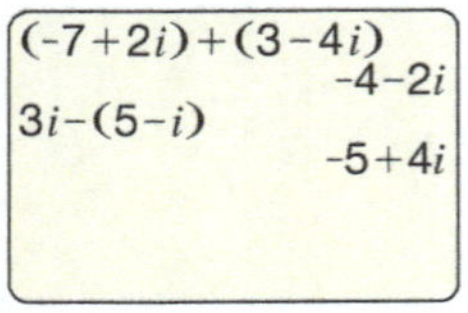

```
(-7+2i)+(3-4i)
            -4-2i
3i-(5-i)
            -5+4i
```

These results are supported in **FIGURE 17.22**.

Now Try Exercises 31, 35

3 Calculating Powers of i

An interesting pattern appears when powers of i are calculated.

$$i^1 = i$$
$$i^2 = -1$$
$$i^3 = i^2 \cdot i = -1 \cdot i = -i$$
$$i^4 = i^2 \cdot i^2 = (-1)(-1) = 1$$
$$i^5 = i^4 \cdot i = (1)i = i$$
$$i^6 = i^4 \cdot i^2 = (1)(-1) = -1$$
$$i^7 = i^4 \cdot i^3 = (1)(-i) = -i$$
$$i^8 = i^4 \cdot i^4 = (1)(1) = 1$$

Note the pattern:
$i = i^5, i^2 = i^6$,
and so on.

The powers of i cycle with the pattern i, -1, $-i$, and 1. These examples suggest the following method for calculating powers of i.

> **POWERS OF i**
>
> The value of i^n can be found by dividing n (a positive integer) by 4. If the remainder is r, then
> $$i^n = i^r.$$
> Note that $i^0 = 1$, $i^1 = i$, $i^2 = -1$, and $i^3 = -i$.

EXAMPLE 4 Calculating powers of i

Evaluate each expression.
(a) i^9 (b) i^{19} (c) i^{40}

Solution
(a) When 9 is divided by 4, the result is 2 with remainder 1. Thus $i^9 = i^1 = i$.
(b) When 19 is divided by 4, the result is 4 with remainder 3. Thus $i^{19} = i^3 = -i$.
(c) When 40 is divided by 4, the result is 10 with remainder 0. Thus $i^{40} = i^0 = 1$.

Now Try Exercises 49, 51, 55

4 Finding Complex Conjugates

The **complex conjugate** of $a + bi$ is $a - bi$. To find the conjugate, we change the sign of the imaginary part b. **TABLE 17.8** contains some complex numbers and their conjugates.

Complex Conjugates

Number	$2 + 5i$	$6 - 3i$	$-2 + 7i$	$-1 - i$	5	$-5i$
Conjugate	$2 - 5i$	$6 + 3i$	$-2 - 7i$	$-1 + i$	5	$5i$

TABLE 17.8

EXAMPLE 5 Finding complex conjugates

Find the complex conjugate of each number.
(a) $5 + 3i$ (b) $-2 - i$ (c) $-4i$ (d) -6

Solution
(a) The conjugate is found by changing the imaginary part of $5 + 3i$ to its opposite. The conjugate is $5 - 3i$.
(b) The conjugate is $-2 + i$.
(c) $-4i$ can be written as $0 - 4i$. The conjugate is $0 + 4i$, or $4i$.
(d) -6 can be written as $-6 + 0i$. The conjugate is $-6 - 0i$, or -6. The conjugate of a real number is simply the same real number.

Now Try Exercises 57, 59, 61, 63

NOTE: The product of a complex number and its conjugate is a *real* number. That is,
$$(a + bi)(a - bi) = a^2 + b^2.$$
For example, $(3 + 4i)(3 - 4i) = 3^2 + 4^2 = 25$, which is a real number. ∎

5 Dividing Complex Numbers

The property of complex conjugates in the NOTE above is used to divide two complex numbers. To convert the quotient $\frac{2 + 3i}{3 - i}$ into standard form $a + bi$, we multiply the numerator and the denominator by the complex conjugate of the *denominator*, which is $3 + i$. The next example illustrates this method.

EXAMPLE 6 Dividing complex numbers

Write each quotient in standard form.

(a) $\dfrac{2 + 3i}{3 - i}$ (b) $\dfrac{4}{2i}$

Solution

(a) Multiply the numerator and denominator by $3 + i$.

$\dfrac{2 + 3i}{3 - i} = \dfrac{(2 + 3i)(3 + i)}{(3 - i)(3 + i)}$ The conjugate of $3 - i$ is $3 + i$.   Multiply by 1.

$= \dfrac{2(3) + (2)(i) + (3i)(3) + (3i)(i)}{(3)(3) + (3)(i) - (i)(3) - (i)(i)}$ Multiply.

$= \dfrac{6 + 2i + 9i + 3i^2}{9 + 3i - 3i - i^2}$ Simplify.

$= \dfrac{6 + 11i + 3(-1)}{9 - (-1)}$ $i^2 = -1$

$= \dfrac{3 + 11i}{10}$ Simplify.

$= \dfrac{3}{10} + \dfrac{11}{10}i$ $\dfrac{a + bi}{c} = \dfrac{a}{c} + \dfrac{b}{c}i$

(b) Multiply the numerator and denominator by $-2i$.

$\dfrac{4}{2i} = \dfrac{(4)(-2i)}{(2i)(-2i)}$ The conjugate of $2i$ is $-2i$.   Multiply by 1.

$= \dfrac{-8i}{-4i^2}$ Simplify.

$= \dfrac{-8i}{-4(-1)}$ $i^2 = -1$

$= \dfrac{-8i}{4}$ Simplify.

$= -2i$ Divide.

These results are supported in **FIGURE 17.23**.

Now Try Exercises 71, 73

Complex Division

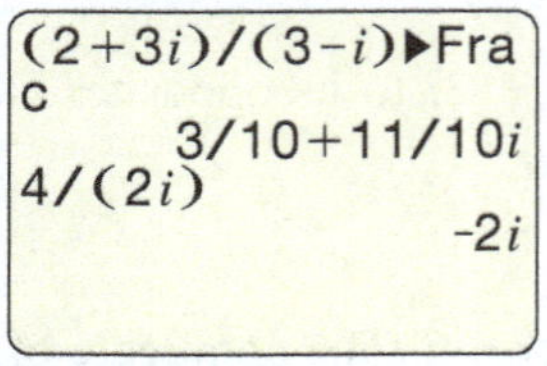

```
(2+3i)/(3-i)▶Fra
c
         3/10+11/10i
4/(2i)
                -2i
```

FIGURE 17.23

17.7 Putting It All Together

CONCEPT	EXPLANATION	EXAMPLES
Complex Numbers	A complex number can be expressed as $a + bi$, where a and b are real numbers. The imaginary unit i satisfies $i = \sqrt{-1}$ and $i^2 = -1$. As a result, we can write $\sqrt{-a} = i\sqrt{a}$ if $a > 0$.	The real part of $5 - 3i$ is 5 and the imaginary part is -3. $\sqrt{-13} = i\sqrt{13}$ and $\sqrt{-9} = 3i$
Addition, Subtraction, and Multiplication	To add (subtract) complex numbers, add (subtract) the real parts and then add (subtract) the imaginary parts.	$(3 + 6i) + (-1 + 2i)$ Sum $= (3 + (-1)) + (6 + 2)i$ $= 2 + 8i$ $(2 - 5i) - (1 + 4i)$ Difference $= (2 - 1) + (-5 - 4)i$ $= 1 - 9i$
	Multiply complex numbers in a manner similar to the way FOIL is used to multiply binomials. Then apply the property $i^2 = -1$.	$(-1 + 2i)(3 + i)$ Product $= (-1)(3) + (-1)(i) + (2i)(3) + (2i)(i)$ $= -3 - i + 6i + 2i^2$ $= -3 + 5i + 2(-1)$ $= -5 + 5i$
Complex Conjugates	The conjugate of $a + bi$ is $a - bi$.	The conjugate of $3 - 5i$ is $3 + 5i$. The conjugate of $2i$ is $-2i$.
Division	To simplify a quotient, multiply the numerator and denominator by the complex conjugate of the *denominator*. Then simplify the expression and write it in standard form as $a + bi$.	$\dfrac{10}{1 + 2i} = \dfrac{10(1 - 2i)}{(1 + 2i)(1 - 2i)}$ Quotient $= \dfrac{10 - 20i}{5}$ $= 2 - 4i$

17.7 Exercises

CONCEPTS AND VOCABULARY

1. Give an example of a complex number that is not a real number.
2. Can you give an example of a real number that is not a complex number? Explain.
3. $\sqrt{-1} =$ _____
4. $i^2 =$ _____
5. $\sqrt{-a} =$ _____, if $a > 0$.
6. The complex conjugate of $10 + 7i$ is _____.
7. The standard form for a complex number is _____.
8. Write $\dfrac{2 + 4i}{2}$ in standard form.
9. The real part of $4 - 5i$ is _____.
10. The imaginary part of $4 - 5i$ is _____.
11. The imaginary part of -7 is _____.
12. The number $7i$ is a(n) _____ imaginary number.

SIMPLIFYING SQUARE ROOTS OF NEGATIVE NUMBERS

Exercises 13–22: Use i to write the expression.

13. $\sqrt{-5}$
14. $\sqrt{-21}$
15. $\sqrt{-100}$
16. $\sqrt{-49}$
17. $\sqrt{-144}$
18. $\sqrt{-64}$

19. $\sqrt{-12}$ **20.** $\sqrt{-8}$
21. $\sqrt{-18}$ **22.** $\sqrt{-48}$

ADDING, SUBTRACTING, AND MULTIPLYING COMPLEX NUMBERS

Exercises 23–46: Write the expression in standard form.

23. $(5 + 3i) + (-2 - 3i)$

24. $(1 - i) + (5 - 7i)$

25. $2i + (-8 + 5i)$ **26.** $-3i + 5i$

27. $(2 - 7i) - (1 + 2i)$ **28.** $(1 + 8i) - (3 + 9i)$

29. $5i - (10 - 2i)$ **30.** $-3(4 - 3i)$

31. $(3 + 2i)(-1 + 5i)$ **32.** $(1 + i) - (1 - i)$

33. $4(5 - 3i)$ **34.** $(1 + 2i)(-6 - i)$

35. $(5 + 4i)(5 - 4i)$ **36.** $(3 + 5i)(3 - 5i)$

37. $(-4i)(5i)$ **38.** $(-6i)(-4i)$

39. $3i + (2 - 3i) - (1 - 5i)$

40. $4 - (5 - 7i) + (3 + 7i)$

41. $(2 + i)^2$ **42.** $(-1 + 2i)^2$

43. $2i(-3 + i)$ **44.** $5i(1 - 9i)$

45. $i(1 + i)^2$ **46.** $2i(1 - i)^2$

Exercises 47 and 48: **Thinking Generally** *Write in standard form.*

47. $(a + 3bi)(a - 3bi)$ **48.** $(a + bi) - (a - bi)$

SIMPLIFYING POWERS OF i

Exercises 49–56: (Refer to Example 4.) Simplify.

49. i^{11} **50.** i^{50} **51.** i^{21}

52. i^{103} **53.** i^{58} **54.** i^{61}

55. i^{64} **56.** i^{28}

FINDING COMPLEX CONJUGATES

Exercises 57–64: Write the complex conjugate.

57. $3 + 4i$ **58.** $1 - 4i$

59. $-6i$ **60.** -10

61. $5 - 4i$ **62.** $7 + 2i$

63. -1 **64.** $19i$

DIVIDING COMPLEX NUMBERS

Exercises 65–78: Write the expression in standard form.

65. $\dfrac{2}{1 + i}$ **66.** $\dfrac{-6}{2 - i}$

67. $\dfrac{3i}{5 - 2i}$ **68.** $\dfrac{-8}{2i}$

69. $\dfrac{8 + 9i}{5 + 2i}$ **70.** $\dfrac{3 - 2i}{1 + 4i}$

71. $\dfrac{5 + 7i}{1 - i}$ **72.** $\dfrac{-7 + 4i}{3 - 2i}$

73. $\dfrac{2 - i}{i}$ **74.** $\dfrac{3 + 2i}{-i}$

75. $\dfrac{1}{i} + \dfrac{1}{2i}$ **76.** $\dfrac{3}{4i} + \dfrac{2}{i}$

77. $\dfrac{1}{-1 + i} - \dfrac{2}{i}$ **78.** $-\dfrac{3}{2i} - \dfrac{2}{1 + i}$

USING ZEROS TO WRITE A FORMULA FOR A FUNCTION

Exercises 79–84: A quadratic function f has the given complex zeros with leading coefficient 1.

(a) *Write the formula for $f(x)$ in factored form.*
(b) *Write the expanded form for $f(x)$ by multiplying the factor form.*

(*Hint: If $-5i$ and $5i$ are the two zeros of $f(x)$, then $f(x)$ can be written in factored form using the formula given by $f(x) = (x + 5i)(x - 5i)$.*)

79. $-i, i$ **80.** $-3i, 3i$

81. $-7i, 7i$ **82.** $-9i, 9i$

83. $-i\sqrt{6}, i\sqrt{6}$ **84.** $-i\sqrt{11}, i\sqrt{11}$

APPLICATIONS INVOLVING COMPLEX NUMBERS

Exercises 85 and 86: **Corrosion in Airplanes** *Corrosion in the metal surface of an airplane can be difficult to detect visually. One test used to locate it involves passing an alternating current through a small area on the plane's surface. If the current varies from one region to another, it may indicate that corrosion is occurring. The impedance Z (or opposition to the flow of electricity) of the metal is related to the voltage V and current I by the equation $Z = \frac{V}{I}$, where Z, V, and I are complex numbers. Calculate Z* for the given values of V and I. (Source: Society for Industrial and Applied Mathematics.)

85. $V = 40 + 70i, I = 2 + 3i$

86. $V = 10 + 20i, I = 3 + 7i$

87. Online Exploration We graph real numbers on a number line. Can we graph complex numbers on a number line? Go online and find out how to graph complex numbers.

88. Online Exploration Go online and find out how to calculate the absolute value of a complex number. If a complex number is graphed, what does the absolute value tell us?

WRITING ABOUT MATHEMATICS

89. A student multiplies $(2 + 3i)(4 - 5i)$ *incorrectly* to obtain $8 - 15i$. What is the student's mistake?

90. A student divides the ratio $\frac{6 - 10i}{3 + 2i}$ *incorrectly* to obtain $2 - 5i$. What is the student's mistake?

SECTION 17.7 Checking Basic Concepts

1. Use i to write each expression.
   (a) $\sqrt{-64}$
   (b) $\sqrt{-17}$

2. Simplify each expression.
   (a) $(2 - 3i) + (1 - i)$
   (b) $4i - (2 + i)$
   (c) $(3 - 2i)(1 + i)$
   (d) $\dfrac{3}{2 - 2i}$

CHAPTER 17 Summary

SECTION 17.1 ■ RADICAL EXPRESSIONS AND FUNCTIONS

Radicals and Radical Notation

Square Root b is a square root of a if $b^2 = a$.

Principal Square Root $\sqrt{a} = b$ if $b^2 = a$ and $b \geq 0$.

Examples: $\sqrt{16} = 4$, $-\sqrt{9} = -3$, and $\pm\sqrt{36} = \pm 6$

Cube Root b is a cube root of a if $b^3 = a$.

Examples: $\sqrt[3]{27} = 3$, $\sqrt[3]{-8} = -2$

nth Root b is an nth root of a if $b^n = a$.

Example: $\sqrt[4]{16} = 2$ because $2^4 = 16$.

NOTE: An *even* root of a *negative* number is not a real number. Also, $\sqrt[n]{a}$ denotes the *principal* nth root. ■

Absolute Value   The expressions $|x|$ and $\sqrt{x^2}$ are equivalent.

Example: $\sqrt{(x + y)^2} = |x + y|$

The Square Root Function The square root function is denoted $f(x) = \sqrt{x}$. Its domain is $\{x \mid x \geq 0\}$ and its graph is shown in the figure.

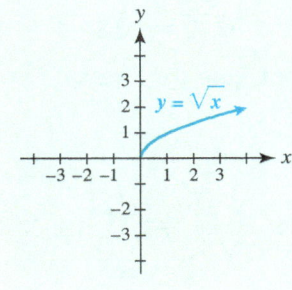

The Cube Root Function The cube root function is denoted $f(x) = \sqrt[3]{x}$. Its domain is all real numbers and its graph is shown in the figure.

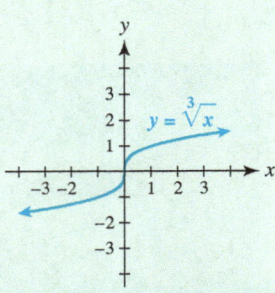

SECTION 17.2 ■ RATIONAL EXPONENTS

The Expression $a^{1/n}$ 　　$a^{1/n} = \sqrt[n]{a}$ if n is an integer greater than 1.

Examples: $5^{1/2} = \sqrt{5}$ 　and　 $64^{1/3} = \sqrt[3]{64} = 4$

The Expression $a^{m/n}$ 　　$a^{m/n} = \sqrt[n]{a^m}$ 　or　 $a^{m/n} = (\sqrt[n]{a})^m$

Examples: $8^{2/3} = \sqrt[3]{8^2} = \sqrt[3]{64} = 4$ 　and
$8^{2/3} = (\sqrt[3]{8})^2 = (2)^2 = 4$

Properties of Exponents

Product Rule　　$a^p a^q = a^{p+q}$ 　　　　$3^{1/7} \cdot 3^{2/7} = 3^{3/7}$

Negative Exponents　　$a^{-p} = \dfrac{1}{a^p},\ \dfrac{1}{a^{-p}} = a^p$ 　　$7^{-1/4} = \dfrac{1}{7^{1/4}},\ \dfrac{1}{7^{-1/4}} = 7^{1/4}$

Negative Exponents for Quotients　　$\left(\dfrac{a}{b}\right)^{-p} = \left(\dfrac{b}{a}\right)^p$ 　　$\left(\dfrac{3}{7}\right)^{-1/3} = \left(\dfrac{7}{3}\right)^{1/3}$

Quotient Rule for Exponents　　$\dfrac{a^p}{a^q} = a^{p-q}$ 　　$\dfrac{5^{4/7}}{5^{1/7}} = 5^{3/7}$

Power Rule for Exponents　　$(a^p)^q = a^{pq}$ 　　$(2^{1/3})^{1/2} = 2^{1/6}$

Power Rule for Products　　$(ab)^p = a^p b^p$ 　　$(5x)^{2/3} = 5^{2/3} x^{2/3}$

Power Rule for Quotients　　$\left(\dfrac{a}{b}\right)^p = \dfrac{a^p}{b^p}$ 　　$\left(\dfrac{5}{3}\right)^{3/4} = \dfrac{5^{3/4}}{3^{3/4}}$

SECTION 17.3 ■ SIMPLIFYING RADICAL EXPRESSIONS

Product Rule for Radical Expressions Provided each expression is defined,
$$\sqrt[n]{a} \cdot \sqrt[n]{b} = \sqrt[n]{a \cdot b}.$$

Example: $\sqrt[3]{3} \cdot \sqrt[3]{9} = \sqrt[3]{27} = 3$

Perfect nth Power An integer a is a perfect nth power if $b^n = a$ for some integer b.

Examples: 25 is a perfect square, 8 is a perfect cube, and 16 is a perfect fourth power.

Simplifying Radicals (nth roots)

STEP 1: Determine the largest perfect nth power factor of the radicand.

STEP 2: Use the product rule to factor out and simplify this perfect nth power.

Examples: $\sqrt{32} = \sqrt{16 \cdot 2} = \sqrt{16} \cdot \sqrt{2} = 4\sqrt{2}$
$\sqrt[3]{32} = \sqrt[3]{8 \cdot 4} = \sqrt[3]{8} \cdot \sqrt[3]{4} = 2\sqrt[3]{4}$

Quotient Rule for Radical Expressions Provided each expression is defined,
$$\sqrt[n]{\dfrac{a}{b}} = \dfrac{\sqrt[n]{a}}{\sqrt[n]{b}}.$$

Example: $\dfrac{\sqrt[3]{24}}{\sqrt[3]{3}} = \sqrt[3]{\dfrac{24}{3}} = \sqrt[3]{8} = 2$

SECTION 17.4 ■ OPERATIONS ON RADICAL EXPRESSIONS

Addition and Subtraction Combine like radicals.

Examples: $2\sqrt[3]{4} + 3\sqrt[3]{4} = 5\sqrt[3]{4}$ and $\sqrt{5} - 2\sqrt{5} = -\sqrt{5}$

Multiplication Sometimes radical expressions can be multiplied like binomials.

Examples: $(4 - \sqrt{2})(2 + \sqrt{2}) = 8 + 4\sqrt{2} - 2\sqrt{2} - 2 = 6 + 2\sqrt{2}$
$(5 - \sqrt{3})(5 + \sqrt{3}) = (5)^2 - (\sqrt{3})^2 = 25 - 3 = 22$ because
$(a - b)(a + b) = a^2 - b^2$.

Rationalizing the Denominator One technique is to multiply the numerator and denominator by the conjugate of the denominator if the denominator is a binomial containing one or more square roots.

Examples: $\dfrac{1}{4 + \sqrt{2}} = \dfrac{1}{(4 + \sqrt{2})} \cdot \dfrac{(4 - \sqrt{2})}{(4 - \sqrt{2})} = \dfrac{4 - \sqrt{2}}{(4)^2 - (\sqrt{2})^2} = \dfrac{4 - \sqrt{2}}{14}$

$\dfrac{4}{\sqrt{7}} = \dfrac{4}{\sqrt{7}} \cdot \dfrac{\sqrt{7}}{\sqrt{7}} = \dfrac{4\sqrt{7}}{7}$

SECTION 17.5 ■ MORE RADICAL FUNCTIONS

Root Functions $f(x) = x^{1/n} = \sqrt[n]{x}$

Odd root functions are defined for all real numbers, and even root functions are defined for nonnegative real numbers.

Examples: $f(x) = x^{1/3} = \sqrt[3]{x}$, $g(x) = x^{1/4} = \sqrt[4]{x}$, and $h(x) = x^{1/5} = \sqrt[5]{x}$

Power Functions If a function can be defined by $f(x) = x^p$, where p is a rational number, then it is a power function.

Examples: $f(x) = x^{4/5}$ and $g(x) = x^{2.3}$

SECTION 17.6 ■ EQUATIONS INVOLVING RADICAL EXPRESSIONS

Power Rule for Solving Radical Equations The solutions to $a = b$ are among the solutions to $a^n = b^n$, where n is a positive integer.

Example: The solutions to the equation $\sqrt{3x + 3} = 2x - 1$ are among the solutions to the equation $3x + 3 = (2x - 1)^2$.

Solving Radical Equations

STEP 1: Isolate a radical term on one side of the equation.

STEP 2: Apply the power rule by raising each side of the equation to the power equal to the index of the isolated radical term.

STEP 3: Solve the equation. If it still contains a radical, repeat Steps 1 and 2.

STEP 4: Check your answers by substituting each result in the *given* equation.

Example: To isolate the radical in $\sqrt{x + 1} + 4 = 6$, subtract 4 from each side to obtain $\sqrt{x + 1} = 2$. Next square each side, which gives $x + 1 = 4$ or $x = 3$. Checking verifies that 3 is a solution.

Pythagorean Theorem If a right triangle has legs a and b with hypotenuse c, then
$$a^2 + b^2 = c^2.$$

Example: If a right triangle has legs 8 and 15, then the hypotenuse equals
$$c = \sqrt{8^2 + 15^2} = \sqrt{289} = 17.$$

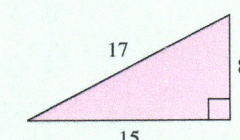

The Distance Formula The distance d between (x_1, y_1) and (x_2, y_2) is
$$d = \sqrt{(x_2 - x_1)^2 + (y_2 - y_1)^2}.$$

continued on next page

Example: The distance between $(-1, 3)$ and $(4, 5)$ is
$$d = \sqrt{(4-(-1))^2 + (5-3)^2} = \sqrt{25 + 4} = \sqrt{29}.$$

Solving the Equation $x^n = k$ Let n be a positive integer.
Take the nth root of each side of $x^n = k$ to obtain $\sqrt[n]{x^n} = \sqrt[n]{k}$.
1. If n is odd, then $\sqrt[n]{x^n} = x$ and the equation becomes $x = \sqrt[n]{k}$.
2. If n is even and $k \geq 0$, then $\sqrt[n]{x^n} = |x|$ and the equation becomes $|x| = \sqrt[n]{k}$.

Examples: $x^3 = -27$ implies that $x = \sqrt[3]{-27} = -3$.
$x^4 = 81$ implies that $|x| = \sqrt[4]{81} = 3$ or $x = \pm 3$.

SECTION 17.7 ■ COMPLEX NUMBERS

Complex Numbers
Imaginary Unit $\quad i = \sqrt{-1}$ and $i^2 = -1$

Standard Form $\quad a + bi$, where a and b are real numbers

Examples: $4 + 3i$, $5 - 6i$, 8, and $-2i$

Real Part $\quad$ The real part of $a + bi$ is a.

Example: The real part of $3 - 2i$ is 3.

Imaginary Part $\quad$ The imaginary part of $a + bi$ is b.

Example: The imaginary part of $2 - i$ is -1.

Arithmetic Operations $\quad$ Arithmetic operations are similar to arithmetic operations on binomials.

Examples:
$(2 + 2i) + (3 - i) = 5 + i$ $\quad$ Sum
$(1 - i) - (1 - 2i) = i$ $\quad$ Difference
$(1 - i)(1 + i) = 1^2 - i^2 = 1 - (-1) = 2$ $\quad$ Product

The complex conjugate of $1 - i$ is $1 + i$, so multiply by $\frac{1+i}{1+i}$.

$\dfrac{2}{1 - i} = \dfrac{2}{1 - i} \cdot \dfrac{1 + i}{1 + i} = \dfrac{2 + 2i}{2} = 1 + i$ $\quad$ Quotient

Powers of i $\quad$ The value of i^n equals i^r, where r is the remainder when n is divided by 4. Note that $i^0 = 1$, $i^1 = i$, $i^2 = -1$, and $i^3 = -i$.

Example: $i^{21} = i^1 = i$ because when 21 is divided by 4, the remainder is 1.

CHAPTER 17 Review Exercises

SECTION 17.1

Exercises 1–12: Simplify the expression.

1. $\sqrt{4}$
2. $\sqrt{36}$
3. $\sqrt{9x^2}$
4. $\sqrt{(x-1)^2}$
5. $\sqrt[3]{-64}$
6. $\sqrt[3]{-125}$
7. $\sqrt[3]{x^6}$
8. $\sqrt[3]{27x^3}$
9. $\sqrt[4]{16}$
10. $\sqrt[5]{-1}$
11. $\sqrt[4]{x^8}$
12. $\sqrt[5]{(x+1)^5}$

SECTION 17.2

Exercises 13–16: Write the expression in radical notation.

13. $14^{1/2}$
14. $(-5)^{1/3}$
15. $\left(\dfrac{x}{y}\right)^{3/2}$
16. $(xy)^{-2/3}$

Exercises 17–20: Evaluate the expression.

17. $(-27)^{2/3}$
18. $16^{1/4}$
19. $16^{3/2}$
20. $81^{3/4}$

Exercises 21–24: Simplify the expression. Assume that all variables are positive.

21. $(z^3)^{2/3}$ **22.** $(x^2 y^4)^{1/2}$

23. $\left(\dfrac{x^2}{y^6}\right)^{3/2}$ **24.** $\left(\dfrac{x^3}{y^6}\right)^{-1/3}$

SECTION 17.3

Exercises 25–40: Simplify the expression. Assume that all variables are positive.

25. $\sqrt{2} \cdot \sqrt{32}$ **26.** $\sqrt[3]{-4} \cdot \sqrt[3]{2}$

27. $\sqrt[3]{x^4} \cdot \sqrt[3]{x^2}$ **28.** $\dfrac{\sqrt{80}}{\sqrt{20}}$

29. $\sqrt[3]{-\dfrac{x}{8}}$ **30.** $\sqrt{\dfrac{1}{3}} \cdot \sqrt{\dfrac{1}{3}}$

31. $\sqrt{48}$ **32.** $\sqrt{54}$

33. $\sqrt[3]{\dfrac{3}{x}} \cdot \sqrt[3]{\dfrac{9}{x^2}}$ **34.** $\sqrt{32 a^3 b^2}$

35. $\sqrt{3xy} \cdot \sqrt{27xy}$ **36.** $\sqrt[3]{-25 z^2} \cdot \sqrt[3]{-5z^2}$

37. $\sqrt{x^2 + 2x + 1}$ **38.** $\sqrt[4]{\dfrac{2a^2}{b}} \cdot \sqrt[4]{\dfrac{8a^3}{b^3}}$

39. $2\sqrt{x} \cdot \sqrt[3]{x}$ **40.** $\sqrt[3]{rt} \cdot \sqrt[4]{r^2 t^4}$

SECTION 17.4

Exercises 41–50: Simplify the expression. Assume that all variables are positive.

41. $3\sqrt{3} + \sqrt{3}$ **42.** $\sqrt[3]{x} + 2\sqrt[3]{x}$

43. $3\sqrt[3]{5} - 6\sqrt[3]{5}$ **44.** $\sqrt[4]{y} - 2\sqrt[4]{y}$

45. $2\sqrt{12} + 7\sqrt{3}$ **46.** $3\sqrt{18} - 2\sqrt{2}$

47. $7\sqrt[3]{16} - \sqrt[3]{2}$ **48.** $\sqrt{4x + 4} + \sqrt{x + 1}$

49. $\sqrt{4x^3} - \sqrt{x}$ **50.** $\sqrt[3]{ab^4} + 2\sqrt[3]{a^4 b}$

Exercises 51–56: Multiply and simplify.

51. $(1 + \sqrt{2})(3 + \sqrt{2})$

52. $(7 - \sqrt{5})(1 + \sqrt{3})$

53. $(3 + \sqrt{6})(3 - \sqrt{6})$

54. $(10 - \sqrt{5})(10 + \sqrt{5})$

55. $(\sqrt{a} + \sqrt{2b})(\sqrt{a} - \sqrt{2b})$

56. $(\sqrt{xy} - 1)(\sqrt{xy} + 2)$

Exercises 57–62: Rationalize the denominator.

57. $\dfrac{4}{\sqrt{5}}$ **58.** $\dfrac{r}{2\sqrt{t}}$

59. $\dfrac{1}{\sqrt{2} + 3}$ **60.** $\dfrac{2}{5 - \sqrt{7}}$

61. $\dfrac{1}{\sqrt{8} - \sqrt{7}}$ **62.** $\dfrac{\sqrt{a} - \sqrt{b}}{\sqrt{a} + \sqrt{b}}$

SECTION 17.5

Exercises 63 and 64: Graph the equation.

63. $y = \sqrt[4]{x}$ **64.** $y = \sqrt[3]{x}$

Exercises 65 and 66: Write $f(x)$ in radical notation and evaluate $f(4)$.

65. $f(x) = x^{1/2}$ **66.** $f(x) = x^{2/7}$

Exercises 67 and 68: Graph the equation. Compare the graph to either $y = \sqrt{x}$ or $y = \sqrt[3]{x}$.

67. $y = \sqrt{x} - 2$ **68.** $y = \sqrt[3]{x - 1}$

Exercises 69–72: Find the domain of f. Write your answer in interval notation.

69. $f(x) = \sqrt{x - 1}$ **70.** $f(x) = \sqrt{6 - 2x}$

71. $f(x) = \sqrt{x^2 + 1}$ **72.** $f(x) = \dfrac{1}{\sqrt{x + 2}}$

SECTION 17.6

Exercises 73–78: Solve. Check your answer.

73. $\sqrt{x + 2} = x$ **74.** $\sqrt{2x - 1} = \sqrt{x + 3}$

75. $\sqrt[3]{x - 1} = 2$ **76.** $\sqrt[3]{3x} = 3$

77. $\sqrt{2x} = x - 4$ **78.** $\sqrt{x} + 1 = \sqrt{x + 2}$

Exercises 79 and 80: Solve graphically. Approximate solutions to the nearest hundredth when appropriate.

79. $\sqrt[3]{2x - 1} = 2$ **80.** $x^{2/3} = 3 - x$

Exercises 81 and 82: A right triangle has legs a and b with hypotenuse c. Find the length of the missing side.

81. $a = 4, b = 7$ **82.** $a = 5, c = 8$

Exercises 83 and 84: Find the exact distance between the given points.

83. $(-2, 3), (2, -2)$ **84.** $(2, -3), (-4, 1)$

Exercises 85–94: Solve.

85. $x^2 = 121$ **86.** $2z^2 = 32$

87. $(x - 1)^2 = 16$ **88.** $x^3 = 64$

89. $(x - 1)^3 = 8$ **90.** $(2x - 1)^3 = 27$

91. $x^4 = 256$ **92.** $x^5 = -1$

93. $(x - 3)^5 = -32$ **94.** $3(x + 1)^4 = 3$

SECTION 17.7

Exercises 95–100: Write the given complex expression in standard form.

95. $(1 - 2i) + (-3 + 2i)$

96. $(1 + 3i) - (3 - i)$

97. $(1 - i)(2 + 3i)$

98. $\dfrac{3 + i}{1 - i}$

99. $\dfrac{i(4 + i)}{2 - 3i}$

100. $(1 - i)^2(1 + i)$

APPLICATIONS

101. Hang Time A football is punted and has a hang time T of 4.6 seconds. Use the formula $T(h) = \frac{1}{2}\sqrt{h}$ to estimate to the nearest foot the height h to which it was kicked.

102. Baseball Diamond The four bases of a baseball diamond form a square that is 90 feet on a side. Find the distance from home plate to second base.

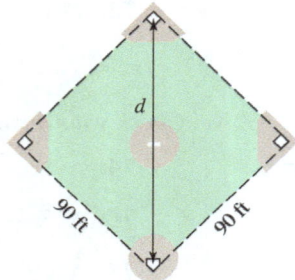

103. Falling Time The time T in seconds for an object to fall from a height of h feet is given by $T = \frac{1}{4}\sqrt{h}$. If a person steps off a 10-foot-high board into a swimming pool, how long is the person in the air?

104. Geometry A cube has sides of length $\sqrt{5}$.
  (a) Find the area of one side of the cube.
  (b) Find the volume of the cube.
  (c) Find the length of the diagonal of one of the sides.
  (d) Find the distance from A to B in the figure.

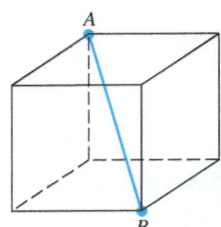

105. Population Growth In 1790, the population of the United States was 4 million, and by 2000, it had grown to 281 million. The *average* annual percentage growth in the population r (expressed as a decimal) can be determined by the polynomial equation $281 = 4(1 + r)^{210}$. Solve this equation for r and interpret the result.

106. Highway Curves If a circular highway curve is banked with a slope of $m = \frac{1}{10}$ (see the accompanying figure) and has a radius of R feet, then the speed limit L in miles per hour for the curve is given by

$$L = \sqrt{3.75R}.$$

(*Source*: N. Garber and L. Hoel, *Traffic and Highway Engineering*.)

  (a) Find the speed limit if R is 500 feet.
  (b) With no banking, the speed limit is given by $L = 1.5\sqrt{R}$. Find the speed limit for a curve with no banking and a radius of 500 feet. How does banking affect the speed limit? Does this result agree with your intuition?

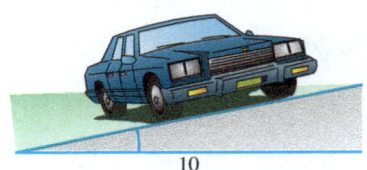

107. Geometry Find the length of a side of a square if the square has an area of 7 square feet.

CHAPTER 17 Test   MyMathLab®, YouTube

Exercises 1–6: Simplify the expression.

1. $\sqrt[3]{-27}$

2. $\sqrt{(z + 1)^2}$

3. $\sqrt{25x^4}$

4. $\sqrt[3]{8z^6}$

5. $\sqrt[4]{16x^4y^5}$; $x > 0, y > 0$

6. $(\sqrt{3} - \sqrt{2})(\sqrt{3} + \sqrt{2})$

Exercises 7 and 8: Write the given expression using radical notation.

7. $7^{2/5}$

8. $\left(\dfrac{x}{y}\right)^{-2/3}$

Exercises 9 and 10: Evaluate the expression by hand.

9. $(-8)^{4/3}$

10. $36^{-3/2}$

Exercises 11 and 12: Use rational exponents to write the expression.

11. $\sqrt[3]{x^4}$

12. $\sqrt{x} \cdot \sqrt[5]{x}$

13. Find the domain of $f(x) = \sqrt{4-x}$. Write your answer in interval notation.

14. Sketch a graph of $y = \sqrt{x+3}$.

Exercises 15–22: Simplify the expression. Assume that all variables are positive.

15. $(2z^{1/2})^3$

16. $\left(\dfrac{y^2}{z^3}\right)^{-1/3}$

17. $\sqrt{3} \cdot \sqrt{27}$

18. $\dfrac{\sqrt{y^3}}{\sqrt{4y}}$

19. $7\sqrt{7} - 3\sqrt{7} + \sqrt{5}$

20. $7\sqrt[3]{x} - \sqrt[3]{x}$

21. $4\sqrt{18} + \sqrt{8}$

22. $\dfrac{\sqrt[3]{32}}{\sqrt[3]{4}}$

23. Solve each equation.
    (a) $\sqrt{x-2} = 5$
    (b) $\sqrt[3]{x+1} = 2$
    (c) $(x-1)^3 = 8$
    (d) $\sqrt{2x+2} = x - 11$

24. Rationalize the denominator.
    (a) $\dfrac{2}{3\sqrt{7}}$
    (b) $\dfrac{1}{1+\sqrt{5}}$

25. Solve $\sqrt{3x-x+1} = \sqrt[3]{x-1}$ graphically. Round solutions to the nearest hundredth.

26. One leg of a right triangle has length 7 and the hypotenuse has length 13. Find the length of the third side.

27. Find the distance between $(-3, 5)$ and $(-1, 7)$.

Exercises 28–31: Write the given complex expression in standard form.

28. $(-5 + i) + (7 - 20i)$

29. $(3i) - (6 - 5i)$

30. $\left(\dfrac{1}{2} - i\right)\left(\dfrac{1}{2} + i\right)$

31. $\dfrac{2i}{5 + 2i}$

32. **Distance to the Horizon** A formula for calculating the distance d in miles that one can see to the horizon on a clear day is approximated by the formula $d = 1.22\sqrt{x}$, where x is the elevation, in feet, of a person.
    (a) If a person climbs a cliff and stops at an elevation of 200 feet, how far can the person see to the horizon?
    (b) How high should the person climb to see 25 miles to the horizon?

33. **Wing Span of a Bird** The wing span L of a bird with weight W can sometimes be modeled by $L = 27.4W^{1/3}$, where L is in inches and W is in pounds. Use this formula to estimate the weight of a bird that has a wing span of 30 inches. (Source: C. Pennycuick, *Newton Rules Biology*.)

CHAPTERS 1–17 Cumulative Review Exercises

1. Evaluate $S = 4\pi r^2$ when $r = 3$.

2. Identify the domain and range of the relation given by $S = \{(-1, 2), (0, 4), (1, 2)\}$.

3. Simplify each expression. Write the result using positive exponents.
    (a) $\left(\dfrac{ab^2}{b^{-1}}\right)^{-3}$
    (b) $\dfrac{(x^2y)^3}{x^2(y^2)^{-3}}$
    (c) $(rt)^2(r^2t)^3$

4. Write 0.00043 in scientific notation.

5. Find $f(3)$ if $f(x) = \dfrac{x}{x-2}$. What is the domain?

6. Find the domain of $f(x) = \sqrt[3]{2x}$.

Exercises 7 and 8: Graph f by hand.

7. $f(x) = 1 - 2x$

8. $f(x) = \sqrt{x+1}$

9. Sketch a graph of $f(x) = x^2 - 4$.
    (a) Identify the domain and range.
    (b) Evaluate $f(-2)$.
    (c) Identify the x-intercepts.
    (d) Solve the equation $f(x) = 0$.

10. Find the slope–intercept form of the line that passes through $(-1, 2)$ and is perpendicular to $y = -2x$.

11. Let f be a linear function. Find a formula for $f(x)$.

x	-2	-1	1	2
$f(x)$	7	4	-2	-5

12. Sketch a graph of a line passing through $(-1, 1)$ with slope $-\dfrac{1}{2}$.

13. Solve $5x - (3 - x) = \dfrac{1}{2}x$.

14. Solve $2x - 5 \leq 4 - x$.

15. Solve $|x - 2| \leq 3$.

16. Solve $-1 \leq 1 - 2x \leq 6$.

17. Solve each system symbolically, if possible.
    (a) $2x - y = 4$
    $x + y = 8$
    (b) $3x - 4y = 2$
    $x - \frac{4}{3}y = 1$

18. Shade the solution set in the xy-plane.
$$x + 2y \leq 2$$
$$-x + 3y \geq 3$$

19. Solve the system.
$$x + 2y - z = 6$$
$$x - 3y + z = -2$$
$$x + y + z = 6$$

20. Calculate $\det \begin{bmatrix} 4 & 2 & -1 \\ 2 & 1 & 0 \\ 0 & -2 & 1 \end{bmatrix}$.

Exercises 21–24: Multiply the expression.

21. $4x(4 - x^3)$
22. $(x - 4)(x + 4)$
23. $(5x + 3)(x - 2)$
24. $(4x + 9)^2$

Exercises 25–30: Factor the expression.

25. $9x^2 - 16$
26. $x^2 - 4x + 4$
27. $15x^3 - 9x^2$
28. $12x^2 - 5x - 3$
29. $r^3 - 1$
30. $x^3 - 3x^2 + 5x - 15$

Exercises 31 and 32: Solve each equation.

31. $x^2 - 3x + 2 = 0$
32. $x^3 = 4x$

Exercises 33 and 34: Simplify the expression.

33. $\dfrac{x^2 + 3x + 2}{x - 3} \div \dfrac{x + 1}{2x - 6}$

34. $\dfrac{2}{x - 1} + \dfrac{5}{x}$

Exercises 35–42: Simplify the given expression. Assume all variables are positive.

35. $\sqrt{36x^2}$
36. $\sqrt[3]{64}$
37. $16^{-3/2}$
38. $\sqrt[4]{625}$
39. $\sqrt{2x} \cdot \sqrt{8x}$
40. $\sqrt{x} \cdot \sqrt[4]{x}$
41. $\dfrac{\sqrt[3]{16x^4}}{\sqrt[3]{2x}}$
42. $4\sqrt{12x} - 2\sqrt{3x}$

43. Multiply $(2x + \sqrt{3})(x - \sqrt{3})$.

44. Find the domain of $f(x) = \sqrt{1 - x}$.

45. Graph $f(x) = 3\sqrt[3]{x}$.

46. Find the distance between $(-2, 3)$ and $(1, 2)$.

47. Write $(1 - i)(2 + 3i)$ in standard form.

48. The lengths of the legs of a right triangle are 5 and 12. What is the length of the hypotenuse?

Exercises 49–52: Solve symbolically.

49. $2\sqrt{x + 3} = x$
50. $\sqrt[3]{x - 1} = 3$
51. $\sqrt{x} + 4 = 2\sqrt{x + 5}$
52. $\frac{1}{3}x^4 = 27$

Exercises 53 and 54: Solve graphically and approximate your answer to the nearest hundredth.

 53. $\sqrt[3]{x^2 - 2} + x = \sqrt{x}$
54. $x - \sqrt[3]{x} = \sqrt{x + 2}$

APPLICATIONS

55. **Calculating Water Flow** The gallons G of water in a tank after t minutes are given by $G(t) = 300 - 15t$. Interpret the y-intercept and the slope of the graph of G as a rate of change.

56. **Distance** A person jogs away from home at 8 miles per hour for 1 hour, rests in a park for 2 hours, and then walks back home at 4 miles per hour. Sketch a graph that models the distance y that the person is from home after x hours.

57. **Investment** A total of $2000 is deposited in two accounts paying 5% and 4% annual interest. If the total interest after 1 year is $93, how much is invested at each rate?

58. **Geometry** A rectangular box with a square base has a volume of 256 cubic inches. If its height is 4 inches less than the length of an edge of the base, find the dimensions of the box.

59. **Height of a Building** A 5-foot 3-inch person casts a shadow that is 7.5 feet while a nearby building casts a 32-foot shadow. Find the height of the building.

60. **Angles in a Triangle** The measure of the largest angle in a triangle is 20° less than the sum of the measures of the two smaller angles. The sum of the measures of the two larger angles is 90° greater than the measure of the smaller angle. Find the measure of each angle.

18 Quadratic Functions and Equations

- **18.1** Quadratic Functions and Their Graphs
- **18.2** Transformations and Translations of Parabolas
- **18.3** Quadratic Equations
- **18.4** The Quadratic Formula
- **18.5** Quadratic Inequalities
- **18.6** Equations in Quadratic Form

What size television should you buy? The farther you sit from your television, the larger it should be. For example, if you sit only 6 feet from the screen, then a 32-inch television would be adequate; if you sit 10 feet from the screen, then a 50-inch television is more appropriate.

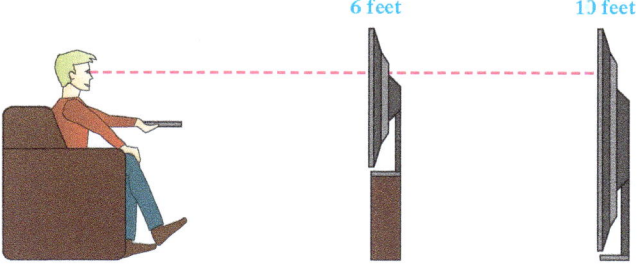

We can use a *quadratic function* to calculate the size S of the television screen needed for a person who sits x feet from the screen. You might want to use S to determine the size of screen that is recommended for you. (See Exercises 99 and 100 in the first section of this chapter.) Quadratic functions occur frequently in applications involving economics, road construction, falling objects, geometry, and modeling real-world data. (*Source: Money*, January 2007, p. 107; hdguru.com.)

18.1 Quadratic Functions and Their Graphs

Objectives

1. Working with Graphs of Quadratic Functions
   - Parabolas
   - The Vertex Formula
   - Graphing Quadratic Functions by Hand
   - Increasing and Decreasing
2. Solving Min-Max Applications
   - Finding Min-Max

NEW VOCABULARY

☐ Quadratic function
☐ Vertex
☐ Axis of symmetry

READING CHECK 1

- How can you identify the vertex and the axis of symmetry on the graph of a parabola?

1 Working with Graphs of Quadratic Functions

PARABOLAS Earlier in this text, we discussed how a quadratic function could be represented by a polynomial of degree 2. We now give an alternative definition of a quadratic function.

> **QUADRATIC FUNCTION**
>
> A **quadratic function** can be written in the form
> $$f(x) = ax^2 + bx + c,$$
> where a, b, and c are constants with $a \neq 0$.

NOTE: The domain of a quadratic function is all real numbers. ■

The graph of *any* quadratic function is a *parabola*. Recall that a parabola is a ∪-shaped graph that opens either upward or downward. The following See the Concept outlines the basic characteristics of parabolas.

See the Concept 1 — **PARABOLAS**

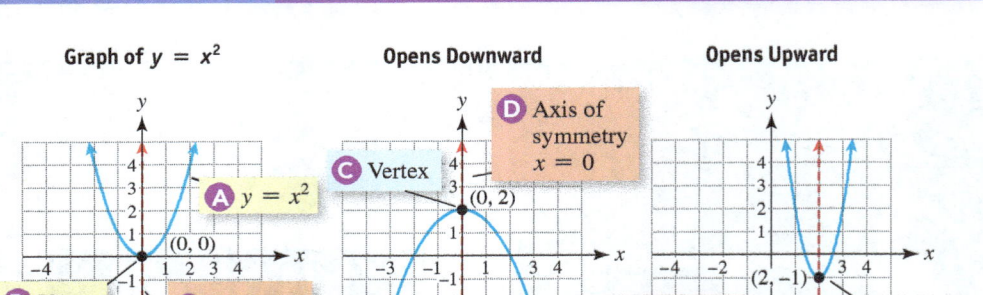

A The graph of the simple quadratic function $f(x) = x^2$ is an important parabola.

B The **vertex** is the *lowest* point on a parabola that opens *upward*.

C The **vertex** is the *highest* point on a parabola that opens *downward*.

D The **axis of symmetry** is a vertical line that passes through the vertex, but it is not part of the parabola. The left and right sides of the parabola would match if we folded the xy-plane along the axis of symmetry.

WATCH VIDEO IN MML.

EXAMPLE 1 Identifying the vertex and the axis of symmetry

Use the graph of the quadratic function to identify the vertex, the axis of symmetry, and whether the parabola opens upward or downward.

(a)

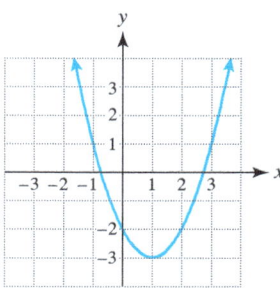

FIGURE 18.1

(b)

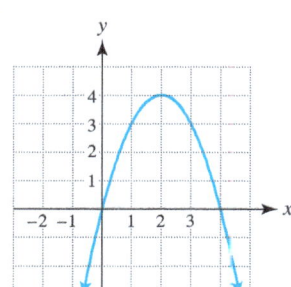

FIGURE 18.2

Solution

(a) The vertex is the lowest point on the graph shown in **FIGURE 18.1**, and its coordinates are $(1, -3)$. The axis of symmetry is the vertical line passing through the vertex, so its equation is $x = 1$. The parabola opens upward.

(b) The vertex is the highest point on the graph shown in **FIGURE 18.2**, and its coordinates are $(2, 4)$. The axis of symmetry is the vertical line passing through the vertex, so its equation is $x = 2$. The parabola opens downward.

Now Try Exercises 25, 27

THE VERTEX FORMULA In order to graph a parabola by hand, it is helpful to know the location of the vertex. The following formula can be used to find the coordinates of the vertex for *any* parabola. This formula can be derived by *completing the square*, a technique discussed in the next section. However, we can also derive the vertex formula by using the fact that a parabola is symmetric, as follows.

The equation for a parabola is $y = ax^2 + bx + c$ and its y-intercept is $(0, c)$. See **FIGURE 18.3**. Because the parabola is symmetric about the axis of symmetry, there must be a second point where the y-value is c.

Finding the x-Value of the Vertex

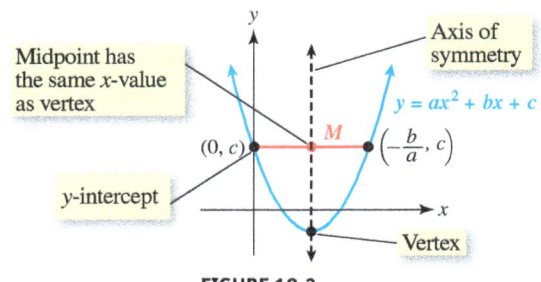

FIGURE 18.3

We can find the two x-values where $y = c$ by solving the following equation.

$$ax^2 + bx + c = c \quad \text{Find } x\text{-values where } y = c.$$
$$ax^2 + bx = 0 \quad \text{Subtract } c \text{ from each side.}$$
$$x(ax + b) = 0 \quad \text{Factor out } x.$$
$$x = 0 \quad \text{or} \quad ax + b = 0 \quad \text{Zero-product property}$$
$$x = 0 \quad \text{or} \quad x = -\frac{b}{a} \quad \text{Solve for } x.$$

1146 CHAPTER 18 QUADRATIC FUNCTIONS AND EQUATIONS

The two points $(0, c)$ and $\left(-\frac{b}{a}, c\right)$ lie on the parabola and the x-coordinate of the vertex is the same as the x-coordinate of the midpoint M of the line segment connecting these two points.

$$x = \frac{0 + \left(-\frac{b}{a}\right)}{2} = -\frac{b}{2a}$$

The x-value of the midpoint is the average of the x-values of the endpoints.

VERTEX FORMULA

The x-coordinate of the vertex of the graph of $y = ax^2 + bx + c$, $a \neq 0$, is given by

$$x = -\frac{b}{2a}.$$

To find the y-coordinate of the vertex, substitute this x-value in $y = ax^2 + bx + c$.

READING CHECK 2

- How do you use the vertex formula to find the coordinates of the vertex?

The following See the Concept shows how the vertex formula can be used to find the vertex (h, k) of a parabola and its axis of symmetry.

See the Concept 2   FINDING THE VERTEX AND AXIS OF SYMMETRY

$$f(x) = ax^2 + bx + c$$

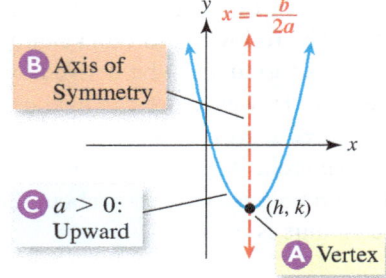
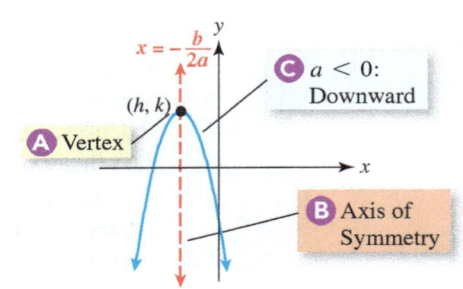

Ⓐ The x-coordinate of the vertex (h, k) is $h = -\frac{b}{2a}$.

The y-coordinate of the vertex (h, k) is $k = f\left(-\frac{b}{2a}\right)$.

Ⓑ The equation of the axis of symmetry is $x = -\frac{b}{2a}$.

Ⓒ The parabola opens upward if $a > 0$, and downward if $a < 0$.

WATCH VIDEO IN MML.

EXAMPLE 2   Finding the vertex of a parabola

Find the vertex for the graph of $f(x) = 2x^2 - 4x + 1$. Support your answer graphically.

Solution
For $f(x) = 2x^2 - 4x + 1$, $a = 2$ and $b = -4$. The x-coordinate of the vertex is

$$x = -\frac{b}{2a} = -\frac{(-4)}{2(2)} = 1. \quad \text{x-coordinate of vertex}$$

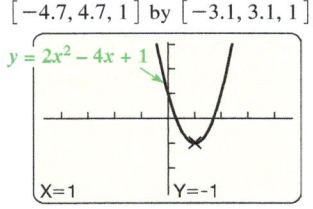

FIGURE 18.4

To find the y-coordinate of the vertex, substitute $x = 1$ in the given formula.

$$f(1) = 2(1)^2 - 4(1) + 1 = -1 \quad \text{y-coordinate of vertex}$$

Thus the vertex is located at $(1, -1)$, which is supported by **FIGURE 18.4**.

Now Try Exercise 29

GRAPHING QUADRATIC FUNCTIONS BY HAND One way to graph a quadratic function without a graphing calculator is to first apply the vertex formula. After the vertex is located, a table of values can be made to locate points on either side of the vertex. This technique is used in the next example to graph three quadratic functions.

EXAMPLE 3 Graphing quadratic functions

Identify the vertex and axis of symmetry on the graph of $y = f(x)$. Graph $y = f(x)$.
(a) $f(x) = x^2 - 1$ **(b)** $f(x) = -(x + 1)^2$ **(c)** $f(x) = x^2 + 4x + 3$

Solution
(a) Begin by applying the vertex formula with $a = 1$ and $b = 0$ to locate the vertex.

$$x = -\frac{b}{2a} = -\frac{0}{2(1)} = 0 \quad \text{x-coordinate of vertex}$$

The y-coordinate of the vertex is found by evaluating $f(0)$.

$$y = f(0) = 0^2 - 1 = -1 \quad \text{y-coordinate of vertex}$$

Thus the coordinates of the vertex are $(0, -1)$. **TABLE 18.1** is made by finding points on either side of the vertex. Plotting these points and connecting a smooth U-shaped graph results in **FIGURE 18.5**. The axis of symmetry is the vertical line passing through the vertex, so its equation is $x = 0$.

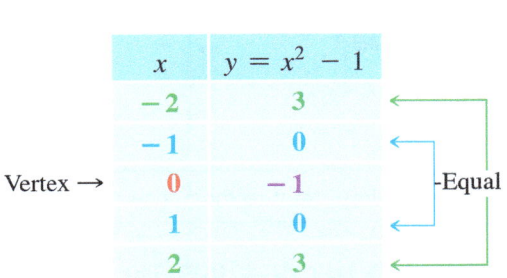

TABLE 18.1

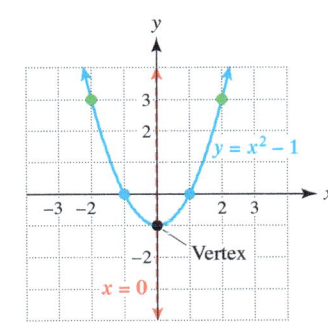

FIGURE 18.5

(b) Before we can apply the vertex formula, we need to determine values for a and b by multiplying the expression for $f(x) = -(x + 1)^2$.

$$-(x + 1)^2 = -(x^2 + 2x + 1) \quad \text{Square the binomial.}$$
$$= -x^2 - 2x - 1 \quad \text{Distribute the negative sign.}$$

Substitute $a = -1$ and $b = -2$ into the vertex formula.

$$x = -\frac{b}{2a} = -\frac{(-2)}{2(-1)} = -1 \quad \text{x-coordinate of vertex}$$

The y-coordinate of the vertex is found by evaluating $f(-1)$.

$$y = f(-1) = -(-1 + 1)^2 = 0 \quad \text{y-coordinate of vertex}$$

Thus the coordinates of the vertex are $(-1, 0)$. **TABLE 18.2** is made by finding points on either side of the vertex. Plotting these points and connecting a smooth ∩-shaped graph result in **FIGURE 18.6**. The axis of symmetry is the vertical line passing through the vertex, so its equation is $x = -1$.

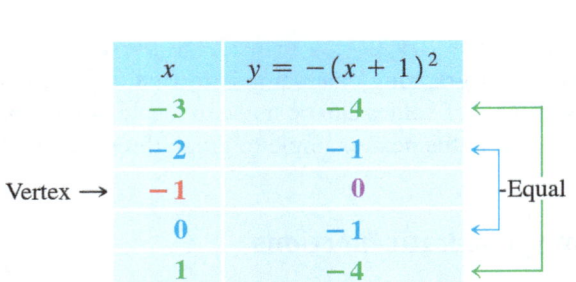

TABLE 18.2

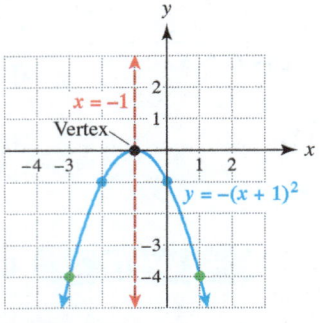

FIGURE 18.6

(c) The graph of $f(x) = x^2 + 4x + 3$ can be found in a manner similar to that used for the graphs in (a) and (b). See **TABLE 18.3** and **FIGURE 18.7**. With $a = 1$ and $b = 4$, the vertex formula can be used to show that the vertex is located at $(-2, -1)$. The equation of the axis of symmetry is $x = -2$.

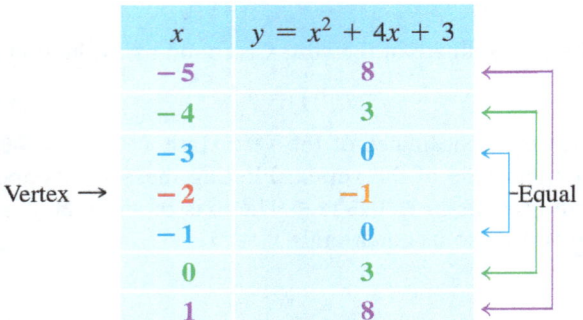

TABLE 18.3

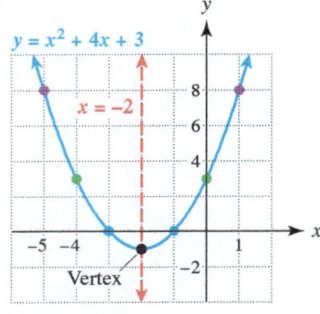

FIGURE 18.7

Now Try Exercises 41, 47, 49

STUDY TIP

The vertex formula is important to memorize because it is often used when graphing parabolas.

INCREASING AND DECREASING The concept of increasing and decreasing is frequently used when modeling data with functions. Suppose that the graph of the equation $y = x^2$ shown in **FIGURE 18.8** represents a valley. If we walk from *left to right* the valley "goes down" and then "goes up." Mathematically, we say that the graph of $y = x^2$ is *decreasing* when $x < 0$ and *increasing* when $x > 0$. In **FIGURE 18.6**, the graph of $y = -(x + 1)^2$ increases when $x < -1$ and decreases when $x > -1$.

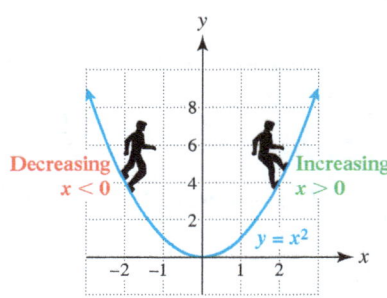

FIGURE 18.8

NOTE: When determining where a graph is increasing and where it is decreasing, we must "walk" along the graph *from left to right*. (We read English from left to right, which might help you remember.) ∎

2 Solving Min–Max Applications

Math in Pizza Eating Context Have you ever noticed that the first piece of pizza tastes really great and gives a person a lot of satisfaction? The second piece may be almost as good. After a few more pieces, there is often a point where the satisfaction starts to level off, and it can even go down as a person starts to eat too much. Eventually, if a person overeats, he or she might regret eating any pizza at all.

In **FIGURE 18.9**, a parabola models the total satisfaction received from eating x slices of pizza. Notice that the satisfaction level increases and then decreases. Although maximum satisfaction would vary with the individual, it always occurs at the vertex. (This is a seemingly simple model, but it is nonetheless used frequently in economics to describe consumer satisfaction from buying x identical items.)

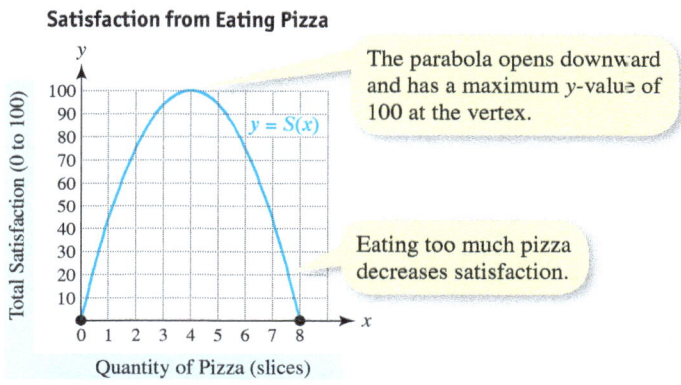

FIGURE 18.9

EXAMPLE 4 Finding maximum satisfaction from pizza

The graph of $S(x) = -6.25x^2 + 50x$ is shown in **FIGURE 18.9**. Use the vertex formula to determine the number of slices of pizza that results in maximum satisfaction. Does this maximum agree with what is shown in the graph of $y = S(x)$?

Solution
The parabola in **FIGURE 18.9** opens downward, so the greatest y-value occurs at the vertex. To find the x-coordinate of the vertex, we can apply the vertex formula. Substitute $a = -6.25$ and $b = 50$ into the vertex formula.

$$x = -\frac{b}{2a} = -\frac{50}{2(-6.25)} = 4 \quad \text{\textit{x}-coordinate of vertex}$$

Maximum satisfaction on a scale of 1 to 100 occurs when **4** slices of pizza are eaten, which agrees with **FIGURE 18.9**. (This number obviously varies from individual to individual.)

Now Try Exercise 83

READING CHECK 3

- How do you determine if a parabola has a maximum y-value or a minimum y-value?

FINDING MIN–MAX When a quadratic function f is used to model real data, the y-coordinate of the vertex represents either a maximum value of $f(x)$, when the parabola opens downward, or a minimum value of $f(x)$, when the parabola opens upward. **FIGURE 18.10** on the next page illustrates maximum and minimum y-values on a parabola.

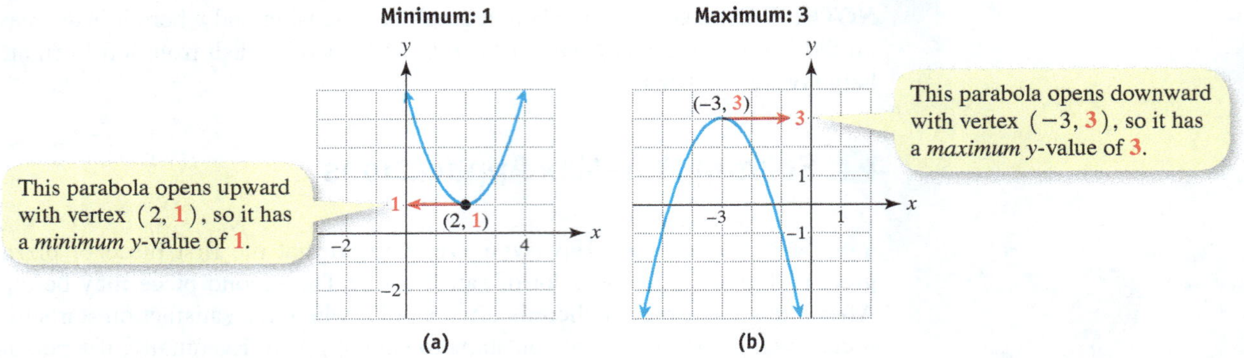

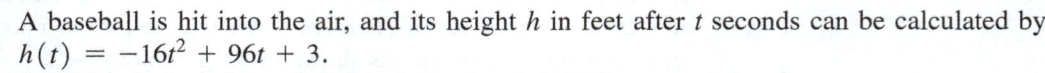

Minimum: 1   Maximum: 3

This parabola opens upward with vertex $(2, 1)$, so it has a *minimum* y-value of 1.

This parabola opens downward with vertex $(-3, 3)$, so it has a *maximum* y-value of 3.

(a)   (b)

FIGURE 18.10

EXAMPLE 5 Finding a minimum y-value

Find the minimum y-value on the graph of $f(x) = x^2 - 4x + 3$.

Solution

To locate the minimum y-value on a parabola, first apply the vertex formula with $a = 1$ and $b = -4$ to find the x-coordinate of the vertex.

$$x = -\frac{b}{2a} = -\frac{(-4)}{2(1)} = 2$$ x-coordinate of vertex

The minimum y-value on the graph is found by evaluating $f(2)$.

$$y = f(2) = 2^2 - 4(2) + 3 = -1$$ y-coordinate of vertex

Thus the minimum y-value on the graph of $y = f(x)$ is -1. This result is supported by **FIGURE 18.11**.

Now Try Exercise 69

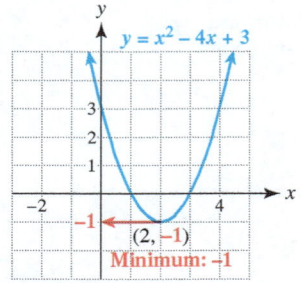

Locating a Minimum y-Value

FIGURE 18.11

🌐 Math in Context (Sports) In the next example, we demonstrate finding a maximum height reached by a baseball.

EXAMPLE 6 Finding maximum height

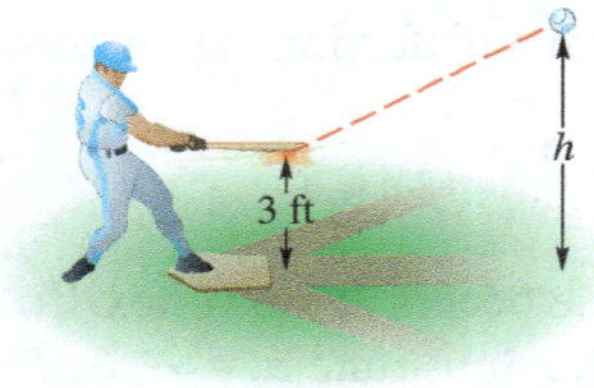

A baseball is hit into the air, and its height h in feet after t seconds can be calculated by $h(t) = -16t^2 + 96t + 3$.
(a) What is the height of the baseball when it is hit?
(b) Determine the maximum height of the baseball.

Solution
(a) The baseball is hit when $t = 0$, so $h(0) = -16(0)^2 + 96(0) + 3 = 3$ feet.
(b) The graph of h opens downward because $a = -16 < 0$. Thus the maximum height of the baseball occurs at the vertex. To find the vertex, we apply the vertex formula with $a = -16$ and $b = 96$ because $h(t) = -16t^2 + 96t + 3$.

$$t = -\frac{b}{2a} = -\frac{96}{2(-16)} = 3 \text{ seconds}$$

The maximum height of the baseball occurs at $t = 3$ seconds and is

$$h(3) = -16(3)^2 + 96(3) + 3 = 147 \text{ feet}.$$

Now Try Exercise 89

Math in Business Context Suppose that a hotel is considering giving a group discount on room rates. The regular price is $80, but for each room rented, the price decreases by $2. On the one hand, if the hotel rents one room, it makes only $78. On the other hand, if the hotel rents 40 rooms, the rooms are all free and the hotel makes nothing. Is there a number of rooms between 1 and 40 that should be rented to maximize the revenue from the group?

In **FIGURE 18.12**, the hotel's revenue is graphed. From the graph, it is apparent that "peak" revenue occurs at the vertex of the parabola.

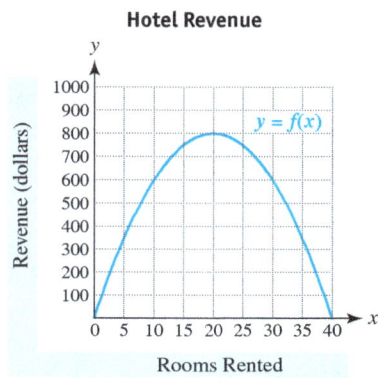

FIGURE 18.12

EXAMPLE 7 Maximizing revenue

A hotel gives the following group discount on room rates. The regular price for a room is $80, but for each room rented the price decreases by $2. A graph of the revenue received from renting x rooms is shown in **FIGURE 18.12**.
(a) Interpret the graph.
(b) What is the maximum revenue? How many rooms should be rented to receive the maximum revenue?
(c) Write a formula for $f(x)$ whose graph is shown in **FIGURE 18.12**.
(d) Use $f(x)$ to determine symbolically the maximum revenue and the number of rooms that should be rented.

Solution
(a) The revenue increases at first, reaches a maximum (which corresponds to the vertex), and then decreases.
(b) In **FIGURE 18.12** the vertex is $(20, 800)$. Thus the maximum revenue of $800 occurs when 20 rooms are rented.
(c) If x rooms are rented, the price for each room is $80 - 2x$. The revenue equals the number of rooms rented times the price of each room. Thus $f(x) = x(80 - 2x)$.
(d) First, multiply $x(80 - 2x)$ to obtain $80x - 2x^2$ and then let $f(x) = -2x^2 + 80x$. The x-coordinate of the vertex is

$$x = -\frac{b}{2a} = -\frac{80}{2(-2)} = 20.$$

The y-coordinate is $f(20) = -2(20)^2 + 80(20) = 800$. These calculations verify our results in part (b).

Now Try Exercise 93

TECHNOLOGY NOTE

Locating a Vertex
Graphing calculators can locate a vertex with the MAXIMUM or MINIMUM utility. The maximum in Example 7 is found in the figure.

$[0, 50, 10]$ by $[0, 1000, 100]$

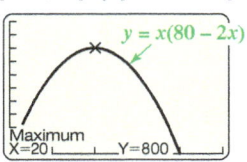

CALCULATOR HELP
To find a minimum or maximum, see Appendix A (page AP-11).

18.1 Putting It All Together

CONCEPT	EXPLANATION	EXAMPLES
Quadratic Function	Can be written as $f(x) = ax^2 + bx + c, a \neq 0$	$f(x) = x^2 + x - 2$ and $g(x) = -2x^2 + 4$ ($b = 0$)
Vertex of a Parabola	The x-coordinate of the vertex for the function $f(x) = ax^2 + bx + c$ with $a \neq 0$ is given by $$x = -\frac{b}{2a}.$$ The y-coordinate of the vertex is found by substituting this x-value in the equation. Hence the vertex is $\left(-\frac{b}{2a}, f\left(-\frac{b}{2a}\right)\right)$.	If $f(x) = -2x^2 + 8x - 7$, then $$x = -\frac{8}{2(-2)} = 2$$ and $f(2) = -2(2)^2 + 8(2) - 7 = 1.$ The vertex is $(2, 1)$. The graph of f opens downward because $a < 0$.
Graph of a Quadratic Function	Its graph is a parabola that opens upward (∪-shaped) if $a > 0$ and downward (∩-shaped) if $a < 0$. The vertex can be used to determine the maximum or minimum y-value.	The graph of $f(x) = -x^2 - 2x$ opens downward because $a = -1 < 0$. The vertex is $(-1, 1)$, so the maximum y-value is 1.

Quadratic functions can have verbal, symbolic, numerical, and graphical representations. For example, the quadratic function represented verbally as *f squares the input x and then subtracts 1* can also be represented as follows.

Symbolic Representation

$f(x) = x^2 - 1$

Numerical Representation

x	$f(x) = x^2 - 1$
-2	3
-1	0
Vertex → 0	-1
1	0
2	3

Equal

Graphical Representation

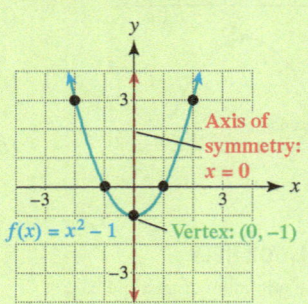

Axis of symmetry: $x = 0$
$f(x) = x^2 - 1$
Vertex: $(0, -1)$

18.1 Exercises

CONCEPTS AND VOCABULARY

1. The graph of a quadratic function is called a(n) _____.

2. If a parabola opens upward, what is the lowest point on the parabola called?

3. If a parabola is symmetric with respect to the y-axis, the y-axis is called the _____.

4. The vertex on the graph of $y = x^2$ is _____.

5. Sketch a parabola that opens downward with a vertex of $(1, 2)$.

6. If $y = ax^2 + bx + c$, the x-coordinate of the vertex is given by $x =$ _____.

7. Any quadratic function can be written in the form $f(x) =$ _____.

8. If a parabola opens downward, the point with the largest y-value is called the _____.

9. (True or False?) The axis of symmetry for the graph of $y = ax^2 + bx + c$ is given by $x = -\frac{b}{2a}$.

10. (True or False?) If the vertex of a parabola is located at (a, b), then the axis of symmetry is given by $x = b$.

11. (True or False?) If a parabola opens downward and its vertex is (a, b), then the minimum y-value on the parabola is b.

12. (True or False?) The graph of $y = -ax^2$ with $a > 0$ opens downward.

EVALUATING QUADRATIC FUNCTIONS

Exercises 13–20: Evaluate the given quadratic function as indicated.

13. Find $f(4)$ if $f(x) = x^2 + 1$.

14. Find $f(3)$ if $f(x) = x^2 - 11$.

15. Find $f(2)$ if $f(x) = x^2 - 6x$.

16. Find $f(-1)$ if $f(x) = x^2 + 5x$.

17. Find $f(-3)$ if $f(x) = 2x^2 + 4x + 1$.

18. Find $f(5)$ if $f(x) = -2x^2 + 7x - 3$.

19. Find $f(8)$ if $f(x) = -3x^2 + 16x + 72$.

20. Find $f(-6)$ if $f(x) = 4x^2 + 11x - 88$.

Exercises 21–24: Use the given graph of f to evaluate the expressions.

21. $f(-2)$ and $f(0)$

22. $f(-2)$ and $f(2)$

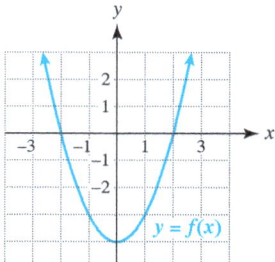 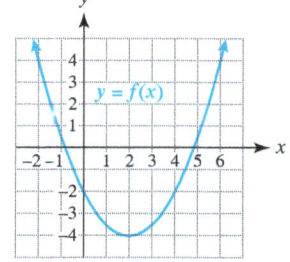

23. $f(-3)$ and $f(1)$

24. $f(-1)$ and $f(2)$

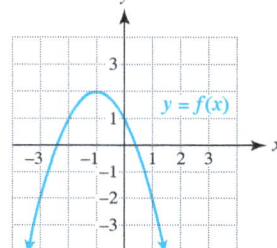 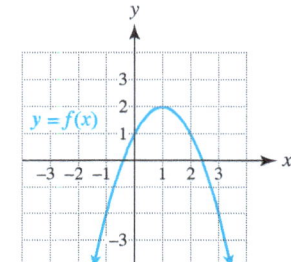

IDENIFYING CHARACTERISTICS OF PARABOLAS

Exercises 25–28: Identify the vertex, axis of symmetry, and whether the parabola opens upward or downward.

25.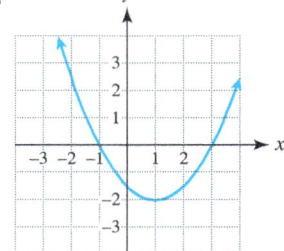

26.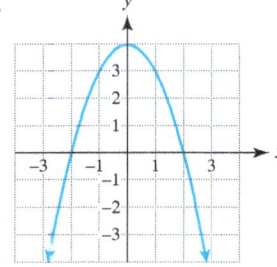

27.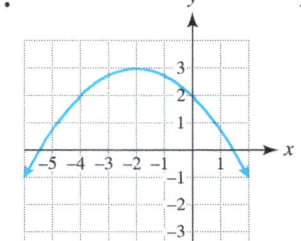

28.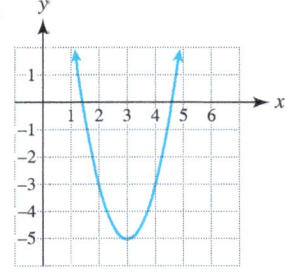

USING THE VERTEX FORMULA

Exercises 29–38: Find the vertex of the parabola.

29. $f(x) = x^2 - 4x - 2$

30. $f(x) = 2x^2 + 6x - 3$

31. $f(x) = -\frac{1}{3}x^2 - 2x + 1$

32. $f(x) = 5 - 4x + x^2$

33. $f(x) = 3 - 2x^2$

34. $f(x) = \frac{1}{4}x^2 - 3x - 2$

35. $f(x) = -0.3x^2 + 0.6x + 1.1$

36. $f(x) = 25 - 10x + 20x^2$

37. $f(x) = 6x - x^2$ 38. $f(x) = x - \frac{1}{2}x^2$

GRAPHING QUADRATIC FUNCTIONS

Exercises 39–58: Do the following for the given $f(x)$.
  (a) Identify the vertex and axis of symmetry on the graph of $y = f(x)$.
  (b) Graph $y = f(x)$.
  (c) Evaluate $f(-2)$ and $f(3)$.

39. $f(x) = \frac{1}{2}x^2$ 40. $f(x) = -3x^2$

41. $f(x) = x^2 - 2$ 42. $f(x) = x^2 - 1$

43. $f(x) = -3x^2 + 1$ 44. $f(x) = \frac{1}{2}x^2 + 2$

45. $f(x) = (x - 1)^2$ 46. $f(x) = (x + 2)^2$

47. $f(x) = -(x + 2)^2$ 48. $f(x) = -(x - 1)^2$

49. $f(x) = x^2 + x - 2$ 50. $f(x) = x^2 - 2x + 2$

51. $f(x) = 2x^2 - 3$ 52. $f(x) = 1 - 2x^2$

53. $f(x) = 2x - x^2$ 54. $f(x) = x^2 + 2x - 8$

55. $f(x) = -2x^2 + 4x - 1$

56. $f(x) = -\frac{1}{2}x^2 + 2x - 3$

57. $f(x) = \frac{1}{4}x^2 - x + 5$ 58. $f(x) = 3 - 6x - 4x^2$

IDENTIFYING INCREASING AND DECREASING INTERVALS

Exercises 59–62: Use the graph of $y = f(x)$ to determine the intervals where f is increasing and where f is decreasing.

59.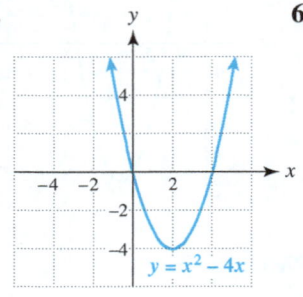

60.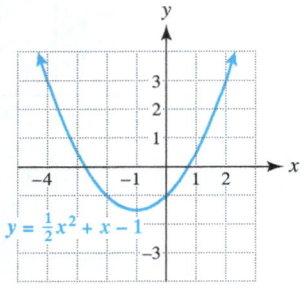

61.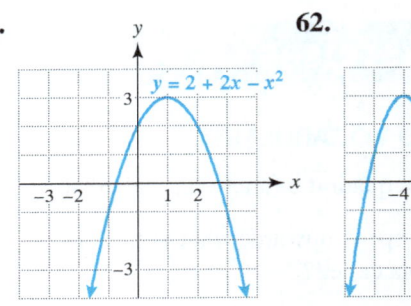

62.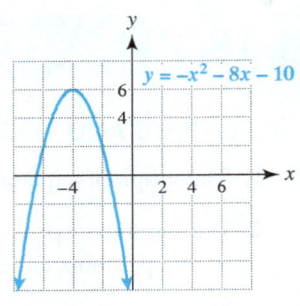

Exercises 63–68: Determine the intervals where $y = f(x)$ is increasing and where it is decreasing.

63. $y = 3x^2 - 4x + 1$ 64. $y = 2x^2 - 5x - 2$

65. $y = -x^2 - 3x$ 66. $y = -2x^2 + 3x - 4$

67. $y = 5 - x - 4x^2$ 68. $y = 3 + x + x^2$

FINDING MAXIMUMS AND MINIMUMS

Exercises 69–74: Find the minimum y-value on the graph of $y = f(x)$.

69. $f(x) = x^2 + 2x - 1$ 70. $f(x) = x^2 + 6x + 2$

71. $f(x) = x^2 - 5x$ 72. $f(x) = x^2 - 3x$

73. $f(x) = 2x^2 + 2x - 3$ 74. $f(x) = 3x^2 - 3x + 7$

Exercises 75–80: Find the maximum y-value on the graph of $y = f(x)$.

75. $f(x) = -x^2 + 2x + 5$

76. $f(x) = -x^2 + 4x - 3$

77. $f(x) = 4x - x^2$

78. $f(x) = 6x - x^2$

79. $f(x) = -2x^2 + x - 5$

80. $f(x) = -5x^2 + 15x - 2$

81. **Numbers** Find two positive numbers whose sum is 20 and whose product is maximum.

82. **Thinking Generally** Find two positive numbers whose sum is k and whose product is maximum.

APPLICATIONS INVOLVING QUADRATIC FUNCTIONS

Exercises 83 and 84: **Eating Pizza** (Refer to Example 4.) Let $S(x)$ denote the satisfaction, on a scale from 0 to 100, from eating x slices of pizza. Find the number of slices that gives the maximum satisfaction.

83. $S(x) = -\frac{100}{9}x^2 + \frac{200}{3}x$

84. $S(x) = -16x^2 + 80x$

Exercises 85–88: **Quadratic Models** *Match the physical situation with the graph (a–d) that models it best.*

85. The height y of a stone thrown from ground level after x seconds

86. The number of people attending a popular movie x weeks after its opening

87. The temperature after x hours in a house when the furnace quits and then a repair person fixes it

88. U.S. population from 1800 to the present

a. 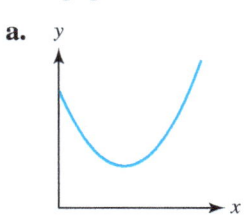   b.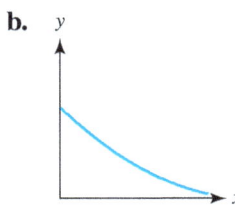

c. 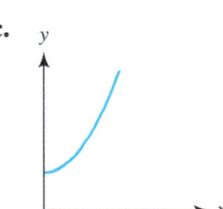   d.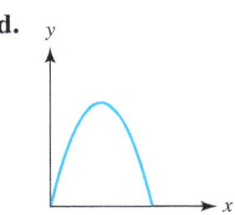

89. **Height Reached by a Baseball** A baseball is hit into the air, and its height h in feet after t seconds is given by $h(t) = -16t^2 + 64t + 2$.
    (a) What is the height of the baseball when it is hit?
    (b) After how many seconds does the baseball reach its maximum height?
    (c) Determine the maximum height of the baseball.

90. **Height Reached by a Golf Ball** A golf ball is hit into the air, and its height h in feet after t seconds is given by $h(t) = -16t^2 + 128t$.
    (a) What is the height of the golf ball when it is hit?
    (b) After how many seconds does the golf ball reach its maximum height?
    (c) Determine the maximum height of the golf ball.

91. **Height Reached by a Baseball** Suppose that a baseball is thrown upward with an initial velocity of 66 feet per second (45 miles per hour) and it is released 6 feet above the ground. Its height h after t seconds is given by
$$h(t) = -16t^2 + 66t + 6.$$
After how many seconds does the baseball reach a maximum height? Estimate this height.

92. **Throwing a Baseball on the Moon** (Refer to Exercise 91.) If the same baseball were thrown the same way on the moon, its height h above the moon's surface after t seconds would be
$$h(t) = -2.55t^2 + 66t + 6.$$
Does the baseball go higher on the moon or on Earth? What is the difference in these two heights?

93. **Concert Tickets** (Refer to Example 7.) An agency is promoting concert tickets by offering a group-discount rate. The regular price is $100 and for each ticket bought the price decreases by $1. (One ticket costs $99, two tickets cost $98 *each*, and so on.)
    (a) A graph of the revenue received from selling x tickets is shown in the figure. Interpret the graph.

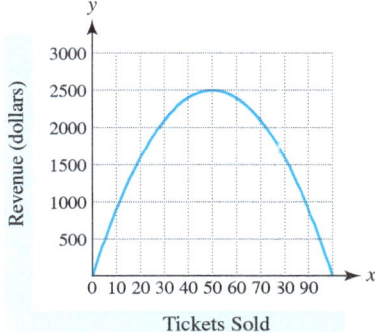

    (b) What is the maximum revenue? How many tickets should be sold to maximize revenue?
    (c) Write a formula for $y = f(x)$ whose graph is shown in the figure.
    (d) Use $f(x)$ to determine symbolically the maximum revenue and the number of tickets that should be sold to maximize revenue.

94. **Maximizing Revenue** The regular price for a round-trip ticket to Las Vegas, Nevada, charged by an airline charter company is $300. For a group rate, the company will reduce the price of each ticket by $1.50 for every passenger on the flight.
    (a) Write a formula $f(x)$ that gives the revenue from selling x tickets.
    (b) Determine how many tickets should be sold to maximize the revenue. What is the maximum revenue?

95. **Monthly Facebook Visitors** The number of *unique* monthly Facebook visitors in millions x years after 2010 can be modeled by
$$V(x) = -0.5x^2 + 10x + 138.$$
    (a) Evaluate $V(1)$, $V(2)$, $V(3)$, and $V(4)$. Interpret your answer.
    (b) Explain why numbers of Facebook visitors can not be modeled by a linear function.

96. **Cell Phone Complexity** People often enjoy having certain features on their cell phones, such as e-mail and the ability to surf the Web. However, as phones become more and more complicated to operate, the benefits from the additional complexity start to decrease. The following parabola models this general situation. Interpret why a parabola is appropriate to model this consumer experience.

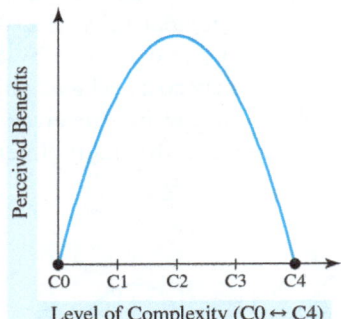

97. **Maximizing Area** A farmer is fencing a rectangular area for cattle and uses a straight portion of a river as one side of the rectangle, as illustrated in the figure. Note that there is no fence along the river. If the farmer has 1200 feet of fence, find the dimensions for the rectangular area that give a maximum area for the cattle.

98. **Maximizing Area** A rectangular pen being constructed for a pet requires 60 feet of fence.
    (a) Write a formula $f(x)$ that can be used to find the total area of the pen if one side of the pen has length x.
    (b) Find the dimensions of the pen that give the largest area. What is the largest area?

Exercises 99 and 100: **Large-Screen Televisions** *(Refer to the introduction to this chapter.) Use the formula*
$$S(x) = -0.227x^2 + 8.155x - 8.8,$$
where $6 \leq x \leq 16$, to estimate the recommended screen size in inches when viewers sit x feet from the screen.

99. $x = 8$ feet   100. $x = 12$ feet

101. **Carbon Emissions** Past and future carbon emissions in billions of metric tons during year x can be modeled by
$$C(x) = \frac{1}{300}x^2 - \frac{199}{15}x + \frac{39{,}619}{3},$$
where $1990 \leq x \leq 2020$. *(Source: U.S. Department of Energy.)*
(a) Evaluate $C(1990)$. Interpret your answer.
(b) Find the expected *increase* in carbon emissions from 1990 to 2020.

102. **Seedling Growth** In a study of the effect of temperature on the growth of melon seedlings, the seedlings were grown at different temperatures, and their heights were measured after a fixed period of time. The findings of this study can be modeled by
$$f(x) = -0.095x^2 + 5.4x - 52.2,$$
where x is the temperature in degrees Celsius and the output $f(x)$ gives the resulting average height in centimeters. *(Source: R. Pearl, "The growth of Cucumis melo seedlings at different temperatures.")*
(a) Graph f in $[20, 40, 5]$ by $[0, 30, 5]$.
(b) Estimate graphically the temperature that resulted in the greatest height for the melon seedlings.
(c) Solve part (b) symbolically.

WRITING ABOUT MATHEMATICS

103. If $f(x) = ax^2 + bx + c$, explain how to find the vertex on the graph of $y = f(x)$.

104. Suppose that a quantity Q is modeled by the formula $Q(x) = ax^2 + bx + c$ with $a < 0$. Explain how to find the x-value that maximizes $Q(x)$. How do you find the maximum value of $Q(x)$?

18.2 Transformations and Translations of Parabolas

Objectives

1. Graphing Basic Transformations of $y = ax^2$
   - The Graph of $y = ax^2, a > 0$
   - The Graph of $y = ax^2, a < 0$
   - Reflections of $y = ax^2$
2. Performing Vertical and Horizontal Translations
3. Applying Vertex Form
   - Graphing with Vertex Form
   - Completing the Square to Find the Vertex
4. Modeling with Quadratic Functions

NEW VOCABULARY

☐ Reflection
☐ Translations
☐ Vertex form
☐ Completing the square

1 Graphing Basic Transformations of $y = ax^2$

THE GRAPH OF $y = ax^2, a > 0$ The graphs of $y_1 = \frac{1}{2}x^2$, $y_2 = x^2$, and $y_3 = 2x^2$ are shown in **FIGURE 18.13** and open upward. Note that $a = \frac{1}{2}$, $a = 1$, and $a = 2$, respectively, and that as the value of a increases, the resulting parabola becomes narrower. When $a > 0$, the graph of $y = ax^2$ never lies *below* the x-axis.

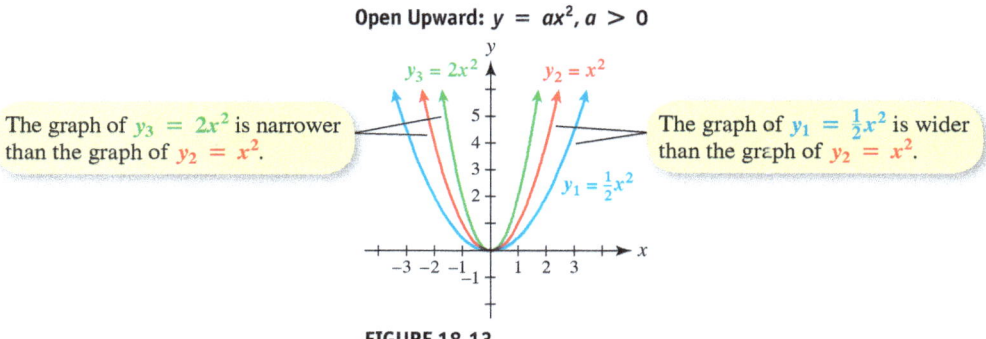

FIGURE 18.13

THE GRAPH OF $y = ax^2, a < 0$ The graphs of $y_4 = -\frac{1}{2}x^2$, $y_5 = -x^2$, and $y_6 = -2x^2$ are shown in **FIGURE 18.14** and open downward. When $a < 0$, the graph of $y = ax^2$ never lies *above* the x-axis.

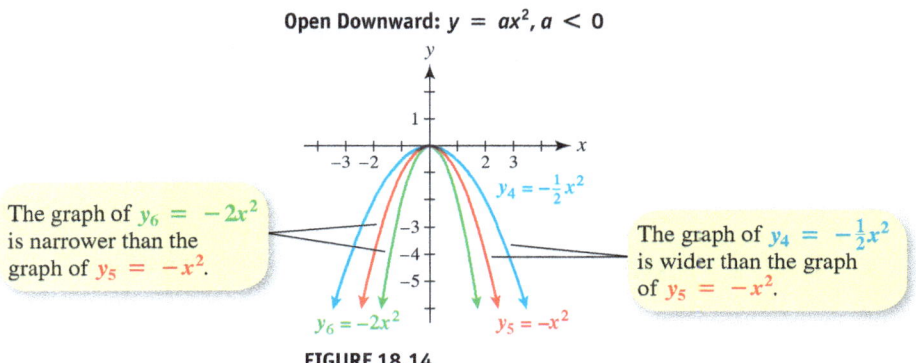

FIGURE 18.14

THE GRAPH OF $y = ax^2$

The graph of $y = ax^2$ is a parabola with the following characteristics.

1. The vertex is $(0, 0)$, and the axis of symmetry is given by $x = 0$.
2. It opens upward if $a > 0$ and opens downward if $a < 0$.
3. It is wider than the graph of $y = x^2$, if $0 < |a| < 1$. It is narrower than the graph of $y = x^2$, if $|a| > 1$.

Reflection Across the x-Axis

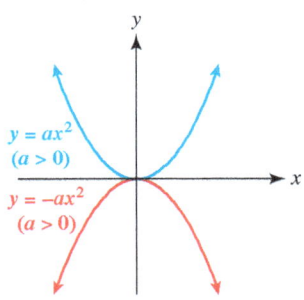

FIGURE 18.15

REFLECTIONS OF $y = ax^2$ The graph of $y_1 = \frac{1}{2}x^2$ in **FIGURE 18.13** can be *transformed* into the graph of $y_4 = -\frac{1}{2}x^2$ in **FIGURE 18.14** by *reflecting* it across the x-axis. In general, the graph of $y = -ax^2$ is a **reflection** of the graph of $y = ax^2$ across the x-axis. That is, if we folded the xy-plane along the x-axis, the two graphs would match. See **FIGURE 18.15**.

EXAMPLE 1 Graphing $y = ax^2$

Compare the graph of $g(x) = -3x^2$ to the graph of $f(x) = x^2$. Then graph both functions in the same xy-plane.

Solution
Both graphs are parabolas. However, the graph of $y = g(x)$ opens downward and is narrower than the graph of $y = f(x)$. These graphs are shown in **FIGURE 18.16**.

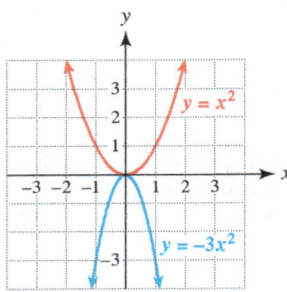

FIGURE 18.16

Now Try Exercise 11

2 Performing Vertical and Horizontal Translations

The graph of $y = x^2$ is a parabola opening upward with vertex $(0, 0)$. Suppose that we graph $y_1 = x^2$, $y_2 = x^2 + 1$, and $y_3 = x^2 - 2$ in the same xy-plane, as calculated in **TABLE 18.4** and shown in **FIGURE 18.17**. All three graphs have the same shape. However, compared to the graph of $y_1 = x^2$, the graph of $y_2 = x^2 + 1$ is shifted *upward* 1 unit and the graph of $y_3 = x^2 - 2$ is shifted *downward* 2 units. Such *shifts* are called **translations** because they do not change the shape of a graph—only its position.

x	$y_2 = x^2 + 1$	$y_1 = x^2$	$y_3 = x^2 - 2$
-2	5	4	2
-1	2	1	-1
0	1	0	-2
1	2	1	-1
2	5	4	2

↑ The x-values do NOT change.

Add 1 to find the y_2-values.

Subtract 2 to find the y_3-values.

TABLE 18.4

Vertical Shifts of $y = x^2$

Shifted up 1 unit
Shifted down 2 units
FIGURE 18.17

Next, suppose that we graph $y_1 = x^2$ and $y_2 = (x - 1)^2$ in the same xy-plane. Compare **TABLE 18.5** and **TABLE 18.6**. Note that the y-values are equal when the x-value for y_2 is 1 unit *larger* than the x-value for y_1. For example, $y_1 = 4$ when $x = -2$ and $y_2 = 4$ when $x = -1$. Thus the graph of $y_2 = (x - 1)^2$ has the same shape as the graph of $y_1 = x^2$ except that it is translated *horizontally to the right* 1 unit, as illustrated in **FIGURE 18.18**.

x	$y_1 = x^2$
-2	4
-1	1
0	0
1	1
2	4

x	$y_2 = (x-1)^2$
-1	4
0	1
1	0
2	1
3	4

Add 1 to find the *x*-values.   The *y*-values do NOT change.

TABLE 18.5   **TABLE 18.6**

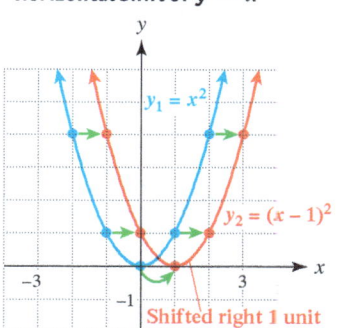

FIGURE 18.18

The graphs $y_1 = x^2$ and $y_2 = (x+2)^2$ are shown in **FIGURE 18.19**. Note that **TABLE 18.7** and **TABLE 18.8** show their *y*-values to be equal when the *x*-value for y_2 is 2 units *smaller* than the *x*-value for y_1. As a result, the graph of $y_2 = (x+2)^2$ has the same shape as the graph of $y_1 = x^2$ except that it is translated *horizontally to the left* 2 units.

x	$y_1 = x^2$
-2	4
-1	1
0	0
1	1
2	4

x	$y_2 = (x+2)^2$
-4	4
-3	1
-2	0
-1	1
0	4

Subtract 2 to find the *x*-values.   The *y*-values do NOT change.

TABLE 18.7   **TABLE 18.8**

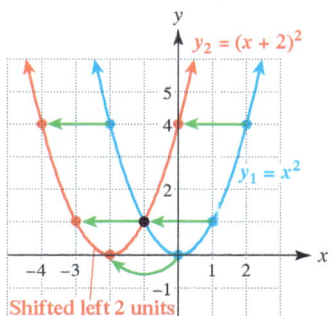

FIGURE 18.19

VERTICAL AND HORIZONTAL TRANSLATIONS OF PARABOLAS

Let h and k be positive numbers.

To graph	*shift the graph of $y = x^2$ by k units*
$y = x^2 + k$	upward.
$y = x^2 - k$	downward.

To graph	*shift the graph of $y = x^2$ by h units*
$y = (x-h)^2$	right.
$y = (x+h)^2$	left.

EXAMPLE 2 Translating the graph $y = x^2$

Graph the equation and identify the vertex. Compare this graph to the graph of $y = x^2$.

(a) $y = x^2 + 2$ (b) $y = (x + 3)^2$ (c) $y = (x - 2)^2 - 3$

Solution
(a) The graph of $y = x^2 + 2$ is similar to the graph of $y = x^2$ except that it is shifted (translated) *upward* 2 units, as shown in **FIGURE 18.20(a)**. The vertex is $(0, 2)$.
(b) The graph of $y = (x + 3)^2$ is similar to the graph of $y = x^2$ except that it is shifted *left* 3 units, as shown in **FIGURE 18.20(b)**. The vertex is $(-3, 0)$.
(c) The graph of $y = (x - 2)^2 - 3$ is similar to the graph of $y = x^2$ except that it is shifted *right* 2 units *and downward* 3 units. See **FIGURE 18.20(c)**. The vertex is $(2, -3)$.

Shifted Up 2 Units   **Shifted Left 3 Units**   **Shifted Right 2 Units and Down 3 Units**

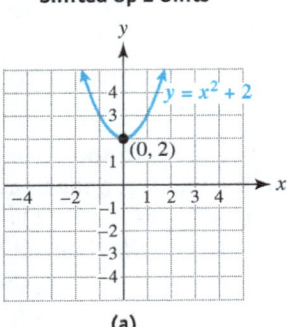
(a)
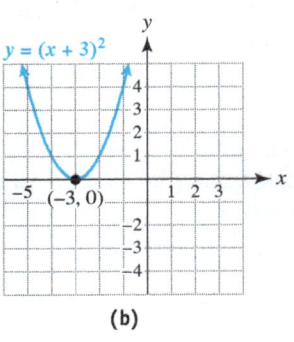
(b)
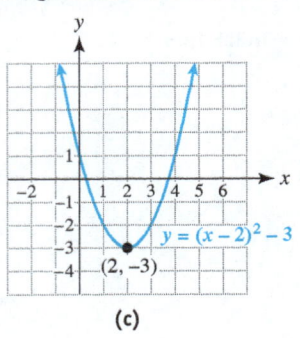
(c)

FIGURE 18.20

Now Try Exercises 23, 27, 35

3 Applying Vertex Form

GRAPHING WITH VERTEX FORM The graphs of $y = 2x^2$ and $y = 2x^2 - 12x + 20$ have *exactly* the same shape, as illustrated in the following See the Concept. However, the vertex for $y = 2x^2$ is $(0, 0)$, whereas the vertex for $y = 2x^2 - 12x + 20$ is $(3, 2)$.

See the Concept 1 FINDING VERTEX FORM

Vertex $(0, 0)$ **Vertex $(3, 2)$**   **Shifting $y = 2x^2$**

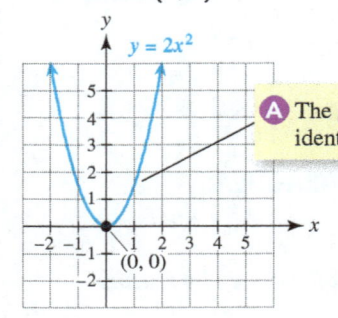
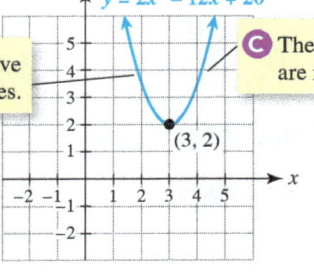
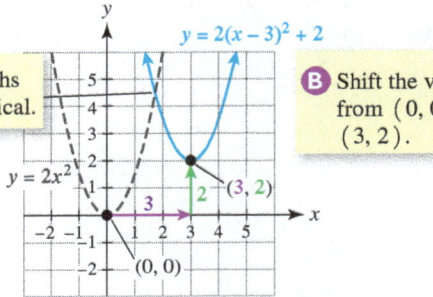

Ⓐ The graphs of $y = 2x^2$ and $y = 2x^2 - 12x + 20$ have identical shapes because $a = 2$ in both equations.

Ⓑ Shifting the graph of $y = 2x^2$ right 3 units and upward 2 units:

Vertex form → $y = 2(x - 3)^2 + 2.$ 2 units upward

3 units right

Ⓒ The graphs of $y = 2x^2 - 12x + 20$ and $y = 2(x - 3)^2 + 2$ are *identical* because they have the same shape and the same vertex $(3, 2)$.

WATCH VIDEO IN MML.

CRITICAL THINKING 1

Expand $y = 2(x - 3)^2 + 2$ and combine like terms. What does it equal?

In general, the equation for any parabola can be written as either $y = ax^2 + bx + c$ or $y = a(x - h)^2 + k$, where the vertex is (h, k). The second form is sometimes called *vertex form*.

READING CHECK 1

- If the equation of a parabola is $y = -3(x + 4)^2 - 5$, identify the vertex.

> **VERTEX FORM**
>
> The **vertex form** of the equation of a parabola with vertex (h, k) is
> $$y = a(x - h)^2 + k,$$
> where $a \neq 0$ is a constant. If $a > 0$, the parabola opens upward; if $a < 0$, the parabola opens downward.

NOTE: The vertex form of the equation of a parabola is sometimes called **standard form for a parabola with a vertical axis.** ■

In the next three examples, we demonstrate graphing parabolas in vertex form, finding their equations, and writing vertex forms of equations.

EXAMPLE 3 Graphing parabolas in vertex form

Compare the graph of $y = f(x)$ to the graph of $y = x^2$. Then sketch a graph of $y = f(x)$ and $y = x^2$ in the same xy-plane.
(a) $f(x) = \frac{1}{2}(x - 5)^2 + 2$
(b) $f(x) = -3(x + 5)^2 - 3$

Solution
(a) Because $a = \frac{1}{2}$, the graph of $y = f(x)$ *opens upward* and is *wider* than the graph of $y = x^2$. **FIGURE 18.21(a)** describes all transformations associated with this graph.
(b) Because $a = -3$, the graph of $y = f(x)$ *opens downward* and is *narrower* than the graph of $y = x^2$. **FIGURE 18.21(b)** describes all transformations associated with this graph.

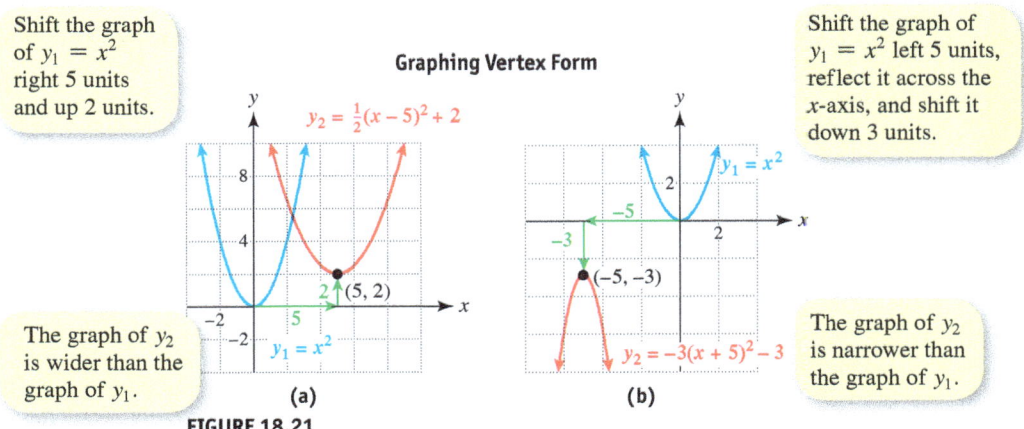

FIGURE 18.21

Now Try Exercises 43, 45

EXAMPLE 4 Finding equations of parabolas

Write the vertex form of a parabola with $a = 2$ and vertex $(-2, 1)$. Then express this equation in the form $y = ax^2 + bx + c$.

Solution
The vertex form of a parabola is given by $y = a(x - h)^2 + k$, where the vertex is (h, k). For $a = 2, h = -2$, and $k = 1$, the equation becomes

$$y = 2(x - (-2))^2 + 1 \quad \text{or} \quad y = 2(x + 2)^2 + 1.$$

To write $y = 2(x + 2)^2 + 1$ in the form $y = ax^2 + bx + c$, do the following.

$$\begin{aligned} y &= 2(x^2 + 4x + 4) + 1 && \text{Multiply } (x + 2)^2 \text{ to get } x^2 + 4x + 4. \\ &= 2x^2 + 8x + 8 + 1 && \text{Distributive property} \\ &= 2x^2 + 8x + 9 && \text{Add.} \end{aligned}$$

The equivalent equation is $y = 2x^2 + 8x + 9$.

Now Try Exercise 49

In the next example, the vertex formula is used to write the equation of a parabola in vertex form.

EXAMPLE 5 Using the vertex formula to write vertex form

Find the vertex on the graph of $y = 3x^2 + 6x + 1$. Write this equation in vertex form.

Solution
We can use the vertex formula, $x = -\frac{b}{2a}$, to find the x-coordinate of the vertex with $a = 3$ and $b = 6$.

$$x = -\frac{b}{2a} = -\frac{6}{2(3)} = -1 \quad \text{x-coordinate of vertex}$$

To find the y-coordinate, let $x = -1$ in $y = 3x^2 + 6x + 1$.

$$y = 3(-1)^2 + 6(-1) + 1 = -2 \quad \text{y-coordinate of vertex}$$

The vertex is $(-1, -2)$. We now find the vertex form with $a = 3, h = -1$, and $k = -2$.

$$\begin{aligned} y &= a(x - h)^2 + k && \text{Vertex form} \\ y &= 3(x - (-1))^2 - 2 && \text{Substitute.} \\ y &= 3(x + 1)^2 - 2 && \text{Simplify.} \end{aligned}$$

The equations $y = 3x^2 + 6x + 1$ and $y = 3(x + 1)^2 - 2$ are equivalent, and their graphs represent the same parabola with vertex $(-1, -2)$.

Now Try Exercise 61

COMPLETING THE SQUARE TO FIND THE VERTEX The vertex of a parabola can be found by using a technique called **completing the square**. To complete the square for the expression $x^2 + 4x$, consider the following See the Concept.

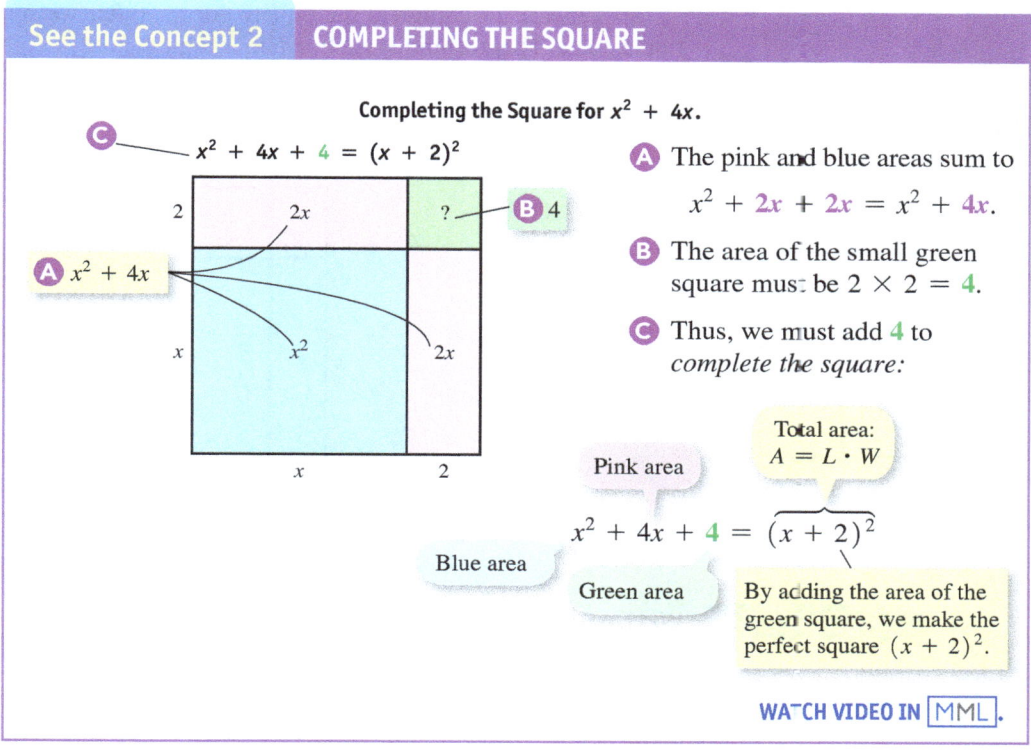

In the next example, we use the above result to complete the square and find the vertex on the graph of $y = x^2 + 4x$.

EXAMPLE 6   Writing vertex form by completing the square

Write the equation $y = x^2 + 4x$ in vertex form by completing the square. Identify the vertex on the graph of the equation.

Solution
As discussed in See the Concept 2, we must add 4 to $x^2 + 4x$ to complete the square. Because we are given an equation, we will add 4 *and* subtract 4 from the right side of the equation in order to keep the equation "balanced."

$$y = x^2 + 4x. \qquad \text{Given equation}$$
$$y = x^2 + 4x + 4 - 4 \qquad \text{Add and subtract 4 on the right.}$$
$$y = (x^2 + 4x + 4) - 4 \qquad \text{Associative property}$$
$$y = (x + 2)^2 - 4 \qquad \text{Perfect square trinomial}$$

The vertex form is $y = (x + 2)^2 - 4$, and the vertex is $(-2, -4)$.

Now Try Exercise 69

NOTE: Adding 4 and subtracting 4 from the right side of the equation in Example 6 is equivalent to adding 0 to the right side, which does not change the equation. ∎

In general, to complete the square for $x^2 + bx$, we must add $\left(\frac{b}{2}\right)^2$, as illustrated in **FIGURE 18.22**. This technique is shown in Example 7.

READING CHECK 2

- Use the figure to *complete the square* for $x^2 + 8x$.

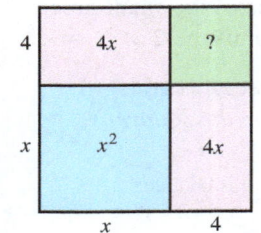

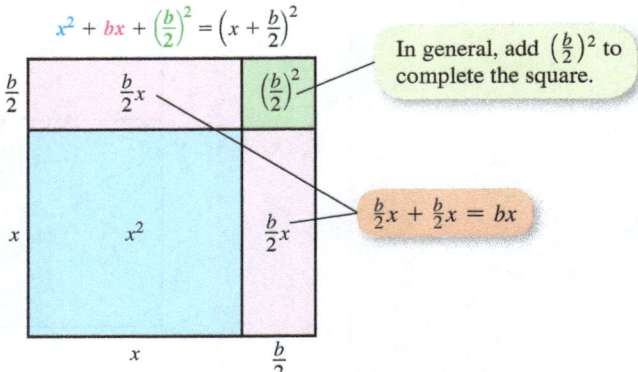

FIGURE 18.22

EXAMPLE 7 Writing vertex form

Write each equation in vertex form. Identify the vertex.
(a) $y = x^2 - 6x - 1$ (b) $y = x^2 + 3x + 4$ (c) $y = 2x^2 + 4x - 1$

Solution
(a) Because $\left(\frac{b}{2}\right)^2 = \left(\frac{-6}{2}\right)^2 = 9$, add *and* subtract 9 on the right side.

$$y = x^2 - 6x - 1 \qquad \text{Given equation}$$
$$= (x^2 - 6x + 9) - 9 - 1 \qquad \text{Add and subtract 9.}$$
$$= (x - 3)^2 - 10 \qquad \text{Perfect square trinomial}$$

The vertex is $(3, -10)$.

(b) Because $\left(\frac{b}{2}\right)^2 = \left(\frac{3}{2}\right)^2 = \frac{9}{4}$, add *and* subtract $\frac{9}{4}$ on the right side.

$$y = x^2 + 3x + 4 \qquad \text{Given equation}$$
$$= \left(x^2 + 3x + \frac{9}{4}\right) - \frac{9}{4} + 4 \qquad \text{Add and subtract } \tfrac{9}{4}.$$
$$= \left(x + \frac{3}{2}\right)^2 + \frac{7}{4} \qquad \text{Perfect square trinomial}$$

The vertex is $\left(-\frac{3}{2}, \frac{7}{4}\right)$.

(c) This equation is slightly different because the leading coefficient is 2 rather than 1. Start by factoring 2 from the first two terms on the right side.

$$y = 2x^2 + 4x - 1 \qquad \text{Given equation}$$
$$= 2(x^2 + 2x) - 1 \qquad \text{Factor out 2.}$$
$$= 2(x^2 + 2x + 1 - 1) - 1 \qquad \left(\tfrac{b}{2}\right)^2 = \left(\tfrac{2}{2}\right)^2 = 1$$
$$= 2(x^2 + 2x + 1) - 2 - 1 \qquad \text{Distributive property: } 2 \cdot (-1)$$
$$= 2(x + 1)^2 - 3 \qquad \text{Perfect square trinomial}$$

Now we have the form $x^2 + bx$ inside the parentheses, so we can complete the square.

The vertex is $(-1, -3)$.

Now Try Exercises 73, 77, 81

4 Modeling with Quadratic Functions

Math in Context *Airline Safety* A taxiway used by an airplane to exit a runway often contains curves. A curve that is too sharp for the speed of the plane is a safety hazard. The scatterplot shown in **FIGURE 18.23** gives an appropriate radius R of a curve designed for an airplane taxiing at x miles per hour. The data are nonlinear because they do not lie on a line. In this subsection we explain how a quadratic function may be used to model such data. (*Source:* FAA.)

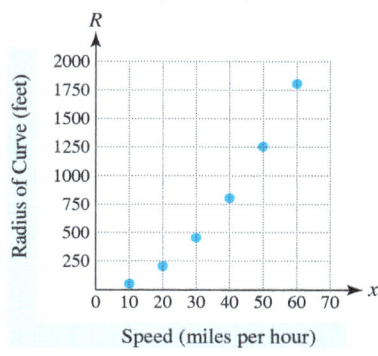

Safe Runway Curve Speed

FIGURE 18.23

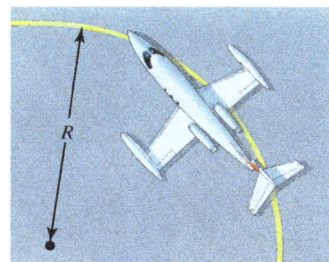

The data shown in **FIGURE 18.23** are listed in **TABLE 18.9**.

Safe Runway Curve Speed

x (mph)	10	20	30	40	50	60
R (ft)	50	200	450	800	1250	1800

Source: Federal Aviation Administration.

TABLE 18.9

A second scatterplot of the data is shown in **FIGURE 18.24**. The data may be modeled by $R(x) = ax^2$ for some value a. To illustrate this relation, graph R for different values of a. In **FIGURES 18.25–18.27**, R has been graphed for $a = 2$, -1, and $\frac{1}{2}$, respectively. When $a > 0$, the parabola opens upward, and when $a < 0$, the parabola opens downward. Larger values of $|a|$ make a parabola narrower, whereas smaller values of $|a|$ make the parabola wider. Through trial and error, $a = \frac{1}{2}$ gives a good fit to the data, so $R(x) = \frac{1}{2}x^2$ models the data.

Finding a so $R(x) = ax^2$ Models the Data

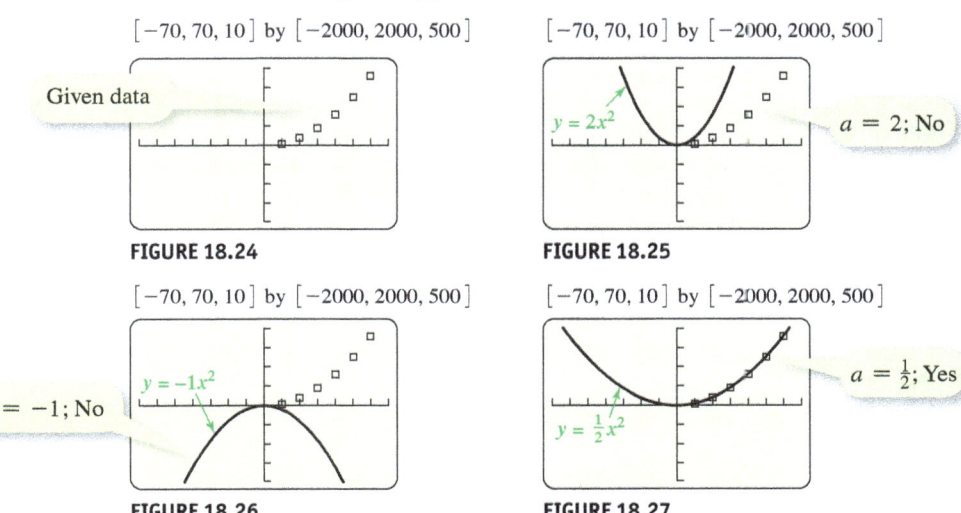

CALCULATOR HELP

To make a scatterplot, see Appendix A (pages AP-3 and AP-4).

FIGURE 18.24 **FIGURE 18.25**

FIGURE 18.26 **FIGURE 18.27**

This value of a can also be found *symbolically*, as demonstrated in the next example.

EXAMPLE 8 Modeling safe runway curve speed

Find a value for the constant a so that $R(x) = ax^2$ models the data in **TABLE 18.9** on the previous page. Check your result by making a table of values for $R(x)$.

Solution
From **TABLE 18.9**, when $x = 10$ miles per hour, the curve radius is 50 feet. Therefore
$$R(10) = 50 \quad \text{or} \quad a(10)^2 = 50.$$

Solving for a gives
$$a = \frac{50}{10^2} = \frac{1}{2}.$$

To be sure that $R(x) = \frac{1}{2}x^2$ is correct, make a table, as shown in **FIGURE 18.28**. Its values agree with those in **TABLE 18.9**.

Now Try Exercise 95

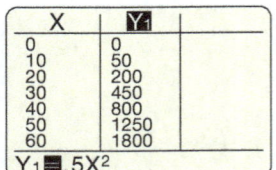

FIGURE 18.28

18.2 Putting It All Together

CONCEPT	EXPLANATION	EXAMPLES
Translations of Parabolas	Compared to the graph $y = x^2$, the graph of $y = x^2 + k$ is shifted vertically k units and the graph of $y = (x - h)^2$ is shifted horizontally h units.	Compared to the graph of $y = x^2$, the graph of $y = x^2 - 4$ is shifted *downward* 4 units. Compared to the graph of $y = x^2$, the graph of $y = (x - 4)^2$ is shifted *right* 4 units and the graph of $y = (x + 4)^2$ is shifted *left* 4 units.
Vertex Form	The vertex form of the equation of a parabola with vertex (h, k) is $$y = a(x - h)^2 + k,$$ where $a \neq 0$ is a constant. If $a > 0$, the parabola opens upward; if $a < 0$, the parabola opens downward.	The graph of $y = 3(x + 2)^2 - 7$ has a vertex of $(-2, -7)$ and opens upward because $3 > 0$.
Completing the Square Method	To complete the square to obtain the vertex form, add *and* subtract $\left(\frac{b}{2}\right)^2$ on the right side of the equation $y = x^2 + bx + c$. Then factor the perfect square trinomial.	If $y = x^2 + 10x - 3$, then add *and* subtract $\left(\frac{b}{2}\right)^2 = \left(\frac{10}{2}\right)^2 = 25$ on the right side of this equation. $y = (x^2 + 10x + 25) - 25 - 3$ $ = (x + 5)^2 - 28$ The vertex is $(-5, -28)$.

18.2 Exercises

CONCEPTS AND VOCABULARY

1. Compared to the graph of $y = x^2$, the graph of $y =$ _____ is shifted upward 2 units.

2. Compared to the graph of $y = x^2$, the graph of $y =$ _____ is shifted to the right 2 units.

3. The vertex of $y = (x - 1)^2 + 2$ is _____.

4. The vertex of $y = (x + 1)^2 - 2$ is _____.

5. A quadratic function f may be written either in the form _____ or _____.

6. The vertex form of a parabola is given by _____ and its vertex is _____.

7. The graph of the equation $y = -x^2$ is a parabola that opens _____.

8. The x-coordinate of the vertex of $y = ax^2 + bx + c$ is $x =$ _____.

9. Compared to the graph of $y = x^2$, the graph of $y = x^2 + k$ with $k > 0$ is shifted k units _____.
   a. upward   b. downward
   c. left   d. right

10. Compared to the graph of $y = x^2$, the graph of $y = (x - k)^2$ with $k > 0$ is shifted k units _____.
    a. upward   b. downward
    c. left   d. right

APPLYING TRANSFORMATIONS OF GRAPHS

Exercises 11–18: Graph $f(x)$. Compare the graph of $y = f(x)$ to the graph of $y = x^2$.

11. $f(x) = -x^2$
12. $f(x) = -2x^2$
13. $f(x) = 2x^2$
14. $f(x) = 3x^2$
15. $f(x) = \frac{1}{4}x^2$
16. $f(x) = \frac{1}{2}x^2$
17. $f(x) = -\frac{1}{2}x^2$
18. $f(x) = -\frac{3}{2}x^2$

RELATING TABLES AND TRANSLATIONS

Exercises 19–22: Complete the table for each translation of $y = x^2$. State what the translation does.

19.
x	−2	−1	0	1	2
$y = x^2$					
$y = x^2 - 3$					

20.
x	−2	−1	0	1	2
$y = x^2$					
$y = x^2 + 5$					

21.
x					
$y = x^2$	4	1	0	1	4
x					
$y = (x - 3)^2$	4	1	0	1	4

22.
x					
$y = x^2$	16	4	0	4	16
x					
$y = (x + 4)^2$	16	4	0	4	16

GRAPHING PARABOLAS

Exercises 23–42: Do the following.

(a) Sketch a graph of the function.
(b) Identify the vertex.
(c) Compare the graph of $y = f(x)$ to the graph of $y = x^2$. (State any transformations used.)

23. $f(x) = x^2 - 4$
24. $f(x) = x^2 - 1$
25. $f(x) = 2x^2 + 1$
26. $f(x) = \frac{1}{2}x^2 + 1$
27. $f(x) = (x - 3)^2$
28. $f(x) = (x + 1)^2$
29. $f(x) = -x^2$
30. $f(x) = -(x + 2)^2$
31. $f(x) = 2 - x^2$
32. $f(x) = (x - 1)^2$
33. $f(x) = (x + 2)^2$
34. $f(x) = (x - 2)^2 - 3$
35. $f(x) = (x + 1)^2 - 2$
36. $f(x) = (x - 3)^2 + 1$
37. $f(x) = (x - 1)^2 + 2$
38. $f(x) = \frac{1}{2}(x + 3)^2 - 3$
39. $f(x) = 2(x - 5)^2 - 4$
40. $f(x) = -3(x + 4)^2 + 5$
41. $f(x) = -\frac{1}{2}(x + 3)^2 + 1$
42. $f(x) = 2(x - 5)^2 + 10$

Exercises 43–46: Compare the graph of $y = f(x)$ to the graph of $y = x^2$. Then sketch a graph of $y = f(x)$ and $y = x^2$ in the same xy-plane.

43. $f(x) = \frac{1}{2}(x - 1)^2 - 2$
44. $f(x) = 2(x + 2)^2 - 1$
45. $f(x) = -2(x + 1)^2 + 3$
46. $f(x) = -\frac{1}{2}(x - 2)^2 + 2$

Exercises 47 and 48: Graph the equation in a window that shows the vertex and all intercepts.

47. $f(x) = -0.4x^2 + 6x - 10$
48. $f(x) = 3x^2 - 40x + 50$

WRITING VERTEX FORM

Exercises 49–52: (Refer to Example 4.) Write the vertex form of a parabola that satisfies the conditions given. Then write the equation in the form $y = ax^2 + bx + c$.

49. Vertex $(3, 4)$ and $a = 3$
50. Vertex $(-1, 3)$ and $a = -5$
51. Vertex $(5, -2)$ and $a = -\frac{1}{2}$
52. Vertex $(-2, -6)$ and $a = \frac{3}{4}$

Exercises 53–56: Write the vertex form of a parabola that satisfies the conditions given. Assume that $a = \pm 1$.

53. Opens upward, vertex $(1, 2)$
54. Opens downward, vertex $(-1, -2)$
55. Opens downward, vertex $(0, -3)$
56. Opens upward, vertex $(5, -4)$

Exercises 57–60: Write the vertex form of the parabola shown in the graph. Assume that $a = \pm 1$.

57.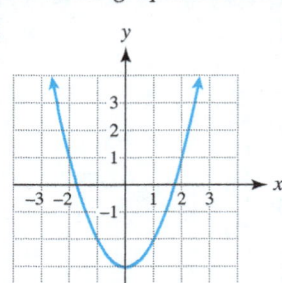

58.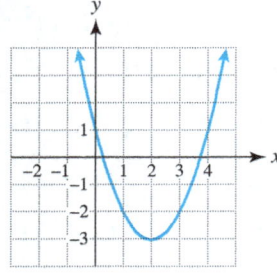

59.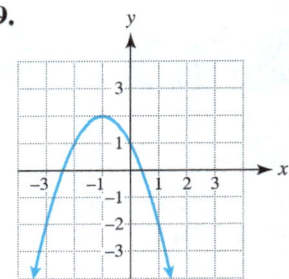

60.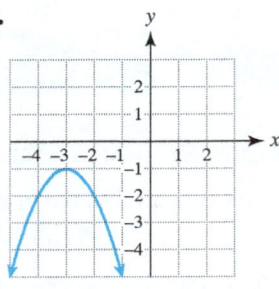

Exercises 61–66: (Refer to Example 5.) Do the following.
(a) Find the vertex on the graph of the equation.
(b) Write the equation in vertex form.

61. $y = 4x^2 - 8x + 5$
62. $y = 3x^2 - 12x + 15$
63. $y = -x^2 - 2x - 3$
64. $y = -x^2 - 4x - 5$
65. $y = -2x^2 - 4x + 1$
66. $y = 2x^2 - 16x + 27$

COMPLETING THE SQUARE

Exercises 67 and 68: Use the given figure to determine what number should be added to the expression to complete the square.

67. $x^2 + 2x$ 68. $x^2 + 6x$

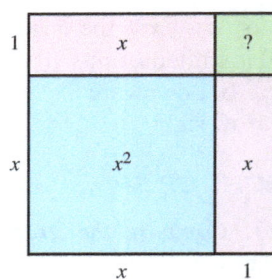 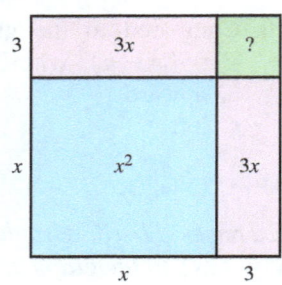

Exercises 69–86: (Refer to Example 7.) Write the equation in vertex form. Identify the vertex.

69. $y = x^2 + 2x$
70. $y = x^2 + 6x$
71. $y = x^2 - 4x$
72. $y = x^2 - 8x$
73. $y = x^2 + 2x - 3$
74. $y = x^2 + 4x + 1$
75. $y = x^2 - 4x + 5$
76. $y = x^2 - 8x + 10$
77. $y = x^2 + 3x - 2$
78. $y = x^2 + 5x - 4$
79. $y = x^2 - 7x + 1$
80. $y = x^2 - 3x + 5$
81. $y = 3x^2 + 6x - 1$
82. $y = 2x^2 + 4x - 9$
83. $y = 2x^2 - 3x$
84. $y = 3x^2 - 7x$
85. $y = -2x^2 - 8x + 5$
86. $y = -3x^2 + 6x + 1$

MODELING DATA

Exercises 87–90: Find a value for the constant a so that $f(x) = ax^2$ models the data. If you are uncertain about your value for a, check it by making a table of values.

87.
x	1	2	3
y	2	8	18

88.
x	−2	0	2
y	6	0	6

89.
x	2	4	6	8
y	1.2	4.8	10.8	19.2

90.
x	5	10	15	20
y	17.5	70	157.5	280

Exercises 91–94: **Modeling Quadratic Data** *Find a quadratic function expressed in vertex form that models the data in the given table. (Hint: Let the first data point in the table be the vertex.)*

91.
x	1	2	3	4
y	−3	−1	5	15

92.
x	−2	−1	0	1	2
y	5	2	−7	−22	−43

93.
x	1980	1990	2000	2010
y	6	56	206	456

94.
x	1990	1995	2000	2005
y	10	60	210	460

95. **Braking Distance** The table lists approximate braking distances D in feet for cars traveling at x miles per hour on dry, level pavement.

x	12	24	36	48
D	12	48	108	192

(a) Make a scatterplot of the data.
(b) Find a function given by $D(x) = ax^2$ that models these data.

96. **Health-Care Costs** The table lists approximate *annual* percent increases in the cost of health insurance premiums between 1992 and 2000.

Year	1992	1994	1996	1998	2000
Increase	11%	4%	1%	4%	11%

Source: Kaiser Family Foundation.

(a) Describe what happened to health-care costs from 1992 to 2000.
(b) Can a linear function model these data? Explain.
(c) What type of function might model these data? Explain.
(d) What might be a good choice for the vertex? Explain.
(e) Find a quadratic function C expressed in vertex form that models the data. (Answers may vary.)
(f) Graph C and the data.

97. **Sub-Saharan Africa** The table lists actual and projected real gross domestic product (GDP) per capita for selected years in sub-Saharan Africa.

Year	1980	1995	2010	2025
Real GDP	$1000	$600	$1000	$2200

Source: IMF WEO, Standard Chartered Research.

(a) Describe what happens to the real GDP.
(b) Can a linear function model these data? Explain.
(c) What type of function might model these data? Explain.
(d) What might be a good choice for the vertex? Explain.
(e) Find a quadratic function C expressed in vertex form that models the data. (Answers may vary.)
(f) Graph C and the data.

98. **Tax Theory and Modeling** If a government wants to generate enough revenue from income taxes, then it is important to assess the correct tax rates. On the one hand, if tax rates are too low, then enough revenue may not be generated. On the other hand, if taxes are too high, people may work less and not earn as much money. Again, enough revenue may not be generated. The following parabolic graph illustrates this phenomenon.

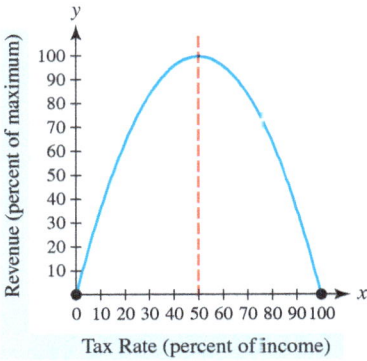

(a) The points $(0, 0)$ and $(100, 0)$ are on this graph. Explain why.
(b) Explain why a linear model is not suitable.
(c) According to *this* graph, what income tax rate maximizes revenue?

WRITING ABOUT MATHEMATICS

99. Explain how to find the vertex of $y = x^2 + bx + c$ by completing the square.

100. If $f(x) = a(x - h)^2 + k$, explain how the values of a, h, and k affect the graph of $y = f(x)$.

101. If $k > 0$, how does the graph of $f(x) = x^2 + k$ compare to the graph of $f(x) = x^2$?

102. A quadratic function can be written in the form $f(x) = ax^2 + bx + c$ or it can be written in the form $f(x) = a(x - h)^2 + k$. Which form is best for identifying the vertex on the graph of $y = f(x)$. Explain.

SECTIONS 18.1 and 18.2 — Checking Basic Concepts

1. Graph each quadratic function. Identify the vertex and axis of symmetry.
   (a) $f(x) = x^2 - 2$
   (b) $f(x) = x^2 - 2x - 2$

2. Compare the graph of $y_1 = 2x^2$ to the graph of $y_2 = -\frac{1}{2}x^2$.

3. Find the maximum y-value on the graph of the equation $y = -3x^2 + 12x - 5$.

4. Sketch a graph of $y = f(x)$. Compare this graph to the graph of $y = x^2$.
   (a) $f(x) = (x - 1)^2 + 2$
   (b) $f(x) = -(x + 3)^2$

5. Write the vertex form for each equation.
   (a) $y = x^2 + 14x - 7$
   (b) $y = 4x^2 + 8x - 2$

18.3 Quadratic Equations

Objectives

1. Learning the Basics of Quadratic Equations
   - Recognizing Quadratic Equations
   - Solving Quadratic Equations (Three Methods)
2. Applying the Square Root Property
3. Completing the Square
4. Solving an Equation for a Variable
5. Solving Applications of Quadratic Equations

NEW VOCABULARY

☐ Quadratic equation
☐ Square root property

1 Learning the Basics of Quadratic Equations

RECOGNIZING QUADRATIC EQUATIONS Any quadratic function f can be represented by $f(x) = ax^2 + bx + c$ with $a \neq 0$.

Quadratic Functions

$$f(x) = 2x^2 - 1, \quad g(x) = -\tfrac{1}{3}x^2 + 2x, \quad h(x) = x^2 + 2x - 1$$

Quadratic functions can be used to write quadratic equations.

Quadratic Equations

$$2x^2 - 1 = 0, \quad -\tfrac{1}{3}x^2 + 2x = 0, \quad x^2 + 2x - 1 = 0$$

QUADRATIC EQUATION

A **quadratic equation** is an equation that can be written as

$$ax^2 + bx + c = 0,$$

where a, b, and c are constants with $a \neq 0$.

READING CHECK 1

- How can you identify a quadratic equation?

The following See the Concepts explains that a quadratic equation can have zero, one, or two real solutions.

See the Concept 1 — SOLUTIONS TO QUADRATIC EQUATIONS

No x-intercepts

A $f(x) = ax^2 + bx + c, \ a < 0$

B $ax^2 + bx + c = 0$ has no solutions.

One x-intercept

A $f(x) = ax^2 + bx + c, \ a > 0$

B $ax^2 + bx + c = 0$ has one solution.

Two x-intercepts

A $f(x) = ax^2 + bx + c, \ a > 0$

B $ax^2 + bx + c = 0$ has two solutions.

A The graph of a quadratic function $f(x) = ax^2 + bx + c$ is either ∪-shaped or ∩-shaped and can intersect the x-axis zero, one, or two times.

B The x-coordinate of an x-intercept corresponds to a real solution to the quadratic equation $ax^2 + bx + c = 0$. Thus a quadratic equation can have zero, one, or two solutions.

WATCH VIDEO IN MML.

NOTE: Some quadratic equations may not have 0 on the right side of the equation, such as $4x^2 = 1$. To solve this quadratic equation graphically, we will sometimes rewrite it as $4x^2 - 1 = 0$ and graph $y = 4x^2 - 1$. Then solutions are the x-coordinates of the x-intercepts, because $y = 0$ on the x-axis. ∎

SOLVING QUADRATIC EQUATIONS (THREE METHODS) We have already solved quadratic equations by factoring, graphing, and constructing tables. In the next example, we apply these three techniques to quadratic equations that have no real solutions, one real solution, and two real solutions.

EXAMPLE 1 Solving quadratic equations

Solve each quadratic equation. Support your results numerically and graphically.
(a) $2x^2 + 1 = 0$ (No real solutions)
(b) $x^2 + 4 = 4x$ (One real solution)
(c) $x^2 - 6x + 8 = 0$ (Two real solutions)

Solution
(a) Symbolic Solution

$$2x^2 + 1 = 0 \quad \text{Given equation}$$
$$2x^2 = -1 \quad \text{Subtract 1 from each side.}$$
$$x^2 = -\frac{1}{2} \quad \text{Divide each side by 2.}$$

This equation has no real-number solutions because $x^2 \geq 0$ for all real numbers x.

Numerical and Graphical Solution On the next page, the points in **TABLE 18.10** for $y = 2x^2 + 1$ are plotted in **FIGURE 18.29** and connected with a parabolic graph. The graph of $y = 2x^2 + 1$ has no x-intercepts, indicating that there are no real solutions.

No Solutions

$y = 2x^2 + 1$

x	y
-2	9
-1	3
0	1
1	3
2	9

TABLE 18.10

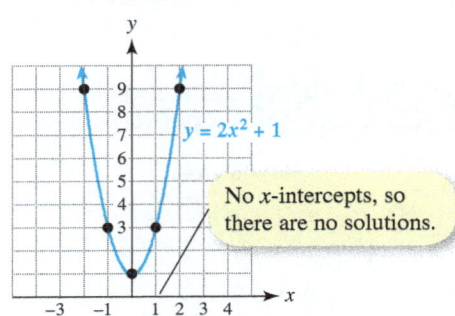

FIGURE 18.29

For all x, $y \neq 0$, so there are no solutions.

No x-intercepts, so there are no solutions.

(b) Symbolic Solution Next, we solve $x^2 + 4 = 4x$.

$$\begin{aligned}
x^2 + 4 &= 4x & &\text{Given equation} \\
x^2 - 4x + 4 &= 0 & &\text{Subtract } 4x \text{ from each side.} \\
(x-2)(x-2) &= 0 & &\text{Factor.} \\
x - 2 = 0 \quad \text{or} \quad x - 2 &= 0 & &\text{Zero-product property} \\
x &= 2 & &\text{There is one solution.}
\end{aligned}$$

Numerical and Graphical Solution Because the given quadratic equation is equivalent to $x^2 - 4x + 4 = 0$, we let $y = x^2 - 4x + 4$. The points in **TABLE 18.11** are plotted in **FIGURE 18.30** and connected with a parabolic graph. The graph of $y = x^2 - 4x + 4$ has one x-intercept, $(2, 0)$. Note that in **TABLE 18.11**, $y = 0$ when $x = 2$, indicating that the equation has one solution.

One Solution

$y = x^2 - 4x + 4$

x	y
0	4
1	1
2	0
3	1
4	4

TABLE 18.11

One solution: 2

FIGURE 18.30

One x-intercept: $(2, 0)$ so there is one solution: 2.

(c) Symbolic Solution Next, we solve $x^2 - 6x + 8 = 0$.

$$\begin{aligned}
x^2 - 6x + 8 &= 0 & &\text{Given equation} \\
(x - 2)(x - 4) &= 0 & &\text{Factor.} \\
x - 2 = 0 \quad \text{or} \quad x - 4 &= 0 & &\text{Zero-product property} \\
x = 2 \quad \text{or} \quad x &= 4 & &\text{There are two solutions.}
\end{aligned}$$

Numerical and Graphical Solution The points in **TABLE 18.12** for $y = x^2 - 6x + 8$ are plotted in **FIGURE 18.31** and connected with a parabolic graph. The graph of $y = x^2 - 6x + 8$ has two x-intercepts, $(2, 0)$ and $(4, 0)$, indicating two solutions. Note that in **TABLE 18.12** $y = 0$ when $x = 2$ or $x = 4$.

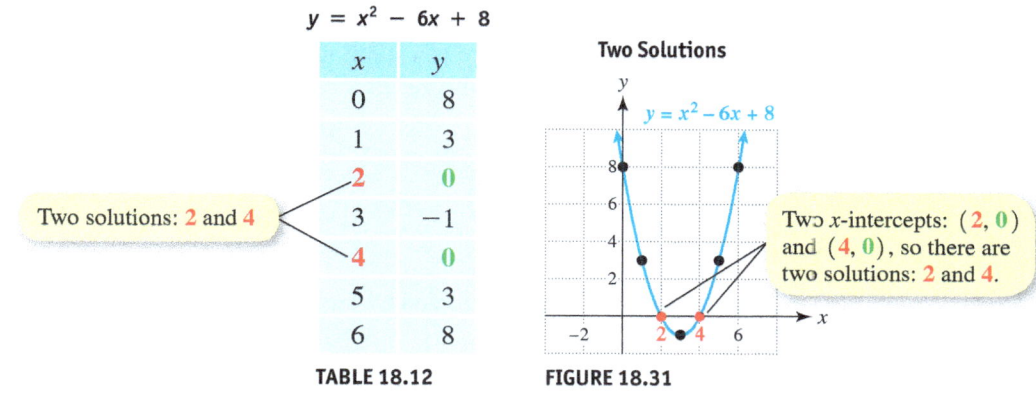

$y = x^2 - 6x + 8$

TABLE 18.12

x	y
0	8
1	3
2	0
3	−1
4	0
5	3
6	8

Two solutions: 2 and 4

Two Solutions

FIGURE 18.31

Two x-intercepts: $(2, 0)$ and $(4, 0)$, so there are two solutions: 2 and 4.

Now Try Exercises 29, 37, 39

Connecting Concepts with Your Life When a person throws a ball into the air, it follows a parabolic path that opens downward. See **FIGURE 18.32**.

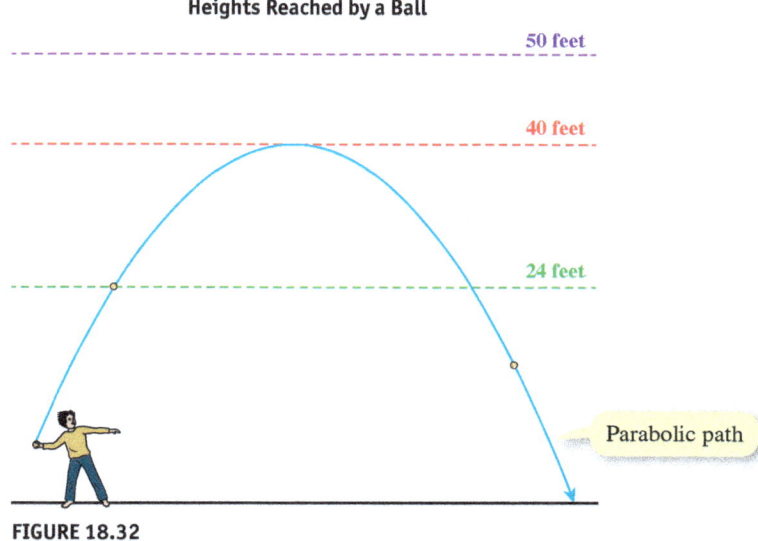

Heights Reached by a Ball

FIGURE 18.32

If the ball reaches a maximum height of 40 feet, then there would be two different times when the ball is 24 feet above the ground because the ball goes up and then down. However, there would be only one time when the ball is 40 feet above the ground, and never a time when the ball is 50 feet above the ground. If the height of the ball h after t seconds is determined by $h(t) = -16t^2 + 48t + 4$, it follows that

$$-16t^2 + 48t + 4 = 24$$

has two solutions because the ball is 24 feet above the ground at two different times,

$$-16t^2 + 48t + 4 = 40$$

has one solution because the ball is 40 feet above the ground at only one point, and

$$-16t^2 + 48t + 4 = 50$$

has no solutions because the ball is never 50 feet above the ground. A quadratic equation can have zero, one, or two solutions in real life. See Exercises 127–130.

READING CHECK 2

- How many solutions can a quadratic equation have?

2 Applying the Square Root Property

The **square root property** is used to solve quadratic equations that have no x-terms. The following is an example of the square root property.

$$x^2 = 25 \text{ is equivalent to } x = \pm 5.$$

The equation $x = \pm 5$ (read "x equals plus or minus 5") indicates that either $x = 5$ or $x = -5$. Each value is a solution because $(5)^2 = 25$ and $(-5)^2 = 25$.

We can derive this result in general for $k \geq 0$.

$$x^2 = k \quad \text{Given quadratic equation}$$
$$\sqrt{x^2} = \sqrt{k} \quad \text{Take the square root of each side.}$$
$$|x| = \sqrt{k} \quad \sqrt{x^2} = |x| \text{ for all } x.$$
$$x = \pm\sqrt{k} \quad |x| = b \text{ implies } x = \pm b, b \geq 0.$$

This result is summarized by the *square root property*.

> **SQUARE ROOT PROPERTY**
>
> Let k be a nonnegative number. Then the solutions to the equation
> $$x^2 = k$$
> are given by $x = \pm\sqrt{k}$. If $k < 0$, then this equation has no real solutions.

EXAMPLE 2 **Using the square root property**

Solve each equation.
(a) $x^2 = 7$ (b) $16x^2 - 9 = 0$ (c) $(x - 4)^2 = 25$

Solution
(a) $x^2 = 7$ is equivalent to $x = \pm\sqrt{7}$ by the square root property. The solutions are $\sqrt{7}$ and $-\sqrt{7}$.

(b)
$$16x^2 - 9 = 0 \quad \text{Given equation}$$
$$16x^2 = 9 \quad \text{Add 9 to each side.}$$
$$x^2 = \frac{9}{16} \quad \text{Divide each side by 16.}$$
$$x = \pm\sqrt{\frac{9}{16}} \quad \text{Square root property}$$
$$x = \pm\frac{3}{4} \quad \text{Simplify.}$$

The solutions are $\frac{3}{4}$ and $-\frac{3}{4}$.

(c)
$$(x - 4)^2 = 25 \quad \text{Given equation}$$
$$(x - 4) = \pm\sqrt{25} \quad \text{Square root property}$$
$$x - 4 = \pm 5 \quad \text{Simplify.}$$
$$x = 4 \pm 5 \quad \text{Add 4 to each side.}$$
$$x = 9 \quad \text{or} \quad x = -1 \quad \text{Evaluate } 4 + 5 \text{ and } 4 - 5.$$

The solutions are 9 and -1.

Now Try Exercises 51, 53, 57

READING CHECK 3

- Would it be more appropriate to use the square root property to solve $x^2 - x + 3 = 0$ or $x^2 - 9 = 0$? Use the square root property to solve the equation that you chose.

Math in Context Falling Object If an object is dropped from a height of h feet, its distance d above the ground after t seconds is given by

$$d(t) = h - 16t^2.$$

This formula can be used to estimate the time it takes for a falling object to hit the ground.

EXAMPLE 3　Modeling a falling object

A toy falls from a window that is 30 feet above the ground. How long does it take for the toy to hit the ground?

Solution
The height h of the window above the ground is 30 feet, so let $d(t) = 30 - 16t^2$. The toy strikes the ground when the distance d above the ground equals 0.

$30 - 16t^2 = 0$	Equation to solve for t
$-16t^2 = -30$	Subtract 30 from each side.
$t^2 = \dfrac{30}{16}$	Divide each side by -16.
$t = \pm\sqrt{\dfrac{30}{16}}$	Square root property
$t = \pm\dfrac{\sqrt{30}}{4}$	Simplify.

Time cannot be negative in this problem, so the appropriate solution is $t = \dfrac{\sqrt{30}}{4} \approx 1.4$. The toy hits the ground after about 1.4 seconds.

Now Try Exercise 125

3 Completing the Square

The *method of completing the square* can be used to solve quadratic equations. Because

$$x^2 + bx + \left(\frac{b}{2}\right)^2 = \left(x + \frac{b}{2}\right)^2,$$

we can solve a quadratic equation in the form $x^2 + bx = d$, where b and d are constants, by adding $\left(\frac{b}{2}\right)^2$ to each side and then factoring the resulting perfect square trinomial. For example, in the equation $x^2 + 6x = 7$, we have $b = 6$, so we add $\left(\frac{6}{2}\right)^2 = 9$ to each side.

READING CHECK 4

- Use the figure to complete the square for $x^2 + 6x$.

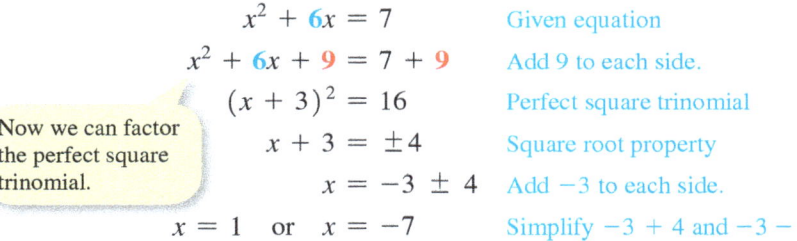

$x^2 + 6x = 7$	Given equation
$x^2 + 6x + 9 = 7 + 9$	Add 9 to each side.
$(x + 3)^2 = 16$	Perfect square trinomial
$x + 3 = \pm 4$	Square root property
$x = -3 \pm 4$	Add -3 to each side.
$x = 1$ or $x = -7$	Simplify $-3 + 4$ and $-3 - 4$.

Now we can factor the perfect square trinomial.

The solutions are 1 and -7. Note that after completing the square, the left side of the equation is a perfect square trinomial. We show how to create one in the next example.

EXAMPLE 4　Creating a perfect square trinomial

Find the term that should be added to $x^2 - 10x$ to form a perfect square trinomial.

Solution
The coefficient of the x-term is -10, so we let $b = -10$. To complete the square, we divide b by 2 and then square the result.

$$\left(\frac{b}{2}\right)^2 = \left(\frac{-10}{2}\right)^2 = 25$$

If we add 25 to $x^2 - 10x$, a perfect square trinomial is formed.

$$x^2 - 10x + 25 = (x - 5)^2$$

Now Try Exercise 77

The following steps can be used to solve any quadratic equation by completing the square.

> **SOLVING A QUADRATIC EQUATION BY COMPLETING THE SQUARE**
>
> **STEP 1:** If necessary, use algebra to write the given equation in the form $x^2 + bx = d$, where the coefficient on x^2 is 1 and the constant term is on the right side of the equation.
>
> **STEP 2:** Add $\left(\frac{b}{2}\right)^2$ to each side of the equation from Step 1.
>
> **STEP 3:** Factor the resulting perfect square trinomial on the left side of the equation.
>
> **STEP 4:** Apply the square root property to solve for x.

EXAMPLE 5 Completing the square when the leading coefficient is 1

Solve the equation $x^2 - 4x + 2 = 0$.

Solution
Start by writing the equation in the form $x^2 + bx = d$.

$$x^2 - 4x + 2 = 0 \qquad \text{Given equation}$$
$$x^2 - 4x = -2 \qquad \text{Step 1: subtract 2.}$$
$$x^2 - 4x + 4 = -2 + 4 \qquad \text{Step 2: add } \left(\tfrac{b}{2}\right)^2 = \left(\tfrac{-4}{2}\right)^2 = 4.$$
$$(x - 2)^2 = 2 \qquad \text{Step 3: perfect square trinomial; add.}$$
$$x - 2 = \pm\sqrt{2} \qquad \text{Step 4: square root property}$$
$$x = 2 \pm \sqrt{2} \qquad \text{Add 2.}$$

The solutions are $2 + \sqrt{2} \approx 3.41$ and $2 - \sqrt{2} \approx 0.59$.

Now Try Exercise 83

EXAMPLE 6 Completing the square when the leading coefficient is not 1

Solve the equation $2x^2 + 7x - 5 = 0$.

Solution

$$2x^2 + 7x - 5 = 0 \qquad \text{Given equation}$$
$$2x^2 + 7x = 5 \qquad \text{Step 1: add 5 to each side.}$$
$$x^2 + \frac{7}{2}x = \frac{5}{2} \qquad \text{Divide each side by 2.}$$
$$x^2 + \frac{7}{2}x + \frac{49}{16} = \frac{5}{2} + \frac{49}{16} \qquad \text{Step 2: add } \left(\tfrac{b}{2}\right)^2 = \left(\tfrac{7}{4}\right)^2 = \tfrac{49}{16}.$$
$$\left(x + \frac{7}{4}\right)^2 = \frac{89}{16} \qquad \text{Step 3: perfect square trinomial; add.}$$
$$x + \frac{7}{4} = \pm\frac{\sqrt{89}}{4} \qquad \text{Step 4: square root property}$$
$$x = -\frac{7}{4} \pm \frac{\sqrt{89}}{4} \qquad \text{Add } -\tfrac{7}{4}.$$
$$x = \frac{-7 \pm \sqrt{89}}{4} \qquad \text{Combine fractions.}$$

Be sure to first write the given equation as $x^2 + bx = d$.

The solutions are $\frac{-7 + \sqrt{89}}{4} \approx 0.61$ and $\frac{-7 - \sqrt{89}}{4} \approx -4.1$.

Now Try Exercise 91

CRITICAL THINKING 1

What happens if you try to solve

$$2x^2 - 13 = 1$$

by completing the square? What method could you use to solve this problem?

4 Solving an Equation for a Variable

We often need to solve an equation or formula for a variable. For example, the formula $V = \frac{1}{3}\pi r^2 h$ calculates the volume of the cone shown in **FIGURE 18.33**. Let's say that we know the volume V is 120 cubic inches and the height h is 15 inches. We can then find the radius of the cone by solving the equation for r.

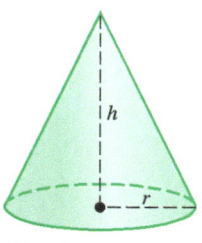

FIGURE 18.33

$$V = \frac{1}{3}\pi r^2 h \qquad \text{Solve the equation for } r.$$

$$3V = \pi r^2 h \qquad \text{Multiply by 3.}$$

$$\frac{3V}{\pi} = r^2 h \qquad \text{Divide by } \pi.$$

$$\frac{3V}{\pi h} = r^2 \qquad \text{Divide by } h.$$

$$r = \pm\sqrt{\frac{3V}{\pi h}} \qquad \text{Square root property; rewrite.}$$

Because $r \geq 0$, we use the positive or *principal square root*. Thus for $V = 120$ cubic inches and $h = 15$ inches,

$$r = \sqrt{\frac{3(120)}{\pi(15)}} = \sqrt{\frac{24}{\pi}} \approx 2.8 \text{ inches.}$$

EXAMPLE 7 Solving equations for variables

Solve each equation for the specified variable.
(a) $s = -\frac{1}{2}gt^2 + h$ for t **(b)** $d^2 = x^2 + y^2$ for y

Solution
(a) Begin by subtracting h from each side of the equation.

$$s = -\frac{1}{2}gt^2 + h \qquad \text{Solve the equation for } t.$$

$$s - h = -\frac{1}{2}gt^2 \qquad \text{Subtract } h.$$

$$-2(s - h) = gt^2 \qquad \text{Multiply by } -2.$$

$$\frac{2h - 2s}{g} = t^2 \qquad \text{Divide by } g; \text{ simplify.}$$

$$t = \pm\sqrt{\frac{2h - 2s}{g}} \qquad \text{Square root property; rewrite.}$$

(b) Begin by subtracting x^2 from each side of the equation.

$$d^2 = x^2 + y^2 \qquad \text{Solve the equation for } y.$$
$$d^2 - x^2 = y^2 \qquad \text{Subtract } x^2.$$
$$y = \pm\sqrt{d^2 - x^2} \qquad \text{Square root property; rewrite.}$$

Now Try Exercises 115, 117

5 Solving Applications of Quadratic Equations

Math in Context — *Airline Safety* In the second section of this chapter, we modeled curves on airport taxiways by using $R(x) = \frac{1}{2}x^2$. In this formula, x represents the airplane's speed in miles per hour, and R represents the radius of the curve in feet. This formula may be used to determine the speed limit for a curve with a radius of **650** feet by solving the following *quadratic equation*. (*Source:* FAA.)

$$\frac{1}{2}x^2 = 650$$

In the next example, we solve this quadratic equation.

EXAMPLE 8 Finding a safe speed limit

Solve $\frac{1}{2}x^2 = 650$ and interpret any solutions.

Solution
Use the square root property to solve this problem.

$$\frac{1}{2}x^2 = 650 \qquad \text{Given equation}$$
$$x^2 = 1300 \qquad \text{Multiply by 2.}$$
$$x = \pm\sqrt{1300} \qquad \text{Square root property}$$

The solutions are $\sqrt{1300} \approx 36$ and $-\sqrt{1300} \approx -36$. The solution of $x \approx 36$ indicates that a safe speed limit for a curve with a radius of 650 feet is about 36 miles per hour. (The negative solution has no physical meaning in this problem.)

Now Try Exercise 123

Math in Context — *Social Network* For many years, MySpace was the largest social network in the United States. However, during 2010 the number of unique monthly visitors fell dramatically from 70 million in January 2010 to 45 million in January 2011. **TABLE 18.13** lists the number of unique visitors V to MySpace in millions x months after January 2010.

Unique MySpace Visitors (millions)

x (months after January 2010)	0	6	12
V (unique monthly visitors)	70	63	45

Source: comScore.
TABLE 18.13

Decreased 7 million Decreased 18 million

The number of unique visitors to MySpace decreased faster in the second half of 2010 than it did in the first half, so a linear function will *not* model these data. Instead, these data can be modeled by the *quadratic function*

$$V(x) = -\frac{25}{144}x^2 + 70.$$

EXAMPLE 9 Modeling the decline in MySpace visitors

Use the equation $V(x) = -\frac{25}{144}x^2 + 70$ given in the above discussion to determine the month when MySpace had 56 million unique visitors.

Solution
To determine when there were 56 million visitors, we must solve

$$-\frac{25}{144}x^2 + 70 = 56.$$

There is no *x*-term in the equation, so we use the square root property.

$$-\frac{25}{144}x^2 + 70 = 56. \qquad \text{Equation to be solved}$$

$$-\frac{25}{144}x^2 = -14 \qquad \text{Subtract 70 from each side.}$$

$$x^2 = \frac{2016}{25} \qquad \text{Multiply each side by } -\frac{144}{25}.$$

$$x = \pm\frac{\sqrt{2016}}{5} \qquad \text{Square root property}$$

Because *x* must be positive, $x = \frac{\sqrt{2016}}{5} \approx 9$. Thus 9 months after January, or in October, MySpace had 56 million unique visitors.

Now Try Exercise 127

18.3 Putting It All Together

CONCEPT	EXPLANATION	EXAMPLES
Square Root Property	If $k \geq 0$, the solutions to the equation $x^2 = k$ are $\pm\sqrt{k}$.	$x^2 = 100$ is equivalent to $x = \pm 10$ and $x^2 = 13$ is equivalent to $x = \pm\sqrt{13}$. $x^2 = -2$ has no real solutions.
Method of Completing the Square	To solve an equation in the form $x^2 + bx = d$, add $\left(\frac{b}{2}\right)^2$ to each side of the equation. Factor the resulting perfect square trinomial and solve for *x* by applying the square root property.	To solve $x^2 + 8x = 3$, add $\left(\frac{8}{2}\right)^2 = 16$ to each side. $x^2 + 8x + 16 = 3 + 16$ Add 16. $(x + 4)^2 = 19$ Perfect square trinomial $x + 4 = \pm\sqrt{19}$ Square root property $x = -4 \pm \sqrt{19}$ Add -4.

18.3 Exercises

CONCEPTS AND VOCABULARY

1. Give an example of a quadratic equation. How many real solutions can a quadratic equation have?

2. Is a quadratic equation a linear equation or a nonlinear equation?

3. Name three symbolic methods that can be used to solve a quadratic equation.

4. Sketch a graph of a quadratic function that has two x-intercepts and opens downward.

5. Sketch a graph of a quadratic function that has no x-intercepts and opens upward.

6. If the graph of $y = ax^2 + bx + c$ intersects the x-axis twice, how many solutions does the equation $ax^2 + bx + c = 0$ have? Explain.

7. Solve $x^2 = 64$. What property did you use?

8. To solve $x^2 + bx = 6$ by completing the square, what value should be added to each side of the equation?

RECOGNIZING QUADRATIC EQUATIONS

Exercises 9–16: Is the given equation quadratic?

9. $x^2 - 3x + 1 = 0$
10. $2x^2 - 3 = 0$
11. $3x + 1 = 0$
12. $x^3 - 3x^2 + x = 0$
13. $-3x^2 + x = 16$
14. $x^2 - 1 = 4x$
15. $x^2 = \sqrt{x} + 1$
16. $\dfrac{1}{x-1} = 5$

SOLVING QUADRATIC EQUATIONS

Exercises 17–20: Approximate to the nearest hundredth.

17. (a) $1 \pm \sqrt{7}$ (b) $-2 \pm \sqrt{11}$
18. (a) $\pm \dfrac{\sqrt{3}}{2}$ (b) $\pm \dfrac{2\sqrt{5}}{7}$
19. (a) $\dfrac{3 \pm \sqrt{13}}{5}$ (b) $\dfrac{-5 \pm \sqrt{6}}{9}$
20. (a) $\dfrac{2}{5} \pm \dfrac{\sqrt{5}}{5}$ (b) $-\dfrac{3}{7} \pm \dfrac{\sqrt{3}}{7}$

Exercises 21–24: A graph of $y = ax^2 + bx + c$ is given. Use this graph to solve $ax^2 + bx + c = 0$, if possible.

21.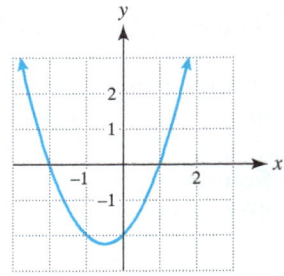
22.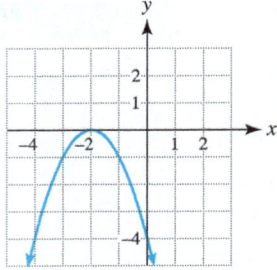
23.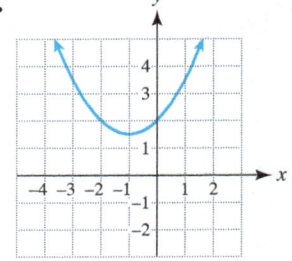
24.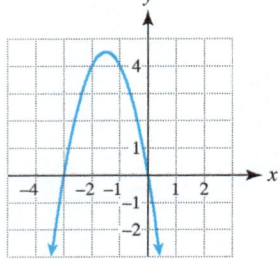

Exercises 25–28: A table of $y = ax^2 + bx + c$ is given. Use this table to solve $ax^2 + bx + c = 0$.

25.

X	Y₁
-3	6
-2	0
-1	-4
0	-6
1	-6
2	-4
3	0

Y₁ = X^2 - X - 6

26.

X	Y₁
-6	0
-4	-16
-2	-24
0	-24
2	-16
4	0
6	24

Y₁ = X^2 + 2X - 24

27.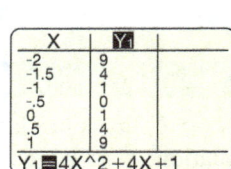

28.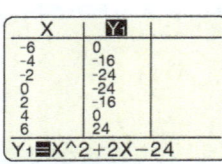

Exercises 29–40: Use each method to solve the equation.
(a) Symbolic   (b) Graphical   (c) Numerical

29. $x^2 - 4x - 5 = 0$
30. $x^2 - x - 6 = 0$
31. $x^2 + 2x = 3$
32. $x^2 + 4x = 5$
33. $x^2 = 9$
34. $x^2 = 4$
35. $4x^2 - 4x = 3$
36. $2x^2 + x = 1$
37. $x^2 + 2x = -1$
38. $4x^2 - 4x + 1 = 0$
39. $x^2 + 2 = 0$
40. $-4x^2 - 1 = 2$

Exercises 41–50: Solve by factoring.

41. $x^2 + 2x - 35 = 0$ **42.** $2x^2 - 7x + 3 = 0$

43. $6x^2 - x - 1 = 0$ **44.** $x^2 + 4x + 6 = -3x$

45. $4x^2 + 13x + 9 = x$ **46.** $9x^2 + 4 = 12x$

47. $25x^2 - 350 = 125x$ **48.** $20x^2 + 150 = 130x$

49. $2(5x^2 + 9) = 27x$ **50.** $15(3x^2 + x) = 10$

USING THE SQUARE ROOT PROPERTY

Exercises 51–62: Use the square root property to solve.

51. $x^2 = 144$ **52.** $4x^2 - 5 = 0$

53. $5x^2 - 64 = 0$ **54.** $3x^2 = 7$

55. $(x + 1)^2 = 25$ **56.** $(x + 4)^2 = 9$

57. $(x - 1)^2 = 64$ **58.** $(x - 3)^2 = 0$

59. $(2x - 1)^2 = 5$ **60.** $(5x + 3)^2 = 7$

61. $10(x - 5)^2 = 50$ **62.** $7(3x + 1)^2 = 14$

Exercises 63–72: **Complex Solutions** *Although the equation $x^2 = -1$ has no real solutions, it does have complex solutions. In the previous chapter, we defined the imaginary unit as $i = \sqrt{-1}$, or $i^2 = -1$. Thus the square root property can be used to solve $x^2 = -1$, to obtain $x = \pm\sqrt{-1}$, or $x = \pm i$. Use the square root property to find the complex solutions to each equation.*

63. $x^2 = -4$ **64.** $x^2 = -9$

65. $x^2 = -3$ **66.** $x^2 = -7$

67. $x^2 = -20$ **68.** $x^2 = -8$

69. $x^2 + 16 = 0$ **70.** $x^2 + 25 = 0$

71. $x^2 + 18 = 0$ **72.** $x^2 + 12 = 0$

COMPLETING THE SQUARE

Exercises 73–76: To solve by completing the square, what value should you add to each side of the equation?

73. $x^2 + 4x = -3$ **74.** $x^2 - 6x = 4$

75. $x^2 - 5x = 4$ **76.** $x^2 + 3x = 1$

Exercises 77–80: (Refer to Example 4.) Find the term that should be added to the expression to form a perfect square trinomial. Write the resulting perfect square trinomial in factored form.

77. $x^2 - 8x$ **78.** $x^2 - 5x$

79. $x^2 + 9x$ **80.** $x^2 + x$

Exercises 81–96: Solve by completing the square.

81. $x^2 - 2x = 24$ **82.** $x^2 - 2x + \frac{1}{2} = 0$

83. $x^2 + 6x - 2 = 0$ **84.** $x^2 - 16x = 5$

85. $x^2 - 3x = 5$ **86.** $x^2 + 5x = 2$

87. $x^2 - 5x + 1 = 0$ **88.** $x^2 - 9x + 7 = 0$

89. $x^2 - 4 = 2x$ **90.** $x^2 + 1 = 7x$

91. $2x^2 - 3x = 4$ **92.** $3x^2 + 6x - 5 = 0$

93. $4x^2 - 8x - 7 = 0$ **94.** $25x^2 - 20x - 1 = 0$

95. $36x^2 + 18x + 1 = 0$ **96.** $12x^2 + 8x - 2 = 0$

Exercises 97–106: Solve by any method.

97. $3x^2 + 12x = 36$ **98.** $6x^2 + 9x = 27$

99. $x^2 + 4x = -2$ **100.** $x^2 + 6x + 3 = 0$

101. $3x^2 - 4 = 2$ **102.** $-2x^2 + 3 = 1$

103. $-6x^2 + 70 = 16x$

104. $-15x^2 + 25x + 10 = 0$

105. $-3x(x - 8) = 6$ **106.** $-2x(4 - x) = 8$

Exercises 107 and 108: **Thinking Generally** *Solve for x. Assume a and c are positive.*

107. $ax^2 - c = 0$ **108.** $ax^2 + bx = 0$

SOLVING EQUATIONS BY MORE THAN ONE METHOD

Exercises 109–114: Solve the quadratic equation
  (a) symbolically,
  (b) graphically, and
  (c) numerically.

109. $x^2 - 3x - 18 = 0$ **110.** $\frac{1}{2}x^2 + 2x - 6 = 0$

111. $x^2 - 8x + 15 = 0$ **112.** $2x^2 + 3 = 7x$

113. $4(x^2 + 35) = 48x$ **114.** $4x(2 - x) = -5$

SOLVING AN EQUATION FOR A VARIABLE

Exercises 115–122: Solve for the specified variable.

115. $x = y^2 - 1$ for y

116. $x = 9y^2$ for y

117. $K = \frac{1}{2}mv^2$ for v

118. $c^2 = a^2 + b^2$ for b

119. $E = \dfrac{k}{r^2}$ for r

120. $W = I^2 R$ for I

121. $LC = \dfrac{1}{(2\pi f)^2}$ for f

122. $F = \dfrac{KmM}{r^2}$ for r

APPLICATIONS INVOLVING QUADRATIC EQUATIONS

123. **Safe Curve Speed** (Refer to Example 8.) Find a safe speed limit x for an airport taxiway curve with the given radius R by using $R = \frac{1}{2}x^2$.
    (a) $R = 450$ feet
    (b) $R = 800$ feet

124. **Braking Distance** The braking distance y in feet that it takes for a car to stop on wet, level pavement can be estimated by $y = \frac{1}{9}x^2$, where x is the speed of the car in miles per hour. Find the speed associated with each braking distance. (Source: L. Haefner, Introduction to Transportation Systems.)
    (a) 25 feet
    (b) 361 feet
    (c) 784 feet

125. **Falling Object** (Refer to Example 3.) How long does it take for a toy to hit the ground if it is dropped out of a window 60 feet above the ground? Does it take twice as long as it takes to fall from a window 30 feet above the ground?

126. **Falling Object** If a metal ball is thrown *downward* with an initial velocity of 22 feet per second (15 mph) from a 100-foot water tower, its height h in feet above the ground after t seconds is modeled by
$$h(t) = -16t^2 - 22t + 100.$$
    (a) Determine symbolically when the height of the ball is 62 feet.
    (b) Support your result in part (a) either graphically or numerically.
    (c) If the ball is thrown *upward* at 22 feet per second, then its height is given by
$$h(t) = -16t^2 + 22t + 100.$$
    Determine when the height of the ball is 80 feet.

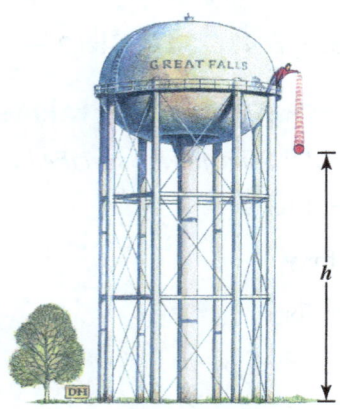

Exercises 127–130: **Height of a Ball** *(Refer to the Connecting Concepts with Your Life discussion in this section.) Suppose that a person throws a ball into the air and the ball's height in feet after t seconds is given by $h(t) = -16t^2 + 48t + 4$. If possible, find the time(s) when the ball reaches the given height.*

127. 24 feet

128. 40 feet

129. 50 feet

130. 4 feet

Exercises 131 and 132: **Television Size** *(Refer to the introduction of this chapter.) The size S of the television screen recommended for a person who sits x feet from the screen ($6 \leq x \leq 16$) is given by*
$$S(x) = -0.227x^2 + 8.155x - 8.8.$$
If a person buys a television set with a size S screen, how far from the screen should the person sit?

131. $S = 42$ inches

132. $S = 50$ inches

133. **Distance** Two athletes start jogging at the same time from the same location. One jogs north at 6 miles per hour while the second jogs east at 8 miles per hour. After how long are the two athletes 20 miles apart?

134. **Geometry** A triangle has an area of 35 square inches, and its base is 3 inches more than its height. Find the base and height of the triangle.

135. **Construction** A rectangular plot of land has an area of 520 square feet and is 6 feet longer than it is wide.
    (a) Write a quadratic equation in the form $ax^2 + bx + c = 0$, whose solution gives the width of the rectangular plot of land.
    (b) Solve the equation. What is the width?

136. **Modeling Motion** The height y in feet of a tennis ball after x seconds is shown in the graph. Estimate when the ball is 25 feet above the ground.

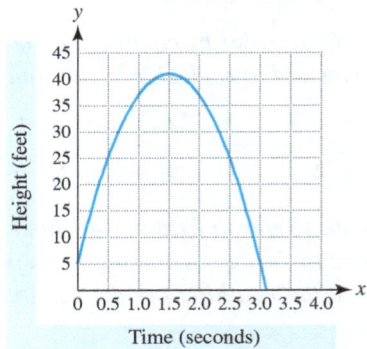

137. **Seedling Growth** The heights of melon seedlings grown at different temperatures are shown in the graph at the top of the next column. At what temperatures are the heights of the seedlings about 22 centimeters?

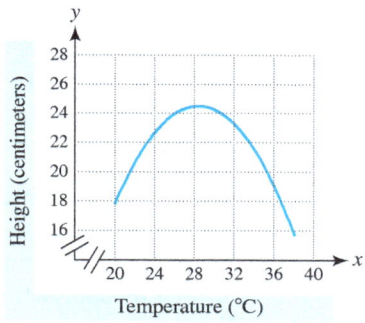

138. U.S. Population The three tables show the population of the United States in millions from 1800 through 2010 for selected years.

Year	1800	1820	1840	1860
Population	5	10	17	31

Year	1880	1900	1920	1940
Population	50	76	106	132

Year	1960	1980	2000	2010
Population	179	226	269	308

Source: U.S. Census Bureau.

(a) Without plotting the data, how do you know that the data are nonlinear?

(b) These data are modeled (approximately) by

$$f(x) = 0.0066(x - 1800)^2 + 5.$$

Find the vertex of the graph of f and interpret it.

(c) Estimate when the U.S. population reached 85 million.

139. Facebook Visitors The number of unique monthly U.S. visitors V in millions to Facebook x years after January 2011 can be modeled by

$$V(x) = -\frac{1}{2}(x - 9)^2 + 188,$$

where $0 \leq x \leq 5$. Determine the year when this number of visitors was 170 million.

140. Federal Debt The federal debt D in trillions of dollars held by foreign and international investors x years after 1970 can be modeled by

$$D(x) = \frac{1}{320}x^2 - \frac{3}{80}x,$$

where $15 \leq x \leq 50$.

(a) Evaluate $D(40)$ and interpret the result.

(b) Determine the year when the federal debt held by foreign and international investors first reached $500 billion ($0.5 trillion).

WRITING ABOUT MATHEMATICS

141. Suppose that you are asked to solve

$$ax^2 + bx + c = 0.$$

Explain how the graph of $y = ax^2 + bx + c$ can be used to find any real solutions to the equation.

142. Explain why a quadratic equation could not have more than two solutions. (*Hint:* Consider the graph of $y = ax^2 + bx + c$.)

Group Activity

Video Games *Exercises 1 and 2: In older video games with two-dimensional graphics, the background is often translated to give the illusion that a character in the game is moving. The simple scene on the left shows a mountain and an airplane. To make it appear that the airplane is flying, the mountain can be translated to the left, as shown in the figure on the right.* (*Source:* C. Pokorny and C. Gerald, *Computer Graphics.*)

1. Suppose that the mountain in the figure on the left is modeled by $f(x) = -0.4x^2 + 4$ and that the airplane is located at the point $(1, 5)$.

   (a) Graph f in $[-4, 4, 1]$ by $[0, 6, 1]$, where the units are kilometers. Plot the point $(1, 5)$ to show the location of the airplane.

   (b) Assume that the airplane is moving horizontally to the right at 0.2 kilometer per second. To give a video game player the illusion that the airplane is moving, graph the image of the mountain and the position of the airplane after 10 seconds.

2. Discuss how you could create the illusion of the airplane moving to the left and gaining altitude as it passes over the mountain. Try to perform a translation of this type. Explain your reasoning.

18.4 The Quadratic Formula

Objectives

1. Solving Quadratic Equations
   - Derivation of the Quadratic Formula
   - Solving Equations with the Quadratic Formula
2. Finding and Applying the Discriminant
3. Using Intercepts to Graph Quadratic Functions
4. Solving Quadratic Equations That Have Complex Solutions

NEW VOCABULARY

☐ Quadratic formula
☐ Discriminant

STUDY TIP

Memorize the quadratic formula.

1 Solving Quadratic Equations

DERIVATION OF THE QUADRATIC FORMULA Thus far, we have solved quadratic equations by factoring, the square root property, and completing the square. In this subsection, we derive a formula that can be used to solve *any* quadratic equation. To do this, we solve the general quadratic equation $ax^2 + bx + c = 0$ for x by completing the square. The resulting formula is called the **quadratic formula**. We assume that $a > 0$ and derive this formula as follows.

$ax^2 + bx + c = 0$	Quadratic equation
$ax^2 + bx = -c$	Subtract c.
$x^2 + \dfrac{b}{a}x = -\dfrac{c}{a}$	Divide by a.
$x^2 + \dfrac{b}{a}x + \dfrac{b^2}{4a^2} = -\dfrac{c}{a} + \dfrac{b^2}{4a^2}$	Add $\left(\dfrac{b/a}{2}\right)^2 = \dfrac{b^2}{4a^2}$.
$\left(x + \dfrac{b}{2a}\right)^2 = -\dfrac{c}{a} + \dfrac{b^2}{4a^2}$	Perfect square trinomial
$\left(x + \dfrac{b}{2a}\right)^2 = -\dfrac{c \cdot 4a}{a \cdot 4a} + \dfrac{b^2}{4a^2}$	Multiply $-\dfrac{c}{a}$ by $\dfrac{4a}{4a}$.
$\left(x + \dfrac{b}{2a}\right)^2 = -\dfrac{4ac}{4a^2} + \dfrac{b^2}{4a^2}$	Simplify.
$\left(x + \dfrac{b}{2a}\right)^2 = \dfrac{-4ac + b^2}{4a^2}$	Add fractions.
$\left(x + \dfrac{b}{2a}\right)^2 = \dfrac{b^2 - 4ac}{4a^2}$	Rewrite.
$x + \dfrac{b}{2a} = \pm\sqrt{\dfrac{b^2 - 4ac}{4a^2}}$	Square root property
$x = -\dfrac{b}{2a} \pm \sqrt{\dfrac{b^2 - 4ac}{4a^2}}$	Add $-\dfrac{b}{2a}$.
$x = -\dfrac{b}{2a} \pm \dfrac{\sqrt{b^2 - 4ac}}{2a}$	Property of square roots
$x = \dfrac{-b \pm \sqrt{b^2 - 4ac}}{2a}$	Combine fractions.

QUADRATIC FORMULA

The solutions to $ax^2 + bx + c = 0$ with $a \neq 0$ are given by

$$x = \dfrac{-b \pm \sqrt{b^2 - 4ac}}{2a}.$$

SOLVING EQUATIONS WITH THE QUADRATIC FORMULA The quadratic formula can be used to solve *any* quadratic equation. It always "works."

EXAMPLE 1   Solving a quadratic equation that has two solutions

Solve the equation $2x^2 - 3x - 1 = 0$. Support your results graphically.

Solution
Symbolic Solution Note that $a = 2$, $b = -3$, and $c = -1$ in $2x^2 - 3x - 1 = 0$.

$$x = \frac{-b \pm \sqrt{b^2 - 4ac}}{2a} \qquad \text{Quadratic formula}$$

$$x = \frac{-(-3) \pm \sqrt{(-3)^2 - 4(2)(-1)}}{2(2)} \qquad \text{Substitute for } a, b, \text{ and } c.$$

$$x = \frac{3 \pm \sqrt{17}}{4} \qquad \text{Simplify.}$$

The solutions are $\frac{3 + \sqrt{17}}{4} \approx 1.78$ and $\frac{3 - \sqrt{17}}{4} \approx -0.28$.

Graphical Solution The graph of $y = 2x^2 - 3x - 1$ is shown in **FIGURE 18.34**. Note that the two x-intercepts correspond to the two solutions for $2x^2 - 3x - 1 = 0$. Estimating from this graph, we see that the solutions are approximately -0.25 and 1.75, which supports our symbolic solution. (You could also use a graphing calculator to find the x-intercepts.)

Two Solutions

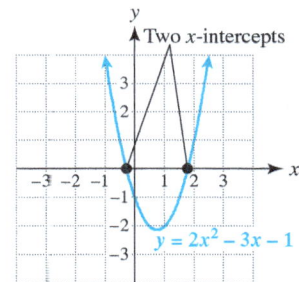

FIGURE 18.34

Now Try Exercise 9

CRITICAL THINKING 1

Use the equation and results from Example 1 to evaluate each expression mentally.

$2\left(\frac{3 + \sqrt{17}}{4}\right)^2 - 3\left(\frac{3 + \sqrt{17}}{4}\right) - 1$ and $2\left(\frac{3 - \sqrt{17}}{4}\right)^2 - 3\left(\frac{3 - \sqrt{17}}{4}\right) - 1$ Both equal 0.

EXAMPLE 2   Solving a quadratic equation that has one solution

Solve the equation $25x^2 + 20x + 4 = 0$. Support your result graphically.

Solution
Symbolic Solution Note that $a = 25$, $b = 20$, and $c = 4$ in $25x^2 + 20x + 4 = 0$.

$$x = \frac{-b \pm \sqrt{b^2 - 4ac}}{2a} \qquad \text{Quadratic formula}$$

$$= \frac{-20 \pm \sqrt{20^2 - 4(25)(4)}}{2(25)} \qquad \text{Substitute for } a, b, \text{ and } c.$$

$$= \frac{-20 \pm \sqrt{0}}{50} \qquad \text{Simplify.}$$

$$= \frac{-20}{50} = -0.4 \qquad \sqrt{0} = 0$$

There is one solution, -0.4.

One Solution

FIGURE 18.35

Graphical Solution The graph of $y = 25x^2 + 20x + 4$ is shown in **FIGURE 18.35**. Note that the one x-intercept, $(-0.4, 0)$, corresponds to the solution to $25x^2 + 20x + 4 = 0$.

Now Try Exercise 11

> **EXAMPLE 3** **Recognizing a quadratic equation that has no real solutions**
>
> Solve the equation $5x^2 - x + 3 = 0$. Support your result graphically.
>
> **Solution**
> **Symbolic Solution** Note that $a = 5, b = -1$, and $c = 3$ in $5x^2 - 1x + 3 = 0$.
>
> $$x = \frac{-b \pm \sqrt{b^2 - 4ac}}{2a} \qquad \text{Quadratic formula}$$
>
> $$= \frac{-(-1) \pm \sqrt{(-1)^2 - 4(5)(3)}}{2(5)} \qquad \text{Substitute for } a, b, \text{ and } c.$$
>
> $$= \frac{1 \pm \sqrt{-59}}{10} \qquad \text{Simplify.}$$
>
> There are *no real solutions* to this equation because $\sqrt{-59}$ *is not a real number*. (Later in this section, we discuss how to find complex solutions to quadratic equations like this one.)
>
> **Graphical Solution** The graph of $y = 5x^2 - x + 3$ is shown in **FIGURE 18.36**. There are no *x*-intercepts, indicating that the equation $5x^2 - x + 3 = 0$ has no real solutions.
>
> **Now Try Exercise 13**

No Real Solutions

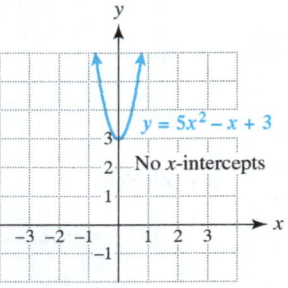

FIGURE 18.36

The following Making Connections shows three ways to solve a quadratic equation.

MAKING CONNECTIONS 1

Three Ways to Solve $x^2 - x - 2 = 0$

Symbolic Solution

Solve $x^2 - x - 2 = 0$.

$$x = \frac{-(-1) \pm \sqrt{(-1)^2 - 4(1)(-2)}}{2(1)}$$

$$x = \frac{1 \pm 3}{2}$$

$$x = 2, -1$$

Solutions are -1 and 2.

Numerical Solution

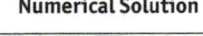

$x^2 - x - 2 = 0$ when
$x = -1$ or 2.

Graphical Solution

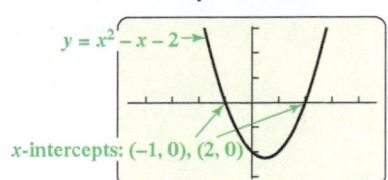

x-intercepts: $(-1, 0), (2, 0)$

The *x*-intercepts are $(-1, 0)$ and $(2, 0)$.

The solutions are -1 and 2.

Each method gives the same solutions.

 Connecting Concepts with Your Life To model the stopping distance of a car, we need to compute two quantities. The first quantity is the *reaction distance*, which is the distance a car travels from the time a driver first recognizes a hazard until the brakes are applied. The second quantity is *braking distance*, which is the distance a car travels after a driver applies the brakes. *Stopping distance* equals the sum of the reaction distance and the braking distance. If a car is traveling x miles per hour, we can estimate the reaction distance in feet as $\frac{11}{3}x$ and the braking distance in feet as $\frac{1}{9}x^2$. See **FIGURE 18.37** at the top of the next page.
(*Source:* L. Haefoer *Introduction to Transportation Systems.*)

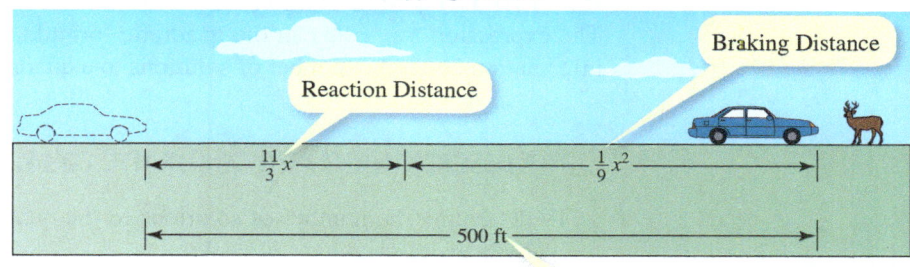

FIGURE 18.37

To estimate the total stopping distance d in feet, add the two expressions to obtain

$$d(x) = \frac{1}{9}x^2 + \frac{11}{3}x.$$

If a car's headlights don't illuminate the road beyond **500** feet, a safe nighttime speed limit x for the car can be determined by solving the quadratic equation

$$\underbrace{\frac{1}{9}x^2}_{\text{Braking Distance}} + \underbrace{\frac{11}{3}x}_{\text{Reaction Distance}} = \underbrace{500}_{\text{Stopping Distance}}. \quad \text{Quadratic equation}$$

EXAMPLE 4 **Modeling stopping distance**

If a car's headlights do not illuminate the road beyond 500 feet, estimate a safe nighttime speed limit x for the car by solving $\frac{1}{9}x^2 + \frac{11}{3}x = 500$.

Solution
Begin by subtracting 500 from each side of the given equation.

$$\frac{1}{9}x^2 + \frac{11}{3}x - 500 = 0 \quad \text{Subtract 500.}$$

To eliminate fractions, multiply each side by the LCD, which is 9. (This step is not necessary, but it makes the problem easier to work.)

$$x^2 + 33x - 4500 = 0 \quad \text{Multiply by 9.}$$

Now let $a = 1$, $b = 33$, and $c = -4500$ in the quadratic formula.

$$\begin{aligned} x &= \frac{-b \pm \sqrt{b^2 - 4ac}}{2a} & \text{Quadratic formula} \\ &= \frac{-33 \pm \sqrt{33^2 - 4(1)(-4500)}}{2(1)} & \text{Substitute for } a, b, \text{ and } c. \\ &= \frac{-33 \pm \sqrt{19{,}089}}{2} & \text{Simplify.} \end{aligned}$$

The solutions are

$$\frac{-33 + \sqrt{19{,}089}}{2} \approx 52.6 \quad \text{and} \quad \frac{-33 - \sqrt{19{,}089}}{2} \approx -85.6.$$

The negative solution has no physical meaning because negative speeds are not possible. The other solution is 52.6, so an appropriate speed limit might be 50 miles per hour.

Now Try Exercise 115

STUDY TIP

The quadratic formula can be used to solve *any* quadratic equation.

2 Finding and Applying the Discriminant

The expression $b^2 - 4ac$ in the quadratic formula is called the **discriminant**. It provides information about the number of solutions to a quadratic equation.

> ### THE DISCRIMINANT AND QUADRATIC EQUATIONS
>
> To determine the number of solutions to the quadratic equation $ax^2 + bx + c = 0$, evaluate the discriminant $b^2 - 4ac$.
>
> 1. If $b^2 - 4ac > 0$, there are two real solutions.
> 2. If $b^2 - 4ac = 0$, there is one real solution.
> 3. If $b^2 - 4ac < 0$, there are no real solutions; there are two complex solutions.

READING CHECK 1

- How many real solutions does $ax^2 + bx + c = 0$ have, if $b^2 - 4ac = 0$?

The following See the Concept shows how the discriminant gives information about the graph of $y = ax^2 + bx + c$ and the quadratic equation $ax^2 + bx + c = 0$.

See the Concept 1 GRAPHS, EQUATIONS, AND THE DISCRIMINANT

$b^2 - 4ac < 0$: No x-intercepts

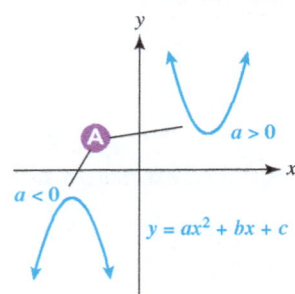

$b^2 - 4ac = 0$: One x-intercept

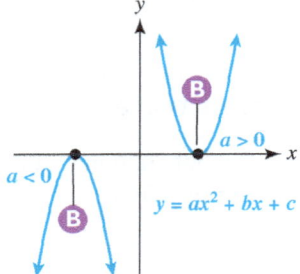

$b^2 - 4ac > 0$: Two x-intercepts

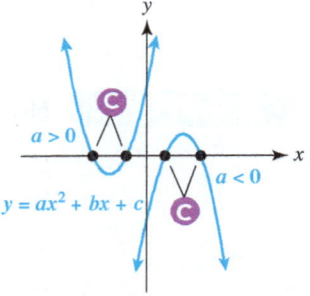

A When $b^2 - 4ac < 0$, the graph of $y = ax^2 + bx + c$ has no x-intercepts and the quadratic equation $ax^2 + bx + c = 0$ has no real solutions.

B When $b^2 - 4ac = 0$, the graph of $y = ax^2 + bx + c$ has one x-intercept and the quadratic equation $ax^2 + bx + c = 0$ has one real solution.

C When $b^2 - 4ac > 0$, the graph of $y = ax^2 + bx + c$ has two x-intercepts and the quadratic equation $ax^2 + bx + c = 0$ has two real solutions.

WATCH VIDEO IN MML.

EXAMPLE 5 Analyzing graphs of quadratic functions

A graph of $f(x) = ax^2 + bx + c$ is shown in **FIGURE 18.38**.
(a) State whether $a > 0$ or $a < 0$.
(b) Solve the equation $ax^2 + bx + c = 0$.
(c) Determine whether the discriminant is positive, negative, or zero.

FIGURE 18.38

Solution
(a) The parabola opens downward, so $a < 0$.
(b) The graph of $f(x) = ax^2 + bx + c$ intersects the x-axis at $(-3, 0)$ and $(2, 0)$. So $f(-3) = 0$ and $f(2) = 0$. The solutions to $ax^2 + bx + c = 0$ are -3 and 2.
(c) There are two solutions, so the discriminant is positive.

Now Try Exercise 43

EXAMPLE 6   Using the discriminant

Use the discriminant to determine the number of solutions to $4x^2 + 25 = 20x$. Then solve the equation, using the quadratic formula.

Solution
Write the equation as $4x^2 - 20x + 25 = 0$ so that $a = 4$, $b = -20$, and $c = 25$. The discriminant evaluates to
$$b^2 - 4ac = (-20)^2 - 4(4)(25) = 0.$$
Thus there is **one real solution**.

$$x = \frac{-b \pm \sqrt{b^2 - 4ac}}{2a} \qquad \text{Quadratic formula}$$

$$= \frac{-(-20) \pm \sqrt{0}}{2(4)} \qquad \text{Substitute.}$$

$$= \frac{20}{8} = 2.5 \qquad \text{Simplify.}$$

The only solution is 2.5.

Now Try Exercise 49(a), (b)

3 Using Intercepts to Graph Quadratic Functions

Instead of using vertex form to graph a quadratic function, we can first plot key points on the parabola to determine its position in the xy-plane. Key points include any intercepts and the vertex. The discriminant can be used to determine the number of x-intercepts. This method is demonstrated in the next example.

EXAMPLE 7   Analyzing the graph of $y = ax^2 + bx + c$

Let $f(x) = -\frac{1}{2}x^2 + x + \frac{3}{2}$.
(a) Does the graph of $y = f(x)$ open upward or downward? Is this graph wider or narrower than the graph of $y = x^2$?
(b) Find the axis of symmetry and the vertex.
(c) Find the discriminant and determine the number of x-intercepts.
(d) Find the y-intercept and any x-intercepts.
(e) Sketch a graph of $y = f(x)$.

Solution
(a) If $f(x) = -\frac{1}{2}x^2 + x + \frac{3}{2}$, then $a = -\frac{1}{2}$, $b = 1$, and $c = \frac{3}{2}$. Because $a = -\frac{1}{2} < 0$, the parabola opens downward. Also, because $0 < |a| < 1$, the graph is wider than the graph of $y = x^2$.

(b) The axis of symmetry is $x = -\dfrac{b}{2a} = -\dfrac{1}{2\left(-\frac{1}{2}\right)} = 1$, or $x = 1$. Because

$$f(1) = -\frac{1}{2}(1)^2 + (1) + \frac{3}{2} = -\frac{1}{2} + 1 + \frac{3}{2} = 2,$$

the vertex is $(1, 2)$.

(c) To find the x-intercepts, we must find the solutions to
$$-\tfrac{1}{2}x^2 + x + \tfrac{3}{2} = 0.$$
Because $a = -\tfrac{1}{2}, b = 1$, and $c = \tfrac{3}{2}$, the discriminant equals
$$b^2 - 4ac = (1)^2 - 4\left(-\tfrac{1}{2}\right)\left(\tfrac{3}{2}\right) = 4 > 0.$$
Thus there are two x-intercepts to plot.

(d) To find the y-intercept, evaluate $f(0)$ because $x = 0$ at the y-intercept.
$$f(0) = -\tfrac{1}{2}(0)^2 + (0) + \tfrac{3}{2} = \tfrac{3}{2}$$
The y-intercept is $\left(0, \tfrac{3}{2}\right)$. Factoring, completing the square, or the quadratic formula can be used to find the x-intercepts. We use factoring and let $y = 0$ to solve for x.

$$\begin{aligned}
-\tfrac{1}{2}x^2 + x + \tfrac{3}{2} &= 0 && \text{Equation to be solved} \\
x^2 - 2x - 3 &= 0 && \text{Multiply by } -2; \text{ clear fractions.} \\
(x + 1)(x - 3) &= 0 && \text{Factor.} \\
x + 1 = 0 \ \text{ or } \ x - 3 &= 0 && \text{Zero-product property} \\
x = -1 \ \text{ or } \ x &= 3 && \text{Solve.}
\end{aligned}$$

The x-intercepts are $(-1, 0)$ and $(3, 0)$

(e) Start by plotting the vertex and intercepts, as shown in **FIGURE 18.39**. Then sketch a smooth, ∩-shaped graph that connects these points.

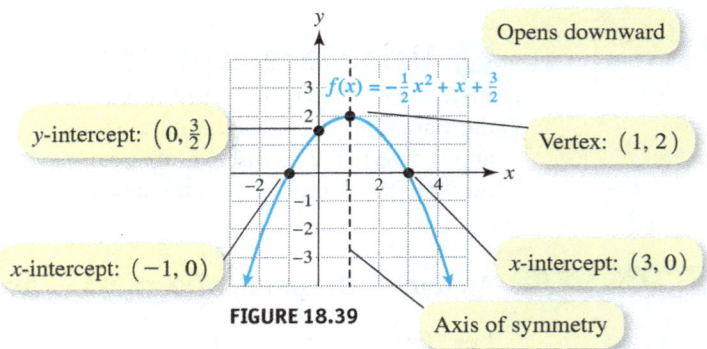

FIGURE 18.39

Now Try Exercise 57

4 Solving Quadratic Equations That Have Complex Solutions

A quadratic equation written in the form $ax^2 + bx + c = 0$ has no real solutions if the discriminant, $b^2 - 4ac$, is negative. For example, the quadratic equation given by $x^2 + 4 = 0$ has $a = 1, b = 0$, and $c = 4$. Its discriminant is
$$b^2 - 4ac = 0^2 - 4(1)(4) = -16 < 0,$$
so this equation has no real solutions. However, if we use complex numbers, we can solve this equation as follows.

$$\begin{aligned}
x^2 + 4 &= 0 && \text{Given equation} \\
x^2 &= -4 && \text{Subtract 4.} \\
x &= \pm\sqrt{-4} && \text{Square root property} \\
x = \sqrt{-4} \ \text{ or } \ x &= -\sqrt{-4} && \text{Meaning of } \pm \\
x = 2i \ \text{ or } \ x &= -2i && \text{The expression } \sqrt{-a}
\end{aligned}$$

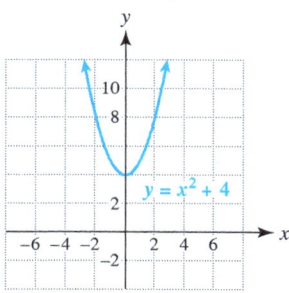

FIGURE 18.40

The solutions are $\pm 2i$. We check each solution to $x^2 + 4 = 0$ as follows.

$$(2i)^2 + 4 = (2)^2 i^2 + 4 = 4(-1) + 4 = 0 \checkmark \quad \text{It checks.}$$
$$(-2i)^2 + 4 = (-2)^2 i^2 + 4 = 4(-1) + 4 = 0 \checkmark \quad \text{It checks.}$$

The fact that the equation $x^2 + 4 = 0$ has only imaginary solutions is apparent from the graph of $y = x^2 + 4$, shown in **FIGURE 18.40**. This parabola does not intersect the x-axis, so the equation $x^2 + 4 = 0$ has no real solutions.

These results can be generalized as follows.

> **THE EQUATION** $x^2 + k = 0$
>
> If $k > 0$, the solutions to $x^2 + k = 0$ are given by $x = \pm i\sqrt{k}$.

NOTE: This result is a form of the *square root property* that includes complex solutions. ∎

EXAMPLE 8 Solving a quadratic equation that has complex solutions

Solve $x^2 + 5 = 0$.

Solution
The solutions are $\pm i\sqrt{5}$. That is, $x = i\sqrt{5}$ or $x = -i\sqrt{5}$.

Now Try Exercise 63

When $b \neq 0$, the preceding method cannot be used. Consider the quadratic equation $2x^2 + x + 3 = 0$, which has $a = 2$, $b = 1$, and $c = 3$. Its discriminant is negative.

$$b^2 - 4ac = 1^2 - 4(2)(3) = -23 < 0$$

This equation has two complex solutions, as demonstrated in the next example.

EXAMPLE 9 Solving a quadratic equation that has complex solutions

Solve $2x^2 + x + 3 = 0$. Write your answer in standard form: $a + bi$.

CALCULATOR HELP

To set your calculator in $a + bi$ mode or to access the imaginary unit i, see Appendix A (page AP-7).

Solution
Let $a = 2$, $b = 1$, and $c = 3$.

$$\begin{aligned}
x &= \frac{-b \pm \sqrt{b^2 - 4ac}}{2a} & &\text{Quadratic formula} \\
&= \frac{-1 \pm \sqrt{1^2 - 4(2)(3)}}{2(2)} & &\text{Substitute for } a, b, \text{ and } c. \\
&= \frac{-1 \pm \sqrt{-23}}{4} & &\text{Simplify.} \\
&= \frac{-1 \pm i\sqrt{23}}{4} & &\sqrt{-23} = i\sqrt{23} \\
&= -\frac{1}{4} \pm i\frac{\sqrt{23}}{4} & &\text{Property of fractions}
\end{aligned}$$

The solutions are $-\frac{1}{4} + i\frac{\sqrt{23}}{4}$ and $-\frac{1}{4} - i\frac{\sqrt{23}}{4}$.

Now Try Exercise 79

CRITICAL THINKING 2

Use the results of Example 9 to evaluate each expression mentally.

$$2\left(-\tfrac{1}{4} + i\tfrac{\sqrt{23}}{4}\right)^2 + \left(-\tfrac{1}{4} + i\tfrac{\sqrt{23}}{4}\right) + 3 \quad \text{and} \quad 2\left(-\tfrac{1}{4} - i\tfrac{\sqrt{23}}{4}\right)^2 + \left(-\tfrac{1}{4} - i\tfrac{\sqrt{23}}{4}\right) + 3$$

Sometimes we can use properties of radicals to simplify a solution to a quadratic equation, as demonstrated in the next example.

EXAMPLE 10 Solving a quadratic equation that has complex solutions

Solve $\tfrac{3}{4}x^2 + 1 = x$. Write your answer in standard form: $a + bi$.

Solution
Begin by subtracting x from each side of the equation and then multiply by 4 to clear fractions. The resulting equation is $3x^2 - 4x + 4 = 0$. Substitute $a = 3$, $b = -4$, and $c = 4$ in the quadratic formula.

$$\begin{aligned}
x &= \frac{-b \pm \sqrt{b^2 - 4ac}}{2a} & &\text{Quadratic formula} \\
&= \frac{-(-4) \pm \sqrt{(-4)^2 - 4(3)(4)}}{2(3)} & &\text{Substitute.} \\
&= \frac{4 \pm \sqrt{-32}}{6} & &\text{Simplify.} \\
&= \frac{4 \pm 4i\sqrt{2}}{6} & &\sqrt{-32} = i\sqrt{32} = i\sqrt{16}\sqrt{2} = 4i\sqrt{2} \\
&= \frac{2}{3} \pm \frac{2}{3}i\sqrt{2} & &\text{Property of fractions; simplify.}
\end{aligned}$$

Now Try Exercise 85

In the next example, we use completing the square to obtain complex solutions.

EXAMPLE 11 Completing the square to find complex solutions

Solve $x(x + 2) = -2$ by completing the square.

Solution
After applying the distributive property, we obtain the equation $x^2 + 2x = -2$. Because $b = 2$, add $\left(\tfrac{b}{2}\right)^2 = \left(\tfrac{2}{2}\right)^2 = 1$ to each side of the equation.

$$\begin{aligned}
x^2 + 2x &= -2 & &\text{Equation to be solved} \\
x^2 + 2x + 1 &= -2 + 1 & &\text{Add 1 to each side.} \\
(x + 1)^2 &= -1 & &\text{Perfect square trinomial; add.} \\
x + 1 &= \pm\sqrt{-1} & &\text{Square root property} \\
x + 1 &= \pm i & &\sqrt{-1} = i, \text{ the imaginary unit} \\
x &= -1 \pm i & &\text{Add } -1 \text{ to each side.}
\end{aligned}$$

The solutions are $-1 + i$ and $-1 - i$.

Now Try Exercise 97

18.4 Putting It All Together

CONCEPT	EXPLANATION	EXAMPLES
Quadratic Formula	The quadratic formula can be used to solve *any* quadratic equation written as $ax^2 + bx + c = 0$. The solutions are given by $$x = \frac{-b \pm \sqrt{b^2 - 4ac}}{2a}.$$	For the equation $$2x^2 - 3x + 1 = 0$$ with $a = 2$, $b = -3$, and $c = 1$, the solutions are $$\frac{-(-3) \pm \sqrt{(-3)^2 - 4(2)(1)}}{2(2)} = \frac{3 \pm \sqrt{1}}{4} = 1, \frac{1}{2}.$$
The Discriminant	The expression $b^2 - 4ac$ is called the discriminant. 1. $b^2 - 4ac > 0$ indicates two real solutions. 2. $b^2 - 4ac = 0$ indicates one real solution. 3. $b^2 - 4ac < 0$ indicates no real solutions; rather, there are two complex solutions.	For the equation $$x^2 + 4x - 1 = 0$$ with $a = 1$, $b = 4$, and $c = -1$, the discriminant is $$b^2 - 4ac = 4^2 - 4(1)(-1) = 20 > 0,$$ indicating two real solutions.
Quadratic Formula and Complex Solutions	If the discriminant is negative $(b^2 - 4ac < 0)$, the two solutions are complex numbers that are not real numbers. If $k > 0$, the solutions to $x^2 + k = 0$ are given by $x = \pm i\sqrt{k}$.	Solve $2x^2 - x + 3 = 0$. $$x = \frac{-(-1) \pm \sqrt{(-1)^2 - 4(2)(3)}}{2(2)}$$ $$= \frac{1 \pm \sqrt{-23}}{4} = \frac{1}{4} \pm i\frac{\sqrt{23}}{4}$$ $x^2 + 7 = 0$ is equivalent to $x = \pm i\sqrt{7}$.

18.4 Exercises

CONCEPTS AND VOCABULARY

1. What is the quadratic formula used for?

2. What basic algebraic technique is used to derive the quadratic formula?

3. Write the discriminant.

4. If the discriminant evaluates to 0, what does that indicate about the quadratic equation?

5. Name four symbolic techniques for solving a quadratic equation.

6. Does every quadratic equation have at least one real solution? Explain.

7. Solve $x^2 - k = 0$, if $k > 0$.

8. Solve $x^2 + k = 0$, if $k > 0$.

USING THE QUADRATIC FORMULA

Exercises 9–14: Use the quadratic formula to solve the equation. Support your result graphically. If there are no real solutions, say so.

9. $2x^2 + 11x - 6 = 0$ 10. $x^2 + 2x - 24 = 0$

11. $-x^2 + 2x - 1 = 0$ 12. $3x^2 - x + 1 = 0$

13. $-2x^2 + x - 1 = 0$ 14. $-x^2 + 4x - 4 = 0$

Exercises 15–32: Solve by using the quadratic formula. If there are no real solutions, say so.

15. $x^2 - 6x - 16 = 0$ 16. $2x^2 - 9x + 7 = 0$

17. $4x^2 - x - 1 = 0$ 18. $-x^2 + 2x + 1 = 0$

19. $-3x^2 + 2x - 1 = 0$ 20. $x^2 + x + 3 = 0$

21. $36x^2 - 36x + 9 = 0$ 22. $4x^2 - 5.6x + 1.96 = 0$

23. $2x(x - 3) = 2$ 24. $x(x + 1) + x = 5$

25. $(x - 1)(x + 1) + 2 = 4x$

26. $\frac{1}{2}(x - 6) = x^2 + 1$ 27. $\frac{1}{2}x(x + 1) = 2x^2 - \frac{3}{2}$

28. $\frac{1}{2}x^2 - \frac{1}{4}x + \frac{1}{2} = x$ 29. $2x(x - 1) = 7$

30. $3x(x - 4) = 4$ 31. $-3x^2 + 10x - 5 = 0$

32. $-2x^2 + 4x - 1 = 0$

Exercises 33–42: Use the quadratic formula to find any x-intercepts on the graph of the function.

33. $f(x) = x^2 - 2x - 1$

34. $f(x) = x^2 + 3x + 1$

35. $f(x) = -2x^2 - x + 3$

36. $f(x) = -3x^2 - x + 4$

37. $f(x) = x^2 + x + 5$

38. $f(x) = 3x^2 - 2x + 5$

39. $f(x) = x^2 + 9$ 40. $f(x) = x^2 + 11$

41. $f(x) = 3x^2 + 4x - 2$

42. $f(x) = 4x^2 - 2x - 3$

IDENTIFYING THE DISCRIMINANT

Exercises 43–48: A graph of $y = ax^2 + bx + c$ is shown.

(a) State whether $a > 0$ or $a < 0$.
(b) Solve $ax^2 + bx + c = 0$, if possible.
(c) Determine whether the discriminant is positive, negative, or zero.

43.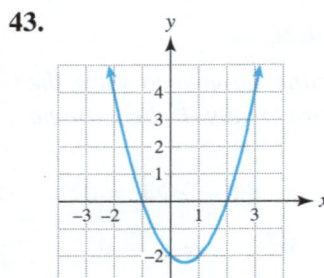

44.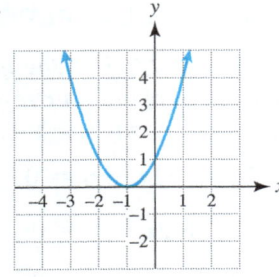

45.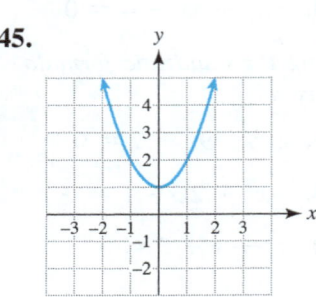

46.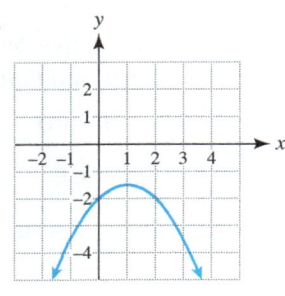

47.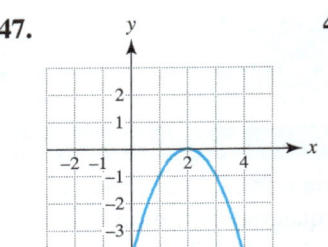

48.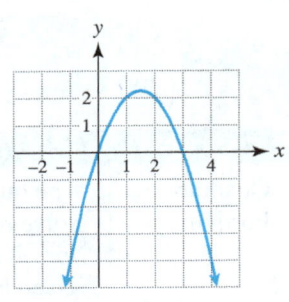

Exercises 49–56: Do the following for the given equation.

(a) Evaluate the discriminant.
(b) How many real solutions are there?
(c) Support your answer for part (b) graphically.

49. $3x^2 + x - 2 = 0$ 50. $5x^2 - 13x + 6 = 0$

51. $x^2 - 4x + 4 = 0$ 52. $\frac{1}{4}x^2 + 4 = 2x$

53. $\frac{1}{2}x^2 + \frac{3}{2}x + 2 = 0$ 54. $x - 3 = 2x^2$

55. $x(x + 3) = 3$

56. $(4x - 1)(x - 3) = -25$

GRAPHING QUADRATIC FUNCTIONS

Exercises 57–62: (Refer to Example 7.) Use the given $f(x)$ to complete the following.

(a) Does the graph of $y = f(x)$ open upward or downward? Is this graph wider, narrower, or the same as the graph of $y = x^2$?
(b) Find the axis of symmetry and the vertex.
(c) Find the discriminant and determine the number of x-intercepts.
(d) Find the y-intercept and any x-intercepts.
(e) Sketch a graph of $y = f(x)$.

57. $f(x) = \frac{1}{2}x^2 + x - \frac{3}{2}$ 58. $f(x) = -x^2 + 4x + 5$

59. $f(x) = 2x - x^2$ 60. $f(x) = x - 2x^2$

61. $f(x) = 2x^2 + 2x - 4$ 62. $f(x) = \frac{1}{2}x^2 - \frac{1}{2}x - 1$

FINDING COMPLEX SOLUTIONS

Exercises 63–94: Solve the equation. Write complex solutions in standard form.

63. $x^2 + 9 = 0$ 64. $x^2 + 16 = 0$

65. $x^2 + 80 = 0$ 66. $x^2 + 20 = 0$

67. $x^2 + \frac{1}{4} = 0$ 68. $x^2 + \frac{9}{4} = 0$

69. $16x^2 + 9 = 0$ 70. $25x^2 + 36 = 0$

71. $x^2 = -6$ 72. $x^2 = -75$

73. $x^2 - 3 = 0$
74. $x^2 - 8 = 0$
75. $x^2 + 2 = 0$
76. $x^2 + 4 = 0$
77. $x^2 - x + 2 = 0$
78. $x^2 + 2x + 3 = 0$
79. $2x^2 + 3x + 4 = 0$
80. $3x^2 - x = 1$
81. $x^2 + 1 = 4x$
82. $3x^2 + 2 = x$
83. $x(x + 1) = -2$
84. $x(x - 4) = -8$
85. $5x^2 + 2x + 4 = 0$
86. $7x^2 - 2x + 4 = 0$
87. $\frac{1}{2}x^2 + \frac{3}{4}x = -1$
88. $-\frac{1}{3}x^2 + x = 2$
89. $x(x + 2) = x - 4$
90. $x - 5 = 2x(2x + 1)$
91. $x(2x - 1) = 1 + x$
92. $2x = x(3 - 4x)$
93. $x^2 = x(1 - x) - 2$
94. $2x^2 = 2x(5 - x) - 8$

COMPLETING THE SQUARE

Exercises 95–100: Solve by completing the square.

95. $x^2 + 2x + 4 = 0$
96. $x^2 - 2x + 2 = 0$
97. $x(x + 4) = -5$
98. $x(8 - x) = 25$
99. $2x^2 - 4x + 6 = 0$
100. $2x^2 + 2x + 1 = 0$

USING THE METHOD OF YOUR CHOICE

Exercises 101–114: Find exact solutions to the quadratic equation, using a method of your choice. Explain why you chose the method you did. Answers may vary.

101. $x^2 - 3x + 2 = 0$
102. $x^2 + 2x + 1 = 0$
103. $0.5x^2 - 1.75x = 1$
104. $\frac{3}{5}x^2 + \frac{9}{10}x = \frac{3}{5}$
105. $x^2 - 5x + 2 = 0$
106. $2x^2 - x - 4 = 0$
107. $2x^2 + x = -8$
108. $4x^2 = 2x - 3$
109. $4x^2 - 1 = 0$
110. $3x^2 = 9$
111. $3x^2 + 6 = 0$
112. $4x^2 + 7 = 0$
113. $9x^2 + 1 = 6x$
114. $10x^2 + 15x = 25$

APPLICATIONS INVOLVING QUADRATIC EQUATIONS

Exercises 115–118: **Modeling Stopping Distance** *(Refer to Example 4.) Use $d = \frac{1}{9}x^2 + \frac{11}{3}x$ to find a safe speed x for the following stopping distances d.*

115. 42 feet
116. 152 feet
117. 390 feet
118. 726 feet

119. **Groupon's Growth** In its early years, Groupon experienced a dramatic increase in value from October 2010 to March 2011. The function
$$G(x) = 0.4x^2 + 1.8x + 6$$
approximates the company's value in billions of dollars x months after October 2010.
(a) Evaluate $G(0)$. Interpret the result.
(b) Determine when Groupon's value was about $15 billion.

120. **Foursquare Users** Foursquare provides a service that allows your friends to know your whereabouts by "checking in." From March 2010 to March 2011, it experienced amazing growth. The function
$$F(x) = \frac{1}{18}x^2 - \frac{1}{12}x + \frac{1}{2}$$
approximates Foursquare users in millions x months after March 2010.
(a) Evaluate $F(3)$. Interpret the result.
(b) Determine symbolically when Foursquare had 4.25 million users.

121. **U.S. AIDS Deaths** In the early years of the HIV/AIDS epidemic, the cumulative numbers in thousands of AIDS deaths from 1984 through 1994 may be modeled by
$$f(x) = 2.39x^2 + 5.04x + 5.1,$$
where $x = 0$ corresponds to 1984, $x = 1$ corresponds to 1985, and so on until $x = 10$ corresponds to 1994. See the accompanying graph. Use the formula for $f(x)$ to estimate the year when the total number of AIDS deaths reached 200 thousand. Compare your result with that shown in the graph.

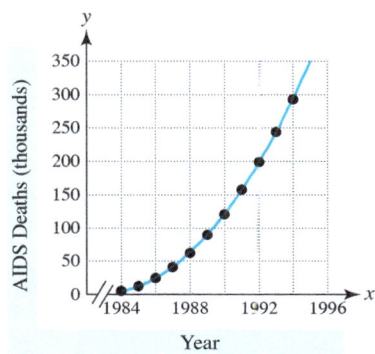

122. **Canoeing** A camper paddles a canoe 2 miles downstream in a river that has a 2-mile-per-hour current. To return to camp, the canoeist travels upstream on

a different branch of the river. It is 4 miles long and has a 1-mile-per-hour current. The total trip (both ways) takes 3 hours. Find the average speed of the canoe in still water. (*Hint:* Time equals distance divided by rate.)

123. **Historical Music Sales** From 1989 to 2009, the global music sales S of both compact discs and tapes in billions of dollars can be modeled by $S(x) = -0.095x^2 + 1.85x + 6$, where x is years after 1989. Estimate symbolically when sales were $12 billion. (*Source:* RIAA, Bain Analysis.)

124. **Monthly Facebook Visitors** The number of *unique* monthly U.S. Facebook visitors in millions x years after 2011 can be modeled by
$$V(x) = -0.5x^2 + 9x + 147,$$
where $0 \leq x \leq 5$. Assuming current trends continue, estimate symbolically when this number was 175 million.

125. **Airplane Speed** A pilot flies 500 miles against a 20-mile-per-hour wind. On the next day, the pilot flies back home with a 10-mile-per-hour tail wind. The total trip (both ways) takes 4 hours. Find the speed of the airplane without a wind.

126. **Distance** Two cars leave an intersection, one traveling south and one traveling east, as shown in the figure. After 1 hour, the two cars are 50 miles apart and the car traveling east has traveled 10 miles farther than the car traveling south. How far did each car travel?

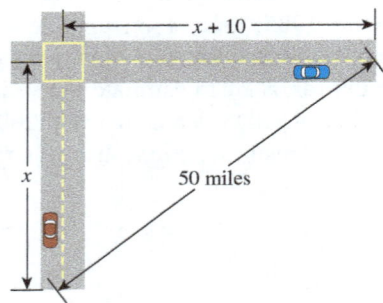

127. **Screen Dimensions** The width of a rectangular computer screen is 3 inches more than its height. If the area of the screen is 154 square inches, find its dimensions
    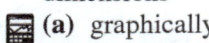 (a) graphically,
    (b) numerically, and
    (c) symbolically.

128. **Sidewalk Dimension** A rectangular flower garden in a park is 30 feet wide and 40 feet long. A sidewalk around the perimeter of the garden is being planned, as shown in the figure. The gardener has enough money to pour 624 square feet of cement sidewalk. Find the width of the sidewalk.

129. **Modeling Water Flow** When water runs out of a hole in a cylindrical container, the height of the water in the container can often be modeled by a quadratic function. The data in the table show the height y in centimeters of water at 30-second intervals in a metal container that has a small hole in it.

Time	0	30	60	90
Height	16	11.9	8.4	5.3

Time	120	150	180
Height	3.1	1.4	0.5

These data are modeled by
$$f(x) = 0.0004x^2 - 0.15x + 16.$$

(a) Explain why a linear function would not be appropriate for modeling these data.
(b) Use the table to estimate the time at which the height was 7 centimeters.
(c) Use the quadratic formula to solve part (b).

130. **Hospitals** The general trend in the number of hospitals in the United States from 1945 through 2000 is shown in the graph and can be modeled by
$$f(x) = -1.38x^2 + 84x + 5865,$$
where $x = 5$ corresponds to 1945, $x = 10$ corresponds to 1950, and so on until $x = 60$ represents 2000.

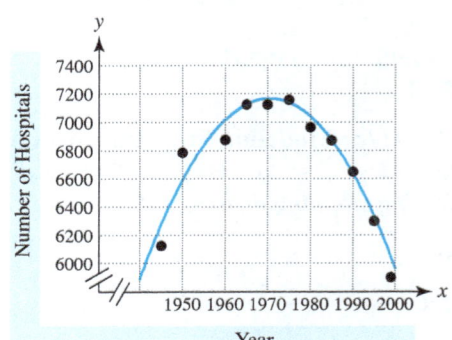

(a) Describe any trends in the numbers of hospitals from 1945 to 2000.
(b) What information does the vertex give?
(c) Use the formula for $f(x)$ to estimate the number of hospitals in 1970. Compare your result with that shown in the graph.
(d) Use the formula for $f(x)$ to estimate the year (or years) when there were 6300 hospitals. Compare your result with that shown in the graph.

WRITING ABOUT MATHEMATICS

131. Explain how the discriminant $b^2 - 4ac$ can be used to determine the number of solutions to a quadratic equation.

132. Let $f(x) = ax^2 + bx + c$ be a quadratic function. If you know the value of $b^2 - 4ac$, what information does this give you about the graph of $y = f(x)$? Explain your answer.

SECTIONS 18.3 and 18.4 — Checking Basic Concepts

1. Solve the quadratic equation $2x^2 - 7x + 3 = 0$ symbolically and graphically.

2. Use the square root property to solve $x^2 = 5$.

3. Complete the square to solve $x^2 - 4x + 1 = 0$.

4. Solve the equation $x^2 + y^2 = 1$ for y.

5. Use the quadratic formula to solve each equation.
(a) $2x^2 = 3x + 1$
(b) $9x^2 - 24x + 16 = 0$

6. Calculate the discriminant for each equation and give the number of *real* solutions.
(a) $x^2 - 5x + 5 = 0$
(b) $2x^2 - 5x + 4 = 0$
(c) $49x^2 - 56x + 16 = 0$

7. Solve each equation.
(a) $x^2 + 5 = 0$
(b) $x^2 + x + 3 = 0$

18.5 Quadratic Inequalities

Objectives
1. Recognizing Quadratic Inequalities
2. Finding Graphical and Numerical Solutions
3. Finding Symbolic Solutions

NEW VOCABULARY
☐ Quadratic inequality
☐ Test value

1 Recognizing Quadratic Inequalities

If the equals sign in a quadratic equation is replaced with $>$, $\geq$, $<$, or $\leq$, a **quadratic inequality** results. Examples of quadratic inequalities include

$$x^2 + 4x - 3 < 0, \quad 5x^2 \geq 5, \quad \text{and} \quad 1 - z \leq z^2.$$

Any quadratic equation can be written as

$$ax^2 + bx + c = 0, \quad a \neq 0, \qquad \text{Quadratic equation}$$

so any quadratic inequality can be written as

$$ax^2 + bx + c > 0, \quad a \neq 0, \qquad \text{Quadratic inequality}$$

where $>$ may be replaced with $\geq$, $<$, or $\leq$.

EXAMPLE 1 Identifying a quadratic inequality

Determine whether the inequality is quadratic.
(a) $5x + x^2 - x^3 \leq 0$ (b) $4 + 5x^2 > 4x^2 + x$

Solution
(a) The inequality $5x + x^2 - x^3 \leq 0$ is not quadratic because it has an x^3-term.
(b) Write the inequality as follows.

$4 + 5x^2 > 4x^2 + x$	Given inequality
$4 + 5x^2 - 4x^2 - x > 0$	Subtract $4x^2$ and x.
$4 + x^2 - x > 0$	Combine like terms.
$x^2 - x + 4 > 0$	Rewrite.

Because the inequality can be written in the form $ax^2 + bx + c > 0$ with $a = 1$, $b = -1$, and $c = 4$, it is a quadratic inequality.

Now Try Exercises 7, 11

READING CHECK 1

Which of the following are quadratic inequalities?
$2x - 5 < 2$
$x^2 - 1 > 0$
$2x^2 + 4x + 5 \leq 0$
$-4x^2 - 5 \geq -7x$
$x < 1$

2 Finding Graphical and Numerical Solutions

Equality often is the boundary between *greater than* and *less than*, so a first step in solving an inequality is to determine the x-values where equality occurs. We begin by using this concept with graphical techniques.

The graph of $y = x^2 - x - 2$ has x-intercepts $(-1, 0)$ and $(2, 0)$. See **FIGURE 18.41**. The solutions to $x^2 - x - 2 = 0$ are given by $x = -1$ or $x = 2$. Between the x-intercepts the graph dips **below** the x-axis and the y-values are **negative**. Thus solutions to $x^2 - x - 2 < 0$ satisfy $-1 < x < 2$. To support this result, we select a **test value**. For example, 0 lies between -1 and 2. If we substitute $x = 0$ in $x^2 - x - 2 < 0$, it results in a true statement.

$0^2 - 0 - 2 < 0$ ✓   True

Visualizing Quadratic Inequalities

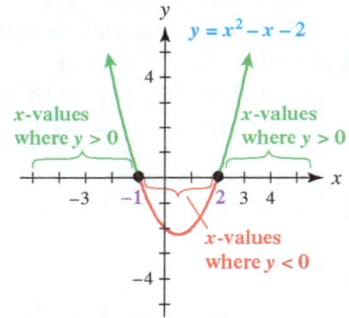

FIGURE 18.41

When $x < -1$ or $x > 2$, the graph lies **above** the x-axis and the y-values are **positive**. Thus the solutions to $x^2 - x - 2 > 0$ satisfy $x < -1$ or $x > 2$. For example, 3 is greater than 2 and -3 is less than -1. Therefore both 3 and -3 are solutions. We can verify this result by substituting 3 and -3 as test values in $x^2 - x - 2 > 0$.

$3^2 - 3 - 2 > 0$ ✓   True
$(-3)^2 - (-3) - 2 > 0$ ✓   True

In the next three examples, we use these concepts to solve quadratic inequalities.

EXAMPLE 2 Solving a quadratic inequality

Make a table of values for $y = x^2 - 3x - 4$ and then sketch the graph. Use the table and graph to solve $x^2 - 3x - 4 \leq 0$. Write your answer in interval notation.

Solution
The points calculated for **TABLE 18.14** are plotted in **FIGURE 18.42** and connected with a smooth U-shaped graph.

Numerical Solution TABLE 18.14 shows that $x^2 - 3x - 4$ equals 0 when $x = -1$ or $x = 4$. Between these values, $x^2 - 3x - 4$ is negative, so the solution set to $x^2 - 3x - 4 \leq 0$ is given by $-1 \leq x \leq 4$ or, in interval notation, $[-1, 4]$.

Graphical Solution In FIGURE 18.42, the graph of $y = x^2 - 3x - 4$ shows that the x-intercepts are $(-1, 0)$ and $(4, 0)$. Between these values, the graph dips **below** the x-axis. Thus the solution set is $[-1, 4]$.

STUDY TIP

Learn how to use a parabola to help solve a quadratic inequality.

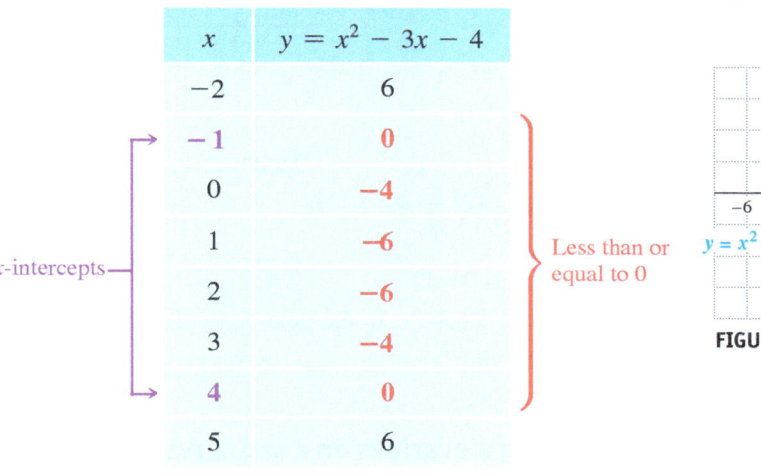

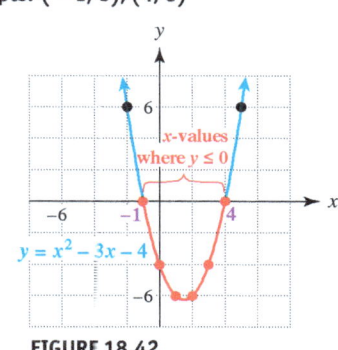

FIGURE 18.42

Now Try Exercises 19, 27

EXAMPLE 3 Solving a quadratic inequality

Solve $x^2 > 1$. Write your answer in interval notation.

Solution
First, we rewrite $x^2 > 1$ as $x^2 - 1 > 0$. The graph of $y = x^2 - 1$ is shown in FIGURE 18.43 with x-intercepts $(-1, 0)$ and $(1, 0)$. The graph lies **above** the x-axis and is shaded green to the left of $x = -1$ or to the right of $x = 1$. Thus the solution set is given by $x < -1$ or $x > 1$, which can be written in interval notation as $(-\infty, -1) \cup (1, \infty)$.

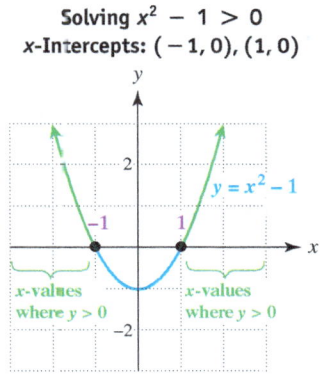

FIGURE 18.43

Now Try Exercise 49

EXAMPLE 4 Solving some special cases

Solve each of the inequalities graphically.
(a) $x^2 + 1 > 0$ (b) $x^2 + 1 < 0$ (c) $(x - 1)^2 \leq 0$

Solution
(a) Because the graph of $y = x^2 + 1$, shown in FIGURE 18.44, is always above the x-axis, $x^2 + 1$ is always greater than 0. The solution set includes all real numbers, or $(-\infty, \infty)$.
(b) Because the graph of $y = x^2 + 1$, shown in FIGURE 18.44, never goes below the x-axis, $x^2 + 1$ is never less than 0. Thus there are no real solutions.

FIGURE 18.44

(c) Because the graph of $y = (x - 1)^2$, shown in **FIGURE 18.45**, never goes below the x-axis, $(x - 1)^2$ is never less than 0. When $x = 1$, $y = 0$, so **1** is the only solution to the inequality $(x - 1)^2 \leq 0$.

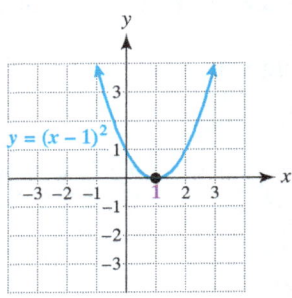

FIGURE 18.45

Now Try Exercises 51, 55, 57

Math in Context (Highway Design) In the next example, we show how quadratic inequalities are used in highway design.

EXAMPLE 5 Determining elevations on a sag curve

Parabolas are frequently used in highway design to model hills and sags (valleys) along a proposed route. Suppose that the elevation E in feet of a sag, or *sag curve*, is given by

$$E(x) = 0.00004x^2 - 0.4x + 2000,$$

where x is the horizontal distance in feet along the sag curve and $0 \leq x \leq 10{,}000$. See **FIGURE 18.46**. Estimate graphically the x-values where the elevation is 1500 feet or less. (*Source:* F. Mannering and W. Kilareski, *Principles of Highway Engineering and Traffic Analysis*.)

FIGURE 18.46

Solution

Graphical Solution We must solve the quadratic inequality

$$0.00004x^2 - 0.4x + 2000 \leq 1500.$$

Let $y_1 = 0.00004x^2 - 0.4x + 2000$ be the sag curve and $y_2 = 1500$ be a line with an elevation of 1500 feet. Their graphs intersect at $x \approx 1464$ and $x \approx 8536$, as shown in **FIGURE 18.47**. The elevation of the road is less than 1500 feet between these x-values. Therefore the elevation is 1500 feet or less when $1464 \leq x \leq 8536$ (approximately).

Determining Where the Elevation Is 1500 Feet

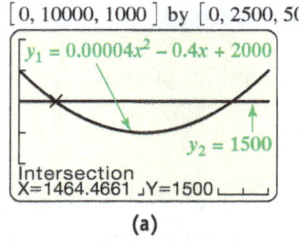   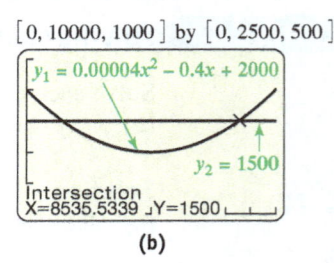

(a)   (b)

FIGURE 18.47

CALCULATOR HELP
To find a point of intersection, see Appendix A (page AP-6).

Now Try Exercise 65

The following chart summarizes visual solutions for several possibilities when solving quadratic equations and inequalities.

Visualizing Solutions to Quadratic Equations and Inequalities

Two x-Intercepts

Case 1: Upward Opening

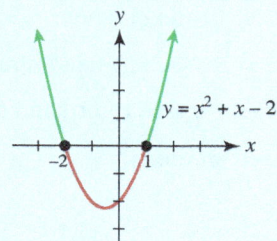

Equation/Inequality	Solutions
$x^2 + x - 2 = 0$	$x = -2, 1$
$x^2 + x - 2 < 0$	$-2 < x < 1$
$x^2 + x - 2 > 0$	$x < -2$ or $x > 1$

Case 2: Downward Opening

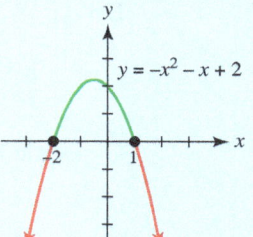

Equation/Inequality	Solutions
$-x^2 - x + 2 = 0$	$x = -2, 1$
$-x^2 - x + 2 < 0$	$x < -2$ or $x > 1$
$-x^2 - x + 2 > 0$	$-2 < x < 1$

One x-Intercept

Case 1: Upward Opening

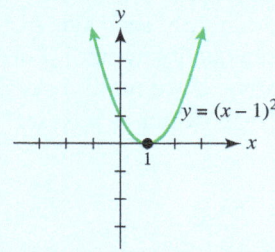

Equation/Inequality	Solutions
$(x - 1)^2 = 0$	$x = 1$
$(x - 1)^2 < 0$	no solutions
$(x - 1)^2 > 0$	$x < 1$ or $x > 1$

Case 2: Downward Opening

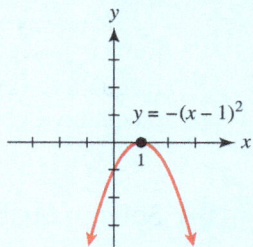

Equation/Inequality	Solutions
$-(x - 1)^2 = 0$	$x = 1$
$-(x - 1)^2 < 0$	$x < 1$ or $x > 1$
$-(x - 1)^2 > 0$	no solutions

No x-Intercepts

Case 1: Upward Opening

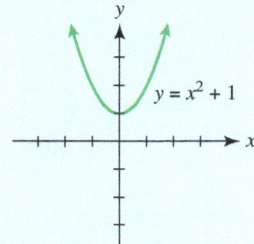

Equation/Inequality	Solutions
$x^2 + 1 = 0$	no solutions
$x^2 + 1 < 0$	no solutions
$x^2 + 1 > 0$	all real numbers

Case 2: Downward Opening

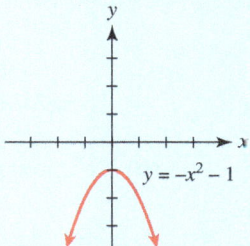

Equation/Inequality	Solutions
$-x^2 - 1 = 0$	no solutions
$-x^2 - 1 < 0$	all real numbers
$-x^2 - 1 > 0$	no solutions

3 Finding Symbolic Solutions

To solve a quadratic inequality symbolically, we first solve the corresponding equality. We can then write the solution to the inequality, using the following method.

SOLUTIONS TO QUADRATIC INEQUALITIES

Let $ax^2 + bx + c = 0$, $a > 0$, have two real solutions p and q, where $p < q$.

$ax^2 + bx + c < 0$ is equivalent to $p < x < q$ (see left-hand figure).

$ax^2 + bx + c > 0$ is equivalent to $x < p$ or $x > q$ (see right-hand figure).

Quadratic inequalities involving $\leq$ or $\geq$ can be solved similarly.

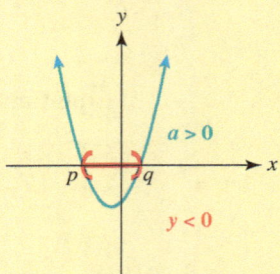
Solutions lie between p and q.

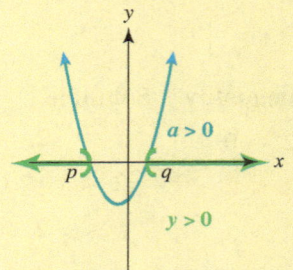
Solutions lie "outside" p and q.

One way to handle the situation where $a < 0$ is to multiply each side of the inequality by -1, in which case we must be sure to *reverse* the inequality symbol. For example, the inequality $-2x^2 + 8 \leq 0$ has $a = -2$, which is negative. If we multiply each side of this inequality by -1, we obtain $2x^2 - 8 \geq 0$ and now $a = 2$, which is positive.

EXAMPLE 6 Solving quadratic inequalities

Solve each inequality symbolically. Write your answer in interval notation.
(a) $6x^2 - 7x - 5 \geq 0$ (b) $x(3 - x) > -18$

Solution
(a) Begin by solving $6x^2 - 7x - 5 = 0$.

$6x^2 - 7x - 5 = 0$ Quadratic equation

$(2x + 1)(3x - 5) = 0$ Factor.

$2x + 1 = 0$ or $3x - 5 = 0$ Zero-product property

$x = -\frac{1}{2}$ or $x = \frac{5}{3}$ Solve.

Therefore solutions to $6x^2 - 7x - 5 \geq 0$ lie "**outside**" these two values and satisfy $x \leq -\frac{1}{2}$ or $x \geq \frac{5}{3}$. In interval notation, the solution set is $\left(-\infty, -\frac{1}{2}\right] \cup \left[\frac{5}{3}, \infty\right)$.

(b) First, rewrite the inequality as follows.

$x(3 - x) > -18$ Given inequality

$3x - x^2 > -18$ Distributive property

$3x - x^2 + 18 > 0$ Add 18.

$-x^2 + 3x + 18 > 0$ Rewrite.

$x^2 - 3x - 18 < 0$ Multiply by -1 because $a < 0$; reverse the inequality symbol.

CRITICAL THINKING 1

The graph of $y = -x^2 + x + 12$ is a parabola with x-intercepts $(-3, 0)$, and $(4, 0)$. Solve $-x^2 + x + 12 < 0$.

Next, solve $x^2 - 3x - 18 = 0$.

$$(x + 3)(x - 6) = 0 \quad \text{Factor.}$$
$$x = -3 \quad \text{or} \quad x = 6 \quad \text{Solve.}$$

Solutions to the inequality $x^2 - 3x - 18 < 0$ lie **between** these two values and satisfy $-3 < x < 6$. In interval notation, the solution set is $(-3, 6)$.

Now Try Exercise 31, 37

EXAMPLE 7 Finding the dimensions of a building

A rectangular building needs to be 7 feet longer than it is wide, as illustrated in **FIGURE 18.48**. The area of the building must be at least 450 square feet. What widths x are possible for this building? Support your results with a table of values.

FIGURE 18.48

Solution

Symbolic Solution If x is the width of the building, $x + 7$ is the length of the building and its area is $x(x + 7)$. The area must be at least 450 square feet, so the inequality $x(x + 7) \geq 450$ must be satisfied.

First solve the following quadratic equation.

$$x(x + 7) = 450 \quad \text{Quadratic equation}$$
$$x^2 + 7x = 450 \quad \text{Distributive property}$$
$$x^2 + 7x - 450 = 0 \quad \text{Subtract 450.}$$
$$x = \frac{-7 \pm \sqrt{7^2 - 4(1)(-450)}}{2(1)} \quad \text{Quadratic formula: } a = 1, b = 7, \text{ and } c = -450$$
$$= \frac{-7 \pm \sqrt{1849}}{2} \quad \text{Simplify.}$$
$$= \frac{-7 \pm 43}{2} \quad \sqrt{1849} = 43$$
$$= 18, -25 \quad \text{Evaluate.}$$

Thus the solutions to $x(x + 7) \geq 450$ are $x \leq -25$ or $x \geq 18$. The width is positive, so the building width must be 18 feet or more.

Numerical Solution A table of values is shown in **FIGURE 18.49**, where $y_1 = x(x + 7)$ equals 450 when $x = 18$. For $x \geq 18$, the area is *at least* 450 square feet.

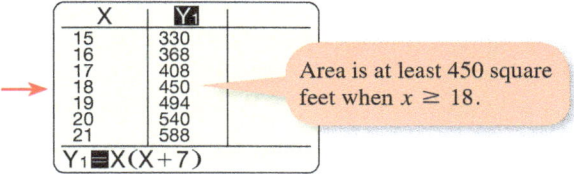

FIGURE 18.49

Now Try Exercise 63

18.5 Putting It All Together

CONCEPT	EXPLANATION
Solving a Quadratic Inequality Symbolically	Let $ax^2 + bx + c = 0$, $a > 0$, have two real solutions p and q, where $p < q$. $ax^2 + bx + c < 0$ is equivalent to $p < x < q$. $ax^2 + bx + c > 0$ is equivalent to $x < p$ or $x > q$. *Examples:* The solutions to $x^2 - 3x + 2 = 0$ are 1 and 2. The solutions to $x^2 - 3x + 2 < 0$ satisfy $1 < x < 2$. The solutions to $x^2 - 3x + 2 > 0$ satisfy $x < 1$ or $x > 2$.
Solving a Quadratic Inequality Graphically	Given $ax^2 + bx + c < 0$ with $a > 0$, graph $y = ax^2 + bx + c$ and locate any x-intercepts. If there are two x-intercepts, then solutions correspond to x-values between the x-intercepts. Solutions to $ax^2 + bx + c > 0$ correspond to x-values "outside" (left or right of) the x-intercepts.
Solving a Quadratic Inequality Numerically	If a quadratic inequality is expressed as $ax^2 + bx + c < 0$ with $a > 0$, then we start by solving $y = ax^2 + bx + c = 0$ with a table. If there are two solutions, then the solutions to the given inequality lie between these values. Solutions to $ax^2 + bx + c > 0$ lie before or after these values in the table.

18.5 Exercises

CONCEPTS AND VOCABULARY

1. How is a quadratic inequality different from a quadratic equation?

2. Do quadratic inequalities typically have two solutions? Explain.

3. Is 3 a solution to $x^2 < 7$?

4. Is 5 a solution to $x^2 \geq 25$?

5. The solutions to $x^2 - 2x - 8 = 0$ are -2 and 4. What are the solutions to $x^2 - 2x - 8 < 0$? Write your answer as an inequality.

6. The solutions to $x^2 + 2x - 3 = 0$ are -3 and 1. What are the solutions to $x^2 + 2x - 3 > 0$? Write your answer as an inequality.

RECOGNIZING QUADRATIC INEQUALITIES AND SOLUTIONS

Exercises 7–12: Is the inequality quadratic?

7. $x^2 + 4x + 5 < 0$
8. $x > x^3 - 5$
9. $x^2 > 19$
10. $x(x - 1) - 2 \geq 0$
11. $4x > 1 - x$
12. $2x(x^2 + 3) < 0$

Exercises 13–18: Is the given value of x a solution?

13. $2x^2 + x - 1 > 0$ $x = 3$
14. $x^2 - 3x + 2 \leq 0$ $x = 2$
15. $x^2 + 2 \leq 0$ $x = 0$
16. $2x(x - 3) \geq 0$ $x = 1$
17. $x^2 - 3x \leq 1$ $x = -3$
18. $4x^2 - 5x + 1 > 30$ $x = -2$

FINDING GRAPHICAL SOLUTIONS

Exercises 19–24: The graph of $y = ax^2 + bx + c$ is given. Solve each equation or inequality.

(a) $ax^2 + bx + c = 0$
(b) $ax^2 + bx + c < 0$
(c) $ax^2 + bx + c > 0$

19.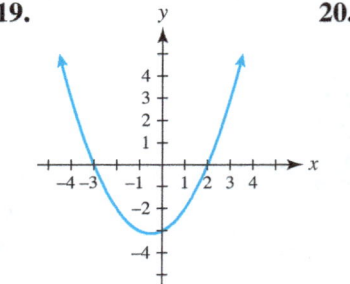

20.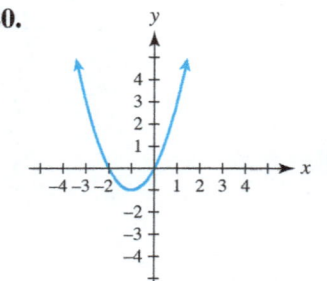

21.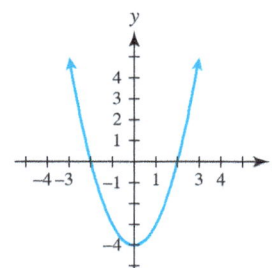
22.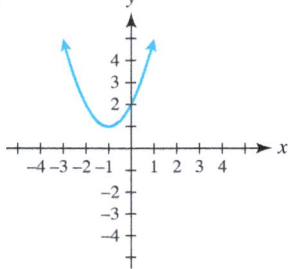

23.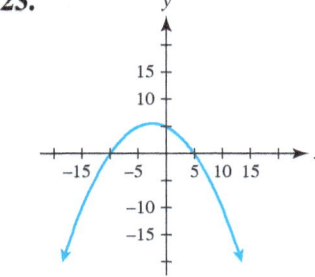
24.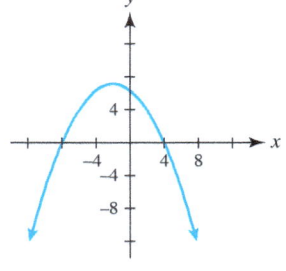

Exercises 25 and 26: *Solve the inequality graphically to the nearest thousandth. Use interval notation.*

25. $\pi x^2 - \sqrt{3}x \leq \frac{3}{11}$ 26. $\sqrt{5}x^2 - \pi^2 x \geq 10.3$

FINDING NUMERICAL SOLUTIONS

Exercises 27–30: *A table for $y = ax^2 + bx + c$ is given. Solve each equation or inequality.*

(a) $ax^2 + bx + c = 0$
(b) $ax^2 + bx + c < 0$
(c) $ax^2 + bx + c > 0$

27. $y = x^2 - 4$

x	-3	-2	-1	0	1	2	3
y	5	0	-3	-4	-3	0	5

28. $y = x^2 - x - 2$

x	-3	-2	-1	0	1	2	3
y	10	4	0	-2	-2	0	4

29. $y = x^2 + 4x$

x	-5	-4	-3	-2	-1	0	1
y	5	0	-3	-4	-3	0	5

30. $y = -2x^2 - 2x + 1.5$

x	-2	-1.5	-1	-0.5	0	0.5	1
y	-2.5	0	1.5	2	1.5	0	-2.5

FINDING SYMBOLIC SOLUTIONS

Exercises 31–40: *Solve the quadratic inequality symbolically. Write your answer in interval notation.*

31. $x^2 + 10x + 21 \leq 0$ 32. $x^2 - 7x - 18 < 0$

33. $3x^2 - 9x + 6 > 0$ 34. $7x^2 + 34x - 5 \geq 0$

35. $x^2 < 10$ 36. $x^2 \geq 64$

37. $x(6 - x) < 0$ 38. $1 - x^2 \leq 0$

39. $x(4 - x) \leq 2$ 40. $2x(1 - x) \geq 2$

USING THE METHOD OF YOUR CHOICE

Exercises 41–44: *Solve the equation in part (a). Use the results to solve the inequalities in parts (b) and (c).*

41. (a) $x^2 - 4 = 0$ 42. (a) $x^2 - 5 = 0$
    (b) $x^2 - 4 < 0$ (b) $x^2 - 5 \leq 0$
    (c) $x^2 - 4 > 0$ (c) $x^2 - 5 \geq 0$

43. (a) $x^2 + x - 1 = 0$
    (b) $x^2 + x - 1 < 0$
    (c) $x^2 + x - 1 > 0$

44. (a) $x^2 + 4x - 5 = 0$
    (b) $x^2 + 4x - 5 \leq 0$
    (c) $x^2 + 4x - 5 \geq 0$

Exercises 45–62: *Solve the inequality by any method. Write your answer in interval notation when appropriate.*

45. $x^2 + 4x + 3 < 0$ 46. $x^2 + x - 2 \leq 0$

47. $2x^2 - x - 15 \geq 0$ 48. $3x^2 - 3x - 6 > 0$

49. $2x^2 \leq 8$ 50. $x^2 < 9$

51. $x^2 > -5$ 52. $-x^2 \geq 1$

53. $-x^2 + 3x > 0$ 54. $-8x^2 - 2x + 1 \leq 0$

55. $x^2 + 2 \leq 0$ 56. $x^2 + 3 \geq -5$

57. $(x - 2)^2 \leq 0$ 58. $(x + 2)^2 \leq 0$

59. $(x + 1)^2 > 0$ 60. $(x - 3)^2 > 0$

61. $x(1 - x) \geq -2$ 62. $x(x - 2) < 3$

APPLICATIONS INVOLVING QUADRATIC EQUATIONS

63. **Dimensions of a Pen** A rectangular pen for a pet is 5 feet longer than it is wide. Give possible values for the width w of the pen if its area must be between 176 and 500 square feet, inclusively.

64. **Dimensions of a Cylinder** The volume of a cylindrical container is given by $V = \pi r^2 h$, where r is its radius and h is its height. See the accompanying figure. If $h = 6$ inches and the volume of the container must be 50 cubic inches or more, estimate to the nearest tenth of an inch possible values for r.

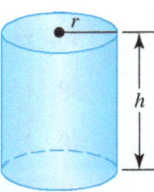

65. **Highway Design** (Refer to Example 5.) The elevation E of a sag curve, in feet, is given by
$$E(x) = 0.0000375x^2 - 0.175x + 1000,$$
where $0 \leq x \leq 4000$.
   (a) Estimate graphically the x-values for which the elevation is 850 feet or less. (*Hint:* Use the window $[0, 4000, 1000]$ by $[500, 1200, 100]$.)
   (b) For which x-values is the elevation 850 feet or more?

66. **Early Cellular Phone Use** The number of cellular subscribers in the United States in thousands from 1985 to 1991 can be modeled by
$$f(x) = 163x^2 - 146x + 205,$$
where x is the year and $x = 0$ corresponds to 1985, $x = 1$ to 1986, and so on. (*Source:* M. Paetsch, *Mobile Communication in the U.S. and Europe.*)
   (a) Write a quadratic inequality whose solution set represents the years when there were 2 million subscribers or more.
   (b) Solve this inequality.

67. **Heart Disease Death Rates** From 1960 to 2010, age-adjusted heart disease rates decreased dramatically. The number of deaths per 100,000 people can be modeled by
$$f(x) = -0.05107x^2 + 194.74x - 184,949,$$
where x is the year, as illustrated in the accompanying figure. (*Source:* Department of Health and Human Services.)

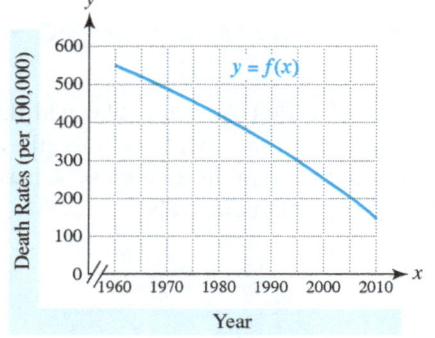

   (a) Evaluate $f(1985)$ using both the formula and the graph. How do your results compare?
   (b) Use the graph to estimate the years when this death rate was 500 or less.
   (c) Solve part (b) by using the quadratic formula.

68. **Accidental Deaths** From 1910 to 2010 the number of accidental deaths per 100,000 people generally decreased and can be modeled by
$$f(x) = -0.001918x^2 + 6.93x - 6156,$$
where x is the year, as shown in the accompanying figure. Note that after 2000, these rates increased and are not modeled by $f(x)$. (*Source:* Department of Health and Human Services.)

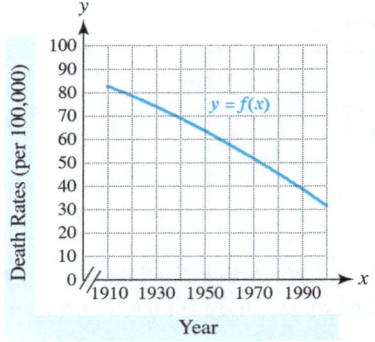

   (a) Evaluate $f(1955)$, using both the formula and the graph. How do your results compare?
   (b) Use the graph to estimate when this death rate was 60 or more.
   (c) Solve part (b) by using the quadratic formula.

WRITING ABOUT MATHEMATICS

69. Consider the inequality $x^2 < 0$. Discuss the solutions to this inequality and explain your reasoning.

70. Explain how the graph of $y = ax^2 + bx + c$ can be used to solve the inequality
$$ax^2 + bx + c > 0$$
when $a < 0$. Assume that the x-intercepts of the graph are $(p, 0)$ and $(q, 0)$ with $p < q$.

18.6 Equations in Quadratic Form

Objectives

1. Solving Higher Degree Polynomial Equations
2. Solving Equations That Have Rational Exponents
3. Solving Equations That Have Complex Solutions

1 Solving Higher Degree Polynomial Equations

Sometimes a fourth-degree polynomial can be factored like a quadratic trinomial, provided it does not have an x-term or an x^3-term. Let's consider the equation $x^4 - 5x^2 + 4 = 0$.

$$x^4 - 5x^2 + 4 = 0 \quad \text{Given equation}$$
$$(x^2)^2 - 5(x^2) + 4 = 0 \quad \text{Properties of exponents}$$

We use the substitution $u = x^2$.

$$u^2 - 5u + 4 = 0 \quad \text{Let } u = x^2.$$
$$(u - 4)(u - 1) = 0 \quad \text{Factor.}$$
$$u - 4 = 0 \quad \text{or} \quad u - 1 = 0 \quad \text{Zero-product property}$$
$$u = 4 \quad \text{or} \quad u = 1 \quad \text{Solve each equation.}$$

Because the given equation $x^4 - 5x^2 + 4 = 0$ uses the variable x, we must give the solutions in terms of x. We substitute x^2 for u in $u = 4$ and $u = 1$ and then solve to obtain the following four solutions.

$$u = 4 \quad \text{or} \quad u = 1 \quad \text{Solutions in terms of } u$$
$$x^2 = 4 \quad \text{or} \quad x^2 = 1 \quad \text{Substitute } x^2 \text{ for } u$$
$$x = \pm 2 \quad \text{or} \quad x = \pm 1 \quad \text{Square root property}$$

The solutions are $-2, -1, 1,$ and 2.

EXAMPLE 1 Solving a sixth-degree equation by substitution

Solve $2x^6 + x^3 = 1$.

Solution

Start by subtracting 1 from each side.

$$2x^6 + x^3 - 1 = 0 \quad \text{Subtract 1.}$$
$$2(x^3)^2 + (x^3) - 1 = 0 \quad \text{Properties of exponents}$$
$$2u^2 + u - 1 = 0 \quad \text{Let } u = x^3.$$
$$(2u - 1)(u + 1) = 0 \quad \text{Factor.}$$
$$2u - 1 = 0 \quad \text{or} \quad u + 1 = 0 \quad \text{Zero-product property}$$
$$u = \frac{1}{2} \quad \text{or} \quad u = -1 \quad \text{Solve.}$$

Now substitute x^3 for u, and solve for x to obtain the following two solutions.

$$x^3 = \frac{1}{2} \quad \text{or} \quad x^3 = -1 \quad \text{Substitute } x^3 \text{ for } u.$$
$$x = \sqrt[3]{\frac{1}{2}} \quad \text{or} \quad x = -1 \quad \text{Take cube root of each side.}$$

Now Try Exercise 3

2 Solving Equations That Have Rational Exponents

Equations that have negative exponents are sometimes reducible to quadratic form. Consider the following example, in which two methods are presented.

EXAMPLE 2 Solving an equation that has negative exponents

Solve $-6m^{-2} + 13m^{-1} + 5 = 0$.

Solution

Method I Use the substitution $u = m^{-1} = \frac{1}{m}$ and $u^2 = m^{-2} = \frac{1}{m^2}$.

$-6m^{-2} + 13m^{-1} + 5 = 0$	Given equation
$-6u^2 + 13u + 5 = 0$	Let $u = m^{-1}$ and $u^2 = m^{-2}$.
$6u^2 - 13u - 5 = 0$	Multiply by -1.
$(2u - 5)(3u + 1) = 0$	Factor.
$2u - 5 = 0$ or $3u + 1 = 0$	Zero-product property
$u = \frac{5}{2}$ or $u = -\frac{1}{3}$	Solve for u.

Because $u = \frac{1}{m}$, $m = \frac{1}{u}$. Thus the solutions are given by $m = \frac{2}{5}$ or $m = -3$.

Method II Another way to solve this equation is to multiply each side by the LCD, m^2.

$-6m^{-2} + 13m^{-1} + 5 = 0$	Given equation
$m^2(-6m^{-2} + 13m^{-1} + 5) = m^2 \cdot 0$	Multiply by m^2.
$-6m^2m^{-2} + 13m^2m^{-1} + 5m^2 = 0$	Distributive property
$-6 + 13m + 5m^2 = 0$	Add exponents.
$5m^2 + 13m - 6 = 0$	Rewrite the equation.
$(5m - 2)(m + 3) = 0$	Factor.
$5m - 2 = 0$ or $m + 3 = 0$	Zero-product property
$m = \frac{2}{5}$ or $m = -3$	Solve.

Now Try Exercise 5

In the next example, we solve an equation that has fractional exponents.

EXAMPLE 3 Solving an equation that has fractional exponents

Solve $x^{2/3} - 2x^{1/3} - 8 = 0$.

Solution
Use the substitution $u = x^{1/3}$.

$x^{2/3} - 2x^{1/3} - 8 = 0$	Given equation
$(x^{1/3})^2 - 2(x^{1/3}) - 8 = 0$	Properties of exponents
$u^2 - 2u - 8 = 0$	Let $u = x^{1/3}$.
$(u - 4)(u + 2) = 0$	Factor.
$u - 4 = 0$ or $u + 2 = 0$	Zero-product property
$u = 4$ or $u = -2$	Solve.

Because $u = x^{1/3}$, $u^3 = (x^{1/3})^3 = x$. Thus $x = 4^3 = 64$ or $x = (-2)^3 = -8$. The solutions are -8 and 64.

Now Try Exercise 13

3 Solving Equations That Have Complex Solutions

Sometimes an equation that is reducible to quadratic form also has complex solutions. This situation is discussed in the next two examples.

EXAMPLE 4 Solving a fourth-degree equation

Find all complex solutions to $x^4 - 1 = 0$.

Solution

$$x^4 - 1 = 0 \quad \text{Given equation}$$
$$(x^2)^2 - 1 = 0 \quad \text{Properties of exponents}$$
$$u^2 - 1 = 0 \quad \text{Let } u = x^2.$$
$$(u - 1)(u + 1) = 0 \quad \text{Factor difference of squares.}$$
$$u - 1 = 0 \quad \text{or} \quad u + 1 = 0 \quad \text{Zero-product property}$$
$$u = 1 \quad \text{or} \quad u = -1 \quad \text{Solve for } u.$$

Now substitute x^2 for u, and solve for x.

$$u = 1 \quad \text{or} \quad u = -1 \quad \text{Solutions in terms of } u$$
$$x^2 = 1 \quad \text{or} \quad x^2 = -1 \quad \text{Let } x^2 = u.$$
$$x = \pm 1 \quad \text{or} \quad x = \pm i \quad \text{Square root property}$$

There are four complex solutions: $-1, 1, -i,$ and i.

Now Try Exercise 25

EXAMPLE 5 Solving a rational equation

Find all complex solutions to $\dfrac{1}{x} + \dfrac{1}{x^2} = -1$.

Solution

This equation is a rational equation. However, if we multiply through by the LCD, x^2, we clear fractions and obtain a quadratic equation with complex solutions.

$$\frac{1}{x} + \frac{1}{x^2} = -1 \quad \text{Given equation}$$
$$\frac{x^2}{x} + \frac{x^2}{x^2} = -1x^2 \quad \text{Multiply each term by } x^2.$$
$$x + 1 = -x^2 \quad \text{Simplify.}$$
$$x^2 + x + 1 = 0 \quad \text{Add } x^2.$$
$$x = \frac{-1 \pm \sqrt{1^2 - 4(1)(1)}}{2(1)} \quad \text{Quadratic formula}$$
$$x = \frac{-1 \pm i\sqrt{3}}{2} \quad \sqrt{-3} = i\sqrt{3}$$
$$x = -\frac{1}{2} \pm \frac{i\sqrt{3}}{2} \quad \frac{a \pm b}{c} = \frac{a}{c} \pm \frac{b}{c}$$

Now Try Exercise 31

18.6 Putting It All Together

CONCEPT	EXPLANATION	EXAMPLES
Higher Degree Polynomial Equations	Let $u = x^n$ for some integer n.	To solve $x^4 - 3x^2 - 4 = 0$, let $u = x^2$. This equation becomes $$u^2 - 3u - 4 = 0.$$
Equations That Have Rational Exponents	Pick a substitution that reduces the equation to quadratic form.	To solve $n^{-2} + 6n^{-1} + 9 = 0$, let $u = n^{-1}$. This equation becomes $$u^2 + 6u + 9 = 0.$$ To solve $6x^{2/5} - 5x^{1/5} - 4 = 0$, let $u = x^{1/5}$. This equation becomes $$6u^2 - 5u - 4 = 0.$$
Equations That Have Complex Solutions	Both polynomial and rational equations can have complex solutions. Use the fact that if $a > 0$, then $\sqrt{-a} = i\sqrt{a}$.	$$1 + \frac{1}{x^2} = 0$$ $$x^2 \cdot \left(1 + \frac{1}{x^2}\right) = 0 \cdot x^2$$ $$x^2 + 1 = 0$$ $$x^2 = -1$$ $$x = \pm i$$

18.6 Exercises

SOLVING EQUATIONS THAT ARE REDUCIBLE TO QUADRATIC FORM

Exercises 1–6: Use the given substitution to solve the equation.

1. $x^4 - 7x^2 + 6 = 0$ $u = x^2$
2. $2k^4 - 7k^2 + 6 = 0$ $u = k^2$
3. $3z^6 + z^3 - 10 = 0$ $u = z^3$
4. $2x^6 + 17x^3 + 8 = 0$ $u = x^3$
5. $4n^{-2} + 17n^{-1} + 15 = 0$ $u = n^{-1}$
6. $m^{-2} + 24 = 10m^{-1}$ $u = m^{-1}$

Exercises 7–24: Solve. Find all real solutions.

7. $x^4 = 8x^2 + 9$
8. $3x^4 = 10x^2 + 8$
9. $3x^6 - 5x^3 - 2 = 0$
10. $6x^6 + 11x^3 + 4 = 0$
11. $2z^{-2} + 11z^{-1} = 40$
12. $z^{-2} - 10z^{-1} + 25 = 0$
13. $x^{2/3} - 2x^{1/3} + 1 = 0$
14. $3x^{2/3} + 18x^{1/3} = 48$
15. $x^{2/5} - 33x^{1/5} = -32$
16. $x^{2/5} - 80x^{1/5} = 81$
17. $x - 13\sqrt{x} + 36 = 0$
18. $x - 17\sqrt{x} + 16 = 0$
19. $z^{1/2} - 2z^{1/4} + 1 = 0$
20. $z^{1/2} - 4z^{1/4} + 4 = 0$
21. $(x + 1)^2 - 5(x + 1) - 14 = 0$
22. $2(x - 5)^2 + 5(x - 5) + 3 = 0$
23. $(x^2 - 1)^2 - 4 = 0$
24. $(x^2 - 9)^2 - 8(x^2 - 9) + 16 = 0$

SOLVING EQUATIONS THAT HAVE COMPLEX SOLUTIONS

Exercises 25–34: Find all complex solutions.

25. $x^4 - 16 = 0$
26. $\frac{1}{3}x^4 - 27 = 0$
27. $x^3 + x = 0$
28. $4x^3 + x = 0$
29. $x^4 - 2 = x^2$
30. $x^4 - 3 = 2x^2$

31. $\dfrac{1}{x} + \dfrac{1}{x^2} = -\dfrac{1}{2}$ 32. $\dfrac{2}{x-1} - \dfrac{1}{x} = -1$

33. $\dfrac{2}{x-2} - \dfrac{1}{x} = -\dfrac{1}{2}$ 34. $\dfrac{1}{x} - \dfrac{1}{x^2} = \dfrac{1}{2}$

WRITING ABOUT MATHEMATICS

35. Explain how to solve $ax^4 - bx^2 + c = 0$. Assume that the left side of the equation factors.

36. Explain what it means for an equation to be reducible to quadratic form.

SECTIONS 18.5 and 18.6 — Checking Basic Concepts

1. Solve the inequality $x^2 - x - 6 > 0$. Write your answer in interval notation.

2. Solve the inequality $3x^2 + 5x + 2 \leq 0$. Write your answer in interval notation.

3. Solve $x^6 + 6x^3 - 16 = 0$.

4. Solve $x^{2/3} - 7x^{1/3} - 8 = 0$.

5. Find all complex solutions to $x^4 + 2x^2 + 1 = 0$.

CHAPTER 18 Summary

SECTION 18.1 ■ QUADRATIC FUNCTIONS AND THEIR GRAPHS

Quadratic Function Any quadratic function f can be written as
$$f(x) = ax^2 + bx + c \quad (a \neq 0).$$

Graph of a Quadratic Function Its graph is a parabola that opens upward if $a > 0$ and opens downward if $a < 0$. The y-intercept is $(0, c)$.

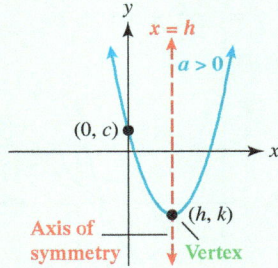 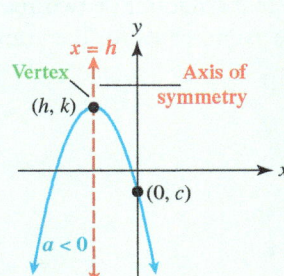

Axis of Symmetry The parabola is symmetric with respect to this vertical line. The axis of symmetry passes through the vertex. Its equation is $x = -\dfrac{b}{2a}$.

Vertex Formula The x-coordinate of the vertex is $-\dfrac{b}{2a}$.

Example: Let $y = x^2 - 4x + 1$ with $a = 1$ and $b = -4$.

$$x = -\dfrac{-4}{2(1)} = 2 \quad \text{and} \quad y = 2^2 - 4(2) + 1 = -3.\text{ The vertex is } (2, -3).$$

Min–Max Applications The graph of a quadratic function $f(x) = ax^2 + bx + c$ is a parabola that has vertex (h, k), where $h = -\dfrac{b}{2a}$ and $k = f\left(-\dfrac{b}{2a}\right)$. If $a > 0$, the parabola opens upward and the minimum y-value is k. If $a < 0$, the parabola opens downward and the maximum y-value is k. These concepts can be used in certain applications to find maximum and minimum values.

SECTION 18.2 ■ TRASFORMATIONS AND TRANSLATIONS OF PARABOLAS

Graph of $y = ax^2$ The graph of $y = ax^2$ is a parabola that has vertex $(0, 0)$. It opens upward if $a > 0$ and opens downward if $a < 0$. It is wider than the graph of $y = x^2$ if $0 < |a| < 1$, and narrower than the graph of $y = x^2$ if $|a| > 1$.

Vertical and Horizontal Translations Let h and k be positive numbers.

To graph	shift the graph of $y = x^2$ by k units
$y = x^2 + k$	upward.
$y = x^2 - k$	downward.

To graph	shift the graph of $y = x^2$ by h units
$y = (x - h)^2$	right.
$y = (x + h)^2$	left.

Example: Compared to $y = x^2$, the graph of $y = (x - 1)^2 + 2$ is translated right 1 unit and upward 2 units.

Vertex Form Any quadratic function can be expressed as $f(x) = a(x - h)^2 + k$. In this form, the point (h, k) is the vertex. A quadratic function can be put in this form by completing the square or by applying the vertex formula.

Example:
$$y = x^2 + 10x - 4 \qquad \text{Given equation}$$
$$= (x^2 + 10x + 25) - 25 - 4 \qquad \left(\frac{b}{2}\right)^2 = \left(\frac{10}{2}\right)^2 = 25; \text{ complete the square.}$$
$$= (x + 5)^2 - 29 \qquad \text{Perfect square trinomial; subtract.}$$

The vertex is $(-5, -29)$.

SECTION 18.3 ■ QUADRATIC EQUATIONS

Quadratic Equations Any quadratic equation can be written as $ax^2 + bx + c = 0$ and can have no real solutions, one real solution, or two real solutions. These solutions correspond to the x-coordinates of the x-intercepts for the graph of $y = ax^2 + bx + c$.

Example:
$$x^2 + x - 2 = 0$$
$$(x + 2)(x - 1) = 0$$
$$x = -2 \quad \text{or} \quad x = 1$$

The solutions are -2 and 1.
See the graph to the right.

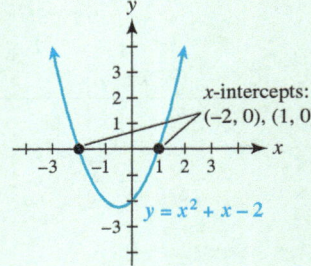

x-intercepts: $(-2, 0), (1, 0)$

$y = x^2 + x - 2$

Square Root Property Let k be a nonnegative number. Then the solutions to the equation $x^2 = k$ are given by $x = \pm\sqrt{k}$. If $k < 0$, then this equation has no real solutions.

Example: $x^2 = 5$ implies that $x = \pm\sqrt{5}$.

Completing the Square Write the equation in the form $x^2 + bx = d$. Complete the square by adding $\left(\frac{b}{2}\right)^2$ to each side of the equation.

Example:
$$x^2 - 8x = 3$$
$$x^2 - 8x + 16 = 3 + 16 \qquad \text{Add } \left(\frac{-8}{2}\right)^2 = 16 \text{ to each side.}$$
$$(x - 4)^2 = 19 \qquad \text{Perfect square trinomial}$$
$$x - 4 = \pm\sqrt{19} \qquad \text{Square root property}$$
$$x = 4 \pm \sqrt{19} \qquad \text{Add 4 to each side.}$$

SECTION 18.4 ■ THE QUADRATIC FORMULA

The Quadratic Formula The solutions to $ax^2 + bx + c = 0$ $(a \neq 0)$ are given by

$$x = \frac{-b \pm \sqrt{b^2 - 4ac}}{2a}.$$

Example: Solve $2x^2 + 3x - 1 = 0$ by letting $a = 2$, $b = 3$, and $c = -1$.

$$x = \frac{-3 \pm \sqrt{3^2 - 4(2)(-1)}}{2(2)} = \frac{-3 \pm \sqrt{17}}{4} \approx 0.28, -1.78$$

The Discriminant The expression $b^2 - 4ac$ is called the discriminant. If $b^2 - 4ac > 0$, there are two real solutions; if $b^2 - 4ac = 0$, there is one real solution; and if $b^2 - 4ac < 0$, there are no real solutions—rather there are two complex solutions.

Example: For $2x^2 + 3x - 1 = 0$, the discriminant is

$$b^2 - 4ac = 3^2 - 4(2)(-1) = 17 > 0.$$

There are two real solutions to this quadratic equation, as shown in the previous example.

Quadratic Equations That Have Complex Solutions A quadratic equation sometimes has no real solutions.

Example: $x^2 + 4 = 0$

$\quad\quad x^2 = -4$ Subtract 4 from each side.

$\quad\quad x = \pm 2i$ Square root property; two complex solutions

SECTION 18.5 ■ QUADRATIC INEQUALITIES

Quadratic Inequalities When the equals sign in a quadratic equation is replaced with $<, >, \leq$, or $\geq$, a quadratic inequality results. For example,

$$3x^2 - x + 1 = 0$$

is a quadratic equation and

$$3x^2 - x + 1 > 0$$

is a quadratic inequality. Like quadratic equations, quadratic inequalities can be solved symbolically, graphically, and numerically. An important first step in solving a quadratic inequality is to solve the corresponding quadratic equation.

Examples: The solutions to $x^2 - 5x - 6 = 0$ are -1 and 6.

        The solutions to $x^2 - 5x - 6 < 0$ satisfy $-1 < x < 6$.

        The solutions to $x^2 - 5x - 6 > 0$ satisfy $x < -1$ or $x > 6$.

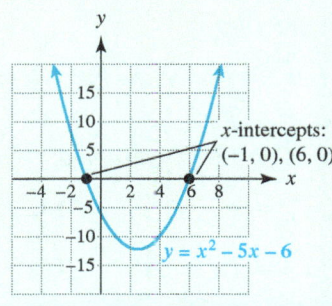

SECTION 18.6 ■ EQUATIONS IN QUADRATIC FORM

Equations Reducible to Quadratic Form An equation that is not quadratic, but can be put into quadratic form by using a substitution, is reducible to quadratic form.

Example: To solve $x^{2/3} - 2x^{1/3} - 15 = 0$, let $u = x^{1/3}$. This equation becomes

$$u^2 - 2u - 15 = 0.$$

Factoring results in $(u + 3)(u - 5) = 0$, so $u = -3$ or $u = 5$.
Because $u = x^{1/3}$, $x = u^3$ and $x = (-3)^3 = -27$ or $x = (5)^3 = 125$.

CHAPTER 18 Review Exercises

SECTION 18.1

Exercises 1 and 2: Identify the vertex, axis of symmetry, and whether the parabola opens upward or downward.

1.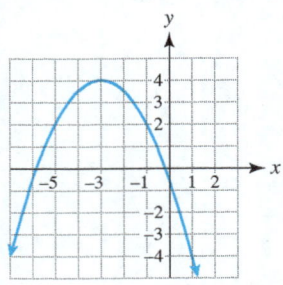
2.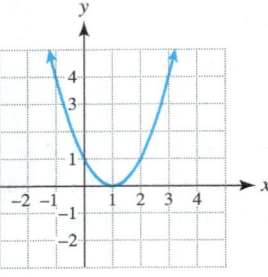

Exercises 3–6: Do the following.
(a) Graph f.
(b) Identify the vertex and axis of symmetry.
(c) Evaluate $f(x)$ at the given value of x.

3. $f(x) = x^2 - 2$, $x = -1$
4. $f(x) = -x^2 + 4x - 3$, $x = 3$
5. $f(x) = -\frac{1}{2}x^2 + x + \frac{3}{2}$, $x = -2$
6. $f(x) = 2x^2 + 8x + 5$, $x = -3$

7. Find the minimum y-value located on the graph of $y = 2x^2 - 6x + 1$.

8. Find the maximum y-value located on the graph of $y = -3x^2 + 2x - 5$.

Exercises 9–12: Find the vertex of the parabola.
9. $f(x) = x^2 - 4x - 2$ 10. $f(x) = 5 - x^2$
11. $f(x) = -\frac{1}{4}x^2 + x + 1$ 12. $f(x) = 2 + 2x + x^2$

SECTION 18.2

Exercises 13–18: Do the following.
(a) Graph f.
(b) Compare the graph of f with the graph of $y = x^2$.

13. $f(x) = x^2 + 2$ 14. $f(x) = 3x^2$
15. $f(x) = (x - 2)^2$ 16. $f(x) = (x + 1)^2 - 3$
17. $f(x) = \frac{1}{2}(x + 1)^2 + 2$
18. $f(x) = 2(x - 1)^2 - 3$

19. Write the vertex form of a parabola with $a = -4$ and vertex $(2, -5)$.

20. Write the vertex form of a parabola that opens downward with vertex $(-4, 6)$. Assume that $a = \pm 1$.

Exercises 21–24: Write the equation in vertex form. Identify the vertex.

21. $y = x^2 + 4x - 7$ 22. $y = x^2 - 7x + 1$
23. $y = 2x^2 - 3x - 8$ 24. $y = 3x^2 + 6x - 2$

Exercises 25 and 26: Find a value for the constant a so that $f(x) = ax^2 - 1$ models the data.

25.
x	1	2	3
$f(x)$	2	11	26

26.
x	-1	0	1
$f(x)$	$-\frac{3}{4}$	-1	$-\frac{3}{4}$

Exercises 27 and 28: Write $f(x)$ in the form given by $f(x) = ax^2 + bx + c$. Identify the y-intercept on the graph of $y = f(x)$.

27. $f(x) = -5(x - 3)^2 + 4$
28. $f(x) = 3(x + 2)^2 - 4$

SECTION 18.3

Exercises 29–32: Use the graph of $y = ax^2 + bx + c$ to solve $ax^2 + bx + c = 0$.

29.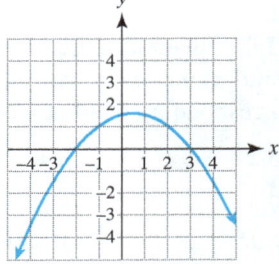
30.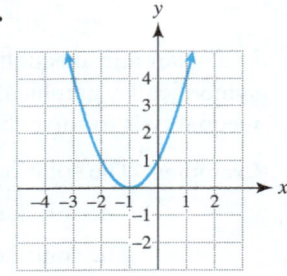
31.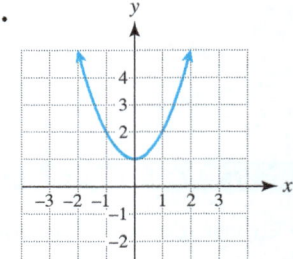
32.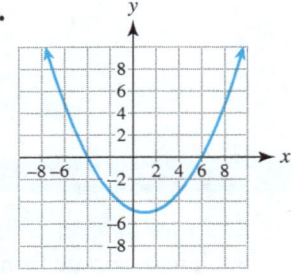

Exercises 33 and 34: A table of $y = ax^2 + bx + c$ is given. Solve $ax^2 + bx + c = 0$.

33.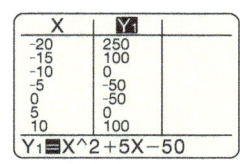

$Y_1 = X^2 + 5X - 50$

34.

$Y_1 = 8X^2 + 2X - 1$

Exercises 35–38: Solve the quadratic equation
(a) graphically and
(b) numerically.

35. $x^2 - 5x - 50 = 0$
36. $\frac{1}{2}x^2 + x - \frac{3}{2} = 0$
37. $\frac{1}{4}x^2 + \frac{1}{2}x = 2$
38. $\frac{1}{2}x + \frac{3}{4} = \frac{1}{4}x^2$

Exercises 39–42: Solve by factoring.

39. $x^2 + x - 20 = 0$
40. $x^2 + 11x + 24 = 0$
41. $15x^2 - 4x - 4 = 0$
42. $7x^2 - 25x + 12 = 0$

Exercises 43–46: Use the square root property to solve.

43. $x^2 = 100$
44. $3x^2 = \frac{1}{3}$
45. $4x^2 - 6 = 0$
46. $5x^2 = x^2 - 4$

Exercises 47–50: Solve by completing the square.

47. $x^2 + 6x = -2$
48. $x^2 - 4x = 6$
49. $x^2 - 2x - 5 = 0$
50. $2x^2 + 6x - 1 = 0$

Exercises 51 and 52: Solve for the specified variable.

51. $F = \dfrac{k}{(R + r)^2}$ for R
52. $2x^2 + 3y^2 = 12$ for y

SECTION 18.4

Exercises 53–58: Use the quadratic formula to solve.

53. $x^2 - 9x + 18 = 0$
54. $x^2 - 24x + 143 = 0$
55. $6x^2 + x = 1$
56. $5x^2 + 1 = 5x$
57. $x(x - 8) = 5$
58. $2x(2 - x) = 3 - 2x$

Exercises 59–64: Solve by any method.

59. $x^2 - 4 = 0$
60. $4x^2 - 1 = 0$
61. $2x^2 + 15 = 11x$
62. $2x^2 + 15 = 13x$
63. $x(5 - x) = 2x + 1$
64. $-2x(x - 1) = x - \frac{1}{2}$

Exercises 65–68: Graphs of $y = ax^2 + bx + c$ are shown at the top of the next column.

(a) State whether $a > 0$ or $a < 0$.
(b) Solve $ax^2 + bx + c = 0$.
(c) Determine whether the discriminant is positive, negative, or zero.

65.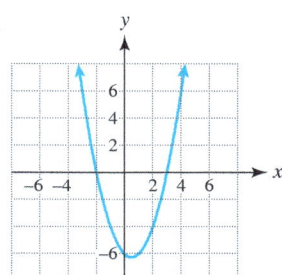

66.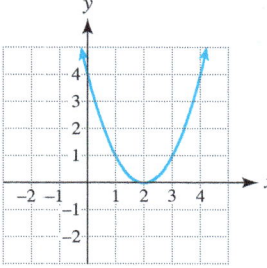

67.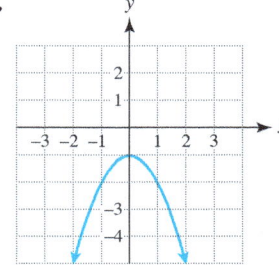

68.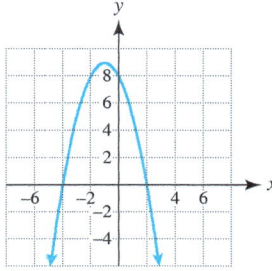

Exercises 69–72: Do the following for the given equation.
(a) Evaluate the discriminant.
(b) How many real solutions are there?
(c) Support your answer for part (b) graphically.

69. $2x^2 - 3x + 1 = 0$
70. $7x^2 + 2x - 5 = 0$
71. $3x^2 + x + 2 = 0$
72. $4.41x^2 - 12.6x + 9 = 0$

Exercises 73–76: Solve. Write any complex solutions in standard form.

73. $x^2 + x + 5 = 0$
74. $2x^2 + 8 = 0$
75. $2x^2 = x - 1$
76. $7x^2 = 2x - 5$

SECTION 18.5

Exercises 77 and 78: The graph of $y = ax^2 + bx + c$ is shown. Solve each equation or inequality.

(a) $ax^2 + bx + c = 0$
(b) $ax^2 + bx + c < 0$
(c) $ax^2 + bx + c > 0$

77.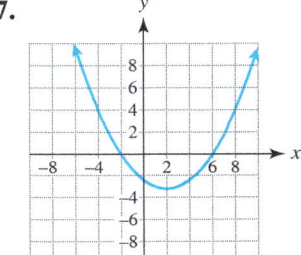

78.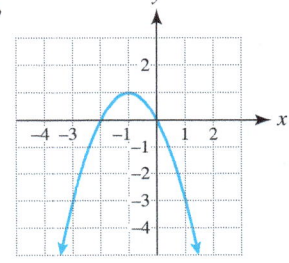

Exercises 79 and 80: A table of $y = ax^2 + bx + c$ is shown. Solve each equation or inequality.

(a) $ax^2 + bx + c = 0$
(b) $ax^2 + bx + c < 0$
(c) $ax^2 + bx + c > 0$

79. $y = x^2 - 16$

x	−6	−4	−2	0	2	4	6
y	20	0	−12	−16	−12	0	20

80. $y = x^2 + x - 2$

x	−3	−2	−1	0	1	2	3
y	4	0	−2	−2	0	4	10

Exercises 81 and 82: Solve the equation in part (a). Use the results to solve the inequalities in parts (b) and (c).

81. (a) $x^2 - 2x - 3 = 0$
(b) $x^2 - 2x - 3 < 0$
(c) $x^2 - 2x - 3 > 0$

82. (a) $2x^2 - 7x - 15 = 0$
(b) $2x^2 - 7x - 15 \leq 0$
(c) $2x^2 - 7x - 15 \geq 0$

Exercises 83–88: Solve the quadratic inequality. Write your answer in interval notation.

83. $x^2 + 4x + 3 \leq 0$ **84.** $5x^2 - 16x + 3 < 0$

85. $6x^2 - 13x + 2 > 0$ **86.** $x^2 \geq 5$

87. $(x - 1)^2 \geq 0$ **88.** $x^2 + 3 < 2$

SECTION 18.6

Exercises 89–92: Solve the equation.

89. $x^4 - 14x^2 + 45 = 0$

90. $2z^{-2} + z^{-1} - 28 = 0$

91. $x^{2/3} - 9x^{1/3} + 8 = 0$

92. $(x - 1)^2 + 2(x - 1) + 1 = 0$

Exercises 93 and 94: Find all complex solutions.

93. $4x^4 + 4x^2 + 1 = 0$ **94.** $\dfrac{1}{x-2} - \dfrac{3}{x} = -1$

APPLICATIONS

95. Construction A rain gutter is being made from a flat sheet of metal so that the cross section of the gutter is a rectangle, as shown in the figure at the top of the next column. The width of the metal sheet is 12 inches.

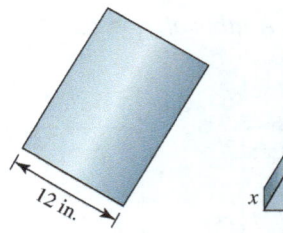
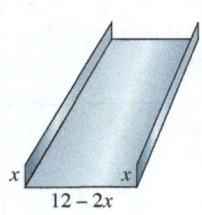

(a) Write a formula $f(x)$ that gives the area of the cross section.
(b) To hold the greatest amount of rainwater, the cross section should have maximum area. Find the dimensions that result in this maximum.

96. Height of a Stone Suppose that a stone is thrown upward with an initial velocity of 44 feet per second (30 miles per hour) and is released 4 feet above the ground. Its height h in feet after t seconds is given by
$$h(t) = -16t^2 + 44t + 4.$$
(a) When does the stone reach a height of 32 feet?
(b) After how many seconds does the stone reach maximum height? Estimate this height.

97. Maximizing Revenue Suppose that hotel rooms cost $90 per night. However, for a group rate the management is considering reducing the cost of a room by $3 for every room rented.
(a) Write a formula $f(x)$ that gives the revenue for x rooms at the group rate.
(b) Graph f in $[0, 30, 5]$ by $[0, 800, 100]$.
(c) How many rooms should be rented to receive revenue of $600?
(d) How many rooms should be rented to maximize revenue?

98. Numbers The product of two numbers is 143. One number is 2 more than the other.
(a) Write a quadratic equation whose solution gives the smaller number x.
(b) Solve the equation.

99. Braking Distance On dry pavement, a safe braking distance d in feet for a car traveling x miles per hour is $d = \frac{x^2}{12}$. For each distance d, find x. (*Source:* F. Mannering, *Principles of Highway Engineering and Traffic Control.*)
(a) $d = 144$ feet
(b) $d = 300$ feet

100. U.S. Energy Consumption From 1950 to 1970, per capita consumption of energy in millions of Btu can be modeled by $f(x) = \frac{1}{4}(x - 1950)^2 + 220$, where x is the year. (*Source:* Department of Energy.)
(a) Find and interpret the vertex.

(b) Graph f in $[1950, 1970, 5]$ by $[200, 350, 25]$. What happened to energy consumption during this time period?

(c) Use f to predict the consumption in 2010. Actual consumption was 321 million Btu. Did f provide a good model for 2010? Explain.

101. Screens A square computer screen has an area of 123 square inches. Approximate its dimensions to the nearest tenth of an inch.

102. Flying a Kite A kite is being flown, as illustrated in the accompanying figure. If 130 feet of string have been let out, find the value of x.

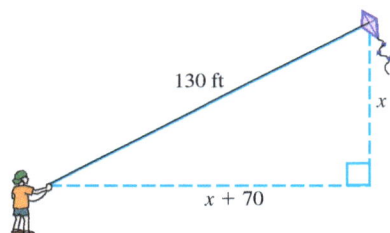

103. Area A uniform strip of grass is to be planted around a rectangular swimming pool, as illustrated in the accompanying figure. The swimming pool is 30 feet wide and 50 feet long. If there is only enough grass seed to cover 250 square feet, estimate the width x that the strip of grass should be.

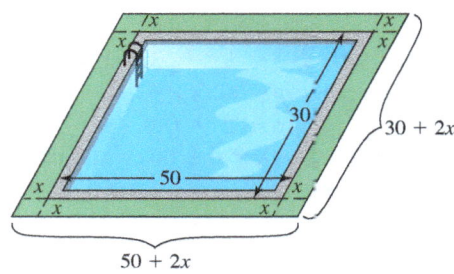

104. Dimensions of a Cone The volume V of a cone is given by $V = \frac{1}{3}\pi r^2 h$, where r is its base radius and h is its height. See the accompanying figure. If $h = 20$ inches and the volume of the cone must be between 750 and 1700 cubic inches, inclusively, estimate, to the nearest tenth of an inch, possible values for r.

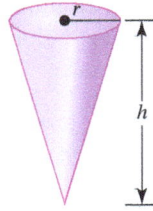

CHAPTER 18 Test   MyMathLab®, YouTube

1. Find the vertex and axis of symmetry for the graph of $f(x) = -\frac{1}{2}x^2 + x + 1$. Evaluate $f(-2)$.

2. Find the minimum y-value located on the graph of $y = x^2 + 3x - 5$.

3. Find the exact value for the constant a so that $f(x) = ax^2 + 2$ models the data in the table.

x	-2	0	2	4
$f(x)$	0	2	0	-6

4. Compare the graph of $y = f(x)$ to the graph of $y = x^2$. Then graph $y = f(x)$.
   (a) $f(x) = (x-1)^2$ (b) $f(x) = x^2 - 2$
   (c) $f(x) = \frac{1}{2}(x-3)^2 + 2$

5. Write $y = x^2 - 6x + 2$ in vertex form. Identify the vertex and axis of symmetry.

6. Use the graph of $f(x) = ax^2 + bx + c$ to solve $ax^2 + bx + c = 0$. Then evaluate $f(1)$.

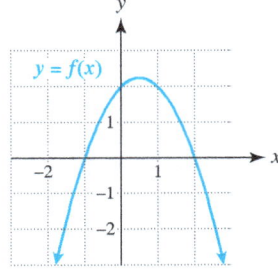

Exercises 7 and 8: Solve the quadratic equation.

7. $3x^2 + 11x - 4 = 0$ 8. $2x^2 = 2 - 6x^2$

9. Solve $x^2 - 8x = 1$ by completing the square.

10. Solve $x(-2x + 3) = -1$ by using the quadratic formula.

11. Solve $9x^2 - 16 = 0$.

12. Solve $F = \dfrac{Gm^2}{r^2}$ for m.

13. A graph of $y = ax^2 + bx + c$ is shown.
    (a) State whether $a > 0$ or $a < 0$.
    (b) Solve $ax^2 + bx + c = 0$.
    (c) Determine whether the discriminant is positive, negative, or zero.

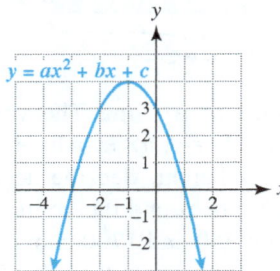

14. Complete the following for $-3x^2 + 4x - 5 = 0$.
    (a) Evaluate the discriminant.
    (b) How many real solutions are there?
    (c) Support your answer for part (b) graphically.

Exercises 15 and 16: The graph of $y = ax^2 + bx + c$ is shown. Solve each equation or inequality.
    (a) $ax^2 + bx + c = 0$
    (b) $ax^2 + bx + c < 0$
    (c) $ax^2 + bx + c > 0$

15.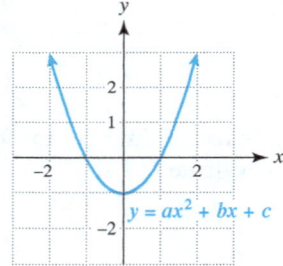

16.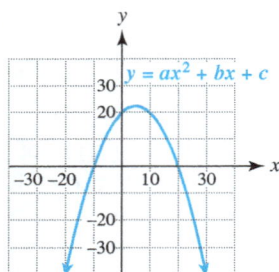

17. Solve the quadratic equation in part (a). Use the result to solve the inequalities in parts (b) and (c) and write your answer in interval notation.
    (a) $8x^2 - 2x - 3 = 0$
    (b) $8x^2 - 2x - 3 \leq 0$
    (c) $8x^2 - 2x - 3 \geq 0$

18. Solve $x^2 + 2x \leq 0$.

19. Solve $x^6 - 3x^3 + 2 = 0$. Find all real solutions.

20. Solve $2x^2 + 4x + 3 = 0$. Write all complex solutions in standard form.

21. Solve $\sqrt{2} - \pi x^2 = 2.12x - 0.5\pi$ graphically. Round your answers to the nearest hundredth.

22. **Braking Distance** On wet pavement, a safe braking distance d in feet for a car traveling x miles per hour is $d = \dfrac{x^2}{9}$. What speed corresponds to a braking distance of 250 feet? (*Source:* F. Mannering, *Principles of Highway Engineering and Traffic Control.*)

23. **Construction** A fence is being constructed along a 20-foot building, as shown in the accompanying figure. No fencing is used along the building.
    (a) If 200 feet of fence are available, find a formula $f(x)$ using only the variable x that gives the area enclosed.
    (b) What value of x gives the greatest area?

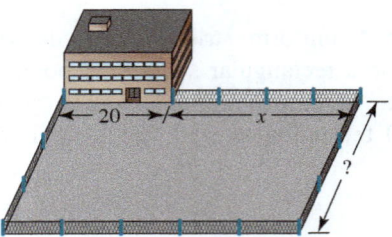

24. **Height of a Stone** Suppose that a stone is thrown upward with an initial velocity of 88 feet per second (60 miles per hour) and is released 8 feet above the ground. Its height h in feet after t seconds is given by
$$h(t) = -16t^2 + 88t + 8.$$
    (a) Graph h in $[0, 6, 1]$ by $[0, 150, 50]$.
    (b) When does the stone strike the ground?
    (c) After how many seconds does the stone reach maximum height? Estimate this height.

CHAPTERS 1–18   Cumulative Review Exercises

1. Evaluate $F = \dfrac{5}{z^2 + 1}$ when $z = -2$.

2. Classify each number as one or more of the following: natural number, whole number, integer, rational number, or irrational number:
$$0.\overline{4},\ \sqrt{7},\ 0,\ -5,\ \sqrt[3]{8},\ -\tfrac{4}{3}.$$

3. Simplify each expression. Write the result using positive exponents.
    (a) $\left(\dfrac{x^2 y^6}{x^{-3}}\right)^2$
    (b) $\dfrac{(xy^{-3})^2}{x(y^{-2})^{-1}}$
    (c) $(a^2 b)^2 (ab^3)^{-4}$

4. Write 9,290,000 in scientific notation.

5. Find $f(-2)$, if $f(x) = \sqrt{2-x}$. Write the domain of f?

6. If $f(2) = 5$, then what point lies on the graph of f?

Exercises 7 and 8: Graph f by hand.

7. $f(x) = x^2 + 2x$

8. $f(x) = |2x - 4|$

9. Find the slope–intercept form of the line that passes through $(4, -1)$ and is parallel to the line passing through $(0, 1)$ and $(-2, 4)$.

10. Find the equation of a vertical line that passes through the point $(-3, 4)$.

11. Solve $2x - 3(x + 2) = 6$.

Exercises 12–14: Solve the inequality. Write your answer in interval notation.

12. $7 - x > 3x$

13. $|3x - 2| \leq 1$

14. $-4 \leq 1 - x < 2$

15. Solve the system.
$$-x - 4y = -3$$
$$5x + y = -4$$

16. Shade the solution set in the xy-plane.
$$3x + y \leq 3$$
$$x - 3y \leq 3$$

17. Solve the system.
$$x + y - z = 3$$
$$x - y + z = 1$$
$$2x - y - z = 1$$

Exercises 18–20: Multiply the expression.

18. $(3x - 2)(2x + 7)$

19. $3xy(x^2 + y^2)$

20. $(\sqrt{x} + 3)(\sqrt{x} - 3)$

Exercises 21 and 22: Factor the expression.

21. $x^3 - x^2 - 2x$

22. $4x^2 - 25$

Exercises 23 and 24: Solve each equation.

23. $x^2 - 3 = 0$

24. $x^2 + 1 = 2x$

Exercises 25 and 26: Simplify the expression.

25. $\dfrac{(x+3)^2}{x+2} \cdot \dfrac{x+2}{2x+6}$

26. $\dfrac{1}{x+2} - \dfrac{1}{x}$

Exercises 27–30: Simplify the expression. Assume all variables are positive.

27. $\sqrt{16x^6}$

28. $16^{-3/2}$

29. $\dfrac{\sqrt[3]{81x}}{\sqrt[3]{3x}}$

30. $\sqrt{8x} + \sqrt{2x}$

31. Graph $f(x) = \sqrt{4x}$.

32. Find the distance between $(-1, 2)$ and $(4, 3)$.

33. Write $\dfrac{3-i}{2+i}$ in standard form.

34. Solve $3\sqrt{x+1} = 2x$.

35. Solve the equation $2x = \sqrt{2.1 - x} + \sqrt[3]{0.1x}$ to the nearest hundredth.

36. Sketch a graph of $f(x) = x^2 - 2x + 3$.
    (a) Find the vertex.
    (b) Evaluate $f(-1)$.
    (c) What is the axis of symmetry?
    (d) Where is f increasing?

37. Write $f(x) = 2x^2 - 4x - 1$ in vertex form.

38. Compare the graph of $f(x) = 4(x + 1)^2 - 2$ to the graph of $y = x^2$.

39. Solve $x^2 + 6x = 2$ by completing the square.

40. Solve $2x^2 - 3x = 1$ by using the quadratic formula.

41. Solve $x(4 - x) = 3$.

42. The graph of $y = ax^2 + bx + c$ is shown. Solve each equation or inequality.
    (a) $ax^2 + bx + c = 0$
    (b) $ax^2 + bx + c \leq 0$

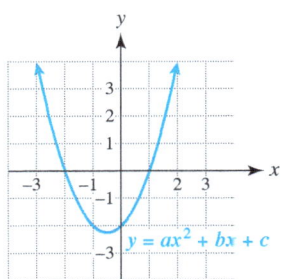

43. Solve $x^2 - 3x + 2 > 0$.

44. Solve $x^4 - 256 = 0$. Find all complex solutions.

Exercises 45–52: **Thinking Generally** *Match the graph (a–h) with its equation. Assume that a, b, and c are positive constants.*

45. $y = ax - b$
46. $y = b$
47. $y = -ax^2 + c$
48. $y = \dfrac{a}{x}$
49. $y = ax^3$
50. $y = |ax + b|$
51. $y = a\sqrt{x}$
52. $y = a\sqrt[3]{x}$

a.
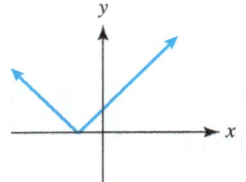

b.
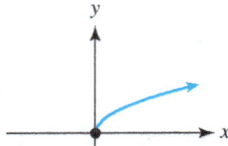

c.
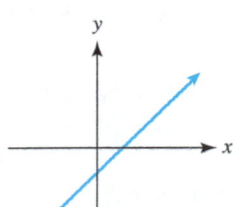

d.
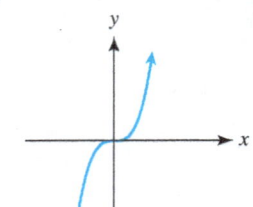

e.
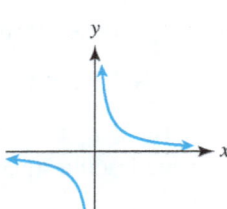

f.
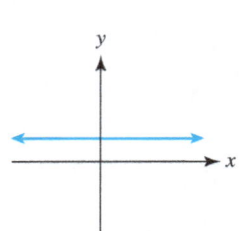

g.
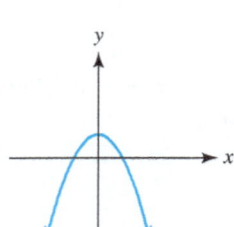

h.
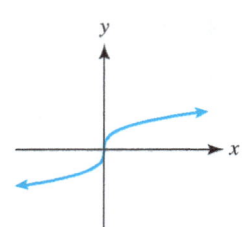

APPLICATIONS

53. Calculating Water Flow Water is being pumped out of a tank. The number of gallons G of water in the tank after t minutes is shown in the figure, where $y = G(t)$.
   (a) Evaluate $G(0)$. Interpret your answer.
   (b) What is the t-intercept? Interpret your answer.
   (c) What is the slope of the graph of G? Interpret your answer.
   (d) Find a formula for $G(t)$.

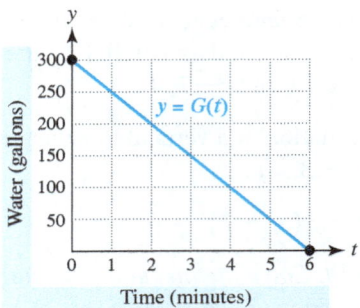

54. Investment Suppose $4000 is deposited in three accounts paying 4%, 5%, and 6% annual interest. The amount invested at 6% is $1000 more than the amount invested at 5%. The interest after 1 year is $216. How much is invested at each rate?

55. Maximizing Area There are 490 feet of fence available to surround the perimeter of a rectangular garden. On one side, there is a 10-foot gate that requires no fencing. What dimensions for the garden give the largest area?

56. Height of a Tree A 6-foot-tall person casts a shadow that is 10 feet long while a nearby tree casts a 55-foot shadow. Find the height of the tree.

19 Exponential and Logarithmic Functions

- 19.1 Composite and Inverse Functions
- 19.2 Exponential Functions
- 19.3 Logarithmic Functions
- 19.4 Properties of Logarithms
- 19.5 Exponential and Logarithmic Equations

When a link is posted on a social network, the majority of hits on the link happen within the first few hours. For example, within the first 3 hours of posting a link on Facebook, the link will receive half of its hits. This pattern can continue over time so the number of hits on the link is decreasing by a factor of $\frac{1}{2}$ every 3 hours. This is sometimes referred to as the half-life of a link. The half-life is even shorter for Twitter links, 2.8 hours, but is much longer, 400 hours, for StumbleUpon. In this chapter, we use exponential functions to model these engagements with links posted on social networks. See Exercises 95–96 in Section 2 of this chapter.

19.1 Composite and Inverse Functions

Objectives

1. Introducing Composition of Functions
   - Symbolic Evaluation
   - Numerical and Graphical Evaluation
2. Recognizing One-to-One Functions
   - Horizontal Line Test
   - Increasing, Decreasing, and One-to-One Functions
3. Finding Inverse Functions
   - Inverse Operations
   - Finding an Inverse
4. Using Tables and Graphs to Express Inverse Functions
   - Tables of Inverse Functions
   - Graphs of Inverse Functions

NEW VOCABULARY

☐ Composite function
☐ Composition
☐ One-to-one correspondence
☐ One-to-one
☐ Horizontal line test
☐ Inverse function

1 Introducing Composition of Functions

Connecting Concepts with Your Life Many tasks in life are performed in *sequence*, such as putting on your socks and then your shoes. These types of situations also occur in mathematics. For example, suppose that we want to calculate the number of ounces in 3 tons. Because there are 2000 pounds in 1 ton, we might first multiply 3 by 2000 to obtain **6000** pounds. There are 16 ounces in a pound, so we could multiply **6000** by 16 to obtain **96,000** ounces. This particular calculation involves a *sequence* of calculations that can be represented by the diagram shown in **FIGURE 19.1**.

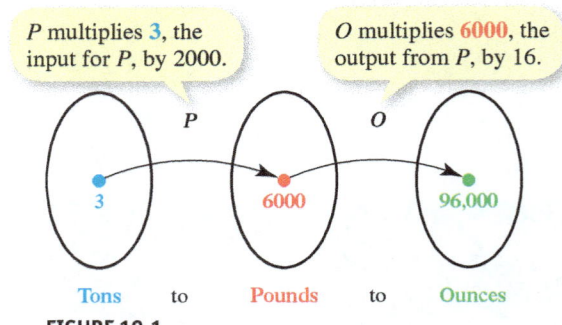

Converting Tons to Ounces

P multiplies **3**, the input for *P*, by 2000.

O multiplies **6000**, the output from *P*, by 16.

Tons   to   Pounds   to   Ounces

FIGURE 19.1

SYMBOLIC EVALUATION The results shown in **FIGURE 19.1** can be calculated by using function notation. Suppose that we let $P(x) = 2000x$ convert x tons to P pounds and also let $O(x) = 16x$ convert x pounds to O ounces. We can calculate the number of ounces in 3 tons by performing the *composition of O and P*. This method can be expressed symbolically as follows.

Composition: $(O \circ P)(3)$ $\quad = O(P(3))$

Input 3
↓
$P(x)$
↓
Output $P(3)$
becomes input for $O(x)$.
↓
$(O \circ P)(x) = O(P(x))$
↓
Output $O(P(3))$

$= O(2000 \cdot 3)$ Let $x = 3$ in $P(x) = 2000x$.
↓
↓ $P(3) = 6000$.
$= O(6000)$ Simplify.
$= 16(6000)$ Let $x = 6000$ in $O(x) = 16x$.
↓
  P first multiplies 3 by 2000 and then *O* multiplies 6000 by 16 to get 96,000. This is written as $(O \circ P)(3)$.
↓
$= 96,000.$ Multiply.

COMPOSITION OF FUNCTIONS

If f and g are functions, then the **composite function** $g \circ f$, or **composition** of g and f, is defined by

$$(g \circ f)(x) = g(f(x)).$$

NOTE: We read $g(f(x))$ as "g of f of x." ∎

READING CHECK 1

- Calculate $(f \circ g)(2)$ and $(g \circ f)(2)$ if $f(x) = x^2$ and $g(x) = 2x - 1$.

> **MAKING CONNECTIONS 1**
>
> **$g \circ f$ vs. $f \circ g$**
>
> The compositions $g \circ f$ and $f \circ g$ represent evaluating functions f and g in different orders. When evaluating $g \circ f$, function f is evaluated first followed by function g, whereas for $f \circ g$, the function g is evaluated first followed by function f. Note that, in general
>
> $$(g \circ f)(x) \neq (f \circ g)(x).$$
>
> That is, *the order in which functions are applied makes a difference*, in the same way that putting on your socks and then your shoes is quite different from putting on your shoes and then your socks. See Example 2(a) and (b).

EXAMPLE 1   **Evaluating composite functions symbolically**

Evaluate $(g \circ f)(2)$ and then find a formula for $(g \circ f)(x)$.
(a) $f(x) = x^3, g(x) = 3x - 2$ (b) $f(x) = 5x, g(x) = x^2 - 3x + 1$
(c) $f(x) = \sqrt{2x}, g(x) = \dfrac{1}{x - 1}$

Solution

(a) $(g \circ f)(2) = g(f(2))$ Composition of functions
$ = g(8)$ $f(2) = 2^3 = 8$
$ = 22$ $g(8) = 3(8) - 2 = 22$

Note that the output from f, which is $f(2)$, becomes the input for g.

$(g \circ f)(x) = g(f(x))$ Composition of functions
$ = g(x^3)$ $f(x) = x^3$
$ = 3x^3 - 2$ Replace x with x^3 in $g(x) = 3x - 2$.

(b) $(g \circ f)(2) = g(f(2))$ Composition of functions
$ = g(10)$ $f(2) = 5(2) = 10$
$ = 71$ $g(10) = 10^2 - 3(10) + 1 = 71$

$(g \circ f)(x) = g(f(x))$ Composition of functions
$ = g(5x)$ $f(x) = 5x$
$ = (5x)^2 - 3(5x) + 1$ Replace x with $5x$ in $g(x) = x^2 - 3x + 1$.
$ = 25x^2 - 15x + 1$ Simplify.

(c) $(g \circ f)(2) = g(f(2))$ Composition of functions
$ = g(2)$ $f(2) = \sqrt{2(2)} = 2$
$ = 1$ $g(2) = \dfrac{1}{2 - 1} = 1$

$(g \circ f)(x) = g(f(x))$ Composition of functions
$ = g(\sqrt{2x})$ $f(x) = \sqrt{2x}$
$ = \dfrac{1}{\sqrt{2x} - 1}$ Replace x with $\sqrt{2x}$ in $g(x) = \dfrac{1}{x - 1}$.

Now Try Exercises 13, 15, 19

NUMERICAL AND GRAPHICAL EVALUATION Composite functions can also be evaluated numerically and graphically, as demonstrated in the next two examples.

EXAMPLE 2 Evaluating composite functions with tables

Use **TABLES 19.1** and **19.2** to evaluate each expression.
(a) $(f \circ g)(2)$ (b) $(g \circ f)(2)$ (c) $(f \circ f)(0)$

x	0	1	2	3
$f(x)$	3	2	0	1

TABLE 19.1

x	0	1	2	3
$g(x)$	1	3	2	0

TABLE 19.2

Solution

(a) $(f \circ g)(2) = f(g(2))$ Composition of functions
$\quad\quad\quad\quad\quad = f(2)$ $g(2) = 2$
$\quad\quad\quad\quad\quad = 0$ $f(2) = 0$

(b) $(g \circ f)(2) = g(f(2))$ Composition of functions
$\quad\quad\quad\quad\quad = g(0)$ $f(2) = 0$
$\quad\quad\quad\quad\quad = 1$ $g(0) = 1$

NOTE: From parts (a) and (b), we see that $(f \circ g)(2) \neq (g \circ f)(2)$. ∎

(c) $(f \circ f)(0) = f(f(0))$ Composition of functions
$\quad\quad\quad\quad\quad = f(3)$ $f(0) = 3$
$\quad\quad\quad\quad\quad = 1$ $f(3) = 1$

Now Try Exercises 23, 25

EXAMPLE 3 Evaluating composite functions graphically

Use **FIGURE 19.2** in the margin to evaluate $(g \circ f)(2)$.

Solution

Because $(g \circ f)(2) = g(f(2))$, start by using **FIGURE 19.2** to evaluate $f(2)$. **FIGURE 19.3(a)** shows that $f(2) = 4$, which becomes the input for g. **FIGURE 19.3(b)** reveals that $g(4) = 2$.

$$(g \circ f)(2) = g(f(2)) = g(4) = 2$$

FIGURE 19.2

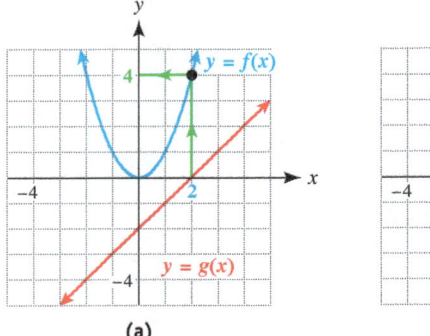

Evaluate: $f(2) = 4$ Evaluate: $g(4) = 2$

(a)   (b)

FIGURE 19.3

Now Try Exercise 29

2 Recognizing One-to-One Functions

Math in Real-World Context In a typical city, there is *usually* a **one-to-one correspondence** between homes and addresses. That is to say that each home has one address, and each address corresponds to one home. We can apply this concept to functions and ask the question, do *different inputs* always result in *different outputs* for every function? The answer is no. Use $f(x) = x^2 + 1$ as an example.

$$f(\underbrace{-2}_{\text{Different Inputs}}) = \underbrace{5 \text{ and } f(2) = 5}_{\text{Same Output}}.$$

However, for $g(x) = 2x$, *different inputs* always result in *different outputs*. For function g, there is a one-to-one correspondence between inputs and outputs. Thus we say that g is a *one-to-one function*, whereas f is not.

ONE-TO-ONE FUNCTION

A function f is **one-to-one** if, for any c and d in the domain of f,

$$c \neq d \quad \text{implies that} \quad f(c) \neq f(d).$$

That is, different inputs always result in different outputs.

HORIZONTAL LINE TEST The following See the Concept shows how to use the horizontal line test to determine if a function is one-to-one.

See the Concept 1   HORIZONTAL LINE TEST

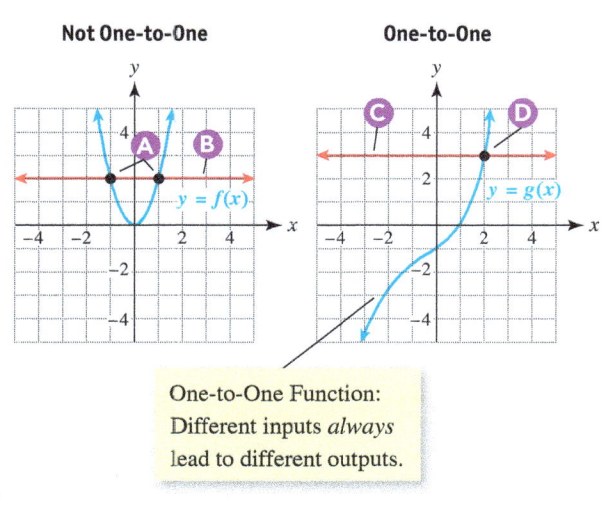

One-to-One Function: Different inputs *always* lead to different outputs.

A The points $(-1, 2)$ and $(1, 2)$ both lie on the graph of f and so $f(-1) = f(1) = 2$. Since different inputs do *not always* lead to different outputs, the function is not one-to-one.

B Two points with different *x*-values and the same *y*-value determine a horizontal line as shown. Thus if any horizontal line intersects the graph of f at more than one point, f is *NOT* one-to-one.

C To determine if a function is one-to-one by its graph, apply the **horizontal line test** and visualize a horizontal line moving up and down over the graph. For a function to be one-to-one, every possible horizontal line can intersect the graph *at most once*.

D Because every horizontal line intersects the graph at most once, g is a one-to-one function.

WATCH VIDEO IN MML.

READING CHECK 2

- How can you determine if a function is one-to-one by looking at its graph?

HORIZONTAL LINE TEST

If every horizontal line intersects the graph of a function f at most once, then f is a one-to-one function.

1226 CHAPTER 19 EXPONENTIAL AND LOGARITHMIC FUNCTIONS

EXAMPLE 4 Using the horizontal line test

Determine whether each graph in **FIGURE 19.4** represents a one-to-one function.

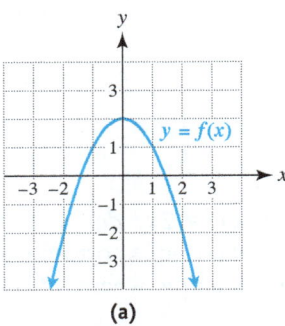 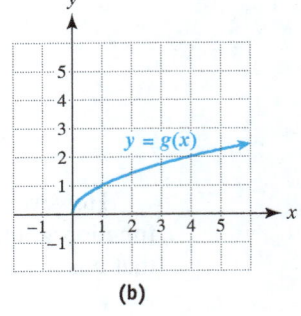

(a)   (b)

FIGURE 19.4

Solution

FIGURE 19.5(a) shows one of many horizontal lines that intersect the graph of $y = f(x)$ twice. Therefore, function f is *not* one-to-one.

Two Points of Intersection: **At Most One Point of**
f **Is Not One-to-One**   **Intersection:** g **Is One-to-One**

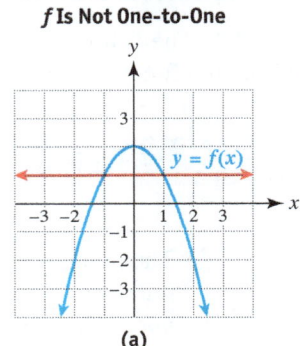 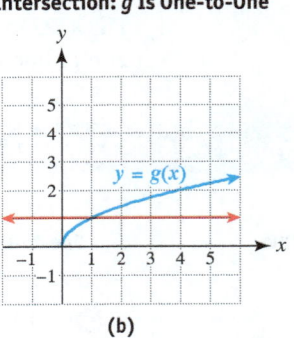

(a)   (b)

FIGURE 19.5

FIGURE 19.5(b) suggests that every horizontal line will intersect the graph of $y = g(x)$ at *most* once. Therefore, function g is one-to-one.

Now Try Exercises 37, 39

STUDY TIP

Be sure that you understand when to use the vertical line test and when to use the horizontal line test.

MAKING CONNECTIONS 2

Vertical and Horizontal Line Tests

The **vertical line test** is used to identify **functions**, whereas the **horizontal line test** is used to identify **one-to-one** functions. For example, consider the graph of $f(x) = x^2$. A vertical line never intersects the graph more than once, so f is a function. A horizontal line can intersect the graph twice, so f is *not* a one-to-one function.

Function; Not One-to-One

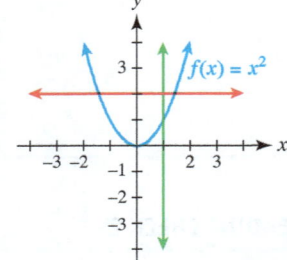

INCREASING, DECREASING, AND ONE-TO-ONE FUNCTIONS When a continuous function *f* is always increasing on its domain, every horizontal line will intersect the graph of *f* at most once. By the horizontal line test, *f* is a one-to-one function. For example, function *f* in Example 4(b) is always increasing on its domain and so it is one-to-one. Similarly, when a continuous function *g* is always decreasing on its domain, *g* is a one-to-one function.

3 Finding Inverse Functions

INVERSE OPERATIONS There are many examples of inverse operations in everyday life.

Connecting Concepts with Your Life Suppose that you walk into a classroom, turn on the lights, and sit down at your desk. How could you undo or reverse these actions? You would stand up from the desk, turn off the lights, and walk out of the classroom. Note that you must not only perform the "inverse" of each action, but you also must do them in the *reverse order*. In mathematics, we undo an arithmetic operation by performing its inverse operation. For example, the inverse operation of addition is subtraction, and the inverse operation of multiplication is division. Adding 5 to *x* and subtracting 5 from *x* are inverse operations.

$$x + 5 - 5 = x$$

Adding 5 and then subtracting 5 gives *x* back.

Similarly, multiplying *x* by 5 and dividing *x* by 5 are inverse operations.

$$\frac{5x}{5} = x$$

Multiplying by 5 and then dividing by 5 gives *x* back.

EXAMPLE 5 **Finding inverse operations**

Give the inverse operations for each statement. Then write a function *f* for the given statement and a function *g* for its inverse operations.
(a) Divide *x* by 3.   (b) Cube *x* and then add 1 to the result.

Solution
(a) The inverse of dividing *x* by 3 is to *multiply x* by 3. Thus

$$f(x) = \frac{x}{3} \quad \text{and} \quad g(x) = 3x.$$

(b) The inverse of cubing a number is taking a cube root, and the inverse of adding 1 is subtracting 1. The inverse operations of "cubing a number and then adding 1" are "subtracting 1 and then taking a cube root." For example, 2 cubed plus 1 is $2^3 + 1 = 9$. For the inverse operations, we *first* subtract 1 from 9 and then take the cube root to obtain 2. That is, $\sqrt[3]{9 - 1} = 2$. When there is more than one operation, we must perform the inverse operations in *reverse order*. Thus

$$f(x) = x^3 + 1 \quad \text{and} \quad g(x) = \sqrt[3]{x - 1}.$$

Now Try Exercises 43, 49

Inverse Functions *f* and *g*

Domain   Range

Range   Domain

The domain of *f* is the range of *g*, and the domain of *g* is the range of *f*.

Functions *f* and *g* in each part of Example 5 are examples of *inverse functions*. Note that in part (a), if $f(x) = \frac{x}{3}$ and $g(x) = 3x$, then

$$f(15) = 5 \quad \text{and} \quad g(5) = 15.$$

If *f* outputs *b* with input *a*, *g* must output *a* with input *b*.

See the diagram to the left.

In general, if *f* and *g* are inverse functions, $f(a) = b$ implies $g(b) = a$. Thus

$$(g \circ f)(a) = g(f(a)) = g(b) = a$$

for any *a* in the domain of *f*, whenever *g* and *f* are inverse functions. The composition of a function with its inverse leaves the input unchanged.

READING CHECK 3

- What happens to the input of the composition of a function with its inverse?

INVERSE FUNCTIONS

Let f be a one-to-one function. Then f^{-1} is the **inverse function** of f if

$$(f^{-1} \circ f)(x) = f^{-1}(f(x)) = x, \quad \text{for every } x \text{ in the domain of } f, \quad \text{and}$$

$$(f \circ f^{-1})(x) = f(f^{-1}(x)) = x, \quad \text{for every } x \text{ in the domain of } f^{-1}.$$

NOTE: In the expression $f^{-1}(x)$, the -1 is *not* an exponent. That is, $f^{-1}(x) \neq \frac{1}{f(x)}$. Rather, if $f(x) = \frac{x}{3}$, then $f^{-1}(x) = 3x$, and if $f(x) = x^3 + 1$, then $f^{-1}(x) = \sqrt[3]{x-1}$. ∎

EXAMPLE 6   Verifying inverses

Verify that $f^{-1}(x) = 3x$ if $f(x) = \frac{x}{3}$.

Solution
We must show that $(f^{-1} \circ f)(x) = x$ and that $(f \circ f^{-1})(x) = x$.

$$(f^{-1} \circ f)(x) = f^{-1}(f(x)) \quad \text{Composition of functions}$$

$$= f^{-1}\left(\frac{x}{3}\right) \quad f(x) = \frac{x}{3}$$

$$= 3\left(\frac{x}{3}\right) \quad f^{-1}(x) = 3x$$

$$= x \checkmark \quad \text{Simplify; it checks.}$$

$$(f \circ f^{-1})(x) = f(f^{-1}(x)) \quad \text{Composition of functions}$$

$$= f(3x) \quad f^{-1}(x) = 3x$$

$$= \frac{3x}{3} \quad f(x) = \frac{x}{3}$$

$$= x \checkmark \quad \text{Simplify; it checks.}$$

Now Try Exercise 51

The definition of inverse functions states that *f must be a one-to-one function*. To understand why, consider **FIGURE 19.6(a)**, where a one-to-one function f is represented by a diagram. To find f^{-1}, the arrows are simply reversed. For example, $f(1) = 3$ implies that $f^{-1}(3) = 1$, so the arrow from 1 to 3 for f must be redrawn in **FIGURE 19.6(b)**, from 3 to 1 for f^{-1}.

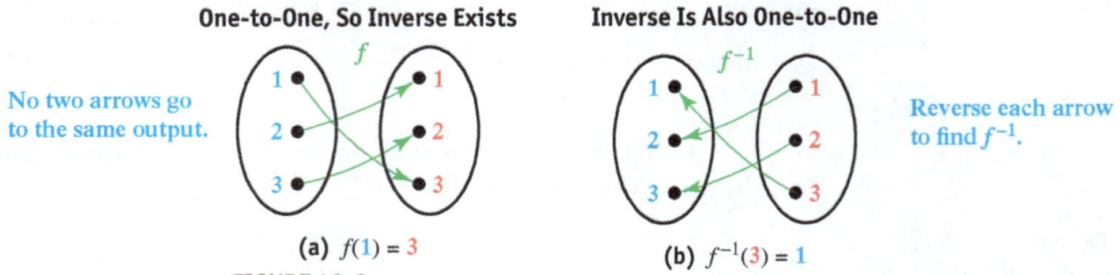

One-to-One, So Inverse Exists   Inverse Is Also One-to-One

No two arrows go to the same output.   Reverse each arrow to find f^{-1}.

(a) $f(1) = 3$ (b) $f^{-1}(3) = 1$

FIGURE 19.6

In **FIGURE 19.7(a)**, function g is *not* one-to-one because two different arrows, from **2** and **3**, both go to **1**. In **FIGURE 19.7(b)**, the arrows are reversed in an attempt to draw the inverse function. However, **two** arrows now go from input **1** to outputs **2** and **3**. This means that *one input has two outputs*, which is not possible if g^{-1} is a function.

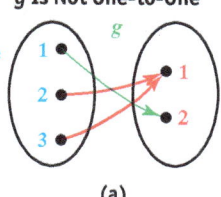
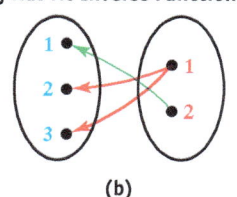

g Is Not One-to-One
Two inputs give the same output: not one-to-one.

g Has No Inverse Function
One input gives two outputs: not a function.

(a)   (b)
FIGURE 19.7

FINDING AN INVERSE The following steps can be used to find the inverse of a function symbolically. Note that for a function f to have an inverse function, f must be one-to-one.

FINDING AN INVERSE FUNCTION

To find f^{-1} for a one-to-one function f, perform the following steps.

STEP 1: Let $y = f(x)$.

STEP 2: Interchange x and y.

STEP 3: Solve the formula for y. The resulting formula is $y = f^{-1}(x)$.

EXAMPLE 7   Finding an inverse function

Find the inverse of each one-to-one function.
(a) $f(x) = 3x - 7$
(b) $g(x) = (x + 2)^3$

Solution
(a) **STEP 1:** Let $y = 3x - 7$.

STEP 2: Interchange x and y to get $x = 3y - 7$.

STEP 3: To solve for y, start by adding 7 to each side.

$$x + 7 = 3y \qquad \text{Add 7 to each side.}$$
$$\frac{x + 7}{3} = y \qquad \text{Divide each side by 3.}$$

Thus $f^{-1}(x) = \frac{x+7}{3}$ or $f^{-1}(x) = \frac{1}{3}x + \frac{7}{3}$.

(b) **STEP 1:** Let $y = (x + 2)^3$.

STEP 2: Interchange x and y to get $x = (y + 2)^3$.

STEP 3: To solve for y, start by taking the cube root of each side.

$$\sqrt[3]{x} = y + 2 \qquad \text{Take the cube root of each side.}$$
$$\sqrt[3]{x} - 2 = y \qquad \text{Subtract 2 from each side.}$$

Thus $g^{-1}(x) = \sqrt[3]{x} - 2$.

Now Try Exercises 63, 71

4 Using Tables and Graphs to Express Inverse Functions

TABLES OF INVERSE FUNCTIONS Inverse functions can be represented with tables. **TABLE 19.3** on the next page shows a table of values for a function f.

A Function f

x	1	2	3	4	5
f(x)	3	6	9	12	15

TABLE 19.3

Because $f(1) = 3$, $f^{-1}(3) = 1$. Similarly, $f(2) = 6$ implies that $f^{-1}(6) = 2$ and so on. **TABLE 19.4** lists values for $f^{-1}(x)$.

The Inverse Function f^{-1}

x	3	6	9	12	15
$f^{-1}(x)$	1	2	3	4	5

← Interchange values

TABLE 19.4

Note that the domain of f is {1, 2, 3, 4, 5} and that the range of f is {3, 6, 9, 12, 15}, whereas the domain of f^{-1} is {3, 6, 9, 12, 15} and the range of f^{-1} is {1, 2, 3, 4, 5}. *The domain of f is the range of f^{-1}, and the range of f is the domain of f^{-1}.* This statement is true in general for a function and its inverse.

GRAPHS OF INVERSE FUNCTIONS If $f(a) = b$, then the point (a, b) lies on the graph of f. This statement also means that $f^{-1}(b) = a$ and that the point (b, a) lies on the graph of f^{-1}. These points are shown in **FIGURE 19.8(a)** with a solid green line segment connecting them. The line $y = x$ is perpendicular to this line segment and divides it into two equal parts. As a result, the graph of f^{-1} can be sketched from the graph of f by reflecting the graph of f across the line $y = x$. See **FIGURE 19.8(b)**.

Reflections Across the Line $y = x$

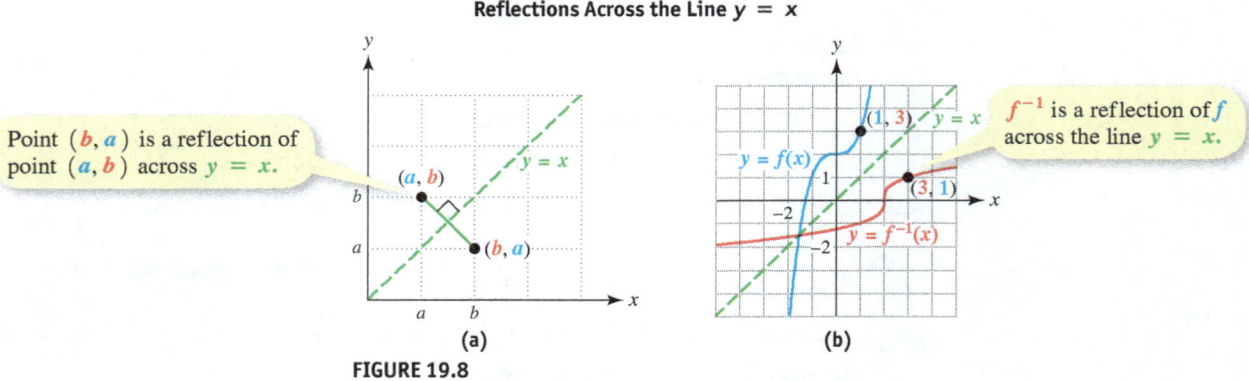

Point (b, a) is a reflection of point (a, b) across $y = x$.

f^{-1} is a reflection of f across the line $y = x$.

FIGURE 19.8

The relationship between the graph of a function and the graph of its inverse function is summarized as follows.

GRAPHS OF FUNCTIONS AND THEIR INVERSES

The graph of f^{-1} is a reflection of the graph of f across the line $y = x$.

19.1 COMPOSITE AND INVERSE FUNCTIONS 1231

EXAMPLE 8 Graphing an inverse function

The graph of $y = f(x)$ is shown in **FIGURE 19.9**. Sketch a graph of $y = f^{-1}(x)$.

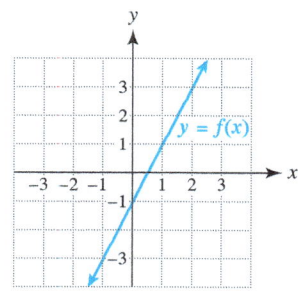

FIGURE 19.9

CRITICAL THINKING 2

The graph of a linear function f passes through the points $(1, 2)$ and $(2, 1)$. What two points does the graph of f^{-1} pass through? Find $f(x)$ and $f^{-1}(x)$.

Solution

The graph of $y = f^{-1}(x)$ is the reflection of the graph of $y = f(x)$ across the line $y = x$ and is shown in **FIGURE 19.10**. Note that the reflection of a line is another line.

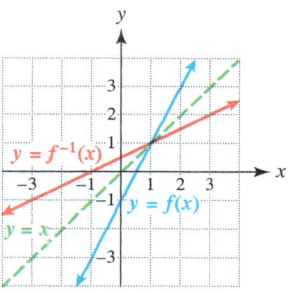

FIGURE 19.10

Now Try Exercise 83

EXAMPLE 9 Graphing an inverse function

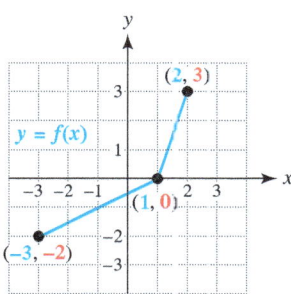

FIGURE 19.11

The line graph shown in **FIGURE 19.11** represents a function f.
(a) Is f a one-to-one function?
(b) Sketch a graph of $y = f^{-1}(x)$.

Solution

(a) Every horizontal line intersects the graph of f at most once. By the horizontal line test, the graph represents a one-to-one function.
(b) The points $(-3, -2)$, $(1, 0)$, and $(2, 3)$ lie on the graph of f. It follows that the points $(-2, -3)$, $(0, 1)$, and $(3, 2)$ lie on the graph of f^{-1}. Plot these three points and then connect them with line segments, as shown in **FIGURE 19.12(a)**. Note that the graph of $y = f^{-1}(x)$ is a reflection of the graph of $y = f(x)$ across the line $y = x$, as shown in **FIGURE 19.12(b)**.

READING CHECK 4

- If we are given the graph of $y = f(x)$, where f is one-to-one, how can we sketch the graph of $y = f^{-1}(x)$?

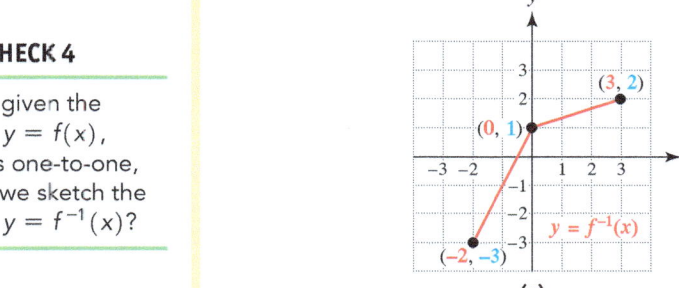 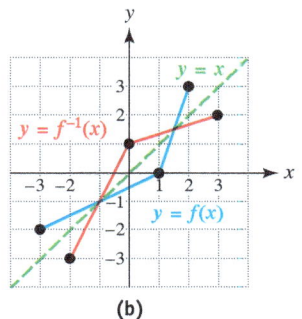

(a) (b)

FIGURE 19.12

Now Try Exercise 81

See the Concept 2: REPRESENTING INVERSE FUNCTIONS

Verbal

f: Multiply x by 2 and add 2.

f^{-1}: Subtract 2 from x and divide by 2.

Use inverse operations in reverse order.

Symbolic

$f(x) = 2x + 2$
$y = 2x + 2$
$x = 2y + 2$
$x - 2 = 2y$
$\dfrac{x-2}{2} = y$

$f^{-1}(x) = \dfrac{x-2}{2}$

Interchange x and y.
Solve for y.

Numerical

x	$f(x)$		x	$f^{-1}(x)$
-2	-2		-2	-2
-1	0		0	-1
0	2		2	0
1	4		4	1

Interchange the x- and y-values.

Graphical

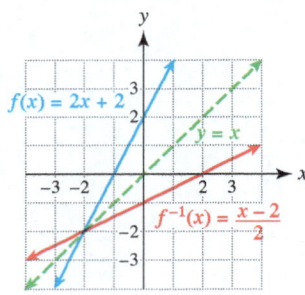

Reflect the graph of f across the line $y = x$.

WATCH VIDEO IN MML.

EXAMPLE 10   Evaluating f and f^{-1} graphically

Use the graph of f in **FIGURE 19.13** to evaluate each expression.
(a) $f(2)$ (b) $f^{-1}(-3)$ (c) $f^{-1}(2)$

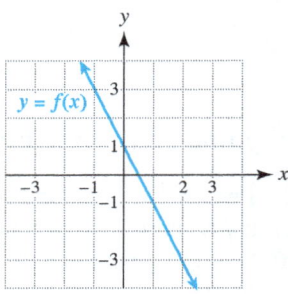

FIGURE 19.13

Solution

(a) To evaluate $f(2)$, find **2** on the x-axis, move downward to the graph of f, and then move left to the y-axis to obtain $f(2) = -3$, as shown in **FIGURE 19.14(a)**.

(b) Start by finding -3 on the y-axis, move right to the graph of f, and then move upward to the x-axis to obtain $f^{-1}(-3) = 2$, as shown in **FIGURE 19.14(b)**. Notice that $f(2) = -3$ from part (a) and $f^{-1}(-3) = 2$ here.

(c) Find **2** on the y-axis, move left to the graph of f, and then move downward to the x-axis. We can see from **FIGURE 19.14(c)** that $f^{-1}(2) = -\tfrac{1}{2}$.

$f(2) = -3$ $f^{-1}(-3) = 2$ $f^{-1}(2) = -\tfrac{1}{2}$

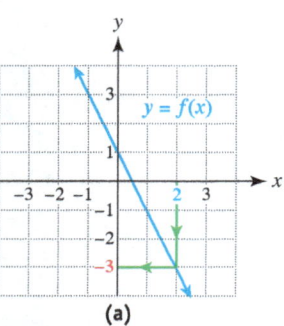 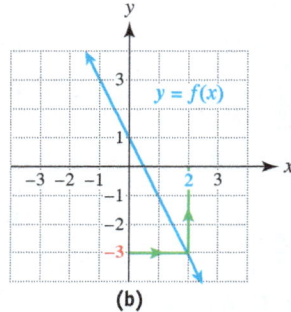 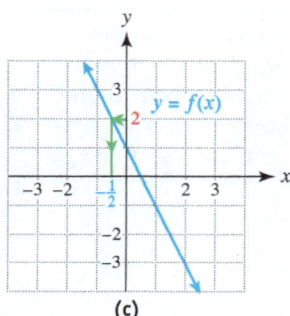

(a)   (b)   (c)

FIGURE 19.14

Now Try Exercise 87

19.1 Putting It All Together

CONCEPT	EXPLANATION	EXAMPLES
Composite Functions	The composition of g and f is $$(g \circ f)(x) = g(f(x)),$$ and represents a *new* function whose name is $g \circ f$.	If $f(x) = 1 - 4x$ and $g(x) = x^3$, then $$(g \circ f)(x) = g(f(x))$$ $$= g(1 - 4x)$$ $$= (1 - 4x)^3.$$
One-to-One Functions	Function f is one-to-one if different inputs always give different outputs.	$f(x) = x^2$ is not one-to-one because $f(-4) = f(4) = 16$, whereas $g(x) = x + 1$ is one-to-one because every input has a unique output.
Horizontal Line Test	This test is used to determine whether a function is one-to-one from its graph: If every horizontal line intersects the graph of a function f at most once, then f is a one-to-one function.	$f(x) = x^2$ is not one-to-one. A horizontal line can intersect its graph more than once. **Not One-to-One** *Fails horizontal line test*
Inverse Functions	f^{-1} will undo the operations performed by f. That is, $$(f^{-1} \circ f)(x) = x \text{ and}$$ $$(f \circ f^{-1})(x) = x.$$	If $f(x) = x^3$, then $f^{-1}(x) = \sqrt[3]{x}$ because cubing a number x and then taking its cube root result in the number x.

19.1 Exercises

CONCEPTS AND VOCABULARY

1. $(g \circ f)(7) = $ _____

2. $(f \circ g)(x) = $ _____

3. Does $(f \circ g)(x)$ always equal $(g \circ f)(x)$?

4. If a function f is one-to-one, then different _____ always result in different _____.

5. If $f(3) = 5$ and $f(7) = 5$, could f be one-to-one?

6. If every horizontal line intersects the graph of f at most once, then f is _____.

7. The inverse operation of subtracting 10 is _____.

8. $(f^{-1} \circ f)(7) = $ _____.

9. If $f(6) = 8$, then f^{-1} (_____) = _____.

10. If $f^{-1}(y) = x$, then f (_____) = _____.

11. For f to have an inverse function, f must be _____.

12. The graph of f^{-1} is a(n) _____ of the graph of f across the line _____.

EVALUATING COMPOSITE FUNCTIONS SYMBOLICALLY

Exercises 13–22: For the given $f(x)$ and $g(x)$, find each of the following.

(a) $(g \circ f)(-2)$ (b) $(f \circ g)(4)$
(c) $(g \circ f)(x)$ (d) $(f \circ g)(x)$

13. $f(x) = x^2$ $g(x) = x + 3$

14. $f(x) = 4x^2$ $g(x) = 5x$

15. $f(x) = 2x$ $g(x) = x^3 - 1$

16. $f(x) = 3x + 1$ $g(x) = x^2 + 4x$

17. $f(x) = \frac{1}{2}x$ $g(x) = |x - 2|$

18. $f(x) = 6x$ $g(x) = \frac{2}{x - 5}$

19. $f(x) = \frac{1}{x}$ $g(x) = 3 - 5x$

20. $f(x) = \sqrt{x + 3}$ $g(x) = x^3 - 3$

21. $f(x) = 2x$ $g(x) = 4x^2 - 2x + 5$

22. $f(x) = 9x - \frac{1}{3x}$ $g(x) = \frac{x}{3}$

EVALUATING COMPOSITE FUNCTIONS NUMERICALLY

Exercises 23–28: For one-to-one functions f and g, evaluate each expression numerically.

x	−2	−1	0	1	2
f(x)	2	1	0	−1	−2

x	−2	−1	0	1	2
g(x)	0	1	−1	2	−2

23. (a) $(f \circ g)(0)$ (b) $(g \circ f)(-1)$

24. (a) $(f \circ g)(1)$ (b) $(g \circ f)(-2)$

25. (a) $(f \circ f)(-1)$ (b) $(g \circ g)(0)$

26. (a) $(g \circ g)(2)$ (b) $(f \circ f)(1)$

27. (a) $(f^{-1} \circ g)(-2)$ (b) $(g^{-1} \circ f)(2)$

28. (a) $(f \circ g^{-1})(1)$ (b) $(g \circ f^{-1})(-2)$

EVALUATING COMPOSITE FUNCTIONS GRAPHICALLY

Exercises 29 and 30: Evaluate each expression graphically.

29. (a) $(f \circ g)(0)$

    (b) $(g \circ f)(1)$

    (c) $(f \circ f)(-1)$

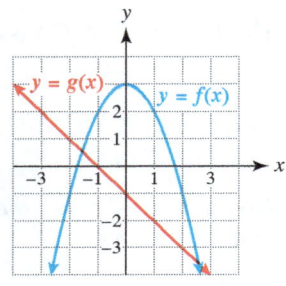

30. (a) $(f \circ g)(1)$

    (b) $(g \circ f)(-2)$

    (c) $(g \circ g)(-2)$

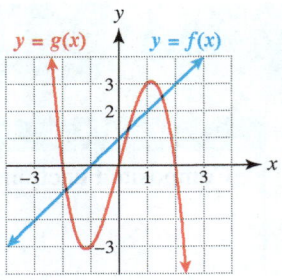

IDENTIFYING ONE-TO-ONE FUNCTIONS

Exercises 31–36: Show that f is not a one-to-one function by finding two inputs that result in the same output. Answers may vary.

31. $f(x) = 5x^2$

32. $f(x) = 4 - x^2$

33. $f(x) = x^4 + 100$

34. $f(x) = \frac{x^2}{x^2 + 1}$

35. $f(x) = x^4 - 3x^2$

36. $f(x) = \sqrt{x^2 - 1}$

APPLYING THE HORIZONTAL LINE TEST

Exercises 37–42: Use the horizontal line test to determine whether the graph represents a one-to-one function.

37.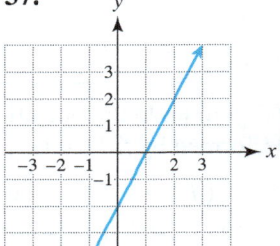

38.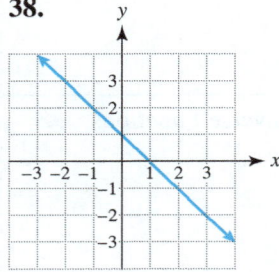

39.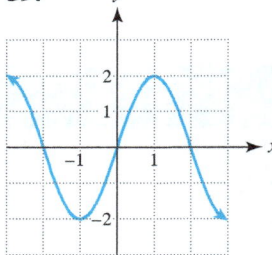

40.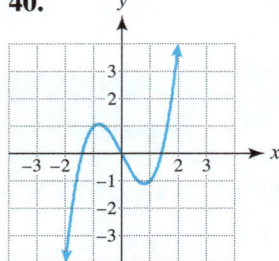

41.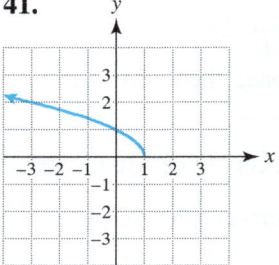

42.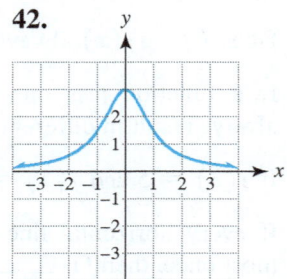

FINDING INVERSE FUNCTIONS

Exercises 43–50: (Refer to Example 5.) Give the inverse operation for the statement. Then write a function f for the given statement and a function g for its inverse.

43. Multiply x by 7.

44. Subtract 10 from x.

45. Add 5 to x and then divide the result by 2.

46. Multiply x by 6 and then add 8 to the result.

47. Multiply x by $\frac{1}{2}$ and then subtract 3 from the result.

48. Divide x by 10 and then add 20 to the result.

49. Cube the sum of x and 5.

50. Take the cube root of x and then subtract 2.

Exercises 51–58: (Refer to Example 6.) Verify that $f(x)$ and $f^{-1}(x)$ are indeed inverse functions.

51. $f(x) = 4x$ $f^{-1}(x) = \frac{x}{4}$

52. $f(x) = \frac{2x}{3}$ $f^{-1}(x) = \frac{3x}{2}$

53. $f(x) = 3x + 5$ $f^{-1}(x) = \frac{x - 5}{3}$

54. $f(x) = x + 7$ $f^{-1}(x) = x - 7$

55. $f(x) = x^3$ $f^{-1}(x) = \sqrt[3]{x}$

56. $f(x) = \sqrt[3]{x - 4}$ $f^{-1}(x) = x^3 + 4$

57. $f(x) = \frac{1}{x}$ $f^{-1}(x) = \frac{1}{x}$

58. $f(x) = \frac{x + 7}{7}$ $f^{-1}(x) = 7x - 7$

Exercises 59–74: (Refer to Example 7.) Find $f^{-1}(x)$.

59. $f(x) = 12x$

60. $f(x) = \frac{3}{4}x$

61. $f(x) = x + 8$

62. $f(x) = x - 3$

63. $f(x) = 5x - 2$

64. $f(x) = 3x + 4$

65. $f(x) = -\frac{1}{2}x + 1$

66. $f(x) = \frac{3}{4}x - \frac{1}{4}$

67. $f(x) = 8 - x$

68. $f(x) = 5 - x$

69. $f(x) = \frac{x + 1}{2}$

70. $f(x) = \frac{3 - x}{5}$

71. $f(x) = \sqrt[3]{2x}$

72. $f(x) = \sqrt[3]{x + 4}$

73. $f(x) = x^3 - 8$

74. $f(x) = (x - 5)^3$

RELATING TABLES AND GRAPHS TO INVERSE FUNCTIONS

Exercises 75–78: Use the table to make a table of values for $f^{-1}(x)$. State the domain and range for f and for f^{-1}.

75.
x	0	1	2	3	4
$f(x)$	0	5	10	15	20

76.
x	−4	−2	0	2	4
$f(x)$	1	2	3	4	5

77.
x	−5	0	5	10	15
$f(x)$	4	2	0	−2	−4

78.
x	0	2	4	6	8
$f(x)$	8	6	4	2	0

Exercises 79–82: (Refer to Example 9.) Use the graph of $y = f(x)$ to sketch a graph of $y = f^{-1}(x)$.

79.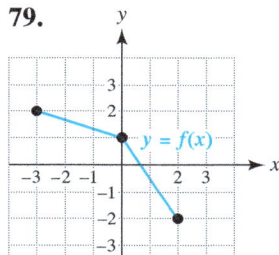

80.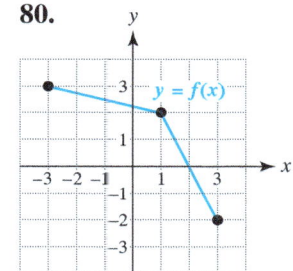

81.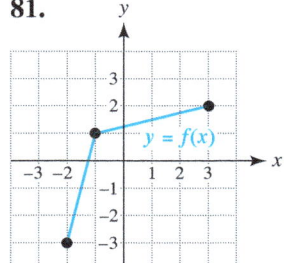

82.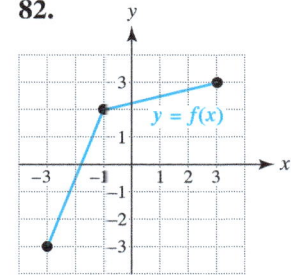

Exercises 83–86: (Refer to Example 8.) Use the graph of $y = f(x)$ to sketch a graph of $y = f^{-1}(x)$. Include the graph of f and the line $y = x$ in your graph.

83.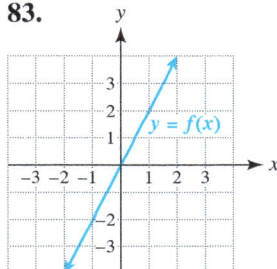

84.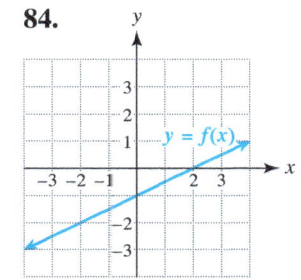

85.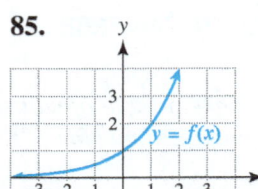

86.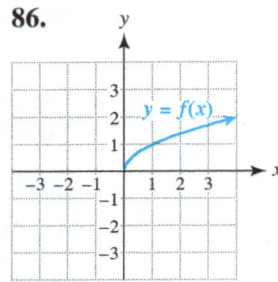

Exercises 87–90: Use the graph of $f(x)$ to evaluate the expression.

87. (a) $f^{-1}(-8)$
(b) $f^{-1}(8)$
(c) $f^{-1}(0)$

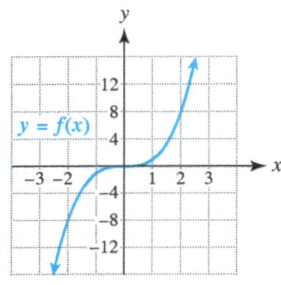

88. (a) $f^{-1}(1)$
(b) $f^{-1}(\frac{1}{2})$
(c) $f^{-1}(2)$

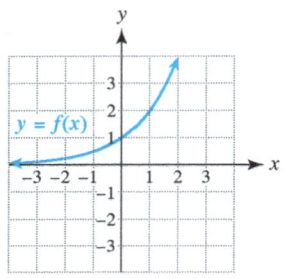

89. (a) $f^{-1}(7)$
(b) $f^{-1}(2)$
(c) $f^{-1}(5)$

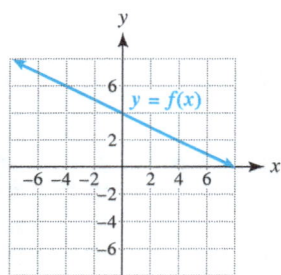

90. (a) $f^{-1}(-5)$
(b) $f^{-1}(3)$
(c) $f^{-1}(5)$

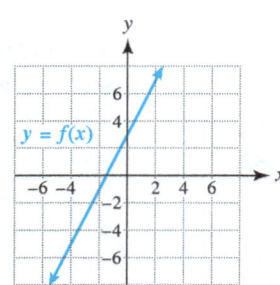

PERFORMING OPERATIONS ON FUNCTIONS

Exercises 91–94: Find each of the following for the given $f(x)$ and $g(x)$.

(a) $(fg)(2)$ (b) $(f-g)(x)$ (c) $(f \circ g)(x)$

91. $f(x) = x^2 - 2, g(x) = x^2 + 2$

92. $f(x) = 2x^2, g(x) = 2x - 1$

93. $f(x) = \dfrac{1}{x}, g(x) = \dfrac{2}{x}$

94. $f(x) = x^3, g(x) = \sqrt[3]{x}$

APPLICATIONS INVOLVING COMPOSITION AND INVERSE FUNCTIONS

95. Circular Wave A stone is dropped in a lake, creating a circular wave. The radius r of the wave in feet after t seconds is $r(t) = 2t$.
(a) The wave's circumference C is $C(r) = 2\pi r$. Evaluate $(C \circ r)(5)$ and interpret your result.
(b) Find $(C \circ r)(t)$.

96. Volume of a Balloon The volume V of a spherical balloon with radius r is given by $V(r) = \frac{4}{3}\pi r^3$. Suppose that the balloon is being inflated so that the radius in inches after t seconds is $r(t) = \sqrt[3]{t}$.
(a) Evaluate $(V \circ r)(3)$ and interpret your result.
(b) Find $(V \circ r)(t)$.

97. College Degree The table lists the percentage P of people 25 or older who have completed 4 or more years of college during year x.

x	1960	1980	2000	2010
$P(x)$	8	16	27	29

Source: U.S Census Bureau.

(a) Evaluate $P(1980)$ and interpret the results.
(b) Make a table for $P^{-1}(x)$.
(c) Evaluate $P^{-1}(16)$.

98. Skin Cancer and Ozone Ozone in the stratosphere filters out most of the harmful ultraviolet (UV) rays from the sun. However, depletion of the ozone layer affects this protection. The formula $U(x) = 1.5x$ calculates the percent increase in UV radiation for an x percent decrease in the thickness of the ozone layer. The formula $C(x) = 3.5x$ calculates the percent increase in skin cancer cases when the UV radiation increases by x percent. (*Source:* R. Turner, D. Pierce, and I. Bateman, *Environmental Economics*).

(a) Evaluate $U(2)$ and $C(3)$ and interpret each result.
(b) Find $(C \circ U)(2)$ and interpret the result.
(c) Find $(C \circ U)(x)$. What does it calculate?

99. **Temperature and Mosquitoes** Temperature can affect the number of mosquitoes observed on a summer night. Graphs of two functions, T and M, are shown. Function T calculates the temperature on a summer evening h hours past midnight, and M calculates the number of mosquitoes observed per 100 square feet when the outside temperature is T.

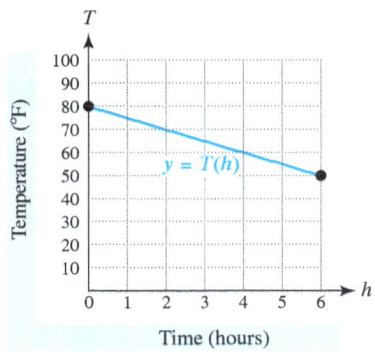

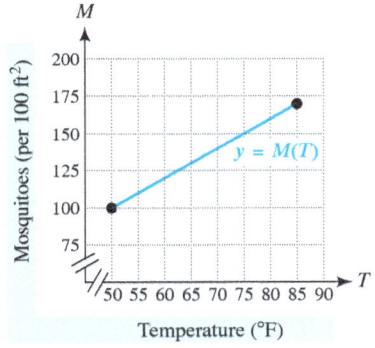

(a) Find $T(1)$ and $M(75)$.
(b) Evaluate $(M \circ T)(1)$ and interpret your result.
(c) What does $(M \circ T)(h)$ calculate?
(d) Find equations for the lines in each graph.
(e) Use your answers from part (d) to write a formula for $(M \circ T)(h)$.

100. **High School Grades** The table lists the percentage P of college freshmen with a high school grade average of A or A– during year x.

x	1970	1980	1990	2000	2010
$P(x)$	20	26	29	43	46

Source: Department of Education.

(a) Evaluate $P(1970)$ and interpret the results.
(b) Make a table for $P^{-1}(x)$.
(c) Evaluate $P^{-1}(43)$.

101. **Temperature** The function given by $f(x) = \frac{9}{5}x + 32$ converts x degrees Celsius to an equivalent temperature in degrees Fahrenheit.

(a) Is f a one-to-one function? Why or why not?
(b) Find $f^{-1}(x)$ and interpret what it calculates.

102. **Feet and Yards** The function $f(x) = 3x$ converts x yards to feet.
(a) Is f a one-to-one function? Why or why not?
(b) Find $f^{-1}(x)$ and interpret what it calculates.

103. **Quarts and Gallons** Write a function f that converts x gallons to quarts. Then find $f^{-1}(x)$ and interpret what it computes.

104. **Wind Speed** The table lists monthly average wind speeds at Hilo, Hawaii, in miles per hour from July through December, where x is the month.

x	July	Aug	Sept	Oct	Nov	Dec
$f(x)$	7	7	7	7	7	7

(a) Is function f one-to-one? Explain.
(b) Does f^{-1} exist?
(c) What happens if you try to make a table for f^{-1}?
(d) Could f be one-to-one if it were computed at a different location? What would have to be true about the monthly average wind speeds?

105. **Online Exploration** Go online and find how many cups there are in one gallon and how many teaspoons there are in one cup.
(a) Write a function C that converts x gallons to cups.
(b) Write a function T that converts x cups to teaspoons.
(c) Write the composite function $(T \circ C)(x)$, where x represents gallons.
(d) Evaluate $(T \circ C)(3)$ and interpret the result.

106. **Online Exploration** Go online and find the number of yuan (Chinese currency) in 1 dollar and the number of rupees (Indian currency) in 1 yuan. Answers may vary over time.
(a) Write a function Y that converts x dollars to yuan.
(b) Write a function R that converts x yuan to rupees.
(c) Write the composite function $(R \circ Y)(x)$, where x represents dollars.
(d) Evaluate $(R \circ Y)(100)$ and interpret the result.

WRITING ABOUT MATHEMATICS

107. Explain the difference between the expressions $(g \circ f)(2)$ and $(f \circ g)(2)$. Are they always equal? If f and g are inverse functions, evaluate $(g \circ f)(2)$ and $(f \circ g)(2)$.

108. Explain what it means for a function to be one-to-one.

19.2 Exponential Functions

Objectives

1. Introducing Exponential Functions
2. Recognizing Graphs of Exponential Functions
   - Exponential and Linear Growth
3. Relating Percent Change and Exponential Functions
   - Percent Change
   - Percent Change and Growth Factor
   - Percent Change and Exponential Functions
4. Computing Compound Interest
   - Interest Paid More Than Once a Year
5. Modeling with Exponential Functions
6. Recognizing the Natural Exponential Function

NEW VOCABULARY

☐ Exponential function with base a and coefficient C
☐ Growth factor
☐ Decay factor
☐ Exponential growth
☐ Exponential decay
☐ Percent change
☐ Compound interest
☐ Base e
☐ Natural exponential function
☐ Continuous growth

1 Introducing Exponential Functions

Math in Biology Context Suppose that an insect population (per acre) doubles each week. **TABLE 19.5** shows the size of the population after x weeks. Note that, as the population of insects becomes larger, the *increase* in population each week becomes greater. The population is increasing by 100%, or doubling, each week. When a quantity increases by a constant percentage (or constant factor) at regular intervals, its growth is *exponential*.

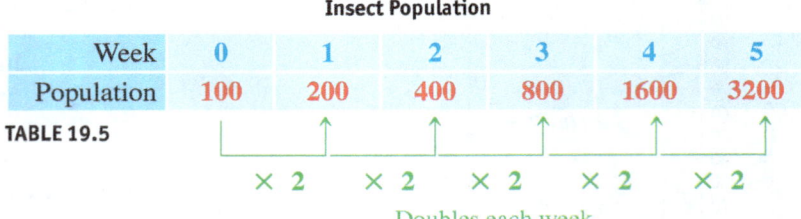

Insect Population

Week	0	1	2	3	4	5
Population	100	200	400	800	1600	3200

TABLE 19.5

Doubles each week (× 2 each step)

We can model the data in **TABLE 19.5** by using the exponential function

$$f(x) = 100(2)^x.$$

For example,

$$f(0) = 100(2)^0 = 100 \cdot 1 = 100,$$
$$f(1) = 100(2)^1 = 100 \cdot 2 = 200,$$
$$f(2) = 100(2)^2 = 100 \cdot 4 = 400,$$

and so on. Note that the exponential function f has a *variable as an exponent*.

> **EXPONENTIAL FUNCTION**
>
> A function represented by
>
> $$f(x) = Ca^x, \quad a > 0 \quad \text{and} \quad a \neq 1,$$
>
> is an **exponential function with base a and coefficient C**. (Unless stated otherwise, we assume that $C > 0$.)

In the formula $f(x) = Ca^x$, a is called the **growth factor** when $a > 1$ and the **decay factor** when $0 < a < 1$. For an exponential function, each time x increases by 1 unit, $f(x)$ increases by a factor of a when $a > 1$ and decreases by a factor of a when $0 < a < 1$. Moreover, because

$$f(0) = Ca^0 = C \cdot 1 = C,$$

the value of C equals the value of $f(x)$ when $x = 0$. That is, $(0, C)$ is the y-intercept. If x represents time, C represents the initial value of f when time equals 0. **FIGURE 19.15** illustrates **exponential growth** and **exponential decay** for $x > 0$.

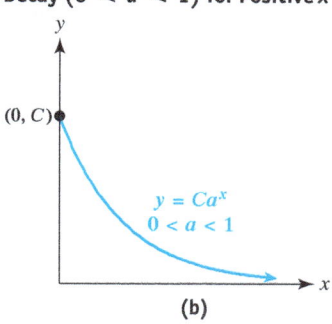

FIGURE 19.15

The set of valid inputs (domain) for an exponential function includes all real numbers. The set of corresponding outputs (range) includes all positive real numbers.

When evaluating an exponential function, we evaluate exponents *before* we multiply. For example, if $f(x) = 4(3)^x$, then

$$f(2) = 4(3)^2 = 4(9) = 36.$$
$$\text{Evaluate exponents first.}$$

The next example illustrates how exponential functions are evaluated.

EXAMPLE 1 Evaluating exponential functions

Evaluate $f(x)$ for the given value of x.
(a) $f(x) = 10(3)^x \quad x = 2$
(b) $f(x) = 5\left(\frac{1}{2}\right)^x \quad x = 3$
(c) $f(x) = \frac{1}{3}(2)^x \quad x = -1$

Solution
(a) $f(2) = 10(3)^2 = 10 \cdot 9 = 90$
(b) $f(3) = 5\left(\frac{1}{2}\right)^3 = 5 \cdot \frac{1}{8} = \frac{5}{8}$
(c) $f(-1) = \frac{1}{3}(2)^{-1} = \frac{1}{3} \cdot \frac{1}{2} = \frac{1}{6}$

NOTE: $f(x)$ is also defined for all *negative* inputs. ∎

Now Try Exercises 11, 13, 15

MAKING CONNECTIONS 1

The Expressions a^{-x} and $\left(\frac{1}{a}\right)^x$

Using properties of exponents, we can write 2^{-x} as

$$2^{-x} = \frac{1}{2^x} = \left(\frac{1}{2}\right)^x.$$

In general, the expressions a^{-x} and $\left(\frac{1}{a}\right)^x$ are equal for all positive values of a.

2 Recognizing Graphs of Exponential Functions

We can graph $f(x) = 2^x$ by first evaluating some points, as in **TABLE 19.6**. If we plot these points and sketch the graph, we obtain **FIGURE 19.16**.

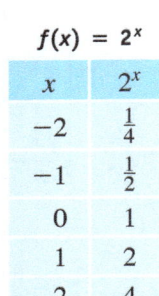

$f(x) = 2^x$

x	2^x
-2	$\frac{1}{4}$
-1	$\frac{1}{2}$
0	1
1	2
2	4

TABLE 19.6

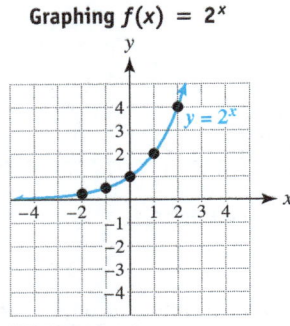

FIGURE 19.16

- This graph is always above the x-axis.
- The graph passes through the point $(0, 1)$.
- Negative x-values give y-values between 0 and 1.
- Positive x-values give y-values greater than 1.

See the Concept 1 — GRAPHING $y = a^x$

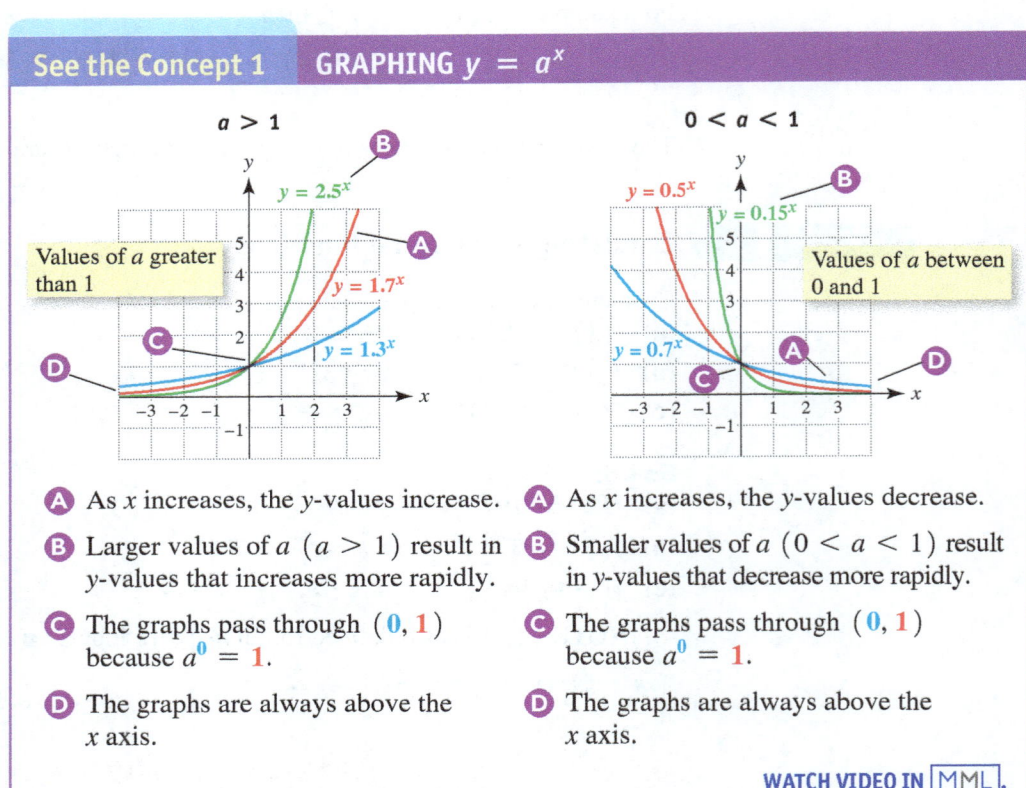

$a > 1$ — Values of a greater than 1

$0 < a < 1$ — Values of a between 0 and 1

A As x increases, the y-values increase.

B Larger values of a ($a > 1$) result in y-values that increases more rapidly.

C The graphs pass through $(0, 1)$ because $a^0 = 1$.

D The graphs are always above the x axis.

A As x increases, the y-values decrease.

B Smaller values of a ($0 < a < 1$) result in y-values that decrease more rapidly.

C The graphs pass through $(0, 1)$ because $a^0 = 1$.

D The graphs are always above the x axis.

WATCH VIDEO IN MML.

EXAMPLE 2 Transforming exponential graphs

Explain how to obtain the graph of $g(x) = 2^{x+1}$ from the graph of $f(x) = 2^x$. Then graph both f and g in the same xy-plane.

Solution
If we replace x with $x + 1$ in the formula $f(x) = 2^x$, we obtain $f(x + 1) = 2^{x+1}$, which is $g(x)$. Thus if we shift the graph of $y = 2^x$ left 1 unit, we obtain the graph of g. See **FIGURE 19.17**.

Now Try Exercise 45

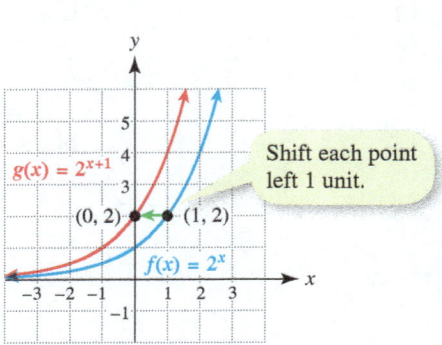

FIGURE 19.17

EXPONENTIAL AND LINEAR GROWTH In the next example, we show the dramatic difference between the outputs of linear and exponential functions.

EXAMPLE 3 Comparing exponential and linear functions

Compare $f(x) = 3^x$ and $g(x) = 3x$ graphically and numerically for $x \geq 0$.

Solution

Graphical Comparison The graphs of $y_1 = 3^x$ and $y_2 = 3x$ are shown in **FIGURE 19.18**. For large values of x, the graph of the exponential function y_1 increases much faster than the graph of the linear function y_2.

Numerical Comparison The tables of $y_1 = 3^x$ and $y_2 = 3x$ are shown in **FIGURE 19.19**. For large values of x, the values for y_1 increase much faster than the values for y_2.

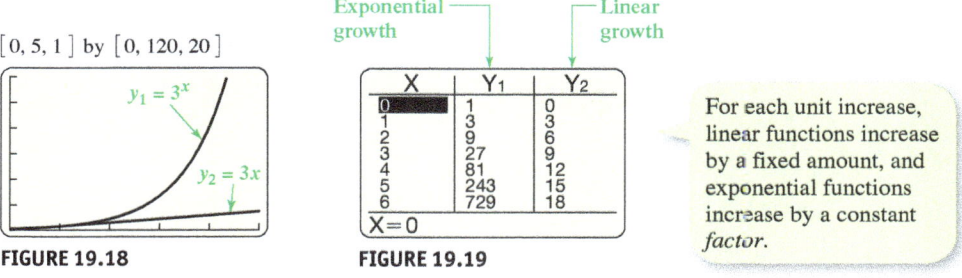

Comparing Exponential and Linear Growth

FIGURE 19.18   FIGURE 19.19

For each unit increase, linear functions increase by a fixed amount, and exponential functions increase by a constant factor.

Now Try Exercise 91

NOTE: The results of Example 3 are true in general: for large enough inputs, exponential functions with $a > 1$ grow far faster than any linear function. ∎

MAKING CONNECTIONS 2

Exponential and Polynomial Functions

The function $f(x) = 2^x$ is an exponential function. The base 2 is a constant and the exponent x is a variable, so $f(3) = 2^3 = 8$.

The function $g(x) = x^2$ is a quadratic (polynomial) function. The base x is a variable and the exponent 2 is a constant, so $g(3) = 3^2 = 9$.

The table clearly shows that the exponential function grows much faster than the quadratic function for large values of x.

	x	0	2	4	6	8	10	12	
Exponential expression →	2^x	1	4	16	64	256	1024	4096	← Exponential growth
Polynomial expression →	x^2	0	4	16	36	64	100	144	← Quadratic growth

---

In the next example, we determine whether a function is linear or exponential.

EXAMPLE 4   Finding linear and exponential functions

For each table, determine whether f is a linear function or an exponential function. Find a formula for f.

(a)
x	0	1	2	3	4
$f(x)$	16	8	4	2	1

(b)
x	0	1	2	3	4
$f(x)$	5	7	9	11	13

(c)
x	0	1	2	3	4
$f(x)$	1	3	9	27	81

(d)
x	0	1	2	3	4
$f(x)$	2	-1	-4	-7	-10

Solution
(a) Each time x increases by 1 unit, $f(x)$ decreases by a factor of $\frac{1}{2}$. Therefore f is an exponential function with a *decay factor* of $\frac{1}{2}$. Because $f(0) = 16$, $C = 16$, so $f(x) = 16\left(\frac{1}{2}\right)^x$. This formula can also be written as $f(x) = 16(2)^{-x}$.
(b) Each time x increases by 1 unit, $f(x)$ increases by 2 units. Therefore f is a linear function, and the slope of its graph is 2. The y-intercept is $(0, 5)$, so $f(x) = 2x + 5$.
(c) Each time x increases by 1 unit, $f(x)$ increases by a factor of 3. Therefore f is an exponential function with a *growth factor* of 3. Because $f(0) = 1$, it follows that $C = 1$ and $f(x) = 1(3)^x$, or $f(x) = 3^x$.
(d) Each time x increases by 1 unit, $f(x)$ **decreases by 3** units. Therefore f is a linear function, and the slope of its graph equals -3. The y-intercept is $(0, 2)$ so $f(x) = -3x + 2$.

Now Try Exercises 47, 49, 51

MAKING CONNECTIONS 3
Linear and Exponential Functions

Linear Function

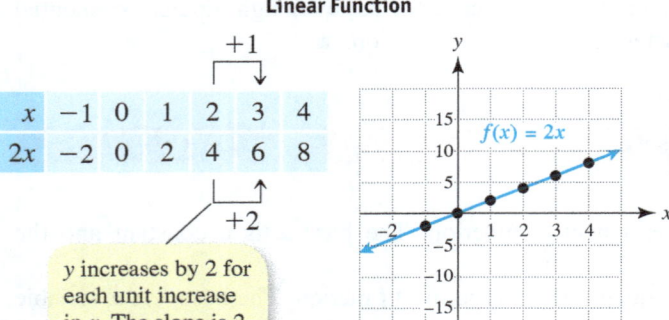

y increases by 2 for each unit increase in x. The slope is 2.

Exponential Function

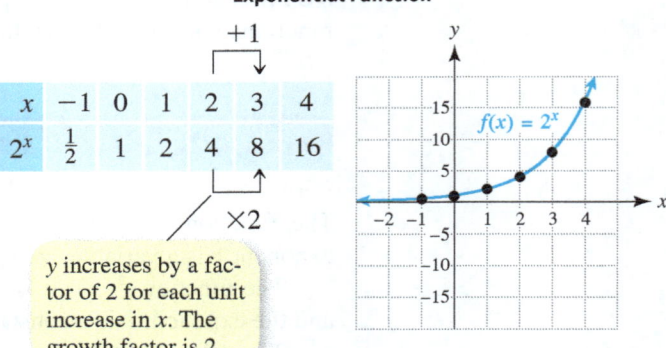

y increases by a factor of 2 for each unit increase in x. The growth factor is 2.

For a *linear function*, given by $f(x) = ax + b$, each time x increases by 1 unit, y increases (or decreases) by a units, where a equals the slope of the graph of f.

For an *exponential function*, given by $f(x) = Ca^x$, each time x increases by 1 unit, y increases by a factor of a when $a > 1$ and decreases by a factor of a when $0 < a < 1$. The constant a equals either the growth factor or the decay factor.

---

3   Relating Percent Change and Exponential Functions

PERCENT CHANGE When an amount A changes to a new amount B, then the **percent change** is calculated by

$$\frac{B - A}{A} \times 100.$$ ← Percent change formula

We multiply the ratio by 100 to change decimal form to percent form.

EXAMPLE 5 **Finding percent change**

Complete the following.
(a) Find the percent change if an account balance increases from $500 to $1000.
(b) Find the percent change if an account balance decreases from $1000 to $500.

Solution
(a) Let $A = 500$ and $B = 1000$.

$$\frac{1000 - 500}{500} \times 100 = \frac{500}{500} \times 100 \qquad \frac{B - A}{A} \times 100$$
$$= 1 \times 100 \qquad \text{Simplify.}$$
$$= 100\% \qquad \text{Multiply.}$$

The percent change (increase) is 100%; that is, the account balance doubles.

(b) Let $A = 1000$ and $B = 500$.

$$\frac{500 - 1000}{1000} \times 100 = -\frac{500}{1000} \times 100 \qquad \frac{B - A}{A} \times 100$$
$$= -\frac{1}{2} \times 100 \qquad \text{Simplify.}$$
$$= -50\% \qquad \text{Multiply.}$$

The percent change (decrease) is -50%.

Now Try Exercise 53

NOTE: In Example 5, the account did *not* increase by 100% and then decrease by 100% to return the account to its initial value of $500. In part (b), the initial amount is $A = \$1000$ and needs to decrease only by a factor of $\frac{1}{2}$, or 50%, to decrease to $500. ∎

PERCENT CHANGE AND GROWTH FACTOR Suppose a savings account A increases from $200 to $600. The percent change, $R\%$, equals

$$\frac{600 - 200}{200} \times 100 = 2 \times 100 = 200\%.$$

Note that the account balance *tripled* from $200 to 600, but the percent change is 200%, *not* 300%. Also, the *increase* in the account balance A is $400 because $600 - \$200 = \$400$. This $400 increase equals 200% of $200. If we let r represent the percent change as a decimal, then

$$200\% \text{ of } \$200 = 2.00 \times \$200 = \$400. \qquad R\% = 200\%; r = 2.00$$

$R\%$ of A = rA = Increase in A

In general, if the *percent increase* in an account balance A is given by r in decimal form, then the *amount of increase* in A is equal to rA and the new balance after the increase is $A + rA$. For example, if $200 increases by 200%, then $r = 2.00$, $A = \$200$, and

$$rA = 2.00(200) = \$400, \qquad \text{Increase in account balance}$$

so the balance increases by $400 as noted above. The new balance for the account is

$$A + rA = \$200 + \$400 = \$600.$$

Initial amount + Increase = Final amount

If we factor out A in the expression $A + rA$, we get

$$A + rA = A(1 + r). \quad \text{Factor out } A.$$

Initial amount + Increase in A = Initial amount × Growth factor

Thus if an account increases from \$200 to \$600, the increase is \$400 and the percent increase is 200%. In addition, the account balance increased by a *growth factor* equal to

$$a = 1 + r = 1 + 2.00 = 3, \quad \text{Growth factor is } a = 3.$$

which means that the account balance tripled.

EXAMPLE 6 Analyzing the decrease in an account balance

An account that contains \$2000 decreases in value by 20%.
(a) Find the decrease in value of the account.
(b) Find the final value of the account.
(c) By what factor a did the account decrease?

Solution
(a) Let $A = 2000$ and $r = -0.20$ (-20% in decimal form). The decrease is

$$rA = -0.20(2000) = -400. \quad \text{Decrease in } A \text{ is } rA.$$

The account decreased in value by \$400.

(b) The final value of the account is

$$A + rA = 2000 + (-400) = \$1600.$$

Initial amount + Decrease = New amount

(c) The account decreased by a factor of

$$a = 1 + r = 1 + (-0.20) = 0.80. \quad \text{Decay factor is } a = 0.80.$$

The account value decreased to 80% of its original value because 80% of \$2000 is \$1600.

Now Try Exercise 57

> **STUDY TIP**
>
> Be sure that you understand how a percent change relates to the growth factor of an exponential function.

NOTE: In general, a positive amount A cannot *decrease* by more than 100% because a 100% decrease would reduce A to 0. However, percent *increases* can be more than 100%. ∎

PERCENT CHANGE AND EXPONENTIAL FUNCTIONS An exponential function results when an *initial value* C is multiplied by a *growth* (or *decay*) *factor* a for each unit increase in x. For example, if the population P of a city is 100,000 people, and the city is growing at 5% per year, then $C = 100,000$ and the growth factor is

$$a = 1 + r = 1 + 0.05 = 1.05.$$

Thus the exponential function

$$P(x) = 100,000(1.05)^x \quad \begin{array}{l} P(x) = Ca^x \text{ with } C = 100,000 \\ \text{and } a = 1.05. \end{array}$$

models the city's population after x years. For example,

$$P(6) = 100,000(1.05)^6 \approx 134,000$$

indicates that after 6 years the city's population has grown to about 134,000 people.

PERCENT CHANGE AND EXPONENTIAL FUNCTIONS

Suppose that an amount A increases or decreases by $R\%$ (or by r expressed in decimal form) for each x-unit increase in time.

1. If $r > 0$, the *growth factor* is $a = 1 + r$ and $a > 1$.
2. If $-1 < r < 0$, the *decay factor* is $a = 1 + r$ and $0 < a < 1$.
3. If the initial amount is C, the amount A after x-units of time is given by
$$A(x) = Ca^x \quad \text{or equivalently,} \quad A(x) = C(1 + r)^x.$$

READING CHECK 1

- If your income increases by 200%, what is the growth factor?

EXAMPLE 7 Analyzing growth of bacteria

Initially, a laboratory culture contains 50,000 bacteria per milliliter and it is increasing in numbers by 20% per hour.
(a) Write the formula for an exponential function B that gives the number of bacteria per milliliter after x hours.
(b) Evaluate $B(3)$ and interpret your result.

Solution
(a) The initial value is $C = 50{,}000$ and the hourly percent increase is 20%, or $r = 0.20$ in decimal form. Thus the growth factor is
$$a = 1 + r = 1 + 0.20 = 1.20. \quad \text{Growth factor} = 1.20 \text{ with } r = 0.20.$$

Because the exponential function can be written as $B(x) = Ca^x$, it follows that
$$B(x) = 50{,}000(1.20)^x.$$

(b) $B(3) = 50{,}000(1.20)^3 = 50{,}000(1.728) = 86{,}400$. Thus after 3 hours, the culture contains 86,400 bacteria per milliliter.

Now Try Exercise 87

4 Computing Compound Interest

Math in Context (Savings) If a total $100 is deposited in a savings account paying 10% annual interest, the interest earned after 1 year equals $100 \times 0.10 = \$10$. The total amount of money in the account after 1 year is $100(1 + 0.10) = \$110$. Each year, the money in the account increases by a growth factor of $a = 1 + r = 1 + 0.10 = 1.10$, so after x years, there will be $100(1.10)^x$ dollars in the account. Thus **compound interest** is an example of exponential growth.

COMPOUND INTEREST

If a total of P dollars is deposited in an account and if interest is paid at the end of each year with an annual rate of interest r, expressed in decimal form, then after t years, the account will contain A dollars, where
$$A = P(1 + r)^t.$$

The growth factor is $(1 + r)$.

NOTE: The compound interest formula takes the form of an exponential function with
$$a = 1 + r.$$ ■

> **EXAMPLE 8** **Calculating compound interest**
>
> A 20-year-old worker deposits $2000 in a retirement account that pays 6% annual interest at the end of each year. How much money will be in the account when the worker is 65 years old? What is the growth factor?
>
> **Solution**
> Here, $P = 2000$, $r = 0.06$, and $t = 45$. The amount in the account after 45 years is
>
> $$A = 2000(1 + 0.06)^{45} \approx \$27{,}529.22, \quad A = P(1 + r)^t$$
>
> which is supported by **FIGURE 19.20**. Each year, the amount of money in the account is multiplied by a factor of $(1 + 0.06)$, so the growth factor is 1.06.

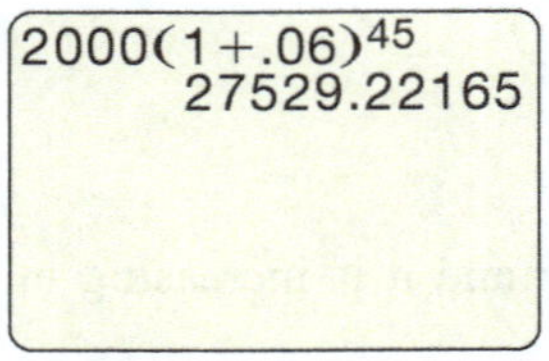

```
2000(1+.06)^45
    27529.22165
```

FIGURE 19.20

Now Try Exercise 67

INTEREST PAID MORE THAN ONCE A YEAR Many times, interest is paid more than once a year. For example, suppose an account gives 8% annual interest that is paid every 3 months, or *quarterly*. It follows that $\frac{1}{4}$ of the annual 8% interest, or 2%, is paid every 3 months. The growth factor is $a = 1 + 0.02 = 1.02$ for each 3-month period. If the initial balance is $1000, then after 5 years, the **quarterly** 2% interest has been paid $4 \cdot 5 = 20$ times, and the account contains

$$\$1000(1.02)^{20} \approx \$1485.95.$$

In general, if a total of P dollars is deposited in an account paying an *annual* interest rate r (in decimal form) and this interest rate is compounded or paid n times per year, then after t years, the account contains A dollars, given by the following formula.

Compound Interest

$$A = P\left(1 + \frac{r}{n}\right)^{nt}$$

Amount after t years — A
Initial deposit — P
Annual interest (decimal) — r
Number of times interest is paid per year — n
Number of years — t

> **EXAMPLE 9** **Calculating compound interest**
>
> Initially, a total of $1500 is deposited in an account paying 6% annual interest, compounded monthly. What is the account balance after 5 years?
>
> **Solution**
> Let $P = 1500$, $r = 0.06$, $n = 12$, and $t = 5$. The balance after 5 years is
>
> $$A = P\left(1 + \frac{r}{n}\right)^{nt} \qquad \text{Interest formula}$$
> $$= 1500\left(1 + \frac{0.06}{12}\right)^{(12 \cdot 5)} \qquad \text{Substitute.}$$
> $$= 1500(1.005)^{60} \qquad \text{Evaluate.}$$
> $$\approx \$2023.28. \qquad \text{Approximate.}$$

Now Try Exercise 75

NOTE: In Example 9, if the annual 6% interest had been paid only once a year, the growth factor would be $a = 1.06$, but interest would have been paid only 5 times. The balance would be $1500(1.06)^5 \approx \$2007.34$, rather than the $2023.28 that resulted from monthly compounding. ■

Math in Context (charity) The Ice Bucket Challenge for ALS became a fundraising phenomenon in August 2014. The next example shows how an exponential function can model donations.

5 Modeling with Exponential Functions

EXAMPLE 10 Modeling donations for the Ice Bucket Challenge for ALS

By August 12th, 2014, there was a cumulative $4 million in donations for the Ice Bucket Challenge for ALS, and this number was increasing at 21.8% per *day* in August.
(a) What was the daily growth factor after August 12?
(b) Write an exponential function $D(x) = Ca^x$ that models the cumulative donations in millions of dollars x days after August 12.
(c) Evaluate $D(16)$ and interpret the result.

Solution
(a) Because $r = 0.218$, the daily growth factor was $a = 1 + 0.218 = 1.218$.
(b) Let $C = 4$ and $a = 1.218$. Then $D(x) = 4(1.218)^x$.
(c) $D(16) = 4(1.218)^{16} \approx 93.8$. In the 16 days following August 12, ALS donations grew from $4 million to approximately $93.8 million.

Now Try Exercise 93

Math in Context (Traffic) Traffic flow at intersections can be modeled by exponential functions whenever traffic patterns occur randomly. In the next example, we model traffic at an intersection by using an exponential function.

EXAMPLE 11 Modeling traffic flow

On average, a particular intersection has 360 vehicles arriving randomly each hour. Traffic engineers use $f(x) = (0.905)^x$ to estimate the likelihood, or probability, that *no* vehicle will enter the intersection within an interval of x seconds. (*Source:* F. Mannering and W. Kilareski, *Principles of Highway Engineering and Traffic Analysis*.)

(a) Compute $f(5)$ and interpret the results.
(b) A graph of $y = f(x)$ is shown in **FIGURE 19.21** Discuss this graph.
(c) Is this function an example of exponential growth or decay?

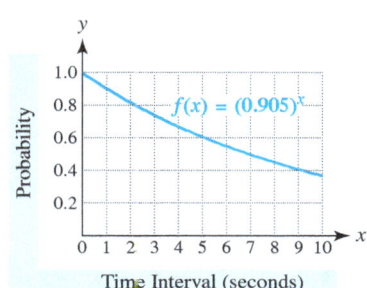

FIGURE 19.21

Solution
(a) The result $f(5) = (0.905)^5 \approx 0.61$ indicates that there is a 61% chance no vehicle will enter the intersection during any particular 5-second interval.
(b) The graph decreases, which means that as the interval of time increases, there is less chance (likelihood) that a car will *not* enter the intersection.
(c) Because the graph is decreasing and $a = 0.905 < 1$, this function is an example of exponential decay.

Now Try Exercise 101

6 Recognizing the Natural Exponential Function

A special type of exponential function that occurs often in applications is called the *natural exponential function*, expressed as $f(x) = e^x$. The **base e** is a special number in mathematics similar to π. The number π is approximately 3.14, whereas the number e is approximately 2.72. The number e is named for the great Swiss mathematician Leonhard Euler (1707–1783). Most calculators have a special key that can be used to compute the natural exponential function.

> **NATURAL EXPONENTIAL FUNCTION**
>
> The function represented by
> $$f(x) = e^x$$
> is the **natural exponential function**, where $e \approx 2.71828183$.

EXAMPLE 12 Sketching a graph of $y = e^x$

Sketch a graph of $y = 2^x$ and $y = 3^x$ in the same xy-plane. Then use these two graphs to sketch a graph of $y = e^x$. How do these graphs compare?

Solution
The graphs of $y = 2^x$, $y = 3^x$, and $y = e^x$ are shown in **FIGURE 19.22**.

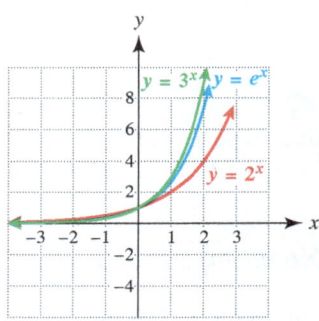

FIGURE 19.22

Because $e \approx 2.718$, the graph of $y = e^x$ lies between the graphs of $y = 2^x$ and $y = 3^x$. It is closer to the graph of $y = 3^x$.

Now Try Exercise 83

🌐 Math in Context (Population) The natural exponential function is frequently used to model **continuous growth** and **continuous decay**. For example, the fact that births and deaths occur throughout the year, not just at one time during the year, must be recognized when population growth or decay is being modeled. If a population P is increasing or decreasing continuously at r percent per year, expressed as a decimal, we can model this population after x years by

$$P = Ce^{rx}, \quad \text{Continuous growth, } r > 0$$
$$\text{Continuous decay, } r < 0$$

where C is the initial population. To evaluate natural exponential functions, we use a calculator, as demonstrated in the next example.

EXAMPLE 13 Modeling population

In 2013, the population of Texas was 26.5 million people and was growing at a continuous rate of 1.8% per year. This population in millions x years after 2013 can be modeled by

$$f(x) = 26.5e^{0.018x}.$$

Estimate the population in 2017.

Solution
Because 2017 is 4 years after 2013, we evaluate $f(4)$ to obtain

$$f(4) = 26.5e^{0.018(4)} \approx 28.5,$$

which is supported by **FIGURE 19.23**. This model estimates the population of Texas to be about 28.5 million in 2017.

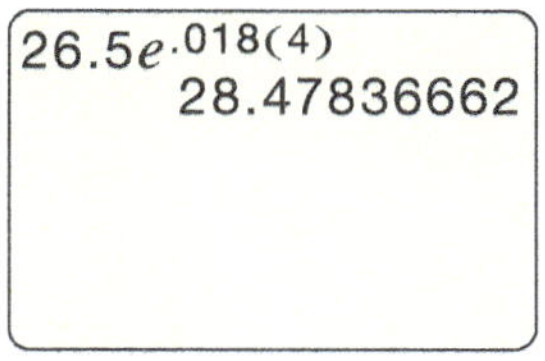

```
26.5e^.018(4)
          28.47836662
```

FIGURE 19.23

CALCULATOR HELP
To evaluate the natural exponential function, see Appendix A (page AP-8).

Now Try Exercise 103

19.2 Putting It All Together

CONCEPT	EXPLANATION	EXAMPLES
Exponential Function	The variable is an exponent. $f(x) = Ca^x$ where C is the initial value when $x = 0$ and a is the growth factor (base). Growth: $a > 1$; Decay: $0 < a < 1$	$f(x) = 3(2)^x$ models exponential **growth**, and $g(x) = 2\left(\frac{1}{3}\right)^x$ models exponential **decay**.
Graphs of Exponential Functions	• Is above the x-axis for all inputs • Increases for $a > 1$ • Decreases for $0 < a < 1$ • The y-intercept is $(0, C)$.	Exponential Functions: $f(x) = Ca^x$ 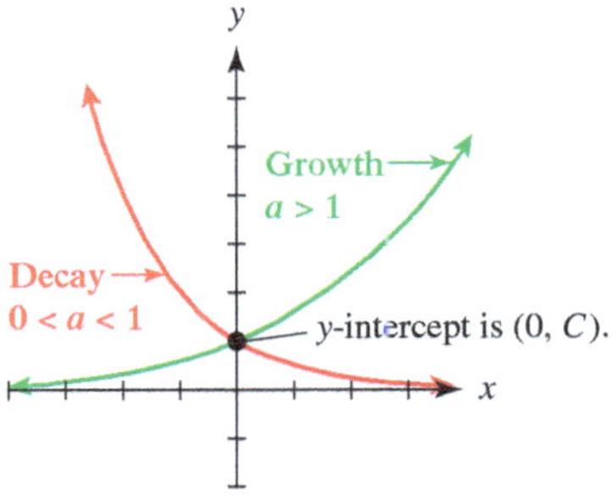 — Growth ($a > 1$), Decay ($0 < a < 1$), y-intercept is $(0, C)$.
Percent Change	An amount changes from A to B: $$\text{Percent change} = \frac{B - A}{A} \times 100.$$	If \$200 increases to \$300, then $$\frac{300 - 200}{200} \times 100 = 50\%$$ is the percent change.

continued on next page

CONCEPT	EXPLANATION	EXAMPLES
Constant Percent Change and Exponential Functions	$f(x) = C(1 + r)^x$ where C is the initial amount and $(1+r)$ is the growth factor, with constant percent change (decimal) r over time x.	If 3000 bacteria increase in number by 12% daily for x days, then $$f(x) = 3000(1.12)^x$$ models these numbers after x days.
Compound Interest	Compounded annually: $$A = P(1 + r)^t$$ Compounded n times per year: $$A = P\left(1 + \frac{r}{n}\right)^{nt}$$ P: Initial deposit A: Final amount r: Annual interest rate (decimal) n: Interest paid n times per year t: Number of years	Depositing $1000 at 3% annual interest for 4 years gives $$A = \$1000(1.03)^4$$ $$\approx 1125.51.$$ Depositing $1000 at 3% annual interest compounded monthly for 4 years gives $$A = \$1000\left(1 + \frac{0.03}{12}\right)^{(12 \cdot 4)}$$ $$\approx \$1127.33.$$
Natural Exponential Function	$f(x) = e^x$ Growth factor, or base, is $e \approx 2.72$.	Using a calculator, $$f(2) = e^2 \approx 7.39.$$

19.2 Exercises MyMathLab

CONCEPTS AND VOCABULARY

1. Give a general formula for an exponential function f.

2. Sketch a graph of an exponential function that illustrates exponential decay.

3. Give the domain and range of an exponential function.

4. Evaluate the expressions 2^x and x^2 for $x = 5$.

5. Approximate e to the nearest thousandth.

6. Evaluate e^2 and π^2 using your calculator.

7. If a quantity y grows exponentially, then for each unit increase in x, y increases by a constant _____.

8. If $f(x) = 1.5^x$, what is the growth factor?

9. If a quantity increases from A to B, then the percent change equals _____.

10. If a quantity increases by 35% each year, then the growth factor a is _____.

EVALUATING EXPONENTIAL FUNCTIONS

Exercises 11–22: Evaluate the exponential function for the given values of x by hand when possible. Approximate answers to the nearest hundredth when appropriate.

11. $f(x) = 3^x$ $x = -2, x = 2$

12. $f(x) = 5^x$ $x = -1, x = 3$

13. $f(x) = 5(2^x)$ $x = 0, x = 5$

14. $f(x) = 3(7^x)$ $x = -2, x = 0$

15. $f(x) = \left(\frac{1}{2}\right)^x$ $x = -2, x = 3$

16. $f(x) = \left(\frac{1}{4}\right)^x$ $x = 0, x = 2$

17. $f(x) = 5(3)^{-x}$ $x = -1, x = 2$

18. $f(x) = 4\left(\frac{3}{7}\right)^x$ $x = 1, x = 4$

19. $f(x) = 1.8^x$ $x = -3, x = 1.5$

20. $f(x) = 0.91^x$ $x = 5.1, x = 10$

21. $f(x) = 3(0.6)^x$ $x = -1, x = 2$

22. $f(x) = 5(4.5)^{-x}$ $x = -2.1, x = 1.9$

Exercises 23 and 24: **Thinking Generally** *For the given exponential function, evaluate $f(0)$ and $f(-1)$.*

23. $f(x) = a^x$

24. $f(x) = (1 + r)^{2x}$

RECOGNIZING GRAPHS OF EXPONENTIAL FUNCTIONS

Exercises 25–28: Match the formula with its graph (a–d). Do not use a calculator.

25. $f(x) = 1.5^x$ **26.** $f(x) = \frac{1}{4}(2^x)$

27. $f(x) = 4\left(\frac{1}{2}\right)^x$ **28.** $f(x) = \left(\frac{1}{3}\right)^x$

a.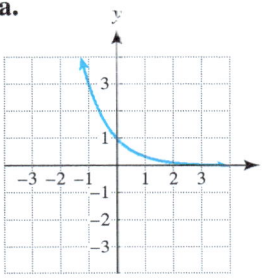

b.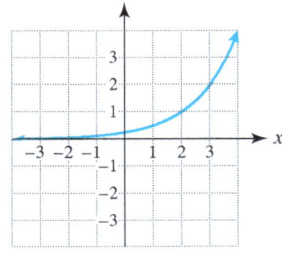

c.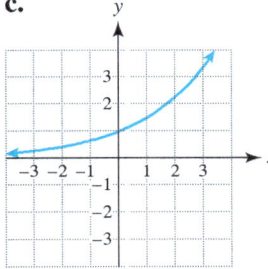

d.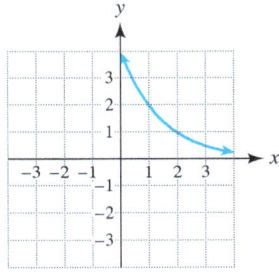

Exercises 29–32: Use the graph of $y = Ca^x$ to determine the constants C and a.

29.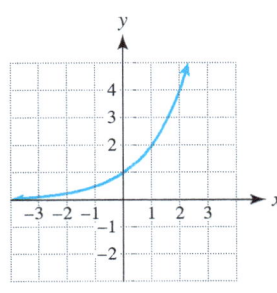

30.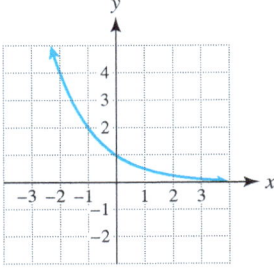

31.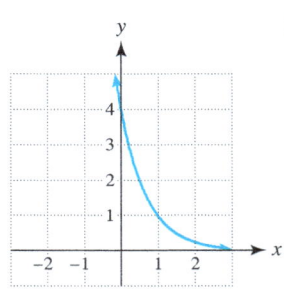

32.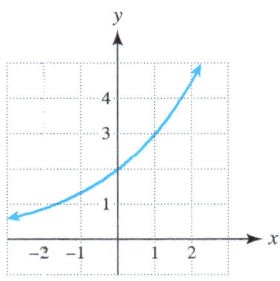

Exercises 33–44: Graph $y = f(x)$. State whether the graph depicts exponential growth or exponential decay.

33. $f(x) = 2^x$ **34.** $f(x) = 3^x$

35. $f(x) = \left(\frac{1}{4}\right)^x$ **36.** $f(x) = \left(\frac{1}{2}\right)^x$

37. $f(x) = 2^{-x}$ **38.** $f(x) = 3^{-x}$

39. $f(x) = 3^x - 1$ **40.** $f(x) = 2^x + 1$

41. $f(x) = 2^{x-1}$ **42.** $f(x) = 2^{x+1}$

43. $f(x) = 4\left(\frac{1}{3}\right)^x$ **44.** $f(x) = 3\left(\frac{1}{2}\right)^x$

Exercises 45–46: Compare the graph of each equation to the graph of $y = f(x)$.

45. $f(x) = 2^x$
(a) $y = 2^{x-1}$
(b) $y = 2^x - 1$
(c) $y = 2^{-x}$
(d) $y = -2^x$

46. $f(x) = e^x$
(a) $y = -e^x$
(b) $y = e^x + 3$
(c) $y = e^{-x}$
(d) $y = e^{x+2}$

COMPARING EXPONENTIAL AND LINEAR GROWTH

Exercises 47–52: (Refer to Example 4.) A table for a function f is given.

(a) *Determine whether function f represents exponential growth, exponential decay, or linear growth.*
(b) *Find a formula for f.*

47.

x	0	1	2	3	4
$f(x)$	64	16	4	1	$\frac{1}{4}$

48.

x	0	1	2	3	4
$f(x)$	$\frac{1}{2}$	1	2	4	8

49.

x	0	1	2	3	4
$f(x)$	8	11	14	17	20

50.

x	-2	-1	0	1	2
$f(x)$	4	2	1	$\frac{1}{2}$	$\frac{1}{4}$

51.

x	-2	-1	0	1	2
$f(x)$	2.56	3.2	4	5	6.25

52.

x	-2	-1	0	1	2
$f(x)$	-6	-2	2	6	10

RELATING PERCENT CHANGE AND EXPONENTIAL FUNCTIONS

Exercises 53–56: For the given amounts A and B, find each of the following.

(a) The percent change if A changes to B
(b) The percent change if B changes to A

53. $A = \$200, B = \$400$

54. $A = \$1.50, B = \$1.00$

55. $A = 150, B = 30$

56. $A = 80, B = 200$

Exercises 57–62: An account contains A dollars and increases or decreases by R%. For each A and R, answer the following.

(a) Find the increase or decrease in the value of the account.
(b) Find the final value of the account.
(c) By what factor did the account value increase or decrease?

57. $A = \$1000, R = 120\%$

58. $A = \$500, R = 230\%$

59. $A = \$650, R = 20\%$

60. $A = \$70, R = 35\%$

61. $A = \$800, R = -10\%$

62. $A = \$950, R = -60\%$

Exercises 63–66: For the given $f(x)$, state the initial value C, the growth or decay factor a, and percent change R for each unit increase in x.

63. $f(x) = 9(1.07)^x$

64. $f(x) = 3(1.351)^x$

65. $f(x) = 1.5(0.45)^x$

66. $f(x) = 0.9^x$

COMPUTING COMPOUND INTEREST

Exercises 67–72: (Refer to Example 8.) The amount P is deposited in an account paying R percent annual interest. Approximate the amount in the account after x years.

67. $P = \$1500 \quad R = 9\% \quad x = 10$ years

68. $P = \$1500 \quad R = 15\% \quad x = 10$ years

69. $P = \$200 \quad R = 20\% \quad x = 50$ years

70. $P = \$5000 \quad R = 8.4\% \quad x = 7$ years

71. $P = \$560 \quad R = 1.4\% \quad x = 25$ years

72. $P = \$750 \quad R = 10\% \quad x = 13$ years

73. Thinking Generally A total of $1000 is deposited in an account paying 8% annual interest for 10 years. If $2000 had been deposited instead of $1000, would there be twice the money in the account after 10 years? Explain.

74. Thinking Generally A total of $500 is deposited in an account paying 5% annual interest for 10 years. If the interest rate had been 10% instead of 5%, would the total interest earned after 10 years be twice as much? Explain.

Exercises 75–78: (Refer to Example 9.) The amount P is deposited in an account giving R% annual interest compounded n times a year. Find the amount A in the account after t years.

75. $P = \$700, R = 4\%, n = 4, t = 3$

76. $P = \$550, R = 3\%, n = 2, t = 5$

77. $P = \$1200, R = 2.5\%, n = 12, t = 7$

78. $P = \$1500, R = 6.5\%, n = 365, t = 20$

EVALUATING AND GRAPHING THE NATURAL EXPONENTIAL FUNCTION

Exercises 79–82: Evaluate $f(x)$ for the given value of x. Approximate answers to the nearest hundredth.

79. $f(x) = e^x \qquad x = 1.2$

80. $f(x) = 2e^x \qquad x = 2$

81. $f(x) = 1 - e^x \qquad x = -2$

82. $f(x) = 4e^{-x} \qquad x = 1.5$

Exercises 83–86: Graph $f(x)$ in $[-4, 4, 1]$ by $[0, 8, 1]$. State whether the graph illustrates exponential growth or exponential decay.

83. $f(x) = e^{0.5x}$
84. $f(x) = e^x + 1$
85. $f(x) = 1.5e^{-0.32x}$
86. $f(x) = 2e^{-x} + 1$

APPLICATIONS OF EXPONENTIAL FUNCTIONS

Exercises 87–90: **Exponential Models** Write the formula for an exponential function, $f(x) = Ca^x$, that models the situation. Evaluate $f(4)$.

87. A sample of 5000 bacteria decreases in number by 25% per week.

88. A sample of 700 insects increases in number by 200% per day.

89. A sample of 50 birds increases in number by 10% per month.

90. A sample of 137 fish decreases in number by 2% per week.

91. **Salary Growth** Suppose your salary is $50,000 per year. Would you rather have a 20% raise each year or a $20 raise each year?

92. **Salary Growth** Suppose your salary is $35,000 per year. Would you rather have a 10% raise each year or a $4000 raise each year? Assume that you will keep this job for 10 years.

93. **Blood Alcohol** Suppose that a person's peak blood alcohol level is 0.07 (grams per 100 mL) and that this level decreases by 40% each hour. (*Source: National Institutes of Health.*)
    (a) What is the hourly decay factor?
    (b) Write an exponential function $B(x) = Ca^x$ that models the blood alcohol level after x hours.
    (c) Evaluate $B(2)$ and interpret the result.

94. **Apple's Revenue** From 2001 to 2015, Apple's first quarter revenue grew at an annual rate of 33.7%, from $1 billion to about $58 billion. (*Source: Apple Corp.*)
    (a) What is the annual growth factor?
    (b) Write an exponential function $R(x) = Ca^x$ that models the revenue in billions of dollars x years after 2001.
    (c) Evaluate $R(14)$ and interpret the result.

95. **Facebook Link Half-Life** (Refer to the chapter opener.) The half-life for a link on Facebook is 3 hours. Write an exponential function F that gives the percentage of hits remaining on a typical Facebook link after t hours. Estimate this percentage after 4 hours.

96. **StumbleUpon Link Half-Life** (Refer to the chapter opener.) The half-life for a link on StumbleUpon is 400 hours. Write an exponential function S that gives the percentage of engagements remaining on a typical StumbleUpon link after t hours. Estimate this percentage after 250 hours.

97. **Tweets per Month** From July 2010 to July 2015, the number of tweets T per month increased dramatically and could be modeled in billions by
$$T(x) = 0.1(2.724)^x,$$
where x represents years after July 2010. (*Source: Twitter.*)
    (a) What is the yearly growth factor?
    (b) Interpret the 0.1 in the formula.
    (c) Evaluate $T(5)$ and interpret the result.

98. **Dating Artifacts** Radioactive carbon-14 is found in all living things and is used to date objects containing organic material. Suppose that an object initially contains C grams of carbon-14. After x years, it will contain A grams, where
$$A = C(0.99988)^x.$$
    (a) Let $C = 10$ and graph A over a 20,000-year period. Is this function an example of exponential growth or decay?
    (b) If $C = 10$, how many grams are left after 5700 years? What fraction of the carbon-14 is left?

99. **E. coli Bacteria** A strain of bacteria that inhabits the intestines of animals is named *Escherichia coli* (*E. coli*). These bacteria are capable of rapid growth and can be dangerous to humans—particularly children. The table shows the results of one study of the growth of *E. coli* bacteria, where concentrations are listed in *thousands* of bacteria per milliliter.

t (minutes)	0	50	100
Concentration	500	1000	2000

t (minutes)	150	200	250
Concentration	4000	8000	16,000

Source: G.S. Stent, *Molecular Biology of Bacterial Viruses.*

(a) Find C and a so that $f(t) = Ca^{t/50}$ models these data.
(b) Use $f(t)$ to estimate the concentration of bacteria after 170 minutes.
(c) Discuss the growth of this strain of bacteria over a 250-minute time period.

100. Swimming Pool Maintenance Chlorine is frequently used to disinfect swimming pools. The chlorine concentration should remain between 1.5 and 2.5 parts per million (ppm). After a warm, sunny day, only 80% of the chlorine may remain in the water, with the other 20% dissipating into the air or combining with other chemicals in the water. (*Source: D. Thomas, Swimming Pool Operator's Handbook.*)
   (a) Let $f(x) = 3(0.8)^x$ model the concentration of chlorine in parts per million after x days. What is the initial concentration of chlorine in the pool?
   (b) If no more chlorine is added, estimate when the chlorine level first drops below 1.6 parts per million.

101. Modeling Traffic Flow (Refer to Example 11.) Construct a table of $f(x) = (0.905)^x$, starting at $x = 0$ and incrementing by 10, until $x = 50$.
   (a) Evaluate $f(0)$ and interpret the result.
   (b) For a time interval of what length is there only a 5% chance that no cars will enter the intersection?

102. Pros and Putts The percentage of putts P from 3 feet to 25 feet made by professional golfers can be modeled by the exponential function
$$P(x) = 99(0.872)^{x-3},$$
where x is the length of the putt.
   (a) Find the percentage of putts made from 3 feet by professionals.
   (b) Evaluate $P(8)$ and interpret the results.
   (c) What is the decay factor? Interpret it.

Exercises 103–106: **Population Growth** (*Refer to Example 13.*) The population P in 2010 for a state is given along with R%, its annual percentage rate of continuous growth.
   (a) Write the formula $f(x) = Pe^{rx}$, where r is in decimal notation, that models the population in millions x years after 2010.
   (b) Estimate the population in 2020.

103. Nevada: $P = 2.7$ million, $R = 1.4$%

104. North Carolina: $P = 9.4$ million, $R = 1.36$%

105. California: $P = 38$ million, $R = 1.02$%

106. Arizona: $P = 6.6$ million, $R = 1.44$%

107. Online Exploration In many states, landlords are mandated to return a tenant's security deposit plus interest. Go online and look up this interest rate for your state. Answers may vary.
   (a) If you are a tenant, calculate the interest you should receive after 1 year. (If you are not, assume the security deposit is $1200.)
   (b) Write a function I that calculates the interest that you should receive after x years. (Does your landlord have to pay compound interest?)

108. Online Exploration Go online and determine how long it takes the fastest growing bacteria to double in number. Answers may vary.
   (a) Suppose that you start with a sample of 4 million such bacteria. Write an exponential function that gives the number N of bacteria in millions after x minutes.
   (b) How many bacteria are there after 2 hours?

WRITING ABOUT MATHEMATICS

109. A student evaluates $f(x) = 4(2)^x$ at $x = 3$ and obtains 512. Did the student evaluate the function correctly? What was the student's error?

110. For a set of data, how can you distinguish between linear growth and exponential growth? Give an example of each type of data.

SECTIONS 19.1 and 19.2 Checking Basic Concepts

1. If $f(x) = 2x^2 + 5x - 1$ and $g(x) = x + 1$, find each expression.
   (a) $(g \circ f)(1)$ (b) $(f \circ g)(x)$

2. Sketch a graph of $f(x) = x^2 - 1$.
   (a) Is f a one-to-one function? Explain.
   (b) Does f have an inverse function?

3. If $f(x) = 4x - 3$, find $f^{-1}(x)$.

4. Evaluate $f(-2)$ if $f(x) = 3(2)^x$.

5. Sketch a graph of $f(x) = \left(\frac{1}{3}\right)^x$.

6. Use the graph of $y = Ca^x$ to determine the constants C and a

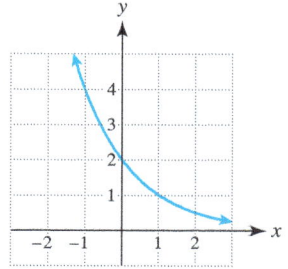

19.3 Logarithmic Functions

Objectives

1. Introducing the Common Logarithmic Function
   • The Graph of the Common Logarithm Function

2. Finding the Inverse of the Common Logarithmic Function

3. Working with Logarithms that Have Other Bases
   • Base 2
   • Base e (The Natural Logarithm)
   • Base a
   • Inverses of Base a Logarithmic Functions

NEW VOCABULARY

☐ Common logarithm of a positive number x
☐ Common logarithmic function
☐ Natural logarithm
☐ Logarithm with base a of a positive number x
☐ Logarithmic function with base a

1 Introducing the Common Logarithmic Function

Math in Measurement Context In applications, measurements can vary greatly in size. TABLE 19.7 lists some examples of objects, with the approximate distances in meters across each.

Sizes of Objects

Object	Distance (meters)
Atom	10^{-9}
Protozoan	10^{-4}
Small Asteroid	10^2
Earth	10^7
Universe	10^{26}

Source: C. Ronan, *The Natural History of the Universe.*
TABLE 19.7

Each distance is listed in the form 10^k for some k. The value of k distinguishes one measurement from another. The *common logarithmic function* or *base-10 logarithmic function*, denoted $\log$ or $\log_{10}$, outputs k if the input x can be written as 10^k for some real number k. For example, $\log 10^{-9} = -9$, $\log 10^2 = 2$, and $\log 10^{1.43} = 1.43$. For any real number k, $\log 10^k = k$. Some values for $\log x$ are given in TABLE 19.8.

Common Logarithms of Powers of 10

x	10^{-4}	10^{-3}	10^{-2}	10^{-1}	10^0	10^1	10^2	10^3	10^4
$\log x$	-4	-3	-2	-1	0	1	2	3	4

TABLE 19.8

Write x as a power of 10.

A common logarithm is the *exponent* on base 10.

We use this information to define the common logarithm.

> **COMMON LOGARITHM**
>
> The **common logarithm of a positive number** x, denoted log x, is calculated as follows. If x is written as $x = 10^k$, then
> $$\log x = k,$$
> where k is a real number. That is, $\log 10^k = k$.
> The function given by
> $$f(x) = \log x$$
> is called the **common logarithmic function**.

NOTE: Previously, we have always used one letter, such as f or g, to name a function. The common logarithm is the *first* function for which we use *three* letters, log, to name it. Thus $f(x)$, $g(x)$, and log (x) all represent functions. Generally, log (x) is written without parentheses as log x. ∎

The equation $\log x = k$ is called logarithmic form and the equation $x = 10^k$ is called exponential form. These forms are *equivalent*. That is, if one equation is true, the other equation must also be true.

Equivalent Logarithmic and Exponential Forms

Logarithmic Form	Exponential Form
$\log 100 = 2$	$10^2 = 100$
$\log 1000 = 3$	$10^3 = 1000$
$\log \frac{1}{10} = -1$	$10^{-1} = \frac{1}{10}$
$\log x = k$	$10^k = x$

General forms

A *logarithm is an exponent*. For example, to evaluate the logarithm log 100, ask the question, "10 to what power equals 100?" The necessary *exponent* equals log 100. That is, $\log 100 = 2$ because $10^2 = 100$.

EXAMPLE 1 Using exponential and logarithmic forms

Change the given equation into the equivalent logarithmic or exponential form.
(a) $\log \frac{1}{100} = -2$ (b) $\log 100{,}000 = 5$ (c) $10^4 = 10{,}000$ (d) $10^{-3} = \frac{1}{1000}$

Solution
(a) We place the exponent -2 on 10 to get $\frac{1}{100}$, so $10^{-2} = \frac{1}{100}$
(b) We place the exponent 5 on 10 to get 100,000, so $10^5 = 100{,}000$
(c) The exponent on 10 is 4, so $\log 10{,}000 = 4$
(d) The exponent on 10 is -3, so $\log \frac{1}{1000} = -3$

Now Try Exercise 13

STEPS USED TO CALCULATE THE COMMON LOGARITHM

STEP 1: Is x positive? If not, then $\log x$ is *undefined*.

STEP 2: Write x as 10^k for some real number k. If this is not possible, then use a calculator to approximate $\log x$.

STEP 3: If $x = 10^k$, then $\log x = k$. That is, $\log 10^k = k$.

EXAMPLE 2   Calculating log x

Evaluate each expression, if possible.
(a) $\log(-4)$ **(b)** $\log 1000$

Solution

(a) STEP 1: Because -4 is negative, $\log -4$ is undefined.

NOTE: Because 10^k is always positive, there is no k such that $10^k = -4$. The common logarithm of *any* negative number is undefined. ∎

(b) STEP 1: Since 1000 is positive, $\log 1000$ is defined.
STEP 2: Because $1000 = 10^3$, we can write $\log 1000$ as $\log 10^3$.
STEP 3: Because $\log 10^k = k$, it follows that
$$\log 1000 = \log 10^3 = 3.$$

Now Try Exercises 29, 33

EXAMPLE 3   Evaluating common logarithms

Evaluate each common logarithm.
(a) $\log 100$ **(b)** $\log \frac{1}{10}$ **(c)** $\log \sqrt{1000}$ **(d)** $\log 45$

Solution
(a) $100 = 10^2$, so $\log 100 = \log 10^2 = 2$
(b) $\frac{1}{10} = 10^{-1}$, so $\log \frac{1}{10} = \log 10^{-1} = -1$
(c) $\log \sqrt{1000} = \log (1000)^{1/2} = \log (10^3)^{1/2} = \log 10^{3/2} = \frac{3}{2}$
(d) It is not obvious how to write 45 as a power of 10. However, we can use a calculator to determine that $\log 45 \approx 1.6532$. Thus $10^{1.6532} \approx 45$.

FIGURE 19.24 supports these answers.

CALCULATOR HELP

To evaluate the common logarithmic function, see Appendix A (page AP-8).

READING CHECK 1

- What does $\log x$ equal?

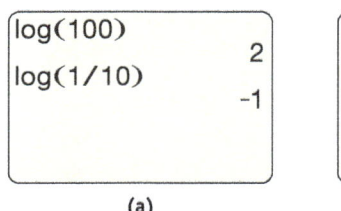

(a)   (b)

FIGURE 19.24

Now Try Exercises 19, 25, 31, 45

THE GRAPH OF THE COMMON LOGARITHM FUNCTION The following See the Concept describes the graph of $y = \log x$.

See the Concept 1 — THE GRAPH OF $y = \log x$

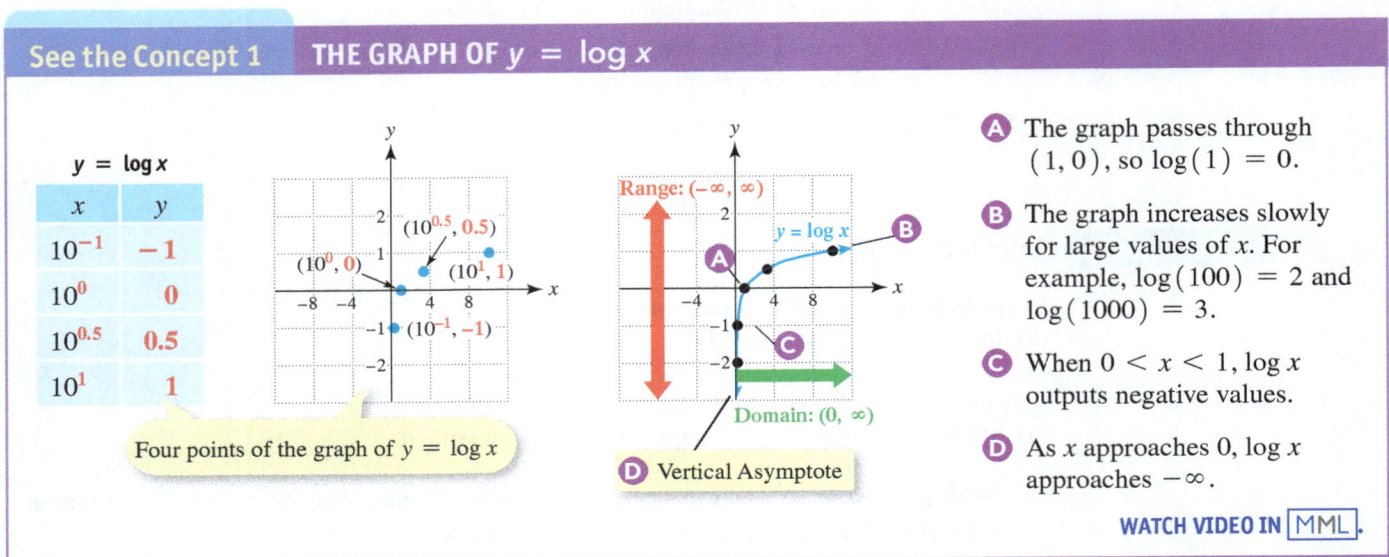

$y = \log x$	
x	y
10^{-1}	-1
10^0	0
$10^{0.5}$	0.5
10^1	1

Four points of the graph of $y = \log x$

A The graph passes through $(1, 0)$, so $\log(1) = 0$.

B The graph increases slowly for large values of x. For example, $\log(100) = 2$ and $\log(1000) = 3$.

C When $0 < x < 1$, $\log x$ outputs negative values.

D As x approaches 0, $\log x$ approaches $-\infty$.

WATCH VIDEO IN MML.

EXAMPLE 4 Graphing logarithmic functions

Graph each function f and compare its graph to $y = \log x$.
(a) $f(x) = \log(x - 2)$ (b) $f(x) = \log(x) + 1$

Solution

(a) Recall that the graph of $y = (x - 2)^2$ is similar to the graph of $y = x^2$, except that it is translated 2 units to the *right*. It follows that the graph of $y = \log(x - 2)$ is similar to the graph of $y = \log x$, except that it is also translated 2 units to the right. This translation is shown in **FIGURE 19.25**. The graph of $y = \log x$ passes through $(1, 0)$, so the graph of $y = \log(x - 2)$ passes through a point that is 2 units to the right, or $(3, 0)$. The vertical asymptote is also translated 2 units to the right to the line $x = 2$.

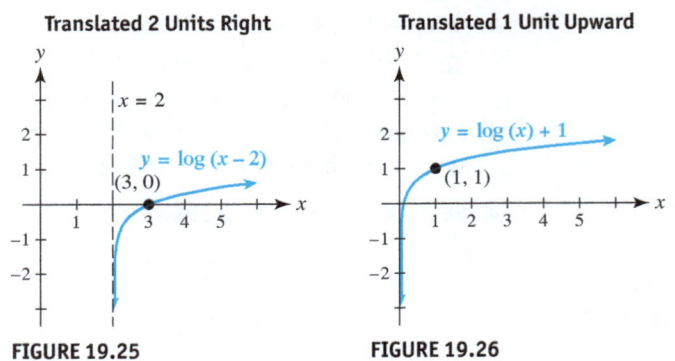

FIGURE 19.25   FIGURE 19.26

NOTE: The graph of $y = \log x$ is undefined when $x < 0$ and the graph of $\log(x - 2)$ in **FIGURE 19.25** is undefined when $x < 2$. ∎

(b) The graph of $y = \log(x) + 1$ is similar to the graph of $y = \log x$, except that it is translated 1 unit *upward*. This graph is shown in **FIGURE 19.26**. Note that the graph of $y = \log(x) + 1$ passes through the point $(1, 1)$.

Now Try Exercises 49, 51

19.3 LOGARITHMIC FUNCTIONS

> **MAKING CONNECTIONS 1**
>
> **The Common Logarithmic Function and the Square Root Function**
> Some similarities of $f(x) = \sqrt{x}$ and $f(x) = \log x$ are outlined in the table below.
>
> | Similarities | $\sqrt{x}$ | $\log x$ |
> |---|---|---|
> | Some calculations are easy. | $\sqrt{100} = 10$ | $\log 1000 = 3$ |
> | Some calculations require a calculator. | $\sqrt{2} \approx 1.41$ | $\log 45 \approx 1.65$ |
> | The implied notation is similar. | $\sqrt{x} = \sqrt[2]{x}$ | $\log x = \log_{10} x$ |
> | The input x can not be negative. | $\sqrt{-3}$ | $\log(-3)$ |
>
> Undefined expressions

2 Finding the Inverse of the Common Logarithmic Function

The graph of $y = \log x$ is a one-to-one function because it passes the horizontal line test. Thus the common logarithmic function has an inverse function. To determine this inverse function for $\log x$, consider **TABLES 19.9** and **19.10**.

$y = 10^x$

x	-2	-1	0	1	2
10^x	10^{-2}	10^{-1}	10^0	10^1	10^2

TABLE 19.9

Inverse functions

$y = \log x$

x	10^{-2}	10^{-1}	10^0	10^1	10^2
$\log x$	-2	-1	0	1	2

TABLE 19.10

If we start with the input **2** for 10^x, we compute 10^2, as shown by (1) in **TABLE 19.9**. If we use 10^2 as the input for $\log x$, we compute **2**, as shown by (2) in **TABLE 19.10**. That is,

$$\log(10^2) = 2. \quad \text{Calculate } 10^x \text{ and then } \log x.$$

Next, suppose we find the logarithm first and then compute a power of 10. If we start with the input 10^2, we compute **2**, as shown by (2) in **TABLE 19.10**. If we use **2** as the input for 10^x, we compute 10^2, as shown by (1) in **TABLE 19.9**. That is,

$$10^{\log(10^2)} = 10^2. \quad \text{Calculate } \log x \text{ and then } 10^x.$$

In general, if $10^a = b$, then $\log b = a$, and if $\log b = a$, then $10^a = b$. Thus the *inverse function* of $f(x) = \log x$ is $f^{-1}(x) = 10^x$. Note that composition of these two functions satisfies the definition of an inverse function.

$$(f \circ f^{-1})(x) = f(f^{-1}(x)) \quad \text{and} \quad (f^{-1} \circ f)(x) = f^{-1}(f(x))$$
$$= f(10^x) \qquad\qquad\qquad\qquad = f^{-1}(\log x)$$
$$= \log 10^x \qquad\qquad\qquad\qquad = 10^{\log x}$$
$$= x \qquad\qquad\qquad\qquad\qquad = x$$

The graphs of $y = \log x$ and $y = 10^x$ are shown in **FIGURE 19.27**. Note that the graph of $y = 10^x$ is a reflection of the graph of $y = \log x$ across the dashed line $y = x$.

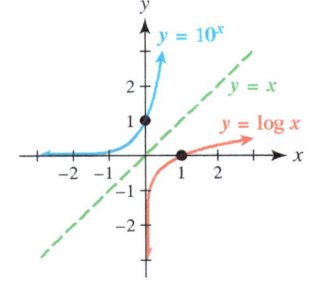

FIGURE 19.27

INVERSE PROPERTIES OF THE COMMON LOGARITHM

The following properties hold for common logarithms.

$$\log 10^x = x, \quad \text{for any real number } x$$
$$10^{\log x} = x, \quad \text{for any positive real number } x$$

EXAMPLE 5 Applying inverse properties

Use inverse properties to simplify each expression.
(a) $\log 10^\pi$ (b) $\log 10^{x^2+1}$ (c) $10^{\log 7}$ (d) $10^{\log 3x}$, $x > 0$

Solution
(a) Because $\log 10^x = x$ for any real number x, $\log 10^\pi = \pi$.
(b) $\log 10^{x^2+1} = x^2 + 1$
(c) Because $10^{\log x} = x$ for any positive real number x, $10^{\log 7} = 7$.
(d) $10^{\log 3x} = 3x$, provided x is a *positive* number.

Now Try Exercises 27, 37, 39, 43

Math in Context *Audiology* Logarithms are used to model quantities that vary greatly in intensity. For example, the human ear is extremely sensitive and able to detect intensities on the eardrum ranging from 10^{-16} watts per square centimeter (w/cm^2) to 10^{-4} w/cm^2, which is usually painful. The next example illustrates modeling sound with logarithms.

EXAMPLE 6 Modeling sound levels

Sound levels in decibels (dB) can be computed by $f(x) = 160 + 10 \log x$, where x is the intensity of the sound in watts per square centimeter. Ordinary conversation has an intensity of 10^{-10} w/cm^2. What decibel level is this? (*Source:* R. Weidner and R. Sells, *Elementary Classical Physics*, Vol. 2.)

Solution
To find the decibel level for ordinary conversation, evaluate $f(10^{-10})$.

$$f(10^{-10}) = 160 + 10 \log (10^{-10}) \quad \text{Substitute } x = 10^{-10}.$$
$$= 160 + 10(-10) \quad \text{Evaluate } \log (10^{-10}).$$
$$= 60 \quad \text{Simplify.}$$

Ordinary conversation corresponds to 60 dB.

Now Try Exercise 111

CRITICAL THINKING 1

If the sound level increases by 10 dB, by what factor does the intensity x increase?

3 Working with Logarithms that Have Other Bases

BASE 2 Common logarithms are base-10 logarithms, but we can define logarithms using other bases. For example, base-2 logarithms are frequently used in computer science. Some values for the base-2 logarithmic function, denoted $f(x) = \log_2 x$, are shown in **TABLE 19.11**. If x can be expressed as $x = 2^k$ for some real number k, then $\log_2 x = \log_2 2^k = k$.

READING CHECK 2

• What does $\log_2 x$ mean?

Base-2 Logarithms of Powers of 2

x	2^{-3}	2^{-2}	2^{-1}	2^0	2^1	2^2	2^3
$\log_2 x$	-3	-2	-1	0	1	2	3

TABLE 19.11

Write x as a power of 2.

A base-2 logarithm is the exponent on base 2.

NOTE: A base-2 logarithm is an *exponent* on base 2. ∎

EXAMPLE 7 Evaluating base-2 logarithms

Simplify each logarithm.
(a) $\log_2 8$ (b) $\log_2 \frac{1}{4}$

Solution
(a) The logarithmic expression $\log_2 8$ represents the exponent on base 2 that gives 8. Because $8 = 2^3$, $\log_2 8 = \log_2 2^3 = 3$.
(b) Because $\frac{1}{4} = \frac{1}{2^2} = 2^{-2}$, $\log_2 \frac{1}{4} = \log_2 2^{-2} = -2$.

Now Try Exercises 87, 89

BASE e (THE NATURAL LOGARITHM) Some values of base-e logarithms are shown in **TABLE 19.12**. A base-e logarithm is referred to as a **natural logarithm** and is denoted either $\log_e x$ or $\ln x$. Natural logarithms are used in many areas including mathematics, science, economics, electronics, and communications.

CALCULATOR HELP
To evaluate the natural logarithmic function, see Appendix A (page AP-8).

Natural Logarithms of Powers of e

x	e^{-3}	e^{-2}	e^{-1}	e^0	e^1	e^2	e^3
$\ln x$	-3	-2	-1	0	1	2	3

TABLE 19.12

NOTE: A natural logarithm is an *exponent* on base e. ∎

To evaluate natural logarithms, we usually use a calculator.

```
ln(10)
        2.302585093
ln(1/2)
        -.6931471806
```

FIGURE 19.28

EXAMPLE 8 Evaluating natural logarithms

Approximate to the nearest hundredth.
(a) $\ln 10$ (b) $\ln \frac{1}{2}$

Solution
(a) **FIGURE 19.28** shows that $\ln 10 \approx 2.30$.
(b) **FIGURE 19.28** shows that $\ln \frac{1}{2} \approx -0.69$.

Now Try Exercises 65, 67

BASE a We now define base-a logarithms.

> **BASE-a LOGARITHMS**
>
> The **logarithm with base a of a positive number x**, denoted $\log_a x$, is calculated as follows. If x is written as $x = a^k$, then
>
> $$\log_a x = k,$$
>
> where $a > 0$, $a \neq 1$, and k is a real number. That is, $\log_a a^k = k$.
> The function given by
>
> $$f(x) = \log_a x$$
>
> is called the **logarithmic function with base a**.

CRITICAL THINKING 2

Evaluate $\log_2 1$.

Remember that *a logarithm is an exponent*. The expression $\log_a x$ equals the exponent k such that $a^k = x$. The graph of $y = \log_a x$ with $a > 1$ is shown in **FIGURE 19.29**. Note that the graph passes through the point $(1, 0)$. Thus $\log_a 1 = 0$. Furthermore, the y-axis is a vertical asymptote and the expression $\log_a x$ is defined only when x is positive.

NOTE: The natural logarithm, $\ln x$, is a base-a logarithm with $a = e$. That is, $\ln x = \log_e x$. ∎

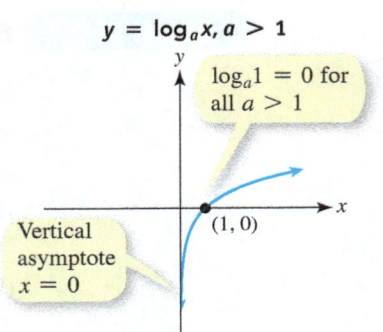

FIGURE 19.29

EXAMPLE 9 **Evaluating base-a logarithms**

Simplify each logarithm.

(a) $\log_5 25$ (b) $\log_4 \frac{1}{64}$ (c) $\log_7 1$ (d) $\log_3 9^{-1}$

Solution

(a) $25 = 5^2$, so $\log_5 25 = \log_5 5^2 = \mathbf{2}$.
(b) $\frac{1}{64} = \frac{1}{4^3} = 4^{-3}$, so $\log_4 \frac{1}{64} = \log_4 4^{-3} = \mathbf{-3}$.
(c) $1 = 7^0$, so $\log_7 1 = \log_7 7^0 = \mathbf{0}$. (The logarithm of 1 is 0, regardless of the base.)
(d) $9^{-1} = (3^2)^{-1} = 3^{-2}$, so $\log_3 9^{-1} = \log_3 3^{-2} = \mathbf{-2}$.

Now Try Exercises 59, 81, 91, 93

INVERSES OF BASE a LOGARITHMIC FUNCTIONS The graph of $y = \log_a x$ in **FIGURE 19.29** passes the horizontal line test, so it is a one-to-one function and it has an inverse. If $f(x) = \log_a x$, then $f^{-1}(x) = a^x$. This is a generalization of the fact that $f(x) = \log x$ and $f^{-1}(x) = 10^x$ represent inverse functions. These and other properties are illustrated for $f(x) = 2^x$ and $g(x) = e^x$ in the following See the Concept.

See the Concept 2   **INVERSES OF EXPONENTIAL FUNCTIONS**

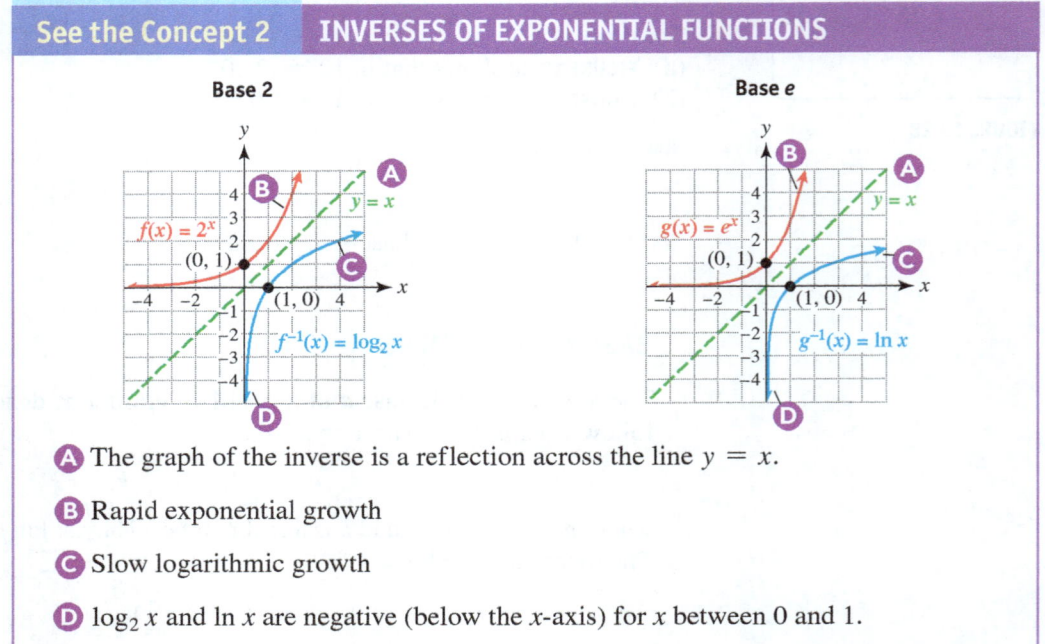

A The graph of the inverse is a reflection across the line $y = x$.

B Rapid exponential growth

C Slow logarithmic growth

D $\log_2 x$ and $\ln x$ are negative (below the x-axis) for x between 0 and 1.

WATCH VIDEO IN MML.

These inverse properties are summarized as follows.

> **INVERSE PROPERTIES OF BASE-a LOGARITHMS**
>
> The following properties hold for logarithms with base a.
>
> $\log_a a^x = x,$ for any real number x
>
> $a^{\log_a x} = x,$ for any positive real number x

NOTE: The inverse function of $f(x) = \ln x$ is $f^{-1}(x) = e^x$.
The inverse function of $f(x) = \log x$ is $f^{-1}(x) = 10^x$. ∎

EXAMPLE 10 Applying inverse properties

Simplify each expression.
(a) $\ln e^{0.5x}$ (b) $e^{\ln 4}$ (c) $2^{\log_2 7x}$ (d) $10^{\log(9x-3)}$

Solution
(a) $\ln e^{0.5x} = 0.5x$ because $\ln e^k = k$ for all k.
(b) $e^{\ln 4} = 4$ because $e^{\ln k} = k$ for all positive k.
(c) $2^{\log_2 7x} = 7x$ for $x > 0$ because $a^{\log_a k} = k$ for all positive k.
(d) $10^{\log(9x-3)} = 9x - 3$ for $x > \frac{1}{3}$ because $10^{\log k} = k$ for all positive k.

Now Try Exercises 41, 61, 63, 95

🌐 Math in Context *Airline Safety* Logarithms occur in many applications, including calculating runway length for airplanes.

EXAMPLE 11 Calculating runway length

There is a mathematical relationship between an airplane's weight x and the runway length required at takeoff. For certain types of airplanes, the minimum runway length L in thousands of feet may be modeled by $L(x) = 1.3 \ln x$, where x is in thousands of pounds.
(*Source:* L. Haefner, *Introduction to Transportation Systems.*)
(a) Estimate the runway length needed for an airplane that weighs 10,000 pounds.
(b) Does a 20,000-pound airplane need twice the runway length that a 10,000-pound airplane needs? Explain.

Solution
(a) Because $L(x) = 1.3 \ln x$, it follows that $L(10) = 1.3 \ln 10 \approx 3$. An airplane that weighs 10,000 pounds requires a runway (at least) 3000 feet long.
(b) Because $L(20) = 1.3 \ln 20 \approx 3.9$, a 20,000-pound airplane does not need twice the runway length needed by a 10,000-pound airplane. Rather, the heavier airplane needs roughly 3900 feet of runway, or only an extra 900 feet.

Now Try Exercise 113

19.3 Putting It All Together

Common logarithms are base-10 logarithms. If a positive number x is written as $x = 10^k$, then $\log x = k$. The value of $\log x$ represents the exponent on the base 10 that gives x.

CONCEPT	EXPLANATION	EXAMPLES
Base-a Logarithms	*Definition:* $\log_a x = k$ means $x = a^k$, where $a > 0$ and $a \neq 1$. *Domain:* all *positive* real numbers *Range:* all real numbers *Graph:* $a > 1$ (shown to the right for $\log x$ and $\ln x$); passes through $(1, 0)$; vertical asymptote: y-axis *Common Logarithm:* base-10 logarithm and denoted $\log x$ *Natural Logarithm:* base-e logarithm, where $e \approx 2.718$, and denoted $\ln x$	$\log 1000 = \log 10^3 = 3$, $\log_2 16 = \log_2 2^4 = 4$, and $\log_3 \frac{1}{81} = \log_3 3^{-4} = -4$
Inverse Properties	The following properties hold for base-a logarithms. $\log_a a^x = x$, for any real number x $a^{\log_a x} = x$, for any positive number x	$\log 10^{7.48} = 7.48$ $2^{\log_2 63} = 63$ $10^{\log 23} = 23$ $\ln e^4 = 4$

19.3 Exercises

CONCEPTS AND VOCABULARY

1. What is the base of the common logarithm?
2. What is the base of the natural logarithm?
3. What are the domain and range of $\log x$?
4. What are the domain and range of 10^x?
5. $\log 10^k = $ _____
6. $\ln e^k = $ _____
7. If $\log x = k$, then $10^k = $ _____.
8. If $x > 0$, then $10^{\log x} = $ _____.
9. What does k equal if $10^k = 5$?
10. What does k equal if $e^k = 5$?
11. $\log 1 = $ _____
12. $\log(-1)$ is _____.

CHANGING EXPONENTIAL AND LOGARITHMIC FORMS

Exercises 13–16: Change the given equation into the equivalent exponential or logarithmic form.

13. (a) $\log 100 = 2$
    (b) $10^{-1} = \frac{1}{10}$
14. (a) $10^5 = 100{,}000$
    (b) $\log \frac{1}{1000} = -3$
15. (a) $10^0 = 1$
    (b) $\log \frac{1}{100{,}000} = -5$
16. (a) $\log 1000 = 3$
    (b) $10^1 = 10$

EVALUATING COMMON LOGARITHMS

Exercises 17 and 18: Complete the table.

17.

x	10^{-5}	10^0	$10^{0.5}$	$10^{2.2}$
$\log x$				

18.

x	10^{-6}	10^{-1}	$10^{\frac{5}{7}}$	10^{π}
$\log x$				

Exercises 19–44: Simplify the expression, if possible.

19. $\log 10^5$
20. $\log 10$
21. $\log 10^{-4}$
22. $\log 10^{-1}$
23. $\log 1$
24. $\log \sqrt[3]{100}$
25. $\log \frac{1}{100}$
26. $\log \frac{1}{10}$
27. $\log 10^{4.7}$
28. $\log 10^{2x+4}$
29. $\log 10{,}000$
30. $\log 100{,}000$
31. $\log \sqrt{10}$
32. $\log 1{,}000{,}000$
33. $\log(-23)$
34. $\log(-8)$
35. $\log 0.001$
36. $\log 0.0001$
37. $10^{\log 2}$
38. $10^{\log 7.5}$
39. $10^{\log x^2}$
40. $10^{\log |x|}$
41. $10^{\log 5}$
42. $\log 10^{\frac{3}{4}}$
43. $\log 10^{2x-7}$
44. $\log 10^{8-4x}$

Exercises 45–48: Evaluate the common logarithm, using a calculator. Round values to the nearest thousandth.

45. $\log 25$
46. $\log 0.501$
47. $\log 1.45$
48. $\log \frac{1}{35}$

GRAPHING COMMON LOGARITHMS

Exercises 49–56: Graph $y = f(x)$. Compare the graph to the graph of $y = \log x$.

49. $f(x) = \log(x) - 1$
50. $f(x) = \log(x) + 2$
51. $f(x) = \log(x + 1)$
52. $f(x) = \log(x + 2)$
53. $f(x) = \log(x - 1)$
54. $f(x) = \log(x - 3)$
55. $f(x) = 2\log x$
56. $f(x) = -\log x$

EVALUATING NATURAL LOGARITHMS

Exercises 57 and 58: Complete the table.

57.

x	$e^{0.5}$	e^{-12}	$e^{\frac{1}{2}}$	$e^{-2.4}$
$\ln x$				

58.

x	e^{-6}	e^{-1}	$e^{\frac{5}{7}}$	e^{π}
$\ln x$				

Exercises 59–64: Simplify the expression.

59. $\ln 1$
60. $\ln e^2$
61. $\ln e^{-5x}$
62. $e^{\ln 2x}$
63. $e^{\ln x^2}$
64. $\ln e^{7x}$

Exercises 65–68: Evaluate the natural logarithm, using a calculator. Round values to the nearest thousandth.

65. $\ln 7$
66. $\ln 126$
67. $\ln \frac{4}{7}$
68. $\ln 0.67$

GRAPHING NATURAL LOGARITHMS

Exercises 69–72: Graph f in $[-4, 4, 1]$ by $[-4, 4, 1]$. Compare this graph to the graph of $y = \ln x$. Identify the domain of f.

69. $f(x) = \ln|x|$
70. $f(x) = \ln(x) - 2$
71. $f(x) = \ln(x + 2)$
72. $f(x) = 2\ln x$

EVALUATING BASE-a LOGARITHMS

Exercises 73 and 74: Complete the table.

73.

x	$5^{\frac{2}{3}}$	5^{-1}	$5^{5.5}$	$5^{-0.9}$
$\log_5 x$				

74.

x	2^{-5}	2^0	$2^{0.5}$	$2^{2.2}$
$\log_2 x$				

Exercises 75–100: Simplify the expression, if possible.

75. $\log_5 5^{6x}$
76. $\log_3 3^3$
77. $\log_2 \sqrt{\frac{1}{8}}$
78. $\log_2 2^6$
79. $\log_2 2^8$
80. $\log_2 2^{-5}$
81. $\log_2 \sqrt{8}$
82. $\log_2 \sqrt{32}$
83. $\log_2 \sqrt[3]{\frac{1}{4}}$
84. $\log_2 \frac{1}{64}$
85. $\log_2(-8)$
86. $\log_2(-7)$
87. $\log_2 4$
88. $\log_2 \frac{1}{32}$
89. $\log_2 \frac{1}{16}$
90. $\log_3 27$
91. $\log_3 \frac{1}{9}$
92. $\log_4 16$
93. $\log_5 5^{-2}$
94. $\log_8 64$
95. $5^{\log_5 17}$
96. $9^{\log_9 73}$
97. $4^{\log_4 (2x)^2}$
98. $b^{\log_b (x-1)}$
99. $5^{\log_5 0.6z}$
100. $7^{\log_7 (x-9)}$

Exercises 101 and 102: Complete the table.

101.

x	$\frac{1}{4}$	$\frac{1}{2}$	1	$\sqrt{2}$	64
$\log_2 x$					

102.

x	$\frac{1}{7}$	1	$\sqrt{7}$	7	49
$\log_7 x$					

GRAPHING BASE-*a* LOGARITHMS

Exercises 103–106: Graph $y = f(x)$.

103. $f(x) = \log_2 x$ **104.** $f(x) = \log_2 (x - 2)$

105. $f(x) = 2 + \log_2 x$ **106.** $f(x) = \log_2 (x + 2)$

Exercises 107–110: Without using a calculator match $f(x)$ with its graph (a–d).

107. $f(x) = \log x$ **108.** $f(x) = \log_3 x$

109. $f(x) = \log_3 (x) + 2$ **110.** $f(x) = \log (x + 1)$

a. 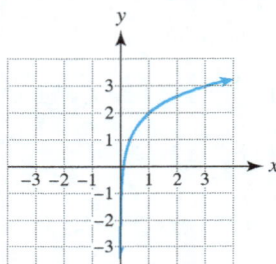 **b.**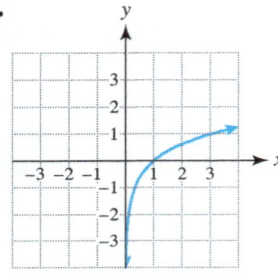

c. 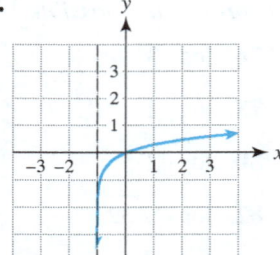 **d.**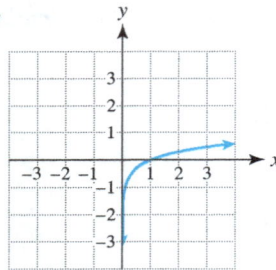

APPLICATIONS OF LOGARITHMS

111. Modeling Sound (Refer to Example 6.) At professional football games in domed stadiums, the decibel level may reach 120. The eardrum usually experiences pain when the intensity of the sound reaches 10^{-4} watts per square centimeter. How many decibels does this quantity represent? Is the noise at a football game likely to hurt some people's eardrums?

112. Hurricanes The barometric air pressure P in inches of mercury at a distance of d miles from the eye of a severe hurricane can sometimes be modeled by the formula $P(d) = 0.48 \ln (d + 1) + 27$. Average air pressure is about 30 inches of mercury. (Source: A. Miller and R. Anthes, *Meteorology*.)

(a) Evaluate $P(0)$ and $P(50)$. Interpret the results.

(b) A graph of $y = P(d)$ is shown in the figure. Describe how the air pressure changes as the distance from the eye of the hurricane increases.

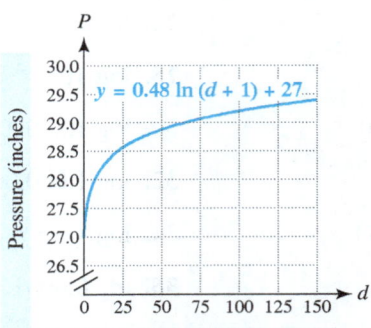

(c) Is the eye of the hurricane a low-pressure area or a high-pressure area?

113. Runway Length (Refer to Example 11.)

(a) A graph of $L(x) = 1.3 \ln x$ is shown in the figure, where $y = L(x)$. As the weight of the plane increases, what can be said about the length of the runway required?

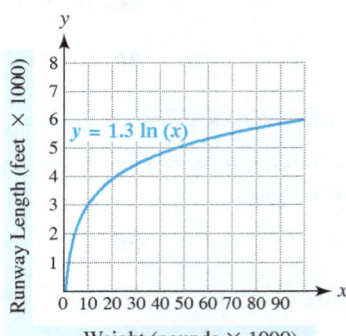

(b) Evaluate $L(50)$ and interpret the result.

114. Growth in Salary Suppose that a person's salary is initially $40,000 and could be determined by either $f(x)$ or $g(x)$, where x represents the number of years of experience.

$$f(x) = 40{,}000(1.1)^x$$
$$g(x) = 40{,}000 \log(10 + x)$$

Would most people prefer that their salaries increase exponentially or logarithmically? Explain.

115. Magnitude of a Star The first stellar brightness scale was developed 2000 years ago by two Greek astronomers, Hipparchus and Ptolemy. The brightest star in the sky was given a magnitude of 1, and the faintest star was given a magnitude of 6. In 1856, this scale was described mathematically by the formula

$$M = 6 - 2.5 \log \frac{I}{I_0},$$

where M is the magnitude of a star with an intensity of I and I_0 is the intensity of the faintest star seen in the sky. (*Source:* M. Zeilik, *Introductory Astronomy and Astrophysics*.)
(a) Find M if $I = 10$ and $I_0 = 1$.
(b) What is the magnitude of a star that is 100 times more intense than the faintest star?
(c) If the intensity of a star increases by a factor of 10, what happens to its magnitude?

116. Population of Urban Regions While less industrialized urban regions of the world are experiencing exponential population growth, industrialized urban regions are experiencing logarithmic population growth. Population in less industrialized urban regions can be modeled by

$$f(x) = 0.338(1.035)^x,$$

whereas the population in industrialized urban regions can be modeled by

$$g(x) = 0.36 + 0.15 \ln(x + 1).$$

In these formulas, the output is in billions of people and x is in years, where $x = 0$ corresponds to 1950, $x = 10$ corresponds to 1960, and so on until $x = 80$ corresponds to 2030. (*Source:* D. Meadows, *Beyond The Limits*.)
(a) Evaluate $f(50)$ and $g(50)$. Interpret the results.
(b) Graph f and g in $[0, 80, 10]$ by $[0, 5, 1]$. Compare the two graphs.
(c) If x increases from 20 to 40, by what factor does $f(x)$ increase? By what factor does $g(x)$ increase?

117. Earthquakes The Richter scale is used to determine the intensity of earthquakes, which corresponds to the amount of energy released. If an earthquake has an intensity of x, its *magnitude*, as computed by the Richter scale, is given by $R(x) = \log \frac{x}{I_0}$, where I_0 is the intensity of a small, measurable earthquake.

(a) On July 26, 1963, an earthquake in Yugoslavia had a magnitude of 6.0 on the Richter scale, and on August 19, 1977, an earthquake in Indonesia measured 8.0. Find the intensity x for each of these earthquakes if $I_0 = 1$.
(b) How many times more intense was the Indonesian earthquake than the Yugoslavian earthquake?

118. Path Loss for Cellular Phones For cellular phones to work throughout a country, large numbers of cellular towers are necessary. How well the signal is propagated throughout a region depends on the location of these towers. One quick way to estimate the strength of a signal at x kilometers is to use the formula

$$D(x) = -121 - 36 \log x.$$

This formula computes the decrease in the signal, using decibels, so it is always negative. For example, $D(1) = -121$ means that at a distance of 1 kilometer, the signal has decreased in strength by 121 decibels. (*Source:* C. Smith, *Practical Cellular & PCS Design*.)

(a) Evaluate $D(3)$ and interpret the result.
(b) Graph D in $[1, 10, 1]$ by $[-160, -120, 10]$.
(c) What happens to the signal as x increases?

WRITING ABOUT MATHEMATICS

119. Explain what $\log_a x$ means and give an example.

120. How would you explain to a student that $\log_a 1 \neq 1$?

Group Activity

Greenhouse Gases Carbon dioxide (CO_2) is a greenhouse gas in the atmosphere that may raise average temperatures on Earth. The burning of fossil fuels could be responsible for the increased levels of carbon dioxide. If current trends continue, future concentrations of atmospheric carbon dioxide in parts per million (ppm) could reach the levels shown in the accompanying table. The CO_2 concentration in the year 2000 was greater than it had been at any time in the previous 160,000 years.

Year	2000	2050	2100	2150	2200
CO_2 (ppm)	364	467	600	769	987

Source: R. Turner, *Environmental Economics.*

(a) Let x be in years, where $x = 0$ corresponds to 2000, $x = 1$ to 2001, and so on. Find values for C and a so that $f(x) = Ca^x$ models the data.
(b) Graph f and the data in the same viewing rectangle.
(c) Use $f(x)$ to estimate graphically the year when the carbon dioxide concentration will be double the preindustrial level of 280 ppm.

19.4 Properties of Logarithms

Objectives

1. Introducing Basic Properties of Logarithms
    - The Product Rule
    - The Quotient Rule
    - The Power Rule
    - Applying More Than One Property
2. Using the Change of Base Formula

NEW VOCABULARY

☐ Change of base formula

1 Introducing Basic Properties of Logarithms

In this subsection, we discuss three important properties of logarithms. The first property is the product rule for logarithms.

THE PRODUCT RULE The product rule for logarithms can be applied when computing a logarithm of a product.

> **PRODUCT RULE FOR LOGARITHMS**
>
> For positive numbers m, n, and $a \neq 1$,
>
> $$\log_a mn = \log_a m + \log_a n.$$

This property is illustrated in **FIGURE 19.30**, which shows that

$$\log 10 = \log (2 \cdot 5) = \log 2 + \log 5.$$

This product rule may be verified by using properties of exponents and the fact that $\log_a a^k = k$ for any real number k. Here, we verify the product property for logarithms. The two other properties presented later can be verified in a similar manner.

If m and n are positive numbers, we can write $m = a^c$ and $n = a^d$ for some real numbers c and d.

$$\log_a mn = \log_a (a^c a^d) = \log_a (a^{c+d}) = c + d \quad \text{and}$$
$$\log_a m + \log_a n = \log_a a^c + \log_a a^d = c + d$$

Thus $\log_a mn = \log_a m + \log_a n$.

```
log(10)
              1
log(2)+log(5)
              1
```

FIGURE 19.30

EXAMPLE 1 Expanding logarithms as sums

Write each expression as a sum of logarithms. Assume that x is positive.
(a) $\log 21$ (b) $\ln 5x$ (c) $\log x^3$

Solution
(a) $\log 21 = \log(3 \cdot 7) = \log 3 + \log 7$ (b) $\ln 5x = \ln(5 \cdot x) = \ln 5 + \ln x$
(c) $\log x^3 = \log(x \cdot x \cdot x) = \log x + \log x + \log x$

Now Try Exercises 13, 15, 17

EXAMPLE 2 Combining logarithms

Write each expression as one logarithm. Assume that x and y are positive.
(a) $\log 5 + \log 6$ (b) $\ln x + \ln xy$ (c) $\log 2x + \log 5x$

Solution
(a) $\log 5 + \log 6 = \log(5 \cdot 6) = \log 30$ (b) $\ln x + \ln xy = \ln(x \cdot xy) = \ln x^2 y$
(c) $\log 2x + \log 5x = \log(2x \cdot 5x) = \log 10x^2$

Now Try Exercises 25, 27, 29

THE QUOTIENT RULE The second property is the quotient rule for logarithms.

> **QUOTIENT RULE FOR LOGARITHMS**
>
> For positive numbers m, n, and $a \neq 1$,
>
> $$\log_a \frac{m}{n} = \log_a m - \log_a n.$$

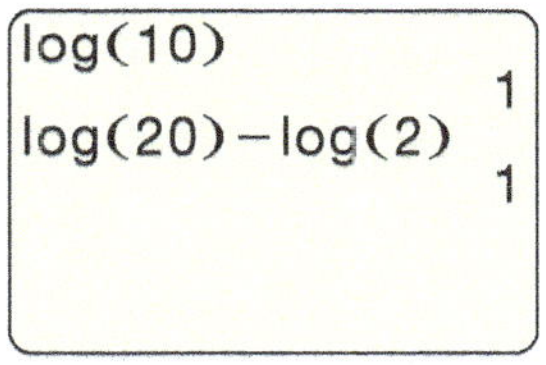

```
log(10)
               1
log(20)−log(2)
               1
```

FIGURE 19.31

This property is illustrated in **FIGURE 19.31**, which shows that

$$\log 10 = \log \frac{20}{2} = \log 20 - \log 2.$$

EXAMPLE 3 Expanding logarithms as differences

Write each expression as a difference of two logarithms. Assume that variables are positive.
(a) $\log \frac{3}{2}$ (b) $\ln \frac{3x}{y}$ (c) $\log_5 \frac{x}{z^4}$

Solution
(a) $\log \frac{3}{2} = \log 3 - \log 2$ (b) $\ln \frac{3x}{y} = \ln 3x - \ln y$

(c) $\log_5 \frac{x}{z^4} = \log_5 x - \log_5 z^4$

Now Try Exercises 19, 21, 23

NOTE: $\log_a(m + n) \neq \log_a m + \log_a n$; $\log_a(m - n) \neq \log_a m - \log_a n$;
$\log_a(mn) \neq \log_a m \cdot \log_a n$; $\log_a\left(\frac{m}{n}\right) \neq \frac{\log_a m}{\log_a n}$

EXAMPLE 4 **Combining logarithms**

Write each expression as one term. Assume that x is positive.
(a) $\log 50 - \log 25$ (b) $\ln x^3 - \ln x$ (c) $\log 15x - \log 5x$

Solution
(a) $\log 50 - \log 25 = \log \dfrac{50}{25} = \log 2$ (b) $\ln x^3 - \ln x = \ln \dfrac{x^3}{x} = \ln x^2$

(c) $\log 15x - \log 5x = \log \dfrac{15x}{5x} = \log 3$

Now Try Exercises 33, 35, 37

READING CHECK 1

- Is $\log(4 + 5)$ the same as $\log 4 + \log 5$? Why or why not?

THE POWER RULE The third property is the power rule for logarithms. To illustrate this rule, we use the following example.

$$\log x^3 = \log(x \cdot x \cdot x) = \log x + \log x + \log x = 3 \log x$$

Thus $\log x^3 = 3 \log x$. This example is generalized in the following rule.

POWER RULE FOR LOGARITHMS

For positive numbers m and $a \neq 1$ and any real number r,

$$\log_a(m^r) = r \log_a m.$$

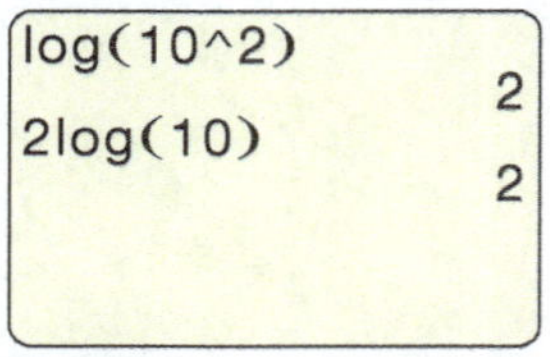

```
log(10^2)
            2
2log(10)
            2
```

FIGURE 19.32

We use a calculator and the equation

$$\log 10^2 = 2 \log 10$$

to illustrate this property. **FIGURE 19.32** shows the result.

EXAMPLE 5 **Applying the power rule**

Rewrite each expression, using the power rule.
(a) $\log 5^6$ (b) $\ln 0.55^{x-1}$ (c) $\log_5 8^{kx}$

Solution
(a) $\log 5^6 = 6 \log 5$ (b) $\ln 0.55^{x-1} = (x - 1) \ln 0.55$
(c) $\log_5 8^{kx} = kx \log_5 8$

Now Try Exercises 39, 41, 47

See the Concept 1 **EXPANDING VERSUS COMBINING**

EXPANDING Ⓐ

$\log 10a = \log 10 + \log a$

$\log \dfrac{x}{5} = \log x - \log 5$

$\log x^3 = 3 \log x$

Ⓑ **COMBINING**

Ⓐ Transforming the left side of the equation into the right side is called **EXPANDING**.

Ⓑ Transforming the right side of the equation into the left side is called **COMBINING**.

WATCH VIDEO IN MML.

APPLYING MORE THAN ONE PROPERTY Sometimes we use more than one property to simplify an expression. We assume that all variables are positive in the next two examples.

EXAMPLE 6 Combining logarithms

Write each expression as the logarithm of a single expression.
(a) $3 \log x + \log x^2$ (b) $2 \ln x - \ln \sqrt{x}$

Solution

(a) $3 \log x + \log x^2 = \log x^3 + \log x^2$ Power rule
$= \log (x^3 \cdot x^2)$ Product rule
$= \log x^5$ Properties of exponents

(b) $2 \ln x - \ln \sqrt{x} = 2 \ln x - \ln x^{1/2}$ $\sqrt{x} = x^{1/2}$
$= \ln x^2 - \ln x^{1/2}$ Power rule
$= \ln \dfrac{x^2}{x^{1/2}}$ Quotient rule
$= \ln x^{3/2}$ Properties of exponents

Now Try Exercises 59, 67

EXAMPLE 7 Expanding logarithms

Write each expression in terms of logarithms of x, y, and z.
(a) $\log \dfrac{x^2 y^3}{\sqrt{z}}$ (b) $\ln \sqrt[3]{\dfrac{xy}{z}}$

Solution

(a) $\log \dfrac{x^2 y^3}{\sqrt{z}} = \log x^2 y^3 - \log \sqrt{z}$ Quotient rule
$= \log x^2 + \log y^3 - \log z^{1/2}$ Product rule; $\sqrt{z} = z^{1/2}$
$= 2 \log x + 3 \log y - \tfrac{1}{2} \log z$ Power rule

(b) $\ln \sqrt[3]{\dfrac{xy}{z}} = \ln \left(\dfrac{xy}{z}\right)^{1/3}$ $\sqrt[3]{m} = m^{1/3}$
$= \tfrac{1}{3} \ln \dfrac{xy}{z}$ Power rule
$= \tfrac{1}{3} (\ln xy - \ln z)$ Quotient rule
$= \tfrac{1}{3} (\ln x + \ln y - \ln z)$ Product rule
$= \tfrac{1}{3} \ln x + \tfrac{1}{3} \ln y - \tfrac{1}{3} \ln z$ Distributive property

Now Try Exercises 73, 75

EXAMPLE 8 Applying properties of logarithms

Using only properties of logarithms and the approximations $\ln 2 \approx 0.7$, $\ln 3 \approx 1.1$, and $\ln 5 \approx 1.6$, find an approximation for each expression.
(a) $\ln 8$ (b) $\ln 15$ (c) $\ln \dfrac{10}{3}$

Solution
Using $\ln 2 \approx 0.7$, $\ln 3 \approx 1.1$, and $\ln 5 \approx 1.6$, we approximate as follows.
(a) $\ln 8 = \ln 2^3 = 3 \ln 2 \approx 3(0.7) = 2.1$
(b) $\ln 15 = \ln(3 \cdot 5) = \ln 3 + \ln 5 \approx 1.1 + 1.6 = 2.7$
(c) $\ln \dfrac{10}{3} = \ln\left(\dfrac{2 \cdot 5}{3}\right) = \ln 2 + \ln 5 - \ln 3 \approx 0.7 + 1.6 - 1.1 = 1.2$

Now Try Exercises 49, 51, 55

READING CHECK 2

- How would you evaluate a logarithm with a base other than 10 or e?

2 Using the Change of Base Formula

Most calculators only have keys to evaluate common and natural logarithms. Occasionally, it is necessary to evaluate a logarithmic function with a base other than 10 or e. In these situations, we use the following **change of base formula**, which we illustrate in the next example.

> **CHANGE OF BASE FORMULA**
>
> Let x and $a \neq 1$ be positive real numbers. Then
>
> $$\log_a x = \frac{\log x}{\log a} \quad \text{or} \quad \log_a x = \frac{\ln x}{\ln a}.$$

EXAMPLE 9 Applying the change of base formula

Approximate $\log_2 14$ to the nearest thousandth.

Solution
Using the change of base formula,

$$\log_2 14 = \frac{\log 14}{\log 2} \approx 3.807 \quad \text{or} \quad \log_2 14 = \frac{\ln 14}{\ln 2} \approx 3.807.$$

FIGURE 19.33 supports these results.

Now Try Exercise 89

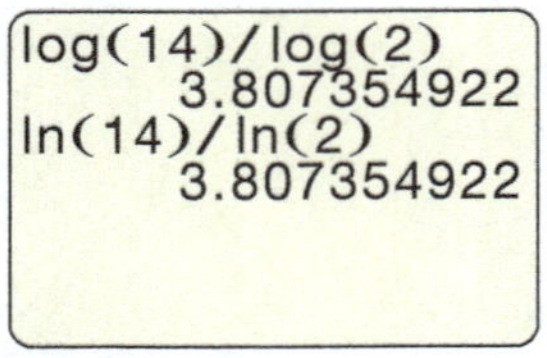

```
log(14)/log(2)
       3.807354922
ln(14)/ln(2)
       3.807354922
```

FIGURE 19.33

19.4 Putting It All Together

CONCEPT	EXPLANATION	EXAMPLES
Properties of Logarithms 1. Product Rule 2. Quotient Rule 3. Power Rule	The following properties hold for positive numbers m, n, and $a \neq 1$ and for any real number r. 1. $\log_a mn = \log_a m + \log_a n$ 2. $\log_a \frac{m}{n} = \log_a m - \log_a n$ 3. $\log_a (m^r) = r \log_a m$	1. $\log(10 \cdot 2) = \log 10 + \log 2$ 2. $\log \frac{45}{6} = \log 45 - \log 6$ 3. $\ln x^6 = 6 \ln x$
Change of Base Formula	Let x and $a \neq 1$ be positive numbers. Then $$\log_a x = \frac{\log x}{\log a} \quad \text{and} \quad \log_a x = \frac{\ln x}{\ln a}.$$	The expression $\log_3 6$ is equivalent to either $$\frac{\log 6}{\log 3} \quad \text{or} \quad \frac{\ln 6}{\ln 3}.$$

19.4 Exercises

NOTE: Assume that all variables are positive and that all expressions are defined in this exercise set. ∎

CONCEPTS AND VOCABULARY

1. $\log 12 = \log 3 + \log (\underline{})$
2. $\ln 5 = \ln 20 - \ln (\underline{})$
3. $\log 8 = (\underline{}) \log 2$
4. $\log mn = \underline{}$
5. $\log \dfrac{m}{n} = \underline{}$
6. $\log (m^r) = \underline{}$
7. Does $\log x + \log y$ equal $\log (x + y)$?
8. Does $\log x - \log y$ equal $\log \left(\dfrac{x}{y}\right)$?
9. Does $\log (xy)$ equal $(\log x)(\log y)$?
10. Does $\log \left(\dfrac{x}{y}\right)$ equal $\dfrac{\log x}{\log y}$?
11. Give the change of base formula.
12. $\log_a 1 = \underline{}$ and $\log_a a = \underline{}$.

APPLYING PROPERTIES OF LOGARITHMS

Exercises 13–18: Write the expression as a sum of two or more logarithms.

13. $\ln (15)$
14. $\log (77)$
15. $\log xy$
16. $\ln 10z$
17. $\log y^2$
18. $\log x^2 y$

Exercises 19–24: Write the expression as a difference of two logarithms.

19. $\log \dfrac{7}{3}$
20. $\ln \dfrac{11}{13}$
21. $\ln \dfrac{x}{y}$
22. $\log \dfrac{2x}{z}$
23. $\log_2 \dfrac{45}{x}$
24. $\log_7 \dfrac{5x}{4z}$

Exercises 25–32: Write the expression as one logarithm.

25. $\log 45 + \log 5$
26. $\log 30 - \log 10$
27. $\ln x + \ln y$
28. $\ln m + \ln n - \ln n$
29. $\ln 7x^2 + \ln 2x$
30. $\ln x + \ln y - \ln z$
31. $\ln x + \ln y^2 - \ln y$
32. $\ln \sqrt{z} - \ln z^3 + \ln y^3$

Exercises 33–38: Write the expression as one term. Evaluate by hand if possible.

33. $\log 20 - \log 4$
34. $\log 900 - \log 9$
35. $\ln x^4 - \ln x^2$
36. $\ln 9x^2 - \ln 3x$
37. $\log 300x - \log 3x$
38. $\log 18x^2 - \log 2x^2$

Exercises 39–48: Rewrite using the power rule.

39. $\log 3^6$
40. $\log x^7$
41. $\ln 2^x$
42. $\ln (0.77)^{x+1}$
43. $\log_2 5^{1/4}$
44. $\log_3 \sqrt{x}$
45. $\log_4 \sqrt[3]{z}$
46. $\log_7 3^\pi$
47. $\log x^{y-1}$
48. $\ln a^{2t}$

Exercises 49–58: (Refer to Example 8.) Using only properties of logarithms and the approximations $\log 2 \approx 0.3$, $\log 5 \approx 0.7$, and $\log 13 \approx 1.1$, find an approximation for the expression.

49. $\log 16$
50. $\log 125$
51. $\log 65$
52. $\log 26$
53. $\log 130$
54. $\log 100$
55. $\log \dfrac{5}{2}$
56. $\log \dfrac{26}{5}$
57. $\log \dfrac{1}{13}$
58. $\log \dfrac{1}{65}$

COMBINING LOGARITHMS

Exercises 59–70: Use properties of logarithms to write the expression as the logarithm of a single expression.

59. $4 \log z - \log z^3$
60. $2 \log_5 y + \log_5 x$
61. $\log x + 2 \log x + 2 \log y$
62. $\log x^2 + 3 \log z - 5 \log y$
63. $\log x - 2 \log \sqrt{x}$
64. $\ln y^2 - 6 \ln \sqrt[3]{y}$
65. $\ln 2^{x+1} - \ln 2$
66. $\ln 8^{1/2} + \ln 2^{1/2}$
67. $\ln \sqrt[3]{x} + \ln \sqrt{x}$
68. $2 \log_3 \sqrt{x} - 3 \log_3 x$
69. $2 \log_a (x + 1) - \log_a (x^2 - 1)$
70. $\log_b (x^2 - 9) - \log_b (x - 3)$

EXPANDING LOGARITHMS

Exercises 71–82: Use properties of logarithms to write the expression in terms of logarithms of x, y, and z.

71. $\log xy^2$

72. $\log \dfrac{x^2}{y^3}$

73. $\ln \dfrac{x^4 y}{z}$

74. $\ln \dfrac{\sqrt{x}}{y}$

75. $\log \dfrac{\sqrt[3]{z}}{\sqrt{y}}$

76. $\log \sqrt{\dfrac{x}{y}}$

77. $\log(x^4 y^3)$

78. $\log(x^2 y^4 z^3)$

79. $\ln \dfrac{1}{y} - \ln \dfrac{1}{x}$

80. $\ln \dfrac{1}{xy}$

81. $\log_4 \sqrt{\dfrac{x^3 y}{z^2}}$

82. $\log_3 \left(\dfrac{x^2 \sqrt{z}}{y^3}\right)$

Exercises 83–86: Graph f and g in the window $[-6, 6, 1]$ by $[-4, 4, 1]$. If the two graphs appear to be identical, prove that they are, using properties of logarithms.

83. $f(x) = \log x^3,\ g(x) = 3 \log x$

84. $f(x) = \ln x + \ln 3,\ g(x) = \ln 3x$

85. $f(x) = \ln(x + 5),\ g(x) = \ln x + \ln 5$

86. $f(x) = \log(x - 2),\ g(x) = \log x - \log 2$

USING THE CHANGE OF BASE FORMULA

Exercises 87–92: Use the change of base formula to approximate each expression to the nearest hundredth.

87. $\log_3 5$

88. $\log_5 12$

89. $\log_2 25$

90. $\log_7 8$

91. $\log_9 102$

92. $\log_6 293$

APPLICATIONS OF LOGARITHMS

93. Modeling Sound The formula $f(x) = 10 \log(10^{16} x)$ can be used to calculate the decibel level of a sound with an intensity x. Use properties of logarithms to simplify this formula to $f(x) = 160 + 10 \log x$.

94. Cellular Phone Technology A formula used to calculate the strength of a signal for a cellular phone is

$$L = 110.7 - 19.1 \log h + 55 \log d,$$

where h is the height of the cellular phone tower and d is the distance the phone is from the tower. Use properties of logarithms to write an expression for L that contains only one logarithm. (*Source:* C. Smith, *Practical Cellular & PCS Design.*)

WRITING ABOUT MATHEMATICS

95. State the three basic properties of logarithms and give an example of each.

96. A student insists that $\log(x - y)$ is equal to the expression $\log x - \log y$. How could you convince the student otherwise?

SECTIONS 19.3 and 19.4 — Checking Basic Concepts

1. Simplify each expression by hand.
   - (a) $\log 10^4$
   - (b) $\ln e^x$
   - (c) $\log_2 \dfrac{1}{8}$
   - (d) $\log_5 \sqrt{5}$

2. Sketch a graph of $f(x) = \log x$.
   - (a) What are the domain and range of f?
   - (b) Evaluate $f(1)$.
   - (c) Can the common logarithm of a positive number be negative? Explain.
   - (d) Can the common logarithm of a negative number be positive? Explain.

3. Write the expression in terms of logarithms of x, y, and z. Assume that variables are positive.
   - (a) $\log xy$
   - (b) $\ln \dfrac{x}{yz}$
   - (c) $\ln x^2$
   - (d) $\log \dfrac{x^2 y^3}{\sqrt{z}}$

4. Write as the logarithm of a single expression.
   - (a) $\log x + \log y$
   - (b) $\ln 2x - 3 \ln y$
   - (c) $2 \log_2 x + 3 \log_2 y - \log_2 z$

19.5 Exponential and Logarithmic Equations

Objectives

1. Recognizing Exponential Equations and Models
   - Changing Exponential Form to Logarithmic Form
   - Solving Exponential Equations
2. Solving Applications of Exponential Equations
3. Recognizing Logarithmic Equations and Models
   - Changing Logarithmic Form to Exponential Form
   - Solving Logarithmic Equations
4. Solving Applications of Logarithmic Equations

1 Recognizing Exponential Equations and Models

Recall that to solve the equation $10 + x = 100$, **we subtract 10 from each side** because addition and subtraction are inverse operations.

$$10 + x - 10 = 100 - 10 \quad \text{Subtract 10 from each side.}$$
$$x = 90$$

Similarly, to solve the equation $10x = 100$, **we divide each side by 10** because multiplication and division are inverse operations.

$$\frac{10x}{10} = \frac{100}{10} \quad \text{Divide each side by 10.}$$
$$x = 10$$

Now suppose that we want to solve the *exponential equation*

$$10^x = 100. \quad \text{Exponential equation}$$

(The exponent is a variable.)

What is new about this type of equation is that the variable x is an *exponent*. The inverse operation of 10^x is $\log x$. Rather than subtracting 10 from each side or dividing each side by 10, **we take the base-10 logarithm of each side**. Doing so results in

$$\log 10^x = \log 100. \quad \text{Take base-10 logarithm of each side.}$$

Because $\log 10^x = x$ for all real numbers x, the equation becomes

$$x = \log 100 \quad \text{or, equivalently,} \quad x = 2.$$

CHANGING EXPONENTIAL FORM TO LOGARITHMIC FORM In order to solve exponential equations, we take a logarithm of each side to change the exponential form of the equation into the logarithmic form. In general, we will show this step, but you may be able to move directly from the exponential form to the logarithmic form. The table shown below gives several examples of exponential equations along with their equivalent logarithmic forms.

Equivalent Forms

Exponential Form	Logarithmic Form	
$10^x = 100$	$x = \log 100$	Take base-10 logarithm of each side.
$2^{3x} = 8$	$3x = \log_2 8$	Take base-2 logarithm of each side.
$e^x = 9$	$x = \ln 9$	Take the natural logarithm of each side.
$8^{4x} = 20$	$4x = \log_8 20$	Take base-8 logarithm of each side.
$a^{2x} = b$	$2x = \log_a b$	Take base-a logarithm of each side.

SOLVING EXPONENTIAL EQUATIONS In the next example, we solve exponential equations by using inverse operations to change them to logarithmic form.

EXAMPLE 1 Solving exponential equations

Solve and approximate to the nearest hundredth.
(a) $10^x = 150$ (b) $e^x = 40$ (c) $2^x = 50$ (d) $0.9^x = 0.5$

Solution

(a)

Change to logarithmic form.

$$10^x = 150 \quad \text{Given equation}$$
$$\log 10^x = \log 150 \quad \text{Take the common logarithm of each side.}$$
$$x = \log 150 \approx 2.18 \quad \text{Inverse property: } \log 10^k = k \text{ for all } k$$

(b) The inverse operation of e^x is $\ln x$, so we take the natural logarithm of each side.

Change to logarithmic form.

$$e^x = 40 \quad \text{Given equation}$$
$$\ln e^x = \ln 40 \quad \text{Take the natural logarithm of each side.}$$
$$x = \ln 40 \approx 3.69 \quad \text{Inverse property: } \ln e^k = k \text{ for all } k$$

(c) The inverse operation of 2^x is $\log_2 x$. Calculators do not usually have a base-2 logarithm key, so we take the common logarithm of each side and then apply the power rule.

$$2^x = 50 \quad \text{Given equation}$$
$$\log 2^x = \log 50 \quad \text{Take the common logarithm of each side.}$$
$$x \log 2 = \log 50 \quad \text{Power rule: } \log m^r = r \log m$$
$$x = \frac{\log 50}{\log 2} \approx 5.64 \quad \text{Divide by } \log 2 \text{ and approximate.}$$

(d) This time, we begin by taking the natural logarithm of each side.

$$0.9^x = 0.5 \quad \text{Given equation}$$
$$\ln 0.9^x = \ln 0.5 \quad \text{Take the natural logarithm of each side.}$$
$$x \ln 0.9 = \ln 0.5 \quad \text{Power rule: } \ln m^r = r \ln m$$
$$x = \frac{\ln 0.5}{\ln 0.9} \approx 6.58 \quad \text{Divide by } \ln 0.9 \text{ and approximate.}$$

Now Try Exercises 21, 23, 33, 35

MAKING CONNECTIONS 1

Logarithms of Quotients and Quotients of Logarithms

The solution in Example 1(c) is $\frac{\log 50}{\log 2}$. Note that

$$\frac{\log 50}{\log 2} \neq \log 50 - \log 2.$$

Rather, $\log 50 - \log 2 = \log \frac{50}{2} = \log 25$ by the quotient rule for logarithms, as shown in the figure.

EXAMPLE 2 Solving exponential equations

Solve each equation and approximate to the nearest hundredth.
(a) $2e^x - 1 = 5$ **(b)** $3^{x-5} = 15$ **(c)** $e^{2x} = e^{x+5}$ **(d)** $3^{2x} = 2^{x+3}$

Solution
(a) Begin by solving for e^x.

$$2e^x - 1 = 5 \quad \text{Given equation}$$
$$2e^x = 6 \quad \text{Add 1 to each side.}$$
$$e^x = 3 \quad \text{Divide each side by 2.}$$

Change to logarithmic form.

$$\ln e^x = \ln 3 \quad \text{Take the natural logarithm.}$$
$$x = \ln 3 \approx 1.10 \quad \text{Inverse property: } \ln e^k = k$$

(b) Start by taking the common logarithm of each side. (We could also take the natural logarithm of each side.)

$$3^{x-5} = 15 \qquad \text{Given equation}$$
$$\log 3^{x-5} = \log 15 \qquad \text{Take the common logarithm of each side.}$$
$$(x - 5) \log 3 = \log 15 \qquad \text{Power rule for logarithms}$$
$$x - 5 = \frac{\log 15}{\log 3} \qquad \text{Divide by log 3.}$$
$$x = \frac{\log 15}{\log 3} + 5 \approx 7.46 \qquad \text{Add 5 to each side and approximate.}$$

(c) For $e^{2x} = e^{x+5}$, the bases are equal, so the exponents must also be equal. To verify this assertion, take the natural logarithm of each side.

$$e^{2x} = e^{x+5} \qquad \text{Given equation}$$
$$\ln e^{2x} = \ln e^{x+5} \qquad \text{Take the natural logarithm.}$$
$$2x = x + 5 \qquad \text{Inverse property: } \ln e^k = k$$
$$x = 5 \qquad \text{Subtract } x.$$

(d) For $3^{2x} = 2^{x+3}$, the bases are not equal. However, we can still solve the equation by taking the common logarithm of each side. A logarithm of any base could be used.

$$3^{2x} = 2^{x+3} \qquad \text{Given equation}$$
$$\log 3^{2x} = \log 2^{x+3} \qquad \text{Take the common logarithm.}$$
$$2x \log 3 = (x + 3) \log 2 \qquad \text{Power rule for logarithms}$$
$$2x \log 3 = x \log 2 + 3 \log 2 \qquad \text{Distributive property}$$
$$2x \log 3 - x \log 2 = 3 \log 2 \qquad \text{Subtract } x \log 2.$$
$$x(2 \log 3 - \log 2) = 3 \log 2 \qquad \text{Factor out } x.$$
$$x = \frac{3 \log 2}{2 \log 3 - \log 2} \qquad \text{Divide by } 2 \log 3 - \log 2.$$
$$x \approx 1.38 \qquad \text{Approximate.}$$

Now Try Exercises 37, 39, 43, 51

EXAMPLE 3 **Solving an exponential equation**

Graphs for $f(x) = 0.2e^x$ and $g(x) = 4$ are shown in **FIGURE 19.34**.
(a) Use the graphs to estimate the solution to the equation $f(x) = g(x)$.
(b) Check your estimate by solving the equation symbolically.

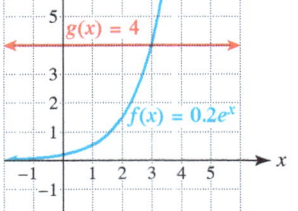

FIGURE 19.34

Solution
(a) The graphs intersect near the point $(3, 4)$. Therefore the solution is given by $x \approx 3$.
(b) We must solve the equation $0.2e^x = 4$.

$$0.2e^x = 4 \qquad \text{Given equation}$$
$$e^x = 20 \qquad \text{Divide each side by 0.2.}$$
$$\ln e^x = \ln 20 \qquad \text{Take the natural logarithm of each side.}$$
$$x = \ln 20 \qquad \text{Inverse property: } \ln e^k = k$$
$$x \approx 2.996 \qquad \text{Approximate.}$$

NOTE: The graphical estimate did not give the *exact* solution of ln 20. ∎

Now Try Exercise 53

2 Solving Applications of Exponential Equations

Math in Savings Context As discussed in Section 2 of this chapter, if a total of $1000 is deposited in a savings account paying 10% annual interest at the end of each year, the amount A in the account after x years is given by

$$A(x) = 1000(1.1)^x.$$

After 10 years, there will be

$$A(10) = 1000(1.1)^{10} \approx \$2593.74$$

in the account. To calculate how long it will take for $4000 to accrue in the account, we need to solve the exponential equation

$$1000(1.1)^x = 4000.$$

We do so in the next example.

EXAMPLE 4 Solving an exponential equation

Solve $1000(1.1)^x = 4000$ symbolically. Give graphical support for your answer.

Solution
Symbolic Solution Begin by dividing each side of the equation by 1000.

$1000(1.1)^x = 4000$	Given equation
$1.1^x = 4$	Divide by 1000.
$\log 1.1^x = \log 4$	Take the common logarithm of each side.
$x \log 1.1 = \log 4$	Power rule for logarithms
$x = \dfrac{\log 4}{\log 1.1} \approx 14.5$	Divide by log 1.1 and approximate.

Interest is paid at the end of the year, so it will take 15 years for $1000 earning 10% annual interest to grow to (at least) $4000.

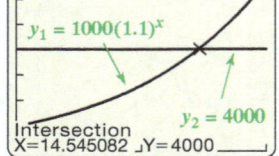

FIGURE 19.35

Graphical Solution Graphical support is shown in **FIGURE 19.35**, where the graphs of $y_1 = 1000(1.1)^x$ and $y_2 = 4000$ intersect when $x \approx 14.5$.

Now Try Exercise 105

Math in Biology Context In the next example, we model the life span of a robin with an exponential function.

EXAMPLE 5 Modeling the life span of a robin

The life spans of 129 robins were monitored over a 4-year period in one study. The formula $f(x) = 10^{-0.42x}$ can be used to calculate the percentage of robins remaining after x years. For example, $f(1) \approx 0.38$ means that after 1 year, 38% of the robins were still alive. (*Source:* D. Lack, *The Life Span of a Robin*.)
(a) Evaluate $f(2)$ and interpret the result.
(b) Determine when 5% of the robins remained.

Solution
(a) $f(2) = 10^{-0.42(2)} \approx \mathbf{0.145}$. After **2** years, about **14.5%** of the robins were still alive.
(b) Use 5% = 0.05 and solve the following equation.

$$10^{-0.42x} = 0.05 \quad \text{Equation to solve}$$
$$\log 10^{-0.42x} = \log 0.05 \quad \text{Take the common logarithm of each side.}$$
$$-0.42x = \log 0.05 \quad \text{Inverse property: } \log 10^k = k$$
$$x = \frac{\log 0.05}{-0.42} \approx 3.1 \quad \text{Divide by } -0.42.$$

After about 3 years, only 5% of the robins were still alive.

Now Try Exercise 107

3 Recognizing Logarithmic Equations and Models

To solve an exponential equation, we use logarithms. To solve a logarithmic equation, we *exponentiate* each side of the equation. To do so, we use the fact that if $x = y$, then $a^x = a^y$ for any positive base a. For example, to solve

$$\log x = 3, \quad \text{Logarithmic equation}$$

we exponentiate each side of the equation, using base 10.

$$10^{\log x} = 10^3 \quad \text{If } a = b, \text{ then } 10^a = 10^b.$$

Because $10^{\log x} = x$ for all positive x, it follows that

$$x = 10^3. \quad \text{Inverse property}$$

To solve logarithmic equations, we frequently use the inverse property

$$a^{\log_a x} = x.$$

Bases must be equal.

Examples of this inverse property include

$$e^{\ln 2k} = 2k, \quad 2^{\log_2 x} = x, \quad \text{and} \quad 10^{\log (x+5)} = x + 5.$$

Both are base-e.   *Both are base-2.*   *Both are base-10.*

CHANGING LOGARITHMIC FORM TO EXPONENTIAL FORM In order to solve logarithmic equations, we exponentiate each side of the equation to change the logarithmic form of the equation into the exponential form. In general, we will show this step, but you may find that you can move directly from the logarithmic form to the exponential form. The table shown below gives several examples of logarithmic equations along with their equivalent exponential forms.

Equivalent Forms

Logarithmic Form	Exponential Form	
$\log x = 3$	$x = 10^3$	Exponentiate each side: base 10.
$\log_2 3x = 5$	$3x = 2^5$	Exponentiate each side: base 2.
$\ln 8x = 6$	$8x = e^6$	Exponentiate each side: base e.
$\log_9 (4 + x) = 5$	$4 + x = 9^5$	Exponentiate each side: base 9.
$\log_a x = b$	$x = a^b$	Exponentiate each side: base a.

SOLVING LOGARITHMIC EQUATIONS In the next example, we solve logarithmic equations by using inverse operations to change them to exponential form.

EXAMPLE 6 Solving logarithmic equations

Solve and approximate solutions to the nearest hundredth when appropriate.
(a) $2 \log x = 4$ (b) $\ln 3x = 5.5$ (c) $\log_2 (x + 4) = 7$

Solution

(a)

$2 \log x = 4$	Given equation
$\log x = 2$	Divide each side by 2.
$10^{\log x} = 10^2$	Exponentiate each side, using base 10.
$x = 100$	Inverse property: $10^{\log k} = k$

Change to exponential form applies to the middle two steps.

(b)

$\ln 3x = 5.5$	Given equation
$e^{\ln 3x} = e^{5.5}$	Exponentiate each side, using base e.
$3x = e^{5.5}$	Inverse property: $e^{\ln k} = k$
$x = \dfrac{e^{5.5}}{3} \approx 81.56$	Divide each side by 3 and approximate.

Change to exponential form applies to the middle two steps.

(c)

$\log_2 (x + 4) = 7$	Given equation
$2^{\log_2 (x+4)} = 2^7$	Exponentiate each side, using base 2.
$x + 4 = 2^7$	Inverse property: $2^{\log_2 k} = k$
$x = 2^7 - 4$	Subtract 4 from each side.
$x = 124$	Simplify.

Change to exponential form applies to the middle two steps.

Now Try Exercises 69, 71, 77

Because the domain of any logarithmic function includes only positive numbers, it is important to check answers, as emphasized in the next example.

EXAMPLE 7 Solving a logarithmic equation

Solve $\log (x + 2) + \log (x - 2) = \log 5$. Check any answers.

Solution

$\log (x + 2) + \log (x - 2) = \log 5$	Given equation
$\log ((x + 2)(x - 2)) = \log 5$	Product rule
$\log (x^2 - 4) = \log 5$	Multiply.
$10^{\log (x^2-4)} = 10^{\log 5}$	Exponentiate using base 10.
$x^2 - 4 = 5$	Inverse properties
$x^2 = 9$	Add 4.
$x = \pm 3$	Square root property

Check each answer.

$\log (3 + 2) + \log (3 - 2) \stackrel{?}{=} \log 5$ $\quad\quad$ $\log (-3 + 2) + \log (-3 - 2) \stackrel{?}{=} \log 5$

$\log 5 + \log 1 \stackrel{?}{=} \log 5$ $\quad\quad\quad\quad\quad\quad\quad$ $\log (-1) + \log (-5) \neq \log 5$ ✗

$\log 5 + 0 \stackrel{?}{=} \log 5$ $\quad\quad\quad\quad\quad\quad\quad\quad\quad\quad\quad$ Undefined

$\log 5 = \log 5$ ✓

Although 3 is a solution, -3 is not, because both $\log (-1)$ and $\log (-5)$ are undefined expressions. *Be sure to check your answers.*

Now Try Exercise 85

4 Solving Applications of Logarithmic Equations

EXAMPLE 8 Modeling bird populations

Near New Guinea, there is a relationship between the number of different species of birds and the size of an island. Larger islands tend to have a greater variety of birds. **TABLE 19.13** lists the number of species of birds y found on islands with an area of x square kilometers.

Numbers of Bird Species on Islands

x (km^2)	0.1	1	10	100	1000
y (species)	10	15	20	25	30

Source: B. Freedman, *Environmental Ecology*.
TABLE 19.13

(a) Find values for the constants a and b so that $y = a + b \log x$ models the data.
(b) Predict the number of bird species on an island of 4000 square kilometers.

Solution
(a) Substitute $x = 1$ and $y = 15$ in the equation to find a.

$$15 = a + b \log 1 \qquad y = a + b \log x$$
$$15 = a + b \cdot 0 \qquad \log 1 = 0$$
$$15 = a \qquad \text{Simplify.}$$

Thus $y = 15 + b \log x$. To find b, substitute $x = 10$ and $y = 20$.

$$20 = 15 + b \log 10 \qquad y = 15 + b \log x$$
$$20 = 15 + b \cdot 1 \qquad \log 10 = 1$$
$$5 = b \qquad \text{Simplify.}$$

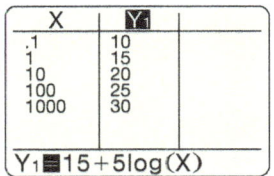

FIGURE 19.36

The data in **TABLE 19.13** are modeled by $y = 15 + 5 \log x$. This result is supported by **FIGURE 19.36**.

(b) To predict the number of species on an island of 4000 square kilometers, let $x = 4000$ and find y.

$$y = 15 + 5 \log 4000 \approx 33$$

The model estimates about 33 different species of birds on this island.

Now Try Exercise 113

EXAMPLE 9 Modeling runway length

For some types of airplanes with weight x, the minimum runway length L required at takeoff is modeled by

$$L(x) = 3 \log x.$$

In this equation, L is measured in thousands of feet and x is measured in thousands of pounds. Estimate the weight of the heaviest airplane that can take off from a runway 5100 feet long. (*Source:* L. Haefner, *Introduction to Transportation Systems*.)

CRITICAL THINKING 1

Previously, we used $L(x) = 1.3 \ln x$ to model runway length. Are $L(x) = 1.3 \ln x$ and $L(x) = 3 \log x$ equivalent formulas? Explain.

Solution

Runway length is measured in thousands of feet, so we must solve the equation $L(x) = 5.1$.

$3 \log x = 5.1$	$L(x) = 5.1$
$\log x = 1.7$	Divide each side by 3.
$10^{\log x} = 10^{1.7}$	Exponentiate each side, using base 10.
$x = 10^{1.7}$	Inverse property: $10^{\log k} = k$
$x \approx 50.1$	Approximate.

The largest airplane that can take off from this runway weighs about 50,000 pounds.

Now Try Exercise 109

19.5 Putting It All Together

TYPE OF EQUATION	PROCEDURE	EXAMPLE	
Exponential	Begin by solving for the exponential expression a^x. Then take a logarithm of each side.	$4e^x + 1 = 9$	Given equation
		$e^x = 2$	Solve for e^x.
		$\ln e^x = \ln 2$	Take the natural logarithm.
		$x = \ln 2$	Inverse property: $\ln e^k = k$
Logarithmic	Begin by solving for the logarithm in the equation. Then exponentiate each side of the equation, using the same base as the logarithm.	$\frac{1}{3} \log 2x = 1$	Given equation
		$\log 2x = 3$	Multiply by 3.
		$10^{\log 2x} = 10^3$	Exponentiate using base 10.
		$2x = 1000$	Inverse property: $10^{\log k} = k$
		$x = 500$	Divide by 2.

19.5 Exercises

MyMathLab

CONCEPTS AND VOCABULARY

1. To solve $x - 5 = 50$, what should be done?

2. To solve $5x = 50$, what should be done?

3. To solve $10^x = 50$, what should be done?

4. To solve $\log x = 5$, what should be done?

5. $\log 10^x = $ _____

6. $10^{\log x} = $ _____

7. $\ln e^{2x} = $ _____

8. $e^{\ln (x+7)} = $ _____

9. Does $\frac{\log 5}{\log 4}$ equal $\log \frac{5}{4}$? Explain.

10. Does $\frac{\log 5}{\log 4}$ equal $\log 5 - \log 4$? Explain.

11. How many solutions are there to $\log x = k$, where k is any real number?

12. How many solutions are there to $10^x = k$, where k is a positive number?

WRITING EQUIVALENT EXPONENTIAL AND LOGARITHMIC FORMS

Exercises 13–16: Change each equation to its equivalent logarithmic form.

13. (a) $6^{3x} = 8$ (b) $e^{7x} = 10$
14. (a) $10^x = 50$ (b) $2^{2x} = 5$
15. (a) $5^{8x} = 12$ (b) $e^x = 3$
16. (a) $a^{3x} = b$ (b) $d^x = c$

Exercises 17–20: Change each equation to its equivalent exponential form.

17. (a) $\log_2 x = 8$ (b) $\ln 2x = 3$
18. (a) $\log(3 + x) = 5$ (b) $\log_9 x = 4$
19. (a) $\log_4 6x = 6$ (b) $\ln(x + 1) = 10$
20. (a) $\log_a(x + 2) = b$ (b) $\log_c 3x = d$

SOLVING EXPONENTIAL EQUATIONS

Exercises 21–52: Solve the given exponential equation. Approximate answers to the nearest hundredth when appropriate.

21. $10^x = 1000$
22. $10^x = 0.01$
23. $2^x = 64$
24. $3^x = 27$
25. $2^{x-3} = 8$
26. $3^{2x} = 81$
27. $4^x + 3 = 259$
28. $3(5^{2x}) = 300$
29. $10^{0.4x} = 124$
30. $0.75^x = 0.25$
31. $e^{-x} = 1$
32. $0.5^{-5x} = 5$
33. $e^x = 25$
34. $e^x = 0.4$
35. $0.4^x = 2$
36. $0.7^x = 0.3$
37. $e^x - 1 = 6$
38. $2e^{4x} = 15$
39. $2(10)^{x+2} = 35$
40. $10^{3x} + 10 = 1500$
41. $3.1^{2x} - 4 = 16$
42. $5.4^{x-1} = 85$
43. $e^{3x} = e^{2x-1}$
44. $e^{x^2} = e^{3x-2}$
45. $5^{4x} = 5^{x^2-5}$
46. $2^{4x} = 2^{x+3}$
47. $e^{2x} \cdot e^x = 10$
48. $10^{x-2} \cdot 10^x = 1000$
49. $e^x = 2^{x+2}$
50. $2^{2x} = 3^{x-1}$
51. $4^{0.5x} = 5^{x+2}$
52. $3^{2x} = 7^{x+1}$

Exercises 53–56: (Refer to Example 3.) The symbolic and graphical representations of f and g are given.

(a) Use the graph to solve $f(x) = g(x)$.
(b) Solve $f(x) = g(x)$ symbolically.

53. $f(x) = 0.2(10^x)$,
    $g(x) = 2$

54. $f(x) = e^x$,
    $g(x) = 7.4$

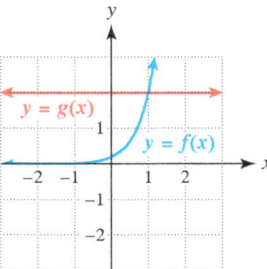
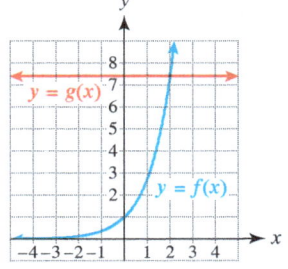

55. $f(x) = 2^{-x}$,
    $g(x) = 4$

56. $f(x) = 0.1(3^x)$,
    $g(x) = 0.9$

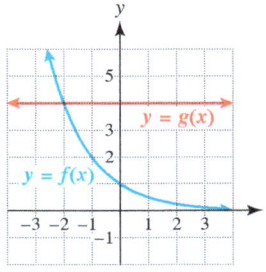
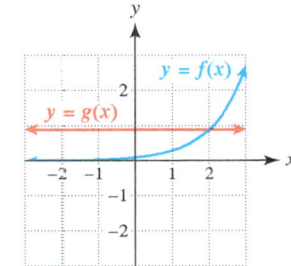

Exercises 57–64: Solve the equation symbolically. Give graphical or numerical support. Approximate answers to the nearest hundredth when appropriate.

57. $10^x = 0.1$
58. $2(10^x) = 2000$
59. $4e^x + 5 = 9$
60. $e^x + 6 = 36$
61. $4^x = 1024$
62. $3^x = 729$
63. $(0.55)^x + 0.55 = 2$
64. $5(0.9)^x = 3$

Exercises 65–68: The given equation cannot be solved symbolically. Find any solutions graphically to the nearest hundredth.

65. $e^x - x = 2$
66. $x \log x = 1$
67. $\ln x = e^{-x}$
68. $10^x - 2 = \log(x + 2)$

SOLVING LOGARITHMIC EQUATIONS

Exercises 69–90: Solve the given equation. Approximate answers to the nearest hundredth when appropriate.

69. $\log x = 2$
70. $\log x = 0.01$
71. $\ln x = 5$
72. $2 \ln x = 4$
73. $\log 2x = 7$
74. $6 \ln 4x = 12$

75. $\log_2 x = 4$
76. $\log_2 x = 32$
77. $\log_2 5x = 2.3$
78. $2 \log_3 4x = 10$
79. $2 \log x + 5 = 7.8$
80. $\ln(x - 1) = 3.3$
81. $5 \ln(2x + 1) = 55$
82. $5 - \log(x + 3) = 2.6$
83. $\log x^2 = \log x$
84. $\ln x^2 = \ln(3x - 2)$
85. $\ln x + \ln(x + 1) = \ln 30$
86. $\log(x - 1) + \log(2x + 1) = \log 14$
87. $\log_3 3x - \log_3(x + 2) = \log_3 2$
88. $\log_4(x^2 - 1) - \log_4(x - 1) = \log_4 6$
89. $\log_2(x - 1) + \log_2(x + 1) = 3$
90. $\log_4(x^2 + 2x + 1) - \log_4(x + 1) = 2$

Exercises 91–96: Solve the equation symbolically. Give graphical or numerical support. Approximate answers to the nearest hundredth.

91. $\log x = 1.6$
92. $\ln x = 2$
93. $\ln(x + 1) = 1$
94. $2 \log(2x + 3) = 8$
95. $17 - 6 \log_3 x = 5$
96. $4 \log_2 x + 7 = 12$

Exercises 97–100: Two functions, f and g, are given.
(a) *Use the graph to solve* $f(x) = g(x)$.
(b) *Solve* $f(x) = g(x)$ *symbolically.*

97. $f(x) = \ln x$,
    $g(x) = 0.7$

98. $f(x) = \log_2 x$,
    $g(x) = 1.6$

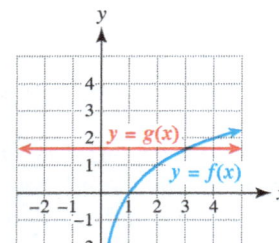

99. $f(x) = 5 \log 2x$,
    $g(x) = 3$

100. $f(x) = 2 \ln(x) - 3$,
     $g(x) = 0.9$

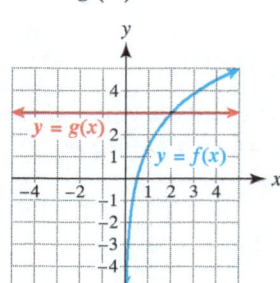
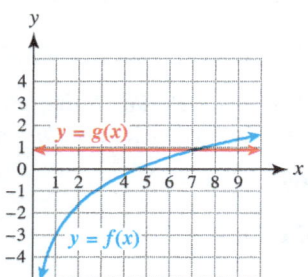

APPLICATIONS OF EXPONENTIAL AND LOGARITHMIC EQUATIONS

101. **Smartwatch Market** The market for smartwatches is increasing dramatically. The annual shipment forecasts in millions can be modeled by $W(x) = 8(1.47)^x$, where x represents years after 2013. (*Source:* BI Intelligence.)
  (a) Evaluate $f(5)$ and interpret the result.
  (b) Determine when the annual shipments of smartwatches will reach 120 million.

102. **Risk of Down Syndrome** The likelihood of a live birth having Down syndrome varies with the age of the mother. The function
$$D(x) = 0.0000816(10^{0.102x})$$
gives the percentage of live births that have Down syndrome when the mother is x years old.
  (a) Evaluate $D(35)$ and interpret the result.
  (b) Determine the mother's age when this percentage reaches 4%.

103. **Bacteria Growth** A sample of 5 million bacteria increases by 15% each hour. To the nearest hour, how long does it take for the sample to reach 20 million?

104. **Bacteria Growth** A sample of 3 million bacteria increases by 125% each day. To the nearest day, how long does it take for the bacteria sample to reach 173 million?

105. **Growth of a Mutual Fund** (Refer to Example 4.) An investor deposits $2000 in a mutual fund that returns 15% at the end of 1 year. Determine the length of time required for the investment to triple its value if the annual rate of return remains the same.

106. **Savings Account** (Refer to Example 4.) If a savings account pays 6% annual interest at the end of each year, how many years will it take for the account to double in value?

107. **Liver Transplants** In the United States, the gap between available organs for liver transplants and people who need them has widened. The number of individuals in thousands waiting for liver transplants can be modeled by
$$f(x) = 120(1.025)^{x-2015},$$
where x is the year. (*Source:* United Network for Organ Sharing.)
  (a) Evaluate $f(2017)$ and interpret the result.
  (b) Determine when the number of individuals waiting for liver transplants will be 139 thousand.

108. **Life Span of a Robin** (Refer to Example 5.) Determine when 50% of the robins in the study were still alive.

109. **Runway Length** (Refer to Example 9.) Determine the weight of the heaviest airplane that can take off from a runway having a length of $\frac{3}{4}$ mile. (*Hint:* 1 mile = 5280 feet.)

110. **Runway Length** (Refer to Example 9.)
    (a) Suppose that an airplane is 10 times heavier than another airplane. How much longer should the runway be for the heavier airplane than for the lighter airplane? (*Hint:* Let the heavier airplane have weight $10x$.)
    (b) If the runway length is increased by 3000 feet, by what factor can the weight of an airplane that uses the runway be increased?

111. **The Decline of Bluefin Tuna** At the end of the 20th century, the number of western Atlantic bluefin tuna declined dramatically. Their numbers in thousands between 1974 and 1991 can be modeled by $f(x) = 230(10)^{-0.055x}$, where x is the number of years after 1974. See the accompanying graph. (*Source:* B. Freedman.)

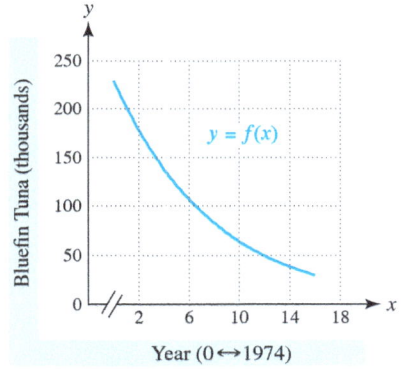

(a) Evaluate $f(1)$ and interpret the result.
(b) Use the graph to estimate the year in which blue-fin tuna numbered 115 thousand.
(c) Solve part (b) symbolically.

112. **Insect Populations** (Refer to Example 8.) The table lists numbers of species of insects y found on islands having areas of x square miles.

x (square miles)	1	2	4
y (species)	1000	1500	2000

x (square miles)	8	16	32
y (species)	2500	3000	3500

(a) Find values for the constants a and b so that $y = a + b \log_2 x$ models the data.
(b) Construct a table for y and verify that your equation models the data.
(c) Estimate the number of species of insects on an island that has an area of 12 square miles.

Exercises 113 and 114: (*Refer to Example 8.*) *Find values for a and b so that* $y = a + b \log x$ *models the data.*

113.
x	0.1	1	10	100
y	22	25	28	31

114.
x	0.01	1	100	1000
y	-10	-2	6	10

115. **Calories Consumed and Land Ownership** In developing countries, there is a relationship between the amount of land a person owns and the average number of calories that person consumes daily. This relationship is modeled by $f(x) = 645 \log(x + 1) + 1925$, where x is the amount of land owned in acres and $0 \le x \le 4$. (*Source:* D. Grigg, *The World Food Problem.*)
    (a) Estimate graphically the number of acres owned by a person consuming 2200 calories per day.
    (b) Solve part (a) symbolically.

116. **Population of Industrialized Urban Regions** The number of people living in industrialized urban regions throughout the world has not grown exponentially. Instead, it has grown logarithmically and is modeled by

$$f(x) = 0.36 + 0.15 \ln(x - 1949).$$

In this formula, the output is billions of people and the input x is the year, where $1950 \le x \le 2030$. (*Source:* D. Meadows, *Beyond The Limits.*)

    (a) Determine graphically when this population may reach 1 billion.
    (b) Solve part (a) symbolically.

117. **Fitbit Sales** The table lists the number of Fitbits sold in millions in year x. (*Source:* Fitbit)

x	2011	2012	2013	2014
y	0.3	1.3	4.5	10.9

(a) Are the data linear or nonlinear? Explain.
(b) The equation $f(x) = 0.345(3.33)^{x-2011}$ may be used to model the data. The growth factor is 3.33. What does this growth factor indicate about Fitbit sales during this period?
(c) If this trend continued, estimate the year when Fitbit sales was 140 million.

118. Greenhouse Gases If current trends continue, concentrations of atmospheric carbon dioxide (CO_2) in parts per million (ppm) are expected to increase. This increase in concentration of CO_2 has been accelerated by burning fossil fuels and deforestation. The exponential equation $y = 364(1.005)^x$ may be used to model CO_2 in parts per million x years after 2000. Estimate the year when the CO_2 concentration could be double the preindustrial level of 280 parts per million. (*Source:* R. Turner, *Environmental Economics.*)

119. Modeling Sound The formula
$$f(x) = 160 + 10 \log x$$
is used to calculate the decibel level of a sound with intensity x measured in watts per square centimeter. The noise level at a basketball game can reach 100 decibels. Find the intensity x of this sound.

120. Loudness of a Sound (Refer to Exercise 119.)
(a) Show that if the intensity of a sound increases by a factor of 10 from x to $10x$, the decibel level increases by 10 decibels. *Hint:* Show that
$$160 + 10 \log 10x = 170 + 10 \log x.$$
(b) Find the increase in decibels if the intensity x increases by a factor of 1000.
(c) Find the increase in the intensity x if the decibel level increases by 20.

121. Hurricanes The barometric air pressure in inches of mercury at a distance of x miles from the eye of a severe hurricane is given by $f(x) = 0.48 \ln(x + 1) + 27$. How far from the eye is the pressure 28 inches of mercury? (*Source:* A. Miller and R. Anthes, *Meteorology.*)

122. Earthquakes The Richter scale is used to determine the intensity of earthquakes, which corresponds to the amount of energy released. If an earthquake has an intensity of x, its magnitude, as computed by the Richter scale, is given by $R(x) = \log \frac{x}{I_0}$, where I_0 is the intensity of a small, measurable earthquake.
(a) If x is 1000 times greater than I_0, how large is this increase on the Richter scale?
(b) If the Richter scale increases from 5 to 8, by what factor does the intensity x increase?

Exercises 123 and 124: **Investment Account** *If a total of x dollars is deposited every 2 weeks (26 times per year) in an account paying an annual interest rate r, expressed in decimal form, the amount A in the account after n years can be approximated by the formula*
$$A = x \left[\frac{(1 + r/26)^{26n} - 1}{(r/26)} \right].$$

123. If a total of $100 is deposited every 2 weeks in an account paying 9% interest, approximate the amount in the account after 10 years.

124. Suppose that your retirement account pays 12% annual interest. Determine how much you should deposit in this account every 2 weeks, in order to have one million dollars at age 65.

WRITING ABOUT MATHEMATICS

125. Explain the basic steps for solving the equation $a(10^x) - b = c$. Then write the solution.

126. Explain the basic steps for solving the equation $a \log 3x = b$. Then write the solution.

SECTION 19.5 Checking Basic Concepts

1. Solve the equation. Approximate answers to the nearest hundredth when appropriate.
   (a) $2(10^x) = 40$ (b) $2^{3x} + 3 = 150$
   (c) $\ln x = 4.1$ (d) $4 \log 2x = 12$

2. Solve $\log(x + 4) + \log(x - 4) = \log 48$. Check the answers.

3. If a total of $500 is deposited in a savings account that pays 3% annual interest at the end of each year, the amount of money A in the account after x years is given by $A = 500(1.03)^x$. Estimate the number of years required for this amount to reach $900.

CHAPTER 19 Summary

SECTION 19.1 ■ COMPOSITE AND INVERSE FUNCTIONS

Composition of Functions If f and g are functions, then the composite function $g \circ f$, or composition of g and f, is defined by $(g \circ f)(x) = g(f(x))$.

Example: If $f(x) = x - 5$ and $g(x) = 2x^2 + 4x - 6$, then $(g \circ f)(x)$ is
$$g(f(x)) = g(x - 5)$$
$$= 2(x - 5)^2 + 4(x - 5) - 6.$$

One-to-One Function A function f is one-to-one if, for any c and d in the domain of f,
$$c \neq d \quad \text{implies that} \quad f(c) \neq f(d).$$
That is, different inputs always result in different outputs.

Example: $f(x) = x^2 + 4$ is *not* one-to-one because $f(-3) = f(3) = 13$. Different inputs; same output

Horizontal Line Test If every horizontal line intersects the graph of a function f at most once, then f is a one-to-one function.

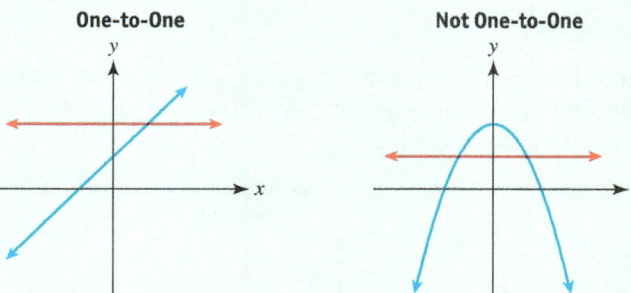

Inverse Functions If f is one-to-one, then f has an inverse function, denoted f^{-1}, that satisfies $(f^{-1} \circ f)(x) = x$ and $(f \circ f^{-1})(x) = x$.

Example: $f(x) = 7x$ and $f^{-1}(x) = \frac{x}{7}$ are inverse functions.

SECTION 19.2 ■ EXPONENTIAL FUNCTIONS

Exponential Function An exponential function is defined by $f(x) = Ca^x$, where $a > 0, C > 0$, and $a \neq 1$. Its domain (set of valid inputs) is all real numbers and its range (outputs) is all positive real numbers. The base is a.

Example: $f(x) = e^x$ is the natural exponential function and $e \approx 2.71828$.

Exponential Growth and Decay When $a > 1$, the graph of $f(x) = Ca^x$ models exponential growth, and when $0 < a < 1$, it models exponential decay. The base a represents either the growth factor or the decay factor. The constant C corresponds to the y-intercept, $(0, C)$.

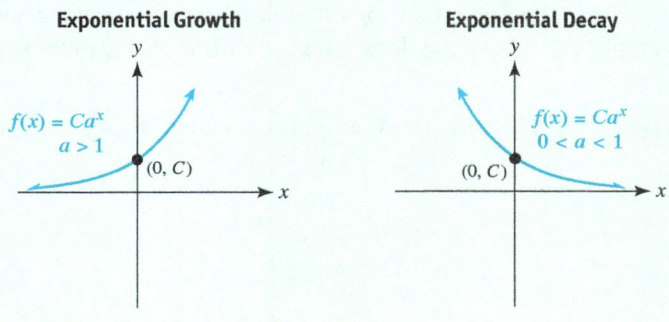

Example: $f(x) = 1.5(2)^x$ is an exponential function with $a = 2$ and $C = 1.5$. It models exponential growth because $a > 1$. The growth factor is 2 because for each unit increase in x, the output from $f(x)$ increases by a *factor* of 2.

Percent Change If an amount A changes to a new amount B, then the percent change is given by

$$\frac{B - A}{A} \times 100.$$

Example: If $500 increases to $700, the percent change is

$$\frac{700 - 500}{500} \times 100 = 40\%.$$

Percent Change and Growth Factor If an initial amount C experiences a constant percent change r (in decimal form) for each x-unit increase in time, then the new amount A is

$$A = C(1 + r)^x,$$

where the growth factor is $a = 1 + r$.

Example: If $800 increases by 4% each year, then after 5 years, the amount is

$$A = 800(1 + 0.04)^5 \approx \$973.32.$$

The growth factor is $a = 1.04$.

Compound Interest If a total of P dollars is deposited in an account paying an *annual* interest rate r (in decimal form) and this interest rate is compounded or paid n times per year, then after t years, the account contains A dollars given by the following.

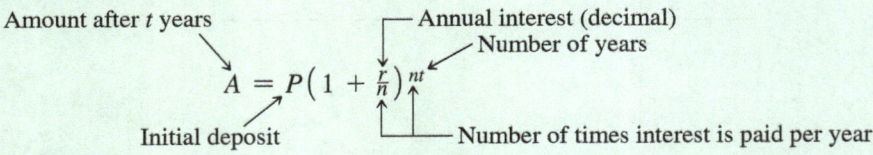

Example: If a total of $1200 is deposited at 6% annual interest, compounded quarterly, then after 8 years, the account contains

$$A = 1200\left(1 + \frac{0.06}{4}\right)^{(4 \cdot 8)} \approx \$1932.39.$$

SECTION 19.3 ■ LOGARITHMIC FUNCTIONS

Base-a Logarithms The logarithm with base a of a positive number x is denoted $\log_a x$. If $\log_a x = b$, then $x = a^b$. That is, $\log_a x$ represents the exponent on base a that results in x.

Examples: $\log_2 16 = 4$ because $16 = 2^4$ and $\log 100 = 2$ because $100 = 10^2$.

Base-a Logarithm

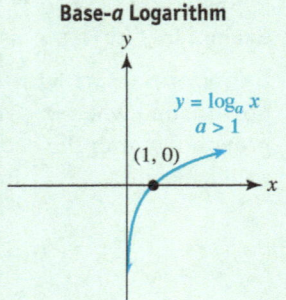

Domain and Range of Logarithmic Functions The domain (set of valid inputs) of a logarithmic function is the set of all positive real numbers and the range (outputs) is the set of real numbers.

Graph of a Logarithmic Function The graph of a logarithmic function passes through $(1, 0)$, as illustrated in the graph. As x becomes large, $\log_a x$ with $a > 1$ grows very slowly.

Exponential and Logarithmic Form The equation $y = a^x$ is equivalent to $\log_a y = x$.

SECTION 19.4 ■ PROPERTIES OF LOGARITHMS

Basic Properties Logarithms have several important properties. For positive numbers m, n, and $a \neq 1$ and any real number r,

1. $\log_a mn = \log_a m + \log_a n$.
2. $\log_a \frac{m}{n} = \log_a m - \log_a n$.
3. $\log_a (m^r) = r \log_a m$.

Examples:
1. $\log 5 + \log 20 = \log (5 \cdot 20) = \log 100 = 2$
2. $\log 100 - \log 5 = \log \frac{100}{5} = \log 20 \approx 1.301$
3. $\ln 2^6 = 6 \ln 2 \approx 4.159$

NOTE: $\log_a 1 = 0$ for any valid base a. Thus $\log 1 = 0$ and $\ln 1 = 0$. ■

Inverse Properties The following inverse properties are important for solving exponential and logarithmic equations.

1. $\log_a a^x = x$, for any real number x
2. $a^{\log_a x} = x$, for any positive number x

Examples:
1. $\log_2 2^\pi = \pi$
2. $10^{\log 2.5} = 2.5$
3. $e^{\ln 5} = 5$
4. $\log 10^7 = 7$

NOTE: If $f(x) = \log x$, then $f^{-1}(x) = 10^x$, and if $g(x) = \ln x$, then $g^{-1}(x) = e^x$. ■

SECTION 19.5 ■ EXPONENTIAL AND LOGARITHMIC EQUATIONS

Solving Equations When solving an exponential equation, we usually take a logarithm of each side. When solving a logarithmic equation, we usually exponentiate each side.

Examples:
1. $2(5)^x = 22$ Exponential equation
   $5^x = 11$ Divide by 2.
   $\log 5^x = \log 11$ Take the common logarithm.
   $x \log 5 = \log 11$ Power rule
   $x = \frac{\log 11}{\log 5}$ Divide by log 5.

2. $\log 2x = 2$ Logarithmic equation
   $10^{\log 2x} = 10^2$ Exponentiate each side.
   $2x = 100$ Inverse properties
   $x = 50$ Divide by 2.

CHAPTER 19 Review Exercises

SECTION 19.1

Exercises 1 and 2: Find the following.
    (a) $(g \circ f)(-2)$ (b) $(f \circ g)(x)$

1. $f(x) = 2x^2 - 4x$, $g(x) = 5x + 1$
2. $f(x) = \sqrt[3]{x - 6}$, $g(x) = 4x^3$

3. Use the two tables to evaluate each expression.
    (a) $(f \circ g)(2)$ (b) $(g \circ f)(1)$

x	0	1	2	3
$f(x)$	3	2	1	0

x	0	1	2	3
$g(x)$	1	2	3	0

4. Use the graph to evaluate each expression.
    (a) $(f \circ g)(-1)$
    (b) $(g \circ f)(2)$
    (c) $(f \circ f)(1)$

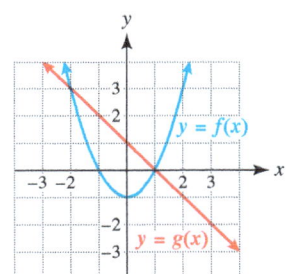

Exercises 5 and 6: Show that f is not one-to-one by finding two inputs that result in the same output. Answers may vary.

5. $f(x) = \dfrac{4}{1+x^2}$ **6.** $f(x) = x^2 - 2x + 1$

Exercises 7 and 8: Use the horizontal line test to determine whether the graph represents a one-to-one function.

7. 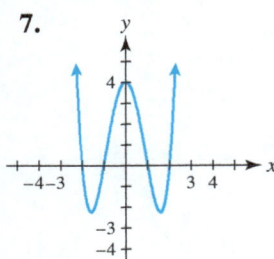   **8.**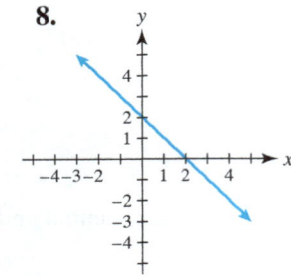

Exercises 9 and 10: Verify that $f(x)$ and $f^{-1}(x)$ are indeed inverse functions.

9. $f(x) = 2x - 9$ $f^{-1}(x) = \dfrac{x+9}{2}$

10. $f(x) = x^3 + 1$ $f^{-1}(x) = \sqrt[3]{x-1}$

Exercises 11–14: Find $f^{-1}(x)$.

11. $f(x) = 5x$ **12.** $f(x) = x - 11$

13. $f(x) = 2x + 7$ **14.** $f(x) = \dfrac{4}{x}$

15. Use the table to make a table of values for $f^{-1}(x)$. What are the domain and range for f^{-1}?

x	0	1	2	3
f(x)	10	8	7	3

16. Use the graph of $y = f(x)$ to sketch a graph of $y = f^{-1}(x)$. Include the graph of f and the line $y = x$.

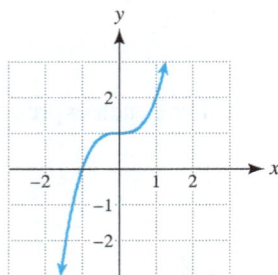

SECTIONS 19.2 AND 19.3

Exercises 17–20: Evaluate the exponential function for the given values of x.

17. $f(x) = 6^x$ $x = -1,\ x = 2$

18. $f(x) = 5(2^{-x})$ $x = 0,\ x = 3$

19. $f(x) = \left(\tfrac{1}{3}\right)^x$ $x = -1,\ x = 4$

20. $f(x) = 3\left(\tfrac{1}{6}\right)^x$ $x = 0,\ x = 1$

Exercises 21–24: Graph f. State whether the graph illustrates exponential growth, exponential decay, or logarithmic growth.

21. $f(x) = 2^x$ **22.** $f(x) = \left(\tfrac{1}{2}\right)^x$

23. $f(x) = \ln(x+1)$ **24.** $f(x) = 3^{-x}$

Exercises 25 and 26: A table for a function f is given.

(a) Determine whether f represents linear or exponential growth.

(b) Find a formula for f.

25.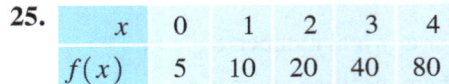

x	0	1	2	3	4
f(x)	5	10	20	40	80

26.

x	0	1	2	3	4
f(x)	5	10	15	20	25

27. Use the graph of $y = Ca^x$ to find C and a.

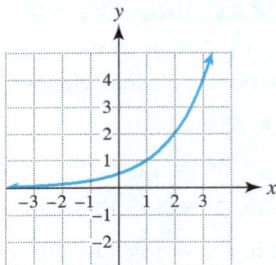

28. Use the graph of $y = k \log_2 x$ to find k.

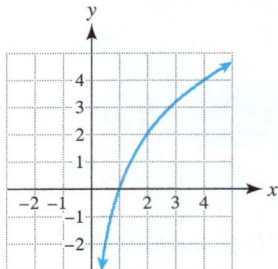

29. Find the percent change if $150 decreases to $120.

30. Find the growth factor if $1500 increases by 7% each year.

Exercises 31 and 32: For the given amounts A and B, find each of the following. Round values to the nearest hundredth of a percent when appropriate.

(a) The percent change if A changes to B

(b) The percent change if B changes to A

31. $A = \$600$, $B = \$1200$

32. $A = \$2.20$, $B = \$1.00$

Exercises 33 and 34: An account contains A dollars and increases or decreases by R%. For each A and R, answer the following.

(a) Find the increase or decrease in value of the account.
(b) Find the final value of the account.
(c) By what factor did the account value increase or decrease?

33. $A = \$500$, $R = 210\%$

34. $A = \$700$, $R = -25\%$

Exercises 35 and 36: **Exponential Models** *Write the formula for an exponential function, $f(x) = Ca^x$, that models the situation. Evaluate $f(2)$.*

35. A city's population of 20,000 decreases in number by 5% per year.

36. A sample of 1500 insects increases in number by 300% per day.

Exercises 37 and 38: If a total of P dollars is deposited in an account that pays R percent annual interest at the end of each year, approximate the amount in the account after x years.

37. $P = \$1200$, $R = 10\%$, $x = 9$ years

38. $P = \$900$, $R = 18\%$, $x = 40$ years

Exercises 39–42: Evaluate $f(x)$ for the given value of x. Approximate answers to the nearest hundredth.

39. $f(x) = 2e^x - 1$ $\qquad x = 5.3$

40. $f(x) = 0.85^x$ $\qquad x = 2.1$

41. $f(x) = 2\log x$ $\qquad x = 55$

42. $f(x) = \ln(2x + 3)$ $\qquad x = 23$

Exercises 43–46: Evaluate the logarithm by hand.

43. $\log 0.001$ $\qquad$ 44. $\log \sqrt{10{,}000}$

45. $\ln e^{-4}$ $\qquad$ 46. $\log_4 16$

Exercises 47–50: Approximate to the nearest thousandth.

47. $\log 65$ $\qquad$ 48. $\ln 0.85$

49. $\ln 120$ $\qquad$ 50. $\log \frac{2}{5}$

Exercises 51–54: Simplify, using inverse properties.

51. $10^{\log 7}$ $\qquad$ 52. $\log_2 2^{5/9}$

53. $\ln e^{6-x}$ $\qquad$ 54. $e^{2 \ln x}$

SECTION 19.4

Exercises 55–60: Write the expression by using sums and differences of logarithms of x, y, and z.

55. $\ln xy$ $\qquad$ 56. $\log \dfrac{x}{y}$

57. $\ln(x^2 y^3)$ $\qquad$ 58. $\log \dfrac{\sqrt{x}}{z^3}$

59. $\log_2 \dfrac{x^2 y}{z}$ $\qquad$ 60. $\log_3 \sqrt[3]{\dfrac{x}{y}}$

Exercises 61–64: Write as the logarithm of one expression.

61. $\log 45 + \log 5 - \log 3$

62. $\log_4 2x + \log_4 5x$

63. $2 \ln x - 3 \ln y$

64. $\log x^4 - \log x^3 + \log y$

Exercises 65–68: Rewrite, using the power rule.

65. $\log 6^3$ $\qquad$ 66. $\ln x^2$

67. $\log_2 5^{2x}$ $\qquad$ 68. $\log_4 0.6^{x+1}$

SECTION 19.5

Exercises 69–78: Solve the given equation. Approximate answers to the nearest hundredth when appropriate.

69. $10^x = 100$ $\qquad$ 70. $2^{2x} = 256$

71. $3e^x + 1 = 28$ $\qquad$ 72. $0.85^x = 0.2$

73. $5 \ln x = 4$ $\qquad$ 74. $\ln 2x = 5$

75. $2 \log x = 80$ $\qquad$ 76. $3 \log x - 5 = 1$

77. $2^{x+4} = 3^x$

78. $\ln(2x + 1) + \ln(x - 5) = \ln 13$

Exercises 79 and 80: Do the following.

(a) Solve $f(x) = g(x)$ graphically.
(b) Solve $f(x) = g(x)$ symbolically.

79. $f(x) = \frac{1}{2}(2^x)$,
    $g(x) = 4$

80. $f(x) = \log_2 2x$,
    $g(x) = 3$

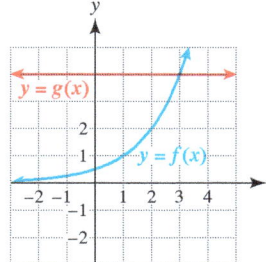
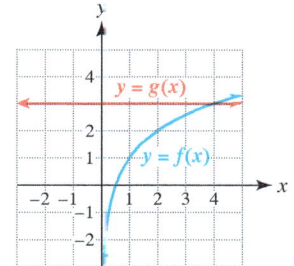

APPLICATIONS

81. Surface Area of a Balloon The surface area S of a spherical balloon with radius r is given by the formula $S(r) = 4\pi r^2$. Suppose that the balloon is being inflated so that its radius in inches after t seconds is $r(t) = \sqrt{2t}$.
(a) Evaluate $(S \circ r)(8)$ and interpret your result.
(b) Find $(S \circ r)(t)$.

82. Sales Tax Suppose that $f(x) = 0.08x$ calculates the sales tax in dollars on an item that costs x dollars.
(a) Is f a one-to-one function? Why?
(b) Find a formula for f^{-1} and interpret what it calculates.

83. Growth of a Mutual Fund An investor deposits $1500 in a mutual fund that returns 11% annually. Determine the time required for the investment to double in value.

84. Modeling Data Find values for the constants a and b so that $y = a + b \log x$ models the data in the table.

x	0.1	1	10	100	1000
y	50	100	150	200	250

85. Modeling Data Find values for the constants C and a so that $y = Ca^x$ models the data in the table.

x	0	1	2	3	4
y	3	6	12	24	48

86. Earthquakes The Richter scale, used to determine the magnitude of earthquakes, is based on the formula $R(x) = \log \frac{x}{I_0}$, where x is the measured intensity. Let $I_0 = 1$. Find the intensity x for an earthquake with $R = 7$.

87. Modeling Population In 2000, the population of Nevada was 2 million and growing continuously at an annual rate of 5.1%. The population of Nevada in millions x years after 2000 can be modeled by
$$f(x) = 2e^{0.051x}.$$
(a) Graph f in the window $[0, 10, 2]$ by $[0, 4, 1]$. Does this function represent exponential growth or decay?
(b) Estimate the population of Nevada in 2010.
(c) Estimate the year when the population of Nevada was 3 million.

88. Modeling Bacteria A colony of bacteria can be modeled by $N(t) = 1000e^{0.0014t}$, where N is measured in bacteria per milliliter and t is in minutes.
(a) Evaluate $N(0)$ and interpret the result.
(b) Estimate how long it takes for N to double.

89. Modeling Wind Speed Wind speeds vary at different heights above the ground. For a particular day, $f(x) = 1.2 \ln(x) + 5$ computes the wind speed in meters per second x meters above the ground, where $x \geq 1$. (*Source:* A. Miller and R. Anthes, *Meteorology*.)
(a) Find the wind speed at a height of 5 meters.
(b) Estimate the height at which the wind speed is 8 meters per second.

CHAPTER 19 Test

1. If $f(x) = 4x^3 - 5x$ and $g(x) = x + 7$, evaluate $(g \circ f)(1)$ and $(f \circ g)(x)$.

2. Use the graph to evaluate each expression.
   (a) $(f \circ g)(-1)$
   (b) $(g \circ f)(1)$

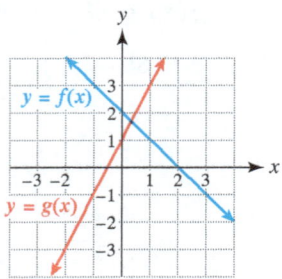

3. Explain why $f(x) = x^2 - 25$ is not a one-to-one function.

4. If $f(x) = 5 - 2x$, find $f^{-1}(x)$.

5. Use the graph of $y = f(x)$ to sketch a graph of $y = f^{-1}(x)$. Include the graph of f and the line $y = x$ in your graph.

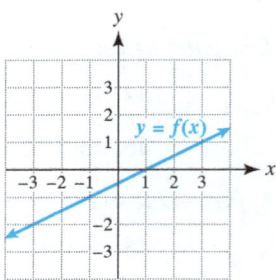

6. Use the table to write a table of values for $f^{-1}(x)$. What are the domain and range of f^{-1}?

x	1	2	3	4
f(x)	8	6	4	2

7. Evaluate $f(x) = 3\left(\tfrac{1}{4}\right)^x$ at $x = 2$.

8. Graph $f(x) = 1.5^{-x}$. State whether the graph of f illustrates exponential growth, exponential decay, or logarithmic growth.

Exercises 9 and 10: A table for a function f is given.
  (a) Determine whether f represents linear or exponential growth.
  (b) Find a formula for $f(x)$.

9. | x | −2 | −1 | 0 | 1 | 2 |
    |--------|------|-----|---|---|----|
    | $f(x)$ | 0.75 | 1.5 | 3 | 6 | 12 |

10. | x | −2 | −1 | 0 | 1 | 2 |
    |--------|----|------|----|-----|---|
    | $f(x)$ | −4 | −2.5 | −1 | 0.5 | 2 |

11. Use the graph of $y = Ca^x$ to find C and a.

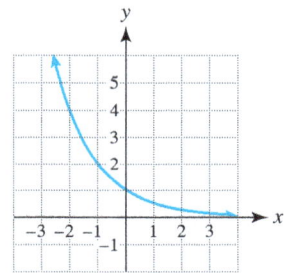

12. Find the percent change if $600 increases to $900.

13. If $1000 increases by 5% per year, what is the growth factor?

14. If a total of $750 is deposited in an account paying 7% annual interest at the end of each year, approximate the amount in the account after 5 years.

15. Let $f(x) = 1.5 \ln(x - 5)$. Approximate $f(21)$ to the nearest hundredth.

16. Evaluate $\log \sqrt{10}$ by hand.

17. Approximate $\log_2 43$ to the nearest thousandth.

18. Graph $f(x) = \log(x - 2)$. Compare this graph to the graph of $y = \log x$.

19. Write $\log \dfrac{x^3 y^2}{\sqrt{z}}$, using sums and differences of logarithms of x, y, and z.

20. Write $4 \ln x - 5 \ln y + \ln z$ as one logarithm.

21. Rewrite $\log 7^{2x}$, using the power rule.

22. Simplify $\ln e^{1-3x}$, using inverse properties.

Exercises 23–26: Solve the given equation. Approximate answers to the nearest hundredth when appropriate.

23. $2e^x = 50$
24. $3(10)^x - 7 = 143$
25. $5 \log x = 9$
26. $3 \ln 5x = 27$

27. **Modeling Data** Find values for constants a and b so that $y = a + b \log x$ models the data.

x	0.01	0.1	1	10	100
y	−1	2	5	8	11

28. **Modeling Bacteria Growth** A sample of bacteria is growing at a rate of 9% per hour and can be modeled by $f(x) = 4(1.09)^x$, where the input x represents elapsed time in hours and the output $f(x)$ is in millions of bacteria.
    (a) What was the initial number of bacteria?
    (b) Evaluate $f(5)$ and interpret the result.
    (c) Does this function represent exponential growth or exponential decay?
    (d) Estimate the elapsed time when there were 8 million bacteria.

29. **Investing** A mutual fund account containing $5000 decreases by 2% per year.
    (a) Write a function A that gives the amount in the account after x years.
    (b) Evaluate $A(3)$ and interpret the result.
    (c) Estimate the number of years for the account to decrease to $4500.

CHAPTERS 1–19 Cumulative Review Exercises

1. Write the number 0.000429 in scientific notation.

2. Classify each real number as one or more of the following: natural number, whole number, integer, rational number, or irrational number.

$$-\tfrac{11}{7},\ -3,\ 0,\ \sqrt{6},\ \pi,\ 5.\overline{18}$$

3. Select the formula that models the data best.

x	−2	−1	0	1	2
y	−7	−5	−3	−1	1

   a. $y = 3x + 1$ **b.** $y = x - 3$ **c.** $y = 2x - 3$

4. State whether the equation illustrates an identity, commutative, associative, or distributive property.

$$(5 - y) + 9 = 9 + (5 - y)$$

Exercises 5–8: Simplify the expression. Write the result with positive exponents.

5. $\left(\dfrac{1}{d^2}\right)^{-2}$

6. $\left(\dfrac{8a^2}{2b^3}\right)^{-3}$

7. $\dfrac{(2x^{-2}y^3)^2}{xy^{-2}}$

8. $\dfrac{x^{-3}y}{4x^2y^{-3}}$

9. Use the graph to express the equation of the line in slope–intercept form.

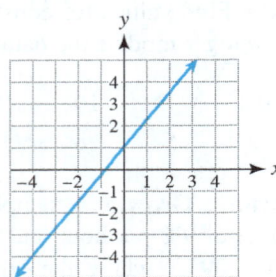

10. Find the domain of $f(x) = \dfrac{10}{x + 3}$.

11. Use the table to write the formula for $f(x) = ax + b$.

x	-2	-1	0	1	2
$f(x)$	-11	-7	-3	1	5

12. Write the equation of the vertical line passing through the point $(4, 7)$.

13. Calculate the slope of the line passing through the points $(4, -1)$ and $(2, -3)$.

14. Sketch the graph of a line passing through the point $(-1, -2)$ with slope $m = 3$.

Exercises 15 and 16: Write the slope–intercept form for a line satisfying the given conditions.

15. Perpendicular to $y = -\tfrac{1}{7}x - 8$, passing through $(1, 1)$

16. Parallel to $y = 3x - 1$, passing through $(0, 5)$

17. Use the graph to solve the equation $y_1 = y_2$.

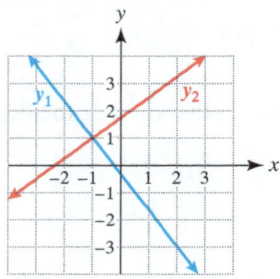

18. Use the table to solve the inequality $y < -4$, where y represents a linear function. Write the solution set as an inequality.

x	-2	-1	0	1	2
y	-24	-14	-4	6	16

Exercises 19–24: Solve the equation or inequality. Write the solutions to the inequalities in interval notation.

19. $\tfrac{2}{3}(x - 3) + 8 = -6$

20. $\tfrac{1}{3}z + 6 < \tfrac{1}{4}z - (5z - 6)$

21. $\left(\dfrac{t + 2}{3}\right) - 10 = \tfrac{1}{3}t - (5t + 8)$

22. $-10 \leq -\tfrac{3}{5}x - 4 < -1$

23. $-2|t - 4| \geq -12$ 24. $\left|\tfrac{1}{2}x - 5\right| = 3$

25. Shade the solution set in the xy-plane.

$$x + y > 3$$
$$2x - y \geq 3$$

26. Evaluate $\det A$ if $A = \begin{bmatrix} -1 & -2 \\ 3 & 4 \end{bmatrix}$.

Exercises 27–30: Solve the system of equations, if possible. Write the solution as an ordered pair or ordered triple where appropriate.

27. $4x - 3y = 1$
    $5x + 2y = 7$

28. $2x - 3y = -2$
    $-6x + 9y = 5$

29. $2x - y + 3z = -2$
    $x + 5y - 2z = -8$
    $-3x - y - 3z = 6$

30. $x + y - z = -1$
    $-x - y - z = -1$
    $x - 2y + z = 1$

Exercises 31–34: Factor completely.

31. $2x^3 - 4x^2 + 2x$ 32. $4a^2 - 25b^2$

33. $8t^3 - 27$ 34. $4a^3 - 2a^2 + 10a - 5$

Exercises 35–38: Solve the equation.

35. $6x^2 - 7x - 10 = 0$ 36. $9x^2 = 4$

37. $x^4 - 2x^3 = 15x^2$ 38. $5x - 10x^2 = 0$

Exercises 39 and 40: Simplify the expression.

39. $\dfrac{x^2 + 5x + 6}{x^2 - 9} \cdot \dfrac{x - 3}{x + 2}$

40. $\dfrac{x^2 - 2x - 8}{x^2 + x - 12} \div \dfrac{(x - 4)^2}{x^2 - 16}$

Exercises 41 and 42: Solve the rational equation. Check your result.

41. $\dfrac{2}{x+2} - \dfrac{1}{x-2} = \dfrac{-3}{x^2-4}$

42. $\dfrac{3y}{y^2+y-2} = \dfrac{1}{y-1} - 2$

43. Solve the equation for J.
$$P = \dfrac{J+2z}{J}$$

44. Simplify the complex fraction.
$$\dfrac{\dfrac{3}{x^2}+x}{x-\dfrac{3}{x^2}}$$

45. Suppose that y varies directly with x. If $y = 15$ when $x = 3$, find y when x is 8.

46. Divide $(3x^3 - 2x - 15)$ by $(x-2)$.

Exercises 47–52: Simplify the expression. Assume that all variables are positive.

47. $\left(\dfrac{x^6}{y^9}\right)^{2/3}$

48. $\sqrt[3]{-x^4} \cdot \sqrt[3]{-x^5}$

49. $\sqrt{5ab} \cdot \sqrt{20ab}$

50. $2\sqrt{24} - \sqrt{54}$

51. $\sqrt[3]{a^5 b^4} + 3\sqrt[3]{a^5 b}$

52. $(5+\sqrt{5})(5-\sqrt{5})$

53. Rationalize the denominator.
$$\dfrac{2}{5-\sqrt{3}}$$

54. Find the domain of f. Write your answer in interval notation.
$$f(x) = \dfrac{3}{\sqrt{x-4}}$$

Exercises 55 and 56: Solve. Check your answer.

55. $2(x+1)^2 = 50$

56. $\sqrt{x+6} = x$

Exercises 57 and 58: Write in standard form.

57. $(-2+3i) - (-5-2i)$

58. $\dfrac{3-i}{1+3i}$

59. Find the vertex on the graph of the function given by $f(x) = 3x^2 - 12x + 13$.

60. Find the maximum y-value on the parabola determined by $y = -2x^2 + 6x - 1$.

Exercises 61–64: Solve the quadratic equation by using the method of your choice.

61. $x^2 - 13x + 40 = 0$

62. $2d^2 - 5 = d$

63. $z^2 - 4z = -2$

64. $x^4 - 10x^2 + 24 = 0$

65. A graph of $y = ax^2 + bx + c$ is shown.
   (a) Solve $ax^2 + bx + c = 0$.
   (b) State whether $a > 0$ or $a < 0$.
   (c) Determine whether the discriminant is positive, negative, or zero.

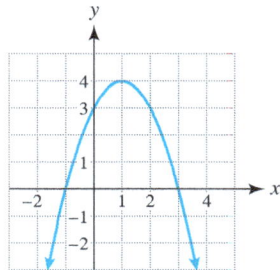

66. Solve $x^2 + 5x - 14 \geq 0$. Write your answer in interval notation.

67. For $f(x) = x^2 - 2$ and $g(x) = 2x + 1$, find each of the following.
   (a) $(f \circ g)(1)$
   (b) $(g \circ f)(x)$

68. Find $f^{-1}(x)$ for $f(x) = \dfrac{3}{x}$.

Exercises 69 and 70: Evaluate without a calculator.

69. $\log_3 81$

70. $e^{\ln 2x}$

71. Write the expression by using sums and differences of logarithms of x and y. Assume x and y are positive.
$$\log \dfrac{\sqrt{x}}{y^2}$$

72. Write $2\ln x + \ln 5x$ as the logarithm of a single expression.

Exercises 73 and 74: Solve the equation. Approximate answers to the nearest hundredth.

73. $6 \log x - 2 = 9$

74. $2^{3x} = 17$

APPLICATIONS

75. **Population Growth** The population P of a community with an annual percentage growth rate r (expressed as a decimal) after t years is given by $P = P_0(1+r)^t$, where P_0 represents the initial population of the community. If a community with an initial population of $P_0 = 12{,}000$ grows to a population of $P = 14{,}600$ in $t = 5$ years, find the annual percentage growth rate r for this community.

76. **Wing Span of a Bird** The wing span L of a bird with weight W can sometimes be modeled by $L = 27.4\sqrt[3]{W}$, where L is in inches and W is in pounds. Estimate the weight of a bird with a wing span of 36 inches. (*Source:* C. Pennycuick, *Newton Rules Biology.*)

77. **Investing for Retirement** A college student invests $8000 in an account that pays interest annually. If the student would like this investment to be worth $1,000,000 in 45 years, what annual interest rate would the account need to pay?

78. **Modeling Wind Speed** Wind speeds are usually measured at heights from 5 to 10 meters above the ground. For a particular day, $f(x) = 1.4 \ln(x) + 7$ computes the wind speed in meters per second x meters above the ground, where $x \geq 1$. (*Source:* A. Miller and R. Anthes, *Meteorology.*)
    (a) Find the wind speed at a height of 8 meters.
    (b) Estimate the height at which the wind speed is 10 meters per second.

20 Conic Sections

- 20.1 **Parabolas and Circles**
- 20.2 **Ellipses and Hyperbolas**
- 20.3 **Nonlinear Systems of Equations and Inequalities**

Throughout history, people have been fascinated with the universe around them and compelled to understand its mysteries. Conic sections, which include parabolas, circles, ellipses, and hyperbolas, have played an important role in gaining this understanding. All celestial objects—including planets, comets, asteroids, and satellites—travel in paths described by conic sections. Today, scientists search the sky for information about the universe with enormous radio telescopes in the shape of parabolic dishes. The Hubble telescope also makes use of a parabolic mirror. As a result, our understanding of the universe has changed dramatically in recent years.

Conic sections have had a profound influence on people's understanding of their world and the cosmos. In this chapter, we introduce you to these age-old curves. (*Source: Historical Topics for the Mathematics Classroom, Thirty-first Yearbook, NCTM.*)

20.1 Parabolas and Circles

Objectives

1. Recognizing Types of Conic Sections
2. Graphing Parabolas with Horizontal Axes of Symmetry and Writing Their Equations
3. Graphing Circles and Writing Their Equations
   - Graphing a Circle
   - Completing the Square to Find the Center and Radius

NEW VOCABULARY

- ☐ Conic sections
- ☐ Circle
- ☐ Radius
- ☐ Center
- ☐ Standard equation of a circle

1 Recognizing Types of Conic Sections

Conic sections are named after the different ways that a plane can intersect a cone. The three basic curves are parabolas, ellipses, and hyperbolas. A circle is a special case of an ellipse. **FIGURE 20.1** shows the three types of conic sections along with an example of the graph in the *xy*-plane associated with each.

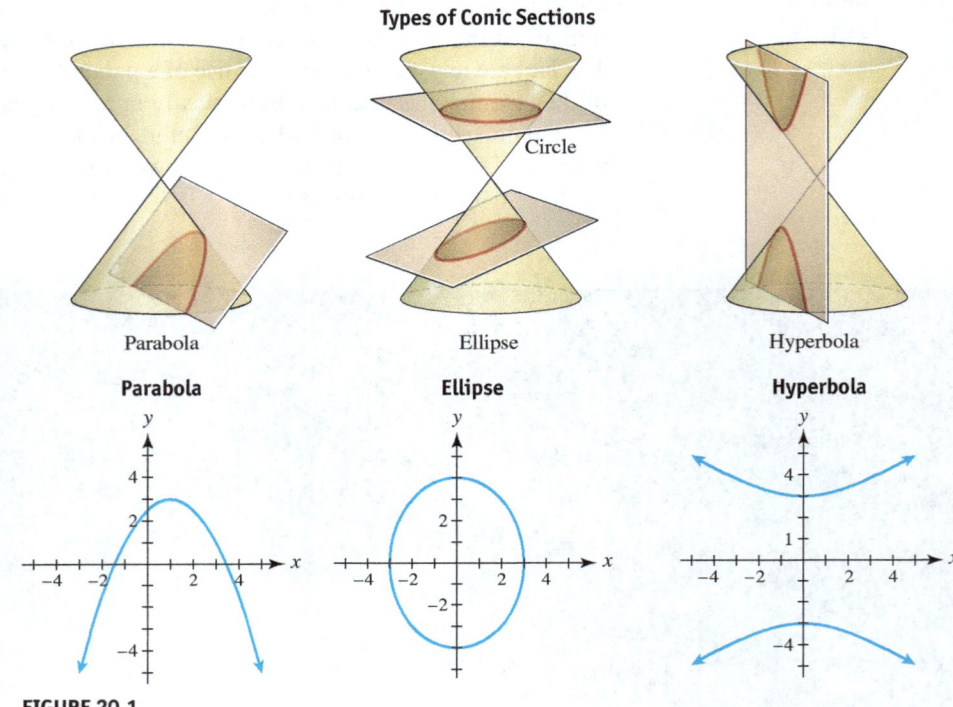

FIGURE 20.1

READING CHECK 1

- State the three basic types of conic sections.

2 Graphing Parabolas with Horizontal Axes of Symmetry and Writing Their Equations

Recall that the *vertex form of a parabola* with a vertical axis of symmetry is

$$y = a(x - h)^2 + k, \quad \text{Vertex form}$$

where (h, k) is the vertex. If $a > 0$, the parabola opens upward; if $a < 0$, the parabola opens downward, as shown in See the Concept 1. The preceding vertex form can also be expressed in the general form

$$y = ax^2 + bx + c. \quad \text{General form}$$

In this form, the *x*-coordinate of the vertex is $x = -\frac{b}{2a}$.

See the Concept 1 — PARABOLAS WITH A VERTICAL AXIS OF SYMMETRY

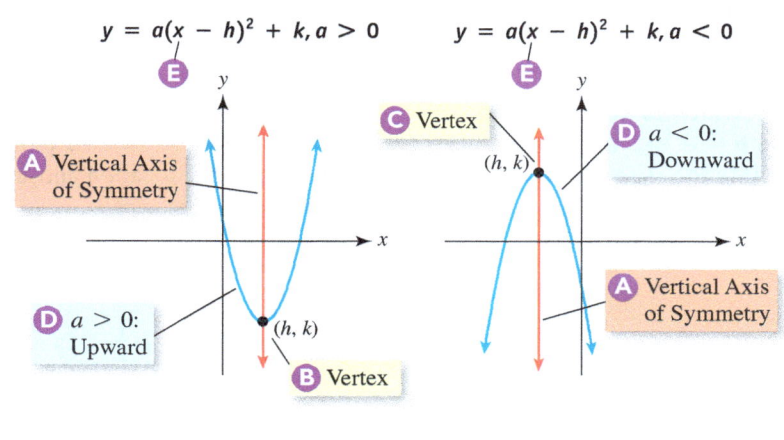

A The graph of this type of parabola *is* a function, with a vertical axis of symmetry.

B The vertex (h, k) is the *lowest* point on a parabola that opens *upward*.

C The vertex (h, k) is the *highest* point on a parabola that opens *downward*.

D Opens upward if $a > 0$, and downward if $a < 0$.

E If the x variable is squared, the axis of symmetry is vertical.

WATCH VIDEO IN MML.

Interchanging the roles of x and y (and also h and k) gives equations for parabolas that open to the right or the left. In this case, their axes of symmetry are horizontal.

PARABOLAS WITH HORIZONTAL AXES OF SYMMETRY

The graph of $x = a(y - k)^2 + h$ is a parabola that opens to the right if $a > 0$ and to the left if $a < 0$. The vertex of the parabola is located at (h, k).

The graph of $x = ay^2 + by + c$ is a parabola opening to the right if $a > 0$ and to the left if $a < 0$. The y-coordinate of its vertex is given by $y = -\frac{b}{2a}$.

READING CHECK 2

- Give vertex form for a parabola with a horizontal axis of symmetry.

See the Concept 2 — PARABOLAS WITH A HORIZONTAL AXIS OF SYMMETRY

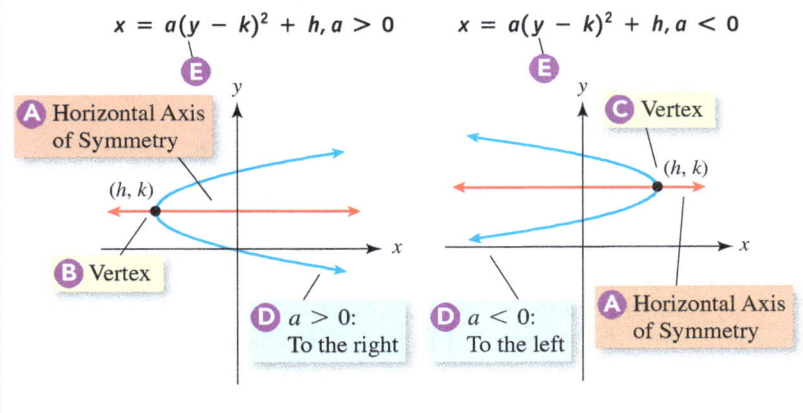

A The graph of this type of parabola *is not* a function. The axis of symmetry is horizontal.

B The vertex (h, k) is the *leftmost* point on a parabola that opens *to the right*.

C The vertex (h, k) is the *rightmost* point on a parabola that opens *to the left*.

D Opens to the right if $a > 0$, and to the left if $a < 0$.

E If the y-variable is squared, the axis of symmetry is horizontal.

WATCH VIDEO IN MML.

EXAMPLE 1 Graphing a parabola

Graph $x = -\frac{1}{2}y^2$. Find its vertex and axis of symmetry.

Solution
The equation can be written in vertex form because $x = -\frac{1}{2}(y - 0)^2 + 0$. The vertex is $(0, 0)$, and because $a = -\frac{1}{2} < 0$, the parabola opens to the left. We can make a table of values, as shown in **TABLE 20.1**, and plot a few points to help determine the location and shape of the graph, as shown in **FIGURE 20.2**. Its axis of symmetry is the x-axis, or $y = 0$.

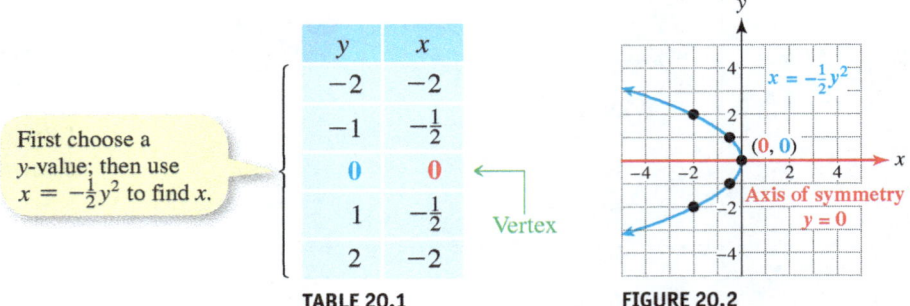

First choose a y-value; then use $x = -\frac{1}{2}y^2$ to find x.

y	x
-2	-2
-1	$-\frac{1}{2}$
0	0 ← Vertex
1	$-\frac{1}{2}$
2	-2

TABLE 20.1

FIGURE 20.2

Now Try Exercise 15

EXAMPLE 2 Graphing a parabola

Graph $x = (y - 3)^2 + 2$. Find its vertex and axis of symmetry.

Solution
Because $h = 2$ and $k = 3$ in the equation $x = a(y - k)^2 + h$, the vertex is $(2, 3)$, and because $a = 1 > 0$, the parabola opens to the right. This parabola has the same shape as $y = x^2$, except that it opens to the right rather than upward. To graph this parabola, we can make a table of values and plot a few points, as shown in **TABLE 20.2** and **FIGURE 20.3**.

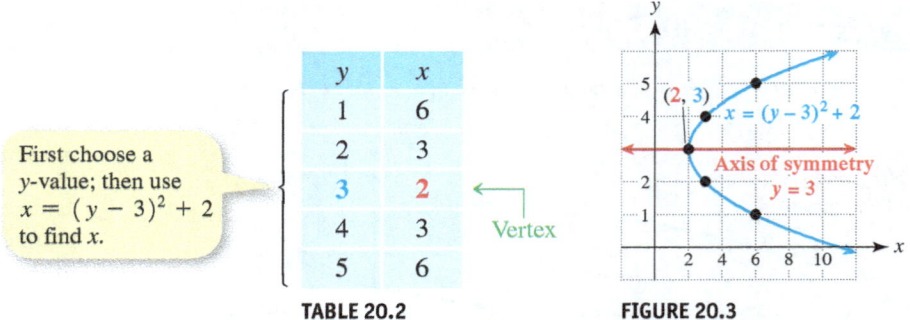

First choose a y-value; then use $x = (y - 3)^2 + 2$ to find x.

y	x
1	6
2	3
3	2 ← Vertex
4	3
5	6

TABLE 20.2

FIGURE 20.3

Both the graph of the parabola and the points from **TABLE 20.2** are shown in **FIGURE 20.3**. There are no y-intercepts and the x-intercept is $(11, 0)$. The axis of symmetry is $y = 3$ because, if we fold the graph on the horizontal line $y = 3$, the two sides match.

NOTE: Sometimes, finding the x- and y-intercepts of the parabola is helpful when you are graphing. To find the x-intercept, let $y = 0$ in $x = (y - 3)^2 + 2$. The x-coordinate of the x-intercept is $x = (0 - 3)^2 + 2 = 11$. To find any y-intercepts, let $x = 0$ in $x = (y - 3)^2 + 2$. Here, $0 = (y - 3)^2 + 2$ means that $(y - 3)^2 = -2$, which has no real solutions, and thus this parabola has no y-intercepts. ∎

Now Try Exercise 23

EXAMPLE 3   Graphing a parabola and finding its vertex

Identify the vertex and then graph each parabola.
(a) $x = -y^2 + 1$ (b) $x = y^2 - 2y - 1$

Solution
(a) If we rewrite $x = -y^2 + 1$ as $x = -(y - 0)^2 + 1$, then $h = 1$ and $k = 0$, so the vertex is $(1, 0)$. By letting $y = 0$ in $x = -y^2 + 1$, we find that the x-coordinate of the x-intercept is $x = -0^2 + 1 = 1$. Similarly, we let $x = 0$ in $x = -y^2 + 1$ to find the y-coordinates of the y-intercepts. The equation $0 = -y^2 + 1$ has solutions -1 and 1.

The parabola opens to the left because $a = -1 < 0$. Additional points given in **TABLE 20.3** help in graphing the parabola shown in **FIGURE 20.4**.

Vertex is $(1, 0)$, so choose y-values on each side of 0.

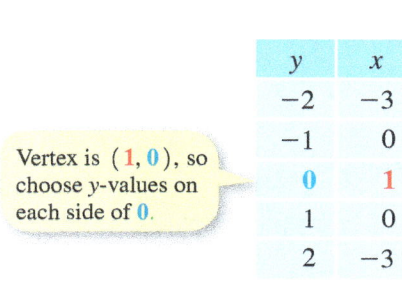

y	x
-2	-3
-1	0
0	1
1	0
2	-3

TABLE 20.3

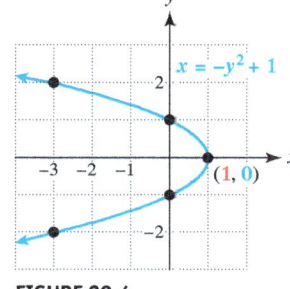

FIGURE 20.4

(b) The y-coordinate of the vertex for the graph of $x = y^2 - 2y - 1$ is given by

$$y = -\frac{b}{2a} = -\frac{-2}{2(1)} = 1.$$

To find the x-coordinate of the vertex, substitute $y = 1$ into $x = y^2 - 2y - 1$.

$$x = (1)^2 - 2(1) - 1 = -2$$

The vertex is $(-2, 1)$. The parabola opens to the right because $a = 1 > 0$. Additional points given in **TABLE 20.4** help in graphing the parabola shown in **FIGURE 20.5**. Note that the y-coordinates of the y-intercepts do not have integer values and that the quadratic formula could be used to find approximations for these ordered pairs. The x-intercept is $(-1, 0)$.

Vertex is $(-2, 1)$, so choose y-values on each side of 1.

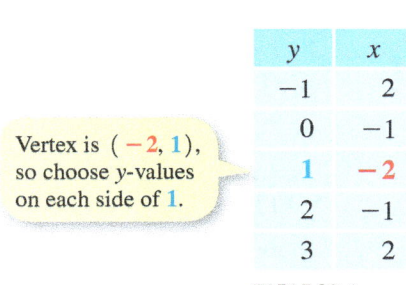

y	x
-1	2
0	-1
1	-2
2	-1
3	2

TABLE 20.4

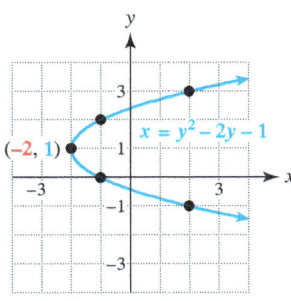

FIGURE 20.5

Now Try Exercises 17, 39

3 Graphing Circles and Writing Their Equations

GRAPHING A CIRCLE A **circle** consists of the set of points in a plane that are the same distance from a fixed point. The fixed distance is called the **radius**, and the fixed point is called the **center**. In **FIGURE 20.6**, all points lying on the blue circle are a distance of 2 units from the center $(2, 1)$. Therefore the radius of the circle equals 2.

Circle

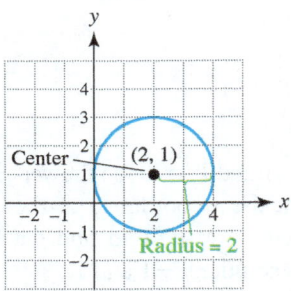

FIGURE 20.6 (repeated)

We can find the equation of the circle shown in **FIGURE 20.6** by using the distance formula. If a point (x, y) lies on the graph of a circle, its distance from the center $(2, 1)$ is 2, and

$$\sqrt{(x-2)^2 + (y-1)^2} = 2. \quad d = \sqrt{(x_2 - x_1)^2 + (y_2 - y_1)^2}$$

Squaring each side gives

$$(x-2)^2 + (y-1)^2 = 2^2.$$

This equation represents the standard equation for a circle with center $(2, 1)$ and radius 2.

> **STANDARD EQUATION OF A CIRCLE**
>
> The **standard equation of a circle** with center (h, k) and radius r is
>
> $$(x-h)^2 + (y-k)^2 = r^2.$$

EXAMPLE 4 Graphing a circle

Graph $x^2 + y^2 = 9$. Find the radius and center.

Solution
The equation $x^2 + y^2 = 9$ can be written in standard form as

$$(x-0)^2 + (y-0)^2 = 3^2.$$

Therefore the center is $(0, 0)$ and the radius is 3. Its graph is shown in **FIGURE 20.7**.

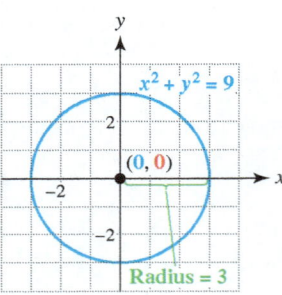

FIGURE 20.7

Now Try Exercise 65

EXAMPLE 5 Graphing a circle

Graph $(x+1)^2 + (y-3)^2 = 4$. Find the radius and center.

Solution
Write the equation as

$$(x - (-1))^2 + (y-3)^2 = 2^2.$$

The center is $(-1, 3)$, and the radius is 2. The circle's graph is shown in **FIGURE 20.8**.

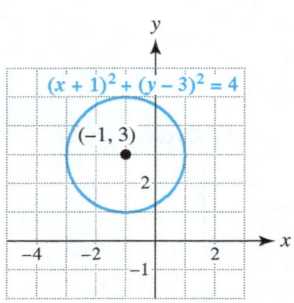

FIGURE 20.8

Now Try Exercise 69

READING CHECK 3

- Give the center and radius of the circle whose equation is $(x-3)^2 + (y+1)^2 = 9$.

TECHNOLOGY NOTE

Graphing Circles

The graph of a circle does not represent a function. One way to graph a circle with a graphing calculator is to solve the equation for y and obtain two equations. One equation gives the upper half of the circle, and the other equation gives the lower half.

For example, to graph $x^2 + y^2 = 4$, begin by solving for y.

$$y^2 = 4 - x^2 \quad \text{Subtract } x^2.$$
$$y = \pm\sqrt{4 - x^2} \quad \text{Square root property}$$

Then graph $y_1 = \sqrt{4 - x^2}$ and $y_2 = -\sqrt{4 - x^2}$. The graph of y_1 is the upper half of the circle, and the graph of y_2 is the lower half of the circle, as shown in **FIGURE 20.9**.

Upper half
$[-4.7, 4.7, 1]$ by $[-3.1, 3.1, 1]$

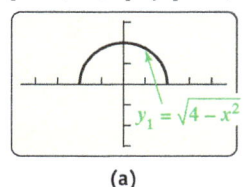

(a)

Lower half
$[-4.7, 4.7, 1]$ by $[-3.1, 3.1, 1]$

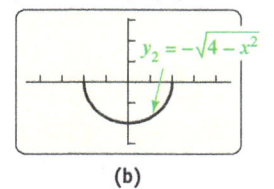

(b)

Both halves
$[-4.7, 4.7, 1]$ by $[-3.1, 3.1, 1]$

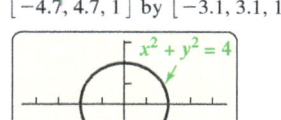

(c)

FIGURE 20.9

$[-4, 4, 1]$ by $[-5, 5, 1]$

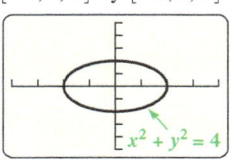

FIGURE 20.10

NOTE: If a circle is not graphed in a *square viewing rectangle*, it will appear to be an oval rather than a circle. In a square viewing rectangle, a circle will appear circular. **FIGURE 20.10** shows the circle graphed in a viewing rectangle that is *not* square. ■

COMPLETING THE SQUARE TO FIND THE CENTER AND RADIUS In the next example, we use the *method of completing the square* to find the center and radius of a circle.

EXAMPLE 6 Finding the center of a circle

Find the center and radius of the circle given by $x^2 + 4x + y^2 - 6y = 5$.

Solution
Begin by writing the equation as

$$(x^2 + 4x + \underline{\quad}) + (y^2 - 6y + \underline{\quad}) = 5.$$

To complete the square, add $\left(\frac{4}{2}\right)^2 = 4$ and $\left(\frac{-6}{2}\right)^2 = 9$ to each side of the equation.

$$(x^2 + 4x + 4) + (y^2 - 6y + 9) = 5 + 4 + 9$$

Factoring each perfect square trinomial yields

$$(x + 2)^2 + (y - 3)^2 = 18.$$

The center is $(-2, 3)$, and because $18 = (\sqrt{18})^2$, the radius is $\sqrt{18}$, or $3\sqrt{2}$.

Now Try Exercise 71

CRITICAL THINKING 1

Does the following equation represent a circle? If so, give its center and radius.

$$x^2 + y^2 + 10y = -32$$

20.1 Putting It All Together

CONCEPT	EXPLANATION	EXAMPLE
Parabola with Horizontal Axis	Vertex form: $x = a(y - k)^2 + h$ If $a > 0$, it opens to the right. If $a < 0$, it opens to the left. The vertex is (h, k). These parabolas may also be expressed as $x = ay^2 + by + c$, where the y-coordinate of the vertex is $y = -\frac{b}{2a}$.	$x = 2(y - 1)^2 + 4$ opens to the right and its vertex is $(4, 1)$. $a > 0$, so opens right.
Standard Equation of a Circle	$(x - h)^2 + (y - k)^2 = r^2$ The radius is r and the center is (h, k).	$(x + 2)^2 + (y - 1)^2 = 16$ has center $(-2, 1)$ and radius 4.

20.1 Exercises

CONCEPTS AND VOCABULARY

1. Name the three general types of conic sections.

2. What is the difference between the parabolas given by $y = ax^2 + bx + c$ and $x = ay^2 + by + c$?

3. If a parabola has a horizontal axis of symmetry, does it represent a function?

4. Sketch a graph of a parabola with a horizontal axis of symmetry.

5. If a parabola has two y-intercepts, does it represent a function? Why or why not?

6. If $x = a(y - k)^2 + h$, what is the vertex?

7. The graph of $x = -y^2$ opens to the _____.

8. The graph of $x = 2y^2 + y - 1$ opens to the _____.

9. The graph of $(x - h)^2 + (y - k)^2 = r^2$ is a(n) _____ with center _____.

10. The graph of $x^2 + y^2 = r^2$ is a circle with center _____ and radius _____.

11. Which of the following (a–d) are the coordinates of the vertex for the parabola given by $x = 4(y + 2)^2 - 3$?
    a. $(-3, 4)$ b. $(3, -2)$
    c. $(-3, 2)$ d. $(-3, -2)$

12. Which of the following (a–d) is the equation of the axis of symmetry for the parabola given by $x = -5(y - 1)^2 + 7$?
    a. $y = 1$ b. $x = -5$
    c. $x = 7$ d. $y = 7$

13. Which of the following (a–d) is the center and radius of the circle given by $x^2 + (y - 2)^2 = 9$?
    a. $(0, -2), r = 3$
    b. $(2, 0), r = 3$
    c. $(0, 2), r = 3$
    d. $(0, 2), r = 9$

14. Which of the following (a–d) is the y-coordinate of the vertex of the parabola given by $x = y^2 - 2y + 3$?
    a. -2 b. 1
    c. 0 d. 3

GRAPHING PARABOLAS AND WRITING THEIR EQUATIONS

Exercises 15–40: Graph the parabola. Find the vertex and axis of symmetry.

15. $x = y^2$
16. $x = -y^2$
17. $x = y^2 + 1$
18. $x = y^2 - 1$
19. $y = x^2 - 1$
20. $y = 2x^2$
21. $x = 2y^2$
22. $x = \frac{1}{4}y^2$
23. $x = (y - 1)^2 + 2$
24. $x = (y - 2)^2 + 1$
25. $y = (x + 2)^2 + 1$
26. $y = (x - 4)^2 + 5$
27. $y = -2(x + 2)^2$
28. $y = (x - 1)^2 - 2$
29. $x = \frac{1}{2}(y + 1)^2 - 3$
30. $x = -2(y + 3)^2 + 1$
31. $x = -3(y - 1)^2$
32. $x = \frac{1}{4}(y + 2)^2 - 3$
33. $y = 2x^2 - x + 1$
34. $y = -x^2 + 2x + 2$
35. $x = -2y^2 + 3y + 2$
36. $x = \frac{1}{2}y^2 + y - 1$
37. $x = 3y^2 + y$
38. $x = -\frac{3}{2}y^2 - 2y + 1$
39. $x = y^2 + 2y + 1$
40. $x = y^2 - 3y - 4$

Exercises 41–44: Use the graph to determine the equation of the parabola. (Hint: Either $a = 1$ or $a = -1$.)

41.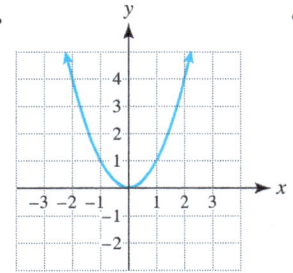

42.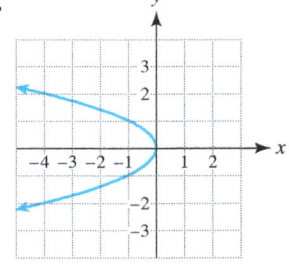

43.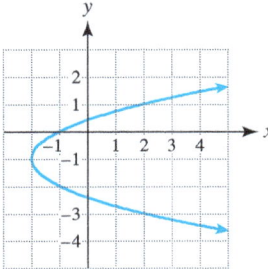

44.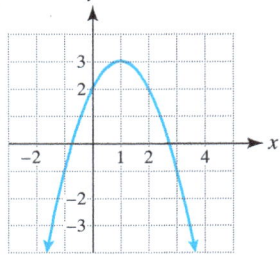

Exercises 45–48: Determine the direction that the parabola opens if it satisfies the given conditions.

45. Passing through $(2, 0)$, $(-2, 0)$, and $(0, -2)$
46. Passing through $(0, -3)$, $(0, 2)$, and $(1, 1)$
47. Vertex $(1, 2)$ passing through the point $(-1, -2)$ with a vertical axis
48. Vertex $(-1, 3)$ passing through the point $(0, 0)$ with a horizontal axis
49. What x-values are possible for the graph of the equation $x = 2y^2$?
50. What y-values are possible for the graph of the equation $x = 2y^2$?
51. **Thinking Generally** How many y-intercepts does a parabola given by
$$x = a(y - k)^2 + h$$
have if $a > 0$ and $h < 0$?
52. **Thinking Generally** Does the graph of the equation $x = ay^2 + by + c$ always have a y-intercept? Explain.
53. What is the x-intercept for the graph of the equation given by
$$x = 3y^2 - y + 1?$$
54. What are the y-intercepts for the graph of the equation given by
$$x = y^2 - 3y + 2?$$

GRAPHING CIRCLES AND WRITING THEIR EQUATIONS

Exercises 55–60: Write the standard equation of the circle with the given radius r and center C.

55. $r = 1$ $\quad C = (0, 0)$
56. $r = 4$ $\quad C = (2, 3)$
57. $r = 3$ $\quad C = (-1, 5)$
58. $r = 5$ $\quad C = (5, -3)$
59. $r = \sqrt{2}$ $\quad C = (-4, -6)$
60. $r = \sqrt{6}$ $\quad C = (0, 4)$

Exercises 61–64: Use the graph to find the standard equation of the circle.

61.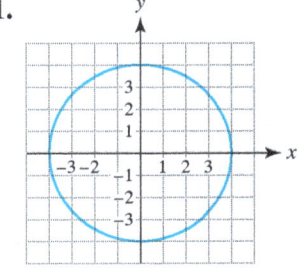

62.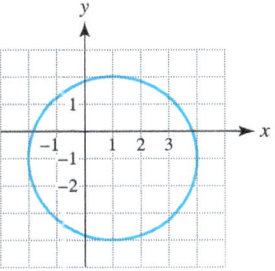

63. 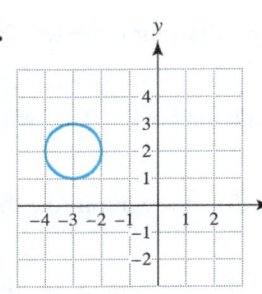　　**64.**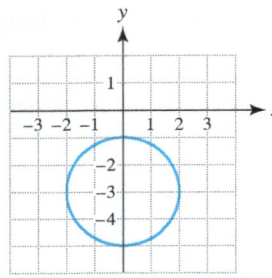

Exercises 65–74: Find the radius and center of the circle. Then graph the circle.

65. $x^2 + y^2 = 4$　　**66.** $x^2 + y^2 = 1$

67. $(x-1)^2 + (y-3)^2 = 9$

68. $(x+2)^2 + (y+1)^2 = 4$

69. $(x+5)^2 + (y-5)^2 = 25$

70. $(x-4)^2 + (y+3)^2 = 16$

71. $x^2 + 6x + y^2 - 2y = -1$

72. $x^2 + y^2 + 12y + 32 = 0$

73. $x^2 + 6x + y^2 - 2y + 3 = 0$

74. $x^2 - 4x + y^2 + 4y = -3$

Exercises 75–78: Use the endpoints of the diameter to find the standard equation of the circle.

75. $(-3, 4), (5, 6)$　　**76.** $(-1, 2), (3, -2)$

77. $(2, 1), (5, 5)$　　**78.** $(3, -2), (1, -4)$

APPLICATIONS OF PARABOLAS

79. Radio Telescopes The Parks radio telescope has the shape of a parabolic dish, as depicted in the figure below. A cross section of this telescope can be modeled by $x = \frac{32}{11,025} y^2$, where $-105 \le y \le 105$; the units are feet. (*Source:* J. Mar, *Structure Technology for Large Radar and Telescope Systems.*)

(a) Graph the cross-sectional shape of the dish in $[-40, 40, 10]$ by $[-120, 120, 20]$.

(b) Find the depth d of the dish.

80. Train Tracks To make a curve safer for trains, parabolic curves are sometimes used instead of circular curves. See the accompanying figures. (*Source:* F. Mannering and W. Kilareski, *Principles of Highway Engineering and Traffic Analysis.*)

(a) Suppose that a curve must pass through the points $(-1, 0)$, $(0, 3)$, and $(0, -3)$, where the units are kilometers. Find an equation for the train tracks in the form
$$x = a(y - h)^2 + k.$$

(b) Find another point that lies on the train tracks.

A parabolic turn　　A circular turn

81. Trajectories of Comets Under certain circumstances, a comet can pass by the sun once and never return. In this situation, the comet may travel in a parabolic path, as illustrated in the figure below. Suppose that a comet's path is given by $x = -2.5y^2$, where the sun is located at $(-0.1, 0)$ and the units are astronomical units (A.U.). One astronomical unit equals 93 million miles. (*Source:* W. Thomson, *Introduction to Space Dynamics.*)

(a) Plot a point for the sun's location and then graph the path of the comet.

(b) Find the distance from the sun to the comet when the comet is located at $(-2.5, 1)$.

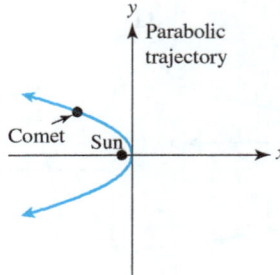

82. Speed of a Comet (Continuation of Exercise 81.) The velocity V in meters per second of a comet traveling in a parabolic trajectory around the sun is given by $V = \frac{k}{\sqrt{D}}$, where D is the comet's distance from the sun in meters and $k = 1.15 \times 10^{10}$.

(a) How does the velocity of the comet change as its distance from the sun changes?

(b) Calculate the velocity of the comet when it is closest to the sun. (*Hint:* 1 mile ≈ 1609 meters.)

WRITING ABOUT MATHEMATICS

83. Suppose that you are given the equation

$$x = a(y - k)^2 + h.$$

(a) Explain how you can determine the direction that the parabola opens.

(b) Explain how to find the axis of symmetry and the vertex.

(c) If the points $(0, 4)$ and $(0, -2)$ lie on the graph of x, what is the axis of symmetry?

(d) Generalize part (c) if $(0, y_1)$ and $(0, y_2)$ lie on the graph of x.

84. Suppose that you are given the vertex of a parabola. Can you determine the axis of symmetry? Explain.

Group Activity

Radio Telescope The U.S. Naval Research Laboratory designed a giant radio telescope weighing 3450 tons. Its parabolic dish has a diameter of 300 feet and a depth of 44 feet, as shown in the figure to the right.
(*Source:* J. Mar, *Structure Technology for Large Radio and Radar Telescope Systems.*)

(a) Determine an equation of the form $x = ay^2$, $a > 0$, that models a cross section of the dish.

(b) Graph your equation in an appropriate viewing rectangle.

20.2 Ellipses and Hyperbolas

Objectives

1. Graphing Ellipses and Writing Their Equations
2. Graphing Hyperbolas and Writing Their Equations

1 Graphing Ellipses and Writing Their Equations

One method used to sketch an ellipse is to tie the ends of a string to two nails driven into a flat board. If a pencil is placed against the string anywhere between the nails, as shown in **FIGURE 20.11**, and is used to draw a curve, the resulting curve is an ellipse. The sum of the distances d_1 and d_2 between the pencil and each of the nails is always fixed by the length of the string. The location of the nails corresponds to the *foci* of the ellipse. An **ellipse** is the set of points in a plane the sum of whose distances from two fixed points is constant. Each fixed point is called a **focus** (plural, foci) of the ellipse.

CRITICAL THINKING 1

What happens to the shape of the ellipse shown in **FIGURE 20.11** as the nails are moved farther apart? What happens to its shape as the nails are moved closer together? When would a circle be formed?

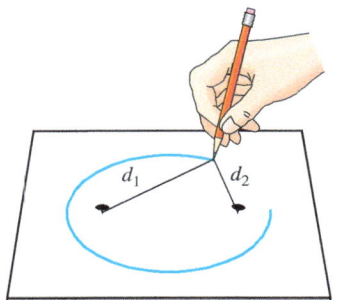

FIGURE 20.11

See the Concept 1: ELLIPSES WITH HORIZONTAL AND VERTICAL MAJOR AXES

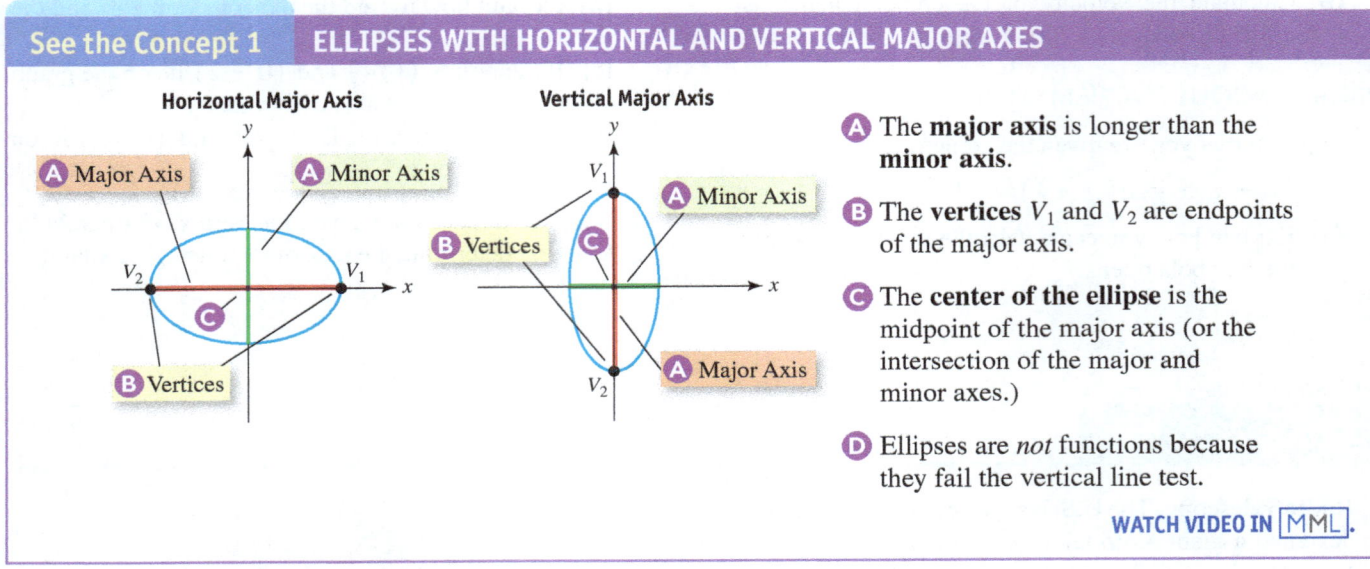

A The **major axis** is longer than the **minor axis**.

B The **vertices** V_1 and V_2 are endpoints of the major axis.

C The **center of the ellipse** is the midpoint of the major axis (or the intersection of the major and minor axes.)

D Ellipses are *not* functions because they fail the vertical line test.

WATCH VIDEO IN MML.

NEW VOCABULARY

- [ ] Ellipse
- [ ] Focus of an ellipse
- [ ] Major/minor axis
- [ ] Vertices of an ellipse
- [ ] Center of an ellipse
- [ ] Hyperbola
- [ ] Focus of a hyperbola
- [ ] Branches
- [ ] Vertices of a hyperbola
- [ ] Transverse axis
- [ ] Fundamental rectangle
- [ ] Asymptotes

Some ellipses can be represented by the following equations.

STANDARD EQUATIONS FOR ELLIPSES CENTERED AT (0, 0)

The ellipse with center at the origin, *horizontal* major axis, and equation

$$\frac{x^2}{a^2} + \frac{y^2}{b^2} = 1, \quad a > b > 0,$$

has vertices $(\pm a, 0)$ and endpoints of the minor axis $(0, \pm b)$.

The ellipse with center at the origin, *vertical* major axis, and equation

$$\frac{x^2}{b^2} + \frac{y^2}{a^2} = 1, \quad a > b > 0,$$

has vertices $(0, \pm a)$ and endpoints of the minor axis $(\pm b, 0)$.

FIGURE 20.12(a) shows an ellipse that has a horizontal major axis; **FIGURE 20.12(b)** shows one that has a vertical major axis. The coordinates of the vertices V_1 and V_2 and endpoints of the minor axis U_1 and U_2 are labeled.

READING CHECK 1

- How can you determine from the standard equation of an ellipse if the major axis is horizontal or vertical?

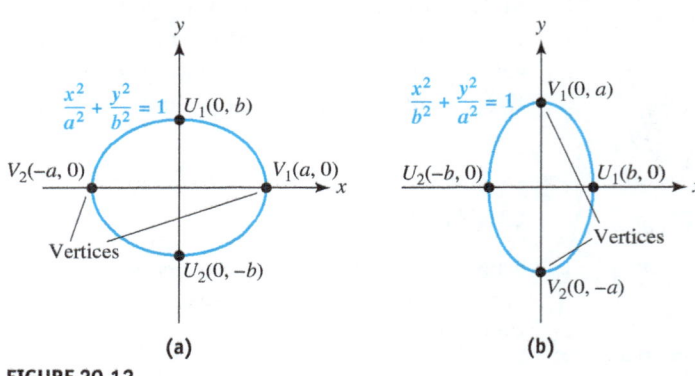

FIGURE 20.12

In the next example, we show how to sketch graphs of ellipses.

EXAMPLE 1 Sketching ellipses

Sketch a graph of each ellipse. Label the vertices and endpoints of the minor axes.

(a) $\dfrac{x^2}{25} + \dfrac{y^2}{4} = 1$ (b) $9x^2 + 4y^2 = 36$

Solution

(a) The equation $\dfrac{x^2}{25} + \dfrac{y^2}{4} = 1$ describes an ellipse with $a^2 = 25$ and $b^2 = 4$. (When you are deciding whether 25 or 4 represents a^2, let a^2 be the larger of the two numbers.) Thus $a = 5$ and $b = 2$, so the ellipse has a horizontal major axis with vertices $(\pm 5, 0)$ and the endpoints of the minor axis are $(0, \pm 2)$. Plot these four points and then sketch the ellipse, as shown in **FIGURE 20.13(a)**.

(b) To write $9x^2 + 4y^2 = 36$ as a standard equation, divide each side by 36.

$$9x^2 + 4y^2 = 36 \quad \text{Given equation}$$

$$\dfrac{9x^2}{36} + \dfrac{4y^2}{36} = \dfrac{36}{36} \quad \text{Divide each side by 36.}$$

$$\dfrac{x^2}{4} + \dfrac{y^2}{9} = 1 \quad \text{Simplify.}$$

This ellipse has a vertical major axis with $a = 3$ and $b = 2$. The vertices are $(0, \pm 3)$, and the endpoints of the minor axis are $(\pm 2, 0)$, as shown in **FIGURE 20.13(b)**.

CRITICAL THINKING 2

Suppose that $a = b$ for an ellipse centered at (0, 0). What can be said about the ellipse? Explain.

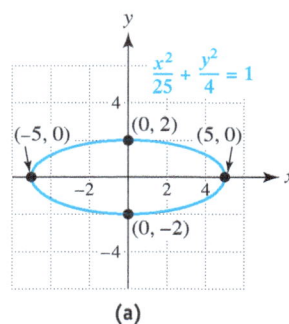
(a)

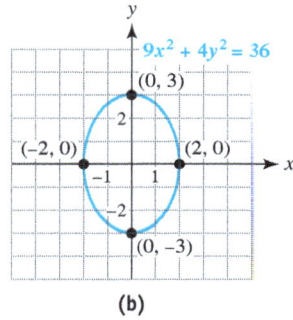
(b)

FIGURE 20.13

Now Try Exercises 15, 21

EXAMPLE 2 Finding the standard equation of an ellipse

Use the graph in **FIGURE 20.14** to determine the standard equation of the ellipse.

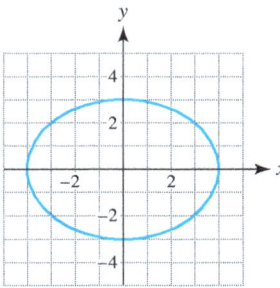

FIGURE 20.14

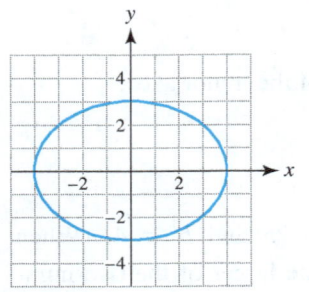

FIGURE 20.14 (repeated)

Solution
The ellipse in **FIGURE 20.14** is centered at $(0, 0)$ with a horizontal major axis. The length of the major axis is 8, so $a = 4$. The length of the minor axis is 6, so $b = 3$. Thus $a^2 = 16$ and $b^2 = 9$, and the standard equation of the ellipse is

$$\frac{x^2}{16} + \frac{y^2}{9} = 1.$$

Now Try Exercise 25

~~Math in Context~~ *Astronomy* Planets travel around the sun in elliptical orbits. Astronomers have measured the values of a and b for each planet. Using this information, we can find the equation of a planet's orbit, as illustrated in the next example.

EXAMPLE 3   Modeling the orbit of Mercury

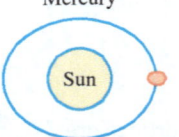

Mercury

Not to Scale

The planet Mercury has one of the least circular orbits of the eight major planets, as illustrated in the margin. For Mercury, $a = 0.387$ and $b = 0.379$. The units are astronomical units (A.U.), where 1 A.U. equals 93 million miles—the distance between Earth and the sun. Graph $\frac{x^2}{a^2} + \frac{y^2}{b^2} = 1$ to model the orbit of Mercury in $[-0.6, 0.6, 0.1]$ by $[-0.4, 0.4, 0.1]$. Then plot the sun at the point $(0.08, 0)$. (*Source*: M. Zeilik, *Introductory Astronomy and Astrophysics*.)

Solution
The orbit of Mercury is given by

$$\frac{x^2}{0.387^2} + \frac{y^2}{0.379^2} = 1. \qquad \frac{x^2}{a^2} + \frac{y^2}{b^2} = 1$$

To graph an ellipse with some graphing calculators, we must solve the equation for y. Doing so results in two equations.

$$\frac{x^2}{0.387^2} + \frac{y^2}{0.379^2} = 1$$

$$\frac{y^2}{0.379^2} = 1 - \frac{x^2}{0.387^2} \qquad \text{Subtract } \frac{x^2}{0.387^2}.$$

$$\frac{y}{0.379} = \pm\sqrt{1 - \frac{x^2}{0.387^2}} \qquad \text{Square root property}$$

$$y = \pm 0.379\sqrt{1 - \frac{x^2}{0.387^2}} \qquad \text{Multiply by 0.379.}$$

The orbit of Mercury can be found by graphing these equations. See **FIGURES 20.15(a)** and **20.15(b)**. The point $(0.08, 0)$ represents the position of the sun in **FIGURE 20.15(b)**.

CRITICAL THINKING 3
Use **FIGURE 20.15** and the information in Example 3 to estimate the minimum and maximum distances that Mercury is from the sun.

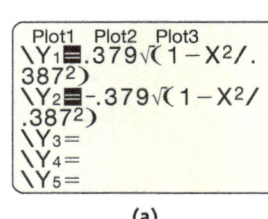

(a)

$[-0.6, 0.6, 0.1]$ by $[-0.4, 0.4, 0.1]$

Orbit of Mercury

Sun

(b)

FIGURE 20.15

Now Try Exercise 49

2 Graphing Hyperbolas and Writing Their Equations

The third type of conic section is the **hyperbola**, which is the set of points in a plane the difference of whose distances from two fixed points is constant. Each fixed point is called a **focus** of the hyperbola.

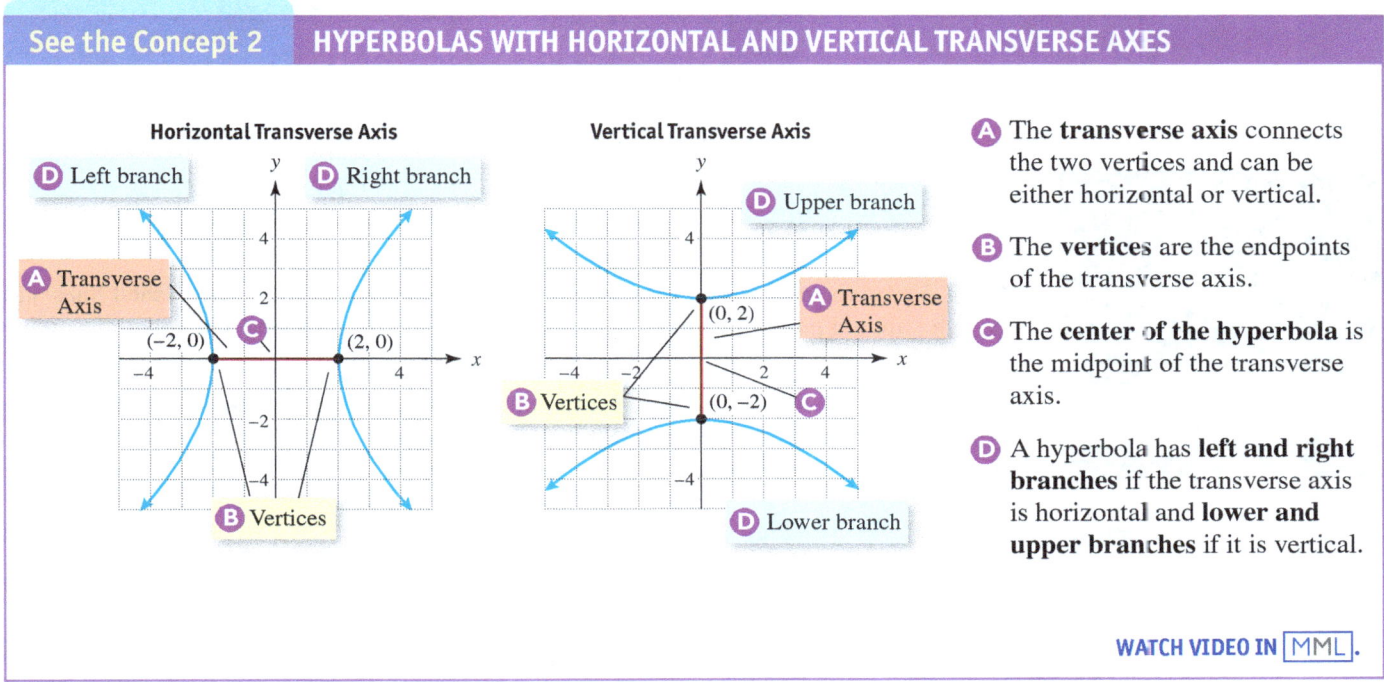

See the Concept 2 — HYPERBOLAS WITH HORIZONTAL AND VERTICAL TRANSVERSE AXES

A The **transverse axis** connects the two vertices and can be either horizontal or vertical.

B The **vertices** are the endpoints of the transverse axis.

C The **center of the hyperbola** is the midpoint of the transverse axis.

D A hyperbola has **left and right branches** if the transverse axis is horizontal and **lower and upper branches** if it is vertical.

WATCH VIDEO IN MML.

READING CHECK 2

- How can you determine from the standard equation of a hyperbola if the transverse axis is horizontal or vertical?

Many hyperbolas can be described by the following equations.

STANDARD EQUATIONS FOR HYPERBOLAS CENTERED AT (0, 0)

The hyperbola with center at the origin, *horizontal* transverse axis, and equation

$$\frac{x^2}{a^2} - \frac{y^2}{b^2} = 1$$

has vertices $(\pm a, 0)$

The hyperbola with center at the origin, *vertical* transverse axis, and equation

$$\frac{y^2}{a^2} - \frac{x^2}{b^2} = 1$$

has vertices $(0, \pm a)$.

Hyperbolas, along with the coordinates of their vertices, are shown in **FIGURE 20.16** on the next page. The two parts of the hyperbola in **FIGURE 20.16(a)** are the *left branch* and *right branch*, whereas in **FIGURE 20.16(b)**, the hyperbola has an *upper branch* and a *lower branch*. The dashed rectangle in each figure is called the **fundamental rectangle**, and its four vertices (corners) are determined by either $(\pm a, \pm b)$ or $(\pm b, \pm a)$. If its diagonals are extended, they correspond to the asymptotes of the hyperbola. The dashed lines $y = \pm \frac{b}{a} x$ and $y = \pm \frac{a}{b} x$ are **asymptotes** for the hyperbolas.

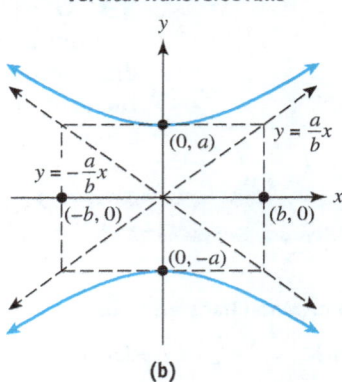

Horizontal Transverse Axis (a)

Vertical Transverse Axis (b)

FIGURE 20.16 Not Functions

NOTE: A hyperbola consists of two solid (blue) curves, or branches. The dashed lines and rectangles are not part of the graph but are used as an aid for sketching the hyperbola. ∎

🌐 Math in Context (Astronomy) One interpretation of an asymptote of a hyperbola can be based on trajectories of comets as they approach the sun. Comets travel in parabolic, elliptic, or hyperbolic trajectories. See **FIGURE 20.17**.

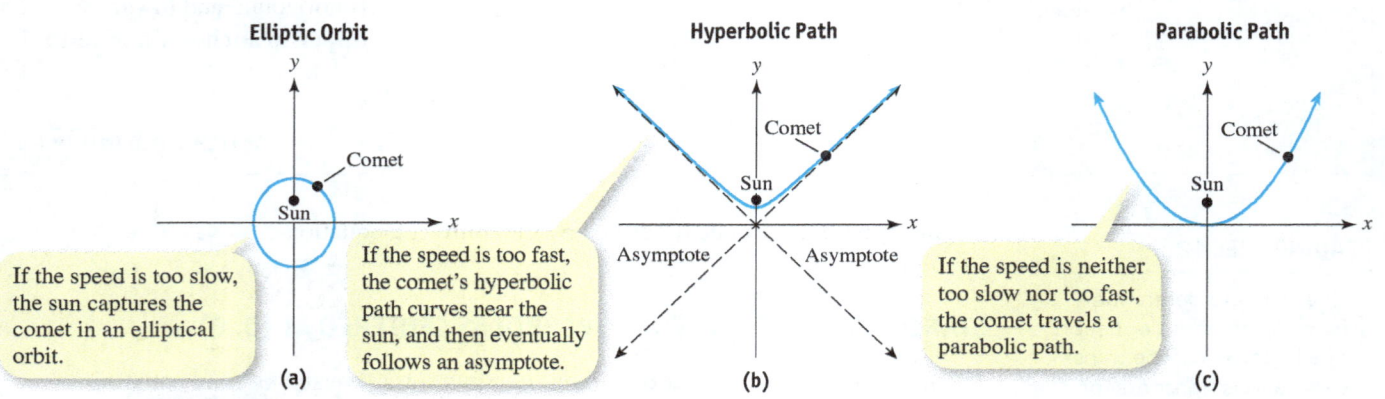

FIGURE 20.17

EXAMPLE 4 **Sketching a hyperbola**

Sketch a graph of $\frac{y^2}{4} - \frac{x^2}{9} = 1$. Label the vertices and show the asymptotes.

Solution

The equation is in standard form with $a^2 = 4$ and $b^2 = 9$, so $a = 2$ and $b = 3$. It has a vertical transverse axis with vertices $(0, -2)$ and $(0, 2)$. The vertices (corners) of the fundamental rectangle are $(\pm 3, \pm 2)$, that is, $(3, 2)$, $(3, -2)$, $(-3, 2)$, and $(-3, -2)$. The asymptotes are the diagonals of this rectangle and are given by $y = \pm \frac{a}{b} x$, or $y = \pm \frac{2}{3} x$. **FIGURE 20.18** shows the hyperbola and these features. The hyperbola has upper and lower branches because the transverse axis is vertical.

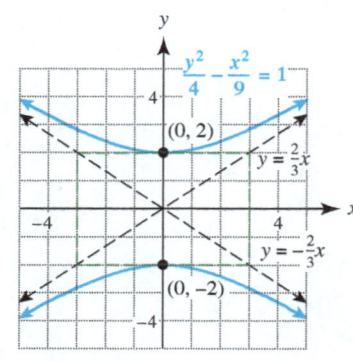

FIGURE 20.18

Now Try Exercise 29

TECHNOLOGY NOTE

Graphing a Hyperbola

The graph of a hyperbola does not represent a function. One way to graph a hyperbola with a graphing calculator is to solve the equation for y and obtain two equations. One equation gives the upper portion of the hyperbola and the other equation gives the lower portion.

For example, to graph $\frac{y^2}{4} - \frac{x^2}{8} = 1$, begin by solving for y.

$$\frac{y^2}{4} = 1 + \frac{x^2}{8} \qquad \text{Add } \tfrac{x^2}{8}.$$

$$y^2 = 4\left(1 + \frac{x^2}{8}\right) \qquad \text{Multiply by 4.}$$

$$y = \pm 2\sqrt{1 + \frac{x^2}{8}} \qquad \text{Square root property}$$

Graph $y_1 = 2\sqrt{1 + x^2/8}$ and $y_2 = -2\sqrt{1 + x^2/8}$, which give the upper and lower branches, respectively. See **FIGURE 20.19**.

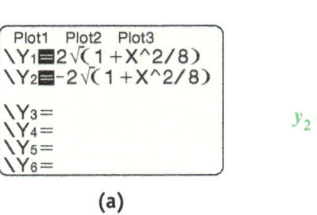

(a)

(b)

FIGURE 20.19

EXAMPLE 5 Determining the standard equation of a hyperbola

Use the graph in **FIGURE 20.20** to determine the standard equation of the hyperbola.

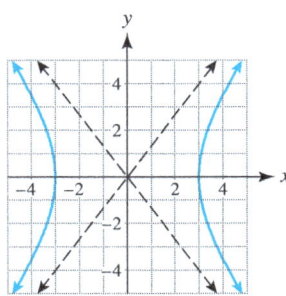

FIGURE 20.20

Solution

The hyperbola has a horizontal transverse axis, so the x^2-term must come first in the equation. The vertices of the hyperbola are $(\pm 3, 0)$, which indicates that $a = 3$ and $a^2 = 9$. The value of b can be found by noting that one of the asymptotes passes through the point $(3, 4)$. This asymptote has the equation $y = \frac{b}{a}x$ or $y = \frac{4}{3}x$, so let $b = 4$ and $b^2 = 16$. The equation of the hyperbola is

$$\frac{x^2}{9} - \frac{y^2}{16} = 1.$$

Now Try Exercise 41

20.2 Putting It All Together

CONCEPT	DESCRIPTION	
Ellipses Centered at $(0,0)$ with $a > b > 0$	**Horizontal Major Axis** Vertices: $(a, 0)$ and $(-a, 0)$ Endpoints of minor axis: $(0, b)$ and $(0, -b)$ $$\frac{x^2}{a^2} + \frac{y^2}{b^2} = 1$$ 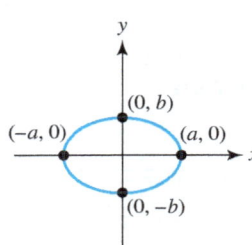	**Vertical Major Axis** Vertices: $(0, a)$ and $(0, -a)$ Endpoints of minor axis: $(-b, 0)$ and $(b, 0)$ $$\frac{x^2}{b^2} + \frac{y^2}{a^2} = 1$$ 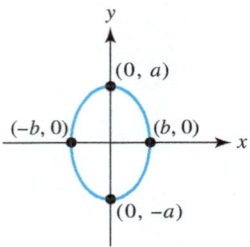
Hyperbolas Centered at $(0,0)$ with $a > 0$ and $b > 0$	**Horizontal Transverse Axis** Vertices: $(a, 0)$ and $(-a, 0)$ Asymptotes: $y = \pm \frac{b}{a}x$ $$\frac{x^2}{a^2} - \frac{y^2}{b^2} = 1$$ 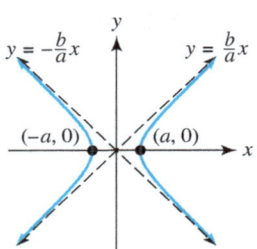	**Vertical Transverse Axis** Vertices: $(0, a)$ and $(0, -a)$ Asymptotes: $y = \pm \frac{a}{b}x$ $$\frac{y^2}{a^2} - \frac{x^2}{b^2} = 1$$ 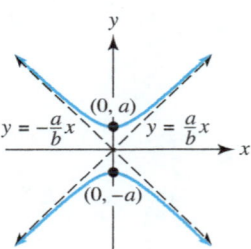

20.2 Exercises

CONCEPTS AND VOCABULARY

1. Sketch an ellipse with a horizontal major axis.

2. Sketch a hyperbola with a vertical transverse axis.

3. The ellipse whose standard equation is $\frac{x^2}{a^2} + \frac{y^2}{b^2} = 1$, $a > b > 0$, has a(n) _____ major axis.

4. The ellipse whose standard equation is $\frac{x^2}{b^2} + \frac{y^2}{a^2} = 1$, $a > b > 0$, has a(n) _____ major axis.

5. What is the maximum number of times that a line can intersect an ellipse?

6. What is the maximum number of times that a parabola can intersect an ellipse?

7. The hyperbola whose equation is $\frac{x^2}{a^2} - \frac{y^2}{b^2} = 1$ has _____ and _____ branches.

8. The hyperbola whose equation is $\frac{y^2}{a^2} - \frac{x^2}{b^2} = 1$ has _____ and _____ branches.

9. How are the asymptotes of a hyperbola related to the fundamental rectangle?

10. Could an ellipse be centered at the origin and have vertices $(4, 0)$ and $(0, -5)$?

11. Which of the following (a–d) represents the coordinates of the vertices on the ellipse $\frac{x^2}{4} + \frac{y^2}{9} = 1$?
    a. $(0, \pm 3)$
    b. $(\pm 2, 0)$
    c. $(\pm 4, 0)$
    d. $(0, \pm 9)$

12. Which of the following (a–d) represents the coordinates of the vertices on the hyperbola $\frac{x^2}{4} - \frac{y^2}{9} = 1$?
    a. $(0, \pm 3)$
    b. $(\pm 2, 0)$
    c. $(\pm 4, 0)$
    d. $(0, \pm 9)$

ELLIPSES AND THIER EQUATIONS

Exercises 13–24: Graph the ellipse. Label the vertices and endpoints of the minor axis.

13. $\frac{x^2}{9} + \frac{y^2}{25} = 1$
14. $\frac{y^2}{9} + \frac{x^2}{25} = 1$
15. $\frac{x^2}{9} + \frac{y^2}{4} = 1$
16. $\frac{x^2}{3} + \frac{y^2}{9} = 1$
17. $x^2 + \frac{y^2}{4} = 1$
18. $\frac{x^2}{9} + y^2 = 1$
19. $\frac{y^2}{5} + \frac{x^2}{7} = 1$
20. $\frac{y^2}{11} + \frac{x^2}{6} = 1$
21. $36x^2 + 4y^2 = 144$
22. $25x^2 + 16y^2 = 400$
23. $6y^2 + 7x^2 = 42$
24. $9x^2 + 5y^2 = 45$

Exercises 25–28: Use the graph to determine the standard equation of the ellipse.

25.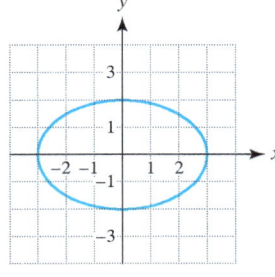
26.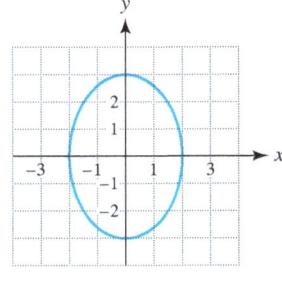
27.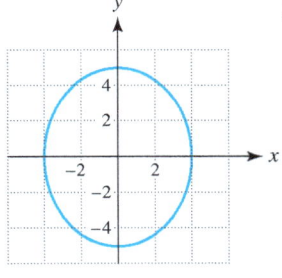
28.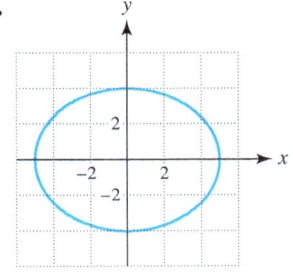

HYPERBOLAS AND THEIR EQUATIONS

Exercises 29–40: Graph the hyperbola. Show the asymptotes and vertices.

29. $\frac{x^2}{4} - \frac{y^2}{9} = 1$
30. $\frac{y^2}{4} - \frac{x^2}{9} = 1$
31. $\frac{y^2}{25} - \frac{x^2}{16} = 1$
32. $\frac{x^2}{25} - \frac{y^2}{16} = 1$
33. $x^2 - y^2 = 1$
34. $y^2 - x^2 = 1$
35. $\frac{x^2}{3} - \frac{y^2}{4} = 1$
36. $\frac{y^2}{5} - \frac{x^2}{8} = 1$
37. $9y^2 - 4x^2 = 36$
38. $36x^2 - 25y^2 = 900$
39. $16x^2 - 4y^2 = 64$
40. $y^2 - 9x^2 = 9$

Exercises 41–44: Use the graph to determine the standard equation of the hyperbola.

41.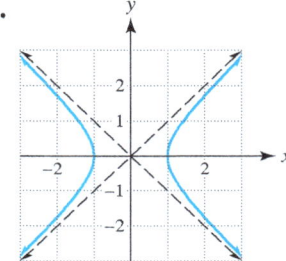
42.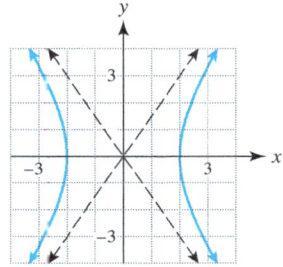
43.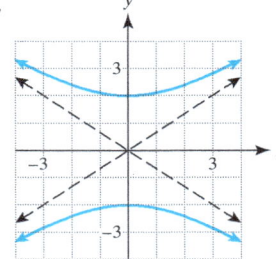
44.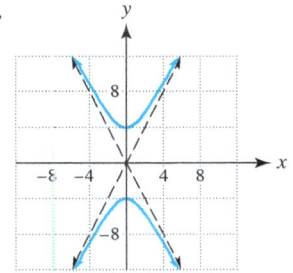

TRANSLATIONS OF ELLIPSES AND HYPERBOLAS

Exercises 45 and 46: Ellipses and hyperbolas can be translated so that they are centered at a point (h, k), rather than at the origin. These techniques are the same as those used for parabolas and circles. To translate a conic section so that it is centered at (h, k) rather than $(0, 0)$, replace x with $(x - h)$ and replace y with $(y - k)$. For example, to center $\frac{x^2}{9} + \frac{y^2}{4} = 1$ at $(-1, 2)$, change its equation to $\frac{(x + 1)^2}{9} + \frac{(y - 2)^2}{4} = 1$. See the accompanying figures on the next page.

continued on next page

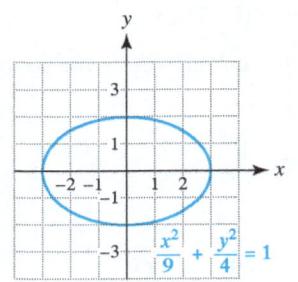

Ellipse centered at (0, 0)

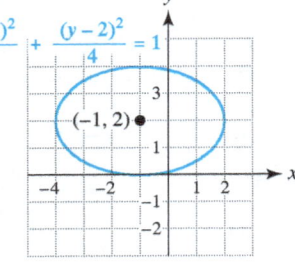

Ellipse centered at (−1, 2)

45. Graph each conic section. For hyperbolas, give the equations of the asymptotes.

(a) $\dfrac{(x-3)^2}{25} + \dfrac{(y-1)^2}{9} = 1$

(b) $\dfrac{(x+1)^2}{4} + \dfrac{(y+2)^2}{16} = 1$

(c) $\dfrac{(x+1)^2}{4} - \dfrac{(y-3)^2}{9} = 1$

(d) $\dfrac{(y-4)^2}{16} - \dfrac{(x+1)^2}{4} = 1$

46. Write the equation of an ellipse that has the following properties.
   (a) Horizontal major axis of length 8, minor axis of length 4, and centered at $(-3, 5)$
   (b) Vertical major axis of length 10, minor axis of length 6, and centered at $(2, -3)$

APPLICATIONS OF ELLIPSES AND HYPERBOLAS

47. Geometry of an Ellipse The area inside an ellipse is given by $A = \pi ab$, and its perimeter can be approximated by

$$P = 2\pi \sqrt{\dfrac{a^2 + b^2}{2}}.$$

Approximate A and P to the nearest hundredth for each ellipse.

(a) $\dfrac{x^2}{16} + \dfrac{y^2}{25} = 1$ (b) $\dfrac{x^2}{7} + \dfrac{y^2}{2} = 1$

48. Geometry of an Ellipse (Refer to the previous exercise.) If $a = b$ in the equation for an ellipse, the ellipse becomes a circle. Let $a = b$ in the formulas for the area and perimeter of an ellipse. Do the equations simplify to the area and perimeter for a circle? Explain.

49. Planet Orbit (Refer to Example 3.) Pluto's orbit is less circular than that of any of the eight planets. For Pluto (now a dwarf planet), $a = 39.44$ A.U. and $b = 38.20$ A.U.

(a) Graph the elliptic orbit of Pluto in the window $[-60, 60, 10]$ by $[-40, 40, 10]$. Plot the point $(9.81, 0)$ to show the position of the sun. Assume that the major axis is horizontal.

(b) Use the information in Exercise 47 to determine how far Pluto travels in one orbit around the sun and approximate the area inside its orbit.

50. Halley's Comet (Refer to Example 3.) The famous Halley's comet travels in an elliptical orbit with $a = 17.95$ and $b = 4.44$ and passes by Earth roughly every 76 years. The most recent pass by Earth was in February 1986. (*Source:* M. Zeilik.)

(a) Graph the orbit of Halley's comet in $[-21, 21, 5]$ by $[-14, 14, 5]$. Assume that the major axis is horizontal and that all units are in astronomical units. Plot a point at $(17.39, 0)$ to represent the position of the sun.

(b) Use the formula in Exercise 47 to estimate how many miles Halley's comet travels in one orbit around the sun.

(c) Estimate the average speed of Halley's comet in miles per hour.

51. Satellite Orbit The orbit of Explorer VII and the outline of Earth's surface are shown in the accompanying figure. This orbit is described by

$$\dfrac{x^2}{4464^2} + \dfrac{y^2}{4462^2} = 1,$$

and the surface of Earth is described by

$$\dfrac{(x - 164)^2}{3960^2} + \dfrac{y^2}{3960^2} = 1.$$

Find the maximum and minimum heights of the satellite above Earth's surface if all units are miles. (*Source:* W. Thomson, *Introduction to Space Dynamics.*)

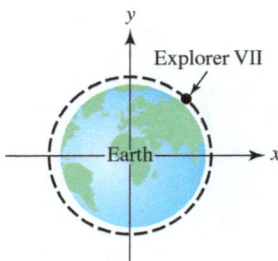

52. Weight Machines Elliptic shapes are used rather than circular shapes in some weight machines. Suppose that the ellipse shown in the figure at the top of the next column is represented by

$$\dfrac{x^2}{16} + \dfrac{y^2}{100} = 1,$$

where the units are inches. Find r_1 and r_2.

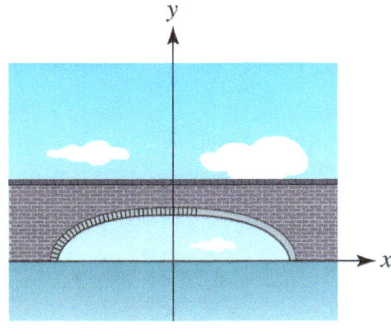

53. Arch Bridge The arch under a bridge is designed as the upper half of an ellipse, as illustrated in the accompanying figure. Its equation is modeled by

$$400x^2 + 10{,}000y^2 = 4{,}000{,}000$$

where the units are feet. Find the height and width of the arch.

54. Thinking Generally Suppose that the population y of a country can be modeled by the upper right branch of the hyperbola

$$\frac{x^2}{a^2} - \frac{y^2}{b^2} = 1,$$

where x represents time in years. What happens to the population after a long period of time?

55. Online Exploration Go online and find one application of ellipses. Write a short paragraph about your findings.

56. Online Exploration Go online and find one application of hyperbolas. Write a short paragraph about your findings.

WRITING ABOUT MATHEMATICS

57. Explain how the values of a and b affect the graph of $\frac{x^2}{a^2} + \frac{y^2}{b^2} = 1$. Assume that $a > b > 0$.

58. Explain how the values of a and b affect the graph of $\frac{x^2}{a^2} - \frac{y^2}{b^2} = 1$. Assume that a and b are positive.

SECTIONS 20.1 and 20.2 — Checking Basic Concepts

1. Graph the parabola $x = (y - 2)^2 + 1$. Find the vertex and axis of symmetry.

2. Find the equation of the circle with center $(1, -2)$ and radius 2. Graph the circle.

3. Find the x- and y-intercepts on the graph of
$$\frac{x^2}{4} + \frac{y^2}{9} = 1.$$

4. Graph the following. Label any vertices and state the type of conic section that it represents.

   (a) $x = y^2$

   (b) $\frac{x^2}{16} + \frac{y^2}{25} = 1$

   (c) $\frac{x^2}{4} - \frac{y^2}{9} = 1$

   (d) $(x - 1)^2 + (y + 2)^2 = 9$

20.3 Nonlinear Systems of Equations and Inequalities

Objectives

1. Solving Nonlinear Systems of Equations
   - Solutions to Nonlinear Systems
   - Method of Substitution
   - Graphical and Symbolic Solutions
2. Solving Nonlinear Systems of Inequalities
   - Recognizing the Solution Set
   - Graphical Solutions

NEW VOCABULARY

☐ Nonlinear system of equations
☐ Method of substitution
☐ Nonlinear system of inequalities

1 Solving Nonlinear Systems of Equations

SOLUTIONS TO NONLINEAR SYSTEMS One way to locate the points at which the line $y = 2x$ intersects the circle $x^2 + y^2 = 5$ is to graph both equations (see **FIGURE 20.21**).

The equation describing the circle is nonlinear. Another way to locate the points of intersection is symbolically, by solving the *nonlinear system of equations*.

$$y = 2x \quad \text{Line}$$
$$x^2 + y^2 = 5 \quad \text{Circle}$$

Linear systems of equations can have no solutions, one solution, or infinitely many solutions. It is possible for a nonlinear system of equations to have *any number* of solutions. **FIGURE 20.21** shows that this nonlinear system of equations has two solutions: $(-1, -2)$ and $(1, 2)$.

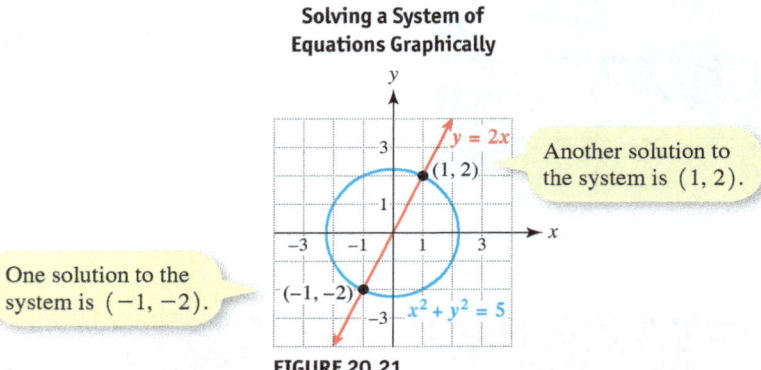

FIGURE 20.21

METHOD OF SUBSTITUTION A symbolic technique for solving a system of equations is the **method of substitution**, which we demonstrate in the next example.

EXAMPLE 1 Solving a nonlinear system of equations symbolically

Solve the following system of equations symbolically. Check any solutions and compare them with **FIGURE 20.21**.

$$y = 2x$$
$$x^2 + y^2 = 5$$

Solution
Substitute $2x$ for y in the second equation $x^2 + y^2 = 5$ and solve for x.

$$x^2 + (2x)^2 = 5 \quad \text{Let } y = 2x \text{ in the second equation.}$$
$$x^2 + 4x^2 = 5 \quad \text{Properties of exponents}$$
$$5x^2 = 5 \quad \text{Combine like terms.}$$
$$x^2 = 1 \quad \text{Divide by 5.}$$
$$x = \pm 1 \quad \text{Square root property}$$

To determine corresponding y-values, substitute $x = \pm 1$ in $y = 2x$; the solutions are $(1, 2)$ and $(-1, -2)$. To check $(1, 2)$, substitute $x = 1$ and $y = 2$ in the given equations.

$$2 \stackrel{?}{=} 2(1) \checkmark \quad \text{True}$$
$$(1)^2 + (2)^2 \stackrel{?}{=} 5 \checkmark \quad \text{True}$$

To check $(-1, -2)$, substitute $x = -1$ and $y = -2$ in the given equations.

$$-2 \stackrel{?}{=} 2(-1) \checkmark \quad \text{True}$$
$$(-1)^2 + (-2)^2 \stackrel{?}{=} 5 \checkmark \quad \text{True}$$

The solutions check and agree with **FIGURE 20.21**.

Now Try Exercise 13

GRAPHICAL AND SYMBOLIC SOLUTIONS In the next example, we solve a nonlinear system of equations graphically and symbolically. The symbolic solution gives the exact solutions.

EXAMPLE 2 Solving a nonlinear system of equations

Solve the nonlinear system of equations graphically and symbolically.

$$x^2 - y = 2$$
$$x^2 + y = 4$$

Solution

Graphical Solution Begin by solving each equation for y.

$$y = x^2 - 2$$
$$y = 4 - x^2$$

Graph $y_1 = x^2 - 2$ and $y_2 = 4 - x^2$. The solutions are approximately $(-1.73, 1)$ and $(1.73, 1)$, as shown in **FIGURE 20.22**. The graphs consist of two parabolas intersecting at two points.

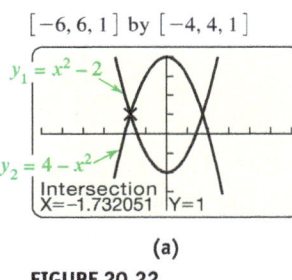
(a)

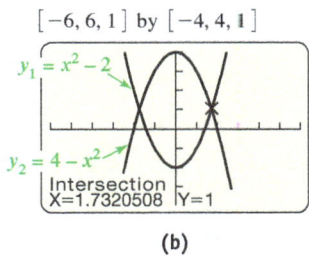
(b)

FIGURE 20.22

Symbolic Solution Solving the first equation for y gives $y = x^2 - 2$. Substitute this expression for y in the second equation and solve for x.

$$x^2 + y = 4 \qquad \text{Second equation}$$
$$x^2 + (x^2 - 2) = 4 \qquad \text{Substitute } y = x^2 - 2.$$
$$2x^2 = 6 \qquad \text{Combine like terms; add 2.}$$
$$x^2 = 3 \qquad \text{Divide by 2.}$$
$$x = \pm\sqrt{3} \qquad \text{Square root property}$$

To determine y, substitute $x = \pm\sqrt{3}$ in $y = x^2 - 2$.

$$y = (\sqrt{3})^2 - 2 = 3 - 2 = 1$$
$$y = (-\sqrt{3})^2 - 2 = 3 - 2 = 1$$

The *exact* solutions are $(\sqrt{3}, 1)$ and $(-\sqrt{3}, 1)$.

Now Try Exercise 29(a), (b)

EXAMPLE 3 Solving a nonlinear system of equations

Solve the nonlinear system of equations symbolically and graphically.

$$x^2 - y^2 = 3$$
$$x^2 + y^2 = 5$$

Solution

Symbolic Solution Instead of using substitution on this nonlinear system of equations, we use elimination. Note that, if we add the two equations, the y-variable will be eliminated.

$$\begin{aligned} x^2 - y^2 &= 3 \quad \text{First equation} \\ x^2 + y^2 &= 5 \quad \text{Second equation} \\ \hline 2x^2 &= 8 \quad \text{Add equations.} \end{aligned}$$

Solving gives $x^2 = 4$, or $x = \pm 2$. To determine y, substitute **4** for x^2 in $x^2 + y^2 = 5$.

$$4 + y^2 = 5 \quad \text{or} \quad y^2 = 1$$

Because $y^2 = 1$, $y = \pm 1$. There are four solutions: $(2, 1)$, $(2, -1)$, $(-2, 1)$, and $(-2, -1)$.

Graphical Solution The graph of the first equation is a hyperbola, and the graph of the second is a circle with radius $\sqrt{5}$. The four points of intersection (solutions) are $(\pm 2, \pm 1)$, as shown in **FIGURE 20.23**.

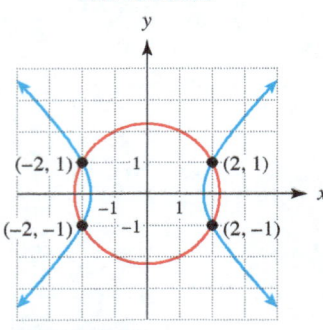

Four Solutions

FIGURE 20.23

Now Try Exercises 23, 25

Cylindrical Container

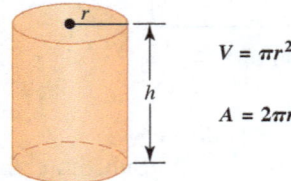

$V = \pi r^2 h$

$A = 2\pi r h$

FIGURE 20.24

🌐 **Math in Context** (Manufacturing) To describe characteristics of curved objects, we often need *nonlinear equations*. The equations of the conic sections discussed in this chapter are but a few examples of nonlinear equations. For instance, cylinders have a curved shape, as illustrated in **FIGURE 20.24**. If the radius of a cylinder is denoted r and its height h, then its volume V is given by the nonlinear equation $V = \pi r^2 h$ and its side area A is given by the nonlinear equation $A = 2\pi r h$.

If we want to manufacture a cylindrical container that holds 35 cubic inches and whose side area is 50 square inches, we need to solve the following nonlinear system of equations.

> Solve to find the r and h that give a volume of 50 in³ and an area of 35 in².

$$\pi r^2 h = 35 \quad \text{The side area is 35 in}^2$$
$$2\pi r h = 50 \quad \text{The volume is 50 in}^3$$

EXAMPLE 4 Modeling the dimensions of a cylindrical container

Find the dimensions of a cylindrical container that has a volume V of 35 cubic inches and a side area A of 50 square inches by solving the following nonlinear system of equations.

$$\pi r^2 h = 35 \quad \text{The side area is 35 in}^2$$
$$2\pi r h = 50 \quad \text{The volume is 50 in}^3$$

Solution

We can find r by solving each equation for h and setting the results equal to each other. This eliminates the variable h.

$$\frac{50}{2\pi r} = \frac{35}{\pi r^2} \qquad h = \tfrac{50}{2\pi r} \text{ and } h = \tfrac{35}{\pi r^2}$$

$$50\pi r^2 = 70\pi r \qquad \text{Eliminate fractions (cross multiply).}$$

$$50\pi r^2 - 70\pi r = 0 \qquad \text{Subtract } 70\pi r.$$

$$10\pi r(5r - 7) = 0 \qquad \text{Factor out } 10\pi r.$$

$$10\pi r = 0 \quad \text{or} \quad 5r - 7 = 0 \qquad \text{Zero-product property}$$

$$r = 0 \quad \text{or} \quad r = \tfrac{7}{5} = 1.4 \qquad \text{Solve.}$$

Because $h = \frac{50}{2\pi r}$, $r = 0$ is not possible, but we can find h by substituting 1.4 for r in the formula.

$$h = \frac{50}{2\pi(1.4)} \approx 5.68$$

A cylindrical container that has a volume of 35 cubic inches and a side area of 50 square inches has a radius of 1.4 inches and a height of about 5.68 inches.

Now Try Exercise 51(b)

READING CHECK 1

- How many solutions can a nonlinear system of equations have?

2 Solving Nonlinear Systems of Inequalities

RECOGNIZING THE SOLUTION SET Previously in this text, we solved systems of *linear* inequalities. A **nonlinear system of inequalities** in two variables can be solved similarly by using graphical techniques. For example, consider the nonlinear system of inequalities

$$y \geq x^2 - 2$$
$$y \leq 4 - x^2.$$

The graph of $y = x^2 - 2$ is a parabola opening upward. The solution set to $y \geq x^2 - 2$ includes all points lying on or above this parabola. See **FIGURE 20.25(a)**. Similarly, the graph of $y = 4 - x^2$ is a parabola opening downward. The solution set to $y \leq 4 - x^2$ includes all points lying on or below this parabola. See **FIGURE 20.25(b)**.

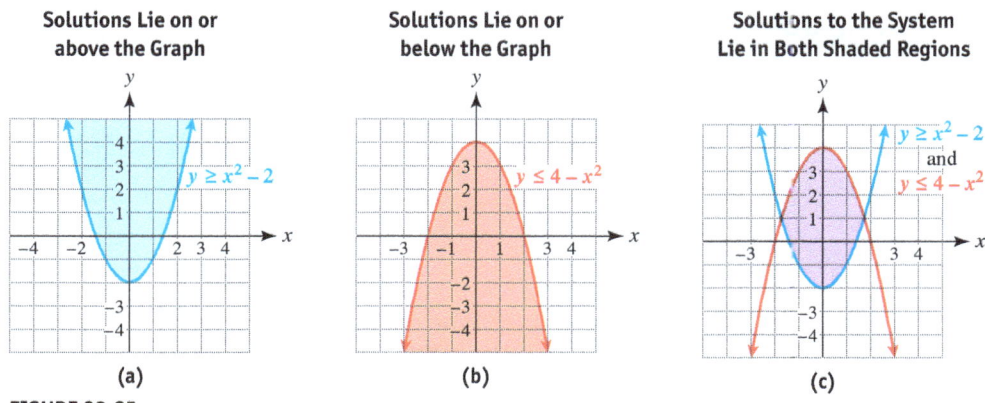

FIGURE 20.25

The solution set for this nonlinear *system* of inequalities includes all points (x, y) in *both* shaded regions. The *intersection* of the shaded regions is shown in **FIGURE 20.25(c)**.

GRAPHICAL SOLUTIONS In the next three examples, graphical solutions are found by shading the solution set in the xy-plane.

EXAMPLE 5   Solving a nonlinear system of inequalities graphically

Shade the solution set for the system of inequalities.

$$\frac{x^2}{4} + \frac{y^2}{9} < 1$$
$$y > 1$$

Solution

The solutions to $\frac{x^2}{4} + \frac{y^2}{9} < 1$ lie *inside* the ellipse $\frac{x^2}{4} + \frac{y^2}{9} = 1$. See **FIGURE 20.26(a)**. Solutions to $y > 1$ lie above the line $y = 1$, as shown in **FIGURE 20.26(b)**. The intersection of these two regions is shown in **FIGURE 20.26(c)**. Any point in this region is a solution.

Solutions Lie inside the Ellipse

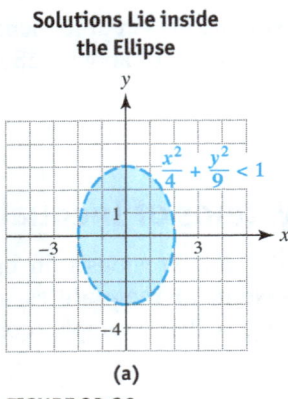

Solutions Lie above the Line

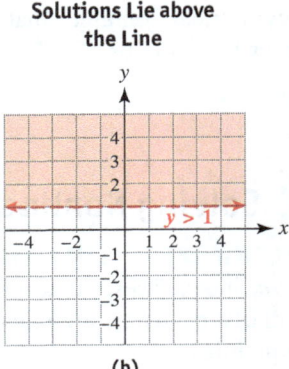

Solutions Lie inside the Ellipse *and* above the Line

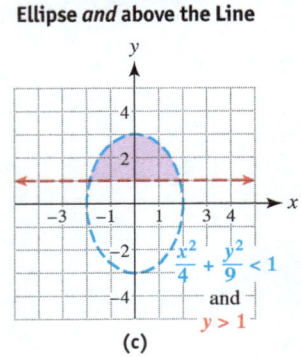

(a) (b) (c)

FIGURE 20.26

For example, the point $(0, 2)$ lies in the shaded region and is a solution to the system. Note that a dashed curve and a dashed line are used when equality is *not* included.

Now Try Exercise 39

EXAMPLE 6   Solving a nonlinear system of inequalities graphically

Shade the solution set for the following system of inequalities.

$$x^2 + y \leq 4$$
$$-x + y \geq 2$$

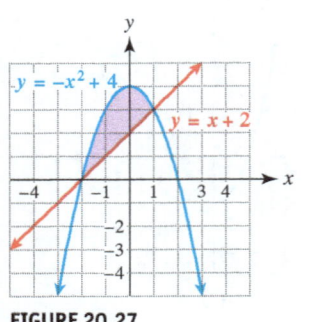

FIGURE 20.27

Solution

The solutions to $x^2 + y \leq 4$ lie *on* or *below* the parabola $y = -x^2 + 4$, and the solutions to $-x + y \geq 2$ lie *on* or *above* the line $y = x + 2$. The appropriate shaded region is shown in **FIGURE 20.27**. Both the parabola and the line are solid because equality is included in both inequalities.

Now Try Exercise 37

In the next example, we use a graphing calculator to shade a region that lies above both graphs, using the "Y$_1$ =" menu. This feature allows us to shade either above or below the graph of a function.

EXAMPLE 7 Solving a system of inequalities with a graphing calculator

Shade the solution set for the following system of inequalities.

$$y \geq x^2 - 2$$
$$y \geq -1 - x$$

Solution

Enter $y_1 = x^2 - 2$ and $y_2 = -1 - x$, as shown in **FIGURE 20.28(a)**. Note that the option to shade above the graphs of Y_1 and Y_2 are selected to the left of Y_1 and Y_2. Then the two inequalities are graphed in **FIGURE 20.28(b)**. The solution set corresponds to the region where there is both vertical and horizontal shading.

CALCULATOR HELP

To shade the solution set to a system of inequalities, see Appendix A (pages AP-6 and AP-7).

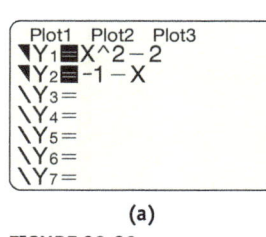
(a)

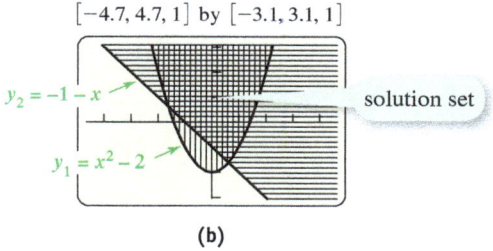
(b)

FIGURE 20.28

Now Try Exercise 49

20.3 Putting It All Together

CONCEPT	EXPLANATION
Nonlinear Systems of Equations in Two Variables	To solve the following system of equations symbolically, using *substitution*, solve the first equation for y to get $y = 5 - x$. $x + y = 5$ First equation $x^2 - y = 1$ Second equation Substitute $5 - x$ for y in the second equation and solve the resulting quadratic equation. (*Elimination* can also be used on this system.) Then $x^2 - (5 - x) = 1$ or $x^2 + x - 6 = 0$. Factoring can be used to determine that the solutions are given by $x = -3$ or $x = 2$. Thus $y = 5 - (-3) = 8$ or $y = 5 - 2 = 3$. The solutions are $(-3, 8)$ and $(2, 3)$. Graphical support is shown in the accompanying figure. **Two Solutions** 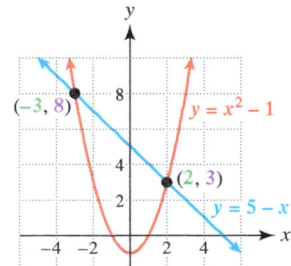 Nonlinear systems of equations can have *any number of solutions*.

continued on next page

1324 CHAPTER 20 CONIC SECTIONS

continued from previous page

CONCEPT	EXPLANATION
Nonlinear Systems of Inequalities in Two Variables	To solve the following system of inequalities graphically, solve each inequality for y. $x + y \leq 5$ or $y \leq 5 - x$ **First inequality** $x^2 - y \leq 1$ or $y \geq x^2 - 1$ **Second inequality** The solutions lie on or below the line and on or above the parabola, as shown in the figure. **Infinitely Many Solutions** The solution set to a nonlinear system of inequalities can be represented by a shaded region in the xy-plane.

20.3 Exercises

CONCEPTS AND VOCABULARY

1. How many solutions can a nonlinear system of equations have?

2. If a nonlinear system of equations has two equations, how many equations does a solution have to satisfy?

3. Determine visually the number of solutions to the following system of equations. Explain your reasoning.

$$y = x$$
$$x^2 + y^2 = 4$$

4. Describe the solution set to $x^2 + y^2 \leq 1$.

5. Does $(-2, -1)$ satisfy $5x^2 - 2y^2 > 18$?

6. Does $(3, 4)$ satisfy $x^2 - 2y \geq 4$?

7. Sketch a parabola and ellipse with four points of intersection.

8. Sketch a line and a hyperbola with two points of intersection.

SOLVING NONLINEAR SYSTEMS OF EQUATIONS

Exercises 9–12: Use the graph to estimate all solutions to the system of equations. Check each solution.

9. $x^2 + y^2 = 10$
$y = 3x$

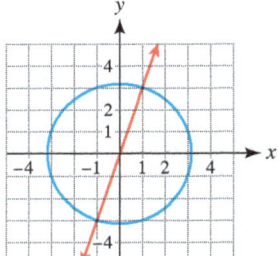

10. $x^2 + 3y^2 = 16$
$y = -x$

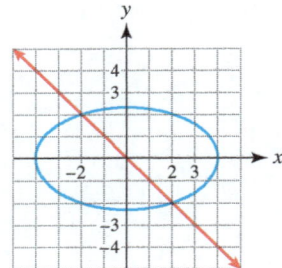

11. $y^2 - x^2 = 1$
$x^2 + 3y^2 = 3$

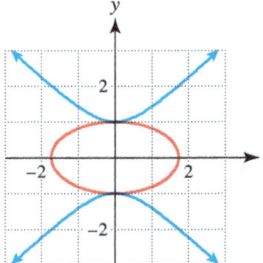

12. $y = 1 - x^2$
$x^2 + y^2 = 1$

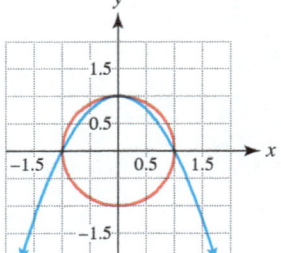

Exercises 13–24: Solve the system of equations symbolically. Check your solutions.

13. $y = 2x$
    $x^2 + y^2 = 45$

14. $y = x$
    $y^2 = 3 - 2x^2$

15. $x + y = 1$
    $x^2 - y^2 = 3$

16. $y - x = -1$
    $y = 2x^2$

17. $y - x^2 = 0$
    $x^2 + y^2 = 6$

18. $x^2 - y^2 = 4$
    $x^2 + y^2 = 4$

19. $3x^2 + 2y^2 = 5$
    $x - y = -2$

20. $x^2 + 2y^2 = 9$
    $2x^2 - y^2 = -2$

21. $x^2 + y^2 = 4$
    $x^2 - 9y^2 = 9$

22. $x^2 + y^2 = 15$
    $9x^2 + 4y^2 = 36$

23. $x^2 + y^2 = 10$
    $2x^2 - y^2 = 17$

24. $2x^2 + 3y^2 = 5$
    $3x^2 - 4y^2 = -1$

Exercises 25–28: Solve the system of equations graphically. Check your solutions.

25. $y = x^2 - 3$
    $2x^2 - y = 1 - 3x$

26. $x + y = 2$
    $x - y^2 = 3$

27. $y - x = -4$
    $x - y^2 = -2$

28. $xy = 1$
    $y = x$

Exercises 29–32: Solve the system of equations
(a) *symbolically,*
(b) *graphically, and*
(c) *numerically.*

29. $y = -2x$
    $x^2 + y = 3$

30. $4x - y = 0$
    $x^3 - y = 0$

31. $xy = 1$
    $x - y = 0$

32. $x^2 + y^2 = 4$
    $y - x = 2$

SOLVING NONLINEAR SYSTEMS OF INEQUALITIES

Exercises 33–36: Shade the solution set in the xy-plane.

33. $y \geq x^2$

34. $y \leq x^2 - 1$

35. $\dfrac{x^2}{4} + \dfrac{y^2}{9} > 1$

36. $x^2 + y^2 \leq 1$

Exercises 37–44: Shade the solution set in the xy-plane. Then use the graph to select one solution.

37. $y > x^2 + 1$
    $y < 3$

38. $y > x^2$
    $y < x + 2$

39. $x^2 + y^2 \leq 1$
    $y < x$

40. $y > x^2 - 2$
    $y \leq 2 - x^2$

41. $x^2 + y^2 \leq 1$
    $(x - 2)^2 + y^2 \leq 1$

42. $x^2 - y \geq 2$
    $(x + 1)^2 + y^2 \leq 4$

43. $x^2 - y^2 \leq 4$
    $x^2 + y^2 \leq 9$

44. $3x + 2y < 6$
    $x^2 + y^2 \leq 16$

Exercises 45 and 46: Match the inequality or system of inequalities with its graph (a or b).

45. $y \leq \dfrac{1}{2}x^2$

46. $y \geq x^2 + 1$
    $y \leq 5$

a.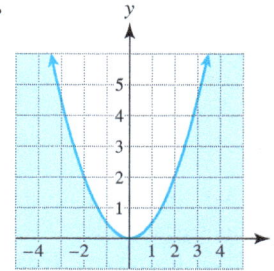

b.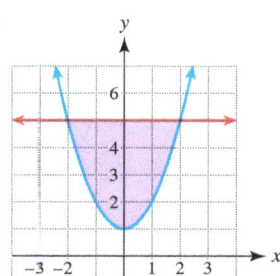

Exercises 47 and 48: Use the graph to write the inequality or system of inequalities.

47.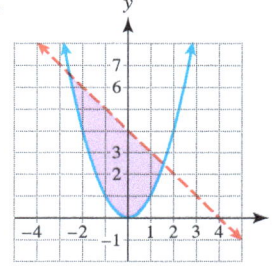

48.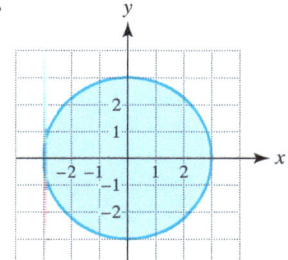

Exercises 49 and 50: Use a graphing calculator to shade the solution set to the system of inequalities.

49. $y \geq x^2 - 1$
    $y \geq 2 - x$

50. $y \leq 4 - x^2$
    $y \leq x^2 - 4$

APPLICATIONS OF NONLINEAR SYSTEMS

51. **Dimensions of a Can** (Refer to Example 4.) Find the dimensions of a cylindrical container with a volume of 40 cubic inches and a side area of 50 square inches (a) graphically and (b) symbolically.

52. **Dimensions of a Can** (Refer to Example 4.) Is it possible to design an aluminum can with volume of 60 cubic inches and side area of 60 square inches? If so, find the dimensions of the can.

53. **Area and Perimeter** The area of a room is 143 square feet, and its perimeter is 48 feet. Let x be the width and y be the length of the room. See the accompanying figure on the next page.
    (a) Write a nonlinear system of equations that models this situation.

continued on next page

(b) Solve the system.

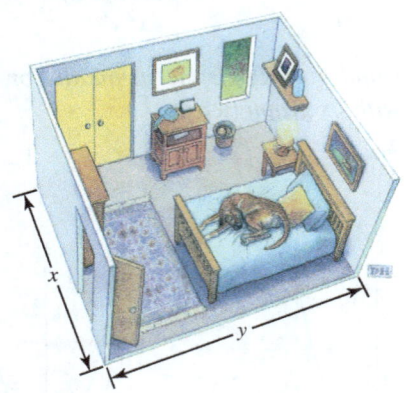

54. Dimensions of a Cone The volume V of a cone is given by $V = \frac{1}{3}\pi r^2 h$, and the surface area S of its side is given by $S = \pi r \sqrt{r^2 + h^2}$, where h is the height and r is the radius of the base. See the accompanying figure.

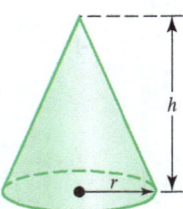

(a) Solve each equation for h.

(b) Estimate r and h graphically for a cone with volume V of 34 cubic feet and surface area S of 52 square feet.

WRITING ABOUT MATHEMATICS

55. A student *incorrectly* changes the following system of inequalities

$$x^2 - y \geq 6$$
$$2x - y \leq -3$$

to

$$y \geq x^2 - 6$$
$$y \leq 2x + 3.$$

The student discovers that $(1, 2)$ satisfies the second system of inequalities but not the first. Explain the student's error.

56. Explain graphically how nonlinear systems of equations can have any number of solutions. Sketch graphs of different systems with zero, one, two, and three solutions.

SECTION 20.3 **Checking Basic Concepts**

1. Solve the following system of equations symbolically and graphically.

$$x^2 - y = 2x$$
$$2x - y = 3$$

2. Determine visually the number of solutions to the following system of equations.

$$y = x^2 - 4$$
$$y = x$$

3. The solution set for a system of inequalities is shown in the accompanying figure to the right.
   (a) Find one ordered pair (x, y) that is a solution and one that is not.
   (b) Write the system of inequalities represented by the graph.

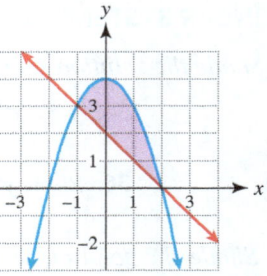

4. Shade the solution set for the following system of inequalities.

$$x^2 + y^2 \leq 4$$
$$y < 1$$

CHAPTER 20 Summary

SECTION 20.1 ■ PARABOLAS AND CIRCLES

Types of Conic Sections There are three basic types of conic sections: parabolas, ellipses, and hyperbolas.

Parabolas A parabola can have a vertical or a horizontal axis of symmetry. Two forms of an equation for a parabola with a *vertical* axis of symmetry are

$$y = ax^2 + bx + c \quad \text{and} \quad y = a(x - h)^2 + k.$$

If $a > 0$, the parabola opens upward, and if $a < 0$, it opens downward. The vertex is located at (h, k). Two forms of an equation for a parabola with a *horizontal* axis of symmetry are

$$x = ay^2 + by + c \quad \text{and} \quad x = a(y - k)^2 + h.$$

If $a > 0$, the parabola opens to the right, and if $a < 0$, it opens to the left. The vertex is located at (h, k).

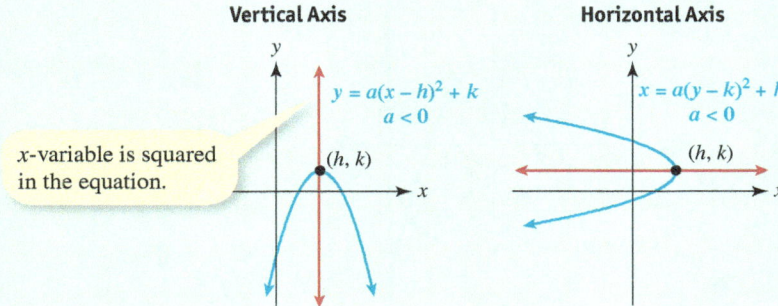

Example: $x = -2(y - 4)^2 + 1$ has vertex $(1, 4)$, has axis of symmetry $y = 4$, and opens to the left.

Circles The standard equation for a circle with center (h, k) and radius r is

$$(x - h)^2 + (y - k)^2 = r^2.$$

Example: $(x + 2)^2 + (y - 3)^2 = 36$ has center $(-2, 3)$ and radius 6.

SECTION 20.2 ■ ELLIPSES AND HYPERBOLAS

Ellipses

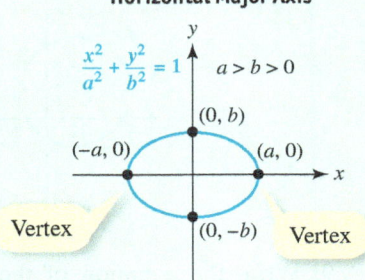
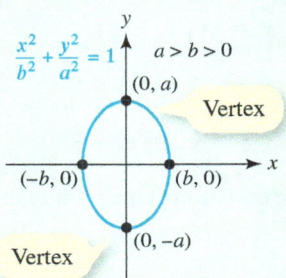

Example: $\frac{x^2}{9} + \frac{y^2}{4} = 1$ is the standard equation for an ellipse with vertices $(\pm 3, 0)$, centered at $(0, 0)$.

Hyperbolas

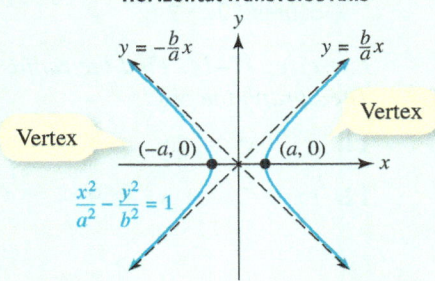
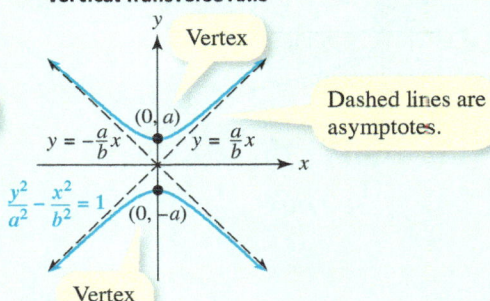

continued on next page

Example: $\frac{x^2}{9} - \frac{y^2}{4} = 1$ is the standard equation for a hyperbola with horizontal transverse axis, vertices $(\pm 3, 0)$, center $(0, 0)$, and asymptotes $y = \pm \frac{2}{3}x$.

SECTION 20.3 ■ NONLINEAR SYSTEMS OF EQUATIONS AND INEQUALITIES

Nonlinear Systems Nonlinear systems of equations can have any number of solutions. The methods of substitution and elimination can often be used to solve a nonlinear system of equations symbolically. Nonlinear systems can also be solved graphically. The solution set for a nonlinear system of two inequalities in two variables is typically a region in the xy-plane. A solution is an ordered pair (x, y) that satisfies both inequalities.

Example: Solve $y \geq x^2 - 2$
$$y \leq 4 - \frac{1}{2}x^2.$$

Nonlinear System of Inequalities

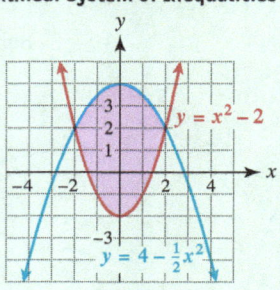

CHAPTER 20   Review Exercises

SECTION 20.1

Exercises 1–6: Graph the parabola. Find the vertex and axis of symmetry.

1. $x = 2y^2$
2. $x = -(y + 1)^2$
3. $x = -2(y - 2)^2$
4. $x = (y + 2)^2 - 1$
5. $x = -3y^2 + 1$
6. $x = \frac{1}{2}y^2 + y - 3$

7. Use the graph to determine the equation of the parabola.

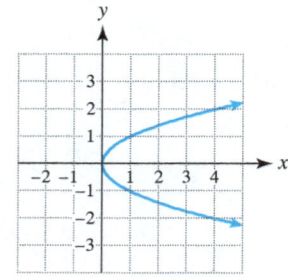

8. Use the graph to find the equation of the circle.

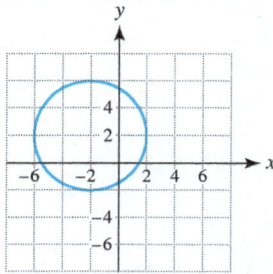

9. Write the equation of the circle with radius 1 and center $(0, 0)$.

10. Write the equation of the circle with radius 4 and center $(2, -3)$.

Exercises 11–14: Find the radius and center of the circle. Then graph the circle.

11. $x^2 + y^2 = 25$
12. $(x - 2)^2 + y^2 = 9$

13. $(x + 3)^2 + (y - 1)^2 = 5$

14. $x^2 - 2x + y^2 + 2y = 7$

SECTION 20.2

Exercises 15–18: Graph the ellipse. Label the vertices and endpoints of the minor axis.

15. $\dfrac{x^2}{4} + \dfrac{y^2}{25} = 1$

16. $x^2 + \dfrac{y^2}{4} = 1$

17. $25x^2 + 20y^2 = 500$

18. $4x^2 + 9y^2 = 36$

19. Use the graph to determine the equation of the ellipse.

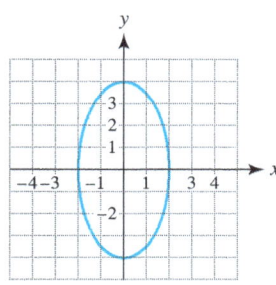

20. Use the graph to find the equation of the hyperbola.

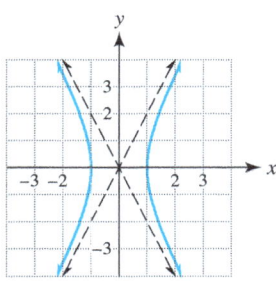

Exercises 21–24: Graph the hyperbola and include the asymptotes in your graph.

21. $\dfrac{x^2}{9} - \dfrac{y^2}{4} = 1$

22. $\dfrac{y^2}{25} - \dfrac{x^2}{16} = 1$

23. $y^2 - x^2 = 1$

24. $25x^2 - 16y^2 = 400$

SECTION 20.3

Exercises 25–28: Use the graph to estimate all solutions to the system of equations. Check each solution.

25. $x^2 + y^2 = 9$
    $x + y = 3$

26. $xy = 2$
    $y = 2x$

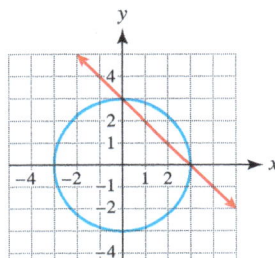
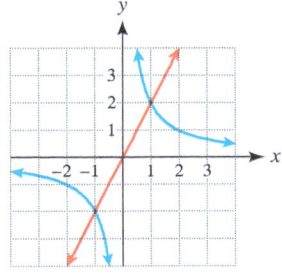

27. $x^2 - y = x$
    $y = x$

28. $x^2 + y^2 = 5$
    $x^2 - y^2 = 3$

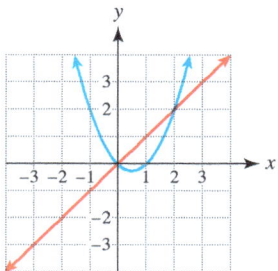
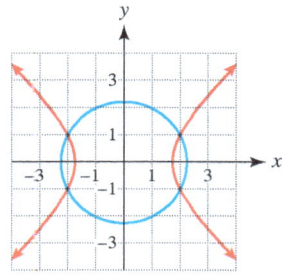

Exercises 29–32: Solve the system of equations. Check your solutions.

29. $y = x$
    $x^2 + y^2 = 32$

30. $x - y = 4$
    $x^2 + y^2 = 16$

31. $y = x^2$
    $2x^2 + y = 3$

32. $y = x^2 + 1$
    $2x^2 - y = 3x - 3$

Exercises 33 and 34: Solve the system graphically.

33. $2x - y = 4$
    $x^2 + y = 4$

34. $x^2 + y = 0$
    $x^2 + y^2 = 2$

Exercises 35 and 36: Solve the system of equations
  (a) *symbolically,*
  (b) *graphically, and*
  (c) *numerically.*

35. $y = x$
    $x^2 + 2y = 8$

36. $y = x^3$
    $x^2 - y = 0$

Exercises 37–44: Shade the solution set in the xy-plane.

37. $y \geq 2x^2$

38. $y < 2x - 3$

39. $y < -x^2$

40. $\dfrac{x^2}{9} + \dfrac{y^2}{16} \leq 1$

41. $y - x^2 \geq 1$
    $y \leq 2$

42. $x^2 + y \leq 4$
    $3x + 2y \geq 6$

43. $y > x^2$
    $y < 4 - x^2$

44. $\dfrac{x^2}{4} + \dfrac{y^2}{9} > 1$
    $x^2 + y^2 < 16$

Exercises 45 and 46: Use the graph to write the system of inequalities.

45.

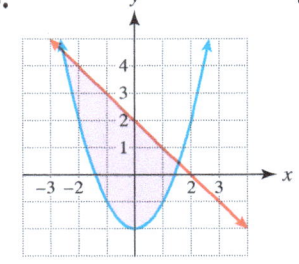

46.

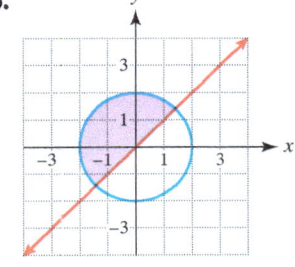

APPLICATIONS

47. Area and Perimeter The area of a desktop is 1000 square inches, and its perimeter is 130 inches. Let x be the width and y be the length of the desktop. See the figure below.
  (a) Write a system of equations that models this situation.
  (b) Solve the system graphically.
  (c) Solve the system symbolically.

48. Numbers The product of two positive numbers is 60, and their difference is 7. Let x be the smaller number and y be the larger number.
  (a) Write a system of equations whose solution gives the two numbers.
  (b) Solve the system graphically.
  (c) Solve the system symbolically.

49. Dimensions of a Container The volume of a cylindrical container is $V = \pi r^2 h$, and its surface area, *excluding* the top and bottom, is $A = 2\pi rh$. Find the dimensions of a container with $A = 100$ square feet and $V = 50$ cubic feet. Is your answer unique?

50. Dimensions of a Container The volume of a cylindrical container is $V = \pi r^2 h$, and its surface area, *including* the top and bottom, is $A = 2\pi rh + 2\pi r^2$. Graphically find the dimensions of a container with $A = 80$ square inches and $V = 35$ cubic inches. Is your answer unique?

51. Geometry of an Ellipse The area inside an ellipse is given by $A = \pi ab$, and its perimeter P can be approximated by
$$P = 2\pi\sqrt{\frac{a^2 + b^2}{2}}.$$
  (a) Graph $\dfrac{x^2}{5} + \dfrac{y^2}{12} = 1$.
  (b) Estimate the ellipse's area and perimeter.

52. Orbit of Mars Mars has an elliptical orbit that is nearly circular, with $a = 1.524$ and $b = 1.517$, where the units are astronomical units (1 A.U. equals 93 million miles). (*Source:* M. Zeilik.)
  (a) Graph the orbit of Mars in $[-3, 3, 1]$ by $[-2, 2, 1]$. Plot the point $(0.15, 0)$ to show the position of the sun. Assume that the major axis is horizontal.
  (b) Use the information in Exercise 51 to estimate how far Mars travels in one orbit around the sun. Approximate the area inside its orbit.

CHAPTER 20 Test

1. Graph the parabola $y = -(x - 1)^2 + 2$. Find the vertex and axis of symmetry.

2. Graph the parabola $x = (y - 4)^2 - 2$. Find the vertex and axis of symmetry.

3. Use the graph to find the equation of the parabola.

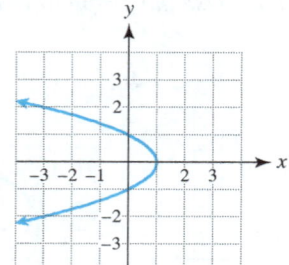

4. Use the graph to find the equation of the circle.

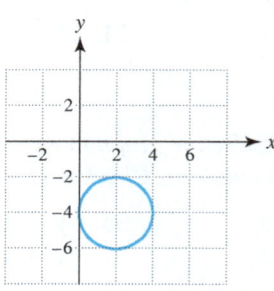

5. Write the equation of the circle with radius 10 and center $(-5, 2)$.

6. Find the radius and center of the circle given by
$$x^2 + 4x + y^2 - 6y = 3.$$
Then graph the circle.

7. Graph the ellipse $\frac{x^2}{16} + \frac{y^2}{49} = 1$. Label the vertices and endpoints of the minor axis.

8. Use the graph to determine the standard equation of the ellipse.

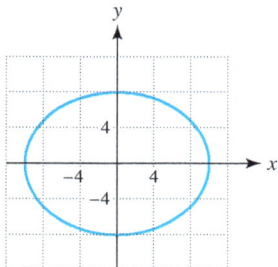

9. Graph the hyperbola $4x^2 - 9y^2 = 36$. Include the asymptotes in your graph.

10. Use the graph to estimate all solutions to the system of equations. Check each solution by substitution in the system of equations.

$$x^2 + y^2 = 16$$
$$x - y = 4$$

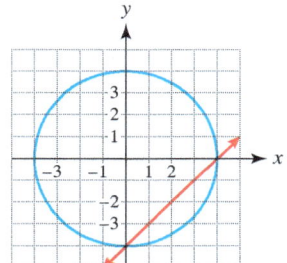

11. Solve the system of equations symbolically.

$$x - y = 3$$
$$x^2 + y^2 = 17$$

12. Solve the system of equations graphically.

$$2x^2 - y = 4$$
$$x^2 + y = 8$$

13. Shade the solution set in the xy-plane.

$$3x + y > 6$$
$$x^2 + y^2 < 25$$

14. Use the graph to write the system of inequalities.

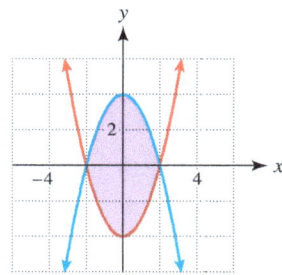

15. **Area and Perimeter** The area of a rectangular swimming pool is 5000 square feet, and its perimeter is 300 feet. Let x be the width and y be the length of the pool.
    (a) Write a nonlinear system of equations that models this situation.
    (b) Solve the system.

16. **Dimensions of a Box** The volume of a rectangular box with a square bottom and open top is $V = x^2y$, and its surface area is $A = x^2 + 4xy$, where x represents its width and length and y represents its height. Estimate graphically the dimensions of a box with $V = 1183$ cubic inches and $A = 702$ square inches. Is your answer unique? (*Hint:* Substitute appropriate values for V and A, and then solve each equation for y.)

17. **Orbit of Uranus** The planet Uranus has an elliptical orbit that is nearly circular, with $a = 19.18$ and $b = 19.16$, where the units are astronomical units (1 A.U. equals 93 million miles). (*Source:* M. Zeilik.)
    (a) Graph the orbit of Uranus in $[-30, 30, 10]$ by $[-20, 20, 10]$. Plot the point $(0.9, 0)$ to show the position of the sun. Assume that the major axis is horizontal.
    (b) Find the minimum distance between Uranus and the sun.

CHAPTERS 1–20 Cumulative Review Exercises

1. Evaluate $K = x^2 + y^2$ when $x = 4$ and $y = -3$.

2. Simplify $\dfrac{(a^{-2}b)^2}{a^{-1}(b^3)^{-2}}$.

3. Write 7.345×10^{-3} in standard (decimal) notation.

4. Find $f(-4)$, if $f(x) = \dfrac{x}{x-4}$. State the domain of f.

5. If the point $(3, 4)$ is on the graph of $y = f(x)$, then $f(\underline{}) = \underline{}$.

Exercises 6 and 7: Graph f by hand.

6. $f(x) = x^2 - x$
7. $f(x) = 1 - 2x$

8. Find the slope–intercept form of the line that passes through the point $(2, -2)$ and is perpendicular to the line given by $y = -\frac{2}{3}x + 1$.

9. Solve $2(1 - x) - 4x = x$.

Exercises 10–12: Solve the inequality. Write your answer in interval notation.

10. $-5 \leq 1 - 2x < 3$ 11. $x^2 - 4 \leq 0$

12. $|1 - x| \geq 2$

13. Solve the system.
$$-2x + y = 1$$
$$5x - y = 2$$

14. Shade the solution set in the *xy*-plane.
$$x + y \leq 4$$
$$x - y \geq 2$$

Exercises 15 and 16: Multiply the expression.

15. $(2x - 1)(x + 5)$ 16. $xy(2x - 3y^2 + 1)$

Exercises 17 and 18: Factor the expression.

17. $6x^2 - 13x - 5$ 18. $x^3 - 4x$

Exercises 19 and 20: Solve the equation.

19. $x^2 + 3x + 2 = 0$ 20. $x^2 + 1 = -3x$

Exercises 21 and 22: Simplify the expression.

21. $\dfrac{x - 2}{x + 2} \div \dfrac{2x - 4}{3x + 6}$ 22. $\dfrac{1}{x + 1} + \dfrac{1}{x - 1}$

Exercises 23–26: Simplify the expression. Assume all variables are positive.

23. $\sqrt{8x^2}$ 24. $8^{2/3}$

25. $\sqrt[3]{2x} \cdot \sqrt[3]{32x^2}$ 26. $3\sqrt{3x} + \sqrt{12x}$

27. Graph $f(x) = -\sqrt{x}$.

28. Find the distance between $(2, -3)$ and $(-2, 0)$.

29. Write $(2 + 3i)(2 - 3i)$ in standard form.

30. Solve $\sqrt{x + 2} = x$.

31. Find the vertex of the graph of $y = x^2 - 6x + 3$.

32. Write $f(x) = x^2 - 2x + 3$ in vertex form.

33. Compare the graph of $f(x) = \sqrt{x - 4}$ to the graph of $y = \sqrt{x}$.

34. Solve $x(3 - x) = 2$.

35. Solve $x^3 + x = 0$. Find all complex solutions.

36. Simplify by hand.
    (a) $\log 10{,}000$ (b) $\log_2 8$
    (c) $\log_3 3^x$ (d) $e^{\ln 6}$
    (e) $\log 2 + \log 50$ (f) $\log_2 24 - \log_2 3$

37. If $f(x) = x^2 + 1$ and $g(x) = 2x$, find the following.
    (a) $(f \circ g)(2)$ (b) $(g \circ f)(x)$

38. If $f(x) = 2 - 3x$, find $f^{-1}(x)$.

39. If a total of $1000 is deposited in an account that pays 5% annual interest, approximate the amount in the account after 6 years.

40. Write $\log \dfrac{x^2 \sqrt{y}}{z^3}$ in terms of logarithms of x, y, and z.

41. Solve $2e^x - 1 = 17$.

42. Solve $3 + \log 4x = 5$.

43. Graph each conic section.
    (a) $x = (y - 1)^2$ (b) $(x - 1)^2 + (y + 1)^2 = 4$
    (c) $\dfrac{x^2}{4} + \dfrac{y^2}{25} = 1$ (d) $4x^2 - 9y^2 = 36$

44. Solve the nonlinear system of equations.
$$x^2 + y^2 = 1$$
$$x^2 + 9y^2 = 9$$

45. Shade the solution set in the *xy*-plane.
$$x^2 + y^2 \leq 4$$
$$x^2 - y \leq 2$$

APPLICATIONS

46. **Calculating Distance** The distance D in miles that a driver of a car is from home after x hours is given by $D(x) = 400 - 50x$.
    (a) Evaluate $D(0)$. Interpret your answer.
    (b) What is the *x*-intercept on the graph of D? Interpret your answer.
    (c) What is the slope of the graph of D? Interpret your answer.

47. **Investment** Suppose a total of $2000 is deposited in three accounts paying 5%, 6%, and 7% annual interest. The amount invested at 6% is $500 more than the amount invested at 5%. The interest after 1 year is $120. How much is invested at each rate?

48. **Area** There are 1200 feet of fence available to surround the perimeter of a rectangular garden. What dimensions for the garden give the largest area?

49. **Population** The population of a city in millions after x years is given by $P = 2e^{0.02x}$. How long will it take for the population to double?

50. **Dimensions of a Can** Find the radius of a cylindrical can with a volume V of 60 cubic inches and a side area S of 50 square inches. (*Hint:* $V = \pi r^2 h$ and $S = 2\pi r h$.)

21 Sequences and Series

21.1 Sequences
21.2 Arithmetic and Geometric Sequences
21.3 Series
21.4 The Binomial Theorem

In this chapter, we present sequences and series, which are essential topics because they are used to model and approximate important quantities. Complicated population growth, such as that experienced by populations of insects and other species over time, can be modeled with *sequences*. Accurate approximations for numbers such as π and e are made with *series*. For example, in about 2000 B.C., the Babylonians thought that π equaled 3.125, whereas at the same time the Chinese thought π equaled 3. By 1700, mathematicians were able to calculate π to 100 decimal places due to the discovery of series. Series are also essential to the solutions of many modern applied mathematics problems. (*Source:* P. Beckman, *A History of Pi.*)

21.1 Sequences

Objectives

1. Finding Terms of a Sequence
2. Recognizing Representations of Sequences
3. Modeling Population Growth with Sequences

NEW VOCABULARY

☐ Terms of a sequence
☐ Finite sequence
☐ Infinite sequence
☐ nth term
☐ General term

1 Finding Terms of a Sequence

🌐 **Math in Context** (Biology) Sequences are *ordered lists*. For example, names listed alphabetically represent a sequence. Listing populations by year is another example of a sequence. **FIGURE 21.1** shows an insect population in thousands per acre over a 6-year period. In mathematics, a sequence is a function for which valid inputs must be natural numbers. For example, we can use a function f to define this sequence by letting $f(1)$ represent the insect population after 1 year, $f(2)$ represent the insect population after 2 years, and in general let $f(n)$ represent the population after n years.

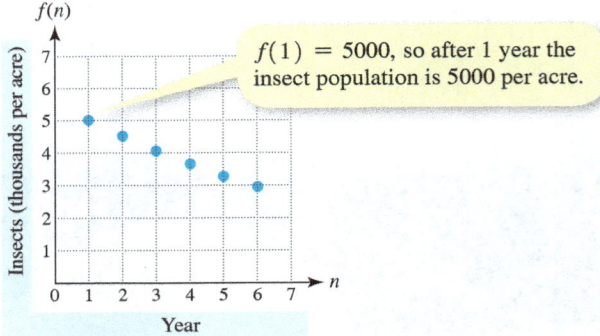

An Insect Population

$f(1) = 5000$, so after 1 year the insect population is 5000 per acre.

FIGURE 21.1

🌐 **Math in Context** (Earnings) Suppose that an individual's starting salary is $40,000 per year and that the person's salary is increased by 10% each year. This situation is modeled by the formula

$$f(n) = 40{,}000(1.10)^n.$$

Symbolic representation

We do not allow the input n to be any real number, but rather limit n to a *natural number* because the individual's salary is constant throughout a particular year. The first five *terms of the sequence* are

$$f(1), f(2), f(3), f(4), f(5).$$

They can be computed as follows.

The first five terms of the sequence are found by evaluating function f for $n = 1, 2, 3, 4, 5$.

$$f(1) = 40{,}000(1.10)^1 = 44{,}000$$
$$f(2) = 40{,}000(1.10)^2 = 48{,}400$$
$$f(3) = 40{,}000(1.10)^3 = 53{,}240$$
$$f(4) = 40{,}000(1.10)^4 = 58{,}564$$
$$f(5) = 40{,}000(1.10)^5 \approx 64{,}420$$

The first five terms of a sequence

This sequence is represented *numerically* in **TABLE 21.1** and *graphically* in **FIGURE 21.2**.

Numerical and Graphical Representations of a Sequence

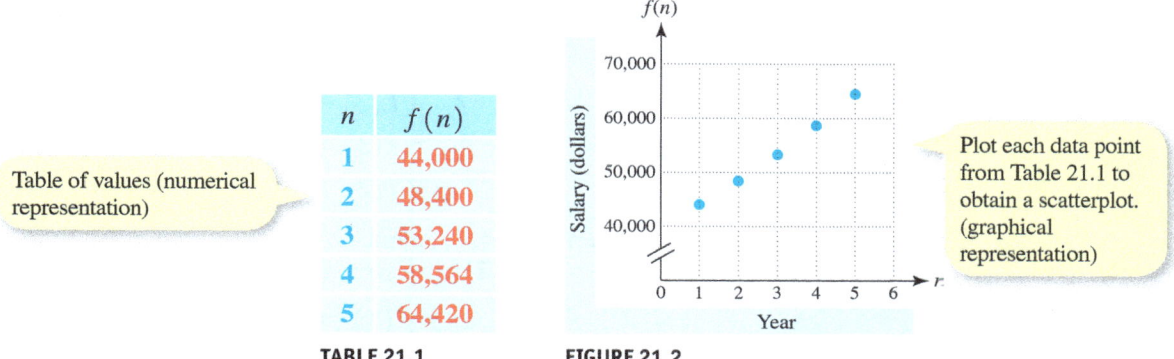

Table of values (numerical representation)

TABLE 21.1

n	$f(n)$
1	44,000
2	48,400
3	53,240
4	58,564
5	64,420

Plot each data point from Table 21.1 to obtain a scatterplot. (graphical representation)

FIGURE 21.2

NOTE: Graphs of sequences are scatterplots. ∎

The preceding sequence is an example of a *finite sequence* of numbers. The even natural numbers,

$$2, 4, 6, 8, 10, 12, 14, \ldots \quad \text{Infinite sequence}$$

are an example of an *infinite sequence* represented by $f(n) = 2n$, where n is a natural number. The three dots, or periods (called an *ellipsis*), indicate that the pattern continues indefinitely.

> **SEQUENCES**
>
> A **finite sequence** is a function whose domain is $D = \{1, 2, 3, \ldots, n\}$ for some fixed natural number n.
> An **infinite sequence** is a function whose domain is the set of natural numbers.

READING CHECK 1

- How do we denote the fourth term of a sequence?

Because sequences are functions, many of the concepts discussed in previous chapters apply to sequences. Instead of letting y represent the output, however, the convention is to write $a_n = f(n)$, where n is a natural number in the domain of the sequence. The *terms* of a sequence are

Terms of a sequence $\quad a_1, a_2, a_3, \ldots, a_n, \ldots$ — nth term

To calculate the terms of a sequence, the first term is given by $a_1 = f(1)$, the second term is given by $a_2 = f(2)$, and so on. The **nth term**, or **general term**, of a sequence is given by $a_n = f(n)$.

EXAMPLE 1 Computing terms of a sequence

Write the first four terms of each sequence for $n = 1, 2, 3,$ and 4.
(a) $f(n) = 2n - 1$ **(b)** $f(n) = 3(-2)^n$ **(c)** $f(n) = \dfrac{n}{n+1}$

Solution

(a) For $a_n = f(n) = 2n - 1$, we write the first four terms as follows.

$$a_1 = f(1) = 2(1) - 1 = 1$$
$$a_2 = f(2) = 2(2) - 1 = 3$$
$$a_3 = f(3) = 2(3) - 1 = 5$$
$$a_4 = f(4) = 2(4) - 1 = 7$$

The first four terms are **1, 3, 5**, and **7**.

(b) For $a_n = f(n) = 3(-2)^n$, we write the first four terms as follows.

$$a_1 = f(1) = 3(-2)^1 = -6$$
$$a_2 = f(2) = 3(-2)^2 = 12$$
$$a_3 = f(3) = 3(-2)^3 = -24$$
$$a_4 = f(4) = 3(-2)^4 = 48$$

The first four terms are **−6, 12, −24**, and **48**.

(c) For $a_n = f(n) = \frac{n}{n+1}$, we write the first four terms as follows.

$$a_1 = f(1) = \frac{1}{1+1} = \frac{1}{2}$$
$$a_2 = f(2) = \frac{2}{2+1} = \frac{2}{3}$$
$$a_3 = f(3) = \frac{3}{3+1} = \frac{3}{4}$$
$$a_4 = f(4) = \frac{4}{4+1} = \frac{4}{5}$$

The first four terms are $\frac{1}{2}, \frac{2}{3}, \frac{3}{4}$, and $\frac{4}{5}$. Note that, although the *input* to a sequence is a natural number, the *output* need not be a natural number.

Now Try Exercises 9, 11, 13

TECHNOLOGY NOTE

Generating Sequences
Many graphing calculators can generate sequences if you change the MODE from function (Func) to sequence (Seq). In **FIGURES 21.3** and **21.4**, the sequences from Example 1 are generated. On some calculators, the sequence utility is found in the LIST OPS menus. The expression

$$\text{seq}(2n - 1, n, 1, 4)$$

represents terms 1 through 4 of the sequence $f(n) = 2n - 1$ with the variable n.

```
seq(2n-1,n,1,4)
       {1 3 5 7}
seq(3(-2)^n,n,1,
4)
     {-6 12 -24 48}
```

FIGURE 21.3

```
seq(n/(n+1),n,1,
4)▶Frac
   {1/2 2/3 3/4 4/...
```

FIGURE 21.4

2 Recognizing Representations of Sequences

Because sequences are functions, they can be represented symbolically, graphically, and numerically. The next two examples illustrate such representations.

EXAMPLE 2 Using a graphical representation

Use **FIGURE 21.5** to write the terms of the sequence.

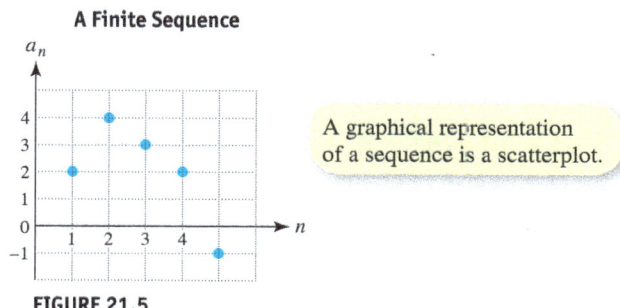

FIGURE 21.5

A graphical representation of a sequence is a scatterplot.

Solution
The points $(1, 2)$, $(2, 4)$, $(3, 3)$, $(4, 2)$, and $(5, -1)$ are shown in the graph. The terms of the sequence are $2, 4, 3, 2,$ and -1.

Now Try Exercise 33

EXAMPLE 3 Representing a sequence

In 2014, the average person in the United States used about 100 gallons of water at home each day. Give symbolic, numerical, and graphical representations for a sequence that models the total amount of water used over a 7-day period. (*Source:* U.S. Geological Survey.)

Solution
Symbolic Representation Let

$$a_n = 100n \text{ for } n = 1, 2, 3, \ldots, 7.$$

Numerical Representation **TABLE 21.2** contains the sequence.

Graphical Representation Plot the points $(1, 100)$, $(2, 200)$, $(3, 300)$, $(4, 400)$, $(5, 500)$, $(6, 600)$, and $(7, 700)$, as shown in **FIGURE 21.6**.

A numerical representation of a sequence is a table of values.

n	a_n
1	100
2	200
3	300
4	400
5	500
6	600
7	700

TABLE 21.2

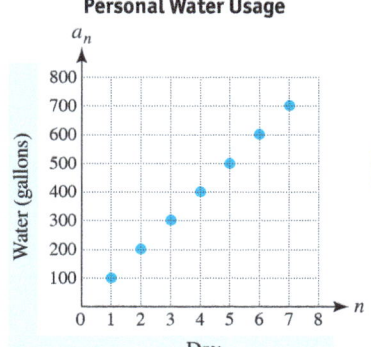

FIGURE 21.6

A graphical representation of a sequence is a scatterplot.

Now Try Exercise 43

3 Modeling Population Growth with Sequences

Math in Context (Entomology) A population model for a species of insect with a life span of 1 year can be described with a sequence. Suppose that each adult female insect produces, on average, r female offspring that survive to reproduce the following year. Let a_n represent the female insect population at the *beginning* of year n. Then the number of female insects is given by

$$a_n = Cr^{n-1},$$

where C is the initial population of female insects. (*Source:* D. Brown and P. Rothery, *Models in Biology*.)

EXAMPLE 4 Modeling numbers of insects

Suppose that the initial population of adult female insects is 500 per acre and that $r = 1.04$. Then the average number of female insects per acre at the beginning of year n is described by

$$a_n = 500(1.04)^{n-1}.$$

Represent the female insect population numerically and graphically for 7 years. Round values to the nearest whole number. Discuss the results. By what percent is the population increasing each year?

Solution
Numerical Representation TABLE 21.3 contains *approximations* for the first 7 terms of the sequence. The insect population increases from 500 to about 633 insects per acre during this time period.

An Insect Population (per acre)

n	1	2	3	4	5	6	7
a_n	500	520	541	562	585	608	633

TABLE 21.3

Graphical Representation Plot the points $(1, 500)$, $(2, 520)$, $(3, 541)$, $(4, 562)$, $(5, 585)$, $(6, 608)$, and $(7, 633)$, as shown in **FIGURE 21.7**. These results indicate that the insect population gradually increases. Because the growth factor is 1.04, the population is increasing by 4% each year.

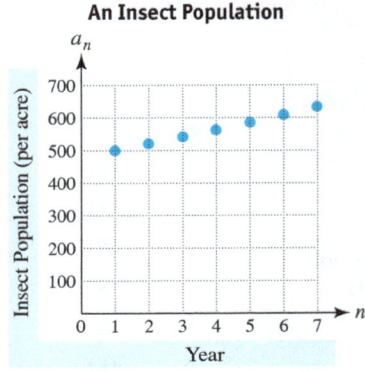

FIGURE 21.7

CRITICAL THINKING 1
Explain how the value of r in Example 4 affects the population of female insects over time. Assume that $r > 0$.

Now Try Exercise 49

TECHNOLOGY NOTE

Graphs and Tables of Sequences

In sequence mode, many graphing calculators are capable of representing sequences graphically and numerically. **FIGURE 21.8(a)** shows how to enter the sequence from Example 4 to produce the table of values shown in **FIGURE 21.8(b)**.

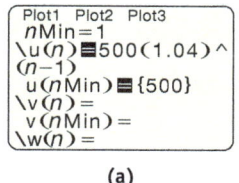

(a)   (b)

FIGURE 21.8

FIGURES 21.9(a) and **(b)** show the set-up used for graphing the sequence from Example 4. The resulting graph is shown in **FIGURE 21.9(c)**.

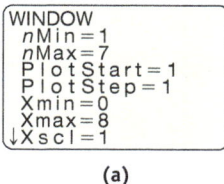 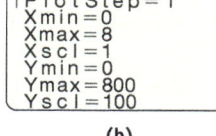 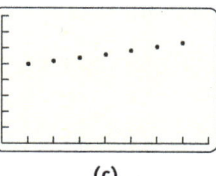

(a)   (b)   (c)

FIGURE 21.9

21.1 Putting It All Together

An infinite sequence is a function whose domain is the set of natural numbers. A finite sequence has the domain $D = \{1, 2, 3, \ldots, n\}$ for some fixed natural number n.

REPRESENTATION	EXAMPLE
Symbolic	$a_n = n - 3$ represents a sequence. The first four terms are $-2, -1, 0,$ and 1: $$a_1 = 1 - 3 = -2, \quad a_2 = 2 - 3 = -1,$$ $$a_3 = 3 - 3 = 0, \quad a_4 = 4 - 3 = 1.$$
Numerical	A numerical representation for $a_n = n - 3$ with $n = 1, 2, 3,$ and 4 is shown in the table. 	n	1	2	3	4	 |---|---|---|---|---| | a_n | -2 | -1 | 0 | 1 | Finite sequence: $-2, -1, 0, 1$
Graphical	For a graphical representation of the first four terms of $a_n = n - 3$, the points $(1, -2)$, $(2, -1)$, $(3, 0)$, and $(4, 1)$ from the previous table are plotted. It is a scatterplot. Finite sequence: $-2, -1, 0, 1$

21.1 Exercises

CONCEPTS AND VOCABULARY

1. Give an example of a finite sequence.

2. Give an example of an infinite sequence.

3. An infinite sequence is a(n) _____ whose domain is the set of _____.

4. An ordered list is a(n) _____.

5. The third term in the sequence $4, -5, 6, -7, 8$ is _____.

6. The graph of a sequence is not a continuous graph but rather a(n) _____.

7. If $f(n)$ represents a sequence, the second term of the sequence is given by _____.

8. If a_n represents a sequence, the fourth term of the sequence is given by _____.

EVALUATING SEQUENCES

Exercises 9–16: Write the first four terms of the sequence for $n = 1, 2, 3,$ and 4.

9. $f(n) = n^2$
10. $f(n) = 3n + 4$
11. $f(n) = \dfrac{1}{n + 5}$
12. $f(n) = 3^n$
13. $f(n) = 5\left(\dfrac{1}{2}\right)^n$
14. $f(n) = n^2 + 2n$
15. $f(n) = 9$
16. $f(n) = (-1)^n$

Exercises 17–24: Write the first three terms of the given sequence for $n = 1, 2,$ and 3.

17. $a_n = n^3$
18. $a_n = 5 - n$
19. $a_n = \dfrac{4n}{3 + n}$
20. $a_n = 3^{-n}$
21. $a_n = 2n^2 + n - 1$
22. $a_n = n^4 - 1$
23. $a_n = -2$
24. $a_n = n^n$

Exercises 25–28: The first five terms of a sequence are shown. Write a formula $f(n)$ that represents the sequence.

25. $1, \dfrac{1}{2}, \dfrac{1}{3}, \dfrac{1}{4}, \dfrac{1}{5}, \dfrac{1}{6}$
26. $\dfrac{1}{2}, \dfrac{2}{3}, \dfrac{3}{4}, \dfrac{4}{5}, \dfrac{5}{6}$
27. $3, 5, 7, 9, 11$
28. $3, 6, 9, 12, 15$

Exercises 29 and 30: **Thinking Generally** *Let b and c be fixed numbers (constants). Find a_1 and a_2.*

29. $a_n = bn + c$

30. $a_n = \dfrac{n + b}{n - c}$, $c \neq 1$ and $c \neq 2$

REPRESENTING SEQUENCES

Exercises 31 and 32: Use the numerical representation to evaluate $\tfrac{1}{2}(a_1 + a_4)$.

31.
n	1	2	3	4	5
a_n	10	8	6	4	2

32.
n	1	2	3	4	5
a_n	-5	0	10	30	60

Exercises 33–36: Write the terms of the sequence.

33.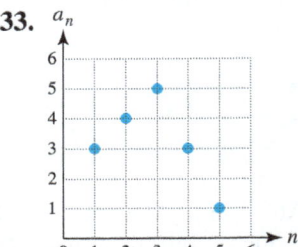
34.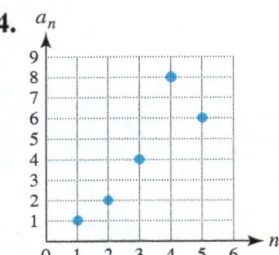
35.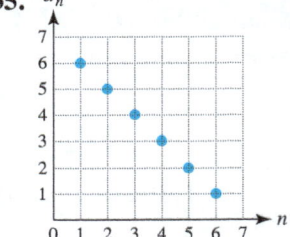
36.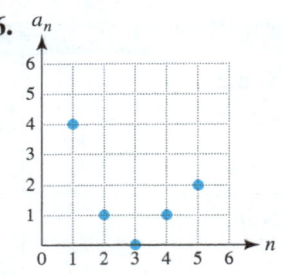

Exercises 37–42: Represent the first seven terms of the sequence numerically and graphically.

37. $a_n = n + 1$
38. $a_n = \tfrac{1}{2}n - \tfrac{1}{2}$
39. $a_n = n^2 - n$
40. $a_n = \tfrac{1}{2}n^2$
41. $a_n = 2^n$
42. $a_n = 2(0.5)^n$

APPLICATIONS OF SEQUENCES

43. **Solid Waste** On average, each U.S. resident generates about 30 pounds of solid waste per week. Give

symbolic, numerical, and graphical representations for a sequence that models the total amount of waste produced over a 7-week period. (*Source:* Environmental Protection Agency.)

44. **Carbon Dioxide Emitters** Because people burn fossil fuels, the United States emits about 5.8 billion metric tons of carbon dioxide per year. (A metric ton is about 2200 pounds.) Give symbolic, numerical, and graphical representations for a sequence that models the total amount of carbon dioxide emitted in billions of metric tons in the United States during a 5-year period. (*Source:* Energy Information Administration.)

45. **Geometry** The lengths of the sides of a sequence of squares are given by 1, 2, 3, and 4, as shown in the figure below. Write sequences that give the following.
    (a) the areas of the squares
    (b) the perimeters of the squares

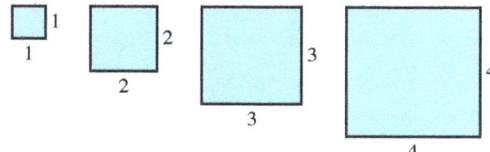

46. **Salaries** An individual's starting salary is $50,000, and the individual receives an increase of 8% per year. Give symbolic, numerical, and graphical representations for this person's salary over 5 years.

47. **Depreciation** Automobiles usually depreciate in value over time. Often, a newer automobile may be worth only 80% of its previous year's value. Suppose that a car is worth $25,000 new.
    (a) How much is it worth after 1 year? After 2 years?
    (b) Write a formula for a sequence that gives the car's value after n years.
    (c) Make a table that shows how much the car was worth each year during the first 7 years.

48. **Falling Object** The distance d that an object falls during *consecutive* seconds is shown in the table. For example, during the third second (from $n = 2$ to $n = 3$), an object falls a distance of 80 feet.

n (seconds)	1	2	3	4	5
d (feet)	16	48	80	112	144

(a) Find values for c and b so that $d = cn + b$ models these data.
(b) How far does an object fall during the sixth second?

49. **Modeling Insect Populations** (Refer to Example 4.) Suppose that the initial population of insects is 2048 per acre and that $r = 0.5$. Use a sequence to represent the insect population over a 7-year period
    (a) symbolically,
    (b) numerically, and
    (c) graphically.

50. **Auditorium Seating** An auditorium has 50 seats in the first row, 55 seats in the second row, 60 seats in the third row, and so on.
    (a) Make a table that shows the number of seats in the first seven rows.
    (b) Write a formula that gives the number of seats in row n.
    (c) How many seats are there in row 23?
    (d) Graph the number of seats in each row for $n = 1, 2, 3, \ldots, 10$.

WRITING ABOUT MATHEMATICS

51. Compare the graph of the function $f(x) = 2x + 1$, where x is a real number, with the graph of the sequence $f(n) = 2n + 1$, where n is a natural number.

52. Explain what a sequence is. Describe the difference between a finite and an infinite sequence.

21.2 Arithmetic and Geometric Sequences

Objectives

1. Recognizing Representations of Arithmetic Sequences
   - Defining Arithmetic Sequences
   - Finding Arithmetic Sequences
2. Recognizing Representations of Geometric Sequences
   - Defining Geometric Sequences
   - Finding Geometric Sequences
3. Solving Applications Involving Sequences

NEW VOCABULARY

☐ Arithmetic sequence
☐ Common difference
☐ Geometric sequence
☐ Common ratio

1 Recognizing Representations of Arithmetic Sequences

DEFINING ARITHMETIC SEQUENCES If a sequence is defined by a linear function, it is an *arithmetic sequence*. For example,

$$f(n) = 2n - 3$$

represents an arithmetic sequence because $f(x) = 2x - 3$ defines a linear function.

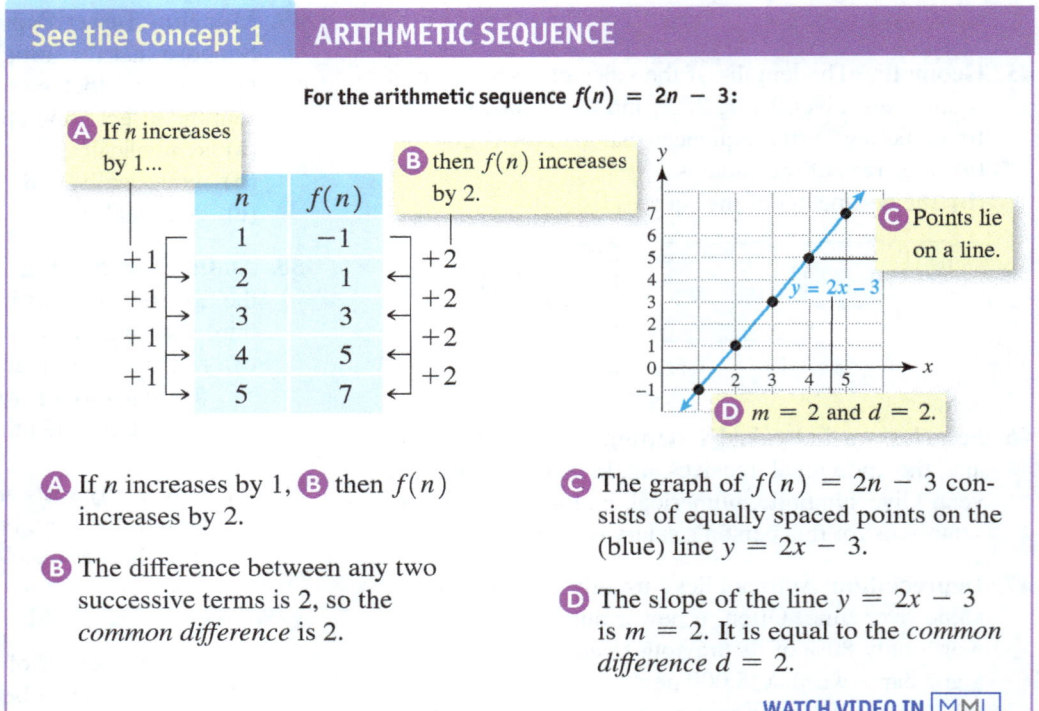

See the Concept 1 — **ARITHMETIC SEQUENCE**

For the arithmetic sequence $f(n) = 2n - 3$:

A If n increases by 1, **B** then $f(n)$ increases by 2.

B The difference between any two successive terms is 2, so the *common difference* is 2.

C The graph of $f(n) = 2n - 3$ consists of equally spaced points on the (blue) line $y = 2x - 3$.

D The slope of the line $y = 2x - 3$ is $m = 2$. It is equal to the *common difference* $d = 2$.

WATCH VIDEO IN MML.

READING CHECK 1

- Given a term in an arithmetic sequence, how can you find the next term and the previous term?

ARITHMETIC SEQUENCE

An **arithmetic sequence** is a linear function given by $a_n = dn + c$ whose domain is the set of natural numbers. The value of d is called the **common difference**.

NOTE: If each term after the first term is obtained by adding a fixed number to the previous term, then the sequence is an arithmetic sequence. The fixed number is the *common difference*. For example, 1, 6, 11, 16, 21, … is an arithmetic sequence because each term (after the first) is found by adding the common difference of **5** to the previous term. That is, $6 = 1 + 5$, $11 = 6 + 5$, $16 = 11 + 5$, and so on. The difference between successive terms is always 5. ∎

EXAMPLE 1 Recognizing arithmetic sequences

Determine whether f is an arithmetic sequence. If it is, identify the common difference d.
(a) $f(n) = 2 - 3n$

(b)

n	$f(n)$
1	10
2	5
3	0
4	-5
5	-10

(c) A graph of f is shown in **FIGURE 21.10**.

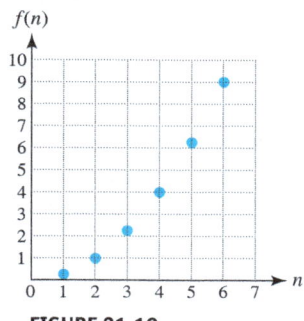

FIGURE 21.10

Solution
(a) This sequence is arithmetic because $f(x) = -3x + 2$ defines a linear function. The common difference is $d = -3$.
(b) The table reveals that each term is found by adding -5 to the previous term. This represents an arithmetic sequence with common difference $d = -5$.
(c) The sequence shown in **FIGURE 21.10** is not an arithmetic sequence because the points are not collinear. That is, the points do not lie on a line and there is no common difference.

Now Try Exercises 11, 17, 25

MAKING CONNECTIONS 1

Common Difference and Slope

The common difference d of an arithmetic sequence equals the slope of the line passing through the collinear points. For example, if $a_n = -2n + 4$, the common difference is -2, and the slope of the line passing through the points on the graph of a_n is also -2, as shown in **FIGURE 21.11**.

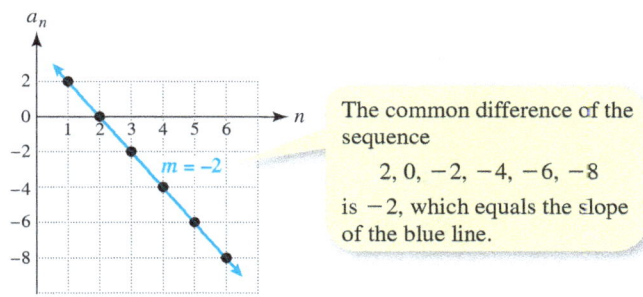

The common difference of the sequence

$$2, 0, -2, -4, -6, -8$$

is -2, which equals the slope of the blue line.

FIGURE 21.11

FINDING ARITHMETIC SEQUENCES In the next example, we write the formula for the general term of an arithmetic sequence.

EXAMPLE 2 **Finding symbolic representations**

Find the general term a_n for each arithmetic sequence.
(a) $a_1 = 3$ and $d = 4$ (b) $a_1 = 3$ and $a_4 = 12$

continued on next page

Solution

(a) Let $a_1 = 3$, $d = 4$, and $a_n = dn + c$. For $d = 4$, we write $a_n = 4n + c$, and to find c, we use $a_1 = 3$.

$$a_1 = 4(1) + c = 3 \quad \text{or} \quad c = -1$$

Thus $a_n = 4n - 1$.

(b) Because $a_1 = 3$ and $a_4 = 12$, the common difference d equals the slope of the line passing through the points $(1, 3)$ and $(4, 12)$, or

$$d = \frac{12 - 3}{4 - 1} = 3.$$

Therefore $a_n = 3n + c$. To find c, we use $a_1 = 3$ and obtain

$$a_1 = 3(1) + c = 3 \quad \text{or} \quad c = 0.$$

Thus $a_n = 3n$.

Now Try Exercises 29, 31

Consider the arithmetic sequence

$$1, 5, 9, 13, 17, 21, 25, 29, \ldots .$$

> The common difference is 4 because the difference between successive terms is 4.

The common difference is $d = 4$, and the first term is $a_1 = 1$. To find the second term, we add d to the first term. To find the third term, we add $2d$ to the first term, and to find the fourth term, we add $3d$ to the first term. That is,

$$\begin{aligned} a_1 &= 1, \\ a_2 &= a_1 + 1d = 1 + 1 \cdot 4 = 5, \\ a_3 &= a_1 + 2d = 1 + 2 \cdot 4 = 9, \\ a_4 &= a_1 + 3d = 1 + 3 \cdot 4 = 13, \end{aligned}$$

> First four terms

and, in general, a_n is determined by

$$a_n = a_1 + (n - 1)d = 1 + (n - 1)4.$$

This result suggests the following formula.

READING CHECK 2

- Write the general term of an arithmetic sequence, given the first term and the common difference.

GENERAL TERM OF AN ARITHMETIC SEQUENCE

The nth term a_n of an arithmetic sequence is given by

$$a_n = a_1 + (n - 1)d,$$

where a_1 is the first term and d is the common difference.

EXAMPLE 3 Finding terms of an arithmetic sequence

If $a_1 = 5$ and $d = 3$, find a_{54}.

Solution
To find a_{54}, apply the formula $a_n = a_1 + (n - 1)d$ with $a_1 = 5$, $n = 54$, and $d = 3$.

$$a_{54} = 5 + (54 - 1)3 = 164$$

Now Try Exercise 35

2 Recognizing Representations of Geometric Sequences

DEFINING GEOMETRIC SEQUENCES The following See the Concept describes a geometric sequence.

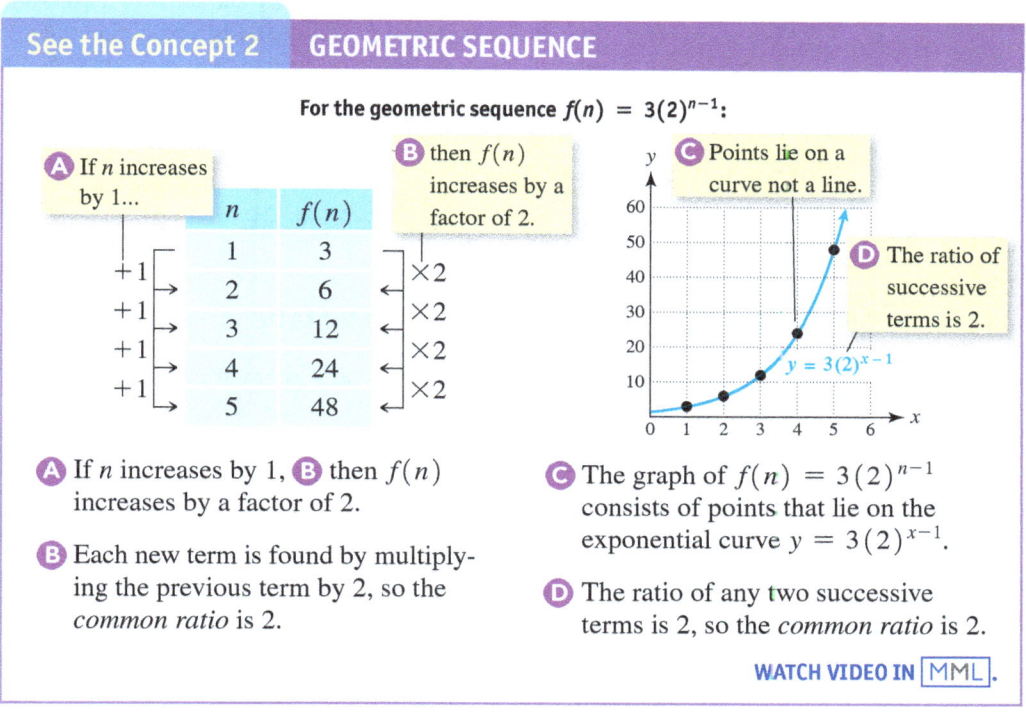

A geometric sequence with a positive common ratio is an exponential function whose domain is the set of natural numbers. Its terms reflect either *exponential growth* or *decay*.

READING CHECK 3

- Given a term in a geometric sequence, how can you find the next term and the previous term?

GEOMETRIC SEQUENCE

A **geometric sequence** is given by $a_n = a_1(r)^{n-1}$, where n is a natural number and r is neither 0 nor 1. The **common ratio** is r, and a_1 is the first term of the sequence.

NOTE: If each term after the first term is obtained by multiplying the previous term by a fixed number, the sequence is a geometric sequence. The fixed number can be either positive or negative, but it cannot be 0 or 1. ∎

EXAMPLE 4 Recognizing geometric sequences

Determine whether f is a geometric sequence. If it is, identify the common ratio.

(a) $f(n) = 2(0.9)^{n-1}$ (c) A graph of f is shown in **FIGURE 21.12**.

(b)

n	$f(n)$
1	8
2	4
3	2
4	1
5	$\frac{1}{2}$

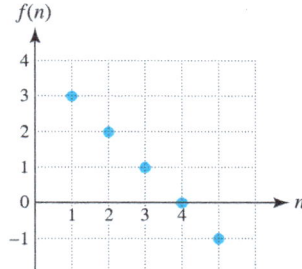

FIGURE 21.12

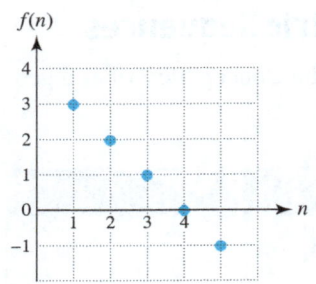

FIGURE 21.12 (repeated)

Solution
(a) This sequence is geometric because $f(x) = 2(0.9)^{x-1}$ defines an exponential function. The common ratio is $r = 0.9$.
(b) The table shows that each successive term 8, 4, 2, 1, $\frac{1}{2}$, is half the previous term. This sequence represents a geometric sequence with a common ratio of $r = \frac{1}{2}$.
(c) The sequence shown in **FIGURE 21.12** (repeated in the margin) is not a geometric sequence because the points are collinear. There is no common ratio.

Now Try Exercises 41, 45, 53

FINDING GEOMETRIC SEQUENCES In the next example, we write the formula for the general term of a geometric sequence.

EXAMPLE 5 Finding symbolic representations

Find a general term a_n for each geometric sequence.
(a) $a_1 = \frac{1}{2}$ and $r = 5$
(b) $a_1 = 2$, $a_3 = 18$, and $r < 0$.

Solution
(a) Let $a_n = a_1(r)^{n-1}$. Because $a_1 = \frac{1}{2}$ and $r = 5$, we can write $a_n = \frac{1}{2}(5)^{n-1}$.
(b) $a_1 = 2$ and $a_3 = 18$, so
$$a_3 = a_1(r)^{3-1}$$
$$18 = 2(r)^2.$$
This equation simplifies to
$$r^2 = 9 \text{ or } r = \pm 3.$$
It is specified that $r < 0$, so $r = -3$ and $a_n = 2(-3)^{n-1}$. Note that the common ratio r can be *negative*. If $r < 0$, successive terms alternate sign (positive or negative).

Now Try Exercises 55, 59

CRITICAL THINKING 1

If we are given a_1 and a_5, can we determine the common ratio of a geometric series? Explain.

EXAMPLE 6 Finding a term of a geometric sequence

If $a_1 = 5$ and $r = 3$, find a_{10}.

Solution
To find a_{10}, apply the formula $a_n = a_1(r)^{n-1}$ with $a_1 = 5$, $r = 3$, and $n = 10$.
$$a_{10} = 5(3)^{10-1} = 5(3)^9 = 98{,}415$$

Now Try Exercise 61

3 Solving Applications Involving Sequences

🌐 *Math in Context* (Public Health) Indoor air pollution has become more hazardous as people spend 80% to 90% of their time in tightly sealed, energy-efficient buildings, which often lack proper ventilation. In the next example, we use a sequence to model proper classroom ventilation.

EXAMPLE 7 Modeling classroom ventilation

Ventilation is an effective means for removing indoor air pollutants. According to the American Society of Heating, Refrigerating, and Air-Conditioning Engineers (ASHRAE), a classroom should have a ventilation rate of 900 cubic feet per hour per person.

(a) Write a sequence that gives the hourly ventilation necessary for 1, 2, 3, 4, and 5 people in a classroom. Is this sequence arithmetic, geometric, or neither?
(b) Write the general term for this sequence. Why is it reasonable to limit the domain to natural numbers?
(c) Find a_{30} and interpret the result.

Solution
(a) One person requires 900 cubic feet of air circulated per hour, two people require 1800, three people 2700, and so on. The first five terms of this sequence are

$$900, 1800, 2700, 3600, 4500.$$

To find the next term, add 900 to the previous term.

This sequence is arithmetic, with a common difference of 900.

(b) The *n*th term equals 900*n*, so we let $a_n = 900n$. Because we cannot have a fraction of a person, limiting the domain to the natural numbers is reasonable.

(c) The result $a_{30} = 900(30) = 27{,}000$ indicates that a classroom with 30 people should have a ventilation rate of 27,000 cubic feet per hour.

Now Try Exercise 65

🌐 *Math in Context* (Chemistry) Chlorine is frequently added to the water to disinfect swimming pools. The chlorine concentration should remain between 1.5 and 2.5 parts per million (ppm). On a warm, sunny day, 30% of the chlorine may dissipate from the water. In the next example, we use a sequence to model the amount of chlorine in a pool at the beginning of each day. (*Source:* D. Thomas, *Swimming Pool Operator's Handbook.*)

EXAMPLE 8 Modeling chlorine in a swimming pool

A swimming pool on a warm, sunny day begins with a high chlorine content of 4 parts per million. (Assume that each day, 30% of the chlorine dissipates.)
(a) Write a sequence that models the amount of chlorine in the pool at the beginning of the first 3 days, assuming that no additional chlorine is added and that the days are warm and sunny. Is this sequence arithmetic, geometric, or neither?
(b) Write the general term for this sequence.
(c) At the beginning of what day does the chlorine first drop below 1.5 parts per million?

Solution
(a) Because 30% of the chlorine dissipates, 70% remains in the water at the beginning of the next day. If the concentration at the beginning of the first day is 4 parts per million, then at the beginning of the second day, it is

$$4 \cdot 0.70 = 2.8 \text{ parts per million},$$

After 1 day, 30% dissipates, or 70% remains.

and at the start of the third day, it is

$$2.8 \cdot 0.70 = 1.96 \text{ parts per million}.$$

The first three terms are 4, 2.8, 1.96. Successive terms are found by multiplying the previous term by 0.7. Thus the sequence is geometric, with common ratio 0.7.

(b) The initial amount is $a_1 = 4$ and the common ratio is $r = 0.7$, so the sequence can be represented by $a_n = 4(0.7)^{n-1}$.

(c) The table shown in **FIGURE 21.13** reveals that $a_4 = 4(0.7)^{4-1} \approx 1.372 < 1.5$. Thus, at the beginning of the fourth day, the chlorine level in the swimming pool drops below the recommended minimum of 1.5 parts per million.

Now Try Exercise 67

n	u(n)
1	4
2	2.8
3	1.96
4	1.372
5	.9604
6	.67228
7	.4706

$u(n) = 4(.7)^{\wedge}(n-1)$

FIGURE 21.13

21.2 Putting It All Together

Arithmetic sequences are linear functions, and geometric sequences with a *positive r* are exponential functions. The inputs for both are limited to the natural numbers. The graph of an arithmetic sequence consists of points that lie on a line, whereas the graph of a geometric sequence (with a positive *r*) consists of points that lie on an exponential curve.

SEQUENCE	FORMULA	EXAMPLE
Arithmetic	$a_n = dn + c$ or $a_n = a_1 + (n-1)d$, where d is the common difference and a_1 is the first term.	If $a_n = 5n + 2$, then the common difference is $d = 5$ and the terms of the sequence are $$7, 12, 17, 22, 27, 32, 37, \ldots.$$ Each term after the first is found by adding 5 to the previous term. The general term can be written as $$a_n = 7 + 5(n-1).$$
Geometric	$a_n = a_1(r)^{n-1}$, where r is the common ratio ($r \neq 0, r \neq 1$) and a_1 is the first term.	If $a_n = 4(-2)^{n-1}$, then the common ratio is $r = -2$ and the first term is $a_1 = 4$. The terms of the sequence are $$4, -8, 16, -32, 64, -128, 256, \ldots.$$ Each term after the first is found by multiplying the previous term by -2.

21.2 Exercises

CONCEPTS AND VOCABULARY

1. An arithmetic sequence is a(n) _____ function.

2. A geometric sequence with $r > 0$ is a(n) _____ function.

3. Give an example of an arithmetic sequence. State the common difference.

4. Give an example of a geometric sequence. State the common ratio.

5. To find successive terms in an arithmetic sequence, _____ the common difference to the _____ term.

6. To find successive terms in a geometric sequence, _____ the previous term by the _____.

7. Find the next term in the arithmetic sequence given by 3, 7, 11, 15. What is the common difference?

8. Find the next term in the geometric sequence given by 2, −4, 8, −16. What is the common ratio?

9. Write the general term a_n for a geometric sequence, using a_1 and r.

10. Write the general term a_n for an arithmetic sequence, using a_1 and d.

RECOGNIZING ARITHMETIC SEQUENCES

Exercises 11–28: (Refer to Example 1.) Determine whether f is an arithmetic sequence. Identify the common difference when possible.

11. $f(n) = 10n - 5$

12. $f(n) = -3n - 5$

13. $f(n) = 6 - n$

14. $f(n) = 6 + \frac{1}{2}n$

15. $f(n) = n^3 + 1$

16. $f(n) = 5\left(\frac{1}{3}\right)^{n-1}$

17.
n	1	2	3	4
$f(n)$	3	6	9	12

18.
n	1	2	3	4
$f(n)$	-7	-5	-3	-1

19.
n	1	2	3	4
$f(n)$	10	7	4	1

20.
n	1	2	3	4
$f(n)$	1	2	4	8

21.
n	1	2	3	4
$f(n)$	-4	0	8	12

22.
n	1	2	3	4
$f(n)$	1	2.5	4	5.5

23.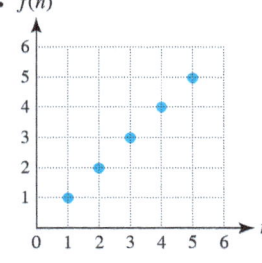

24.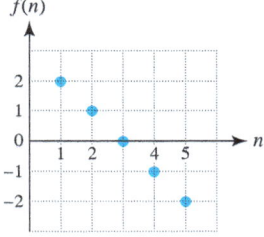

25.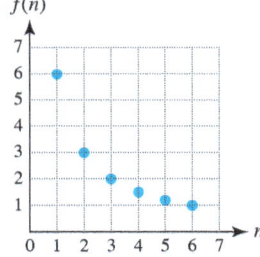

26.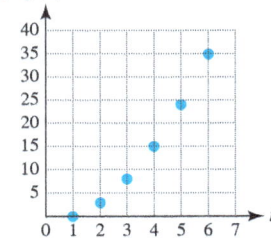

27.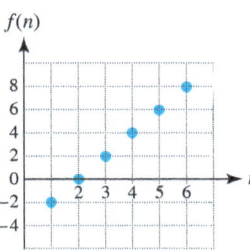

28.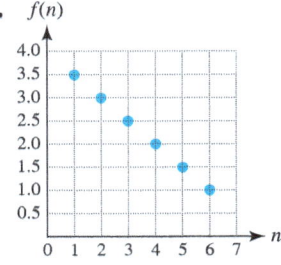

FINDING ARITHMETIC SEQUENCES

Exercises 29–34: (Refer to Example 2.) Find the general term a_n for the arithmetic sequence.

29. $a_1 = 7$ and $d = -2$

30. $a_1 = 5$ and $a_2 = 9$

31. $a_1 = -2$ and $a_3 = 6$

32. $a_2 = 7$ and $a_3 = 10$

33. $a_8 = 16$ and $a_{12} = 8$

34. $a_3 = 7$ and $d = -5$

Exercises 35–38: (Refer to Example 3.)

35. If $a_1 = -3$ and $d = 2$, find a_{32}.

36. If $a_1 = 2$ and $d = -3$, find a_{19}.

37. If $a_1 = -3$ and $a_2 = 0$, find a_9.

38. If $a_3 = -3$ and $d = 4$, find a_{62}.

RECOGNIZING GEOMETRIC SEQUENCES

Exercises 39–54: (Refer to Example 4.) Determine whether f is a geometric sequence. Identify the common ratio when possible.

39. $f(n) = 3^n$

40. $f(n) = 2(4)^n$

41. $f(n) = \frac{2}{3}(0.8)^{n-1}$

42. $f(n) = 7 - 3n$

43. $f(n) = 2(n-1)^2$

44. $f(n) = 2\left(-\frac{3}{4}\right)^{n-1}$

45.
n	1	2	3	4
$f(n)$	2	4	8	16

46.
n	1	2	3	4
$f(n)$	-6	3	-1.5	0.75

47.
n	1	2	3	4
$f(n)$	1	4	9	16

48.
n	1	2	3	4
$f(n)$	7	4	-1	-8

49.
n	1	2	3	4
$f(n)$	2	8	32	128

50.
n	1	2	3	4
$f(n)$	1	$\frac{1}{2}$	$\frac{1}{4}$	$\frac{1}{8}$

51. 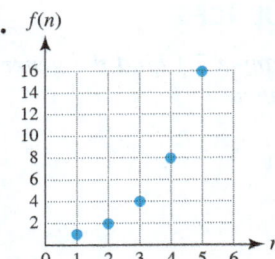 **52.**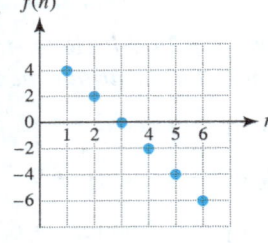

53. 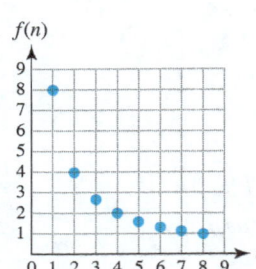 **54.**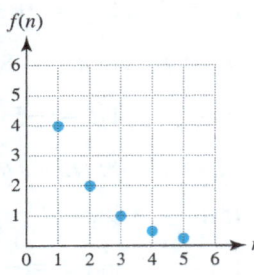

FINDING GEOMETRIC SEQUENCES

Exercises 55–60: (Refer to Example 5.) Find the general term a_n for the geometric sequence.

55. $a_1 = 1.5$ and $r = 4$ **56.** $a_1 = 3$ and $r = \frac{1}{4}$

57. $a_1 = -3$ and $a_2 = 6$ **58.** $a_1 = 2$ and $a_4 = 54$

59. $a_1 = 1$, $a_3 = 16$, and $r > 0$

60. $a_2 = 3$, $a_4 = 12$, and $r < 0$

Exercises 61–64: (Refer to Example 6.)

61. If $a_1 = 2$ and $r = 3$, find a_8.

62. If $a_1 = 4$ and $a_2 = 2$, find a_9.

63. If $a_1 = -1$ and $a_2 = 3$, find a_6.

64. If $a_3 = 5$ and $r = -3$, find a_7.

APPLICATIONS OF ARITHMETIC AND GEOMETRIC SEQUENCES

65. Room Ventilation (Refer to Example 7.) In areas that allow smoking, the ventilation rate should be 3000 cubic feet per hour per person. (Source: ASHRAE.)
 (a) Write a sequence that gives the hourly ventilation necessary for 1, 2, 3, 4, and 5 people in a smoking area. Is this sequence arithmetic, geometric, or neither?
 (b) Write the general term for this sequence.
 (c) Find a_{20} and interpret the result.
 (d) Give a graphical representation for eight terms of this sequence, using $n = 1, 2, 3, \ldots, 8$. Are the points collinear?

66. Salary Suppose that an employee receives a $2000 raise each year and that the sequence a_n models the employee's salary after n years. Is this sequence arithmetic, geometric, or neither? Explain.

67. Chlorine in Swimming Pools (Refer to Example 8.) Suppose that the water in a swimming pool initially has a chlorine content of 3 parts per million and that 20% of the chlorine dissipates each day.
 (a) If no additional chlorine is added, write the general term for a sequence that gives the chlorine concentration at the beginning of each day.
 (b) Give a graphical representation for this sequence, using $n = 1, 2, 3, \ldots, 8$. Are the points collinear? Is this sequence arithmetic, geometric, or neither?

68. Salary Suppose that an employee receives a 7% increase in salary each year and that the sequence a_n models the employee's salary after n years. Is this sequence arithmetic, geometric, or neither? Explain.

69. Bouncing Ball A tennis ball bounces back to 85% of the height from which it was dropped and then to 85% of the height of each successive bounce.
 (a) Write the general term for a sequence a_n that gives the maximum height of the ball on the nth bounce. Let $a_1 = 5$ feet.
 (b) Is the sequence arithmetic or geometric? Explain.
 (c) Find a_8 and interpret the result.

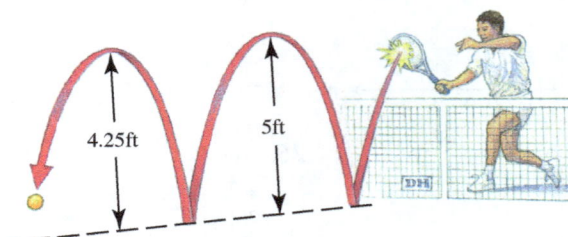

70. Falling Object The total distance D_n that an object falls in n seconds is shown in the table. Is the sequence arithmetic, geometric, or neither? Explain your reasoning.

n (seconds)	1	2	3	4	5
D_n (feet)	16	64	144	256	400

71. Theater Seating A theater has 40 seats in the first row, 42 seats in the second row, 44 seats in the third row, and so on.
 (a) Can the number of seats in each row be modeled by an arithmetic or geometric sequence? Explain.

(b) Write the general term for a sequence a_n that gives the number of seats in row n.
(c) How many seats are there in row 20?

72. **Appreciation of Lake Property** A certain type of lake property in northern Minnesota is increasing in value by 15% per year. Let the sequence a_n give the value of this type of lake property at the beginning of year n.
    (a) Is a_n arithmetic, geometric, or neither? Explain your reasoning.
    (b) Write the general term a_n for this sequence if $a_1 = \$100,000$.
    (c) Find a_7 and interpret the result.
    (d) Give a graph for a_n, where $n = 1, 2, 3, \ldots, 10$.

Exercises 73 and 74: **Recursive Sequences** *Some sequences are not defined by a formula for a_n. Instead, they are defined recursively. With a recursive formula, you must find terms a_1 through a_{n-1} before you can find a_n. For example, let*

$$a_1 = 2$$
$$a_n = a_{n-1} + 3, \quad \text{for } n \geq 2.$$

To find $a_2, a_3,$ and a_4, we let $n = 2, 3, 4$.

$$a_2 = a_1 + 3 = 2 + 3 = 5$$
$$a_3 = a_2 + 3 = 5 + 3 = 8$$
$$a_4 = a_3 + 3 = 8 + 3 = 11$$

The first four terms of the sequence are 2, 5, 8, 11.

73. **Fibonacci Sequence** The Fibonacci sequence dates back to 1202 and is one of the most famous sequences in mathematics. It can be defined recursively as follows.

$$a_1 = 1, \quad a_2 = 1$$
$$a_n = a_{n-1} + a_{n-2}, \quad \text{for } n \geq 3$$

Find the first 12 terms of this sequence.

74. **Insect Populations** Frequently, the population of a particular insect does not continue to grow indefinitely. Instead, its population grows rapidly at first and then levels off because of competition for limited resources. In one study, the behavior of the winter moth was modeled with a sequence similar to the following, where a_n gives the population density in thousands per acre during year n.

$$a_1 = 1$$
$$a_n = 2.85a_{n-1} - 0.19a_{n-1}^2, \quad n \geq 2$$

(*Source:* G. Varley and G. Gradwell, "Population models for the winter moth.")

(a) Make a table for $n = 1, 2, 3, \ldots, 7$. Describe what happens to the population density of the winter moth.
(b) Graph the sequence for $n = 1, 2, 3, \ldots, 20$. Discuss the graph.

NOTE: Many graphing calculators are capable of generating tables and graphs for a recursive sequence. ■

WRITING ABOUT MATHEMATICS

75. If you are given a table of values for a sequence, how can you determine if the sequence is geometric? Give an example.

76. If you are given a graph of a sequence, how can you determine if the sequence is arithmetic? Give an example.

SECTIONS 21.1 and 21.2 — **Checking Basic Concepts**

1. Write the first four terms of the sequence defined by $a_n = \dfrac{n}{n+4}$.

2. Represent the sequence $a_n = n + 1$ numerically and graphically for $n = 1, 2, 3, 4, 5$.

3. Use the table to determine whether the sequence is arithmetic or geometric. Write the general term for the sequence.

    (a)
n	1	2	3	4	5
a_n	−2	1	4	7	10

    (b)
n	1	2	3	4	5
a_n	3	−6	12	−24	48

4. Determine the general term a_n for an arithmetic sequence with $a_1 = 5$ and $d = 2$.

5. Determine the general term a_n for a geometric sequence with $a_1 = 5$ and $r = 2$.

21.3 Series

Objectives
1. Introducing Series
2. Working with Arithmetic Series
3. Working with Geometric Series
4. Using Summation Notation

NEW VOCABULARY
- ☐ Finite series
- ☐ Arithmetic series
- ☐ Sum of the first n terms of an arithmetic series
- ☐ Geometric series
- ☐ Sum of the first n terms of a geometric series
- ☐ Annuity
- ☐ Summation notation
- ☐ Index of summation
- ☐ Lower limit
- ☐ Upper limit

1 Introducing Series

Although the terms *sequence* and *series* are sometimes used interchangeably in everyday life, they represent different mathematical concepts. A sequence is an *ordered list* of numbers, whereas a series is a *summation of the terms* in a sequence.

Suppose that a person has a starting salary of $30,000 per year and receives a $2000 raise each year. Then the *sequence*

$$30{,}000,\ 32{,}000,\ 34{,}000,\ 36{,}000,\ 38{,}000$$
Sequence

A sequence is an ordered list.

lists these salaries over a 5-year period. The total amount earned is given by the *series*

$$30{,}000\ +\ 32{,}000\ +\ 34{,}000\ +\ 36{,}000\ +\ 38{,}000,$$
Series

A series is the sum of the terms of a sequence.

whose sum is $170,000.

See the Concept 1 — SEQUENCES AND SERIES

For the first seven odd integers:

A Sequence — $1, 3, 5, 7, 9, 11, 13$

A A sequence is an ordered list of terms.

B Series — $1 + 3 + 5 + 7 + 9 + 11 + 13$

B A series is the sum of the terms in a sequence.

C Sum of the series — 49

C When we evaluate the sum in **B**, we say that we are finding the "sum of series."

WATCH VIDEO IN MML.

We now define the concept of a series.

FINITE SERIES

A **finite series** is an expression of the form

$$a_1 + a_2 + a_3 + \cdots + a_n.$$

EXAMPLE 1 Twitter worldwide advertising revenues

TABLE 21.4 presents a sequence a_n that gives Twitter worldwide advertising revenues in millions of dollars n years after 2011. For example, $a_3 = 1095$ indicates that in 2014, advertising revenues were $1095 million.

Twitter Advertising

n	1	2	3	4	5
a_n	269	595	1095	1692	2444

TABLE 21.4
Source: Statista.

(a) Write a series whose sum represents the total worldwide advertising revenues for Twitter from 2012 to 2016. Find the sum of the series.

(b) Interpret $a_1 + a_2 + a_3 + a_4 + a_5$.

Solution

(a) The required series and sum are given by

$$269 + 595 + 1095 + 1692 + 2444 = 6095.$$

(b) Revenues from 2012 to 2016 were $6095 million, or $6.095 billion. This series represents total revenues for 5 years from 2012 to 2016.

Now Try Exercise 45

READING CHECK 1

- How do you distinguish between a sequence and a series?

2 Working with Arithmetic Series

Summing the terms of an arithmetic sequence results in an **arithmetic series**. For example, $a_n = 2n - 1$ for $n = 1, 2, 3, \ldots, 7$ defines the arithmetic sequence

$$\underbrace{1, 3, 5, 7, 9, 11, 13.}_{\text{Arithmetic sequence}}$$

The corresponding arithmetic *series* is

$$\underbrace{1 + 3 + 5 + 7 + 9 + 11 + 13,}_{\text{Arithmetic series}}$$

> An arithmetic series is the sum of the terms of an arithmetic sequence.

whose sum is 49. The following formula gives the sum of the first n terms of an arithmetic sequence. Note that the sum of the first n terms of a sequence is a finite series.

READING CHECK 2

- Use a formula to find the sum of $1 + 5 + 9 + 13 + 17 + 21$. Check your answer by adding the terms.

SUM OF THE FIRST n TERMS OF AN ARITHMETIC SEQUENCE

The **sum of the first n terms of an arithmetic sequence**, denoted S_n, is found by averaging the first and nth terms and then multiplying by n. That is,

$$S_n = a_1 + a_2 + a_3 + \cdots + a_n = n\left(\frac{a_1 + a_n}{2}\right).$$

The series $1 + 3 + 5 + 7 + 9 + 11 + 13$ consists of 7 terms, where the first term is 1 and the last term is 13. Substituting in the formula gives

$$S_7 = 7\left(\frac{1 + 13}{2}\right) = 49,$$

which agrees with the sum obtained by adding the 7 terms.

Because $a_n = a_1 + (n - 1)d$ for an arithmetic sequence, S_n can also be written

$$S_n = n\left(\frac{a_1 + a_n}{2}\right)$$

$$= \frac{n}{2}(a_1 + a_n)$$

$$= \frac{n}{2}(a_1 + a_1 + (n - 1)d)$$

$$= \frac{n}{2}(2a_1 + (n - 1)d).$$

EXAMPLE 2 Finding the sum of the terms of an arithmetic sequence

Suppose that a person has a starting annual salary of $30,000 and receives a $1500 raise each year. Calculate the total amount earned after 10 years.

Solution
The sequence that gives the salary during year n is given by
$$a_n = 30{,}000 + 1500(n - 1).$$
One way to calculate the sum of the first 10 terms, denoted S_{10}, is to find a_1 and a_{10}.
$$a_1 = 30{,}000 + 1500(1 - 1) = \mathbf{30{,}000} \quad \text{Salary after first year}$$
$$a_{10} = 30{,}000 + 1500(10 - 1) = \mathbf{43{,}500} \quad \text{Salary after tenth year}$$
Thus the total amount earned during this 10-year period is
$$S_{10} = 10\left(\frac{a_1 + a_{10}}{2}\right)$$
$$= 10\left(\frac{30{,}000 + 43{,}500}{2}\right)$$
$$= \$367{,}500.$$
This sum can also be found with the second formula by letting $d = 1500$.
$$S_n = \frac{n}{2}(2a_1 + (n - 1)d)$$
$$= \frac{10}{2}(2 \cdot 30{,}000 + (10 - 1)1500)$$
$$= 5(60{,}000 + 9 \cdot 1500)$$
$$= \$367{,}500.$$

NOTE: $\underbrace{30{,}000 + 31{,}500 + 33{,}000 + 34{,}500 + \cdots + 43{,}500}_{10 \text{ terms}} = 367{,}500$ ∎

Now Try Exercise 53

EXAMPLE 3 Finding the sum of the terms of an arithmetic sequence

Use a formula to find the sum of the series $2 + 4 + 6 + \cdots + 100$.

Solution
The first term is $a_1 = 2$, and the common difference is $d = 2$. This series represents the even numbers from 2 to 100, so the number of terms is $n = 50$. Using the formula
$$S_n = n\left(\frac{a_1 + a_n}{2}\right), \quad \text{Sum formula}$$
we obtain
$$S_{50} = 50\left(\frac{2 + 100}{2}\right) = 2550.$$

Now Try Exercise 13

TECHNOLOGY NOTE

Sum of the First n Terms of a Sequence
The "seq(" utility, found on some calculators under the LIST OPS menus, generates a sequence. The "sum(" utility, found on some calculators under the LIST MATH menus, calculates the sum of the sequence inside the parentheses. To verify the result in Example 2, let $a_n = 30{,}000 + 1500(n - 1)$. The value 367,500 for S_{10} is shown in **FIGURE 21.14(a)**. The result found in Example 3 is shown in **FIGURE 21.14(b)**.

Example 2

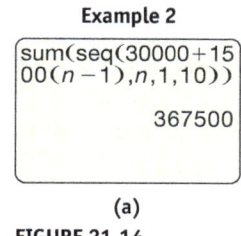

Example 3

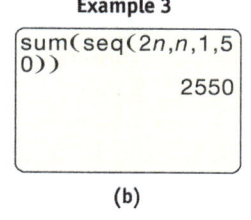

(a)   (b)
FIGURE 21.14

3 Working with Geometric Series

A **geometric series** is the sum of the terms of a geometric sequence. For example,

$$\underbrace{1, 2, 4, 8, 16, 32}_{\text{Geometric sequence}}$$

is a geometric sequence with $a_1 = 1$ and $r = 2$. Then

$$\underbrace{1 + 2 + 4 + 8 + 16 + 32}_{\text{Geometric series}}$$

> A geometric series is the sum of the terms of a geometric sequence.

is a geometric series. We can use the following formula to sum the first n terms of a geometric sequence.

READING CHECK 3

- Use a formula to find the sum

   $1 + 4 + 16 + 64 + 256$.

   Check your answer by adding the terms.

> **SUM OF THE FIRST n TERMS OF A GEOMETRIC SEQUENCE**
>
> If its first term is a_1 and its common ratio is r, then the **sum of the first n terms of a geometric sequence** is given by
>
> $$S_n = a_1 \left(\frac{1 - r^n}{1 - r} \right),$$
>
> provided $r \neq 1$.

EXAMPLE 4   Finding the sum of the terms of a geometric sequence

Find each sum.
(a) $1 + 2 + 4 + 8 + 16 + 32 + 64 + 128 + 256$
(b) $\frac{1}{2} - \frac{1}{4} + \frac{1}{8} - \frac{1}{16} + \frac{1}{32}$

Solution
(a) This series is geometric, with $n = 9$, $a_1 = 1$, and $r = 2$, so

$$S_9 = 1\left(\frac{1 - 2^9}{1 - 2}\right) = 511. \qquad S_n = a_1\left(\frac{1-r^n}{1-r}\right)$$

(b) This series is geometric, with $n = 5$, $a_1 = \frac{1}{2} = 0.5$, and $r = -\frac{1}{2} = -0.5$, so

$$S_5 = 0.5\left(\frac{1 - (-0.5)^5}{1 - (-0.5)}\right) = \frac{11}{32} = 0.34375.$$

Now Try Exercises 19, 21

🌐 *Math in Context* (Investment) A sum of money from which regular payments are made is called an **annuity**. An annuity may be purchased with a lump sum deposit or by deposits made at various intervals. Suppose that a total of $1000 is deposited at the end of each year in an annuity account that pays an annual interest rate I expressed as a decimal. At the end of the first year, the account contains $1000. At the end of the second year, $1000 is deposited again. In addition, the first deposit of $1000 would have received interest during the second year. Therefore the value of the annuity after 2 years is

$$1000 + 1000(1 + I).$$

After 3 years, the balance is

$$1000 + 1000(1 + I) + 1000(1 + I)^2,$$

and after n years, this amount is given by

$$1000 + 1000(1 + I) + 1000(1 + I)^2 + \cdots + 1000(1 + I)^{n-1}.$$

This series is a geometric series with its first term $a_1 = 1000$ and the common ratio $r = (1 + I)$. The sum of the first n terms is given by

$$S_n = a_1\left(\frac{1 - (1 + I)^n}{1 - (1 + I)}\right) = a_1\left(\frac{(1 + I)^n - 1}{I}\right).$$

EXAMPLE 5 **Finding the future value of an annuity**

Suppose that a 20-year-old worker deposits $1000 into an annuity account at the end of each year. If the interest rate is 5%, find the future value of the annuity when the worker is 65 years old.

Solution
Let $a_1 = 1000$, $I = 0.05$, and $n = 45$. The future value of the annuity is

$$S_n = a_1\left(\frac{(1 + I)^n - 1}{I}\right)$$

$$= 1000\left(\frac{(1 + 0.05)^{45} - 1}{0.05}\right)$$

$$\approx \$159{,}700.$$

Now Try Exercise 41

4 Using Summation Notation

Summation notation is used to write series efficiently. The symbol Σ, the uppercase Greek letter *sigma*, is used to indicate a sum.

SUMMATION NOTATION

$$\sum_{k=1}^{n} a_k = a_1 + a_2 + a_3 + \cdots + a_n$$

The letter k is called the **index of summation**. The numbers 1 and n represent the subscripts of the first and last terms in the series. They are called the **lower limit** and **upper limit** of the summation, respectively.

EXAMPLE 6 Using summation notation

Find each sum.

(a) $\sum_{k=1}^{5} k^2$ (b) $\sum_{k=1}^{4} 5$ (c) $\sum_{k=3}^{6} (2k - 5)$

Solution

(a) $\sum_{k=1}^{5} k^2 = 1^2 + 2^2 + 3^2 + 4^2 + 5^2 = 55$

(b) $\sum_{k=1}^{4} 5 = 5 + 5 + 5 + 5 = 20$

(c) $\sum_{k=3}^{6} (2k - 5) = \underbrace{(2(3) - 5)}_{k=3} + \underbrace{(2(4) - 5)}_{k=4} + \underbrace{(2(5) - 5)}_{k=5} + \underbrace{(2(6) - 5)}_{k=6}$

$= 1 + 3 + 5 + 7 = 16$

Now Try Exercises 27, 29, 33

Summation notation is used frequently in statistics. The next example demonstrates how averages can be expressed in summation notation.

EXAMPLE 7 Applying summation notation

Express the average of the n numbers $x_1, x_2, x_3, \ldots, x_n$ in summation notation.

Solution
The average of n numbers can be written as

$$\frac{x_1 + x_2 + x_3 + \cdots + x_n}{n}.$$

This expression is equivalent to $\frac{1}{n}\left(\sum_{k=1}^{n} x_k\right)$.

Now Try Exercise 35

NOTE: $\sum_{k=1}^{n} x_k$ is equivalent to $\sum_{k=1}^{n} x_k$. ∎

🌐 *Math in Context* (Public Health) Sometimes it is necessary to have air pass through more than one filter in order to clean the air properly. In the next example, we use a series to describe this situation.

EXAMPLE 8 Modeling air filtration

Suppose that an air filter removes 90% of the impurities entering it.
(a) Find a series that represents the amount of impurities removed by a sequence of n air filters. Express this answer in summation notation.
(b) How many air filters would be necessary to remove 99.99% of the impurities?

Impurities Passing through Air Filters

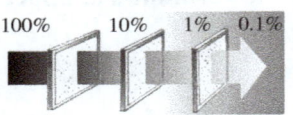

FIGURE 21.15

Solution

(a) The first filter removes 90% of the impurities, so 10%, or 0.1, passes through it. Of the 0.1 that passes through the first filter, 90% is removed by the second filter, while 10% of 10%, or 0.01, passes through. Then, 10% of 0.01, or 0.001, passes through the third filter. FIGURE 21.15 depicts these results, from which we can establish a pattern. If we let 100%, or 1, represent the amount of impurities entering the first air filter, the amount *removed* by n filters equals

$$(0.9)(1) + (0.9)(0.1) + (0.9)(0.01) + (0.9)(0.001) + \cdots + (0.9)(0.1)^{n-1}.$$

In summation notation, we write this series as $\sum_{k=1}^{n} 0.9(0.1)^{k-1}$.

(b) To remove 99.99%, or 0.9999, of the impurities requires 4 air filters, because

$$\sum_{k=1}^{4} 0.9(0.1)^{k-1} = (0.9)(1) + (0.9)(0.1) + (0.9)(0.01) + (0.9)(0.001)$$

$$= 0.9 + 0.09 + 0.009 + 0.0009$$

$$= 0.9999.$$

Now Try Exercise 47

21.3 Putting It All Together

A finite sequence is an ordered list such as

$$a_1, a_2, a_3, a_4, a_5, \ldots, a_n.$$

A finite series is the summation of the terms of a sequence and can be expressed as

$$a_1 + a_2 + a_3 + a_4 + a_5 + \cdots + a_n.$$

SERIES	DESCRIPTION	EXAMPLE
Finite Arithmetic	$a_1 + a_2 + a_3 + \cdots + a_n$, where $a_n = dn + c$ or $a_n = a_1 + (n-1)d$. The sum of the first n terms is $$S_n = n\left(\frac{a_1 + a_n}{2}\right)$$ or $$S_n = \frac{n}{2}(2a_1 + (n-1)d),$$ where a_1 is the first term and d is the common difference.	The series $$4 + 7 + 10 + 13 + 16 + 19 + 22$$ is obtained from the sequence $$a_n = 3n + 1 \quad \text{or} \quad a_n = 4 + 3(n-1).$$ Its sum is $$S_7 = 7\left(\frac{4+22}{2}\right) = 91 \quad \text{or}$$ $$S_7 = \frac{7}{2}(2 \cdot 4 + (7-1)3) = 91.$$
Finite Geometric	$a_1 + a_2 + a_3 + \cdots + a_n$, where $a_n = a_1(r)^{n-1}$ for nonzero constants a_1 and r. The sum of the first n terms is $$S_n = a_1\left(\frac{1-r^n}{1-r}\right),$$ where a_1 is the first term and r is the common ratio ($r \neq 1$).	The series $$3 + 6 + 12 + 24 + 48 + 96$$ has $n = 6$, $a_1 = 3$, and $r = 2$. Its sum is $$S_6 = 3\left(\frac{1-2^6}{1-2}\right) = 189.$$

21.3 Exercises

CONCEPTS AND VOCABULARY

1. The summation of the terms of a sequence is called a(n) ____.

2. Find the sum of the series $1 + 2 + 3 + 4$.

3. The series $1 + 3 + 5 + 7 + 9$ is an example of a(n) ____ series.

4. The series $1 + 3 + 9 + 27 + 81$ is an example of a(n) ____ series.

5. If $a_1 + a_2 + a_3 + \cdots + a_n$ is an arithmetic series, its sum is $S_n =$ ____.

6. If $a_1 + a_2 + a_3 + \cdots + a_n$ is a geometric series with the common ratio $r \neq 1$, its sum is $S_n =$ ____.

7. The symbol Σ is used to indicate a(n) ____.

8. Write $\sum_{k=1}^{4} a_k$ as a sum.

9. $\sum_{n=1}^{5} a_1 + (n-1)d$ is an example of a(n) ____ series.

10. $\sum_{n=1}^{4} a_1 r^{n-1}$ is an example of a(n) ____ series.

WORKING WITH ARITHMETIC SERIES

Exercises 11–18: Find the sum by using a formula.

11. $3 + 5 + 7 + 9 + 11 + 13$

12. $7.5 + 6 + 4.5 + 3 + 1.5 + 0 + (-1.5)$

13. $1 + 2 + 3 + 4 + \cdots + 40$

14. $1 + 3 + 5 + 7 + \cdots + 99$

15. $-7 + (-4) + (-1) + 2 + 5$

16. $89 + 84 + 79 + 74 + 69 + 64 + 59 + 54$

17. $-1 + (-4) + (-7) + (-10) + \cdots + (-22)$

18. $-2 + 3 + 8 + 13 + 18 + \cdots + 33$

WORKING WITH GEOMETRIC SERIES

Exercises 19–26: Find the sum by using a formula.

19. $3 + 9 + 27 + 81 + 243 + 729 + 2187$

20. $2 - 1 + \frac{1}{2} - \frac{1}{4} + \frac{1}{8} - \frac{1}{16} + \frac{1}{32}$

21. $1 - 2 + 4 - 8 + 16 - 32 + 64 - 128$

22. $2 + \frac{1}{2} + \frac{1}{8} + \frac{1}{32} + \frac{1}{128} + \frac{1}{512}$

23. $0.5 + 1.5 + 4.5 + 13.5 + 40.5 + 121.5$

24. $0.6 + 0.3 + 0.15 + 0.075 + 0.0375$

25. $5 + 20 + 80 + 320 + \cdots + 5120$

26. $90 + 30 + 10 + \frac{10}{3} + \frac{10}{9} + \frac{10}{27}$

USING SUMMATION NOTATION

Exercises 27–34: Write the terms of the series and find their sum.

27. $\sum_{k=1}^{4} 2k$

28. $\sum_{k=1}^{6} (k - 1)$

29. $\sum_{k=1}^{8} 4$

30. $\sum_{k=2}^{6} (5 - 2k)$

31. $\sum_{k=1}^{7} k^2$

32. $\sum_{k=1}^{4} 5(2)^{k-1}$

33. $\sum_{k=4}^{5} (k^2 - k)$

34. $\sum_{k=1}^{4} \log k$

Exercises 35–38: Write in summation notation.

35. $1^4 + 2^4 + 3^4 + 4^4 + 5^4 + 6^4$

36. $1 + \frac{1}{5^1} + \frac{1}{5^2} + \frac{1}{5^3} + \frac{1}{5^4}$

37. $1 + \frac{1}{2^2} + \frac{1}{3^2} + \frac{1}{4^2} + \frac{1}{5^2}$

38. $1 + \frac{1}{10} + \frac{1}{100} + \frac{1}{1000} + \frac{1}{10{,}000}$

39. Verify that $\sum_{k=1}^{n} k = \frac{n(n+1)}{2}$ by using a formula for the sum of the first n terms of an arithmetic series.

40. Use Exercise 39 to find the sum of the series $\sum_{k=1}^{200} k$.

APPLICATIONS OF ARITHMETIC AND GEOMETRIC SEQUENCES

Exercises 41–44: Annuities (Refer to Example 5.) Find the future value of the annuity.

41. $a_1 = \$2000 \quad I = 0.08 \quad n = 20$

42. $a_1 = \$500 \quad I = 0.15 \quad n = 10$

43. $a_1 = \$10{,}000 \quad I = 0.11 \quad n = 5$

44. $a_1 = \$3000 \quad I = 0.19 \quad n = 45$

45. **Female CEO Compensation** The following table lists the compensation in millions of dollars for the six top-paid female CEOs of S&P 500 companies in 2015. (*Source: Statista.*)

continued on next page

Company	Xerox	Oracle	TJX
Compensation	22.2	37.7	28.7

Company	Yahoo!	Lockheed Martin	Pepsico
Compensation	42.1	33.7	22.5

(a) Write a series whose sum is the total compensation in millions for the six top-paid female CEOs.
(b) Find the sum of the series in millions of dollars.

46. Global Social Networks The following table lists the number of active users in millions of the top six global social networks in 2015. (*Source:* Statista.)

Company	Facebook Messenger	Whats App	QQ
Active Users	700	800	832

Company	Facebook	QZone	WeChat
Active Users	1490	668	549

(a) Write a series whose sum is the total number of active users in millions for the top six global social networks.
(b) Find the sum of the series in millions.

47. Air Filtration (Refer to Example 8.) Suppose that an air filter removes 80% of the impurities entering it.
(a) Find a series that represents the amount of impurities removed by a sequence of n air filters. Express the answer in summation notation.
(b) How many filters would be necessary to remove 96% of the impurities?

48. Air Filtration Suppose that an air filter removes 70% of the impurities entering it.
(a) Find a series that represents the amount of impurities removed by a sequence of n air filters. Express the answer in summation notation.
(b) How many filters would be necessary to remove 97.3% of the impurities?

49. Area A sequence of smaller squares is formed by connecting the midpoints of the sides of a larger square as shown in the figure.

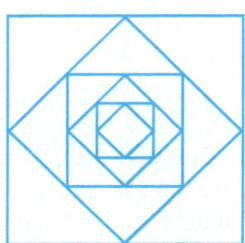

(a) If the area of the largest square is 1 square unit, determine the first five terms of a sequence that describes the area of each successive square.
(b) Use a formula to sum the areas of the first 10 squares.

50. Perimeter (Refer to the previous exercise.) Use a formula to find the sum of the perimeters of the first 10 squares.

51. Stacking Logs A stack of logs is made in layers, with one log less in each layer, as shown in the accompanying figure. If the top layer has 6 logs and the bottom layer has 14 logs, what is the total number of logs in the pile? Use a formula to find this sum.

52. Stacking Logs (Refer to the previous exercise.) Suppose that a stack of logs has 15 logs in the top layer and a total of 10 layers. How many logs are in the stack?

53. Salaries (Refer to Example 2.) Suppose that an individual's starting salary is $35,000 per year and that the individual receives a $2000 raise each year. Find the total amount earned over 20 years.

54. Salaries Suppose that an individual's starting salary is $35,000 per year and that the individual receives a 10% raise each year. Find the total amount earned over 20 years.

55. Bouncing Ball A tennis ball first bounces to 75% of the height from which it was dropped and then to 75% of the height of each successive bounce. If it is dropped from a height of 10 feet, find the distance it *falls* between the fourth and fifth bounce.

56. Bouncing Ball A tennis ball first bounces to 75% of the height from which it was dropped and then to 75% of the height of each successive bounce. If it is dropped from a height of 10 feet, find the *total* distance it travels before it reaches its fifth bounce. (*Hint:* Make a sketch.)

57. Infinite Series The sum S of an infinite geometric series can be found if its common ratio r satisfies $|r| < 1$. It is given by

$$S = \frac{a_1}{1 - r}.$$

(If $|r| \geq 1$, this sum does not exist.) For example, the infinite geometric series

$$1 + \frac{1}{2} + \frac{1}{4} + \frac{1}{8} + \frac{1}{16} + \cdots$$

has $a_1 = 1$ and $r = \frac{1}{2}$. Therefore its sum S equals

$$S = \frac{1}{1 - \frac{1}{2}} = 2.$$

You might want to add terms of this series to see how increasing the number of terms results in a sum closer to 2. Find the sum of each infinite geometric series.

(a) $2 - 1 + \frac{1}{2} - \frac{1}{4} + \frac{1}{8} - \frac{1}{16} + \cdots$
(b) $1 + \frac{1}{3} + \frac{1}{9} + \frac{1}{27} + \frac{1}{81} + \cdots$
(c) $0.1 + 0.01 + 0.001 + 0.0001 + \cdots$
(d) $0.12 + 0.0012 + 0.000012 + 0.00000012 + \cdots$

58. Online Exploration Go online and find a series that can be used to calculate π. Use the first five terms of your series to estimate π.

59. Online Exploration Go online and find a series that can be used to calculate the number e. Use the first five terms of your series to estimate e.

WRITING ABOUT MATHEMATICS

60. Discuss the difference between a sequence and a series. Give an example of each.

61. Suppose that an arithmetic series has $a_1 = 1$ and a common difference of $d = 2$, whereas a geometric series has $a_1 = 1$ and a common ratio of $r = 2$. Discuss how their sums compare as the number of terms n becomes large. (*Hint:* Calculate each sum for $n = 10, 20,$ and 30.)

Group Activity

Calculating π The quest for an accurate estimation for π is a fascinating story covering thousands of years. Because π is an irrational number, it cannot be represented exactly by a fraction. Its decimal expansion neither repeats nor has a pattern. The ability to compute π was essential to the development of societies because π appears in formulas used in construction, surveying, and geometry. In early historical records, π was given the value of 3. Later the Egyptians used a value of

$$\frac{256}{81} \approx 3.1605.$$

Not until the discovery of the series was an exceedingly accurate decimal approximation of π possible. In 1989, π was computed to 1,073,740,000 digits, which required 100 hours of supercomputer time. Why would anyone want to compute π to so many decimal places? One practical reason is to test electrical circuits in new computers. If a computer has a small defect in its hardware, there is a good chance that an error will appear after it has performed trillions of arithmetic calculations during the computation of π. The series

$$\frac{\pi^4}{90} \approx \frac{1}{1^4} + \frac{1}{2^4} + \frac{1}{3^4} + \frac{1}{4^4} + \cdots + \frac{1}{n^4}$$

gives an estimate of π, where larger values of n give better approximations. (*Source:* P. Beckmann, *A History of Pi*.)

(a) Approximate π by finding the sum of the first four terms.
(b) Use a calculator to approximate π by summing the first 50 terms. Compare the result to the actual value of π.

21.4 The Binomial Theorem

Objectives

1. Introducing Pascal's Triangle
2. Writing Factorial Notation
3. Finding Binomial Coefficients
4. Using the Binomial Theorem
   - Expanding a Binomial
   - Finding a Particular Term of $(a + b)^n$

NEW VOCABULARY

☐ Pascal's triangle
☐ Factorial notation
☐ Binomial coefficient
☐ Binomial theorem

1 Introducing Pascal's Triangle

We demonstrate how to expand expressions of the form $(a + b)^n$, where n is a natural number. These expressions occur in statistics, finite mathematics, computer science, and calculus. The first method that we discuss is Pascal's triangle. Expanding $(a + b)^n$ for increasing values of n gives the following results.

Expanding Binomials

$(a + b)^0 = 1$

$(a + b)^1 = 1a + 1b$

$(a + b)^2 = 1a^2 + 2ab + 1b^2$

$(a + b)^3 = 1a^3 + 3a^2b + 3ab^2 + 1b^3$

$(a + b)^4 = 1a^4 + 4a^3b + 6a^2b^2 + 4ab^3 + 1b^4$

$(a + b)^5 = 1a^5 + 5a^4b + 10a^3b^2 + 10a^2b^3 + 5ab^4 + 1b^5$

Note that $(a + b)^1$ has two terms, starting with a and ending with b; $(a + b)^2$ has three terms, starting with a^2 and ending with b^2; and, in general, $(a + b)^n$ has $n + 1$ terms, starting with a^n and ending with b^n. From left to right, the exponent on a decreases by 1 each successive term, and the exponent on b increases by 1 each successive term.

The triangle formed by the red numbers is called **Pascal's triangle**. This triangle consists of 1s along the sides, and each element inside the triangle is the sum of the two numbers above it, as shown in **FIGURE 21.16**. Pascal's triangle is usually written without variables and can be extended to include as many rows as needed.

Pascal's Triangle

```
          1
         1 1
        1 2 1
       1 3 3 1
      1 4 6 4 1
     1 5 10 10 5 1
```

FIGURE 21.16

READING CHECK 1

- Explain how to find the numbers in Pascal's triangle.

We can use this triangle to expand $(a + b)^n$, where n is a natural number. For example, the expression $(m + n)^4$ consists of five terms written as

$$(m + n)^4 = \_\_m^4 + \_\_m^3n^1 + \_\_m^2n^2 + \_\_m^1n^3 + \_\_n^4.$$

Because there are five terms, the coefficients can be found in the fifth row of Pascal's triangle, which is

$$1\ 4\ 6\ 4\ 1.$$

Thus

$$(m + n)^4 = \underline{1}\,m^4 + \underline{4}\,m^3n^1 + \underline{6}\,m^2n^2 + \underline{4}\,m^1n^3 + \underline{1}\,n^4$$
$$= m^4 + 4m^3n + 6m^2n^2 + 4mn^3 + n^4.$$

EXAMPLE 1 Expanding a binomial

Expand each binomial, using Pascal's triangle.
(a) $(x + 2)^5$ (b) $(2m - n)^3$

Solution
(a) To find the coefficients, use the sixth row (1 5 10 10 5 1) in Pascal's triangle.
$$(x + 2)^5 = \underline{1}x^5 + \underline{5}x^4 \cdot 2^1 + \underline{10}x^3 \cdot 2^2 + \underline{10}x^2 \cdot 2^3 + \underline{5}x^1 \cdot 2^4 + \underline{1}(2^5)$$
$$= x^5 + 10x^4 + 40x^3 + 80x^2 + 80x + 32$$

(b) To find the coefficients, use the fourth row (1 3 3 1) in Pascal's triangle.
$$(2m - n)^3 = \underline{1}(2m)^3 + \underline{3}(2m)^2(-n)^1 + \underline{3}(2m)^1(-n)^2 + \underline{1}(-n)^3$$
$$= 8m^3 - 12m^2n + 6mn^2 - n^3$$

Now Try Exercises 13, 15

2 Writing Factorial Notation

An alternative to Pascal's triangle is the binomial theorem, which uses **factorial notation**.

---

n FACTORIAL (n!)

For any positive integer n,
$$n! = 1 \cdot 2 \cdot 3 \cdot \cdots \cdot n.$$

We also define $0! = 1$.

NOTE: Because multiplication is commutative, n factorial can also be defined as
$$n! = n \cdot (n - 1) \cdot (n - 2) \cdot \cdots \cdot 2 \cdot 1. \blacksquare$$

Examples include the following.
$$0! = 1$$
$$1! = 1$$
$$2! = 1 \cdot 2 = 2$$
$$3! = 1 \cdot 2 \cdot 3 = 6$$
$$4! = 1 \cdot 2 \cdot 3 \cdot 4 = 24$$
$$5! = 1 \cdot 2 \cdot 3 \cdot 4 \cdot 5 = 120$$

FIGURE 21.17 supports these results. On some calculators, factorial (!) can be accessed in the MATH PRB menus.

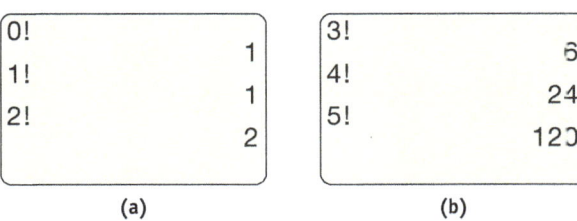

FIGURE 21.17

EXAMPLE 2 Evaluating factorial expressions

Simplify the expression.

(a) $\dfrac{5!}{3!2!}$ (b) $\dfrac{4!}{4!0!}$

Solution

(a) $\dfrac{5!}{3!2!} = \dfrac{1 \cdot 2 \cdot 3 \cdot 4 \cdot 5}{(1 \cdot 2 \cdot 3)(1 \cdot 2)} = \dfrac{120}{6 \cdot 2} = 10$ (b) $0! = 1$, so $\dfrac{4!}{4!0!} = \dfrac{4!}{4!(1)} = \dfrac{4!}{4!} = 1$

Now Try Exercises 25, 27

3 Finding Binomial Coefficients

The expression $_nC_r$ represents a *binomial coefficient* that can be used to calculate the numbers in Pascal's triangle.

> **BINOMIAL COEFFICIENT $_nC_r$**
>
> For n and r nonnegative integers, $n \geq r$,
>
> $$_nC_r = \dfrac{n!}{(n-r)!\,r!}$$
>
> is a **binomial coefficient**.

Values of $_nC_r$ for $r = 0, 1, 2, \ldots, n$ correspond to the $n + 1$ numbers in row $n + 1$ of Pascal's triangle.

EXAMPLE 3 Calculating $_nC_r$

Calculate $_3C_r$ for $r = 0, 1, 2, 3$ by hand. Check your results on a calculator. Compare these numbers with the fourth row in Pascal's triangle.

Solution

$$_3C_0 = \dfrac{3!}{(3-0)!0!} = \dfrac{6}{6 \cdot 1} = 1 \qquad _3C_1 = \dfrac{3!}{(3-1)!1!} = \dfrac{6}{2 \cdot 1} = 3$$

$$_3C_2 = \dfrac{3!}{(3-2)!2!} = \dfrac{6}{1 \cdot 2} = 3 \qquad _3C_3 = \dfrac{3!}{(3-3)!3!} = \dfrac{6}{1 \cdot 6} = 1$$

These results are supported in **FIGURE 21.18**. The fourth row of Pascal's triangle is

$$1 \quad 3 \quad 3 \quad 1,$$

which agrees with the calculated values for $_3C_r$ when $r = 0, 1, 2, 3$. On some calculators, the MATH PRB menus are used to calculate $_nC_r$.

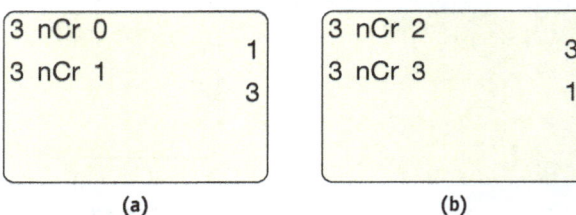

FIGURE 21.18

Now Try Exercises 29, 33

4 Using the Binomial Theorem

EXPANDING A BINOMIAL The binomial coefficients can be used to expand expressions of the form $(a + b)^n$. To do so, we use the **binomial theorem**.

> **BINOMIAL THEOREM**
>
> For any positive integer n and any numbers a and b,
> $$(a + b)^n = {}_nC_0 a^n + {}_nC_1 a^{n-1}b^1 + \cdots + {}_nC_{n-1}a^1 b^{n-1} + {}_nC_n b^n.$$

For example, using the results of Example 3, we write

$$\begin{aligned}(a + b)^3 &= {}_3C_0 a^3 + {}_3C_1 a^2 b^1 + {}_3C_2 a^1 b^2 + {}_3C_3 b^3\\ &= 1a^3 + 3a^2 b + 3ab^2 + 1b^3\\ &= a^3 + 3a^2 b + 3ab^2 + b^3.\end{aligned}$$

EXAMPLE 4 Expanding a binomial

Use the binomial theorem to expand each expression.
(a) $(x + y)^5$ (b) $(3 - 2x)^4$

Solution
(a) The coefficients are calculated as follows.

$${}_5C_0 = \frac{5!}{(5-0)!0!} = 1, \quad {}_5C_1 = \frac{5!}{(5-1)!1!} = 5, \quad {}_5C_2 = \frac{5!}{(5-2)!2!} = 10$$

$${}_5C_3 = \frac{5!}{(5-3)!3!} = 10, \quad {}_5C_4 = \frac{5!}{(5-4)!4!} = 5, \quad {}_5C_5 = \frac{5!}{(5-5)!5!} = 1$$

Using the binomial theorem, we arrive at the following result.

$$\begin{aligned}(x + y)^5 &= {}_5C_0 x^5 + {}_5C_1 x^4 y^1 + {}_5C_2 x^3 y^2 + {}_5C_3 x^2 y^3 + {}_5C_4 x y^4 + {}_5C_5 y^5\\ &= 1x^5 + 5x^4 y + 10x^3 y^2 + 10x^2 y^3 + 5xy^4 + 1y^5\\ &= x^5 + 5x^4 y + 10x^3 y^2 + 10x^2 y^3 + 5xy^4 + y^5\end{aligned}$$

(b) The coefficients are calculated as follows.

$${}_4C_0 = \frac{4!}{(4-0)!0!} = 1, \quad {}_4C_1 = \frac{4!}{(4-1)!1!} = 4, \quad {}_4C_2 = \frac{4!}{(4-2)!2!} = 6,$$

$${}_4C_3 = \frac{4!}{(4-3)!3!} = 4, \quad {}_4C_4 = \frac{4!}{(4-4)!4!} = 1$$

Using the binomial theorem with $a = 3$ and $b = (-2x)$, we arrive at the following result.

$$\begin{aligned}(3 - 2x)^4 &= {}_4C_0 (3)^4 + {}_4C_1 (3)^3 (-2x) + {}_4C_2 (3)^2 (-2x)^2\\ &\quad + {}_4C_3 (3)(-2x)^3 + {}_4C_4 (-2x)^4\\ &= 1(81) + 4(27)(-2x) + 6(9)(4x^2) + 4(3)(-8x^3) + 1(16x^4)\\ &= 81 - 216x + 216x^2 - 96x^3 + 16x^4\end{aligned}$$

Now Try Exercises 41, 43

FINDING A PARTICULAR TERM OF $(a + b)^n$ The binomial theorem gives *all* of the terms of $(a + b)^n$. However, we can find a particular term by noting that the $(r + 1)$st term in the binomial expansion for $(a + b)^n$ is given by the formula

$$_nC_r a^{n-r} b^r,$$

for $0 \leq r \leq n$. The next example shows how to use this formula to find the $(r + 1)$st term of $(a + b)^n$.

EXAMPLE 5 Finding the *k*th term in a binomial expansion

Find the third term of $(x - y)^5$.

Solution

In this example, the $(r + 1)$st term is the *third* term in the expansion of $(x - y)^5$. That is, $r + 1 = 3$, or $r = 2$. Also, the exponent in the expression is $n = 5$. To get this binomial into the form $(a + b)^n$, we note that the first term in the binomial is $a = x$ and that the second term in the binomial is $b = -y$. Substituting the values for r, n, a, and b in the formula $_nC_r a^{n-r} b^r$ for the $(r + 1)$st term yields

$$_5C_2 (x)^{5-2} (-y)^2 = 10x^3 y^2.$$

The third term in the binomial expansion of $(x - y)^5$ is $10x^3 y^2$.

Now Try Exercise 55

21.4 Putting It All Together

CONCEPT	EXPLANATION	EXAMPLE
Pascal's Triangle	1 1 1 1 2 1 **1 3 3 1** 1 4 6 4 1 1 5 10 10 5 1	$(a + b)^3 = 1a^3 + 3a^2 b + 3ab^2 + 1b^3$ (Row 4) To expand $(a + b)^n$, use row $n + 1$ in the triangle.
Factorial Notation	The expression $n!$ equals $1 \cdot 2 \cdot 3 \cdot \ldots \cdot n$.	$5! = 1 \cdot 2 \cdot 3 \cdot 4 \cdot 5 = 120$ $0! = 1$
Binomial Coefficient $_nC_r$	$_nC_r = \dfrac{n!}{(n - r)! r!}$	$_6C_4 = \dfrac{6!}{(6 - 4)! 4!} = \dfrac{6!}{2! 4!} = \dfrac{720}{2 \cdot 24} = 15$
Binomial Theorem	$(a + b)^n = {_nC_0} a^n + {_nC_1} a^{n-1} b^1 + \cdots$ $\qquad + {_nC_{n-1}} a^1 b^{n-1} + {_nC_n} b^n$	$(a + b)^4 = {_4C_0} a^4 + {_4C_1} a^3 b + {_4C_2} a^2 b^2$ $\qquad + {_4C_3} ab^3 + {_4C_4} b^4$ $= 1a^4 + 4a^3 b + 6a^2 b^2 + 4ab^3 + 1b^4$ $= a^4 + 4a^3 b + 6a^2 b^2 + 4ab^3 + b^4$

21.4 Exercises

CONCEPTS AND VOCABULARY

1. How many terms result from expanding $(a + b)^4$?
2. How many terms result from expanding $(a + b)^n$?
3. To find the coefficients for the expansion of $(a + b)^3$, what row of Pascal's triangle do you use?
4. Write down the first 5 rows of Pascal's triangle.
5. $4! = $ _____
6. $1 \cdot 2 \cdot 3 \cdot 4 \cdot 5 \cdot 6 = $ _____
7. $_nC_r = $ _____
8. $(a + b)^2 = $ _____
9. Thinking Generally $\frac{n!}{(n-1)!} = $ _____
10. Thinking Generally $_nC_n = $ _____

USING PASCAL'S TRIANGLE

Exercises 11–18: Use Pascal's triangle to expand the given expression.

11. $(x + y)^3$
12. $(x + y)^4$
13. $(2x + 1)^4$
14. $(2x - 1)^4$
15. $(a - b)^5$
16. $(3x + 2y)^3$
17. $(x^2 + 1)^3$
18. $\left(\frac{1}{2} - x^2\right)^5$

EVALUATING FACTORIAL NOTATION

Exercises 19–28: Evaluate the expression.

19. $3!$
20. $6!$
21. $7!$
22. $0!$
23. $\frac{4!}{3!}$
24. $\frac{6!}{3!}$
25. $\frac{2!}{0!}$
26. $\frac{5!}{1!}$
27. $\frac{5!}{2!3!}$
28. $\frac{6!}{4!2!}$

FINDING BINOMIAL COEFFICIENTS

Exercises 29–34: Evaluate the binomial coefficient by hand.

29. $_5C_4$
30. $_3C_1$
31. $_6C_5$
32. $_2C_2$
33. $_4C_0$
34. $_4C_3$

Exercises 35–40: Evaluate the binomial coefficient with a calculator.

35. $_{12}C_7$
36. $_{13}C_8$
37. $_9C_5$
38. $_{25}C_4$
39. $_{19}C_{11}$
40. $_{10}C_6$

EXPANDING USING THE BINOMIAL THEOREM

Exercises 41–52: Use the binomial theorem to expand the expression.

41. $(m + n)^3$
42. $(m + n)^5$
43. $(x - y)^4$
44. $(1 - 3x)^4$
45. $(2a + 1)^3$
46. $(x^2 - 1)^3$
47. $(x + 2)^5$
48. $(a - 3)^5$
49. $(3 + 2m)^4$
50. $(m - 3n)^3$
51. $(2x - y)^3$
52. $(2a + 3b)^4$

FINDING A PARTICULAR TERM

Exercises 53–58: The $(r + 1)$st term of the expression $(a + b)^n$, $0 \leq r \leq n$, is given by $_nC_r a^{n-r}b^r$. Find the specified term. Refer to Example 5.

53. The first term of $(a + b)^8$
54. The second term of $(a - b)^{10}$
55. The fourth term of $(x + y)^7$
56. The sixth term of $(a + b)^9$
57. The first term of $(2m + n)^9$
58. The eighth term of $(2a - b)^8$

WRITING ABOUT MATHEMATICS

59. Explain how to find the numbers in Pascal's triangle.
60. Compare the expansion of $(a + b)^n$ to the expansion of $(a - b)^n$. Give an example.

SECTIONS 21.3 and 21.4 — Checking Basic Concepts

1. Determine whether the series is arithmetic or geometric.
   (a) $\frac{1}{2} + \frac{1}{4} + \frac{1}{8} + \cdots + \frac{1}{256}$
   (b) $\frac{1}{2} + \frac{5}{2} + \frac{9}{2} + \frac{13}{2} + \frac{17}{2}$

2. Use a formula to find the sum.
   $$4 + 8 + 12 + \cdots + 48.$$

3. Use a formula to find the sum.
   $$1 - 2 + 4 - 8 + 16 - 32 + 64 - 128 + 256 - 512.$$

4. Use Pascal's triangle to expand $(x - y)^4$.

5. Use the binomial theorem to expand $(x + 2)^3$.

CHAPTER 21 Summary

SECTION 21.1 ■ SEQUENCES

Sequences An *infinite sequence* is a function whose domain is the natural numbers. A *finite sequence* is a function whose domain is $D = \{1, 2, 3, \ldots, n\}$ for some natural number n.

Example: $a_n = 2n$ is a symbolic representation of the even natural numbers. The first six terms, 2, 4, 6, 8, 10, 12, of this sequence are represented numerically and graphically in the table and figure.

Numerical Representation

n	a_n
1	2
2	4
3	6
4	8
5	10
6	12

Graphical Representation

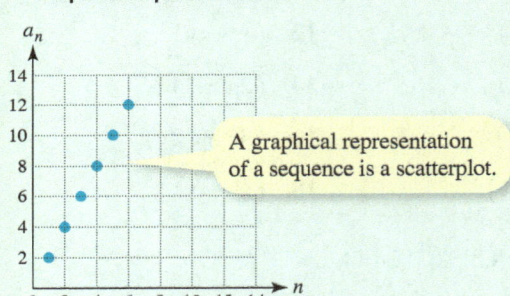

A graphical representation of a sequence is a scatterplot.

SECTION 21.2 ■ ARITHMETIC AND GEOMETRIC SEQUENCES

Two common types of sequences are arithmetic and geometric.

Arithmetic Sequence An arithmetic sequence is determined by a linear function of the form $f(n) = dn + c$ or $f(n) = a_1 + (n - 1)d$. Successive terms in an arithmetic sequence are found by adding the common difference d to the previous term.

Example: The sequence 1, 3, 5, 7, 9, 11, . . . is an arithmetic sequence with its first term $a_1 = 1$, common difference $d = 2$, and general term $a_n = 2n - 1$.

Geometric Sequence The general term for a geometric sequence is given by $f(n) = a_1 r^{n-1}$. Successive terms in a geometric sequence are found by multiplying the previous term by the common ratio r.

Example: The sequence 3, 6, 12, 24, 48, . . . is a geometric sequence with its first term $a_1 = 3$, common ratio $r = 2$, and general term $a_n = 3(2)^{n-1}$.

SECTION 21.3 ■ SERIES

Series A series results when the terms of a sequence are summed. The series associated with the sequence 2, 4, 6, 8, 10 is

$$2 + 4 + 6 + 8 + 10,$$

and its sum equals 30. An arithmetic series results when the terms of an arithmetic sequence are summed, and a geometric series results when the terms of a geometric sequence are summed. In this section, we discussed formulas for finding sums of arithmetic and geometric series. See Putting It All Together for Section 3 of this chapter.

Summation Notation Summation notation can be used to write series efficiently.

Example: $1^2 + 2^2 + 3^2 + 4^2 + 5^2 = \sum_{k=1}^{5} k^2$.

SECTION 21.4 ■ THE BINOMIAL THEOREM

Pascal's triangle may be used to find the coefficients for the expansion of $(a + b)^n$, where n is a natural number.

$$\begin{array}{ccccccccc}
 & & & & 1 & & & & \\
 & & & 1 & & 1 & & & \\
 & & 1 & & 2 & & 1 & & \\
 & 1 & & 3 & & 3 & & 1 & \\
1 & & 4 & & 6 & & 4 & & 1 \\
\end{array}$$
$$1 \quad 5 \quad 10 \quad 10 \quad 5 \quad 1$$

Example: To expand $(x + y)^4$, use the fifth row of Pascal's triangle.

$$(x + y)^4 = \mathbf{1}x^4 + \mathbf{4}x^3y + \mathbf{6}x^2y^2 + \mathbf{4}xy^3 + \mathbf{1}y^4$$
$$= x^4 + 4x^3y + 6x^2y^2 + 4xy^3 + y^4$$

The binomial theorem can also be used to expand powers of binomials. See Putting It All Together for Section 4 of this chapter.

CHAPTER 21 Review Exercises

SECTION 21.1

Exercises 1–4: Write the first four terms of the sequence for $n = 1, 2, 3,$ and 4.

1. $f(n) = n^3$
2. $f(n) = 5 - 2n$
3. $f(n) = \dfrac{2n}{n^2 + 1}$
4. $f(n) = (-2)^n$

Exercises 5 and 6: Write the terms of the sequence.

5.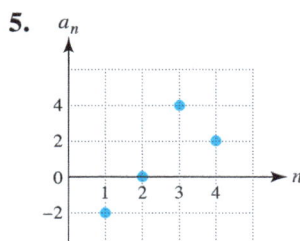

6.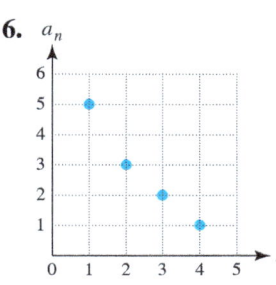

Exercises 7–10: Represent the first seven terms of the sequence numerically and graphically.

7. $a_n = 2n$
8. $a_n = n^2 - 4$
9. $a_n = 4\left(\frac{1}{2}\right)^n$
10. $a_n = \sqrt{n}$

SECTION 21.2

Exercises 11–18: Determine whether f is an arithmetic sequence. Identify the common difference when possible.

11. $f(n) = 5n - 1$
12. $f(n) = 4 - n^2$
13. $f(n) = 2^n$
14. $f(n) = 4 - \frac{1}{3}n$

15.
n	1	2	3	4
$f(n)$	20	17	14	11

16.
n	1	2	3	4
$f(n)$	-3	0	6	12

17. 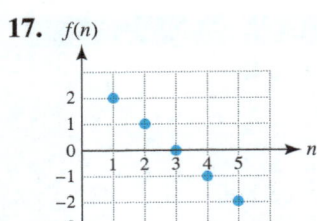   **18.**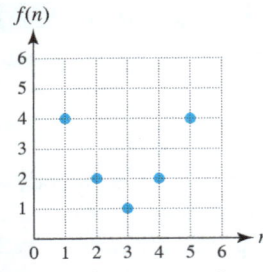

Exercises 19 and 20: Find the general term a_n for the arithmetic sequence.

19. $a_1 = -3$ and $d = 4$ **20.** $a_1 = 2$ and $a_2 = -3$

Exercises 21–28: Determine whether f is a geometric sequence. Identify the common ratio when possible.

21. $f(n) = 2(4)^n$ **22.** $f(n) = 2n^4$

23. $f(n) = 1 - 2n$ **24.** $f(n) = 5(0.7)^n$

25.

n	1	2	3	4
$f(n)$	5	4	3	1

26.

n	1	2	3	4
$f(n)$	27	−9	3	−1

27. 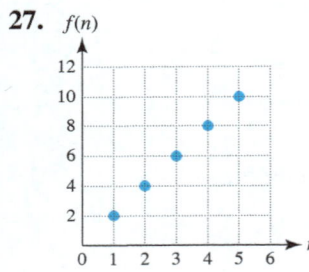   **28.**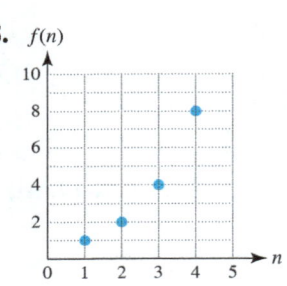

Exercises 29 and 30: Find the general term a_n for the geometric sequence.

29. $a_1 = 5$ and $r = 0.9$ **30.** $a_1 = 2$ and $a_2 = 8$

SECTION 21.3

Exercises 31–34: Find the sum, using a formula.

31. $4 + 9 + 14 + 19 + 24 + 29 + 34 + 39 + 44$

32. $4.5 + 3.0 + 1.5 + 0 - 1.5$

33. $1 - 4 + 16 - 64 + \cdots + 4096$

34. $1 + \frac{1}{2} + \frac{1}{4} + \frac{1}{8} + \frac{1}{16} + \cdots + \frac{1}{256}$

Exercises 35–38: Write the terms of the series.

35. $\sum_{k=1}^{5} 2k + 1$ **36.** $\sum_{k=1}^{4} \frac{1}{k+1}$

37. $\sum_{k=1}^{4} k^3$ **38.** $\sum_{k=2}^{7} (1 - k)$

Exercises 39–42: Write the series in summation notation.

39. $1 + 2 + 3 + \cdots + 20$

40. $1 + \frac{1}{2} + \frac{1}{3} + \cdots + \frac{1}{20}$

41. $\frac{1}{2} + \frac{2}{3} + \frac{3}{4} + \cdots + \frac{9}{10}$

42. $1^2 + 2^2 + 3^2 + 4^2 + 5^2 + 6^2 + 7^2$

SECTION 21.4

Exercises 43–46: Use Pascal's triangle to expand the given expression.

43. $(x + 4)^3$ **44.** $(2x + 1)^4$

45. $(x - y)^5$

46. $(a - 1)^6$

Exercises 47–50: Evaluate the expression.

47. $3!$ **48.** $\dfrac{5!}{3!2!}$

49. $_6C_3$ **50.** $_4C_3$

Exercises 51–54: Use the binomial theorem to expand the given expression.

51. $(m + 2)^4$ **52.** $(a + b)^5$

53. $(x - 3y)^4$ **54.** $(3x - 2)^3$

APPLICATIONS

55. Salaries An individual's starting salary is $45,000, and the individual receives a 10% raise each year. Give symbolic, numerical, and graphical representations for this person's salary over 7 years. What type of sequence is it?

56. Salaries An individual's starting salary is $45,000, and the individual receives an increase of $5000 each year. Give symbolic, numerical, and graphical representations for this person's salary over 7 years. What type of sequence is it?

57. Rain Forests Rain forests are defined as forests that grow in regions that receive more than 70 inches of rain each year. The world is losing an estimated 30 million acres of rain forests annually. Give symbolic, numerical, and graphical representations for a sequence that models the total number of acres (in millions) lost over a 7-year period. (*Source: Scientific American 2009.*)

58. Prison Escapes In 1995, there were 12,000 inmates that escaped from state institutions. Since then these escapes have declined, on average, by 10% per year. (*Source:* Bureau of Justice Statistics.)

(a) Write a sequence a_n that models the number of escapes, where $n = 1$ corresponds to 1995, $n = 2$ to 1996, and so on.

(b) Is a_n arithmetic, geometric, or neither? Explain.

(c) Find a_{19} and interpret the result.

CHAPTER 21 Test MyMathLab® YouTube

1. Write the first four terms of the given sequence for $n = 1, 2, 3,$ and 4.
$$f(n) = \frac{n^2}{n+1}$$

2. Use the graph to write the terms of the sequence.

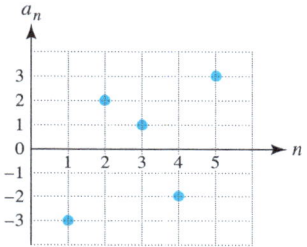

3. List the first seven terms of $a_n = n^2 - n$ in a table. Let $n = 1, 2, \ldots, 7$.

4. Expand the expression $(2x - 1)^4$.

Exercises 5 and 6: Determine whether the sequence is arithmetic or geometric. Identify either the common difference or the common ratio.

5. $f(n) = 7 - 3n$

6.
n	1	2	3	4
$f(n)$	-2	4	-8	16

7. Find the general term a_n for the arithmetic sequence if $a_1 = 2$ and $d = -3$.

8. Find the general term a_n for the geometric sequence if $a_1 = 2$ and $a_3 = 4.5$. Assume $r > 0$.

Exercises 9 and 10: Determine whether $a_n = f(n)$ is a geometric sequence. Identify the common ratio when possible.

9. $f(n) = -3(2.5)^n$

10.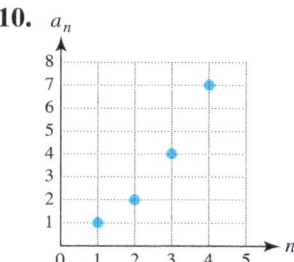

Exercises 11 and 12: Find the sum, using a formula.

11. $-1 + 2 + 5 + 8 + 11 + 14 + 17 + 20 + 23$

12. $1 - \frac{2}{3} + \frac{4}{9} - \frac{8}{27} + \frac{16}{81} - \frac{32}{243} + \frac{64}{729}$

13. Write the terms of the series $\sum_{k=2}^{7} 3k$.

14. Write the series $1^3 + 2^3 + 3^3 + \cdots + 60^3$ in summation notation.

15. Evaluate $\frac{7!}{4!\,3!}$.

16. Evaluate $_5C_3$.

17. **Auditorium Seating** An auditorium has 50 seats in the first row, 57 seats in the second row, 64 seats in the third row, and so on. Use a formula to find the total number of seats in the first 45 rows.

18. **Mobile Addicts** A mobile addict is someone who opens apps on a smartphone more than 60 times a day. In 2014, there were 176 million mobile addicts worldwide, and this number was growing by 59% per year. Give symbolic, numerical, and graphical representations for the number of mobile addicts in millions over a 5-year period, starting in 2014. What type of sequence is it? (*Source:* Business Insider.)

19. **Tent Caterpillars** Large numbers of tent caterpillars can defoliate trees and ruin crops. After they mature, they spin a cocoon and develop into moths that lay eggs. Suppose that an initial population of 2000 tent caterpillars doubles every 5 days.

continued on next page

(a) Write a formula for a_n that models the number of tent caterpillars after $n - 1$ five-day time periods. (Hint: $a_1 = 2000$, $a_2 = 4000$, and $a_3 = 8000$.)

(b) Is a_n arithmetic, geometric, or neither? Explain your reasoning.

(c) Find a_6 and interpret the result.

(d) Give a graph for a_n, where $n = 1, 2, 3, 4, 5, 6$.

CHAPTERS 1–21 Cumulative Review Exercises

1. State whether the equation illustrates an identity, commutative, associative, or distributive property.

$$29(102) = 29(100) + 29(2)$$

2. Identify the domain and range of the relation given by $S = \{(-6, 5), (-2, 1), (0, 3), (2, 0)\}$.

Exercises 3–6: Simplify the expression. Write the result using positive exponents.

3. $\dfrac{x^{-2}y^3}{(3xy^{-2})^3}$

4. $\left(\dfrac{3b}{6a^2}\right)^{-4}$

5. $\left(\dfrac{1}{z^2}\right)^{-5}$

6. $\dfrac{8x^{-3}y^2}{4x^3y^{-1}}$

7. Find the domain of $f(x) = \dfrac{-5}{x - 8}$.

8. Use the table to write the formula for $f(x) = ax + b$.

x	-2	-1	0	1	2
$f(x)$	5	3	1	-1	-3

9. Write the equation of the horizontal line that passes through the point $(2, 3)$.

10. Find the slope and the y-intercept of the graph of $f(x) = -3x + 5$.

Exercises 11 and 12: Write the slope-intercept form for a line satisfying the given conditions.

11. Perpendicular to the line $y = -\frac{2}{3}x - 4$, and passing through $(1, 4)$

12. Parallel to $y = 2x - 7$, passing through $(5, 2)$

Exercises 13–18: Solve the equation or inequality. Write the solutions to inequalities in interval notation.

13. $\frac{2}{5}(x - 4) = -12$

14. $\frac{2}{5}z + \frac{1}{4}z > 2 - (z - 1)$

15. $-3|t - 5| \leq -18$

16. $|4 + \frac{2}{3}x| = 6$

17. $\frac{1}{4}t - (2t + 5) + 6 = \dfrac{t + 3}{4}$

18. $-3 \leq \frac{2}{3}x + 5 < 11$

19. Shade the solution set in the xy-plane.

$$x - y < 4$$
$$x + 2y \geq 7$$

Exercises 20 and 21: Solve the system of equations. Write the solution as an ordered pair or ordered triple where appropriate.

20. $\begin{aligned} x - 2y &= 1 \\ -2x + 7y &= 4 \end{aligned}$

21. $\begin{aligned} x + y + z &= 5 \\ -2x - y + z &= -10 \\ x + 2y + 8z &= 1 \end{aligned}$

22. Evaluate det A.

$$A = \begin{bmatrix} 4 & -3 \\ 3 & 2 \end{bmatrix}$$

Exercises 23 and 24: Multiply the expressions.

23. $2x^3(4x^4 - 3x^3 + 5)$

24. $(2z - 7)(3z + 4)$

Exercises 25 and 26: Factor completely.

25. $4x^2 - 9y^2$

26. $2a^3 - a^2 + 8a - 4$

Exercises 27 and 28: Solve the equation.

27. $4x^2 - x - 3 = 0$

28. $x^4 - 10x^3 = -24x^2$

Exercises 29 and 30: Simplify the expression.

29. $\dfrac{x^2 - 7x + 10}{x^2 - 25} \cdot \dfrac{x + 5}{x + 1}$

30. $\dfrac{x^2 + 7x + 12}{x^2 - 9} \div \dfrac{x^2 - 5x + 6}{(x - 3)^2}$

Exercises 31 and 32: Solve the rational equation. Check your result.

31. $\dfrac{2}{x+5} = \dfrac{-3}{x^2-25} + \dfrac{1}{x-5}$

32. $\dfrac{2y}{y^2-3y+2} = \dfrac{1}{y-2} + 2$

33. Solve the equation for W.
$$R = \dfrac{3C - 2W}{5}$$

34. Simplify the complex fraction.
$$\dfrac{\dfrac{1}{x^2} + \dfrac{2}{x}}{\dfrac{1}{x^2} - \dfrac{4}{x}}$$

Exercises 35 and 36: Simplify the expression. Assume that all variables are positive.

35. $\sqrt[3]{x^4 y^4} - 2\sqrt[3]{xy}$

36. $(4 + \sqrt{2})(4 - \sqrt{2})$

Exercises 37 and 38: Solve. Check your answers.

37. $8(x-3)^2 = 200$

38. $3\sqrt{2x+6} = 6x$

Exercises 39 and 40: Write the complex expression in standard form.

39. $(-3 + i)(-4 - 2i)$

40. $\dfrac{2 - 6i}{1 + 2i}$

41. Find the minimum y-value located on the graph of $y = 3x^2 + 8x + 5$.

42. Write the equation $y = 2x^2 + 8x + 17$ in vertex form and identify the vertex.

Exercises 43 and 44: Solve by using the method of your choice. Write any complex solutions in standard form.

43. $x^2 - 4x + 13 = 0$

44. $z^2 - 4z = 32$

45. Solve the quadratic inequality. Write your answer in interval notation.
$$x^2 + 2x - 3 < 0$$

46. For $f(x) = x^2 + 1$ and $g(x) = 3x - 2$, find each of the following.
(a) $(f \circ g)(-2)$
(b) $(g \circ f)(x)$

47. Find $f^{-1}(x)$ for the one-to-one function
$$f(x) = \dfrac{3x + 1}{2}.$$

48. Write $\ln(x^3 \sqrt{y})$ by using sums and differences of logarithms of x and y.

49. Write $2 \log x - \log 4xy$ as one logarithm. Assume x and y are positive.

Exercises 50 and 51: Solve the equation. Approximate answers to the nearest hundredth.

50. $8 \log x + 3 = 17$ **51.** $4^{2x} = 5$

52. Graph the parabola $x = (y-3)^2 + 1$. Find the vertex and the axis of symmetry.

53. Find the center and the radius of the circle whose equation is $x^2 - 6x + y^2 + 2y = -6$.

Exercises 54 and 55: Graph the ellipse or hyperbola. Label the vertices and the endpoints of the minor axis on the ellipse. Show the asymptotes on the hyperbola.

54. $\dfrac{x^2}{4} + \dfrac{y^2}{9} = 1$ **55.** $\dfrac{x^2}{16} - \dfrac{y^2}{4} = 1$

Exercises 56 and 57: Use the graph to determine the equation of the ellipse or hyperbola.

56. 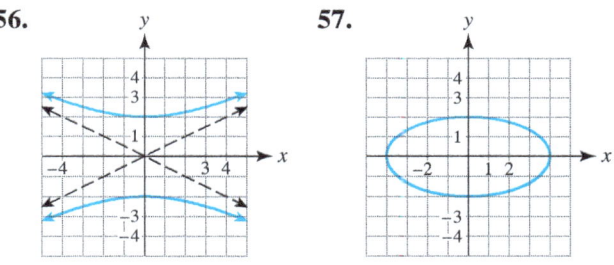 **57.**

58. Solve the system of equations. Check your solutions.
$$y = x^2 + 1$$
$$x^2 + 2y = 5$$

59. Shade the solution set in the xy-plane.
$$y \geq x^2 - 2$$
$$y \leq -x$$

Exercises 60 and 61: Determine whether f is an arithmetic or a geometric sequence. If it is arithmetic, find the common difference. If it is geometric, find the common ratio.

60. $f(n) = 5 - 2n$ **61.** $f(n) = 3(0.2)^n$

62. Find the general term a_n for the arithmetic sequence where $a_1 = 2$ and $a_2 = 5$.

63. Find the general term a_n for the geometric sequence where $a_1 = 4$ and $a_2 = 12$.

Exercises 64 and 65: Find the sum using a formula.

64. $3 + 7 + 11 + 15 + 19 + \cdots + 35$

65. $1 - 2 + 4 - 8 + 16 - \cdots + 1024$

Exercises 66 and 67: Expand the binomial expression.

66. $(2x + 3)^4$ **67.** $(2a - 5b)^3$

APPLICATIONS

68. Radius of a Circle If a circle has an area of A square units, its radius r is given by $r = \sqrt{\frac{A}{\pi}}$. Approximate the radius of a circle with an area of 14 square inches to the nearest hundredth of an inch.

69. Exercise and Fluid Consumption When a person exercises, the total amount of fluid he or she will need that day increases depending on the person's weight and the duration of the exercise. To determine the number of ounces of fluid needed, divide the person's weight by 2 and then add 0.4 ounces for every minute of exercise. *(Source: Runner's World.)*
   (a) Write a function that gives the fluid requirements for a person weighing 170 pounds who exercises for x minutes a day.
   (b) If an athlete who exercises for 90 minutes requires 130 ounces of fluid, find the athlete's weight.

70. Airplane Speed An airplane travels 1080 miles into the wind in 3 hours. The return trip with the wind takes 2.7 hours. Find the average speed of the airplane and the average wind speed.

71. Working Together Suppose that one person can weed a garden in 60 minutes and a second person can weed the same garden in 90 minutes. How long would it take these two people to weed the garden if they worked together?

72. Numbers The product of two positive numbers is 96. If the larger number is subtracted from 3 times the smaller number, the result is 12. Let x be the smaller number and let y be the larger number.
   (a) Write a system of equations for this situation.
   (b) What are the two numbers?

73. Marching Band A band is marching in a triangular formation so that 1 person is in the first row, 3 people are in the second row, 5 people are in the third row, and so on. Use a formula to find the total number of musicians in the marching band if the last row contains 23 people.

APPENDIX A
Using the Graphing Calculator

1 Overview of the Appendix

This appendix provides instruction for the TI-83, TI-83 Plus, and TI-84 Plus graphing calculators that may be used in conjunction with this textbook. It includes specific keystrokes needed to work several examples from the text. Students are advised to consult the *Graphing Calculator Guidebook* provided by the manufacturer.

2 Entering Mathematical Expressions

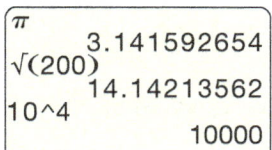

FIGURE A.1

EVALUATING π To evaluate π, use the following keystrokes, as shown in the first and second lines of **FIGURE A.1**. (Do *not* use 3.14 or $\frac{22}{7}$ for π.)

$$\boxed{\text{2nd}}\,\boxed{\wedge[\pi]}\,\boxed{\text{ENTER}}$$

EVALUATING A SQUARE ROOT To evaluate a square root, such as $\sqrt{200}$, use the following keystrokes, as shown in the third and fourth lines of **FIGURE A.1**.

$$\boxed{\text{2nd}}\,\boxed{x^2[\sqrt{\,}]}\,\boxed{2}\,\boxed{0}\,\boxed{0}\,\boxed{)}\,\boxed{\text{ENTER}}$$

EVALUATING AN EXPONENTIAL EXPRESSION To evaluate an exponential expression, such as 10^4, use the following keystrokes, as shown in the last two lines of **FIGURE A.1**.

$$\boxed{1}\,\boxed{0}\,\boxed{\wedge}\,\boxed{4}\,\boxed{\text{ENTER}}$$

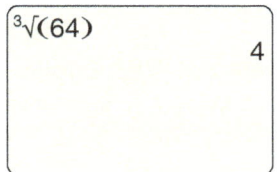

FIGURE A.2

EVALUATING A CUBE ROOT To evaluate a cube root, such as $\sqrt[3]{64}$, use the following keystrokes, as shown in **FIGURE A.2**.

$$\boxed{\text{MATH}}\,\boxed{4}\,\boxed{6}\,\boxed{4}\,\boxed{)}\,\boxed{\text{ENTER}}$$

> **SUMMARY: ENTERING MATHEMATICAL EXPRESSIONS**
>
> To access the *number* π, use $\boxed{\text{2nd}}\,\boxed{\wedge[\pi]}$.
>
> To evaluate a *square root*, use $\boxed{\text{2nd}}\,\boxed{x^2[\sqrt{\,}]}$.
>
> To evaluate an *exponential expression*, use the $\boxed{\wedge}$ key. To square a number, the $\boxed{x^2}$ key can also be used.
>
> To evaluate a *cube root*, use $\boxed{\text{MATH}}\,\boxed{4}$.

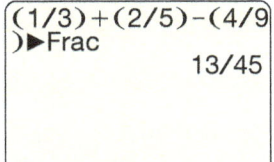

FIGURE A.3

3 Expressing Answers as Fractions

To evaluate $\frac{1}{3} + \frac{2}{5} - \frac{4}{9}$ in fraction form, use the following keystrokes, as shown in **FIGURE A.3**.

> **SUMMARY: EXPRESSING ANSWERS AS FRACTIONS**
>
> Enter the arithmetic expression. To access the "Frac" feature, use the keystrokes
> MATH 1. Then press ENTER.

FIGURE A.4

4 Displaying Numbers in Scientific Notation

To display numbers in scientific notation, set the graphing calculator in scientific mode (Sci) by using the following keystrokes. See **FIGURE A.4**. (These keystrokes assume that the calculator is starting from normal mode.)

In scientific mode, we can display the numbers 5432 and 0.00001234 in scientific notation, as shown in **FIGURE A.5**.

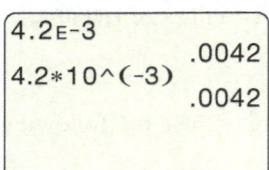

FIGURE A.5

> **SUMMARY: SETTING SCIENTIFIC MODE**
>
> If your calculator is in normal mode, it can be set in scientific mode by pressing
>
>
>
> These keystrokes return the graphing calculator to the home screen.

5 Entering Numbers in Scientific Notation

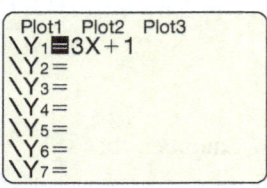

FIGURE A.6

Numbers can be entered in scientific notation. For example, to enter 4.2×10^{-3} in scientific notation, use the following keystrokes. (Be sure to use the negation key (–) rather than the subtraction key.)

This number can also be entered using the following keystrokes. See **FIGURE A.6**.

> **SUMMARY: ENTERING NUMBERS IN SCIENTIFIC NOTATION**
>
> One way to enter a number in scientific notation is to use the keystrokes
>
> 2nd ,[EE]
>
> to access an exponent (EE) of 10.

6 Making a Table

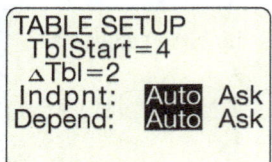

FIGURE A.7

To make a table of values for $y = 3x + 1$ starting at $x = 4$ and incrementing by 2, begin by pressing Y= and then entering the formula $Y_1 = 3X + 1$, as shown in **FIGURE A.7**. (See Entering a Formula on page AP-4.) To set the table parameters, press the following keys. See **FIGURE A.8**.

FIGURE A.8

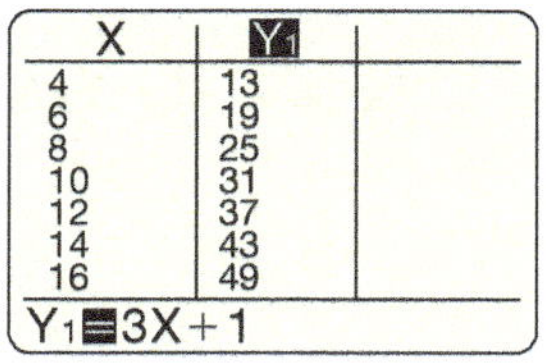

FIGURE A.9

These keystrokes specify a table that starts at $x = 4$ and increments the x-values by 2. Therefore, the values of Y_1 at $x = 4, 6, 8, \ldots$ appear in the table. To create this table, press the following keys.

(2nd) (GRAPH [TABLE])

We can scroll through x- and y-values by using the arrow keys. See **FIGURE A.9**. Note that there is no first or last x-value in the table.

SUMMARY: MAKING A TABLE

1. Enter the formula for the equation using (Y =).
2. Press (2nd) (WINDOW [TBLSET]) to set the starting x-value and the increment between x-values appearing in the table.
3. Create the table by pressing (2nd) (GRAPH [TABLE]).

7 Setting the Viewing Rectangle (Window)

There are at least two ways to set the standard viewing rectangle of $[-10, 10, 1]$ by $[-10, 10, 1]$. The first involves pressing (ZOOM) followed by (6). See **FIGURE A.10**. The second method for setting the standard viewing rectangle is to press (WINDOW) and enter the following keystrokes. See **FIGURE A.11**.

((–)) (1) (0) (ENTER) (1) (0) (ENTER) (1) (ENTER)

((–)) (1) (0) (ENTER) (1) (0) (ENTER) (1) (ENTER)

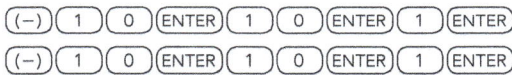

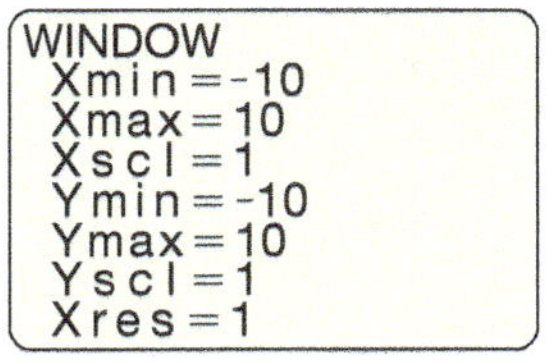

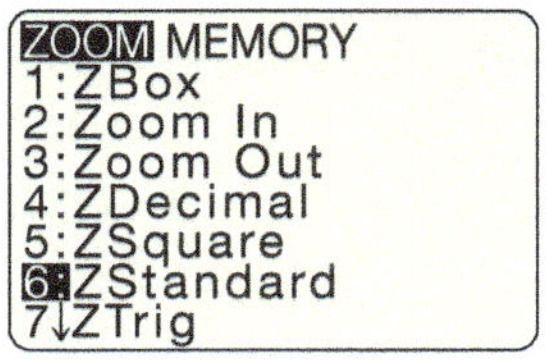

FIGURE A.10

FIGURE A.11

(Be sure to use the negation key $(-)$ rather than the subtraction key.) Other viewing rectangles can be set in a similar manner by pressing (WINDOW) and entering the appropriate values. To see the viewing rectangle, press (GRAPH).

SUMMARY: SETTING THE VIEWING RECTANGLE

To set the standard viewing rectangle, press (ZOOM) (6). To set any viewing rectangle, press (WINDOW) and enter the necessary values. To see the viewing rectangle, press (GRAPH).

NOTE: You do not need to change "Xres" from 1. ■

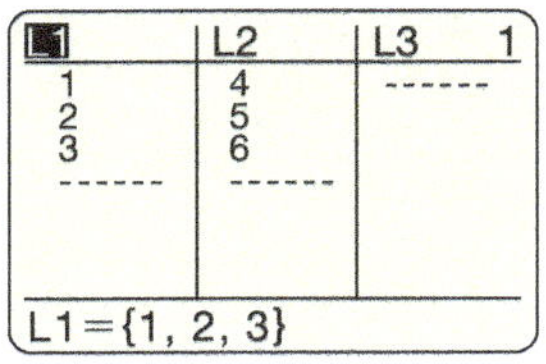

FIGURE A.12

FIGURE A.13

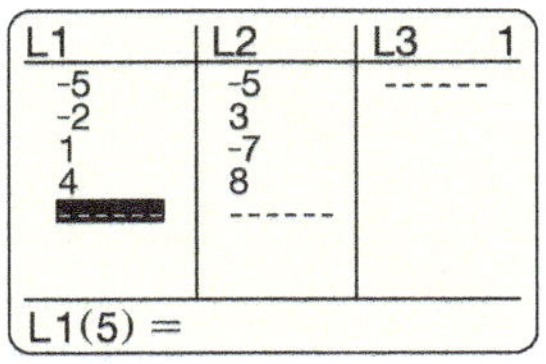

8 Making a Scatterplot or a Line Graph

To make a scatterplot with the points $(-5, -5)$, $(-2, 3)$, $(1, -7)$, and $(4, 8)$, begin by following these steps.

1. Press (STAT) followed by (1).
2. If list L1 is not empty, use the arrow keys to place the cursor on L1, as shown in **FIGURE A.12**. Then press (CLEAR) followed by (ENTER). This deletes all elements in the list. Similarly, if L2 is not empty, clear the list.
3. Input each x-value into list L1 followed by (ENTER). Input each y-value into list L2 followed by (ENTER). See **FIGURE A.13**.

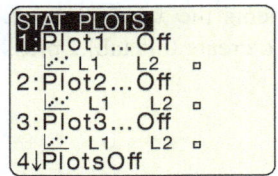

FIGURE A.14

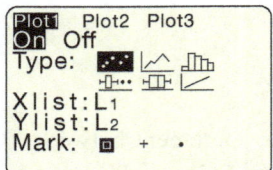

FIGURE A.15

$[-10, 10, 1]$ by $[-10, 10, 1]$

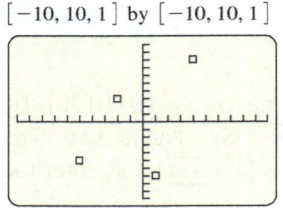

FIGURE A.16

It is essential that both lists have the same number of values—otherwise, an error message appears when a scatterplot is attempted. Before these four points can be plotted, "STAT PLOT" must be turned on. It is accessed by pressing

$$\boxed{\text{2nd}}\;\boxed{\text{Y = [STAT PLOT]}},$$

as shown in **FIGURE A.14**.

There are three possible "STAT PLOTS," numbered 1, 2, and 3. Any one of the three can be selected. The first plot can be selected by pressing $\boxed{1}$. Next, place the cursor over "On" and press $\boxed{\text{ENTER}}$ to turn "Plot1" on. There are six types of plots that can be selected. The first type is a *scatterplot* and the second type is a *line graph*, so place the cursor over the first type of plot and press $\boxed{\text{ENTER}}$ to select a scatterplot. (To make the line graph, place the cursor over the second type of plot and press $\boxed{\text{ENTER}}$.) The *x*-values are stored in list L1, so select L1 for "Xlist" by pressing $\boxed{\text{2nd}}\;\boxed{1}$. Similarly, press $\boxed{\text{2nd}}\;\boxed{2}$ for the "Ylist," since the *y*-values are stored in list L2. Finally, there are three styles of marks that can be used to show data points in the graph. We will usually use the first because it is largest and shows up the best. Make the screen appear as in **FIGURE A.15**. Before plotting the four data points, be sure to set an appropriate viewing rectangle. (See Setting the Viewing Rectangle on the previous page.) Then press $\boxed{\text{GRAPH}}$. The data points appear as in **FIGURE A.16**.

REMARK 1: A fast way to set the viewing rectangle for any scatterplot is to select the "ZOOMSTAT" feature by pressing $\boxed{\text{ZOOM}}\;\boxed{9}$. This feature automatically scales the viewing rectangle so that all data points are shown.

REMARK 2: If an equation has been entered into the $\boxed{\text{Y =}}$ menu and selected, it will be graphed with the data. This feature is used frequently to model data.

> **SUMMARY: MAKING A SCATTERPLOT OR A LINE GRAPH**
>
> The following are basic steps necessary to make either a scatterplot or a line graph.
>
> 1. Use $\boxed{\text{STAT}}\;\boxed{1}$ to access lists L1 and L2.
> 2. If list L1 is not empty, place the cursor on L1 and press $\boxed{\text{CLEAR}}\;\boxed{\text{ENTER}}$. Repeat for list L2, if it is not empty.
> 3. Enter the *x*-values into list L1 and the *y*-values into list L2.
> 4. Use $\boxed{\text{2nd}}\;\boxed{\text{Y = [STAT PLOT]}}$ to select appropriate parameters for the scatterplot or line graph.
> 5. Set an appropriate viewing rectangle. Press $\boxed{\text{GRAPH}}$. Otherwise, press $\boxed{\text{ZOOM}}\;\boxed{9}$. This feature automatically sets the viewing rectangle and plots the data.
>
> **NOTE:** $\boxed{\text{ZOOM}}\;\boxed{9}$ *cannot* be used to set a viewing rectangle for the graph of an equation. ∎

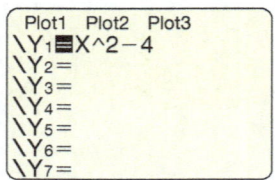

FIGURE A.17

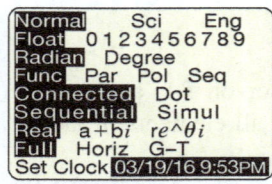

FIGURE A.18

9 Entering a Formula

To enter a formula, press $\boxed{\text{Y =}}$. For example, use the following keystrokes after "$Y_1 =$" to enter $y = x^2 - 4$. See **FIGURE A.17**.

$$\boxed{\text{Y =}}\;\boxed{\text{CLEAR}}\;\boxed{X, T, \theta, n}\;\boxed{\wedge}\;\boxed{2}\;\boxed{-}\;\boxed{4}$$

Note that there is a built-in key to enter the variable X. If "$Y_1 =$" does not appear after pressing $\boxed{\text{Y =}}$, press $\boxed{\text{MODE}}$ and make sure the calculator is set in *function mode*, denoted "Func". See **FIGURE A.18**.

SUMMARY: ENTERING A FORMULA

To enter a formula, press (Y=). To delete a formula, press (CLEAR).

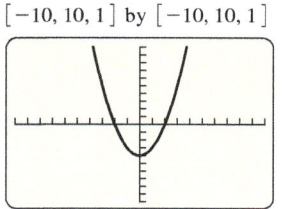

$[-10, 10, 1]$ by $[-10, 10, 1]$

FIGURE A.19

10 Graphing an Equation

To graph an equation, such as $y = x^2 - 4$, start by pressing (Y=) and enter $Y_1 = X^2 - 4$. If there is an equation already entered, remove it by pressing (CLEAR). The equals signs in "$Y_1 =$" should be in reverse video (a dark rectangle surrounding a white equals sign), which indicates that the equation will be graphed. If the equals sign is not in reverse video, place the cursor over it and press (ENTER). Set an appropriate viewing rectangle and then press (GRAPH). The graph will appear in the specified viewing rectangle. See **FIGURES A.17** and **A.19**.

SUMMARY: GRAPHING AN EQUATION

1. Use the (Y=) menu to enter the formula.
2. Use the (WINDOW) menu to set an appropriate viewing rectangle.
3. Press (GRAPH).

11 Graphing a Vertical Line

Set an appropriate window (or viewing rectangle). Then return to the home screen by pressing

(2nd)(MODE [QUIT]).

To graph a vertical line, such as $x = -4$, press

(2nd)(PRGM [DRAW])(4)((-))(4).

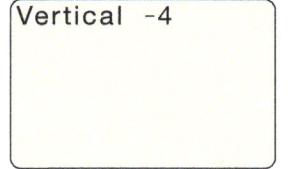

FIGURE A.20

See **FIGURE A.20**. Pressing (ENTER) will make the vertical line appear, as shown in **FIGURE A.21**.

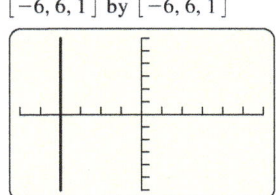

$[-6, 6, 1]$ by $[-6, 6, 1]$

FIGURE A.21

SUMMARY: GRAPHING THE VERTICAL LINE $x = h$

1. Set an appropriate window by pressing (WINDOW).
2. Return to the home screen by pressing (2nd)(MODE [QUIT]).
3. Draw a vertical line by pressing (2nd)(PRGM [DRAW])(4)(h)(ENTER).

12 Squaring a Viewing Rectangle

In a square viewing rectangle the graph of $y = x$ is a line that makes a 45° angle with the positive x-axis, a circle appears circular, and all sides of a square have the same length. An approximate square viewing rectangle can be set if the distance along the x-axis is 1.5 times the distance along the y-axis. Examples of viewing rectangles that are (approximately) square include

$$[-6, 6, 1] \text{ by } [-4, 4, 1] \quad \text{and} \quad [-9, 9, 1] \text{ by } [-6, 6, 1].$$

Square viewing rectangles can be set automatically by pressing either

(ZOOM)(4)   or   (ZOOM)(5).

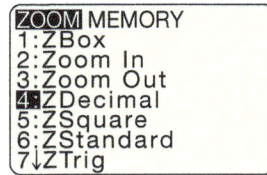

FIGURE A.22

ZOOM 4 provides a *decimal window*, which is discussed on page AP-11. See **FIGURE A.22**.

SUMMARY: SQUARING A VIEWING RECTANGLE

Either (ZOOM) 4 or (ZOOM) 5 may be used to produce a square viewing rectangle. An (approximately) square viewing rectangle has the form

$$[-1.5k, 1.5k, 1] \text{ by } [-k, k, 1],$$

where k is a positive number.

13 Locating a Point of Intersection

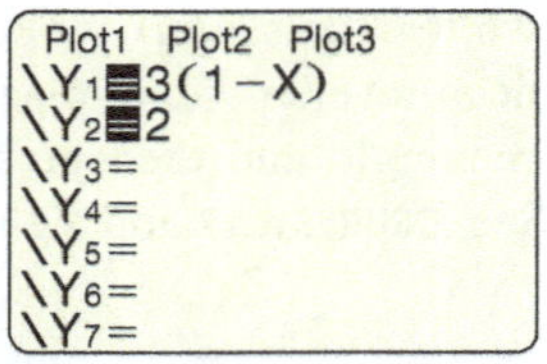

To find the point of intersection for the graphs of

$$y_1 = 3(1 - x) \quad \text{and} \quad y_2 = 2,$$

start by entering Y_1 and Y_2, as shown in **FIGURE A.23**. Set the window, and graph both equations. Then press the following keys to find the intersection point.

$$\boxed{\text{2nd}}\ \boxed{\text{TRACE [CALC]}}\ \boxed{5}$$

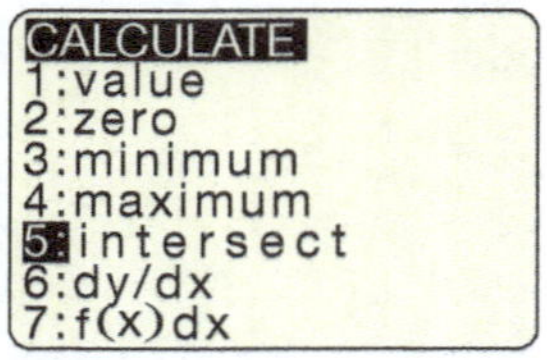

See **FIGURE A.24**, where the "intersect" utility is being selected. The calculator prompts for the first curve, as shown in **FIGURE A.25**. Use the arrow keys to locate the cursor near the point of intersection and press (ENTER). Repeat these steps for the second curve. Finally, we are prompted for a guess. For each of the three prompts, place the free-moving cursor near the point of intersection and press (ENTER). The approximate coordinates of the point of intersection will be shown.

FIGURE A.23

FIGURE A.24

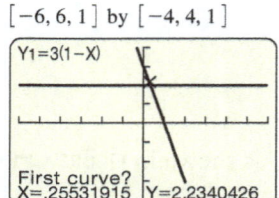

FIGURE A.25

SUMMARY: FINDING A POINT OF INTERSECTION

1. Graph the two equations in an appropriate viewing rectangle.
2. Press (2nd) (TRACE [CALC]) (5).
3. Use the arrow keys to select an approximate location for the point of intersection. Press (ENTER) to make the three selections for "First curve?", "Second curve?", and "Guess?". (Note that if the cursor is near the point of intersection, you usually do not need to move the cursor for each selection. Just press (ENTER) three times.)

14 Shading Inequalities

To shade the solution set for one or more linear inequalities such as $2x + y \leq 5$ and $-2x + y \geq 1$, begin by solving each inequality for y to obtain $y \leq 5 - 2x$ and $y \geq 2x + 1$. Then let $Y_1 = 5 - 2X$ and $Y_2 = 2X + 1$, as shown in **FIGURE A.26**. Position the cursor to the left of Y_1 and press (ENTER) three times. The triangle that appears indicates that the calculator will shade the region below the graph of Y_1. Next locate the cursor to the left of Y_2 and press (ENTER) twice. This triangle indicates that the calculator will shade the region above the graph of Y_2. After setting the viewing rectangle to $[-15, 15, 5]$ by $[-10, 10, 5]$, press (GRAPH). The result is shown in **FIGURE A.27**.

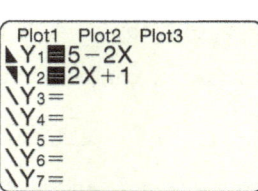

FIGURE A.26 **FIGURE A.27**

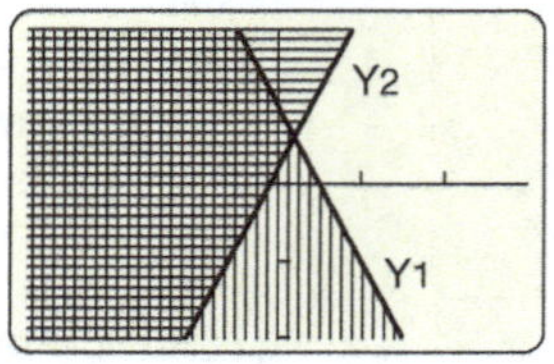

SUMMARY: SHADING ONE OR MORE INEQUALITIES

1. Solve each inequality for y.
2. Enter each formula as Y_1 and Y_2 in the (Y=) menu.
3. Locate the cursor to the left of Y_1 and press (ENTER) two or three times to shade either above or below the graph of Y_1. Repeat for Y_2.
4. Set an appropriate viewing rectangle.
5. Press (GRAPH).

NOTE: The "Shade" utility in the DRAW menu can also be used to shade the region *between* two graphs. ∎

FIGURE A.28

15 Setting $a + bi$ Mode

To evaluate expressions containing square roots of negative numbers, such as $\sqrt{-25}$, set your calculator in $a + bi$ mode by using the following keystrokes.

See **FIGURES A.28** and **A.29**.

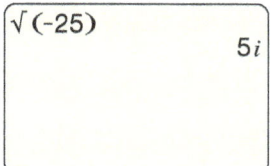

FIGURE A.29

SUMMARY: SETTING $a + bi$ MODE

1. Press (MODE).
2. Move the cursor to the seventh line, highlight $a + bi$, and press (ENTER).
3. Press (2nd)(MODE [QUIT]) and return to the home screen.

16 Evaluating Complex Arithmetic

Complex arithmetic can be performed much like other arithmetic expressions. This is done by entering

to obtain the imaginary unit i from the home screen. For example, to find the sum $(-2 + 3i) + (4 - 6i)$, perform the following keystrokes on the home screen.

FIGURE A.30

The result is shown in **FIGURE A.30**. Other complex arithmetic operations are done similarly.

SUMMARY: EVALUATING COMPLEX ARITHMETIC

Enter a complex expression in the same way as you would any arithmetic expression. To obtain the complex number i, use (2nd)(. [i]).

FIGURE A.31

17 Evaluating Other Mathematical Expressions

EVALUATING OTHER ROOTS To evaluate a fifth root, such as $\sqrt[5]{23}$, use the following keystrokes, as shown in **FIGURE A.31**.

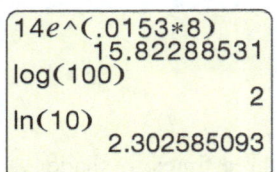

FIGURE A.32

EVALUATING THE NATURAL EXPONENTIAL FUNCTION To evaluate $14e^{0.0153(8)}$, use the following keystrokes, as shown in the first and second lines of **FIGURE A.32**.

EVALUATING THE COMMON LOGARITHMIC FUNCTION To evaluate $\log(100)$, use the following keystrokes, as shown in the third and fourth lines of **FIGURE A.32**.

EVALUATING THE NATURAL LOGARITHMIC FUNCTION To evaluate $\ln(10)$, use the following keystrokes, as shown in the last two lines of **FIGURE A.32**.

SUMMARY: OTHER MATHEMATICAL EXPRESSIONS

To evaluate a *kth root*, use (k)(MATH)(5).

To access the *natural exponential function*, use (2nd)(LN [eˣ]).

To access the *common logarithmic function*, use (LOG).

To access the *natural logarithmic function*, use (LN).

18 Accessing the Absolute Value

To graph $y_1 = |x - 50|$, begin by entering $Y_1 = \text{abs}(X - 50)$. The absolute value (abs) is accessed by pressing

(MATH)(▷)(1).

See **FIGURE A.33**.

SUMMARY: ACCESSING THE ABSOLUTE VALUE

1. Press (MATH).
2. Position the cursor over "NUM".
3. Press (1) to select the absolute value.

19 Entering the Elements of a Matrix

The elements of the augmented matrix A given by

$$A = \begin{bmatrix} 1 & 1 & 2 & | & 1 \\ -1 & 0 & 1 & | & -2 \\ 2 & 1 & 5 & | & -1 \end{bmatrix}$$

can be entered by using the following keystrokes on the TI-83 Plus or TI-84 Plus to define a matrix A with dimension 3×4. (*Note:* On the TI-83 the matrix menu is found by pressing (MATRX).)

Input the 12 elements of the matrix A, row by row. Finish each entry by pressing (ENTER). See **FIGURE A.34**. After these elements have been entered, press

(2nd)(MODE [QUIT])

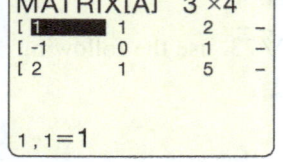

FIGURE A.34

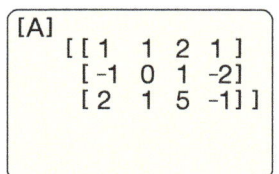

[A]
[[1   1   2   1]
 [-1   0   1 -2]
 [2   1   5 -1]]

FIGURE A.35

to return to the home screen. To display the matrix A, press

$\boxed{\text{2nd}}\boxed{x^{-1}\text{ [MATRIX]}}\boxed{1}\boxed{\text{ENTER}}$.

See **FIGURE A.35**.

SUMMARY: ENTERING THE ELEMENTS OF A MATRIX A

1. Begin by accessing the matrix A by pressing $\boxed{\text{2nd}}\boxed{x^{-1}\text{ [MATRIX]}}\boxed{\triangleright}\boxed{\triangleright}\boxed{1}$.
2. Enter the dimension of A by pressing $\boxed{m}\boxed{\text{ENTER}}\boxed{n}\boxed{\text{ENTER}}$, where the dimension of the matrix is $m \times n$.
3. Input each element of the matrix, row by row. Finish each entry by pressing $\boxed{\text{ENTER}}$. Use $\boxed{\text{2nd}}\boxed{\text{MODE [QUIT]}}$ to return to the home screen.

NOTE: On the TI-83, replace the keystrokes $\boxed{\text{2nd}}\boxed{x^{-1}\text{ [MATRIX]}}$ with $\boxed{\text{MATRX}}$. ∎

20 Finding Reduced Row–Echelon Form

To find the reduced row–echelon form of matrix A (entered above in **FIGURE A.35**), use the following keystrokes from the home screen on the TI-83 Plus or TI-84 Plus.

$\boxed{\text{2nd}}\boxed{x^{-1}\text{ [MATRIX]}}\boxed{\triangleright}\boxed{\text{ALPHA}}\boxed{\text{APPS [B]}}\boxed{\text{2nd}}\boxed{x^{-1}\text{ [MATRIX]}}\boxed{1}\boxed{)}\boxed{\text{ENTER}}$

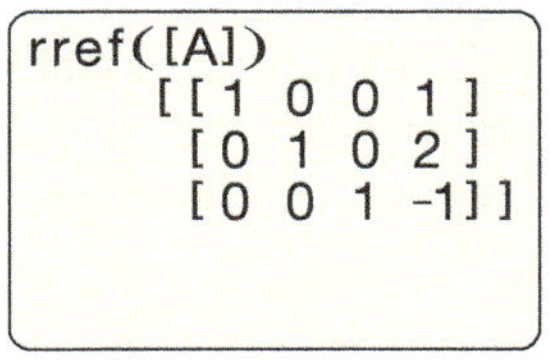

rref([A])
[[1 0 0   1]
 [0 1 0   2]
 [0 0 1 -1]]

FIGURE A.36

The resulting matrix is shown in **FIGURE A.36**. On the TI-83 graphing calculator, use the following keystrokes to find the reduced row–echelon form.

$\boxed{\text{MATRX}}\boxed{\triangleright}\boxed{\text{ALPHA}}\boxed{\text{MATRX [B]}}\boxed{\text{MATRX}}\boxed{1}\boxed{)}\boxed{\text{ENTER}}$

SUMMARY: FINDING THE REDUCED ROW–ECHELON FORM OF A MATRIX

1. To make rref([A]) appear on the home screen, use the following keystrokes for the TI-83 Plus or TI-84 Plus graphing calculator.

   $\boxed{\text{2nd}}\boxed{x^{-1}\text{ [MATRIX]}}\boxed{\triangleright}\boxed{\text{ALPHA}}\boxed{\text{APPS [B]}}\boxed{\text{2nd}}\boxed{x^{-1}\text{ [MATRIX]}}\boxed{1}\boxed{)}$

2. Press $\boxed{\text{ENTER}}$ to calculate the reduced row–echelon form.
3. Use arrow keys to access elements that do not appear on the screen.

NOTE: On the TI-83, replace the keystrokes $\boxed{\text{2nd}}\boxed{x^{-1}\text{ [MATRIX]}}$ with $\boxed{\text{MATRX}}$ and $\boxed{\text{APPS [B]}}$ with $\boxed{\text{MATRX [B]}}$. ∎

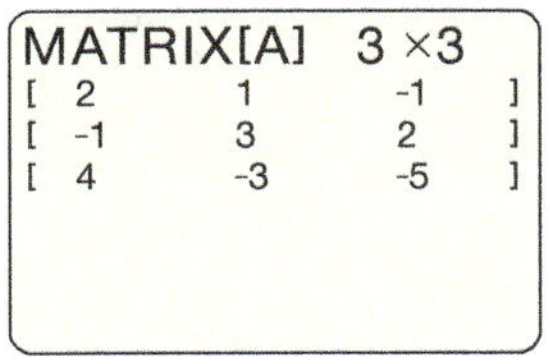

MATRIX[A] 3 ×3
[2   1   -1   ]
[-1   3   2   ]
[4   -3   -5   ]

FIGURE A.37

21 Evaluating a Determinant

To evaluate the determinant of matrix A given by

$$A = \begin{bmatrix} 2 & 1 & -1 \\ -1 & 3 & 2 \\ 4 & -3 & -5 \end{bmatrix},$$

start by entering the 9 elements of the 3×3 matrix, as shown in **FIGURE A.37**. To compute det A, perform the following keystrokes from the home screen.

det([A])
                -6

FIGURE A.38

$\boxed{\text{2nd}}\boxed{x^{-1}\text{ [MATRIX]}}\boxed{\triangleright}\boxed{1}\boxed{\text{2nd}}\boxed{x^{-1}\text{ [MATRIX]}}\boxed{1}\boxed{)}\boxed{\text{ENTER}}$

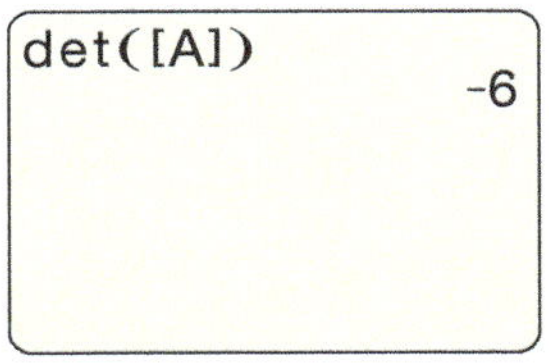

The results are shown in **FIGURE A.38**.

> **SUMMARY: EVALUATING A DETERMINANT OF A MATRIX**
>
> 1. Enter the dimension and elements of the matrix A.
> 2. Return to the home screen by pressing (2nd)(MODE [QUIT]).
> 3. On the TI-83 Plus or TI-84 Plus, perform the following keystrokes.
>
>   (2nd)(x^{-1} [MATRIX]) (▷) (1) (2nd)(x^{-1} [MATRIX]) (1) ()) (ENTER)
>
> **NOTE:** On the TI-83, replace the keystrokes (2nd)(x^{-1} [MATRIX]) with (MATRX). ■

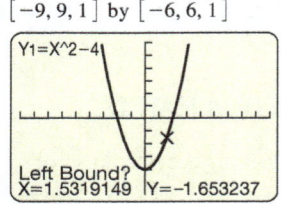

FIGURE A.39

22 Locating an *x*-Intercept or Zero

To locate an *x*-intercept or *zero* of $f(x) = x^2 - 4$, start by entering $Y_1 = X^2 - 4$ into the (Y=) menu. Set the viewing rectangle to $[-9, 9, 1]$ by $[-6, 6, 1]$ and graph Y_1. Afterwards, press the following keys to invoke the zero finder. See **FIGURE A.39**.

(2nd)(TRACE [CALC]) (2)

The graphing calculator prompts for a left bound. Use the arrow keys to set the cursor to the left of the *x*-intercept and press (ENTER). The graphing calculator then prompts for a right bound. Set the cursor to the right of the *x*-intercept and press (ENTER). Finally, the graphing calculator prompts for a guess. Set the cursor roughly at the *x*-intercept and press (ENTER). See **FIGURES A.40–A.42**. The calculator then approximates the *x*-intercept or zero automatically, as shown in **FIGURE A.43**. The zero of -2 can be found similarly.

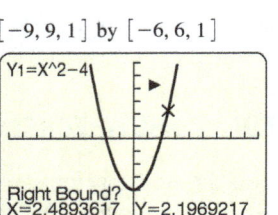

FIGURE A.40 **FIGURE A.41** **FIGURE A.42** **FIGURE A.43**

> **SUMMARY: LOCATING AN *x*-INTERCEPT OR ZERO**
>
> 1. Graph the function in an appropriate viewing rectangle.
> 2. Press (2nd)(TRACE [CALC])(2).
> 3. Select the left and right bounds, followed by a guess. Press (ENTER) after each selection. The calculator then approximates the *x*-intercept or zero.

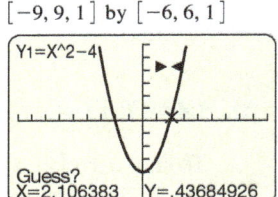

FIGURE A.44

23 Setting Connected or Dot Mode

To set your graphing calculator in dot mode, press (MODE), position the cursor over "Dot," and press (ENTER). See **FIGURE A.44**. Graphs will now appear in dot mode rather than connected mode.

> **SUMMARY: SETTING CONNECTED OR DOT MODE**
>
> 1. Press (MODE).
> 2. Position the cursor over "Connected" or "Dot". Press (ENTER).

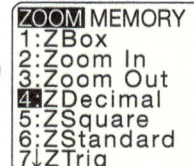

FIGURE A.45

24 Setting a Decimal Window

With a decimal window, the cursor stops on convenient x-values. In the decimal window $[-9.4, 9.4, 1]$ by $[-6.2, 6.2, 1]$ the cursor stops on x-values that are multiples of 0.2. If we reduce the viewing rectangle to $[-4.7, 4.7, 1]$ by $[-3.1, 3.1, 1]$, the cursor stops on x-values that are multiples of 0.1. To set this smaller window automatically, press ZOOM 4. See **FIGURE A.45**. Decimal windows are also useful when graphing rational functions with asymptotes in connected mode.

SUMMARY: SETTING A DECIMAL WINDOW

1. Press ZOOM 4 to set the viewing rectangle $[-4.7, 4.7, 1]$ by $[-3.1, 3.1, 1]$.
2. A larger decimal window is $[-9.4, 9.4, 1]$ by $[-6.2, 6.2, 1]$.

25 Finding Maximum and Minimum Values

To find a minimum y-value (or vertex) on the graph of $f(x) = 1.5x^2 - 6x + 4$, start by entering $Y_1 = 1.5X\text{^}2 - 6X + 4$ from the Y = menu. Set the viewing rectangle and then perform the following keystrokes to find the minimum y-value.

2nd TRACE [CALC] 3

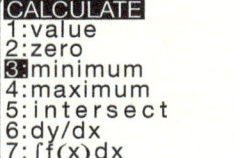

FIGURE A.46

See **FIGURE A.46**.

The calculator prompts for a left bound. Use the arrow keys to position the cursor left of the vertex and press ENTER. Similarly, position the cursor to the right of the vertex for the right bound and press ENTER. Finally, the graphing calculator asks for a guess between the left and right bounds. Place the cursor near the vertex and press ENTER. See **FIGURES A.47–A.49**. The minimum value is shown in **FIGURE A.50**.

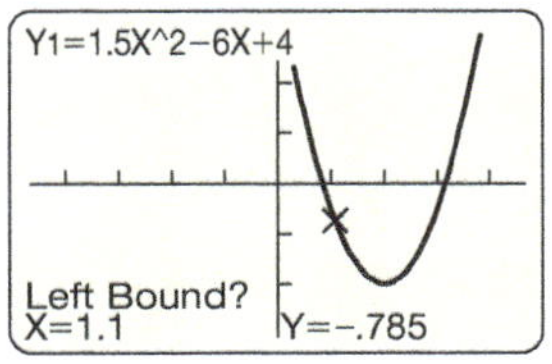

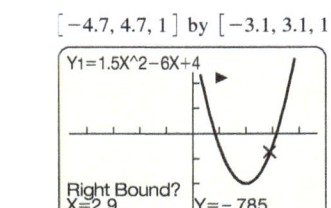

$[-4.7, 4.7, 1]$ by $[-3.1, 3.1, 1]$

FIGURE A.47

$[-4.7, 4.7, 1]$ by $[-3.1, 3.1, 1]$

FIGURE A.48

$[-4.7, 4.7, 1]$ by $[-3.1, 3.1, 1]$

FIGURE A.49

$[-4.7, 4.7, 1]$ by $[-3.1, 3.1, 1]$

FIGURE A.50

$[0, 50, 10]$ by $[0, 1000, 100]$

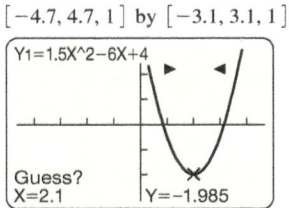

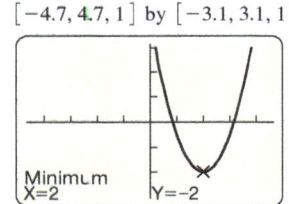

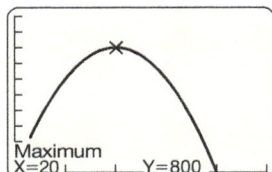

FIGURE A.51

A maximum of the function f on an interval can be found similarly, except enter

2nd TRACE [CALC] 4.

The calculator prompts for left and right bounds, followed by a guess. Press ENTER after the cursor has been located appropriately for each prompt. The graphing calculator will display the maximum y-value. An example is shown in **FIGURE A.51**, where $f(x) = x(80 - 2x)$.

SUMMARY: FINDING MAXIMUM AND MINIMUM VALUES

1. Graph the function in an appropriate viewing rectangle.
2. Press 2nd TRACE [CALC] 3 to find a minimum y-value.
3. Press 2nd TRACE [CALC] 4 to find a maximum y-value.
4. Use the arrow keys to locate the left and right x-bounds, followed by a guess. Press ENTER to select each position of the cursor.

APPENDIX B
Sets

NEW VOCABULARY

- ☐ Set
- ☐ Elements
- ☐ Empty (null) set
- ☐ Universal set
- ☐ Subset
- ☐ Equal sets
- ☐ Venn diagrams
- ☐ Complement
- ☐ Union
- ☐ Intersection
- ☐ Finite/infinite set

1 Determining the Elements of a Set

A **set** is a collection of things, and the members of a set are called **elements**. A set can be described by listing its elements between braces. For example, the set W containing the *weekdays* is

$$\text{Set} \to W = \{\underbrace{\text{Monday, Tuesday, Wednesday, Thursday, Friday}}_{\text{Elements}}\}.$$

This set has 5 elements. For example, Monday *is an element of* W, which is denoted

$$\text{Monday} \in W.$$

However, Sunday *is not an element of* W, which is denoted

$$\text{Sunday} \notin W.$$

If a set contains no elements, then it is called the **empty set** or **null set**. The empty set is denoted $\emptyset$, or $\{\ \}$. For example, the set Z that contains the names of U.S. states starting with the letter Z is the empty set. That is, $Z = \emptyset$ or $Z = \{\ \}$.

NOTE: Do *not* write the empty set as $\{\emptyset\}$. ∎

EXAMPLE 1   Listing the elements of a set

Use set notation to list the elements of each set S described.
(a) The natural numbers from 1 to 12 that are odd
(b) The days of the week that start with the letter T
(c) The last names of U.S. presidents in office from 2005 to 2015

Solution
(a) The list of the natural numbers from 1 to 12 is

$$1, 2, 3, 4, 5, 6, 7, 8, 9, 10, 11, \text{ and } 12.$$

The set of odd natural numbers from this list is

$$S = \{1, 3, 5, 7, 9, 11\}.$$

(b) $S = \{\text{Tuesday, Thursday}\}$
(c) $S = \{\text{Bush, Obama}\}$

Now Try Exercises 1, 3, 5

EXAMPLE 2   Determining the elements of sets

Use $\in$ or $\notin$ to make each statement true.
(a) 5 ____ $\{1, 2, 3, 4, 5, 6\}$
(b) -2 ____ $\{-4, 0, 2, 4, 6\}$
(c) $\frac{1}{2}$ ____ $\{0, 0.5, 1.0, 1.5, 2.0\}$

Solution
(a) Because 5 is an element of $\{1, 2, 3, 4, 5, 6\}$, we write

$$5 \in \{1, 2, 3, 4, 5, 6\}.$$

(b) Because -2 is not an element of $\{-4, 0, 2, 4, 6\}$, we write
$$-2 \notin \{-4, 0, 2, 4, 6\}.$$

(c) Because $\frac{1}{2} = 0.5$, we write
$$\frac{1}{2} \in \{0, 0.5, 1.0, 1.5, 2.0\}.$$

Now Try Exercises 11, 13, 15

2 Determining the Universal Set

When discussing sets, we assume that there is a *universal set*. The **universal set** contains all elements under consideration. For example, if the universal set U is all days of the week, then the set S containing the days that start with the letter S is
$$S = \{\text{Sunday, Saturday}\}.$$

However, if the universal set U is only the weekdays, then the set S containing the days starting with the letter S is

the *empty set*, or $S = \{\ \}$.

EXAMPLE 3 Using different universal sets

Determine the set O of odd integers that belong to each universal set U.
(a) $U = \{1, 2, 3, 4, 5, 6, 7, 8, 9, 10\}$
(b) $U = \{1, 6, 11, 16, 21, 26\}$
(c) $U = \{1, 2, 3, 4, \ldots\}$

Solution
(a) The odd integers in U are 1, 3, 5, 7, and 9, so
$$O = \{1, 3, 5, 7, 9\}.$$
(b) The odd integers in $U = \{1, 6, 11, 16, 21, 26\}$ are 1, 11, and 21. Thus
$$O = \{1, 11, 21\}.$$
(c) The three dots in $\{1, 2, 3, 4, \ldots\}$ indicate that U contains all natural numbers. Thus the set O contains all odd natural numbers, or
$$O = \{1, 3, 5, 7, 9, \ldots\}.$$

Now Try Exercises 21, 23

3 Determining Subsets

If every element in a set B is contained in a set A, then we say that B is a **subset** of A, denoted $B \subseteq A$. For example, if $A = \{1, 2, 3, 4\}$ and $B = \{2, 4\}$, then $B \subseteq A$ because every element in B belongs to A. However, A is not a subset of B, denoted $A \not\subseteq B$, because the elements 1 and 3 are in A but *not* in B. The symbol $\not\subseteq$ is read "is not a subset of."

If every element in set A is in set B and every element in set B is in set A, then A and B are **equal sets**, denoted $A = B$. Note that if $A \subseteq B$ and $B \subseteq A$, then $A = B$.

EXAMPLE 4 Determining subsets

Let $A = \{a, b, c, d, e\}$, $B = \{b, c, d\}$, $C = \{b, e\}$, and $D = \{e, b\}$. Determine whether each statement is true or false.
(a) $A \subseteq B$ (b) $B \subseteq A$ (c) $C = D$ (d) $C \not\subseteq A$ (e) $\emptyset \subseteq B$ (f) $A \subseteq A$

Solution

(a) False; the elements **a** and **e** in A are not in B, so A is *not* a subset of B.
(b) True; every element in B is in A, so B is a subset of A.
(c) True; although the elements in C and D are listed in a different order, they contain exactly the same elements, so C and D are equal.
(d) False; every element in C is in A, so C *is* a subset of A.
(e) True; the empty set, or null set, is a subset of *every* set.
(f) True; every element in A is in A, so A is a subset of itself.

NOTE: Every set is a subset of itself. ∎

Now Try Exercises 29, 31, 33, 35

4 Using Venn Diagrams and Finding Complements

Venn diagrams are often used to depict relationships among sets. A large rectangle typically represents the universal set, and subsets of the universal set are represented by regions within the universal set. In **FIGURE B.1**, the universal set U is represented by everything inside the large rectangle. The set A is represented by the red circular region within this rectangle because A is a subset of U.

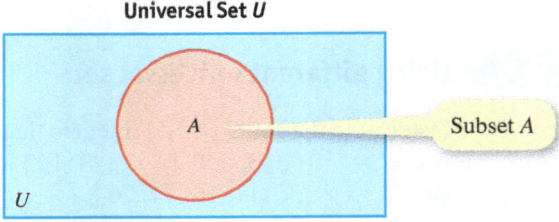

FIGURE B.1 $A \subseteq U$

The **complement** of a set A, denoted A', is the set containing all elements in the universal set that are *not* in A. That is, if $a \notin A$, then $a \in A'$. For example, if

$$U = \{1, 2, 3, 4, 5, 6\} \quad \text{and} \quad A = \{1, 2, 3\},$$

then

$$A' = \{4, 5, 6\}$$

because the elements 4, 5, and 6 are found in U but not in A. This situation is illustrated by the Venn diagram in **FIGURE B.2**. The red region is A, and the blue region is A'. Together, the red and blue regions make up the universal set U.

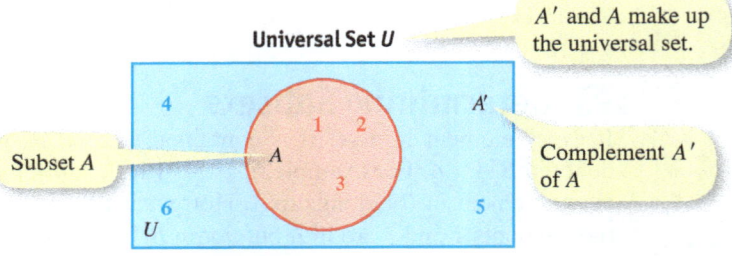

FIGURE B.2 A and A'

NOTE: Every element in U must be in either A or A', but not both. ∎

EXAMPLE 5 Determining complements

Let the universal set be $U = \{\text{red, blue, yellow, green, black, white}\}$, and let two subsets of U be $A = \{\text{red, blue, yellow}\}$ and $B = \{\text{black, white}\}$. Find each of the following.
(a) A' (b) B' (c) U'

Solution
(a) The elements in U that are not in A are in $A' = \{\text{green, black, white}\}$.
(b) $B' = \{\text{red, blue, yellow, green}\}$
(c) Because the universal set U contains every element under consideration, the complement of U, or U', is empty. That is, $U' = \{\ \}$ or $\varnothing$.

Now Try Exercises 73, 75, 81

5 Finding Unions and Intersections

Although we do not perform arithmetic operations, such as multiplication or division, on sets, we can find the *union* or *intersection* of two or more sets. The **union** of two sets A and B, denoted $A \cup B$ and read "A union B," is the set containing any element that can be found in *either* set A *or* set B. If an element is in both A and B, then this element is listed only once in $A \cup B$. For example, if

$$A = \{1, 2, 3, 4\} \quad \text{and} \quad B = \{3, 4, 5, 6\},$$

then

$$A \cup B = \{1, 2, 3, 4, 5, 6\}. \quad \text{Elements in set } A \text{ or in set } B$$

Note that elements 3 and 4 are in both A and B, but are listed only once in $A \cup B$. A Venn diagram of this situation is shown in **FIGURE B.3**. The region that represents the union is shaded blue.

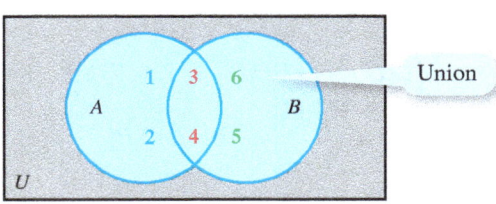

FIGURE B.3

The **intersection** of two sets A and B, denoted $A \cap B$ and read "A intersect B," is the set containing elements that can be found in *both* set A *and* set B. For example, if

$$A = \{1, 2, 3, 4\} \quad \text{and} \quad B = \{3, 4, 5, 6\},$$

then

$$A \cap B = \{3, 4\}. \quad \text{Elements in both set } A \text{ and set } B$$

Note that elements 3 and 4 belong to *both A and B*. This situation is illustrated by the Venn diagram in **FIGURE B.4** on the next page. The purple region containing elements 3 and 4 represents the intersection of A and B.

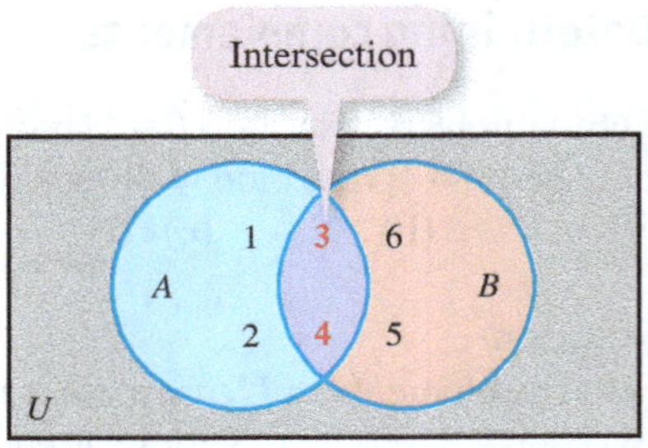

FIGURE B.4

> **EXAMPLE 6** Finding unions and intersections of sets
>
> Let $A = \{a, b, c, x, z\}$, $B = \{a, x, y\}$, and $C = \{c, x, y, z\}$. Find each set.
> (a) $A \cup B$ (b) $B \cap C$ (c) $A \cup C$ (d) $A \cap B \cap C$
>
> **Solution**
> (a) The union contains the elements belonging to either A or B or both. Thus
> $$A \cup B = \{a, b, c, x, y, z\}.$$
> (b) The intersection contains the elements belonging to both B and C. Thus
> $$B \cap C = \{x, y\}.$$
> (c) $A \cup C = \{a, b, c, x, y, z\}$
> (d) For an element to be in the intersection of three sets, the element must belong to each set. The only element belonging to A and B and C is x. Thus $A \cap B \cap C = \{x\}$.
>
> Now Try Exercises 63, 65, 69, 79

> **EXAMPLE 7** Using set operations
>
> Let $U = \{1, 2, 3, 4, 5, 6, 7, 8\}$, $A = \{2, 4, 6, 8\}$, $B = \{1, 3, 5, 6\}$, and $C = \{5, 6, 7, 8\}$. Find each set.
> (a) $A \cap B$ (b) $B \cap B'$ (c) $B \cup C$ (d) $A \cup C'$
>
> **Solution**
> (a) $A \cap B = \{6\}$
> (b) Because $B = \{1, 3, 5, 6\}$ and $B' = \{2, 4, 7, 8\}$, $B \cap B' = \{\ \}$.
>
>   **NOTE:** The intersection of a set and its complement is always the empty set. ∎
> (c) $B \cup C = \{1, 3, 5, 6, 7, 8\}$
> (d) Because $A = \{2, 4, 6, 8\}$ and $C' = \{1, 2, 3, 4\}$, $A \cup C' = \{1, 2, 3, 4, 6, 8\}$.
>
> Now Try Exercises 67, 71, 77, 83

6 Recognizing Finite and Infinite Sets

Sets can either have a finite number of elements or infinitely many elements. The elements of a **finite set** can be listed explicitly, whereas the elements of an **infinite set** cannot be listed because there are infinitely many elements. For example, the set of integers S from -3 to 3 is a finite set with 7 elements because

$$S = \{-3, -2, -1, 0, 1, 2, 3\}. \quad \text{Finite set}$$

In contrast, the set of natural numbers,

$$N = \{1, 2, 3, 4, \ldots\}, \quad \text{Infinite set}$$

is an infinite set because the list continues without end.

EXAMPLE 8 Recognizing finite and infinite sets

Identify each set as finite or infinite.
(a) The set of integers
(b) The set of natural numbers between 1 and 5, inclusive
(c) The set of rational numbers between 1 and 5, inclusive

Solution
(a) There are infinitely many integers, so the set of integers is infinite.
(b) The set of natural numbers between 1 and 5, inclusive, is $\{1, 2, 3, 4, 5\}$. Because we can list each element of this set, it is a finite set.
(c) Because there are infinitely many rational numbers (fractions) between 1 and 5, this set is an infinite set.

Now Try Exercises 87, 89

B Exercises

FINDING ELEMENTS OF A SET

Exercises 1–10: Use set notation to list the elements of the set described.

1. The natural numbers less than 8
2. The natural numbers greater than 4 and less than or equal to 10
3. The days of the week starting with the letter S
4. U.S. states whose names start with the letter A
5. The letters of the alphabet from A to G
6. The letters of the alphabet from D to M
7. Dogs that can fly under their own power
8. People having an income of $10 trillion in 2017
9. The even integers greater than -2 and less than 11
10. The odd integers less than 21 and greater than 12

Exercises 11–20: Insert $\in$ or $\notin$ to make the statement true.

11. 10 _____ $\{5, 10, 15, 20\}$
12. 7 _____ $\{1, 3, 5, 7, 9, 11\}$
13. 5 _____ $\{2, 4, 6, 8, 10\}$
14. -1 _____ $\{0, 1, 2, 3, 4, 5\}$
15. $\frac{1}{4}$ _____ $\{0, 0.25, 0.5, 0.75, 1\}$
16. $\frac{2}{5}$ _____ $\{0, 0.2, 0.4, 0.6, 0.8, 1\}$
17. $\frac{1}{3}$ _____ $\{0, 0.33, 0.67, 1\}$
18. $\frac{4}{2}$ _____ $\{1, 3, 4, 5, 7\}$
19. red _____ $\{\text{blue, green, red, yellow}\}$
20. M _____ $\{M, T, W, R, F\}$

Exercises 21–24: Determine the set E of even integers that belong to the given universal set U.

21. $U = \{1, 2, 3, 4, 5, 6, 7, 8, 9, 10\}$
22. $U = \{-3, -2, -1, 0, 1, 2, 3\}$
23. $U = \{1, 2, 3, 4, \ldots\}$
24. $U = \{1, 3, 5, 7, 9, 11, \ldots\}$

Exercises 25–28: Determine the set A of words starting with the letter A that belong to the given universal set U.

25. $U = \{\text{Apple, Orange, Pear, Apricot}\}$
26. $U = \{\text{Apple, Orange}\}$
27. $U = \{\text{Calculus, Algebra, Geometry}\}$
28. $U = \{\text{Calculus, Algebra, Geometry, Arithmetic}\}$

FINDING SUBSETS

Exercises 29–36: Let $A = \{1, 2, 3, 4\}$, $B = \{3, 4, 5, 6\}$, and $C = \{3, 6\}$. Determine whether the statement is true or false.

29. $A \subseteq B$
30. $B \subseteq A$
31. $C \subseteq B$
32. $\emptyset \subseteq C$
33. $C \subseteq C$
34. $A \not\subseteq B$
35. $B \not\subseteq C$
36. $C \subseteq A$

Exercises 37–44: Let $A = \{a, b, c, d, e\}$, $B = \{b, d\}$, and $C = \{a, c, e\}$. Determine whether the statement is true or false.

37. $B \subseteq A$
38. $C \subseteq A$
39. $B \not\subseteq C$
40. $C = B$
41. $C \subseteq \emptyset$
42. $C \not\subseteq B$
43. $A \subseteq A$
44. $C \not\subseteq C$

SKETCHING VENN DIAGRAMS

Exercises 45–52: Let A and B be two sets and U be the universal set. Sketch a Venn diagram and shade the region that illustrates the given set.

45. $A \cup B$
46. $A \cap B$
47. A'
48. B'
49. $A \cap U$
50. $B \cup U$
51. $(A \cap B)'$
52. $(A \cup B)'$

FINDING UNIONS, INTERSECTIONS, AND COMPLEMENTS

Exercises 53–62: Write the expression in terms of one set.

53. $\{1, 2, 3\} \cup \{3, 4\}$
54. $\{1, 2, 3\} \cup \{6\}$
55. $\{a, b, c\} \cap \{a, b, d\}$
56. $\{x, y, z\} \cap \{a, b, c\}$
57. $\{a, b, c\} \cup \{\ \}$
58. $\{a, b, c\} \cap \emptyset$
59. $\{a, b\} \cup \{c, b\} \cup \{d, a\}$
60. $\{1, 3, 5\} \cup \{4, 5\} \cup \{4, 5, 6\}$
61. $\{4, 5, 8, 9\} \cap \{3, 5, 8, 9\} \cap \{4, 5, 8\}$
62. $\{1, 2, 5\} \cap \{2, 1\} \cap \{1, 2, 3\}$

Exercises 63–86: Let
$$U = \{1, 2, 3, 4, 5, 6, 7, 8, 9, 10\},$$
$$A = \{1, 3, 5, 7, 9\}, B = \{2, 4, 6, 8, 10\},$$
$$C = \{3, 4, 5, 6, 7\}, \text{ and } D = \{4, 5, 6, 9\}.$$

Use set notation to list the elements in the given set.

63. $A \cup B$
64. $A \cap B$
65. $C \cap D$
66. $A \cup D$
67. $U \cap \emptyset$
68. $\emptyset \cup U$
69. $A \cup \emptyset$
70. $\emptyset \cap B$
71. $D \cap D'$
72. $A \cup A$
73. U'
74. A'
75. B'
76. D'
77. $A \cup D'$
78. $A' \cap B'$
79. $A \cap B \cap C$
80. $B \cup C \cup D$
81. C'
82. $D' \cup D$
83. $B \cap C$
84. $B \cap C \cap D$
85. $C \cup C'$
86. $D' \cap B \cap C$

RECOGNIZING FINITE AND INFINITE SETS

Exercises 87–92: Determine whether the given set is finite or infinite.

87. The set of whole numbers
88. The set of real numbers
89. The set of natural numbers less than 1000
90. The set of natural numbers greater than 4 and less than 20
91. The days of the week
92. The names of the students in your class

APPENDIX C
Linear Programming

NEW VOCABULARY

☐ Objective function
☐ Constraint
☐ Feasible solutions
☐ Linear programming problem
☐ Optimal value
☐ Vertex

1 Understanding Objective Functions and Constraints

Math in Business Context Suppose that a small business sells candy for $3 per pound and freshly ground coffee for $5 per pound. All inventory is sold by the end of the day. The revenue R collected in dollars is given by

$$R = 3x + 5y,$$ *Objective function*

where x is the pounds of candy sold and y is the pounds of coffee sold. For example, if the business sells **80** pounds of candy and **40** pounds of coffee during a day, then its revenue is

$$R = 3(80) + 5(40) = \$440.$$

The function $R = 3x + 5y$ is called an **objective function**.

Suppose also that the company cannot package more than 150 pounds of candy and coffee per day. Then the inequality

$$x + y \leq 150$$ *Constraint*

represents a **constraint** on the objective function, which limits the company's revenue for any one day. A goal of this business might be to maximize the objective function

$$R = 3x + 5y,$$ Revenue equation

subject to the following constraints.

$$x + y \leq 150 \quad \text{Constraint on production}$$
$$x \geq 0, y \geq 0 \quad \text{Variables cannot be negative.}$$ *Constraints*

Note that the constraints $x \geq 0$ and $y \geq 0$ are included because the number of pounds of candy or coffee cannot be negative. The problem that we have described is called a *linear programming problem*. Before learning how to solve a linear programming problem, we need to discuss the region of *feasible solutions*.

2 Finding the Region of Feasible Solutions

The constraints for a linear programming problem consist of linear inequalities. These inequalities are satisfied by some points in the xy-plane but not by others. The set of solutions to these constraints is called the **feasible solutions**. For example, the *region of feasible solutions* to the constraints for the business just described is shaded in **FIGURE C.1**.

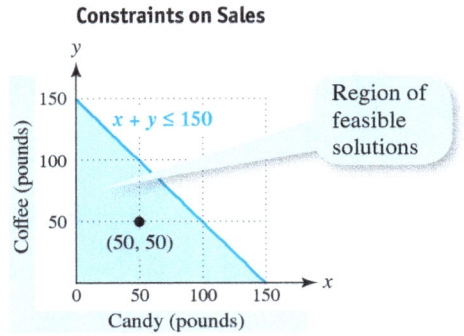

FIGURE C.1

The point $(50, 50)$ lies in the shaded region and represents the business selling 50 pounds of candy and 50 pounds of coffee. In the next example, we shade the region of feasible solutions to a set of constraints.

EXAMPLE 1 Finding the region of feasible solutions

Shade the region of feasible solutions for the following constraints.

$$x + 2y \leq 30$$
$$2x + y \leq 30$$
$$x \geq 0, y \geq 0$$

Solution

The feasible solutions are the ordered pairs (x, y) that *satisfy all four* inequalities. They lie on or below the lines $x + 2y = 30$ and $2x + y = 30$, and on or above the line $y = 0$ and on or to the right of $x = 0$, as shown in **FIGURE C.2**. Note that the inequalities $x \geq 0$ and $y \geq 0$ restrict the region of feasible solutions to quadrant I (including the x- and y-axes.)

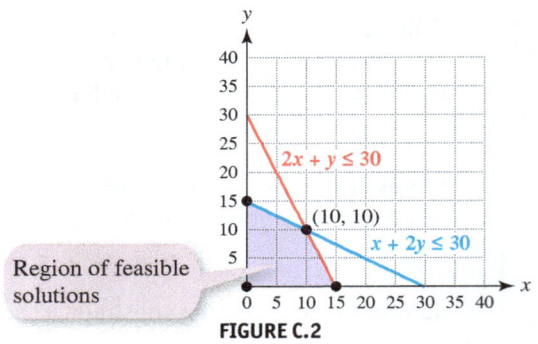

FIGURE C.2

Now Try Exercise 15

STUDY TIP

Be sure that you can graph inequalities (see Example 1) before you attempt to solve a linear programming problem.

3 ▶ Solving Linear Programming Problems

A **linear programming problem** consists of an *objective function* and a system of linear inequalities called *constraints*. The solution set for the system of linear inequalities is called the *region of feasible solutions*. The objective function describes a quantity that is to be optimized. The **optimal value** for a linear programming problem often results in maximum revenue or minimum cost.

When the system of constraints has only two variables, the boundary of the region of feasible solutions often consists of line segments intersecting at points called *vertices* (plural of **vertex**). To solve a linear programming problem, we use the *fundamental theorem of linear programming*.

FUNDAMENTAL THEOREM OF LINEAR PROGRAMMING

If the optimal value for a linear programming problem exists, then it occurs at a vertex of the region of feasible solutions.

The fundamental theorem of linear programming is used to solve the following linear programming problem. A justification of this theorem is given after Example 2.

EXAMPLE 2 **Maximizing an objective function**

Maximize the objective function $R = 2x + 3y$ subject to

$$x + 2y \leq 30$$
$$2x + y \leq 30$$
$$x \geq 0, y \geq 0.$$

Solution
The region of feasible solutions is shaded in **FIGURE C.2** from Example 1. Note that the vertices on the boundary of feasible solutions are $(0, 0)$, $(15, 0)$, $(10, 10)$, and $(0, 15)$. To find the maximum value of R, substitute each vertex in the formula for R. The maximum value of R is 50 when $x = 10$ and $y = 10$. See **TABLE C.1**.

Vertex	$R = 2x + 3y$
$(0, 0)$	$2(0) + 3(0) = 0$
$(15, 0)$	$2(15) + 3(0) = 30$
$(10, 10)$	$2(10) + 3(10) = 50$
$(0, 15)$	$2(0) + 3(15) = 45$

TABLE C.1

Maximum R is 50.

Now Try Exercise 33

The following steps are helpful in solving linear programming word problems.

STEPS FOR SOLVING A LINEAR PROGRAMMING WORD PROBLEM

STEP 1: Read the problem carefully. Consider making a table.

STEP 2: Write the objective function and all the constraints.

STEP 3: Sketch a graph of the region of feasible solutions. Identify all vertices.

STEP 4: Evaluate the objective function at each vertex. A maximum (or a minimum) occurs at a vertex.

NOTE: If the region is unbounded, a maximum (or minimum) may not exist. ∎

JUSTIFICATION OF THE FUNDAMENTAL THEOREM To better understand the fundamental theorem of linear programming, consider the following example. Suppose that we want to maximize profit $P = 3x + 7y$ subject to the following four constraints:

$$x \geq 1, \quad x \leq 5, \quad y \geq x, \quad \text{and} \quad y \leq 6.$$

The corresponding region of feasible solutions is shown in **FIGURE C.3** on the next page.

Each value of P determines a unique line. For example, if $P = 70$, then the equation for P becomes $3x + 7y = 70$. The resulting line, shown in **FIGURE C.4** on the next page, does not intersect the region of feasible solutions. Thus there are no values for x and y that lie in this region and result in a profit of 70. **FIGURE C.4** also shows the lines that result from letting $P = 0, 10,$ and 30. If $P = 10$, then the line intersects the region of feasible solutions only at the vertex $(1, 1)$. This means that if $x = 1$ and $y = 1$, then $P = 3(1) + 7(1) = 10$. If $P = 30$, then the line $3x + 7y = 30$ intersects the region of feasibility infinitely many

CRITICAL THINKING

What is the minimum value for P subject to the given constraints?

times. However, it appears that values greater than 30 are possible for P. In **FIGURE C.5**, lines are drawn for $P = 57, 63$, and 70. Notice that there are no points of intersection for $P = 63$ or $P = 70$, but there is one vertex in the region of feasible solutions at $(5, 6)$ that gives $P = 57$. Thus the maximum value of P is 57, and this maximum occurs at a vertex of the region of feasible solutions. The fundamental theorem of linear programming generalizes this result.

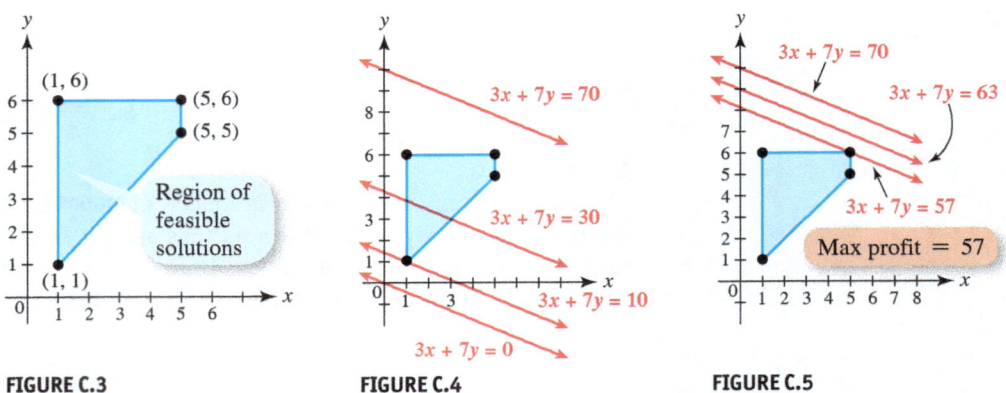

FIGURE C.3 **FIGURE C.4** **FIGURE C.5**

EXAMPLE 3   Minimizing the cost of vitamins

A breeder is mixing two different vitamins, Brand X and Brand Y, into pet food. Each serving of pet food should contain at least 60 units of vitamin A and 30 units of vitamin C. Brand X costs 80 cents per ounce and Brand Y costs 50 cents per ounce. Each ounce of Brand X contains 15 units of vitamin A and 10 units of vitamin C, whereas each ounce of Brand Y contains 20 units of vitamin A and 5 units of vitamin C. Determine how much of each brand of vitamin should be mixed to produce a minimum cost per serving.

Solution

STEP 1: Begin by listing the information, as illustrated in **TABLE C.2**.

Brand	Amount	Vitamin A	Vitamin C	Cost
X	x	15	10	80 cents
Y	y	20	5	50 cents
Minimum		60	30	

TABLE C.2

STEP 2: If x ounces of Brand X are purchased at 80 cents per ounce and if y ounces of Brand Y are purchased at 50 cents per ounce, then the total cost C is given by $C = 80x + 50y$. Because each ounce of Brand X contains 15 units of vitamin A and each ounce of Brand Y contains 20 units of vitamin A, the total number of units of vitamin A is $15x + 20y$. If each serving of pet food must contain at least 60 units of vitamin A, the constraint is $15x + 20y \geq 60$. Similarly, because each serving requires at least 30 units of vitamin C, $10x + 5y \geq 30$. The linear programming problem then becomes the following.

Minimize: $C = 80x + 50y$ Cost (in cents)
Subject to: $15x + 20y \geq 60$ Vitamin A constraint
             $10x + 5y \geq 30$ Vitamin C constraint
             $x \geq 0, y \geq 0$ Amounts cannot be negative.

STEP 3: The region containing the feasible solutions is shown in **FIGURE C.6**.

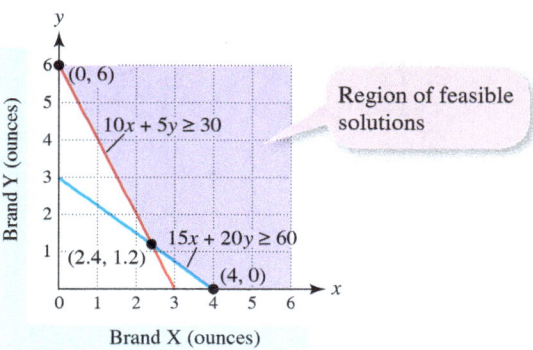

FIGURE C.6

NOTE: To determine the vertex $(2.4, 1.2)$, solve the system of equations

$$15x + 20y = 60$$
$$10x + 5y = 30$$

by using elimination. ∎

STEP 4: In **FIGURE C.6**, the vertices are $(0, 6)$, $(2.4, 1.2)$, and $(4, 0)$. Evaluate the objective function $C = 80x + 50y$ at each vertex, as shown in **TABLE C.3**.

Vertex	$C = 80x + 50y$
$(0, 6)$	$80(0) + 50(6) = 300$
$(2.4, 1.2)$	$80(2.4) + 50(1.2) = 252$ — Minimum cost (cents)
$(4, 0)$	$80(4) + 50(0) = 320$

TABLE C.3

The minimum cost occurs when **2.4** ounces of Brand X and **1.2** ounces of Brand Y are mixed, at a cost of **$2.52** per serving.

Now Try Exercise 47

C Exercises

MyMathLab®

CONCEPTS AND VOCABULARY

1. A procedure used in business to optimize quantities such as cost and profit is called _____.

2. In linear programming, the function to be optimized is called the _____ function.

3. The region in the xy-plane that satisfies the constraints is called the region of _____.

4. In linear programming, constraints typically consist of a system of _____.

5. If the optimal value for a linear programming problem exists, then it occurs at a(n) _____ of the region of feasible solutions.

6. To find the optimal value in a linear programming problem, substitute each vertex into the _____ function.

FINDING REGIONS OF FEASIBLE SOLUTIONS

Exercises 7–20: Shade the region of feasible solutions for the following constraints.

7. $x + y \leq 3$
   $x \geq 0, y \geq 0$

8. $2x + y \leq 4$
   $x \geq 0, y \geq 0$

9. $4x + 3y \leq 12$
   $x \geq 0, y \geq 0$

10. $5x + 3y \leq 15$
    $x \geq 0, y \geq 0$

11. $x \leq 5$
    $y \leq 2$
    $x \geq 0, y \geq 0$

12. $x \leq 3$
    $y \leq 4$
    $x \geq 1, y \geq 1$

13. $x + y \leq 5$
    $x + y \geq 2$
    $x \geq 0, y \geq 0$

14. $2x + y \leq 6$
    $x + y \geq 3$
    $x \geq 0, y \geq 0$

15. $3x + 2y \leq 6$
    $2x + 3y \leq 6$
    $x \geq 0, y \geq 0$

16. $5x + 3y \leq 30$
    $3x + 5y \leq 30$
    $x \geq 0, y \geq 0$

17. $x + y \leq 3$
    $x + 3y \geq 3$
    $x \geq 0, y \geq 0$

18. $3x + y \geq 6$
    $x + 2y \geq 6$
    $x \geq 0, y \geq 0$

19. $x + 2y \geq 4$
    $3x + 2y \geq 6$
    $x \geq 0, y \geq 0$

20. $4x + 3y \geq 12$
    $3x + 4y \geq 12$
    $x \geq 0, y \geq 0$

SOLVING LINEAR PROGRAMMING PROBLEMS

Exercises 21–24: Find the maximum of R on the region of feasible solutions shown in the figure.

21. $R = 4x + 5y$

22. $R = 2x + 3y$

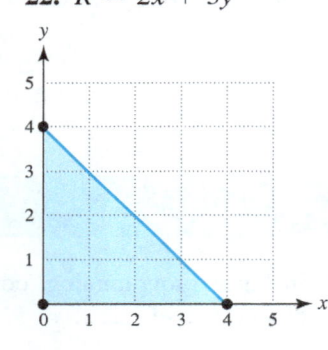

23. $R = x + 3y$

24. $R = 12x + 9y$

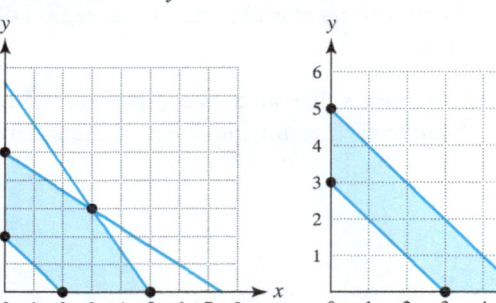

Exercises 25–28: Use the figures in Exercises 21–24 to complete Exercises 25–28, respectively, to minimize C.

25. $C = 2x + 3y$

26. $C = 3x + y$

27. $C = 5x + y$

28. $C = 2x + 7y$

Exercises 29–36: Maximize the objective function R, subject to the given constraints.

29. $R = 3x + 5y$
    $x + y \leq 150$
    $x \geq 0, y \geq 0$

30. $R = 6x + 5y$
    $x + y \leq 8$
    $x \geq 0, y \geq 0$

31. $R = 3x + 2y$
    $2x + y \leq 6$
    $x \geq 0, y \geq 0$

32. $R = x + 3y$
    $x + 2y \leq 4$
    $x \geq 0, y \geq 0$

33. $R = 12x + 9y$
    $3x + y \leq 6$
    $x + 3y \leq 6$
    $x \geq 0, y \geq 0$

34. $R = 10x + 30y$
    $x + 3y \leq 12$
    $3x + y \leq 12$
    $x \geq 0, y \geq 0$

35. $R = 4x + 5y$
    $x + y \geq 2$
    $x + 2y \leq 4$
    $x \geq 0, y \geq 0$

36. $R = 3x + 7y$
    $x + y \geq 1$
    $3x + y \leq 3$
    $x \geq 0, y \geq 0$

Exercises 37–42: Minimize the objective function C, subject to the given constraints.

37. $C = x + 2y$
    $x \leq 3, y \leq 2$
    $x \geq 0, y \geq 0$

38. $C = 3x + y$
    $x \leq 5, y \leq 3$
    $x \geq 1, y \geq 1$

39. $C = 8x + 15y$
    $x + y \geq 4$
    $x \geq 0, y \geq 0$

40. $C = x + 2y$
    $3x + 4y \geq 12$
    $x \geq 0, y \geq 0$

41. $C = 30x + 40y$
    $2x + y \leq 6$
    $x + y \geq 2$
    $x \geq 0, y \geq 0$

42. $C = 50x + 70y$
    $2x + 3y \leq 6$
    $x + y \geq 1$
    $x \geq 0, y \geq 0$

APPLICATIONS OF LINEAR PROGRAMMING

Exercises 43–50: Solve the linear programming problem.

43. **Maximizing Revenue** A small business sells candy for $4 per pound and coffee for $6 per pound. The business can package and sell at most a total of 100 pounds of candy and coffee per day, but at least 20 pounds of candy must be sold each day. Determine how many pounds of candy and coffee need to be sold each day to maximize revenue.

44. **Maximizing Revenue** An organic foods store sells almonds for $10 per pound and flax seed for $5 per pound. The business can package and sell, at most, a total of 50 pounds of almonds and flax seed per day, but it must sell at least 15 pounds of flax seed per day. Determine the maximum daily revenue.

45. **Minimizing Cost** It costs a business $50 to make a graphing calculator and $20 to make a scientific calculator. Each week, the company must produce at least 90 calculators. At least twice as many scientific calculators must be made as graphing calculators. Determine the minimum weekly cost.

46. **Minimizing Cost** It costs a business $20 to make one compact disc player and $10 to make one radio. Each week, the company must make a combined total of at least 50 compact disc players and radios. At least as many compact disc players as radios must be manufactured. Determine how many compact disc players and radios should be made to minimize weekly costs.

47. **Vitamin Cost** A pet owner is mixing two different brands of vitamins, Brand X and Brand Y, into pet food. Brand X costs 90 cents per ounce and Brand Y costs 60 cents per ounce. Each serving is a mixture of the two brands and should contain at least 40 units of vitamin A and 30 units of vitamin C. Each ounce of Brand X contains 20 units of vitamin A and 10 units of vitamin C, whereas each ounce of Brand Y contains 10 units of vitamin A and 10 units of vitamin C. Determine how much of each brand of vitamin should be mixed to produce a minimum cost per serving.

48. **Pet Food Cost** A pet owner is buying two brands of food, X and Y, for his animals. Each serving of the mixture of the two foods should contain at least 60 grams of protein and 40 grams of fat. Brand X costs 75 cents per unit and Brand Y costs 50 cents per unit. Each unit of Brand X contains 20 grams of protein and 10 grams of fat, whereas each unit of Brand Y contains 10 grams of protein and 10 grams of fat. Determine how much of each brand should be bought to obtain a minimum cost per serving.

49. **Raising Animals** A breeder can raise no more than 50 hamsters and mice but no more than 20 hamsters. If she sells the hamsters for $15 each and the mice for $10 each, find the maximum revenue.

50. **Maximizing Profit** A business manufactures two parts, X and Y. Machines A and B are needed to make each part. To make part X, machine A is needed 3 hours and machine B is needed 1 hour. To make part Y, machine A is needed 1 hour and machine B is needed 2 hours. Machine A is available 60 hours per week and machine B is available 50 hours per week. The profit from part X is $300 and the profit from part Y is $250. How many parts of each type should be made to maximize weekly profit?

APPENDIX D
Synthetic Division

NEW VOCABULARY

☐ Synthetic division

1 Performing Synthetic Division

A shortcut called **synthetic division** can be used to divide $x - k$, where k is a constant, into a polynomial. For example, to divide $x - 2$ into $3x^3 - 8x^2 + 7x - 6$, we do the following (with the equivalent long division shown at the right).

Synthetic Division

$$\begin{array}{r|rrrr} 2 & 3 & -8 & 7 & -6 \\ & & 6 & -4 & 6 \\ \hline & 3 & -2 & 3 & 0 \end{array}$$

Add to find row 3.

Long Division of Polynomials

$$\begin{array}{r} 3x^2 - 2x + 3 \\ x - 2 \overline{\smash{\big)}\, 3x^3 - 8x^2 + 7x - 6} \\ \underline{3x^3 - 6x^2} \\ -2x^2 + 7x \\ \underline{-2x^2 + 4x} \\ 3x - 6 \\ \underline{3x - 6} \\ 0 \end{array}$$

Note that the blue numbers in the expression for long division correspond to the third row in synthetic division. The remainder is 0, which is the last number in the third row. The quotient is $3x^2 - 2x + 3$. Its coefficients are 3, -2, and 3 and are located in the third row. To divide $x - 2$ into $3x^3 - 8x^2 + 7x - 6$ with synthetic division, use the following steps.

STEP 1: In the top row, write 2 (the value of k) on the left and then write the coefficients of the dividend $3x^3 - 8x^2 + 7x - 6$.

STEP 2: (a) Copy the leading coefficient 3 of $3x^3 - 8x^2 + 7x - 6$ into the third row and multiply it by 2 (the value of k). Write the result 6 in the second row below -8. Add -8 and 6 in the second column to obtain the -2 in the third row.
(b) Repeat the process by multiplying -2 by 2 and place the result -4 below 7. Then add 7 and -4 to obtain 3.
(c) Multiply 3 by 2 and place the result 6 below the -6. Adding 6 and -6 gives 0.

STEP 3: The last number in the third row is 0, which is the remainder. The other numbers in the third row are the coefficients of the quotient, which is $3x^2 - 2x + 3$.

EXAMPLE 1 Performing synthetic division

Use synthetic division to divide $x^4 - 5x^3 + 9x^2 - 10x + 3$ by $x - 3$.

Solution
Because the divisor is $x - 3$, the value of k is 3.

$$\begin{array}{r|rrrrr} 3 & 1 & -5 & 9 & -10 & 3 \\ & & 3 & -6 & 9 & -3 \\ \hline & 1 & -2 & 3 & -1 & 0 \end{array}$$

The quotient is $x^3 - 2x^2 + 3x - 1$ and the remainder is 0. This result is expressed by

$$\frac{x^4 - 5x^3 + 9x^2 - 10x + 3}{x - 3} = x^3 - 2x^2 + 3x - 1 + \frac{0}{x - 3}, \text{ or}$$

$$\frac{x^4 - 5x^3 + 9x^2 - 10x + 3}{x - 3} = x^3 - 2x^2 + 3x - 1.$$

Now Try Exercise 3

EXAMPLE 2 **Performing synthetic division**

Use synthetic division to divide $2x^3 - x + 5$ by $x + 1$.

Solution
Write $2x^3 - x + 5$ as $2x^3 + 0x^2 - x + 5$. The divisor $x + 1$ can be written as

$$x + 1 = x - (-1),$$

so we let $k = -1$.

$$\begin{array}{r|rrrr} -1 & 2 & 0 & -1 & 5 \\ & & -2 & 2 & -1 \\ \hline & 2 & -2 & 1 & 4 \end{array}$$

The remainder is 4, and the quotient is $2x^2 - 2x + 1$. This result can also be expressed as

$$\frac{2x^3 - x + 5}{x + 1} = 2x^2 - 2x + 1 + \frac{4}{x + 1}.$$

Now Try Exercise 13

D Exercises

PERFORMING SYNTHETIC DIVISION

Exercises 1–16: Use synthetic division to divide.

1. $\dfrac{x^2 + 3x - 1}{x - 1}$

2. $\dfrac{2x^2 + x - 1}{x - 3}$

3. $(3x^2 - 22x + 7) \div (x - 7)$

4. $(5x^2 + 29x - 6) \div (x + 6)$

5. $\dfrac{x^3 + 7x^2 + 14x + 8}{x + 4}$

6. $\dfrac{2x^3 + 3x^2 + 2x + 4}{x + 1}$

7. $\dfrac{2x^3 + x^2 - 1}{x - 2}$

8. $\dfrac{x^3 + x - 2}{x + 3}$

9. $(x^3 - 2x^2 - 2x + 4) \div (x - 4)$

10. $(2x^4 + 3x^2 - 4) \div (x + 2)$

11. $(x^3 - 3x^2 - 8x - 10) \div (x - 5)$

12. $(x^4 - 1) \div (x + 1)$

13. $(2x^4 - x) \div (x + 2)$

14. $(2x^3 + x - 1) \div (x - 3)$

15. $(b^4 - 1) \div (b - 1)$

16. $(a^2 + a) \div (a + 2.5)$

APPENDIX E
Using a Calculator

1 Overview of the Appendix

This appendix offers instruction for performing arithmetic on common scientific calculators as well as the TI-83, TI-83 Plus, and TI-84 Plus graphing calculators. For more detailed instructions on the use of a specific calculator, students are advised to consult the user manual provided by the manufacturer.

2 Operations on Whole Numbers

ADDING WHOLE NUMBERS To find a sum such as $328 + 4169$ on a calculator, use the following keystrokes.

Scientific Calculator: (3)(2)(8)(+)(4)(1)(6)(9)(=)

The result shown in the display will be [4497].

Graphing Calculator: (3)(2)(8)(+)(4)(1)(6)(9)(ENTER)

The result is shown in the first and second lines of **FIGURE E.1**.

FIGURE E.1

328+4169
            4497
4039−372
            3667

SUBTRACTING WHOLE NUMBERS A difference such as $4039 - 372$ can be found on a calculator by using the following keystrokes.

Scientific Calculator: (4)(0)(3)(9)(−)(3)(7)(2)(=)

The result shown in the display will be [3667].

Graphing Calculator: (4)(0)(3)(9)(−)(3)(7)(2)(ENTER)

The result is shown in the third and fourth lines of **FIGURE E.1**.

MULTIPLYING WHOLE NUMBERS The following keystrokes can be used to find a product such as 386×73 on a calculator.

Scientific Calculator: (3)(8)(6)(×)(7)(3)(=)

The result shown in the display will be [28178].

Graphing Calculator: (3)(8)(6)(×)(7)(3)(ENTER)

The result is shown in the first and second lines of **FIGURE E.2**.

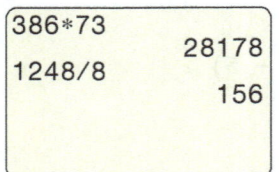

FIGURE E.2

386*73
         28178
1248/8
           156

DIVIDING WHOLE NUMBERS To find a quotient such as $1248 \div 8$ on a calculator, use the following keystrokes.

Scientific Calculator: (1)(2)(4)(8)(÷)(8)(=)

The result shown in the display will be [156].

Graphing Calculator: (1)(2)(4)(8)(÷)(8)(ENTER)

The result is shown in the third and fourth lines of **FIGURE E.2**.

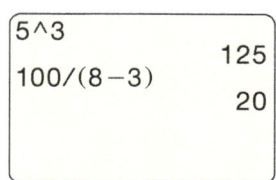

FIGURE E.3

EVALUATING EXPONENTS On a scientific calculator, an exponent is entered using the y^x key. On a graphing calculator, an exponent is entered using the $\wedge$ key. To evaluate an exponential expression such as 5^3 on a calculator, use the following keystrokes.

Scientific Calculator: (5)(y^x)(3)(=)

The result shown in the display will be $\boxed{125}$.

Graphing Calculator: (5)($\wedge$)(3)(ENTER)

The result is shown in the first and second lines of **FIGURE E.3**.

USING PARENTHESES An expression such as $100 \div (8 - 3)$ can be evaluated on a calculator by using the following keystrokes.

Scientific Calculator: (1)(0)(0)(÷)(( )(8)(−)(3)( ))(=)

The result shown in the display will be $\boxed{20}$.

Graphing Calculator: (1)(0)(0)(÷)(( )(8)(−)(3)( ))(ENTER)

The result is shown in the third and fourth lines of **FIGURE E.3**.

3 Operations on Integers

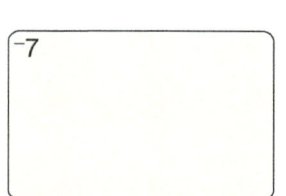

FIGURE E.4

ENTERING NEGATIVE NUMBERS On a scientific calculator, a negative integer is entered using the $+/-$ key. On a graphing calculator, a negative integer is entered using the $(-)$ key. For example, -7 is entered in a calculator as follows.

Scientific Calculator: (7)($+/-$)

The result shown in the display will be $\boxed{-7}$.

Graphing Calculator: ($(-)$)(7)

The result is shown in **FIGURE E.4**.

FINDING ABSOLUTE VALUE On some scientific calculators, an absolute value can be found by pressing the (abs) key or a key with "abs" displayed directly above it. On a graphing calculator, absolute value is found under the "NUM" menu after pressing the (MATH) key. For example, to find $|-3|$ with a calculator, use the following keystrokes.

Scientific Calculator: (abs)(3)($+/-$)(=) or (2nd)([abs])(3)($+/-$)(=)

The result shown in the display will be $\boxed{3}$.

Graphing Calculator: (MATH)(▶)(1)($(-)$)(3)( ))(ENTER)

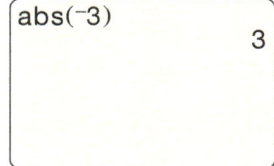

FIGURE E.5

The result is shown in **FIGURE E.5**.

ADDING INTEGERS The following calculator keystrokes can be used to find a sum such as $14 + (-9)$.

Scientific Calculator: (1)(4)(+)(9)($+/-$)(=)

The result shown in the display will be $\boxed{5}$.

Graphing Calculator: (1)(4)(+)(( )($(-)$)(9)( ))(ENTER)

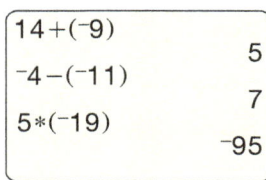

FIGURE E.6

The result is shown in the first and second lines of **FIGURE E.6**.

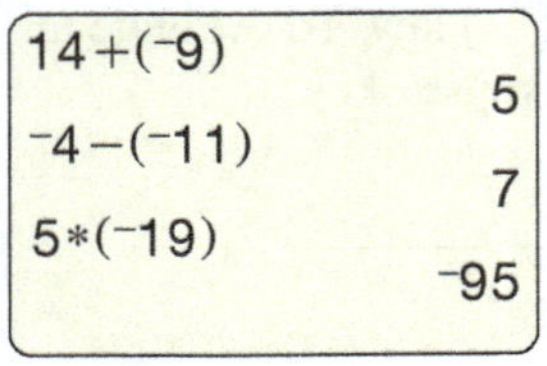

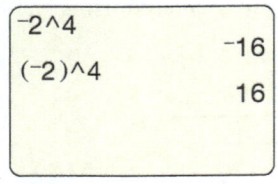

FIGURE E.6 (Repeated)

SUBTRACTING INTEGERS To find a difference such as $-4 - (-11)$ on a calculator, use the following keystrokes.

Scientific Calculator: [4] [+/−] [−] [1] [1] [+/−] [=]

The result shown in the display will be ⬚ 7 .

Graphing Calculator: [(−)] [4] [−] [(] [(−)] [1] [1] [)] [ENTER]

The result is shown in the third and fourth lines of **FIGURE E.6**.

MULTIPLYING INTEGERS A product such as $5 \cdot (-19)$ can be found on a calculator by using the following keystrokes.

Scientific Calculator: [5] [×] [1] [9] [+/−] [=]

The result shown in the display will be ⬚ −95 .

Graphing Calculator: [5] [×] [(] [(−)] [1] [9] [)] [ENTER]

The result is shown in the fifth and sixth lines of **FIGURE E.6**.

EVALUATING EXPONENTS ON INTEGERS When using a calculator to evaluate exponential expressions such as -2^4 and $(-2)^4$, it is important to apply the order of operations agreement correctly. To evaluate the expression -2^4 on a calculator, use the following keystrokes.

Scientific Calculator: [2] [y^x] [4] [=] [+/−]

The result shown in the display will be ⬚ −16 .

Graphing Calculator: [(−)] [2] [^] [4] [ENTER]

The result is shown in the first and second lines of **FIGURE E.7**.

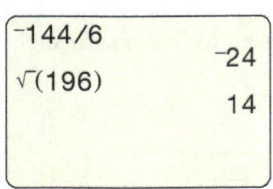

FIGURE E.7

To evaluate the expression $(-2)^4$ on a calculator, use the following keystrokes.

Scientific Calculator: [2] [+/−] [y^x] [4] [=]

The result shown in the display will be ⬚ 16 .

Graphing Calculator: [(] [(−)] [2] [)] [^] [4] [ENTER]

The result is shown in the third and fourth lines of **FIGURE E.7**.

DIVIDING INTEGERS The following keystrokes can be used to find a quotient such as $-144 \div 6$ on a calculator.

Scientific Calculator: [1] [4] [4] [+/−] [÷] [6] [=]

The result shown in the display will be ⬚ −24 .

Graphing Calculator: [(−)] [1] [4] [4] [÷] [6] [ENTER]

The result is shown in the first and second lines of **FIGURE E.8**.

FINDING A SQUARE ROOT To find a square root such as $\sqrt{196}$ on a calculator, use the following keystrokes.

Scientific Calculator: [1] [9] [6] [√]

The result shown in the display will be ⬚ 14 .

Graphing Calculator: [2nd] [x^2 [√]] [1] [9] [6] [)] [ENTER]

The result is shown in the third and fourth lines of **FIGURE E.8**.

FIGURE E.8

4 Operations on Fractions

SIMPLIFYING FRACTIONS On some scientific calculators, a fraction can be simplified using the $\boxed{a^{b/c}}$ key. Such calculators often display the symbol (⌐) to indicate a fraction bar. On a graphing calculator, a fraction feature is found under the "MATH" menu after pressing the $\boxed{\text{MATH}}$ key. For example, the fraction $\frac{84}{120}$ can be simplified to $\frac{7}{10}$ with a calculator by using the following keystrokes.

Scientific Calculator: $\boxed{8}\boxed{4}\boxed{a^{b/c}}\boxed{1}\boxed{2}\boxed{0}\boxed{=}$

The result shown in the display will be $\boxed{7\lrcorner 10}$.

Graphing Calculator: $\boxed{8}\boxed{4}\boxed{\div}\boxed{1}\boxed{2}\boxed{0}\boxed{\text{MATH}}\boxed{1}\boxed{\text{ENTER}}$

The result is shown in the first and second lines of **FIGURE E.9**.

```
84/120►Frac
           7/10
(-1/6)*(5/8)►Fra
c
          -5/48
```

FIGURE E.9

MULTIPLYING FRACTIONS A product such as $-\frac{1}{6} \cdot \frac{5}{8}$ can be found on a calculator by using the following keystrokes.

Scientific Calculator: $\boxed{1}\boxed{a^{b/c}}\boxed{6}\boxed{+/-}\boxed{\times}\boxed{5}\boxed{a^{b/c}}\boxed{8}\boxed{=}$

The result shown in the display will be $\boxed{-5\lrcorner 48}$.

Graphing Calculator: $\boxed{(}\boxed{(-)}\boxed{1}\boxed{\div}\boxed{6}\boxed{)}\boxed{\times}\boxed{(}\boxed{5}\boxed{\div}\boxed{8}\boxed{)}\boxed{\text{MATH}}\boxed{1}\boxed{\text{ENTER}}$

The result is shown in the third, fourth, and fifth lines of **FIGURE E.9**.

EVALUATING EXPONENTS ON FRACTIONS On some scientific calculators, the keys $\boxed{\text{2nd}}$ and $\boxed{\text{F◄►D}}$ are used to express an answer in fraction form. The following keystrokes can be used when evaluating an exponential expression such as $\left(-\frac{2}{3}\right)^3$ with a calculator.

Scientific Calculator: $\boxed{2}\boxed{a^{b/c}}\boxed{3}\boxed{+/-}\boxed{y^x}\boxed{3}\boxed{=}\boxed{\text{2nd}}\boxed{\text{F◄►D}}$

The result shown in the display will be $\boxed{-8\lrcorner 27}$.

Graphing Calculator: $\boxed{(}\boxed{(-)}\boxed{2}\boxed{\div}\boxed{3}\boxed{)}\boxed{\wedge}\boxed{3}\boxed{\text{MATH}}\boxed{1}\boxed{\text{ENTER}}$

The result is shown in the first and second lines of **FIGURE E.10**.

```
(-2/3)^3►Frac
          -8/27
√(45/80)►Frac
            3/4
```

FIGURE E.10

FINDING A SQUARE ROOT OF A FRACTION To find a square root such as $\sqrt{\frac{45}{80}}$ on a calculator, use the following keystrokes.

Scientific Calculator: $\boxed{4}\boxed{5}\boxed{a^{b/c}}\boxed{8}\boxed{0}\boxed{\sqrt{}}\boxed{\text{2nd}}\boxed{\text{F◄►D}}$

The result shown in the display will be $\boxed{3\lrcorner 4}$.

Graphing Calculator: $\boxed{\text{2nd}}\boxed{x^2[\sqrt{}]}\boxed{4}\boxed{5}\boxed{\div}\boxed{8}\boxed{0}\boxed{)}\boxed{\text{MATH}}\boxed{\cdot}\boxed{\text{ENTER}}$

The result is shown in the third and fourth lines of **FIGURE E.10**.

DIVIDING FRACTIONS A quotient such as $-\frac{5}{6} \div \frac{5}{4}$ can be found on a calculator by using the following keystrokes.

Scientific Calculator: $\boxed{5}\boxed{a^{b/c}}\boxed{6}\boxed{+/-}\boxed{\div}\boxed{5}\boxed{a^{b/c}}\boxed{4}\boxed{=}$

The result shown in the display will be $\boxed{-2\lrcorner 3}$.

Graphing Calculator: $\boxed{(}\boxed{(-)}\boxed{5}\boxed{\div}\boxed{6}\boxed{)}\boxed{\div}\boxed{(}\boxed{5}\boxed{\div}\boxed{4}\boxed{)}\boxed{\text{MATH}}\boxed{1}\boxed{\text{ENTER}}$

The result is shown in the first, second, and third lines of **FIGURE E.11**.

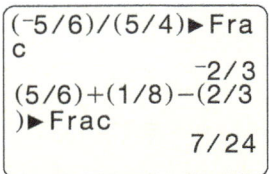

```
(-5/6)/(5/4)►Fra
c
          -2/3
(5/6)+(1/8)-(2/3
)►Frac
          7/24
```

FIGURE E.11

AP-32 APPENDIX E USING A CALCULATOR

ADDING OR SUBTRACTING FRACTIONS To evaluate expressions involving sums and differences such as $\frac{5}{6} + \frac{1}{8} - \frac{2}{3}$ with a calculator, use the following keystrokes.

Scientific Calculator: [5] [a b/c] [6] [+] [1] [a b/c] [8] [−] [2] [a b/c] [3] [=]

The result shown in the display will be ⟨ 7⌐24 ⟩.

Graphing Calculator: [(] [5] [÷] [6] [)] [+] [(] [1] [÷] [8] [)] [−] [(] [2] [÷] [3] [)]
[MATH] [1] [ENTER]

The result is shown in the fourth, fifth, and sixth lines of **FIGURE E.11**.

Figure E.11 (Repeated)
```
(-5/6)/(5/4)▶Fra
c
                -2/3
(5/6)+(1/8)-(2/3
)▶Frac
                7/24
```

5 Operations on Decimals

ADDING OR SUBTRACTING DECIMALS To evaluate expressions involving sums and differences such as $4.3 + 6.7 - (-1.2)$ with a calculator, use the following keystrokes.

Scientific Calculator: [4] [.] [3] [+] [6] [.] [7] [−] [1] [.] [2] [+/−] [=]

The result shown in the display will be ⟨ 12.2 ⟩.

Graphing Calculator: [4] [.] [3] [+] [6] [.] [7] [−] [(] [(−)] [1] [.] [2] [)] [ENTER]

The result is shown in the first and second lines of **FIGURE E.12**.

MULTIPLYING DECIMALS The following keystrokes can be used to find a product such as $(-8.2)(-0.51)$ on a calculator.

Scientific Calculator: [8] [.] [2] [+/−] [×] [0] [.] [5] [1] [+/−] [=]

The result shown in the display will be ⟨ 4.182 ⟩.

Graphing Calculator: [(] [(−)] [8] [.] [2] [)] [(] [(−)] [0] [.] [5] [1] [)] [ENTER]

The result is shown in the third and fourth lines of **FIGURE E.12**.

Figure E.12
```
4.3+6.7-(-1.2)
                12.2
(-8.2)(-0.51)
               4.182
```

DIVIDING DECIMALS A quotient such as $3.72 \div 5$ can be found on a calculator by using the following keystrokes.

Scientific Calculator: [3] [.] [7] [2] [÷] [5] [=]

The result shown in the display will be ⟨ 0.744 ⟩.

Graphing Calculator: [3] [.] [7] [2] [÷] [5] [ENTER]

The result is shown in the first and second lines of **FIGURE E.13**.

ESTIMATING A SQUARE ROOT To estimate a square root such as $\sqrt{11}$ on a calculator, use the following keystrokes.

Scientific Calculator: [1] [1] [√]

The result shown in the display will be ⟨ 3.31662479 ⟩.

Graphing Calculator: [2nd] [x^2 [√]] [1] [1] [)] [ENTER]

The result is shown in the third and fourth lines of **FIGURE E.13**.

Figure E.13
```
3.72/5
                .744
√(11)
          3.31662479
```

6 Financial Math

USING THE SIMPLE INTEREST FORMULA The formula $I = Prt$ can be used to find simple interest. For example, when $P = \$9520$, $r = 3\%$, and $t = 5$ years, the following keystrokes can be used to find the simple interest.

Scientific Calculator: ⑨⑤②⓪×⓪.⓪③×⑤=

The result shown in the display will be ☐ 1428 ☐.

Graphing Calculator: ⑨⑤②⓪×⓪.⓪③×⑤ ENTER

The result is shown in the first and second lines of **FIGURE E.14**.

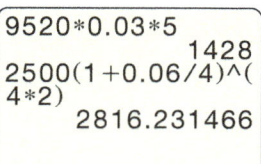

```
9520*0.03*5
            1428
2500(1+0.06/4)^(
4*2)
      2816.231466
```

Figure E.14

USING THE COMPOUND INTEREST FORMULA The final amount in an account that pays compound interest is given by the formula

$$A = P\left(1 + \frac{r}{n}\right)^{nt}.$$

It is important to use parentheses correctly when working with this formula. For example, when $P = \$2500$, $r = 6\%$, $n = 4$, and $t = 2$ years, the following keystrokes can be used to find the final amount in the account.

Scientific Calculator: ②⑤⓪⓪×(①+⓪.⓪⑥÷④) y^x
(④×②)=

The result shown in the display will be ☐ 2816.231467 ☐.

Graphing Calculator: ②⑤⓪⓪(①+⓪.⓪⑥÷④)^
(④×②) ENTER

The result is shown in the third, fourth, and fifth lines of **FIGURE E.14**.

7 Scientific Notation

USING SCIENTIFIC NOTATION On some scientific calculators, scientific notation can be entered into the calculator by pressing the EE key. On a graphing calculator, the EE feature is accessed by pressing the 2nd and . keys. For example, to find a quotient such as $(8.4 \times 10^{-5}) \div (2.1 \times 10^2)$ with a calculator, use the following keystrokes.

Scientific Calculator: ⑧.④ EE ⑤ +/− ÷ ②.① EE ② =

The result shown in the display will be ☐ 0 0000004 ☐.

Graphing Calculator: (⑧.④ 2nd [EE] (−)⑤) ÷ (②.①
2nd [EE] ②) ENTER

The result is shown (in scientific notation) in **FIGURE E.15**.

```
(8.4E-5)/(2.1E2)
             4E-7
```

FIGURE E.15

Answers to Selected Exercises

1 Whole Numbers

SECTION 1.1 (pp. 11–15)

1. counting **2.** 0 **3.** periods **4.** standard **5.** place value
6. trillions **7.** word **8.** expanded **9.** graph **10.** bar graph and line graph **11.** thousands **13.** ten-millions
15. hundreds **17.** ones **19.** hundred-millions **21.** 7
23. 8 **25.** 3 **27.** 2 **29.** 4 **31.** Four hundred seventy-two thousand, five hundred **33.** Ninety-three thousand, two hundred six **35.** One thousand, six hundred fifty-one
37. 2055 **39.** 599,616,423 **41.** 39,410,000
43. 83,000,600,012 **45.** 342,563 **47.** 7,905,377
49. 2,000,000 + 500,000 + 10,000 + 30 + 6
51. 600 + 20 + 9 **53.** 600,000 + 3000 + 100 + 30 + 8
55. ; (a) < (b) > (c) >
57. ; (a) > (b) < (c) <
59. ; (a) > (b) > (c) >
61. > **63.** < **65.** > **67.** < **69.** < **71.** India
73. France **75.** Mackenzie **77.** 2575 km **79.** 2014
81. $315,000,000 **83.** Minimum wage increased.
85. 1990 to 2010 **87.** Website Design **89.** 4
91. Power Tools **93.** Power Tools **95.** 2006 **97.** Males
99. 2010 **101.** Public **103.** 1000 + 100 + 20 + 4
105. 34,359,738,378 **107.** Four hundred twenty-three billion
109. The truck driver

SECTION 1.2 (pp. 26–29)

1. addends **2.** sum **3.** Yes **4.** commutative
5. associative **6.** identity **7.** addition
8. minuend; subtrahend **9.** difference **10.** No
11. identity **12.** subtraction **13.** solution **14.** solutions
15. 28 **17.** 599 **19.** 885 **21.** 7868 **23.** 7872
25. 27,064 **27.** 81,827 **29.** 1,064,159 **31.** 8967
33. 88,051 **35.** Commutative property
37. Identity property **39.** Associative property **41.** 66
43. 70 **45.** 13 **47.** 431 **49.** 1422 **51.** 151
53. 2316 **55.** 32,905 **57.** 31,433 **59.** 101,027
61. 2618 **63.** 17,685 **65.** 10,876 **67.** 22 + 57; $79
69. 793 − 54; 739 photos **71.** 62 − 19; 43 eggs
73. 1200 + 300; 1500 patients **75.** 645 − 3; 642 DVDs
77. 39 + 71; 110 Web pages **79.** 9 **81.** 3 **83.** 20
85. 27 **87.** 151 **89.** 27 **91.** 529 **93.** 56 ft **95.** 35 cm
97. 54 mi **99.** 78 in. **101.** 12 **103.** 43 in. **105.** 622
107. 130 pounds **109.** 17 **111.** 1102 pounds
113. 2000 **115.** 6,000,000

Checking Basic Concepts 1.1 & 1.2 (p. 29)

1. (a) ten-thousands **(b)** hundreds
2. 70,000 + 4000 + 200 + 90 + 3 **3** 48,239,610
4.
5. (a) > **(b)** < **6. (a)** 4317 **(b)** 210,530
7. (a) 8659 **(b)** 600,884
8. (a) 97 − 45; 52 **(b)** 106 + 73; 179
9. (a) 5 **(b)** 29 **10.** 108 cm

SECTION 1.3 (pp. 42–45)

1. addition **2.** factors **3.** product **4.** commutative
5. associative **6.** identity **7.** zero **8.** distributive
9. multiplication **10.** subtraction **11.** dividend; divisor
12. quotient **13.** identity **14.** 0; undefined
15. long division **16.** division **17.** 1 square unit **18.** area
19. Associative property **21.** Commutative property
23. Identity property **25.** Distributive property
27. Zero property **29.** 5 · 6 + 5 · 9 **31.** 4 · 8 − 4 · 1
33. 6 · 3 − 2 · 3 **35.** 7 **37.** 0 **39.** 54 **41.** 336
43. 1812 **45.** 1704 **47.** 2408 **49.** 51,625 **51.** 283,504
53. 117,066 **55.** 21,000 **57.** 680,000 **59.** 1,500,000
61. 9 **63.** 1 **65.** 11 **67.** Undefined **69.** 0 **71.** 1
73. 12 **75.** 961 r2 **77.** 80 r7 **79.** 200 **81.** Undefined
83. 126 r67 **85.** 65 **87.** 62 r735 **89.** 14 · 3; 42 square feet
91. 5 · 15; $75 **93.** 126 ÷ 7; 18 miles per gallon
95. 75 ÷ 15; 5 days **97.** 3 **99.** 5 **101.** 50 **103.** 6
105. 40 square inches **107.** 289 square miles
109. 8100 square feet **111.** 3 **113.** $30 million
115. 33 hours **117.** 8 squares; 32 squares **119.** 5000
121. 0 calories **123.** 240,000 pixels **125.** 250
127. (a) 15 feet **(b)** 50 feet **129.** 13 **131.** 3; $2

SECTION 1.4 (pp. 53–56)

1. exponential notation **2.** 4; 7 **3.** 2 **4.** 3 **5.** 9 **6.** 10^7
7. variable **8.** algebraic expression **9.** equation
10. formula **11.** $P = 2l + 2w$ **12.** $A = s^2$ **13.** evaluate
14. variable **15.** expression **16.** equation **17.** 8^3 **19.** 2^5
21. $2^3 \cdot 5^2$ **23.** $5^3 \cdot 7^3$ **25.** 7^2 **27.** 4^9 **29.** 2^3 **31.** 3^5
33. 81 **35.** 32 **37.** 256 **39.** 216 **41.** 1000 **43.** 8,000,000
45. 3000 **47.** $A - 5$ **49.** $6 \cdot G$ **51.** $S + 10$ **53.** $P \div 2$
55. 20 **57.** 0 **59.** 65 **61.** 46 **63.** 10 **65.** 4 **67.** 6

69. 16 **71.** 16 inches **73.** 15 square inches **75.** 69 yards
77. $6M$, where M represents monthly income
79. $p \div 3$, where p represents the number of pizza slices
81. $a + h$, where a represents age and h represents heart rate
83. Yes **85.** No **87.** No **89.** 10 **91.** 14 **93.** 12 **95.** 8
97. 10 feet **99.** 40 inches **101. (a)** $C = 6p$
(b) $C = 8d$ **(c)** painting: 1440; clearing: 1920
103. 19,767 **105.** 3; $1 **107.** $20 **109.** 2 **111.** 12

Checking Basic Concepts 1. 3 & 1.4 (p. 56)

1. (a) 286 **(b)** 0 **(c)** 207 **(d)** 64,668 **2. (a)** 35
(b) 140 r5 **(c)** Undefined **(d)** 171 **3. (a)** 12 **(b)** 72
4. 120,000 **5. (a)** 7^4 **(b)** $2^2 \cdot 8^2$ **6. (a)** 40 **(b)** 300,000
7. (a) 13 **(b)** 30 **8.** Yes **9. (a)** 8 **(b)** 12 **10.** 44 inches
11. 7 **12.** 800 square feet

SECTION 1.5 (pp. 63–64)

1. rounding **2.** right **3.** estimation **4.** approximation
5. highest **6.** perfect **7.** square root **8.** radical; radicand
9. 7 **10.** is approximately equal to **11.** 700 **13.** 59,000
15. 80 **17.** 900 **19.** 380,000 **21.** 54,200 **23.** 780
25. 30,000,000 **27.** 3000 **29.** 90,000,000 **31.** 70
33. 20,000 **35.** 1500 **37.** 400 **39.** 3000 **41.** 8200
43. 4,900,000 **45.** 20 **47.** 280,000 **49.** 250 **51.** 5
53. 11 **55.** 19 **57.** 25 **59.** 2000 hours **61.** 180 years
63. 9000 **65.** 12,000 feet **67.** 1210 **69.** 220 million
71. 2003

SECTION 1.6 (pp. 69–70)

1. order of operations **2.** Parentheses; radical **3.** before
4. subtraction **5.** the quantity **6.** addition **7.** 24
9. 16 **11.** 21 **13.** 14 **15.** 73 **17.** 25 **19.** 30 **21.** 19
23. 180 **25.** 256 **27.** 86 **29.** 0 **31.** 4 **33.** 10
35. 9 **37.** 67 **39.** 8 **41.** 12 **43.** 22
45. $(14 - 12) \cdot 5 - 10$ **47.** $(36 - 6^2) \div (5 - 1)$
49. $(32 \div 4^2 - 2) \cdot 9$ **51.** 8 **53.** 7 **55.** 41 **57.** 1
59. 21 **61.** 19 **63.** $5 + 12$; 17 **65.** $21 - 9$; 12
67. $7^2 + 4$; 53 **69.** $(6 + 5) \cdot 9$; 99 **71.** $(6 + 2) \cdot 5$; 40
73. $2^3 \cdot 3^2$; 72 **75.** $\sqrt{16} + 9$; 13 **77.** $7 \cdot 3 - 2$; 19
79. 177 bpm **81.** 8000 watts **83.** 8 thousand per acre
85. 40 °C

Checking Basic Concepts 1. 5 & 1.6 (p. 71)

1. 45,000 **2. (a)** 1200 **(b)** 4600 **3.** 6000
4. (a) 9 **(b)** 13 **5. (a)** 13 **(b)** 12 **6.** 2 **7.** 9
8. $(4 + 2) \cdot 3$; 18 **9.** 8400 **10.** 20 feet

SECTION 1.7 (pp. 77–78)

1. Expression **2.** Equation **3.** term **4.** coefficient
5. like **6.** unlike **7.** Expression **9.** Equation
11. Equation **13.** Expression **15.** Like **17.** Unlike
19. Like **21.** Unlike **23.** $11x$ **25.** $7yz$ **27.** Not possible
29. $16ab$ **31.** $7x + 9$ **33.** $11y + 5$ **35.** $5a + 4$
37. $8z - 14$ **39.** $3x + 2$ **41.** $5x + 9$ **43.** $3ab + 4y$
45. (a) $7x = 4x + 15$ **(b)** 5 checks in both equations.
47. (a) $2x + 11 = 3x + 6$ **(b)** 5 checks in both equations.
49. $6x - 2x = 36$, where x is the number of inches.
51. $14 = x - 9$, where x is her score. **53.** $4x = 28$, where x is her shoe size. **55.** $x = 3x - 8$, where x is the score.
57. $53,000 **59.** 4 gallons **61.** 8480 **63.** 12
65. 4 **67.** $21 **69.** 3 inches

Checking Basic Concepts 1. 7 (p. 79)

1. (a) Equation **(b)** Expression **2. (a)** Like **(b)** Unlike
3. (a) $5pq$ **(b)** Not possible **4. (a)** $2x + 1$ **(b)** $9x + 3$
5. $x - 7 = 23$, where x is his age. **6.** 4 feet

CHAPTER 1 REVIEW (pp. 83–86)

1. hundreds **2.** hundred-thousands **3.** 1 **4.** 0
5. Forty-eight thousand, three hundred nine **6.** Thirty-seven
7. $600 + 70 + 3$ **8.** $60,000 + 1000 + 4$ **9.** 58,345
10. **11.** $>$ **12.** $<$
13. 35 **14.** 1125 **15.** 6005 **16.** 11,886 **17.** 766
18. 2134 **19.** 32,827 **20.** 32,708 **21.** $83 - 21$; 62
22. $103 + 48$; 151 **23.** 22 **24.** 11 **25.** $5 \cdot 4 + 5 \cdot 2$
26. $7 \cdot 8 - 7 \cdot 5$ **27.** 0 **28.** 99 **29.** 79,636
30. 367,392 **31.** 21 **32.** 67 **33.** 83 r6 **34.** Undefined
35. $66 \div 11$; 6 **36.** 26×7; 182 **37.** 5 **38.** 3 **39.** 8^5
40. 9^3 **41.** 49 **42.** 125 **43.** 400 **44.** 900,000 **45.** 54
46. 216 **47.** 35 feet **48.** 120 square inches **49.** 10
50. 3 **51.** 11 **52.** 27 **53.** 6 **54.** 28 **55.** 160
56. 980,000 **57.** 50,000 **58.** 400,000 **59.** 1300
60. 1400 **61.** 16 **62.** 11 **63.** 35 **64.** 31 **65.** 23
66. 52 **67.** 8 **68.** 42 **69.** 10 **70.** 13 **71.** 31 **72.** 21
73. $9 - (2 + 6)$; 1 **74.** $4 \cdot 3 - 1$; 11 **75.** Expression
76. Equation **77.** Unlike **78.** Like **79.** $18x$ **80.** $10b$
81. $13mn$ **82.** Not possible **83.** $4y + 7$ **84.** $10z - 6$
85. $15a + 4$ **86.** $5y + 9$ **87.** $x - 7 = 64$, where x is his height. **88.** $14 = x + 12$, where x is her score.
89. Stamp price increased. **90.** 22 cents **91.** 3
92. 88 square inches **93.** 59 bpm **94.** 68 bpm **95.** 64 cm
96. 6; $1 **97.** $40 **98.** 10 feet **99.** 25 °C **100.** 5
101. About $3 billion **102.** 2008 **103.** 2013
104. About $3.5 billion **105.** 8096 **106.** 80 minutes

CHAPTER 1 TEST (p. 86)

1. Ten-thousands **2.** $7000 + 300 + 40 + 1$ **3.** 78,000
4. $>$ **5.** 4341 **6.** 6026 **7.** 130,059 **8.** 514 r38
9. 3^4 **10.** 200 **11.** 13 **12.** 6 **13.** 19 **14.** 5 **15.** 12
16. 11 **17.** 16 **18.** 9 **19.** $5p^2$ **20.** Not possible
21. Zero Property **22.**
23. $8x + 2$ **24.** $4y + 13$ **25.** 56 mi **26.** 6; $8
27. $1268 **28.** $C = 14h$

2 Integers

Section 2.1 (pp. 95–97)

1. positive **2.** negative **3.** opposite **4.** 0 **5.** a
6. integers **7.** origin **8.** absolute value **9.** -3
10. $+17$ or 17 **11.** -7 **13.** 43 **15.** 237 **17.** $-93{,}000$
19. -8 **21.** 26 **23.** 0 **25.** 23 **27.** -5 **29.** 1
31. number line from -5 to 5 with points at $-3, 0, 2, 4$
33. number line from -20 to 20 with points at $-12, -4, 8, 16$
35. number line from -100 to 100 with points at $-87, 5, 76$
37. $>$ **39.** $>$ **41.** $<$ **43.** $<$ **45.** $>$ **47.** $<$ **49.** 10
51. 0 **53.** 18 **55.** 87 **57.** -2 **59.** -19 **61.** 0
63. $>$ **65.** $>$ **67.** $=$ **69.** $<$ **71.** -282 **73.** $19{,}340$
75. Floors that are above ground **77. (a)** Romania
(b) Malta and Tonga **79.** $-\$1745$ **81.** $-\$200$
83. (a) Pacific **(b)** Arctic **(c)** Southern **(d)** Indian
85. $-\$4000$ **87.** 400 videos

Section 2.2 (pp. 103–105)

1. absolute values **2.** Negative **3.** Positive **4.** Negative
5. additive inverses or opposites **6.** 0 **7.** right; left
8. $\cap$; $\cup$ **9.** 12 **11.** -12 **13.** 41 **15.** -42 **17.** -15
19. 23 **21.** 0 **23.** 39 **25.** -1 **27.** -66 **29.** -143
31. -16 **33.** 13 **35.** -93 **37.** Commutative
39. Inverse **41.** Associative **43.** Identity **45.** -4
47. -17 **49.** 0 **51.** 64 **53.** -45 **55.** -4 **57.** 4
59. -5 **61.** 0 **63.** 4 **65.** -2 **67.** -9 **69.** 5 **71.** $49\,°F$
73. 13 yards **75.** -1019 feet, or 1019 feet below ground level **77.** $-\$320$ million **79.** $\$2489$

Checking Basic Concepts 2.1 & 2.2 (p. 105)

1. (a) -23 **(b)** 16 **2. (a)** 52 **(b)** -9
3. number line from -5 to 5 with points at $-3, 0, 2, 4$ **4. (a)** $>$ **(b)** $>$
5. (a) 17 **(b)** 31 **6. (a)** 8 **(b)** -35 **(c)** -21 **(d)** 83
7. (a) 4 **(b)** -6 **8. (a)** 2 **(b)** -5 **9.** $-\$420$
10. 2405 feet

Section 2.3 (pp. 110–112)

1. opposite **2.** (-7) **3.** 9 **4.** The commutative and associative properties **5.** addition **6.** No
7. subtraction symbol **8.** True **9.** 6 **11.** -5 **13.** -15
15. -42 **17.** 27 **19.** 29 **21.** -5 **23.** 8 **25.** 34
27. 52 **29.** 9 **31.** 45 **33.** -67 **35.** -2 **37.** 13
39. 4 **41.** -31 **43.** 14 **45.** -4 **47.** 3 **49.** -6
51. 0 **53.** -5 **55.** -2 **57.** 10 **59.** 7 **61.** $87\,°F$
63. $14{,}776$ feet **65.** 15 feet **67.** $\$189{,}400$ **69.** $\$176$
71. $-10\,°F$

Section 2.4 (pp. 118–119)

1. positive **2.** negative **3.** identity **4.** distributive
5. positive **6.** negative **7.** positive **8.** negative
9. $\sqrt{4}$ **10.** $-\sqrt{4}$ **11.** -12 **13.** 40 **15.** -18 **17.** 170
19. 0 **21.** -42 **23.** -150 **25.** Associative **27.** Zero
29. Commutative **31.** Distributive **33.** Identity **35.** -36
37. 30 **39.** -21 **41.** -90 **43.** 0 **45.** -80 **47.** 400
49. -3 **51.** 5 **53.** 12 **55.** -2 **57.** Undefined **59.** 1
61. 0 **63.** -6 **65.** $-5, 5$ **67.** $-9, 9$ **69.** No integer square roots **71.** 0 **73.** 4 **75.** -6 **77.** 10
79. Not an integer **81.** -1 **83.** -21 **85.** -10 **87.** -5
89. 25 **91.** 10 **93.** Not an integer **95.** $-44\,°C$
97. $5 \cdot (-107) = -535$; 535 feet deep
99. $-300 \div 12 = -25$; 25 fewer prisoners
101. $5 \cdot (-29) = -145$; $\$145$

Checking Basic Concepts 2.3 & 2.4 (p. 120)

1. (a) -34 **(b)** -14 **(c)** 33 **(d)** 15 **2. (a)** -8 **(b)** 9
3. (a) -8 **(b)** -5 **4.** 4 **5. (a)** -44 **(b)** 39
(c) -48 **(d)** 50 **6.** -48 **7. (a)** -4 **(b)** -5
(c) 9 **(d)** -25 **8. (a)** 8 **(b)** -4 **9.** $\$165$
10. $3 \cdot (-23) = -69$; 69 feet deep

Section 2.5 (pp. 125–127)

1. order of operations **2.** innermost **3.** parentheses
4. average **5.** -10 **7.** 3 **9.** 27 **11.** -1 **13.** -2
15. 9 **17.** 39 **19.** 30 **21.** 0 **23.** 8 **25.** 50 **27.** -13
29. 3 **31.** -15 **33.** -8 **35.** 30 **37.** 1 **39.** 16
41. $-20 + 10 \cdot (14 - 12)$ **43.** $-5^2 \div (3 + 2) + 5$
45. $(32 \div 4^2 - 2) \cdot 9$ **47.** $(16 - 4^2) \div (4 - 9)$
49. -8 **51.** 16 **53.** -81 **55.** -1 **57.** -27
59. $-1{,}000{,}000$ **61.** $10{,}000$ **63.** -1 **65.** 0 **67.** 12
69. 11 **71.** -5 **73.** -6 **75.** 29 **77.** -1 **79.** -16
81. 0 **83.** $-20\,°C$ **85.** $5\,°F$ **87.** $-\$1000$ **89.** $-6\,°F$

Section 2.6 (pp. 133–135)

1. equation **2.** solution **3.** variable **4.** check **5.** Yes
7. No **9.** No **11.** No **13.** No **15.** Yes **17.** Yes
19. No **21.** -15 **23.** -5 **25.** 8 **27.** -5 **29.** 5 **31.** 7
33. Table values: $0, 1, 2, 3, 4$; solution: -1
35. Table values: $-1, 2, 5, 8, 11$; solution: 1
37. Table values: $-7, -1, 5, 11, 17$; solution: 2
39. Table values: $4, 3, 2, 1, 0$; solution: -6
41. -1 **43.** -2 **45.** 2 **47.** 0 **49.** 7 **51.** 3 **53.** 3
55. 2012 **57.** Table values: $120, 60, 40, 30, 24$; solution: 4 gallons **59.** 2012

Checking Basic Concepts 2.5 & 2.6 (p. 136)

1. (a) 23 **(b)** 0 **(c)** -14 **2. (a)** 8 **(b)** -11 **(c)** 6
3. No **4.** Yes **5. (a)** -5 **(b)** 6
6. Table values: $9, 7, 5, 3, 1$; solution: -1 **7.** $-58\,°F$

Chapter 2 Review (pp. 140–142)

1. -19 **2.** 52 **3.** 31 **4.** -2
5. number line from -5 to 5 with points at $-3, -1, 2, 4$

6. [number line from -5 to 5 with points at -4, -2, 1, 3]
7. < 8. > 9. > 10. < 11. -6 12. -1 13. 0
14. 12 15. > 16. = 17. Sunday 18. Increase
19. -1 20. -15 21. -51 22. 22 23. 53 24. -18
25. -46 26. 19 27. 5 28. -5 29. Inverse
30. Associative 31. Commutative 32. Identity 33. -3
34. 4 35. -4 36. 2 37. 19 38. -32 39. -30
40. -18 41. -17 42. -22 43. 7 44. -22 45. -21
46. -12 47. -49 48. 160 49. 3 50. -2 51. -4
52. 5 53. -3 54. 9 55. -6 56. 90 57. -24
58. 1 59. -15 60. 3 61. -90 62. 48
63. Associative 64. Identity 65. Zero 66. Distributive
67. -4 68. Not an integer 69. -8 70. -45 71. -5
72. Not an integer 73. -49 74. 49 75. -22 76. 8
77. 1 78. -4 79. 21 80. -5 81. $-10 + 5 \cdot (8 - 6)$
82. $14 - (16 - 3) - 1$ 83. $(7 - 11) \cdot 4^2 + 64$
84. $-3^2 \div (5 - 2) + 3$ 85. 0 86. 6 87. -20
88. -5 89. Yes 90. Yes 91. No 92. No 93. -11
94. -3 95. 6 96. 20 97. Table values: -11, -8, -5, -2, 1; solution: 2 98. Table values: 11, 9, 7, 5, 3; solution: -1 99. 6 100. 5 101. (a) Syria (b) Serbia
102. $592 103. 2013 104. $-260{,}000 \div 13 = -20{,}000$; a 20,000 decrease each year 105. -25 °C 106. 2017
107. -3 °F 108. $8 \cdot (-19) = -152$; $152
109. -$7 million 110. 3 °C 111. -$64 112. $850

Chapter 2 Test (p. 143)

1. [number line from -5 to 5 with points at -3, -1, 2, 4] 2. 5 3. > 4. =
5. -5 6. -50 7. 84 8. -7 9. 1 10. -84
11. -121 12. -10 13. -8 14. -7 15. 8
16. -4 17. 1 18. -4 19. Yes 20. No 21. -6
22. 9 23. Table values: 13, 9, 5, 1, -3; solution: 2
24. 7 25. $132,900 26. Table values: 70, 51, 32, 13, -6; solution: 4 miles

Chapters 1–2 Cumulative Review (p. 144)

1. 8 2. $30{,}000 + 2000 + 10$ 3. 6064 4. 11,105
5. 6417 6. 81 r27 7. 7^3 8. 19 9. 33,000,000
10. 1500 11. 9 12. 39 13. $5x - 5$ 14. <
15. -17 16. 5 17. -10 18. 100 19. -1
20. 4 21. -80 22. Zero property 23. 8
24. 1 25. No 26. Table values: -26, -19, -12, -5, 2; solution: 1 27. $979 28. -4 °F 29. 177 bpm
30. 40 cm

3 Algebraic Expressions and Linear Equations

Section 3.1 (pp. 152–153)

1. term 2. coefficient 3. unlike 4. like
5. commutative; associative 6. evaluate 7. $7y$
9. $-6m$ 11. $10x^3$ 13. $4b - 8$ 15. 7 17. $8x + 16$
19. $2m - 4$ 21. $4x + 8$ 23. $2y - 18$

25. $3n^2 - 7n + 3$ 27. 0 29. $7a$ 31. $4x + y$
33. $-2y + 9$ 35. $-15m - 3$ 37. $8a - 7$
39. $-2x^2 + 3x - 1$ 41. $10m + 13$ 43. $y - 5$ 45. 0
47. $4x - 12$ 49. $-2t$ 51. $10x$ 53. $-40n$ 55. $-27p$
57. 0 59. $-8y - 24$ 61. $-63x + 7$ 63. 0
65. $12a + 1$ 67. $2b + 10$ 69. $10x - 1$ 71. $y + 3$
73. $3m - 1$ 75. $11x$ 77. $6t - 28$ 79. $9x + 8$
81. $-7a + 38$
83. (a) $16x + 14x$ (b) $30x$ (c) 150 square feet
85. (a) $3(x + 2)$ (b) $3x + 6$ (c) 21 square units

Section 3.2 (pp. 158–159)

1. − 2. · 3. = 4. ÷ 5. + 6. define
7. $2p$, where p represents the ticket price
9. $h - 8$, where h represents his height
11. $5 + g$, where g represents the number of gallons
13. $t \div 6$, where t represents the number of toys
15. $7(a + 6)$, where a represents her age
17. $9(a - 6)$, where a represents his age
19. $3n - 8$, where n represents the number
21. $x \div (-10) = 9$ 23. $64x = -256$ 25. $x - 3 = 12$
27. $x - 12 = x \div 2$ 29. $2(x + 6) = -20$
31. $x - 30 = 135$ 33. $2x - 12 = 6$
35. $x + 10 = 3x - 6$ 37. $(x + 2) \div 2 = 13$
39. $2x + 2(x + 5) = 70$ 41. $4x = x^3$

Checking Basic Concepts 3.1 & 3.2 (p. 159)

1. $-6y + 5$
2. (a) $a + 9$ (b) $-7y + 7$ (c) $-5m - 3$ (d) $-5x + 6$
3. $s + 5$, where s represents her score 4. $x - 9 = 15$
5. $x + 10{,}000 = 76{,}300$ 6. $3x + 45 = 798$

Section 3.3 (pp. 166–167)

1. solution 2. variable 3. equivalent 4. addition
5. subtraction 6. multiplication 7. division 8. variable
9. Yes 11. No 13. Yes 15. Yes 17. No
19. Equivalent 21. Not equivalent 23. Equivalent
25. Not equivalent 27. Not equivalent 29. -7 31. 7
33. -3 35. 17 37. -49 39. -13 41. 10 43. 6
45. 4 47. 1 49. 6 51. -10 53. 7 55. 27
57. 48 59. -35 61. -17 63. -8 65. 4 67. -48
69. 14 71. 9 73. 60 75. -8 77. 19 79. -88
81. 5 inches 83. 3 years

Section 3.4 (pp. 177–180)

1. linear 2. -2; 0 3. 1 4. numerically
5. symbolically 6. given 7. Linear; $a = 3, b = 7$
9. Not linear 11. Linear; $a = 4, b = -9$ 13. Not linear
15. Not linear 17. Linear; $a = 1, b = -1$
19. Linear; $a = 6, b = 15$ 21. -4 23. -7 25. 3
27. 11 29. 5 31. 2 33. 5 35. -24 37. -3
39. -3 41. 4 43. 15 45. -1 47. -12 49. -6
51. 16 53. 0 55. -13 57. 2 59. 6 61. 7
63. -2 65. 17 67. 6 69. 1 71. -22 73. 8
75. Table values: -4, -3, -2, -1, 0; solution: 2

77. Table values: 11, 9, 7, 5, 3; solution: 0
79. Table values: −12, −9, −6, −3, 0; solution: 2
81. −2   **83.** 0   **85.** 1   **87.** 8
89. (a) 6 miles   (b) 6 miles   (c) 6 miles
91. (a) 4 hours   (b) 4 hours   (c) 4 hours

Checking Basic Concepts 3. 3 & 3.4 (p. 180)

1. Yes   **2.** No   **3.** (a) 7   (b) −5   (c) −32   (d) −56
4. (a) Linear; $a = 4, b = -3$ (b) Not linear
(c) Not linear   (d) Linear; $a = -2, b = -8$
5. (a) 3   (b) 2   (c) −3   (d) 1
6. Table values: 9, 6, 3, 0, −3; solution: −1
7. −2   **8.** 5 inches   **9.** 4 miles

Section 3.5 (pp. 186–188)

1. understanding; variable   **2.** equation   **3.** solve; original
4. check   **5.** 24   **7.** −3   **9.** 3   **11.** −5   **13.** −2
15. 8   **17.** 10   **19.** −2   **21.** 165 pounds
23. 9 quadrillion BTU   **25.** 8   **27.** 24   **29.** 24
31. 3181 million   **33.** Length: 19 inches; width: 12 inches
35. 26 feet, 35 feet, 41 feet   **37.** 2003: 150,000; 2010: 98,000
39. Mother: 49; daughter: 17   **41.** Prius: 50 mpg; Fit: 36 mpg   **43.** 2004

Checking Basic Concepts 3. 5 (p. 189)

1. 6   **2.** −3   **3.** 0   **4.** 5900 megawatts

CHAPTER 3 REVIEW (pp. 191–193)

1. $9y$ **2.** $-17n$ **3.** 0   **4.** $8b - 8$ **5.** $9a - 8$
6. $-4y + 5$ **7.** $-35p$ **8.** $15w - 5$ **9.** $2b + 21$
10. $-3x - 15$ **11.** $-5t - 26$ **12.** $-5m + 1$
13. $2p$, where p represents the number of pancakes
14. $3a$, where a represents her age
15. $5(s + 3)$, where s represents her score
16. $2n - 8$, where n represents the number
17. $(p + 3) \div 5$, where p represents the price
18. $3n + 4$, where n represents the number
19. $5x = 45$ **20.** $x + 7 = 14$ **21.** $9 = x - 11$
22. $x - 10 = 3x$ **23.** $2(x + 4) = -12$
24. $(x + 14) \div 3 = 7$ **25.** Yes   **26.** No   **27.** No
28. Yes   **29.** Equivalent   **30.** Equivalent
31. Not equivalent   **32.** Not equivalent   **33.** 15   **34.** 2
35. 3   **36.** −1   **37.** −6   **38.** 36   **39.** 35   **40.** −72
41. 1   **42.** −7   **43.** −45   **44.** −38
45. Linear; $a = 2, b = -6$ **46.** Linear; $a = 6, b = -10$
47. Not linear   **48.** Not linear   **49.** −7   **50.** 3
51. 4   **52.** 2   **53.** −7   **54.** −1   **55.** −13   **56.** 2
57. −2   **58.** 5   **59.** −2   **60.** 5   **61.** 12   **62.** −5
63. Table values: −9, −7, −5, −3, −1; solution: −1
64. Table values: 12, 9, 6, 3, 0; solution: 2   **65.** −2
66. −8   **67.** 18   **68.** −2   **69.** −3   **70.** 5   **71.** 10
72. −3   **73.** (a) $8x + 6x$ (b) $14x$ (c) 112 windows
74. $x + 23 = 3x - 7$ **75.** 16 inches   **76.** 4 miles
77. 43   **78.** Length: 22 inches; width: 15 inches
79. *Gunsmoke:* 20 yr; *Lassie:* 17 yr   **80.** 15 inches
81. $(6x - 2) + x = 12$ **82.** Mexico: 2; Sweden: 10
83. 2010   **84.** 4 inches

CHAPTER 3 TEST (p. 194)

1. $3x + 4$ **2.** $-3n - 7$ **3.** $24y$ **4.** $-2m + 13$ **5.** 14
6. $2w - 7$ **7.** $x + 8 = 3x$ **8.** $5(x + 3) = -60$ **9.** 56
10. −2   **11.** −7   **12.** −14   **13.** −180   **14.** −24
15. Linear; $a = 6, b = -13$ **16.** Not linear   **17.** 3
18. −1   **19.** −7   **20.** −9
21. Table values: 7, 4, 1, −2, −5; solution: 2
22. 4   **23.** −32   **24.** 3   **25.** $x + 15 = 139$
26. Length: 19 inches; width: 11 inches
27. AK: 20; SD: 54

CHAPTERS 1–3 CUMULATIVE REVIEW (p. 195)

1. 0   **2.** 5^4 **3.** 80,000   **4.** 8000   **5.** 11   **6.** 19   **7.** $2x + 2$
8. =   **9.** 5   **10.** −8   **11.** 0   **12.** −1   **13.** Yes
14. Table values: −15, −11, −7, −3, 1; solution: 1
15. $-5w - 3$ **16.** $7y - 9$ **17.** $10 = x + 7$
18. $3(x + 6) = -18$ **19.** −9   **20.** −9   **21.** −1   **22.** 0
23. 9   **24.** −5   **25.** 8; $4   **26.** 56 in.   **27.** −2
28. $200,000   **29.** $1400 + x = 1550$
30. AR: 14 million acres; CA: 25 million acres

4 Fractions

SECTION 4.1 (pp. 207–209)

1. fraction   **2.** numerator; denominator   **3.** rational
4. proper   **5.** improper   **6.** mixed number
7. Numerator: 6; Denominator: 13   **9.** Numerator: 12; Denominator: 5   **11.** Numerator: x; Denominator: y
13. Numerator: $3p$; Denominator: 14   **15.** $\frac{1}{6}$ **17.** $\frac{5}{8}$ **19.** $\frac{13}{16}$
21. $\frac{1}{8}$ **23.** 1   **25.** 0   **27.** −13   **29.** Undefined   **31.** 1
33. 53   **35.** $\frac{13}{4}$; $3\frac{1}{4}$ **37.** $\frac{17}{10}$; $1\frac{7}{10}$ **39.** $\frac{43}{8}$; $5\frac{3}{8}$ **41.** $5\frac{1}{2}$
43. $2\frac{5}{6}$ **45.** −4   **47.** $11\frac{3}{8}$ **49.** $-7\frac{2}{5}$ **51.** $\frac{23}{4}$ **53.** $-\frac{26}{3}$
55. $\frac{77}{8}$ **57.** $-\frac{107}{3}$ **59.** $\frac{561}{5}$

61. [number line from 0 to 1, point between]
63. [number line from 0 to 2, point near 1]
65. [number line from −2 to 0, point near −1]
67. [number line from 0 to 2, point between 1 and 2]
69. [number line from −5 to 0, point near −3]

71. $\frac{4}{50}$ **73.** $\frac{243}{365}$ **75.** $\frac{4}{7}$ **77.** $\frac{39}{134}$ **79.** $\frac{35}{68}$

SECTION 4.2 (pp. 220–222)

1. factor   **2.** divisible   **3.** 5   **4.** 3   **5.** prime
6. composite   **7.** prime   **8.** factor   **9.** equivalent
10. greatest common factor   **11.** both   **12.** lowest terms
13. cross   **14.** rational   **15.** No   **17.** Yes   **19.** Yes

21. No **23.** Yes **25.** Yes **27.** No **29.** Yes
31. Yes **33.** Prime **35.** Composite **37.** Neither
39. Composite **41.** $2 \cdot 2 \cdot 2 \cdot 2$ **43.** $3 \cdot 3 \cdot 5$
45. $2 \cdot 2 \cdot 5 \cdot 7$ **47.** $3 \cdot 7 \cdot 11$ **49.** $2 \cdot 13 \cdot 17$ **51.** 21
53. 42 **55.** 5 **57.** 3 **59.** 20 **61.** 4 **63.** 2 **65.** 10
67. 24 **69.** 4 **71.** 16 **73.** $\frac{2}{3}$ **75.** $\frac{2}{5}$ **77.** $-\frac{6}{11}$ **79.** $\frac{3}{10}$
81. $-\frac{4}{9}$ **83.** < **85.** < **87.** > **89.** = **91.** Yes
93. No **95.** Yes **97.** $2x$ **99.** $7a^2$ **101.** 50 **103.** $8a^2c$
105. $\frac{x}{3}$ **107.** $-\frac{3}{10y}$ **109.** $\frac{1}{2}$ **111.** $-2xy$ **113.** $\frac{5x}{9y^2}$
115. $\frac{1}{3}$ **117.** Gen X **119.** 43 **121.** 1970: $\frac{2}{25}$; 2016: $\frac{8}{25}$

Checking Basic Concepts 4.1 & 4.2 (p. 223)

1. Numerator: 5; Denominator: 18 **2.** $\frac{11}{4}$; $2\frac{3}{4}$ **3.** $3\frac{4}{5}$
4. $\frac{45}{7}$ **5.** Yes **6.** (a) Composite (b) Prime
7. (a) $2 \cdot 5 \cdot 5$ (b) $3 \cdot 5 \cdot 7$ **8.** 30 **9.** (a) $\frac{5}{8}$ (b) $-\frac{4x}{7}$
10. $\frac{6}{25}$

SECTION 4.3 (pp. 233–235)

1. numerators; denominators **2.** fractions **3.** multiply
4. numerator; denominator **5.** reciprocals
6. multiplication **7.** $\frac{3}{20}$ **9.** $\frac{22}{27}$ **11.** $-\frac{10}{9}$ **13.** $-\frac{63}{20}$
15. $\frac{1}{24}$ **17.** 6 **19.** $-\frac{9}{2}$ **21.** $\frac{1}{5}$ **23.** $-\frac{4}{3}$ **25.** $\frac{4}{21}$ **27.** $\frac{1}{10}$
29. -3 **31.** $\frac{4}{15}$ **33.** $-\frac{15}{22}$ **35.** $\frac{11}{18}$ **37.** $\frac{1}{4}$ **39.** $\frac{5x}{2y}$
41. $\frac{8p^2}{15q^2}$ **43.** $\frac{3ab^2}{2}$ **45.** $\frac{9x^2}{2y}$ **47.** $-\frac{2u^3}{3v^2}$ **49.** $-\frac{7y}{3x}$ **51.** $\frac{1}{6x}$
53. 4 **55.** $\frac{x}{5y}$ **57.** $\frac{1}{16}$ **59.** $\frac{9}{25}$ **61.** $-\frac{27}{64}$ **63.** $\frac{64}{121}$ **65.** $\frac{3}{5}$
67. $\frac{1}{8}$ **69.** 4 **71.** $\frac{2}{3}$ **73.** $\frac{5}{3}$ **75.** -12 **77.** $\frac{1}{15}$ **79.** 1
81. $\frac{5}{3}$ **83.** $-\frac{20}{3}$ **85.** $-\frac{24}{5}$ **87.** 6 **89.** $-\frac{5}{3}$ **91.** $\frac{5}{9}$
93. $-\frac{9}{22}$ **95.** $\frac{7}{26}$ **97.** $\frac{xy}{2}$ **99.** $-9b$ **101.** $-\frac{5p}{6q}$ **103.** $9xy^2$
105. 1 **107.** $\frac{5}{6}$ **109.** -3 **111.** $\frac{70}{3}$ **113.** 44 million
115. 81 **117.** $480 **119.** $\frac{1}{3}$ cup **121.** $\frac{1}{12}$
123. 20 square inches **125.** $\frac{91}{2}$ square miles

SECTION 4.4 (pp. 240–243)

1. like; common **2.** numerators **3.** denominator
4. numerators **5.** denominator **6.** fractions **7.** $\frac{3}{5}$
9. $-\frac{5}{9}$ **11.** $\frac{24}{25}$ **13.** $\frac{10}{13}$ **15.** $\frac{4}{7}$ **17.** $\frac{2}{3}$ **19.** -2 **21.** 0
23. 1 **25.** $\frac{1}{2} + \frac{1}{3} + \frac{1}{4}$ **27.** $\frac{1}{10} + \frac{1}{2} + \frac{1}{2}$ **29.** Less than 1
31. Greater than 1 **33.** $\frac{2}{7}$ **35.** $\frac{8}{9}$ **37.** $-\frac{12}{17}$ **39.** $-\frac{5}{11}$
41. $\frac{1}{4}$ **43.** 0 **45.** $-\frac{1}{5}$ **47.** $-\frac{1}{3}$ **49.** 1 **51.** $\frac{1}{3}$ **53.** $-\frac{1}{2}$
55. $\frac{2}{3}$ **57.** $\frac{1}{15}$ **59.** $-\frac{8}{15}$ **61.** -1 **63.** $\frac{7x}{9y}$ **65.** $-\frac{4}{5m^2}$
67. $\frac{7x - 3w}{y}$ **69.** $\frac{1}{d}$ **71.** $\frac{4y}{x^2}$ **73.** $-\frac{k^2}{3c}$ **75.** $3x^2$
77. $\frac{3}{4}$ inch **79.** $\frac{1}{125}$ **81.** $\frac{13}{2500}$ **83.** $\frac{2}{5}$ **85.** $\frac{1}{5}$ **87.** $\frac{3}{5}$

Checking Basic Concepts 4.3 & 4.4 (p. 243)

1. (a) $\frac{4}{3}$ (b) -5 (c) $\frac{4}{5}$ (d) $-\frac{3}{4}$ (e) $\frac{1}{12y}$ (f) -3
(g) $-10a^2$ (h) $\frac{3x}{2y}$ **2.** (a) $\frac{4}{25}$ (b) 5 **3.** (a) $\frac{12}{19}$ (b) $\frac{1}{3}$
(c) $\frac{2}{7}$ (d) -1 (e) $4x^2$ (f) $\frac{m^2}{4n^3}$ **4.** 39

SECTION 4.5 (pp. 251–253)

1. least **2.** divisible **3.** listing **4.** prime factorization
5. maximum **6.** LCM **7.** 20 **9.** 60 **11.** 12 **13.** 80
15. 90 **17.** 30 **19.** 180 **21.** 135 **23.** 1296 **25.** $9x$
27. $24y^3$ **29.** $24a^3b^2$ **31.** $12xyz$ **33.** 36 **35.** 28
37. 120 **39.** 60 **41.** $6xy^2$ **43.** $16mn^2$ **45.** abc
47. $\frac{9}{12}; \frac{2}{12}$ **49.** $\frac{22}{36}; \frac{27}{36}$ **51.** $\frac{15}{30}; \frac{25}{30}; \frac{27}{30}$ **53.** $\frac{11}{20}$ **55.** $-\frac{7}{40}$
57. $-\frac{13}{9}$ **59.** $\frac{1}{6}$ **61.** $\frac{2}{3}$ **63.** $-\frac{7}{6}$ **65.** 1 **67.** $\frac{3}{4}$ **69.** $\frac{5}{2}$
71. $\frac{19x}{12}$ **73.** $\frac{m^2}{3}$ **75.** $\frac{3x + 20y}{48}$ **77.** $\frac{20y - 3}{24y^2}$ **79.** $\frac{9x^2 + 25y^2}{15xy}$
81. $\frac{37}{100}$ **83.** $\frac{3}{50}$ **85.** $\frac{5}{12}$ **87.** $\frac{5}{8}$ **89.** 4 ft **91.** $\frac{37}{18}$ in.
93. 7 yd

SECTION 4.6 (pp. 261–262)

1. $\frac{1}{2}$ **2.** less **3.** improper fractions **4.** estimate
5. vertically **6.** regroup **7.** $\frac{21}{5}$ **9.** $-\frac{52}{15}$ **11.** $\frac{89}{10}$
13. $-\frac{61}{4}$ **15.** 8 **17.** -14 **19.** 11 **21.** -10 **23.** 10
25. $39\frac{1}{2}$ **27.** $7\frac{1}{24}$ **29.** $6\frac{3}{4}$ **31.** $-33\frac{4}{9}$ **33.** $-1\frac{1}{8}$
35. $-9\frac{9}{10}$ **37.** $13\frac{9}{19}$ **39.** $4\frac{23}{35}$ **41.** $11\frac{7}{12}$ **43.** $7\frac{8}{21}$
45. $5\frac{7}{24}$ **47.** $14\frac{13}{24}$ **49.** 4 **51.** $11\frac{6}{11}$ **53.** $-13\frac{19}{21}$
55. $14\frac{7}{8}$ **57.** $-23\frac{2}{3}$ **59.** $2\frac{13}{15}$ **61.** 10 pieces; $4\frac{1}{2}$ inches
63. $5\frac{1}{3}$ cups **65.** About 676 thousand **67.** $2\frac{3}{8}$ inches
69. $\frac{7}{12}$ gallon **71.** $16\frac{2}{3}$ square feet **73.** $57\frac{19}{25}$ square inches

Checking Basic Concepts 4.5 & 4.6 (p. 262)

1. (a) 60 (b) $20x^2$ **2.** (a) 72 (b) $30x^2y$
3. (a) $\frac{7}{24}$ (b) $-\frac{1}{20}$ (c) $\frac{33}{40}$ (d) $-\frac{x}{24}$
4. (a) $-\frac{1}{2}$ (b) $6\frac{17}{20}$ (c) $30\frac{3}{4}$ (d) $5\frac{1}{3}$ **5.** $\frac{13}{24}$

SECTION 4.7 (pp. 268–269)

1. complex **2.** reciprocal **3.** I **4.** II **5.** $\frac{5}{14}$ **7.** 9 **9.** $\frac{16}{5}$
11. $\frac{1}{12}$ **13.** $\frac{25}{2xy}$ **15.** 12 **17.** $\frac{5}{3}$ **19.** $\frac{4}{5}$ **21.** 5 **23.** $\frac{17}{22}$
25. $\frac{16x}{21}$ **27.** $\frac{21x}{y}$ **29.** $\frac{1}{5}$ **31.** $\frac{11}{8}$ **33.** $\frac{2}{7}$ **35.** $\frac{4}{9}$ **37.** $\frac{3x}{4}$
39. 8 **41.** 3 **43.** $\frac{3}{4}$ **45.** $-\frac{17}{32}$ **47.** $\frac{1}{10}$ **49.** -11
51. $-\frac{9}{16}$ **53.** $\frac{1}{6}$ **55.** $\frac{29}{30}$ **57.** $3\frac{1}{8}$ mph **59.** $\frac{3}{4}$ hour
61. 28 °C

SECTION 4.8 (pp. 281–283)

1. addition **2.** multiplication **3.** distributive
4. LCD **5.** $\frac{13}{20}$ **7.** $-\frac{2}{15}$ **9.** $-\frac{7}{12}$ **11.** $\frac{2}{5}$ **13.** 14
15. $-\frac{15}{2}$ **17.** $\frac{9}{4}$ **19.** $-\frac{3}{10}$ **21.** 11 **23.** $\frac{2}{9}$ **25.** $-\frac{5}{12}$
27. $\frac{9}{20}$ **29.** $\frac{1}{9}$ **31.** $-\frac{6}{7}$ **33.** $\frac{1}{6}$ **35.** 2 **37.** $-\frac{11}{16}$
39. 6 **41.** $\frac{12}{7}$ **43.** $\frac{4}{5}$ **45.** $\frac{7}{20}$ **47.** $\frac{1}{3}$
49. Table values: $-\frac{8}{3}, -2, -\frac{4}{3}, -\frac{2}{3}, 0$; solution: 9
51. Table values: $\frac{23}{5}, \frac{19}{5}, 3, \frac{11}{5}, \frac{7}{5}$; solution: -1
53. Table values: $-\frac{7}{3}, -\frac{4}{3}, -\frac{1}{3}, \frac{2}{3}, \frac{5}{3}$; solution: 4
55. -2 **57.** 4 **59.** 1 **61.** $-\frac{11}{4}$ **63.** $\frac{21}{2}$ **65.** $-\frac{34}{3}$
67. $-\frac{7}{8}$ **69.** $-\frac{1}{45}$ **71.** $\frac{2}{5}$ **73.** $\frac{7}{4}$ feet **75.** $\frac{10}{3}$ inches

77. 8 °C   **79.** 3 minutes   **81. (a)** 2004   **(b)** 2004   **(c)** 2004   **(d)** The solutions are the same.

Checking Basic Concepts 4.7 & 4.8 (p. 284)

1. (a) $\frac{3}{2}$ **(b)** $\frac{w}{27}$ **(c)** $\frac{11}{6}$ **(d)** $\frac{2x}{3}$ **2. (a)** $\frac{7}{24}$ **(b)** $\frac{1}{4}$
3. (a) $\frac{5}{8}$ **(b)** -30 **(c)** 6   **(d)** 1
4. Table values: 6, 5, 4, 3, 2; solution: -7 **5.** -2
6. 74 °F

CHAPTER 4 REVIEW (pp. 291–293)

1. (a) Numerator: 7; Denominator: 18
(b) Numerator: x; Denominator: 5
2. (a) $\frac{2}{5}$ **(b)** $\frac{7}{8}$ **3. (a)** 0   **(b)** -2 **(c)** 1   **(d)** Undefined
4. (a) $\frac{13}{10}$; $1\frac{3}{10}$ **(b)** $\frac{17}{6}$; $2\frac{5}{6}$ **5. (a)** $6\frac{1}{3}$ **(b)** $-2\frac{4}{5}$
6. (a) $-\frac{32}{5}$ **(b)** $\frac{70}{9}$
7. (a) number line with point at -3 (range -5 to 0)
(b) number line with point at $\frac{1}{2}$ (range 0 to 2)
8. (a) No   **(b)** Yes   **9. (a)** Composite   **(b)** Prime
(c) Neither   **(d)** Composite   **10. (a)** $2 \cdot 2 \cdot 2 \cdot 5$
(b) $2 \cdot 5 \cdot 11$ **11. (a)** 11   **(b)** 18   **12. (a)** 6   **(b)** 8
13. (a) $-\frac{4}{13}$ **(b)** $\frac{8}{11}$ **14. (a)** $<$ **(b)** $>$ **15. (a)** $3x$
(b) $2xy^2$ **16. (a)** $\frac{4x}{5}$ **(b)** $-3x^2$ **17.** $\frac{25}{48}$ **18.** $\frac{5}{12}$ **19.** -3
20. $\frac{8}{21}$ **21.** $\frac{3x}{10w}$ **22.** $-\frac{9y}{4x}$ **23. (a)** $-\frac{1}{64}$ **(b)** $\frac{7}{10}$
24. (a) $-\frac{9}{4}$ **(b)** 10   **25.** $\frac{2}{15}$ **26.** $-\frac{18}{5}$ **27.** $\frac{xy}{8}$ **28.** $\frac{3p^2}{8q}$
29. -20 **30.** $\frac{5}{24}$ **31.** 1   **32.** $\frac{4}{5}$ **33.** 2   **34.** $-\frac{5}{8}$ **35.** $\frac{y^2}{4x}$
36. 0   **37. (a)** 60   **(b)** 48   **(c)** $12x^2$ **(d)** $12mn$
38. (a) 48   **(b)** $24y^2$ **39.** $\frac{3}{18}$; $\frac{16}{18}$ **40.** $\frac{35}{42}$; $\frac{12}{42}$; $\frac{9}{42}$ **41.** $\frac{11}{12}$
42. $-\frac{17}{40}$ **43.** $\frac{2}{3}$ **44.** $\frac{4}{9}$ **45.** $\frac{7y^2}{3}$ **46.** $\frac{11}{8m}$ **47.** -4 **48.** 5
49. $13\frac{1}{12}$ **50.** $-21\frac{1}{2}$ **51.** $2\frac{14}{15}$ **52.** $5\frac{5}{9}$ **53.** $-13\frac{3}{4}$
54. $4\frac{1}{2}$ **55.** $2\frac{2}{3}$ **56.** $-8\frac{2}{9}$ **57.** $\frac{2}{3}$ **58.** $\frac{5}{2}$ **59.** $2x$ **60.** $\frac{9y}{25}$
61. $\frac{2}{3}$ **62.** $\frac{5}{4}$ **63.** $\frac{23}{24}$ **64.** -72 **65.** $\frac{3}{14}$ **66.** 2   **67.** $\frac{7}{12}$
68. $-\frac{10}{3}$ **69.** -2 **70.** -3 **71.** Table values: $-\frac{23}{5}$, $-\frac{21}{5}$, $-\frac{19}{5}$, $-\frac{17}{5}$, -3; solution: 5   **72.** Table values: $\frac{1}{2}$, 2, $\frac{7}{2}$, 5, $\frac{13}{2}$; solution: 3   **73.** -2 **74.** 4   **75.** $\frac{93}{100}$ **76.** Red class
77. $\frac{1}{2}$ cup   **78.** $\frac{4}{15}$ **79.** $\frac{7}{10}$ **80.** $25\frac{1}{4}$ inches   **81.** $132
82. $4\frac{4}{5}$ hours

CHAPTER 4 TEST (pp. 293–294)

1. $\frac{19}{6}$; $3\frac{1}{6}$ **2.** $\frac{33}{7}$ **3.** $7\frac{1}{3}$ **4. (a)** Composite   **(b)** Neither
(c) Prime   **(d)** Composite   **5.** $2 \cdot 2 \cdot 2 \cdot 3 \cdot 5$
6. (a) 5   **(b)** 27   **7. (a)** 8   **(b)** $7x$ **8. (a)** $-\frac{3}{4}$ **(b)** $4x$
9. (a) $-\frac{3}{2}$ **(b)** $\frac{7}{20}$ **(c)** $\frac{5x}{7y}$ **(d)** $-\frac{8xy}{3}$ **10. (a)** $\frac{9}{11}$ **(b)** $\frac{25}{36}$
11. (a) $\frac{7}{12}$ **(b)** $-\frac{7}{30}$ **(c)** $-\frac{23}{15}$ **(d)** $\frac{4w}{x}$ **12. (a)** $11\frac{5}{12}$
(b) $-22\frac{2}{5}$ **13. (a)** $\frac{25}{26}$ **(b)** 1   **14. (a)** -16 **(b)** $\frac{3}{4}$ **(c)** 5
(d) $\frac{1}{2}$ **15.** Table values: $-\frac{13}{6}$, $-\frac{5}{6}$, $\frac{1}{2}$, $\frac{11}{6}$, $\frac{19}{6}$; solution: 2
16. 3   **17.** $672   **18.** 2013

CHAPTERS 1–4 CUMULATIVE REVIEW (pp. 294–295)

1. 36,285   **2.** $15 \cdot 4$; 60   **3.** 222 r15   **4.** 5   **5.** 3700   **6.** 6
7. $3x - 1$ **8.** 4   **9.** Associative   **10.** 6   **11.** -60 **12.** 1
13. Table values: -11, -8, -5, -2, 1; solution: 0
14. $2y + 5$ **15.** $7 = x - 13$ **16.** $2(x + 3) = -14$
17. No   **18.** No   **19.** -2 **20.** -3 **21.** $\frac{37}{10}$; $3\frac{7}{10}$
22. $2 \cdot 3 \cdot 3 \cdot 11$ **23.** $-\frac{3}{4}$ **24.** $\frac{1}{6}$ **25.** 0   **26.** $\frac{7}{15}$
27. $8\frac{23}{24}$ **28.** $\frac{13}{30}$ **29.** $\frac{1}{3}$ **30.** $\frac{11}{4}$ **31.** $40   **32.** 2014
33. Length: 9 inches; width: 4 inches   **34.** $\frac{2}{5}$

5 Decimals

Section 5.1 (pp. 305–307)

1. decimal   **2.** decimal; decimals   **3.** and   **4.** 10
5. 10   **6.** unequal   **7.** fifty-six hundredths
9. seven and one hundred sixteen thousandths
11. negative fifty-eight and seven tenths
13. negative two and one thousand three millionths
15. five hundred one and twelve ten-thousandths
17. One hundred twenty-nine and 68/100
19. $\frac{3}{10}$ **21.** $-\frac{1}{25}$ **23.** $\frac{17}{20}$ **25.** $-8\frac{1}{5}$ **27.** $12\frac{3}{4}$
29. $23\frac{41}{200}$ **31.** $-1\frac{7}{250}$ **33.** $6\frac{41}{80}$
35. number line with point near 8.5 (range 8.0 to 9.0)
37. number line with point at 26.76 (range 26.70 to 26.80)
39. number line with point at 0.315 (range 0.310 to 0.320)
41. number line with point near -2.1 (range -3.0 to -2.0)
43. number line with point near -5.79 (range -5.80 to -5.70)
45. $<$ **47.** $<$ **49.** $>$ **51.** $<$ **53.** $>$ **55.** $<$ **57.** $<$
59. $>$ **61.** 0.4   **63.** 52.01   **65.** -7.0094 **67.** 9.003
69. -1.106021 **71.** 5.73829   **73.** $4   **75.** $143.30
77. $20   **79.** one thousand four hundred fifty-three and seventy-one hundredths   **81.** $2\frac{9}{20}$ **83.** Usain Bolt
85. Illinois; Illinois: $2.07; Florida $2.07   **87.** $2.06
89. $\frac{4}{125}$

Section 5.2 (pp. 313–315)

1. estimate   **2.** rounded   **3.** decimal points   **4.** vertically
5. $20 + 400 = 420$ **7.** $1500 - 300 = 1200$
9. $690 - 90 = 600$ **11.** $1650 + 350 = 2000$
13. $-300 + 70 = -230$ **15.** $-40 - 130 = -170$
17. Estimate: 10; Actual: 10.388
19. Estimate: 670; Actual: 672.899
21. Estimate: 830; Actual: 833.62
23. Estimate: 760; Actual: 758.259
25. Estimate: 6550; Actual: 6547.18
27. Estimate: 170; Actual: 173.48
29. Estimate: 700; Actual: 700.857

31. Estimate: 160; Actual: 162.7
33. Estimate: −200; Actual: −198.37
35. Estimate: −690; Actual: −695.61
37. Estimate: 1300; Actual: 1298.94
39. Estimate: 60; Actual: 56.32   41. 134.207   43. 800.96
45. −3154.29   47. −61.692   49. 1235.899   51. 5006.555
53. Greater than 1   55. Less than 1   57. Less than 0
59. Greater than 0   61. 2.598   63. 1.473   65. $2.78
67. $4.14   69. 11.03$y$ 71. 9.6w − 5   73. 1.8n^2
75. −2.6p^2 + 13.6   77. 1.6x + 2.7y 79. 6.3 million tons
81. $19.52   83. 116 thousand   85. 1980
87. 0.92 thousand gallons   89. 296.9 miles
91. 40.2 inches   93. 40.16 feet   95. 15.5 inches
97. $867.85   99. 5.969 hours

Checking Basic Concepts 5.1 & 5.2 (p. 316)

1. twenty-three and ninety-seven thousandths
2. (a) $-5\frac{3}{5}$ (b) $\frac{13}{25}$
3. <number line: 34.20 34.22 34.24 34.26 34.28 34.30>
4. (a) <   (b) >   5. 0.278   6. $9
7. (a) 150 + 20 = 170   (b) 7000 − 200 = 6800
8. (a) 36.02   (b) 434.59   (c) −71.66   (d) 282.62
9. (a) 5.04x (b) 3.5y^2 + 2   10. $8.07

Section 5.3 (pp. 328–330)

1. estimate   2. sum   3. right   4. above   5. right
6. 2   7. left   8. denominator; numerator   9. 2 · 27 = 54
11. 12 ÷ 4 = 3   13. 500 · 5 = 2500   15. 87 ÷ 1 = 87
17. 123.3   19. 2.94   21. 3.953   23. 15.96   25. 34.4344
27. −43.8   29. −7.02   31. 2.091   33. 124.89
35. −467.9   37. 410   39. −9.8   41. 34,498   43. 9.4
45. 2.97   47. −12.9   49. 4.$\overline{6}$ 51. 3.7$\overline{3}$ 53. 1.38
55. 14.75   57. 86.25   59. −0.376   61. −600
63. 1.779   65. −0.634   67. 78.94   69. −0.0076
71. 0.0001   73. 0.25   75. 0.375   77. −0.24
79. 0.$\overline{3}$ 81. 0.2$\overline{6}$ 83. −0.365   85. 3.2   87. −9.5$\overline{3}$
89. 17.$\overline{6}$ 91. 49.68   93. −2.3364   95. 42.2
97. −0.05   99. 3.51x + 21.6   101. −20.4y − 18.36
103. 0.8x 105. −80w 107. 578.5 calories   109. 51 mpg
111. $2,584.4 million   113. 4.9%   115. $4.68
117. $264.12   119. $459,334.80   121. $512.46

Section 5.4 (pp. 337–339)

1. irrational   2. real   3. real   4. $\sqrt{36}$ 5. Guess-and-check
6. Babylonian   7. (a) 3, $\sqrt{4}$ (b) 3, 0, $\sqrt{4}$ (c) 3, 0, $\sqrt{4}$
(d) $-\frac{5}{8}$, 3, 0, $\sqrt{4}$ (e) $\sqrt{5}$ 9. (a) $\frac{9}{3}$ (b) 0, $\frac{9}{3}$ (c) 0, $\frac{9}{3}$
(d) 0, $\frac{9}{3}$, $-1.\overline{2}$, 6.4   (e) $\sqrt{10}$ 11. 2.2   13. 3.9   15. 9.1
17. 3.74   19. 6.48   21. 9.95   23. 2.45   25. 4.24
27. 8.83   29. 1.4142   31. 5.4772   33. 7.4162   35. 9.8
37. 5.15   39. 13.9   41. 1.25   43. 16.8$\overline{3}$ 45. 3   47. 7.18
49. 10.61   51. −9.1   53. 9   55. −10.$\overline{6}$ 57. 32 feet
59. $144.1 million per year   61. 46.28 square inches
63. 16.35 square miles   65. 29.5 mpg   67. 3.095 million
69. 6.93 feet

Checking Basic Concepts 5.3 & 5.4 (p. 339)

1. (a) 7.56   (b) −244.3   (c) 26.3   (d) −19.096
2. (a) 3.56   (b) 9.1$\overline{5}$ (c) 0.635   (d) −0.4
3. (a) 0.45   (b) 0.4$\overline{6}$ (c) −3.125   (d) 9.8$\overline{3}$
4. (a) 19x + 7.5   (b) −70y 5. 4.9   6. 4.47
7. (a) 6.7   (b) 1.$\overline{8}$ 8. −6.3   9. $1169.62
10. 35 feet

Section 5.5 (pp. 347–350)

1. Symbolic, numerical, and graphical   2. decimals
3. decimal places   4. numerically   5. 16.5   7. −12.4
9. 3.5   11. 4.9   13. 3.2   15. 6.8   17. −0.$\overline{3}$ 19. −3.05
21. 3.7   23. 18.7   25. −40.25   27. −0.5   29. 44.8
31. Table values: −2.5, 2.3, 7.1, 11.9, 16.7; solution: 2
33. Table values: 9.5, 5.6, 1.7, −2.2, −6.1; solution: −2
35. Table values: 11, 11.7, 12.4, 13.1, 13.8; solution: 2.1
37. −5   39. 3   41. 3   43. 9.15   45. 3   47. 8.5
49. −1.2   51. 1150   53. 6.5 feet   55. 5.7 inches
57. 2014   59. 14 °C   61. (a) 2014   (b) 2014   (c) 2014
(d) The solutions are the same.

Section 5.6 (pp. 359–363)

1. circle   2. radius   3. diameter   4. radius
5. circumference   6. π 7. semicircle   8. right
9. legs   10. hypotenuse   11. mean   12. median
13. mode   14. weighted   15. 12π; 37.68 inches
17. 50π; 157 yards   19. 0.6π; 1.884 inches
21. 3π; 9.42 feet   23. 25π; 78.5 square feet
25. 0.16π; 0.5024 square yards   27. π; 3.14 square inches
29. 2.25π; 7.065 square feet   31. 84 square inches
33. 114.24 square yards   35. 178.46 square feet
37. 35.44 square yards   39. 10 inches   41. 8 yards
43. 6.6 feet   45. 10.8 yards   47. 69.8   49. 29.2   51. 3.4
53. 40.5   55. 27.65   57. 1   59. 4.7   61. No mode
63. Two modes: 5 and 9   65. 3.25   67. 3.8
69. 1293.68 feet   71. 103,918.5 square feet   73. 17 feet
75. 23 seasons   77. 19 seasons   79. 17 seasons

Checking Basic Concepts 5.5 & 5.6 (p. 363)

1. (a) 3.2   (b) 18.1   (c) 8.$\overline{36}$
2. Table values: −6.1, −2.7, 0.7, 4.1, 7.5; solution: −2
3. 4   4. $C \approx$ 43.96 inches; $A \approx$ 153.86 square inches
5. 7 inches   6. Mean: 7.875; median: 7.5; mode: 6
7. 2.625   8. 2020

Chapter 5 Review (pp. 369–371)

1. seventy-six hundredths
2. negative five and two hundred six thousandths
3. $-\frac{2}{25}$ 4. $37\frac{1}{4}$
5. <number line: 7.0 7.2 7.4 7.6 7.8 8.0>
6. <number line: −5.20 −5.18 −5.16 −5.14 −5.12 −5.10>
7. >   8. <   9. −4.0083   10. 3591.01   11. $12
12. $41.81   13. Estimate: 250; Actual: 250.145

14. Estimate: 660; Actual: 659.98
15. Estimate: 670; Actual: 669.979
16. Estimate: 840; Actual: 840.07
17. Estimate: 290; Actual: 290.34
18. Estimate: 180; Actual: 177.8
19. $10.77y$ **20.** $145.7x^3$ **21.** $7q - 1.1$ **22.** $7.41n + 4b$
23. $3.6x + 4.7y$ **24.** $0.3a - 3.5b + 0.9$
25. $4 \cdot 30 = 120$ **26.** $200 \div 10 = 20$ **27.** -1.56
28. -1.89 **29.** 194.6 **30.** 16.3 **31.** 13.56 **32.** 6.256
33. 17.475 **34.** 92.46 **35.** 140 **36.** 6.8125 **37.** -13.8
38. 10.504 **39.** -0.44 **40.** $0.4\overline{3}$ **41.** $8.1\overline{3}$ **42.** 64.85
43. $7.7x + 17.6$ **44.** $-1.5y + 0.6$ **45.** $0.6x$ **46.** $-60n$
47. (a) $2, \sqrt{9}$ **(b)** $2, 0, \sqrt{9}$ **(c)** $2, 0, \sqrt{9}$
(d) $\frac{5}{8}, 2, 0, \sqrt{9}$ **(e)** $-\sqrt{7}$ **48. (a)** 7 **(b)** 7 **(c)** 7
(d) $7, -\frac{2}{5}, 3.\overline{6}, -7.4$ **(e)** $\sqrt{5}$ **49.** 3.61 **50.** 4.58
51. 2.65 **52.** 5.29 **53.** 6.775 **54.** 1 **55.** 8.25 **56.** 3
57. 4.9 **58.** 9.5 **59.** -32.3 **60.** 9.2 **61.** 16.9
62. 5.6 **63.** 3.1 **64.** 1.25 **65.** -9.9 **66.** 5 **67.** 4.8
68. 0.125 **69.** -30.2 **70.** 2.2
71. Table values: $-7.5, -1.6, 4.3, 10.2, 16.1$; solution: 3
72. Table values: $9, 9.8, 10.6, 11.4, 12.2$; solution: 0.6
73. -5 **74.** 3 **75.** 18π; 56.52 inches **76.** 22π; 69.08 feet
77. 81π; 254.34 square feet **78.** 64π; 200.96 square inches
79. 64.26 square yards **80.** 42.74 square feet **81.** 20 feet
82. 10.9 inches **83.** Mean: 4.8; median: 4.7; mode: 4.7
84. Mean: 12.4; median: 12.5; mode: 15 **85.** 2.8 **86.** 1.25
87. Bacterium A **88.** 470.4 miles **89.** 189 calories
90. 72 feet **91.** $94.1°F$ **92.** 24 feet **93.** $\$26.37$ **94.** $\frac{38}{125}$
95. 104 bpm **96.** $\$20.15$ **97.** 540.08 feet **98.** 493

Chapter 5 Test (pp. 371–372)

1. $-\frac{17}{20}$ **2.** $13\frac{5}{8}$ **3.** $<$ **4.** 91.581 **5.** 130.347
6. 3708.23 **7.** 22.685 **8.** 413.8 **9.** 69.24 **10.** 48
11. $15x + 1$ **12.** $-50x$ **13.** $17.1\overline{6}$ **14.** -0.275
15. (a) $9, \sqrt{4}$ **(b)** $9, \sqrt{4}, 0$ **(c)** $9, \sqrt{4}, 0$
(d) $9, -\frac{2}{3}, \sqrt{4}, 0$ **(e)** π **16.** 4.58 **17.** 4.5 **18.** 3.1
19. 4.1 **20.** 7.5 **21.** Table values: $-7.4, -6.7, -6,$
$-5.3, -4.6$; solution: -0.5 **22.** 2 **23.** $C \approx 18.84$ inches;
$A \approx 28.26$ square inches **24.** 6.9 inches **25.** Mean: 11.9;
median: 12.2; mode: 12.6 **26.** 2.75 **27.** 65.5 feet
28. 62 years

Chapters 1–5 Cumulative Review (pp. 372–373)

1. $60,000 + 1000 + 5$ **2.** $48,000$ **3.** 8 **4.** 1500 **5.** 11
6. $=$ **7.** -7 **8.** Commutative **9.** $4x + 8$ **10.** -2
11. Table values: $-14, -9, -4, 1, 6$; solution: -1 **12.** No
13. $x - 12 = -5$ **14.** $2(x + 4) = -10$ **15.** 2 **16.** 4
17. -2 **18.** Yes **19.** 2 **20.** $\frac{23}{10}; 2\frac{3}{10}$ **21.** $-\frac{3}{2}$ **22.** $\frac{11}{40}$
23. $0.2\overline{6}$ **24.** -10.7 **25.** 6.71 **26.** $<$ **27.** 14.76
28. 3.2 **29.** 64 inches **30.** $\$2862$

6 Ratios, Proportions, and Measurement

Section 6.1 (pp. 381–383)

1. ratio **2.** 1 **3.** denominator; numerator **4.** rate **5.** unit
6. pricing **7.** $\frac{1}{3}$ **9.** $\frac{3}{8}$ **11.** $\frac{4}{5}$ **13.** $\frac{9}{10}$ **15.** $\frac{2}{5}$ **17.** $\frac{10}{3}$
19. $\frac{5}{4}$ **21.** $\frac{7}{6}$ **23.** $\frac{3}{1} = 3$ **25.** $\frac{0.625}{1} = 0.625$
27. $\frac{5.5}{1} = 5.5$ **29.** $\frac{1}{1} = 1$ **31.** $\frac{2 \text{ inches}}{3 \text{ hours}}$ **33.** $\frac{13 \text{ dollars}}{2 \text{ hours}}$
35. $\frac{12 \text{ seats}}{1 \text{ row}}$ **37.** $\frac{1 \text{ copier}}{17 \text{ employees}}$ **39.** $\frac{2 \text{ slices}}{1 \text{ person}}$ **41.** $\$8.75$/hr
43. 25.6 mi/gal **45.** 0.8 in./hr **47.** $\$2.25$/drink
49. 62 beats/min **51.** $\$0.42$/oz **53.** $\$3.95$/lb
55. $\$0.00425$/lb **57.** $\frac{32}{35}$ **59.** $\frac{2}{3}$ **61.** $\frac{18}{5}$ **63.** $\frac{19}{10}$
65. (a) $\frac{1}{3}$ **(b)** $\frac{11}{250}$ **(c)** $\frac{3}{41}$ **67.** $\frac{3}{197}$ **69.** 0.72; yes
71. University: 43.25; community college: 32.5; there
are 32.5 students for each instructor. **73.** Receptionist: 66
words/min; office manager: 63.5 words/min; the receptionist
75. Large: $\$0.22$/oz; small: $\$0.205$/oz; the small jar
77. Generic ($\$0.256$/pill vs. $\$0.365$/pill)
79. (a) Beets: $\$0.14$/oz; mints: $\$0.35$/oz **(b)** We are not
comparing size options for the same product.

Section 6.2 (pp. 392–394)

1. proportion **2.** $\frac{a}{b} = \frac{c}{d}$ **3.** cross products **4.** equal
5. similar **6.** proportional **7.** Yes **9.** No **11.** Yes
13. Yes **15.** No **17.** Yes **19.** No **21.** No **23.** 35
25. 6.3 **27.** $\frac{1}{3}$ **29.** 1 **31.** -12 **33.** 19.2 **35.** 3.6
37. 6.9 **39.** $\frac{2}{3}$ **41.** $-\frac{3}{5}$ **43.** $-\frac{3}{4}$ **45.** 14 inches
47. 1.6 yards **49.** $\frac{1}{4}$ inch **51.** 140 inches **53.** 14 days
55. 12 inches **57.** $\$1.14$ **59.** 4 students **61.** 288 slices
63. $2\frac{1}{2}$ cups **65.** 3500 songs **67.** $10,500$ people
69. 240 freshman **71.** 3.6 inches of rain **73.** 25 meters
75. 143 feet **77.** 0.6 inch

Checking Basic Concepts 6.1 & 6.2 (p. 395)

1. (a) $\frac{4}{9}$ **(b)** $\frac{16}{3}$ **2. (a)** $\frac{2 \text{ inches}}{5 \text{ hours}}$ **(b)** $\frac{291 \text{ miles}}{4 \text{ hours}}$
3. (a) $\$9.25$/hr **(b)** 1.5 in./hr **4. (a)** $\$2.95$/lb **(b)** $\$0.18$/oz
5. (a) Yes **(b)** No **6. (a)** -30 **(b)** 1 **7.** 3 inches
8. Large: $\$0.07$/oz; small: $\$0.095$/oz; the large drink
9. $1\frac{2}{3}$ **10.** 84 feet

Section 6.3 (pp. 401–402)

1. length **2.** unit **3.** denominator **4.** numerator **5.** area
6. capacity **7.** volume **8.** capacity; volume **9.** weight
10. ounce **11.** 4 ft **13.** $\frac{1}{2}$ or 0.5 mi **15.** 24 in. **17.** 29 yd
19. 6336 ft **21.** 1584 in. **23.** $\frac{1}{12}$ or $0.08\overline{3}$ ft **25.** 1440 in^2
27. $30,976$ yd^2 **29.** 36 yd^2 **31.** $\frac{1}{100}$ or 0.01 mi^2 **33.** 6 c
35. 1 pt **37.** 41 qt **39.** 13 pt **41.** $6\frac{1}{4}$ or 6.25 pt
43. 6 pt **45.** $3\frac{3}{4}$ or 3.75 gal **47.** 1 c **49.** 80 c **51.** 25 qt
53. $27\frac{1}{2}$ or 27.5 lb **55.** 122 oz **57.** 8 oz **59.** 20 lb
61. 55 T **63.** $11\frac{1}{4}$ or 11.25 T **65.** $1\frac{1}{4}$ or 1.25 pt
67. 110 yd **69.** 11 stones **71. (a)** 225 ft^2 **(b)** 25 yd^2
(c) $\$800$ **73.** 16 servings **75.** 56 pieces; 1 foot
77. 500 containers

Section 6.4 (pp. 409–411)

1. 1000 **2.** deci **3.** meter **4.** Decimeter **5.** liter
6. mL **7.** gram **8.** Hectogram **9.** kiloliter
11. centimeter **13.** dekameter **15.** milligram
17. 34 dm **19.** 1210 hm **21.** 4.5 dm **23.** 0.01459 km
25. 60 cm **27.** 2.5 dm **29.** 40 cm^2 **31.** 0.07 km^2
33. 0.012 hm^2 **35.** 2500 m^2 **37.** 51,000 dl **39.** 9000 L
41. 130 hl **43.** 9 mL **45.** 120,000 dl **47.** 0.0005 dl
49. 0.0038 hg **51.** 0.095 dg **53.** 4.5 g **55.** 570 cg
57. 8.793 kg **59.** 0.55 cg **61.** 0.5 L **63.** 23 kg
65. (a) 1250 m^2 (b) 12.5 dam^2 **67.** 5.5 t
69. 25 servings **71.** 125 pieces **73.** 75 tourists

Checking Basic Concepts 6.3 & 6.4 (p. 411)

1. (a) 6 yd (b) 12 in. **2.** 1152 in^2
3. (a) 36 oz (b) $5\frac{1}{2}$ or 5.5 gal **4.** (a) 4 lb (b) 23 lb
5. (a) 4.7 m (b) 0.3 hm **6.** 30,000 cm^2
7. (a) 5800 mL (b) 3.25 dal **8.** (a) 83 kg (b) 0.62 cg
9. 24 servings **10.** 50 lengths

Section 6.5 (pp. 418–421)

1. 2.54 **2.** 1.09 **3.** 0.62 **4.** 1.06 **5.** 29.57 **6.** 2.2
7. 28.35 **8.** Fahrenheit; Celsius **9.** 43.18 cm
11. 220.18 m **13.** 135.48 km **15.** (a) 63.43 yd
(b) 63.22 yd **17.** (a) 2.79 mi (b) 2.79 mi
19. 8.26 m **21.** 111.55 ft **23.** 9.84 in. **25.** 4.84 km
27. 9.81 ft **29.** 473.12 mL **31.** 52.83 L **33.** 19.88 gal
35. 7.55 L **37.** 4.04 c **39.** 14.53 gal **41.** 14.81 oz
43. 134.64 lb **45.** 3827.25 mg **47.** 91.52 oz
49. 659.09 g **51.** 2909.09 mg **53.** 25 °C **55.** 37.4 °F
57. $-26.\overline{1}$ °C **59.** 212 °F **61.** 36.9 °C **63.** 101.9 °F
65. 37.0 °C **67.** 68 °F (actual: 68 °F)
69. -23 °C (actual: $-23.\overline{3}$ °C) **71.** 10 °C (actual: 10 °C)
73. 591 mL **75.** 50 lb **77.** 115.7 °F
79. Yes (about 60.007 ft) **81.** 3 L **83.** No (about 18.46 ft)
85. 5.44 by 2.64 in. **87.** -140 °C **89.** 3.9 kg

Section 6.6 (pp. 424–425)

1. days **2.** 60; 60 **3.** speed **4.** distance; time
5. 315 min **7.** 1.5 hr **9.** 14,760 sec **11.** 420 hr
13. 1.15 d **15.** 0.5 yr **17.** 102 ft/sec **19.** 12 m/d
21. 0.2 m/sec **23.** 22.4 in./wk **25.** 44 ft/sec
27. 420,000 m/hr **29.** 4000 yd/hr **31.** 10.8 km/hr
33. 36 ft/sec **35.** 57 mi/hr **37.** 180 mi/hr
39. 70 mi/hr **41.** No (48.4 km/hr)

Checking Basic Concepts 6.5 & 6.6 (p. 425)

1. (a) 6.35 m (b) 1636.8 ft (c) 1182.8 mL (d) 3.18 pt
(e) 2.75 lb (f) 20 kg **2.** (a) 23 °F (b) 70 °C
3. 90 cm/sec **4.** 18 in./d **5.** 2.9 lb **6.** No (49.6 mi/hr)

Chapter 6 Review (pp. 428–431)

1. $\frac{1}{4}$ **2.** $\frac{4}{11}$ **3.** $\frac{1}{6}$ **4.** $\frac{8}{25}$ **5.** $\frac{0.8}{1}$ = 0.8 **6.** $\frac{4.\overline{6}}{1}$ = $4.\overline{6}$
7. $\frac{3\text{ miles}}{16\text{ minutes}}$ **8.** $\frac{1\text{ laptop}}{4\text{ students}}$ **9.** \$7.35/hr **10.** 74 beats/min
11. \$0.143/lb **12.** \$0.49/L **13.** Yes **14.** No **15.** No
16. Yes **17.** -16 **18.** -1 **19.** 3 **20.** $-\frac{7}{6}$ **21.** 7 inches
22. 5 inches **23.** 150 in. **24.** 9504 ft **25.** $\frac{3}{5}$ or 0.6 mi
26. 4 yd **27.** 30 yd^2 **28.** 1,115,136 ft^2 **29.** 56 qt
30. 3.5 qt **31.** 17 c **32.** 44 c **33.** 120 oz **34.** 24.5 lb
35. 8 lb **36.** 41 T **37.** milligram **38.** dekaliter
39. 75 cm **40.** 0.114 dam **41.** 126 hm **42.** 1.9 dm
43. 7.8 m^2 **44.** 850 m^2 **45.** 525.3 kl **46.** 7.5 mL
47. 7.69 dal **48.** 0.001 dl **49.** 8.7 dg **50.** 28 kg
51. 77 cg **52.** 2.1 g **53.** (a) 7.20 km (b) 7.19 km
54. (a) 1.55 mi (b) 1.55 mi **55.** 5.59 m **56.** 272.80 yd
57. 11.29 km **58.** 274.32 cm **59.** 26.5 gal **60.** 94.62 mL
61. 38.68 L **62.** 10.40 gal **63.** 14.33 lb **64.** 200,000 mg
65. 0.13 oz **66.** 477.27 g **67.** 48.2 °F **68.** -25 °C
69. 36.8 °C **70.** 98.5 °F **71.** -35 °C (actual: $-34.\overline{4}$ °C)
72. 77 °F (actual: 77 °F) **73.** 72 hr **74.** 576 min
75. 4.7 hr **76.** 2.5 wk **77.** 7500 m/hr **78.** 2.5 ft/wk
79. 40 in./hr **80.** 480,000 m/hr **81.** 121 km/hr
82. 392 ft/sec **83.** University: 51.6; college: 38.2; there are 38.2 students for each instructor. **84.** 8 students
85. 337 bales; 80 in. **86.** 20 servings **87.** 233 lb
88. Yes (56.45 km/hr) **89.** 117 ft
90. Large: \$0.195/oz; small: \$0.205/oz; the large coffee drink **91.** 22,700 g **92.** More **93.** 50 °C
94. -297.4 °F

Chapter 6 Test (pp. 431–432)

1. $\frac{1}{3}$ **2.** 0.125 mi/min **3.** \$0.215/oz **4.** No **5.** 27
6. $-\frac{1}{42}$ **7.** 2.1 ft **8.** 234 in. **9.** 3.75 ft^2 **10.** 68 pt
11. 17.5 T **12.** 0.0154 m **13.** 0.54 km^2 **14.** 75 mL
15. 2.4 hg **16.** 0.32 km **17.** 1.82 c **18.** 1136.36 g
19. 59 °F **20.** 21 °C (actual: $21.\overline{1}$ °C) **21.** 28.5 d
22. 259,200 sec **23.** 300 ft/hr **24.** 3226 m/d
25. Large: \$0.13/oz; small: \$0.148/oz; the large fruit smoothie **26.** 8 servings **27.** Tuesday

Chapters 1–6 Cumulative Review (pp. 432–433)

1. Thirty-four thousand two hundred six **2.** Unlike
3. [number line from 0 to 5 with points at 2 and 4]
4. 730,000 **5.** < **6.** -16 **7.** -60 **8.** 1 **9.** $4x$ **10.** 4
11. Table values: $-9, -7, -5, -3, -1$; solution: 2
12. 9 **13.** $\frac{23}{10}$; $2\frac{3}{10}$ **14.** (a) Yes (b) No **15.** $\frac{2}{9}$ **16.** $\frac{7}{10}$
17. 594.03 **18.** 64 **19.** 100π; 314 square feet **20.** 3.00
21. 58 mi/hr **22.** -20 **23.** 80 bpm **24.** 2010 **25.** 4 in.
26. $29\frac{3}{4}$ in. **27.** 739 **28.** 27 in.

7 Percent and Probability

Section 7.1 (pp. 440–442)

1. percent **2.** % **3.** $\frac{x}{100}$ **4.** $0.01x$ **5.** left **6.** 100%
7. right **8.** circle graph **9.** $\frac{7}{25}$ **11.** $\frac{3}{80}$ **13.** $\frac{11}{20}$
15. $\frac{1}{25}$ **17.** $\frac{3}{40}$ **19.** $\frac{29}{25}$ or $1\frac{4}{25}$ **21.** $\frac{33}{400}$ **23.** $\frac{1}{3}$
25. 0.58 **27.** 0.095 **29.** 1.73 **31.** 0.06 **33.** 0.0025
35. 0.003 **37.** 1.16 **39.** 0.084 **41.** 70% **43.** 22.5%
45. 550% **47.** 18% **49.** $66.\overline{6}$% **51.** $58.\overline{3}$%
53. 125% **55.** 81% **57.** 1% **59.** 160% **61.** 7.2%
63. 299% **65.** 0.5% **67.** 4.01%
69.

Percent	Decimal	Fraction
80%	0.8	$\frac{4}{5}$
30%	0.3	$\frac{3}{10}$
65%	0.65	$\frac{13}{20}$
22.5%	0.225	$\frac{9}{40}$
260%	2.6	$\frac{13}{5}$ or $2\frac{3}{5}$
40%	0.4	$\frac{2}{5}$

71. (a) Recycling **(b)** $\frac{1}{4}$ **(c)** 0.33 **73. (a)** 10% **(b)** $\frac{8}{25}$
(c) 0.05 **75.** $2.\overline{3}$% **77.** $\frac{3}{5}$ **79.** 110%; Sweden has more cell phones than people. **81.** 0.236 **83.** $\frac{4}{25}$ **85.** 60%

Section 7.2 (pp. 447–448)

1. whole; part **2.** of **3.** multiply; equals
4. percent · whole = part **5.** problem **6.** decimal
7. 15% of 120 is 18 **9.** 150% of 42 is 63
11. 2.5% of 760 is 19 **13.** 80% of 5 is 4
15. 85% of 20 is 17 **17.** $0.18 \cdot 40 = 7.2$
19. $1.4 \cdot 35 = 49$ **21.** $0.64 \cdot 75 = 48$
23. $0.125 \cdot 24 = 3$ **25.** $0.625 \cdot 24 = 15$
27. $0.68 \cdot x = 17$ **29.** $x \cdot 95 = 39.9$
31. $1.04 \cdot 70 = x$ **33.** $x \cdot 60 = 48$
35. $x \cdot 9 = 3$ **37.** 3 **39.** $33.\overline{3}$% **41.** 6 **43.** 748.6
45. 60 **47.** 150 **49.** $66.\overline{6}$% **51.** 3550 **53.** 17.5%
55. 27.5 **57.** 20% **59.** 76 **61.** 170 beats per minute
63. 17%

Checking Basic Concepts 7.1 & 7.2 (p. 449)

1. (a) $\frac{19}{25}$ **(b)** $\frac{1}{15}$ **(c)** $\frac{12}{125}$ **2. (a)** 0.065 **(b)** 2.14
(c) 0.028 **3. (a)** 12% **(b)** 350% **(c)** 87.5%
4. (a) 29% **(b)** 7.3% **(c)** 297% **5. (a)** 350% of 14 is 49
(b) 44% of 25 is 11 **6. (a)** $1.8 \cdot 30 = 54$
(b) $0.25 \cdot 36 = 9$ **7. (a)** $0.75 \cdot x = 64$
(b) $x \cdot 25 = 5$ **8. (a)** 20% **(b)** 45 **(c)** 130.8 **9.** $\frac{51}{100}$
10. (a) Upper class **(b)** $\frac{8}{25}$ **(c)** 0.2

Section 7.3 (pp. 454–455)

1. proportion **2.** equal **3.** percent **4.** $\frac{\text{percent}}{100} = \frac{\text{part}}{\text{whole}}$
5. $\frac{24}{100} = \frac{16.8}{70}$ **7.** $\frac{180}{100} = \frac{45}{25}$ **9.** $\frac{40}{100} = \frac{34}{85}$ **11.** $\frac{12.5}{100} = \frac{8}{64}$
13. $\frac{62.5}{100} = \frac{25}{40}$ **15.** $\frac{48}{100} = \frac{19}{x}$ **17.** $\frac{x}{100} = \frac{38.4}{114}$
19. $\frac{x}{100} = \frac{24}{80}$ **21.** $\frac{42.5}{100} = \frac{39}{x}$ **23.** $\frac{99}{100} = \frac{297}{x}$ **25.** 18
27. $66.\overline{6}$% **29.** 400% **31.** 27 **33.** 40 **35.** 25 **37.** 130
39. 35 **41.** $33.\overline{3}$% **43.** 32.5 **45.** 20% **47.** 23.5%
49. 8,717 **51.** 17%

Section 7.4 (pp. 463–465)

1. *percent* **2.** sales **3.** total price **4.** discount
5. sale price **6.** commission **7.** gross pay
8. withholdings **9.** net pay **10.** increase; decrease
11. $279 **13.** $1.4 billion **15.** $1.9345 million
17. 185 pounds **19.** 20,000 **21.** $70,000 **23.** 7%
25. 65% **27.** $26.\overline{6}$% **29.** $4.16; $68.16 **31.** $11.16; $159.96 **33.** 4% **35.** $160 **37.** $140; $420
39. $53.55; $22.95 **41.** 18% **43.** $35 **45.** $11,400
47. $4370 **49.** $8700 **51.** 6% **53.** $177.60; $562.40
55. 12.5% **57.** $3900 **59.** 20% increase **61.** 45%
63. 12% decrease

Checking Basic Concepts 7.3 & 7.4 (p. 465)

1. (a) $\frac{64}{100} = \frac{144}{x}$ **(b)** $\frac{x}{100} = \frac{80}{60}$ **(c)** $\frac{99}{100} = \frac{x}{70}$
2. (a) 200% **(b)** 15 **(c)** 25 **(d)** $16.\overline{6}$% **3.** 15 milligrams
4. 3800 **5.** $7.15; $137.15 **6.** $12.16; $3.04 **7.** $5112
8. 2% decrease

Section 7.5 (pp. 471–473)

1. principal **2.** interest **3.** interest rate **4.** simple
5. annual **6.** decimal; years **7.** principal; interest
8. compound **9.** annual percentage rate **10.** decimal; years **11.** $32 **13.** $21 **15.** $520 **17.** $131.25
19. $1 **21.** $14,000 **23.** $729.60 **25.** $2320
27. $922.50 **29.** 4 years **31.** 6% **33.** $7000
35. $1856.03 **37.** $1348.35 **39.** $275.38 **41.** $6087.60
43. $111,177.59 **45.** $5493.05 **47.** $21.44

Section 7.6 (pp. 478–479)

1. experiment **2.** outcome **3.** event **4.** probability
5. percent chance **6.** impossible event **7.** certain event
8. complement **9.** True, false **11.** 1, 2, 3, 4, 5, 6, 7, 8
13. Monday, Tuesday, Wednesday, Thursday, Friday
15. Certain; 1 **17.** Impossible; 0 **19.** Certain; 1
21. 50% **23.** 25% **25.** 40% **27.** $\frac{1}{3}$ **29.** $\frac{1}{2}$ **31.** $\frac{3}{5}$
33. $\frac{1}{52}$ **35.** $\frac{3}{26}$ **37.** $\frac{1}{2}$ **39.** $\frac{1}{12}$ **41.** $\frac{1}{3}$ **43.** 20%
45. $\frac{3}{10}$ **47.** 40%

Checking Basic Concepts 7.5 & 7.6 (p. 479)

1. $2.50 **2.** $1227 **3.** 9 months **4.** $25,776.80
5. 0 **6.** $\frac{3}{52}$ **7.** 80% **8.** $\frac{2}{11}$

Chapter 7 Review (pp. 483–486)

1. $\frac{17}{25}$ **2.** $\frac{9}{5}$ or $1\frac{4}{5}$ **3.** $\frac{1}{16}$ **4.** $\frac{9}{125}$ **5.** 0.29 **6.** 0.004
7. 6.25 **8.** 0.3575 **9.** 68% **10.** 320% **11.** $77.\overline{7}$%
12. 31.25% **13.** 16% **14.** 4.9% **15.** 702% **16.** 3.05%
17. Tuition **18.** $\frac{3}{25}$ **19.** 130% of 40 is 52

20. 75% of 12 is 9 **21.** 0.775 · 80 = 62
22. 0.50 · 184 = 92 **23.** 0.75 · x = 46.5
24. x · 72 = 48 **25.** 13.6 **26.** 35 **27.** 33.$\overline{3}$%
28. 60% **29.** $\frac{240}{100} = \frac{96}{40}$ **30.** $\frac{80}{100} = \frac{28}{35}$
31. $\frac{165}{100} = \frac{x}{20}$ **32.** $\frac{x}{100} = \frac{3}{9}$ **33.** 7.8 **34.** 66.$\overline{6}$%
35. 50% **36.** 4400 **37.** 2.8 grams **38.** 3120 **39.** 4%
40. 35% **41.** $10.89; $208.89 **42.** 4% **43.** $76; $304
44. 17% **45.** $187.50 **46.** $555 **47.** 12%
48. $196.80; $623.20 **49.** 22.$\overline{2}$% **50.** $2280
51. 33.$\overline{3}$% increase **52.** 20% decrease **53.** $7.20
54. $9276 **55.** $2520 **56.** $26,880 **57.** $6660
58. $1245 **59.** $3900 **60.** 9 months **61.** $114,840.86
62. $8933.82 **63.** $58,533.43 **64.** $1512.22
65. Red, black, blue **66.** 2, 3, 5, 7, 11, 13, 17, 19
67. Certain; 1 **68.** Impossible; 0 **69.** 25%
70. 37.5% **71.** $\frac{1}{3}$ **72.** $\frac{4}{5}$ **73.** $\frac{1}{13}$ **74.** $\frac{5}{26}$ **75.** 25%
76. $\frac{3}{11}$ **77.** 66% **78.** $\frac{2}{5}$ **79.** 0.428 **80.** 62.5%

Chapter 7 Test (p. 486)

1. $\frac{11}{20}$ **2.** $\frac{13}{5}$ or $2\frac{3}{5}$ **3.** 95% **4.** 7.8% **5.** x · 160 = 32
6. $\frac{12}{100} = \frac{x}{90}$ **7.** 140 **8.** 33.$\overline{3}$% **9.** 95% **10.** 800
11. $31.50 **12.** $26,000 **13.** 3% **14.** $2391.24
15. 2, 4, 6, 8, 10 **16.** Impossible; 0 **17.** $\frac{1}{2}$ **18.** 62%
19. 140 pounds **20.** $64,000 **21.** $17.45; $366.45
22. $11.55; $7.70 **23.** $4620 **24.** 15% increase

Chapters 1–7 Cumulative Review (pp. 487–488)

1. 46,391 **2.** 658,000 **3.** 11 **4.** 9w − 10 **5.** −21
6. −72 **7.** −2 **8.** 1 **9.** −7t − 8
10. 2(x + 5) = −14 **11.** −16 **12.** Table values: 12, 9, 6, 3, 0; solution: 1 **13.** 3 · 5 · 7 **14.** −$\frac{1}{14}$
15. $6\frac{1}{3}$ **16.** $\frac{47}{48}$ **17.** 468.05 **18.** (a) 9 (b) 9 (c) 9
(d) 9, −$\frac{1}{5}$, 4.$\overline{6}$, −2.1 (e) $\sqrt{7}$ **19.** 6 **20.** 49π; 153.86 square feet **21.** 13,728 ft **22.** 8.9 mL **23.** −35 °C
24. 300,000 m/hr **25.** 1.37 **26.** 33.$\overline{3}$% **27.** 126
28. $\frac{3}{26}$ **29.** 177 bpm **30.** 2012 **31.** 3 miles
32. $26\frac{1}{4}$ inches **33.** 521 **34.** Large: $0.175/oz; small: $0.185/oz; the large fruit drink **35.** $9.50; $28.50
36. 12% decrease

8 Geometry

Section 8.1 (pp. 497–499)

1. point **2.** line **3.** line segment **4.** ray **5.** angle
6. vertex **7.** right **8.** straight **9.** acute
10. obtuse **11.** congruent **12.** complementary
13. supplementary **14.** parallel **15.** intersecting
16. vertical **17.** adjacent **18.** perpendicular
19. corresponding **20.** alternate interior **21.** alternate exterior **22.** supplementary **23.** Line AB, or $\overleftrightarrow{AB}$
25. ∠m **27.** Ray CD, or $\overrightarrow{CD}$ **29.** Point K
31. Line segment PQ, or $\overline{PQ}$ **33.** (a) N (b) $\overrightarrow{NQ}$ and $\overrightarrow{NR}$
(c) ∠PNQ, ∠QNR, and ∠PNR (d) The naming does not clearly define a single angle. **35.** Right **37.** Acute
39. Obtuse **41.** Straight **43.** 166° **45.** 109° **47.** 81°
49. m∠a = 49°; m∠b = 131°; m∠d = 131°
51. m∠j = 96°; m∠l = 96°; m∠m = 84°
53. Congruent **55.** Congruent **57.** Supplementary
59. Supplementary **61.** Congruent **63.** Congruent
65. m∠b = 119°; m∠g = 119°; m∠h = 61°
67. m∠a = 92°; m∠c = 88°; m∠h = 92°

Section 8.2 (pp. 505–507)

1. acute **2.** obtuse **3.** right **4.** scalene **5.** isosceles
6. equilateral **7.** 180° **8.** congruent **9.** ASA **10.** SAS
11. SSS **12.** AAA **13.** Right **15.** Acute **17.** Obtuse
19. Acute **21.** Obtuse **23.** Right **25.** Isosceles
27. Equilateral **29.** Scalene **31.** Equilateral **33.** Scalene
35. Isosceles **37.** 51° **39.** 34° **41.** 118° **43.** 60°
45. 15° **47.** 36° **49.** 18° **51.** 50° **53.** SSS **55.** SAS
57. ASA **59.** Congruent; ASA **61.** Not congruent
63. Congruent; SSS

Checking Basic Concepts 8.1 & 8.2 (p. 507)

1. (a) ∠n (b) Ray AB, or $\overrightarrow{AB}$ **2.** (a) Right (b) Acute
3. 71° **4.** 66° **5.** m∠x = 19°; m∠w = 161°; m∠z = 19°
6. m∠d = 41°; m∠e = 41°; m∠g = 139°
7. (a) Obtuse (b) Right **8.** (a) Equilateral (b) Scalene
9. 25° **10.** Congruent; SAS

Section 8.3 (pp. 513–515)

1. polygon **2.** hexagon **3.** heptagon **4.** regular
5. (n − 2) · 180° **6.** $\frac{(n-2) \cdot 180°}{n}$ **7.** circle; center
8. radius **9.** diameter **10.** 2r; $\frac{d}{2}$ **11.** It has a curved side.
13. The sides intersect. **15.** It is not closed. **17.** Pentagon
19. Heptagon **21.** Triangle **23.** Quadrilateral
25. Hexagon **27.** Octagon **29.** 1080° **31.** 360°
33. 900° **35.** 120° **37.** 135° **39.** 108°
41. Parallelogram **43.** Rhombus, parallelogram
45. Trapezoid **47.** Rectangle, parallelogram **49.** Kite
51. 9 m **53.** $\frac{2}{9}$ ft **55.** 0.15 cm **57.** 16 mi **59.** 9 ft
61. 5 m **63.** 135° **65.** 540°

Section 8.4 (pp. 520–522)

1. perimeter **2.** sides **3.** circumference **4.** C = πd
5. C = 2πr **6.** composite **7.** 35 cm **9.** 17.3 ft
11. 23 km **13.** $2\frac{3}{8}$ or $\frac{19}{8}$ m **15.** 21 m **17.** 416 ft
19. $\frac{7}{2}$ cm **21.** 108 mi **23.** 262 in. **25.** 320 m
27. (a) 12π m (b) 37.68 m **29.** (a) 10π in. (b) 31.4 in.
31. (a) 2.5π ft (b) 7.85 ft **33.** (a) 2.8π cm (b) 8.792 cm
35. (a) $\frac{7}{11}$π in. (b) 2 in. **37.** (a) π km (b) $\frac{22}{7}$ km
39. (a) $\frac{5}{4}$π m (b) $\frac{55}{14}$ m **41.** (a) 7π mi (b) 22 mi
43. 18.28 in. **45.** 278.5 m **47.** 41.4 mi **49.** 1040 feet
51. 4740 km **53.** 400 m

Checking Basic Concepts 8.3 & 8.4 (p. 522)

1. (a) Octagon (b) Quadrilateral 2. (a) 540° (b) 900°
3. (a) 60° (b) 120° 4. (a) Kite (b) Rhombus, parallelogram 5. 21 m 6. 11.4 in. 7. $10\frac{1}{2}$ or $\frac{21}{2}$ in.
8. 31.5 cm 9. 72 ft 10. 9.42 m 11. $\frac{8}{7}$ ft 12. 144.25 m
13. 19 in. 14. 64 ft

Section 8.5 (pp. 528–530)

1. square 2. cubic 3. square 4. False 5. volume
6. surface area 7. 4900 m² 9. 15 in² 11. 6.25 ft²
13. 450 cm² 15. 94.5 km² 17. 113.0 in² 19. 132.7 m²
21. 1.5 km² 23. 7539.1 ft² 25. 343 in³ 27. 42 ft³
29. 3014.4 cm³ 31. 37.7 in³ 33. 270 ft³ 35. 33.5 cm³
37. 216 in² 39. 136 ft² 41. 326.6 cm² 43. 254.0 in²
45. 57 ft² 47. 615.4 cm² 49. 188.4 ft³ 51. 277.5 in²
53. 7 ft² 55. 188.4 in²

Checking Basic Concepts 8.5 (p. 530)

1. (a) 20 m² (b) 0.6 ft² 2. (a) 706.5 ft² (b) 452.2 in²
3. (a) 2744 km³ (b) 565.2 ft³
4. (a) 25.1 m² (b) 45,216 cm²

Chapter 8 Review (pp. 535–537)

1. Line XY, or $\overleftrightarrow{XY}$ 2. Ray PQ, or $\overrightarrow{PQ}$
3. Line segment AB, or $\overline{AB}$ 4. Point T 5. Acute
6. Obtuse 7. Right 8. Straight 9. 66° 10. 147°
11. $m\angle c = 76°$; $m\angle f = 104°$; $m\angle e = 76°$
12. $m\angle a = 115°$; $m\angle b = 65°$; $m\angle c = 65°$
13. Supplementary 14. Congruent 15. Congruent
16. Supplementary 17. Acute 18. Obtuse 19. Right
20. Obtuse 21. Scalene 22. Equilateral 23. Isosceles
24. Scalene 25. 84° 26. 94° 27. 18° 28. 40° 29. ASA
30. SSS 31. Not congruent 32. Congruent; SAS
33. Quadrilateral 34. Hexagon 35. Pentagon
36. Octagon 37. 720° 38. 540° 39. 60° 40. 135°
41. Rhombus, parallelogram 42. Kite 43. Trapezoid
44. Rectangle, parallelogram 45. 14 m 46. $\frac{4}{9}$ ft
47. 10.6 in. 48. 1 cm 49. 104 in. 50. $\frac{11}{8}$ in.
51. 4 m 52. 52.5 in. 53. (a) 5π ft (b) 15.7 ft
54. (a) 13π m (b) 40.82 m 55. (a) 3π km (b) $\frac{66}{7}$ km
56. (a) $\frac{14}{11}\pi$ cm (b) 4 cm 57. 110 ft 58. 95.98 m
59. 25.3 m² 60. 37.5 ft² 61. 49 ft² 62. 15 cm²
63. 254.3 cm² 64. 0.8 ft² 65. 125 ft³ 66. 3052.1 cm³
67. 50.2 in³ 68. 256.4 m³ 69. 207 in³ 70. 641.8 m²
71. 100.5 cm² 72. 146 ft² 73. 15 ft 74. 129.92 in.
75. 16.3 m² 76. 628 in³

Chapter 8 Test (pp. 537–538)

1. $\angle ABC$ or $\angle B$ 2. Obtuse 3. 19° 4. 62°
5. $m\angle a = 108°$; $m\angle f = 72°$; $m\angle h = 108°$ 6. Acute
7. Isosceles 8. 99° 9. Congruent; ASA 10. Pentagon
11. (a) 1080° (b) 135° 12. Parallelogram 13. $\frac{4}{9}$ ft
14. 38 cm 15. (a) 40π ft (b) 125.6 ft 16. 140 cm
17. (a) 2025 m² (b) 216 in² 18. 28.3 ft²

19. (a) 33.5 in³ (b) 336 ft³
20. (a) 1536 cm² (b) 207.2 ft²

Chapters 1–8 Cumulative Review (pp. 538–539)

1. $10,000 + 8000 + 20$ 2. 9^4 3. 74,000 4. 27
5. (number line from −5 to 5 with points at −4, −2, 1, 3) 6. 4 7. 28
8. Table values: $-12, -7, -2, 3, 8$; solution: 1 9. Yes
10. -2 11. 1 12. $4 = x + 9$ 13. $\frac{17}{6}, 2\frac{5}{6}$
14. (a) Composite (b) Neither (c) Prime (d) Composite
15. -10 16. $-\frac{7}{10}$ 17. 3.3 in. 18. 341.57 19. 2.4
20. (a) $8, \sqrt{9}$ (b) $8, \sqrt{9}, 0$ (c) $8, \sqrt{9}, 0$
(d) $8, -\frac{1}{5}, \sqrt{9}, 0$ (e) π 21. 2.9 ft 22. $0.23/oz
23. 60 pt 24. 2100 ft/hr 25. $x \cdot 80 = 16$ 26. 4%
27. 22% 28. 800 29. $m\angle a = 109°$; $m\angle b = 71°$; $m\angle h = 109°$ 30. 105° 31. (a) 1080° (b) 135°
32. 425.2 cm² 33. 18 feet 34. $x + 17 = 2x - 6$
35. 79.7 °F 36. 11.57; $189.57

9 Linear Equations and Inequalities in One Variable

Section 9.1 (pp. 552–554)

1. constant 2. $ax + b = 0$ 3. Exactly one
4. numerically 5. Addition, multiplication 6. LCD
7. None 8. Infinitely many 9. Yes; $a = 3, b = -7$
11. Yes; $a = \frac{1}{2}, b = 0$ 13. No 15. No
17. Yes; $a = 1.1, b = -0.9$ 19. Yes; $a = 2, b = -6$
21. No

23.

x	−1	0	1	2	3
$x - 3$	−4	−3	−2	−1	0

; 2

25.

x	0	1	2	3	4
$-3x + 7$	7	4	1	−2	−5

; 2

27.

x	−2	−1	0	1	2
$4 - 2x$	8	6	4	2	0

; −1

29. $\frac{3}{11}$ 31. 23 33. 28 35. $-\frac{11}{5}$ 37. $-\frac{7}{2}$ 39. 3 41. $\frac{9}{4}$
43. $-\frac{5}{2}$ 45. $\frac{10}{3}$ 47. $\frac{9}{8}$ 49. 1 51. 1 53. -0.055 55. 8
57. $\frac{1}{3}$ 59. $-\frac{b}{a}$ 61. No solutions 63. One solution
65. Infinitely many solutions 67. No solutions
69. Infinitely many solutions 71. $y = -3x + 2$
73. $y = -\frac{4}{3}x + 4$ 75. $w = \frac{A}{l}$ 77. $h = \frac{V}{\pi r^2}$
79. $a = \frac{2A}{h} - b$ 81. $w = \frac{V}{lh}$ 83. $b = 2s - a - c$
85. $b = a - c$ 87. $a = \frac{cd}{b - d}$ 89. 15 in.

91. (a)

Hours (x)	0	1	2	3	4
Distance (D)	4	12	20	28	36

(b) $D = 8x + 4$ (c) 28 miles; yes
(d) 2.25 hours; the bicyclist is 22 miles from home after 2 hours and 15 minutes.
93. 2009 95. 2010 97. 2013

Section 9.2 (pp. 561–564)

1. Check your solution. 2. + 3. = 4. $n+1$ and $n+2$
5. rt 6. distance; time 7. $2+x=12; 10$
9. $2x+7=9; 1$ 11. $7-x=-2; 9$ 13. $\frac{x}{2}=17; 34$
15. 31, 32, 33 17. 34 19. 8 21. 21 23. 7 years old
25. 210 pounds 27. 12 billionaires 29. 14 reptile species 31. *Les Mis*: $2.6 billion; *Phantom*: $5.6 billion
33. 306 ascents 35. 24 inches 37. Length: 19 inches; width: 12 inches 39. Facebook: 198 million users; Twitter: 49 million users 41. 26 hours 43. $d=8$ miles
45. $r=20$ feet per second 47. $t=5$ hours 49. 60 mph
51. $\frac{3}{8}$ hour 53. 0.6 hr at 5 mph; 0.5 hr at 8 mph 55. 54 mph and 60 mph 57. 60 mph 59. $\frac{60}{7} \approx 8.6$ mph
61. 30 ounces 63. $1500 at 6%; $2500 at 5% 65. 5.5 lb of the $12 brand and 2.5 lb of the $7 brand 67. 8 mph: 54 min, 9 mph: 18 min 69. Student: 56 tickets, Adult: 95 tickets
71. 6 gallons 73. 10 grams

Checking Basic Concepts 9.1 & 9.2 (p. 564)

1. (a) No (b) Yes
2.
x	3	3.5	4	4.5	5
$4x-3$	9	11	13	15	17

; 4

3. (a) 18 (b) $\frac{1}{6}$ (c) $2.\overline{6}$ or $\frac{8}{3}$ (d) 3
4. (a) One solution (b) Infinitely many solutions (c) No solutions
5. (a) $D=300-75x$ (b) $0=300-75x$ (c) 4 hours
6. $-32, -31, -30$ 7. 6.5 hr 8. $5000 at 6%; $3000 at 7%

Section 9.3 (pp. 575–579)

1. $<, \leq, >, \geq$ 2. greater than; less than 3. solution
4. equivalent 5. True 6. True 7. number line
8. > 9. < 10. > 11. Yes 13. No 15. Yes 17. No

19. [number line graph]
21. [number line graph]
23. [number line graph]
25. [number line graph]
27. $x<0$ 29. $x \leq 3$ 31. $x \geq 10$
33. $x > -2$ 35. $x < 1$
37.
x	1	2	3	4	5
$-2x+6$	4	2	0	-2	-4

; $x \geq 3$

39.
x	-3	-2	-1	0	1
$5-x$	8	7	6	5	4
$x+7$	4	5	6	7	8

; $x < -1$

41. $[6, \infty)$ 43. $(-2, \infty)$
45. $(-\infty, 7]$
47. $x > 3$; [number line graph]
49. $y \geq -2$; [number line graph]
51. $z > 8$; [number line graph]
53. $t \leq -5$; [number line graph]
55. $x < 5$; [number line graph]
57. $t \leq -2$; [number line graph]
59. $y > -\frac{3}{20}$; [number line graph]
61. $z \geq -\frac{14}{3}$; [number line graph]
63. $x<7; (-\infty, 7)$ 65. $x \leq -\frac{4}{3}; (-\infty, -\frac{4}{3}]$
67. $x > -\frac{39}{2}; (-\frac{39}{2}, \infty)$ 69. $x \leq \frac{3}{2}; (-\infty, \frac{3}{2}]$
71. $x > \frac{1}{2}; (\frac{1}{2}, \infty)$ 73. $x \leq \frac{5}{4}; (-\infty, \frac{5}{4}]$
75. $x > -\frac{8}{3}; (-\frac{8}{3}, \infty)$ 77. $x \leq -\frac{1}{3}; (-\infty, -\frac{1}{3}]$
79. $x<5; (-\infty, 5)$ 81. $x \leq -\frac{3}{2}; (-\infty, -\frac{3}{2}]$
83. $x > -\frac{9}{8}; (-\frac{9}{8}, \infty)$ 85. $x \leq \frac{4}{7}; (-\infty, \frac{4}{7}]$
87. $x < \frac{21}{11}; (-\infty, \frac{21}{11})$ 89. $\{x | x > 1\}$
91. $\{x | x \geq -7\}$ 93. $\{x | x < 6\}$ 95. $x > 60$
97. $x \geq 21$ 99. $x > 40{,}000$ 101. $x \leq 70$
103. Less than 10 feet 105. 86 or more
107. 4.5 hours 109. 4 days
111. (a) $C = 1.5x + 2000$ (b) $R = 12x$
(c) $P = 10.5x - 2000$ (d) 191 or more compact discs
113. (a) After 48 minutes (b) After more than 48 minutes
115. Altitudes more than 4.5 miles

Checking Basic Concepts 9.3 (p. 579)

1. [number line graph]
2. $x \leq 1$ 3. Yes
4.
x	-2	-1	0	1	2
$5-2x$	9	7	5	3	1

; $x \geq -1$

5. (a) $x > 3$ (b) $x \geq -35$ (c) $x \leq -1$
6. $x \leq 12$ 7. More than 13 inches

Chapter 9 Review (pp. 581–583)

1. Yes; $a=-4, b=1$ 2. No 3. 2 4. 22 5. $\frac{27}{5}$
6. 7 7. -7 8. $-\frac{2}{3}$ 9. $\frac{9}{4}$ 10. $\frac{11}{14}$ 11. $\frac{7}{8}$ 12. 0
13. No solutions 14. Infinitely many solutions
15. Infinitely many solutions 16. One solution
17.
x	0.5	1.0	1.5	2.0	2.5
$-2x+3$	2	1	0	-1	-2

; 1.5

18.
x	-2	-1	0	1	2
$-(x+1)+3$	4	3	2	1	0

; 0

19. $y = 3x - 5$ 20. $y = -x + 8$ 21. $y = \frac{z}{2x}$
22. $b = 3S - a - c$ 23. $b = \frac{12T - 4a}{3}$ 24. $c = \frac{ab}{d-b}$
25. $6x = 72; 12$ 26. $x + 18 = -23; -41$

27. $2x - 5 = x + 4; 9$ **28.** $x + 4 = 3x; 2$ **29.** 9
30. $-52, -51, -50$ **31.** $d = 24$ mi **32.** $d = 3850$ ft
33. $r = 25$ yards per second **34.** $t = \frac{25}{3}$ hr

35.

36.

37.

38.

39. $x < 3$ **40.** $x \geq -1$ **41.** Yes **42.** No

43.

x	0	1	2	3	4	$x < 2$
5 − x	5	4	3	2	1	

44.

x	1	1.5	2	2.5	3	$x \leq 2.5$
2x − 5	−3	−2	−1	0	1	

45. $x > 3$ **46.** $x \geq -5$ **47.** $x \leq -1$ **48.** $x < \frac{23}{3}$
49. $x \leq \frac{1}{9}$ **50.** $x \geq \frac{9}{4}$ **51.** $x < 50$ **52.** $x \leq 45{,}000$
53. $x \geq 16$ **54.** $x < 1995$ or $x \leq 1994$

55. (a)

Hours (x)	1	2	3	4	5
Distance (D)	40	30	20	10	0

(b) $D = 50 - 10x$ (c) 20 miles; yes
(d) 3 hours or less or from noon to 3:00 P.M.
56. $\frac{1}{6}$ hr, or 10 min
57. 50 mL **58.** $500 at 3%; $800 at 5%
59. 25 inches or less **60.** 74 or more **61.** 6 hr
62. (a) $C = 85x + 150{,}000$ (b) $R = 225x$
(c) $P = 140x - 150{,}000$ (d) 1071 or fewer

Chapter 9 Test (p. 583)

1. -6 **2.** $\frac{5}{2}$ **3.** $-\frac{16}{11}$ **4.** $\frac{25}{6}$ **5.** No solutions
6. Infinitely many solutions
7. $x = \frac{y - z}{3y}$ **8.** $x = \frac{20R - 4y}{5}$
9.

x	0	1	2	3	4	; 3
6−2x	6	4	2	0	−2	

10. $x + (-7) = 6; 13$ **11.** $2x + 6 = x - 7; -13$
12. 111, 112, 113 **13.** $r = 50$ ft/sec
14. $x > 0$ **15.** $x \leq 6$ **16.** $x > -\frac{1}{7}$
17. (a) $S = 2x + 5$ (b) 21 in. (c) 17.5 in. **18.** 2000 mL

Chapters 1-9 Cumulative Review (pp. 583–584)

1. 3 **2.** 9^5 **3.** 1300 **4.** 4 **5.** No **6.** Table values:
11, 8, 5, 2, −1; solution: 2 **7.** $11 = x + 4$ **8.** -2
9. 0 **10.** 0.35 **11.** $\frac{17}{20}$ **12.** $\frac{1}{6}$ **13.** $<$ **14.** 57 mi/hr
15. -20 **16.** $7.5d$ **17.** 12.5% **18.** 105 **19.** $\frac{1}{2}$ **20.** 1
21. Infinitely many **22.** No solutions **23.** 29, 30, 31
24. 65 mph **25.** $x = \frac{a + 4}{3y}$ **26.** $x = 3A - y - z$

27. $x < 1$

28. $x \leq 2$

29. $\frac{1}{3}$ **30.** 2009 **31.** 560 messages **32.** $8.01; $186.01
33. 300 mL **34.** 272 billion kilowatt-hours

10 Graphing Equations

Section 10.1 (pp. 591–593)

1. xy-plane **2.** origin **3.** $(0, 0)$ **4.** 4 **5.** False **6.** $x; y$
7. scatterplot **8.** line **9.** $(-2, -2), (-2, 2), (0, 0), (2, 2)$
11. $(-1, 0), (0, -3), (0, 2), (2, 0)$
13. $(1, 3)$: I; $(0, -3)$: None; $(-2, 2)$: II
15. $(C, 6)$: None; $(8, -4)$: IV; $(-6, -6)$: III

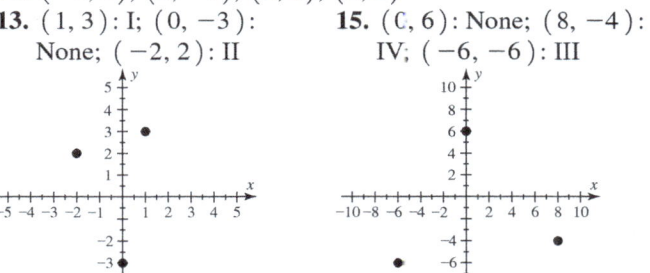

17. (a) I (b) III **19.** (a) None (b) I
21. (a) II (b) IV **23.** I and III

25. **27.**

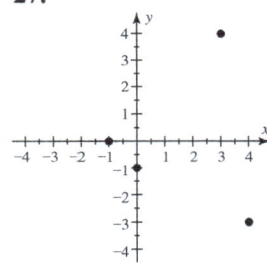

29. **31.**

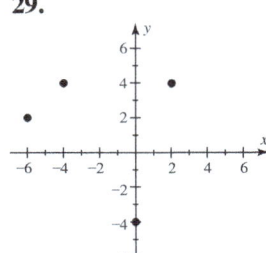 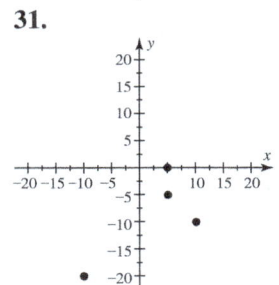

33.

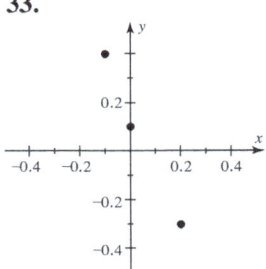

35. (1960, 484), (1980, 632), (2000, 430), (2010, 325); in 1960, there were 484 billion cigarettes consumed in the United States (answers may vary slightly).

37. 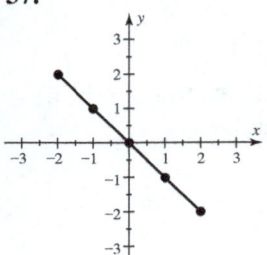   **39.**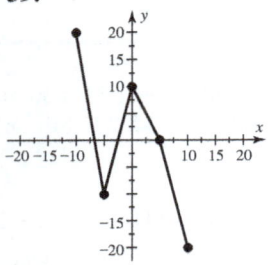

29.

x	−3	0	3	6
y	−9	0	9	18

31.

x	8	6	4	2
y	−2	0	2	4

33.

x	y
−8	−2
−4	0
0	2
4	4

35.

x	y
$-\frac{1}{2}$	−2
$-\frac{1}{4}$	−1
0	0
$\frac{1}{4}$	1

41. 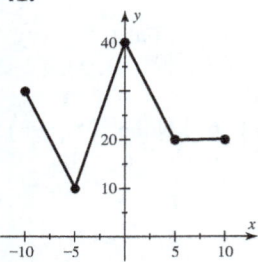   **43. (a)**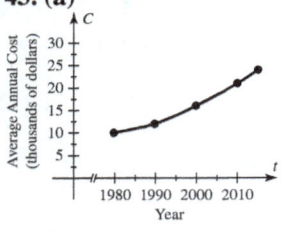

(b) The cost increased.

37. They must be multiples of 5.

39.

x	−1	0	1
y	2	0	−2

Table values may vary.

41.

x	0	1	2
y	3	2	1

Table values may vary.

45. (a) 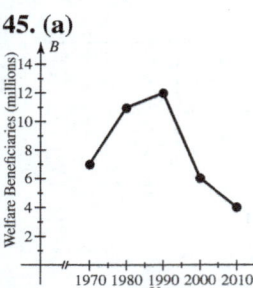   **47. (a)**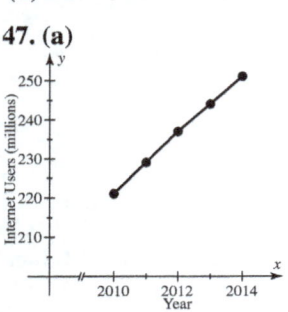

(b) The number of welfare beneficiaries increased and then decreased.

(b) The number of Internet users is increasing.

49. (a) The rate decreased.   **(b)** 9   **(c)** About −70%

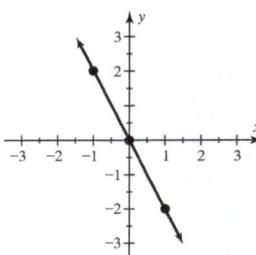   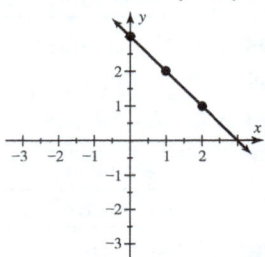

43.

x	−2	0	2
y	3	2	1

Table values may vary.

45.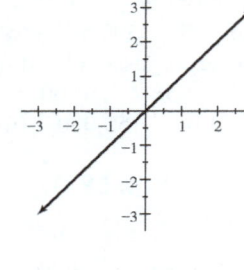

Section 10.2 (pp. 601–604)

1. True **2.** False **3.** ordered pair **4.** linear **5.** False
6. True **7.** graph **8.** line **9.** False **10.** False **11.** Yes
13. No **15.** No **17.** Yes **19.** No

21.

x	−2	−1	0	1	2
y	−8	−4	0	4	8

23.

x	6	3	0	−3	−9
y	−2	0	2	4	8

25.

x	y
−8	−4
−4	0
0	4
4	8
8	12

27.

x	y
−6	−4
0	−2
6	0
12	2
18	4

47. 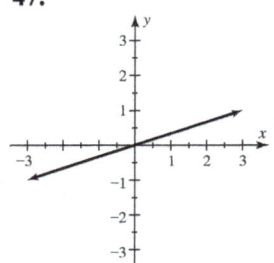   **49.**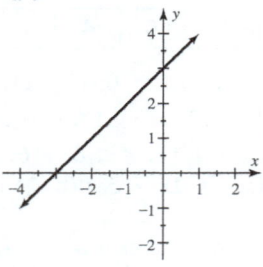

51. 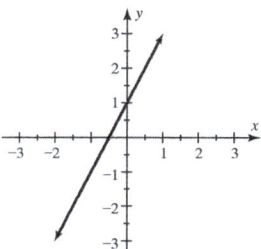 **53.**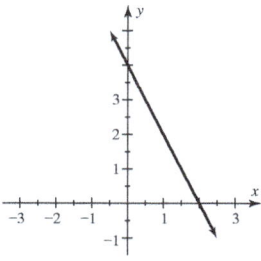

71.
x	y
−2	3
−1	0
0	−1
1	0
2	3

73.
x	y
−2	8
−1	2
0	0
1	2
2	8

55. 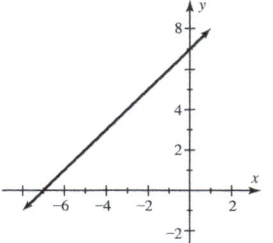 **57.**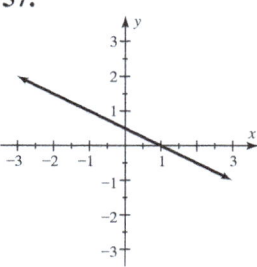

59. 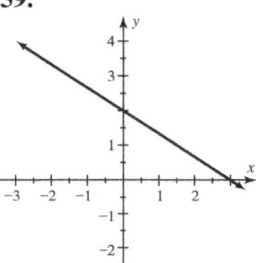 **61.**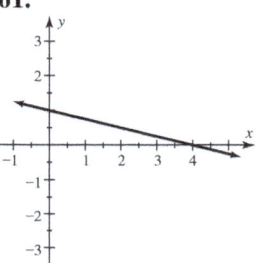

75.
x	y
−3	2
−1	0
0	−1
1	0
3	2

77.
x	y
0	0
1	1
4	2
9	3

63. 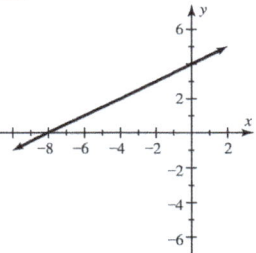 **65.**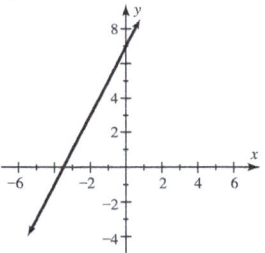

67. 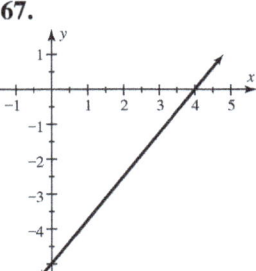 **69.**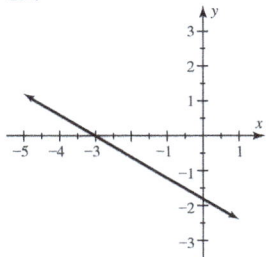

79. (a)
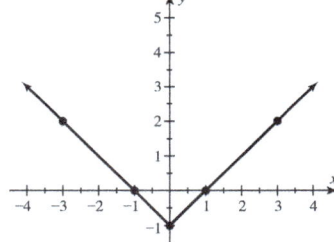
t	2010	2020	2030	2040	2050
P	13.2	15.0	16.7	18.5	20.3

(b) 2030

81. (a)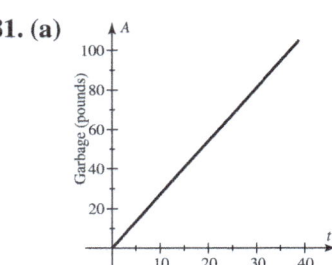

(b) About 37 days

A-18 ANSWERS TO SELECTED EXERCISES

83. (a) 68.3; 24.3; In 2008, Internet Explorer had 68.3% of the desktop web-browser market, and in 2014, it fell to 24.3%.

(b) 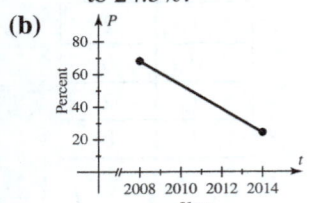 (c) 2010

Checking Basic Concepts 10.1 & 10.2 (p. 604)

1. $(-2, 2)$, II; $(-1, -2)$, III; $(1, 3)$, I; $(3, 0)$, none

2. **3.**

4. Yes

5.

x	-2	-1	0	1	2
y	5	3	1	-1	-3

6. (a) (b)

7. (a) 1; 9. After one month, the global share was 1%; after 12 months, the global share was 9%.

(b) (c) About 5 months

Section 10.3 (pp. 612–615)

1. x-intercept **2.** 0 **3.** y-intercept **4.** 0 **5.** horizontal
6. $y = b$ **7.** vertical **8.** $x = k$ **9.** $(3, 0); (0, -2)$
11. $(0, 0); (0, 0)$ **13.** $(-2, 0), (2, 0); (0, 4)$
15. $(1, 0); (0, 1)$

17.

x	-2	-1	0	1	2
y	0	1	2	3	4

$(-2, 0); (0, 2)$

19.

x	-4	-2	0	2	4
y	-6	-4	-2	0	2

$(2, 0); (0, -2)$

21. x-int: $(3, 0)$; y-int: $(0, -2)$

23. x-int: $(6, 0)$; y-int: $(0, -2)$

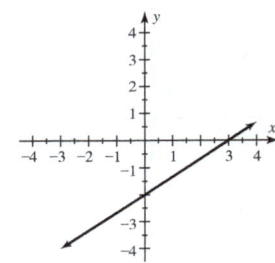 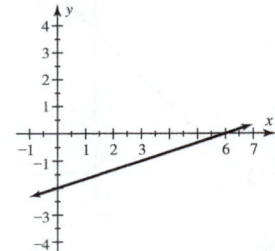

25. x-int: $(-1, 0)$; y-int: $(0, 6)$

27. x-int: $(7, 0)$; y-int: $(0, 3)$

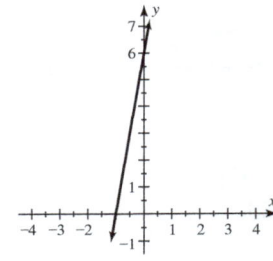 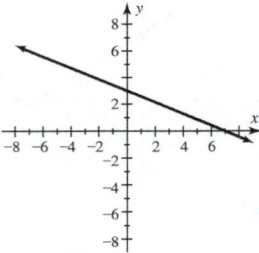

29. x-int: $(4, 0)$; y-int: $(0, -3)$

31. x-int: $(4, 0)$; y-int: $(0, -2)$

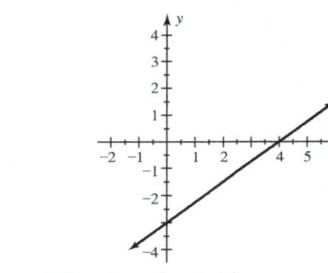 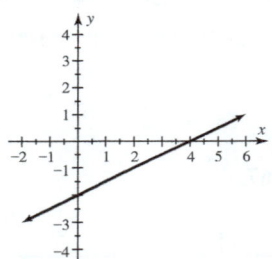

33. x-int: $(-4, 0)$; y-int: $(0, 3)$

35. x-int: $(3, 0)$; y-int: $(0, 2)$

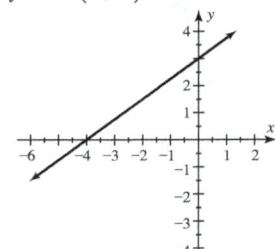 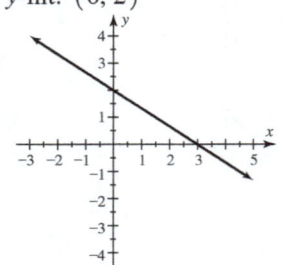

37. x-int: $(-2, 0)$; y-int: $(0, 5)$

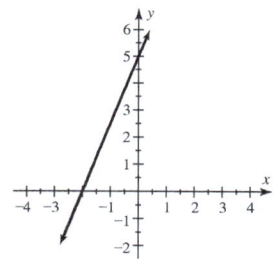

39. x-int: $\left(\frac{C}{A}, 0\right)$; y-int: $\left(0, \frac{C}{B}\right)$ **41.** $y = 1$ **43.** $x = -6$

45. (a) 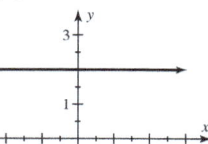 **(b)**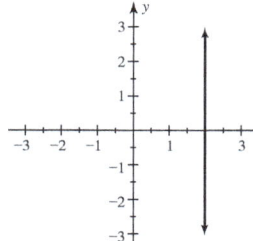

47. (a) 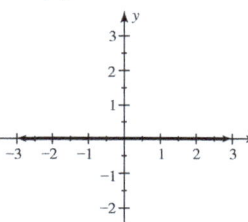 **(b)**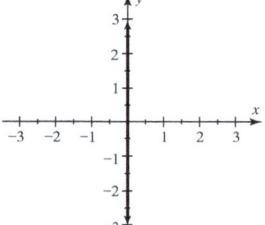

49. (a) 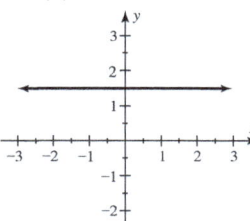 **(b)**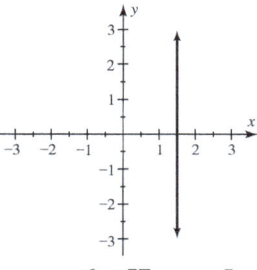

51. $y = 4$ **53.** $x = -1$ **55.** $y = -6$ **57.** $x = 5$
59. $y = 2; x = 1$ **61.** $y = -45; x = 20$
63. $y = 5; x = 0$ **65.** $x = -1$ **67.** $y = -\frac{5}{6}$
69. $x = 4$ **71.** $x = -\frac{2}{3}$ **73.** $y = 0$

75. (a) y-int: $(0, 200)$; x-int: $(4, 0)$ **(b)** The driver was initially 200 miles from home; the driver arrived home after 4 hours.

77. (a) y-int: $(0, 128)$; t-int: $(4, 0)$ **(b)** The initial velocity was 128 ft/sec; the velocity after 4 seconds was 0.

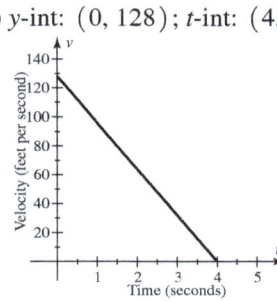

79. (a) y-int: $(0, 2000)$; x-int: $(4, 0)$ **(b)** The pool initially contained 2000 gallons; the pool was empty after 4 hours.

Section 10.4 (pp. 625–629)

1. True **2.** True **3.** rise; run **4.** horizontal **5.** vertical
6. $\frac{y_2 - y_1}{x_2 - x_1}$ **7.** positive **8.** negative **9.** rate **10.** Toward
11. Positive **13.** Zero **15.** Negative **17.** Undefined
19. 0; the rise always equals 0.
21. 1; the graph rises 1 unit for each unit of run.
23. 2; the graph rises 2 units for each unit of run.
25. Undefined; the run always equals 0.
27. $-\frac{3}{2}$; the graph falls 3 units for each 2 units of run.
29. -1; the graph falls 1 unit for each unit of run.

31. 2; 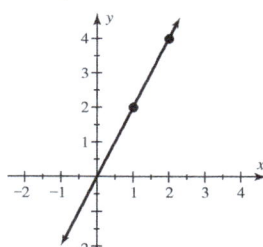 **33.** Undefined;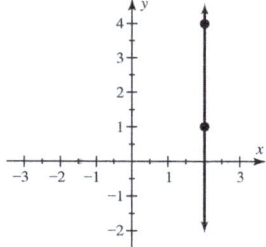

35. $-\frac{2}{3}$; 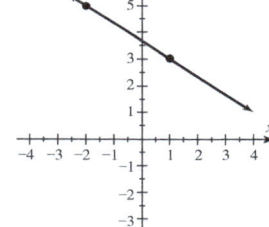 **37.** 0;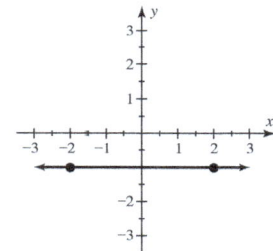

39. 1 **41.** $-\frac{6}{7}$ **43.** 0 **45.** Undefined **47.** $\frac{13}{20}$
49. $\frac{9}{100}$ **51.** $-\frac{5}{7}$ **53.** 49

55. 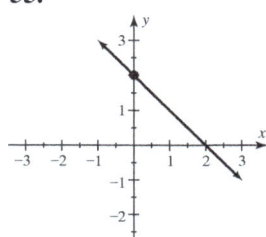 **57.**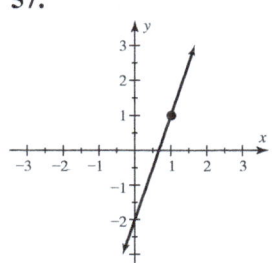

59. 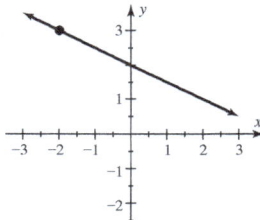 **61.**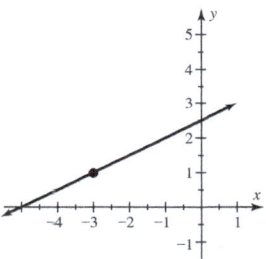

63.

x	0	1	2	3
y	−4	−2	0	2

65.

x	1	2	3	4
y	4	1	−2	−5

67.

x	−4	−2	0	2
y	0	3	6	9

69. c.   **71.** b.   **73. (a)** $m_1 = 1000; m_2 = -1000$
(b) $m_1 = 1000$: Water is being added to the pool at a rate of 1000 gallons per hour. $m_2 = -1000$: Water is being removed from the pool at a rate of 1000 gallons per hour.
(c) Initially, the pool contained 2000 gallons of water. Over the first 3 hours, water was pumped into the pool at a rate of 1000 gallons per hour. For the next 2 hours, water was pumped out of the pool at a rate of 1000 gallons per hour.
75. (a) $m_1 = 50; m_2 = 0; m_3 = -50$ **(b)** $m_1 = 50$: The car is moving away from home at a rate of 50 mph. $m_2 = 0$: The car is not moving. $m_3 = -50$: The car is moving toward home at a rate of 50 mph. **(c)** Initially, the car is at home. Over the first 2 hours, the car travels away from home at a rate of 50 mph. Then the car is parked for 1 hour. Finally, the car travels toward home at a rate of 50 mph.
77. m^3/min

79. 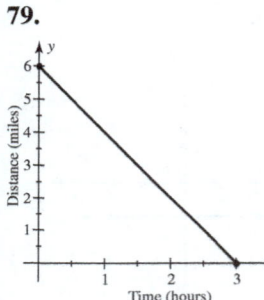   **81.**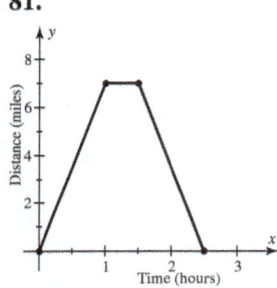

83. (a) 76,000,000 **(b)** 281,000,000 **(c)** 2,050,000/yr
85. (a) 40,800 **(b)** The celebrity gained 40,800 followers each year, on average.
87. (a) 25 **(b)** The revenue is $25 per flash drive.
89. (a)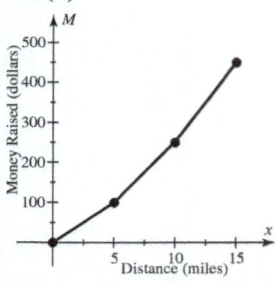

(b) $m_1 = 20; m_2 = 30; m_3 = 40$ **(c)** $m_1 = 20$: Each mile between 0 and 5 miles is worth $20 per mile. $m_2 = 30$: Each mile between 5 and 10 miles is worth $30 per mile. $m_3 = 40$: Each mile between 10 and 15 miles is worth $40 per mile.
91. (a) 750 **(b)** Median family income increased, on average, by $750/yr over the period.
93. (a) Toward **(b)** 10 ft **(c)** 2 ft/min **(d)** 5

Checking Basic Concepts 10.3 & 10.4 (pp. 629–630)

1. $(-2, 0); (0, 3)$
2.

x	−2	−1	0	1	2
y	−6	−4	−2	0	2

$(1, 0); (0, -2)$
3. (a) x-int: $(6, 0)$; y-int: $(0, -3)$ **(b)** y-int: $(0, 2)$

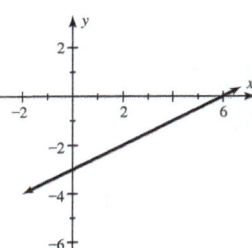   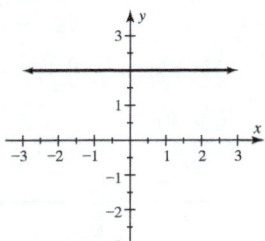

(c) x-int: $(-1, 0)$

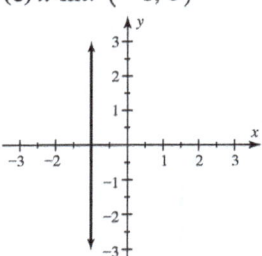

4. $y = 4; x = -2$ **5. (a)** $\frac{3}{4}$ **(b)** 0 **(c)** Undefined
6. $-\frac{1}{2}$
7.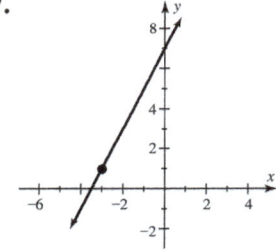

8. (a) $m_1 = 0; m_2 = 2; m_3 = -\frac{2}{3}$ **(b)** $m_1 = 0$: The depth is not changing. $m_2 = 2$: The depth increased at a rate of 2 feet per hour. $m_3 = -\frac{2}{3}$: The depth decreased at a rate of $\frac{2}{3}$ foot per hour. **(c)** Initially, the pond had a depth of 5 feet. For the first hour, there was no change in the depth of the pond. For the next hour, the depth of the pond increased at a rate of 2 feet per hour to a depth of 7 feet. Finally, the depth of the pond decreased for 3 hours at a rate of $\frac{2}{3}$ foot per hour until it was 5 feet deep.

Section 10.5 (pp. 638–641)

1. $y = mx + b$ **2.** slope **3.** y-coordinate of the y-intercept
4. origin **5.** parallel **6.** slope **7.** perpendicular
8. reciprocals **9.** f. **10.** d. **11.** a. **12.** b. **13.** e.
14. c. **15.** $y = x - 1$ **17.** $y = -2x + 1$ **19.** $y = -2x$
21. $y = \frac{3}{4}x + 2$

23. $y = x + 2$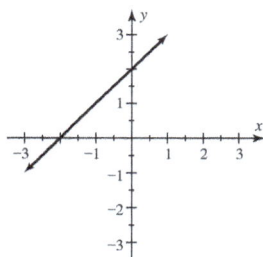

25. $y = 2x - 1$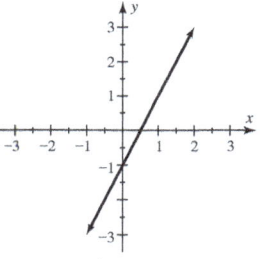

27. $y = -\frac{1}{2}x - 2$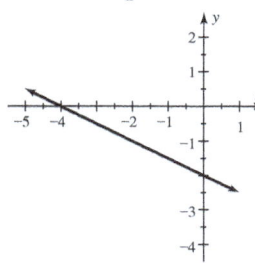

29. $y = \frac{1}{3}x$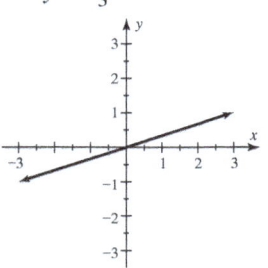

31. $y = 3$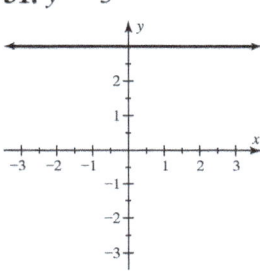

33. (a) $y = -x + 4$ (b) $-1; (0, 4)$
35. (a) $y = -2x + 4$ (b) $-2; (0, 4)$
37. (a) $y = \frac{1}{2}x + 2$ (b) $\frac{1}{2}; (0, 2)$
39. (a) $y = \frac{2}{3}x - 2$ (b) $\frac{2}{3}; (0, -2)$
41. (a) $y = \frac{1}{4}x + \frac{3}{2}$ (b) $\frac{1}{4}; (0, \frac{3}{2})$
43. (a) $y = -\frac{1}{3}x + \frac{2}{3}$ (b) $-\frac{1}{3}; (0, \frac{2}{3})$

45.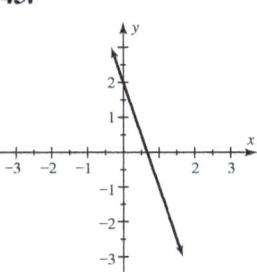

47.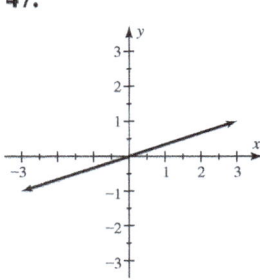

49.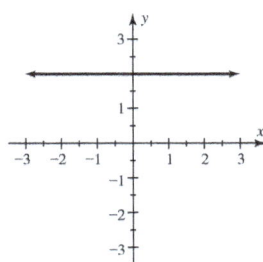

51.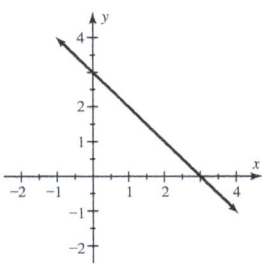

53.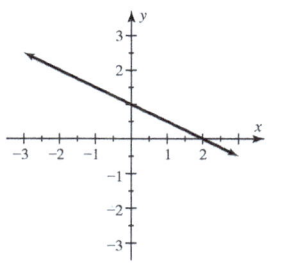

55.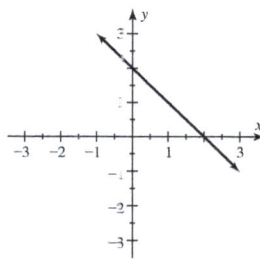

57. $y = 2x + 2$ **59.** $y = x - 2$ **61.** $-\frac{1}{5}$ **63.** $\frac{5}{4}$
65. 4 **67.** $y = \frac{4}{7}x + 3$ **69.** $y = 3x$ **71.** $y = -\frac{1}{2}x + \frac{5}{2}$
73. $y = 2x$ **75.** $y = \frac{1}{3}x + \frac{1}{3}$ **77.** (a) $25 (b) 25 cents
(c) $(0, 25)$; the fixed cost of renting the car is $25 (d) 0.25;
the cost per mile of driving the car **79.** (a) $7.45
(b) $C = 0.07x + 3.95$ (c) 67 min
81. (a) $m = 0.35; b = 164.3$ (b) The fixed cost of owning the car for one month

Section 10.6 (pp. 649–652)

1. True **2.** True **3.** $y = mx + b$
4. $y - y_1 = m(x - x_1)$ or $y = m(x - x_1) + y_1$
5. 1; 3 **6.** distributive **7.** Yes; every nonvertical line has exactly one slope and one y-intercept. **8.** No; it depends on the point used. **9.** No **11.** Yes **13.** $y - 3 = -2(x + 2)$
15. $y - 2 = \frac{3}{4}(x - 1)$ **17.** $y + 1 = -\frac{1}{2}(x - 3)$
19. $y - 1 = 4(x + 3)$ **21.** $y + 3 = \frac{1}{2}(x + 5)$
23. $y - 30 = 1.5(x - 2010)$ **25.** $y - 4 = \frac{7}{3}(x - 2)$
27. $y = \frac{3}{5}(x - 5)$ **29.** $y - 15 = 5(x - 2003)$
31. $y = 3x - 2$ **33.** $y = \frac{1}{3}x$ **35.** $y = \frac{2}{3}x + \frac{1}{12}$
37. $y = -2x + 9$ **39.** $y = \frac{3}{5}x - 2$ **41.** $y = -16x - 19$
43. $y = -2x + 5$ **45.** $y = -x + 1$ **47.** $y = -\frac{1}{9}x + \frac{1}{3}$
49. $y = 2x - 7$ **51.** $y = 2x - 15$ **53.** $y = \frac{1}{2}x + 5$
55. $y = 2x - 3$ **57.** $y = -2x - 1$ **59.** $y = \frac{1}{2}x - \frac{5}{2}$
61. $y = 2x - 5$ **63.** $y = -\frac{1}{2}x - \frac{3}{2}$
65. $(0, -mx_1 + y_1)$. **67.** (a) $-5\,°F$ per hour
(b) $T = -5x + 45$; the temperature is decreasing at a rate of $5\,°F$ per hour. (c) $(9, 0)$; at 9 A.M. the temperature was $0\,°F$.
(d)

(e) At 4 A.M. the temperature was $25\,°F$.

69. (a) Entering; 300 gallons (b) (0, 100); initially the tank contains 100 gallons. (c) $y = 50x + 100$; the amount of water is increasing at a rate of 50 gallons per minute. (d) 8 minutes **71.** (a) Toward (b) After 1 hour, the person is 250 miles from home. After 4 hours, the person is 100 miles from home. (c) 50 mph (d) $y = -50x + 300$; the car is traveling toward home at 50 mph.
73. (a) In 2009, there were 700 million mobile Internet users. (b) $y - 700 = 200(x - 2009)$ (c) 1500 million; 1500 million (d) The number of mobile Internet users increased on average by 200 million per year.
75. (a) $y - 1567 = -165(x - 2012)$ (b) 1237
77. Cigarette consumption decreased on average by 10.33 billion cigarettes per year.
79. (a) $y = 0.009375x - 18.1875$ (b) Sea level is rising 0.009375 meter per year. (c) 0.75 meter

Checking Basic Concepts 10.5 & 10.6 (p. 653)

1. $y = -3x + 1$ **2.** $y = \frac{4}{5}x - 4; \frac{4}{5}; (0, -4)$
3.

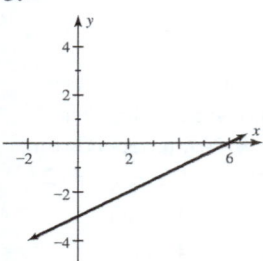

4. (a) $y = 3x - 2$ (b) $y = -\frac{3}{2}x$ (c) $y = -\frac{7}{3}x - \frac{5}{3}$
5. $y - 3 = -2(x + 1)$ **6.** $y = -2x + 1$
7. $y = 2x + 3$ **8.** (a) $y = -12x + 48$ (b) 12 mph (c) 4 P.M. (d) 48 miles   **9.** (a) 5 inches (b) 2 inches per hour (c) (0, 5); five inches of snow fell before noon (d) 2; the average rate of snowfall was 2 inches per hour.

Section 10.7 (pp. 658–661)

1. linear   **2.** exact   **3.** approximate   **4.** constant
5. constant rate of change   **6.** initial amount   **7.** f.
8. d.   **9.** a.   **10.** c.   **11.** e.   **12.** b.   **13.** Yes   **15.** No
17. No   **19.** Exact; $y = 2x - 2$ **21.** Approximate; $y = 2x + 2$ **23.** Approximate; $y = 2$
25. $y = 5x + 40$ **27.** $y = -20x - 5$ **29.** $y = 8$
31. (a) Yes   (b)

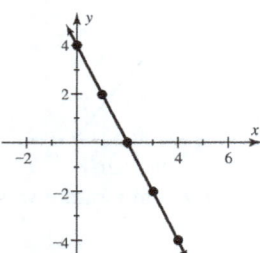

(c) $y = -2x + 4$

33. (a) No   (b)

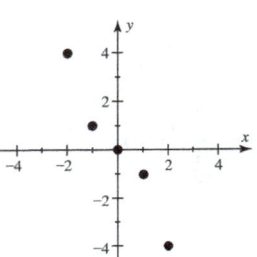
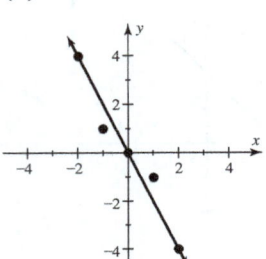

(c) $y = -2x$

35. (a) Yes   (b)

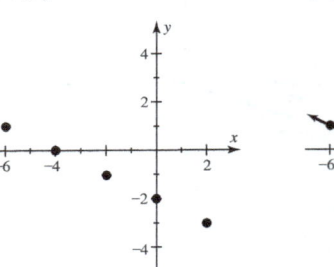
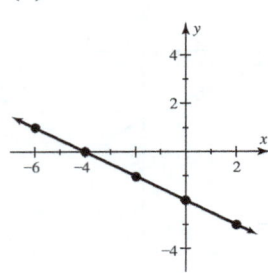

(c) $y = -\frac{1}{2}x - 2$

37. (a) No   (b)

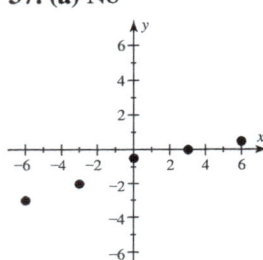
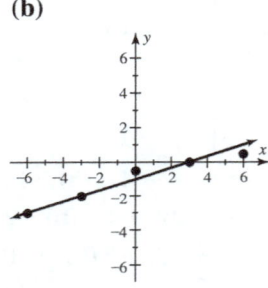

(c) $y = \frac{1}{3}x - 1$
39. $g = 5t + 200$, where g represents gallons of water and t represents time in minutes.
41. $d = 6t + 5$, where d represents distance in miles and t represents time in hours.
43. $p = 8t + 200$, where p represents total pay in dollars and t represents time in hours.
45. $r = t + 5$, where r represents total number of roofs shingled and t represents time in days.
47. (a) $-\frac{2}{45}$ acre per year   (b) $A = -\frac{2}{45}t + 5$
49. (a) $y = \frac{19}{6}x - 6155.5$ (b) About 232,000
51. (a)   (b)

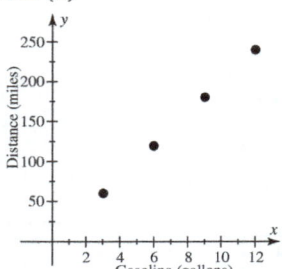
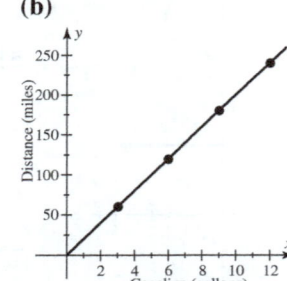

(c) 20; the mileage is 20 miles per gallon.
(d) $y = 20x$ (e) 140 miles

53.

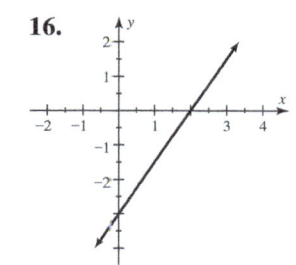

55. $y = 1106x - 2{,}196{,}295$; $(0, -2{,}196{,}295)$; no

Checking Basic Concepts 10.7 (p. 661)

1. Yes   **2.** Approximate; $y = x - 1$
3. (a) $y = 10x + 50$ **(b)** $y = -2x + 200$
4. (a)   **(b)**

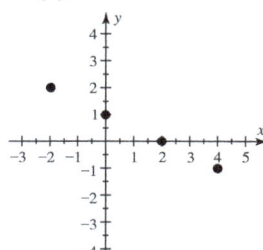   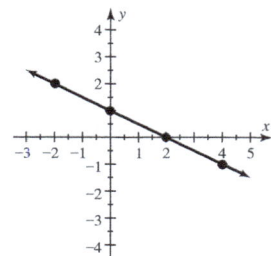

(c) $y = -\frac{1}{2}x + 1$

5. (a) $T = 0.075x$ **(b)** $5.4°F$

Chapter 10 Review (pp. 665–669)

1. $(-2, 0)$: none; $(-1, 2)$: II; $(0, 0)$: none; $(1, -2)$: IV; $(1, 3)$: I
2.

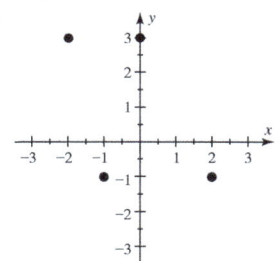

3. (a) II   **(b)** IV   **4. (a)** None   **(b)** III
5.   **6.**

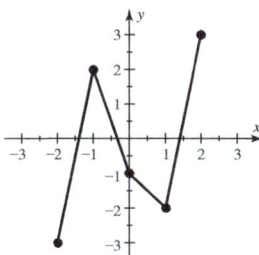   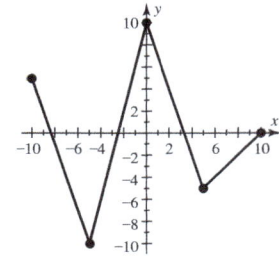

7. Yes   **8.** No
9.

x	-2	-1	0	1	2
y	6	3	0	-3	-6

10.

x	4	3	2.5	2	1
y	-3	-1	0	1	3

11.

x	-2	0	2	4
y	-4	2	8	14

12.

x	-1	-3	-5	-7
y	1	2	3	4

13.   **14.**

15.   **16.**

17. $(3, 0)$; $(0, -2)$ **18.** $(-2, 0)$ and $(2, 0)$; $(0, -4)$

19.

x	-2	-1	0	1	2
y	4	3	2	1	0

$(2, 0)$; $(0, 2)$

20.

x	-4	-2	0	2	4
y	-4	-3	-2	-1	0

$(4, 0)$; $(0, -2)$

21. x-int: $(3, 0)$;   **22.** x-int: $(1, 0)$;
y-int: $(0, -2)$ y-int: $(0, -5)$

23. x-int: $(4, 0)$;   **24.** x-int: $(2, 0)$;
y-int: $(0, -2)$ y-int: $(0, 3)$

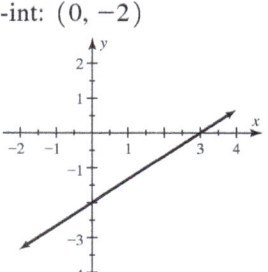

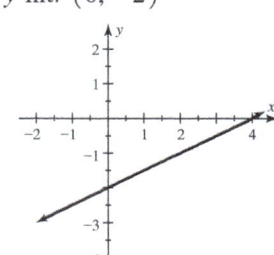   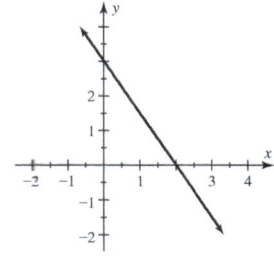

25. $y = 1$ **26.** $x = 3$
27. (a) 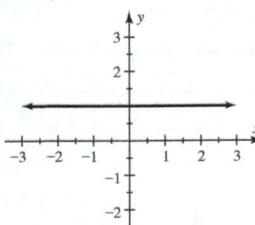 **(b)**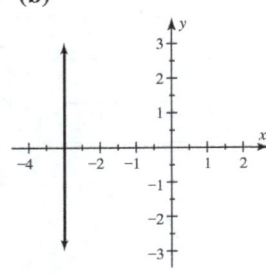

28. $x = -1; y = 1$ **29.** $y = 3; x = -2$
30. $x = 4$ **31.** $y = -5$ **32. (a)** y-int: $(0, 90)$; x-int: $(3, 0)$
(b) The driver is initially 90 miles from home; the driver arrives home after 3 hours. **33.** 2 **34.** $-\frac{2}{5}$
35. 0 **36.** Undefined **37.** 3 **38.** $-\frac{1}{2}$

39. (a) 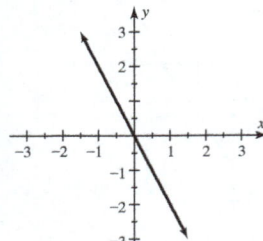 **40. (a)**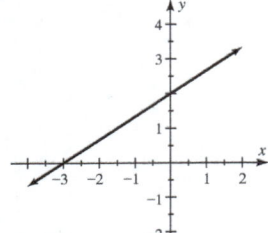

(b) $-2; (0, 0)$ **(b)** $\frac{2}{3}; (0, 2)$
41. 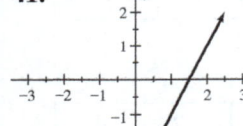 **42.**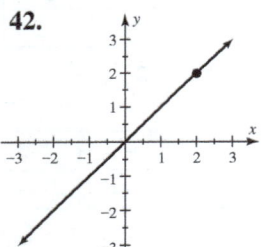

43.
x	0	1	2	3
y	1	$\frac{3}{2}$	2	$\frac{5}{2}$

44. Positive **45.** $y = x + 1$ **46.** $y = -2x + 2$
47. $y = 2x - 2$ **48.** $y = -\frac{3}{4}x + 3$

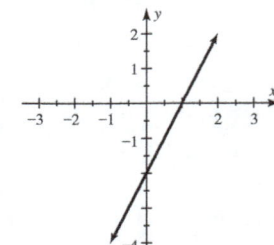 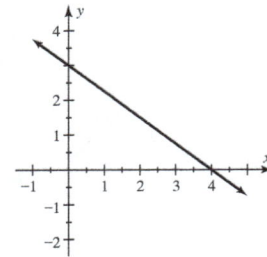

49. (a) $y = -x + 3$ **(b)** $-1; (0, 3)$
50. (a) $y = \frac{5}{6}x - 5$ **(b)** $\frac{5}{6}; (0, -5)$

51. 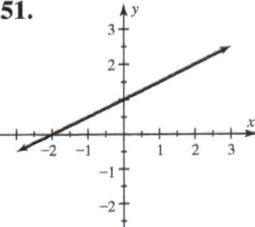 **52.**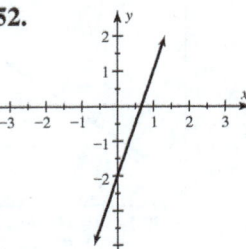

53. 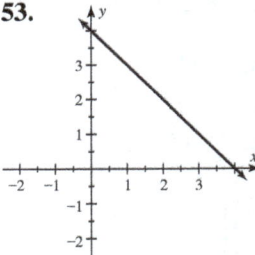 **54.**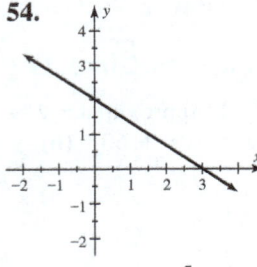

55. $y = 5x - 5$ **56.** $y = -2x$ **57.** $y = -\frac{5}{6}x + 2$
58. $y = -2x - 3$ **59.** $y = \frac{2}{3}x - 2$ **60.** $y = -\frac{1}{5}x - 2$
61. Yes **62.** No **63.** $y - 2 = 5(x - 1)$
64. $y + 30 = 3(x - 20)$ **65.** $y = \frac{4}{3}(x - 3)$
66. $y = 2\left(x - \frac{1}{2}\right)$ **67.** $y - 7 = 2(x - 5)$
68. $y = -\frac{2}{3}(x + 1)$ **69.** $y = 3x + 5$
70. $y = -\frac{1}{4}x + 3$ **71.** No
72. Approximate; $y = -x + 5$ **73.** $y = -2x + 40$
74. $y = 20x + 200$ **75.** $y = 50$ **76.** $y = 5x - 20$

77. (a) Yes **(b)**

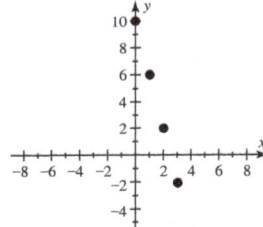 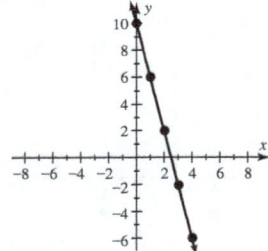

(c) $y = -4x + 10$
78. (a) No **(b)**

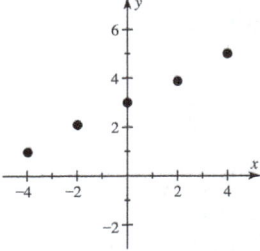 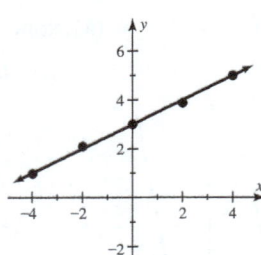

(c) $y = \frac{1}{2}x + 3$

79. (a)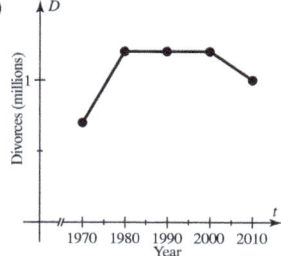

(b) The number of divorces increased significantly between 1970 and 1980, remained unchanged from 1980 to 2000, and then decreased.
80. (a) $G = 100t$
(b)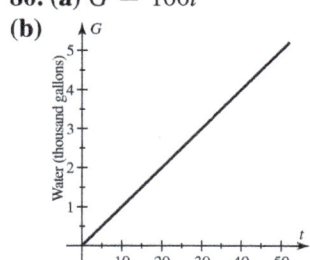

(c) 50 days
81. (a)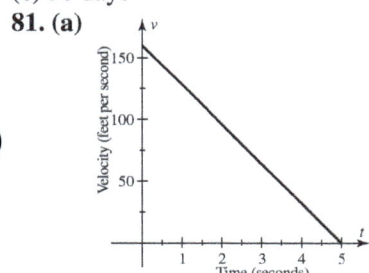

(b) v-int: $(0, 160)$; t-int: $(5, 0)$; the initial velocity was 160 ft/sec, and the velocity after 5 seconds was 0.
82. (a) $m_1 = 0$; $m_2 = -1500$; $m_3 = 500$ **(b)** $m_1 = 0$: The population remained unchanged. $m_2 = -1500$: The population decreased at a rate of 1500 insects per week. $m_3 = 500$: The population increased at a rate of 500 insects per week. **(c)** For the first week, the population did not change from its initial value of 4000. Over the next two weeks, the population decreased at a rate of 1500 insects per week until it reached 1000. Finally, the population increased at a rate of 500 per week for two weeks, reaching 2000.
83.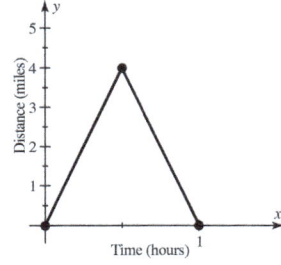

84. (a) $\frac{2}{3}$ **(b)** Revenue increased, on average, by about $0.67 billion/yr
85. (a) $35 **(b)** 20¢ **(c)** $(0, 35)$; the fixed cost of renting the car **(d)** 0.2; the cost for each mile driven
86. (a) Toward; the slope is negative. **(b)** After 1 hour, the car is 200 miles from home; after 3 hours, the car is 100 miles from home. **(c)** $y = -50x + 250$; the car is moving toward home at 50 mph. **(d)** 150 miles; 150 miles
87. (a)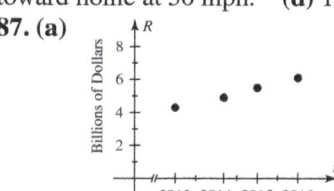

(b) $R = 0.6t - 1203.5$; Revenue increased by $0.6 billion/yr.
(c) Yes **(d)** $7.3 billion
88. (a) **(b)**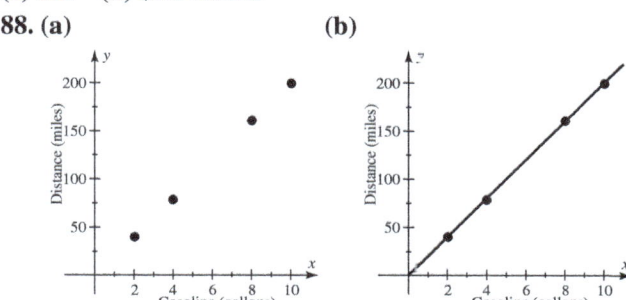

(c) 20; the mileage is 20 miles per gallon.
(d) $y = 20x$; no **(e)** About 180 miles

Chapter 10 Test (pp. 669–670)

1. $(-2, -2)$: III; $(-2, 1)$: II; $(0, -2)$: none; $(1, 0)$: none; $(2, 3)$: I; $(3, -1)$: IV
2.

3.

x	-2	-1	0	1	2
y	-8	-6	-4	-2	0

$(2, 0)$; $(0, -4)$
4. Yes
5.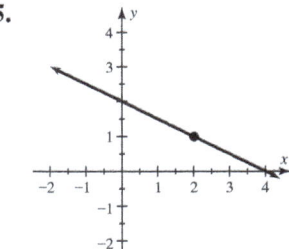

6. x-int: $(3,0)$; y-int: $(0,-5)$

7.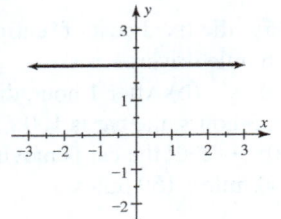

8.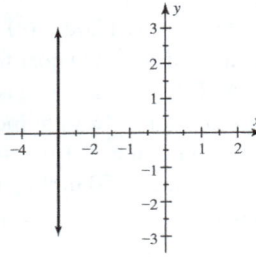

9.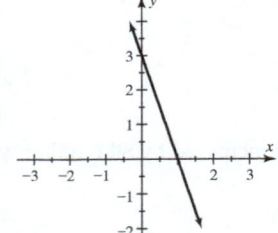

10.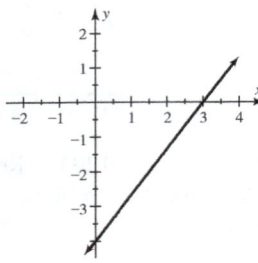

11. $y = -x + 2$ **12.** $y = 2x + \frac{1}{2}$; 2; $(0, \frac{1}{2})$
13. $y = -5$; $x = 1$ **14.** $-\frac{2}{9}$ **15.** $y = -\frac{4}{3}x - 5$
16. $y = 3x - 11$ **17.** $y = -3x + 5$ **18.** $y = -\frac{1}{2}x$
19. $y = \frac{1}{2}x + 5$ **20.** Yes, $y = 3x - 2$
21. $y - 7 = -3(x + 2)$ **22.** Approximate; $y = x - 1$
23.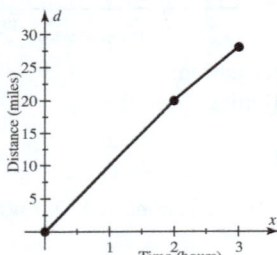

24. (a) $m_1 = 2$; $m_2 = -9$; $m_3 = 2$; $m_4 = 5$
(b) $m_1 = 2$: The population increased at a rate of 2000 fish per year. $m_2 = -9$: The population decreased at a rate of 9000 fish per year. $m_3 = 2$: The population increased at a rate of 2000 fish per year. $m_4 = 5$: The population increased at a rate of 5000 fish per year.
(c) For the first year, the population increased from an initial value of 8000 to 10,000 at a rate of 2000 fish per year. During the second year, the population dropped dramatically to 1000 at a rate of 9000 fish per year. Over the third year, the population grew to 3000 at a rate of 2000 fish per year. Finally, over the fourth year, the population grew at a rate of 5000 fish per year to reach 8000.
25. $N = 100x + 2000$

Chapters 1–10 Cumulative Review (pp. 670–671)

1. Composite; $40 = 2 \cdot 2 \cdot 2 \cdot 5$ **2.** Prime
3. $n + 10$ **4.** $n^2 - 2$ **5.** $\frac{2}{3}$ **6.** $\frac{17}{30}$ **7.** 14
8. -9 **9.** Rational **10.** Irrational **11.** $7x + 1$
12. $x - 4$ **13.** 2 **14.** 0 **15.** 720 ft² **16.** 20 ft²

17.

x	-2	-1	0	1	2
y	10	8	6	4	2

;1

18. $2n + 2 = n - 5$; -7 **19.** 8 hours
20. $x < 2$ **21.** $\{x \mid x > 0\}$ **22.** $\{x \mid x \leq \frac{1}{2}\}$
23. $(-2, -3)$: III; $(0, 3)$: none; $(2, -2)$: IV; $(2, 1)$: I
24.

x	-2	-1	0	1	2
y	3	2.5	2	1.5	1

25.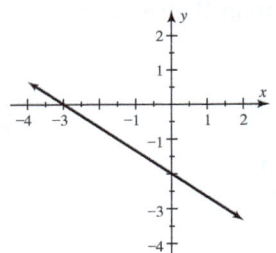

26.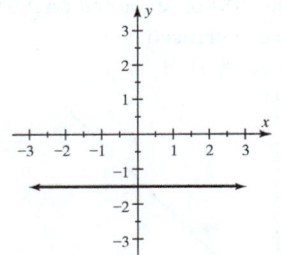

27. x-int: $(-10, 0)$; y-int: $(0, 8)$ **28.** $x = \frac{3}{2}$; $y = -2$
29. $y = 3x - 3$
30. $y = \frac{3}{5}x - 3$;

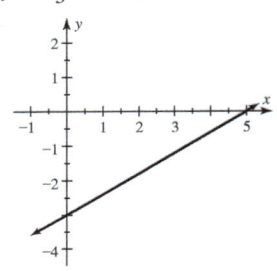

31. $y = -\frac{2}{3}x - 3$ **32.** $y = -2x + 1$
33. $y = 5x + 100$ **34.** $I = 5000x + 20,000$
35. $C = 8x$ **36.** $\frac{49}{100}$ **37.** $1000 at 3%; $2000 at 4%
38. (a) $85 **(b)** $25 **(c)** 30¢

11 Systems of Linear Equations in Two Variables

Section 11.1 (pp. 682–685)

1. ordered **2.** intersection-of-graphs **3.** no solutions; one solution; infinitely many **4.** consistent **5.** inconsistent
6. independent **7.** dependent **8.** table **9.** the same
10. intercepts **11.** 1 **13.** 2 **15.** -2 **17.** -1
19.

x	0	1	2	3
$y = x + 3$	3	4	5	6

;1

21.

x	-4	-2	2	4
$y = 3 - 3x$	15	9	-3	-9

;2

23.

x	-3	0	3	6
$2x + 3y = 6$	4	2	0	-2

;3

25. One; consistent; independent
27. Infinitely many; consistent; dependent
29. None; inconsistent   **31.** $(1, 1)$ **33.** $(2, -3)$
35. $\left(\frac{1}{5}, \frac{4}{5}\right)$ **37.** $(2, 1)$ **39.** $(3, 2)$ **41.** $(-1, 1)$
43. $(2, 4)$ **45.** $(3, 1)$
47.

x	0	1	2	3
$y = x + 2$	2	3	4	5
$y = 4 - x$	4	3	2	1

; $(1, 3)$

49. (a) $(-1, 1)$ **(b)** $(-1, 1)$ **51. (a)** $(3, 1)$ **(b)** $(3, 1)$
53. (a) $(1, 3)$ **(b)** $(1, 3)$ **55.** $(-1, -2)$ **57.** $(1, -1)$
59. $(1, 2)$ **61.** $(3, 2)$ **63.** $(3, 1)$ **65.** $(1, 2)$
67. (a) $x + y = 4, x - y = 0$ **(b)** 2, 2
69. (a) $2x + y = 7, x - y = 2$ **(b)** 3, 1
71. (a) $x = 3y, x - y = 4$ **(b)** 6, 2
73. (a) $C = 0.5x + 50$ **(b)** 60 mi   **(c)** 60 mi
75. (a) $x + y = 60, x = 2y$ **(b)** $(40, 20)$; in 2013, digital downloads accounted for 40% of all music revenue and streaming subscriptions accounted for 20%.
77. 180 e-readers at $120 each
79. (a) $x - y = 4; 2x + 2y = 28$ where x is length, y is width   **(b)** $(9, 5)$; the rectangle is 9 in. × 5 in.
81. 23; yes   **83.** 23; yes

Section 11.2 (pp. 692–695)

1. exact   **2.** parentheses   **3.** It has infinitely many solutions.
4. It has no solutions.   **5.** $(3, 6)$ **7.** $(2, 1)$ **9.** $(-1, 0)$
11. $(3, 0)$ **13.** $\left(\frac{1}{2}, 0\right)$ **15.** $(0, 4)$ **17.** $(1, 2)$
19. $(6, -4)$ **21.** $(4, -3)$ **23.** $\left(1, -\frac{1}{2}\right)$ **25.** $(1, 3)$
27. $(0, -2)$ **29.** $(2, 2)$ **31.** $(-3, 5)$ **33.** No solutions
35. Infinitely many   **37.** $(4, 1)$ **39.** Infinitely many
41. No solutions   **43.** $\left(-\frac{14}{33}, -\frac{37}{33}\right)$ **45.** $(0, 0)$
47. No solutions   **49.** No solutions
51. They are a single line.
53. (a) $L - W = 10, 2L + 2W = 72$ **(b)** $(23, 13)$
55. (a) $x = \frac{1}{2}y; x + y = 90$ **(b)** $(30, 60)$ **(c)** $(30, 60)$
57. (a) $x - y = 30, y = 0.864x$ **(b)** $(220.59, 190.59)$
59. 94 ft × 50 ft   **61.** 17 and 53   **63.** 10 mph; 2 mph
65. $3.\overline{3}$ L of 20% solution, $6.\overline{6}$ L of 50% solution
67. (a) $x + y = 15, 3x + 4y = 53$ **(b)** $(7, 8)$
69. (a) $x + y = 48, 12x + 5y = 450$ **(b)** $(30, 18)$
(c) Tread climber: 360 cal, bicycle: 90 cal
71. Superior: 32,000 mi²; Michigan: 22,000 mi²

Checking Basic Concepts 11.1 & 11.2 (p. 695)

1. (a) -2 **(b)** 6   **2.** $(4, 2)$ **3.** $(2, 1)$
4. (a) No solutions   **(b)** $(1, -1)$
(c) Infinitely many solutions
5. (a) $x + y = 300, 150x + 200y = 55,000$
(b) $(100, 200)$

Section 11.3 (pp. 703–706)

1. Substitution; elimination   **2.** addition   **3.** =   **4.** =
5. It has infinitely many solutions.   **6.** It has no solutions.

7. $(1, 1)$ **9.** $(-2, 1)$ **11.** No solutions   **13.** Infinitely many
15. $(6, 1)$ **17.** $(-1, 4)$ **19.** $(2, 4)$ **21.** $(2, 1)$
23. $(4, -1)$ **25.** $(-4, 1)$ **27.** $\left(\frac{4}{5}, 1\right)$ **29.** $(-2, -5)$
31. $(1, 0)$ **33.** $(2, 2)$ **35.** $\left(\frac{4}{5}, \frac{9}{5}\right)$ **37.** $(3, 2)$
39. $(0, 1)$ **41.** $(2, 1)$ **43.** $(4, -3)$ **45.** $(1, 0)$
47. Infinitely many; 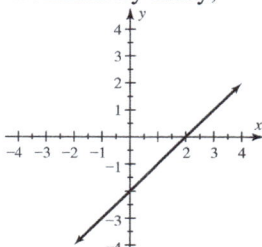   **49.** One;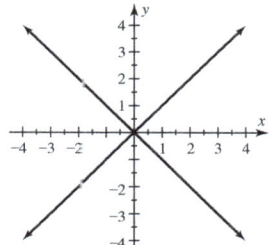

51. No solutions; 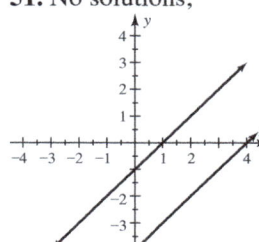   **53.** No solutions;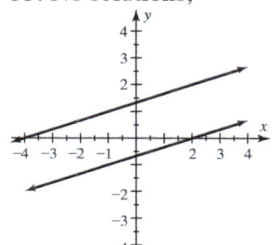

55. Infinitely many;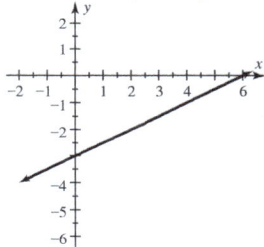

57. healthcare: 11.7 million; ncaa: 8.7 million
59. Bicycle: 18 min; stair climber: 12 min
61. Current: 2 mph; boat: 6 mph
63. $2000 at 3%; $3000 at 5%   **65.** $-43, 26$
67. (a) 20 in. × 40 in.   **(b)** 20 in. × 40 in.

Section 11.4 (pp. 714–717)

1. inequality   **2.** ordered pair   **3.** All points below and including the line $y = k$ **4.** All points to the right of the line $x = k$.   **5.** All points above and including the line $y = x$
6. test   **7.** dashed   **8.** solid   **9.** $Ax + By = C$
10. both inequalities   **11.** intersect   **12.** No   **13.** Yes
15. No   **17.** No   **19.** Yes   **21.** Yes   **23.** No   **25.** $x > 1$
27. $y \geq 2$ **29.** $y < x$ **31.** $-x + y \leq 1$
33. 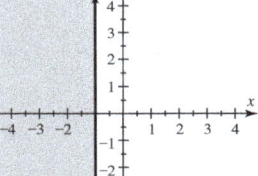   **35.**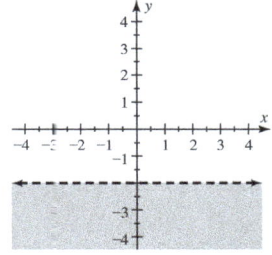

A-28 ANSWERS TO SELECTED EXERCISES

37. 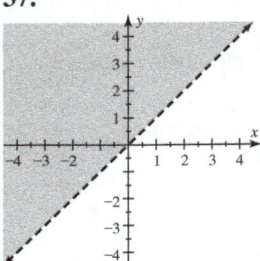 **39.**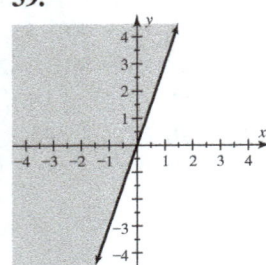

41. 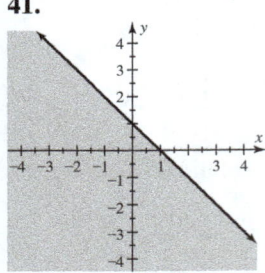 **43.**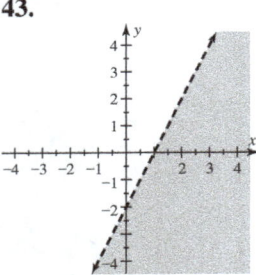

45. Yes **47.** No **49.** Yes
51. The region containing $(1, 2)$
53. The region containing $(1, 0)$
55. 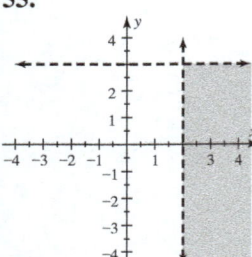 **57.**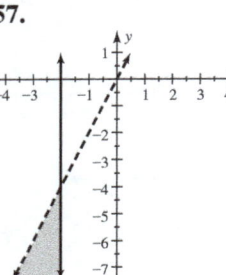

59. 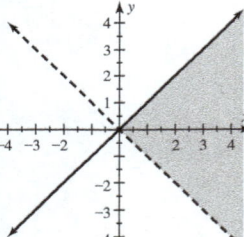 **61.**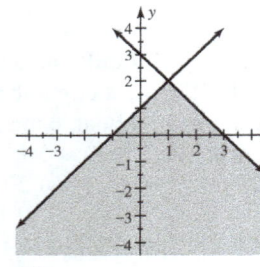

63. 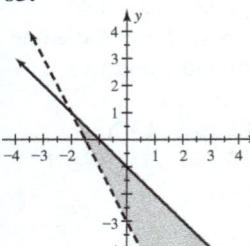 **65.**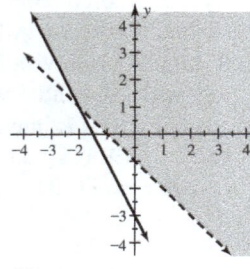

67. 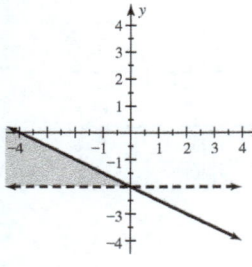 **69.**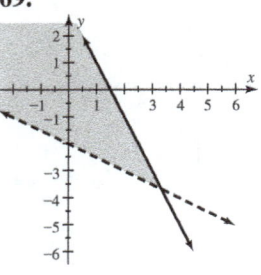

71. 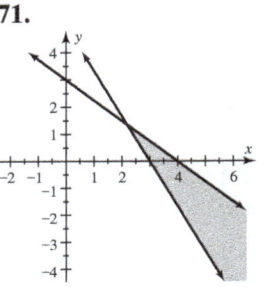 **73.**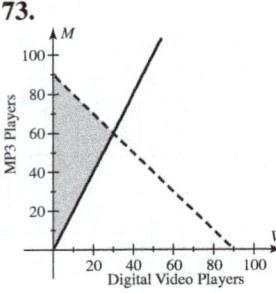

75. (a) 200 bpm; 150 bpm
(b)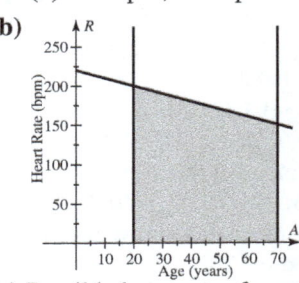

(c) Possible heart rates for ages 20 to 70 **77.** 150 to 200 lb
79. (a) 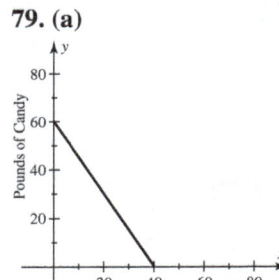 **(b)**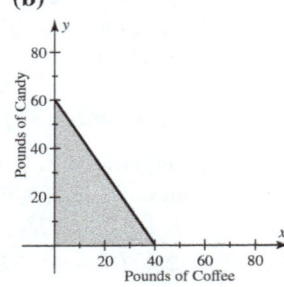

81. Wetter and cooler regions **83.** $7P - 5T \geq -70$
85. $7P - 5T \leq -70$
$35P - 3T \geq 140$

Checking Basic Concepts 11.3 & 11.4 (p. 717)

1. $(1, 1)$ **2. (a)** $(0, -1)$; one **(b)** No solutions
(c) Infinitely many **3.** Infinitely many
4. (a) 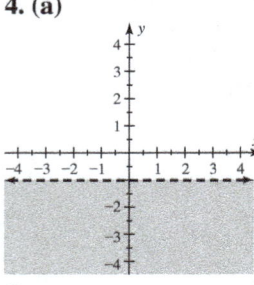 **(b)**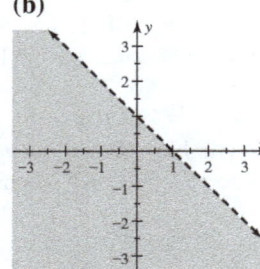

5.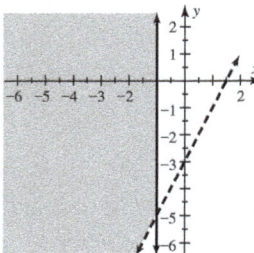

6. (a) $x + y = 11.1, x - y = 5.7$ **(b)** $(8.4, 2.7)$

Chapter 11 Review (pp. 720–723)

1. 3 2. 2 3. (a) None (b) Inconsistent
4. (a) One (b) Consistent; independent
5. (a) Infinitely many (b) Consistent; dependent
6. (a) None (b) Inconsistent 7. $(1, 2)$
8. $(5, 2)$ 9. $(4, 3)$ 10. $(2, -4)$ 11. $(2, 2)$
12. $(1, 2)$ 13. $(2, 6)$ 14. $(1, 1)$
15. $(4, -3)$ 16. $(1, 2)$ 17. $(1, 1)$
18. $(1, 2)$ 19. $(1, 1)$ 20. $(-2, -1)$
21. $(2, 6)$ 22. $(2, -10)$ 23. $(-2, 1)$
24. $(-4, -4)$ 25. No solutions
26. No solutions 27. Infinitely many 28. $(1, 1)$
29. $(2, 1)$ 30. $(-1, 2)$ 31. $(11, -1)$ 32. $(1, 0)$
33. $\left(\frac{9}{4}, \frac{7}{4}\right)$ 34. $\left(\frac{5}{2}, 1\right)$ 35. $(5, -7)$ 36. $(8, 2)$
37. $(-2, 3)$ 38. $(1, 0)$ 39. $(1, 3)$ 40. $(2, -1)$
41. Infinitely many 42. Infinitely many 43. No solutions
44. One 45. Yes 46. No 47. No 48. Yes 49. $y > 1$
50. $y \le 2x + 1$

51.
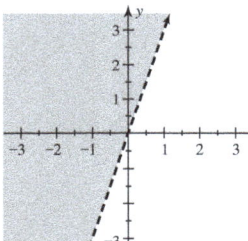

52.
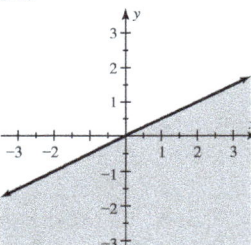

53.
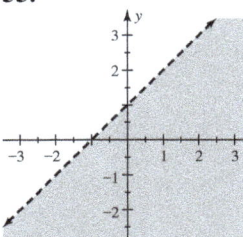

54.
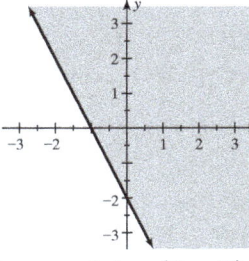

55.
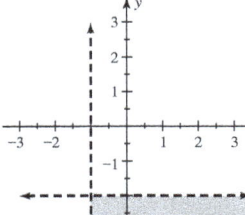

56.
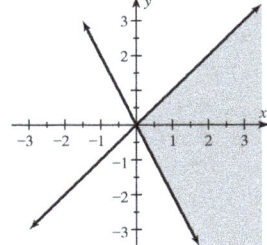

57. Yes 58. No 59. The region containing $(2, -2)$
60. The region containing $(1, 3)$

61.
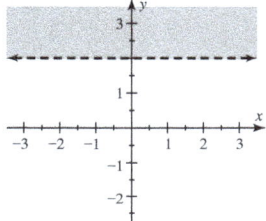

62.

63.
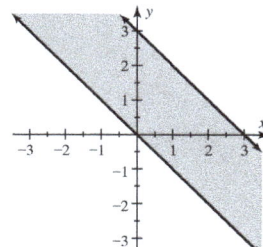

64.
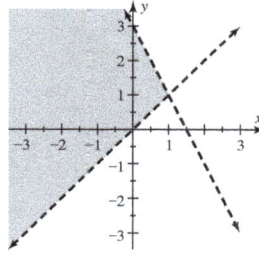

65.
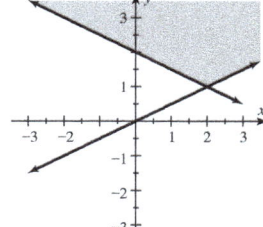

66.
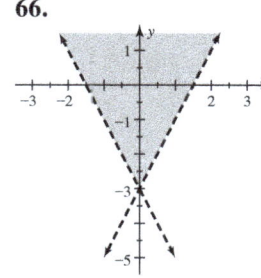

67. 3100 deaths in 1912; 33,790 deaths in 2012
68. Display: $6.8 billion; native: $5.0 billion
69. (a) $C = 0.2x + 40$ (b) 250 mi (c) 250 mi
70. 75°, 105° 71. (a) $2x + y = 180$, $2x - y = 40$
(b) $(55, 70)$ (c) $(55, 70)$ 72. 20 ft × 24 ft
73. (a) $x + y = 10$, $80x + 120y = 920$; x is $80 rooms, y is $120 rooms. (b) $(7, 3)$ 74. 7 lb of $2 candy; 11 lb of $3 candy 75. Bicycle: 35 min; stair climber: 25 min
76. 2 mph 77. (a) 16 ft × 24 ft (answers may vary slightly)
(b) 16 ft × 24 ft

78.
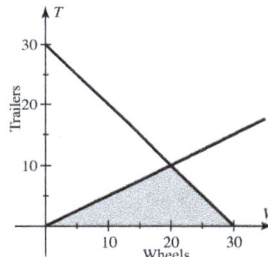

Chapter 11 Test (pp. 723–724)

1. $(1, 2)$ 2. $(-2, -1)$ 3. $(-1, -2)$ 4. $(-2, 3)$
5. $(1, 3)$ 6. (a) $(2, 1)$; one; consistent (b) No solutions; zero; inconsistent 7. (a) Infinitely many (b) Consistent; dependent 8. (a) None (b) Inconsistent 9. $(-3, 4)$
10. No solutions 11. Infinitely many solutions
12. $(-1, 3)$ 13. $y \le -\frac{1}{2}x$ 14. Yes

15.
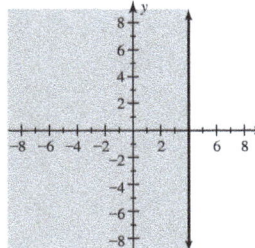

16.
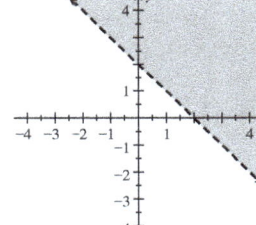

17. 4

18. 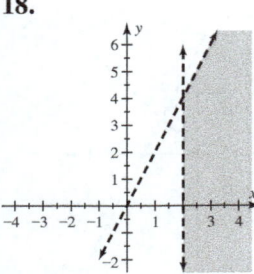 **19.** 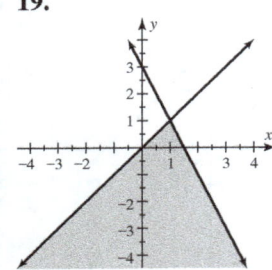 **25.** 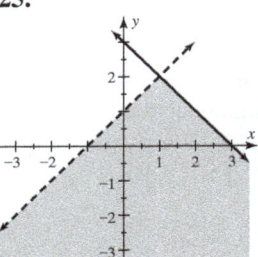 **26.**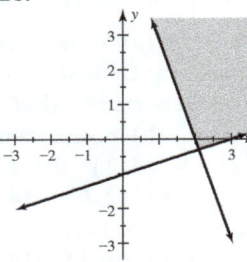

20. 3.4 million in 2012; 6.1 million in 2013
21. 7.5 L of 20% solution and 2.5 L of 60% solution
22. $\frac{2}{3}$ hr at 6 mph; $\frac{1}{3}$ hr at 9 mph

27. $F = 16.5R$ **28.** 94 °C **29.** $245 **30.** $158.05
31. $G = -3x + 30$ **32.** $1200 at 5%; $1200 at 6%

Chapters 1–11 Cumulative Review (pp. 724–725)

1. $2 \cdot 2 \cdot 2 \cdot 3 \cdot 5$ **2. (a)** 4 **(b)** 8 **3. (a)** Rational
(b) Irrational **4. (a)** < **(b)** > **5. (a)** $4x^2$ **(b)** $5x - 2$
6. $\frac{2}{9}$ **7.** $\frac{1}{2}$ **8.** No solutions **9.** 11, 12, 13, 14
10. 432 in^2 or 3 ft^2 **11.** $x = \frac{W + 7y}{3}$ **12.** $x \geq 1$
13.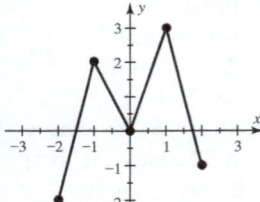

14. (a) 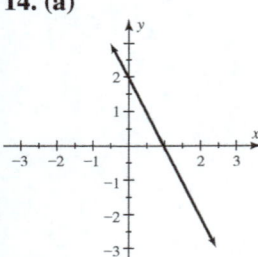 **(b)**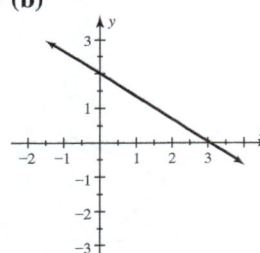

15. $(2, 0); (0, -8)$
16. (a) 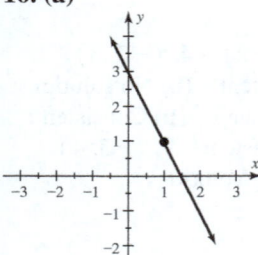 **(b)**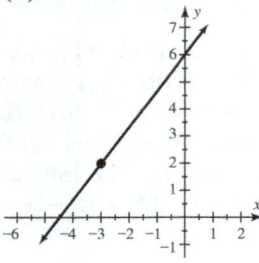

17. (a) $y = -\frac{1}{2}x - 1$ **(b)** $y = 4x - 3$
18. $y = -\frac{1}{4}x + 3$ **19.** $y = -3x + 4$ **20.** $(4, 4)$
21. $(1, -3)$ **22.** No solutions **23.** Infinitely many solutions
24. $(-1, 2)$

12 Polynomials and Exponents

Section 12.1 (pp. 734–735)

1. base; exponent **2.** 1 **3.** a^{m+n} **4.** a^{mn} **5.** $a^n b^n$ **6.** $\frac{a^n}{b^n}$
7. 64 **9.** −8 **11.** −8 **13.** 11 **15.** $\frac{1}{2}$ **17.** 1 **19.** 5 **21.** 1
23. 3^3 or 27 **25.** 4^8 or 65,536 **27.** x^9 **29.** x^6 **31.** $20x^7$
33. $-3x^3 y^4$ **35.** 2^6 or 64 **37.** n^{12} **39.** x^7 **41.** $49b^2$
43. $a^3 b^3$ **45.** 1 **47.** $-64b^6$ **49.** $x^{14} y^{21}$ **51.** $\frac{1}{27}$
53. $\frac{a^5}{b^5}$ **55.** $\frac{(x-y)^3}{27}$ **57.** $\frac{25}{(a+b)^2}$ **59.** $x^{12} y^9$ **61.** $a^{10} b^8$
63. $\frac{8x^3}{125}$ **65.** $\frac{27x^6}{125 y^{12}}$ **67.** $(x+y)^4$ **69.** $(a+b)^5$ **71.** 6
73. $a^3 + 2ab^2$ **75.** $12a^5 + 6a^3 b$ **77.** $x^3 y^3 - x^2 y^4$
79. $a = 3, b = 1, m = 1, n = 0$ (answers may vary)
81. $10x^4$ **83.** $9\pi x^4$ **85.** $8x^3$ **87.** $1157.63
89. (a) 3^2 **(b)** $72,000

Section 12.2 (pp. 744–746)

1. monomial **2.** polynomial **3.** degree **4.** degree
5. binomial **6.** trinomial **7.** like **8.** like **9.** opposite
10. vertically **11.** 2; 3 **13.** 2; −1 **15.** 2; −5 **17.** 0; 6
19. Yes; 1; 1; 1 **21.** Yes; 3; 1; 2 **23.** No **25.** No
27. Yes; 2; 3; 6 **29.** Yes; x **31.** Yes; $-5x^3$ **33.** No
35. Yes; $2ab$ **37.** $-x + 9$ **39.** $4x^2 + 8x + 1$
41. $4y^3 + 3y^2 + 4y - 9$ **43.** $4xy + 1$ **45.** $2a^2 b^3$
47. $9x^2 + x - 6$ **49.** $x^2 - 7x - 1$ **51.** $-5x^2$
53. $-3a^2 + a - 4$ **55.** $2t^2 + 3t - 4$ **57.** $4x - 2$
59. $-3x^2 + 5x + 2$ **61.** $z^3 - 6z^2 - 6z - 1$
63. $3xy + 2x^2 y^2$ **65.** $-a^3 b$ **67.** $-x^2 - 5x - 4$
69. $-2x^3 - 6x - 2$ **71.** 9, 9 **73.** −93, 3
75. (a) 200 bpm **(b)** 100 bpm **(c)** It decreases quickly at first, then more slowly. **77.** $2z^2$; 200 in^2
79. $2x^2 + x^2$ or $3x^2$; 108 ft^2 **81.** $\pi x^2 + \pi y^2$; 13π ft^2
83. (a) $m_1 \approx 0.077$; $m_2 \approx 0.083$; $m_3 \approx 0.077$. A line is reasonable but not exact. **(b)** For the given years, its estimates are reasonable.

Checking Basic Concepts 12.1 & 12.2 (p. 746)

1. (a) −25 **(b)** 1 **2. (a)** 10^8 **(b)** $-12x^7$ **(c)** $a^6 b^2$
(d) $\frac{x^4}{z^{12}}$ **3. (a)** 1 **(b)** $9x^{14}$ **(c)** $10a^5 - 14a^2$
4. 3; 2; 4 **5. (a)** $2w^2 h$ **(b)** 2880 in^3 **6. (a)** $3a^2 + 6$
(b) $2z^3 + 7z - 8$ **(c)** $4xy$

Section 12.3 (pp. 752–754)

1. The product rule 2. Distributive 3. term; term
4. vertically 5. x^7 7. $-12a^2$ 9. $20x^5$ 11. $4x^2y^3$
13. $-12x^3y^2$ 15. $3x + 12$ 17. $-45x - 5$ 19. $4z - z^2$
21. $-5y - 3y^2$ 23. $15x^3 - 12x$ 25. $6x^3 - 6x^2$
27. $-32t^2 - 8t - 8$ 29. $-5n^4 + n^3 - 2n^2$
31. $x^2y + xy^2$ 33. $x^4y - x^3y^2$ 35. $-a^4b + 2ab^4$
37. $x^2 + 3x$ 39. $x^2 + 4x + 4$ 41. $x^2 + 9x + 18$
43. $x^2 + 8x + 15$ 45. $x^2 - 17x + 72$
47. $6z^2 - 19z + 10$ 49. $64b^2 - 1$ 51. $10y^2 - 3y - 7$
53. $5 - 13a + 6a^2$ 55. $1 - 9x^2$ 57. $x^3 - x^2 + x - 1$
59. $4x^3 - 3x^2 + 16x - 12$ 61. $2n^3 + n^2 + 6n + 3$
63. $m^3 + 4m^2 + 4m + 1$ 65. $6x^3 - 7x^2 + 14x - 8$
67. $x^3 + 1$ 69. $4b^4 + 3b^3 + 19b^2 + 9b + 21$
71. $x^3 - x^2 - 5x + 2$ 73. $a^3 - 8$
75. $6x^4 - 2x^3 + 5x^2 - x + 1$ 77. $m \cdot n$
79. (a) $h^3 - 2h^2 - 8h$ (b) 14,175 in^3
81. (a) $50x - x^2$ (b) 625 83. $6x^2 + 12x + 6$
85. (a) $64t - 16t^2$ (b) 64; 64 (c) Yes; yes

Section 12.4 (pp. 759–761)

1. $a^2 - b^2$ 2. $a^2 + 2ab + b^2$ 3. $a^2 - 2ab + b^2$
4. $(a + b)^2$ 5. False 6. False 7. $x^2 - 9$ 9. $16x^2 - 1$
11. $1 - 4a^2$ 13. $4x^2 - 9y^2$ 15. $a^2b^2 - 25$ 17. $a^4 - b^4$
19. $x^6 - y^6$ 21. 9999 23. 391 25. 9900
27. $a^2 - 4a + 4$ 29. $4x^2 + 12x + 9$
31. $9b^2 + 30b + 25$ 33. $\frac{9}{16}a^2 - 6a + 16$
35. $1 - 2b + b^2$ 37. $25 + 10y^3 + y^6$
39. $a^4 + 2a^2b + b^2$ 41. $a^3 + 3a^2 + 3a + 1$
43. $x^3 - 6x^2 + 12x - 8$ 45. $8x^3 + 12x^2 + 6x + 1$
47. $216u^3 - 108u^2 + 18u - 1$ 49. $20x + 36$
51. $x^2 + 2x - 35$ 53. $9x^2 - 30x + 25$
55. $25x^2 + 35x + 12$ 57. $16b^2 - 25$
59. $-20x^3 + 35x^2 - 10x$ 61. $64 - 48a + 12a^2 - a^3$
63. $x^3 + 6x^2 + 9x$ 65. $x^4 - 5x^2 + 4$ 67. $a^{2n} - b^{2n}$
69. (a) $x^2 + 4x + 4$ (b) $x^2 + 4x + 4$
71. (a) $4x^2 + 12x + 9$ (b) $4x^2 + 12x + 9$
73. (a) $6x^2 + 60x + 150$ (b) $x^3 + 15x^2 + 75x + 125$
75. (a) $1 + 2x + x^2$ (b) 1.21; the money increases by 1.21 times in 2 years if the interest rate is 10%.
77. (a) $1 - 2x + x^2$ (b) 0.25; if the chance of rain on each day is 50%, then there is a 25% chance that it will not rain on either day.
79. (a) $32z + 256$ (b) 2176; the area of an 8-foot-wide sidewalk around a 60 × 60 foot pool is 2176 ft^2.

Checking Basic Concepts 12.3 & 12.4 (p. 762)

1. (a) $-15x^3y^5$ (b) $-6x + 4x^2$ (c) $3a^3b - 6a^2b^2 + 3ab^3$
2. (a) $4x^2 + 9x - 9$ (b) $2x^4 - 2$ (c) $x^3 + y^3$
3. (a) $25x^2 - 4$ (b) $x^2 + 6x + 9$ (c) $4 - 28x + 49x^2$
(d) $t^3 + 6t^2 + 12t + 8$ 4. (a) $m^2 + 10m + 25$
(b) $m^2 + 10m + 25$

Section 12.5 (pp. 770–772)

1. $\frac{1}{a^n}$ 2. a^n 3. a^{m-n} 4. $\frac{b^m}{a^n}$ 5. $\left(\frac{b}{a}\right)^n$ 6. $1 \le b < 10$
7. (a) $\frac{1}{4}$ (b) 9 9. (a) 2 (b) $\frac{1}{1000}$ 11. (a) $\frac{1}{81}$ (b) $\frac{1}{8}$
13. (a) $\frac{1}{576}$ (b) 64 15. (a) 1 (b) 64 17. (a) 25 (b) $\frac{49}{4}$
19. (a) $\frac{1}{x}$ (b) $\frac{1}{a^4}$ 21. (a) $\frac{1}{x^2}$ (b) $\frac{1}{a^8}$ 23. (a) $\frac{y^3}{x^3}$ (b) $\frac{1}{x^3y^3}$
25. (a) $\frac{1}{16t^4}$ (b) $\frac{1}{(x+1)^7}$ 27. (a) a^8 (b) $\frac{1}{r^2t^6}$ 29. (a) $\frac{b^2}{a^4}$
(b) x^2 31. (a) a^{13} (b) $\frac{2}{z^3}$ 33. (a) $-\frac{2y^3}{3x^2}$ (b) $\frac{1}{x^3}$
35. (a) $2b$ (b) $\frac{a^3}{b^3}$ 37. (a) $\frac{x^7}{3y^8}$ (b) $\frac{4a^5}{b^6}$ 39. (a) y^5 (b) $2t^3$
41. (a) $\frac{3a^{10}}{8}$ (b) $\frac{1}{32b^9}$ 43. (a) x^2y^2 (b) a^6b^3
45. (a) $\frac{n^4}{72m^{11}}$ (b) $\frac{16x^{11}}{y^{13}}$ 47. (a) $\frac{b^2}{a^2}$ (b) $\frac{4v}{u}$
49. (a) $\frac{4}{9a^6b^6}$ (b) $\frac{16m^{14}}{25n^2}$ 51. $\frac{a^n}{a^m} = a^{n-m} = a^{-(m-n)} = \frac{1}{a^{m-n}}$
53. Thousand 55. Billion 57. Hundredth 59. -0.002
61. 45,000 63. 0.008 65. 0.000456 67. 39,000,000
69. $-500,000$ 71. 2×10^3 73. -5.67×10^2
75. 1.2×10^7 77. 4×10^{-3} 79. 8.95×10^{-4}
81. -5×10^{-2} 83. 1,500,000 85. -1.5 87. 2000
89. 0.002 91. (a) About 5.859×10^{12} mi
(b) About 2.5×10^{13} mi 93. About 9.2×10^9 mi
95. 1×10^{100} 97. (a) 1.5684×10^{13} (b) About $49,476

Section 12.6 (pp. 778–779)

1. $\frac{a}{d} + \frac{b}{d}$ 2. $\frac{a}{d} + \frac{b}{d} - \frac{c}{d}$ 3. term 4. False 5. False
6. 9; 4; 1 7. $x + 1$; $2x^2 - 2x + 1$; 4 8. $0x^2$ 9. $2x$
11. $z^3 + z^2$ 13. $\frac{a^2}{2} - 3$ 15. $\frac{1}{3y^2} + \frac{2}{y}$ 17. $\frac{4}{x} - 7x^2$
19. $\frac{2}{y} + \frac{1}{y^2}$ 21. $3x^3 - 1 + \frac{2}{x}$ 23. $4y^2 - 1 + \frac{2}{y}$
25. $3m^2 - 2m + 4$ 27. $2x + 1 + \frac{3}{x-2}$ 29. $x + 1$
31. $x^2 + 1 + \frac{-1}{x-1}$ 33. $x^2 + 3x - 1$
35. $x^2 - x + 2 + \frac{1}{4x+1}$ 37. $x^2 + 2x + 3 + \frac{8}{x-2}$
39. $3x^2 + 3x + 3 + \frac{5}{x-1}$ 41. $x + 3 + \frac{-x-2}{x^2+1}$
43. $x + 1$ 45. $x^2 - 2x + 4$ 47. They are the same.
49. $4x$
51. $x + 2$;

Checking Basic Concepts 12.5 & 12.6 (p. 779)

1. (a) $\frac{1}{81}$ (b) $\frac{1}{2x^7}$ (c) $\frac{b^8}{16a^2}$ 2. (a) z^5 (b) $\frac{y^6}{x^3}$ (c) $\frac{x^6}{27}$
3. (a) 4.5×10^4 (b) 2.34×10^{-4} (c) 1×10^{-2}
4. (a) 47,100 (b) 0.006 5. $5a - 3$
6. $3x + 2$; R: -2 7. $x^2 + 2x + 1$; R: $x + 1$
8. (a) 9.3×10^7 (b) 500 sec (8 min 20 sec)

Chapter 12 Review (pp. 783–785)

1. 125 2. -81 3. 4 4. 11 5. -5 6. 1 7. 6^5
8. 10^{12} 9. z^9 10. y^6 11. $30x^9$ 12. a^4b^4 13. 2^{10}
14. m^{20} 15. a^3b^3 16. x^8y^{12} 17. x^7y^{11} 18. 1
19. $(r-t)^9$ 20. $(a+b)^6$ 21. $\frac{9}{(x-y)^2}$ 22. $\frac{(x+y)^3}{8}$
23. $6x^3 - 10x^2$ 24. $12x^2 + 3x^4$ 25. 7; 6 26. 5; -1
27. Yes; 1; 1; 1 28. Yes; 4; 1; 3 29. Yes; 3; 2; 2
30. No 31. $5x^2 - x + 3$ 32. $-6x^2 + 3x + 7$
33. $3x + 4$ 34. $-2x^2 - 13$ 35. $-2x^2 + 9x + 5$

36. $3x^2 - 2x - 6$ **37.** $2a^3 - a^2 + 7a$ **38.** $-2x + 12$
39. $5y^2 - 3xy$ **40.** $4xy$ **41.** $-x^5$ **42.** $-r^3t^4$
43. $-6t + 15$ **44.** $2y - 12y^2$ **45.** $18x^5 + 30x^4$
46. $-x^3 + 2x^2 - 9x$ **47.** $-a^3b + 2a^2b^2 - ab^3$
48. $a^2 + 3a - 10$ **49.** $8x^2 + 13x - 6$
50. $-2x^2 + 3x - 1$ **51.** $2y^3 + y^2 + 2y + 1$
52. $2y^4 - y^2 - 1$ **53.** $z^3 + 1$ **54.** $4z^3 - 15z^2 + 13z - 3$
55. $z^2 + z$ **56.** $2x^2 + 4x$ **57.** $z^2 - 4$ **58.** $25z^2 - 81$
59. $1 - 9y^2$ **60.** $25x^2 - 16y^2$ **61.** $r^2t^2 - 1$
62. $4m^4 - n^4$ **63.** $x^2 + 2x + 1$ **64.** $16x^2 + 24x + 9$
65. $y^2 - 6y + 9$ **66.** $4y^2 - 20y + 25$
67. $16 + 8a + a^2$ **68.** $16 - 8a + a^2$
69. $x^4 + 2x^2y^2 + y^4$ **70.** $x^2y^2 - 4xy + 4$
71. $z^3 + 15z^2 + 75z + 125$ **72.** $8z^3 - 12z^2 + 6z - 1$
73. 3599 **74.** 396 **75.** $\frac{1}{9}$ **76.** $\frac{1}{9}$ **77.** 4 **78.** $\frac{1}{1000}$
79. 36 **80.** $\frac{1}{25}$ **81.** $\frac{9}{16}$ **82.** 1 **83.** $\frac{1}{z^2}$ **84.** $\frac{1}{y^4}$ **85.** $\frac{1}{a^2}$
86. $\frac{1}{x^5}$ **87.** $\frac{1}{4t^2}$ **88.** $\frac{1}{a^3b^6}$ **89.** $\frac{1}{y^3}$ **90.** x^4 **91.** $\frac{2}{x^3}$ **92.** $\frac{2x^4}{3y^3}$
93. $\frac{a^5}{b^5}$ **94.** $4t^4$ **95.** $\frac{n^7}{72m^{12}}$ **96.** $\frac{9x^{10}}{y^{10}}$ **97.** $\frac{y^2}{x^2}$ **98.** $\frac{2v}{3u}$
99. 600 **100.** 52,400 **101.** -0.0037 **102.** -0.06234
103. 1×10^4 **104.** 5.61×10^7 **105.** 5.4×10^{-5}
106. 1×10^{-3} **107.** 24,000,000 **108.** 0.2 **109.** $\frac{5x}{3} + 1$
110. $3b^2 - 2 + \frac{1}{b^2}$ **111.** $3x + 2 + \frac{4}{x-1}$
112. $3x - 4 + \frac{6}{3x+2}$ **113.** $x^2 - 3x - 1$
114. $x^2 + \frac{-1}{2x-1}$ **115.** $x - 1 + \frac{-2x+2}{x^2+1}$
116. $x^2 + 2x + 5$ **117. (a)** 60 bpm **(b)** 110 bpm
(c) It increases. **118.** $6xy$; 72 ft² **119.** $10z^2 + 25z$; 510 in²
120. x^4y^2 **121.** 2.64×10^9 hours
122. $\frac{4}{3}\pi x^3 + 8\pi x^2 + 16\pi x + \frac{32}{3}\pi$ **123. (a)** $96t - 16t^2$
(b) 128; after 2 seconds the ball is 128 ft high.
124. (a) $600L - L^2$ **(b)** 27,500; a rectangular building with a perimeter of 1200 ft and a side of length 50 ft has an area of 27,500 ft².
125. (a) $x^2 + 10x + 25$ **(b)** $x^2 + 10x + 25$
126. (a) $16x$ **(b)** 1600

Chapter 12 Test (p. 786)

1. (a) -1 **(b)** -81 **2. (a)** -6 **(b)** $\frac{1}{64}$ **(c)** 8 **(d)** -3
3. 3; 2; 3 **4.** $x^3 - 4x + 8$ **5.** $4x + 6$ **6.** $-3y^3 - 2y + 1$
7. $x^2 + x - 7$ **8.** $4a^3 + 2ab$ **9.** $24y^{11}$ **10.** a^5b^8 **11.** x^4
12. $\frac{a^3}{b^6}$ **13.** $a^3b - ab^3$ **14.** $\frac{4}{9a^4b^6}$ **15.** $\frac{2y^3}{x}$ **16.** $\frac{16}{(a+b)^4}$
17. $12x^5 - 18x^3 + 3x^2$ **18.** $2z^2 - 2z - 12$
19. $49y^4 - 9$ **20.** $9x^2 - 12x + 4$
21. $m^3 + 9m^2 + 27m + 27$ **22.** $y^3 - y + 6$ **23.** 6396
24. 0.0061 **25.** 5.41×10^3 **26.** $3x - 2 + \frac{1}{x}$
27. $x^2 - x + 1 + \frac{-1}{x+2}$ **28. (a)** $20t$ **(b)** $2t + 2000$
(c) $18t - 2000$; profit from selling t tickets
29. $12x^2$; 1200 ft² **30.** $3x^3 + 27x^2 + 54x$
31. (a) $88t - 16t^2$ **(b)** 120; after 3 seconds, the ball is 120 ft high.

Chapters 1–12 Cumulative Review (p. 787)

1. (a) 8 **(b)** 8 **2. (a)** 29 **(b)** $\frac{3}{2}$ **3. (a)** 7 **(b)** No solutions
4. (a) Infinitely many solutions **(b)** -2 **5.** 68 mph
6. (a) $\frac{21}{50}$ **(b)** $\frac{19}{250}$

7. 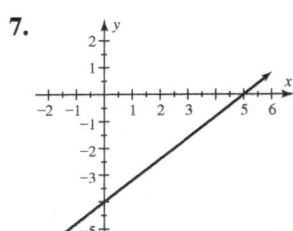 **8.**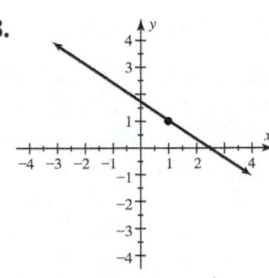

9. $y = -2x + 1$ **10.** $(2, 0); (0, -3)$ **11.** $y = \frac{1}{2}x - 4$
12. $y = 3x + 1$ **13.** No solutions **14.** $(2, -1)$
15. Infinitely many solutions **16.** $(0, -6)$

17. 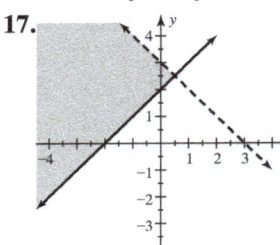 **18.**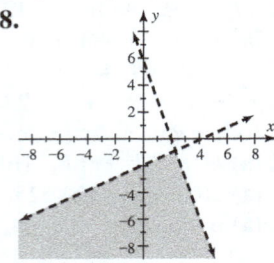

19. (a) $15x^5$ **(b)** $x^{10}y^7$ **20. (a)** $2x^2 - x + 3$
(b) $12a^3 + a - 5$ **21. (a)** $2x^2 - 11x - 21$
(b) $y^3 - 10y - 3$ **(c)** $16x^2 - 49$ **(d)** $25a^2 + 30a + 9$
22. (a) $\frac{1}{x}$ **(b)** $\frac{x^9}{8}$ **(c)** $\frac{x^4}{2y^2}$ **(d)** $\frac{x^7}{y^8}$ **23.** 2.4×10^{10}
24. 0.000000471 **25. (a)** $4x^2 - 1$ **(b)** $2x - 5 + \frac{1}{x+3}$
26. -25% **27.** 200 mL **28. (a)** $2x^2 + 10x$
(b) $x^2 + 7x + 10$ **(c)** $2x^2 + 4x$ **(d)** $10x^2 + 42x + 20$

13 Factoring Polynomials and Solving Equations

Section 13.1 (pp. 796–798)

1. factor **2.** a **3.** multiplying **4.** prime; multiplication
5. greatest common factor (GCF) **6.** grouping
7. $1, 2, x, 2x$ **8.** $2x$
9. $2(x + 2)$;

	2x	4
2		
	x	2

11. $z(z + 4)$;

	z^2	$4z$
z		
	z	4

13. $3y(y + 4)$;

	$3y^2$	$12y$
3y		
	y	4

15. $3x(x + 3)$ **17.** $2y^2(2y - 1)$ **19.** $2z(z^2 + 4z - 2)$
21. $3xy(2x - y)$ **23.** $6x$; $6x(1 - 3x)$
25. $4y^2$; $4y^2(2y - 3)$ **27.** $3z$; $3z(2z^2 + z + 3)$
29. x^2; $x^2(x^2 - 5x - 4)$

31. $5y^2$; $5y^2(y^3 + 2y^2 - 3y + 2)$ 33. x; $x(y + z)$
35. ab; $ab(b - a)$ 37. $5x^2y^3$; $5x^2y^3(y + 2x)$
39. ab; $ab(a + b + 1)$ 41. $(x - 2)(x + 1)$
43. $(z + 4)(z + 5)$ 45. $2x(2x^2 - 1)(x - 5)$
47. $(x^2 + 3)(x + 2)$ 49. $(y^2 + 1)(2y + 1)$
51. $(5z^2 + 2)(3z - 1)$ 53. $(4t^2 + 3)(t - 5)$
55. $(3r^2 - 2)(3r + 2)$ 57. $(7x^2 - 2)(x + 3)$
59. $(y^2 - 2)(2y - 7)$ 61. $(z^2 - 7)(z - 4)$
63. $(x^3 + 2)(2x - 3)$ 65. $(x + y)(a + b)$
67. $3(x^2 + 1)(x + 2)$ 69. $2y(3y^2 - 1)(y - 4)$
71. $x^2(x^2 - 3)(x + 2)$ 73. $2x^2(x^2 - 3)(2x + 1)$
75. $xy(x + y)(x - 2)$ 77. $2x^2y^2(x + 2)(y - x)$
79. $a\left(x^2 + \frac{b}{a}x + \frac{c}{a}\right)$ 81. (a) $16t$ (b) $16t(5 - t)$
83. $2x^2 - 4x$; $2x(x - 2)$ 85. $8y^2 - xy$; $y(8y - x)$
87. (a) 168 in^3 (b) $4x(x^2 - 15x + 50)$

Section 13.2 (pp. 804–806)

1. 1 2. $x^2 + (m + n)x + mn$; $m + n$; mn
3. mn; $m + n$ 4. prime 5. $x^2 + 7x + 10$
7. $x^2 - 9x + 18$ 9. $x^2 + x - 56$ 11. 4, 7 13. 3, −10
15. −5, 10 17. −7, −4 19. $(x + 1)(x + 2)$
21. $(y + 2)(y + 2)$ 23. Prime 25. $(x + 3)(x + 5)$
27. $(m + 4)(m + 9)$ 29. $(n + 10)(n + 10)$
31. $(x - 1)(x - 5)$ 33. $(y - 3)(y - 4)$
35. $(z - 5)(z - 8)$ 37. $(a - 7)(a - 9)$ 39. Prime
41. $(b - 5)(b - 25)$ 43. $(x - 5)(x + 18)$
45. $(m - 5)(m + 9)$ 47. Prime
49. $(n - 10)(n + 20)$ 51. $(x - 1)(x + 23)$
53. $(a - 4)(a + 8)$ 55. $(b + 4)(b - 5)$ 57. Prime
59. $(x + 8)(x - 9)$ 61. $(y + 2)(y - 17)$
63. $(z + 6)(z - 11)$ 65. $5(x - 4)(x + 2)$
67. $-3(m + 4)(m - 1)$ 69. $y(y - 2)(y - 5)$
71. $-x(x + 5)(x - 3)$ 73. $3a(a + 1)(a + 6)$
75. $-2x(x^2 - 3x + 4)$ 77. $2m^2(m - 7)(m + 2)$
79. $-3x^2(x + 1)(x - 2)$ 81. $(x + 1)(x + 5)$
83. $(x - 1)(x - 3)$ 85. $(6 - x)(2 + x)$
87. $(8 + x)(4 - x)$ 89. $(x + 1)(x + k)$
91. $x + 1$; 93. $x + 1, x + 2$;

95. $x + 1$ 97. $(x + 2)(x + 6)$ 99. $(x + 3)(x + 2)$
101. $(x - 5)(x - 6)$ 103. $(x + 6)(x - 2)$

Checking Basic Concepts 13.1 & 13.2 (p. 806)

1. $4x$ 2. $6z^2(2z - 3)$ 3. (a) $(6y + 5)(y - 2)$
(b) $(x^2 + 5)(2x + 1)$ (c) $4(z^2 + 1)(z - 3)$
4. (a) $(x + 2)(x + 4)$ (b) $(x - 7)(x + 6)$ (c) Prime
(d) $4a(a + 2)(a + 3)$ 5. $(x + 5)(x + 5)$

Section 13.3 (pp. 813–814)

1. ac; b 2. ax^2; c 3. +; + 4. +; − 5. −; −
6. −; + 7. 3; x 8. x; $4x$ 9. 1; 3 10. 1; $2x$
11. $3x$; 1 12. $6x$; x 13. 5; 3 14. $6x$; 8
15. $(x + 3)(2x + 1)$ 17. Prime 19. $(x + 1)(3x + 1)$
21. $(2x + 3)(3x + 1)$ 23. $(x - 2)(5x - 1)$
25. $(y - 1)(2y - 5)$ 27. Prime 29. $(z - 5)(7z - 2)$
31. $(t - 3)(3t + 2)$ 33. $(3r + 2)(5r - 3)$
35. $(3m - 4)(8m + 3)$ 37. $(5x - 1)(5x + 2)$
39. $(x + 2)(6x - 1)$ 41. Prime
43. $(3n + 1)(7n - 1)$ 45. $(2y + 3)(7y + 1)$
47. $(4z - 3)(7z - 1)$ 49. $(3x - 2)(10x - 3)$
51. Prime 53. $(2t + 3)(9t - 2)$
55. $3(2a - 1)(2a + 3)$ 57. $z(2z + 3)(5z + 2)$
59. $3x(4x - 3)(2x - 1)$ 61. $2x^2(4x^2 - 3x + 1)$
63. $7x^2(2x + 1)(2x + 3)$ 65. $(3x + 1)(x + k)$
67. $(7x + 1)(x + 2)$ 69. $(2x - 1)(x - 2)$
71. $-(4x + 3)(2x - 1)$ 73. $-(x + 5)(2x - 3)$
75. $-(x - 3)(5x + 1)$
77. $3x + 2$ by $2x + 1$;

	$2x$	$6x^2$	$4x$
	1	$3x$	2
		$3x$	2

79. $(2x + 1)(x + 3)$

Section 13.4 (pp. 820–821)

1. $(a - b)(a + b)$ 2. $6x$; $7y$ 3. False 4. False
5. $(a + b)^2$ 6. $(a - b)^2$ 7. $6x$ 8. $20rt$
9. $(a + b)(a^2 - ab + b^2)$ 10. $(a - b)(a^2 + ab + b^2)$
11. $2x$; $3y$ 12. x; 1 13. −; + 14. +; −
15. $(x - 1)(x + 1)$ 17. $(z - 10)(z + 10)$
19. $(2y - 1)(2y + 1)$ 21. $(6z - 5)(6z + 5)$
23. $(3 - x)(3 + x)$ 25. $(1 - 3y)(1 + 3y)$
27. $(2a - 3b)(2a + 3b)$ 29. $(6m - 5n)(6m + 5n)$
31. $(9r - 7t)(9r + 7t)$ 33. $(x + 4)^2$ 35. Not possible
37. $(x - 3)^2$ 39. $(3y + 1)^2$ 41. $(2z - 1)^2$
43. Not possible 45. $(3x + 5)^2$ 47. $(2a - 9)^2$
49. $(x + y)^2$ 51. $(r - 5t)^2$ 53. Not possible
55. $(z + 1)(z^2 - z + 1)$ 57. $(x + 4)(x^2 - 4x + 16)$
59. $(y - 2)(y^2 + 2y + 4)$ 61. $(n - 1)(n^2 + n + 1)$

63. $(2x + 1)(4x^2 - 2x + 1)$
65. $(m - 4n)(m^2 + 4mn + 16n^2)$
67. $(2x + 5y)(4x^2 - 10xy + 25y^2)$
69. $4(x - 2)(x + 2)$ 71. $2(y - 7)^2$
73. $5(z + 2)(z^2 - 2z + 4)$ 75. $xy(x - y)(x + y)$
77. $2m(m^2 - 5m + 9)$
79. $7x^2(10x - 3y)(10x + 3y)$
81. $2(2a + b)(4a^2 - 2ab + b^2)$ 83. $4b^2(b + 3)^2$
85. $4(5r - 2t)(25r^2 + 10rt + 4t^2)$
87. $2x + 3$;

	$2x$	3
$2x$	$4x^2$	$6x$
3	$6x$	9

89. $(x - 5)(x + 5)$

Checking Basic Concepts 13.3 & 13.4 (p. 821)

1. (a) $(x - 4)(2x + 3)$ (b) $(2x + 7)(3x - 2)$
2. (a) Prime (b) $2y(3y + 1)(y - 2)$
3. $(3x + 2)(x + 3)$ 4. (a) $(z - 8)(z + 8)$
(b) $(3r - 2t)(3r + 2t)$ 5. (a) $(x + 6)^2$
(b) $(3a - 2b)^2$ 6. (a) $(m - 3)(m^2 + 3m + 9)$
(b) $(5n + 3)(25n^2 - 15n + 9)$
7. (a) $4(2x - 1)(2x + 1)$ (b) $3y(y + 2)(y^2 - 2y + 4)$

Section 13.5 (pp. 826–827)

1. Greatest common factor 2. GCF 3. Grouping
4. squares; cubes; cubes 5. No; a sum of squares cannot be factored. 6. Yes; a sum of cubes can be factored.
7. square; grouping; FOIL 8. completely 9. $2(2x - 1)$
11. $2(y^2 - 2y + 2)$ 13. $(z - 2)(z + 2)$
15. $(a + 2)(a^2 - 2a + 4)$ 17. $(2b - 3)^2$
19. Not possible 21. $(x^2 + 5)(x - 1)$
23. $(y - 4)(y - 1)$ 25. $(x - 3)(x + 3)(x + 4)$
27. $8(a - 2)(a^2 + 2a + 4)$
29. $2x(3x^2 + 1)(2x - 3)$
31. $2t(3t + 2)(9t^2 - 6t + 4)$
33. $2(r - 1)(r + 1)(r + 3)$ 35. $3z^2(2z + 3)(z - 5)$
37. $2b^2(3b - 1)(2b - 1)$ 39. $6z(1 - 2z)(1 + 2z)$
41. $3y(x - 5)^2$ 43. $(3m - 2n)(9m^2 + 6mn + 4n^2)$
45. $3(x - 2)(x + 2)(x - 1)(x^2 + x + 1)$
47. $(5a + 3)(a - 6)$ 49. $3r(t + 5)(t + 6)$
51. $(3b^2 + 4)(3b + 2)$ 53. $2n(3n^2 + n - 5)$
55. $4(x - 3y)(x + 3y)$ 57. $2a(a - 4)^2$
59. $4x(2y + 1)(4y^2 - 2y + 1)$
61. $2b(2b - 1)(2b + 1)(b + 3)$ 63. $3x + 1$

Section 13.6 (pp. 833–835)

1. 0; 0 2. No; one side of the equation must be zero.
3. $2x = 0$; $x + 6 = 0$ 4. Subtract $4x$ from each side. 5. Apply the zero-product property. 6. solving
7. zero 8. 2 9. $ax^2 + bx + c = 0$ with $a \neq 0$
10. $x^2 - 6x + 1 = 0$ 11. descending 12. 0 13. 0
15. $-8, 0$ 17. $1, 2$ 19. $-\frac{5}{6}, 7$ 21. $\frac{1}{3}, \frac{3}{7}$ 23. $0, 5, 8$
25. $0, 1$ 27. $0, 5$ 29. $-\frac{3}{2}, 0$ 31. $-1, 1$ 33. $-\frac{1}{2}, \frac{1}{2}$
35. $-2, -1$ 37. $5, 7$ 39. $-2, \frac{1}{2}$ 41. $-\frac{5}{2}, -\frac{2}{3}$
43. $-5, 5$ 45. $0, 5$ 47. $-3, 0$ 49. $-1, 6$
51. $-\frac{7}{2}, 2$ 53. $-2, 1$ 55. $-3, \frac{1}{2}$ 57. $-\frac{1}{2}$
59. $-1, -\frac{2}{3}$ 61. 12 ft 63. 4 65. 4
67. (a) 6 sec
(b)

Time (t)	0	1	2	3	4	5	6
Height (h)	0	80	128	144	128	80	0

; 3 sec

69. (a) 81.8 ft; 327.3 ft; when the speed doubles, the braking distance quadruples. (b) About 19 mph
(c) About 19 mph; yes 71. 40 by 50 pixels

Checking Basic Concepts 13.5 & 13.6 (p. 835)

1. (a) $9(a^2 - 2a + 3)$ (b) $7x(y^2 + 4)$
2. (a) $2z^2(3z - 2)(z - 4)$ (b) $2r^2(t - 3)(t + 3)$
3. (a) $4x(3x - 2)^2$ (b) $3(2b - 3)(4b^2 + 6b + 9)$
4. (a) $0, \frac{3}{2}$ (b) $-1, \frac{3}{5}$ 5. $-3, 1$ 6. After $\frac{11}{2}$ sec

Section 13.7 (pp. 839–841)

1. GCF 2. zero-product 3. factors
4. $(x^2 - y^2)^2$ 5. $(z^2 + 1)(z^2 + 2)$
6. No; $x^4 - 1 = (x - 1)(x + 1)(x^2 + 1)$
7. Yes; $x^2 + 1$ cannot be factored further.
8. Subtract x from each side. 9. Factor out x^2. 10. One
11. $5(x - 3)(x + 2)$ 13. $-4(y + 2)(y + 6)$
15. $-10(z + 5)(2z + 1)$ 17. $-4(t + 3)(7t - 5)$
19. $r(r - 1)(r + 1)$ 21. $3x(x - 2)(x + 3)$
23. $12z(2z - 1)(3z + 2)$ 25. $x^2(x - 2)(x + 2)$
27. $t^2(t - 1)(t + 2)$ 29. $(x^2 - 3)(x^2 - 2)$
31. $(x^2 + 3)(2x^2 + 1)$ 33. $(y^2 + 3)^2$
35. $(x^2 - 3)(x^2 + 3)$ 37. $(x - 3)(x + 3)(x^2 + 9)$
39. $z^3(z + 1)^2$ 41. $(x + y)(2x - y)$
43. $(a + b)^2(a - b)^2$ 45. $x(x + y)(x - y)$
47. $x(2x + y)^2$ 49. (a) $x(x - 2)(x + 2)$ (b) $-2, 0, 2$
51. (a) $2y(y - 6)(y + 3)$ (b) $-3, 0, 6$
53. (a) $(x^2 + 4)(x - 1)$ (b) 1 55. $-8, -3$
57. $-1, 3$ 59. $-1, 0, 4$ 61. $-6, 0, 4$ 63. $-6, 0, 6$
65. $-7, 0, 1$ 67. $-3, -2, 2, 3$ 69. $-1, 1$
71. $-3, 0, 3$ 73. $-1, 1, 2$ 75. 5

77. (a) $x < 7.5$ in. because the width is 15 in.
(b) $300 - 4x^2$ **(c)** 2.5 in. **79.** 8 **81.** 18.6 trillion ft^3
83. Factoring is very difficult (answers may vary).

Checking Basic Concepts 13.7 (p. 841)

1. (a) $3(x - 4)(x + 2)$ **(b)** $-5(2y + 1)(y - 1)$
2. (a) $(z^2 - 5)(z^2 + 5)$ **(b)** $7(t - 1)(t + 1)(t^2 + 1)$
3. (a) $(x - 2)^2(x + 2)^2$ **(b)** $y(y + 10)(2y - 3)$
4. $-4, 0, 3$ **5.** 3

Chapter 13 Review (pp. 844–846)

1. 5; $5(x - 3)$ **2.** y; $y(y + 2)$ **3.** $4z^2$; $4z^2(2z - 1)$
4. $3x^2$; $3x^2(2x^2 + x - 4)$ **5.** $3y$; $3y(3x + 5z^2)$
6. a^2b^2; $a^2b^2(b + a)$ **7.** $(x - 3)(x + 2)$
8. $y(y + 3)(x - 5)$ **9.** $(z^2 + 5)(z - 2)$
10. $(t^2 + 8)(t + 1)$ **11.** $(x^2 + 6)(x - 3)$
12. $(x - y)(a + b)$ **13.** $x^2(x^2 - 2)(x + 3)$
14. $2y(y^2 + 1)(y + 3)$ **15.** 4, 5 **16.** $-3, 7$
17. $-9, -4$ **18.** $-25, 4$ **19.** $(x - 4)(x + 3)$
20. $(x + 4)(x + 6)$ **21.** $(x - 2)(x + 8)$
22. $(x - 7)(x + 6)$ **23.** Prime **24.** Prime
25. $(x - 1)(x + 3)$ **26.** $(x + 10)(x + 12)$
27. $2x(x - 2)(x + 5)$ **28.** $x^2(x + 4)(x - 7)$
29. $(2 - x)(5 - x)$ **30.** $(6 - x)(4 + x)$
31. $-2(x + 5)(x - 3)$ **32.** $-x(x + 10)(x - 1)$
33. $(3x - 1)(3x + 2)$ **34.** $(x - 1)(2x + 5)$
35. $(x + 3)(3x + 5)$ **36.** $(5x - 1)(7x + 1)$
37. Prime **38.** Prime **39.** $(3x + 1)(8x - 5)$
40. $(x + 9)(4x - 3)$ **41.** $3x(2x + 7)(2x + 1)$
42. $2x^2(x + 3)(4x - 5)$ **43.** $(3 - 2x)(4 + x)$
44. $(1 - 2x)(1 + 5x)$ **45.** $(z - 2)(z + 2)$
46. $(3z - 8)(3z + 8)$ **47.** $(6 - y)(6 + y)$
48. $(10a - 9b)(10a + 9b)$ **49.** $(x + 7)^2$
50. $(x - 5)^2$ **51.** $(2x - 3)^2$ **52.** $(3x + 8)^2$
53. $(2t - 1)(4t^2 + 2t + 1)$
54. $(3r + 2t)(9r^2 - 6rt + 4t^2)$
55. $2x(x - 5)(x + 5)$ **56.** $3(2x + 3)(4x^2 - 6x + 9)$
57. $2x(x + 7)^2$ **58.** $2x(x - 4)(x^2 + 4x + 16)$
59. $4(3x - 2)$ **60.** $3x^2(2x + 3)$
61. $3(3y^2 - 2y + 2)$ **62.** $y(z - 3)(z + 3)$
63. $x(x - 2)(x + 2)(x + 7)$ **64.** $3x(2x + 3)^2$
65. $3a(b - 2)(b^2 + 2b + 4)$ **66.** $5x(x^2 + 4)$

67. $6x(2x - y)(2x + y)$ **68.** $y(x + 3)(x^2 - 3x + 9)$
69. $m = 0$ or $n = 0$ **70.** 0 **71.** $-9, \frac{3}{4}$ **72.** $-\frac{6}{5}, \frac{1}{4}$
73. 0, 1, 2 **74.** 0, 7 **75.** $-8, 8$ **76.** $-7, -2$
77. $-2, 3$ **78.** $-\frac{3}{2}, \frac{2}{5}$ **79.** $-4, 18$ **80.** $-2, \frac{5}{2}$
81. $5(x - 5)(x + 2)$ **82.** $-3(x - 3)(x + 5)$
83. $y(y - 2)(y + 2)$ **84.** $3y(y - 1)(y + 3)$
85. $2z^2(z + 2)(z + 5)$ **86.** $8z^2(z - 2)(z + 2)$
87. $(x^2 - 3)^2$ **88.** $(x - 3)(x + 3)(2x^2 + 3)$
89. $(a + 5b)^2$ **90.** $x(x + y)(x - y)$ **91.** $-\frac{1}{2}, 5$
92. $0, \frac{5}{2}, 3$ **93.** $-5, 0, 5$ **94.** $0, 3, 4$ **95.** $-2, 2$
96. $-4, 4$ **97.** -4 **98.** $-1, 1$
99. $3x + 7$; **100.** $x + 1$ by $x + 5$;

	$3x$	7
$3x$	$9x^2$	$21x$
7	$21x$	49

	x	5
x	x^2	$5x$
1	x	5

101. $x + 1$ **102.** $(x + 1)(x + 3)$
103. $(2x + 3)(x + 6)$ **104.** 2 **105.** 8 ft by 15 ft
106. After 2 sec and 3 sec **107. (a)** 390 ft **(b)** 15 mph
(c) 15 mph; yes **108. (a)** $10,000 **(b)** $50 or $150
(c) $50 or $150; yes **109. (a)** $2000 - 4x^2$ **(b)** 5 in.
110. $x + 2$

Chapter 13 Test (pp. 846–847)

1. $4xy$; $4xy(x - 5y + 3)$ **2.** $3a^2b^2$; $3a^2b^2(3a + 1)$
3. $(y + z)(a + b)$ **4.** $(x^2 - 5)(3x + 1)$
5. $(y - 2)(y + 6)$ **6.** $(2x + 5)^2$ **7.** $(z - 4)(4z - 3)$
8. $(t - 7)(2t - 3)$ **9.** Prime **10.** Prime
11. $3x(x + 1)(2x - 1)$ **12.** $2(z - 3)(z + 3)(z^2 + 3)$
13. $4y(3y - 5)(3y + 5)$
14. $7x(x + 2)(x^2 - 2x + 4)$ **15.** $a^2(4a + 3)^2$
16. $2(b - 2)(b + 2)(b^2 + 4)$ **17.** $-4, 4$ **18.** $-4, 5$
19. $\frac{4}{3}$ **20.** $-6, 11$ **21.** $-3, 0, 3$ **22.** $-2, -1, 1, 2$
23. $3x + 5$ **24.** $(x + 2)(x + 3)$ **25. (a)** 275 ft
(b) 33 mph **26.** 1 sec and 2 sec

Chapters 1–13 Cumulative Review (pp. 847–848)

1. $\frac{3}{7}$ **2.** $\frac{7}{10}$ **3.** 17 **4.** $-\frac{7}{2}$
5. 1;

x	-2	-1	0	1	2
$2x + 3$	-1	1	3	5	7

6. $3n - 5 = n - 7$; -1 **7.** $\frac{57}{1000}$; 0.057
8. $W = \frac{P - 2L}{2}$ **9.** $z > 2$

A-36 ANSWERS TO SELECTED EXERCISES

10. 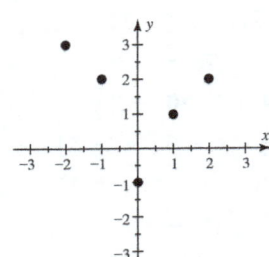   **11.** *x*-int: $\left(\frac{2}{3}, 0\right)$; *y*-int: $(0, -2)$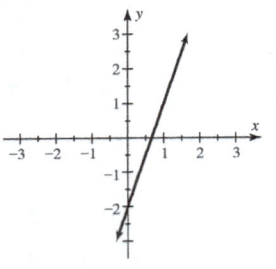

12. *y*-int: $(0, -2)$

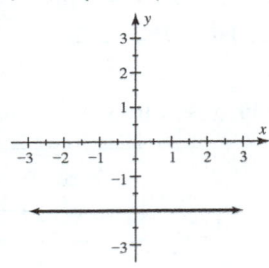

13. $y = -\frac{3}{2}x + \frac{7}{2}$ **14.** $y = \frac{4}{3}x + \frac{11}{3}$
15. $(-1, 0)$; $(0, -2)$; $y = -2x - 2$ **16.** $(1, 2)$
17. $(1, -1)$ **18.** $(-2, 5)$

19. 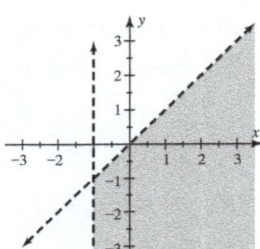   **20.**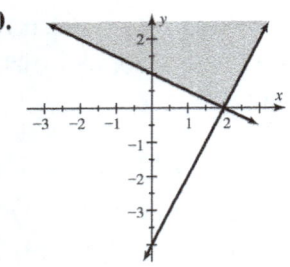

21. -16 **22.** 1   **23.** $\frac{x^{10}}{y^4}$ **24.** $-14x^5 + 21x^4$ **25.** $\frac{1}{a^2}$
26. $\frac{1}{4t^6}$ **27.** $\frac{1}{x^2y^5}$ **28.** $32x^5y^{10}$ **29.** $2x^2 + 4x$
30. $3x + \frac{-4x + 1}{x^2 + 1}$ **31.** $(x - 4)(x + 7)$
32. $(2y + 3)(3y - 4)$ **33.** $(5x - 2y)(5x + 2y)$
34. $(8x - 1)^2$ **35.** $(3t - 2)(9t^2 + 6t + 4)$
36. $-4(x - 3)(x + 2)$ **37.** $-5, 0, 5$ **38.** $-2, 1$
39. $-\frac{7}{2}, 0, \frac{7}{2}$ **40.** $-3, 3$ **41. (a)** $10x + 8x$ or $18x$
(b) 50 min   **42.** $6\frac{11}{12}$ mi   **43.** 25 min skiing; 35 min running
44. (a) $C = 0.25x + 20$ **(b)** 320 mi
45. (a) $2x + y = 180$; $2x - y = 20$ **(b)** $(50, 80)$, the angles are 50°, 50°, and 80°.   **46.** $18xy$; 108 yd²
47. $x + 6$ **48. (a)** 4 sec   **(b)** After 1 sec and 3 sec

14 Rational Expressions

Section 14.1 (pp. 857–860)

1. $\frac{P}{Q}$; polynomials   **2.** Yes; both x and $2x^2 + 1$ are polynomials.   **3.** denominator   **4.** a **5.** $\frac{a}{b}$ **6.** rational   **7.** $-\frac{3}{7}$
9. $-\frac{4}{9}$ **11.** $-\frac{1}{4}$ **13.** Undefined   **15.** Undefined   **17.** -1
19. 1

21.

x	-2	-1	0	1	2
$\frac{x}{x+1}$	2	—	0	$\frac{1}{2}$	$\frac{2}{3}$

23.

x	-2	-1	0	1	2
$\frac{3x}{2x^2+1}$	$-\frac{2}{3}$	-1	0	1	$\frac{2}{3}$

25. 0   **27.** 3   **29.** $-\frac{4}{5}$ **31.** None   **33.** $-5, 5$ **35.** $-3, -2$
37. $1, \frac{5}{2}$ **39.** $\frac{2}{3}$ **41.** $\frac{1}{2}$ **43.** $-\frac{2}{5}$ **45.** $-\frac{1}{3}$ **47. (a)** $\frac{1}{2}$ **(b)** $\frac{1}{2}$
49. (a) -1 **(b)** -1 **51.** $\frac{1}{2x^2}$ **53.** $\frac{4y}{3x}$ **55.** $\frac{1}{2}$ **57.** $\frac{3}{5}$
59. $\frac{x+1}{x+6}$ **61.** $\frac{5y+3}{y+2}$ **63.** -1 **65.** -1 **67.** $-\frac{1}{3}$ **69.** $-\frac{1}{2}$
71. 1   **73.** $-\frac{3x+5}{3x-5}$ **75.** $\frac{n-1}{n-5}$ **77.** $\frac{x}{6}$ **79.** $\frac{z-2}{z-3}$
81. $\frac{x+4}{3x+2}$ **83.** $\frac{1}{3x-2}$ **85.** 1   **87.** $-\frac{1}{2}$ **89.** -1
91. (a) Equation   **(b)** 6   **93. (a)** Expression   **(b)** $\frac{1}{x+1}$
95. (a) Expression   **(b)** $x - 2$ **97. (a)** Equation   **(b)** 8

99. (a)

x	-4	-3	-2	-1	0	1	2	3	4
y	-0.2	-0.25	$-0.\overline{3}$	-0.5	-1	—	1	0.5	$0.\overline{3}$

(b) 1

(c)–(e)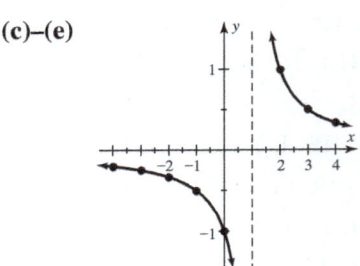

101. (a)

x	-4	-3	-2	-1	0	1	2	3	4
y	$1.\overline{3}$	1.5	2	—	0	0.5	$0.\overline{6}$	0.75	0.8

(b) -1 **(c)–(e)**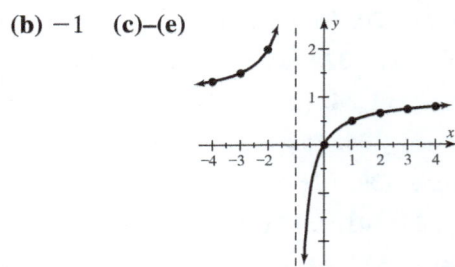

103. (a) $\frac{1}{2}$; when traffic arrives at a rate of 3 vehicles per minute, the average wait is $\frac{1}{2}$ minute.

(b)

x	2	4	4.5	4.9	4.99
T	$\frac{1}{3}$	1	2	10	100

As x nears 5 vehicles per minute, a small increase in x increases the wait dramatically.

105. (a)

x	0	12	36	72
P	0.5	4.83	6.07	6.5

(b) 50   **(c)** The population increased quickly at first, but then leveled off.
107. (a) 6 hr   **(b)** $\frac{M}{60}$ **109. (a)** $x = 5$ **(b)** As the average rate nears 5 cars per minute, a small increase in x increases the wait dramatically.   **111.** $\frac{1}{2}$ **113. (a)** $\frac{3}{n}$ **(b)** $\frac{n-3}{n}$; $\frac{97}{100}$; there is a 97% chance that a winning ball will not be drawn.

Section 14.2 (pp. 865–867)

1. numerators; denominators   **2.** reciprocal   **3.** $\frac{AC}{BD}$ **4.** $\frac{AD}{BC}$
5. $\frac{2}{5}$ **7.** $\frac{12}{7}$ **9.** $\frac{2}{3}$ **11.** $\frac{2}{11}$ **13.** 4   **15.** $\frac{8}{27}$ **17.** 10   **19.** $\frac{12}{5}$

21. 1   **23.** $\frac{z+1}{z+4}$ **25.** $\frac{2}{3}$ **27.** $\frac{x+2}{x-2}$ **29.** $\frac{8(x+1)}{x^2}$
31. $\frac{x-3}{x}$ **33.** $\frac{z+3}{z-7}$ **35.** 1   **37.** $(t+1)(t+2)$
39. $\frac{x(x+4)}{x^2+4}$ **41.** $\frac{z-1}{z+2}$ **43.** $\frac{y}{y-1}$ **45.** $\frac{x+1}{x-3}$ **47.** $\frac{x-3}{x-1}$
49. $\frac{2}{2x+3}$ **51.** -2 **53.** $\frac{z-1}{z+1}$ **55.** $\frac{y+2}{2y+1}$ **57.** $\frac{4(t-1)}{t^2+1}$
59. $\frac{y-3}{y-5}$ **61.** $2x$ **63.** $\frac{2z+1}{z+5}$ **65.** $\frac{1}{t+6}$ **67.** $\frac{2a+3b}{a+b}$
69. $x-y$ **71.** 1   **73.** (a) $D = \frac{30}{x}$ (b) 300 ft; 75 ft; stopping distance on dry pavement is one-fourth as long.
75. (a) $\frac{1}{n+1}$ (b) $\frac{1}{100}$ **77.** (a) $R = \frac{350(x-3)^2 + 50}{260(x-3)^2 + 150}$
(b) Six years after startup, Google's revenues were about 1.6 times Facebook's revenues.   **79.** (a) $54; it costs $54, on average, to make each camera when 5000 are produced.
(b) $T = 50x + 20{,}000$ (c) 270,000; it costs $270,000 to make 5000 cameras.

Checking Basic Concepts 14.1 & 14.2 (p. 867)

1. Undefined; $\frac{3}{8}$ **2.** (a) $\frac{2x}{5y}$ (b) 5   (c) $\frac{x+2}{x+4}$
3. (a) $\frac{4}{9}$ (b) $\frac{2}{x-1}$ **4.** (a) $\frac{5z}{6}$ (b) $x+1$
5. (a)

x	0.5	1.0	1.5	1.9
T	$\frac{2}{3}$	1	2	10

(b) As x nears 2 customers per minute, a small increase in x increases the wait dramatically.

Section 14.3 (pp. 873–875)

1. numerators; denominators   **2.** numerators; denominators
3. $\frac{A+B}{C}$ **4.** $\frac{A-B}{C}$ **5.** 1   **7.** $\frac{6}{5}$ **9.** 1   **11.** $\frac{3}{7}$ **13.** $\frac{1}{2}$
15. $\frac{1}{2}$ **17.** $\frac{2}{3}$ **19.** $\frac{3}{x}$ **21.** $\frac{1}{2}$ **23.** 3   **25.** 1   **27.** $\frac{4z}{4z+3}$
29. 2   **31.** 1   **33.** $\frac{1}{x-1}$ **35.** $z-1$ **37.** 2   **39.** $y-1$
41. $\frac{x}{2}$ **43.** $z-2$ **45.** $x-3$ **47.** $\frac{7n}{2n^2-n+5}$ **49.** $\frac{6}{x+3}$
51. $\frac{9}{ab}$ **53.** $\frac{1}{x+y}$ **55.** $\frac{10}{x-y}$ **57.** 0   **59.** $\frac{b}{2a}$ **61.** 1
63. $a-b$ **65.** $\frac{1}{x+2}$ **67.** $\frac{1}{3x-5}$ **69.** $2x-3y$
71. It equals 7.   **73.** (a) $\frac{14}{n+1}$ (b) $\frac{7}{50}$; when there are 100 batteries, there are 7 chances in 50 that a defective battery is chosen.   **75.** $C = \frac{7x + 1{,}500{,}000}{x}$

Section 14.4 (pp. 882–884)

1. Examples include 36 and 54 (answers may vary).
2. xy **3.** $\frac{3}{3}$ **4.** $\frac{x+1}{x+1}$ **5.** 12   **7.** 6   **9.** 30   **11.** 72
13. $12x$ **15.** $10x^2$ **17.** $x(x+1)$ **19.** $(2x+1)(x+3)$
21. $x(x-1)(x+1)$ **23.** $(x-8)^2(x+1)$
25. $(2x-1)(2x+1)$ **27.** $(x-1)(x+1)$
29. $(2x+3)(x+2)(x+3)$ **31.** $3y(y+2)(y-1)$
33. $\frac{13}{10}$ **35.** $\frac{2}{9}$ **37.** $\frac{14}{25}$ **39.** $\frac{9}{20}$ **41.** $\frac{3}{9}$ **43.** $\frac{15}{21}$ **45.** $\frac{2x^2}{8x^3}$
47. $\frac{x-2}{x^2-4}$ **49.** $\frac{x}{x^2+x}$ **51.** $\frac{2x^2+2x}{x^2+2x+1}$ **53.** $\frac{13}{12x}$
55. $\frac{5z-7}{z^3}$ **57.** $\frac{y-x}{xy}$ **59.** $\frac{a^2+b^2}{ab}$ **61.** $\frac{7}{2(x+2)}$
63. $\frac{t+2}{t(t-2)}$ **65.** $\frac{n^2+4n+5}{(n-1)(n+1)}$ **67.** $-\frac{3}{x-3}$ **69.** 0
71. $\frac{6x-4}{(x-1)^2}$ **73.** $\frac{3}{2y-1}$ **75.** $\frac{x-1}{x(x+2)}$ **77.** $\frac{3x+5}{(x-2)(x+2)}$
79. $\frac{x+9}{x(x-3)(x+3)}$ **81.** $\frac{x^2+4x-4}{x(x-2)(x+2)}$ **83.** $\frac{2x+2}{(x+2)^2}$
85. $\frac{x^2+2x-1}{(x+1)(x+2)(x+3)}$ **87.** $-\frac{2b}{(a+b)(a-b)}$
89. 0   **91.** $\frac{6}{x-2}$ **93.** 0   **95.** $\frac{3}{x}$ **97.** $\frac{x^3-x+2}{(x-1)(x+1)}$

99. $\frac{3x^2+x+4}{(x-1)(x+1)}$ **101.** $\frac{1}{75}$; $\frac{1}{75}$; yes
103. $\frac{D-F}{FD}$ **105.** 60 years

Checking Basic Concepts 14.3 & 14.4 (p. 884)

1. (a) 1   (b) $\frac{2-x}{3x}$ (c) z **2.** (a) $15x$ (b) $4x(x+1)$
(c) $(x+1)(x-1)$ **3.** (a) $\frac{6x+5}{x(x+1)}$ (b) $\frac{4}{x-3}$
(c) $-\frac{x+4}{2x+1}$ **4.** $\frac{a^2+b^2}{(a-b)(a+b)}$

Section 14.5 (pp. 891–892)

1. $\frac{1}{2} \cdot \frac{4}{3} = \frac{2}{3}$ **2.** $\frac{a}{b} \cdot \frac{d}{c} = \frac{ad}{bc}$ **3.** fractions   **4.** $\dfrac{\frac{x}{2}}{\frac{1}{x-1}}$
5. $\dfrac{\frac{a}{b}}{\frac{c}{d}}$ **6.** $x(x+2)$ **7.** 30   **9.** $(x-1)(x+1)$
11. $x(2x-1)(2x+1)$ **13.** $\frac{4}{5}$ **15.** $\frac{10}{7}$ **17.** $\frac{9}{14}$ **19.** $\frac{1}{2}$
21. $\frac{3y}{x}$ **23.** 3   **25.** $\frac{p+1}{p+2}$ **27.** $\frac{5}{z}$ **29.** $\frac{y}{y-3}$ **31.** $\frac{4x-1}{4x+1}$
33. $\frac{x^2}{3}$ **35.** 1   **37.** $\frac{n+m}{n-m}$ **39.** $\frac{2x+y}{2x-y}$ **41.** $\frac{1+b}{1-a}$
43. $\frac{q+2}{q}$ **45.** $2x+3$ **47.** $\frac{2x}{(x+1)(2x-1)(2x+1)}$
49. $\frac{1}{ab}$ **51.** $\frac{ab}{a+b}$ **53.** $\dfrac{P\left(1+\frac{r}{26}\right)^{52} - P}{\frac{r}{26}}$
55. $R = \frac{ST}{S+T}$

Section 14.6 (pp. 903–907)

1. rational   **2.** $\frac{2x+5}{3x}$; $\frac{2x+5}{3x} = 9$ (answers may vary)
3. $ad = bc$; b; d **4.** Yes   **5.** $12x$ **6.** $V; T$ **7.** $\frac{3}{2}$ **9.** $\frac{5}{2}$
11. 0   **13.** $\frac{1}{4}$ **15.** $\frac{10}{11}$ **17.** 3   **19.** 25   **21.** 5   **23.** $\frac{5}{16}$
25. 1   **27.** No solutions (extraneous: $-\frac{1}{2}$)   **29.** $\frac{1}{2}$
31. -1 **33.** $-\frac{3}{2}$, -1 **35.** 4   **37.** 4   **39.** -6 **41.** 6
43. -3 (extraneous: 1)   **45.** 1 (extraneous: -2)   **47.** $\frac{7}{12}$
49. $\frac{1}{6}$, 5   **51.** 3   **53.** No solutions (extraneous: 1)   **55.** 3
57. No solutions (extraneous: -1)   **59.** No solutions
61. -2 **63.** Expression; 1; 1   **65.** Equation; 2
67. Equation; 4   **69.** Expression; $x+2$; 4   **71.** $-\frac{1}{2}$; $\frac{1}{2}$
73. -1; $\frac{1}{2}$ **75.** (a) -3, 1   (b) -3, 1   **77.** (a) -1
(b) -1 **79.** (a) 2   (b) 2   **81.** (a) -2, 2   (b) -2, 2
83. 0.300, 1.100   **85.** 1.084   **87.** $a = \frac{F}{m}$ **89.** $r = \frac{V}{I} - R$
91. $b = \frac{2A}{h}$ **93.** $z = \frac{15}{k-3}$ **95.** $b = \frac{aT}{a-T}$
97. $x = \frac{ky}{3y+2k}$ **99.** 9 cars per minute
101. $\frac{12}{7} \approx 1.7$ hours   **103.** $\frac{8}{3} \approx 2.7$ days
105. 20 mph; 18 mph   **107.** (a) 120; the braking distance is 120 feet when the slope of the road is -0.05.   (b) -0.1
109. 15 mph   **111.** 200 mph   **113.** 10 mph running; 5 mph jogging   **115.** (a) 1; when cars leave the ramp at a rate of 4 vehicles per minute, the wait is 1 minute.   (b) As more cars try to exit, the waiting time increases; yes
(c) $4.\overline{6}$ vehicles/min

Checking Basic Concepts 14.5 & 14.6 (p. 907)

1. (a) $\frac{5}{6}$ (b) $\frac{1}{9x^2}$ (c) $\frac{b-a}{b+a}$ (d) $\frac{r+t}{2rt}$ **2.** (a) $\frac{1}{5}$
(b) -4 (c) 2   (d) $-2, \frac{1}{2}$ (e) -3 (extraneous: 1)
3. (a) $x = \frac{2(b+3y)}{a}$ (b) $m = \frac{k}{2k-1}$ **4.** (a) 300; when the slope of the hill is 0.1, the braking distance is 300 feet.
(b) 0.3; the braking distance is 200 feet when the slope of the road is 0.3.

Section 14.7 (pp. 917–921)

1. A statement that two ratios are equal **2.** $\frac{5}{6} = \frac{x}{7}$
3. It doubles. **4.** It is halved. **5.** constant **6.** constant
7. Directly; if the number being fed doubles, the bill will double. **8.** Inversely; doubling the number of painters will halve the time. **9.** inverse **10.** direct **11.** 15 **13.** 21
15. 48 **17.** $\frac{21}{8}$ **19.** $-4, 4$ **21.** $-\frac{7}{2}, \frac{7}{2}$ **23.** $b = \frac{ad}{c}$
25. (a) $\frac{5}{8} = \frac{9}{x}$ (b) $\frac{72}{5}$ **27.** (a) $\frac{4}{8} = \frac{10}{x}$ (b) 20
29. (a) $\frac{98}{7} = \frac{x}{11}$ (b) $154 **31.** (a) $\frac{3}{750} = \frac{7}{x}$ (b) 1750
33. (a) 2 (b) 12 **35.** (a) $\frac{3}{2}$ (b) 9 **37.** (a) $-\frac{15}{2}$
(b) -45 **39.** (a) 24 (b) 3 **41.** (a) 40 (b) 5
43. (a) 400 (b) 50 **45.** (a) $k = 0.25$
(b) $z = 8.75$ **47.** (a) $k = 11$ (b) $z = 385$
49. (a) $k = 10$ (b) $z = 350$
51. (a) Direct (b) $y = \frac{3}{2}x$ **53.** (a) Inverse (b) $y = \frac{36}{x}$
(c) 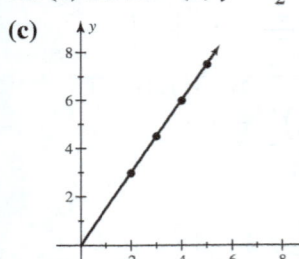 (c)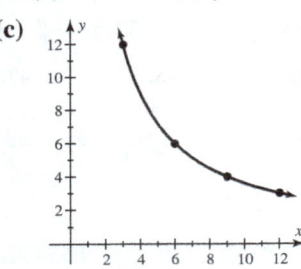

55. (a) Neither (b) NA (c) NA **57.** Direct; 2
59. Neither **61.** Inverse; 8 **63.** 47.6 minutes
65. 1.625 inches **67.** About 2123 lb **69.** 18.75 ft
71. (a) Direct; the ratios $\frac{R}{W}$ always equal 0.012.
(b) $R = 0.012W$
73. (a) Direct; the ratios $\frac{G}{A}$ always equal 27.
(b) $G = 27A$

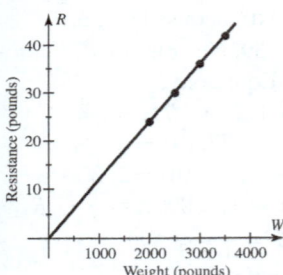 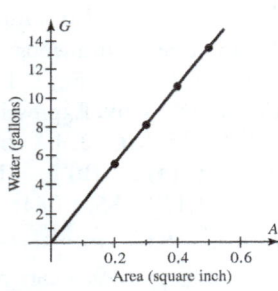

(c) 38.4 pounds

(c) For each square-inch increase in the cross-sectional area of the hose, the flow increases by 27 gallons per minute.

75. (a) $F = \frac{1200}{L}$ (b) 80 pounds
77. (a) Direct (b) $y = -19x$ (c) Negative; for each 1-mile increase in altitude, the temperature decreases by 19° F. (d) 47.5° F decrease **79.** 1.8 ohms
81. $z = 0.5104x^2y^3$ **83.** About 133 pounds **85.** About 35.2 pounds **87.** 750 **89.** 1200 lb per square inch

Checking Basic Concepts 14.7 (p. 921)

1. (a) $\frac{18}{5}$ (b) $\frac{15}{4}$ **2.** (a) $\frac{4}{6} = \frac{8}{x}$; 12 (b) $\frac{2}{148} = \frac{5}{x}$; 370 minutes
3. 60; 6 **4.** (a) Direct; the ratios $\frac{y}{x}$ always equal $\frac{3}{2}$; $\frac{3}{2}$.
(b) Inverse; the products xy always equal 24; 24. **5.** $160

Chapter 14 Review (pp. 925–928)

1. $-\frac{3}{5}$ **2.** -3 **3.** Undefined **4.** Undefined
5.

x	-2	-1	0	1	2
$\frac{3x}{x-1}$	2	$\frac{3}{2}$	0	—	6

6. $-2, 2$ **7.** $\frac{5y^3}{3x^2}$ **8.** $x - 6$ **9.** $\frac{x+3}{x+1}$ **10.** $\frac{x+4}{x+1}$
11. (a) Expression (b) $\frac{1}{x-3}$ **12.** (a) Equation (b) 18
13. 2 **14.** $\frac{1}{x+5}$ **15.** $\frac{1}{z+3}$ **16.** $\frac{x}{x-2}$ **17.** $\frac{5}{6}$
18. $\frac{8}{x(x+1)}$ **19.** $\frac{1}{2}$ **20.** 1 **21.** $\frac{10}{x+10}$ **22.** $\frac{1}{x-1}$ **23.** 1
24. 1 **25.** $\frac{1}{x+1}$ **26.** $\frac{2}{x-5}$ **27.** $\frac{2}{xy}$ **28.** $\frac{x}{y}$ **29.** $15x$
30. $10x^2$ **31.** $x(x-5)$ **32.** $10x^2(x-1)$ **33.** $\frac{9}{24}$
34. $\frac{16}{12x}$ **35.** $\frac{3x^2+6x}{x^2-4}$ **36.** $\frac{2x^2+4x}{2x^2+x-6}$ **37.** $\frac{19}{24}$ **38.** $\frac{7}{4x}$
39. $-\frac{1}{9x}$ **40.** $\frac{4x+3}{x(x-1)}$ **41.** $\frac{2x}{(x-1)(x+1)}$ **42.** $\frac{8-9x}{6x^2}$
43. $\frac{2x-7}{6x}$ **44.** $-\frac{1}{(x-1)(x+1)}$ **45.** $\frac{5y-x}{(x-y)(x+y)}$
46. -1 **47.** $\frac{33}{28}$ **48.** $\frac{7}{10}$ **49.** $\frac{n}{2}$ **50.** $\frac{3(p+1)}{p-1}$ **51.** $\frac{1}{3}$
52. $\frac{2-x}{2+x}$ **53.** $\frac{1}{x^2}$ **54.** $x + 3$ **55.** $\frac{20}{7}$ **56.** $\frac{8}{3}$ **57.** $\frac{1}{5}$
58. -5 **59.** 4 **60.** -2 **61.** 5 **62.** 7 **63.** $-2, 5$
64. $-3, 3$ **65.** 8 **66.** No solutions
67. No solutions (extraneous: -1)
68. $-\frac{3}{2}$ (extraneous: 0) **69.** -3 **70.** -12
71. (a) Equation (b) $-2, 2$ **72.** (a) Expression
(b) $x + 3$ **73.** $b = \frac{2ac}{3a-c}$ **74.** $x = \frac{y}{y-1}$ **75.** 2
76. $\frac{15}{7}$ **77.** (a) $\frac{6}{x} = \frac{13}{20}$ (b) $\frac{120}{13}$ **78.** (a) $\frac{341}{11} = \frac{x}{8}$
(b) $248 **79.** (a) 4 (b) 20 **80.** (a) 3 (b) 15 **81.** (a) 10
(b) 2 **82.** (a) 21 (b) $\frac{21}{5}$ **83.** $k = 3$ **84.** $z = 720$
85. (a) Inverse (b) $y = \frac{60}{x}$ **86.** (a) Direct (b) $y = 3x$
(c) 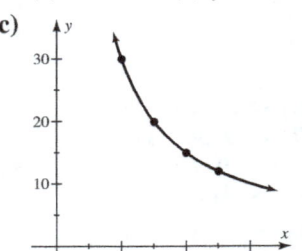 (c)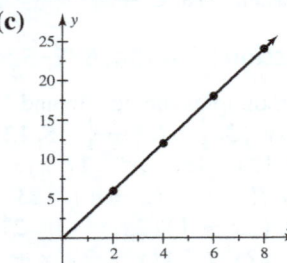

87. Direct; $\frac{1}{2}$ **88.** Inverse; 12 **89.** (a) $\frac{1}{5} = 0.2$; when the average rate of arrival is 10 cars per minute, the average wait is 0.2 minute, or 12 seconds.

(b)

x	5	10	13	14	14.9
T	$\frac{1}{10}$	$\frac{1}{5}$	$\frac{1}{2}$	1	10

(c) It increases dramatically. **90.** 60 mph
91. $\frac{800}{13} \approx 61.5$ hours **92.** 10 mph and 12 mph

93. 8 mph   **94.** About 33.3 feet   **95.** 200 vehicles
96. $468   **97.** 36 pounds   **98.** 1.6 inches
99. 750 to 2250 seconds, or 12.5 to 37.5 minutes
100. About 771 lb   **101.** 6493.8 watts   **102.** 702 pounds

Chapter 14 Test (p. 929)

1. $\frac{9}{5}$ **2.** -2 **3.** $x + 5$ **4.** $x - 5$ **5.** 3   **6.** 2   **7.** $\frac{x-1}{10x}$
8. $\frac{6}{x(x+3)}$ **9.** $\frac{4x+1}{x+4}$ **10.** $\frac{t+7}{2t-3}$ **11.** $6x^2(x-1)$
12. $\frac{4x-4}{7x^2-7x}$ **13.** $-\frac{1}{y+1}$ **14.** $\frac{x^2y-x+y}{xy^2}$ **15.** $\frac{b}{15}$
16. $\frac{p}{p-2}$ **17.** $\frac{35}{2}$ **18.** 3   **19.** 1   **20.** $-1, 2$ **21.** $\frac{2}{7}$
22. -12 **23.** No solutions (extraneous: $\frac{1}{2}$)   **24.** 0 (extraneous: 5)   **25.** $x = \frac{2+5y}{3y}$ **26.** $b = \frac{a}{a-1}$ **27. (a)** $\frac{7}{2}$
(b) 21   **28.** Inversely; 32   **29.** 24 hours   **30.** 67.5 feet
31. $\frac{16}{5} = 3.2$; when the average arrival rate is 24 people per hour, there are about 3 people in line, on average.

Chapters 1–14 Cumulative Review (pp. 930–931)

1. $24\pi \approx 75.4$ **2.** $2x - 2$ **3.** $\frac{2}{5}$ **4.** $\frac{3}{4}$ **5.** $2x + 2$
6. $y - 11$ **7.** -1 **8.** $x \leq \frac{1}{4}$
9. x-intercept: $(3, 0)$ **10.** x-intercept: $(1, 0)$
    y-intercept: $(0, -2)$ y-intercept: none

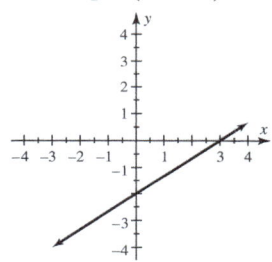

11. $y = 3x + 5$

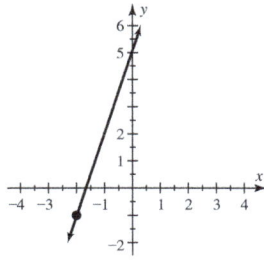

12. $y = 2x - 1$ **13.** $y = -\frac{2}{3}x + \frac{1}{3}$ **14.** $y = \frac{2}{3}x + \frac{8}{3}$
15. $(1, -2)$ **16.** $(2, -8)$ **17.** Infinitely many solutions
18. No solutions   **19.** $15z^8$ **20.** a^3b^3
21. $10y^2 - 11y - 6$ **22.** $x^4 - 2x^2y^2 + y^4$ **23.** $\frac{1}{27x^6}$
24. $\frac{2}{x^2}$ **25.** 1.23×10^{-3} **26.** $2x + 1 + \frac{4}{x-1}$
27. $(3-x)(2+5x)$ **28.** $(3z-2)(3z+2)$
29. $(t+8)^2$ **30.** $x(x-4)(x+4)$ **31.** $-7, 2$
32. $-2, 0, 2$ **33.** $\frac{1}{12x}$ **34.** $\frac{1}{2x+4}$ **35.** $\frac{x+2}{x-2}$
36. $x = \frac{z+2y}{3}$ **37.** $\frac{7}{12}$ **38.** $-1, 2$ **39.** 5.5
40. Inverse; $y = \frac{20}{x}$ **41. (a)** $21x$ **(b)** 90 min

42. 35 min running, 25 min walking
43. (a) $2x + y = 180, 2x - y = 32$ **(b)** $(53, 74)$ or $53°$, $53°, 74°$ **44.** About 500 lb

15 Introduction to Functions

Section 15.1 (pp. 947–952)

1. function   **2.** y equals f of x **3.** symbolic   **4.** numerical
5. domain   **6.** range   **7.** one   **8.** False   **9.** True
10. $(3, 4)$; 3; 6   **11.** (a, b) **12.** d **13.** 1   **14.** equal
15. Yes   **16.** Yes   **17.** No   **18.** No   **19.** Yes   **20.** No
21. $-6; -2$ **23.** $0; \frac{3}{2}$ **25.** $25; \frac{9}{4}$ **27.** $3; 3$ **29.** $13; -22$
31. $-\frac{1}{2}; \frac{2}{5}$ **33. (a)** $I(x) = 36x$ **(b)** $I(10) = 360$; there are 360 inches in 10 yards.   **35. (a)** $M(x) = \frac{x}{5280}$
(b) $M(10) = \frac{10}{5280} \approx 0.0019$; There is $\frac{10}{5280}$ mile in 10 feet.
37. (a) $A(x) = 43,560x$ **(b)** $A(10) = 435,600$; there are 435,600 square feet in 10 acres.
39. $f = \{(1, 3), (2, -4), (3, 0)\}$;
$D = \{1, 2, 3\}; R = \{-4, 0, 3\}$
41. $f = \{(a, b), (c, d), (e, a), (d, b)\}$;
$D = \{a, c, d, e\}; R = \{a, b, d\}$

43. 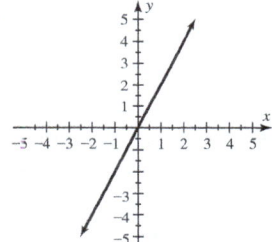   **45.**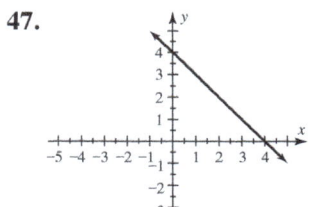

47. 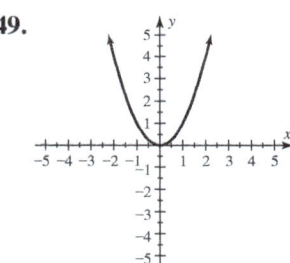   **49.**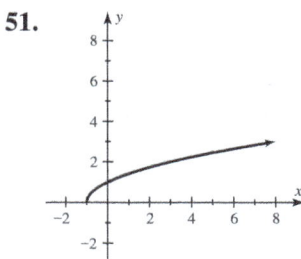

51.

53. $3; -1$ **55.** $0; 2$ **57.** $-4; -3$ **59.** $5.5; 3.7$
61. 26.9; in 1990, average fuel efficiency was 26.9 mpg.

63. Numerical:

x	-3	-2	-1	0	1	2	3
y = f(x)	2	3	4	5	6	7	8

Graphical:

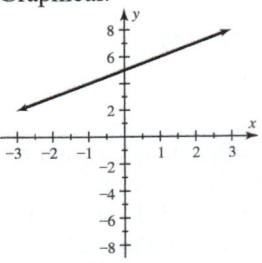

Symbolic: $y = x + 5$

65. Numerical:

x	-3	-2	-1	0	1	2	3
y = f(x)	-17	-12	-7	-2	3	8	13

Graphical:

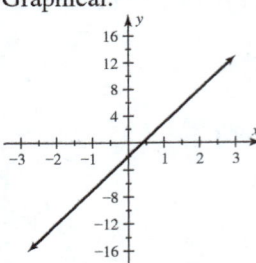

Symbolic: $y = 5x - 2$

67. Subtract $\frac{1}{2}$ from the input x to obtain the output y.
69. Divide the input x by 3 to obtain the output y.
71. Subtract 1 from the input x and then take the square root to obtain the output y.
73. $f(x) = 0.60x$

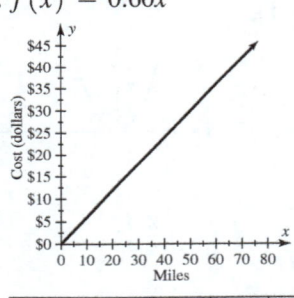

Miles	10	20	30	40	50	60	70
Cost	$6	$12	$18	$24	$30	$36	$42

75. 1400; in 2014, there were 1400 billion, or 1.4 trillion Google searches.
77. $D: -2 \leq x \leq 2; R: 0 \leq y \leq 2$
79. $D: -2 \leq x \leq 4; R: -2 \leq y \leq 2$
81. $D:$ All real numbers; $R: y \geq -1$
83. $D: -3 \leq x \leq 3; R: -3 \leq y \leq 2$
85. $D = \{1, 2, 3, 4\}; R = \{5, 6, 7\}$
87. All real numbers **89.** All real numbers **91.** $x \neq 5$
93. All real numbers **95.** $x \geq 1$ **97.** All real numbers
99. $x \neq 0$ **101. (a)** 1488, 1488 whales were sighted in 2015. **(b)** $D = \{2011, 2012, 2013, 2014, 2015\}$; $R = \{1054, 1126, 1331, 1488, 1607\}$ **(c)** Decreased from the first to the second year and then increased every year after that. **103.** $D = \{1, 2, 3, \ldots, 20\}$; $R = \{200, 400, 600, \ldots, 4000\}$

105. Relation; $D = \{2, 3\}; R = \{-4, 3, 5\}$
107. Relation; Function; $D = \{-7, 3, 4\}; R = \{5\}$
109. Neither **111.** Relation; Function; $D = \{-5, -2, 0, 3\}$; $R = \{-5, -2, 0, 3\}$
113. No **115.** Yes **117. (a)** 0.2 **(b)** Yes. Each month has one average amount of precipitation. **(c)** 2, 3, 7, 11
119. Yes. D: All real numbers; R: All real numbers
121. No **123.** Yes. $D: -4 \leq x \leq 4; R: 0 \leq y \leq 4$
125. Yes. D: All real numbers; $R: y = 3$ **127.** No
129. Yes. $D = \{-6, -4, 2, 4\}; R = \{-4, 2\}$ **131.** Yes
133. No **135.** The person walks away from home, then turns around and walks back a little more slowly.
137.

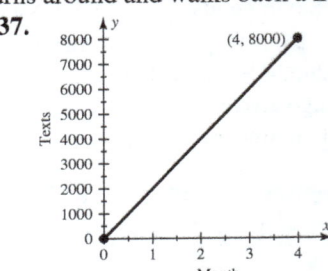

139. 10; 10
$[-10, 10, 1]$ by $[-10, 10, 1]$

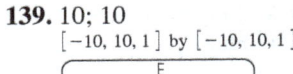

141. 10; 5
$[0, 100, 10]$ by $[-50, 50, 10]$

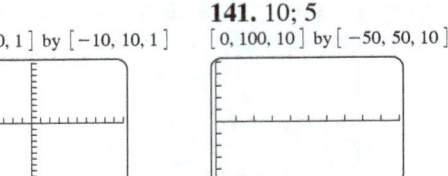

143. 16; 5
$[1980, 1995, 1]$ by $[12000, 16000, 1000]$

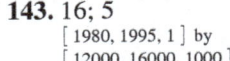

145.
$[-6, 6, 1]$ by $[-6, 6, 1]$

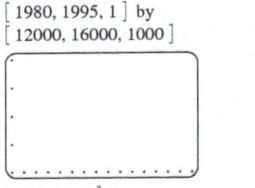

147. $[-30, 30, 5]$ by $[-50, 50, 5]$

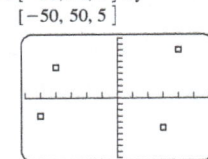

149. $[-200, 200, 50]$ by $[-250, 250, 50]$

151. Numerical

Graphical
$[-10, 10, 1]$ by $[-10, 10, 1]$

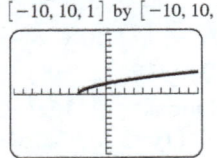

153. Numerical

Graphical (Dot Mode)
$[-10, 10, 1]$ by $[-10, 10, 1]$

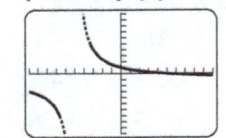

Section 15.2 (pp. 964–968)

1. $mx + b$ **2.** b **3.** line **4.** horizontal **5.** 7 **6.** 0
7. True **8.** False **9.** Carpet costs $2 per square foot. Ten square feet of carpet costs $20.

10. The rate at which water is leaving the tank is 4 gallons per minute. After 5 minutes, the tank contains 80 gallons of water.
11. Yes; $m = \frac{1}{2}, b = -6$ **13.** No
15. Yes; $m = 0, b = -9$ **17.** Yes; $m = -9, b = 0$
19. Yes **21.** No **23.** Yes; $f(x) = 3x - 6$ **25.** Yes; $f(x) = -\frac{3}{2}x + 3$ **27.** No **29.** Yes; $f(x) = 2x$
31. Yes; $f(x) = -4$ **33.** $-16; 20$ **35.** $\frac{17}{3}; 2$
37. $-22; -22$ **39.** $-2; 0$ **41.** $-1; -4$ **43.** $1; 1$
45. $f(x) = 6x; 18$ **47.** $f(x) = \frac{x}{6} - \frac{1}{2}; 0$ **49.** d. **51.** b.
53. 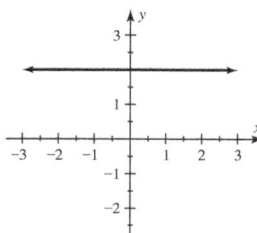 **55.**
57. 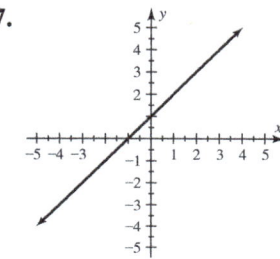 **59.**
61.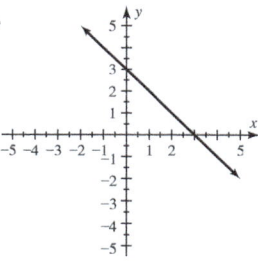
63. $f(x) = \frac{1}{16}x$ **65.** $f(t) = 65t$ **67.** $f(x) = 24$ **69.** a
71. (a) f multiplies the input x by -2 and then adds 1 to obtain the output y.
(b)

x	-2	0	2
$y = f(x)$	5	1	-3

(c)

73. (a) f multiplies the input x by $\frac{1}{2}$ and then subtracts 1 to obtain the output y.
(b)

x	-2	0	2
$y = f(x)$	-2	-1	0

(c)

75. (a) 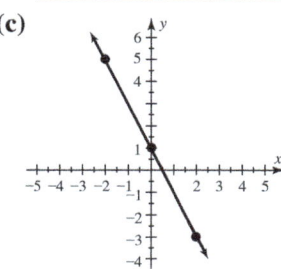 **(b)** $[-6, 6, 1]$ by $[-4, 4, 1]$

77. (a) 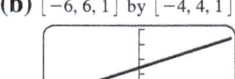 **(b)** $[-6, 6, 1]$ by $[-4, 4, 1]$

79. b. **81.** c. **83. (a)** $G(x) = \frac{X}{E}$ **(b)** $C(x) = \frac{3x}{E}$
85. -1 **87.** $(-1, 0)$ **89.** $\left(-\frac{1}{2}, \frac{1}{2}\right)$ **91.** $(20, 0)$
93. $(-8, -1)$ **95.** $\left(-1, -\frac{1}{6}\right)$ **97.** $(0.2, 0.25)$
99. $(2005, 9)$ **101.** $(2a, 2b)$
103. (a) $f(x) = -2x + 5; 1$
(b) 1 **(c)** Equal **105. (a)** $f(x) = \frac{2}{5}x + \frac{1}{5}; 1$
(b) 1 **(c)** Equal **107.** 79.8 years
109. 256.5 million **111.** $50,850
113. (a) Symbolic: $f(x) = 70$
Graphical:

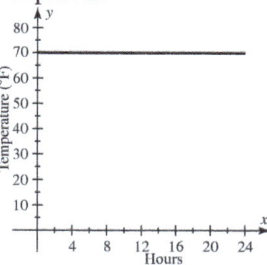

(b)

Hours	0	4	8	12	16	20	24
Temp. (F)	70	70	70	70	70	70	70

(c) Constant
115. $D(t) = 60t + 50$
117. (a) $K(x) = 130x$ **(b)** $A(x) = 5x$
(c) 47,450; 1825; On average, someone between ages 18 and 24 sends 47,450 texts in 1 year, while someone over 65 sends 1825 texts. **119. (a)** 1.6 million **(b)** 0.1 million
(c) $f(x) = 0.1x + 1.6$ **(d)** 2.2 million
121. (a) 189 million **(b)** In 2010, there were about 145 million users. **(c)** Users increased, on average, by 11 million per year. **123. (a)** $V(T) = 0.5T + 137$
(b) 162 cm³ **125.** $f(x) = 40x$; about 2.92 pounds
127. (a) 43% **(b)** 3% **(c)** $P(x) = 3x + 43$ **(d)** 55%

Checking Basic Concepts 15.1 & 15.2 (p. 969)

1. Symbolic: $f(x) = x^2 - 1$
Graphical:

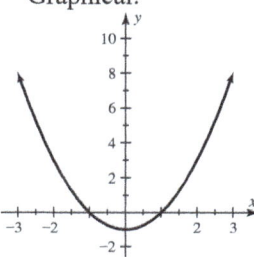

2. (a) D: $-3 \leq x \leq 3$; R: $-4 \leq y \leq 4$ **(b)** 0; 4
(c) No. The graph is not a line. **3. (a)** Yes **(b)** No
(c) Yes **(d)** Yes

4. $f(-2) = 10$

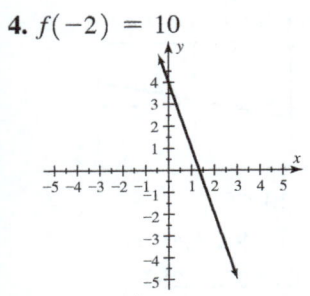

5. $f(x) = \frac{1}{2}x - 1$ **6. (a)** 36.7: In 2015, the median age was about 37 years. **(b)** 0.2: The median age is increasing by 0.2 year each year. 27.7: In 1970, the median age was 27.7 years. **7.** $(1, -1)$

Section 15.3 (pp. 978–982)

1. $x > 1$ and $x \leq 7$ (answers may vary)
2. $x \leq 3$ or $x > 5$ (answers may vary)
3. No **4.** Yes **5.** Yes
6. Numerically, graphically, symbolically
7. Yes, no **8.** No, yes **9.** No, yes **10.** Yes, no
11. No, yes **12.** Yes, yes **13.** $[2, 10]$ **15.** $(5, 8]$
17. $(-\infty, 4)$ **19.** $(-2, \infty)$ **21.** $[-2, 5)$
23. $(-8, 8]$ **25.** $(3, \infty)$ **27.** $(-\infty, -2] \cup [4, \infty)$
29. $(-\infty, 1) \cup [5, \infty)$ **31.** $(-3, 5]$ **33.** $(-\infty, -2)$
35. $(-\infty, 4)$ **37.** $(-\infty, 1) \cup (2, \infty)$
39. $\{x \mid -1 \leq x \leq 3\}$

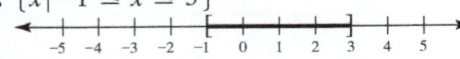

41. $\{x \mid -2 < x < 2.5\}$

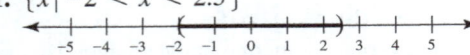

43. $\{x \mid x > 3\}$

45. $\{x \mid x \leq -1 \text{ or } x \geq 2\}$

47. All real numbers

49. $[-6, 7]$ **51.** $(\frac{13}{4}, \infty)$ **53.** $(-\infty, \infty)$
55. $(-\infty, -\frac{11}{5}) \cup (-1, \infty)$ **57.** No solutions
59. $[-6, 1)$ **61.** $[-\frac{1}{4}, \frac{11}{8})$ **63.** $[-9, 3]$
65. $[-4, -\frac{1}{4})$ **67.** $(-1, 1]$ **69.** $[-2, 1]$ **71.** $(0, 2)$
73. $(9, 21]$ **75.** $[\frac{12}{5}, \frac{22}{5}]$ **77.** $(-\frac{13}{3}, 1)$
79. $[0, 12]$ **81.** $[-1, 2]$ **83.** $(-1, 2)$ **85.** $[-3, 1]$
87. $(-\infty, -2) \cup (0, \infty)$
89. (a) Toward, because distance is decreasing
(b) 4 hours, 2 hours **(c)** From 2 to 4 hours
(d) During the first 2 hours
91. (a) 2 **(b)** 4 **(c)** $\{x \mid 2 \leq x \leq 4\}$
(d) $\{x \mid 0 \leq x < 2\}$ **93.** $[1, 4]$
95. $(-\infty, -2) \cup (0, \infty)$ **97.** $[2002, 2010]$ **99.** $[1, 3]$
101. $(-\infty, \infty)$ **103.** $(-\infty, 1) \cup [3, \infty)$
105. $(c - b, d - b]$

107.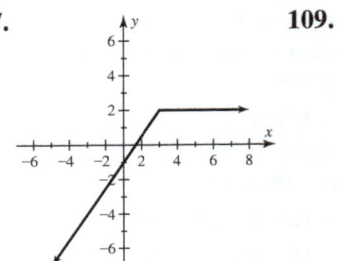

109.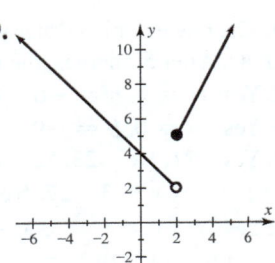

111. (a) $-5 \leq x \leq 5$ **(b)** 2; 3; 6
(c)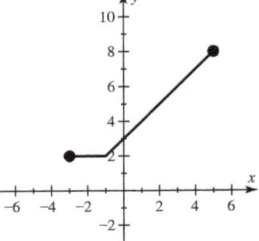

(d) $2 \leq y \leq 8$
113. (a) $-1 \leq x \leq 2$ **(b)** undefined; 0; undefined
(c)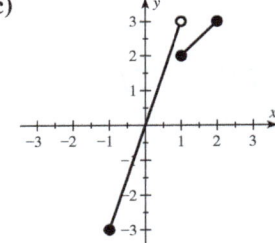

(d) $-3 \leq y \leq 3$
115. (a) $-3 \leq x \leq 3$ **(b)** $-2; 1; -1$
(c)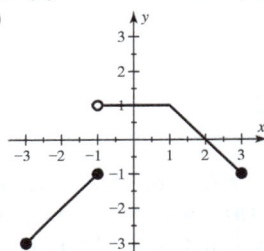

(d) $-3 \leq y \leq 1$
117. (a) From 2013 to 2017 **(b)** From 2013 to 2017
(c) From 2013 to 2017 **119.** From 0.5 to 1.5 miles
121. From $5.\overline{6}$ to 9 feet **123.** From 1.5 to 2.5 miles
125. (a) $M(x) = 25x + 250$ **(b)** From 2014 to 2018
127. (a) maximum: 55 mph; minimum: 30 mph **(b)** About 12 miles **(c)** $f(4) = 40; f(12) = 30; f(18) = 55$

Section 15.4 (pp. 995–1000)

1. domain **2.** range **3.** $(-\infty, \infty)$
4. $(-\infty, 5) \cup (5, \infty)$ **5.** absolute value **6.** exponent
7. 2 **8.** 1 **9.** rational **10.** $-\frac{1}{2}$ **11.** c. **12.** a.
13. $D: (-\infty, \infty); R: (-\infty, \infty)$
15. $D: (-\infty, \infty); R: (-\infty, \infty)$
17. $D: (-\infty, \infty); R: [2, \infty)$

19. $D: (-\infty, \infty); R: (-\infty, 0]$
21. $D: [-1, \infty); R: [0, \infty)$
23. $D: (-\infty, \infty); R: [0, \infty)$
25. $(-\infty, 1) \cup (1, \infty)$
27. $(-\infty, 2) \cup (2, \infty)$
29. $(-\infty, -2) \cup (-2, 2) \cup (2, \infty)$
31. $(-\infty, 0) \cup (0, 2) \cup (2, \infty)$
33. $(-\infty, 1) \cup (1, \infty)$
35. $(-\infty, -1) \cup (-1, 3) \cup (3, \infty)$
37. $D: (-\infty, \infty); R: (-\infty, \infty)$
39. $D: [-2, 2]; R: [-2, 2]$ **41.** $D: [-2, 3]; R: [-2, 2]$
43. Yes; 1; linear **45.** Yes; 3; cubic **47.** No
49. Yes; 2; quadratic **51.** No **53.** Yes; 4; fourth degree
55. 12; 0 **57.** 1; $\frac{11}{4}$ **59.** 0; 6 **61.** -14; 14 **63.** -6; 9
65. $\frac{1}{11}; -\frac{1}{7}$ **67.** $-\frac{5}{6}$; undefined **69.** $\frac{5}{6}$; undefined
71. 1; -1 **73.** $-2; -2$ **75.** $-4; 0$ **77.** -1; undefined
79.

x	-2	-1	0	1	2
$f(x) = \frac{1}{x-1}$	$-\frac{1}{3}$	$-\frac{1}{2}$	-1	—	1

; undefined

81. 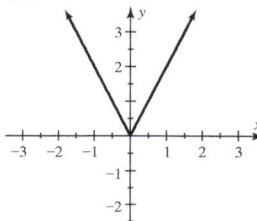 **83.**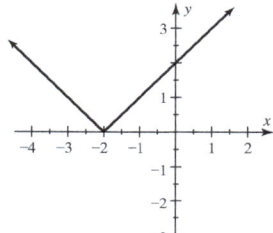

85. 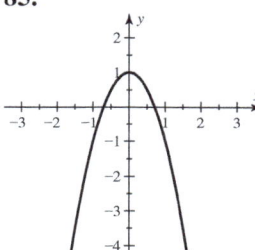 **87.**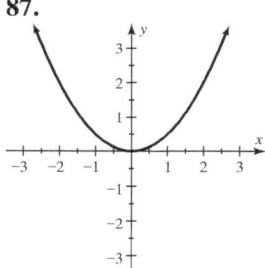

89. 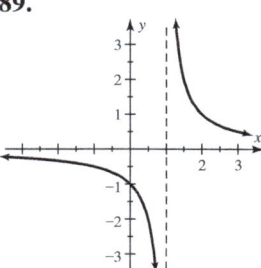 **91.**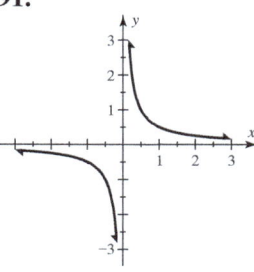

93. 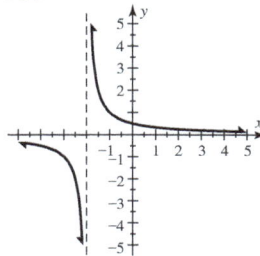 **95.**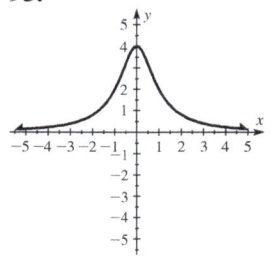

97. **99.**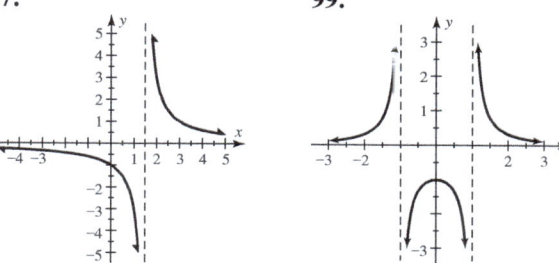

101. (a) 19 **(b)** -9 **(c)** 150 **(d)** 0
103. (a) 41 **(b)** -21 **(c)** 900 **(d)** Undefined
105. (a) $2x + 3$ **(b)** -1 **(c)** $x^2 + 3x + 2$ **(d)** $\frac{x+1}{x+2}$
107. (a) $x^2 - x + 1$ **(b)** $1 - x - x^2$ **(c)** $x^2 - x^3$
(d) $\frac{1-x}{x^2}$ **109.** $a^2 - 2a$ **111.** b **113.** d
115. $f(-2) \approx 5; f(1) \approx 0$
117. $f(-1) \approx -1; f(1) \approx 1; f(2) \approx -2$
119. $f(-3) = -63; f(1) = 3; f(4) = 10$
121. $D: [-3, 3]$

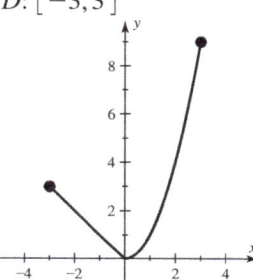

123. c. **125.** d. **127. (a)** No; answers may vary.
(b) Yes **(c)** No; $0 \le t \le 10$
129. (a) 1; when cars are leaving the lot at a rate of 4 vehicles per minute, the average wait is 1 minute.
(b) As more cars try to exit, the waiting time increases; yes. **(c)** $4.\overline{6}$ vehicles per minute
131. (a) 75; the braking distance is 75 feet when the uphill grade is 0.05. **(b)** 0.15
133. (a) 123 thousand, which is close to the actual value.
(b) 478.6 thousand, which is too high; AIDS deaths did not continue to rise as rapidly as the model predicts.
135. About 45% and 22%; after 1 day (3 days), students remember 45% (22%) of what they have learned.
137. (a) 130; it costs $130 thousand to make 100 notebook computers. **(b)** The y-intercept for C is $(0, 100)$. The company has $100 thousand in fixed costs even if it makes 0 computers. The y-intercept for R is $(0, 0)$. If the company sells 0 computers, its revenue is $0. **(c)** $P(x) = 0.45x - 100$
(d) 223 or more

Checking Basic Concepts 15.3 & 15.4 (p. 1000)

1. (a) Yes **(b)** No
2. (a) $[-3, 1]$ **(b)** $(-\infty, -1] \cup [3, \infty)$ **(c)** $[-\frac{8}{3}, \frac{8}{3})$
3. (a) $(-\infty, \infty)$ **(b)** $(-\infty, 1) \cup (1, \infty)$ **(c)** $[0, \infty)$
4. (a) $D: [-2, 1]; R: [-3, 1]$ **(b)** 1; -3

5.

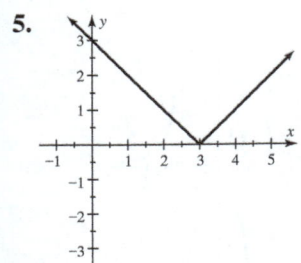

Section 15.5 (pp. 1011–1014)

1. $|3x + 2| = 6$ (answers may vary)
2. $|2x - 1| \leq 17$ (answers may vary)
3. Yes   **4.** Yes   **5.** Yes
6. No, it is equivalent to $-3 < x < 3$.   **7.** 2   **8.** 0
9. No, yes   **10.** Yes, no   **11.** No, yes   **12.** No, yes
13. Yes, yes   **14.** No, yes

15. 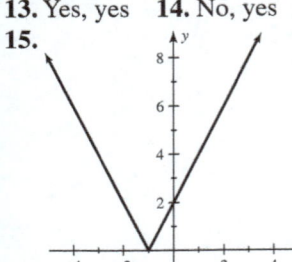   **17.**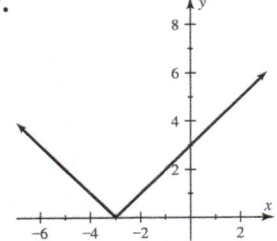

19.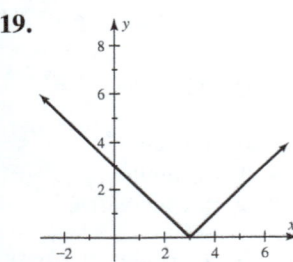

21. $-7, 7$ **23.** No solutions
25. $-\frac{9}{4}, \frac{9}{4}$ **27.** $-4, 4$ **29.** $-6, 5$ **31.** $1, 2$ **33.** $\frac{3}{4}$
35. $-8, 12$ **37.** No solutions   **39.** $-15, 18$ **41.** $-1, \frac{1}{3}$
43. $-5, \frac{3}{5}$ **45.** -6 **47.** $-3, 3$ **49.** $(-3, 3)$
51. $(-\infty, -3) \cup (3, \infty)$
53. (a) $-4, 4$ (b) $(-4, 4)$ (c) $(-\infty, -4) \cup (4, \infty)$
55. (a) $\frac{1}{2}, 2$ (b) $[\frac{1}{2}, 2]$ (c) $(-\infty, \frac{1}{2}] \cup [2, \infty)$
57. $[-3, 3]$ **59.** $(-\infty, -4) \cup (4, \infty)$
61. No solutions   **63.** $(-\infty, 0) \cup (0, \infty)$
65. $(-\infty, -\frac{7}{2}) \cup (\frac{7}{2}, \infty)$ **67.** $(-3, 5)$
69. $(-\infty, -9] \cup [-1, \infty)$ **71.** $[\frac{5}{6}, \frac{11}{6}]$ **73.** $[-10, 14]$
75. No solutions   **77.** $(-\infty, \infty)$ **79.** No solutions
81. $(-\infty, \infty)$ **83.** $(-\infty, -13] \cup [17, \infty)$
85. $[0.9, 1.1]$ **87.** $(-\infty, 9.5) \cup (10.5, \infty)$
89. (a) $-1, 3$ (b) $(-1, 3)$ (c) $(-\infty, -1) \cup (3, \infty)$
91. $0, 4$ **93.** (a) $-1, 0$ (b) $[-1, 0]$
(c) $(-\infty, -1] \cup [0, \infty)$ **95.** $(-\infty, -1] \cup [1, \infty)$
97. $[-2, 4]$ **99.** $(-\infty, 1) \cup (3, \infty)$ **101.** $(2, 4.\overline{6})$
103. $(-\infty, \infty)$ **105.** $\{x \mid -3 \leq x \leq 3\}$
107. $\{x \mid x < 2 \text{ or } x > 3\}$ **109.** $|x| \leq 4$
111. $|y| > 2$ **113.** $|2x + 1| \leq 0.3$ **115.** $|\pi x| \geq 7$
117. two   **119.** (a) $\{T \mid 19 \leq T \leq 67\}$

(b) Monthly average temperatures vary from 19 °F to 67 °F.
121. (a) $\{T \mid -26 \leq T \leq 46\}$ (b) Monthly average temperatures vary from -26 °F to 46 °F.
123. (a) About 19,058 feet   (b) Africa and Europe
(c) South America, North America, Africa, Europe, and Antarctica   (d) $|E - A| \leq 5000$
125. $\{d \mid 2.498 \leq d \leq 2.502\}$; the diameter can vary from 2.498 to 2.502 inches.
127. $|d - 3.8| \leq 0.03$
129. Values between 19 and 21, exclusive

Checking Basic Concepts 15.5 (p. 1014)

1. $-\frac{28}{3}, 12$ **2.** (a) $-\frac{2}{3}, \frac{14}{3}$ (b) $(-\frac{2}{3}, \frac{14}{3})$
(c) $(-\infty, -\frac{2}{3}) \cup (\frac{14}{3}, \infty)$
3. $(0, 6)$; $(-\infty, 0] \cup [6, \infty)$
4. (a) $1, 3$ (b) $[1, 3]$ (c) $(-\infty, 1] \cup [3, \infty)$

Chapter 15 Review (pp. 1018–1022)

1. $-7; 0$ **2.** $-22; 2$ **3.** $-2; 1$ **4.** $5; 5$
5. (a) $P(q) = 2q$ (b) $P(5) = 10$; there are 10 pints in 5 quarts.   **6.** (a) $f(x) = 4x - 3$
(b) $f(5) = 17$; three less than four times 5 is 17.
7. $(3, -2)$ **8.** $4; -6$

9. 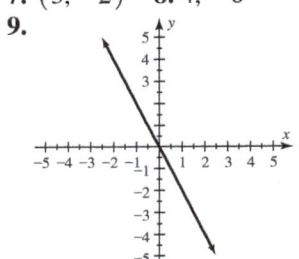   **10.**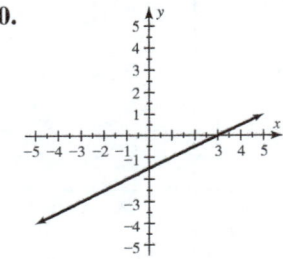

11. 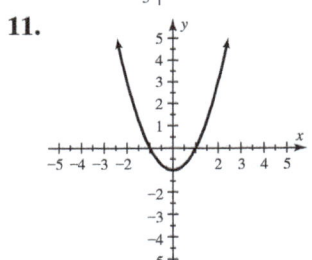   **12.**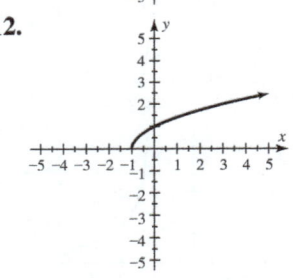

13. $1; 4$ **14.** $1; -2$ **15.** $7; -1$
16. Numerical:

x	-3	-2	-1	0	1	2	3
$y = f(x)$	-11	-8	-5	-2	1	4	7

Symbolic: $f(x) = 3x - 2$
Graphical:

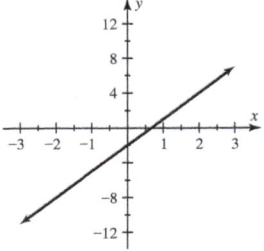

17. D: All real numbers; R: $y \leq 4$
18. D: $-4 \leq x \leq 4$; R: $-4 \leq y \leq 0$
19. Yes 20. No
21. $D = \{-3, -1, 2, 4\}$; $R = \{-1, 3, 4\}$; yes
22. $D = \{-1, 0, 1, 2\}$; $R = \{-2, 2, 3, 4, 5\}$; no
23. All real numbers 24. $x \geq 0$ 25. $x \neq 0$
26. All real numbers 27. $x \leq 5$ 28. $x \neq -2$
29. All real numbers 30. All real numbers
31. No 32. Yes 33. Yes; $m = -4$, $b = 5$
34. Yes; $m = -1$, $b = 7$ 35. No
36. Yes; $m = 0$, $b = 6$ 37. Yes; $f(x) = \frac{3}{2}x - 3$
38. No 39. 1 40. $f(-2) = -3$; $f(1) = 0$
41. 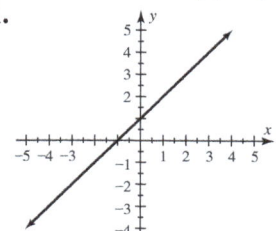 42.
43. 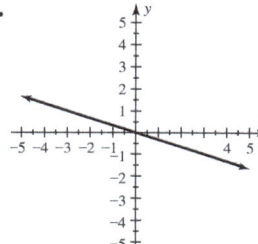 44.
45. $H(x) = 24x$; $H(2) = 48$, there are 48 hours in 2 days.
46. (a) 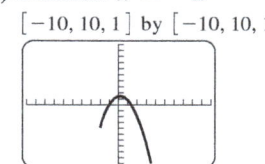 (b) Domain: $x \geq -2$

$[-10, 10, 1]$ by $[-10, 10, 1]$

47. $(0.5, -3)$ 48. $\left(\frac{5}{12}, \frac{3}{8}\right)$
49. $[-2, 2]$
50. $(-\infty, -3]$
51. $\left(-\infty, \frac{4}{5}\right] \cup (2, \infty)$
52. $(-\infty, \infty)$
53. $[-2, 1]$ 54. (a) -4 (b) 2 (c) $[-4, 2]$
(d) $(-\infty, 2)$ 55. (a) 2 (b) $(2, \infty)$ (c) $(-\infty, 2)$
56. (a) 4 (b) 2 (c) $(2, 4)$
57. $\left[-3, \frac{2}{3}\right]$ 58. $(-6, 45]$ 59. $\left(-\infty, \frac{7}{2}\right)$
60. $[1.8, \infty)$ 61. $(-3, 4)$ 62. $(-\infty, 4) \cup (10, \infty)$
63. $(-5, 5)$ 64. $[8, 28]$ 65. $(-9, 21)$ 66. $[88, 138)$
67. 4, 2 68. -10, undefined

69. $D = (-\infty, \infty)$; $R = [0, \infty)$
70. $D = (-\infty, \infty)$; $R = [0, \infty)$
71. $(-\infty, 4) \cup (4, \infty)$
72. $D = [-3, 1]$; $R = [-3, 6]$ 73. Yes; 2; quadratic
74. Yes; 1; linear 75. Yes; 3; cubic
76. No 77. 11; 2 78. $-\frac{4}{5}$; undefined
79. 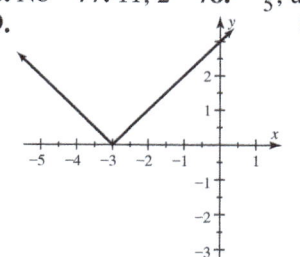 80.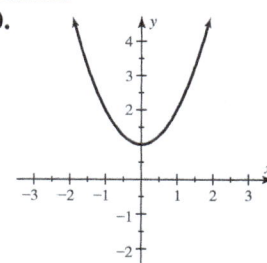
81. 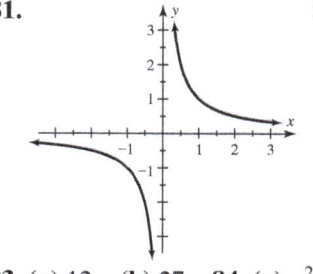 82.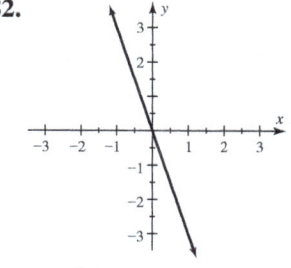

83. (a) 12 (b) 27 84. (a) $x^2 - x$ (b) $x + 1$
85. No, no 86. No, yes 87. No, yes 88. No, yes
89. (a) 0, 4 (b) $(0, 4)$ (c) $(-\infty, 0) \cup (4, \infty)$
90. (a) $-3, 1$ (b) $[-3, 1]$ (c) $(-\infty, -3] \cup [1, \infty)$
91. $-22, 22$ 92. 1, 8 93. $-26, 42$ 94. $-\frac{23}{3}, \frac{25}{3}$
95. 0, 2 96. $-3, \frac{9}{5}$ 97. (a) $-8, 6$ (b) $[-8, 6]$
(c) $(-\infty, -8] \cup [6, \infty)$
98. (a) $-\frac{5}{2}, \frac{7}{2}$ (b) $\left[-\frac{5}{2}, \frac{7}{2}\right]$ (c) $\left(-\infty, -\frac{5}{2}\right] \cup \left[\frac{7}{2}, \infty\right)$
99. $(-\infty, -3) \cup (3, \infty)$ 100. $(-4, 4)$ 101. $[-3, 4]$
102. $\left(-\infty, -\frac{5}{2}\right] \cup [5, \infty)$ 103. $[4.4, 4.6]$
104. $\left[\frac{3}{13}, \frac{7}{13}\right]$ 105. $(-\infty, \infty)$ 106. $\frac{3}{2}$
107. $(-\infty, -1.5] \cup [1.5, \infty)$ 108. $[-2, 6]$
109. $|x| \leq 0.05$ 110. $|5x - 1| > 4$
111. (a) About 25.1 (b) Decreased
$[1885, 1965, 10]$ by $[22, 26, 1]$

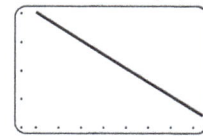

(c) -0.0492; the median age decreased by about 0.0492 year per year.
112. (a) 2.1 million (b) The number of marriages each year did not change.
113. (a) $f(x) = 8x$ (b) 8 (c) The total fat increases at the rate of 8 grams per cup.
114. (a)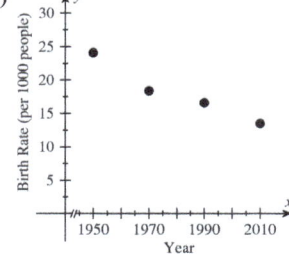

(b) $f(x) = -0.1675x + 350$
(c) About 15 per 1000 people (answers may vary)
115. (a) $f(2010) = 128$; in 2010, there were 128 days with an unhealthy air quality in Los Angeles.
(b) $D = \{2006, 2008, 2010, 2012, 2014\}$;
$R = \{120, 122, 128, 145, 148\}$ **(c)** It decreased.
116. (a) Linear

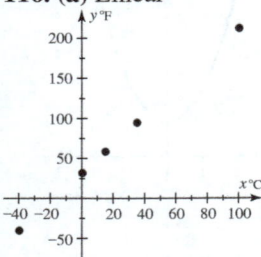

(b) $f(x) = \frac{9}{5}x + 32$; a 1 °C change equals a $\frac{9}{5}$ °F change.
(c) 68 °F **117. (a)** 3 hours **(b)** 2 hours and 4 hours
(c) Between 2 and 4 hours, exclusive **(d)** 20 miles per hour
118. (a) $f(x) = -0.05x + 1.3$ **(b)** About 1.15 million
119. (a) $|A - 3.9| \le 1.7$ **(b)** $2.2 \le A \le 5.6$
120. Values between 32.2 and 37.8, exclusive

Chapter 15 Test (pp. 1023–1024)

1. 46; (4, 46)
2. $C(x) = 4x$; $C(5) = 20$, 5 pounds of candy costs $20.
3. (a) **(b)**

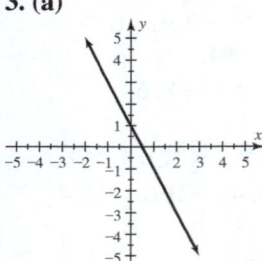 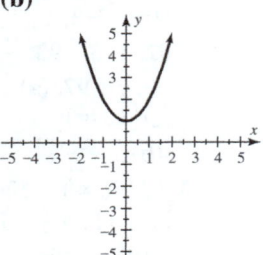

(c) **(d)**

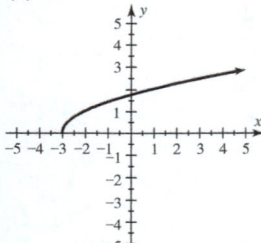 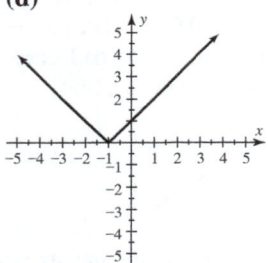

4. 0, −3; D: $-3 \le x \le 3$; R: $-3 \le y \le 0$
5. Symbolic: $f(x) = x^2 - 5$
Numerical:

x	−3	−2	−1	0	1	2	3
$y = f(x)$	4	−1	−4	−5	−4	−1	4

Graphical:

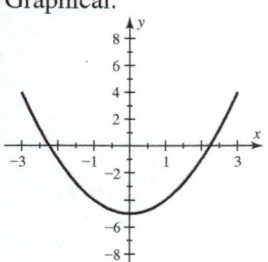

6. No, it fails the vertical line test.
7. (a) $D = \{-2, -1, 0, 5\}$ **(b)** All real numbers
(c) $x \ge -4$ **(d)** All real numbers **(e)** $x \ne 5$
8. Yes; $f(x) = -8x + 6$
9.
$\xleftarrow{\quad\quad\quad}|\!\!-\!\!\!|\xrightarrow{\quad\quad\quad}$
$-5\ -4\ -3\ -2\ -1\ \ 0\ \ 1\ \ 2\ \ 3\ \ 4\ \ 5$
10. $(-\infty, -2) \cup (1, \infty)$ **11. (a)** −5 **(b)** 5
(c) $[-5, 5]$ **(d)** $(-\infty, 5)$ **12.** $(-8, 0)$
13. $-12, 24$ **14. (a)** $[-5, 5]$ **(b)** $(-\infty, 0) \cup (0, \infty)$
15. Yes; 3; cubic **16.** $\frac{8}{7}$; $(-\infty, 5) \cup (5, \infty)$
17. (a) 9 **(b)** $2x^3 + 2x$
18. (a) $f(x) = 0.4x + 75$ **(b)** 35 minutes
19. (a) Yes; $P(0) = 100, P(20) = 134, P(30) = 150,$
$P(50) = 180$ **(b)** Probably not because the race is over after 50 seconds; $[0, 50]$.
20. (a) $x = 25$
$[0, 25, 5]$ by $[0, 2, 0.5]$

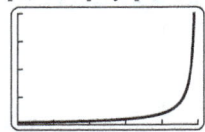

(b) 24 vehicles/min

Chapters 1–15 Cumulative Review (pp. 1024–1025)

1. $2^3 \cdot 3 \cdot 5$ **2.** $3n - 4 = n; 2$ **3.** $-\frac{1}{6}$ **4.** 6
5. 15 **6.** 7.5% **7.** $a = \frac{3A + b}{2}$ **8.** $\{t | t > 2\}$
9. **10.**

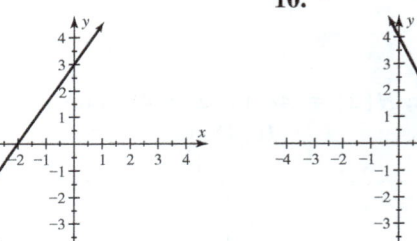

11. $y = -\frac{1}{2}x + 2$ **12.** $(1, 1)$
13. **14.**

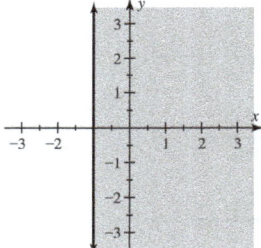

15.

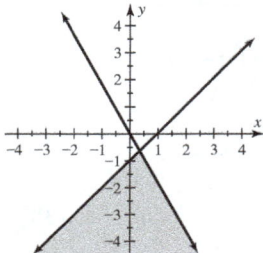

16. $-x^2 + 2x - 9$ **17.** $\frac{1}{x}$ **18.** $6b$ **19.** $24t^3$
20. $\frac{1}{4x^2y^2}$ **21.** $2x^5 - 8x^4 - 10x^2$ **22.** $10x^2 - 33x - 7$

23. $y^2 - 9$ **24.** $x^2 - 8xy + 16y^2$ **25.** 25,000
26. 2.8×10^{-2} **27.** $3x^2 - 2x + 4$
28. $3x^2 + 5x + 5 + \frac{6}{x-1}$ **29.** $5x^2y^2(2y - 3x)$
30. $(x - 1)(x + 1)(x + 3)$ **31.** $(z - 1)(2z + 3)$
32. $(4x - 5)(4x + 5)$ **33.** $(a - 2)(a^2 + 2a + 4)$
34. $(z^2 + 1)(z^2 + 6)$ **35.** $-5, 0$ **36.** 0
37. $-3, \frac{1}{2}$ **38.** $-1, 0, 1$ **39.** $x + 2$ **40.** $-\frac{5}{3}$; undefined
41. $2x - 4$ **42.** $\frac{3x + 7}{(x + 1)(x + 3)}$ **43.** -8 **44.** 10 **45.** 12
46.

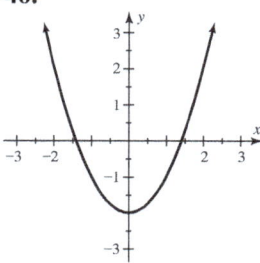

$D: (-\infty, \infty); R: [-2, \infty)$
47. $(-\infty, 5) \cup (8, \infty)$ **48.** $-1, 5$ **49.** $[-1, 5]$
50. $(-\infty, 2) \cup (6, \infty)$ **51.** $d = 72t$
52. (a) $y = \frac{1}{4}x + 2$ **(b)** $\frac{1}{4}$
(c) Rain is falling at a rate of $\frac{1}{4}$ inch per hour.
(d) 3 inches **53.** $2100 at 5%, $2900 at 7% **54.** 104 feet
55. (a) $20x$ **(b)** 50 min **56.** $1650

16 Systems of Linear Equations

Section 16.1 (pp. 1036–1038)

1. No; three planes cannot intersect at exactly 2 points.
2. $x + y + z = 5, 2x - 3y + z = 7, x + 2y - 4z = 2$
(answers may vary) **3.** Yes **4.** No; a solution must be an ordered triple. **5.** Two **6.** Three **7.** no **8.** Infinitely many **9.** $(1, 2, 3)$ **11.** $(-1, 1, 2)$ **13.** $(3, -1, 1)$
15. $\left(\frac{17}{2}, -\frac{3}{2}, 2\right)$ **17.** $(5, -2, -2)$ **19.** $(11, 2, 2)$
21. $(1, -1, 2)$ **23.** $(0, -3, 2)$ **25.** $(-1, 3, -2)$
27. $(-1, 2, 4)$ **29.** $(8, 5, 2)$ **31.** No solutions
33. $(4 - z, 2, z)$ **35.** $(3, -3, 0)$ **37.** $\left(-\frac{3}{2}, 5, -10\right)$
39. $\left(\frac{3}{2}, 1, -\frac{1}{2}\right)$ **41.** No solutions **43.** $(4 - z, 1, z)$
45. No solutions **47. (a)** $x + 2y + 4z = 10$,
$x + 4y + 6z = 15, 3y + 2z = 6$ **(b)** $(2, 1, 1.5)$;
a hamburger costs $2, fries $1, and a soft drink $1.50.
49. (a) $x + y + z = 180, x - z = 55, x - y - z = -10$
(b) $x = 85°, y = 65°,$ and $z = 30°$ **(c)** These values check. **51.** 110°, 50°, 20°
53. (a) $a + 600b + 4c = 525, a + 400b + 2c = 365,$
$a + 900b + 5c = 805$ **(b)** $a = 5, b = 1, c = -20,$
$F = 5 + A - 20W$ **(c)** 445 fawns
55. (a) $N + P + K = 80, N + P - K = 8, 9P - K = 0$
(b) $(40, 4, 36)$; 40 pounds nitrogen, 4 pounds phosphorus, 36 pounds potassium **57.** $7500 at 4%, $8500 at 5%, and $14,000 at 7.5%

Section 16.2 (pp. 1047–1050)

1. A rectangular array of numbers
2. $\begin{bmatrix} 2 & 1 & 3 \\ 0 & -4 & 2 \end{bmatrix}$ is 2×3 (answers may vary).
3. $\begin{bmatrix} 1 & 3 & | & 10 \\ 2 & -6 & | & 4 \end{bmatrix}$ is 2×3 (answers may vary).
4. 3×4 **5.** $\begin{bmatrix} 1 & 0 & | & -3 \\ 0 & 1 & | & 5 \end{bmatrix}$ (answers may vary)
6. $4, 2, -3$ **7.** 3×3 **9.** 3×2
11. $\begin{bmatrix} 1 & -3 & | & 1 \\ -1 & 3 & | & -1 \end{bmatrix}$ **13.** $\begin{bmatrix} 2 & -1 & 2 & | & -4 \\ 1 & -2 & 0 & | & 2 \\ -1 & 1 & -2 & | & -6 \end{bmatrix}$
15. $x + 2y = -6, 5x - y = 4$ **17.** $x - y + 2z = 6,$
$2x + y - 2z = 1, -x + 2y - z = 3$
19. $x = 4, y = -2, z = 7$
21. $\begin{bmatrix} 0 & 1 & 1 & 1 \\ 1 & 0 & 1 & 0 \\ 0 & 0 & 0 & 1 \\ 1 & 0 & 1 & 0 \end{bmatrix}$ **23.** $\begin{bmatrix} 0 & 1 & 0 & 0 \\ 0 & 0 & 0 & 0 \\ 1 & 1 & 0 & 1 \\ 0 & 1 & 0 & 0 \end{bmatrix}$ **25.** $(1, 3)$
27. $\left(-\frac{3}{2}, 2\right)$ **29.** $\left(\frac{7}{2}, -\frac{3}{2}\right)$ **31.** $\left(\frac{7}{2}, \frac{3}{2}\right)$ **33.** $(1, 2, 3)$
35. $(3, -3, 3)$ **37.** $(-1, 1, 0)$ **39.** $(1, 1, 1)$
41. $(-3, 2, 2)$ **43.** $(-7, 5)$ **45.** $(1, 2, 3)$
47. $(1, 0.5, -3)$ **49.** $(0.5, 0.25, -1)$
51. $(0.5, -0.2, 1.7)$ **53.** Dependent **55.** Inconsistent
57. Dependent **59.** $\left(\frac{1}{a}, \frac{1}{b}, \frac{2}{ab}\right)$ **61.** 214 pounds
63. (a) $a + 16b + 26c = 80$
$a + 28b + 45c = 344$
$a + 31b + 54c = 416$
(b) $a \approx -272.9, b \approx 19.8,$ and $c \approx 1.4$ **(c)** About 216 lb
65. $\frac{1}{2}$ hour at 5 miles per hour, 1 hour at 6 miles per hour, and $\frac{1}{2}$ hour at 8 miles per hour **67.** $500 at 3%, $1000 at 4%, and $1500 at 6% **69.** Answers may vary.

71. (a) $\begin{bmatrix} 1 & 19 & 57.5 & 32 & | & 125 \\ 1 & 26 & 65 & 42 & | & 316 \\ 1 & 30 & 72 & 48 & | & 436 \\ 1 & 30.5 & 75 & 54 & | & 514 \end{bmatrix}$

(b) $a \approx -552.272, b \approx 8.733, c \approx 2.859, d \approx 10.843$

Checking Basic Concepts 16.1 & 16.2 (p. 1051)

1. $(1, 3, -1)$ **2.** $(1, 2, 3)$ **3.** $(-2, 2, -1)$ **4.** $200 at 1%; $400 at 2%; $900 at 4%

Section 16.3 (pp. 1056–1057)

1. square **2.** number **3.** system of linear equations **4.** 0
5. -2 **7.** -53 **9.** -323 **11.** -5 **13.** -36 **15.** -42
17. -50 **19.** 0 **21.** -3555 **23.** -7466.5 **25.** abc
27. 15 square feet **29.** 52 square feet **31.** 25.5 square feet
33. $(2, -2)$ **35.** $\left(-\frac{29}{11}, -\frac{34}{11}\right)$ **37.** $(-5, 7)$ **39.** $(1, 3, 2)$
41. $(1, -1, 1)$ **43.** $(-1, 0, 4)$

Checking Basic Concepts 16.3 (p. 1058)

1. (a) -1 **(b)** 17 **2.** $(-4, 6)$ **3.** 21 square units

Chapter 16 Review (pp. 1059–1061)

1. Yes **2.** $(1, -1, 2)$ **3.** $(-4, 3, 2)$ **4.** $(-1, 5, -3)$
5. $(-1, 3, 2)$ **6.** $(1, 1, 1)$ **7.** No solutions
8. $(2 - z, 2z - 1, z)$

9. $\begin{bmatrix} 1 & 1 & 1 & | & -6 \\ 1 & 2 & 1 & | & -8 \\ 0 & 1 & 1 & | & -5 \end{bmatrix}; (-1, -2, -3)$

10. $\begin{bmatrix} 1 & 1 & 1 & | & -3 \\ -1 & 1 & 0 & | & 5 \\ 0 & 1 & 1 & | & -1 \end{bmatrix}; (-2, 3, -4)$

11. $\begin{bmatrix} 1 & 2 & -1 & | & 1 \\ -1 & 1 & -2 & | & 5 \\ 0 & 2 & 1 & | & 10 \end{bmatrix}; (-5, 4, 2)$

12. $\begin{bmatrix} 2 & 2 & -2 & | & -14 \\ -2 & -3 & 2 & | & 12 \\ 1 & 1 & -4 & | & -22 \end{bmatrix}; (-4, 2, 5)$

13. $(-7, 4, 2)$ **14.** $(5.4, 2.1, 9.7)$ **15.** -8 **16.** 30
17. 89 **18.** 130 **19.** $181{,}845$ **20.** 67.688 **21.** 46 square feet **22.** 128 square feet **23.** $(2, -1)$ **24.** $(-5, 3)$
25. 5489 in 1994; 4641 in 2004 **26.** \$8 tickets: 285; \$12 tickets: 195 **27. (a)** $m + 3c + 5b = 14$, $m + 2c + 4b = 11$, $c + 3b = 5$ **(b)** Malts: \$3; cones: \$2; bars: \$1 **28.** $100°, 65°$, and $15°$ **29.** 2 pounds of \$1.50 candy, 4 pounds of \$2 candy, 6 pounds of \$2.50 candy
30. (a) $a + 202b + 63c = 40$, $a + 365b + 70c = 50$, $a + 446b + 77c = 55$ **(b)** $a \approx 27.134; b \approx 0.061$, $c \approx 0.009$ **(c)** About 46 inches

Chapter 16 Test (p. 1061)

1. No. Three planes cannot intersect at exactly three points.
2. $(1, -2, -2)$ **3.** $(-4, 2, -5)$ **4.** No solutions
5. $(1, 2, 3)$ **6.** Infinitely many solutions: $(1, z, z)$

7. (a) $\begin{bmatrix} 2 & -4 & | & -10 \\ -3 & -2 & | & 7 \end{bmatrix}$ **(b)** $(-3, 1)$

8. (a) $\begin{bmatrix} 1 & 1 & 1 & | & 2 \\ 1 & -1 & -1 & | & 3 \\ 2 & 2 & 1 & | & 6 \end{bmatrix}$ **(b)** $\left(\frac{5}{2}, \frac{3}{2}, -2\right)$

9. 6 **10.** 114 **11.** $\left(-\frac{47}{2}, -\frac{83}{2}\right)$ **12.** 6 mph: 30 min; 7 mph: 12 min; 9 mph: 18 min **13.** $85°, 60°$, and $35°$
14. (a) $a + 20b + 25c = 168$; $a + 24b + 40c = 270$; $a + 30b + 50c = 405$
(b) $a = -270, b = 20.1, c = 1.44$ **(c)** About 316 lb

Chapters 1–16 Cumulative Review (pp. 1062–1064)

1. $2^3 \cdot 3^2 \cdot 5$ **2.** $2n + 7 = n - 2; -9$ **3.** $\frac{23}{21}$ **4.** $-\frac{1}{6}$
5. 18 **6.** $\frac{1}{10}$ **7.** $\frac{6}{5}$ **8.** 2 **9.** $\frac{31}{25}; 1.24$ **10.** 6 miles
11. $t < \frac{4}{7}$

12.

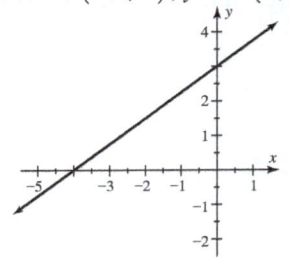

13. x-int: $(-4, 0)$; y-int: $(0, 3)$

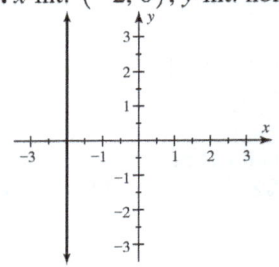

14. x-int: $(-2, 0)$; y-int: none

15. x-int: $(2, 0)$; y-int: $(0, 4)$; $y = -2x + 4$
16. $y = -2x + 1$ **17.** $y = \frac{3}{5}x + \frac{34}{5}$ **18.** $y = 2x + 3$
19. $P = 500x + 4000$ **20.** $(7.5, 10)$

21. **22.**

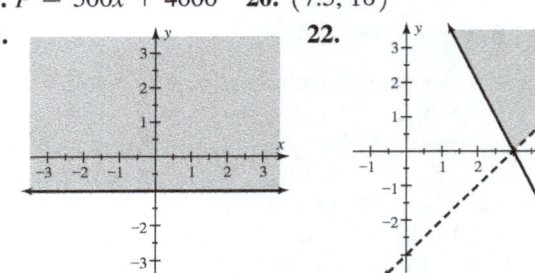

23. -76 **24.** $24t^4$ **25.** 1 **26.** $\frac{1}{4t^6}$ **27.** $2ab^3$ **28.** $\frac{a^6}{8b^6}$
29. $2a^4 - 4a^3 + 6a^2$ **30.** $5x^2 - 34x - 7$
31. $4x^2 + 12xy + 9y^2$ **32.** $a^2 - b^2$ **33.** $2x^2 - 4x + 3$
34. $x^3 - 4x^2 + 3x - 2 + \frac{1}{x-5}$ **35.** $5ab^2(2 - 5a^2b^3)$
36. $(y - 3)(y^2 + 2)$ **37.** $(2z + 3)(3z - 1)$
38. $(2z - 3)(2z + 3)$ **39.** $(2y - 5)^2$
40. $(a - 3)(a^2 + 3a + 9)$ **41.** $(z^2 - 3)(4z^2 - 5)$
42. $ab(a + b)(2a - b)$ **43.** $-2, 1$ **44.** $-\frac{1}{3}, \frac{3}{2}$
45. $x - 4$ **46.** $\frac{x-6}{3x-4}$ **47.** $2(x - 1)$ or $2x - 2$ **48.** 1
49. 8 **50.** $\frac{17}{12}$ **51.** 10 **52.** 13

ANSWERS TO SELECTED EXERCISES   A-49

53. $D = (-\infty, \infty); R = [-2, \infty)$

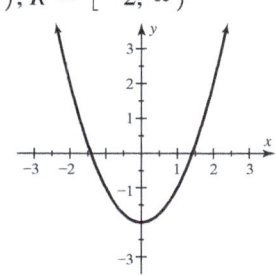

54. $2, -2$ **55.** $(3, \infty)$ **56.** $(-\infty, 0]$
57. $(-1, 5]$

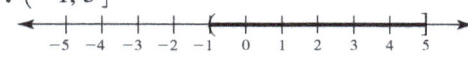

58. $\left(-\infty, -\frac{3}{2}\right) \cup [3, \infty)$

59. $[-2, 4]$ **60.** $[-16, 10]$ **61. (a)** $-3, 1$ **(b)** $[-3, 1]$
(c) $(-\infty, -3] \cup [1, \infty)$ **62.** $-6, 18$
63. $(-\infty, -6) \cup \left(\frac{8}{3}, \infty\right)$ **64.** $[4, 7]$ **65.** $(1, 1, 2)$

66. $\begin{bmatrix} 1 & 1 & -1 & | & 4 \\ -1 & -1 & -1 & | & 0 \\ 1 & -2 & 1 & | & -9 \end{bmatrix}; (-1, 3, -2)$

67.

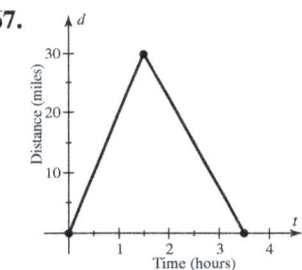

68. 3 children and 5 adults   **69.** $\frac{6}{7}$ hour   **70. (a)** 112 feet
(b) 5.5 seconds   **71.** 61.25 feet
72. (a) $4b + 3f + 4m = 23; b + 2f + m = 7;$
$3b + f + 2m = 13$ **(b)** Burger: \$2; fries: \$1; malt: \$3

17 Radical Expressions and Functions

Section 17.1 (pp. 1074–1077)

1. ± 3 **2.** 3   **3.** 2   **4.** Yes   **5.** b   **6.** $|x|$
7. Not a real number   **8.** -3 **9.** d.   **10.** a.

11.

12.

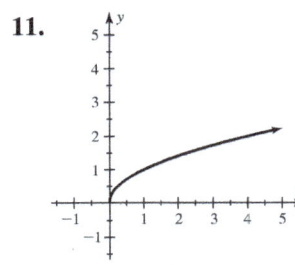

13. $\{x | x \geq 0\}$ **14.** All real numbers   **15.** 3   **17.** 0.6
19. $\frac{4}{5}$ **21.** x **23.** 3   **25.** -4 **27.** $\frac{2}{3}$ **29.** $-x^3$ **31.** $3x$
33. 3   **35.** -3 **37.** Not a real number   **39.** 3.32
41. 1.71   **43.** -1.48 **45.** 4   **47.** $|y|$ **49.** $|x - 5|$
51. $|x - 1|$ **53.** $|y|$ **55.** $|x^3|$ **57.** x **59.** 3, not a real number   **61.** $\sqrt{6}$, not a real number   **63.** $\sqrt{20}$ or $2\sqrt{5}, \sqrt{6}$ **65.** 1, 2   **67.** $-3, 2$ **69.** $-1, \sqrt[3]{-6}$ or $-\sqrt[3]{6}$
71. 4   **73.** 5   **75.** $[-2, \infty)$ **77.** $[2, \infty)$ **79.** $[2, \infty)$
81. $(-\infty, 1]$ **83.** $\left(-\infty, \frac{8}{5}\right]$ **85.** $(-\infty, \infty)$ **87.** $\left(-\frac{1}{2}, \infty\right)$
89. $[-2, \infty)$ **91.** $[0, \infty)$

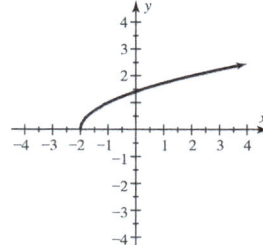   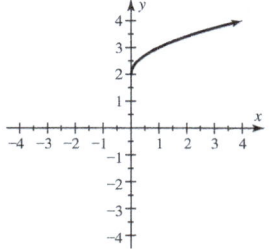

Shifted 2 units left   Shifted 2 units upward

93. $(-\infty, \infty)$

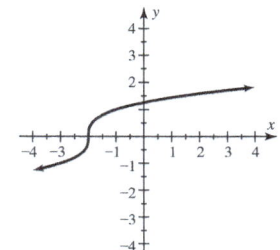

Shifted 2 units left

95.

x	$\sqrt{x} + 1$
-1	—
0	1
1	2
4	3
9	4

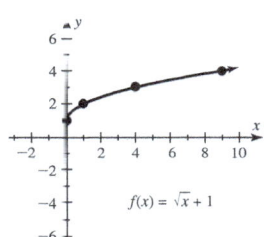

97.

x	$\sqrt{3x}$
-1	—
0	0
$\frac{1}{3}$	1
$\frac{4}{3}$	2
3	3

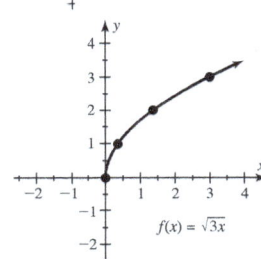

99.

x	$2\sqrt[3]{x}$
-8	-4
-1	-2
0	0
1	2
8	4

101.

x	$\sqrt[3]{x-1}$
-7	-2
0	-1
1	0
2	1
9	2

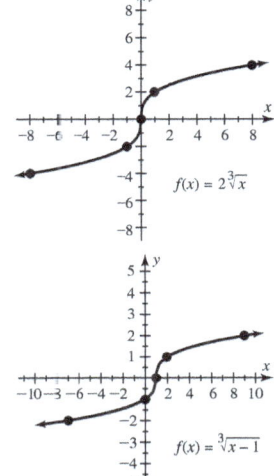

103. $A = 6$ **105.** 1 sec **107.** 122 mi
109. About $14 thousand; no
111. (a) $P(25,000) = \$71{,}246$; if \$25,000 is spent on equipment per worker, each worker will produce about \$71,246 worth of goods.

(b)
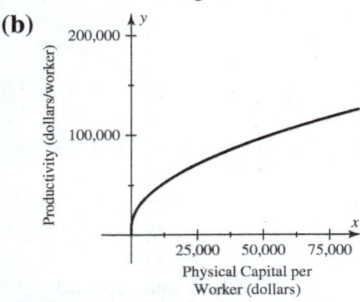

(c) \$26,197; \$20,102; an additional \$25,000 is spent on equipment per worker, but productivity levels off. There is a point where the business starts to lose money.
113. (a) 50, about 82, about 95 **(b)** In 2012, about 95% of the soybeans that were planted were genetically engineered. **(c)** For 1990, $x = -10$ and the equation is undefined for negative values of x.

Section 17.2 (pp. 1083–1085)

1. 2 **2.** 2 **3.** $\frac{1}{2}$ **4.** $\frac{1}{2}$ **5.** $x^{1/2}$ **6.** $a^{4/3}$ **7.** $\sqrt[n]{a}$
8. $(\sqrt[n]{a})^m$ or $\sqrt[n]{a^m}$ **9.** c. **10.** f. **11.** g. **12.** h. **13.** d.
14. e. **15.** a. **16.** b. **17.** 1.74 **19.** 1.71 **21.** 3.74
23. 0.55 **25.** $\sqrt{7}$ **27.** $\sqrt[3]{a}$ **29.** $\sqrt[6]{x^5}$ **31.** $\sqrt{x+y}$
33. $\frac{1}{\sqrt[3]{b^2}}$ **35.** $t^{1/2}$ **37.** $(x+1)^{1/3}$ **39.** $(x+1)^{-1/2}$
41. $(a^2-b^2)^{1/2}$ **43.** $x^{-7/3}$ **45.** $\sqrt{9}$; 3 **47.** $\sqrt[3]{8}$; 2
49. $\sqrt{\frac{4}{9}}$; $\frac{2}{3}$ **51.** $\sqrt[3]{(-8)^2}$ or $(\sqrt[3]{-8})^2$; 4 **53.** $\sqrt[3]{8}$; 2
55. $\frac{1}{\sqrt[4]{16^3}}$ or $\frac{1}{(\sqrt[4]{16})^3}$; $\frac{1}{8}$ **57.** $\frac{1}{(\sqrt[4]{4})^3}$; $\frac{1}{8}$ **59.** $\sqrt[4]{z}$ **61.** $\frac{1}{\sqrt[5]{y^2}}$
63. $\sqrt[3]{3x}$ **65.** x **67.** $2x^{2/3}$ **69.** $\frac{7}{x^{1/6}}$ **71.** x^3 **73.** xy^2
75. $y^{13/6}$ **77.** $\frac{x^4}{9}$ **79.** $\frac{y^3}{x}$ **81.** $y^{1/4}$ **83.** $\frac{1}{a^{2/3}}$ **85.** $\frac{1}{k^2}$
87. $b^{3/4}$ **89.** p^2+p **91.** $x^{5/6}-x$ **93.** $\frac{3}{x^{1/6}}$ **95.** b^3
97. (a)

x	0	20	40	60	80
$A(x)$	0	27%	37%	44%	50%

(b) The abandonment rate levels off. The longer a person watches a video, the more likely he or she will continue to watch. **99.** About 1 step/sec **101. (a)** 400 in^2
(b) $A = 100W^{2/3}$ **103. (a)** 42.4 cm; 44.7 cm
(b) First year **105.** Answers may vary.

Checking Basic Concepts 17.1 & 17.2 (p. 1085)

1. (a) ± 7 **(b)** 7 **2. (a)** -2 **(b)** -3 **3. (a)** $\sqrt{x^3}$ or $(\sqrt{x})^3$ **(b)** $\sqrt[3]{x^2}$ or $(\sqrt[3]{x})^2$ **(c)** $\frac{1}{\sqrt[5]{x^2}}$ or $\frac{1}{(\sqrt[5]{x})^2}$
4. $|x-1|$ **5. (a)** 3 **(b)** 5 **(c)** $\left(\frac{12}{7}\right)^{7/12}$ or about 1.37.

Section 17.3 (pp. 1092–1094)

1. Yes **2.** No **3.** $\sqrt[3]{ab}$ **4.** $\frac{a}{b}$ **5.** $\frac{a}{b}$ **6.** $\frac{1}{3}$ **7.** No
8. Yes **9.** No, since $1^3 \neq 3$ **10.** Yes; $4^3 = 64$
11. 3 **13.** 10 **15.** 4 **17.** $\frac{3}{5}$ **19.** $\frac{1}{4}$ **21.** $\frac{2}{3}$ **23.** x^3
25. $\frac{\sqrt[3]{7}}{3}$ **27.** $\frac{\sqrt[4]{x}}{3}$ **29.** $\frac{x}{9}$ **31.** $\frac{x}{4}$ **33.** 3 **35.** 4 **37.** 3
39. -2 **41.** $2y$ **43.** 3 **45.** $2x^2$ **47.** $-a^2\sqrt[3]{5}$
49. $2x\sqrt[4]{y}$ **51.** $6x^3$ **53.** $4x^2y^3$ **55.** $\frac{3}{2}$ **57.** $\sqrt[3]{12ab}$
59. $5\sqrt{z}$ **61.** $\frac{1}{a}$ **63.** $\sqrt{x^2-16}$ **65.** $\sqrt[3]{a^3+1}$
67. $\sqrt{x+1}$ **69.** 10 **71.** 2 **73.** 3 **75.** $10\sqrt{2}$ **77.** $3\sqrt[3]{3}$
79. $2\sqrt{2}$ **81.** $-2\sqrt[5]{2}$ **83.** $b^2\sqrt{b}$ **85.** $2n\sqrt{2n}$
87. $2ab^2\sqrt{3b}$ **89.** $-5xy\sqrt[3]{xy^2}$ **91.** $5\sqrt[3]{5t^2}$ **93.** $\frac{3t}{r\sqrt[4]{5r^3}}$
95. $\frac{3\sqrt[3]{x^2}}{y}$ **97.** $\frac{7a\sqrt{a}}{9}$
99. $\left(\sqrt[mn]{a^m b^m}\right)^n = (a^m b^m)^{n/mn}$
$= (a^m b^m)^{1/m}$
$= a^{m/m} b^{m/m}$
$= ab$
101. $\sqrt[6]{3^5}$ **103.** $2\sqrt[12]{2^5}$ **105.** $3\sqrt[12]{3^{11}}$ **107.** $x\sqrt[12]{x}$
109. $\sqrt[12]{r^{11}t^7}$ **111. (a)** $S = 5\sqrt{M}$ **(b)** 50 miles per hour
113. (a) 12; there is a 12% price markup 4 days prior to game day. **(b)** $y = 6\sqrt{x}$ **(c)** 12; yes **115. (a)** $\frac{9}{2}$ sec
(b) $T = \frac{\sqrt{h}}{2}$

Section 17.4 (pp. 1104–1106)

1. $2\sqrt{a}$ **2.** $3\sqrt[3]{b}$ **3.** like
4. Yes; $4\sqrt{15} - 3\sqrt{15} = \sqrt{15}$
5. No **6.** $\frac{\sqrt{7}}{\sqrt{7}}$ **7.** $\sqrt{t}+5$ **8.** $\frac{5+\sqrt{2}}{5+\sqrt{2}}$
9. Not possible **11.** $\sqrt{7}, 2\sqrt{7}, 3\sqrt{7}$ **13.** $2\sqrt[3]{2}, -3\sqrt[3]{2}$
15. Not possible **17.** $2\sqrt[3]{xy}, xy\sqrt[3]{xy}$ **19.** $9\sqrt{3}$ **21.** $6\sqrt[3]{5}$
23. Not possible **25.** Not possible **27.** Not possible
29. $5\sqrt[3]{2}$ **31.** $8\sqrt{2}$ **33.** $6\sqrt{11}$ **35.** $2\sqrt{x}-\sqrt{y}$
37. $2\sqrt[3]{z}$ **39.** $-5\sqrt[3]{6}$ **41.** y^2-y **43.** $9\sqrt{7}$ **45.** $6\sqrt[4]{3}$
47. $7\sqrt{x}$ **49.** $8\sqrt{2k}$ **51.** $-2\sqrt{11}$ **53.** $5\sqrt[3]{2}-\sqrt{2}$
55. $-\sqrt[3]{xy}$ **57.** $3\sqrt{x+2}$ **59.** $\sqrt{x+2}$
61. $(x-1)\sqrt{x+1}$ **63.** $4x\sqrt{x}$ **65.** $\frac{\sqrt[3]{7x}}{6}$ **67.** $\frac{3\sqrt{3}}{2}$
69. $\frac{71\sqrt{2}}{10}$ **71.** $2\sqrt{2}$ **73.** $(5x-1)\sqrt[4]{x}$
75. $(8x+2)\sqrt{x}$ or $2\sqrt{x}(4x+1)$
77. $(3ab-1)\sqrt[3]{ab}$ **79.** $(n-2)\sqrt[3]{n}$
81. $(f+g)(x) = 3\sqrt{x}+1$; $(f-g)(x) = 7\sqrt{x}-5$
83. $(f+g)(x) = 4\sqrt[3]{x}$; $(f-g)(x) = 2$
85. $x - \sqrt{x} - 6$ **87.** 2 **89.** 119 **91.** $x-64$
93. $ab - c$ **95.** $x + \sqrt{x} - 56$ **97.** $\frac{\sqrt{7}}{7}$ **99.** $\frac{4\sqrt{3}}{3}$
101. $\frac{\sqrt{5}}{3}$ **103.** $\frac{\sqrt{3b}}{6}$ **105.** $\frac{t\sqrt{r}}{2r}$ **107.** $\frac{3+\sqrt{2}}{7}$
109. $\sqrt{10} - 2\sqrt{2}$ **111.** $\frac{11-4\sqrt{7}}{3}$ **113.** $\sqrt{7}+\sqrt{6}$
115. $\frac{z+3\sqrt{z}}{z-9}$ **117.** $\frac{a+2\sqrt{ab}+b}{a-b}$ **119.** $\sqrt{x+1}+\sqrt{x}$
121. $\frac{3\sqrt[3]{x^2}}{x}$ **123.** $\frac{\sqrt[3]{x}}{x}$ **125.** $12\sqrt{3} \approx 20.8$ cm **127.** $2\sqrt{6}$
129. $120\sqrt{2}$ ft **131.** $\sqrt{2x}$ ft **133.** 850 ft^3/sec
135. $P = 325$; there were about 325 thousand adults incarcerated in state or federal prisons for drug law violations in the year 2000. $J = 105$; there were about 105 thousand adults incarcerated in jails for drug law violations in the year 2000.
137. $T = 90\sqrt{x} + 70$

Checking Basic Concepts 17.3 & 17.4 (p. 1106)

1. (a) $\frac{1}{8}$ **(b)** 10 **(c)** $-2x\sqrt[3]{xy}$ **(d)** $\frac{4b^2}{5}$ **2.** $\sqrt[6]{7^5}$
3. (a) 6 **(b)** 3 **(c)** $6x^3$ **4. (a)** $7\sqrt{6}+\sqrt{7}$
(b) $5\sqrt[3]{x}$ **(c)** $\sqrt{x}$ **5. (a)** $(y-x)\sqrt[3]{xy}$ **(b)** 14
6. $\frac{\sqrt{6}}{2}$ **7.** $\frac{\sqrt{5}+1}{2}$

Section 17.5 (pp. 1111–1114)

1. 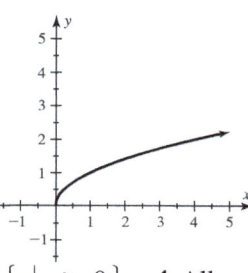   **2.**

3. $\{x \mid x \geq 0\}$ **4.** All real numbers   **5.** $f(x) = x^p$, p is rational.   **6.** $f(x) = \sqrt[n]{x}$, where n is an integer greater than 1.   **7.** $\{x \mid x \geq 0\}$ **8.** All real numbers
9. $\sqrt{3} \approx 1.73$; not a real number   **11.** $\sqrt[4]{6} \approx 1.57$; not a real number   **13.** $\sqrt[5]{13} \approx 1.67$; 1   **15.** $\sqrt[3]{6} \approx 1.82$; -1
17. $f(x) = \sqrt{x}$ **19.** $f(x) = \sqrt[3]{x^2}$ **21.** $f(x) = \frac{1}{\sqrt[5]{x}}$
23. 32; 55.90   **25.** $-\frac{1}{128} \approx -0.01$; 0.04
27. 4; not possible   **29.** 4; 4
31. $[0, \infty)$ **33.** $(-\infty, \infty)$

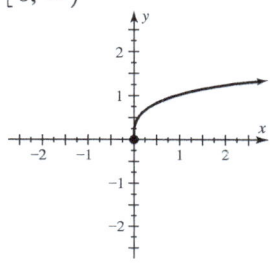   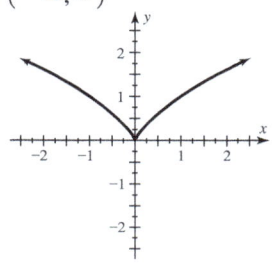

35. $g(x)$ is greater.   **37.** $f(x)$ is greater.
$[0, 6, 1]$ by $[0, 6, 1]$ $[0, 6, 1]$ by $[0, 6, 1]$

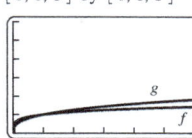

39. $x^p > x^q$ **41. (a)** 6   **(b)** $\sqrt{2x}$ **(c)** $4|x|$
(d) 2   **43.** b.   **45.** c.   **47.** d.   **49.** b.
51. (a) $f(x) = (x + 2)(x - 2)$ **(b)** $f(x) = x^2 - 4$
53. (a) $f(x) = (x + \sqrt{7})(x - \sqrt{7})$
(b) $f(x) = x^2 - 7$
55. (a) $f(x) = (x + \sqrt{6})(x - \sqrt{6})$
(b) $f(x) = x^2 - 6$ **57.** 2877 in^2
59. About 35%   **61.** No, for $x \geq 10$, it is less than double.
63. (a) 6 yr   **(b)** The twin in the spaceship will be 4 years younger than the twin on Earth.
65. (a) $[0, 5, 1]$ by $[0, 0.5, 0.1]$ **(b)** $k \approx 0.16$

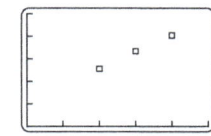

(c) $[0, 5, 1]$ by $[0, 0.5, 0.1]$; yes   **(d)** 0.295 m^2

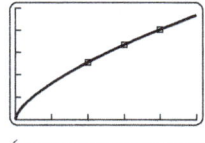

(answers may vary)

67. (a) $k \approx 0.91$ **(b)** $[0, 15, 0.1]$ by $[0, 1, 0.1]$

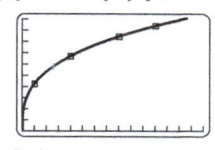

It increases.
(c) About 0.808 m   **(d)** About 0.788; a bird weighing 0.65 kg has a wing span of about 0.788 m.

Section 17.6 (pp. 1123–1127)

1. Square each side.   **2.** Cube each side.   **3.** Yes
4. Check them in the given equation.   **5.** Finding an unknown side of a right triangle (answers may vary)   **6.** 5
7. $d = \sqrt{(x_2 - x_1)^2 + (y_2 - y_1)^2}$ **8.** $x^{1/2} + x^{3/4} = 2$
9. 2   **11.** x **13.** $2x + 1$ **15.** $5x^2$ **17.** 64   **19.** 81   **21.** 6
23. 48   **25.** 6   **27.** 3   **29.** 27   **31.** -2 **33.** 15   **35.** $\frac{1}{2}$
37. 0, 1   **39.** 2   **41.** -4 **43.** 9   **45.** 9   **47.** ± 7
49. ± 10 **51.** $-5, 3$ **53.** $\frac{1}{3}, \frac{11}{3}$ **55.** 4   **57.** -4
59. 1   **61.** $\frac{7}{5}$ **63.** ± 2 **65.** $\sqrt[5]{12}$ **67.** $-5, 1$ **69.** 3
71. 1.88   **73.** $-1, 0.70$ **75.** 1.79   **77.** No Solutions
79. (a–c) 16   **81. (a–c)** 11   **83.** $L = \frac{8T^2}{\pi^2}$ **85.** $A = \pi r^2$
87. Yes   **89.** Yes   **91.** Yes   **93.** No   **95.** $\sqrt{32} = 4\sqrt{2}$
97. 7   **99.** $c = 5$ **101.** $b = \sqrt{61}$ **103.** $a = 14$
105. $\sqrt{20} = 2\sqrt{5}$ **107.** $\sqrt{4000} = 20\sqrt{10}$ **109.** $\sqrt{89}$
111. 5   **113.** 4   **115.** 2, 122   **117.** $W = 8$ pounds
119. About 81 sec   **121.** About 3 miles   **123.** About 269 feet
125. 19 inches   **127.** $h \approx 16.3$ inches, $d \approx 33.3$ inches
129. (a) 121 feet   **(b)** About 336 feet   **131. (a)** About 38 miles per hour; the vehicle involved in the accident was traveling about 38 miles per hour.   **(b)** About 453 feet
133. (a) It increases by a factor of 8.   **(b)** $v = \sqrt[3]{\frac{W}{3.8}}$
(c) 20 mph   **135.** $a\sqrt{2}$ **137. (a)** 30x
(b) $d = \sqrt{(1000 - x)^2 + 500^2}$
(c) $50\sqrt{(1000 - x)^2 + 500^2}$
(d) $30x + 50\sqrt{(1000 - x)^2 + 500^2}$
(e) $[0, 1000, 100]$ by $[40000, 60000, 5000]$

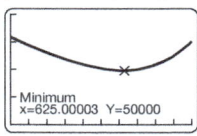

Checking Basic Concepts 17.5 & 17.6 (p. 1127)

1. (a)   **(b)**

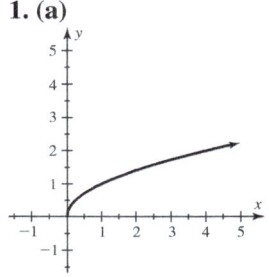   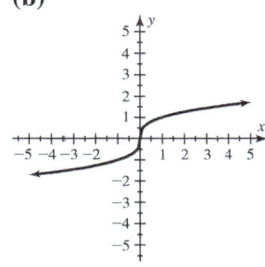

$f(-1)$ is not a real number. $f(-1) = -1$

(c)

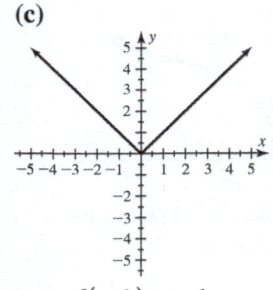

$f(-1) = 1$

2. 3.2 **3.** $[4, \infty)$ **4. (a)** 4 **(b)** 28 **(c)** 3 **5.** 13
6. 9.6 in. **7.** −3, 1

Section 17.7 (pp. 1133–1135)

1. $2 + 3i$ (answers may vary) **2.** No; any real number a can be written as $a + 0i$. **3.** i **4.** −1 **5.** $i\sqrt{a}$
6. $10 - 7i$ **7.** $a + bi$ **8.** $1 + 2i$ **9.** 4 **10.** −5 **11.** 0
12. pure **13.** $i\sqrt{5}$ **15.** $10i$ **17.** $12i$ **19.** $2i\sqrt{3}$
21. $3i\sqrt{2}$ **23.** 3 **25.** $-8 + 7i$ **27.** $1 - 9i$ **29.** $-10 + 7i$
31. $-13 + 13i$ **33.** $20 - 12i$ **35.** 41 **37.** 20 **39.** $1 + 5i$
41. $3 + 4i$ **43.** $-2 - 6i$ **45.** −2 **47.** $a^2 + 9b^2$
49. $-i$ **51.** i **53.** −1 **55.** 1 **57.** $3 - 4i$ **59.** $6i$
61. $5 + 4i$ **63.** −1 **65.** $1 - i$ **67.** $-\frac{6}{29} + \frac{15}{29}i$ **69.** $2 + i$
71. $-1 + 6i$ **73.** $-1 - 2i$ **75.** $-\frac{3}{2}i$ **77.** $-\frac{1}{2} + \frac{3}{2}i$
79. (a) $f(x) = (x + i)(x - i)$ **(b)** $f(x) = x^2 + 1$
81. (a) $f(x) = (x + 7i)(x - 7i)$ **(b)** $f(x) = x^2 + 49$
83. (a) $f(x) = (x + i\sqrt{6})(x - i\sqrt{6})$
(b) $f(x) = x^2 + 6$ **85.** $\frac{290}{13} + \frac{20}{13}i$ **87.** They are graphed using a real axis and an imaginary axis.

Checking Basic Concepts 17.7 (p. 1135)

1. (a) $8i$ **(b)** $i\sqrt{17}$ **2. (a)** $3 - 4i$ **(b)** $-2 + 3i$
(c) $5 + i$ **(d)** $\frac{3}{4} + \frac{3}{4}i$

Chapter 17 Review (pp. 1138–1140)

1. 2 **2.** 6 **3.** $3|x|$ **4.** $|x - 1|$ **5.** −4 **6.** −5 **7.** x^2
8. $3x$ **9.** 2 **10.** −1 **11.** x^2 **12.** $x + 1$ **13.** $\sqrt{14}$
14. $\sqrt[3]{-5}$ **15.** $\left(\sqrt{\frac{x}{y}}\right)^3$ or $\sqrt{\left(\frac{x}{y}\right)^3}$ **16.** $\frac{1}{\sqrt[3]{(xy)^2}}$ or $\frac{1}{(\sqrt[3]{xy})^2}$
17. 9 **18.** 2 **19.** 64 **20.** 27 **21.** z^2 **22.** xy^2 **23.** $\frac{x^3}{y^9}$
24. $\frac{y^2}{x}$ **25.** 8 **26.** −2 **27.** x^2 **28.** 2 **29.** $-\frac{\sqrt[3]{x}}{2}$ **30.** $\frac{1}{3}$
31. $4\sqrt{3}$ **32.** $3\sqrt{6}$ **33.** $\frac{3}{x}$ **34.** $4ab\sqrt{2a}$ **35.** $9xy$
36. $5z\sqrt[3]{z}$ **37.** $x + 1$ **38.** $\frac{2a\sqrt[4]{a}}{b}$ **39.** $2\sqrt[6]{x^5}$
40. $\sqrt[6]{r^5t^8}$ or $t\sqrt[6]{r^5t^2}$ **41.** $4\sqrt{3}$ **42.** $3\sqrt[3]{x}$ **43.** $-3\sqrt[3]{5}$
44. $-\sqrt[4]{y}$ **45.** $11\sqrt{3}$ **46.** $7\sqrt{2}$ **47.** $13\sqrt[3]{2}$
48. $3\sqrt{x + 1}$ **49.** $(2x - 1)\sqrt{x}$ **50.** $(b + 2a)\sqrt[3]{ab}$
51. $5 + 4\sqrt{2}$ **52.** $7 + 7\sqrt{3} - \sqrt{5} - \sqrt{15}$ **53.** 3
54. 95 **55.** $a - 2b$ **56.** $xy + \sqrt{xy} - 2$ **57.** $\frac{4\sqrt{5}}{5}$
58. $\frac{r\sqrt{t}}{2t}$ **59.** $\frac{3 - \sqrt{2}}{7}$ **60.** $\frac{5 + \sqrt{7}}{9}$
61. $\sqrt{8} + \sqrt{7}$ **62.** $\frac{a - 2\sqrt{ab} + b}{a - b}$

63.

64.

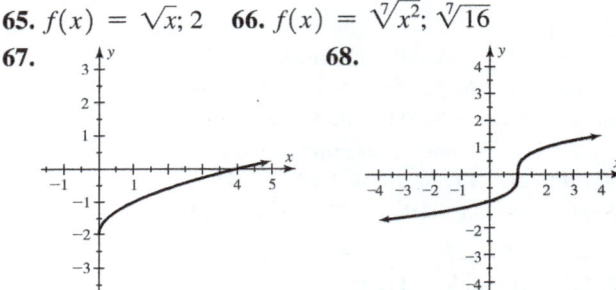

65. $f(x) = \sqrt{x}$; 2 **66.** $f(x) = \sqrt[7]{x^2}$; $\sqrt[7]{16}$
67.

68.

Shifted 2 units downward   Shifted 1 unit to the right

69. $[1, \infty)$ **70.** $(-\infty, 3]$ **71.** $(-\infty, \infty)$ **72.** $(-2, \infty)$
73. 2 **74.** 4 **75.** 9 **76.** 9 **77.** 8 **78.** $\frac{1}{4}$ **79.** 4.5
80. 1.62 **81.** $c = \sqrt{65}$ **82.** $b = \sqrt{39}$ **83.** $\sqrt{41}$
84. $\sqrt{52} = 2\sqrt{13}$ **85.** ±11 **86.** ±4 **87.** −3, 5
88. 4 **89.** 3 **90.** 2 **91.** ±4 **92.** −1 **93.** 1
94. −2, 0 **95.** −2 **96.** $-2 + 4i$ **97.** $5 + i$ **98.** $1 + 2i$
99. $-\frac{14}{13} + \frac{5}{13}i$ **100.** $2 - 2i$ **101.** About 85 feet
102. $\sqrt{16,200} = 90\sqrt{2} \approx 127.3$ feet
103. About 0.79 second **104. (a)** 5 square units
(b) $5\sqrt{5}$ cubic units **(c)** $\sqrt{10}$ units **(d)** $\sqrt{15}$ units
105. $r = \sqrt[210]{\frac{281}{4}} - 1 \approx 0.02$; from 1790 through 2000, the average annual percent growth rate was about 2%.
106. (a) About 43 miles per hour **(b)** About 34 miles per hour; a steeper bank allows for a higher speed limit; yes
107. $\sqrt{7} \approx 2.65$ feet

Chapter 17 Test (pp. 1140–1141)

1. −3 **2.** $|z + 1|$ **3.** $5x^2$ **4.** $2z^2$ **5.** $2xy\sqrt[4]{y}$ **6.** 1
7. $\sqrt[5]{7^2}$ or $(\sqrt[5]{7})^2$ **8.** $\sqrt{\left(\frac{y}{x}\right)^2}$ or $\left(\sqrt[3]{\frac{y}{x}}\right)^2$ **9.** 16
10. $\frac{1}{216}$ **11.** $x^{4/3}$ **12.** $x^{7/10}$ **13.** $(-\infty, 4]$
14.

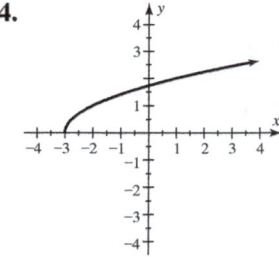

15. $8z^{3/2}$ **16.** $\frac{z}{y^{2/3}}$ **17.** 9 **18.** $\frac{y}{2}$ **19.** $4\sqrt{7} + \sqrt{5}$
20. $6\sqrt[3]{x}$ **21.** $14\sqrt{2}$ **22.** 2 **23. (a)** 27 **(b)** 7 **(c)** 3
(d) 17 **24. (a)** $\frac{2\sqrt{7}}{21}$ **(b)** $\frac{-1 + \sqrt{5}}{4}$ **25.** 2.63
26. $\sqrt{120} \approx 10.95$ **27.** $\sqrt{8} = 2\sqrt{2}$ **28.** $2 - 19i$
29. $-6 + 8i$ **30.** $\frac{5}{4}$ **31.** $\frac{4}{29} + \frac{10}{29}i$ **32. (a)** About 17.25 mi
(b) About 420 ft **33.** 1.31 pounds

Chapters 1–17 Cumulative Review (pp. 1141–1142)

1. 36π **2.** $D = \{-1, 0, 1\}, R = \{2, 4\}$ **3. (a)** $\frac{1}{a^3b^9}$
(b) x^4y^9 **(c)** r^8t^5 **4.** 4.3×10^{-4} **5.** $3; x \neq 2$
6. All real numbers

7. 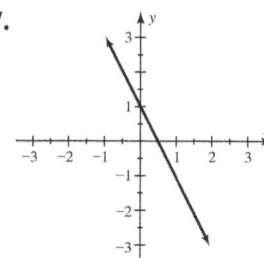 **8.**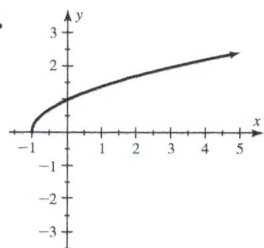

9.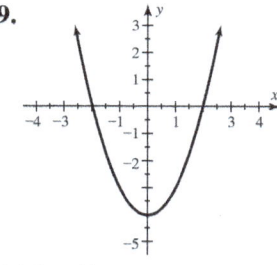

(a) D: all real numbers; R: $y \geq -4$ **(b)** 0
(c) $(-2, 0), (2, 0)$ **(d)** $-2, 2$ **10.** $y = \frac{1}{2}x + \frac{5}{2}$
11. $f(x) = 1 - 3x$

12.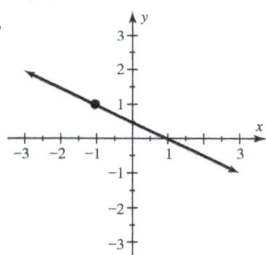

13. $\frac{6}{11}$ **14.** $(-\infty, 3]$ **15.** $[-1, 5]$ **16.** $\left[-\frac{5}{2}, 1\right]$
17. (a) $(4, 4)$ **(b)** No solutions

18.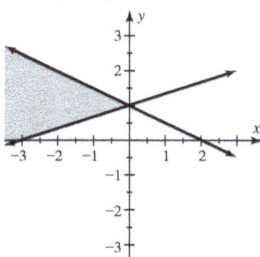

19. $(3, 2, 1)$ **20.** 4 **21.** $-4x^4 + 16x$ **22.** $x^2 - 16$
23. $5x^2 - 7x - 6$ **24.** $16x^2 + 72x + 81$
25. $(3x - 4)(3x + 4)$ **26.** $(x - 2)^2$ **27.** $3x^2(5x - 3)$
28. $(4x - 3)(3x + 1)$ **29.** $(r - 1)(r^2 + r + 1)$
30. $(x - 3)(x^2 + 5)$ **31.** $1, 2$ **32.** $-2, 0, 2$ **33.** $2x + 4$
34. $\frac{7x - 5}{x^2 - x}$ **35.** $6x$ **36.** 4 **37.** $\frac{1}{64}$ **38.** 5 **39.** $4x$
40. $x^{3/4}$ or $\sqrt[4]{x^3}$ **41.** $2x$ **42.** $6\sqrt{3x}$ **43.** $2x^2 - \sqrt{3}x - 3$
44. $(-\infty, 1]$

45.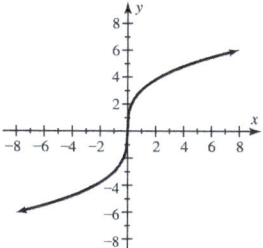

46. $\sqrt{10}$ **47.** $5 + i$ **48.** 13 **49.** 6 **50.** 28 **51.** $\frac{4}{9}, 4$
52. ± 3 **53.** 1.41 **54.** 4.06 **55.** The tank initially contains 300 gallons of water. Water is leaving the tank at 15 gal/min.

56.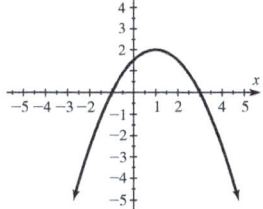

57. $1300 at 5%, $700 at 4% **58.** 8 in. by 8 in. by 4 in.
59. 22.4 feet **60.** 45°, 55°, 80°

18 Quadratic Functions and Equations

Section 18.1 (pp. 1153–1156)

1. parabola **2.** The vertex **3.** axis of symmetry **4.** $(0, 0)$

5.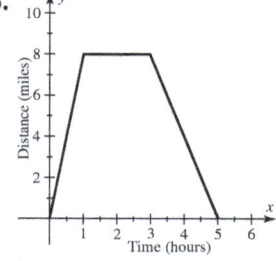

(answers may vary)
6. $-\frac{b}{2a}$ **7.** $ax^2 + bx + c$ with $a \neq 0$ **8.** vertex
9. True **10.** False **11.** False **12.** True **13.** 17
15. -8 **17.** 7 **19.** 8 **21.** $0, -4$ **23.** $-2, -2$
25. $(1, -2); x = 1$; upward **27.** $(-2, 3); x = -2$; downward **29.** $(2, -6)$ **31.** $(-3, 4)$
33. $(0, 3)$ **35.** $(1, 1.4)$ **37.** $(3, 9)$
39. (a) $(0, 0); x = 0$ **41. (a)** $(0, -2); x = 0$
(b) 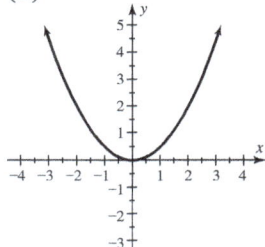 **(b)**

(c) 2; 4.5 **(c)** 2; 7

43. (a) $(0, 1); x = 0$
(b)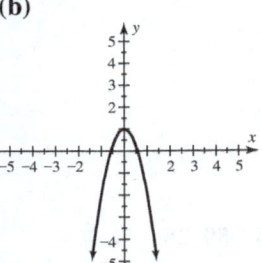
(c) $-11; -26$

45. (a) $(1, 0); x = 1$
(b)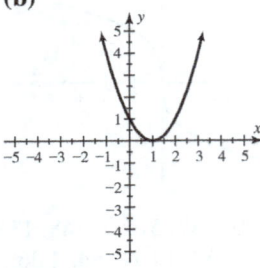
(c) $9; 4$

47. (a) $(-2, 0); x = -2$
(b)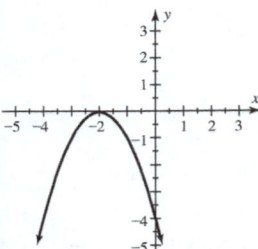
(c) $0; -25$

49.
(a) $(-0.5, -2.25); x = -0.5$
(b)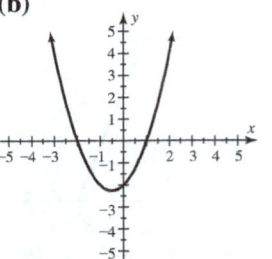
(c) $0; 10$

51. (a) $(0, -3); x = 0$
(b)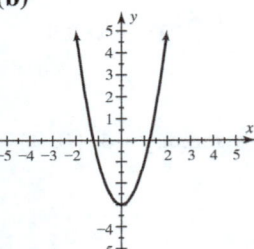
(c) $5; 15$

53. (a) $(1, 1); x = 1$
(b)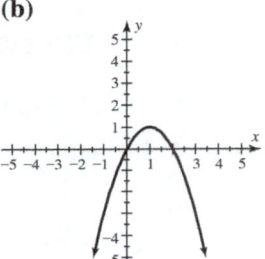
(c) $-8; -3$

55. (a) $(1, 1); x = 1$
(b)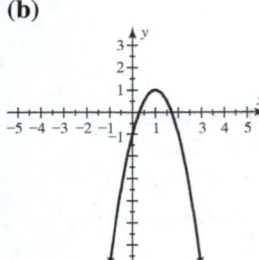
(c) $-17; -7$

57. (a) $(2, 4); x = 2$
(b)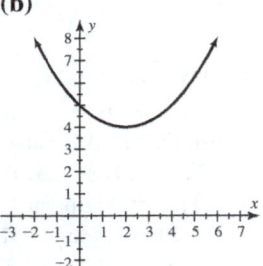
(c) $8; 4.25$

59. Incr: $x > 2$, decr: $x < 2$ **61.** Incr: $x < 1$, decr: $x > 1$
63. Incr: $x > \frac{2}{3}$, decr: $x < \frac{2}{3}$
65. Incr: $x < -\frac{3}{2}$, decr: $x > -\frac{3}{2}$
67. Incr: $x < -\frac{1}{8}$, decr: $x > -\frac{1}{8}$
69. -2 **71.** $-\frac{25}{4}$ **73.** $-\frac{7}{2}$ **75.** 6 **77.** 4
79. $-\frac{39}{8}$ **81.** 10, 10 **83.** 3 slices **85.** d. **87.** a.
89. (a) 2 feet (b) 2 seconds (c) 66 feet
91. $\frac{66}{32} \approx 2$ seconds; about 74 feet **93.** (a) The revenue increases at first up to 50 tickets and then it decreases.
(b) $2500; 50 (c) $f(x) = x(100 - x)$ (d) $2500; 50

95. (a) $V(1) = 147.5$, $V(2) = 156$, $V(3) = 163.5$, $V(4) = 170$; in 2011, there were 147.5 million unique Facebook visitors in one month. Other values can be interpreted similarly. (b) A linear function does not model these data because these three increases are not equal (or nearly equal).
97. 300 feet by 600 feet **99.** 42 in. **101.** (a) 6; in 1990, emissions were 6 billion metric tons. (b) 3 billion metric tons.

Section 18.2 (pp. 1167–1170)

1. $x^2 + 2$ **2.** $(x - 2)^2$ **3.** $(1, 2)$ **4.** $(-1, -2)$
5. $f(x) = ax^2 + bx + c; f(x) = a(x - h)^2 + k$
6. $y = a(x - h)^2 + k; (h, k)$ **7.** downward **8.** $-\frac{b}{2a}$
9. a. **10.** d.

11.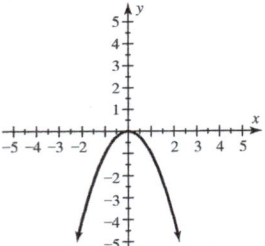
Reflected across the x-axis

13.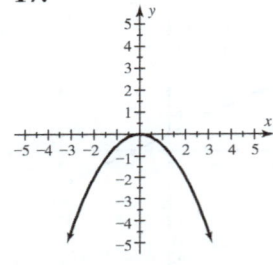
Narrower

15.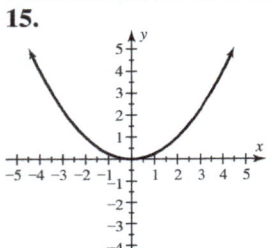
Wider

17.
Reflected across the x-axis and wider

19.

x	-2	-1	0	1	2
$y = x^2$	4	1	0	1	4
$y = x^2 - 3$	1	-2	-3	-2	1

Shifted 3 units downward

21.

x	-2	-1	0	1	2
$y = x^2$	4	1	0	1	4

x	1	2	3	4	5
$y = (x-3)^2$	4	1	0	1	4

Shifted 3 units right

23. (a)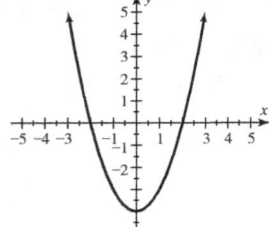
(b) $(0, -4)$
(c) Down 4 units

25. (a)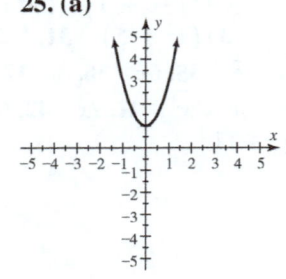
(b) $(0, 1)$
(c) Narrower and up 1 unit

27. (a)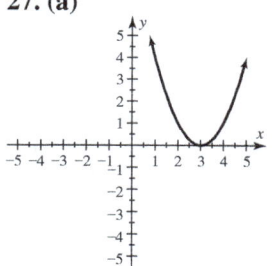
(b) $(3, 0)$
(c) Right 3 units

29. (a)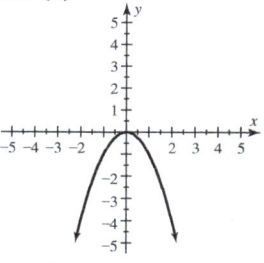
(b) $(0, 0)$
(c) Reflected across the x-axis

31. (a)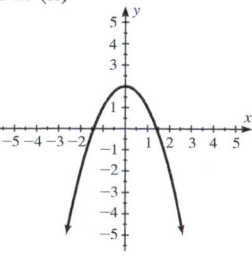
(b) $(0, 2)$
(c) Reflected across the x-axis and up 2 units

33. (a)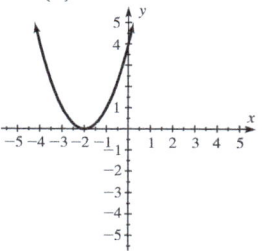
(b) $(-2, 0)$
(c) Left 2 units

35. (a)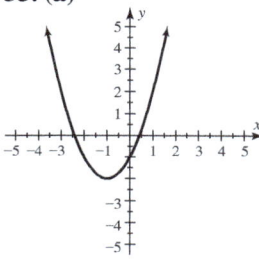
(b) $(-1, -2)$
(c) Left 1 unit and down 2 units

37. (a)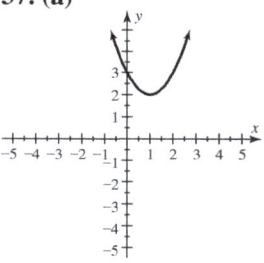
(b) $(1, 2)$
(c) Right 1 unit and up 2 units

39. (a)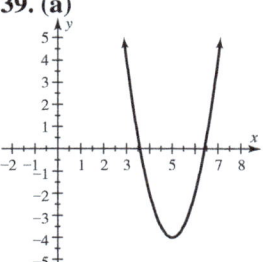
(b) $(5, -4)$
(c) Narrower, right 5 units and down 4 units

41. (a)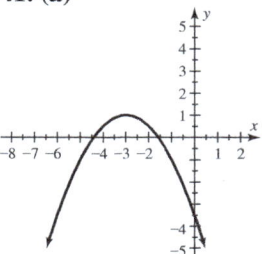
(b) $(-3, 1)$
(c) Wider, reflected across the x-axis, left 3 units and up 1 unit

43. Translated 1 unit right, 2 units downward, and is wider
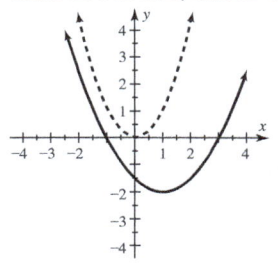

45. Translated 1 unit left, 3 units upward, opens downward, and is narrower
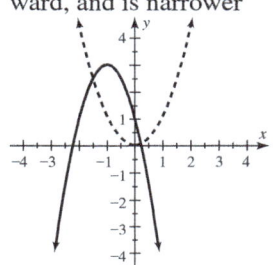

47. $[-20, 20, 2]$ by $[-20, 20, 2]$
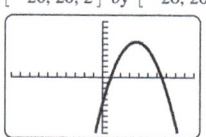

49. $y = 3(x - 3)^2 + 4;\ y = 3x^2 - 18x + 31$
51. $y = -\frac{1}{2}(x - 5)^2 - 2;\ y = -\frac{1}{2}x^2 + 5x - \frac{29}{2}$
53. $y = (x - 1)^2 + 2$ **55.** $y = -(x - 0)^2 - 3$
57. $y = (x - 0)^2 - 3$ **59.** $y = -(x + 1)^2 + 2$
61. (a) $(1, 1)$ **(b)** $y = 4(x - 1)^2 + 1$ **63. (a)** $(-1, -2)$
(b) $y = -(x + 1)^2 - 2$ **65. (a)** $(-1, 3)$
(b) $y = -2(x + 1)^2 + 3$ **67.** 1
69. $y = (x + 1)^2 - 1;\ (-1, -1)$
71. $y = (x - 2)^2 - 4;\ (2, -4)$
73. $y = (x + 1)^2 - 4;\ (-1, -4)$
75. $y = (x - 2)^2 + 1;\ (2, 1)$
77. $y = \left(x + \frac{3}{2}\right)^2 - \frac{17}{4};\ \left(-\frac{3}{2}, -\frac{17}{4}\right)$
79. $y = \left(x - \frac{7}{2}\right)^2 - \frac{45}{4};\ \left(\frac{7}{2}, -\frac{45}{4}\right)$
81. $y = 3(x + 1)^2 - 4;\ (-1, -4)$
83. $y = 2\left(x - \frac{3}{4}\right)^2 - \frac{9}{8};\ \left(\frac{3}{4}, -\frac{9}{8}\right)$
85. $y = -2(x + 2)^2 + 13;\ (-2, 13)$ **87.** $a = 2$
89. $a = 0.3$ **91.** $y = 2(x - 1)^2 - 3$
93. $y = 0.5(x - 1980)^2 + 6$

95. (a) 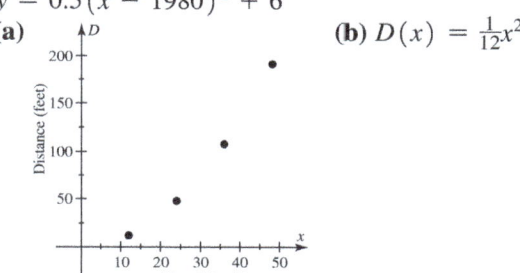 **(b)** $D(x) = \frac{1}{12}x^2$

97. (a) Decreases and then increases
(b) No, because the data decrease and then increase.
(c) Quadratic; it can model data that decrease and increase.
(d) $(1995, 600)$; it has the minimum y-value.
(e) $C(x) = 1.8(x - 1995)^2 + 600$
(f) $[1970, 2030, 10]$ by $[0, 2500, 500]$
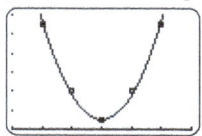

Checking Basic Concepts 18.1 & 18.2 (p. 1170)

1. (a) 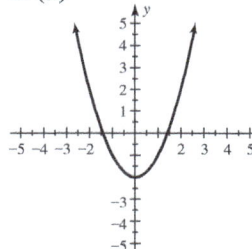 **(b)**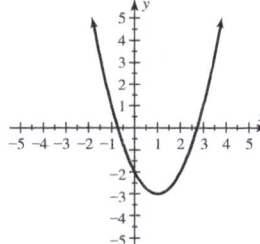

$(0, -2);\ x = 0$ $\qquad\qquad (1, -3);\ x = 1$

2. y_1 opens upward, whereas y_2 opens downward, y_1 is narrower than y_2.   **3.** 7
4. (a) 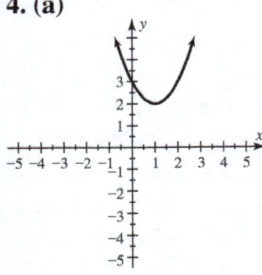   **(b)**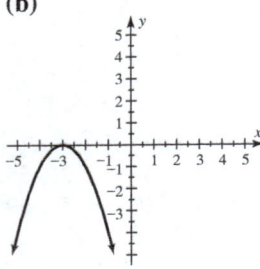

1 unit right, 2 units up   Reflected across the x-axis, 3 units left

5. (a) $y = (x + 7)^2 - 56$ **(b)** $y = 4(x + 1)^2 - 6$

Section 18.3 (pp. 1180–1183)

1. $x^2 + 3x - 2 = 0$ (answers may vary); it can have 0, 1, or 2 solutions   **2.** Nonlinear   **3.** Factoring, square root property, completing the square
4. 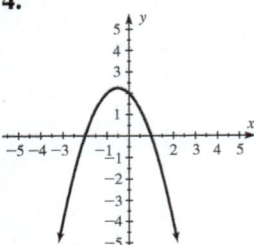   **5.**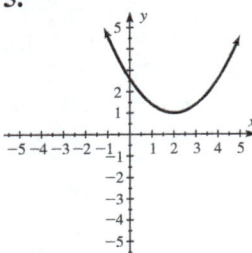

(answers may vary)   (answers may vary)
6. Two; the solutions are the x-coordinates of the x-intercepts.
7. ± 8; the square root property (answers may vary)
8. $\left(\frac{b}{2}\right)^2$ **9.** Yes   **11.** No   **13.** Yes   **15.** No
17. (a) 3.65, −1.65   **(b)** 1.32, −5.32   **19. (a)** 1.32, −0.12
(b) −0.28, −0.83   **21.** −2, 1   **23.** No real solutions
25. −2, 3   **27.** −0.5   **29.** −1, 5   **31.** −3, 1
33. −3, 3   **35.** $-\frac{1}{2}, \frac{3}{2}$ **37.** −1   **39.** No real solutions
41. −7, 5   **43.** $-\frac{1}{3}, \frac{1}{2}$ **45.** $-\frac{3}{2}$ **47.** −2, 7   **49.** $\frac{6}{5}, \frac{3}{2}$
51. ± 12 **53.** $\pm \frac{8}{\sqrt{5}}$ or $\pm \frac{8\sqrt{5}}{5}$ **55.** −6, 4
57. −7, 9   **59.** $\frac{1 \pm \sqrt{5}}{2}$ **61.** $5 \pm \sqrt{5}$
63. $\pm 2i$ **65.** $\pm i\sqrt{3}$ **67.** $\pm 2i\sqrt{5}$ **69.** $\pm 4i$
71. $\pm 3i\sqrt{2}$ **73.** 4   **75.** $\frac{25}{4}$ **77.** 16; $(x - 4)^2$
79. $\frac{81}{4}$; $\left(x + \frac{9}{2}\right)^2$ **81.** −4, 6   **83.** $-3 \pm \sqrt{11}$
85. $\frac{3 \pm \sqrt{29}}{2}$ **87.** $\frac{5 \pm \sqrt{21}}{2}$ **89.** $1 \pm \sqrt{5}$
91. $\frac{3 \pm \sqrt{41}}{4}$ **93.** $\frac{2 \pm \sqrt{11}}{2}$ **95.** $\frac{-3 \pm \sqrt{5}}{12}$
97. −6, 2   **99.** $-2 \pm \sqrt{2}$ **101.** $\pm \sqrt{2}$ **103.** $-5, \frac{7}{3}$
105. $4 \pm \sqrt{14}$ **107.** $\pm \sqrt{\frac{c}{a}}$ **109.** −3, 6   **111.** 3, 5
113. 5, 7   **115.** $y = \pm \sqrt{x + 1}$ **117.** $v = \pm \sqrt{\frac{2K}{m}}$
119. $r = \pm \sqrt{\frac{k}{E}}$ **121.** $f = \pm \frac{1}{2\pi \sqrt{LC}}$ **123. (a)** 30 miles per hour   **(b)** 40 miles per hour   **125.** About 1.9 seconds; no
127. 0.5 sec, 2.5 sec   **129.** Does not happen
131. About 8 feet   **133.** 2 hours
135. (a) $x^2 + 6x - 520 = 0$ **(b)** −26 or 20; 20 feet
137. About 23 °C and 34 °C   **139.** 2014

Section 18.4 (pp. 1193–1197)

1. To solve quadratic equations that are written in the form $ax^2 + bx + c = 0$ **2.** Completing the square
3. $b^2 - 4ac$ **4.** One solution   **5.** Factoring, square root property, completing the square, and the quadratic formula
6. No; not when $b^2 - 4ac < 0$ **7.** $\pm \sqrt{k}$ **8.** $\pm i\sqrt{k}$
9. $-6, \frac{1}{2}$ **11.** 1   **13.** No real solutions   **15.** −2, 8
17. $\frac{1 \pm \sqrt{17}}{8}$ **19.** No real solutions   **21.** $\frac{1}{2}$ **23.** $\frac{3 \pm \sqrt{13}}{2}$
25. $2 \pm \sqrt{3}$ **27.** $\frac{1 \pm \sqrt{37}}{6}$ **29.** $\frac{1 \pm \sqrt{15}}{2}$ **31.** $\frac{5 \pm \sqrt{10}}{3}$
33. $(1 \pm \sqrt{2}, 0)$ **35.** $\left(-\frac{3}{2}, 0\right), (1, 0)$ **37.** None
39. None   **41.** $\left(\frac{-2 \pm \sqrt{10}}{3}, 0\right)$ **43. (a)** $a > 0$
(b) −1, 2   **(c)** Positive   **45. (a)** $a > 0$ **(b)** No real solutions   **(c)** Negative   **47. (a)** $a < 0$ **(b)** 2   **(c)** Zero
49. (a) 25   **(b)** 2   **51. (a)** 0   **(b)** 1   **53. (a)** $-\frac{7}{4}$ **(b)** 0
55. (a) 21   **(b)** 2
57. (a) Upward; wider   **59. (a)** Downward; the same
(b) $x = -1$; $(-1, -2)$ **(b)** $x = 1$; $(1, 1)$
(c) 4, 2   **(c)** 4, 2
(d) y-int: $\left(0, -\frac{3}{2}\right)$;   **(d)** y-int: $(0, 0)$; x-int:
x-int: $(-3, 0), (1, 0)$ $(0, 0), (2, 0)$
(e)   **(e)**

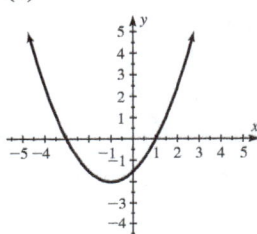

61. (a) Upward; narrower
(b) $x = -\frac{1}{2}$; $\left(-\frac{1}{2}, -\frac{9}{2}\right)$
(c) 36, 2
(d) y-int: $(0, -4)$; x-int: $(-2, 0), (1, 0)$
(e)

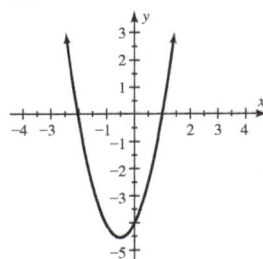

63. $\pm 3i$ **65.** $\pm 4i\sqrt{5}$ **67.** $\pm \frac{1}{2}i$ **69.** $\pm \frac{3}{4}i$ **71.** $\pm i\sqrt{6}$
73. $\pm \sqrt{3}$ **75.** $\pm i\sqrt{2}$ **77.** $\frac{1}{2} \pm i\frac{\sqrt{7}}{2}$ **79.** $-\frac{3}{4} \pm i\frac{\sqrt{23}}{4}$
81. $2 \pm \sqrt{3}$ **83.** $-\frac{1}{2} \pm i\frac{\sqrt{7}}{2}$ **85.** $-\frac{1}{5} \pm i\frac{\sqrt{19}}{5}$
87. $-\frac{3}{4} \pm i\frac{\sqrt{23}}{4}$ **89.** $-\frac{1}{2} \pm i\frac{\sqrt{15}}{2}$ **91.** $\frac{1 \pm \sqrt{3}}{2}$
93. $\frac{1}{4} \pm i\frac{\sqrt{15}}{4}$ **95.** $-1 \pm i\sqrt{3}$ **97.** $-2 \pm i$
99. $1 \pm i\sqrt{2}$ **101.** 1, 2   **103.** $-\frac{1}{2}, 4$ **105.** $\frac{5 \pm \sqrt{17}}{2}$
107. $-\frac{1}{4} \pm \frac{3}{4}i\sqrt{7}$ **109.** $\pm \frac{1}{2}$ **111.** $\pm i\sqrt{2}$ **113.** $\frac{1}{3}$
115. 9 miles per hour   **117.** 45 miles per hour

119. (a) 6; in October 2010, Groupon's value was $6 billion. **(b)** January 2011 **121.** $x \approx 8.04$, or about 1992; this agrees with the graph. **123.** About 1993 and 2004 ($x \approx 4.11, 15.36$) **125.** $130 + 5\sqrt{634} \approx 256$ mph **127. (a)–(c)** 11 in. by 14 in. **129. (a)** The rate of change is not constant. **(b)** 75 seconds (answers may vary) **(c)** 75 seconds

Checking Basic Concepts 18.3 & 18.4 (p. 1197)

1. $\frac{1}{2}, 3$ **2.** $\pm\sqrt{5}$ **3.** $2 \pm \sqrt{3}$ **4.** $y = \pm\sqrt{1 - x^2}$ **5. (a)** $\frac{3 \pm \sqrt{17}}{4}$ **(b)** $\frac{4}{3}$ **6. (a)** 5; two real solutions **(b)** -7; no real solutions **(c)** 0; one real solution **7. (a)** $\pm i\sqrt{5}$ **(b)** $-\frac{1}{2} \pm i\frac{\sqrt{11}}{2}$

Section 18.5 (pp. 1204–1206)

1. It has an inequality symbol rather than an equals sign. **2.** No, they often have infinitely many. **3.** No **4.** Yes **5.** $-2 < x < 4$ **6.** $x < -3$ or $x > 1$ **7.** Yes **9.** Yes **11.** No **13.** Yes **15.** No **17.** No **19. (a)** $-3, 2$ **(b)** $-3 < x < 2$ **(c)** $x < -3$ or $x > 2$ **21. (a)** $-2, 2$ **(b)** $-2 < x < 2$ **(c)** $x < -2$ or $x > 2$ **23. (a)** $-10, 5$ **(b)** $x < -10$ or $x > 5$ **(c)** $-10 < x < 5$ **25.** $[-0.128, 0.679]$ **27. (a)** $-2, 2$ **(b)** $-2 < x < 2$ **(c)** $x < -2$ or $x > 2$ **29. (a)** $-4, 0$ **(b)** $-4 < x < 0$ **(c)** $x < -4$ or $x > 0$ **31.** $[-7, -3]$ **33.** $(-\infty, 1) \cup (2, \infty)$ **35.** $(-\sqrt{10}, \sqrt{10})$ **37.** $(-\infty, 0) \cup (6, \infty)$ **39.** $(-\infty, 2 - \sqrt{2}] \cup [2 + \sqrt{2}, \infty)$ **41. (a)** $-2, 2$ **(b)** $-2 < x < 2$ **(c)** $x < -2$ or $x > 2$ **43. (a)** $\frac{-1 \pm \sqrt{5}}{2}$ **(b)** $\frac{-1 - \sqrt{5}}{2} < x < \frac{-1 + \sqrt{5}}{2}$ **(c)** $x < \frac{-1 - \sqrt{5}}{2}$ or $x > \frac{-1 + \sqrt{5}}{2}$ **45.** $(-3, -1)$ **47.** $(-\infty, -2.5] \cup [3, \infty)$ **49.** $[-2, 2]$ **51.** $(-\infty, \infty)$ **53.** $(0, 3)$ **55.** No real solutions **57.** 2 **59.** $(-\infty, -1) \cup (-1, \infty)$ **61.** $[-1, 2]$ **63.** From 11 feet to 20 feet **65. (a)** From 1131 feet to 3535 feet (approximately) **(b)** Before 1131 feet or after 3535 feet (approximately) **67. (a)** About 383; they agree (approx.). **(b)** About 1969 or after **(c)** About 1969 or after

Section 18.6 (pp. 1210–1211)

1. $\pm 1, \pm\sqrt{6}$ **3.** $-\sqrt[3]{2}, \sqrt[3]{\frac{5}{3}}$ **5.** $-\frac{4}{5}, -\frac{1}{3}$ **7.** $-3, 3$ **9.** $-\sqrt[3]{\frac{1}{3}}, \sqrt[3]{2}$ **11.** $-\frac{1}{8}, \frac{2}{5}$ **13.** 1 **15.** $1, 32^5 = 33{,}554{,}432$ **17.** 16, 81 **19.** 1 **21.** $-3, 6$ **23.** $-\sqrt{3}, \sqrt{3}$ **25.** $\pm 2, \pm 2i$ **27.** $0, \pm i$ **29.** $\pm\sqrt{2}, \pm i$ **31.** $-1 \pm i$ **33.** $\pm 2i$

Checking Basic Concepts 18.5 & 18.6 (p. 1211)

1. $(-\infty, -2) \cup (3, \infty)$ **2.** $\left[-1, -\frac{2}{3}\right]$ **3.** $-2, \sqrt[3]{2}$ **4.** $-1, 8^3 = 512$ **5.** $\pm i$

Chapter 18 Review (pp. 1214–1217)

1. $(-3, 4)$; $x = -3$; downward **2.** $(1, 0)$; $x = 1$; upward **3. (a)**

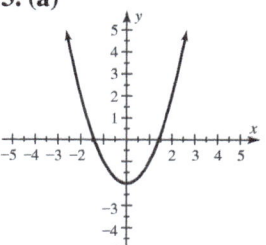

(b) $(0, -2)$; $x = 0$ **(c)** -1

4. (a)

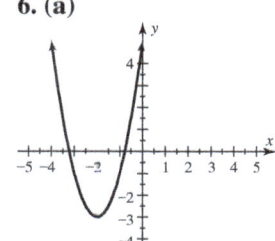

(b) $(2, 1)$; $x = 2$ **(c)** 0

5. (a)

(b) $(1, 2)$; $x = 1$ **(c)** -2.5

6. (a)

(b) $(-2, -3)$; $x = -2$ **(c)** -1

7. $-\frac{7}{2}$ **8.** $-\frac{14}{3}$ **9.** $(2, -6)$ **10.** $(0, 5)$ **11.** $(2, 2)$ **12.** $(-1, 1)$

13. (a)

(b) Up 2 units

14. (a)

(b) Narrower

15. (a)

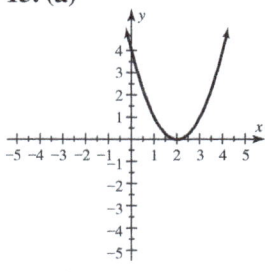

(b) Right 2 units

16. (a)

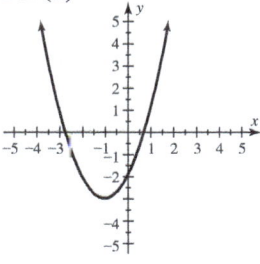

(b) Left 1 unit, down 3 units

17. (a)

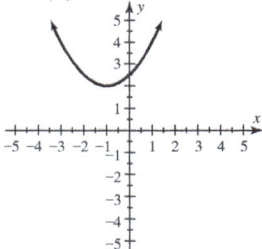

(b) Wider, left 1 unit, up 2 units

18. (a)

(b) Narrower, right 1 unit, down 3 units

19. $y = -4(x - 2)^2 - 5$ **20.** $y = -(x + 4)^2 + 6$
21. $y = (x + 2)^2 - 11; (-2, -11)$
22. $y = \left(x - \frac{7}{2}\right)^2 - \frac{45}{4}; \left(\frac{7}{2}, -\frac{45}{4}\right)$
23. $y = 2\left(x - \frac{3}{4}\right)^2 - \frac{73}{8}; \left(\frac{3}{4}, -\frac{73}{8}\right)$
24. $y = 3(x + 1)^2 - 5; (-1, -5)$ **25.** $a = 3$
26. $a = \frac{1}{4}$ **27.** $f(x) = -5x^2 + 30x - 41; (0, -41)$
28. $f(x) = 3x^2 + 12x + 8; (0, 8)$ **29.** $-2, 3$
30. -1 **31.** No real solutions   **32.** $-4, 6$ **33.** $-10, 5$
34. $-0.5, 0.25$ **35.** $-5, 10$ **36.** $-3, 1$ **37.** $-4, 2$
38. $-1, 3$ **39.** $-5, 4$ **40.** $-8, -3$ **41.** $-\frac{2}{5}, \frac{2}{3}$
42. $\frac{4}{7}, 3$ **43.** ± 10 **44.** $\pm \frac{1}{3}$ **45.** $\pm \frac{\sqrt{6}}{2}$
46. No real solutions   **47.** $-3 \pm \sqrt{7}$ **48.** $2 \pm \sqrt{10}$
49. $1 \pm \sqrt{6}$ **50.** $\frac{-3 \pm \sqrt{11}}{2}$ **51.** $R = -r \pm \sqrt{\frac{k}{F}}$
52. $y = \pm\sqrt{\frac{12 - 2x^2}{3}}$ **53.** $3, 6$ **54.** $11, 13$ **55.** $-\frac{1}{2}, \frac{1}{3}$
56. $\frac{5 \pm \sqrt{5}}{10}$ **57.** $4 \pm \sqrt{21}$ **58.** $\frac{3 \pm \sqrt{3}}{2}$ **59.** ± 2
60. $\pm \frac{1}{2}$ **61.** $\frac{5}{2}, 3$ **62.** $\frac{3}{2}, 5$ **63.** $\frac{3 \pm \sqrt{5}}{2}$
64. $\frac{1 \pm \sqrt{5}}{4}$ **65. (a)** $a > 0$ **(b)** $-2, 3$ **(c)** Positive
66. (a) $a > 0$ **(b)** 2 **(c)** Zero   **67. (a)** $a < 0$
(b) No real solutions   **(c)** Negative   **68. (a)** $a < 0$
(b) $-4, 2$ **(c)** Positive   **69. (a)** 1 **(b)** 2 **70. (a)** 144
(b) 2 **71. (a)** -23 **(b)** 0 **72. (a)** 0 **(b)** 1
73. $-\frac{1}{2} \pm i\frac{\sqrt{19}}{2}$ **74.** $\pm 2i$ **75.** $\frac{1}{4} \pm i\frac{\sqrt{7}}{4}$ **76.** $\frac{1}{7} \pm i\frac{\sqrt{34}}{7}$
77. (a) $-2, 6$ **(b)** $-2 < x < 6$ **(c)** $x < -2$ or $x > 6$
78. (a) $-2, 0$ **(b)** $x < -2$ or $x > 0$ **(c)** $-2 < x < 0$
79. (a) $-4, 4$ **(b)** $-4 < x < 4$ **(c)** $x < -4$ or $x > 4$
80. (a) $-2, 1$ **(b)** $-2 < x < 1$ **(c)** $x < -2$ or $x > 1$
81. (a) $-1, 3$ **(b)** $-1 < x < 3$ **(c)** $x < -1$ or $x > 3$
82. (a) $-\frac{3}{2}, 5$ **(b)** $-\frac{3}{2} \le x \le 5$ **(c)** $x \le -\frac{3}{2}$ or $x \ge 5$
83. $[-3, -1]$ **84.** $\left(\frac{1}{5}, 3\right)$ **85.** $\left(-\infty, \frac{1}{6}\right) \cup (2, \infty)$
86. $(-\infty, -\sqrt{5}] \cup [\sqrt{5}, \infty)$ **87.** $(-\infty, \infty)$
88. No solutions   **89.** $\pm\sqrt{5}, \pm 3$ **90.** $-\frac{1}{4}, \frac{2}{7}$
91. $1, 512$ **92.** 0 **93.** $\pm i\frac{\sqrt{2}}{2}$ **94.** $2 \pm i\sqrt{2}$
95. (a) $f(x) = x(12 - 2x)$ **(b)** 6 inches by 3 inches
96. (a) After 1 second and 1.75 seconds   **(b)** 1.375 seconds; 34.25 feet   **97. (a)** $f(x) = x(90 - 3x)$
(b) $[0, 30, 5]$ by $[0, 800, 100]$

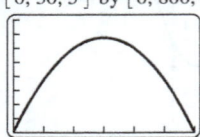

(c) 10 or 20 rooms   **(d)** 15 rooms
98. (a) $x(x + 2) = 143$ **(b)** $x = -13$ or $x = 11$; the numbers are -13 and -11 or 11 and 13.
99. (a) $\sqrt{1728} \approx 41.6$ miles per hour   **(b)** 60 miles per hour
100. (a) $(1950, 220)$; in 1950, the per capita consumption was at a low of 220 million Btu.
(b) $[1950, 1970, 5]$ by $[200, 350, 25]$

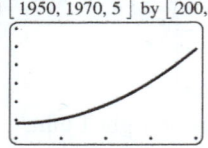

It increased.

(c) $f(2010) = 1120$; no; the trend represented by this model did not continue after 1970.   **101.** About 11.1 inches by 11.1 inches   **102.** 50 feet   **103.** About 1.5 feet
104. About 6.0 to 9.0 inches

Chapter 18 Test (pp. 1217–1218)

1. $\left(1, \frac{3}{2}\right); x = 1; -3$ **2.** $-\frac{29}{4}$ **3.** $a = -\frac{1}{2}$
4. (a) Same as $y = x^2$ except shifted 1 unit right   **(b)** Same as $y = x^2$ except shifted 2 units downward

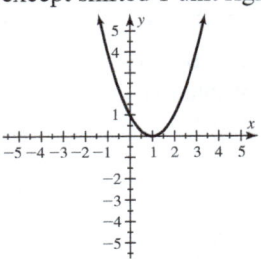   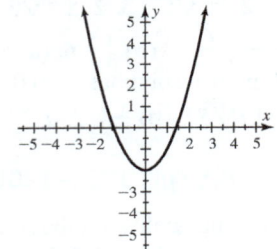

(c) Same as $y = x^2$ except it is wider, shifted right 3 units, and shifted upward 2 units

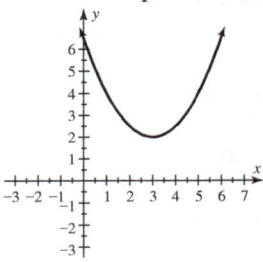

5. $y = (x - 3)^2 - 7; (3, -7); x = 3$ **6.** $-1, 2; 2$
7. $-4, \frac{1}{3}$ **8.** $-\frac{1}{2}, \frac{1}{2}$ **9.** $4 \pm \sqrt{17}$ **10.** $\frac{3 \pm \sqrt{17}}{4}$
11. $\pm \frac{4}{3}$ **12.** $m = \pm\sqrt{\frac{Fr^2}{G}}$ **13. (a)** $a < 0$ **(b)** $-3, 1$
(c) Positive   **14. (a)** -44 **(b)** No real solutions
(c) The graph of $y = -3x^2 + 4x - 5$ does not intersect the x-axis.   **15. (a)** $-1, 1$ **(b)** $-1 < x < 1$
(c) $x < -1$ or $x > 1$ **16. (a)** $-10, 20$
(b) $x < -10$ or $x > 20$ **(c)** $-10 < x < 20$
17. (a) $-\frac{1}{2}, \frac{3}{4}$ **(b)** $\left[-\frac{1}{2}, \frac{3}{4}\right]$ **(c)** $\left(-\infty, -\frac{1}{2}\right] \cup \left[\frac{3}{4}, \infty\right)$
18. $[-2, 0]$ **19.** $\sqrt[3]{2}, 1$ **20.** $-1 \pm i\frac{\sqrt{2}}{2}$
21. $-1.37, 0.69$ **22.** $\sqrt{2250} \approx 47.4$ miles per hour
23. (a) $f(x) = (x + 20)(90 - x)$ **(b)** 35
24. (a) $[0, 6, 1]$ by $[0, 150, 50]$

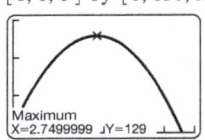

(b) After about 5.6 seconds
(c) 2.75 seconds; 129 feet

Chapters 1–18 Cumulative Review (pp. 1218–1220)

1. 1 **2.** Natural: $\sqrt[3]{8}$; whole: $0, \sqrt[3]{8}$; integer: $0, -5, \sqrt[3]{8}$; rational: $0.\overline{4}, 0, -5, \sqrt[3]{8}, -\frac{4}{3}$; irrational: $\sqrt{7}$ **3. (a)** $x^{10}y^{12}$
(b) $\frac{x}{y^8}$ **(c)** $\frac{1}{b^{10}}$ **4.** 9.29×10^6 **5.** $2; x \le 2$ **6.** $(2, 5)$

7. 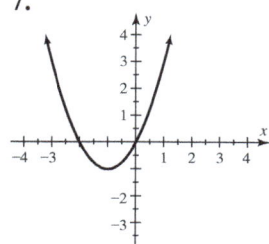 **8.**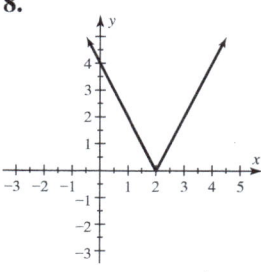

9. $y = -\frac{3}{2}x + 5$ **10.** $x = -3$ **11.** -12 **12.** $\left(-\infty, \frac{7}{4}\right)$
13. $\left[\frac{1}{3}, 1\right]$ **14.** $(-1, 5]$ **15.** $(-1, 1)$
16.

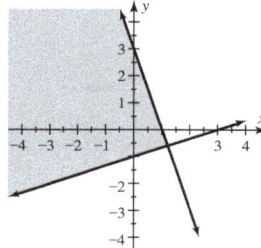

17. $(2, 2, 1)$ **18.** $6x^2 + 17x - 14$ **19.** $3x^3y + 3xy^3$
20. $x - 9$ **21.** $x(x + 1)(x - 2)$
22. $(2x + 5)(2x - 5)$ **23.** $\pm\sqrt{3}$ **24.** 1 **25.** $\frac{x + 3}{2}$
26. $-\frac{2}{x(x + 2)}$ **27.** $4x^3$ **28.** $\frac{1}{64}$ **29.** 3 **30.** $3\sqrt{2x}$
31.

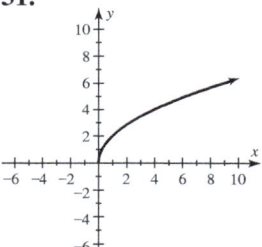

32. $\sqrt{26}$ **33.** $1 - i$ **34.** 3 **35.** 0.79
36.

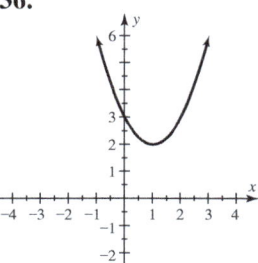

(a) $(1, 2)$ (b) 6 (c) $x = 1$ (d) $x > 1$
37. $f(x) = 2(x - 1)^2 - 3$ **38.** Shifted 1 unit left, 2 units downward; narrower **39.** $-3 \pm \sqrt{11}$ **40.** $\frac{3 \pm \sqrt{17}}{4}$
41. 1, 3 **42.** (a) $-2, 1$ (b) $-2 \leq x \leq 1$
43. $x < 1$ or $x > 2$ **44.** $\pm 4, \pm 4i$ **45.** c. **46.** f. **47.** g.
48. e. **49.** d. **50.** a. **51.** b. **52.** h. **53.** (a) 300; initially, the tank holds 300 gal. (b) $(6, 0)$; after 6 minutes, the tank is empty. (c) -50; water is pumped out at 50 gal/min.
(d) $G(t) = 300 - 50t$ **54.** $2200 at 6%, $1200 at 5%, $600 at 4% **55.** 125 ft by 125 ft **56.** 33 feet

19 Exponential and Logarithmic Functions

Section 19.1 (pp. 1233–1237)

1. $g(f(7))$ **2.** $f(g(x))$ **3.** No **4.** inputs; outputs
5. No **6.** one-to-one **7.** adding 10 **8.** 7 **9.** 8; 6
10. $x; y$ **11.** one-to-one **12.** reflection; $y = x$
13. (a) 7 (b) 49 (c) $(g \circ f)(x) = x^2 + 3$
(d) $(f \circ g)(x) = (x + 3)^2$ **15.** (a) -65 (b) 126
(c) $(g \circ f)(x) = 8x^3 - 1$ (d) $(f \circ g)(x) = 2x^3 - 2$
17. (a) 3 (b) 1 (c) $(g \circ f)(x) = \frac{1}{2}x - 2$
(d) $(f \circ g)(x) = \frac{1}{2}|x - 2|$ **19.** (a) $\frac{11}{2}$ (b) $-\frac{1}{17}$
(c) $(g \circ f)(x) = 3 - \frac{5}{x}$ (d) $(f \circ g)(x) = \frac{1}{3 - 5x}$
21. (a) 77 (b) 122 (c) $(g \circ f)(x) = 16x^2 - 4x + 5$
(d) $(f \circ g)(x) = 8x^2 - 4x + 10$ **23.** (a) 1 (b) 2
25. (a) -1 (b) 1 **27.** (a) 0 (b) 2 **29.** (a) 2 (b) -3
(c) -1 **31.** $f(1) = f(-1) = 5$
33. $f(1) = f(-1) = 101$ **35.** $f(2) = f(-2) = 4$
37. Yes **39.** No **41.** Yes **43.** Divide x by 7;
$f(x) = 7x; g(x) = \frac{x}{7}$ **45.** Multiply x by 2, then subtract 5;
$f(x) = \frac{x + 5}{2}; g(x) = 2x - 5$ **47.** Add 3 to x and multiply the result by 2; $f(x) = \frac{1}{2}x - 3; g(x) = 2(x + 3)$
49. Take the cube root of x and subtract 5;
$f(x) = (x + 5)^3; g(x) = \sqrt[3]{x} - 5$
51. $(f \circ f^{-1})(x) = 4\left(\frac{x}{4}\right) = x; (f^{-1} \circ f)(x) = \frac{4x}{4} = x$
53.–57. Show $(f \circ f^{-1})(x) = (f^{-1} \circ f)(x) = x$. See the answer to Exercise 51 above. **59.** $f^{-1}(x) = \frac{x}{12}$
61. $f^{-1}(x) = x - 8$ **63.** $f^{-1}(x) = \frac{x + 2}{5}$
65. $f^{-1}(x) = -2(x - 1)$ **67.** $f^{-1}(x) = 8 - x$
69. $f^{-1}(x) = 2x - 1$ **71.** $f^{-1}(x) = \frac{x^3}{2}$
73. $f^{-1}(x) = \sqrt[3]{x + 8}$
75.

x	0	5	10	15	20
$f^{-1}(x)$	0	1	2	3	4

Domain of f = range of f^{-1} = $\{0, 1, 2, 3, 4\}$
Range of f = domain of f^{-1} = $\{0, 5, 10, 15, 20\}$

77.

x	4	2	0	-2	-4
$f^{-1}(x)$	-5	0	5	10	15

Domain of f = range of f^{-1} = $\{-5, 0, 5, 10, 15\}$
Range of f = domain of f^{-1} = $\{-4, -2, 0, 2, 4\}$

79. 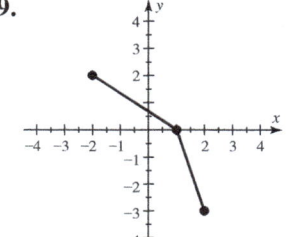 **81.**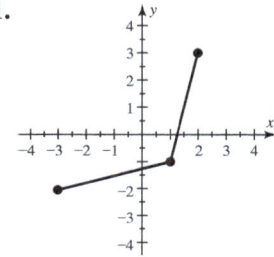

83. 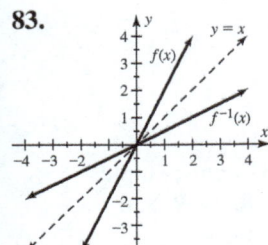 85.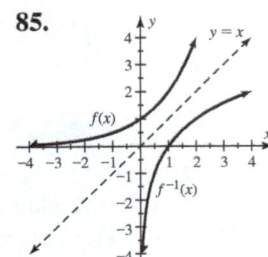

87. (a) −2 (b) 2 (c) 0   89. (a) −6 (b) 4 (c) −2
91. (a) 12 (b) −4 (c) $x^4 + 4x^2 + 2$ 93. (a) $\frac{1}{2}$
(b) $-\frac{1}{x}$ (c) $\frac{x}{2}$ 95. (a) 20π; after 5 seconds, the wave has a circumference of $20\pi \approx 62.8$ feet. (b) $(C \circ r)(t) = 4\pi t$
97. (a) 16; in 1980, 16% of people 25 or older had completed four or more years of college.

(b)

x	8	16	27	29
$P^{-1}(x)$	1960	1980	2000	2010

(c) 1980

99. (a) 75°; 150   (b) 150; one hour after midnight, there are 150 mosquitoes per 100 square feet.   (c) The number of mosquitoes per 100 square feet, h hours after midnight
(d) $T(h) = -5h + 80$; $M(T) = 2T$
(e) $(M \circ T)(h) = -10h + 160$
101. (a) Yes, different inputs result in different outputs.
(b) $f^{-1}(x) = \frac{5}{9}(x - 32)$ converts x degrees Fahrenheit to an equivalent temperature in degrees Celsius.
103. $f(x) = 4x$; $f^{-1}(x) = \frac{x}{4}$ converts x quarts to gallons.
105. (a) $C(x) = 16x$ (b) $T(x) = 48x$
(c) $(T \circ C)(x) = 768x$ (d) 2304; there are 2304 tsp in 3 gal.

Section 19.2 (pp. 1250–1254)

1. $f(x) = Ca^x$
2. Answers may vary.

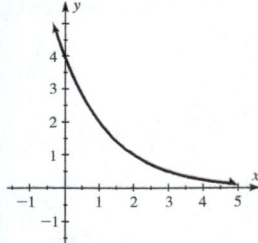

3. D: all real numbers; R: all positive real numbers
4. 32; 25   5. 2.718   6. 7.389; 9.870 (approximately)
7. factor   8. 1.5   9. $\frac{B - A}{A} \times 100$ 10. 1.35
11. $\frac{1}{9}$; 9   13. 5; 160   15. 4; $\frac{1}{8}$ 17. 15; $\frac{5}{9}$
19. 0.17; 2.41   21. 5; 1.08   23. 1; $\frac{1}{a}$ 25. c.   27. d.
29. $C = 1, a = 2$ 31. $C = 4, a = \frac{1}{4}$
33. 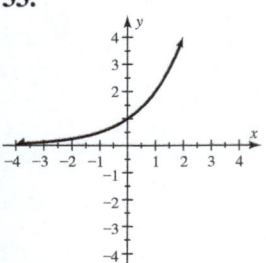   35.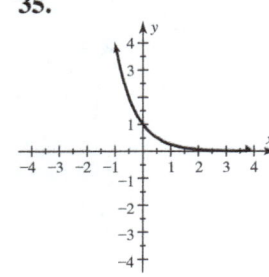

Growth   Decay

37. 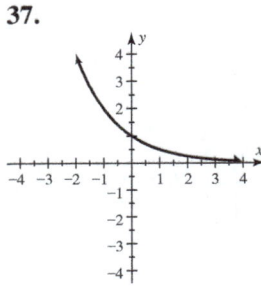   39.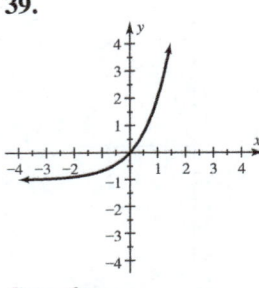

Decay   Growth

41. 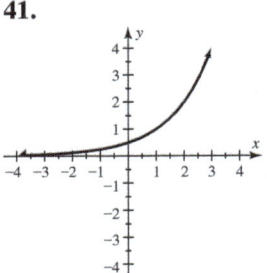   43.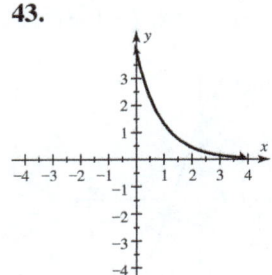

Growth   Decay

45. (a) 1 unit right   (b) 1 unit down   (c) Reflected across the y-axis   (d) Reflected across the x-axis
47. (a) Exponential decay   (b) $f(x) = 64\left(\frac{1}{4}\right)^x$
49. (a) Linear growth   (b) $f(x) = 3x + 8$
51. (a) Exponential growth   (b) $f(x) = 4(1.25)^x$
53. (a) 100%   (b) −50%   55. (a) −80%   (b) 400%
57. (a) $1200   (b) $2200   (c) 2.2
59. (a) $130   (b) $780   (c) 1.2
61. (a) −$80   (b) $720   (c) 0.9
63. $C = 9, a = 1.07, R = 7\%$
65. $C = 1.5, a = 0.45, R = -55\%$ 67. $3551.05
69. $1,820,087.63   71. $792.75   73. Yes; this is equivalent to having two accounts, each containing $1000 initially.
75. $788.78   77. $1429.24   79. 3.32   81. 0.86
83. $[-4, 4, 1]$ by $[0, 8, 1]$ 85. $[-4, 4, 1]$ by $[0, 8, 1]$

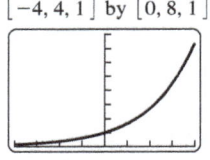   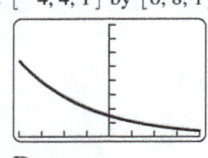

Growth   Decay
87. $f(x) = 5000(0.75)^x$; $f(4) \approx 1582$
89. $f(x) = 50(1.1)^x$; $f(4) \approx 73$ 91. 20% is much better
93. (a) 0.6   (b) $B(x) = 0.07(0.6)^x$ (c) 0.0252; after 2 hours, blood alcohol is 0.0252 g/100 mL
95. $F(t) = 100\left(\frac{1}{2}\right)^{t/3}$, 40%   97. (a) 2.724
(b) In July 2010, there were about 0.1 billion tweets per month.
(c) About 15; after 5 years, there were 15 billion tweets per month.   99. (a) $C = 500, a = 2$ (b) About 5278 thousand per milliliter   (c) The growth is exponential and doubles every 50 seconds.

101.

X	Y1
0	1
10	.36854
20	.13582
30	.05006
40	.01845
50	.0068
60	.00251

Y1■(0.905)^X

(a) 1; the probability that no vehicle will enter the intersection during a period of 0 seconds is 1 or 100%.
(b) About 30 seconds
103. (a) $f(x) = 2.7e^{0.014x}$ **(b)** 3.1 million
105. (a) $f(x) = 38e^{0.0102x}$ **(b)** 42 million
107. Answers may vary.

Checking Basic Concepts 19.1 & 19.2 (p. 1255)

1. (a) 7 **(b)** $(f \circ g)(x) = 2x^2 + 9x + 6$
2.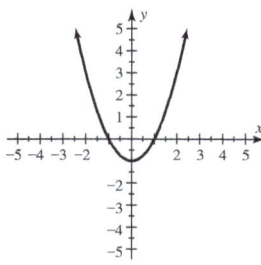

(a) No, it does not pass the horizontal line test. **(b)** No
3. $f^{-1}(x) = \frac{x+3}{4}$ **4.** $\frac{3}{4}$
5.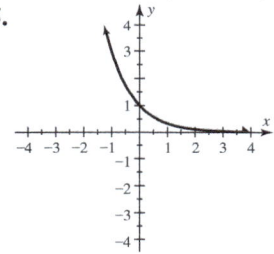

6. $C = 2; a = \frac{1}{2}$

Section 19.3 (pp. 1264–1267)

1. 10 **2.** e **3.** $D = \{x \mid x > 0\}$; R: all real numbers
4. D: all real numbers; $R = \{x \mid x > 0\}$ **5.** k **6.** k
7. x **8.** x **9.** log 5 **10.** ln 5 **11.** 0 **12.** undefined
13. (a) $10^2 = 100$ **(b)** $\log \frac{1}{10} = -1$
15. (a) log 1 = 0 **(b)** $10^{-5} = \frac{1}{100{,}000}$

17.

x	10^{-5}	10^0	$10^{0.5}$	$10^{2.2}$
log x	-5	0	0.5	2.2

19. 5 **21.** -4 **23.** 0 **25.** -2 **27.** 4.7 **29.** 4 **31.** $\frac{1}{2}$
33. Undefined **35.** -3 **37.** 2 **39.** $x^2, x \neq 0$ **41.** 5
43. $2x - 7$ **45.** 1.398 **47.** 0.161
49.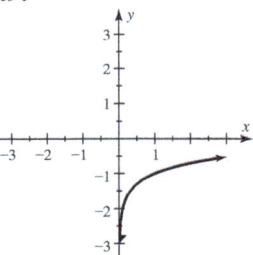

1 unit downward

51.

1 unit to the left

53.

1 unit to the right

55.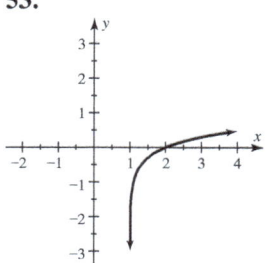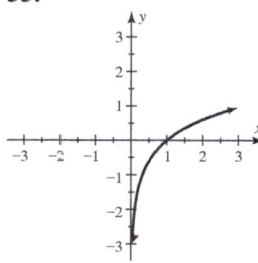

Increases faster

57.

x	$e^{0.5}$	e^{-12}	$e^{\frac{1}{2}}$	$e^{-2.4}$
ln x	0.5	-12	$\frac{1}{2}$	-2.4

59. 0 **61.** $-5x$ **63.** $x^2, x \neq 0$ **65.** 1.946 **67.** -0.560
69. $[-4, 4, 1]$ by $[-4, 4, 1]$ **71.** $[-4, 4, 1]$ by $[-4, 4, 1]$

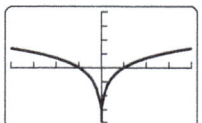 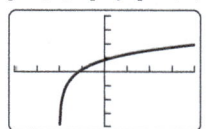

Reflected across the y-axis together with the graph of $y = \ln x$; $D = \{x \mid x \neq 0\}$

2 units to the left; $D = \{x \mid x > -2\}$

73.

x	$5^{\frac{2}{3}}$	5^{-1}	$5^{5.5}$	$5^{-0.9}$
$\log_5 x$	$\frac{2}{3}$	-1	5.5	-0.9

75. $6x$ **77.** $-\frac{3}{2}$ **79.** 8 **81.** $\frac{3}{2}$ **83.** $-\frac{2}{3}$ **85.** Undefined
87. 2 **89.** -4 **91.** -2 **93.** -2 **95.** 17
97. $(2x)^2, x \neq 0$ **99.** $0.6z, z > 0$

101.

x	1/4	1/2	1	$\sqrt{2}$	64
$\log_2 x$	-2	-1	0	1/2	6

103. 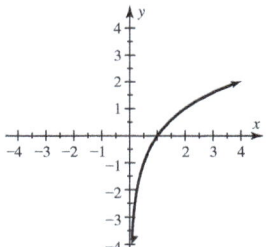 **105.**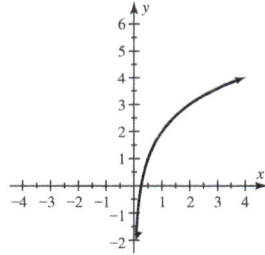

107. d. **109.** a. **111.** 120 dB; yes **113. (a)** It increases, but it doesn't double when the weight of the plane doubles. **(b)** About 5.086; a 50,000-pound airplane needs a runway 5086 feet long. **115. (a)** 3.5 **(b)** 1 **(c)** It decreases by 2.5.
117. (a) 10^6; 10^8 **(b)** 100 times

Section 19.4 (pp. 1273–1274)

1. 4 **2.** 4 **3.** 3 **4.** log m + log n **5.** log m − log n
6. $r \log m$ **7.** No **8.** Yes **9.** No **10.** No
11. $\log_a x = \frac{\log x}{\log a}$ or $\log_a x = \frac{\ln x}{\ln a}$ **12.** 0; 1
13. ln 3 + ln 5 **15.** log x + log y **17.** log y + log y
19. log 7 − log 3 **21.** ln x − ln y **23.** $\log_2 45 - \log_2 x$
25. log 225 **27.** ln xy **29.** ln $14x^3$ **31.** ln xy **33.** log 5
35. ln x^2 **37.** 2 **39.** 6 log 3 **41.** x ln 2 **43.** $\frac{1}{4} \log_2 5$
45. $\frac{1}{3} \log_4 z$ **47.** $(y - 1) \log x$ **49.** 1.2 **51.** 1.8

53. 2.1   **55.** 0.4   **57.** −1.1   **59.** $\log z$ **61.** $\log x^3 y^2$
63. 0   **65.** $\ln 2^x$ **67.** $\ln x^{5/6}$ **69.** $\log_a \frac{x+1}{x-1}$
71. $\log x + 2 \log y$ **73.** $4 \ln x + \ln y - \ln z$
75. $\frac{1}{3} \log z - \frac{1}{2} \log y$ **77.** $4 \log x + 3 \log y$
79. $\ln x - \ln y$ **81.** $\frac{3}{2} \log_4 x + \frac{1}{2} \log_4 y - \log_4 z$
83. $[-6, 6, 1]$ by $[-4, 4, 1]$ $[-6, 6, 1]$ by $[-4, 4, 1]$

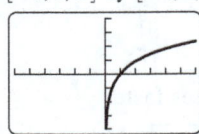 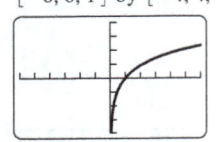

By the power rule, $\log x^3 = 3 \log x$
85. $[-6, 6, 1]$ by $[-4, 4, 1]$ $[-6, 6, 1]$ by $[-4, 4, 1]$

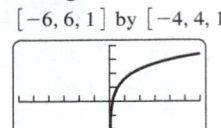

Not the same
87. 1.46   **89.** 4.64   **91.** 2.10
93. $10 \log(10^{16} x) = 10(\log 10^{16} + \log x) = 10(16 + \log x) = 160 + 10 \log x$

Checking Basic Concepts 19.3 & 19.4 (p. 1274)

1. (a) 4   (b) x (c) −3   (d) $\frac{1}{2}$
2.

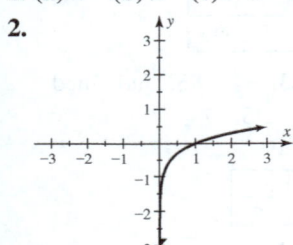

(a) $D = \{x \mid x > 0\}$; R: all real numbers   (b) 0
(c) Yes; for example, $\log \frac{1}{10} = -1$.   (d) No; negative numbers are not in the domain of $\log x$.
3. (a) $\log x + \log y$ (b) $\ln x - \ln y - \ln z$
(c) $2 \ln x$ (d) $2 \log x + 3 \log y - \frac{1}{2} \log z$
4. (a) $\log xy$ (b) $\ln \frac{2x}{y^3}$ (c) $\log_2 \frac{x^2 y^3}{z}$

Section 19.5 (pp. 1282–1286)

1. Add 5 to each side.   **2.** Divide each side by 5.
3. Take the common logarithm of each side.
4. Exponentiate each side using base 10.   **5.** x
6. $x, x > 0$ **7.** $2x$ **8.** $x + 7, x > -7$
9. No; $\log \frac{5}{4} = \log 5 - \log 4$
10. No; $\log 5 - \log 4 = \log \frac{5}{4}$ **11.** 1   **12.** 1
13. (a) $3x = \log_6 8$ (b) $7x = \ln 10$
15. (a) $8x = \log_5 12$ (b) $x = \ln 3$
17. (a) $2^8 = x$ (b) $e^3 = 2x$
19. (a) $4^6 = 6x$ (b) $e^{10} = x + 1$
21. 3   **23.** 6   **25.** 6   **27.** 4   **29.** $\frac{\log 124}{0.4} \approx 5.23$
31. 0   **33.** $\ln 25 \approx 3.22$ **35.** $\frac{\ln 2}{\ln 0.4} \approx -0.76$
37. $\ln 7 \approx 1.95$ **39.** $\log \frac{35}{2} - 2 \approx -0.76$

41. $\frac{\log 20}{2 \log 3.1} \approx 1.32$ **43.** −1   **45.** −1, 5
47. $\frac{\ln 10}{3} \approx 0.77$ **49.** $\frac{2 \ln 2}{1 - \ln 2} \approx 4.52$
51. $\frac{2 \log 5}{0.5 \log 4 - \log 5} \approx -3.51$ **53.** (a) 1   (b) 1
55. (a) −2   (b) −2   **57.** −1   **59.** 0   **61.** 5
63. $\frac{\log 1.45}{\log 0.55} \approx -0.62$ **65.** −1.84, 1.15   **67.** 1.31
69. 100   **71.** $e^5 \approx 148.41$ **73.** 5,000,000   **75.** 16
77. $\frac{2^{2.3}}{5} \approx 0.98$ **79.** $10^{1.4} \approx 25.12$
81. $\frac{e^{11} - 1}{2} \approx 29{,}936.57$ **83.** 1   **85.** 5   **87.** 4   **89.** 3
91. $10^{1.6} \approx 39.81$ **93.** $e - 1 \approx 1.72$ **95.** 9
97. (a) 2   (b) $e^{0.7} \approx 2.01$
99. (a) 2   (b) $\frac{1}{2}(10^{0.6}) \approx 1.99$
101. (a) About 54.9; in 2018, there will be an annual shipment of 54.9 million smartwatches.   (b) 2020
103. 10 hours   **105.** 8 years   **107.** (a) About 126; in 2017, about 126 thousand people were waiting for liver transplants.   (b) Around 2021   **109.** About 20,893 pounds
111. (a) About 203; in 1975, there were about 203 thousand bluefin tuna.   (b) 1979   (c) 1979
113. $a = 25$, $b = 3$ **115.** (a) About 1.67 acres
(b) About 1.67 acres   **117.** (a) Nonlinear; they do not increase at a constant rate.   (b) Each year, Fitbit sales increased by a factor of 3.33.   (c) 2016
119. 10^{-6} watts/square centimeter   **121.** About 7 miles
123. About $42,055.97

Checking Basic Concepts 19.5 (p. 1286)

1. (a) $\log 20 \approx 1.30$ (b) $\frac{\log 147}{3 \log 2} \approx 2.40$
(c) $e^{4.1} \approx 60.34$ (d) 500   **2.** 8   **3.** 20 years

Chapter 19 Review (pp. 1289–1292)

1. (a) 81   (b) $(f \circ g)(x) = 50x^2 - 2$ **2.** (a) −32
(b) $(f \circ g)(x) = \sqrt[3]{4x^3 - 6}$ **3.** (a) 0   (b) 3
4. (a) 3   (b) −2   (c) −1   **5.** $f(1) = f(-1) = 2$
6. $f(0) = f(2) = 1$ **7.** No   **8.** Yes
9. $(f \circ f^{-1})(x) = 2\left(\frac{x+9}{2}\right) - 9 = x$
$(f^{-1} \circ f)(x) = \frac{(2x-9)+9}{2} = x$
10. $(f \circ f^{-1}) = (\sqrt[3]{x-1})^3 + 1 = x$
$(f^{-1} \circ f)(x) = \sqrt[3]{(x^3+1)-1} = x$
11. $f^{-1}(x) = \frac{x}{5}$ **12.** $f^{-1}(x) = x + 11$
13. $f^{-1}(x) = \frac{x-7}{2}$ **14.** $f^{-1}(x) = \frac{4}{x}$
15.

x	10	8	7	3
$f^{-1}(x)$	0	1	2	3

$D = \{3, 7, 8, 10\}$; $R = \{0, 1, 2, 3\}$
16.

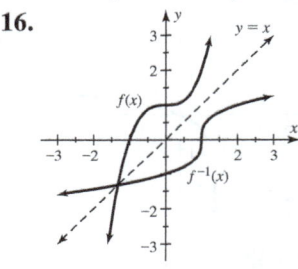

17. $\frac{1}{6}$; 36 18. 5; $\frac{5}{8}$ 19. 3; $\frac{1}{81}$ 20. 3; $\frac{1}{2}$
21. 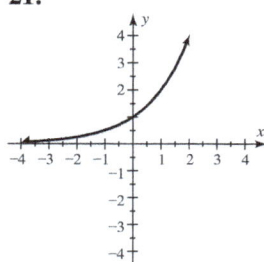 22.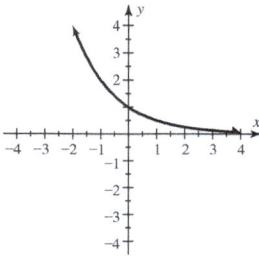

Exponential growth   Exponential decay

23. 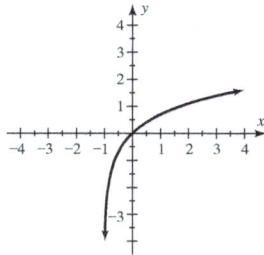 24.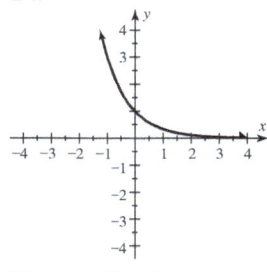

Logarithmic growth   Exponential decay

25. (a) Exponential growth (b) $f(x) = 5(2)^x$
26. (a) Linear growth (b) $f(x) = 5x + 5$
27. $C = \frac{1}{2}$, $a = 2$ 28. $k = 2$ 29. -20% 30. 1.07
31. (a) 100% (b) -50% 32. (a) $-54.\overline{54}\%$ (b) 120%
33. (a) $1050 (b) $1550 (c) 3.1 34. (a) $-\$175$
(b) $525 (c) 0.75 35. $f(x) = 20{,}000(0.95)^x$;
$f(2) = 18{,}050$ 36. $f(x) = 1500(4)^x$; $f(2) = 24{,}000$
37. $2829.54 38. $675,340.51 39. 399.67 40. 0.71
41. 3.48 42. 3.89 43. -3 44. 2 45. -4 46. 2
47. 1.813 48. -0.163 49. 4.787 50. -0.398 51. 7
52. $\frac{5}{9}$ 53. $6 - x$ 54. x^2, $x > 0$ 55. $\ln x + \ln y$
56. $\log x - \log y$ 57. $2 \ln x + 3 \ln y$
58. $\frac{1}{2} \log x - 3 \log z$ 59. $2 \log_2 x + \log_2 y - \log_2 z$
60. $\frac{1}{3} \log_3 x - \frac{1}{3} \log_3 y$ 61. $\log 75$ 62. $\log_4 (10x^2)$
63. $\ln \frac{x^2}{y^3}$ 64. $\log xy$ 65. $3 \log 6$ 66. $2 \ln x$
67. $2x \log_2 5$ 68. $(x + 1) \log_4 0.6$ 69. 2 70. 4
71. $\ln 9 \approx 2.20$ 72. $\frac{\log 0.2}{\log 0.85} \approx 9.90$ 73. $e^{0.8} \approx 2.23$
74. $\frac{1}{2}e^5 \approx 74.21$ 75. 10^{40} 76. 100
77. $\frac{4 \log 2}{\log 3 - \log 2} \approx 6.84$ 78. 6 79. (a) 3 (b) 3
80. (a) 4 (b) 4 81. (a) 64π; after 8 seconds, the balloon has a surface area of $64\pi \approx 201$ in^2.
(b) $(S \circ r)(t) = 8\pi t$ 82. (a) Yes, different inputs result in different outputs. (b) $f^{-1}(x) = \frac{x}{0.08}$ calculates the cost of an item whose sales tax is x dollars. 83. 7 years
84. $a = 100$, $b = 50$ 85. $C = 3$, $a = 2$ 86. 10^7
87. (a) $[0, 10, 2]$ by $[0, 4, 1]$

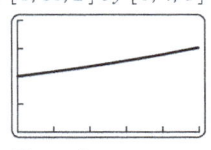

Growth
(b) About 3.3 million (c) 2008 88. (a) 1000; there were 1000 bacteria/mL initially. (b) About 495.11 minutes
89. (a) About 6.93 meters/second (b) 12.18 meters

Chapter 19 Test (pp. 1292–1293)

1. 6; $(f \circ g)(x) = 4(x + 7)^3 - 5(x + 7)$
2. (a) 3 (b) 3 3. $f(-5) = f(5) = 0$
(answers may vary) 4. $f^{-1}(x) = \frac{5 - x}{2}$
5.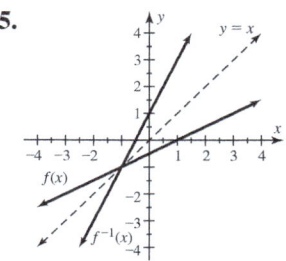

6.
x	8	6	4	2
$f^{-1}(x)$	1	2	3	4

$D = \{2, 4, 6, 8\}$; $R = \{1, 2, 3, 4\}$
7. $\frac{3}{16}$
8.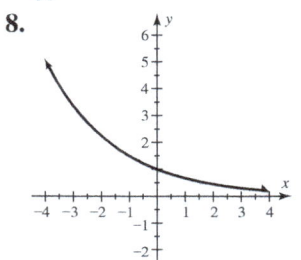

Exponential decay
9. (a) Exponential growth (b) $f(x) = 3(2)^x$
10. (a) Linear growth (b) $f(x) = 1.5x - 1$
11. $C = 1$, $a = \frac{1}{2}$ 12. 50% 13. 1.05 14. $1051.91
15. 4.16 16. $\frac{1}{2}$ 17. 5.426
18.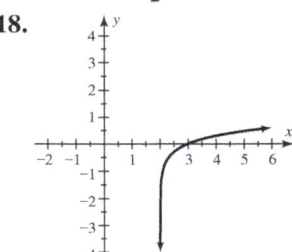

Shifted to the right 2 units
19. $3 \log x + 2 \log y - \frac{1}{2} \log z$ 20. $\ln \frac{x^4 z}{y^5}$ 21. $2x \log 7$
22. $1 - 3x$ 23. $\ln 25 \approx 3.22$ 24. $\log 50 \approx 1.70$
25. $10^{1.8} \approx 63.10$ 26. $\frac{1}{5}e^9 \approx 1620.62$
27. $a = 5$, $b = 3$ 28. (a) 4 million (b) About 6.15; after 5 hours, there were about 6.15 million bacteria.
(c) Growth (d) After 8 hours
29. (a) $A(x) = 5000(0.98)^x$ (b) $A(3) \approx 4705.96$; after 3 years, the account contains $4705.96. (c) About 5

Chapters 1–19 Cumulative Review (pp. 1293–1296)

1. 4.29×10^{-4} 2. Natural: none; whole: 0; integer: $-3, 0$; rational: $-\frac{11}{7}, -3, 0, 5.\overline{18}$; irrational: $\sqrt{6}, \pi$ 3. c.

4. Commutative 5. d^4 **6.** $\frac{b^9}{64a^6}$ **7.** $\frac{4y^8}{x^5}$ **8.** $\frac{y^4}{4x^5}$
9. $y = \frac{5}{4}x + 1$ **10.** $\{x \mid x \neq -3\}$ **11.** $f(x) = 4x - 3$
12. $x = 4$ **13.** 1
14.

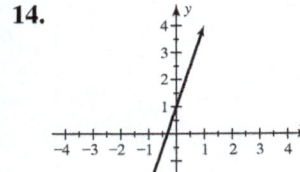

4.

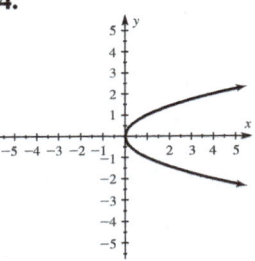

5. No; it does not pass the vertical line test. **6.** (h, k)
7. left **8.** right **9.** circle; (h, k) **10.** $(0, 0); r$
11. d. **12.** a. **13.** c. **14.** b.
15. **17.**

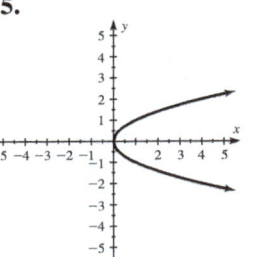 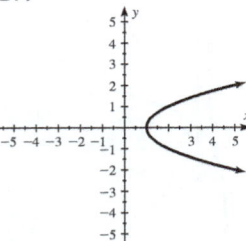

$(0, 0); y = 0$ $(1, 0); y = 0$
19. **21.**

15. $y = 7x - 6$ **16.** $y = 3x + 5$ **17.** -1 **18.** $x < 0$
19. -18 **20.** $(-\infty, 0)$ **21.** $\frac{4}{15}$ **22.** $(-5, 10]$
23. $[-2, 10]$ **24.** 4, 16
25.

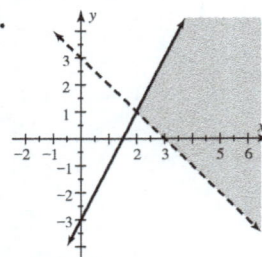

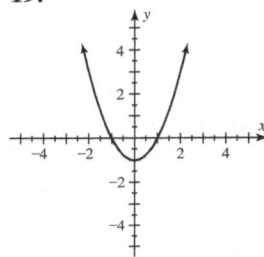 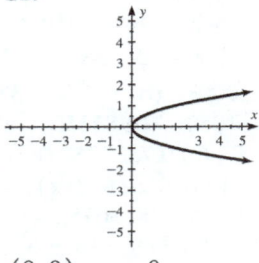

$(0, -1); x = 0$ $(0, 0); y = 0$
23. **25.**

26. 2 **27.** $(1, 1)$ **28.** No solutions **29.** $(-4, 0, 2)$
30. $(0, 0, 1)$ **31.** $2x(x - 1)^2$ **32.** $(2a - 5b)(2a + 5b)$
33. $(2t - 3)(4t^2 + 6t + 9)$ **34.** $(2a^2 + 5)(2a - 1)$
35. $-\frac{5}{6}, 2$ **36.** $-\frac{2}{3}, \frac{2}{3}$ **37.** $-3, 0, 5$ **38.** $0, \frac{1}{2}$ **39.** 1
40. $\frac{x+2}{x-3}$ **41.** 3 **42.** -3 **43.** $J = \frac{2z}{P-1}$ **44.** $\frac{x^3+3}{x^3-3}$
45. 40 **46.** $3x^2 + 6x + 10 + \frac{5}{x-2}$ **47.** $\frac{x^4}{y^6}$ **48.** x^3
49. $10ab$ **50.** $\sqrt{6}$ **51.** $(b + 3)a\sqrt[3]{a^2b}$ **52.** 20
53. $\frac{5 + \sqrt{3}}{11}$ **54.** $(4, \infty)$ **55.** $-6, 4$ **56.** 3 **57.** $3 + 5i$
58. $-i$ **59.** $(2, 1)$ **60.** $\frac{7}{2}$ **61.** $5, 8$ **62.** $\frac{1 \pm \sqrt{41}}{4}$
63. $2 \pm \sqrt{2}$ **64.** $\pm 2, \pm \sqrt{6}$ **65. (a)** $-1, 3$ **(b)** $a < 0$
(c) Positive **66.** $(-\infty, -7] \cup [2, \infty)$ **67. (a)** 7
(b) $(g \circ f)(x) = 2x^2 - 3$ **68.** $f^{-1}(x) = \frac{3}{x}$ **69.** 4
70. $2x, x > 0$ **71.** $\frac{1}{2} \log x - 2 \log y$ **72.** $\ln(5x^3)$
73. $10^{11/6} \approx 68.13$ **74.** $\frac{\log 17}{3 \log 2} \approx 1.36$ **75.** 4%
76. 2.27 pounds **77.** About 11.3%
78. (a) 9.91 meters per second **(b)** 8.52 meters

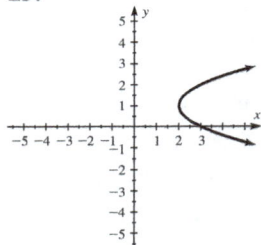 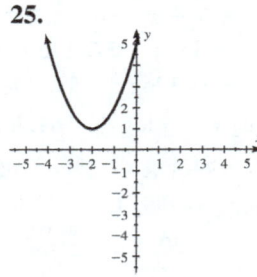

$(2, 1); y = 1$ $(-2, 1); x = -2$
27. **29.**

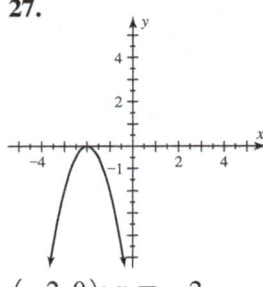 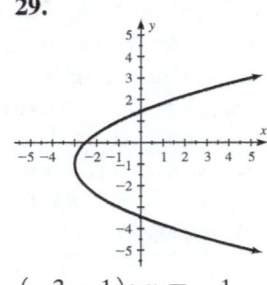

$(-2, 0); x = -2$ $(-3, -1); y = -1$

20 Conic Sections

Section 20.1 (pp. 1304–1307)

1. Parabola, ellipse, hyperbola **2.** The axes of symmetry are vertical and horizontal, respectively. **3.** No

31.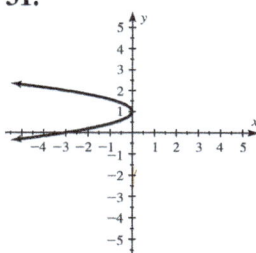
$(0, 1); y = 1$

33.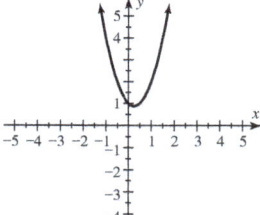
$\left(\frac{1}{4}, \frac{7}{8}\right); x = \frac{1}{4}$

35.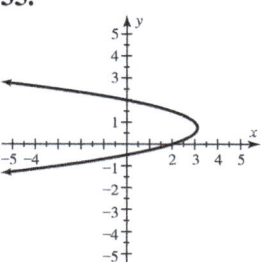
$\left(\frac{25}{8}, \frac{3}{4}\right); y = \frac{3}{4}$

37.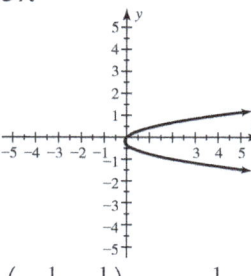
$\left(-\frac{1}{12}, -\frac{1}{6}\right); y = -\frac{1}{6}$

39.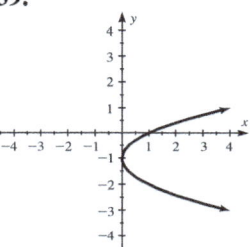
$(0, -1); y = -1$

41. $y = x^2$ **43.** $x = (y + 1)^2 - 2$ **45.** Upward
47. Downward **49.** $x \geq 0$ **51.** Two **53.** $(1, 0)$
55. $x^2 + y^2 = 1$ **57.** $(x + 1)^2 + (y - 5)^2 = 9$
59. $(x + 4)^2 + (y + 6)^2 = 2$ **61.** $x^2 + y^2 = 16$
63. $(x + 3)^2 + (y - 2)^2 = 1$
65. $2; (0, 0)$ **67.** $3; (1, 3)$
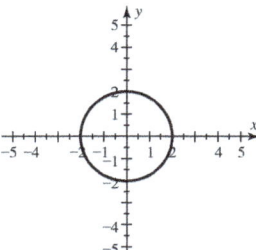 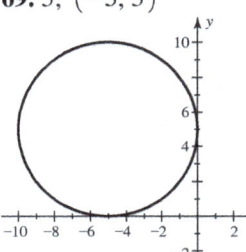

69. $5; (-5, 5)$ **71.** $3; (-3, 1)$

73. $\sqrt{7}; (-3, 1)$
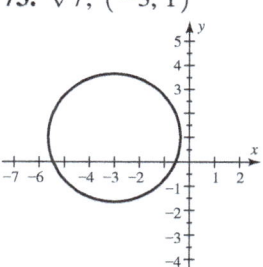

75. $(x - 1)^2 + (y - 5)^2 = 17$
77. $\left(x - \frac{7}{2}\right)^2 + (y - 3)^2 = \frac{25}{4}$
79. (a)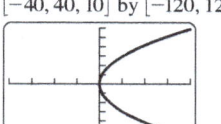
$[-40, 40, 10]$ by $[-120, 120, 20]$
(b) 32 ft

81. (a)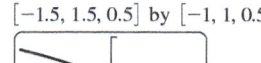
$[-1.5, 1.5, 0.5]$ by $[-1, 1, 0.5]$
(b) 2.6 A.U., or 241,800,000 miles

Section 20.2 (pp. 1314–1317)

1.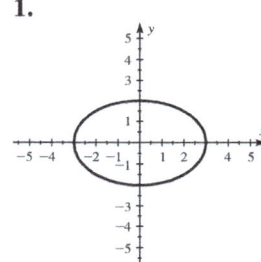
(answers may vary)

2.
(answers may vary)

3. horizontal **4.** vertical **5.** 2 **6.** 4 **7.** left; right
8. lower; upper **9.** They are the diagonals extended.
10. No **11.** a. **12.** b.

13.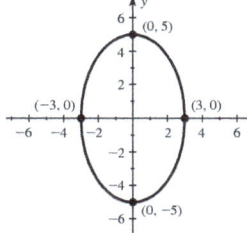

15.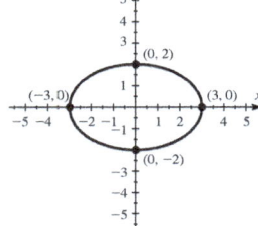

17.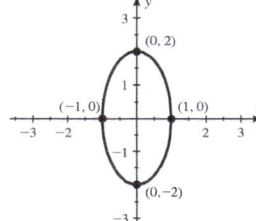

19.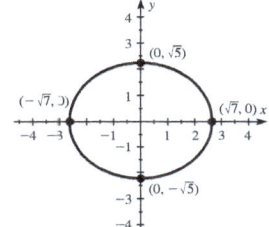

21. 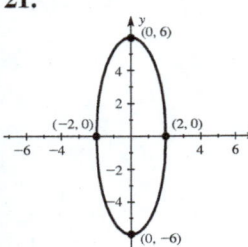 **23.**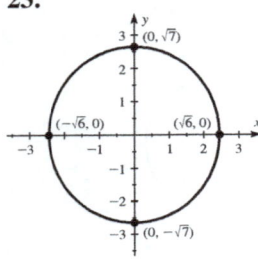

25. $\frac{x^2}{9} + \frac{y^2}{4} = 1$ **27.** $\frac{y^2}{25} + \frac{x^2}{16} = 1$

29. 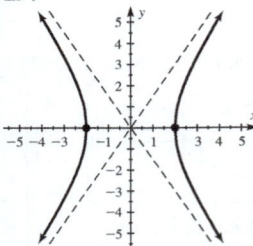 **31.**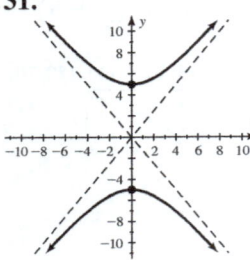

33. 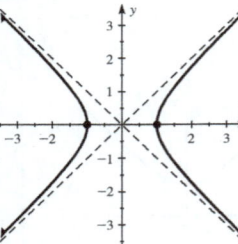 **35.**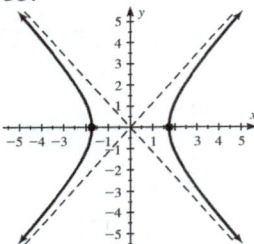

37. 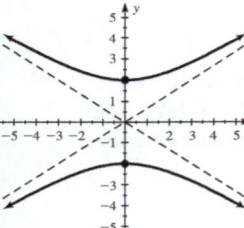 **39.**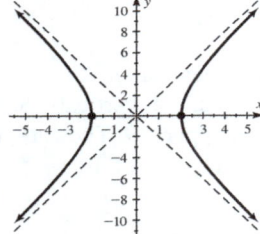

41. $x^2 - y^2 = 1$ **43.** $\frac{y^2}{4} - \frac{x^2}{9} = 1$

45. (a) 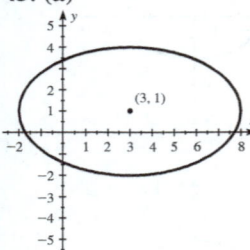 **(b)**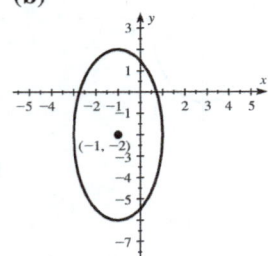

(c) 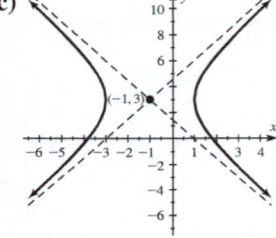 **(d)**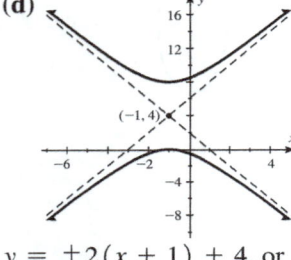

$y = \pm\frac{3}{2}(x+1) + 3$, or $y = \frac{3}{2}x + \frac{9}{2}, y = -\frac{3}{2}x + \frac{3}{2}$

$y = \pm 2(x+1) + 4$, or $y = 2x + 6, y = 2x + 2$

47. (a) $A \approx 62.83; P \approx 28.45$
(b) $A \approx 11.75; P \approx 13.33$
49. (a) $[-60, 60, 10]$ by $[-40, 40, 10]$

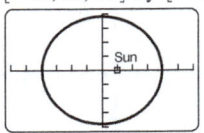

(b) $P \approx 243.9$ A.U., or about 2.27×10^{10} miles; $A \approx 4733$ square A.U., or about 4.09×10^{19} square miles
51. Maximum: 668 miles; minimum: 340 miles
53. Height: 20 feet; width: 200 feet
55. Answers may vary.

Checking Basic Concepts 20.1 & 20.2 (p. 1317)

1. 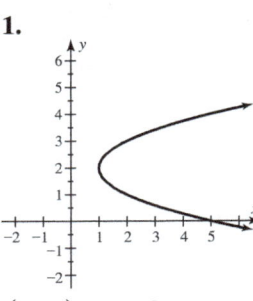 **2.** $(x-1)^2 + (y+2)^2 = 4$

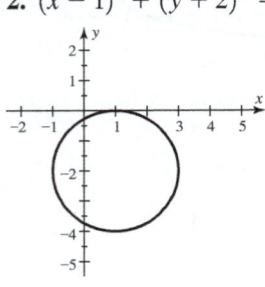

$(1, 2); y = 2$
3. x-intercepts: $(\pm 2, 0)$; y-intercepts: $(0, \pm 3)$
4. (a) 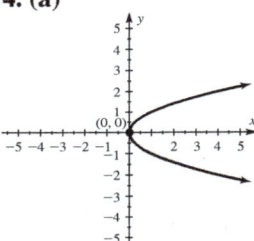 **(b)**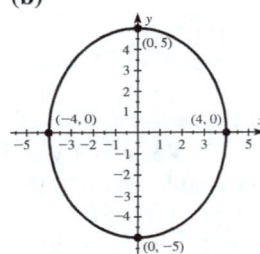

Parabola   Ellipse

(c) 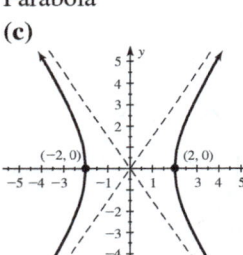 **(d)**

Hyperbola   Circle (and ellipse)

Section 20.3 (pp. 1324–1326)

1. Any number **2.** Two **3.** Two; the line intersects the circle twice. **4.** All points inside and including a circle of radius 1 centered at the origin **5.** No **6.** No

7. 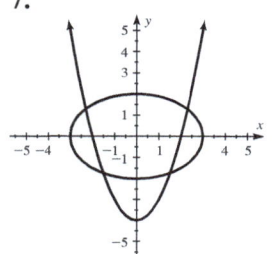   **8.**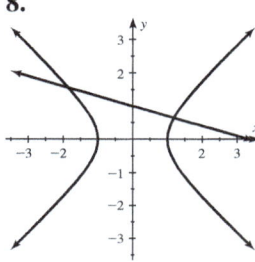

(answers may vary)   (answers may vary)

9. $(1, 3), (-1, -3)$ **11.** $(0, -1), (0, 1)$
13. $(3, 6), (-3, -6)$ **15.** $(2, -1)$ **17.** $(-\sqrt{2}, 2), (\sqrt{2}, 2)$
19. $\left(-\frac{3}{5}, \frac{7}{5}\right), (-1, 1)$ **21.** No solutions   **23.** $(\pm 3, \pm 1)$
25. $(-1, -2), (-2, 1)$ **27.** $(7, 3), (2, -2)$
29. $(-1, 2), (3, -6)$ **31.** $(-1, -1), (1, 1)$

33. 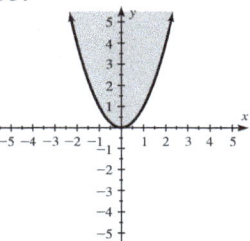   **35.**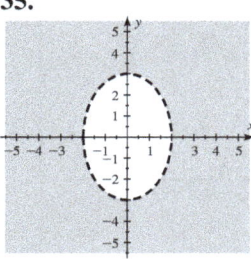

37. 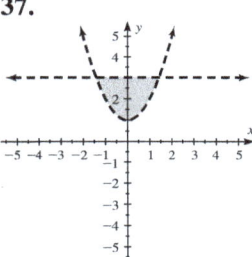   **39.**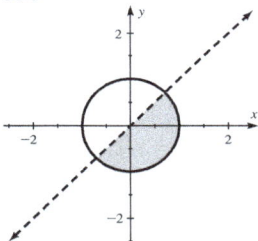

$(0, 2)$ (answers may vary) $\left(\frac{1}{2}, -\frac{1}{2}\right)$ (answers may vary)

41. 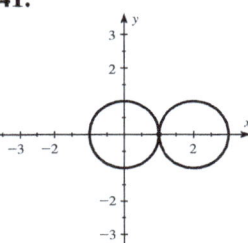   **43.**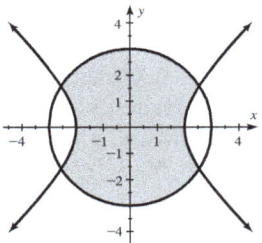

$(1, 0)$ $(0, 0)$ (answers may vary)

45. a.   **47.** $y \geq x^2; y < 4 - x$
49. $[-10, 10, 1]$ by $[-10, 10, 1]$

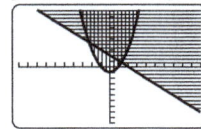

51. $r = 1.6$ inches; $h \approx 4.97$ inches
53. (a) $xy = 143, 2x + 2y = 48$
(b) $x = 11$ ft, $y = 13$ ft

Checking Basic Concepts 20.3 (p. 1326)

1. $(1, -1), (3, 3)$ **2.** Two   **3. (a)** $(0, 3), (4, 4)$
(answers may vary)   **(b)** $y \geq 2 - x$ and $y \leq 4 - x^2$

4.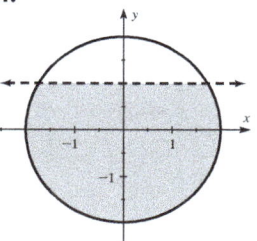

Chapter 20 Review (pp. 1328–1330)

1. 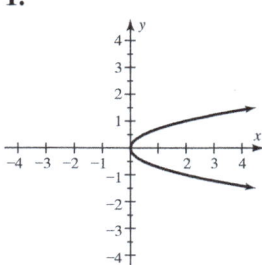   **2.**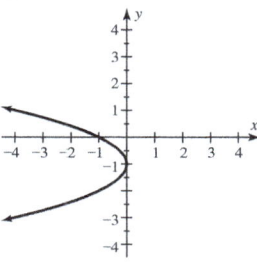

$(0, 0); y = 0$ $(0, -1); y = -1$

3. 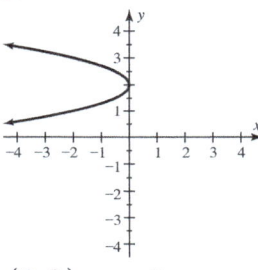   **4.**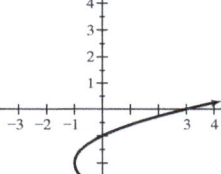

$(0, 2); y = 2$ $(-1, -2); y = -2$

5. 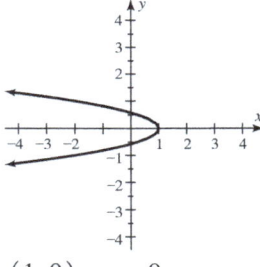   **6.**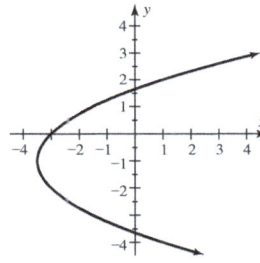

$(1, 0); y = 0$ $\left(-\frac{7}{2}, -1\right); y = -1$

7. $x = y^2$ **8.** $(x + 2)^2 + (y - 2)^2 = 16$
9. $x^2 + y^2 = 1$ **10.** $(x - 2)^2 + (y + 3)^2 = 16$
11. 5; $(0, 0)$ **12.** 3; $(2, 0)$

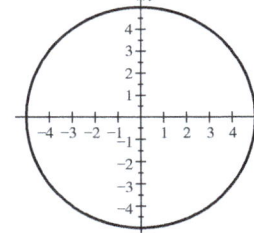   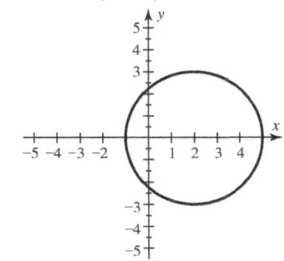

13. $\sqrt{5}$; $(-3, 1)$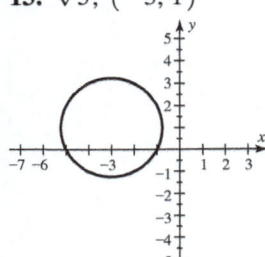

14. 3; $(1, -1)$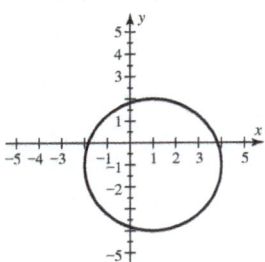

15.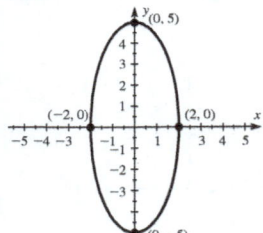

16.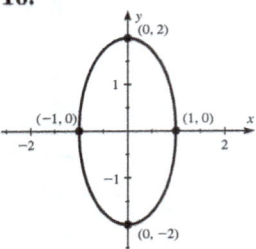

17.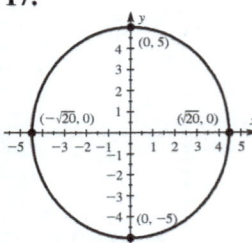

18.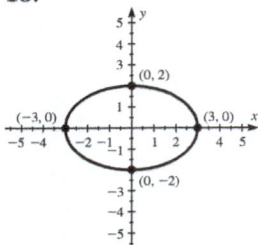

19. $\frac{y^2}{16} + \frac{x^2}{4} = 1$ **20.** $x^2 - \frac{y^2}{4} = 1$

21.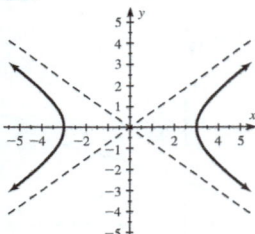

22.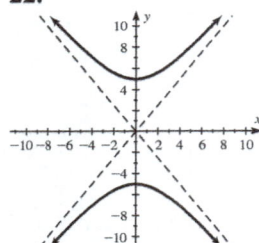

23.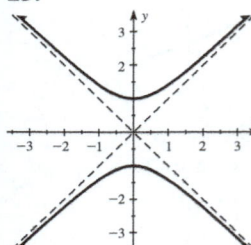

24.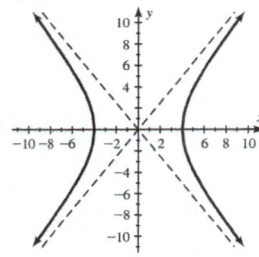

25. $(0, 3), (3, 0)$ **26.** $(-1, -2), (1, 2)$
27. $(0, 0), (2, 2)$
28. $(-2, -1), (-2, 1), (2, -1), (2, 1)$
29. $(-4, -4), (4, 4)$ **30.** $(0, -4), (4, 0)$
31. $(-1, 1), (1, 1)$ **32.** $(1, 2), (2, 5)$
33. $(-4, -12), (2, 0)$ **34.** $(-1, -1), (1, -1)$
35. $(2, 2), (-4, -4)$ **36.** $(1, 1), (0, 0)$

37.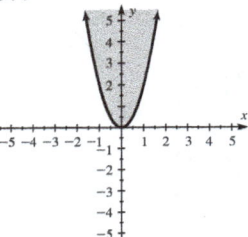

38.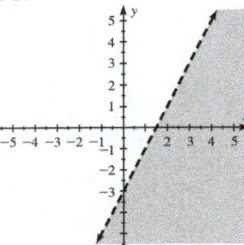

39.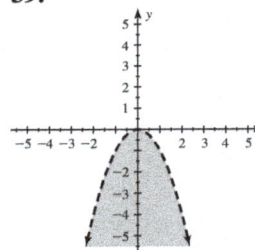

40.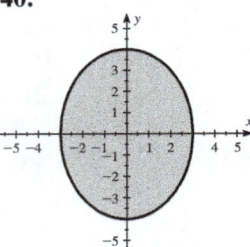

41.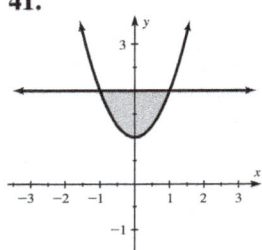

42.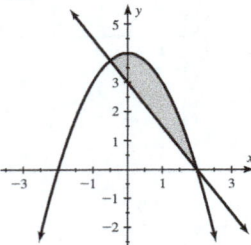

43.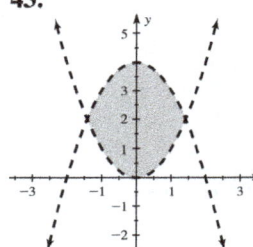

44.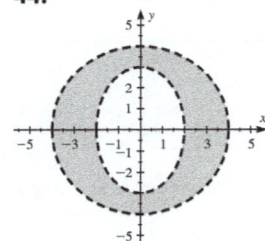

45. $y \geq x^2 - 2, y \leq 2 - x$ **46.** $y \geq x, x^2 + y^2 \leq 4$
47. (a) $xy = 1000, 2x + 2y = 130$
(b) $x = 25$ inches; $y = 40$ inches
(c) $x = 25$ inches; $y = 40$ inches
48. (a) $xy = 60, y - x = 7$ (b) $x = 5; y = 12$
(c) $x = 5; y = 12$ **49.** $r = 1$ foot, $h \approx 15.92$ feet; yes
50. Either $r \approx 0.94$ inches, $h \approx 12.60$ inches or $r \approx 3.00$ inches, $h \approx 1.23$ inches; no
51. (a) $[-7.5, 7.5, 1]$ by $[-5, 5, 1]$

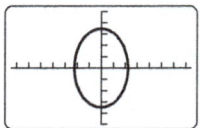

(b) $A \approx 24.33, P \approx 18.32$
52. (a) $[-3, 3, 1]$ by $[-2, 2, 1]$

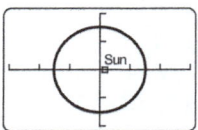

(b) $P \approx 9.55$ A.U., or about 8.9×10^8 miles; $A \approx 7.26$ square A.U., or about 6.3×10^{16} square miles

Chapter 20 Test (pp. 1330–1331)

1. 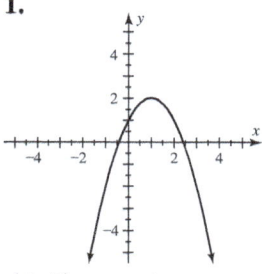 **2.**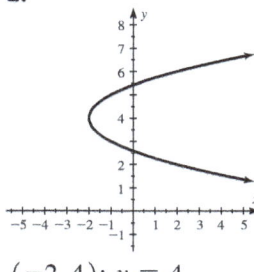

$(1, 2);\ x = 1$ $(-2, 4);\ y = 4$

3. $x = -y^2 + 1$ **4.** $(x - 2)^2 + (y + 4)^2 = 4$
5. $(x + 5)^2 + (y - 2)^2 = 100$
6. $r = 4$, center $= (-2, 3)$ **7.**

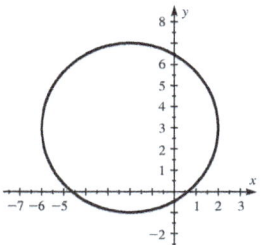   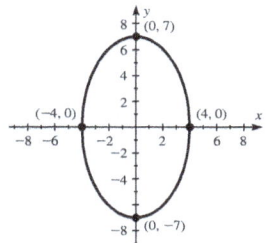

8. $\frac{x^2}{100} + \frac{y^2}{64} = 1$
9.

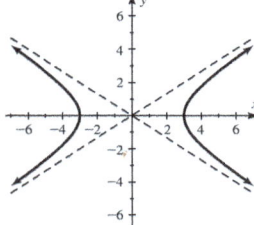

10. $(0, -4), (4, 0)$ **11.** $(-1, -4), (4, 1)$
12. $(-2, 4), (2, 4)$
13.

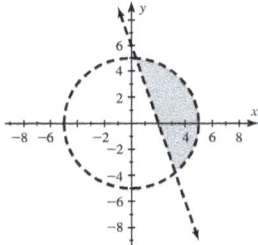

14. $y \leq 4 - x^2,\ y \geq x^2 - 4$
15. (a) $xy = 5000,\ 2x + 2y = 300$
(b) 50 feet by 100 feet
16. Either $x \approx 22.08$ in., $y \approx 2.43$ in. or $x \approx 7.29$ in., $y \approx 22.24$ in.; no
17. (a) $[-30, 30, 10]$ by $[-20, 20, 10]$

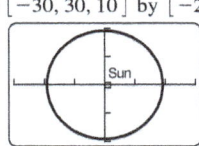

(b) 18.28 A.U., or about 1,700,040,000 miles

Chapters 1–20 Cumulative Review (pp. 1331–1332)

1. 25   **2.** $\frac{b^8}{a^3}$ **3.** 0.007345   **4.** $\frac{1}{2};\ x \neq 4$ **5.** 3; 4
6.   **7.**

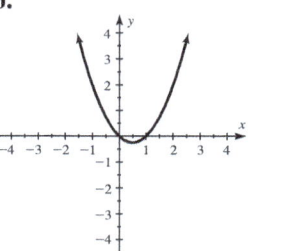   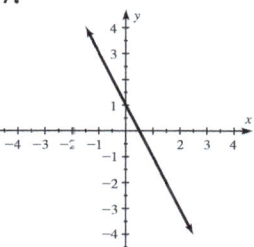

8. $y = \frac{3}{2}x - 5$ **9.** $\frac{2}{7}$ **10.** $(-1, 3]$ **11.** $[-2, 2]$
12. $(-\infty, -1] \cup [3, \infty)$ **13.** $(1, 2)$
14.

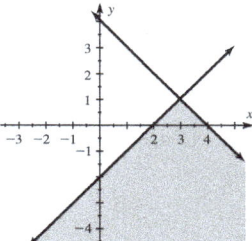

15. $2x^2 + 9x - 5$ **16.** $2x^2y - 3xy^3 + xy$
17. $(3x + 1)(2x - 5)$ **18.** $x(x + 2)(x - 2)$ **19.** $-2, -1$
20. $\frac{-3 \pm \sqrt{5}}{2}$ **21.** $\frac{3}{2}$ **22.** $\frac{2x}{x^2 - 1}$ **23.** $2x\sqrt{2}$ **24.** 4
25. $4x$ **26.** $5\sqrt{3x}$
27.

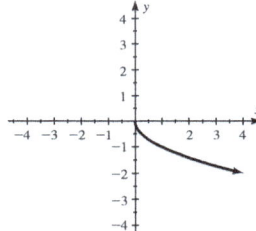

28. 5   **29.** 13   **30.** 2   **31.** $(3, -6)$
32. $f(x) = (x - 1)^2 + 2$ **33.** Shifted 4 units right
34. 1, 2   **35.** 0, $\pm i$ **36. (a)** 4   **(b)** 3   **(c)** x **(d)** 6   **(e)** 2
(f) 3   **37. (a)** 17   **(b)** $2x^2 + 2$ **38.** $\frac{2}{3} - \frac{1}{3}x$
39. $1340.10   **40.** $2 \log x + \frac{1}{2} \log y - 3 \log z$ **41.** $\ln 9$
42. 25
43. (a)   **(b)**

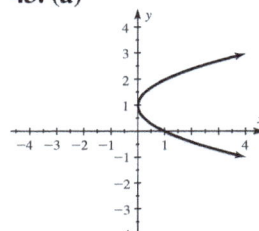   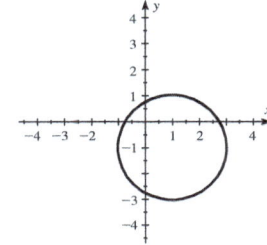

(c) 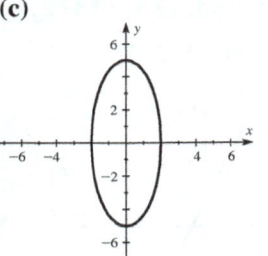 (d)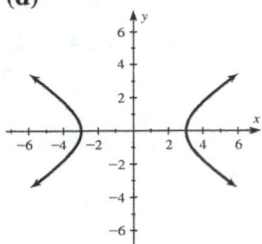

44. $(0, -1), (0, 1)$

45.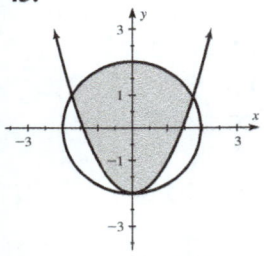

46. (a) 400; initially, the driver is 400 miles from home.
(b) $(8, 0)$; after 8 hours, the driver arrives at home.
(c) -50; the driver is traveling 50 mph toward home.
47. $500 at 5%, $1000 at 6%, $500 at 7%
48. 300 ft by 300 ft   **49.** $50 \ln 2 \approx 34.7$ yr
50. 2.4 in.

21 Sequences and Series

Section 21.1 (pp. 1340–1341)

1. 1, 2, 3, 4 (answers may vary)   **2.** 1, 3, 5, 7, … (answers may vary)   **3.** function; natural numbers   **4.** sequence
5. 6   **6.** scatterplot   **7.** $f(2)$ **8.** a_4 **9.** 1, 4, 9, 16
11. $\frac{1}{6}, \frac{1}{7}, \frac{1}{8}, \frac{1}{9}$ **13.** $\frac{5}{2}, \frac{5}{4}, \frac{5}{8}, \frac{5}{16}$ **15.** 9, 9, 9, 9   **17.** 1, 8, 27
19. $1, \frac{8}{5}, 2$ **21.** 2, 9, 20   **23.** $-2, -2, -2$ **25.** $f(n) = \frac{1}{n}$
27. $f(n) = 2n + 1$ **29.** $b + c; 2b + c$
31. 7   **33.** 3, 4, 5, 3, 1   **35.** 6, 5, 4, 3, 2, 1
37.

n	1	2	3	4	5	6	7
a_n	2	3	4	5	6	7	8

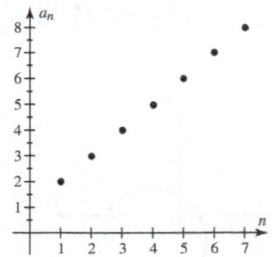

39.

n	1	2	3	4	5	6	7
a_n	0	2	6	12	20	30	42

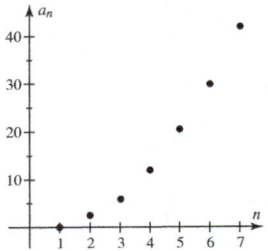

41.

n	1	2	3	4	5	6	7
a_n	2	4	8	16	32	64	128

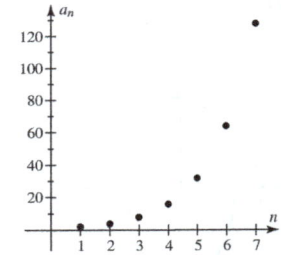

43. $a_n = 30n$ for $n = 1, 2, 3, \ldots, 7$

n	1	2	3	4	5	6	7
a_n	30	60	90	120	150	180	210

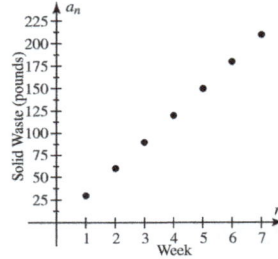

45. (a) 1, 4, 9, 16   (b) 4, 8, 12, 16
47. (a) $20,000; $16,000   (b) $a_n = 25,000(0.8)^n$
(c)

n	1	2	3	4	5	6	7
a_n	20,000	16,000	12,800	10,240	8192	6553.6	5242.9

49. (a) $a_n = 2048(0.5)^{n-1}$, for $n = 1, 2, 3, \ldots, 7$
(b)

n	1	2	3	4	5	6	7
a_n	2048	1024	512	256	128	64	32

(c)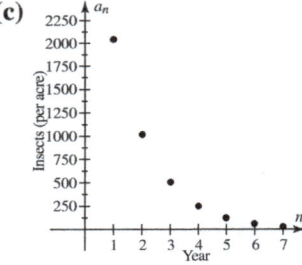

Section 21.2 (pp. 1348–1351)

1. linear   **2.** exponential   **3.** $a_n = 3n + 1$; 3 (answers may vary)   **4.** $a_n = 5(2)^{n-1}$; 2 (answers may vary)
5. add; previous   **6.** multiply; common ratio   **7.** 19; 4
8. 32; -2 **9.** $a_n = a_1(r)^{n-1}$ **10.** $a_n = a_1 + (n-1)d$
11. Yes; 10   **13.** Yes; -1 **15.** No   **17.** Yes; 3
19. Yes; -3 **21.** No   **23.** Yes; 1   **25.** No   **27.** Yes; 2
29. $a_n = -2n + 9$ **31.** $a_n = 4n - 6$

33. $a_n = -2n + 32$ **35.** 59 **37.** 21 **39.** Yes; 3
41. Yes; 0.8 **43.** No **45.** Yes; 2 **47.** No **49.** Yes; 4
51. Yes; 2 **53.** No **55.** $a_n = 1.5(4)^{n-1}$
57. $a_n = -3(-2)^{n-1}$ **59.** $a_n = 1(4)^{n-1}$ **61.** 4374
63. 243 **65.** (a) 3000, 6000, 9000, 12,000, 15,000; arithmetic
(b) $a_n = 3000n$ (c) 60,000; when there are 20 people, the ventilation rate should be 60,000 cubic feet per hour.
(d)

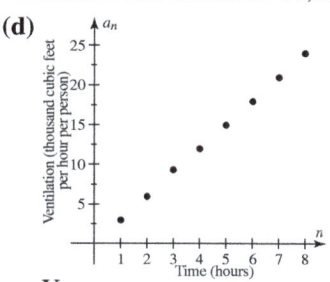

Yes

67. (a) $a_n = 3(0.8)^{n-1}$
(b) No; geometric

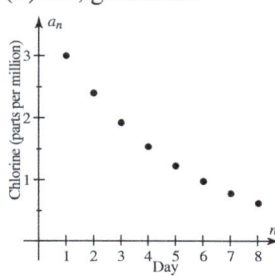

69. (a) $a_n = 5(0.85)^{n-1}$ (b) Geometric; the common ratio is 0.85. (c) About 1.6; on the 8th bounce, the ball reaches a maximum height of about 1.6 ft.
71. (a) Arithmetic; the common difference is 2.
(b) $a_n = 40 + 2(n - 1)$ or $a_n = 38 + 2n$ (c) 78
73. 1, 1, 2, 3, 5, 8, 13, 21, 34, 55, 89, 144

Checking Basic Concepts 21.1 & 21.2 (p. 1351)

1. $\frac{1}{5}, \frac{1}{3}, \frac{3}{7}, \frac{1}{2}$
2.

n	1	2	3	4	5
a_n	2	3	4	5	6

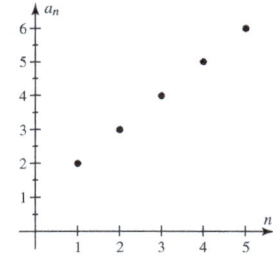

3. (a) Arithmetic; $a_n = 3n - 5$ (b) Geometric; $a_n = 3(-2)^{n-1}$ **4.** $a_n = 2n + 3$ **5.** $a_n = 5(2)^{n-1}$

Section 21.3 (pp. 1359–1361)

1. series **2.** 10 **3.** arithmetic **4.** geometric
5. $n\left(\frac{a_1 + a_n}{2}\right)$ or $\frac{n}{2}(2a_1 + (n-1)d)$ **6.** $a_1\left(\frac{1 - r^n}{1 - r}\right)$
7. sum **8.** $a_1 + a_2 + a_3 + a_4$ **9.** arithmetic
10. geometric **11.** 48 **13.** 820 **15.** -5 **17.** -92

19. 3279 **21.** -85 **23.** 182 **25.** 6825
27. $2 + 4 + 6 + 8$; 20
29. $4 + 4 + 4 + 4 + 4 + 4 + 4 - 4$; 32
31. $1 + 4 + 9 + 16 + 25 + 36 + 49$; 140
33. $12 + 20$; 32 **35.** $\sum_{k=1}^{6} k^4$ **37.** $\sum_{k=1}^{5} \frac{1}{k^2}$
39. $\sum_{k=1}^{n} k = n\left(\frac{a_1 + a_n}{2}\right) = n\left(\frac{1 + n}{2}\right) = \frac{n(n + 1)}{2}$
41. $91,523.93 **43.** $62,278.01
45. (a) $22.2 + 37.7 + 28.7 + 42.1 + 33.7 + 22.5$
(b) 186.9 **47.** (a) $\sum_{k=1}^{n} 0.8(0.2)^{k-1}$ (b) 2
49. (a) $1, \frac{1}{2}, \frac{1}{4}, \frac{1}{8}, \frac{1}{16}$ (b) $\frac{1023}{512}$ **51.** 90 logs **53.** $1,080,000
55. About 3.16 feet **57.** (a) $\frac{4}{3}$ (b) $\frac{3}{2}$ (c) $0.\overline{1} = \frac{1}{9}$
(d) $0.\overline{12} = \frac{4}{33}$ **59.** Answers may vary.

Section 21.4 (p. 1367)

1. 5 **2.** $n + 1$ **3.** 4
4.
```
          1
        1   1
      1   2   1
    1   3   3   1
  1   4   6   4   1
```
5. 24 **6.** $6! = 720$ **7.** $\frac{n!}{(n-r)!\,r!}$ **8.** $a^2 + 2ab + b^2$
9. n **10.** 1 **11.** $x^3 + 3x^2y + 3xy^2 + y^3$
13. $16x^4 + 32x^3 + 24x^2 + 8x + 1$
15. $a^5 - 5a^4b + 10a^3b^2 - 10a^2b^3 + 5ab^4 - b^5$
17. $x^6 + 3x^4 + 3x^2 + 1$ **19.** 6 **21.** 5040 **23.** 4
25. 2 **27.** 10 **29.** 5 **31.** 6 **33.** 1 **35.** 792 **37.** 126
39. 75,582 **41.** $m^3 + 3m^2n + 3mn^2 + n^3$
43. $x^4 - 4x^3y + 6x^2y^2 - 4xy^3 + y^4$
45. $8a^3 + 12a^2 + 6a + 1$
47. $x^5 + 10x^4 + 40x^3 + 80x^2 + 80x + 32$
49. $81 + 216m + 216m^2 + 96m^3 + 16m^4$
51. $8x^3 - 12x^2y + 6xy^2 - y^3$ **53.** a^8 **55.** $35x^4y^3$
57. $512m^9$

Checking Basic Concepts 21.3 & 21.4 (p. 1368)

1. (a) Geometric (b) Arithmetic **2.** 312
3. -341 **4.** $x^4 - 4x^3y + 6x^2y^2 - 4xy^3 + y^4$
5. $x^3 + 6x^2 + 12x + 8$

Chapter 21 Review (pp. 1369–1371)

1. 1, 8, 27, 64 **2.** 3, 1, -1, -3 **3.** $1, \frac{4}{5}, \frac{3}{5}, \frac{8}{17}$
4. $-2, 4, -8, 16$ **5.** $-2, 0, 4, 2$ **6.** 5, 3, 2, 1
7.

n	1	2	3	4	5	6	7
a_n	2	4	6	8	10	12	14

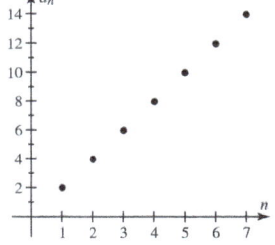

8.

n	1	2	3	4	5	6	7
a_n	−3	0	5	12	21	32	45

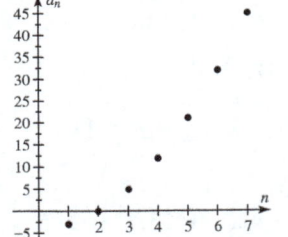

9.

n	1	2	3	4	5	6	7
a_n	2	1	0.5	0.25	0.125	0.0625	0.0313

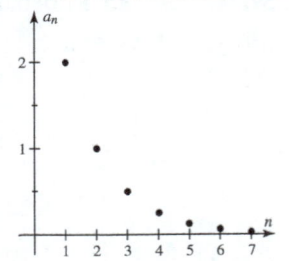

10.

n	1	2	3	4	5	6	7
a_n	1	1.4142	1.7321	2	2.2361	2.4495	2.6458

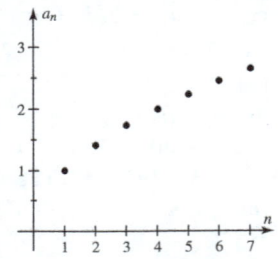

11. Yes; 5 **12.** No **13.** No **14.** Yes; $-\frac{1}{3}$ **15.** Yes; −3
16. No **17.** Yes; −1 **18.** No **19.** $a_n = 4n - 7$
20. $a_n = -5n + 7$ **21.** Yes; 4 **22.** No **23.** No
24. Yes; 0.7 **25.** No **26.** Yes; $-\frac{1}{3}$ **27.** No **28.** Yes; 2
29. $a_n = 5(0.9)^{n-1}$ **30.** $a_n = 2(4)^{n-1}$ **31.** 216 **32.** 7.5
33. 3277 **34.** $\frac{511}{256}$ **35.** $3 + 5 + 7 + 9 + 11$
36. $\frac{1}{2} + \frac{1}{3} + \frac{1}{4} + \frac{1}{5}$ **37.** $1 + 8 + 27 + 64$
38. $-1 + (-2) + (-3) + (-4) + (-5) + (-6)$
39. $\sum_{k=1}^{20} k$ **40.** $\sum_{k=1}^{20} \frac{1}{k}$ **41.** $\sum_{k=1}^{9} \frac{k}{k+1}$ **42.** $\sum_{k=1}^{7} k^2$
43. $x^3 + 12x^2 + 48x + 64$
44. $16x^4 + 32x^3 + 24x^2 + 8x + 1$
45. $x^5 - 5x^4y + 10x^3y^2 - 10x^2y^3 + 5xy^4 - y^5$
46. $a^6 - 6a^5 + 15a^4 - 20a^3 + 15a^2 - 6a + 1$
47. 6 **48.** 10 **49.** 20 **50.** 4
51. $m^4 + 8m^3 + 24m^2 + 32m + 16$
52. $a^5 + 5a^4b + 10a^3b^2 + 10a^2b^3 + 5ab^4 + b^5$
53. $x^4 - 12x^3y + 54x^2y^2 - 108xy^3 + 81y^4$
54. $27x^3 - 54x^2 + 36x - 8$

55. $a_n = 45,000(1.1)^{n-1}$ for $n = 1, 2, 3, \ldots, 7$; geometric

n	1	2	3	4	5	6	7
a_n	45,000	49,500	54,450	59,895	65,885	72,473	79,720

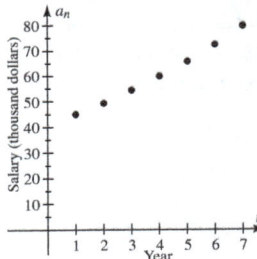

56. $a_n = 45,000 + 5000(n - 1)$ for $n = 1, 2, 3, \ldots, 7$; arithmetic

n	1	2	3	4	5	6	7
a_n	45,000	50,000	55,000	60,000	65,000	70,000	75,000

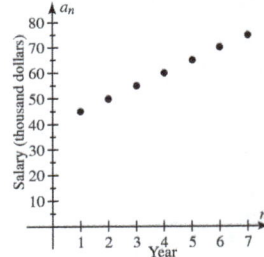

57. $a_n = 30n$ for $n = 1, 2, 3, \ldots, 7$

n	1	2	3	4	5	6	7
a_n	30	60	90	120	150	180	210

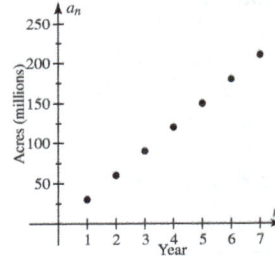

58. (a) $a_n = 12,000(0.9)^{n-1}$ (b) Geometric; the common ratio is 0.9. (c) About 1800; in 2013, there were about 1800 escapes.

Chapter 21 Test (pp. 1371–1372)

1. $\frac{1}{2}, \frac{4}{3}, \frac{9}{4}, \frac{16}{5}$ **2.** −3, 2, 1, −2, 3
3.

n	1	2	3	4	5	6	7
a_n	0	2	6	12	20	30	42

4. $16x^4 - 32x^3 + 24x^2 - 8x + 1$
5. Arithmetic; −3 **6.** Geometric; −2
7. $a_n = 2 - 3(n - 1)$ or $a_n = 5 - 3n$
8. $a_n = 2(1.5)^{n-1}$ **9.** Yes; 2.5 **10.** No **11.** 99
12. $\frac{463}{729}$ **13.** $6 + 9 + 12 + 15 + 18 + 21$
14. $\sum_{k=1}^{60} k^3$ **15.** 35 **16.** 10 **17.** 9180

18. $a_n = 176(1.59)^{n-1}$ for $n = 1, 2, 3, 4, 5$; geometric

n	1	2	3	4	5
a_n	176	279.84	444.95	707.46	1124.87

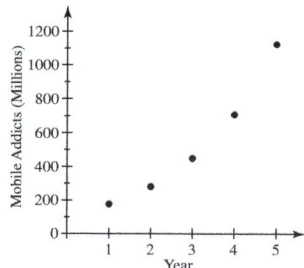

19. (a) $a_n = 2000(2)^{n-1}$ **(b)** Geometric; the common ratio is 2. **(c)** 64,000; after 25 days, there are 64,000 caterpillars.
(d)

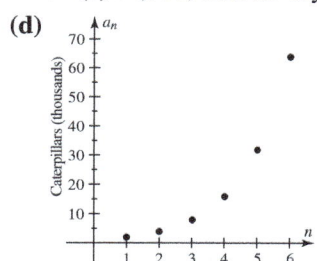

Chapters 1–21 Cumulative Review (pp. 1372–1374)

1. Distributive **2.** $D = \{-6, -2, 0, 2\}$; $R = \{0, 1, 3, 5\}$
3. $\frac{y^9}{27x^5}$ **4.** $\frac{16a^8}{b^4}$ **5.** z^{10} **6.** $\frac{2y^3}{x^6}$ **7.** $D = \{x | x \neq 8\}$
8. $f(x) = -2x + 1$ **9.** $y = 3$ **10.** $-3; 5$
11. $y = \frac{3}{2}x + \frac{5}{2}$ **12.** $y = 2x - 8$ **13.** -26
14. $\left(\frac{20}{11}, \infty\right)$ **15.** $(-\infty, -1] \cup [11, \infty)$ **16.** $-15, 3$
17. $\frac{1}{8}$ **18.** $[-12, 9)$
19.

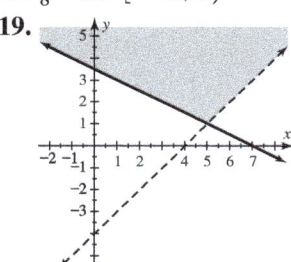

20. $(5, 2)$ **21.** $(3, 3, -1)$ **22.** 17
23. $8x^7 - 6x^6 + 10x^3$ **24.** $6z^2 - 13z - 28$
25. $(2x - 3y)(2x + 3y)$ **26.** $(a^2 + 4)(2a - 1)$
27. $-\frac{3}{4}, 1$ **28.** $0, 4, 6$ **29.** $\frac{x-2}{x+1}$ **30.** $\frac{x+4}{x-2}$ **31.** 12
32. $\frac{1}{2}, 3$ **33.** $W = \frac{3C - 5R}{2}$ **34.** $\frac{1 + 2x}{1 - 4x}$
35. $(xy - 2)\sqrt[3]{xy}$ **36.** 14 **37.** $-2, 8$ **38.** $\frac{3}{2}$
39. $14 + 2i$ **40.** $-2 - 2i$ **41.** $-\frac{1}{3}$
42. $y = 2(x + 2)^2 + 9; (-2, 9)$ **43.** $2 \pm 3i$
44. $-4, 8$ **45.** $(-3, 1)$ **46. (a)** 65
(b) $(g \circ f)(x) = 3x^2 + 1$ **47.** $f^{-1}(x) = \frac{2x - 1}{3}$
48. $3 \ln x + \frac{1}{2} \ln y$ **49.** $\log \frac{x}{4y}$ **50.** 56.23 **51.** 0.58

52.

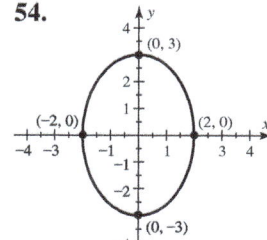

$(1, 3); y = 3$
53. $(3, -1); 2$
54. 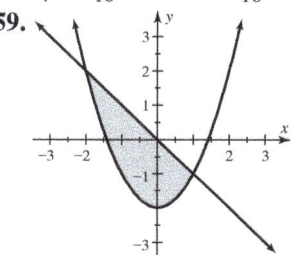 **55.**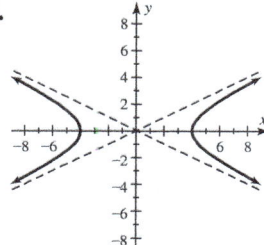

56. $\frac{y^2}{4} - \frac{x^2}{16} = 1$ **57.** $\frac{x^2}{16} + \frac{y^2}{4} = 1$ **58.** $(1, 2), (-1, 2)$
59.

60. Arithmetic; -2 **61.** Geometric; 0.2
62. $a_n = 3n - 1$ **63.** $a_n = 4(3)^{n-1}$ **64.** 171
65. 683 **66.** $16x^4 + 96x^3 + 216x^2 + 216x + 81$
67. $8a^3 - 60a^2b + 150ab^2 - 125b^3$
68. $\sqrt{\frac{14}{\pi}} \approx 2.11$ inches
69. (a) $f(x) = 0.4x + 85$ **(b)** 188 pounds
70. Airplane: 380 mph; wind: 20 mph
71. 36 minutes **72. (a)** $xy = 96, 3x - y = 12$
(b) 8 and 12 **73.** 144

B Sets

(pp. AP-17–AP-18)

1. $\{1, 2, 3, 4, 5, 6, 7\}$ **3.** $\{\text{Sunday, Saturday}\}$
5. $\{A, B, C, D, E, F, G\}$ **7.** $\varnothing$ **9.** $\{0, 2, 4, 6, 8, 10\}$
11. $\in$ **13.** $\notin$ **15.** $\in$ **17.** $\notin$ **19.** $\in$
21. $E = \{2, 4, 6, 8, 10\}$ **23.** $E = \{2, 4, 6, 8, 10, \ldots\}$
25. $A = \{\text{Apple, Apricot}\}$ **27.** $A = \{\text{Algebra}\}$
29. False **31.** True **33.** True **35.** True **37.** True
39. True **41.** False **43.** True
45. 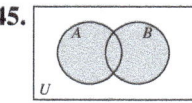 **47.**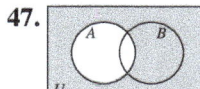

49. 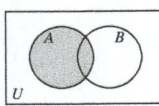 **51.**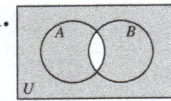

53. $\{1, 2, 3, 4\}$ **55.** $\{a, b\}$ **57.** $\{a, b, c\}$
59. $\{a, b, c, d\}$ **61.** $\{5, 8\}$
63. $\{1, 2, 3, 4, 5, 6, 7, 8, 9, 10\}$ **65.** $\{4, 5, 6\}$
67. $\varnothing$ **69.** $\{1, 3, 5, 7, 9\}$ **71.** $\varnothing$ **73.** $\varnothing$
75. $\{1, 3, 5, 7, 9\}$
77. $\{1, 2, 3, 5, 7, 8, 9, 10\}$ **79.** $\varnothing$ **81.** $\{1, 2, 8, 9, 10\}$
83. $\{4, 6\}$ **85.** $\{1, 2, 3, 4, 5, 6, 7, 8, 9, 10\}$ **87.** Infinite
89. Finite **91.** Finite

C Linear Programming

(pp. AP-23–AP-25)

1. linear programming
3. feasible solutions **5.** vertex

7. 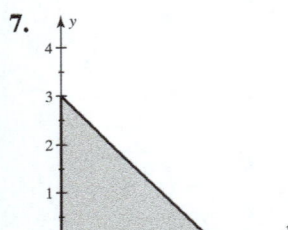 **9.**

11. 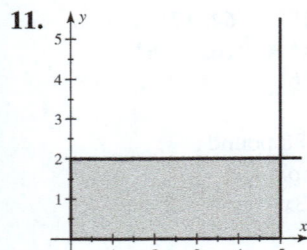 **13.**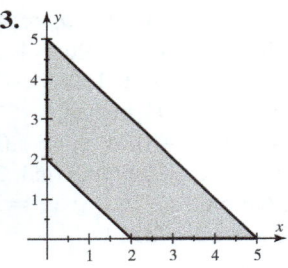

15. 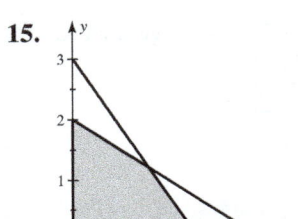 **17.**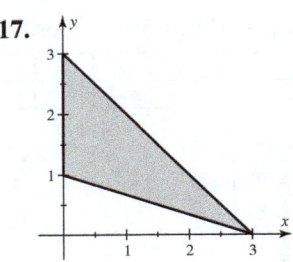

19.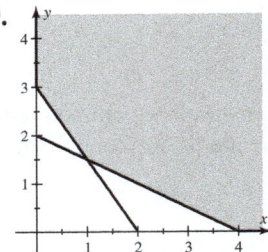

21. $R = 31$ **23.** $R = 15$ **25.** $C = 5$ **27.** $C = 2$
29. $R = 750$ **31.** $R = 12$ **33.** $R = 31.5$ **35.** $R = 16$
37. $C = 0$ **39.** $C = 32$ **41.** $C = 60$
43. 20 pounds of candy, 80 pounds of coffee **45.** $1800
47. 1 ounce Brand X, 2 ounces Brand Y **49.** $600

D Synthetic Division

(p. AP-27)

1. $x + 4 + \frac{3}{x - 1}$ **3.** $3x - 1$ **5.** $x^2 + 3x + 2$
7. $2x^2 + 5x + 10 + \frac{19}{x - 2}$ **9.** $x^2 + 2x + 6 + \frac{28}{x - 4}$
11. $x^2 + 2x + 2$ **13.** $2x^3 - 4x^2 + 8x - 17 + \frac{34}{x + 2}$
15. $b^3 + b^2 + b + 1$

Glossary

absolute value A nonnegative number, written $|a|$, that is equal to the distance of a from the origin on the number line.

absolute value equation An equation that contains an absolute value.

absolute value function The function defined by $f(x) = |x|$.

absolute value inequality An inequality that contains an absolute value.

acute angle An angle that measures between 0° and 90°.

acute triangle A triangle with only acute angles.

addends In an addition problem, the two numbers that are added.

addition property of equality If a, b, and c are real numbers, then $a = b$ is equivalent to $a + c = b + c$.

additive identity The number 0.

additive inverse (opposite) The additive inverse, or opposite, of a number a is $-a$.

adjacency matrix A matrix used to represent a map showing distances between cities or a social network.

adjacent angles Supplementary pairs of angles formed by intersecting lines.

algebraic expression An expression consisting of numbers, variables, operation symbols, such as $+$, $-$, $\times$, and $\div$, and grouping symbols, such as parentheses.

alternate exterior angles When two parallel lines are cut by a transversal, alternate exterior angles are outside of the parallel lines on opposite sides of the transversal, and are *not* adjacent angles.

alternate interior angles When two parallel lines are cut by a transversal, alternate interior angles are between the parallel lines on opposite sides of the transversal, and are *not* adjacent angles.

angle-side-angle (ASA) Two triangles are congruent if two angles and the included side of one triangle are congruent to two angles and the included side of the other triangle.

annual interest rate An interest rate that is applied once per year.

annual percentage rate (APR) The interest rate usually used in the computation of compound interest.

annuity A sum of money from which regular payments are made.

approximately equal The symbol $\approx$ indicates that two quantities are nearly equal.

approximation An answer found by estimating that is usually not exactly accurate.

area The number of square units that are needed to cover a region.

arithmetic mean (mean) The arithmetic mean of a list of numbers is the sum of the numbers divided by the number of numbers in the list.

arithmetic sequence A linear function given by $a_n = dn + c$ whose domain is the set of natural numbers.

arithmetic series The sum of the terms of an arithmetic sequence.

associative property for addition For any real numbers a, b, and c, $(a + b) + c = a + (b + c)$.

associative property for multiplication For any real numbers a, b, and c, $(a \cdot b) \cdot c = a \cdot (b \cdot c)$.

asymptotes of a hyperbola The two lines determined by the diagonals of the hyperbola's fundamental rectangle.

augmented matrix A matrix used to represent a system of linear equations; a vertical line is positioned in the matrix where the equals signs occur in the system of equations.

average The result of adding up the numbers of a set and then dividing the sum by the number of elements in the set.

axis of symmetry of a parabola The line passing through the vertex of the parabola that divides the parabola into two symmetric parts.

bar graph A graph with rectangular bars, where the length of each bar represents a data value.

base The value of b in the expression b^n.

base-10 number system The decimal system that uses digits 0-9 to represent all numbers.

base e If the base of an exponential expression is e (approximately 2.72), then we say this expression has base e.

base unit In the metric system of measurement, the meter, liter, and gram are the base units used to measure length, capacity, and mass, respectively.

bases (of a trapezoid) The parallel sides of a trapezoid.

basic complex fraction A complex fraction where both the numerator and denominator are single fractions.

basic percent equation An equation that is written in the form $percent \cdot whole = part$.

basic percent statement form A percent statement written in the form *a percent of the whole is a part*.

basic principle of fractions When simplifying fractions, the principle that states $\frac{a \cdot c}{b \cdot c} = \frac{a}{b}$.

basic rational equation A rational equation that has a single rational expression on each side of the equals sign.

binary operation An operation that requires two numbers to calculate an answer.

binomial A polynomial with two terms.

binomial coefficient The expression ${}_nC_r = \frac{n!}{(n-r)!r!}$, where n and r are nonnegative integers, $n \geq r$, that can be used to calculate the numbers in Pascal's triangle.

binomial theorem A theorem that provides a formula to expand expressions of the form $(a + b)^n$.

braces { } Symbols used to enclose the elements of a set.

branches A hyperbola has two branches, a left branch and a right branch, or an upper branch and a lower branch.

byte A unit of computer memory, capable of storing one letter of the alphabet.

capacity A measure of the amount of a substance that a container can hold.

Celsius Units used in the metric system to measure temperature.

center of a circle The point that is a fixed distance from all the points on a circle.

center of an ellipse The midpoint of the major axis.

certain event An event that is certain to occur.

change of base formula A formula used to evaluate a logarithm of one base by using a logarithm of a different base and given by $\log_a x = \frac{\log x}{\log a}$ or $\log_a x = \frac{\ln x}{\ln a}$.

circle The set of points in a plane that are the same distance from a fixed point.

circle graph A circle with shaded regions that visually represent percentages.

circular cylinder A solid with two parallel ends that are identical circles and a curved side.

circumference The perimeter of a circle.

coefficient The numeric constant of a term.

coefficient of a monomial The number in a monomial.

commission A percent of total sales, usually paid to the salesperson.

common denominator A denominator that is the same in two or more fractions.

common difference The value of d in an arithmetic sequence, $a_n = dn + c$.

common logarithmic function The function given by $f(x) = \log x$.

common logarithm of a positive number x Denoted $\log x$, it may be calculated as follows: If x is expressed as $x = 10^k$, then $\log x = k$, where k is a real number. That is, $\log 10^k = k$.

common multiple A number that two or more numbers will divide into evenly.

common ratio The value of r in a geometric sequence, $a_n = a_1(r)^{n-1}$.

commutative property for addition For any real numbers a and b, $a + b = b + a$.

commutative property for multiplication For any real numbers a and b, $a \cdot b = b \cdot a$.

complement The set containing all elements in the universal set that are not in A, denoted A'.

complement of an event All outcomes of an experiment that are *not* part of the given event.

complementary angles Two angles whose measures sum to 90°.

completely factored When the coefficient of a term is written as the product of prime numbers and any powers of variables are written as repeated multiplications.

completing the square method An important technique in mathematics that involves adding a constant to a binomial so that a perfect square trinomial results.

complex conjugate The complex conjugate of $a + bi$ is $a - bi$.

complex fraction A rational expression that contains fractions in its numerator, denominator, or both.

complex number A complex number can be written in standard form as $a + bi$, where a and b are real numbers and i is the imaginary unit.

composite function If f and g are functions, then g of f, or the composition of g and f, is defined by $(g \circ f)(x) = g(f(x))$ and is read "g of f of x."

composite number A natural number greater than 1 that is not a prime number.

composite region (figure) A region that consists of more than one geometric shape.

composition Replacing a variable with an algebraic expression in function notation is called composition of functions; when functions are applied to a variable in sequence.

compound inequality Two inequalities joined by the word *and* or the word *or*.

compound interest A type of interest paid at the end of each year using the formula $A = P(1 + r)^t$, where P is the amount deposited, r is the annual interest rate in decimal form, and t is the number of years for the account to contain A dollars.

cone A solid that tapers smoothly from a circular base to a single point.

conic section The curve formed by the intersection of a plane and a cone.

congruent angles Angles with the same measure.

congruent triangles Triangles in which the measures of corresponding angles are equal and corresponding sides have the same length.

conjugate The conjugate of $a + b$ is $a - b$.

consistent system with dependent equations A system of linear equations with infinitely many solutions.

consistent system with independent equations A system of linear equations with exactly one solution.

constant function A linear function with $m = 0$ that can be written as $f(x) = b$.

constant of proportionality (constant of variation) In the equation $y = kx$, the nonzero number k.

constraint In linear programming, an inequality which limits the objective function.

continuous growth Growth in a quantity that is directly proportional to the amount present.

contradiction An equation that is always false regardless of the values of any variables.

coordinates The numbers in an ordered pair.

corresponding angles When two parallel lines are cut by a transversal, corresponding angles are in the same relative position with respect to the parallel lines.

Cramer's rule A method that uses determinants to solve linear systems of equations.

cross product For the equation $\frac{a}{b} = \frac{c}{d}$ the cross products are ad and bc.

cube A solid object that has exactly six identical square faces.

cube root The number b is a cube root of a if $b^3 = a$.

cube root function The function defined by $f(x) = \sqrt[3]{x}$.

cubic function A function f of degree 3 represented by $f(x) = ax^3 + bx^2 + cx + d$, where a, b, c, and d are constants and $a \neq 0$.

decay factor The value of a in an exponential function, $f(x) = Ca^x$, when $0 < a < 1$.

decimal notation Notation used to represent a number with an integer part and a fractional part separated by a decimal point.

decimal number A number written in decimal notation.

decimal places The place values of the digits in a decimal number; also the number of digits to the right of the decimal point.

decimal point The dot in a decimal number that separates the integer part and the fractional part.

defining a variable Specifically stating what a variable represents.

degree A degree (°) is $\frac{1}{360}$ of a revolution.

degree of a monomial The sum of the exponents of the variables.

degree of a polynomial The degree of the term (or monomial) with highest degree.

denominator The bottom number in a fraction.

dependent equations Equations in a linear system that have infinitely many solutions.

dependent variable The variable that represents the output of a function.

descending order A polynomial in one variable is written in descending order if the exponents on the variable decrease as the terms are written from left to right.

determinant A real number associated with a square matrix.

diagrammatic representation A function represented by a diagram.

diameter The distance across a circle on a straight line through its center.

difference The answer to a subtraction problem.

difference of two cubes An expression in the form $a^3 - b^3$, which can be factored as $(a - b)(a^2 + ab + b^2)$.

difference of two squares An expression in the form $a^2 - b^2$, which can be factored as $(a - b)(a + b)$.

dimension of a matrix The size expressed in number of rows and columns. For example, if a matrix has m rows and n columns, its dimension is $m \times n$ (m by n).

directly proportional (varies directly) A quantity y is directly proportional to x if there is a nonzero number k such that $y = kx$.

discount A percent of the original price that is subtracted from the original price.

discriminant The expression $b^2 - 4ac$ in the quadratic formula.

distance The distance d between the points (x_1, y_1) and (x_2, y_2) in the xy-plane is $d = \sqrt{(x_2 - x_1)^2 + (y_2 - y_1)^2}$.

distributive properties For any real numbers a, b, and c, $a(b + c) = ab + ac$ and $a(b - c) = ab - ac$.

dividend In a division problem, the number being divided.

divisible One whole number is divisible by a second whole number if their quotient has remainder 0.

division sign The mathematical symbol that separates the dividend and divisor.

divisor In a division problem, the number being divided *into* the dividend.

domain The set of all x-values of the ordered pairs in a function.

element of a matrix Each number in a matrix.

elements of a set The members of a set.

elimination method A symbolic method used to solve a system of equations that is based on the property that if "equals are added to equals the results are equal."

ellipse The set of points in a plane the sum of whose distances from two fixed points is constant.

empty set (null set) A set that contains no elements.

equal sets If every element in a set A is in set B and every element in set B is in set A, then A and B are equal sets, denoted $A = B$.

equation A mathematical statement that two algebraic expressions are equal.

equation of a line Point-slope form and slope-intercept form are examples of an equation of a line.

equilateral triangle A triangle with three sides of the same length.

equivalent equations Equations that have the same solution set.

equivalent fractions Fractions that name the same number.

estimation A rough calculation used to find a reasonably accurate answer.

evaluate To find the value of an expression by replacing any variables with given values.

even root The nth root, $\sqrt[n]{a}$, where n is even.

event Any outcome or group of outcomes of an experiment.

expanded form of a whole number A whole number written as a sum of the numbers represented by each digit in the whole number.

expansion of a determinant by minors A method of finding a 3×3 determinant by using determinants of 2×2 matrices.

experiment An activity with an observable result.

exponent The value of n in the expression b^n.

exponential decay When $0 < a < 1$, the graph of $f(x) = Ca^x$ models exponential decay.

exponential equation An equation that has a variable as an exponent.

exponential expression An expression that has an exponent.

exponential function with base a and coefficient C A function represented by $f(x) = Ca^x$, where $a > 0$, $C > 0$, and $a \neq 1$.

exponential growth When $a > 1$, the graph of $f(x) = Ca^x$ models exponential growth.

exponential notation A notation involving exponents that can be used to represent repeated multiplication.

extraneous solution A solution that does not satisfy the given equation.

factor tree A visual diagram used to find the prime factorization of a number.

factorial notation $n! = 1 \cdot 2 \cdot 3 \cdot \cdots \cdot n$ for any positive integer.

factoring a polynomial The process of writing a polynomial as a product of lower degree polynomials.

factoring by grouping A technique that uses the associative and distributive properties by grouping four terms of a polynomial in such a way that the polynomial can be factored even though its greatest common factor is 1.

factors In a multiplication problem, the two numbers multiplied.

Fahrenheit Units used in the U.S. system to measure temperature.

feasible solutions In linear programming, the set of solutions that satisfy the constraints.

finite sequence A function with domain $D = \{1, 2, 3, \ldots, n\}$ for some fixed natural number n.

finite series A series that contains a finite number of terms, and that can be expressed in the form $a_1 + a_2 + a_3 + \cdots + a_n$ for some n.

finite set A set whose elements can be listed explicitly.

focus (plural: **foci**) A fixed point used to determine the points that form a parabola, an ellipse, or a hyperbola.

FOIL A method for multiplying two binomials $(A + B)$ and $(C + D)$. Multiply **F**irst terms AC, **O**utside terms AD, **I**nside terms BC, and **L**ast terms BD; then combine like terms.

formula A special type of equation used to calculate one quantity from given values of other quantities.

fraction A number that can be used to describe a portion of a whole.

function A set of ordered pairs (x, y), where each x-value corresponds to exactly one y-value.

function notation The notation $y = f(x)$, where the input x produces output y.

fundamental rectangle The rectangle of a hyperbola whose four vertices are determined by either $(\pm a, \pm b)$ or $(\pm b, \pm a)$, where $\frac{x^2}{a^2} - \frac{y^2}{b^2} = 1$ or $\frac{y^2}{a^2} - \frac{x^2}{b^2} = 1$.

Gauss–Jordan elimination A method used to solve a linear system in which matrix row transformations are applied to an augmented matrix.

general term (nth term) of a sequence a_n, where n is a natural number in the domain of a sequence $a_n = f(n)$.

geometric sequence An exponential function given by $a_n = a_1(r)^{n-1}$, where n is a natural number and $r \neq 0$ or 1.

geometric series The sum of the terms of a geometric sequence.

Grade point average (GPA) If the variables a, b, c, d, and f represent the *number of credits* earned with a grade of A, B, C, D, and F, respectively, then the grade point average (GPA) is given by
$$\text{GPA} = \frac{4a + 3b + 2c + 1d + 0f}{a + b + c + d + f}.$$

graph of a linear equation A straight line in the xy-plane.

graph of a whole number A dot placed on a number line at a whole number's position.

graphical representation A graph of a function.

graphical solution A solution to an equation obtained by graphing.

greater than If a real number b is located to the right of a real number a on the number line, we say that b is greater than a, and write $b > a$.

greater than or equal to If a real number a is greater than or equal to b, denoted $a \geq b$, then either $a > b$ or $a = b$ is true.

greatest common factor (GCF) for numbers The largest number that divides evenly into two or more given numbers.

greatest common factor (GCF) of a polynomial The term with the highest degree and greatest coefficient that is a factor of all terms in the polynomial.

gross pay Salary or hourly wages before withholdings have been subtracted.

growth factor The value of a in the exponential function, $f(x) = Ca^x$, when $a > 1$.

half-life The time it takes for an amount to decay to half its original amount.

heptagon A polygon with seven sides.

hexagon A polygon with six sides.

horizontal axis A horizontal number line that divides a graphing area into upper and lower regions; it intersects the vertical axis at 0.

horizontal line test If every horizontal line intersects the graph of a function f at most once, then f is a one-to-one function.

hyperbola The set of points in a plane the difference of whose distances from two fixed points is constant.

hypotenuse The longest side of a right triangle.

identity An equation that is always true regardless of the values of any variables.

identity properties of division If $a \neq 0$, then $\frac{a}{a} = 1$; for any number a, $\frac{a}{1} = a$.

identity property of multiplication If any number a is multiplied by 1, the result is a, that is, $a \cdot 1 = 1 \cdot a = a$.

identity property of 1 If any number a is multiplied by 1, the result is a, that is, $a \cdot 1 = 1 \cdot a = a$.

identity property for addition If 0 is added to any real number a, the result is a, that is, $a + 0 = 0 + a = a$.

imaginary number A complex number $a + bi$ with $b \neq 0$.

imaginary part The value of b in the complex number $a + bi$.

imaginary unit A number denoted i whose properties are $i = \sqrt{-1}$ and $i^2 = -1$.

impossible event An event that cannot possibly occur.

improper fraction A fraction whose numerator is greater than or equal to its denominator in absolute value.

inch, foot, yard, mile The U.S. units of measurement for length.

inconsistent system A system of linear equations that has no solution.

independent equations Equations in a linear system that have different graphs.

independent variable The variable that represents the input of a function.

index The value of n in the expression $\sqrt[n]{a}$.

index of summation The variable k in the expression $\sum_{k=1}^{n}$.

inequality When the equals sign in an equation is replaced with any one of the symbols $<, \leq, >,$ or $\geq$, an inequality results.

infinite sequence A function whose domain is the set of natural numbers.

infinite set A set with infinitely many elements.

infinity Values that increase without bound.

input An element of the domain of a function.

integers A set of numbers including natural numbers, their opposites, and 0, or $\ldots, -3, -2, -1, 0, 1, 2, 3, \ldots$.

intercept form A linear equation in the form $\frac{x}{a} + \frac{y}{b} = 1$.

interest A fee for the use of someone's money, which is usually a percent of the principal.

interest rate The percent used to calculate the interest on an investment or loan.

intersecting lines Two lines that cross at a point.

intersection The set containing elements that belong to *both A and B*, denoted $A \cap B$ and read "A intersect B."

intersection-of-graphs method A graphical technique for solving two equations.

interval notation A notation for number line graphs that eliminates the need to draw the entire line.

inverse function If f is a one-to-one function, then f^{-1} is the inverse function of f, if $(f^{-1} \circ f)(x) = f^{-1}(f(x)) = x$ for every x in the domain of f, and $(f \circ f^{-1})(x) = f(f^{-1}(x)) = x$ for every x in the domain of f^{-1}.

inverse property for addition For any number a, $a + (-a) = 0$.

inversely proportional (varies inversely) A quantity y is inversely proportional to x if there is a nonzero number k such that $y = \frac{k}{x}$.

irrational numbers Real numbers that cannot be expressed as fractions, such as π or $\sqrt{2}$.

isosceles triangle A triangle with at least two sides of the same length.

joint variation A quantity z varies jointly with x and y if there is a nonzero number k such that $z = kxy$.

key (on a pictograph) The portion of a pictograph that gives the meaning of one picture or symbol in the graph.

kite A quadrilateral with two pairs of adjacent sides that are equal in length.

leading coefficient In a polynomial of one variable, the coefficient of the monomial with highest degree.

least common denominator (LCD) The LCD of two or more fractions is the smallest number that is divisible by every denominator.

least common multiple (LCM) The smallest number that two or more numbers will divide into evenly.

legs of a right triangle The two shorter sides of a right triangle.

less than If a real number a is located to the left of a real number b on the number line, we say that a is less than b and write $a < b$.

less than or equal to If a real number a is less than or equal to b, denoted $a \leq b$, then either $a < b$ or $a = b$ is true.

like fractions Fractions with the same denominator.

like radicals Radicals that have the same index and the same radicand.

like terms Two terms, or monomials, that contain the same variables raised to the same powers.

linear equation An equation that can be written in the form $ax + b = 0$, where $a \neq 0$.

linear equation in two variables An equation that can be written in the form $Ax + By = C$, where A, B, and C are fixed numbers and A and B are not both equal to 0.

linear function A function f represented by $f(x) = mx + b$, where m and b are constants.

linear inequality A linear inequality results whenever the equals sign in a linear equation is replaced with any one of the symbols $<, \leq, >,$ or $\geq$.

linear inequality in two variables When the equals sign in a linear equation in two variables is replaced with $<, \leq, >,$ or $\geq$, a linear inequality in two variables results.

linear polynomial A polynomial of degree 1 that can be written as $ax + b$, where $a \neq 0$.

linear programming problem A problem consisting of an objective function and a system of linear inequalities called constraints.

linear system in three variables A system of three equations in which each equation can be written in the form $ax + by + cz = d$; an ordered triple (x, y, z) is a solution to the system of equations if the values for x, y, and z make *all three* equations true.

line graph The resulting graph when consecutive data points in a scatterplot are connected with straight line segments.

listing method A method for finding the LCM that involves listing multiples of the given numbers.

logarithm with base a of a positive number x Denoted $\log_a x$, it may be calculated as follows: If x can be expressed as $x = a^k$, then $\log_a x = k$, where $a > 0$, $a \neq 1$, and k is a real number. That is, $\log_a a^k = k$.

logarithmic function with base a The function represented by $f(x) = \log_a x$.

logistic function A function used to model growth of a population.

lower limit In summation notation, the number representing the subscript of the first term of the series.

lowest terms A fraction is in lowest terms if its numerator and denominator have no factors in common.

main diagonal In an augmented matrix, the diagonal set of numbers from the upper left of the matrix to the lower right.

major axis The longer axis of an ellipse, which connects the vertices.

mass A measure of the amount of matter in an object.

matrix A rectangular array of numbers.

matrix row transformations Operations performed on rows of an augmented matrix that result in an equivalent system of linear equations.

mean (arithmetic mean) The mean of a list of numbers is the sum of the numbers divided by the number of numbers in the list.

measures of central tendency Measures of the location of the "middle" of a list of numbers.

median In an ordered list of numbers, the median is the middle number in a list with an odd number of values, or it is the mean of the two middle numbers in a list with an even number of values.

meter, liter, gram The meter, liter, and gram are metric units of length, capacity, and mass, respectively.

method of substitution A symbolic method for solving a system of equations in which one equation is solved for one of the variables and then the result is substituted into the other equation.

minor axis The shorter axis of an ellipse.

minors The 2×2 matrices that are used to find a determinant of a 3×3 matrix.

minuend The number a in the difference $a - b$.

mixed number An integer written with a proper fraction.

mode The value that occurs most often in a list of numbers. A data set can have more than one mode or no mode.

monomial A number, a variable, or a product of numbers and variables raised to natural number powers.

multiplication A fast way to perform repeated addition.

multiplication property of equality If a, b, and c are real numbers with $c \neq 0$, then $a = b$ is equivalent to $ac = bc$.

multiplication sign The mathematical symbol that separates two factors.

multiplicative identity The number 1.

multiplicative inverse (reciprocal) The multiplicative inverse of a nonzero number a is $\frac{1}{a}$.

name of function In the function given by $f(x)$, we call the function f.

natural exponential function The function represented by $f(x) = e^x$, where $e \approx 2.71828$.

natural logarithm The base-e logarithm, denoted either $\log_e x$ or $\ln x$.

natural numbers The set of (counting) numbers expressed as 1, 2, 3, 4, 5, 6, ...

negative infinity Values that decrease without bound.

negative number A number that is less than zero.

negative reciprocals Slopes of two lines satisfy $m_1 = -\frac{1}{m_2}$, or $m_1 \cdot m_2 = -1$. These lines are perpendicular.

negative slope On a graph, the slope of a line that falls from left to right.

negative square root of a Denoted $-\sqrt{a}$.

net pay The amount remaining after withholdings have been subtracted from gross pay.

nonlinear data If data points do not lie on a (straight) line, the data are nonlinear.

nonlinear equation An equation that is not a linear equation; its graph is not a straight line.

nonlinear function A function that is *not* a linear function; its graph is not a straight line.

nonlinear system of equations Two or more equations at least one of which is nonlinear.

nonlinear system of inequalities Two or more inequalities at least one of which is nonlinear.

non real complex number A complex number $a + bi$ with $b \neq 0$; sometimes called an imaginary number.

nth root The number b is an nth root of a if $b^n = a$, where n is a positive integer.

nth term (general term) of a sequence See general term (nth term) of a sequence.

null set (empty set) A set that contains no elements.

number line A horizontal line marked with evenly spaced tick marks that is often used to graph numbers.

numerator The top number in a fraction.

numerical representation A table of values for a function.

numerical solution A solution often obtained by using a table of values.

objective function The given function to be optimized in a linear programming problem.

obtuse angle An angle that measures between 90° and 180°.

obtuse triangle A triangle with one obtuse angle.

octagon A polygon with eight sides.

odd root The nth root, $\sqrt[n]{a}$, where n is odd.

one to one (correspondence) When two sets have the same number of elements, so that each element in one set can be paired exactly once with an element in the second set; for functions, when different inputs always result in different outputs.

one-to-one function A function f in which for any c and d in the domain of f, $c \neq d$ implies that $f(c) \neq f(d)$. That is, different inputs always result in different outputs.

opposite (additive inverse) The opposite, or additive inverse, of a number a is $-a$.

opposite of a polynomial The polynomial obtained by negating each term in a given polynomial.

optimal value In linear programming, the value that maximizes or minimizes the objective function.

order of operations agreement A set of rules used to ensure consistency in the evaluation of mathematical expressions.

ordered pair A pair of numbers written in parentheses (x, y), in which the order of the numbers is important.

ordered triple Can be expressed as (x, y, z), where x, y, and z are numbers and represent a solution to a linear system in three variables.

origin On the number line, the point associated with the real number 0; in the xy-plane, the point where the axes intersect, $(0, 0)$.

ounce, cup, pint, quart, gallon The U.S. units of measurement for capacity.

ounce, pound, ton The U.S. units of measurement for weight.

outcome A result of an experiment.

output An element of the range of a function.

parabola The $\cup$-shaped graph of a quadratic function that opens either upward or downward.

parallel lines Two or more lines in the same plane that never intersect; they have the same slope.

parallelogram A quadrilateral with two pairs of parallel sides.

partial dividend The number formed by starting at the left end of a dividend and selecting the fewest digits that give a number that is greater than the divisor.

Pascal's triangle A triangle made up of numbers in which there are 1s along the sides and each element inside the triangle is the sum of the two numbers above it.

pentagon A polygon with five sides.

percent A ratio with a denominator of 100.

percent chance A probability written as a percent.

percent change If a quantity changes from x to y, then the percent change is $\frac{y - x}{x} \times 100$.

percent decrease A negative percent change.

percent increase A positive percent change.

percent problem A question asking us to find the percent, whole, or part in a percent statement.

perfect cube An integer with an integer cube root.

perfect nth power The value of a if there exists an integer b such that $b^n = a$.

perfect square An integer with an integer square root.

perfect square trinomial A trinomial that can be factored as the square of a binomial, for example, $a^2 + 2ab + b^2 = (a + b)^2$.

perimeter The distance around an enclosed region.

period A group of digits in a whole number.

perpendicular lines Two lines in a plane that intersect to form a right ($90°$) angle.

pictograph A graph that uses a picture or symbol to represent a quantity visually.

piecewise-defined function A function that is defined by using different formulas on different intervals of its domain.

piecewise-defined linear function A piecewise-defined function in which every piece is a linear function.

place value The position of a digit in a number that is written in standard form.

plane A flat surface that continues without end.

plane geometry Geometry that focuses on figures with two or fewer dimensions.

plotting Graphing points in the xy-plane.

point–slope form The line with slope m passing through the point (x_1, y_1), given by the equation $y - y_1 = m(x - x_1)$ or, equivalently, $y = m(x - x_1) + y_1$.

polygon A closed plane figure determined by three or more line segments.

polynomial The sum of one or more monomials.

polynomial functions of one variable Functions that are defined by a polynomial in one variable.

polynomials in one variable Polynomials that contain one variable.

positive number A number that is greater than zero.

positive slope On a graph, the slope of a line that rises from left to right.

power function A function that can be represented by $f(x) = x^p$, where p is a rational number.

prime factorization A number written as a product of prime numbers.

prime factorization method A method for finding the LCM or GCF that uses the prime factorizations of the given numbers.

prime number A natural number greater than 1 that has *only* itself and 1 as natural number factors.

prime polynomial A polynomial with integer coefficients that cannot be factored by using integer coefficients.

principal An amount of money that is initially borrowed or invested.

principal nth root of a Denoted $\sqrt[n]{a}$.

principal square root The square root of a that is nonnegative, denoted $\sqrt{a}$.

probability A real number between 0 and 1, inclusive. A probability of 0 indicates that an event is impossible, whereas a probability of 1 indicates that an event is certain.

product The answer to a multiplication problem.

proper fraction A fraction whose numerator is less than its denominator in absolute value.

proportion A statement that two ratios are equal.

pure imaginary number A complex number $a + bi$ with $a = 0$ and $b \neq 0$.

Pythagorean theorem If a right triangle has legs a and b with hypotenuse c, then $a^2 + b^2 = c^2$.

quadrants The four regions determined by the xy-plane.

quadratic equation An equation that can be written in the form $ax^2 + bx + c = 0$, where a, b, and c are constants, with $a \neq 0$.

quadratic formula The solutions to the quadratic equation, $ax^2 + bx + c = 0$, $a \neq 0$, are $\frac{-b \pm \sqrt{b^2 - 4ac}}{2a}$.

quadratic function A function f represented by the equation $f(x) = ax^2 + bx + c$, where a, b, and c are constants, with $a \neq 0$.

quadratic inequality If the equals sign in a quadratic equation is replaced with $>$, $\geq$, $<$, or $\leq$, a quadratic inequality results.

quadratic polynomial A polynomial of degree 2 that can be written as $ax^2 + bx + c$, with $a \neq 0$.

quadrilateral A polygon with four sides.

quotient The answer to a division problem.

radical expression An expression that contains a radical sign.

radical sign The symbol $\sqrt{}$ or $\sqrt[n]{}$ for some positive integer n.

radicand The expression under the radical sign.

radius The fixed distance between the center and any point on a circle.

range The set of all y-values of the ordered pairs in a function.

rate A ratio used to compare different kinds of quantities.

rate of change Slope can be interpreted as a rate of change. It indicates how fast the graph of a line is changing.

ratio A comparison of two quantities, expressed as a quotient.

rational equation An equation that contains one or more rational expressions.

rational expression A polynomial divided by a nonzero polynomial.

rational function A function defined by $f(x) = \frac{p(x)}{q(x)}$, where $p(x)$ and $q(x)$ are polynomials and the domain of f includes all x-values such that $q(x) \neq 0$.

rational number Any number that can be expressed as the ratio of two integers $\frac{p}{q}$, where $q \neq 0$; a fraction.

rationalizing the denominator The process of removing radicals from a denominator so that the denominator contains only rational numbers.

real numbers All rational and irrational numbers; any number that can be represented by decimal numbers.

real part The value of a in the complex number $a + bi$.

reciprocal (multiplicative inverse) The reciprocal of a nonzero number a is $\frac{1}{a}$.

rectangle A quadrilateral in which all angles measure $90°$.

rectangular coordinate system (xy-plane) The xy-plane used to plot points and graph data.

reduced row–echelon form A matrix form for representing a system of linear equations in which there are 1s on the main diagonal with 0s above and below each 1.

reflection If the point (x, y) is on the graph of a function, then $(x, -y)$ is on the graph of its reflection across the x-axis.

regular polygon A polygon in which all sides have equal length and all angles have equal measure.

rectangular prism A solid object that has exactly six rectangular faces.

relation A set of ordered pairs.

remainder The amount left over when the quotient of two whole numbers is not a whole number.

repeat bar A bar written above a group of digits in a decimal number to indicate that the digits repeat.

repeating decimal A decimal number with digits to the right of the decimal point that continue without end in a repeating pattern.

rhombus A quadrilateral with four sides of equal length.

right angle An angle that measures 90°.

right triangle A triangle with one right angle.

rise The change in y between two points on a line, that is, $y_2 - y_1$.

root function In the power function $f(x) = x^p$, if $p = \frac{1}{n}$, where $n \geq 2$ is an integer, then f is also a root function, which is given by $f(x) = \sqrt[n]{x}$.

rounding a number Approximating a number to a given level of accuracy.

run The change in x between two points on a line, that is, $x_2 - x_1$.

sale price The result obtained by subtracting the discount from the original price.

sales tax A percent of the original price that is added to the original price.

scalene triangle A triangle with no sides of the same length.

scatterplot A graph of distinct points plotted in the xy-plane.

scientific notation A real number a written as $b \times 10^n$, where $1 \leq |b| < 10$ and n is an integer.

second, minute, hour, day, week, month, year Units of time used in both the U.S. and metric systems of measurement.

semicircle Half of a circle.

set A collection of things.

set-builder notation Notation to describe a set of numbers without having to list all of the elements. For example, $\{x|x > 5\}$ is read as "the set of all real numbers x such that x is greater than 5."

side-angle-side (SAS) Two triangles are congruent if two sides and the included angle of one triangle are congruent to two sides and the included angle of the other triangle.

side-side-side (SSS) Two triangles are congruent if three sides of one triangle are congruent to three sides of the other triangle.

sides of an angle The two rays that form an angle.

sides of a polygon The line segments that form a polygon.

signed numbers Positive numbers, negative numbers, and zero.

similar figures Two figures in which the measures of corresponding angles are equal and the measures of corresponding sides are proportional.

simple interest Interest that is based only on the original principal.

slope The ratio of the change in y (rise) to the change in x (run) along a line. The slope m of a line passing through the points (x_1, y_1) and (x_2, y_2) is $m = \frac{y_2 - y_1}{x_2 - x_1}$, where $x_1 \neq x_2$.

slope–intercept form The line with slope m and y-intercept $(0, b)$ is given by $y = mx + b$.

solution Each value of the variable that makes the equation true.

solution set The set of all solutions to an equation.

solution to a system In a system of two equations in two variables, an ordered pair, (x, y), that makes *both* equations true.

solving an equation Finding all of the solutions to an equation.

speed A rate that gives a distance traveled in an amount of time.

sphere A ball-shaped solid.

spider chart A chart used to represent data visually. It often resembles a spider web and is useful for displaying data that are divided into several categories.

square A regular quadrilateral.

square-based pyramid A solid with a square base and four triangular sides that meet at a single point.

square matrix A matrix in which the number of rows and the number of columns are equal.

square root The number b is a square root of a number a if $b^2 = a$.

square root function The function given by $f(x) = \sqrt{x}$, where $x \geq 0$.

square root property If k is a nonnegative number, then the solutions to the equation $x^2 = k$ are given by $x = \pm\sqrt{k}$. If $k < 0$, then this equation has no real solutions.

1 square unit A measure of area represented by a square that measures 1 unit on each side.

standard equation of a circle The standard equation of a circle with center (h, k) and radius r is $(x - h)^2 + (y - k)^2 = r^2$.

standard form (of a linear equation in two variables) The form $Ax + By = C$, where A, B, and C are constants, with A and B not both 0.

standard form of a complex number $a + bi$, where a and b are real numbers.

standard form of a quadratic equation The equation given by $ax^2 + bx + c = 0$, where $a \neq 0$.

standard form of a whole number A whole number written in digits with periods separated by commas.

straight angle An angle that measures 180°.

standard viewing rectangle of a graphing calculator Xmin $= -10$, Xmax $= 10$, Xscl $= 1$, Ymin $= -10$, Ymax $= 10$, and Yscl $= 1$, denoted $[-10, 10, 1]$ by $[-10, 10, 1]$.

subscript The symbol x_1 has a subscript of 1 and is read "x sub one" or "x one".

subset If every element in a set B is contained in a set A, then we say that B is a subset of A, denoted $B \subseteq A$.

subtrahend The number b in the difference $a - b$.

sum The answer to an addition problem.

sum of the first n terms of an arithmetic sequence Denoted S_n, is found by averaging the first and nth terms and then multiplying by n.

sum of the first n terms of a geometric sequence Given by $S_n = a_1\left(\frac{1 - r^n}{1 - r}\right)$, if its first term is a_1 and its common ratio is r, provided $r \neq 1$.

sum of two cubes An expression in the form $a^3 + b^3$, which can be factored as $(a + b)(a^2 - ab + b^2)$.

summation notation Notation in which the uppercase Greek letter sigma represents the sum, for example,
$$\sum_{k=1}^{n} a_k = a_1 + a_2 + a_3 + \cdots + a_n.$$

supplementary angles Two angles whose measures sum to 180°.

surface area A measure of the amount of exposed area for a given solid.

symbolic representation Representing a function with a formula; for example, $f(x) = x^2 - 2x$.

symbolic solution A solution to an equation obtained by using properties of equations; the resulting solution set is exact.

synthetic division A shortcut that can be used to divide $x - k$, where k is a number, into a polynomial.

system of linear equations in two variables A system of equations in which each equation can be written as $Ax + By = C$.

system of linear inequalities in two variables Two or more linear inequalities to be solved at the same time, the solution to which must satisfy each inequality.

table A structure used to display information visually in a rectangular array.

table of values An organized way to display the inputs and outputs of a function; a numerical representation.

temperature A measure of how warm or cool an object or environment is.

term A number, a variable, or a product of numbers and variables raised to powers.

terms of a sequence $a_1, a_2, a_3, \ldots a_n, \ldots$ where the first term is $a_1 = f(1)$, the second term is $a_2 = f(2)$, and so on.

test point When graphing the solution set to an inequality, a point chosen to determine which region of the xy-plane to include in the solution set.

test value A real number chosen to determine the solution set to an inequality.

three-part inequality A compound inequality written in the form $a < x < b$, where $\leq$ may replace $<$.

total amount paid The sum of the original price and the sales tax.

total value (of an investment) The sum of the principal and the interest.

translation The shifting of a graph upward, downward, to the right, or to the left in such a way that the shape of the graph stays the same.

transversal A line that intersects two other lines in the same plane.

transverse axis In a hyperbola, the line segment that connects the vertices.

trapezoid A quadrilateral with one pair of parallel sides.

triangle A polygon with three sides.

trinomial A polynomial with three terms.

unary operation An operation that requires only one number.

undefined slope The slope of a line that is vertical.

union Denoted $A \cup B$ and read "A union B," it is the set containing any element that can be found in *either A or B*.

undefined An expression is undefined when it involves division by zero.

unit fraction A fraction that is equivalent to 1.

unit pricing A ratio used to compare pricing.

unit rate Rates expressed with a denominator of 1.

unit ratio A ratio expressed with a denominator of 1.

universal set A set that contains all elements under consideration.

unlike terms Terms that are not like terms.

upper limit In summation notation, the number representing the subscript of the last term of the series.

variable A symbol, such as x, y, or z, used to represent any unknown quantity.

varies directly A quantity y varies directly with x if there is a nonzero number k such that $y = kx$.

varies inversely A quantity y varies inversely with x if there is a nonzero number k such that $y = \frac{k}{x}$.

varies jointly A quantity z varies jointly with x and y if there is a nonzero number k such that $z = kxy$.

Venn diagrams Diagrams used to depict relationships between sets.

verbal representation A description, in words, of what a function computes.

vertex The lowest point on the graph of a parabola that opens upward or the highest point on the graph of a parabola that opens downward.

vertex form of a parabola The vertex form of a parabola with vertex (h, k) is $y = a(x - h)^2 + k$, where $a \neq 0$ is a constant.

vertex of an angle The common endpoint of the two rays that form an angle.

vertex of a polygon A common endpoint of two sides of a polygon.

vertical angles Congruent pairs of angles formed by intersecting lines.

vertical asymptote A vertical asymptote typically occurs in the graph of a rational function when the denominator of a rational expression equals 0 but the numerator does not equal 0; it can be represented by a vertical line in the graph of a rational function.

vertical axis A vertical number line that divides a graphing area into left and right regions; it intersects the horizontal axis at 0.

vertical line test If every vertical line intersects a graph at no more than one point, then the graph represents a function.

vertices of an ellipse The endpoints of the major axis.

vertices of a hyperbola The endpoints of the transverse axis.

viewing rectangle (window) On a graphing calculator, the window that determines the x- and y-values shown in the graph.

volume A measure of the amount of a substance.

weight A measure of the force on an object due to gravity.

weighted mean A process for finding the mean of a list of numbers in which each number is multiplied by the number of times it occurs in the list.

whole numbers The set of numbers 0, 1, 2, 3, 4, 5, ...

withholdings Taxes, insurance, and other deductions that are subtracted from gross pay.

word form of a whole number A whole number written in words.

***x*-axis** The horizontal axis in the *xy*-plane.

***x*-coordinate** The first value in an ordered pair.

***x*-intercept** A point where a graph intersects the *x*-axis.

Xmax Regarding the viewing rectangle of a graphing calculator, Xmax is the maximum *x*-value along the *x*-axis.

Xmin Regarding the viewing rectangle of a graphing calculator, Xmin is the minimum *x*-value along the *x*-axis.

Xscl Regarding the viewing rectangle of a graphing calculator, the distance between consecutive tick marks on the *x*-axis.

***xy*-plane (rectangular coordinate system)** The system used to plot points and graph data.

***y*-axis** The vertical axis in the *xy*-plane.

***y*-coordinate** The second value in an ordered pair.

***y*-intercept** A point where a graph intersects the *y*-axis.

Ymax Regarding the viewing rectangle of a graphing calculator, Ymax is the maximum *y*-value along the *y*-axis.

Ymin Regarding the viewing rectangle of a graphing calculator, Ymin is the minimum *y*-value along the *y*-axis.

Yscl Regarding the viewing rectangle of a graphing calculator, the distance between consecutive tick marks on the *y*-axis.

zero of a polynomial An *x*-value that results in 0 when it is substituted into a polynomial; for example, the zeros of $x^2 - 4$ are 2 and -2.

zero-product property If the product of two numbers is 0, then at least one of the numbers must be 0. That is, $ab = 0$ implies $a = 0$ or $b = 0$ (or both).

zero property of addition If 0 is added to any real number a, the result is a, that is, $a + 0 = 0 + a = a$.

zero property of multiplication If any real number a is multiplied by 0, the result is 0, that is, $a \cdot 0 = 0 \cdot a = 0$.

zero properties of division If $a \neq 0$, then $\frac{0}{a} = 0$; for any number a, $\frac{a}{0}$ is undefined.

zero slope The slope of a line that is horizontal.

Photo Credits

Cover, Oleg_P/Shutterstock *First endsheet*: **upper right,** Jakubzak/Fotolia **upper left,** Wendy Rockswold **middle right,** Dotshock/Shutterstock **lower left,** Wendy Rockswold **lower right,** Wendy Rockswold *Second endsheet*: **upper left,** Tommaso Lizzul/Fotolia **upper right,** IQoncept/Shutterstock **middle left,** Jules Selmes/Pearson Education, Inc. **lower right,** Wendy Rockswold **lower left,** Little_Desire/Shutterstock **iii,** Wendy Rockswold **iii,** Terry A. Krieger **xxv,** Pearson Education, Inc.

Chapter 1 **1,** Monkey Business Images/Shutterstock **7,** Smileus/Shutterstock **24,** Wendy Rockswold **41,** Wendy Rockswold **45,** Andrea Danti/Shutterstock **51,** Jupiterimages/Stockbyte/Getty Images **52,** Wendy Rockswold

Chapter 2 **87,** Martin Valigursky/Fotolia **88,** Jakubzak/Fotolia

Chapter 3 **145,** Wachirakl/Fotolia

Chapter 4 **196,** Sdecoret/Shutterstock **235,** Wendy Rockswold **242,** VanHart/Shutterstock **250,** Leksele/Shutterstock

Chapter 5 **296,** PCN Black/PCN Photography/Alamy Stock Photo **306,** Stuart Monk/Shutterstock **325,** Michael Shake/Fotolia **355,** Donald Sawvel/Shutterstock

Chapter 6 **374,** ZUMA Press Inc/Alamy Stock Photo **382,** Trubach/Shutterstock **387 (left),** Reddogs/Shutterstock **387 (right),** Aleksandrs Kobilanskis/Shutterstock **399 (left),** Monchai Tudsamalee/123RF **399 (middle left),** Hurst Photo/Shutterstock **399 (middle right),** Doomu/Fotolia **399 (right),** Bestvc/Fotolia **422,** Wendy Rockswold

Chapter 7 **434,** Mircea BEZERGHEANU/Shutterstock

Chapter 8 **489,** An Qi/Alamy Stock Photo **515 (top),** Digital Vision/Getty Images **515 (middle),** Wendy Rockswold **515 (bottom),** John Foxx/Stockbyte/Getty Images **530,** Luciano Mortula/Shutterstock

Chapter 9 **540,** Wendy Rockswold **545,** Wendy Rockswold **554,** VikOl/Shutterstock **562,** Peter Zaharov/Shutterstock **568,** Wendy Rockswold **573,** Wendy Rockswold

Chapter 10 **585,** Wendy Rockswold **587,** Lculig/123RF **641,** Wendy Rockswold **660,** Graeme Shannon/Shutterstock

Chapter 11 **672,** Sergey Nivens/Fotolia **678,** Little_Desire/Shutterstock **679,** Jules Selmes/Pearson Education, Inc **685,** Nikola Volrábová/123RF **689,** Gary Arbach/123RF **701,** Wendy Rockswold **723,** Wendy Rockswold

Chapter 12 **726,** lzf/Shutterstock **736,** T-Design/Shutterstock **757,** Areipa.lt/Shutterstock **767,** Fusebulb/Shutterstock **772,** Viktar Malyshchyts/Shutterstock **773,** Wendy Rockswold

Chapter 13 **788,** Minicel73/Fotolia **831,** Tony Bowler/Shutterstock

Chapter 14 **849,** Bart Everett/Shutterstock **854,** Wendy Rockswold **855,** Wendy Rockswold **860,** Jfergusonphotos/Fotolia **863,** Wendy Rockswold **872,** Wendy Rockswold **908,** Wendy Rockswold **920,** Wendy Rockswold **921 (left),** Wendy Rockswold **921 (right),** Wendy Rockswold **928,** Wendy Rockswold

Chapter 15 **932,** Wendy Rockswold **944,** Mikhail P./Shutterstock **950,** Wendy Rockswold **957,** Dmitry Nikolaev/Shutterstock **958,** Wendy Rockswold **959,** Szaffy/E+/Getty Images **976,** Minerva Studio/Shutterstock **986,** Dotshock/Shutterstock **991,** Chanpipat/Shutterstock **992,** Wendy Rockswold

Chapter 16 **1026,** Betty Shelton/123RF **1029,** Sascha Burkard/Fotolia **1038,** Pictureguy56/123RF **1040,** Goodluz/Shutterstock **1054,** Celeborn/Shutterstock

Chapter 17 **1065,** Wendy Rockswold **1067,** Wendy Rockswold **1079,** Georgejmclittle/Fotolia **1081,** Martin Harvey/Digital Vision/Gettyimages **1085,** Wendy Rockswold **1088,** Janenet/Fotolia **1098,** Baloncici/Shutterstock **1106,** Wendy Rockswold **1118,** Wendy Rockswold **1122,** Wendy Rockswold **1125 (top),** Di Studio/Fotolia **1125 (bottom),** Wendy Rockswold **1135,** Wendy Rockswold

Chapter 18 **1143,** Toria/Shutterstock **1149,** Tommaso Lizzul/Fotolia **1156,** Levent Konuk/Shutterstock

Chapter 19 **1221,** IQoncept/Shutterstock **1236,** Jim Barber/Shutterstock **1247,** Wendy Rockswold **1254,** Holbox/Shutterstock686 **1278,** ChrisHill/Shutterstock **1281,** Froemic/Fotolia

Chapter 20 **1297,** Cardens Design/Shutterstock **1312,** MarcelClemens/Shutterstock

Chapter 21 **1333,** Wendy Rockswold

Index

$a + bi$ mode, 1128–1129, AP-7
Absolute value, 91–92, 94, 1069, AP-8, AP-29
Absolute value equations, 1001–1005, 1010, 1011
Absolute value functions, 984, 994
Absolute value inequalities, 1005–1010, 1011
Acute angles, 492, 496
Acute triangles, 500, 504
Addends, 15, 25
Addition
    of algebraic expressions, 148
    application problem-solving involving, 23–25
    of complex numbers, 1129–1130, 1133
    of decimal numbers, 308–309, 313, AP-32
    equations involving, 22–23
    of fractions, 236–237, 240, 246–249, 251, 868, AP-32
    of fractions, with unlike denominators, 877
    of integers, 97–101, 102, 107, AP-29
    mental calculations for, 18–19
    of mixed numbers, 256, 258–259, 260
    of polynomials, 737–740, 743
    properties of, 17–18, 25
    of radical expressions, 1095–1097, 1103
    of rational expressions, 238–239, 240, 249, 869–870, 873
    regrouping in, 16–17
    of signed decimal numbers, 309–310
    visual, of integers, 100–101
    of whole numbers, 15–17, 25, AP-28
    words associated with, 19, 25
Addition properties for integers, 99–100, 103
Addition property of equality, 161–163, 165, 270–271
Addition property of inequalities, 568–570, 575
Additive inverse, 89, 94, 99, 740. *See also* Opposites
Adjacent angles, 494
Algebraic expressions
    adding, 148
    combining like terms in, 146–147
    compared to equations, 53, 71, 77
    containing integers, 123, 125
    for converting temperatures, 415–416
    decimal numbers in, 311, 324–325, 334–335
    defined, 48, 53
    evaluating, 49
    multiplying, 150
    opposites of, 148–150, 152
    order of operations in, 65–67, 68
    simplifying, 71–73, 77, 146–147, 151, 152
    subtracting, 148–150
    translating words to, 50, 53, 154–155
Alternate exterior angles, 495
Alternate interior angles, 495
Angles
    defined, 490, 496
    naming, 491
    in regular polygons, 509
    right, 353, 492, 496
    types of, 491–495
Angle-side-angle (ASA) property, 503, 505
Annual percentage rate (APR), 469
Applications. *See* Problem solving
Approximation, 58, 62, 1067
Area
    of circles, 351–352, 359, 525–526, 528
    of composite regions, 352–353
    defined, 40
    formulas for, 50
    metric units of, 405–406, 409
    of parallelograms, 523–524
    of plane figures, 528
    of polygons, 523–525, 528
    of rectangles, 40, 42, 523
    of squares, 523
    surface, 526–528
    of trapezoids, 277, 523–525
    of triangles, 231–232, 233, 523
    U.S. units of, 397–398, 400
Arithmetic mean, 355–356
Arithmetic operations
    on decimals, AP-32
    on functions, 992–993, 994
    on graphing calculators, AP-7, AP-28–AP-32
    on integers, AP-29–AP-30
    with mixed numbers, 255–260
    order of, 727
    performing, on functions, 992–993, 994
    on scientific calculators, AP-28–AP-32
Arithmetic sequences, 1342–1344, 1348, 1353–1355
Arithmetic series, 1353–1355, 1358
Arithmetic symbols, words associated with, 50, 154–155, 157, 555
Associative property
    for addition, 17–18, 99
    for multiplication, 31, 113
Asymptotes, 987, 990, 992, 1311, 1314
Atanasoff, John, 1026
Augmented matrices (matrix), 1039, 1047
Average rates of change, 622–623
Average(s), 123–124, 125
Axes (axis)
    of ellipses, 1308, 1314
    horizontal, 586, 590
    of symmetry, 1144–1147, 1152, 1298–1301, 1304
    transverse, 1311, 1312, 1314
    vertical, 586, 590

Babylonian algorithm, 332–333, 337
Bar graphs, 7, 11
Base
    in exponential expressions, 46, 727, 733
    of a trapezoid, 277
Base units, of metric system, 403
Base-a logarithms, 1261–1263, 1264
Base-e, 1248, 1261, 1264
Base-10 logarithms, 1255–1257
Base-10 number system, 47
Base-2 logarithms, 1260–1261
Basic percent equation, 444, 447
Basic percent statement form, 443, 447
Basic principle of fractions, 213–214, 225, 851
Binomial(s)
    coefficients, 1364–1365, 1366
    cubing, 758, 759
    defined, 737
    expanding powers of, 1362–1366
    factoring, 793–795, 814–817, 819, 822, 823, 825
    factors, signs in, 802, 810
    multiplying, 749–750, 798
    squaring, 756–757, 759
    theorem, 1365–1366
Braces, use of, AP-12
Brackets, use of, 565, 568
Branches, of hyperbolas, 1311–1312

Calculators. *See* Graphing calculators; Scientific calculators
Canceling, 216

Capacity
  converting between U.S and metric units of, 413–414, 418
  defined, 398
  metric units of, 406–407, 409
  U.S. units of, 398–399, 400
Celsius temperatures, 414–417
Centers
  of circles, 511, 1301–1302, 1303
  of ellipses, 1308
  of hyperbolas, 1311
Centers, of circles, 350
Centigrams, 407–408
Centiliters, 407
Centimeters, 404
Certain events, 474, 478
Change of base formula, 1272
Charts, spider, 8–9
Circle graphs, 438, 440
Circles, 350–352, 358
  area of, 525–526, 528
  centers of, 1301–1302, 1303
  circumference of, 518–519, 520
  defined, 511, 513
  as ellipses, 1298
  on graphing calculators, 1303
  radius and diameter of, 511–512
  standard equation of, 1302, 1304
Circular cylinders, 527
Circumference, 351, 358, 518–519, 520
Coefficients
  binomial, 1364–1365, 1366
  C, in exponential functions, 1238
  defined, 72, 146
  leading, 798–799, 1176
  of monomials, 736, 743
  signs of, and factoring, 802, 810–812
Combining like terms, 72–73, 146–147, 152
Commas, use of, 2, 568
Commission, 459–460, 462
Common (like) denominators, 236–240, 868–871. *See also* Least common denominator (LCD)
Common difference, 1342, 1343, 1348
Common factors
  greatest. *see* Greatest common factor (GCF)
  in polynomials, 789–792, 796, 822, 835, 838
Common logarithmic function, 1255–1260
Common multiples, 244
Common ratio, 1345, 1348
Commutative property
  for addition, 17, 99
  for multiplication, 30–31, 113
Comparison
  of decimal numbers, 301–302, 304
  of fractions, 217–218
  of integers, 90–91
  of whole numbers, 6, 10
Complementary angles, 492–493, 497

Complements
  of events, 476–477, 478
  of sets, AP-14–AP-15
Completing the square method
  to find center and radius of a circle, 1303
  to find complex solutions, 1192, 1193
  to find vertex coordinates in parabolas, 1145, 1162–1164, 1166
  to solve quadratic equations, 1175–1176, 1179, 1193
Complex conjugates, 1131–1132, 1133
Complex fractions, 263–267, 884–890
  basic, 885, 890
  defined, 885, 890
  methods for simplifying, 885–890, 891
  and order of operations, 266–267
  simplifying, 263–266
Complex numbers, 1128–1132, 1133
  addition and subtraction of, 1129–1130, 1133
  dividing, 1132, 1133
  multiplying, 1130, 1133
  non-real, 1128
Complex solutions
  completing the square method for, 1192, 1193
  equations with, 1190–1192, 1193, 1209, 1210
Composite figures, 519
Composite functions, 1222–1224, 1233
Composite numbers, 211–212, 213, 219
Composite regions, 352–353
Compound inequalities, 969–977, 978
  containing "and," 971, 978
  containing "or," 972, 978
  defined, 969, 978
  solutions to, 976
  as three-part inequalities, 972–975
Compound interest, 468–470, 471, 1245–1247, 1250, AP-33
Conditional equations, 548
Cones, 527
Congruent angles, 492–493, 496, 497
Congruent triangles, 502–504, 505
Conic sections, 1298–1303
Conjugates, 1101–1102, 1103, 1131–1132, 1133
Connected mode, 992, AP-10
Consistent systems of linear equations, 676
Constant function, 960, 963
Constant of proportionality, 909, 912, 916
Constraints, AP-19, AP-20
Continuous growth/decay, 1248
Contradiction equations, 548
Conversions
  within metric system of units, 403–409
  of temperatures, 415–417
  of units of speed, 422–423
  of units of time, 421–422
  between U.S and metric systems, 412–418
  within U.S. system of units, 395–400

Corresponding angles, 495
Counting numbers. *See* Natural numbers
Cramer, Gabriel, 1054
Cramer's rule, 1054–1055, 1056
Cross product method, 217–218, 220
Cross product rule, 384, 391
Cross products, 452, 454
Cross-multiplication, 893–894
Cube root functions, 1073–1074, 1107
Cube roots, 1067–1068, 1117–1118, AP-1
Cubes
  factoring, 816–817, 819, 823–824
  perfect, 1087
Cubes (geometric shape), 527
Cubic functions, 985
Cubing, 47
Cups, 398–399

Data analysis, 914–915
Data modeling
  exact and approximate, 654
  with exponential functions, 1247–1249
  increasing and decreasing in, 1148–1149
  with linear equations, 653–657, 658
  with linear functions, 957–958, 963
  with linear systems, 1029, 1033
  with quadratic functions, 1165–1166
  using sequences, 1338
Days, 421
Decay factor, 1238
Decimals, 297–327
  adding, 308–310, 313
  in algebraic expressions, 311, 324–325
  application problem-solving involving, 312, 325–326, 345–346
  arithmetic operations on, AP-32
  clearing, 341–343, 346
  clearing, from equations, 546–548
  comparing, 301–302, 304
  dividing, 319–322, 327
  estimating, 307–308, 313, 316–317, 326
  fractions as, 323–324, 327
  as fractions or mixed numbers, 299–300, 304
  mixed numbers as, 324
  multiplying, 316–319, 326
  on number lines, 300–301, 304
  as percents, 437, 439
  percents as, 435–436, 439
  repeating, 320–321
  rounding, 302–303, 304
  signed, 309–310, 318
  solving equations containing, 340–344, 346–347
  subtracting, 308, 309–310, 313, AP-32
  in word form, 298–299, 304
Decimal notation, 297, 304
Decimal places, 297, 298
Decimal points, 297
Decimal windows, 992, AP-11

Decimeters, 404
Defining variables, 154, 157
Degree(s)
   in angles, 491, 496
   of monomials, 736, 743
   of polynomials, 737
Dekagrams, 407–408
Dekaliters, 407
Dekameters, 404
Denominators
   common (like), 236–240, 868–871
   in complex fractions, 885–890, 891
   defined, 197–198
   least common. *see* Least common denominator (LCD)
   rationalizing, 1100–1103
   unlike, 246–250, 877–879, 881
Dependent variables, 933, 946
Determinants, 1052–1054, 1055, AP-9–AP-10
Diagrammatic representation of functions, 937, 947
Diameter, 350, 511–512, 513
Difference(s)
   common, 1342, 1343, 1348
   defined, 19, 25
   estimating, 59, 313
   estimating decimal, 307–308
   factoring, of cubes, 816–817, 819
   factoring, of squares, 814–815, 820
   logarithms as, 1269
   power rules and, 732
   of rational expressions, 870–872, 873
Digits, 2
Dimension of a matrix, 1039
Direct variation, 909–911, 914, 916
Discounts, 459, 462
Discriminant, 1188–1189, 1193
Distance formula, 557–558, 1120–1121, 1123
Distributive properties
   application of, 747
   for combining like terms, 146–147
   defined, 752
   for multiplication, 32–34, 113
   solving linear equations by applying, 545–546
Dividend, in division, 35, 37, 42
Divisibility tests (divisible by), 209–211, 219
Division
   application problem-solving involving, 41, 116–117
   of complex numbers, 1132, 1133
   of decimal numbers, 319–322, 327, AP-32
   defined, 35
   equations involving, 39–40
   of fractions, AP-31, 228–230, 233, 862
   of integers, 115, 117, AP-30
   long, 37–39, 42, 323–324
   of mixed numbers, 258
   by monomials, 773–774, 777
   of polynomials, 773–778
   by polynomials, 775–777, 778
   properties for, 36–37, 42
   of rational expressions, 230, 233, 863–864, 865
   synthetic, AP-26–AP-27
   of whole numbers, 35–36, 42, AP-28
   words associated with, 39, 42
Divisor, 35, 42
Domain(s)
   of functions, 933, 939–941, 946
   in interval notation, 983–984, 994
   of rational functions, 988
   of square root functions, 1071–1072
Dot, use of, 49
Dot mode, 992, AP-10
Double negative rule, 89

Elements
   of matrices, 1039, AP-8–AP-9
   of sets, AP-12–AP-13
Elimination method
   application problem-solving using, 701–702
   Gauss-Jordan, 1041–1043
   multiplying before applying, 697–699
   of solving nonlinear system of equations, 1320
   of solving systems of linear equations, 696–701, 703, 1030, 1036
Ellipses, 1298, 1307–1310, 1314
   application problem-solving involving, 1310
   on graphing calculators, 1310
   graphs of, 1307–1309
   standard equation for, 1308, 1309
Ellipsis, 2
Empty sets, AP-12
Equal sets, AP-13
Equality, properties of, 160–165, 270–275, 280
Equals sign ($=$), 155
Equations. *See also* Formulas
   absolute value, 1001–1005, 1010, 1011
   compared to expressions, 53, 71, 77, 831
   conditional, 548
   containing complex numbers, 1209–1210
   contradiction, 548
   defined, 22, 26, 127
   dependent, 676, 1043
   equivalent, 160–161, 165
   exponential, 1275–1278, 1282
   graphical solutions to, 131–132, 133
   on graphing calculators, 598, AP-5
   guess-and-check method for, 128–129, 133
   higher-degree polynomial, 1207–1209, 1210
   for horizontal lines, 607–608
   identity, 548
   independent, 676
   with integer solutions, 127–128
   involving addition and subtraction, 22–23
   involving multiplication and division, 39–40
   involving percents, 444–447
   involving variables, 51–52
   linear. *see* Linear equations
   of lines, 642–648
   logarithmic, 1279–1282
   polynomial, 837–838, 839
   power rules for, 1114, 1123
   quadratic. *see* Quadratic equations
   radical. *see* Radical equations
   rational. *see* Rational equations
   solving, for variables, 899–900, 1177–1178
   symbolic solutions to, 270–275, 280, 340–343, 346
   systems of. *see* Nonlinear systems of equations; Systems of linear equations
   tables of values for solving, 129–130
   translating sentences or words into, 75, 155–157, 555–556
   for vertical lines, 608–610
   with infinitely many solutions, 548–549
   with no solutions, 548–549
Equilateral triangles, 500–501, 504
Equivalent equations, 160–161, 165
Equivalent fractions, 213–214, 217–218, 219, 323
Estimation
   defined, 58, 62
   of sums and differences, 58–60, 237, 307–308, 313
   using graphs, 60–61, 62
Euler, Leonhard, 1248
Evaluation
   compared to simplification, 151
   defined, 48–49
Even root functions, 1107, 1111
Even roots, 1068
Events
   certain, 474, 478
   complements of, 476–477, 478
   defined, 473, 477
   impossible, 474, 478
   probability of, 475–476
Expanded form
   defined, 4–5, 10
   and distributive properties, 34
   and regrouping, 17, 21
Expansion by minors, of a determinant, 1052
Experiments, 473–474, 477
Exponents
   application problem-solving involving, 733
   on calculators, 728, 763, AP-1, AP-30, AP-29, AP-31

Exponents (*continued*)
 defined, 46
 fractions as, 1208
 logarithms and, 1256, 1275–1282
 negative integer, 762–766, 769, 1207–1208
 rational, 1077–1082, 1083, 1207–1208, 1210
 rational numbers as, 1077–1082, 1083
 rules of, 729–733, 763–766
 zero, 728–729, 733
Exponential decay, 1238, 1241–1242
Exponential equations, 1275–1278, 1282
Exponential expressions, 89, 120–121, 727, AP-1, AP-31
Exponential form, 1256, 1275–1280
Exponential functions, 1238–1250
 compound interest and, 1245–1247
 data modeling with, 1247–1249
 graphs of, 1239–1242, 1249
 growth and decay factors in, 1238
 inverses of, 1262
 linear functions versus, 1241–1242
 natural, 1248, 1250
 percent change and, 1244–1245, 1250
 polynomial functions and, 1241
Exponential growth, 1238, 1241–1242
Exponential notation, 46, 53
Expressions
 algebraic. *see* Algebraic expressions
 compared to equations, 53, 71, 77, 831
 exponential. *see* Exponential expressions
 numerical, 334
 radical. *see* Radical expressions
 rational. *see* Rational expressions
Extraneous solutions, 895–896, 1114–1117

Factor(s)
 binomial, signs in, 802, 810–812
 common. *see* Common factors
 decay, 1238
 defined, 30, 41, 789
 divisibility tests for, 209–211, 219
 greatest common, 214–215, 220
 growth, 1238, 1243–1245
 more than two integer, 114
 repeated, 46
Factor trees, 212
Factorial notation, 1363, 1366
Factoring
 binomials, 793–795, 814–817, 819, 823–824, 825
 with FOIL in reverse, 810–812
 by grouping, 793–795, 796, 806–809, 825
 higher-degree polynomials, 835–837, 839
 perfect square trinomials, 815–816
 polynomials, 789–792, 815–819, 822–825, 835–837
 solving quadratic equations by, 828–830

 trinomials, 798–803, 804
 using FOIL in reverse, 826
 visual, using rectangles, 803–804
Fahrenheit temperatures, 414–417
Feasible solutions, AP-19–AP-20
Fifth root functions, 1107
Figures
 composite, 519
 geometric, 490–491
 plane, 523–528
 similar, 386–391, 392
Financial math, AP-33
Finite sequences, 1335, 1358
Finite series, 1352, 1358
Finite sets, AP-16–AP-17
Focus (foci)
 of ellipses, 1307
 of hyperbolas, 1311
FOIL method
 for binomial multiplication, 749, 750
 reverse, in factoring trinomials, 810–812, 826
Foot (feet), 395
Formula(s). *See also* Equations; Symbolic representation
 for area, 523, 528
 change of base, 1272
 for converting temperatures, 415–416
 defined, 49, 53
 distance, 557–558, 1120–1121, 1123
 evaluating, 49–50
 fractions in, 231
 geometric, 49–50
 grade point average (GPA), 358
 on graphing calculators, AP-4–AP-5
 midpoint, 960–963
 quadratic, 1184–1186, 1190–1192, 1193
 solving, for variables, 549–551, 899–900
 translating words to, 50, 53
 vertex, 1145–1147, 1162
 volume and surface area, 527
Fourth root functions, 1107
Fractional parts. *See* Decimal places
Fractions, 197–205
 adding, 236–237, 240, 246–249, 251, 868, AP-32
 application problem-solving involving, 230–231, 239–240, 249–250, 277–278
 basic principle of, 213–214, 225
 on calculators, 868, AP-1–AP-2, AP-31–AP-32
 clearing, 273–274, 275, 546–547
 comparing, 217–218
 complex. *see* Complex fractions
 decimal numbers as, 299–300, 304, 323–324, 327
 defined, 197, 206
 dividing, 228–230, 233, 862, AP-31
 equivalent, 213–214, 217–218, 219, 323

 as exponents, 1208
 exponents on, AP-31
 in formulas, 231
 graphs of, 204–205, 206
 like, 236–240
 multiplying, 223–227, 232, 861, AP-31
 negative signs in, 853–854
 as percents, 437, 439
 percents as, 435–436, 439
 proper and improper, 200–204, 206
 properties for, 200, 206
 raising, to a power, 227
 rates as, 377–378
 rational numbers as, 199
 ratios as, 375–376
 represented by shading, 198
 simplifying, 214–216, 851, AP-31
 solving equations containing, 270–280
 square roots of, AP-31, 228, 233
 subtracting, 237–238, 240, 246–249, 251, 868, AP-32
 unit, 395, 400
 with unlike denominators, 877
Function(s), 933–946
 absolute value, 984, 994
 arithmetic operations on, 992–993, 994
 basic concepts about, 933–934
 composition of, 1222–1224, 1233
 constant, 960, 963
 cube root, 1073–1074, 1107
 cubic, 985
 defined, 939–940, 946
 domains of, 933, 939–941, 946, 1288
 exponential, 1238–1250
 identifying, 942–944
 inverse, 1227–1232, 1233, 1259
 linear, 953–963, 985, 1241–1242
 logarithmic, 1255–1263
 naming, 933
 nonlinear, 942, 983–994
 objective, AP-19–AP-21
 one-to-one, 1224–1227, 1233
 order of evaluation for, 1223
 ordered pairs as, 939, 942
 piecewise-defined, 976–977
 polynomial, 1241
 polynomial, of one variable, 985–986, 994
 power, 1108–1110, 1111
 quadratic, 985, 1143–1152, 1165–1166, 1188–1190
 rational, 987–992, 994
 representations of, 934–942, 947, 953–957
 root, 1069–1072, 1074, 1107–1108, 1111
 step, 976
 vertical line test for, 943–944
Function notation, 934, 1222
Fundamental rectangles, 1311
Fundamental theorem of linear programming, AP-20–AP-22

Gallons, 398–399
Gauss, Carl Friedrich, 1041
Gauss-Jordan elimination, 1041–1043
General terms, of sequences, 1335, 1344
Geometric figures, 490–491
Geometric sequences, 1345–1346, 1348, 1355–1356
Geometric series, 1355–1356, 1358
Grade point average (GPA), 357–358, 359
Grams, 403, 407–408
Graphical representation. *See* Graphs
Graphical solutions
    to decimal equations, 276–277, 280, 344, 347
    to integer equations, 131–132, 133
    to linear equations, 172–174, 177
    to nonlinear systems of inequalities, 1321–1322
Graphing calculators, 944–946, AP-1–AP-11, AP-28–AP-33
    $a + bi$ mode on, AP-7
    absolute value on, AP-8
    arithmetic operations on, AP-7, AP-28–AP-32
    asymptotes on, 992, AP-11
    binomial coefficients on, 1364
    circles on, 1303
    complex arithmetic on, AP-7
    decimal windows on, 992, AP-11
    determinants on, 1053, AP-9–AP-10
    dot mode on, 992, AP-10
    entering expressions on, AP-1, AP-7–AP-8
    equations on, 598, AP-5
    exponents on, 728, 763, AP-1, AP-29
    factorial notation on, 1363
    financial math on, AP-33
    formulas on, AP-4–AP-5
    fractions on, 868, AP-1–AP-2
    hyperbola on, 1313
    imaginary unit i on, AP-7
    inequalities on, 708, AP-6–AP-7
    integers on, AP-29–AP-30
    intersection points on, 675, AP-6
    line graphs on, 589, AP-3–AP-4
    linear functions on, 956–957
    logarithmic functions on, AP-8
    matrices on, 1045, 1053, AP-8–AP-10
    min-max values on, AP-11
    natural exponential function on, AP-8
    negative numbers on, AP-29
    parentheses on, AP-29
    pi on, AP-1
    rational exponents on, 1079
    reduced row-echelon form on, AP-9
    scatterplots on, 589, 945–946, AP-3–AP-4
    scientific notation on, 767, AP-2, AP-33
    sequences on, 1336, 1339, 1355
    solution checking on, 680
    square roots of negative numbers on, 1070, AP-7
    square roots on, 1067
    sum of sequences on, 1355
    systems of inequalities on, 1323
    systems of linear equations on, 1045
    tables of values on, 855
    tables on, 543, AP-2–AP-3
    vertex (vertices) on, 1151
    vertical lines on, AP-5
    viewing rectangle on, 944–945, AP-3, AP-5–AP-6
    zero finder on, AP-10
Graphs
    bar, 7, 11
    choosing scale for, 588
    circle, 438, 440
    of circles, 1301–1303
    composite function, 1224
    for converting temperatures, 417
    of ellipses, 1307–1309
    estimation using, 60–61, 62
    of exponential functions, 1239–1242, 1249
    of fractions, 204–205, 206
    of functions, 934–936, 938, 943–944, 947
    horizontal line, 607–608
    of hyperbolas, 1312–1313, 1314
    of integers, 90
    of inverse functions, 1230–1232
    line. *see* Line graphs
    of linear functions, 953, 955–957, 963
    of linear inequalities, 707–709
    of logarithmic functions, 1258, 1264
    of mixed numbers, 205, 206
    of nonlinear equations, 600
    of nonlinear functions, 942
    number line, 565–566, 575
    of ordered pairs, 588–589
    of parabolas, 1157–1162, 1298–1301
    of polynomials, 736
    power function, 1109–1110
    of proper and improper fractions, 204–205, 206
    of quadratic equations, 1147–1148, 1185–1186
    of quadratic functions, 1144–1152, 1172–1173, 1188–1190
    of quadratic inequalities, 1198–1201, 1204
    of radical equations, 898, 1117, 1118–1119
    rectangular coordinate system for, 586
    root function, 1070, 1073
    scatterplot, 586, 588–589, 591, 945–946, AP-3–AP-4
    of sequences, 1337, 1339
    and solution sets, 600
    of systems of linear equations, 673–677, 681, 699, 700
    of systems of linear inequalities, 710–711
    tables versus, 586
    variation, 911, 913
    vertex form and, 1160–1162
    vertical line, 608–610, 612, AP-5
    of whole numbers, 5–6
Greater than ($>$), 6, 90–91
Greatest common factor (GCF)
    defined, 220, 796
    factoring out, 791–792, 795, 803, 809–810, 818–819, 825
    finding, 214–215
    for rational expressions, 218–219
Gross pay, 460
Grouping. *See also* Regrouping
    factoring by, 793–795, 796, 806–809, 825
    nested, 66–67, 122
    symbols, 121–122
Growth factors, 1238, 1243–1245
Guess-and-check method
    for equations, 128–129, 133
    for square roots, 332, 337

Hectograms, 407–408
Hectoliters, 407
Hectometers, 404
Heptagons, 508
Hexagons, 508
Higher-degree polynomials, 835–837, 839, 1207–1209, 1210
Horizontal axis, 172
Horizontal lines, 607–608, 610, 611
Horizontal line test, 1225–1227, 1233
Hours, 421
Hyperbolas
    defined, 1298
    on graphing calculators, 1313
    graphs of, 1312–1313, 1314
    standard equation for, 1311, 1313
Hypotenuse, 353

Identity equations, 548
Identity properties
    for addition, 18, 25, 99
    for division, 36
    for fractions, 200, 206
    for multiplication, 31, 113
    for subtraction, 21–22, 25
Imaginary numbers, 1128
Imaginary parts, of complex numbers, 1128
Imaginary unit i, 1128–1131, AP-7
Impossible events, 474, 478
Improper fractions, 200–204
    defined, 200
    as mixed numbers, 201–204, 206
    mixed numbers as, 254, 260
    represented by shading, 201
Inch(es), 395
Inconsistent systems of linear equations, 676, 1034, 1043
Independent variables, 933, 946

Index(es)
  of radicals, 1068, 1089
  of summation, 1357
Inequalities
  absolute value, 1005–1010, 1011
  addition property of, 568–570, 575
  compound, 969–977, 978
  on graphing calculators, 708, AP-6–AP-7
  in interval notations, 973
  linear. *see* Linear inequalities
  multiplication property of, 570–571, 575
  nonlinear systems of, 1321–1323, 1324
  quadratic, 1197–1203, 1204
  symbols of, 565, 573
  systems of linear, 710–714
  three-part, 972–975, 978
  translating words to, 572–573
Infinite sequences, 1335
Infinite sets, AP-16–AP-17
Infinity, in interval notation, 568, 973
Input variables. *See* Domains; Independent variables
Integers, 89–94. *See also* Numbers
  adding, 97–102, 103, 107, AP-29
  in algebraic expressions, 123, 125
  application problem-solving involving, 92–93, 102
  arithmetic operations on, AP-29–AP-30
  consecutive, 556
  defined, 89, 94
  dividing, 115, 117, AP-30
  in equations, 127–132
  graphs of, 90
  multiplying, 112–114, 117, AP-30
  negative, as exponents, 762–766, 769, 1207–1208
  order of operations with, 121
  square roots of, 116, 117
  subtracting, 106–109, 110, AP-30
  visual addition of, 100–101
Integer pairs, 798–800
Intercepts, 604–607, 611, 1189–1190, 1201
Interest
  compound, 468–470, 471, AP-33
  defined, 466, 471
  simple, 466–467, 469, 471, AP-33
Interest rate, 466, 471
Intersecting lines, 493–494, 497
Intersection
  point of, 675, AP-6
  of sets, 970, AP-15–AP-16
Intersection-of-graph method, 673
Interval notation
  compound inequalities in, 972–973, 978
  domain and range in, 983–984, 994
  infinity in, 568, 973
  solution sets in, 567–568, 575
Inverse functions, 1227–1232, 1233, 1259

Inverse properties, 99, 1259–1260, 1262–1263, 1264
Inverse variation, 912–914, 916
Investments, 467
Irrational numbers, 330–331, 336
Isosceles triangles, 500–501, 504

Joint variation, 915, 916

Kilograms, 407–408
Kiloliters, 407
Kilometers, 404
Kites (geometric shape), 511
Kowa, Sei, 1052
kth root, AP-8

Leading coefficients, 798–799, 1176
Least common denominator (LCD)
  and clearing decimals, 548
  defined, 251
  finding, 246–250
  rational equations solving and, 894–896
  in rational expressions, 877–880, 881
  simplifying complex fractions using, 888–890, 891
Least common multiples (LCMs), 244–245, 251, 875–877, 881
Legs, of triangles, 353
Leibniz, Gottfried, 1052
Length
  converting between U.S and metric units of, 412–413, 418
  metric units of, 403–405, 409
  U.S. units of, 395–396, 400
Less than ($<$), 6, 90–91
Like (common) denominators, 236–240, 868–871
Like fractions, 236–240
Like radicals, 1095, 1097, 1103
Like signs
  adding integers with, 97–98, 103
  of products, 112–113
  of quotients, 115
Like terms, 71–73, 152
  combining, 146–147
  in monomials, 738–739, 743
Line(s)
  data modeling using, 653–657, 658
  defined, 490, 496
  equations of, 642–648
  graphing of, using intercepts, 606–607
  parallel, 634, 637
  perpendicular, 634–636, 638
  point-slope form of, 642–646
  slope of. *see* Slope
  slope-intercept form of, 630–637, 644–646
  standard form of, 611
  vertical, 608–610, 612, AP-5
Line graphs, 7–8, 589–590
  defined, 591
  function of, 11, 586
  on graphing calculators, AP-3–AP-4

Line segments, 490, 496
Line tests
  horizontal, 1225–1227, 1233
  vertical, 943–944, 947, 1226
Linear equations, 541–551. *See also* Equations
  application problem-solving involving, 545
  clearing fractions and decimals to solve, 546–548
  data modeling with, 653–657, 658
  distributive property for solving, 545–546
  numerical solutions to, 543, 551
  in one variable, 541–542
    application problem-solving involving, 181–182
    methods for solving, 168–176
  symbolic solutions to, 169–171, 176, 543–544, 551
  systems of. *see* Systems of linear equations
  in two variables
    graphs of, 597–600, 606–607
    ordered pairs as solutions to, 594–595, 601
    standard form for, 599, 601, 611
    systems of, 673–703, 710–713, 714, 1054–1055, 1056
    tables of solutions for, 595–597, 601
Linear functions, 953–963, 985. *See also* Functions
  data modeling with, 953–954, 963
  defined, 954, 963
  exponential functions versus, 1241–1242
  identifying, 954–955
  midpoints and, 962–963
  piecewise-defined, 976
  rate of change for, 953–954, 963
  representations of, 953–957
Linear inequalities, 565–575
  application problem-solving involving, 573–574
  defined, 574
  graphing, 565–566
  graphs of, 707–709
  solutions and solution sets to, 565–568, 571–572, 575, 706–709
  systems of. *see* Systems of linear inequalities
  tables of solutions for, 567
  test points for, 707
  in two variables, 706–709
Linear models, 653–654, 657. *See also* Data modeling
Linear programming problems, AP-19–AP-23
  fundamental theorem for, AP-21
  steps for solving, AP-21

Listing method, 215, 244
Liters, 403, 406–407
Loan periods, 468
Loans, 467
Logarithmic equations, 1279–1282
Logarithmic form, 1256, 1275–1280
Logarithmic functions
    with base-a, 1261–1263
    change of base formula for, 1272
    domain and range of, 1288
    on graphing calculators, AP-8
    graphs of, 1258, 1264
    inverse of, 1259–1260
    square root function and, 1259
Logarithms
    base-a, 1261–1263, 1264
    with bases other than 10, 1260–1261
    common, 1255–1257, 1264
    expanding versus combining, 1270–1271
    exponents and, 1256, 1275–1282
    inverse, 1259–1260, 1262–1263, 1264
    natural, 1261, 1264
    properties of, 1268–1271, 1272
    as sums and differences, 1269
Long division, 37–39, 42, 323–324
Lower limits, of summations, 1357
Lowest terms
    rational expressions in, 851
    simplifying fractions to, 215–216, 220

Main diagonal, of a matrix, 1039
Major axis, of ellipses, 1308
Mass
    converting between U.S and metric units of, 414, 418
    defined, 407
    metric units of, 407–408, 409
Matrices (matrix), 1039–1046, 1047
    augmented, 1039, 1047
    determinants of, 1052–1054, 1055
    dimensions of, 1039
    on graphing calculators, 1045, 1053, AP-8–AP-10
    row transformations in, 1041–1043
    square, 1039
Mean
    arithmetic, 355–356, 359
    weighted, 357
Measurement
    of angles, 494, 495–496
    converting between U.S and metric systems of, 412–418
    metric system of, 403–409
    U.S. system of, 395–400
Measure(s)
    of central tendency, 355
    of one angle, 510
Median, 356, 359
Meters, 403–405
Method of substitution. *See* Substitution method

Metric system, of measurement, 403–409
    base units of, 403
    conversion to U.S. system of measurement, 412–418
    prefixes used in, 403, 409
Mid-point formula, 960–963
Miles, 395
Milligrams, 407–408
Milliliters, 406–407
Millimeters, 404
Min-max values, 1149–1151, AP-11
Minor axis, of ellipses, 1308
Minors, in matrices, 1052
Minuend, 19, 25
Minus signs, use of, 89
Minutes, 421
Mixed numbers
    arithmetic operations with, 255–260
    as decimal numbers, 324
    decimal numbers as, 299–300, 304
    defined, 201, 206
    graphs of, 205, 206
    as improper fractions, 254, 260
    improper fractions as, 201–204, 206
    rounding, 255, 260
Mixture problems, 559–560
Mode, 357, 359
Modeling. *See* Data modeling
Monomials
    coefficients of, 736
    defined, 736, 743
    degrees of, 736
    dividing by, 773–774, 777
    evaluating expressions containing, 741–742
    multiplying, 746–749
    properties of, 736, 743
Month, 421
Multiplication
    of algebraic expressions, 150
    application problem-solving involving, 40–41, 116–117
    of binomials, 798
    of complex numbers, 1130, 1133
    cross, 893–894
    of decimal numbers, 316–319, 326, AP-32
    equations involving, 39–40
    of fractions, 223–227, 232, 861, AP-31
    of integers, 112–114, 117, AP-30
    of larger numbers, 33
    of mixed numbers, 257, 260
    of monomials, 746–749
    of numbers ending in zero, 34
    of polynomials, 746–751, 752
    properties for, 30–34, 42
    properties for integers, 113, 117
    of radical expressions, 1086–1087, 1099–1100, 1103
    of rational expressions, 226–227, 232, 862–863, 864
    regrouping in, 33

    of whole numbers, 30–34, 41, AP-28
    words associated with, 34–35, 42
Multiplication property
    of equality, 164–165, 271–273
    of inequalities, 570–571, 575
Multiplication sign $(-)$, 30
Multiplicative inverses, 228–229

Names, of functions, 933
Natural exponential function, 1248, 1250, AP-8
Natural logarithm, 1261, 1264
Natural numbers, 2, 10
Negative infinity, 973
Negative integer exponents, 762–766, 769, 1207–1208
Negative numbers
    adding, 97–98, 103
    defined, 88
    entering, on calculators, AP-29
    rational, 199
Negative reciprocals, 635–636
Negative signs, 853–854
Negative slope, 616–617, 624
Nested grouping, 56–67, 122
Net pay, 460, 462
Nonlinear data, 736, 835–837
Nonlinear equations
    graphs of, 600
    identifying, 542
    systems of, 1318–1321, 1323
Nonlinear functions, 942, 983–994
Nonlinear systems of equations, 1318–1321, 1323
    elimination method to solve, 1320
    graphical solutions to, 1320
    substitution method to solve, 1318–1319
Nonlinear systems of inequalities, 1321–1323, 1324
    graphical solutions to, 1321–1322
    on graphing calculators, 1323
    solution sets to, 1321–1322
Non-real complex numbers, 1128
nth power, perfect, 1087, 1092
nth roots, 1121–1122, 1123
    principal, 1068, 1074
    of real numbers, 1068–1069, 1074
    simplifying, 1087, 1092
nth terms, of sequences, 1335, 1344
Null sets, AP-12
Numbers. *See also* Integers
    complex. *see* Complex numbers
    composite, 211–212, 213, 219
    consecutive, 556
    decimal. *see* Decimals
    imaginary, 1128
    irrational, 330–331, 336
    mixed. *see* Mixed numbers
    natural, 2, 10
    negative, 88, 97–98, 199, AP-29
    positive, 88, 97–98

Numbers (*continued*)
   prime, 211–212, 219
   pure imaginary, 1128
   rational. *see* Rational numbers
   real. *see* Real numbers
   signed, 88–89, 94, 309–310, 318
   whole. *see* Whole numbers
Number line(s)
   decimal numbers on, 300–301, 304
   graphs, 575
   integer addition on, 100–101, 103
   integer subtraction on, 107–108, 110
   midpoint formula on, 961
   origin on, 90
   solution sets on, 565–566
   three-part inequalities on, 972
   whole numbers on, 5–6, 10
Number problems, 556–557
Numerators
   in complex fractions, 885–890, 891
   identifying, 197–198
Numerical expressions, 334
Numerical representation. *See also* Ordered pairs; Tables
   of functions, 934–936, 947
   of inverse functions, 1232
   of linear functions, 953, 955, 963
   of nonlinear functions, 942
   of sequences, 1337, 1339
Numerical solutions
   to equations containing decimals, 343–344, 347
   to equations containing fractions, 276, 280
   to linear equations, 171–172, 177, 543, 551
   for linear inequalities, 567
   to quadratic equations, 1171–1173
   to quadratic inequalities, 1198–1199, 1204
   to systems of linear equations, 674–675, 677–678, 681, 700

Objective function, AP-19–AP-21
Obtuse angles, 492, 496
Obtuse triangles, 500, 504
Octagons, 508
Odd root functions, 1107, 1111
Odd roots, 1068
1 square unit, 40
One-to-one functions, 1224–1227, 1233
Operations. *See* Arithmetic operations; Order of operations agreement
Opposites. *See also* Additive inverse
   and absolute value, 91
   of algebraic expressions, 148–150, 152
   defined, 88, 94
   finding, 89
   integer subtraction using, 106–107
   of polynomials, 740, 743
Optimal value, AP-20

Order of operations agreement
   applying, 65–67, 68, 121, 124
   for complex fractions, 266–267
   for decimal numbers, 333–334
   to evaluate fractions, 727
Ordered pairs
   as functions, 939, 942
   graphs of, 588–589
   plotting and reading, 587, 590
   as solutions, 594–595, 601
Ordered triples, 1027, 1035
Origin, on number lines, 90
Ounces
   fluid, 398–399, 400
   as unit of weight, 399–400
Outcomes, 473–474, 477
Output variables. *See* Dependent variables; Domains

Parabolas, 1144–1151, 1152, 1298–1301, 1304
   graphs of, 1157–1162, 1298–1301
   with horizontal axis of symmetry, 1299–1301, 1304
   min-max application problem-solving using, 1149–1151
   transformations of, 1157–1158
   translations of, 1158–1160, 1166
   vertex form of, 1298–1299, 1304
   vertex of, 1144–1147, 1152
   with vertical axis of symmetry, 1144–1146, 1152, 1298–1299
Parallel lines
   cut by transversals, 494–495, 497
   defined, 493, 497
   and slope, 634, 637
Parallelograms, 511, 523–524
Parentheses, use of, AP-29, 565, 568
Partial dividend, 37
Pascal's Triangle, 1362–1363, 1366
Pentagon (geometric shape), 508
Percents
   application problem-solving involving, 439, 456–462
   as decimals, 435–436, 439
   decimals as, 437, 439
   defined, 435, 439
   as fractions, 435–436, 439
   fractions as, 437, 439
   problems involving, 444–447, 451–454
   sales tax as, 457–458
Percent chance, 474, 477
Percent change, 461–462, 1242–1245, 1249, 1250
Percent equations, 444–446, 447
Percent statements
   form of, 443, 447
   translating, to proportions, 450–451, 454
Perfect cubes, 1087
Perfect nth power, 1087
Perfect square trinomials, 815–816, 819, 823, 1175

Perfect squares, 61, 62, 1087
Perimeter
   of circles. *see* Circumference
   of composite figures, 519
   defined, 23–24, 26
   formulas for, 50
   of polygons, 516–517, 520
Periods, in whole numbers, 2, 298
Perpendicular lines, 493, 497, 634–636, 638
$Pi\,(\pi)$, 330, 1248, AP-1
Piecewise-defined function, 976–977, 978
Piecewise-defined linear function, 976
Pints, 398–399
Place value, 2–3, 10
Plane (geometric), 490
Plane figures
   area of, 523–526, 528
   volume and surface area of, 526–528
Plane geometry, 490
Plotting points. *See* Ordered pairs; Scatterplots
Plus signs, use of, 5, 89
Points, in plane geometry, 490, 496
Point-slope form, 642–646, 649
Polya, George, 74, 555
Polygons
   angles in, 509–510
   area of, 523–525
   defined, 508, 512
   names of, 508, 513
   perimeter of, 516–517, 520
   properties of, 508–511
   regular, 509, 513
Polynomials, 736–743, 788–826. *See also* Binomials; Trinomials
   adding, 737–740, 743
   application problem-solving involving, 757
   defined, 737, 743
   degrees of, 737
   dividing, by monomials, 773–775, 777
   dividing by, 775–777
   evaluating expressions containing, 741–742, 743
   factoring, 789–792, 815–819, 822–825, 835–837
   four-term, 793–794, 796, 823
   graphs of, 736
   higher-degree, 835–838, 839, 1207–1209, 1210
   least common multiples of, 875–877, 881
   multiplying, 746–751, 752
   in one variable, 737
   opposites of, 740, 743
   prime, 802, 808–809
   properties of, 737
   quadratic, 828–829
   subtracting, 740–741, 743
   terms in, number of, 823
   zeros of, 828
Polynomial functions, 985–986, 994, 1241

Positive numbers
 adding, 97–98
 defined, 88
Positive slope, 616–617, 624
Pounds, 399–400
Powers
 of imaginary units, 1130–1131
 raising, to a power, 730–731, 734
 raising fractions to, 227, 233
 solving equations involving, 1081–1082, 1083, 1114, 1123
 of 10
  clearing decimals with, 341–343
  common logarithms of, 1255
  with decimals, 318–319, 322, 323, 327
  defined, 47, 53
  in scientific notation, 767
Power functions, 1108–1110, 1111
Power rules
 for exponents, 730–732, 734
 for logarithms, 1270, 1272
 for radical equations, 1114, 1123
Prefixes, metric, 403, 409
Pricing, unit, 379–380
Prime factorization, 212–213, 215, 219
Prime factorization method, 244–245, 251
Prime numbers, 211–212, 219
Prime polynomials, 802, 808–809
Principal, 466, 471
Principal nth roots, 1068, 1074
Probability, 474–476, 477, 856
Problem solving
 choosing best method for, 186
 linear equation, 181–182
 min-max applications, 1149–1151
 steps for, 74–75, 181, 186, 555
 translating words or sentences when, 75, 154–157, 555–556, 572–573
 using estimation, 59–60
 word problems, 555–561, AP-21–AP-22
Products
 cross, 217–218, 220
 defined, 30, 41
 estimating, 59, 316–317, 326
 with more than two factors, 114, 117
 raising, to a power, 731, 734
 of rational expressions, 862–863
 signs of, 112–113, 114, 117
 of sums and differences, 755–756, 759
Product rule
 for exponents, 729–730, 734, 1081–1082, 1083
 for logarithms, 1268, 1272
 for radical expressions, 1086–1087, 1092
Proper fractions, 200–201
Proportion, 383–386
 application problem-solving involving, 389–391
 and cross products, 384–385, 452

defined, 383, 391, 449–450, 454, 916
in similar figures, 387
solving, 907–909
translating percent statements to, 450–451, 454
Pure imaginary numbers, 1128
Pyramids, 527
Pythagoras, 353
Pythagorean theorem, 353–354, 359, 1120, 1123

Quadrants, 586, 590
Quadratic equations
 application problem-solving involving, 831–832, 1178–1179
 basics of, 1170–1173
 completing the square with, 1175–1176, 1179
 complex numbers and, 1190–1192, 1193
 discriminants and, 1188–1189
 graphs of, 1147–1148, 1172–1173, 1185–1186
 numerical solutions to, 1171–1173
 quadratic formula and, 1184–1186
 solving, by factoring, 828–830, 833
 square root property for, 1173–1174, 1179
 standard form of, 828–829
 symbolic solutions to, 1171–1172
Quadratic formula, 1184–1192, 1193
 derivation of, 1184
 discriminant in, 1188–1189
 solving equations using, 1184–1186
Quadratic functions
 data modeling using, 1165–1166
 graphs of, 1144–1151, 1152, 1188–1190
 min-max values using, 1149–1151
 as polynomial function of one variable, 985
 uses of, 1143
Quadratic inequalities, 1197–1203, 1204
 graphs of, 1198–1201, 1204
 numerical solutions to, 1198–1199, 1204
 symbolic solutions to, 1202–1203, 1204
Quadratic polynomials, 828–829
Quadrilaterals, 508, 510–511, 513
Quarts, 398–399
Quotients
 defined, 35, 42
 estimating, 59, 316–317, 326
 of logarithms and logarithms of quotients, 1276
 negative exponents and, 765–766, 770
 raising, to a power, 731–732, 734
 of rational expressions, 863–864
 signs of, 115, 117
 simplification of, 763–765, 1090

Quotient rule
 for exponents, 764–765, 766, 770, 1081–1082, 1083
 for logarithms, 1269, 1272
 for radical expressions, 1089–1090, 1092

Radical equations
 containing cube roots, 1117–1118
 extraneous solutions to, 1114–1117
 graphs of, 1117, 1118–1119
 solutions to, 1119
 squaring twice in, 1117
Radical expressions, 1066–1069, 1095–1103
 adding, 1095–1097, 1103
 defined, 1066
 equations containing. see Radical equations
 multiplying, 1086–1087, 1099–1100, 1103
 order of operations in, 66
 product rule for, 1086, 1092
 quotient rule for, 1089–1090, 1092
 simplifying, 1087–1091, 1092
 square roots as, 61
 subtracting, 1097–1099, 1103
Radical sign, 61, 62, 1066
Radicals, like, 1095, 1097, 1103
Radicand, 61, 62, 1066
Radius (radii)
 defined, 350, 513, 1301
 finding the, 511–512, 1301–1302, 1303
Range
 of functions, 933, 940, 946
 in interval notation, 983–984, 994
Rate of change
 constant, 657, 953
 for linear functions, 953–954, 963
 slope as, 617, 621–622, 624
Rate(s)
 defined, 377, 380
 as fractions, 377–378
 speed as a, 422
 unit, 378–379, 380
Ratio(s)
 common, 1345, 1348
 proportions and, 907–908, 912
Rational equations, 893–902, 903
 application problem-solving involving, 901–902
 basic, 893–894, 903
 defined, 893
 extraneous solutions to, 895–896
 graphs of, 898
 methods for solving, 893–896, 903
 versus rational expressions, 853, 896–898
 symbolic solutions to, 898–899
Rational exponents, 1077–1082, 1083, 1207–1208, 1210

Rational expressions, 220, 850–881
  adding, 238–239, 240, 249, 869–870, 873, 877–879, 881
  application problem-solving involving, 854–855, 880–881
  basic principle of, 852, 857
  as complex fractions, 884–890
  containing unlike denominators, 877–879, 881
  defined, 850, 857
  dividing, 230, 233, 863–864, 865
  least common denominator in, 877–880, 881
  multiplying, 226–227, 232, 862–863, 864
  negative signs and, 853–854
  versus rational equations, 853, 896–898
  simplifying, 218–219, 851–852
  subtracting, 238–239, 240, 249, 870–872, 873, 877–879, 881
  undefined, 850–851, 857
Rational functions, 987–992, 994
Rational numbers
  defined, 199, 206
  as exponents, 1077–1082, 1083
  in scientific notation, 768
Ratio(s)
  defined, 375, 380
  as fractions, 375–376
  in similar figures, 387
  unit, 377, 380
Rays, 490, 496
Real numbers, 330–331
  application problem-solving involving, 334–335
  defined, 330, 336
  nth roots of, 1068–1069, 1074
Real parts, of complex numbers, 1128
Reciprocals, 228–229, 635–636
Rectangles
  defined, 511
  formulas for, 50, 523
  fundamental, 1311
  using, for visual factoring, 803–804
Rectangular coordinate system, 586, 590
Rectangular prism, 527
Reduced row-echelon form, 1041–1043, 1047, AP-9
Reflections, in graphs, 1157
Regrouping
  in addition, 16–17
  and expanded form, 17, 21
  in multiplication, 33
  in subtraction, 20–21
Relations, in functions, 939, 940
Remainders, 36, 42
Repeat bars, 321
Repeating decimals, 320–321
Rhombus, 511
Right angles, 353, 492, 496
Right triangles, 353, 500, 504
Rise, 616–618, 624
Root functions, 1107–1108, 1111

Rounding
  decimal numbers, 302–303, 304
  defined, 62
  mixed numbers, 255, 260
  whole numbers, 57–58
Run, 616–618, 624

Sale prices, 459, 462
Sales tax, 457–458, 462
Scalene triangles, 500–501, 504
Scatterplots, 586, 588–589, 591, 945–946, AP-3–AP-4
Scientific calculators
  arithmetic operations on, AP-28
  exponent evaluation on, AP-29
  financial math on, AP-33
  integers on, AP-30
  negative numbers on, AP-29
  parentheses on, AP-29
  scientific notation on, AP-33
Scientific notation, 767–769, 770, AP-2, AP-33
Seconds, 421
Semicircle, 352
Sentences and equations, 75, 155–157, 555–556
Sequences
  application problem-solving involving, 1346–1347
  arithmetic, 1342–1344, 1353–1355
  data modeling using, 1338
  defined, 1334
  finite and infinite, 1335
  geometric, 1345–1346, 1355–1356
  on graphing calculators, 1336, 1339, 1355
  representations of, 1337, 1339
  versus series, 1352
  terms of, 1335–1336, 1344–1345, 1346
Series
  arithmetic, 1353–1355, 1358
  finite, 1352, 1358
  geometric, 1355–1356, 1358
  versus sequences, 1352
  summation notation for, 1356–1357
Set-builder notation, 571–572, 575
Sets
  complements of, AP-14–AP-15
  elements of, AP-12–AP-13
  empty (null), AP-12
  equal, AP-13
  finite and infinite, AP-16–AP-17
  intersection of, AP-15–AP-16
  of numbers, 1128
  subsets of, AP-13–AP-14
  unions of, 971, AP-15–AP-16
  universal, AP-13
Shading
  equivalent fractions represented by, 213
  fractions represented by, 198
  improper fractions represented by, 201

  inequalities on graphing calculator, AP-6–AP-7
Side-angle-side (SAS) property, 503, 505
Sides
  of angles, 491, 496
  of polygons, 508
Side-side-side (SSS) property, 503, 505
Sigma (Σ), 1356–1357
Signs
  in binomial factors, 802, 810
  choice of, when factoring, 812
Signed numbers, 88–89, 94, 309–310, 318
Similar figures, 386–391, 392
Simple interest, AP-33, 466–467, 469, 471
Simplifying
  algebraic expressions, 71–73, 77
  compared to evaluation, 151
  complex fractions, 263–267, 885–890, 891
  before division of fractions, 230, 233
  fractions, AP-31, 851
  before multiplication of fractions, 225–226, 232
  nth root, 1087, 1092
  quotients, 763–765, 1090
  radical expressions, 1087–1091, 1092
  rational expressions, 218–219, 851–852
  using arithmetic properties, 73
Slope, 615–624
  calculation of, 617–620, 624
  common difference and, 1343
  of parallel lines, 634
  of perpendicular lines, 635, 636–637
  as rate of change, 617, 621–623, 624
  types of, 616–617, 624
Slope-intercept form, 630–633, 637, 644–646
Solutions. *See also* Solution sets
  checking, 51–52, 74, 127–128, 133, 160, 680
  complex, 1190–1192, 1193, 1209, 1210
  to compound inequalities, 976
  defined, 22, 26, 127
  and equivalent equations, 160, 165
  extraneous, 895–896, 1114–1117
  feasible, AP-19–AP-20
  graphical. *see* Graphical solutions
  guess-and-check method for, 128–129, 133
  to inequalities, 565
  in interval notations, 972–973
  to linear equations, 548–549, 551
  to linear inequalities, 565–568, 571–572, 575, 706–709
  numerical. *see* Numerical solutions
  ordered pairs as, 594–595, 601, 675
  symbolic. *see* Symbolic solutions
  to systems of equations, 688–689, 700–701, 703, 1034–1035
  tables of, 129–130, 133, 595–597, 601

Solution sets. *See also* Solutions
   defined, 565
   graphing, 565–566
   and graphs, 600
   in interval notation, 567–568, 575
   to linear inequalities, 565–568, 571–572, 575, 706–709
   to nonlinear systems of inequalities, 1321–1322
   on number lines, 565–566
   in set-builder notation, 571–572
   to systems of linear equations, 707–708, 710–711, 714
   to systems of linear inequalities, 710–711, 714
Speed, 422–423
Spheres, 527
Spider charts, 8–9
Square(s)
   completing the, 1162–1164, 1166, 1175–1176, 1179, 1192
   concept of, 47
   factoring, 814–815, 819, 823
   perfect, 815–816, 819, 1087
Square(s) (geometric shape)
   defined, 511
   formulas for, 50, 523
   units, 40
Square foot (feet), 397
Square matrix, 1039
Square root(s)
   approximating, 331–333, 337, 1067
   defined, 61, 1066
   finding, 61–62
   of fractions, 228, 233, AP-31
   functions, 1069–1072, 1074, 1107
   on graphing calculators, 1067, AP-1, AP-7, AP-30, AP-32
   of integers, 116, 117
   of negative numbers, 1067, 1070, 1129, AP-7
   principal, 1066
   property, 1173–1174, 1179
   on scientific calculators, AP-30, AP-32
Square viewing rectangle, 1303, AP-5–AP-6
Square yards, 397
Square-based pyramids, 527
Standard form
   for complex numbers, 1128
   for linear equations in two variables, 599, 601, 611
   for lines, 611
   for a parabola, 1161
   for quadratic equations, 828–829
   for whole numbers, 2, 10
Standard viewing rectangle, 945
STAT PLOT, AP-4
Step function, 976
Straight angles, 492, 496
Subsets, AP-13–AP-14

Substitution method, 686–691, 692
   application problem-solving using, 689–691
   with higher degree polynomial equations, 1207
   with infinitely many solutions, 688–689
   with linear systems, 1030, 1036
   with no solutions, 688–689
   with nonlinear systems of equations, 1318–1319
   steps for using, 686
Subtraction
   of algebraic expressions, 148–150
   application problem-solving involving, 23–25, 109–110
   of complex numbers, 1129–1130, 1133
   of decimal numbers, 308, 309, 313, AP-32
   equations involving, 22–23
   of fractions, 237–238, 240, 246–249, 251, 868, 877, AP-32
   of integers, 106–109, 110, AP-30
   of mixed numbers, 256–257, 259, 260
   of polynomials, 740–741, 743
   properties of, 21–22, 25
   of radical expressions, 1097–1099, 1103
   of rational expressions, 238–239, 240, 249, 870–872, 873
   regrouping in, 20–21
   of signed decimal numbers, 310
   visual methods of, 107–109, 110
   of whole numbers, 19–21, 25, AP-28
   words associated with, 22, 25
Subtrahend, 19, 25
Summation notation, 1356–1357
Sum(s)
   of the angle measures, 501–502, 504, 509–510
   defined, 15, 25
   estimating, 59, 237, 307–308, 310, 313
   factoring, of cubes, 816–817, 819, 823–824
   factoring, of squares, 815
   of first n terms of arithmetic sequences, 1353–1355
   of first n terms of geometric sequences, 1355–1356
   logarithms as, 1269
   power rules and, 732
   of rational expressions, 869–870, 873
Supplementary angles, 492–493, 497
Surface area, 526–528
Symbolic expressions. *See* Algebraic expressions
Symbolic representation. *See also* Formulas
   of functions, 934, 936, 937
   of inverse functions, 1232
   of linear functions, 947, 954, 963

   of nonlinear functions, 942
   of sequences, 1337, 1339, 1343–1344, 1346
Symbolic solutions
   to equations containing decimals, 340–343, 346
   to equations containing fractions, 270–275, 280
   to linear equations, 169–171, 176, 543–544, 551
   to quadratic equations, 1171–1172
   to quadratic inequalities, 1202–1203, 1204
   to rational equations, 899
   to systems of linear equations, 686–687, 699, 700
Symbols
   arithmetic, 50, 154, 155, 555
   grouping, 121–122
   of inequalities, 565, 573
   infinity, 973
   integer addition using, 101–102, 103
   integer subtraction using, 108–109, 110
   with multiple purposes, 89
   square root, 1066
   translating words to, 67–68, 155, 157
   union, 971
Synthetic division, AP-26–AP-27
Systems of linear equations, 673–703, 1026–1047
   application problem-solving involving, 678–679
   consistent, 675
   Cramer's rule for, 1054–1055, 1056
   determinants for solving, 1052–1054, 1055
   elimination method to solve, 696–702, 703, 1030, 1036
   Gauss-Jordan elimination method for, 1041–1043
   graphing calculators and, 1045
   graphs of, 673–677, 681, 699, 700
   inconsistent, 676, 1034, 1043
   with infinitely many solutions, 700–701, 703, 1034–1035
   matrix solutions for, 1039–1047
   with no solutions, 700–701, 703, 1034
   numerical solutions to, 674–675, 677–678, 681, 700
   solution sets to, 707–708, 710–711, 714
   solutions to, 675–678, 681, 714, 1027–1028, 1035
   substitution method to solve, 686–691, 692, 1030, 1036
   symbolic solutions to, 686–687, 699, 700
   in three variables, 1027–1035
   in two variables, 673–703, 1054–1055, 1056
   types of, 675–676, 682, 1027

Systems of linear inequalities, 710–714
  application problem-solving involving, 712–713
  graphs of, 710–711
  solution sets to, 710–711, 714
  solutions to, 706–711, 714
  in two variables, 710–713, 714
Systems of nonlinear equations, 1318–1321, 1323
Systems of nonlinear inequalities, 1321–1323, 1324

Tables, 133, 171–172, 177. *See also* Numerical solutions
  for converting temperatures, 416–417
  defined, 9, 11
  for equations, 129–130
  to find intercepts, 605
  to find slope-intercept form, 645–646
  on graphing calculators, 543, 855, AP-2–AP-3
  versus graphs, 586
  for inequalities, 567
  multiplication, 30
  of sequences, 1339
  and slope calculation, 620–621
  with two variables, 595–597, 601
Temperature, 414–417, 418
Term(s)
  defined, 71, 146
  like, 71–73, 738–739, 743
  lowest, 215–216, 220, 851
  of polynomials, 737
  of sequences, 1335–1336, 1344–1345, 1346
  unlike, 72
  used in plane geometry, 490
Test points, 707
Three-part inequalities, 972–975, 978
Time
  and simple interest, 468
  units of, 421–422, 423
Tons, 399–400
Total amount paid, 457–458, 462
Total value, 467, 469
Transformations, of parabolas, 1157–1158
Translations, of parabolas, 1158–1160, 1166
Transversals, 494–495
Transverse axis, 1311, 1312, 1314
Trapezoids
  area of, 277, 523–525, 528
  defined, 511
Triangles, 500–505
  application problem-solving involving, 355
  area of, 231–232, 523
  classifying, 500
  formulas for, 50, 1056
  as polygon, 508
  properties of, 502–504
  and Pythagorean theorem, 353–354
  types of, 500–501, 504
Trinomials
  defined, 737
  factoring, 798–803, 804, 815–816, 826
    by grouping, 806–809, 826
    using FOIL in reverse, 810–812, 826
    perfect square, 815–816, 819, 822, 823, 1175
    signs of the binomial factors based on form of, 802, 810–812

Undefined slope, 616–617, 624
Unions, of sets, 971, AP-15–AP-16
Unit analysis, 412
Unit fractions, 395, 400, 412
Unit pricing, 379–380
Unit rates, 378–379, 380
Unit ratios, 377, 380
Universal sets, AP-13
Unlike denominators, 246–250, 877–879, 881
Unlike signs
  adding integers with, 98–99, 103
  of products, 112–113
  of quotients, 115
Unlike terms, 72, 146
Upper limits, in summation notation, 1357
U.S. system of measurement
  conversion to metric system, 412–418
  units of measurement in, 395–400

Variables
  defined, 48, 53
  defining, 154, 157
  dependent, 933, 946
  in formulas, 549–551
  independent, 933, 946
  order of operations with, 67, 68
  solving for, 899–900, 1177–1178
Variation, 909–916
  application problem-solving involving, 909–910, 912–913, 915
  constant of, 909, 912
  data analysis involving, 914–915
  direct, 909–911, 916
  inverse, 912–914, 916
  joint, 915, 916
Venn diagrams, AP-14
Verbal representation
  of functions, 934, 936, 947
  of linear functions, 954, 963
Vertex (vertices)
  of angles, 491, 496
  completing the square to find, 1162–1163
  of ellipses, 1308
  on graphing calculators, 1151
  of hyperbolas, 1311
  of parabolas, 1144–1147, 1152, 1298–1299
  of polygons, 508
  of the region of feasible solutions, AP-20
Vertex form, 1160–1164, 1166, 1298–1299, 1304
Vertex formula, 1145–1148, 1162
Vertical angles, 494
Vertical asymptotes, 987, 990, 992
Vertical axis, 172
Vertical line test, 943–944, 947, 1226
Vertical lines, 608–610, 612, AP-5
Viewing rectangles, 944–945, 1303, AP-3, AP-5–AP-6
Visualization method, 1006–1007
Volume
  converting between U.S and metric units of, 413–414, 418
  of geometric solids, 526–527, 528
  metric units of, 406–407, 409
  U.S. units of, 398–399, 400

Week, 421
Weight
  U.S. units of, 399–400
  converting between U.S and metric units of, 414, 418
  defined, 407
Weighted mean, 357
Whole numbers
  adding, 15–17, 25, AP-28
  comparison of, 6, 10
  defined, 2, 10
  dividing, AP-28, 35–36
  divisibility tests for, 209–211, 219
  in expanded form, 4–5
  graphing, 5–6
  multiplying, AP-28, 30–34, 41
  perfect squares of, 61
  periods in, 2
  rounding, 57–58
  in standard form, 2–3
  subtracting, 19–21, 25, AP-28
  in word form, 3–4
Windows, on graphing calculators. *See* Viewing rectangles
Withholdings, 460
Word form, 3, 10
Word problems, 555, 561, AP-21–AP-22. *See also* Problem solving
Words
  associated with
    addition, 19, 25
    arithmetic symbols, 50, 154, 155, 555
    division, 39, 42
    multiplication, 34–35, 42
    subtraction, 22, 25
  translating
    to equations, 75, 155–157, 555–556
    to expressions, 50, 53, 154–155
    to formulas, 50, 53

to inequalities, 572–573
to symbols, 67–68, 154–155, 157
as verbal representation of functions, 934
writing decimals in, 298–299, 304

x-axis, 586, 590
x-coordinate, 588
x-intercepts, 604–605, 611, AP-10
Xmax, 944
Xmin, 944
Xscl, 944
xy-plane, 586, 590, 961

Yards, 395
y-axis, 586, 590
y-coordinate, 588
Year, 421
y-intercepts, 604–605, 611
Ymax, 944
Ymin, 944
Yscl, 944

Zero(s)
multiplying numbers ending in, 34
of polynomials, 828
values less than, 88

Zero exponents, 728–729, 733
Zero finder, on graphing calculator, AP-10
Zero properties
for division, 36–37
for fractions, 200, 206
for multiplication, 31, 113
Zero slope, 616–617, 624
Zero-product property, 827–828, 833
ZOOMSTAT, AP-4

Formulas from Geometry

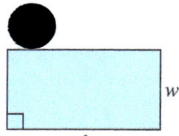
Rectangle
$A = lw$
$P = 2l + 2w$

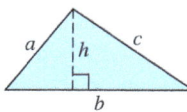
Triangle
$A = \frac{1}{2}bh$
$P = a + b + c$

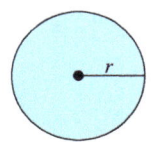
Circle
$A = \pi r^2$
$C = 2\pi r$

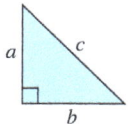
Pythagorean Theorem
$a^2 + b^2 = c^2$

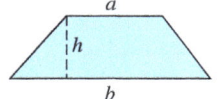
Trapezoid
$A = \frac{1}{2}(a+b)h$

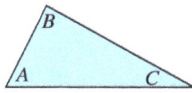
Sum of the Angles in a Triangle
$A + B + C = 180°$

Rectangular Prism
$V = lwh$
$S = 2lw + 2lh + 2wh$

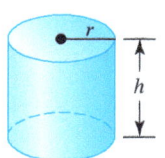
Circular Cylinder
$V = \pi r^2 h$
$S = 2\pi rh + 2\pi r^2$

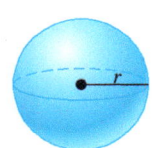
Sphere
$V = \frac{4}{3}\pi r^3$
$S = 4\pi r^2$

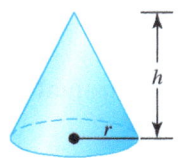
Cone
$V = \frac{1}{3}\pi r^2 h$
$S = \pi r\sqrt{r^2 + h^2} + \pi r^2$

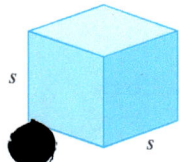
Cube
$V = s^3$
$S = 6s^2$

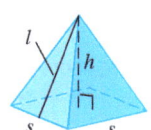
Square-Based Prism
$V = \frac{1}{3}s^2 h$
$S = s^2 + 2sl$

Formulas from Algebra

Basic Properties
$a + b = b + a$
$ab = ba$
$(a + b) + c = a + (b + c)$
$(ab)c = a(bc)$
$a(b + c) = ab + ac$
$\dfrac{a}{c} + \dfrac{b}{c} = \dfrac{a + b}{c}$
$\dfrac{a}{b} \times \dfrac{c}{d} = \dfrac{ac}{bd}$
$\dfrac{a}{b} \div \dfrac{c}{d} = \dfrac{ad}{bc}$

Exponents
$a^m a^n = a^{m+n}$
$(a^m)^n = a^{mn}$
$(ab)^n = a^n b^n$
$\left(\dfrac{a}{b}\right)^n = \dfrac{a^n}{b^n}$
$\dfrac{a^m}{a^n} = a^{m-n}$
$a^{-n} = \dfrac{1}{a^n}$
$\dfrac{a^{-m}}{a^{-n}} = \dfrac{a^n}{a^m}$
$\left(\dfrac{a}{b}\right)^{-n} = \left(\dfrac{b}{a}\right)^n$

Logarithms
$\log_a x = k$ means $x = a^k$
$\log_a 1 = 0$ and $\log_a a = 1$
$\log_a m + \log_a n = \log_a mn$
$\log_a m - \log_a n = \log_a \dfrac{m}{n}$
$\log_a m^r = r \log_a m$
$\log_a a^x = x$
$a^{\log_a x} = x$ $(x > 0)$
$\log_a x = \dfrac{\log_b x}{\log_b a}$ (change of base)

Radicals
$a^{1/n} = \sqrt[n]{a}$
$a^{m/n} = \sqrt[n]{a^m}$
$a^{m/n} = (\sqrt[n]{a})^m$
$\sqrt[n]{ab} = \sqrt[n]{a} \cdot \sqrt[n]{b}$
$\sqrt[n]{\dfrac{a}{b}} = \dfrac{\sqrt[n]{a}}{\sqrt[n]{b}}$

Special Factoring
$a^2 - b^2 = (a - b)(a + b)$
$a^2 + 2ab + b^2 = (a + b)^2$
$a^2 - 2ab + b^2 = (a - b)^2$
$a^3 + b^3 = (a + b)(a^2 - ab + b^2)$
$a^3 - b^3 = (a - b)(a^2 + ab + b^2)$

Constants
$\pi \approx 3.141593$
$e \approx 2.718282$
$i = \sqrt{-1}$
$i^2 = -1$

Distance Formula
$d = \sqrt{(x_2 - x_1)^2 + (y_2 - y_1)^2}$

Midpoint Formula
$M = \left(\dfrac{x_1 + x_2}{2}, \dfrac{y_1 + y_2}{2}\right)$

Interval Notation
$x < 4$: $(-\infty, 4)$
$1 < x \leq 6$: $(1, 6]$
$x \geq 5$: $[5, \infty)$

Slope of a Line
$m = \dfrac{\text{rise}}{\text{run}} = \dfrac{y_2 - y_1}{x_2 - x_1}$ $(x_1 \neq x_2)$

Point–Slope Form
$y - y_1 = m(x - x_1)$, or
$y = m(x - x_1) + y_1$

Distance, rate, and time
$d = rt$

Vertical Line
$x = k$

Horizontal Line
$y = b$

Slope of Parallel Lines
$m_1 = m_2$

Slope of Perpendicular Lines
$m_1 \cdot m_2 = -1$, or
$m_1 = -\dfrac{1}{m_2}$

Quadratic Formula
$x = \dfrac{-b \pm \sqrt{b^2 - 4ac}}{2a}$

Vertex Formula
$x = -\dfrac{b}{2a}$

Vertex (Standard) Form
$y = a(x - h)^2 + k$

Slope–Intercept Form
$y = mx + b$

Standard Equation of a Circle
$(x - h)^2 + (y - k)^2 = r^2$